Brief Contents

University Physics

FOR THE LIFE SCIENCES

University Physics

FOR THE LIFE SCIENCES

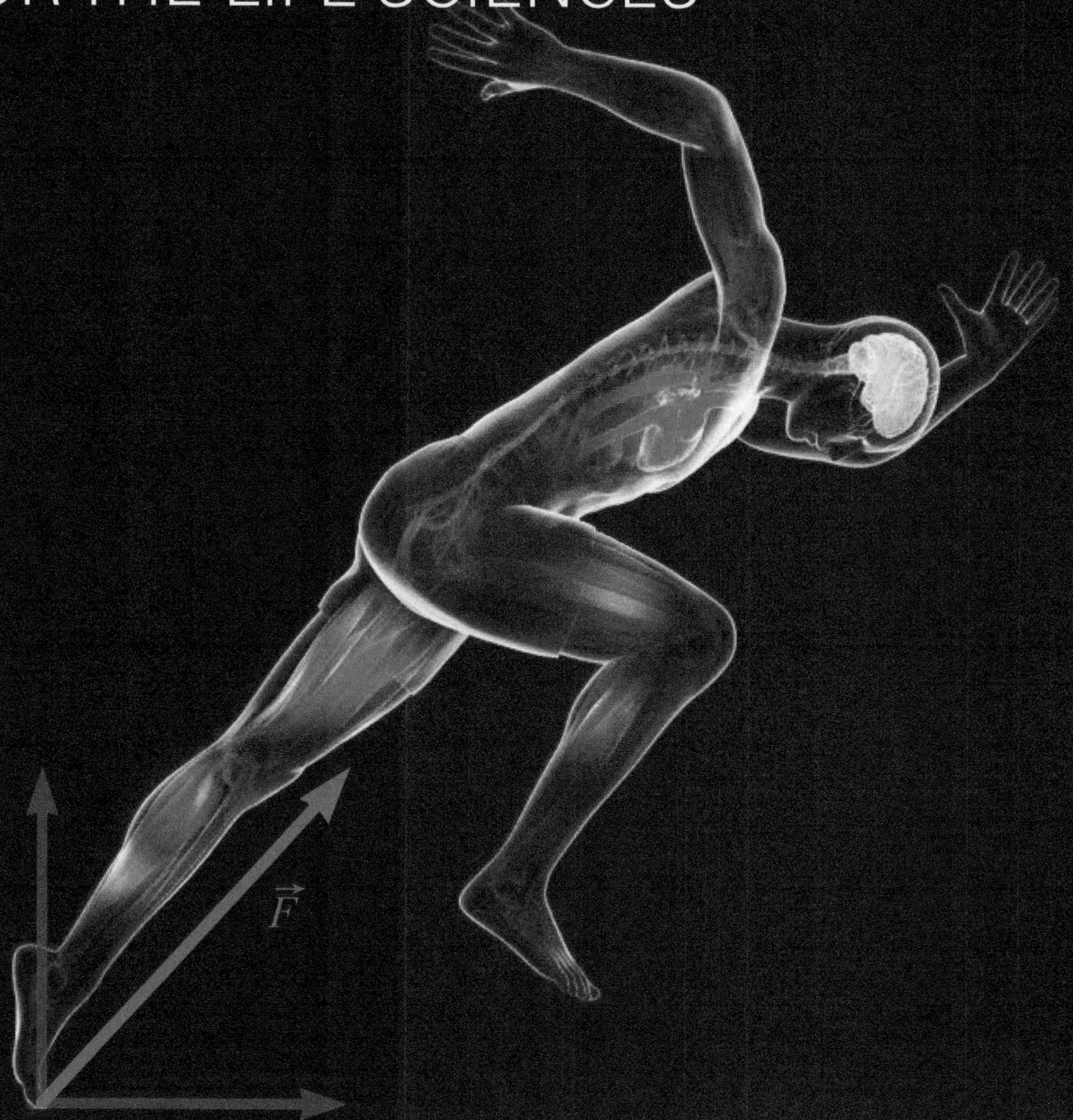

$\vec{F}$

RANDALL D. KNIGHT
California Polytechnic State University, San Luis Obispo

BRIAN JONES
Colorado State University

STUART FIELD
Colorado State University

With contributions by Catherine Crouch, Swarthmore College

Content Development
Director HE Content Management Science & Health Sciences: Jeanne Zalesky
Senior Analyst HE Global Content Strategy–Physical Sciences: Deborah Harden
Senior Content Developer, Science (Physics/Astronomy): David Hoogewerff

Content Management
Manager of Content Development–HE Science: Matt Walker
Development Editor: Edward Dodd

Content Production
Director, Production & Digital Studio, Science/ECS: Katherine Foley
Producer–Science/ECS: Kristen Sanchez
Managing Producer: Kristen Flathman
Senior Content Producer: Martha Steele
Content Producer, Media: Keri Rand
Copyeditor: Carol Reitz
Proofreader: Joanna Dinsmore
Art and Design Director: Mark Ong/Side By Side Studios
Interior/Cover Designer: Preston Thomas/Cadence Design Studio

Product Management
Product Manager Science–Physical Sciences: Darien Estes

Product Marketing
Senior Product Marketing Manager, Physical & Geological Sciences: Candice Madden

Rights and Permissions
Rights & Permissions Management: Ben Ferrini
Photo Researcher: Paul James Tan

Please contact https://support.pearson.com/getsupport/s/ with any queries on this content.

Cover Image by SCIEPRO, Science Photo Library

Copyright © 2022 by Pearson Education, Inc. or its affiliates, 221 River Street, Hoboken, NJ 07030. All Rights Reserved. Manufactured in the United States of America. This publication is protected by copyright, and permission should be obtained from the publisher prior to any prohibited reproduction, storage in a retrieval system, or transmission in any form or by any means, electronic, mechanical, photocopying, recording, or otherwise. For information regarding permissions, request forms, and the appropriate contacts within the Pearson Education Global Rights and Permissions department, please visit www.pearsoned.com/permissions/.

Acknowledgments of third-party content appear on the appropriate page within the text or on pages C-1 and C-2, which constitutes an extension of this copyright page.

PEARSON, ALWAYS LEARNING, and Mastering™ Physics are exclusive trademarks owned by Pearson Education, Inc. or its affiliates in the U.S. and/or other countries.

Unless otherwise indicated herein, any third-party trademarks, logos, or icons that may appear in this work are the property of their respective owners, and any references to third-party trademarks, logos, icons, or other trade dress are for demonstrative or descriptive purposes only. Such references are not intended to imply any sponsorship, endorsement, authorization, or promotion of Pearson's products by the owners of such marks, or any relationship between the owner and Pearson Education, Inc., or its affiliates, authors, licensees, or distributors.

Library of Congress Cataloging-in-Publication Data

Names: Knight, Randall Dewey, author. | Jones, Brian, author. | Field, Stuart, author.
Title: University physics for the life sciences / Randall D. Knight, California Polytechnic State University, San Luis Obispo, Brian Jones, Colorado State University, Stuart Field, Colorado State University; with contributions by Catherine Crouch, Swarthmore College.
Description: Hoboken: Pearson, 2021. | Includes index. | Summary: "University Physics for the Life Sciences has been written in response to the growing call for an introductory physics course explicitly designed for the needs and interests of life science students anticipating a career in biology, medicine, or a health-related field"—Provided by publisher.
Identifiers: LCCN 2020053801 | ISBN 9780135822180 (hardcover) | ISBN 9780135821053 (ebook)
Subjects: LCSH: Physics—Textbooks.
Classification: LCC QC23.2 .K658 2021 | DDC 530—dc23
LC record available at https://lccn.loc.gov/2020053801

Cataloging-in-Publication Data is on file at the Library of Congress.

11 2024

Access Code Card
ISBN-10: 0-135-82129-0
ISBN-13: 978-0-135-82129-9

Rental
ISBN-10: 0-135-82218-1
ISBN-13: 978-0-135-82218-0

About the Authors

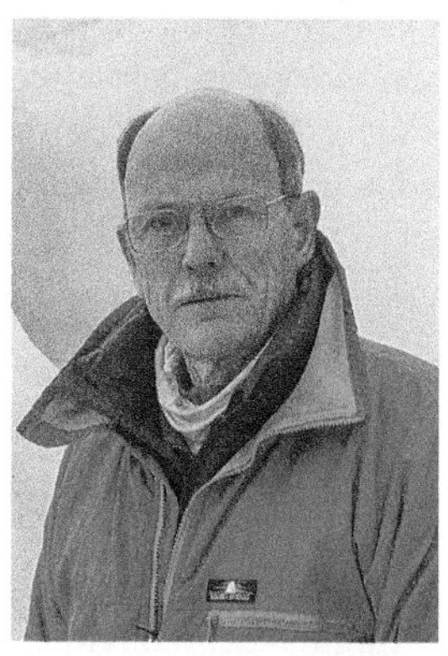

Randy Knight taught introductory physics for thirty-two years at Ohio State University and California Polytechnic State University, where he is Professor Emeritus of Physics. Professor Knight received a PhD in physics from the University of California, Berkeley, and was a postdoctoral fellow at the Harvard-Smithsonian Center for Astrophysics before joining the faculty at Ohio State University. A growing awareness of the importance of research in physics education led first to *Physics for Scientists and Engineers: A Strategic Approach* and later to *College Physics: A Strategic Approach.* Professor Knight's research interests are in the fields of laser spectroscopy and environmental science. When he's not in front of a computer, you can find Randy hiking, traveling, playing the piano, or spending time with his wife Sally and their five cats.

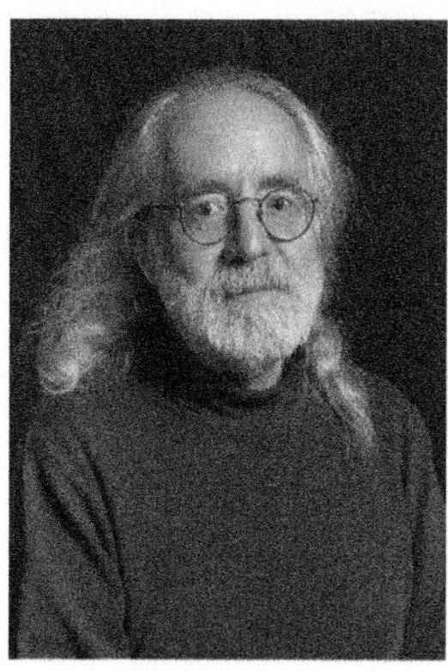

Brian Jones has won several teaching awards at Colorado State University during his thirty years teaching in the Department of Physics. His teaching focus in recent years has been the College Physics class, including writing problems for the MCAT exam and helping students review for this test. In 2011, Brian was awarded the Robert A. Millikan Medal of the American Association of Physics Teachers for his work as director of the Little Shop of Physics, a hands-on science outreach program. He is actively exploring the effectiveness of methods of informal science education and how to extend these lessons to the college classroom. Brian has been invited to give workshops on techniques of science instruction throughout the United States and in Belize, Chile, Ethiopia, Azerbaijan, Mexico, Slovenia, Norway, Namibia, and Uganda. Brian and his wife Carol have dozens of fruit trees and bushes in their yard, including an apple tree that was propagated from a tree in Isaac Newton's garden.

Stuart Field has been interested in science and technology his whole life. While in school he built telescopes, electronic circuits, and computers. After attending Stanford University, he earned a Ph.D. at the University of Chicago, where he studied the properties of materials at ultralow temperatures. After completing a postdoctoral position at the Massachusetts Institute of Technology, he held a faculty position at the University of Michigan. Currently at Colorado State University, Stuart teaches a variety of physics courses, including algebra-based introductory physics, and was an early and enthusiastic adopter of Knight's *Physics for Scientists and Engineers*. Stuart maintains an active research program in the area of superconductivity. Stuart enjoys Colorado's great outdoors, where he is an avid mountain biker; he also plays in local ice hockey leagues.

Contributing author **Catherine Hirshfeld Crouch** is Professor of Physics at Swarthmore College, where she has taught since 2003. Dr. Crouch's work developing and evaluating curriculum for introductory physics for life science students has been used by faculty around the country and has been supported by the National Science Foundation. She earned her PhD at Harvard University in experimental condensed matter physics, and then remained at Harvard in a dual postdoctoral fellowship in materials physics and physics education with Eric Mazur, including developing and evaluating pedagogical best practices for undergraduate physics. She has published numerous peer-reviewed research articles in physics education and experimental physics, and has involved dozens of Swarthmore undergraduate students in her work. She is married to Andy Crouch and they have two young adult children, Timothy and Amy.

To the Student

If you're taking a physics course that uses this text, chances are that you intend a career in medicine or the life sciences. What are you expected to learn in physics that's relevant to your future profession?

Understanding physics is essential to a mastery of the life sciences for two key reasons:

- Physics and physical laws underlie all physiological processes, from the exchange of gases as you breathe to the propagation of nerve impulses.
- Many of the modern technologies used in biology and medicine, from fluorescent microscopy to radiation therapy, are based on physics concepts.

Because of this critical role, physics is a major component of the MCAT.

Biological systems are also physical systems, and a deep knowledge of biology requires understanding how the laws of physics apply to and sometimes constrain biological processes. One of our goals in this text is to build on the science you've learned in biology and chemistry to provide a solid understanding of the physical basis of biology and medicine.

Another important goal is to help you develop your quantitative reasoning skills. Quantitative reasoning is more than simply doing calculations. It is important to be able to do calculations, but our primary focus will be to discover and use *patterns* and *relationships* that occur in nature. Right away, in Chapter 1, we'll present evidence showing that there's a quantitative relationship between a mammal's mass and its metabolic rate. That is, knowing the metabolic rate of a mouse allows you to predict the metabolic rate of an elephant. Making and testing predictions are at the heart of what science and medicine are all about. Physics, the most quantitative of the sciences, is a great place to practice these skills.

Physics and biology are both sciences. They share many similarities, but learning physics requires a different approach than learning biology. In physics, exams will rarely test your ability to simply recall information. Instead, the emphasis will be on learning procedures and skills that, on exams, you will need to apply to new situations and new problems.

You may be nervous about the amount of mathematics used in physics. This is common, but be reassured that you can do it! The math we'll use is overwhelmingly the algebra, geometry, and trigonometry you learned in high school. You may be a bit rusty (see Appendix A for a review of the math we'll be using), and you almost certainly will understand this math better after using it in physics, but our many years of teaching experience find that nearly all students can handle the math.

This text does use some calculus, and your instructor will decide how much or how little of that to include. Many of the ideas of physics—how fast things happen, how things accumulate—are expressed most naturally in the language of calculus. We'll introduce the ideas gently and show you how calculus can be an important thinking and reasoning tool. In fact, many students find they understand calculus best after using it in physics. It's important to become comfortable with calculus because it is increasingly used as a quantitative tool in the life sciences.

How To Learn Physics

There's no single strategy for learning physics that works for everyone, but we can make a few suggestions that will help most students succeed:

- **Read all of each chapter!** This might seem obvious, but we know that many students focus their study on the worked examples. The worked examples are important and helpful, but to succeed on exams you will have to apply these ideas to completely new problems. To do so, you need to understand the underlying principles and logic that are explained in the body of the chapter.
- **Use the chapter summaries.** The chapter summaries are designed to help you see the big picture of how the pieces fit together. That said, the summaries are not a substitute for reading the chapter; their purpose is to help you consolidate your knowledge *after* you've read the chapter. Notice that there are also *part summaries* at the end of each of the text's six parts.
- **Actively participate in class.** Take notes, answer questions, and participate in discussions. There is ample evidence that *active participation* is far more effective for learning science than passive listening.
- **Apply what you've learned.** Give adequate time and attention to the assigned homework questions and problems. Much of your learning occurs while wrestling with problems. We encourage you to form a study group with two or three classmates. At the same time, make sure you fully understand how each problem is solved and are not simply borrowing someone else's solution.
- **Solve new problems as you study for exams.** Questions and problems on physics exams will be *entirely new problems,* not simply variations on problems you solved for homework. Your instructor wants you to demonstrate that you understand the physics by being able to apply it in new situations. Do review the solutions to worked examples and homework problems, focusing on the underlying reasoning rather than the calculations, but don't stop there. A much better use of time is to practice solving additional end-of-chapter problems while, as much as possible, referring only to the chapter summaries.

Our sincere wish is that you'll find your study of physics to be a rewarding experience that helps you succeed in your chosen field by enhancing your understanding of biology and medicine. Many of our students report this was their experience!

To the Instructor

University Physics for the Life Sciences has been written in response to the growing call for an introductory physics course explicitly designed for the needs and interests of life science students anticipating a career in biology, medicine, or a health-related field. The need for such a course has been recognized within the physics education community as well as by biological and medical professional societies. The *Conference on Introductory Physics for the Life Sciences Report* (American Association of Physics Teachers, 2014, available at compadre.org/ipls/) provides background information and makes many recommendations that have guided the development of this text.

This new text is based on Knight *Physics for Scientists and Engineers* (4th edition, 2017) and Knight, Jones, Field *College Physics* (4th edition, 2019). As such, it is a research-based text based on decades of studies into how students learn physics and the challenges they face. It continues the engaging, student-oriented writing style for which the earlier books are known. At the same time, we have fully rethought the content, ordering, examples, and end-of-chapter problems to ensure that this text matches the needs of the intended audience.

Objectives

Our goals in writing this textbook have been:

- To produce a textbook that recognizes and meets the needs of students in the life sciences.
- To integrate proven techniques from physics education research and cognitive psychology into the classroom in a way that accommodates a range of teaching and learning styles.
- To help students develop conceptual and quantitative reasoning skills that will be important in their professional lives.
- To prepare students to succeed on the Chemical and Physical Foundations of Biological Systems portion of the MCAT exam.

Content and Organization

Why develop a new textbook? What is needed to best meet the needs of life science students? The purpose of this text is to prepare students to grasp and apply physics content as needed to their discipline of choice—biology, biochemistry, and/or health sciences. However, the introductory physics course taken by most life science students has for decades covered pretty much the same topics as those taught in the course for engineering and physics majors but with somewhat less mathematics. Few of the examples or end-of-chapter problems deal with living systems. Such a course does not help life science students see the relevance of physics to their discipline.

Many topics of biological importance are missing in a standard introductory physics textbook. These include viscosity, surface tension, diffusion, osmosis, and electrostatics in salt water. Applications such as imaging, whether in the form of fluorescence microscopy or scanning electron microscopy, are barely touched on. A physics course designed for life science students must be grounded in the fundamental laws of physics, a goal to which this text remains firmly committed. But how those laws are applied to the life sciences, and the examples that are explored, differ significantly from their application to engineering and physics.

To endeavor to connect physics to the life sciences, we have added many topics that are important for biologists and physicians. To make time for these, we've scaled back some topics that are important for physicists and engineers but much less so for students in the life sciences. There's less emphasis on standard force-and-motion problems; circular and rotational motion has been de-emphasized (and the text has been written to allow instructors to omit rotation entirely); some aspects of electricity and magnetism have been reduced; and relativity is omitted. After careful consideration, and consultation with experts in biology and physics education, we've made the choice that these topics are less relevant to the audience than the new content that needed to be added.

The most significant change is in the treatment of energy and thermodynamics. Energy and entropy are crucial to all living systems, and introductory physics could play a key role to help students understand these ideas. However, physicists and biologists approach these topics in very different ways.

The standard physics approach that emphasizes conservation of mechanical energy provides little insight into biological systems, where mechanical energy is almost never conserved. Further, biologists need to understand not how work is performed by a heat engine but how useful work can be extracted from a chemical reaction. Biologists describe energy use in terms of enthalpy and Gibbs free energy—concepts from chemistry—rather than heat and entropy. A presentation of energy and entropy must connect to and elucidate reaction dynamics, enthalpy, and free energy if it is to help students see the relevance of physics to biology.

Thus we've developed a new unit on energy and thermodynamics that provides a coherent development of energy ideas, from work and kinetic energy through the laws of thermodynamics. Students bring a knowledge of atoms and molecules to the course, so a kinetic-theory perspective is emphasized. Molecular energy diagrams and the Boltzmann factor are used to understand what happens in a chemical reaction, and ideas about randomness lead not only to entropy but also to Gibbs free energy and what that tells us about the energetics of reactions.

This text does use simple calculus, but more lightly than in the calculus-based introductory course for physicists and engineers. Calculus is now a required course for biology majors at many universities, it is increasingly used as a quantitative analysis tool in biological research, and many medical schools expect at least a semester of calculus. Few results depend on calculus, and it can easily be sidestepped if an instructor desires an algebra-based course. Similarly, there are topics where the instructor could supplement the text with a somewhat more rigorous use of calculus if his or her students have the necessary math background.

Although this text is oriented toward the life sciences, it assumes no background in biology on the part of the instructor. Examples and problems are self-contained. A basic familiarity with chemistry and chemical reactions is assumed.

Key Features

Many of the key features of this textbook are grounded in physics education research.

- **Annotated figures,** now seen in many textbooks, were introduced to physics in the first edition of Knight's *Physics for Scientists and Engineers*. Research shows that the "instructor's voice" greatly increases students' ability to understand the many figures and graphs used in physics.
- **Stop to Think Questions** throughout the chapters are based on documented student misconceptions.
- **NOTES** throughout the chapters call students' attention to concepts or procedures known to cause difficulty.
- **Tactics Boxes** and **Problem-Solving Strategies** help students develop good problem-solving skills.
- **Chapter Summaries** are explicitly hierarchical in design to help students connect the ideas and see the big picture.

Instructor Resources

A variety of resources are available to help instructors teach more effectively and efficiently. Most can be downloaded from the Instructor Resources area of Mastering™ Physics.

- **Ready-To-Go Teaching Modules** are an online instructor's guide. Each chapter contains background information on what is known from physics education research about student misconceptions and difficulties, suggested teaching strategies, suggested lecture demonstrations, and suggested pre- and post-class assignments.
- **Mastering Physics** is Pearson's online homework system through which the instructor can assign pre-class reading quizzes, tutorials that help students solve a problem with hints and wrong-answer feedback, direct-measurement videos, and end-of-chapter questions and problems. Instructors can devise their own assignments or utilize pre-built assignments that have been designed with a good balance of problem types and difficulties.
- **PowerPoint Lecture Slides** can be modified by the instructor but provide an excellent starting point for class preparation. The lecture slides include QuickCheck questions.
- **QuickCheck "Clicker Questions"** are conceptual questions, based on known student misconceptions. They are designed to be used as part of an active-learning teaching strategy. The Ready-To-Go teaching modules provide information on the effective use of QuickCheck questions.
- The **Instructor's Solution Manual** is available in both Word and PDF formats. We do require that solutions for student use be posted only on a secure course website.

Instructional Package

University Physics for the Life Sciences provides an integrated teaching and learning package of support material for students and instructors.

NOTE For convenience, instructor supplements can be downloaded from the Instructor Resources area of Mastering Physics.

Supplement	Print	Online	Instructor or Student Supplement	Description
Mastering Physics with Pearson eText		✓	Instructor and Student Supplement	This product features all of the resources of Mastering Physics in addition to the new Pearson eText 2.0. Now available on smartphones and tablets, Pearson eText 2.0 comprises the full text, including videos and other rich media.
Instructor's Solutions Manual		✓	Instructor Supplement	This comprehensive solutions manual contains complete solutions to all end-of-chapter questions and problems.
TestGen Test Bank		✓	Instructor Supplement	The Test Bank contains more than 2,000 high-quality problems, with a range of multiple-choice, true/false, short answer, and regular homework-type questions. Test files are provided in both TestGen® and Word format.
Instructor's Resource Materials	✓	✓	Instructor Supplement	All art, photos, and tables from the book are available in JPEG format and as modifiable PowerPoints™. In addition, instructors can access lecture outlines as well as "clicker" questions in PowerPoint format, editable content for key features, and all the instructor's resources listed above.
Ready-to-Go Teaching Modules		✓	Instructor Supplement	Ready-to-Go Teaching Modules provide instructors with easy-to-use tools for teaching the toughest topics in physics. Created by the authors and designed to be used before, during, and after class, these modules demonstrate how to effectively use all the book, media, and assessment resources that accompany *University Physics for the Life Sciences.*

Acknowledgments

We have relied on conversations with and the written publication of many members of the physics education community and those involved in the Introductory Physics for Life Sciences movement. Those who may recognize their influence include the late Lillian McDermott and members of the Physics Education Research Group at the University of Washington, Edward "Joe" Redish and members of the Physics Education Research Group at the University of Maryland, Ben Dreyfus, Ben Geller, Bob Hilborn, Dawn Meredith, and the late Steve Vogel. Ben Geller of Swarthmore College provided useful insights into teaching thermodynamics and contributed to the Ready-To-Go Teaching Modules.

We are especially grateful to our contributing author Catherine Crouch at Swarthmore College for many new ideas and examples, based on her experience developing a course for life science majors, and for a detailed review of the entire manuscript.

Thanks to Christopher Porter, The Ohio State University, for the difficult task of writing the *Instructor's Solutions Manual*; to Charlie Hibbard for accuracy checking every figure and worked example in the text; to Elizabeth Holden, University of Wisconsin-Platteville, for putting together the lecture slides; and to Jason Harlow, University of Toronto, for updating the QuickCheck "clicker" questions.

We especially want to thank Director HE Content Management Science & Health Sciences, Jeanne Zalesky; Product Manager Science–Physical Sciences, Darien Estes; Senior Analyst HE Global Content Strategy–Physical Sciences, Deborah Harden; Senior Development Editor, Alice Houston; Development Editor, Edward Dodd; Senior Content Producer, Martha Steele; Senior Associate Content Analyst Physical Science, Pan-Science, Harry Misthos; and all the other staff at Pearson for their enthusiasm and hard work on this project. It has been very much a team effort.

Thanks to Margaret McConnell, Project Manager, and the composition team at Integra for the production of the text; Carol Reitz for her fastidious copyediting; Joanna Dinsmore for her precise proofreading; and Jan Troutt and Tim Brummett at Troutt Visual Services for their attention to detail in the rendering and revising of the art.

And, last but not least, we each want to thank our wives for their encouragement and patience.

Randy Knight
California Polytechnic State University.

Brian Jones
Colorado State University.

Stuart Field
Colorado State University

Reviewers and Classroom Testers

Ward Beyermann, *University of California–Riverside*
Jim Buchholz, *California Baptist University*
David Buehrle, *University of Maryland*
Robert Clare, *University of California–Riverside*
Carl Covatto, *Arizona State University*
Nicholas Darnton, *Georgia Tech*
Jason Deibel, *Wright State University*
Deborah Hemingway, *University of Maryland*
David Joffe, *Kennesaw State University*
Lisa Lapidus, *Michigan State University*
Eric Rowley, *Wright State University*
Josh Samani, *University of California–Los Angeles*
Kazumi Tolich, *University of Washington*
Luc Wille, *Florida Atlantic University*
Xian Wu, *University of Connecticut*

Detailed Contents

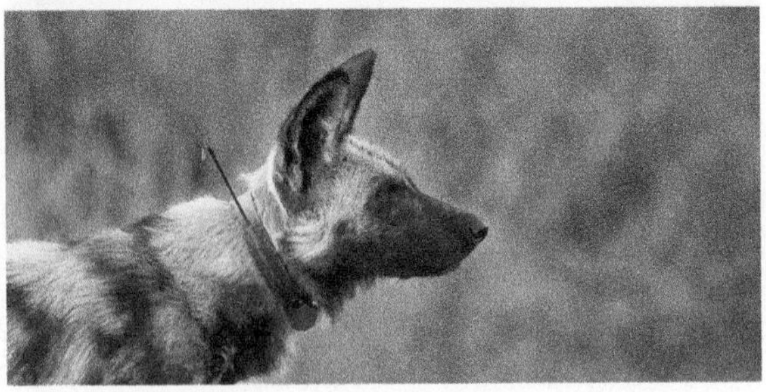

PART I

Force and Motion

Elite athletes push the human body's physical limits. What forces act on this sprinter as she accelerates? How much force can her muscles, tendons, and bones endure? How much air can flow into her lungs? How rapidly can blood be pumped through her veins? These are physics questions that help us understand human performance, questions we'll address in Part I.

OVERVIEW

The Science of Physics

Physics is the foundational science that underlies biology, chemistry, earth science, and all other fields that attempt to understand our natural world. Physicists couple careful experimentation with theoretical insights to build powerful and predictive models of how the world works. A key aspect of physics is that it is a *unifying* discipline: A relatively small number of key concepts can explain a vast array of natural phenomena. In this text, we have organized the chapters into parts according to six of these unifying principles. Each of the six parts opens with an overview that gives you a look ahead, a glimpse of where your journey will take you in the next few chapters. It's easy to lose sight of the big picture while you're busy negotiating the terrain of each chapter. In Part I, the big picture is, in a word, *change*.

Why Things Change

Simple observations of the world around you show that most things change. Some changes, such as aging, are biological. Others, such as the burning of gasoline in your car, are chemical. In Part I, we will look at changes that involve *motion* of one form or another—from running and jumping to swimming microorganisms.

There are two big questions we must tackle to study how things change by moving:

- **How do we describe motion?** How should we measure or characterize the motion if we want to analyze it quantitatively?
- **How do we explain motion?** Why do objects have the particular motion they do? When you toss a ball upward, why does it go up and then come back down rather than keep going up? What are the "laws of nature" that allow us to predict an object's motion?

Two key concepts that will help answer these questions are *force* (the "cause") and *acceleration* (the "effect"). Our basic tools will be three laws of motion worked out by Isaac Newton to relate force and motion. We will use Newton's laws to explore a wide range of problems—from how a sprinter accelerates to how blood flows through the circulatory system. As you learn to solve problems dealing with motion, you will be learning techniques that you can apply throughout this text.

Using Models

Another key aspect of physics is the importance of models. Suppose we want to analyze a ball moving through the air. Is it necessary to analyze the way the atoms in the ball are connected? Or the details of how the ball is spinning? Or the small drag force it experiences as it moves? These are interesting questions, of course. But if our task is to understand the motion of the ball, we need to simplify!

We can conduct a perfectly fine analysis of the ball's motion by treating the ball as a single particle moving through the air. This is a *model* of the situation. A model is a simplified description of reality that is used to reduce the complexity of a problem so it can be analyzed and understood. Both physicists and biologists make extensive use of models to simplify complex situations, and in Part I you'll begin to learn where and how models are employed and assessed. Learning how to simplify a situation is the essence of successful problem solving.

1 Physics for the Life Sciences

Magnetic resonance imaging (MRI) is just one of many ways that physics has contributed to biology and medicine. We'll look at how images like this are created in Chapter 26.

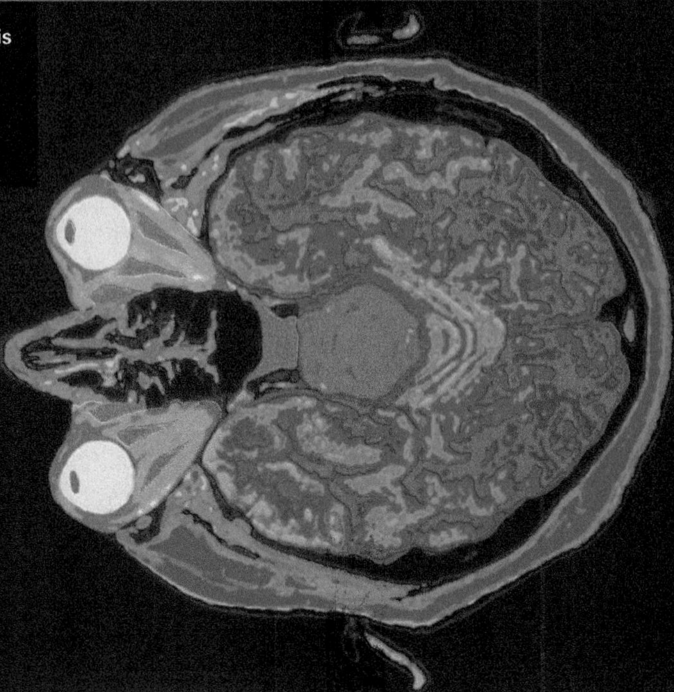

LOOKING AHEAD

Chapter Previews

Each chapter starts with a preview outlining the major topics of the chapter and how they are relevant to the life sciences.

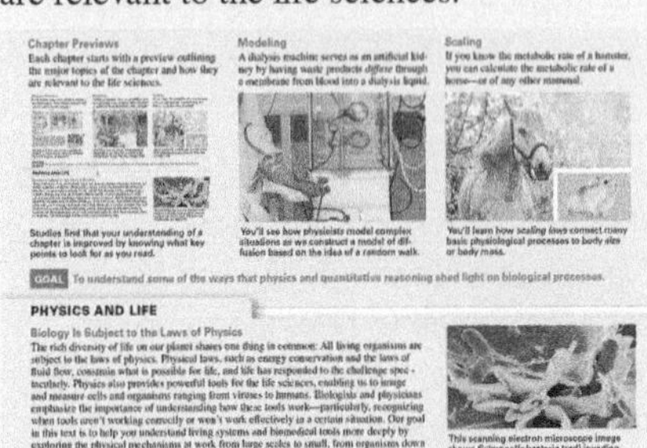

Studies find that your understanding of a chapter is improved by knowing what key points to look for as you read.

Modeling

A dialysis machine serves as an artificial kidney by having waste products *diffuse* through a membrane from blood into a dialysis liquid.

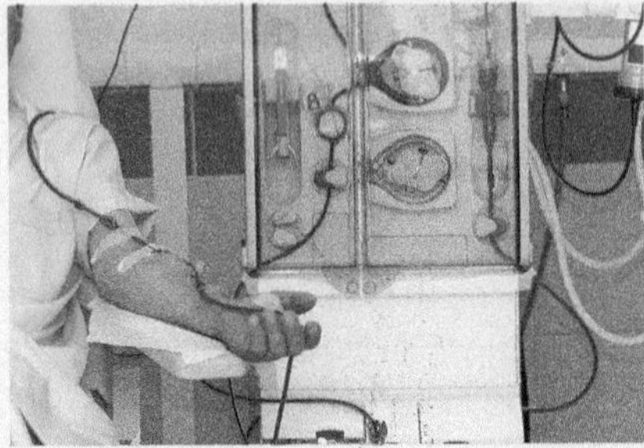

You'll see how physicists model complex situations as we construct a model of diffusion based on the idea of a random walk.

Scaling

If you know the metabolic rate of a hamster, you can calculate the metabolic rate of a horse—or of any other mammal.

You'll learn how *scaling laws* connect many basic physiological processes to body size or body mass.

GOAL **To understand some of the ways that physics and quantitative reasoning shed light on biological processes.**

PHYSICS AND LIFE

Biology Is Subject to the Laws of Physics

The rich diversity of life on our planet shares one thing in common: All living organisms are subject to the laws of physics. Physical laws, such as energy conservation and the laws of fluid flow, constrain what is possible for life, and life has responded to the challenge spectacularly. Physics also provides powerful tools for the life sciences, enabling us to image and measure cells and organisms ranging from viruses to humans. Biologists and physicians emphasize the importance of understanding how these tools work—particularly, recognizing when tools aren't working correctly or won't work effectively in a certain situation. Our goal in this text is to help you understand living systems and biomedical tools more deeply by exploring the physical mechanisms at work from large scales to small, from organisms down to molecules. This first chapter will introduce you to several key ideas.

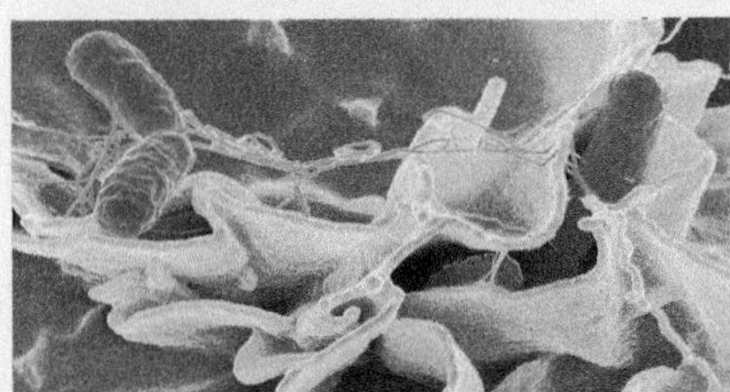

This scanning electron microscope image shows *Salmonella* bacteria (red) invading immune cells. You'll learn about electron microscopes in Chapter 28.

1.1 Why Physics?

Why take physics? Does knowing about projectiles, pendulums, or magnetic fields help you understand biological systems? Do doctors or microbiologists or ecologists think about or use the principles of physics?

Actually, the answer to these questions is Yes. The existence of separate biology, chemistry, and physics departments may make it seem like these are distinct sciences, but that couldn't be further from the truth. Our overarching goal in this text is to help you discover the central importance of physics to the life sciences. Biological systems are part of the physical world, and biological processes are physical processes, so the laws of physics can help you understand a great deal about biology, biochemistry, and medicine. Let's look at some examples:

Physics in biology

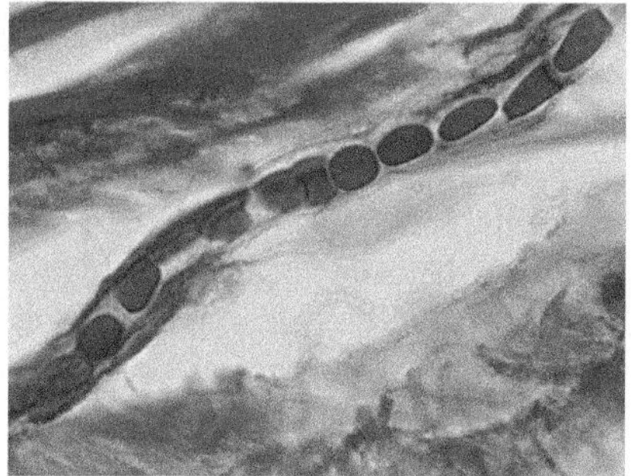

The circulatory system, from the heart to the capillaries, is governed by fluid dynamics. You'll study the physics of circulation in Chapter 9.

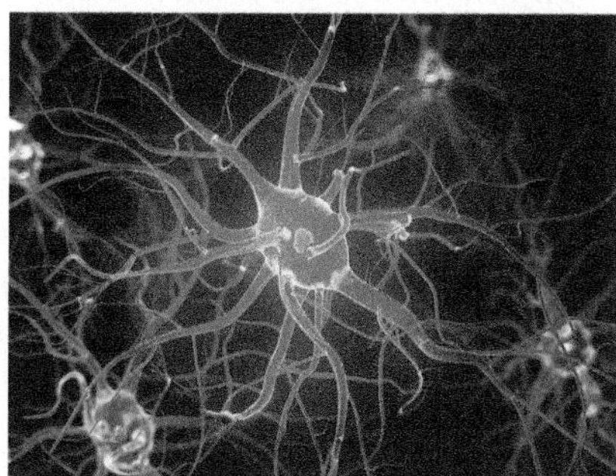

Neurons signal each other via electric pulses that travel along axons. This is a topic we'll visit in Chapter 25.

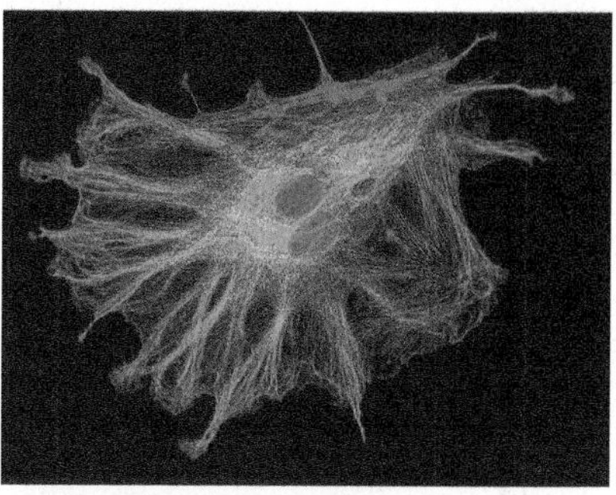

The light-emitting properties of the green fluorescent protein are used to visualize cell structure. Fluorescence is covered in Chapter 29.

Physics in biomechanics

The physics of locomotion affects how fast an animal can run. Locomotion is the topic of several chapters in Part I.

Physics helps us understand the limits of athletic performance. Chapter 11 looks at how the body uses energy.

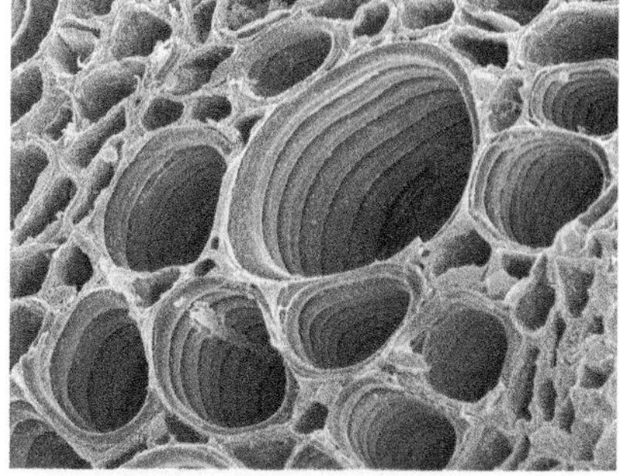

And physics explains how sap rises in trees through the xylem. This somewhat counter-intuitive fluid flow is described in Chapter 9.

Physics in medicine

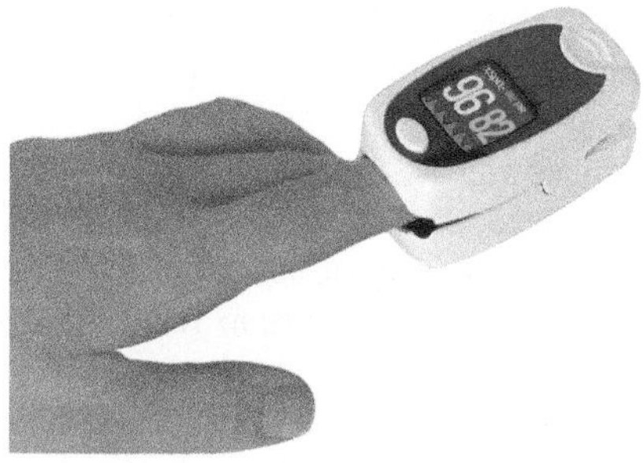

Pulse oximetry measures blood oxygen with a clip-on device. Chapter 29 discusses how blood absorbs different colors of light.

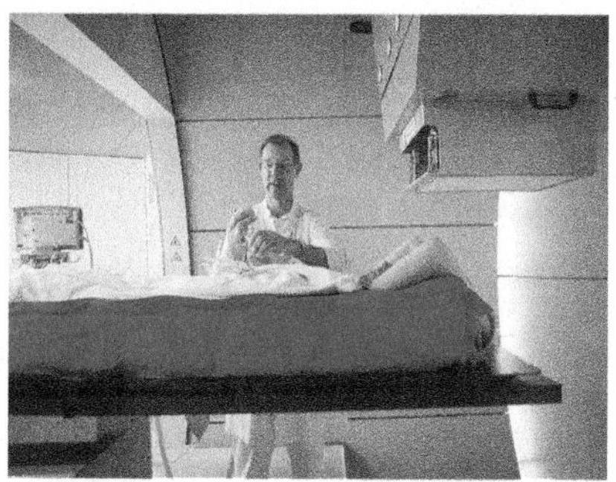

A patient prepares to have cancer treated by proton irradiation. Radiation and radiation therapy are topics in Chapter 30.

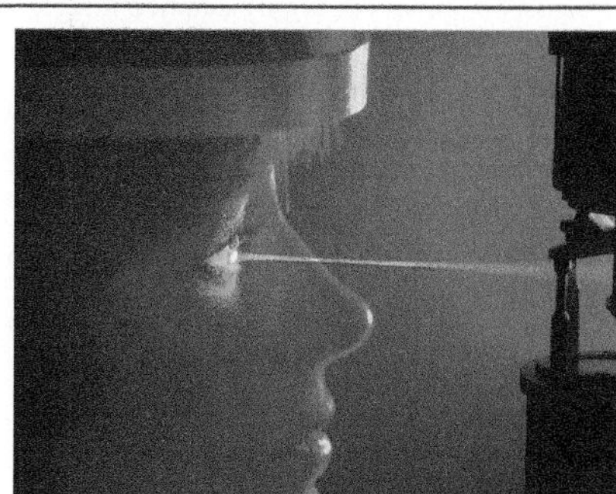

Lasers are used for high-precision eye surgery of both the cornea and the retina. Chapter 20 covers the optics of the human eye.

These are among the many applications of physics to the life sciences that you'll learn about in this book. That said, this is a physics textbook, not a biology textbook. We need to develop the underlying principles of physics before we can explore the applications, so we will often start with rolling balls, oscillating springs, and other simple systems that illustrate the physical principles. Then, after laying the foundations, we will move on to see how these ideas apply to biology and medicine. As authors, our sincere hope is that this course will help you see why many things in biology happen as they do.

Physics and Biology

TABLE 1.1 Characteristics of biology and physics

Biology	Physics
Irreducibly complex systems	Simple models
More qualitative	More quantitative
Focus on specific examples	Focus on broad principles

Physicists and biologists are scientists: They share many common views of what science is and how science operates. At the same time, as TABLE 1.1 illustrates, physicists and biologists often see the world in rather different ways. Your physics course will be less stressful and more productive if you're aware of these differences.

Physics tries to get at the essence of a process by identifying broad principles, such as energy conservation, and applying them to systems that have been greatly simplified by stripping away superfluous details. This approach is less common in biology, where systems often can't be simplified without throwing out details that are essential to biological function. You may have taken biology courses in which you were expected to memorize a great deal of information; there's much less to memorize in physics. Instead, most physicists agree that students demonstrate their understanding of physics by using general principles to solve unfamiliar problems.

With that in mind, we can establish four large-scale goals for this text. By the end, you should be able to:

- Recognize and use the principles of physics to explain physical phenomena.
- Understand the importance of models in physics.
- Reason quantitatively.
- Apply the principles of physics to biological systems.

The examples in this text will help you at each step along the way, and the end-of-chapter problems will provide many opportunities for practice.

Mathematics

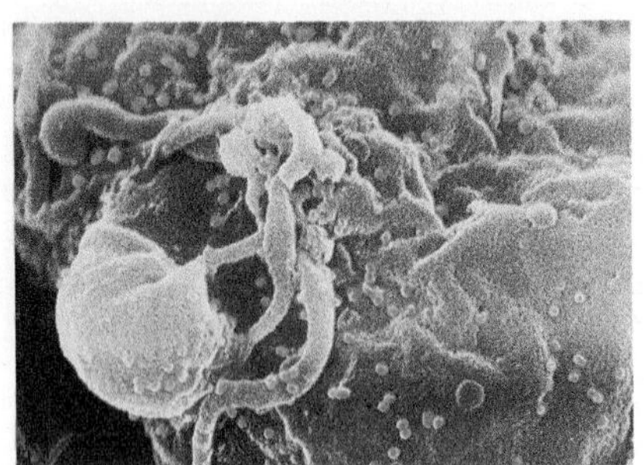

Seeing the details A *scanning electron microscope* produces highly detailed images of biological structures only a few nanometers in size. The level of detail in this image of a budding HIV virus can provide new understanding of how the virus spreads and how it might be stopped. Physics has been at the forefront of developing a wide variety of imaging technologies.

Physics is the most quantitative of all the sciences. As physics has developed, the laws of physics have come to be stated as mathematical equations. These equations can be used to make quantitative, testable predictions about nature, whether it's the orbit of a satellite or the pressure needed to pump blood through capillaries. This approach to science has proven to be extremely powerful; much of modern biomedical technology—from electron microscopes to radiation therapy—depends on the equations of physics.

But math is used in physics for more than simply doing calculations. The equations of physics tell a story; they're a shorthand way to describe how different concepts are related to one another. So, while we can use the ideal-gas law to calculate the pressure in a container of gas, it's more useful to recognize that the pressure of a gas in a rigid container (one that has a constant volume) increases in exactly the same proportion as the absolute temperature. Consequently, doubling the temperature causes the pressure to double. The ideal-gas law expresses a set of deep ideas about how gases behave.

So, yes, we will often use equations to calculate values. But, more fundamentally, physics is about using math to reason and to analyze; that is, math is a *thinking tool* as much as it is a calculation tool. This may be a new way of using math for you, but—with some practice and experience—we think you'll come to recognize the power of this way of thinking.

The math used in this book is mostly math you already know: algebra, geometry, and some trigonometry. You might need some review (see Appendix A), but we're confident that you can handle the math. Physics does use many *symbols* to represent quantities, such as F for force, p for pressure, and E for energy, so many of our equations—like the ideal-gas law—are algebraic equations that show how quantities are related to one another. It's customary to use Greek letters to represent some quantities, but we'll let you know what those letters are when we introduce them.

We will, in some chapters, use a little bit of the calculus of derivatives and integrals. Don't panic! Our interest, once again, is not so much in doing calculations as in understanding how different physical quantities are related to one another. And we'll remind you, at the appropriate times, what derivatives and integrals are all about. In fact, most students feel that they come to understand calculus much better after studying physics because physics provides a natural context for illustrating why calculus is useful.

1.2 Models and Modeling

The real world is messy and complicated. A well-established procedure in physics is to brush aside many of the real-world details in order to discern *patterns* that occur over and over. For example, an object oscillating back and forth on a spring, a swinging pendulum, a vibrating guitar string, a sound wave, and an atom jiggling in a crystal seem very different—yet perhaps they are not really so different. Each is an example of a system oscillating around an equilibrium position. If we focus on understanding the properties of a very simple oscillating system, we'll find that we understand quite a bit about the many real-world manifestations of oscillations.

Stripping away the details to focus on essential features is a process called *modeling*. A **model** is a simplified picture of reality, but one that still captures the essence of what we want to study. Thus "mass on a spring" is a model of almost all oscillating systems. Models allow us to make sense of complex situations by providing a framework for thinking about them.

A memorable quote attributed to Albert Einstein is that physics "should be as simple as possible—but not simpler." We want to use the simplest model that allows us to understand the phenomenon we're studying, but we can't make the model so simple that essential features of the phenomenon are lost. That's somewhat of a problem in this text because understanding the *physics* of a biological system is not the same as understanding the biology. Our models of biological systems may seem to throw out much of the relevant biology, but keep in mind that our goal is to understand the ways in which the *physical* properties of the system have a meaningful effect on aspects of the biology.

We'll develop and use many models throughout this textbook; they'll be one of our most important thinking tools. These models will be of two types:

- *Descriptive models*: What are the essential characteristics and properties of a phenomenon? How do we describe it in the simplest possible terms? For example, we will often model a cell as a water-filled sphere. This omits all the details about what's inside, but for some purposes, such as estimating a cell's mass or volume, we don't need to know what's inside.
- *Explanatory models*: Why do things happen as they do? What causes what? Explanatory models have predictive power, allowing us to test—against experimental data—whether a model provides an adequate explanation of an observed phenomenon. A spring-like model of molecular bonds will allow us to explain many of the thermal properties of materials.

Biologists also use models. A biological model, often qualitative rather than quantitative, is a simplified representation of the structure and function of a biological system or a biological process. The *cell model* is a simplified presentation of an immensely complex system, but it allows you to think logically about the key pieces and processes of a cell. Mice, fruit flies, and nematodes are important *model organisms* that lend themselves to the study of particular biological questions without unnecessary complications.

The eyes have it We study a *model organism* not to learn details about the organism but because the organism lends itself to the understanding of broad biological principles. *Drosophila melanogaster*, a common fruit fly, has provided immeasurable insights into genetics for more than a century.

At the same time, biology and medicine are becoming increasingly quantitative, with models more and more like those constructed by physicists. For example:

- Mathematical models of enzyme kinetics provide quantitative predictions of complex biochemical pathways.
- Neural network models improve our understanding of both real brains and artificial intelligence.
- Epidemiological models increase our knowledge of how disease spreads.
- Global climate models that illustrate the earth's climate and how it is changing depend on a complex interplay between living (e.g., photosynthesis) and nonliving (e.g., solar radiation) processes.

These models skip over many details in order to provide a big-picture understanding of the system. That's the purpose of a model. This is not to say that details are unimportant—most scientists spend most of their careers studying the details—but that we don't want to let the trees obscure our view of the forest.

1.3 Case Study: Modeling Diffusion

Diffusion is one of the most important physical processes in biology. Oxygen diffuses from your lungs to your blood, and neurotransmitters diffuse across the synapses that connect one neuron to the next. We'll study diffusion extensively in Chapter 13, but as a case study let's see how a physicist might *model* diffusion and conclude when diffusion has biological significance. That is, our goal is to create a model that makes testable predictions of *how far* molecules can diffuse and *how long* it takes them to do so.

NOTE ▸ The analysis in this section is more complex than you are expected to do on your own. However, it is expected that you will follow the reasoning and be able to answer questions about the procedure and the results. ◂

FIGURE 1.1 The possible motion of one molecule as it collides with other molecules.

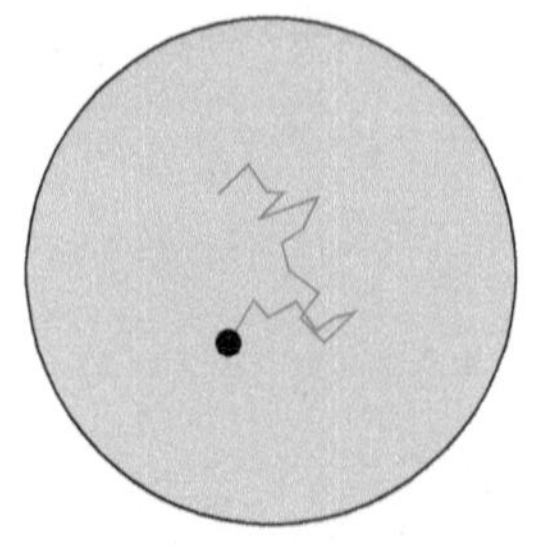

On a microscopic scale, molecules are constantly jostling around and colliding with one another, an atomic-level motion we'll later associate with thermal energy. Any one molecule moves only a short distance before a collision sends it off in a different direction. Its trajectory, if we could see it, might look something like FIGURE 1.1: lots of short, straight segments of apparently random lengths in what appear to be random directions.

This is a chaotic and complex motion. The essence of modeling is to make simplifying assumptions, so how might we begin? The photographs in FIGURE 1.2 provide a clue. The left photo shows blue dye carefully deposited as a thin layer in a test tube of agar gel. The right photo is the same test tube one week later. It's a slow process, but the dye has diffused both up and down in a symmetrical pattern. This suggests that we start not with the complex three-dimensional motion of a molecule but with the simpler case of diffusion along a one-dimensional linear axis.

FIGURE 1.2 Vertical diffusion of blue dye.

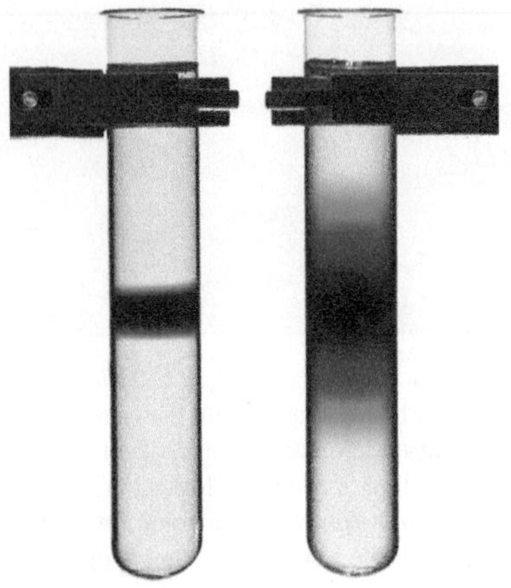

A sidewalk is a linear axis. Suppose you stand at one location on an east-west sidewalk and flip a coin. If it's heads, you take one step to the east; if tails, one step to the west. Then you toss the coin again, and again, and again, each time randomly taking a step to the east or to the west. You would be engaged in what physicists and mathematicians call a one-dimensional **random walk**. Because collisions randomly redirect molecules, let's see what we can learn by modeling molecular motion as a random walk.

NOTE ▸ Variations on the random-walk model are used in applications ranging from protein folding and genetic drift to predicting share prices on the stock market. ◂

In particular, let's imagine a molecule that starts at the origin, $x = 0$, and then takes a random walk along the x-axis. That is, the molecule, at regular intervals, randomly takes a step whose length we'll call d in either the $+x$-direction or the $-x$-direction. At each step there's a 50% chance of going either way. To help make this clear, FIGURE 1.3 shows what might be the first 10 steps of the molecule. The first two steps are to the right, the third back to the left, and so on. The molecule seems to wander aimlessly—that's the essence of random motion—and after 10 steps its position is $x_{10} = -2d$.

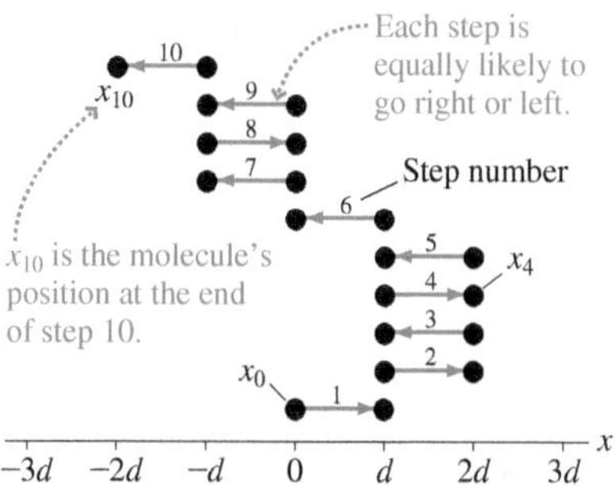

FIGURE 1.3 The first 10 steps of one possible random walk.

Now diffusion involves not one molecule but many, so imagine that we have a very large number of molecules that each move in one dimension along the x-axis. Assume that each molecule starts at the origin and then undergoes a random walk. What can we say about the collective behavior of a large number of random-walking molecules?

This is a problem that can be worked out exactly by using statistics, but the mathematical manipulations are a bit tricky. Instead, we'll explore the model as a computer simulation, one that you could do yourself in a spreadsheet. Suppose you put a zero in a spreadsheet cell to show a molecule's starting position. In the cell to the right, you use the spreadsheet's random-number generator—a digital coin flip—to either add d to the initial position (a step to the right) or subtract d from the initial position (a step to the left). Then you do the same thing in the next cell to the right, and then the next cell to the right of that, and so on, each time adding or subtracting d from the previous cell with a 50% chance of each. The 101st cell will be the molecule's position after taking 100 random steps.

Now you can add a second row of cells for a second molecule, a third row for a third molecule, and so on. It might take a big spreadsheet, but you can have as many molecules and as many steps as you wish. We've used exactly this procedure to simulate the random walks of 1000 molecules for 100 steps each. FIGURE 1.4 gives the result by using dots to show the final positions of each of these molecules. Note that all molecules must be at an even multiple of d after an even number of steps, so the possible positions after 100 steps are 0, $\pm 2d$, $\pm 4d$, and so on. You can see that the first molecule—the top row—ends up at $x_{100} = 4d$ while the 453rd molecule in the 453rd row ends at $x_{100} = 24d$.

FIGURE 1.4 The result of 1000 molecules following a random walk for 100 steps.

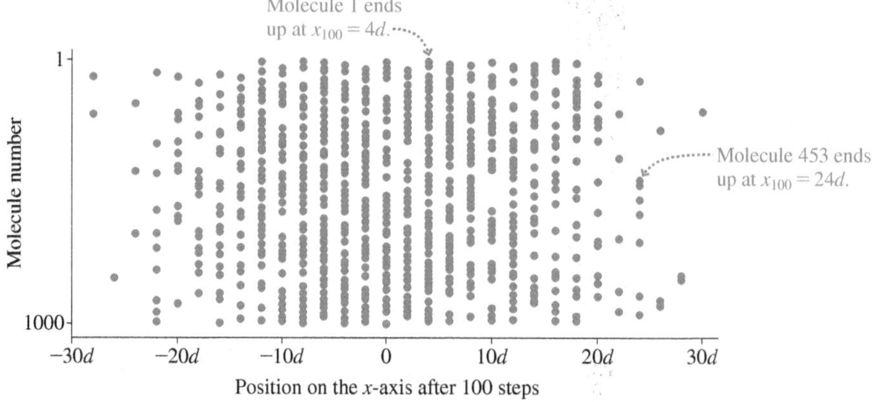

The motion may be random, but the collective motion of many molecules reveals a pattern. Figure 1.4 shows that roughly half the molecules end up to the right of the origin and half to the left. That's exactly what we would expect from randomly taking steps to the right or left. Mathematically, we can say that the *average position* of all 1000 molecules is $x_{avg} = 0$. That is, the center of this array of dots is at the origin. You can see this in the second photo of Figure 1.2: The dye has spread, but it has done so symmetrically so that the center of the dye, its average position, is still in the middle of the test tube.

The center may not have moved, but the molecules are *spreading out,* which is exactly what diffusion is. However, the spreading is perhaps not as much as you might have expected. After 100 steps, a molecule could have reached $x = 100d$, but you can see from Figure 1.4 that, in fact, most are less than $10d$ from the origin. Even the most adventuresome molecule—one in a thousand—has moved out a distance of only $30d$. A moment's thought tells why. Any molecule's position after 100 steps is determined by 100 coin tosses. Although you don't expect to get exactly 50 heads and 50 tails, you do expect the number of heads to be fairly similar to the number of tails. In the same way, a molecule won't move very far from the origin if the number of steps it takes to the right is fairly similar to the number it takes to the left. Reaching $x = 100d$ would require tossing 100 heads in a row, an outcome that is extraordinarily unlikely.

FIGURE 1.5 A histogram of the 1000 molecular positions in Figure 1.4.

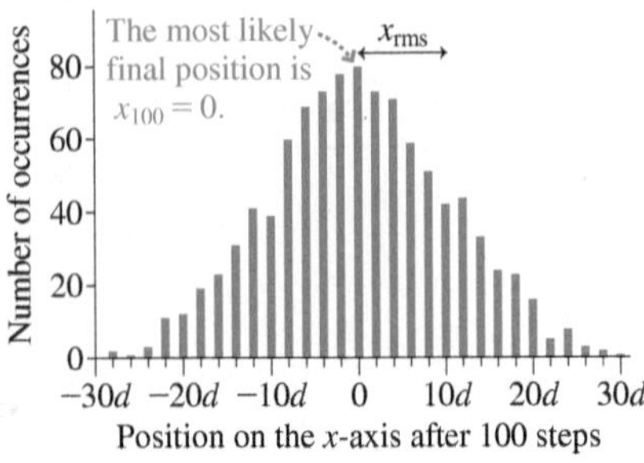

We can analyze the 1000 molecular positions in Figure 1.4 by creating the *histogram* of FIGURE 1.5. A histogram is a bar chart that shows how many molecules end up at each position. The histogram is a *bell curve,* showing that $x_{100} = 0$ is the most likely final position, occurring 80 times out of the 1000 molecules, and that final positions becomes less likely as you move away from the origin. The computer-simulation bell curve is a bit ragged with only 1000 molecules, but, as you can imagine, it would become a very smooth curve if we could run the simulation with billions of molecules.

Measuring the Spread Due to Diffusion

Diffusing molecules spread out, and that's exactly what the molecules in our model are doing. How can we measure this? What quantity describes the amount of spreading? Averaging the positions of all the molecules results in zero, as we've seen, so the average position doesn't tell us anything about spreading. You might think of averaging the *absolute values* of the positions; those are all positive, and their average—which is not zero—really would give some information about spreading. However, it turns out that working with absolute values presents mathematical difficulties.

FIGURE 1.6 A graph showing how x^2 changes for one molecule and, on average, for 1000 molecules during the first 100 steps of a random walk.

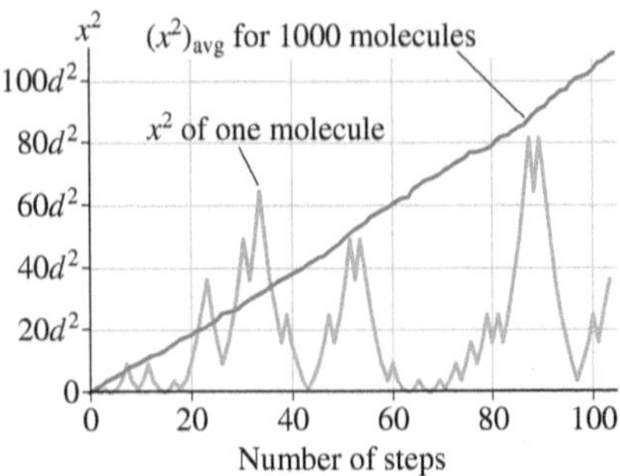

Rather than averaging absolute values, let's average the *squares* of the positions of each of the molecules. That is, we square the positions of all 1000 molecules, add the squares, and then divide by 1000 to find the average. We'll use the notation $(x^2)_{avg}$ to indicate this average of the squares of the positions. Squares are positive, so $(x^2)_{avg}$ is not zero. But is it useful? To find out, FIGURE 1.6 graphs x^2 during the 100 steps of our computer simulation both for one molecule and for $(x^2)_{avg}$ when averaged over all 1000 molecules.

The graph for one molecule is always positive, as it must be, but not very useful. It looks like the molecule managed to walk out to $x = \pm 8d$ in 33 steps, making $x^2 = 64d^2$, then returned to the origin after 43 steps, got as far as $x = \pm 9d$ after 87 and 89 steps, but ended back at $x = \pm 6d$. All in all, it's pretty much what you might expect for one molecule engaged in a random walk.

But *averaging* the squares of the positions of a large number of molecules tells a different story. You can see that $(x^2)_{avg}$ increases linearly with the number of steps, and the average becomes a better and better straight line if we increase the number of molecules in the average. Furthermore, it appears that $(x^2)_{avg} \approx 50d^2$ after 50 steps and $(x^2)_{avg} \approx 100d^2$ after 100 steps. That is, our computer simulation suggests that $(x^2)_{avg} = nd^2$ after n steps. And, indeed, a rigorous statistical analysis proves that this is true.

The square root of $(x^2)_{avg}$ is called the **root-mean-square distance** x_{rms}, often called the *rms distance*. Taking the square root of $(x^2)_{avg} = nd^2$ gives

$$x_{rms} = \sqrt{(x^2)_{avg}} = \sqrt{n}\,d \tag{1.1}$$

The root mean square—the square *root* of the *mean* (i.e., the average) of the *squares*—is a useful way of describing the spread of a set of values that are symmetrical about zero. We'll revisit this concept at several points throughout this text. For our random-walking molecules, the root-mean-square distance x_{rms} after n steps is simply the step size d multiplied by the square root of n. Thus $x_{rms} = \sqrt{16}\,d = 4d$ after 16 steps and $x_{rms} = \sqrt{100}\,d = 10d$ after 100 steps, as our simulations showed.

The root-mean-square distance is shown on the histogram of Figure 1.5, and you can see that x_{rms} is a reasonably good answer to the question: What is the typical distance that the molecules have spread after n steps?

It can be shown, using the tools of probability and statistics, that x_{rms} is the *standard deviation* of the histogram, a term you may have encountered in statistics. One can show that, on average, 68% of all the molecules have positions between $-x_{rms}$ and $+x_{rms}$, while only 5% have traveled farther than $2x_{rms}$. Of our 1000 molecules that took 100 steps and have $x_{rms} = 10d$, we would expect 680 to have traveled no farther than $10d$ from the origin and only 50 to have exceeded $20d$. This is a good description of the results shown in Figures 1.4 and 1.5. Thus x_{rms} appears to be quite a good indicator of about how much spreading has occurred after n steps. We can say that x_{rms} is the *diffusion distance*.

EXAMPLE 1.1 Diffusive spreading

What is the diffusion distance as a fraction of the maximum possible travel distance after 10^2, 10^6, and 10^{12} steps?

PREPARE The maximum possible travel distance increases linearly with the number of steps n, but the diffusion distance increases more slowly, with the square root of n.

SOLVE If a molecule moves in the same direction for every step, it will travel distance $x_{max} = nd$ in n steps. This is the maximum possible travel distance. The diffusion distance is $x_{diff} = x_{rms} = \sqrt{n}d$. Thus

$$\frac{x_{diff}}{x_{max}} = \frac{\sqrt{n}d}{nd} = \frac{1}{\sqrt{n}}$$

For the different values of n we can compute

$$\frac{x_{diff}}{x_{max}} = \begin{cases} 10^{-1} & \text{for } n = 10^2 \\ 10^{-3} & \text{for } n = 10^6 \\ 10^{-6} & \text{for } n = 10^{12} \end{cases}$$

ASSESS Compared to how far a molecule *could* travel, the diffusion distance rapidly decreases as the number of steps increases. Diffusion is 10% of the maximum distance after 100 steps but only one-millionth of the maximum distance after 10^{12} steps, a number that is comparable to the number of collisions a molecule undergoes each second. Having lots of collisions does *not* mean rapid spreading.

NOTE ▶ Worked examples in this text will follow a three-part problem-solving strategy: Prepare, Solve, and Assess. These examples are intended to illustrate good problem-solving procedures, and we strongly encourage you to use the same approach in your own work. We'll look at this problem-solving strategy in more detail in Chapter 3. ◀

Connecting to the Real World

Our analysis of the random walk has revealed some interesting similarities to what is known about diffusion, but how can we judge whether this model is an accurate description of diffusion? What quantitative prediction can we make to connect this model to what happens in the physical world?

Two things that we can easily measure are distance and time, so let's work out the relationship between the diffusion distance and the length of time that the system spends diffusing. We can then compare measurements to the predictions of this model to see how good the model is. First, let's represent the time interval between steps of the random walk by the symbol Δt, where Δ is an uppercase Greek delta. As you'll see in Chapter 2, we use the symbol t to represent a specific instant of time and Δt to represent an *interval* of time. If the molecules start moving at time $t = 0$, then at a later time t they will have taken $n = t/\Delta t$ steps. If, for example, a step is taken every $\Delta t = \frac{1}{10}$ s, there will have been 100 steps at $t = 10$ s. We can write x_{rms} in terms of t and Δt as

$$x_{rms} = \sqrt{n}d = \sqrt{\frac{t}{\Delta t}}\,d = \sqrt{\frac{td^2}{\Delta t}} = \sqrt{vdt}$$

For the last step we used the fact that $d/\Delta t$ = distance/time, as in miles per hour or meters per second, is the molecule's *speed*. Throughout this text we'll use the symbol v (from velocity) to represent speed, with $v = d/\Delta t$.

Let's define the **diffusion coefficient** D to be

$$D = \frac{1}{2}\frac{d^2}{\Delta t} = \frac{1}{2}vd \tag{1.2}$$

The diffusion coefficient connects our model to the physical world because it depends on the speed v of molecules and the distance d traveled between collisions. A factor of $\frac{1}{2}$ is included in the definition to avoid a factor of 2 in an important equation that you'll meet in Chapter 13. Thus our model predicts that the root-mean-square distance should increase with time as

$$x_{\text{rms}} = \sqrt{2Dt} \tag{1.3}$$

The real world is three dimensional, not a straight line. That turns out to be an easy extension of our model. Suppose each step is $\pm d$ along the x-axis *and* $\pm d$ along the y-axis *and* $\pm d$ along the z-axis. In this 3D world, a molecule's distance from the starting point is $r = \sqrt{x^2 + y^2 + z^2}$. Averaging the squares, as we did above, gives

$$(r^2)_{\text{avg}} = (x^2)_{\text{avg}} + (y^2)_{\text{avg}} + (z^2)_{\text{avg}}$$

The x-axis does not differ from the y-axis or the z-axis, so *averages* should be the same along each axis. Thus

$$(r^2)_{\text{avg}} = (x^2)_{\text{avg}} + (y^2)_{\text{avg}} + (z^2)_{\text{avg}} = 2Dt + 2Dt + 2Dt = 6Dt \tag{1.4}$$

If we define the three-dimensional root-mean-square distance as $r_{\text{rms}} = \sqrt{(r^2)_{\text{avg}}}$, then

$$r_{\text{rms}} = \sqrt{6Dt} \tag{1.5}$$

This is our model's prediction for how far molecules will diffuse in three dimensions in time t.

Equation 1.5 is a quantitative, testable prediction of our model. It says that the distance molecules diffuse should increase with the *square root* of the elapsed time. This is not what your intuition suggests. If you walk or run or drive at a steady speed, you go twice as far in twice the time. But because $\sqrt{2} = 1.41$, our prediction is that molecules will diffuse only 1.41 times as far in 20 s as they do in 10 s. It will take 40 s to diffuse twice as far as in 10 s. In fact, this prediction is confirmed by countless experiments.

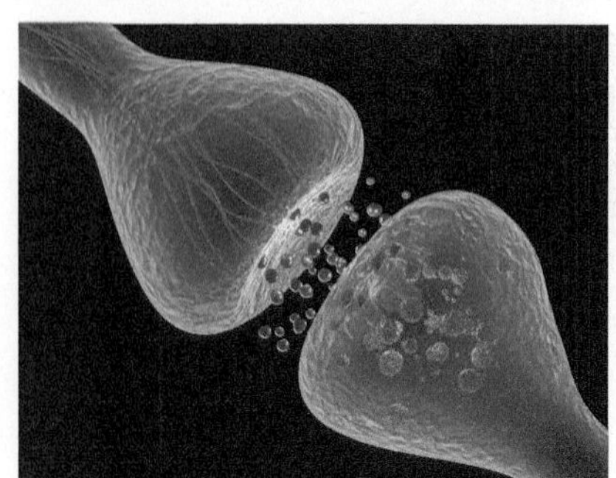

Crossing the gap Nerve impulses traveling from neuron to neuron have to cross the *synaptic cleft* between the axon terminal of one neuron and the dendrite receptor of the next. When the sending terminal on the axon is electrically activated, it releases chemicals called *neurotransmitters* that simply diffuse across the gap. The gap is so narrow, approximately 20 nm in width, that the diffusing neurotransmitters reach the other side and stimulate a response in less than a microsecond.

To really test our model, we would also like to predict a numerical value for the diffusion coefficient D. What do we know that will help us at least estimate D? *Estimating* is a valuable scientific skill, one we'll spend a fair amount of time on in this text, so let's look at how to approach an estimation.

Our random-walk model is a simplification of the idea that molecules are constantly colliding and changing direction, so let's think about what is known about molecular motions and collisions. For one thing, we can infer that molecules in a liquid are pretty much touching each other. Liquids, unlike gases, are essentially *incompressible*—the molecules can't get any closer together. The distance a molecule can move between collisions is *at most* the diameter of a molecule, probably less. You learned in chemistry that the size of small molecules such as oxygen and water is about 0.1 nm, where $1 \text{ nm} = 1 \text{ nanometer} = 10^{-9} \text{ m}$. So let's estimate that the random-walk step size, the distance between collisions, is $d \approx 0.05 \text{ nm} = 5 \times 10^{-11} \text{ m}$.

How fast do molecules move? We can estimate molecular speeds by thinking about sound waves. Sound travels through a medium because the molecules run into each other, propagating a sound wave forward. We'll look at the details when we study sound, but it seems reasonable that the sound speed is probably fairly similar to the speed of a molecule. We could look up the speed of sound in a reference book or on the Internet, but we can also estimate it from an experience that many of you have had. You may have learned that the distance to a lightning strike can be determined by counting the seconds from seeing the lightning to hearing the thunder, then dividing by 5 to get the distance in miles. The implication is that the speed of sound is about

$\frac{1}{5}$ mi/s. A mile is about 1600 m, so $\frac{1}{5}$ mi/s = 320 m/s. Consequently, our second estimate is that the speed with which molecules travel between collisions is $v \approx 320$ m/s.

We have assumed that molecular speeds in liquids are not greatly different from molecular speeds in air. You'll learn when we study waves that the speed of sound in water is about four times the speed in air. That's a big difference for some applications, such as whether ultrasound imaging works better in air or water, but it's not a large discrepancy when our goal is merely a rough estimate of the diffusion coefficient.

With that, we can now predict that the diffusion coefficient for small molecules diffusing through a liquid like water should be approximately

$$D = \tfrac{1}{2}vd \approx \tfrac{1}{2}(320 \text{ m/s})(5.0 \times 10^{-11} \text{ m}) \approx 8 \times 10^{-9} \text{ m}^2\text{/s} \qquad (1.6)$$

Notice the rather unusual units for a diffusion coefficient; these arise because of the square-root relationship between diffusion distance and time.

Also notice that the value is given to only one significant figure because the numbers that went into this calculation are only estimates; they are not precisely known. This is called an **order-of-magnitude estimate,** meaning that we wouldn't be surprised to be off by a factor of three or four but that our estimate is almost certainly within a factor of ten of the real value. That may seem a poor outcome, but initially we had no idea what the value of a diffusion coefficient might be. To have determined by simple reasoning that D is approximately 10^{-8} m²/s (rounding the value of D in Equation 1.6 to the nearest power of 10), rather than 10^{-10} m²/s or 10^{-6} m²/s, is an important increase in our understanding of diffusion, one that can be improved only by careful laboratory measurements.

EXAMPLE 1.2 **Diffusion times**

Estimate how long it takes small molecules such as oxygen to diffuse 10 μm (about the diameter of a cell) and 10 cm (roughly the size of a small animal) through water, the primary component of intercellular fluid. Note that 1 μm = 1 micrometer or 1 micron = 10^{-6} m.

PREPARE The root-mean-square distance r_{rms} tells us how far diffusion carries the molecules in time t. We'll use our estimated $D \approx 8 \times 10^{-9}$ m²/s as the diffusion coefficient.

SOLVE The root-mean-square distance is $r_{rms} = \sqrt{6Dt}$. Solving for the time needed for molecules to diffuse distance r_{rms} gives

$$t = \frac{(r_{rms})^2}{6D}$$

Diffusing across a cell requires $r_{rms} = 10\ \mu\text{m} = 1 \times 10^{-5}$ m. Thus

$$t_{cell} = \frac{(1.0 \times 10^{-5} \text{ m})^2}{6(8 \times 10^{-9} \text{ m}^2\text{/s})} \approx 2 \times 10^{-3} \text{ s} = 2 \text{ ms}$$

But diffusing through a small animal requires $r_{rms} = 10$ cm = 0.1 m. In this case,

$$t_{cell} = \frac{(0.1 \text{ m})^2}{6(8 \times 10^{-9} \text{ m}^2\text{/s})} \approx 2 \times 10^5 \text{ s} \approx 2 \text{ days}$$

ASSESS Diffusion of oxygen across a cellular distance of a few μm is very rapid. Oxygen in your blood is easily transported throughout cells by diffusion as long as no cell is more than a couple of cells away from a capillary. Unicellular organisms such as amoebas can use diffusion to passively harvest the oxygen from their environment. But diffusion across larger distances is impractically slow. A small animal, or one of your organs, is roughly 10^4 times larger than a cell. Because diffusion times depend on the *square* of the distance, diffusion across 10 cm takes 10^8 times as long as diffusion across a cell. The second photo in Figure 1.2, the dye diffusion, was taken *a week* after the experiment started, and the diffusion distance is only a few centimeters.

We should feel pretty satisfied. Starting with a very simplified idea about what happens in molecular collisions and using just a few basic ideas about the physical world, we've put together a model of diffusion that has strong explanatory power. Our estimated value of D for small molecules turns out to be about a factor of four too large, as you'll see in Chapter 13, but, as we might say colloquially, it's certainly in the ballpark.

Furthermore, this result tells us why an amoeba doesn't have lungs. A creature of its size doesn't need them! But you do. Diffusion is inadequate to transport oxygen and other molecules throughout an organism of your size, so you need an actively powered (i.e., using metabolic energy) circulatory system. The laws of physics have real consequences for organisms.

Our goal in this section was to demonstrate how physicists model a complex situation and how physics can increase your understanding of biological systems. The mathematics we used was not especially complex, but we certainly emphasized quantitative reasoning. Graphs were also a useful tool to shape our thinking. These are the types of reasoning skills that we will help you develop throughout this textbook.

STOP TO THINK 1.1 Our estimate of the diffusion coefficient D is for the diffusion of small molecules through water. The same type of reasoning predicts that the diffusion coefficient for diffusion through air is _______ the diffusion coefficient for diffusion through water.

A. Smaller than
B. About the same as
C. Larger than

NOTE ► Each chapter will have several *Stop to Think* questions. They are designed to see if you have processed and understood what you've read. The answers are given at the end of the chapter, but you should make a serious effort to think about these questions before turning to the answers. They are an important part of learning. A wrong answer should spur you to review the section before going on. ◄

FIGURE 1.7 Comparing a shrew and an elephant.

1.4 Proportional Reasoning: Scaling Laws in Biology

The smallest terrestrial mammal is the Etruscan shrew, with a length of 4 cm and mass of a mere 2 g. The largest is the 7-m-long, 6000 kg African elephant. If we were to enlarge a shrew by a factor of 175, giving it a 7 m length, would it look pretty much like a furry elephant?

We can't carry out the experiment, but we can simulate it with photographs. FIGURE 1.7 shows a shrew and an elephant with, at least photographically, the same body length. We could say that the shrew has been *scaled* to the size of the elephant. In some regards, the answer to our question is Yes. The scaled-up shrew is a bit chubbier than the elephant, but the height, the ear size, and other features are all about the same. But what's with the legs? Compared to the elephant, the legs of the scaled-up shrew are twigs rather than tree trunks.

This is not an accident; there are underlying physical reasons why large animals are not just scaled-up versions of small animals. It all has to do with **scaling laws,** which are regularities in how the physical characteristics of organisms—from bone size to heart rate—depend on the organism's size.

Proportionality

FIGURE 1.8 A graph showing linear proportionality.

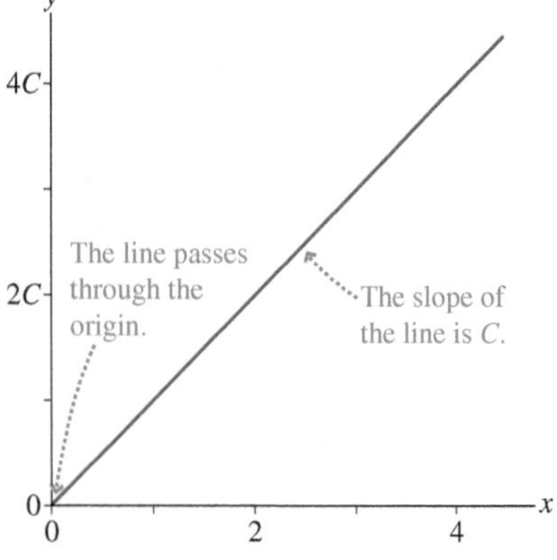

You often hear it said that one quantity is *proportional* to another. What does this mean? We say that y is **linearly proportional** to x if

$$y = Cx \tag{1.7}$$

where C is a constant called the **proportionality constant.** This is sometimes written $y \propto x$, where the symbol $\propto$ means "is proportional to." As FIGURE 1.8 shows, a graph of y versus x is a straight line *passing through the origin* with slope C.

NOTE ► A graph of "a versus b" means that a is graphed on the vertical axis and b on the horizontal axis. Saying "graph a versus b" is a shorthand way of saying "graph a as a function of b," indicating that b is the independent variable on the horizontal axis. ◄

For example, the mass of an object is related to its volume by $m = \rho V$, where ρ (the Greek letter rho—we use a lot of Greek letters in physics so as to have enough symbols for our needs) is the object's density. Thus mass is linearly proportional to volume with, in this case, density being the proportionality constant.

Proportionality allows us to use ratios to draw conclusions without needing to know the proportionality constant. Suppose x has an initial value x_1, and thus y has the initial value $y_1 = Cx_1$. If we change x to x_2, then y changes to $y_2 = Cx_2$. The ratio of y_2 to y_1 is

$$\frac{y_2}{y_1} = \frac{Cx_2}{Cx_1} = \frac{x_2}{x_1} \tag{1.8}$$

That is, the ratio of y_2 to y_1 is exactly the same as the ratio of x_2 to x_1, meaning that x and y scale up or down by the *same factor*. If you double an object's volume, its mass doubles. If you decrease an object's volume by a factor of three, its mass decreases by a factor of three. This type of reasoning is called **ratio reasoning.**

NOTE ▶ Linear proportionality is more specific than a *linear relationship*. The linear relationship $y = Cx + B$ also has a straight-line graph, but the graph has a y-intercept B; the line does not pass through the origin. Ratio reasoning does *not* work with a linear relationship because the constants don't cancel. A linear proportionality is a special case of a linear relationship whose graph passes through the origin ($B = 0$). ◀

Now consider the surface area of a sphere: $A = 4\pi r^2$. This is also a proportionality, but in this case we would say, "The area is proportional to the *square* of the radius" or $A \propto r^2$. Proportionalities don't have to be linear (i.e., dependent on the first power of x); any relationship in which one quantity is a constant times another quantity is a proportionality. And ratio reasoning works with any proportionality because the constant cancels.

EXAMPLE 1.3 The surface area of a cell

A spherical type A cell has a surface area of 2 μm^2. The diameter of a type B cell is three times that of a type A cell. What is the surface area of a type B cell?

PREPARE We could use the surface-area formula to find the radius of a type A cell, multiple it by 3, and then calculate the surface area of a type B cell. That's a lot of calculation, though, with the possibility for making mistakes. Instead, let's use ratio reasoning.

SOLVE The ratio of the surface areas is

$$\frac{A_B}{A_A} = \frac{4\pi r_B^2}{4\pi r_A^2} = \left(\frac{r_B}{r_A}\right)^2$$

The proportionality constant 4π cancels, and we find that the ratio of the areas is the *square* of the ratio of the radii. If $r_B/r_A = 3$, then $A_B = 3^2 \times A_A = 9A_A = 18\ \mu\text{m}^2$.

ASSESS This is what we mean by *reasoning* with math rather than seeing math as simply a calculation tool. With practice, this is a problem you can solve in your head! "Let's see, area is proportional to the square of the radius. If the radius increases by a factor of 3, the area will increase by a factor of $3^2 = 9$."

EXAMPLE 1.4 Time to work

For a person or object that moves at a steady speed, the time needed to travel a specified distance is *inversely* proportional to the speed; that is, $\Delta t \propto 1/v$, where v is the symbol for speed. Angela gets to work in 15 minutes if she walks at a steady speed. How long will it take if she rides her bicycle and cycles three times as fast as she walks?

PREPARE We don't know either the distance to work or Angela's walking speed, but we don't need either if we use ratio reasoning.

SOLVE Because $\Delta t = C/v$, with C being some unknown constant, we have

$$\frac{\Delta t_{\text{cycle}}}{\Delta t_{\text{walk}}} = \frac{C/v_{\text{cycle}}}{C/v_{\text{walk}}} = \frac{v_{\text{walk}}}{v_{\text{cycle}}}$$

The ratio of the times is the *inverse* of the ratio of the speeds. We know that $v_{\text{cycle}} = 3v_{\text{walk}}$, so

$$\frac{\Delta t_{\text{cycle}}}{\Delta t_{\text{walk}}} = \frac{v_{\text{walk}}}{3v_{\text{walk}}} = \frac{1}{3}$$

Thus $\Delta t_{\text{cycle}} = \frac{1}{3}\Delta t_{\text{walk}} = 5$ min.

ASSESS It makes sense that it takes one-third as long if Angela goes three times as fast.

Proportionality helps us understand why the elephant's legs are much thicker than the legs of the scaled-up shrew. An object's volume V depends on the *cube* of some linear dimension l, such as a radius or an edge length. The exact formula for volume depends on the specific shape, but for any object—including animals—$V \propto l^3$. For a given density, mass is proportional to volume and, because all mammals have about the same density, mass also obeys $M \propto l^3$. We say that mass *scales* with the cube of the linear dimension. If we scale up an animal by a factor of five, quintupling the size of the animal in all three dimensions, its mass increases by a factor of $5^3 = 125$.

An animal's legs have to support its weight. An animal that's 125 times more massive needs bones that are 125 times stronger so as not to break under the load. You will see in Chapter 6 that the strength of bones is proportional to their cross-section area, but that's a problem for large animals. Areas, whether those of circles, squares, or any other shape, scale as the square of the linear dimension: $A \propto l^2$. In scaling up an animal by a factor of five, areas and bone strength increase by only a factor of $5^2 = 25$. In other words, bone strength is not keeping up with the animal's increase in weight. And to get from a shrew to an elephant we have to scale not by a factor of five but by a factor of 175!

An animal's mass scales as l^3, where l is its linear dimension, but its bone strength scales as only l^2. The weight of an animal that is simply scaled up would quickly outstrip its leg bones' ability to support that weight, so a simple scaling up in size doesn't work. To support the weight, the cross-section area of the bones must also scale as l^3. This means that the bone diameter has to scale as $l^{3/2}$, faster than the rate at which the length of the animal increases. That's why the diameter of the elephant's tree-trunk-like legs is a much larger fraction of its body length than the twig-like legs of the shrew.

Allometry

It may seem unlikely that there are mathematically expressible "laws of biology." The great core principles of biology—such as evolution, structure-function relationships, and the connections between an organism and its genetic information—don't readily lend themselves to being expressed in equations. But consider FIGURE 1.9, a graph of *basal metabolic rate* (the rate at which a resting animal uses energy, abbreviated BMR) versus mass m for a large number of mammals, ranging from the smallest (the 2 g Etruscan shrew) to the largest (a 1.5×10^8 g blue whale). Recall

FIGURE 1.9 Basal metabolic rate as a function of mass for mammals.

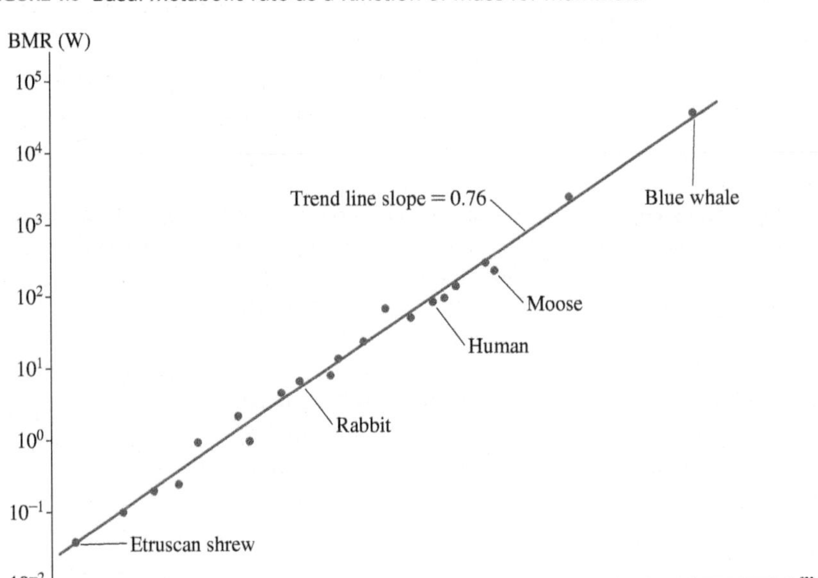

that "BMR versus mass" graphs BMR on the vertical axis and mass—the independent variable—on the horizontal axis.

NOTE ▶ Notice that graph axis labels tell us the units that are being used. Mass is measured in grams (g), while BMR is measured in watts (W). Metabolic rate is energy used per second, and the watt—just as used with your appliances to show the rate of energy use—is the metric unit. ◀

The data points fall just about perfectly along a straight line, suggesting that there *is* some kind of law relating a mammal's BMR to its mass. The mass of the blue whale is nearly 10^8 times larger than the mass of the shrew, so this apparent law holds over an extremely wide range of masses. When two numbers differ by a factor of ≈ 10, we say that they differ by an **order of magnitude.** Thus the range of masses spans eight orders of magnitude. Similarly, the BMR values shown on the vertical axis span about six orders of magnitude.

But you may have noticed that Figure 1.9 is not the type of graph you usually work with. You're familiar with graphs where the tick marks on the axis represent constant *intervals*. That is, the tick marks might be labeled 0, 1, 2, 3, . . . with an interval of 1. Or 0, 20, 40, 60, . . . with an interval of 20. But the horizontal-axis tick marks in Figure 1.9 are labeled 10^0, 10^1, 10^2, 10^3, ... That is, each tick mark is a constant *factor* of 10 larger than the preceding tick mark. The same is true on the vertical axis.

Figure 1.9, called a **log-log graph,** graphs not BMR versus mass but the logarithm of BMR versus the logarithm of mass—that is, log(BMR) versus log(m). Log-log graphs are widely used in science, especially—as here—when the data span many orders of magnitude. Data that fall on a straight line on a log-log graph are said to follow a *scaling law*. We need to review some properties of logarithms to see what this means.

Any positive number a can be written as a power of 10 in the form $a = 10^b$. Simple cases are $1 = 10^0$, $10 = 10^1$, and $100 = 10^2$. You can use your calculator to find that $\sqrt{10} = 3.16$, so $3.16 = 10^{1/2} = 10^{0.50}$. We define the *logarithm* of a as follows:

$$\text{If } a = 10^b \text{ then } \log(a) = b \tag{1.9}$$

Thus $\log(1) = 0$, $\log(10) = 1$, $\log(100) = 2$, and $\log(3.16) = 0.50$. You can use your calculator to find that $\log(25) = 1.40$, which means that $25 = 10^{1.40}$.

NOTE ▶ We're using *base-10 logarithms,* also called *common logarithms*. You may also be familiar with *natural logarithms*. We'll see those later, but data analysis is usually done with common logarithms. ◀

Because $10^b \cdot 10^d = 10^{b+d}$, you should be able to convince yourself that $\log(a \cdot c) = \log(a) + \log(c)$. This is an important property of logarithms. And because a^n is $a \cdot a \cdot a \cdot \cdots \cdot a$, multiplied n times, the logarithm of a^n is $\log(a)$ added n times; that is, $\log(a^n) = n\log(a)$.

Log-log graphs are widely used in science. It's important to know how to read them, but the axis labels can be confusing. FIGURE 1.10 shows a portion of the horizontal axis from Figure 1.9 with two different ways of labeling the axis. The labels above the axis—powers of 10—are what Figure 1.9 shows, but they are *not* what is graphed. What's actually graphed is the logarithms of these powers of 10, which are the labels shown below the axis. That is, the tick mark labeled 10^3 actually represents the value 3.0, the logarithm of the label. So the axis really is increasing 0, 1, 2, 3, . . . , with a constant interval between the tick marks, but the tick marks are labeled, instead, with the powers of 10. Notice how the midpoints between the tick marks are about 3, 30, 300, ... This follows from the fact that their logarithms are about 0.5, 1.5, 2.5, ...

This method of labeling, while potentially confusing until you get used to it, is quite useful. The top labels in Figure 1.10 tell the values of the masses, which is what

FIGURE 1.10 Two ways to label a logarithmic axis.

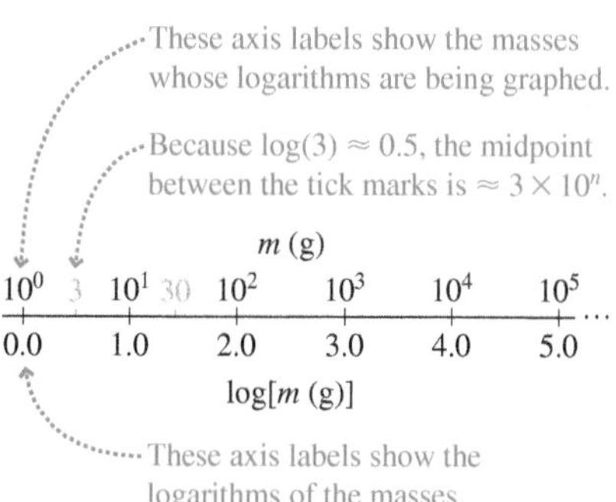
These axis labels show the masses whose logarithms are being graphed.
Because log(3) ≈ 0.5, the midpoint between the tick marks is ≈ 3 × 10ⁿ.
m (g)
10⁰ 3 10¹ 30 10² 10³ 10⁴ 10⁵
0.0 1.0 2.0 3.0 4.0 5.0
log[m (g)]
These axis labels show the logarithms of the masses.

you really want to know. The bottom labels tell only the logarithms of the masses, which are harder to interpret.

Returning to Figure 1.9, we see that the *trend line*—the straight line that best "fits" the data—has a slope of 0.76. Slope, you will recall, is the rise-over-run ratio. But the rise and run of what? Because this is a graph of log(BMR) versus log(m), the slope is measured as the "rise" in log(BMR) divided by the "run" in log(m). That is, the slope is determined from the unseen logarithm labels 0, 1, 2, 3, . . . rather than the power-of-ten labels used in the graph.

Let's now imagine that some quantity Y (the BMR in this example) depends on another quantity X (the mass) via the proportionality

$$Y = CX^r \tag{1.10}$$

where C is a constant. We say that Y scales as the rth power of X.

We take the logarithm of both sides of Equation 1.10 and use the properties of logarithms:

$$\log Y = \log(CX^r) = \log C + \log(X^r) = r\log X + \log C \tag{1.11}$$

To interpret this equation, recall that a linear equation of the form $y = mx + b$ graphs as a straight line with slope m and y-intercept b. By defining $y = \log Y$ and $x = \log X$, we see that Equation 1.11 is actually a linear equation. That is, a graph of $\log Y$ versus $\log X$ is a straight line with slope r. Conversely, and importantly, **if a graph of $\log Y$ versus $\log X$ is a straight line, then Y and X obey a scaling law like Equation 1.10 and the value of the exponent r is given by the slope of the line.**

NOTE ▸ Spreadsheets and other graphing software that you might use to analyze data usually have a "logarithmic axis" option. If you choose that option, the computer calculates and plots the logarithms of the data for you (taking logarithms is not something you have to do yourself) and then labels the axes using the power-of-ten notation shown above the axis in Figure 1.10. Using the logarithmic axis option allows you to look for functional relationships that are scaling laws. ◂

Figure 1.9 graphed the logarithm of the basal metabolic rate against the logarithm of the mass for a large number of mammals. We had no reason to suspect that these two quantities are related in any particular way, but the graph turned out to be a straight line with a slope of almost exactly $\frac{3}{4}$. Consequently, we've discovered a scaling law telling us that a mammal's BMR scales as the $\frac{3}{4}$th power of its mass m; that is, $\text{BMR} = Cm^{3/4}$ or, as we would usually write, $\text{BMR} \propto m^{3/4}$. We could use the graph's y-intercept to determine the value of the constant C, but the constant is usually less interesting than the exponent.

This result has broad generality across all mammals. It is interesting that birds and insects also have metabolic rates that scale as the $\frac{3}{4}$th power of mass. This scaling law really seems to be a law of biology. Why? Scaling laws, especially ones that span many orders of magnitude, tell us that there's some underlying regularity in the physics of the organism.

A simple model relating an organism's metabolic rate to its size is based on the fact that metabolism generates heat and that heat has to be dissipated. Heat is dissipated through an organism's surface via heat convection and radiation, so this model predicts that BMR should be proportional to the organism's surface area. We've seen that surface area scales as $A \propto l^2$, where l is the organism's linear size, and thus BMR should scale with the size of an organism as $\text{BMR} \propto l^2$. The organism's mass is proportional to its volume, so mass scales as $m \propto l^3$. Combining these, by eliminating l, we see that this model predicts that BMR should scale with the organism's mass m as $\text{BMR} \propto m^{2/3}$. That is, a geometric model in which the organism's BMR is determined by how quickly it can dissipate heat through its surface predicts a scaling law, but one in which BMR should scale as the $\frac{2}{3}$rd power of mass.

We found instead that BMR scales as the $\frac{3}{4}$th power of mass. A model must make *testable predictions,* and we must reject a model if its predictions fail. The geometric model of BMR fails; even though it seemed logical, there's something wrong with its assumptions.

The study of scaling laws in biology is called **allometry.** It is the study of how physiological processes scale with body size. It turns out that *quarter-power scaling laws* are everywhere in biology. A log-log graph of heart rate, respiration, life span, blood-vessel diameter, or any number of other physiological features versus mass tends to be a straight line with a slope that is an integer multiple of $\frac{1}{4}$—that is, $\frac{1}{4}$ or $\frac{2}{4}$ or $\frac{3}{4}$. This calls for an explanation, but, as we just saw, the explanation involves not simply an organism's size. Only recently has a better explanation been found. Organisms are composed of *networks* that move energy, metabolites, and information from central reservoirs (heart, lungs, brain) to localized points of use. A network model of organisms—more complex than we can delve into here—successfully predicts quarter-power scaling. The significant conclusion, learned from scaling laws, is that an organism's metabolic rate is determined not by how efficiently it can dissipate heat through its surface but by how efficiently its networks can deliver fuel and oxygen to its cells.

NOTE ▸ We should point out that quarter-power scaling remains somewhat controversial. Some biologists maintain that the $\frac{2}{3}$-power scaling predicted by the surface-area model is a better description of the admittedly imperfect data. This is an active field of biological research. ◂

EXAMPLE 1.5 Heart rates of hamsters and horses

FIGURE 1.11 is a log-log graph of heart rate R versus body mass m for mammals. A linear trend line follows the data well, showing that heart rate scales with body mass.

a. What is the scaling law for heart rate as a function of body mass?

b. A 1200 kg horse has a resting heart rate of 38 beats/min. According to the scaling law, what should be the resting heart rate of an 80 g hamster?

FIGURE 1.11 Heart rate as a function of mass for mammals.

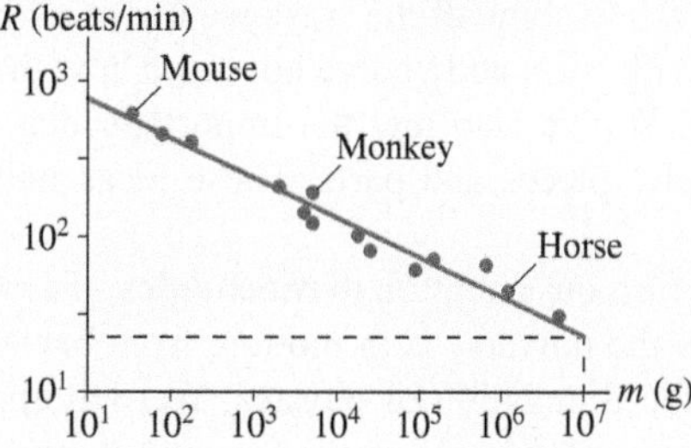

PREPARE The scaling law is $R = Cm^r$, where the exponent r is the slope of the graph. This is a log-log graph, so the actual values of the tick marks along the axes are the logarithms of the mass and heart rate values shown. That is, the tick marks on the horizontal axis actually run from 1 to 7, while those on the vertical axis run from 1 to 3. These are the values we need to measure the slope of the line.

SOLVE

a. We can determine the scaling law by measuring the slope of the trend line. Recall that the slope of a line is "rise over run." This is a descending line, so the rise, and thus the slope, is negative. For the full trend line, the run goes from the first tick mark to the last. The values at these points are not the 10^1 and 10^7 displayed but, instead, the logarithms of these values: 1 and 7. Thus run $= 7.0 - 1.0 = 6.0$. The labeled tick marks on the vertical axis have the actual values 1, 2, and 3, so the trend line starts at 2.8 and ends at 1.3, which gives rise $= 1.3 - 2.8 = -1.5$. Thus the slope is

$$r = \text{slope} = \frac{\text{rise}}{\text{run}} = \frac{-1.5}{6.0} = -0.25 = -\frac{1}{4}$$

The data show that heart rate scales as the $-\frac{1}{4}$th power of body mass or, equivalently, inversely with the $\frac{1}{4}$th power of body mass. Mathematically, $R = C/m^{1/4}$.

b. We could use the graph to determine the value of the proportionality constant C, but it isn't necessary. It's easier to use ratio reasoning to write

$$\frac{R_{\text{hamster}}}{R_{\text{horse}}} = \frac{C/(m_{\text{hamster}})^{1/4}}{C/(m_{\text{horse}})^{1/4}} = \left(\frac{m_{\text{horse}}}{m_{\text{hamster}}}\right)^{1/4}$$

Using the given values of the masses, with the hamster's mass converted to kg, we get

$$R_{\text{hamster}} = R_{\text{horse}}\left(\frac{m_{\text{horse}}}{m_{\text{hamster}}}\right)^{1/4} = (38 \text{ beats/min})\left(\frac{1200 \text{ kg}}{0.080 \text{ kg}}\right)^{1/4}$$

$$= 420 \text{ beats/min}$$

ASSESS Heart rate is another example that follows the quarter-power scaling law. Reported heart rates for hamsters range from 280 to 500 beats/min. A horse's mass is 15,000 times that of a hamster, but a simple scaling law has allowed us to use data for horses to make an accurate prediction of the heart rate of a hamster.

EXAMPLE 1.6 **Heartbeats in a lifetime**

As Example 1.5 showed, the heart rate of mammals scales inversely as the $\frac{1}{4}$th power of body mass. A log-log graph of average lifetime versus mass is a straight line with a slope of $\frac{1}{4}$, so lifetime scales as the $\frac{1}{4}$th power of mass. How does the number of heartbeats in a lifetime scale with mass?

PREPARE We're given two scaling laws: one for heart rate versus mass and one for lifetime versus mass. We need to combine these to see how the number of heartbeats scales with mass.

SOLVE We can symbolize an animal's heart rate by R and its lifetime by T. If R is measured in beats per minute and T in minutes, the number N of heartbeats in a lifetime is

$$N = RT$$

This is exactly the same as calculating the number of miles driven by multiplying the speed in miles per hour by the elapsed time in hours. If $R \propto m^{-1/4}$ and $T \propto m^{1/4}$, then

$$N \propto m^{-1/4} \cdot m^{1/4} = m^0 = 1$$

That is, the number of heartbeats in a lifetime *does not depend on an animal's mass*; it is the *same for all mammals*. It turns out to be $\approx 1.5 \times 10^9$ heartbeats.

ASSESS Small animals have fast heart rates and short lifetimes. Large animals have slow heart rates but long lifetimes—at least when lifetimes are measured in years. But when lifetimes are measured in heartbeats, all mammals have about the same lifetime. For humans, $R \approx 70$ beats/min gives $T \approx 40$ years. Indeed, that was about the average human lifespan prior to the development of modern medicine.

STOP TO THINK 1.2 The strength of the electric force exerted by one ion on another scales inversely as the square of the distance between the ions. Suppose the force between two ions is 1.00 nN (1.00 nanonewton), where (as you'll learn in Chapter 4) the *newton* is the metric unit of force. If the distance between the ions doubles, the force will be

A. 0.25 nN
B. 0.50 nN
C. 1.00 nN
D. 2.00 nN
E. 4.00 nN

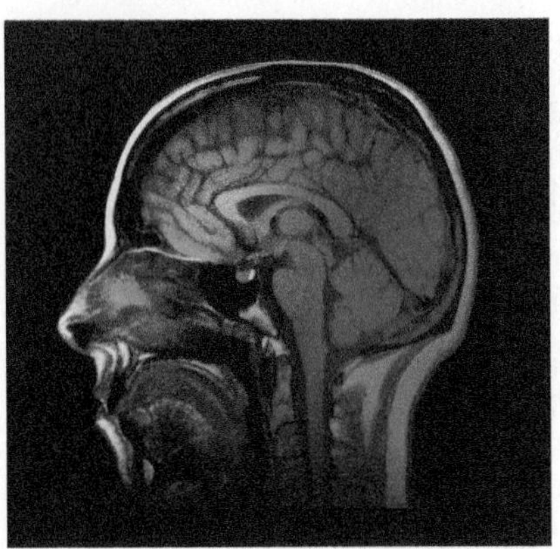

One step at a time Our goal is eventually to be able to say something meaningful about the hydraulics of blood flow, the electricity of nerve conduction, the optics of the eye, and how MRI images like this are made. There are many steps along the way, and we'll explore them one at a time.

1.5 Where Do We Go from Here?

This first chapter has introduced you to some of the ways we use numbers, quantitative reasoning, and scaling laws in physics, and you've now seen how these tools can shed light on biological systems. You've also met the important idea of *modeling* a complex system by using simpler pieces and parts. These ideas will be with us throughout the book.

For the next eight chapters we turn our attention to **mechanics,** the study of force and motion. Almost everything in the universe is in motion, from bacteria to galaxies, so understanding motion is the foundation of physics. We will apply what we learn to topics ranging from the biomechanics of moving organisms to blood flow in the circulatory system. Let's get started.

SUMMARY

GOAL To understand some of the ways that physics and quantitative reasoning shed light on biological processes.

IMPORTANT CONCEPTS

Models

A **model** is a simplified picture of reality, but one that still captures the essence of what we want to study. Models can be

- Descriptive: What are the essential characteristics and properties?
- Explanatory: Why do things happen as they do?

Physicists look for models that make quantitative predictions.

Mathematics

Physics uses math in two different ways:

- To do calculations.
- To reason and analyze by discovering relationships between measurable quantities.

Math is a thinking tool as much as it is a calculation tool.

Diffusion and the Random Walk

Diffusion can be modeled as a **random walk** in which molecules move left or right with equal probability.

The **root-mean-square distance** r_{rms} measures the spreading of diffusion. After time t,

$$r_{rms} = \sqrt{6Dt}$$

where D is the **diffusion coefficient.** The amount of diffusive spreading increases with the square root of the elapsed time.

Scaling and Proportional Reasoning

A quantity y is **linearly proportional** to another quantity x if

$$y = Cx$$

A graph of y versus x is a straight line *passing through the origin* with slope C.

If a quantity Y depends on another quantity X via the proportionality

$$Y = CX^r$$

then we say that Y *scales* as the rth power of X. A graph of log Y versus log X is a straight line with slope r.

LEARNING OBJECTIVES

After studying this chapter, you should be able to:

- Distinguish between descriptive and explanatory models. *Problems 1.6, 1.30–1.33*
- Reason about diffusion using the random-walk model. *Conceptual Questions 1.2–1.4; Problems 1.2, 1.4, 1.6, 1.7*
- Solve proportionality problems using ratio reasoning. *Conceptual Questions 1.5, 1.6; Problems 1.8, 1.9, 1.11, 1.12, 1.16*
- Use and interpret log-log graphs. *Multiple-Choice Question 1.9; Problems 1.13, 1.14*

STOP TO THINK ANSWERS

Stop to Think 1.1: C. The molecular speed v is likely to be about the same, but the step size d is much larger because gas molecules are much farther apart.

Stop to Think 1.2: A. Force F scaling inversely as the square of distance d means that $F = C/d^2$. If d doubles, d^2 increases by a factor of 4. Because d^2 is in the denominator, F is reduced to $\frac{1}{4}$ of its original value.

QUESTIONS

Conceptual Questions

1. If molecules diffuse distance d in 1 s, how far will they diffuse in 4 s?
2. Xenon atoms are heavier and, on average, slower than helium atoms at the same temperature. Do you expect the diffusion coefficient for xenon through air to be larger than, smaller than, or about the same as the diffusion coefficient for helium through air? Explain.
3. Do you expect the diffusion coefficient of a gas through hot water to be larger than, smaller than, or about the same as its diffusion coefficient through cold water? Explain.
4. The elephant trunk snake and the sea snake live in water. Both have lungs and breathe, but both can also exchange gases through their skin. The elephant trunk snake can be up to 2 m long, three times the maximum length of a sea snake. The sea snake exchanges about 35% of its carbon dioxide through its skin, compared to only 10% for the elephant trunk snake. Why might you expect this result?

Problem difficulty is labeled as | (straightforward) to ||||| (challenging). Problems labeled INT integrate significant material from earlier chapters; BIO are of biological or medical interest; CALC require calculus to solve.

5. A spherical Type A cell has volume V. The diameter of a Type B cell is twice that of a Type A cell. What is the volume of a Type B cell?
6. The period of a pendulum is proportional to the square root of the length of the pendulum. Suppose a 1-m-long pendulum has a period of 1 s. What is the length of a pendulum that has a period of 2 s?

Multiple-Choice Questions

7. | The largest source of naturally occurring radiation exposure for humans is radon, a radioactive inert gas that diffuses out of the soil into the atmosphere after being produced by the decay of minute quantities of uranium in the soil. Radon has a short half-life, and it has effectively decayed completely after 20 days. Radon produced deep below the earth's surface can't make it into the atmosphere, but radon in the top layers of soil can. For soil with a radon diffusion coefficient of 1×10^{-6} m^2/s, typical of relatively dry sandy or loamy soil, roughly what is the thickness of the soil layer from which we can expect radon to diffuse into the atmosphere? Note that this problem is best modeled as one-dimensional diffusion.
   A. 6 mm B. 3 cm
   C. 2 m D. 3 m
8. || For mammals, the diameter of the aorta scales with the BIO mass of the animal as $d \propto m^{3/8}$. A 70 kg human has an aorta diameter of about 2 cm. What diameter aorta would you expect for a 5000 kg African elephant?
   A. 0.4 cm B. 5 cm
   C. 10 cm D. 50 cm
9. || The slope of the trend line in the log-log graph of Figure Q1.9 is
   A. 1 B. 2
   C. 3 D. 1000

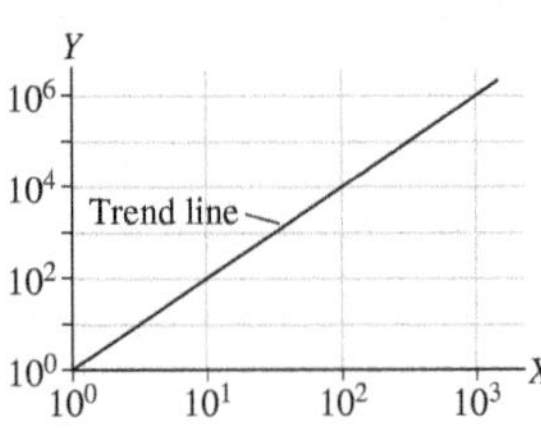

FIGURE Q1.9

10. || If a mass oscillates on a spring, the frequency of oscillation—the number of cycles per second—is inversely proportional to the square root of the mass. Suppose a 200 g mass oscillates with a frequency of 2.0 cycles per second. What would be the frequency of a 400 g mass oscillating on the same spring?
    A. 1.0 cycle/second B. 1.4 cycles/second
    C. 2.8 cycles/second D. 4.0 cycles/second

PROBLEMS

Section 1.3 Case Study: Modeling Diffusion

1. | 1000 random walkers, all starting from $x = 0$, take steps of length d. After two steps, how many walkers do you expect to find at (a) $x = 0$, (b) $x = d$, and (c) $x = 2d$?
2. | A large group of randomly walking molecules has $(x^2)_{avg} = 100\ \mu m^2$ after 10^6 steps.
   a. What is the step size in nm?
   b. What is x_{avg} after these 10^6 steps?
3. || A large group of randomly walking molecules takes 2.0 μm steps along a linear axis every 2.0 ms.
   a. What will the root-mean-square distance be after 20 s?
   b. To diffuse the same distance in 10 s with the same step size, what must the time between steps be?
   c. To diffuse the same distance in 10 s with the same time between steps, what must the step size be?
4. || A random walker takes 50 cm steps along a linear axis every 1.0 s. On average, how long will it take the walker to get (a) 5.0 m and (b) 50 m from her starting point?
5. | It takes a sample of molecules 10 s to diffuse 1.0 mm. How long will it take to diffuse 2.0 mm?
6. || You'll learn in Chapter 13 that the *mean free path* (the average distance between collisions) of small molecules in air is about 250 nm.
   a. Estimate the diffusion coefficient for a small molecule in air.
   b. About how long does it take molecules to diffuse 3.0 m in still air?
   c. If you open a bottle of a strong-smelling substance, you can soon smell it throughout the room. This is frequently cited as evidence of diffusion. Based on your answer to part b, is this a valid assertion?
7. | The diffusion coefficient for glucose in water is about BIO 7×10^{-10} m^2/s. Diffusion through cell membranes is reported to be similar. Estimate the time it takes glucose molecules to travel by diffusion
   a. 8 nm across a cell membrane.
   b. 1 mm from a photosynthetic cell on the surface of a leaf to a vein (phloem) that can transport them to the rest of the plant.

Section 1.4 Proportional Reasoning: Scaling Laws in Biology

8. | The period T of a pendulum—the time for one complete oscillation—is proportional to the square root of the pendulum's length L. If a 1.0-m-long pendulum has a period of 4.0 s, what is the period of a 2.0-m-long pendulum?
9. | The distance molecules spread by diffusion is proportional to the square root of the time over which spreading occurs. Oxygen molecules diffuse about 10 μm through water in 2 ms. How far do they diffuse in 20 ms?
10. ||| The average speed of molecules is inversely proportional BIO to the square root of their mass. A small molecule such as nitrogen or oxygen has molecular mass ≈ 30 Da $= 30$ u, where the dalton (Da) and the atomic mass unit (u) are two different names for the same thing. A typical protein has molecular mass ≈ 100 kDa.
    a. Estimate the diffusion coefficient of a protein through water. (Because molecules in a liquid are essentially touching, is the distance a protein moves between collisions likely to differ much from our estimate of how far a small molecule moves between collisions?)
    b. About how long does it take a protein to diffuse across a 10-μm-diameter cell?

11. || For mammals, resting heart rate scales as mass to the $-\frac{1}{4}$th
BIO power; that is, $R \propto m^{-1/4}$. The heart rate of a 5000 kg elephant is 30 bpm (beats per minute). What do you predict for the heart rate of a 5.0 kg monkey?

12. || A well-known result from ecology and evolution is the *species-*
BIO *area relationship*, which says that the number of species S living in a confined area A scales as A^z, where the value of the exponent z depends on the nature of the confined area. For oceanic islands, which have a clear boundary and well-defined area, $z \approx 0.33$. Suppose island B is $\frac{1}{10}$th the size of island A. If island A has 300 species, how many species do you predict for island B?

13. || What scaling law is implied by the trend line in Figure P1.13? That is, Y scales as what power of X?

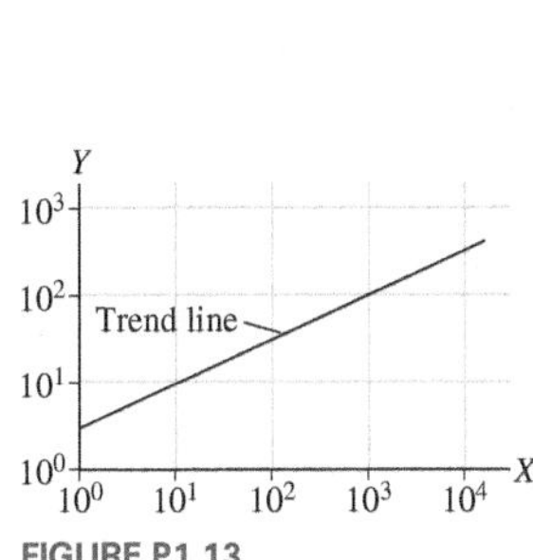

FIGURE P1.13

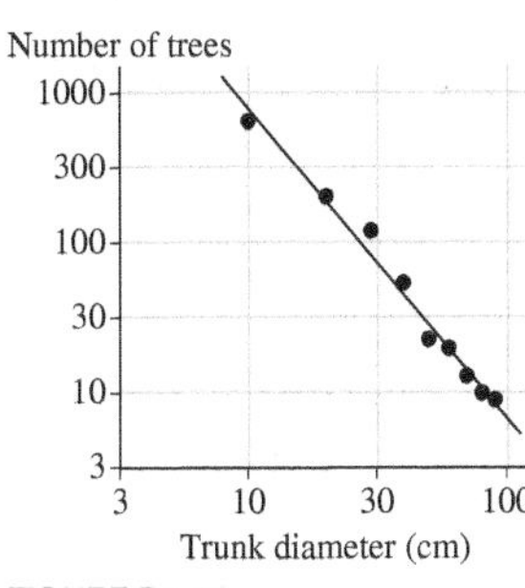

FIGURE P1.14

14. || Scaling laws apply not just to organisms but also to entire
BIO ecosystems. Figure P1.14 shows the number of trees in a 5 acre section of tropical forest with the trunk diameters shown on the horizontal axis. You can see that the forest structure is not random but follows a scaling law. In fact, this scaling law was shown to persist for over 30 years even though individual trees in this plot grew and died. Determine the scaling law by reading the graph carefully and calculating the slope of the trend line. The number of trees scales with what power of the trunk diameter? Round your answer to the nearest quarter-power.

15. || The total leaf area A of a leafy plant scales with the total
BIO plant's mass m as $A \propto m^{3/4}$. If you replace one large tree with two smaller trees, each with half the mass of the original tree, by what factor do you increase the total leaf area?

16. || The average distance d that a Galapagos tortoise travels during
BIO a day scales with the tortoise's mass m as $d \propto m^{2/3}$. If a 50 kg tortoise travels 300 m in a day, how far does a 200 kg tortoise travel?

17. || The flying speed of birds in steady flight scales as the $-\frac{1}{4}$th
BIO power of the bird's mass; that is, $v \propto m^{-1/4}$. Canada geese (5 kg) fly at a speed of 60 km/h. How fast do you expect a 1 kg red-tailed hawk to fly?

General Problems

18. || Microscopy finds that a yeast cell typically contains 40 mi-
BIO tochondria. Mitochondria can be modeled as 1-μm-diameter, 2-μm-long cylinders. Yeast cells are 5-μm-diameter spheres. Estimate the percentage of the cell volume that is filled with mitochondria. See Appendix A for a review of calculating the volume and surface area of different shapes.

19. || The roundworm *C. elegans* is a widely used model organism
BIO in biology. The adult has exactly 959 cells, all of which are visible through a microscope. An adult can be modeled as a 0.8-mm-long, 0.1-mm-diameter cylinder.
a. If we assume the cells are roughly identical in size, estimate the volume of a cell in μm^3.
b. If the cells are modeled as cubes that stack to form the organism, what is the edge length of a cell in μm?

20. || A healthy adult has 6×10^6 red blood cells per μL of blood.
BIO a. An adult has 5 L of blood. Estimate the number of red blood cells in an adult.
b. A laboratory measurement finds 150 μg of hemoglobin per μL of blood. Hemoglobin, the blood protein that transports oxygen, has a molecular weight of 64 kDa or 64,000 u, where 1 dalton (Da) = 1 atomic mass unit (u) = 1.66×10^{-24} g. Estimate the number of hemoglobin proteins in one red blood cell.

21. || RNA polymerase is an enzyme that moves along strands of
BIO DNA, reads the genetic code, and creates a matching strand of mRNA (messenger RNA). RNA polymerase transcribes DNA at the rate of about 60 bp/s (base pairs per s). DNA has a length of 0.34 nm per base pair. An average gene in the human genome is about 12,000 bp in length.
a. What is the speed in nm/s at which RNA polymerase moves along a strand of DNA?
b. How long does it take to transcribe a gene of average length and produce one mRNA molecule?

22. ||| Agnes Pockels, in experiments conducted around 1890, was the first scientist to make a reasonable measurement of the size of a molecule. She placed a 1.0 mg drop of olive oil (density 0.92 g/cm^3) on the surface of a large pool of water and allowed it to spread out to its maximum size. This produced an "oil slick" with an area of 8500 cm^2. Using this information and making appropriate assumptions about the state of the oil when it has spread to maximum size, estimate the diameter in nm of a molecule of olive oil. Your answer will be somewhat too large due to the limitations of this technique, but Pockels's result was the first estimate of molecular size.

23. || There's been considerable interest in recent years about the
BIO human *microbiome,* the microbes that live in and on our bodies. It has been estimated that the number of microbial cells you carry with you is three times the number of your own human cells. On average, a mammalian cell can be modeled as a 12-μm-diameter sphere, while a bacterial cell is a 1-μm-diameter sphere. All cells have densities approximately that of water, 1000 kg/m^3.
a. About how many human cells does a 70 kg adult have?
b. What fraction of this adult's weight is due to their microbiome?

24. || The human genome consists of approximately 3 billion
BIO (3×10^9) *base pairs,* corresponding to 6 billion individual *nucleotides,* the basic building blocks of DNA. X-ray crystallography finds that each nucleotide along a strand of DNA can be modeled as a cube 0.3 nm on each side.
a. If all the nucleotides were arranged end to end along one strand of DNA, how long would the strand be?
b. It's hard to imagine that much material fitting into a cell nucleus, which is typically 5 μm in diameter. Suppose 6 billion nucleotides were packed tightly together to form a sphere. What would the diameter of the sphere be? DNA is not packed that tightly, but you should see that fitting the DNA into the nucleus is not an issue.

25. || A human head can be modeled as a 15-cm-diameter sphere,
BIO with the scalp covering approximately 25%. On average, adults have 100,000 scalp hairs. We want to estimate the distance between two neighboring hairs.
a. What is the density of hairs, measured as hairs per cm^2?
b. Suppose the hairs are arranged in a perfect rectangular grid. What is the distance between two neighboring hairs in this grid? Real hairs aren't arranged this regularly, but your answer is a good estimate of the typical spacing between neighboring hairs.

26. || The volume of air in your lungs changes as you breathe, but a BIO typical value, corresponding to mid-breath, is $1\text{ L} = 1000\text{ cm}^3$.
    a. Suppose you model each lung as a sphere. Approximately what is the diameter of a lung?
    b. What would be the total lung surface area?
    c. In fact, the total lung surface area is approximately 70 m^2 because the lungs are not empty spheres but are subdivided into an estimated 300 million *alveoli*, small sacs from which gases diffuse into and out of capillaries. Approximately what is the diameter in μm of an alveolus?
27. || A study of South American mammals found that the number BIO of species in a large area of land scales as the average mass of individuals of the species to the $-\frac{3}{4}$th power; that is, $S \propto m^{-3/4}$, where S is the number of distinct species (not the number of individuals). Species like sloths and monkeys have masses of roughly 5 kg. Smaller mammals, such as guinea pigs and chinchillas, have masses of roughly 0.5 kg. What is the expected ratio of the number of 0.5 kg species to the number of 5 kg species?
28. || The world population is approaching 8 billion people. Suppose we all lined up front to back, touching each other.
    a. Estimate, to one significant figure, how long the line would stretch.
    b. The earth's radius is 6.4×10^6 m. To one significant figure, how many times would the line of people wrap around the earth's equator?
29. || Loamy soil contains an estimated 10^9 bacteria per cm^3 of BIO soil. Do these many bacteria affect the weight of the soil? A typical bacterium can be modeled as a 1-μm-diameter sphere that has the density of water (1000 kg/m^3). The soil itself has a density of 1300 kg/m^3. To one significant figure, what fraction of the soil's weight is due to bacteria?

MCAT-Style Passage Problems

The Microbiome

There has been a good deal of interest in recent years in determining scaling laws that apply across widely disparate species. The rapid evolution of tools for analyzing the DNA of microbes that exist in and on animals has led to the discovery of scaling laws that relate the number of microbes and the diversity of microbe species to the mass of the host animal. The total number of microbes N that an animal harbors seems to scale with the animal's mass m as $N \propto m^{1.07}$. The diversity of gut microbe species also seems to scale with the animal's mass. For one specific measure of diversity that we can call D, diversity scales with the animal's mass m as $D \propto m^{0.34}$.

30. The relationship between animal mass and number of microbes given in the passage is consistent with which of the following general principles?
    A. Most microbes live on the surface of animals.
    B. Most microbes live in the gut, and the size of the gut is approximately proportional to the mass of the animal.
    C. Most of the microbes that live in and on animals have yet to be discovered.
    D. Microbes exact a cost on their host animal, and larger animals pay a higher cost.
31. Cows and elephants are both herbivores. The mass of a typical dairy cow is 500 kg; the mass of a typical African elephant is 5000 kg. By what approximate factor do you expect the diversity of an elephant's gut to exceed that of a cow?
    A. 2
    B. 3
    C. 10
    D. 12
32. If the above rules hold true across the lifetime of an organism as it grows, how will the number and diversity of gut microbes change?
    A. The number of microbes and their diversity will both increase, but diversity will increase faster.
    B. The number of microbes and their diversity will both increase, but the number will increase faster.
    C. The number of microbes will increase; the diversity will decrease.
    D. The diversity of microbes will increase; the number will decrease.
33. The number of cells in an animal's body is proportional to the animal's mass. If the above rules hold true across the lifetime of an animal, how will the ratio of microbial cells to animal cells change?
    A. As the animal grows, there will be fewer microbial cells per animal cell.
    B. As the animal grows, the ratio of microbial cells to animal cells will not change.
    C. As the animal grows, there will be more microbial cells per animal cell.

2 Describing Motion

A falcon moves in a graceful arc through the air.

LOOKING AHEAD

Describing Motion

This series of images of a skier clearly shows his motion. Such visual depictions are a good first step in describing motion.

In this chapter, you'll learn to make *motion diagrams* that provide a simplified view of the motion of an object.

Velocity and Acceleration

The winner of a sprint accelerates quickly out of the block, then maintains a large acceleration until reaching the highest velocity.

You'll see how the concepts of *velocity* and *acceleration* are essential to understanding how an object moves as it does.

Numbers and Units

Quantitative descriptions involve numbers, and numbers require units. This speedometer gives speed in mph and km/h.

You'll learn the units used in physics, and you'll learn to convert between these and more familiar units.

GOAL To introduce the fundamental concepts of motion.

PHYSICS AND LIFE

Movement Is Essential to Life

Movement is essential to the survival of all animals, from microorganisms to mammals. Plants, though not ambulatory, also move in response to external stimuli and forces, and they equip their seeds to travel long distances while falling or floating. Biological processes depend on the motion of molecules and fluids within and between cells. In addition, motion is a useful analogy, both conceptually and mathematically, for processes such as the spread of diseases. Thus we begin our study of physics by establishing the basic ideas of motion. This chapter and the next will focus on how to describe motion with pictures, graphs, and equations. Chapters 4 and 5 will then analyze how forces and interactions give rise to and determine motion.

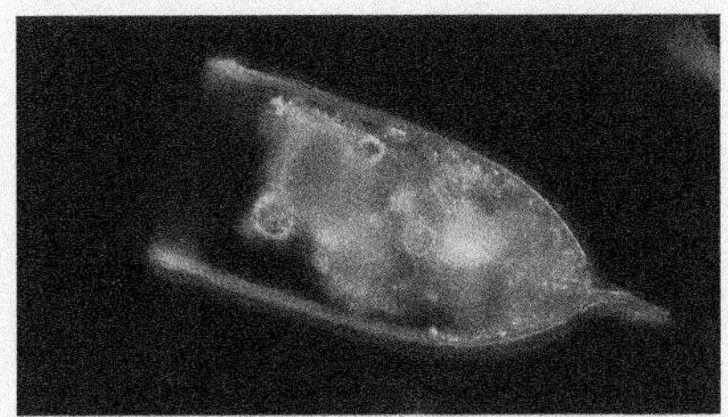

The word "plankton" is derived from a Greek root meaning *drifter*. These microorganisms move by drifting with the ocean currents.

2.1 Motion Diagrams

Motion is an essential capability for life. It is also an essential characteristic of many physical processes, and so it is a theme that will appear in one form or another throughout this entire text. Although we all have intuition about motion, based on our experiences, some of the important aspects of motion turn out to be rather subtle. So rather than jumping immediately into a lot of mathematics and calculations, we'll begin by *visualizing* motion and becoming familiar with the *concepts* needed to describe a moving object, whether it is a molecule, an organism, or an inanimate object.

FIGURE 2.1 Four basic types of motion.

Linear motion

Circular motion

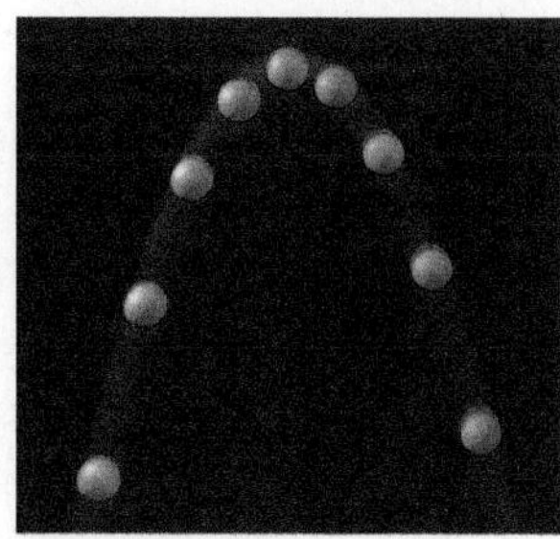

Projectile motion

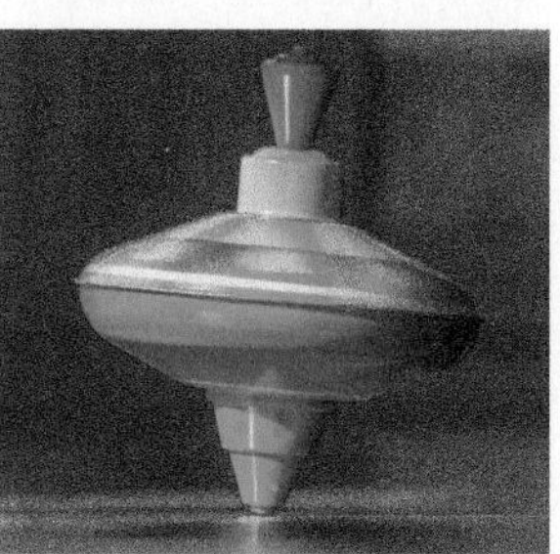

Rotational motion

Motion is simply the change of an object's position with time. FIGURE 2.1 shows four basic types of motion that we will study in this book. The first three—linear, circular, and projectile motion—in which the object moves through space are called **translational motion.** The path along which the object moves, whether straight or curved, is called the object's **trajectory.** Rotational motion is somewhat different because there's movement but the object as a whole doesn't change position. We'll defer rotational motion until later and, for now, focus on translational motion.

FIGURE 2.2 Four frames from a video.

Making a Motion Diagram

An easy way to study motion is to make a video of a moving object. A video camera or the camera in your phone takes images at a fixed rate, typically 30 every second. Each separate image is called a *frame*. As an example, FIGURE 2.2 shows four frames from a video of a car going past. Not surprisingly, the car is in a different position in each frame.

Suppose we edit the video by layering the frames on top of each other, creating the composite image shown in FIGURE 2.3. This edited image, showing an object's position at several *equally spaced instants of time,* is called a **motion diagram.** As the examples below show, we can define concepts such as constant speed, speeding up, and slowing down in terms of how an object appears in a motion diagram.

FIGURE 2.3 A motion diagram of the car shows all the frames simultaneously.

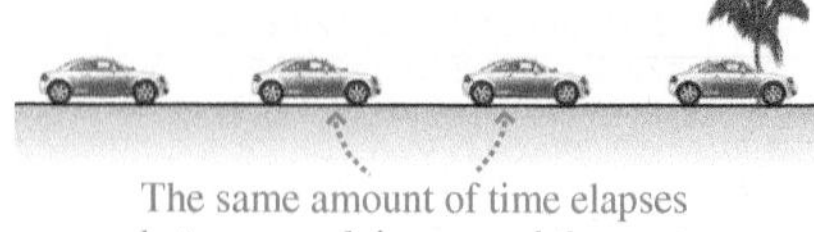

The same amount of time elapses
between each image and the next.

NOTE ► It's important to keep the camera in a *fixed position* as the object moves by. Don't "pan" it to track the moving object. ◄

Examples of motion diagrams

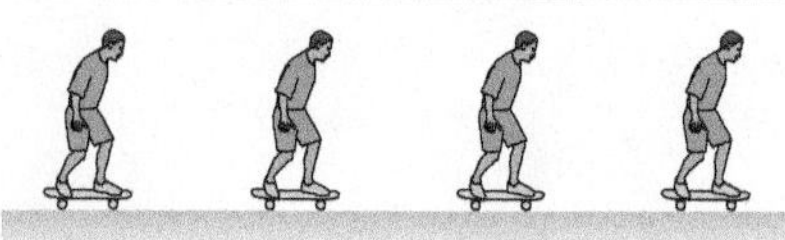

Images that are *equally spaced* indicate an object moving with *constant speed.*

An *increasing distance* between the images shows that the object is *speeding up*.

A *decreasing distance* between the images shows that the object is *slowing down*.

The Particle Model

For many types of motion, from swimming bacteria to accelerating rockets, the motion of the object *as a whole* is not influenced by the details of the object's size and shape. All we really need to keep track of is the motion of a single point on the object, so we can treat the object *as if* all its mass were concentrated into this single point. An object that can be represented as a mass at a single point in space is called a **particle.** A particle has no size, no shape, and no distinction between top and bottom or between front and back.

If we model an object as a particle, we can represent the object in each frame of a motion diagram as a simple dot rather than having to draw a full picture. FIGURE 2.4 shows how much simpler motion diagrams appear when the object is represented as a particle. Note that the dots have been numbered 0, 1, 2, . . . to tell the sequence in which the frames were exposed.

FIGURE 2.4 Using the particle model to draw a motion diagram.

Motion diagram of a car stopping

Same motion diagram using the particle model

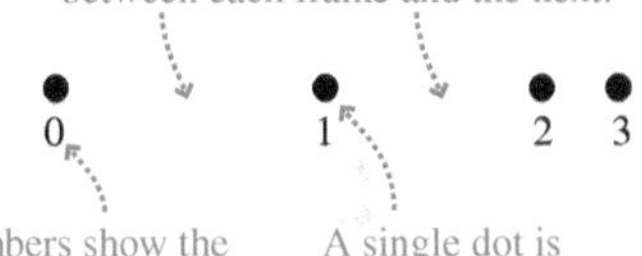

The same amount of time elapses between each frame and the next.
0
1
2
3
Numbers show the order in which the frames were taken.
A single dot is used to represent the object.

Treating an object as a particle is, of course, a simplification of reality—but that's what modeling is all about, as you learned in Chapter 1. The **particle model** of motion is a simplification in which we treat a moving object as if all of its mass were concentrated at a single point. The particle model is an excellent approximation of reality for the translational motion of animals, bicycles, cars, and similar objects.

Of course, not everything can be modeled as a particle. Whether or not a model is useful depends on the question we seek to answer. Some types of human and animal motion can't be understood without considering the separate motions of the limbs, a description that goes beyond the particle model. We'll need to develop new models when we get to new types of motion, but the particle model will serve us well throughout Part I of this book.

STOP TO THINK 2.1 Which car is going faster, A or B? Assume there are equal intervals of time between the frames of both videos.

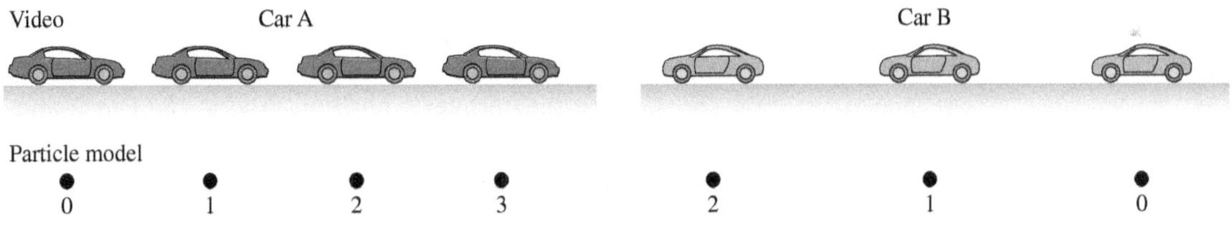

NOTE ► Each chapter will have several *Stop to Think* questions. These questions are designed to see if you've understood the basic ideas that have been presented. The answers are given at the end of the chapter, but you should make a serious effort to think about these questions before turning to the answers. ◄

2.2 Position, Time, and Displacement

Our everyday descriptions of motion are rather hazy and imprecise. A science of motion needs a set of well-defined concepts, giving us an appropriate language to describe motion. The two most basic facts about a moving object are *where* it is (i.e., its **position**) and *when* the object is at that position (i.e., the **time**). Position measurements can be made by laying a coordinate-system grid over a motion diagram. You can then measure the (x, y) coordinates of each point in the motion diagram. Of course, the world does not come with a coordinate system attached. A coordinate system is an artificial grid that *you* place over a problem in order to analyze the motion. You place the origin of your coordinate system wherever it is useful for you; different observers of a moving object might all choose to use different origins.

FIGURE 2.5 Coordinate systems for a race and a falling rock.

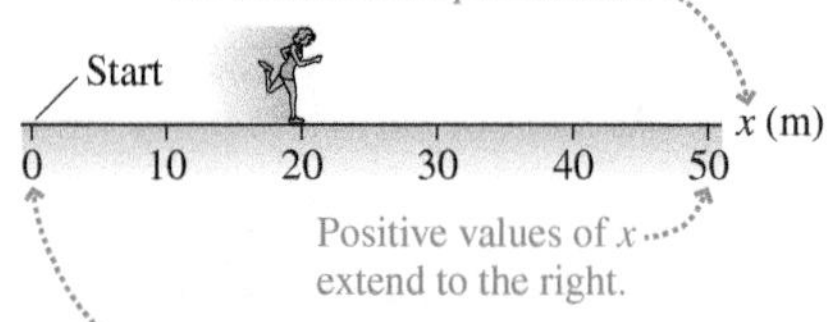
This is the coordinate used to represent positions along the axis. The units of x are shown in the parentheses.
Start
x (m)
0 10 20 30 40 50
Positive values of x extend to the right.
The start of the race is a natural choice for the origin.

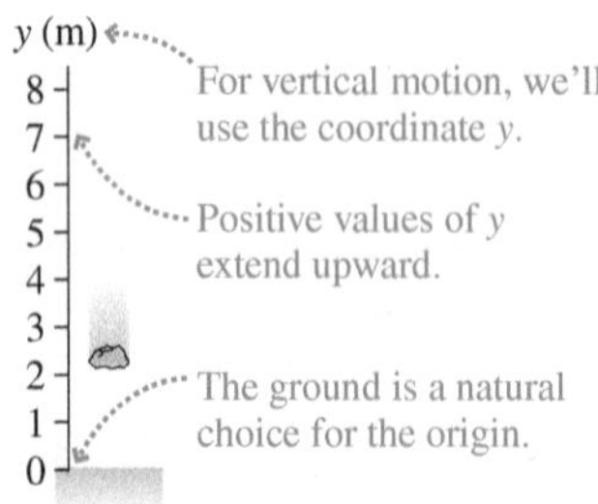
y (m)
For vertical motion, we'll use the coordinate y.
Positive values of y extend upward.
The ground is a natural choice for the origin.
8 7 6 5 4 3 2 1 0

To illustrate, FIGURE 2.5 shows how we can set up coordinate systems for a sprinter and a falling rock. In math, a horizontal axis is usually labeled the x-axis, so we use the symbol x to represent the sprinter's position. Similarly, we use the y-axis for vertical motion and represent the rock's position with the symbol y. Using symbols to represent position (and, later, other quantities) allows us to work with these concepts mathematically. We place the origin of the sprinter's coordinate system at the start of the race and the origin of the rock's coordinate system at the ground, but that is a *choice* rather than a requirement. You're free to place the origin at whatever position makes it easiest for you to solve a problem.

We also make the choice that the positive end of a horizontal axis (the x-axis) extends to the right and the positive end of a vertical axis (the y-axis) extends upward. We could define axes in other ways, but this particular *convention* allows us all to agree about which quantities are positive and which negative.

Time, in a sense, is also a coordinate system, although you may never have thought of time this way. You can pick an arbitrary point in the motion and label it "$t = 0$ seconds." This is simply the instant you decide to start your clock or stopwatch, so it is the origin of your time coordinate. Different observers might choose to start their clocks at different moments. A video frame labeled "$t = 4$ seconds" was taken 4 seconds after you started your clock.

We typically choose $t = 0$ to represent the "beginning" of a problem or situation, but the object may have been moving before then. Those earlier instants would be measured as negative times, just as objects on the x-axis to the left of the origin have negative values of position. Negative numbers are not to be avoided; they simply locate an event in space or time *relative to an origin.*

Displacement: A Change of Position

FIGURE 2.6 Motion diagrams for the sprinter and the falling rock, each illustrating one displacement vector.

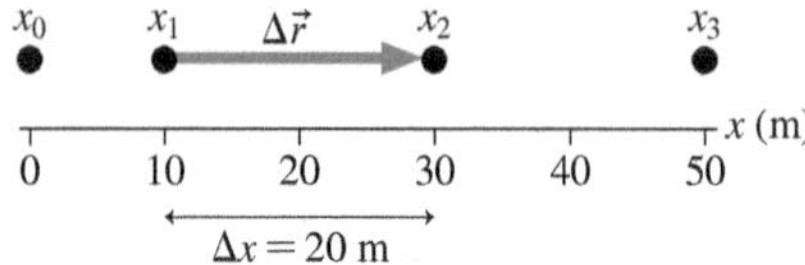
Sprinter
x_0 x_1 $\Delta\vec{r}$ x_2 x_3
x (m)
0 10 20 30 40 50
$\Delta x = 20$ m

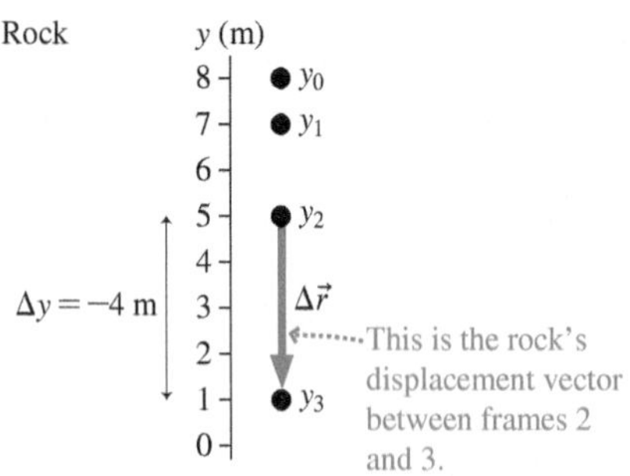
Rock
y (m)
y_0
y_1
y_2
$\Delta\vec{r}$
y_3
$\Delta y = -4$ m
This is the rock's displacement vector between frames 2 and 3.

We began by saying that motion is the change in an object's position with time, but how do we measure a change of position? A motion diagram is the perfect tool. FIGURE 2.6 shows motion diagrams for the sprinter and the falling rock. Notice how we've used subscripts to label the positions, starting with x_0 and y_0 at time $t = 0$ s. At later times, you can see that the sprinter's position is $x_1 = 10$ m in frame 1 and $x_2 = 30$ m in frame 2. In moving from one position to another, she has *changed* position by

$$\Delta x = x_2 - x_1 = 20 \text{ m}$$

This change of position is called **displacement;** in this case, her **displacement** is 20 m. Similarly, the rock's displacement between frame 2 and frame 3 is

$$\Delta y = y_3 - y_2 = -4 \text{ m}$$

This is a negative displacement because the value of y, which measures the rock's position, is decreasing as the rock falls.

NOTE ▸ The Greek letter delta (Δ) is used in math and science to indicate the *change* in a quantity—in this case, the change in an object's position. The quantity Δx is always the *later* (or final) value of x minus the *earlier* (or initial) value: $\Delta x = x_{\text{later}} - x_{\text{earlier}}$. This means that Δx can be positive or negative; the sign is important. Δx is a *single* symbol. You cannot cancel out or remove the Δ. ◂

A different way to indicate displacement is by drawing an arrow between the dots of the motion diagram, as we did in Figure 2.6. This arrow is called the **displacement vector** $\Delta\vec{r}$. The symbol $\vec{r}$ is used in math and physics to represent an object's *position vector,* an arrow drawn from the origin of the coordinate system to the object's location, and so $\Delta\vec{r}$ indicates a *change of position*. We won't have any need to use position vectors, but the displacement vector $\Delta\vec{r}$ will appear in a number of different contexts.

Scalars and Vectors

Some physical quantities, such as time, mass, and temperature, can be described completely by a single number with a unit. For example, the mass of an object is 6 kg and its temperature is 30°C. A single number (with a unit) that describes a physical quantity is called a **scalar.** A scalar can be positive, negative, or zero.

Many other quantities, however, have a directional aspect and cannot be described by a single number. To describe the motion of a car, for example, you must specify not only how fast it is moving, but also the *direction* in which it is moving. A quantity having both a *size* (the "How far?" or "How fast?") and a *direction* (the "Which way?") is called a **vector.** The size or length of a vector is called its *magnitude.* Many quantities in physics are vectors, and we'll look at vectors more thoroughly in Chapter 4. All we need for now is a little basic information.

We indicate a vector by drawing an arrow over the letter that represents the quantity. Thus $\vec{a}$ and $\vec{A}$ are symbols for vectors, whereas a and A, without the arrows, are symbols for scalars. In handwritten work you must draw arrows over all symbols that represent vectors. This may seem strange until you get used to it, but it is very important because we will often use both a and $\vec{a}$, or both A and $\vec{A}$, in the same problem, and they mean different things! Note that the arrow over the symbol always points to the right, regardless of which direction the actual vector points. Thus we write $\vec{a}$ or $\vec{A}$, never $\overleftarrow{a}$ or $\overleftarrow{A}$.

To specify a vector, we have to give both its magnitude—which is always positive—and its direction. For example, you could write the sprinter's displacement vector between the first and second frames as $\Delta\vec{r}_{\text{sprinter}} = (20\text{ m, right})$ and that of the rock between the third and fourth frames as $\Delta\vec{r}_{\text{rock}} = (4\text{ m, down})$. The specification "down" gives the same directional information as the minus sign in Δy.

FIGURE 2.7 Motion diagrams with the displacement vectors.

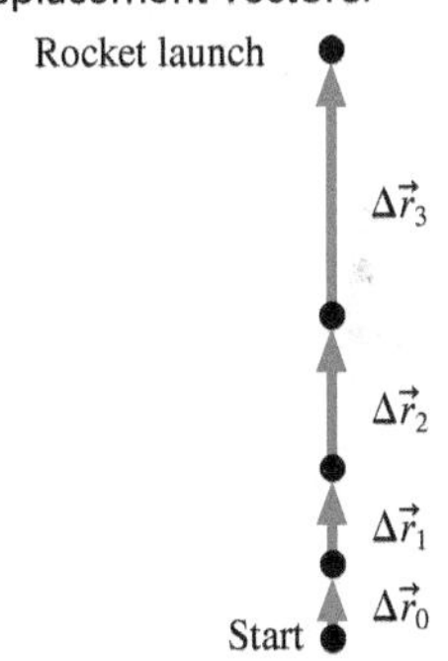

Motion Diagrams with Displacement Vectors

The first step in analyzing a motion diagram is to determine *all* of the displacement vectors, which, as you saw in Figure 2.6, are simply the arrows connecting each dot to the next. Label each arrow with a *vector* symbol $\Delta\vec{r}_n$, starting with $n = 0$. FIGURE 2.7 shows the idea with motion diagrams of a rocket launch and a car coming to a stop. It's a good idea to write "Start" or "Stop" beside a dot where an object starts from rest or ends at rest.

Now we can conclude, more precisely than before, that, as time proceeds:

- An object is speeding up if its displacement vectors are increasing in length.
- An object is slowing down if its displacement vectors are decreasing in length.

EXAMPLE 2.1 Headfirst into the snow

Alice is sliding along a smooth, icy road on her sled when she suddenly runs headfirst into a large, very soft snowbank that gradually brings her to a halt. Draw a motion diagram for Alice. Show and label all displacement vectors.

PREPARE The details of Alice and the sled—their size, shape, color, and so on—are not relevant to understanding their overall motion. So we can model Alice and the sled as one particle.

SOLVE FIGURE 2.8 shows a motion diagram. The problem statement suggests that the sled's speed is very nearly constant until it hits the snowbank. Thus the displacement vectors are of equal length as Alice slides along the icy road. She begins slowing when she hits the snowbank, so the displacement vectors then get shorter until the sled stops. We're told that her stop is gradual, so we want the vector lengths to get shorter gradually rather than suddenly.

FIGURE 2.8 The motion diagram of Alice and the sled.

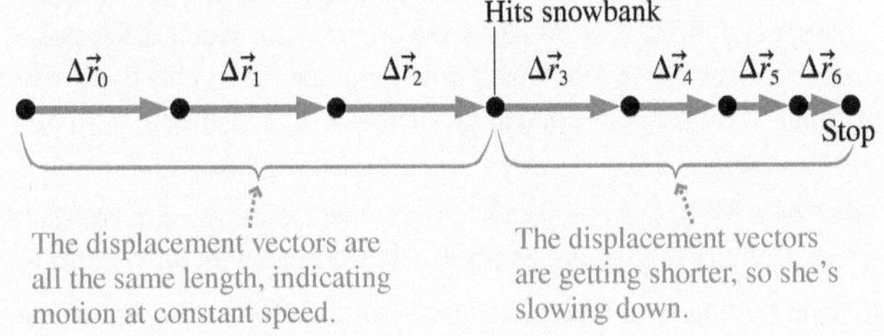

Time Interval

It's also useful to consider a *change* in time. For example, the clock readings of two frames of a video might be t_1 and t_2. The specific values are arbitrary because they are timed relative to an arbitrary instant that you chose to call $t = 0$. But the **time interval** $\Delta t = t_2 - t_1$ is *not* arbitrary. It represents the elapsed time for the object to move from one position to the next.

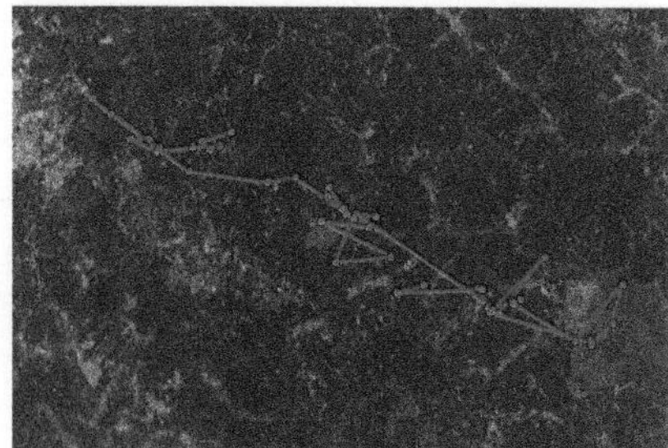

Tracking elephants Conservation scientists track the movements of elephants and other animals using GPS collars that record their positions and other data at regular, preset intervals of time. The data for this Indian elephant clearly show periods of slow and rapid motion. This particular study was observing how elephants move between forest areas and agricultural land, a critical issue for conservation because elephants are more likely to be killed if they destroy crops.

The time interval $\Delta t = t_b - t_a$ measures the elapsed time as an object moves from position a at time t_a to position b at time t_b. The value of Δt is independent of the specific clock used to measure the times.

To summarize the main idea of this section, we have added coordinate systems and clocks to our motion diagrams in order to measure *when* each frame was exposed and *where* the object was located at that time. Different observers of the motion may choose different coordinate systems and different clocks. However, it's especially important to notice that all observers find the *same* values for the displacement vectors $\Delta\vec{r}$ and the time intervals Δt because these are independent of the specific coordinate system used to measure them.

STOP TO THINK 2.2 Sarah starts at a positive position along the x-axis. She then undergoes a displacement to the left. Her final position

A. Is positive. B. Is negative. C. Could be either positive or negative.

2.3 Velocity

It's no surprise that, during a given time interval, a speeding bullet travels farther than a speeding snail. To extend our study of motion so that we can compare the bullet to the snail, we need a way to measure how fast or how slowly an object moves.

You're familiar with the idea of **speed,** defined as the ratio

$$\text{speed } v = \frac{\text{distance traveled}}{\text{time interval spent traveling}} = \frac{d}{\Delta t} \tag{2.1}$$

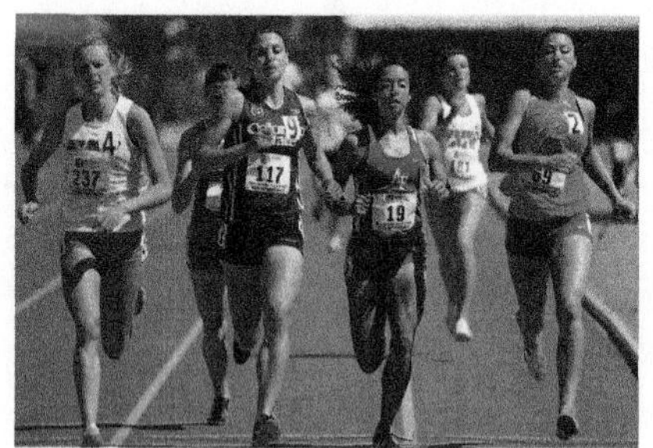

The victory goes to the runner with the highest average speed.

If you drive 15 miles (mi) in 30 minutes $\left(\frac{1}{2}\text{ h}\right)$, your speed is

$$v = \frac{15\text{ mi}}{\frac{1}{2}\text{ h}} = 30\text{ mph} \tag{2.2}$$

The symbol for speed, v, comes from its close connection to velocity, a connection we'll look at shortly.

EXAMPLE 2.2 Transcribing DNA

A strand of messenger RNA (mRNA), which contains the information to synthesize one protein, is transcribed from DNA by the enzyme RNA polymerase. Molecular biology experiments show that an RNA polymerase molecule moves along a DNA molecule at the rate of roughly 60 nt/s, where nt stands for *nucleotide,* the "letters" of the DNA code. X-ray crystallography studies have found that nucleotides are strung along DNA at 3.0 nt/nm, where 1 nm = 1 nanometer = 10^{-9} m. What is the speed with which RNA polymerase moves along DNA, and how long does it take to transcribe the mRNA molecule needed to synthesize a typical protein with 300 amino acids?

PREPARE We'll assume that RNA polymerase moves at a constant speed. Cellular motion is often best characterized with units of nm/s.

SOLVE 60 nt/s is in a sense a speed, but in DNA units per second rather than distance per second. To express this in nm/s is an exercise in unit conversion. We'll look at unit conversion more carefully later in this chapter; you've done it in other science classes. The speed is

$$v = 60\,\frac{\text{nt}}{\text{s}} \times \frac{1\text{ nm}}{3.0\text{ nt}} = 20\text{ nm/s}$$

You learn in biology that three sequential nucleotides code for one amino acid, so 900 nucleotides have to be transcribed to mRNA to build a protein with 300 amino acids. At 3.0 nt/nm, this requires a 300 nm length of DNA. The RNA polymerase molecule will move 300 nm in

$$\Delta t = \frac{d}{v} = \frac{300\text{ nm}}{20\text{ nm/s}} = 15\text{ s}$$

ASSESS A major point of this example is that simple calculations can provide insights into cellular processes. Here we find that DNA transcription is actually pretty slow, with each gene able to generate only about four mRNA molecules per minute. (Several different genes on a single DNA molecule can be doing this simultaneously.) Combined with additional information about the overall rate of protein synthesis, this example reveals that each cell must have a large number of ribosomes, the organelles that carry out protein synthesis from the mRNA transcripts, to compensate for the slow transcription rate.

Although the concept of speed is widely used in our day-to-day lives, it is not a sufficient basis for a science of motion. To see why, imagine you're trying to land a jet plane on an aircraft carrier. It matters a great deal to you whether the aircraft carrier is moving at 20 mph (miles per hour) to the north or 20 mph to the east. Simply knowing that the ship's speed is 20 mph is not enough information—you also need to know the direction in which it is moving.

It's the displacement $\Delta\vec{r}$, a vector quantity, that tells us not only the distance traveled by a moving object, but also the *direction* of motion. Consequently, a more useful ratio than $d/\Delta t$ is the ratio $\Delta\vec{r}/\Delta t$. In addition to measuring how fast an object moves, this ratio is a vector that points in the direction of motion.

It is convenient to give this ratio a name. We call it the **velocity,** and it has the symbol $\vec{v}$. **The velocity of an object during the time interval Δt, in which the object undergoes a displacement $\Delta\vec{r}$, is the vector**

$$\text{velocity } \vec{v} = \frac{\text{displacement}}{\text{time interval}} = \frac{\Delta\vec{r}}{\Delta t} \tag{2.3}$$

Velocity of an object

FIGURE 2.9 The displacement vectors and velocities of ships A and B.

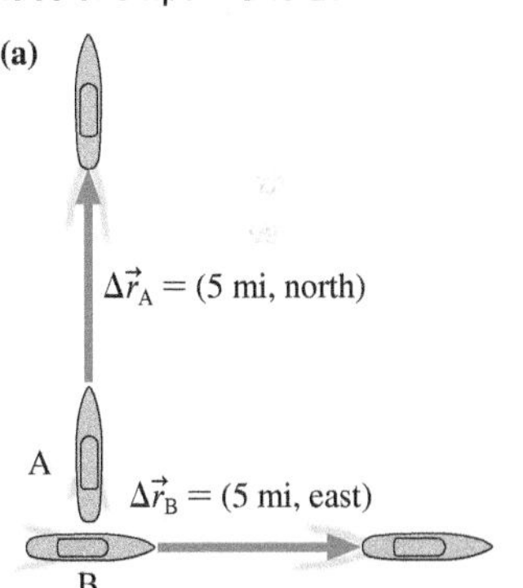

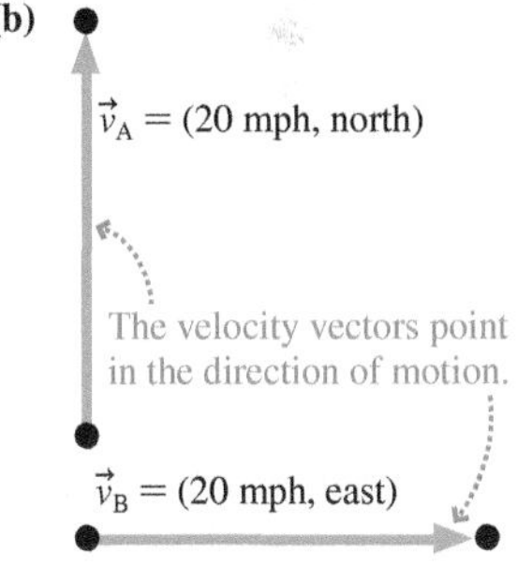

An object's velocity vector points in the same direction as the displacement vector $\Delta\vec{r}$. This is the direction of motion. The object's speed, which lacks directional information, is the magnitude of the velocity vector.

NOTE ▶ Biology uses terms that can be quite complex, challenging your memory, but they are very precise. In contrast, physics terms are often common, everyday words that are given an uncommon meaning. For example, in everyday language we do not make a distinction between speed and velocity, but in physics *the distinction is very important.* In particular, speed is simply "How fast?" whereas velocity is "How fast, and in which direction?" As we go along we will be giving other words more specific meanings in physics than they have in everyday language. ◀

As an example, FIGURE 2.9a shows two ships that move 5 miles in 15 minutes. Using Equation 2.3 with $\Delta t = 0.25$ h, we find

$$\begin{aligned} \vec{v}_A &= (20 \text{ mph, north}) \\ \vec{v}_B &= (20 \text{ mph, east}) \end{aligned} \tag{2.4}$$

Both ships have a speed of 20 mph, but their velocities differ. Notice how the velocity *vectors* in FIGURE 2.9b point in the direction of motion.

NOTE ▶ Our goal in this chapter is to *visualize* motion with motion diagrams. Strictly speaking, the vector we have defined in Equation 2.3, and the vector we will show on motion diagrams, is the *average* velocity $\vec{v}_{avg}$. Our definitions and symbols, which somewhat blur the distinction between average and instantaneous quantities, are adequate for visualization purposes, but they're not the final word. We will refine these definitions in Chapter 3, where our goal will be to develop the mathematics of motion. ◀

The "Per" in Miles Per Hour

The units for speed and velocity are a unit of distance, such as feet, meters, or miles, divided by a unit of time, such as seconds or hours. Thus we could measure velocity in units of mi/h (or mph) or m/s, pronounced "miles *per* hour" or "meters *per* second." The word "per" will often arise in physics when we consider the ratio of two quantities. What do we mean, exactly, by "per"?

If a car moves with a speed of 23 m/s, we mean that it travels 23 meters *for each* second of elapsed time. The word "per" thus associates the number of units in the numerator (23 m) with *one* unit of the denominator (1 s). We'll see many other examples of this idea as the text progresses. You may already know a bit about *density;* for example, the density of gold is 19.3 g/cm^3 ("grams *per* cubic centimeter"). This means that there are 19.3 grams of gold *for each* cubic centimeter of the metal. Gold is much denser than water, which you may recall has a density of only 1.0 g/cm^3.

FIGURE 2.10 Motion diagram of the tortoise racing the hare.

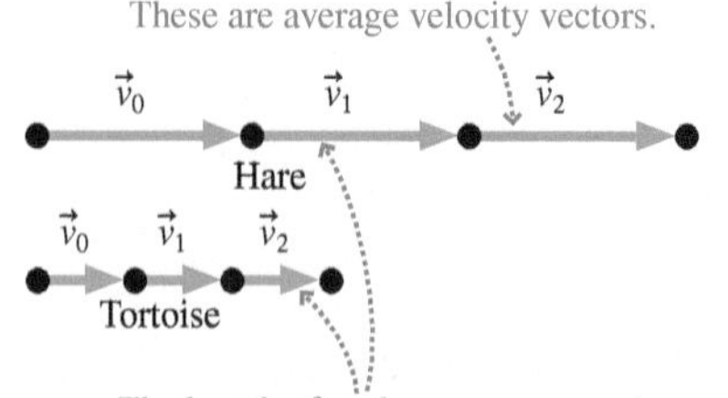
These are average velocity vectors.
$\vec{v}_0$
$\vec{v}_1$
$\vec{v}_2$
Hare
$\vec{v}_0$
$\vec{v}_1$
$\vec{v}_2$
Tortoise
The length of each arrow represents the average speed. The hare moves faster than the tortoise.

Motion Diagrams with Velocity Vectors

The velocity vector points in the same direction as the displacement $\Delta\vec{r}$, and the length of $\vec{v}$ is directly proportional to the length of $\Delta\vec{r}$. Consequently, the vectors connecting each dot of a motion diagram to the next, which we previously labeled as displacements, could equally well be identified as velocity vectors.

This idea is illustrated in FIGURE 2.10, which shows four frames from the motion diagram of a tortoise racing a hare. The vectors connecting the dots are now labeled as velocity vectors $\vec{v}$. **The length of a velocity vector represents the average speed with which the object moves between the two points.** Longer velocity vectors indicate faster motion. You can see that the hare moves faster than the tortoise.

Notice that the hare's velocity vectors do not change; each has the same length and direction. We say the hare is moving with *constant velocity.* The tortoise is also moving with its own constant velocity.

EXAMPLE 2.3 Accelerating up a hill

The light turns green and a car accelerates, starting from rest, up a 10° hill. Draw a motion diagram showing the car's velocity.

PREPARE We can use the particle model to represent the car as a dot.

SOLVE The car's motion takes place along a straight line, but the line is neither horizontal nor vertical. A motion diagram should show the object moving with the correct orientation—in this case, at an angle of 10°. FIGURE 2.11 shows several frames of the motion diagram, where we see the car speeding up. The car starts from rest, so the first arrow is drawn as short as possible and the first dot is labeled "Start." The displacement vectors have been drawn from each dot to the next, but then they are identified and labeled as average velocity vectors $\vec{v}$.

FIGURE 2.11 Motion diagram of a car accelerating up a hill.

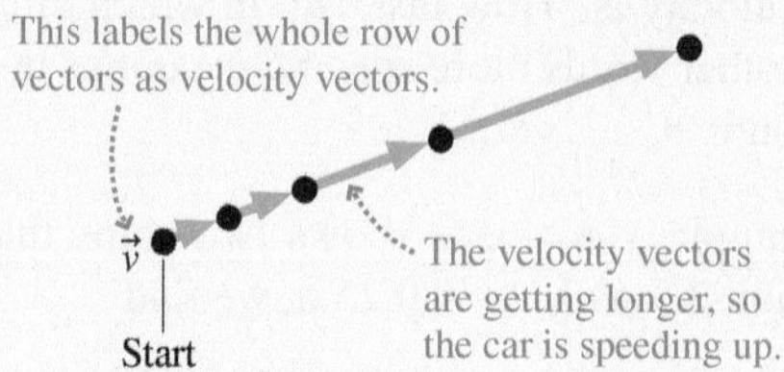
This labels the whole row of vectors as velocity vectors.
$\vec{v}$
Start
The velocity vectors are getting longer, so the car is speeding up.

EXAMPLE 2.4 A rolling soccer ball

Marcos kicks a soccer ball. It rolls along the ground until stopped by Jose. Draw a motion diagram of the ball.

PREPARE This example is typical of how many problems in science and engineering are worded. The problem does not give a clear statement of where the motion begins or ends. Are we interested in the motion of the ball just during the time it is rolling between Marcos and Jose? What about the motion *as* Marcos kicks it (ball rapidly speeding up) or *as* Jose stops it (ball rapidly slowing down)? The point is that *you* will often be called on to make a *reasonable interpretation* of a problem statement. In this problem, the details of kicking and stopping the ball are complex. The motion of the ball across the ground is easier to describe, and it's a motion you might expect to learn about in a physics class. So our *interpretation* is that the motion diagram should start as the ball leaves Marcos's foot (ball already moving) and should end the instant it touches Jose's foot (ball still moving). In between, the ball will slow down a little. We will model the ball as a particle.

SOLVE With this interpretation in mind, FIGURE 2.12 shows the motion diagram of the ball. Notice how, in contrast to the car of Figure 2.11, the ball is already moving as the motion diagram video begins. As before, the average velocity vectors are found by connecting the dots. You can see that the average velocity vectors get slightly shorter as the ball slows. Each $\vec{v}$ is different, so this is *not* constant-velocity motion.

FIGURE 2.12 Motion diagram of a soccer ball rolling from Marcos to Jose.

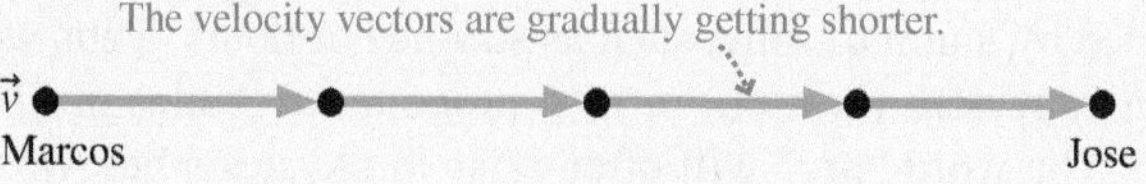
The velocity vectors are gradually getting shorter.
$\vec{v}$
Marcos
Jose

STOP TO THINK 2.3 A particle moves from position 1 to position 2 during the time interval Δt. Which vector shows the particle's average velocity?

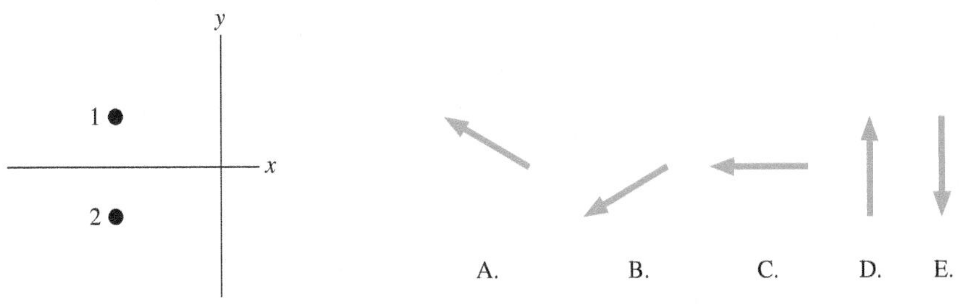

2.4 Acceleration

Position, time, and velocity are important concepts, and at first glance they might appear to be sufficient to describe motion. But that is not the case. Sometimes an object's velocity is constant, as it was in Figure 2.10. More often, an object's velocity changes as it moves, as in Figures 2.11 and 2.12. We need one more motion concept—*acceleration*—to describe a *change* in the velocity.

Because velocity is a vector, it can change in two possible ways:

1. The magnitude can change, indicating a change in speed; or
2. The direction can change, indicating that the object has changed direction.

We will concentrate for now on the first case, a change in speed. The car accelerating up a hill in Figure 2.11 was an example in which the magnitude of the velocity vector changes but not the direction. We'll return to the second case in Chapter 7.

When we wanted to measure changes in position, the ratio $\Delta\vec{r}/\Delta t$ was useful. This ratio is the *rate of change of position.* By analogy, consider an object whose velocity changes from $\vec{v}_a$ to $\vec{v}_b$ during the time interval Δt. The quantity $\Delta\vec{v} = \vec{v}_b - \vec{v}_a$ is the change of velocity, and thus the ratio $\Delta\vec{v}/\Delta t$ is the *rate of change of velocity.* It has a large magnitude for objects that speed up quickly and a small magnitude for objects that speed up slowly.

The ratio $\Delta\vec{v}/\Delta t$ is called the **acceleration,** and its symbol is $\vec{a}$. **The acceleration of an object during the time interval Δt, in which the object's velocity changes by $\Delta\vec{v}$, is**

$$\text{acceleration } \vec{a} = \frac{\text{velocity change}}{\text{time interval}} = \frac{\Delta\vec{v}}{\Delta t} \tag{2.5}$$

Acceleration of an object

The acceleration vector points in the same direction as the vector $\Delta\vec{v}$.

Acceleration is a fairly abstract concept, and it's surprisingly easy to confuse velocity and acceleration. Yet it is essential to develop a good intuition about acceleration because it will be a key concept for understanding why objects move as they do. Motion diagrams will be an important tool for developing that intuition.

Pressed into your seat The Audi TT accelerates from 0 to 60 mph in 6 s. Our bodies are able to directly sense acceleration through the vestibular system in our inner ears. However, we don't have a direct sense of velocity; we recognize that we're moving only by visually observing that we are moving relative to other objects or, sometimes, by sensing the movement of air or water on our skin. Our physical sensations of movement come primarily from sensing acceleration.

Finding the Acceleration Vectors on a Motion Diagram

Perhaps the most important use of a motion diagram is in determining the acceleration vector $\vec{a}$ at each point in the motion. From its definition, Equation 2.5, we see that $\vec{a}$ **points in the same direction as $\Delta\vec{v}$,** the change of velocity, so we need to find the direction of $\Delta\vec{v}$. To do so, we can rewrite the definition $\Delta\vec{v} = \vec{v}_b - \vec{v}_a$ as $\vec{v}_b = \vec{v}_a + \Delta\vec{v}$. That is, the later velocity $\vec{v}_b$ is the earlier velocity $\vec{v}_a$ plus $\Delta\vec{v}$. This is a *vector addition* problem: What vector must we add to $\vec{v}_a$ to get to $\vec{v}_b$? We'll explore vector addition in Chapter 4, but when all the vectors point in the same (or opposite) direction, as they do here, then vector addition simply lengthens or shortens the arrows. Tactics Box 2.1 shows how.

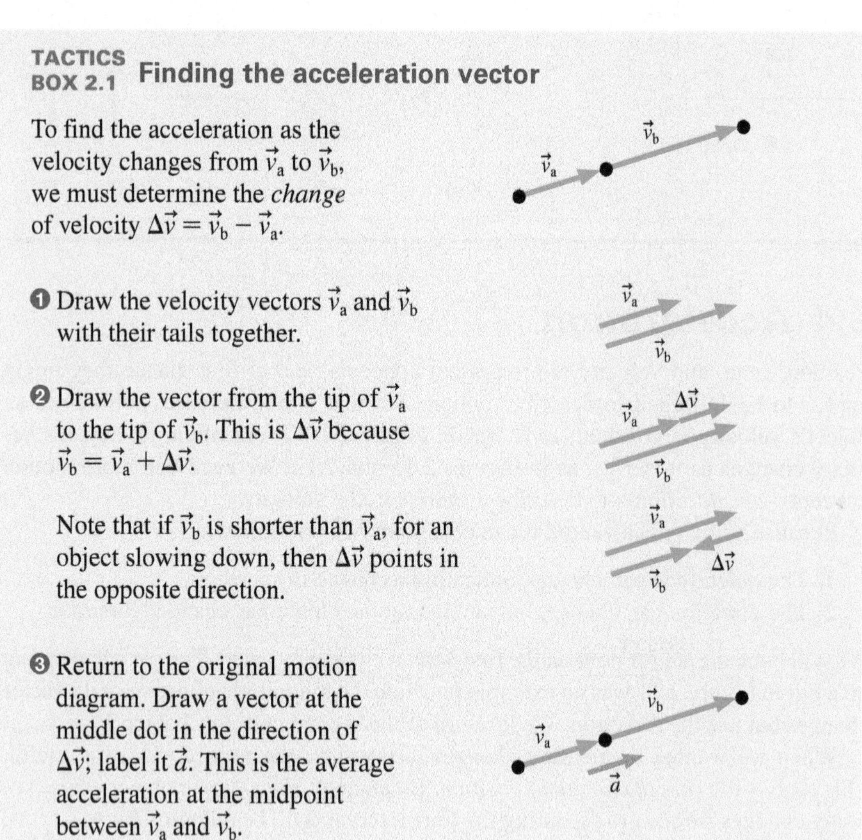
TACTICS BOX 2.1 Finding the acceleration vector
To find the acceleration as the velocity changes from $\vec{v}_a$ to $\vec{v}_b$, we must determine the *change* of velocity $\Delta\vec{v} = \vec{v}_b - \vec{v}_a$.
$\vec{v}_a$
$\vec{v}_b$
❶ Draw the velocity vectors $\vec{v}_a$ and $\vec{v}_b$ with their tails together.
$\vec{v}_a$
$\vec{v}_b$
❷ Draw the vector from the tip of $\vec{v}_a$ to the tip of $\vec{v}_b$. This is $\Delta\vec{v}$ because $\vec{v}_b = \vec{v}_a + \Delta\vec{v}$.
$\Delta\vec{v}$
$\vec{v}_a$
$\vec{v}_b$
Note that if $\vec{v}_b$ is shorter than $\vec{v}_a$, for an object slowing down, then $\Delta\vec{v}$ points in the opposite direction.
$\vec{v}_a$
$\Delta\vec{v}$
$\vec{v}_b$
❸ Return to the original motion diagram. Draw a vector at the middle dot in the direction of $\Delta\vec{v}$; label it $\vec{a}$. This is the average acceleration at the midpoint between $\vec{v}_a$ and $\vec{v}_b$.
$\vec{v}_a$
$\vec{v}_b$
$\vec{a}$

Notice that the acceleration vector goes beside the middle dot, not beside the velocity vectors. This is because each acceleration vector is determined by the *difference* between the *two* velocity vectors on either side of a dot. The length of $\vec{a}$ does not have to be the exact length of $\Delta\vec{v}$; it is the direction of $\vec{a}$ that is most important.

The procedure of Tactics Box 2.1 can be repeated to find $\vec{a}$ at each point in the motion diagram. Note that we cannot determine $\vec{a}$ at the first and last points because we have only one velocity vector and can't find $\Delta\vec{v}$.

The Complete Motion Diagram

Tactics Boxes to help you accomplish specific tasks will appear throughout this text. The next one summarizes how to draw a motion diagram.

TACTICS BOX 2.2 Drawing a motion diagram

Determine whether it is appropriate to model the moving object as a particle. Make simplifying assumptions when interpreting the problem statement. Then

❶ Show the position of the object in each frame of the video as a dot. Use five or six dots for simple motions to make the motion clear without overcrowding the picture. More complex motions will need a finer level of detail with more dots. The diagram should show the motion changing *gradually* from one frame to the next.

❷ Draw the velocity vectors by connecting each dot in the motion diagram to the next with a vector arrow. There is *one* velocity vector linking each *two* position dots. Label the velocity vectors $\vec{v}$.

❸ Use Tactics Box 2.1 to find the acceleration vectors. There is *one* acceleration vector linking each *two* velocity vectors. Each acceleration vector is drawn at the dot between the two velocity vectors it links. Use $\vec{0}$ to indicate a point at which the acceleration is zero. Label the acceleration vectors $\vec{a}$.

STOP TO THINK 2.4 A particle undergoes acceleration $\vec{a}$ while moving from point 1 to point 2. Which of the choices shows the velocity vector $\vec{v}_2$ as the particle moves from point 2 to point 3?

2 $\vec{v}_1$ 1
$\vec{a}$

A: 3 $\vec{v}_2$ 2
B: 3 $\vec{v}_2$ 2
C: 2 $\vec{v}_2$ 3
D: 2 $\vec{v}_2$ 3

Examples of Motion Diagrams

Let's look at some examples of the full strategy for drawing motion diagrams.

EXAMPLE 2.5 A hungry cheetah

Cheetahs are renowned for their rapid acceleration and top speed. But a cheetah can maintain that high speed for only a few seconds. If a hungry cheetah begins its charge from too far away, its prey animal will escape if it manages to stay ahead of the cheetah until the cheetah tires. Draw the motion diagram of a cheetah that accelerates quickly, runs at top speed for a short distance, and then slows after failing to catch its dinner.

PREPARE The cheetah is small in comparison with the distance covered, so it's reasonable to model the cheetah as a particle. We'll assume that the motion is along a straight line.

SOLVE FIGURE 2.13 is a complete motion diagram of the cheetah. The numbered circles in the insets use the steps of Tactics Box 2.1 to find the acceleration vector $\vec{a}$ during two segments of the motion. All the other acceleration vectors during those segments will be similar. There's no acceleration during the constant-speed segment of the run because there's no change of velocity. Notice that we've explicitly written $\vec{a} = \vec{0}$ where the velocity is constant.

ASSESS The acceleration and velocity vectors point in the *same direction* when the cheetah is speeding up. The acceleration is acting to make the velocity vectors longer. The acceleration and velocity vectors point in *opposite directions* when the cheetah is slowing down. In this case, the acceleration is acting to make the velocity vectors shorter.

FIGURE 2.13 Motion diagram of a cheetah.

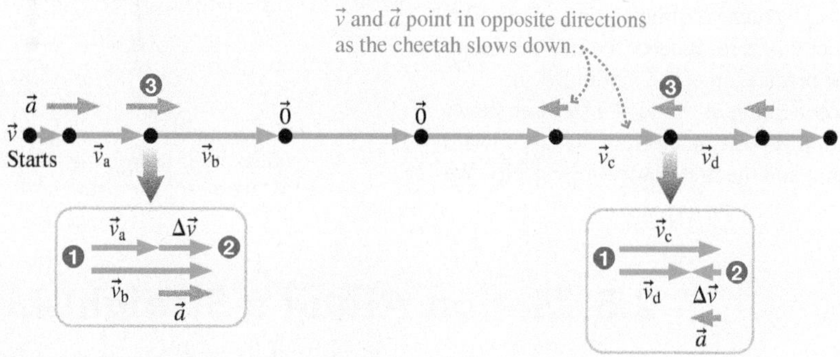

Our observations at the end of Example 2.5 are always true for motion in a straight line. **For motion along a line:**

- **An object is speeding up if and only if $\vec{v}$ and $\vec{a}$ point in the same direction.**
- **An object is slowing down if and only if $\vec{v}$ and $\vec{a}$ point in opposite directions.**
- **An object's velocity is constant if and only if $\vec{a} = \vec{0}$.**

NOTE ▸ In everyday language, we use the word "accelerate" to mean *speed up* and the word "decelerate" to mean *slow down.* But speeding up and slowing down are both changes in the velocity and consequently, by our definition, *both* are accelerations. In physics, "acceleration" refers to changing the velocity, no matter what the change is, and not just to speeding up. ◂

EXAMPLE 2.6 Tossing a ball

Draw the motion diagram of a ball tossed straight up in the air.

PREPARE This problem calls for some interpretation. Should we include the toss itself, or only the motion after the ball is released? What about catching it? It appears that this problem is really concerned with the ball's motion through the air. Consequently, we begin the motion diagram at the instant that the tosser releases the ball and end the diagram at the instant the ball touches his hand. We will consider neither the toss nor the catch. And, of course, we will model the ball as a particle.

SOLVE We have a slight difficulty here because the ball retraces its route as it falls. A literal motion diagram would show the upward motion and downward motion on top of each other, leading to confusion. We can avoid this difficulty by horizontally separating the upward motion and downward motion diagrams. This will not affect our conclusions because it does not change any of the vectors. FIGURE 2.14 shows the motion diagram drawn this way. Notice that the very top dot is shown twice—as the end point of the upward motion and the beginning point of the downward motion.

The ball slows down as it rises. You've learned that the acceleration vectors point opposite the velocity vectors for an object that is slowing down along a line, and they are shown accordingly. Similarly, $\vec{a}$ and $\vec{v}$ point in the same direction as the falling ball speeds up. Notice something interesting: The acceleration vectors point downward both while the ball is rising *and* while it is falling. Both "speeding up" and "slowing down" occur with the *same* acceleration vector. This is an important conclusion, one worth pausing to think about.

Now let's look at the top point on the ball's trajectory. The velocity vectors point upward but are getting shorter as the ball approaches the top. As the ball starts to fall, the velocity vectors point downward and are getting longer. There must be a moment—just an instant as $\vec{v}$ switches from pointing up to pointing down—when the velocity is zero. Indeed, the ball's velocity *is* zero for an instant at the precise top of the motion!

But what about the acceleration at the top? The inset shows how the average acceleration is determined from the last upward velocity before the top point and the first downward velocity. We find that the acceleration at the top is pointing downward, just as it does elsewhere in the motion.

Many people expect the acceleration to be zero at the highest point. But the velocity at the top point *is* changing—from up to down. If the velocity is changing, there *must* be an acceleration. A downward-pointing acceleration vector is needed to turn the velocity vector from up to down. Another way to think about this is to note that zero acceleration would mean no change of velocity. When the ball reached zero velocity at the top, it would hang there and not fall if the acceleration were also zero!

FIGURE 2.14 Motion diagram of a ball tossed straight up in the air.

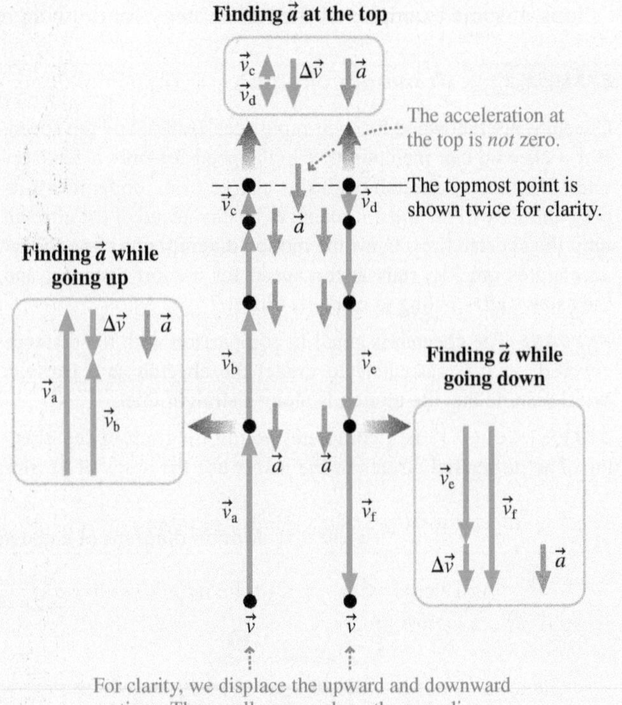

2.5 Motion Along a Straight Line

An object's motion can be described in terms of three fundamental quantities: its position $\vec{r}$, velocity $\vec{v}$, and acceleration $\vec{a}$. These are vectors, but for motion along a straight line, the vectors are restricted to point only "forward" or "backward." Consequently, we can describe one-dimensional motion with the simpler quantities x, v_x, and a_x (or y, v_y, and a_y). However, we need to give each of these quantities an explicit *sign,* positive or negative, to indicate whether the position, velocity, or acceleration vector points right or left (or, for vertical motion, up or down).

NOTE ▶ The subscripts on v and a are important. They indicate that this is really a vector quantity in which we're using a positive or negative sign to show direction. The quantities v and a without a subscript represent only the magnitude of the velocity (i.e., the speed) and the magnitude of the acceleration. ◀

Determining the Signs of Position, Velocity, and Acceleration

Position, velocity, and acceleration are measured with respect to a coordinate system, a grid or axis that *you* impose on a problem to analyze the motion. We will use an x-axis to describe horizontal motion and a y-axis for vertical motion. A coordinate axis has two essential features:

1. An origin to define zero; and
2. An x or y label (with units) at the positive end of the axis.

As mentioned earlier, we will follow the convention that **the positive end of an x-axis is to the right and the positive end of a y-axis is up.** The signs of position, velocity, and acceleration are based on this convention.

TACTICS BOX 2.3 Determining the sign of the position, velocity, and acceleration

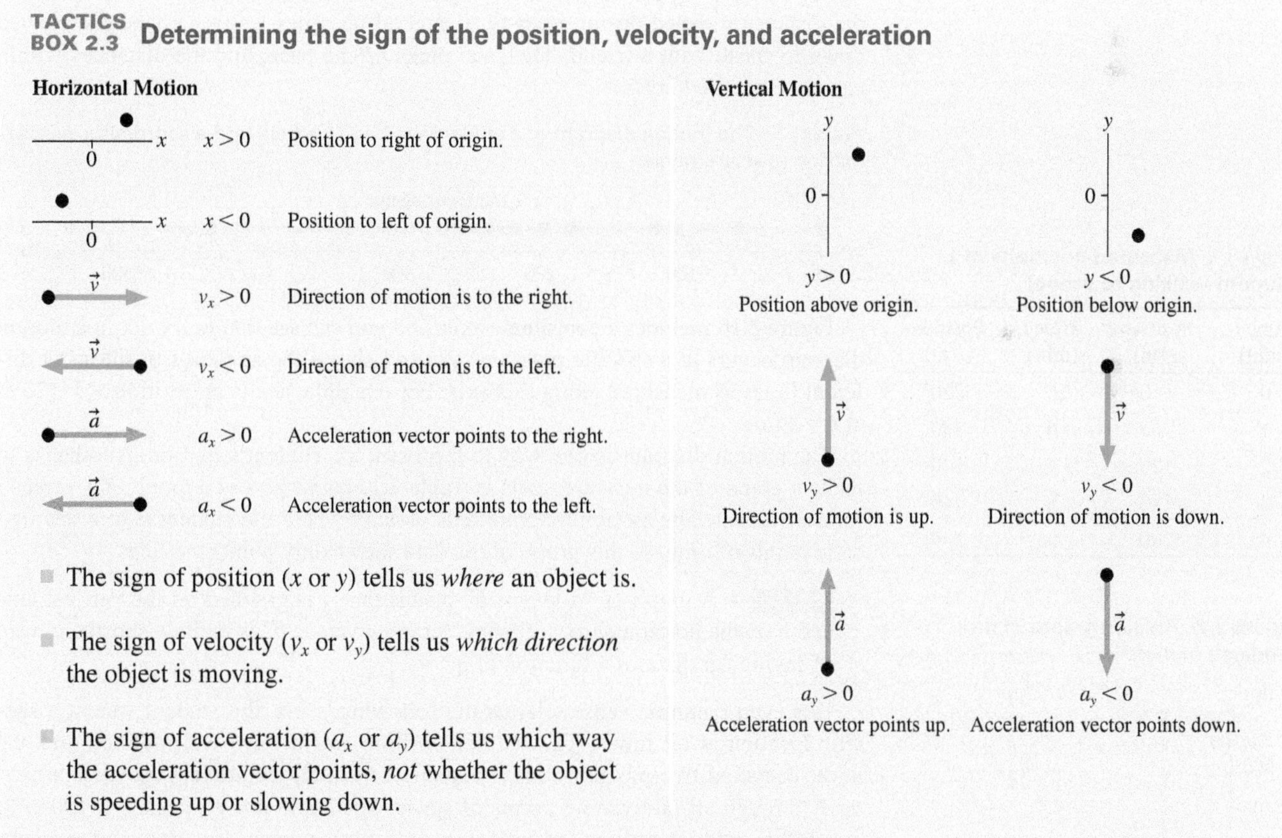
Horizontal Motion
$x > 0$ Position to right of origin.
$x < 0$ Position to left of origin.
$v_x > 0$ Direction of motion is to the right.
$v_x < 0$ Direction of motion is to the left.
$a_x > 0$ Acceleration vector points to the right.
$a_x < 0$ Acceleration vector points to the left.
Vertical Motion
$y > 0$ Position above origin.
$y < 0$ Position below origin.
$v_y > 0$ Direction of motion is up.
$v_y < 0$ Direction of motion is down.
$a_y > 0$ Acceleration vector points up.
$a_y < 0$ Acceleration vector points down.
The sign of position (x or y) tells us *where* an object is.
The sign of velocity (v_x or v_y) tells us *which direction* the object is moving.
The sign of acceleration (a_x or a_y) tells us which way the acceleration vector points, *not* whether the object is speeding up or slowing down.

Acceleration is where things get a bit tricky. A natural tendency is to think that a positive value of a_x or a_y describes an object that is speeding up while a negative value describes an object that is slowing down (decelerating). However, this interpretation *does not work.*

Acceleration is defined as $\vec{a} = \Delta\vec{v}/\Delta t$. The direction of $\vec{a}$ can be determined by using a motion diagram to find the direction of $\Delta\vec{v}$. The one-dimensional acceleration a_x (or a_y) is then positive if the vector $\vec{a}$ points to the right (or up), negative if $\vec{a}$ points to the left (or down).

FIGURE 2.15 One of these objects is speeding up, the other slowing down, but they both have a positive acceleration a_x.

(a) Speeding up to the right

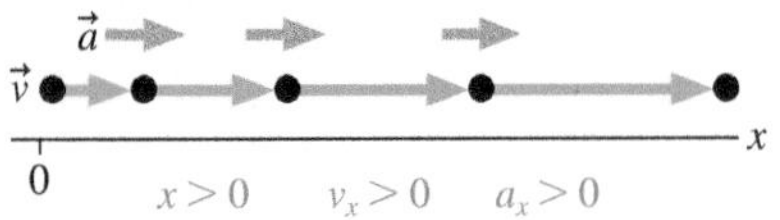

(b) Slowing down to the left

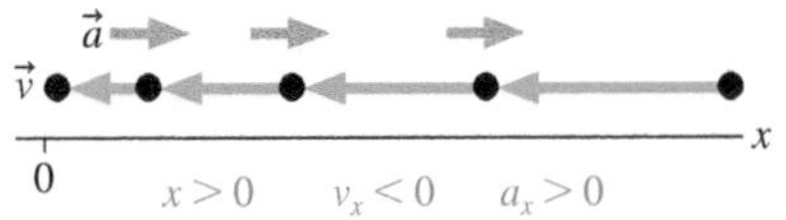

FIGURE 2.15 shows that this method for determining the sign of a does not conform to the simple idea of speeding up and slowing down. The object in Figure 2.15a has a positive acceleration ($a_x > 0$) not because it is speeding up (although it is) but because the vector $\vec{a}$ points in the positive direction. Compare this with the motion diagram of Figure 2.15b. Here the object is slowing down, but it has a positive acceleration ($a_x > 0$) because $\vec{a}$ still points to the right.

In the previous section, we found that an object is speeding up if $\vec{v}$ and $\vec{a}$ point in the same direction, slowing down if they point in opposite directions. For one-dimensional motion this rule becomes:

- An object is speeding up if and only if v_x and a_x have the same sign.
- An object is slowing down if and only if v_x and a_x have opposite signs.
- An object's velocity is constant if and only if $a_x = 0$.

Notice how the first two of these rules are at work in Figure 2.15.

Position Graphs

FIGURE 2.16 is a motion diagram, made at 1 frame per minute, of a student walking to school. You can see that she leaves home at a time we choose to call $t = 0$ min and makes steady progress for a while. Beginning at $t = 3$ min there is a period where the distance traveled during each time interval becomes less—perhaps she slowed down to speak with a friend. Then she picks up the pace, and the distances within each interval are longer.

FIGURE 2.16 The motion diagram of a student walking to school and a coordinate axis for making measurements.

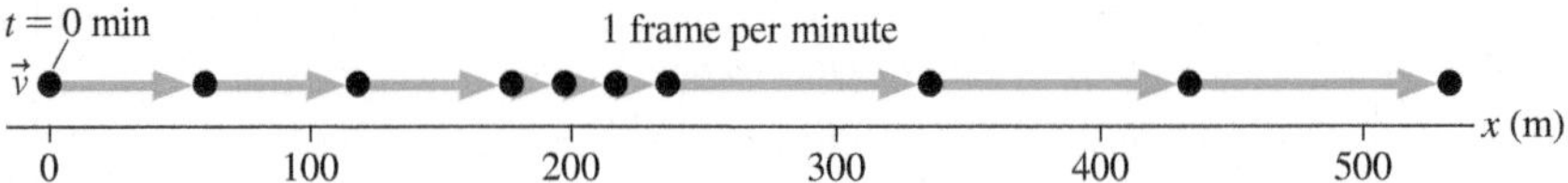

Figure 2.16 includes a coordinate axis, and you can see that every dot in a motion diagram occurs at a specific position. TABLE 2.1 shows the student's positions at different times as measured along this axis. For example, she is at position $x = 120$ m at $t = 2$ min.

TABLE 2.1 **Measured positions of a student walking to school**

Time t (min)	Position x (m)	Time t (min)	Position x (m)
0	0	5	220
1	60	6	240
2	120	7	340
3	180	8	440
4	200	9	540

The motion diagram is one way to represent the student's motion. Another is to make a graph of the measurements in Table 2.1. FIGURE 2.17a is a graph of x versus t for the student. The motion diagram tells us only where the student is at a few discrete points of time, so this graph of the data shows only points, no lines.

NOTE ▸ A graph of "a versus b" means that a is graphed on the vertical axis and b on the horizontal axis. Saying "graph a versus b" is really a shorthand way of saying "graph a as a function of b." ◂

FIGURE 2.17 Position graphs of the student's motion.

(a)

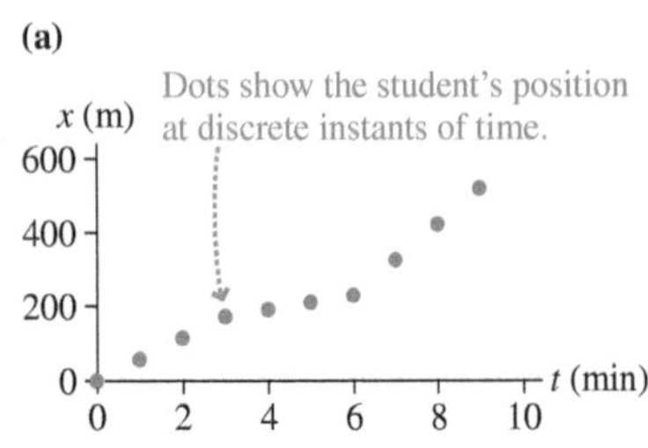

(b)

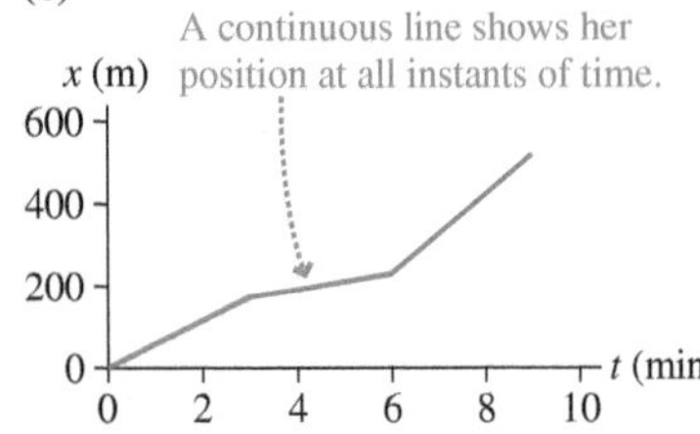

However, common sense tells us the following. First, the student was at a specific location at all times. That is, her position was always well defined and she never occupied two positions simultaneously. Second, the student moved *continuously* through all intervening points of space. She could not go from $x = 100$ m to $x = 200$ m without passing through every point in between. It is thus quite reasonable to believe that her motion can be shown as a continuous line passing through the measured points, as shown in FIGURE 2.17b. A continuous line or curve that shows an object's position as a function of time, either x versus t or y versus t, is called a **position graph.**

NOTE ▸ A graph is *not* a "picture" of the motion. The student is walking along a straight line, but the graph itself is not a straight line. Further, we've graphed her position on the vertical axis even though her motion is horizontal. Graphs are *abstract representations* of motion. We will place significant emphasis on the process of interpreting graphs, and many of the exercises and problems will give you a chance to practice these skills. ◂

EXAMPLE 2.7 Interpreting a position graph

The graph in FIGURE 2.18a represents the motion of a car along a straight road. Describe the motion of the car.

PREPARE We'll model the car as a particle with a precise position at each instant.

FIGURE 2.18 Position graph of a car.

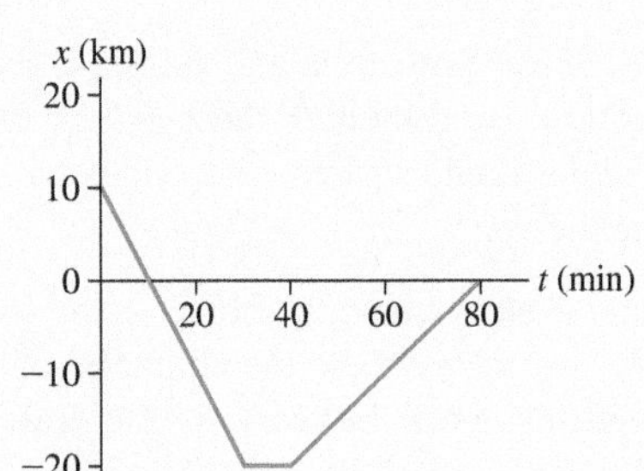
(a)
x (km)
20
10
0
−10
−20
20
40
60
80
t (min)

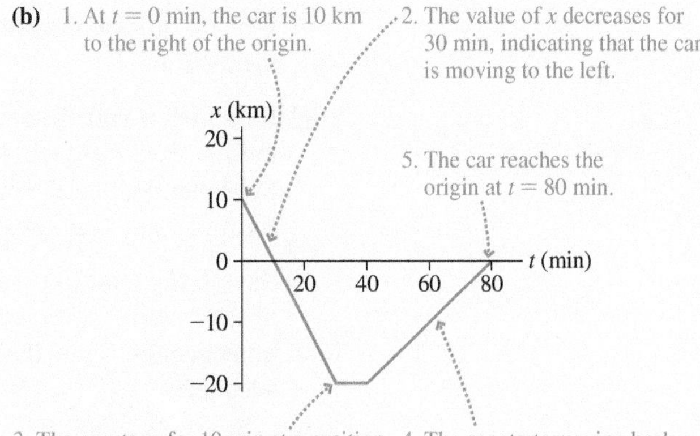
(b)
1. At t = 0 min, the car is 10 km to the right of the origin.
2. The value of x decreases for 30 min, indicating that the car is moving to the left.
5. The car reaches the origin at t = 80 min.
x (km)
20
10
0
−10
−20
20
40
60
80
t (min)
3. The car stops for 10 min at a position 20 km to the left of the origin.
4. The car starts moving back to the right at t = 40 min.

SOLVE As FIGURE 2.18b shows, the position graph represents a car that travels to the left for 30 minutes, stops for 10 minutes, then travels back to the right for 40 minutes.

Symbols and Mathematics

Physics is not mathematics. Math problems are clearly stated, such as "What is $2 + 2$?" Physics is about the world around us, and to describe that world we must use language. Now, language is wonderful—we couldn't communicate without it—but language can sometimes be imprecise or ambiguous.

The challenge when reading a physics problem is to translate the words into symbols that can be manipulated, calculated, and graphed. **The translation from words to symbols is the heart of problem solving in physics.** This is the point where ambiguous words and phrases must be clarified, where the imprecise must be made precise, and where you arrive at an understanding of exactly what the question is asking.

Mathematics is a language that allows us to specify quantitative relationships between quantities. In mathematical equations, concepts such as position and velocity are represented by symbols. Some symbols are logical; others are historical and may seem arbitrary. Furthermore, it's common for biologists, chemists, and physicists to use different symbols for the same ideas, so it's important to always be sure you know what underlying idea is represented by a given symbol! This book will use the symbols that are most common in physics. We'll always explain them when we introduce them, but thereafter it's up to you to learn to read this language with facility.

We will use subscripts on symbols, such as x_3, to designate a particular point in the problem. Physicists usually label the starting point of the problem with the subscript "0," not the subscript "1" that you might expect. When using subscripts, make sure that all symbols referring to the same point in the problem have the *same numerical subscript.* To have the same point in a problem characterized by position x_1 but velocity v_{2x} is guaranteed to lead to confusion!

The Pictorial Representation

It's often the case that the first step in solving a physics problem is to "draw a picture." The purpose of drawing a picture is to aid you in the words-to-symbols translation. Complex problems have far more information than you can keep in your head at one time. Think of a picture as a "memory extension," helping you organize and keep track of vital information.

Although any picture is better than none, there really is a *method* for drawing pictures that will help you be a better problem solver. It is called the **pictorial representation** of the problem. We'll add other pictorial representations as we go along, but the following procedure is appropriate for motion problems.

TACTICS BOX 2.4 Drawing a pictorial representation

1. **Draw a motion diagram.** The motion diagram develops your intuition for the motion.
2. **Establish a coordinate system.** Select your axes and origin to match the motion. For one-dimensional motion, choose either the x-axis or the y-axis to be parallel to the motion. The coordinate system determines whether the signs of v and a are positive or negative.
3. **Sketch the situation.** Not just any sketch. Show the object at the *beginning* of the motion, at the *end,* and at any point where the character of the motion changes. Show the object, not just a dot, but very simple drawings are adequate.
4. **Define symbols.** On the sketch, define symbols representing quantities such as position, velocity, acceleration, and time. Use subscripts 0, 1, 2, ... to distinguish the points in the motion that you identified in step 3. *Every* variable used later in the mathematical solution should be defined on the sketch. Some will have known values, others are initially unknown, but all should be given symbolic names.
5. **List known information.** Make a table of the quantities whose values you can determine from the problem statement or that can be found quickly with simple geometry or unit conversions. Some quantities are implied by the problem, rather than explicitly given. Others are determined by your choice of coordinate system.
6. **Identify the desired unknowns.** What quantity or quantities will allow you to answer the question? These should have been defined as symbols in step 4. Don't list every unknown, only the one or two needed to answer the question.

It's not an overstatement to say that a well-done pictorial representation of the problem will take you halfway to the solution. The following example illustrates how to construct a pictorial representation for a problem that is typical of problems you will see in the next few chapters.

EXAMPLE 2.8 Drawing a pictorial representation

Draw a pictorial representation for the following problem: A rocket sled accelerates horizontally at 50 m/s^2 for 5.0 s, then coasts for 3.0 s. What is the total distance traveled?

PREPARE We'll follow the steps of Tactics Box 2.4. Assume that the motion is to the right.

SOLVE FIGURE 2.19 is the pictorial representation. The motion diagram shows an acceleration phase followed by a coasting phase. Because the motion is horizontal, the appropriate coordinate system is an x-axis. We've chosen to place the origin at the starting point. The motion has a beginning, an end, and a point where the motion changes from accelerating to coasting, and these are the three sled positions sketched in the figure. The quantities x, v_x, and t are needed at each of three *points,* so these have been defined on the sketch and distinguished by subscripts 0, 1, and 2. Accelerations are associated with *intervals* between the points, so only two accelerations are defined. Values for three quantities are given in the problem statement, although we need to use the motion diagram, where we find that $\vec{a}$ points to the right, to know that $a_{0x} = +50$ m/s^2 rather than -50 m/s^2. The values $x_0 = 0$ m and $t_0 = 0$ s are choices we made when setting up the coordinate system. The value $v_{0x} = 0$ m/s is part of our *interpretation* of the problem. Finally, we identify x_2 as the quantity that will answer the question. We now understand quite a bit about the problem and would be ready to start a quantitative analysis.

FIGURE 2.19 A pictorial representation.

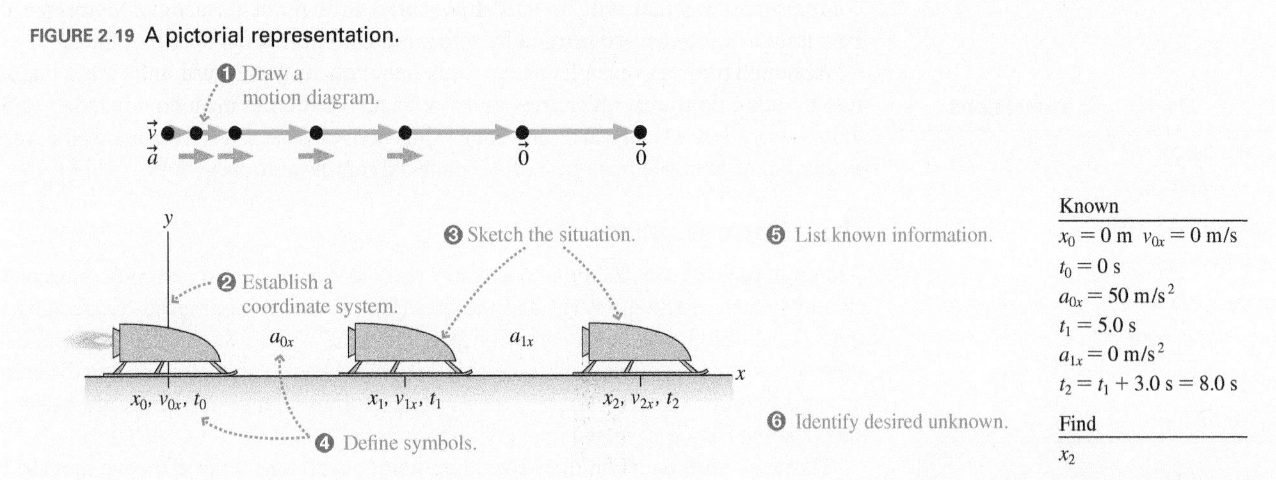

We didn't *solve* the problem; that is not the purpose of the pictorial representation. The pictorial representation is a systematic way to go about interpreting a problem and getting ready for a mathematical solution. Although you might be able to solve this problem without a pictorial representation if you've studied physics, there will be many problems in this course where you'll find a pictorial representation essential to avoid making errors. Developing good problem-solving skills at the beginning will make them second nature later.

Visualization A new building requires careful planning. The architect's visualization and drawings have to be complete before the detailed procedures of construction get under way. The same is true for solving problems in physics.

Types of Representations

A picture is one way to *represent* your knowledge of a situation. You could also represent your knowledge using words, graphs, or equations. Each **representation of knowledge** gives us a different perspective on the problem. The more tools you have for thinking about a complex problem, the more likely you are to solve it.

There are four representations of knowledge that we will use over and over:

1. The *verbal* representation. A problem statement, in words, is a verbal representation of knowledge. So is an explanation that you write.
2. The *pictorial* representation. The pictorial representation, which we've just presented, is the most literal depiction of the situation.
3. The *graphical* representation. We will make extensive use of graphs.
4. The *mathematical* representation. Equations that can be used to find the numerical values of specific quantities are the mathematical representation.

NOTE ▶ The mathematical representation is only one of many. Much of physics is more about thinking and reasoning than it is about solving equations. ◀

2.6 Units and Significant Figures

Science is based upon experimental measurements, and measurements require *units.* The system of units used in physics is called *le Système Internationale d'Unités.* These are commonly referred to as **SI units.** All of the quantities needed to understand motion can be expressed in terms of the three basic SI units shown in TABLE 2.2. Other quantities can be expressed as a combination of these basic units. Velocity, expressed in meters per second or m/s, is a ratio of the length unit to the time unit.

We will have many occasions to use lengths, times, and masses that are either much less or much greater than the standards of 1 meter, 1 second, and 1 kilogram. We will do so by using *prefixes* to denote various powers of 10. TABLE 2.3 lists the

TABLE 2.2 **The basic SI units**

Quantity	Unit	Abbreviation
time	second	s
length	meter	m
mass	kilogram	kg

TABLE 2.3 **Common prefixes**

Prefix	Power of 10	Abbreviation
giga-	10^9	G
mega-	10^6	M
kilo-	10^3	k
centi-	10^{-2}	c
milli-	10^{-3}	m
micro-	10^{-6}	μ
nano-	10^{-9}	n
pico-	10^{-12}	p

common prefixes that will be used frequently throughout this book. Memorize it! Few things in science are learned by rote memory, but this list is one of them.

Although prefixes make it easier to talk about quantities, the SI units are seconds, meters, and kilograms. Quantities given with prefixed units must be converted to SI units before any calculations are done. Unit conversions are best done at the very beginning of a problem, as part of the pictorial representation.

TABLE 2.4 **Useful unit conversions**

1 in = 2.54 cm
1 mi = 1.609 km
1 mph = 0.447 m/s
1 m = 39.37 in
1 km = 0.621 mi
1 m/s = 2.24 mph
1 $m^3 = 10^3$ L
1 $m^3 = 10^6\ cm^3$

Unit failure In 1999, the $125 million Mars Climate Orbiter burned up in the Martian atmosphere instead of entering a safe orbit. The problem was faulty units! The engineering team supplied data in English units, but the navigation team assumed that the data were in SI units.

Unit Conversions

Although physics uses SI units, chemistry and biology frequently employ other metric units, such as the gram (g) and the liter (L), as well as nonmetric or specialized units. In addition, we cannot entirely forget that the United States still uses English units. Thus it's very important to be able to convert back and forth between different systems of units. Unit conversions are a critical skill in science and medicine. TABLE 2.4 shows some frequently used conversion factors.

There are various techniques for doing unit conversions. One effective method is to write the conversion factor as a ratio equal to one. For example, using information in Tables 2.3 and 2.4, we have

$$\frac{10^{-6}\ \text{m}}{1\ \mu\text{m}} = 1 \quad \text{and} \quad \frac{2.54\ \text{cm}}{1\ \text{in}} = 1$$

Because multiplying any expression by 1 does not change its value, these ratios are easily used for conversions. To convert 3.5 μm to meters we compute

$$3.5\ \cancel{\mu\text{m}} \times \frac{10^{-6}\ \text{m}}{1\ \cancel{\mu\text{m}}} = 3.5 \times 10^{-6}\ \text{m}$$

Similarly, the conversion of 2 feet to meters is

$$2.00\ \cancel{\text{ft}} \times \frac{12\ \cancel{\text{in}}}{1\ \cancel{\text{ft}}} \times \frac{2.54\ \cancel{\text{cm}}}{1\ \cancel{\text{in}}} \times \frac{10^{-2}\ \text{m}}{1\ \cancel{\text{cm}}} = 0.610\ \text{m}$$

Notice how units in the numerator and in the denominator cancel until only the desired units remain at the end. You can continue this process of multiplying by 1 as many times as necessary to complete all the conversions.

EXAMPLE 2.9 Finding the density of glucose syrup

A lab assignment asks students to measure the density of glucose syrup. One group of students carefully fills a 10.0 mL graduated cylinder with the syrup and then places it on a scale, recording a reading of 40.1 g. They had previously determined that the mass of an empty cylinder is 25.8 g. What is the density of glucose syrup in SI units?

PREPARE In chemistry and biology, density—mass per unit volume—is often given in the units g/cm^3. However, the SI unit of density is kg/m^3. You may recall that 1 mL = 1 cm^3. An important conversion factor, found in Table 2.4, is 1 m^3 = 10^6 cm^3.

SOLVE The mass of the syrup is 40.1 g − 25.8 g = 14.3 g, so its density is

$$\frac{14.3\ \text{g}}{10.0\ \text{cm}^3} = 1.43\ \text{g/cm}^3$$

This is a little larger than the density of water, 1.00 g/cm^3. The conversion to SI units then takes two steps:

$$1.43\ \frac{\text{g}}{\text{cm}^3} \times \frac{1\ \text{kg}}{10^3\ \text{g}} \times \frac{10^6\ \text{cm}^3}{1\ \text{m}^3} = 1430\ \text{kg/m}^3$$

The density of water in SI units is 1000 kg/m^3, a value worth memorizing to use as a comparison.

ASSESS Thick sugar syrups are denser than water, so 1430 kg/m^3 seems reasonable. Measuring in grams and milliliters may be practical in the laboratory, but quantities need to be expressed in SI units for many kinds of calculations. Examples and homework throughout this text use a wide variety of units to give you real-world practice.

Assessment

As we get further into problem solving, you will need to decide whether or not the answer to a problem “makes sense.” To determine this, at least until you have more experience with SI units, you may need to convert from SI units back to the English units in which you think. But this conversion does not need to be very accurate. For example, if you are working a problem about blood flow and reach an answer of 0.35 m/s, all you really want to know is whether or not this is a realistic speed for blood flowing through an artery. That requires a "quick and dirty" conversion, not a conversion of great accuracy.

TABLE 2.5 shows several approximate conversion factors that can be used to assess the answer to a problem. Using 1 m ≈ 3 ft, you find that 0.35 m/s is roughly 1 ft/s, a reasonable speed for arterial blood flow. But an answer of 35 m/s, which you might get after making a calculation error, would be an unreasonable 100 ft/s. Practice with these conversion factors will allow you to develop intuition for metric units; you'll probably find it useful to memorize a few of them.

NOTE ▸ These approximate conversion factors are accurate to only one significant figure. This is sufficient to assess the answer to a problem, but do *not* use the conversion factors from Table 2.5 for converting English units to SI units at the start of a problem. Use Table 2.4. ◂

TABLE 2.5 Approximate conversion factors. Use these for assessment, not in problem solving.

1 cm ≈ $\frac{1}{2}$ in
10 cm ≈ 4 in
1 m ≈ 1 yard
1 m ≈ 3 feet
1 km ≈ 0.6 mile
1 m/s ≈ 2 mph

Significant Figures

When we measure any quantity, such as the length of a bone or the weight of a specimen, we can do so with only a certain *precision.* If you make a measurement with the ruler shown in FIGURE 2.20, you probably can't be more accurate than about ±1 mm, so the ruler has a precision of 1 mm. The digital calipers shown can make a measurement to within ±0.01 mm, so it has a precision of 0.01 mm. The precision of a measurement can also be affected by the skill or judgment of the person performing the measurement. A stopwatch might have a precision of 0.001 s, but, due to your reaction time, your measurement of the time of a sprinter would be much less precise.

It is important that your measurement be reported in a way that reflects its actual precision. Suppose you use a ruler to measure the length of a particular frog. You judge that you can make this measurement with a precision of about 1 mm, or 0.1 cm. In this case, the frog's length should be reported as, say, 6.2 cm. We interpret this to mean that the actual value falls between 6.15 cm and 6.25 cm and thus rounds to 6.2 cm. If you reported the frog's length as simply 6 cm, you would be saying less than you know; you would be withholding information. If you reported the number as 6.213 cm, however, anyone reviewing your work would interpret this to mean that the actual length falls between 6.2125 cm and 6.2135 cm, a precision of 0.001 cm. In this case, you would be claiming to have more information than you really possessed.

The way to state your knowledge precisely is through the proper use of **significant figures.** You can think of a significant figure as being a digit that is reliably known. A number such as 6.2 cm has *two* significant figures because the next decimal place—the one-hundredths—is not reliably known. As FIGURE 2.21 shows, the best way to determine how many significant figures a number has is to write it in scientific notation.

What about numbers like 320 m and 20 kg? Whole numbers with trailing zeros are ambiguous unless written in scientific notation. Even so, writing 2.0×10^1 kg is

FIGURE 2.20 The precision of a measurement depends on the instrument used to make it.

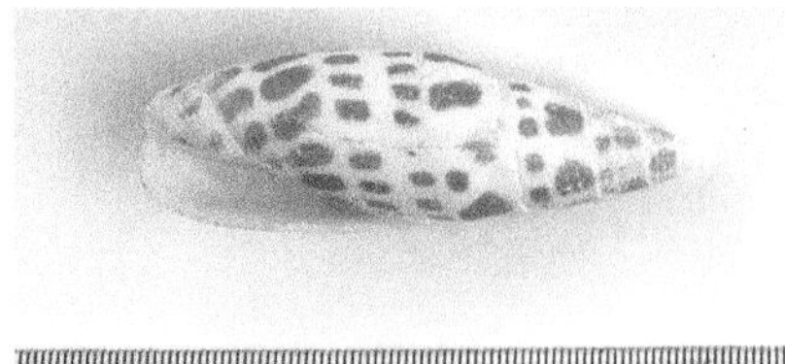
8
9
10
11
12
13
14
15
16
17
This ruler has a precision of 1 mm.

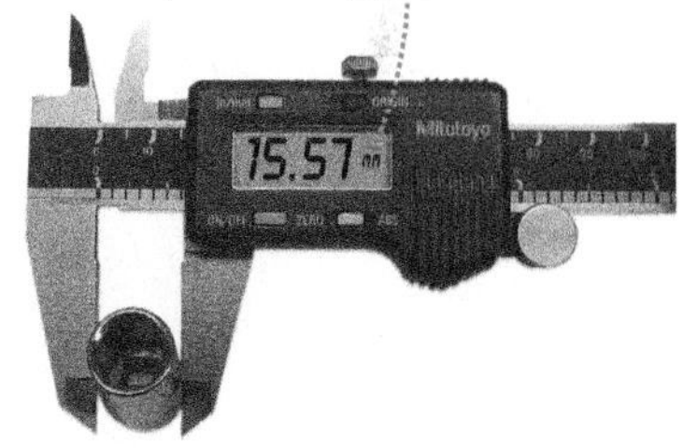
These calipers have a precision of 0.01 mm.
15.57

FIGURE 2.21 Determining significant figures.

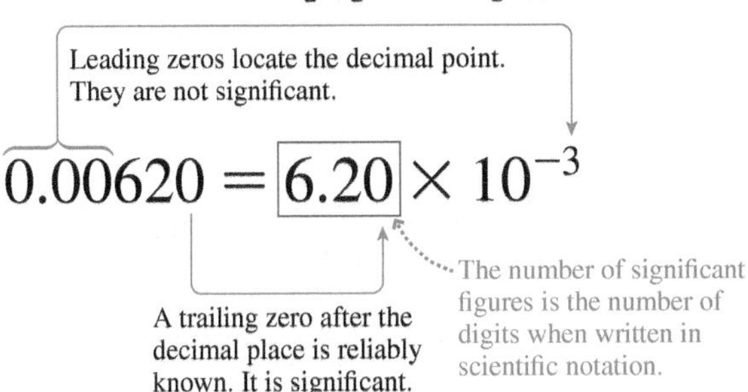
Leading zeros locate the decimal point. They are not significant.
0.00620 = 6.20 × 10^-3
A trailing zero after the decimal place is reliably known. It is significant.
The number of significant figures is the number of digits when written in scientific notation.

- The number of significant figures ≠ the number of decimal places.
- Changing units shifts the decimal point but does not change the number of significant figures.

tedious, and few practicing scientists would do so. In this textbook, we'll adopt the rule that **whole numbers always have at least two significant figures,** even if one of those is a trailing zero. By this rule, 320 m, 20 kg, and 8000 s each have two significant figures, but 8050 s would have three.

Calculations with numbers follow the "weakest link" rule. The saying, which you probably know, is that "a chain is only as strong as its weakest link." If nine out of ten links in a chain can support a 1000 pound weight, that strength is meaningless if the tenth link can support only 200 pounds. Nine out of the ten numbers used in a calculation might be known with a precision of 0.01%; but if the tenth number is poorly known, with a precision of only 10%, then the result of the calculation cannot possibly be more precise than 10%.

TACTICS BOX 2.5 Using significant figures

1. When multiplying or dividing several numbers, or taking roots, the number of significant figures in the answer should match the number of significant figures of the *least* precisely known number used in the calculation.
2. When adding or subtracting several numbers, the number of decimal places in the answer should match the *smallest* number of decimal places of any number used in the calculation.
3. Exact numbers are perfectly known and do not affect the number of significant figures an answer should have. Examples of exact numbers are the 2 and the π in the formula $C = 2\pi r$ for the circumference of a circle.
4. It is acceptable to keep one or two extra digits during intermediate steps of a calculation, to minimize rounding error, as long as the final answer is reported with the proper number of significant figures.
5. For examples and problems in this text, **the appropriate number of significant figures for the answer is determined by the data provided.** Whole numbers with trailing zeros, such as 20 kg, are interpreted as having at least two significant figures.

NOTE ▶ Be careful! Many calculators have a default setting that shows two decimal places, such as 5.23. This is dangerous. If you need to calculate 5.23/58.5, your calculator will show 0.09 and it is all too easy to write that down as an answer. By doing so, you have reduced a calculation of two numbers having three significant figures to an answer with only one significant figure. The proper result of this division is 0.0894 or 8.94×10^{-2}. You will avoid this error if you keep your calculator set to display numbers in *scientific notation* with two decimal places. ◀

EXAMPLE 2.10 Using significant figures

An object consists of two pieces. The mass of one piece has been measured to be 6.47 kg. The volume of the second piece, which is made of aluminum, has been measured to be $4.44 \times 10^{-4}\ \text{m}^3$. A handbook lists the density of aluminum as $2.7 \times 10^3\ \text{kg/m}^3$. What is the total mass of the object?

SOLVE First, we calculate the mass of the second piece:

$$m = (4.44 \times 10^{-4}\ \text{m}^3)(2.7 \times 10^3\ \text{kg/m}^3)$$
$$= 1.199\ \text{kg} = 1.2\ \text{kg}$$

The number of significant figures of a product must match that of the *least* precisely known number, which is the two-significant-figure density of aluminum. Now we add the two masses:

$$\begin{array}{r} 6.47\ \text{kg} \\ +\ 1.2\ \text{kg} \\ \hline 7.7\ \text{kg} \end{array}$$

The sum is 7.67 kg, but the hundredths place is not reliable because the second mass has no reliable information about this digit. Thus we must round to the one decimal place of the 1.2 kg. The best we can say, with reliability, is that the total mass is 7.7 kg.

Significant figures are the way scientists and physicians communicate how accurate a measurement is, and thus they are extremely important in determining what kind of decisions can be made from that measurement. For this reason, it is important to understand how to properly use significant figures.

Orders of Magnitude and Estimating

Precise calculations are appropriate when we have precise data, but there are many times when a very rough estimate is sufficient. Suppose you see a rock fall off a cliff and would like to know how fast it was going when it hit the ground. By doing a mental comparison with the speeds of familiar objects, such as cars and bicycles, you might judge that the rock was traveling at "about" 20 mph.

This is a one-significant-figure estimate. With some luck, you can distinguish 20 mph from either 10 mph or 30 mph, but you certainly cannot distinguish 20 mph from 21 mph. A one-significant-figure estimate or calculation, such as this, is called an *order-of-magnitude estimate.* An order-of-magnitude estimate is indicated by the symbol ~, which indicates even less precision than the "approximately equal" symbol ≈. You would say that the speed of the rock is $v \sim 20$ mph.

Order-of-magnitude estimates can be extremely useful in biology. In Example 2.2, for instance, order-of-magnitude data on the speed of DNA transcription indicated the need for a very large number of ribosomes. Other simple estimates can support—or contradict—the plausibility of a particular hypotheses. Because we have little direct personal experience with the microscopic world, however, many biologists find it useful to memorize a few benchmark numbers, such as the characteristic sizes of a bacterium, a nucleus, and a molecular bond.

TABLES 2.6 and 2.7 have information that will be useful for doing estimates.

TABLE 2.6 **Some approximate lengths**

	Length (m)
Distance across campus	1000
Length of a football field	100
Length of a classroom	10
Length of your arm	1
Width of a textbook	0.1
Length of a fingernail	0.01
Diameter of a mammalian cell	10^{-5}
Diameter of a bacterium	10^{-6}

TABLE 2.7 **Some approximate masses**

	Mass (kg)
Small car	1000
Large human	100
Medium-size dog	10
Science textbook	1
Apple	0.1
Pencil	0.01
Raisin	0.001
Mass of a mammalian cell	10^{-12}
Mass of a bacterium	10^{-15}

EXAMPLE 2.11 Estimating the number of proteins

The single-cell *E. coli* bacterium is widely used in biological and medical research. Cell analysis has found that its water content is ≈70% and that proteins make up ≈50% of the dry mass. A typical protein consists of 300 amino acids, and the average molecular mass of an amino acid is 100 daltons or, equivalently, 100 atomic mass units (amu). Use this information to estimate the number of proteins in an *E. coli* bacterium. A useful conversion factor is 1 dalton = 1 amu = 1.7×10^{-27} kg.

SOLVE Table 2.7 shows that the mass of a bacterium is $\approx 10^{-15}$ kg. 30% of that is dry mass, and 50% of the dry mass is protein, so the total protein mass of a cell is

$$M_{\text{protein}} \approx 0.5 \times 0.3 \times 10^{-15}\text{ kg} = 1.5 \times 10^{-16}\text{ kg}$$

On average, each protein contains 300 amino acids with, on average, each having a molecular mass of 100 daltons. Thus the mass in kg of a typical protein is

$$m_{1\text{ protein}} \approx 300\text{ amino acids} \times \frac{100\text{ daltons}}{1\text{ amino acid}} \times \frac{1.7 \times 10^{-27}\text{ kg}}{1\text{ dalton}}$$

$$= 5.1 \times 10^{-23}\text{ kg}$$

The number of proteins is the total protein mass divided by the mass per protein:

$$N_{\text{protein}} = \frac{M_{\text{protein}}}{m_{1\text{ protein}}} \approx \frac{1.5 \times 10^{-16}\text{ kg}}{5.1 \times 10^{-23}\text{ kg}} = 3 \times 10^{6}\text{ proteins}$$

We kept two significant figures during each stage of the calculation but then, because this is an order-of-magnitude estimate, rounded to one significant figure at the end.

ASSESS Despite its small size, a single *E. coli* cell contains several million proteins. The mass of a mammalian cell is 1000 times larger, so we would expect to find several billion proteins in a mammalian cell. An end-of-chapter problem will let you continue this analysis to estimate the average distance between two protein molecules in a cell.

STOP TO THINK 2.5 Rank in order, from the most to the least, the number of significant figures in the following numbers. For example, if B has more than C, C has the same number as A, and A has more than D, you could give your answer as B > C = A > D.

A. 82 B. 0.0052 C. 0.430 D. 4.321×10^{-10}

SUMMARY

GOAL To introduce the fundamental concepts of motion.

IMPORTANT CONCEPTS

Motion Diagrams

The **particle model** represents a moving object as if all its mass were concentrated at a single point. Using this model, we can represent motion with a **motion diagram.** A motion diagram is an especially important tool for finding acceleration vectors.

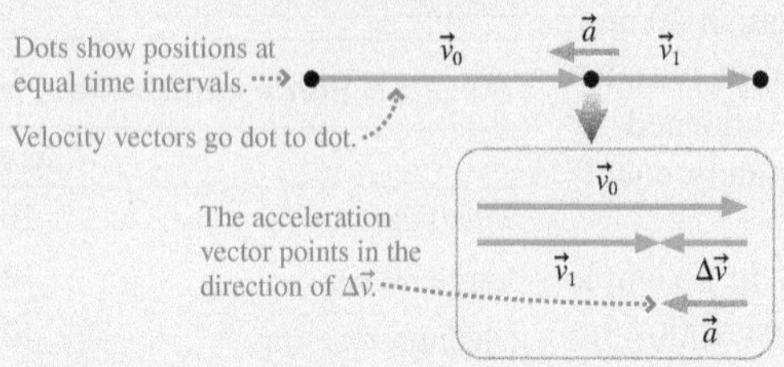

Describing Motion

To create a **pictorial representation:**

1. Draw a motion diagram.
2. Establish coordinates.
3. Sketch the situation.
4. Define symbols.
5. List knowns.
6. Identify desired unknown.

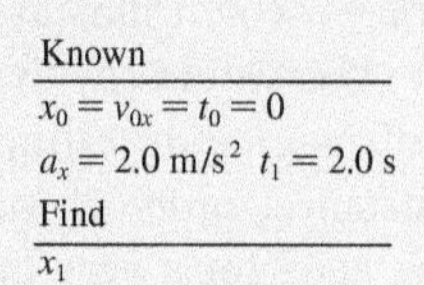

a_x; x; 0; x_0, v_{0x}, t_0; x_1, v_{1x}, t_1

Known
$x_0 = v_{0x} = t_0 = 0$
$a_x = 2.0\text{ m/s}^2$ $t_1 = 2.0\text{ s}$
Find
x_1

Motion Along a Line

Position locates an object with respect to a chosen coordinate system. Change in position is the **displacement** $\Delta\vec{r}$.

Velocity is the rate of change of position: $\vec{v} = \Delta\vec{r}/\Delta t$.

Acceleration is the rate of change of velocity: $\vec{a} = \Delta\vec{v}/\Delta t$.

- **Speeding up:** $\vec{v}$ and $\vec{a}$ point in the same direction, v_x and a_x have the same sign.
- **Slowing down:** $\vec{v}$ and $\vec{a}$ point in opposite directions, v_x and a_x have opposite signs.
- **Constant speed:** $\vec{a} = \vec{0}$, $a_x = 0$.

Acceleration a_x (a_y) is positive if $\vec{a}$ points right (up).

Acceleration a_x (a_y) is negative if $\vec{a}$ points left (down).

A **position graph** is an abstract representation of the motion that shows position on the vertical axis and time on the horizontal axis.

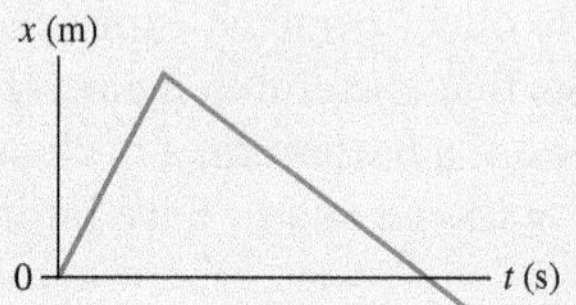

Units

Every measurement of a quantity must include a unit. The system of units used in physics is the **SI system.** Common SI units include:

- Length: meters (m)
- Time: seconds (s)
- Mass: kilograms (kg)

APPLICATIONS

Working with Numbers

In scientific notation, a number is expressed as a decimal number between 1 and 10 multiplied by a power of ten. In scientific notation, the diameter of the earth is 1.27×10^7 m.

A prefix can be used before a unit to indicate a multiple of 10 or 1/10. Thus we can write the diameter of the earth as 12,700 km, where the k in km denotes 1000.

We can perform a **unit conversion** to convert the diameter of the earth to a different unit, such as miles. We do so by multiplying by a conversion factor equal to 1, such as $1 = 1$ mi/1.61 km.

Significant figures are reliably known digits. The number of significant figures for:

- Multiplication, division, and powers is set by the value with the fewest significant figures.
- Addition and subtraction is set by the value with the smallest number of decimal places.

The appropriate number of significant figures in a calculation is determined by the accuracy of the data provided.

An **order-of-magnitude estimate** is an estimate that has an accuracy of about one significant figure. Such estimates are often based on everyday experience.

LEARNING OBJECTIVES After studying this chapter, you should be able to:

- Draw and interpret motion diagrams to represent motion. *Conceptual Questions 2.2, 2.3; Problems 2.1, 2.2, 2.15, 2.16*
- Describe motion in terms of position, displacement, and time. *Conceptual Questions 2.7, 2.9; Problems 2.3–2.6*
- Determine and show velocity vectors on a motion diagram. *Conceptual Questions 2.2, 2.3; Problems 2.11, 2.34–2.36, 2.38*
- Determine the acceleration vector for an object in linear motion. *Conceptual Questions 2.10, 2.11; Problems 2.13, 2.14, 2.36–2.38*
- Draw a position graph and a pictorial representation. *Problems 2.17, 2.18*
- Use scientific notation. *Multiple-Choice Question 2.24*
- Express quantities with the appropriate units and the proper number of significant figures. *Problems 2.24–2.27*
- Perform unit conversions. *Problems 2.21–2.23, 2.28, 2.29*

STOP TO THINK ANSWERS

Stop to Think 2.1: B. The images of B are farther apart, so it travels a larger distance than does A during the same intervals of time.

Stop to Think 2.2: C. Depending on her initial positive position and how far she moves in the negative direction, she could end up on either side of the origin.

Stop to Think 2.3: E. The average velocity vector is found by connecting one dot in the motion diagram to the next.

Stop to Think 2.4: B. $\vec{v}_2 = \vec{v}_1 + \Delta\vec{v}$, and $\Delta\vec{v}$ points in the direction of $\vec{a}$.

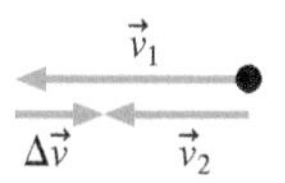

$\vec{v}_1$ $\Delta\vec{v}$ $\vec{v}_2$

Stop to Think 2.5: D > C > B = A.

QUESTIONS

Conceptual Questions

1. A softball player slides into second base. Use the particle model to draw a motion diagram of the player from the time he begins to slide until he reaches the base. Show and label the velocity and acceleration vectors.
2. A car travels to the left at a steady speed for a few seconds, then brakes for a stop sign. Use the particle model to draw a motion diagram of the car for the entire motion described here. Show and label the velocity and acceleration vectors.
3. BIO The bush baby, a small African mammal, is a remarkable jumper. Although only about 8 inches long, it can jump, from a standing start, straight up to a height of over 7 feet! Use the particle model to draw a motion diagram for a bush baby's jump, from its start until it reaches its highest point. Show and label the velocity and acceleration vectors.
4. Your roommate drops a tennis ball from a third-story balcony. It hits the sidewalk and bounces as high as the second story. Draw a motion diagram, using the particle model, showing the ball's velocity and acceleration vectors from the time it is released until it reaches the maximum height on its bounce.
5. A ball is dropped from the roof of a tall building and students in a physics class are asked to sketch a motion diagram for this situation. A student submits the diagram shown in Figure Q2.5. Is the diagram correct? Explain.

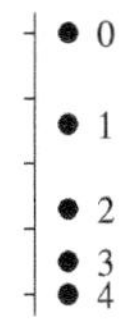

0 1 2 3 4

FIGURE Q2.5

6. Give an example of a trip you might take in your car for which the distance traveled as measured on your car's odometer is not equal to the displacement between your initial and final positions.
7. Two friends watch a jogger complete a 400 m lap around the track in 100 s. One of the friends states, "The jogger's velocity was 4 m/s during this lap." The second friend objects, saying, "No, the jogger's speed was 4 m/s." Who is correct? Justify your answer.
8. The motion of a skateboard along a horizontal axis is observed for 5 s. The initial position of the skateboard is negative with respect to a chosen origin, and its velocity throughout the 5 s is also negative. At the end of the observation time, is the skateboard closer to or farther from the origin than initially? Explain.
9. You are standing on a straight stretch of road and watching the motion of a bicycle; you choose your position as the origin. At one instant, the position of the bicycle is negative and its velocity is positive. Is the bicycle getting closer to you or farther away? Explain.
10. Does the object represented in Figure Q2.10 have a positive or negative value of a_x? Explain.

FIGURE Q2.10 $\vec{v}$

11. Does the object represented in Figure Q2.11 have a positive or negative value of a_y? Explain.

$\vec{v}$

FIGURE Q2.11

12. Determine the signs (positive, negative, or zero) of the position, velocity, and acceleration for the particle in Figure Q2.12.

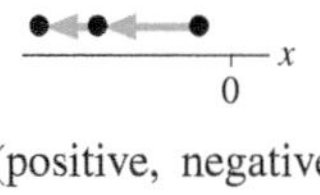

x 0

FIGURE Q2.12

13. Determine the signs (positive, negative, or zero) of the position, velocity, and acceleration for the particle in Figure Q2.13.

y 0

FIGURE Q2.13

Problem difficulty is labeled as | (straightforward) to ||||| (challenging). Problems labeled INT integrate significant material from earlier chapters; BIO are of biological or medical interest; CALC require calculus to solve.

14. Your friend Travis claims to have set the new world speed record for riding a unicycle. His top speed, he says, was 55 m/s. Do you believe him? Explain.

Multiple-Choice Questions

15. | Which of the following motions could be described by the motion diagram of Figure Q2.15?
   A. A hockey puck sliding across smooth ice.
   B. A cyclist braking to a stop.
   C. A sprinter starting a race.
   D. A ball bouncing off a wall.

FIGURE Q2.15 (dots labeled 5 4 3 2 1 0)

16. | Which of the following motions is described by the motion diagram of Figure Q2.16?
   A. An ice skater gliding across the ice.
   B. An airplane braking to a stop after landing.
   C. A car pulling away from a stop sign.
   D. A pool ball bouncing off a cushion and reversing direction.

FIGURE Q2.16 (dots labeled 0 1 2 3 4 5)

17. || Weddell seals make holes in sea ice so that they can swim down to forage on the ocean floor below. Measurements for one seal showed that it dived straight down from such an opening, reaching a depth of 0.30 km in a time of 5.0 min. What was the speed of the diving seal?
BIO
   A. 0.60 m/s B. 6.0 m/s
   C. 1.0 m/s D. 10 m/s
18. || Hicham El Guerrouj of Morocco holds the world record in the 1500 m running race. He ran the final 400 m in a time of 51.9 s. What was his average speed in mph over the last 400 m?
   A. 14.2 mph B. 15.5 mph
   C. 17.2 mph D. 23.9 mph
19. | You throw a rock upward. The rock is moving upward, but it is slowing down. If we define the ground as the origin, the position y of the rock is ____ and the velocity v_y of the rock is ____.
   A. positive, positive B. positive, negative
   C. negative, positive D. negative, negative
20. | You drop a rock into a well. It is moving downward and speeding up. The velocity v_y of the rock is ____ and the acceleration a_y of the rock is ____.
   A. positive, positive B. positive, negative
   C. negative, positive D. negative, negative
21. | Compute 3.24 m + 0.532 m to the correct number of significant figures.
   A. 3.7 m B. 3.77 m
   C. 3.772 m D. 3.7720 m
22. || An American football field is 109.7 m long and 48.8 m wide. To the correct number of significant figures, what is its area?
   A. 5353 m^2 B. 5353.4 m^2
   C. 5350 m^2 D. 5400 m^2
23. | The earth formed 4.57×10^9 years ago. What is this time in seconds?
   A. 1.67×10^{12} s B. 4.01×10^{13} s
   C. 2.40×10^{15} s D. 1.44×10^{17} s
24. ||| An object's average density ρ is defined as the ratio of its mass to its volume: $\rho = M/V$. The earth's mass is 5.94×10^{24} kg, and its volume is 1.08×10^{12} km^3. What is the earth's average density?
   A. 5.50×10^3 kg/m^3 B. 5.50×10^6 kg/m^3
   C. 5.50×10^9 kg/m^3 D. 5.50×10^{12} kg/m^3

PROBLEMS

Section 2.1 Motion Diagrams

1. | A bicycle slowly speeds up from a stop sign, travels a short distance at steady speed, then quickly brakes to a halt for a red light. Use the particle model to draw a motion diagram of the bicycle. Number the dots in order, starting from zero.
2. | Amanda has just entered an elevator. The elevator rises and stops at the third floor. Use the particle model to draw a motion diagram of Amanda during her entire ride on the elevator. Number the dots in order, starting from zero. (Be sure to consider how the elevator speeds up and slows down.)

Section 2.2 Position, Time, and Displacement

3. | Figure P2.3 shows Sue along the straight-line path between her home and the cinema. What is Sue's position x if
   a. Her home is the origin?
   b. The cinema is the origin?

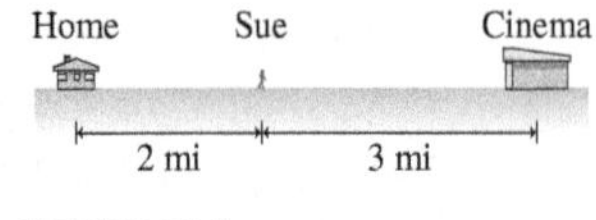

FIGURE P2.3

4. | Figure P2.3 shows Sue along the straight-line path between her home and the cinema. Now Sue walks home. What is Sue's displacement Δx if
   a. Her home is the origin?
   b. The cinema is the origin?
5. | Kendra observes a paramecium as it swims along the edge of a ruler. The paramecium starts at the 65 mm mark at 9:11:15 A.M. and ends at the 42 mm mark at 9:12:05 A.M. What are the displacement Δx in mm and the time interval Δt in s for the paramecium's swim?
BIO
6. | A car travels along a straight east-west road. A coordinate system is established on the road, with x increasing to the east. The car ends up 14 mi west of the origin, which is defined as the intersection with Mulberry Road. If the car's displacement was −23 mi, what side of Mulberry Road did the car start on? How far from the intersection was the car at the start?

Section 2.3 Velocity

7. | A security guard walks at a steady pace, traveling 110 m in one trip around the perimeter of a building. It takes him 240 s to make this trip. What is his speed?

8. || List the following items in order of decreasing speed, from greatest to least: (i) A wind-up toy car that moves 0.15 m in 2.5 s. (ii) A soccer ball that rolls 2.3 m in 0.55 s. (iii) A bicycle that travels 0.60 m in 0.075 s. (iv) A cat that runs 8.0 m in 2.0 s.

9. || BIO In Michael Johnson's 1999 world-record 400 m sprint, he ran the first 100 m in 11.20 s; then he reached the 200 m mark after a total time of 21.32 s had elapsed, reached the 300 m mark after 31.76 s, and finished in 43.18 s.
   a. During what 100 m segment was his speed the highest?
   b. During this segment, what was his speed in m/s?

10. || BIO Kinesin is a motor protein that moves cargo within cells in a series of rapid 8-nm-long steps. Each step takes 50 μs, and there is an average 15 ms delay between steps. To one significant figure, what is the average speed of a kinesin motor protein?

11. || Figure P2.11 shows the motion diagram for a horse galloping in one direction along a straight path. Not every dot is labeled, but the dots are at equally spaced instants of time. What is the horse's velocity v_x
   a. During the first 10 seconds of its gallop?
   b. During the interval from 30 s to 40 s?
   c. During the interval from 50 s to 70 s?

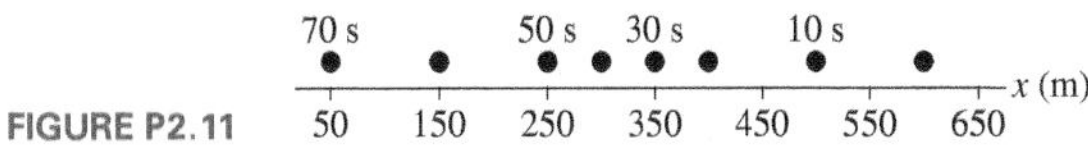

FIGURE P2.11

12. | A dog trots from $x = -12$ m to $x = 8$ m in 10 s. What is its velocity v_x?

Section 2.4 Acceleration

13. || Figure P2.13 shows two dots of a motion diagram and vector $\vec{v}_1$. Copy this figure, then add dot 3 and the next velocity vector $\vec{v}_2$ if the acceleration vector $\vec{a}$ at dot 2 (a) points up and (b) points down.

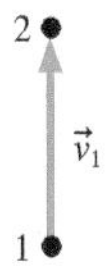

2 $\vec{v}_1$ 1

FIGURE P2.13

14. || Figure P2.14 shows two dots of a motion diagram and vector $\vec{v}_2$. Copy this figure, then add dot 4 and the next velocity vector $\vec{v}_3$ if the acceleration vector $\vec{a}$ at dot 3 (a) points right and (b) points left.

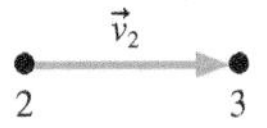

$\vec{v}_2$ 2 3

FIGURE P2.14

15. | A speed skater accelerates from rest and then keeps skating at a constant speed. Draw a complete motion diagram of the skater.

16. | A roof tile falls straight down from a two-story building. It lands in a swimming pool and settles gently to the bottom. Draw a complete motion diagram of the tile.

Section 2.5 Motion Along a Straight Line

17. || Figure P2.17 shows the motion diagram of a drag racer. The camera took one frame every 2 s.

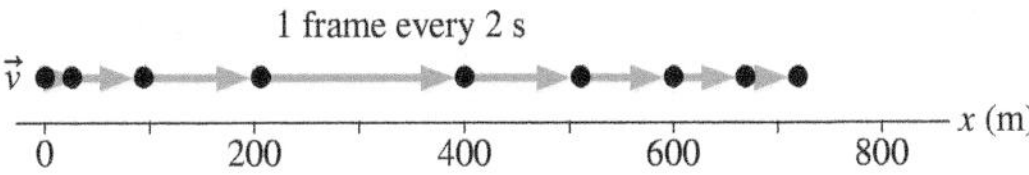

FIGURE P2.17

   a. Measure the x-value of the racer at each dot. List your data in a table similar to Table 2.1, showing each position and the time at which it occurred.
   b. Make a position graph for the drag racer. Because you have data only at certain instants, your graph should consist of dots that are not connected together.

18. | Write a short description of the motion of a real object for which Figure P2.18 would be a realistic position graph.

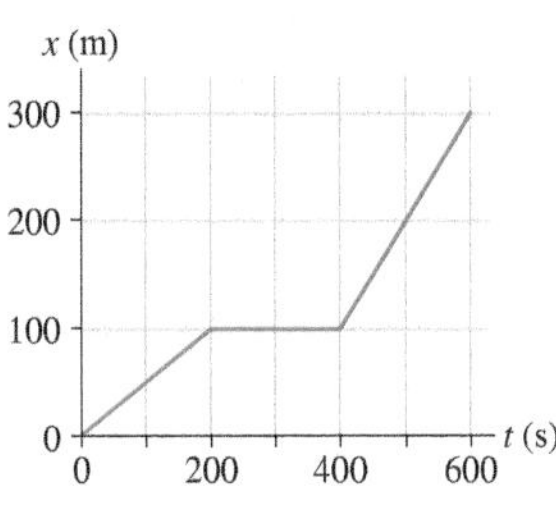

FIGURE P2.18

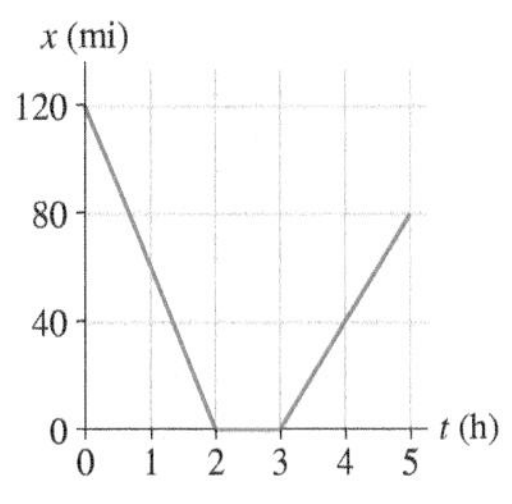

FIGURE P2.19

19. | Write a short description of the motion of a real object for which Figure P2.19 would be a realistic position graph.

20. || What are the signs (positive or negative) of the (a) position y, (b) velocity v_y, and (c) acceleration a_y for the particle in Figure P2.20?

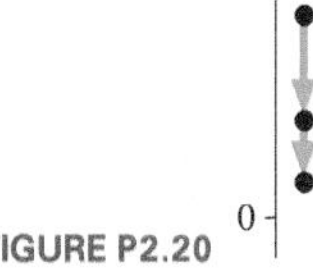

FIGURE P2.20

Section 2.6 Units and Significant Figures

21. | Convert the following to SI base units:
   a. 9.12 μs b. 3.42 km
   c. 44 cm/ms d. 80 km/h

22. | Convert the following to SI units:
   a. 8.0 in b. 66 ft/s c. 60 mph

23. | Convert the following to SI units:
   a. 1.0 hour b. 1.0 day c. 1.0 year

24. | How many significant figures does each of the following numbers have?
   a. 6.21 b. 62.1
   c. 0.620 d. 0.062

25. | How many significant figures does each of the following numbers have?
   a. 0.621 b. 0.006200
   c. 1.0621 d. 6.21×10^3

26. | Compute the following numbers to three significant figures.
   a. 33.3×25.4 b. $33.3 - 25.4$
   c. $\sqrt{33.3}$ d. $333.3 \div 25.4$

27. | Perform the following calculations with the correct number of significant figures.
   a. 159.31×204.6 b. $5.1125 + 0.67 + 3.2$
   c. $7.662 - 7.425$ d. $16.5/3.45$

28. | Using the approximate conversion factors in Table 2.5, convert the following SI units to English units to one significant figure.
   a. 30 cm b. 25 m/s
   c. 5 km d. 0.5 cm

29. | Using the approximate conversion factors in Table 2.5, convert the following to SI units *without* using your calculator.
   a. 20 ft b. 60 mi
   c. 60 mph d. 8 in

30. ‖ If you make multiple measurements of your height, you are likely to find that the results vary by nearly half an inch in either direction due to measurement error and actual variations in height. You are slightly shorter in the evening, after gravity has compressed and reshaped your spine over the course of a day. One measurement of a man's height is 6 feet and 1 inch. Express his height in meters, using the appropriate number of significant figures.
BIO

31. ||| Estimate the average speed, in m/s, with which the hair on your head grows. Make this estimate from your own experience noting, for instance, how often you cut your hair and how much you trim. Express your result in scientific notation.
BIO

32. ||| The nematode worm *C. elegans* has been a model organism for biological studies since the 1960s. Interestingly, each worm has exactly 959 cells. How big is each cell? The organism is reasonably well described as a 80-μm-diameter, 1000-μm-long cylinder. To find an estimate, suppose the cells are identical cubes with edge length L stacked side by side. Make an order-of-magnitude estimate, to one significant figure, of the cell edge length in μm.
BIO

33. ‖ An average adult human can be modeled as a 70 kg, 140-cm-tall cylinder with a diameter of 25 cm. Use information in Tables 2.6 and 2.7 to make an order-of-magnitude estimate, to one significant figure, of the number of cells in a human body by considering (a) a person's mass and (b) a person's size. Should you be concerned that your two estimates are not exactly the same?
BIO

General Problems

Problems 34 through 38 are motion problems similar to those you will learn to solve in Chapter 3. For now, simply *interpret* the problems by drawing (a) a motion diagram and (b) a pictorial representation. **Do *not* solve these problems** or do any mathematics.

34. ‖ A Porsche accelerates from a stoplight at 5.0 m/s^2 for five seconds, then coasts for three more seconds. How far has it traveled?

35. ‖ Billy drops a watermelon from the top of a three-story building, 10 m above the sidewalk. How fast is the watermelon going when it hits?

36. ‖ A speed skater moving across frictionless ice at 8.0 m/s hits a 5.0-m-wide patch of rough ice. She slows steadily, then continues on at 6.0 m/s. What is her acceleration on the rough ice?

37. ‖ The giant eland, an African antelope, is an exceptional jumper, able to leap 1.5 m off the ground. To jump this high, with what speed must the eland leave the ground?
BIO

38. ‖ A motorist is traveling at 20 m/s. He is 60 m from a stop light when he sees it turn yellow. His reaction time, before stepping on the brake, is 0.50 s. What steady deceleration while braking will bring him to a stop right at the light?

Problems 39 through 41 show a motion diagram. For each, write a one or two sentence "short story" about a real object that has this motion. Your stories should mention people or objects by name and say what they are doing. Problems 34 through 38 are examples of motion short stories.

39. |

40. |

41. |

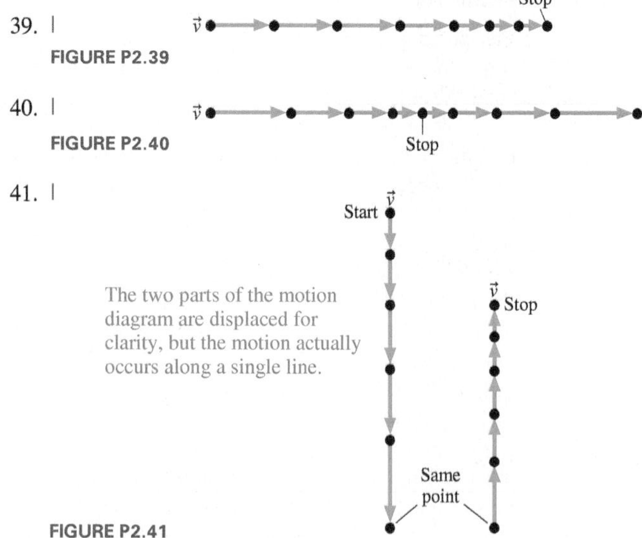
Stop
$\vec{v}$
Stop
Start
$\vec{v}$
The two parts of the motion diagram are displaced for clarity, but the motion actually occurs along a single line.
Stop
Same point

FIGURE P2.39

FIGURE P2.40

FIGURE P2.41

42. ||| On a highway trip, Joseph drives the first 25 miles at 55 mph, and the next 15 miles at 70 mph. What is his average speed for this trip?

43. ||| Evan is just leaving his house to visit his grandmother. Normally, the trip takes him 25 minutes on the freeway, going 55 mph. But tonight he's running 5 minutes late. How fast will he need to drive on the freeway to make up the 5 minutes?

44. ||| Gretchen runs the first 4.0 km of a race at 5.0 m/s. Then a stiff wind comes up, so she runs the last 1.0 km at only 4.0 m/s. If she later ran the same course again, what constant speed would let her finish in the same time as in the first race?

45. ‖ The end of Hubbard Glacier in Alaska advances by an average of 105 feet per year. What is the speed of advance of the glacier in m/s?

46. ‖ The Greenland shark is thought to be the longest-living vertebrate on earth. Early estimates of its maximum age were based on the fact that such sharks grow in length only about a centimeter per year, and yet an adult shark can reach a length of 15 feet. Estimate how long a 15 foot shark might have lived. (A newborn shark is about 1 foot long.)
BIO

47. ||| The winner of the 2016 Keystone (Colorado) Uphill/Downhill mountain bike race finished in a total time of 47 minutes and 25 seconds. The uphill leg was 4.6 miles long, and on this leg his average speed was 8.75 mph. The downhill leg was 6.9 miles. What was his average speed on this leg?

48. ||| Shannon decides to check the accuracy of her speedometer. She adjusts her speed to read exactly 70 mph on her speedometer and holds this steady, measuring the time between successive mile markers separated by exactly 1.00 mile. If she measures a time of 54 s, is her speedometer accurate? If not, is the speed it shows too high or too low?

49. ‖ Motor neurons in mammals transmit signals from the brain to skeletal muscles at approximately 25 m/s. Estimate how much time in ms (10^{-3} s) it takes for a signal to get from your brain to your hand.
BIO

50. ‖ Kinesin is a motor protein found in eukaryotic cells that carries organelles throughout the cell. Kinesin molecules, powered by the hydrolysis of ATP, "walk" along microtubules in discrete steps. Biologists can directly observe the motion of the kinesin by attaching a small (50 nm) bead to the molecule, then tracking the bead
BIO

with reflected laser light. Figure P2.50 is a simplified graph of the data from one such experiment, showing the position of the bead versus time. You see a series of steps that, on this time scale, seem to be instantaneous, each followed by a plateau of variable length while the protein remains stationary.

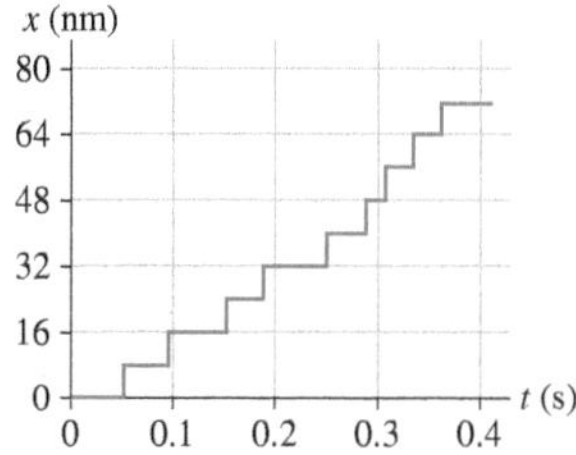

FIGURE P2.50

a. What is the length in nm of each step?
b. Use the total time and the total number of steps to calculate the average time interval between steps.
c. What is the average speed in nm/s of the kinesin protein measured in this experiment?

51. | The bacterium *Escherichia coli* (or *E. coli*) is a single-celled organism that lives in the gut of healthy humans and animals. Its body shape can be modeled as a 2-μm-long cylinder with a 1 μm diameter, and it has a mass of 1×10^{-12} g. Its chromosome consists of a single double-stranded chain of DNA 700 times longer than its body length. The bacterium moves at a constant speed of 20 μm/s, though not always in the same direction. Answer the following questions about *E. coli* using SI base units (unless specifically requested otherwise) and correct significant figures. BIO

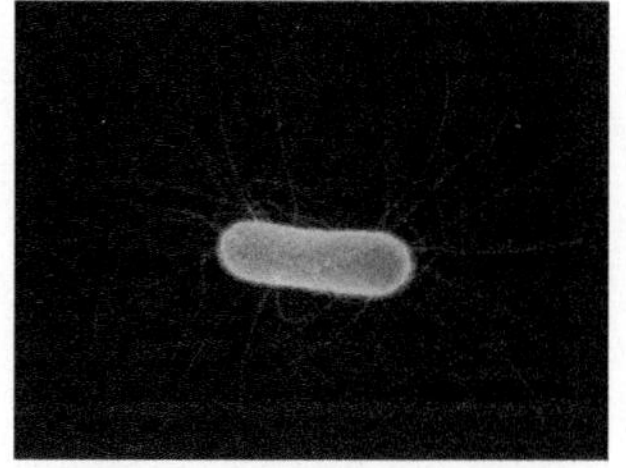

a. What is its length?
b. What is its diameter?
c. What is its mass?
d. What is the length of its DNA, in millimeters?
e. If the organism were to move along a straight path, how many meters would it travel in one day?

52. | The bacterium *Escherichia coli* (or *E. coli*) is a single-celled organism that lives in the gut of healthy humans and animals. When grown in a uniform medium rich in salts and amino acids, it swims along zig-zag paths at a constant speed changing direction at varying time intervals. Figure P2.52 shows the positions of an *E. coli* as it moves from point A to point J. Each segment of the motion can be identified by two letters, such as segment BC. During which segments, if any, does the bacterium have the same BIO

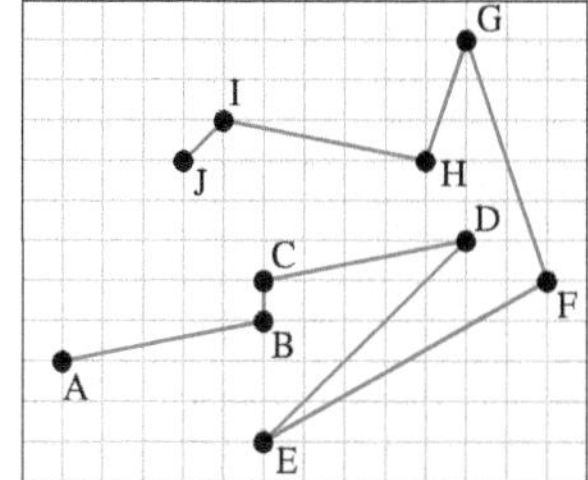

FIGURE P2.52

a. Displacement? b. Speed? c. Velocity?

53. ||| A blood count finds typically 5×10^6 red blood cells per μL of blood. An adult has 5 L of blood. BIO
a. How many red blood cells does an adult have?
b. Laboratory measurements find 150 μg of hemoglobin per μL of blood. Hemoglobin is a protein with a mass of 64,000 daltons or, equivalently, 64,000 atomic mass units (amu). Make an order-of-magnitude estimate, to one significant figure, of the number of hemoglobin molecules in each red blood cell. A useful conversion factor is 1 dalton = 1 amu = 1.7×10^{-27} kg.

54. ||| A yeast cell is a 5.0-μm-diameter sphere. Electron microscopy finds that there are typically 40 mitochondria in a yeast cell, each a 0.75-μm-diameter sphere. What percent of the volume of a yeast cell is filled with mitochondria? BIO

55. ||| Example 2.11 found that a single *E. coli* bacterium contains 3×10^6 proteins. The bacterium can be modeled as a 1-μm-diameter, 2-μm-long cylinder. Estimate the spacing between protein molecules. To do so, suppose that each protein sits at the center of a sphere of radius r and that the total volume of all these spheres is the volume of the bacterium. Then the spacing between two proteins is $2r$, the distance from the center of one sphere to the center of an adjacent but touching sphere. This is an *estimate,* not a precise calculation, but it will give the right order of magnitude for the spacing between protein molecules. BIO
a. To one significant figure, what is the spacing in nm between protein molecules in *E. coli*?
b. Protein molecules themselves can be modeled as 5-nm-diameter spheres. Based on this and your answer to part a, do the proteins in *E. coli* form a dense fluid (molecules close together relative to their size) or a dilute fluid (molecules far apart compared to their size)?

56. || A regulation soccer field for international play is a rectangle with a length between 100 m and 110 m and a width between 64 m and 75 m. What are the smallest and largest areas that the field could be?

57. || An intravenous saline drip has 9.0 g of sodium chloride per liter of water. By definition, 1 mL = 1 cm^3. Express the salt concentration in kg/m^3. BIO

58. || The quantity called *mass density* is the mass per unit volume of a substance. What are the mass densities in SI base units of the following objects?
a. A 215 cm^3 solid with a mass of 0.0179 kg.
b. 95 cm^3 of a liquid with a mass of 77 g.

MCAT-Style Passage Problems

Growth Speed

The images of trees in Figure P2.59 come from a catalog advertising fast-growing trees. If we mark the position of the top of the tree in the successive years, as shown in the graph in the figure, we obtain a motion diagram much like ones we have seen for other kinds of motion. The motion isn't steady, of course. In some months the tree grows rapidly; in other months, quite slowly. We can see, though, that the average speed of growth is fairly constant for the first few years.

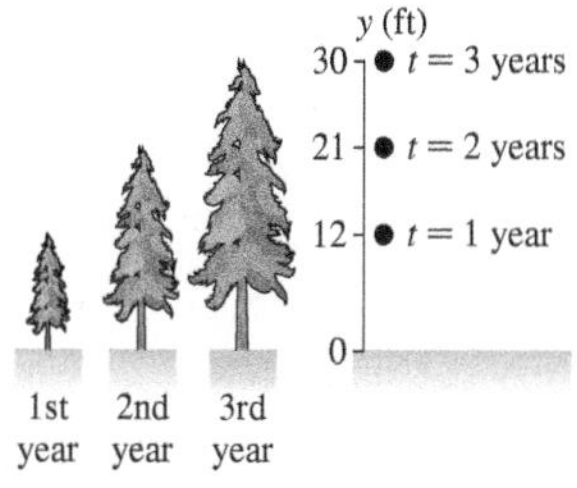

FIGURE P2.59

59. What is the tree's speed of growth, in feet per year, from $t = 1$ yr to $t = 3$ yr?
A. 12 ft/yr B. 9 ft/yr C. 6 ft/yr D. 3 ft/yr

60. What is this speed in m/s?
A. 9×10^{-8} m/s B. 3×10^{-9} m/s
C. 5×10^{-6} m/s D. 2×10^{-6} m/s

61. At the end of year 3, a rope is tied to the very top of the tree to steady it. This rope is staked into the ground 15 feet away from the tree. What angle does the rope make with the ground?
A. 63° B. 60°
C. 30° D. 27°

3 Motion Along a Line

A world-class sprinter has a peak acceleration of about 8 m/s^2.

LOOKING AHEAD

Uniform Motion

Successive images of the Segway rider are the same distance apart, so his velocity is constant. This is called *uniform motion.*

You'll learn to describe motion in terms of quantities such as distance and velocity, an important first step in analyzing motion.

Acceleration

A cheetah is capable of running at very high speeds. Equally important, it is capable of a rapid *change* in speed—a large *acceleration.*

You'll use the concept of acceleration to solve problems of changing velocity, such as races or predators chasing prey.

Free Fall

After the diver jumps, his motion is determined by gravity alone. Motion under the influence of only gravity is called *free fall.*

How long will it take this diver to reach the water? This is the type of free-fall problem you'll learn to solve.

GOAL To describe and analyze motion along a line.

PHYSICS AND LIFE

Animals Speed Up and Slow Down as They Move

Speeding up, slowing down, or moving at a steady speed. Running, hopping, or flying. Creatures and objects can move in many different ways. Under what conditions can one animal catch another, an athlete win a race, a large bird or airplane manage to take off, or a seed or spore travel far enough from its parent plant so that it can grow? Quantifying motion will help us answer questions such as these. It will also give us the necessary tools to later determine the metabolic demands of various types of motion and to understand the tremendous variety of ways animals utilize energy resources to achieve speed, agility, and endurance.

A click beetle can launch itself with an amazing acceleration of more than 400*g*.

3.1 Uniform Motion

Chapter 2 introduced the basic concepts of motion and focused on how to visualize it. That was an important first step, but now we want to go further and solve quantitative problems about motion. The mathematical description of motion is called **kinematics,** our topic for this chapter.

The simplest possible motion is motion along a straight line at a constant, unvarying speed. We call this **uniform motion.** Because velocity is the combination of speed and direction, **uniform motion is motion with constant velocity.**

FIGURE 3.1 shows the motion diagram of an object in uniform motion. For example, this might be you riding your bicycle along a straight line at a perfectly steady 5 m/s (≈10 mph). Notice how all the displacements are exactly the same; this is a characteristic of uniform motion.

FIGURE 3.1 Motion diagram and position graph for uniform motion.

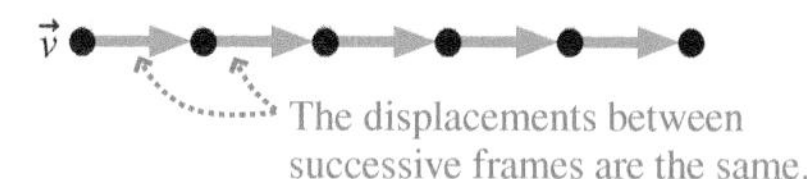

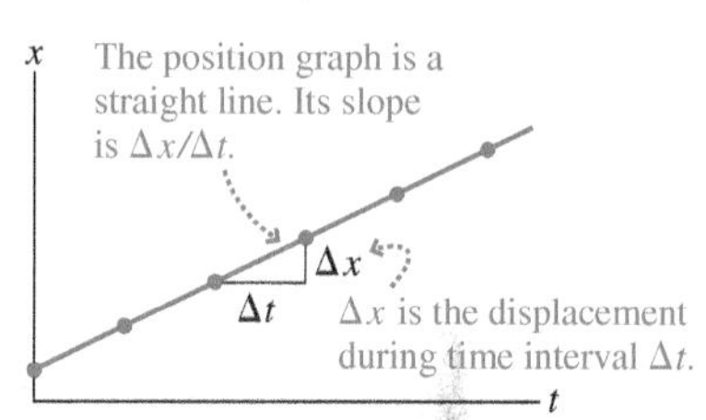

If we make a position graph—remember that position is graphed on the *vertical* axis—it's a straight line. In fact, an alternative definition is that **an object's motion is uniform if and only if its position graph is a straight line.**

◀◀ SECTION 2.3 defined an object's velocity as $\Delta\vec{r}/\Delta t$. For one-dimensional motion, this is simply $\Delta x/\Delta t$ (for horizontal motion) or $\Delta y/\Delta t$ (for vertical motion). Recall that Δx is the object's displacement during the time interval Δt. You can see in Figure 3.1 that Δx and Δt are, respectively, the "rise" and "run" of the position graph. Because rise over run is the slope of a line,

$$v_x = \frac{\Delta x}{\Delta t} \left(\text{or } v_y = \frac{\Delta y}{\Delta t}\right) = \text{slope of the position graph} \tag{3.1}$$

That is, **the velocity is the slope of the position graph.** Velocity has units of "length per time," such as "miles per hour." The SI unit of velocity is meters per second, abbreviated m/s.

An object's **speed** v is how fast it's going, independent of direction. This is simply $v = |v_x|$ or $v = |v_y|$, the magnitude or absolute value of the object's velocity. Although we will use the concept of speed from time to time, our mathematical analysis of motion is based on velocity, not speed. The subscript in v_x or v_y is an essential part of the notation, reminding us that, even in one dimension, the velocity is a vector.

EXAMPLE 3.1 Relating a velocity graph to a position graph

FIGURE 3.2 is the position graph of a car.

a. Draw the car's velocity graph.
b. Describe the car's motion.

PREPARE We can model the car as a particle, with a well-defined position at each instant of time. Figure 3.2 is a graphical representation of the car's motion.

SOLVE

a. The car's position graph is a sequence of three straight lines. Each of these straight lines represents uniform motion at a constant velocity. We can determine the car's velocity during each interval of time by measuring the slope of the line.

The position graph starts out sloping downward—a negative slope. Although the car moves a distance of 4.0 m during the first 2.0 s, its *displacement* is

$$\Delta x = x_{\text{at } 2.0\text{ s}} - x_{\text{at } 0.0\text{ s}} = -4.0\text{ m} - 0.0\text{ m} = -4.0\text{ m}$$

FIGURE 3.2 Position graph.

FIGURE 3.3 The corresponding velocity graph.

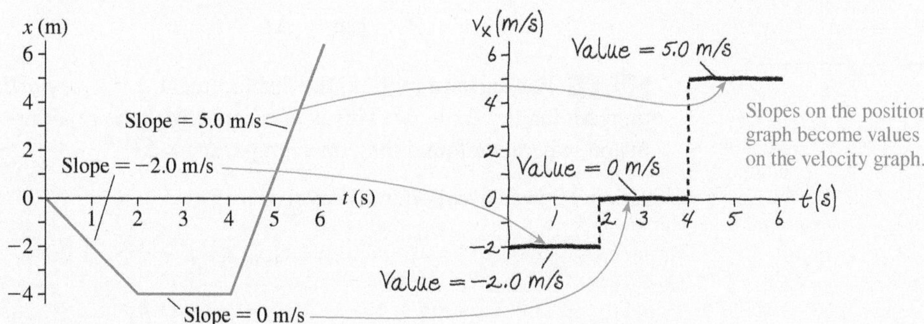

Continued

The time interval for this displacement is $\Delta t = 2.0$ s, so the velocity during this interval is

$$v_x = \frac{\Delta x}{\Delta t} = \frac{-4.0\text{ m}}{2.0\text{ s}} = -2.0\text{ m/s}$$

The car's position does not change from $t = 2$ s to $t = 4$ s ($\Delta x = 0$), so $v_x = 0$. Finally, the displacement between $t = 4$ s and $t = 6$ s is $\Delta x = 10.0$ m. Thus the velocity during this interval is

$$v_x = \frac{10.0\text{ m}}{2.0\text{ s}} = 5.0\text{ m/s}$$

These velocities are shown on the velocity graph of FIGURE 3.3 on the previous page.

b. The car backs up for 2 s at 2.0 m/s, sits at rest for 2 s, then drives forward at 5.0 m/s for at least 2 s. We can't tell from the graph what happens for $t > 6$ s.

ASSESS The velocity graph and the position graph look completely different. The *value* of the velocity graph at any instant of time equals the *slope* of the position graph.

Example 3.1 brought out a few points that are worth emphasizing.

TACTICS BOX 3.1 Interpreting position graphs

1. Steeper slopes correspond to faster speeds.
2. Negative slopes correspond to negative velocities and, hence, to motion in the negative direction (usually to the left or down).
3. The slope is a ratio of intervals, $\Delta x / \Delta t$, not a ratio of coordinates. That is, the slope is *not* simply x / t.

NOTE ▶ If you were to use a ruler to measure the rise and the run of the graph, you could compute the actual slope of the line as drawn on the page. That is not the slope to which we are referring when we equate the velocity with the slope of the line. Instead, we find the slope by measuring the rise and run using the scales along the axes. The "rise" Δx is some number of meters; the "run" Δt is some number of seconds. The rise and run include units, and the ratio of these units gives the units of the slope. ◀

The Equations of Uniform Motion

The physics and mathematics of motion are the same regardless of whether an object moves along the x-axis or the y-axis. We will write equations using x, v_x, and a_x, but the same equations apply to vertical motion if we use y, v_y, and a_y.

FIGURE 3.4 The velocity is found from the slope of the position graph.

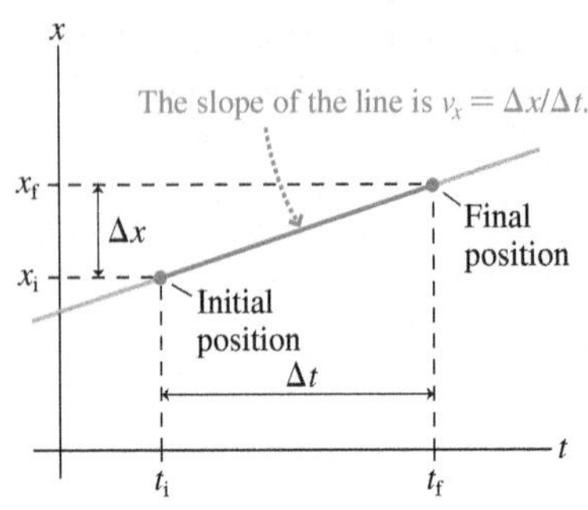

Consider an object in uniform motion along the x-axis with the linear position graph shown in FIGURE 3.4. The object's **initial position** is x_i at time t_i. The term *initial position,* designated with subscript i, refers to the starting point of our analysis or the starting point in a problem; the object may or may not have been in motion prior to t_i. At a later time t_f, the ending point of our analysis, the object's **final position,** denoted by f, is x_f.

The object's velocity v_x along the x-axis can be determined by finding the slope of the graph:

$$v_x = \frac{\text{rise}}{\text{run}} = \frac{\Delta x}{\Delta t} = \frac{x_f - x_i}{t_f - t_i} \tag{3.2}$$

NOTE For uniform motion, the displacement Δx is *proportional to* the elapsed time Δt; that is, $\Delta x \propto \Delta t$. This is in contrast with the random-walk motion of diffusion, where we found that, on average, $\Delta x \propto \Delta t^{1/2}$.

Equation 3.2 is easily rearranged to give

$$x_f = x_i + v_x\,\Delta t \tag{3.3}$$

Position of an object in uniform motion

Equation 3.3 tells us that the object's position increases linearly as the elapsed time Δt increases—exactly as we see in the straight-line position graph.

EXAMPLE 3.2 Lunch in Cleveland?

Bob leaves home in Chicago at 9:00 A.M. and drives east at 60 mph. Susan, 400 miles to the east in Pittsburgh, leaves at the same time and travels west at 40 mph. Where will they meet for lunch?

PREPARE Here is a problem where, for the first time, we can really put all the pieces of our problem-solving strategy into play. To begin, we'll model Bob's and Susan's cars as being in uniform motion. Their real motion is certainly more complex, but over a long drive it's reasonable to approximate their motion as constant speed along a straight line.

FIGURE 3.5 shows the pictorial representation. To show a figure like one *you* would draw, the pictorial representation has been done in a "pencil-sketch" style. We will include pencil-sketch figures in nearly every chapter to illustrate what a good problem solver would draw. In this sketch, the equal spacings of the dots in the motion diagram indicate that the motion is uniform. In evaluating the given information, we recognize that the starting time of 9:00 A.M. is not relevant to the problem. Consequently, the initial time is chosen as simply $t_0 = 0$ h. Bob and Susan are traveling in opposite directions, hence one of the velocities must be a negative number. We have chosen a coordinate system in which Bob starts at the origin and moves to the right (east) while Susan is moving to the left (west). Thus Susan has the negative velocity.

Notice how we've assigned position, velocity, and time symbols to each point in the motion. Pay special attention to how subscripts are used to distinguish different points in the problem and to distinguish Bob's symbols from Susan's. Notice that we've used the numerical subscripts 0 and 1 rather than the generic initial i and final f. One reason is that you're less likely to make mistakes when you use numerical subscripts. Another is that numerical subscripts will be essential in more complex problems where you need to work with more than two positions.

One purpose of the pictorial representation is to establish what we need to find. Bob and Susan meet when they have the same position at the same time t_1. Thus we want to find $(x_1)_B$ at the time when $(x_1)_B = (x_1)_S$. Notice that $(x_1)_B$ and $(x_1)_S$ are Bob's and Susan's *positions,* which are equal when they meet, not the distances they have traveled.

SOLVE The goal of the mathematical representation is to proceed from the pictorial representation to a mathematical solution of the problem. We can begin by using Equation 3.3 to find Bob's and Susan's positions at time t_1 when they meet:

$$(x_1)_B = (x_0)_B + (v_x)_B(t_1 - t_0) = (v_x)_B t_1$$

$$(x_1)_S = (x_0)_S + (v_x)_S(t_1 - t_0) = (x_0)_S + (v_x)_S t_1$$

Notice two things. First, we started by writing the *full* statement of Equation 3.3, with $\Delta t = t_1 - t_0$. Only then did we simplify by dropping those terms known to be zero. You're less likely to make accidental errors if you follow this procedure. Second, we replaced the generic subscripts i and f with the specific symbols 0 and 1 that we defined in the pictorial representation. This is also good problem-solving technique.

The condition that Bob and Susan meet is

$$(x_1)_B = (x_1)_S$$

By equating the right-hand sides of the above equations, we get

$$(v_x)_B t_1 = (x_0)_S + (v_x)_S t_1$$

Solving for t_1 we find that they meet at time

$$t_1 = \frac{(x_0)_S}{(v_x)_B - (v_x)_S} = \frac{400 \text{ miles}}{60 \text{ mph} - (-40) \text{ mph}} = 4.0 \text{ hours}$$

Finally, inserting this time back into the equation for $(x_1)_B$ gives

$$(x_1)_B = \left(60 \frac{\text{miles}}{\text{hour}}\right) \times (4.0 \text{ hours}) = 240 \text{ miles}$$

As noted in Chapter 2, this textbook will assume that all data are good to at least two significant figures, even when one of those is a trailing zero. So 400 miles, 60 mph, and 40 mph each have two significant figures, and consequently we've calculated results to two significant figures.

While 240 miles is a number, it is not yet the answer to the question. The phrase "240 miles" by itself does not say anything meaningful. Because this is the value of Bob's *position,* and Bob was driving east, the answer to the question is They meet 240 miles east of Chicago.

ASSESS Before stopping, we should check whether or not this answer seems reasonable. We certainly expected an answer between 0 miles and 400 miles. We also know that Bob is driving faster than Susan, so we expect that their meeting point will be *more* than halfway from Chicago to Pittsburgh. Our assessment tells us that 240 miles is a reasonable answer.

FIGURE 3.5 Pictorial representation for Example 3.2.

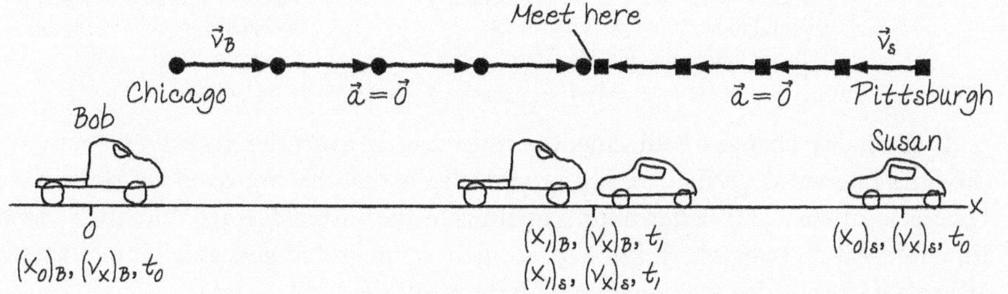

Known
$(x_0)_B = 0$ mi $(v_x)_B = 60$ mph
$(x_0)_S = 400$ mi $(v_x)_S = -40$ mph
$t_0 = 0$ h t_1 is when $(x_1)_B = (x_1)_S$

Find
$(x_1)_B$

FIGURE 3.6 Position graphs for Bob and Susan.

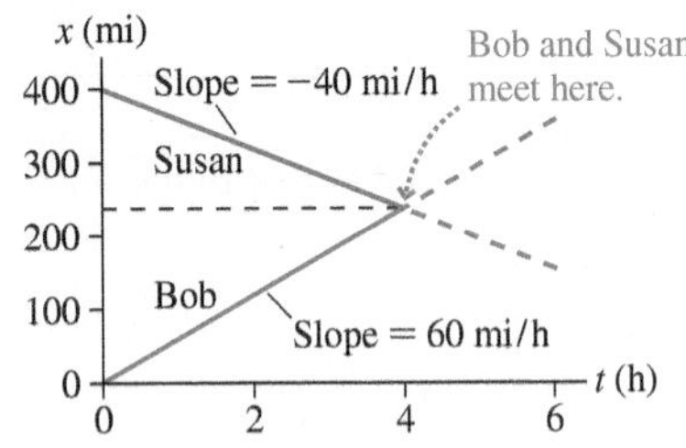

It is instructive to look at this example from a graphical perspective. FIGURE 3.6 shows position graphs for Bob and Susan. Notice the negative slope for Susan's graph, indicating her negative velocity. The point of interest is the intersection of the two lines; this is where Bob and Susan have the same position at the same time. Our method of solution, in which we equated $(x_1)_B$ and $(x_1)_S$, is really just solving the mathematical problem of finding the intersection of two lines. This procedure is useful for many problems in which there are two moving objects.

STOP TO THINK 3.1 Which position graph represents the motion shown in the motion diagram?

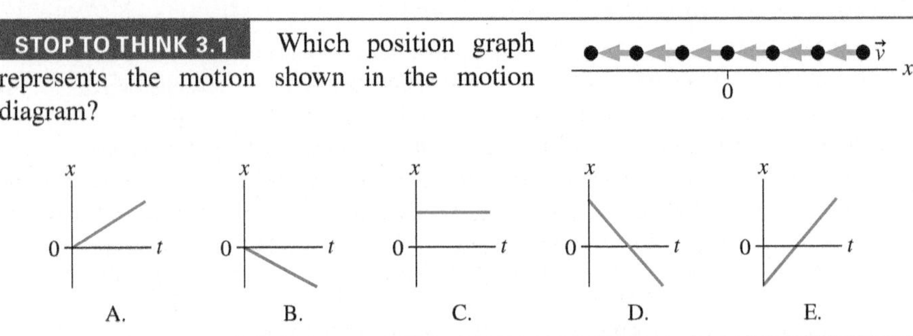

FIGURE 3.7 Motion diagram and position graph of a car speeding up.

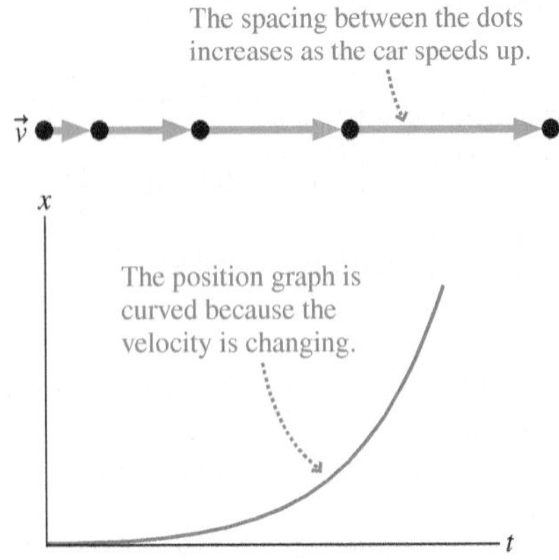

3.2 Instantaneous Velocity

Uniform motion is simple, but objects and organisms rarely travel for long with a constant velocity. Far more common is a velocity that changes with time. For example, FIGURE 3.7 shows the motion diagram and position graph of a car speeding up after the light turns green. Notice how the velocity vectors increase in length, causing the graph to curve upward as the car's displacements get larger and larger.

If you were to watch the car's speedometer, you would see it increase from 0 mph to 10 mph to 20 mph and so on. At any instant of time, the speedometer tells you how fast the car is going *at that instant*. If we include directional information, we can define an object's **instantaneous velocity**—speed and direction—as its velocity at a single instant of time.

For uniform motion, the slope of the straight-line position graph is the object's velocity. FIGURE 3.8 shows that there's a similar connection between instantaneous velocity and the slope of a curved position graph.

FIGURE 3.8 Instantaneous velocity at time t is the slope of the tangent to the curve at that instant.

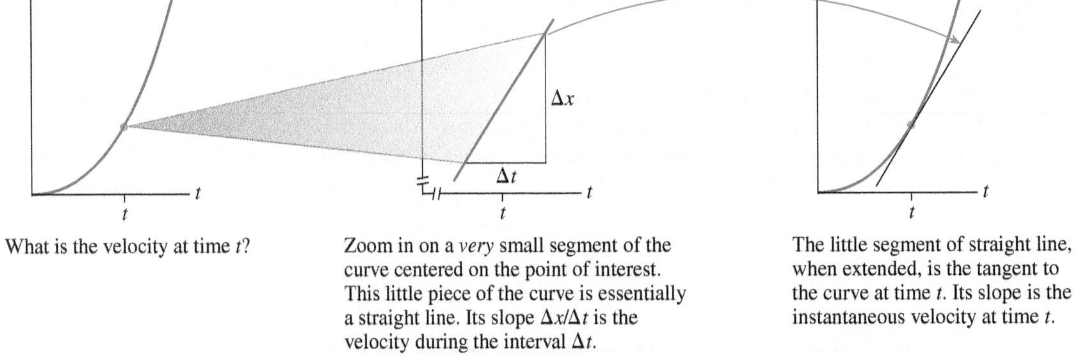

x
t
Δx
Δt
What is the velocity at time t?
Zoom in on a very small segment of the curve centered on the point of interest. This little piece of the curve is essentially a straight line. Its slope Δx/Δt is the velocity during the interval Δt.
The little segment of straight line, when extended, is the tangent to the curve at time t. Its slope is the instantaneous velocity at time t.

If a velocity changes with time, the ratio $\Delta x/\Delta t$ gives the *average* velocity over the time interval Δt. What we see graphically is that the rise-over-run ratio $\Delta x/\Delta t$ becomes a better and better approximation to the instantaneous velocity v_x as the time interval Δt over which the displacement is measured gets smaller and smaller. We can state this idea mathematically in terms of the limit $\Delta t \rightarrow 0$:

$$v_x = \lim_{\Delta t \to 0} \frac{\Delta x}{\Delta t} = \frac{dx}{dt} \qquad (3.4)$$

Instantaneous velocity, the derivative of position with respect to time, is the rate of change of position.

As Δt continues to get smaller, the ratio $\Delta x/\Delta t$ reaches a constant or *limiting* value. That is, **the instantaneous velocity at time *t* is the velocity during a time interval Δt, centered on *t*, as Δt approaches zero.** In calculus, this limit is called *the derivative of x with respect to t*, and it is denoted dx/dt.

Graphically, $\Delta x/\Delta t$ is the slope of a straight line. As Δt gets smaller (i.e., more and more magnification), the straight line becomes a better and better approximation of the curve *at that one point.* In the limit $\Delta t \to 0$, the straight line is tangent to the curve. As Figure 3.8 shows, **the instantaneous velocity at time *t* is the slope of the line that is tangent to the position graph at time *t*.** That is,

$$v_x = \text{slope of the position graph at time } t \qquad (3.5)$$

The steeper the slope, the larger the magnitude of the velocity.

EXAMPLE 3.3 Finding velocity from position graphically

FIGURE 3.9 shows the position graph of an elevator.

a. At which labeled point or points does the elevator have minimum velocity?
b. At which point or points does the elevator have maximum velocity?
c. Sketch an approximate velocity graph for the elevator.

FIGURE 3.9 Position graph.

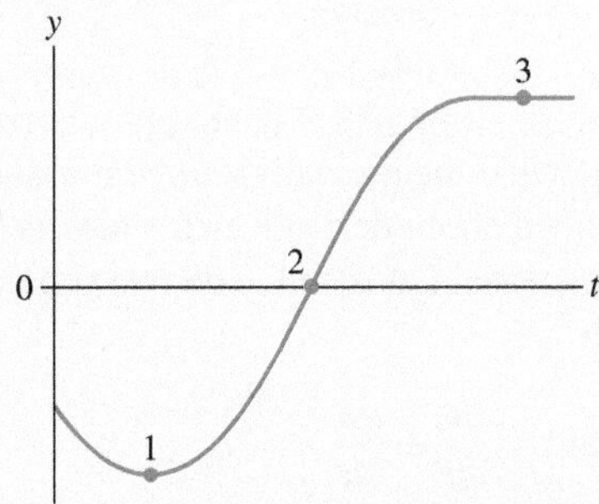

PREPARE We can model the elevator as a particle. The position graph of Figure 3.9 is the graphical representation. The slope of the position graph changes, so we know that the elevator's velocity changes.

SOLVE

a. At any instant, an object's velocity is the slope of its position graph. FIGURE 3.10a shows that the elevator has the least velocity—no velocity at all!—at points 1 and 3 where the slope is zero. At point 1, the velocity is only instantaneously zero. At point 3, the elevator has actually stopped and remains at rest.
b. The elevator has maximum velocity at 2, the point of steepest slope.
c. Although we cannot find an exact velocity graph, we can see that the slope, and hence v_y, is initially negative, becomes zero at point 1, rises to a maximum value at point 2, decreases back to zero a little before point 3, then remains at zero thereafter. Thus FIGURE 3.10b shows, at least approximately, the elevator's velocity graph.

FIGURE 3.10 The velocity graph is found from the slope of the position graph.

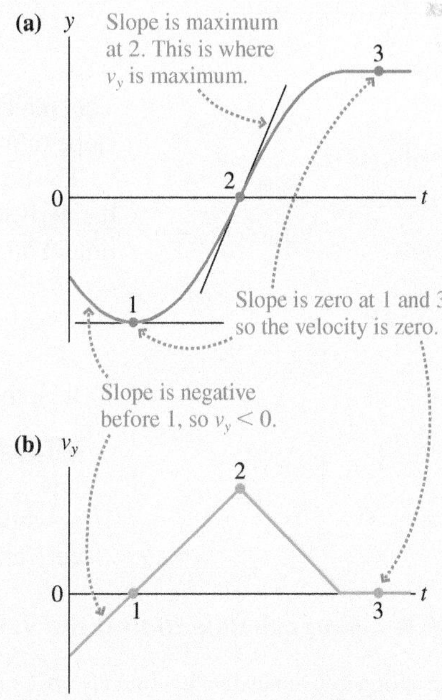

ASSESS Once again, the shape of the velocity graph is very different from the shape of the position graph. You must transfer *slope* information from the position graph to *value* information on the velocity graph.

A Little Calculus: Derivatives

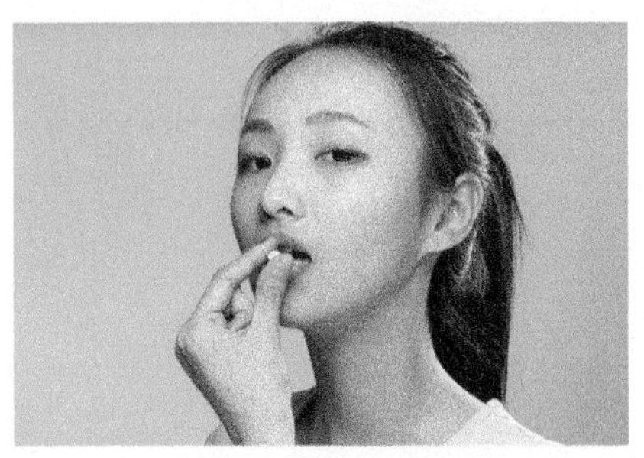

Take your medicine Doctors and pharmacologists use calculus to determine how the bloodstream concentration of a medicine initially increases with time, then decreases as the medicine is eliminated from the body.

Calculus is the branch of mathematics that deals with instantaneous quantities. In other words, it provides us with the tools for evaluating limits such as the one in Equation 3.4.

The notation dx/dt is called *the derivative of x with respect to t*, and Equation 3.4 defines it as the limiting value of a ratio. As Figure 3.8 showed, dx/dt can be interpreted graphically as the slope of the line that is tangent to the position graph.

The most common functions we will use in Part I of this book are powers and polynomials. (Exponential functions will be considered separately in Section 3.8.) Consider the function $u(t) = ct^n$, where c and n are constants. The symbol u is a "dummy name" to represent any function of time, such as $x(t)$ or $y(t)$. The following result is proven in calculus:

$$\text{The derivative of } u = ct^n \text{ is } \frac{du}{dt} = nct^{n-1} \tag{3.6}$$

For example, suppose the position of a particle as a function of time is $x(t) = 2t^2$ m, where t is in s. We can find the particle's velocity $v_x = dx/dt$ by using Equation 3.6 with $c = 2$ and $n = 2$ to calculate

$$v_x = \frac{dx}{dt} = 2 \cdot 2t^{2-1} = 4t$$

This is an expression for the particle's velocity as a function of time.

FIGURE 3.11 shows the particle's position and velocity graphs. It is critically important to understand the relationship between these two graphs. The *value* of the velocity graph at any instant of time, which we can read directly off the vertical axis, is the *slope* of the position graph at that same time. This is illustrated at $t = 3$ s.

FIGURE 3.11 Position graph and the corresponding velocity graph.

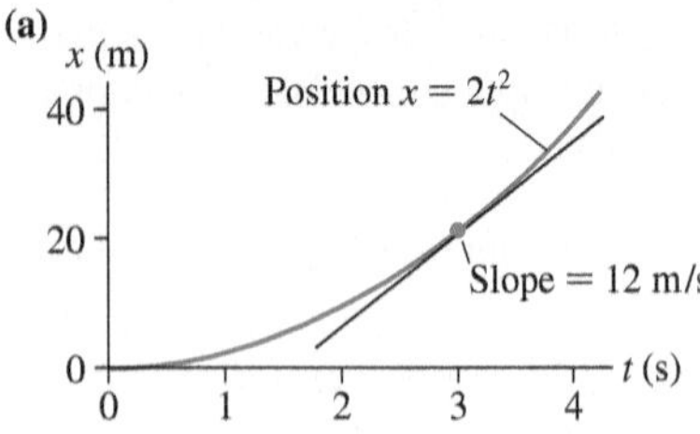

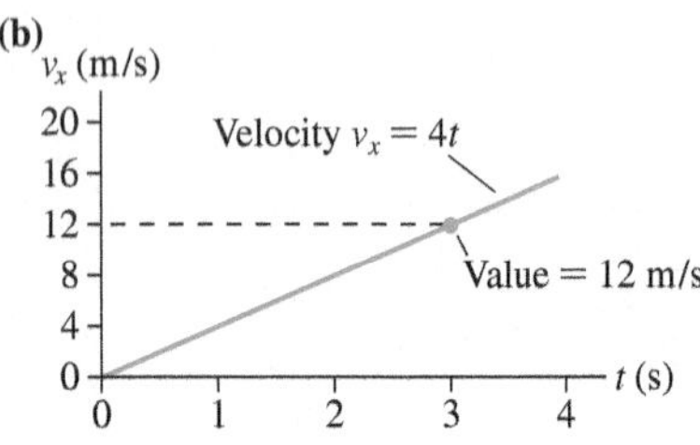

A value that doesn't change with time, such as the position of an object at rest, can be represented by the function $u = c =$ constant. That is, the exponent of t^n is $n = 0$. You can see from Equation 3.6 that the derivative of a constant is zero. That is,

$$\frac{du}{dt} = 0 \text{ if } u = c = \text{constant} \tag{3.7}$$

This makes sense. The graph of the function $u = c$ is simply a horizontal line. The slope of a horizontal line—which is what the derivative du/dt measures—is zero.

The only other information we need about derivatives for now is how to evaluate the derivative of the sum of two functions. Let u and w be two separate functions of time. You will learn in calculus that

$$\frac{d}{dt}(u + w) = \frac{du}{dt} + \frac{dw}{dt} \tag{3.8}$$

That is, the derivative of a sum is the sum of the derivatives.

NOTE ▶ You may have learned in calculus to take the derivative dy/dx, where y is a function of x. The derivatives we use in physics are the same; only the notation is different. We're interested in how quantities change with time, so our derivatives are with respect to t instead of x. ◀

EXAMPLE 3.4 Using calculus to find the velocity

A particle's position is given by the function $x(t) = (-t^3 + 3t)$ m, where t is in s.

a. What are the particle's position and velocity at $t = 2$ s?
b. Draw graphs of x and v_x during the interval $-3 \text{ s} \le t \le 3$ s.
c. Draw a motion diagram to illustrate this motion.

PREPARE We can find the particle's velocity by taking the time derivative of its position.

SOLVE

a. We can compute the position directly from the function x:

$$x(\text{at } t = 2 \text{ s}) = -(2)^3 + (3)(2) = -8 + 6 = -2 \text{ m}$$

The velocity is $v_x = dx/dt$. The function for x is the sum of two polynomials, so

$$v_x = \frac{dx}{dt} = \frac{d}{dt}(-t^3 + 3t) = \frac{d}{dt}(-t^3) + \frac{d}{dt}(3t)$$

The first derivative is a power with $c = -1$ and $n = 3$; the second has $c = 3$ and $n = 1$. Using Equation 3.6, we have

$$v_x = (-3t^2 + 3) \text{ m/s}$$

where t is in s. Evaluating the velocity at $t = 2$ s gives

$$v_x(\text{at } t = 2 \text{ s}) = -3(2)^2 + 3 = -9 \text{ m/s}$$

The negative sign indicates that the particle, at this instant of time, is moving to the *left* at a speed of 9 m/s.

b. FIGURE 3.12 shows the position graph and the velocity graph. You can make graphs like these with a graphing calculator or graphing software. The slope of the position graph at $t = 2$ s is -9 m/s; this becomes the *value* that is graphed for the velocity at $t = 2$ s.

c. Finally, we can interpret the graphs in Figure 3.12 to draw the motion diagram shown in FIGURE 3.13.

- The particle is initially to the right of the origin ($x > 0$ at $t = -3$ s) but moving to the left ($v_x < 0$). Its *speed* is slowing ($v = |v_x|$ is decreasing), so the velocity vector arrows are getting shorter.
- The particle passes the origin $x = 0$ m at $t \approx -1.5$ s, but it is still moving to the left.
- The position reaches a minimum at $t = -1$ s; the particle is as far left as it is going. The velocity is *instantaneously* $v_x = 0$ m/s as the particle reverses direction.
- The particle moves back to the right between $t = -1$ s and $t = 1$ s ($v_x > 0$).
- The particle turns around again at $t = 1$ s and begins moving back to the left ($v_x < 0$). It keeps speeding up, then disappears off to the left.

ASSESS A point in the motion where a particle reverses direction is called a **turning point.** It is a point where the velocity is instantaneously zero while the position is a maximum or minimum. This particle has two turning points, at $t = -1$ s and again at $t = +1$ s. We will see many other examples of turning points.

FIGURE 3.12 Position and velocity graphs.

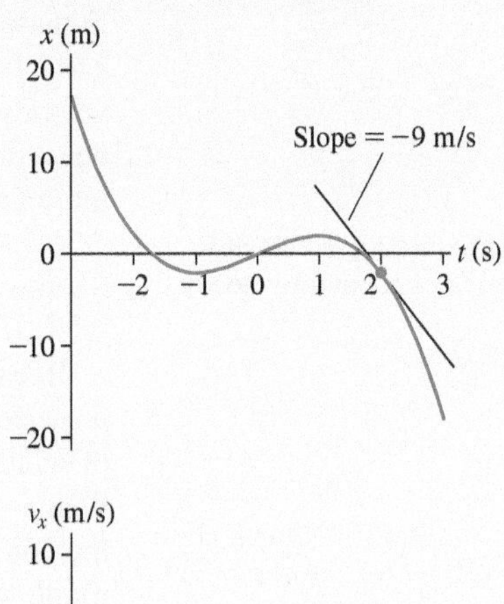

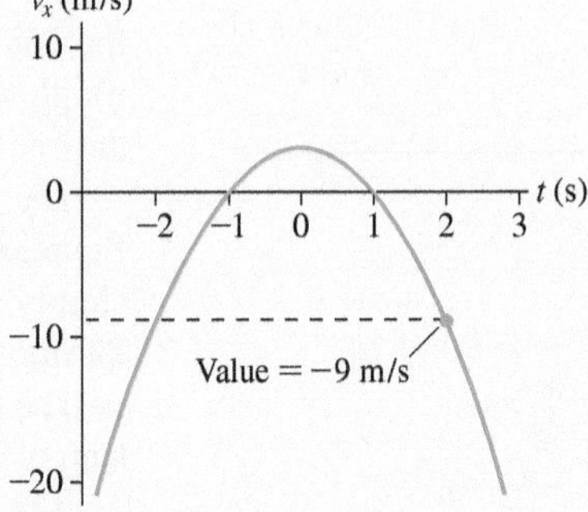

FIGURE 3.13 Motion diagram for Example 3.4.

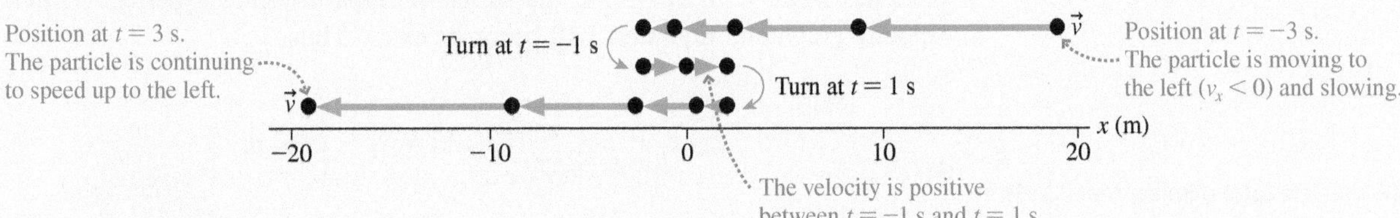

STOP TO THINK 3.2 Which velocity graph goes with the position graph on the left?

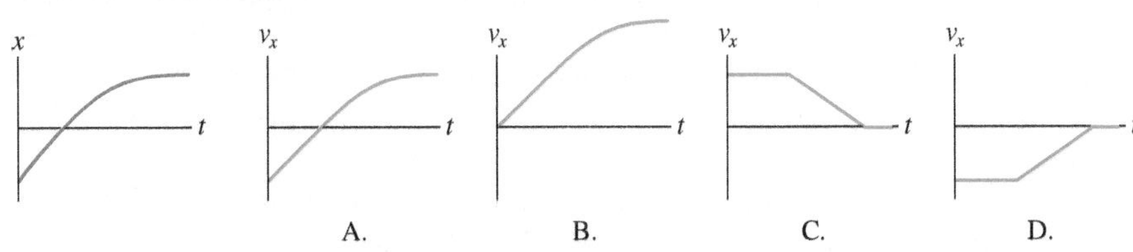

3.3 Finding Position from Velocity

Equation 3.4 allows us to find the instantaneous velocity v_x if we know the position x as a function of time. But what about the reverse problem? Can we use the object's velocity to calculate its position at some future time t? Equation 3.3, $x_f = x_i + v_x \Delta t$, does this for the case of uniform motion with a constant velocity. We need to find a more general expression that is valid when v_x is not constant.

FIGURE 3.14 Approximating a velocity graph with a series of constant-velocity steps.

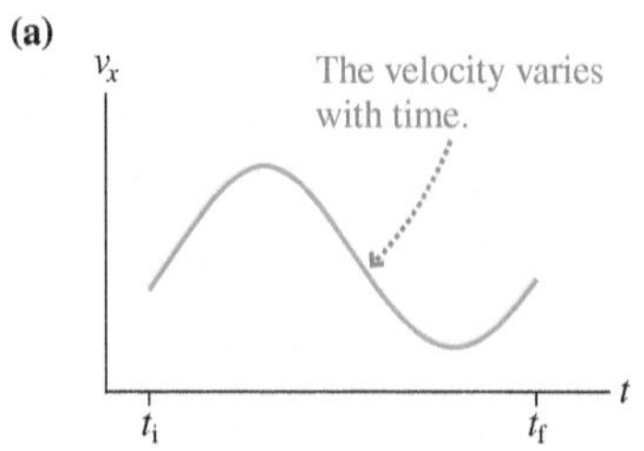
(a)
v_x
The velocity varies with time.
t
t_i
t_f

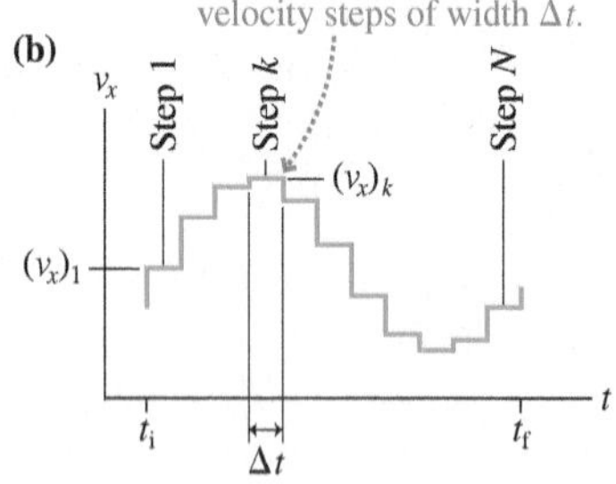
The velocity curve is approximated by constant-velocity steps of width Δt.
(b)
v_x
Step 1
Step k
Step N
$(v_x)_k$
$(v_x)_1$
t
t_i
Δt
t_f

FIGURE 3.14a is a velocity graph for an object whose velocity varies with time. Suppose we know the object's position to be x_i at an initial time t_i. Our goal is to find its final position x_f at a later time t_f.

Because we know how to handle constant velocities, using Equation 3.3, let's *approximate* the velocity function of Figure 3.14a as a series of constant-velocity steps of width Δt. This is illustrated in FIGURE 3.14b. During the first step, from time t_i to time $t_i + \Delta t$, the velocity has the constant value $(v_x)_1$. The velocity during step k has the constant value $(v_x)_k$. Although the approximation shown in the figure is rather rough, with only 11 steps, we can easily imagine that it could be made as accurate as desired by having more and more ever-narrower steps.

The velocity during each step is constant (uniform motion), so we can apply Equation 3.3 to each step. The object's displacement Δx_1 during the first step is simply $\Delta x_1 = (v_x)_1 \Delta t$. The displacement during the second step $\Delta x_2 = (v_x)_2 \Delta t$, and during step k the displacement is $\Delta x_k = (v_x)_k \Delta t$.

The total displacement of the object between t_i and t_f can be approximated as the sum of all the individual displacements during each of the N constant-velocity steps. That is,

$$\Delta x = x_f - x_i \approx \Delta x_1 + \Delta x_2 + \cdots + \Delta x_N = \sum_{k=1}^{N} (v_x)_k \Delta t \tag{3.9}$$

where Σ (Greek sigma) is the symbol for summation. With a simple rearrangement, the particle's final position is

$$x_f \approx x_i + \sum_{k=1}^{N} (v_x)_k \Delta t \tag{3.10}$$

Our goal was to use the object's velocity to find its final position x_f. Equation 3.10 nearly reaches that goal, but Equation 3.10 is only approximate because the constant-velocity steps are only an approximation of the true velocity graph. But if we now let $\Delta t \rightarrow 0$, each step's width approaches zero while the total number of steps N approaches infinity. In this limit, the series of steps becomes a perfect replica of the velocity graph and Equation 3.10 becomes exact. Thus

$$x_f = x_i + \lim_{\Delta t \to 0} \sum_{k=1}^{N} (v_x)_k \Delta t = x_i + \int_{t_i}^{t_f} v_x \, dt \tag{3.11}$$

Displacement, the integral of velocity, is the area under the velocity curve.

FIGURE 3.15 The total displacement Δx is the "area under the curve."

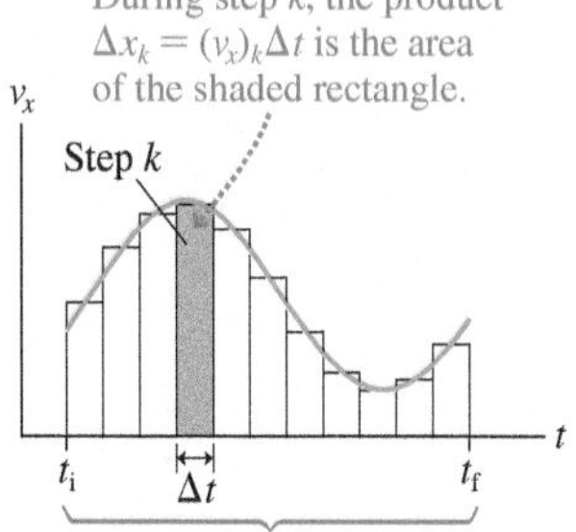
During step k, the product $\Delta x_k = (v_x)_k \Delta t$ is the area of the shaded rectangle.
v_x
Step k
t
t_i
Δt
t_f
During the interval t_i to t_f, the total displacement Δx is the "area under the curve."

The expression on the right is read, "the integral of $v_x \, dt$ from t_i to t_f." Equation 3.11 is the result that we were seeking. It allows us to predict an object's position x_f at a future time t_f.

We can give Equation 3.11 an important geometric interpretation. FIGURE 3.15 shows step k in the approximation of the velocity graph as a tall, thin rectangle of height $(v_x)_k$ and width Δt. The product $\Delta x_k = (v_x)_k \Delta t$ is the area (base $\times$ height) of this small rectangle. The sum in Equation 3.11 adds up all of these rectangular areas to give the total area enclosed between the t-axis and the tops of the steps. The limit of this sum as $\Delta t \rightarrow 0$ is the total area enclosed between the t-axis and the velocity curve. This is called the "area under the curve." Thus a graphical interpretation of Equation 3.11 is

$$x_f = x_i + \text{area under the velocity curve } v_x \text{ between } t_i \text{ and } t_f \tag{3.12}$$

NOTE ▶ Wait a minute! The displacement $\Delta x = x_f - x_i$ is a length. How can a length equal an area? Recall earlier, when we found that the velocity is the slope of the position graph, we emphasized the need to use the units of the quantities on each axis. The same idea applies here. The area of each small rectangle is $v_x\,\Delta t$, but v_x is some number of meters per second while Δt is some number of seconds. Hence their product, the area under the curve, has units of meters. ◀

EXAMPLE 3.5 The displacement of a cyclist

FIGURE 3.16 shows the velocity graph of a bicycle after the stoplight turns green. How far does the cyclist move during the first 6.0 s?

FIGURE 3.16 Velocity graph for a bicycle.

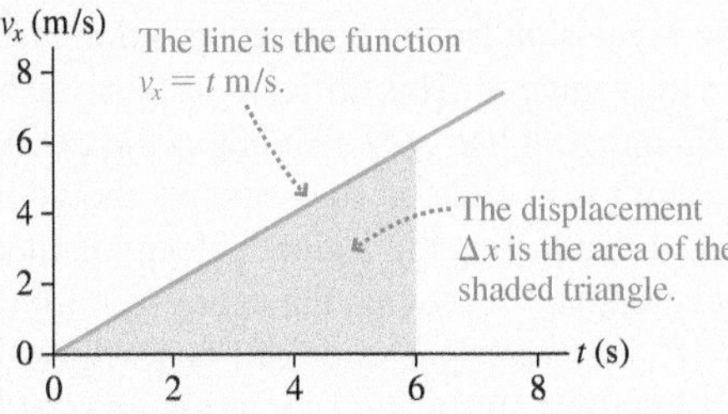

PREPARE We can model the bicycle as a particle with a well-defined position at all times. Figure 3.16 is the graphical representation. The question "How far?" indicates that we need to find a displacement Δx rather than a position x.

SOLVE According to Equation 3.12, the cyclist's displacement $\Delta x = x_f - x_i$ between $t = 0$ s and $t = 6.0$ s is the area under the curve from $t = 0$ s to $t = 6.0$ s. The curve in this case is an angled line, so the area is that of a triangle:

$$\begin{aligned}\Delta x &= \text{area of triangle between } t = 0 \text{ s and } t = 6.0 \text{ s}\\ &= \tfrac{1}{2} \times \text{base} \times \text{height}\\ &= \tfrac{1}{2} \times 6.0 \text{ s} \times 6.0 \text{ m/s} = 18 \text{ m}\end{aligned}$$

The cyclist moves 18 m during the first 6.0 seconds.

ASSESS The "area" is a product of s and m/s, so Δx has the proper units of m. Using 1 m ≈ 3 ft, we find that the cyclist has moved about 54 ft, or 18 yards, in the first 6 s. This seems quite reasonable for a bicycle, which gives us confidence in our answer.

A Little More Calculus: Integrals

Taking the derivative of a function is equivalent to finding the slope of a graph of the function. Similarly, evaluating an integral is equivalent to finding the area under a graph of the function. The graphical method is very important for building intuition about motion but is limited in its practical application. Just as derivatives of standard functions can be evaluated and tabulated, so can integrals.

The integral in Equation 3.11 is called a *definite integral* because there are two definite boundaries to the area we want to find. These boundaries are called the lower (t_i) and upper (t_f) *limits of integration.* For the important function $u(t) = ct^n$, the essential result from calculus is that

$$\int_{t_i}^{t_f} u\,dt = \int_{t_i}^{t_f} ct^n\,dt = \frac{ct^{n+1}}{n+1}\bigg|_{t_i}^{t_f} = \frac{ct_f^{\,n+1}}{n+1} - \frac{ct_i^{\,n+1}}{n+1} \qquad (n \neq -1) \qquad (3.13)$$

The vertical bar in the third step with subscript t_i and superscript t_f is a shorthand notation from calculus that means—as seen in the last step—the integral evaluated at the upper limit t_f *minus* the integral evaluated at the lower limit t_i. You also need to know that for two functions u and w,

$$\int_{t_i}^{t_f} (u + w)\,dt = \int_{t_i}^{t_f} u\,dt + \int_{t_i}^{t_f} w\,dt \qquad (3.14)$$

That is, the integral of a sum is equal to the sum of the integrals.

EXAMPLE 3.6 Using calculus to find the position of a cyclist

Use calculus to solve Example 3.5 for a bicycle after the stoplight turns green.

PREPARE The cyclist's velocity increases linearly from 0 m/s to 8 m/s in 8 s. As Figure 3.16 shows, the line in the graph is the function $v_x = t$ m/s, where t is in s. Let the bicycle's initial position be $x_i = 0$ m.

Continued

SOLVE We can find the position x at time t by using Equation 3.11:

$$x = x_i + \int_0^t v_x\,dt = 0\text{ m} + \int_0^t t\,dt$$

The integral is a function of the form $u = ct^n$ with $c = 1$ and $n = 1$. We can evaluate the integral by using Equation 3.13:

$$\int_0^t t\,dt = \frac{t^2}{2}\bigg|_0^t = \tfrac{1}{2}(t^2 - 0^2) = \tfrac{1}{2}t^2$$

Thus the bicycle's position at time t is

$$x = \tfrac{1}{2}t^2\text{ m}$$

where t is in s. This is a general result for the position at *any* time t for which the cyclist's velocity continues to increase linearly. At $t = 6.0$ s, the cyclist has reached

$$x(\text{at } t = 6.0\text{ s}) = \tfrac{1}{2}(6.0)^2\text{ m} = 18\text{ m}$$

ASSESS This agrees with the answer we found using a graphical approach.

It may seem troubling that the expression for the bicycle's position in Example 3.6 seems to equate a distance with a time squared. This difficulty is resolved by recognizing that the factor $\frac{1}{2}$ is not a pure number; it has units. We began the example by writing the velocity as $v_x = t$ m/s, where t is in s. To be rigorous, we should have written $v_x = (1\text{ m/s}^2)t$, which makes it clear that having t in s gives velocity in m/s. If you follow the 1 m/s^2 through the calculation, you'll see that the expression for the bicycle's position is actually $x = \left(\frac{1}{2}\text{ m/s}^2\right)t^2$, so inserting a value of t in s will correctly give a position in m. However, explicitly writing out $v_x = (1\text{ m/s}^2)t$ is awkward and rarely done. Instead, keep in mind that the constants in algebraic expressions often have units that are not explicitly shown.

STOP TO THINK 3.3 Which position graph corresponds to the velocity graph on the left? The particle's position at $t_i = 0$ s is $x_i = -10$ m.

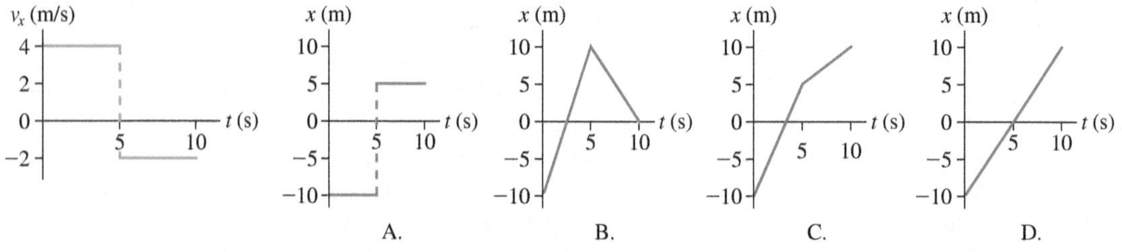

3.4 Constant Acceleration

We need one more major concept to describe one-dimensional motion: acceleration. Acceleration will turn out to be the linchpin that connects motion and forces. You'll see in Chapter 4 that we can determine an object's acceleration by knowing the forces acting on it, a set of principles called *Newton's laws*.

Let's conduct a race between a Volkswagen Beetle and a Porsche to see which can reach a speed of 30 m/s (≈60 mph) in the least amount of time. Both cars are equipped with computers that will record the speedometer reading 10 times each second. This gives a nearly continuous record of the *instantaneous* velocity of each car. TABLE 3.1 shows some of the data. The velocity graphs, based on these data, are shown in FIGURE 3.17.

TABLE 3.1 **Velocities of a Porsche and a Volkswagen Beetle**

t (s)	v_{Porsche} (m/s)	v_{VW} (m/s)
0.0	0.0	0.0
0.1	0.5	0.2
0.2	1.0	0.4
0.3	1.5	0.6
⋮	⋮	⋮

How can we describe the difference in performance of the two cars? It is not that one has a different velocity from the other; both achieve every velocity between 0 and 30 m/s. The distinction is how long it took each to *change* its velocity from 0 to 30 m/s. The Porsche changed velocity quickly, in 6.0 s, while the VW needed 15 s to make the same velocity change. Because the Porsche had a velocity change $\Delta v_x = 30$ m/s during a time interval $\Delta t = 6.0$ s, the *rate* at which its velocity changed was

$$\text{rate of velocity change} = \frac{\Delta v_x}{\Delta t} = \frac{30\text{ m/s}}{6.0\text{ s}} = 5.0\text{ (m/s)/s} \tag{3.15}$$

Notice the units. They are units of "velocity per second." A velocity changing at the rate of 5.0 "meters per second per second" means that the velocity increases by 5.0 m/s during the first second, by another 5.0 m/s during the next second, and so on. In fact, the velocity will increase by 5.0 m/s during any second in which it is changing at the rate of 5.0 (m/s)/s.

Chapter 2 introduced *acceleration* as "the rate of change of velocity." That is, acceleration measures how quickly or slowly an object's velocity changes. In parallel with our treatment of velocity, let's define the acceleration a_x during the time interval Δt to be

$$a_x = \frac{\Delta v_x}{\Delta t} \tag{3.16}$$

Equations 3.15 and 3.16 show that the Porsche had the rather large acceleration of 5.0 (m/s)/s.

Because Δv_x and Δt are the "rise" and "run" of a velocity graph, we see that a_x can be interpreted graphically as the *slope* of a straight-line velocity graph. In other words,

$$a_x = \text{slope of the velocity graph} \tag{3.17}$$

Figure 3.17 uses this idea to show that the VW's average acceleration is

$$(a_{VW})_x = \frac{\Delta v_x}{\Delta t} = \frac{10 \text{ m/s}}{5.0 \text{ s}} = 2.0 \text{ (m/s)/s}$$

This is less than the acceleration of the Porsche.

NOTE ▶ It is customary to abbreviate the acceleration unit (m/s)/s as m/s^2. For example, the Porsche has an acceleration of 5.0 m/s^2. We will use this notation, but keep in mind the *meaning* of the notation as "(meters per second) per second." ◀

FIGURE 3.17 Velocity graphs for the Porsche and the VW Beetle.

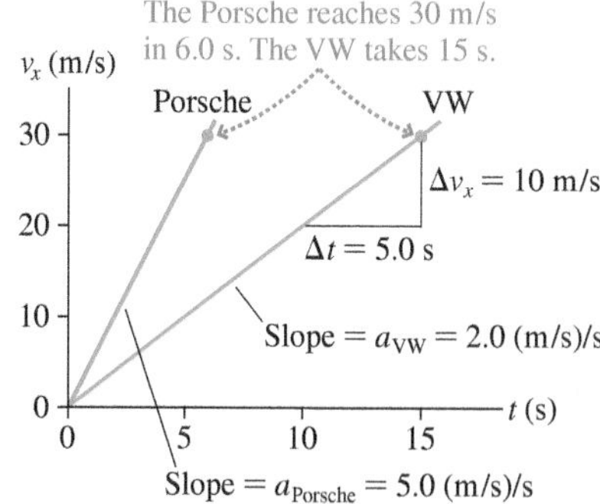

Getting up to speed A bird must have a minimum speed to fly. Generally, the larger the bird, the faster the takeoff speed. Small birds can get moving fast enough to fly with a vigorous jump, but larger birds need a running start. These swans must accelerate for a long distance to reach the speed they need to fly, so they make a frenzied dash across the frozen surface of a pond.

EXAMPLE 3.7 **Animal acceleration**

Lions, like most predators, are capable of very rapid starts. Measurements show that a lion's initial acceleration, starting from rest, can be as high as 9.5 m/s^2. How long does it take a lion to go from rest to 10 mph, the speed of a fairly accomplished distance runner?

PREPARE The lion's speed increases by 9.5 m/s every second. We need the runner's speed in m/s; then we can use Equation 3.16 to calculate how long it takes the lion to reach this speed.

SOLVE The final velocity the lion must reach is

$$(v_x)_f = 10 \frac{\text{mi}}{\text{h}} \times \frac{0.447 \text{ m/s}}{1 \text{ mph}} = 4.5 \text{ m/s}$$

We took the conversion factor from Table 2.4. By definition, the *change* of velocity is $\Delta v_x = (v_x)_f - (v_x)_i$. The lion starts from rest, so its initial velocity $(v_x)_i$ is 0 m/s. Thus a final velocity of $(v_x)_f = 4.5$ m/s means that the lion changes velocity by $\Delta v_x = 4.5$ m/s. Knowing that the acceleration is $a_x = 9.5$ m/s^2, we can now use Equation 3.16 to find that the lion matches the runner's speed in

$$\Delta t = \frac{\Delta v_x}{a_x} = \frac{4.5 \text{ m/s}}{9.5 \text{ m/s}^2} = 0.47 \text{ s}$$

ASSESS The approximate conversion 1 m/s ≈ 2 mph tells us that the runner's speed is ≈5 m/s. The lion increases its speed by 9.5 m/s every second, so it's reasonable (if a bit intimidating) that it reaches the runner's speed in ≈0.5 s.

STOP TO THINK 3.4 A particle moves with the velocity graph shown here. At which labeled point is the magnitude of the acceleration the greatest?

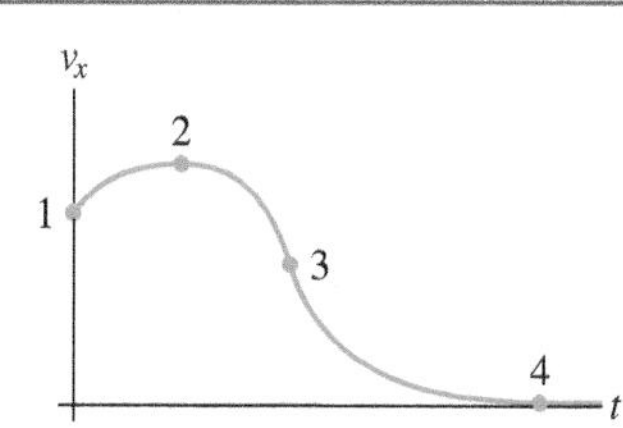

The Sign of the Acceleration

In everyday language, "to accelerate" means to speed up. We noted in Chapter 2 that physics uses the term a bit differently. *Any* change in velocity, either speeding up or slowing down, is an acceleration. It's a natural tendency to think that a positive value of a_x describes an object that is speeding up, while a negative value means the object is slowing. As logical as that seems, *it is not correct.*

It's true that there are many situations where the sign doesn't matter—we simply want to know the magnitude of the acceleration $a = |a_x|$, a positive number that tells us the rate at which the speed is changing. However, you'll learn in the next chapter that Newton's laws of motion, the laws that govern how an object moves in response to forces acting on it, depend on the idea that acceleration is a vector—it has a direction. For motion along a line, the direction is given by the sign of a_x. FIGURE 3.18 shows that a_x is positive when the vector $\vec{a}$ points to the right, negative when $\vec{a}$ points to the left. This agrees with the sign of the slope of the velocity graph. Similarly, a_y is positive when the vector $\vec{a}$ points up, negative when $\vec{a}$ points down. But we must repeat, for emphasis, that **positive and negative accelerations do *not* correspond to speeding up and slowing down.**

FIGURE 3.18 Determining the sign of the acceleration.

The object is moving to the right and speeding up. Its acceleration vector points to the right, so $a_x > 0$.

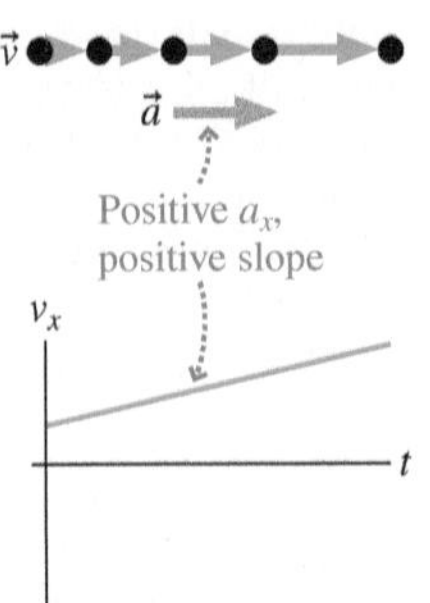

The object is moving to the left and slowing down. Its acceleration vector points to the right, so $a_x > 0$ even though it is slowing.

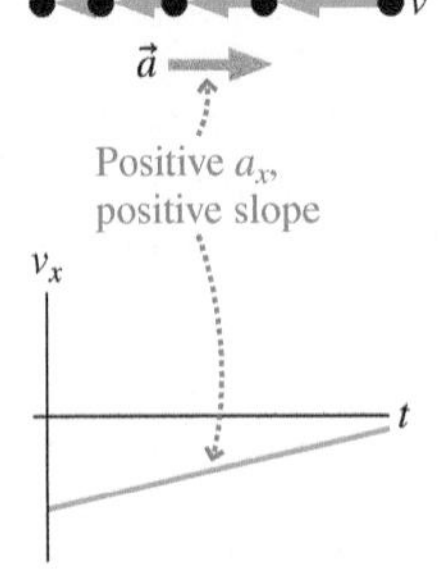

The object is moving to the right and slowing down. Its acceleration vector points to the left, so $a_x < 0$.

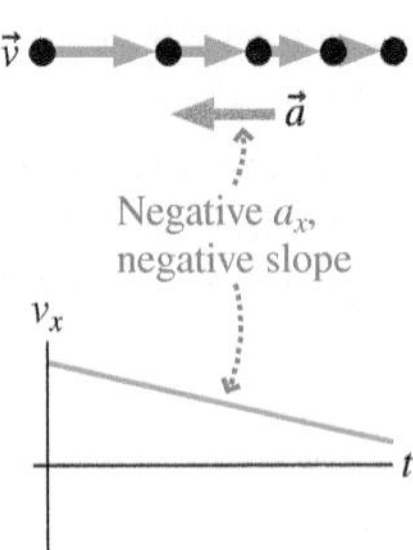

The object is moving to the left and speeding up. Its acceleration vector points to the left, so $a_x < 0$ even though it is getting faster.

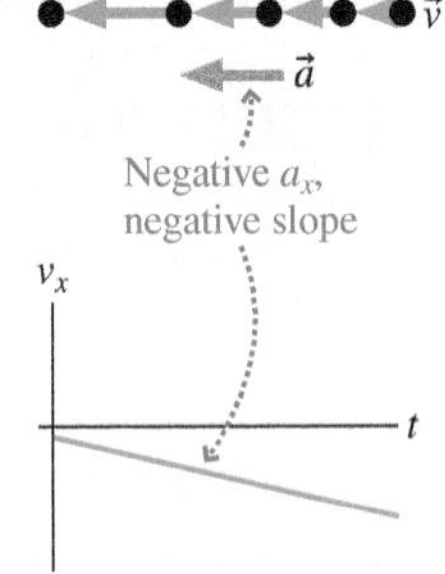

EXAMPLE 3.8 Running the court

A basketball player starts at the left end of the court and moves to the right with the velocity shown in FIGURE 3.19. Draw a motion diagram and an acceleration graph for the basketball player.

FIGURE 3.19 Velocity graph for the basketball player of Example 3.8.

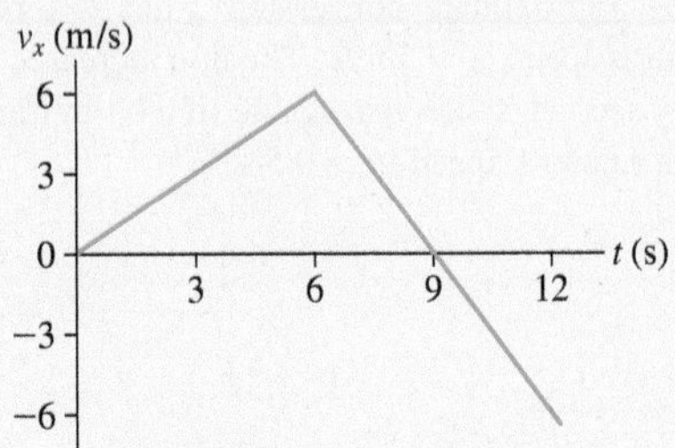

PREPARE Looking at the graph, we can see that the velocity is positive (motion to the right) and increasing for the first 6 s, so the velocity arrows in the motion diagram are to the right and getting longer. From $t = 6$ s to 9 s the motion is still to the right (v_x is still positive), but the arrows are getting shorter because v_x is decreasing. There's a turning point at $t = 9$ s, when $v_x = 0$ m/s, and after that the motion is to the left (v_x is negative) and getting faster.

SOLVE The motion diagram of FIGURE 3.20a shows the velocity and the acceleration vectors. Acceleration is the slope of the velocity graph. For the first 6 s, the slope has the constant value

$$a_x = \frac{\Delta v_x}{\Delta t} = \frac{6.0\ \text{m/s}}{6.0\ \text{s}} = 1.0\ \text{m/s}^2$$

FIGURE 3.20 Motion diagram and acceleration graph for Example 3.8.

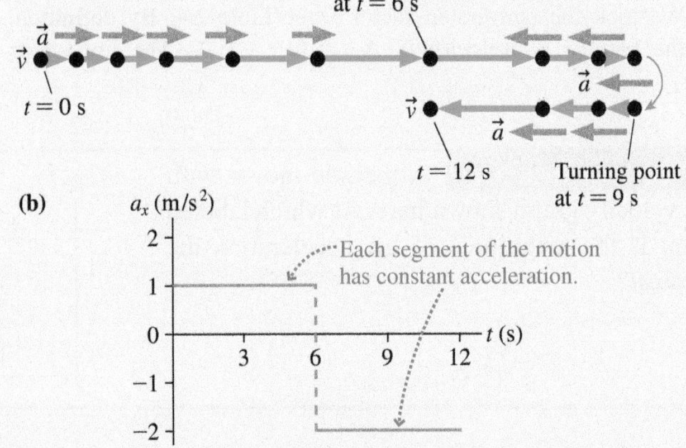

The velocity then decreases by 12 m/s during the 6 s interval from $t = 6$ s to $t = 12$ s, so

$$a_x = \frac{\Delta v_x}{\Delta t} = \frac{-12 \text{ m/s}}{6.0 \text{ s}} = -2.0 \text{ m/s}^2$$

The acceleration graph for these 12 s is shown in FIGURE 3.20b. Notice that there is no change in the acceleration at $t = 9$ s, the turning point.

ASSESS The *sign* of a_x does *not* tell us whether the object is speeding up or slowing down. The basketball player is slowing down from $t = 6$ s to $t = 9$ s, then speeding up from $t = 9$ s to $t = 12$ s. Nonetheless, his acceleration is negative during this entire interval because his acceleration vector, as seen in the motion diagram, always points to the left.

The Kinematic Equations of Constant Acceleration

Constant acceleration is not the only kind of motion; it's not even the most common kind of motion in the biological world. Nonetheless, constant acceleration is a useful model for many types of motion, biological and otherwise. In this section we will use the relationships among position, velocity, and acceleration to derive some useful results that we can use to analyze motion with constant acceleration. We'll extend these ideas to motion with changing acceleration in Section 3.8.

Consider an object whose acceleration a_x remains constant during the time interval $\Delta t = t_f - t_i$. At the beginning of this interval, at time t_i, the object has initial velocity $(v_x)_i$ and initial position x_i. Note that t_i is often zero, but it does not have to be. We would like to predict the object's final position x_f and final velocity $(v_x)_f$ at time t_f.

The object's velocity is changing because the object is accelerating. FIGURE 3.21a shows the acceleration graph, a horizontal line between t_i and t_f. It is not hard to find the object's velocity $(v_x)_f$ at a later time t_f. By definition,

$$a_x = \frac{\Delta v_x}{\Delta t} = \frac{(v_x)_f - (v_x)_i}{\Delta t} \tag{3.18}$$

which is easily rearranged to give

$$(v_x)_f = (v_x)_i + a_x \Delta t \tag{3.19}$$

The velocity graph, shown in FIGURE 3.21b, is a straight line that starts at $(v_x)_i$ and has slope a_x.

As you learned in the last section, the object's final position is

$$x_f = x_i + \text{area under the velocity curve } v_x \text{ between } t_i \text{ and } t_f \tag{3.20}$$

The shaded area in Figure 3.21b can be subdivided into a rectangle of area $(v_x)_i \Delta t$ and a triangle of area $\frac{1}{2}(a_x \Delta t)(\Delta t) = \frac{1}{2}a_x(\Delta t)^2$. Adding these gives

Final and initial position (m) — Time interval (s)

$$x_f = x_i + (v_x)_i \Delta t + \tfrac{1}{2}a_x(\Delta t)^2 \tag{3.21}$$

Initial velocity (m/s) — Acceleration (m/s²)

where $\Delta t = t_f - t_i$ is the elapsed time. The quadratic dependence on Δt causes the position graph for constant-acceleration motion to have a parabolic shape.

Equations 3.19 and 3.21 are two of the basic kinematic equations for motion with *constant* acceleration. They allow us to predict an object's position and velocity at a future instant of time. We need one more equation to complete our set, a direct relationship between position and velocity. We first use Equation 3.19 to write $\Delta t = ((v_x)_f - (v_x)_i)/a_x$. Then we substitute this into Equation 3.21 to get

$$x_f = x_i + (v_x)_i\left(\frac{(v_x)_f - (v_x)_i}{a_x}\right) + \tfrac{1}{2}a_x\left(\frac{(v_x)_f - (v_x)_i}{a_x}\right)^2 \tag{3.22}$$

FIGURE 3.21 Acceleration and velocity graphs for constant acceleration.

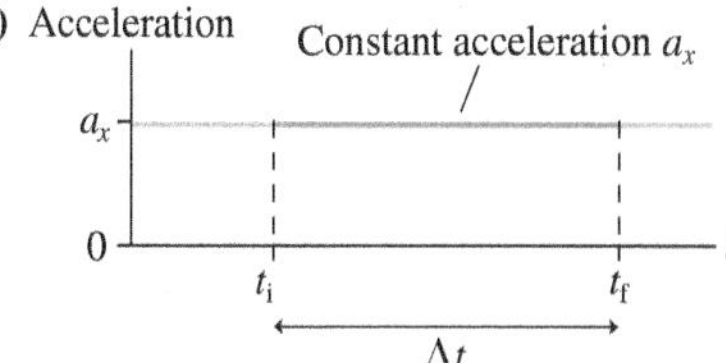

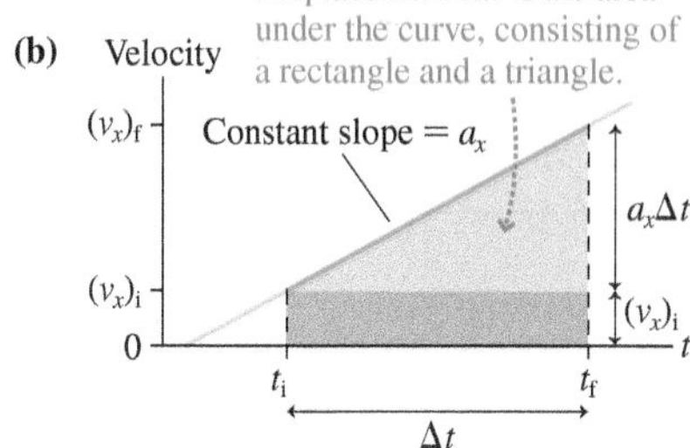

Dinner at a distance A chameleon's tongue is a powerful tool for catching prey. Certain species can extend the tongue to a distance of over 1 ft in less than 0.1 s! An analysis of videos of the motion of the chameleon tongue reveals that the tongue has a period of rapid acceleration, over 500 m/s², followed by a period of constant velocity.

With a bit of algebra, this is rearranged to read

Final and initial velocity (m/s)

$$(v_x)_f^2 = (v_x)_i^2 + 2a_x \Delta x \tag{3.23}$$

Acceleration (m/s^2) Change in position (m)

where $\Delta x = x_f - x_i$ is the *displacement* (not the distance!). Equation 3.23 is the last of the three kinematic equations for motion with constant acceleration. Note that Equations 3.19, 3.21, and 3.23 are not independent equations—each can be derived from the others—but are simply different ways of grouping the kinematic variables.

The Constant-Acceleration Model

In reality, few objects move with a perfectly constant acceleration. Even so, it is often reasonable to *model* their acceleration as being constant. It's an appropriate model when the acceleration changes little over the time period during which we wish to analyze the motion or when an order-of-magnitude estimate is sufficient. In this text, we'll usually model objects such as runners, cars, and rockets as having constant acceleration even though their actual acceleration is often more complicated.

One of the goals of this text is to help you learn a *strategy* for solving physics problems. You've seen that our worked examples have three parts: Prepare, Solve, and Assess. This is how expert problem solvers proceed, and the examples model good problem-solving techniques that we hope you will mimic. The Prepare step is especially important because thoughtful analysis and visualization of the problem are essential to a successful solution. We will present a number of specific *problem-solving strategies* at appropriate places in the text. The constant-acceleration model is the basis for our first problem-solving strategy.

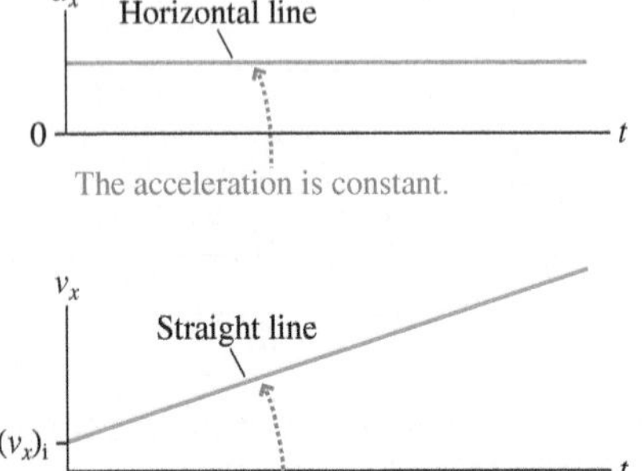

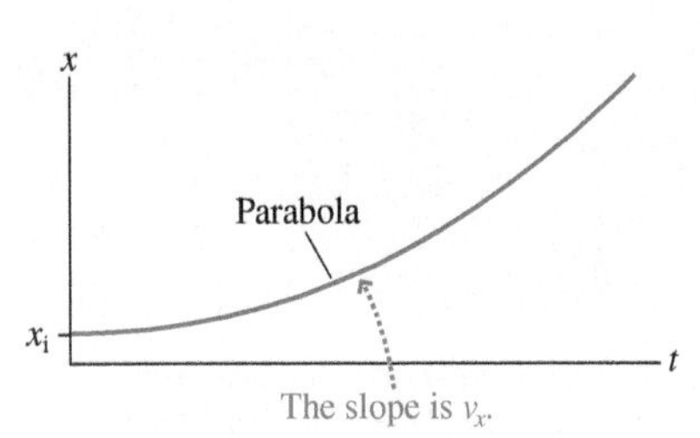

A graphical representation of motion with constant acceleration.

PROBLEM-SOLVING STRATEGY 3.1 Motion with constant acceleration

PREPARE Model the object as having constant acceleration. Use different representations of the information in the problem.

- Draw a *pictorial representation.* This helps you assess the information you are given and starts the process of translating the problem into symbols.
- Use a *graphical representation* if it is appropriate for the problem.
- Go back and forth between these two representations as needed.

SOLVE Develop a mathematical representation using the three kinematic equations:

$$(v_x)_f = (v_x)_i + a_x \Delta t \tag{3.19}$$

$$x_f = x_i + (v_x)_i \Delta t + \tfrac{1}{2} a_x (\Delta t)^2 \tag{3.21}$$

$$(v_x)_f^2 = (v_x)_i^2 + 2a_x \Delta x \tag{3.23}$$

- It is customary to use y as the symbol for vertical motion.
- Replace i and f with numerical subscripts defined in the pictorial representation.
- These equations also describe uniform motion if you set $a_x = 0$.

ASSESS Check that your result has the correct units and significant figures, is reasonable, and answers the question.

NOTE ► The three-step procedure—Prepare, Solve, Assess—will be followed in all the problem-solving strategies in this book. These are key steps in developing good problem-solving skills. ◄

STOP TO THINK 3.5 Which velocity graph or graphs go with the acceleration graph on the left? The particle is initially moving to the right.

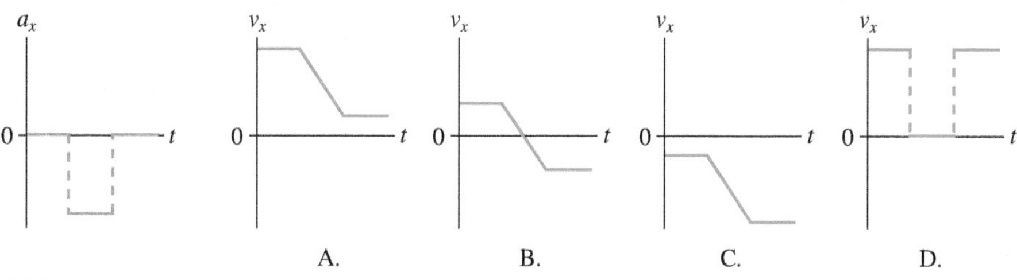

3.5 Solving Kinematics Problems

We've been developing many concepts and tools. Now it's time to put them all together to solve some kinematics problems of motion along a straight line.

Notice that we place great emphasis on drawing the pictorial representation. A common tendency of students is to read a problem and immediately ask "What equation do I use?" That approach works only for the simplest one-step problems. Problems throughout this text require higher-order problem-solving skills, and for that the Prepare step of the problem-solving strategy is critical. Expert problem solvers put *most* of their effort into modeling the situation and drawing the pictorial representation; that's where the critical thinking occurs. It's safe to say that a well-drawn pictorial representation takes you halfway to a solution and lays out a clear road map for the remaining steps. We strongly urge you to study these examples not for *Which equation is used?* but *What procedure is followed?* and then apply the same approach to your own problem solving.

EXAMPLE 3.9 Kinematics of a rocket launch

A rocket is launched straight up with a constant acceleration of $18\ \mathrm{m/s^2}$. After 150 s, how fast is the rocket moving and how far has it traveled?

PREPARE We are given the acceleration and the time interval, suggesting that this is a constant-acceleration problem. FIGURE 3.22 is a pictorial representation showing the whole problem in a nutshell. Notice how the pictorial representation shows the axis, identifies the important points of the motion, and defines the variables, using subscripts 0 and 1 to distinguish the two points in the motion. Finally, we include a list of values that gives the known and unknown quantities. We have now taken the statement of the problem in words and made it much more precise. The pictorial representation contains everything we need to know about the problem.

FIGURE 3.22 Pictorial representation of the rocket launch.

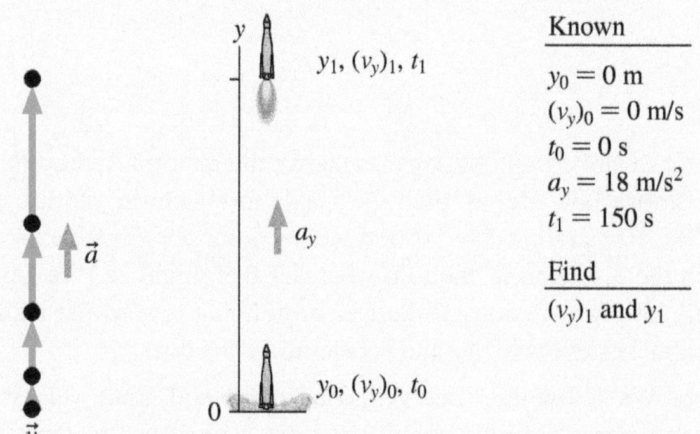

SOLVE Our first task is to find the final velocity. Our list of values includes the initial velocity, the acceleration, and the time interval, so we can use the first kinematic equation of Problem-Solving Strategy 3.1 to find the final velocity:

$$(v_y)_1 = (v_y)_0 + a_y\,\Delta t = 0\ \mathrm{m/s} + (18\ \mathrm{m/s^2})(150\ \mathrm{s})$$
$$= 2700\ \mathrm{m/s}$$

The distance traveled is found using the second equation of Problem-Solving Strategy 3.1:

$$y_1 = y_0 + (v_y)_0\,\Delta t + \tfrac{1}{2}a_y(\Delta t)^2$$
$$= 0\ \mathrm{m} + (0\ \mathrm{m/s})(150\ \mathrm{s}) + \tfrac{1}{2}(18\ \mathrm{m/s^2})(150\ \mathrm{s})^2$$
$$= 2.0 \times 10^5\ \mathrm{m} = 200\ \mathrm{km}$$

ASSESS The acceleration is very large, and it goes on for a long time, so the large final velocity and large distance traveled seem reasonable.

EXAMPLE 3.10 The motion of a rocket sled

Rocket-powered sleds, known as *rocket sleds,* are used to test equipment under extreme accelerations. Suppose that a rocket sled's engines fire for 5.0 s, boosting the sled to a speed of 250 m/s. The sled then deploys a braking parachute, slowing by 3.0 m/s per second until it stops. What is the total distance traveled?

PREPARE We're not given the sled's initial acceleration, while the rockets are firing, but rocket sleds are aerodynamically shaped to minimize air resistance and so it seems reasonable to model the sled as a particle undergoing constant acceleration. FIGURE 3.23 shows the pictorial representation. We've made the reasonable assumptions that the sled starts from rest and that the braking parachute is deployed just as the rocket burn ends. There are three points of interest in this problem: the start, the change from propulsion to braking, and the stop. Each of these points has been assigned a position, velocity, and time. Here, with three distinct points in the motion, it's essential to use numerical subscripts 0, 1, and 2 rather than the generic subscripts i and f. Accelerations are associated not with specific points in the motion but with the intervals between the points, so acceleration $(a_x)_0$ is the acceleration between points 0 and 1 while acceleration $(a_x)_1$ is the acceleration between points 1 and 2. The acceleration vector $\vec{a}_1$ points to the left, so $(a_x)_1$ is negative. The sled stops at the end point, so $(v_x)_2 = 0$ m/s.

SOLVE We know how long the rocket burn lasts and the velocity at the end of the burn. Because we're modeling the sled as having uniform acceleration, we can use the first kinematic equation of Problem-Solving Strategy 3.1 to write

$$(v_x)_1 = (v_x)_0 + (a_x)_0(t_1 - t_0) = (a_x)_0 t_1$$

We started with the complete equation, then simplified by noting which terms were zero. Solving for the boost-phase acceleration, we have

$$(a_x)_0 = \frac{(v_x)_1}{t_1} = \frac{250 \text{ m/s}}{5.0 \text{ s}} = 50 \text{ m/s}^2$$

Notice that we worked algebraically until the last step—a hallmark of good problem-solving technique that minimizes the chances of calculation errors. Also, in accord with the significant figure rules of Chapter 2, 50 m/s^2 is considered to have two significant figures.

Now we have enough information to find out how far the sled travels while the rockets are firing. The second kinematic equation of Problem-Solving Strategy 3.1 is

$$x_1 = x_0 + (v_x)_0(t_1 - t_0) + \tfrac{1}{2}(a_x)_0(t_1 - t_0)^2 = \tfrac{1}{2}(a_x)_0 t_1^2$$
$$= \tfrac{1}{2}(50 \text{ m/s}^2)(5.0 \text{ s})^2 = 625 \text{ m}$$

The braking phase is a little different because we don't know how long it lasts. But we do know both the initial and final velocities, so we can use the third kinematic equation of Problem-Solving Strategy 3.1:

$$(v_x)_2^2 = (v_x)_1^2 + 2(a_x)_1 \Delta x = (v_x)_1^2 + 2(a_x)_1(x_2 - x_1)$$

Notice that Δx is *not* x_2; it's the displacement $(x_2 - x_1)$ during the braking phase. We can now solve for x_2:

$$x_2 = x_1 + \frac{(v_x)_2^2 - (v_x)_1^2}{2(a_x)_1}$$
$$= 625 \text{ m} + \frac{0 - (250 \text{ m/s})^2}{2(-3.0 \text{ m/s}^2)} = 11{,}000 \text{ m}$$

We kept three significant figures for x_1 at an intermediate stage of the calculation but rounded to two significant figures at the end.

ASSESS The total distance is 11 km ≈ 7 mi. That's large but believable. Using the approximate conversion factor 1 m/s ≈ 2 mph from Table 2.5, we see that the top speed is ≈500 mph. It will take a long distance for the sled to gradually stop from such a high speed.

FIGURE 3.23 Pictorial representation of the rocket sled.

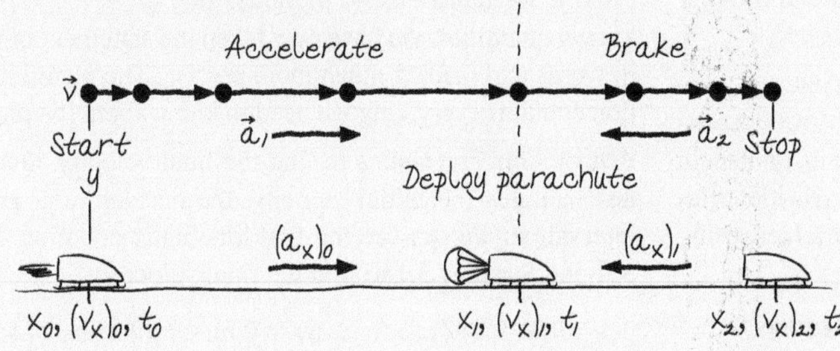

EXAMPLE 3.11 Jumping fleas

The mechanics of how fleas jump so strongly has fascinated biologists for more than a hundred years. A flea jumps by rapidly extending its hind legs. The jumps can be at any angle, but an angle of 20° above horizontal is typical. High-speed videos of one species show that the flea's body reaches a takeoff speed of 1.3 m/s over a distance of 1.0 mm. How long does the jump last?

PREPARE We can model the takeoff as constant-acceleration motion over a distance of 1.0 mm, a model that conforms to the data reasonably well. FIGURE 3.24 shows the pictorial representation. Notice how we've tilted the x-axis to match the motion of the flea. For straight-line motion we need our coordinate axis to match the direction of motion, whatever that might be. We have set up the problem so that the flea is jumping toward the right, which makes the velocity and acceleration positive.

SOLVE We know the flea's displacement and final velocity. However, there's no kinematic equation we can use to directly calculate the duration of the jump. We first need to find the flea's

FIGURE 3.24 Pictorial representation of a jumping flea.

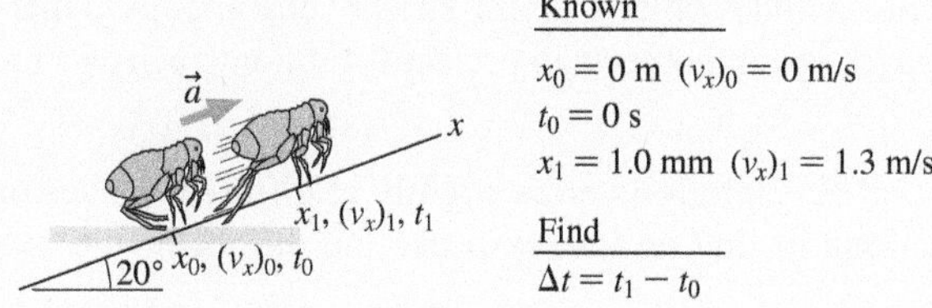

Known
$x_0 = 0$ m $(v_x)_0 = 0$ m/s
$t_0 = 0$ s
$x_1 = 1.0$ mm $(v_x)_1 = 1.3$ m/s
Find
$\Delta t = t_1 - t_0$
$\vec{a}$
x
$x_1, (v_x)_1, t_1$
$20°$ $x_0, (v_x)_0, t_0$

acceleration, and then we can compute Δt. From the third kinematic equation in Problem-Solving Strategy 3.1 we find

$$a_x = \frac{(v_x)_1^2 - (v_x)_0^2}{2\Delta x} = \frac{(1.3 \text{ m/s})^2 - (0 \text{ m/s})^2}{2(0.0010 \text{ m})} = 845 \text{ m/s}^2$$

This is a staggeringly large acceleration but possible for such a small animal. Knowing the acceleration, we can then use the first kinematic equation to calculate the duration of the jump:

$$\Delta t = \frac{(v_x)_1 - (v_x)_0}{a_x} = \frac{1.3 \text{ m/s}}{845 \text{ m/s}^2} = 0.0015 \text{ s} = 1.5 \text{ ms}$$

ASSESS The angled coordinate axis does not affect the kinematics but will, when we return to this problem later, affect the forces of the jump. It's hard to assess how reasonable our answer is other than to note that flea jumps seem instantaneous to the human eye, so the duration of the jump is certainly very short.

3.6 Free Fall

If you drop an apple and a feather, you know what will happen. The apple quickly strikes the ground, and the feather drifts slowly downward and lands some time later. But if you do this experiment in a vacuum, the result is strikingly different: Both the apple and the feather experience the exact same acceleration, undergo the exact same motion, and strike the ground at the same time.

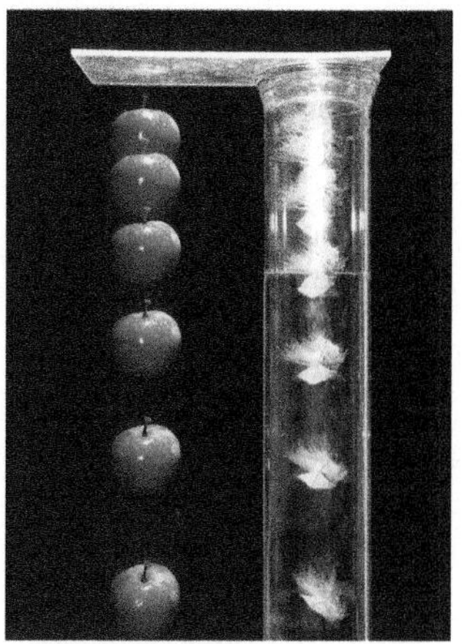

Free-falling In a vacuum, the apple and feather fall at the same rate and hit the ground at the same time.

Being in a vacuum removes air resistance, so objects are acted upon by only one force—gravity. If an object moves under the influence of gravity only, and no other forces, we call the resulting motion **free fall.** Many experiments have shown that **all objects in free fall, regardless of their mass, have the same acceleration.** Thus, if you drop two objects and they are both in free fall, they hit the ground at the same time.

Air resistance is a small factor when a dense, heavy object like a hammer or a rock falls through air, so we make only a slight error in treating a heavy object *as if* it were in free fall. Motion with air resistance is a problem we will study in Chapter 5. Until then, we will restrict our attention to situations in which air resistance is very small, and thus we will model the falling objects as being in free fall.

FIGURE 3.25a shows the motion diagram for an object that was released from rest and falls freely. Since the acceleration is the same for all objects, the diagram and graph would be the same for a falling baseball or a falling boulder! FIGURE 3.25b shows the object's velocity graph. The velocity changes at a steady rate. The slope of the velocity graph is the free-fall acceleration $a_{\text{free fall}}$.

Instead of dropping the object, suppose we throw it upward. What happens then? You know that the object will move up and that its speed will decrease as it rises. This is illustrated in the motion diagram of FIGURE 3.25c, which shows a surprising result: Even though the object is moving up, its acceleration still points down. In fact, **the free-fall acceleration always points down,** no matter what direction an object is moving.

FIGURE 3.25 Motion of an object in free fall.

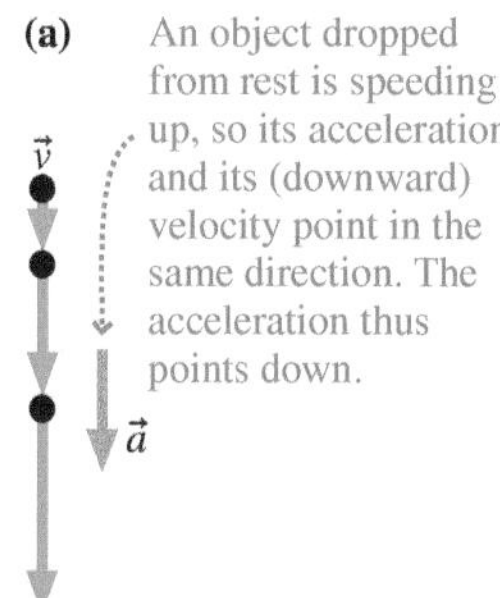

(a)
An object dropped from rest is speeding up, so its acceleration and its (downward) velocity point in the same direction. The acceleration thus points down.
$\vec{v}$
$\vec{a}$

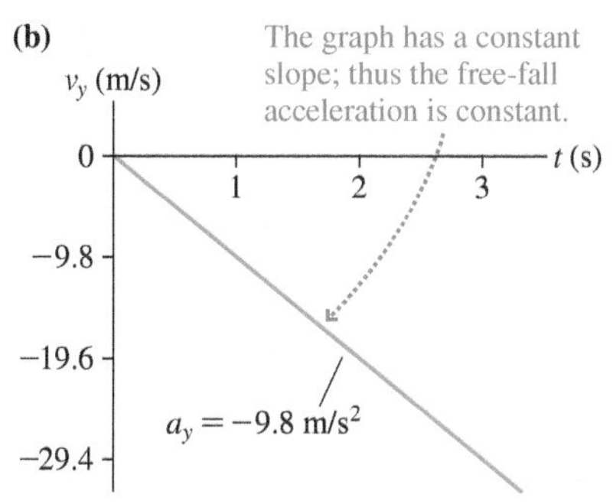

(b)
v_y (m/s)
The graph has a constant slope; thus the free-fall acceleration is constant.
0
1
2
3
t (s)
−9.8
−19.6
−29.4
$a_y = -9.8$ m/s^2

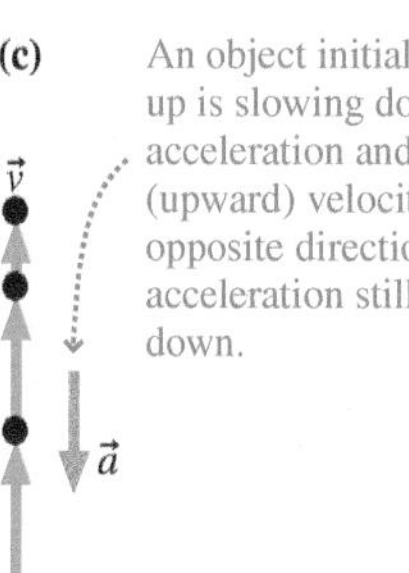

(c)
An object initially moving up is slowing down, so its acceleration and its (upward) velocity point in opposite directions. The acceleration still points down.
$\vec{v}$
$\vec{a}$

Free falling We can't tell from the photo if these children are moving up or down. Either way, they are in free fall—and so are accelerating downward at 9.8 m/s^2.

NOTE ▸ Despite the name, free fall is not restricted to objects that are literally falling. Any object moving under the influence of gravity only, and no other forces, is in free fall. This includes objects falling straight down, objects that have been tossed or shot straight up, and objects in projectile motion (such as a basketball free throw). ◂

The value of the free-fall acceleration varies slightly at different places on the earth, but for the calculations in this text we will use the the following average value:

$$\vec{a}_{\text{free fall}} = (9.80 \text{ m/s}^2, \text{vertically downward}) \tag{3.24}$$

Standard value for the acceleration of an object in free fall

The magnitude of the **free-fall acceleration** has the special symbol g:

$$g = 9.80 \text{ m/s}^2$$

Several points about free fall are worthy of note:

- g, by definition, is *always* positive. **There will never be a problem that uses a negative value for g.**
- The velocity graph in Figure 3.25b has a negative slope. Even though a falling object speeds up, it has *negative* acceleration. Thus g is *not* the object's acceleration, simply the *magnitude* of the acceleration. The one-dimensional acceleration is

$$a_y = -g$$

- Because free fall is motion with constant acceleration, we can use the kinematic equations for constant acceleration with $a_y = -g$.
- Once an object is acted upon by only the force of gravity, it is in free fall with $a_y = -9.8$ m/s^2. It doesn't matter how the object entered free fall: Once in the air, a football that was punted straight up has the *same* acceleration of $-g$ as a stone dropped off a bridge.
- g is not called "gravity." Gravity is a force, not an acceleration. g is the *free-fall acceleration.* It is also, in some books, called the *acceleration of gravity.*
- $g = 9.80$ m/s^2 only on earth. Other planets have different values of g. You will learn in Chapter 5 how to determine g for other planets.
- We will sometimes compute the acceleration of objects not in free fall in units of g. An acceleration of 9.8 m/s^2 is an acceleration of 1g; an acceleration of 19.6 m/s^2 is 2g. Generally, we can compute

$$\text{acceleration (in units of } g\text{, or } g\text{'s)} = \frac{\text{acceleration (in units of m/s}^2)}{9.8 \text{ m/s}^2} \tag{3.25}$$

EXAMPLE 3.12 Analyzing a rock's fall

A heavy rock is dropped from rest at the top of a cliff and falls 100 m before hitting the ground. How long does the rock take to fall to the ground, and what is its velocity when it hits?

PREPARE We will ignore air resistance for the heavy rock and model the motion as constant acceleration with $a_y = -g$. FIGURE 3.26 shows a pictorial representation with all necessary data. We have placed the origin at the ground, so $y_0 = 100$ m.

SOLVE The first question in the problem statement involves a relationship between time and distance. Using $(v_y)_0 = 0$ m/s and $t_0 = 0$ s, we find

$$y_1 = y_0 + (v_y)_0\Delta t + \tfrac{1}{2}a_y(\Delta t)^2 = y_0 - \tfrac{1}{2}g(\Delta t)^2 = y_0 - \tfrac{1}{2}g{t_1}^2$$

Solving for t_1 gives

$$t_1 = \sqrt{\frac{2(y_0 - y_1)}{g}} = \sqrt{\frac{2(100 \text{ m} - 0 \text{ m})}{9.80 \text{ m/s}^2}} = 4.5 \text{ s}$$

Now that we know the fall time, we can find $(v_y)_1$:

$$(v_y)_1 = (v_y)_0 - g\,\Delta t = -gt_1 = -(9.80 \text{ m/s}^2)(4.5 \text{ s})$$
$$= -44 \text{ m/s}$$

The final velocity is negative because the motion is downward. The impact *speed* is 44 m/s.

FIGURE 3.26 Pictorial representation of a falling rock.

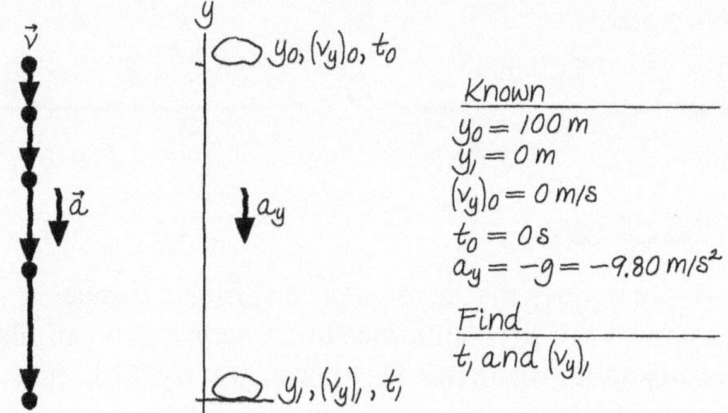

ASSESS Are the answers reasonable? Well, 100 m is about 300 feet, which is about the height of a 30-floor building. How long does it take an object to fall 30 floors? Four or five seconds seems pretty reasonable. How fast would the object be going at the bottom? Using the approximate conversion factor 1 m/s ≈ 2 mph, we find that 44 m/s ≈ 90 mph. That also seems like a pretty reasonable speed for something that has fallen 30 floors. Suppose we had made a mistake. If we had misplaced a decimal point, we could have calculated a speed of 440 m/s, or about 900 mph! This is clearly *not* reasonable. The point of assessing your answer is not to prove that it's right but to rule out answers that are obviously wrong.

EXAMPLE 3.13 Finding the height of a leap

The springbok is an antelope found in southern Africa that gets its name from its remarkable jumping ability. When a springbok is startled, it will leap straight up into the air—a maneuver called a "pronk." A particular springbok goes into a crouch to perform a pronk. It then extends its legs forcefully, accelerating at 35 m/s^2 for 0.70 m as its legs straighten. Legs fully extended, it leaves the ground and rises into the air.

a. At what speed does the springbok leave the ground?
b. How high does it go?

PREPARE The springbok changes shape, but we can still model it as a particle if we focus on the movement of the body, treating the legs as separate objects that push on the body. This is not too bad an approximation because the legs are only a relatively small part of the mass of the animal. We then have a two-part problem. In the first phase of its motion, the springbok accelerates upward, reaching some maximum speed just as it leaves the ground. The problem statement indicates that it's a constant acceleration. As soon as it does so, the springbok is subject to only the force of gravity, so it is in free fall. FIGURE 3.27 is the pictorial representation, where we've identified three distinct phases in the motion.

FIGURE 3.27 Pictorial representation of the springbok's leap.

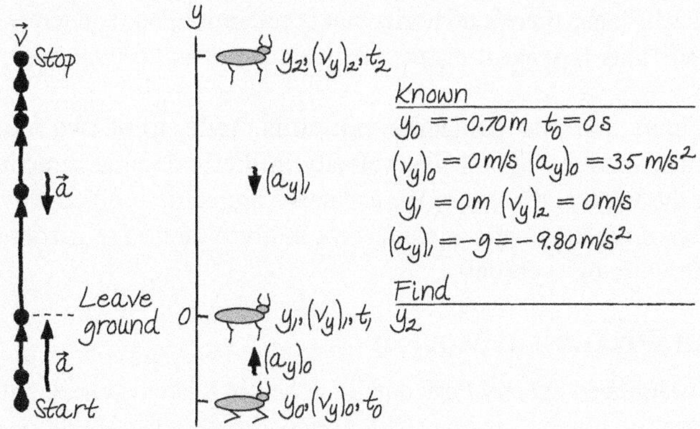

We've placed the origin at point 1, where the springbok leaves the ground, so that y_2 is the height we seek. This makes the initial position y_0 negative. A negative position doesn't mean that the springbok is below ground, merely beneath the point we've chosen for the origin. Notice that the highest point is a turning point of the motion, with zero instantaneous velocity: $(v_y)_2 = 0$ m/s. This was not explicitly stated but is part of our interpretation of the problem.

SOLVE

a. For the first phase, pushing off the ground, we have information about displacement, initial velocity, and acceleration, but we don't know anything about the time interval. The third equation in Problem-Solving Strategy 3.1 is perfect for this type of situation. We can use it to solve for the velocity with which the springbok lifts off the ground:

$$(v_y)_1^2 = (v_y)_0^2 + 2(a_y)_0 \Delta y$$
$$= (0 \text{ m/s})^2 + 2(35 \text{ m/s}^2)(0.70 \text{ m}) = 49 \text{ m}^2/\text{s}^2$$
$$(v_y)_1 = \sqrt{49 \text{ m}^2/\text{s}^2} = 7.0 \text{ m/s}$$

The springbok leaves the ground with a speed of 7.0 m/s.

b. Now we are ready for the second phase of the motion, the vertical motion after leaving the ground. The displacement-velocity equation is again appropriate because again we don't know the time. Because $y_1 = 0$, the springbok's displacement is $\Delta y = y_2 - y_1 = y_2$, the height of the leap. From part a, the initial velocity is $(v_y)_1 = 7.0$ m/s, and the final velocity is $(v_y)_2 = 0$. This is free-fall motion, with $(a_y)_1 = -g$; thus

$$(v_y)_2^2 = 0 = (v_y)_1^2 - 2g\Delta y = (v_y)_1^2 - 2gy_2$$

which gives

$$(v_y)_1^2 = 2gy_2$$

Solving for y_2, we get a jump height of

$$y_2 = \frac{(7.0 \text{ m/s})^2}{2(9.8 \text{ m/s}^2)} = 2.5 \text{ m}$$

ASSESS 2.5 m is a remarkable leap—a bit over 8 ft—but these animals are known for their jumping ability, so this seems reasonable.

STOP TO THINK 3.6 A volcano ejects a chunk of rock straight up at a velocity of $v_y = 30$ m/s. Ignoring air resistance, what will be the velocity v_y of the rock when it falls back into the volcano's crater?

A. > 30 m/s B. 30 m/s C. 0 m/s D. −30 m/s E. < −30 m/s

3.7 Projectile Motion

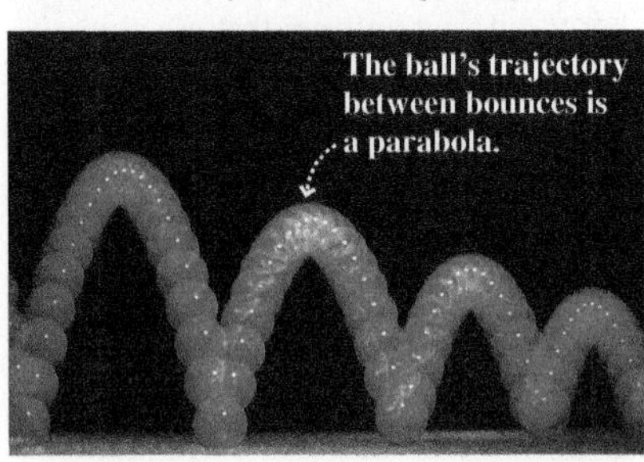

The ball's trajectory
between bounces is
a parabola.

FIGURE 3.28 A parabolic trajectory.

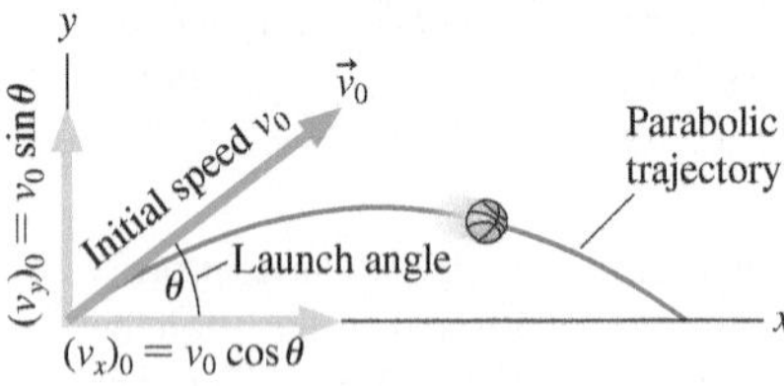

FIGURE 3.29 A projectile launched with initial velocity $\vec{v}_0$.

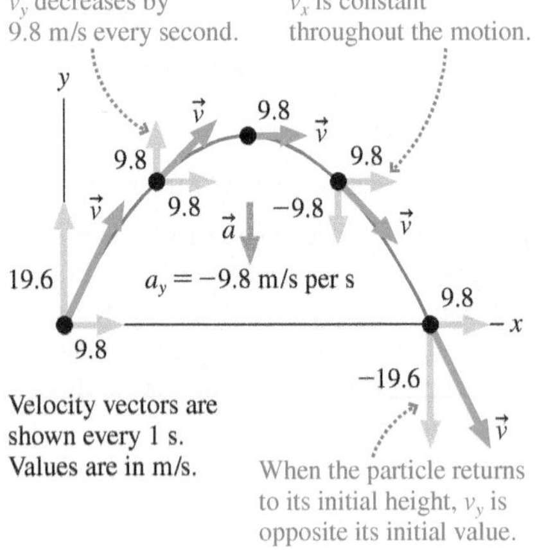

$a_y = -9.8$ m/s per s

shown every 1 s.
Values are in m/s.

to its initial height, v_y is
opposite its initial value.

FIGURE 3.30 The velocity and acceleration vectors of a projectile.

Baseballs and tennis balls flying through the air, Olympic divers, and daredevils shot from cannons all exhibit what we call **projectile motion.** A *projectile* is **an object that moves in two dimensions under the influence of only gravity.** Projectile motion is motion along a curved line, rather than a straight line, but it's an extension of free-fall motion and worthy of a short digression. We will continue to neglect the influence of air resistance, leading to results that are a good approximation of reality for relatively heavy objects moving relatively slowly over relatively short distances. Projectiles follow a *parabolic trajectory* like the one seen in FIGURE 3.28.

The start of a projectile's motion, be it thrown by hand or shot from a gun, is called the *launch,* and the angle θ of the initial velocity $\vec{v}_0$ above the horizontal (i.e., above the x-axis) is called the **launch angle.** This is illustrated in FIGURE 3.29. We'll deal with vectors in Chapter 4, but you can see that the initial velocity vector $\vec{v}_0$ can be broken into a horizontal component $(v_x)_0$ and a vertical component $(v_y)_0$. You learned in trigonometry that the adjacent and opposite sides of a right triangle are the hypotenuse multiplied by cosine and sine; thus

$$\begin{aligned}(v_x)_0 &= v_0\cos\theta \\ (v_y)_0 &= v_0\sin\theta\end{aligned} \tag{3.26}$$

where v_0 is the initial speed.

NOTE ▶ A projectile launched at an angle *below* the horizontal (such as a ball thrown downward from the roof of a building) has *negative* values for θ and $(v_y)_0$. However, the *speed* v_0 is always positive. ◀

Gravity acts downward, and we know that objects released from rest fall straight down, not sideways. Hence a projectile has no horizontal acceleration, while its vertical acceleration is simply that of free fall. Thus

$$\begin{aligned}a_x &= 0 \\ a_y &= -g\end{aligned} \tag{3.27}$$

In other words, **the vertical component of acceleration a_y is just the familiar $-g$ of free fall, while the horizontal component a_x is zero. Projectiles are in free fall.**

To see how these conditions influence the motion, FIGURE 3.30 shows a projectile launched from $(x_0, y_0) = (0\text{ m}, 0\text{ m})$ with $(v_x)_0 = 9.8$ m/s and $(v_y)_0 = 19.6$ m/s. The value of v_x never changes because there's no horizontal acceleration, but v_y decreases by 9.8 m/s every second. This is what it *means* to accelerate at $a_y = -9.8\text{ m/s}^2 = (-9.8\text{ m/s})$ per second.

You can see from Figure 3.30 that **projectile motion is made up of two independent motions:** uniform motion at constant velocity in the horizontal direction and free-fall motion in the vertical direction. We can solve projectile problems by a *simultaneous* application of the kinematic equations for uniform motion (horizontal) and constant-acceleration motion (vertical).

Reasoning About Projectile Motion

Suppose a heavy ball is launched exactly horizontally at height h above a horizontal field. At the exact instant that the ball is launched, a second ball is simply dropped from height h. Which ball hits the ground first?

It may seem hard to believe, but—if air resistance is neglected—the balls hit the ground *simultaneously*. They do so because the horizontal and vertical components of projectile motion are independent of each other. The initial horizontal velocity of the first ball has *no* influence over its vertical motion. Neither ball has any initial motion in the vertical direction, so both fall distance h in the same amount of time. You can see this in FIGURE 3.31.

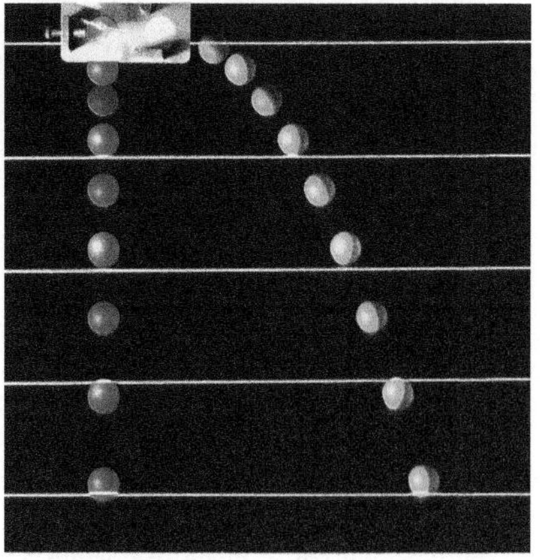

FIGURE 3.31 A projectile launched horizontally falls in the same time as a projectile that is released from rest.

FIGURE 3.32a shows a useful way to think about the trajectory of a projectile. Without gravity, a projectile would follow a straight line. Because of gravity, the particle at time t has "fallen" a distance $\frac{1}{2}gt^2$ below this line. The separation grows as $\frac{1}{2}gt^2$, giving the trajectory its parabolic shape.

Use this idea to think about the following "classic" problem in physics:

> A hungry bow-and-arrow hunter in the jungle wants to shoot down a coconut that is hanging from the branch of a tree. He points his arrow directly at the coconut, but as luck would have it, the coconut falls from the branch at the *exact* instant the hunter releases the string. Does the arrow hit the coconut?

You might think that the arrow will miss the falling coconut, but it doesn't. Although the arrow travels very fast, it follows a slightly curved parabolic trajectory, not a straight line. Had the coconut stayed on the tree, the arrow would have curved under its target as gravity caused it to fall a distance $\frac{1}{2}gt^2$ below the straight line. But $\frac{1}{2}gt^2$ is also the distance the coconut falls while the arrow is in flight. Thus, as FIGURE 3.32b shows, the arrow and the coconut fall the same distance and meet at the same point!

FIGURE 3.32 A projectile follows a parabolic trajectory because it "falls" a distance $\frac{1}{2}gt^2$ below a straight-line trajectory.

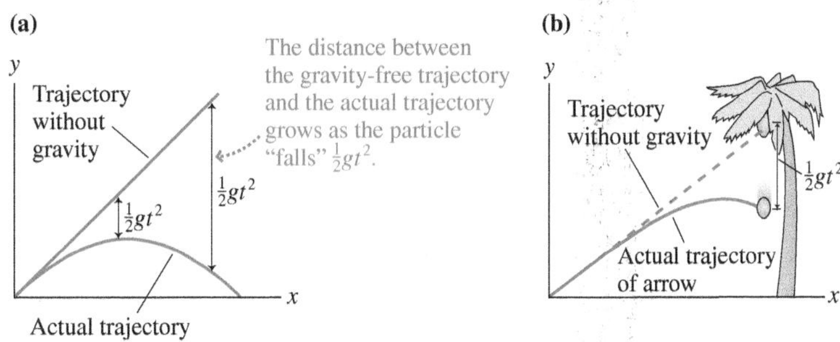

Solving Projectile Motion Problems

PROBLEM-SOLVING STRATEGY 3.2 Projectile motion problems

PREPARE Establish a coordinate system with the x-axis horizontal and the y-axis vertical. Define symbols and identify what the problem is trying to find. For a launch at angle θ, the initial velocity components are $(v_x)_i = v_0 \cos\theta$ and $(v_y)_i = v_0 \sin\theta$.

SOLVE The acceleration is known: $a_x = 0$ and $a_y = -g$. The kinematic equations are

Horizontal	Vertical
$x_f = x_i + (v_x)_i \Delta t$	$y_f = y_i + (v_y)_i \Delta t - \frac{1}{2}g(\Delta t)^2$
$(v_x)_f = (v_x)_i = \text{constant}$	$(v_y)_f = (v_y)_i - g\Delta t$

Δt is the same for the horizontal and vertical components of the motion. Find Δt from one component, then use that value for the other component.

ASSESS Check that your result has correct units and significant figures, is reasonable, and answers the question.

EXAMPLE 3.14 **Leaping lemurs**

Lemurs, primates found only on the island of Madagascar, are known for their ability to jump from tree branch to tree branch. Suppose a lemur wants to leap to a branch 5.0 m to the side of and 2.5 m lower than where it is sitting. If it pushes off horizontally, with what speed does it need to leave its current branch in order to reach its target?

PREPARE The lemur is small compared to the distance it needs to travel. If we ignore air resistance, a fairly good approximation, we can model the lemur as a projectile in free fall. The pictorial representation of FIGURE 3.33 is very important because there are a large number of quantities to keep track of. The position of the origin of the coordinate system is our choice. We could place the origin at the lemur's starting position, but then y_1 would be negative. Instead, we position the origin level with the ending position so that all quantities are positive. The statement that the lemur leaps horizontally means that $(v_x)_0 = v_0$ and $(v_y)_0 = 0$ m/s.

FIGURE 3.33 Pictorial representation of a leaping lemur.

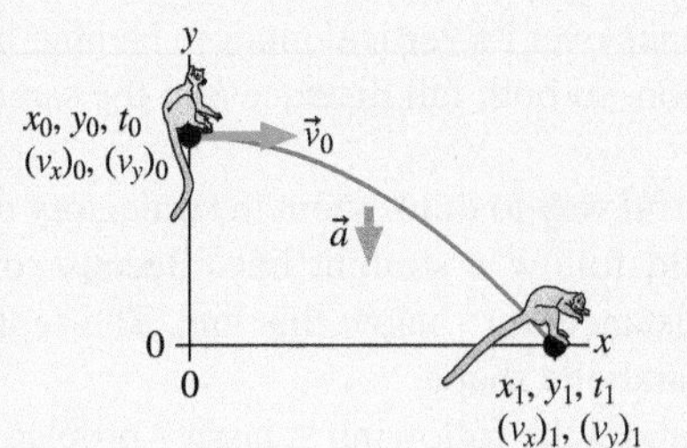

Known	
$x_0 = 0$ m	$(v_x)_0 = v_0$
$y_0 = 2.5$ m	$(v_y)_0 = 0$ m/s
$t_0 = 0$ s	
$x_1 = 5.0$ m	$y_1 = 0$ m
$a_x = 0$ m/s^2	$a_y = -g$
Find	
v_0	

SOLVE Each point on the trajectory has *x*- and *y*-components of position and velocity but only *one* value of time. That is, the time needed to move horizontally to x_1 is the *same* time needed to fall vertically to y_1. **Although the horizontal and vertical motions are independent, they are connected through the time *t*.** This is a critical observation for solving projectile-motion problems. The kinematics equations from Problem-Solving Strategy 3.2 are

$$x_1 = x_0 + (v_x)_0\,\Delta t = v_0 t_1$$
$$y_1 = y_0 + (v_y)_0\,\Delta t - \tfrac{1}{2}g(\Delta t)^2 = y_0 - \tfrac{1}{2}g t_1^2$$

where we use $\Delta t = t_1$ because $t_0 = 0$ s. The *y*-equation allows us to find the time needed to fall 2.5 m to the lower branch:

$$t_1 = \sqrt{\frac{2y_0}{g}} = \sqrt{\frac{2(2.5\text{ m})}{9.8\text{ m/s}^2}} = 0.71\text{ s}$$

We can then solve the *x*-equation for the launch speed v_0. Using the value of t_1, we find

$$v_0 = \frac{x_1}{t_1} = \frac{5.0\text{ m}}{0.71\text{ s}} = 7.0\text{ m/s}$$

ASSESS We can use the approximate conversion factor 1 m/s ≈ 2 mph to see that 7 m/s ≈ 14 mph. That's fast for a leap but not unreasonable for an animal noted for its leaping ability.

EXAMPLE 3.15 **Jumping frog contest**

Frogs, with their long, strong legs, are excellent jumpers. And thanks to the good folks of Calaveras County, California, who have a jumping frog contest every year in honor of a Mark Twain story, we have very good data on how far a determined frog can jump.

High-speed cameras show that a good jumper goes into a crouch, then rapidly extends his legs by typically 15 cm during a 65 ms push off, leaving the ground at a 30° angle. How far does this frog leap?

PREPARE We model the push off as linear motion with uniform acceleration. A bullfrog is fairly heavy and dense, so we ignore air resistance and model the leap as projectile motion. This is a two-part problem: linear acceleration followed by projectile motion. A key observation is that **the final velocity for pushing off the ground becomes the initial velocity of the projectile motion.** FIGURE 3.34 shows a separate pictorial representation for each part. Notice that we've used different coordinate systems for the two parts; coordinate systems are our choice, and for each part of the motion we've chosen the coordinate system that makes the problem easiest to solve.

SOLVE While pushing off, the frog travels 15 cm = 0.15 m in 65 ms = 0.065 s. We could find his speed at the end of pushing off if we knew the acceleration. Because the initial velocity is zero, we can find the acceleration from the position-acceleration-time kinematic equation:

$$x_1 = x_0 + (v_x)_0\Delta t + \tfrac{1}{2}a_x(\Delta t)^2 = \tfrac{1}{2}a_x(\Delta t)^2$$

$$a_x = \frac{2x_1}{(\Delta t)^2} = \frac{2(0.15\text{ m})}{(0.065\text{ s})^2} = 71\text{ m/s}^2$$

This is a substantial acceleration, but it doesn't last long. At the end of the 65 ms push off, the frog's velocity is

$$(v_x)_1 = (v_x)_0 + a_x\Delta t = (71\text{ m/s}^2)(0.065\text{ s}) = 4.62\text{ m/s}$$

We'll keep an extra significant figure here to avoid round-off error in the second half of the problem.

The end of the push off is the beginning of the projectile motion, so the second part of the problem is to find the distance of a projectile launched with velocity $\vec{v}_0 = (4.62\text{ m/s}, 30°)$. The initial *x*- and *y*-components of the launch velocity are

$$(v_x)_0 = v_0\cos\theta \qquad (v_y)_0 = v_0\sin\theta$$

The kinematic equations of projectile motion, with $a_x = 0$ and $a_y = -g$, are

$$x_1 = x_0 + (v_x)_0\,\Delta t = (v_0\cos\theta)\Delta t \qquad y_1 = y_0 + (v_y)_0\,\Delta t - \tfrac{1}{2}g(\Delta t)^2 = (v_0\sin\theta)\Delta t - \tfrac{1}{2}g(\Delta t)^2$$

We can find the time of flight from the vertical equation by setting $y_1 = 0$:

$$0 = (v_0\sin\theta)\Delta t - \tfrac{1}{2}g(\Delta t)^2 = \left(v_0\sin\theta - \tfrac{1}{2}g\,\Delta t\right)\Delta t$$

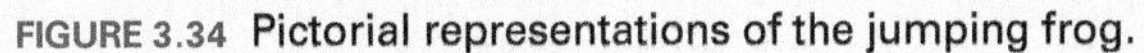

FIGURE 3.34 Pictorial representations of the jumping frog.

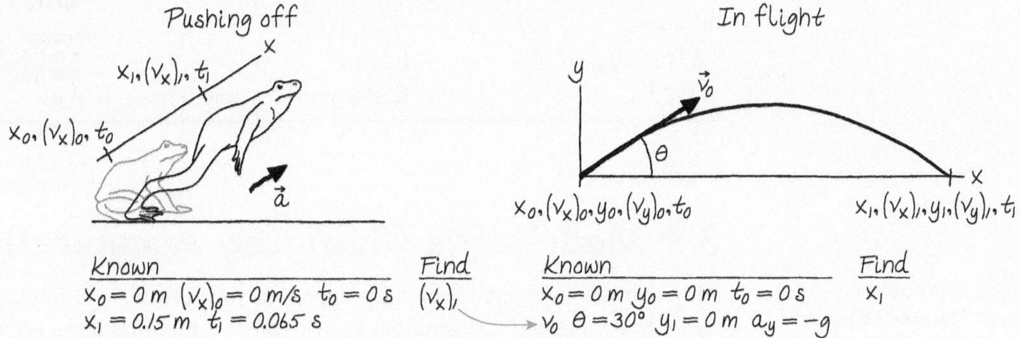

and thus

$$\Delta t = 0 \quad \text{or} \quad \Delta t = \frac{2v_0 \sin\theta}{g}$$

Both are legitimate solutions. The first corresponds to the instant when $y = 0$ at the launch, the second to when $y = 0$ as the frog hits the ground. Clearly, we want the second solution. Substituting this expression for Δt into the equation for x_1 gives

$$x_1 = (v_0 \cos\theta)\frac{2v_0 \sin\theta}{g} = \frac{2v_0^2 \sin\theta \cos\theta}{g}$$

We can simplify this result with the trigonometric identity $2\sin\theta\cos\theta = \sin(2\theta)$. Thus the distance traveled by the frog is

$$x_1 = \frac{v_0^2 \sin(2\theta)}{g}$$

Using $v_0 = 4.62$ m/s and $\theta = 30°$, we find that the frog leaps a distance of 1.9 m.

ASSESS 1.9 m is about 6 feet, or about 10 times the frog's body length. That's amazing, but true. Jumps of 2.2 m have been seen in the lab. And the Calaveras County record holder, Rosie the Ribeter, covered 6.5 m—21 feet—in three jumps!

The distance a projectile travels is called its *range*. As Example 3.15 found, a projectile that lands at the same elevation from which it was launched has

$$\text{range} = \frac{v_0^2 \sin(2\theta)}{g} \tag{3.28}$$

If the projectile can be launched at the same speed at all angles—which is not always true—and if there's no air resistance, then the maximum range occurs for $\theta = 45°$, where $\sin(2\theta) = 1$. But there's more that we can learn from this equation. Because $\sin(180° - x) = \sin x$, it follows that $\sin(2(90° - \theta)) = \sin(2\theta)$. Consequently, a projectile launched either at angle θ *or* at angle $(90° - \theta)$ will travel the same distance *over level ground*. FIGURE 3.35 shows several trajectories of projectiles launched with the same initial speed.

FIGURE 3.35 Trajectories of a projectile launched with a speed of 99 m/s.

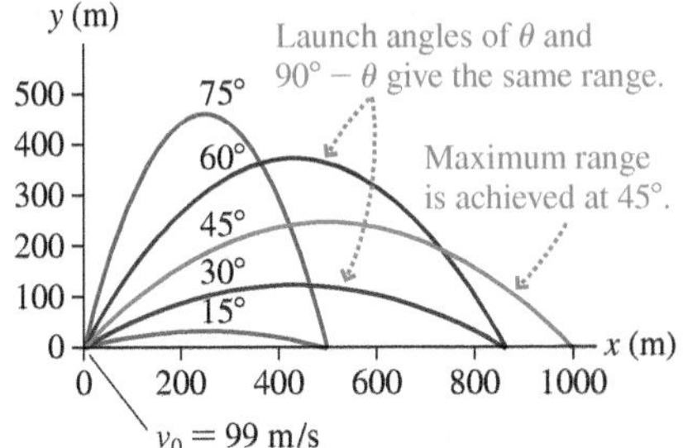

NOTE ▶ Equation 3.28 is *not* a general result. It applies *only* in situations where the projectile lands at the same elevation from which it was fired. ◀

It is interesting that human athletes who throw an object for distance, whether a baseball or a javelin, throw at an angle well under the 45° that a physicist might predict. The reason lies in the biomechanics of throwing. You can throw a ball horizontally at high speed, but it's very difficult to throw a ball straight up. Human arms and shoulders aren't designed to do that. Your maximum throwing speed decreases as the angle of the throw increases, and a faster throw at a smaller angle can outdistance a slower throw at 45°. Different throwing sports require somewhat different biomechanics, but elite athletes in many sports throw at an angle in the 30°–35° range. Long jumpers launch at an even smaller angle, typically 20°–25°, because the run-up speed carries into the launch speed only for jumps at fairly shallow angles.

Air resistance is also a factor for real-world projectiles. Shorter flight times are less influenced by air resistance than longer flight times, and that tends to favor smaller launch angles over larger launch angles. For example, a batted baseball travels the maximum distance if it is hit at an angle of about 40°.

A long long jump A 45° angle gives the greatest range for a projectile, so why do long jumpers take off at a much shallower angle? The main reason is that athletes can jump faster at smaller angles. The gain from the faster speed outweighs the effect of the smaller angle.

STOP TO THINK 3.7 A 50 g marble rolls off a table and hits 2 m from the base of the table. A 100 g marble rolls off the same table with the same speed. It lands at distance

A. Less than 1 m. B. 1 m. C. Between 1 m and 2 m.
D. 2 m. E. Between 2 m and 4 m. F. 4 m.

3.8 Modeling a Changing Acceleration

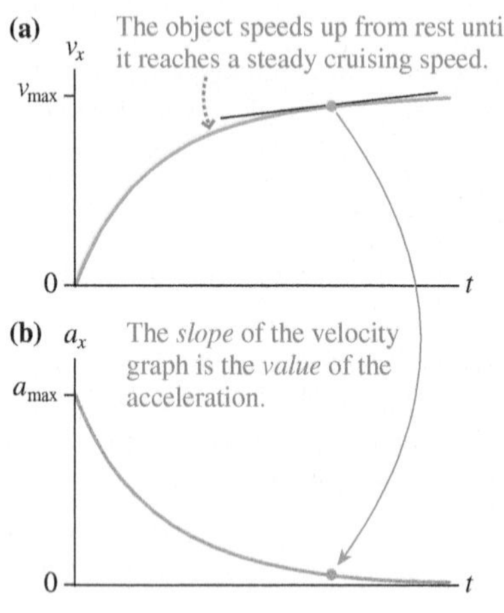

(a)
v_x
The object speeds up from rest until it reaches a steady cruising speed.
v_{max}
0
t
(b) a_x
The *slope* of the velocity graph is the *value* of the acceleration.
a_{max}
0
t

FIGURE 3.36 Realistic velocity and acceleration graphs that could describe a runner or a car.

Although the constant-acceleration model is very useful, real moving objects only rarely have constant acceleration. FIGURE 3.36a is a more realistic velocity graph that could describe a runner beginning a race or a car leaving a stop sign. The graph is not a straight line, so this is *not* motion with constant acceleration.

We can define an instantaneous acceleration much as we defined the instantaneous velocity. The instantaneous velocity at time t is the slope of the position graph at that time or, mathematically, the derivative of the position with respect to time. By analogy: The **instantaneous acceleration a_x is the slope of the line that is tangent to the velocity curve at time t.** Mathematically, this is

$$a_x = \frac{dv_x}{dt} = \text{slope of the velocity graph at time } t \tag{3.29}$$

FIGURE 3.36b applies this idea by showing the object's acceleration graph. At each instant of time, the *value* of the object's acceleration is the *slope* of its velocity graph. The initially steep slope indicates a large initial acceleration. The acceleration decreases to zero as the object reaches its maximum speed.

The reverse problem—to find the velocity v_x if we know the acceleration a_x at all instants of time—is also important. Again, with analogy to velocity and position, we have

$$(v_x)_f = (v_x)_i + \int_{t_i}^{t_f} a_x\, dt \tag{3.30}$$

The graphical interpretation of Equation 3.30 is

$$(v_x)_f = (v_x)_i + \text{area under the acceleration curve } a_x \text{ between } t_i \text{ and } t_f \tag{3.31}$$

EXAMPLE 3.16 Finding velocity from acceleration

FIGURE 3.37 shows the acceleration graph for a particle with an initial velocity of 10 m/s. What is the particle's velocity at $t = 8$ s?

PREPARE Figure 3.37 is a graphical representation of the motion. The acceleration changes, so to find the velocity we will need to integrate—which we can do graphically by finding the area under the curve.

FIGURE 3.37 Acceleration graph for Example 3.16.

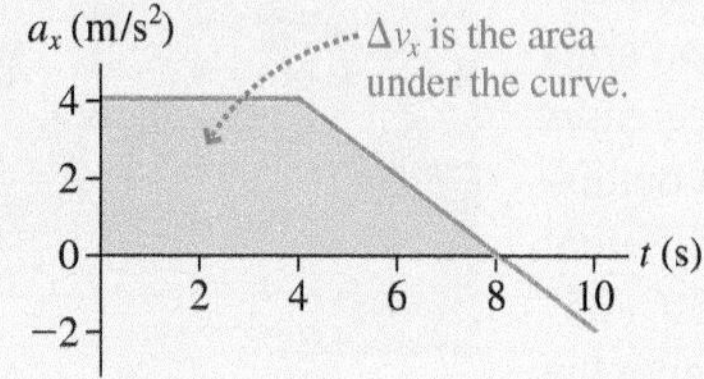

a_x (m/s²)
Δv_x is the area under the curve.
4
2
0
−2
2
4
6
8
10
t (s)

SOLVE The change in velocity is found as the area under the acceleration curve:

$$(v_x)_f = (v_x)_i + \text{area under the acceleration curve } a_x \text{ between } t_i \text{ and } t_f$$

The area under the curve between $t_i = 0$ s and $t_f = 8$ s can be subdivided into a rectangle $(0\text{ s} \le t \le 4\text{ s})$ and a triangle $(4\text{ s} \le t \le 8\text{ s})$. These areas are easily computed. Thus

$$v_x(\text{at } t = 8\text{ s}) = 10\text{ m/s} + (4\text{ (m/s)/s})(4\text{ s}) + \tfrac{1}{2}(4\text{ (m/s)/s})(4\text{ s}) = 34\text{ m/s}$$

Modeling Motion with Exponential Functions

Whether it's a racing car or a human sprinter, any vehicle or animal running an all-out sprint starts from rest, accelerates to faster and faster speeds, and eventually levels out at a top speed. That is, acceleration doesn't go on forever. It turns out that all-out sprints can be modeled quite accurately with *exponential functions*—a result that can be justified from biomechanics.

An exponential function has the form

$$u = Ae^{-t/\tau} = A\exp(-t/\tau) \tag{3.32}$$

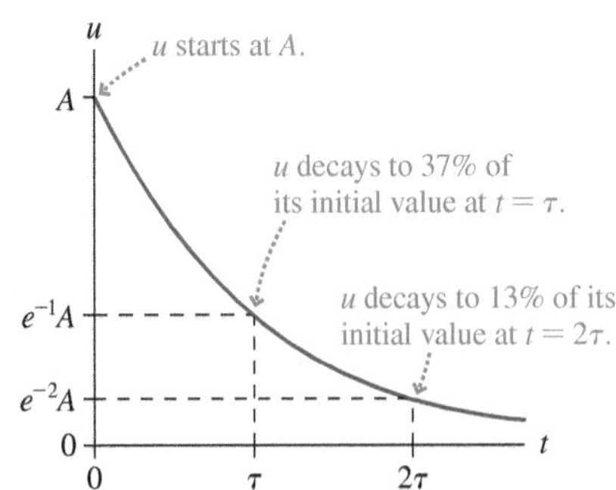

FIGURE 3.38 The exponential function.

where $e = 2.71828\ldots$ is the base of natural logarithms in the same way that 10 is the base of ordinary logarithms. FIGURE 3.38 is a graph of u versus t, showing what is called *exponential decay*. The graph starts from $u(t = 0) = A$ because $e^0 = 1$, and then asymptotically approaches zero for very large values of t.

The constant τ (Greek tau) is called the **time constant.** When $t = \tau$, $u = e^{-1}A = 0.37A$. When $t = 2\tau$, $u = e^{-2}A = 0.13A$. We will see numerous examples of exponential decay in this text, and for all of them **the quantity decays to 37% of its initial value after one time constant has elapsed.** Exponential decay is a common mathematical model; once you've learned the properties of exponential functions, you will immediately know how to apply this knowledge to a new situation.

The acceleration curve of Figure 3.36b looks very much like an exponential decay, so let's try modeling a sprinter's acceleration as

$$a_x = a_{max}\, e^{-t/\tau} \tag{3.33}$$

Then what about velocity? The exponential function has important properties that make it useful in modeling: (1) The derivative of an exponential is an exponential, and (2) the integral of an exponential is an exponential. Specifically,

$$\frac{d}{dt}(e^{-t/\tau}) = -\frac{1}{\tau}e^{-t/\tau} \text{ and } \int e^{-t/\tau}dt = -\tau\, e^{-t/\tau} \tag{3.34}$$

Differentiating or integrating gives back the same exponential, simply multiplied by a (negative) constant.

Starting with our exponential model of acceleration, we can use Equations 3.30 and 3.34 to find the sprinter's velocity. The sprinter starts from rest, so $(v_x)_i = 0$ and thus

$$v_x = \int_0^t a_x dt = a_{max}\int_0^t e^{-t/\tau}dt = -a_{max}\tau e^{-t/\tau}\Big|_0^t \tag{3.35}$$
$$= -a_{max}\tau(e^{-t/\tau} - e^0)$$
$$= v_{max}(1 - e^{-t/\tau})$$

where in the last step we used $e^0 = 1$.

We also wrote $v_{max} = a_{max}\tau$, and that needs a justification. After a very long time, $t \gg \tau$, the exponential function approaches zero: $e^{-t/\tau} \rightarrow 0$. That's what *decay* means. As the exponential approaches zero, the quantity $(1 - e^{-t/\tau})$ approaches one. Thus the constant in front of the parentheses in Equation 3.35 is the sprinter's top speed, for $t \gg \tau$, and that is v_{max}. FIGURE 3.39 shows the exponential acceleration and velocity graphs. At $t = \tau$, the acceleration has decreased to 37% of its maximum value while the velocity has increased to 100% − 37% = 63% of its maximum. A homework problem will let you compute the distance covered.

FIGURE 3.39 A sprinter's velocity and acceleration can be modeled with exponential functions.

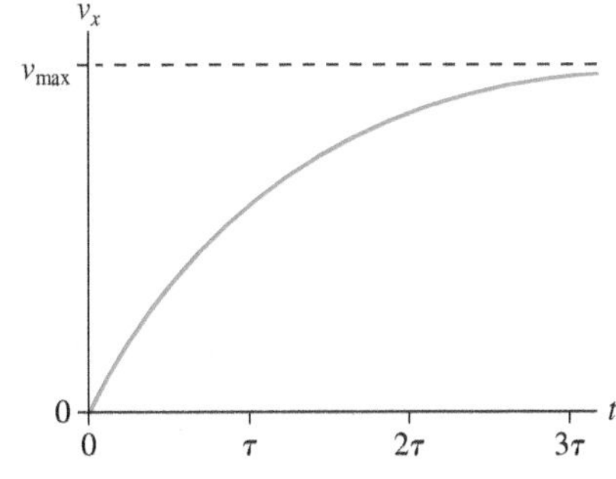

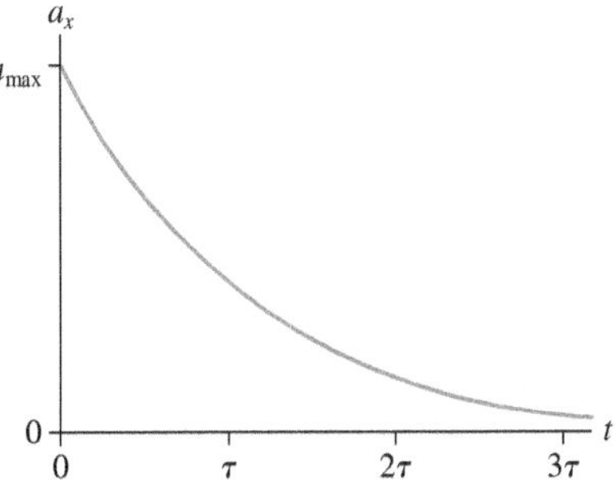

These graphs have the qualitative features we already identified: They show an initial rapid increase in speed and then a leveling off at maximum speed as the acceleration drops to zero. But is this actually the right mathematical model? A comparison with field data verifies that Equations 3.33 and 3.35 give very good quantitative agreement for real sprinters, each of whom can be characterized by specific values for v_{max} and τ. For example, Usain Bolt's record-setting 100 m dash in 2009 can be modeled with $v_{max} = 12.1$ m/s and $\tau = 1.32$ s. With these values, Equation 3.35 can be used to predict his race time to within 0.01 s of the actual result.

EXAMPLE 3.17 **Horse race**

A Quarter Horse is a breed that excels in fast sprints and short races—typically a quarter-mile. A champion Quarter Horse has an initial acceleration of 5.7 m/s^2. Its top speed is an impressive 86 km/h (≈55 mph). What are the speed and the magnitude of the acceleration of a sprinting Quarter Horse after 2.0 s and after 8.0 s?

PREPARE Real animals don't accelerate with a constant acceleration. We'll model the horse's velocity and acceleration with exponential functions—a much closer approximation to reality.

SOLVE The horse's initial acceleration is the maximum acceleration: $a_{max} = 5.7$ m/s^2. For a model in which velocity and acceleration change exponentially, the maximum speed is related to the maximum acceleration and the time constant by $v_{max} = a_{max}\tau$. First we need to convert the maximum speed to SI units:

$$v_{max} = 86\,\frac{\text{km}}{\text{h}} \times \frac{1000\text{ m}}{1\text{ km}} \times \frac{1\text{ h}}{60\text{ min}} \times \frac{1\text{ min}}{60\text{ s}} = 24\text{ m/s}$$

We can then find the time constant:

$$\tau = \frac{v_{max}}{a_{max}} = \frac{24\text{ m/s}}{5.7\text{ m/s}^2} = 4.2\text{ s}$$

Thus 4.2 s after starting a race, the horse's acceleration has decreased to 37% of its initial value and its speed has increased to 63% of its final value. We can calculate specific values using

$$v = |v_x| = (24\text{ m/s})(1 - e^{-t/4.2\text{ s}})$$
$$a = |a_x| = (5.7\text{ m/s}^2)\,e^{-t/4.2\text{ s}}$$

Thus we find:

At $t = 2.0$ s:	$v = 9.1$ m/s	$a = 3.5$ m/s^2
At $t = 8.0$ s:	$v = 20$ m/s	$a = 0.8$ m/s^2

ASSESS We see big changes within the first 2 s, and the horse is close to full speed after 8 s. If you've ever watched a horse race, the horses do seem to be at full speed within about 10 s, so our answers seem reasonable.

Example 3.17 raises an interesting question: Who is the winner in a race between a horse and a man? The surprising answer is "It depends." Specifically, the winner depends on the length of the race.

FIGURE 3.40 Velocity and position graphs for a sprint between a man and a horse.

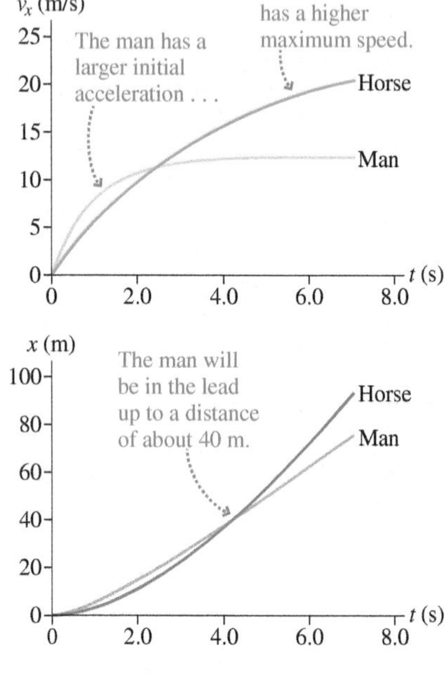

Some animals are capable of high speed; others are capable of great acceleration. Horses can run much faster than humans, but, when starting from rest, humans are capable of much greater initial acceleration. **FIGURE 3.40** shows velocity and position graphs for an elite male sprinter and a thoroughbred racehorse. The horse's maximum speed is about twice that of the man, but the man's initial acceleration—the slope of the velocity graph at early times—is greater than that of the horse. As the second graph shows, a man could win a short race, but for a longer race the horse's higher maximum speed will put it in the lead. The men's world-record time for the mile is a bit under 4 min, but a horse can easily run a mile in less than 2 min.

STOP TO THINK 3.8 Rank in order, from most positive to least positive, the accelerations at points 1 to 3.

A. $a_1 > a_2 > a_3$
B. $a_3 > a_1 > a_2$
C. $a_3 > a_2 > a_1$
D. $a_2 > a_1 > a_3$

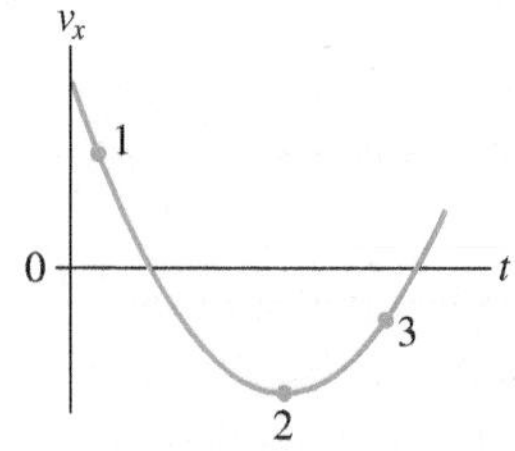

Integrated Examples

Each of the remaining chapters will end with a longer, more complex Integrated Example. These examples will require a synthesis of ideas, often from multiple chapters. In all of them, pay close attention to the *reasoning,* not so much to the equations. When you are faced with solving your own problems on homework and exams, getting the reasoning right will lead you to the proper equations. Equation hunting without correct reasoning rarely yields a correct answer.

CHAPTER 3 INTEGRATED EXAMPLE Speed versus endurance

Cheetahs have the highest top speed of any land animal, but they usually fail in their attempts to catch their prey because their endurance is limited. They can maintain their maximum speed of 30 m/s for only about 15 s before they need to stop.

Thomson's gazelles, their preferred prey, have a lower top speed than cheetahs, but they can maintain this speed for a few minutes. When a cheetah goes after a gazelle, success or failure is a simple matter of kinematics: Is the cheetah's high speed enough to allow it to reach its prey before the cheetah runs out of steam? The following problem uses realistic data for such a chase.

A cheetah has spotted a gazelle. The cheetah leaps into action, reaching its top speed of 30 m/s in a few seconds. At this instant, the gazelle, 160 m from the running cheetah, notices the danger and heads directly away. The gazelle accelerates at 4.5 m/s^2 for 6.0 s, then continues running at a constant speed. After reaching its maximum speed, the cheetah can continue running for only 15 s. Does the cheetah catch the gazelle, or does the gazelle escape?

PREPARE The question asks, "Does the cheetah catch the gazelle?" Our most challenging task is to translate these words into a mathematical problem that we can solve using the techniques of this chapter. For a problem of this complexity, it will be particularly important to prepare a complete pictorial representation. The pictorial representation lays out all the relevant information in a concise visual form, helping to guide your mathematical solution.

This example consists of two related problems, the motion of the cheetah and the motion of the gazelle, for which we'll use the subscripts C and G. Let's take our starting time, $t_0 = 0$ s, as the instant that the gazelle notices the cheetah and begins to run. We'll take the position of the cheetah at this instant as the origin of our coordinate system, so $x_{0C} = 0$ m and $x_{0G} = 160$ m—the gazelle is 160 m away when it notices the cheetah. The gazelle accelerates until time t_1, and the cheetah has to give up at time t_2. We've used this information to draw the pictorial representation in FIGURE 3.41.

With a clear picture of the situation, we can now rephrase the problem this way: Compute the position of the cheetah and the position of the gazelle at $t_2 = 15$ s, the time when the cheetah needs to break off the chase. If $x_{2G} > x_{2C}$, then the gazelle stays out in front and escapes. If $x_{2G} \leq x_{2C}$, the cheetah wins the race—and gets its dinner.

SOLVE The cheetah is in uniform motion for the entire duration of the problem, so

$$x_{2C} = x_{0C} + (v_x)_{0C}\,\Delta t = 0\text{ m} + (30\text{ m/s})(15\text{ s}) = 450\text{ m}$$

The gazelle's motion has two phases: one of constant acceleration and then one of constant velocity. We can solve for the position and the velocity at t_1, the end of the first phase, using the first and second kinematic equations of Problem-Solving Strategy 3.1. Let's find the velocity first:

$$(v_x)_{1G} = (v_x)_{0G} + (a_x)_G\,\Delta t = 0\text{ m/s} + (4.5\text{ m/s}^2)(6.0\text{ s}) = 27\text{ m/s}$$

With its head start of 160 m, the gazelle's position at t_1 is

$$x_{1G} = x_{0G} + (v_x)_{0G}\,\Delta t + \tfrac{1}{2}(a_x)_G(\Delta t)^2$$
$$= 160\text{ m} + 0 + \tfrac{1}{2}(4.5\text{ m/s}^2)(6.0\text{ s})^2 = 240\text{ m}$$

From t_1 to t_2, a total of 9.0 s, the gazelle moves at a constant speed, so we can use the uniform motion equation to find its final position. Be sure to notice that the gazelle starts this phase of the motion at $x_{1G} = 240$ m.

$$x_{2G} = x_{1G} + (v_x)_{1G}\,\Delta t = 240 + (27\text{ m/s})(9.0\text{ s}) = 480\text{ m}$$

x_{2C} is 450 m; x_{2G} is 480 m. The gazelle is 30 m ahead of the cheetah when the cheetah has to break off the chase, so the gazelle escapes.

ASSESS Does our solution make sense? Let's look at the final result. The numbers in the problem statement are realistic, so we expect our results to mirror real life. The speed for the gazelle is close to that of the cheetah, which seems reasonable for two animals known for their speed. And the result is the most common occurrence—the chase is close, but the gazelle gets away.

FIGURE 3.41 Pictorial representation for the cheetah and for the gazelle.

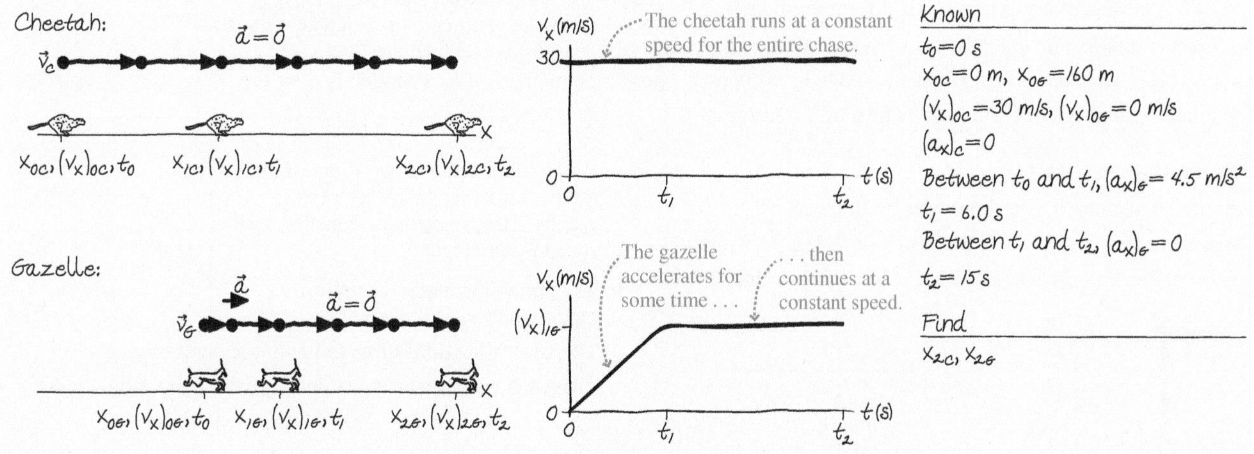

SUMMARY

GOAL To describe and analyze motion along a line.

GENERAL PRINCIPLES

Kinematic Equations

Kinematics describes motion in terms of position, velocity, and acceleration.

Acceleration: $a_x = dv_x/dt$ = slope of the velocity graph

Velocity: $v_x = dx/dt$ = slope of the position graph

$$(v_x)_f = (v_x)_i + \int_{t_i}^{t_f} a_x\,dt = (v_x)_i + \begin{cases}\text{area under the accelera-}\\ \text{tion curve from } t_i \text{ to } t_f.\end{cases}$$

Position: $$x_f = x_i + \int_{t_i}^{t_f} v_x\,dt = x_i + \begin{cases}\text{area under the velocity}\\ \text{curve from } t_i \text{ to } t_f.\end{cases}$$

Solving Constant-Acceleration Problems

PREPARE Model the object as having constant acceleration. Draw a pictorial representation.

SOLVE The kinematic equations for constant acceleration are

$$(v_x)_f = (v_x)_i + a_x\Delta t$$

$$x_f = x_i + (v_x)_i\,\Delta t + \frac{1}{2}a_x(\Delta t)^2$$

$$(v_x)_f^2 = (v_x)_i^2 + 2a_x\Delta x$$

Uniform motion has $a_x = 0$ and $x_f = x_i + v_x\,\Delta t$.

ASSESS Is the result reasonable? Does it have proper units and significant figures?

IMPORTANT CONCEPTS

Position, velocity, and acceleration are related graphically.

- The slope of the position graph is the value on the velocity graph.
- The slope of the velocity graph is the value on the acceleration graph
- x is a maximum or minimum at a turning point, and $v_x = 0$.

The sign of a_x indicates which way $\vec{a}$ points, *not* whether the object is speeding up or slowing down.

- $a_x > 0$ if $\vec{a}$ points to the right or up.
- $a_x < 0$ if $\vec{a}$ points to the left or down.

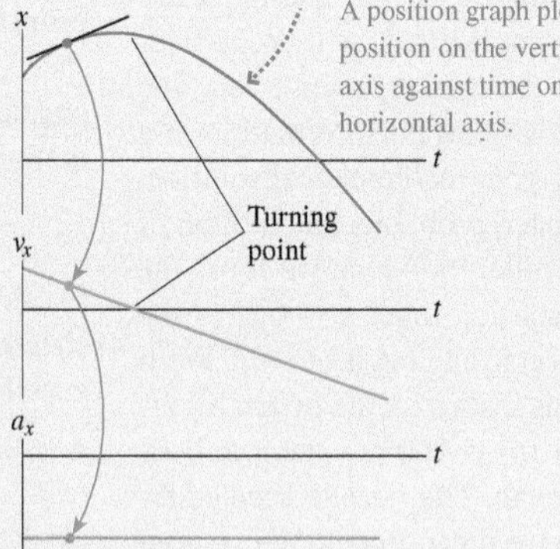

Displacement is the area under the velocity curve.

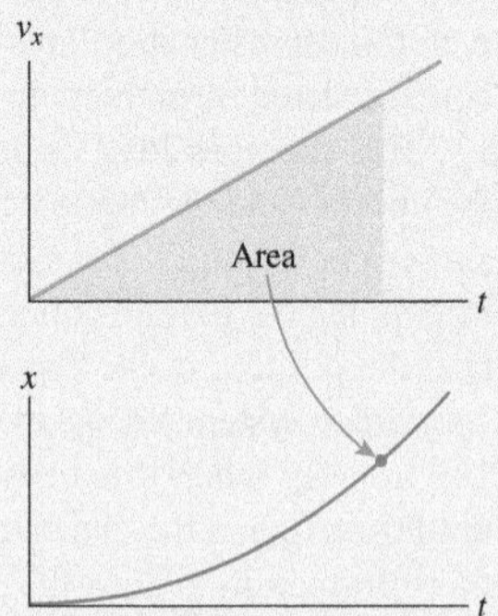

APPLICATIONS

Changing Velocity

The sign of v_x indicates the direction of motion.

- $v_x > 0$ is motion to the right; $v_y > 0$ is upward motion
- $v_x < 0$ is motion to the left; $v_y < 0$ is downward motion

An object is *speeding up* if and only if v_x and a_x have the same sign.

An object is *slowing down* if and only if v_x and a_x have opposite signs.

The direction and sign of $\vec{a}$ are found with a motion diagram.

Free fall is motion under the influence of only gravity. It is a special case of constant-acceleration motion. The acceleration vector always points straight down, whether an object is moving up or down, with magnitude $g = 9.80\text{ m/s}^2$. The y-component of acceleration is $a_y = -g$.

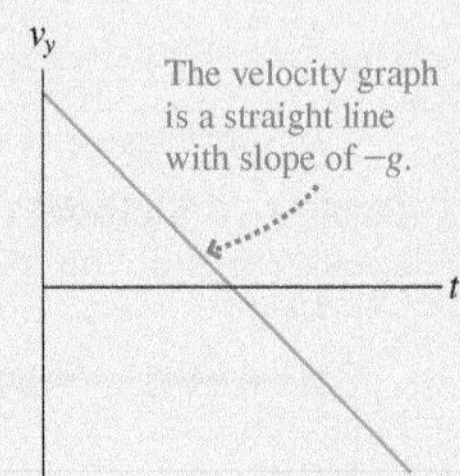

Pictorial Representation

- A motion diagram to determine acceleration vectors.
- A sketch of important points in the motion that defines symbols.
- A list of known values and of what you're trying to find.

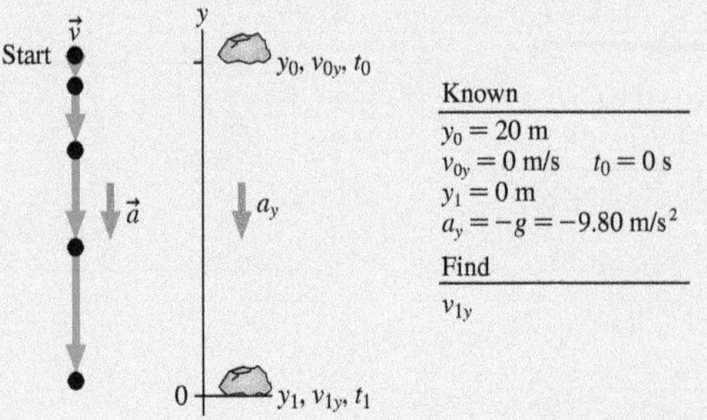

Projectile motion is two-dimensional motion under the influence of only gravity. The projectile is launched with speed v_0 and angle θ.

The horizontal motion is uniform motion with $v_x = v_0\cos\theta$.

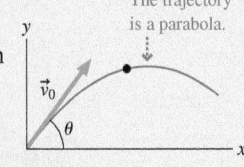

The trajectory is a parabola.

The vertical motion is free fall with acceleration $a_y = -g$.

The x and y kinematic equations have the *same* value for Δt.

LEARNING OBJECTIVES After studying this chapter, you should be able to:

- Solve problems involving objects in uniform motion. *Conceptual Questions 3.3, 3.5; Problems 3.1–3.3, 3.5, 3.6*
- Interpret position and velocity graphs. *Conceptual Questions 3.7, 3.9; Problems 3.8–3.10*
- Calculate and work with instantaneous velocity. *Conceptual Questions 3.4, 3.6; Problems 3.7, 3.19*
- Relate position, velocity, and acceleration using simple calculus. *Problems 3.11, 3.44, 3.45*
- Determine and interpret the sign of acceleration. *Conceptual Question 3.10; Problems 3.15, 3.16*
- Solve problems of motion with constant acceleration. *Conceptual Question 3.11; Problems 3.12, 3.14, 3.19, 3.20, 3.22*
- Solve problems for objects in free fall. *Conceptual Questions 3.12, 3.13; Problems 3.24, 3.25, 3.27–3.29*
- Solve problems about projectiles that follow parabolic trajectories. *Conceptual Question 3.14; Problems 3.33, 3.35, 3.36, 3.38, 3.39*
- Model the motion of an object with a changing acceleration. *Conceptual Question 3.8; Problems 3.41–3.43*

STOP TO THINK ANSWERS

Stop to Think 3.1: D. The particle starts with positive x and moves to negative x.

Stop to Think 3.2: C. The velocity is the slope of the position graph. The slope is positive and constant until the position graph crosses the axis, then positive but decreasing, and finally zero when the position graph is horizontal.

Stop to Think 3.3: B. A constant positive v_x corresponds to a linearly increasing x, starting from $x_i = -10$ m. The constant negative v_x then corresponds to a linearly decreasing x.

Stop to Think 3.4: 3. Acceleration is the slope of the velocity graph. The largest magnitude of the slope is at point 3.

Stop to Think 3.5: A and B. The velocity is constant while $a = 0$: it decreases linearly while a is negative. Graphs A, B, and C all have the same acceleration, but only graphs A and B have a positive initial velocity that represents a particle moving to the right.

Stop to Think 3.6: D. The final velocity will have the same *magnitude* as the initial velocity, but the velocity is negative because the rock will be moving downward.

Stop to Think 3.7: D. A projectile's downward acceleration does not depend on its mass. The second marble has the same initial velocity and the same acceleration, so it follows the same trajectory and lands at the same position.

Stop to Think 3.8: C. Acceleration is the slope of the graph. The slope is zero at 2. Although the graph is steepest at 1, the slope at that point is negative, and so $a_1 < a_2$. Only 3 has a positive slope, so $a_3 > a_2$.

QUESTIONS

Conceptual Questions

1. Write a very short "story" about an animal or object that could have the position graph of Figure Q3.1. Make specific reference to information you obtain from the graph.

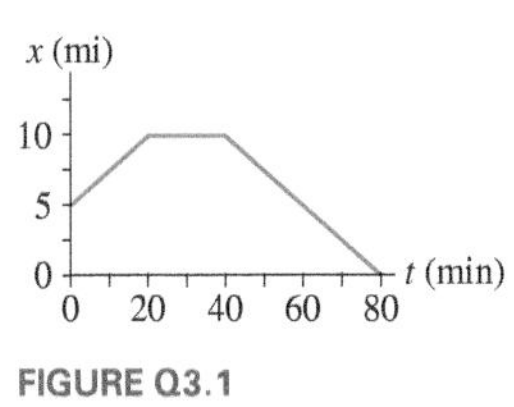

FIGURE Q3.1

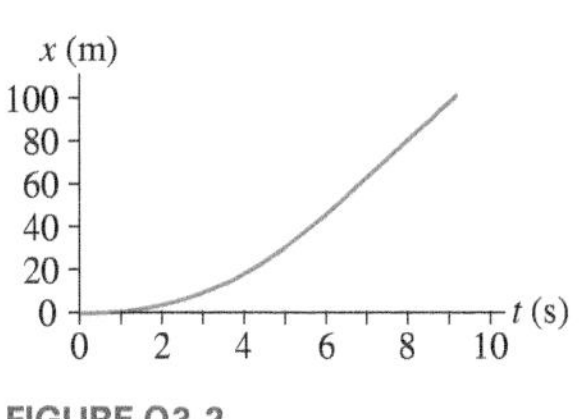

FIGURE Q3.2

2. Write a very short "story" about an animal or object that could have the position graph of Figure Q3.2. Make specific reference to information you obtain from the graph.
3. Figure Q3.3 shows a position graph for the motion of objects A and B as they move along the same axis.
   a. At the instant $t = 1$ s, is the speed of A greater than, less than, or equal to the speed of B? Explain.
   b. Do objects A and B ever have the *same* speed? If so, at what time or times? Explain.

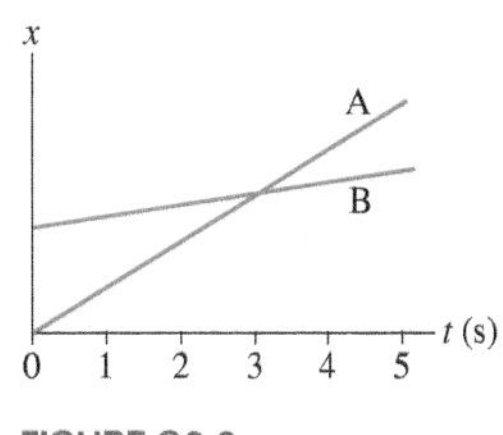

FIGURE Q3.3

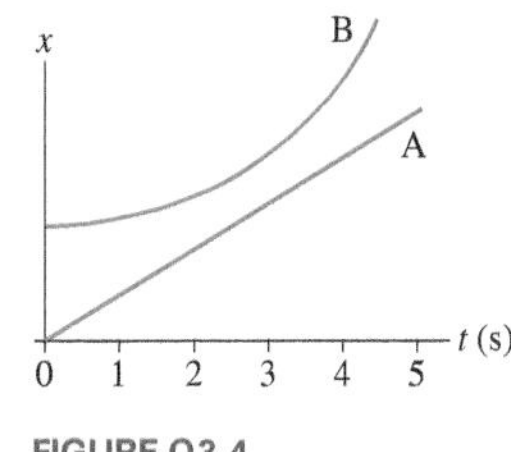

FIGURE Q3.4

4. Figure Q3.4 shows a position graph for the motion of objects A and B as they move along the same axis.
   a. At the instant $t = 1$ s, is the speed of A greater than, less than, or equal to the speed of B? Explain.
   b. Do objects A and B ever have the *same* speed? If so, at what time or times? Explain.

Problem difficulty is labeled as | (straightforward) to ||||| (challenging). Problems labeled INT integrate significant material from earlier chapters; BIO are of biological or medical interest; CALC require calculus to solve.

5. You are driving down the road at a constant speed. Another car going a bit faster catches up with you and passes you. Draw a position graph for both vehicles on the same set of axes, and note the point on the graph where the other vehicle passes you.
6. Figure Q3.6 shows the position graph for a moving object. At which numbered point or points:
   a. Is the object moving the fastest?
   b. Is the object moving to the left?
   c. Is the object speeding up?
   d. Is the object turning around?

FIGURE Q3.6

7. Figure Q3.7 is the velocity graph for an object moving along the x-axis.
   a. During which segment(s) is the velocity constant?
   b. During which segment(s) is the object speeding up?
   c. During which segment(s) is the object slowing down?
   d. During which segment(s) is the object standing still?
   e. During which segment(s) is the object moving to the right?

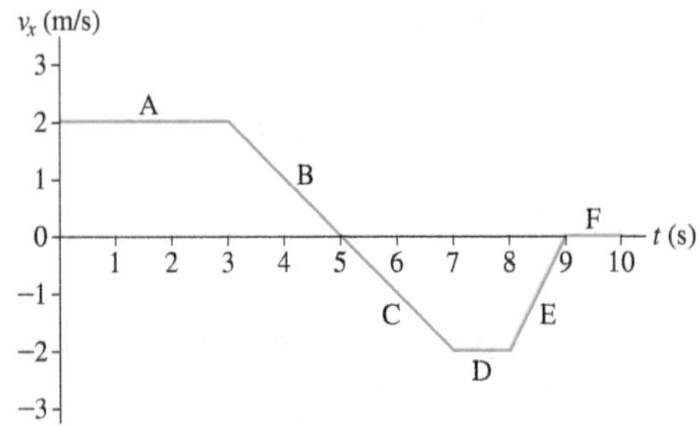

FIGURE Q3.7

8. BIO Figure Q3.8 shows growth rings in a tree's trunk. The wide and narrow rings correspond to years of fast and slow growth. Think of the rings as a motion diagram for the tree's growth. If we define an axis as shown, with x measured out from the center of the tree, use the appearance of the rings to sketch a velocity graph for the radial growth of the tree.

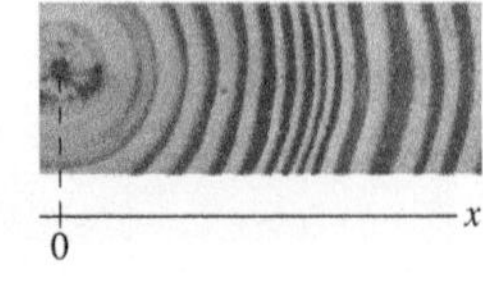

FIGURE Q3.8

9. Figure Q3.9 is a velocity graph for two objects, A and B. Students Victoria, Wes, and Zach are asked to tell stories that correspond to the motion. Victoria says, "I think A and B are two balls; ball A is thrown upward and ball B is thrown downward." "I'm not so sure," says Wes. "I think they are two airplanes going the same direction on parallel runways. Plane A is taking off; plane B is landing." Zach replies, "No, they are two cars on the highway that pass each other while going in opposite directions." Which of the students, if any, is correct? Explain.

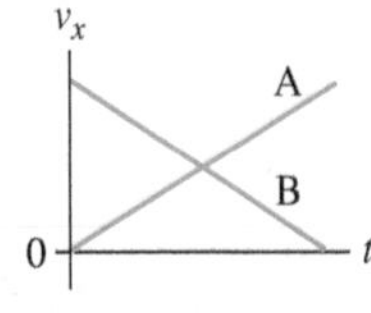

FIGURE Q3.9

10. a. Give an example of a vertical motion with a positive velocity and a negative acceleration.
    b. Give an example of a vertical motion with a negative velocity and a negative acceleration.
11. BIO Certain animals are capable of running at great speeds; other animals are capable of tremendous accelerations. Speculate on which would be more beneficial to a predator—large maximum speed or large acceleration.
12. Sketch a velocity graph for a rock that is thrown straight upward, from the instant it leaves the hand until the instant it hits the ground.
13. A ball is thrown straight up into the air. At each of the following instants, is the ball's acceleration a_y equal to g, $-g$, 0, $<g$, or $>g$?
    a. Just after leaving your hand.
    b. At the very top (maximum height).
    c. Just before hitting the ground.
14. A projectile is launched at an angle of 30°.
    a. Is there any point on the trajectory where $\vec{v}$ and $\vec{a}$ are parallel to each other? If so, where?
    b. Is there any point where $\vec{v}$ and $\vec{a}$ are perpendicular to each other? If so, where?

Multiple-Choice Questions

15. | Figure Q3.15 shows the position graph of a bicycle traveling on a straight road. At which labeled instant is the speed of the bicycle greatest?

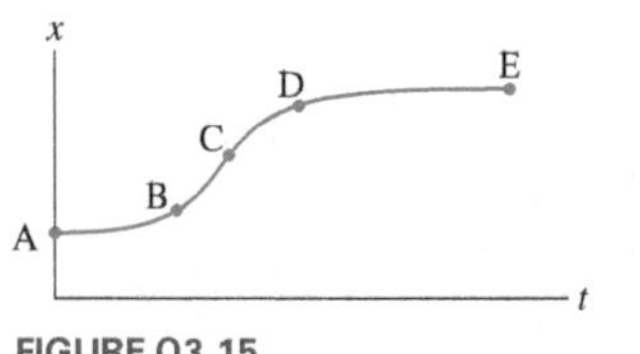

FIGURE Q3.15

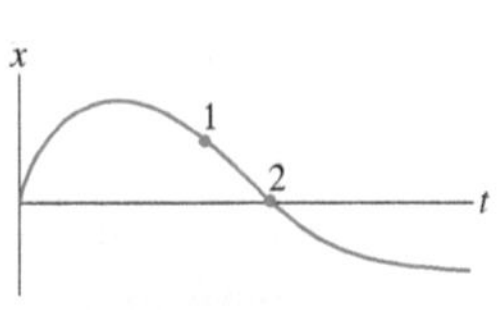

FIGURE Q3.16

16. || Figure Q3.16 shows the position graph of a car traveling on a straight road. The velocity at instant 1 is ________ and the velocity at instant 2 is ________.
    A. positive, negative
    B. positive, positive
    C. negative, negative
    D. negative, zero
    E. positive, zero
17. | Figure Q3.17 shows an object's position graph. What is the velocity of the object at $t = 6$ s?
    A. 0.67 m/s
    B. 0.83 m/s
    C. 3.3 m/s
    D. 4.2 m/s
    E. 25 m/s

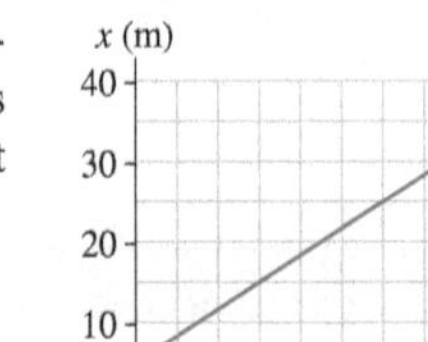

FIGURE Q3.17

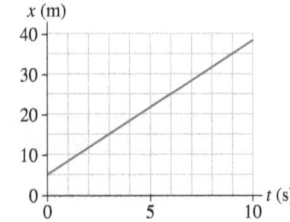

18. || Figure Q3.18 shows a motion diagram with the clock reading (in seconds) shown at each position. From $t = 9$ s to $t = 15$ s the object is at the same position. After that, it returns along the same track. The positions of the dots for $t \geq 16$ s are offset for clarity. Which graph best represents the object's *velocity*?

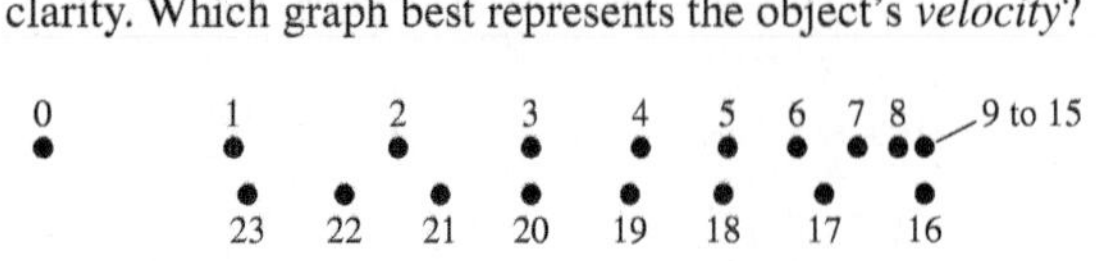

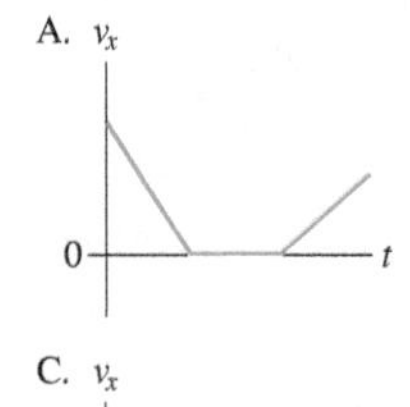

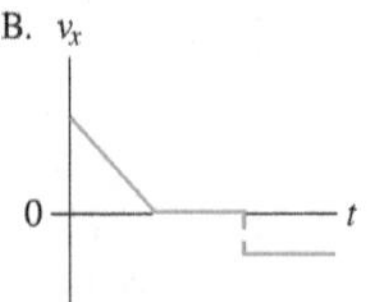

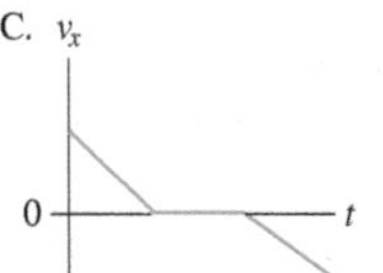

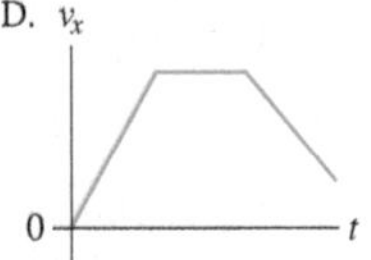

FIGURE Q3.18

19. | The following options describe the motion of four cars A–D. Which car has the largest acceleration?
 A. Goes from 0 m/s to 10 m/s in 5.0 s
 B. Goes from 0 m/s to 5.0 m/s in 2.0 s
 C. Goes from 0 m/s to 20 m/s in 7.0 s
 D. Goes from 0 m/s to 3.0 m/s in 1.0 s

20. | A car is traveling at $v_x = 20$ m/s. The driver applies the brakes, and the car slows with $a_x = -4.0$ m/s^2. What is the stopping distance?
 A. 5.0 m B. 25 m C. 40 m D. 50 m

21. || Velocity graphs for three running dogs are shown in Figure Q3.21. At $t = 5.0$ s, which dog has run the farthest distance?
 A. Rudy B. Spot
 C. Tess D. All have run the same distance.

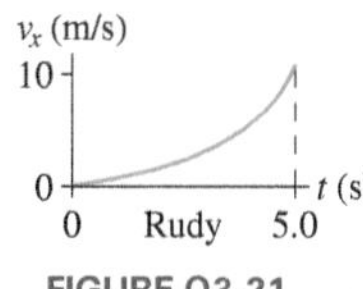

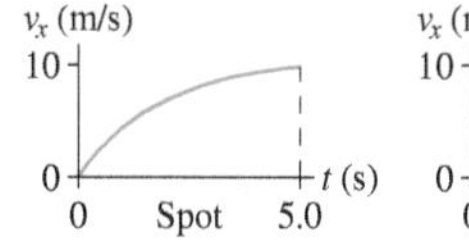

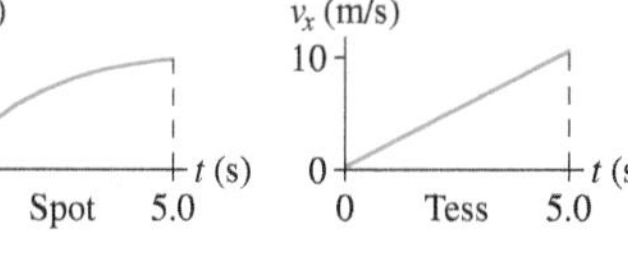

FIGURE Q3.21

22. || Which of the three dogs in Figure Q3.21 had the largest initial acceleration?
 A. Rudy B. Spot
 C. Tess D. All have the same initial acceleration.

23. || Figure Q3.23 is a simplified velocity graph for a lion chasing a gazelle. The lion has a greater initial acceleration, but the gazelle escapes if it hasn't been caught at the instant its speed exceeds the lion's speed. At this instant, how much farther has the lion run than the gazelle?
BIO
 A. 6 m B. 12 m C. 24 m D. 36 m

FIGURE Q3.23

24. | The velocity graph for a car driving down a straight road is shown in Figure Q3.24. What is the acceleration of the car during the period shown?
 A. 1.0 m/s^2
 B. 2.5 m/s^2
 C. 3.8 m/s^2
 D. 5.0 m/s^2

FIGURE Q3.24

25. || The velocity graph for a car driving down a straight road is shown in Figure Q3.24. How far does the car travel during the time interval from $t = 0$ s to $t = 4.0$ s?
 A. 10 m B. 20 m
 C. 40 m D. 60 m

26. | A ball thrown at an initial angle of 37.0° and initial velocity of 23.0 m/s reaches a maximum height h, as shown in Figure Q3.26. With what initial speed must a ball be thrown *straight up* to reach the same maximum height h?
 A. 13.8 m/s B. 17.3 m/s
 C. 18.4 m/s D. 23.0 m/s

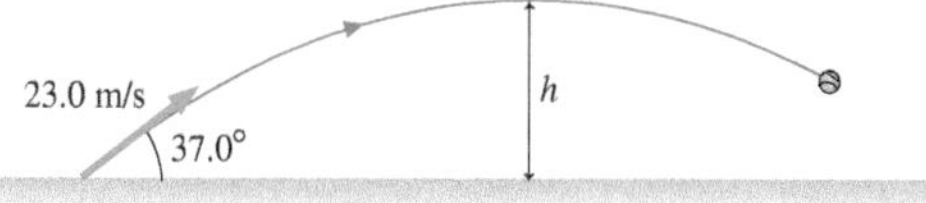

FIGURE Q3.26

27. || Liam throws a water balloon horizontally at 8.2 m/s out of a window 10 m from the ground. How far from the base of the building does the balloon land?
 A. 4.2 m B. 8.0 m C. 12 m
 D. 15 m E. 18 m

PROBLEMS

Section 3.1 Uniform Motion

1. || In major league baseball, the pitcher's mound is 60 feet from the batter. If a pitcher throws a 95 mph fastball, how much time elapses from when the ball leaves the pitcher's hand until the ball reaches the batter?

2. | Alan leaves Los Angeles at 8:00 AM to drive to San Francisco, 400 mi away. He travels at a steady 50 mph. Beth leaves Los Angeles at 9:00 AM and drives a steady 60 mph.
 a. Who gets to San Francisco first?
 b. How long does the first to arrive have to wait for the second?

3. ||| Julie drives 100 mi to Grandmother's house. On the way to Grandmother's, Julie drives half the distance at 40 mph and half the distance at 60 mph. On her return trip, she drives half the time at 40 mph and half the time at 60 mph.
 a. What is Julie's average speed on the way to Grandmother's house?
 b. What is her average speed on the return trip?

4. | Richard is driving home to visit his parents. 125 mi of the trip are on the interstate highway where the speed limit is 65 mph. Normally Richard drives at the speed limit, but today he is running late and decides to take his chances by driving at 70 mph. How many minutes does he save?

5. || In an 8.00 km race, one runner runs at a steady 11.0 km/h and another runs at 14.0 km/h. How far from the finish line is the slower runner when the faster runner finishes the race?

6. | While running a marathon, a long-distance runner uses a stopwatch to time herself over a distance of 100 m. She finds that she runs this distance in 18 s. Answer the following by considering ratios, without computing her velocity.
 a. If she maintains her speed, how much time will it take her to run the next 400 m?
 b. How long will it take her to run a mile at this speed?

Section 3.2 Instantaneous Velocity

Section 3.3 Finding Position from Velocity

7. || Figure P3.7 shows actual data from Usain Bolt's 2009 word-record run in the 100 m sprint. From this graph, estimate his top speed in m/s and in mph.

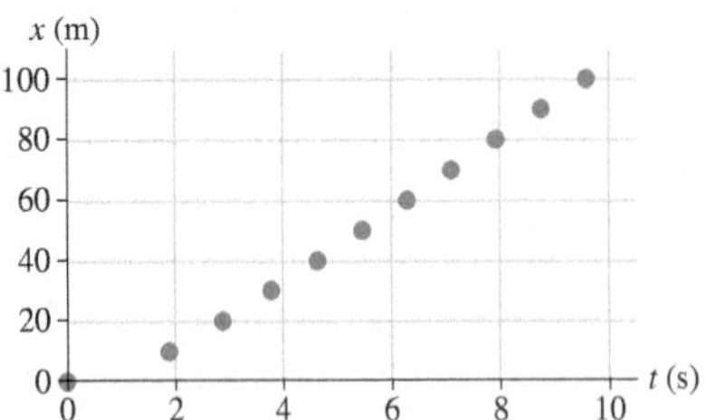

FIGURE P3.7

8. | Figure P3.8 shows the position graph of a particle.
   a. Draw the particle's velocity graph for the interval $0\text{ s} \leq t \leq 4\text{ s}$.
   b. Does this particle have a turning point or points? If so, at what time or times?

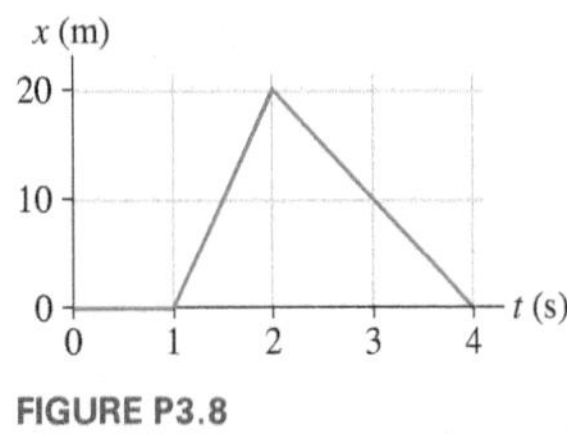

FIGURE P3.8

9. || A somewhat idealized graph of the speed of the blood in the ascending aorta during one beat of the heart appears as in Figure P3.9.
   a. Approximately how far, in cm, does the blood move during one beat?
   b. It's approximately 30 cm from your heart to your brain. How many heartbeats does it take to move blood from your heart to your brain?

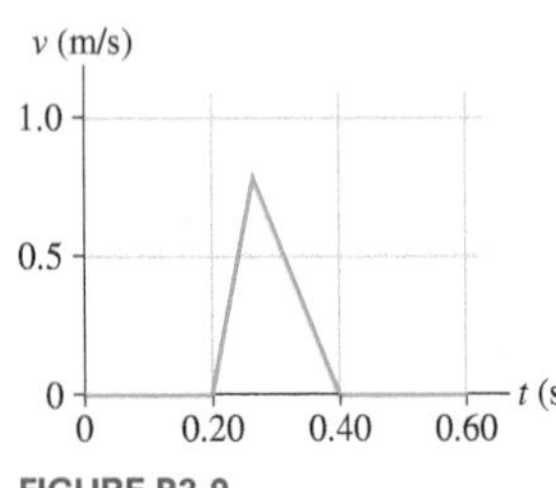

FIGURE P3.9

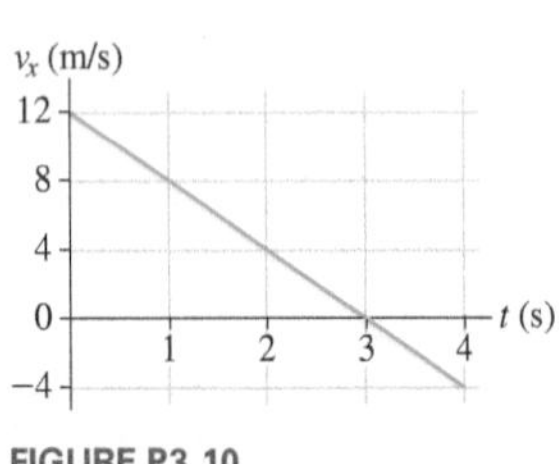

FIGURE P3.10

10. ||| A car starts from $x_i = 10$ m at $t_i = 0$ s and moves with the velocity graph shown in Figure P3.10.
    a. What is the car's position at $t = 2$ s, 3 s, and 4 s?
    b. Does this car ever change direction? If so, at what time?

11. ||| A particle's position at time t is given by $x = 2t^2 - 8t$ m,
CALC where t is in s. What is the particle's velocity at (a) $t = 1.0$ s, (b) $t = 2.0$ s, and (c) $t = 3.0$ s?

Section 3.4 Constant Acceleration

Section 3.5 Solving Kinematics Problems

12. || Figure P3.12 shows the velocity graph of a bicycle. Draw the bicycle's acceleration graph for the interval $0\text{ s} \leq t \leq 4\text{ s}$. Give both axes an appropriate numerical scale.

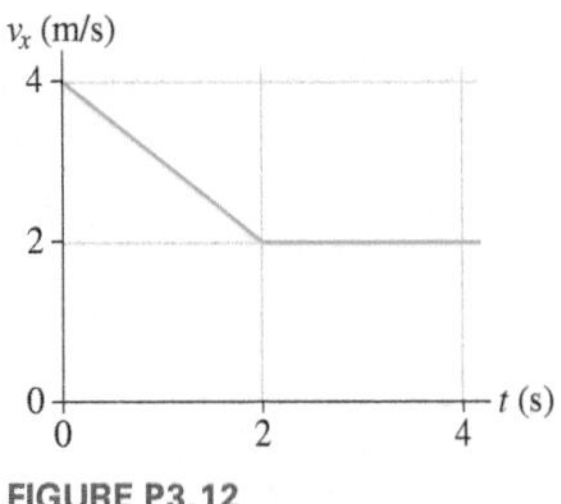

FIGURE P3.12

13. || Figure P3.13 is the velocity graph of a toy train. Its position at $t = 0$ s is $x = 2.0$ m.
    a. What is the train's acceleration at $t = 3.0$ s?
    b. What is the train's position at $t = 6.0$ s?

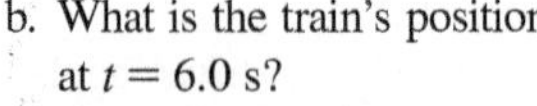

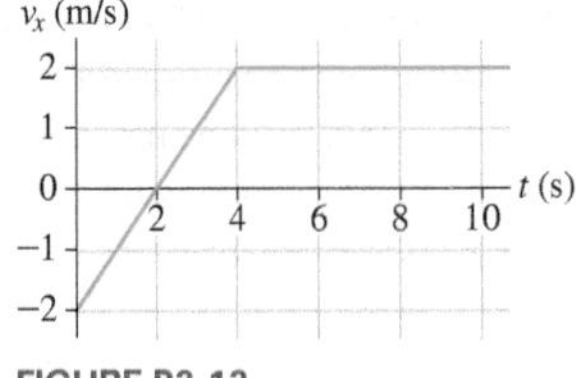

FIGURE P3.13

14. || Figure P3.14 shows a velocity graph for a particle moving along the x-axis. At $t = 0$ s, assume that $x = 0$ m.
    a. What are the particle's position, velocity, and acceleration at $t = 1.0$ s?
    b. What are the particle's position, velocity, and acceleration at $t = 3.0$ s?

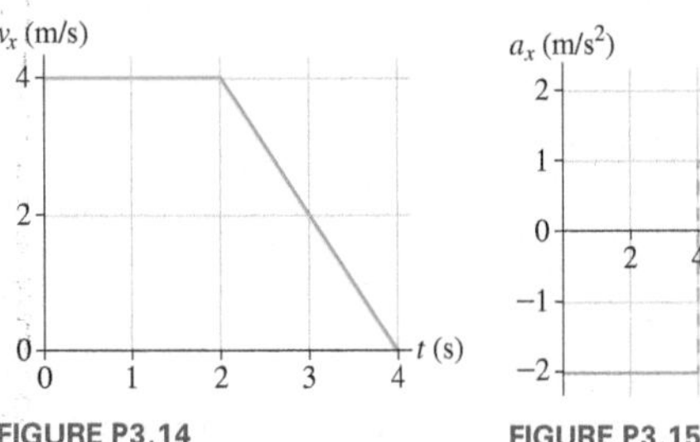

FIGURE P3.14

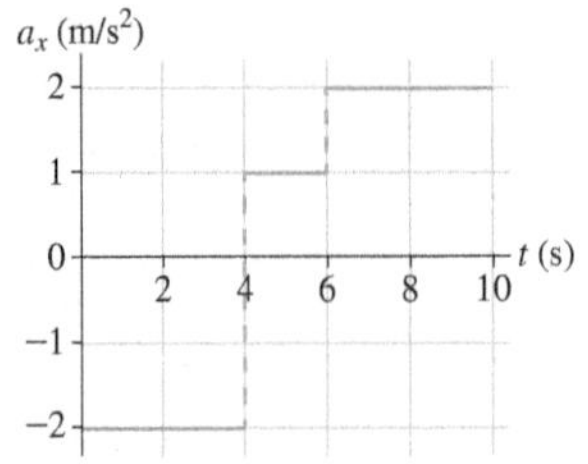

FIGURE P3.15

15. ||| An object has the acceleration graph shown in Figure P3.15. Its velocity at $t = 0$ s is $v_x = 2.0$ m/s. Draw the object's velocity graph.

16. || Figure P3.9 showed data for the speed of blood in the aorta.
BIO Determine the magnitude of the acceleration for both phases, speeding up and slowing down.

17. | Figure P3.17 is a somewhat simplified velocity graph for Olympic sprinter Usain Bolt starting a 100 m dash. Estimate his acceleration during each of the intervals A, B, and C.

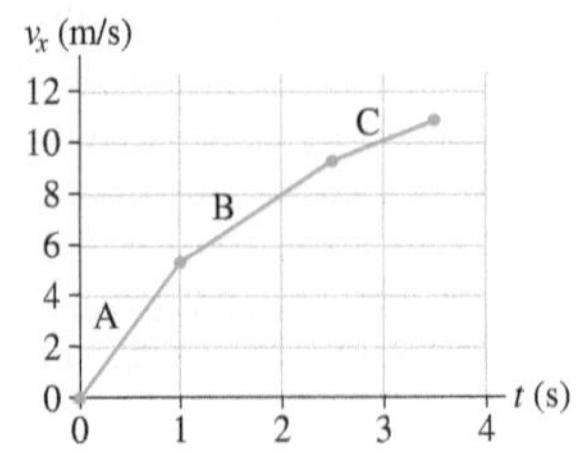

FIGURE P3.17

18. | Small frogs that are good
BIO jumpers are capable of remarkable accelerations. One species reaches a takeoff speed of 3.7 m/s in 60 ms. What is the frog's acceleration during the jump?

19. | A Thomson's gazelle can reach a speed of 13 m/s in 3.0 s. A
BIO lion can reach a speed of 9.5 m/s in 1.0 s. A trout can reach a speed of 2.8 m/s in 0.12 s. Which animal has the largest acceleration?

20. || When striking, the pike, a
BIO predatory fish, can accelerate from rest to a speed of 4.0 m/s in 0.11 s.
    a. What is the acceleration of the pike during this strike?
    b. How far does the pike move during this strike?

21. || When you sneeze, the air in your lungs accelerates from rest
BIO to 150 km/h in approximately 0.50 s. What is the acceleration of the air in m/s^2?

22. ||| A simple model for a person running the 100 m dash is to assume the sprinter runs with constant acceleration until reaching top speed, then maintains that speed through the finish line. If a sprinter reaches his top speed of 11.2 m/s in 2.14 s, what will be his total time?

Section 3.6 Free Fall

23. || Here's an interesting challenge you
BIO can give to a friend. Hold a \$1 (or larger!) bill by an upper corner. Have a friend prepare to pinch a lower corner, putting her fingers near but not touching the bill. Tell her to try to catch the bill when you drop it by simply closing her fingers. This seems like it should be easy, but it's not. After she sees that you have released the bill, it will take her about 0.25 s to react and close her fingers—which is not fast enough to catch the bill. How much time does it take for the bill to fall beyond her grasp? The length of a bill is 16 cm.

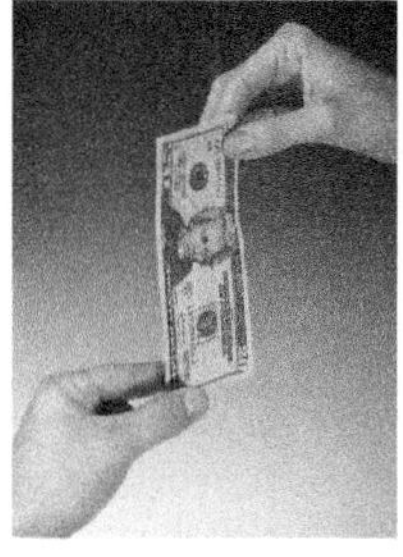

24. || In the preceding problem we saw that a person's reac-
BIO tion time is generally not quick enough to allow the person to catch a \$1 bill dropped between the fingers. The 16 cm length of the bill passes through a student's fingers before she can grab it if she has a typical 0.25 s reaction time. How long would a bill need to be for her to have a good chance of catching it?

25. | A gannet is a seabird that fishes by diving from a great height.
BIO If a gannet hits the water at 32 m/s (which they do), what height did it dive from? Assume that the gannet was motionless before starting its dive.

26. ||| Scientists explored how hoverflies detect motion by testing
BIO their response to a free-fall condition. Hoverflies were dropped from rest in a 40-cm-tall enclosure. Air resistance was not significant, and the flies could easily withstand a crash landing. The illumination in the enclosure and the patterning of the walls were adjusted between trials. Hoverflies dropped in darkness were generally not able to detect that they were falling in time to avoid crashing into the floor, while hoverflies dropped in a lighted enclosure with striped walls were generally able to avoid this fate. The findings imply that hoverflies rely on a visual rather than a kinesthetic sense to detect the condition of free fall. Suppose a hoverfly detects that it is falling 150 ms after being dropped, a typical time, and then starts beating its wings.
   a. How far has it fallen after 150 ms?
   b. What subsequent vertical acceleration is needed to avoid a crash landing?

27. ||| Steelhead trout migrate upriver to spawn. Occasionally
BIO they need to leap up small waterfalls to continue their journey. Fortunately, steelhead are remarkable jumpers, capable of leaving the water at a speed of 8.0 m/s. What is the maximum height that a steelhead can jump?

28. || Excellent human jumpers can leap straight up to a height of
BIO 110 cm off the ground. To reach this height, with what speed would a person need to leave the ground?

29. || A football is kicked straight up into the air; it hits the ground 5.2 s later.
   a. What was the greatest height reached by the ball? Assume it is kicked from ground level.
   b. With what speed did it leave the kicker's foot?

30. ||| In an action movie, the villain is rescued from the ocean by grabbing onto the ladder hanging from a helicopter. He is so intent on gripping the ladder that he lets go of his briefcase of counterfeit money when he is 130 m above the water. If the briefcase hits the water 6.0 s later, what was the speed at which the helicopter was ascending?

31. || Spud Webb was, at 5 ft 6 in, one of the shortest basketball
BIO players to play in the NBA. But he had an amazing vertical leap; he could jump to a height of 1.1 m off the ground, so he could easily dunk a basketball. For such a leap, what was his "hang time"—the time spent in the air after leaving the ground and before touching down again?

Section 3.7 Projectile Motion

32. || A physics student on Planet Exidor throws a ball, and it follows the parabolic trajectory shown in Figure P3.32. The ball's position is shown at 1.0 s intervals until $t = 3.0$ s. At $t = 1.0$ s, the ball's velocity has components $v_x = 2.0$ m/s, $v_y = 2.0$ m/s.

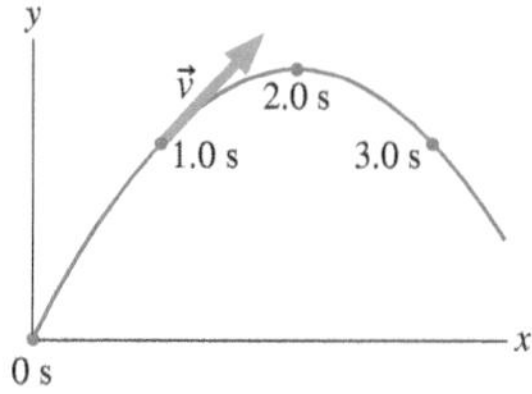

FIGURE P3.32

   a. Determine the x- and y-components of the ball's velocity at $t = 0.0$ s, 2.0 s, and 3.0 s.
   b. What is the value of g on Planet Exidor?
   c. What was the ball's launch angle?

33. || A ball with a horizontal speed of 1.25 m/s rolls off a bench 1.00 m above the floor.
   a. How long will it take the ball to hit the floor?
   b. How far from a point on the floor directly below the edge of the bench will the ball land?

34. || A pipe discharges storm water into a creek. Water flows horizontally out of the pipe at 1.5 m/s, and the end of the pipe is 2.5 m above the creek. How far out from the end of the pipe is the point where the stream of water meets the creek?

35. ||| A running mountain lion can make a leap 10.0 m long, reach-
BIO ing a maximum height of 3.0 m.
   a. What is the speed of the mountain lion just as it leaves the ground?
   b. At what angle does it leave the ground?

36. ||| A rifle is aimed horizontally at a target 50 m away. The bullet hits the target 2.0 cm below the aim point.
   a. What was the bullet's flight time?
   b. What was the bullet's speed as it left the barrel?

37. ||| A gray kangaroo can bound across a flat stretch of ground
BIO with each jump carrying it 10 m from the takeoff point. If the kangaroo leaves the ground at a 20° angle, what are its (a) takeoff speed and (b) horizontal speed?

38. ||| On the Apollo 14 mission to the moon, astronaut Alan Shepard hit a golf ball with a golf club improvised from a tool. The free-fall acceleration on the moon is 1/6 of its value on earth. Suppose he hit the ball with a speed of 25 m/s at an angle 30° above the horizontal.
   a. How long was the ball in flight?
   b. How far did it travel?
   c. Ignoring air resistance, how much farther would it travel on the moon than on earth?

39. || A soccer player takes a free kick from a spot that is 20 m from the goal. The ball leaves his foot at an angle of 32°, and it eventually hits the crossbar of the goal, which is 2.4 m from the ground. At what speed did the ball leave his foot?

40. ||| Sharpshooters are insects that suck copious quantities of BIO xylem fluid from plants. Over the course of a day, a sharpshooter may ingest an amount of liquid equal to 300 times its mass, extracting nutrients and then excreting the remaining liquid. Some sharpshooters rid themselves of this liquid waste by collecting small droplets that they fling, quite vigorously, at regular intervals. Suppose a sharpshooter accelerates a droplet at $100g$ through a 500 μm distance, launching the droplet at a 60° angle above horizontal. (These are typical numbers.) If we ignore air resistance and the droplet lands at the same height from which it was launched, how far from the sharpshooter does it touch down?

Section 3.8 Modeling a Changing Acceleration

41. || Figure P3.41 shows the acceleration graph of a particle moving along the x-axis. Its initial velocity is $(v_x)_0 = 8.0$ m/s at $t_0 = 0$ s. What is the particle's velocity at $t = 4.0$ s?

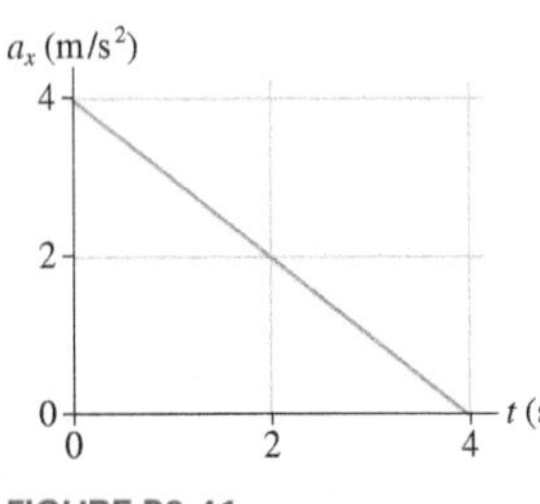

FIGURE P3.41

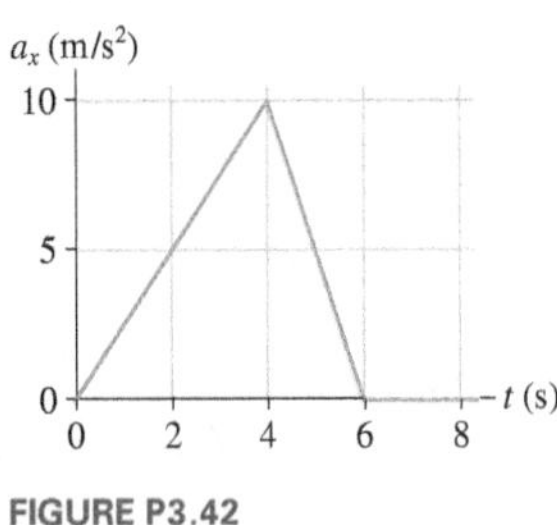

FIGURE P3.42

42. || Figure P3.42 shows the acceleration graph for a particle that starts from rest at $t = 0$ s. What is the particle's velocity at $t = 6$ s?

43. ||| Archerfish are known for BIO their ability to take down insect prey by precisely shooting water jets, but they can also leap from the surface of the water to capture prey directly. A 10-cm-long archerfish begins such a leap with its body vertical in the water, using its fins and undulating its body to propel itself sharply upward. Its acceleration steadily decreases as it leaves the water because less of its body and fewer of its fins are in contact with the water. The graph of Figure P3.43 is somewhat idealized data for the acceleration of an archerfish from the instant it begins a leap until the instant it leaves the water.
   a. With what speed does the fish leave the water?
   b. What height above the water does it reach?

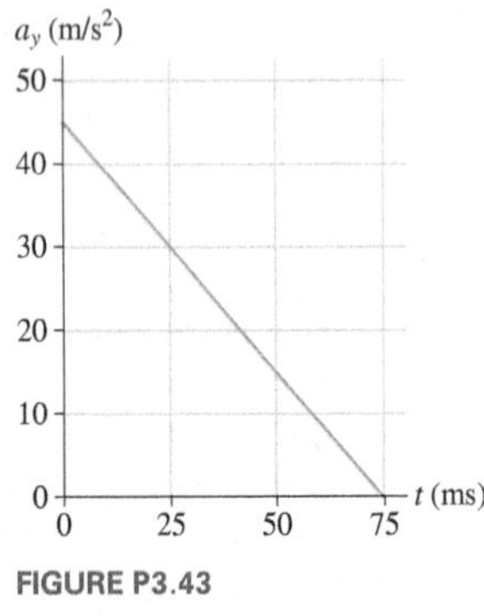

FIGURE P3.43

44. ||| A particle moving along the x-axis has its position de- CALC scribed by the function $x = (2t^3 + 2t + 1)$ m, where t is in s. At $t = 2$ s what are the particle's (a) position, (b) velocity, and (c) acceleration?

45. ||| A particle moving along the x-axis has its velocity described CALC by the function $v_x = 2t^2$ m/s, where t is in s. Its initial position is $x_0 = 1$ m at $t_0 = 0$ s. At $t = 1$ s what are the particle's (a) position, (b) velocity, and (c) acceleration?

General Problems

46. || Figure P3.46 shows the motion diagram, made at two frames of film per second, of a ball rolling along a track. The track has a 3.0-m-long sticky section.
   a. Use the scale to determine the positions of the center of the ball. Place your data in a table, similar to Table 2.1, showing each position and the instant of time at which it occurred.
   b. Make a graph of x versus t for the ball. Because you have data only at certain instants of time, your graph should consist of dots that are not connected together.
   c. What is the *change* in the ball's position from $t = 0$ s to $t = 1.0$ s?
   d. What is the *change* in the ball's position from $t = 2.0$ s to $t = 4.0$ s?
   e. What is the ball's velocity before reaching the sticky section?
   f. What is the ball's velocity after passing the sticky section?
   g. Determine the ball's acceleration on the sticky section of the track.

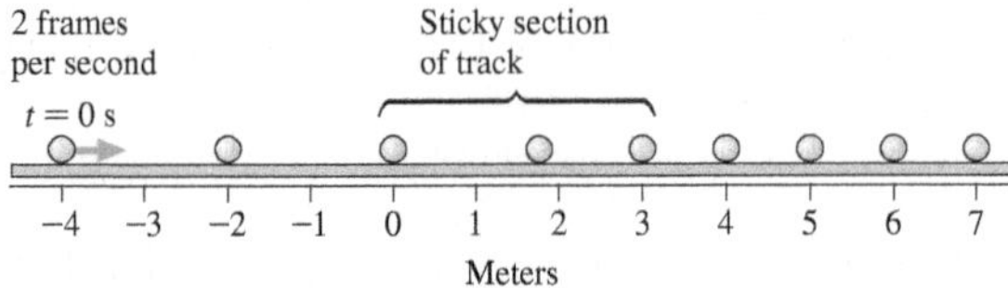

FIGURE P3.46

47. || A truck driver has a shipment of apples to deliver to a destination 440 miles away. The trip usually takes him 8 hours. Today he finds himself daydreaming and realizes 120 miles into his trip that he is running 15 minutes later than his usual pace at this point. At what speed must he drive for the remainder of the trip to complete the trip in the usual amount of time?

48. ||| In a 5000 m race, the athletes run $12\frac{1}{2}$ laps; each lap is 400 m. Kara runs the race at a constant pace and finishes in 17.5 min. Hannah runs the race in a blistering 15.3 min, so fast that she actually passes Kara during the race. How many laps has Hannah run when she passes Kara? Give your answer to the nearest tenth of a lap.

49. || The takeoff speed for an Airbus A320 jetliner is 80 m/s. Velocity data measured during takeoff are as shown in the table.
   a. What is the jetliner's acceleration during takeoff, in m/s² and in g's?
   b. At what time do the wheels leave the ground?
   c. For safety reasons, in case of an aborted takeoff, the length of the runway must be three times the takeoff distance. What is the minimum length runway this aircraft can use?

t (s)	v_x (m/s)
0	0
10	23
20	46
30	69

50. ||| The waterwheel is a car- BIO nivorous aquatic plant that closes a pair of vanes to trap insects in between. The motion is quite speedy. The tip of one vane moves 1.5 mm in a mere 20 ms as the vanes snap shut. Smoothed data for the velocity of the tip of one vane are shown in Figure P3.50. What is v_{max}, the maximum speed of the tip of a vane?

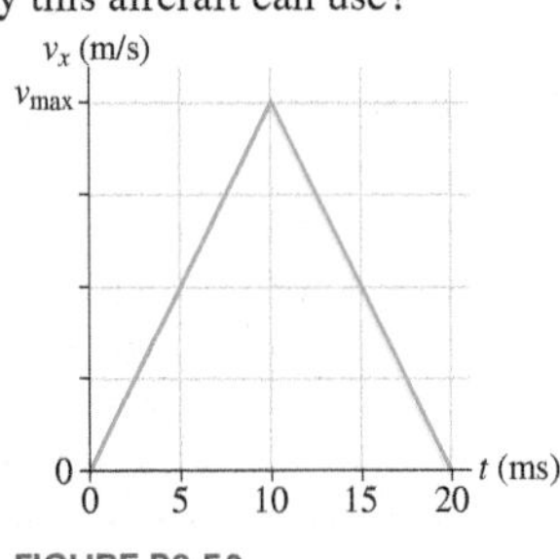

FIGURE P3.50

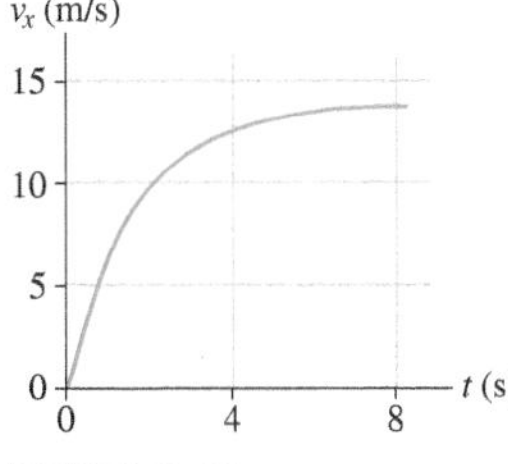

FIGURE P3.51

51. ||| Actual velocity data for a lion pursuing prey are shown in Figure P3.51. Estimate:
BIO
   a. The initial acceleration of the lion.
   b. The acceleration of the lion at 4 s.
   c. The distance traveled by the lion between 0 s and 8 s.

52. ||| The position of a particle is given by the function $x = (2t^3 - 6t^2 + 12)$ m, where t is in s.
CALC
   a. At what time does the particle reach its minimum velocity? What is $(v_x)_{min}$?
   b. At what time is the acceleration zero?

53. |||| A particle's velocity is described by the function $v_x = kt^2$ m/s, where k is a constant and t is in s. The particle's position at $t_0 = 0$ s is $x_0 = -9.0$ m. At $t_1 = 3.0$ s, the particle is at $x_1 = 9.0$ m. Determine the value of the constant k. Be sure to include the proper units.
CALC

54. ||| A particle's acceleration is described by the function $a_x = (10 - t)$ m/s^2, where t is in s. Its initial conditions are $x_0 = 0$ m and $(v_x)_0 = 0$ m/s at $t = 0$ s.
CALC
   a. At what time is the velocity again zero?
   b. What is the particle's position at that time?

55. ||| You are driving to the grocery store at 20 m/s. You are 110 m from an intersection when the traffic light turns red. Assume that your reaction time is 0.70 s and that your car brakes with constant acceleration.
   a. How far are you from the intersection when you begin to apply the brakes?
   b. What magnitude braking acceleration will bring you to rest right at the intersection?
   c. How long does it take you to stop?

56. ||| A cross-country skier is skiing along at a zippy 8.0 m/s. She stops pushing and simply glides along, slowing to a reduced speed of 6.0 m/s after gliding for 5.0 m. What is the magnitude of her acceleration as she slows?

57. || A small propeller airplane can comfortably achieve a high enough speed to take off on a runway that is 1/4 mile long. A large, fully loaded passenger jet has about the same acceleration from rest, but it needs to achieve twice the speed to take off. What is the minimum runway length that will serve? **Hint:** You can solve this problem using ratios without having any additional information.

58. ||| Chameleons catch insects with their tongues, which they can rapidly extend to great lengths. In a typical strike, the chameleon's tongue accelerates at a remarkable 250 m/s^2 for 20 ms, then travels at constant speed for another 30 ms. During this total time of 50 ms, 1/20 of a second, how far does the tongue reach?
BIO

59. || A parent and a three-year-old child are on a walk in the woods. The parent doesn't notice that the child is lagging behind, being deep in thought about the physics of the surrounding world, until suddenly hearing, "Look!" The parent turns to see the child 20 m behind, pointing to a beehive and about to reach into it. After an amazingly short 0.25 s reaction time, the parent accelerates toward the child at 2.3 m/s^2 for 3.0 s, then maintains top speed until grabbing the child away from the beehive. How much time elapses between the child's exclamation and rescue?

60. ||| You're driving down the highway late one night at 20 m/s when a deer steps onto the road 35 m in front of you. Your reaction time before stepping on the brakes is 0.50 s, and the maximum deceleration of your car is 10 m/s^2.
   a. How much distance is between you and the deer when you come to a stop?
   b. What is the maximum speed you could have and still not hit the deer?

61. || A car is traveling at a steady 80 km/h in a 50 km/h zone. A police motorcycle takes off at the instant the car passes it, accelerating at a steady 8.0 m/s^2.
   a. How much time elapses before the motorcycle is moving as fast as the car?
   b. How far is the motorcycle from the car when it reaches this speed?

62. || The velocity graph for the vertical jump of a green leafhopper, a small insect, is shown in Figure P3.62. This insect is unusual because it jumps with nearly constant acceleration.
BIO
   a. Estimate the leafhopper's acceleration.
   b. About how far does it move during this phase of its jump?

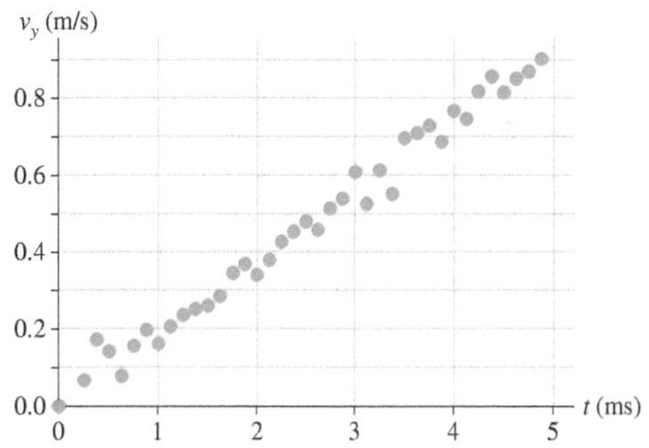

FIGURE P3.62

63. ||| When you blink your eye, it takes a mere 0.024 s for your top eyelid to go from rest, with your eye open, to fully closed.
BIO
   a. Use a ruler and a mirror to measure the distance your top eyelid moves during a blink.
   b. What is the magnitude of the eyelid's acceleration, assuming it is constant?
   c. What is the upper eyelid's speed as it hits the bottom eyelid?

64. ||| A bush baby, an African primate, is capable of a remarkable vertical leap. The bush baby goes into a crouch and extends its legs, pushing upward for a distance of 0.16 m. After this upward acceleration, the bush baby leaves the ground and travels upward for 2.3 m. What is the acceleration during the pushing-off phase? Give your answer in m/s^2 and in g's.
BIO

65. ||| Certain insects can achieve seemingly impossible accelerations while jumping. The click beetle accelerates at an astonishing 400g over a distance of 0.60 cm as it rapidly bends its thorax, making the "click" that gives it its name.
BIO
   a. Assuming the beetle jumps straight up, at what speed does it leave the ground?
   b. How much time is required for the beetle to reach this speed?
   c. Ignoring air resistance, how high would it go?

66. | A test of canine locomotion found that a young dog, starting
BIO from rest, could reach a speed of 2.7 m/s in a distance of 1.0 m. Assume that the acceleration is constant over this short distance.
a. What is the magnitude of the dog's acceleration?
b. What is the dog's speed when it reaches the 1.0 m mark?

67. || Researchers have compiled excellent data for zebras being
BIO pursued by lions. Under real-world conditions, a stationary zebra that spots a predator accelerates at 4.0 m/s^2 until reaching a speed of 11 m/s, then continues running at a steady speed. How much distance does a pursued zebra cover during the first 10 s of a run?

68. || Lions are capable of remarkable acceleration, but under
BIO realistic pursuit conditions they accelerate at a more modest 5.5 m/s^2 until reaching a steady speed of 14 m/s. Under these conditions, how long would it take a lion to run a 100 m dash?

69. || A student standing on the ground throws a ball straight up. The ball leaves the student's hand with a speed of 15 m/s when the hand is 2.0 m above the ground. How long is the ball in the air before it hits the ground? (The student moves her hand out of the way.)

70. ||| A rock is tossed straight up with a speed of 20 m/s. When it returns, it falls into a hole 10 m deep.
a. What is the rock's velocity as it hits the bottom of the hole?
b. How long is the rock in the air, from the instant it is released until it hits the bottom of the hole?

71. ||| Haley is driving down a straight highway at 75 mph. A construction sign warns that the speed limit will drop to 55 mph in 0.50 mi. What is the magnitude of the constant acceleration (in m/s) that will bring Haley to this lower speed in the distance available?

72. |||| A car starts from rest at a stop sign. It accelerates at 2.0 m/s^2 for 6.0 seconds, coasts for 2.0 s, and then slows down at a rate of 1.5 m/s^2 for the next stop sign. How far apart are the stop signs?

73. ||| Heather and Jerry are standing on a bridge 50 m above a river. Heather throws a rock straight down with a speed of 20 m/s. Jerry, at exactly the same instant of time, throws a rock straight up with the same speed. Ignore air resistance.
a. How much time elapses between the first splash and the second splash?
b. Which rock has the faster speed as it hits the water?

74. || A Thomson's gazelle can run at very high speeds, but its
BIO acceleration is relatively modest. A reasonable model for the sprint of a gazelle assumes an acceleration of 4.2 m/s^2 for 6.5 s, after which the gazelle continues at a steady speed.
a. What is the gazelle's top speed?
b. A human would win a very short race with a gazelle. The best time for a 30 m sprint for a human runner is 3.6 s. How much time would the gazelle take for a 30 m race?
c. A gazelle would win a longer race. The best time for a 200 m sprint for a human runner is 19.3 s. How much time would the gazelle take for a 200 m race?

75. ||| A pole-vaulter is nearly motionless as he clears the bar, set 4.2 m above the ground. He then falls onto a thick pad. The top of the pad is 80 cm above the ground, and it compresses by 50 cm as he comes to rest. What is his acceleration as he comes to rest on the pad?

76. ||| The longest recorded pass in an NFL game traveled 83 yards in the air from the quarterback to the receiver. Assuming that the pass was thrown at the optimal 45° angle, what was the speed at which the ball left the quarterback's hand?

77. ||| A spring-loaded gun, fired vertically, shoots a marble 6.0 m straight up in the air. What is the marble's range if it is fired horizontally from 1.5 m above the ground?

78. || Small-plane pilots regularly compete in "message drop" competitions, dropping heavy weights (for which air resistance can be ignored) from their low-flying planes and scoring points for having the weights land close to a target. A plane 60 m above the ground is flying directly toward a target at 45 m/s.
a. At what distance from the target should the pilot drop the weight?
b. The pilot looks down at the weight after she drops it. Where is the plane located at the instant the weight hits the ground. Is it not yet over the target, directly over the target, or past the target?

79. ||| An empirical finding (i.e., one deduced from data, not derived
BIO from a theory) is that a baseball player's fastest throwing speed is

$$v = v_{max}\sqrt{1 - \frac{\theta}{95^{\circ}}}$$

where θ, the angle of the throw above horizontal, is measured in degrees. This makes sense. A player's fastest throw is v_{max} at a horizontal $\theta = 0°$. A standing player can barely throw a ball straight up ($\theta = 90°$), and throwing a ball backward more than about 5° beyond vertical is essentially impossible. Consider a professional player who can throw a baseball at $v_{max} = 40$ m/s.
a. At what angle of throw can the player achieve maximum range? Assume that the ball is thrown and caught at the same height. It's not possible to find an exact solution, but you can easily use your calculator to solve the problem numerically. Calculate the distance of the throw for a 30° angle, which is likely less than the optimal angle, then repeat at angles of 31°, 32°, . . . until you find the maximum distance.
b. What is the maximum throwing distance?
c. At this angle, what is the player's throwing speed as a percent of v_{max}?

80. ||| Carl Lewis, a renowned Olympic sprinter in the 1980s and 1990s, ran a 100 m dash that can be accurately modeled with exponential functions using $v_{max} = 11.8$ m/s and $\tau = 1.45$ s. What were his speed and acceleration at (a) $t = 0$ s, (b) $t = 3.00$ s, and (c) $t = 6.00$ s?

81. |||| Example 3.17 modeled the sprint of a Quarter Horse that has
BIO a maximum speed of 24 m/s and a maximum acceleration of
CALC 5.7 m/s^2. How far has this horse run at (a) $t = 2.0$ s, (b) $t = 4.0$ s, and (c) $t = 8.0$ s? You need to use integration to solve this problem.

82. |||| Microorganisms that live in water have to swim constantly
BIO to keep moving against the drag of the water's viscosity. If they
CALC stop actively swimming, their velocity decreases exponentially with a very short time constant: $v_x = v_{swim}e^{-t/\tau}$, where t is the elapsed time since the animal stopped swimming and started coasting. For example, *Paramecium,* a unicellular inhabitant of freshwater ponds, swims at about 200 μm/s and stops with a time constant of a mere 0.5 ms. If a *Paramecium* suddenly stops swimming, what are (i) its speed in μm/s and (ii) the magnitude of its deceleration in m/s^2 at (a) $t = 0$ ms, (b) $t = 0.5$ ms, and (c) $t = 1$ ms? The data are known to only one significant figure, so your answers should have one significant figure.

MCAT-Style Passage Problems

Free Fall on Different Worlds

Objects in free fall on the earth have acceleration $a_y = -9.8\ \text{m/s}^2$. On the moon, free-fall acceleration is approximately 1/6 of the acceleration on earth. This changes the scale of problems involving free fall. For instance, suppose you jump straight upward, leaving the ground with velocity $(v_x)_i$ and then steadily slowing until reaching zero velocity at your highest point. Because your initial velocity is determined mostly by the strength of your leg muscles, we can assume your initial velocity would be the same on the moon. But considering the final equation in Problem-Solving Strategy 3.1 we can see that, with a smaller free-fall acceleration, your maximum height would be greater. The following questions ask you to think about how certain athletic feats might be performed in this reduced-gravity environment.

83. If an astronaut can jump straight up to a height of 0.50 m on earth, how high could he jump on the moon?
    A. 1.2 m B. 3.0 m C. 3.6 m D. 18 m

84. On the earth, an astronaut can safely jump to the ground from a height of 1.0 m; her velocity when reaching the ground is slow enough to not cause injury. From what height could the astronaut safely jump to the ground on the moon?
    A. 2.4 m
    B. 6.0 m
    C. 7.2 m
    D. 36 m

85. On the earth, an astronaut throws a ball straight upward; it stays in the air for a total time of 3.0 s before reaching the ground again. If a ball were to be thrown upward with the same initial speed on the moon, how much time would pass before it hit the ground?
    A. 7.3 s
    B. 18 s
    C. 44 s
    D. 108 s

4 Force and Motion

This swimmer propels herself through the water with powerful thrusts from her arms and legs.

LOOKING AHEAD

Force

A *force* is a push or pull. It is an interaction between an *agent* (here the thumb) and the *object* (the syringe).

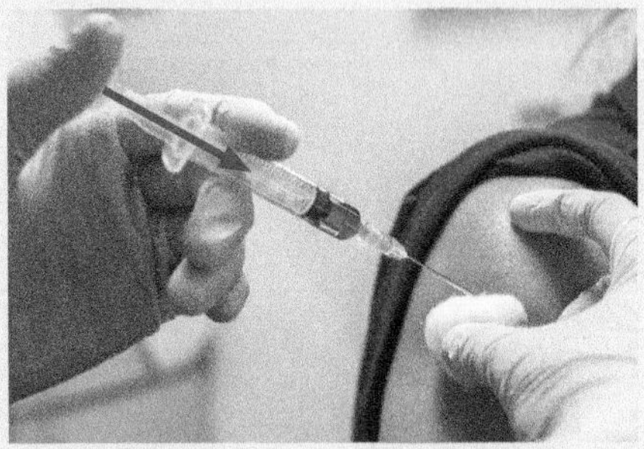

In this chapter, you'll learn how to identify forces and show them on a visual tool called a *free-body diagram.*

Equilibrium

Objects at rest are in *mechanical equilibrium.* Forces are necessary to keep an object in equilibrium, but the net force is zero.

This gymnast is in equilibrium because all the forces acting on her are exactly balanced. This is *Newton's first law.*

Force and Motion

A net force causes an object to *accelerate.* The forward acceleration of the sled requires a net force in the forward direction.

A larger acceleration requires a larger force. You'll learn that this connection between force and motion is *Newton's second law.*

GOAL To establish the connection between force and motion.

PHYSICS AND LIFE

Organisms Use Forces

Organisms use forces to support themselves and to move through their environment. In this chapter we begin to examine both how forces give rise to movement and how forces must balance for an object to remain still. Surprisingly, forces are *not* required to sustain movement in an ideal, frictionless world. However, organisms almost always live in environments with dissipative forces that oppose motion, such as friction and air resistance, which require the organism to generate a compensating force to maintain its motion. As we will explore further in future chapters, the forces needed for support and locomotion are one important reason organisms that live in water have very different body structures than those that live on land.

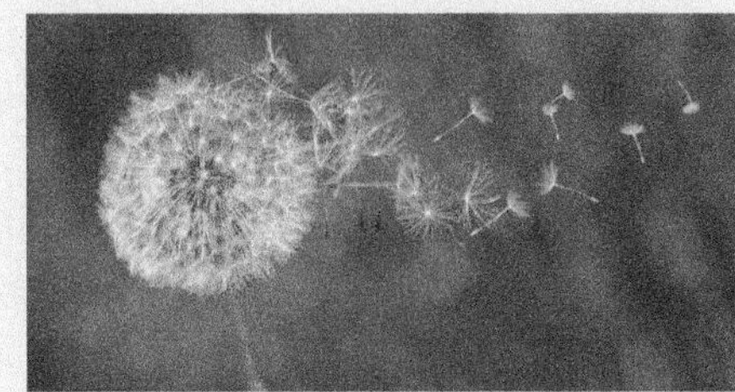

A dandelion's seeds float on the wind because the drag force of air resistance is large enough to balance the seeds' weight.

4.1 Motion and Forces: Newton's First Law

You have now learned how to *describe* motion using pictures, graphs, and equations. That's certainly an important skill. But we've not addressed the more fundamental question of *why* things move as they do. What enables a cheetah to accelerate, a bird to fly, or a bacterium's flagellum to rotate? In this chapter, we will turn our attention to the *causes* of motion. This topic is called **dynamics,** which joins with kinematics to form the subject of **mechanics,** the general science of motion. We'll begin our study of dynamics qualitatively, to introduce the concepts, then add quantitative detail as we develop the ideas.

Interstellar coasting A nearly perfect example of Newton's first law is the pair of Voyager space probes launched in 1977. Both spacecraft long ago ran out of fuel and are now coasting through the frictionless vacuum of space. Although not entirely free of influence from the sun's gravity, they are now so far from the sun and other stars that gravitational influences are very nearly zero, and the probes will continue their motion for billions of years.

What Causes Motion?

Let's start with a basic question: Do you need to keep pushing on something—to keep applying a force—to keep it going? Your daily experience might suggest that the answer is Yes. If you slide your textbook across the desk and then stop pushing, the book will quickly come to rest. Other objects will continue to move for a longer time: A hockey puck sliding across the ice keeps going for a long time, but it, too, comes to rest at some point.

FIGURE 4.1 shows a series of experiments in which a sled has been given a push across a horizontal surface. The sled traveling across smooth snow soon comes to rest because of friction between the snow and the sled's runners. The friction is much less on the slick ice of Figure 4.1b, so the sled could slide for quite a distance before stopping. Now, *imagine* the situation in Figure 4.1c, where the sled slides on idealized *frictionless* ice. In this case, the sled, once started in its motion, would continue in motion forever, moving in a straight line with no loss of speed.

FIGURE 4.1 A sled sliding on increasingly smooth surfaces.

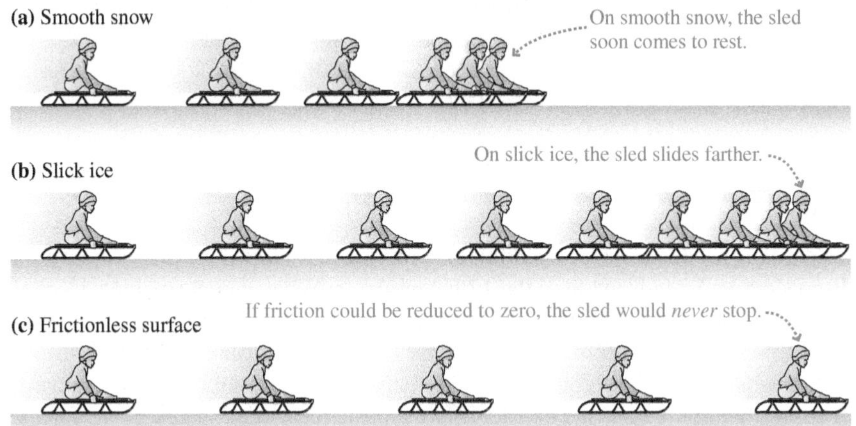

(a) Smooth snow
On smooth snow, the sled soon comes to rest.
(b) Slick ice
On slick ice, the sled slides farther.
(c) Frictionless surface
If friction could be reduced to zero, the sled would *never* stop.

In the absence of friction, a moving sled will stay in motion without anything or anyone pushing it. Our intuition suggests that this is not true because we have no direct experience with frictionless motion, but careful experiments—notably by Galileo in the years around 1600—verify that this is, in fact, the way the world works. It's also true that a sled sitting at rest will stay at rest if nothing pushes it; it won't start moving on its own.

It was Isaac Newton who, in 1666 at the age of 23, built on the experimental work of Galileo by developing a *theory* to explain how forces affect the motion of objects. Newton's theory consists of three *laws of motion*. The first law generalizes Galileo's observations:

Newton's first law Consider an object that has no forces acting on it. If it is at rest, it will remain at rest. If it is moving, it will continue to move in a straight line at a constant speed.

The surprising and counterintuitive implication of Newton's first law is that **no force is needed to maintain an object's motion!** Something had to initiate the movement—and we'll get to that—but an object that is in uniform motion (including being at rest, which is uniform motion with zero velocity) will continue unchanged, indefinitely, as long as no force acts on it. Nothing has to keep pushing on the object to sustain the motion. This resistance to change—to stay at rest or to keep moving in a straight line—is called **inertia.**

As an important application of Newton's first law, consider the crash test of FIGURE 4.2. As the car contacts the wall, the wall exerts a force on the car and the car begins to slow. But the wall is a force on the *car,* not on the dummy. In accordance with Newton's first law, the unbelted dummy continues to move straight ahead at its original speed. Sooner or later, a force will act to bring the dummy to rest. The only questions are when and how large the force will be. In the case shown, the dummy comes to rest in a short, violent collision with the dashboard of the stopped car. Seatbelts and air bags slow a dummy—or a driver—at a much lower rate and provide a much gentler stop.

FIGURE 4.2 Newton's first law tells us: Wear your seatbelts!

At the instant of impact, the car and the dummy are moving at the same speed.

The car slows as it hits, but the dummy continues at the same speed . . .

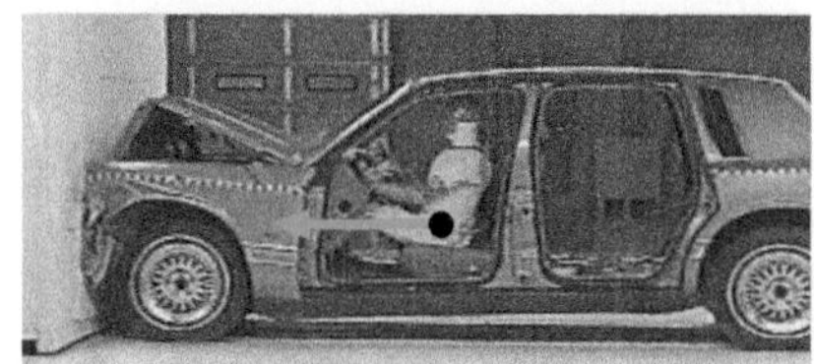

. . . until it hits the now-stationary dashboard. Ouch!

4.2 Force

No cause is needed for a moving object to remain in motion. The proper question, as discovered by Isaac Newton in the 17th century, is: What causes an object to *change* its motion? Why does a moving object stop? Why does a stationary object start moving? Newton's reply is that **a *force* is what causes an object to change its motion.** But what is a force? The concept of force is best introduced by looking at examples of some common forces and considering the basic properties shared by all forces.

What is a force?

A force is a push or a pull.

Our commonsense idea of a **force** is that it is a *push* or a *pull*. This is a good starting point, and we will refine this idea as we go along. Notice our careful choice of words: We refer to "*a* force" rather than simply "force." We want to think of a force as a very specific *action,* so that we can talk about a single force or perhaps about two or three individual forces that we can clearly distinguish—hence the concrete idea of "a force" acting on an object.

A force acts on an object.

Implicit in our concept of force is that **a force acts on an object**. In other words, pushes and pulls are applied *to* something—an **object.** From the object's perspective, it has a force *exerted* on it. Forces do not exist in isolation from the object that experiences them.

Agent

A force requires an agent.

Every force has an **agent,** something that acts or pushes or pulls; that is, a force has a specific, identifiable *cause*. As you throw a ball, it is your hand, while in contact with the ball, that is the agent or the cause of the force exerted on the ball. *If* a force is being exerted on an object, you must be able to identify a specific cause (i.e., the agent) of that force. Conversely, a force is not exerted on an object *unless* you can identify a specific cause or agent. Note that an agent can be an inert, inanimate object such as a tabletop or a wall. Such agents are the cause of many common forces.

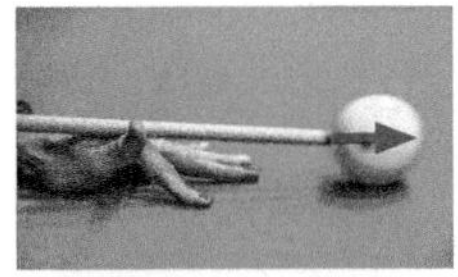

A force is a vector.

If you push an object, you can push either gently or very hard. Similarly, you can push either left or right, up or down. To quantify a push, we need to specify both a magnitude *and* a direction. It should thus come as no surprise that a force is a **vector** quantity. The general symbol for a force is the vector symbol $\vec{F}$. The size or strength of such a force is its magnitude F.

A force can be either a contact force ...

There are two basic classes of forces, depending on whether the agent touches the object or not. **Contact forces** are forces that act on an object by touching it at a point of contact. The bat must touch the ball to hit it. A string must be tied to an object to pull it. The majority of forces that we will examine are contact forces.

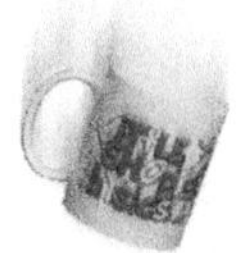

... or a long-range force.

Long-range forces are forces that act on an object without physical contact. Magnetism is an example of a long-range force. You have undoubtedly held a magnet over a paper clip and seen the paper clip leap up to the magnet. A coffee cup released from your hand is pulled to the earth by the long-range force of gravity.

NOTE ► In the particle model, objects cannot exert a force on themselves. A force will always have an agent or cause external to the object. ◄

Force Vectors

We can use a simple diagram to visualize how forces are exerted on objects. Because we are using the particle model, in which objects are treated as points, the process of drawing a force vector is straightforward:

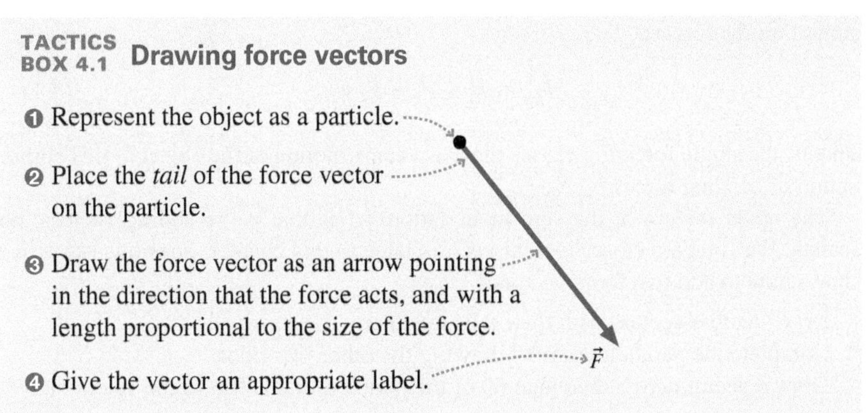
TACTICS BOX 4.1 Drawing force vectors
❶ Represent the object as a particle.
❷ Place the *tail* of the force vector on the particle.
❸ Draw the force vector as an arrow pointing in the direction that the force acts, and with a length proportional to the size of the force.
❹ Give the vector an appropriate label.
$\vec{F}$

Step 2 may seem contrary to what a "push" should do (it may look as if the force arrow is *pulling* the object rather than *pushing* it), but moving a vector does not change it as long as the length and angle do not change. The vector $\vec{F}$ is the same regardless of whether the tail or the tip is placed on the particle. Our reason for using the tail will become clear when we consider how to combine several forces.

FIGURE 4.3 Three force vectors.

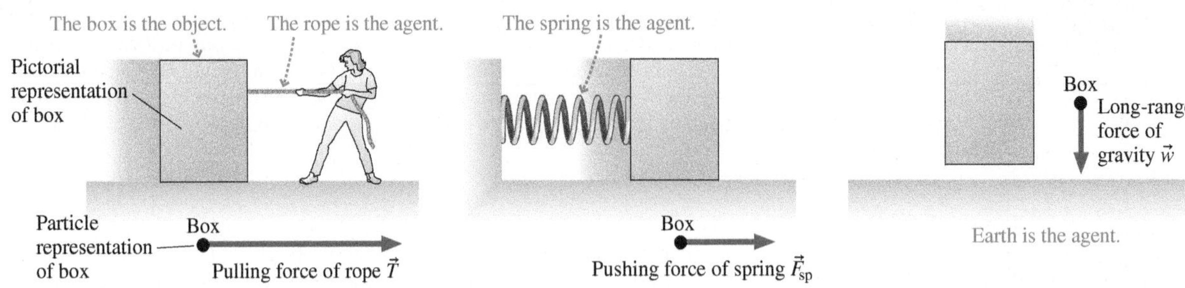

FIGURE 4.3 shows three examples of force vectors. One is a pull, one a push, and one a long-range force, but in all three the *tail* of the force vector is placed on the particle that represents the object. Although the generic symbol for a force is $\vec{F}$, as the figure shows we often use special symbols for certain forces that arise frequently, such as $\vec{T}$ for the tension force in a rope, $\vec{F}_{sp}$ for the force of a spring, and $\vec{w}$ for the force of gravity (an object's *weight*).

NOTE ▸ You might think that the person is the agent in the left frame of Figure 4.3. However, the person is not touching the box. The box responds to the tension in the rope, a contact force, regardless of how that tension is established. The person *would* be an agent if we were to examine the rope as a moving object. We'll need Newton's third law, a Chapter 5 topic, to fully analyze how the box, the rope, and the person all interact. For now, we'll focus on understanding how an individual object moves. ◂

Combining Forces

FIGURE 4.4 Two forces applied to a box.

(a)

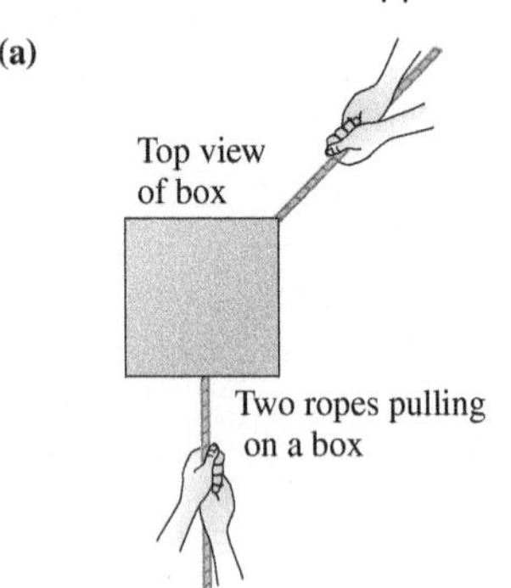

(b)

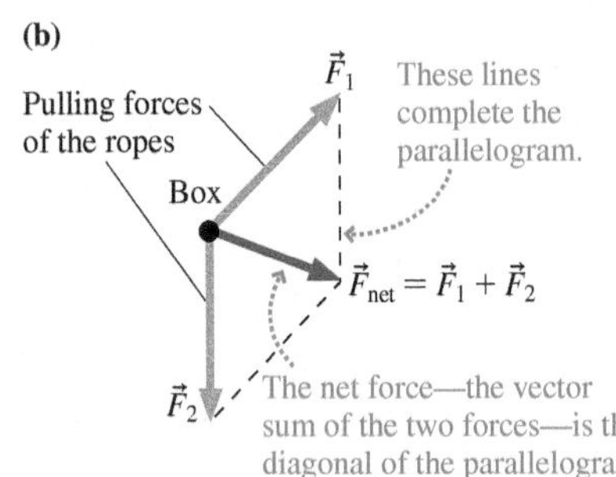

FIGURE 4.4a shows a top view of a box being pulled by two ropes, each exerting a force on the box. How will the box respond? Experiments show that when several forces $\vec{F}_1, \vec{F}_2, \vec{F}_3, \ldots$ are exerted on an object, they combine to form a **net force** that is the sum of all the forces:

$$\vec{F}_{net} = \vec{F}_1 + \vec{F}_2 + \vec{F}_3 + \cdots \tag{4.1}$$

That is, the single force $\vec{F}_{net}$ causes the exact same motion of the object as this combination of original forces $\vec{F}_1, \vec{F}_2, \vec{F}_3, \ldots$.

The novel feature of the sum in Equation 4.1 is that we're adding vectors, not scalars. We'll take a closer look at vectors later in this chapter. For now, FIGURE 4.4b shows how to add two force vectors:

- Draw the two vectors with their tails together.
- Complete the parallelogram by drawing the other two sides.
- Draw a vector across the diagonal of the parallelogram. This is the vector sum of the two forces.
- If there are more than two forces, repeat these steps to add another vector to the sum already obtained.

Almost all realistic situations involve multiple forces, so being able to determine the net force is a critical skill.

NOTE ▸ It is important to realize that the net force $\vec{F}_{net}$ is not a new force acting *in addition* to the original forces $\vec{F}_1, \vec{F}_2, \vec{F}_3, \ldots$. It's simply the combination of all the original forces. ◂

STOP TO THINK 4.1 Two of the three forces exerted on an object are shown. The net force points directly to the left. Which vector could be the missing third force?

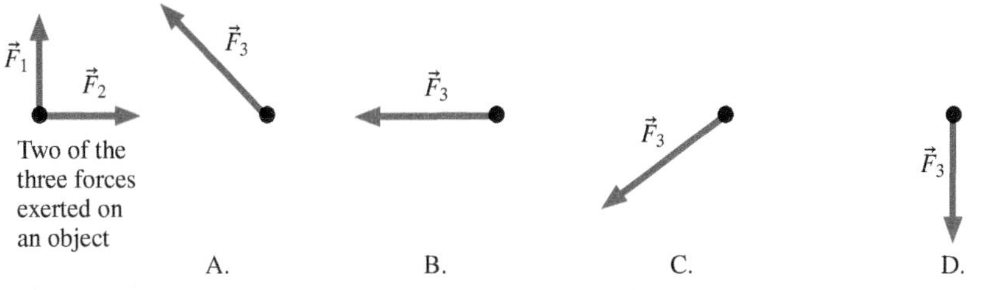
$\vec{F}_1$
$\vec{F}_2$
Two of the three forces exerted on an object
$\vec{F}_3$
A.
B.
C.
D.

4.3 A Short Catalog of Forces

There are many forces we will deal with over and over. This section will introduce you to some of them and to the symbols we use to represent them.

NOTE ▶ Be aware that other textbooks may use different symbols for these forces. ◀

Weight Force

A falling rock is pulled toward the earth by the long-range force of gravity. Gravity is what keeps you in your chair, keeps the planets in their orbits around the sun, and shapes the large-scale structure of the universe. For now we'll concentrate on objects on or near the surface of the earth (or other planet).

The gravitational pull of the earth on an object on or near the surface of the earth is called the **weight force** or, more informally, simply *weight.* Our symbol for the weight force is $\vec{w}$. Weight is the only long-range force we will encounter in the next few chapters. The agent for the weight force is the *entire earth* pulling on an object. The weight force is in some ways the simplest force we'll study. As FIGURE 4.5 shows, **an object's weight force vector always points vertically downward,** no matter how the object is moving.

FIGURE 4.5 The weight force always points vertically downward.

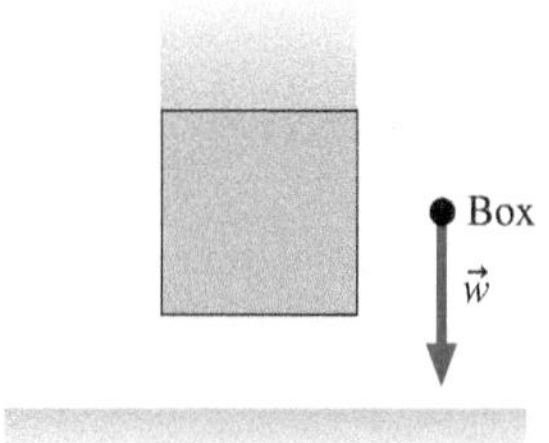
The weight force pulls the box down.
Box
$\vec{w}$

NOTE ▶ We often refer to "the weight" of an object. This is an informal expression for w, the magnitude of the weight force exerted on the object. Note that **weight is not the same thing as mass.** We will briefly examine mass later in the chapter, and we'll explore the connection between weight and mass in Chapter 5. ◀

A spring in his step When the flexible blade of this athlete's prosthesis hits the ground, it compresses just like an ordinary spring.

Spring Force

Springs exert one of the most basic contact forces. A spring can either push (when compressed) or pull (when stretched). FIGURE 4.6 shows the **spring force.** In both cases, pushing and pulling, the tail of the force vector is placed on the particle in the force diagram. There is no special symbol for a spring force, so we simply use a subscript label: $\vec{F}_{sp}$.

FIGURE 4.6 The spring force is parallel to the spring.

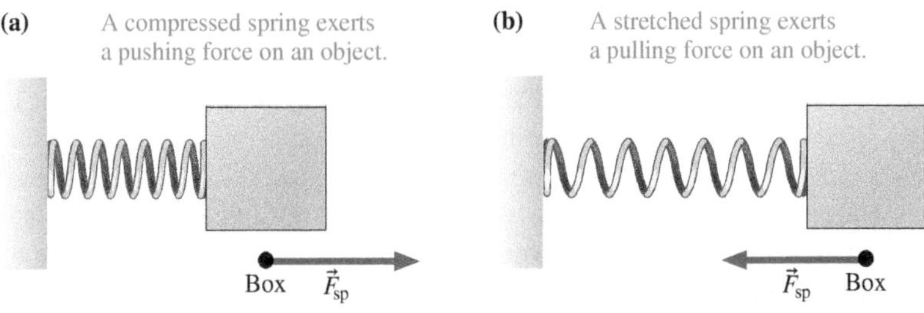
(a)
A compressed spring exerts a pushing force on an object.
Box $\vec{F}_{sp}$
(b)
A stretched spring exerts a pulling force on an object.
$\vec{F}_{sp}$ Box

Although you may think of a spring as a metal coil that can be stretched or compressed, this is only one type of spring. Hold a ruler, or any other thin piece of wood or metal, by the ends and bend it slightly. When you let go, it "springs" back to its original shape. This is just as much a spring as is a metal coil.

FIGURE 4.7 Tension is parallel to the rope.

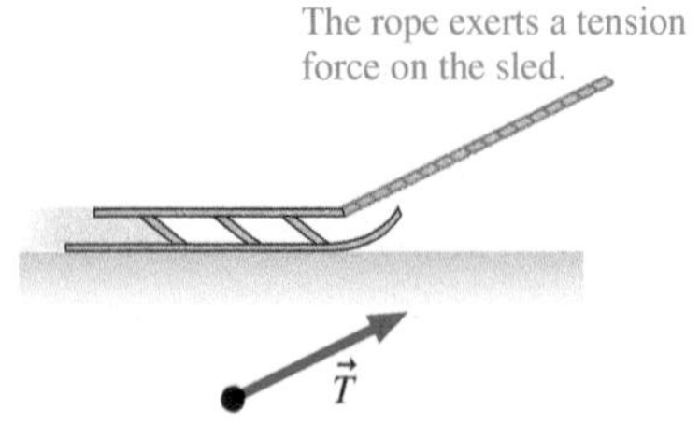
The rope exerts a tension force on the sled.
$\vec{T}$

Tension Force

When a string or rope or wire pulls on an object, it exerts a contact force that we call the **tension force,** represented by $\vec{T}$. **The direction of the tension force is always in the direction of the string or rope,** as you can see in FIGURE 4.7. When we speak of "the tension" in a string, this is an informal expression for T, the size or magnitude of the tension force. Note that the tension force can only *pull* in the direction of the string; if you try to *push* with a string, it will go slack and be unable to exert a force.

We can think about the tension force using a microscopic picture. If you were to use a very powerful microscope to look inside a rope, you would "see" that it is made of *atoms* joined together by *molecular bonds.* Molecular bonds are not rigid connections between the atoms. They are more accurately thought of as tiny *springs* holding the atoms together, as in FIGURE 4.8. Pulling on the ends of a string or rope stretches the spring-like molecular bonds ever so slightly. The tension within a rope and the tension force experienced by an object at the end of the rope are really the net spring force exerted by billions and billions of microscopic springs.

FIGURE 4.8 An atomic model of tension.

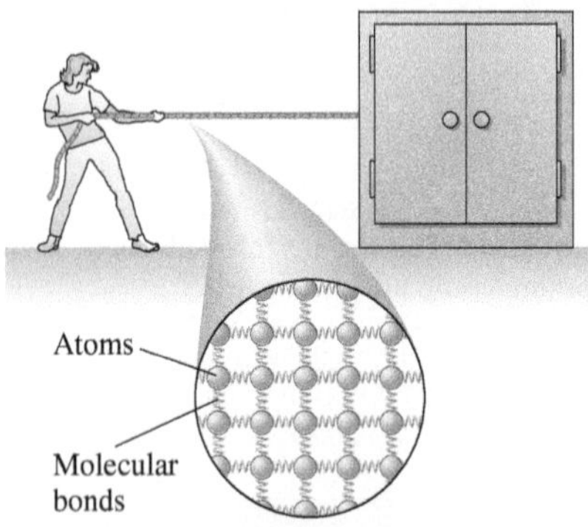
Atoms
Molecular bonds

This atomic-level view of tension introduces a new idea: a microscopic **atomic model** for understanding the behavior and properties of **macroscopic** (i.e., containing many atoms) objects. We will frequently use atomic models to obtain a deeper understanding of our observations.

The atomic model of tension also helps to explain one of the basic properties of ropes and strings. When you pull on a rope tied to a heavy box, the rope in turn exerts a tension force on the box. If you pull harder, the tension force on the box becomes greater. How does the box "know" that you are pulling harder on the other end of the rope? According to our atomic model, when you pull harder on the rope, its microscopic springs stretch a bit more, increasing the spring force they exert on each other—and on the box they're attached to.

FIGURE 4.9 An atomic model of the normal force exerted by a table.

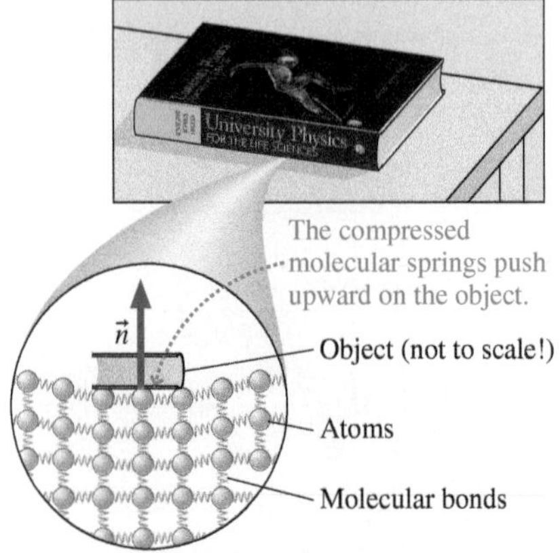
University Physics
The compressed molecular springs push upward on the object.
$\vec{n}$
Object (not to scale!)
Atoms
Molecular bonds

Normal Force

If you sit on a bed, the springs in the mattress compress and, as a consequence of the compression, exert an upward force on you. Stiffer springs would show less compression but would still exert an upward force. The compression of extremely stiff springs might be measurable only by sensitive instruments. Nonetheless, the springs would compress ever so slightly and exert an upward spring force on you.

FIGURE 4.9 shows a book resting on top of a sturdy table. The table may not visibly flex or sag, but—just as you do to the bed—the book compresses the molecular "springs" in the table. The compression is very small, but it is not zero. As a consequence, the compressed molecular springs *push upward* on the book. We say that "the table" exerts the upward force, but it is important to understand that the pushing is *really* done by molecular springs. Similarly, an object resting on the ground compresses the molecular springs holding the ground together and, as a consequence, the ground pushes up on the object.

FIGURE 4.10 The wall pushes outward against your hand.

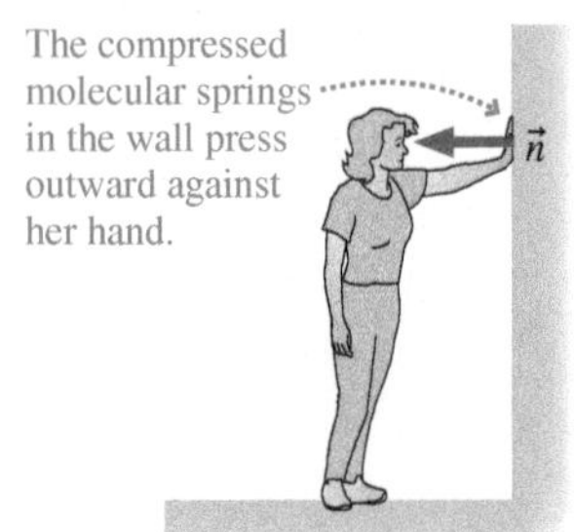
The compressed molecular springs in the wall press outward against her hand.
$\vec{n}$

We can extend this idea. Suppose you place your hand on a wall and lean against it, as shown in FIGURE 4.10. Does the wall exert a force on your hand? As you lean, you compress the molecular springs in the wall and, as a consequence, they push outward *against* your hand. So the answer is Yes, the wall does exert a force on you. It's not hard to see this if you examine your hand as you lean: You can see that your hand is slightly deformed, and becomes more so the harder you lean. This deformation is direct evidence of the force that the wall exerts on your hand. Consider also what would happen if the wall suddenly vanished. Without the wall there to push against you, you would topple forward.

The force the table surface exerts is vertical, while the force the wall exerts is horizontal. In all cases, the force exerted on an object that is pressing against a surface is in a direction *perpendicular* to the surface. Mathematicians refer to a line that is perpendicular to a surface as being *normal* to the surface. In keeping with this terminology, we define the **normal force** as the force exerted by a surface (the agent) against an object that is pressing against the surface. The symbol for the normal force is $\vec{n}$.

We're not using the word "normal" to imply that the force is an "ordinary" force or to distinguish it from an "abnormal force." A surface exerts a force *perpendicular* (i.e., normal) to itself as the molecular springs press *outward.* FIGURE 4.11 shows an object on an inclined surface, a common situation. Notice how the normal force $\vec{n}$ is perpendicular to the surface.

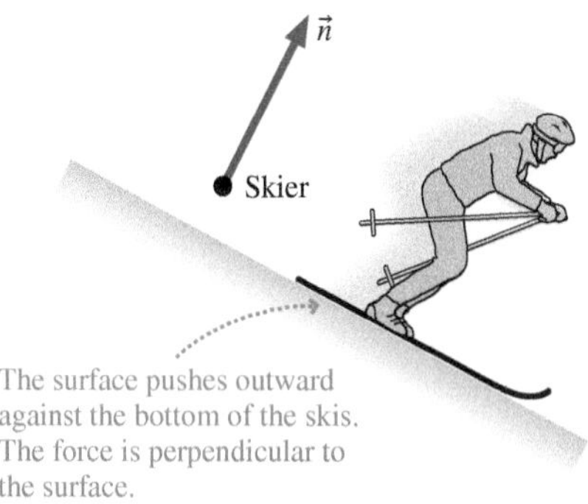
$\vec{n}$
Skier
The surface pushes outward against the bottom of the skis. The force is perpendicular to the surface.

FIGURE 4.11 The normal force is perpendicular to the surface.

The normal force is a very real force arising from the very real compression of molecular bonds. It is in essence just a spring force, but one exerted by a vast number of microscopic springs acting at once. The normal force is responsible for the "solidness" of solids. It is what prevents you from passing right through the chair you are sitting in and what causes the pain and the lump if you bang your head into a door. Your head can then tell you that the force exerted on it by the door was very real!

Friction Force

You've certainly observed that a rolling or sliding object, if not pushed or propelled, slows down and eventually stops. You've probably discovered that you can slide better across a sheet of ice than across asphalt. And you also know that most objects stay in place on a table without sliding off even if the table is tilted a bit. The force responsible for these sorts of behavior is the **friction force** or, informally, simply *friction.* The friction force is represented by the symbol $\vec{f}$.

The friction force, like the normal force, is exerted by a surface. Unlike the normal force, however, **the friction force is always *parallel* to the surface,** not perpendicular to it. (In many cases, a surface will exert *both* a normal and a friction force.) On a microscopic level, friction arises as atoms from the object and atoms on the surface run into each other. The rougher the surface is, the more these atoms are forced into close proximity and, as a result, the larger the friction force. We will develop a simple model of friction in the next chapter that will be sufficient for our needs. For now, it is useful to distinguish between two kinds of friction:

- *Kinetic friction,* denoted $\vec{f}_k$, acts as an object *slides* across a surface. Kinetic friction is a force that always "opposes the motion," meaning that the friction force $\vec{f}_k$ on a sliding object points in the direction opposite to the direction of the object's motion.
- *Static friction,* denoted $\vec{f}_s$, is the force that keeps an object "stuck" on a surface and prevents its motion relative to the surface. Finding the direction of $\vec{f}_s$ is a little trickier than finding the direction of $\vec{f}_k$. The static friction force points opposite the direction in which the object *would* move if there were no friction; that is, it points in the direction necessary to *prevent* motion.

FIGURE 4.12 shows examples of kinetic and static friction.

FIGURE 4.12 Kinetic and static friction are parallel to the surface.

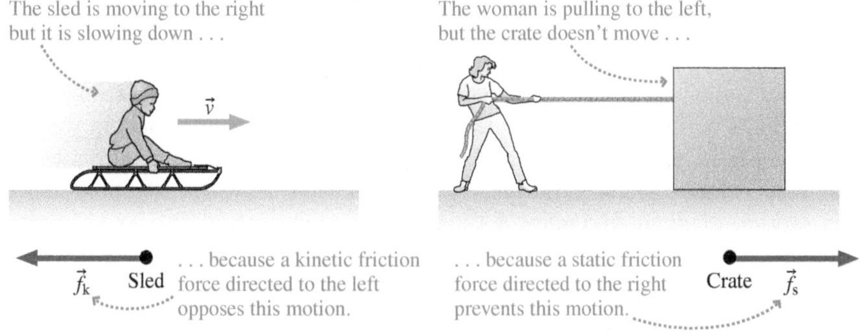
The sled is moving to the right but it is slowing down . . .
$\vec{v}$
$\vec{f}_k$
Sled
. . . because a kinetic friction force directed to the left opposes this motion.
The woman is pulling to the left, but the crate doesn't move . . .
. . . because a static friction force directed to the right prevents this motion.
Crate
$\vec{f}_s$

Drag Force

Friction at a surface is one example of a *resistive force,* a force that opposes or resists motion. Resistive forces are also experienced by objects moving through *fluids*—gases (like air) and liquids (like water). This kind of resistive force—the force of a fluid on a moving object—is called the **drag force** and is symbolized as $\vec{D}$. Like kinetic friction, **the drag force points opposite the direction of motion.** FIGURE 4.13 shows an example of drag.

FIGURE 4.13 Air resistance is an example of drag.

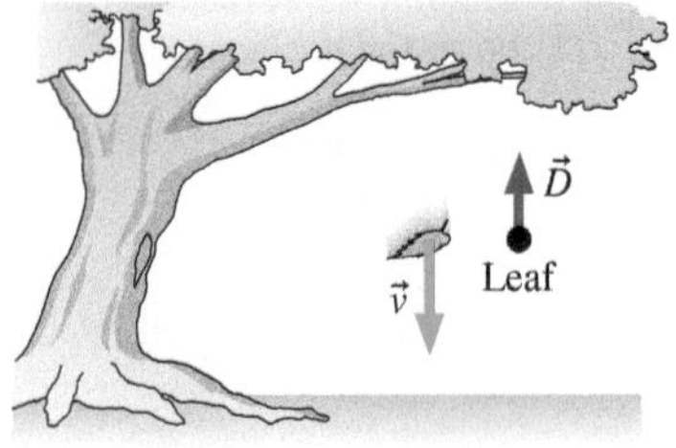
Air resistance is a significant force on falling leaves. It points opposite the direction of motion.
$\vec{D}$
$\vec{v}$
Leaf

Drag can be a large force for objects moving at high speeds or in dense fluids. Hold your arm out the window as you ride in a car and feel how hard the air pushes against your arm. Note also how the air resistance against your arm increases rapidly as the car's speed increases. For a small particle moving in water, such as a swimming *Paramecium,* drag can be the dominant force.

On the other hand, for objects that are heavy and compact, moving in air, and with a speed that is not too great, the drag force of air resistance is fairly small. Under these conditions, **you can neglect air resistance unless a problem explicitly asks you to include it.** The error introduced into calculations by this approximation is generally pretty small.

Thrust Force

A jet airplane obviously has a force that propels it forward; likewise for the rocket in FIGURE 4.14. This force, the **thrust force,** occurs when a jet or rocket engine expels gas molecules at high speed. Thrust is a contact force, with the exhaust gas being the agent that pushes on the engine. We will study the process by which thrust is generated in Chapter 5. For now, we need only consider that **thrust is a force opposite the direction in which the exhaust gas is expelled.** There's no special symbol for the thrust force, so we use $\vec{F}_{\text{thrust}}$.

FIGURE 4.14 The thrust force on a rocket is opposite the direction of the expelled gases.

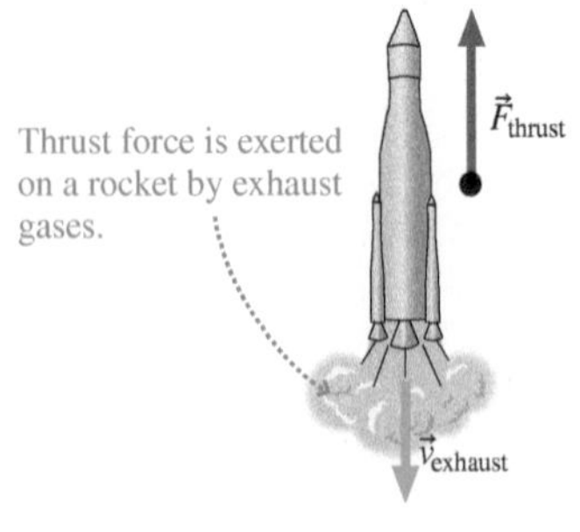
Thrust force is exerted on a rocket by exhaust gases.
$\vec{F}_{\text{thrust}}$
$\vec{v}_{\text{exhaust}}$

Electric and Magnetic Forces

Electricity and magnetism, like gravity, exert long-range forces. The forces of electricity and magnetism act on charged particles. We will study electric and magnetic forces in detail in Part VI of this text. These forces—and the forces inside the nucleus, which we will also see later in the text—won't be important for the dynamics problems we consider in the next several chapters.

STOP TO THINK 4.2 A frog is resting on a slope. What can you say about the friction force acting on the frog?

A. There is no friction force.
B. There is a kinetic friction force directed up the slope.
C. There is a static friction force directed up the slope.
D. There is a kinetic friction force directed down the slope.
E. There is a static friction force directed down the slope.

It's not just rocket science Rockets are propelled by thrust, but many animals are as well. Scallops are shellfish with no feet and no fins, but they can escape from predators or move to new territory by using a form of jet propulsion. A scallop forcibly ejects water from the rear of its shell, resulting in a thrust force that moves it forward.

4.4 What Do Forces Do?

How does an object move when a force is exerted on it? The only way to answer this question is to do experiments. FIGURE 4.15 shows an experiment—one you might do in your physics laboratory—in which a glider on an air track is pulled by a string. The horizontal force pulling the glider is the tension force of the string. A falling weight keeps the tension constant while a motion sensor tracks the glider and allows us to determine its velocity and acceleration.

The motion diagram of Figure 4.15 shows the outcome of this experiment: The glider *accelerates*. That's not too surprising. Far more important is that the acceleration vectors are all the same length; that is, **an object pulled with a constant force moves with a *constant* acceleration.** This finding could not have been

FIGURE 4.15 An experiment to measure the motion of a glider that is pulled by a constant tension force.

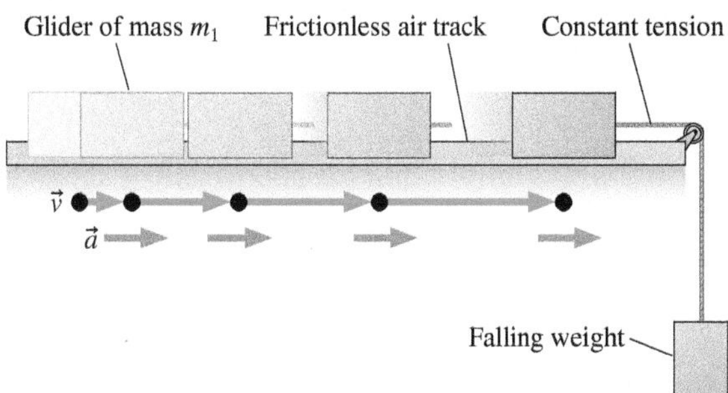

FIGURE 4.16 A graph of how a glider of constant mass is accelerated by different forces.

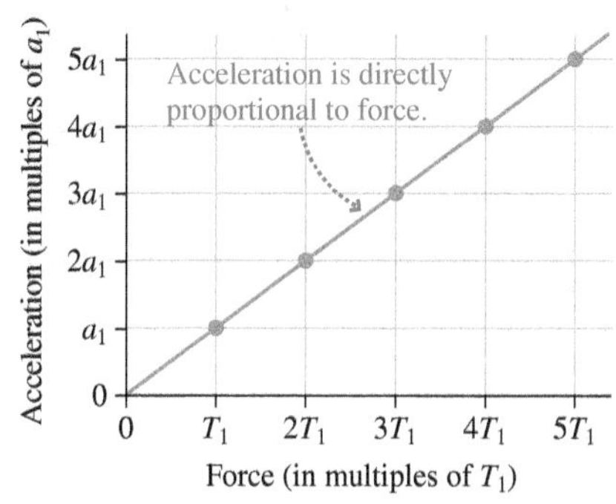

anticipated. It's conceivable that the glider would speed up for a while and then move at a steady speed. Or speed up with an acceleration that steadily decreases rather than being constant. But that's not what happens. The experimental evidence is that the glider moves *with a constant acceleration* for as long as we pull it with a constant force.

What happens if we increase the force? We can re-run the experiment with more tension by increasing the size of the falling weight. FIGURE 4.16 is a graph of the results. We've not yet introduced units of force, so the horizontal force axis simply shows multiples of the tension force T_1 used in our first experiment. Likewise, the vertical acceleration axis shows acceleration as multiples of the initial acceleration a_1. The graph shows that **an object's acceleration is directly proportional to the force acting on it.** If we double the force acting on an object, its acceleration doubles.

A final question is: For a given force, how does an object's acceleration depend on its mass? We can find out by increasing the mass of the glider *without changing the tension.* FIGURE 4.17 shows that doubling the mass from an initial m_1 halves the acceleration, quadrupling the mass results in an acceleration only one-fourth of the initial acceleration a_1, and so on. Thus our final important result is that, for a fixed amount of force, **an object's acceleration is inversely proportional to its mass.**

You're familiar with this idea: It's much harder to get your car rolling by pushing it than to get your bicycle rolling. A larger mass has more *resistance* to acceleration or, equivalently, more *inertia.*

FIGURE 4.17 A graph of how gliders of different mass are accelerated by the same force.

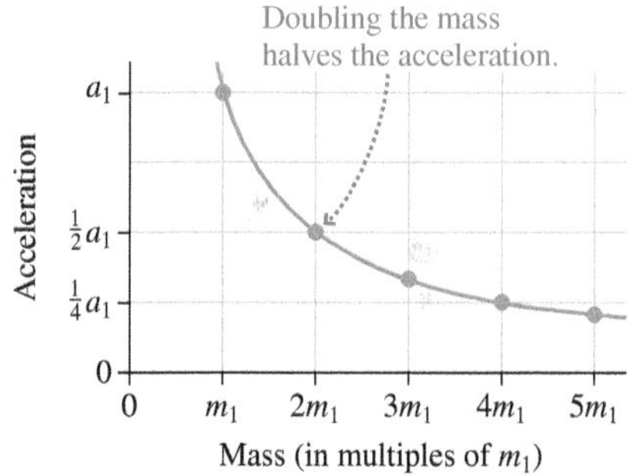

EXAMPLE 4.1 Finding the mass of a glider

When a 1.0 kg air-track glider is pulled with a constant tension force, it accelerates at 3.0 m/s^2. When a second glider is pulled with the same tension force, it accelerates at 5.0 m/s^2. What is the mass of the second glider?

PREPARE Because acceleration is inversely proportional to mass, we can use ratio reasoning to solve this problem. We use m to represent the unknown mass of the second glider.

SOLVE A quantity y is said to be *inversely proportional* to another quantity x if $y = A/x$, where A is a proportionality constant. This relationship is sometimes written as $y \propto 1/x$. If we have two values of x—say, x_1 and x_2—and corresponding y-values

$$y_1 = \frac{A}{x_1} \quad \text{and} \quad y_2 = \frac{A}{x_2}$$

then the *ratio* of the y-values is

$$\frac{y_2}{y_1} = \frac{A/x_2}{A/x_1} = \frac{x_1}{x_2}$$

That is, the ratio of the y-values is the inverse of the ratio of the x-values. This *ratio reasoning* allows us to solve some problems without needing to know the value of A. Because acceleration is inversely proportional to mass, we can write the ratios as

$$\frac{3.0\ \text{m/s}^2}{5.0\ \text{m/s}^2} = \frac{m}{1.0\ \text{kg}}$$

and thus

$$m = \frac{3.0\ \text{m/s}^2}{5.0\ \text{m/s}^2} \times (1.0\ \text{kg}) = 0.60\ \text{kg}$$

ASSESS With the same force applied, the second glider had a *larger* acceleration than the 1.0 kg glider. It makes sense, then, that its mass—its resistance to acceleration—is *less* than 1.0 kg.

Feel the difference Because of its high sugar content, a can of regular soda has a mass about 4% greater than that of a can of diet soda. If you try to judge which can is more massive by simply holding one in each hand, this small difference is almost impossible to detect. If you *move* the cans up and down, however, the difference becomes subtly but noticeably apparent: People evidently are more sensitive to how the mass of each can resists acceleration than they are to the cans' weights alone.

STOP TO THINK 4.3 An air-track glider accelerates at 2 m/s^2 when pulled with tension force T. Suppose a second glider with twice the mass is pulled with tension force $2T$. What is the acceleration of the second glider?

A. 1 m/s^2 B. 2 m/s^2 C. 4 m/s^2 D. 8 m/s^2 E. 16 m/s^2

4.5 Newton's Second Law

We can now summarize the results of our experiments. What we've seen is that **a force causes an object to accelerate. The acceleration a is directly proportional to the force F and inversely proportional to the mass m.** We can express both these relationships in equation form as

$$a = \frac{F}{m} \tag{4.2}$$

Note that if we double the size of the force F, the acceleration a will double, as we found experimentally. And if we quadruple the mass m, the acceleration will be only one-fourth as large, again agreeing with our experiments.

Equation 4.2 tells us the magnitude of an object's acceleration in terms of its mass and the force applied. But our experiments also had another important finding: The *direction* of the acceleration was the same as the direction of the force. We can express this fact by writing Equation 4.2 in *vector* form as

$$\vec{a} = \frac{\vec{F}}{m} \tag{4.3}$$

Finally, our experiment was limited to looking at an object's response to a *single* applied force acting in a single direction. Realistically, an object is likely to be subjected to several distinct forces $\vec{F}_1, \vec{F}_2, \vec{F}_3, \ldots$ that may point in different directions. What happens then? Experiments show that the acceleration of the object is determined by the *net force* acting on it. Recall from Figure 4.4 and Equation 4.1 that the net force is the *vector sum* of all forces acting on the object. So if several forces are acting, we use the net force in Equation 4.4.

Newton was the first to recognize these connections between force and motion. This relationship is known today as **Newton's second law.**

Footsteps As you walk, it takes muscular effort—forces—to accelerate your foot as it lifts up and then again to decelerate your foot before it touches down. This continual start-stop effort, which requires metabolic energy, makes walking a rather inefficient mode of locomotion. Cycling, where your feet are in continuous motion, is more efficient. We'll return to this topic in Chapter 11.

Newton's second law An object of mass m subjected to forces $\vec{F}_1, \vec{F}_2, \vec{F}_3, \ldots$ will undergo an acceleration $\vec{a}$ given by

$$\vec{a} = \frac{\vec{F}_{\text{net}}}{m} \tag{4.4}$$

where the net force $\vec{F}_{\text{net}} = \vec{F}_1 + \vec{F}_2 + \vec{F}_3 + \cdots$ is the vector sum of all forces acting on the object. **The acceleration vector $\vec{a}$ points in the same direction as the net force vector $\vec{F}_{\text{net}}$.**

While some relationships are found to apply only in special circumstances, others seem to have universal applicability. Those equations that appear to apply at all times and under all conditions have come to be called "laws of nature." Newton's first and second laws are laws of nature; you will meet others as we go through this text.

We can rewrite Newton's second law in the form

$$\vec{F}_{\text{net}} = m\vec{a} \tag{4.5}$$

which is how you'll see it presented in many textbooks and how, in practice, we'll often use the second law. Equations 4.4 and 4.5 are mathematically equivalent, but

Equation 4.4 better describes the central idea of Newtonian mechanics: A force applied to an object causes the object to accelerate, and the acceleration is in the direction of the net force. It is worth noting that **an object responds only to forces acting on it *at this instant*.** The object has no memory of forces that may have been exerted at earlier times.

NOTE ▶ When several forces act on an object, be careful not to think that the strongest force "overcomes" the others to determine the motion on its own. It is $\vec{F}_{net}$, the sum of *all* the forces, that determines the acceleration $\vec{a}$. ◀

CONCEPTUAL EXAMPLE 4.2 **Acceleration of a wind-blown basketball**

You drop a basketball while a stiff breeze is blowing to the right. In what direction does the ball accelerate?

REASON The breeze pushes the basketball to the right while gravity pulls straight down on it. Thus FIGURE 4.18a shows two forces acting on the ball. Newton's second law tells us that the direction of the acceleration is the same as the direction of the net force $\vec{F}_{net}$. In FIGURE 4.18b we find $\vec{F}_{net}$ by vector addition of $\vec{w}$ and $\vec{F}_{breeze}$. We see that $\vec{F}_{net}$ and therefore $\vec{a}$ point downward and to the right.

ASSESS This makes sense on the basis of your experience. Weight pulls the ball down, and the wind pushes the ball to the right. The net result is an acceleration down and to the right.

FIGURE 4.18 A basketball falling in a strong breeze.

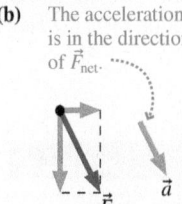

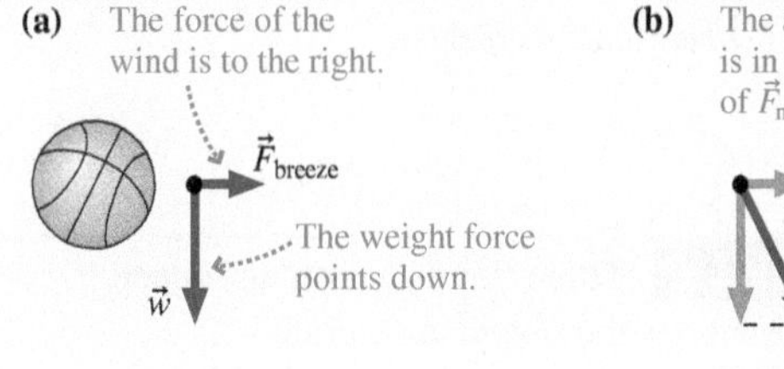

The Unit of Force

Because $\vec{F}_{net} = m\vec{a}$, the units of force must be the unit of mass (kg) multiplied by the unit of acceleration (m/s^2); thus the units of force are kg · m/s^2. This unit of force is called the **newton**:

$$1 \text{ newton} = 1 \text{ N} = 1 \frac{\text{kg} \cdot \text{m}}{\text{s}^2}$$

The abbreviation for the newton is N. TABLE 4.1 lists some typical forces in newtons.

The unit of force in the English system is the *pound* (abbreviated lb). Although the definition of the pound has varied, it is now defined in terms of the newton:

$$1 \text{ pound} = 1 \text{ lb} = 4.45 \text{ N}$$

You very likely associate pounds with kilograms rather than with newtons. Everyday language often confuses the ideas of mass and weight, but we're going to need to make a clear distinction between them. We'll have more to say about this in the next chapter.

TABLE 4.1 **Approximate magnitude of some typical forces**

Force	Approximate magnitude (newtons)
Weight of a U.S. nickel	0.05
Weight of ¼ cup of sugar	0.5
Weight of a 1 pound object	5
Weight of a typical house cat	50
Weight of a 110 pound person	500
Propulsion force of a car	5000
Thrust force of a small jet engine	50,000
Pulling force of a locomotive	500,000

EXAMPLE 4.3 **Racing down the runway**

A Boeing 737—a small, short-range jet with a mass of 51,000 kg—sits at rest at the start of a runway. The pilot turns the pair of jet engines to full throttle, and the thrust accelerates the plane down the runway. After traveling 940 m, the plane reaches its takeoff speed of 70 m/s and leaves the ground. What is the thrust of each engine?

PREPARE If we assume that the plane undergoes a constant acceleration (a reasonable assumption), we can use kinematics to find the magnitude of that acceleration. Then we will use Newton's second law to find the force—the thrust—that produced this acceleration. FIGURE 4.19 is a pictorial representation of the airplane's motion.

FIGURE 4.19 Pictorial representation of the accelerating airplane.

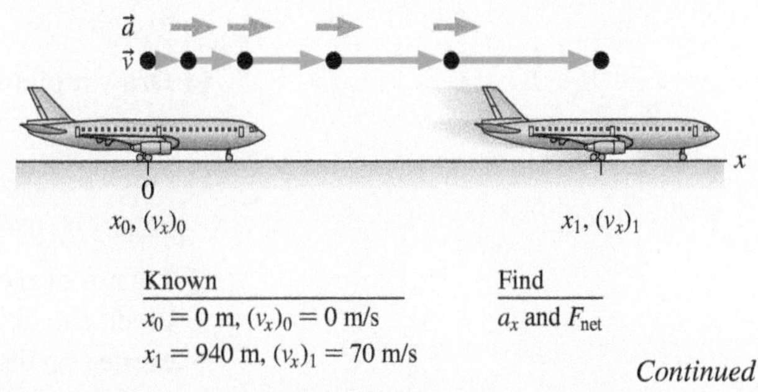

Continued

SOLVE We don't know how much time it takes the plane to reach its takeoff speed, but we do know that it travels a distance of 940 m. We can use the kinematic equation

$$(v_x)_1^2 = (v_x)_0^2 + 2a_x \Delta x$$

to find the acceleration. The displacement is $\Delta x = x_1 - x_0 = 940$ m, and the initial velocity is 0. Thus

$$a_x = \frac{(v_x)_1^2}{2\,\Delta x} = \frac{(70\text{ m/s})^2}{2(940\text{ m})} = 2.61\text{ m/s}^2$$

We've kept an extra significant figure because this isn't our final result—we are asked to solve for the thrust. We complete the solution by using Newton's second law:

$$F = ma_x = (51{,}000\text{ kg})(2.61\text{ m/s}^2) = 133{,}000\text{ N}$$

The thrust of each engine is half of this total force:

$$\text{thrust of one engine} = 67{,}000\text{ N} = 67\text{ kN}$$

ASSESS An acceleration of about $\frac{1}{4}g$ seems reasonable for an airplane: It's zippy, but it's not a thrill ride. And the final value we find for the thrust of each engine is close to the value given in Table 4.1. This gives us confidence that our final result makes good physical sense.

STOP TO THINK 4.4 Three forces act on an object. In which direction does the object accelerate?

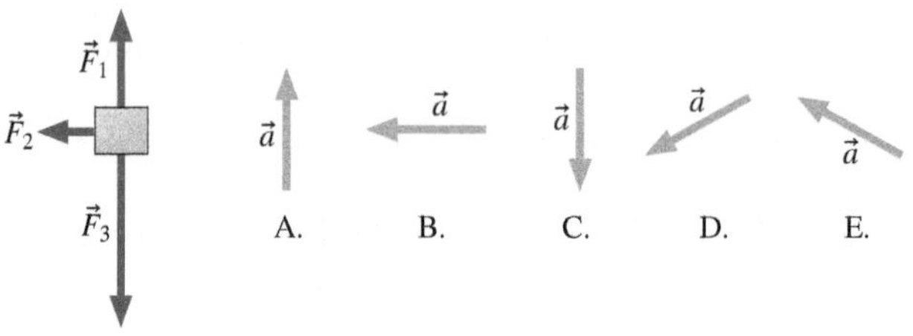

4.6 Free-Body Diagrams

In a typical physics problem, an object is being pushed and pulled in various directions by several distinct forces. Some forces are given explicitly, while others are only implied. In order to proceed, it is necessary to determine all the forces that act on the object. It is also necessary to avoid including forces that do not really exist. Now that you have learned the properties of forces and seen a catalog of typical forces, we can develop a step-by-step method for identifying each force in a problem. A list of the most common forces we'll come across in the next few chapters is given in TABLE 4.2.

TABLE 4.2 **Common forces and their notation**

Force	Notation
General force	$\vec{F}$
Weight	$\vec{w}$
Spring force	$\vec{F}_{sp}$
Tension	$\vec{T}$
Normal force	$\vec{n}$
Static friction	$\vec{f}_s$
Kinetic friction	$\vec{f}_k$
Drag	$\vec{D}$
Thrust	$\vec{F}_{thrust}$

NOTE ▸ Occasionally, you'll see labels for forces that aren't included in Table 4.2. For instance, if you push a book across a table with your hand, we might simply refer to the force you apply as $\vec{F}_{hand}$. ◂

Tactics Box 4.2 will help you correctly identify the forces acting on an object. It's followed by two examples that explicitly illustrate the steps of the Tactics Box.

TACTICS BOX 4.2 Identifying forces

1. **Identify the object of interest.** This is the object whose motion you wish to study.
2. **Draw a picture of the situation.** Show the object of interest and all other objects—such as ropes, springs, and surfaces—that touch it.
3. **Draw a closed curve around the object.** Only the object of interest is inside the curve; everything else is outside.
4. **Locate every point on the boundary of this curve where other objects touch the object of interest.** These are the points where *contact forces* are exerted on the object.

5. **Name and label each contact force acting on the object.** There is at least one force at each point of contact; there may be more than one. When necessary, use subscripts to distinguish forces of the same type.

6. **Name and label each long-range force acting on the object.** For now, the only long-range force we'll consider is weight, the force of gravity.

CONCEPTUAL EXAMPLE 4.4 **Identifying forces on a bungee jumper**

A bungee jumper has leapt off a bridge and is nearing the bottom of her fall. What forces are being exerted on the bungee jumper?

REASON FIGURE 4.20 Forces on a bungee jumper.

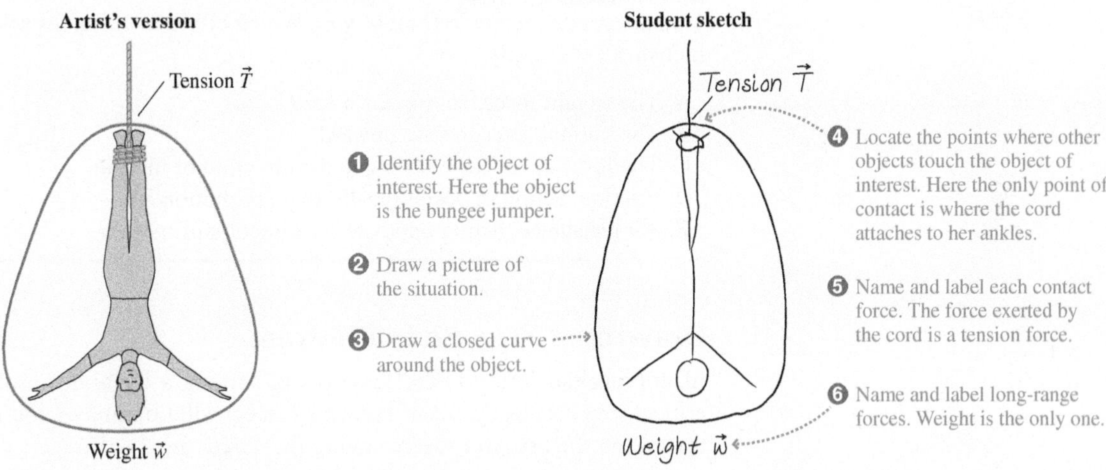

CONCEPTUAL EXAMPLE 4.5 **Identifying forces on a skier**

A skier is being towed up a snow-covered hill by a tow rope. What forces are being exerted on the skier?

REASON FIGURE 4.21 Forces on a skier.

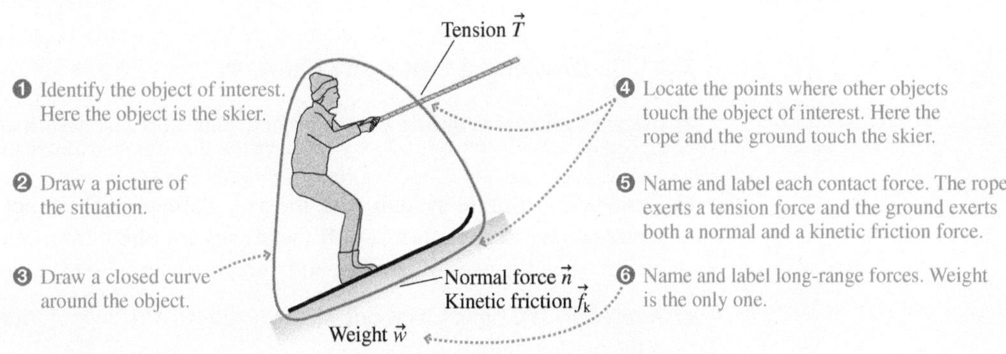

NOTE ► You might have expected two friction forces and two normal forces in Example 4.5, one on each ski. Keep in mind, however, that we're working within the particle model, which represents the skier by a single point. A particle has only one contact with the ground, so there is a single normal force and a single friction force. The particle model is valid if we want to analyze the motion of the skier as a whole, but we would have to go beyond the particle model to find out what happens to each ski. ◄

CONCEPTUAL EXAMPLE 4.6 **Identifying forces on a rocket**

A rocket is flying upward through the air, high above the ground. Air resistance is not negligible. What forces are being exerted on the rocket?

REASON

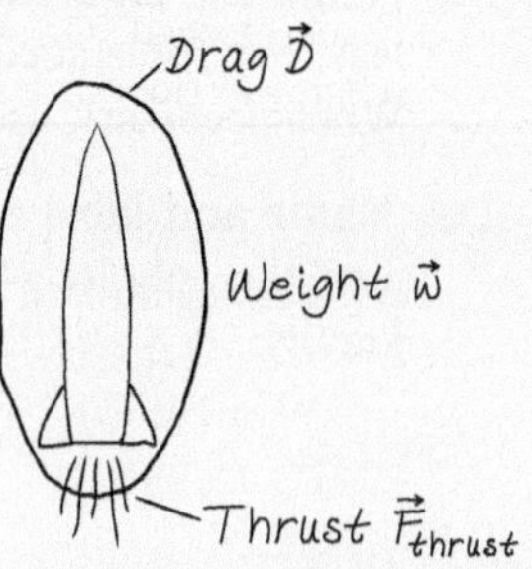

FIGURE 4.22 Forces on a rocket.

STOP TO THINK 4.5 You've just kicked a rock, and it is now sliding across the ground about 2 meters in front of you. Which of these are forces acting on the rock? Include all that apply.

A. The weight force, acting downward
B. The normal force, acting upward
C. The force of the kick, acting in the direction of motion
D. Friction, acting opposite the direction of motion
E. Air resistance, acting opposite the direction of motion

Drawing a Free-Body Diagram

Identifying the forces is the first step in solving a dynamics problem, and it's a critical step because you can't reach a correct solution unless you're working with the correct forces. But just knowing the forces isn't sufficient; you also have to get them all pointing in the right directions. So as a second step we assemble all of the information about the forces that act on an object (or "body") into a single diagram called a **free-body diagram.** A free-body diagram represents the object as a particle and shows *all* of the forces that act on the object. Learning how to draw a correct free-body diagram is a very important skill, one that will become a critical part of our approach to solving motion problems. For dynamics problems, the free-body diagram is an essential part of the *pictorial representation.*

TACTICS BOX 4.3 **Drawing a free-body diagram**

❶ **Identify all forces acting on the object.** This step was described in Tactics Box 4.2.

❷ **Draw a coordinate system.** Use the axes defined in your pictorial representation (see Tactics Box 2.4). If those axes are tilted, for motion along an incline, then the axes of the free-body diagram should be similarly tilted.

❸ **Represent the object as a dot at the origin of the coordinate axes.** This is the particle model.

❹ **Draw vectors representing each of the identified forces.** This was described in Tactics Box 4.1. Be sure to label each force vector.

❺ **Draw and label the *net force* vector $\vec{F}_{net}$.** Draw this vector beside the diagram, not on the particle. Then check that $\vec{F}_{net}$ points in the same direction as the acceleration vector $\vec{a}$ on your motion diagram. Or, if appropriate, write $\vec{F}_{net} = \vec{0}$.

EXAMPLE 4.7 Forces on an elevator

An elevator, suspended by a cable, speeds up as it moves upward from the ground floor. Draw a free-body diagram of the elevator.

PREPARE We can model the elevator as a particle.

SOLVE We'll follow the steps of Tactics Box 4.3. Notice that the elevator is moving upward with an increasing speed. That means its acceleration vector points upward, so $\vec{F}_{net}$ must point upward as well. This will be true if the magnitude of the (upward) tension force $\vec{T}$ is greater than that of the (downward) weight force $\vec{w}$. We have shown this in the free-body diagram by drawing the tension vector longer than the weight vector.

ASSESS Let's take a look at our picture and see if it makes sense. The coordinate axes, with a vertical y-axis, are the ones we would use in a pictorial representation of the motion, so we've chosen the correct axes. And, as noted, the tension force is drawn longer than the weight, which indicates an upward net force and hence an upward acceleration.

FIGURE 4.23 Free-body diagram of an elevator accelerating upward.

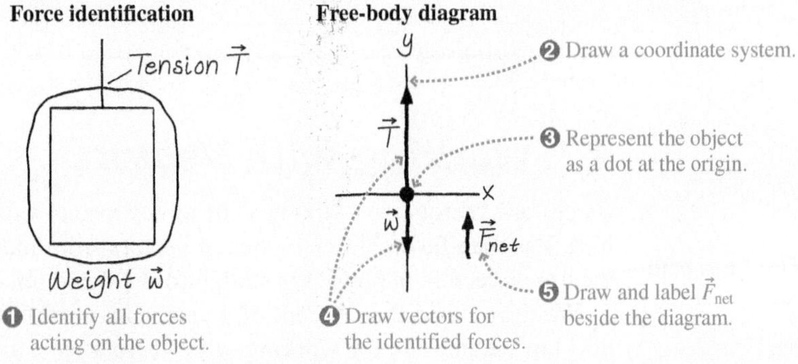

EXAMPLE 4.8 Forces on a towed skier

A tow rope pulls a skier up a snow-covered hill at a constant speed. Draw a full pictorial representation of the skier.

PREPARE We can model the skier as a particle. A full pictorial representation consists of both a motion diagram and a free-body diagram.

SOLVE If we were doing a kinematics problem, the pictorial representation would use a coordinate system with the x-axis parallel to the slope, so we use these same tilted coordinate axes for the free-body diagram. The full pictorial representation is shown in FIGURE 4.24.

ASSESS We have shown $\vec{T}$ pulling parallel to the slope and $\vec{f}_k$, which opposes the direction of motion, pointing down the slope. The normal force $\vec{n}$ is perpendicular to the surface and thus along the y-axis. Finally, and this is important, the weight $\vec{w}$ is *vertically* downward, *not* along the negative y-axis.

The skier moves in a straight line with constant speed, so $\vec{a} = \vec{0}$. Newton's second law then tells us that $\vec{F}_{net} = m\vec{a} = \vec{0}$. Thus we have drawn the vectors such that the forces add to zero. We'll learn more about how to do this in Chapter 5.

FIGURE 4.24 Pictorial representation for a skier being towed at a constant speed.

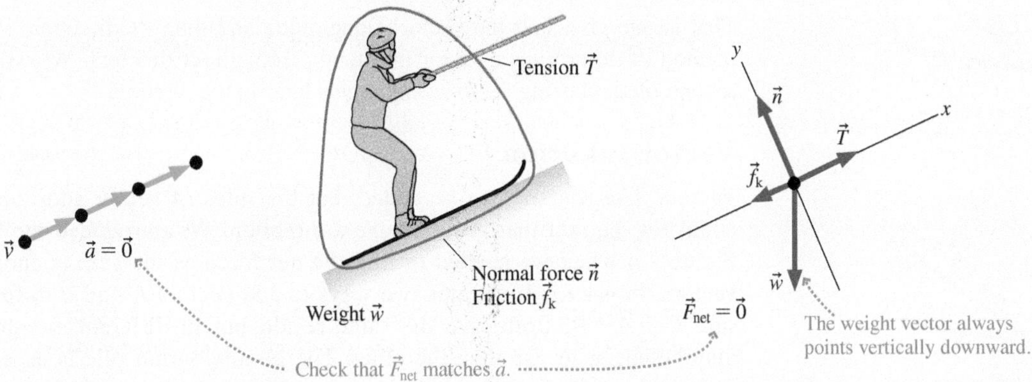

Free-body diagrams will be our major tool for the next several chapters. Careful practice will pay immediate benefits in the next chapter. Indeed, it is fair to say that a problem is more than half solved when you correctly complete the free-body diagram.

STOP TO THINK 4.6 An elevator suspended by a cable is moving upward and slowing to a stop. Which free-body diagram is correct?

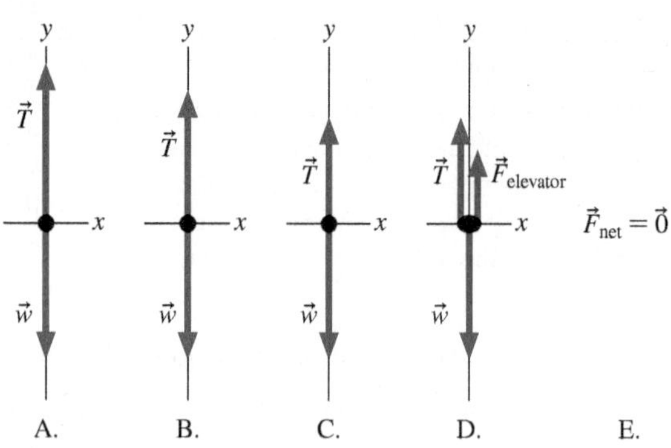

4.7 Working with Vectors

Forces are vectors, so working with forces means working with vectors. You may have less than favorable memories of vectors from math classes, but don't despair; the use of vectors in physics is straightforward and not difficult.

FIGURE 4.25 The velocity vector $\vec{v}$ has both a magnitude and a direction.

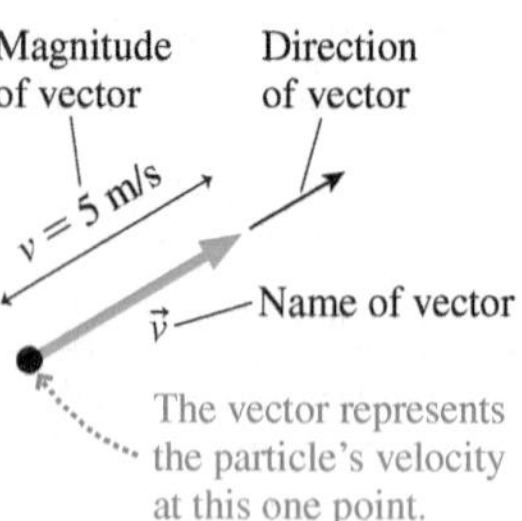

We introduced the concept of a vector in ◂◂ SECTION 2.2. In this section we will develop techniques for working with vectors as a tool for analyzing forces and motion. As you learned, a vector is a quantity with both a size (magnitude) and a direction. FIGURE 4.25 shows how to represent a particle's velocity as a vector $\vec{v}$. The particle's speed at this point is 5 m/s *and* it is moving in the direction indicated by the arrow. Recall that the magnitude of a vector is represented by the letter symbol of the vector, but without an arrow. In this case, the particle's speed—the magnitude of the velocity vector $\vec{v}$—is $v = 5$ m/s. The magnitude of a vector, a *scalar* quantity, cannot be a negative number.

NOTE ▸ Although the vector arrow is drawn across the page, from its tail to its tip, this arrow does *not* indicate that the vector "stretches" across this distance. Instead, the arrow tells us the value of the vector quantity only at the one point where the tail of the vector is placed. ◂

To describe a vector, we must specify both its magnitude and its direction. For example, we might write the velocity vector of Figure 4.25 as

$$\vec{v} = (5 \text{ m/s}, 30° \text{ above horizontal})$$

That is, we give first the vector's magnitude and then its direction. We'll use this method of describing a vector frequently throughout this text. We will introduce a second method using *vector components* later in this section.

FIGURE 4.26 Two vectors can be added using the tip-to-tail rule or the parallelogram rule.

(a)

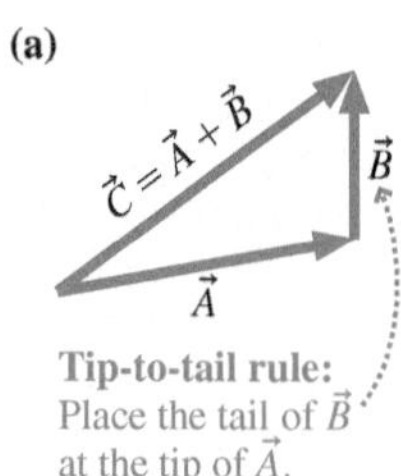

Tip-to-tail rule: Place the tail of $\vec{B}$ at the tip of $\vec{A}$.

(b)

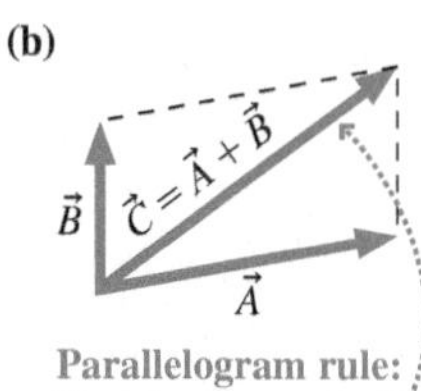

Parallelogram rule: Find the diagonal of the parallelogram formed by $\vec{A}$ and $\vec{B}$.

Vector Addition

Vectors, like scalars, can be added, but the rules of vector addition have to account for the fact that vectors have a direction. We introduced the basic idea in Figure 4.4 when we needed to find the net force as the sum of individual force vectors. FIGURE 4.26 illustrates two ways to add vectors $\vec{A}$ and $\vec{B}$ to find the vector sum $\vec{C} = \vec{A} + \vec{B}$. Both give the same result, but in different circumstances one may be easier to use than the other. The parallelogram rule is especially useful for working with free-body diagrams where the vectors are drawn with their tails together. Note that vector addition is commutative: $\vec{A} + \vec{B} = \vec{B} + \vec{A}$. You can add vectors in any order you wish.

Suppose acceleration $\vec{a}_2$ is twice acceleration $\vec{a}_1$. We write this as $\vec{a}_2 = 2\vec{a}$ and draw vector $\vec{a}_2$ with twice the length of $\vec{a}_1$ *in the same direction*. That is, multiplication of a vector by a positive constant simply stretches or compresses the vector without changing its direction. FIGURE 4.27 shows this for vectors $\vec{E}$ and $\vec{F} = c\vec{E}$.

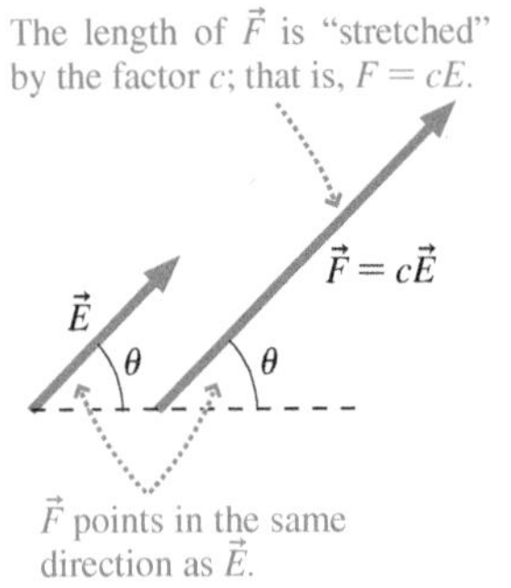

FIGURE 4.27 Multiplication of a vector by a positive scalar.

The length of $\vec{F}$ is "stretched" by the factor c; that is, $F = cE$.

$\vec{E}$ θ θ $\vec{F} = c\vec{E}$

$\vec{F}$ points in the same direction as $\vec{E}$.

Vector Subtraction

How do we *subtract* vector $\vec{P}$ from vector $\vec{Q}$? In arithmetic we can write $5 - 3 = 5 + (-3)$; that is, subtraction becomes addition once we introduce the idea of negative numbers. Similarly, we can write the vector subtraction $\vec{P} - \vec{Q}$ as $\vec{P} + (-\vec{Q})$ if we define the negative of a vector as

$$-Q = (Q, \text{direction opposite } \vec{Q}) \tag{4.6}$$

That is, $-\vec{Q}$ has the same magnitude as $\vec{Q}$ but points in the opposite direction.

This makes sense. FIGURE 4.28 uses the tip-to-tail rule to show that adding vector $-\vec{R}$ to vector $\vec{R}$ brings us back to our starting point, a vector that has zero length or magnitude. This vector is known as the **zero vector,** denoted $\vec{0}$. This is the graphical depiction of the mathematical statement $\vec{R} + (-\vec{R}) = \vec{R} - \vec{R} = \vec{0}$. Note that the direction of the zero vector is not defined; you cannot describe the direction of an arrow of zero length!

FIGURE 4.28 The negative of a vector.

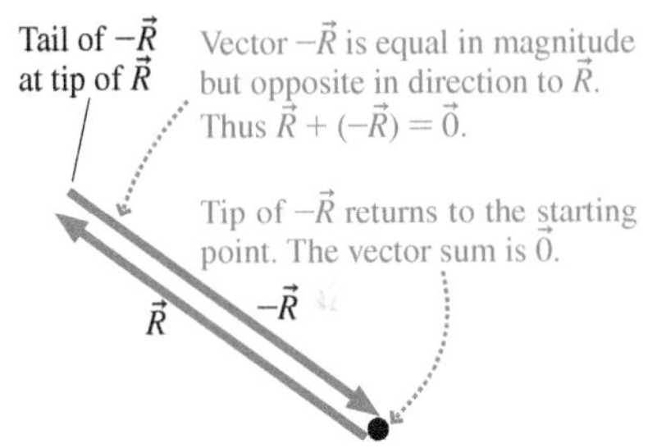
Tail of $-\vec{R}$ at tip of $\vec{R}$
Vector $-\vec{R}$ is equal in magnitude but opposite in direction to $\vec{R}$. Thus $\vec{R} + (-\vec{R}) = \vec{0}$.
Tip of $-\vec{R}$ returns to the starting point. The vector sum is $\vec{0}$.
$\vec{R}$
$-\vec{R}$

FIGURE 4.29 shows that we can now use the rules of vector addition to perform the vector subtraction $\vec{P} - \vec{Q}$. We do so by first drawing $\vec{P}$, then $-\vec{Q}$, then using the tip-to-tail addition rule to find $\vec{P} + (-\vec{Q})$.

FIGURE 4.29 Graphical vector subtraction.

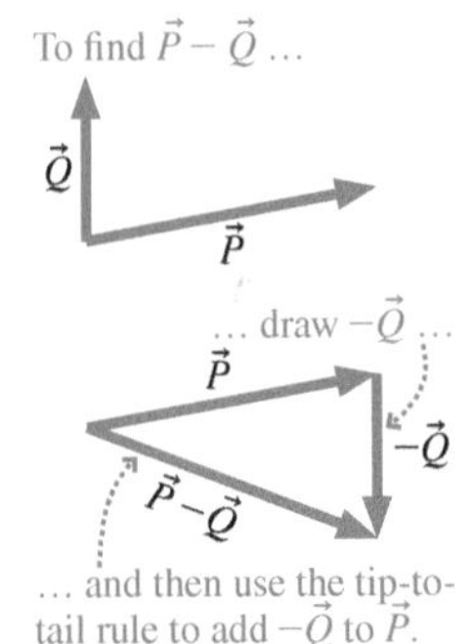
To find $\vec{P} - \vec{Q}$ ...
$\vec{Q}$
$\vec{P}$
... draw $-\vec{Q}$ ...
$\vec{P}$
$-\vec{Q}$
$\vec{P} - \vec{Q}$
... and then use the tip-to-tail rule to add $-\vec{Q}$ to $\vec{P}$.

Vector Components

FIGURE 4.30 shows a vector $\vec{A}$ and an xy-coordinate system that we've chosen. Once the directions of the axes are known, we can define two new vectors $\vec{A}_x$ and $\vec{A}_y$ *parallel to the axes* such that $\vec{A}$ is the vector sum:

$$\vec{A} = \vec{A}_x + \vec{A}_y \tag{4.7}$$

In essence, we have "broken" vector $\vec{A}$ into two perpendicular vectors that are parallel to the coordinate axes. We say that we have **decomposed,** or **resolved,** vector $\vec{A}$ into its component vectors.

NOTE ▶ It is not necessary for the tail of $\vec{A}$ to be at the origin. All we need to know is the *orientation* of the coordinate system so that we can draw $\vec{A}_x$ and $\vec{A}_y$ parallel to the axes. ◀

Suppose we have a vector $\vec{A}$ that has been decomposed into vectors $\vec{A}_x$ and $\vec{A}_y$ parallel to the coordinate axes. We can describe each of these vectors with a single number called the **component** that gives its length and direction. The *x-component* and *y-component* of vector $\vec{A}$, denoted A_x and A_y, are determined as follows:

FIGURE 4.30 Vectors $\vec{A}_x$ and $\vec{A}_y$ are drawn parallel to the coordinate axes such that $\vec{A} = \vec{A}_x + \vec{A}_y$.

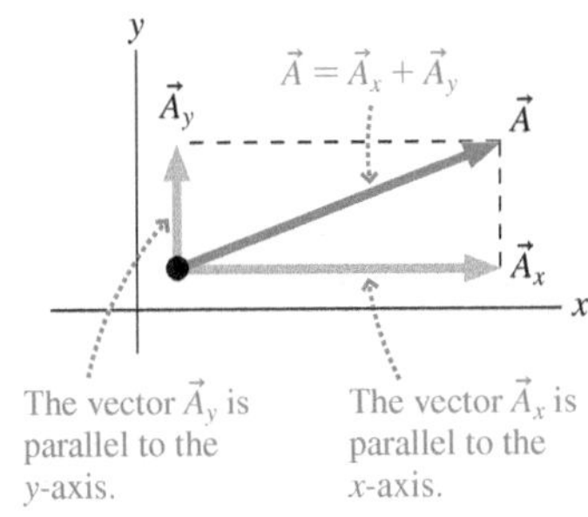
y
$\vec{A} = \vec{A}_x + \vec{A}_y$
$\vec{A}_y$
$\vec{A}$
$\vec{A}_x$
x
The vector $\vec{A}_y$ is parallel to the y-axis.
The vector $\vec{A}_x$ is parallel to the x-axis.

TACTICS BOX 4.4 Determining the components of a vector

1. The absolute value $|A_x|$ of the x-component A_x is the magnitude of the component vector $\vec{A}_x$.
2. The *sign* of A_x is positive if $\vec{A}_x$ points in the positive x-direction, negative if $\vec{A}_x$ points in the negative x-direction.
3. The y-component A_y is determined similarly.

FIGURE 4.31 Determining the components of a vector.

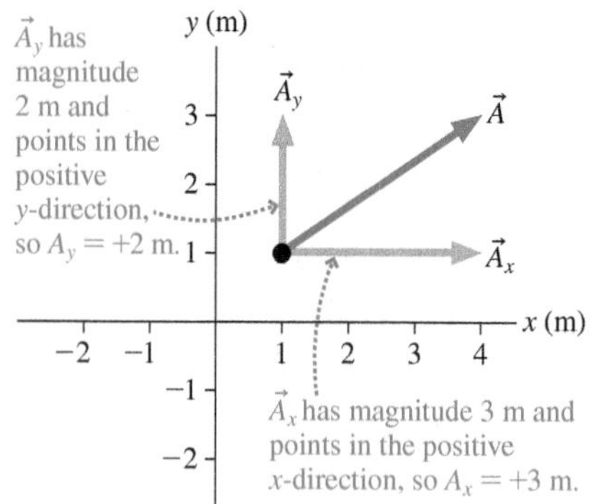

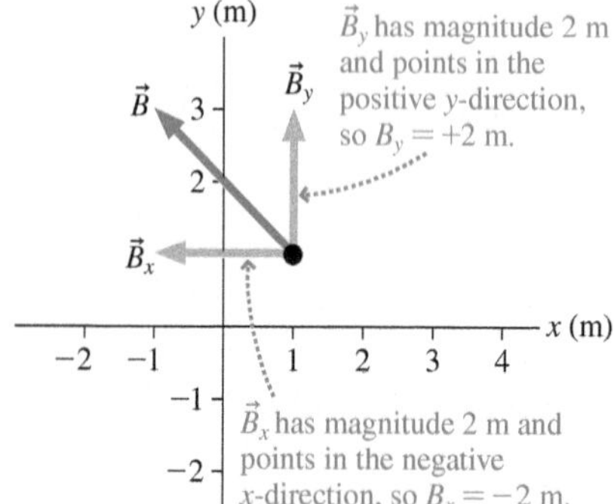

In other words, the component A_x tells us two things: how big $\vec{A}_x$ is and toward which end of the axis $\vec{A}_x$ points. FIGURE 4.31 shows two examples of determining the components of a vector.

Much of physics is expressed in the language of vectors. We will frequently need to decompose a vector into its components or to "reassemble" a vector from its components, moving back and forth between the graphical and the component representations of a vector. Let's start with the problem of decomposing a vector into its x- and y-components. FIGURE 4.32a shows a vector $\vec{A}$ at an angle θ above horizontal. It is *essential* to use a picture or diagram such as this to define the angle you are using to describe a vector's direction. $\vec{A}$ points to the right and up, so Tactics Box 4.4 tells us that the components A_x and A_y are both positive.

We can find the components using trigonometry, as illustrated in FIGURE 4.32b. For this case, we find that

$$A_x = A\cos\theta \qquad A_y = A\sin\theta \tag{4.8}$$

where A is the magnitude, or length, of $\vec{A}$. These equations convert the length and angle description of vector $\vec{A}$ into the vector's components, but they are correct *only* if θ is measured from horizontal.

NOTE ▸ It may have been a while since you used trigonometry to relate the sides of a right triangle. The Trigonometry Review at the bottom of the page summarizes the definitions and uses of sines, cosines, and tangents. ◂

FIGURE 4.32 Finding the components of a vector.

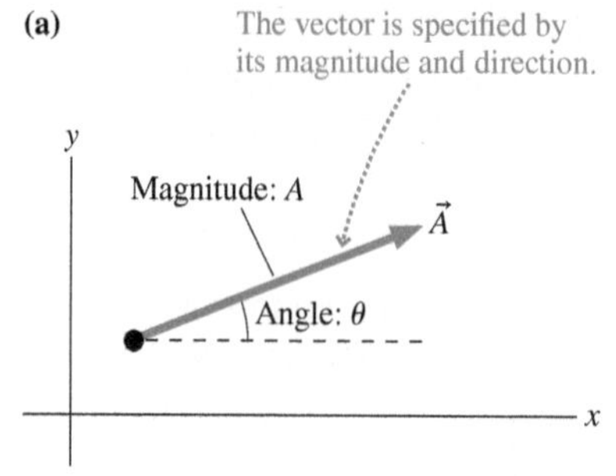

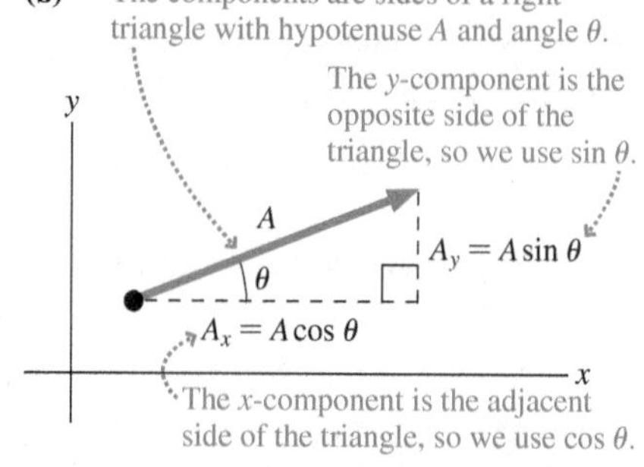

Trigonometry Review

We specify the sides of a right triangle in relation to one of the angles.

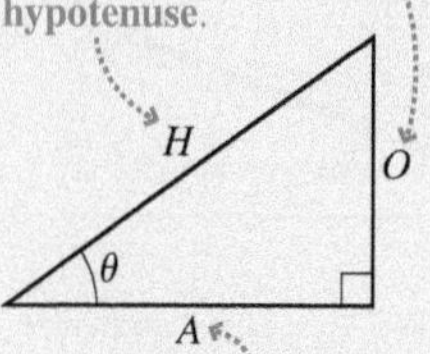

This is the side opposite to angle θ.

This is the side adjacent to angle θ.

The three sides are related by the *Pythagorean theorem*:

$$H = \sqrt{A^2 + O^2}$$

The sine, cosine, and tangent of angle θ are defined as ratios of the side lengths.

$$\sin\theta = \frac{O}{H}$$

$$\cos\theta = \frac{A}{H}$$

$$\tan\theta = \frac{O}{A}$$

We can rearrange these equations in useful ways:

$$O = H\sin\theta$$

$$A = H\cos\theta$$

Given the length of the hypotenuse and one angle, we can find the side lengths.

y is opposite to the angle; use the sine formula.

20 cm

30°

y

x

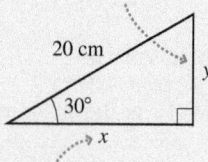

x is adjacent to the angle; use the cosine formula.

$$x = (20\text{ cm})\cos(30°) = 17\text{ cm}$$

$$y = (20\text{ cm})\sin(30°) = 10\text{ cm}$$

Inverse trig functions let us use lengths to find angles.

$$\theta = \sin^{-1}\left(\frac{O}{H}\right)$$

$$\theta = \cos^{-1}\left(\frac{A}{H}\right)$$

$$\theta = \tan^{-1}\left(\frac{O}{A}\right)$$

The inverse functions are also called arcsin, arccos, and arctan.

If we are given the lengths of the triangle's sides, we can find angles.

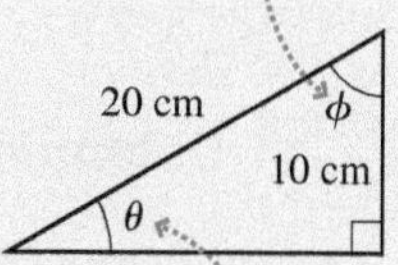

φ is adjacent to the 10 cm side; use the cos⁻¹ formula.
20 cm
φ
10 cm
θ
θ is opposite to the 10 cm side; use the sin⁻¹ formula.

$$\phi = \cos^{-1}\left(\frac{10\text{ cm}}{20\text{ cm}}\right) = 60°$$

$$\theta = \sin^{-1}\left(\frac{10\text{ cm}}{20\text{ cm}}\right) = 30°$$

Alternatively, if we are given the x- and y-components of a vector, we can determine the length and angle of the vector, as shown in FIGURE 4.33. Because A in Figure 4.33 is the hypotenuse of a right triangle, its length is given by the Pythagorean theorem:

$$A = \sqrt{A_x^2 + A_y^2} \tag{4.9}$$

Similarly, the tangent of angle θ is the ratio of the opposite side to the adjacent side, so

$$\theta = \tan^{-1}\left(\frac{A_y}{A_x}\right) \tag{4.10}$$

Equations 4.9 and 4.10 can be thought of as the "inverse" of Equations 4.8.

Vectors don't always point to the right and up as in Figure 4.32, and the angle defining their direction may not always be measured from the x-axis. For example, FIGURE 4.34 shows vector $\vec{C}$ pointing down and to the right. In this case, the component vector $\vec{C}_y$ is pointing *down*, in the negative y-direction, so the y-component C_y is a *negative* number. In addition, the angle ϕ is drawn measured from the y-axis, so the components of $\vec{C}$ are

$$\begin{aligned} C_x &= C \sin \phi \\ C_y &= -C \cos \phi \end{aligned} \tag{4.11}$$

The roles of sine and cosine are reversed from those in Equations 4.8 because the angle ϕ is measured with respect to vertical, not horizontal.

NOTE ▶ Whether the x- and y-components use the sine or cosine depends on how you define the vector's angle. As noted previously, you *must* draw a diagram to define the angle that you use, and you must be sure to refer to the diagram when computing components. Don't use Equations 4.8 or 4.11 as general rules—they aren't! They appear as they do because of how we defined the angles. ◀

When we determine the direction of a vector from its components, we must consider the signs of the components. Finding the angle of vector $\vec{C}$ in Figure 4.34 requires the length of C_y *without* the minus sign, so vector $\vec{C}$ has direction

$$\phi = \tan^{-1}\left(\frac{C_x}{|C_y|}\right) \tag{4.12}$$

Notice that the roles of x and y differ from those in Equation 4.10.

FIGURE 4.33 Specifying a vector from its components.

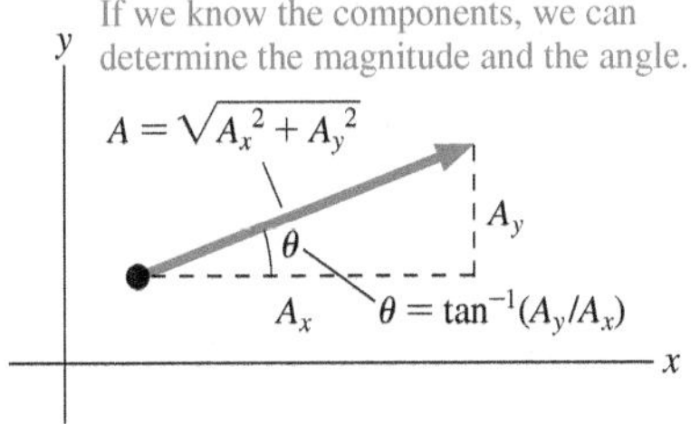

FIGURE 4.34 Relationships for a vector with a negative component.

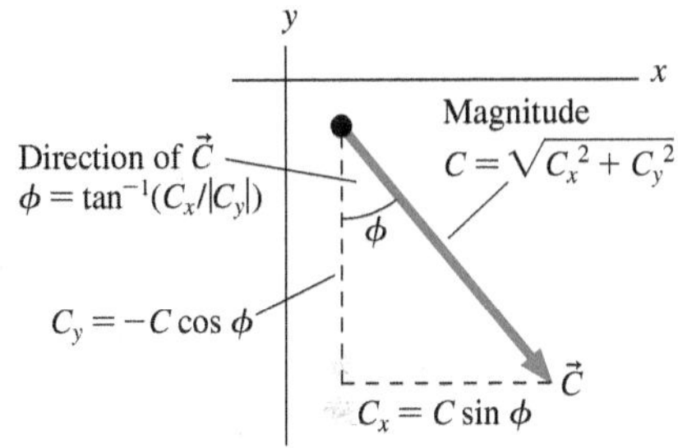

EXAMPLE 4.9 Finding the components of an acceleration vector

Seen from above, a hummingbird's acceleration is $\vec{a} = 6.0$ m/s^2, 30° south of west. What are the components of the acceleration vector in a coordinate system with the x-axis pointing east?

PREPARE Making a sketch is crucial to setting up this problem. FIGURE 4.35 shows the original vector $\vec{a}$ decomposed into vectors parallel to the axes.

FIGURE 4.35 The components of the acceleration vector.

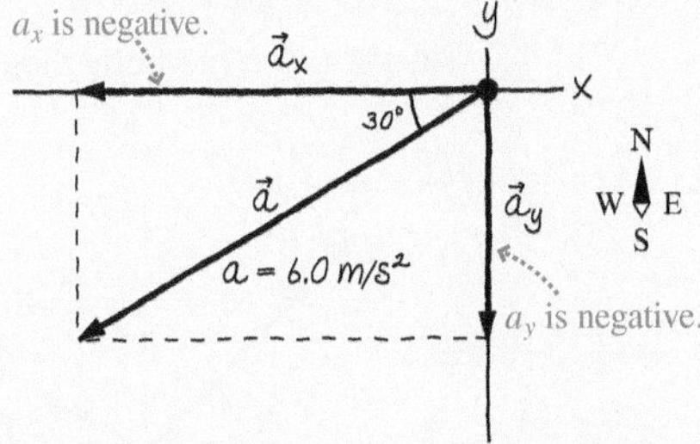

SOLVE The acceleration vector $\vec{a} = (6.0$ m/s^2, 30° below the negative x-axis) points to the left (negative x-direction) and down (negative y-direction), so the components a_x and a_y are both negative:

$$a_x = -a\cos 30° = -(6.0\text{ m/s}^2)\cos 30° = -5.2\text{ m/s}^2$$
$$a_y = -a\sin 30° = -(6.0\text{ m/s}^2)\sin 30° = -3.0\text{ m/s}^2$$

ASSESS The magnitude of the y-component is smaller than that of the x-component, as seems to be the case in Figure 4.35, a good check on our work. The units of a_x and a_y are the same as the units of vector $\vec{a}$. Notice that we had to insert the components' minus signs manually after observing that the vector points down and to the left.

STOP TO THINK 4.7 What are the x- and y-components C_x and C_y of vector $\vec{C}$?

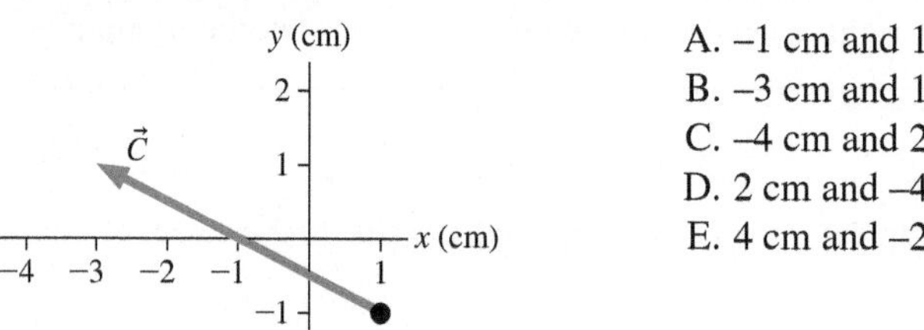

A. −1 cm and 1 cm
B. −3 cm and 1 cm
C. −4 cm and 2 cm
D. 2 cm and −4 cm
E. 4 cm and −2 cm

Working with Components

You've seen how to add and subtract vectors graphically. You can also use components to work with vectors. To illustrate, let's look at the vector sum $\vec{C} = \vec{A} + \vec{B}$ for the vectors shown in FIGURE 4.36. You can see that the component vectors of $\vec{C}$ are the sums of the component vectors of $\vec{A}$ and $\vec{B}$. The same is true of the components: $C_x = A_x + B_x$ and $C_y = A_y + B_y$.

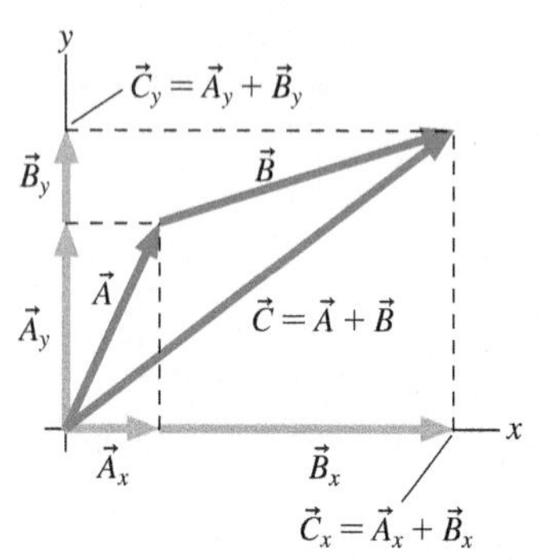

FIGURE 4.36 Using components to add vectors.

In general, if $\vec{D} = \vec{A} + \vec{B} + \vec{C} + \cdots$, then the x- and y-components of vector $\vec{D}$ are

$$D_x = A_x + B_x + C_x + \cdots$$
$$D_y = A_y + B_y + C_y + \cdots \qquad (4.13)$$

The next few chapters will make frequent use of *vector equations*. For example, the net force on a car skidding to a stop is the vector sum of the normal force, the weight force, and the friction force. We write this as

$$\vec{F}_{net} = \vec{n} + \vec{w} + \vec{f} \qquad (4.14)$$

Equation 4.14 is really just a shorthand way of writing the two simultaneous equations:

$$(F_{net})_x = n_x + w_x + f_x$$
$$(F_{net})_y = n_y + w_y + f_y \qquad (4.15)$$

In other words, a vector equation is interpreted as meaning: Equate the x-components on both sides of the equals sign, then equate the y-components. Vector notation allows us to write these two equations in a more compact form.

EXAMPLE 4.10 Finding the net force

FIGURE 4.37a is a free-body diagram showing three forces acting on an object. What is the net force $\vec{F}_{net} = \vec{F}_1 + \vec{F}_2 + \vec{F}_3$? Give your answer as a magnitude and direction.

PREPARE Figure 4.37a shows the forces, which we need to decompose into components.

FIGURE 4.37 Free-body diagram of three forces.

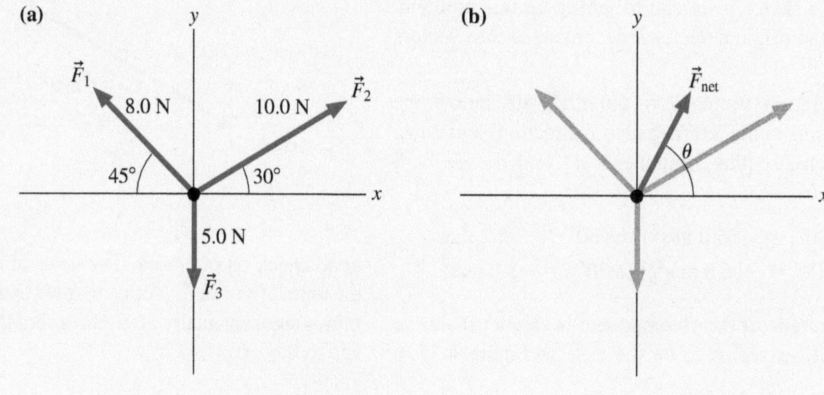

SOLVE The vector equation $\vec{F}_{net} = \vec{F}_1 + \vec{F}_2 + \vec{F}_3$ is really two simultaneous equations:

$$(F_{net})_x = F_{1x} + F_{2x} + F_{3x}$$
$$(F_{net})_y = F_{1y} + F_{2y} + F_{3y}$$

We determine the components using the magnitudes and angles. Thus

$$F_{1x} = -(8.0\text{ N})\cos 45° = -5.66\text{ N} \quad F_{1y} = (8.0\text{ N})\sin 45° = 5.66\text{ N}$$
$$F_{2x} = (10.0\text{ N})\cos 30° = 8.66\text{ N} \quad F_{2y} = (10.0\text{ N})\sin 30° = 5.00\text{ N}$$
$$F_{3x} = 0\text{ N} \quad F_{3y} = -5.00\text{ N}$$

The minus sign of F_{1x} is critical, and it appears not from some formula but because we recognized—from the figure—that the x-component of $\vec{F}_1$ points to the left. This is why actually drawing the free-body diagram is essential to solving dynamics problems. We can now combine the pieces by adding the columns, finding

$$(F_{net})_x = 3.00\text{ N} \qquad (F_{net})_y = 5.66\text{ N}$$

To find the magnitude and direction of $\vec{F}_{net}$, FIGURE 4.37b uses the components to draw the vector and to define a directional angle θ. The magnitude is the hypotenuse of the triangle:

$$F_{net} = |\vec{F}_{net}| = \sqrt{(3.00\text{ N})^2 + (5.66\text{ N})^2} = 6.4\text{ N}$$

The two components are the opposite and adjacent sides of a triangle, whose ratio is the tangent of the angle, so

$$\theta = \tan^{-1}\left(\frac{(F_{net})_y}{(F_{net})_x}\right) = \tan^{-1}\left(\frac{5.66\text{ N}}{3.00\text{ N}}\right) = 62°$$

Thus the net force is $\vec{F}_{net} = (6.4\text{ N}, 62°\text{ above the } +x\text{-axis})$.

ASSESS From the free-body diagram, it appears that the sum of the three vectors probably points to the right and up and that the magnitude will be similar to that of the three individual forces. Our calculation agrees, so we have good reason to believe the result is correct.

Tilted Axes

Although we are used to having the x-axis horizontal, there is no requirement that it has to be that way. Finding components with tilted axes is no harder than what we have done so far. Vector $\vec{C}$ in FIGURE 4.38 can be decomposed into component vectors $\vec{C}_x$ and $\vec{C}_y$, with $C_x = C\cos\theta$ and $C_y = C\sin\theta$. This will be useful when we need to find the components of a vector parallel to and perpendicular to a tilted surface. Doing so is easy if we let the x-axis be along the surface while the y-axis is perpendicular to the surface.

FIGURE 4.38 A coordinate system with tilted axes.

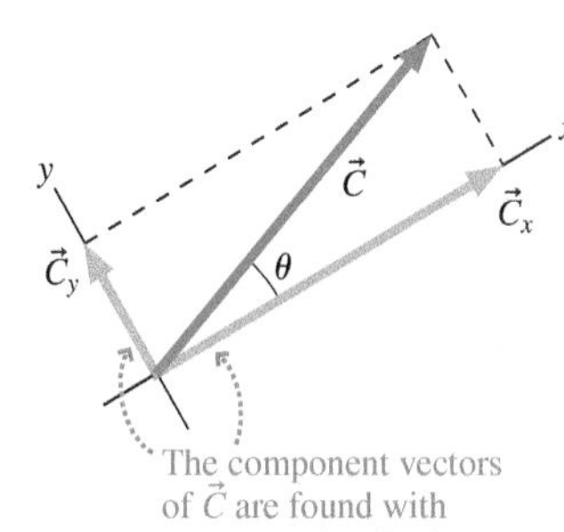
y
x
$\vec{C}$
$\vec{C}_x$
$\vec{C}_y$
θ
The component vectors of $\vec{C}$ are found with respect to the tilted axes.

4.8 Using Newton's Laws

We will conclude this chapter with some examples that illustrate using Newton's laws to solve problems. Chapter 5 will then look in more detail at some of the particular forces that are encountered in and by living systems.

It's vital to keep in mind that Newton's second law, $\vec{F} = m\vec{a}$, is a *vector* equation. Directions matter. In practice, we work with Newton's second law by using vector components and writing

$$\begin{aligned}(F_{net})_x &= F_{1x} + F_{2x} + F_{3x} + \cdots = \sum_i (F_i)_x = ma_x \\ (F_{net})_y &= F_{1y} + F_{2y} + F_{3y} + \cdots = \sum_i (F_i)_y = ma_y\end{aligned} \tag{4.16}$$

where Σ (Greek sigma) stands for "the sum of." In other words, Newton's second law for motion in a plane (which is all that we'll consider) is really two simultaneous equations: one for acceleration along the x-axis and one for acceleration along the y-axis. This is easier than it may sound, as the examples that follow will show; it's mostly a matter of paying careful attention to identifying all the forces and finding their x- and y-components.

We summarize these ideas in Problem-Solving Strategy 4.1.

PROBLEM-SOLVING STRATEGY 4.1 **Newtonian mechanics**

PREPARE Model the object as a particle. Make other simplifications depending on what kinds of forces are acting. Then draw a pictorial representation.

- Show important points in the motion with a sketch, establish a coordinate system, define symbols, and identify what the problem is trying to find. Align one coordinate axis with the direction of motion.
- Use a motion diagram to determine the object's acceleration vector $\vec{a}$. The acceleration is zero for an object in uniform motion.
- Identify all forces acting on the object *at this instant* and show them on a free-body diagram.
- It's OK to go back and forth between these steps as you visualize the situation.

SOLVE The mathematical representation is based on Newton's second law:

$$(F_{\text{net}})_x = \sum F_x \quad \text{and} \quad (F_{\text{net}})_y = \sum F_y$$

The forces are "read" directly from the free-body diagram. Depending on the problem,

- Solve for the unknown forces in an equilibrium problem with zero acceleration;
- Solve for the acceleration, then use kinematics to find velocities and positions; or
- Use kinematics to determine the acceleration, then solve for unknown forces.

ASSESS Check that your result has correct units and significant figures, is reasonable, and answers the question.

Mechanical Equilibrium

An important application of Newton's laws, especially for biological systems, is in analyzing the forces that act on an object at rest. For example, what forces are applied to your ankle by your bones and ligaments when you are standing at rest? What forces are needed to keep a tall tree from falling over in a strong wind? An object at rest has no acceleration ($\vec{a} = \vec{0}$) and thus the net force acting on it must be zero ($\vec{F}_{\text{net}} = \vec{0}$). It is said to be in **mechanical equilibrium.** For an extended object, like a tree, being in mechanical equilibrium requires not only no acceleration but also no rotation. We'll return to this topic in Chapter 6 after we introduce the idea of *torque,* the rotational equivalent of force. For now we'll focus on equilibrium situations where the object can be modeled as a particle that doesn't rotate.

EXAMPLE 4.11 **Readying a wrecking ball**

A wrecking ball weighing 2500 N hangs from a cable. Prior to swinging, it is pulled back to a 20° angle by a second, horizontal cable. What is the tension in the horizontal cable?

FIGURE 4.39 Pictorial representation of a wrecking ball just before release.

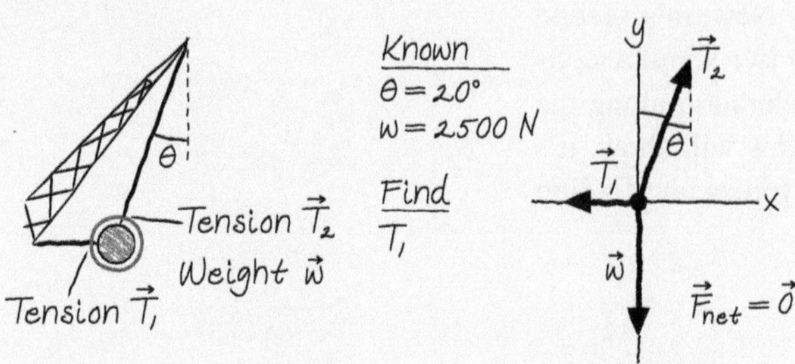

PREPARE We can model the wrecking ball as a particle. The ball is at rest, so it is in mechanical equilibrium. In **FIGURE 4.39**, we start by identifying all the forces acting on the ball: a tension force from each cable and the ball's weight. We've used different symbols $\vec{T}_1$ and $\vec{T}_2$ for the two different tension forces. We then construct a free-body diagram for these three forces, noting that $\vec{F}_{\text{net}} = m\vec{a} = \vec{0}$. We're looking for the magnitude T_1 of the tension force $\vec{T}_1$ in the horizontal cable.

SOLVE The requirement of equilibrium is $\vec{F}_{\text{net}} = m\vec{a} = \vec{0}$. In component form, we have the two equations:

$$\sum F_x = T_{1x} + T_{2x} + w_x = ma_x = 0$$

$$\sum F_y = T_{1y} + T_{2y} + w_y = ma_y = 0$$

You might have been tempted to write $-T_{1x}$ in the first equation because force $\vec{T}_1$ points to the left. However, the net force is the *sum* of the individual forces. The fact that $\vec{T}_1$ points to the left will be taken into account when we *evaluate* the components.

Now we're ready to write the components of each force vector in terms of the magnitudes and directions of those vectors. With practice you'll learn to read the components directly off the free-body diagram, but to begin it's worthwhile to organize the components into a table.

Force	Name of x-component	Value of x-component	Name of y-component	Value of y-component
$\vec{T}_1$	T_{1x}	$-T_1$	T_{1y}	0
$\vec{T}_2$	T_{2x}	$T_2 \sin\theta$	T_{2y}	$T_2 \cos\theta$
$\vec{w}$	w_x	0	w_y	$-w$

We see from the free-body diagram that $\vec{T}_1$ points along the negative x-axis, so $T_{1x} = -T_1$ and $T_{1y} = 0$. Any negative signs enter here as we evaluate the components, not when we write Newton's second law. We need to be careful with geometry and trigonometry as we find the components of $\vec{T}_2$. First, notice that the vector's angle from the y-axis is the same as the cable's angle from vertical. Then, recalling that the side of a triangle adjacent to the angle is related to the cosine, we see that the vertical (y) component of $\vec{T}_2$ is $T_2 \cos\theta$. Similarly, the horizontal (x) component is $T_2 \sin\theta$. The weight vector points straight down, so its y-component is $-w$. With these components, Newton's second law now becomes

$$-T_1 + T_2 \sin\theta + 0 = 0 \quad \text{and} \quad 0 + T_2 \cos\theta - w = 0$$

We can rewrite these equations as

$$T_1 = T_2 \sin\theta \quad \text{and} \quad T_2 \cos\theta = w$$

These are two simultaneous equations with two unknowns: T_1 and T_2. To eliminate T_2 from the two equations, we solve the second equation for T_2, giving $T_2 = w/\cos\theta$. Inserting this result into the first equation gives

$$T_1 = \frac{w}{\cos\theta} \sin\theta = \frac{\sin\theta}{\cos\theta} w = w \tan\theta = (2500\text{ N}) \tan 20° = 910\text{ N}$$

where we made use of the fact that $\tan\theta = \sin\theta/\cos\theta$.

ASSESS It seems reasonable that the force needed to pull the ball to a fairly modest angle is less than the ball's weight.

EXAMPLE 4.12 Towing a skier

A tow rope pulls a skier with a weight of 800 N (≈180 lb) up a 15° slope at constant speed. The friction force on the skier is 90 N. What are the magnitudes of the tension force and the normal force?

PREPARE In Example 4.8 we analyzed the forces on the skier and drew a free-body diagram, which is redrawn in FIGURE 4.40 with a table of relevant data. We've drawn a tilted coordinate system with the x-axis parallel to the direction of motion. Notice that the angle θ between the weight vector $\vec{w}$ and the negative y-axis is the same as the angle of the slope; this situation occurs in many problems. This is an equilibrium problem, even though the skier is not at rest, because the acceleration is zero and thus the net force on the skier must be $\vec{F}_{\text{net}} = \vec{0}$.

SOLVE The free-body diagram shows four forces acting on the skier. Newton's second law with $\vec{a} = \vec{0}$ is

$$\sum F_x = T_x + n_x + w_x + (f_k)_x = ma_x = 0$$

$$\sum F_y = T_y + n_y + w_y + (f_k)_y = ma_y = 0$$

FIGURE 4.40 Pictorial representation of a skier being towed up a hill.

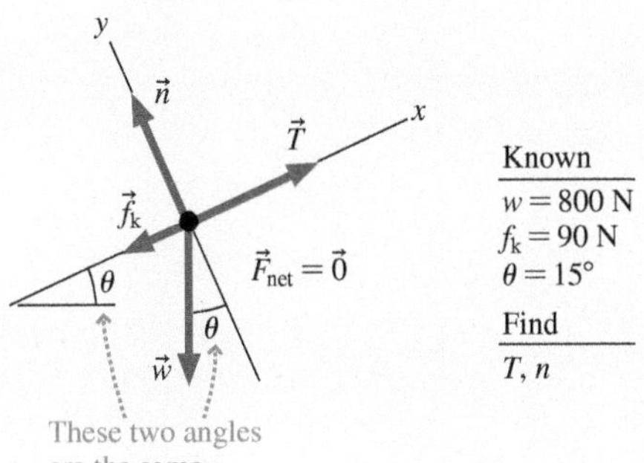

We can again find the components directly from the free-body diagram. Some are zero.

$$T_x = T \qquad T_y = 0$$
$$n_x = 0 \qquad n_y = n$$
$$w_x = -w \sin\theta \qquad w_y = -w \cos\theta$$
$$(f_k)_x = -f_k \qquad (f_k)_y = 0$$

The weight force points down and to the left in this coordinate system, so both components are negative.

We can use these components to simplify Newton's second law:

$$T - w \sin\theta - f_k = 0$$

$$n - w \cos\theta = 0$$

We can solve these two equations directly for the magnitudes of the tension force and the normal force:

$$T = w \sin\theta + f_k = (800\text{ N}) \sin 15° + 90\text{ N} = 300\text{ N}$$

$$n = w \cos\theta = (800\text{ N}) \cos 15° = 770\text{ N}$$

ASSESS Because of friction, a 90 N tension force would be needed to pull the skier over level snow. Pulling the skier up a slope requires even more tension, so our calculation of 300 N seems reasonable. The free-body diagram shows that the normal force is balanced by the y-component of the weight, so we expect the magnitude of the normal force to be slightly less than the weight. Our calculation finds that this is true.

Accelerated Motion

Mechanical equilibrium is important, but Newton's laws are especially important for understanding why objects *accelerate* in response to forces. We'll look at one example now, then many more in the next chapter. Notice how even examples of straight-line motion require that we analyze forces in two dimensions, with some forces having components perpendicular to the motion. That's why free-body diagrams are so important and why we need to work with vectors. Part of our strategy for using Newton's second law successfully is to **align one of the coordinate axes with the direction of motion.** That choice makes one component of the acceleration (either a_x or a_y) zero, which greatly simplifies the math.

EXAMPLE 4.13 Towing a car

A car with a mass of 1500 kg is being towed by a rope held at a 20° angle to the horizontal. A friction force of 320 N opposes the car's motion. What is the tension in the rope if the car goes from rest to 12 m/s in 10 s?

PREPARE We'll model the car as a particle. FIGURE 4.41 is a pictorial representation of the problem showing that the car's acceleration is directed to the right. Consequently, we've chosen a coordinate system in which the car accelerates along a horizontal x-axis. Force identification finds three contact forces in addition to the long-range force of gravity. These forces are shown on the free-body diagram, which also shows that there is a net force directed to the right, in the same direction as the acceleration.

SOLVE Newton's second law in component form is

$$\sum F_x = n_x + T_x + f_x + w_x = ma_x$$

$$\sum F_y = n_y + T_y + f_y + w_y = ma_y = 0$$

We've written the equations as sums, as we did with equilibrium problems, but now we have an acceleration along the x-axis. However, our choice of coordinate system dictates that $a_y = 0$ because there's no motion along the y-axis. We can now determine the components of the forces just by "reading" the free-body diagram. Three forces are along an axis and have only one nonzero component; trigonometry is needed only to find the x- and y-components of the tension. Thus Newton's second law becomes

$$T\cos\theta - f = ma_x$$

$$n + T\sin\theta - w = 0$$

Because the car speeds up from rest to 12 m/s in 10 s, we can use kinematics to find the acceleration:

$$a_x = \frac{\Delta v_x}{\Delta t} = \frac{(v_x)_1 - (v_x)_0}{t_1 - t_0} = \frac{(12\text{ m/s}) - (0\text{ m/s})}{(10\text{ s}) - (0\text{ s})} = 1.2\text{ m/s}^2$$

We can now use the first Newton's-law equation above to solve for the tension. We have

$$T = \frac{ma_x + f}{\cos\theta} = \frac{(1500\text{ kg})(1.2\text{ m/s}^2) + 320\text{ N}}{\cos 20°} = 2300\text{ N}$$

ASSESS The tension is about half of what Table 4.2 showed as a typical force needed to propel a car. That seems reasonable. Although we did not need the y-equation to solve this problem, that's unusual; most of the problems you will solve in Chapter 5 require both equations.

FIGURE 4.41 Pictorial representation of a car being towed.

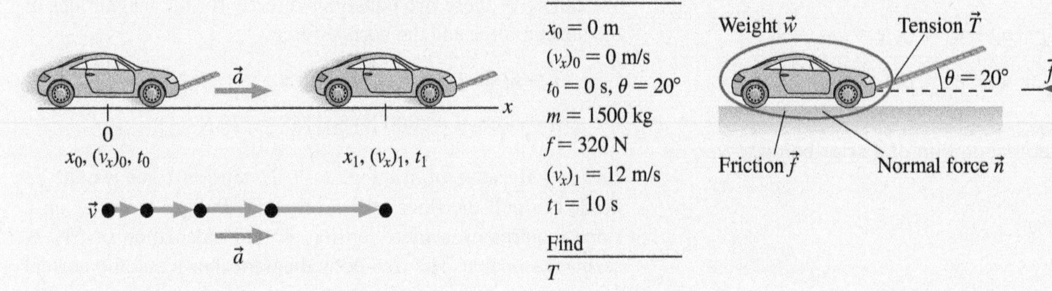

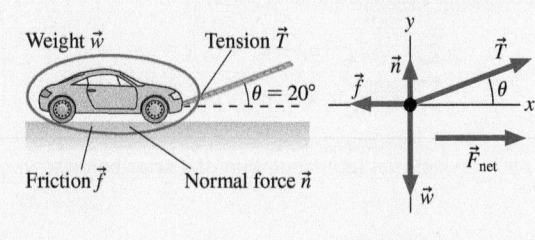

CHAPTER 4 INTEGRATED EXAMPLE Finding the force on the kneecap

The concept of torque, which we will study in Chapter 6, reveals how the structure of the kneecap (patella) enables the quadriceps muscle to move the femur. For now, we'll use a simple model to look at the forces in this system. The kneecap is attached by a tendon to the quadriceps muscle. This tendon pulls at a 10° angle relative to the femur, the bone of your upper leg. The patella is also attached to your lower leg (tibia) by a tendon that pulls parallel to the leg. To balance these forces, the end of your femur pushes outward on the patella. Bending your knee increases the tension in the tendons, and both have a tension of 60 N when the knee is bent to make a 70° angle between the upper and lower leg. What force does the femur exert on the kneecap in this position?

PREPARE We can model the kneecap as a particle in mechanical equilibrium. FIGURE 4.42 shows how to draw a pictorial representation. We've chosen to align the x-axis with the femur. The three forces—shown on the free-body diagram—are labeled $\vec{T}_1$ and $\vec{T}_2$ for the tensions and $\vec{F}$ for the femur's push. We've *defined* angle θ to indicate the direction of the femur's force on the kneecap.

SOLVE This is an equilibrium problem, with three forces on the kneecap that must sum to zero. For $\vec{a} = \vec{0}$, Newton's second law, written in component form, is

$$(F_{\text{net}})_x = \sum_i (F_i)_x = T_{1x} + T_{2x} + F_x = 0$$

$$(F_{\text{net}})_y = \sum_i (F_i)_y = T_{1y} + T_{2y} + F_y = 0$$

The components of the force vectors can be evaluated directly from the free-body diagram:

$$T_{1x} = -T_1 \cos 10° \qquad T_{1y} = T_1 \sin 10°$$

$$T_{2x} = -T_2 \cos 70° \qquad T_{2y} = -T_2 \sin 70°$$

$$F_x = F\cos\theta \qquad F_y = F\sin\theta$$

This is where signs enter, with T_{1x} being assigned a negative value because $\vec{T}_1$ points to the left. Similarly, $\vec{T}_2$ points both to the left and down, so both T_{2x} and T_{2y} are negative. With these components, Newton's second law becomes

$$-T_1 \cos 10° - T_2 \cos 70° + F\cos\theta = 0$$

$$T_1 \sin 10° - T_2 \sin 70° + F\sin\theta = 0$$

These are two simultaneous equations for the two unknowns F and θ. We will encounter equations of this form on many occasions, so make a note of the method of solution. First, rewrite the two equations as

$$F\cos\theta = T_1 \cos 10° + T_2 \cos 70°$$

$$F\sin\theta = -T_1 \sin 10° + T_2 \sin 70°$$

Next, divide the second equation by the first to eliminate F:

$$\frac{F\sin\theta}{F\cos\theta} = \tan\theta = \frac{-T_1 \sin 10° + T_2 \sin 70°}{T_1 \cos 10° + T_2 \cos 70°}$$

Then solve for θ:

$$\theta = \tan^{-1}\left(\frac{-T_1 \sin 10° + T_2 \sin 70°}{T_1 \cos 10° + T_2 \cos 70°}\right)$$

$$= \tan^{-1}\left(\frac{-(60\text{ N})\sin 10° + (60\text{ N})\sin 70°}{(60\text{ N})\cos 10° + (60\text{ N})\cos 70°}\right) = 30°$$

Finally, use θ to find F:

$$F = \frac{T_1 \cos 10° + T_2 \cos 70°}{\cos\theta}$$

$$= \frac{(60\text{ N})\cos 10° + (60\text{ N})\cos 70°}{\cos 30°} = 92\text{ N}$$

The question asked What force? and force is a vector, so we must specify both the magnitude and the direction. With the knee in this position, the femur exerts a force $\vec{F}$ = (92 N, 30° above the femur) on the kneecap.

ASSESS The magnitude of the force would be 0 N if the leg were straight, 120 N if the knee could be bent 180° so that the two tendons pull in parallel. The knee is closer to fully bent than to straight, so we would expect a femur force between 60 N and 120 N. Thus the calculated magnitude of 92 N seems reasonable.

FIGURE 4.42 Pictorial representation of the kneecap in equilibrium.

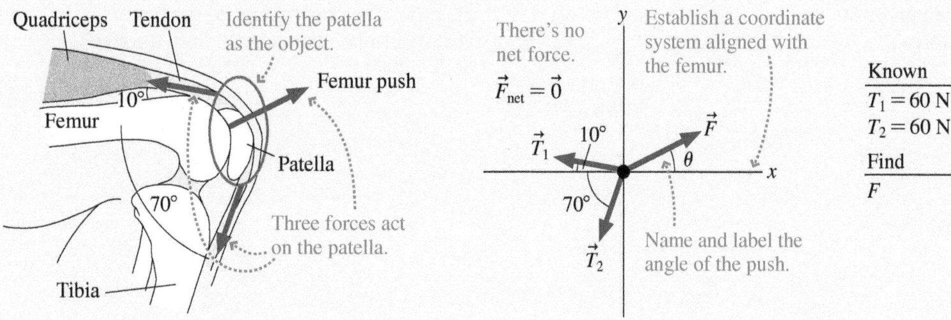

SUMMARY

GOAL To establish the connection between force and motion.

GENERAL PRINCIPLES

Newton's First Law

Consider an object with no force acting on it. If it is at rest, it will remain at rest. If it is in motion, it will continue to move in a straight line at a constant speed.

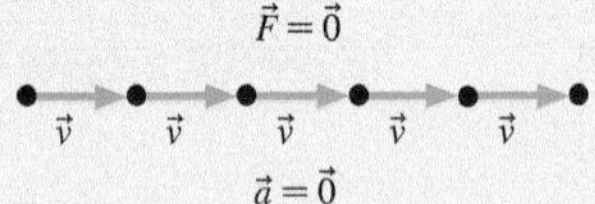

The first law tells us that no "cause" is needed for motion. An object that experiences no force will experience no acceleration. Without a force, an object in uniform motion will continue that motion forever.

Newton's Second Law

If an object with mass m experiences a net force, it will undergo acceleration

$$\vec{a} = \frac{\vec{F}_{\text{net}}}{m}$$

where $\vec{F}_{\text{net}} = \vec{F}_1 + \vec{F}_2 + \vec{F}_3 + \cdots$ is the vector sum of all the individual forces acting on the object.

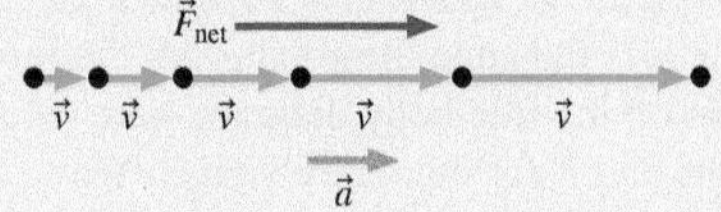

The second law tells us the connection between force and motion.

IMPORTANT CONCEPTS

Force is a push or a pull on an object.

- Force is a vector, with a magnitude and a direction.
- A force requires an agent.
- A force is either a contact force or a long-range force.

The SI unit of force is the **newton** (N). A 1 N force will cause a 1 kg mass to accelerate at 1 m/s^2.

Net force is the vector sum of all the forces acting on an object.

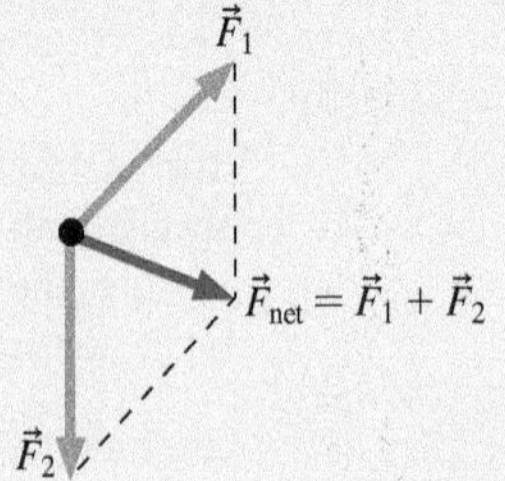

A **vector** can be decomposed into x- and y-components. The signs depend on the direction the vector points.

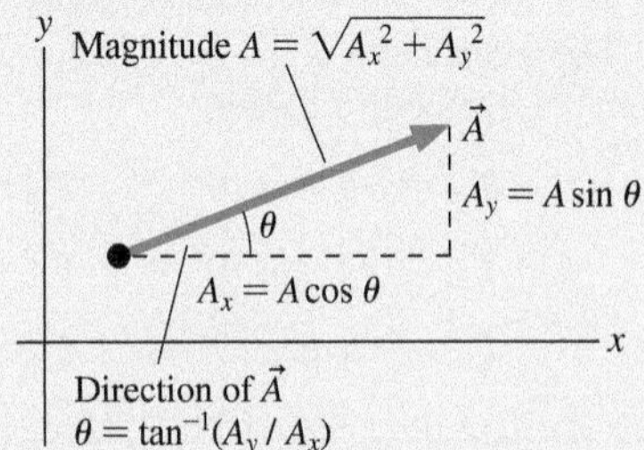

APPLICATIONS

Identifying Forces

Forces are identified by locating the points where other objects touch the object of interest. These are the points where contact forces are exerted. In addition, objects are acted on by the long-range weight force.

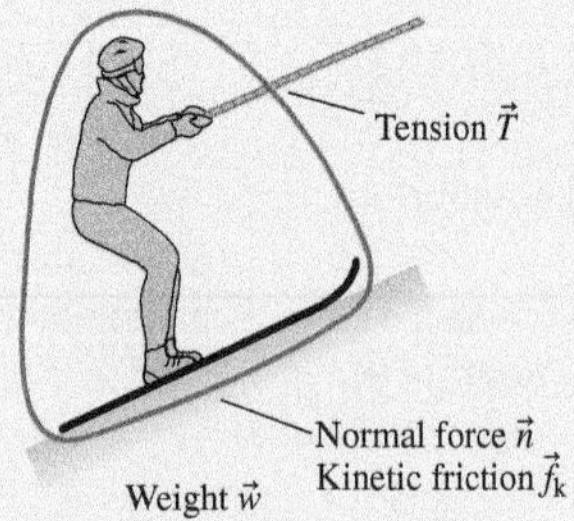

Free-Body Diagrams

A **free-body diagram** represents the object as a particle at the origin of a coordinate system. Force vectors are drawn with their tails on the particle. The net force vector is drawn beside the diagram. The components of the force vectors can be read directly from the free-body diagram.

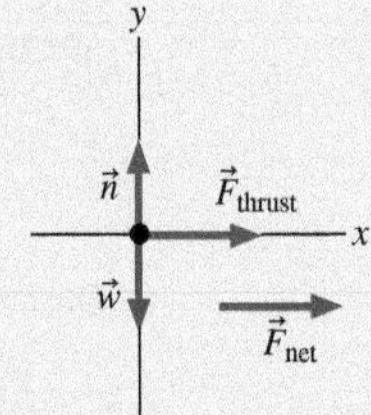

LEARNING OBJECTIVES After studying this chapter, you should be able to:

- Reason with Newton's first law. *Conceptual Questions 4.1, 4.3, 4.4, 4.12; Problem 4.1*
- Recognize and identify the forces acting on an object. *Conceptual Questions 4.4, 4.7, 4.8; Problems 4.7–4.10*
- Understand the connection between force and motion. *Conceptual Questions 4.2, 4.10, 4.13; Problems 4.11–4.15*
- Draw and analyze a free-body diagram. *Conceptual Questions 4.6, 4.7; Problems 4.27, 4.28, 4.31–4.35*
- Work with vectors and use vector math. *Conceptual Question 4.13; Problems 4.4–4.6, 4.37, 4.42–4.45*
- Solve equilibrium and dynamics problems using Newton's second law. *Conceptual Questions 4.10, 4.11; Problems 4.48, 4.50, 4.51, 4.53*

STOP TO THINK ANSWERS

Stop to Think 4.1: C.

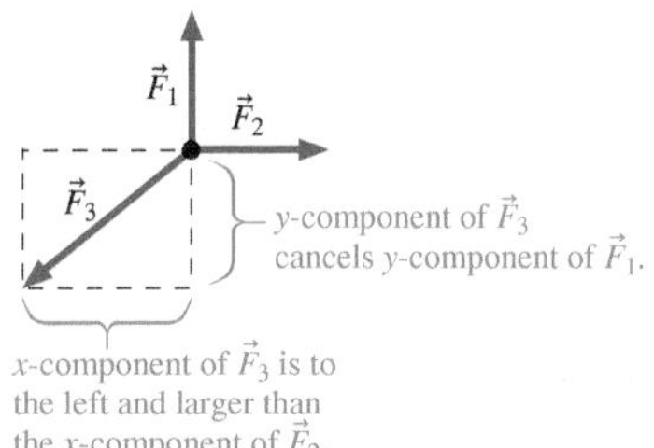
$\vec{F}_1$
$\vec{F}_2$
$\vec{F}_3$
y-component of $\vec{F}_3$ cancels y-component of $\vec{F}_1$.
x-component of $\vec{F}_3$ is to the left and larger than the x-component of $\vec{F}_2$.

Stop to Think 4.2: C. The frog isn't moving, so a static friction force is keeping it at rest. If there was no friction, the weight force would cause the frog to slide down the slope. The static friction force opposes this, so it must be directed up the slope.

Stop to Think 4.3: B. Acceleration is proportional to force, so doubling the tension doubles the acceleration of the original glider from 2 m/s^2 to 4 m/s^2. But acceleration is also inversely proportional to mass. Doubling the mass cuts the acceleration in half, back to 2 m/s^2.

Stop to Think 4.4: D.

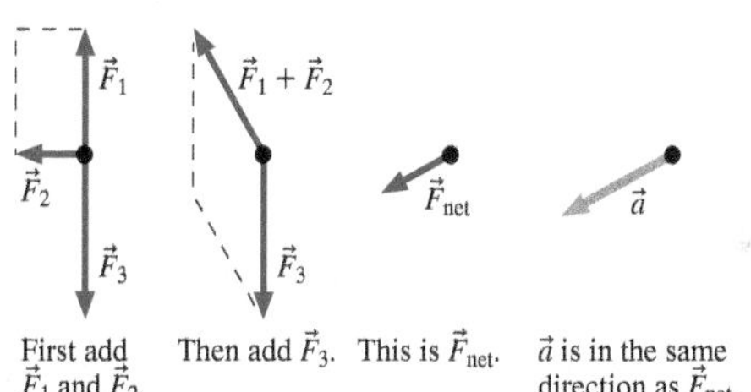
$\vec{F}_1$
$\vec{F}_2$
$\vec{F}_3$
$\vec{F}_1 + \vec{F}_2$
$\vec{F}_3$
$\vec{F}_{net}$
$\vec{a}$
First add $\vec{F}_1$ and $\vec{F}_2$.
Then add $\vec{F}_3$.
This is $\vec{F}_{net}$.
$\vec{a}$ is in the same direction as $\vec{F}_{net}$.

Stop to Think 4.5: A, B, and D. Friction and the normal force are the only contact forces. Nothing is touching the rock to provide a "force of the kick." We've agreed to ignore air resistance unless a problem specifically calls for it.

Stop to Think 4.6: C. The acceleration vector points downward as the elevator slows. $\vec{F}_{net}$ points in the same direction as $\vec{a}$, so $\vec{F}_{net}$ also points downward. This will be true if the tension is less than the weight: $T < w$.

Stop to Think 4.7: C. The vector stretches 4 units to the left and 2 units upward.

QUESTIONS

Conceptual Questions

1. If an object is not moving, does that mean that there are no forces acting on it? Explain.
2. If you know all of the forces acting on a moving object, can you tell in which direction the object is moving? If the answer is Yes, explain how. If the answer is No, give an example.
3. When your car accelerates away from a stop sign, you feel like you're being pushed back into your seat. Can you identify the force that is pushing you back? If not, why do you feel like you're being pushed back?
4. Three arrows are shot horizontally. They have left the bow and are traveling parallel to the ground as shown in Figure Q4.4. Air resistance is negligible. Rank in order, from largest to smallest, the magnitudes of the *horizontal* forces F_1, F_2, and F_3 acting on the arrows. Some may be equal. State your reasoning.

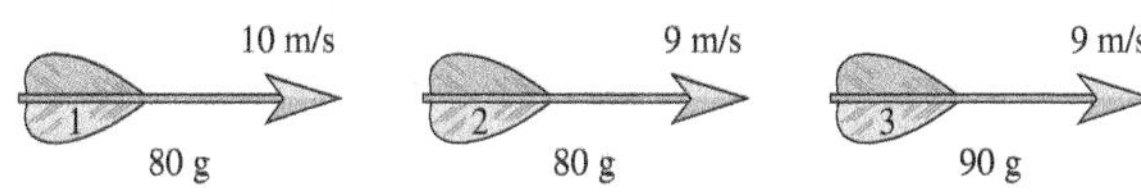
10 m/s
1
80 g
9 m/s
2
80 g
9 m/s
3
90 g

FIGURE Q4.4

Problem difficulty is labeled as | (straightforward) to ||||| (challenging). Problems labeled INT integrate significant material from earlier chapters; BIO are of biological or medical interest; CALC require calculus to solve.

5. Here's a great everyday use of the physics described in this chapter. If you are trying to get ketchup out of the bottle, the best way to do it is to turn the bottle upside down and give the bottle a sharp *upward* smack, forcing the bottle rapidly upward. Think about what subsequently happens to the ketchup, which is initially at rest, and use Newton's first law to explain why this technique is so successful.

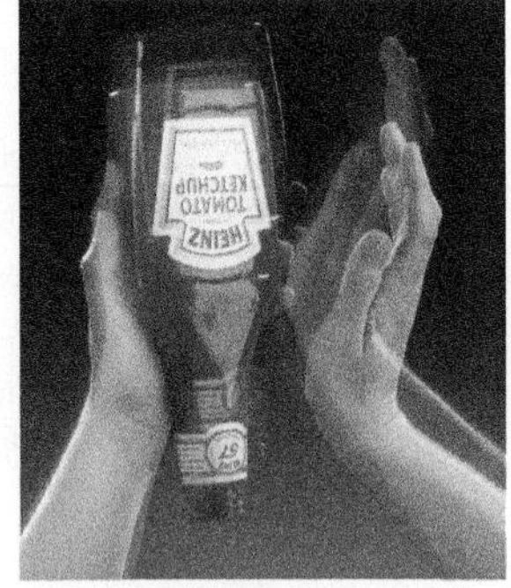

6. An elevator suspended by a cable is descending at constant velocity. How many force vectors would be shown on a free-body diagram? Name them.
7. A compressed spring is pushing a block across a rough horizontal table. How many force vectors would be shown on a free-body diagram? Name them.
8. You toss a ball straight up in the air. Immediately after you let go of it, what force or forces are acting on the ball? For each force you name, (a) state whether it is a contact force or a long-range force and (b) identify the agent of the force.
9. a. Give an example of the motion of an object in which the frictional force on the object is directed opposite to the motion.
   b. Give an example of the motion of an object in which the frictional force on the object is in the same direction as the motion.
10. Suppose you are an astronaut in deep space, far from any source of gravity. You have two objects that look identical, but one has a large mass and the other a small mass. How can you tell the difference between the two?
11. Jonathan accelerates away from a stop sign. His eight-year-old daughter sits in the passenger seat. On whom does the back of the seat exert a greater force? Or are the forces the same?
12. Figure Q4.12 shows a hollow tube forming three-quarters of a circle. It is lying flat on a table. A ball is shot through the tube at high speed. As the ball emerges from the other end, does it follow path A, path B, or path C? Explain.

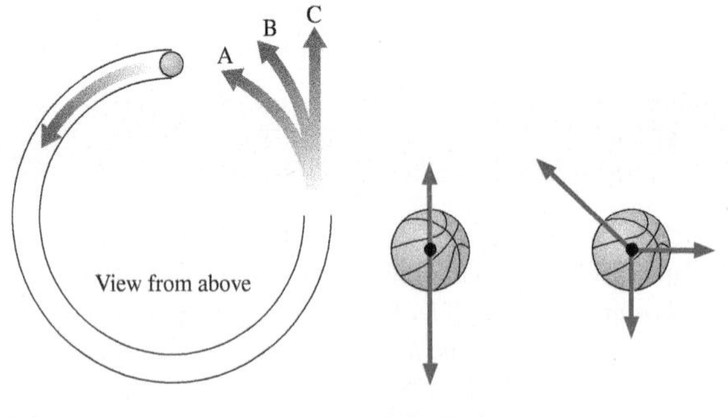

FIGURE Q4.12 FIGURE Q4.13

13. Which, if either, of the basketballs in Figure Q4.13 are in equilibrium? Explain.

Multiple-Choice Questions

14. | A block has acceleration a when pulled by a string. If two identical blocks are glued together and pulled with twice the original force, their acceleration will be

    A. $\frac{1}{4}a$ B. $\frac{1}{2}a$ C. a D. $2a$ E. $4a$

15. | A 5.0 kg block has an acceleration of 0.20 m/s^2 when a force is exerted on it. A second block has an acceleration of 0.10 m/s^2 when subject to the same force. What is the mass of the second block?
    A. 10 kg
    B. 5.0 kg
    C. 2.5 kg
    D. 7.5 kg
16. | Tennis balls experience a large drag force. A tennis ball is hit so that it goes straight up and then comes back down. The direction of the drag force is
    A. Always up.
    B. Up and then down.
    C. Always down.
    D. Down and then up.
17. || A group of students is making model cars that will be propelled by model rocket engines. These engines provide a nearly constant thrust force. The cars are light—most of the weight comes from the rocket engine—and friction and drag are very small. As the engine fires, it uses fuel, so it is much lighter at the end of the run than at the start. A student ignites the engine in a car, and the car accelerates. As the fuel burns and the car continues to speed up, the magnitude of the acceleration will
    A. Increase.
    B. Stay the same.
    C. Decrease.
18. | A person gives a box a shove so that it slides up a ramp, then reverses its motion and slides down. The direction of the force of friction is
    A. Always down the ramp.
    B. Up the ramp and then down the ramp.
    C. Always down the ramp.
    D. Down the ramp and then up the ramp.
19. || Craig is trying to push a heavy crate up a wooden ramp, but the crate won't budge. The direction of the static friction force acting on the crate is
    A. Up the ramp.
    B. Down the ramp.
    C. The static friction force is zero.
    D. There is not enough information to tell.
20. | As shown in the chapter, scallops use jet propulsion to move
BIO from one place to another. Their shells make them denser than water, so they normally rest on the ocean floor. If a scallop wishes to remain stationary, hovering a fixed distance above the ocean floor, it must eject water _____ so that the thrust force on the scallop is _____.
    A. upward, upward
    B. upward, downward
    C. downward, upward
    D. downward, downward
21. | Figure Q4.21 shows block A sitting on top of block B. A constant force $\vec{F}$ is exerted on block B, causing block B to accelerate to the right. Block A rides on block B without slipping. Which statement is true?

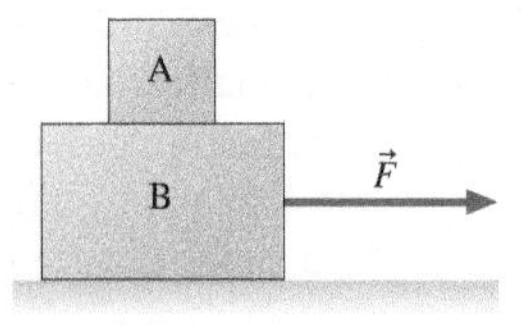

FIGURE Q4.21

    A. Block B exerts a friction force on block A, directed to the left.
    B. Block B exerts a friction force on block A, directed to the right.
    C. Block B does not exert a friction force on block A.

22. || Which combination of the vectors shown in Figure Q4.22 has the largest magnitude?
A. $\vec{A} + \vec{B} + \vec{C}$
B. $\vec{B} + \vec{A} - \vec{C}$
C. $\vec{A} - \vec{B} + \vec{C}$
D. $\vec{B} - \vec{A} - \vec{C}$

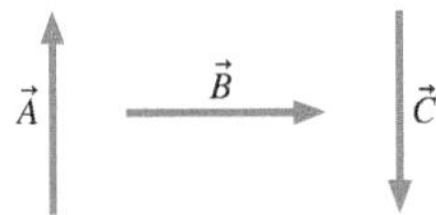

FIGURE Q4.22

23. || Two vectors appear as in Figure Q4.23. Which combination points directly to the left?
A. $\vec{P} + \vec{Q}$
B. $\vec{P} - \vec{Q}$
C. $\vec{Q} - \vec{P}$
D. $-\vec{Q} - \vec{P}$

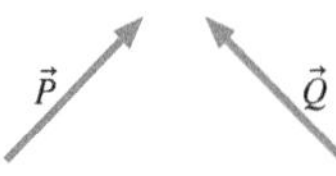

FIGURE Q4.23

PROBLEMS

Section 4.1 Motion and Forces: Newton's First Law

1. | Whiplash injuries during an automobile accident are caused by the inertia of the head. If someone is wearing a seatbelt, her body will tend to move with the car seat. However, her head is free to move until the neck restrains it, causing damage to the neck. Brain damage can also occur.
BIO
Figure P4.1 shows two sequences of head and neck motion for a passenger in an auto accident. One corresponds to a head-on collision, the other to a rear-end collision. Which is which? Explain.

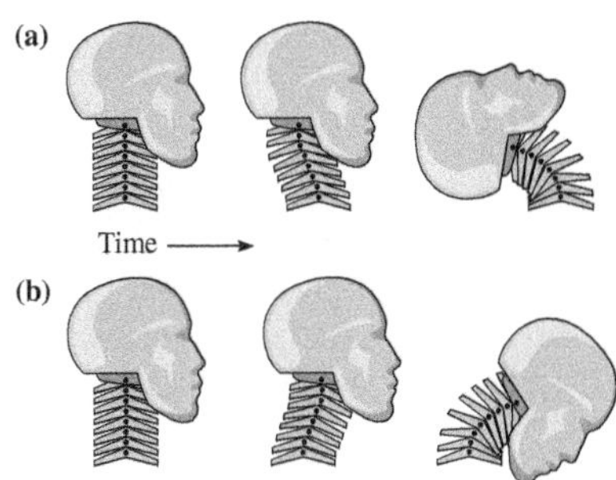

FIGURE P4.1

2. | An automobile has a head-on collision. A passenger in the car experiences a compression injury to the brain. Is this injury most likely to be in the front or rear portion of the brain? Explain.
BIO

3. | In a head-on collision, an infant is much safer in a child safety seat when the seat is installed facing the rear of the car. Explain.
BIO

Problems 4 through 6 show two forces acting on an object at rest. Redraw the diagram, then add a third force that results in a net force of zero. Label the new force $\vec{F}_3$.

4. ||
5. ||
6. ||

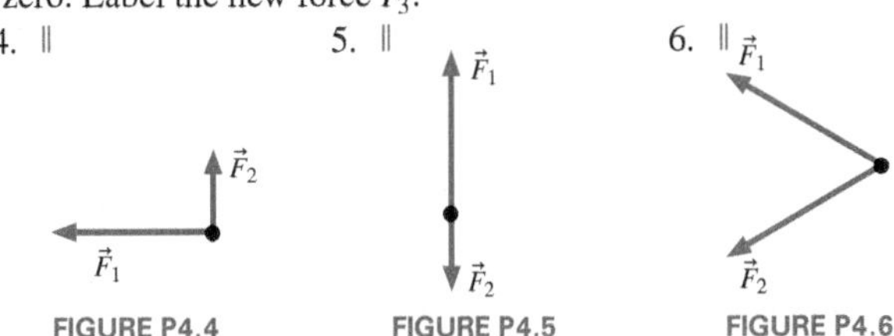

FIGURE P4.4 FIGURE P4.5 FIGURE P4.6

Section 4.2 Force

Section 4.3 A Short Catalog of Forces

7. || A car is parked on a steep hill. Identify the forces on the car.
8. | A baseball player is sliding into second base. Identify the forces on the baseball player.
9. || A jet plane is speeding down the runway during takeoff. Air resistance is not negligible. Identify the forces on the jet.
10. || An arrow has just been shot from a bow and is now traveling horizontally. Air resistance is not negligible. Identify the forces on the arrow.

Section 4.4 What Do Forces Do?

11. | A constant force applied to object A causes it to accelerate at 5 m/s^2. The same force applied to object B causes an acceleration of 3 m/s^2. Applied to object C, it causes an acceleration of 8 m/s^2.
a. Which object has the largest mass?
b. Which object has the smallest mass?
c. What is the ratio of mass A to mass B (m_A/m_B)?
12. || A compact car has a maximum acceleration of 4.0 m/s^2 when it carries only the driver and has a total mass of 1200 kg. What is its maximum acceleration after picking up four passengers and their luggage, adding an additional 400 kg of mass?
13. | A constant force is applied to an object, causing the object to accelerate at 8.0 m/s^2. What will the acceleration be if
a. The force is doubled?
b. The object's mass is doubled?
c. The force and the object's mass are both doubled?
d. The force is doubled and the object's mass is halved?
14. ||| Figure P4.14 shows acceleration graphs for two objects pulled by rubber bands. Each rubber band exerts the same tension force. What is the mass ratio m_1/m_2?

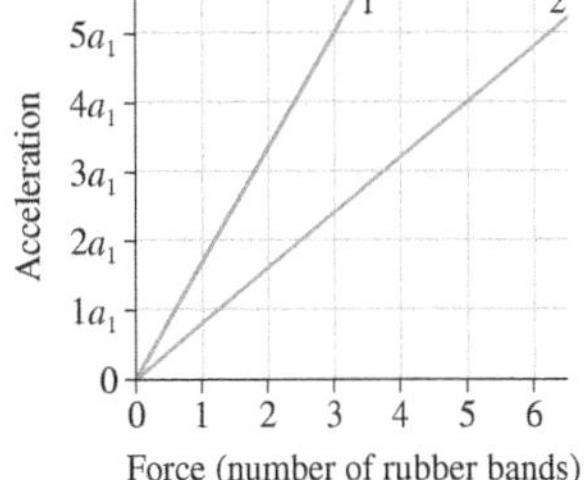

FIGURE P4.14

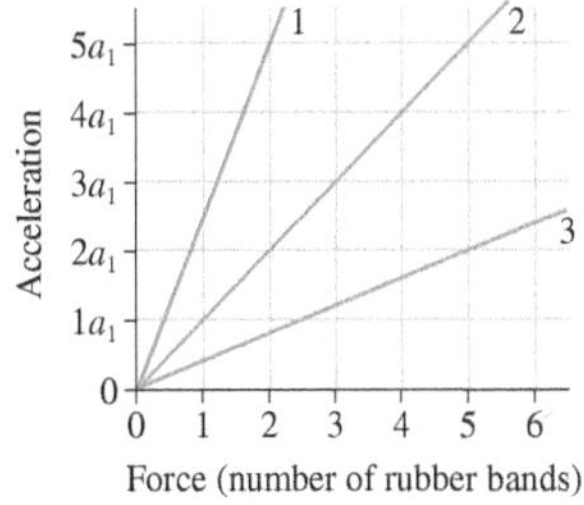

FIGURE P4.15

15. ||| Figure P4.15 shows an acceleration graph for three objects pulled by rubber bands. Each rubber band exerts the same tension force. The mass of object 2 is 0.20 kg. What are the masses of objects 1 and 3? Explain your reasoning.
16. ||| A man pulling an empty wagon causes it to accelerate at 1.4 m/s^2. What will the acceleration be if he pulls with the same force when the wagon contains a child whose mass is three times that of the wagon?

Section 4.5 Newton's Second Law

17. | Redraw the motion diagram shown in Figure P4.17, then INT draw a vector beside it to show the direction of the net force acting on the object. Explain your reasoning.

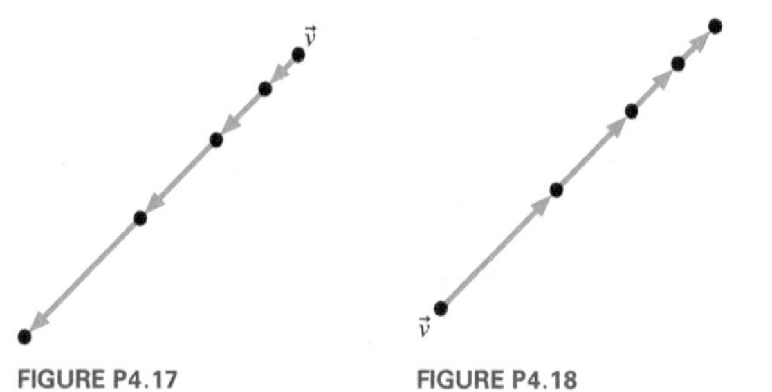

FIGURE P4.17 FIGURE P4.18

18. | Redraw the motion diagram shown in Figure P4.18, then INT draw a vector beside it to show the direction of the net force acting on the object. Explain your reasoning.

19. || Figure P4.19 shows an acceleration graph for a 200 g object. What force values go in the blanks on the horizontal scale?

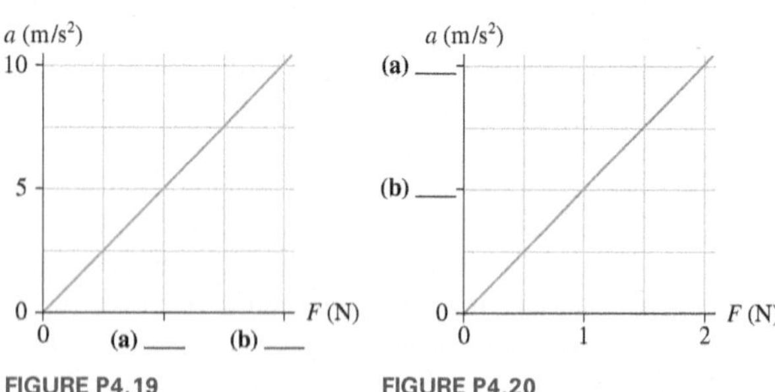

FIGURE P4.19 FIGURE P4.20

20. || Figure P4.20 shows an acceleration graph for a 500 g object. What acceleration values go in the blanks on the vertical scale?

21. || Figure P4.21 shows the acceleration of objects of different mass that experience the same force. What is the magnitude of the force?

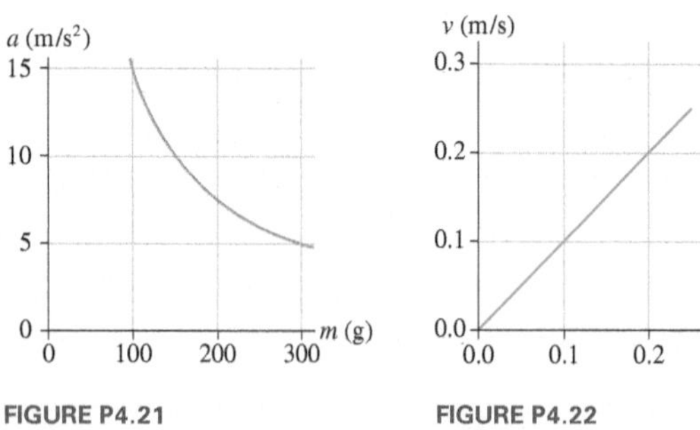

FIGURE P4.21 FIGURE P4.22

22. || Scallops eject water from their shells to provide a thrust BIO force. The graph shows a smoothed graph of actual data for the INT initial motion of a 25 g scallop speeding up to escape a predator. What is the magnitude of the net force needed to achieve this motion?

23. || The Lamborghini Huracán has an initial acceleration of $0.75g$. Its mass, with driver, is 1510 kg. If an 80 kg passenger rode along, what would the car's acceleration be?

24. || In tee-ball, young players use a bat to hit a stationary ball off INT a stand. The 140 g ball has about the same mass as a baseball, but it is larger and softer. In one hit, the ball leaves the bat at 12 m/s after being in contact with the bat for 2.0 ms. Assume constant acceleration during the hit.
    a. What is the acceleration of the ball?
    b. What is the net force on the ball during the hit?

25. || A 55 kg ice skater is gliding along at 3.5 m/s. Five seconds INT later her speed has dropped to 2.9 m/s. What is the magnitude of the kinetic friction acting on her skates?

26. || The *head injury criterion* (HIC) is used to assess the likelihood BIO of head injuries arising from various types of collisions; an HIC greater than about 1000 s is likely to result in severe injuries or even death. The criterion can be written as HIC $= (a_{avg}/g)^{2.5}\Delta t$, where a_{avg} is the average acceleration during the time Δt that the head is being accelerated, and g is the free-fall acceleration. Figure P4.26 shows a simplified graph of the net force on a crash dummy's 4.5 kg head as it hits the airbag during a automobile collision. What is the HIC in this collision?

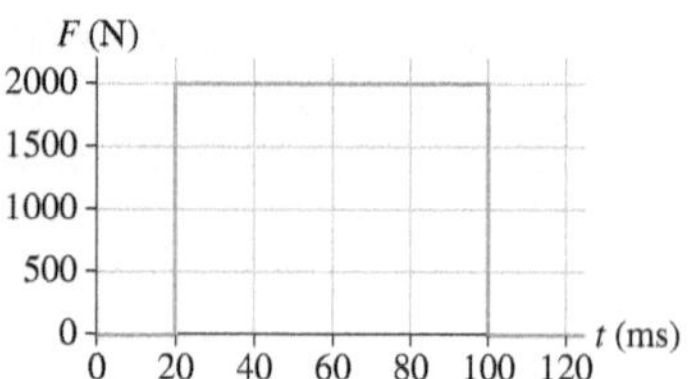

FIGURE P4.26

Section 4.6 Free-Body Diagrams

Problems 27 and 28 show a free-body diagram. For each problem, (a) redraw the free-body diagram and (b) write a short description of a real object for which this is the correct free-body diagram. Use the situations described in Conceptual Examples 4.4, 4.5, and 4.6 as models of what a description should be like.

27. | 28. |

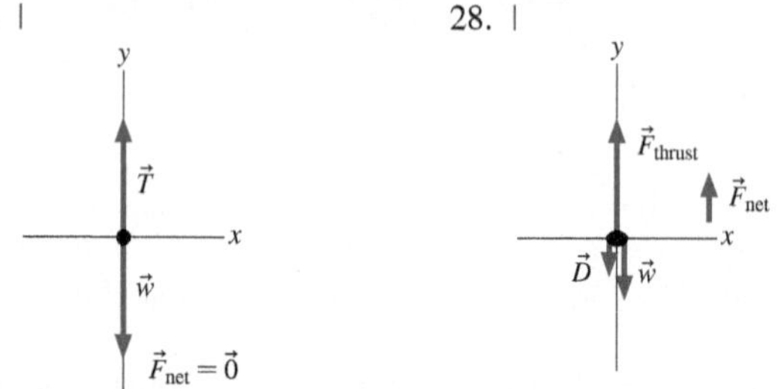

FIGURE P4.27 FIGURE P4.28

29. ||| A student draws the flawed free-body diagram shown in Figure P4.29 to represent the forces acting on a car traveling at constant speed on a level road. Identify the errors in the diagram, then draw a correct free-body diagram for this situation.

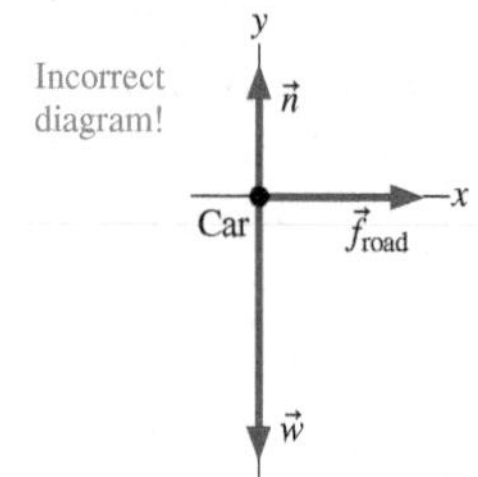

FIGURE P4.29

30. ||| A student draws the flawed free-body diagram shown in Figure P4.30 to represent the forces acting on a golf ball that is traveling upward and to the right a very short time after being hit off the tee. Air resistance is assumed to be relevant. Identify the errors in the diagram, then draw a correct free-body diagram for this situation.

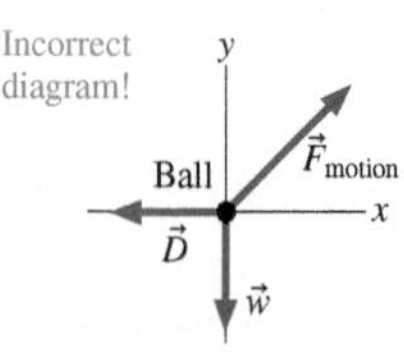

FIGURE P4.30

Problems 31 through 36 describe a situation. For each problem, identify all the forces acting on the object and draw a free-body diagram of the object.

31. | Your car is accelerating from a stop.
32. | Your car is skidding to a stop from a high speed.
33. | Your car is parked on a steep hill.
34. | An ascending elevator, hanging from a cable, is coming to a stop.
35. | A box is being dragged across the floor at a constant speed by a rope pulling horizontally on it. Friction is not negligible.
36. ||| A cannonball has just been shot out of a cannon aimed 45° above the horizontal. Drag cannot be neglected.

Section 4.7 Working with Vectors

37. || Trace the vectors in Figure P4.37 onto your paper. Then use graphical methods to draw the vectors (a) $\vec{A} + \vec{B}$ and (b) $\vec{A} - \vec{B}$.

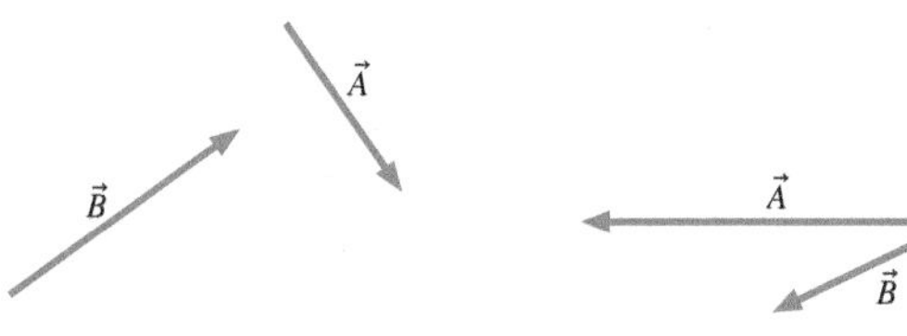

FIGURE P4.37 FIGURE P4.38

38. || Trace the vectors in Figure P4.38 onto your paper. Then use graphical methods to draw the vectors (a) $\vec{A} + \vec{B}$ and (b) $\vec{A} - \vec{B}$.
39. || Trace the vectors in Figure P4.39 onto your paper. Then draw the vector $\vec{C}$ such that $\vec{A} + \vec{B} + \vec{C} = 0$.

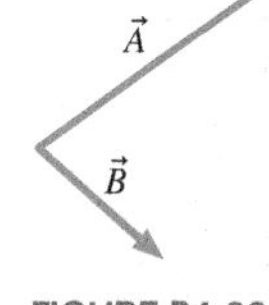

FIGURE P4.39

40. || Two vectors $\vec{A}$ and $\vec{B}$ are at right angles to each other. The magnitude of $\vec{A}$ is 1. What should be the length of $\vec{B}$ so that the magnitude of their vector sum is 2?
41. || A position vector with magnitude 10 m points to the right and up. Its x-component is 6.0 m. What is the value of its y-component?
42. ||| A velocity vector 40° above the positive x-axis has a y-component of 10 m/s. What is the value of its x-component?
43. || A cannon tilted upward at 30° fires a cannonball with a speed of 100 m/s. At that instant, what is the component of the cannonball's velocity parallel to the ground?
44. || Draw each of the following vectors, then find its x- and y-components.
    a. $\vec{d} = (100 \text{ m}, 45° \text{ below } +x\text{-axis})$
    b. $\vec{v} = (300 \text{ m/s}, 20° \text{ above } +x\text{-axis})$
    c. $\vec{a} = (5.0 \text{ m/s}^2, -y\text{-direction})$
45. || Draw each of the following vectors, then find its x- and y-components.
    a. $\vec{d} = (2.0 \text{ km}, 30° \text{ left of } +y\text{-axis})$
    b. $\vec{v} = (5.0 \text{ cm/s}, -x\text{-direction})$
    c. $\vec{a} = (10 \text{ m/s}^2, 40° \text{ left of } -y\text{-axis})$
46. || Each of the following vectors is given in terms of its x- and y-components. Draw the vector, label an angle that specifies the vector's direction, then find the vector's magnitude and direction.
    a. $v_x = 20$ m/s, $v_y = 40$ m/s
    b. $a_x = 2.0$ m/s², $a_y = -6.0$ m/s²
47. || Each of the following vectors is given in terms of its x- and y-components. Draw the vector, label an angle that specifies the vector's direction, then find the vector's magnitude and direction.
    a. $v_x = 10$ m/s, $v_y = 30$ m/s
    b. $a_x = 20$ m/s², $a_y = 10$ m/s²

Section 4.8 Using Newton's Laws

48. || The three ropes in Figure P4.48 are tied to a small, very light ring. Two of the ropes are anchored to walls at right angles, and the third rope pulls as shown. What are T_1 and T_2, the magnitudes of the tension forces in the first two ropes?

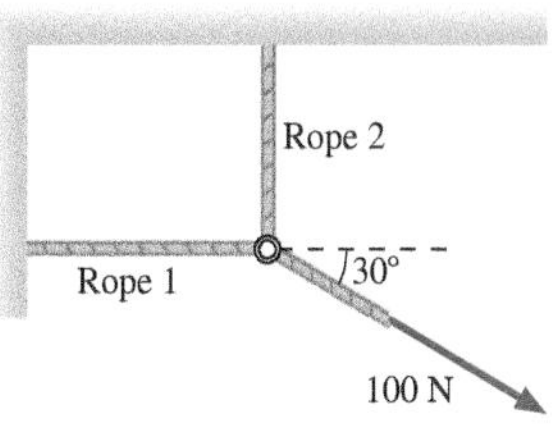

FIGURE P4.48

49. ||| A construction crew would like to support a 1000 kg steel beam with two angled ropes as shown in Figure P4.49. Their rope can support a maximum tension of 5600 N. Is this rope strong enough to do the job?

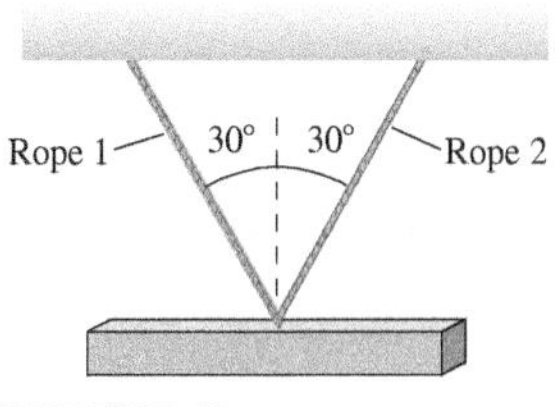

FIGURE P4.49

50. || An early submersible craft for deep-sea exploration was raised and lowered by a cable from a ship. When the craft was stationary, the tension in the cable was 6000 N. When the craft was lowered or raised at a steady rate, the motion through the water added an 1800 N drag force.
    a. What was the tension in the cable when the craft was being lowered to the seafloor?
    b. What was the tension in the cable when the craft was being raised from the seafloor?
51. || The forces in Figure P4.51 are acting on a 2.0 kg object. What is a_x, the x-component of the object's acceleration?

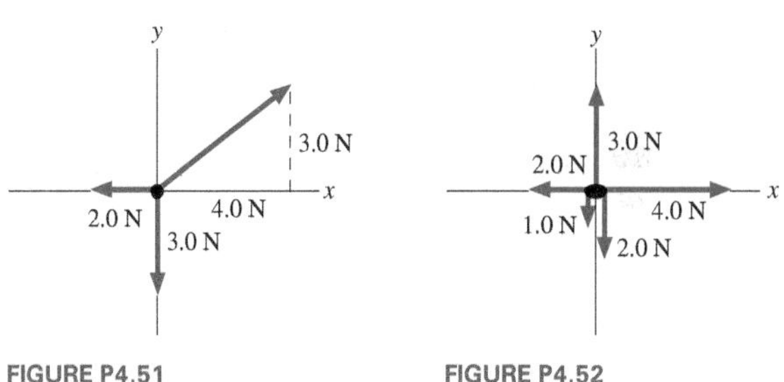

FIGURE P4.51 FIGURE P4.52

52. | The forces in Figure P4.52 are acting on a 2.0 kg object. Find the values of a_x and a_y, the x- and y-components of the object's acceleration.
53. | A horizontal rope is tied to a 50 kg box on frictionless ice. What is the tension in the rope if
    a. The box is at rest?
    b. The box moves at a steady 5.0 m/s?
    c. The box has $v_x = 5.0$ m/s and $a_x = 5.0$ m/s²?

General Problems

Problems 54 through 60 describe a situation. For each problem, draw a motion diagram, a force identification diagram, and a free-body diagram.

54. || An elevator, suspended by a single cable, has just left the tenth floor and is speeding up as it descends toward the ground floor.
55. ||| A rocket is being launched straight up. Air resistance is not negligible.
56. ||| A jet plane is speeding down the runway during takeoff. Air resistance is not negligible.
57. || You've slammed on the brakes and your car is skidding to a stop while going down a 20° hill.
58. || A cave explorer is being lowered into a vertical shaft by a rope. He is slowing down as he nears the bottom.

59. ||| A spring-loaded gun shoots a plastic ball. The trigger has just been pulled and the ball is starting to move down the barrel. The barrel is horizontal.
60. || A person on a bridge throws a rock straight down toward the water. The rock has just been released.

61. ||| Figure P4.61 shows vectors $\vec{A}$ and $\vec{B}$. What are the components C_x and C_y of vector $\vec{C}$ such that $\vec{A} + \vec{B} + \vec{C} = \vec{0}$?

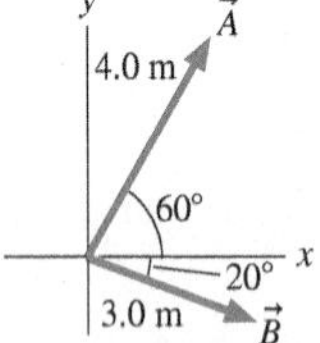

FIGURE P4.61

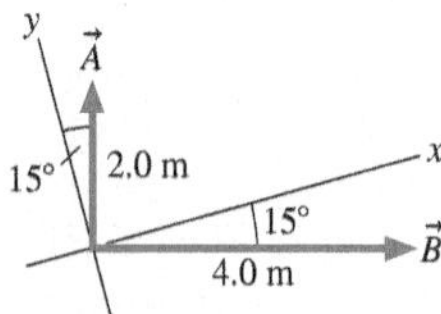

FIGURE P4.62

62. ||| Figure P4.62 shows vectors $\vec{A}$ and $\vec{B}$. What are the components D_x and D_y of vector $\vec{D} = 2\vec{A} + 3\vec{B}$?
63. || Bethany, who weighs 560 N, lies in a hammock suspended by ropes tied to two trees. One rope makes an angle of 45° with the ground; the other makes an angle of 30°. Find the tension in each of the ropes.
64. || A 65 kg student is walking on a slackline, a length of webbing stretched between two trees. The line stretches and so has a noticeable sag, as shown in Figure P4.64. At the point where his foot touches the line, the rope applies a tension force in each direction, as shown. What is the tension in the line?

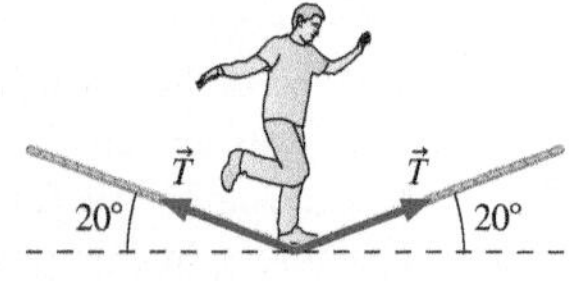

FIGURE P4.64

65. || A football coach sits on a sled while two of his players build their strength by dragging the sled across the field with ropes. The friction force on the sled is 1000 N, the players have equal pulls, and the angle between the two ropes is 20°. How hard must each player pull to drag the coach at a steady 2.0 m/s? Assume that the ropes are parallel to the ground.
66. || A construction worker with a weight of 850 N stands on a roof that is sloped at 20°. What is the magnitude of the normal force of the roof on the worker?
67. || A 50 kg box hangs from a rope. What is the tension in the rope if
    a. The box is at rest?
    b. The box has $v_y = 5.0$ m/s and is speeding up at 5.0 m/s^2?
68. || A car has a mass of 1500 kg. If the driver applies the brakes while on a gravel road, the maximum friction force that the tires can provide without skidding is 7200 N. If the car is moving at 20 m/s, what is the shortest distance in which the car can stop safely?
69. || A study of bluefish locomotion found that their tail motion produces an average thrust of 0.65 N. Fish are streamlined, so drag can be neglected for fish swimming at slow speeds. Suppose a 1.1 kg bluefish that is coasting horizontally at 0.45 m/s suddenly begins tail motion. How fast will it be moving after swimming 1.0 m?
    BIO INT
70. ||| Researchers have measured the acceleration of racing greyhounds as a function of their velocity; a simplified version of their results is shown in Figure P4.70. The acceleration at low speeds is constant and is limited by the fact that any greater acceleration would result in the dog pitching forward because of the force acting on its hind legs during its power stroke. At higher speeds, the dog's acceleration is limited by the maximum power its muscles can provide.
    BIO INT
    a. What is the agent of the force that causes the dog to accelerate?
    b. What is the forward force on a 32 kg dog during its initial acceleration?
    c. How far has the dog run when its speed reaches 4 m/s?

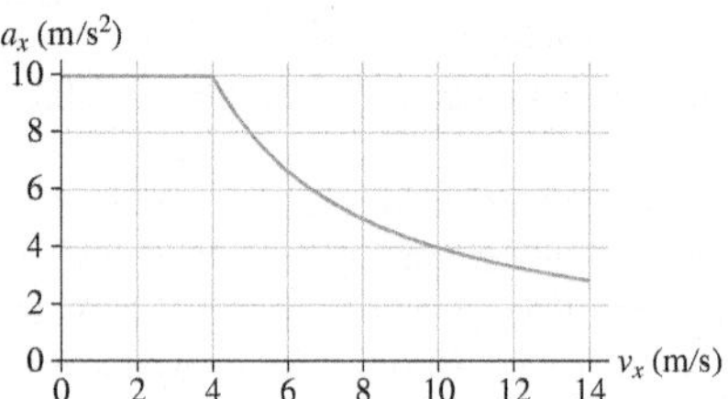

FIGURE P4.70

71. ||| The fastest pitched baseball was clocked at 47 m/s. Model the throw as being caused by a constant, horizontal force over a distance of 1.0 m. What was the force of the throw during this record-setting pitch? A baseball has a mass of 150 g.
    BIO INT
72. ||| The trap-jaw ant, found throughout tropical South America, catches its prey by closing its mandibles *extremely* fast on its victim. Figure P4.72, based on high-speed video, shows the speed of one mandible jaw versus time in *micro*seconds. The speed increases rapidly as the mandibles are closing. They reach full closure at $t = 80\ \mu s$, followed by a rapid deceleration as mechanical structures in the joint stop their motion. Our goal is to determine how much "bite force" the mandibles can exert, and that occurs during the closure phase. We'll model the mandibles as particles—not a great model in this situation but sufficient for an estimate.
    BIO INT
    a. What is the maximum acceleration of a mandible during the closure phase?
    b. A muscle, similar to your jaw muscle, pulls the mandibles closed. What is the maximum force exerted by this muscle? This is the force that could be applied to prey held in the mandibles. The mass of a trap-jaw ant mandible was found to be 130 μg. Be careful with conversions; the SI unit of mass is kg, not g.
    c. The ant's mass is 14 mg. What is the ant's mandible force as a multiple of its weight?

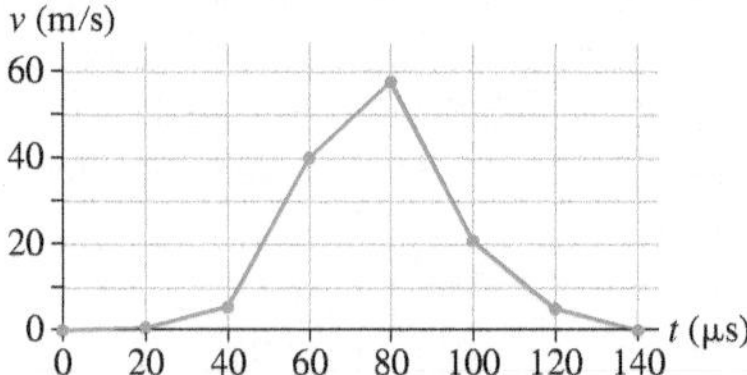

FIGURE P4.72

73. || The jumping ability of the African desert locust was measured by placing the insect on a *force plate*, a platform that can accurately measure the force that acts on it. When the 0.50 g locust jumped straight up, the upward force was measured to follow the curve in Figure P4.73. Notice that the vertical scale is in millinewtons. What was the locust's maximum upward acceleration?
    BIO

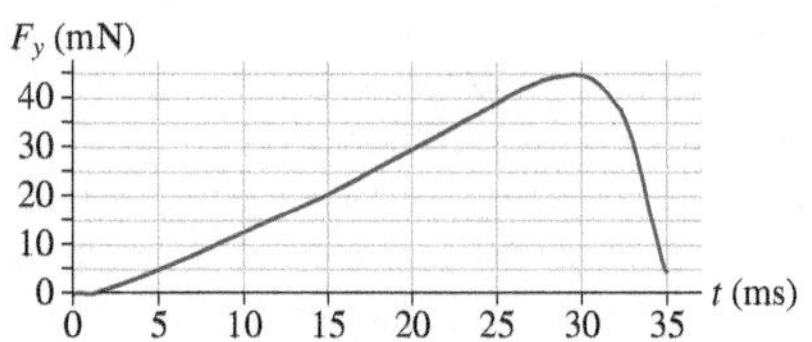

FIGURE P4.73

MCAT-Style Passage Problems

A Simple Solution for a Stuck Car

If your car is stuck in the mud and you don't have a winch to pull it out, you can use a piece of rope and a tree to do the trick. First, you tie one end of the rope to your car and the other to a tree, then pull as hard as you can on the middle of the rope, as shown in Figure P4.74a. This technique applies a force to the car much larger than the force that you can apply directly. To see why the car experiences such a large force, look at the forces acting on the center point of the rope, as shown in Figure P4.74b. The sum of the forces is zero, thus the tension is much greater than the force you apply. It is this tension force that acts on the car and, with luck, pulls it free.

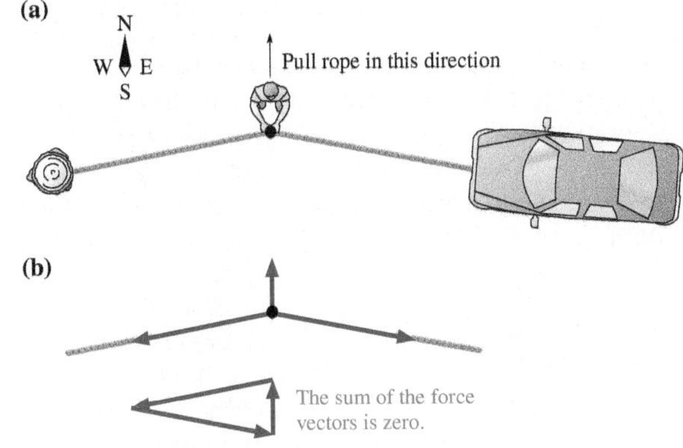
(a)
N
W
E
S
Pull rope in this direction
(b)
The sum of the force vectors is zero.

FIGURE P4.74

74. The sum of the three forces acting on the center point of the rope is assumed to be zero because
    A. This point has a very small mass.
    B. Tension forces in a rope always cancel.
    C. This point is not accelerating.
    D. The angle of deflection is very small.
75. When you are pulling on the rope as shown, what is the approximate direction of the tension force on the tree?
    A. North
    B. South
    C. East
    D. West
76. Assume that you are pulling on the rope but the car is not moving. What is the approximate direction of the force of the mud on the car?
    A. North
    B. South
    C. East
    D. West
77. Suppose your efforts work, and the car begins to slowly move forward out of the mud at a steady speed. As it does so, the force of the rope on the car is
    A. Zero.
    B. Less than the force of the mud on the car.
    C. Equal to the force of the mud on the car.
    D. Greater than the force of the mud on the car.

5 Interacting Systems

Flying squirrels don't really fly. Instead, the drag force on the parachute-like membrane that stretches from wrist to ankle allows them to glide for long distances.

LOOKING AHEAD

Working with Forces

A skydiver's maximum speed is determined by a balance between the weight force and the drag force due to the air.

You'll learn to solve dynamics problems by using mathematical expressions for the various forces that you met in Chapter 4.

Springs

A kangaroo's hop is powered by spring-like tendons in its back legs. The tendons stretch as the kangaroo lands, then launch it back into the air.

Many biological materials are *elastic.* A spring is the model for elastic forces, and you'll see that the force of a spring is given by a simple law.

Interactions

The hammer exerts a downward force on the nail. Surprisingly, the nail exerts an equal upward force on the hammer.

You'll learn how to identify and work with *interaction pairs* of forces and to reason with Newton's third law.

GOAL To use Newton's laws to solve problems of force and motion.

PHYSICS AND LIFE

Evolutionary Niches Are Shaped by the Physical Environment

Living organisms evolve and adapt to make the most of their interactions with their environment. Plants and animals that live in water have profoundly different body plans and structures than those that live on land—in large part because of the different forces and stresses they experience. Microorganisms that swim through liquids face very different challenges than do larger organisms. They also need to attach to and detach from surfaces, and biologists are just starting to learn how single cells are stimulated and influenced by their attachments to the surrounding matrix. In this chapter, we will explore more deeply the many ways in which objects interact with each other and their surroundings, with a goal of understanding the constraints that the physical environment places on the creatures that live in it.

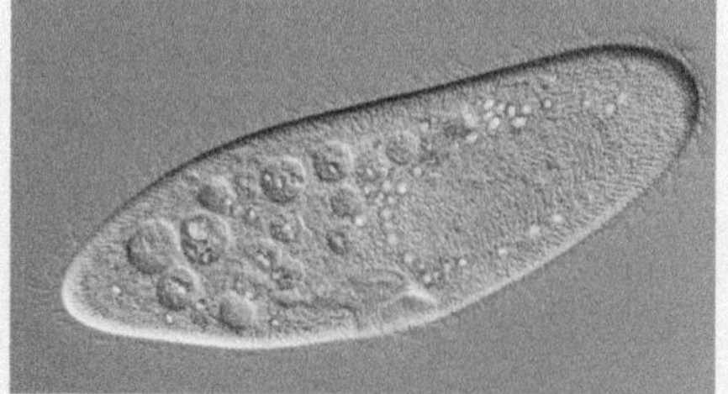

The motion of a swimming *Paramecium* is dominated by the viscous drag forces that it experiences.

5.1 Systems and Interactions

Chapters 2, 3, and 4 have focused on developing tools and fundamental concepts. You've learned to describe motion, to use pictorial and mathematical representations, and to work with vectors. You've also learned that Newton's second law—the link between force and motion—is the key idea that *explains* why objects move as they do. With that background, we can now turn our attention to the physics of specific types of forces that occur in nature. We'll be especially interested in forces that affect living systems, but that's not our exclusive focus; it's also important to understand the motions of balls, cars, and airplanes.

One important idea is that forces are **interactions.** Consider the hammer hitting the nail in FIGURE 5.1a. The hammer exerts a force on the nail as it drives the nail forward. At the same time, the nail exerts a force on the hammer, a force that would dent the hammer if it were made of a very soft material. The hammer and nail *interact.*

In fact, any time an object A pushes or pulls on another object B, B pushes or pulls back on A. If you're sitting, the chair pushes up on you (the normal force) as you press down on the chair. If you're rock climbing, the rope pulls up on you as you pull down on the rope. The two objects *interact* with each other and influence each other. We can describe the interaction by saying that *for every action there is an equal but opposite reaction,* a deep and powerful idea that is often referred to as *Newton's third law.* We will return later in the chapter to examine the implications of Newton's third law.

To start, though, we're going to look at situations where we can focus on how just one of the interacting objects responds to the interaction. For example, we normally want to know what happens to the nail, not to the hammer. We usually want to know how friction with the road slows a car rather than what the car does to the road. To do so, let's define the **system** as the object whose motion we wish to analyze and the **environment** as those objects external to the system. This is a *conceptual framework* for thinking about interactions. FIGURE 5.1b looks again at the hammer and the nail, identifying the nail as the system and the hammer as part of the environment.

Our Chapter 4 discussion of forces said that forces always have agents—something exerting the force. For the moment, we're placing all agents in the environment, and thus all forces that act on the system originate in the environment. The key point for now is that even the simplest force is an interaction between the system and the environment.

FIGURE 5.1 The hammer and the nail interact with each other.

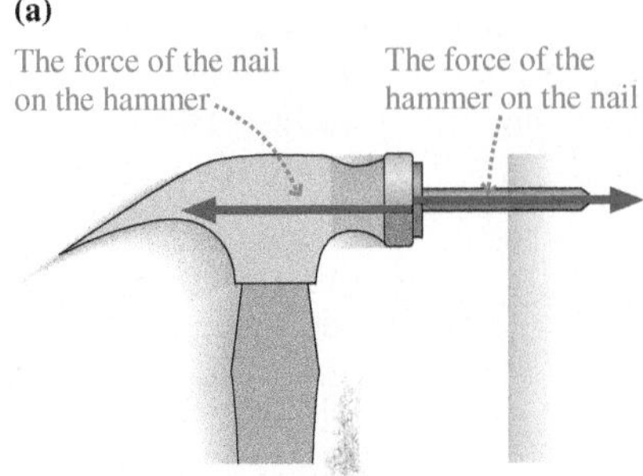

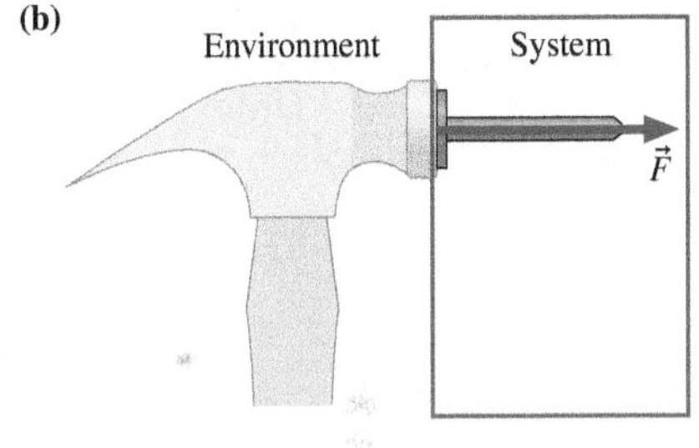

NOTE ▸ Distinguishing between system and environment is an approach that we'll use throughout this text in a wide variety of contexts. In all cases, the system is what we want to understand; we might want to know the system's motion, its thermal properties, or its electric properties. Except in the rare case of an *isolated system,* the system interacts with and is influenced by external factors. The agents of those external factors are what make up the environment. Our use of the word "environment" is quite similar to how it's used in biology to mean the surroundings in which an organism lives and with which it interacts. (In chemistry, "surroundings" is the term often used for this concept.) ◂

5.2 Mass and Weight

When the doctor asks what you weigh, what does she really mean? In everyday speech people rarely distinguish between the terms "weight" and "mass," but in physics understanding the difference is of critical importance.

Mass, you'll recall from Chapter 4, is a quantity that describes an object's inertia, its tendency to resist being accelerated. Loosely speaking, it also describes the amount of matter in an object. Mass, measured in kilograms, is an intrinsic property of an object; it has the same value wherever the object may be and whatever forces might be acting on it.

Weight, however, is a *force.* Specifically, it is the gravitational force exerted on an object by a planet. Thus weight is an *interaction* between an object and the entire planet. Weight is a vector, not a scalar, and the vector's direction is always straight down. Weight is measured in newtons.

FIGURE 5.2 The free-body diagram of an object in free fall.

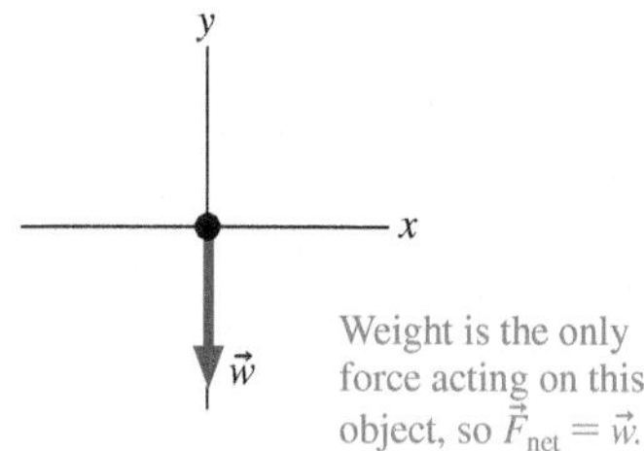

y
x
$\vec{w}$
Weight is the only force acting on this object, so $\vec{F}_{\text{net}} = \vec{w}$.

Mass and weight are not the same thing, but they are related. FIGURE 5.2 shows the free-body diagram of an object in free fall. The *only* force acting on this object is its weight $\vec{w}$, the downward pull of gravity. The object is in free fall, so, as we saw in ◀◀ SECTION 3.6, the acceleration is vertical, with $a_y = -g$, where g is the free-fall acceleration of 9.80 m/s^2. Newton's second law for this object is thus

$$\sum F_y = -w = -mg$$

which tells us that

$$w = mg \tag{5.1}$$

The magnitude of the weight force, which we call simply "the weight," is directly proportional to the mass, with g as the constant of proportionality.

NOTE ▶ Although we derived the relationship between mass and weight for an object in free fall, the weight of an object is *independent* of its state of motion. Equation 5.1 holds for an object at rest on a table, sliding horizontally, or moving in any other way. ◀

Weight loss On the moon, astronaut John Young jumped 2 feet straight up, despite his spacesuit that weighs 180 pounds on earth. On the moon, where $g = 1.6$ m/s^2, he and his suit together weighed only 60 pounds.

Because an object's weight depends on g, and the value of g varies from planet to planet, weight is not a fixed, constant property of an object. The value of g at the surface of the moon is about one-sixth its earthly value, so an object on the moon would have only one-sixth its weight on earth. The object's weight on Jupiter would be greater than its weight on earth. Its mass, however, would be the same. The amount of matter has not changed, only the gravitational force exerted on that matter.

So, when the doctor asks what you weigh, she really wants to know your *mass.* That's the amount of matter in your body. You can't really "lose weight" by going to the moon, even though you would weigh less there!

We need to make a clarification here. It's likely that you give your weight in pounds, but the pound is the unit of *force,* not mass, in the English system. (We noted in Chapter 4 that the pound is defined as 1 lb = 4.45 N.) You might then "convert" this to kilograms. But a kilogram is a unit of mass, not a unit of force. An object that weighs 1 pound, meaning $w = mg = 4.45$ N, has a mass of

$$m = \frac{w}{g} = \frac{4.45\ \text{N}}{9.80\ \text{m/s}^2} = 0.454\ \text{kg} = 454\ \text{g}$$

This calculation is different from converting, for instance, feet to meters. Both feet and meters are units of length, and it's always true that 1 m = 3.28 ft. When you "convert" from pounds to kilograms, you are determining the mass that has a certain weight—two fundamentally different quantities—and this calculation depends on the value of g. But we are usually working on or close to the earth, where we assume that $g = 9.80$ m/s^2. In this case, a given mass always corresponds to the same weight, and we can use the relationships listed in TABLE 5.1.

TABLE 5.1 **Mass, weight, and force**

Conversion between force units:
1 pound = 4.45 N
1 N = 0.225 pound
Correspondence between mass and weight on earth:
1 kg ↔ 2.20 lb
1 lb ↔ 0.454 kg = 454 g

EXAMPLE 5.1 Typical masses and weights

What are the weight, in N, and the mass, in kg, of a 90 pound gymnast, a 150 pound professor, and a 240 pound football player?

PREPARE We can use the conversions and correspondences in Table 5.1.

SOLVE We use the correspondence between mass and weight just as we use the conversion factor between different forces:

$$w_{\text{gymnast}} = 90\ \text{lb} \times \frac{4.45\ \text{N}}{1\ \text{lb}} = 400\ \text{N} \qquad m_{\text{gymnast}} = 90\ \text{lb} \times \frac{0.454\ \text{kg}}{1\ \text{lb}} = 41\ \text{kg}$$

$$w_{\text{prof}} = 150\ \text{lb} \times \frac{4.45\ \text{N}}{1\ \text{lb}} = 670\ \text{N} \qquad m_{\text{prof}} = 150\ \text{lb} \times \frac{0.454\ \text{kg}}{1\ \text{lb}} = 68\ \text{kg}$$

$$w_{\text{player}} = 240\ \text{lb} \times \frac{4.45\ \text{N}}{1\ \text{lb}} = 1070\ \text{N} \qquad m_{\text{player}} = 240\ \text{lb} \times \frac{0.454\ \text{kg}}{1\ \text{lb}} = 110\ \text{kg}$$

ASSESS We can use the information in this problem to assess the results of future problems. If you get an answer of 1000 N, you now know that this is approximately the weight of a football player, which can help with your assessment.

Apparent Weight

The weight of an object is the force of gravity on that object. You may never have thought about it, but gravity is not a force that you can feel or sense directly. Your *sensation* of weight—how heavy you *feel*—is due to *contact forces* supporting you and the response of your *vestibular system* to any motion. (The vestibular system is the sensory system in the inner ear that provides your brain with information about motion and spatial orientation.) As you read this, your sensation of weight is due to the normal force exerted on you by the chair in which you are sitting. The chair's surface touches you and activates nerve endings in your skin. You sense the magnitude of this force, and this is your sensation of weight. When you stand, you feel the contact force of the floor pushing against your feet. If you are hanging from a rope, you feel the friction force between the rope and your hands.

Let's define your **apparent weight** w_{app} in terms of the force you feel:

$$w_{app} = \text{magnitude of supporting contact forces} \quad (5.2)$$

Definition of apparent weight

If you are in equilibrium, your weight and apparent weight are generally the same. But this is not the case if you are accelerating. For example, you feel "heavy" when an elevator you are riding in suddenly accelerates upward, and you feel lighter than normal as the upward-moving elevator brakes to a halt. Your true weight $w = mg$ has not changed during these events, but your *sensation* of your weight has.

Let's look at the details for this case. Imagine a man standing in an elevator as it accelerates upward. As FIGURE 5.3 shows, the only forces acting on the man are the upward normal force of the floor holding him up and the downward weight force. Because the man has an acceleration $\vec{a}$, Newton's second law tells us that there must be a net force acting on him in the direction of $\vec{a}$.

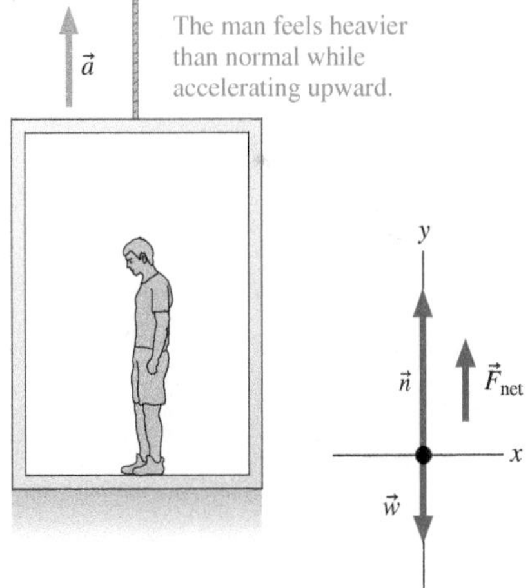

FIGURE 5.3 A man in an accelerating elevator.

Looking at the free-body diagram in Figure 5.3, we see that the y-component of Newton's second law is

$$\sum F_y = n_y + w_y = n - w = ma_y \quad (5.3)$$

where m is the man's mass. Solving Equation 5.3 for n gives

$$n = w + ma_y \quad (5.4)$$

The normal force is the contact force supporting the man, so, given the definition of Equation 5.2, we can rewrite Equation 5.4 as

$$w_{app} = w + ma_y$$

For the case shown in Figure 5.3, where the elevator is accelerating upward, $a_y = +a$ and thus $w_{app} = w + ma$. The apparent weight is greater than the man's weight w, so he feels heavier than normal. If, instead, the elevator has a downward acceleration, then $a_y = -a$ and $w_{app} = w - ma$. The apparent weight is less than the man's weight, and he feels lighter than normal.

Apparent weight isn't just a sensation, though. You can measure it with a scale. When you stand on a bathroom scale, the scale reading is the upward force of the scale on you. If you aren't accelerating (usually a pretty good assumption!), the upward force of the scale on you equals your weight, so the scale reading is your true weight. But if you are accelerating, this correspondence may not hold. In the example above, if the man were standing on a bathroom scale in the elevator, the contact force supporting him would be the upward force of the scale. If the elevator is accelerating upward, the scale shows an increased weight: The scale reading is equal to w_{app}.

Apparent weight can be measured by a scale, so it's no surprise that apparent weight has real, physical implications. Astronauts are nearly crushed by their apparent weight during a rocket launch when a is much greater than g. This occurs because the internal contact forces in the body have to be much larger than normal to accelerate the astronaut's limbs and organs. Much of the thrill of amusement park rides, such as roller coasters, comes from rapid changes in your apparent weight.

EXAMPLE 5.2 **Finding a rider's apparent weight in an elevator**

Anjay's mass is 70 kg. He is standing on a scale in an elevator that is moving at 5.0 m/s. As the elevator slows to a stop, the scale reads 750 N. Before it stopped, was the elevator moving up or down? How long did the elevator take to come to rest?

PREPARE The scale reading as the elevator comes to rest, 750 N, is Anjay's apparent weight. Anjay's actual weight is

$$w = mg = (70\text{ kg})(9.80\text{ m/s}^2) = 686\text{ N}$$

This is an intermediate step in the calculation, so we are keeping an extra significant figure.

SOLVE Because Anjay's apparent weight w_{app} is greater than his actual weight w, his acceleration $a_y = (w_{app} - w)/m$ is *positive.* You learned in Chapter 3 that if an object is slowing, as is the case here, then velocity and acceleration have opposite signs. Consequently, the elevator's velocity must be *negative*—it is moving down.

We can use kinematics to find the stopping time. For this step we need the acceleration, which is

$$a_y = \frac{w_{app} - w}{m} = \frac{750\text{ N} - 686\text{ N}}{70\text{ kg}} = 0.91\text{ m/s}^2$$

Then we use the kinematic equation

$$(v_y)_f = (v_y)_i + a_y \Delta t$$

The elevator is initially moving downward, so $(v_y)_i = -5.0$ m/s, and it then comes to a halt, so $(v_y)_f = 0$. We know the acceleration, so the time interval is

$$\Delta t = \frac{(v_y)_f - (v_y)_i}{a_y} = \frac{0 - (-5.0\text{ m/s})}{0.91\text{ m/s}^2} = 5.5\text{ s}$$

ASSESS Think back to your experiences riding elevators. If the elevator is moving downward and then comes to rest, you "feel heavy." This gives us confidence that our analysis of the motion is correct. And 5.0 m/s is a pretty fast elevator: At this speed, the elevator will be passing more than one floor per second. If you've been in a fast elevator in a tall building, you know that 5.5 s is reasonable for the time it takes for the elevator to slow to a stop.

A different physical situation in which our sensation of weight is altered is when we are submerged in water. You may have experienced feeling lighter when you stand in a deep pool. This change in your sensation also arises because there are forces acting in addition to gravity, but this time those forces are exerted by the water; the water itself supports you, leaving a sensation of very little weight. The supportive force from the water is called the *buoyant force,* a topic we'll explore in Chapter 9. This reduced apparent weight is the reason that many water-dwelling animals and plants have very soft bodies; they don't require as strong a skeleton to maintain their body plan.

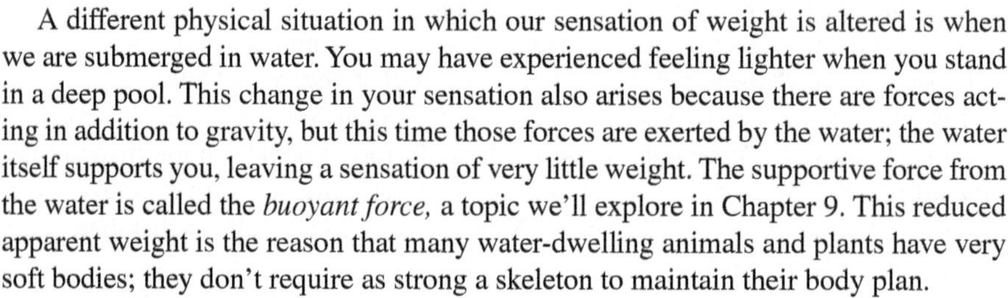

A weightless experience As we learned in Chapter 3, objects undergoing projectile motion are in free fall. This specially adapted plane flies in the same parabolic trajectory as would a projectile with no air resistance. Objects inside, such as these passengers, thus move along a free-fall trajectory. They feel weightless, and they float with respect to the plane's interior until the plane resumes normal flight, up to 30 seconds later.

Weightlessness

Let's return to the elevator example. Suppose Anjay's apparent weight was 0 N—meaning that the scale read *zero.* Could this happen? Recall that Anjay's apparent weight is the contact force supporting him, so we are saying that the upward force from the scale is zero. If we use $w_{app} = 0$ in the equation for the apparent acceleration, we find

$$0 = w + ma_y = mg + ma_y$$

so that $a_y = -g = -9.8\text{ m/s}^2$. This is a case we've seen before—it is free fall! This means that **a person (or any object) in free fall has zero apparent weight.**

Think about this carefully. Imagine a man inside an elevator that is in free fall (a frightening case, for sure!). This man is also in free fall, so he is falling at the exact same rate as the elevator. From the man's perspective, he is "floating" inside the elevator. A small push off the floor would send him floating toward the top of the elevator. He is what we call *weightless.*

"Weightless" does *not* mean "no weight." An object that is **weightless** has no *apparent* weight. The distinction is significant. The man's weight is still mg because gravity is still pulling down on him, but he has no *sensation* of weight as he free falls. The term "weightless" is in a sense a poor one because it implies that objects have no weight. As we see, that is not the case.

You've seen videos of astronauts and various objects floating inside the International Space Station as it orbits the earth. If an astronaut tries to stand on a scale, it does not exert any force against her feet and reads zero. She is said to be weightless. But if the criterion to be weightless is to be in free fall, and if astronauts orbiting the earth are weightless, does this mean that they are in free fall?

They are! Earth's gravity has not disappeared in space—gravity is what holds the space station in orbit—so astronauts are moving under the influence of only gravity, which is the definition of being in free fall. An astronaut, like a projectile, is continuously falling—she is, after all, following a curved trajectory—but she doesn't "fall out" of orbit because her large speed allows the curvature of her trajectory to match the curvature of the earth's surface. Thus astronauts "fall" without getting closer to the earth.

STOP TO THINK 5.1 You're bouncing up and down on a trampoline. After you have left the trampoline and are moving upward, your apparent weight is

A. More than your true weight.
B. Less than your true weight.
C. Equal to your true weight.
D. Zero.

5.3 Interactions with Surfaces

Interactions with surfaces are very common and very important. The locomotion of any animal or vehicle is an interaction with the ground. Static friction, a surface interaction, *prevents* your foot from slipping as you walk up a hill. *Osteoarthritis* is a consequence of increased surface interactions between the bones in a joint as the smooth cartilage wears away. We can divide surface interactions into two basic types: those perpendicular to the surface (normal forces) and those parallel to the surface (friction forces).

The Normal Force

In Chapter 4 we saw that an object at rest on a table is subject to an upward force due to the table. This force is called the *normal force* because it is always directed normal, or perpendicular, to the surface of contact. As we saw, the normal force has its origin in the atomic "springs" that make up the surface. The harder the object bears down on the surface, the more these springs are compressed and the harder they push back. Thus the normal force *adjusts* itself so that the object stays on the surface without penetrating it. This fact is key in solving for the normal force.

EXAMPLE 5.3 **Normal force on a pressed book**

A 1.2 kg book lies on a table. You press down on the book from above with a force of 15 N. What is the normal force acting on the book from the table below?

PREPARE We'll let the book be the system. We need to identify the forces acting on the book and prepare a free-body diagram showing these forces. These steps are illustrated in FIGURE 5.4.

FIGURE 5.4 Finding the normal force on a book pressed from above.

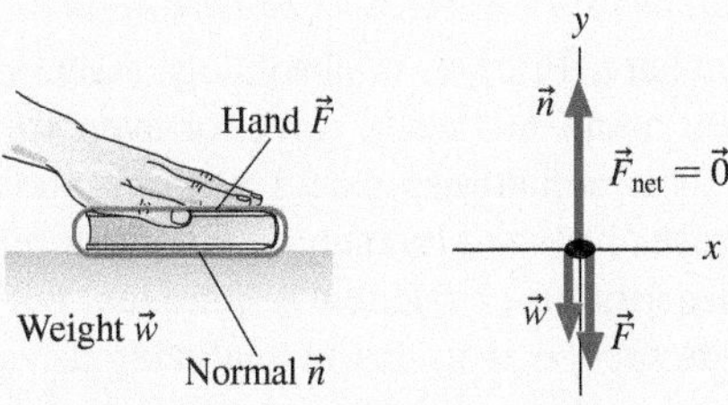

SOLVE Because the book is in equilibrium, the net force on it must be zero. The only forces acting are in the y-direction, so Newton's second law is

$$\sum F_y = n_y + w_y + F_y = n - w - F = ma_y = 0$$

We learned in the last section that the weight force is $w = mg$. The weight of the book is thus

$$w = mg = (1.2\text{ kg})(9.8\text{ m/s}^2) = 12\text{ N}$$

With this information, we see that the normal force exerted by the table is

$$n = F + w = 15\text{ N} + 12\text{ N} = 27\text{ N}$$

ASSESS The magnitude of the normal force is *larger* than the weight of the book. From the table's perspective, the extra force from the hand pushes the book further into the atomic springs of the table. These springs then push back harder, giving a normal force that is greater than the weight of the book.

NOTE ▸ You'll see numerous problems in which an upward normal force and a downward weight force are the only vertical forces and so, if there's no vertical motion, $n = w = mg$. But this is a special case, not a rule. The weight and the normal force are different forces that, in general, are not equal. ◂

EXAMPLE 5.4 **Acceleration of a downhill skier**

A skier slides down a steep 27° slope. On a slope this steep, friction is much smaller than the other forces acting on the skier and can be ignored. What is the skier's acceleration?

PREPARE An object moving on a ramp or incline is a common situation. Part of the problem-solving strategy for using Newton's laws is to choose a coordinate system that has one axis along the direction of motion. We'll let the x-axis point down the slope. This simplifies the analysis because the skier does not move in the y-direction at all, which makes $a_y = 0$. FIGURE 5.5 is a pictorial representation. The skier, who is circled, is the system of interest. The normal force, by definition, is perpendicular to the surface, so it points along the y-axis. Be careful with the weight force. It points vertically *straight down,* regardless of the coordinate system. In fact, the angle between the weight vector and the y-axis is exactly the same as the angle of the slope, here designated θ. Don't make the mistake of thinking that the weight force must lie on the y-axis!

SOLVE We can use Newton's second law in component form to find the skier's acceleration:

$$\sum F_x = w_x + n_x = ma_x$$

$$\sum F_y = w_y + n_y = ma_y$$

FIGURE 5.5 Pictorial representation of a downhill skier.

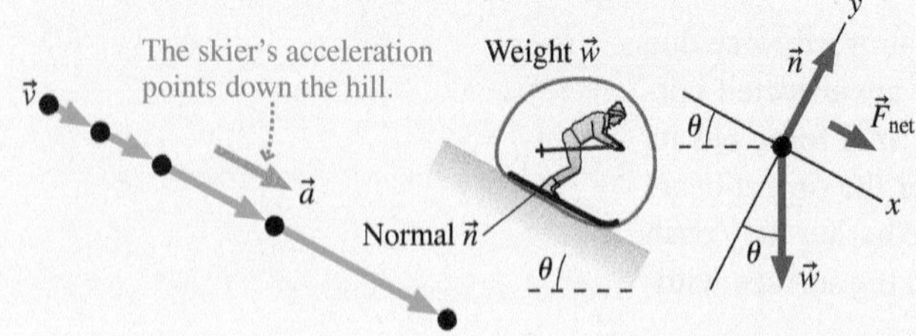

Because $\vec{n}$ points directly in the positive y-direction, $n_y = n$ and $n_x = 0$. We can use the fact that the angle between $\vec{w}$ and the negative y-axis is the same as the slope angle θ to write $w_x = w\sin\theta = mg\sin\theta$ and $w_y = -w\cos\theta = -mg\cos\theta$, where we used $w = mg$. Be sure you understand where these expressions come from, including the signs, because you will need to do similar analyses when you solve problems.

With these components in hand, Newton's second law becomes

$$\sum F_x = w_x + n_x = mg\sin\theta = ma_x$$

$$\sum F_y = w_y + n_y = -mg\cos\theta + n = ma_y = 0$$

In the second equation we used the fact that $a_y = 0$. The m cancels in the first of these equations, leaving us with

$$a_x = g\sin\theta$$

We can use this to calculate the skier's acceleration:

$$a_x = g\sin\theta = (9.8\text{ m/s}^2)\sin 27° = 4.4\text{ m/s}^2$$

ASSESS Our result shows that when $\theta = 0$, so that the slope is horizontal, the skier's acceleration is zero, as it should be. Further, when $\theta = 90°$ (a vertical slope), his acceleration is g, which makes sense because he's in free fall when $\theta = 90°$. Notice that the mass canceled out, so we didn't need to know the skier's mass. It turned out that we didn't need the y-equation for this problem, but we will use it in similar problems as soon as we include friction.

STOP TO THINK 5.2 A mountain biker is climbing a steep 20° slope at a constant speed. The cyclist and bike have a combined weight of 800 N. What can you say about the magnitude of the normal force of the ground on the bike?

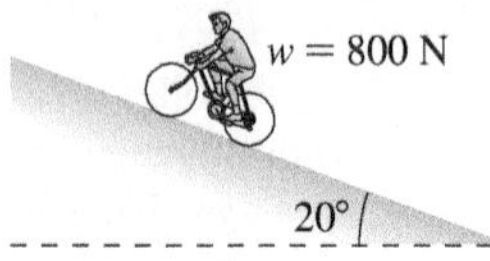

A. $n > 800$ N
B. $n = 800$ N
C. $n < 800$ N

FIGURE 5.6 Static friction keeps an object from slipping.

(a) Force identification

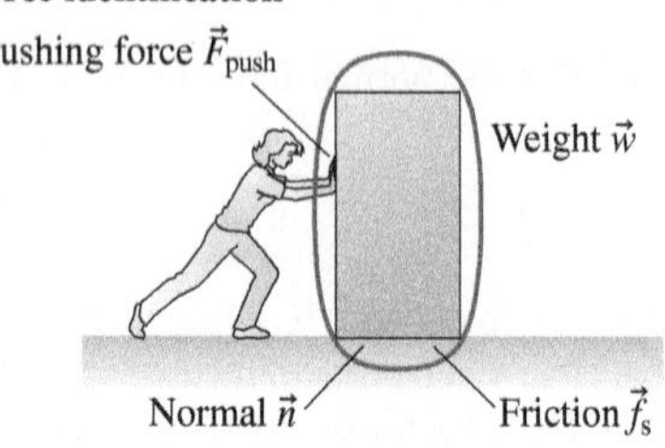

(b) Free-body diagram

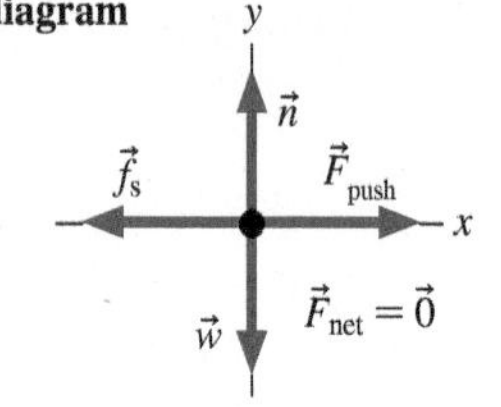

Static Friction

In everyday life, friction is everywhere. Friction is absolutely essential for many things we do. Without friction you could not walk, drive, or even sit down (you would slide right off the chair!). It is sometimes useful to think about idealized frictionless surfaces, but in most cases we want to analyze situations in which friction plays a crucial role. A full description of friction is complex, but the simple model of friction we discuss in this section is useful in analyzing a wide range of situations.

We'll begin with static friction $\vec{f}_s$, the force that a surface exerts on an object to keep it from slipping across that surface. Consider the woman pushing on the box in FIGURE 5.6a. Because the box is not moving with respect to the floor, the woman's push to the right must be balanced by a static friction force $\vec{f}_s$ pointing to the left. This is the general rule for finding the *direction* of $\vec{f}_s$: Decide which way the object *would* move if there were no friction. The static friction force $\vec{f}_s$ then points in the opposite direction, to prevent motion relative to the surface.

Determining the *magnitude* of $\vec{f}_s$ is a bit trickier. The box, at rest, is in equilibrium with no net force, and the free-body diagram of FIGURE 5.6b shows that the only horizontal forces are friction and the pushing force. These have to balance, so $f_s = F_{push}$. In other words, **the harder the woman pushes, the harder the static friction force from the floor pushes back.** This is illustrated in FIGURE 5.7a and b. If she reduces her pushing force, the friction force will automatically be reduced to match. Static friction acts in *response* to an applied force.

But there's clearly a limit to how big $\vec{f}_s$ can get. If the woman pushes hard enough, the box will slip and start to move across the floor. In other words, the static friction force has a *maximum* possible magnitude $f_{s\,max}$, as illustrated in FIGURE 5.7c. Experiments with friction show that, in many circumstances, $f_{s\,max}$ is proportional to the magnitude of the normal force between the surface and the object; that is,

$$f_{s\,max} = \mu_s n \tag{5.5}$$

where μ_s is called the **coefficient of static friction.** The coefficient is a number that depends on the materials from which the object and the surface are made. The higher the coefficient of static friction, the greater the "stickiness" between the object and the surface, and the harder it is to make the object slip. TABLE 5.2 lists some approximate values of coefficients of friction.

NOTE ▸ Equation 5.5 does not say $f_s = \mu_s n$. The value of f_s depends on the force or forces that static friction has to balance to keep the object from moving. It can have any value from zero up to, but not exceeding, $\mu_s n$; that is, $f_s \le \mu_s n$. ◂

So our rules for static friction are:

- The direction of static friction is such as to oppose motion.
- The magnitude f_s of static friction adjusts itself so that the net force is zero and the object doesn't move.
- The magnitude of static friction cannot exceed the maximum value $f_{s\,max}$ given by Equation 5.5. If the friction force needed to keep the object stationary is greater than $f_{s\,max}$, the object slips and starts to move.

FIGURE 5.7 Static friction acts in response to an applied force.

(a) Pushing gently: friction pushes back gently.

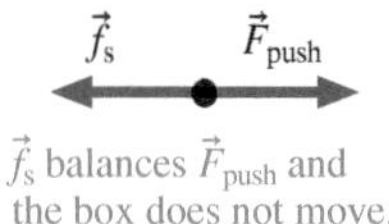

$\vec{f}_s$ balances $\vec{F}_{push}$ and the box does not move.

(b) Pushing harder: friction pushes back harder.

$\vec{f}_s$ grows as $\vec{F}_{push}$ increases, but the two still cancel and the box remains at rest.

(c) Pushing harder still: $\vec{f}_s$ is now pushing back as hard as it can.

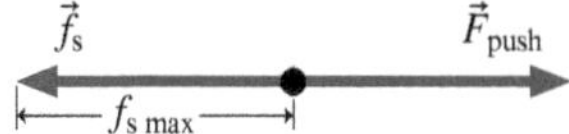

Now the magnitude of $\vec{f}_s$ has reached its maximum value $f_{s\,max}$. If $\vec{F}_{push}$ gets any bigger, the forces will *not* cancel and the box will start to accelerate.

TABLE 5.2 **Coefficients of friction**

Material	Static μ_s	Kinetic μ_k	Rolling μ_r
Rubber on concrete	1.00	0.80	0.02
Steel on steel (dry)	0.80	0.60	0.002
Steel on steel (lubricated)	0.10	0.05	
Wood on wood	0.50	0.20	
Wood on snow	0.12	0.06	
Ice on ice	0.10	0.03	

EXAMPLE 5.5 Walking on ice

As you may know from experience, you have to take very short steps to walk on a slippery surface such as ice. Why? The biomechanics of locomotion is complex, but we can use a simple model to understand some essential ideas. FIGURE 5.8 shows a person in midstride, with the legs at equal angles θ and both feet on the surface. What is the maximum leg angle θ_{max} for walking without slipping on concrete ($\mu_s = 1.0$) and on ice ($\mu_s = 0.15$)?

FIGURE 5.8 Walking without slipping depends on static friction.

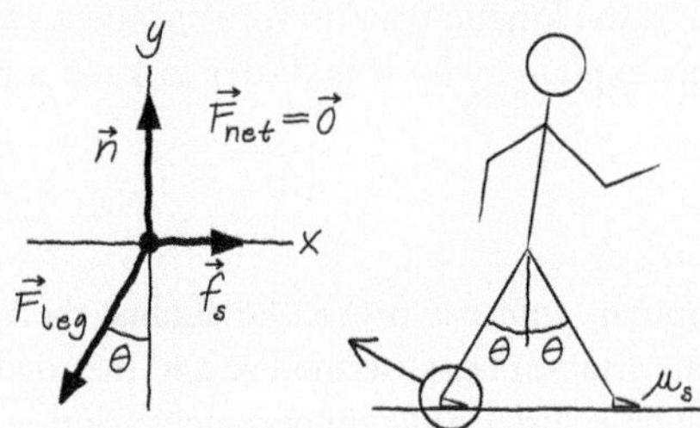

PREPARE Static friction has a maximum value, and an object—in this case, the person's foot—will slip if the friction force needed to keep the object stationary exceeds $f_{s\,max}$. The goal is to find the angle at which this occurs. Our system, the object we wish to focus on, is a foot. We've chosen the rear foot, but either will do. The foot's free-body diagram shows the normal force of the surface pushing up and static friction pointing to the right. Static friction acts to *prevent* slipping; the foot would slip to the left were it not for a friction force to the right.

The foot itself has negligible weight, but we need to account for the weight of the body that the foot and leg are holding up. A stretched rope, in tension, pulls on whatever is attached to each end. The legs in our model are similar, but they are in *compression;* thus they *push* on whatever is attached to each end—in this case, the torso and the foot. The leg is the agent, so we've labeled this force $\vec{F}_{leg}$. It's exerted in the direction of the leg, at angle θ, and its magnitude is proportional to the person's weight.

Continued

SOLVE The foot is temporarily at rest when it's planted on the ground in midstride. This is an equilibrium situation with $\vec{F}_{net} = \vec{0}$; hence Newton's second law for the foot is

$$\sum F_x = (f_s)_x + n_x + (F_{leg})_x = f_s - F_{leg}\sin\theta = 0$$

$$\sum F_y = (f_s)_y + n_y + (F_{leg})_y = n - F_{leg}\cos\theta = 0$$

We've again written the *sum* of all the force components along each axis, then used the free-body diagram to see that some components are zero and both components of $\vec{F}_{leg}$ are negative.

We see from the x-component equation that the friction force needed to prevent slipping is

$$f_s = F_{leg}\sin\theta$$

We also know, from our model of static friction, that slipping can be prevented only if $f_s \leq f_{s\,max} = \mu_s n$. The y-component equation yields $n = F_{leg}\cos\theta$, so the no-slipping condition is

$$f_s = F_{leg}\sin\theta \leq \mu_s n = \mu_s(F_{leg}\cos\theta)$$

Dividing both sides by $F_{leg}\cos\theta$, we find

$$\frac{\sin\theta}{\cos\theta} = \tan\theta \leq \mu_s$$

F_{leg} canceled, which means that we need to know only the angle of the leg force, not its magnitude. (As a practice exercise you might want to convince yourself that an analysis of the torso finds $F_{leg} = mg/2\cos\theta$.) The person is stable in this position as long as $\tan\theta \leq \mu_s$, so the maximum leg angle is

$$\theta_{max} = \tan^{-1}(\mu_s) = \begin{cases} 45^\circ & \mu_s = 1.0 \text{ on concrete} \\ 8.5^\circ & \mu_s = 0.15 \text{ on ice} \end{cases}$$

ASSESS The results tell us that you can spread your legs far apart on a rough surface without slipping, but not on a slick surface such as ice. To confirm this with a simple experiment, see how far you can spread your legs without slipping on a smooth floor, first while wearing rubber-soled shoes and then wearing your socks.

Kinetic Friction

Once the box starts to slide, as in FIGURE 5.9, the static friction force is replaced by a kinetic (or sliding) friction force $\vec{f}_k$. Kinetic friction is in some ways simpler than static friction: The direction of $\vec{f}_k$ is always opposite to the direction in which an object slides across the surface, and experiments show that kinetic friction, unlike static friction, has a nearly *constant* magnitude, given by

$$f_k = \mu_k n \tag{5.6}$$

where μ_k is called the **coefficient of kinetic friction.** Equation 5.6 also shows that kinetic friction, like static friction, is proportional to the magnitude of the normal force n. Notice that **the magnitude of the kinetic friction force does not depend on how fast the object is sliding.**

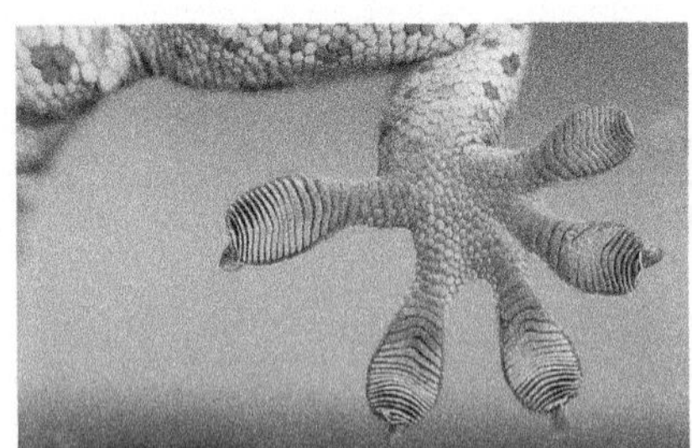

Geckos stick to walls, but their feet aren't sticky A gecko can cling to a smooth vertical wall or even hang upside down from the ceiling. It does this with a force of *adhesion,* not of static friction. Gecko foot pads are covered by millions of *setae,* each of which branches into hundreds of nanometer-sized hairs. These tiny hairs are in such close contact with the atoms of a surface that they experience a *van der Waals attraction,* a weak intermolecular force related to the hydrogen bonding that shapes proteins and other macromolecules. These interactions pull the gecko to the surface with a force many times larger than its weight.

FIGURE 5.9 The kinetic friction force is *opposite* to the direction of motion.

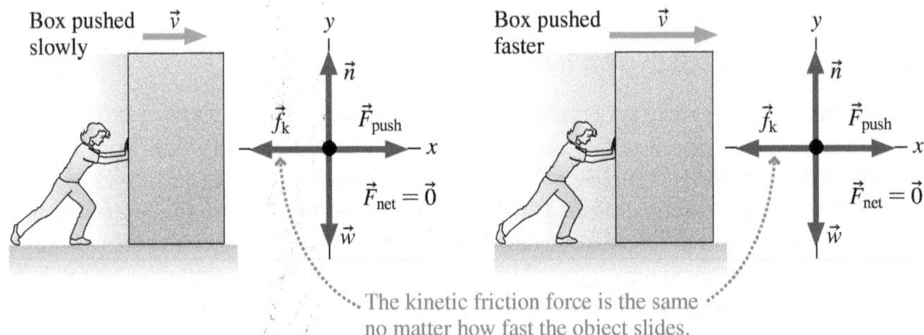

Table 5.2 includes approximate values of μ_k. Notice that, for any pair of materials, μ_k is always less than μ_s. That is, the kinetic friction force is always less than the *maximum* static friction force. This explains why it is easier to keep a box moving than it is to start it moving.

FIGURE 5.10 Rolling friction is also opposite the direction of motion.

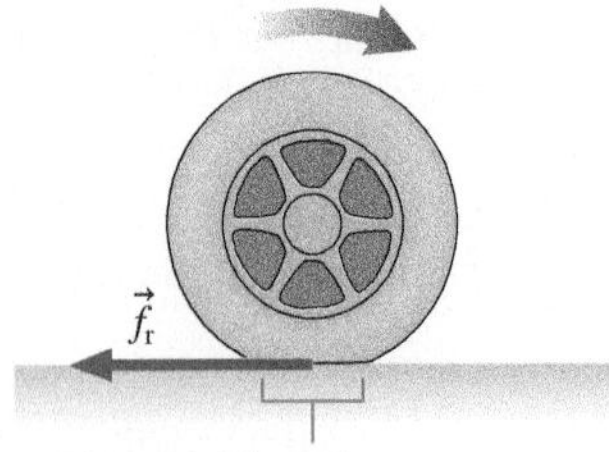

Rolling Friction

If you slam on the brakes hard enough, your car tires slide against the road surface and leave skid marks. This is kinetic friction because the tire and the road are *sliding* against each other. A wheel *rolling* on a surface also experiences friction, but not kinetic friction: As FIGURE 5.10 shows, the portion of the wheel that contacts the surface is stationary with respect to the surface, not sliding.

The interaction between a rolling wheel and the road involves adhesion between and deformation of surfaces and can be quite complicated, but in many cases we can treat it like another type of friction force that opposes the motion, one defined by a **coefficient of rolling friction** μ_r:

$$f_r = \mu_r n \tag{5.7}$$

Rolling friction acts very much like kinetic friction, but values of μ_r (see Table 5.2) are much smaller than values of μ_k. It's easier to roll something on wheels than to slide it!

STOP TO THINK 5.3 Rank in order, from largest to smallest, the size of the friction forces $\vec{f}_A$ to $\vec{f}_E$ in the five different situations (one or more friction forces could be zero). The box and the floor are made of the same materials in all situations.

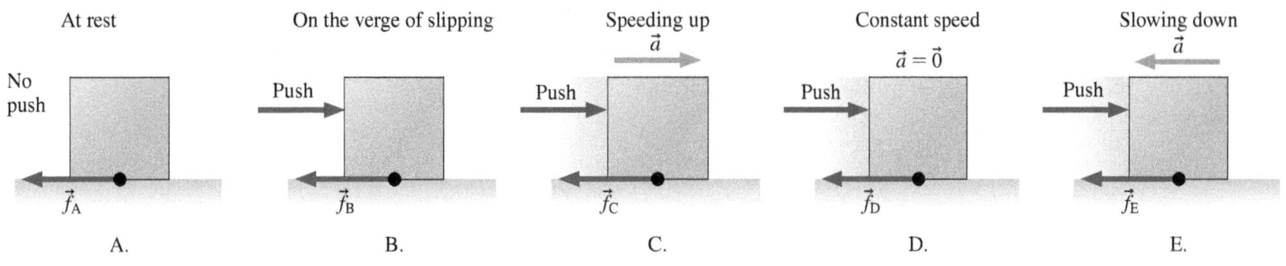

Working with Friction Forces

These ideas can be summarized in a *model* of friction:

$$\begin{aligned}
&\text{Static: } \vec{f}_s = (\text{magnitude} \le f_{s\,max} = \mu_s n, \text{ direction as necessary to prevent motion})\\
&\text{Kinetic: } \vec{f}_k = (\mu_k n, \text{ direction opposite the motion})\\
&\text{Rolling: } \vec{f}_r = (\mu_r n, \text{ direction opposite the motion})
\end{aligned} \tag{5.8}$$

Push or pull

Friction

Motion is *relative to the surface.*

To skid or not to skid If you brake as hard as you can without skidding, the force that stops your car is the static friction force between your tires and the road. This force is bigger than the kinetic friction force, so if you skid, not only can you lose control of your car but you also end up taking a longer distance to stop. Antilock braking systems brake the car as hard as possible without skidding, stopping the car in the shortest possible distance while also retaining control.

Here "motion" means "motion relative to the surface." The maximum value of static friction $f_{s\,max} = \mu_s n$ occurs when the object slips and begins to move.

NOTE ▶ Equations 5.8 are a "model" of friction, not a "law" of friction. These equations provide a reasonably accurate, but not perfect, description of how friction forces act under many, but not all, circumstances. They are a simplification of reality that works reasonably well, which is what we mean by a "model." They are not a "law of nature" on a level with Newton's laws. This model fails for extremely smooth surfaces, like microscope slides, and surfaces that have an ordered microscale structure, such as gecko feet, where intermolecular forces are more important than friction forces. ◀

EXAMPLE 5.6 Finding the force to slide a sofa

Carol wants to move her 32 kg sofa to a different room in the house. She places "sofa sliders" under the feet of the sofa. These are slippery disks with a coefficient of kinetic friction of 0.080 on Carol's hardwood floor. She then pushes the sofa at a steady 0.40 m/s across the floor. How much force does she apply to the sofa?

PREPARE We'll let the sofa be the system and assume that it slides to the right. In this case, a kinetic friction force $\vec{f}_k$ opposes

Continued

the motion by pointing to the left. In FIGURE 5.11 we identify the forces acting on the sofa and construct a free-body diagram. We've noted that the net force is zero because the sofa moves at constant speed.

FIGURE 5.11 Forces on a sofa being pushed across a floor.

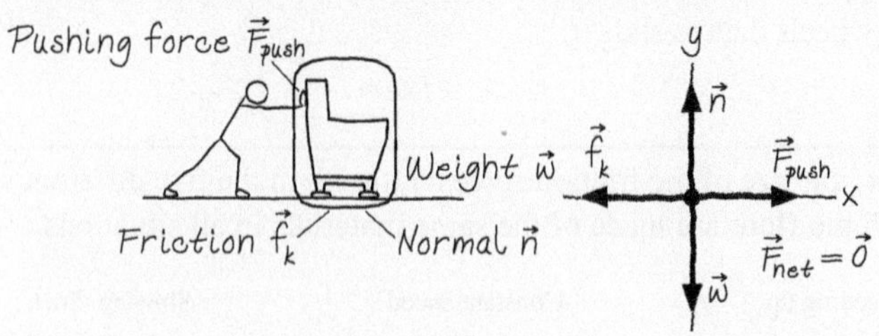

SOLVE The sofa is not accelerating, so both the x- and y-components of the net force must be zero:

$$\sum F_x = n_x + w_x + (F_{push})_x + (f_k)_x = 0 + 0 + F_{push} - f_k = 0$$

$$\sum F_y = n_y + w_y + (F_{push})_y + (f_k)_y = n - w + 0 + 0 = 0$$

In the first equation, the x-component of $\vec{f}_k$ is equal to $-f_k$ because $\vec{f}_k$ is directed to the left. Similarly, in the second equation, $w_y = -w$ because the weight force points down.

From the first equation, we see that Carol's pushing force is $F_{push} = f_k$. To evaluate this, we need f_k. Here we can use our model for kinetic friction:

$$f_k = \mu_k n$$

We can find the normal force n from the vertical equation, $n - w = 0$. The weight force is $w = mg$, so we have

$$n = mg$$

With this, we find that Carol's pushing force is

$$F_{push} = f_k = \mu_k n = \mu_k mg$$
$$= (0.080)(32\,\text{kg})(9.80\,\text{m/s}^2) = 25\,\text{N}$$

ASSESS The speed with which Carol pushes the sofa does not enter into the answer. This makes sense because the kinetic friction force doesn't depend on speed. The final result of 25 N is a rather small force—only about $5\frac{1}{2}$ pounds—but we expect this because Carol has used slippery disks to move the sofa. Notice that here we *did* need both components of Newton's second law even though the motion is entirely horizontal. This is typical of problems that involve friction.

EXAMPLE 5.7 A sliding dog

A 12 kg dog running across a smooth floor at 4.0 m/s tries to stop but, as is often the case, ends up sliding. If the dog slides 180 cm, what is the coefficient of kinetic friction between her paws and the floor?

PREPARE The dog is the system, which we will model as a particle sliding with constant acceleration. Sliding is associated with kinetic friction; in this case, it's the only horizontal force acting on the dog. Our model of kinetic friction is $f_k = \mu_k n$. To find μ_k, we'll need to use kinematics to find the dog's acceleration while stopping, then Newton's second law to determine f_k. FIGURE 5.12 is a pictorial representation of the situation.

FIGURE 5.12 Pictorial representation of a dog sliding to a stop.

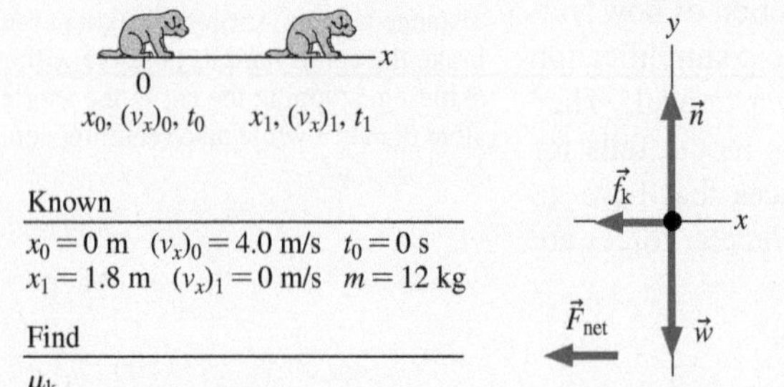

SOLVE The motion is entirely along the x-axis, so $a_y = 0$. However, we do need the y-component equation of Newton's second law to find n. The second-law equations are

$$\sum F_x = (f_k)_x = -f_k = ma_x$$

$$\sum F_y = n_y + w_y = n - w = n - mg = 0$$

We're gaining more experience with Newton's second law, so in this case we omitted the force components that, from the free-body diagram, we know are zero. We also used $w = mg$ for the dog's weight.

The two equations tell us that $f_k = -ma_x$ and $n = mg$. Using these in the model of kinetic friction, we find

$$f_k = -ma_x = \mu_k n = \mu_k mg$$

and thus

$$\mu_k = -\frac{a_x}{g}$$

We don't actually need to know the dog's mass, but we do need to use kinematics to find a_x. We know the distance of the slide but not the time, so the appropriate kinematic equation is

$$(v_x)_1^2 = 0 = (v_x)_0^2 + 2a_x \Delta x$$

Solving for a_x gives

$$a_x = -\frac{(v_x)_0^2}{2(x_1 - x_0)} = -\frac{(4.0\,\text{m/s})^2}{2(1.8\,\text{m})} = -4.44\,\text{m/s}^2$$

The acceleration is negative because the acceleration vector points to the left. Knowing the acceleration, we can now find

$$\mu_k = -\frac{(-4.44\,\text{m/s}^2)}{9.80\,\text{m/s}^2} = 0.45$$

ASSESS By comparing our answer to the values of μ_k in Table 5.2, we see that 0.45 is a reasonable coefficient of kinetic friction for paws on a smooth floor.

Causes of Friction

It is worth taking a brief pause to look at the *causes* of friction. All surfaces, even those that are quite smooth to the touch, are very rough on a microscopic scale. When two objects are placed in contact, they do not make a smooth fit. Instead, as FIGURE 5.13 shows, the high points on one surface become jammed against the high points on the other surface, while the low points are not in contact at all. Only a very small fraction (typically 10^{-4}) of the surface area is in actual contact. The amount of contact depends on how hard the surfaces are pushed together, which is why friction forces are proportional to n.

For an object to slip, you must push it hard enough to overcome the forces exerted at these contact points. Once the two surfaces are sliding against each other, their high points undergo constant collisions, deformations, and even brief bonding that lead to the resistive force of kinetic friction.

FIGURE 5.13 A microscopic view of friction.

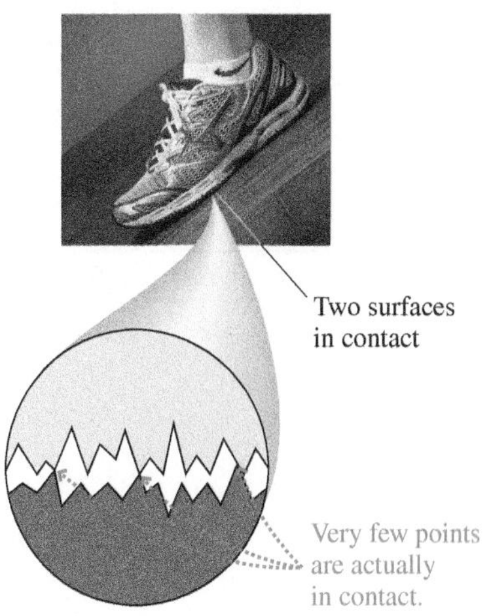
Two surfaces in contact
Very few points are actually in contact.

5.4 Drag

Drag is a force that opposes or retards the motion of an object as it moves through a *fluid*—a substance that flows, such as air or water. You experience drag forces due to the air—sometimes called *air resistance*—whenever you jog, bicycle, or drive your car. In biology, drag forces are extremely important for microorganisms moving in either air or water. The drag force $\vec{D}$:

- Is opposite to the direction of motion, as shown in FIGURE 5.14.
- Increases in magnitude as the object's speed increases.

FIGURE 5.14 The drag force is opposite the direction of motion.

$\vec{D}$

NOTE ▸ Drag is *not* a constant force because it depends on the object's speed. Thus constant-acceleration kinematics can *not* be applied to problems that involve drag. ◂

Drag is more complex than friction because there are two rather different physical causes of drag. For a relatively large object, such as a flying bird, most of the drag—the force that resists the bird's motion—occurs because the moving bird has to push away the air in front of it. This drag is said to be caused by *inertial forces,* forces that arise because a fluid's mass makes the fluid difficult to move. In contrast, the drag on a small unicellular organism swimming through a liquid is mostly due to *viscous forces* from the fluid molecules sticking to one another and to the organism.

Fortunately, there are many situations where only one of these causes of drag is significant; that is, the drag is almost entirely due to inertial forces or almost entirely due to viscous forces. Fairly simply models will allow us to characterize drag in these cases. We can tell which case applies—inertial drag or viscous drag—by examining a quantity called the *Reynolds number.*

Reynolds Number

For an object moving through a fluid, the **Reynolds number** is defined as the ratio of inertial forces to viscous forces acting on the object. The Reynolds number is high if inertial forces dominate, low if viscous forces dominate. Two important properties of a fluid are its density ρ (Greek rho) and its **viscosity** η (Greek eta). For an object of size L moving through a fluid with speed v, the Reynolds number is

$$Re = \frac{\text{inertial forces}}{\text{viscous forces}} = \frac{\rho v L}{\eta} \tag{5.9}$$

Reynolds number for an object moving through a fluid

NOTE ▸ The symbol for the Reynolds number is Re. It is the only symbol in this text that uses two letters. ◂

We won't attempt to justify this expression for the Reynolds number, but it makes sense. More force is needed to deflect a denser fluid or to deflect the fluid around an object that is larger or moving faster, and these quantities all appear in the numerator. Likewise, the size of any viscous forces is related to the fluid's viscosity, which we see in the denominator.

Notice that the Reynolds number, a ratio of forces, is a pure number with no units; it is said to be *dimensionless*. When Re is high, inertial drag is dominant; when Re is low, viscous drag is more important.

In SI units, density is measured in kg/m^3 and viscosity in $Pa \cdot s$. Pa is the abbreviation for *pascal*, the SI unit of pressure, which is defined as $1\ Pa = 1\ N/m^2$. We'll have a lot more to say about pressure in later chapters. TABLE 5.3 lists the values of density and viscosity for some typical fluids. Viscosity is *very* dependent on temperature—think how quickly honey loses viscosity as you heat it—so the values shown are not appropriate at temperatures other than those listed.

TABLE 5.3 Density and viscosity

Fluid	ρ (kg/m^3)	η (Pa·s)
Air (20°C)	1.2	1.8×10^{-5}
Ethyl alcohol (20°C)	790	1.3×10^{-3}
Olive oil (20°C)	910	8.4×10^{-2}
Water (20°C)	1000	1.0×10^{-3}
Water (40°C)	1000	7.0×10^{-4}
Honey (20°C)	1400	10
Honey (40°C)	1400	1.7

The object's size L is a characteristic or typical size of the object, which could be a height, width, or diameter. This seems a bit odd; for an object that is rectangular, the value of Re depends on whether you select the width or the height for L. However, in practice, how you define L makes very little difference. As you'll see, we want to know whether Re is "high" or "low," values that differ by factors of 1000 or more. A difference of a factor of 2 or 3 will not affect our judgment about the size of Re. Said another way, our use of the Reynolds number needs only approximate values, not precise calculations.

As a simple example, consider a salmon swimming at a typical speed of 0.5 m/s through 20°C water. Seen head-on, an adult salmon is typically 15 cm tall and 10 cm wide, so we'll use 15 cm as the salmon's size L. The density and viscosity of water are found in Table 5.3. The Reynolds number for the salmon's motion is

$$Re = \frac{\rho v L}{\eta} = \frac{(1000\ \text{kg/m}^3)(0.5\ \text{m/s})(0.15\ \text{m})}{1.0 \times 10^{-3}\ \text{Pa} \cdot \text{s}} = 75{,}000$$

As you'll see, this is considered a very high Reynolds number.

TABLE 5.4 Drag coefficients

Object	C_D
Commercial airliner	0.024
Swimming fish	0.15
Toyota Prius	0.24
Bird in a dive	0.35
Pitched baseball	0.35
Racing cyclist	0.88
Running person	1.2

Drag at High Reynolds Number

A Reynolds number greater than about 1000 is considered high. A high Reynolds number means, as we've seen, that the drag arises primarily from inertia, with viscosity playing a minor role. In this case, drag occurs because the moving object has to push the fluid out of the way. The Reynolds number is high for most ordinary objects—balls, birds, cars—moving through air at ordinary speeds. Re is also high for fish and larger objects moving through water.

For high Reynolds numbers, the drag force for motion through a fluid at speed v is

$$\vec{D} = \left(\tfrac{1}{2} C_D \rho A v^2,\ \text{direction opposite the motion}\right) \tag{5.10}$$

Drag force for high Reynolds number, $Re > 1000$

Here ρ is again the density of the fluid, A is the cross-section area of the object (in m^2) as defined below, and the dimensionless **drag coefficient** C_D depends on the object's shape. More streamlined, aerodynamic shapes have lower values of C_D. TABLE 5.4 lists the drag coefficients for some common moving objects.

Notice that, for a high Reynolds number, **the magnitude of the drag force is proportional to the *square* of the object's speed.** If the speed doubles, drag increases by a factor of 4.

The cross-section area A is the two-dimensional projection that you see when an object is coming toward you. If you were to shine a spotlight at the object as it approaches you, A is the area of the object's shadow on a screen behind it. Many objects can be modeled as spheres or cylinders, so FIGURE 5.15 shows cross-section areas and drag coefficients for a sphere and for two different orientations of a cylinder.

FIGURE 5.15 Cross-section areas and drag coefficients for a sphere and a cylinder.

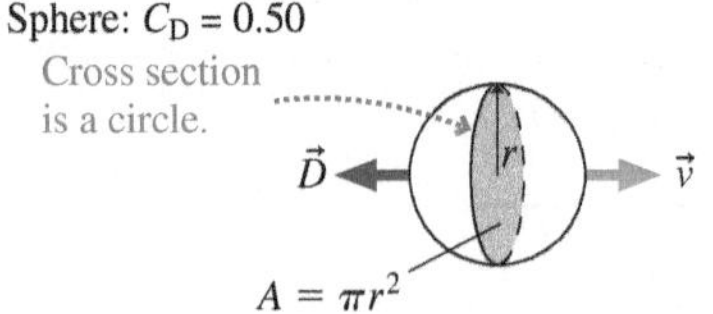

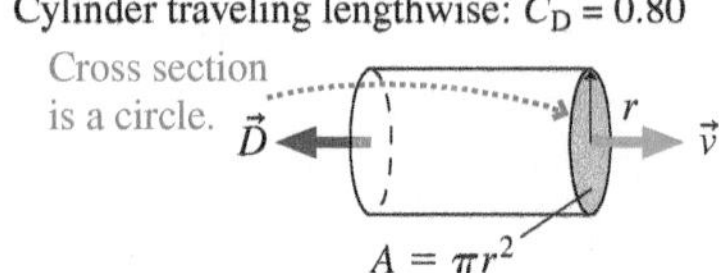

Cylinder traveling sideways: $C_D = 1.1$

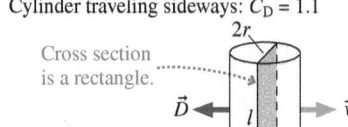

Cross section is a rectangle.

$2r$, l, $\vec{D}$, $\vec{v}$

$A = 2rl$

EXAMPLE 5.8 **The drag coefficient of a swimming penguin**

Biologists have measured the drag coefficient of a swimming penguin by observing the rate at which a penguin's speed decreases in its glide phase, when it's not actively swimming and is slowing down. In one study, a gliding 4.8 kg Gentoo penguin was found to be decelerating at 0.52 m/s^2 when its speed was 1.60 m/s. If its frontal area is 0.020 m^2, what is the penguin's drag coefficient?

PREPARE When the penguin is gliding horizontally, the only force acting to slow it down is the drag force. Using Newton's second law, we will find the drag force from the penguin's acceleration and mass. Once the drag force is known, we will use Equation 5.10 to find the drag coefficient. Note that the density ρ is that of the fluid through which the object moves, not the object's density.

SOLVE Assume that the penguin is moving to the right along the x-axis. Then the drag force points left and Newton's second law is

$$(F_{\text{net}})_x = D_x = -D = ma_x$$

The acceleration a_x is negative for an object that is slowing while moving to the right. Thus the magnitude of the drag force is

$$D = -ma_x = -(4.8\text{ kg})(-0.52\text{ m/s}^2) = 2.5\text{ N}$$

Finally, from Equation 5.10 we can solve for the drag coefficient. We have

$$C_D = \frac{2D}{\rho A v^2} = \frac{2(2.5\text{ N})}{(1000\text{ kg/m}^3)(0.020\text{ m}^2)(1.60\text{ m/s})^2} = 0.098$$

ASSESS This drag coefficient is quite a bit better than that of the Toyota Prius in Table 5.4 and slightly better than that of a fish. This is reasonable given that penguins have a highly adapted, streamlined shape.

Terminal Speed

Suppose an object starting from rest is pushed or pulled through a fluid by a constant applied force $\vec{F}_{\text{applied}}$. Initially the speed is low and the drag force is small, as shown in FIGURE 5.16a, so the net force causes the object to speed up. But the drag force increases as the speed increases, and eventually the object reaches a speed at which the drag force has exactly the same magnitude as $\vec{F}_{\text{applied}}$. Now the net force is zero, as shown in FIGURE 5.16b, so the object can no longer accelerate but will maintain this steady speed for as long as the force is applied. The steady, unchanging speed at which drag exactly counterbalances an applied force is called the object's **terminal speed.**

Terminal speed while falling, where gravity is the applied force, is just one example. A fish reaches its maximum speed—also a terminal speed—when the drag of the water is equal and opposite to the propulsion force the fish generates by the lateral motion of its tail: $\vec{D} = \vec{F}_{\text{prop}}$. In the case of an object fired at an initial speed *greater* than the terminal speed, the very large drag force slows the object until it is moving at the terminal speed.

FIGURE 5.16 An object reaches its terminal speed when the drag force exactly balances the applied force.

(a) At low speeds, D is small and the object accelerates.

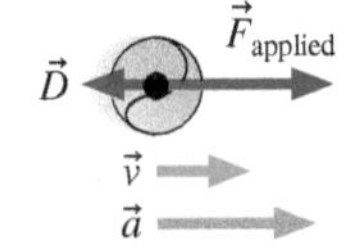

(b) Eventually, v reaches a value such that $D = F_{\text{applied}}$. Then the net force is zero and the object moves at a constant terminal speed v_{term}.

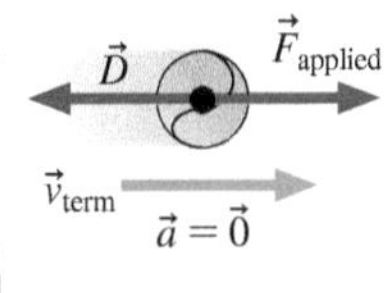

EXAMPLE 5.9 **The terminal speeds of a man and a mouse**

A 75 kg skydiver and his 20 g pet mouse falling after jumping from a plane are shown in FIGURE 5.17. Find the terminal speed of each.

PREPARE Gravity is a constant force pulling down on a falling object, while drag due to the air is an upward force. The terminal speed is reached when the drag on each object is equal to the object's weight mg. We will model both the man and the mouse as cylinders falling sideways.

FIGURE 5.17 A falling skydiver and mouse, and their cross-section areas.

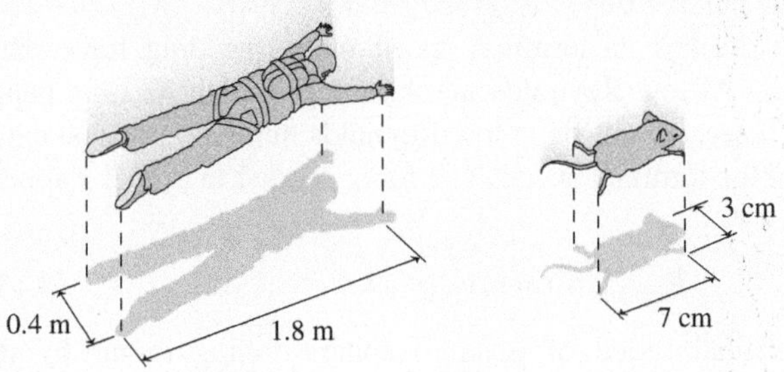

SOLVE From Figure 5.15 we see that $C_D = 1.1$ for a cylinder falling sideways and that a cylinder's cross-section area as seen from the side is $A = 2rl$. With the dimensions given, we can calculate $A_{\text{man}} = 0.72\text{ m}^2$ and $A_{\text{mouse}} = 2.1 \times 10^{-3}\text{ m}^2$. We assume that the two skydivers are at sufficiently low altitude that we can use the sea-level value of the density of air.

The terminal speed is reached when $D = mg$, or

$$\frac{1}{2}C_D \rho A (v_{\text{term}})^2 = mg$$

from which we can find the terminal velocity as

$$v_{\text{term}} = \sqrt{\frac{2mg}{C_D \rho A}}$$

Continued

Thus our two skydivers have terminal speeds

$$v_{man} = \sqrt{\frac{2(75\text{ kg})(9.8\text{ m/s}^2)}{(1.1)(1.2\text{ kg/m}^3)(0.72\text{ m}^2)}} = 39\text{ m/s}$$

$$v_{mouse} = \sqrt{\frac{2(0.020\text{ kg})(9.8\text{ m/s}^2)}{(1.1)(1.2\text{ kg/m}^3)(2.1 \times 10^{-3}\text{ m}^2)}} = 12\text{ m/s}$$

ASSESS 39 m/s is about 90 mph. Reported terminal speeds for skydivers falling in the prone position are in the 100–120 mph range, so our simple model of the fall gives a result that is close but a bit too low. We've probably overestimated both ρ, because skydivers are at a high enough altitude that the air density is lower than at sea level, and A, because their legs are actually spread apart and allow air to flow between them. A more realistic, but also more complex, model would give a better prediction. The mouse, though, falls at a much more modest 12 m/s ≈ 25 mph. Small animals can usually survive a fall from *any* height because their terminal speed is not terribly fast. Many tree-dwelling animals, such as the flying squirrel in the photograph at the beginning of the chapter, can extend flaps of skin to increase their area and thus fall at an even slower speed.

NOTE ▶ The drag force is usually much less than the weight force for heavier objects, such as balls or bicycles, moving at not-too-fast speeds through air. We typically neglect drag in our model of these situations because doing so simplifies the analysis while causing only a very small error in the calculations. ◀

Drag at Low Reynolds Number

A 10-μm-diameter spore, released from a fungus, settles to the ground at about 20 mm/s. It's not hard to calculate that the Reynolds number for the falling spore is $Re \approx 0.01$. This is an example of motion at low Reynolds number, which we define to be $Re < 1$. The drag force for low Reynolds number is almost entirely a viscous force.

For low Reynolds number, the drag force for motion through a fluid at speed v is

$$\vec{D} = (bv, \text{ direction opposite the motion}) \tag{5.11}$$

where b is a constant that depends on the size and shape of the object and on the fluid's viscosity. For low Reynolds number, **the magnitude of the drag force is proportional to the object's speed.** If the speed doubles, drag also increases by a factor of 2. This is in contrast to the quadratic dependence on speed at high Reynolds number.

NOTE ▶ Drag is more complex for intermediate values of Reynolds number, $1 < Re < 1000$. We don't consider those situations in this text. ◀

Viscous drag is mostly applicable to very small objects, many of which can be reasonably modeled as spheres. That's fortunate because only for spheres does theory give us any guidance for the coefficient b. For a spherical object of radius r moving in a fluid with viscosity η, it can be shown that $b = 6\pi\eta r$. Thus

$$\vec{D}_{sphere} = (6\pi\eta r v, \text{ direction opposite the motion}) \tag{5.12}$$

Stokes' law for low Reynolds number, $Re < 1$

This expression for the viscous drag on a sphere is called **Stokes' law.** Notice that viscous drag at low Reynolds number depends on the fluid's viscosity, whereas inertial drag at high Reynolds number depends on the fluid's density.

As before, an object reaches its terminal speed when the drag force exactly balances an applied force. At low Reynolds number, unlike high *Re*, this happens almost instantly, as you'll see, so motion at low Reynolds number is almost entirely constant-speed motion at the terminal speed. The force needed to propel a sphere at its terminal speed is

$$F_{applied} = 6\pi\eta r v_{term} \tag{5.13}$$

We can find the terminal speed of a small sphere falling in air by using $F_{applied} = mg$, as we did for high Reynolds number, but this approach doesn't quite work for objects falling in liquids. The reason is that the upward buoyant force of the liquid—a topic we'll study in Chapter 9—cannot be neglected. We'll return to this after we introduce buoyancy.

EXAMPLE 5.10 **Measuring the mass of a pollen grain**

Pollen grains are very light. In one experiment to determine their mass, researchers dropped grains inside a clear glass cylinder and then watched their motion with a microscope. A 40-μm-diameter pollen grain was observed to fall at a rate of 5.3 cm/s. What is the mass in nanograms of this grain?

PREPARE We will model the grain as a sphere. At terminal speed, the force of gravity on the grain is equal to the drag force. This is a very small object moving quite slowly, so its Reynolds number is very low and the drag is viscous drag. We'll assume the air in the cylinder was at a room temperature of 20°C and use the value for air viscosity in Table 5.3. In SI units, the terminal speed is $v_{term} = 0.053$ m/s.

SOLVE We set the force of gravity equal to the viscous drag force to get

$$mg = 6\pi\eta r v_{term}$$

We can then solve for the mass as

$$m = \frac{6\pi\eta r v_{term}}{g}$$
$$= \frac{6\pi(1.8 \times 10^{-5}\ \text{Pa}\cdot\text{s})(20 \times 10^{-6}\ \text{m})(0.053\ \text{m/s})}{9.8\ \text{m/s}^2}$$
$$= 3.7 \times 10^{-11}\ \text{kg}$$

Converting this result to nanograms (ng) gives

$$m = 3.7 \times 10^{-11}\ \text{kg} \times \frac{1000\ \text{g}}{\text{kg}} \times \frac{1\ \text{ng}}{10^{-9}\ \text{g}} = 37\ \text{ng}$$

ASSESS This extremely tiny mass is hard to assess on its own. It's likely that a pollen grain's density is near that of water, and the mass of a 40-μm-diameter sphere of water can be calculated—volume times density—to be about 30 ng. So our answer seems reasonable.

Life at Low Reynolds Number

From protozoa swimming in pond water to bacteria moving in intercellular fluid (essentially water), biology is filled with examples of motion at extremely low Reynolds number. For example, a 1-μm-diameter bacterium moving at a typical speed of 30 μm/s has $Re \approx 3 \times 10^{-5}$. Life at such low Reynolds number is very much at odds with our everyday experience.

The motion of your body and other everyday objects is at high Reynolds number, where inertia plays a central role. When you apply a force to an object, it takes time—due to inertia—for it to gain speed. When you remove the force, drag forces don't cause the object to stop immediately; because of its inertia it takes some time for the object to coast to a stop. Newton's second law, which relates force and motion, is key to understanding motion at high Reynolds number.

Motion at extremely low Reynolds number is dramatically different. After a force is applied, the object reaches its terminal speed v_0 *almost instantaneously* and then moves at constant speed. When the force is removed, the object comes to a dead halt *almost instantaneously*. Viscosity is so much more important than inertia that there's no coasting. The only way a tiny organism can move at all is with continuous propulsion; that's what the hair-like beating cilia and rotating flagella of microorganisms do.

Suppose a small spherical object of mass m and radius r is moving through a viscous liquid at speed v_0 when the applied force suddenly vanishes, leaving only drag. How long does it take the object to stop, and how far does it move while stopping? It is interesting that this motion is described with *exponential functions,* a type of motion that we modeled in ◀◀ SECTION 3.8.

Assume the object is moving to the right, so the x-components of the drag force and the acceleration are negative. From Stokes' law, the acceleration is

$$a_x = \frac{(F_{drag})_x}{m} = -\frac{6\pi\eta r v_x}{m}$$

By definition, the object's acceleration is $a_x = dv_x/dt$. Thus the motion is described by the equation

$$\frac{dv_x}{dt} = -\frac{6\pi\eta r v_x}{m} \tag{5.14}$$

This looks complicated, but we can find the velocity by integration.

First, we rearrange Equation 5.14 to get all the velocity terms on one side:

$$\frac{dv_x}{v_x} = -\frac{6\pi\eta r}{m}\,dt$$

The fraction on the right is a group of constants, so it is constant. Now we integrate both sides from an initial speed v_0 at time $t_0 = 0$ to a later speed v at time t:

$$\int_{v_0}^{v} \frac{dv_x}{v_x} = -\frac{6\pi\eta r}{m}\int_0^t dt = -\frac{6\pi\eta r}{m}t = -\frac{t}{\tau} \tag{5.15}$$

In Equation 5.15 we were able to take the constants outside the time integration. We then grouped them together and defined the *time constant* τ of the motion to be

$$\tau = \frac{m}{6\pi\eta r} \tag{5.16}$$

It's not obvious, but by using the units of viscosity (Pa • s), one can show that the units of τ are s, as they should be for a time. We'll see that the time constant is a *characteristic time* for how long it takes the object to come to a halt.

To finish the integration, recall from calculus that $\int dx/x = \ln x$, where $\ln x$ is the *natural logarithm* of x. In this case,

$$\int_{v_0}^{v} \frac{dv_x}{v_x} = \ln v_x\Big|_{v_0}^{v} = \ln v - \ln v_0 = \ln\left(\frac{v}{v_0}\right) \tag{5.17}$$

Thus

$$\ln\left(\frac{v}{v_0}\right) = -\frac{t}{\tau} \tag{5.18}$$

To solve for v, we also need to recall that the logarithmic function and the exponential function are inverses of each other, which means that $\ln(e^x) = x$ and $e^{\ln x} = x$. Applying the latter of these to Equation 5.18, we find

$$e^{\ln(v/v_0)} = \frac{v}{v_0} = e^{-t/\tau} \tag{5.19}$$

If we multiply through by v_0, we find that **the object's speed decreases exponentially with time:**

$$v = v_0 e^{-t/\tau} \tag{5.20}$$

This is shown in FIGURE 5.18.

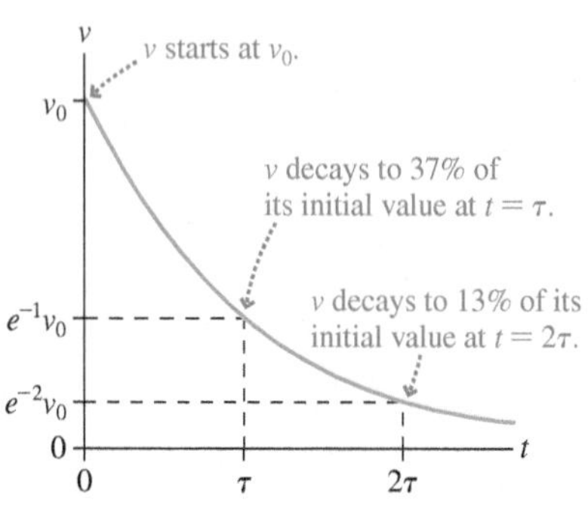
v
v starts at v_0.
v_0
v decays to 37% of its initial value at $t = \tau$.
$e^{-1}v_0$
v decays to 13% of its initial value at $t = 2\tau$.
$e^{-2}v_0$
0
0
τ
2τ
t

FIGURE 5.18 Speed of an object coasting to a halt at low Reynolds number with time constant τ.

Because the object's speed decays exponentially, starting from v_0, it will have decreased to $e^{-1}v_0 = 0.37v_0$ at $t = \tau$ and to $e^{-2}v_0 = 0.13v_0$ at $t = 2\tau$. As a practical matter, we could say that the motion has pretty well ceased at $t = 2\tau$, so we define a *stopping time*

$$\Delta t_{\text{stop}} = 2\tau = \frac{m}{3\pi\eta r} \tag{5.21}$$

We'll omit the details, but an integration of $v_x = dx/dt = v_0 \exp(-t/\tau)$ would find that the total distance the object traveled while stopping is

$$\Delta x_{\text{stop}} = v_0\tau = \tfrac{1}{2}v_0\,\Delta t_{\text{stop}} \tag{5.22}$$

NOTE ▸ This same analysis, with an integration leading to an exponential decay, will reappear in several other situations in this text. ◂

EXAMPLE 5.11 **A *Paramecium* coasts to a stop**

Paramecia, common unicellular inhabitants of freshwater ponds, are about 100 μm in size and swim at about 200 μm/s. If a *Paramecium* in 20°C water suddenly stops swimming, how long does it take to come to a stop and how far does it coast while stopping?

PREPARE A *Paramecium*'s swimming speed is its terminal speed in response to the pushing force generated by beating its cilia. Although *Paramecia* are somewhat elongated, we'll simplify the situation by modeling the *Paramecium* as a 100-μm-diameter sphere. The density of a *Paramecium* is essentially that of water, so we can calculate its mass from the density of water and the volume of a sphere.

SOLVE The mass is $m = \frac{4}{3}\pi r^3 \rho$, where ρ is the density of water. If we use this expression for the *Paramecium*'s mass in Equation 5.21, then the time it takes to come to a stop is

$$\Delta t_{\text{stop}} = \frac{\frac{4}{3}\pi r^3 \rho}{3\pi \eta r} = \frac{4r^2\rho}{9\eta} = \frac{4(50 \times 10^{-6}\text{ m})^2(1000\text{ kg/m}^3)}{9(1.0 \times 10^{-3}\text{ Pa}\cdot\text{s})}$$
$$= 1.1 \times 10^{-3}\text{ s} \approx 1\text{ ms}$$

The density and viscosity of water are taken from Table 5.3. During this time, the *Paramecium* travels distance

$$\Delta x_{\text{stop}} = \tfrac{1}{2} v_0 \,\Delta t_{\text{stop}} = \tfrac{1}{2}(200 \times 10^{-6}\text{ m/s})(1.1 \times 10^{-3}\text{ s})$$
$$= 1.1 \times 10^{-7}\text{ m} \approx 0.1\ \mu\text{m}$$

ASSESS Stopping in 1 ms over a distance of 0.1 μm, which is only $\frac{1}{1000}$ the diameter of the organism, is, for all practical purposes, an instantaneous stop. This agrees with our assertion about motion at low Reynolds number.

You can swim through water because your hands are able to push the water away from you, but swimming is much more challenging at very low Reynolds number. If you tried to swim through an extremely viscous fluid, like honey, your back-and-forth strokes would cause you to rock back and forth but there would be no net motion.

Microorganisms that need to move at very low Reynolds number had to evolve a different form of locomotion, their flagella and cilia. FIGURE 5.19 shows a simple model of a flagellum like the one a bacterium uses; it's basically a rotating cork-screw. The rotary motion—rather than the back-and-forth stroking motion of human swimming—enables the flagellum to apply a continuous push to the fluid. Thus the fluid, by Newton's third law, applies a continuous pushing force to the organism, and, as we've just seen, the organism almost instantaneously reaches its terminal speed and continues swimming at that constant speed.

FIGURE 5.19 A rotating flagellum exerts a continuous, propeller-like push on the fluid. In response, the fluid exerts a forward propulsion force on the organism.

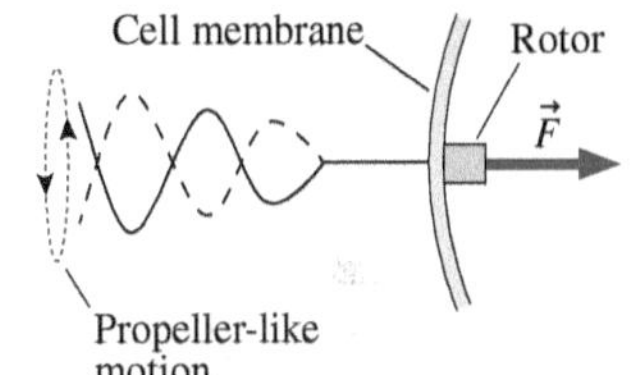

Protozoa are propelled by cilia. The beating motion of cilia is more complex than the motion of a rotating flagellum, but the end result is the same: a constant push against the fluid. The essential point is that the physics of life at low Reynolds number dictates evolutionary constraints not faced by larger organisms.

STOP TO THINK 5.4 *Chlamydomonas* are unicellular green algae that use a pair of flagella to swim at a speed of 200 μm/s. The graph shows the speed of one alga after it ceases swimming and coasts to a halt. What is the time constant of the motion?

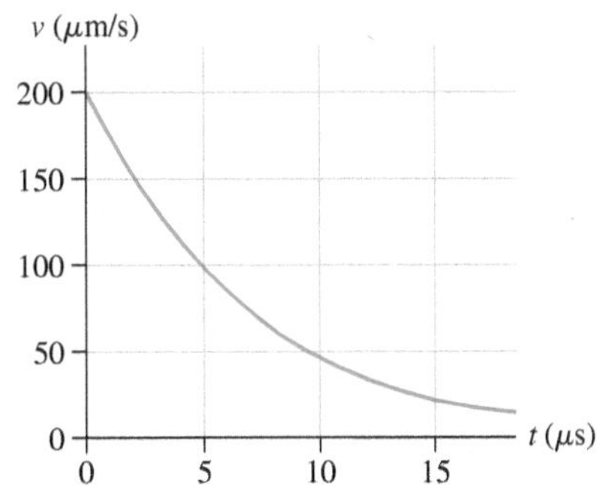

A. 5 μs
B. 7 μs
C. 10 μs
D. 14 μs

5.5 Springs and Elastic Forces

If you stretch a rubber band, a force tries to pull the rubber band back to its equilibrium, or unstretched, length. A force that restores a system to an equilibrium position is called a **restoring force.** Objects that exert restoring forces are called **elastic.** The most basic examples of elasticity are things like springs and rubber bands, but other examples of elasticity and restoring forces abound. For example, the steel beams flex slightly as you drive your car over a bridge, but they are restored to equilibrium after your car passes by. Biological elasticity occurs at the molecular level (e.g., nucleic acids and proteins) and at the organismal level (e.g., tendons, blood vessels, and even bones). Nearly everything that stretches, compresses, flexes, bends, or twists exhibits a restoring force and can be called elastic.

We're going to use a simple spring as our *model* of elasticity. Suppose you have a spring whose **equilibrium length** is L_0. This is the length of the spring when it is neither pushing nor pulling. If you stretch (or compress) the spring, how hard does it pull (or push) back? Measurements show that

- The force is *opposite the displacement.* This is what we *mean* by a restoring force.
- If you don't stretch or compress the spring too much, the force is *proportional to the displacement from equilibrium.* The farther you push or pull, the larger the force.

FIGURE 5.20 Properties of a spring.

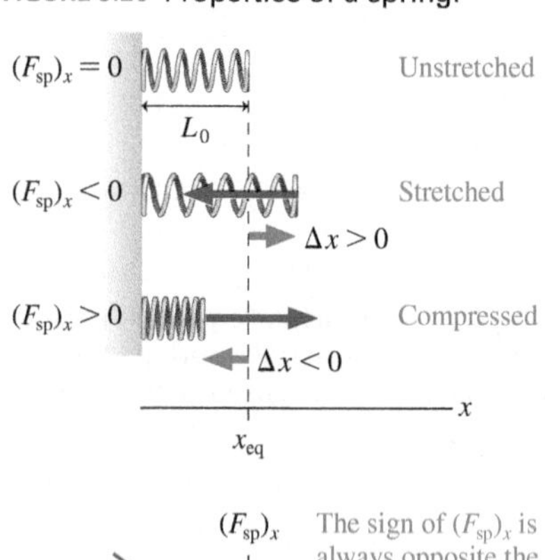

FIGURE 5.20 shows a horizontal spring exerting force $\vec{F}_{sp}$. The free end of the spring, when it is neither stretched nor compressed, is at position x_{eq}. Note that x_{eq} is usually *not* the same as the spring's equilibrium length L_0; that will be true only if the origin of the x-axis is at the wall. When the spring is stretched, the **spring displacement** $\Delta x = x - x_{eq}$ is positive while $(F_{sp})_x$, the x-component of the restoring force, is negative. Similarly, compressing the spring makes $\Delta x < 0$ and $(F_{sp})_x > 0$. The graph of force versus displacement is a straight line with negative slope, showing that the spring force is proportional to but *opposite* the displacement.

The equation of the straight-line graph passing through the origin is

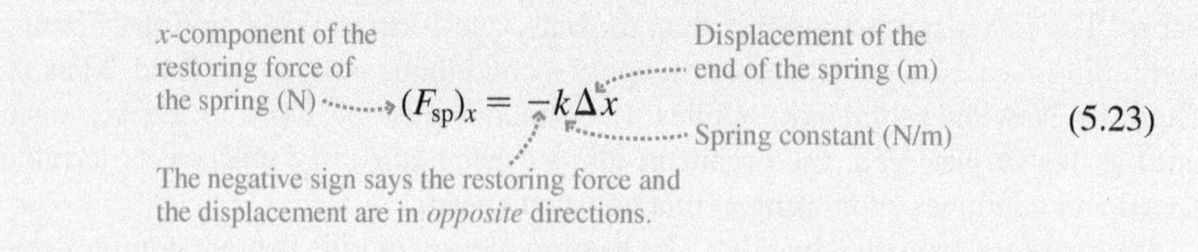

$$(F_{sp})_x = -k\,\Delta x \tag{5.23}$$

For a vertical spring with displacement Δy we would write $(F_{sp})_y = -k\,\Delta y$.

The minus sign is the mathematical indication of a *restoring* force, and the constant k—the absolute value of the slope of the line—is called the **spring constant** of the spring. The units of the spring constant are N/m. This relationship between the force and displacement of a spring was discovered by Robert Hooke, a contemporary (and sometimes bitter rival) of Newton. **Hooke's law** is not a true "law of nature," in the sense that Newton's laws are, but is actually just a *model* of a restoring force. It works well for *small* displacements from equilibrium, but fails for any real spring that is compressed or stretched too far. A hypothetical massless spring for which Hooke's law is true at all displacements is called an **ideal spring.** We'll consider only ideal springs in this chapter.

NOTE ► The force does not depend on the spring's physical length L but, instead, on the *displacement* Δx of the end of the spring. ◄

The spring constant k is a property that characterizes a spring, just as mass m characterizes a particle. For a given spring, k is a constant—it does not change as the spring is stretched or compressed. If k is large, it takes a large pull to cause a significant stretch, and we call the spring a "stiff" spring. A spring with small k can be stretched with very little force, and we call it a "soft" spring.

Many biological materials, such as tendons and bones, behave much like springs when they are only slightly stretched or compressed. By modeling them as springs, we can use an appropriate spring constant to characterize their elastic properties. We'll return to this idea in the next chapter.

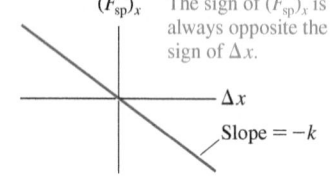

EXAMPLE 5.12 **A spring in your step**

Several companies manufacture running shoes that have steel springs in the heel pads to launch a runner back into the air after a heel strike. When a 70 kg runner rocks back on his heels so that all of his weight is supported by the heel springs, the springs each compress by 1.2 mm. When he is running hard and hits the ground on one heel, the force compressing the spring is 5.0 times his weight.

a. What is the spring constant k of a heel spring?
b. By how much does a spring compress during a heel strike?

PREPARE When the spring in the heels is compressed, it produces a restoring force that pushes upward on the runner's foot. When the runner rocks back on his heels, we can assume that half of the runner's weight is on each heel. The pictorial representation of **FIGURE 5.21** shows the situation. We know the force that is compressing the spring and how much it compresses, so we will use those values to find the spring constant. With the spring constant in hand, we will find the compression in part b.

SOLVE

a. The runner rocking back on his heels is in equilibrium, with no net force, so for each foot

$$\sum F_y = (F_{sp})_y + w_y = (F_{sp})_y - \tfrac{1}{2}mg = 0$$

FIGURE 5.21 Pictorial representation of the heel spring.

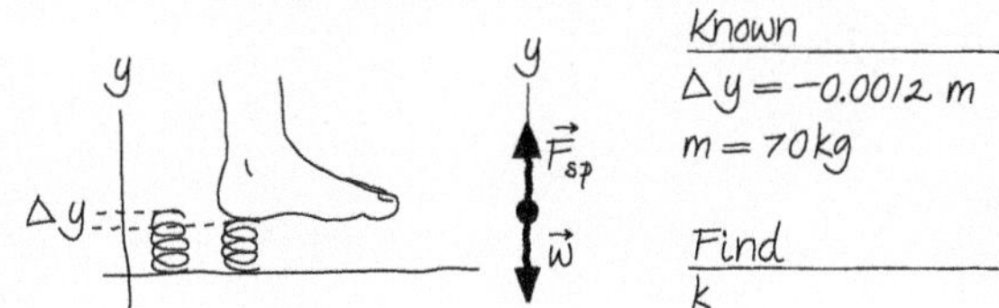

where we use $w = \tfrac{1}{2}mg$ because only half his weight rests on each spring. The force exerted by the spring is $(F_{sp})_y = -k\,\Delta y$. Notice that $\Delta y = y_f - y_i = -1.2$ mm is *negative* because the displacement is downward. Using this expression for the spring force in the second-law equation gives

$$-k\,\Delta y - \tfrac{1}{2}mg = 0$$

We can now solve for the spring constant:

$$k = -\frac{mg}{2\,\Delta y} = -\frac{(70\text{ kg})(9.8\text{ m/s}^2)}{2(-0.0012\text{ m})} = 2.9 \times 10^5\text{ N/m}$$

b. The restoring force during a heel strike is 5.0 times the runner's weight. Now we have

$$(F_{sp})_y = -k\,\Delta y = 5mg$$

$$\Delta y = -\frac{5mg}{k} = -0.012\text{ m} = -12\text{ mm}$$

The downward displacement is negative, but the answer to By how much? is that the spring is compressed by 12 mm.

ASSESS The spring is quite stiff, approximately 300,000 N/m. This is reasonable; the spring compresses only a little under the large force of the runner's weight. The force of the heel strike is 5 times the runner's weight, or 10 times the original force that was half the runner's weight. This gives a compression that is 10 times the original compression, 12 mm instead of 1.2 mm. This is reasonable because the spring's restoring force is linearly proportional to the displacement.

EXAMPLE 5.13 **The spring constant of DNA**

When uncoiled, a double-helix DNA molecule is surprisingly stretchy. Single-molecule experiments with *optical tweezers* are able to measure the elastic properties of DNA and other biomolecules. One such experiment, shown in **FIGURE 5.22**, used a 3700-base-pair (bp) length of DNA. One end of the molecule was attached to a glass slide, the other to a tiny 1-μm-diameter polystyrene bead that was held at the focus of a laser beam. (Optical tweezers will be discussed in Chapter 20.) The optical tweezers moved the bead a few nanometers at a time, slowly stretching the molecule while measuring its restoring force. The graph in Figure 5.22 shows the results; notice that small biomechanical forces are measured in piconewtons (1 pN = 1 piconewton = 10^{-12} N). What are (a) the spring constant of DNA for small displacements and (b) the fractional change in length for which Hooke's law is valid? DNA has 3.0 bp/nm.

FIGURE 5.22 An optical tweezers experiment to measure the elastic properties of DNA.

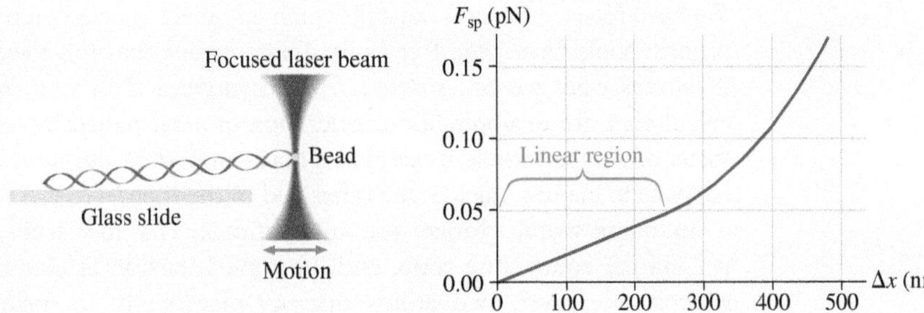

Continued

PREPARE The data show the *magnitude* of the restoring force, so this graph has a positive slope rather than the negative slope of Figure 5.20, which graphed the x-component of the force. We can see that DNA is not an ideal spring because the graph is not a straight line. However, the graph is linear over the first ≈250 nm of stretching, so Hooke's law is valid for these small displacements.

SOLVE

a. The magnitude of the restoring force is $F_{sp} = k\,\Delta x$, which tells us that the spring constant is the slope of the force graph over its linear region. You can see from the graph that the force increases to 0.050 pN when the molecule has been stretched by 250 nm, so the slope, the spring constant, is

$$k = \frac{0.050 \text{ pN}}{250 \text{ nm}} = \frac{0.050 \text{ pN}}{0.25\ \mu\text{m}} = 0.20 \text{ pN}/\mu\text{m}$$

These are appropriate units for biological work, where piconewton forces and micrometer distances are typical. We can compare the strand of DNA to macroscopic springs by converting to standard SI units:

$$k = 0.20 \frac{\text{pN}}{\mu\text{m}} \times \frac{10^{-12} \text{ N}}{1 \text{ pN}} \times \frac{1\ \mu\text{m}}{10^{-6} \text{ m}} = 2.0 \times 10^{-7} \text{ N/m}$$

A macroscopic length of a single DNA molecule would be an extremely soft spring!

b. This particular DNA molecule, with 3700 base pairs, obeys Hooke's law for displacements of up to ≈250 nm. Over what range of displacements would Hooke's law be valid for a molecule with, say, 500 base pairs or 5000 base pairs? It seems plausible that any DNA molecule, regardless of overall length, will stay on the linear part of the graph if stretched by the same *fraction* of its length because that keeps the amount of stretch per base pair the same. Thus a more general result, applicable to any DNA molecule, is to specify the maximum displacement as a fraction of the molecule's length. Given that DNA has 3.0 bp/nm, a characteristic of all DNA, the length of this particular DNA molecule was

$$L = 3700 \text{ bp} \times \frac{1 \text{ nm}}{3.0 \text{ bp}} = 1230 \text{ nm}$$

We've kept an extra significant figure for this calculation to avoid round-off errors. A displacement of 250 nm is a fractional change in length of

$$\frac{\Delta x}{L} = \frac{250 \text{ nm}}{1230 \text{ nm}} = 0.20 = 20\%$$

This indicates that a strand of DNA of any length will obey Hooke's law and act as an ideal spring if stretched to no more than 20% of its relaxed length. A stretch of more than 20% leads to nonideal behavior that is not described by Hooke's law.

ASSESS Cell structure and cellular processes such as mitosis depend on the mechanical properties of the cellular materials. Elasticity is one such property, but certainly not the only one. We'll take a closer look at the mechanical properties of biological materials in the next chapter.

5.6 Newton's Third Law

FIGURE 5.23 An interaction pair of forces.

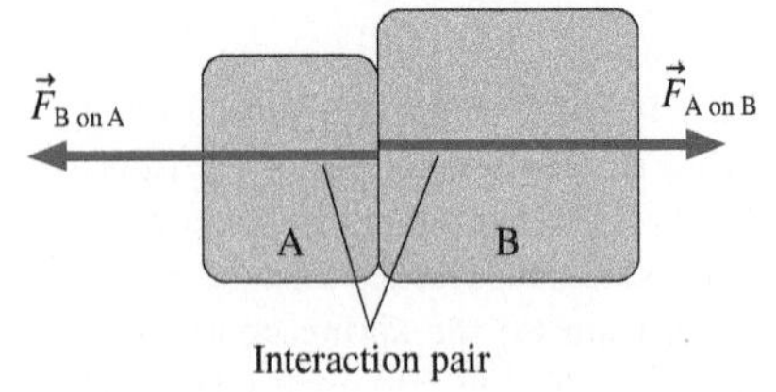

This chapter opened with a hammer hitting a nail; we noted that the hammer pushes on the nail while the nail pushes back on the hammer. In general, any time one object A exerts a force $\vec{F}_{\text{A on B}}$ on another object B, then object B exerts a force $\vec{F}_{\text{B on A}}$ *in the opposite direction* on object A. This pair of forces, shown in FIGURE 5.23, is called an **interaction pair.** Notice the very explicit subscripts on the force vectors. The first letter is the *agent;* the second letter is the object on which the force acts. $\vec{F}_{\text{A on B}}$ is a force exerted *by* A *on* B.

NOTE ▶ The two members of an interaction pair of forces (1) act on different objects and (2) act in opposite directions. ◀

Objects, Systems, and the Environment

We have focused on situations in which an object is acted on by external forces; our primary tools have been free-body diagrams and Newton's second law. FIGURE 5.24a illustrates what we call *single-particle dynamics.* This is a powerful approach for calculating, for example, the acceleration of a car pulled by a rope. But the narrow focus of single-particle dynamics ignores important questions. For example: What happens to the tow truck at the other end of the rope?

In other words, forces are interactions. The tow truck and the car *interact* via the connecting rope, and the rope's tension affects both of them. At the microscopic level, two atoms *interact* electrically to form a molecular bond.

FIGURE 5.24 Single-particle dynamics and a model of interacting objects.

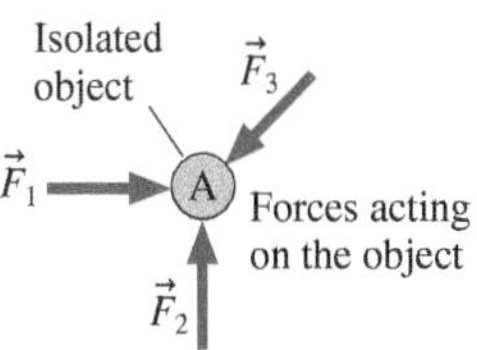

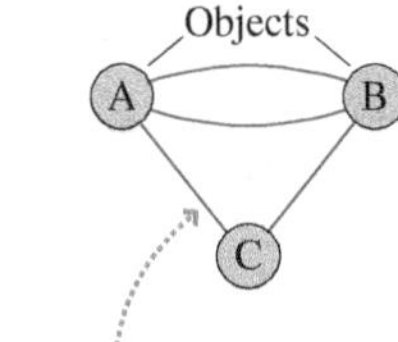

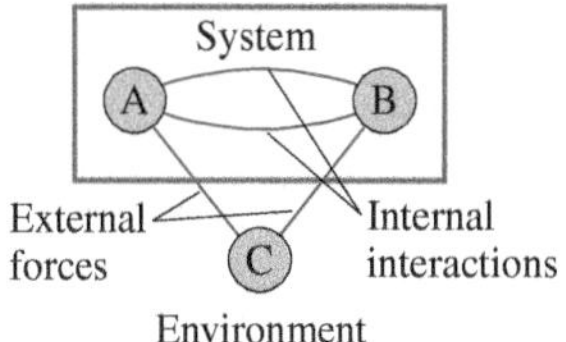

Bonding and chemical energy cannot be understood from a single-particle perspective; we need to understand the interactions themselves, and that requires a broader perspective.

As an example, FIGURE 5.24b shows three objects interacting via interaction pairs of forces. We might need to know how all three objects move as a result of these interactions. On the other hand, we'll often be interested in the motion of some of the objects, say objects A and B, but not of others. For example, objects A and B might be the hammer and the nail, while object C is the earth. The earth interacts with both the hammer and the nail via gravity, but in a practical sense the earth remains "at rest" while the hammer and nail move. Let's define the *system,* as we did earlier, to be those objects whose motion we want to analyze and the *environment* as objects external to the system.

FIGURE 5.24c is a new kind of diagram, an **interaction diagram,** in which we've enclosed the objects of the system in a box and represented interactions as lines connecting objects. This is a rather abstract, schematic diagram, but it captures the essence of the interactions. Notice that interactions between objects inside the system are referred to as **internal interactions** while interactions with objects in the environment are called **external forces.** For the hammer and nail, the impact is an internal interaction whereas the gravitational force on each—an interaction with the earth—is an external force.

NOTE ▸ *Every* force is one member of an interaction pair, so there is no such thing as a true "external force." What we call an external force is simply an interaction between an object we've chosen to place inside the system and an object outside the system. ◂

Propulsion

A system with an internal source of energy, such as glucose or gasoline, can use that energy to drive itself forward. We call this **propulsion.** Propulsion is an important aspect not only of walking or running but also of the forward motion of swimmers, cars, and rockets. Understanding propulsion, which can be somewhat counterintuitive, depends on thinking about interaction pairs of forces.

If you try to walk across a frictionless floor, your foot slips and slides *backward.* In order for you to walk, friction is needed so that your foot *sticks* to the floor as you straighten your leg, moving your body forward. The friction that prevents slipping is *static* friction. Static friction, you will recall, acts in the direction that prevents slipping. The static friction force has to point in the *forward* direction to prevent your foot from slipping backward. It is this forward-directed static friction force that propels you forward! The force of your foot on the floor, the other half of the interaction pair, is in the opposite direction.

You can walk forward, whereas a crate sitting on the ground cannot, because you have an *internal source of energy* that allows you to straighten your leg by pushing

Jetting around An octopus propels itself by expelling a jet of water. The animal pushes on the water and—by Newton's third law—the water pushes on the animal. The animal accelerates due to the force of the water, a propulsive force.

FIGURE 5.25 Examples of propulsion.

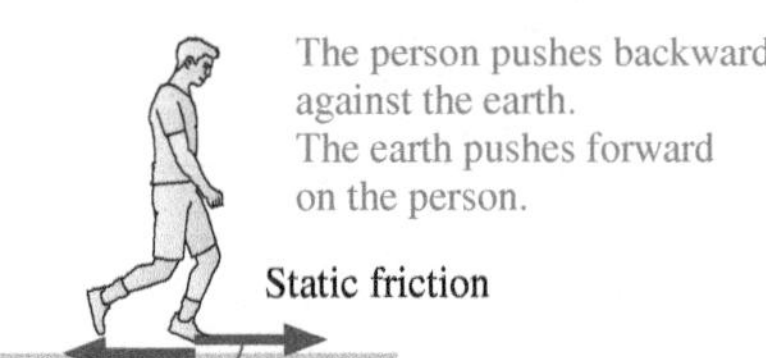
The person pushes backward against the earth.
The earth pushes forward on the person.
Static friction
The car pushes backward against the earth.
The earth pushes forward on the car.
Static friction

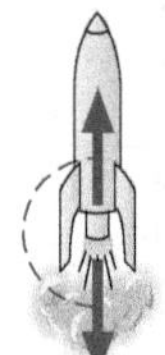
The rocket pushes the hot gases backward.
The gases push the rocket forward.
Thrust

backward against the surface. In essence, you walk by pushing the earth away from you. The earth's surface responds by pushing you forward. These are static friction forces. In contrast, all the crate can do is slide, so *kinetic* friction opposes the motion of the crate. FIGURE 5.25 shows examples of propulsion.

We'll illustrate these new ideas with a concrete example.

CONCEPTUAL EXAMPLE 5.14 Pushing a crate

FIGURE 5.26 shows a person pushing a large crate across a rough surface. Identify all interactions, show them on an interaction diagram, then draw free-body diagrams of the person and the crate.

FIGURE 5.26 A person pushes a crate across a rough floor.

REASON The interaction diagram of FIGURE 5.27 starts by representing every object as a circle in the correct position but separated from all other objects. The person and the crate are obvious objects. The earth is also an object that both exerts and experiences forces, but it's necessary to distinguish between the surface, which exerts contact forces, and the entire earth, which exerts the long-range weight force.

Figure 5.27 also identifies the various interactions. Some, like the pushing interaction between the person and the crate, are fairly obvious. The interactions with the earth are a little trickier. Gravity, a long-range force, is an interaction between each object and the earth as a whole. Friction forces and normal forces are contact interactions between each object and the earth's surface. These are two different interactions, so two interaction lines connect the crate to the surface and the person to the surface. Finally, we've enclosed the person and crate in a box labeled System. These are the objects whose motion we wish to analyze.

NOTE ► Interactions are between two *different* objects. None of the interactions are between an object and itself. ◄

The objects in the system—the crate and the person—are what move, so we need a free-body diagram for each of them. FIGURE 5.28 correctly places the crate's free-body diagram to the right of the person's free-body diagram. Each object has three interaction lines that cross the system boundary and thus represent external forces. These are the weight force due to the entire earth, the upward normal force from the surface, and a friction force from the surface. We can use familiar labels such as $\vec{n}_P$ and $\vec{f}_C$, but **it's very important to distinguish different forces with subscripts.** There's now more than one normal force. If you call both simply $\vec{n}$, you're almost certain to make mistakes when you start writing out the second-law equations.

The directions of the normal forces and the weight forces are clear, but we have to be careful with friction. Friction force $\vec{f}_C$ is a kinetic friction force acting on the crate as it slides across the surface, so it points left, opposite the motion. But what about friction between the person and the surface? It is tempting to draw force $\vec{f}_P$ pointing to the left. After all, friction forces are supposed to be in the direction opposite the motion. But if we did so, the person would have two forces to the left, $\vec{F}_{C \text{ on } P}$ and $\vec{f}_P$, and none to the

FIGURE 5.27 The interaction diagram.

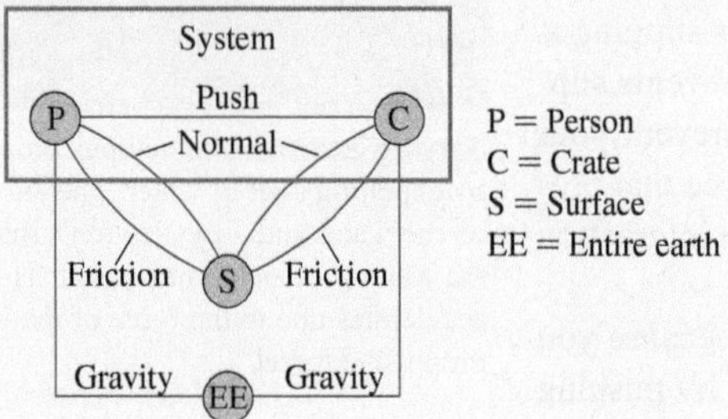
System
Push
Normal
P
C
S
EE
Friction
Friction
Gravity
Gravity
P = Person
C = Crate
S = Surface
EE = Entire earth

FIGURE 5.28 Free-body diagrams of the person and the crate.

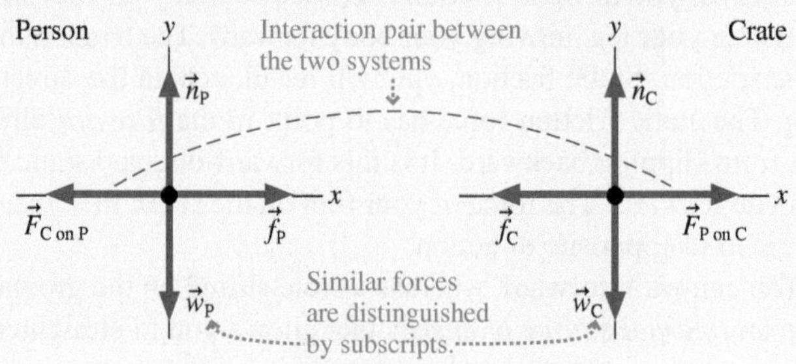
Person
Crate
Interaction pair between the two systems
y
x
$\vec{n}_P$
$\vec{n}_C$
$\vec{F}_{C \text{ on } P}$
$\vec{f}_P$
$\vec{f}_C$
$\vec{F}_{P \text{ on } C}$
$\vec{w}_P$
$\vec{w}_C$
Similar forces are distinguished by subscripts.

right, causing the person to accelerate *backward*! That is clearly not what happens, so what is wrong?

The friction force on the person is a *propulsion force*. It's a static friction force because the person's foot is planted on the ground, not sliding across the surface. The person pushes backward against the surface with force $\vec{f}_{\text{P on S}}$, and in turn the surface pushes forward on the person with force $\vec{f}_{\text{S on P}}$, which we've shortened to $\vec{f}_{\text{P}}$. This is the force that moves the person to the right.

Finally, we have one internal interaction. The crate is pushed with force $\vec{F}_{\text{P on C}}$. If A pushes or pulls on B, then B pushes or pulls back on A, so the reaction to force $\vec{F}_{\text{P on C}}$ is $\vec{F}_{\text{C on P}}$, the crate pushing back against the person's hands. Force $\vec{F}_{\text{P on C}}$ is a force exerted on the crate, so it's shown on the crate's free-body diagram. Force $\vec{F}_{\text{C on P}}$ is exerted *on the person,* so it is drawn on the person's free-body diagram.

ASSESS This has been a long example, but it illustrates how interacting objects are analyzed. It is especially important to notice that **the two forces of an interaction diagram never occur on the same free-body diagram.** We've drawn $\vec{F}_{\text{P on C}}$ as a force on the crate and $\vec{F}_{\text{C on P}}$ as a force on the person, then connected them with a dashed line to show that they are an interaction pair.

Applying the Third Law

Newton was the first to recognize how the two members of an interaction pair of forces are related to each other. Today we know this as **Newton's third law:**

Newton's third law Every force occurs as one member of an interaction pair of forces.

- The two members of an interaction pair act on two *different* objects.
- The two members of an interaction pair are equal in magnitude but opposite in direction: $\vec{F}_{\text{A on B}} = -\vec{F}_{\text{B on A}}$.

We deduced most of the third law above, where we found that the two members of an interaction pair are always opposite in direction. But the most significant portion of the third law, which is by no means obvious, is that the two members of an interaction pair have *equal* magnitudes. That is, $F_{\text{A on B}} = F_{\text{B on A}}$. This is the quantitative relationship that will allow you to solve problems of interacting objects.

Newton's third law is frequently stated as "For every action there is an equal but opposite reaction." While this is indeed a catchy phrase, it lacks the preciseness of our preferred version. For one thing, it's usually not possible to identify one force as the action and the other as the reaction; it's simply a pair of forces. In addition, this phrase fails to capture an essential feature of interaction pairs—that they each act on a *different* object.

NOTE ▶ Newton's third law extends and completes our concept of *force*. We can now recognize force as an *interaction* between objects rather than as some "thing" with an independent existence of its own. The concept of an interaction will become increasingly important as we begin to study the laws of energy and momentum. ◀

Newton's third law is easy to state but harder to grasp. For example, consider what happens when you release a ball. Not surprisingly, it falls down. But if the ball and the earth exert equal and opposite forces on each other, as Newton's third law alleges, why doesn't the earth "fall up" to meet the ball?

The key to understanding this and many similar puzzles is that Newton's third law equates the size of two forces, not two accelerations. The acceleration depends not only on the force but also on the mass, as Newton's second law states. **In an interaction between two objects of different mass, the lighter mass will have a larger acceleration even though the forces exerted on the two objects are equal.** The earth is almost infinitely massive compared to the falling ball, so—with equal forces—the earth's acceleration is infinitely small in comparison to that of the ball.

CONCEPTUAL EXAMPLE 5.15 The forces on accelerating boxes

The hand shown in FIGURE 5.29 pushes boxes A and B to the right across a frictionless table. The mass of B is larger than the mass of A.

a. Draw an interaction diagram and free-body diagrams of A, B, and the hand H, showing only the *horizontal* forces. Connect interaction pairs with dashed lines.
b. Rank in order, from largest to smallest, the horizontal forces shown on your free-body diagrams.

FIGURE 5.29 Hand H pushes boxes A and B.

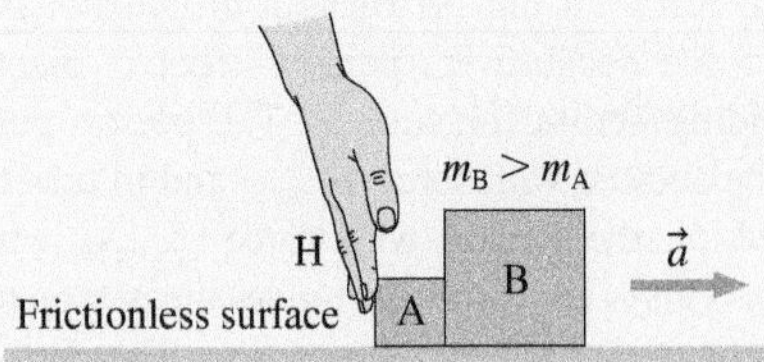

REASON

a. This somewhat artificial example illustrates the reasoning process in using Newton's second and third laws together to understand an interaction. There are five important steps:
   - Define the system.
   - Draw an interaction diagram to identify the forces and interactions.
   - Show only the horizontal forces and interactions on free-body diagrams, *using a separate diagram for each object.*
   - Apply Newton's third law to any interaction pair of forces, one on each free-body diagram.
   - Apply Newton's second law to each object.

FIGURE 5.30 shows the first three steps. We've defined the hand and the two boxes to be the system and identified the external forces from the environment. Boxes A and B each experience a downward weight force as they interact with the entire earth and an upward normal force from the surface. The hand is driven by an external agent that we can call the arm. Internally, hand H pushes on box A, and A pushes back on H, so $\vec{F}_{H \text{ on } A}$ and $\vec{F}_{A \text{ on } H}$ are an interaction pair. Similarly, A and B push on each other. **The hand does not touch box B, so there is no interaction between H and B.** Altogether there are five horizontal forces and two interaction pairs. Notice that each force is shown on the free-body diagram of the object that it acts *on*.

b. According to Newton's third law, $F_{A \text{ on } H} = F_{H \text{ on } A}$ and $F_{A \text{ on } B} = F_{B \text{ on } A}$. But the third law is not our only tool. The boxes are *accelerating* to the right, because there's no friction, so Newton's *second* law tells us that box A must have a net force to the right. Consequently, $F_{H \text{ on } A} > F_{B \text{ on } A}$. Similarly, $F_{\text{arm on } H} > F_{A \text{ on } H}$ is needed to accelerate the hand. Thus

$$F_{\text{arm on } H} > F_{A \text{ on } H} = F_{H \text{ on } A} > F_{B \text{ on } A} = F_{A \text{ on } B}$$

ASSESS You might have expected $F_{A \text{ on } B}$ to be larger than $F_{H \text{ on } A}$ because $m_B > m_A$. It's true that the *net* force on B is larger than the *net* force on A, but we have to reason more closely to judge the individual forces. Notice how we used both the second and the third laws to answer this question.

FIGURE 5.30 The interactions between the hand and the boxes.

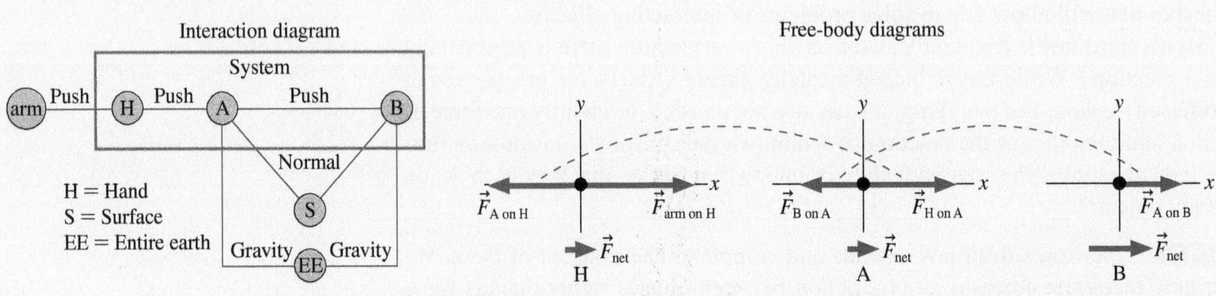

STOP TO THINK 5.5 A small car is pushing a larger truck that has a dead battery. The mass of the truck is larger than the mass of the car. Which of the following statements is true?

A. The car exerts a force on the truck, but the truck doesn't exert a force on the car.
B. The car exerts a larger force on the truck than the truck exerts on the car.
C. The car exerts the same amount of force on the truck as the truck exerts on the car.
D. The truck exerts a larger force on the car than the car exerts on the truck.
E. The truck exerts a force on the car, but the car doesn't exert a force on the truck.

Tension Revisited

Many objects are connected by strings, ropes, tendons, and so on. In single-particle dynamics, we defined "tension" as the force exerted on an object by a rope or string. Now we need to think more carefully about the string itself. Just what do we mean when we talk about the tension "in" a string?

FIGURE 5.31 shows a heavy safe hanging from a rope, placing the rope under tension. If you cut the rope, the safe and the lower portion of the rope will fall. Thus there must be a force *within* the rope by which the upper portion of the rope pulls upward on the lower portion to prevent it from falling.

Chapter 4 introduced an atomic-level model in which tension is due to the stretching of spring-like molecular bonds. Stretched springs exert pulling forces, and the combined pulling force of billions of stretched molecular springs in a string or rope is what we call *tension.*

An important aspect of tension is that it pulls equally *in both directions.* To gain a mental picture, imagine holding your arms outstretched and having two friends pull on them. You'll remain at rest—but "in tension"—as long as they pull with equal strength in opposite directions. But if one lets go, analogous to the breaking of molecular bonds if a rope breaks or is cut, you'll fly off in the other direction!

FIGURE 5.31 Tension forces within the rope are due to stretching the spring-like molecular bonds.

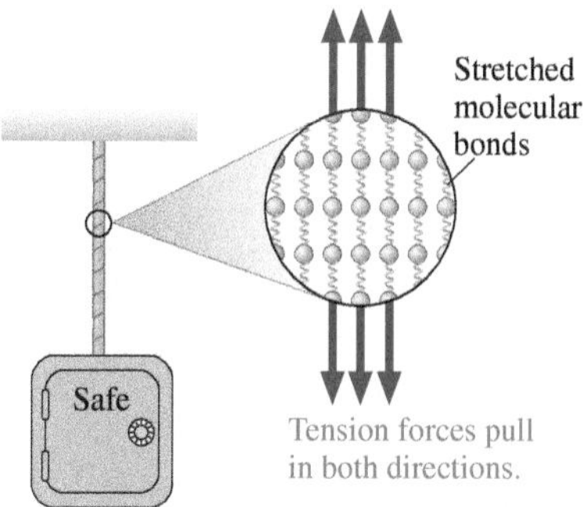

CONCEPTUAL EXAMPLE 5.16 **Pulling a rope**

FIGURE 5.32a shows a student pulling horizontally with a 100 N force on a rope that is attached to a wall. In FIGURE 5.32b, two students in a tug-of-war pull on opposite ends of a rope with 100 N each. Is the tension in the second rope larger than, smaller than, or the same as that in the first rope?

FIGURE 5.32 Which rope has a larger tension?

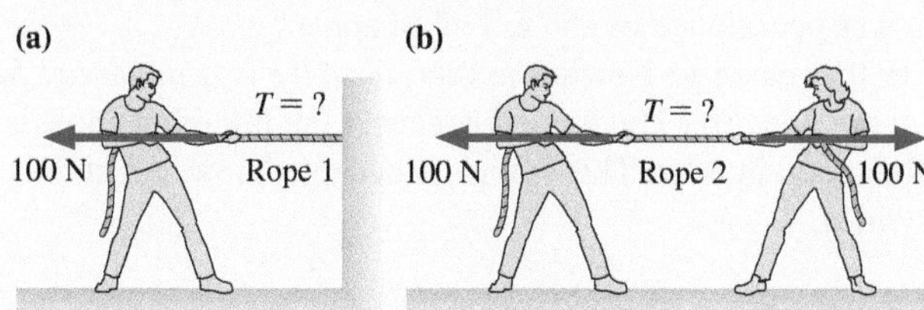

REASON Surely pulling on a rope from both ends causes more tension than pulling on one end. Right? Before jumping to conclusions, let's analyze the situation carefully.

FIGURE 5.33a shows the first student, the rope, and the wall as separate, interacting objects. Force $\vec{F}_{S \text{ on } R}$ is the student pulling on the rope, so it has magnitude 100 N. Forces $\vec{F}_{S \text{ on } R}$ and $\vec{F}_{R \text{ on } S}$ are an interaction pair and must have equal magnitudes. Similarly for forces $\vec{F}_{W \text{ on } R}$ and $\vec{F}_{R \text{ on } W}$. Finally, because the rope is in equilibrium, *force* $\vec{F}_{W \text{ on } R}$ *has to balance force* $\vec{F}_{S \text{ on } R}$. Thus

$$F_{R \text{ on } W} = F_{W \text{ on } R} = F_{S \text{ on } R} = F_{R \text{ on } S} = 100 \text{ N}$$

The first and third equalities are Newton's third law; the second equality follows from Newton's second law for the rope with $a_R = 0$.

Forces $\vec{F}_{R \text{ on } S}$ and $\vec{F}_{R \text{ on } W}$ are the pulling forces exerted by the rope and are what we *mean* by "the tension in the rope." Thus the tension in the first rope is 100 N.

FIGURE 5.33b repeats the analysis for the rope pulled by two students. Each student pulls with 100 N, so $F_{S1 \text{ on } R} = 100$ N and $F_{S2 \text{ on } R} = 100$ N. Just as before, there are two interaction pairs and the rope is in static equilibrium. Thus

$$F_{R \text{ on } S2} = F_{S2 \text{ on } R} = F_{S1 \text{ on } R} = F_{R \text{ on } S1} = 100 \text{ N}$$

The tension in the rope—the pulling forces $\vec{F}_{R \text{ on } S1}$ and $\vec{F}_{R \text{ on } S2}$—is still 100 N!

You may have *assumed* that the student on the right in Figure 5.32b is doing something to the rope that the wall in Figure 5.32a does not do. But our analysis finds that the wall, just like the student, pulls to the right with 100 N. The rope doesn't care whether it's pulled by a wall or a hand. It experiences the same forces in both cases, so the rope's tension is the same in both.

ASSESS Ropes and strings exert forces at *both* ends. The force with which they pull—and thus the force pulling on them at each end—*is* the tension in the rope. Tension is not the sum of the pulling forces.

FIGURE 5.33 Analysis of tension forces.

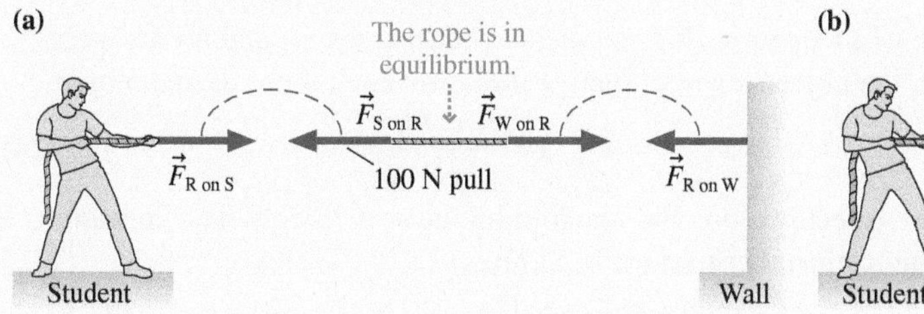

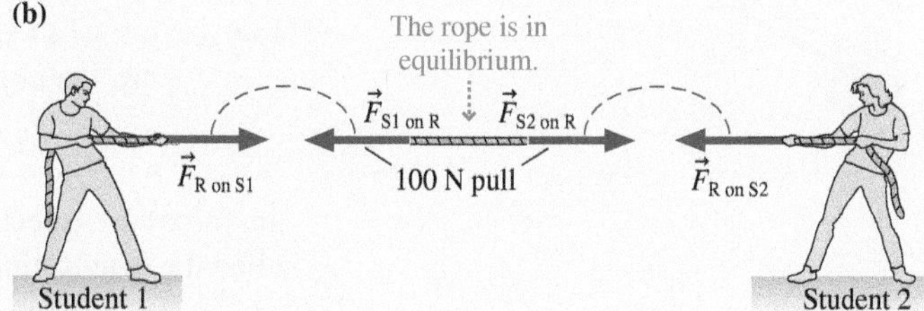

FIGURE 5.34 The massless string approximation allows objects A and B to act *as if* they are directly interacting.

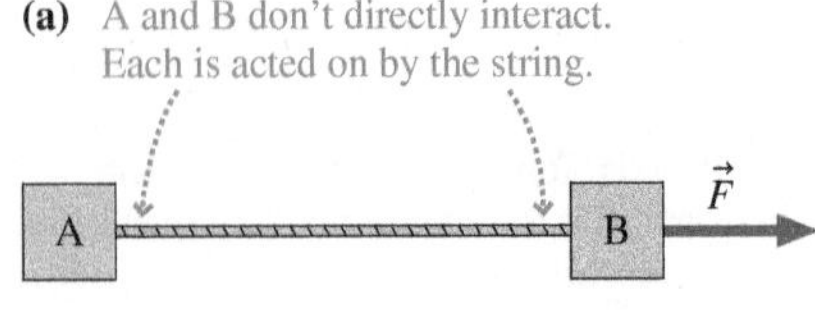
(a) A and B don't directly interact.
Each is acted on by the string.
A
B
$\vec{F}$

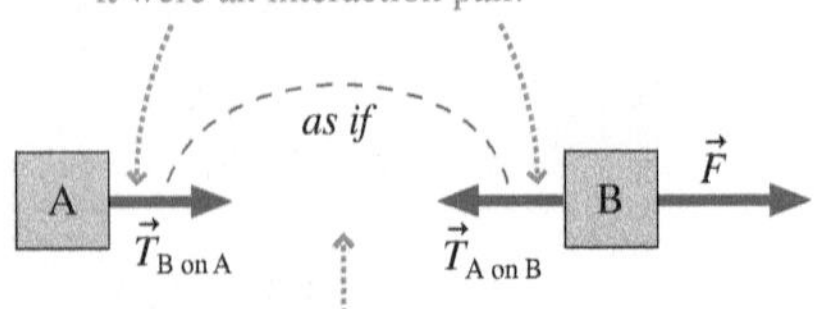
(b) This pair of forces acts as if it were an interaction pair.
as if
A
B
$\vec{F}$
$\vec{T}_{\text{B on A}}$
$\vec{T}_{\text{A on B}}$
We can omit the string if we assume it is massless.

FIGURE 5.34a shows objects A and B connected by a string or rope under tension. Each is acted on by the tension in the string. If the mass of the string is much less than the masses of the objects, which is often the case, we can adopt the *massless string approximation*. Because it takes no net force to accelerate a massless string, the tension force of the string pulling forward on A is exactly the same as the tension force pulling backward on B. Although A and B really interact with the string, not directly with each other, a massless string allows us to treat the forces $\vec{T}_{\text{A on B}}$ and $\vec{T}_{\text{B on A}}$ in FIGURE 5.34b *as if* they are an interaction pair of forces. That is, a massless string simply transmits a force from A to B without changing the magnitude of that force. For problems in this book, you can assume that strings and ropes are massless unless stated otherwise.

STOP TO THINK 5.6 Blocks A and B are connected by massless string 2 and pulled across a frictionless table by massless string 1. B has a larger mass than A. Is the tension T_2 in string 2 larger than, smaller than, or equal to the tension T_1 in string 1?

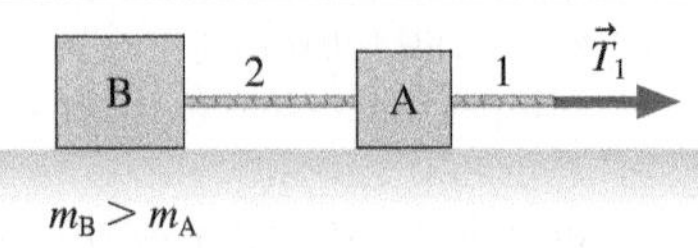
B
2
A
1
$\vec{T}_1$
$m_B > m_A$

5.7 Newton's Law of Gravity

Gravity enters our everyday lives as the force responsible for the weight of objects on or near the surface of a planet. But gravity has more far-reaching implications as the force that holds the solar system together. Our current understanding of the force of gravity begins with Isaac Newton. The popular image of Newton coming to a key realization about gravity after an apple fell on his head is at least close to the truth: Newton himself said that the "notion of gravitation" came to him as he "sat in a contemplative mood" and "was occasioned by the fall of an apple."

The important notion that came to Newton is this: *Gravity is a universal force that affects all objects in the universe.* The force that causes the fall of an apple is the same force that keeps the moon in orbit. This is widely accepted now, but at the time this was a revolutionary idea.

Gravity Obeys an Inverse-Square Law

Newton proposed that *every* object in the universe attracts *every other* object with a force that has the following properties:

1. The force is inversely proportional to the square of the distance between the objects.
2. The force is directly proportional to the product of the masses of the two objects.

FIGURE 5.35 shows two spherical objects with masses m_1 and m_2 separated by distance r. Each object exerts an attractive force on the other, a force that we call the **gravitational force.** These two forces form an interaction pair, so $\vec{F}_{1\text{ on }2}$ is equal in magnitude and opposite in direction to $\vec{F}_{2\text{ on }1}$. The magnitude of the forces is given by Newton's law of gravity.

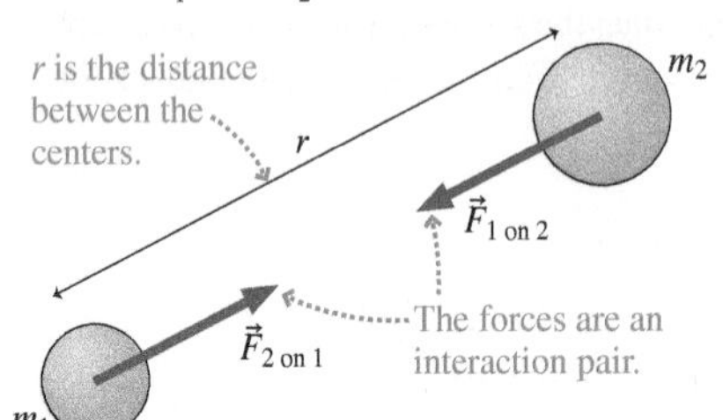
r is the distance between the centers.
r
m_2
$\vec{F}_{1\text{ on }2}$
$\vec{F}_{2\text{ on }1}$
The forces are an interaction pair.
m_1

FIGURE 5.35 The gravitational forces on masses m_1 and m_2.

Newton's law of gravity If two objects with masses m_1 and m_2 are a distance r apart, the objects exert attractive forces on each other of magnitude

$$F_{1\text{ on }2} = F_{2\text{ on }1} = \frac{Gm_1m_2}{r^2} \tag{5.24}$$

The forces are directed along the line joining the two objects. The constant G is called the **gravitational constant.** In SI units,

$$G = 6.67 \times 10^{-11}\ \text{N}\cdot\text{m}^2/\text{kg}^2$$

NOTE ▶ Strictly speaking, Newton's law of gravity applies to *particles* with masses m_1 and m_2. However, it can be shown that the law also applies to the force between two spherical objects if r is the distance between their centers. ◀

As the distance r between two objects increases, the gravitational force between them decreases. Because the distance appears squared in the denominator, Newton's law of gravity is what we call an **inverse-square law.** Doubling the distance between two masses causes the force between them to decrease by a factor of 4.

CONCEPTUAL EXAMPLE 5.17 The gravitational force between two spheres

The gravitational force between two giant lead spheres is 0.010 N when the centers of the spheres are 20 m apart. What is the distance between their centers when the gravitational force between them is 0.160 N?

REASON We can solve this problem without knowing the masses of the two spheres. The key is to consider the ratios of forces and distances. Gravity is an inverse-square relationship; the force is related to the inverse square of the distance. The force *increases* by a factor of $(0.160\text{ N})/(0.010\text{ N}) = 16$, so the distance must *decrease* by a factor of $\sqrt{16} = 4$. The distance is thus $(20\text{ m})/4 = 5.0\text{ m}$.

ASSESS This type of ratio reasoning is a very good way to get a quick handle on the solution to a problem.

EXAMPLE 5.18 Finding the gravitational force between two people

You are seated in your physics class next to another student 0.60 m away. Estimate the magnitude of the gravitational force between you. Assume that you each have a mass of 65 kg.

PREPARE We model each of you as a sphere. This is not a particularly good model, but it will do for making an estimate. We then take 0.60 m as the distance between your centers.

SOLVE The gravitational force is given by Equation 5.24:

$$F_{\text{you on student}} = \frac{Gm_{\text{you}}m_{\text{student}}}{r^2}$$

$$= \frac{(6.67 \times 10^{-11}\text{ N} \cdot \text{m}^2/\text{kg}^2)(65\text{ kg})(65\text{ kg})}{(0.60\text{ m})^2}$$

$$= 7.8 \times 10^{-7}\text{ N}$$

ASSESS The force is quite small, roughly the weight of one hair on your head. This seems reasonable; you don't normally sense this attractive force!

There is a gravitational force between all objects in the universe, but the gravitational force between two ordinary-sized objects is extremely small. Only when one (or both) of the masses is exceptionally large does the force of gravity become important. The downward force of the earth on you—your weight—is large because the earth has an enormous mass. And the attraction is mutual: By Newton's third law, you exert an equal upward force on the earth. However, the large mass of the earth makes the *effect* of this force on the earth negligible.

EXAMPLE 5.19 The gravitational force of the earth on the moon

What is the magnitude of the gravitational force of the earth on the moon? If the moon were to stop revolving around the earth, what would be its acceleration toward the earth? The masses of the earth and the moon are, respectively, 5.98×10^{24} kg and 7.36×10^{22} kg. The distance between their centers is 384,000 km.

PREPARE The earth and moon are spheres, and the force between them is given by Newton's law of gravity. In SI units, the distance between them is $r = 384{,}000\text{ km} = 3.84 \times 10^5\text{ km} = 3.84 \times 10^8\text{ m}$.

SOLVE The earth's gravitational force on the moon is

$$F_{\text{earth on moon}} = \frac{GM_{\text{earth}}M_{\text{moon}}}{r^2}$$

$$= \frac{(6.67 \times 10^{-11}\text{ N} \cdot \text{m}^2/\text{kg}^2)(5.98 \times 10^{24}\text{ kg})(7.36 \times 10^{22}\text{ kg})}{(3.84 \times 10^8\text{ m})^2}$$

$$= 1.99 \times 10^{20}\text{ N}$$

This is a very large force, but it's acting on a very large mass. We can use Newton's second law to find that the magnitude of the moon's acceleration toward the earth would be

$$a = \frac{F_{\text{earth on moon}}}{M_{\text{moon}}} = 2.7 \times 10^{-3}\text{ m/s}^2$$

ASSESS The moon is much farther from the center of the earth than you are while standing on the surface, so it's not surprising that the moon's acceleration would be much less than the free-fall acceleration when you drop a ball. And this actually *is* the moon's acceleration, only it's a *centripetal acceleration* of circular motion instead of a free-fall acceleration straight toward the earth—a topic we will look at in Chapter 7.

Big *G* and Little *g*

The force of gravitational attraction between the earth and you is responsible for your weight. If you traveled to another planet, your *mass* would be the same but your *weight* would differ. Indeed, when astronauts ventured to the moon, television images showed them walking—and even jumping and skipping—with ease, even though they were wearing life-support systems with a mass greater than 80 kg.

FIGURE 5.36 An astronaut weighing a mass on another planet.

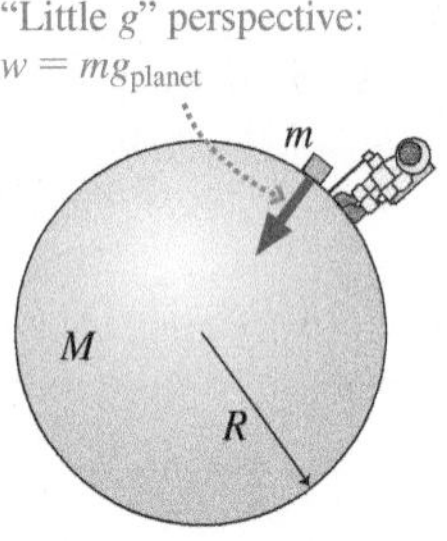

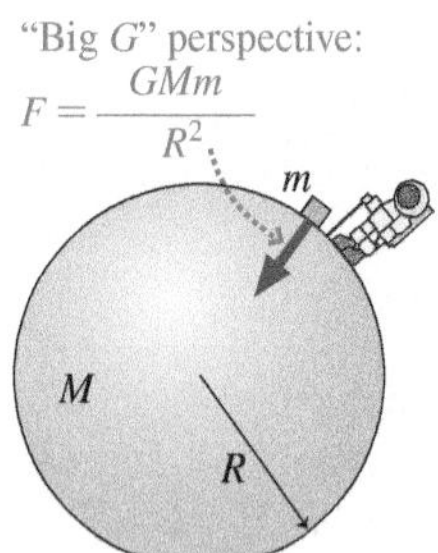

FIGURE 5.36 shows an astronaut on another planet weighing a rock of mass m. When we compute the weight of an object on the surface of the earth, we use the formula $w = mg$. We can do the same calculation for any planet as long as we use the value of g for that planet:

$$w = mg_{\text{planet}} \tag{5.25}$$

This is the "little g" perspective.

But we can also take a "big G" perspective. The weight of the rock comes from the gravitational attraction of the planet, which has mass M, and we can compute this weight using Equation 5.24. The distance r is the planet's radius R. Thus

$$F = \frac{GMm}{R^2} \tag{5.26}$$

Because Equations 5.25 and 5.26 are two names and two expressions for the same force, we can equate the right-hand sides to find that

$$g_{\text{planet}} = \frac{GM}{R^2} \tag{5.27}$$

Free-fall acceleration on the surface of a planet

NOTE ▸ We will use uppercase R and M to represent the large mass and radius of a star or planet. ◂

Using the earth's mass and radius, we find $g = 9.83$ m/s^2. This is our familiar free-fall acceleration at the surface, with the slight difference from the measured 9.80 m/s^2 being a small correction due to the earth's rotation. If we use values for the mass and the radius of the moon, we compute $g_{\text{moon}} = 1.62$ m/s^2. This means that an object would weigh less on the moon than it would on the earth. A 70 kg astronaut wearing an 80 kg spacesuit would weigh more than 330 lb on the earth but only 54 lb on the moon.

Walking on the moon You will learn in Chapter 7 that a person's maximum walking speed depends on the value of g. The low lunar gravity makes walking very easy but also makes it very slow: The maximum walking speed on the moon is about 1 m/s—a very gentle stroll! Walking at a reasonable pace was difficult for the Apollo astronauts, but the reduced weight made jumping quite easy. Videos from the surface of the moon often showed the astronauts getting from place to place by hopping or skipping—not for fun, but for speed and efficiency.

Equation 5.27 gives g at the surface of a planet. More generally, imagine an object at distance $r > R$ from the center of a planet. Its free-fall acceleration at this distance is

$$g = \frac{GM}{r^2} \tag{5.28}$$

This more general result agrees with Equation 5.27 if $r = R$, but it allows us to determine the "local" free-fall acceleration at distances $r > R$.

As you're flying in a jet airplane at an altitude of about 10 km, the free-fall acceleration is about 0.3% less than on the ground. At the height of the International Space Station, about 400 km, Equation 5.28 gives $g = 8.7$ m/s^2, about 13% less than the free-fall acceleration on the earth's surface. This value of g, only slightly less than the ground-level value, emphasizes the point that an object in orbit is not "weightless" due to the absence of gravity, but rather because it is in free fall.

At the distance of the moon, earth's free-fall acceleration is 2.7×10^{-3} m/s^2, exactly the acceleration we found for the moon in Example 5.19. Make sure you understand why.

Gravity on a Grand Scale

Although relatively weak, gravity is a long-range force. No matter how far apart two objects may be, there is a gravitational attraction between them. Consequently, gravity is the most ubiquitous force in the universe. It not only keeps your feet on the ground, but also is at work on a much larger scale. The Milky Way galaxy, the collection of stars of which our sun is a part, is held together by gravity. But why doesn't the attractive force of gravity simply pull all of the stars together?

The reason is that all of the stars in the galaxy are in orbit around the center of the galaxy. The gravitational attraction keeps the stars moving in orbits around the center of the galaxy rather than falling inward, much as the planets orbit the sun rather than falling into the sun. In the nearly 5 billion years that our solar system has existed, it has orbited the center of the galaxy approximately 20 times.

The galaxy as a whole doesn't rotate as a fixed object, though. All of the stars in the galaxy are different distances from the galaxy's center, and so orbit with different periods. As the stars orbit, their relative positions shift. Stars that are relatively near neighbors now could be on opposite sides of the galaxy at some later time.

Home sweet home A spiral galaxy, similar to our Milky Way galaxy.

STOP TO THINK 5.7 Rank in order, from largest to smallest, the free-fall accelerations on the surfaces of the following planets.

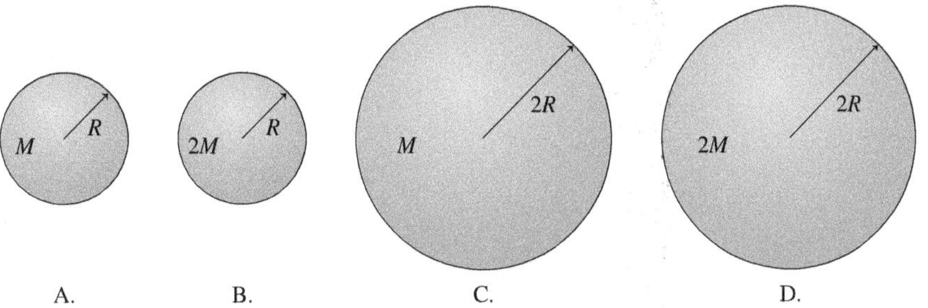

CHAPTER 5 INTEGRATED EXAMPLE **Stopping distances of a skidding car**

You may have noticed "Steep Hill" signs on highways to warn drivers that stopping a car is more difficult when going downhill. This example will give you a sense of how much extra stopping distance is needed. Suppose a 1500 kg car is traveling at a speed of 30 m/s when the driver slams on the brakes and skids to a halt. Determine the stopping distance if the car is traveling up a 10° slope, down a 10° slope, or on a level road.

PREPARE We'll treat the car as a particle and use the model of kinetic friction. We can use Newton's second law to find the car's acceleration and then apply kinematics to calculate the stopping distance. We want to solve the problem only once, not three separate times, so we will leave the slope angle θ unspecified until the end.

FIGURE 5.37 (next page) shows the pictorial representation. We show the car sliding uphill, but this representation works equally well for a level or downhill slide if we let θ be zero or negative, respectively. We use a tilted coordinate system so that the motion is along the x-axis. The car *skids* to a halt, so we use the coefficient of *kinetic* friction for rubber on concrete from Table 5.2.

SOLVE Newton's second law and the model of kinetic friction are

$$\sum F_x = n_x + w_x + (f_k)_x = 0 - mg\sin\theta - f_k = ma_x$$

$$\sum F_y = n_y + w_y + (f_k)_y = n - mg\cos\theta + 0 = ma_y = 0$$

$$f_k = \mu_k n$$

We've written the first two equations by "reading" the motion diagram and the free-body diagram. Notice that both components of the weight vector $\vec{w}$ are negative. $a_y = 0$ because the motion is entirely along the x-axis.

The second equation gives $n = mg\cos\theta$. Using this in the friction model, we find $f_k = \mu_k mg\cos\theta$. Inserting this result back into the first equation then gives

$$ma_x = -mg\sin\theta - \mu_k mg\cos\theta = -mg(\sin\theta + \mu_k\cos\theta)$$

$$a_x = -g(\sin\theta + \mu_k\cos\theta)$$

Continued

This is a constant acceleration. Constant-acceleration kinematics gives

$$(v_x)_1^2 = 0 = (v_x)_0^2 + 2a_x(x_1 - x_0) = (v_x)_0^2 + 2a_x x_1$$

which we can solve for the stopping distance x_1:

$$x_1 = -\frac{(v_x)_0^2}{2a_x} = \frac{(v_x)_0^2}{2g(\sin\theta + \mu_k \cos\theta)}$$

Notice how the minus sign in the expression for a_x canceled the minus sign in the expression for x_1. Evaluating our result at the three different angles gives the stopping distances:

$$x_1 = \begin{cases} 48 \text{ m} & \theta = 10^\circ \quad \text{uphill} \\ 57 \text{ m} & \theta = 0^\circ \quad \text{level} \\ 75 \text{ m} & \theta = -10^\circ \quad \text{downhill} \end{cases}$$

The implications are clear about the danger of driving downhill too fast!

ASSESS 30 m/s ≈ 60 mph and 57 m ≈ 180 feet on a level surface. These are similar to the stopping distances you learned when you got your driver's license, so the results seem reasonable. Additional confirmation comes from noting that the expression for a_x becomes $a_x = -g$ at a slope angle θ of 90°. Such an angle would represent falling straight down, rather than skidding, so the acceleration should, as we find, be the free-fall acceleration.

FIGURE 5.37 Pictorial representation of a skidding car.

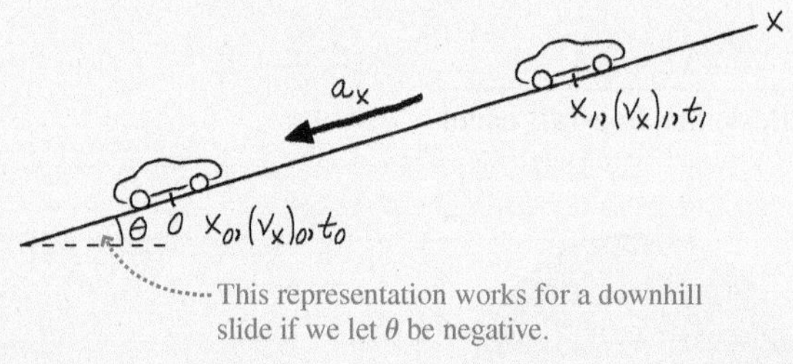

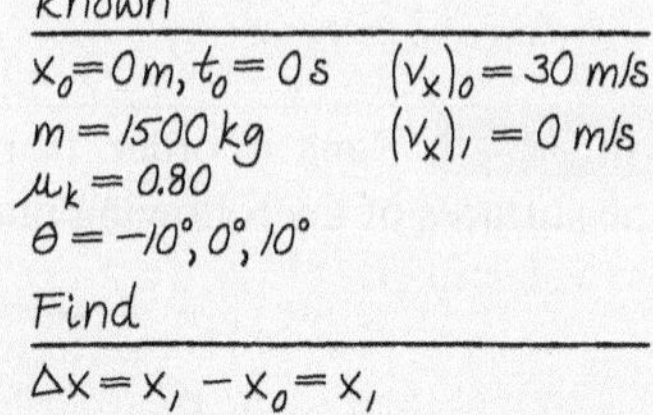

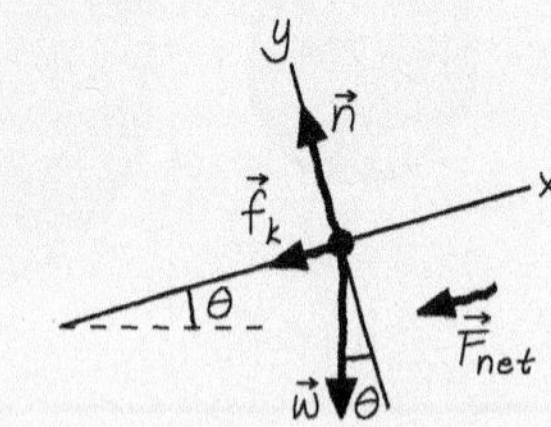

SUMMARY

GOAL To use Newton's laws to solve problems of force and motion.

GENERAL PRINCIPLES

A Problem-Solving Strategy

A three-part strategy applies to both equilibrium and dynamics problems.

PREPARE Model the object as a particle and make simplifying assumptions.

- Translate words into symbols.
- Draw a sketch to define the situation.
- Draw a motion diagram.
- Identify forces.
- Draw a free-body diagram.

(Go back and forth between these steps as needed.)

SOLVE Use Newton's second law:

$$\vec{F}_{\text{net}} = \sum_i \vec{F}_i = m\vec{a}$$

"Read" the vectors from the free-body diagram. Use kinematics to find velocities and positions.

ASSESS Is the result reasonable? Does it have correct units and significant figures?

Newton's Third Law

Every force occurs as one member of an **interaction pair** of forces. The two members of an interaction pair:

- Act on two *different* objects.
- Are equal in magnitude but opposite in direction:

$$\vec{F}_{\text{A on B}} = -\vec{F}_{\text{B on A}}$$

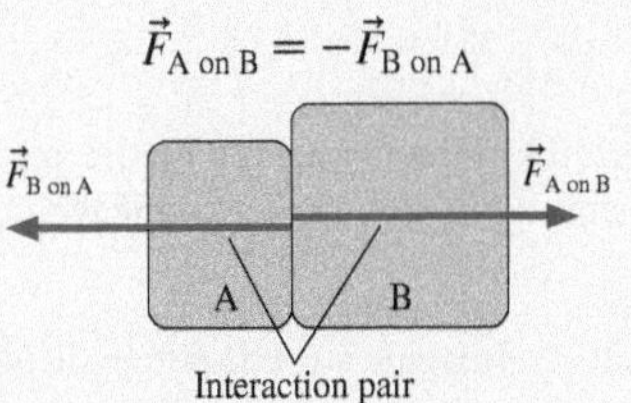

Objects whose motion is of interest are the **system.**

External forces originate in the **environment.**

Interactions can be shown on an **interaction diagram.**

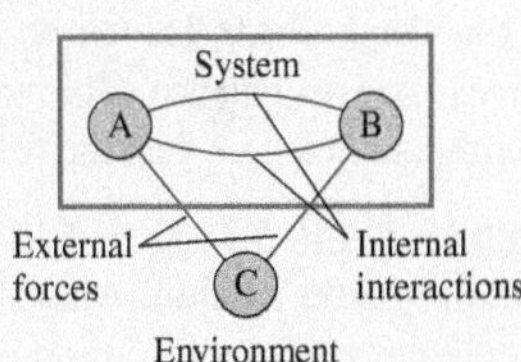

IMPORTANT CONCEPTS

Three important forces

Weight: $\vec{w} = (mg, \text{downward})$

Friction: $\vec{f}_s = (0 \text{ to } \mu_s n$, direction as necessary to prevent motion)

$\vec{f}_k = (\mu_k n$, direction opposite to the motion)

$\vec{f}_r = (\mu_r n$, direction opposite to the motion)

Drag is directed opposite to the motion:

$D = \frac{1}{2}C_D \rho A v^2$, high *Re*

$D = 6\pi\eta r v$, sphere at low *Re*

Using the second law

Newton's second law is a vector equation. In component form, it is the two simultaneous equations

$$(F_{\text{net}})_x = \sum F_x = ma_x$$
$$(F_{\text{net}})_y = \sum F_y = ma_y$$

The acceleration is zero

- In equilibrium problems.
- Along an axis perpendicular to the direction of motion

Interacting objects

When two objects interact, you need to draw two separate free-body diagrams.

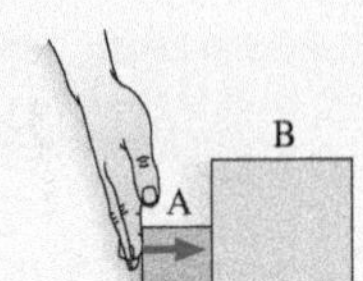

Interaction pairs of forces have equal magnitude and opposite directions.

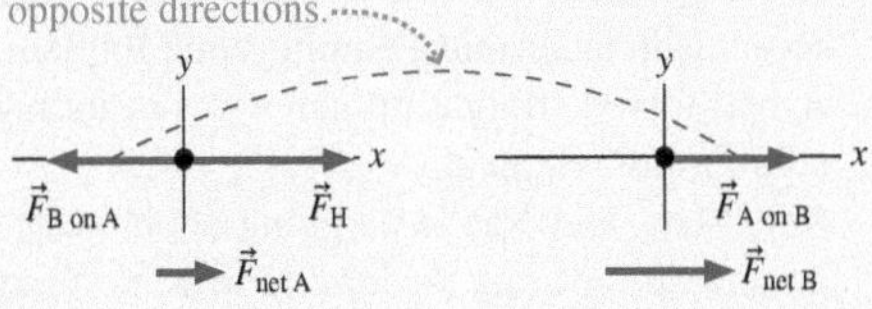

APPLICATIONS

Apparent weight is the magnitude of the contact force supporting an object. It is what a scale reads, and it is your sensation of weight.

An object reaches **terminal speed** when the drag force exactly balances an applied force.

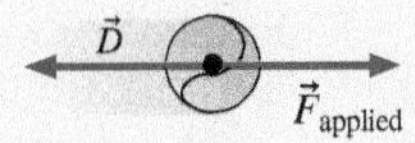

Newton's law of gravity

Two objects with masses m_1 and m_2 that are distance r apart exert an attractive gravitational force on each other of magnitude

$$F_{1 \text{ on } 2} = F_{2 \text{ on } 1} = \frac{Gm_1m_2}{r^2}$$

where $G = 6.67 \times 10^{-11}\ \text{N} \cdot \text{m}^2/\text{kg}^2$ is the gravitational constant. The law of gravity is an **inverse-square force law.**

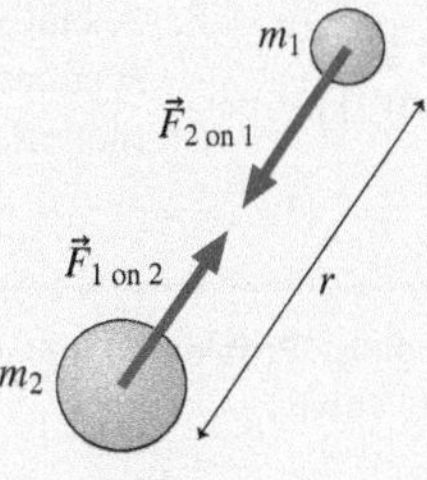

Springs and strings

The tension force in a string and the restoring force of a spring are due to the stretching of molecular bonds.

Hooke's law for the restoring force of an ideal spring is

$$(F_{sp})_x = -k\Delta x$$

where k is the **spring constant.**

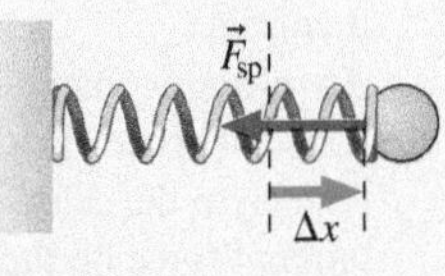

Tension in a massless string or rope pulls equally in both directions.

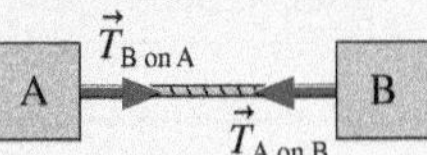

LEARNING OBJECTIVES After studying this chapter, you should be able to:

- Solve dynamics problems using free-body diagrams and Newton's second law. *Conceptual Questions 5.3–5.5; Problems 5.7, 5.10, 5.33*
- Work with and distinguish among mass, weight, and apparent weight. *Conceptual Question 5.6; Problems 5.1, 5.2, 5.4–5.6*
- Determine static, kinetic, and rolling friction. *Conceptual Questions 5.5, 5.9; Problems 5.8–5.10, 5.12, 5.14*
- Use two models of drag to solve problems of motion through a fluid. *Conceptual Questions 5.8, 5.10; Problems 5.18–5.21*
- Calculate the force exerted by a spring. *Conceptual Question 5.12; Problems 5.24–5.28*
- Identify forces on and solve problems about interacting objects. *Conceptual Question 5.11; Problems 5.30–5.34*
- Calculate the gravitational force between two massive bodies. *Conceptual Question 5.13; Problems 5.36–5.39*

STOP TO THINK ANSWERS

Stop to Think 5.1: D. When you are in the air, there is no contact force supporting you, so your apparent weight is zero: You are weightless.

Stop to Think 5.2: C. Imagine a y-axis perpendicular to the slope. The only forces along the y-axis are the normal force and the y-component of the weight force. There's no net force in the y-direction, so these two forces have to balance. The weight force points straight down, so its y-component is less than the full weight w. Thus $n < w$.

Stop to Think 5.3: $f_B > f_C = f_D = f_E > f_A$. Situations C, D, and E are all kinetic friction, which does not depend on either velocity or acceleration. Kinetic friction is less than the maximum static friction that is exerted in B. $f_A = 0$ because no friction is needed to keep the object at rest.

Stop to Think 5.4: B. At $t = \tau$ the curve has decreased to $e^{-1} \approx 37\%$ of its initial height.

Stop to Think 5.5: C. Newton's third law says that the force of A on B is *equal* and opposite to the force of B on A. This is always true. The mass of the objects isn't relevant.

Stop to Think 5.6: Less than. Both blocks are accelerating to the right because there is no friction. Block A is pulled to the right by tension $\vec{T}_1$ and to the left by tension $\vec{T}_2$. Acceleration to the right requires a net force to the right, so $T_1 > T_2$ or, equivalently, $T_2 < T_1$.

Stop to Think 5.7: B > A > D > C. The free-fall acceleration is proportional to the mass, but inversely proportional to the square of the radius.

QUESTIONS

Conceptual Questions

1. An object is subject to two forces that do not point in opposite directions. Is it possible to choose their magnitudes so that the object is in mechanical equilibrium? Explain.
2. A ball tossed straight up has $v = 0$ at its highest point. Is it in equilibrium? Explain.
3. Kat, Matt, and Nat are arguing about why a physics book on a table doesn't fall. According to Kat, "Gravity pulls down on it, but the table is in the way so it can't fall." "Nonsense," says Matt. "An upward force simply overcomes the downward force to prevent it from falling." "But what about Newton's first law?" counters Nat. "It's not moving, so there can't be any forces acting on it." None of the statements is exactly correct. Who comes closest, and how would you change his or her statement to make it correct?
4. An elevator, hanging from a single cable, moves upward at constant speed. Friction and air resistance are negligible. Is the tension in the cable greater than, less than, or equal to the gravitational force on the elevator? Explain. Include a free-body diagram as part of your explanation.
5. Boxes A and B in Figure Q5.5 both remain at rest. Is the friction force on A larger than, smaller than, or equal to the friction force on B? Explain.

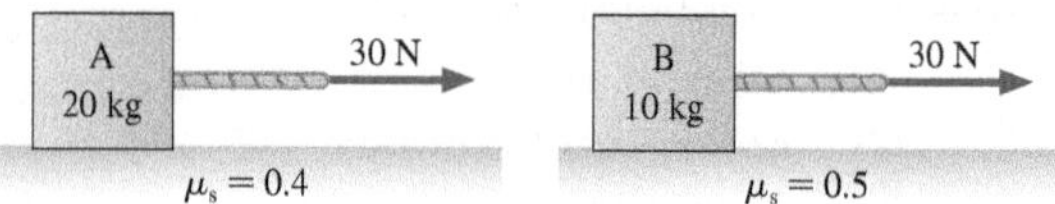

FIGURE Q5.5

6. An astronaut takes his bathroom scale to the moon and then stands on it. Is the reading of the scale his true weight? Explain.
7. An astronaut orbiting the earth is handed two balls that have identical outward appearances. However, one is hollow while the other is filled with lead. How can the astronaut determine which is which? Cutting or altering the balls is not allowed.
8. A ball is thrown straight up. Taking the drag force of air into account, does it take longer for the ball to travel to the top of its motion or for it to fall back down again?

Problem difficulty is labeled as | (straightforward) to ||||| (challenging). Problems labeled INT integrate significant material from earlier chapters; BIO are of biological or medical interest; CALC require calculus to solve.

9. Suppose you are holding a box in front of you and away from your body by squeezing the sides, as shown in Figure Q5.9. Draw a free-body diagram showing all of the forces on the box. What is the force that is holding the box up, the force that is opposite the weight force?

FIGURE Q5.9

10. Three objects move through the air as shown in Figure Q5.10. Rank in order, from largest to smallest, the three drag forces D_1, D_2, and D_3. Some may be equal. Give your answer in the form $A < B = C$ and state your reasoning.

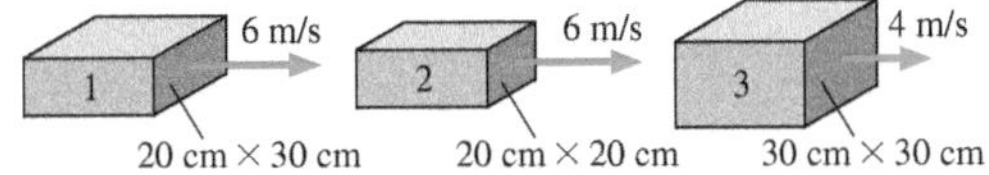

FIGURE Q5.10

11. The hand in Figure Q5.11 is pushing on the back of block A. Blocks A and B, with $m_B > m_A$, are connected by a massless string and slide on a frictionless surface. Is the force of the string on B larger than, smaller than, or equal to the force of the hand on A? Explain.

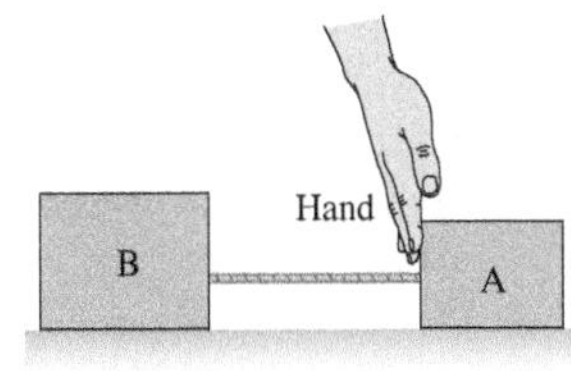

FIGURE Q5.11

12. The left end of a spring is attached to a wall. When Bob pulls on the right end with a 200 N force, he stretches the spring by 20 cm. The same spring is then used for a tug-of-war between Bob and Carlos. Each pulls on his end of the spring with a 200 N force. How far does the spring stretch? Explain.
13. Is the earth's gravitational force on the sun larger than, smaller than, or equal to the sun's gravitational force on the earth? Explain.

Multiple-Choice Questions

14. || The wood block in Figure Q5.14 is at rest on a wood ramp. In which direction is the static friction force on block 1?
    A. Up the slope.
    B. Down the slope.
    C. The friction force is zero.
    D. There's not enough information to tell.

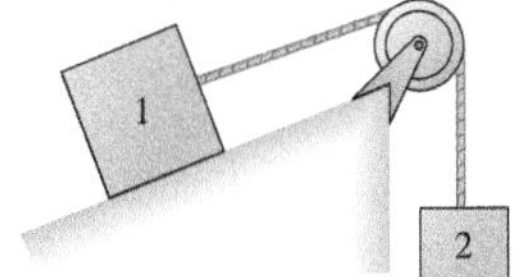

FIGURE Q5.14

15. | A 5.0 kg dog sits on the floor of an elevator that is accelerating *downward* at 1.20 m/s^2.
    a. What is the magnitude of the normal force of the elevator floor on the dog?
    A. 34 N B. 43 N C. 49 N
    D. 55 N E. 74 N
    b. What is the magnitude of the force of the dog on the elevator floor?
    A. 4.2 N B. 49 N C. 55 N
    D. 43 N E. 74 N
16. || A 2.0 kg ball is suspended by two light strings as shown in Figure Q5.16. What is the tension T in the angled string?
    A. 9.5 N B. 15 N
    C. 20 N D. 26 N
    E. 30 N

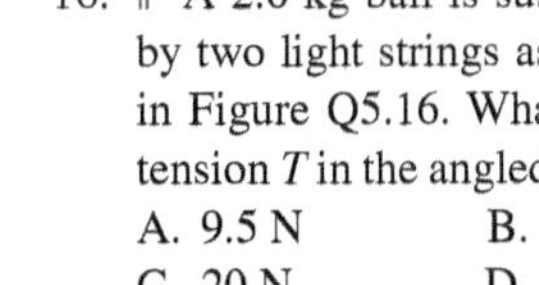
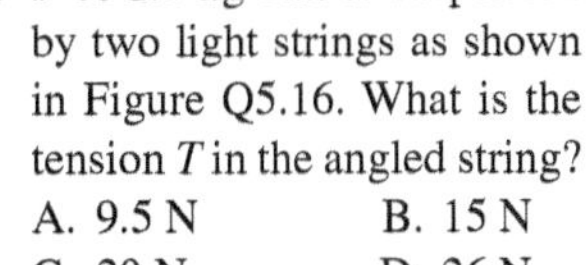
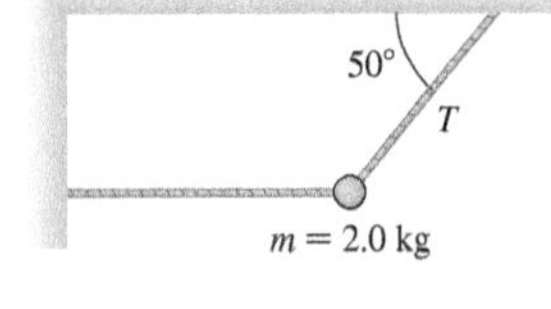

FIGURE Q5.16

17. | Eric has a mass of 62 kg. He is standing on a scale in an elevator that is accelerating downward at 1.7 m/s^2. What is the reading on the scale?
    A. 0 N B. 500 N C. 600 N D. 700 N
18. | A football player at practice pushes a 60 kg blocking sled across the field at a constant speed. The coefficient of kinetic friction between the grass and the sled is 0.30. How much force must the player apply to the sled?
    A. 18 N B. 60 N C. 180 N D. 600 N
19. | Two football players are pushing a 60 kg blocking sled across the field at a constant speed of 2.0 m/s. The coefficient of kinetic friction between the grass and the sled is 0.30. Once they stop pushing, how far will the sled slide before coming to rest?
    A. 0.20 m B. 0.68 m C. 1.0 m D. 6.6 m
20. || A 2000 kg car is parked on a 30° slope. The coefficients of static and kinetic friction are 0.90 and 0.40, respectively. What is the magnitude of the friction force on the car?
    A. 6800 N B. 9800 N C. 15,000 N D. 20,000 N
21. || A truck is traveling at 30 m/s on a slippery road. The driver slams on the brakes and the truck starts to skid. If the coefficient of kinetic friction between the tires and the road is 0.20, how far will the truck skid before stopping?
    A. 230 m B. 300 m C. 450 m D. 680 m
22. || Suppose one night the radius of the earth doubled but its mass stayed the same. What would be an approximate new value for the free-fall acceleration at the surface of the earth?
    A. 2.5 m/s^2 B. 5.0 m/s^2 C. 10 m/s^2 D. 20 m/s^2
23. || An elastic climbing rope is designed to stretch. Suppose a 10.0-m-long rope stretches to 10.5 m when supporting a 60 kg climber. What will the rope's length be when supporting a 75 kg climber?
    A. 10.6 m B. 12.5 m C. 13.3 m D. 20.0 m

PROBLEMS

Section 5.1 Systems and Interactions

Section 5.2 Mass and Weight

1. | A woman has a mass of 55.0 kg.
    a. What is her weight on earth?
    b. What are her mass and her weight on the moon, where $g = 1.62$ m/s^2?
2. || The acceleration of the spacecraft in which the Apollo astronauts took off from the moon was 3.4 m/s^2. On the moon, $g = 1.6$ m/s^2. What was the apparent weight of a 75 kg astronaut during takeoff?
3. || a. How much force does an 80 kg astronaut exert on their chair while sitting at rest on the launch pad on earth?
    b. How much force does the astronaut exert on their chair while accelerating straight up at 10 m/s^2?

4. || Riders on the Power Tower are launched skyward with an acceleration of 4.0g, after which they experience a period of free fall. What is a 60 kg rider's apparent weight
   a. During the launch?
   b. During the period of free fall?

5. || Zach, whose mass is 80 kg, is in an elevator descending at 10 m/s. The elevator takes 3.0 s to brake to a stop at the first floor.
   a. What is Zach's apparent weight before the elevator starts braking?
   b. What is Zach's apparent weight while the elevator is braking?
6. || Figure P5.6 shows the velocity graph of a 75 kg passenger in an elevator. What is the passenger's apparent weight at $t = 1.0$ s? At 5.0 s? At 9.0 s?

INT

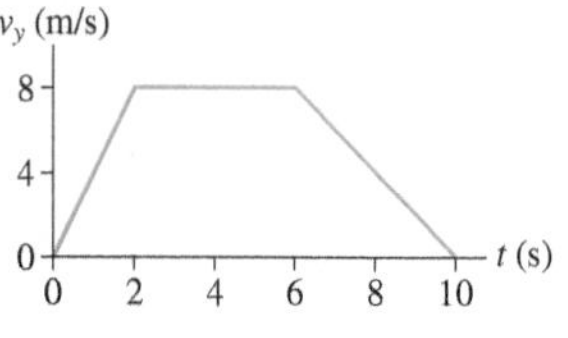

FIGURE P5.6

Section 5.3 Interactions with Surfaces

7. | Mountain goats can easily scale slopes angled at 60° from horizontal. What are the normal force and the static friction force acting on a mountain goat that weighs 900 N and is standing on such a slope?

BIO

8. || A 23 kg child goes down a straight slide inclined 35° above horizontal. The coefficient of kinetic friction between her pants and the slide is 0.25.
   a. What is the magnitude of the normal force acting on the child?
   b. What is the magnitude of the friction force acting on the child?
9. ||| A crate pushed along the floor with velocity $\vec{v}_0$ slides a distance d after the pushing force is removed.
   a. If the mass of the crate is doubled but the initial velocity is not changed, what distance does the crate slide before stopping? Explain.
   b. If the initial velocity of the crate is doubled to $2\vec{v}_0$ but the mass is not changed, what distance does the crate slide before stopping? Explain.
10. || Two workers are sliding a 300 kg crate across the floor at constant speed. One worker pushes forward on the crate with a force of 380 N while the other pulls in the same direction with a force of 350 N using a rope connected to the crate. What is the crate's coefficient of kinetic friction on the floor?
11. ||| A 4000 kg truck is parked on a 7.0° slope. How big is the friction force on the truck?
12. || A 1000 kg car traveling at a speed of 40 m/s skids to a halt on wet concrete where $\mu_k = 0.60$. How long are the skid marks?
13. ||| A 10 kg crate is placed on a horizontal conveyor belt. The materials are such that $\mu_s = 0.50$ and $\mu_k = 0.30$.
    a. Draw a free-body diagram showing all the forces on the crate if the conveyer belt runs at constant speed.
    b. Draw a free-body diagram showing all the forces on the crate if the conveyer belt is speeding up.
    c. What is the maximum acceleration the belt can have without the crate slipping?
    d. If the acceleration of the belt exceeds the value determined in part c, what is the acceleration of the crate?
14. || The rolling resistance for steel on steel is quite low; the coefficient of rolling friction is typically $\mu_r = 0.002$. Suppose a 180,000 kg locomotive is rolling at 10 m/s (just over 20 mph) on level rails. If the engineer disengages the engine, how much time will it take the locomotive to coast to a stop? How far will the locomotive move during this time?
15. ||| A rubber-wheeled 50 kg cart rolls down a 15° concrete incline. What is the cart's acceleration if rolling friction is (a) neglected and (b) included?

Section 5.4 Drag

16. || To one significant figure, what is the Reynolds number of (a) a 3.0-cm-diameter ball moving in air at 15 m/s, (b) a 30-cm-diameter cylindrical dolphin swimming in water at 15 m/s, (c) a 1.0-mm-diameter sand grain falling in air at 15 m/s, and (d) a 20-μm-diameter diatom sinking in water at 15 mm/s? Assume a temperature of 20°C for each.
17. ||| Oceanographers use submerged sonar systems, towed by a cable from a ship, to map the ocean floor. In addition to their downward weight, there are buoyant forces and forces from the flowing water that allow them to travel in a horizontal path. One such submersible has a cross-section area of 1.3 m^2, a drag coefficient of 1.2, and, when towed at 5.1 m/s, the tow cable makes an angle of 30° with the horizontal. What is the tension in the cable?
18. || At its widest point, the diameter of a bottlenose dolphin is 0.50 m. Bottlenose dolphins are particularly sleek, having a drag coefficient of only about 0.090.
    a. What is the drag force acting on such a dolphin swimming at 7.5 m/s?
    b. Using the dolphin's diameter as its characteristic length, what is the Reynolds number as it swims at this speed in 20° C water?

BIO

19. |||| The most dangerous particles in polluted air are those with diameters less than 2.5 μm because they can penetrate deeply into the lungs. A 15-cm-tall closed container holds a sample of polluted air containing many spherical particles with a diameter of 2.5 μm and a mass of 1.4×10^{-14} kg. How long does it take for all of the particles to settle to the bottom of the container?

BIO

20. ||| The air is less dense at higher elevations, so skydivers reach a high terminal speed. The highest recorded speed for a skydiver was achieved in a jump from a height of 39,000 m. At this elevation, the density of the air is only 4.3% of the surface density. Use the data from Example 5.9 to estimate the terminal speed of a skydiver at this elevation.
21. ||| Running on a treadmill is slightly easier than running outside because there is no drag force to work against. Suppose a 60 kg runner completes a 5.0 km race in 18 minutes. Use the cross-section area estimate of Example 5.9 to determine the drag force on the runner during the race. What is this force as a fraction of the runner's weight?

BIO

22. ||| What is the magnitude of the acceleration of a skydiver who is currently falling at one-half his eventual terminal speed?
23. |||| A small, 250-μm-diameter bead is fired horizontally into 20°C olive oil at a speed of 0.25 m/s. Its stopping distance in the olive oil is 11 μm. What is the mass of the bead in μg?

Section 5.5 Springs and Elastic Forces

24. | A scale used to weigh fish consists of a spring hung from a support. The spring's equilibrium length is 10.0 cm. When a 4.0 kg fish is suspended from the end of the spring, it stretches to a length of 12.4 cm.
   a. What is the spring constant k for this spring?
   b. If an 8.0 kg fish is suspended from the spring, what will be the length of the spring?
25. || A spring has an unstretched length of 10 cm. It exerts a restoring force F when stretched to a length of 11 cm.
   a. For what total stretched length of the spring is its restoring force $3F$?
   b. At what compressed length is the restoring force $2F$?
26. || A passenger railroad car has a total of 8 wheels. Springs on each wheel compress—slightly—when the car is loaded. Ratings for the car give the stiffness per wheel (the spring constant, treating the entire spring assembly as a single spring) as 2.8×10^7 N/m. When 30 passengers, each with average mass 80 kg, board the car, how much does the car move down on its spring suspension? Assume that each wheel supports 1/8 the weight of the car.
27. || One end of a 10-cm-long spring is attached to the ceiling. When a 2.0 kg mass is hung from the other end, the spring stretches to a length of 15 cm.
   a. What is the spring constant?
   b. How long is the spring when a 3.0 kg mass is suspended from it?
28. | You need to make a spring scale to measure the mass of objects hung from it. You want each 1.0 cm length along the scale to correspond to a mass difference of 0.10 kg. What should be the value of the spring constant?

Section 5.6 Newton's Third Law

For Problems 29 through 31:
   a. Draw an interaction diagram.
   b. Identify the "system" on your interaction diagram.
   c. Draw a free-body diagram for each object in the system. Use dashed lines to connect members of an interaction pair.

29. | A soccer ball and a bowling ball have a head-on collision at this instant. Rolling friction is negligible.
30. | A weightlifter stands up at constant speed from a squatting position while holding a heavy barbell across her shoulders.
31. | A steel cable with mass is lifting a girder. The girder is speeding up.
32. || A 1000 kg car pushes a 2000 kg truck that has a dead battery. When the driver steps on the accelerator, the drive wheels of the car push backward against the ground with a force of 4500 N.
   a. What is the magnitude of the force of the car on the truck?
   b. What is the magnitude of the force of the truck on the car?
33. ||| An 80 kg spacewalking astronaut pushes off a 640 kg satellite, exerting a 100 N force for the 0.50 s it takes him to straighten his arms. How far apart are the astronaut and the satellite after 1.0 min?
34. |||| Blocks with masses of 1.0 kg, 2.0 kg, and 3.0 kg are lined up in a row on a frictionless table. All three are pushed forward by a 12 N force applied to the 1.0 kg block. How much force does the 2.0 kg block exert on (a) the 3.0 kg block and (b) the 1.0 kg block?
35. ||| The 1.0 kg block in Figure P5.35 is tied to the wall with a rope. It sits on top of the 2.0 kg block. The lower block is pulled to the right with a tension force of 20 N. The coefficient of kinetic friction at both the lower and upper surfaces of the 2.0 kg block is $\mu_k = 0.40$.
   a. What is the tension in the rope attached to the wall?
   b. What is the acceleration of the 2.0 kg block?

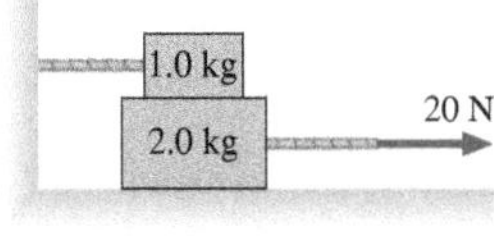

FIGURE P5.35

Section 5.7 Newton's Law of Gravity

36. || The gravitational force of a star on an orbiting planet 1 is F_1. Planet 2, which is twice as massive as planet 1 and orbits at twice the distance from the star, experiences gravitational force F_2. What is the ratio F_2/F_1? You can ignore the gravitational force between the two planets.
37. ||| The centers of a 10 kg lead ball and a 100 g lead ball are separated by 10 cm.
   a. What gravitational force does each exert on the other?
   b. What is the ratio of this gravitational force to the weight of the 100 g ball?
38. || The free-fall acceleration at the surface of planet 1 is 20 m/s^2. The radius and the mass of planet 2 are twice those of planet 1. What is the free-fall acceleration on planet 2?
39. || In recent years, astronomers have found planets orbiting nearby stars that are quite different from planets in our solar system. Kepler-12b has a diameter that is 1.7 times that of Jupiter, but a mass that is only 0.43 that of Jupiter. The mass and radius of Jupiter are 1.90×10^{27} kg and 6.99×10^4 km. What is the value of g on this large, but low-density, world?

General Problems

40. || Dana has a sports medal suspended by a long ribbon from her rearview mirror. As she accelerates onto the highway, she notices that the medal is hanging at an angle of 10° from the vertical.
   a. Does the medal lean toward or away from the windshield? Explain.
   b. What is her acceleration?
41. || Figure P5.41 shows the velocity graph of a 2.0 kg object as it moves along the x-axis. What is the x-component of the net force acting on this object at $t = 1$ s? At 4 s? At 7 s? (INT)

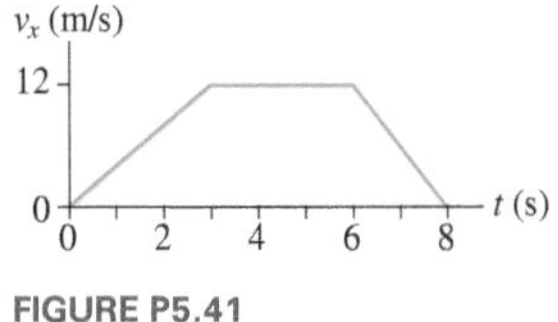

FIGURE P5.41

42. | Your forehead can withstand a force of about 6.0 kN before fracturing, while your cheekbone can only withstand about 1.3 kN. (BIO)
   a. If a 140 g baseball strikes your head at 30 m/s and stops in 0.0015 s, what is the magnitude of the ball's acceleration?
   b. What is the magnitude of the force that stops the baseball?
   c. What force does the baseball apply to your head? Explain.
   d. Are you in danger of a fracture if the ball hits you in the forehead? In the cheek?
43. || A fisherman has caught a very large, 5.0 kg fish from a dock that is 2.0 m above the water. He is using lightweight fishing line that will break under a tension of 54 N or more. He is eager to get the fish to the dock in the shortest possible time. If the fish is at rest at the water's surface, what's the least amount of time in which the fisherman can raise the fish to the dock without losing it?

44. || Riders on the Tower of Doom, an amusement park ride, experience 2.0 s of free fall, after which they are slowed to a stop in 0.50 s. What is a 65 kg rider's apparent weight as the ride is coming to rest? By what factor does this exceed her actual weight?

45. || A rocket headed to the International Space Station takes 8.0 s to reach a speed of 160 km/h. During this phase, what is the apparent weight of a 72 kg astronaut?

46. ||| Woodpeckers are amazing birds. A skull that dissipates
BIO energy allows them to forcefully drive their beak into wood approximately 20 times per second without sustaining brain injury. One study of woodpeckers found that the beak strikes a tree trunk at 3.6 m/s, experiences a nearly constant force for a mere 2.0 ms, and then rebounds at the same 3.6 m/s.
   a. What is the magnitude of the acceleration of the woodpecker's head?
   b. How far in mm does its beak penetrate into the wood?
   c. A woodpecker's head has a mass of 9.0 g. What is the magnitude of the force on a woodpecker head?

47. || Seat belts and air bags save lives by reducing the forces ex-
BIO erted on the driver and passengers in an automobile collision. Cars are designed with a "crumple zone" in the front of the car. In the event of an impact, the passenger compartment decelerates over a distance of about 1 m as the front of the car crumples. An occupant restrained by seat belts and air bags decelerates with the car. By contrast, an unrestrained occupant keeps moving forward with no loss of speed (Newton's first law!) until hitting the dashboard or windshield, as you saw in Figure 4.2. These are unyielding surfaces, and the unfortunate occupant then decelerates over a distance of only about 5 mm.
   a. A 60 kg person is in a head-on collision. The car's speed at impact is 15 m/s. Estimate the net force on the person if they are wearing a seat belt and if the air bag deploys.
   b. Estimate the net force that ultimately stops the person if they are not restrained by a seat belt or air bag.
   c. How do these two forces compare to the person's weight?

48. ||| Corey, whose mass is 95 kg, stands on a bathroom scale in an elevator. The scale reads 830 N for the first 3.0 s after the elevator starts to move, then 930 N for the next 3.0 s. What is the elevator's velocity 6.0 s after starting?

49. ||| You've always wondered about the acceleration of the elevators in the 101-story-tall Empire State Building. One day, while visiting New York, you take your bathroom scale into the elevator and stand on it. The scale reads 150 lb as the door closes. The reading varies between 120 lb and 170 lb as the elevator makes its upward journey.
   a. What is the magnitude of the acceleration as the elevator starts upward?
   b. What is the magnitude of the acceleration as the elevator brakes to a stop?

50. ||| A 5.0 kg object initially at rest at the origin is subjected to the time-varying force shown in Figure P5.50. What is the object's velocity at $t = 6$ s?

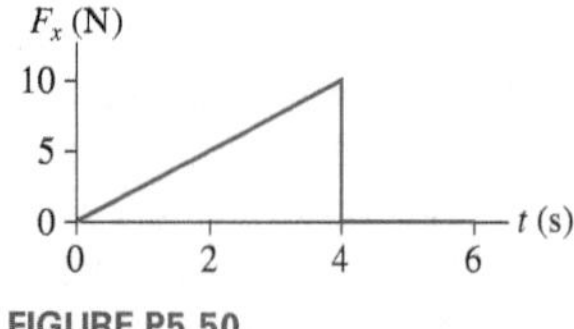

FIGURE P5.50

51. |||| A 50 g particle that can move along the x-axis experiences
CALC the net force $F_x = 2.0t^2$ N, where t is in s. The particle is at rest at $t = 0$ s. What is the particle's speed at $t = 2.0$ s?

52. |||| A 50 g particle that can move along the x-axis experiences
CALC the net force $F_x = 2.0e^{-t/1.0\,\text{s}}$ N, where t is in s. The particle is at rest at $t = 0$ s. What is the particle's speed at $t = 2.0$ s?

53. || An impala is an African antelope capable of a remarkable
BIO vertical leap. In one recorded leap, a 45 kg impala went into a deep crouch, pushed straight up for 0.21 s, and reached a height of 2.5 m above the ground. To achieve this vertical leap, with what force did the impala push down on the ground? What is the ratio of this force to the antelope's weight?

54. |||| The standing vertical jump is a good test of an athlete's
BIO strength and fitness. The athlete goes into a deep crouch, then extends his legs rapidly; when his legs are fully extended, he leaves the ground and rises to his highest height. It is the force of the ground on the athlete during the extension phase that accelerates the athlete to the final speed with which he leaves the ground. A good jumper can exert a force on the ground equal to twice his weight. If his crouch is 60 cm deep, how far off the ground does he rise?

55. ||| A 1.0 kg wood block is pressed against a vertical wood wall by a 12 N force as shown in Figure P5.55. If the block is initially at rest, will it move upward, move downward, or stay at rest?

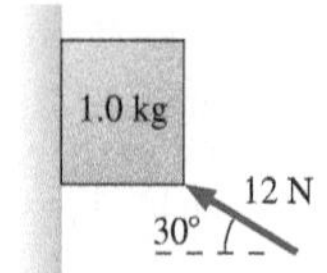

FIGURE P5.55

56. ||| Researchers test canine locomotion by
BIO measuring the *ground reaction force* (the vector sum of the normal force and the propulsion force) as dogs push down and backward with both rear legs. In one trial, a dog generated an instantaneous ground reaction force that was 1.7 times the dog's weight and angled 15° from vertical. What were the magnitudes of the dog's horizontal and vertical accelerations at that moment?

57. |||| A simple model shows how drawing a bow across a violin string causes the string to vibrate. As the bow moves across the string, static friction between the bow and the string pulls the string along with the bow. At some point, the tension pulling the string back exceeds the maximum static friction force and the string snaps back. This process repeats cyclically, causing the string's vibration. Assume the tension in a 0.33-m-long violin string is 50 N, and the coefficient of static friction between the bow and the string is $\mu_s = 0.80$. If the normal force of the bow on the string is 0.75 N, how far can the string be pulled before it slips if the string is bowed at its center?

58. || Josh starts his sled at the top of a 3.0-m-high hill that has a constant slope of 25°. After reaching the bottom, he slides across a horizontal patch of snow. Ignore friction on the hill, but assume that the coefficient of kinetic friction between his sled and the horizontal patch of snow is 0.050. How far from the base of the hill does he end up?

59. ||| An Airbus A320 jetliner has a takeoff mass of 75,000 kg. It reaches its takeoff speed of 82 m/s (180 mph) in 35 s. What is the thrust of the engines? You can neglect air resistance but not rolling friction.

60. ||| A 4.0 g cricket sits at rest on a board. Out of curiosity, you slowly lift one end of the board so that the cricket is sitting on a steeper and steeper slope. Finally, the cricket begins to slide when the board is at an angle of 40° above horizontal. What is the coefficient of static friction between the cricket's feet and the board?

61. || A 6.5-cm-diameter ball has a terminal speed of 26 m/s. What is the ball's mass?

62. ||| A 70 kg bicyclist is coasting down a long hill with a 3.5° slope. She's moving quite rapidly, so air drag is important. Her cross-section area is 0.32 m^2 and her drag coefficient is 0.88. What speed does she eventually reach, in mph?

63. ||| BIO Gannets are large sea birds (2.5 kg) known for folding their wings and plunging straight into the ocean. Seen from the front, the cross section of a gannet with folded wings can be approximated as a $15\,\text{cm} \times 17\,\text{cm}$ rectangle. Its drag coefficient is a modestly streamlined 0.30.
   a. Gannets cruise over the ocean at a height of about 25 m, searching for fish. If a gannet folds its wings and dives from this height, with what speed will it reach the water? Drag is not significant for this part of the motion.
   b. Drag is very significant as the gannet enters the water. What is the initial drag force while the bird is still moving at its impact speed? Give your answer both in newtons and as a multiple of the bird's weight. The density of seawater is $1030\,\text{kg/m}^3$.
   c. A gannet is built to withstand this high impact force, and it quickly decelerates to its terminal speed. What is the terminal speed of a gannet in seawater?
64. ||| BIO Dandelion seeds have a plume made of thin filaments that create a very large drag coefficient due to subtle aerodynamics as the air moves through and past the plume. The slower terminal speed and longer time in the air provide greater dispersal than would be possible for simple spherical seeds. A 0.63 mg seed has a plume with a cross-section area of 13 mm^2, and it falls with a terminal speed of 40 cm/s.
   a. What is the drag coefficient for this seed? Assume an air density of $1.2\,\text{kg/m}^3$. Be careful when you convert the data to SI units.
   b. How many times larger is your answer than the drag coefficient of a simple sphere?
65. |||| BIO A microorganism swimming through water at a speed of 150 μm/s suddenly stops swimming. Its speed drops to 75 μm/s in 1.5 ms. What is the total distance in μm it travels while stopping?
66. || Small particulates can be removed from the emissions of a coal-fired power plant by *electrostatic precipitation*. The particles are given a small electric charge that draws them toward oppositely charged plates, where they stick. Consider a spherical particulate with a diameter of 1.0 μm. The electric force on this particle is 2.0×10^{-13} N. What is the speed of such a particle? (The electric force is much greater than the particle's weight, which can be ignored.)
67. || BIO Researchers often use *force plates* to measure the forces that people exert against the floor during movement. A force plate works like a bathroom scale, but it keeps a record of how the reading changes with time. Figure P5.67 shows the data from a force plate as a woman jumps straight up and then lands.

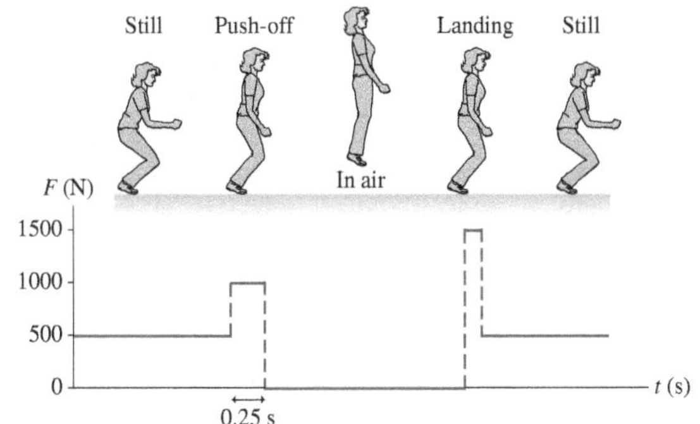

FIGURE P5.67

   a. What was the vertical component of her acceleration during push-off?
   b. What was the vertical component of her acceleration while in the air?
   c. What was the vertical component of her acceleration during the landing?
   d. What was her speed as her feet left the force plate?
   e. How high did she jump?
68. ||| BIO You probably think of wet surfaces as being slippery. Surprisingly, the opposite is true for human skin, as you can demonstrate by sliding a dry versus a slightly damp fingertip along a smooth surface such as a desktop. Researchers have found that the static coefficient of friction between dry skin and steel is 0.27, while that between damp skin and steel can be as high as 1.4. Suppose a man holds a steel rod vertically in his hand, exerting a 400 N grip force on the rod. What is the heaviest rod he can hold without slipping if
   a. His hands are dry?
   b. His hands are wet?
69. ||| BIO A person with compromised pinch strength in his fingers can only exert a normal force of 6.0 N to either side of a pinch-held object, such as the book shown in Figure P5.69. What is the greatest mass book he can hold onto vertically before it slips out of his fingers? The coefficient of static friction of the surface between the fingers and the book cover is 0.80.

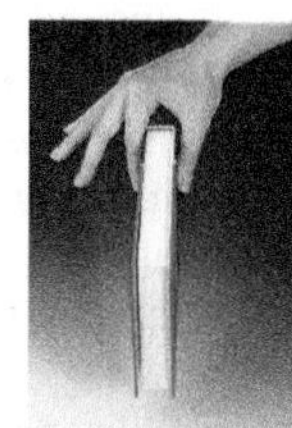

FIGURE P5.69

70. || It's possible for a determined group of people to pull an aircraft. Drag is negligible at low speeds, and the only force impeding motion is the rolling friction of the rubber tires on the concrete runway.

In 2000, a team of 60 British police officers set a world record by pulling a Boeing 747, with a mass of 200,000 kg, a distance of 100 m in 53 s. The plane started at rest. Estimate the force with which each officer pulled on the plane, assuming constant pulling force and constant acceleration.
71. ||| Two blocks are at rest on a frictionless incline, as shown in Figure P5.71. What are the tensions in the two strings?

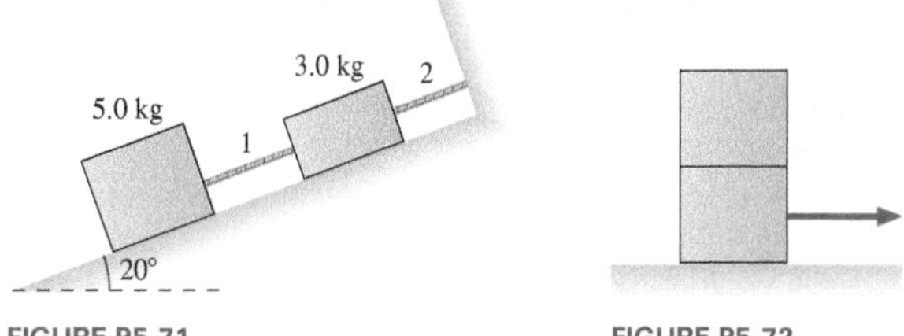

FIGURE P5.71

FIGURE P5.72

72. || Two identical 2.0 kg blocks are stacked as shown in Figure P5.72. The bottom block is free to slide on a frictionless surface. The coefficient of static friction between the blocks is 0.35. What is the maximum horizontal force that can be applied to the lower block without the upper block slipping?
73. |||| A Federation starship (2.0×10^6 kg) uses its tractor beam to pull a shuttlecraft (2.0×10^4 kg) aboard from a distance of 10 km away. The tractor beam exerts a constant force of 4.0×10^4 N on the shuttlecraft. Both spacecraft are initially at rest. How far does the starship move as it pulls the shuttlecraft aboard?

74. |||| A fish can be modeled as a cylinder of radius r and mass
BIO m swimming in water with velocity v_x. The drag on the fish is
CALC characterized by a high Reynolds number. If you write Newton's second law for the fish, using $F_x = ma_x = mdv_x/dt$ and the expression for drag (remember that the drag force points opposite the direction of motion), you find that $dv_x/dt = -cv_x^2$, where c is a constant that depends on m, r, the water's density ρ, and the coefficient of drag C_D.
   a. Rewrite this equation as $dv_x/v_x^2 = -cdt$, then integrate it to find an expression giving the fish's velocity as a function of time. The initial conditions, which you'll need to determine the integration constant, are $v_x = v_0$ at $t = 0$.
   b. Suppose a 12 kg, 20-cm-diameter tuna with a drag coefficient of 0.15 is swimming at 4.0 m/s in seawater. If it stops actively swimming, how long will it take for its speed to decrease to 50% of its initial value? The density of seawater is 1030 kg/m^3.

75. ||| Two identical, side-by-side springs with spring constant 240 N/m support a 2.00 kg hanging box. Each spring supports the same weight. By how much is each spring stretched?

76. | A 5.0 kg mass hanging from a spring scale is slowly lowered onto a vertical spring, as shown in Figure P5.76. The scale reads in newtons.
   a. What does the spring scale read just before the mass touches the lower spring?
   b. The scale reads 20 N when the lower spring has been compressed by 2.0 cm. What is the value of the spring constant for the lower spring?
   c. At what compression distance will the scale read zero?

FIGURE P5.76

77. ||| A 25 kg child bounces on a pogo stick. The pogo stick has a spring with spring constant 2.0×10^4 N/m. When the child makes a nice big bounce, she finds that at the bottom of the bounce she is accelerating *upward* at 9.8 m/s^2. How much is the spring compressed?

78. || The Achilles tendon is the strongest tendon in the human body.
BIO On average, the Achilles tendon of an adult is 150 mm in length and 20 mm^2 in cross section. Surprisingly, it is quite elastic; it can store and release energy like a spring, a topic we'll return to in Chapter 10. But it's a very stiff spring, with a spring constant of approximately 2.5×10^6 N/m. The force on the Achilles tendon while running can be 7 times the runner's weight. By how many mm does a 65 kg runner's Achilles tendon stretch?

79. |||| A sensitive gravimeter at a mountain observatory finds that the free-fall acceleration is 0.0075 m/s^2 less than that at sea level. What is the observatory's altitude? The radius of the earth is 6.37×10^6 m.

80. |||| In 2014, the Rosetta space probe reached the comet Churyumov–Gerasimenko. Although the comet's core is actually far from spherical, in this problem we'll model it as a sphere with a mass of 1.0×10^{13} kg and a radius of 1.6 km. If a rock were dropped from a height of 1.0 m above the comet's surface, how long would it take to hit the surface?

81. ||| A 20 kg sphere is at the origin and a 10 kg sphere is at $(x, y) = (20 \text{ cm}, 0 \text{ cm})$. At what point or points could you place a small mass such that the net gravitational force on it due to the spheres is zero?

MCAT-Style Passage Problems

Sliding on the Ice

In the winter sport of curling, players give a 20 kg stone a push across a sheet of ice. The stone moves approximately 40 m before coming to rest. The final position of the stone, in principle, only depends on the initial speed at which it is launched and the force of friction between the ice and the stone, but team members can use brooms to sweep the ice in front of the stone to adjust its speed and trajectory a bit; they must do this without touching the stone. Judicious sweeping can lengthen the travel of the stone by 3 m.

82. A curler pushes a stone to a speed of 3.0 m/s over a time of 2.0 s. Ignoring the force of friction, how much force must the curler apply to the stone to bring it up to speed?
   A. 3.0 N B. 15 N
   C. 30 N D. 150 N

83. Suppose the stone is launched with a speed of 3 m/s and travels 40 m before coming to rest. What is the *approximate* magnitude of the friction force on the stone?
   A. 0 N B. 2 N
   C. 20 N D. 200 N

84. Suppose the stone's mass is increased to 40 kg, but it is launched at the same 3 m/s. Which one of the following is true?
   A. The stone would now travel a longer distance before coming to rest.
   B. The stone would now travel a shorter distance before coming to rest.
   C. The coefficient of friction would now be greater.
   D. The force of friction would now be greater.

Jumping for Joy

A study of jumping uses *a force plate* that measures and records the magnitude of the vertical force pressing down on it. According to Newton's third law, this magnitude is the same as the magnitude of the normal force pushing up on the subject. An athlete stands in a crouch and then pushes off vigorously, leaves the ground, comes back down and lands on the force plate, and finally stands at rest. Data recorded by the force plate during the jump are shown in the graph of Figure P5.85.

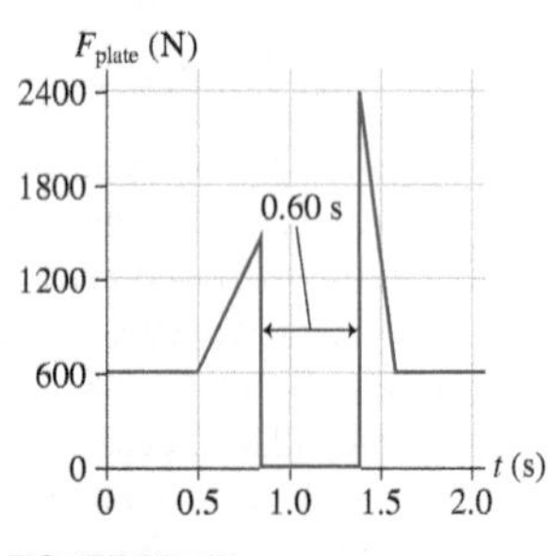

FIGURE P5.85

85. What is the approximate mass of the athlete?
   A. 50 kg B. 60 kg
   C. 70 kg D. 80 kg

86. During the push-off phase, approximately what is the athlete's maximum acceleration?
   A. 1.5g B. 2.0g
   C. 2.5g D. 3.0g

87. Approximately what maximum height does the athlete reach?
   A. 15 cm B. 30 cm
   C. 45 cm D. 60 cm

88. The magnitude of the maximum acceleration on landing is ________ times greater than the magnitude of the maximum acceleration when pushing off.
   A. 1.5 B. 2.0
   C. 2.5 D. 3.0

6 Equilibrium and Elasticity

A dancer's foot withstands enormous stresses as she balances on her toes.

LOOKING AHEAD

Torque

Torque, a force applied through a *moment arm,* tries to turn or rotate an object. Torque is the rotational equivalent of force.

You'll learn how to calculate the torque that the rocks on the right and the left exert about the balance point.

Static Equilibrium

As this cyclist balances on his back tire, the net force *and* the net torque on the bicycle must be zero.

You'll find out how to use Newton's laws to analyze extended objects that are in *static equilibrium.*

Elasticity

All materials have some "give"; we say that they are *elastic.* If you pull on them, they stretch, and at some point they break.

Is silk as strong as steel? You'll learn to think about what this question means, and how to answer it.

GOAL To learn about the static equilibrium of extended objects and the elastic properties of materials.

PHYSICS AND LIFE

Organisms Need Balance and Stability

Balance and locomotion are of critical importance to animals. Bones, muscles, and tendons work together to generate the forces that produce movement or stillness. In this chapter we'll investigate the forces required for an animal to maintain balance and the limits on those forces imposed by the properties of the materials involved. Dramatic injuries, such as a ruptured tendon or a broken bone, can result when those limits are exceeded. We will find that most biomaterials (even many of those, like bone, that we think of as hard) are elastic—they can compress, stretch, and bend—and we can understand the properties of such materials by extending the ideas we previously developed for springs. Their springiness allows these materials to store or dissipate energy when needed.

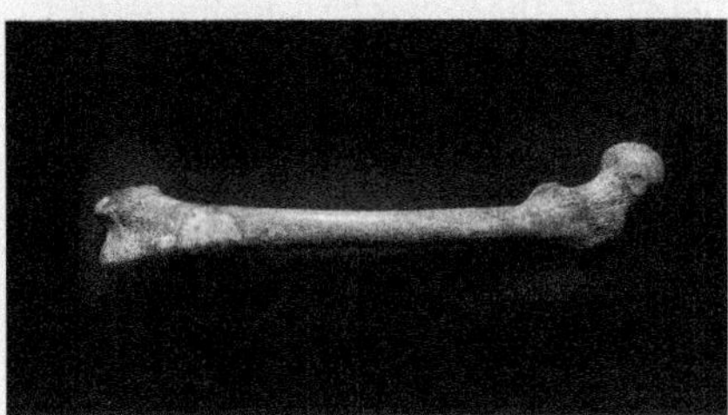

The elastic properties of bones determine whether they'll support your weight or whether they'll break upon impact.

6.1 Extended Objects

$\vec{T}$
$\vec{F}$
$\vec{w}$

Finding equilibrium This climber is in equilibrium. What does that say about the forces acting on her?

It surely hasn't escaped your notice that real objects are not particles. The particle model was a good starting point that helps explain many real-world applications. For example, the velocity and acceleration of a sprinter and the drag force on a swimming paramecium can be understood by modeling the objects as particles. If you throw a ball, we can model the ball as a particle that accelerates due to the force of your hand. But suppose your interest is not the motion of the ball but the biomechanics of throwing. How do forces exerted by your muscles enable you to rotate and extend your arm in a throwing motion? Could understanding those forces allow you to optimize a throw? What mechanical limitations might lead to injury? We're not going to be able to answer questions like these by modeling your arm as a particle.

Your arm, like all real objects, is an **extended object**—an object with a size and shape that can't be ignored. And, like your arm, an extended object can *rotate*. We're actually going to begin our study of extended objects by considering the forces that *prevent* an object from rotating. For example, how can you stand on two feet, or even one, without falling over—a rotation—while most animals need four feet for stability? What forces allow you to hold your arm out, parallel to the floor, without gravity pulling it down? Then, in the next chapter, we'll develop the ideas needed to understand rotating objects.

Chapter 4 introduced the idea of *mechanical equilibrium* for objects at rest. For a particle, the sole requirement for mechanical equilibrium is that there be no net force: $\vec{F}_{net} = \vec{0}$. All the applied forces must balance. However, an extended object in mechanical equilibrium has more requirements; not only is it not accelerating, it also is not rotating. We'll begin by introducing the important concept of *torque* in order to establish a no-rotation condition.

Extended objects have an additional complication that we can't overlook. Some objects, such as steel and to a somewhat lesser extent bone, are *rigid;* they maintain their size and shape. In contrast, materials like rubber and most biomaterials *deform* when forces are applied to them. Indeed, deformation is essential to their function. Perfectly rigid tree branches would snap off in a strong wind or even under their own weight. Your arteries would not be able to withstand the pressure increases during a heartbeat if they couldn't expand a little with each beat.

FIGURE 6.1 A steel beam is well modeled as a rigid body.

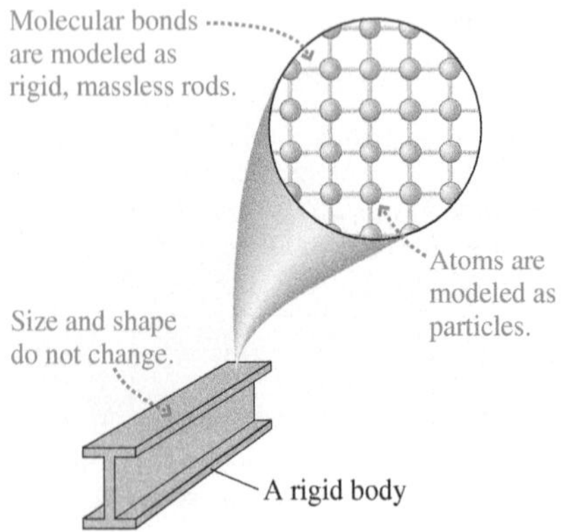
Molecular bonds are modeled as rigid, massless rods.
Atoms are modeled as particles.
Size and shape do not change.
A rigid body

We'll return to the elastic properties of materials later in the chapter. For now, let's focus on extended objects whose size and shape do not change. We call such an object a **rigid body.** For example, the steel beam in FIGURE 6.1 is an excellent approximation of a rigid body. No real object can be perfectly rigid; even a steel beam bends if it is loaded too heavily. Thus treating an object as if it's perfectly rigid is another model, the **rigid-body model.** The rigid-body model applies well to materials such as steel, wood, and bone under many circumstances, and for most of this chapter we will assume that those circumstances hold. We'll look at the limitations of the rigid-body model when we return to the elasticity of materials at the end of the chapter.

6.2 Torque

FIGURE 6.2 The four forces are the same strength, but they have different effects on the swinging door.

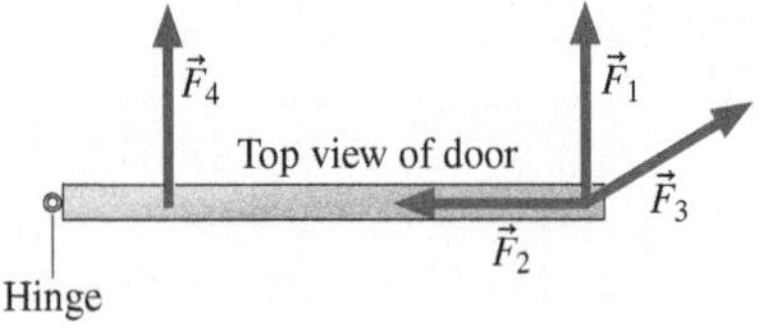
$\vec{F}_4$
$\vec{F}_1$
Top view of door
$\vec{F}_3$
$\vec{F}_2$
Hinge

Consider the common experience of pushing open a heavy door. FIGURE 6.2 is a top view of a door that is hinged on the left. Four forces are shown, all of equal strength. Which of these will be most effective at opening the door?

Force $\vec{F}_1$ will open the door, but force $\vec{F}_2$, which pushes straight toward the hinge, will not. Force $\vec{F}_3$ will open the door, but not as effectively as $\vec{F}_1$. What about $\vec{F}_4$? It is perpendicular to the door and it has the same magnitude as $\vec{F}_1$, but you know from experience that pushing close to the hinge is not as effective as pushing at the outer edge of the door.

The ability of a force to cause a rotation thus depends on three factors:

1. The magnitude F of the force
2. The distance r from the pivot point—the axis about which the object can rotate—to the point at which the force is applied
3. The angle at which the force is applied

We can incorporate these three factors into a single quantity called the **torque** τ (Greek tau). Loosely speaking, torque measures the effectiveness of a force at causing an object to rotate about a pivot point. **Torque is the rotational equivalent of force.** In Figure 6.2, for instance, the torque τ_1 due to $\vec{F}_1$ is greater than τ_4 due to $\vec{F}_4$.

To make these ideas specific, FIGURE 6.3a shows a force $\vec{F}$ applied at one point on a wrench that's loosening a nut. The force is applied at distance r from the pivot point; its direction is specified by the angle ϕ (Greek phi), measured from the *radial line* that extends outward from the pivot point.

We saw in Figure 6.2 that force $\vec{F}_1$, which was directed perpendicular to the door, was effective in opening it, but force $\vec{F}_2$, directed toward the hinges, was not—in fact, such a force does not open the door at all. This suggests that only the component of force *perpendicular* to the radial line causes rotation. To use this idea, FIGURE 6.3b breaks the force on the wrench into two component vectors: $\vec{F}_\perp$ directed perpendicular to the radial line, and $\vec{F}_\parallel$ directed parallel to it. Because $\vec{F}_\parallel$ points either directly toward or away from the pivot point, it has no effect on the nut's rotation and thus contributes nothing to the torque. Only $\vec{F}_\perp$ tends to cause rotation of the nut, so it is this component of the force that determines the torque.

FIGURE 6.3 Force $\vec{F}$ exerts a torque about the pivot point.

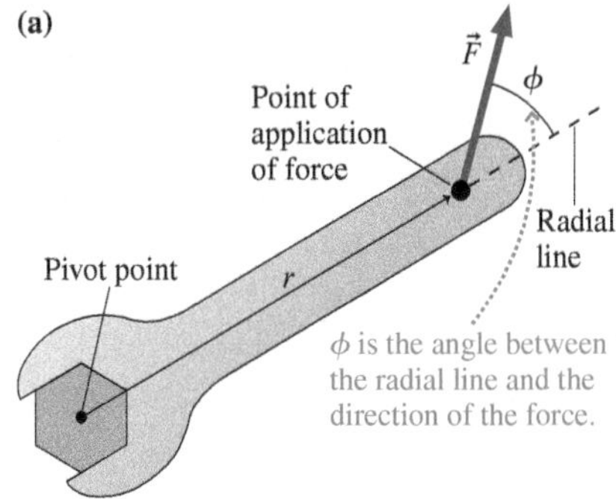
(a)
$\vec{F}$
ϕ
Point of application of force
Radial line
Pivot point
r
ϕ is the angle between the radial line and the direction of the force.

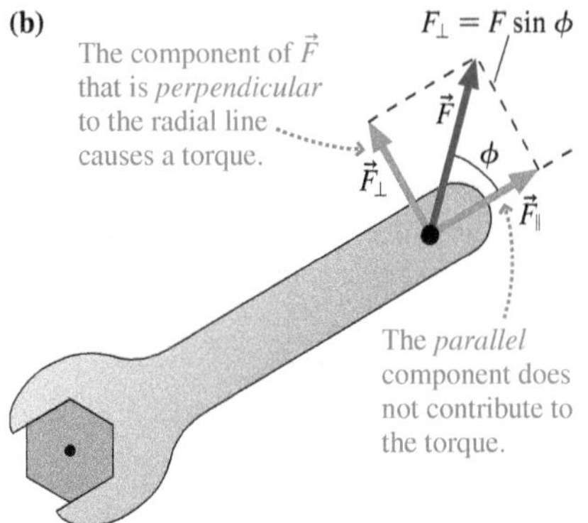
(b)
$F_\perp = F \sin\phi$
The component of $\vec{F}$ that is *perpendicular* to the radial line causes a torque.
$\vec{F}$
ϕ
$\vec{F}_\perp$
$\vec{F}_\parallel$
The *parallel* component does not contribute to the torque.

NOTE ▸ The perpendicular component $\vec{F}_\perp$ is pronounced "F perpendicular" and the parallel component $\vec{F}_\parallel$ is "F parallel." ◂

A force applied at a greater distance r from the pivot point has a greater effect on rotation, so we expect a larger value of r to give a greater torque. We also saw that only $\vec{F}_\perp$ contributes to the torque. Both these observations are contained in our first expression for torque:

$$\tau = rF_\perp \tag{6.1}$$

Torque due to a force with perpendicular component $F_\perp$ acting at a distance r from the pivot point

From this equation, we see that the SI unit of torque is the newton-meter, abbreviated N · m. In English units, torque is measured in foot-pounds.

FIGURE 6.4 shows an alternative way to calculate torque. The dashed line that is in the direction of the force is called the *line of action*. The perpendicular distance from this line to the pivot point is called the **moment arm** (or *lever arm*) $r_\perp$. You can see from the figure that $r_\perp = r\sin\phi$. Further, Figure 6.3b showed that $F_\perp = F\sin\phi$. We can then write Equation 6.1 as $\tau = rF\sin\phi = F(r\sin\phi) = Fr_\perp$. Thus an equivalent expression for the torque is

$$\tau = r_\perp F \tag{6.2}$$

Torque due to a force F with moment arm $r_\perp$

Equations 6.1 and 6.2 are two different ways to calculate the torque. You can use the one that is more convenient, depending on the problem.

FIGURE 6.4 You can also calculate torque in terms of the moment arm between the pivot point and the line of action.

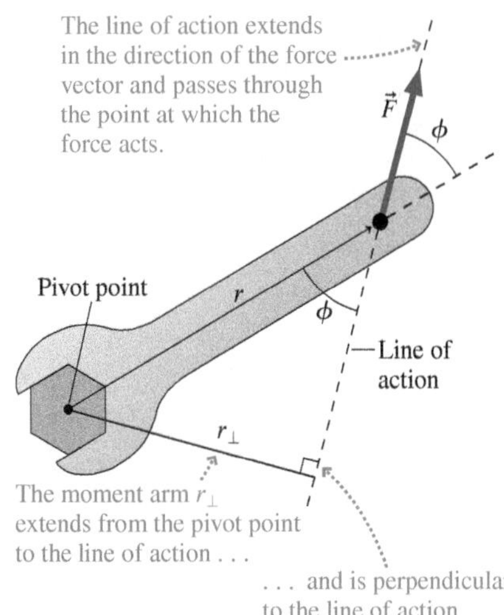
The line of action extends in the direction of the force vector and passes through the point at which the force acts.
$\vec{F}$
ϕ
Pivot point
r
ϕ
Line of action
$r_\perp$
The moment arm $r_\perp$ extends from the pivot point to the line of action . . .
. . . and is perpendicular to the line of action.

Different torques Different wheelchair designs give different results. A basketball player needs to start and stop quickly, which requires large torques on the wheel. To make the torque as large as possible, the handrim—the outside wheel that she actually grabs—is almost as big as the wheel itself. The racer needs to move continuously at high speed, so his wheels spin much faster. To allow his hands to keep up, his handrim is much smaller than the chair's wheel, making its speed correspondingly lower. The smaller radius means, however, that the torque he can apply is lower as well.

CONCEPTUAL EXAMPLE 6.1 Pedaling a bike

It's hard to get going if you try to start your bike with one pedal at its highest point. Why? Where should the pedal be for the quickest and easiest start?

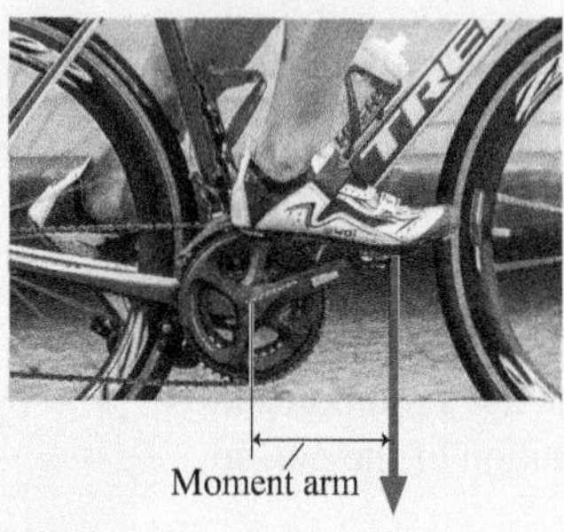

REASON Your leg presses almost straight down on the pedal. When the pedal is at the top, the force is directed almost straight toward the pivot point, so it causes only a small torque. The moment arm, which is the horizontal distance between the pivot point and the vertical line along which the force acts, is very small when the pedal is near the top (exactly zero when exactly at the top), so the torque is very small. You achieve the maximum torque, and thus the quickest acceleration, if one pedal is all the way forward because then the moment arm is as large as it can be.

ASSESS If you've ever ridden a bike, this agrees with your experience. Getting the quickest start depends not so much on how hard you press down but on where the pedal is when you press it.

NOTE ▸ Torque differs from force in a very important way. Torque is calculated or measured *about a particular point.* To say that a torque is 20 N·m is meaningless without specifying the point about which the torque is calculated. Torque can be calculated about any point, but its value depends on the point chosen because this choice determines r and ϕ. In practice, we usually calculate torques about a hinge, pivot, or axle. ◂

EXAMPLE 6.2 Torque to open a stuck door

Ryan is trying to open a stuck door. He pushes it at a point 0.75 m from the hinges with a 240 N force directed 20° away from being perpendicular to the door. There's a natural pivot point, the hinge. What torque does Ryan exert about the hinge? How could he exert more torque?

PREPARE We start by drawing the pictorial representation of FIGURE 6.5. The figure shows the door, identifies the pivot point (the hinge, in this case), draws the radial line, and shows where and how the force is applied.

FIGURE 6.5 Ryan's force exerts a torque on the door.

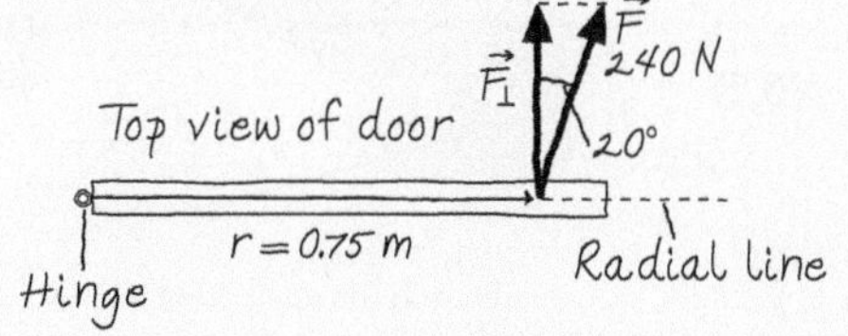

SOLVE As you can see in the figure, the component of $\vec{F}$ that is perpendicular to the radial line is $F_\perp = F\cos 20° = 226$ N. The distance from the hinge to the point at which the force is applied is $r = 0.75$ m. We can now find the torque on the door from Equation 6.1:

$$\tau = rF_\perp = (0.75 \text{ m})(226 \text{ N}) = 170 \text{ N}\cdot\text{m}$$

The torque depends on how hard Ryan pushes, where he pushes, and at what angle. If he wants to exert more torque, he could push at a point a bit farther out from the hinge, or he could push exactly perpendicular to the door. Or he could simply push harder!

ASSESS As you'll see by doing more problems, 170 N·m is a significant torque, but this makes sense if you are trying to free a stuck door.

Equations 6.1 and 6.2 give only the magnitude of the torque. But torque, like a force component, has a sign. We will follow the convention that **a torque that tends to rotate the object in a counterclockwise direction is positive, while a torque that tends to rotate the object in a clockwise direction is negative.**

FIGURE 6.6 summarizes what we've said about the magnitude and the sign of a torque. Notice that a force pushing straight toward the pivot point or pulling straight out from the pivot point exerts *no* torque.

NOTE ▶ When calculating a torque, you must supply the appropriate sign by observing the direction in which the torque acts. ◀

FIGURE 6.6 Signs and strengths of the torque.

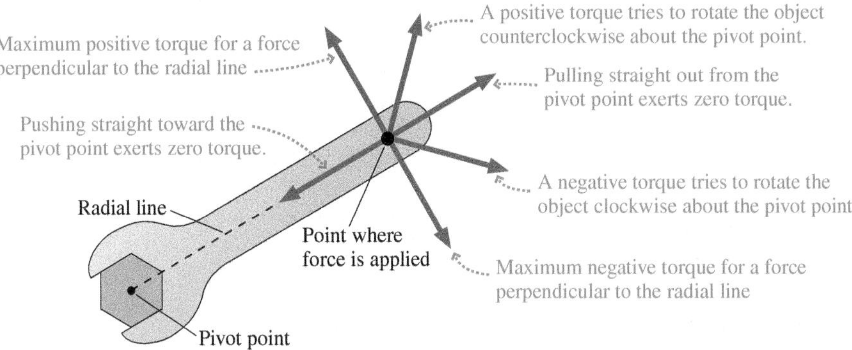
A positive torque tries to rotate the object counterclockwise about the pivot point.
Maximum positive torque for a force perpendicular to the radial line
Pulling straight out from the pivot point exerts zero torque.
Pushing straight toward the pivot point exerts zero torque.
A negative torque tries to rotate the object clockwise about the pivot point
Radial line
Point where force is applied
Maximum negative torque for a force perpendicular to the radial line
Pivot point

Net Torque

FIGURE 6.7 shows the forces acting on the crankset of a bicycle. Forces $\vec{F}_1$ and $\vec{F}_2$ are due to the rider pushing on the pedals, and $\vec{F}_3$ and $\vec{F}_4$ are tension forces from the chain. The crankset is free to rotate about a fixed axle, but the axle prevents it from having any translational motion with respect to the bike frame. It does so by exerting force $\vec{F}_{axle}$ on the object to balance the other forces and keep $\vec{F}_{net} = \vec{0}$.

Forces $\vec{F}_1$, $\vec{F}_2$, $\vec{F}_3$, and $\vec{F}_4$ exert torques τ_1, τ_2, τ_3, and τ_4 on the crank (measured about the axle), but $\vec{F}_{axle}$ does *not* exert a torque because it is applied at the pivot point—the axle—and so has zero moment arm. The *net* torque about the axle is the sum of the torques due to the *applied* forces:

$$\tau_{net} = \tau_1 + \tau_2 + \tau_3 + \tau_4 + \cdots = \sum \tau \tag{6.3}$$

Keep in mind that torque can be positive or negative, so the net torque can be zero even when several torques are applied.

FIGURE 6.7 The forces exert a net torque about the pivot point.

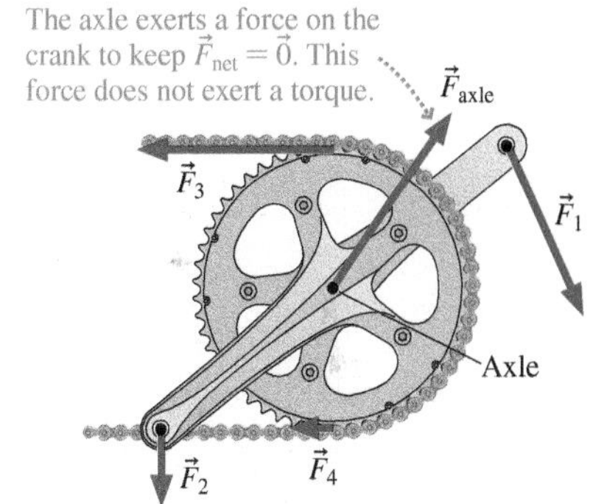
The axle exerts a force on the crank to keep $\vec{F}_{net} = \vec{0}$. This force does not exert a torque.
$\vec{F}_{axle}$
$\vec{F}_3$
$\vec{F}_1$
Axle
$\vec{F}_2$
$\vec{F}_4$

EXAMPLE 6.3 Torque at the gym

Exercise equipment uses a complex system of cables, weights, and springs to provide a controlled, steady tension. We can illustrate the basic ideas with the simple machine shown in FIGURE 6.8. This device to exercise your shoulder and upper arm consists of a pivoted, lightweight, 1.0-m-long bar with a horizontal cable attached at the midpoint. To use it, you set the cable tension and then, while gripping the end of the bar, quickly rotate the bar from vertical to horizontal. To rotate the bar in this fashion requires a net torque of 10 N · m about the pivot point. What is your pushing force when the bar is 60° above horizontal, as shown, if you've set the cable tension at 250 N?

PREPARE We model the bar as a rigid body. The pictorial representation has been drawn with the bar rotating counterclockwise, so the net torque of 10 N · m is positive. Your push, which tends to rotate the bar counterclockwise, creates a positive torque; the cable tension acts in the opposite direction and so exerts a negative torque. We're told that the bar is "lightweight," so we'll assume that any torque due to the weight of the bar can be neglected.

FIGURE 6.8 Pictorial representation of an exercise bar being rotated.

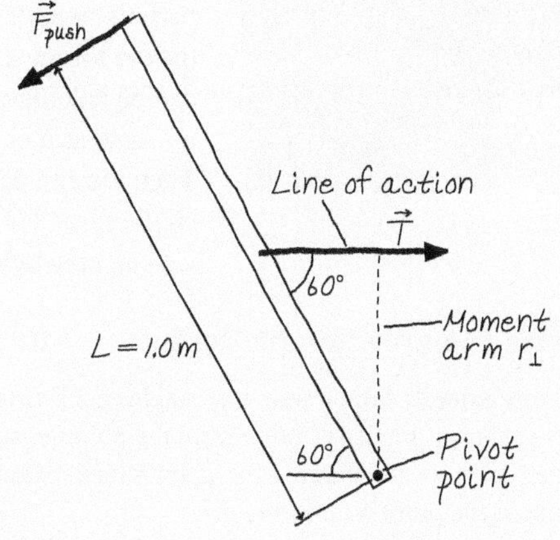
$\vec{F}_{push}$
Line of action
$\vec{T}$
60°
Moment arm $r_\perp$
$L = 1.0$ m
60°
Pivot point

Continued

SOLVE The net torque is $\tau_{net} = \tau_{push} + \tau_{tension} = 10\ \text{N}\cdot\text{m}$. The torque you apply is $\tau_{push} = rF_{\perp}$. Your pushing force is perpendicular to the radial line, so $F_{\perp} = F_{push}$. The point at which $\vec{F}_{push}$ acts is at distance $r = L$ from the pivot, where L is the length of the bar, so the torque due to your push is $\tau_{push} = rF_{\perp} = LF_{push}$. To calculate the torque due to the tension, we see that the moment arm—the perpendicular distance from the pivot point to the line of action—is the side of a right triangle opposite the 60° angle. The hypotenuse is half the bar length, $\frac{1}{2}L$, so

$$r_{\perp} = \tfrac{1}{2}L \sin 60° = 0.433\ \text{m}$$

Thus, from Equation 6.2, the torque due to the tension is

$$\tau_{tension} = -r_{\perp}T = -(0.433\ \text{m})(250\ \text{N}) = -108\ \text{N}\cdot\text{m}$$

Notice that *we* supplied the minus sign after observing that the tension would rotate the bar clockwise; the sign does not appear automatically. (Similarly, we previously noted that the torque you exert is positive.) Knowing the net torque, we can now write

$$\tau_{push} = LF_{push} = \tau_{net} - \tau_{tension} = 10\ \text{N}\cdot\text{m} - (-108\ \text{N}\cdot\text{m}) = 118\ \text{N}\cdot\text{m}$$

With one last step, we find that your pushing force is

$$F_{push} = \frac{\tau_{push}}{L} = \frac{118\ \text{N}\cdot\text{m}}{1.0\ \text{m}} = 120\ \text{N}$$

We rounded the final answer to two significant figures.

ASSESS By using $w = mg$, we see that 120 N is roughly the weight of a 12 kg mass. 12 kg corresponds to a little more than 25 lb. You can probably pick up a 25 lb weight with one arm, but it takes some effort. You would likely apply about the same effort while working out, so a pushing force of 120 N seems reasonable.

STOP TO THINK 6.1 A wheel turns freely on an axle at the center. Which one of the forces shown in the figure will provide the largest positive torque about the axle?

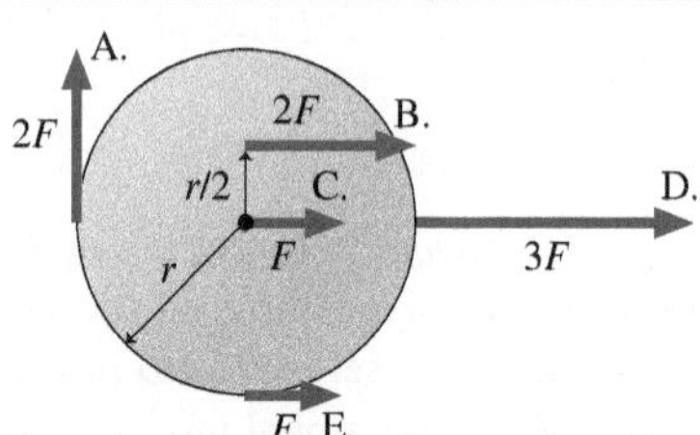

FIGURE 6.9 The center of gravity is the point where the weight appears to act.

(a) Gravity exerts a force and a torque on each particle that makes up the gymnast. Rotation axis

(b) The weight force provides a torque about the rotation axis.

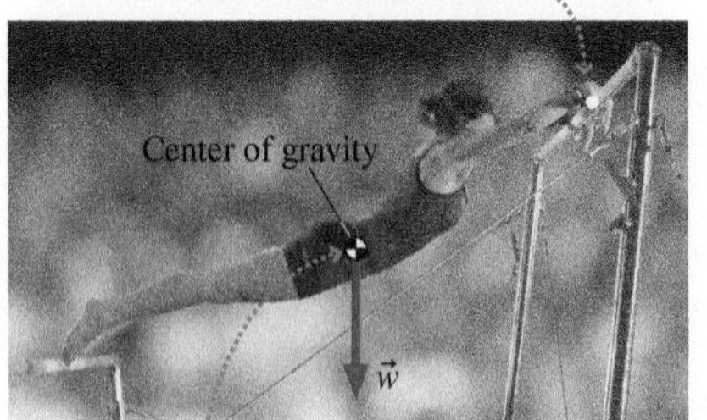

The gymnast responds *as if* her entire weight acts at her center of gravity.

6.3 Gravitational Torque and the Center of Gravity

As the gymnast in FIGURE 6.9 pivots around the bar, a torque due to the force of gravity causes her to rotate toward a vertical position. A falling tree and a car hood slamming shut are other examples where gravity exerts a torque on an object. Stationary objects can also experience a torque due to gravity. A diving board experiences a gravitational torque about its fixed end. It doesn't rotate because of a counteracting torque provided by forces from the base at its fixed end.

We've learned how to calculate the torque due to a single force acting on an object. But gravity doesn't act at a single point on an object. It pulls downward on *every particle* that makes up the object, as shown for the gymnast in Figure 6.9a, and so each particle experiences a small torque. The gravitational torque on the object as a whole is the *net* torque exerted on all the particles. We won't prove it, but the gravitational torque can be calculated by assuming that the net force of gravity—that is, the object's weight $\vec{w}$—acts at a single special point of the object called its **center of gravity** (symbol ⊗). Then we can calculate the torque due to gravity by the methods we learned earlier for a single force ($\vec{w}$) acting at a single point (the center of gravity). Figure 6.9b shows how we can consider the gymnast's weight as acting at her center of gravity.

EXAMPLE 6.4 The gravitational torque on a flagpole

A 3.2 kg flagpole extends from a wall at an angle of 25° from the horizontal. Its center of gravity is 1.6 m from the point where the pole is attached to the wall. What is the gravitational torque on the flagpole about the point of attachment?

PREPARE In all the problems in this chapter, a pictorial representation is crucial to help you understand the details, so we will start with the diagram in FIGURE 6.10. We'll note dimensions, draw the

FIGURE 6.10 Pictorial representation of the flagpole.

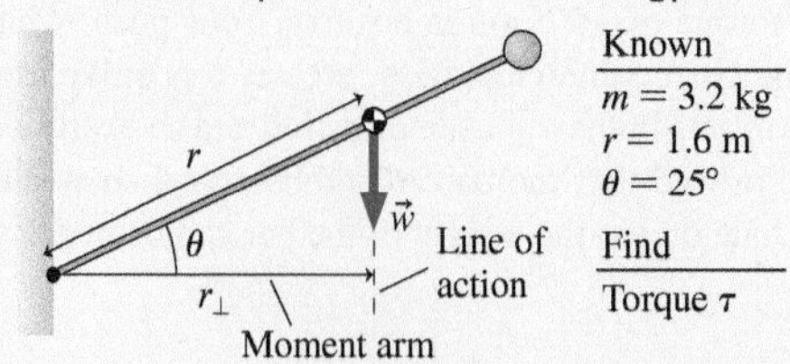

weight vector at the proper point, and choose one of the approaches we've seen for calculating the torque. For the purpose of calculating gravitational torque, we can consider the entire weight of the pole as acting at the center of gravity. Because the moment arm $r_\perp$ is simple to visualize here, we'll use Equation 6.2 for the torque.

SOLVE From Figure 6.10, we see that the moment arm is $r_\perp = (1.6\text{ m})\cos 25° = 1.45\text{ m}$. Thus the gravitational torque on the flagpole, about the point where it attaches to the wall, is

$$\tau = -r_\perp w = -r_\perp mg = -(1.45\text{ m})(3.2\text{ kg})(9.8\text{ m/s}^2)$$
$$= -45\text{ N}\cdot\text{m}$$

We inserted the minus sign because the torque tries to rotate the pole in a clockwise direction. If the pole were attached to the wall by a hinge, the gravitational torque would cause the pole to fall. However, the actual rigid connection provides a counteracting (positive) torque to the pole that prevents this. The net torque is zero.

ASSESS A torque is a product of a force and a distance. The force is the weight of the pole, just over 30 N; the distance is the moment arm, just shy of 1.5 m. So 45 N·m is a reasonable result.

If you hold a ruler by its end so that it is free to pivot, you know that it will quickly rotate so that it hangs straight down. FIGURE 6.11a explains this result in terms of the center of gravity and gravitational torque. The center of gravity of the ruler lies at its center. If the center of gravity is directly below the pivot point, there is no gravitational torque and the ruler will stay put. If you rotate the ruler to the side, the resulting gravitational torque will quickly pull it back until the center of gravity is again below the pivot point. We've shown this for a ruler, but this is a general principle: **An object that is free to rotate about a pivot will come to rest with the center of gravity below the pivot point.**

If the center of gravity lies directly *above* the pivot point, as in FIGURE 6.11b, there is no torque due to the object's weight and it can remain balanced. However, if the object is even slightly displaced to either side, the gravitational torque will no longer be zero and the object will begin to rotate. This question of *balance*—the behavior of an object whose center of gravity lies above the pivot point—will be explored later in this chapter.

FIGURE 6.11 Gravity acts on a ruler.

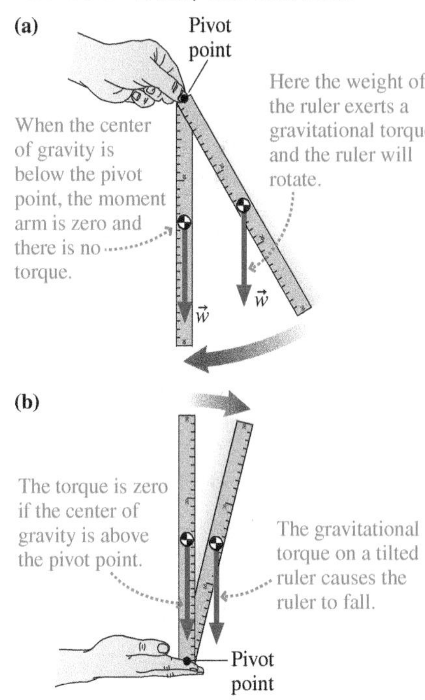
(a)
Pivot point
When the center of gravity is below the pivot point, the moment arm is zero and there is no torque.
Here the weight of the ruler exerts a gravitational torque and the ruler will rotate.
$\vec{w}$
$\vec{w}$
(b)
The torque is zero if the center of gravity is above the pivot point.
The gravitational torque on a tilted ruler causes the ruler to fall.
Pivot point

Calculating the Position of the Center of Gravity

You have to know where an object's center of gravity is before you can calculate the gravitational torque or know whether the object is balanced. For a highly symmetrical object, such as a wheel, the center of gravity is at the center of the object. But for irregularly shaped objects, we need a way to calculate the position of the center of gravity.

Consider the dumbbell shown in FIGURE 6.12. If we slide the triangular pivot back and forth until the dumbbell balances, then the torque due to gravity must be zero and thus the pivot point must be directly under the center of gravity at position x_{cg}. We can calculate the gravitational torque directly by calculating and summing the torques about this point due to the two individual weights. Gravity acts on weight 1 with moment arm r_1, so the torque about the pivot point at position x_{cg} is

FIGURE 6.12 Finding the center of gravity of a dumbbell.

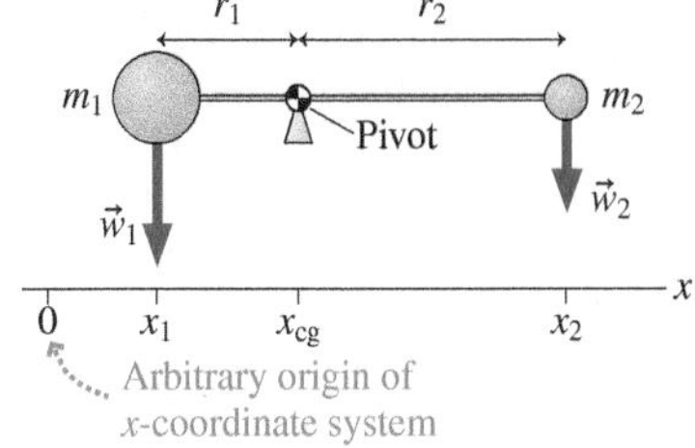
r_1
r_2
m_1
m_2
Pivot
$\vec{w}_1$
$\vec{w}_2$
x
0
x_1
x_{cg}
x_2
Arbitrary origin of x-coordinate system

$$\tau_1 = r_1 w_1 = (x_{cg} - x_1)m_1 g$$

Similarly, the torque due to weight 2 is

$$\tau_2 = -r_2 w_2 = -(x_2 - x_{cg})m_2 g$$

This torque is negative because it tends to rotate the dumbbell in a clockwise direction. We've just argued that the net torque must be zero because the pivot is directly under the center of gravity, so

$$\tau_{net} = 0 = \tau_1 + \tau_2 = (x_{cg} - x_1)m_1 g - (x_2 - x_{cg})m_2 g$$

We can solve this equation for the position of the center of gravity x_{cg}:

$$x_{cg} = \frac{x_1 m_1 + x_2 m_2}{m_1 + m_2} \tag{6.4}$$

The following Tactics Box shows how Equation 6.4 can be generalized to find the center of gravity of *any* number of particles. If the particles don't all lie along the x-axis, then we'll also need to find the y-coordinate of the center of gravity.

TACTICS BOX 6.1 Finding the center of gravity

1. Choose an origin for your coordinate system. You can choose any convenient point as the origin. Model each discrete part of the object as a particle.
2. Determine the coordinates (x_1, y_1), (x_2, y_2), (x_3, y_3), . . . for the particles of masses $m_1, m_2, m_3, \ldots$, respectively.
3. The x-coordinate of the center of gravity is

$$x_{cg} = \frac{x_1 m_1 + x_2 m_2 + x_3 m_3 + \cdots}{m_1 + m_2 + m_3 + \cdots} = \frac{1}{M}\sum_i x_i m_i \tag{6.5}$$

where we use a capital M for the total mass of an extended object.

4. Similarly, the y-coordinate of the center of gravity is

$$y_{cg} = \frac{y_1 m_1 + y_2 m_2 + y_3 m_3 + \cdots}{m_1 + m_2 + m_3 + \cdots} = \frac{1}{M}\sum_i y_i m_i \tag{6.6}$$

Because the center of gravity depends on products such as $x_1 m_1$, objects with large masses count more heavily than objects with small masses. Consequently, **the center of gravity tends to lie closer to the heavier objects or particles** that make up the entire object.

NOTE ▶ In other books or contexts you may see the term "center of mass." For our purposes, there's no difference between "center of mass" and "center of gravity." ◀

EXAMPLE 6.5 Balancing a seesaw

Emma and Noah are playing on a seesaw. Noah, mass 22 kg, is seated 1.6 m away from the pivot point. Emma, mass 25 kg, wants to sit at a point that will make the seesaw exactly balance on its pivot. How far away from the pivot must she sit? Assume that the center of gravity of the seesaw itself is directly over the pivot.

PREPARE In the pictorial representation of FIGURE 6.13 we've defined coordinates so that the pivot point is at $x = 0$ m. Noah is 1.6 m from the pivot, so $x_N = -1.6$ m. The seesaw will be balanced if the system's center of gravity is directly over the pivot. The center of gravity of the seesaw itself is already over the pivot, so we only need to consider the children.

FIGURE 6.13 Children balance on a seesaw.

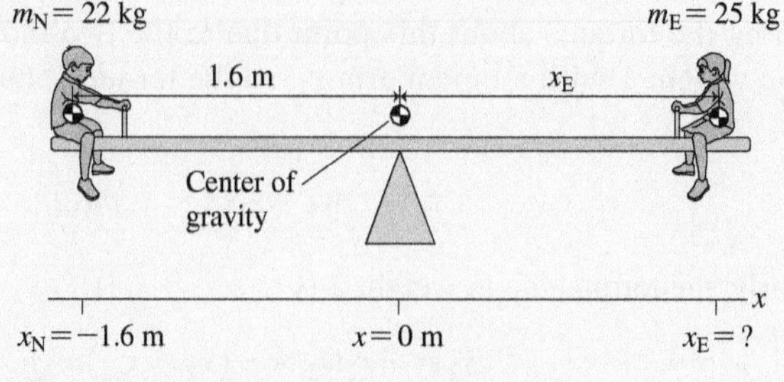

SOLVE There are two masses, the masses of Noah and Emma, and the position of the center of gravity is at the pivot, so we can simplify Equation 6.5 for the position of the center of gravity:

$$x_{cg} = 0 = \frac{x_N m_N + x_E m_E}{m_N + m_E}$$

If the fraction on the right is equal to zero, the numerator must be zero, so

$$x_N m_N + x_E m_E = 0$$

We can solve this for Emma's position:

$$x_E m_E = -x_N m_N$$

$$x_E = -x_N\left(\frac{m_N}{m_E}\right) = -(-1.6\text{ m})\frac{22\text{ kg}}{25\text{ kg}} = 1.4\text{ m}$$

Emma should sit 1.4 m from the pivot.

ASSESS If you've ever played on a seesaw, you know that to achieve balance the heavier person needs to sit closer to the pivot, so our answer makes sense.

EXAMPLE 6.6 Finding the center of gravity of a gymnast

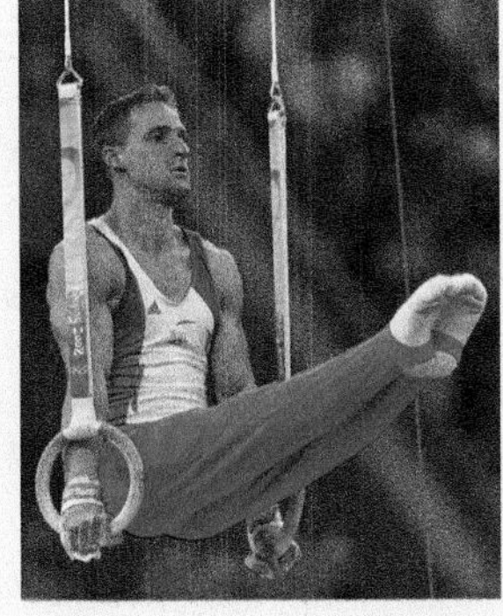

The gymnast in the photo, demonstrating the pike position, has his hands slightly in front of his hips. He's not rotating, so his hands—the pivot points—must be directly below his center of gravity. FIGURE 6.14 shows how we can consider his body to consist of two segments whose masses and center-of-gravity positions are shown. For a composite object with distinct pieces, each with a known mass and center of gravity, you can locate the entire object's center of gravity by using Equations 6.5 and 6.6 with the masses and center-of-gravity coordinates of the individual pieces. Where is the gymnast's center of gravity relative to his hips?

FIGURE 6.14 Centers of gravity of two segments of a gymnast.

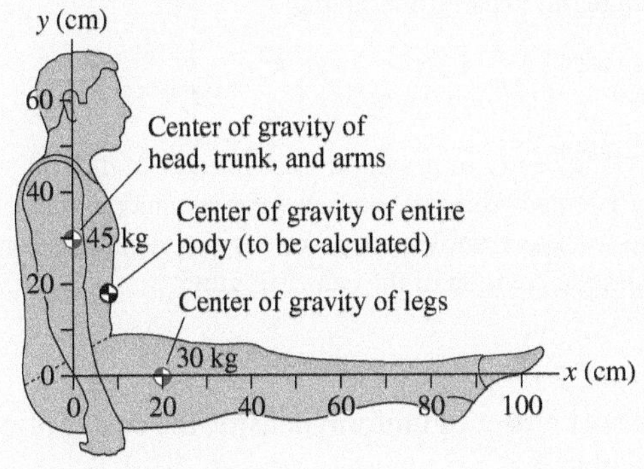

PREPARE The hips, where the upper body and lower body pivot, are a natural choice for the origin of a coordinate system. From Figure 6.14 we can read the x- and y-coordinates of the center of gravity of each segment:

$$x_{\text{trunk}} = 0\text{ cm} \qquad y_{\text{trunk}} = 30\text{ cm}$$

$$x_{\text{legs}} = 20\text{ cm} \qquad y_{\text{legs}} = 0\text{ cm}$$

SOLVE The x- and y-coordinates of the center of gravity are given by Equations 6.5 and 6.6:

$$x_{\text{cg}} = \frac{x_{\text{trunk}}m_{\text{trunk}} + x_{\text{legs}}m_{\text{legs}}}{m_{\text{trunk}} + m_{\text{legs}}}$$

$$= \frac{(0\text{ cm})(45\text{ kg}) + (20\text{ cm})(30\text{ kg})}{45\text{ kg} + 30\text{ kg}} = 8\text{ cm}$$

and

$$y_{\text{cg}} = \frac{y_{\text{trunk}}m_{\text{trunk}} + y_{\text{legs}}m_{\text{legs}}}{m_{\text{trunk}} + m_{\text{legs}}}$$

$$= \frac{(30\text{ cm})(45\text{ kg}) + (0\text{ cm})(30\text{ kg})}{45\text{ kg} + 30\text{ kg}} = 18\text{ cm}$$

ASSESS The gymnast's center of gravity in this position is 18 cm above and 8 cm in front of his hips, just about on the surface of his abdomen. That's consistent with where we see his hands in the photo. Notice that the center of gravity is closer to the heavier trunk segment than to the lighter legs.

Finding the Center of Gravity by Integration

Equations 6.5 and 6.6 work well for an object composed of a few discrete parts but not so well for an extended object like the one in FIGURE 6.15. To find the center of gravity of an extended object, let's imagine dividing it into many small cells or boxes, each with the same very small mass Δm. We will number the cells 1, 2, 3, ... Cell i has coordinates (x_i, y_i) and mass $m_i = \Delta m$. The center-of-gravity coordinates are then

$$x_{\text{cg}} = \frac{1}{M}\sum_i x_i\,\Delta m \qquad \text{and} \qquad y_{\text{cg}} = \frac{1}{M}\sum_i y_i\,\Delta m$$

Suppose we let the cells become smaller and smaller, with the total number of cells increasing. As each cell becomes infinitesimally small, we can replace Δm with

FIGURE 6.15 Calculating the center of gravity of an extended object.

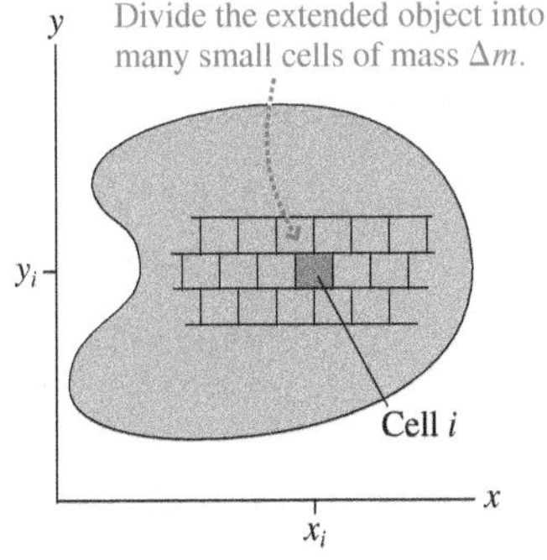

dm and the sum by an integral. Converting a sum to an integral in this way is a procedure that you've seen in calculus. Then

$$x_{cg} = \frac{1}{M}\int x\,dm \quad \text{and} \quad y_{cg} = \frac{1}{M}\int y\,dm \tag{6.7}$$

Equations 6.7 are a formal definition of the center of gravity, but they are *not* ready to integrate in this form. First, integrals are carried out over *coordinates,* not over masses. Before we can integrate, we must replace dm by an equivalent expression involving a coordinate differential such as dx or dy. Second, no limits of integration have been specified. The procedure for using Equations 6.7 is best shown with an example.

EXAMPLE 6.7 The center of gravity of a rod

Find the center of gravity of a thin, uniform rod of length L and mass M.

PREPARE FIGURE 6.16 shows the rod. We've chosen a coordinate system such that the rod lies along the x-axis from 0 to L. Because the rod is "thin," we'll assume that $y_{cg} = 0$.

FIGURE 6.16 Finding the center of gravity of a long, thin rod.

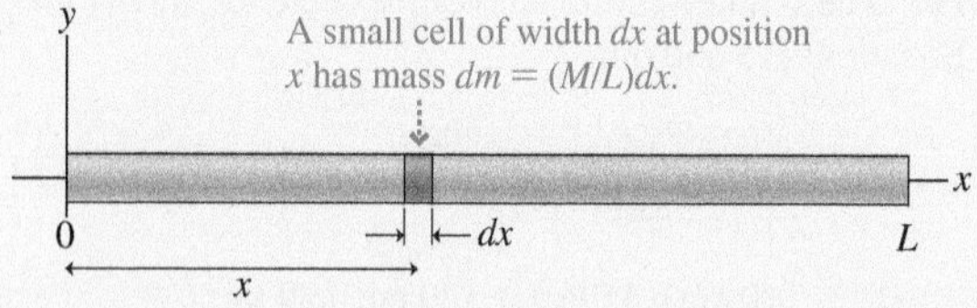

SOLVE Our task is to find x_{cg}, which lies somewhere on the x-axis. To do this, we divide the rod into many small cells of mass dm. One such cell, at position x, is shown. The cell's width is dx. Because the rod is *uniform,* the mass of this little cell is the *same fraction* of the total mass M that dx is of the total length L. That is,

$$\frac{dm}{M} = \frac{dx}{L}$$

Consequently, we can express dm in terms of the coordinate differential dx as

$$dm = \frac{M}{L}dx$$

NOTE ▶ The change of variables from dm to the differential of a coordinate is *the* key step in calculating the center of gravity. ◀

With this expression for dm, Equation 6.7 for x_{cg} becomes

$$x_{cg} = \frac{1}{M}\left(\frac{M}{L}\int x\,dx\right) = \frac{1}{L}\int_0^L x\,dx$$

where in the last step we've noted that summing "all the mass in the rod" means integrating from $x = 0$ to $x = L$. This is a straightforward integral to carry out, giving

$$x_{cg} = \frac{1}{L}\left[\frac{x^2}{2}\right]_0^L = \frac{1}{L}\left[\frac{L^2}{2} - 0\right] = \tfrac{1}{2}L$$

ASSESS The center of gravity is at the center of the rod, as you probably guessed. But we could use this technique with objects that have a more complex shape or varying density where you can't guess the position of the center of gravity.

NOTE ▶ For any symmetrical object of uniform density, the center of gravity is at the physical center of the object. ◀

STOP TO THINK 6.2 A baseball bat is cut into two pieces at its center of gravity. Which end is heavier?

A. The handle end (left end)
B. The hitting end (right end)
C. The two ends weigh the same.

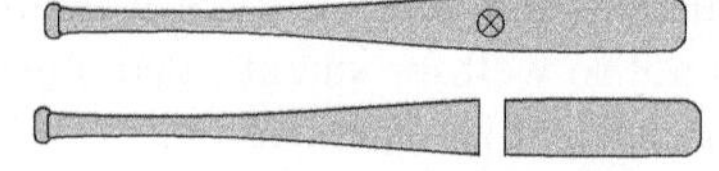

6.4 Static Equilibrium

We have now spent several chapters studying motion and its causes. In many disciplines, it is just as important to understand the conditions under which objects do *not* move. Buildings and dams must be designed such that they remain practically motionless, even when huge forces act on them. And joints in the body must sustain large forces when the body is supporting heavy loads, as in holding or carrying heavy objects.

Recall from ◀◀ SECTION 4.7 that an object at rest is in *equilibrium.* As long as an object can be modeled as a *particle,* the condition necessary for equilibrium is that the net force $\vec{F}_{net}$ on the particle is zero, as in FIGURE 6.17a, where the two forces applied to the particle balance and the particle can remain at rest.

Now we've moved beyond the particle model to study extended objects that can rotate. Consider, for example, the object in FIGURE 6.17b. In this case both the net force is zero and, because the two forces act along the same line, the net torque is zero; the object is in equilibrium with no tendency to move or to rotate. But what about the object in FIGURE 6.17c? The net force is still zero, but this time the object begins to rotate because the two forces exert a net torque. For an extended object, $\vec{F}_{net} = \vec{0}$ is not by itself enough to ensure equilibrium. There is a second condition for equilibrium of an extended object: The net torque τ_{net} on the object must also be zero.

Extended objects at rest are said to be in **static equilibrium.** If we write the net force in component form, the conditions for static equilibrium are

$$\left.\begin{array}{l}\sum F_x = 0 \\ \sum F_y = 0\end{array}\right\} \text{ No net force}$$
$$\sum \tau = 0 \quad \text{No net torque} \tag{6.8}$$

Conditions for static equilibrium of an extended object

FIGURE 6.17 An object with no net force acting on it may still be out of equilibrium.

(a) When the net force on a particle is zero, the particle is in equilibrium.

(b) Both the net force and the net torque are zero, so the object is in equilibrium.

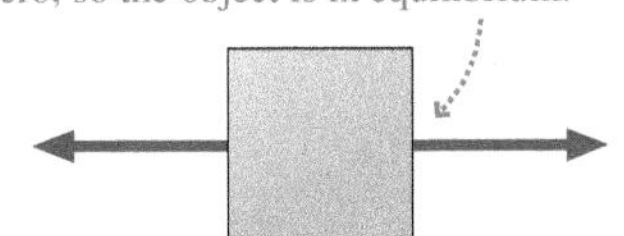

(c) The net force is still zero, but the net torque is *not* zero. The object will rotate and is not in equilibrium.

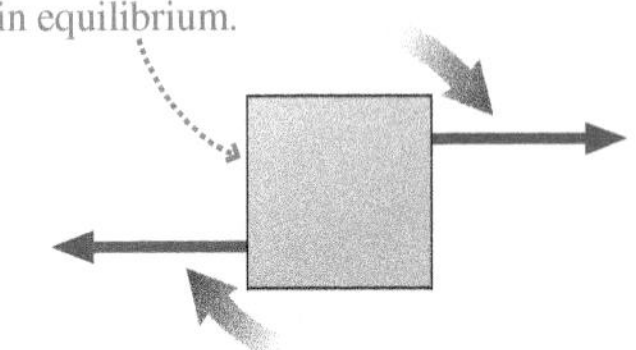

Let's look at an example to see how we can use these conditions to analyze a physical situation.

EXAMPLE 6.8 Lifting a load

A participant in a "strongman" competition uses an old-fashioned device to move a large load with only muscle power. A 1 ton (910 kg) bucket of rocks is suspended from a sturdy, lightweight beam 0.85 m from a pivot. The man lifts the beam at its end, 3.6 m from the pivot, and holds it steady. How much force must the man apply? What is the force on the beam from the pivot?

PREPARE This is a static equilibrium problem—nothing is moving. The first step in our solution is to find an object to focus on. The way we've drawn the diagram in FIGURE 6.18, with the highlight color on the beam, hints at our choice. The object of interest in this case is the beam, not the man or the bucket. The beam has different forces acting on it at different points, and it's not going anywhere. For the beam, the net force and the net torque must both be zero.

Our next step is to draw the pictorial representation of FIGURE 6.19, in which we clearly identify both the forces that act on the beam and where they act. The pivot, like the man, is holding up one end of the beam, so it exerts an upward force at the right end. Although the beam is the object of interest, we'll assume that its weight can be neglected in comparison with the 1 ton bucket of rocks. We will measure the distances from the pivot, which is the natural point about which to compute torques. The force of the bucket on the beam is equal to the weight of the bucket:

$$F_b = w_b = m_b g = (910\text{ kg})(9.8\text{ m/s}^2) = 8900\text{ N}$$

This information has been included in the pictorial representation.

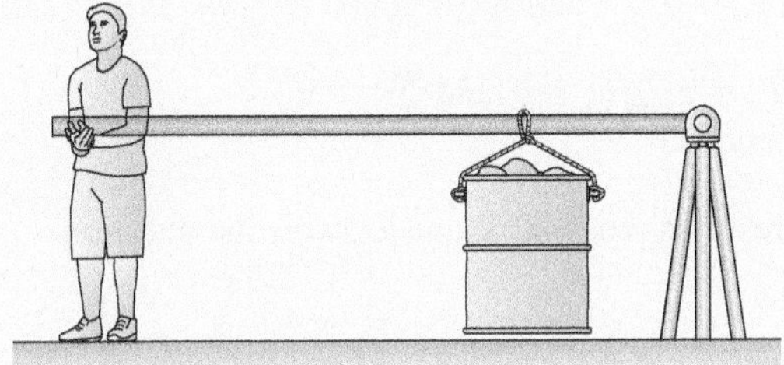

FIGURE 6.18 Using a beam to lift a load.

SOLVE The first criterion for equilibrium is that there is no net force. There are no forces acting horizontally, so we consider the condition $\sum F_y = 0$, which gives

$$\sum F_y = F_m + F_p - F_b = 0$$

This clearly isn't enough information to solve the problem; there are two unknown forces in this equation. But we have another condition, $\sum \tau = 0$. All the forces act perpendicular to the beam, so we can compute the torques as $\tau = rF_\perp$, where $F_\perp$ is the magnitude of each force. There is no torque from the force of the pivot because this force acts right at the pivot. The force of the bucket tries to rotate the beam counterclockwise, while the force of the man tries to rotate the beam clockwise. Thus the torque equation is

$$\sum \tau = F_b d_b - F_m d_m = 0$$

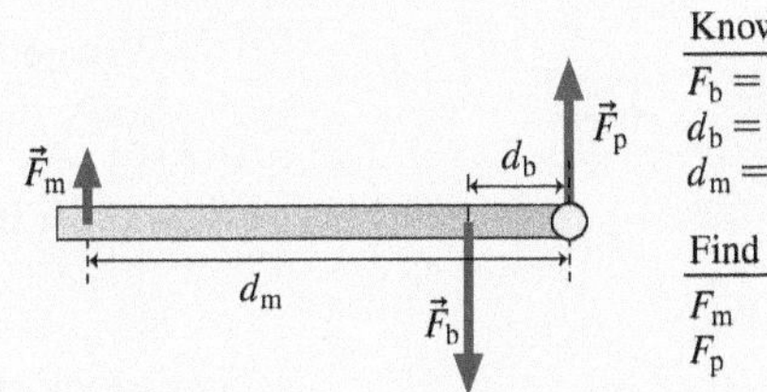

FIGURE 6.19 The forces acting on the beam.

Continued

We know the value of every quantity in this equation except F_m, so we can calculate the force exerted by the man:

$$F_m = F_b\frac{d_b}{d_m} = (8900\text{ N})\frac{0.85\text{ N}}{3.6\text{ m}} = 2100\text{ N}$$

With this, we can now revisit the equation for forces to compute the force of the pivot:

$$F_p = F_b - F_m = 8900\text{ N} - 2100\text{ N} = 6800\text{ N}$$

ASSESS The force the man must apply is 2100 N, the weight of about 200 kg. That's a lot of force, but it is a strongman competition so this makes sense.

STOP TO THINK 6.3 Which of these objects is in static equilibrium?

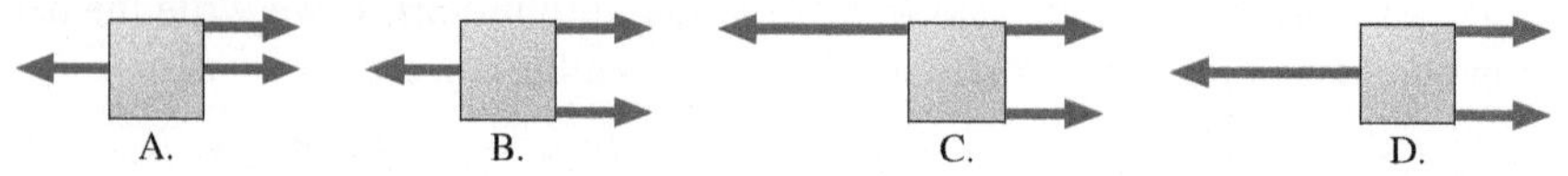
A.
B.
C.
D.

Choosing the Pivot Point

Earlier, we saw that the value of the torque depends on the choice of the pivot point. In Example 6.8, there was a natural choice for the pivot point, but we could have chosen another point. Would that choice have affected the result of our calculation?

Consider the hammer shown in FIGURE 6.20, supported on a pegboard by two pegs A and B. Because the hammer is in static equilibrium, the net torque around the pivot point at peg A must be zero: The clockwise torque due to the weight $\vec{w}$ is exactly balanced by the counterclockwise torque due to the force $\vec{n}_B$ of peg B. (Recall that the torque due to $\vec{n}_A$ is zero because here $\vec{n}_A$ acts at the pivot point A.) But if instead we take B as the pivot point, the net torque is still zero. The counterclockwise torque due to $\vec{w}$ (with a large force but small moment arm) balances the clockwise torque due to $\vec{n}_A$ (with a small force but large moment arm). Indeed, **for an object in static equilibrium, the net torque about *every* point must be zero.** This means you can pick *any* point you wish as a pivot point for calculating the torque.

FIGURE 6.20 A hammer resting on two pegs.

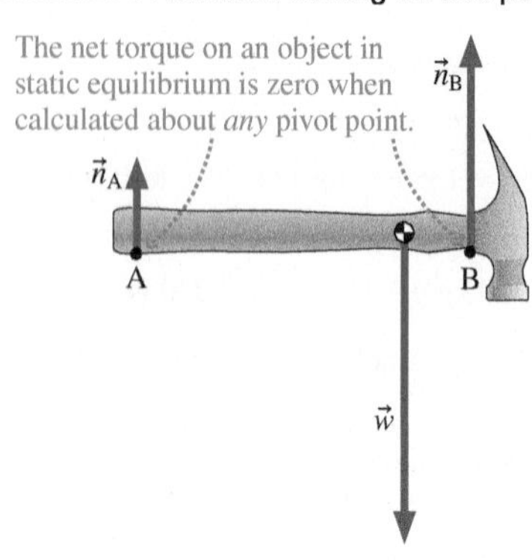
The net torque on an object in static equilibrium is zero when calculated about *any* pivot point.
$\vec{n}_B$
$\vec{n}_A$
A
B
$\vec{w}$

Although any choice of a pivot point will work, some choices are better because they simplify the calculations. It's often good to choose a point at which several forces act because the torques exerted by those forces will be zero.

PROBLEM-SOLVING STRATEGY 6.1 Static equilibrium problems

PREPARE Begin by making decisions:

- Model the object of interest as a rigid body.
- Draw a pictorial representation showing all forces and where they act.
- Decide on a pivot point about which to compute torques.
- Determine the distances from the pivot point to the points where forces act.

SOLVE The mathematical steps are based on the conditions:

$$\vec{F}_{net} = \vec{0} \quad \text{and} \quad \tau_{net} = 0$$

- Write equations for $\sum F_x = 0$, $\sum F_y = 0$, and $\sum \tau = 0$.
- Solve the resulting equations.

ASSESS Check that your result is reasonable and answers the question.

Let's look at some examples.

EXAMPLE 6.9 Finding the center of gravity of the human body

The human body's center of gravity can be determined using a device called a *reaction board*. In one such measurement, illustrated in FIGURE 6.21, a 60 kg woman lies on a 2.5-m-long, 6.0 kg reaction board with her feet over a fixed support at one end of the board. A scale under the other end reads 250 N. What is the distance from the woman's feet to her center of gravity?

FIGURE 6.21 Using a reaction board to measure a subject's center of gravity.

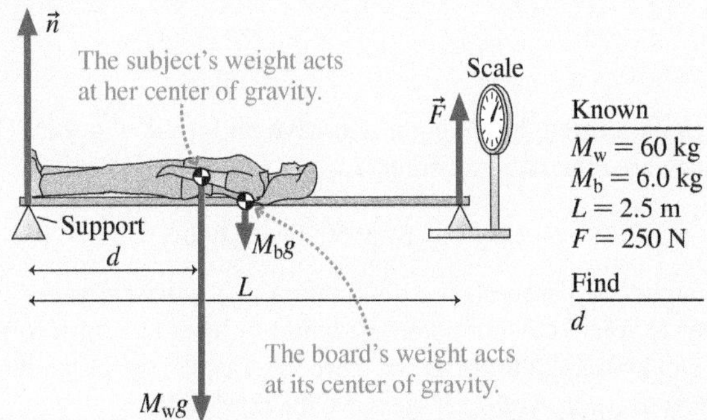

PREPARE The problem is about the woman and the position of her center of gravity, but it makes sense to take the woman and the board, considered together, as the object of interest. As the figure shows, there is an upward force from the scale, a downward weight force acting on the board (at the board's center of gravity, its midpoint), and a downward weight force acting on the woman at her center of gravity. We know these forces, but a fourth force—the upward force from the support—is unknown, so it makes sense to choose the pivot point here.

SOLVE Because the board and woman are in static equilibrium, the net force and net torque on them must be zero. The force equation reads

$$\sum F_y = n - M_b g - M_w g + F = 0$$

and the torque equation about the support on the left gives

$$\sum \tau = -\frac{L}{2} M_b g - d M_w g + LF = 0$$

In this case, the force equation isn't needed because we can solve the torque equation for d:

$$d = \frac{LF - \frac{1}{2} L M_b g}{M_w g} = \frac{(2.5 \text{ m})(250 \text{ N}) - \frac{1}{2}(2.5 \text{ m})(6.0 \text{ kg})(9.8 \text{ m/s}^2)}{(60 \text{ kg})(9.8 \text{ m/s}^2)}$$

$$= 0.94 \text{ m}$$

ASSESS If the woman is 1.70 m (5 ft 7 in) tall, her center of gravity is $(0.94 \text{ m})/(1.70 \text{ m}) = 55\%$ of her height, or a little more than halfway up her body. This seems reasonable and this is, in fact, a typical value for women. The center of gravity for men tends to be a bit higher in the body.

EXAMPLE 6.10 Walking the plank

Adrienne (50 kg) and Bo (90 kg) are playing on a 100 kg rigid plank resting on the supports seen in FIGURE 6.22. If Adrienne stands on the left end, can Bo walk all the way to the right end without the plank tipping over? If not, how far can he get past the support on the right?

FIGURE 6.22 Adrienne and Bo on the plank.

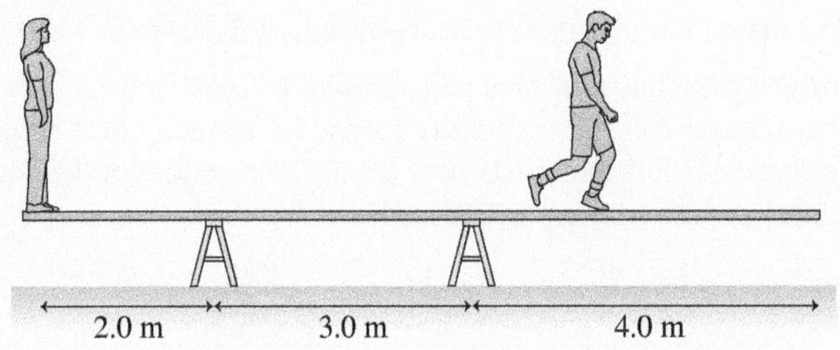

FIGURE 6.23 A pictorial representation of the forces on the plank.

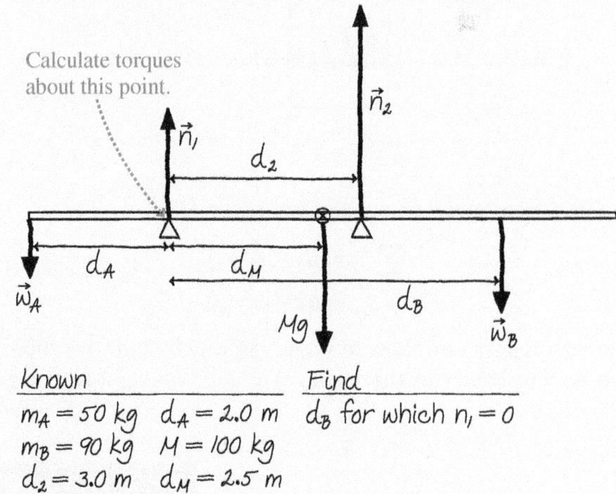

PREPARE We model the plank as a uniform rigid body with its center of gravity at the center. FIGURE 6.23 shows the forces acting on the plank. Both supports exert upward forces. $\vec{w}_A$ and $\vec{w}_B$ are Adrienne's and Bo's weights pushing down on the board.

SOLVE Because the plank is resting on the supports, not held down, forces $\vec{n}_1$ and $\vec{n}_2$ must point upward. (The supports could pull down if the plank were nailed to them, but that's not the case here.) Force $\vec{n}_1$ will decrease as Bo moves to the right, and the tipping point occurs when $n_1 = 0$. The plank remains in static equilibrium right up to the tipping point, so both the net force and the net torque on it are zero. The force equation is

$$\sum F_y = n_1 + n_2 - w_A - w_B - Mg$$
$$= n_1 + n_2 - m_A g - m_B g - Mg = 0$$

We can again choose any point we wish for calculating torque. Let's use the support on the left. Adrienne and the support on the right exert positive torques about this point; the other forces exert negative torques. Force $\vec{n}_1$ exerts no torque because it acts at the pivot point. Thus the torque equation is

$$\tau_{net} = d_A m_A g - d_B m_B g - d_M M g + d_2 n_2 = 0$$

Continued

At the tipping point, where $n_1 = 0$, the force equation gives $n_2 = (m_A + m_B + M)g$. Substituting this into the torque equation and then solving for Bo's position give

$$d_B = \frac{d_A m_A - d_M M + d_2(m_A + m_B + M)}{m_B} = 6.3 \text{ m}$$

Bo doesn't quite make it to the end. The plank tips when he's 6.3 m past the left support, our pivot point, and thus 3.3 m past the support on the right.

ASSESS We could have solved this problem somewhat more simply had we chosen the support on the right for calculating the torques. However, you might not recognize the "best" point for calculating the torques in a problem. The point of this example is that it doesn't matter which point you choose.

EXAMPLE 6.11 **Will the ladder slip?**

A 3.0-m-long ladder leans against a frictionless wall at an angle of 60°. What is the minimum value of μ_s, the coefficient of static friction with the ground, that prevents the ladder from slipping?

PREPARE The ladder is a rigid rod of length L. To not slip, it must remain in static equilibrium with $\vec{F}_{net} = \vec{0}$ and $\tau_{net} = 0$. **FIGURE 6.24** shows the ladder and the forces acting on it.

FIGURE 6.24 A ladder in static equilibrium.

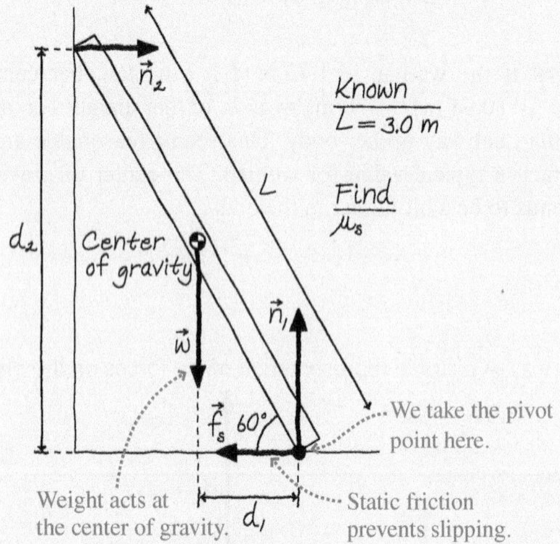

SOLVE The x- and y-components of $\vec{F}_{net} = \vec{0}$ are

$$\sum F_x = n_2 - f_s = 0$$

$$\sum F_y = n_1 - w = 0$$

The net torque is zero about *any* point, so which should we choose? The bottom corner of the ladder is a good choice because two forces pass through this point and have no torque about it. The torque about the bottom corner is

$$\tau_{net} = d_1 w - d_2 n_2 = \tfrac{1}{2}(L\cos 60°)w - (L\sin 60°)n_2 = 0$$

The signs are based on the observation that w would cause the ladder to rotate counterclockwise while $\vec{n}_2$ would cause it to rotate clockwise. All together, we have three equations in the three unknowns n_1, n_2, and f_s. If we solve the third for n_2,

$$n_2 = \frac{\tfrac{1}{2}(L\cos 60°)w}{L\sin 60°} = \frac{w}{2\tan 60°}$$

we can then substitute this into the first to find

$$f_s = \frac{w}{2\tan 60°}$$

This is the static friction force needed to keep the ladder from slipping. But we know there's an upper limit to static friction. Our model of friction is $f_s \le f_{s\,max} = \mu_s n_1$. We can find n_1 from the second equation: $n_1 = w$. Using this, the model of static friction tells us that

$$f_s \le \mu_s w$$

Comparing these two expressions for f_s, we see that μ_s must obey

$$\mu_s \ge \frac{1}{2\tan 60°} = 0.29$$

Thus the minimum value of the coefficient of static friction is 0.29.

ASSESS You know from experience that you can lean a ladder or other object against a wall if the ground is "rough," but it slips if the surface is too smooth. 0.29 is a "medium" value for the coefficient of static friction, which is reasonable.

STOP TO THINK 6.4 What does the scale read?

A. 500 N
B. 1000 N
C. 2000 N
D. 4000 N

Massless rod

Scale

1000 N

6.5 Stability and Balance

If you tilt a box up on one edge by a small amount and let go, it falls back down. If you tilt it too much, it falls over. And if you tilt it "just right," you can get the box to balance on its edge. What determines these three possible outcomes?

FIGURE 6.25 illustrates the idea with a car, but the results are general and apply in many situations. An extended object has a *base of support* on which it rests when in static equilibrium. If you tilt the object, one edge of the base of support becomes a pivot point. As long as the object's center of gravity remains over the base of support, torque due to gravity will rotate the object back toward its stable equilibrium position; we say that the object is **stable.** This is the situation in Figure 6.25b.

FIGURE 6.25 A car—or any object—will fall over when tilted too far.

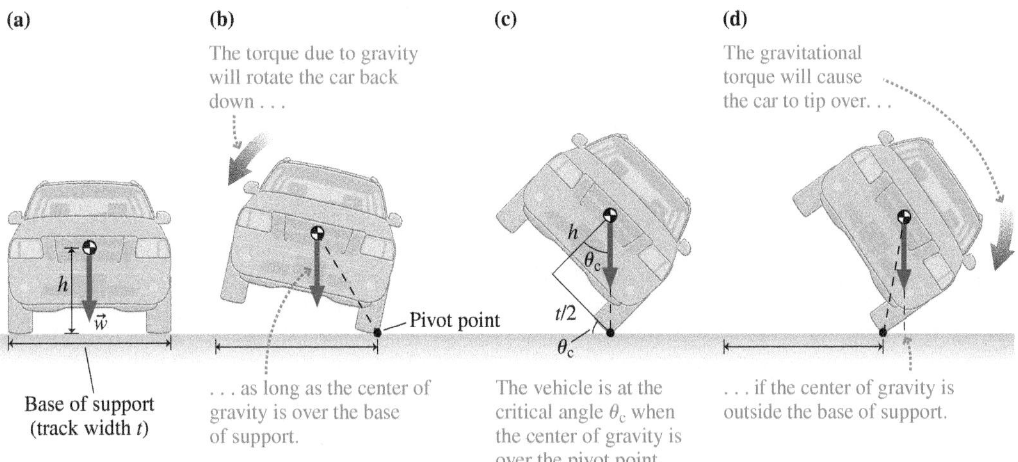
(a)
(b)
(c)
(d)
The torque due to gravity will rotate the car back down . . .
The gravitational torque will cause the car to tip over. . .
h
w
Pivot point
t/2
θc
Base of support (track width t)
. . . as long as the center of gravity is over the base of support.
The vehicle is at the critical angle θc when the center of gravity is over the pivot point.
. . . if the center of gravity is outside the base of support.

A *critical angle* θ_c is reached when the center of gravity is directly over the pivot point, as in Figure 6.25c. This is the point of balance, with no net torque. If the car continues to tip, the center of gravity moves outside the base of support, as in Figure 6.25d. Now, the gravitational torque causes a rotation in the opposite direction and the car rolls over; it is **unstable.** If an accident causes a vehicle to pivot up onto two wheels, it will roll back to an upright position as long as $\theta < \theta_c$, but it will roll over if $\theta > \theta_c$.

For vehicles, the distance between the tires—the base of support—is called the track width t. The height of the center of gravity is h. You can see from Figure 6.25c that the ratio of the two, the height-to-width ratio, determines when the critical angle is reached. An analysis of the triangle in the third image in Figure 6.25 shows how the critical angle depends on the track width t and the height of the center of gravity h:

$$\theta_c = \tan^{-1}\left(\frac{\frac{1}{2}t}{h}\right)$$

Decreasing the height of the center of gravity or increasing the width of the base of support will increase the critical angle, leading to a more stable situation. The same argument can be made for any object, leading to the general rule that **a wider base of support and/or a lower center of gravity improves stability.**

CONCEPTUAL EXAMPLE 6.12 Understanding ladder safety

A man is standing on a ladder to paint a wall. He (unwisely) leans out from the side of the ladder, as in FIGURE 6.26 (next page). If he leans too far, the ladder will tip to the side. Explain why and when the ladder will tip.

REASON For stability, the combined center of gravity of the man and the ladder must be over the ladder's base. When the man leans to the side, his center of gravity shifts. This changes the position of the combined center of gravity quite a bit; the man has more mass than the ladder, so the combined center of gravity is closer to his center of gravity than that of the ladder. FIGURE 6.27 (next page) shows that the situation is at a literal tipping point—if the man moves any farther to the left, the center of gravity will be to the left of the ladder's supports, and the ladder will tip.

ASSESS This result makes sense to anyone who has used a ladder for such work; there is a very narrow lean angle for safe use. The design of the ladder, with a wider base, enhances safety, but only up to a point.

Continued

FIGURE 6.26 Unsafe lean on a ladder.

FIGURE 6.27 Analyzing the unsafe lean.

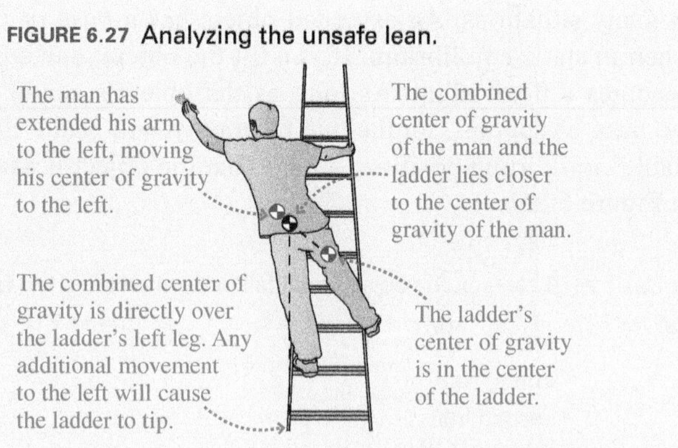
The man has extended his arm to the left, moving his center of gravity to the left.
The combined center of gravity of the man and the ladder lies closer to the center of gravity of the man.
The combined center of gravity is directly over the ladder's left leg. Any additional movement to the left will cause the ladder to tip.
The ladder's center of gravity is in the center of the ladder.

Balancing a soda can Try to balance a soda can—full or empty—on the narrow bevel at the bottom. It can't be done because, either full or empty, the center of gravity is near the center of the can. If the can is tilted enough to sit on the bevel, the center of gravity lies far outside this small base of support. But if you put about 2 ounces (60 ml) of water in an empty can, the center of gravity will be right over the bevel and the can will balance.

FIGURE 6.28 Standing on tiptoes.

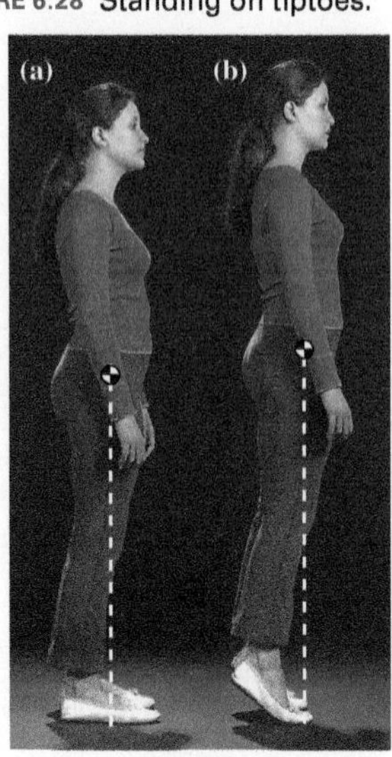
(a)
(b)

Stability and Balance of the Human Body

A tripod or three-legged stool is stable as long its center of gravity remains over the base of support—the triangle defined by the three points where the legs touch the ground. The three-point stance used by athletes in several sports is especially stable not only for being like a tripod but also for lowering the center of gravity. A four-legged quadruped is even more stable, and a quadruped can easily lift one foot off the ground, becoming a tripod, without fear of falling over.

The stability of bipeds is much more challenging because the base of support—the area of and between the two feet—is small. One solution, adopted by birds, is to have very wide feet relative to their body size. The human body, in contrast, has fairly small feet. Our stability is due to our brain's ability to constantly adjust our stance, no matter what we're doing, to keep our center of gravity over the quite small base of support provided by the soles of our feet. To maintain stability, we unconsciously adjust the positions of our arms and legs. This is seen dramatically when something upsets our balance, risking a fall, and we instinctively thrust out an arm or leg to change our center of gravity.

FIGURE 6.28a shows the body in its normal standing position. Individuals vary, but the human center of gravity while standing is roughly at the height of the hips. A standing posture has the center of gravity well centered over the base of support (the feet), ensuring stability. If the woman were now to stand on tiptoes *without* otherwise adjusting her body position, her center of gravity would fall behind the base of support, which is now the balls of the feet, and she would fall backward.

To prevent this, as shown in FIGURE 6.28b, the body naturally leans forward, regaining stability by moving the center of gravity over the balls of the feet. Try this: Stand facing a wall with your toes touching the base of the wall. Now try standing on your toes. Your body can't move forward to keep your center of gravity over your toes, so you can't do it!

Changing your body shape and posture when you're not standing straight requires effort to maintain your stability. When you bend over, your torso moves forward while, simultaneously, your rear end moves backward just the right distance to keep your center of gravity over your feet. The gymnast in FIGURE 6.29 is a more extreme example. To balance with only her hands on the beam, her center of gravity must be directly above her hands.

This may seem surprising because almost none of her body is located along the dashed line. In fact, if we were to calculate the location of her center of gravity by adding up the contributions of all the parts, we would find that her center of gravity is outside her body. (Indeed, the center of gravity of a complex shape is often outside the object; a doughnut's center of gravity is in the center of the hole.)

However, if you look at how her mass is distributed, you can imagine that the gravitational torque on her extended leg offsets the gravitational torque on the rest of her body. The positioning of her legs is crucial to getting her center of gravity in exactly the right location. It takes long practice to learn to perform this remarkable feat of balance.

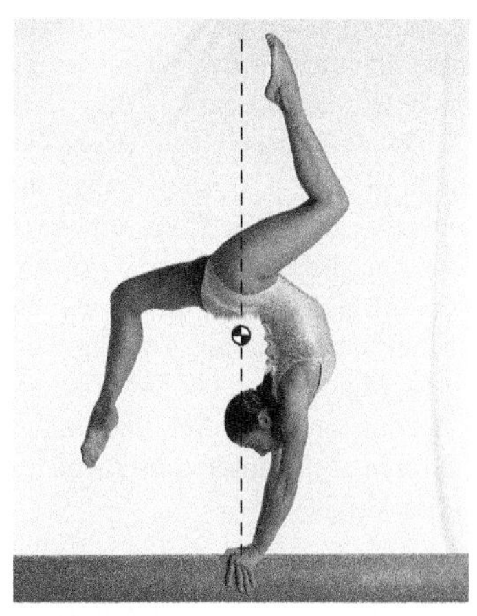

FIGURE 6.29 Stability requires the gymnast's center of gravity to be directly above her hands.

STOP TO THINK 6.5 Rank in order, from least stable to most stable, the three objects shown in the figure. The positions of their centers of gravity are marked. (For the centers of gravity to be positioned like this, the objects must have a nonuniform composition.)

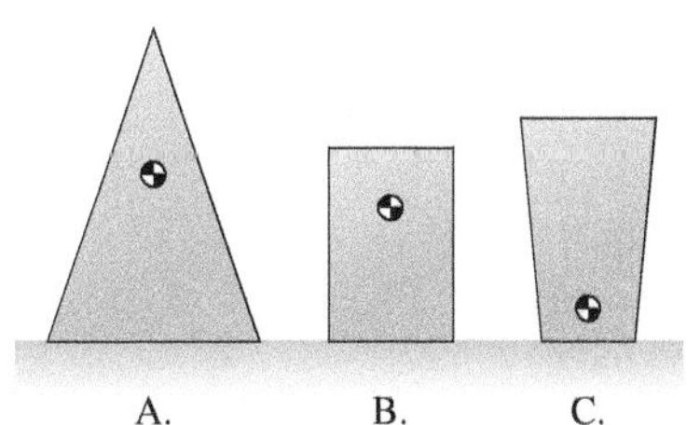

6.6 Forces and Torques in the Body

It's unlikely that you can crack a nut with the force of your bare hand, so you use a nutcracker. Why? Because the nutcracker provides you with what we call *mechanical advantage.* FIGURE 6.30a shows the forces on the handle from your hand, the nut, and—in the opposite direction—the rivet. The handle is in static equilibrium, so these forces sum to zero. But static equilibrium also requires that the net torque be zero. Suppose we take the rivet to be the pivot point. Zero net torque about the pivot point tells us that the force of the nut on the handle must be much larger than the force of your hand because the force of the nut acts much closer to the pivot point and has a smaller moment arm.

According to Newton's third law, the force of the nutcracker on the nut is equal but opposite to the force of the nut on the handle. Thus, as FIGURE 6.30b shows, the nutcracker amplifies or magnifies the force you're able to bring to bear. A typical nutcracker can exert a force 5 or 6 times larger than what you could achieve with your bare hand.

The nutcracker is an example of a *lever,* which you also see in tools such as bottle openers, pliers, and pry bars. A lever, shown in FIGURE 6.31, is a rigid bar that pivots at a support point called a *fulcrum.* If you exert effort to push or pull one end of the lever with force $\vec{F}_{\text{effort}}$, then the other end of the lever exerts a force $\vec{F}_{\text{on load}}$ on a load. In turn, according to Newton's third law, the load exerts force $\vec{F}_{\text{on lever}} = -\vec{F}_{\text{on load}}$ on the lever, a force equal in magnitude but opposite in direction.

Zero net torque about the fulcrum requires that $r_{\text{effort}} F_{\text{effort}} = r_{\text{load}} F_{\text{on lever}}$, where r_{effort} and r_{load} are the distances from the fulcrum. But $F_{\text{on load}} = F_{\text{on lever}}$, so

$$F_{\text{on load}} = \frac{r_{\text{effort}}}{r_{\text{load}}} F_{\text{effort}} = MA \times F_{\text{effort}} \tag{6.9}$$

where $MA = r_{\text{effort}}/r_{\text{load}}$, called the **mechanical advantage,** is simply the ratio of the lever arms. For example, a nutcracker has a mechanical advantage of 5 or 6 because your hand is placed 5 or 6 times farther from the pivot point than the nut is. It's often useful to think of mechanical advantage as a *force amplification factor* because a long lever arm allows you to greatly amplify your effort.

This idea is relevant because many of the joints in your body function as levers. It is interesting, though, that many have $MA < 1$, which means a force *reduction.* That may seem counterproductive, but we'll find that levers have uses other than simply transferring force.

FIGURE 6.30 The operation of a nutcracker.

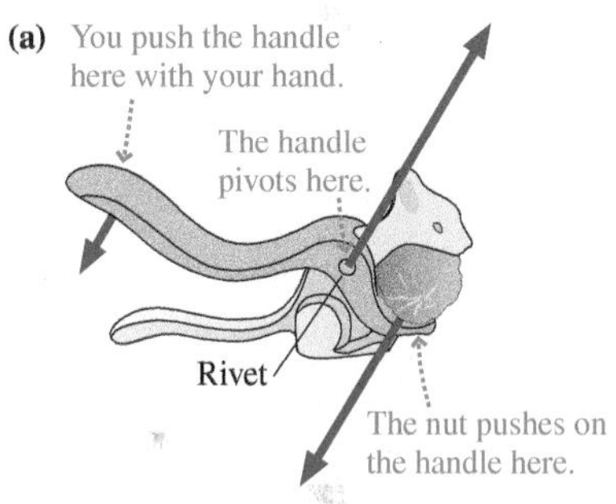

(a) You push the handle here with your hand.
The handle pivots here.
Rivet
The nut pushes on the handle here.

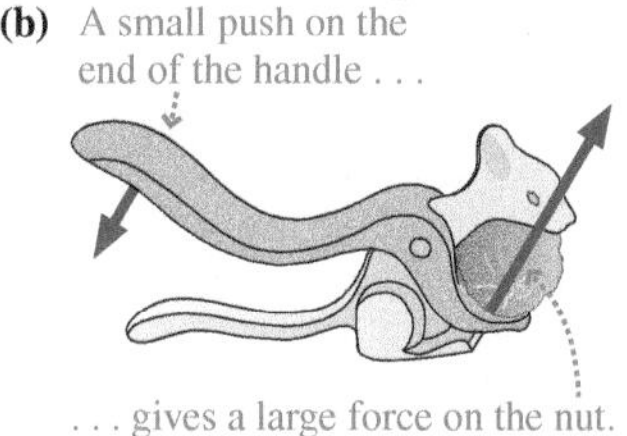

(b) A small push on the end of the handle . . .
. . . gives a large force on the nut.

FIGURE 6.31 A lever.

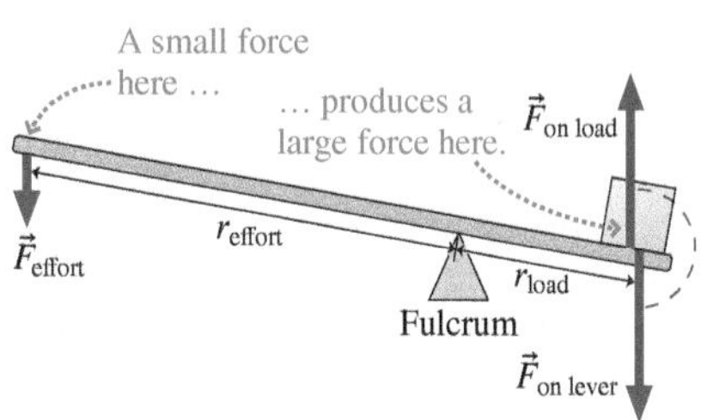

EXAMPLE 6.13 Forces in the ankle joint

Your foot and ankle are complex, but let's make a simple model of the foot as a rigid bar that rotates about a pivot point in the ankle. In other words, let's model the foot as a lever. This simple model will help us explore the operation of the ankle joint.

When you stand on tiptoe, your foot pivots about your ankle. FIGURE 6.32 shows that there's an upward force on your toes from the floor, an upward force on the heel from your Achilles tendon, and a downward force at the pivot point from the leg bone. At the same time, your toes push downward on the floor. Suppose a 61 kg woman stands on one foot, on tiptoe. Measurements of her foot are shown in the figure. What is the tension in her Achilles tendon? What is the magnitude of the force at the ankle joint? And how do these forces compare to her weight?

FIGURE 6.32 Forces on the foot when standing on tiptoe.

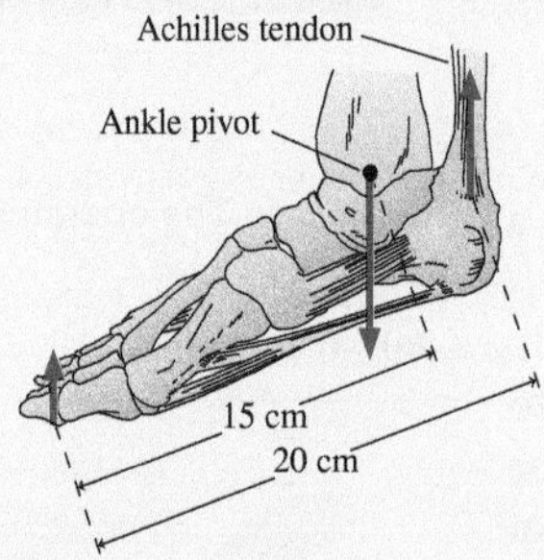

PREPARE FIGURE 6.33 is our model of the foot. We've assumed that the Achilles tendon pulls vertically, so the leg force applied at the ankle joint must also be vertical. What we see is a lever, similar to the nutcracker. In this case, the tension in the Achilles tendon pulling up on the heel (the effort) causes the toes to push down against the ground, so the force on "the load" is the force $\vec{F}_{\text{on floor}}$. The two lever-arm distances are $r_{\text{effort}} = 5$ cm and $r_{\text{load}} = 15$ cm.

SOLVE We can calculate that the mechanical advantage of this lever is $MA = r_{\text{effort}}/r_{\text{load}} = \frac{1}{3}$, a value less than 1. Thus Equation 6.9 tells us that

$$F_{\text{on floor}} = MA \times F_{\text{effort}} = \tfrac{1}{3}F_{\text{tendon}}$$

FIGURE 6.33 A simplified model of the foot.

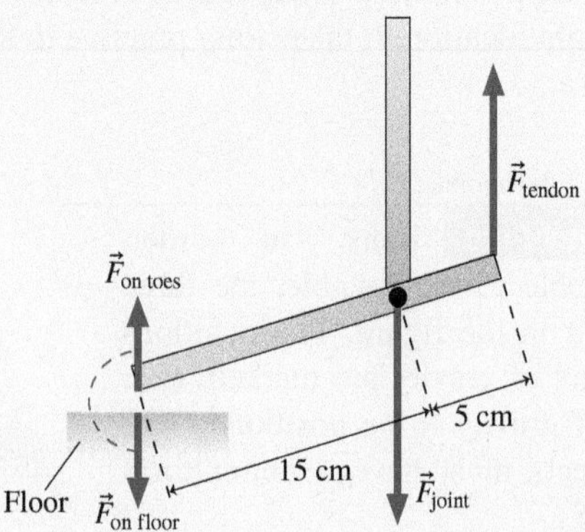

Now $\vec{F}_{\text{on floor}}$ and $\vec{F}_{\text{on toes}}$ are an interaction pair with equal magnitudes, and $\vec{F}_{\text{on toes}}$ is really just the normal force of the floor pushing up. It's supporting her weight, so the magnitude is $F_{\text{on toes}} = mg = (61\text{ kg})(9.8\text{ m/s}^2) = 600\text{ N}$. Thus $F_{\text{on floor}} = 600$ N and the tension in the tendon is

$$F_{\text{tendon}} = 3F_{\text{on floor}} = 1800\text{ N}$$

This is three times the woman's weight. We can then find the force acting at the ankle joint, the pivot point, from the requirement of no net force in the vertical direction:

$$F_{\text{joint}} = F_{\text{tendon}} + F_{\text{on floor}} = 2400\text{ N}$$

This is four times the woman's weight.

ASSESS Standing on tiptoes places a great deal of stress on your Achilles tendon and ankle, so forces several times the woman's weight are not surprising. You might have noticed that the answer did not depend on the angle of the foot. That does seem surprising, but it's a consequence of our model of the foot being too simple. A real foot also pivots at the base of the toes, which stretches the muscles and tendons that run along the sole of your foot and requires an increasing effort from the Achilles tendon. Nonetheless, our model gives us a good sense of the forces involved.

EXAMPLE 6.14 Analyzing forces and torques in the elbow

In weightlifting, where the applied forces are large, holding the body in static equilibrium requires muscle forces that are quite large indeed. In the strict curl event, a standing athlete lifts a barbell by moving only his forearms, which pivot at the elbow. Record lifts are in the range of 110 kg. FIGURE 6.34 shows the arm bones and the main lifting muscle when the forearm is horizontal. What is the tension in the tendon at this point in performing a 110 kg strict curl?

PREPARE We can model the arm, as shown in FIGURE 6.35, as two rigid rods connected by a hinge. We'll ignore the arm's weight because it is so much smaller than the barbell. The tendon does pull at a slight angle, but we'll model it as being vertical. The forearm is in motion, but the motion is modest and so we will treat this as a static equilibrium problem. The force of the barbell on one arm is the weight of half the barbell's mass:

$$F_{\text{barbell}} = \left(\frac{110\text{ kg}}{2}\right)g = 540\text{ N}$$

The force in the joint acts at the pivot, so it does not contribute a torque.

SOLVE The tension in the tendon tries to rotate the arm counterclockwise, so it produces a positive torque. The torque due to the barbell, which tries to rotate the arm in a clockwise direction, is negative. Thus the zero-net-torque requirement is

$$\tau_{\text{net}} = 0 = r_{\text{tendon}}F_{\text{tendon}} - r_{\text{barbell}}F_{\text{barbell}}$$

FIGURE 6.34 The arm lifting a barbell.

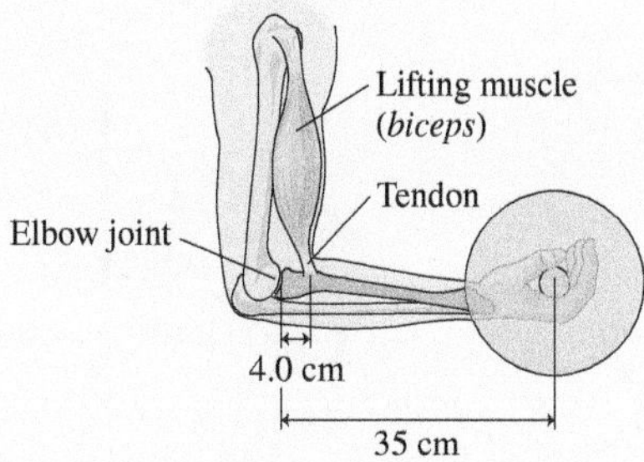

FIGURE 6.35 A simplified model of the arm and weight.

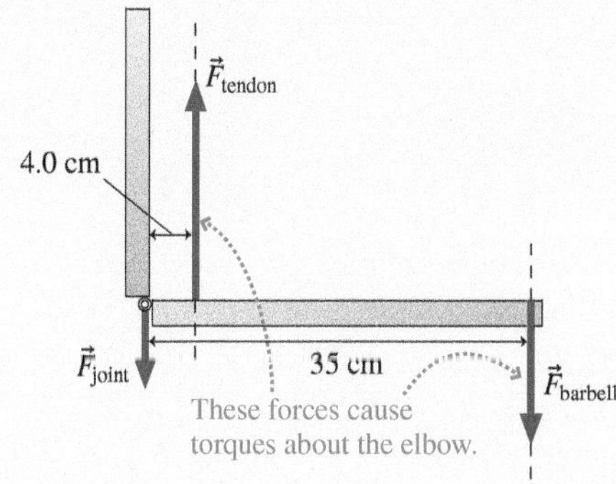

We can solve this equation for the tension in the tendon. The distances appear in a ratio, so the units cancel; there is no need for unit conversion:

$$F_{\text{tendon}} = (540\ \text{N})\frac{35\ \text{cm}}{4.0\ \text{cm}} = 4700\ \text{N}$$

The tendon tension comes from the muscles, which must provide a force nearly 9 times the weight lifted! For this lift, the biceps in the arms are pulling with a combined force of about 1 ton, which makes this impressive lift seem even more amazing.

Notice that this is another lever, one with the fulcrum at the end (the elbow) rather than in the middle. The mechanical advantage is a small $MA = (4.0\ \text{cm})/(35\ \text{cm}) = 0.11$ and thus $F_{\text{barbell}} = 0.11 \times F_{\text{tendon}}$.

ASSESS The large value for the tendon tension makes sense, given the problem statement. In fact, as we'll see in the next section, this tension is a significant fraction of what the tendon can support before failure. The lift is possible, but it's nearing the limit of what the tissues of the body can do.

Strength Versus Speed and Range of Motion

A nutcracker, with mechanical advantage $MA > 1$, amplifies the force your hand alone can exert. But two examples of levers in the body found a mechanical advantage $MA < 1$. That may seem surprising, but levers have uses other than force amplification.

If you look back at the lever in Figure 6.31, which has $MA > 1$, you'll notice that a large *displacement* of the lever at the end where you push causes only a small displacement of the load. That is, you amplify the force on the load by the factor MA by reducing the distance the load moves. We could say that the load has a limited *range of motion.*

Conversely, placing the fulcrum close to the end where you push, which makes $MA < 1$, allows a fairly small displacement on your end to produce a large displacement—a large range of motion—of the load. The motion amplification is simply the *inverse* of the mechanical advantage, $1/MA$. An increased range of motion is obtained by reducing the force on the load.

Thus a lever can be used either to amplify force at the expense of motion *or* to amplify motion at the expense of force. A lever with $MA > 1$ is used to increase force by limiting the distance through which the load moves. A lever with $MA < 1$ and thus $1/MA > 1$ is used to increase range of motion by limiting the force available to lift a load.

Your elbow is not designed to act like a nutcracker but, instead, to give your hand a large range of motion with only a small contraction or elongation of your biceps. Thus we would expect the lever of your forearm, with its fulcrum at the elbow, to have $MA < 1$. And, indeed, Example 6.14 found $MA = 0.11$. The inverse, $1/MA = 8.8$, is how much a biceps contraction is amplified at the hand. The price to be paid is that the tension in the biceps and tendon has to be 8.8 times larger than the weight of anything you lift. Your ankle joint is similar; you want a relatively small change in the length of your calf muscle to cause a fairly large range of motion of your toes, and this requires $MA < 1$.

Let's see how this applies to the motion of the jaw. A typical human bite force is 1200 N at the second molars, a force that is probably greater than the person's weight. The *masseter muscle* that provides most of the force to close your jaw isn't a particularly large muscle, but, as **FIGURE 6.36** (next page) shows, its attachment is quite favorable for providing large forces.

The line of action of the muscular force is about 5 cm from the pivot point, while the second molars are about 7 cm away. This is another lever in which the

Making light work of moving The tendon force in Example 6.14 is very large because the weight is supported much farther from the elbow than the point where the tendon attaches. If the weight is supported closer to the elbow, as demonstrated by the two people in the picture, then the downward torque of the weight is much less and the necessary tendon and muscle force is much reduced.

FIGURE 6.36 Jaw motion in a human and a dog.

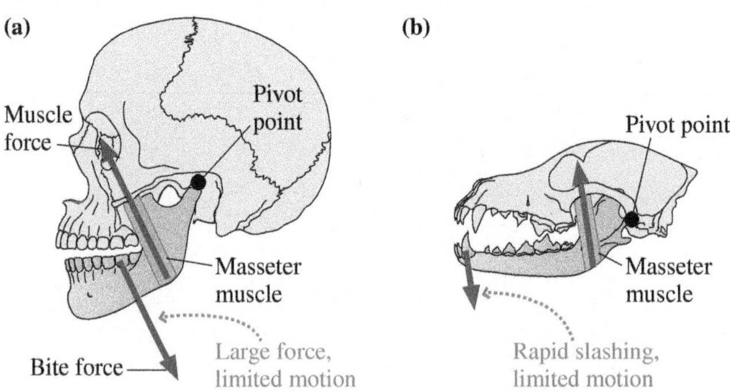

fulcrum, like your elbow joint, is at the end, and this lever has mechanical advantage $MA = 5/7$. The force you can apply to grinding food with your molars, which is key to the human diet, is only slightly smaller than the full force of the muscle. There's a trade-off, though. You may be capable of great bite strength, but your jaw has limited range of motion and you have limited bite speed.

Dogs and other predators have different needs. The prominent canine teeth at the front of their jaw are much farther from the pivot point than the masseter muscle, with a much smaller mechanical advantage, so the canine teeth have a wide range of motion (dogs and cats can open their mouths much wider than we can) and are well adapted for rapid, slashing bites. The trade-off in this case is less bite force, although that's compensated for by teeth that are much sharper than ours.

6.7 Elasticity and Deformation

A hammer is made of steel, not plastic or rubber. The reason is pretty obvious: Steel is extremely rigid and does not give appreciably under the force of an impact, whereas plastic and rubber dent, flex, bend, and possibly break. We've been modeling extended objects as rigid bodies, but no object is perfectly rigid and many—especially most biological materials—aren't even close. Real objects undergo *deformation* in response to applied forces, and that is the topic of this final section in the chapter.

From macromolecules to cells to entire organisms, the abilities of biological structures to accomplish their functions are determined not only by their biochemical properties but just as much, and sometimes more, by their mechanical properties. The actin and myosin molecules in muscle fibers have to stretch—but not too much. Sessile organisms in the tidal zone have to bend with the ever-changing waves while maintaining a firm grip on the substrate. Your arteries have to flex with the pressure of each heartbeat. Bones have to withstand impacts. These structures can do their jobs only if they are built from materials that have the necessary mechanical properties.

Biological materials with extraordinary properties have evolved to accomplish all of these needs and more. One common feature of nearly all biomaterials is that they are typically elastic rather than rigid—they deform in response to applied forces. Although you might expect deformation to be a bad thing, in fact it's often functional, storing energy for later use or providing a way to dissipate energy safely. (We'll return to look at energy storage in Chapter 11.) So in this final section we go beyond the rigid-body model to explore the principles of how various kinds of materials deform.

We'll first develop the core concepts by looking at homogeneous materials like metal and concrete, which have fairly simple properties. We will then extend our discussion to include biomaterials.

Stress and Strain

Suppose you clamp one end of a solid rod while using a strong machine to pull on the other with force $\vec{F}$. FIGURE 6.37a shows the experimental arrangement. We usually think of solids as being, well, solid. But any material, be it plastic, concrete, or steel, will stretch as the spring-like molecular bonds expand.

FIGURE 6.37b shows actual data for the force needed to stretch a 1.0-cm-diameter, 1.0-m-long steel rod. The vertical scale is in kilonewtons, so you can see, not surprisingly, that it takes a very large force, 16,000 N, to stretch the rod by a mere 1.0 mm. But it does stretch! Steel, like a very stiff spring, is what we call **elastic.** Our pulling stretches the spring-like molecular bonds, and then they return to their equilibrium lengths after the load is removed.

Notice that the force is linearly proportional to the change in length, just as in an ideal spring. That is, the force needed to stretch the rod by ΔL is

$$F = k\,\Delta L \qquad (6.10)$$

where k is the slope of the graph. You'll recognize Equation 6.10 as none other than the relationship between force and extension for a spring—Hooke's law. Objects that we think of as rigid have very large values of k, meaning that they are very difficult to stretch. An object with a small value of k is easily stretched, like a rubber band, but snaps back to its original shape. Such objects are called **pliant.**

The difficulty with Equation 6.10 is that the proportionality constant k depends both on the composition of the rod—whether it is, say, steel or copper—and on the rod's length and cross-section area. It would be useful to characterize the elastic properties of steel in general, or copper in general, without needing to know the dimensions of a specific rod.

We can meet this goal by thinking about Hooke's law at the atomic scale. The elasticity of a material is directly related to the spring constant of the molecular bonds between neighboring atoms. As FIGURE 6.38 shows, the force pulling each bond is proportional to the quantity F/A. This force causes each bond to stretch by an amount proportional to $\Delta L/L$. We don't know what the proportionality constants are, but we don't need to. Hooke's law applied to a molecular bond tells us that the force pulling on a bond is proportional to the amount that the bond stretches. Thus F/A must be proportional to $\Delta L/L$. We can write their proportionality as

$$\frac{F}{A} = Y\left(\frac{\Delta L}{L}\right) \qquad (6.11)$$

The ratio of force to cross-section area is called **stress**. The ratio of the change in length to the original length is called **strain**.

The proportionality constant Y is called **Young's modulus** or, sometimes, the *modulus of elasticity*. It is directly related to the spring constant of the molecular bonds, so it is a property of the material from which the object is made. Basically, Young's modulus characterizes a material's stiffness or rigidity, with larger values indicating a stiffer material.

A comparison of Equations 6.10 and 6.11 shows that Young's modulus can be written as $Y = kL/A$. This is not a definition of Young's modulus but simply an expression for making an experimental determination of the value of Young's modulus. This k is the spring constant of the rod seen in Figure 6.37a, easily measured as the slope of the graph in Figure 6.37b.

The quantity F/A, where A is the cross-section area, is called **tensile stress,** measured in N/m^2. It is represented by the symbol σ (Greek sigma). The quantity $\Delta L/L$, the fractional increase in the length, is called **strain.** Strain, represented by the symbol ε (Greek epsilon), has no units; we say that strain is dimensionless. The numerical values of strain are usually very small because solids cannot be stretched very much before reaching the breaking point. The terms are easily confused; not only do "stress" and "strain" sound similar, but they have similar meanings in everyday language. You need to pay careful attention to which is which as you practice using these concepts.

With these definitions, Equation 6.11 can be written

$$\sigma = Y\varepsilon \qquad (6.12)$$

stress = Y × strain

Because strain is dimensionless, Young's modulus Y has the same dimensions as stress, namely N/m^2. TABLE 6.1 (next page) gives values of Young's modulus for several common materials. (Table 6.1 shows additional material properties that we'll use later in this section.) Large values of Y characterize materials that are stiff and rigid.

FIGURE 6.37 Stretching a rod.

(a)

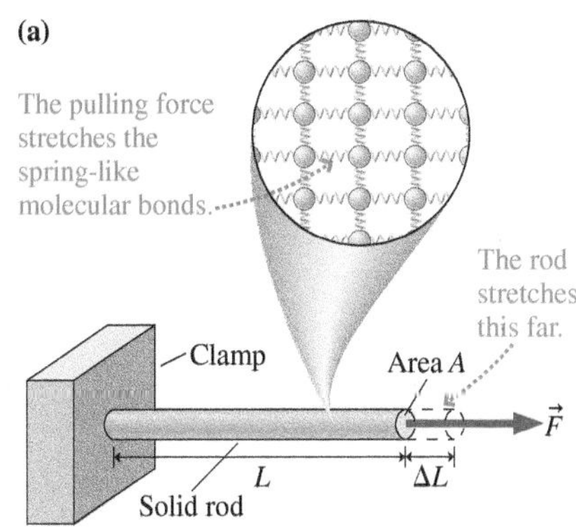

(b) Data for the stretch of a 1.0-m-long, 1.0-cm-diameter steel rod. The error bars are too small to see.

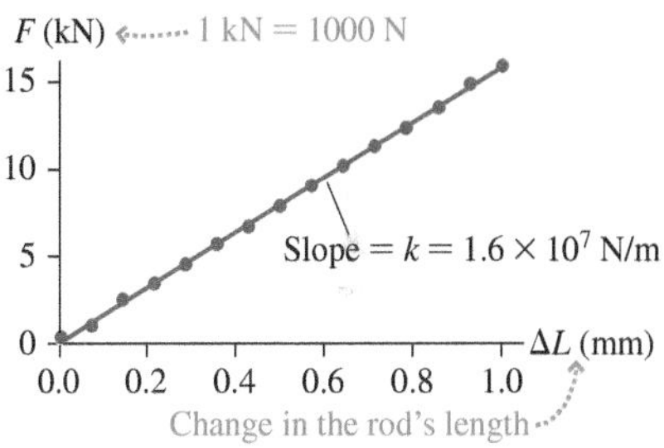

FIGURE 6.38 A material's elasticity is directly related to the spring constant of the molecular bonds.

The number of bonds is proportional to area A. If the rod is pulled with force F, the force pulling on each bond is proportional to F/A.

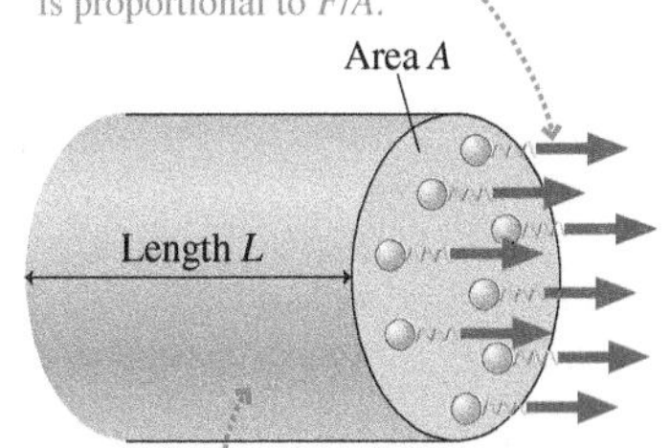

The number of bonds along the rod is proportional to length L. If the rod stretches by ΔL, the stretch of each bond is proportional to $\Delta L/L$.

TABLE 6.1 Elastic properties of various materials

Substance	Young's modulus (10^6 N/m^2)	Shear modulus (10^6 N/m^2)	Tensile strength (10^6 N/m^2)
Steel (structural)	200,000	79,000	1000
Copper	110,000	45,000	220
Concrete (typical)	30,000	21,000	5
Compact bone	18,000	5000	140
Wood (Douglas fir)	10,000	700	75
Dragline spider silk	10,000	2000	1000
Tendon (collagen)	1500		100
Nuchal ligament (elastin)	1		2

"Softer" materials, at least relatively speaking, have smaller values of Y. In general, biomaterials are softer than metals or engineered materials.

NOTE ▸ You will learn in Chapter 9 that *pressure* is also defined as F/A. The SI unit of pressure, the *pascal,* is 1 Pa = 1 N/m^2. Consequently, some books and papers measure stress and Young's modulus in Pa (or MPa or GPa) rather than N/m^2. Despite the similar definitions, pressure and stress differ in that pressure exerts its force in all directions while stress is force applied along an axis. For that reason, using N/m^2 helps remind us what stress measures. ◂

We introduced Young's modulus by considering how materials stretch. But materials also can be compressed, and Equation 6.12 and Young's modulus also apply to the compression of materials. In that case, the stress is *compressive stress.* Bones and the structural elements in buildings are compressed by the load they bear. Concrete is often compressed, as in columns that support highway overpasses, but rarely stretched.

NOTE ▸ Whether the rod is stretched or compressed, Equation 6.12 is valid only in the linear region of the graph in Figure 6.37b. ◂

EXAMPLE 6.15 Stretching a wire

A 2.0-m-long, 1.0-mm-diameter wire is suspended from the ceiling. Hanging a 4.5 kg mass from the wire stretches the wire's length by 1.0 mm. What is Young's modulus for this wire? Can you identify the material?

PREPARE The hanging mass creates tensile stress in the wire.

SOLVE The force pulling on the wire, which is simply the weight of the hanging mass, produces tensile stress

$$\sigma = \frac{F}{A} = \frac{mg}{\pi r^2} = \frac{(4.5\text{ kg})(9.80\text{ m/s}^2)}{\pi(5.0 \times 10^{-4}\text{ m})^2} = 5.6 \times 10^7\text{ N/m}^2$$

The resulting stretch of 1.0 mm is a strain of $\varepsilon = \Delta L/L =$ (1.0 mm)/(2000 mm) $= 5.0 \times 10^{-4}$. Thus Young's modulus for the wire is

$$Y = \frac{\sigma}{\varepsilon} = 1.1 \times 10^{11}\text{ N/m}^2 = 110{,}000 \times 10^6\text{ N/m}^2$$

Referring to Table 6.1, we see that the wire is made of copper.

STOP TO THINK 6.6 A 10 kg mass is hung from a 1-m-long cable, causing the cable to stretch by 2 mm. Suppose a 10 kg mass is hung from a 2 m length of the same cable. By how much does the cable stretch?

A. 0.5 mm B. 1 mm C. 2 mm D. 3 mm E. 4 mm

Shear

A linear stretch or compression is not the only way an object can deform. The object in FIGURE 6.39 is subject to equal but opposite forces, like a wire under tension, but in this case the forces are offset from each other. A perfectly rigid body would rotate in response to the torque caused by these forces. With a deformable object, however, offset forces can cause deformation; such deformation is called **shear.**

Suppose force F is applied to the top and bottom surfaces, which have surface area a. The quantity F/a is called the **shear stress.** It is measured in N/m^2, just like tensile stress. However, the relevant area for shear is *parallel* to the force, rather than a cross section perpendicular to the force, which is why we've labeled it a rather than A. We can also define the **shear strain** $\Delta l/t$, where t is the object's thickness. Notice that $\Delta l/t = \tan\gamma$, where γ (Greek gamma) is the angle through which the side of the object is tilted.

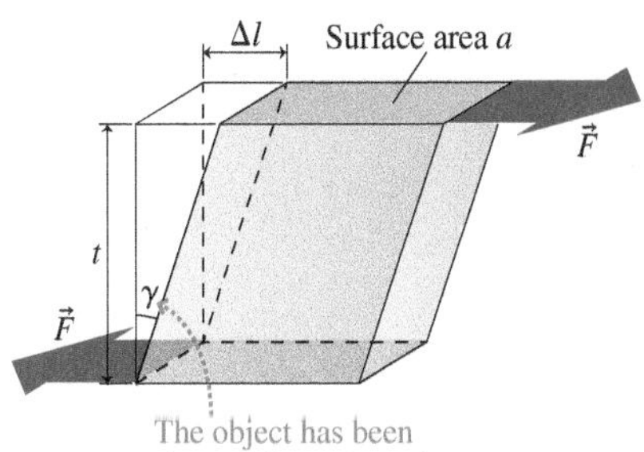

FIGURE 6.39 Offset opposing forces cause a shear deformation.

Once again, stress is directly proportional to strain, so we can write

$$\frac{F}{a} = S\frac{\Delta l}{t} \tag{6.13}$$

where S is the **shear modulus,** a material property that characterizes its propensity for shear deformation. Table 6.1 gives values of the shear modulus where they are known. Most, with one exception, are only a little less than Young's modulus. Wood, the exception, is a biomaterial with a layered structure. A piece of wood has very strong resistance to compression along the grain, but the layers of wood can fairly easily slide against each other, so it's not difficult to shear wood.

As a quick example, suppose 10,000 N forces are applied to opposite faces of a $1\text{ cm} \times 1\text{ cm} \times 1\text{ cm}$ cube of copper, a fairly soft metal. 10,000 N is a large force, approximately the weight of a 1000 kg mass. Through what angle is the cube sheared? The face of the cube has an area of $1\text{ cm}^2 = 10^{-4}\text{ m}^2$, so the shear stress is $F/a = 10^8\text{ N/m}^2$. Table 6.1 gives the shear modulus of copper as $45 \times 10^9\text{ N/m}^2$, so the shear strain is

$$\Delta l/t = \tan\gamma = (F/a)/S = 0.0022$$

Thus the deformation angle is $\gamma = \tan^{-1}(0.0022) = 0.1°$, a tiny deformation you would not even notice.

Whether it's tension or shear, it takes a large force to have much impact on a metal or on engineered materials. But biomaterials are usually softer, so commonly encountered forces can cause significant deformation. We'll look at examples later in this section.

Beyond the Elastic Limit

Stress is directly proportional to strain, with Young's modulus Y or shear modulus S being the proportionality constant, only if the strain is fairly small—that is, for fairly small deformations. This linear relationship breaks down when strain and deformation become too large. FIGURE 6.40 shows complete stress-strain curves for two quite different materials: steel and rubber.

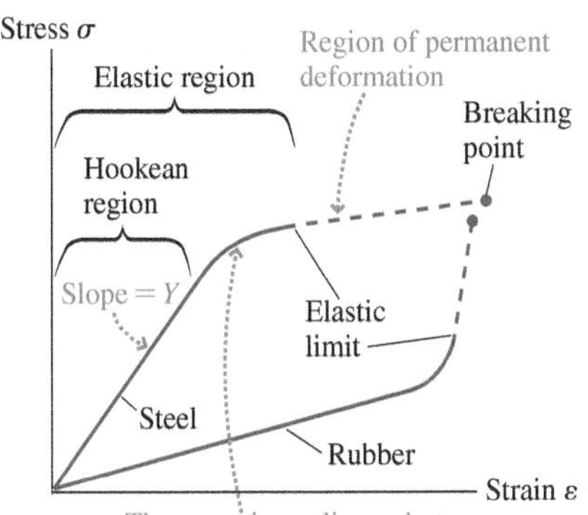

FIGURE 6.40 Stress-strain curves for steel and rubber. The scales (not shown) are not the same for the two materials.

As you can see, each graph begins with a linear region, called the *Hookean region,* where stress is directly proportional to strain and thus the material acts like an ideal spring that obeys Hooke's law. Only in this region can we use Equation 6.11 for the tensile stress. The *elastic region,* ending at the **elastic limit,** is the full range of strains for which the material will return to its original shape when the forces are removed; strain beyond the elastic limit results in a permanent deformation.

Finally, any material will break if you pull too hard. The tensile stress that causes a material to rupture is called the **tensile strength.** Note that it's stress—force divided by the cross-section area—that matters, not simply the applied force. A material can sustain very large forces if they are spread over a large enough area. Table 6.1 gives some typical values of tensile strength. Materials can also fail under compression at a maximum stress called the *compressive strength.*

NOTE ▶ A material's breaking point occurs far beyond the Hookean region of the stress-strain curve. Thus you cannot use Young's modulus to calculate the strain at the breaking point. ◀

But why are the curves for steel and rubber so different? If you stretch a metal, as when you tighten a steel guitar string, you're directly stretching the bonds between atoms. Molecular bonds become weaker as they lengthen (the spring constant decreases), and eventually a bond begins to rapidly increase in length in response to little additional force. You may have noticed that a piece of metal gets "soft" if you bend it too far. That's what we see in a stress-strain curve that flattens out: a large increase in strain (elongation) with only a small increase of stress. This is the expected behavior of any material, such as metal or glass, where the microscopic structure is one of strong bonds between neighboring atoms.

That's not how rubber is constructed. Rubber is a polymer. In its relaxed state, the long-chain molecules are folded and coiled, with only hydrogen bonds holding the polymer in the folded configuration. When you begin to stretch rubber, you're unfolding and uncoiling the polymers. Hydrogen bonds are much weaker than the covalent bonds between carbon atoms along the polymer backbone, so it takes little energy to break them; consequently rubber is very pliant. But it's a different story once the polymers are uncoiled because further stretching requires lengthening the carbon-carbon bonds. That's much more difficult, so rubber gets stiffer rather than softer as it's stretched, which we see in the increasing steepness of the stress-strain curve. You've probably noticed that it takes almost no effort to start stretching a rubber band, but eventually the rubber band gets much stiffer and—until it breaks—refuses to stretch any farther. Many biomaterials are polymers, so it makes sense that their mechanical properties are more like rubber than steel.

Case Studies of Biological Materials

We've noted that the functions of biological structures depend on their mechanical properties. The microscopic structure of biological materials, especially the presence of long fibers, makes them more complex than metals. In many cases, their mechanical properties depend on whether forces are applied along or perpendicular to the fiber direction. Think of how a plant stem is strong along the axis of the stem, due to cellulose fibers, but has little resistance to bending or kinking.

A comprehensive study of the mechanical properties of biological materials and structures is an advanced subject, but we can begin to appreciate some of the issues with three case studies.

FIGURE 6.41 Stress-strain curves for steel and spider silk.

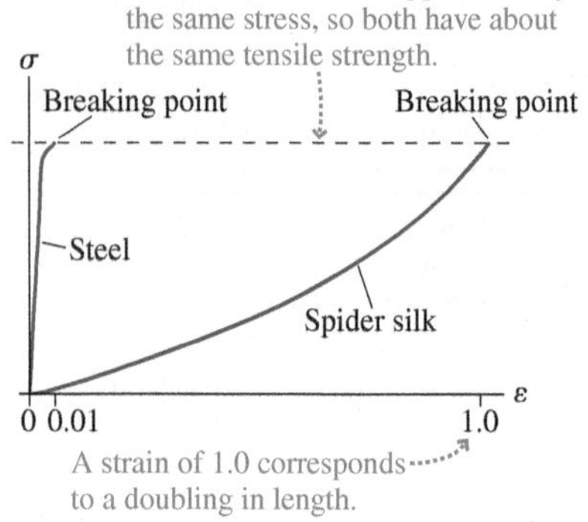

Spider silk: It's sometimes said that silk is stronger than steel. What does that mean? And is it true? Nature has many producers of silk fibers, from spiders to silkworm moths, but the champion fiber in terms of mechanical properties is the dragline silk of orb-weaving spiders. Suppose we take equal lengths of steel wire and dragline silk, slowly stretch them, and measure the force and displacement of each until it breaks. **FIGURE 6.41** shows the resulting stress-strain curves.

Initially, in the linear Hookean region, silk has a much smaller slope and thus a much smaller Young's modulus. Referring back to Table 6.1, you can see that $Y_{\text{silk}} \approx \frac{1}{20} Y_{\text{steel}}$. If the same stress is applied to both, the silk will stretch about 20 times as far as the steel. In other words, steel is much *stiffer* than spider silk.

Interestingly, though, we find that steel and silk ultimately fail at about the same stress. That is, they have about the same tensile strength, $\approx 1 \times 10^9\ \text{N/m}^2$, and in this sense spider silk is "as strong as steel." The steel listed in Table 6.1 is structural steel, the kind used in construction. *High-tensile steel* used in machine tools has a tensile strength about 3 times larger, as does the synthetic fiber Kevlar used in bulletproof vests, so it's not correct to say that spider silk is "stronger than steel" or stronger than any manufactured product. But for a natural product, made at room temperature by spiders digesting small insects, it comes very close!

EXAMPLE 6.16 Breaking wires

A strand of dragline silk is 6.0 μm in diameter, about 10% the diameter of a human hair. What is the largest mass that can be supported by a "wire" of dragline silk and by a very fine copper wire of the same diameter?

PREPARE The largest mass that can be supported depends on the material's cross-section area and tensile strength, but not its length. The tensile strengths of silk and copper are given in Table 6.1 as 1000×10^6 N/m^2 and 220×10^6 N/m^2, respectively.

SOLVE Stress is $\sigma = F/A$, and the tensile strength is σ_{max}. Thus the maximum force, one that is on the verge of breaking the wire, is $F_{max} = A\sigma_{max}$. The silk and the copper wires have the same cross-section area, $A = \pi r^2 = 2.8 \times 10^{-11}$ m^2. Thus

$$F_{max} = (2.8 \times 10^{-11} \text{ m}^2)\sigma_{max} = \begin{cases} 0.028 \text{ N} & \text{silk} \\ 0.0062 \text{ N} & \text{copper} \end{cases}$$

In this case, the force stretching the wire is the weight of the hanging mass, so $F_{max} = m_{max}g$. The largest mass that can be supported (that brings the wire to the breaking point) is

$$m_{max} = \frac{F_{max}}{g} = \begin{cases} 2.9 \text{ g} & \text{silk} \\ 0.63 \text{ g} & \text{copper} \end{cases}$$

ASSESS The mass of a penny is 2.5 g. A strand of spider silk could hold up a penny without breaking, but an equal-diameter copper wire would fail.

Figure 6.41 showed the response of silk to being stretched. FIGURE 6.42 tells a different story. This graph shows the response of silk as the stress is increased (but remaining well below the breaking point) and then reduced, allowing the fiber to contract back to its initial length. These are called the *loading* and *unloading* curves. Unlike an ideal spring, whose unloading curve exactly retraces the loading curve, silk has a "memory" of having been stressed. When the stress is first reduced, after reaching its maximum, there's very little reduction in strain. That is, a silk fiber initially retains nearly its full elongation even though the pulling force is decreasing. The fiber snaps back only after the stress has been significantly reduced. Many other biological materials behave similarly.

FIGURE 6.42 Loading and unloading curves for spider silk.

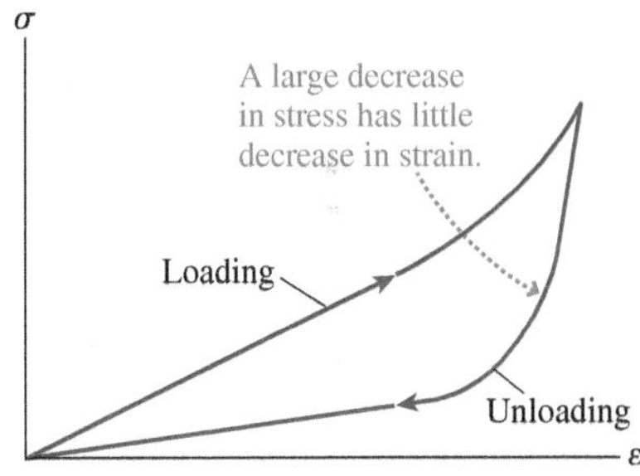

A stretched spring, as you know, stores energy—a topic we'll explore in Chapter 11. An ideal spring gives back all the stored energy as it returns to its initial length. You'll later see how the stress-strain curve can be used to calculate the stored energy. The fact that silk's unloading curve falls below its loading curve means that the energy stored while stretching is *not* all returned upon contraction; much of the stretching energy is converted to *thermal energy* that slightly raises the temperature of the silk.

The fraction of energy that is returned is called the material's **resilience.** Spider silk has a resilience of about 0.35, meaning that only 35% of the energy stored when silk stretches is returned as it contracts. But that's just what a spider needs! A spider web has to dissipate the energy of a flying insect, not store up that energy and then fling the insect back out. Some biological structures need to store and then release energy; others, such as a spider web, need to dissipate energy. Organisms have evolved biomaterials that suit their functions.

Arteries: If your arteries were very stiff, like steel pipes, they would experience a huge surge of pressure each time your heart beats. Indeed, *atherosclerosis* (sometimes called *hardening of the arteries*) is accompanied by an increase in systolic blood pressure, the larger of the two blood-pressure numbers that corresponds to a heart contraction. Arteries need to be rather soft and elastic so that an expansion with each heartbeat can act as a shock absorber. At the same time, an artery can't be *too* elastic without having an increased risk of aneurysm.

Artery walls have a complex structure, but their mechanical properties can be understood by modeling them as a composite of the proteins *elastin* and *collagen.* Collagen occurs in almost pure form in the tendons that anchor muscle to bone. It is a stiff, strong, fibrous material with a resilience of about 0.9, meaning that nearly all the energy stored when a tendon stretches is returned when the tendon contracts. That's an important property of a material that's essential for locomotion.

Elastin is the body's rubber. It is a soft, pliant, easily stretched material. The nuchal ligament of grazing mammals, such as cows, is nearly pure elastin. This rubbery ligament, which runs from the rear of the skull along the top of the neck to the thoracic vertebrae, helps support the head and seems to act as a shock absorber.

FIGURE 6.43 shows simplified views of an artery wall. The wall's relaxed state has kinked and coiled collagen fibers embedded in a soft elastin matrix. As a result, a

FIGURE 6.43 Cross sections of an artery wall.

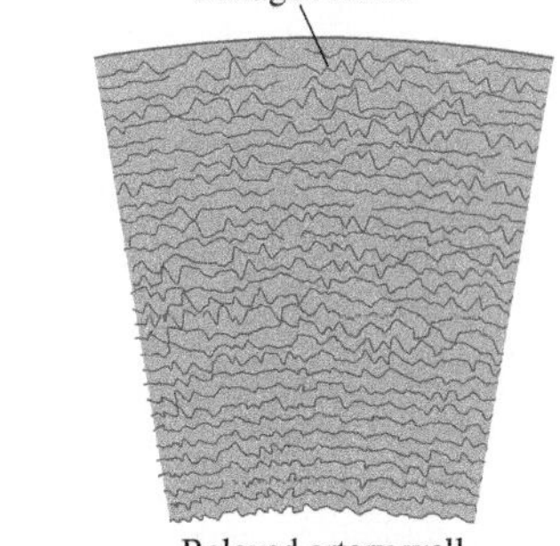

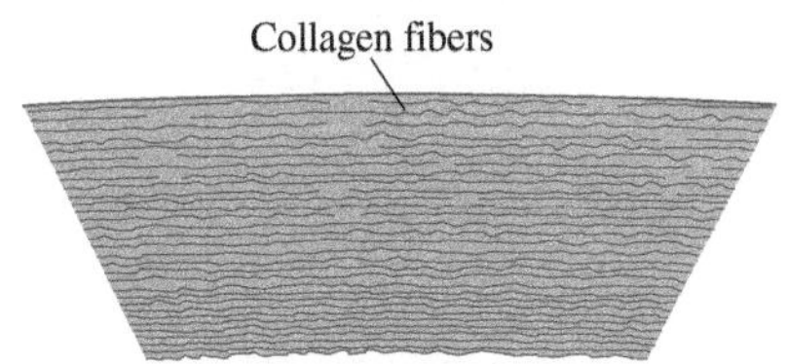

healthy artery's response to stress is very much like that of a rubber band: an easy initial stretch of the elastin, followed by an increasing resistance to stretch as the collagen fibers straighten out. The elastin allows an artery to expand to buffer the increased pressure of a heart contraction, while the collagen provides a safety net that prevents runaway expansion.

Your skin is a similar elastin-collagen composite; it's quite elastic in response to small stresses but highly resistant to large deformations. Unfortunately for all of us, skin loses its ability to produce elastin as we age. Sags and wrinkles are evidence that skin has lost much of its elasticity.

FIGURE 6.44 Cross section of a long bone.

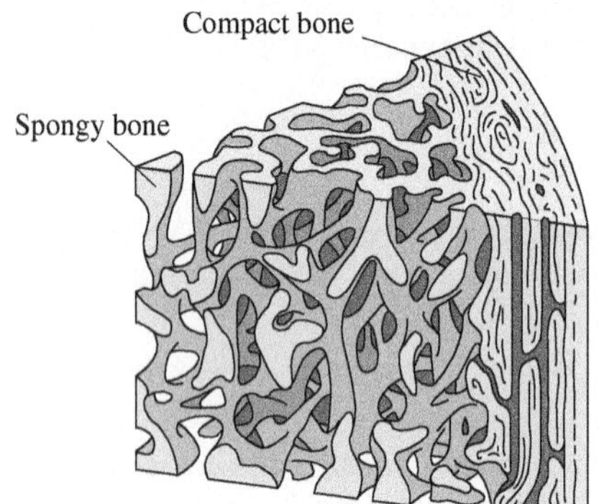

Bone: Bone is the ultimate structural material in vertebrates. Our skeletons give us shape, support us against the forces of gravity, and withstand the high stresses of walking and running. Most of the bones in our bodies are made of two different kinds of bony materials: a shell of dense and rigid *compact bone* on the exterior and porous and flexible *spongy bone* (also called *cancellous bone*) on the interior. FIGURE 6.44 shows a cross section of a typical bone. The spongy bone serves essential physiological functions but contributes essentially nothing to a bone's mechanical properties. The mechanical properties of bone can be modeled as those of a hollow cylinder of compact bone.

EXAMPLE 6.17 Breaking a leg

Do you risk breaking a leg during normal activities of running and jumping? To find out, an adult human femur can be modeled as a 480-mm-long hollow cylinder with a 2.4 cm outer diameter and a 1.2 cm inner diameter. The compressive strength of compact bone is $180 \times 10^6\ \text{N/m}^2$, somewhat greater than the tensile strength shown in Table 6.1. If an 85 kg (190 lb) man stands on one leg, what is the stress on his femur as a fraction of the bone's compressive strength? By how much does the bone compress? What if the man runs, where measurements show that the stress of impact is about 7 times the runner's weight?

PREPARE Compressive strength, analogous to tensile strength, is the compressive stress at which an object fails. Muscle has little rigidity, so the weight of a man standing on one leg is carried by his femur.

SOLVE Stress is $\sigma = F/A$, and the compressive strength is σ_{max}. The applied force is the man's weight (minus the weight of his lower leg, which we'll ignore): $w = mg = 833\ \text{N}$. The cross-section area of the bone is the area of the outer circle minus that of the inner circle:

$$A = \pi r_{\text{out}}^2 - \pi r_{\text{in}}^2 = \pi((1.2\ \text{cm})^2 - (0.6\ \text{cm})^2)$$

$$= 3.4\ \text{cm}^2 \times \frac{1\ \text{m}^2}{10^4\ \text{cm}^2} = 3.4 \times 10^{-4}\ \text{m}^2$$

Thus the stress on the femur of standing is

$$\sigma = \frac{F}{A} = \frac{833\ \text{N}}{3.4 \times 10^{-4}\ \text{m}^2} = 2.45 \times 10^6\ \text{N/m}^2$$

This seems like a very large stress, but as a fraction of the compressive strength it is only

$$\frac{2.45 \times 10^6\ \text{N/m}^2}{180 \times 10^6\ \text{N/m}^2} = 0.014 = 1.4\%$$

The mechanical design of the femur has a very large margin of safety for activities such as standing.

A stress this far from the failure point is almost certainly within the Hookean region of the stress-strain curve, so we can use Young's modulus, taken from Table 6.1, to compute the strain:

$$\varepsilon = \frac{\Delta L}{L} = \frac{\sigma}{Y} = \frac{2.45 \times 10^6\ \text{N/m}^2}{18{,}000 \times 10^6\ \text{N/m}^2} = 1.4 \times 10^{-4}$$

This is a very small strain. The bone is compressed by

$$\Delta L = (1.4 \times 10^{-4})(480\ \text{mm}) = 0.065\ \text{mm}$$

That's about the diameter of a human hair. An increase in the stress by a factor of 7 while running increases the compression by a factor of 7 to nearly 0.5 mm, no longer negligible. But the stress is still only about 10% of the compressive strength, nowhere near failure.

ASSESS Bones are hard, so only a hair-width compression while standing is not surprising. And bones must have a reasonable safety margin to withstand an occasional high impact without breaking. Even so, a fall from a height or the sudden impact in a car collision could exceed the compressive strength, and that's one way in which bones are broken. *Osteoporosis,* a condition in which bones become more porous, not only makes bones weaker but also thins the wall of compact bone. A smaller cross-section area means higher stress during simple events such as walking, and higher stress combined with less compressive strength creates a heightened risk for bone fractures.

Bones can break from compressive failure, but most fractures occur when bones are bent too far. FIGURE 6.45 shows a long beam (a bone or a board) as it's bent by a force perpendicular to the long axis. There are methods for calculating the curvature of the bend, but we'll limit ourselves to observing that one side of the object is in tension, the other in compression. Example 6.17 pointed out that the tensile strength of bone—the tension stress at which bone fails—is less than the compressive strength. If a bone is bent far enough, it will fail on the outside edge of the bend when the stress at the outer edge reaches the tensile strength. The energy released when the stressed bone breaks then drives a crack—a bone fracture—well into or even through the bone.

The lesson of these three case studies is that knowledge of a biological structure's mechanical properties is essential for understanding both how it functions and how it might fail. Many injuries and accidents, from pulled muscles to broken bones, are material failures where an applied stress exceeds the material's tensile strength or compressive strength.

FIGURE 6.45 A bent object is in tension on one side, in compression on the other.

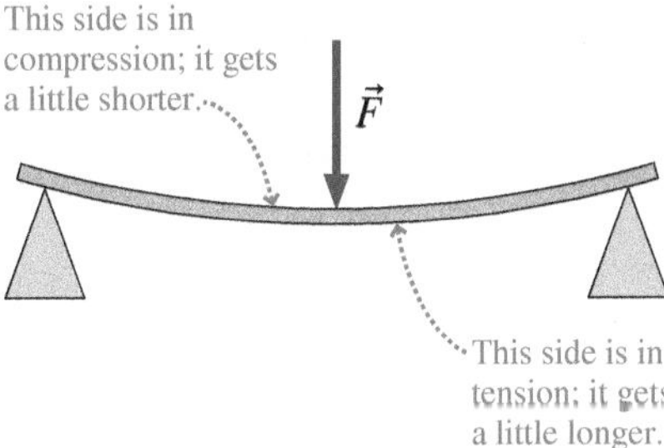
This side is in compression; it gets a little shorter.
$\vec{F}$
This side is in tension; it gets a little longer.

CHAPTER 6 INTEGRATED EXAMPLE Elevator cable stretch

The steel cables that hold elevators stretch only a very small fraction of their length, but in a tall building this small fractional change can add up to a noticeable stretch. This example uses realistic numbers for such an elevator to make this point. The 2300 kg car of a high-speed elevator in a tall building is supported by six 1.27-cm-diameter cables. Steel cables are slightly more pliant than structural steel, with a Young's modulus of 1.0×10^{11} N/m^2. When the elevator is on the bottom floor, the cables rise 90 m up the shaft to the motor above.

On a busy morning, the elevator is on the bottom floor and fills up with 20 people who have a total mass of 1500 kg. The elevator then accelerates upward at 2.3 m/s^2 until it reaches its cruising speed. How much do the cables stretch due to the weight of the car alone? How much additional stretch occurs when the passengers get in but before the car leaves? And what is the total stretch of the cables while the elevator is accelerating? In all cases, you can ignore the mass of the cables.

PREPARE Although there are six cables, we can imagine them combined into one cable with a cross-section area A six times that of each individual cable. Each cable has radius 0.00635 m and cross-section area $\pi r^2 = 1.27 \times 10^{-4}$ m^2. Multiplying by 6 gives $A = 7.62 \times 10^{-4}$ m^2.

FIGURE 6.46 shows the details. The upward force $\vec{F}$ is the tension of the cable, which is due to the stretch of the cable. The stress-strain relationship for the cable is $\sigma = (F/A) = Y\epsilon = Y(\Delta L/L)$, where L is the cable length and Y is Young's modulus. We can solve this equation for the stretch of the cable in response to a known force F:

$$\Delta L = \frac{LF}{YA}$$

The elevator is in static equilibrium for the first two questions, but the third requires a net force because the elevator is accelerating.

SOLVE When the elevator is at rest, the net force is zero and so the tension in the cable is equal to the weight suspended from it. For the first two questions, the forces are

$$F_1 = m_{\text{car}}g = (2300\text{ kg})(9.8\text{ m/s}^2) = 22{,}500\text{ N}$$

$$F_2 = (m_{\text{car}} + m_{\text{passengers}})g = (2300\text{ kg} + 1500\text{ kg})(9.8\text{ m/s}^2) = 37{,}200\text{ N}$$

These forces stretch the cable by

$$\Delta L_1 = \frac{(90\text{ m})(22{,}500\text{ N})}{(1.0 \times 10^{11}\text{ N/m}^2)(7.62 \times 10^{-4}\text{ m}^2)} = 0.027\text{ m} = 2.7\text{ cm}$$

$$\Delta L_2 = \frac{(90\text{ m})(37{,}200\text{ N})}{(1.0 \times 10^{11}\text{ N/m}^2)(7.62 \times 10^{-4}\text{ m}^2)} = 0.044\text{ m} = 4.4\text{ cm}$$

FIGURE 6.46 Details of the cable stretch and the forces acting on the elevator.

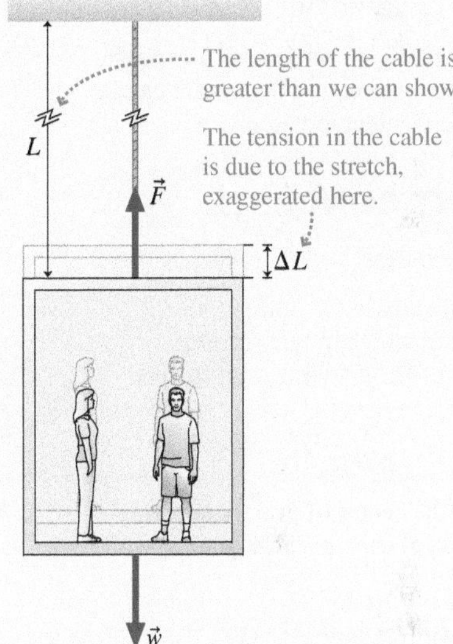
The length of the cable is greater than we can show.
L
The tension in the cable is due to the stretch, exaggerated here.
$\vec{F}$
ΔL
$\vec{w}$

The additional stretch when the passengers board is

$$4.4\text{ cm} - 2.7\text{ cm} = 1.7\text{ cm}$$

When the elevator is accelerating upward, the tension in the cables must increase. Newton's second law for the vertical motion is

$$\Sigma F_y = F - w = ma_y$$

The cable tension is

$$F_3 = w + ma_y = mg + ma_y = (3800\text{ kg})(9.8\text{ m/s}^2 + 2.3\text{ m/s}^2) = 46{,}000\text{ N}$$

Thus right at the start of the motion, when the full 90 m of cable is still deployed, we find that the cables are stretched by

$$\Delta L_3 = \frac{(90\text{ m})(46{,}000\text{ N})}{(1.0 \times 10^{11}\text{ N/m}^2)(7.62 \times 10^{-4}\text{ m}^2)} = 0.054\text{ m} = 5.4\text{ cm}$$

ASSESS When the passengers enter the car, the cable stretches by 1.7 cm, or about two-thirds of an inch. This is large enough to notice (as we might expect given the problem statement) but not large enough to cause concern to the passengers. The total stretch for a fully loaded elevator accelerating upward is 5.4 cm, just greater than 2 inches. This is not unreasonable for cables that are nearly as long as a football field; the fractional change in length is still quite small.

SUMMARY

GOAL To learn about the static equilibrium of extended objects and the elastic properties of materials.

GENERAL PRINCIPLES

Static Equilibrium

An object in **static equilibrium** must have no net force on it and no net torque.

Mathematically, we express this as

$$\sum F_x = 0$$
$$\sum F_y = 0$$
$$\sum \tau = 0$$

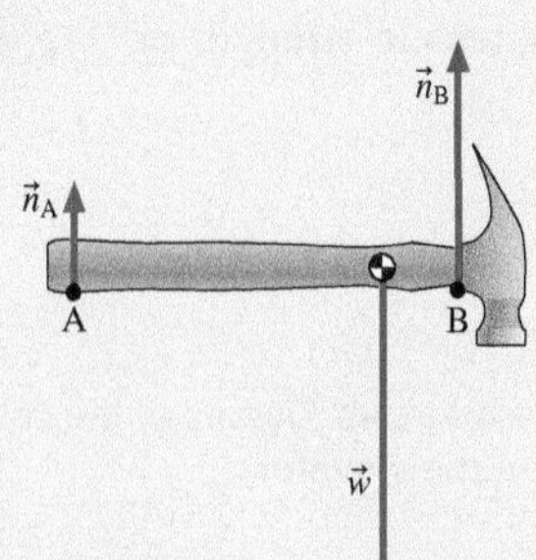

$\vec{n}_B$
$\vec{n}_A$
A
B
$\vec{w}$

Because the net torque is zero about *any* point, the pivot point for calculating the torque can be chosen at any convenient location.

Torque

Torque, the rotational equivalent of force, is the ability of a force to cause a rotation. There are two different ways to calculate torque.

Method 1: $\tau = rF_\perp$ Method 2: $\tau = r_\perp F$

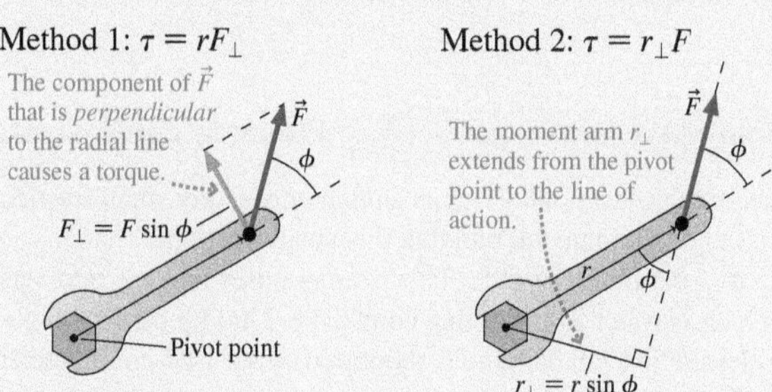

The component of $\vec{F}$ that is *perpendicular* to the radial line causes a torque.
$\vec{F}$
ϕ
$F_\perp = F \sin\phi$
Pivot point
The moment arm $r_\perp$ extends from the pivot point to the line of action.
$\vec{F}$
ϕ
r
$r_\perp = r \sin\phi$

Torque is positive if it would cause a counterclockwise rotation.

IMPORTANT CONCEPTS

Center of gravity

The **center of gravity** is the point of an extended object at which gravity can be considered to act. The **gravitational torque** about a pivot is

$$\tau = r_\perp w = r_\perp Mg$$

The position of the center of gravity depends on the distance x_i of each particle of mass m_i from the origin:

$$x_{cg} = \frac{x_1m_1 + x_2m_2 + x_3m_3 + \cdots}{m_1 + m_2 + m_3 + \cdots} = \frac{1}{M}\sum_i x_im_i$$

A similar equation is used to find y_{cg}.

Elasticity

Elasticity describes how a solid material responds when stretched, compressed, or sheared. For a small linear deformation where Hooke's law remains valid:

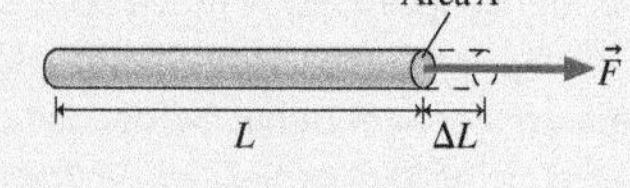

Area A
$\vec{F}$
L
ΔL

$$(F/A) = Y(\Delta L/L)$$

Tensile stress, Young's modulus, Strain

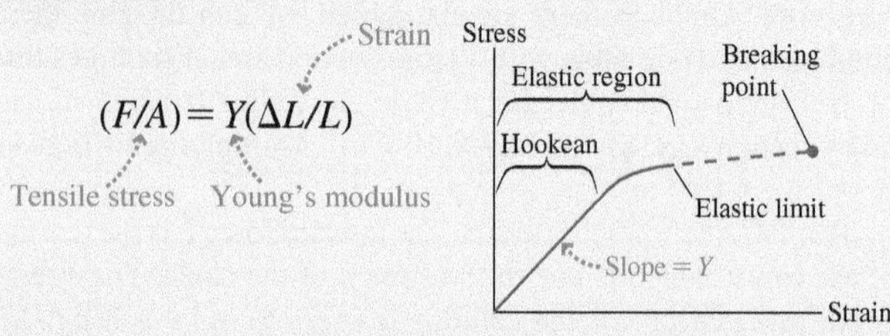

Center of gravity
Pivot point
$r_\perp$
$\vec{w}$
Stress
Elastic region
Breaking point
Hookean
Elastic limit
Slope = Y
Strain

If stretched too far, an object will permanently deform and finally break. The stress at which an object fails is its **tensile strength.**

APPLICATIONS

Stability and balance

An object is **stable** if its center of gravity is over its base of support; otherwise, it is **unstable.**

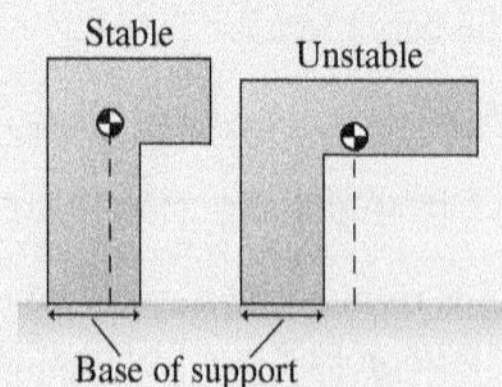

Stable
Unstable
Base of support

If an object is tipped, it will reach the limit of its stability when its center of gravity is over the edge of the base. This defines the **critical angle** θ_c.

Greater stability is possible with a lower center of gravity or a broader base of support.

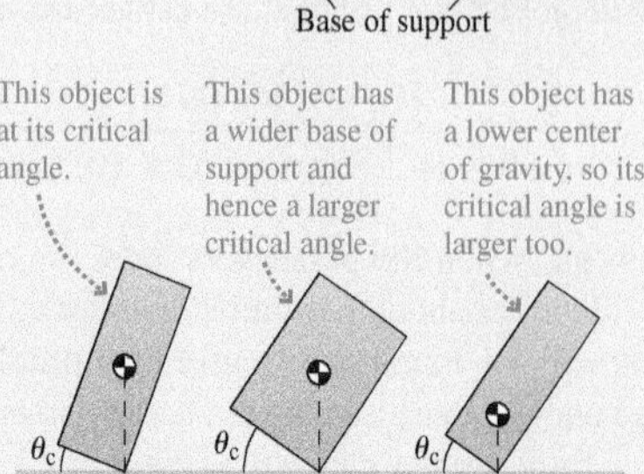

This object is at its critical angle.
This object has a wider base of support and hence a larger critical angle.
This object has a lower center of gravity, so its critical angle is larger too.
θ_c
θ_c
θ_c

Forces and torques in the body

Muscles and tendons apply the forces and torques needed to maintain static equilibrium. These forces and the corresponding stresses may be quite large.

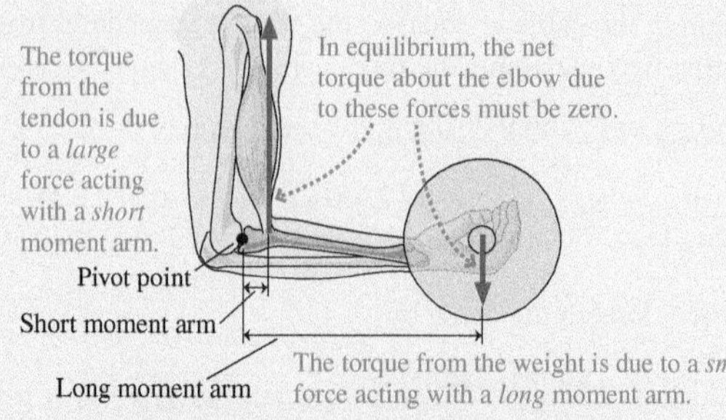

The torque from the tendon is due to a *large* force acting with a *short* moment arm.
In equilibrium, the net torque about the elbow due to these forces must be zero.
Pivot point
Short moment arm
Long moment arm
The torque from the weight is due to a *small* force acting with a *long* moment arm.

The materials of which the body is made, from muscles and bones to skin and arteries, have mechanical properties and mechanical limits that place physical limits on functionality.

LEARNING OBJECTIVES After studying this chapter, you should be able to:

- Calculate the torque about a pivot point. *Conceptual Questions 6.2, 6.3; Problems 6.1–6.3, 6.6, 6.7*
- Determine an object's center of gravity and calculate the gravitational torque about a pivot point. *Conceptual Questions 6.1, 6.4; Problems 6.8–6.10, 6.13, 6.14*
- Solve static equilibrium problems using force and torque. *Conceptual Questions 6.5, 6.7; Problems 6.15, 6.16, 6.18, 6.21, 6.22*
- Determine an object's stability. *Problems 6.24, 6.25, 6.27–6.29*
- Solve problems about forces and torques in the human body. *Conceptual Question 6.12; Problems 6.11, 6.32–6.34, 6.36*
- Calculate the deformation of an object using Young's modulus. *Conceptual Questions 6.9, 6.11; Problems 6.40, 6.41, 6.44, 6.45, 6.47*
- Interpret stress-strain curves of biological materials. *Problems 6.46, 6.48, 6.49*

STOP TO THINK ANSWERS

Stop to Think 6.1: E. Forces D and C act at or in line with the pivot point and provide no torque. Of the others, only E tries to rotate the wheel counterclockwise, so it is the only choice that gives a positive nonzero torque.

Stop to Think 6.2: B. The center of the hitting end is *closer* to the center of gravity, so it must have *more* mass.

Stop to Think 6.3: D. Only object D has both zero net force and zero net torque.

Stop to Think 6.4: C. The scale exerts an upward force at half the distance from the pivot point as the weight's downward force. To exert an equal but opposite torque for static equilibrium requires twice the force.

Stop to Think 6.5: B, A, C. The critical angle θ_c, shown in the figure, measures how far the object can be tipped before falling. B has the smallest critical angle, followed by A, then C.

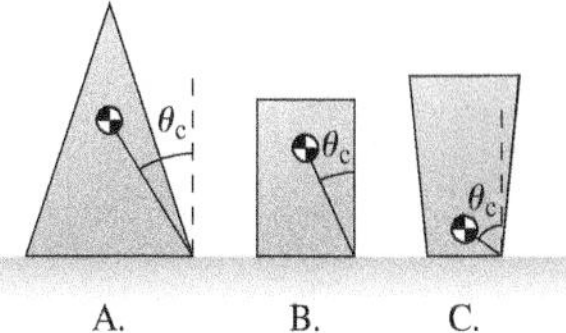

Stop to Think 6.6: E. The cables have the same diameter, and the force is the same, so the stress is the same in both cases. This means that the strain, $\Delta L/L$, is the same. The 2 m cable will experience twice the change in length of the 1 m cable.

QUESTIONS

Conceptual Questions

1. The solid spheres in Figure Q6.1 are made of the same material. Is the center of gravity at point A, B, or C? Explain.

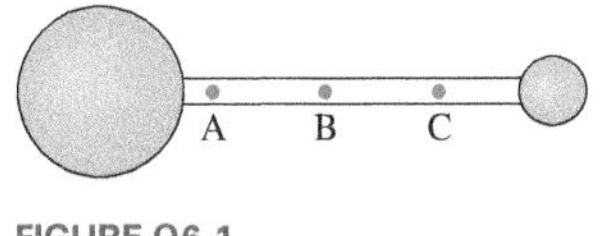

FIGURE Q6.1

2. Four equal-strength forces are applied to a bar that can rotate about a pivot. Rank in order, from largest to smallest, the torques τ_A to τ_D about the pivot.

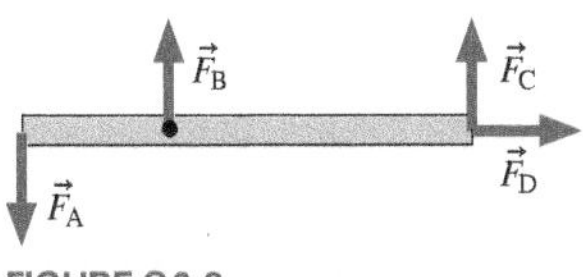

FIGURE Q6.2

3. If you are using a wrench to loosen a very stubborn nut, you can make the job easier by using a "cheater pipe." This is a piece of pipe that slides over the handle of the wrench, as shown in Figure Q6.3, making it effectively much longer. Explain why this would help you loosen the nut.

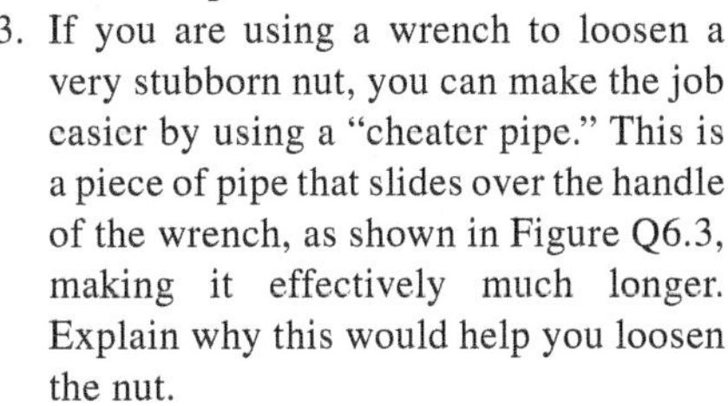

4. If you stand with your back and heels against a wall, and bend forward at the waist, you will topple forward. Explain why.

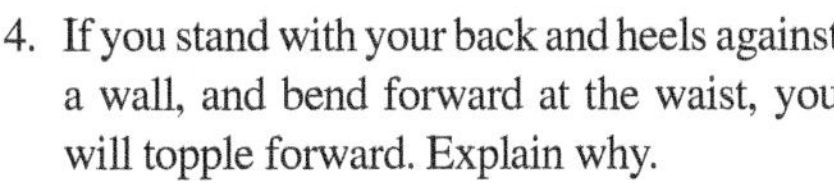
FIGURE Q6.3

5. Figure Q6.5 shows a steel angle bracket with equal-length arms. Which point is the center of gravity?

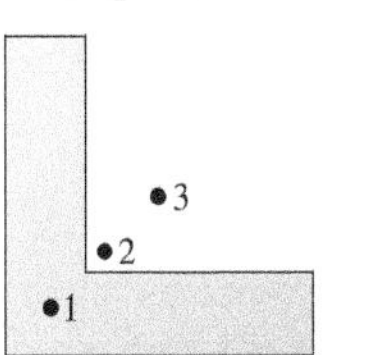

FIGURE Q6.5

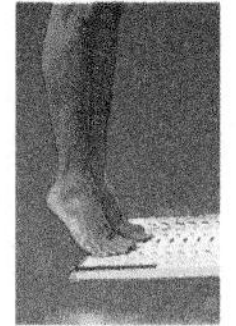
FIGURE Q6.6

6. As divers stand on tiptoes on the edge of a diving platform, in preparation for a high dive, as shown in Figure Q6.6, they usually extend their arms in front of them. Why do they do this?
7. You must lean quite far forward as you rise from a chair (try it!). Explain why.
8. Figure Q6.8 shows a wheelbarrow. The wheelbarrow together with its load have a weight of 300 N. Their center of gravity is shown. How much upward force must you apply at the handles to move the wheelbarrow?

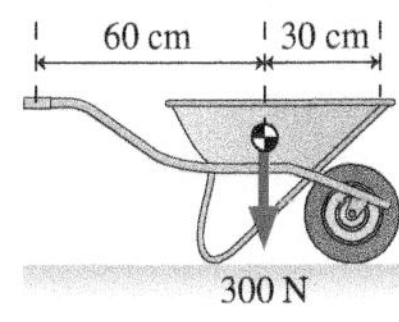

FIGURE Q6.8

9. A wire is stretched to its breaking point by a 5000 N force. A longer wire made of the same material has the same diameter. Is the force that will stretch it right to its breaking point larger than, smaller than, or equal to 5000 N? Explain.

Problem difficulty is labeled as I (straightforward) to IIIII (challenging). Problems labeled INT integrate significant material from earlier chapters; BIO are of biological or medical interest; CALC require calculus to solve.

10. A 100-cm-long wire is under 1×10^7 N/m^2 of stress when stretched to a length of 101 cm. At what length will the stress be 3×10^7 N/m^2?
11. The maximum force that wire A can withstand without breaking is 2000 N. Wire B is made of the same material as wire A but has twice the diameter. What is the maximum force that wire B can withstand?
12. You can apply a much larger bite force with your molars, in the back of your mouth, than with your incisors, at the front of your mouth. Explain why this is so. BIO

Multiple-Choice Questions

13. | A nut needs to be tightened with a wrench. Which force shown in Figure Q6.13 will apply the greatest torque to the nut?

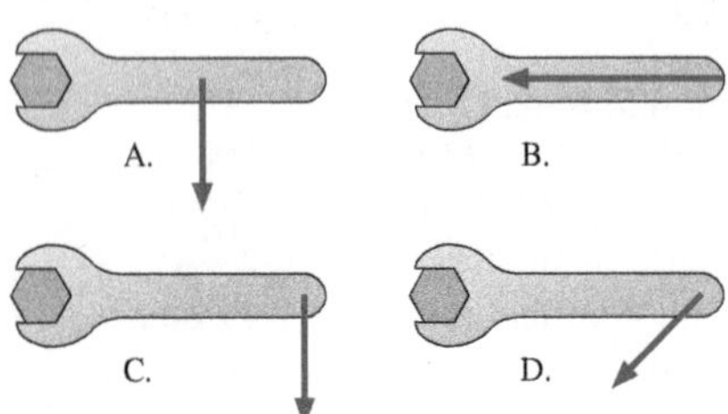

FIGURE Q6.13

14. | Specifications require the oil filter for your car to be tightened with a torque of 30 N · m. You are using a 15-cm-long oil filter wrench, and you apply a force at the very end of the wrench in the direction that produces maximum torque. How much force should you apply?
A. 2000 N B. 400 N C. 200 N D. 30 N
15. | A machine part is made up of two pieces, with centers of gravity shown in Figure Q6.15. Which point could be the center of gravity of the entire part?

FIGURE Q6.15

16. || Two children hold opposite ends of a lightweight, 1.8-m-long horizontal pole with a water bucket hanging from it. The older child supports twice as much weight as the younger child. How far is the bucket from the older child?
A. 0.3 m B. 0.6 m C. 0.9 m D. 1.2 m

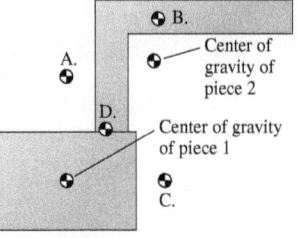

17. || The uniform rod in Figure Q6.17 has a weight of 14.0 N. What is the magnitude of the normal force of the surface on the left end of the rod?
A. 7 N
B. 14 N
C. 28 N
D. It can't be determined.

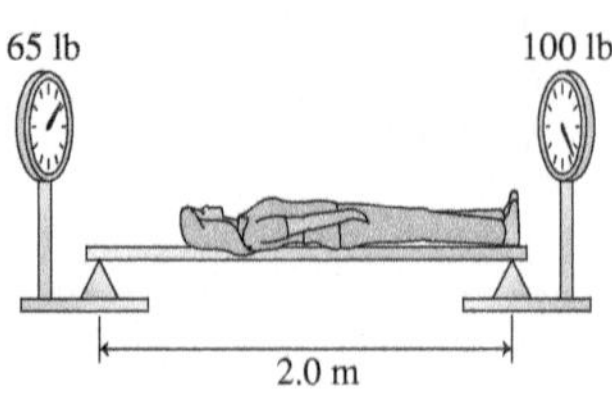

FIGURE Q6.17

18. | A student lies on a very light, rigid board with a scale under each end. Her feet are directly over one scale, and her body is positioned as shown in Figure Q6.18. The two scales read the values shown in the figure. What is the student's weight?
A. 65 lb B. 75 lb C. 100 lb D. 165 lb

FIGURE Q6.18

19. | For the student in Figure Q6.18, approximately how far from her feet is her center of gravity?
A. 0.6 m B. 0.8 m
C. 1.0 m D. 1.2 m
20. || Some climbing ropes are designed to noticeably stretch under a load to lessen the forces in a fall. A weight is attached to a 10 m length of climbing rope, which then stretches by 20 cm. Now this single rope is replaced by a doubled rope—two pieces of rope next to each other. How much does the doubled rope stretch?
A. 5.0 cm B. 10 cm
C. 20 cm D. 40 cm
21. || Some climbing ropes are designed to noticeably stretch under a load to lessen the forces in a fall. A 10 m length of climbing rope is supporting a climber; this causes the rope to stretch by 60 cm. If the climber is supported by a 20 m length of the same rope, by how much does the rope stretch?
A. 30 cm B. 60 cm
C. 90 cm D. 120 cm
22. || You have a heavy piece of equipment hanging from a 1.0-mm-diameter wire. Your supervisor asks that the length of the wire be doubled without changing how far the wire stretches. What diameter must the new wire have?
A. 1.0 mm B. 1.4 mm
C. 2.0 mm D. 4.0 mm

PROBLEMS

Section 6.1 Extended Objects

Section 6.2 Torque

1. | What is the net torque about the axle on the pulley in Figure P6.1?
2. || The tune-up specifications of a car call for the spark plugs to be tightened to a torque of 38 N · m. You plan to tighten the plugs by pulling on the end of a 25-cm-long wrench. Because of the cramped space under the hood, you'll need to pull at an angle of 120° with respect to the wrench shaft. With what force must you pull?

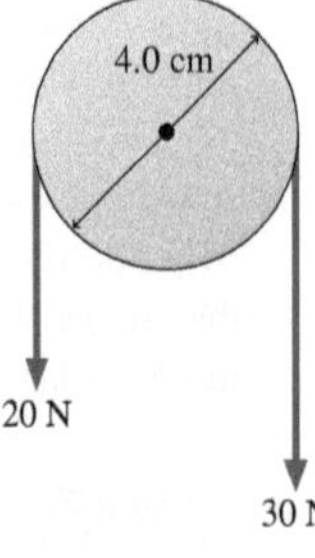

FIGURE P6.1

3. || In Figure P6.3, force $\vec{F}_2$ acts half as far from the pivot as $\vec{F}_1$. What magnitude of $\vec{F}_2$ causes the net torque on the rod to be zero?

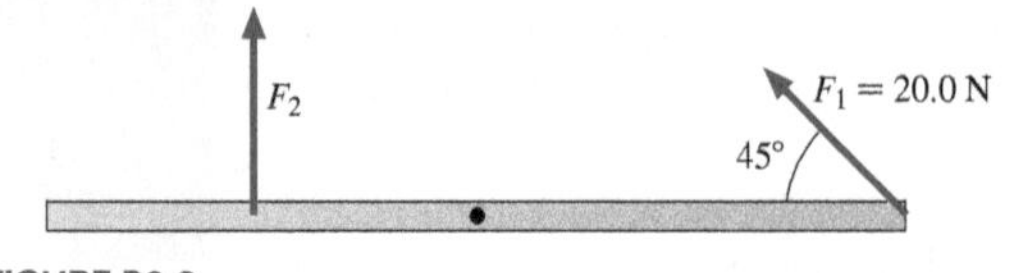

FIGURE P6.3

4. || What is the net torque on the bar shown in Figure P6.4, about the axis indicated by the dot?

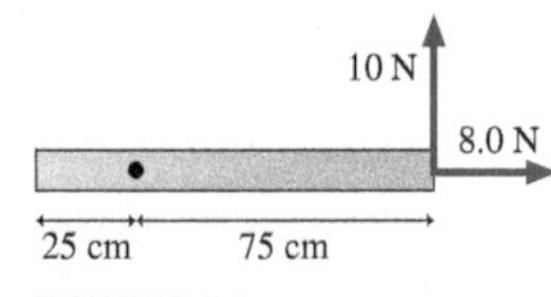

FIGURE P6.4

5. || A typical jar that has been tightened to a reasonable degree requires 2.0 N · m to open. If you grab a 7.0-cm-diameter jar lid with one hand so that your thumb and fingers exert equal magnitude forces on opposite sides of the lid, as in Figure P6.5, what is the magnitude of each of the forces?

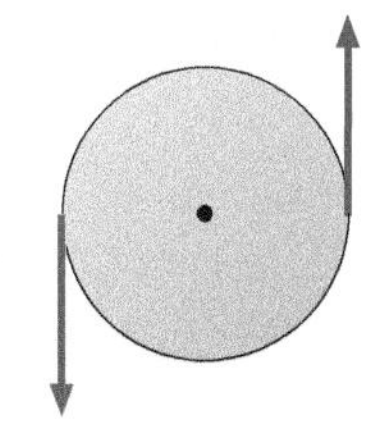

FIGURE P6.5

6. || What is the net torque on the bar shown in Figure P6.6, about the axis indicated by the dot?

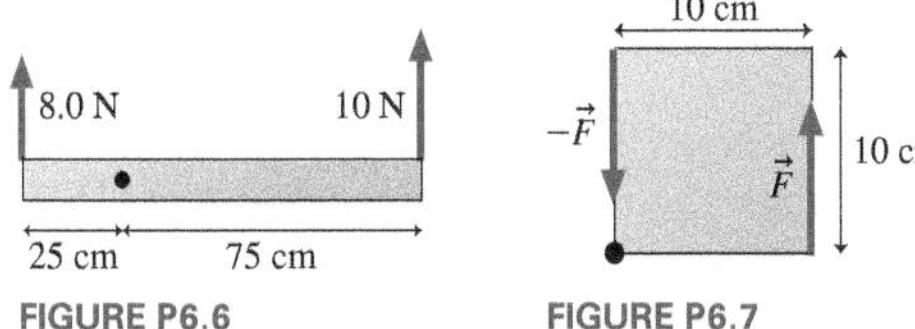

FIGURE P6.6

FIGURE P6.7

7. || In Figure P6.7, what magnitude force provides 5.0 N · m net torque about the axle?

Section 6.3 Gravitational Torque and the Center of Gravity

8. || When you stand quietly, you pivot back and forth
BIO a very small amount about your ankles. It's easier to maintain stability if you tip slightly forward, so that your center of gravity is slightly in front of your ankles. The 68 kg man shown in Figure P6.8 is 1.8 m tall; his center of gravity is 1.0 m above the ground and, during quiet standing, is 5.5 cm in front of his ankles. What is the magnitude of the gravitational torque about his ankles?

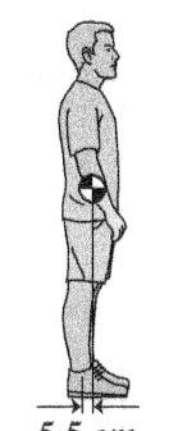

FIGURE P6.8

9. ||| The 2.0 kg, uniform, horizontal rod in Figure P6.9 is seen from the side. What is the gravitational torque about the point shown?

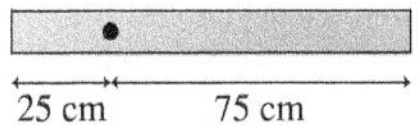

25 cm 75 cm

FIGURE P6.9

10. ||| A 4.00-m-long, 500 kg steel beam extends horizontally from the point where it has been bolted to the framework of a new building under construction. A 70.0 kg construction worker stands at the far end of the beam. What is the magnitude of the gravitational torque about the point where the beam is bolted into place?

11. ||| An athlete at the gym holds a 3.0 kg steel ball in his hand.
BIO His arm is 70 cm long and has a mass of 4.0 kg with the center of gravity 30 cm from the shoulder. What is the magnitude of the gravitational torque about his shoulder if he holds his arm
 a. Straight out to his side, parallel to the floor?
 b. Straight, but 45° below horizontal?

12. || The three masses shown in Figure P6.12 are connected by massless, rigid rods. What are the coordinates of the center of gravity?

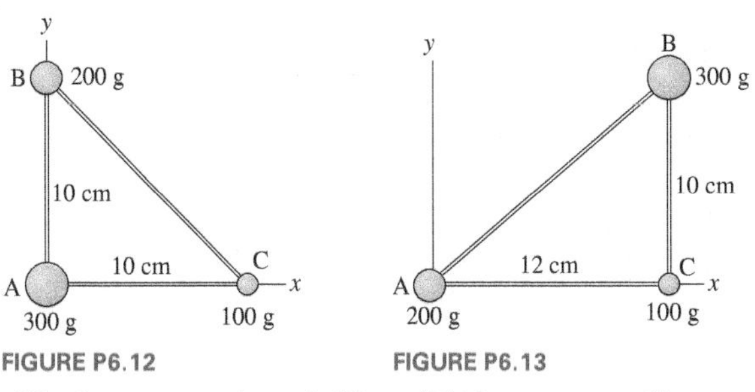

FIGURE P6.12

FIGURE P6.13

13. || The three masses shown in Figure P6.13 are connected by massless, rigid rods. What are the coordinates of the center of gravity?

14. ||| Figure P6.14 shows two thin beams joined at right angles. The vertical beam is 15.0 kg and 1.00 m long and the horizontal beam is 25.0 kg and 2.00 m long.

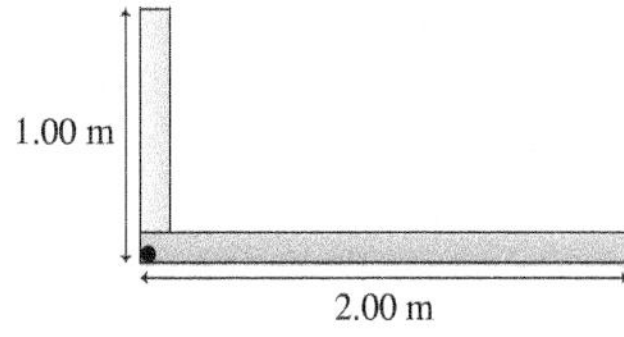

FIGURE P6.14

 a. Find the center of gravity of the two joined beams. Express your answer in the form (x, y), taking the origin at the corner where the beams join.
 b. Calculate the gravitational torque on the joined beams about an axis through the corner. The beams are seen from the side.

Section 6.4 Static Equilibrium

15. || A 64 kg student stands on a very light, rigid board that rests on a bathroom scale at each end, as shown in Figure P6.15. What is the reading on each of the scales?

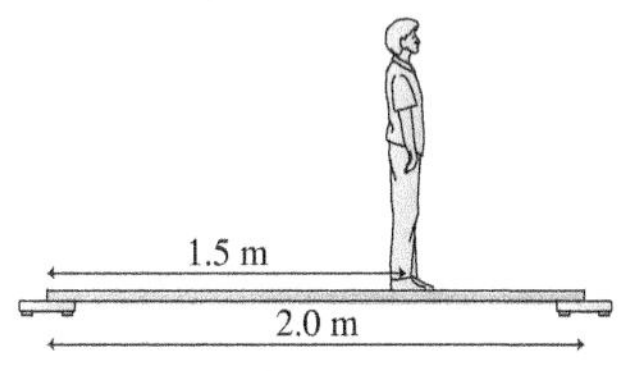

FIGURE P6.15

16. ||| You're carrying a 3.6-m-long, 25 kg pole to a construction site when you decide to stop for a rest. You place one end of the pole on a fence post and hold the other end of the pole 85 cm from its tip. How much force must you exert to keep the pole motionless in a horizontal position?

17. | A typical horse weighs 5000 N. The
BIO distance between the front and rear hooves and the distance from the rear hooves to the center of gravity for a typical horse are shown in Figure P6.17. What fraction of the horse's weight is borne by the front hooves?

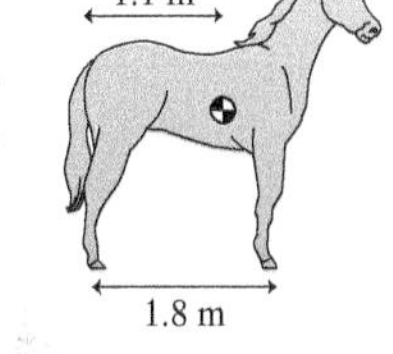

FIGURE P6.17

18. || A vendor hangs an 8.0 kg sign in front of his shop with a cable held away from the building by a lightweight pole. The pole is free to pivot about the end where it touches the wall, as shown in Figure P6.18. What is the tension in the cable?

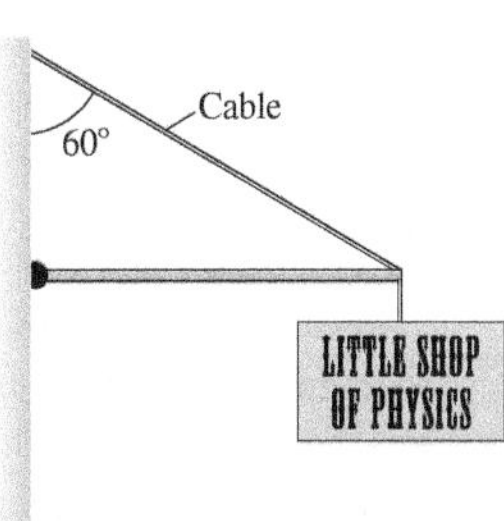

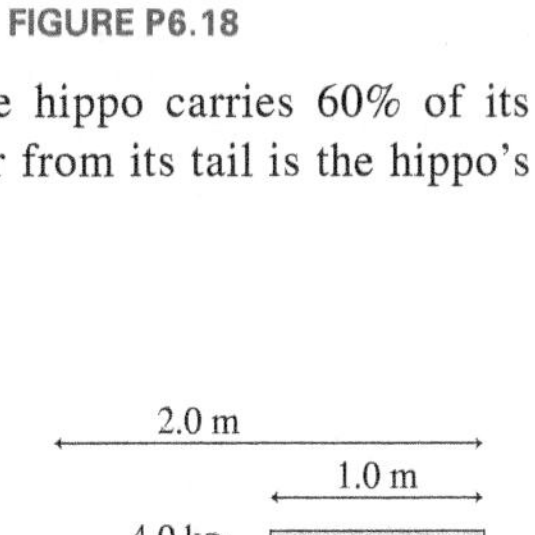

FIGURE P6.18

19. || A hippo's body is 4.0 m
BIO long with front and rear feet located as in Figure P6.19. The hippo carries 60% of its weight on its front feet. How far from its tail is the hippo's center of gravity?

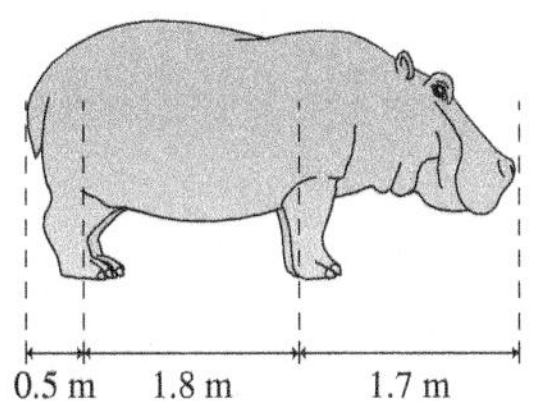

FIGURE P6.19

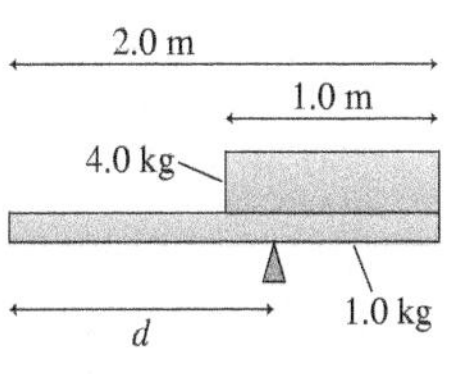

FIGURE P6.20

20. ||| The two objects in Figure P6.20 are balanced on the pivot. What is distance d?

21. ||| A bicycle mechanic is checking a road bike's chain. He applies a 45 N force to a pedal at the angle shown in Figure P6.21 while keeping the wheel from rotating. The pedal is 17 cm from the center of the crank; the gear has a diameter of 16 cm. What is the tension in the chain?

FIGURE P6.21

22. || A 60 kg diver stands at the end of a 30 kg springboard, as shown in Figure P6.22. The board is attached to a hinge at the left end but simply rests on the right support. What is the magnitude of the vertical force exerted by the hinge on the board?

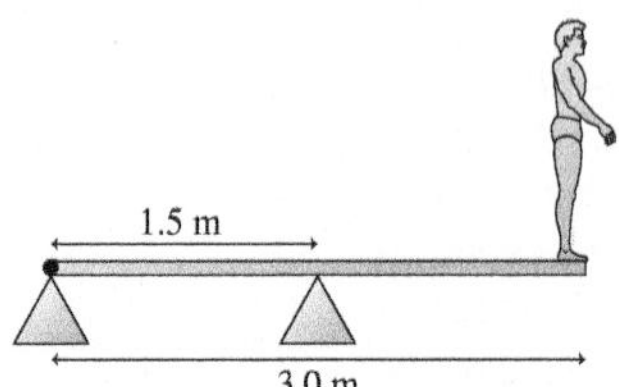

FIGURE P6.22

23. ||| A 5.0 kg cat and a 2.0 kg bowl of tuna fish are at opposite ends of the 4.0-m-long seesaw of Figure P6.23. How far to the left of the pivot must a 4.0 kg cat stand to keep the seesaw balanced?

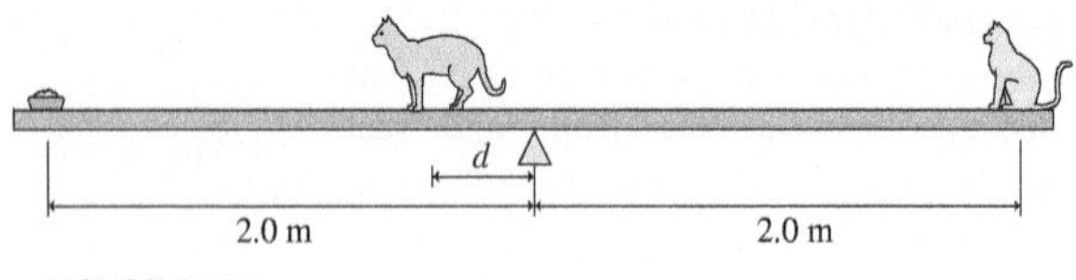

FIGURE P6.23

Section 6.5 Stability and Balance

24. | A standard four-drawer filing cabinet is 52 inches high and 15 inches wide. If it is evenly loaded, the center of gravity is at the center of the cabinet. A worker is tilting a filing cabinet to the side to clean under it. To what angle can he tilt the cabinet before it tips over?

25. | A double-decker London bus might be in danger of rolling over in a highway accident, but at the low speeds of its urban environment, it's plenty stable. The track width is 2.05 m. With no passengers, the height of the center of gravity is 1.45 m, rising to 1.73 m when the bus is loaded to capacity. What are the critical angles for both the unloaded and loaded bus?

26. | The stability of a vehicle is often rated by the *static stability factor,* which is one-half the track width divided by the height of the center of gravity above the road. A typical SUV has a static stability factor of 1.2. What is the critical angle?

27. | A magazine rack has a center of gravity 16 cm above the floor, as shown in Figure P6.27. Through what maximum angle, in degrees, can the rack be tilted without falling over?

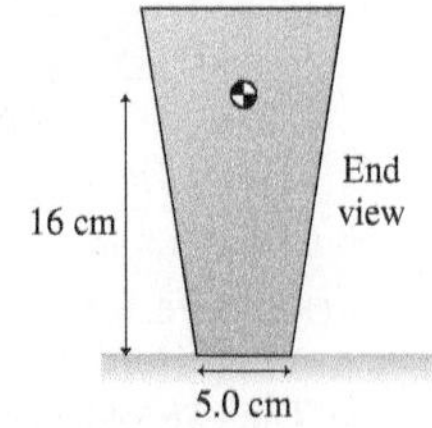

FIGURE P6.27

28. || A car manufacturer claims that you can drive its new vehicle across a hill with a 47° slope before the vehicle starts to tip. If the vehicle is 2.0 m wide, how high is its center of gravity?

29. || A thin 2.00 kg box rests on a 6.00 kg board that hangs over the end of a table, as shown in Figure P6.29. How far can the center of the box be from the end of the table before the board begins to tilt?

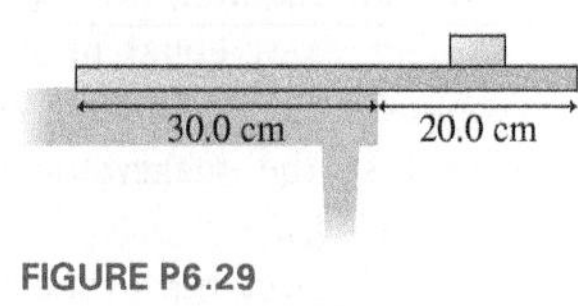

FIGURE P6.29

Section 6.6 Forces and Torques in the Body

30. || You need to pry up a 150 kg rock in your garden. To do so, you grab a 2.0-m-long steel bar to use as a lever and a round log that can be used as a fulcrum. Your largest downward force, if you lean on one end of the bar with almost all your weight, is 650 N. How far from the rock should you place the fulcrum so that your largest force is just sufficient to lift the rock?

31. || A woman is pushing a load in a wheelbarrow, as in Figure P6.31. The combined mass of the wheelbarrow and the load is 110 kg, with a center of gravity 0.25 m behind the axle. The woman supports the wheelbarrow at the handles, 1.1 m behind the axle.
   a. What is the force required to support the wheelbarrow?
   b. What fraction of the weight of the wheelbarrow and the load does this force represent?

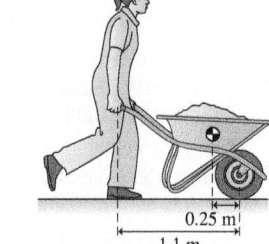

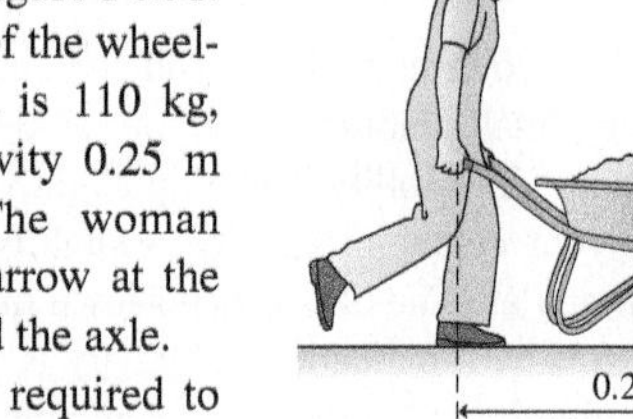

FIGURE P6.31

32. || Figure P6.32 shows the operation of a garlic press. The lower part of the press is held steady, and the upper handle is pushed down, thereby crushing a garlic clove through a screen. Approximate distances are shown in the figure. If the user exerts a 12 N force on the upper handle, estimate the force on the clove.

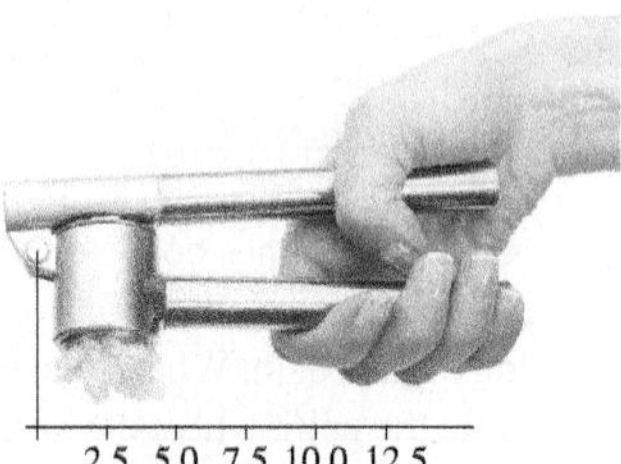

FIGURE P6.32

33. || Hold your upper arm vertical and your lower arm horizontal with your hand palm-down on a table, as shown in Figure P6.33. If you now push down on the table, you'll feel that your triceps muscle has contracted and is trying to pivot your lower arm about the elbow joint. If a person with the arm dimensions shown pushes down hard with a 90 N force (about 20 lb), what force must the triceps muscle provide? You can ignore the mass of the arm and hand in your calculation.
BIO

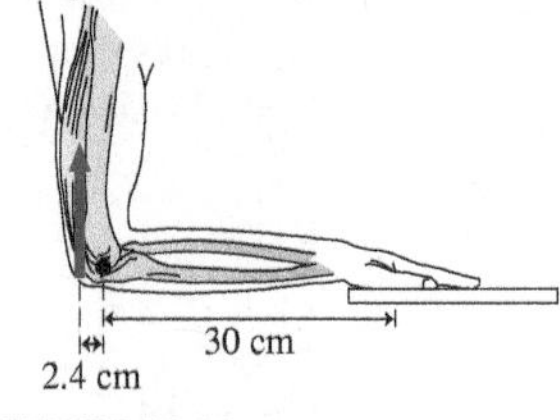

FIGURE P6.33

34. || If you stand on one foot while holding your other leg up behind you, your muscles apply a force to hold your leg in this raised position. We can model this situation as in Figure P6.34. The leg pivots at the knee joint, and the force that holds the leg up is provided by a tendon attached to the lower leg as shown. Assume that the lower leg and the foot have a combined mass of 4.0 kg, and that their combined center of gravity is at the center of the lower leg.
BIO
   a. How much force must the tendon exert to keep the leg in this position? Assume that the tendon pulls vertically.
   b. As you hold your leg in this position, the upper leg exerts a force on the lower leg at the knee joint. What are the magnitude and direction of this force?

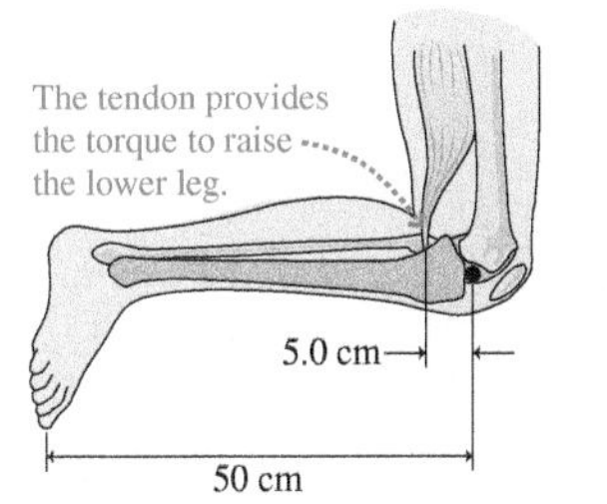

FIGURE P6.34

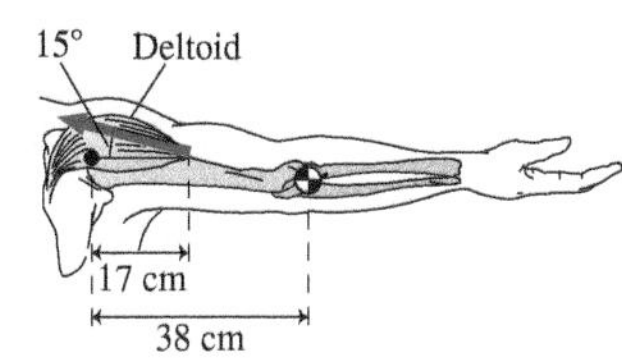

FIGURE P6.35

35. || If you hold your arm outstretched with palm upward, as in Figure P6.35, the force to keep your arm from falling comes from your deltoid muscle. The arm of a typical person has mass 4.0 kg and the distances and angles shown in the figure.
BIO
   a. What force must the deltoid muscle provide to keep the arm in this position?
   b. By what factor does this force exceed the weight of the arm?

36. || Dogs, like many animals, stand and walk on their toes. Figure P6.36 shows the rear foot and leg of a 20 kg dog along with the relevant distances and angles. The Achilles tendon pulls on the heel at an angle 15° from vertical. What is the tension in the tendon if the dog is supporting one-fourth of its weight on each foot?
BIO

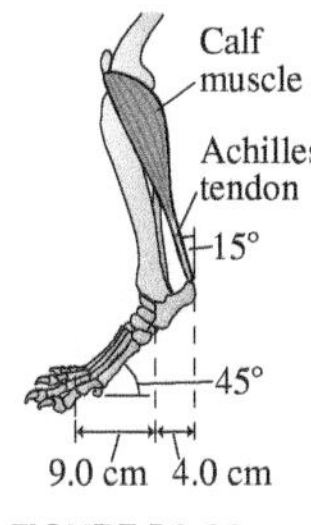

FIGURE P6.36

Section 6.7 Elasticity and Deformation

37. || Figure P6.37 shows the force needed to stretch a 2.0-m-long, 2.0-mm-diameter wire.
   a. What is the strain when the wire has been stretched by 5.0 cm?
   b. What is the stress when the wire has been stretched by 5.0 cm?
   c. What is Young's modulus for this wire?

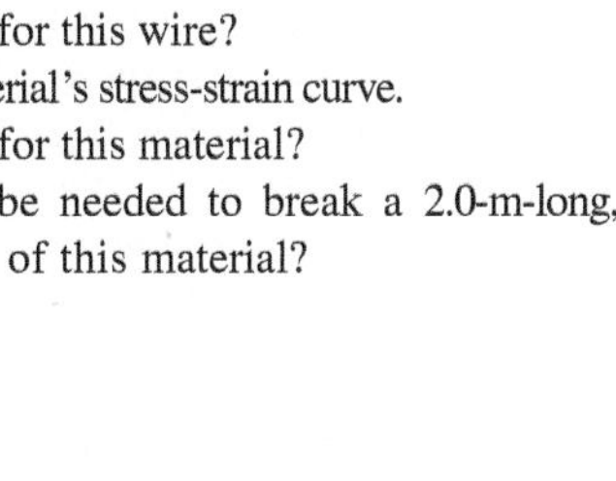

FIGURE P6.37

38. || Figure P6.38 shows a material's stress-strain curve.
   a. What is Young's modulus for this material?
   b. How much force would be needed to break a 2.0-m-long, 1.0×1.0 mm square wire of this material?

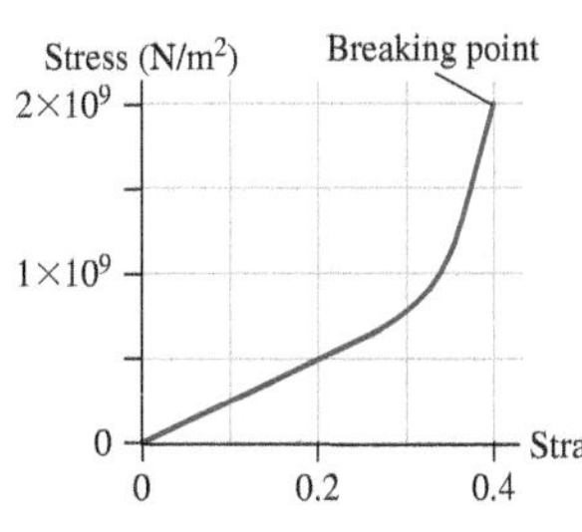

FIGURE P6.38

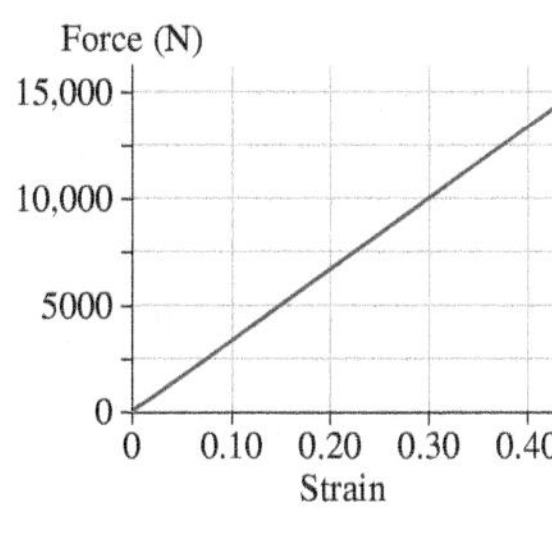

FIGURE P6.39

39. ||| Dynamic climbing ropes are designed to be quite pliant, allowing a falling climber to slow down over a long distance. The graph in Figure P6.39 shows force-versus-strain data for an 11-mm-diameter climbing rope. What is the Young's modulus for this rope?

40. || A force stretches a wire by 1.0 mm.
   a. A second wire of the same material has the same cross section and twice the length. How far will it be stretched by the same force?
   b. A third wire of the same material has the same length and twice the diameter as the first. How far will it be stretched by the same force?

41. || To meet a certain specification, an 11-mm-diameter rope must experience a maximum elongation of 5.0% when supporting a 150 kg load. What is the minimum Young's modulus?

42. || A 70 kg mountain climber dangling in a crevasse stretches a 50-m-long, 1.0-cm-diameter rope by 8.0 cm. What is Young's modulus for the rope?

43. ||| What hanging mass will stretch a 2.0-m-long, 0.50-mm-diameter steel wire by 1.0 mm?

44. ||| An 80-cm-long, 1.0-mm-diameter steel guitar string must be tightened to a tension of 2.0 kN by turning the tuning screw. By how much is the string stretched?

45. |||| A three-legged wooden bar stool made out of solid Douglas fir has legs that are 2.0 cm in diameter. When a 75 kg man sits on the stool, by what percent does the length of the legs decrease? Assume, for simplicity, that the stool's legs are vertical and that each bears the same load.

46. || One species of jellyfish uses the energy from a folded-collagen spring to fire a microscopic needle, called a stylet, with enough speed to penetrate a crustacean shell, after which it injects venom into the crustacean. High-speed video showed that the impact of the 75-nm-diameter stylet generated a stress of 7.0×10^9 N/m^2, similar to the stress of a bullet impact. What force does the stylet apply to the crustacean?
BIO

47. || A floor can be dented by the application of too much stress. Who is more likely to dent your floor: a 50 kg woman in high-heeled shoes (assume a 6.0-mm-diameter heel) standing on one heel or a 5000 kg African elephant (40-cm-diameter foot pads) standing on all four legs?

48. || The Achilles tendon connects the muscles in your calf to the back of your foot. When you are sprinting, your Achilles tendon alternately stretches, as you bring your weight down onto your forward foot, and contracts to push you off the ground. A 70 kg runner has an Achilles tendon that is 15 cm long with a cross-section area of 1.1 cm^2.
BIO
   a. By how much will the runner's Achilles tendon stretch if the maximum force on it is 8.0 times his weight, a typical value while running?
   b. What is the strain of the tendon at its maximum stretch?

49. || The tendon known as the anterior cruciate ligament (ACL) BIO in a knee is made of a material that has a Young's modulus of 5.0×10^7 N/m^2. One athlete's ligament is 25 mm long and 9.0 mm in diameter.
   a. How much will this athlete's ACL stretch if a force of 400 N is applied to it?
   b. What is the maximum force this athlete's ACL can withstand without breaking if the ligament's tensile strength is 1.5×10^7 N/m^2?

50. || Hydrogels are materials that are easily deformed. One experiment created hydrogel blocks that were 5.0 mm tall and had a 20 mm × 20 mm base. With the base adhered to a horizontal surface, a sticky 20-mm-wide blade pulled the top surface of the hydrogel parallel to the surface, causing a shear deformation. A 5.5 N force, weighing only slightly more than a pound, pulled the top surface 4.0 mm to the side. What was the shear modulus for this hydrogel?

General Problems

51. |||| The 20-cm-diameter disk in Figure P6.51 can rotate on an axle through its center. What is the net torque about the axle?

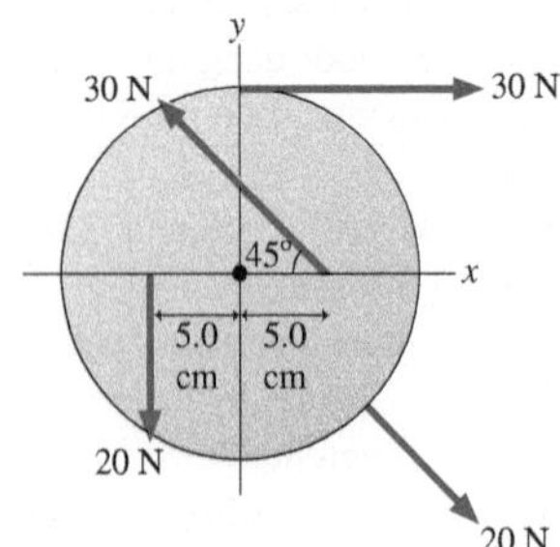

FIGURE P6.51

52. |||| A combination lock has a 1.0-cm-diameter knob that is INT part of the dial you turn to unlock the lock. To turn that knob, you grip it between your thumb and forefinger with a force of 0.60 N as you twist your wrist. Suppose the coefficient of static friction between the knob and your fingers is only 0.12 because some oil accidentally got onto the knob. What is the most torque you can exert on the knob without having it slip between your fingers?

53. |||| A 70 kg man's arm, including the hand, can be modeled as a BIO 75-cm-long uniform cylinder with a mass of 3.5 kg. In raising both his arms, from hanging down to straight up, by how much does he raise his center of gravity?

54. ||| We can model a pine tree in the forest as having a compact BIO INT canopy at the top of a relatively bare trunk. Wind blowing on the top of the tree exerts a horizontal force, and thus a torque that can topple the tree if there is no opposing torque. Suppose a tree's canopy presents an area of 9.0 m^2 to the wind centered at a height of 7.0 m above the ground. (These are reasonable values for forest trees.) If the wind blows at 6.5 m/s:
   a. What is the magnitude of the drag force of the wind on the canopy? Assume a drag coefficient of 0.50.
   b. What torque does this force exert on the tree, measured about the point where the trunk meets the ground?

55. |||| A long rod of length L and cross-section area A has a variable CALC density given by $\rho = \rho_0 x/L$, where x is measured from one end of the rod. That is, the rod fades out to nothing ($\rho \rightarrow 0$) at one end and reaches maximum density ρ_0 at the other. Imagine a very thin slice of the rod at position x having width dx and cross-section area A. This slice has volume $dV = A\,dx$ and mass $dm = \rho\,dV = (\rho_0 x/L)(A\,dx) = (\rho_0 A/L)x\,dx$.
   a. Find an expression for ρ_0 in terms of M, L, and A. Do this by integrating dm over the length of the rod.
   b. Find the position x_{cg} of the rod's center of gravity.

56. |||| An 800 g steel plate has the shape of the isosceles triangle CALC shown in Figure P6.56. What are the x- and y-coordinates of the center of mass?
Hint: Divide the triangle into vertical strips of width dx, then relate the mass dm of a strip at position x to the values of x and dx.

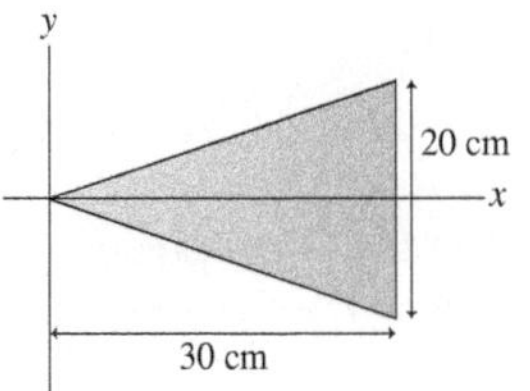

FIGURE P6.56

57. |||| A 3.0-m-long rigid beam with a mass of 100 kg is supported at each end, as shown in Figure P6.57. An 80 kg student stands 2.0 m from support 1. How much upward force does each support exert on the beam?

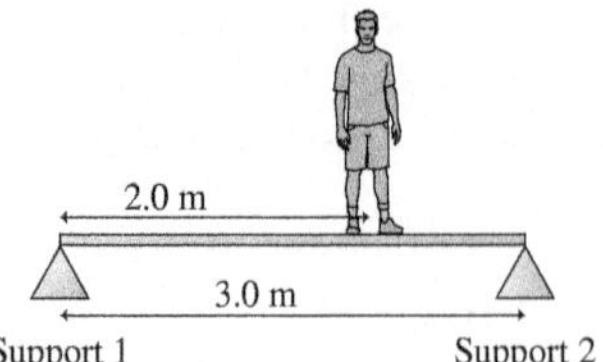

FIGURE P6.57

58. ||| An 80 kg construction worker sits down 2.0 m from the end of a 1450 kg steel beam to eat his lunch, as shown in Figure P6.58. The cable supporting the beam is rated at 15,000 N. Should the worker be worried?

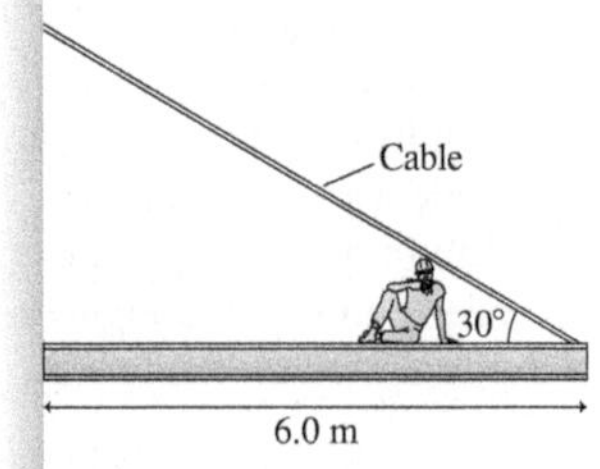

FIGURE P6.58

59. ||| A man is attempting to raise a 7.5-m-long, 28 kg flagpole that has a hinge at the base by pulling on a rope attached to the top of the pole, as shown in Figure P6.59. With what force does the man have to pull on the rope to hold the pole motionless in this position?

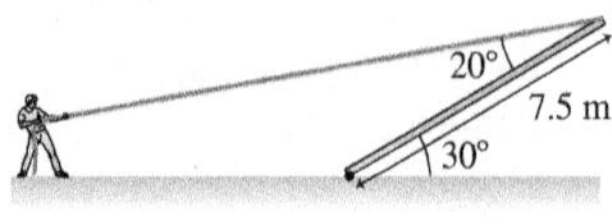

FIGURE P6.59

60. |||| A 40 kg, 5.0-m-long beam is supported by, but not attached to, the two posts in Figure P6.60. A 20 kg boy starts walking along the beam. How close can he get to the right end of the beam without it tipping?

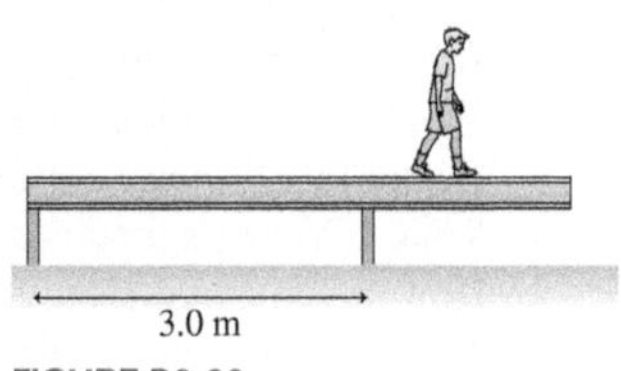

FIGURE P6.60

61. ||| Figure P6.61 shows a lightweight plank supported at its right end by a 7.0-mm-diameter rope with a tensile strength of $6.0 \times 10^7\ N/m^2$.
   a. What is the maximum force that the rope can support?
   b. What is the greatest distance, measured from the pivot, that the center of gravity of an 800 kg piece of heavy machinery can be placed without snapping the rope?

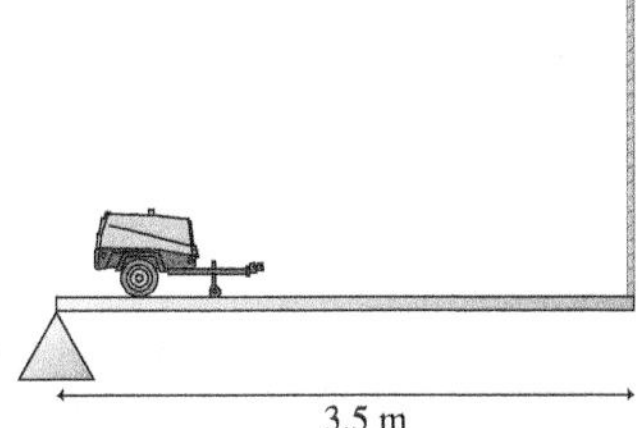

FIGURE P6.61

62. ||| There is a disk of cartilage between each pair of vertebrae in your spine. Suppose a disk is 0.50 cm thick and 4.0 cm in diameter. If this disk supports half the weight of a 65 kg person, by what fraction of its thickness does the disk compress?
BIO

63. || Orb spiders make silk with a diameter of 12 μm.
BIO
   a. A large orb spider has a mass of 0.50 g. If this spider suspends itself from a single 12-cm-long strand of silk, by how much will the silk stretch?
   b. What is the maximum weight that a single thread of this silk could support?

64. ||| Larger animals have sturdier bones than smaller animals. A mouse's skeleton is only a few percent of its body weight, compared to 16% for an elephant. To see why this must be so, recall, from Example 6.17, that the stress on the femur for a man standing on one leg is 1.4% of the bone's compressive strength. Suppose we scale this man up by a factor of 10 in all dimensions, keeping the same body proportions. Use the data for Example 6.17 to compute the following.
BIO
   a. Both the inside and outside diameter of the femur, the region of compact bone, will increase by a factor of 10. What will be the new cross-section area?
   b. The man's body will increase by a factor of 10 in each dimension. What will be his new mass?
   c. If the scaled-up man now stands on one leg, what fraction of the compressive strength is the stress on the femur?

65. ||| The main muscles that hold your head upright attach to your spine in back of the point where your head pivots on your neck. Figure P6.65 shows typical numbers for the distance from the pivot to the muscle attachment point and the distance from the pivot to the center of gravity of the head. The muscles pull down to keep your head upright. If the muscle relaxes—if, for instance, you doze in one of your classes besides Physics—your head tips forward. In the questions that follow, assume that your head has a mass of 4.8 kg, and that you maintain the relative angle between your head and your spine.
BIO
   a. With the head held level, as in Figure P6.65, what muscle force is needed to keep a 4.8 kg head upright?
   b. Suppose you lie on your stomach on a bench with your head extending over the edge. What muscle force is needed to keep your head aligned with your spine?
   c. If you tip your body backward, you will reach a point where no muscle force is needed to keep your head upright. For the distances given in Figure P6.65, at what angle from vertical does this balance occur?

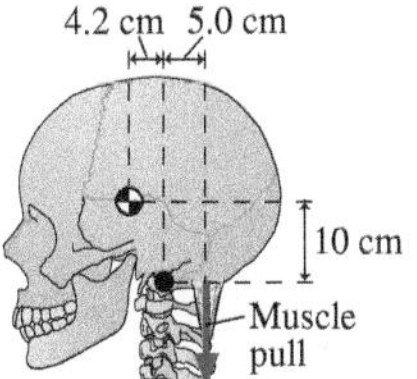

FIGURE P6.65

66. || The nuchal ligament is a very large, very elastic ligament that helps support the weight of a horse's head. It attaches to the back of the skull and to a point low on the cervical spine. The strain in the ligament varies from 0.20 to 0.60 as the horse moves its head up and down.
BIO
   a. The ligament has a length of 75 cm when the strain is 0.40. What is the unstretched length of the ligament?
   b. What is the range of tension forces if the cross-section area of the ligament is 6.0 cm^2?
   c. A horse's head is 12% of its body mass, and 500 kg is a typical mass for a horse. What is the largest ligament force as a fraction of the weight of the horse's head?

67. ||| A woman weighing 580 N does a pushup from her knees, as shown in Figure P6.67. What are the normal forces of the floor on (a) each of her hands and (b) each of her knees?
BIO

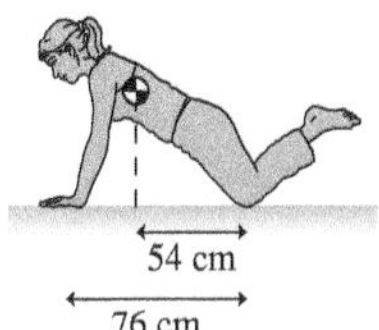

FIGURE P6.67

68. |||| When you bend over, a series of large muscles, the erector spinae, pull on your spine to hold you up. Figure P6.68 shows a simplified model of the spine as a rod of length L that pivots at its lower end. In this model, the center of gravity of the 320 N weight of the upper torso is at the center of the spine. The 160 N weight of the head and arms acts at the top of the spine. The erector spinae muscles are modeled as a single muscle that acts at an 12° angle to the spine. Suppose the person in Figure P6.68 bends over to an angle of 30° from the horizontal.
BIO
   a. What is the tension in the erector muscle?
   Hint: Align your x-axis with the axis of the spine.
   b. A force from the pelvic girdle acts on the base of the spine. What is the component of this force in the direction of the spine? (This large force is the cause of many back injuries).

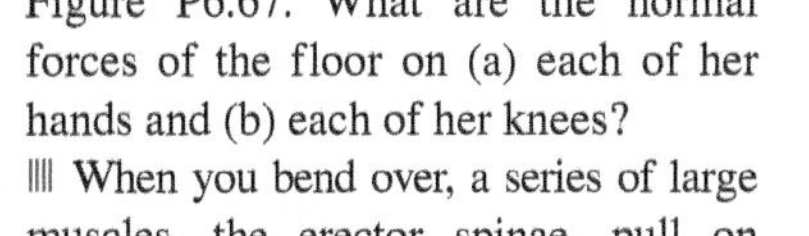

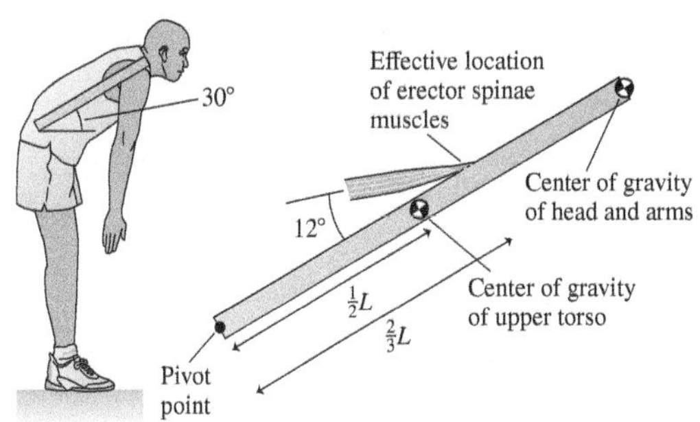

FIGURE P6.68

MCAT-Style Passage Problems

Elasticity of a Ligament

The nuchal ligament in a horse supports the weight of the horse's head. This ligament is much more elastic than a typical ligament, stretching from 15% to 45% longer than its resting length as a horse's head moves up and down while it runs. This stretch of the ligament stores energy, making locomotion more efficient. Measurements on a segment of ligament show a linear stress-versus-strain relationship until the strain approaches 0.80. Smoothed data for the stretch are shown in Figure P6.69.

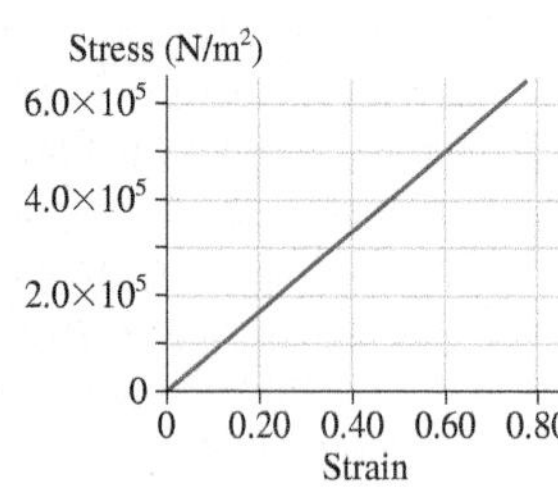

FIGURE P6.69

69. What is the approximate Young's modulus for the ligament?
A. 4×10^5 N/m^2 B. 8×10^5 N/m^2
C. 3×10^6 N/m^2 D. 6×10^6 N/m^2
70. The segment of ligament tested has a resting length of 40 mm. How long is the ligament at a strain of 0.60?
A. 46 mm B. 52 mm C. 58 mm D. 64 mm
71. Suppose the ligament has a circular cross section. For a certain ligament, an investigator measures the restoring force at a strain of 0.40. If the ligament is replaced with one that has twice the diameter, by what factor does the restoring force increase?
A. 1.4 B. 2 C. 4 D. 8

Sprinting Forces

Excellent sprinters have more than just strong muscles working for them; they also tend to have biomechanics that gives them a slight edge. Figure P6.72a shows a simplified view of the human ankle joint. The foot is tipped forward, with the ground force on the toes and the moment arms noted. The Achilles tendon attaches to the calf muscle, which isn't shown. You might expect that sprinters would have a higher mechanical advantage than nonsprinters because excellence in sprinting entails producing a large force when accelerating down the track, but the opposite is true. In one study with size-matched runners, the elite sprinters had an average L_1 of 130 mm and an average L_2 of 52 mm, giving them a reduced mechanical advantage compared to the nonsprinters, who had an average L_1 of 120 mm and L_2 of 58 mm. This apparent contradiction is resolved by noting that muscles don't produce a constant force when they contract; instead, the maximum possible force decreases as the contraction speed increases, as shown in Figure P6.72b. A lower mechanical advantage, which would normally be a disadvantage, provides a lower muscle contraction speed, and the increased muscle force can compensate to give an overall increase in the force on the ground.

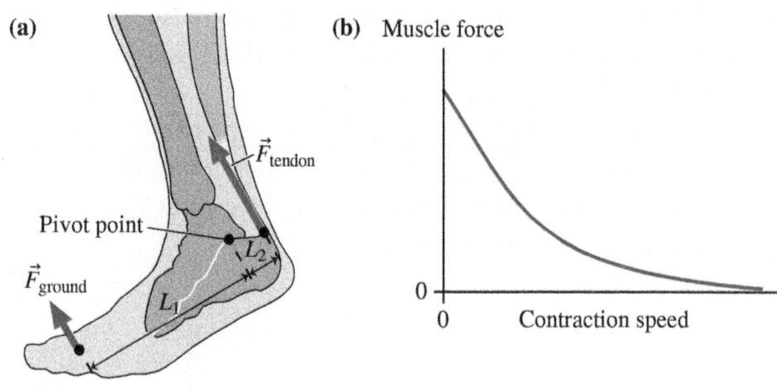

FIGURE P6.72

72. A 50 kg sprinter pushes off with the toes of one foot, exerting a force on the track that is 1.5 times his weight. For the data given in the passage, what approximate Achilles tendon force does this require? Consider the foot to be in equilibrium for the purpose of making an estimate, although technically it is not.
A. 500 N B. 750 N C. 1500 N D. 1900 N
73. Given the information in the passage, we expect that, as a sprinter gains speed,
A. The force provided by his calf muscles will increase.
B. The force provided by his calf muscles will stay the same.
C. The force provided by his calf muscles will decrease.
74. Suppose a 50 kg sprinter and a 50 kg nonsprinter are each doing an exercise in which they stand flat on one foot and then raise the heel of that foot off the ground. What can you say about the relative muscle force required for this exercise?
A. The muscle force is greater for the nonsprinter.
B. The muscle force is the same for the sprinter and the nonsprinter.
C. The muscle force is greater for the sprinter.
75. The range of motion when flexing the foot at the ankle is limited by the range of stretch of the calf muscles, which is approximately the same for both sprinters and nonsprinters. Which would you expect to have a greater range of motion of their toes?
A. Nonsprinters will have a greater range of motion.
B. Sprinters and nonsprinters will have the same range of motion.
C. Sprinters will have a greater range of motion.

7 Circular and Rotational Motion

The wind exerts a torque on these windmills, causing them to rotate. In doing so, they convert the wind's energy to electric energy.

LOOKING AHEAD

Uniform Circular Motion

These fairgoers are enjoying a ride in which they move in *uniform circular motion,* which is circular motion at a constant speed.

You'll find that the speed of an object in circular motion is related to the *period* and *frequency* of the motion.

Centripetal Acceleration

The extreme acceleration in this high-speed centrifuge is used to separate the components of blood: red blood cells, platelets, and plasma.

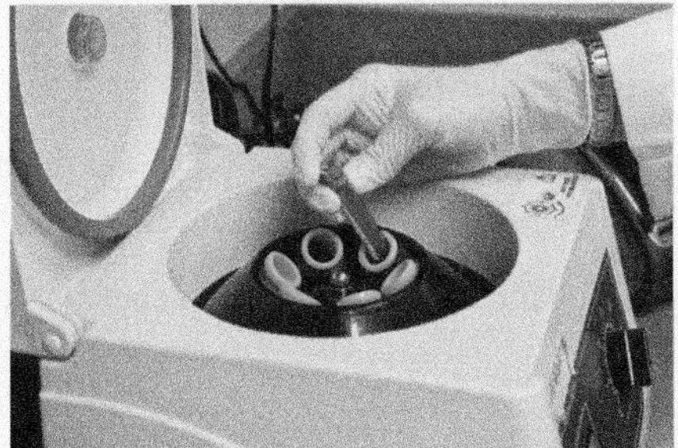

You'll learn that circular motion requires an acceleration directed toward the center of the circle. We'll call this a *centripetal acceleration.*

Rotational Motion

The rotation of a potter's wheel is driven by the *torque* of an electric motor and slowed by the frictional torque of the potter's hands.

You'll learn a rotational version of Newton's second law that relates an object's *angular acceleration* to the applied torque.

GOAL To apply Newton's laws to problems of circular and rotational motion.

PHYSICS AND LIFE

Much of Biomechanics Is About Rotational Motion

Your maximum walking speed is determined by your center of gravity moving in a circular arc as your body pivots over your forward foot. The maximum speed with which you can pivot or turn a corner is determined by the force of static friction. Your maximum throwing speed is determined by the torques on your shoulder as it rotates. An object in motion tries to go in a straight line—that's Newton's first law—so understanding circular and rotational motion is all about the forces that have to be applied to move an object in a circular trajectory. These are the forces that, for example, separate samples by density in an ultracentrifuge. In this chapter, we'll explore the forces and torques that are required for circular and rotational motion and then apply our knowledge to a host of applications.

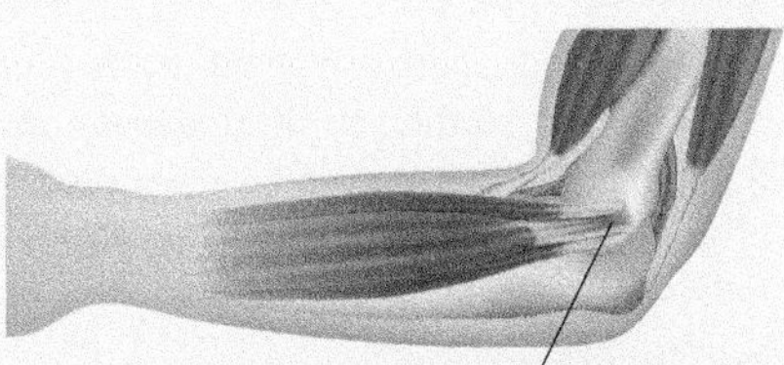

In many sports, your muscles need to exert large torques to hit or throw an object. The stress of this exertion can lead to injury, such as tennis elbow.

7.1 Uniform Circular Motion

The London Eye Ferris wheel

The 32 cars on the London Eye Ferris wheel move at a stately 0.5 m/s in a vertical circle that has a diameter of 130 m (430 ft). Motion in a circle at constant speed is called **uniform circular motion.** Other examples are a point on the tire of a bicycle that is coasting at steady speed, the elbow of a spinning figure skater, and simply your hand when your arm swings in an arc at steady speed.

Although an object needn't complete a full circle to be in uniform circular motion, we'll focus on objects that complete multiple full circles of motion, one after another. Since the motion is uniform, each time around the circle is just a repeat of the one before, so the motion is *periodic*. The time interval it takes an object to go around a circle one time, completing one revolution (abbreviated rev), is called the **period** of the motion. In this book, we will represent the period by the symbol T. Be alert to context so as not to confuse this with tension.

Rather than stating the time for one revolution, we can specify circular motion by its **frequency,** the number of revolutions per second, for which we use the symbol f. An object with a period of one-half second completes 2 revolutions each second. Similarly, an object can make 10 revolutions in 1 s if its period is one-tenth of a second. This shows that frequency is the inverse of the period:

$$f = \frac{1}{T} \tag{7.1}$$

Although frequency is often expressed as "revolutions per second," *revolutions* are not true units but merely the counting of events. Thus the SI unit of frequency is simply inverse seconds, or s^{-1}.

NOTE ▶ In practice, frequency is often given in revolutions per minute, or rpm. Values in rpm usually need to be converted to SI units of s^{-1} before calculations are done. ◀

FIGURE 7.1 Relating period, radius, and speed.

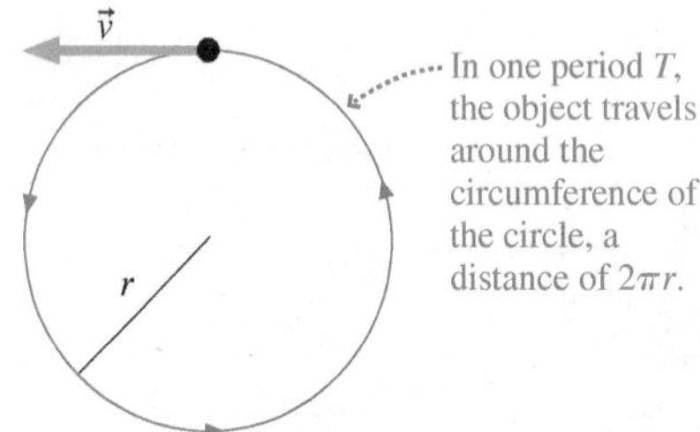
$\vec{v}$
In one period T, the object travels around the circumference of the circle, a distance of $2\pi r$.
r

FIGURE 7.1 shows an object moving at a constant speed in a circular path of radius r. We know the time for one revolution—one period T—and we know the distance traveled—one circumference. Speed is distance divided by time, so the speed of uniform circular motion is

$$v = \frac{1 \text{ circumference}}{1 \text{ period}} = \frac{2\pi r}{T} \tag{7.2}$$

Given Equation 7.1 relating frequency and period, we can also write the speed as

$$v = 2\pi f r \tag{7.3}$$

EXAMPLE 7.1 A spinning table saw blade

The circular blade of a table saw is 25 cm in diameter and spins at 3600 rpm. How much time is required for one revolution? What is the speed of a tooth at the edge of the blade?

PREPARE Each tooth of the saw blade is undergoing uniform circular motion, so the period, frequency, and speed are related by Equations 7.1–7.3. Before we get started, we need to do a couple of unit conversions. The diameter of the blade is 0.25 m, so its radius is 0.125 m. The frequency is given in rpm; we need to convert this to s^{-1}:

$$f = 3600 \frac{\text{rev}}{\text{min}} \times \frac{1 \text{ min}}{60 \text{ s}} = 60 \frac{\text{rev}}{\text{s}} = 60 \text{ s}^{-1}$$

SOLVE The time for one revolution is the period, given by Equation 7.1:

$$T = \frac{1}{f} = \frac{1}{60 \text{ s}^{-1}} = 0.017 \text{ s}$$

The speed of the tooth is given by Equation 7.3:

$$v = 2\pi f r = 2\pi (60 \text{ s}^{-1})(0.125 \text{ m}) = 47 \text{ m/s}$$

ASSESS The speed of the tooth is extremely high: 47 m/s ≈ 100 mph! Still, this seems plausible for a high-speed saw.

Centripetal Acceleration

An object in uniform circular motion has a constant speed, but it does *not* have a constant velocity. Velocity is a vector that indicates both the speed and the direction of motion. Although its speed may be constant, an object moving in uniform circular motion has a constantly changing direction. Consequently, and perhaps surprisingly, *an object in uniform circular motion is accelerating*!

◀◀ SECTION 2.4 introduced acceleration as the rate of change of velocity, and we noted that velocity can either change speed or change direction. We've focused thus far on linear motion, where acceleration is due to a change of speed, but we now want to look at the second possibility. Acceleration is $\vec{a} = \Delta\vec{v}/\Delta t$, so the acceleration vector points in the direction of $\Delta\vec{v}$. Tactics Box 2.1 showed how to find $\Delta\vec{v}$. for two parallel velocity vectors of a motion diagram by drawing them with their tails together, then drawing the connecting vector. This procedure works equally well for circular motion, and FIGURE 7.2 follows the same steps to find the direction of the acceleration at one point on the circle.

At this point, on the right edge of the circle, the acceleration vector points to the left—toward the center of the circle. That result is not specific to this point; no matter which point of the motion diagram we choose, **the acceleration vector $\vec{a}$ points to the center of the circle.** This acceleration of circular motion is called **centripetal acceleration,** from a Greek root meaning "center seeking." Centripetal acceleration is not a new kind of acceleration; all we're doing is *naming* an acceleration that corresponds to a particular type of motion.

The motion diagram tells us the direction of $\vec{a}$, but it doesn't give us a value for the magnitude a. During one revolution, the vector $\vec{v}$ turns through a complete circle. Imagine keeping the tail of $\vec{v}$ fixed while the tip traces out a circle of radius v. The circumference of this circle, $2\pi v$, is the magnitude of $\Delta\vec{v}$ over one complete revolution. The time required for one revolution in one period is $\Delta t = T = 2\pi r/v$. Combining these expressions, we find that the magnitude of the centripetal acceleration is

$$a = \frac{\Delta v}{\Delta t} = \frac{2\pi v}{2\pi r/v} = \frac{v^2}{r} \tag{7.4}$$

In vector notation, we can write

$$\vec{a} = \left(\frac{v^2}{r}, \text{ toward center of circle}\right) \tag{7.5}$$

Centripetal acceleration for uniform circular motion

FIGURE 7.2 Finding the acceleration of an object in uniform motion.

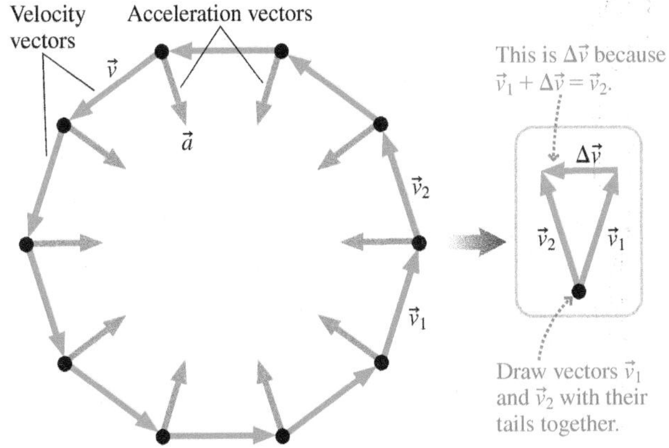
Velocity vectors
Acceleration vectors
$\vec{v}$
$\vec{a}$
$\vec{v}_2$
$\vec{v}_1$
This is $\Delta\vec{v}$ because $\vec{v}_1 + \Delta\vec{v} = \vec{v}_2$.
$\Delta\vec{v}$
$\vec{v}_2$
$\vec{v}_1$
Draw vectors $\vec{v}_1$ and $\vec{v}_2$ with their tails together.
The acceleration vector points in the same direction as $\Delta\vec{v}$, toward the center of the circle.
$\vec{a}$

NOTE ▸ Centripetal acceleration has a constant magnitude, but it's not a constant acceleration because the direction is constantly changing. **Thus the constant-acceleration kinematic equations of Chapter 3 cannot be used for circular motion; they do not apply.** ◂

EXAMPLE 7.2 **Acceleration and period of a carnival ride**

In the Quasar carnival ride, passengers travel in a horizontal 5.0-m-radius circle. For safe operation, the maximum sustained acceleration that riders may experience is 20 m/s^2, approximately twice the free-fall acceleration. How fast are the riders moving when the Quasar operates at maximum acceleration, and what is the period of the ride?

PREPARE The riders are in uniform circular motion. The pictorial representation of FIGURE 7.3 shows a top view of the motion of the ride.

FIGURE 7.3 Pictorial representation for the Quasar carnival ride.

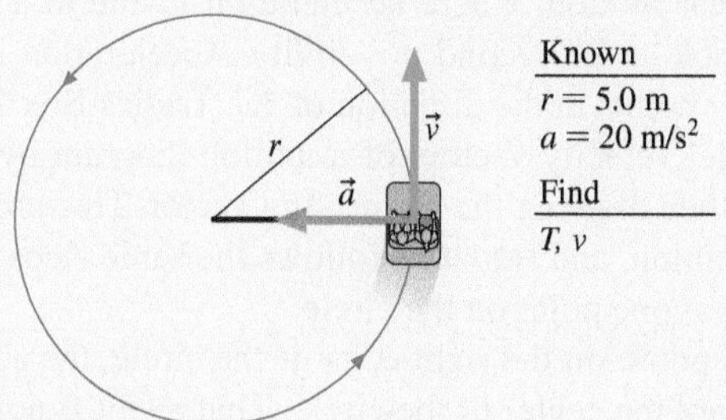

SOLVE We can use Equation 7.5 for the centripetal acceleration to find the speed:

$$v = \sqrt{ar} = \sqrt{(20\text{ m/s}^2)(5.0\text{ m})} = 10\text{ m/s}$$

Then, with the speed and radius given, we can use Equation 7.2 to find the period:

$$T = \frac{2\pi r}{v} = \frac{2\pi(5.0\text{ m})}{10\text{ m/s}} = 3.1\text{ s}$$

ASSESS One rotation in just over 3 seconds seems reasonable for a pretty zippy carnival ride. (The period for this particular ride is actually 3.7 s, so it runs a bit slower than the maximum safe speed.)

STOP TO THINK 7.1 Rank in order, from largest to smallest, the period of the motion of particles A to D.

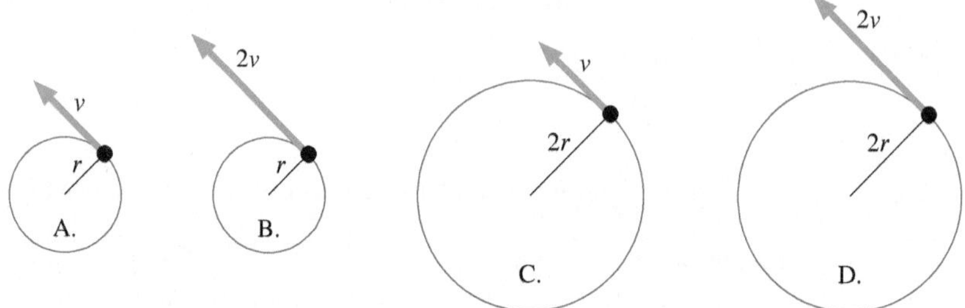

7.2 Dynamics of Uniform Circular Motion

The teeth on a spinning saw blade and the riders on a circular carnival ride are accelerating. Consequently, according to Newton's second law, they must have a *net force* acting on them.

We've already determined the acceleration of a particle in uniform circular motion—the centripetal acceleration of Equation 7.5. Newton's second law tells us what the net force must be to cause this acceleration:

$$\vec{F}_{\text{net}} = m\vec{a} = \left(\frac{mv^2}{r}, \text{ toward center of circle}\right) \tag{7.6}$$

Net force producing the centripetal acceleration of uniform circular motion

In other words, **a particle of mass m moving at constant speed v around a circle of radius r must always have a net force of magnitude mv^2/r pointing toward the center of the circle,** as in FIGURE 7.4. It is this net force that causes the centripetal acceleration of circular motion. Without such a net force, the particle would move off in a straight line tangent to the circle.

The force described by Equation 7.6 is not a *new* kind of force. The net force will be due to one or more of our familiar forces, such as tension, friction, or the normal force. Equation 7.6 simply tells us how the net force needs to act—how strongly and in which direction—to cause the particle to move with speed v in a circle of radius r.

In each example of circular motion that we will consider in this chapter, a physical force or a combination of forces directed toward the center produces the necessary acceleration.

FIGURE 7.4 Net force for circular motion.

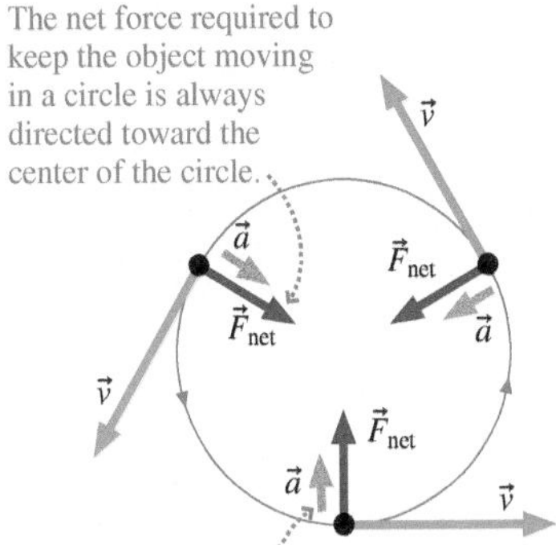
The net force required to keep the object moving in a circle is always directed toward the center of the circle.
$\vec{v}$
$\vec{a}$
$\vec{F}_{net}$
The net force causes a centripetal acceleration.

CONCEPTUAL EXAMPLE 7.3 Tarzan swings

Tarzan swings through the jungle on a vine. At the very bottom of a swing, is the tension in the vine less than, greater than, or equal to Tarzan's weight?

REASON A free-body diagram of Tarzan would show an upward tension force from the vine and a downward weight force. These are vertical forces, but Tarzan's motion at the bottom of the swing is horizontal. It seems like the tension and the weight force should balance—but they don't. Although Tarzan's motion is horizontal *at this instant,* he is not moving along a straight line. He is undergoing circular motion with a constantly changing direction, and thus he is accelerating. A swinging motion is not uniform circular motion; Tarzan speeds up, then slows down. However, the bottom of the arc is a point where, at that instant, only the direction of the velocity vector is changing, not its magnitude. Thus the same motion-diagram analysis that we did in Figure 7.2 finds that Tarzan has a centripetal acceleration directed upward toward the center of the circle. This is shown in FIGURE 7.5. An upward acceleration requires a *net* force in the upward direction, so the vine tension must be greater than Tarzan's weight.

FIGURE 7.5 Tarzan swinging on a vine.

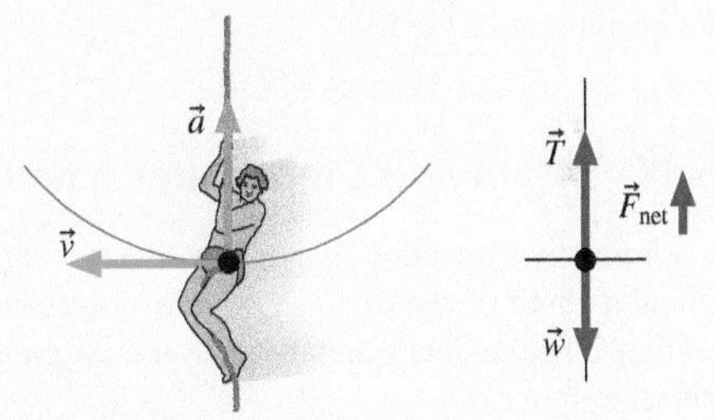
$\vec{a}$
$\vec{v}$
$\vec{T}$
$\vec{F}_{net}$
$\vec{w}$

ASSESS You may have noticed that while riding on a swing you feel a bit "heavier" at the bottom of the arc. This is because the upward force of the seat of the swing is greater than your weight, creating the net upward force and the centripetal acceleration of your circular motion.

Solving Circular Dynamics Problems

The problem-solving strategy for circular dynamics problems is similar to the strategy you have been using to solve dynamics problems for motion along a line.

PROBLEM-SOLVING STRATEGY 7.1 Circular dynamics problems

Circular motion involves an acceleration and thus a net force.

PREPARE Model the object as a particle in uniform circular motion. Draw a pictorial representation in which you sketch the motion, define the axes, define symbols, and identify what the problem is trying to find. There are two common situations:

- If the motion is in a horizontal plane, like a tabletop, draw the free-body diagram with the circle viewed edge-on. Choose the x-axis to point from the object toward the center of the circle; the y-axis is then perpendicular to the plane of the circle.
- If the motion is in a vertical plane, like a Ferris wheel, draw the free-body diagram with the circle viewed face-on. Choose the x-axis to point from the object toward the center of the circle; the y-axis is then tangent to the circle.

Continued

Hurling the heavy hammer Scottish games involve feats of strength. Here, a man is throwing a 30 lb hammer for distance. He starts by swinging the hammer rapidly in a circle. You can see from how he is leaning that he is providing a large force directed toward the center to produce the necessary centripetal acceleration. When the hammer is heading in the right direction, the man lets go. With no force directed toward the center, the hammer will stop going in a circle and fly in the chosen direction across the field.

SOLVE Newton's second law for uniform circular motion, $\vec{F}_{\text{net}} = (mv^2/r$, toward center of circle), is a vector equation. In the coordinate system just described, with the x-axis pointing toward the center of the circle,

$$\sum F_x = \frac{mv^2}{r} \quad \text{and} \quad \sum F_y = 0$$

That is, the net force toward the center of the circle has magnitude mv^2/r while the net force perpendicular to the circle is zero. The components of the forces are found directly from the free-body diagram. Depending on the problem, either:

- Use the net force to determine the speed v, then use circular kinematics to find frequencies or other details of the motion.
- Use circular kinematics to determine the speed v, then solve for unknown forces.

ASSESS Make sure your net force points toward the center of the circle. Check that your result has the correct units, is reasonable, and answers the question.

EXAMPLE 7.4 Finding the maximum speed for a car to turn a corner

What is the maximum speed with which a 1500 kg car can make a turn around a curve of radius 10 m on a level (unbanked) road without sliding? This radius is about what you might expect at a major intersection in a city.

PREPARE The car doesn't complete a full circle, but it's in uniform circular motion for a quarter of a circle while turning. Thus we'll model the car as a particle in uniform circular motion. FIGURE 7.6 is a pictorial representation. The issue we must address is *how* a car turns a corner. If the road were frictionless, like a very icy road, turning the steering wheel would have no effect; the car would slide straight ahead.

The force that prevents sliding is *static* friction. We've seen that static friction is the cause of propulsion, but in that case static friction parallel or opposite to the direction of motion causes an object to speed up or slow down. This car has a steady speed but a changing direction, and that requires a force not parallel to the motion but *perpendicular* to the motion. Thus, as the figure shows, static friction with the road pushes *sideways* on the tire, toward the center of the circle. But that's exactly the force needed for circular motion. Maximum turning speed will be reached when the static friction force reaches its maximum possible value, $f_{\text{s max}} = \mu_s n$. If a car enters the curve at a higher speed, static friction cannot provide the necessary centripetal acceleration and the car will slide.

Because the motion is in a horizontal plane, we've drawn the free-body diagram from behind the car, looking edge-on at the circle. The x-axis points toward the center of the circle, and the y-axis is perpendicular to the plane of motion.

SOLVE The only force in the x-direction, toward the center of the circle, is static friction. Newton's second law along the x-axis is

$$\sum F_x = f_s = \frac{mv^2}{r}$$

There's no motion along the y-axis, so Newton's second law in the y-direction is

$$\sum F_y = n - w = ma_y = 0$$

Thus $n = w = mg$.

The car will slide when the static friction force reaches its maximum value. Recall from Chapter 5 that this maximum force is given by

$$f_{\text{s max}} = \mu_s n = \mu_s mg$$

The maximum speed occurs when the static friction force reaches its maximum value, or when

$$f_{\text{s max}} = \frac{m(v_{\text{max}})^2}{r}$$

FIGURE 7.6 Pictorial representation of a car turning a corner.

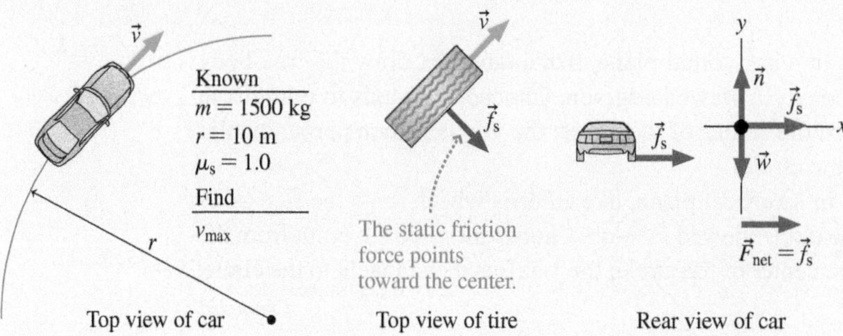

Using the known value of $f_{s\,max}$, we find

$$\frac{m(v_{max})^2}{r} = f_{s\,max} = \mu_s mg$$

Rearranging, we get

$$(v_{max})^2 = \mu_s gr$$

For rubber tires on pavement, we find from Table 5.2 that $\mu_s = 1.0$. We then have

$$v_{max} = \sqrt{\mu_s gr} = \sqrt{(1.0)(9.8\ \text{m/s}^2)(10\ \text{m})} = 9.9\ \text{m/s}$$

ASSESS 9.9 m/s ≈ 20 mph, which seems like a reasonable upper limit for the speed at which a car can turn a city corner without sliding. There are two other things to note about the solution:

- The car's mass canceled out. The maximum speed *does not* depend on the mass of the vehicle, though this may seem surprising.
- The final expression for v_{max} *does* depend on the coefficient of friction and the radius of the turn. Both of these factors make sense. You know, from experience, that the speed at which you can take a turn decreases if μ_s is less (the road is wet or icy) or if r is smaller (the turn is tighter).

Because v_{max} depends on μ_s and because μ_s depends on road conditions, the maximum safe speed through turns can vary dramatically. A car that easily handles a curve in dry weather can suddenly slide out of control when the pavement is wet. Icy conditions are even worse. If you lower the value of the coefficient of friction in Example 7.4 from 1.0 (dry pavement) to 0.1 (icy pavement), the maximum speed for the turn goes down to 3.1 m/s—about 6 mph!

Highway curves—and also racetracks—are *banked* by a raised outside edge that allows cars to turn at a higher speed. Banking allows the normal force of the road to provide some, or even all, of the centripetal acceleration, as you'll see in the next example.

A banked turn on a racetrack.

EXAMPLE 7.5 Finding a car's speed on a banked turn

A highway curve of radius 70 m is banked at a 15° angle. At what speed can a car take this curve without assistance from friction?

PREPARE After our discussion of the role of friction in turning corners, it seems surprising to suggest that a car can make the turn without friction. But it can if the curve is banked. The purpose of banking becomes clear if you look at the free-body diagram in FIGURE 7.7. The normal force is perpendicular to the road, so tilting the road causes $\vec{n}$ to have a component toward the center of the circle. **The horizontal component of $\vec{n}$ causes the centripetal acceleration needed to make the turn.** Although the car is tilted, it is still moving in a horizontal circle. Thus, following Problem-Solving Strategy 7.1, we've again chosen the x-axis to be horizontal and pointing toward the center of the circle.

FIGURE 7.7 Pictorial representation for the car on a banked turn.

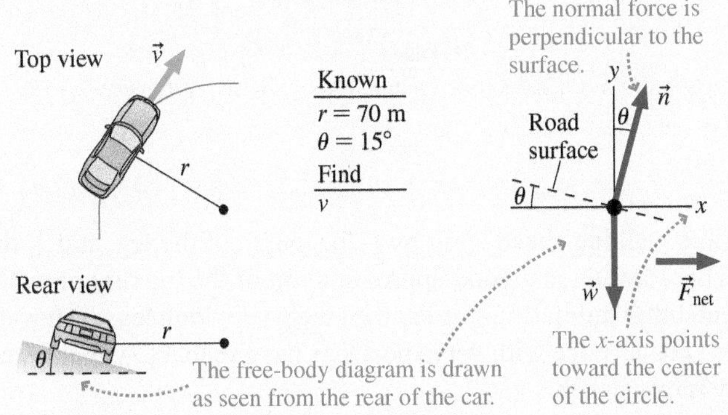

SOLVE Without friction, $n_x = n\sin\theta$ is the only component of force toward the center of the circle. It is this inward component of the normal force on the car that causes it to turn the corner. Newton's second law is

$$\sum F_x = n\sin\theta = \frac{mv^2}{r}$$

$$\sum F_y = n\cos\theta - w = 0$$

where θ is the angle at which the road is banked, and we've assumed that the car is traveling at the correct speed v. From the y-equation,

$$n = \frac{w}{\cos\theta} = \frac{mg}{\cos\theta}$$

Substituting this into the x-equation and solving for v give

$$\left(\frac{mg}{\cos\theta}\right)\sin\theta = mg\tan\theta = \frac{mv^2}{r}$$

$$v = \sqrt{rg\tan\theta} = 14\ \text{m/s}$$

ASSESS This is ≈30 mph, a reasonable speed. Only at this exact speed can the turn be negotiated without reliance on friction forces.

The friction force provides the necessary centripetal acceleration not only for cars turning corners, but also for bicycles. The cyclists in FIGURE 7.8 (next page) are going through a tight turn; you can tell by how they lean. The road exerts both a vertical normal force and a horizontal friction force on their tires. Notice that the cyclists lean at an angle where the vector sum of the road forces is pointing toward their center of gravity. That means there's no torque about the center of gravity, which keeps them balanced as they round the curve.

FIGURE 7.8 Road forces on a cyclist leaning into a turn.

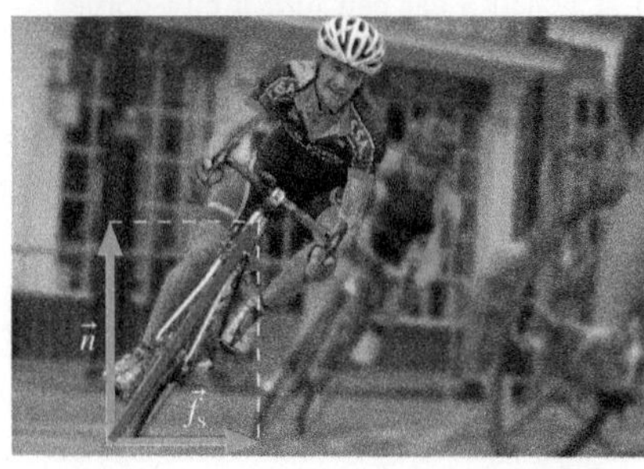

STOP TO THINK 7.2 A car is rolling over the top of a hill at constant speed v. At this instant,

A. $n > w$.
B. $n < w$.
C. $n = w$.
D. We can't tell about n without knowing v.

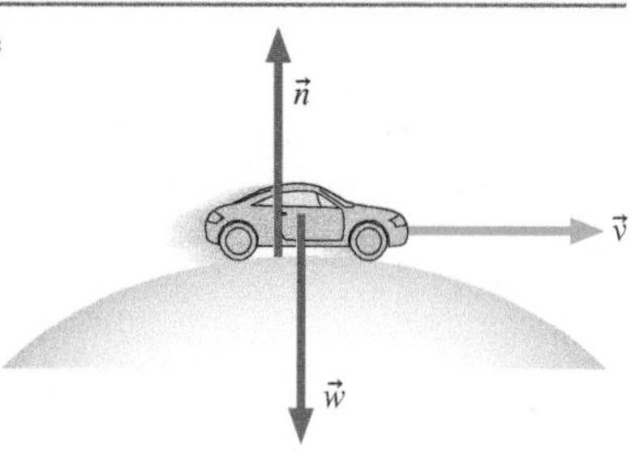

Maximum Walking Speed

Humans and other two-legged animals have two basic gaits: walking and running. At slow speeds, you walk. When you need to go faster, you run. Why don't you just walk faster? There is an upper limit to the speed of walking, and this limit is set by the physics of circular motion.

Think about the motion of your body as you take a walking stride. You put one foot forward, then push off with your rear foot. Your body pivots over your front foot, and you bring your rear foot forward to take the next stride. As you can see in FIGURE 7.9a, the path that your body takes during this stride is the arc of a circle. **In a walking gait, your body is in circular motion as you pivot on your forward foot.**

A net force toward the center of the circle is required for this circular motion. FIGURE 7.9b shows the forces acting on the woman's body during the midpoint of the stride: her weight, directed down, and the normal force of the ground, directed up. Newton's second law for the x-axis is

$$\sum F_x = w - n = \frac{mv^2}{r}$$

Notice the unusual x-axis. Problem-Solving Strategy 7.1 says the x-axis should point toward the center of the circle, so here the x-axis points down. That defines the positive direction of the axis, so in this case $w_x = +w$, in contrast to the more typical problem in which weight lies on the negative y-axis with $w_y = -w$. The normal force is in the $-x$-direction; thus $n_x = -n$.

The net force must point toward the center of the circle, or, in this case, down. In order for the net force to point down, the normal force must be *less* than her weight. Your body tries to "lift off" as it pivots over your foot, decreasing the normal force exerted on you by the ground. The normal force becomes smaller as you walk faster, but n cannot be less than zero. Thus the maximum possible walking speed v_{max} occurs when $n = 0$. Setting $n = 0$ in Newton's second law gives

$$w = mg = \frac{m(v_{max})^2}{r}$$

Thus

$$v_{max} = \sqrt{gr} \tag{7.7}$$

The maximum possible walking speed is set by r, the length of the leg, and g, the free-fall acceleration. This formula is a good approximation of the maximum walking speed for humans and other animals. Giraffes, with their very long legs, can walk at high speeds. Animals such as mice with very short legs have such a low maximum walking speed that they rarely use this gait.

For humans, the length of the leg is approximately 0.7 m, giving $v_{max} \approx$ 2.6 m/s $\approx$ 6 mph. You *can* walk this fast, though it becomes energetically unfavorable to walk at speeds above 4 mph. Most people make a transition to a running gait at about this speed.

FIGURE 7.9 Analysis of a walking stride.

(a) Walking stride

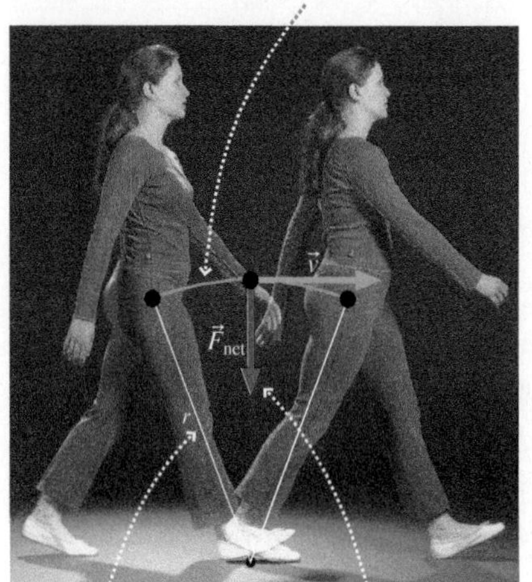
During each stride, her hip undergoes circular motion.
The radius of the circular motion is the length of the leg from the foot to the hip.
The circular motion requires a force directed toward the center of the circle.

(b) Forces in the stride

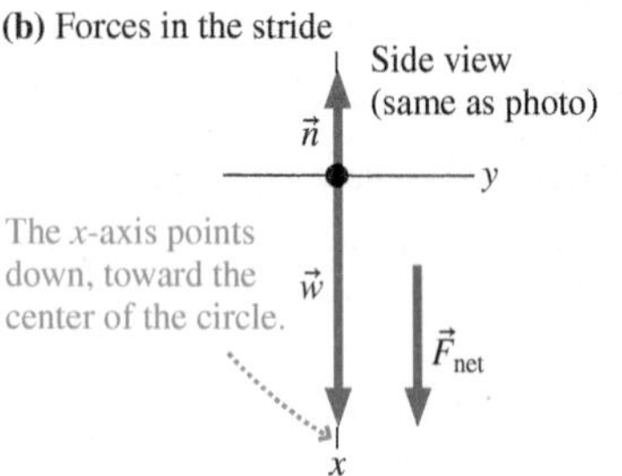
Side view (same as photo)
The x-axis points down, toward the center of the circle.
y
x

Centrifugal Force?

FIGURE 7.10 shows a carnival ride that spins the riders around inside a large cylinder. The people are "stuck" to the inside wall of the cylinder! As you probably know from

FIGURE 7.10 Inside the Gravitron, a rotating circular room.

experience, the riders *feel* that they are being pushed outward, into the wall. But our analysis has found that an object in circular motion must have an *inward* force to create the centripetal acceleration. How can we explain this apparent difference?

If you are a passenger in a car that turns a corner quickly, you may feel "thrown" by some mysterious force against the door. But is there really such a force? FIGURE 7.11 shows a bird's-eye view of you riding in a car as it makes a left turn. Just before the turn, you were moving in a straight line so, according to Newton's first law, as the car begins to turn, you want to continue moving along that same line. However, the door moves into your path and so runs into you! You feel the force of the door because it is this force, pushing *inward* toward the center of the curve, that is causing you to turn the corner. But you were not "thrown" into the door; the door ran into you.

A "force" that *seems* to push an object to the outside of a circle is called a *centrifugal force.* Despite having a name, there really is no such force. What you feel is your body trying to move ahead in a straight line (which would take you away from the center of the circle) as outside forces act to turn you in a circle. The only real forces, those that appear on free-body diagrams, are the ones pushing inward toward the center. **A centrifugal force will never appear on a free-body diagram and never be included in Newton's laws.**

With this in mind, let's revisit the rotating carnival ride. A person watching from above would see the riders in the cylinder moving in a circle with the walls providing the inward force that causes their centripetal acceleration. The riders *feel* as if they're being pushed outward because their natural tendency to move in a straight line is being resisted by the wall of the cylinder, which keeps getting in the way. But feelings aren't forces. The only actual force is the contact force of the cylinder wall pushing *inward.*

FIGURE 7.11 Bird's-eye view of a passenger in a car turning a corner.

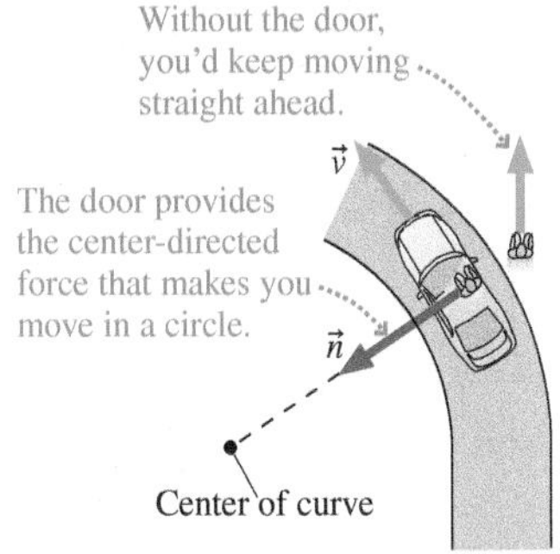
Without the door, you'd keep moving straight ahead.
$\vec{v}$
The door provides the center-directed force that makes you move in a circle.
$\vec{n}$
Center of curve

FIGURE 7.12 The operation of a centrifuge.

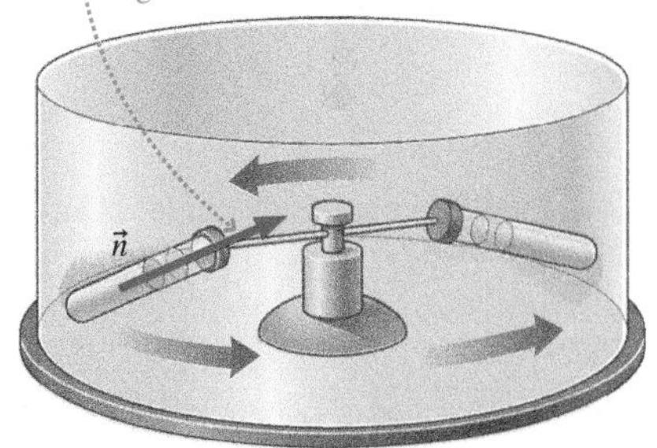
The high centripetal acceleration requires a large normal force, which leads to a large effective gravitational acceleration.
$\vec{n}$

The Centrifuge

The *centrifuge,* FIGURE 7.12, is an important instrument used in biology and biochemistry to separate liquids and small particles (such as macromolecules or organelles) that have different densities. In principle it's not necessary to use an instrument to do this; mixtures of oil and water spontaneously separate, as do cream and milk, with the less dense liquid floating on the more dense liquid. However, separation is very slow if the liquids have been thoroughly mixed.

Small particles suspended in liquid in a test tube also separate by density—given enough time—because more dense particles sink through the liquid faster than less dense particles of similar size. For microscopic particles, such as cell components, the settling time under even ideal conditions may be days or weeks. Under real-world conditions, in which diffusion and temperature variations cause some mixing, microscopic particles may never settle out. We need to drastically speed up the separation process to make it useful, and that's what a centrifuge does by creating an extremely strong "artificial gravity."

The settling speed of particles in a liquid is determined both by viscous drag (which you studied in Chapter 5) and by buoyancy in the liquid (a topic we'll take up in Chapter 9). As you'll learn there, a particle's terminal speed—how fast it settles—is proportional to $(\rho_{\text{particle}} - \rho_{\text{liquid}})g$, where ρ_{particle} and ρ_{liquid} are the densities of the particle and the liquid. If the liquid's density increases with depth, the particle will fall only to the point where its density matches that of the liquid. And, not surprisingly, the terminal speed is proportional to g; settling would happen more quickly on a planet with stronger gravity.

If we place the test tube in a centrifuge and spin it at speed v, the particle is in uniform circular motion with centripetal acceleration v^2/r, so from the particle's perspective g has been replaced by an *effective gravitational acceleration* $g_{\text{eff}} = v^2/r$. The same settling and sedimentation processes occur, only much faster if $g_{\text{eff}} \gg g$. If g_{eff} is large enough, as in an ultracentrifuge spinning at 75,000 rpm or higher, then settling rates are faster than diffusion and convection, allowing even the smallest cellular components to be separated.

Human centrifuge If you spin your arm rapidly in a vertical circle, the motion will produce an effect like that in a centrifuge. The motion will assist outbound blood flow in your arteries and retard inbound blood flow in your veins. There will be a buildup of fluid in your hand that you will be able to see (and feel!) quite easily. Although this exercise is harmless, it gives you an idea of how large accelerations can interfere with blood flow.

EXAMPLE 7.6 **Analyzing the ultracentrifuge**

A 15-cm-diameter ultracentrifuge produces an extraordinarily large $g_{eff} = 750{,}000g$, where g is the free-fall acceleration due to gravity. What is its frequency in rpm? What is the apparent weight of a sample with a mass of 1.5 g?

PREPARE The acceleration in SI units is

$$g_{eff} = 750{,}000(9.80 \text{ m/s}^2) = 7.35 \times 10^6 \text{ m/s}^2$$

The radius is half the diameter, or $r = 7.5$ cm $= 0.075$ m.

SOLVE The effective gravitational acceleration is $g_{eff} = v^2/r$, so the speed of the sample is

$$v = \sqrt{g_{eff}r} = \sqrt{(7.35 \times 10^6 \text{ m/s}^2)(0.075 \text{ m})} = 742 \text{ m/s}$$

The relationship between speed and frequency, Equation 7.3, is $v = 2\pi fr$. We can use this to find the frequency:

$$f = \frac{v}{2\pi r} = \frac{742 \text{ m/s}}{2\pi(0.075 \text{ m})} = 1570 \text{ rev/s}$$

Converting to rpm, we find

$$f = 1570 \frac{\text{rev}}{\text{s}} \times \frac{60 \text{ s}}{1 \text{ min}} = 94{,}000 \text{ rpm}$$

The sample's apparent weight is determined by the effective gravitational acceleration:

$$w_{app} = mg_{eff} = (1.5 \times 10^{-3} \text{ kg})(7.35 \times 10^6 \text{ m/s}^2) = 11{,}000 \text{ N}$$

The 1.5 g sample has an apparent weight of about 2500 lb!

ASSESS Commercial ultracentrifuges spin at 100,000 rpm, or even higher, so our answer is plausible. The sample speed is so large, exceeding the speed of sound in air, that all the air has to be evacuated from an ultracentrifuge. From Newton's third law, the sample pulls back on the axle of the centrifuge with its effective weight, ≈2500 lb, yet the rotors are not especially strong. To deal with this, a centrifuge has to be carefully balanced by loading it with several tubes whose forces on the axle balance one another.

7.3 The Rotation of a Rigid Body

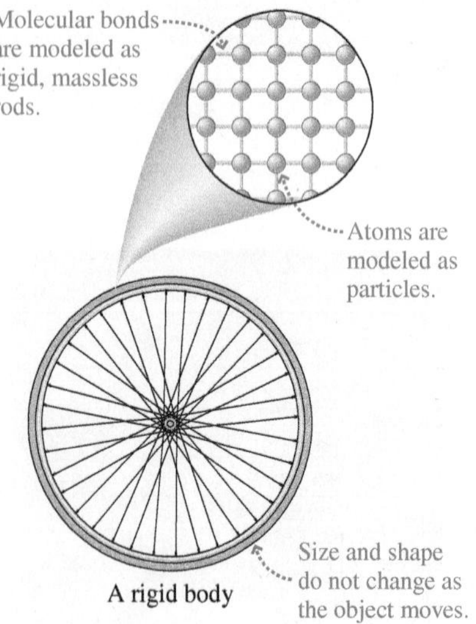
Molecular bonds are modeled as rigid, massless rods.
Atoms are modeled as particles.
Size and shape do not change as the object moves.
A rigid body

FIGURE 7.13 The rigid-body model of an extended object.

We are surrounded by objects that rotate—from bicycle wheels to bacterial flagella. Much of biomechanics deals with how muscular forces rotate your bones about a joint or your entire body about a pivot point. Rotational motion is an extension of circular motion, but we'll need to develop some new concepts to describe the rotation of solid objects.

Chapter 6 introduced the idea of a *rigid body,* an extended object whose size and shape do not change as it moves. For example, a bicycle wheel can be thought of as a rigid body. **FIGURE 7.13** shows a rigid body as a collection of atoms held together by the rigid "massless rods" of molecular bonds.

Real molecular bonds are, of course, not perfectly rigid. That's why an object seemingly as rigid as a bicycle wheel can flex and bend. Thus Figure 7.13 is really a simplified *model* of an extended object, the **rigid-body model.** The rigid-body model is a very good approximation for objects that are stiff, but it is not an appropriate model for entire organisms that flex and bend as they move. Even so, a flexible organism can sometimes be modeled as a group of connected rigid bodies.

FIGURE 7.14 illustrates the three basic types of motion of a rigid body: **translational motion, rotational motion,** and **combination motion.** We've already studied translational motion of a rigid body using the particle model. If a rigid body doesn't rotate,

FIGURE 7.14 Three basic types of motion of a rigid body.

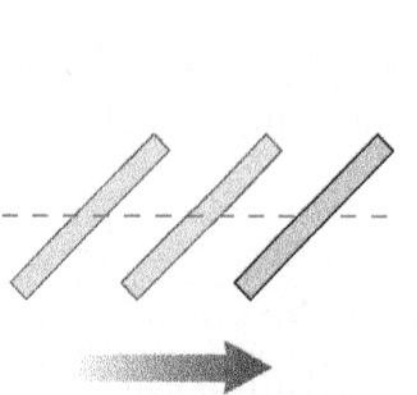
Translational motion:
The object as a whole moves along a trajectory but does not rotate.

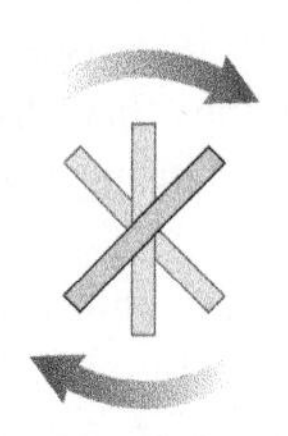
Rotational motion:
The object rotates about a fixed point. Every point on the object moves in a circle.

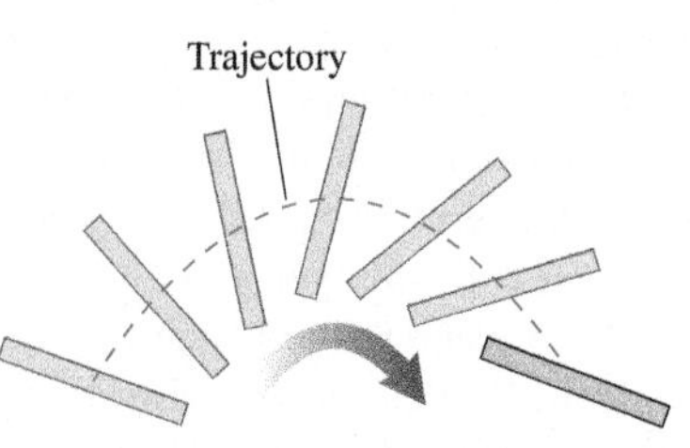
Trajectory
Combination motion:
An object rotates as it moves along a trajectory.

this model is often adequate for describing its motion. The rotational motion of a rigid body will be the main focus of this section and the next. We'll also discuss an important case of combination motion—that of a *rolling* object—later in this chapter.

Angular Position and Velocity

Our first task is to find a language to describe rotational motion. We used position, velocity, and acceleration to understand motion along a line. Are there equivalent descriptions of rotation?

FIGURE 7.15 shows a wheel rotating on an axle. Suppose we designate one point on the wheel, at radius r from the axle, as a *reference point*. The **angular position** of the wheel is the reference point's angle θ from the positive x-axis. We'll follow the mathematical convention that angle θ is positive when measured *counterclockwise* from the x-axis.

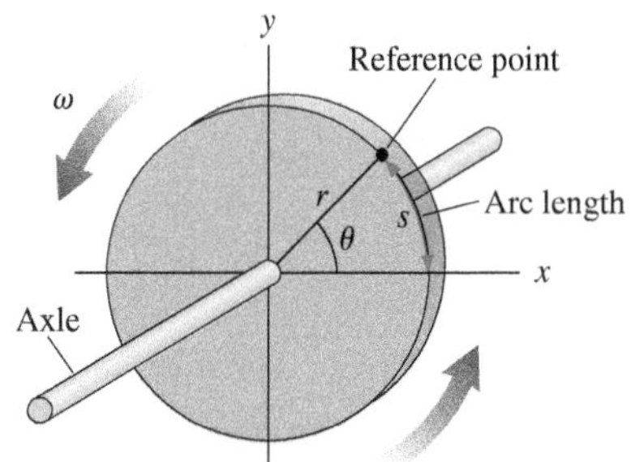

FIGURE 7.15 The angular position of a rotating object is the angle of a reference point on the object.

Although θ could be measured in degrees or revolutions, scientists and mathematicians usually measure θ in the angular unit of *radians*. Figure 7.15 shows the **arc length** s, the distance the reference point has moved along its circular path from the x-axis. The angular unit of **radians** is defined such that

$$\theta(\text{in radians}) = \frac{s}{r} \tag{7.8}$$

The radian is the SI unit of angle. An angle of 1 rad has an arc length s exactly equal to the radius r.

The arc length all the way around a circle is the circle's circumference $2\pi r$. Thus the angle of a full circle is

$$\theta_{\text{full circle}} = \frac{2\pi r}{r} = 2\pi \text{ rad}$$

This is the basis for the well-known conversion

$$1 \text{ rev} = 360° = 2\pi \text{ rad}$$

We can use this conversion to find that

$$1 \text{ rad} = 1 \text{ rad} \times \frac{360°}{2\pi \text{ rad}} = 57.3°$$

As a rough approximation, 1 rad $\approx$ 60°. We'll often specify angles in degrees, but keep in mind that the SI unit is the radian. For example, the common right angle of 90° is $\pi/2$ rad.

An important consequence of Equation 7.8 is that the arc length spanning angle θ is

$$s = r\theta \text{ if } \theta \text{ is in rad} \tag{7.9}$$

This is a result we'll use often, but it is valid *only* if θ is measured in radians, not degrees.

For linear motion, we defined an object's velocity to be $v_x = dx/dt$, the rate at which its position changes with time. By analogy, let's define the **angular velocity** of a rotating object to be

$$\omega = \frac{d\theta}{dt} \tag{7.10}$$

Angular velocity of a rotating object

The symbol ω is a lowercase Greek omega, *not,* in spite of the resemblance, the familiar English letter w. A rotating object's angular velocity is the rate at which its angular position changes with time. Angular velocity describes the rotation of the *entire object;* that is, every point on the object is moving in a circle with the same angular velocity. The larger the value of ω, the more rapidly the object is rotating.

The SI unit of angular velocity is rad/s, the rotational equivalent of m/s. But we can say more. A rotating object has a period $T = 1/f$, just like an object in circular motion. The object makes one rotation with $\Delta\theta = 2\pi$ rad during one period with $\Delta t = T$, so the angular velocity is

$$\omega = \frac{2\pi \text{ rad}}{T} = (2\pi \text{ rad})f \qquad (7.11)$$

Equation 7.11 connects a rotating object's angular velocity ω in rad/s to the circular frequency f in rev/s of points on the object.

We also found, in Equation 7.2, that the speed of a point in uniform circular motion is $v = 2\pi r/T$. Combining this with Equation 7.11, we find that a point on a rotating object that is distance r from the axis of rotation moves with speed

$$v = \omega r \qquad (7.12)$$

Speed of a point on a rotating object

EXAMPLE 7.7 **Rotation of a bicycle crank**

The *crank* on your bicycle is the arm that connects the pedal to the axle. The standard length for the crank is 17 cm. If you pedal at 75 rpm, a typical pedaling cadence, what are the angular velocity of the crank and the speed of your foot?

PREPARE We need to convert the circular frequency of your feet to rev/s and then use Equations 7.11 and 7.12.

SOLVE The frequency with which the crank rotates is

$$f = 75\,\frac{\text{rev}}{\text{min}} \times \frac{1 \text{ min}}{60 \text{ s}} = 1.25 \text{ rev/s}$$

Thus the crank's angular velocity—a constant angular velocity in this instance—is

$$\omega = (2\pi \text{ rad})f = (2\pi \text{ rad})(1.25 \text{ rev/s}) = 7.9 \text{ rad/s}$$

Remember that rev is not a true unit, simply a reminder of what we're counting, so it drops out when we no longer need it. Your foot is on the pedal at distance $r = 17$ cm $= 0.17$ m from the axis of rotation, so it moves with speed

$$v = \omega r = (7.9 \text{ rad/s})(0.17 \text{ m}) = 1.3 \text{ m/s}$$

ASSESS If you picture the circle your foot makes while pedaling, its circumference seems to be about a meter. You're pedaling at a little more than 1 rev/s, so a speed of a little more than 1 m/s seems reasonable.

NOTE ▶ The units of angles are often troublesome. Unlike the kilogram or the second, for which we have standards, the *radian,* similar to the *revolution,* is really just a name to remind us that we're dealing with an angle. Consequently, the radian unit sometimes appears or disappears without warning, which seems rather mysterious until you get used to it. The speed of your foot is not an angular quantity, so the rad in rad/s dropped out of the calculation. ◀

FIGURE 7.16 Positive and negative angular velocities.

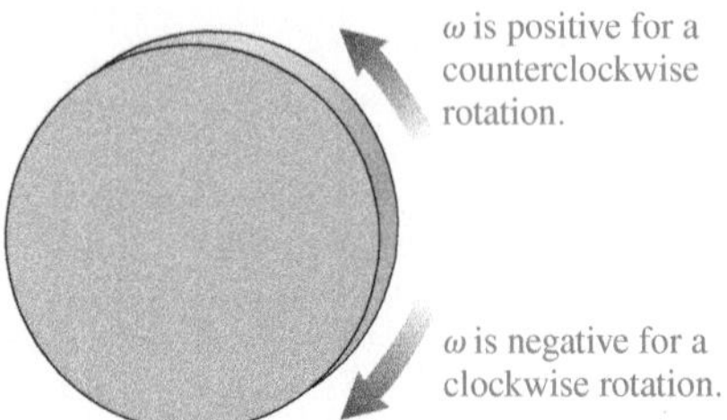
ω is positive for a counterclockwise rotation.
ω is negative for a clockwise rotation.

The velocity v_x of linear motion uses a sign, positive or negative, to indicate direction—motion to the right or motion to the left. Angular velocity also has a sign to tell us whether an object is rotating clockwise (cw) or counterclockwise (ccw). The convention that we'll follow, shown in FIGURE 7.16, is based on the fact that θ is measured counterclockwise from the x-axis. The absolute value of ω, if you don't care about the rotation direction, is the *angular speed.*

Angular Acceleration

A bicycle wheel rotates faster and faster, with increasing angular velocity, as you speed up. Just as we needed the concept of acceleration to understand the changing velocity of motion along a line, we need an *angular acceleration* to describe the changing angular velocity of rotation.

FIGURE 7.17 The signs of angular velocity and angular acceleration. The rotation is speeding up if ω and α have the same sign, slowing down if they have opposite signs.

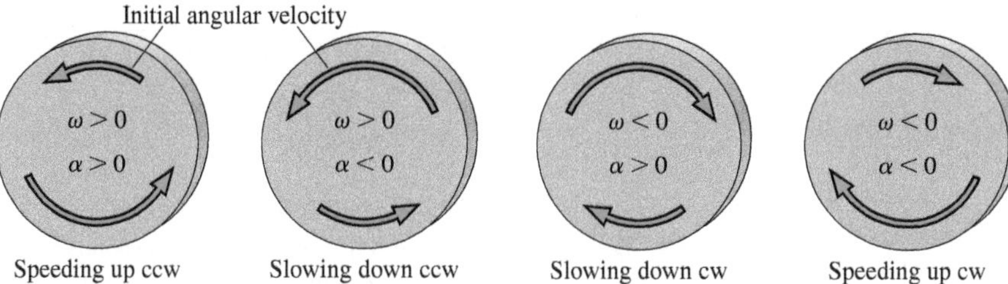

Chapter 3 defined an object's *linear* acceleration as $a_x = dv_x/dt$, the rate at which the velocity changes. By analogy, we now define the **angular acceleration** to be

$$\alpha = \frac{d\omega}{dt} \tag{7.13}$$

Angular acceleration of a rotating object

We use the symbol α (Greek alpha) for angular acceleration. Angular acceleration is the rate at which a rotating object's angular velocity changes. The units of angular acceleration are rad/s^2.

Be careful with the sign of α! You learned in Chapter 3 that positive and negative values of a_x don't mean simply speeding up and slowing down. Instead, an object is speeding up if v_x and a_x have the same sign, slowing down if they have opposite signs. **FIGURE 7.17** shows that the same idea applies to rotational motion.

NOTE ▶ Don't confuse angular acceleration with centripetal acceleration. Angular acceleration is how rapidly the angular velocity of a rotating object is changing. Centripetal acceleration, a vector that points to the center of the circle, is the changing-direction acceleration of a particle in circular motion. The individual particles of a rotating object undergo centripetal acceleration even if the object's angular velocity is constant. ◀

Because angular velocity and angular acceleration are defined exactly the same as linear velocity and acceleration, **the kinematic equations for rotational motion with constant angular acceleration are exactly the same as the kinematic equations for linear motion with constant acceleration.** This is shown in **TABLE 7.1**.

TABLE 7.1 Rotational and linear kinematics for constant acceleration

Rotational kinematics	Linear kinematics
$\omega_f = \omega_i + \alpha\,\Delta t$	$(v_x)_f = (v_x)_i + a_x\,\Delta t$
$\Delta\theta = \omega_i\,\Delta t + \frac{1}{2}\alpha(\Delta t)^2$	$\Delta x = (v_x)_i\,\Delta t + \frac{1}{2}a_x(\Delta t)^2$
${\omega_f}^2 = {\omega_i}^2 + 2\alpha\,\Delta\theta$	$(v_x)_f^2 = (v_x)_i^2 + 2a_x\,\Delta x$

EXAMPLE 7.8 Spinning up a Blu-ray disc

A Blu-ray disc spins up from rest to a final angular velocity of 5400 rpm in 2.00 s. What is the angular acceleration of the disc? At the end of 2.00 s, how many revolutions has the disc made?

PREPARE This problem is clearly analogous to the linear motion problems we saw in Chapter 3; something starts from rest and then moves with constant angular acceleration for a fixed interval of time. We will solve this problem by referring to Table 7.1 and using equations that correspond to the equations we'd use to solve a similar linear motion problem.

SOLVE The initial angular velocity is $\omega_i = 0$ rad/s. The final angular velocity is $\omega_f = 5400$ rpm. In rad/s this is

$$\omega_f = \frac{5400\text{ rev}}{\text{min}} \times \frac{1\text{ min}}{60\text{ s}} \times \frac{2\pi\text{ rad}}{1\text{ rev}} = 565\text{ rad/s}$$

Continued

Thus the angular acceleration is

$$\alpha = \frac{\Delta\omega}{\Delta t} = \frac{565 \text{ rad/s} - 0 \text{ rad/s}}{2.00 \text{ s}} = 282.5 \text{ rad/s}^2$$

We've kept an extra significant figure for later calculations, but we'll report the result as $\alpha = 283 \text{ rad/s}^2$.

Next, we use the equation for the angular displacement—analogous to the equation for linear displacement—during the period of constant angular acceleration to determine the angle through which the disc moves:

$$\Delta\theta = \omega_i \Delta t + \tfrac{1}{2}\alpha(\Delta t)^2$$
$$= (0 \text{ rad/s})(2.00 \text{ s}) + \tfrac{1}{2}(282.5 \text{ rad/s}^2)(2.00 \text{ s})^2$$
$$= 565 \text{ rad}$$

Each revolution corresponds to an angular displacement of 2π, so we have

$$\text{number of revolutions} = \frac{565 \text{ rad}}{2\pi \text{ rad/revolution}}$$
$$= 90 \text{ revolutions}$$

The disc completes 90 revolutions during the first 2.00 seconds.

ASSESS The disc spins up to 5400 rpm, which corresponds to 90 rev/s. It would undergo 180 revolutions if it spun at full speed for 2 s. But since it starts up from rest, it makes sense that it completes only half this number of revolutions.

STOP TO THINK 7.3 A ball at one end of a massless but rigid rod swings in a horizontal circle about a pivot at the other end at a steady rate of twice per second. Which of the following quantities are zero? Which are constant but not zero?

A. Speed
B. Frequency
C. Velocity
D. Angular velocity
E. Centripetal acceleration
F. Angular acceleration

7.4 Rotational Dynamics and Moment of Inertia

◀◀ SECTION 6.2, which is well worth a quick review, introduced torque as the rotational equivalent of force. An object in static equilibrium is *not* rotating, so a requirement of static equilibrium is that the net torque be zero. But what if the net torque on an object isn't zero? You know that an object responds to a net force by accelerating; that's Newton's second law. As you'll see, **an object responds to a net torque by rotating with an angular acceleration.**

To see where the connection between torque and angular acceleration comes from, let's start by looking at a single particle. FIGURE 7.18 shows a particle of mass m connected to a pivot point by a rigid but massless rod of length r. A constant-magnitude force perpendicular to the rod causes the particle to speed up with acceleration $a = F/m$ as it moves around the circle. (This *speeding up* acceleration tangential to the circle is in addition to any centripetal acceleration directed toward the center of the circle.) After a time interval Δt, the particle's speed will have increased by $\Delta v = a\Delta t = (F/m)\Delta t$.

FIGURE 7.18 A force of magnitude F exerts a torque on the particle and causes an angular acceleration.

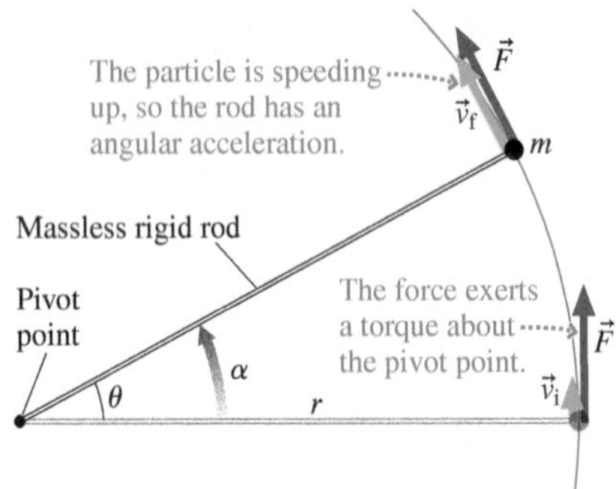
The particle is speeding up, so the rod has an angular acceleration.
$\vec{F}$
$\vec{v}_f$
m
Massless rigid rod
Pivot point
The force exerts a torque about the pivot point.
$\vec{F}$
α
θ
r
$\vec{v}_i$

The particle also has an angular velocity because the angle θ is increasing with time. We found, in Equation 7.12, that $v = \omega r$, so an increase of speed corresponds to an angular velocity increase of

$$\Delta\omega = \frac{\Delta v}{r} = \frac{F}{mr}\Delta t$$

Thus the particle, speeding up under the influence of force F, has angular acceleration

$$\alpha = \frac{d\omega}{dt} = \frac{\Delta\omega}{\Delta t} = \frac{F}{mr} \tag{7.14}$$

where we used the average angular acceleration $\Delta\omega/\Delta t$ because this is a constant angular acceleration.

The force in Figure 7.18 exerts a torque about the pivot point. Because the force is always perpendicular to the radial line, the torque is $\tau = rF_\perp = rF$. Substituting $F = \tau/r$ into Equation 7.14, we find that a torque on the particle causes it to have angular acceleration

$$\alpha = \frac{\tau}{mr^2} \tag{7.15}$$

Now all that remains is to extend this idea from a single particle to a rigid body.

Newton's Second Law for Rotation

FIGURE 7.19 shows a rigid body that undergoes rotation about a fixed and unmoving axis. According to the rigid-body model, we can think of the object as consisting of particles with masses $m_1, m_2, m_3, \ldots$ at fixed distances $r_1, r_2, r_3, \ldots$ from the axis. Suppose forces $\vec{F}_1, \vec{F}_2, \vec{F}_3, \ldots$ act on these particles. These forces exert torques around the rotation axis, so the object will undergo an angular acceleration α. Because all the particles that make up the object rotate together, each particle has this *same* angular acceleration α. Rearranging Equation 7.15 slightly, we can write the torques on the particles as

$$\tau_1 = m_1 r_1^2 \alpha \qquad \tau_2 = m_2 r_2^2 \alpha \qquad \tau_3 = m_3 r_3^2 \alpha$$

and so on for every particle in the object. If we add up all these torques, the *net* torque on the object is

$$\begin{aligned}\tau_{net} &= \tau_1 + \tau_2 + \tau_3 + \cdots = m_1 r_1^2 \alpha + m_2 r_2^2 \alpha + m_3 r_3^2 \alpha + \cdots \\ &= \alpha(m_1 r_1^2 + m_2 r_2^2 + m_3 r_3^2 + \cdots) = \alpha \sum m_i r_i^2\end{aligned} \tag{7.16}$$

In factoring α out of the sum, we're making explicit use of the fact that every particle in a rotating rigid body has the *same* angular acceleration α.

The quantity $\sum mr^2$ in Equation 7.16, which is the proportionality constant between angular acceleration and net torque, is called the object's **moment of inertia** I:

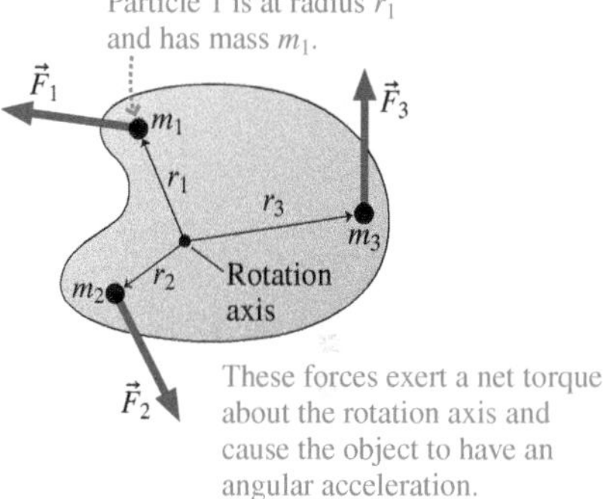
Particle 1 is at radius r_1 and has mass m_1.
$\vec{F}_1$
m_1
r_1
r_3
$\vec{F}_3$
m_3
m_2
r_2
Rotation axis
$\vec{F}_2$
These forces exert a net torque about the rotation axis and cause the object to have an angular acceleration.

FIGURE 7.19 The forces on a rigid body exert a torque about the rotation axis.

$$I = m_1 r_1^2 + m_2 r_2^2 + m_3 r_3^2 + \cdots = \sum m_i r_i^2 \tag{7.17}$$

Moment of inertia for rotation around a specified axis

The unit of moment of inertia is mass times distance squared, or kg · m^2. An object's moment of inertia, like torque, *depends on the axis of rotation.* Once the axis is specified, allowing the values of $r_1, r_2, r_3, \ldots$ to be determined, the moment of inertia *about that axis* can be calculated from Equation 7.17.

NOTE ▸ The word "moment" in "moment of inertia" and "moment arm" has nothing to do with time. It stems from the Latin *momentum,* meaning "motion." ◂

Substituting the moment of inertia into Equation 7.16 puts the final piece of the puzzle into place, giving us the fundamental equation for rigid-body dynamics:

Newton's second law for rotation An object that experiences a net torque τ_{net} about the axis of rotation undergoes an angular acceleration

$$\alpha = \frac{\tau_{net}}{I} \tag{7.18}$$

where I is the moment of inertia of the object *about the rotation axis.*

Spinning a sphere The moment of inertia depends on the mass and also on where the mass is located. The granite ball in the photo has a mass of 8200 kg. It's also physically large, with much of the mass far from the center. For both of these reasons, the moment of inertia is extremely large. The sphere can freely spin on a thin layer of pressurized water. Even though the girl exerts a large torque, the extremely large moment of inertia means that the angular acceleration is very small.

TABLE 7.2 **Rotational and linear dynamics**

Rotational dynamics		Linear dynamics	
Torque (N · m)	τ_{net}	force (N)	$\vec{F}_{net}$
Moment of inertia (kg · m²)	I	mass (kg)	m
Angular acceleration (rad/s²)	α	acceleration (m/s²)	$\vec{a}$
Newton's second law	$\alpha = \tau_{net}/I$	Newton's second law	$\vec{a} = \vec{F}_{net}/m$

In practice we often write $\tau_{net} = I\alpha$, but Equation 7.18 better conveys the idea that **a net torque is the cause of angular acceleration.** In the absence of a net torque ($\tau_{net} = 0$), the object has zero angular acceleration α, so it either does not rotate ($\omega = 0$) or rotates with *constant* angular velocity ($\omega =$ constant).

TABLE 7.2 summarizes the analogies between linear and rotational dynamics.

Interpreting the Moment of Inertia

Before rushing to calculate moments of inertia, let's get a better understanding of its meaning. First, notice that **moment of inertia is the rotational equivalent of mass.** It plays the same role in Equation 7.18 as does mass m in the now-familiar $\vec{a} = \vec{F}_{net}/m$. Recall that objects with larger mass have a larger *inertia,* meaning that they're harder to accelerate. Similarly, an object with a larger moment of inertia is harder to get rotating: It takes a larger torque to spin up an object that has a larger moment of inertia than an object with a smaller moment of inertia. The fact that "moment of inertia" retains the word "inertia" reminds us of this.

FIGURE 7.20 Moment of inertia depends on both the mass and how the mass is distributed.

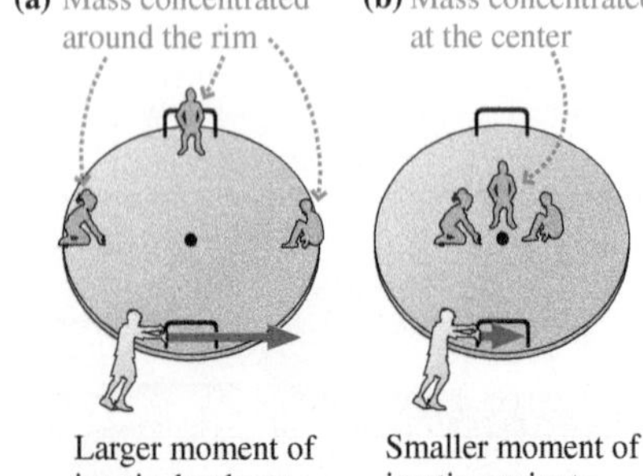

But why does the moment of inertia depend on the distances r from the rotation axis? Think about trying to start a merry-go-round from rest, as shown in FIGURE 7.20. By pushing on the rim of the merry-go-round, you exert a torque on it, and its angular velocity begins to increase. As you probably know from experience, it's much easier to push if your friends sit near the axle than if they sit at the rim. This is because having the mass near the center, with smaller values of r, lowers the moment of inertia.

Thus an object's moment of inertia depends not only on the object's mass but also on *how the mass is distributed* around the rotation axis. This is of great importance to bicycle racers. Every time a cyclist accelerates, she has to "spin up" the wheels and tires. The larger the moment of inertia, the more effort it takes and the smaller her acceleration. For this reason, racers use the lightest possible tires, and they put those tires on wheels that have been designed to keep the mass as close as possible to the center without sacrificing the necessary strength and rigidity.

NOTE ▶ The moment of inertia, unlike mass, is not simply a property of the object. The moment of inertia also depends on the location of the axis of rotation because that determines the values of $r_1, r_2, r_3, \ldots$. An object's moment of inertia for rotation around one axis differs from its moment of inertia for rotation around a different axis. ◀

The Moments of Inertia of Common Shapes

Newton's second law for rotational motion is easy to write, but we can't make use of it without knowing an object's moment of inertia. Unlike mass, we can't measure moment of inertia by putting an object on a scale. And although we can guess that the center of gravity of a symmetrical object is at the physical center of the object, we can *not* guess the moment of inertia of even a simple object. A short list of common moments of inertia is given in TABLE 7.3. We use a capital M for the total mass of an extended object.

We can make some general observations about the moments of inertia in Table 7.3. For example, the moment of inertia of a solid cylinder is less than that of a cylindrical hoop because the mass is concentrated nearer the center.

TABLE 7.3 **Moments of inertia of objects with uniform density and total mass *M***

Object and axis	Picture	I	Object and axis	Picture	I
Thin rod (of any cross section), about center	L	$\frac{1}{12}ML^2$	Cylinder or disk, about center	R	$\frac{1}{2}MR^2$
Thin rod (of any cross section), about end	L	$\frac{1}{3}ML^2$	Cylindrical hoop, about center	R	MR^2
Plane or slab, about center	a, b	$\frac{1}{12}Ma^2$	Solid sphere, about diameter	R	$\frac{2}{5}MR^2$
Plane or slab, about edge	a, b	$\frac{1}{3}Ma^2$	Spherical shell, about diameter	R	$\frac{2}{3}MR^2$

In the same way we can see why a slab rotated about its center has a lower moment of inertia than the same slab rotated about its edge: In the latter case, some of the mass is twice as far from the axis as the farthest mass in the former case. Those particles contribute *four times* as much to the moment of inertia, leading to an overall larger moment of inertia for the slab rotated about its edge.

EXAMPLE 7.9 The point of the pole

The photo shows a high-wire walker striding confidently along a steel cable. If he tips slightly to the side, a gravitational torque will start to rotate his body even farther—which is clearly not good. To help him balance, he increases his moment of inertia by carrying a long, heavy pole. Now the gravitational torque will cause a smaller angular acceleration, giving him more time to recover his balance.

Suppose the man is 1.8 m tall with a mass of 88 kg. The pole is 9.1 m long, with a mass of 20 kg. What is the moment of inertia of his body? Of the pole?

PREPARE Let's model both the man's body and the pole as rods. The difference is the location of the pivot point. The man's body pivots about the end of the rod, his feet, so the moment of inertia is $\frac{1}{3}ML^2$, where M is his mass and L is his height. The pole pivots about the center, so the moment of inertia is $\frac{1}{12}ML^2$, where M is the mass of the pole and L its length.

SOLVE The moment of inertia of the man's body is

$$I_{\text{man}} = \tfrac{1}{3}ML^2 = \tfrac{1}{3}(88\text{ kg})(1.8\text{ m})^2 = 95\text{ kg}\cdot\text{m}^2$$

The moment of inertia of the pole is

$$I_{\text{pole}} = \tfrac{1}{12}ML^2 = \tfrac{1}{12}(20\text{ kg})(9.1\text{ m})^2 = 140\text{ kg}\cdot\text{m}^2$$

ASSESS Although the pole's mass is less than one-quarter of the man's mass, its moment of inertia is larger. The formula for moment of inertia includes the square of a length, so the pole's greater length more than makes up for its smaller mass. Holding the pole significantly increases the man's moment of inertia; the moment of inertia of the man holding the pole is 2.5 times that of the man alone. This results in a significant decrease in angular acceleration, providing much greater stability. The pole is flexible, and this provides an additional benefit. The downward sag of the pole lowers the position of the overall center of gravity, thus decreasing gravitational torque; this improves stability as well.

STOP TO THINK 7.4 Four very lightweight disks of equal radii each have three identical heavy marbles glued to them as shown. Rank in order, from largest to smallest, the moments of inertia of the disks about the indicated axis.

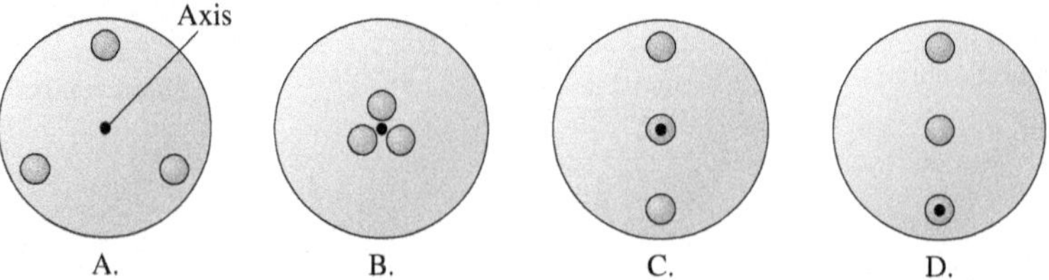

Using Newton's Second Law for Rotation

In this section we'll look at two examples of rotational dynamics for rigid bodies that rotate about a *fixed axis*. The restriction to a fixed axis avoids complications that arise for an object undergoing a combination of rotational and translational motion.

EXAMPLE 7.10 Starting an airplane engine

The engine in a small airplane is specified to have a torque of 120 N · m. This engine drives a 2.0-m-long, 40 kg propeller. On start-up, how long does it take the propeller to reach 200 rpm?

PREPARE The propeller can be modeled as a rigid rod that rotates about its center. The engine exerts a torque on the propeller, as shown in FIGURE 7.21. The specified angular velocity is

$$\omega_f = 200\,\frac{\text{rev}}{\text{min}} \times \frac{1\text{ min}}{60\text{ s}} \times \frac{2\pi\text{ rad}}{1\text{ rev}} = 20.9\text{ rad/s}$$

FIGURE 7.21 A rotating airplane propeller.

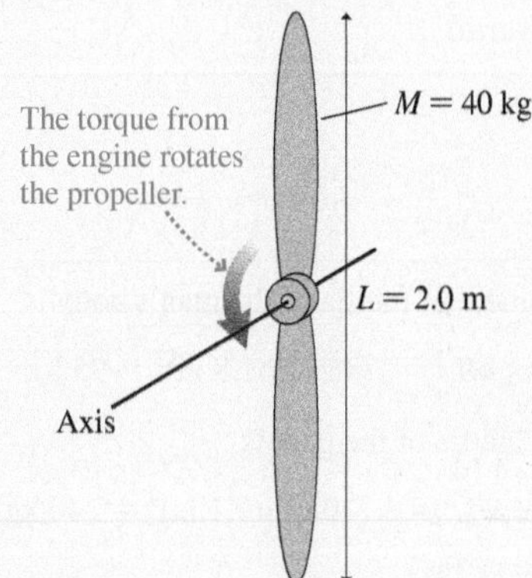

SOLVE The moment of inertia of a rod rotating about its center is found from Table 7.3:

$$I = \tfrac{1}{12}ML^2 = \tfrac{1}{12}(40\text{ kg})(2.0\text{ m})^2 = 13.33\text{ kg}\cdot\text{m}^2$$

The 120 N · m torque of the engine causes an angular acceleration

$$\alpha = \frac{\tau}{I} = \frac{120\text{ N}\cdot\text{m}}{13.33\text{ kg}\cdot\text{m}^2} = 9.0\text{ rad/s}^2$$

This is a constant acceleration because the torque is constant. Thus we can use constant-acceleration kinematics to find that the time needed to reach $\omega_f = 20.9$ rad/s is

$$\Delta t = \frac{\Delta\omega}{\alpha} = \frac{\omega_f - \omega_i}{\alpha} = \frac{20.9\text{ rad/s} - 0\text{ rad/s}}{9.0\text{ rad/s}^2} = 2.3\text{ s}$$

ASSESS It takes quite a few seconds for an airplane propeller to reach full speed, so a little over 2 s to reach a modest 200 rpm seems about right. We did assume a constant angular acceleration, and that's a reasonable approximation for the first few seconds while the propeller is still turning slowly. Eventually, air resistance and friction will cause opposing torques and the angular acceleration will decrease. At full speed, the opposing torque due to air resistance and friction cancels the torque of the engine. Then $\tau_{net} = 0$ and the propeller turns at *constant* angular velocity with no angular acceleration.

EXAMPLE 7.11 Biomechanics of pitching

A professional baseball pitcher's motion is one of the fastest and most stressful in sports. High-speed video allows the biomechanics of a throw to be studied in great detail. During the main acceleration phase, torque at the shoulder joint rotates the arm through an angle of approximately 60° in 0.060 s. The actual motion is complex, involving twisting the arm and moving the elbow, but we can understand some of the essential biomechanics by modeling the arm as a rigid rod of mass m and length L pivoted at the shoulder. The mass of the arm is not equally distributed, being concentrated in the upper arm, so we'll model the shape of the arm as a cone.

You can look up the moment of inertia of a cone. For a cone whose length is much greater than its radius, and for rotation about the base of the cone, $I = \frac{1}{10}ML^2$. What torque is needed if the arm's mass and length have typical values of 5.0 kg and 70 cm? With what speed is the wrist moving at the end of the throw?

PREPARE FIGURE 7.22 shows a pictorial representation of the throw. We'll ignore the mass of the ball because it's much less than the mass of the arm. We're assuming that this phase of the motion starts from rest. We'll also assume that the angular acceleration is constant during the overall angular displacement of $\Delta\theta = 60° = 1.1$ rad.

SOLVE We can use the second rotational kinematics equation in Table 7.1 to find the arm's angular acceleration:

$$\alpha = \frac{2\,\Delta\theta}{(\Delta t)^2} = \frac{2(1.1\text{ rad})}{(0.060\text{ s})^2} = 610\text{ rad/s}^2$$

We need to evaluate the moment of inertia before we can calculate the torque:

$$I = \tfrac{1}{10}ML^2 = \tfrac{1}{10}(5.0\text{ kg})(0.70\text{ m})^2 = 0.25\text{ kg}\cdot\text{m}^2$$

Thus, from Newton's second law for rotational motion, the torque that causes this angular acceleration is

$$\tau = I\alpha = (0.25\text{ kg}\cdot\text{m}^2)(610\text{ rad/s}^2) = 150\text{ N}\cdot\text{m}$$

FIGURE 7.22 A simple model of throwing a baseball.

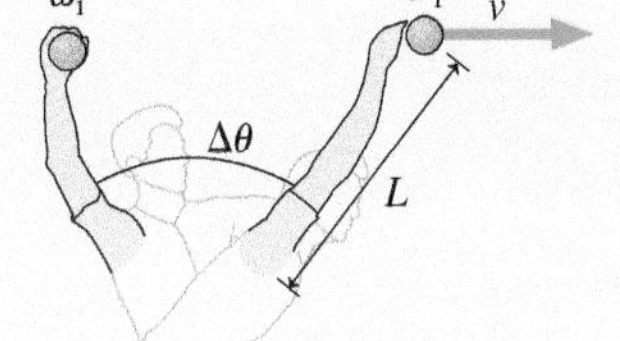

Known
$\Delta\theta = 60°$ $\quad \Delta t = 0.060$ s
$\omega_i = 0$ rad/s
$M = 5.0$ kg $\quad L = 70$ cm

Find
τ, v

Using the angular acceleration in the first rotational kinematics equation in Table 7.1, we find that the final angular velocity is

$$\omega_f = \omega_i + \alpha\,\Delta t = (610\text{ rad/s}^2)(0.060\text{ s}) = 37\text{ rad/s}$$

Thus the speed of the wrist, which is pretty much at the end of the arm, is

$$v = \omega_f r = (37\text{ rad/s})(0.70\text{ m}) = 26\text{ m/s}$$

ASSESS 26 m/s ≈ 55 mph. That seems slow, since a major-league pitcher can throw at nearly 100 mph. Something else must be going on. In fact, the full speed of an actual throw is greater than the wrist speed because an additional acceleration is imparted by a separate motion of the wrist at the end of the arm motion.

How much force is required to generate the 150 N · m torque responsible for the wrist speed? This torque is generated by several shoulder muscles, all of which attach to the upper arm bone (the *humerus*) very near the joint, providing moment arms of only a few cm. If one muscle, for example, exerts 40 N · m of the torque through a moment arm of only 4 cm, the force required from that muscle is ≈1000 N, which is nearly 150% of the pitcher's weight. These enormous muscular forces sometimes cause severe shoulder injuries; to prevent such injuries, strict limits are normally placed on the number of throws a pitcher makes during a game.

A full biomechanical analysis would model the upper arm, the lower arm, and the hand as three separate rigid bodies connected by three joints—the shoulder, the elbow, and the wrist—with additional torque generated by twisting of the entire body. That's a very complex calculation, as you can imagine, requiring computer simulation. Even so, each part of the problem involves torques, moments of inertia, and angular accelerations, much as we've done in this simple model of a throw.

Calculating the Moment of Inertia

Moments of inertia for common shapes are tabulated and can be found online. Occasionally, for an unusual shape, it's necessary to calculate the moment of inertia.

Equation 7.17 defines the moment of inertia as a sum over all the particles in the system. As we did for calculating the center of gravity, we can replace the individual particles with cells 1, 2, 3, … of mass Δm. Then the moment of inertia summation can be converted to an integration:

$$I = \sum_i r_i^2\,\Delta m \xrightarrow[\Delta m \to 0]{} \int r^2\,dm \qquad (7.19)$$

where r is the distance from the rotation axis. If we let the rotation axis be the z-axis, then we can write the moment of inertia as

$$I = \int (x^2 + y^2)\,dm \qquad \text{(rotation about the } z\text{-axis)} \qquad (7.20)$$

NOTE ▶ You *must* replace dm by an equivalent expression involving a coordinate differential such as dx or dy before you can carry out the integration. ◀

You can use any coordinate system to calculate the coordinates x_{cm} and y_{cm} of the center of mass. But the moment of inertia is defined for rotation about a particular axis, and r is measured from that axis. Thus the coordinate system used for moment-of-inertia calculations *must* have its origin at the pivot point. We'll do two examples to illustrate the idea.

EXAMPLE 7.12 Moment of inertia of a rod about a pivot at one end

Find the moment of inertia of a thin, uniform rod of length L and mass M that rotates about a pivot at one end.

PREPARE An object's moment of inertia depends on the axis of rotation. In this case, the rotation axis is at the end of the rod. FIGURE 7.23 defines an x-axis with the origin at the pivot point.

FIGURE 7.23 Setting up the integral to find the moment of inertia of a rod.

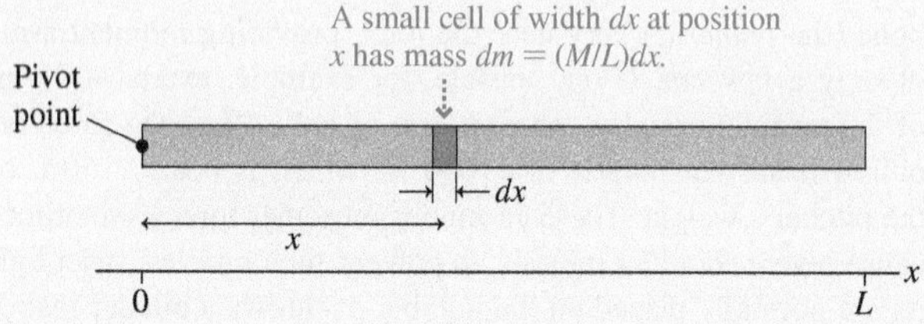
A small cell of width dx at position x has mass dm = (M/L)dx.
Pivot point
dx
x
0
L
x

SOLVE Because the rod is thin, we can assume that $y \approx 0$ for all points on the rod. Thus

$$I = \int x^2 \, dm$$

The small amount of mass dm in the small length dx is $dm = (M/L)\,dx$, as we found in Example 6.7 in Chapter 6. The rod extends from $x = 0$ to $x = L$, so the moment of inertia about one end is

$$I_{end} = \frac{M}{L}\int_0^L x^2\,dx = \frac{M}{L}\left[\frac{x^3}{3}\right]_0^L = \tfrac{1}{3}ML^2$$

ASSESS The moment of inertia involves a product of the total mass M with the *square* of a length, in this case L. All moments of inertia have a similar form, although the fraction in front will vary. This is the result shown earlier in Table 7.3.

EXAMPLE 7.13 Moment of inertia of a circular disk about an axis through the center

Find the moment of inertia of a circular disk of radius R and mass M that rotates on an axis passing through its center.

PREPARE FIGURE 7.24 shows the disk and defines distance r from the axis.

SOLVE This is a situation of great practical importance. To solve this problem, we need to use a two-dimensional integration scheme that you learned in calculus. Rather than dividing the disk into little boxes, let's divide it into narrow *rings* of mass dm. Figure 7.24 shows one such ring, of radius r and width dr. Let dA represent the area of this ring. The mass dm in this ring is the same fraction of the total mass M as dA is of the total area A. That is,

$$\frac{dm}{M} = \frac{dA}{A}$$

FIGURE 7.24 Setting up the integral to find the moment of inertia of a disk.

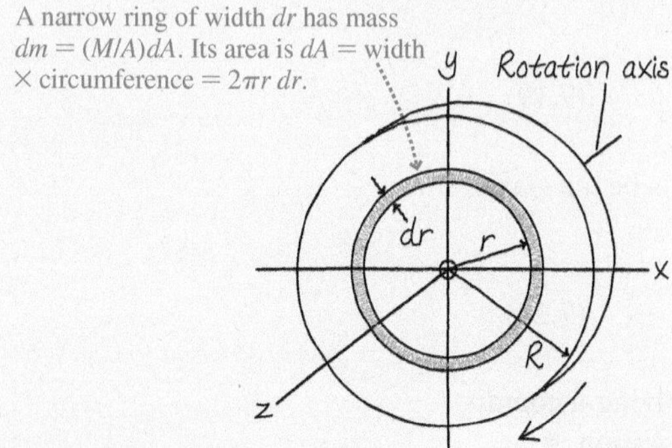
A narrow ring of width dr has mass dm = (M/A)dA. Its area is dA = width × circumference = 2πr dr.
y
Rotation axis
dr
r
R
x
z

Thus the mass in the small area dA is

$$dm = \frac{M}{A}\,dA$$

This is the reasoning we used to find the center of mass of the rod in Example 6.7, only now we're using it in two dimensions.

The total area of the disk is $A = \pi R^2$, but what is dA? If we imagine unrolling the little ring, it would form a long, thin rectangle of length $2\pi r$ and height dr. Thus the *area* of this little ring is $dA = 2\pi r\,dr$. With this information we can write

$$dm = \frac{M}{\pi R^2}(2\pi r\,dr) = \frac{2M}{R^2}r\,dr$$

Now we have an expression for dm in terms of a coordinate differential dr, so we can proceed to carry out the integration for I. Using Equation 7.19, we find

$$I_{disk} = \int r^2\,dm = \int r^2\left(\frac{2M}{R^2}r\,dr\right) = \frac{2M}{R^2}\int_0^R r^3\,dr$$

where in the last step we have used the fact that the disk extends from $r = 0$ to $r = R$. Performing the integration gives

$$I_{disk} = \frac{2M}{R^2}\left[\frac{r^4}{4}\right]_0^R = \tfrac{1}{2}MR^2$$

ASSESS Once again, the moment of inertia involves a product of the total mass M with the *square* of a length, in this case R.

7.5 Rolling Motion

Rolling is a *combination motion* in which an object rotates about an axis that is moving along a straight-line trajectory. For example, FIGURE 7.25 is a time-exposure photo of a rolling wheel with one lightbulb on the axis and a second lightbulb at the edge. The axis light moves straight ahead, but the edge light follows a curve called a *cycloid.*

To understand rolling motion, consider FIGURE 7.26, which shows a round object—a wheel or a sphere—that rolls forward, *without slipping,* exactly one revolution. The point initially at the bottom follows the blue curve to the top and then back to the bottom. The overall position of the object is measured by the position x_{cg} of the object's center of gravity. Because the object doesn't slip, in one revolution the center of gravity moves forward exactly one circumference, so that $\Delta x_{cg} = 2\pi R$. The time for the object to turn one revolution is its period T, so we can compute the speed of the object's center of gravity as

$$v_{cg} = \frac{\Delta x_{cg}}{T} = \frac{2\pi R}{T} \tag{7.21}$$

FIGURE 7.25 The trajectories of the center of a wheel and of a point on the rim are seen in a time-exposure photograph.

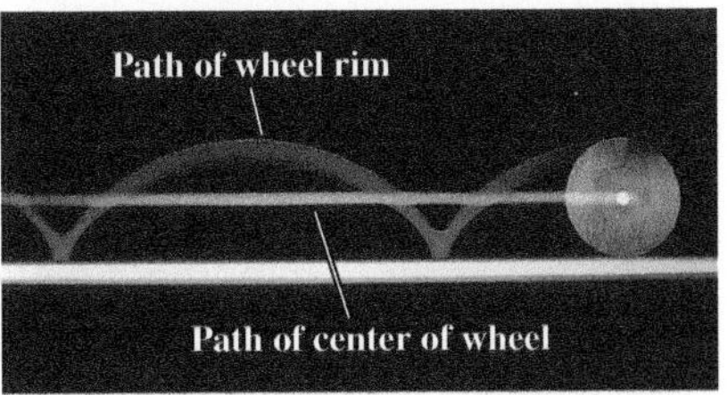

FIGURE 7.26 An object rolling through one revolution.

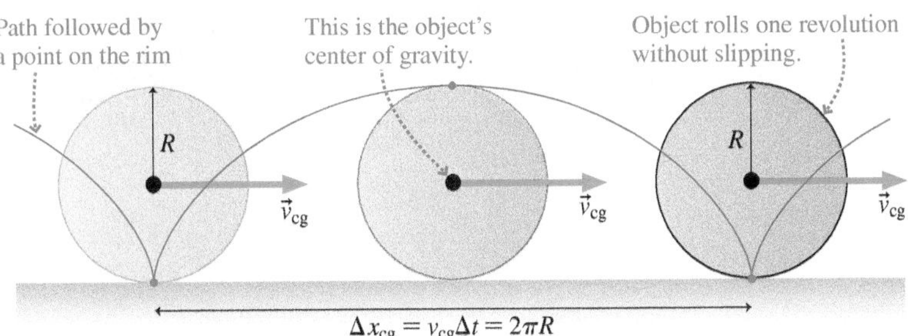

But $2\pi/T$ is the angular velocity ω, which leads to

$$v_{cg} = \omega R \tag{7.22}$$

Equation 7.22 is the **rolling constraint,** the basic link between translation and rotation for objects that roll without slipping.

We can find the velocity for any point on a rolling object by adding the velocity of that point when the object is in pure translation, without rolling, to the velocity of the point when the object is in pure rotation, without translating. FIGURE 7.27 shows how the velocity vectors at the top, center, and bottom of a rotating wheel are found in this way.

Thus the point at the top of the wheel has a forward speed of v_{cg} due to its translational motion plus a forward speed of $\omega R = v_{cg}$ due to its rotational motion. The speed of a point at the top of a wheel is then $v_{top} = 2\omega R = 2v_{cg}$, or *twice* the speed of its center. On the other hand, the point at the bottom of the wheel, where it touches the ground, still has a forward speed of v_{cg} due to its translational motion. But its

Ancient movers The great stone *moai* of Easter Island were moved as far as 16 km from a quarry to their final positions. Archeologists believe that one possible method of moving these 14 ton statues was to place them on rollers. One disadvantage of this method is that the statues, placed on top of the rollers, move twice as fast as the rollers themselves. Thus rollers are continuously left behind and have to be carried back to the front and reinserted. Sadly, the indiscriminate cutting of trees for moving *moai* may have hastened the demise of this island civilization.

FIGURE 7.27 Rolling is a combination of translation and rotation.

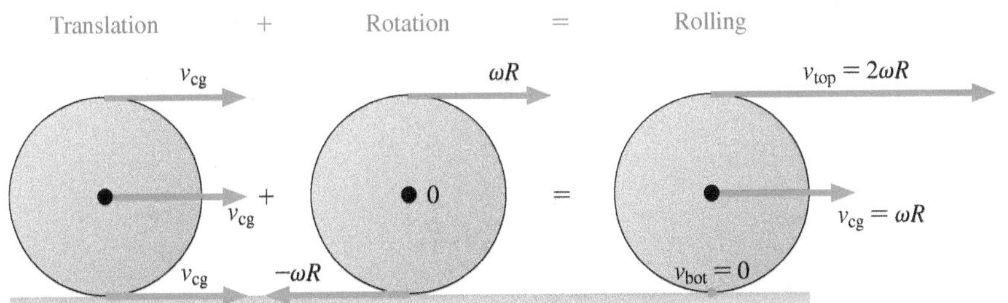

velocity due to rotation points *backward* with a magnitude of $\omega R = v_{cg}$. Adding these, we find that the velocity of this lowest point is $v_{bot} = 0$. In other words, **the point on the bottom of a rolling object is instantaneously at rest.**

Although this seems surprising, it is really what we mean by "rolling without slipping." If the bottom point had a velocity, it would be moving horizontally relative to the surface. In other words, it would be slipping or sliding across the surface. To roll without slipping, the bottom point, the point touching the surface, must be at rest.

EXAMPLE 7.14 Spinning your wheels

You are driving down the highway at 65 mph, which is 29 m/s. Your tires have a radius of 0.30 m.

a. How many times per second does each tire rotate?
b. What is the speed of a point at the top of a tire, relative to the ground?

PREPARE We can use Equation 7.22 to find the angular velocity of the wheel given the speed of the car. If we know this, we can find the frequency of the rotation. Figure 7.27 shows us how to find the speed of a point at the top of a tire.

SOLVE

a. The car is moving at 29 m/s, so a rearrangement of Equation 7.22 gives the angular velocity of the tire as

$$\omega = \frac{v_{cg}}{R} = \frac{29 \text{ m/s}}{0.30 \text{ m}} = 97 \text{ rad/s}$$

The frequency is

$$f = \frac{\omega}{2\pi \text{ rad}} = \frac{97 \text{ rad/s}}{2\pi \text{ rad}} = 15 \text{ rev/s}$$

b. The velocity of a point on the top of the tire is the sum of the velocity of the car and the velocity of the rotational motion of the wheel. Figure 7.27 shows us that this is twice the speed of the car, or 58 m/s.

ASSESS The tire is spinning 15 times per second. That's a high rate, but it's reasonable because you know that the wheels on rapidly moving cars spin too fast for you to perceive the motion. If you drive for an hour at highway speeds, your tires rotate more than 50,000 times!

STOP TO THINK 7.5 A wheel rolls without slipping. Which is the correct velocity vector for point P on the wheel?

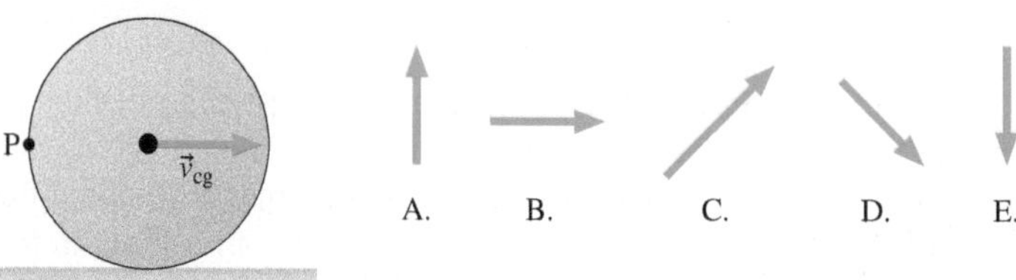

CHAPTER 7 INTEGRATED EXAMPLE Predator and prey

A cheetah can outrun a gazelle if both are moving in a straight line, but real predator-prey chases are far more complex. A gazelle fleeing from a cheetah will run in a straight line until the cheetah is close, then start making evasive maneuvers. The cheetah will swerve to follow its prey, but the gazelle may escape, even though running at a slower speed, if it can make a tighter turn than the cheetah—that is, a turn with a smaller radius.

This example will explore a simplified model of such a chase, shown in FIGURE 7.28, that nonetheless preserves the key physics. The gazelle and cheetah are running to the right at speeds v_{gaz}

FIGURE 7.28 A gazelle trying to outmaneuver a cheetah.

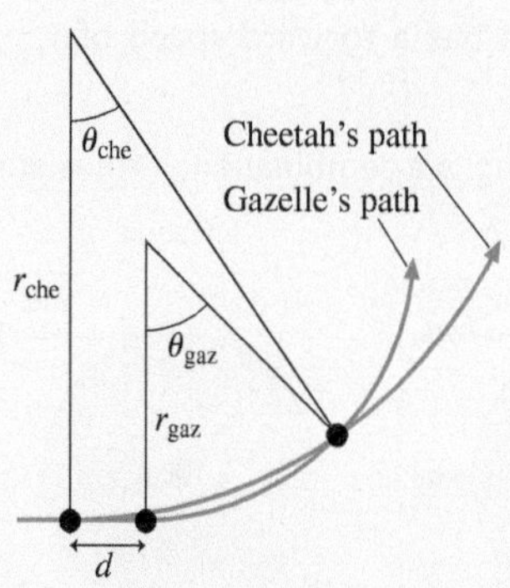

and v_{che}. When the cheetah has closed to distance d, the gazelle, without loss of speed, suddenly turns onto a circular path of radius r_{gaz}. The cheetah responds by veering onto its own circular path of radius $r_{che} > r_{gaz}$. To keep our model simple, we'll assume that the cheetah's reaction time is zero.

This is a life-and-death situation, so we'll assume that each is turning as tightly as it can manage at that speed. Their paths eventually cross, as shown in the figure, but the gazelle will escape if it can reach the intersection point before the cheetah; it is then inside the cheetah's circle, and the cheetah cannot turn tightly enough to catch it without slowing down.

There are two requirements for the gazelle to escape. A full analysis of this model, which is more complex than we will undertake here, finds that the gazelle must have a larger centripetal acceleration than the cheetah. But that alone is not sufficient. If the gazelle turns too soon, the cheetah will get to the intersection point first and catch the gazelle. Thus the gazelle must wait until the cheetah is very close before it swerves.

Animals, like runners or cars, have a maximum possible centripetal acceleration set by the interaction of their feet with the ground. Thus there's a trade-off between speed and turning: To turn at a higher speed requires a larger turning radius. Although cheetahs can run at speeds in excess of 25 m/s, videos of their hunts show that they often reduce their speed as they close in on a prey in anticipation of needing to make sudden sharp turns. In one field study that used accelerometers in collars, a cheetah's maximum centripetal acceleration during a hunt was measured to be $0.50g$, and a cheetah with this acceleration could turn through a 45° angle in 2.0 s.

a. Assume a gazelle is running at 80% of the cheetah's speed. What is the maximum radius of the turn that, in this model, will allow it to just barely escape?
b. Suppose that r_{gaz} is 90% of the value of part a, which ensures that the gazelle's centripetal acceleration is larger than the cheetah's. We'll skip the geometry details, but one can show, for two specific values of d—the cheetah-gazelle separation distance at which the gazelle swerves—that the angles of the intersection (defined in Figure 7.28) are:

For $d = 5.0$ m, $\theta_{gaz} = 47.6°$ and $\theta_{che} = 35.8°$
For $d = 15$ m, $\theta_{gaz} = 117°$ and $\theta_{che} = 80.9°$

The first of these is the situation illustrated in Figure 7.28. For each of these cases, does the gazelle escape?

PREPARE We assume that both animals maintain a constant speed. Even so, they are accelerating when in circular motion with a centripetal acceleration of magnitude $a = v^2/r$. To have the possibility of escape, the gazelle's centripetal acceleration must exceed that of the cheetah.

SOLVE

a. We want to find the turning radius for which the gazelle just barely escapes. We are given the cheetah's acceleration and information about its angular speed, but the gazelle's speed is given in terms of the cheetah's. Thus we'll start by finding the cheetah's speed and turning radius, then use them to find the gazelle's speed.

The field study found $a_{che} = (v_{che})^2/r_{che} = 0.50g = g/2$. In addition, the cheetah could turn through a 45° angle in $\Delta t = 2.0$ s. 45° is one-eighth of a circle, so the distance traveled during Δt is one-eighth of the circumference: $(2\pi r_{che})/8 = \pi r_{che}/4$. Thus, symbolically, the cheetah's speed is

$$v_{che} = \frac{\text{distance}}{\text{time}} = \frac{\pi r_{che}/4}{\Delta t} = \frac{\pi r_{che}}{4\Delta t}$$

Substituting this into the expression for the centripetal acceleration gives

$$\frac{(v_{che})^2}{r_{che}} = \frac{(\pi r_{che}/4\Delta t)^2}{r_{che}} = \left(\frac{\pi}{4\Delta t}\right)^2 r_{che} = g/2$$

We can solve this equation for r_{che}, then use the result to calculate v_{che}. Doing so gives

$$r_{che} = 31.8 \text{ m} \quad v_{che} = 12.5 \text{ m/s}$$

If the gazelle runs at 80% of the cheetah's speed, then $v_{gaz} = 10.0$ m/s. We already know that the cheetah's centripetal acceleration is $g/2$, and the gazelle's acceleration must exceed this to allow any possibility of escape. The requirement $a_{gaz} = (v_{gaz})^2/r_{gaz} > g/2$ leads to

$$r_{gaz} < \frac{2(v_{gaz})^2}{g} = \frac{2(10.0 \text{ m/s})^2}{9.80 \text{ m/s}^2} = 20.4 \text{ m}$$

To escape, the gazelle's turn radius must be no larger than 20.4 m.

b. The gazelle will escape if it passes through the intersection point first. To determine this, we need to find how long it takes each animal to reach the intersection. The time the gazelle needs to reach the intersection at speed v_{gaz} is $\Delta t_{gaz} = l_{gaz}/v_{gaz}$, where l_{gaz} is the length of the gazelle's circular arc. The gazelle would travel a full circumference $2\pi r_{gaz}$ if it turned through a 360° circle. By turning through the smaller angle θ_{gaz}, it travels just a fraction of the circumference: $l_{gaz} = (\theta_{gaz}/360°)2\pi r_{gaz}$. Thus the gazelle's travel time is

$$\Delta t_{gaz} = \frac{l_{gaz}}{v_{gaz}} = \left(\frac{\theta_{gaz}}{360°}\right)\frac{2\pi r_{gaz}}{v_{gaz}}$$

Remember that r_{gaz} is 90% of 20.4 m. The analysis for the cheetah is exactly the same, leading to

$$\Delta t_{che} = \frac{l_{che}}{v_{che}} = \left(\frac{\theta_{che}}{360°}\right)\frac{2\pi r_{che}}{v_{che}}$$

We know the speeds and radii of the turns for both animals, and the problem statement gives the angles of the intersection point for two values of d, the gazelle's distance ahead when it starts to swerve. Thus we can calculate:

For $d = 5.0$ m, $\Delta t_{gaz} = 1.53$ s and $\Delta t_{che} = 1.59$ s
For $d = 15$ m, $\Delta t_{gaz} = 3.75$ s and $\Delta t_{che} = 3.59$ s

A gazelle that waits until the cheetah is only 5.0 m behind before turning manages to just beat the cheetah through the intersection point to escape. But a gazelle that swerves too soon, with the cheetah still 15 m behind, isn't so lucky.

ASSESS We indeed find that the gazelle's greater centripetal acceleration, and its ability to turn with a smaller radius, can make up for its slower speed as long as the evasive swerve is timed correctly. A model in which predator and prey are running in uniform circular motion is clearly oversimplified, but it manages to capture the salient features of a hunt in which a slower but more rapidly turning prey is able to elude capture.

SUMMARY

GOAL To apply Newton's laws to problems of circular and rotational motion.

GENERAL PRINCIPLES

Newton's Second Law for Circular Motion

An object in **uniform circular motion** requires a net force toward the center of the circle to provide the necessary **centripetal acceleration:**

$$\vec{F}_{\text{net}} = \left(\frac{mv^2}{r}, \text{toward center of circle}\right)$$

If we choose the x-axis to point toward the center of the circle, Newton's second law is

$$\sum F_x = \frac{mv^2}{r} \qquad \sum F_y = 0$$

Newton's Second Law for Rotational Motion

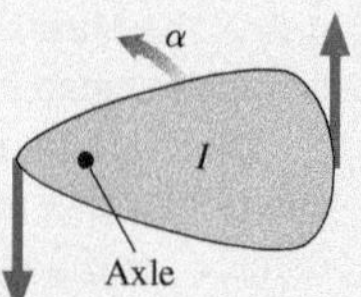

If a net torque acts on an extended object, the object will experience an angular acceleration given by

$$\alpha = \frac{\tau_{\text{net}}}{I}$$

where I is the object's moment of inertia.

This is Newton's second law for rotational motion, analogous to $\vec{a}_{\text{net}} = \vec{F}_{\text{net}}/m$ for linear motion.

IMPORTANT CONCEPTS

Kinematics of circular motion

For an object moving in a circle of radius r at constant speed v:

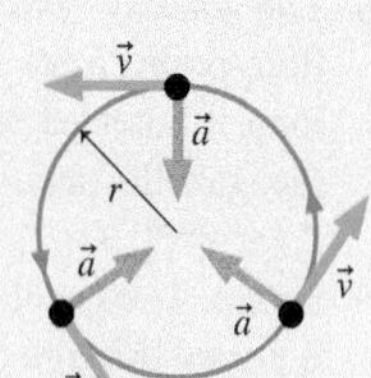

- **Period** T is the time for one revolution.
- **Frequency** is the number of revolutions per second. It is related to the period by $f = 1/T$.
- Frequency and period are connected to the speed by

$$v = 2\pi fr = 2\pi r/T$$

- The centripetal acceleration toward the center of the circle has magnitude

$$a = \frac{v^2}{r}$$

Kinematics of rotational motion

Rotational motion is described by:

- **Angular velocity** $\omega = d\theta/dt$ rad/s
- **Angular acceleration** $\alpha = d\omega/dt$ rad/s^2
- Angles are measured in *radians:*

$$1 \text{ rev} = 360° = 2\pi \text{ rad}$$

- The speed of a point at radius r is $v = \omega r$.
- The kinematic equations for constant angular acceleration are

$$\omega_f = \omega_i + \alpha\,\Delta t$$
$$\Delta\theta = \omega_i\,\Delta t + \tfrac{1}{2}\alpha(\Delta t)^2$$
$$\omega_f^{\,2} = \omega_i^{\,2} + 2\alpha\Delta\theta$$

These are analogous to the kinematic equations for constant acceleration along a straight line.

APPLICATIONS

Rigid-body model

The object is modeled as particle-like atoms connected by massless, rigid rods. The object's size and shape do not change as it moves.

Forces in circular motion

Circular motion requires a net force pointing to the center. At the top or bottom of a circle, the normal force or tension is *not* equal to the object's weight.

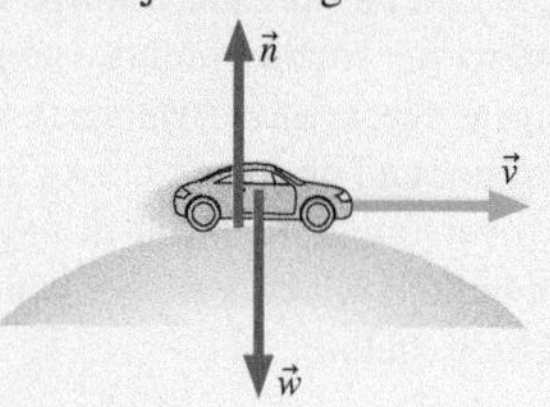

Moment of inertia

Moment of inertia is the rotational equivalent of mass. The moment of inertia is smaller if mass is concentrated closer to the axis of rotation.

Moments of inertia of common shapes:

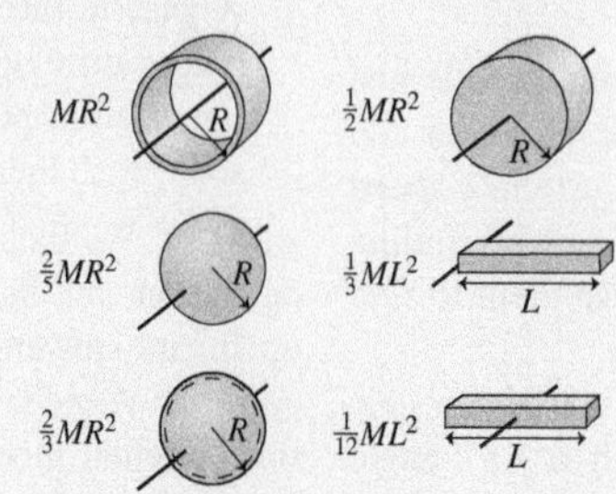

Rolling motion

For an object that rolls without slipping, with angular velocity ω, the center-of-gravity speed is

$$v_{\text{cg}} = \omega R$$

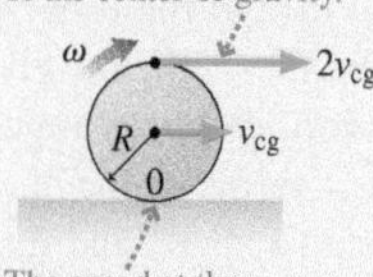

LEARNING OBJECTIVES After studying this chapter, you should be able to:

- Calculate the period, frequency, and speed of an object in uniform circular motion. *Conceptual Question 7.1; Problems 7.1, 7.3–7.5, 7.9*
- Calculate and interpret the centripetal acceleration of an object in uniform circular motion. *Problems 7.7, 7.8, 7.11, 7.12, 7.23*
- Solve dynamics problems for objects in uniform circular motion. *Conceptual Questions 7.3–7.5; Problems 7.15, 7.16, 7.19, 7.21, 7.24*
- Calculate the angular velocity and angular acceleration of a rotating rigid body. *Conceptual Questions 7.2, 7.6; Problems 7.29, 7.32–7.34, 7.36*
- Determine and interpret an object's moment of inertia. *Conceptual Questions 7.7, 7.8, 7.10; Problems 7.37–7.39*
- Solve rotational dynamics problems. *Problems 7.41–7.45*
- Analyze rolling motion. *Problems 7.46–7.49*

STOP TO THINK ANSWERS

Stop to Think 7.1: C > D = A > B. Rearranging Equation 7.2 gives $T = 2\pi r/v$. For the cases shown, speed is either v or $2v$; the radius of the circular path is r or $2r$. Going around a circle of radius r at a speed v takes the same time as going around a circle of radius $2r$ at a speed $2v$. It's twice the distance at twice the speed.

Stop to Think 7.2: B. The car is moving in a circle, so there must be a net force toward the center of the circle. The center of the circle is below the car, so the net force must point downward. This can be true only if $w > n$. This makes sense; $n < w$, so the apparent weight is less than the true weight. The riders in the car "feel light"; if you've driven over a rise like this, you know that this is what you feel.

Stop to Think 7.3: F (zero); A, B, and D (constant). Velocity and centripetal acceleration are both vectors that change direction. There is no angular acceleration. Speed, frequency, and angular velocity, which are proportional to one another, are all constant.

Stop to Think 7.4: $I_D > I_A > I_C > I_B$. The moments of inertia are $I_B \approx 0$, $I_C = 2mr^2$, $I_A = 3mr^2$, and $I_D = mr^2 + m(2r)^2 = 5mr^2$.

Stop to Think 7.5: C. The velocity of P is the vector sum of $\vec{v}$ directed to the right and an upward velocity of the same magnitude due to the rotation of the wheel.

QUESTIONS

Conceptual Questions

1. In uniform circular motion, which of the following quantities are constant: speed, instantaneous velocity, the tangential component of velocity, the radial component of acceleration? Which of these quantities are zero throughout the motion?
2. Figure Q7.2 shows three points on a steadily rotating wheel.
   a. Rank in order, from largest to smallest, the angular velocities ω_1, ω_2, and ω_3 of these points. Explain.
   b. Rank in order, from largest to smallest, the speeds v_1, v_2, and v_3 of these points. Explain.

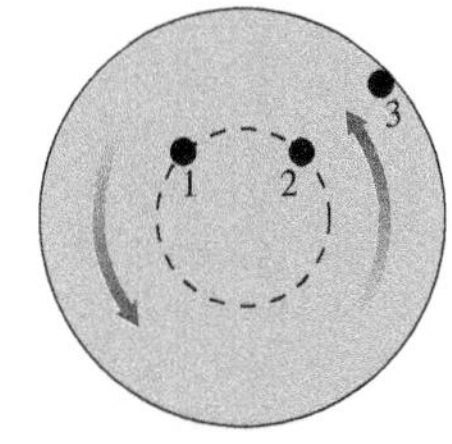

FIGURE Q7.2

3. A car runs out of gas while driving down a hill. It rolls through the valley and starts up the other side. At the very bottom of the valley, which of the free-body diagrams in Figure Q7.3 is correct? The car is moving to the right, and drag and rolling friction are negligible.

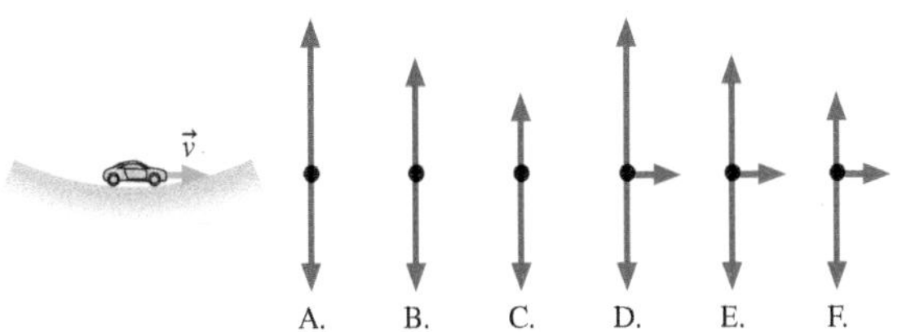

FIGURE Q7.3

4. Playground swings move through an arc of a circle. When you are on a swing, and at the lowest point of your motion, is your apparent weight greater than, less than, or equal to your true weight? Explain.
5. Assume the earth is a perfect sphere rotating about an axis that passes through the north and south poles. Would your apparent weight when standing on the equator be larger than, smaller than, or equal to your apparent weight when standing at the north pole? Explain.
6. Figure Q7.6 shows four pulleys, each with a heavy and a light block strung over it. The blocks' velocities are shown. What are the signs (+ or −) of the angular velocity and angular acceleration of the pulley in each case?

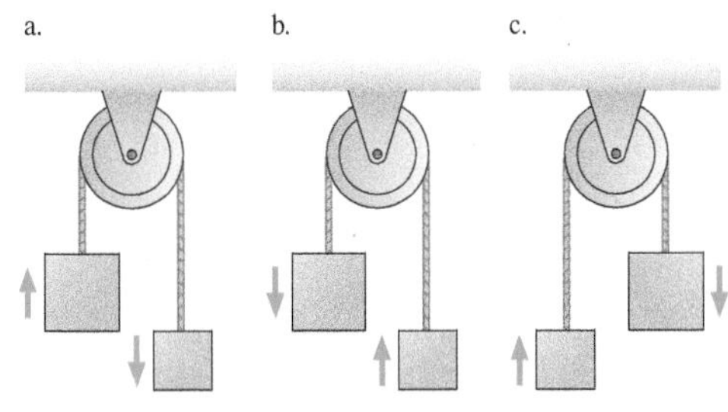

FIGURE Q7.6

7. Must an object be rotating to have a moment of inertia? Explain.
8. You have two solid steel spheres. Sphere B has twice the radius of sphere A. By what *factor* does the moment of inertia I_B of sphere B exceed the moment of inertia I_A of sphere A?

Problem difficulty is labeled as I (straightforward) to IIIII (challenging). Problems labeled INT integrate significant material from earlier chapters; BIO are of biological or medical interest; CALC require calculus to solve.

9. The professor hands you two spheres. They have the same mass, the same radius, and the same exterior surface. The professor claims that one is a solid sphere and the other is hollow. Can you determine which is which without cutting them open? If so, how? If not, why not?
10. The solid cylinder and cylindrical shell in Figure Q7.10 have the same mass, same radius, and turn on frictionless, horizontal axles. (The cylindrical shell has lightweight spokes connecting the shell to the axle.) A rope is wrapped around each cylinder and tied to a block. The blocks have the same mass and are held the same height above the ground. Both blocks are released simultaneously. Which hits the ground first? Or is it a tie? Explain.

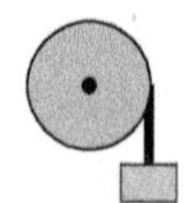
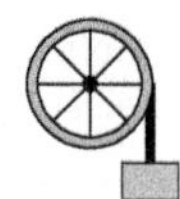

FIGURE Q7.10

11. A diver in the pike position (legs straight, hands on ankles) usually makes only one or one-and-a-half rotations. To make two or three rotations, the diver goes into a tuck position (knees bent, body curled up tight). Why?

Multiple-Choice Questions

12. | The cylindrical space station in Figure Q7.12, 200 m in diameter, rotates in order to provide artificial gravity of g for the occupants. How much time does the station take to complete one rotation?
A. 3 s B. 20 s
C. 28 s D. 32 s

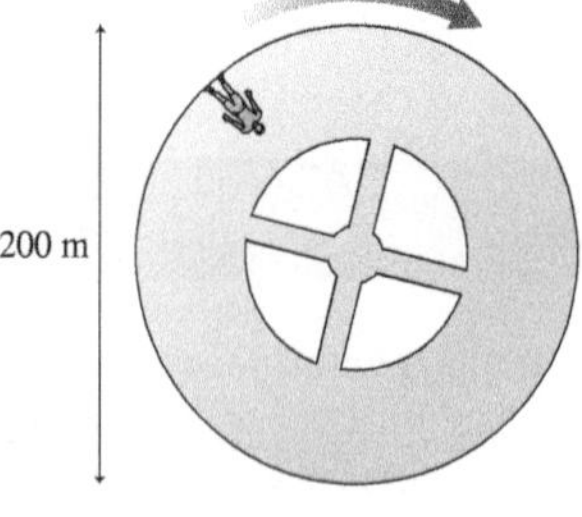

FIGURE Q7.12

13. | Your roommate is working on her bicycle and has the bike upside down. She spins the 60-cm-diameter wheel, and you notice that a pebble stuck in the tread goes around three times every second. Approximately what is the pebble's acceleration?
A. 1 m/s^2 B. 20 m/s^2
C. 110 m/s^2 D. 350 m/s^2
14. | A ball on a string moves around a complete circle, once a second, on a frictionless, horizontal table. The tension in the string is measured to be 6.0 N. What would the tension be if the ball went around in only half a second?
A. 1.5 N B. 3.0 N
C. 12 N D. 24 N
15. | As seen from above, a car rounds the curved path shown in Figure Q7.15 at a constant speed. Which vector best represents the net force acting on the car?

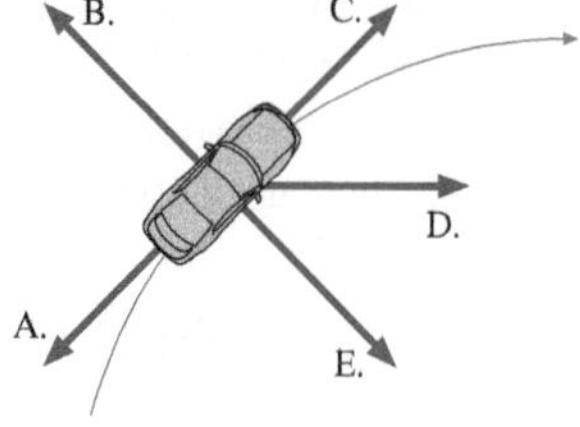

FIGURE Q7.15

16. || On a snowy day, when the coefficient of friction μ_s between a car's tires and the road is 0.50, the maximum speed that the car can go around a curve is 20 mph. What is the maximum speed at which the car can take the same curve on a sunny day when $\mu_s = 1.0$?
A. 20 mph B. 24 mph C. 28 mph
D. 33 mph E. 40 mph
17. | A runner going in circles wants to make a tighter turn with half the initial radius. By what factor must the runner's speed change so that the tighter turn can be made with the same friction force?
A. The runner's speed must be halved.
B. The runner's speed must decrease by a factor of $\sqrt{2}$.
C. The runner's speed must stay the same.
D. The runner's speed must increase by a factor of $\sqrt{2}$.
E. The runner's speed must double.

Questions 18 through 20 concern the spin of a figure skater at the end of a performance. An initial very rapid spin slows to about 120 rpm when the skater stretches her arms out. Suppose the skater's hands are 80 cm from the rotation axis.

18. || What is the approximate angular speed of the spinning skater?
A. 6 rad/s B. 13 rad/s C. 20 rad/s D. 750 rad/s
19. || What is the approximate speed of the skater's hand?
A. 5 m/s B. 10 m/s C. 20 m/s D. 100 m/s
20. || What is the approximate centripetal acceleration of the skater's hand?
A. 12 m/s^2 B. 50 m/s^2 C. 130 m/s^2 D. 210 m/s^2
21. || An audio CD has a mass of 15 g and a diameter of 120 mm. What is its moment of inertia about an axis through its center, perpendicular to the disk?
A. 2.7×10^{-5} kg $\cdot$ m^2 B. 5.4×10^{-5} kg $\cdot$ m^2
C. 1.1×10^{-4} kg $\cdot$ m^2 D. 2.2×10^{-4} kg $\cdot$ m^2
22. | Two horizontal rods are each held up by vertical strings tied to their ends. Rod A has length L and mass M; rod B has length $2L$ and mass $2M$. Each rod then has one of its supporting strings cut, causing the rod to begin pivoting about the end that is still tied up. Which rod has a larger initial angular acceleration?
A. Rod A
B. Rod B
C. The initial angular acceleration is the same for both.

PROBLEMS

Section 7.1 Uniform Circular Motion

1. || A 5.0-m-diameter merry-go-round is turning with a 4.0 s period. What is the speed of a child on the rim?
2. | The earth, with a radius of 6.4×10^6 m, rotates on its axis once a day. What is the speed of a person standing on the equator?
3. | A vinyl record rotates at $33\frac{1}{3}$ rpm.
a. What is its frequency in rev/s?
b. What is its period in seconds?
4. | A typical hard disk in a computer spins at 5400 rpm.
a. What is the frequency in rev/s?
b. What is the period in seconds?
5. | In the hammer throw, one of the earliest Olympic events, a heavy ball attached to a chain is swung several times in a circle, as the athlete spins, and then released. For 2016 Olympic champion Anita Wlodarczyk, the ball revolved in a circle of radius 2.1 m and her final rotation took only 0.43 s.
a. What was the frequency of this final rotation in rev/s?
b. What was the speed of the ball?
c. What was the ball's acceleration?
6. ||| The radius of the earth's very nearly circular orbit around the sun is 1.50×10^{11} m. Find the magnitude of the earth's (a) velocity and (b) centripetal acceleration as it travels around the sun. Assume a year of 365 days.

7. | Modern wind turbines are larger than they appear, and despite their apparently lazy motion, the speed of the blades tips can be quite high—many times higher than the wind speed. A typical modern turbine has blades 56 m long that spin at 13 rpm. At the tip of a blade, what are (a) the speed and (b) the centripetal acceleration?

8. | The California sea lion is capable of making extremely fast,
BIO tight turns while swimming underwater. In one study, scientists observed a sea lion making a circular turn with a radius of 0.35 m while swimming at 4.2 m/s.
   a. What is the sea lion's centripetal acceleration, in units of g?
   b. What percentage is this acceleration of that of an F-15 fighter jet's maximum centripetal acceleration of $9g$?
9. || Baseball pitching machines are used to fire baseballs toward a batter for hitting practice. In one kind of machine, the ball is fed between two wheels that are rapidly rotating in opposite directions; as the ball is pulled in between the wheels, it rapidly accelerates up to the speed of the wheels' rims, at which point it is ejected. For a machine with 35-cm-diameter wheels, what rotational frequency (in rpm) do the wheels need to pitch a 90 mph fastball?

10. | Bumblebees are skilled aerialists, able to fly with confidence
BIO around and through the leaves and stems of plants. In one test of bumblebee aerial navigation, bees in level flight flew at a constant 0.35 m/s, turning right and left as they navigated an obstacle-filled track. While turning, the bees maintained a reasonably constant centripetal acceleration of4.0 m/s^2.
   a. What is the radius of curvature for such a turn?
   b. How much time is required for a bee to execute a 90° turn?
11. || A typical running track is an oval with 74-m-diameter half circles at each end. A runner going once around the track covers a distance of 400 m. Suppose a runner, moving at a constant speed, goes once around the track in 1 min 40 s. What is her centripetal acceleration during the turn at each end of the track?
12. || Suppose you are running on a rather slippery surface. You know that you can successfully make a turn of radius 10 m when running at 4.0 m/s, but you are right on the edge of slipping. If you want to make a turn with 1.5 times greater speed, by what factor must you increase the radius of your turn?

Section 7.2 Dynamics of Uniform Circular Motion

13. || In short-track speed skating, the track has 16-m-diameter semicircles at each end. Assume that a 65 kg skater goes around the turn at a constant 12 m/s.
   a. What is the horizontal force on the skater?
   b. What is the ratio of this force to the skater's weight?

14. || In addition to their remarkable top speeds, cheetahs have im-
BIO pressive cornering abilities. In one study, the sideways force on the feet of a 60 kg cheetah, with its claws dug into the ground, was found to be 960 N.
   a. What was the cheetah's centripetal acceleration?
   b. What was the radius of the turn if the cheetah was running at 20 m/s?
15. || A cyclist is rounding a 20-m-radius curve at 12 m/s. What is the minimum possible coefficient of static friction between the bike tires and the ground?
16. || A highway curve of radius 500 m is designed for traffic moving at a speed of 90 km/h. What is the correct banking angle of the road?
17. || A fast pitch softball player does
BIO a "windmill" pitch, illustrated in Figure P7.17, moving her hand through a circular arc to pitch a ball at 70 mph. The 0.19 kg ball is 60 cm from the pivot point at her shoulder. The ball reaches its maximum speed at the lowest point of the circle, where it is released.
   a. At the bottom of the circle, just before the ball leaves her hand, what is its centripetal acceleration?
   b. What are the magnitude and direction of the force her hand exerts on the ball at this point?

FIGURE P7.17

18. | A wind turbine has 12,000 kg blades that are 38 m long. The blades spin at 22 rpm. If we model a blade as a point mass at the midpoint of the blade, what is the inward force necessary to provide each blade's centripetal acceleration?
19. ||| You're driving your pickup truck around a curve that has a radius of 20 m. How fast can you drive around this curve before a steel toolbox slides on the steel bed of the truck?
20. || Gibbons, small Asian apes, move by *brachiation,* swinging
BIO below a handhold to move forward to the next handhold. A 9.0 kg gibbon has an arm length (hand to shoulder) of 0.60 m. We can model its motion as that of a point mass swinging at the end of a 0.60-m-long, massless rod. At the lowest point of its swing, the gibbon is moving at 3.5 m/s. What upward force must a branch provide to support the swinging gibbon?
21. ||| The spin cycle of a clothes washer extracts the water in clothing by greatly increasing the water's apparent weight so that it is efficiently squeezed through the clothes and out the holes in the drum. In a top loader's spin cycle, the 45-cm-diameter drum spins at 1200 rpm around a vertical axis. What is the apparent weight of a 1.0 g drop of water?
22. || To withstand "g-forces" of up to 10 g's, caused by suddenly
BIO pulling out of a steep dive, fighter jet pilots train on a "human centrifuge." 10 g's is an acceleration of 98 m/s^2. If the length of the centrifuge arm is 12 m, at what speed is the rider moving when she experiences 10 g's?
23. || A typical laboratory centrifuge rotates at 4000 rpm. Test
BIO tubes have to be placed into a centrifuge very carefully because
INT of the very large accelerations.
   a. What is the acceleration at the end of a test tube that is 10 cm from the axis of rotation?
   b. For comparison, what is the magnitude of the acceleration a test tube would experience if stopped in a 1.0-ms-long encounter with a hard floor after falling from a height of 1.0 m?

24. ||| The passengers in a roller coaster car feel 50% heavier than their true weight as the car goes through a dip with a 30 m radius of curvature. What is the car's speed at the bottom of the dip?

25. ||| You're driving your new sports car at 75 mph over the top of a hill that has a radius of curvature of 525 m. What fraction of your normal weight is your apparent weight as you crest the hill?

Section 7.3 The Rotation of a Rigid Body

26. | A child on a merry-go-round takes 3.0 s to go around once. What is his angular displacement during a 1.0 s time interval? Give your answer in both radians and degrees.

27. || *Chlamydomonas,* algae known as *green yeast,* is a popular organism for lab study. These cells move through their watery environment by waving a pair of flagella. Normally the flagella wave in synchrony, moving the cell smoothly. However, some individuals have only a single flagellum, and its asymmetric wave causes the cell body to rotate as it moves. Figure P7.27 shows microscope data for the body rotation of a single-flagellum cell. How many rotations does the cell complete in 10 s?
BIO

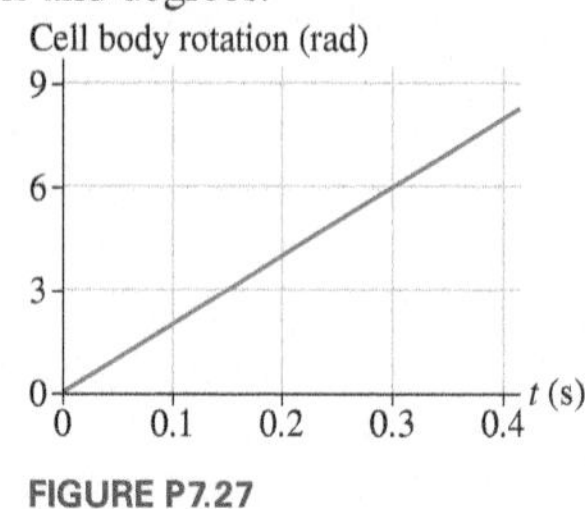

FIGURE P7.27

28. | The 1.00-cm-long second hand on a watch rotates smoothly.
    a. What is its angular velocity?
    b. What is the speed of the tip of the hand?

29. || A vinyl record rotates on a turntable at 45 rpm. What are (a) the angular speed in rad/s and (b) the period of the motion?

30. ||| The earth's radius is about 4000 miles. Kampala, the capital of Uganda, and Singapore are both nearly on the equator. The distance between them is 5000 miles.
    a. Through what angle do you turn, relative to the earth, if you fly from Kampala to Singapore? Give your answer in both radians and degrees.
    b. The flight from Kampala to Singapore takes 9 hours. What is the plane's angular speed relative to the earth?

31. ||| A turntable rotates counterclockwise at 78 rpm. A speck of dust on the turntable is at $\theta = 0.45$ rad at $t = 0$ s. What is the angle of the speck at $t = 8.0$ s? Your answer should be between 0 and 2π rad.

32. || Figure P7.32 shows the angular position of a potter's wheel.
    a. What is the angular displacement of the wheel between $t = 5$ s and $t = 15$ s?
    b. What is the angular velocity of the wheel at $t = 15$ s?
    c. What is the maximum speed of a point on the outside of the wheel, 15 cm from the axle?

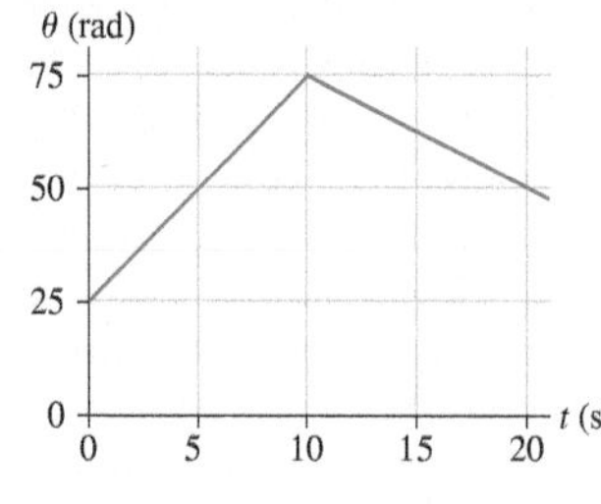

FIGURE P7.32

33. ||| To throw a discus, the thrower holds it with a fully outstretched arm. Starting from rest, he begins to turn with a constant angular acceleration, releasing the discus after making one complete revolution. The diameter of the circle in which the discus moves is about 1.8 m. If the thrower takes 1.0 s to complete one revolution, starting from rest, what will be the speed of the discus at release?

34. || The bigclaw snapping shrimp shown in Figure P7.34 is aptly named—it has one big claw that snaps shut with remarkable speed. The part of the claw that moves rotates through a 90° angle in 1.3 ms. Assume that the claw is 1.5 cm long and that it closes with constant angular acceleration.
BIO
    a. What is the angular acceleration in rad/s^2?
    b. What is the final angular speed of the claw?
    c. How fast is the tip of the claw moving at the end of its motion?

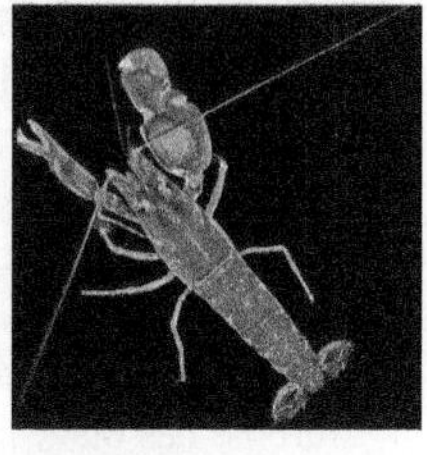
FIGURE P7.34

35. || A computer hard disk starts from rest, then speeds up with an angular acceleration of 190 rad/s^2 until it reaches its final angular speed of 7200 rpm. How many revolutions has the disk made 10.0 s after it starts up?

36. || The crankshaft in a race car goes from rest to 3000 rpm in 2.0 s.
    a. What is the crankshaft's angular acceleration?
    b. How many revolutions does it make while reaching 3000 rpm?

Section 7.4 Rotational Dynamics and Moment of Inertia

37. || A regulation table tennis ball is a thin spherical shell 40 mm in diameter with a mass of 2.7 g. What is its moment of inertia about an axis that passes through its center?

38. || A solid cylinder with a radius of 4.0 cm has the same mass as a solid sphere of radius R. If the cylinder and sphere have the same moment of inertia about their centers, what is the sphere's radius?

39. || A bicycle rim has a diameter of 0.65 m and a moment of inertia, measured about its center, of 0.19 kg · m^2. What is the mass of the rim?

40. | A small grinding wheel has a moment of inertia of 4.0×10^{-5} kg · m^2. What net torque must be applied to the wheel for its angular acceleration to be 150 rad/s^2?

41. | An object's moment of inertia is 2.0 kg · m^2. Its angular velocity is increasing at the rate of 4.0 rad/s per second. What is the net torque on the object?

42. ||| The lightweight wheel on a road bike has a moment of inertia of 0.097 kg · m^2. A mechanic, checking the alignment of the wheel, gives it a quick spin; it completes 5 rotations in 2.0 s. To bring the wheel to rest, the mechanic gently applies the disk brakes, which squeeze pads against a metal disk connected to the wheel. The pads touch the disk 7.1 cm from the axle, and the wheel slows down and stops in 1.5 s. What is the magnitude of the friction force on the disk?

43. ||| A 200 g, 20-cm-diameter plastic disk is spun on an axle through its center by an electric motor. What torque must the motor supply to take the disk from 0 to 1800 rpm in 4.0 s?

44. || The engine in a small airplane is specified to have a torque of 500 N · m. This engine drives a 2.0-m-long, 40 kg single-blade propeller. On startup, how long does it take the propeller to reach 2000 rpm?

45. ||| If you lift the front wheel of a poorly maintained bicycle off the ground and then start it spinning at 0.72 rev/s, friction in the bearings causes the wheel to stop in just 12 s. If the moment of inertia of the wheel about its axle is 0.30 kg · m^2, what is the magnitude of the frictional torque?

Section 7.5 Rolling Motion

46. | A bicycle with 0.80-m-diameter tires is coasting on a level road at 5.6 m/s. A small blue dot has been painted on the tread of the rear tire.
    a. What is the angular speed of the tires?
    b. What is the speed of the blue dot when it is 0.80 m above the road?
    c. What is the speed of the blue dot when it is 0.40 m above the road?
47. || A typical road bike wheel has a diameter of 70 cm including the tire. In a time trial, when a cyclist is racing along at 12 m/s:
    a. How fast is a point at the top of the tire moving?
    b. How fast, in rpm, are the wheels spinning?
48. || A car tire is 60 cm in diameter. The car is traveling at a speed of 20 m/s.
    a. What is the tire's angular velocity, in rpm?
    b. What is the speed of a point at the top edge of the tire?
    c. What is the speed of a point at the bottom edge of the tire?
49. || The top point on a 64-cm-diameter rolling bicycle wheel has an instantaneous speed of 14 m/s. How many revolutions will the wheel make in 10 s?

General Problems

50. || Figure P7.50 shows the angular velocity graph for a particle moving in a circle. How many revolutions does the object make during the first 4 s?

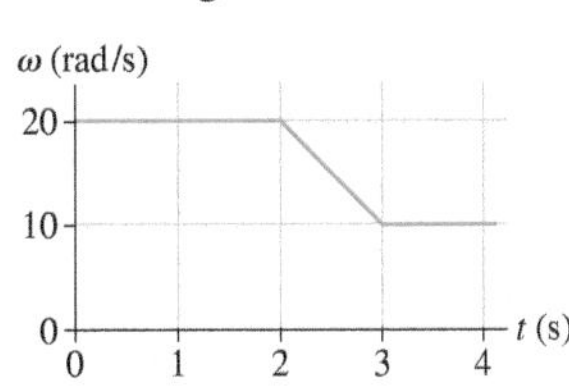

FIGURE P7.50

FIGURE P7.51

51. | Figure P7.51 shows the angular position graph for a particle moving in a circle. What is the particle's angular velocity at (a) $t = 1$ s, (b) $t = 4$ s, and (c) $t = 7$ s?
52. | The graph in Figure P7.52 shows the angular velocity of the crankshaft in a car. Draw a graph of the angular acceleration versus time. Include appropriate numerical scales on both axes.

INT

FIGURE P7.52

53. || How fast must a plane fly along the earth's equator so that the sun stands still relative to the passengers? In which direction must the plane fly, east to west or west to east? Give your answer in both km/h and mph. The radius of the earth is 6400 km.
54. || Peregrine falcons are known for their maneuvering ability. In a tight circular turn, a falcon can attain a centripetal acceleration 1.5 times the free-fall acceleration. What is the radius of the turn if the falcon is flying at 25 m/s?

BIO

55. || A turbine is spinning at 3800 rpm. Friction in the bearings is so low that it takes 10 min to coast to a stop. How many revolutions does the turbine make while stopping?
56. ||| The angular velocity of a process control motor is $\omega = \left(20 - \frac{1}{2}t^2\right)$ rad/s, where t is in seconds.

CALC

    a. At what time does the motor reverse direction?
    b. Through what angle does the motor turn between $t = 0$ s and the instant at which it reverses direction?
57. ||| A painted tooth on a spinning gear has angular acceleration $\alpha = (20 - t)$ rad/s^2, where t is in s. Its initial angular velocity, at $t = 0$ s, is 300 rpm. What is the tooth's angular velocity in rpm at $t = 20$ s?

CALC

58. ||| Communications satellites are placed in a circular orbit where they stay directly over a fixed point on the equator as the earth rotates. These are called *geosynchronous orbits.* The radius of the earth is 6.37×10^6 m, and the altitude of a geosynchronous orbit is 3.58×10^7 m ($\approx$22,000 miles). What are (a) the speed and (b) the magnitude of the acceleration of a satellite in a geosynchronous orbit?
59. || It is well known that runners run more slowly around a curved track than a straight one. One hypothesis to explain this is that the *total* force from the track on a runner's feet—the magnitude of the *vector* sum of the normal force (that has average value mg to counteract gravity) and the inward-directed friction force that causes the runner's centripetal acceleration—is greater when running around a curve than on a straight track. Runners compensate for this greater force by increasing the time their feet are in contact with the ground, which slows them down. For a sprinter running at 10 m/s around a curved track of radius 20 m, how much greater (as a percentage) is the average total force on their feet compared to when they are running in a straight line?

BIO

60. || Astronauts in the International Space Station must work out every day to counteract the effects of weightlessness. Researchers have investigated if riding a stationary bicycle while experiencing artificial gravity from a rotating platform gives any additional cardiovascular benefit. What frequency of rotation, in rpm, is required to give an acceleration of $1.4g$ to an astronaut's feet if her feet are 1.1 m from the platform's rotational axis?

BIO

61. || Biologists have studied the running ability of the northern quoll, a marsupial indigenous to Australia. In one set of experiments, they studied the maximum speed that quolls could run around a smooth circular track without slipping. One quoll was running at 2.8 m/s around a curve with a radius of 1.2 m when it started to slip. What was the coefficient of static friction between the quoll's feet and the track in this trial?

BIO

62. || You are driving your car through a roundabout that has a radius of 9.0 m. Your physics textbook is lying on the seat next to you. What is the fastest speed at which you can go around the curve without the book sliding? The coefficient of static friction between the book and the seat is 0.30.
63. || In the Wall of Death carnival attraction, stunt motorcyclists ride around the inside of a large, 10-m-diameter wooden cylinder that has *vertical* walls. The coefficient of static friction between the riders' tires and the wall is 0.90. What is the minimum speed at which the motorcyclists can ride without slipping down the wall?

64. ||| In the swing carousel amusement park ride, riders sit in chairs that are attached by a chain to a large rotating drum. As the carousel turns, the riders move in a large circle with the chains tilted out from the vertical. In one such carousel, the riders move in a 16.5-m-radius circle and take 8.3 s to complete one revolution. What is the angle of the chains, as measured from the vertical?

65. || A 5.0 g coin is placed 15 cm from the center of a turntable. INT The coin has static and kinetic coefficients of friction with the turntable surface of $\mu_s = 0.80$ and $\mu_k = 0.50$. The turntable speeds up slowly, starting from rest. At what frequency in rpm does the coin slide off?

66. ||| In an old-fashioned amusement park ride, passengers stand inside a 3.0-m-tall, 5.0-m-diameter hollow steel cylinder with their backs against the wall. The cylinder begins to rotate about a vertical axis. Then the floor on which the passengers are standing suddenly drops away! If all goes well, the passengers will "stick" to the wall and not slide. Clothing has a static coefficient of friction against steel in the range 0.60 to 1.0 and a kinetic coefficient in the range 0.40 to 0.70. What is the minimum rotational frequency, in rpm, for which the ride is safe?

67. || The ultracentrifuge is an important tool for separating and BIO INT analyzing proteins in biological research. Because of the enormous centripetal accelerations that can be achieved, the apparatus must be carefully balanced so that each sample is matched by another on the opposite side of the rotor shaft. Any difference in mass of the opposing samples will cause a net force in the horizontal plane on the shaft of the rotor; this force can actually be large enough to destroy the centrifuge. Suppose that a scientist makes a slight error in sample preparation, and one sample has a mass 10 mg greater than the opposing sample. If the samples are 10 cm from the axis of the rotor and the ultracentrifuge spins at 70,000 rpm, what is the magnitude of the net force on the rotor due to the unbalanced samples?

68. ||| Suppose the moon were held in its orbit not by gravity but by a massless cable attached to the center of the earth. What would be the tension in the cable? The moon is 3.8×10^8 m from the earth, has a mass of 7.4×10^{22} kg, and orbits with a period of 27.3 days.

69. |||| Two wires are tied to the 300 g sphere shown in Figure P7.69. The sphere revolves in a horizontal circle at a constant speed of 7.5 m/s. What is the tension in each of the wires?

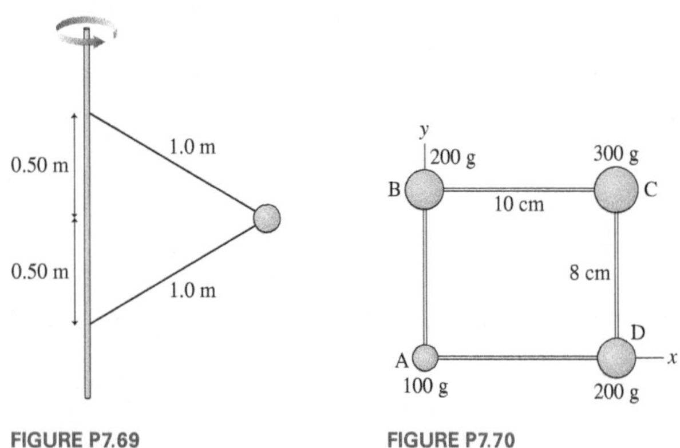

FIGURE P7.69 FIGURE P7.70

70. ||| The four masses shown in Figure P7.70 are connected by INT massless, rigid rods.

a. Find the coordinates of the center of gravity.

b. Find the moment of inertia about an axis that passes through mass A and is perpendicular to the page.

71. |||| A 1.0 kg ball and a 2.0 kg ball are connected by a 1.0-m-long rigid, massless rod. The rod is rotating clockwise about its center of mass at 20 rpm. What net torque will bring the balls to a halt in 5.0 s?

72. ||| Calculate by direct integration the moment of inertia for a CALC thin rod of mass M and length L about an axis located distance d from one end. Confirm that your answer agrees with Table 7.3 when $d = 0$ and when $d = L/2$.

73. |||| a. A cylinder of mass M and radius R has a cylindrical hole CALC of radius r centered on the axis. Find an expression for the moment of inertia of the cylinder. You should confirm that your answer agrees with Table 7.3 when $r = 0$ (a solid disk) and when $r = R$ (a cylindrical hoop).

b. A 7.5 kg, 25-cm-diameter cylinder has a 10-cm-diameter hole. What torque is needed to spin up the cylinder to 90 rpm in 5.0 s, starting from rest?

74. ||| Calculate the moment of inertia of a thin, uniform rod of CALC length L and mass M that rotates about a pivot at distance $L/3$ from one end.

75. || Flywheels are large, massive wheels used to store energy. They can be spun up slowly, then the wheel's energy can be released quickly to accomplish a task that demands high power. An industrial flywheel has a 1.5 m diameter and a mass of 250 kg. A motor spins up the flywheel with a constant torque of 50 N · m. How long does it take the flywheel to reach its top angular speed of 1200 rpm?

76. ||| Sprinters must move their BIO legs very rapidly to achieve high speeds. Figure P7.76 shows data for the knee angle of a sprinter during one complete stride early in a 100 m dash. The angle is that of the lower leg relative to the upper leg, ranging from 0° if the leg is completely straight to 180° if the lower leg is completely bent back against the thigh. We will focus on the first part of the motion—between 0 s and 0.05 s. Consider only the motion of the lower leg relative to the upper leg, ignoring the fact that the entire leg is also rotating about the hip.

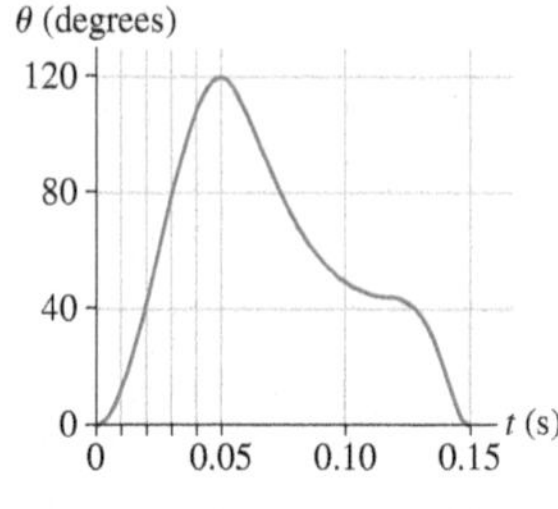

FIGURE P7.76

a. Use the graph to estimate the angular velocity in rad/s of the lower leg relative to the upper leg at 0.025 s.

b. What is the average angular acceleration during the first 0.025 s?

c. The lower leg can be modeled as a 50-cm-long, 4.8 kg rod pivoted about one end. What is the lower leg's moment of inertia?

d. What torque is required to provide the lower leg's angular acceleration?

77. || A trap-jaw ant has mandibles that can snap shut with some BIO INT force, as you might expect from its name. The formidable snap is good for more than capturing prey. When an ant snaps its jaws against the ground, the resulting force can launch the ant into the air. Here are typical data: An ant rotates its mandible, of length 1.30 mm and mass 130 μg (which we can model as a uniform rod rotated about its end), at a high angular speed. As the tip strikes the ground, it undergoes an angular acceleration of magnitude 3.5×10^8 rad/s^2. If we assume that the tip of the mandible hits perpendicular to the ground, what is the force on the tip? How does this compare to the weight of a 12 mg ant?

78. ||| A chef sharpens a knife by pushing it with a constant force INT against the rim of a grindstone. The 30-cm-diameter stone is spinning at 200 rpm and has a mass of 28 kg. The coefficient of kinetic friction between the knife and the stone is 0.20. If the stone slows steadily to 180 rpm in 10 s of grinding, what is the force with which the chef presses the knife against the stone?

79. ||||| A toy top with a spool of diameter 5.0 cm has a moment of inertia of 3.0×10^{-5} kg·m^2 about its rotation axis. To get the top spinning, its string is pulled with a tension of 0.30 N. How long does it take for the top to complete the first five revolutions? The string is long enough that it is wrapped around the top more than five turns.

80. ||||| A 2.0-m-long slab of concrete is supported by rollers, as shown in Figure P7.80. If the slab is pushed to the right, it will move off the supporting rollers, one by one, on the left side. How far can the slab be moved before its center of gravity is to the right of the contact point with the rightmost roller—at which point the slab begins to tip?

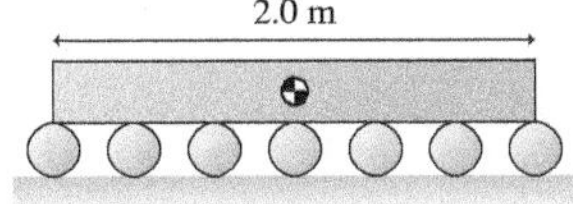

FIGURE P7.80

MCAT-Style Passage Problems

The Bunchberry

The bunchberry flower has the fastest-moving parts ever seen in a plant. Initially, the stamens are held by the petals in a bent position, storing energy like a coiled spring. As the petals release, the tips of the stamens fly up and quickly release a burst of pollen. Figure P7.81 shows the details of the motion. The tips of the stamens act like a catapult, flipping through a 60° angle; the times on the earlier photos show that this happens in just 0.30 ms. We can model a stamen tip as a 1.0-mm-long, 10 μg rigid rod with a 10 μg anther sac at one end and a pivot point at the opposite end. Though an oversimplification, we will model the motion by assuming the angular acceleration is constant throughout the motion.

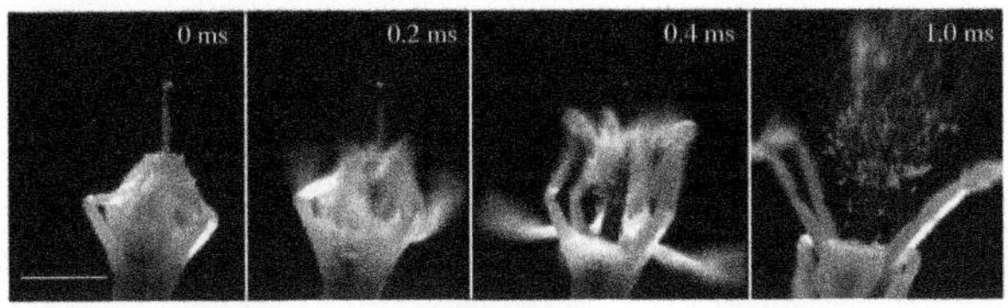

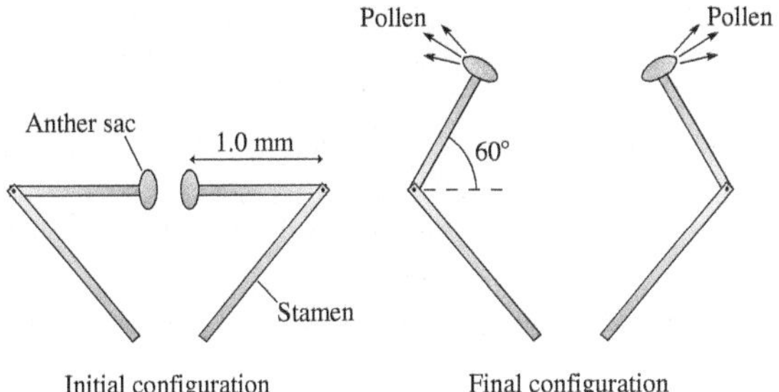

FIGURE P7.81

81. What is the angular acceleration of the anther sac during the motion?
A. 3.5×10^3 rad/s^2
B. 7.0×10^3 rad/s^2
C. 1.2×10^7 rad/s^2
D. 2.3×10^7 rad/s^2

82. What is the speed of the anther sac as it releases its pollen?
A. 3.5 m/s
B. 7.0 m/s
C. 10 m/s
D. 14 m/s

83. How large is the "straightening torque"? (You can omit gravitational forces from your calculation; the gravitational torque is much less than this.)
A. 2.3×10^{-7} N·m
B. 3.1×10^{-7} N·m
C. 2.3×10^{-5} N·m
D. 3.1×10^{-5} N·m

Lions and Zebras

Lions are capable of higher bursts of speed than the zebras that they pursue, but real chases in the wild generally aren't straight-line sprints. A zebra being chased by a lion runs in straight-line segments punctuated by rapid changes of direction. Data from chases in the wild show that both lions and zebras in an active chase run at speeds lower than they are capable of; the slower speeds allow for tighter turns. An extensive study found that a lion chasing a zebra maintained a constant speed of about 14 m/s, alternating between straight-line runs and turns with a centripetal acceleration of 6.5 m/s^2. Typical values for the zebras were a bit lower—a constant speed of 11 m/s and turns with an acceleration of 5.5 m/s^2.

84. For these values, which animal is capable of a tighter turn—that is, a circular turn with a smaller radius?
A. The zebra has a tighter turn.
B. The lion has a tighter turn.
C. The zebra and the lion have very nearly the same turn radius.

85. For these values, which animal is capable of a faster turn—that is, turning a 90° angle in less time?
A. The zebra has a faster turn.
B. The lion has a faster turn.
C. The zebra and the lion turn in very nearly the same amount of time.

86. Approximately what is the net force of the ground on a 150 kg lion as it makes a turn?
A. 1000 N
B. 1500 N
C. 1800 N
D. 2400 N

8 Momentum

These rams butt heads at high speeds in a ritual to assert their dominance. Thick skulls and a spongy cover protect their brains during these high-impact collisions.

LOOKING AHEAD

Impulse

This golf club delivers an *impulse* to the ball as the club strikes it. The size of the impulse determines the momentum of the ball as it flies away.

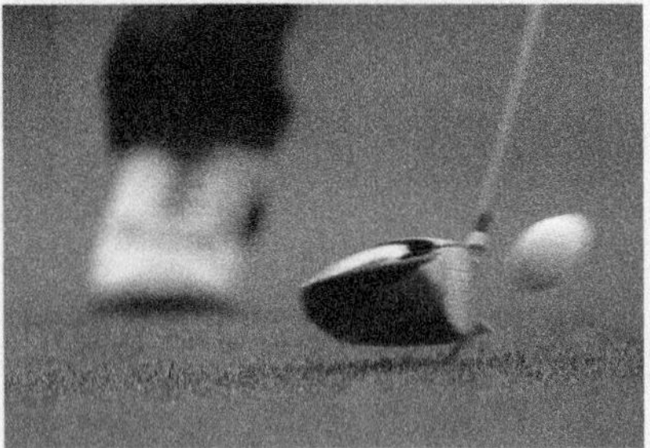

You'll learn that a longer-lasting, stronger force delivers a larger impulse to an object.

Conservation of Momentum

The heavy curling stone keeps moving in a straight line at nearly constant speed. It has a large amount of *momentum,* a tendency to keep going.

You'll see that the total momentum of an *isolated system* does not change. It is *conserved,* an important new idea.

Angular Momentum

The spinning diver also has momentum, but in this case it's *angular momentum,* a tendency to keep rotating.

You'll learn a new *before-and-after* problem-solving strategy for problems of momentum and angular momentum.

GOAL To learn about impulse, momentum, and a new problem-solving strategy based on conservation laws.

PHYSICS AND LIFE

Conservation Laws Constrain What an Organism Can Do

When two objects collide, whether animals or atoms, their subsequent motion is determined by their momentum. Momentum and a related idea, impulse, help us understand how protective sports equipment and their biological analogs can be designed to limit the damage from a collision. Momentum and impulse also explain how the collisions of molecules with the walls of a container produce the pressure exerted by a fluid. An essential idea is that momentum, along with energy and angular momentum, is *conserved;* they are never used up, just moved around. Conservation laws place limits on what an organism can do. Motion and other behaviors that would violate a conservation law simply can't occur. Conservation of momentum allows us to analyze situations that otherwise would be extremely difficult to understand, such as an ice skater using arm movements to speed up a spin or an aquatic organism using jet propulsion to flee from a predator.

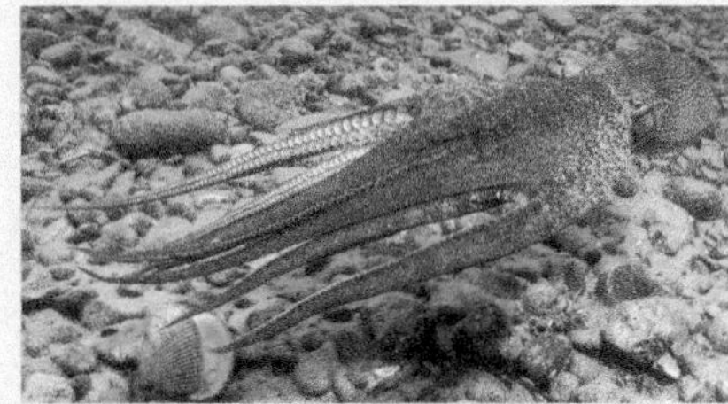

This octopus propels itself in one direction by squirting a jet of water in the opposite direction. The total momentum of the octopus and the water is conserved.

8.1 Momentum and Impulse

An important task in many sports is to cause a sudden large change in the velocity of our bodies (e.g., jumping) or of an object (e.g., by throwing it). Sudden collisions between athletes sometimes lead to serious injuries. Forces are very much involved in these events, but a direct application of Newton's laws is not always the best way to understand them. In this chapter, we'll develop a different way to look at interactions based on the concept of *momentum.* We'll also introduce the idea of a *conservation law,* one of the most important concepts in science.

A **collision** is a short-duration interaction between two objects that strike each other. The collision between a tennis ball and a racket, or a soccer ball and your foot, may seem instantaneous to your eye, but high-speed photography reveals that the side of the ball is significantly flattened during the collision. It takes time to compress the ball, and more time for the ball to re-expand as it leaves the racket or bat.

A tennis ball collides with a racket. Notice that the right side of the ball is flattened.

The duration of a collision depends on the materials from which the objects are made, but 1 to 10 ms (0.001 to 0.010 s) is fairly typical. This is the time during which the two objects are in contact with each other. The harder the objects, the shorter the contact time. A collision between two steel balls lasts less than 200 microseconds.

FIGURE 8.1 shows an object colliding with a wall. The object approaches with an initial horizontal velocity $(v_x)_i$, experiences a force of duration Δt, and leaves with final velocity $(v_x)_f$. Notice that the object, as in the photo above, *deforms* during the collision. A particle cannot be deformed, so we cannot model colliding objects as particles. Instead, we model a colliding object as an *elastic object* that compresses and then expands, much like a spring. Indeed, that's exactly what happens at the microscopic level during a collision: Molecular bonds compress, store elastic energy, then re-expand as the object rebounds.

FIGURE 8.1 A collision.

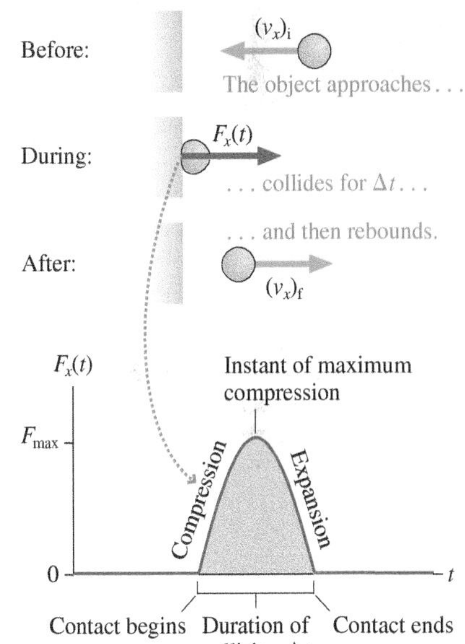

The force of a collision is usually very large in comparison to other forces exerted on the object. A large force exerted for a small interval of time is called an **impulsive force.** The graph of Figure 8.1 shows how a typical impulsive force behaves, rapidly growing to a maximum at the instant of maximum compression, then decreasing back to zero. The force is zero before contact begins and after contact ends. Because an impulsive force is a function of time, we will write it as $F_x(t)$.

NOTE ▶ Both v_x and F_x are components of vectors and thus have *signs* indicating which way the vectors point. ◀

We can use Newton's second law to find how the object's velocity changes as a result of the collision. Acceleration in one dimension is $a_x = dv_x/dt$, so the second law is

$$ma_x = m\frac{dv_x}{dt} = F_x(t)$$

After multiplying both sides by dt, we can write the second law as

$$m\,dv_x = F_x(t)\,dt \tag{8.1}$$

The force is nonzero only during an interval of time from t_i to $t_f = t_i + \Delta t$, so let's integrate Equation 8.1 over this interval. The velocity changes from $(v_x)_i$ to $(v_x)_f$ during the collision; thus

$$m\int_{v_i}^{v_f} dv_x = m(v_x)_f - m(v_x)_i = \int_{t_i}^{t_f} F_x(t)\,dt \tag{8.2}$$

We need some new tools to help us make sense of Equation 8.2.

Momentum

The product of an object's mass and velocity is called the *momentum* of the object:

$$\textbf{momentum} = \vec{p} = m\vec{v} \tag{8.3}$$

Momentum of an object of mass m and velocity $\vec{v}$

FIGURE 8.2 The momentum $\vec{p}$ can be decomposed into x- and y-components.

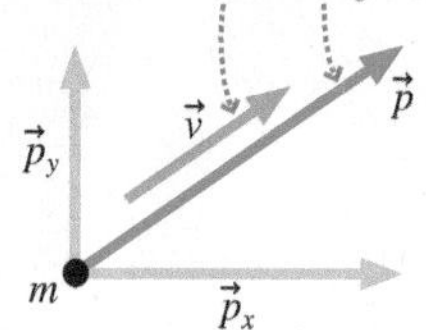
Momentum is a vector pointing in the same direction as the object's velocity.
v
p
p_y
m
p_x

The units of momentum are kg • m/s.

Momentum $\vec{p}$ is a vector. Specifically, momentum $\vec{p}$ points in the same direction as the velocity vector $\vec{v}$ because multiplying a vector by a scalar doesn't change the direction of the vector. FIGURE 8.2 shows that $\vec{p}$, like any vector, can be decomposed into x- and y-components. Equation 8.3, which is a vector equation, is a shorthand way to write the simultaneous equations

$$p_x = mv_x$$
$$p_y = mv_y$$

NOTE ▸ One of the most common errors in momentum problems is a failure to use the appropriate signs. **The momentum component p_x has the same sign as v_x.** Momentum is *negative* for an object moving to the left (on the x-axis) or down (on the y-axis). ◂

Legging it A frog making a jump wants to gain as much momentum as possible before leaving the ground. This means that it wants the greatest impulse $J = F_{avg}\,\Delta t$ delivered to it by the ground. There is a maximum force that muscles can exert, limiting F_{avg}. But the time interval Δt during which the force is exerted can be greatly increased by having long legs that can push against the ground for a longer amount of time. Many animals that are good jumpers have particularly long legs.

An object can have a large momentum by having either a small mass but a large velocity or a small velocity but a large mass. For example, a 5.5 kg (12 lb) bowling ball rolling at a modest 2 m/s has momentum of magnitude $p = (5.5\text{ kg})(2\text{ m/s}) = 11\text{ kg}\cdot\text{m/s}$. This is almost exactly the same momentum as a 9 g bullet fired from a high-speed rifle at 1200 m/s.

Newton actually formulated his second law in terms of momentum rather than acceleration:

$$\vec{F} = m\vec{a} = m\frac{d\vec{v}}{dt} = \frac{d(m\vec{v})}{dt} = \frac{d\vec{p}}{dt} \tag{8.4}$$

This statement of the second law, saying that force is the rate of change of momentum, is more general than our earlier version $\vec{F} = m\vec{a}$. It allows for the possibility that the mass of the object might change, such as a jetting squid or octopus losing mass as it expels water.

Returning to Equation 8.2, you can see that $m(v_x)_i$ and $m(v_x)_f$ are $(p_x)_i$ and $(p_x)_f$, the x-component of the object's momentum before and after the collision. Further, $(p_x)_f - (p_x)_i$ is Δp_x, the *change* in the object's momentum. In terms of momentum, Equation 8.2 is

$$\Delta p_x = (p_x)_f - (p_x)_i = \int_{t_i}^{t_f} F_x(t)\,dt \tag{8.5}$$

Now we need to examine the right-hand side of Equation 8.5.

FIGURE 8.3 Looking at the impulse graphically.

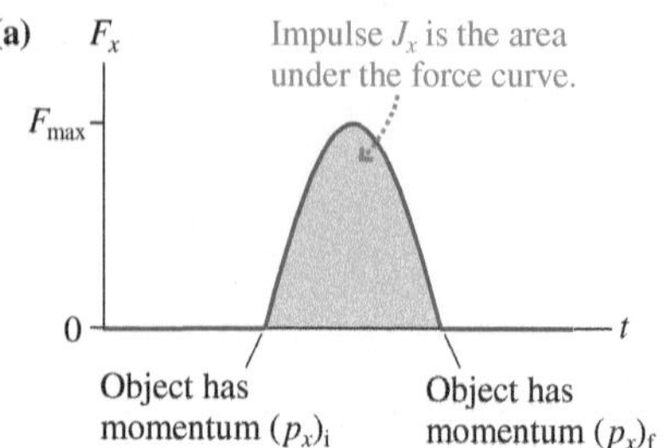
(a)
F_x
Impulse J_x is the area under the force curve.
F_{max}
0
t
Object has momentum $(p_x)_i$
Object has momentum $(p_x)_f$

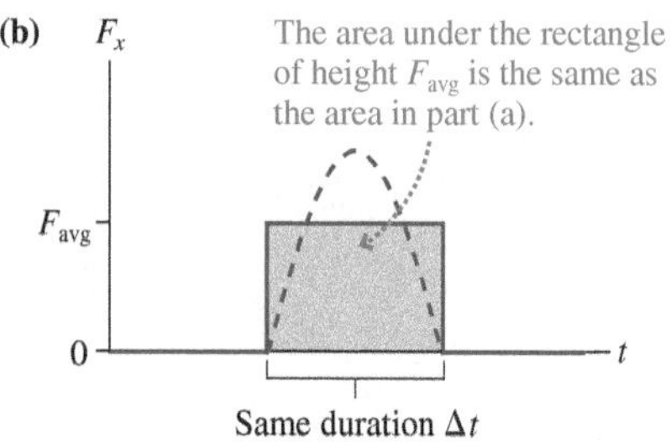
(b)
F_x
The area under the rectangle of height F_{avg} is the same as the area in part (a).
F_{avg}
0
t
Same duration Δt

Impulse

Equation 8.5 tells us that the object's change in momentum is related to the time integral of the force. Let's define a quantity J_x called the *impulse* to be

$$\textbf{impulse} = J_x = \int_{t_i}^{t_f} F_x(t)\,dt \tag{8.6}$$
$$= \text{area under the } F_x(t) \text{ curve between } t_i \text{ and } t_f$$

Impulse caused by a force

Strictly speaking, impulse has units of N • s, but you should be able to show that N • s are equivalent to kg • m/s, the units of momentum.

The interpretation of the integral in Equation 8.6 as an area under a curve is especially important. FIGURE 8.3a portrays the impulse graphically. Because the force changes in a complicated way during a collision, it is often useful to describe the collision in terms of an *average* force F_{avg}. As FIGURE 8.3b shows, F_{avg} is the height of a rectangle that has the same area, and thus the same impulse, as the real force curve. The impulse exerted during the collision is

$$J_x = F_{avg}\,\Delta t \tag{8.7}$$

Equation 8.2, which we found by integrating Newton's second law, can now be rewritten in terms of impulse and momentum as

$$\Delta p_x = J_x \tag{8.8}$$

The momentum principle

This result, called the **momentum principle,** says that **an impulse delivered to an object changes the object's momentum.** The momentum $(p_x)_f$ "after" an interaction, such as a collision or an explosion, is equal to the momentum $(p_x)_i$ "before" the interaction *plus* the impulse that arises from the interaction:

$$(p_x)_f = (p_x)_i + J_x \tag{8.9}$$

FIGURE 8.4 illustrates the momentum principle for a rubber ball bouncing off a wall. **Notice the signs; they are very important.** The ball is initially traveling toward the right, so $(v_x)_i$ and $(p_x)_i$ are positive. After the bounce, $(v_x)_f$ and $(p_x)_f$ are negative. The force *on the ball* is toward the left, so F_x is also negative. The graphs show how the force and the momentum change with time.

Although the interaction is very complex, the impulse—the area under the force graph—is all we need to know to find the ball's velocity as it rebounds from the wall. The final momentum is

$$(p_x)_f = (p_x)_i + J_x = (p_x)_i + \text{area under the force curve} \tag{8.10}$$

and the final velocity is $(v_x)_f = (p_x)_f/m$. In this example, the area has a negative value.

FIGURE 8.4 The momentum principle helps us understand a rubber ball bouncing off a wall.

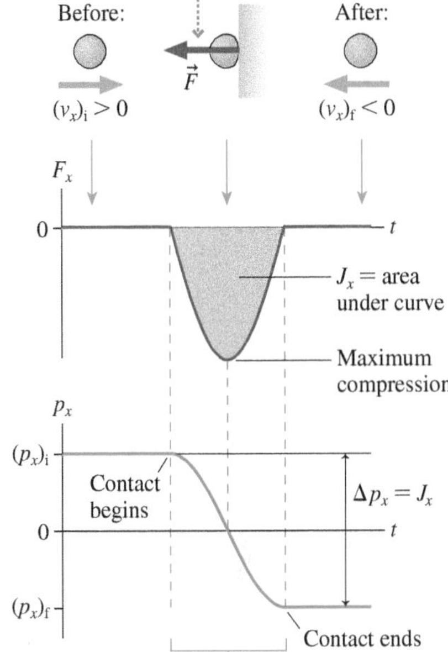

STOP TO THINK 8.1 The cart's change of momentum is

A. −30 kg • m/s
B. −20 kg • m/s
C. 0 kg • m/s
D. 10 kg • m/s
E. 20 kg • m/s
F. 30 kg • m/s

Before: 10 kg, 2 m/s
After: 1 m/s

Momentum Bar Charts

The momentum principle tells us that **impulse transfers momentum to an object.** If an object moving along the x-axis initially has 2 kg • m/s of momentum, a 1 kg • m/s impulse delivered to the object increases its momentum to 3 kg • m/s. That is, $(p_x)_f = (p_x)_i + J_x$.

We can represent this "momentum accounting" with a **momentum bar chart,** a tool for visualizing momentum changes. For example, the bar chart of FIGURE 8.5 represents the ball colliding with a wall in Figure 8.4. You can see that four units of negative impulse (impulse directed to the left) add to an initial two units of momentum to give two negative units of final momentum.

FIGURE 8.5 A momentum bar chart for the ball in Figure 8.4.

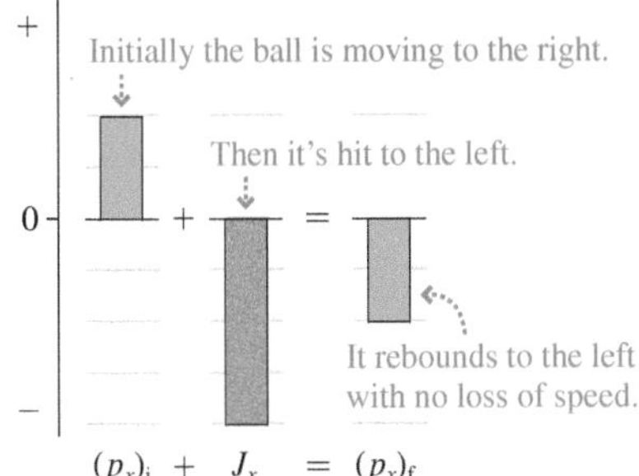

NOTE ▸ The vertical scale of a momentum bar chart has no numbers; it can be adjusted to match any problem. However, be sure that all bars in a given problem use a consistent scale. ◂

Reducing Impact

You might have seen, or even participated in, an egg-tossing contest. A pair of contestants toss an egg back and forth, taking a step away from each other after each successful catch. The winners are the pair who can toss an egg the farthest distance without breaking it. The winning strategy, as you might guess, is to pull your hands inward, toward your body, while catching the egg so that it slows down over the longest possible time and distance.

The momentum principle explains why. The egg undergoes a momentum change Δp_x as it stops when your hands apply an impulse $J_x = F_{avg}\,\Delta t$. Because $\Delta p_x = J_x$, the average force you apply to the egg to stop it in a time interval Δt is

$$F_{avg} = \frac{\Delta p_x}{\Delta t} \tag{8.11}$$

If you lengthen the time of the catch by moving your hands, you lessen the force that's needed and thus lessen the chance of breaking the egg.

This inverse relationship between force and time has many applications. Your car is designed with *crumple zones* to increase the duration of a head-on collision so as to reduce the impact force on passengers. Seat belts and air bags also bring occupants to rest over a longer interval of time. Similarly, a bike helmet is filled with a crushable material that, in an accident, brings the rider's head to rest more slowly than in a direct impact with the road. Lengthening collision times reduces impact forces and saves lives.

Some animals have evolved similar safeguards. The opening photograph in this chapter shows two male Rocky Mountain bighorn sheep engaged in a head-butting contest that seems almost certain to injure, if not kill, both participants. However, thin, flexible bones in their sinus cavities and a thick, spongy mass between a double-wall cranium stretch out the collision time during which the brain has to come to rest to roughly 0.25 s. The impacts, while dramatic, are not forceful enough to cause brain injury. Similarly, woodpeckers have spongy, elastic bones in their skulls that spread out the impact of hammering on tree trunks.

Solving Impulse and Momentum Problems

Pictorial representations are an important problem-solving tool. The pictorial representations you've been using are oriented toward Newton's laws and a subsequent kinematics analysis. Momentum problems and, soon, energy problems are better served with a new representation, a **before-and-after pictorial representation.** A before-and-after pictorial representation is composed of *two* drawings, labeled "Before" and "After," that show the objects *before* they interact and again *after* the interaction. Let's look at an example.

EXAMPLE 8.1 Hitting a baseball

A 150 g baseball is thrown to the left with a speed of 20 m/s. It is hit straight back toward the pitcher at a speed of 40 m/s. The interaction force between the ball and the bat is shown in FIGURE 8.6. What *maximum* force F_{max} does the bat exert on the ball? What is the *average* force of the bat on the ball?

FIGURE 8.6 The interaction force between the baseball and the bat.

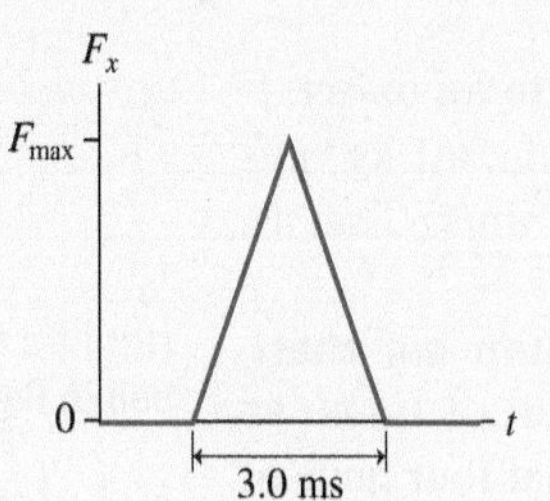

PREPARE We'll model the baseball as an elastic object and the interaction as a collision. We assume that the motion is along the x-axis. FIGURE 8.7 is a before-and-after pictorial representation. Before-and-after pictures are usually simpler than the pictures we used for dynamics problems, so listing the known information right on the sketch is adequate. Notice that we converted the statements about *speeds* into information about *velocities*, with $(v_x)_i$ negative.

SOLVE The momentum principle is

$$\Delta p_x = J_x = \text{area under the force curve}$$

FIGURE 8.7 A before-and-after pictorial representation.

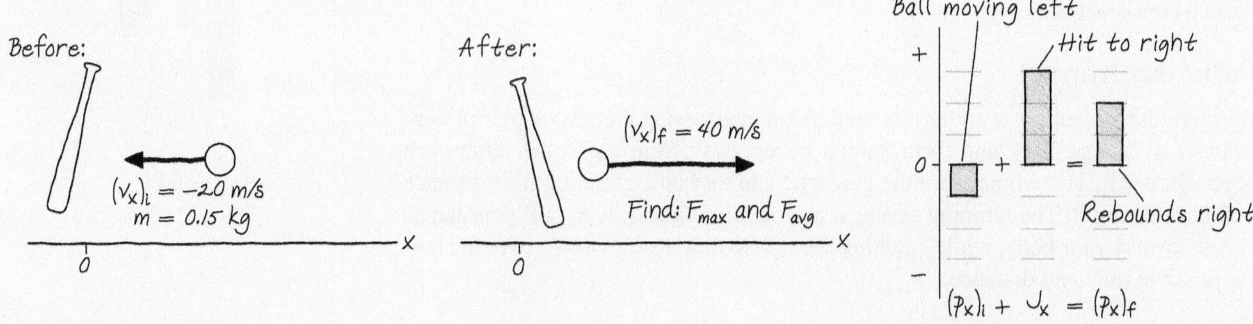

We know the velocities before and after the collision, so we can calculate the ball's momenta:

$$(p_x)_i = m(v_x)_i = (0.15\text{ kg})(-20\text{ m/s}) = -3.0\text{ kg}\cdot\text{m/s}$$
$$(p_x)_f = m(v_x)_f = (0.15\text{ kg})(40\text{ m/s}) = 6.0\text{ kg}\cdot\text{m/s}$$

Thus the *change* in momentum is

$$\Delta p_x = (p_x)_f - (p_x)_i = 9.0\text{ kg}\cdot\text{m/s}$$

The force curve is a triangle with height F_{max} and width 3.0 ms. The area under the curve is

$$J_x = \text{area} = \tfrac{1}{2}(F_{max})(0.0030\text{ s}) = (F_{max})(0.0015\text{ s})$$

Using this information in the momentum principle, we have

$$9.0\text{ kg}\cdot\text{m/s} = (F_{max})(0.0015\text{ s})$$

Thus the *maximum* force is

$$F_{max} = \frac{9.0\text{ kg}\cdot\text{m/s}}{0.0015\text{ s}} = 6000\text{ N}$$

The *average* force, which depends on the collision duration $\Delta t = 0.0030$ s, has the smaller value:

$$F_{avg} = \frac{J_x}{\Delta t} = \frac{\Delta p_x}{\Delta t} = \frac{9.0\text{ kg}\cdot\text{m/s}}{0.0030\text{ s}} = 3000\text{ N}$$

ASSESS F_{max} is a large force, but quite typical of the impulsive forces during collisions. The main thing to focus on is our new perspective: An impulse changes the momentum of an object.

Other forces often act on an object during a collision or other brief interaction. In Example 8.1, for instance, the baseball is also acted on by gravity. Usually these other forces are *much* smaller than the interaction forces. The 1.5 N weight of the ball is vastly less than the 3000 N average force of the bat on the ball. We can reasonably neglect these small forces *during* the brief time of the impulsive force by using what is called the **impulse approximation.**

When we use the impulse approximation, $(p_x)_i$ and $(p_x)_f$ (and $(v_x)_i$ and $(v_x)_f$) are the momenta (and velocities) *immediately* before and *immediately* after the collision. For example, the velocities in Example 8.1 are those of the ball just before and after it collides with the bat. We could then do a follow-up problem, including gravity and drag, to find the ball's speed a second later as the second baseman catches it. We'll look at some two-part examples later in the chapter.

A spiny cushion The spines of a hedgehog obviously help protect it from predators. But they serve another function as well. If a hedgehog falls from a tree—a not uncommon occurrence—it simply rolls itself into a ball before it lands. Its thick spines then cushion the blow by increasing the time it takes for the animal to come to rest. Indeed, hedgehogs have been observed to fall out of trees on purpose to get to the ground!

STOP TO THINK 8.2 A 10 g rubber ball and a 10 g clay ball are thrown at a wall with equal speeds. The rubber ball bounces, the clay ball sticks. Which ball delivers a larger impulse to the wall?

A. The clay ball delivers a larger impulse because it sticks.
B. The rubber ball delivers a larger impulse because it bounces.
C. They deliver equal impulses because they have equal momenta.
D. Neither delivers an impulse to the wall because the wall doesn't move.

8.2 Conservation of Momentum

The momentum principle was derived from Newton's second law and is really just an alternative way of looking at single-particle dynamics. To discover the real power of momentum for problem solving, we need also to invoke Newton's third law, which will lead us to one of the most important principles in physics: conservation of momentum.

FIGURE 8.8 shows two objects with initial velocities $(v_{1x})_i$ and $(v_{2x})_i$. The objects collide, then bounce apart with final velocities $(v_{1x})_f$ and $(v_{2x})_f$. The forces during the collision, as the objects are interacting, are the interaction pair $\vec{F}_{1\text{ on }2}$ and $\vec{F}_{2\text{ on }1}$. We'll continue to assume that the motion is one dimensional along the x-axis.

NOTE ▸ The notation, with all the subscripts, may seem excessive. But there are two objects, and each has an initial and a final velocity, so we need to distinguish among four different velocities. ◂

Newton's third law tells us that the interaction forces $\vec{F}_{1\text{ on }2}$ and $\vec{F}_{2\text{ on }1}$ are equal in magnitude but opposite in direction. We saw in Equation 8.4 that Newton's second

FIGURE 8.8 A collision between two objects.

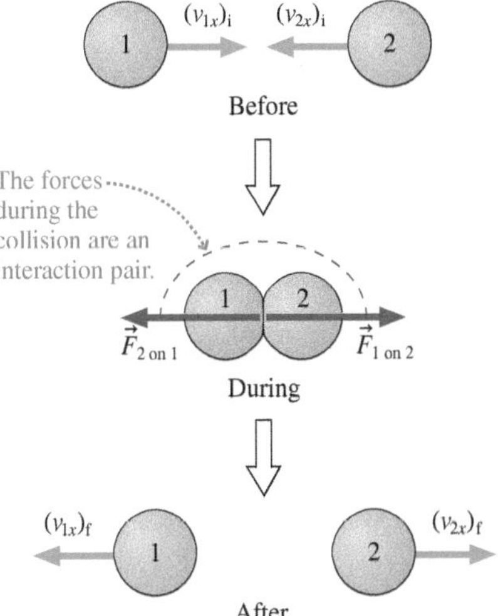

law can be written in terms of momentum as $\vec{F} = d\vec{p}/dt$. Thus the x-component of the third law is

$$(F_x)_{1 \text{ on } 2} = \frac{dp_{2x}}{dt} = -(F_x)_{2 \text{ on } 1} = -\frac{dp_{1x}}{dt} \tag{8.12}$$

The sum $p_{1x} + p_{2x}$ is the *total momentum* of the two objects along the x-axis. Because $dp_{2x}/dt = -dp_{1x}/dt$, we have

$$\frac{d}{dt}(p_{1x} + p_{2x}) = \frac{dp_{1x}}{dt} + \frac{dp_{2x}}{dt} = 0 \tag{8.13}$$

You know from calculus that if the time derivative of $p_{1x} + p_{2x}$ is zero, it must be the case that

$$p_{1x} + p_{2x} = \text{constant} \tag{8.14}$$

In other words, the total momentum $p_{1x} + p_{2x}$ is *not changed* by the collision; it has the same value *after* the collision as it did *before* the collision. We can write this as

$$(p_{1x})_\text{f} + (p_{2x})_\text{f} = (p_{1x})_\text{i} + (p_{2x})_\text{i} \tag{8.15}$$

Rocketing ahead The total momentum of the rocket + gases system is conserved, so the rocket accelerates forward as the gases are expelled backward.

We say that the total momentum $p_{1x} + p_{2x}$ is **conserved,** which means it is unchanged by the interaction. Furthermore, this conservation is independent of the interaction force; we don't need to know anything about forces $\vec{F}_{1 \text{ on } 2}$ and $\vec{F}_{2 \text{ on } 1}$ to make use of Equation 8.15.

As an example, **FIGURE 8.9** is a before-and-after pictorial representation of two train cars colliding and coupling. Equation 8.15 relates the momenta of the cars after the collision to their momenta before the collision:

$$m_1(v_{1x})_\text{f} + m_2(v_{2x})_\text{f} = m_1(v_{1x})_\text{i} + m_2(v_{2x})_\text{i}$$

Initially, car 1 is moving with velocity $(v_{1x})_\text{i} = v_\text{i}$ while car 2 is at rest. Afterward, they roll together with the common final velocity v_f. Furthermore, car 2 is twice as heavy as car 1, with $m_1 = m$ and $m_2 = 2m$. With this information, the momentum equation is

$$mv_\text{f} + 2mv_\text{f} = mv_\text{i} + 0$$

The mass cancels, and we find that the train cars' final speed is $v_\text{f} = \frac{1}{3}v_\text{i}$. That is, we can predict that the final speed is exactly one-third the initial speed of car 1 without knowing anything about the complex interaction between the two cars as they collide.

FIGURE 8.9 Two colliding train cars.

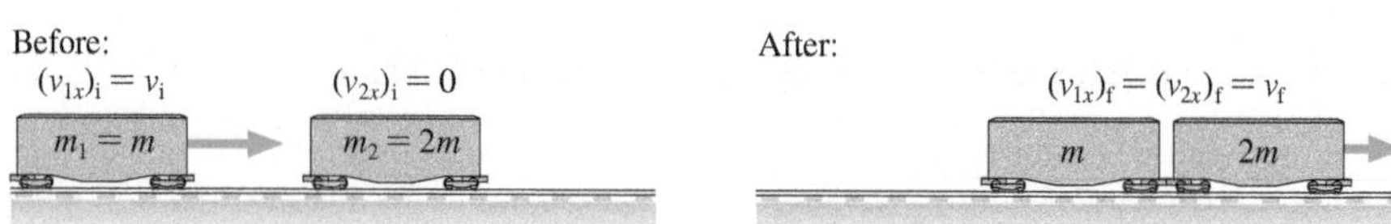

Total Momentum

Equation 8.15 shows that the quantity $p_{1x} + p_{2x}$ is constant; it has the same value after a two-body interaction as it did before. However, the analysis leading to Equation 8.15 was for the specific instance of a collision between two objects. We'll skip the mathematical details, but it's probably not surprising that we can extend this idea to a larger system.

Suppose we have a system of N objects interacting both with each other and perhaps with external forces from the environment. We define the **total momentum** $\vec{P}$ of the system to be

$$\vec{P} = \vec{p}_1 + \vec{p}_2 + \vec{p}_3 + \cdots + \vec{p}_N \tag{8.16}$$

That is, the total momentum is the vector sum of all the individual momenta. Note that **total momentum is a property of the system,** not of any individual object. This is our first encounter with a *system property,* but it won't be the last.

The result of the full analysis is that

$$\frac{d\vec{P}}{dt} = \vec{F}_{ext} \tag{8.17}$$

where $\vec{F}_{ext}$ is the net force exerted on the system by agents *outside* the system. But this is just Newton's second law for the system as a whole. That is, **the rate of change of the total momentum of the system is equal to the net force applied to the system.**

Equation 8.17 tells us that any interactions between objects *inside* the system do not change the total momentum of the system. That was the implication of Equation 8.15 for two objects, but the principle is general. Only external forces change the total momentum. In fact, we've been using this idea all along as an assumption of the particle model. When we model balls and cars as particles, we assume that the internal forces between the atoms—the interactions that hold the object together—do not affect the motion of the object as a whole.

Isolated Systems

Consider a system on which there are *no* external forces or in which the external forces are balanced and add to zero. We call this an **isolated system.** An isolated system is not influenced or affected by external forces from the environment.

For an isolated system, Equation 8.17 is simply

$$\frac{d\vec{P}}{dt} = \vec{0} \quad \text{(isolated system)} \tag{8.18}$$

In other words, **the *total* momentum of an isolated system is conserved; it does not change.** The total momentum $\vec{P}$ remains constant, *regardless* of whatever interactions are going on *inside* the system. The importance of this result is sufficient to elevate it to a law of nature, alongside Newton's laws.

> **Law of conservation of momentum** The total momentum $\vec{P}$ of an isolated system is a constant. Interactions within the system do not change the system's total momentum. Mathematically, the law of conservation of momentum is
>
> $$\vec{P}_f = \vec{P}_i \tag{8.19}$$

The total momentum *after* an interaction is equal to the total momentum *before* the interaction. Because Equation 8.19 is a vector equation, the equality is true for each of the components of the momentum vector. That is,

$$\begin{aligned}
&\text{x-component:}\quad \overbrace{(p_{1x})_f + (p_{2x})_f + (p_{3x})_f + \cdots}^{\text{Final momentum}} = \overbrace{(p_{1x})_i + (p_{2x})_i + (p_{3x})_i + \cdots}^{\text{Initial momentum}}\\
&\qquad\qquad\qquad\text{Object 1}\quad\text{Object 2}\quad\text{Object 3}\\
&\text{y-component:}\quad (p_{1y})_f + (p_{2y})_f + (p_{3y})_f + \cdots = (p_{1y})_i + (p_{2y})_i + (p_{3y})_i + \cdots
\end{aligned} \tag{8.20}$$

The x-equation is an extension of Equation 8.15 to N interacting objects.

NOTE ► It is worth emphasizing the critical role of Newton's third law. The law of conservation of momentum is a direct consequence of the fact that interactions within an isolated system are interaction pairs. ◄

A Strategy for Conservation of Momentum Problems

PROBLEM-SOLVING STRATEGY 8.1 **Conservation of momentum**

PREPARE Clearly define the *system.*

- If possible, choose a system that is isolated ($\vec{F}_{\text{net}} = \vec{0}$) or within which the interactions are sufficiently short and intense that you can ignore external forces for the duration of the interaction (the impulse approximation). Momentum is conserved.
- If it's not possible to choose an isolated system, try to divide the problem into parts such that momentum is conserved during one segment of the motion. Other segments of the motion can be analyzed using Newton's laws or, as you'll learn in upcoming chapters, conservation of energy.
- Draw a before-and-after pictorial representation. Establish a coordinate system. Define symbols that will be used in the problem, list known values, and identify what you're trying to find.

SOLVE The mathematical representation is based on the law of conservation of momentum: $\vec{P}_{\text{f}} = \vec{P}_{\text{i}}$. In component form, this is

$$(p_{1x})_{\text{f}} + (p_{2x})_{\text{f}} + (p_{3x})_{\text{f}} + \cdots = (p_{1x})_{\text{i}} + (p_{2x})_{\text{i}} + (p_{3x})_{\text{i}} + \cdots$$
$$(p_{1y})_{\text{f}} + (p_{2y})_{\text{f}} + (p_{3y})_{\text{f}} + \cdots = (p_{1y})_{\text{i}} + (p_{2y})_{\text{i}} + (p_{3y})_{\text{i}} + \cdots$$

ASSESS Check that your result has correct units and significant figures, is reasonable, and answers the question.

EXAMPLE 8.2 **The speed of a squid**

Squid rely on jet propulsion when a rapid escape is necessary. A 1.5 kg squid at rest slowly pulls 100 g of water into its mantle, then, with a strong muscular contraction, ejects this water at a remarkable 45 m/s. How fast is the squid moving immediately after this ejection?

PREPARE Let's define the system to consist of the squid and the water it has pulled into its mantle; the surrounding ocean water is the environment. This is *not* an isolated system because the drag force of the water is an external force that will slow the squid. However, the drag force will not have a significant influence during the very brief time when the water is being ejected, so we can model the system as being isolated for the short duration of the ejection. The total momentum of an isolated system is conserved; this allows us to find the squid's speed *immediately* after the ejection. The motion takes place along a line, so we'll establish a coordinate system with the squid moving in the positive x-direction and the ejected water in the negative x-direction. FIGURE 8.10 is a before-and-after pictorial representation of the situation. Momentum depends on velocity, not speed, so notice that the final velocity of the water has a negative value. The initial total momentum is $\vec{P}_{\text{i}} = \vec{0}$ because both the squid and the water are at rest. Consequently, the total momentum will be $\vec{P}_{\text{f}} = \vec{0}$ *after* the water has been ejected.

FIGURE 8.10 Before-and-after pictorial representation of a jetting squid.

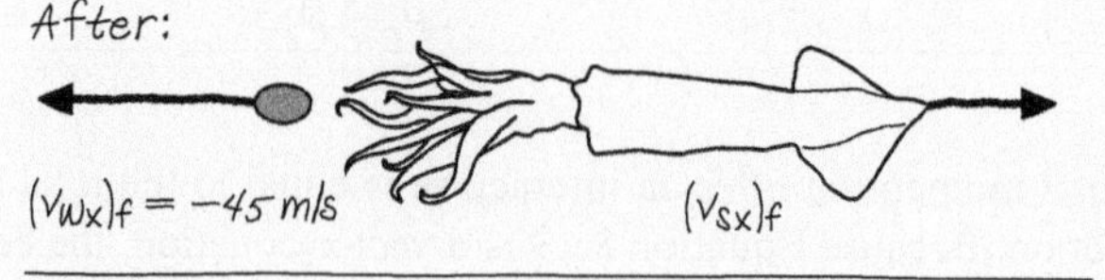

SOLVE The motion is linear, so we need to consider only the x-components of momentum. The initial total momentum is zero because the squid and the water are at rest. We can write the final momentum of the squid as $m_{\text{S}}(v_{\text{S}x})_{\text{f}}$ and that of the ejected water as $m_{\text{W}}(v_{\text{W}x})_{\text{f}}$. Thus the law of conservation of momentum is

$$(P_x)_{\text{f}} = m_{\text{S}}(v_{\text{S}x})_{\text{f}} + m_{\text{W}}(v_{\text{W}x})_{\text{f}} = (P_x)_{\text{i}} = 0$$

Solving for $(v_{\text{S}x})_{\text{f}}$, we find

$$(v_{\text{S}x})_{\text{f}} = -\frac{m_{\text{W}}}{m_{\text{S}}}(v_{\text{W}x})_{\text{f}} = -\frac{0.10\text{ kg}}{1.5\text{ kg}}(-45\text{ m/s}) = 3.0\text{ m/s}$$

The squid jets away with a speed of 3.0 m/s. Both the squid and the water are now moving, but their total momentum is still zero because the x-component of the water's momentum is negative.

ASSESS Nature documentaries show that squids can jet very quickly over short distances, so 3.0 m/s $\approx$ 6 mph seems reasonable. We were able to use momentum conservation to find the squid's speed immediately after the water ejection, but for longer periods of time, the squid + water system is not isolated and its momentum is not conserved. We would need to use Newton's second law and the drag force of the water to find out how the squid subsequently slows down. You must be careful to apply conservation laws only to isolated systems.

EXAMPLE 8.3 Rolling away

Bob sees a stationary cart 8.0 m in front of him. He decides to run to the cart as fast as he can, jump on, and roll down the street. Bob has a mass of 75 kg and the cart's mass is 25 kg. If Bob accelerates at a steady 1.0 m/s^2, what is the cart's speed just after Bob jumps on?

PREPARE This is a two-part problem. First Bob accelerates across the ground. Then Bob lands on and sticks to the cart, a "collision" between Bob and the cart. The interaction forces between Bob and the cart (i.e., friction) act only over the fraction of a second it takes Bob's feet to become stuck to the cart. We can model the system Bob + cart as an isolated system during the brief interval of the "collision," and thus the total momentum of Bob + cart is conserved during this interaction. But the system Bob + cart is *not* an isolated system for the entire problem because Bob's initial acceleration has nothing to do with the cart.

Our strategy is to divide the problem into an *acceleration* part, which we can analyze using kinematics, and a *collision* part, which we can analyze with momentum conservation. The pictorial representation of FIGURE 8.11 includes information about both parts. With three points in time, we've used the numerical subscripts 0, 1, and 2 to distinguish these points. Notice that Bob's velocity $(v_{Bx})_1$ at the end of his run is his "before" velocity for the collision.

SOLVE The first part of the mathematical representation is kinematics. We don't know how long Bob accelerates, but we do know his acceleration and the distance. Thus

$$(v_{Bx})_1^2 = (v_{Bx})_0^2 + 2a_x\,\Delta x = 2a_x x_1$$

His velocity after accelerating for 8.0 m is

$$(v_{Bx})_1 = \sqrt{2a_x x_1} = 4.0\text{ m/s}$$

The second part of the problem, the collision, uses conservation of momentum: $(P_x)_2 = (P_x)_1$. Equation 8.20 is

$$m_B(v_{Bx})_2 + m_C(v_{Cx})_2 = m_B(v_{Bx})_1 + m_C(v_{Cx})_1 = m_B(v_{Bx})_1$$

where we've used $(v_{Cx})_1 = 0$ m/s because the cart starts at rest. In this problem, Bob and the cart move together at the end with a common velocity, so we can replace both $(v_{Bx})_2$ and $(v_{Cx})_2$ with simply $(v_x)_2$. Solving for $(v_x)_2$, we find

$$(v_x)_2 = \frac{m_B}{m_B + m_C}(v_{Bx})_1 = \frac{75\text{ kg}}{100\text{ kg}} \times 4.0\text{ m/s} = 3.0\text{ m/s}$$

The cart's speed is 3.0 m/s immediately after Bob jumps on.

ASSESS Bob's final speed is a bit less than the top speed of his run, which makes sense.

FIGURE 8.11 Pictorial representation of Bob and the cart.

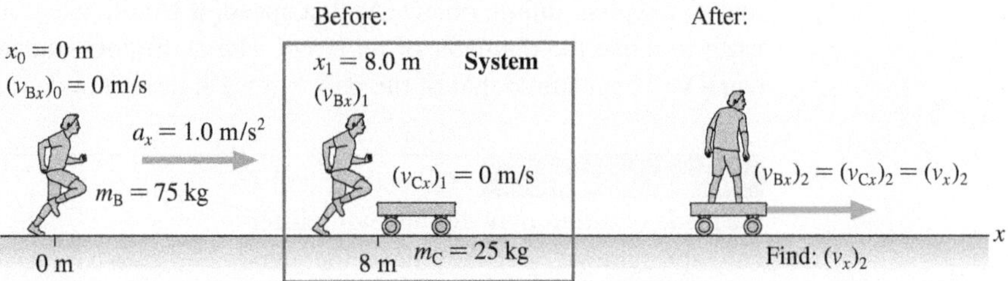

Notice how easy this was! No forces, no need to use kinematic equations. Conservation laws are powerful tools for making before-and-after comparisons, and we will use them extensively throughout this text. At the same time, it's important to recognize when a conservation law is *not* the appropriate tool. If we want to find how far Bob slides across the cart before sticking to it or how long the slide takes, then we need to use Newton's laws. Newton's laws deal with the *details* of the interaction, whereas conservation laws only compare the initial and final situations. Each has its place, depending on the question asked, and we'll look at how the wording of a problem provides clues for selecting the right tool.

Choosing the System

The first step in the conservation of momentum problem-solving strategy is to clearly define the system. **The goal is to choose a system whose momentum will be conserved.** Even then, it is the *total* momentum of the system that is conserved, not the momenta of the individual objects within the system.

As an example, consider what happens if you drop a rubber ball and let it bounce off a hard floor. Is momentum conserved? You might be tempted to answer yes because the ball's rebound speed is very nearly equal to its impact speed. But there are two errors in this reasoning.

First, momentum depends on *velocity,* not speed. The ball's velocity and momentum vectors reverse direction during the collision. Even if their magnitudes are equal, the ball's momentum after the collision is *not* equal to its momentum before the collision.

FIGURE 8.12 Whether or not momentum is conserved as a ball falls to earth depends on your choice of the system.

(a)

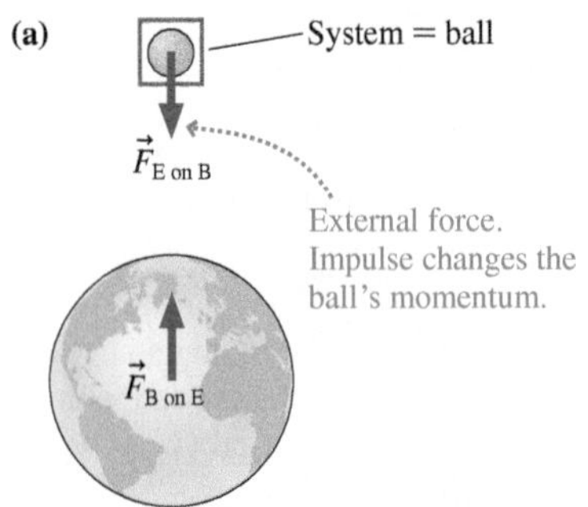

(b)

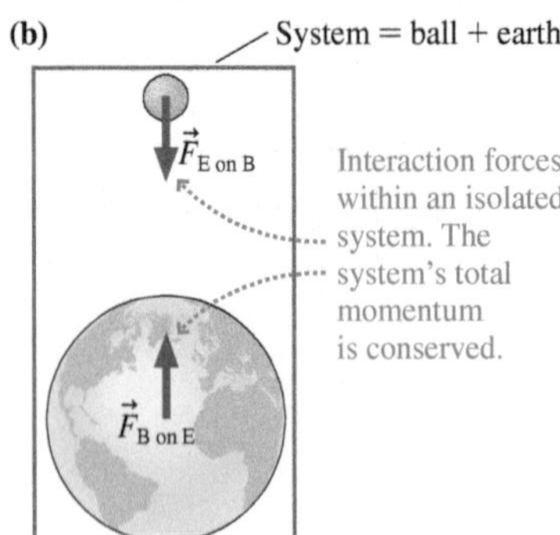

But more important, we haven't defined the system. The momentum of what? Whether or not momentum is conserved depends on the system. FIGURE 8.12 shows two different choices of systems. In Figure 8.12a, where the ball itself is chosen as the system, the gravitational force of the earth on the ball is an external force. This force causes the ball to accelerate toward the earth, changing the ball's momentum. When the ball hits, the force of the floor on the ball is also an external force. The impulse of $\vec{F}_{\text{floor on ball}}$ changes the ball's momentum from "down" to "up" as the ball bounces. The momentum of this system is most definitely *not* conserved.

Figure 8.12b shows a different choice. Here the system is ball + earth. Now the gravitational forces and the impulsive forces of the collision are interactions *within* the system. This is an isolated system, so the *total* momentum $\vec{P} = \vec{p}_{\text{ball}} + \vec{p}_{\text{earth}}$ is conserved.

In fact, the total momentum (in this reference frame) is $\vec{P} = \vec{0}$ because both the ball and the earth are initially at rest. The ball accelerates toward the earth after you release it, while the earth—due to Newton's third law—accelerates toward the ball in such a way that their individual momenta are always equal but opposite.

Why don't we notice the earth "leaping up" toward us each time we drop something? Because the earth's mass is enormous compared to everyday objects—roughly 10^{25} times larger. Momentum is the product of mass and velocity, so the earth would need an "upward" speed of only about 10^{-25} m/s to match the momentum of a typical falling object. At that speed, it would take 300 million years for the earth to move the diameter of an atom! The earth does, indeed, have a momentum equal and opposite to that of the ball, but we'll never notice it.

STOP TO THINK 8.3 Objects A and C are made of different materials, with different "springiness," but they have the same mass and are initially at rest. When ball B collides with object A, the ball ends up at rest. When ball B is thrown with the same speed and collides with object C, the ball rebounds to the left. Compare the speeds of A and C after the collisions. Is v_A greater than, equal to, or less than v_C?

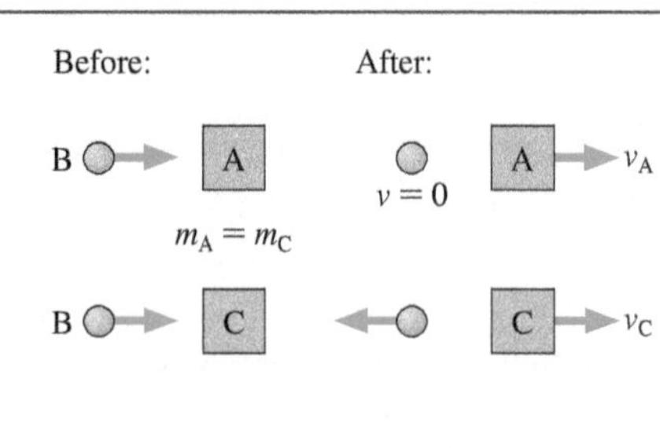

8.3 Collisions and Explosions

Collisions, with two or more objects running together, and *explosions,* with two or more objects flying apart, are very much the opposites of each other. What they have in common is that the objects form an isolated system, so the system's total momentum is conserved.

FIGURE 8.13 An inelastic collision.

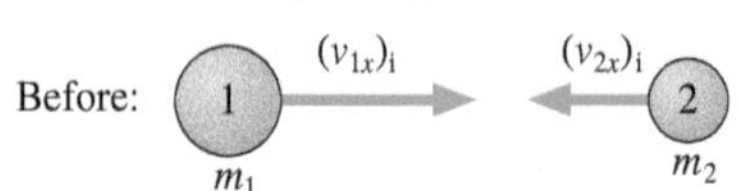

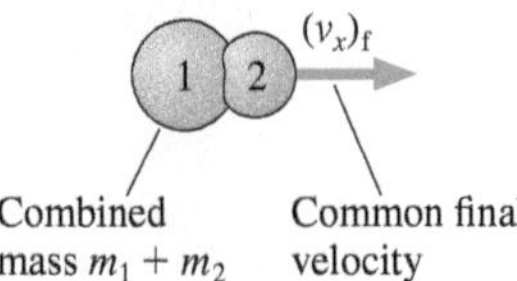

Inelastic Collisions

A collision in which the two objects stick together and move with a common final velocity is called a **perfectly inelastic collision.** Examples include railroad cars coupling, balls of clay running into each other, and a dart hitting a dart board. As FIGURE 8.13 shows, the key to analyzing a perfectly inelastic collision is the fact that **the two objects have a common final velocity.**

EXAMPLE 8.4 **An inelastic glider collision**

In a laboratory experiment, a 200 g air-track glider and a 400 g air-track glider are pushed toward each other from opposite ends of the track. The gliders have Velcro tabs on the front and will stick together when they collide. The 200 g glider is pushed with an initial speed of 3.0 m/s. The collision causes it to reverse direction at 0.40 m/s. What was the initial speed of the 400 g glider?

PREPARE We define the system to be the two gliders. This is an isolated system, so its total momentum is conserved in the collision. The gliders stick together, so this is a perfectly inelastic collision.

FIGURE 8.14 shows a pictorial representation. We've chosen to let the 200 g glider (glider 1) start out moving to the right, so $(v_{1x})_i$ is a positive 3.0 m/s. The gliders move to the left after the collision, so their common final velocity is $(v_x)_f = -0.40$ m/s.

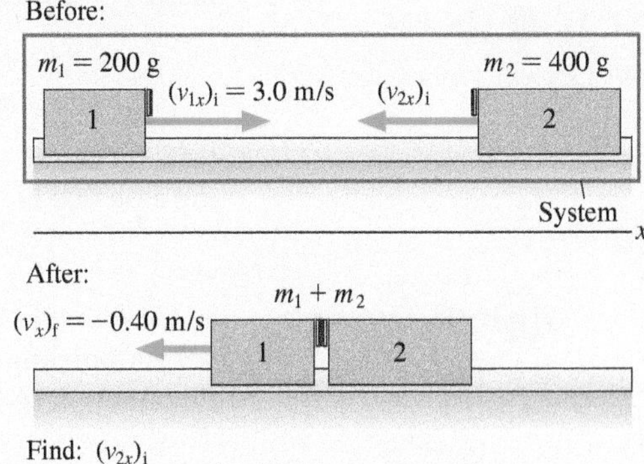

FIGURE 8.14 The before-and-after pictorial representation of an inelastic collision.

SOLVE The law of conservation of momentum, $(P_x)_f = (P_x)_i$, is

$$(m_1 + m_2)(v_x)_f = m_1(v_{1x})_i + m_2(v_{2x})_i$$

where we made use of the fact that the combined mass $m_1 + m_2$ moves together after the collision. We can easily solve for the initial velocity of the 400 g glider:

$$(v_{2x})_i = \frac{(m_1 + m_2)(v_x)_f - m_1(v_{1x})_i}{m_2}$$

$$= \frac{(0.60\text{ kg})(-0.40\text{ m/s}) - (0.20\text{ kg})(3.0\text{ m/s})}{0.40\text{ kg}}$$

$$= -2.1\text{ m/s}$$

The negative sign indicates that the 400 g glider started out moving to the left. The initial *speed* of the glider, which we were asked to find, is 2.1 m/s.

ASSESS The post-collision speed is very small, so the gliders must have started with nearly equal momenta in opposite directions. Glider 2 has twice the mass of glider 1, so its initial speed must have been roughly half that of glider 1. A speed of 2.1 m/s is a bit more than half, but close, and therefore a reasonable answer.

STOP TO THINK 8.4 The two objects are both moving to the right. Object 1 catches up with object 2 and collides with it. The objects stick together and continue on with speed v_f. Which of these statements is true?

A. v_f is greater than v_1.
B. $v_f = v_1$
C. v_f is greater than v_2 but less than v_1.
D. $v_f = v_2$
E. v_f is less than v_2.
F. Can't tell without knowing the masses.

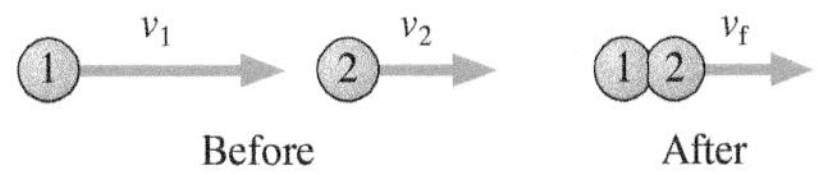

v_1 v_2 v_f
1 2 12
Before After

Elastic Collisions

A ball of clay doesn't bounce when it hits the floor; its kinetic energy of motion is *dissipated* as an increase in thermal energy, raising the temperature of the ball and the floor. In contrast, the **perfectly elastic collision** of FIGURE 8.15 is a perfect bounce in which kinetic energy is conserved. A perfectly elastic collision is an idealization, like a frictionless surface, but collisions between two very hard objects, such as two billiard balls or two steel balls, come close to being perfectly elastic.

The analysis of perfectly elastic collisions requires both the law of conservation of momentum and the law of conservation of energy; we will take up the latter in Chapter 11. Now we'll present the results without proof for the special case in which *ball 2 is initially at rest:*

FIGURE 8.15 A perfectly elastic collision.

Before:
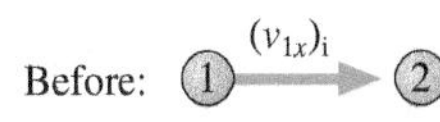

During:
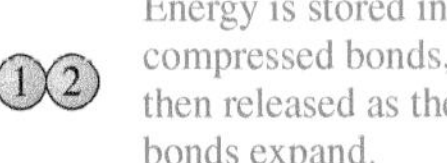

After:
1 2
$(v_{1x})_f$ $(v_{2x})_f$

$$(v_{1x})_f = \frac{m_1 - m_2}{m_1 + m_2}(v_{1x})_i \qquad (v_{2x})_f = \frac{2m_1}{m_1 + m_2}(v_{1x})_i \tag{8.21}$$

Perfectly elastic collision with ball 2 initially at rest

FIGURE 8.16 Three special elastic collisions.

Case a: $m_1 = m_2$

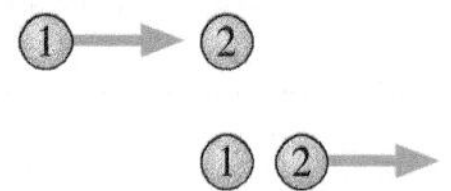

Ball 1 stops. Ball 2 goes forward with $(v_{2x})_f = (v_{1x})_i$.

Case b: $m_1 \gg m_2$

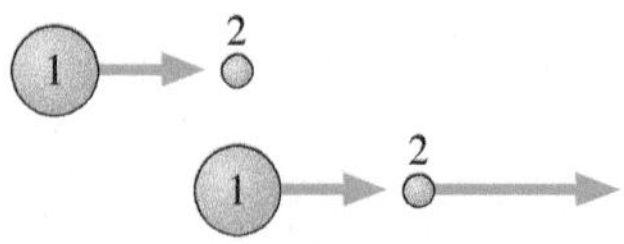

Ball 1 hardly slows down. Ball 2 is knocked forward at $(v_{2x})_f \approx 2(v_{1x})_i$.

Case c: $m_1 \ll m_2$

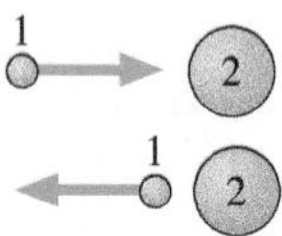

Ball 1 bounces off ball 2 with almost no loss of speed. Ball 2 hardly moves.

Equations 8.21 allow us to compute the final velocity of each ball. These equations are a little difficult to interpret, so let us look at the three special cases shown in **FIGURE 8.16**.

Case a: $m_1 = m_2$. This is the case of one billiard ball striking another of equal mass. For this case, Equations 8.16 give

$$(v_{1x})_f = 0 \qquad (v_{2x})_f = (v_{1x})_i$$

Case b: $m_1 \gg m_2$. This is the case of a bowling ball running into a Ping-Pong ball. We do not want an exact solution here, but an approximate solution for the limiting case that $m_1 \to \infty$. Equations 8.16 in this limit give

$$(v_{1x})_f \approx (v_{1x})_i \qquad (v_{2x})_f \approx 2(v_{1x})_i$$

Case c: $m_1 \ll m_2$. Now we have the reverse case of a Ping-Pong ball colliding with a bowling ball. Here we are interested in the limit $m_1 \to 0$, in which case Equations 8.16 become

$$(v_{1x})_f \approx -(v_{1x})_i \qquad (v_{2x})_f \approx 0$$

These cases agree well with our expectations and give us confidence that Equations 8.16 accurately describe a perfectly elastic collision.

NOTE ▸ Most collisions are *partially elastic*—somewhere between perfectly elastic and perfectly inelastic. However, the math for partially elastic collisions is beyond the scope of this textbook, as is the math for perfectly elastic collisions in which both objects are moving. We'll limit our examples and problems to the two ideal collisions. ◂

One important situation in which a collision really is perfectly elastic, with no dissipation of energy, is a collision between two atoms or molecules. The reasons are rooted in quantum physics, but you can see that it must be true by thinking about a gas. The molecules in a gas undergo uncountable billions of collisions every second. If even the tiniest bit of energy were dissipated in these collisions, the molecules would gradually slow down until eventually all motion would cease. That doesn't happen, so the collisions must be perfectly elastic.

EXAMPLE 8.5 Molecular collisions

You'll learn in a few chapters that a nitrogen molecule in room-temperature air is likely to be moving at a speed of approximately 500 m/s. Suppose that a nitrogen molecule (N_2) at this speed has a head-on collision with an oxygen molecule (O_2) that, although this is extremely unlikely, happens to be at rest. What are the speed and direction of each molecule after the collision?

PREPARE The two molecules form an isolated system, at least for the short interval of time during their collision, so the interaction is a perfectly elastic collision that conserves both momentum and energy. Equations 8.21 are for ball 2 at rest, so we'll designate the nitrogen molecule ball 1 with $(v_{1x})_i = 500$ m/s. The masses of nitrogen and oxygen (ball 2) are found by doubling the atomic masses listed in the periodic table. This gives 28 u and 32 u, respectively, where 1 u = 1 *atomic mass unit,* equivalent to 1 dalton. The equations use the *ratios* of masses, so we can leave both masses in u and do not have to convert them to kg.

SOLVE From Equations 8.21, we find:

$$(v_{1x})_f = \frac{28\text{ u} - 32\text{ u}}{28\text{ u} + 32\text{ u}} \times 500\text{ m/s} = -33\text{ m/s}$$

$$(v_{2x})_f = \frac{2(28\text{ u})}{28\text{ u} + 32\text{ u}} \times 500\text{ m/s} = 470\text{ m/s}$$

These are velocities, not speeds, so a negative value indicates motion in the opposite direction. We see that the nitrogen molecule rebounds at a very slow 33 m/s while the oxygen molecule is knocked forward at very nearly the nitrogen's initial speed.

ASSESS The molecules are very similar in mass, so this is close to being elastic collision Case a in which the initially moving molecule stops completely. Because oxygen is slightly heavier than nitrogen, the nitrogen molecule does rebound slowly while the oxygen molecule doesn't need quite as much speed to conserve momentum.

Explosions

An **explosion,** where the particles of the system move apart from each other after a brief, intense interaction, is the opposite of a collision. The explosive forces, which could be from an expanding spring or from expanding hot gases, are *internal* forces. If the system is isolated, its total momentum during the explosion will be conserved.

EXAMPLE 8.6 **Radioactivity**

A ^{238}U uranium nucleus is radioactive. It spontaneously disintegrates into a small fragment that is ejected with a measured speed of 1.50×10^7 m/s and a "daughter nucleus" that recoils with a measured speed of 2.56×10^5 m/s. What are the atomic masses of the ejected fragment and the daughter nucleus?

PREPARE The notation ^{238}U indicates the isotope of uranium with an atomic mass of 238 u. The nucleus contains 92 protons (uranium is atomic number 92) and 146 neutrons. The disintegration of a nucleus is, in essence, an explosion. Only *internal* nuclear forces are involved, so the total momentum is conserved in the decay.

FIGURE 8.17 shows the pictorial representation. The mass of the daughter nucleus is m_1 and that of the ejected fragment is m_2. Notice that we converted the speed information to velocity information, giving $(v_{1x})_f$ and $(v_{2x})_f$ opposite signs.

SOLVE The nucleus was initially at rest, hence the total momentum is zero. The momentum after the decay is still zero if the two pieces fly apart in opposite directions with momenta equal in magnitude but opposite in sign. That is,

$$(P_x)_f = m_1(v_{1x})_f + m_2(v_{2x})_f = (P_x)_i = 0$$

Although we know both final velocities, this is not enough information to find the two unknown masses. However, we also have another conservation law, conservation of mass, that requires

$$m_1 + m_2 = 238 \text{ u}$$

Combining these two conservation laws gives

$$m_1(v_{1x})_f + (238 \text{ u} - m_1)(v_{2x})_f = 0$$

The mass of the daughter nucleus is

$$m_1 = \frac{(v_{2x})_f}{(v_{2x})_f - (v_{1x})_f} \times 238 \text{ u}$$
$$= \frac{1.50 \times 10^7 \text{ m/s}}{(1.50 \times 10^7 \text{ m/s}) - (-2.56 \times 10^5 \text{ m/s})} \times 238 \text{ u}$$
$$= 234 \text{ u}$$

With m_1 known, the mass of the ejected fragment is $m_2 = 238 - \text{m}_1 = 4$ u.

ASSESS All we learn from a momentum analysis is the masses. Chemical analysis shows that the daughter nucleus is the element thorium, atomic number 90, with two fewer protons than uranium. The ejected fragment carried away two protons as part of its mass of 4 u, so it must be a particle with two protons and two neutrons. This is the nucleus of a helium atom, ^{4}He, which in nuclear physics is called an *alpha particle* α. Thus the radioactive decay of ^{238}U can be written as $^{238}\text{U} \rightarrow {}^{234}\text{Th} + \alpha$.

FIGURE 8.17 Before-and-after pictorial representation of the decay of a ^{238}U nucleus.

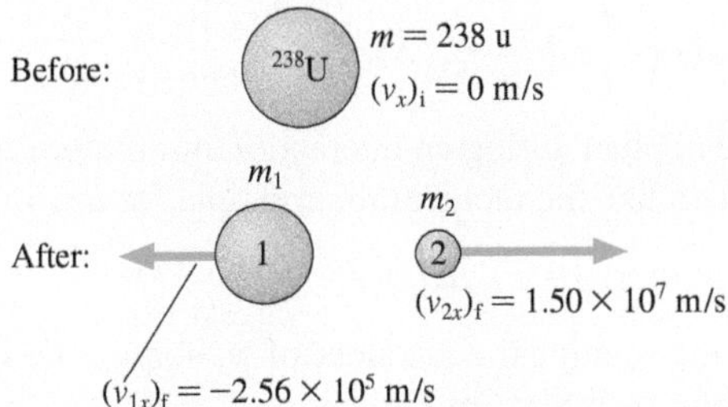

Example 8.6 would be a very difficult problem to solve using Newton's laws, but it turns out to be straightforward when approached from the before-and-after perspective of momentum conservation. The key to using conservation laws is selecting an appropriate system in which the complicated interactions—whether spring forces or nuclear forces—are internal forces that don't change the total momentum.

Much the same reasoning explains how a rocket or jet aircraft accelerates. FIGURE 8.18 shows a rocket with a parcel of fuel on board. Burning converts the fuel to hot gases that are expelled from the rocket motor. If we choose rocket + gases to be the system, the burning and expulsion are both internal forces. There are no other forces, so the total momentum of the rocket + gases system must be conserved. The rocket gains forward velocity and momentum as the exhaust gases are shot out the back, but the *total* momentum of the system remains zero.

Many people have a hard time understanding how a rocket can accelerate in the vacuum of space because there is nothing to "push against." In fact, the rocket doesn't push against anything *external,* only against the gases that it pushes out the back. In return, in accordance with Newton's third law, the gases push forward on the rocket.

FIGURE 8.18 Rocket propulsion is an example of conservation of momentum.

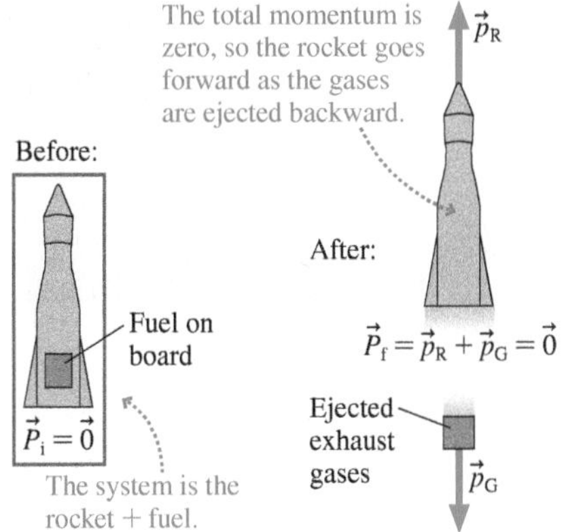

STOP TO THINK 8.5 An explosion in a rigid pipe shoots out three pieces. A 6 g piece comes out the right end. A 4 g piece comes out the left end with twice the speed of the 6 g piece. From which end, left or right, does the third piece emerge?

8.4 Angular Momentum

For a single object, the law of conservation of momentum is an alternative way of stating Newton's first law. Rather than saying that an object will continue to move in a straight line at constant speed unless acted on by a net force, we can say that the momentum of an isolated object is conserved. Both express the idea that an object moving in a straight line tends to "keep going" unless something acts on it to change its motion.

We found rotational analogs for velocity (angular velocity) and force (torque). Is there a rotational analog for momentum? Momentum itself isn't conserved for an object that is moving in a circle. Momentum is a vector, and the momentum of an object in circular motion changes as the direction of motion changes. Nonetheless, a ball that is moving in a circle at the end of a string tends to "keep going" in a circular path. And a spinning bicycle wheel would keep turning if it were not for friction. The quantity that expresses this idea for circular motion is called *angular momentum.*

FIGURE 8.19 The boy is applying a torque to the merry-go-round, increasing its angular momentum.

FIGURE 8.19 shows a boy pushing a merry-go-round. By exerting a torque he's giving the merry-go-round an angular acceleration and increasing its angular velocity. Angular acceleration, you will recall, is $\alpha = d\omega/dt$, and Newton's second law for the rotational motion of an object with moment of inertia I is $\alpha = \tau/I$. Combining these two expressions gives $I\,d\omega = \tau\,dt$. Integrating then results in

$$I\,\Delta\omega = \int \tau\,dt = \tau_{avg}\,\Delta t \qquad (8.22)$$

where τ_{avg} is the average torque applied during an interaction of duration Δt.

Equation 8.22 looks very much like the momentum principle for *linear* motion:

$$\Delta p_x = m\,\Delta v_x = F_{avg}\,\Delta t = J_x \qquad (8.23)$$

The quantity $I\omega$ appears to be the rotational equivalent of $p_x = mv_x$, the linear momentum, so we define the **angular momentum** L to be

$$L = I\omega \qquad (8.24)$$

Angular momentum of an object with moment of inertia I rotating at angular velocity ω

The SI units of angular momentum are those of moment of inertia times angular velocity, or kg • m²/s.

Just as an object in linear motion can have a large momentum by having either a large mass or a high speed, a rotating object can have a large angular momentum by having a large moment of inertia or a large angular velocity. The merry-go-round in Figure 8.19 has a larger angular momentum if it's spinning fast than if it's spinning slowly. Also, the merry-go-round (large I) has a much larger angular momentum than a toy top (small I) spinning with the same angular velocity.

The eye of a hurricane As air masses from the slowly rotating outer zones of a hurricane are drawn toward the low-pressure center, their moment of inertia about the center decreases. Because the angular momentum of these air masses is conserved, their speed must *increase* as they approach the center, leading to the high wind speeds near the center of the storm.

Conservation of Angular Momentum

Having now defined angular momentum, we can write Equation 8.22 as

$$\Delta L = \tau_{avg}\,\Delta t = \text{angular impulse} \qquad (8.25)$$

in exact analogy with Equation 8.23 for linear momentum. Equation 8.25 tells us that an object's change in angular momentum is equal to the *angular impulse* applied to the object by a torque. If there's no torque, a rotating object will "keep going" by continuing to rotate with constant angular velocity.

For a system of more than one rotating object, we define the *total angular momentum* to be the sum of the angular momenta of the objects in the system: $L_{tot} = L_1 + L_2 + \cdots$. Notice that L has the same sign as the angular velocity ω, so two objects rotating in opposite directions have angular momenta with opposite signs.

If the system is isolated, with no *external* torques, then there can be no change in the total angular momentum. We can state this conclusion as a law:

Law of conservation of angular momentum The total angular momentum L_{tot} of an isolated system is a constant. Interactions within the system do not change the total angular momentum. Mathematically, the law of conservation of angular momentum is

$$(L_{tot})_f = (L_{tot})_i \quad (8.26)$$

In terms of the individual objects in the system, Equation 8.26 is

Final (f) and initial (i) moment of inertia and angular velocity of object 1

$$(I_1)_f(\omega_1)_f + (I_2)_f(\omega_2)_f + \cdots = (I_1)_i(\omega_1)_i + (I_2)_i(\omega_2)_i + \cdots \quad (8.27)$$

Final (f) and initial (i) moment of inertia and angular velocity of object 2

We will look at problems involving multiple objects in which the angular momentum before and after an interaction is the same. The problem-solving strategy will be the same as the one we used for problems of linear momentum.

Varying Moment of Inertia

There is one aspect of angular momentum that is not analogous to linear momentum. As we've noted, for an isolated object, the *linear* momentum $\vec{p} = m\vec{v}$ is constant. The object's mass can't change, so the velocity can't change either. The object keeps moving at a constant speed. For an isolated object, the *angular* momentum $L = I\omega$ is constant as well. The object's mass can't change, but it can experience a change in its moment of inertia because the *distribution* of mass can change. (Changing the mass distribution requires *internal* forces, but we don't need to know what they are because internal forces don't exert a torque.) As a result, **a rotating object can experience a change in angular velocity.**

The classic example is a spinning figure skater. A skater in a spin on the ice experiences very little friction. The normal force balances her weight force, so she is, to a good approximation, an isolated system. If she starts the spin with her arms and legs far out from the axis of her spin, she'll start with a large moment of inertia. As she pulls her arms and legs in, her moment of inertia decreases, so her angular velocity increases. FIGURE 8.20 illustrates this idea. By dramatically reducing their moment of inertia, world-class skaters can increase the angular speed of a spin by a factor of 5 or more.

FIGURE 8.20 A spinning figure skater.

CONCEPTUAL EXAMPLE 8.7 Diving into the pool

A diver performing a forward flip—or several—goes into a tuck position, then opens up before entering the water. Why?

REASON FIGURE 8.21 (next page) shows a diver at several points in the dive. As she pushes off, the upward normal force from the diving board is larger than her downward weight force. The net upward force gives the diver height, which is important, but that's not all. If you jump straight up, the normal force points toward your center of gravity and exerts no torque; you go straight up without rotating. But the diver has angled forward slightly during the push-off, so the normal force points behind her center of gravity and does exert a torque. Consequently, she both rises *and* begins to rotate about her center of gravity.

Once clear of the board, the diver's linear momentum is not conserved—her weight is an external force—but she experiences no torque because the weight force acts at her center of gravity. Thus she is an isolated system and her angular momentum *is* conserved. Initially the diver's rotation speed is fairly small. However, going into the tuck position significantly decreases her

Continued

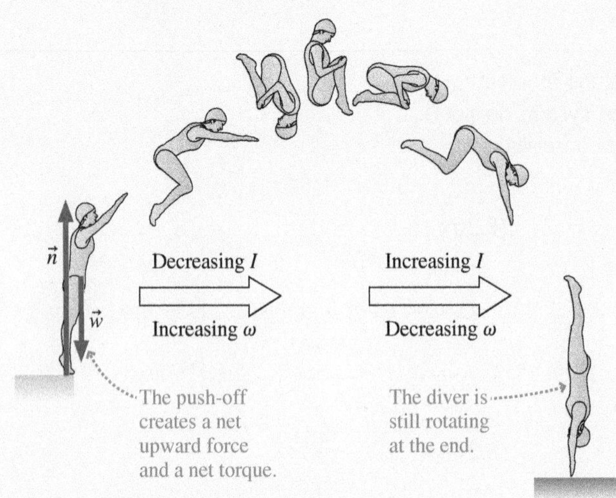

FIGURE 8.21 Motion of a diver.

moment of inertia. To compensate, and to keep angular momentum constant, her angular velocity must increase significantly. The faster spin allows her to complete a rotation—or two or three—during the time she's in the air. Opening up at the end increases her moment of inertia and decreases her angular velocity, but not to zero. Divers are still rotating as they enter the water, and it's very easy, if their timing isn't perfect, to over-rotate and not enter vertically.

ASSESS Our analysis makes sense. This is the same strategy the figure skater uses, only while in free fall. Divers can have a less compact position in a dive that requires only one turn, but they have to squeeze into a very tight ball when performing multiple flips in order to achieve the highest angular velocity.

EXAMPLE 8.8 Period of a merry-go-round

Juan, whose mass is 36 kg, stands at the center of a 200 kg merry-go-round that is rotating once every 2.5 s. While it is rotating, Juan walks out to the edge of the merry-go-round, 2.0 m from its center. What is the rotational period of the merry-go-round when Juan gets to the edge?

PREPARE We take the system to be Juan + merry-go-round and assume frictionless bearings. This is an isolated system with no external torque, so the angular momentum of the system will be conserved. As shown in the pictorial representation of FIGURE 8.22, we model the merry-go-round as a uniform disk. From Table 7.3, the moment of inertia of a disk is $I_{\text{disk}} = \frac{1}{2}MR^2$. If we model Juan as a particle of mass m, his moment of inertia is zero when he is at the center, but it increases to mR^2 when he reaches the edge.

FIGURE 8.22 Pictorial representation of the merry-go-round.

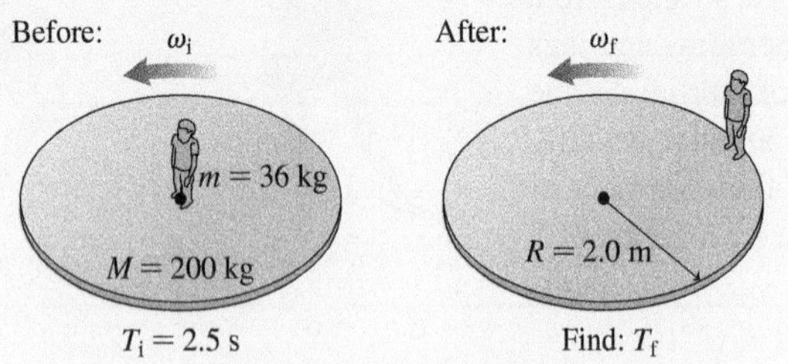

SOLVE The mathematical statement of the law of conservation of angular momentum is Equation 8.27. The initial angular momentum is

$$L_i = (I_{\text{Juan}})_i(\omega_{\text{Juan}})_i + (I_{\text{disk}})_i(\omega_{\text{disk}})_i = 0 \cdot \omega_i + \tfrac{1}{2}MR^2\omega_i = \tfrac{1}{2}MR^2\omega_i$$

Here we have used the fact that both Juan and the disk have the same initial angular velocity, which we have called ω_i. Similarly, the final angular momentum is

$$L_f = (I_{\text{Juan}})_f\,\omega_f + (I_{\text{disk}})_f\,\omega_f = mR^2\omega_f + \tfrac{1}{2}MR^2\omega_f = (mR^2 + \tfrac{1}{2}MR^2)\,\omega_f$$

where ω_f is the common final angular velocity of both Juan and the disk.

The law of conservation of angular momentum states that $L_f = L_i$, so that

$$(mR^2 + \tfrac{1}{2}MR^2)\,\omega_f = \tfrac{1}{2}MR^2\omega_i$$

Canceling the R^2 terms from both sides and solving for ω_f give

$$\omega_f = \left(\frac{M}{M + 2m}\right)\omega_i$$

The initial angular velocity is related to the initial period of rotation T_i by

$$\omega_i = \frac{2\pi}{T_i} = \frac{2\pi}{2.5\text{ s}} = 2.51\text{ rad/s}$$

Thus the final angular velocity is

$$\omega_f = \left(\frac{200\text{ kg}}{200\text{ kg} + 2(36\text{ kg})}\right)(2.51\text{ rad/s}) = 1.85\text{ rad/s}$$

When Juan reaches the edge, the period of the merry-go-round has increased to

$$T_f = \frac{2\pi}{\omega_f} = \frac{2\pi}{1.85\text{ rad/s}} = 3.4\text{ s}$$

ASSESS The merry-go-round rotates *more slowly* after Juan moves out to the edge. This makes sense because if the system's moment of inertia increases, as it does when Juan moves out, the angular velocity must decrease to keep the angular momentum constant.

Solving Example 8.8 using Newton's laws would be quite difficult. We would have to deal with internal forces, such as Juan's feet against the merry-go-round, and other complications. Problems like this show the true power of conservation laws to solve problems. As long as we are concerned only about the before and after states, we can use conservation laws to make a connection between the two without needing to account for, or even know, the details of the interactions.

STOP TO THINK 8.6 The left figure shows two boys of equal mass standing halfway to the edge on a turntable that is freely rotating at angular velocity ω_i. They then walk to the positions shown in the right figure. The final angular velocity ω_f is

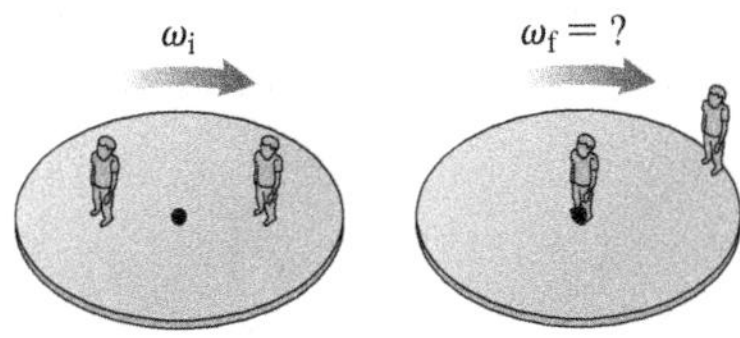

A. Greater than ω_i. B. Less than ω_i. C. Equal to ω_i.

CHAPTER 8 INTEGRATED EXAMPLE Aerial firefighting

A forest fire is easiest to attack when it's just getting started. In remote locations, this often means using airplanes to rapidly deliver large quantities of water and fire suppressant to the blaze.

The "Superscooper" is an amphibious aircraft that can pick up a 6000 kg load of water by skimming over the surface of a river or lake and scooping water directly into its storage tanks. An empty Superscooper has a mass of 13,000 kg as it approaches the water's surface at a speed of 45 m/s.

a. It takes the plane 12 s to pick up a full load of water. What is its speed immediately after picking up the water?
b. What is the magnitude of the average force of the water on the plane?
c. The plane then flies over the fire zone at 50 m/s. It releases water by opening doors in the belly of the plane, allowing the water to fall straight down with respect to the plane. What is the plane's speed after dropping the water if it takes 5.0 s to do so?

PREPARE This is a momentum problem, but we have to choose the system with care in order to use Problem-Solving Strategy 8.1. The plane alone is not an isolated system because, as it scoops up the water, the water exerts a large drag force on the plane. Instead, we'll choose the plane *and* the water to be the system; then the complicated forces between the plane and the water are *internal* forces that do not change the total momentum of the plane + water system. The plane is also acted on by the thrust of the propellers and the drag of the air, but these are balanced to give no net force while the plane is flying at constant speed. Thus the plane + water system can be modeled as an isolated system during the brief intervals while water is being scooped or dropped. With these decisions made, FIGURE 8.23 shows a before-and-after pictorial representation of the problem.

SOLVE

a. The x-component of the law of conservation of momentum is

$$(P_x)_f = (P_x)_i$$

or

$$(m_P + m_W)(v_x)_f = m_P(v_{Px})_i + m_W(v_{Wx})_i = m_P(v_{Px})_i + 0$$

Here we've used the facts that the initial velocity of the water is zero and that the final situation, as in an inelastic collision, has the combined mass of the plane and water moving with the same velocity $(v_x)_f$. Solving for $(v_x)_f$, we find

FIGURE 8.23 Pictorial representation of the plane and water.

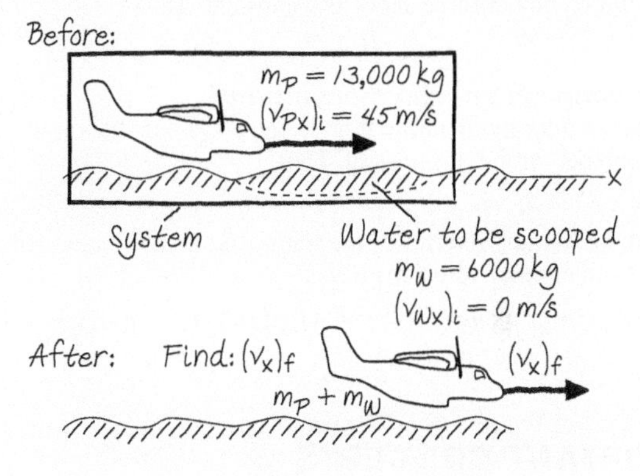

$$(v_x)_f = \frac{m_P(v_{Px})_i}{m_P + m_W} = \frac{(13{,}000\ \text{kg})(45\ \text{m/s})}{(13{,}000\ \text{kg}) + (6000\ \text{kg})} = 31\ \text{m/s}$$

b. The average force of the water on the plane during a time interval Δt can be calculated if we know the impulse of the water on the plane: $F_{avg} = J_x/\Delta t$. We can find the impulse by applying the momentum principle $J_x = \Delta p_x$, where $\Delta p_x = m_P \Delta v_x$ is the change in the plane's momentum. We can calculate

$$J_x = m_P \Delta v_x = m_P[(v_x)_f - (v_{Px})_i]$$
$$= (13{,}000\ \text{kg})(31\ \text{m/s} - 45\ \text{m/s}) = -1.8 \times 10^5\ \text{kg} \cdot \text{m/s}$$

Thus the average force is

$$(F_{avg})_x = \frac{J_x}{\Delta t} = \frac{-1.8 \times 10^5\ \text{kg} \cdot \text{m/s}}{12\ \text{s}} = -15{,}000\ \text{N}$$

This is the x-component of the force vector, which is negative because we've chosen a coordinate system in which the plane is flying to the right. The magnitude of the force is 15,000 N.

c. Because the water drops straight down *relative to the plane,* it has the same x-component of velocity immediately after being dropped as before being dropped. That is, simply opening the doors doesn't cause the water to speed up or slow down horizontally, so the water's horizontal momentum doesn't change upon being dropped. Because the total momentum of the plane + water system is conserved, the momentum of the plane doesn't change either. The plane's speed after the drop is still 50 m/s. The water does quickly slow after experiencing drag from the air, as you can see in the photo, but that's after the water has stopped interacting with the plane.

ASSESS The mass of the water is nearly half that of the plane, so the significant decrease in the plane's velocity as it scoops up the water is reasonable. The force of the water on the plane is large, but is still only about 10% of the plane's weight, $mg \approx 130{,}000$ N, so the answer seems to be reasonable.

SUMMARY

GOAL To learn about impulse, momentum, and a new problem-solving strategy based on conservation laws.

GENERAL PRINCIPLES

Conservation Laws

When a quantity *before* an interaction is the same *after* the interaction, we say that the quantity is **conserved.**

Conservation of momentum
The total momentum $\vec{P} = \vec{p}_1 + \vec{p}_2 + \cdots$ of an **isolated system**—one on which no net force acts—is a constant. Thus

$$\vec{P}_f = \vec{P}_i$$

Conservation of angular momentum
The total angular momentum of an isolated system, one on which there is no net torque, is constant. Thus

$$L_f = L_i$$

This can be written in terms of the initial and final moments of inertia I and angular velocities ω as

$$(I_1)_f(\omega_1)_f + (I_2)_f(\omega_2)_f + \cdots = (I_1)_i(\omega_1)_i + (I_2)_i(\omega_2)_i + \cdots$$

Solving Momentum Conservation Problems

PREPARE Choose an isolated system or a system that is isolated during at least part of the problem. Draw a before-and-after pictorial representation. Define symbols, list known values, and determine what you need to find to answer the question.

SOLVE Write the **law of conservation of momentum** in terms of vector components:

$$(p_{1x})_f + (p_{2x})_f + \cdots = (p_{1x})_i + (p_{2x})_i + \cdots$$
$$(p_{1y})_f + (p_{2y})_f + \cdots = (p_{1y})_i + (p_{2y})_i + \cdots$$

In terms of masses and velocities, this is

$$m_1(v_{1x})_f + m_2(v_{2x})_f + \cdots = m_1(v_{1x})_i + m_2(v_{2x})_i + \cdots$$
$$m_1(v_{1y})_f + m_2(v_{2y})_f + \cdots = m_1(v_{1y})_i + m_2(v_{2y})_i + \cdots$$

ASSESS Is the result reasonable? Does it have correct units and significant figures?

IMPORTANT CONCEPTS

Momentum is $\vec{p} = m\vec{v}$.

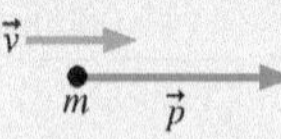

Impulse is $J_x =$ area under the force curve.

Impulse and momentum are related by the **momentum principle:**

$$\Delta p_x = J_x$$

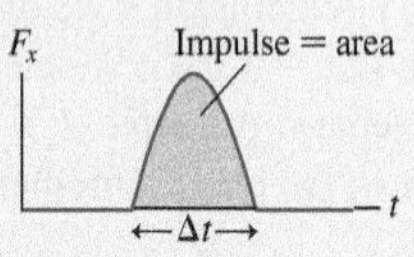

This is an alternative statement of Newton's second law.

Angular momentum $L = I\omega$ is the rotational analog of linear momentum.

Before-and-after pictorial representation

- Define the system.
- Use two drawings to show the system *before* and *after* the interaction.
- List known information and identify what you are trying to find.

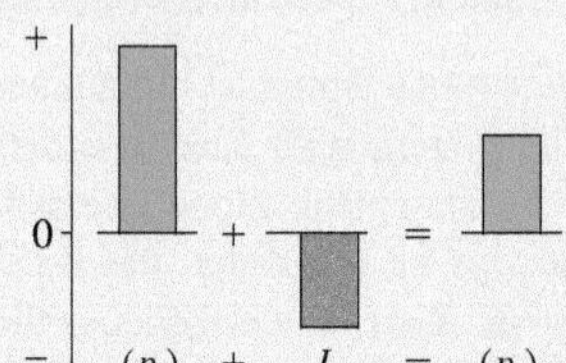

Momentum bar charts

A momentum bar chart displays the momentum principle $(p_x)_i + J_x = (p_x)_f$ in graphical form.

APPLICATIONS

Collisions

In a **perfectly inelastic collision,** two objects stick together and move with a common final velocity.

In a **perfectly elastic collision,** the objects bounce apart and conserve energy as well as momentum.

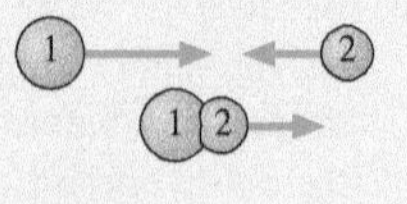

Explosions

In an **explosion,** objects fly apart from each other but their total momentum is conserved.

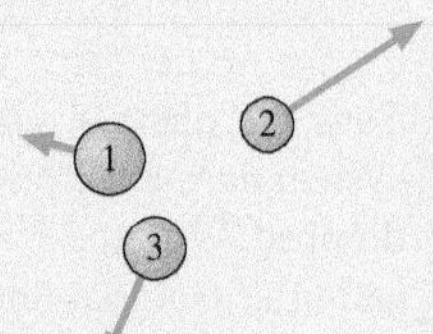

In jet propulsion, whether rockets or squid, the momentum of the object + exhaust system is conserved.

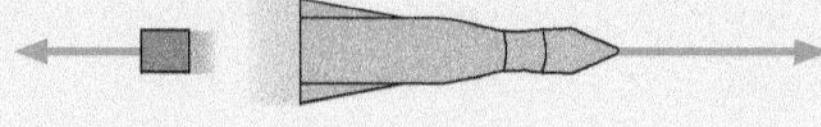

LEARNING OBJECTIVES After studying this chapter, you should be able to:

- Calculate the momentum of and impulse on an object. *Conceptual Questions 8.1, 8.6; Problems 8.1, 8.2, 8.5, 8.6, 8.8, 8.14*
- Solve impulse and momentum problems. *Conceptual Questions 8.2, 8.4, 8.5; Problems 8.7–8.9, 8.11, 8.12, 8.15*
- Use Problem-Solving Strategy 8.1 to identify isolated systems and solve momentum conservation problems. *Conceptual Questions 8.7, 8.8; Problems 8.16, 8.18–8.21*
- Apply the law of conservation of momentum to collisions and explosions. *Conceptual Questions 8.9–8.11; Problems 8.25–8.27, 8.30, 8.34, 8.35*
- Understand and use angular momentum and its law of conservation. *Conceptual Questions 8.12, 8.13; Problems 8.36–8.40*

STOP TO THINK ANSWERS

Stop to Think 8.1: F. The cart is initially moving in the negative *x*-direction, so $(p_x)_i = -20\text{ kg}\cdot\text{m/s}$. After it bounces, $(p_x)_f = +10\text{ kg}\cdot\text{m/s}$. Thus $\Delta p = (10\text{ kg}\cdot\text{m/s}) - (-20\text{ kg}\cdot\text{m/s}) = 30\text{ kg}\cdot\text{m/s}$.

Stop to Think 8.2: B. The clay ball goes from $(v_x)_i = v$ to $(v_x)_f = 0$, so $J_{\text{clay}} = \Delta p_x = -mv$. The rubber ball rebounds, going from $(v_x)_i = v$ to $(v_x)_f = -v$ (same speed, opposite direction). Thus $J_{\text{rubber}} = \Delta p_x = -2mv$. The rubber ball has a larger momentum change, and this requires a larger impulse.

Stop to Think 8.3: Less than. The ball's momentum $m_B v_B$ is the same in both cases. Momentum is conserved, so the *total* momentum is the same after both collisions. The ball that rebounds from C has *negative* momentum, so C must have a larger momentum than A.

Stop to Think 8.4: C. Momentum conservation requires $(m_1 + m_2) \times v_f = m_1 v_1 + m_2 v_2$. Because $v_1 > v_2$, it must be that $(m_1 + m_2) \times v_f = m_1 v_1 + m_2 v_2 > m_1 v_2 + m_2 v_2 = (m_1 + m_2) v_2$. Thus $v_f > v_2$. Similarly, $v_2 < v_1$, so $(m_1 + m_2) v_f = m_1 v_1 + m_2 v_2 < m_1 v_1 + m_2 v_1 = (m_1 + m_2) v_1$. Thus $v_f < v_1$. The collision causes m_1 to slow down and m_2 to speed up.

Stop to Think 8.5: Right end. The pieces started at rest, so the total momentum of the system is zero. It's an isolated system, so the total momentum after the explosion is still zero. The 6 g piece has momentum $6v$. The 4 g piece, with velocity $-2v$, has momentum $-8v$. The combined momentum of these two pieces is $-2v$. In order for *P* to be zero, the third piece must have a *positive* momentum $(+2v)$ and thus a positive velocity.

Stop to Think 8.6: B. Angular momentum $L = I\omega$ is conserved. Both boys have mass *m* and initially stand distance *R*/2 from the axis. Thus the initial moment of inertia is $I_i = I_{\text{disk}} + 2 \times m(R/2)^2 = I_{\text{disk}} + \frac{1}{2}mR^2$. The final moment of inertia is $I_f = I_{\text{disk}} + 0 + mR^2$, because the boy standing at the axis contributes nothing to the moment of inertia. Because $I_f > I_i$ we must have $\omega_f < \omega_i$.

QUESTIONS

Conceptual Questions

1. Rank in order, from largest to smallest, the momenta $(p_x)_A$ to $(p_x)_E$ of the objects in Figure Q8.1.

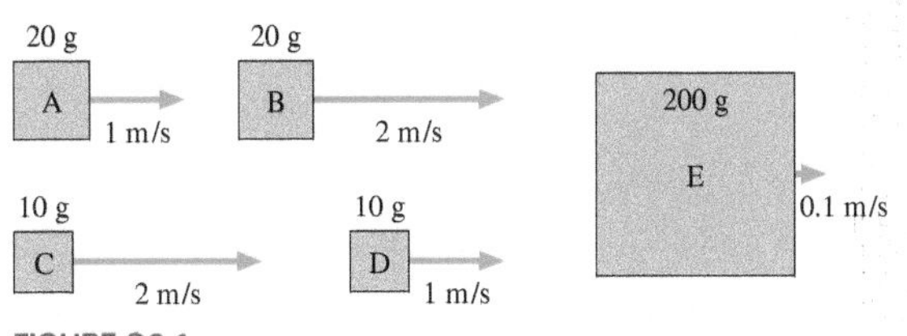

FIGURE Q8.1

2. A 2 kg object is moving to the right with a speed of 1 m/s. What are the object's speed and direction after receiving an impulse of (a) 4 N · s and (b) −4 N · s?
3. A 0.2 kg plastic cart and a 20 kg lead cart can both roll without friction on a horizontal surface. Equal forces are used to push both carts forward for a time of 1 s, starting from rest. After the force is removed at $t = 1$ s, is the momentum of the plastic cart greater than, less than, or equal to the momentum of the lead cart? Explain.
4. Angie, Bai, and Carlos are discussing a physics problem in which two identical bullets are fired with equal speeds at equal-mass wood and steel blocks resting on a frictionless table. One bullet bounces off the steel block while the second becomes embedded in the wood block. "All the masses and speeds are the same," says Angie, "so I think the blocks will have equal speeds after the collisions." "But what about momentum?" asks Bai. "The bullet hitting the wood block transfers all its momentum and energy to the block, so the wood block should end up going faster than the steel block." "I think the bounce is an important factor," replies Carlos. "The steel block will be faster because the bullet bounces off it and goes back the other direction." Which of these three do you agree with, and why?
5. It feels better to catch a hard ball while wearing a padded glove than to catch it bare handed. Use the ideas of impulse and momentum to explain why.
6. When you leap down from a high perch, you have a gentler landing with less force if you bend your knees as you land. Use the ideas of impulse and momentum to explain why.
7. A golf club continues forward after hitting the golf ball. Is momentum conserved in the collision? Explain, making sure you are careful to identify "the system."

Problem difficulty is labeled as | (straightforward) to ||||| (challenging). Problems labeled INT integrate significant material from earlier chapters; BIO are of biological or medical interest; CALC require calculus to solve.

8. Suppose a rubber ball collides head-on with a more massive steel ball traveling in the opposite direction with equal speed. Which ball, if either, receives the larger impulse? Explain.
9. Two objects collide, one of which was initially moving and the other initially at rest.
   a. Is it possible for *both* objects to be at rest after the collision? Give an example in which this happens, or explain why it can't happen.
   b. Is it possible for *one* object to be at rest after the collision? Give an example in which this happens, or explain why it can't happen.
10. Two ice skaters initially at rest, Paula and Ricardo, push off from each other. Ricardo weighs more than Paula.
    a. Which skater, if either, has the greater momentum after the push-off? Explain.
    b. Which skater, if either, has the greater speed after the push-off? Explain.
11. To win a prize at the county fair, you're trying to knock down a bowling pin by hitting it with a thrown object. Should you choose to throw a rubber ball, which will bounce off the pin, or a beanbag, which will strike the pin and not bounce? Assume the ball and beanbag have equal size and weight, and that you throw both with the same speed. Explain.
12. If the polar ice caps melt due to global warming, water now locked up near the earth's rotation axis will spread out around the surface of the globe. Will the length of a day increase, decrease, or be unchanged?
13. Is the angular momentum of disk A in Figure Q8.13 larger than, smaller than, or equal to the angular momentum of disk B? Explain.

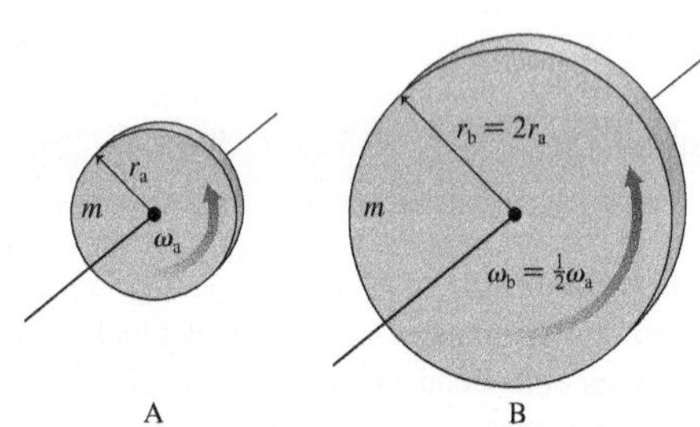

FIGURE Q8.13

Multiple-Choice Questions

14. | Model rocket engines are rated by their thrust force and by the impulse they provide. You can use this information to determine the time interval for which the engines fire. Two rocket engines provide the same impulse. The first engine provides 6 N of thrust for 2 s; the second provides 4 N of thrust. For how long does this second engine fire?
    A. 1 s B. 2 s C. 3 s D. 4 s
15. || In Figure Q8.15, what value of F_{max} gives an impulse of 6.0 N · s?
    A. 750 N
    B. 1000 N
    C. 1500 N
    D. 3000 N

FIGURE Q8.15

16. | Curling is a sport played with 20 kg stones that slide across an ice surface. Suppose a curling stone sliding at 1 m/s strikes another, stationary stone and comes to rest in 2 ms. Approximately how much force is there on the stone during the impact?
    A. 200 N B. 1000 N
    C. 2000 N D. 10,000 N
17. | Two friends are sitting in a stationary canoe. At $t = 3.0$ s the person at the front tosses a sack to the person in the rear, who catches the sack 0.2 s later. Which plot in Figure Q8.17 shows the velocity of the boat as a function of time? Positive velocity is forward, negative velocity is backward. Neglect any drag force on the canoe from the water.

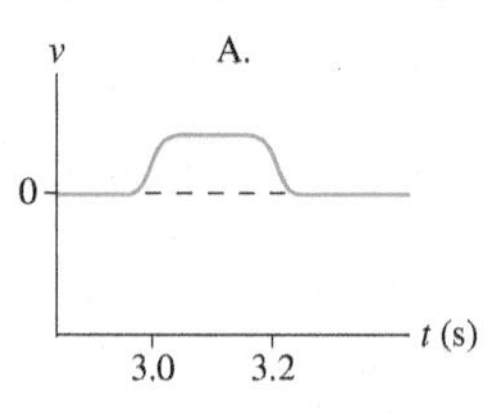

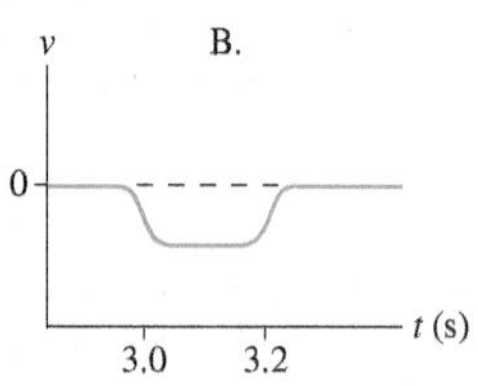

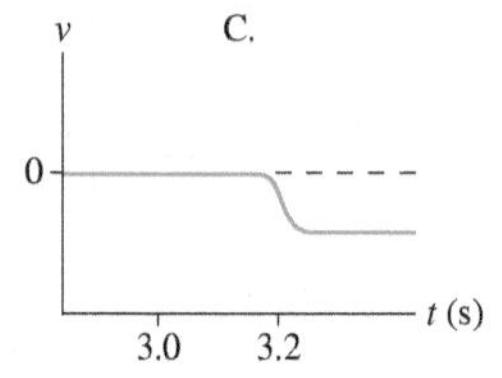

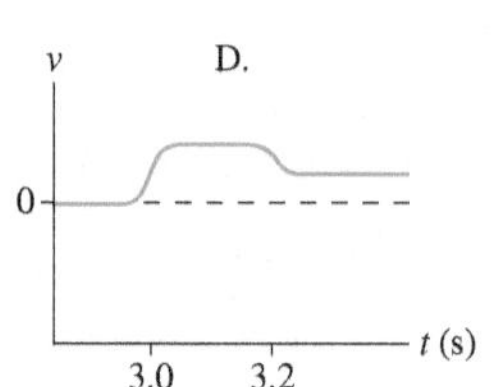

FIGURE Q8.17

18. | Two balls are hung from cords. The first ball, of mass 1.0 kg, is pulled to the side and released, reaching a speed of 2.0 m/s at the bottom of its arc. Then, as shown in Figure Q8.18, it hits and sticks to another ball. The speed of the pair just after the collision is 1.2 m/s. What is the mass of the second ball?

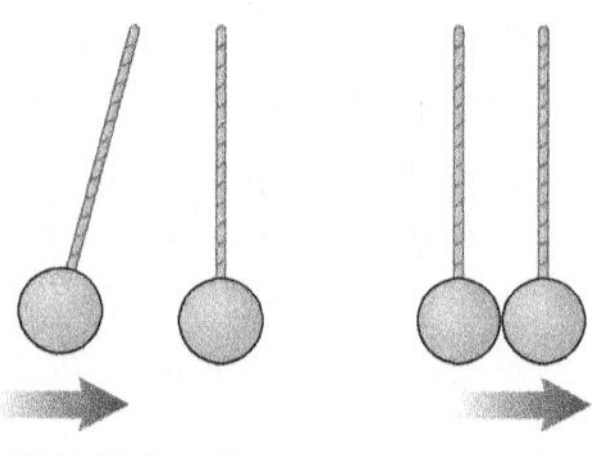
FIGURE Q8.18

    A. 0.67 kg B. 2.0 kg C. 1.7 kg D. 1.0 kg
19. | Figure Q8.19 shows two blocks sliding on a frictionless surface. Eventually the smaller block catches up with the larger one, collides with it, and sticks. What is the speed of the two blocks after the collision?
    A. $v_i/2$ B. $4v_i/5$ C. v_i D. $5v_i/4$ E. $2v_i$

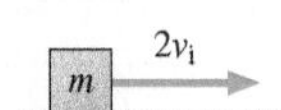

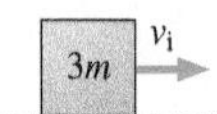

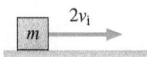

FIGURE Q8.19

20. | A 4.0-m-diameter playground merry-go-round, with a moment of inertia of 400 kg · m^2, is freely rotating with an angular velocity of 2.0 rad/s. Raj, whose mass is 80 kg, runs on the ground around the outer edge of the merry-go-round in the opposite direction to its rotation. Still moving, he jumps directly onto the rim of the merry-go-round, bringing it (and himself) to a halt. How fast was Raj running when he jumped on?
    A. 2.0 m/s B. 4.0 m/s C. 5.0 m/s
    D. 7.5 m/s E. 10 m/s
21. || A disk rotates freely on a vertical axis with an angular velocity of 30 rpm. An identical disk rotates above it in the same direction about the same axis, but without touching the lower disk, at 20 rpm. The upper disk then drops onto the lower disk. After a short time, because of friction, they rotate together. The final angular velocity of the disks is
    A. 50 rpm B. 40 rpm C. 25 rpm
    D. 20 rpm E. 10 rpm

PROBLEMS

Section 8.1 Momentum and Impulse

1. | At what speed do a bicycle and its rider, with a combined mass of 100 kg, have the same momentum as a 1500 kg car traveling at 5.0 m/s?
2. | What is the magnitude of the momentum of
   a. A 3000 kg truck traveling at 15 m/s?
   b. A 200 g baseball thrown at 40 m/s?
3. | A 57 g tennis ball is served at 45 m/s. If the ball started from rest, what impulse was applied to the ball by the racket?
4. || Model rocket engines are rated by the impulse that they deliver when they fire. A particular engine is rated to deliver an impulse of 3.5 kg • m/s. The engine powers a 120 g rocket, including the mass of the engine. What is the final speed of the rocket once the engine has fired? (Ignore the change in mass as the engine fires and ignore the weight force during the short duration firing of the engine.)
5. || What impulse does the force shown in Figure P8.5 exert on a 250 g particle?

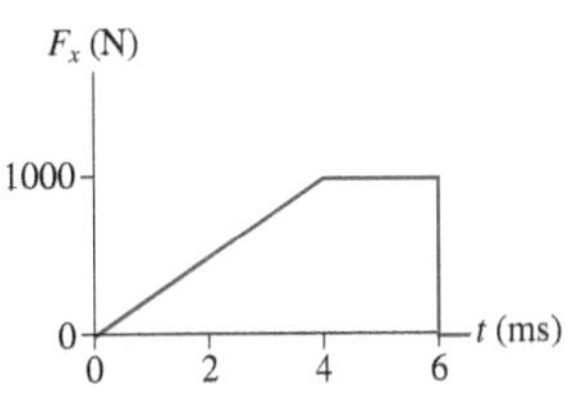

FIGURE P8.5

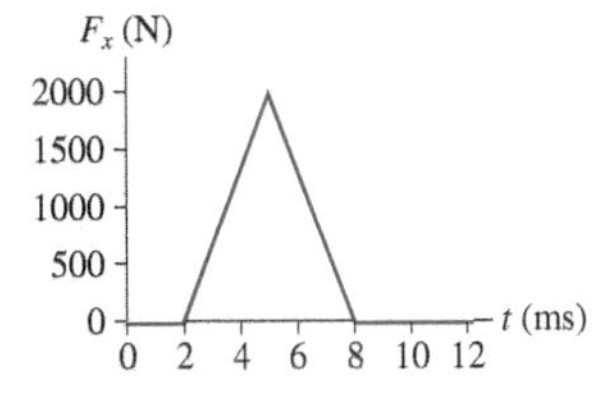

FIGURE P8.6

6. || What is the impulse on a 3.0 kg particle that experiences the force shown in Figure P8.6?
7. || Figure P8.7 is an incomplete momentum bar chart for a collision that lasts 10 ms. What are the magnitude and direction of the average collision force exerted on the object?

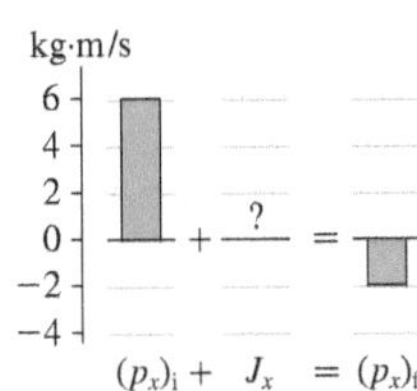

FIGURE P8.7

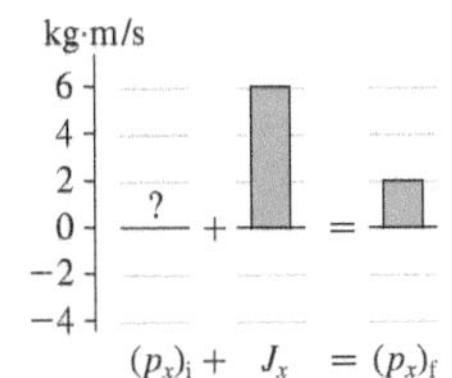

FIGURE P8.8

8. || Figure P8.8 is an incomplete momentum bar chart for a 50 g particle that experiences an impulse lasting 10 ms. What were the speed and direction of the particle before the impulse?
9. || A 2.0 kg object is moving to the right with a speed of 1.0 m/s when it experiences the force shown in Figure P8.9. What are the object's speed and direction after the force ends?

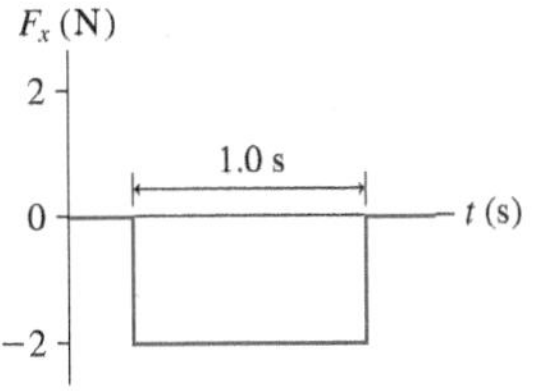

FIGURE P8.9

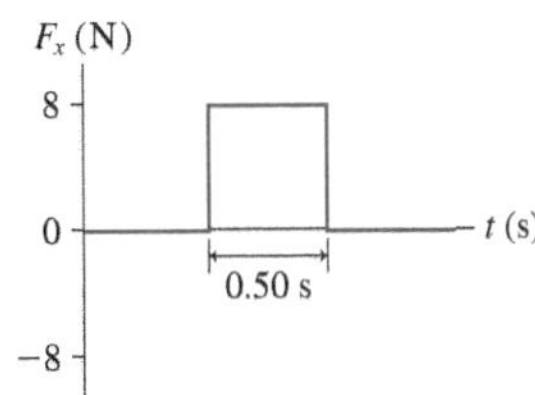

FIGURE P8.10

10. || A 2.0 kg object is moving to the right with a speed of 1.0 m/s when it experiences the force shown in Figure P8.10. What are the object's speed and direction after the force ends?
11. ||| A 250 g ball collides with a wall. Figure P8.11 shows the ball's velocity and the force exerted on the ball by the wall. What is $(v_x)_f$, the ball's rebound velocity?

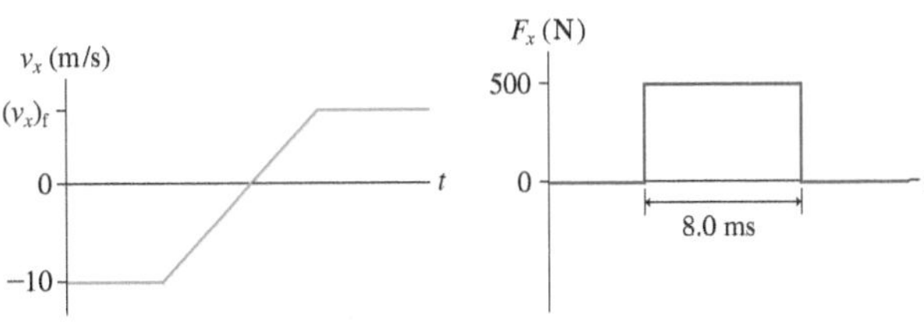

FIGURE P8.11

12. ||| A 600 g air-track glider collides with a spring at one end of the track. Figure P8.12 shows the glider's velocity and the force exerted on the glider by the spring. How long is the glider in contact with the spring?

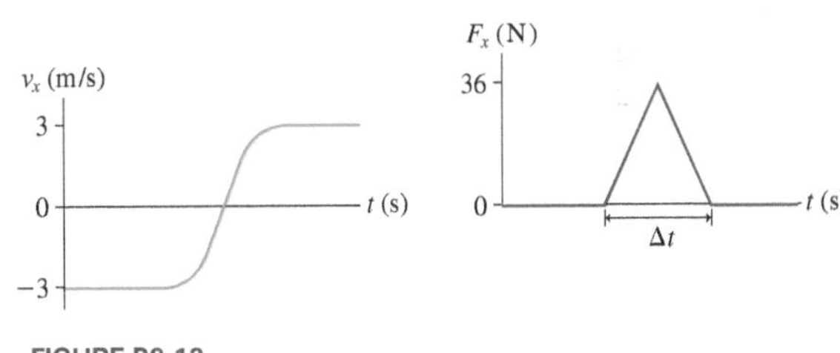

FIGURE P8.12

13. || The flowers of the bunchberry plant open with astonishing force and speed, causing the pollen grains to be ejected out of the flower in a mere 0.30 ms at an acceleration of 2.5×10^4 m/s^2. If the acceleration is constant, what impulse is delivered to a pollen grain with a mass of 1.0×10^{-7} g? BIO
14. || Ferns spread spores instead of seeds, and some ferns eject the spores at surprisingly high speeds. One species accelerates 1.4 μg spores to a 4.5 m/s ejection speed in a time of 1.0 ms. What impulse is provided to the spores? What is the average force on a spore? BIO
15. ||| Researchers studying the possible effects of "heading" a soccer ball—hitting it with the head—use a force plate to measure the interaction force between a ball and a hard surface. Figure P8.15 shows smoothed data of the force when a 430 g soccer ball is fired horizontally at the force plate with a speed of 16 m/s. With what speed does the ball rebound from the plate? BIO

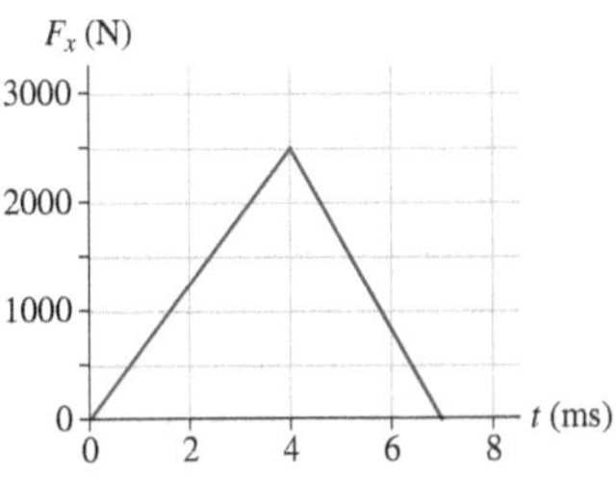

FIGURE P8.15

Section 8.2 Conservation of Momentum

16. | A 5000 kg open train car is rolling on frictionless rails at 22 m/s when it starts pouring rain. A few minutes later, the car's speed is 20 m/s. What mass of water has collected in the car?
17. | A 10,000 kg railroad car is rolling at 2.0 m/s when a 4000 kg load of gravel is suddenly dropped in. What is the car's speed just after the gravel is loaded?

18. || A 10-m-long glider with a mass of 680 kg (including the passengers) is gliding horizontally through the air at 30 m/s when a 60 kg skydiver drops out by releasing his grip on the glider. What is the glider's velocity just after the skydiver lets go?

19. | Three identical train cars, coupled together, are rolling east at speed v_0. A fourth car traveling east at $2v_0$ catches up with the three and couples to make a four-car train. A moment later, the train cars hit a fifth car that was at rest on the tracks, and it couples to make a five-car train. What is the speed of the five-car train?

20. || Aaliyah (36 kg) and Diego (47 kg) are bouncing on a trampoline. Just as Aaliyah reaches the high point of her bounce, Diego is moving upward past her at 4.1 m/s. At that instant he grabs hold of her. What is their speed just after he grabs her?

21. || If you free the cork in a highly pressurized champagne bottle, the resulting launch of the cork will, in principle, cause the bottle to recoil. A filled champagne bottle has a mass of 1.8 kg. The cork has a mass of 7.5 g and is launched at 20 m/s. If the bottle could move freely, with what speed would it recoil? Is this something you are likely to notice?

Section 8.3 Collisions and Explosions

22. || A 300 g bird flying along at 6.0 m/s sees a 10 g insect heading straight toward it at a speed of 30 m/s. The bird opens its mouth wide and enjoys a nice lunch. What is the bird's speed immediately after swallowing?

23. || The parking brake on a 2000 kg Cadillac has failed, and it is rolling slowly, at 1.0 mph, toward a group of small children. Seeing the situation, you realize you have just enough time to drive your 1000 kg Volkswagen head-on into the Cadillac and save the children. With what speed should you impact the Cadillac to bring it to a halt?

24. || A small, 100 g cart is moving at 1.20 m/s on a frictionless track when it collides with a larger, 1.00 kg cart at rest. After the collision, the small cart recoils at 0.850 m/s. What is the speed of the large cart after the collision?

25. || A 71 kg baseball player jumps straight up to catch a hard-hit ball. If the 140 g ball is moving horizontally at 45 m/s, and the catch is made when the ballplayer is at the highest point of his leap, what is his speed immediately after stopping the ball?

26. || Peregrine falcons, which can dive at 200 mph (90 m/s), grab prey birds from the air. The impact usually kills the prey. Suppose a 480 g falcon diving at 75 m/s strikes a 240 g pigeon, grabbing it in her talons. We can assume that the slow-flying pigeon is stationary. The collision between the birds lasts 15 ms.
BIO
a. What is the final speed of the falcon and pigeon?
b. What is the average impact force on the pigeon?

27. || A 50 g ball of clay traveling at 15 m/s hits and sticks to a 1.0 kg brick sitting at rest on a frictionless surface. What is the speed of the brick after the collision?

28. || Fred (mass 60 kg) is running with the football at a speed of 6.0 m/s when he is met head-on by Brutus (mass 120 kg), who is moving at 4.0 m/s. Brutus grabs Fred in a tight grip, and they fall to the ground. Which way do they slide, and how far? The coefficient of kinetic friction between football uniforms and Astroturf is 0.30.
INT

29. || A 50 g marble moving at 2.0 m/s strikes a 20 g marble at rest. What is the speed of each marble immediately after the collision?

30. || A proton is traveling to the right at 2.0×10^7 m/s. It has a head-on perfectly elastic collision with a carbon atom at rest. The mass of the carbon atom is 12 times the mass of the proton. What are the speed and direction of each after the collision?

31. | A 50 kg archer, standing on frictionless ice, shoots a 100 g arrow at a speed of 100 m/s. What is the recoil speed of the archer?

32. ||| A 70.0 kg football player is gliding across very smooth ice at 2.00 m/s. He throws a 0.450 kg football straight forward. What is the player's speed afterward if the ball is thrown at
a. 15.0 m/s relative to the ground?
b. 15.0 m/s relative to the player?

33. ||| Dan is gliding on his skateboard at 4.0 m/s. He suddenly jumps backward off the skateboard, kicking the skateboard forward at 8.0 m/s. How fast is Dan going as his feet hit the ground? Dan's mass is 50 kg and the skateboard's mass is 5.0 kg.

34. || Two ice skaters, with masses of 50 kg and 75 kg, are at the center of a 60-m-diameter circular rink. The skaters push off against each other and glide to opposite edges of the rink. If the heavier skater reaches the edge in 20 s, how long does the lighter skater take to reach the edge?

35. || A ball of mass m and another ball of mass $3m$ are placed inside a smooth metal tube with a massless spring compressed between them. When the spring is released, the heavier ball flies out of one end of the tube with speed v_0. With what speed does the lighter ball emerge from the other end?

Section 8.4 Angular Momentum

36. || What is the angular momentum about the axle of the 500 g rotating bar in Figure P8.36?

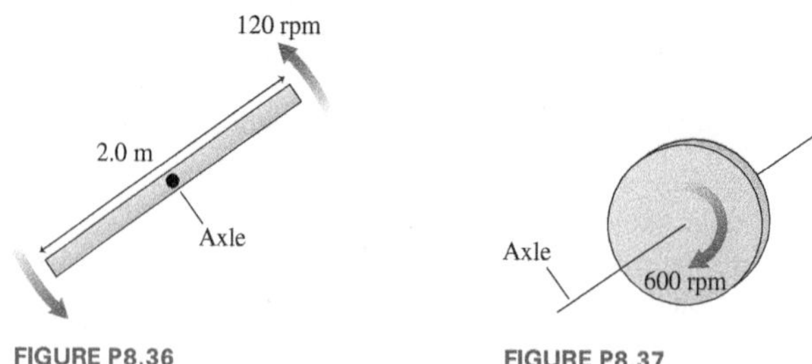

FIGURE P8.36 FIGURE P8.37

37. |||| What is the angular momentum about the axle of the 2.0 kg, 4.0-cm-diameter rotating disk in Figure P8.37?

38. ||| How fast, in rpm, would a 5.0 kg, 22-cm-diameter bowling ball have to spin to have an angular momentum of 0.23 kg · m²/s?

39. | Divers change their body position in midair while rotating about their center of gravity. In one dive, the diver leaves the board with her body nearly straight, then tucks into a somersault position. If the moment of inertia of the diver in a straight position is 14 kg · m² and in a tucked position is 4.0 kg · m², by what factor does her angular speed increase?
BIO

40. || Ice skaters often end their performances with spin turns, where they spin very fast about their center of gravity with their arms folded in and legs together. Upon ending, their arms extend outward, proclaiming their finish. Not quite as noticeably, one leg goes out as well. Suppose that the moment of inertia of a skater with arms out and one leg extended is 3.2 kg · m² and for arms and legs in is 0.80 kg · m². If she starts out spinning at 5.0 rev/s, what is her angular speed (in rev/s) when her arms and one leg open outward?
BIO

General Problems

41. ||| A tennis player swings her 1000 g racket with a speed of 10 m/s. She hits a 60 g tennis ball that was approaching her at a speed of 20 m/s. The ball rebounds at 40 m/s.

a. How fast is her racket moving immediately after the impact? You can ignore the interaction of the racket with her hand for the brief duration of the collision.

b. If the tennis ball and racket are in contact for 10 ms, what is the average force that the racket exerts on the ball? How does this compare to the gravitational force on the ball?

42. ||| A 200 g ball is dropped from a height of 2.0 m, bounces on
INT a hard floor, and rebounds to a height of 1.5 m. Figure P8.42 shows the impulse received from the floor. What maximum force does the floor exert on the ball?

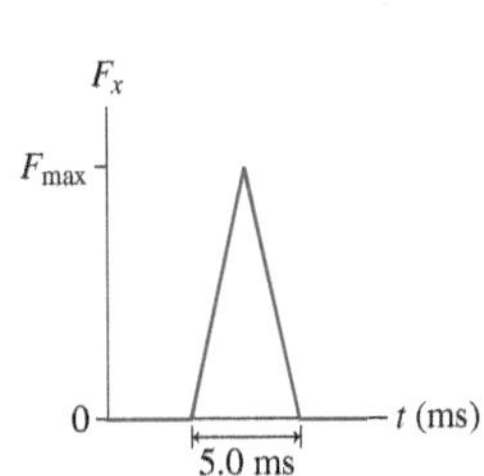

FIGURE P8.42

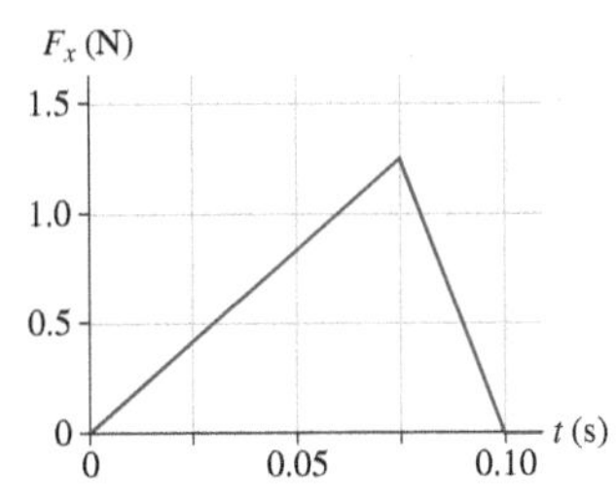

FIGURE P8.43

43. ||| Researchers exploring the jumping strategies of frogs used a
BIO force sensor to measure the force exerted on a fixed perch by a 22 g Cuban tree frog as it leapt. The smoothed and somewhat simplified data are shown in Figure P8.43. What was the frog's leaping speed?

44. ||| Air-track gliders with masses 300 g, 400 g, and 200 g are lined up and held in place with lightweight springs compressed between them. All three are released at once. The 200 g glider flies off to the right while the 300 g glider goes left. Their position graphs, as measured by motion detectors, are shown in Figure P8.44. What are the direction (right or left) and speed of the 400 g glider that was in the middle?

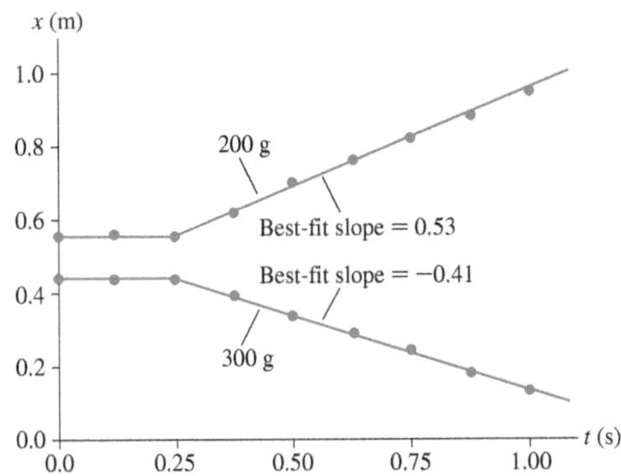

FIGURE P8.44

45. ||| An object of mass m is at rest at $t = 0$. Its momentum for $t > 0$
CALC is given by $p_x = 6t^2$ kg·m/s, where t is in s. Find an expression for $F_x(t)$, the force exerted on the object as a function of time.

46. ||| A 500 g particle has velocity $v_x = -5.0$ m/s at $t = -2$ s.
CALC Force $F_x = (4 - t^2)$ N, where t is in s, is exerted on the particle between $t = -2$ s and $t = 2$ s. This force increases from 0 N at $t = -2$ s to 4 N at $t = 0$ s and then back to 0 N at $t = 2$ s. What is the particle's velocity at $t = 2$ s?

47. || When firefighters use a fire hose to deliver water to a fire, it takes a great deal of force to hold the hose steady. A typical fire hose delivers 20 kg of water per second. The water moves at 6.0 m/s in the hose and then speeds up to 36 m/s as it exits the nozzle. How much force must a firefighter exert on the hose to keep it from recoiling?

48. || One measure of heart health in patients who have congestive
BIO heart failure is the force with which the left ventricle pushes on blood that it is ejecting. An average force of less than 0.20 N is associated with serious impairment. Ultrasound measurements show a patient's heart ejecting 17 g of blood at a speed of 0.73 m/s in a time of 0.10 s. What average ejection force results from these numbers?

49. || Male bighorn sheep (120 kg) engage in head-butting contests
BIO to establish dominance. High-speed photography has shown that the impact speed is typically 6.0 m/s and that the collision, which brings both rams to a halt, lasts approximately 0.25 s. Suppose the collision force increases linearly with time until it reaches a maximum F_{max}, then decreases linearly to zero at $t = 0.25$ s. What is F_{max}?

50. || Sports doctors are very concerned about brain injuries in con-
BIO tact sports. Brain injuries can occur if the head accelerates or decelerates at more than $50g$, with injuries becoming serious or even life-threatening at $150g$. Padded helmets are designed to reduce forces by lengthening the contact time. Suppose two 95 kg football players each running at 4.5 m/s have a (illegal) head-to-head collision that brings both of them to a halt. A real collision of this sort is complex, but we can model it as a simple collision in which the force is constant for the duration of the collision.

a. What magnitude force would decelerate each player at $75g$?

b. What is the minimum collision duration in ms for which their head accelerations would not exceed $75g$?

c. For players who are not wearing helmets, a true head-to-head collision has a duration of approximately 2.0 ms. In such a situation, what would be the head deceleration in g's of these two players?

51. || A 0.010 g water strider moves forward with a quick flick of
BIO its legs, pushing water backward to enable it to acquire forward momentum. The legs push for 0.010 s, and surface tension limits the horizontal force to a maximum of 5.0×10^{-4} N. What is the maximum speed a motionless water strider can achieve in one stroke?

52. || Most geologists believe that the dinosaurs became extinct 65 million years ago when a large comet or asteroid struck the earth, throwing up so much dust that the sun was blocked out for a period of many months. Suppose an asteroid with a diameter of 2.0 km and a mass of 1.0×10^{13} kg hits the earth (6.0×10^{24} kg) with an impact speed of 4.0×10^4 m/s.

a. What is the earth's recoil speed after such a collision? (Use a reference frame in which the earth was initially at rest.)

b. What percentage is this of the earth's speed around the sun? The earth orbits the sun at a distance of 1.5×10^{11} m.

53. || Squids rely on jet propulsion to move around. A 1.50 kg
BIO squid (including the mass of water inside the squid) drifting at 0.40 m/s suddenly ejects 0.100 kg of water to get itself moving at 2.50 m/s. If drag is ignored over the small interval of time needed to expel the water (the impulse approximation), what is the water's ejection speed relative to the squid?

54. || You've probably seen cartoons in which a person is blown
BIO backward by a sneeze. Could this actually happen? Suppose a 60 kg woman sneezes rather violently while standing motionless on a frictionless surface, ejecting 3.0 g of air and water vapor at a speed of 40 m/s. These are typical values for a sneeze. If the air is ejected horizontally, how fast is the woman moving after the sneeze?

55. ||| a. A bullet of mass m is fired into a block of mass M that is at
INT rest. The block, with the bullet embedded, slides distance d across a horizontal surface. The coefficient of kinetic friction is μ_k. Find an expression for the bullet's speed v_{bullet}.
b. What is the speed of a 10 g bullet that, when fired into a 10 kg stationary wood block, causes the block to slide 5.0 cm across a wood table? The coefficient of kinetic friction for wood on wood was given in Table 5.2 as 0.20.

56. |||| You are part of a search-
INT and-rescue mission that has been called out to look for a lost explorer. You've found the missing explorer, but, as Figure P8.56 shows, you're separated from him by a 200-m-high cliff and a 30-m-wide raging river. To save his life, you need to get a 5.0 kg package of emergency supplies across the river. Unfortunately, you can't throw the package hard enough to make it across. Fortunately, you happen to have a 1.0 kg rocket intended for launching flares. Improvising quickly, you attach a sharpened stick to the front of the rocket, so that it will impale itself into the package of supplies, then fire the rocket at ground level toward the supplies. What minimum speed must the rocket have just before impact in order to save the explorer's life?

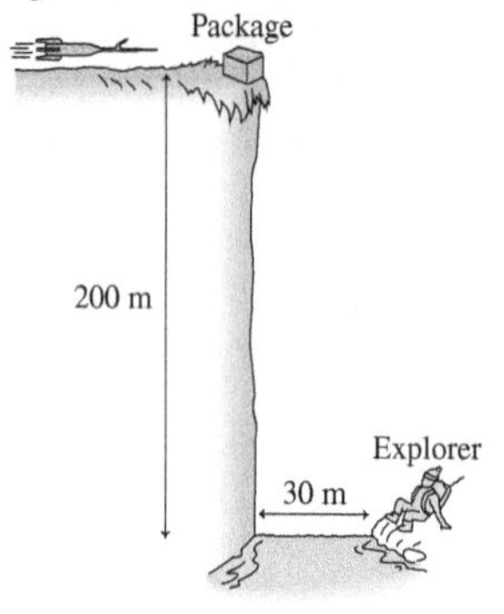

FIGURE P8.56

57. || In a ballistics test, a 25 g bullet traveling horizontally at 1200 m/s goes through a 30-cm-thick 350 kg stationary target and emerges with a speed of 900 m/s. The target is free to slide on a smooth horizontal surface. What is the target's speed just after the bullet emerges?

58. || Two 500 g blocks of wood are 2.0 m apart on a frictionless table. A 10 g bullet is fired at 400 m/s toward the blocks. It passes all the way through the first block, then embeds itself in the second block. The speed of the first block immediately afterward is 6.0 m/s. What is the speed of the second block after the bullet stops in it?

59. ||| A typical raindrop is much more massive than a mosquito
BIO and much faster than a mosquito flies. How does a mosquito survive the impact? Recent research has found that the collision of a falling raindrop with a mosquito is a perfectly inelastic collision. That is, the mosquito is "swept up" by the raindrop and ends up traveling along with the raindrop. Once the relative speed between the mosquito and the raindrop is zero, the mosquito is able to detach itself from the drop and fly away.
a. A hovering mosquito is hit by a raindrop that is 40 times as massive and falling at 8.2 m/s, a typical raindrop speed. How fast is the raindrop, with the attached mosquito, falling immediately afterward if the collision is perfectly inelastic?
b. Because a raindrop is "soft" and deformable, the collision duration is a relatively long 8.0 ms. What is the mosquito's average acceleration, in g's, during the collision? The peak acceleration is roughly twice the value you found, but the mosquito's rigid exoskeleton allows it to survive accelerations of this magnitude. In contrast, humans cannot survive an acceleration of more than about 10g.

60. ||| Ann (mass 50 kg) is standing at the left end of a 15-m-long, 500 kg cart that has frictionless wheels and rolls on a frictionless track. Initially both Ann and the cart are at rest. Suddenly, Ann starts running along the cart at a speed of 5.0 m/s relative to the cart. How far will Ann have run *relative to the ground* when she reaches the right end of the cart?

61. ||| A neon atom (mass 20 u) traveling at 350 m/s has a head-on collision with a helium atom (mass 4 u). The collision causes each atom to reverse direction with no change of speed. What is the speed of the helium atom?

62. |||| A 100 g granite cube slides down a 40° frictionless ramp. At
INT the bottom, just as it exits onto a horizontal table, it collides with a 200 g steel cube at rest. How high above the table should the granite cube be released to give the steel cube a speed of 150 cm/s?

63. || A 30 ton rail car and a 90 ton rail car, initially at rest, are connected together with a giant but massless compressed spring between them. When released, the 30 ton car is pushed away at a speed of 4.0 m/s relative to the 90 ton car. What is the speed of the 30 ton car relative to the ground?

64. || A proton (mass 1 u) is shot toward an unknown target nucleus at a speed of 2.50×10^6 m/s. The proton rebounds with its speed reduced by 25% while the target nucleus acquires a speed of 3.12×10^5 m/s. What is the mass, in atomic mass units, of the target nucleus?

65. || The nucleus of the polonium isotope ^{214}Po (mass 214 u) is radioactive and decays by emitting an alpha particle (a helium nucleus with mass 4 u). Laboratory experiments measure the speed of the alpha particle to be 1.92×10^7 m/s. Assuming the polonium nucleus was initially at rest, what is the recoil speed of the nucleus that remains after the decay?

66. |||| A spaceship of mass 2.0×10^6 kg is cruising at a speed of 5.0×10^6 m/s when the antimatter reactor fails, blowing the ship into three pieces. One section, having a mass of 5.0×10^5 kg, is blown straight backward with a speed of 2.0×10^6 m/s. A second piece, with mass 8.0×10^5 kg, continues forward at 1.0×10^6 m/s. What are the direction and speed of the third piece?

67. ||| Figure P8.67 shows a 100 g
INT puck revolving at 100 rpm on a 20-cm-radius circle on a frictionless table. A string attached to the puck passes through a hole in the middle of the table. The end of the string below the table is then slowly pulled down until the puck is revolving in a 10-cm-radius circle. How many revolutions per minute does the puck make at this new radius?

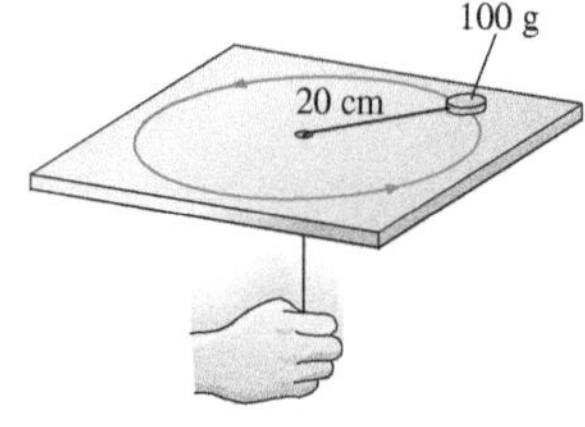

FIGURE P8.67

68. || A 2.0 kg, 20-cm-diameter turntable rotates at 100 rpm on frictionless bearings. Two 500 g blocks fall from above, hit the turntable simultaneously at opposite ends of a diameter, and stick. What is the turntable's angular speed, in rpm, just after this event?

69. ||| Disk A, with a mass of 2.0 kg and a radius of 40 cm, rotates clockwise about a frictionless vertical axle at 30 rev/s. Disk B, also 2.0 kg but with a radius of 20 cm, rotates counterclockwise about that same axle, but at a greater height than disk A, at 30 rev/s. Disk B slides down the axle until it lands on top of disk A, after which they rotate together. After the collision, what is their common angular speed (in rev/s) and in which direction do they rotate?

MCAT-Style Passage Problems

Hitting a Golf Ball

Consider a golf club hitting a golf ball. To a good approximation, we can model this as a collision between the rapidly moving head of the golf club and the stationary golf ball, ignoring the shaft of the club and the golfer.

A golf ball has a mass of 46 g. Suppose a 200 g club head is moving at a speed of 40 m/s just before striking the golf ball. After the collision, the golf ball's speed is 60 m/s.

70. What is the momentum of the club + ball system right before the collision?
    A. 1.8 kg · m/s
    B. 8.0 kg · m/s
    C. 3220 kg · m/s
    D. 8000 kg · m/s
71. Immediately after the collision, the momentum of the club + ball system will be
    A. Less than before the collision.
    B. The same as before the collision.
    C. Greater than before the collision.
72. A manufacturer makes a golf ball that compresses more than a traditional golf ball when struck by a club. How will this affect the average force during the collision?
    A. The force will decrease.
    B. The force will not be affected.
    C. The force will increase.
73. By approximately how much does the club head slow down as a result of hitting the ball?
    A. 4 m/s
    B. 6 m/s
    C. 14 m/s
    D. 26 m/s

9 Fluids

Kayaking is all about fluids, from the flow of the river to the buoyancy of the boat.

LOOKING AHEAD

Pressure

Pressure varies with elevation, so an accurate blood-pressure measurement requires the cuff to be level with your heart.

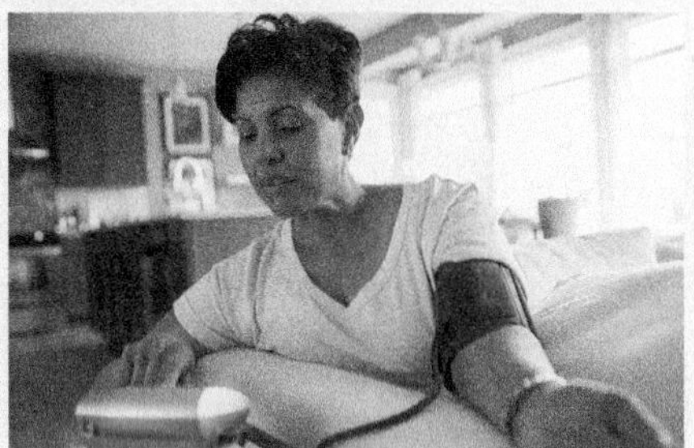

You'll learn about the different units with which *pressure* is measured and how to calculate pressure at different depths in a liquid.

Buoyancy

Majestic icebergs are a beautiful sight in the Arctic Ocean. But they are a danger to ships because nearly 90% of an iceberg is underwater.

You'll learn how to use *Archimedes' principle* to find the *buoyant force* on an object that is floating on or submerged in a fluid.

Flow

A smoke stream shows that the air flowing past a cyclist changes from smooth *laminar flow* to chaotic *turbulent flow*.

You'll explore how fluids flow and the laws that govern the flow of both *ideal fluids* and *viscous fluids.*

GOAL To understand the static and dynamic properties of fluids.

PHYSICS AND LIFE

Organisms Depend on Fluids

Cells and organisms are filled with and surrounded by fluids, whether air or water. Larger plants and animals have developed circulatory or vascular systems that use fluids to transport vital nutrients and dissolved gases. Understanding the physics of moving fluids is essential for knowing how these systems work and what to do when they degrade and fail. In addition, the properties of fluids are of critical importance in determining how organisms move, what body structures can be supported, and the size limits of organisms. The physics of fluids makes often spectacular demands of organisms, and an equally spectacular variety of creatures have evolved in response.

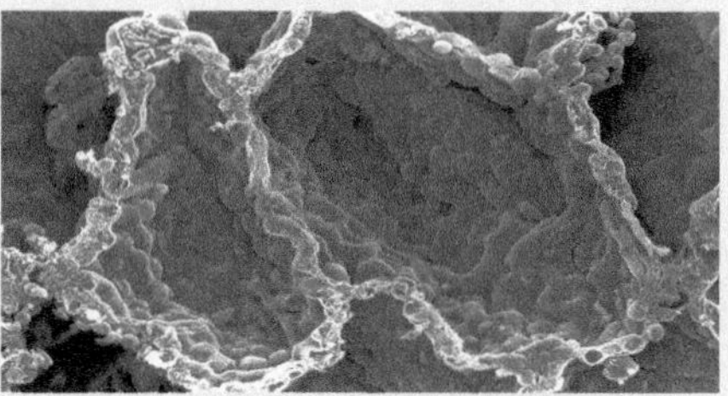

The proper functioning of *alveoli* in the lungs depends on the smooth flow of air into and out of the lungs and on surface tension to control the pressure.

9.1 Properties of Fluids

Life depends on fluids. Blood flows through our bodies, cellular processes operate within the intracellular fluid, we breathe air, and aquatic organisms live in water. Quite simply, a **fluid** is a substance that flows. Liquids and gases may seem quite different, but both are fluids, and their similarities are often more important than their differences.

The detailed behavior of gases and especially liquids can be complex. Fortunately, we can capture the essential characteristics of gases and liquids in simple molecular models. A gas, as shown in FIGURE 9.1a, is a system in which each molecule moves freely through space until, on occasion, it collides with another molecule or the wall. A gas is *compressible* because there's empty space between the molecules.

A liquid, shown in FIGURE 9.1b, is a system in which the molecules are weakly bound to each other, though less strongly than in a solid. The weak bonds keep the molecules close together but allow them to move around. Liquids, like solids but unlike gases, are essentially *incompressible;* the molecules are about as close together as they can get.

FIGURE 9.1 Molecular models of gases and liquids.

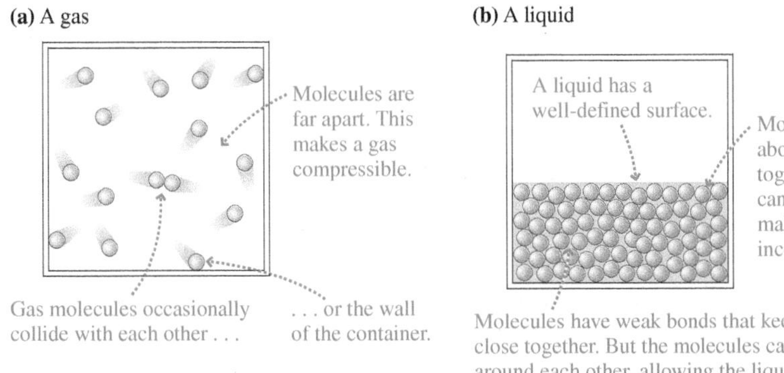

(a) A gas
Molecules are far apart. This makes a gas compressible.
Gas molecules occasionally collide with each other . . .
. . . or the wall of the container.
(b) A liquid
A liquid has a well-defined surface.
Molecules are about as close together as they can get. This makes a liquid incompressible.
Molecules have weak bonds that keep them close together. But the molecules can slide around each other, allowing the liquid to flow.

Volume and Density

One important parameter that characterizes a macroscopic system is its volume V, the amount of space the system occupies. The SI unit of volume is m^3. Nonetheless, both cm^3 and liters (L) are widely used metric units of volume. Pay careful attention to units in problems that use volume; in many cases you will need to convert volume units to m^3 before doing calculations.

While it is true that $1\text{ m} = 100\text{ cm}$, it is *not* true that $1\text{ m}^3 = 100\text{ cm}^3$. FIGURE 9.2 shows that the volume conversion factor is $1\text{ m}^3 = 10^6\text{ cm}^3$. A liter is 1000 cm^3, so $1\text{ m}^3 = 10^3\text{ L}$. A milliliter (1 mL) is the same as 1 cm^3.

FIGURE 9.2 There are 10^6 cm^3 in 1 m^3.

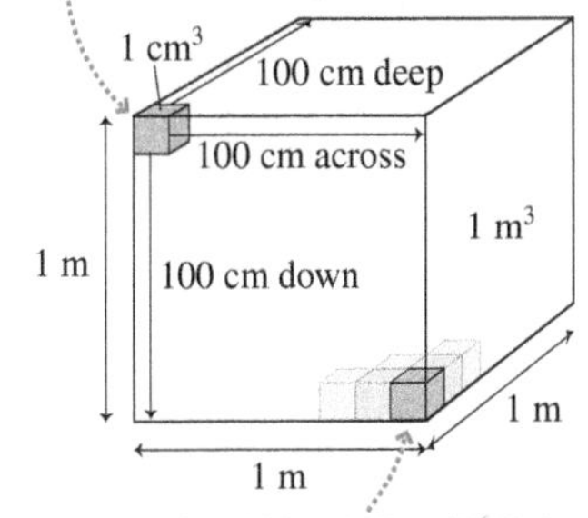

Subdivide the 1 m × 1 m × 1 m cube into little cubes 1 cm on a side. You will get 100 subdivisions along each edge.

100 cm across
100 cm down

A system is also characterized by its *density*. Suppose you have several blocks of copper, each of different size. Each block has a different mass m and a different volume V. Nonetheless, all the blocks are copper, so there should be some quantity that has the *same* value for all the blocks, telling us, "This is copper, not some other material." The most important such parameter is the *ratio* of mass to volume, which we call the **mass density** ρ (lowercase Greek rho):

$$\rho = \frac{m}{V} \quad \text{(mass density)} \tag{9.1}$$

Conversely, an object of density ρ has mass $m = \rho V$.

The SI units of mass density are kg/m^3. Nonetheless, units of g/cm^3 are widely used. You will often need to convert these to SI units. You must convert both the grams to kilograms and the cubic centimeters to cubic meters. The net result is the conversion factor

$$1\text{ g/cm}^3 = 1000\text{ kg/m}^3$$

TABLE 9.1 **Densities at 1 atm pressure**

Substance	ρ (kg/m^3)
Helium gas (20°C)	0.166
Air (0°C)	1.29
Air (20°C)	1.20
Air (37°C)	1.14
Gasoline	680
Ethyl alcohol	790
Oil (typical)	900
Water	1000
Seawater	1030
Blood (whole)	1060
Glycerin	1260
Mercury	13,600

The mass density is usually called simply "the density" if there is no danger of confusion. However, we will meet other types of density as we go along, and sometimes it is important to be explicit about which density you are using. TABLE 9.1 provides a short list of mass densities of various fluids. Notice the enormous difference between the densities of gases and liquids. Gases have lower densities because the molecules in gases are farther apart than in liquids.

STOP TO THINK 9.1 A piece of glass is broken into two pieces of different size. Rank in order, from largest to smallest, the mass densities of pieces A, B, and C.

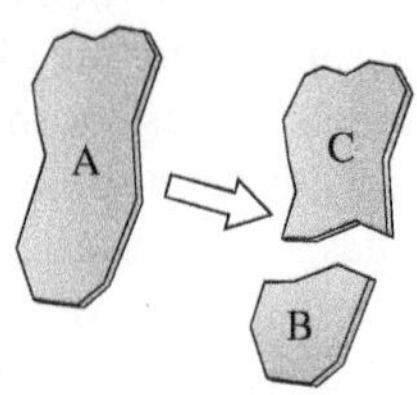

9.2 Pressure

"Pressure" is a word we all know and use. You probably have a commonsense idea of what pressure is. For example, you feel the effects of varying pressure against your eardrums when you swim underwater or take off in an airplane. Cans of whipped cream are "pressurized" to make the contents squirt out when you press the nozzle. It's hard to open a "vacuum sealed" jar of jelly the first time, but easy after the seal is broken.

FIGURE 9.3 Water pressure pushes the water *sideways,* out of the holes.

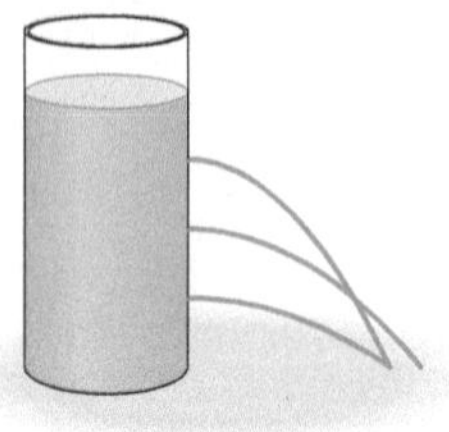

You've probably seen water squirting out of a hole in the side of a container, as in FIGURE 9.3. Notice that the water emerges at greater speed from a hole at greater depth. And you've probably felt the air squirting out of a hole in a bicycle tire or inflatable air mattress. These observations suggest that

- "Something" pushes the water or air *sideways,* out of the hole.
- In a liquid, the "something" is larger at greater depths.

FIGURE 9.4 shows a fluid—either a liquid or a gas—pressing against a small area A with force $\vec{F}$. This is the force that pushes the fluid out of a hole. In the absence of a hole, $\vec{F}$ pushes against the wall of the container. Let's define the **pressure** at this point in the fluid to be the ratio of the force to the area on which the force is exerted:

$$p = \frac{F}{A} \tag{9.2}$$

FIGURE 9.4 The fluid presses against area A with force $\vec{F}$.

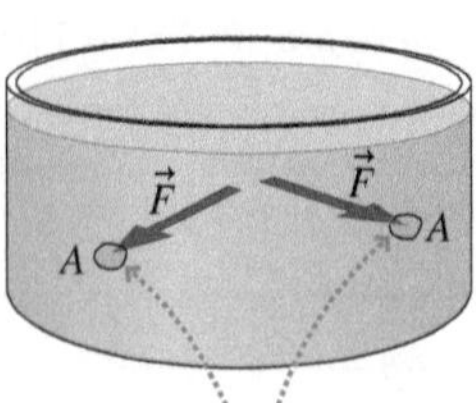

Notice that pressure is a scalar, not a vector. You can see, from Equation 9.2, that a fluid exerts a force of magnitude

$$F = pA \tag{9.3}$$

on a surface of area A. The force is *perpendicular* to the surface.

NOTE ► Pressure itself is *not* a force, even though we sometimes talk informally about "the force exerted by the pressure." The correct statement is that the *fluid* exerts a force on a surface. ◄

From its definition, pressure has units of N/m^2. The SI unit of pressure is the **pascal,** defined as

$$1 \text{ pascal} = 1 \text{ Pa} = 1 \text{ N/m}^2$$

This unit is named for the 17th-century French scientist Blaise Pascal, the inventor of the syringe. Large pressures are often given in kilopascals, where 1 kPa = 10^3 Pa, or megapascals, where 1 MPa = 10^6 Pa.

FIGURE 9.5 Learning about pressure.

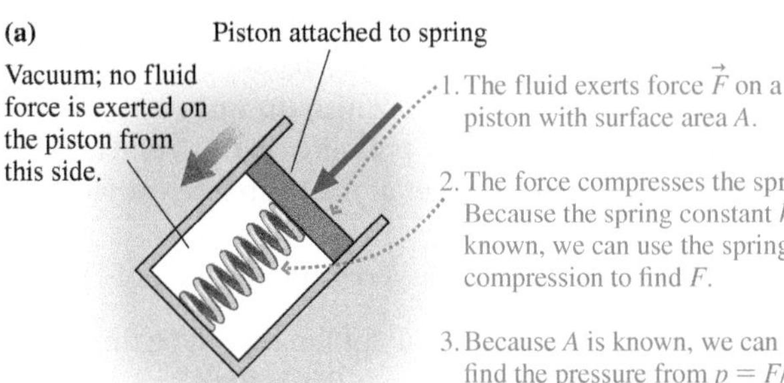

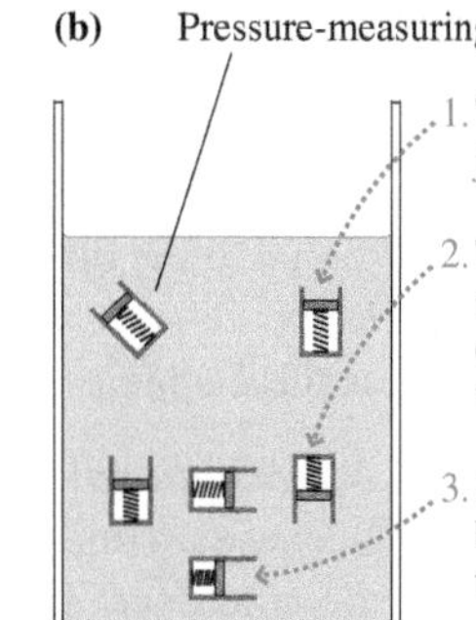

(a) Piston attached to spring
Vacuum; no fluid force is exerted on the piston from this side.

2. The force compresses the spring. Because the spring constant k is known, we can use the spring's compression to find F.
3. Because A is known, we can find the pressure from $p = F/A$.

(b) Pressure-measuring device in fluid

FIGURE 9.5a shows a simple spring-loaded device we could use to measure pressure. Once we've built such a device, we can place it in various liquids and gases to learn about pressure. FIGURE 9.5b shows what we can learn from a series of simple experiments.

The first statement in Figure 9.5b is especially important. Pressure exists at *all* points within a fluid, not just at the walls of the container. You may recall that tension exists at *all* points in a string, not only at its ends where it is tied to an object. We understood tension as the different parts of the string *pulling* against each other. Pressure is an analogous idea, except that the different parts of a fluid are *pushing* against each other.

NOTE ▸ Stress, which you studied in Chapter 6, is also defined as F/A. And while we used N/m^2 as the units of stress, some texts and tables use Pa. The distinction between stress and pressure is that stress is exerted in a particular direction but pressure is homogeneous, felt equally in all directions. ◂

Causes of Pressure

Gases and liquids are both fluids, but they have some important differences. Liquids are nearly incompressible; gases are highly compressible. The molecules in a liquid attract each other via molecular bonds; the molecules in a gas do not interact other than through collisions. These differences affect the origins of pressure in gases and liquids.

Imagine that you have two sealed jars, each containing a small amount of mercury and nothing else. All the air has been removed from the jars. Suppose you take the two jars into orbit on the space station, where they are weightless. One jar you keep cool, so that the mercury is a liquid. The other you heat until the mercury becomes a gas. What can we say about the pressure in these two jars?

As FIGURE 9.6 shows, molecular bonds hold the liquid mercury together. It might quiver like Jello, but it remains a cohesive drop floating in the center of the jar. The liquid drop exerts no forces on the walls, so there's *no* pressure in the jar containing the liquid. (If we actually did this experiment, a very small fraction of the mercury would be in the vapor phase and create what is called *vapor pressure.*)

The gas is different. The gas molecules collide with the wall of the container, and each bounce exerts a force on the wall. The force from any one collision is incredibly small, but there are an extraordinarily large number of collisions every second. These collisions cause the gas to have a pressure. We will do the calculation in Chapter 13.

FIGURE 9.7 shows the jars back on earth. Because of gravity, the liquid now fills the bottom of the jar and exerts a force on the bottom and the sides. Liquid mercury is incompressible, so the volume of liquid in Figure 9.7 is the same as in

FIGURE 9.6 A liquid and a gas in a weightless environment.

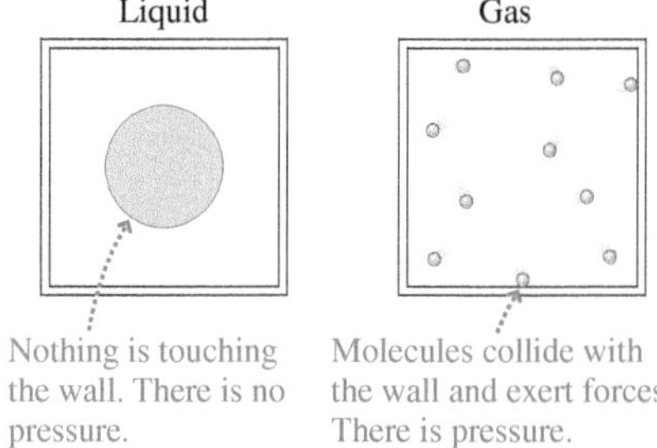

Liquid
Gas
Nothing is touching the wall. There is no pressure.
Molecules collide with the wall and exert forces. There is pressure.

FIGURE 9.7 Gravity affects the pressure of the fluids.

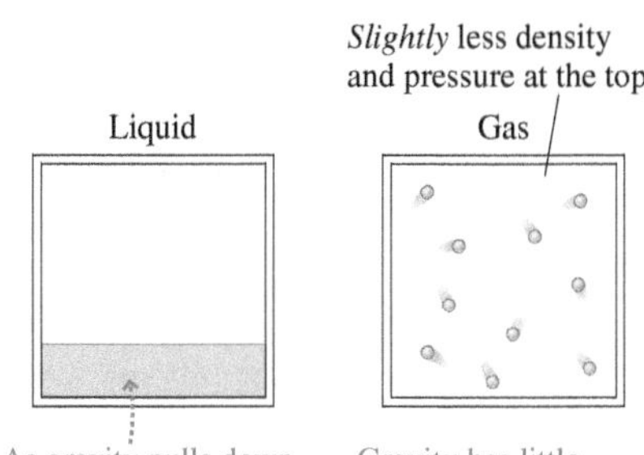

Slightly less density and pressure at the top
Liquid
Gas
As gravity pulls down, the liquid exerts a force on the bottom and sides of its container.
Gravity has little effect on the pressure of the gas.

Figure 9.6. There is still no pressure on the top of the jar (other than the very small vapor pressure).

At first glance, the situation in the gas-filled jar seems unchanged from Figure 9.6. However, the earth's gravitational pull causes the gas density to be *slightly* more at the bottom of the jar than at the top. Because the pressure due to collisions is proportional to the density, the pressure is *slightly* larger at the bottom of the jar than at the top.

Thus there appear to be two contributions to the pressure in a container of fluid:

1. A *gravitational contribution* that arises from gravity pulling down on the fluid. Because a fluid can flow, forces are exerted on both the bottom and sides of the container. The gravitational contribution depends on the strength of the gravitational force.
2. A *thermal contribution* due to the collisions of freely moving molecules with the walls. The thermal contribution depends on the absolute temperature of the gas.

A detailed analysis finds that these two contributions are not entirely independent of each other, but the distinction is useful for a basic understanding of pressure. Let's see how these two contributions apply to different situations.

Pressure in Gases

The pressure in a laboratory-size container of gas is due almost entirely to the thermal contribution. A container would have to be ≈ 100 m tall for gravity to cause the pressure at the top to be even 1% less than the pressure at the bottom. Laboratory-size containers are much less than 100 m tall, so we can quite reasonably assume that the pressure has the *same* value at all points in a laboratory-size container of gas.

Decreasing the number of molecules in a container decreases the gas pressure simply because there are fewer collisions with the walls. If a container is completely empty, with no atoms or molecules, then the pressure is $p = 0$ Pa. This is a *perfect vacuum.* No perfect vacuum exists in nature, not even in the most remote depths of outer space, because it is impossible to completely remove every atom from a region of space. In practice, a **vacuum** is an enclosed space in which $p \ll 1$ atm. Using $p = 0$ Pa is then a very good approximation.

Atmospheric Pressure

FIGURE 9.8 The pressure and density decrease with increasing height in the atmosphere.

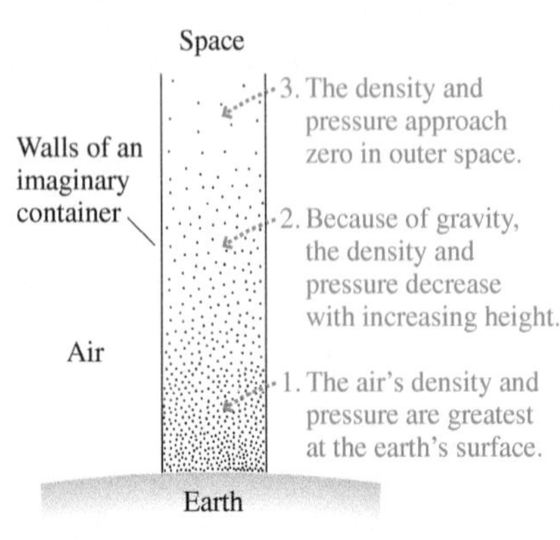

The earth's atmosphere is *not* a laboratory-size container. The height of the atmosphere is such that the gravitational contribution to pressure *is* important. As FIGURE 9.8 shows, the density of air slowly decreases with increasing height until approaching zero in the vacuum of outer space. Consequently, the pressure of the air, what we call the *atmospheric pressure* p_{atmos}, decreases with height. The air pressure is less in Denver (elevation 1600 m) than in Miami (sea level).

The atmospheric pressure *at sea level* varies slightly with the weather, but the global average sea-level pressure is 101,000 Pa. Consequently, we define the **standard atmosphere** as

$$1 \text{ standard atmosphere} = 1 \text{ atm} = 101{,}000 \text{ Pa} = 101 \text{ kPa}$$

The standard atmosphere, usually referred to simply as "atmospheres," is a commonly used unit of pressure, especially in chemistry. But it is not an SI unit, so for physics problems you often must convert atmospheres to pascals before doing calculations with pressure. Again, pay attention to the units of other quantities you're working with to see what unit conversions are needed.

NOTE ► Unless you happen to live right at sea level, the atmospheric pressure around you is somewhat less than 1 atm. Pressure experiments use a barometer to determine the actual atmospheric pressure. For simplicity, this textbook will always assume that the pressure of the air is $p_{atmos} = 1$ atm unless stated otherwise. ◄

Given that the pressure of the air at sea level is 101 kPa, you might wonder why the weight of the air doesn't crush your forearm when you rest it on a table. Your forearm has a surface area of $\approx 200 \text{ cm}^2 = 0.02 \text{ m}^2$, so the force of the air pressing against it is ≈ 2000 N (≈ 450 pounds). How can you even lift your arm?

The reason, as FIGURE 9.9 shows, is that a fluid exerts pressure forces in *all* directions. There *is* a downward force of ≈ 2000 N on your forearm, but the air underneath your arm exerts an upward force of the same magnitude. The *net* force is very close to zero.

But, you say, there isn't any air under my arm if I rest it on a table. Actually, there is. There would be a *vacuum* under your arm if there were no air. Imagine placing your arm on the top of a large vacuum cleaner suction tube. What happens? You feel a downward force as the vacuum cleaner "tries to suck your arm in." However, the downward force you feel is not a *pulling* force from the vacuum cleaner. It is the *pushing* force of the air above your arm *when the air beneath your arm is removed and cannot push back.* Air molecules do not have hooks! They have no ability to "pull" on your arm. The air can only push.

Vacuum cleaners, suction cups, and other similar devices are powerful examples of how strong atmospheric pressure forces can be *if* the air is removed from one side of an object so as to produce an unbalanced force. The fact that we are *surrounded* by the fluid allows us to move around in the air, just as we swim underwater, oblivious of these strong forces. Similarly, the fluid pressure inside your body must be very close to 1 atm to prevent you from imploding or exploding!

Thick windows The deeper you dive in the ocean, the greater the pressure. At the 4500 m depths the research submarine *Alvin* reaches, the pressure is greater than 450 atm! The pilots certainly want to look outside, but putting a window in a craft at this depth is no mean feat. In the photo, a technician inspects the window that will fill a viewport. The tapered shape will have its larger face toward the sea. The water pressure will push the window firmly into its conical seat, making a tight seal.

FIGURE 9.9 Pressure forces in a fluid push with equal strength in all directions.

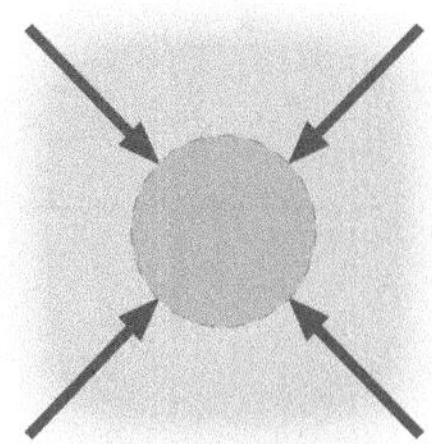

The forces of a fluid push in *all* directions.

EXAMPLE 9.1 A suction cup

A 10.0-cm-diameter suction cup is pushed against a smooth ceiling. What is the maximum mass of an object that can be suspended from the suction cup without pulling it off the ceiling? The mass of the suction cup is negligible.

PREPARE Pushing the suction cup against the ceiling pushes the air out. We'll assume that the volume enclosed between the suction cup and the ceiling is a perfect vacuum with $p = 0$ Pa. We'll also assume that the pressure in the room is 1 atm.

FIGURE 9.10 shows a free-body diagram of the suction cup stuck to the ceiling. The downward normal force of the ceiling is distributed around the rim of the suction cup, but in the particle model we can show this as a single force vector.

FIGURE 9.10 A suction cup is held to the ceiling by air pressure pushing upward on the bottom.

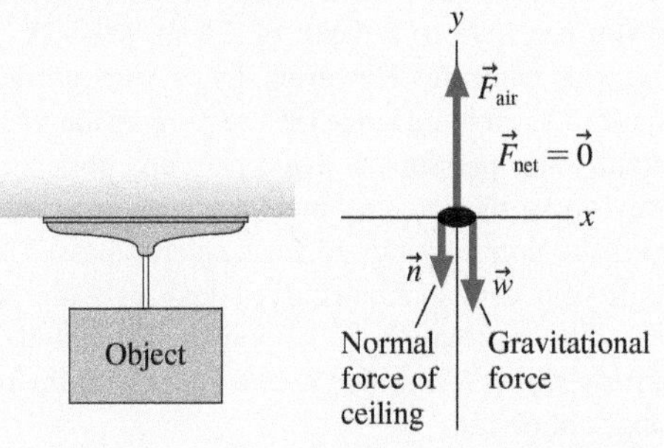

SOLVE The suction cup remains stuck to the ceiling, in static equilibrium, as long as $F_{air} = n + w$. The magnitude of the upward force exerted by the air is

$$F_{air} = pA = p\pi r^2 = (101{,}000 \text{ Pa})\pi(0.050 \text{ m})^2 = 793 \text{ N}$$

There is no downward force from the air in this case because there is no air inside the cup. Increasing the hanging mass decreases the normal force n by an equal amount. The maximum weight has been reached when n is reduced to zero. Thus

$$w_{max} = mg = F_{air} = 793 \text{ N}$$

$$m = \frac{793 \text{ N}}{g} = 81 \text{ kg}$$

ASSESS The suction cup can support a mass of up to 81 kg if all the air is pushed out, leaving a perfect vacuum inside. A real suction cup won't achieve a perfect vacuum, but suction cups can hold substantial weight. In the same way, the suckers of cephalopods, such as octopus and squid, enable them to grip rocks and prey.

Pressure in Liquids

Gravity causes a liquid to fill the bottom of a container. We'd like to determine the pressure at depth d below the surface of the liquid. We will assume that the liquid is at rest; flowing liquids will be considered later in this chapter.

The shaded cylinder of liquid in FIGURE 9.11 extends from the surface to depth d. This cylinder, like the rest of the liquid, is in static equilibrium with $\vec{F}_{net} = \vec{0}$. Three forces act on this cylinder: the weight mg of the liquid in the cylinder, a downward force p_0A due to the pressure p_0 at the surface of the liquid, and an upward force pA due to the liquid beneath the cylinder pushing up on the bottom of the cylinder. This third force is a consequence of our earlier observation that different parts of a fluid push against each other. Pressure p, which is what we're trying to find, is the pressure at the bottom of the cylinder.

FIGURE 9.11 Measuring the pressure at depth d in a liquid.

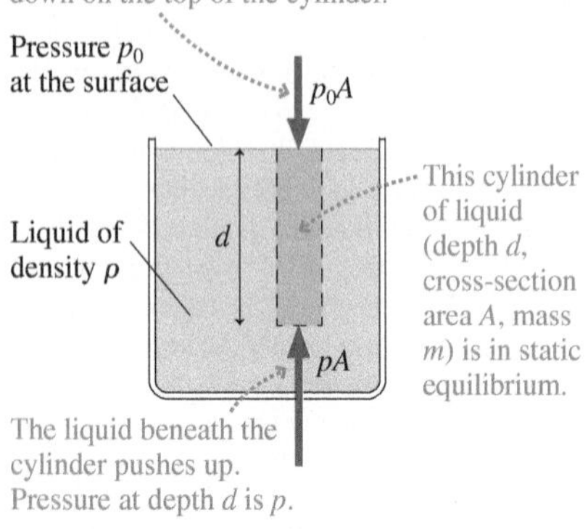
Whatever is above the liquid pushes down on the top of the cylinder.
Pressure p_0 at the surface
p_0A
This cylinder of liquid (depth d, cross-section area A, mass m) is in static equilibrium.
Liquid of density ρ
d
pA
The liquid beneath the cylinder pushes up. Pressure at depth d is p.

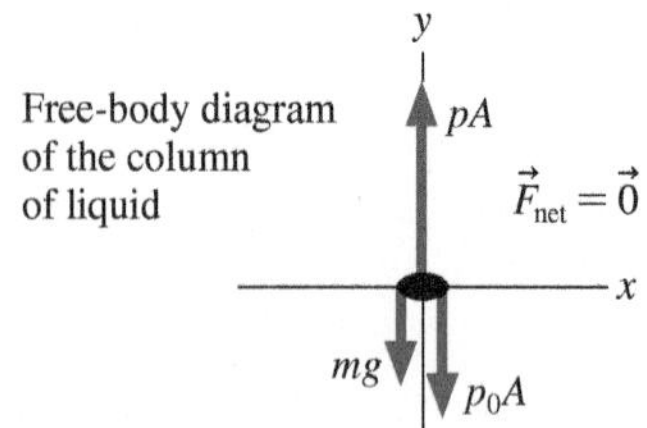
y
Free-body diagram of the column of liquid
pA
$\vec{F}_{net} = \vec{0}$
x
mg
p_0A

The upward force balances the two downward forces, so

$$pA = p_0A + mg \tag{9.4}$$

The liquid is a cylinder of cross-section area A and height d. Its volume is $V = Ad$ and its mass is $m = \rho V = \rho Ad$. Substituting this expression for the mass of the liquid into Equation 9.4, we find that the area A cancels from all terms. The pressure at depth d in a liquid is

$$p = p_0 + \rho gd \tag{9.5}$$

Hydrostatic pressure of a liquid at depth d

where ρ is the liquid's density. Because the fluid is at rest, the pressure given by Equation 9.5 is called the **hydrostatic pressure.** The fact that g appears in Equation 9.5 reminds us that pressure in a liquid is due to gravity.

As expected, $p = p_0$ at the surface, where $d = 0$. Pressure p_0 is often due to the air or other gas above the liquid. $p_0 = 1$ atm $= 101$ kPa for a liquid that is open to the air. However, p_0 can also be the pressure due to a piston or a closed surface pushing down on the top of the liquid.

NOTE ▶ Equation 9.5 assumes that the fluid is *incompressible;* that is, its density ρ doesn't increase with depth. This is an excellent assumption for liquids, but not a good one for a gas. Equation 9.5 should not be used to calculate the pressure in a gas. ◀

EXAMPLE 9.2 Pressure on an eardrum

An ocean snorkeler dives to a depth of 5.5 m. What is the pressure at this depth? What is the net force on her eardrum if the pressure inside her ear is 1.0 atm? An adult's eardrum has a typical diameter of 8.0 mm.

PREPARE You've very likely felt pressure on your eardrum when you swim underwater. The water pressure is hydrostatic pressure that increases with depth. As a result, the water exerts force $F = pA$ on a surface of area A. The pressure at the surface of the water is assumed to be $p_0 = 1.0$ atm $= 1.01 \times 10^5$ Pa due to the air. The density of seawater, $\rho = 1030$ kg/m^3, is found in Table 9.1.

SOLVE From Equation 9.5, the pressure at depth $d = 5.5$ m is

$$p = p_0 + \rho gd = 1.01 \times 10^5 \text{ Pa} + (1030 \text{ kg/m}^3)(9.80 \text{ m/s}^2)(5.5 \text{ m})$$
$$= 1.6 \times 10^5 \text{ Pa} = 160 \text{ kPa}$$

The area of the eardrum is $A = \pi r^2 = \pi(0.0040 \text{ m})^2 = 5.0 \times 10^{-5}$ m^2. Thus the force of the water on the eardrum is

$$F_{water} = pA = (1.6 \times 10^5 \text{ Pa})(5.0 \times 10^{-5} \text{ m}^2) = 8.0 \text{ N}$$

But that's only the force pushing inward. The 1.0 atm pressure inside the ear pushes outward with force

$$F_{inner\ ear} = pA = (1.01 \times 10^5 \text{ Pa})(5.0 \times 10^{-5} \text{ m}^2) = 5.1 \text{ N}$$

As a result, the *net* force on the eardrum, pushing inward, is

$$F_{net} = F_{water} - F_{inner\ ear} = 2.9 \text{ N}$$

ASSESS This is not an extremely large force, but it makes sense because you can dive to a depth of 5.5 m, or ≈ 18 ft, without rupturing your eardrum. However, divers who go deeper than ≈ 5 m have to learn to equalize the pressure inside their ear with the external water pressure to avoid pain and possible ear damage. (They do so by pinching their nose and blowing gently to force open the normally closed Eustachian tubes.) The internal pressure of your ears spontaneously equalizes with the external pressure—your ears may "pop" as you change altitude—but this is a relatively slow process that doesn't help with the rapid pressure changes of diving.

The hydrostatic pressure in a liquid depends only on the depth and the pressure at the surface. This observation has some important implications. FIGURE 9.12a shows two connected tubes. It's certainly true that the larger volume of liquid in the wide tube weighs more than the liquid in the narrow tube. You might think that this extra weight would push the liquid in the narrow tube higher than in the wide tube. But it doesn't. If d_1 were larger than d_2, then, according to the hydrostatic pressure equation, the pressure at the bottom of the narrow tube would be higher than the pressure at the bottom of the wide tube. This *pressure difference* would cause the liquid to *flow* from right to left until the heights were equal.

Thus a first conclusion: **A connected liquid in hydrostatic equilibrium rises to the same height in all open regions of the container.**

FIGURE 9.12b shows two connected tubes of different shape. The conical tube holds more liquid above the dotted line, so you might think that $p_1 > p_2$. But it isn't. Both points are at the same depth, thus $p_1 = p_2$. If p_1 were larger than p_2, the pressure at the bottom of the left tube would be larger than the pressure at the bottom of the right tube. This would cause the liquid to flow until the pressures were equal.

Thus a second conclusion: **The pressure is the same at all points on a horizontal line through a connected liquid in hydrostatic equilibrium.**

NOTE ▶ Both of these conclusions are restricted to liquids in hydrostatic equilibrium. The situation is different for flowing fluids, as we'll see later in the chapter. ◀

FIGURE 9.12 Some properties of a liquid in hydrostatic equilibrium are not what you might expect.

(a)

(b)

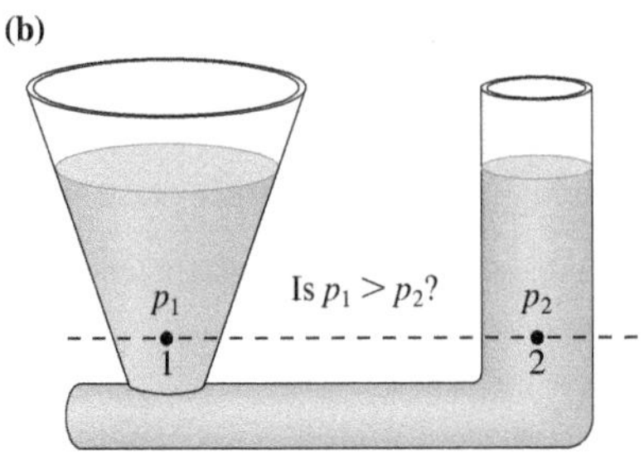

EXAMPLE 9.3 Pressure in a closed tube

Water fills the tube shown in FIGURE 9.13. What is the pressure at the top of the closed tube?

PREPARE This is a liquid in hydrostatic equilibrium. The closed tube is not an open region of the container, so the water cannot rise to an equal height. Nevertheless, the pressure is still the same at all points on a horizontal line. In particular, the pressure at the top of the closed tube equals the pressure in the open tube at the height of the dashed line. Assume $p_0 = 1.00$ atm.

FIGURE 9.13 A water-filled tube.

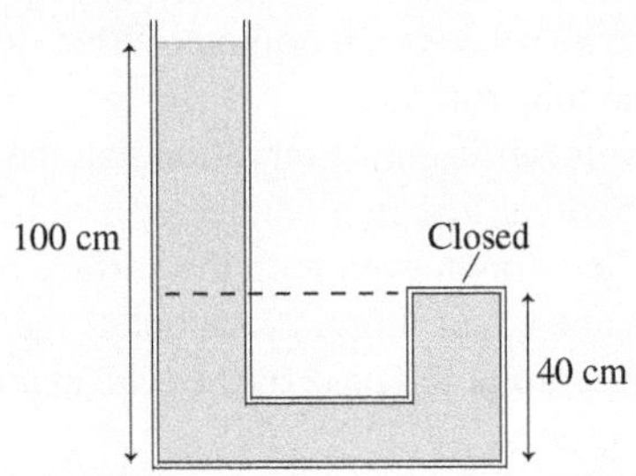

SOLVE A point 40 cm above the bottom of the open tube is at a depth of 60 cm. The pressure at this depth is

$$p = p_0 + \rho g d$$
$$= 1.01 \times 10^5 \text{ Pa} + (1000 \text{ kg/m}^3)(9.80 \text{ m/s}^2)(0.60 \text{ m})$$
$$= 1.07 \times 10^5 \text{ Pa} = 1.06 \text{ atm}$$

This is the pressure at the top of the closed tube.

ASSESS The water in the open tube *pushes* the water in the closed tube up against the top of the tube, which is why the pressure is greater than 1 atm.

STOP TO THINK 9.2 Water is slowly poured into the container until the water level has risen into tubes A, B, and C. The water doesn't overflow from any of the tubes. How do the water depths in the three columns compare to each other?

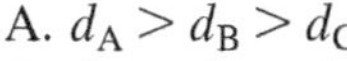

Water

A B C

A. $d_A > d_B > d_C$
B. $d_A < d_B < d_C$
C. $d_A = d_B = d_C$
D. $d_A = d_C > d_B$
E. $d_A = d_C < d_B$

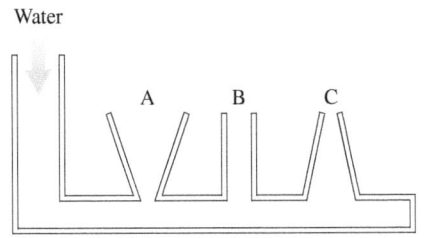

Gauge Pressure

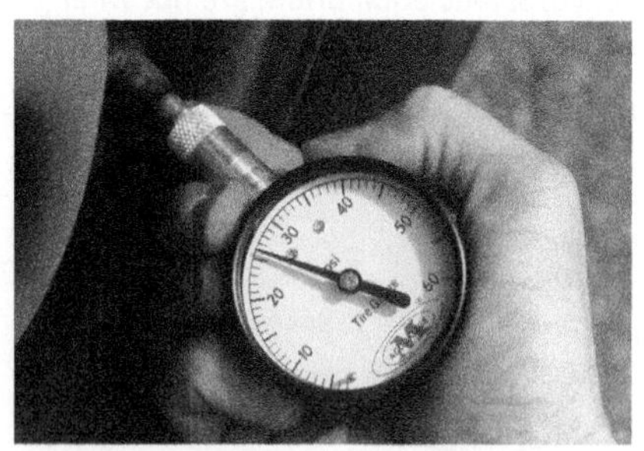

A tire-pressure gauge reads the gauge pressure p_g, not the absolute pressure p.

The pressure in a fluid is measured with a *pressure gauge*. The fluid pushes against some sort of spring, and the spring's displacement is registered by a pointer on a dial.

Blood-pressure cuffs and many pressure gauges, such as tire gauges and the gauges on air tanks, measure not the actual or absolute pressure p but what is called **gauge pressure.** The gauge pressure, denoted p_g, is the pressure *in excess* of 1 atm. That is,

$$p_g = p - 1 \text{ atm} \tag{9.6}$$

You must add 1 atm = 101 kPa to the reading of a pressure gauge to find the absolute pressure p that you need for doing many types of calculations: $p = p_g + 1$ atm.

EXAMPLE 9.4 An underwater pressure gauge

An underwater pressure gauge reads 60 kPa. What is its depth?

PREPARE The gauge reads gauge pressure, not absolute pressure.

SOLVE The hydrostatic pressure at depth d, with $p_0 = 1$ atm, is $p = 1 \text{ atm} + \rho g d$. Thus the gauge pressure is

$$p_g = p - 1 \text{ atm} = (1 \text{ atm} + \rho g d) - 1 \text{ atm} = \rho g d$$

The term $\rho g d$ is the pressure *in excess* of atmospheric pressure and thus *is* the gauge pressure. Solving for d, we find

$$d = \frac{60{,}000 \text{ Pa}}{(1000 \text{ kg/m}^3)(9.80 \text{ m/s}^2)} = 6.1 \text{ m}$$

ASSESS A depth of a few meters seems reasonable for an underwater gauge.

Barometers and Pumps

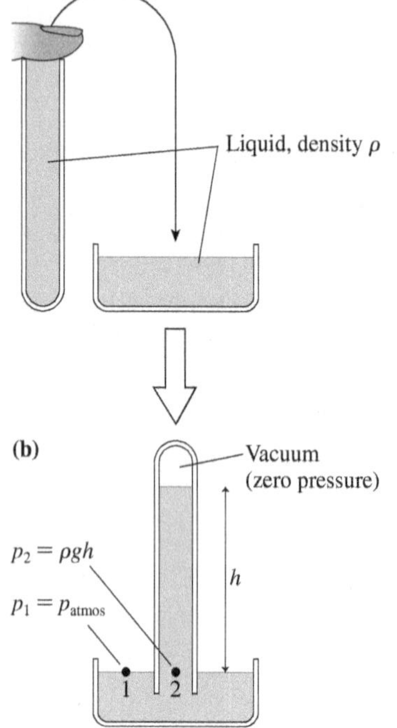

FIGURE 9.14 A barometer.

A traditional but still important pressure-measuring instrument is the **barometer,** which is used to measure the atmospheric pressure p_{atmos}. FIGURE 9.14a shows a glass tube, sealed at the bottom, that has been completely filled with a liquid. If we temporarily seal the top end, we can invert the tube and place it in a beaker of the same liquid. When the temporary seal is removed, some, but not all, of the liquid runs out, leaving a liquid column in the tube that is a height h above the surface of the liquid in the beaker. This device, shown in FIGURE 9.14b, is a barometer. What does it measure? And why doesn't *all* the liquid in the tube run out?

The key to understanding the barometer is our observation that the pressure is the same at any two points along a horizontal line in a connected fluid. Points 1 and 2 in Figure 9.14b are on a horizontal line drawn even with the surface of the liquid, so the pressure at these two points must be equal. Liquid runs out of the tube only until a balance is reached between the pressure at the base of the tube and the pressure of the air.

You can think of a barometer as rather like a seesaw. If the pressure of the atmosphere increases, it presses down on the liquid in the beaker. This forces liquid up the tube until the pressures at points 1 and 2 are equal. If the atmospheric pressure falls, liquid has to flow out of the tube to keep the pressures equal at these two points.

The pressure at point 2 is the pressure due to the weight of the liquid in the tube plus the pressure of the gas above the liquid. But in this case there is no gas above the liquid! Because the tube had been completely full of liquid when it was inverted, the space left behind when the liquid ran out is a vacuum (ignoring a very slight *vapor pressure* of the liquid, negligible except in extremely precise measurements). Thus pressure p_2 is simply $p_2 = \rho g h$.

Equating p_1 and p_2 gives

$$p_{atmos} = \rho g h \tag{9.7}$$

Thus we can measure the atmosphere's pressure by measuring the height of the liquid column in a barometer.

The average air pressure at sea level causes a column of mercury in a mercury barometer to stand 760 mm above the surface. Knowing that the density of mercury is 13,600 kg/m^3 (at 0°C), we can use Equation 9.7 to find that the average atmospheric pressure is

$$p_{\text{atmos}} = \rho_{\text{Hg}} g h = (13{,}600\ \text{kg/m}^3)(9.80\ \text{m/s}^2)(0.760\ \text{m})$$
$$= 1.01 \times 10^5\ \text{Pa} = 101\ \text{kPa}$$

This is the value given earlier as "one standard atmosphere."

Because of the historical importance of barometers, it is still common in some areas of science and medicine to report pressure—especially blood pressure—in terms of the height of liquid in a mercury barometer. The conversion factor is 1 atm = 760 millimeters of mercury (abbreviated "mm Hg") = 101 kPa. Or, equivalently, 1 mm Hg = 133 Pa.

The actual barometric pressure varies slightly from day to day as the weather changes. Weather systems are called *high-pressure systems* or *low-pressure systems,* depending on whether the local sea-level pressure is higher or lower than one standard atmosphere. Higher pressure is usually associated with fair weather, while lower pressure portends rain.

Mercury is 13.6 times denser than water, so a barometer that uses water rather than mercury would have to stand 13.6 times higher at 10.3 m, or $\approx$34 ft. A 34-foot-tall barometer is not very practical, but this idea has a very practical application.

FIGURE 9.15 shows a hand pump lifting water up a height h. The pump works by reducing the air pressure above the water, after which the water is not "sucked" upward—recall our discussion of suction cups—but, instead, is *pushed* upward by the atmosphere pressing down on the water below. It's basically a barometer. If the pressure p_0 above the column of water is reduced only slightly below atmospheric pressure, as is the case with most hand pumps, then the atmosphere can push the water up only a small distance. But suppose you have a powerful pump that can completely remove the air above the water, creating a vacuum. Then the water will rise to the height given by Equation 9.7, or 10.3 m.

That's the limit. You can't use atmospheric pressure to pump water to a height of more than 10.3 m. But many trees are taller than 10 m, sometimes even 100 m. Clearly atmospheric pressure cannot lift water this high, so how does water get from the roots to the top of a tree? We'll return to this question in Section 9.4.

FIGURE 9.15 A simple water pump.

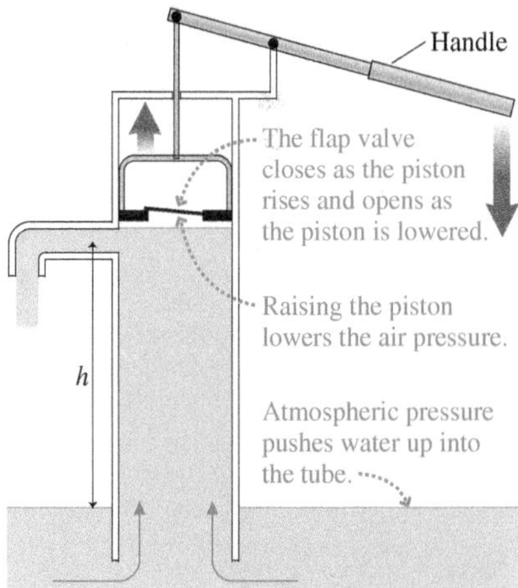

Handle
The flap valve closes as the piston rises and opens as the piston is lowered.
Raising the piston lowers the air pressure.
h
Atmospheric pressure pushes water up into the tube.

Pressure Units

In practice, pressure is measured in a number of different units. This plethora of units and abbreviations has arisen historically as scientists and engineers working on different subjects (liquids, high-pressure gases, low-pressure gases, weather, etc.) developed what seemed to them the most convenient units. These units continue in use through tradition, so it is necessary to convert back and forth among them. TABLE 9.2 gives the basic conversions.

TABLE 9.2 **Pressure units**

Unit	Abbreviation	Conversion to Pa	Uses
pascal	Pa		SI unit: 1 Pa = 1 N/m^2 used in most calculations
atmosphere	atm	1 atm = 101 kPa	general
millimeters of mercury	mm Hg	1 mm Hg = 133 Pa	gases, blood pressure, and barometric pressure
pounds per square inch	psi	1 psi = 6.89 kPa	U.S. engineering and industry

STOP TO THINK 9.3 A U-shaped tube is open to the atmosphere on both ends. Water is poured into the tube, followed by oil, which floats on the water because it is less dense than water. Which figure shows the correct equilibrium configuration?

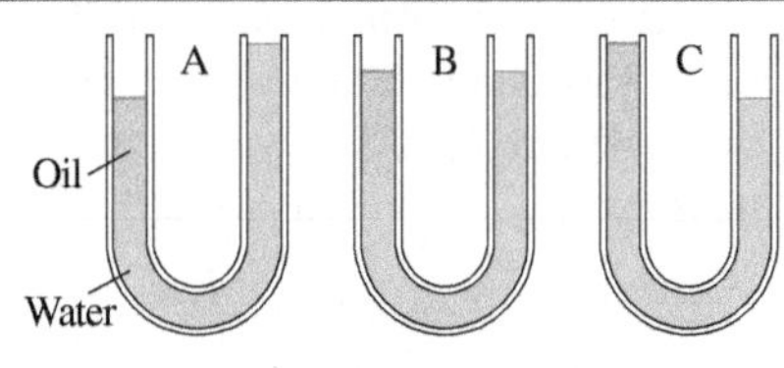
A
B
C
Oil
Water

9.3 Buoyancy

Wood floats on the surface of a lake. A penny with a mass of a few grams sinks, but a massive steel barge floats. How can we understand these diverse phenomena?

An air mattress floats effortlessly on the surface of a swimming pool. If you've ever tried to push an air mattress underwater, you know it is nearly impossible. As you push down, the water pushes up. This upward force of a fluid is called the **buoyant force.**

The basic reason for the buoyant force is straightforward. FIGURE 9.16 shows a cylinder submerged in a liquid. The pressure in the liquid increases with depth, so the pressure at the bottom of the cylinder is greater than at the top. Both cylinder ends have equal area, so force $\vec{F}_{up}$ is greater than force $\vec{F}_{down}$. (Remember that pressure forces push in *all* directions.) Consequently, the pressure in the liquid exerts a *net upward force* on the cylinder of magnitude $F_{net} = F_{up} - F_{down}$. This is the buoyant force.

FIGURE 9.16 The buoyant force arises because the fluid pressure at the bottom of the cylinder is greater than that at the top.

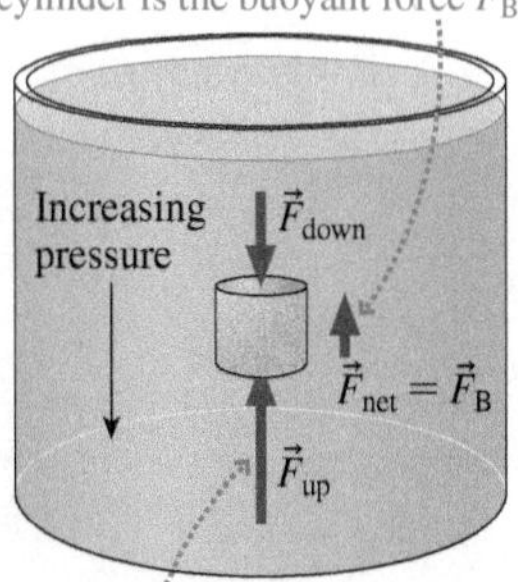
The net force of the fluid on the cylinder is the buoyant force $\vec{F}_B$.
Increasing pressure
$\vec{F}_{down}$
$\vec{F}_{net} = \vec{F}_B$
$\vec{F}_{up}$
$F_{up} > F_{down}$ because the pressure is greater at the bottom. Hence the fluid exerts a net upward force.

The submerged cylinder illustrates the idea in a simple way, but the result is not limited to cylinders or to liquids. Suppose we isolate a parcel of fluid of arbitrary shape and volume by drawing an imaginary boundary around it, as shown in FIGURE 9.17a. This parcel is in static equilibrium. Consequently, the parcel's weight force pulling it down must be balanced by an upward force. The upward force, which is exerted on this parcel of fluid by the surrounding fluid, is the buoyant force $\vec{F}_B$. The buoyant force matches the weight of the fluid: $F_B = w$.

Now imagine that we remove this parcel of fluid and instantaneously replace it with an object having exactly the same shape and size, as shown in FIGURE 9.17b. Because the buoyant force is exerted by the *surrounding* fluid, and the surrounding fluid hasn't changed, the buoyant force on this new object is *exactly the same* as the buoyant force on the parcel of fluid that we removed.

When an object (or a portion of an object) is immersed in a fluid, it *displaces* fluid that would otherwise fill that region of space. The displaced fluid's volume is exactly the volume of the portion of the object that is immersed in the fluid. Figure 9.17 leads us to conclude that the magnitude of the upward buoyant force matches the weight of this displaced fluid.

This idea was first recognized by the ancient Greek mathematician and scientist Archimedes, and today we know it as *Archimedes' principle:*

FIGURE 9.17 The buoyant force on an object is the same as the buoyant force on an equal volume of fluid.

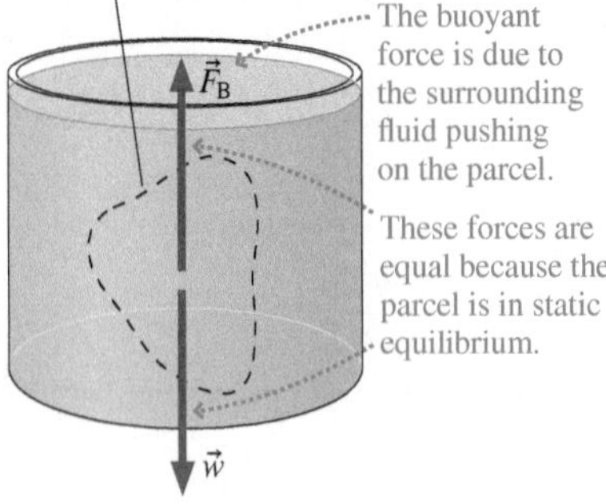
(a) Imaginary boundary around a parcel of fluid
$\vec{F}_B$
The buoyant force is due to the surrounding fluid pushing on the parcel.
These forces are equal because the parcel is in static equilibrium.
$\vec{w}$

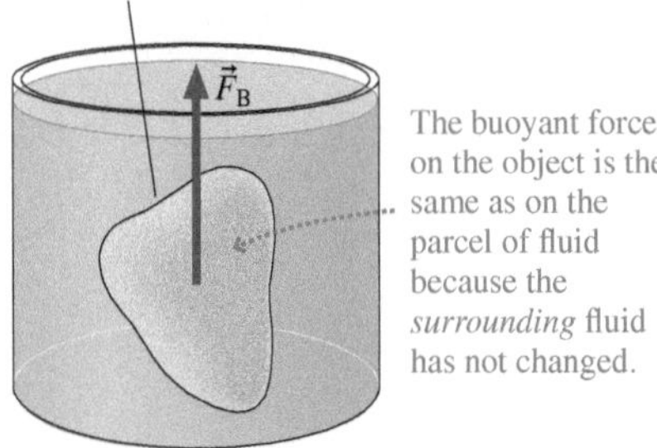
(b) Real object with same size and shape as the parcel of fluid
$\vec{F}_B$
The buoyant force on the object is the same as on the parcel of fluid because the *surrounding* fluid has not changed.

Archimedes' principle A fluid exerts an upward buoyant force $\vec{F}_B$ on an object immersed in or floating on the fluid. The magnitude of the buoyant force equals the weight of the fluid displaced by the object.

Suppose the fluid has density ρ_f and the object displaces a volume V_f of fluid. The mass of the displaced fluid is then $m_f = \rho_f V_f$ and so its weight is $w_f = \rho_f V_f g$. Thus Archimedes' principle in equation form is

$$F_B = \rho_f V_f g \tag{9.8}$$

Buoyant force on an object displacing volume V_f of fluid of density ρ_f

It is important to distinguish the density and volume of the displaced fluid from the density and volume of the object. To do so, we use subscript f for the fluid and o for the object. If the object is completely submerged, the volume of fluid displaced is the same as the volume of the object, so $V_f = V_o$.

NOTE ► An object's weight w_o, the force of gravity, is unaffected by being submerged. The fact that you "feel lighter" in the water is due to the buoyant force of the water, not to any change in your weight. ◄

EXAMPLE 9.5 Is the crown gold?

Legend has it that Archimedes was asked by King Hiero of Syracuse to determine whether a crown was pure gold or had been adulterated with a less valuable metal by an unscrupulous goldsmith. It was this problem that led Archimedes to the principle that bears his name. In a modern version of his method, a crown weighing 8.30 N is suspended underwater from a string. The tension in the string is measured to be 7.81 N. Is the crown pure gold?

PREPARE To discover whether the crown is pure gold, we will proceed as Archimedes did—by determining its density ρ_o and comparing that to the known density of gold. **FIGURE 9.18** shows the forces acting on the crown. In addition to the familiar tension and weight forces, the water exerts an upward buoyant force on the crown. The magnitude of the buoyant force is given by Archimedes' principle.

FIGURE 9.18 The forces acting on the submerged crown.

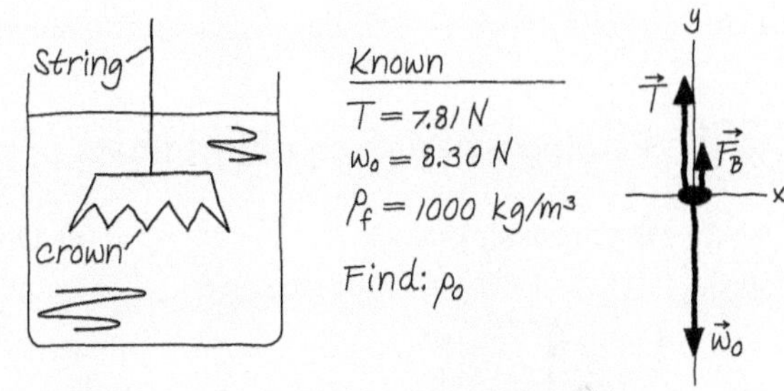

SOLVE Because the crown is in equilibrium, its acceleration and the net force on it are zero. Newton's second law then reads

$$\sum F_y = F_B + T - w_o = 0$$

from which the buoyant force is

$$F_B = w_o - T = 8.30\text{ N} - 7.81\text{ N} = 0.49\text{ N}$$

According to Archimedes' principle, $F_B = \rho_f V_f g$, where V_f is the volume of the fluid displaced. Here, where the crown is completely submerged, the volume of the fluid displaced is equal to the volume V_o of the crown. The crown's weight is $w_o = m_o g = \rho_o V_o g$, so its volume is

$$V_o = \frac{w_o}{\rho_o g}$$

Inserting this volume into Archimedes' principle gives

$$F_B = \rho_f V_o g = \rho_f\left(\frac{w_o}{\rho_o g}\right)g = \frac{\rho_f}{\rho_o} w_o$$

or, solving for ρ_o,

$$\rho_o = \frac{\rho_f w_o}{F_B} = \frac{(1000\text{ kg/m}^3)(8.30\text{ N})}{0.49\text{ N}} = 17{,}000\text{ kg/m}^3$$

The crown's density is considerably lower than that of pure gold, which is 19,300 kg/m³. The crown is not pure gold.

ASSESS The buoyant force is much less than the crown's weight. This tells us that the rather massive crown displaces relatively little water and thus must be made of a material much denser than water, as we found.

STOP TO THINK 9.4 The five blocks have the masses and volumes shown. Using what you learned in Figure 9.17, rank in order, from greatest to least, the buoyant forces on the blocks.

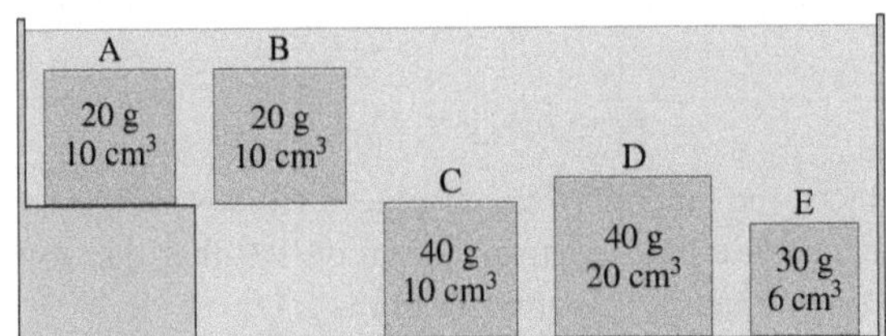

Float or Sink?

If we *hold* an object underwater and then release it, it rises to the surface, sinks, or remains "hanging" in the water. How can we predict which it will do? Whether it heads for the surface or the bottom depends on whether the upward buoyant force F_B on the object is larger or smaller than the downward weight force w_o.

The magnitude of the buoyant force is $\rho_f V_f g$. The weight of a uniform object, such as a block of steel, is simply $\rho_o V_o g$. But a nonuniform object, such as a fish, may have pieces of varying density. If we define the **average density** to be $\rho_{avg} = m_o/V_o$, the weight of a nonuniform object can be written as $w_o = \rho_{avg} V_o g$.

Comparing $\rho_f V_f g$ to $\rho_{avg} V_o g$, and noting that $V_f = V_o$ for an object that is fully submerged, we see that an object floats or sinks depending on whether the fluid density ρ_f is larger or smaller than the object's average density ρ_{avg}. If the densities are equal, the object is in equilibrium and hangs motionless. This is called **neutral buoyancy.** These conditions are summarized in Tactics Box 9.1.

TACTICS BOX 9.1 Finding whether an object floats or sinks

❶ Object sinks

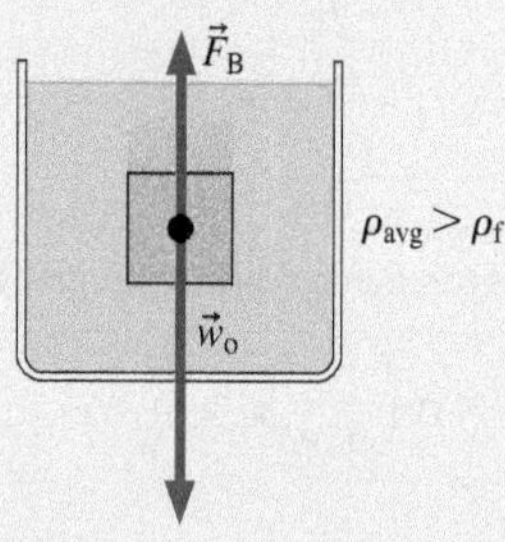

An object sinks if it weighs more than the fluid it displaces—that is, if its average density is greater than the density of the fluid.

❷ Object floats

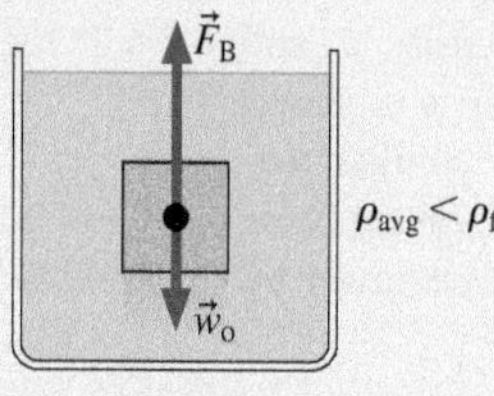

An object rises to the surface if it weighs less than the fluid it displaces—that is, if its average density is less than the density of the fluid.

❸ Object has neutral buoyancy

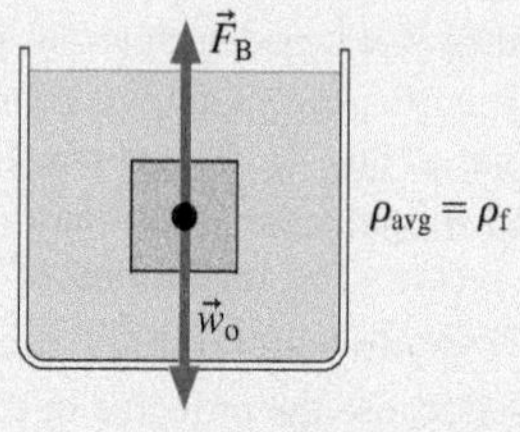

An object hangs motionless if it weighs exactly the same as the fluid it displaces—that is, if its average density equals the density of the fluid.

Steel is denser than water, so a chunk of steel sinks. Oil is less dense than water, so oil floats on water. Scuba divers use weighted belts to adjust their average density to match the density of the water, so they maintain their depth in the water without exerting effort.

If we release a block of wood underwater, the net upward force causes the block to shoot to the surface. Then what? To understand floating, let's begin with a *uniform* object such as the block shown in FIGURE 9.19. This object contains nothing tricky, like indentations or voids. Because it's floating, it must be the case that $\rho_o < \rho_f$.

FIGURE 9.19 A floating object is in static equilibrium.

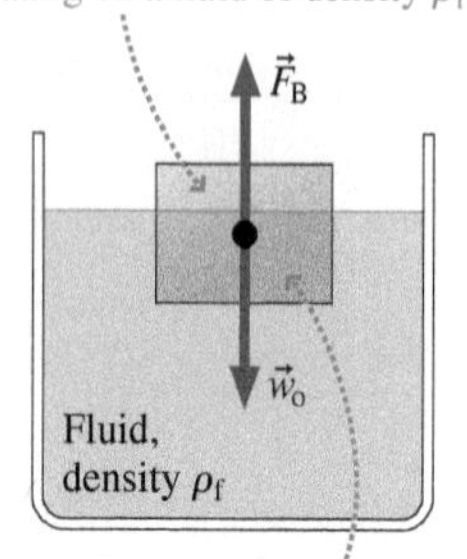

Now that the object is floating, it's in equilibrium. Thus, the upward buoyant force, given by Archimedes' principle, exactly balances the downward weight of the object; that is,

$$F_B = \rho_f V_f g = w_o = \rho_o V_o g \tag{9.9}$$

For a floating object, the volume of the displaced fluid is *not* the same as the volume of the object. In fact, we can see from Equation 9.9 that the volume of fluid displaced by a floating object of uniform density is

$$V_f = \frac{\rho_o}{\rho_f} V_o \tag{9.10}$$

which is *less* than V_o because $\rho_o < \rho_f$.

NOTE ► Equation 9.10 applies only to *uniform* objects. It does not apply to boats, hollow spheres, or other objects of nonuniform composition. ◄

CONCEPTUAL EXAMPLE 9.6 **Which has the greater buoyant force?**

A block of iron sinks to the bottom of a vessel of water while a block of wood of the *same size* floats. On which is the buoyant force greater?

REASON The buoyant force is equal to the volume of water displaced. The iron block is completely submerged, so it displaces a volume of water equal to its own volume. The wood block floats, so it displaces only the fraction of its volume that is underwater, which is *less* than its own volume. The buoyant force on the iron block is therefore greater than on the wood block.

ASSESS This result may seem counterintuitive, but remember that the iron block sinks because of its high density, while the wood block floats because of its low density. A smaller buoyant force is sufficient to keep it floating.

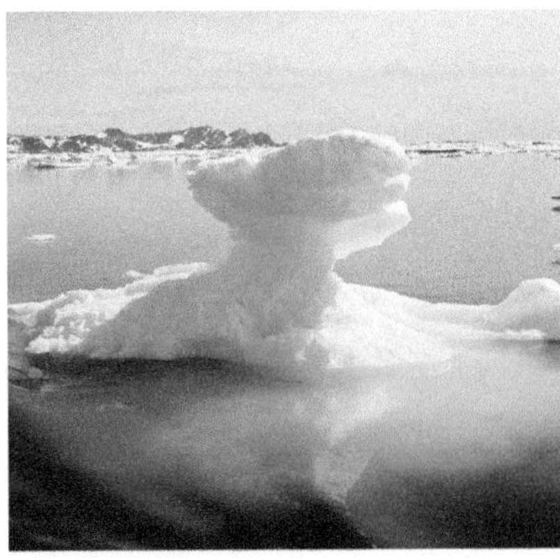

Hidden depths Most icebergs break off glaciers and are freshwater ice with a density of 917 kg/m^3. The density of seawater is 1030 kg/m^3. Thus

$$V_f = \frac{917\ \text{kg/m}^3}{1030\ \text{kg/m}^3} V_o = 0.89V_o$$

V_f, the volume of the displaced water, is also the volume of the iceberg that is underwater. The saying "90% of an iceberg is underwater" is correct!

EXAMPLE 9.7 **Measuring the density of an unknown liquid**

You need to determine the density of an unknown liquid. You notice that a wooden block floats in this liquid with 4.6 cm of the side of the block submerged. When the block is placed in water, it also floats but with 5.8 cm submerged. What is the density of the unknown liquid?

PREPARE It's useful to draw a picture to visualize the situation. We don't know the cross-section area of the block, so we simply label it A. FIGURE 9.20 shows the block and submerged lengths h_u in the unknown liquid and h_w in water.

FIGURE 9.20 A wooden block floating in two liquids.

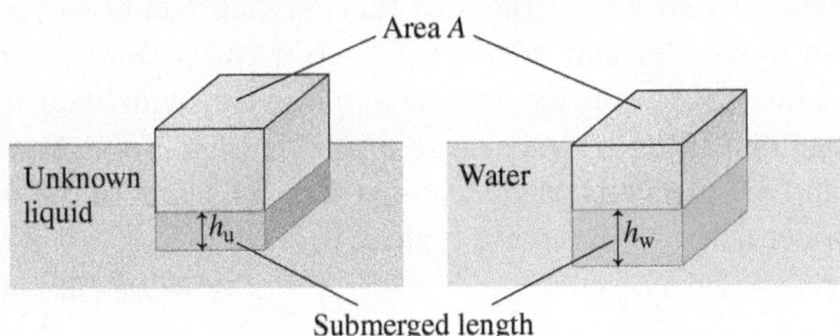

SOLVE The block displaces volume $V_u = Ah_u$ of the unknown liquid. Thus

$$V_u = Ah_u = \frac{\rho_o}{\rho_u} V_o$$

Similarly, the block displaces volume $V_w = Ah_w$ of the water, leading to

$$V_w = Ah_w = \frac{\rho_o}{\rho_w} V_o$$

Because there are two fluids, we've used subscripts w for water and u for the unknown in place of the fluid subscript f. The product $\rho_o V_o$ appears in both equations. In the first $\rho_o V_o = \rho_u A h_u$, and in the second $\rho_o V_o = \rho_w A h_w$. Equating the right-hand sides gives

$$\rho_u A h_u = \rho_w A h_w$$

The area A cancels, and the density of the unknown liquid is

$$\rho_u = \frac{h_w}{h_u} \rho_w = \frac{5.8\ \text{cm}}{4.6\ \text{cm}} \times 1000\ \text{kg/m}^3 = 1300\ \text{kg/m}^3$$

Comparison with Table 9.1 shows that the unknown liquid is likely to be glycerin.

ASSESS The object floats slightly higher in the unknown liquid than in water, so we expect a density slightly greater than that of water, as we found.

Floating your boat Steel is much denser than water, so how can a boat float? A boat is essentially a hollow shell, but it displaces a three-dimensional volume of water. The water is less dense, but the volume of displaced water is much larger than the volume of steel. As a boat settles into water, it sinks until the weight of the displaced water exactly matches the boat's weight. The upward buoyant force then balances the weight and the boat floats in equilibrium.

Buoyancy and Bodies

Masters of density Manatees live in fresh water and must dive for their food. Too much fat would make them less dense than water, which would increase the energy needed to dive. A typical body fat percentage for a manatee is about 7%, comparable to elite human athletes. Manatees can adjust the amount of residual air in their lungs to achieve nearly neutral buoyancy. They can also move air from one lung to the other to roll, and from the top of the lungs to the bottom to tip forward for a dive.

Different components of your body have different densities. Of the different body components, only fat has a density lower than that of water. An average college student's body is about 20% fat, with the balance a mix of water, muscle, blood, and bone. With this typical fat percentage, the average density of the body is about $1050\ \text{kg/m}^3$. Submerged in water, a person with this density would sink. But if the person takes a deep breath, she reduces her average density to approximately $990\ \text{kg/m}^3$ and will float. This is a common experience for most people: After taking a deep breath, you float; if you exhale as completely as possible, you sink.

In Example 9.5, we saw that the density of an object could be determined by weighing it both underwater and in air. The same thing can be done with a person. Since fat has a lower density than muscle or bone, a lower average body density implies a greater proportion of body fat. To determine density, a person is first weighed in air and then lowered completely into water, asked to exhale as completely as possible, and weighed again. It's then a reasonably straightforward calculation to determine the percentage of body fat. Empirical formulas take into account the air space that remains in the body and expected percentages of the other elements to give an estimate of the amount of fat in a person's body.

Aquatic animals also have to deal with buoyancy. The body of a typical fish is similar to yours. It is made of muscle, skin, bone, and fat, with an average density of about $1050\ \text{kg/m}^3$. This is greater than the density of fresh water or of seawater, and fish have no lungs filled with air to compensate. If this was all there was to the story, fish would have a tendency to sink, so they would need to expend energy to stay at the same depth.

For this reason, all fish have structures that provide additional buoyancy. The most common, in freshwater and saltwater fish, is a *swim bladder,* a gas-filled sac inside the body. The fish controls the volume of gas in the swim bladder, and thereby adjusts its average density to be very close to the density of the water in which it swims. For ocean fish, the swim bladder might take up 1–3% of the volume of the body; for freshwater fish, 5–7% is more typical.

EXAMPLE 9.8 How big a bladder?

With its swim bladder deflated, a 9.5 kg yellowfin tuna has an average density of $1050\ \text{kg/m}^3$. It is swimming near the surface of the ocean. What volume of gas in mL must the fish have in its swim bladder to achieve neutral buoyancy?

PREPARE For neutral buoyancy, the average density of the fish must equal the density of seawater. The density of seawater is given in Table 9.1 as $1030\ \text{kg/m}^3$.

SOLVE The average density of the fish is calculated by dividing its mass by its *total* volume—the volume of the fish itself plus the volume of air in its swim bladder. (We'll ignore the small mass of the air in the bladder.) We can use the fish's density and mass to find the volume of just the fish, with its swim bladder deflated:

$$V_{\text{fish}} = \frac{m}{\rho} = \frac{9.5\ \text{kg}}{1050\ \text{kg/m}^3} = 0.00905\ \text{m}^3$$

With its bladder inflated, the fish's average density, which we'll equate to the seawater density, is thus

$$\rho_{\text{avg}} = \frac{9.5\ \text{kg}}{0.00905\ \text{m}^3 + V_{\text{bladder}}} = \rho_{\text{sea}} = 1030\ \text{kg/m}^3$$

Solving for V_{bladder}, we find

$$V_{\text{bladder}} = 0.00017\ \text{m}^3 \times \frac{1000\ \text{L}}{1\ \text{m}^3}$$

$$= 0.17\ \text{L} = 170\ \text{mL}$$

ASSESS This is about 2% of the fish's volume, which is typical for an ocean fish, so our result seems reasonable. Fish typically maintain a slight negative buoyancy, so real data for tuna show a volume slightly less than this.

Sharks don't have swim bladders; neither do marine mammals. Sharks and many marine mammals have areas of low-density fats that reduce their average density. Sperm whales have a great deal of low-density oil in an organ in their heads. There is evidence that these deep-diving whales adjust the temperature of the oil to change its density. They cool the oil to make them negatively buoyant at the start of a dive; they then warm the oil during the dive so they achieve positive buoyancy when it is time to resurface.

STOP TO THINK 9.5 An ice cube is floating in a glass of water that is filled to the brim. When the ice cube melts, the water level will

A. Fall. B. Stay the same. C. Rise, causing the water to spill.

Terminal Speed in a Liquid

In ◀◀ SECTION 5.4, where we looked at the drag force on objects moving through fluids, we calculated the terminal speed of an object falling in air but not that of an object sinking in a liquid. The reason is that the buoyant force of air can usually be neglected, but the buoyant force of a liquid cannot.

FIGURE 9.21 shows three forces acting on an object that is sinking in a liquid. In addition to the downward weight force there are two upward forces: the drag force $\vec{F}_{drag}$, opposing the downward motion, and a buoyant force $\vec{F}_B$ from the displaced fluid. When the object has reached terminal speed and is no longer accelerating, these three forces must balance. Thus the drag force is

$$F_{\text{drag at term speed}} = w - F_B = \rho_o V_o g - \rho_f V_f g = (\rho_o - \rho_f) V_o g \quad (9.11)$$

where, in the last step, we used $V_f = V_o$ for an object that is submerged. There's often little difference between ρ_o and ρ_f for objects falling in liquids, so the terminal speed is much less than the speed in air.

The key idea of Section 5.4 was that the drag force depends on the Reynolds number *Re*, the ratio of inertial forces to viscous forces. At high Reynolds number, $Re > 1000$, the drag is inertial and depends on the square of v: $F_{drag} = \frac{1}{2} C_D \rho_f A v^2$. At low Reynolds number, $Re < 1$, the drag on a sphere of radius r is viscous and is linear in v: $F_{drag} = 6\pi\eta_f r v$, where η_f is the viscosity of the fluid. As a rough rule of thumb, objects smaller than about 100 μm fall in water with viscous drag while objects larger than a few mm experience inertial drag. There's no simple expression for drag for in-between cases. Equation 9.11 can be solved for the terminal speed v_{term} after the correct model for F_{drag} has been identified.

FIGURE 9.21 The forces on an object sinking in a liquid.

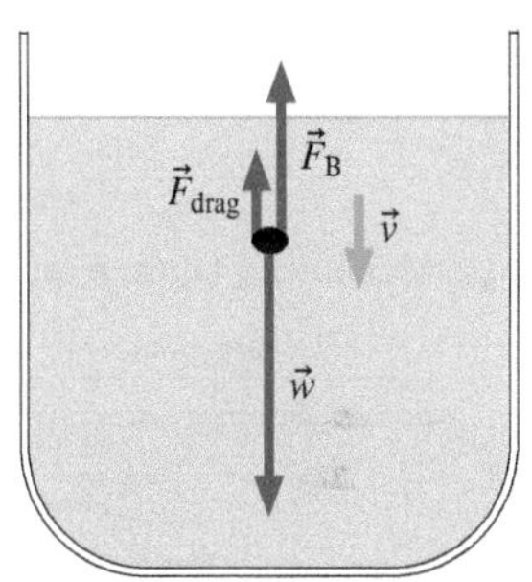

EXAMPLE 9.9 **Ocean sediment**

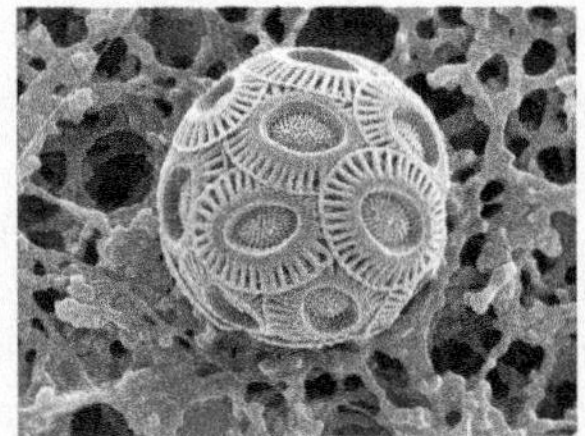

Sediment accumulates on the deep ocean floor as microorganisms die and then very slowly settle to the bottom. *Coccolithophores,* unicellular phytoplankton, are a major contributor to the sediment. These microorganisms construct a nearly spherical exoskeleton of calcium carbonate, and it is the exoskeleton shell and its enclosed water that settle after the cell dies. Thick deposits of coccolithophore exoskeletons, laid down over tens of thousands of years, form chalk. The average mid-ocean depth is 4.0 km. How long does it take a 30-μm-diameter, 24 ng coccolithophore to fall 4.0 km in seawater? The viscosity of seawater at deep-ocean temperatures is 1.7×10^{-3} Pa · s.

PREPARE The density ρ_o in Equation 9.11 is the coccolithophore's *average* density, which we can find from the mass and volume. Because the diameter is so small, this is a case of low Reynolds number for which we can use Stokes' law in conjunction with Equation 9.11.

SOLVE The volume of the coccolithophore is

$$V_o = \tfrac{4}{3}\pi r^3 = \tfrac{4}{3}\pi(15 \times 10^{-6}\ \text{m})^3 = 1.41 \times 10^{-14}\ \text{m}^3$$

We need mass in kg; 24 ng is 2.4×10^{-8} g, which is 2.4×10^{-11} kg. Thus the object's average density is

$$\rho_o = \frac{m}{V_o} = \frac{2.4 \times 10^{-11}\ \text{kg}}{1.41 \times 10^{-14}\ \text{m}^3} = 1700\ \text{kg/m}^3$$

The density is greater than that of seawater, so the coccolithophore will sink.

Equation 9.11, with Stokes' law for the drag force, is

$$F_{\text{drag at term speed}} = 6\pi\eta r v_{term} = (\rho_o - \rho_f)V_o g = \tfrac{4}{3}\pi r^3(\rho_o - \rho_f)g$$

We inserted the expression for the volume of a sphere in order to simplify the final result. Solving for the terminal speed gives

$$v_{term} = \frac{2r^2(\rho_o - \rho_f)g}{9\eta}$$

$$= \frac{2(15 \times 10^{-6}\ \text{m})^2(1700\ \text{kg/m}^3 - 1030\ \text{kg/m}^3)(9.8\ \text{m/s}^2)}{9(1.7 \times 10^{-3}\ \text{Pa}\cdot\text{s})}$$

$$= 1.9 \times 10^{-4}\ \text{m/s} = 0.19\ \text{mm/s}$$

Continued

At this speed, the time it takes to fall 4.0 km = 4000 m is

$$\Delta t = \frac{4000\text{ m}}{1.9 \times 10^{-4}\text{ m/s}} = 2.1 \times 10^7\text{ s}$$

$$= (2.1 \times 10^7\text{ s}) \times \frac{1\text{ h}}{3600\text{ s}} \times \frac{1\text{ day}}{24\text{ h}} = 240\text{ days}$$

ASSESS The terminal speed is an extremely slow ≈ 0.2 mm/s. But 30 μm is only about half the diameter of a human hair, and it's not surprising that such a small object would sink very slowly in water. As a result, it takes roughly 8 months for a coccolithophore to reach the bottom of the abyss.

9.4 Surface Tension and Capillary Action

Bubbles and water drops assume a spherical shape, insects walk across water, and liquid rises in a capillary tube. These are all manifestations of a *surface tension* present at the interface between a liquid and a gas. Many biological processes, from the functioning of alveoli in the lungs to the extraction of water in the soil by roots, are governed by surface tension.

FIGURE 9.22 Measuring surface tension.

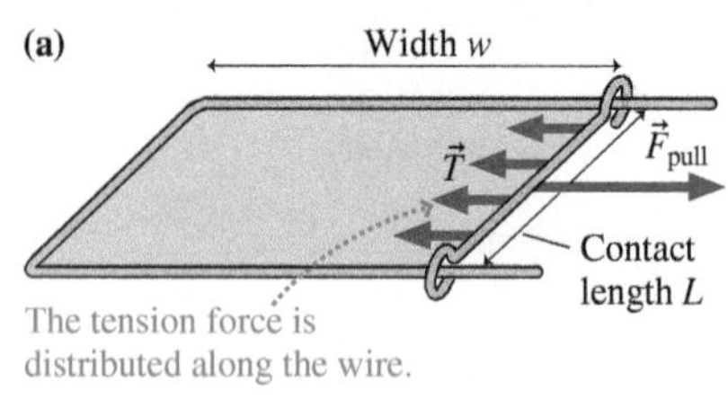

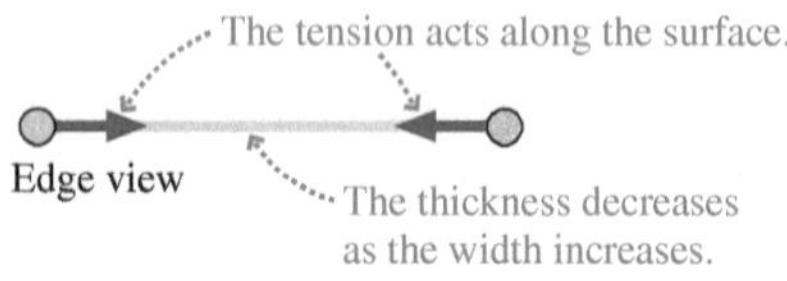

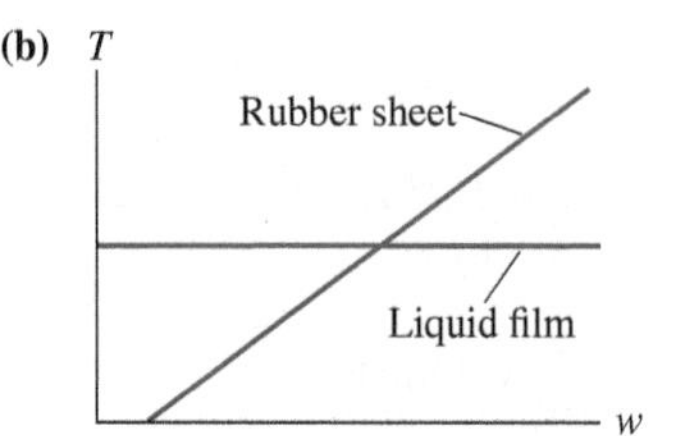

FIGURE 9.22a shows an experiment in which a thin film of liquid, such as soapy water, is stretched between a U-shaped frame and a sliding wire. It is interesting that a pulling force $\vec{F}_{\text{pull}}$ is needed simply to hold the wire in place; if you let go, the wire will snap back to the left as the surface area of the liquid is reduced. It would appear that the liquid pulls on the wire with a tension-like force $\vec{T}$, rather like a stretched spring. The forces are balanced when the wire is held in equilibrium, so measuring the pulling force is a way of determining the tension T exerted by the liquid.

Further investigation finds that this is an unusual tension force. If the wire were to pull on a thin sheet of rubber, the tension would increase linearly with the horizontal width w of the sheet; that is, more stretching requires more force. But for a liquid film, as shown in the graph of FIGURE 9.22b, *the tension is independent of the width;* a 10-cm-wide film pulls on the wire with the same force as a 1-cm-wide film.

Why the difference? Stretching a rubber sheet stretches the spring-like intermolecular bonds; that takes more and more force. In contrast, extending a liquid film causes molecules to move from the interior volume to the surface, *creating new surface* while the thickness of the film decreases. A wider film has more molecules on the surface, but the average distance between molecules has not changed and so the weak bonding forces between them—the source of the tension force—does not change.

Thus the tension force is exerted at the liquid-air interface—it's a *surface* tension—not within the bulk of the liquid. A microscopic view of a liquid would show that molecules experience both repulsive forces due to collisions with neighbors and attractive forces binding molecules weakly to each other. The attractive forces that hold a liquid together are called **cohesive forces.** All these forces are balanced for molecules in the interior of a liquid, but molecules right at the surface experience an asymmetry because interactions with liquid molecules are not matched by interactions with the more sparse gas molecules. The molecular details are complex, but the net result of this asymmetry is a "pulling together" of surface molecules that we observe as the surface tension.

The tension force may be independent of the film's width w, but it is directly proportional to the sliding wire's length L because a longer wire is touched and pulled on by more surface molecules. By factoring out the contact length, we define the **surface tension** γ (lowercase Greek gamma) as

$$\gamma = \frac{\text{tension provided by the surface}}{\text{contact length}} = \frac{T}{L} \tag{9.12}$$

Surface tension, despite its name, is not a tension and does not have units of newtons. The actual tension T of a liquid surface is distributed across the line of contact, and what we call surface tension is the tension *per meter of contact*. Its units are N/m.

Put another way, the actual pulling tension exerted along a length L of a surface with surface tension γ is $T = \gamma L$.

NOTE ▶ The tension force is tangent to the surface of the liquid at the line of contact. ◀

Unlike tension in a rope or spring, which is a response to external forces, **surface tension is a property of the liquid.** Surface tension, like density or viscosity, can be tabulated, and TABLE 9.3 does so for some typical liquids. The commonly used units are mN/m, which need to be converted to N/m when used in calculations. The size of the surface tension is an indication of the strength of the cohesive forces that hold the liquid together. Water has larger surface tension than most liquids because the cohesive hydrogen bonds between water molecules are fairly strong.

Adding a surfactant such as soap or detergent reduces the cohesive forces, which decreases the surface tension significantly. You'll see in a few pages that this is of supreme importance to mammalian lungs. Surface tension also decreases with increasing temperature; this is part of the reason hot water cleans things better than cold water.

TABLE 9.3 **Surface tension of liquids (temperature is 20°C unless otherwise indicated)**

Liquid	Surface tension (mN/m)
Water (20°C)	73
Water (40°C)	70
Water (80°C)	63
Blood	58
Blood (37°C)	52
Olive oil	32
Soapy water (typical)	25
Ethanol	22

EXAMPLE 9.10 Surface tension supports a weight

A U-shaped frame like the one in Figure 9.22 is oriented vertically. A film of soapy water across the frame is held down by the weight of a 10-cm-long sliding wire. What is the wire's mass?

PREPARE A film of soapy water has *two* surfaces, and each exerts an upward force $T = \gamma L$ on a wire of length L. The surface tension of soapy water is given in Table 9.3 as $\gamma = 25\ \text{mN/m} = 0.025\ \text{N/m}$.

SOLVE The net upward surface tension force $2T$ has to balance the downward weight force mg along a contact length $L = 10\ \text{cm} = 0.10\ \text{m}$. Thus

$$m = \frac{2T}{g} = \frac{2\gamma L}{g} = \frac{2(0.025\ \text{N/m})(0.10\ \text{m})}{9.8\ \text{m/s}^2}$$
$$= 5.1 \times 10^{-4}\ \text{kg} = 0.51\ \text{g}$$

ASSESS You may have seen a film of soapy water on a frame made of copper wires. Half a gram is roughly the mass of a 10 cm length of 1-mm-diameter copper wire, so our answer agrees with experience.

Water striders are insects known for their ability to walk on water. The feet of a strider are hydrophobic due to thousands of microscopic hairs that enable the feet to push the water down into a dimple without breaking through the surface. The upward component of the surface tension around the curved dimple then supports the weight of the strider. The details are somewhat complex, but we can understand the basic idea with a simpler example of a needle floating on water.

A water strider is supported by the water's surface tension.

EXAMPLE 9.11 Floating a needle on water

A 4.0-cm-long needle will float if it is placed gently on the surface of 20°C water. What is the maximum mass of a needle that can float? What is this needle's diameter if it is made of stainless steel with density 7900 kg/m^3?

PREPARE FIGURE 9.23 is a pictorial representation and a free-body diagram showing a cross section of the needle. We'll assume that the needle is a uniform cylinder without any taper. You might think this is an example of buoyancy, but the needle is not

FIGURE 9.23 Pictorial representation and free-body diagram for a needle floating on water.

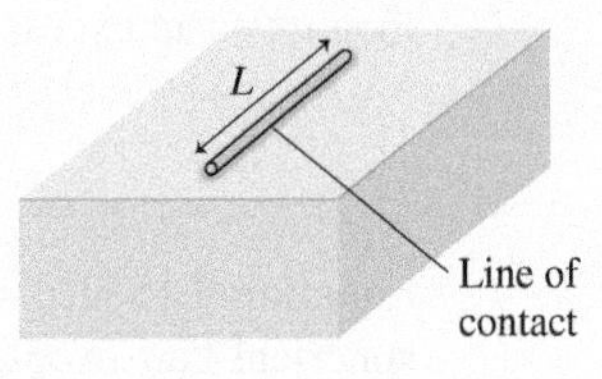

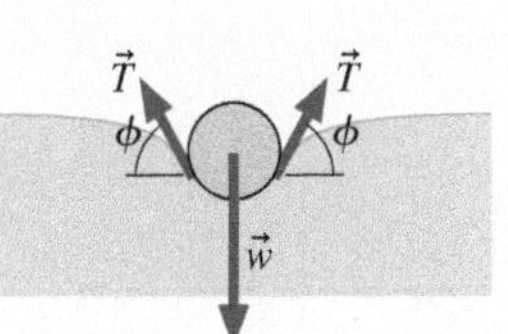

Continued

submerged in the water (i.e., the water level follows the outline of the needle rather than covering any of it), so the buoyant force is zero. Instead, the needle depresses the surface of the water to make a cylindrical dimple. This creates a tension force tangent to the surface at the line of contact, so the water pulls up on the needle at an angle ϕ above horizontal. The net force must be zero for the needle to float.

SOLVE The water tension acts on *both* sides of the needle with vertical component $T\sin\phi$. The tension force is distributed along each edge of the needle with total magnitude $T = \gamma L$, where L is the needle length. Equilibrium requires

$$w = mg = 2T\sin\phi = 2\gamma L\sin\phi$$

A heavier needle sinks deeper into the water, increasing the angle ϕ. But ϕ can't exceed 90°, at which point $\sin\phi_{max} = 1$, so the maximum mass that will float is

$$m_{max} = \frac{2\gamma L\sin\phi_{max}}{g} = \frac{2\gamma L}{g} = \frac{2(0.073\ \text{N/m})(0.040\ \text{m})}{9.8\ \text{m/s}^2}$$
$$= 6.0\times10^{-4}\ \text{kg} = 0.60\ \text{g}$$

The mass of a cylinder of density ρ is $m = \rho V = \pi r^2 L\rho$. Thus the maximum radius is

$$r_{max} = \sqrt{\frac{m_{max}}{\pi L\rho}} = \sqrt{\frac{6.0\times10^{-4}\ \text{kg}}{\pi(0.040\ \text{m})(7900\ \text{kg/m}^3)}}$$
$$= 7.8\times10^{-4}\ \text{m} = 0.78\ \text{mm}$$

Our calculation finds that the maximum diameter of a floating stainless steel needle is 1.6 mm.

ASSESS Sewing needles are typically 0.5 mm to 1 mm in diameter, so our answer is consistent with the fact that a sewing needle does float on water.

Bubbles and Droplets

Surface tension pulls an isolated bubble or liquid droplet into the shape of a sphere. But why don't the tension forces shrink the size down to zero? Because the inward pulling tension increases the *pressure* inside the sphere, sometimes by a surprising amount, until a balance is reached between pressure forces and surface tension forces.

Imagine that we could cut a bubble or droplet of radius R into halves, as shown in FIGURE 9.24, with pressure p_{in} inside and pressure p_{out} outside. Because of the pressures, the left hemisphere experiences forces $\vec{F}_R = (p_{out}A$, to the right) and $\vec{F}_L = (p_{in}A$, to the left), where A is the sphere's circular cross-section area $A = \pi R^2$. In addition to the pressure forces, the surface tension that holds the sphere together acts tangent to the hemisphere along the line of contact, which in this case, being the circumference of the sphere, is of length $L = 2\pi R$.

FIGURE 9.24 The pressure inside a bubble must be larger than the exterior pressure to balance the surface tension forces.

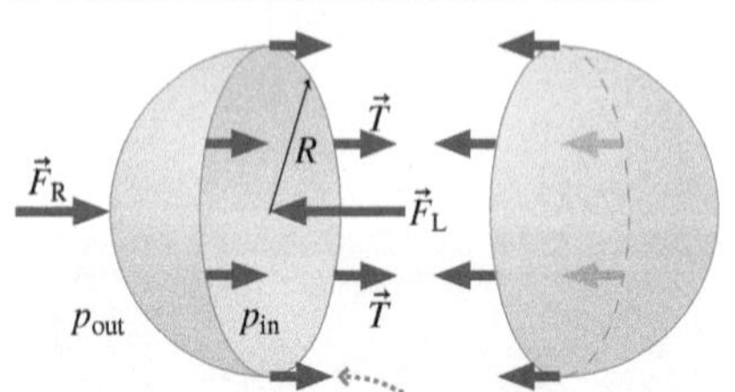

First consider a bubble created from a thin liquid film, like a soap bubble, with air both inside and outside. This bubble has *two* surfaces, an inner and an outer surface, so the net tension force is $T = 2\gamma L = 4\pi\gamma R$. The equilibrium condition for the left hemisphere is

$$T + F_R - F_L = 4\pi\gamma R + \pi R^2 p_{out} - \pi R^2 p_{in} = 0 \tag{9.13}$$

Thus there is a pressure *difference* between the interior and the exterior:

$$\Delta p_{bubble} = p_{in} - p_{out} = \frac{4\gamma}{R} \tag{9.14}$$

If the outside pressure is atmospheric pressure, then Δp_{bubble} is the gauge pressure inside the bubble. You can calculate that the excess pressure inside a 2-cm-diameter soap bubble is a very modest 10 Pa.

A spherical droplet of liquid has only a single surface separating it from the air, so the net tension force is a smaller $T = \gamma L = 2\pi\gamma R$. The same equilibrium condition then gives an excess pressure of

$$\Delta p_{drop} = p_{in} - p_{out} = \frac{2\gamma}{R} \tag{9.15}$$

Equation 9.15 also applies to air bubbles in a liquid because they can be considered spherical "air droplets" within the liquid. It's straightforward to calculate that the gauge pressure for air bubbles in 20°C water is approximately 30 Pa, 3000 Pa, and 300,000 Pa for bubbles of diameters 1 cm, 100 μm, and 1 μm, respectively. The first may be insignificant, but the last is a very large difference of 3 atm between the inside and outside of the bubble.

Equations 9.14 and 9.15, sometimes called the **law of Laplace,** have numerous biological and medical implications. We'll look at one example. At the end of the air pathway through the lungs are approximately 500 million *alveoli,* which are roughly spherical sacs a fraction of a millimeter in diameter. Gas exchange to and from the blood takes place by diffusion across the walls of the alveoli. Physically, the alveoli are very much like gas bubbles in the watery interstitial fluid.

According to Equation 9.15, the excess pressure needed to inflate an alveolus is proportional to the surface tension and inversely proportional to its radius. That might cause problems. For example, because the surface tension of water is large, excessive effort would be required to expand the alveoli during inhalation—like trying to blow up an especially stiff balloon. Also, a reduction in the number of molecules inside the alveoli during exhalation would create an imbalance between tension forces and pressure forces, allowing the strong surface tension to collapse the alveoli. Finally, if one alveolus became larger than the others, it would require less pressure for further inflation, so that one alveolus would inflate more at the expense of others.

These problems are avoided in healthy lungs by a *pulmonary surfactant,* a lipid-protein complex that is absorbed into the air-water interface at the surface of the alveoli. A *surfactant* is a compound that lowers the surface tension of a liquid by disrupting the attractive intermolecular forces; soaps and detergents are the most familiar surfactants. The lowered surface tension makes it much easier for the alveoli to expand during inhalation.

The pulmonary surfactant has a second important property. We know that the surface tension of a pure substance, such as water, is independent of the surface area. That's no longer true after a surfactant has been added. When an alveolus expands, the increasing surface area dilutes the surfactant and thus it has less influence. Consequently, surface tension *increases* as an alveolus expands and decreases as it contracts and the surfactant becomes more concentrated. The decreasing surface tension as the alveoli contract during exhalation is what prevents alveolar collapse.

Premature infants often lack sufficient pulmonary surfactant to breathe on their own, a condition known as *neonatal respiratory distress syndrome.* It can be treated by introducing an artificial surfactant directly into the lungs.

STOP TO THINK 9.6 Two soap bubbles of different diameters are connected through a small tube. What will happen?

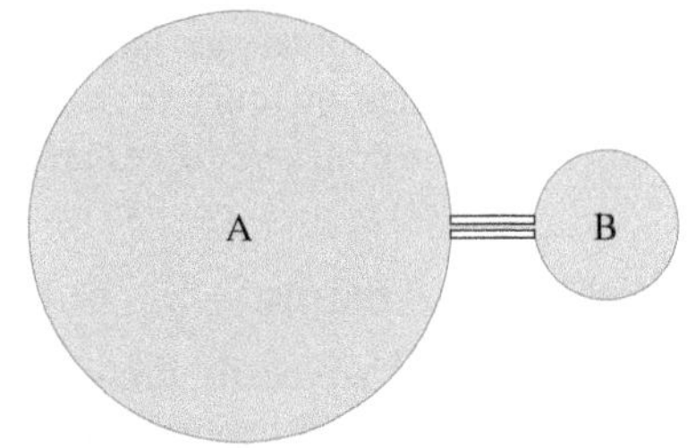

A. Nothing.
B. Bubble B will shrink to nothing and bubble A will expand.
C. Bubble A will shrink to nothing and bubble B will expand.
D. Bubble A will shrink and bubble B will expand until they are the same size.

Wetting and Contact Angle

The water drop in FIGURE 9.25a (next page) has beaded up into an almost perfect sphere on the waxy surface of a leaf. The same water drop would spread out to a flat puddle on a clean glass surface, especially if a small amount of detergent were added.

When a liquid touches a solid surface, as in FIGURE 9.25b, there's a *line of contact* where the liquid, the solid, and the air all meet. The shape of the drop depends not only on the cohesive surface tension along the liquid-air interface but also on

TABLE 9.4 Contact angles

Interface	Contact angle
Mercury-glass	130°
Water-Teflon	108°
Water-polyvinyl chloride	85°
Water-glass (typical)	30°
Blood-glass (typical)	25°
Water-clean glass	≈0°

FIGURE 9.25 A liquid meets a solid surface at a contact angle determined by the balance between cohesive and adhesive forces.

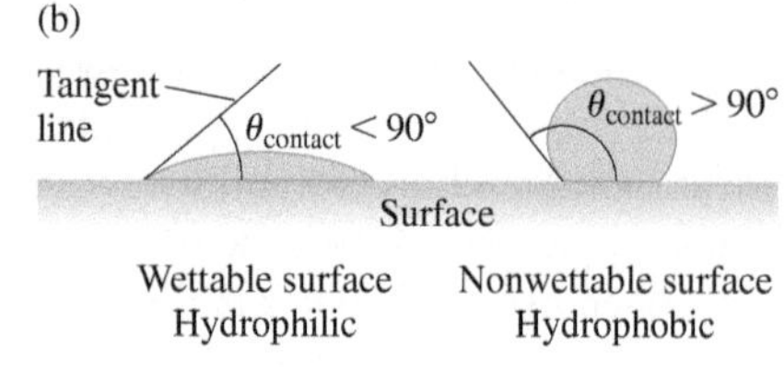

adhesive forces that attract the liquid to the solid. The equilibrium situation can be characterized by the **contact angle** $\theta_{contact}$ between the surface and the tangent to the drop at the line of contact.

If the adhesive forces that pull liquid molecules toward the surface are larger than the cohesive forces within the liquid, then the drop spreads out and the contact angle is smaller than 90°. The surface is said to be *wettable* or *hydrophilic*. Adding a surfactant causes the drop to spread even more, better wetting the surface with a smaller contact angle. Wettability is an important parameter in the interaction of biological fluids with surfaces, whether that is blood serum interacting with a microscope slide or saliva interacting with dental enamel.

Conversely, the liquid pulls itself into a near-spherical shape with a contact angle larger than 90° when the cohesive forces are larger than the adhesive forces. This surface is said to be *nonwettable* or *hydrophobic*. The waxy leaf surface in Figure 9.25a is very hydrophobic, as are the surfaces of nonstick cookware.

TABLE 9.4 lists a few contact angles. In practice, the contact angle is very sensitive to surface roughness and cleanliness, as all but the cleanest surfaces have small amounts of other molecules on them. Water spreads into a thin sheet with $\theta_{contact} \approx 0°$ on very clean, very smooth glass, which indicates that the adhesive forces between water and glass are much stronger than the already strong cohesive forces of water. But in more typical cases, such as raindrops on your windshield, water on glass forms flattened drops with $\theta_{contact} \approx 30°$.

FIGURE 9.26 The shape of a meniscus at the surface of a liquid depends on the balance between cohesive and adhesive forces.

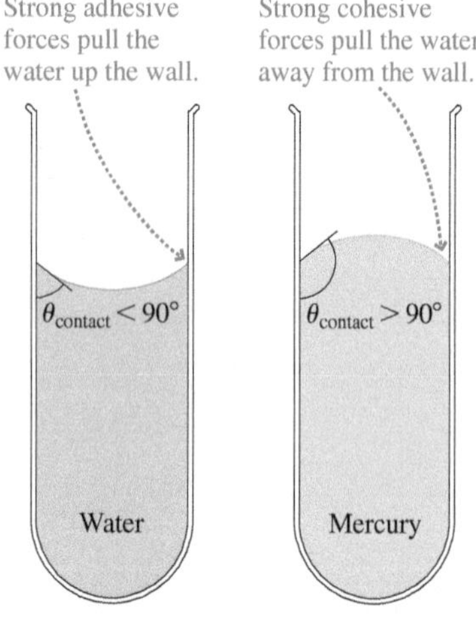

The adhesive forces at the liquid-surface interface are responsible for the curved *meniscus* of a liquid in a vertical tube. As FIGURE 9.26 shows, the meniscus is concave downward when the contact angle with the wall is smaller than 90°, which means that the adhesive forces are larger than the cohesive forces. A convex upward meniscus, seen when mercury is in a glass tube, forms when the adhesive forces are smaller than the cohesive forces and $\theta_{contact} > 90°$.

Capillary Action

FIGURE 9.27 The rise or fall of liquid in a narrow capillary tube depends on whether adhesive forces between the liquid and the wall are stronger or weaker than cohesive forces within the liquid.

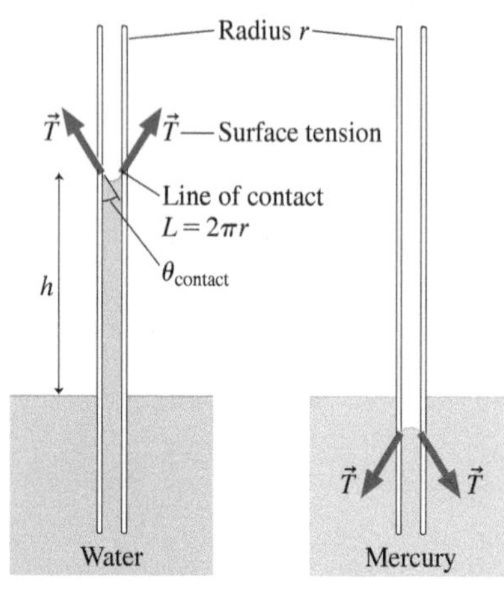

Water and blood are easily drawn into a capillary tube, seemingly defying gravity. This **capillary action** occurs when the adhesive forces between the liquid and the wall of the tube exceed the cohesive forces inside the liquid—the same conditions that create a concave meniscus. In this case, as shown on the left in FIGURE 9.27, the adhesive forces between the liquid and the wall pull the liquid up the wall of a vertical tube until the surface tension at the line of contact with the wall balances the weight of the raised column of liquid. We can use the equilibrium condition to determine the height of the liquid.

The meniscus touches the wall of the tube around a line of contact with length equal to the tube's circumference, so $L = 2\pi r$. The surface tension force is $T = \gamma L = 2\pi\gamma r$, but it's angled outward at the contact angle $\theta_{contact}$. Thus the vertical component of the surface tension, which has to balance the weight of the liquid, is $2\pi\gamma r\cos\theta_{contact}$. The mass of the liquid is $m = \rho V$, and its volume—ignoring a small correction due to the curvature of the meniscus—is that of a cylinder: $V = \pi r^2 h$. Equating the upward and downward forces on the raised column of liquid gives

$$2\pi\gamma r\cos\theta_{\text{contact}} = \rho(\pi r^2 h)g \qquad (9.16)$$

Consequently, liquid in a capillary tube rises to a height of

$$h = \frac{2\gamma\cos\theta_{\text{contact}}}{\rho g r} \qquad (9.17)$$

EXAMPLE 9.12 **Using a capillary tube**

A capillary tube used to sample blood from a finger prick has an inner diameter of 0.80 mm. To what height will the blood rise in a vertical capillary tube?

PREPARE Equation 9.17 predicts the height. The density, surface tension, and contact angle of blood can be found in Tables 9.1, 9.3, and 9.4, respectively. The surface tension at a body temperature of 37°C is the correct value to use.

SOLVE The height will be

$$h = \frac{2\gamma\cos\theta_{\text{contact}}}{\rho g r} = \frac{2(0.052\ \text{N/m})\cos 25°}{(1060\ \text{kg/m}^3)(9.8\ \text{m/s}^2)(0.40\times10^{-3}\ \text{m})}$$

$$= 0.023\ \text{m} = 23\ \text{mm}$$

ASSESS 23 mm is a bit less than an inch. Blood draws are able to fill an 80-mm-long tube because the tube is held horizontally, which reduces the influence of gravity.

Capillary action plays an important role in many phenomena. *Wicking,* such as the movement of water through a paper towel or molten wax through a candle wick, is capillary action through small-diameter pores. Similarly, pores through soil help bring water up from deeper layers into the root zones of plants. And the *lacrimal ducts* in the corners of the eyes use capillary action to drain constantly produced tear fluid from the eyes into the nasal cavity.

Sap Flow in Trees

Trees have a complex hydraulic system of small-diameter tubes, the *xylem,* that transport water and nutrients from the roots to the leaves, where it is used for photosynthesis and evaporated during transpiration. Water can ascend more than 100 m in the tallest trees. How?

Your first thought might be that water is lifted to the leaves by capillary action. Capillary action is important in small plants, but it can't be the mechanism for lifting water in trees. The diameters of xylem vessels are typically a few tens of micrometers. Knowing this, we can use Equation 9.17 to find that capillary action cannot lift water more than about a meter.

Maybe transpiration lowers the pressure and "sucks" water up the xylem. We looked at a pressure-reduction pump in Section 9.2—which is really the atmosphere *pushing* liquid up rather than it being pulled—and found that the maximum possible lift is 10.3 m. That's not nearly enough. Nor is any root pressure sufficient to push water up to a height of many meters.

Instead, the vertical transport of water in trees depends on the strong cohesive forces—due to hydrogen bonds—that give water a large surface tension. Because of cohesion, water and other liquids can, under the right conditions, be placed *in tension* like a rope. Water gets to the leaves not by being pushed from below but by being *pulled* from above.

Water evaporation takes place through pores—*stomata*—on the underside of leaves. The amount of evaporation can be large: several liters per minute for a large tree. Because water molecules exert attractive forces on each other, molecules leaving the pores pull and tug on the water molecules that remain in the leaf. Those, in turn, pull and tug on molecules in the xylem, and so on all the way down to the roots.

FIGURE 9.28 The weight of a column of water can be supported by tensile forces.

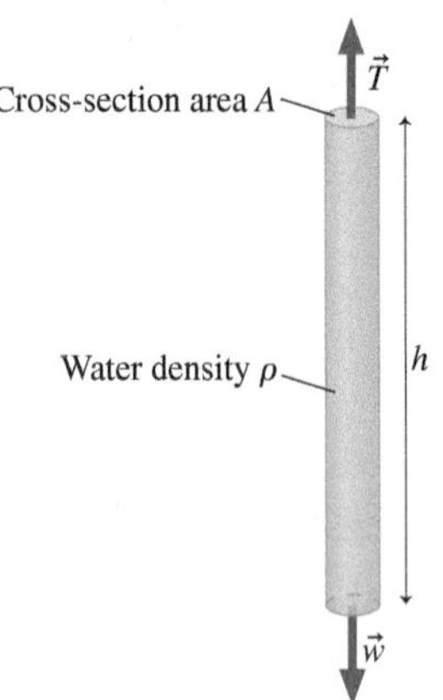

As a result, the entire water column is in tension, just as pulling one end of a rope places the entire rope in tension.

FIGURE 9.28 shows a tension force pulling up on a water column of height h and cross-section area A. In equilibrium, the tension has to balance the weight of the water $w = mg = \rho Ahg$. Thus the *stress* at the top of the water column is

$$\text{stress} = \frac{T}{A} = \frac{mg}{A} = \rho gh \tag{9.18}$$

The stress for a 100-m-tall tree is easily calculated to be $1 \times 10^6\ \text{N/m}^2$. We previously used N/m^2 as the units for stress, but we noted that these units are equivalent to Pa. So we can say that the tensile stress needed to support a 100-m-long column of water is 1 MPa $\approx$ 10 atm.

In fact, the water column is not just hanging in equilibrium but is *moving* upward through the xylem. We'll look later in the chapter at the forces that are needed to keep a viscous fluid moving through a pipe, but for trees the additional force needed to move the fluid is approximately the same as the force needed to support the weight of the fluid, so the actual tensile stress in the xylem of a 100-m-tall tree is more like 2 MPa $\approx$ 20 atm. This tension, due to the cohesive forces in the water, actively *pulls* the water up the tree.

Can water actually do this without the water column breaking? You certainly can't pick up a glass of water by pulling upward on the top surface. But very narrow tubes of water, supported by the hydrophilic surface of the xylem, can withstand a surprising amount of tension. Recall that *tensile strength* is the tensile stress that causes a material to fail; metals have tensile strengths of several hundred MPa. Experiments show that the tensile strength of a narrow tube of water is $\approx$30 MPa, nothing like that of metal but more than adequate to withstand the tensions needed to move water up the tallest trees.

Equation 9.18 looks very similar to the equation $p = p_0 + \rho gd$ for the hydrostatic pressure at depth d. In this case, the quantity ρgh is the hydrostatic pressure due to the weight of a column of water of height h. If the pressure at the top of the xylem is $p_0 \approx 1$ atm, then the water pressure in the xylem at the base of a 100-m-tall tree should be $\approx$11 atm. If the pressure in the xylem were that large, cutting through the bark of a tree into the xylem would cause sap to come spraying out. But sap only oozes from a cut, which suggests that the pressure in the xylem near ground level is $\approx$1 atm.

If that's so, and if the hydrostatic pressure difference between the bottom and top of a 100-m-tall tree is $\approx$10 atm, then the pressure at the top would have to be a *negative* -9 atm! Allowing for the additional pressure difference needed to support the flow of the moving sap, we would have to conclude that the water pressure within the xylem at the top of a tall tree is ≈ -20 atm.

We defined pressure by looking at repulsive interactions—collisions of molecules with each other and with the walls of the container—that cause fluids to *push* on things. However, now we have a situation where the attractive interactions are more important and the fluid *pulls* on things. We've called that a tension force, but you could think of tension as being a negative pressure. This cannot happen in a gas; the interactions of gas molecules are exclusively repulsive, so gas pressures are always positive. But under the right circumstances a liquid can be in tension, and in that sense it has a negative pressure.

You might think that the idea of negative pressure is contrary to the basic concept of pressure. Many physicists agree and prefer to describe fluid flow through the xylem in terms of tensile stress in the liquid. However, plant physiologists routinely explain liquid transport in terms of negative pressures, saying that transpiration at the leaves creates a negative pressure at the top of the tree. These are different perspectives, not competing explanations, and both give the same results for quantities such as flow rates. Just think of negative pressure as an alternative way of describing tension.

Two more features of flow through trees are worth noting. First, the entire system is powered by the sun, through evaporation, and requires no expenditure of metabolic energy. Second, bubbles are the enemy of pulling a liquid under tension. A bubble, cavity, or embolism breaks the water column and stops the flow just as effectively as cutting a rope. The narrow diameters of the xylem vessels may well be an evolutionary response to this danger because bubbles don't readily form in narrow tubes that have hydrophilic walls. But freezing is another story, and freeze-thaw cycles tend to introduce bubbles that prevent the xylem vessels from reactivating in the spring. Instead, trees in climates with cold winters grow a new set of xylem vessels each spring just under the bark; these become the *growth rings* that you see across the end of a log.

9.5 Fluids in Motion

We've focused thus far on fluid statics, but it's time to turn our attention to fluids in motion—blood flowing through your veins or air moving into and out of your lungs. Fluid flow is a complex subject. Many aspects of fluid flow, especially turbulence and the formation of eddies, are still not well understood and are areas of current research. We can avoid most of the complexity by making certain assumptions. In our discussion, we'll assume:

- **Fluids are incompressible.** This is a very good assumption for liquids, but it also holds reasonably well for a gas moving at less than about 30% of the speed of sound. Thus air moving at speeds less than 100 m/s (≈220 mph) can be considered incompressible.
- **The flow is steady.** That is, the fluid velocity at each point in the fluid is constant; it does not fluctuate or change with time. Flow under these conditions is called **laminar flow,** and it is distinguished from **turbulent flow.**

FIGURE 9.29 Rising smoke changes from laminar flow to turbulent flow.

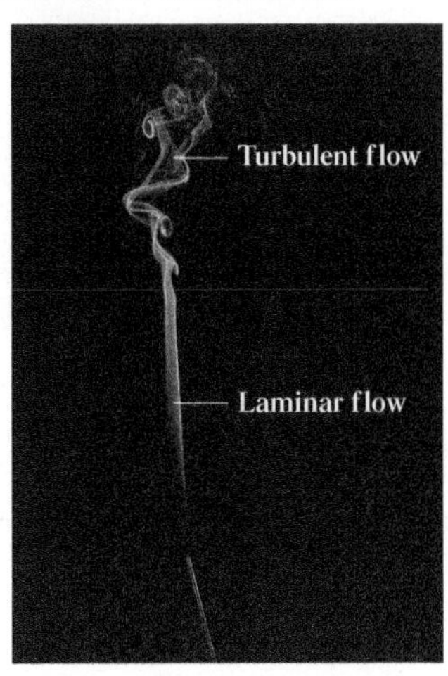

The rising smoke in FIGURE 9.29 begins as laminar flow, recognizable by its smooth contours, but at some point it undergoes a transition to turbulent flow. Our model of fluids can be applied to the laminar flow, but not to the turbulent flow. This limitation isn't as limiting as it might seem. The motion of blood in your circulatory system, for instance, is always laminar if things are working as they should—turbulence spells trouble. Turbulence in blood vessels makes a very particular sound in a stethoscope, which allows a physician listening to your heart and lungs to detect potentially serious problems in a routine physical exam.

The Equation of Continuity

Consider a fluid flowing through a tube—oil through a pipe or blood through an artery. If the tube's diameter changes, as happens in FIGURE 9.30, what happens to the speed of the fluid?

FIGURE 9.30 Flow speed changes through a tapered tube.

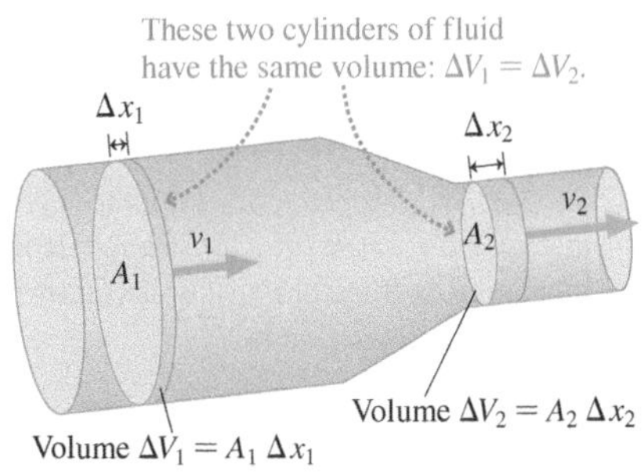

When you squeeze a toothpaste tube, the volume of toothpaste that emerges matches the amount by which you reduce the volume of the tube. An *incompressible* fluid flowing through a rigid tube or pipe acts the same way. Fluid is neither created nor destroyed within the tube, and there's no place to store any extra fluid introduced into the tube. If fluid of volume V enters the tube during some interval of time Δt, then an equal volume of fluid must leave the tube.

To see the implications of this idea, suppose all the molecules of the fluid in Figure 9.30 are moving to the right with speed v_1 at a point where the cross-section area is A_1. Farther along the tube, where the cross-section area is A_2, their speed is v_2. During an interval of time Δt, the molecules in the wider section move forward a distance $\Delta x_1 = v_1\,\Delta t$ and "vacate" a volume $\Delta V_1 = A_1\,\Delta x_1$. At the same time, molecules in the narrower section move $\Delta x_2 = v_2\,\Delta t$ and fill a new volume $\Delta V_2 = A_2\,\Delta x_2$. Because the fluid is incompressible, the volumes ΔV_1 and ΔV_2 must be equal; that is,

$$\Delta V_1 = A_1\,\Delta x_1 = A_1 v_1\,\Delta t = \Delta V_2 = A_2\,\Delta x_2 = A_2 v_2\,\Delta t \qquad (9.19)$$

Dividing both sides of the equation by Δt gives the **equation of continuity:**

$$v_1 A_1 = v_2 A_2 \tag{9.20}$$

The equation of continuity relating the speed v of an incompressible fluid to the cross-section area A of the tube in which it flows

FIGURE 9.31 The speed of the water is inversely proportional to the cross-section area of the stream.

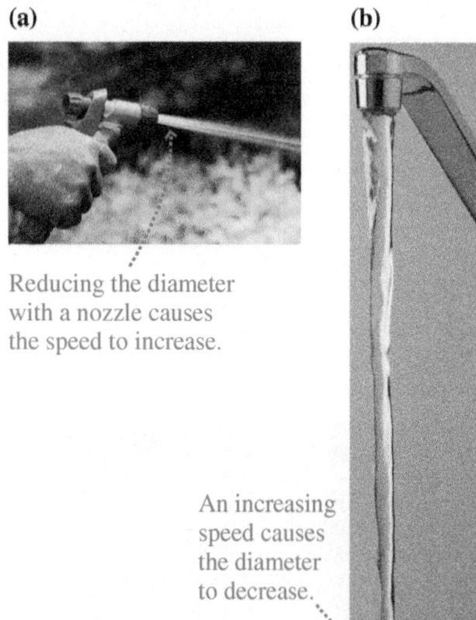

(a)
(b)
Reducing the diameter with a nozzle causes the speed to increase.
An increasing speed causes the diameter to decrease.

Equation 9.20 says that **the volume of an incompressible fluid entering one part of a tube or pipe must be matched by an equal volume leaving downstream.**

An important consequence of the equation of continuity is that **flow is faster in narrower parts of a tube, slower in wider parts.** You're familiar with this conclusion from many everyday observations. The garden hose shown in FIGURE 9.31a squirts water farther after you put a nozzle on it. This is because the narrower opening of the nozzle gives the water a higher exit speed. Water flowing from the faucet shown in FIGURE 9.31b picks up speed as it falls. As a result, the flow tube "necks down" to a smaller diameter.

The *rate* at which fluid flows through the tube—volume per second—is $\Delta V/\Delta t$. This is called the **volume flow rate** Q. We can see from Equation 9.19 that

$$Q = \frac{\Delta V}{\Delta t} = vA \tag{9.21}$$

Volume flow rate for liquid moving at speed v through a tube of cross-section area A

The SI units of Q are m^3/s, although in practice Q may be measured in cm^3/s, liters per minute, or, in the United States, gallons per minute and cubic feet per minute. Another way to express the meaning of the equation of continuity is to say that **the volume flow rate is constant at all points in a tube.**

EXAMPLE 9.13 The physics of breathing

On average, a resting adult male breathes 12 times per minute and inhales a total of 8.0 L of air. The breathing cycle is not symmetrical: 40% of the time is spent in inhalation, 60% in exhalation. The average diameter of an adult male's trachea—the windpipe—is 1.8 cm. The trachea branches into two bronchi as it descends into the lungs. What is the average air speed in the trachea during a resting inhalation? What diameter bronchi will maintain this air speed?

PREPARE The breathing data can be expressed as a volume flow rate, from which we can calculate the flow speed. The flow rates through the trachea and the bronchi have to be the same, and they are connected by the equation of continuity.

SOLVE At 12 breaths per minute, one cycle takes 5.0 s. 40% of that, or 2.0 s, is the time for one inhalation. The air volume of one inhalation is 8.0 L/12 = 0.67 L. Thus the average volume flow rate during inhalation is

$$Q = \frac{\Delta V}{\Delta t} = \frac{0.67\text{ L}}{2.0\text{ s}} \times \frac{1\text{ m}^3}{1000\text{ L}} = 3.3 \times 10^{-4}\text{ m}^3/\text{s}$$

The flow rate through cross-section area A is $Q = vA = \pi r^2 v$, so we can calculate that the average flow speed in the trachea is

$$v_{trac} = \frac{Q}{\pi (r_{trac})^2} = \frac{3.3 \times 10^{-4}\text{ m}^3/\text{s}}{\pi (0.0090\text{ m})^2} = 1.3\text{ m/s}$$

Continuity requires the air to flow through the bronchi at the same *total* rate, but there are two bronchi and thus

$$Q_{bron} = 2v_{bron}A_{bron} = 2\pi (r_{bron})^2 v_{bron}$$
$$= Q_{trac} = v_{trac}A_{trac} = \pi (r_{trac})^2 v_{trac}$$

If there's no loss of air speed as the air moves from the trachea to the bronchi—an assumption we'll justify in the next section—then $v_{bron} = v_{trac}$ and thus

$$r_{bron} = \frac{r_{trac}}{\sqrt{2}} = \frac{0.90\text{ cm}}{\sqrt{2}} = 0.64\text{ cm}$$

The bronchi diameter will be twice this, or 1.3 cm.

ASSESS An air speed of 1.3 m/s is modest—a fast walking pace—and is in good agreement with measurements. This is an average, with the instantaneous speed varying from 0 m/s to approximately 2.5 m/s. The average diameter of an adult male's bronchus is 1.2 cm, quite close to our prediction. The small discrepancy could be because our model of the respiratory system is a bit too simplified. But because we're using averages, not the measurements of a specific individual, it might be that the discrepancy is simply due to the limitations of the data.

Representing Fluid Flow: Streamlines and Fluid Elements

Representing the flow of fluid is more complicated than representing the motion of a point particle because fluid flow is the collective motion of a vast number of particles. FIGURE 9.32 gives us an idea of one possible fluid-flow representation. Here smoke is being used to help engineers visualize the airflow around a car in a wind tunnel. The smoothness of the flow tells us this is laminar flow. But notice also how the individual smoke trails retain their identity. They don't cross or get mixed together. Each smoke trail represents a *streamline* in the fluid.

Imagine that we could inject a tiny colored drop of water into a stream of water undergoing laminar flow. Because the flow is steady and the water is incompressible, this colored drop would maintain its identity as it flowed along. The path or trajectory followed by this "particle of fluid" is called a **streamline.** Smoke particles mixed with the air allow us to see the streamlines in the wind-tunnel photograph of Figure 9.32. FIGURE 9.33 illustrates three important properties of streamlines.

FIGURE 9.32 Streamlines in the laminar airflow around a car in a wind tunnel.

Streamline

FIGURE 9.33 Particles in a fluid move along streamlines.

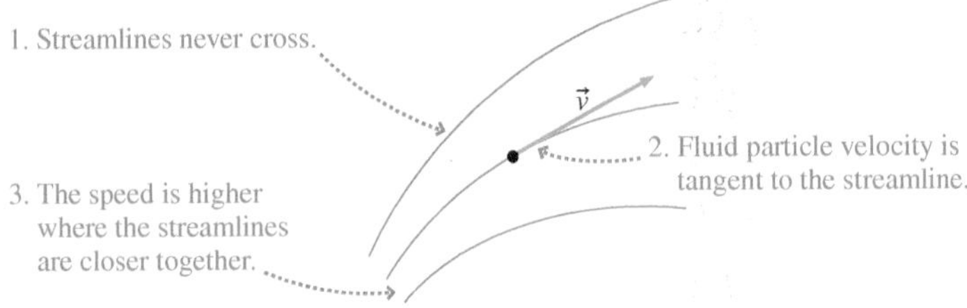
1. Streamlines never cross.
$\vec{v}$
2. Fluid particle velocity is tangent to the streamline.
3. The speed is higher where the streamlines are closer together.

STOP TO THINK 9.7 The figure shows volume flow rates (in cm^3/s) for all but one tube. What is the volume flow rate through the unmarked tube? Is the flow direction in or out?

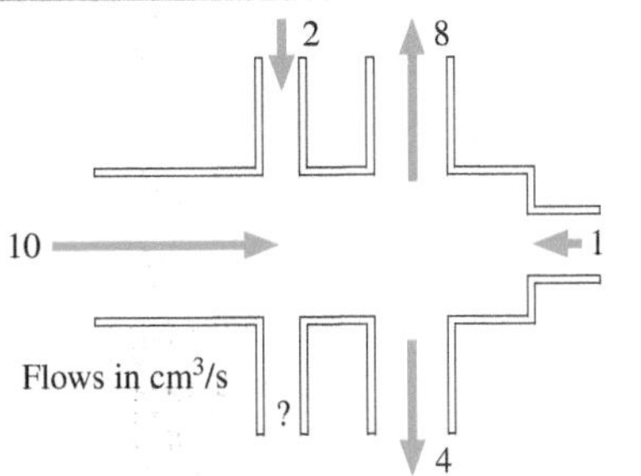
2
8
10
1
Flows in cm^3/s
?
4

9.6 Ideal Fluid Dynamics

The equation of continuity describes a moving fluid but doesn't tell us anything about *why* the fluid is in motion. It is equivalent to the kinematic equations that describe the motion of particles but don't explain their motion. Consider a fluid moving through the tapered tube shown in FIGURE 9.34. The equation of continuity tells us that the fluid speed is increasing from left to right, which is indicated by the streamlines getting closer together; that is, the fluid is *accelerating*. Thus we again need to turn to Newton's second law, the law that relates acceleration to force and mass.

FIGURE 9.35 shows a thin section of fluid within the tube that has length Δx. This section, with cross-section area A, has mass $m = \rho V = \rho A \Delta x$. Further, this section of fluid is accelerating, so it must be experiencing a net force. We're assuming a horizontal motion, which rules out gravity, and there are no external forces. Instead, this section of fluid is pushed on by the surrounding fluid—that is, by *pressure forces*. In particular, a net force to the right indicates that the pressure p_1 on the left is larger than the pressure p_2 on the right, so this section of fluid is accelerated by the *pressure difference* across it.

FIGURE 9.34 A fluid accelerates as it moves through a tapered tube.

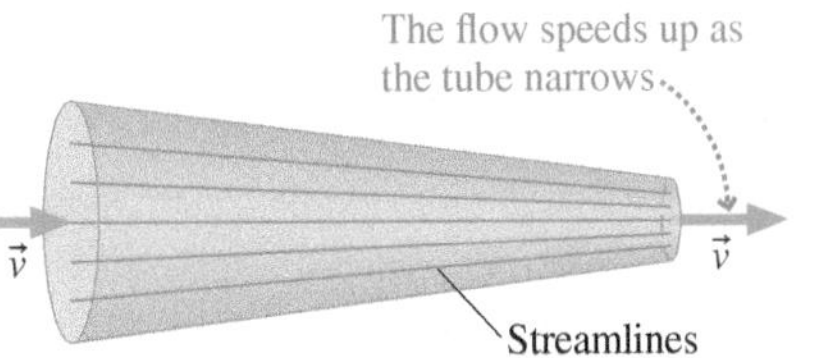
The flow speeds up as the tube narrows.
$\vec{v}$
$\vec{v}$
Streamlines

FIGURE 9.35 A section of the fluid experiences a net force and accelerates due to a pressure difference across it.

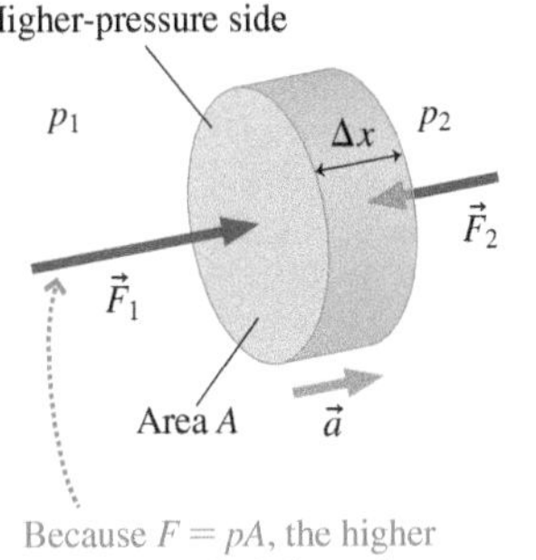
Higher-pressure side
p_1
Δx
p_2
$\vec{F}_2$
$\vec{F}_1$
Area A
$\vec{a}$
Because $F = pA$, the higher pressure on the left results in a net force to the right.

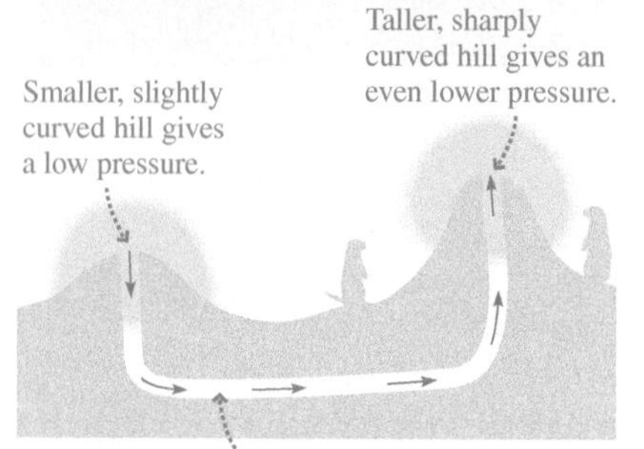

Taller, sharply curved hill gives an even lower pressure.
Smaller, slightly curved hill gives a low pressure.
The pressure difference between the two ends pushes air through the burrow.

Nature's air conditioning Prairie dogs ventilate their underground burrows with aerodynamic forces and pressures. The two entrances to their burrows are surrounded by mounds, one higher than the other. When the wind blows across these mounds, the pressure is reduced at the top. The taller mound, with its greater curvature, has the lower pressure of the two entrances. Air then is pushed through the burrow toward this lower-pressure side.

Suppose there are no viscous forces and thus no resistance to flow. This is a strong supposition—like assuming that the motion of solid objects is frictionless—and we'll return to this assumption later, after we've further developed the ideas of fluid dynamics, to see the conditions under which it is justified. An incompressible fluid that flows with no viscous forces is called an **ideal fluid,** and the *ideal-fluid model* is the first of two models of fluid dynamics that we'll explore. Flow without viscosity is sometimes called *inviscid flow.*

With no viscous forces to consider, Newton's second law for the section of fluid in Figure 9.35 is

$$(F_{\text{net}})_x = F_1 - F_2 = p_1A - p_2A = ma_x = (\rho A\Delta x)a_x \tag{9.22}$$

One of the equations of constant-acceleration kinematics, for acceleration through a displacement Δx, was $v_2^{\,2} = v_1^{\,2} + 2a_x\Delta x$. Using this for a_x in Equation 9.22 gives

$$p_1A - p_2A = (\rho A\Delta x)a_x = \rho A(a_x\Delta x) = \tfrac{1}{2}\rho Av_2^{\,2} - \tfrac{1}{2}\rho Av_1^{\,2}$$

The cross-section area A cancels, and a regrouping of the terms leaves us with

$$p_2 + \tfrac{1}{2}\rho v_2^{\,2} = p_1 + \tfrac{1}{2}\rho v_1^{\,2} \tag{9.23}$$

Bernoulli's equation relating the speed and pressure at two points on a streamline

The main implication of Equation 9.23 is that **the pressure is higher at a point on a streamline where the fluid is moving slower, lower where the fluid is moving faster.** This property of moving fluids was discovered by the 18th-century Swiss scientist Daniel Bernoulli, and thus Equation 9.23 is called **Bernoulli's equation.** Although we used a thin section of fluid to derive it, it applies to any two points *on the same streamline.*

Notice that no pressure difference is needed to keep an ideal fluid flowing along at constant speed. This is just Newton's first law. With no viscosity, just as with no friction, an object—in this case the fluid—needs no net force to continue moving in a straight line at constant speed. A pressure difference is needed only to *change* the fluid's velocity. Example 9.13 assumed that the air flow speed in the lung's bronchi is the same as in the trachea. That assumption is justified because there's almost no pressure difference between the bronchi and the trachea and thus almost no acceleration that would change the air speed.

NOTE ► If you've taken physics before, you may have noticed that the term $\frac{1}{2}\rho v^2$ in Bernoulli's equation looks a lot like the expression $\frac{1}{2}mv^2$ for kinetic energy. (If not, you'll learn this in the next chapter.) In fact, Bernoulli's equation can be interpreted as a statement about energy, saying that the pressure forces change the fluid's kinetic energy. This is not true for a viscous fluid because viscous forces, like friction, dissipate kinetic energy, converting it into thermal energy and an increased temperature. ◄

FIGURE 9.36 The pressure is lower where wind blows faster across the top of a hill.

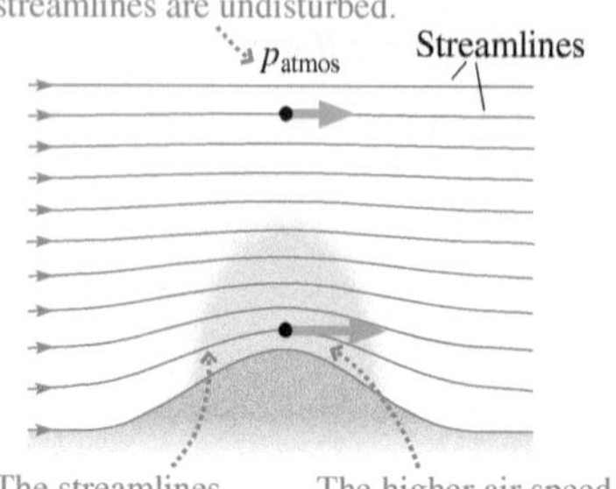

To see how Bernoulli's equation can be applied, consider air flowing over a hill, as shown in FIGURE 9.36. Far to the left, before the hill, the wind blows at a constant speed and so the streamlines are equally spaced. But the streamlines have to bunch together as the air goes over the hill, showing that the air speeds up. According to Bernoulli's equation, there must be a region of lower pressure near the top of the hill where the air is moving the fastest. Prairie dogs have learned to use this effect to ventilate their burrows.

EXAMPLE 9.14 Pressure loss at a constriction

Water flows through a 12-mm-diameter pipe at 4.0 m/s and a gauge pressure of 150 kPa. An imperfection in a 50-cm-long section of the pipe has caused scale to build up on the pipe wall, which constricts the diameter to 8.0 mm. Assume that water can be modeled as a nonviscous fluid when flowing through a relatively large diameter pipe. What is the gauge pressure in this section of the pipe?

PREPARE We're assuming that the water is nonviscous, so we can use the ideal-fluid model and Bernoulli's equation. We'll also need the equation of continuity. FIGURE 9.37 is a pictorial representation of the situation. Bernoulli's equation relates two points on a streamline, which we'll call points 1 and 2; they are indicated with dots.

SOLVE We can use Bernoulli's equation to find pressure p_2, but first we need to know the flow speed v_2 in the constriction.

FIGURE 9.37 Pictorial representation of water flowing through a constricted pipe.

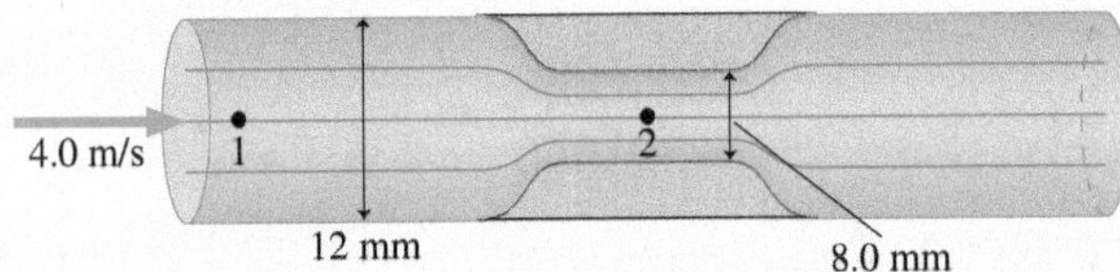

The flow speed can be found from the equation of continuity, Equation 9.20:

$$v_1A_1 = v_1\pi r_1^2 = v_2A_2 = v_2\pi r_2^2$$

$$v_2 = \left(\frac{r_1}{r_2}\right)^2 v_1 = \left(\frac{6.0\text{ mm}}{4.0\text{ mm}}\right)^2(4.0\text{ m/s}) = 9.0\text{ m/s}$$

Constricting the cross-section area has forced the water to speed up. Now we can use Bernoulli's equation, Equation 9.23, to calculate

$$\begin{aligned} p_2 &= p_1 + \tfrac{1}{2}\rho(v_1^2 - v_2^2) \\ &= 1.5 \times 10^5\text{ Pa} + \tfrac{1}{2}(1000\text{ kg/m}^3)[(4.0\text{ m/s})^2 - (9.0\text{ m/s})^2] \\ &= 1.2 \times 10^4\text{ Pa} = 120\text{ kPa} \end{aligned}$$

We were able to leave the pressures as gauge pressures because the conversions to absolute pressures—adding the same term on both sides of the equation—cancel.

ASSESS It's not surprising that the water has to speed up to pass through a narrower section of pipe. The fact that the pressure decreases is less obvious, but in this case it takes a pressure difference of 30 kPa, or $\approx$0.3 atm, to accelerate the water to the higher speed. The pressure returns to its initial 150 kPa on the far side of the constriction where the water has slowed to its initial speed.

The conclusion of Example 9.14 has real implications for blood flow through arteries. Plaque formation on the walls of arteries—the disease of *atherosclerosis*—causes a drop in blood pressure at the site of the narrowing. The technique of *Doppler ultrasound,* which we'll examine more closely in Chapter 16, uses sound waves to measure the speed of flowing blood. FIGURE 9.38 shows a longitudinal section of a carotid artery that contains sizable plaques; yellow indicates a higher blood speed than red, showing the increased speed in the constricted section of the artery. Bernoulli's equation can then be used to compute the pressure drop at the constriction. In severe cases, the pressure loss allows the vessel walls to contract, which squeezes the artery even more tightly closed.

FIGURE 9.38 A Doppler ultrasound image of a carotid artery occluded by plaque. The blood speed through the artery is color coded, with yellow representing a higher speed than red.

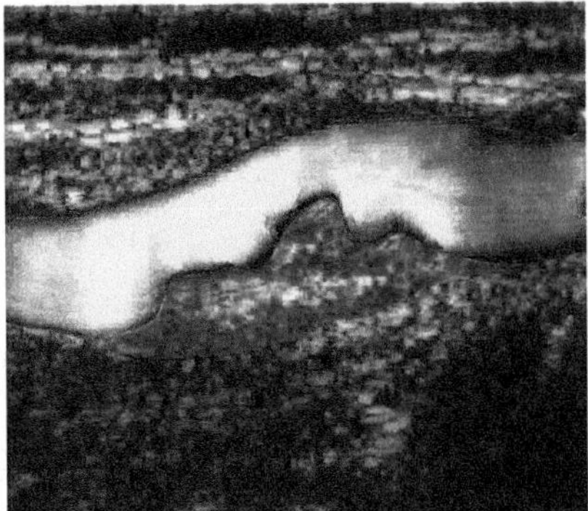

STOP TO THINK 9.8 Water flows through a pipe whose diameter changes. You can't see inside the pipe, but you can observe the height of the water in three vertical glass tubes attached to the top of the pipe. Rank in order, from highest to lowest, the water speeds at points 1, 2, and 3.

1 2 3

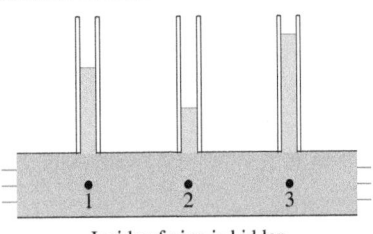

Inside of pipe is hidden

A. $v_1 > v_2 > v_3$
B. $v_1 = v_2 = v_3$
C. $v_2 > v_1 > v_3$
D. $v_3 > v_1 > v_2$

EXAMPLE 9.15 Using the difference in blood speeds to compute a blood-pressure drop

If the aortic valve, through which blood exits the left ventricle, becomes narrowed, a condition called *aortic stenosis,* it can lead to significant health problems. One of the diagnostic criteria is a drop in pressure as the blood traverses the valve. If this pressure drop exceeds 40 mm Hg, a diagnosis of severe aortic stenosis is warranted. Clinicians can use Doppler echocardiography, a form of ultrasound, to measure the speeds of the blood upstream and downstream of the valve. The speeds can then be used to compute a corresponding pressure drop. In one patient, as the left ventricle is ejecting blood, measurements show speeds of 1.1 m/s upstream of the valve and 3.5 m/s just past the valve. Does the patient meet the criterion for severe aortic stenosis?

PREPARE Blood pressure, which we'll look at more closely in Section 9.8, is measured in mm Hg. We saw earlier that 1 mm Hg = 133 Pa. Assume that blood's viscosity in the aorta and the other large blood vessels can be neglected, in which case Bernoulli's equation applies to this situation. (A more complete model of blood flow would include its viscosity, but the ideal-fluid model is surprisingly good for flows over this short distance.)

SOLVE The drop in pressure is

$$\Delta p = p_1 - p_2 = \tfrac{1}{2}\rho(v_2^2 - v_1^2)$$
$$= \tfrac{1}{2}(1050\ \text{kg/m}^3)[(3.5\ \text{m/s})^2 - (1.1\ \text{m/s})^2]$$
$$= 5800\ \text{Pa} = 44\ \text{mm Hg}$$

This is above the threshold, so a positive diagnosis is warranted.

ASSESS The clinical data show a significant increase in speed and a corresponding drop in pressure as the blood exits the valve, so it's no surprise that this person has a serious condition. In a normal heart, there is some increase in speed as the blood exits the aortic valve. This causes a reduction in pressure, but normally the pressure increases again as the flow slows down on entering the aorta. If the increase in speed is too great, as it is for this patient, the flow may cross the boundary between laminar and turbulent flow, with negative consequences.

9.7 Viscous Fluid Dynamics

Bernoulli's equation is for an ideal fluid, one that has no friction-like viscous forces. That's a reasonable assumption in many situations, but there are many other circumstances where the fluid's viscosity cannot be overlooked. Thus *viscous flow* is our second important model of fluid dynamics.

Viscosity η, a fluid's resistance to flow, was introduced in ◀◀ SECTION 5.4 when we looked at the drag force on an object moving through a stationary fluid. Now we want to analyze a moving fluid in a stationary tube. We'll restrict our analysis to tubes that have circular cross sections. TABLE 9.5 gives the values of η for some common fluids. Note that viscosity decreases *very* rapidly with an increase in temperature.

TABLE 9.5 Viscosities of fluids

Fluid	η (Pa · s)
Air (20°C)	1.8×10^{-5}
Water (20°C)	1.0×10^{-3}
Water (60°C)	4.7×10^{-4}
Milk (20°C)	3.0×10^{-3}
Whole blood (37°C)	3.5×10^{-3}
Olive oil (20°C)	8.4×10^{-2}
Motor oil (20°C)	2.4×10^{-1}
Motor oil (100°C)	8.0×10^{-3}

Viscosity has a profound effect on how a fluid moves through a tube. FIGURE 9.39a shows that in an ideal fluid, which has no viscosity, all the fluid particles move with the same speed v, the speed that appears in Bernoulli's equation. For a viscous fluid, seen in FIGURE 9.39b, the fluid moves fastest in the center of the tube. The speed decreases away from the center until it reaches zero on the walls of the tube. That is, **the layer of fluid in contact with the wall of the tube does not move at all.** Whether it is water moving through pipes or blood through arteries, the fact that the fluid at the outer edges "lingers" and barely moves allows deposits to build up on the inside walls of a tube.

FIGURE 9.39 Viscosity alters the velocity profile of a fluid flowing through a tube.

(a) Ideal fluid

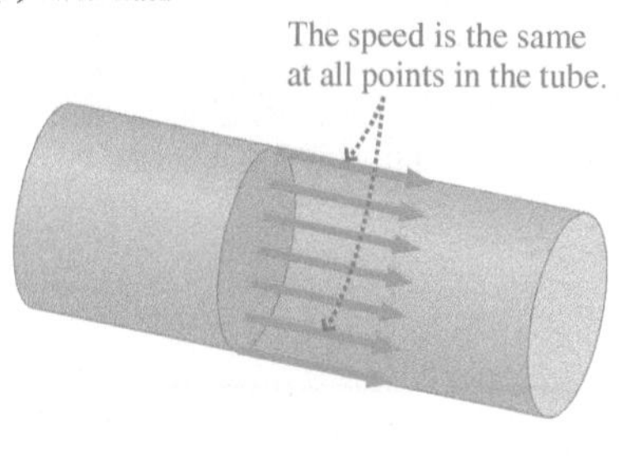
The speed is the same
at all points in the tube.

(b) Viscous fluid

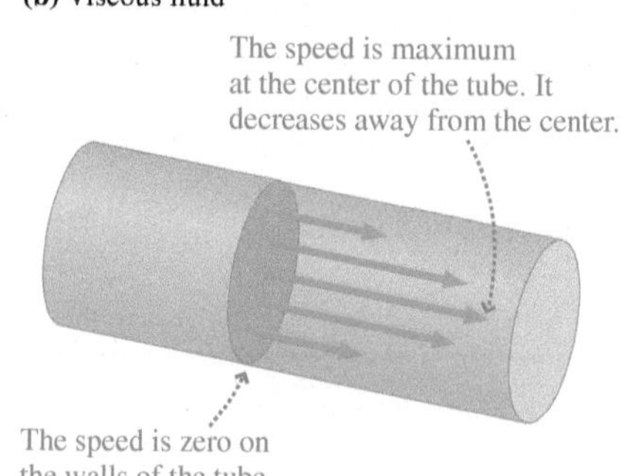
The speed is maximum
at the center of the tube. It
decreases away from the center.
The speed is zero on
the walls of the tube.

We can't characterize the flow of a viscous fluid by a single speed v, but we can define an *average* flow speed v_{avg}. The volume flow rate $Q = \Delta V/\Delta t$ is still a well-defined quantity, so, in parallel with $Q = vA$ for an ideal fluid, we define the average flow speed of a viscous fluid by $v_{avg} = Q/A$.

FIGURE 9.40 shows fluid flowing smoothly (i.e., laminar flow) with average speed v_{avg} through a horizontal tube that has constant diameter R. If this is an ideal fluid, then we learn from Bernoulli's equation that $\Delta p = p_1 - p_2 = 0$. That is, there's no net force on this segment of fluid; it simply "coasts" through the tube at a constant speed with no change in pressure. It's equivalent to a puck gliding at a constant speed across a frictionless surface with no applied force.

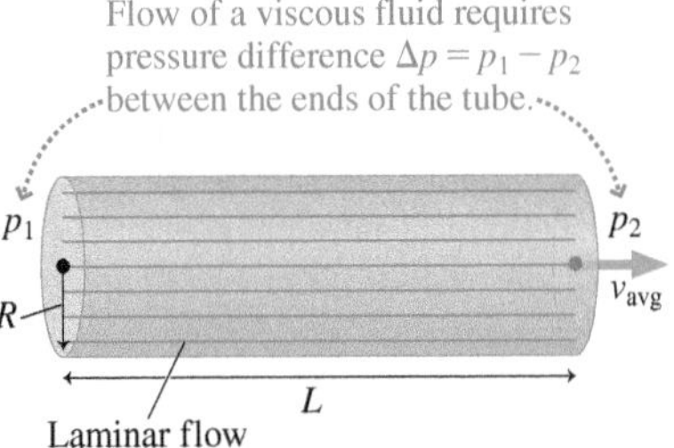
Flow of a viscous fluid requires pressure difference $\Delta p = p_1 - p_2$ between the ends of the tube.
p_1
p_2
v_{avg}
R
L
Laminar flow

FIGURE 9.40 A pressure gradient is needed to keep a viscous fluid flowing.

But if there's friction, something has to apply a steady force—equal to and opposite the friction force—to keep the puck moving at a steady speed. Likewise, something has to apply a steady force—equal to and opposite the viscous drag—to push a viscous fluid through a tube at a steady average speed. That "something" is the pressure difference Δp between the ends of the tube; that is, **a pressure difference is needed to keep a viscous fluid flowing.** The pressure force on the segment of fluid in the tube of Figure 9.40 is $F_{press} = A\Delta p$, where A is the tube's cross-section area.

◀◀ SECTION 5.4 found that the drag on a sphere of radius r moving through a fluid with viscosity η is $F_{drag} = 6\pi\eta vr$. We see that drag depends linearly on the viscosity, speed, and distance over which viscous forces act. Thus it would seem likely that the viscous drag on a segment of fluid flowing through a tube of length L is $F_{drag} = c\eta v_{avg}L$, where c is an unknown constant. Delving deeper into the fundamentals of fluid mechanics than we will go in this text can show that $c = 8\pi$ and thus $F_{drag} = 8\pi\eta v_{avg}L$.

Steady motion of the fluid occurs when the pressure force balances the viscous drag: $F_{press} = F_{drag}$. Using $A = \pi R^2$ for a circular pipe with radius R, we find that the required pressure difference to move the fluid at speed v_{avg} is

$$\Delta p = \frac{F_{press}}{A} = \frac{F_{drag}}{\pi R^2} = \frac{8\eta v_{avg}L}{R^2} \tag{9.24}$$

A larger pressure difference is needed for a more viscous fluid, a longer tube, or a narrower tube.

If we solve Equation 9.24 for v_{avg}, we find that a pressure difference Δp across a tube of length L causes a fluid to flow with average speed

$$v_{avg} = \frac{R^2}{8\eta}\frac{\Delta p}{L} \tag{9.25}$$

Viscosity is in the denominator, which tells us, not surprisingly, that a more viscous fluid flows more slowly.

The more practical quantity, because it's easily measured, is the volume flow rate Q. Because $Q = v_{avg}A = \pi R^2 v_{avg}$, we find

$$Q = \frac{\pi R^4}{8\eta}\frac{\Delta p}{L} \tag{9.26}$$

Poiseuille's equation for viscous flow in a pipe with a pressure gradient

This is called **Poiseuille's equation** (also sometimes called the *Poiseuille-Hagan equation*) for viscous flow, named for a 19th-century French scientist who studied fluids and was especially interested in blood flow.

The quantity $\Delta p/L$ is called the **pressure gradient.** It is the pressure change per unit length or, equivalently, the slope of a pressure graph. A large pressure change over a small distance is a large pressure gradient. We can interpret Poiseuille's equation as saying that **a pressure gradient drives a fluid flow.** This is our first introduction to the idea that **gradients drive flows,** but it won't be our last.

One surprising consequence of Poiseuille's equation is the very strong dependence of the flow on the tube's radius: The flow rate is proportional to the *fourth* power of R. Fairly small changes in the radius have a large effect on the flow rate.

EXAMPLE 9.16 Blood flow through a clogged artery

Atherosclerosis is a condition in which arteries are narrowed by the buildup of fatty plaques on the inner wall of the artery. The process restricts blood flow to organs, including the heart, and can have serious health consequences. Suppose the diameter of an artery is reduced by 25%, a not untypical value. By what percent is the blood flow reduced if the pressure gradient along the artery stays constant? By what factor would the pressure difference established by the heart have to increase to keep the blood flow rate constant?

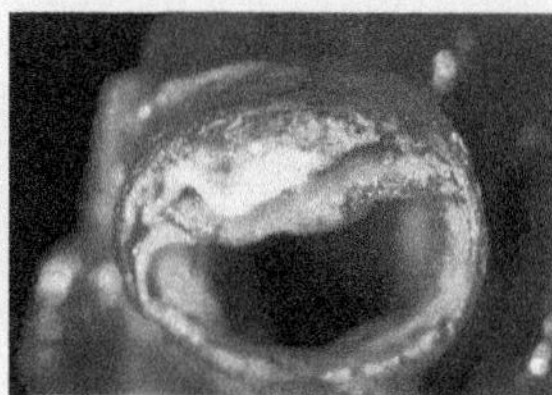
Cross section of an artery clogged with plaque.

PREPARE We'll model the blood flow through the artery as viscous flow through a tube with a constant diameter.

SOLVE Suppose the flow rate is Q_1 through an unclogged artery of radius R_1 and Q_2 through a clogged artery with a smaller radius $R_2 = 0.75R_1$. From Poiseuille's equation, Equation 9.26, we find

$$\frac{Q_2}{Q_1} = \frac{R_2{}^4}{R_1{}^4} = (0.75)^4 = 0.32$$

A 25% reduction in diameter leads to a huge 68% reduction in flow if the pressure gradient is unchanged. Conversely, for a constant flow rate Q, the pressure difference Δp is inversely proportional to R^4. Thus

$$\frac{\Delta p_2}{\Delta p_1} = \frac{(1/R_2)^4}{(1/R_1)^4} = \left(\frac{1}{0.75}\right)^4 = 3.2$$

Blood pressure would have to increase by a factor of 3.2 to maintain the flow rate.

ASSESS The fourth-power dependence on R has profound consequences for the flow of a viscous fluid. Physiologically, the heart cannot greatly increase the amount of pressure it generates, so atherosclerosis typically causes a significant reduction of blood flow. It is the leading cause of death in the United States.

EXAMPLE 9.17 Blood pressure in the veins

Blood samples are taken from veins rather than arteries because the pressure is much lower as blood returns to the heart. Blood donors who give whole blood have a 6.0-mm-long, 16-gauge needle (1.2 mm inner diameter) needle inserted into a vein in their arm. Blood flows through the needle, then through 75 cm of 3.6-mm-diameter tubing to the collection bag. What is the venous blood pressure of a patient who makes a 500 mL donation in 8.0 min?

PREPARE Blood flow through narrow-diameter tubes and needles is viscous flow. Table 9.5 gives the viscosity of whole blood as 3.5×10^{-3} Pa · s. Blood is incompressible, so the volume flow rates through the needle and through the tube are the same. There are two pressure decreases, one across the needle and a second across the tubing. Their sum gives the total pressure decrease, which is the blood pressure in the vein relative to atmospheric pressure.

SOLVE The volume flow rate in Poiseuille's equation is in SI units of m^3/s, so we need to begin with some unit conversions:

$$Q = \frac{500\text{ mL}}{8.0\text{ min}} \times \frac{1\text{ L}}{1000\text{ mL}} \times \frac{1\text{ m}^3}{1000\text{ L}} \times \frac{1\text{ min}}{60\text{ s}} = 1.0 \times 10^{-6}\text{ m}^3\text{/s}$$

The pressure decrease needed across the needle to drive this flow is

$$\Delta p_{\text{needle}} = \frac{8\eta LQ}{\pi R^4} = \frac{8(3.5 \times 10^{-3}\text{ Pa}\cdot\text{s})(0.0060\text{ m})(1.0 \times 10^{-6}\text{ m}^3\text{/s})}{\pi(0.0006\text{ m})^4} = 410\text{ Pa}$$

Similarly, the pressure decrease needed across the tubing is

$$\Delta p_{\text{tubing}} = \frac{8\eta LQ}{\pi R^4} = \frac{8(3.5 \times 10^{-3}\text{ Pa}\cdot\text{s})(0.75\text{ m})(1.0 \times 10^{-6}\text{ m}^3\text{/s})}{\pi(0.0018\text{ m})^4} = 640\text{ Pa}$$

The total pressure decrease, which is the pressure in the vein, is

$$p_{\text{vein}} = \Delta p_{\text{needle}} + \Delta p_{\text{tubing}} = 1050\text{ Pa} \times \frac{760\text{ mm Hg}}{101{,}000\text{ Pa}} = 7.9\text{ mm Hg}$$

ASSESS We'll look at blood pressure in the next section, but you may already know that the baseline blood pressure at the output of the heart, the second (diastolic) number in a blood-pressure reading, is typically 75 mm Hg. The pressure decreases as the blood flows through the narrow arterioles and capillaries, and it is, indeed, only a few mm Hg by the time it is collected in the veins for return to the heart. Our calculation is consistent with measured values. Blood could be collected in $\approx\frac{1}{10}$ the time from a major artery with ≈ 10 times the pressure, but at significantly higher risk.

Boundary Layers

The wind may be blowing hard enough to shake the leaves on trees, but a leaf lying flat on the ground is undisturbed. The current in a river or the surge through a tidal pool may be strong enough to knock you off your feet, yet small aquatic organisms live happily on and around the rocks at the bottom. These observations suggest that the flow speed of the fluid is much reduced near the edge of the flow—a *boundary*.

A key attribute of viscous flow through a tube is that the flow speed goes to zero at the wall of the tube. This is sometimes called the *no-slip condition*. The same idea applies to any laminar viscous flow (i.e., steady flow without turbulence): The flow speed goes to zero at a boundary.

FIGURE 9.41 shows a fluid of density ρ flowing over a flat surface. Far from the surface, the fluid moves uniformly with the *free-stream speed* v_{free}. At the surface, the speed is zero. Thus there's a transition zone, a **boundary layer,** in which the flow speed changes rapidly. Objects or organisms that reside in the boundary layer experience less flow, and thus less drag, than objects or organisms in the free stream. This is especially true toward the bottom of the boundary layer.

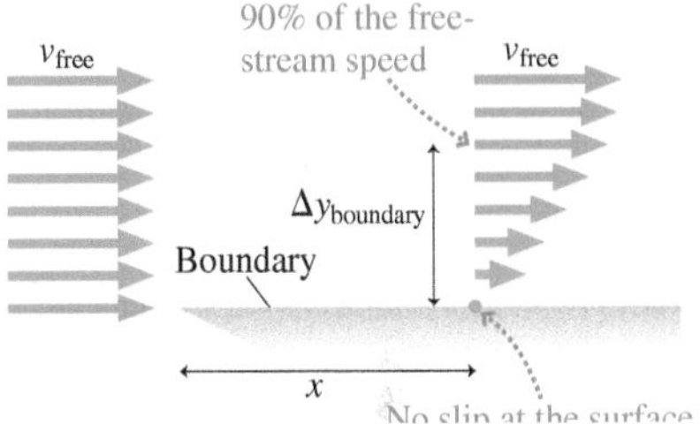

FIGURE 9.41 The boundary layer is where the flow speed changes from zero to the free-stream speed.

We define the thickness of the boundary layer $\Delta y_{boundary}$ to be the distance over which the speed increases from zero to 90% of the free-stream speed. The result of a fairly complex analysis is

$$\Delta y_{boundary} \approx 3\sqrt{\frac{\eta x}{\rho v_{free}}} \tag{9.27}$$

where x is the distance from the edge or lip where the boundary begins. The boundary layer thickness increases with this distance, but not rapidly because it depends on only the square root of x.

Notice that the free-stream speed appears in the denominator. Somewhat surprisingly, a faster free-stream speed has a *thinner* boundary layer. That is, a faster flow has a much larger *velocity gradient* not only because there's a larger speed difference between the edge and the free steam but also because this change happens over a smaller distance.

The boundary layer thickness depends on x, so it doesn't have a single, well-defined value. Nonetheless, for wind blowing at 5 m/s ($\approx$12 mph), the boundary layer thickness from a few to 100 m behind an edge or lip is a few cm. For a typical river current or ocean surge, the boundary layer thickness from 1 to a few meters behind an edge or lip is a few mm.

This finding is especially important for small organisms that live at or near the surface. Terrestrial organisms no more than a few cm tall hardly know the wind is blowing; they spend their time in a quiet refuge with very little air motion. Aquatic organisms are somewhat more constrained, but staying below a few mm in height substantially reduces the drag from water flowing past. Sessile organisms such as mussels live mostly in the boundary layer, so the drag forces that try to pull them from the rocks are less than you might expect based on your experiences higher in the water column.

Turbulence

Our analysis of both ideal and viscous fluids has assumed that the flow is *laminar*, or smooth, with fluid particles that move along gradually curving streamlines. In laminar flow, pressure and velocity change gradually in predictable ways. The layers in a fluid—and *laminar* means "arranged in layers"—remain distinct and do not mix.

Not all fluid flow is laminar. Many flowing fluids exhibit **turbulence,** which is pretty much the opposite of laminar flow. Turbulent flow is characterized by chaotic and erratic changes in pressure and velocity. Whirlpool-like eddies and vortices are

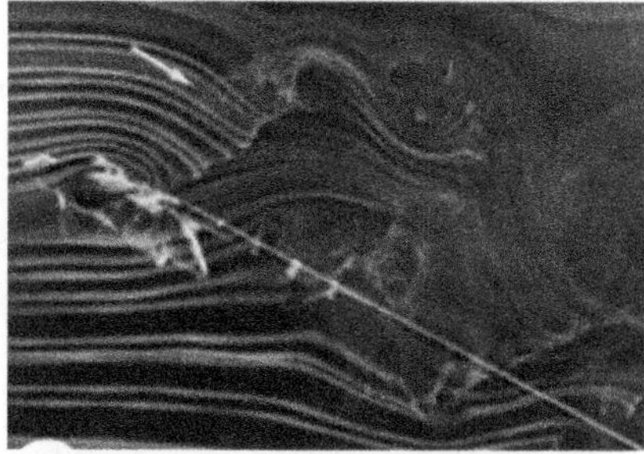

A turbulent encounter Bird and insect wings generate turbulence because the wings move through the air rapidly, at high Reynolds number. In an ingenious experiment, researchers attached a hawk moth to a lightweight support and let it "fly" in a wind tunnel where smoke particles show the airflow over the wings. The flow is laminar as it crosses the moth, then breaks up into turbulent eddies and vortices. It was found that the vortices created by the leading edges of wings produce additional lift that enables the moth to hover and maneuver.

common, and the fluid is well mixed. The photo of Figure 9.29 showed the dramatic difference when a flow transitions from laminar to turbulent.

What determines whether a fluid has nicely behaved laminar flow or irregular and unpredictable turbulent flow? The Reynolds number! Chapter 5 introduced the Reynolds number, $Re = \rho vL/\eta$, as the ratio of inertial forces to viscous forces. The "characteristic size" L is the diameter $D = 2R$ both for liquid flowing through a circular tube, the subject of this section, and for a sphere moving through a fluid, the situation we considered in Chapter 5.

Viscous fluids, like honey, tend to flow very smoothly, so we expect laminar flow for low Reynolds number. Flow at high Reynolds number, where inertia dominates, is like a stampede of thousands of closely packed cattle: It is smooth as long as every cow maintains absolutely the same velocity, but the slightest stumble or the slightest irregularity causes the flow to break up into a chaotic mess. So, as the Reynolds number increases, there comes a point where the slightest irregularity—even just molecular diffusion—will cause a flow to become turbulent.

There is no satisfactory theory of turbulence—it remains an unsolved problem of classical physics—and thus no way to predict the range of Reynolds numbers that characterize laminar or turbulent flow. Nonetheless, experimental measurements have established the relevant Reynolds numbers for many situations of practical importance. FIGURE 9.42 illustrates laminar and turbulent flows for fluid flow through a circular tube and for a sphere moving through a fluid.

We see that Poiseuille's equation for flow through a circular tube, which assumes laminar flow, is valid only when $Re < 2000$. That's true in many practical situations, but this condition can be violated for flows at high speeds or in larger-diameter tubes. Fluid flow becomes fully turbulent for $Re > 4000$. In between is a complicated transition region where flows exhibit both laminar and turbulent characteristics.

You should recall, from Chapter 5, that an object moving through a fluid experiences *viscous drag* when $Re < 1$ and *inertial drag* when $Re > 1000$. Now we can see why. For low Reynolds number, the fluid flow around a sphere is laminar; the streamlines part, then come back together, leaving no trace of the sphere's passage. In contrast, at high Reynolds number, the sphere leaves a fully turbulent *wake* behind it. You may have observed this with boats; water moves smoothly around a boat that is gliding slowly through the water, but a fast boat leaves a turbulent wake behind. The inertial drag of high Reynolds number, which is proportional to the square of the speed, occurs when the object creates a turbulent wake.

The transition from laminar to turbulent flow has numerous implications in medicine. Blood flow is normally laminar, but some pathologies disrupt the flow

FIGURE 9.42 Laminar and turbulent fluid flows.

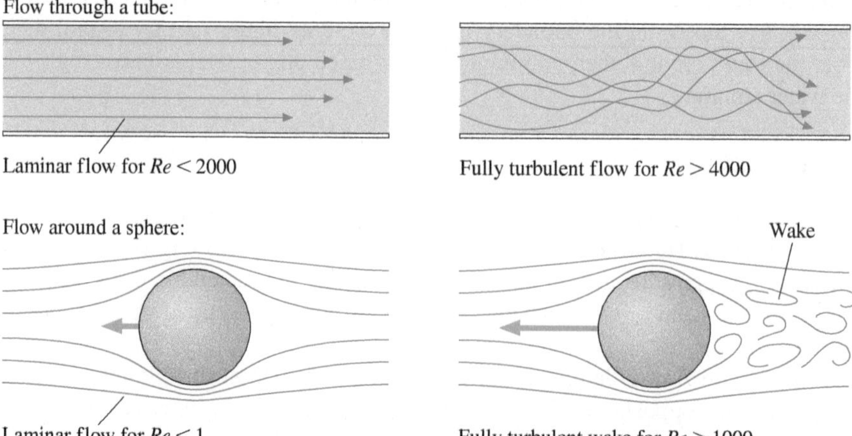

and create turbulence. Thus detection and recognition of turbulence are important in the diagnosis of disease. The passage of air through the respiratory system is laminar at modest respiratory rates but can become turbulent as the air speed increases at elevated respiratory rates. Turbulent flow is less efficient than laminar flow, which means that a given pressure gradient generates a smaller volume flow rate.

Bernoulli's equation is somewhat of a paradox. A nonviscous fluid would seem to have an infinitely high Reynolds number, so you would expect any flow to be turbulent. Yet Bernoulli's equation for nonviscous fluids applies only to laminar flow. The reason is that Bernoulli's equation assumes an *ideal fluid* that can maintain laminar flow for any value of *Re*. No real fluid can do this, so using Bernoulli's equation—which does have many important applications—requires establishing that the flow really is laminar. Bernoulli's equation works best for very streamlined shapes that restrain or delay the onset of turbulence.

STOP TO THINK 9.9 A viscous fluid flows through the pipe shown. The three marked segments are of equal length. Across which segment is the pressure gradient the largest?

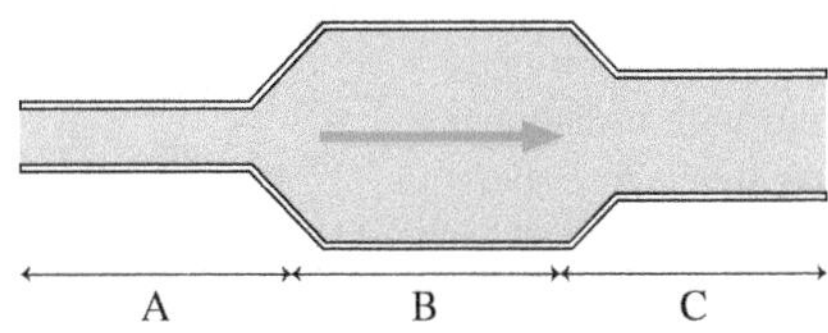

9.8 The Circulatory System

Once every second, more or less, your heart beats, and the contraction of the chambers of your heart sends blood flowing through your body's arteries, capillaries, and veins. In this section, we'll use the principles that we've learned in earlier sections to explain and explore the motion of blood through your circulatory system.

We'll focus on the motion of blood through your body, from your heart's left ventricle to the right atrium; we won't examine the pulmonary circulation through your lungs. As usual, we'll make a simplified model of the system. We'll assume, as we've done before, that the fluid is incompressible (a very good approximation) and that the flow is laminar. In the larger vessels, we'll be able to assume that the effect of viscosity is negligible; in the smaller vessels, it isn't.

FIGURE 9.43 Blood pressure during one cycle of a heartbeat.

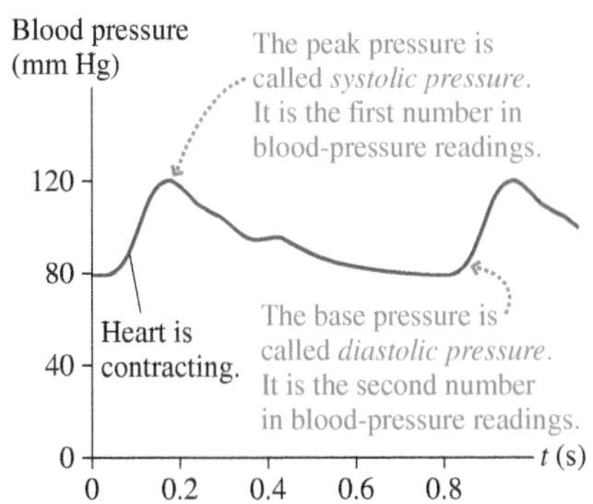

Blood Pressure

When your heart muscle contracts, the pressure of the blood increases, and blood leaves the left ventricle and moves into the large artery called the aorta. FIGURE 9.43 is a pressure graph showing how the blood pressure at the heart changes during a bit more than one cycle of the heartbeat. Blood pressure is a gauge pressure, the pressure in excess of the pressure of the atmosphere, and it is traditionally measured in mm Hg. The graph shows a fairly typical blood pressure for a healthy young adult, with a *systolic* (peak) pressure of 120 mm Hg and a *diastolic* (base) pressure of 80 mm Hg—that is, a blood pressure of "120 over 80."

Blood pressure that is too high or too low can cause health problems. For this reason, a visit to the doctor usually starts with a measurement of blood pressure. Although automated blood-pressure measurements are becoming more common, the traditional procedure, shown in FIGURE 9.44, is still used. A cuff is placed around the upper arm, then a nurse pumps air into the cuff while using a stethoscope to listen to the flow of blood in a large artery and reading a gauge that shows the pressure in the cuff. The nurse first increases the cuff pressure until the cuff squeezes the artery shut and cuts off the blood flow. He or she then slowly lets air leak out of

FIGURE 9.44 Measuring blood pressure.

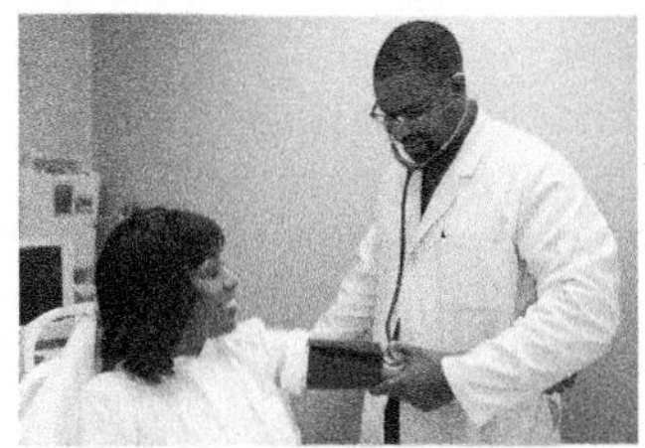

the cuff, listening for telltale sounds of blood flow in your artery. When the cuff pressure drops below the systolic pressure, the pressure pulse during each beat of your heart forces the artery to open briefly and a squirt of blood goes through. The nurse records the pressure when he or she hears the blood start to flow. This is your systolic pressure. This pulsing of the blood through your artery lasts until the cuff pressure reaches the diastolic pressure, when the artery remains open continuously and the blood flows smoothly. This transition is easily heard in the stethoscope, and the nurse records your diastolic pressure.

The clinically important value of blood pressure is the pressure at the heart, but we've noted that blood pressure is measured in an artery in the arm. Are these two pressures the same? We know that there can be a difference in pressure across a tube or pipe due to the effects of viscosity; there can also be a difference in pressure at different points in a fluid due to differences in elevation.

EXAMPLE 9.18 How much does blood pressure change?

A woman is lying down with her arms horizontal. Along her upper arm, her brachial artery stretches 20 cm with a diameter of 4.0 mm. Blood moves through the artery at an average speed of 9.0 cm/s. What is the pressure drop along the 20 cm length?

PREPARE Any pressure difference along the artery isn't due to a height difference because the artery is horizontal. Therefore, any pressure difference must be due to the viscosity of the blood, which is given in Table 9.5 as 3.5×10^{-3} Pa · s. We'll model the artery as a 2.0-mm-radius, 20-cm-long tube. Our equations use SI units, so we'll need to use the conversion factor 1 mm Hg = 133 Pa to get the appropriate blood-pressure units at the end.

SOLVE Equation 9.24 gives the pressure difference needed to push a viscous fluid through a tube:

$$\Delta p = \frac{8\eta v_{avg} L}{R^2}$$

$$= \frac{8(3.5 \times 10^{-3}\ \text{Pa}\cdot\text{s})(0.20\ \text{m})(0.090\ \text{m/s})}{(0.0020\ \text{m})^2}$$

$$= 126\ \text{Pa} \times \frac{1\ \text{mm Hg}}{133\ \text{Pa}} = 0.95\ \text{mm Hg}$$

ASSESS We know that, clinically, the blood pressure in the arm is taken to be the same as the blood pressure at the heart, so we'd expect a very small difference in pressure along large arteries such as the brachial artery. And, indeed, the number that we found is a small fraction of the average blood pressure—less than 1%—and certainly small compared to the rapid fluctuations in blood pressure that occur naturally. The very small effect of viscosity is the reason blood flowing through the larger arteries can be modeled as an ideal fluid that obeys Bernoulli's equation.

There is only a minimal change in pressure due to viscous losses along your arteries. But there can certainly be differences in pressure due to differences in elevation. The blood in your arteries is a connected fluid, so the pressures are lower at greater heights. In the blood pressure measurement shown in Figure 9.43, the cuff is placed on the arm at the level of the heart so that there is no difference in pressure. How much error is introduced if this is not the case?

EXAMPLE 9.19 Computing the change in blood pressure with height

A patient is lying in bed. Suppose she raises her arm for a blood-pressure measurement, with the cuff placed 15 cm (about 6 inches) above her heart. How much of an error does this introduce in the measurement? Ignore any pressure drop due to the viscosity of the blood.

PREPARE A contracting heart generates pressure that moves the blood through the circulatory system, but that's in addition to a steady hydrostatic pressure. The hydrostatic pressure is $p = p_0 + \rho g d$, so the pressure difference between two points that differ in elevation by Δd is $\Delta p = \rho g \Delta d$.

SOLVE The pressure at the cuff is lower than the pressure at the heart by

$$\Delta p = \rho g \Delta d = (1050\ \text{kg/m}^3)(9.8\ \text{m/s}^2)(0.15\ \text{m})$$

$$= 1540\ \text{Pa} \approx 12\ \text{mm Hg}$$

ASSESS This is a significant difference—more than 10 times the difference in pressure due to viscous losses. To avoid introducing such an error, the blood-pressure cuff must be placed at the level of the heart.

Blood-pressure decreases with increasing height above the heart. When you are upright, the blood pressure in your brain is noticeably lower than the pressure at your heart. For animals with long necks, this difference in pressure is correspondingly greater.

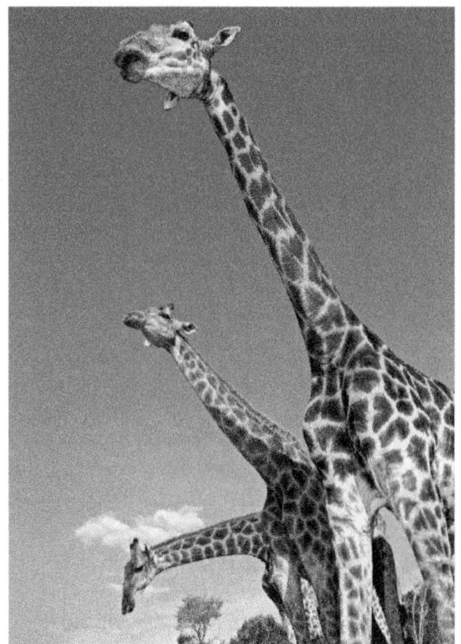

Extreme blood pressure A giraffe's head can be more than 2 meters above its heart. If the blood pressure in the brain is greater than zero (as it must be), the pressure at the heart must be quite high. A typical blood pressure for a resting giraffe is 240/160 mm Hg—a number that would raise serious alarms in a human!

The Arteries and Capillaries

In the human body, blood pumped from the heart to the body starts its journey in a single large artery, the aorta. The flow then branches into smaller blood vessels, the large arteries that feed the head, the trunk, and the limbs. These branch into still smaller arteries, which then branch into a network of much smaller arterioles, which branch further into the capillaries. FIGURE 9.45 shows a schematic outline of the circulatory system and provides some relevant numbers. The pressure in the vessels, shown graphically, assumes that the person is lying down so that there is no pressure change due to differences in elevation.

The numbers in the table are averages from different sources, representing variations among individuals of different sizes, ages, and levels of physical conditioning. The numbers are given to one significant figure; we really can't claim more precision. Nonetheless, a few broad trends emerge:

- As we've seen, there is only a very small change in pressure across the larger arteries. The pressure begins to drop only when the blood enters the smaller arteries and, even more so, the arterioles, where viscosity starts to have a significant effect. Most of the pressure drop occurs in these smaller vessels. Changes in the size of the small arteries and arterioles have significant effects on blood flow and on blood pressure.
- As blood moves from the aorta to the arteries and then to the arterioles, the diameter of the individual vessels decreases but the total cross-section area of all of the vessels increases. There is a high degree of branching; going from the aorta to the arterioles, the area of an individual vessel decreases by a factor of 1,000,000, but the area of all of the vessels considered together increases by a factor of nearly 200. This implies that the single aorta eventually branches to 200,000,000 arterioles!
- As the total cross-section area for the flow increases, the continuity equation indicates that the flow speed must decrease. As the flow branches from a large vessel to a number of smaller vessels, the area of the branches exceeds that of the initial vessel, so the speed decreases.

FIGURE 9.45 Schematic overview of blood flow in the circulatory system.

	Left ventricle	Aorta	Arteries	Arterioles	Capillaries	Venules	Veins	Vena cava	Right atrium
Diameter (cm)		2	0.5	0.002	0.0009	0.003	0.5	3	
Total cross-section area (cm^2)		3	20	500	4000	3000	80	7	
Approximate number		1	100	2×10^8	6×10^9	4×10^8	400	1	

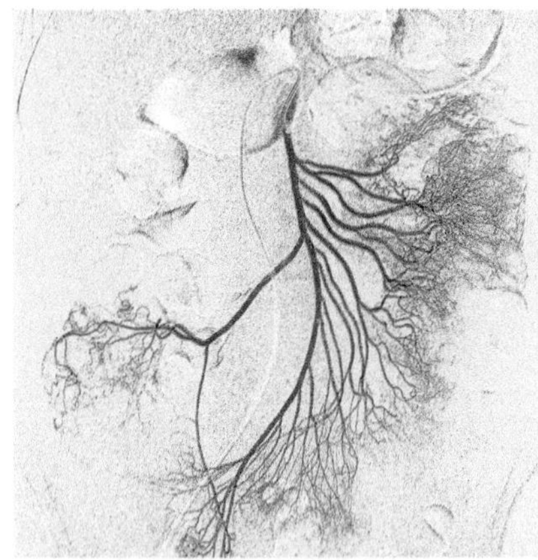

Divided ever finer This preserved section of blood vessels shows the tremendous increase in number and in total area as blood vessels branch from large arteries to arterioles. One large artery gives rise to thousands of smaller vessels.

Measurements show that the adult heart pumps 5.0 L of blood per minute. Thus the volume flow rate is

$$Q = \frac{5.0 \text{ liters}}{1 \text{ minute}} = \frac{0.0050 \text{ m}^3}{60 \text{ s}} = 8.3 \times 10^{-5} \text{ m}^3/\text{s}$$

All of this blood goes through the aorta, for which Figure 9.45 shows a diameter of 2 cm. The average speed of the flow can be computed using Equation 9.21:

$$v_{\text{avg}} = \frac{Q}{A} = \frac{8.3 \times 10^{-5} \text{ m}^3/\text{s}}{\pi(0.010 \text{ m})^2} = 0.26 \text{ m/s}$$

This is a reasonable number for the average flow speed. As the vessels branch and the total area increases, the speed decreases, as we've discussed.

This decrease in speed is important for the primary purpose of blood—delivering oxygen to and removing carbon dioxide from cells in the body. These gases diffuse across the thin walls of the tiniest vessels, the capillaries. The low speed of blood flow in these tiny, thin vessels allows more time for this diffusion to take place. Once gas exchange has occurred, the blood is collected by a venule to be returned to the heart.

EXAMPLE 9.20 **Determining the details of the capillaries**

Figure 9.45 gives typical values for the diameter of a capillary as well as the total cross-section area of all the capillaries together. A typical capillary has a length of 1.0 mm. If the heart is pumping blood at a typical 5 liters per minute, how much time does it take for a red blood cell to move through a capillary?

PREPARE Given the volume flow rate and the cross-section area, we can find the flow speed in the capillaries, which we will use to find the time a blood cell takes to traverse a capillary. The total cross-section area of all of the capillaries is given in Figure 9.45 as $4000 \text{ cm}^2 = 0.4 \text{ m}^2$. We showed previously that a flow rate of 5 liters per minute corresponds to $8.3 \times 10^{-5} \text{ m}^3/\text{s}$.

SOLVE We can find the speed of blood through the capillaries using Equation 9.21:

$$v = \frac{Q}{A} = \frac{8.3 \times 10^{-5} \text{ m}^3/\text{s}}{0.4 \text{ m}^2} = 2.1 \times 10^{-4} \text{ m/s}$$

At this speed, it takes a blood cell a time

$$\Delta t = \frac{\Delta x}{v} = \frac{0.0010 \text{ m}}{2.1 \times 10^{-4} \text{ m/s}} = 4.8 \text{ s} \approx 5 \text{ s}$$

to traverse a capillary.

ASSESS The relatively long time for the blood to traverse the capillary makes sense; there must be sufficient time for diffusion to occur.

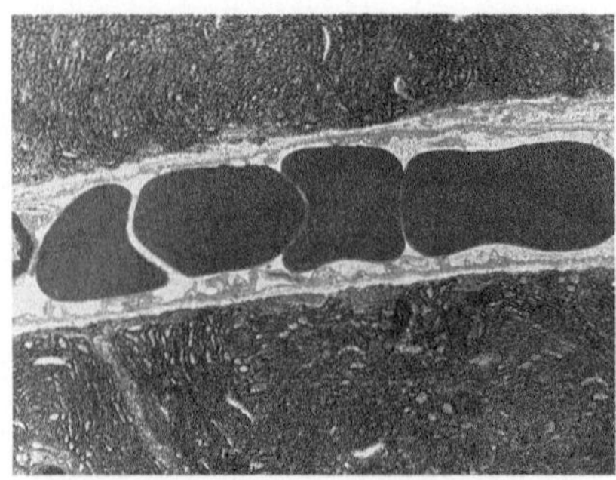

Single file, please The capillaries are about as small as they can be because red blood cells can barely fit through. The red blood cells move through one after another as shown in the photo; they even deform to fit the contours of the capillary. This puts the cells close to the walls of the capillary, which enhances diffusion. This ordering actually reduces the viscosity of the blood and so reduces the pressure needed to move blood through these vessels.

We've assumed that the flow in the circulatory system is laminar. Is that a valid assumption? With the numbers we now have for the aorta, we can calculate that the Reynolds number is

$$Re = \frac{\rho v D}{\eta} = \frac{(1050 \text{ kg/m}^3)(0.26 \text{ m/s})(0.020 \text{ m})}{3.5 \times 10^{-3} \text{ Pa} \cdot \text{s}} = 1600$$

This is in the range for laminar flow, which we saw requires $Re < 2000$, but it's not too far below the threshold for turbulence to develop. If there is narrowing in the large arteries, speeds increase, and the combination of these two factors can put Re above the critical value, leading to turbulent flow. A physician can detect turbulent flow from its characteristic sound. This is one thing a physician is screening for when listening to your heart; the telltale sound of turbulent flow is a sign of trouble.

The Veins

The pressure difference along the arteries keeps the blood moving. But this doesn't work for the veins. Figure 9.45 shows that, when the blood reaches the veins, the pressure is quite low. How then does the blood make its way back to the heart? The veins must have other structures that keep the blood moving.

Consider the veins in your lower legs and feet. When you are lying down, the gauge pressure in these veins is nearly zero. When you are upright, this pressure rises. If you are walking or exercising, the venous pressure rises to about 30 mm Hg. What would the pressure be if the veins were simple tubes?

Walking improves the circulation A horse has no muscles in its lower legs or hoofs that aid in blood return to the heart. Instead, the normal force of the ground compresses a network of veins in the hoof, pushing blood upward to where valves in the veins prevent return flow. When a horse walks, the changing forces on each hoof make this system work as a pump, efficiently returning blood to the heart.

EXAMPLE 9.21 What is the pressure in the veins?

A woman's feet are 1.3 m below her heart. If we assume her veins are simple tubes, what is the blood pressure in the veins of her feet?

PREPARE When the blood returns to the right atrium, the gauge pressure is approximately zero. If the veins from the feet to the heart are simple tubes, then the pressure below the heart is a hydrostatic pressure.

SOLVE The increase in gauge pressure at depth d is

$$p = \rho g d = (1050\ \text{kg/m}^3)(9.8\ \text{m/s}^2)(1.3\ \text{m})$$

$$= 13.4\ \text{kPa} = 100\ \text{mm Hg}$$

ASSESS The hydrostatic pressure would be far above what is observed, so the veins can't be simple tubes; there must be more to the story.

The veins of the legs, and other veins in the body, have one-way valves. The contraction of muscles surrounding the veins moves the blood by squeezing the veins, which causes the blood to squirt upward. The valves ensure that this motion is toward the heart, and they prevent the blood from moving back down the veins when the muscles relax. If you sit or stand for a long time, this system can't do its job, and blood can collect in your legs. (Your veins are quite expandable; they change size to accommodate changes in blood volume and blood flow.) At some point, you'll want to get up and move around, using your leg muscles to push the blood back toward the heart.

CHAPTER 9 INTEGRATED EXAMPLE An intravenous transfusion

A patient in the hospital often receives fluids via an intravenous (IV) infusion. A bag of the fluid is held at a fixed height above the patient's body. The fluid then travels down a large-diameter, flexible tube to a catheter—a short tube with a small diameter—inserted into the patient's vein.

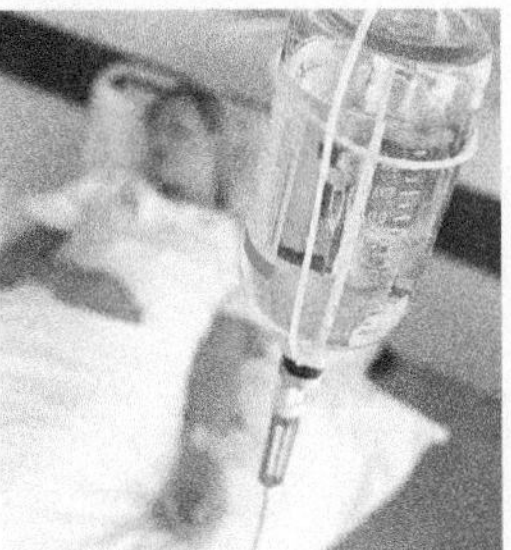

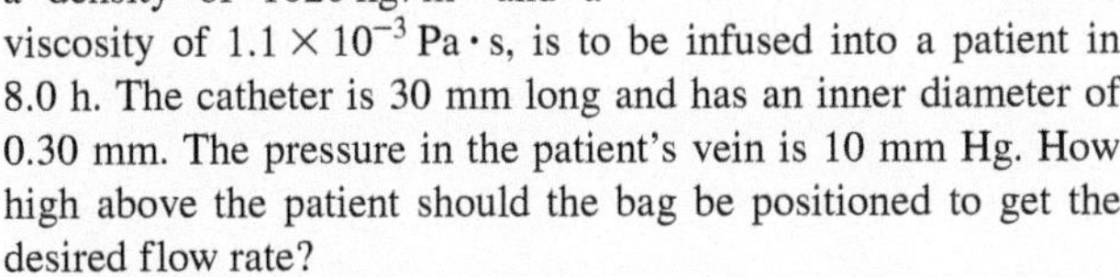

1.0 L of saline solution, with a density of $1020\ \text{kg/m}^3$ and a viscosity of $1.1 \times 10^{-3}\ \text{Pa} \cdot \text{s}$, is to be infused into a patient in 8.0 h. The catheter is 30 mm long and has an inner diameter of 0.30 mm. The pressure in the patient's vein is 10 mm Hg. How high above the patient should the bag be positioned to get the desired flow rate?

PREPARE We're concerned with the flow of a viscous fluid. According to Poiseuille's equation, the flow rate depends inversely on the fourth power of a tube's radius. The tube from the elevated bag to the catheter has a large diameter, while the diameter of the catheter is small. Thus, we expect the flow rate to be determined entirely by the flow through the narrow catheter. We will determine the pressure difference across the catheter that is necessary to provide the needed flow. Then we will use this to determine the height at which the bag must be placed. FIGURE 9.46 shows a sketch of the situation, defines variables, and lists the known information.

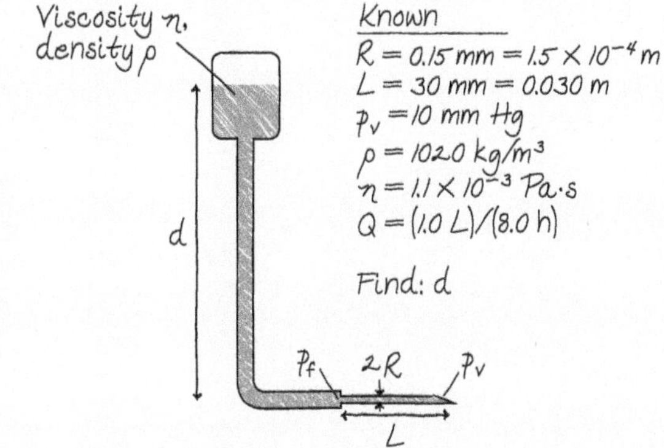

FIGURE 9.46 Pictorial representation of an IV transfusion.

SOLVE The desired volume flow rate is

$$Q = \frac{\Delta V}{\Delta t} = \frac{1.0\ \text{L}}{8.0\ \text{h}} = 0.125\ \text{L/h}$$

Converting to SI units using $1.0\ \text{L} = 1.0 \times 10^{-3}\ \text{m}^3$, we have

$$Q = 0.125\ \frac{\text{L}}{\text{h}} \times \frac{1.0 \times 10^{-3}\ \text{m}^3}{\text{L}} \times \frac{1\ \text{h}}{3600\ \text{s}} = 3.47 \times 10^{-8}\ \text{m}^3/\text{s}$$

Continued

Poiseuille's equation for viscous fluid flow is

$$Q = \frac{\pi R^4 \Delta p}{8\eta L}$$

Thus the pressure difference needed between the ends of the catheter tube to produce the desired flow rate Q is

$$\Delta p = \frac{8\eta L Q}{\pi R^4}$$
$$= \frac{8(1.1 \times 10^{-3}\ \text{Pa}\cdot\text{s})(0.030\ \text{m})(3.47 \times 10^{-8}\ \text{m}^3/\text{s})}{\pi(1.5 \times 10^{-4}\ \text{m})^4}$$
$$= 5760\ \text{Pa}$$

We know the pressure on the end of the catheter in the patient's vein is $p_v = 10$ mm Hg or, converting to SI units using Table 9.2,

$$p_v = 10\ \text{mm Hg} \times \frac{101 \times 10^3\ \text{Pa}}{760\ \text{mm Hg}} = 1330\ \text{Pa}$$

This is a gauge pressure, the pressure in excess of 1 atm. Thus the pressure on the fluid side of the catheter, which we can call p_f, must be

$$p_f = p_v + \Delta p = 1330\ \text{Pa} + 5760\ \text{Pa} = 7090\ \text{Pa}$$

This pressure, like the vein pressure, is a gauge pressure, so the absolute pressure in the fluid at the entrance to the catheter is $p = 1\ \text{atm} + 7090\ \text{Pa}$. This is a hydrostatic pressure, and thus

$$p = p_0 + \rho g d = 1\ \text{atm} + \rho g d$$

We see that $\rho g d = 7090$ Pa. Solving for d, we find the required elevation of the bag above the patient's arm:

$$d = \frac{p_f}{\rho g} = \frac{7090\ \text{Pa}}{(1020\ \text{kg/m}^3)(9.8\ \text{m/s}^2)} = 0.71\ \text{m}$$

ASSESS This height of almost a meter seems reasonable for the height of an IV bag. In practice, the bag can be raised or lowered to adjust the fluid flow rate.

SUMMARY

GOAL To understand the static and dynamic properties of fluids.

GENERAL PRINCIPLES

Fluid Statics

Gases

- Freely moving particles
- Compressible
- Pressure is due to collisions.
- Pressure is constant in a laboratory-size container.

Liquids

- Loosely bound particles
- Incompressible
- Pressure is due to gravity.
- The **hydrostatic pressure** at depth d is

$$p = p_0 + \rho g d$$

- The pressure is the same at all points on a horizontal line through a liquid in hydrostatic equilibrium.

Fluid Dynamics

We assume that fluids are incompressible and the flow is laminar along streamlines. In addition, an **ideal fluid** is nonviscous.

The **volume flow rate** is

$$Q = \frac{\Delta V}{\Delta t} = v_1 A_1 = v_2 A_2$$

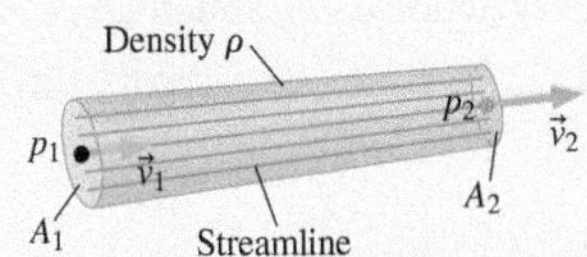

Bernoulli's equation is an application of Newton's second law to ideal fluids:

$$p_1 + \tfrac{1}{2}\rho v_1^2 = p_2 + \tfrac{1}{2}\rho v_2^2$$

Poiseuille's equation governs the flow of viscous fluids through a tube:

$$Q = v_{avg} A = \frac{\pi R^4 \Delta p}{8\eta \; L}$$

where $\Delta p/L$ is the **pressure gradient.**

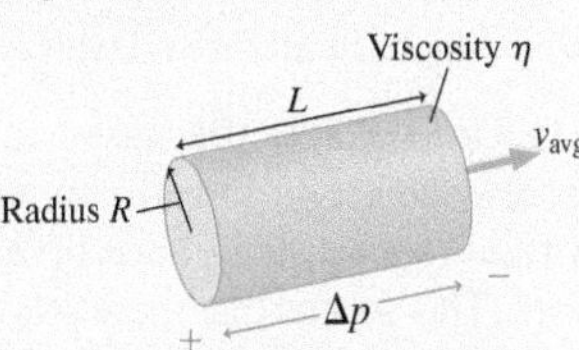

IMPORTANT CONCEPTS

Density $\rho = m/V$, where m is mass and V is volume.

Pressure $p = F/A$, where F is the force magnitude and A is the area on which the force acts.

- Pressure exists at all points in a fluid.
- Pressure pushes equally in all directions.
- Gauge pressure $p_g = p - 1$ atm.

Viscosity η is the property of a fluid that makes it resist flowing.

Streamlines are the paths of individual fluid particles.

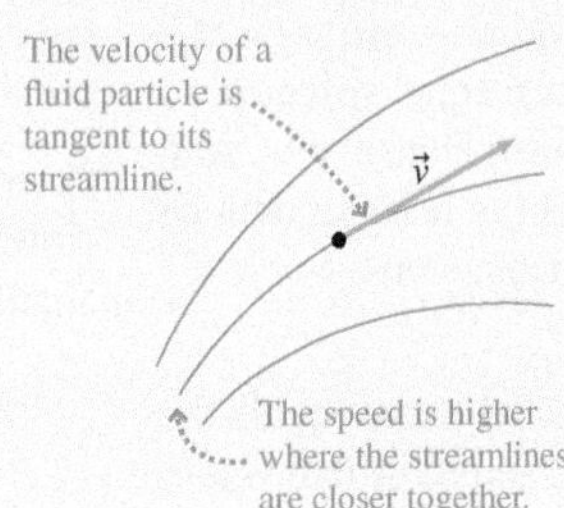

Surface tension γ is a property of a liquid. At the surface, the tension force along a line of contact of length L is

$$T = \gamma L$$

APPLICATIONS

Buoyancy is the upward force of a fluid on an object immersed in the fluid.

Archimedes' principle: The magnitude of the buoyant force equals the weight of the fluid displaced by the object.

Sink: $\rho_{avg} > \rho_f$ $\quad F_B < w_o$

Float: $\rho_{avg} < \rho_f$ $\quad F_B > w_o$

Neutrally buoyant: $\rho_{avg} = \rho_f$ $\quad F_B = w_o$

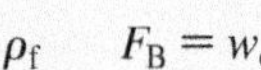

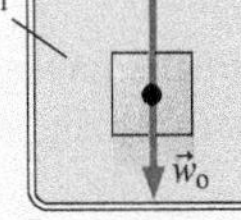

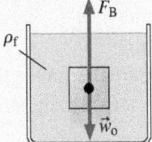

Bubbles and drops

Surface tension causes a pressure difference between the inside and outside of a bubble or drop. For a bubble or drop of radius R, the **law of Laplace** is

$$\Delta p = \begin{cases} 2\gamma/R & \text{drop} \\ 4\gamma/R & \text{bubble} \end{cases}$$

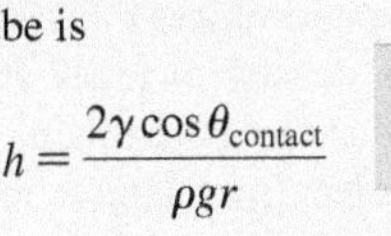

Capillary action

Wetting and capillary action occur when the adhesive force between a liquid and a surface differs from the cohesive force of the liquid. The height of rise in a capillary tube is

$$h = \frac{2\gamma \cos\theta_{contact}}{\rho g r}$$

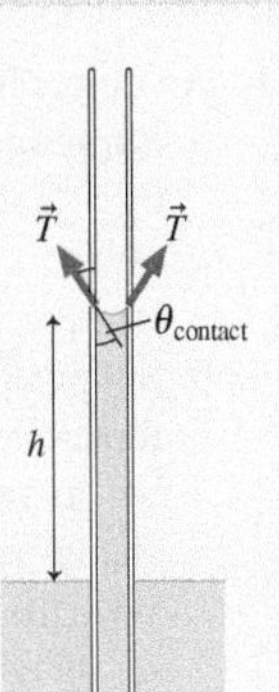

LEARNING OBJECTIVES After studying this chapter, you should be able to:

- Interpret the mass density of an object. *Conceptual Question 9.1; Problems 9.1–9.4*
- Calculate and interpret the hydrostatic pressure in a liquid. *Conceptual Questions 9.2, 9.4, 9.6; Problems 9.7, 9.8, 9.10, 9.12, 9.13*
- Solve buoyancy problems using Archimedes' principle. *Conceptual Questions 9.8, 9.9; Problems 9.15–9.17, 9.21, 9.22*
- Calculate the force of surface tension and apply it to bubbles and capillary action. *Conceptual Question 9.13; Problems 9.26–9.31*
- Relate the fluid speed at two different points in a flow. *Conceptual Question 9.14; Problems 9.32–9.35*
- Solve problems about the motion of an ideal fluid. *Conceptual Questions 9.14, 9.15; Problems 9.36–9.39*
- Determine the motion of viscous fluids. *Conceptual Question 9.16; Problems 9.41–9.45*
- Understand the motion of blood in the circulatory system. *Problems 9.47–9.51*

STOP TO THINK ANSWERS

Stop to Think 9.1: $\rho_A = \rho_B = \rho_C$. Density depends only on what the object is made of, not how big the pieces are.

Stop to Think 9.2: C. These are all open tubes, so the liquid rises to the same height in all three despite their different shapes.

Stop to Think 9.3: C. Because points 1 and 2 are at the same height and are connected by the same fluid (water), they must be at the same pressure. This means that these two points can support the same weight of fluid above them. The oil is less dense than water, so this means a taller column of oil can be supported on the left side than of water on the right side.

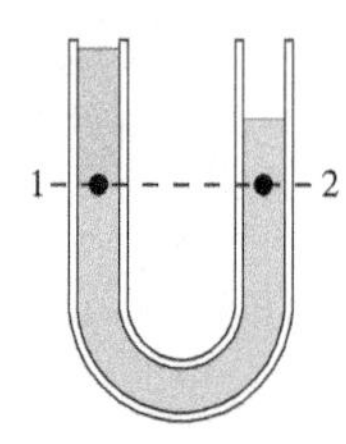

Stop to Think 9.4: $(F_B)_D > (F_B)_A = (F_B)_B = (F_B)_C > (F_B)_E$. The buoyant force on an object is equal to the weight of the fluid it displaces. Block D has the greatest volume and thus displaces the greatest amount of fluid. Blocks A, B, and C have equal volumes, so they displace equal amounts of fluids, but less than D displaces. And block E has the smallest volume and displaces the least amount of fluid. The buoyant force on an object does not depend on the object's mass, or its position in the fluid.

Stop to Think 9.5: B. The weight of the displaced water equals the weight of the ice cube. When the ice cube melts and turns into water, that amount of water will exactly fill the volume that the ice cube is now displacing.

Stop to Think 9.6: B. The pressure in a bubble is inversely proportional to its radius, so the smaller bubble B has the higher pressure. The air inside will flow from the higher pressure of B to the lower pressure of A. The pressure inside B will increase as it shrinks, driving the process even faster until B disappears.

Stop to Think 9.7: 1 cm^3/s out. The fluid is incompressible, so the sum of what flows in must match the sum of what flows out. 13 cm^3/s is known to be flowing in while 12 cm^3/s flows out. An additional 1 cm^3/s must flow out to achieve balance.

Stop to Think 9.8: C. There's no vertical motion, so the pressure at points 1, 2, and 3 can be found as the hydrostatic pressure that depends on the height of the liquid above these points. We can see from the heights that $p_2 < p_1 < p_3$. According to Bernoulli's equation, the pressure is lowest where the flow speed is highest.

Stop to Think 9.9: A. All three segments have the same volume flow rate Q. According to Poiseuille's equation, the segment with the smallest radius R has the greatest pressure gradient $\Delta p/L$.

QUESTIONS

Conceptual Questions

1. An object has density ρ.
   a. Suppose each of the object's three dimensions is increased by a factor of 2 without changing the material of which the object is made. Will the density change? If so, by what factor? Explain.
   b. Suppose each of the object's three dimensions is increased by a factor of 2 without changing the object's mass. Will the density change? If so, by what factor? Explain.
2. Rank in order, from largest to smallest, the pressures at 1, 2, and 3 in Figure Q9.2. Explain.

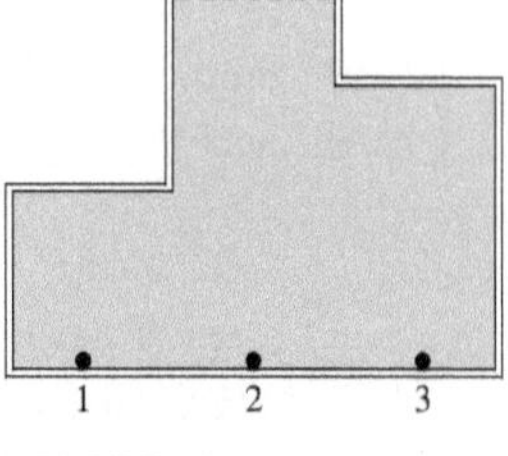

FIGURE Q9.2

3. Figure Q9.3 shows two rectangular tanks, A and B, full of water. They have equal depths and equal thicknesses (the dimension into the page) but different widths.
   a. Compare the forces the water exerts on the bottoms of the tanks. Is F_A larger than, smaller than, or equal to F_B? Explain.
   b. Compare the forces the water exerts on the sides of the tanks. Is F_A larger than, smaller than, or equal to F_B? Explain.

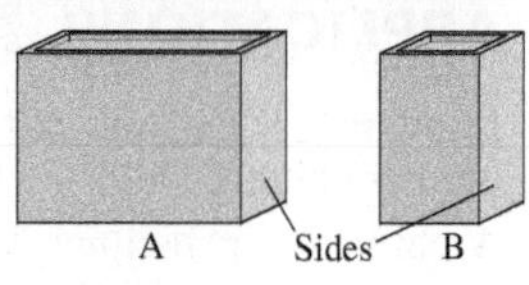

FIGURE Q9.3

Problem difficulty is labeled as | (straightforward) to ||||| (challenging). Problems labeled INT integrate significant material from earlier chapters; BIO are of biological or medical interest; CALC require calculus to solve.

4. In Figure Q9.4, is p_1 larger than, smaller than, or equal to p_2? Explain.

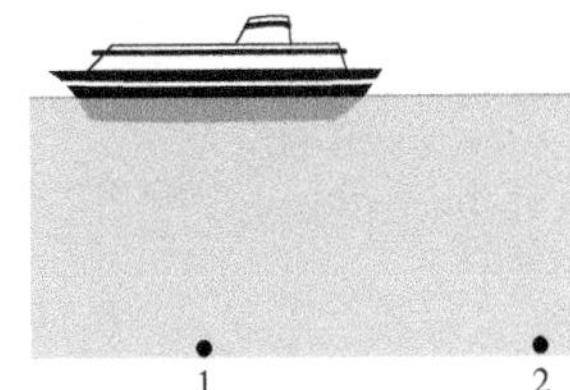

FIGURE Q9.4

5. A gas cylinder has two pressure gauges, one that shows absolute pressure and one that shows gauge pressure. Near sea level, the difference between the two readings is 100 kPa. If the cylinder is carried to the top of a mountain, will the difference be smaller than, larger than, or still 100 kPa? Explain.
6. Water expands when heated. Suppose a beaker of water is heated from 10°C to 90°C. Does the pressure at the bottom of the beaker increase, decrease, or stay the same? Explain.
7. Rank in order, from largest to smallest, the densities of the floating blocks A, B, and C in Figure Q9.7. Explain.

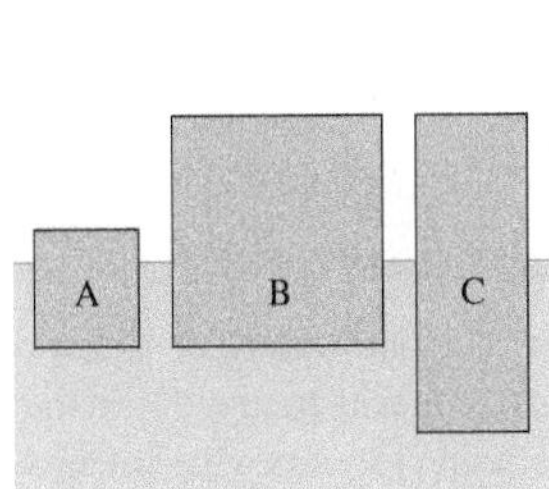

FIGURE Q9.7

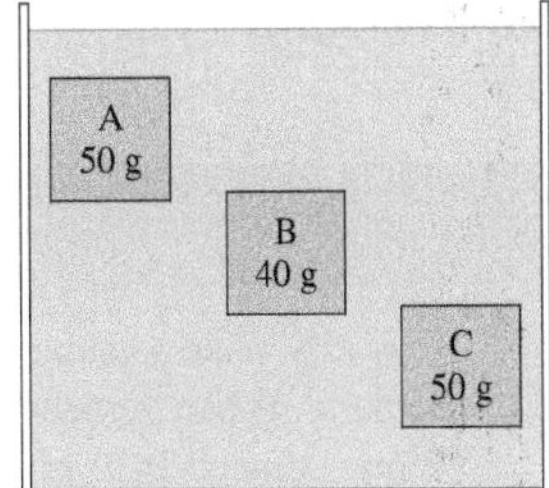

FIGURE Q9.8

8. Blocks A, B, and C in Figure Q9.8 have the same volume. Rank in order, from largest to smallest, the sizes of the buoyant forces F_A, F_B, and F_C on A, B, and C. Explain.
9. Blocks A, B, and C in Figure Q9.8 have the same density. Rank in order, from largest to smallest, the sizes of the buoyant forces F_A, F_B, and F_C on A, B, and C. Explain.
10. The two identical beakers in Figure Q9.10 are filled to the same height with water. Beaker B has a plastic sphere floating in it. Which beaker, with all its contents, weighs more? Or are they equal? Explain.

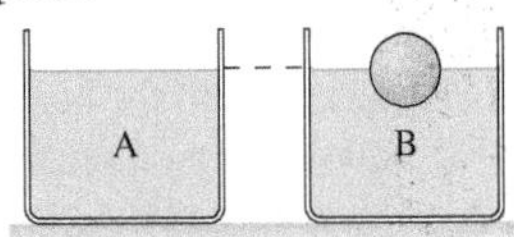

FIGURE Q9.10

11. BIO Freshwater fish tend to have larger swim bladders than saltwater fish. Explain why you would expect this to be true.
12. BIO A higher level of hemoglobin in the blood increases the blood's density. This is the basis for a simple test that can be used to see if a prospective blood donor has a high enough hemoglobin level to donate safely. A drop of blood is placed in a copper sulfate solution, and the time for the drop to sink to the bottom is measured. If this time is too long, the hemoglobin level is too low. Explain how this test works.
13. Figure Q9.13 shows two vertical loops in which the tension of a liquid supports a 2 g mass. Which liquid has the larger surface tension? Or are they equal?

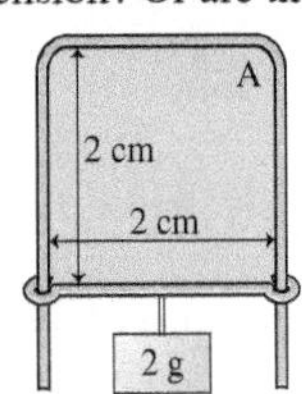

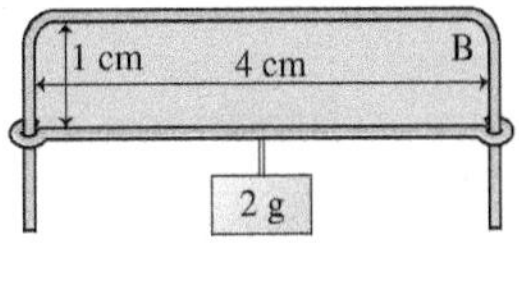

FIGURE Q9.13

14. A liquid with negligible viscosity flows through the pipe shown in Figure Q9.14. This is an over-head view.
    a. Rank in order, from largest to smallest, the flow speeds v_1 to v_4 at points 1 to 4. Explain.
    b. Rank in order, from largest to smallest, the pressures p_1 to p_4 at points 1 to 4. Explain.

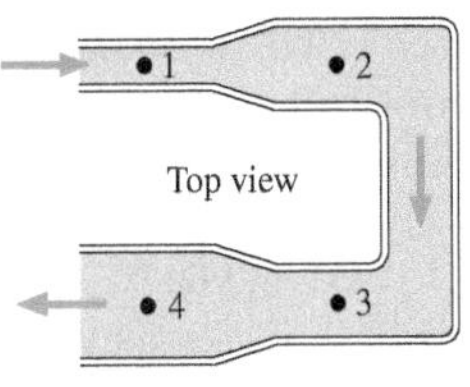

FIGURE Q9.14

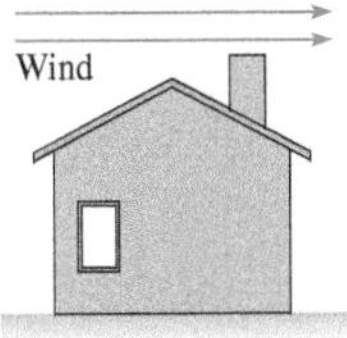

FIGURE Q9.15

15. Wind blows over the house in Figure Q9.15. A window on the ground floor is open. Is there an airflow through the house? If so, does the air flow in the window and out the chimney, or in the chimney and out the window? Explain.
16. Oil flows through an $\frac{1}{8}$-inch-diameter tube at the rate of 4 quarts/min. What will be the flow rate through a $\frac{1}{4}$-inch-diameter tube of equal length if the pressure difference is unchanged?

Multiple-Choice Questions

17. | Figure Q9.17 shows a 100 g block of copper (density $8900\ \text{kg/m}^3$) and a 100 g block of aluminum (density $2700\ \text{kg/m}^3$) connected by a massless string that runs over two massless, frictionless pulleys. The two blocks exactly balance, since they have the same mass. Now suppose that the whole system is submerged in water. What will happen?
    A. The copper block will fall, the aluminum block will rise.
    B. The aluminum block will fall, the copper block will rise.
    C. Nothing will change.
    D. Both blocks will rise.

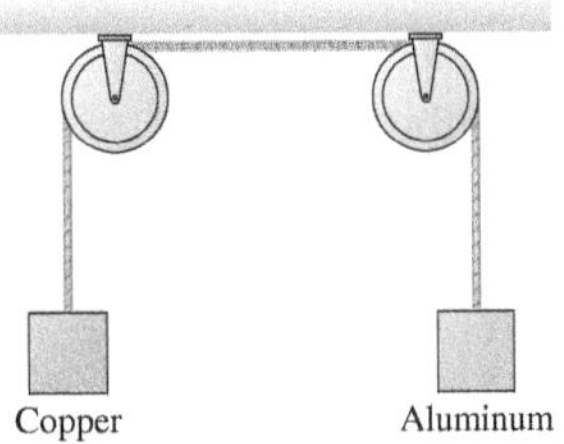

FIGURE Q9.17

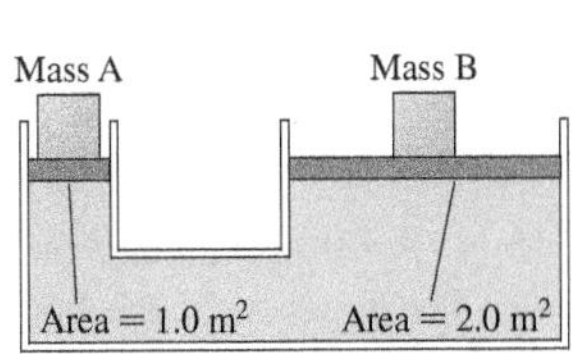

FIGURE Q9.18

18. | Masses A and B rest on very light pistons that enclose a fluid, as shown in Figure Q9.18. There is no friction between the pistons and the cylinders they fit inside. Which of the following is true?
    A. Mass A is greater.
    B. Mass B is greater.
    C. Mass A and mass B are the same.
19. BIO | If you dive underwater, you notice an uncomfortable pressure on your eardrums due to the increased pressure. The human eardrum has an area of about 70 mm^2 ($7 \times 10^{-5}\ \text{m}^2$), and it can sustain a force of about 7 N without rupturing. If your body had no means of balancing the extra pressure (which, in reality, it does), what would be the maximum depth you could dive without rupturing your eardrum?
    A. 0.3 m B. 1 m C. 3 m D. 10 m

20. | A large beaker of water is filled to its rim with water. A block of wood is then carefully lowered into the beaker until the block is floating. In this process, some water is pushed over the edge and collects in a tray. The weight of the water in the tray is
A. Greater than the weight of the block.
B. Less than the weight of the block.
C. Equal to the weight of the block.

21. | The density of a typical hippo's body is 1030 kg/m^3, so it
BIO will sink in fresh water. What is the buoyant force on a submerged 1500 kg hippo?
A. 14,000 N B. 14,300 N
C. 14,600 N D. 14,900 N

22. | An object floats in water, with 75% of its volume submerged. What is its approximate density?
A. 250 kg/m^3 B. 750 kg/m^3
C. 1000 kg/m^3 D. 1250 kg/m^3

23. || The gauge pressure inside a $100\text{-}\mu\text{m}$-diameter air bubble in benzene is 8.5 mm Hg. What is the surface tension of benzene?
A. 56 mN/m B. 28 mN/m
C. 14 mN/m D. 7 mN/m

24. | A syringe is being used to squirt water as shown in Figure Q9.24. The water is ejected from the nozzle at 10 m/s. At what speed is the plunger of the syringe being depressed?

FIGURE Q9.24

A. 0.01 m/s B. 0.1 m/s
C. 1 m/s D. 10 m/s

25. | Water flows through a 4.0-cm-diameter horizontal pipe at a speed of 1.3 m/s. The pipe then narrows down to a diameter of 2.0 cm. Ignoring viscosity, what is the pressure difference between the wide and narrow sections of the pipe?
A. 850 Pa B. 3400 Pa C. 9300 Pa
D. 12,700 Pa E. 13,500 Pa

26. || A 15-m-long garden hose has an inner diameter of 2.5 cm. One end is connected to a spigot; 20°C water flows from the other end at a rate of 1.2 L/s. What is the gauge pressure at the spigot end of the hose?
A. 1900 Pa B. 2700 Pa C. 4200 Pa
D. 5800 Pa E. 7300 Pa

PROBLEMS

Section 9.1 Properties of Fluids

1. || What is the volume in mL of 55 g of a liquid with density 1100 kg/m^3?
2. | A 75 kg college student's body has a density of 1050 kg/m^2. What is the student's volume?
3. | Fat cells in humans are composed almost entirely of pure
BIO triglycerides with an average density of about 900 kg/m^3. If 20% of the mass of a 70 kg student's body is fat (a typical value), what is the total volume of the fat in the student's body?
4. || a. 50 g of gasoline are mixed with 50 g of water. What is the average density of the mixture?
b. 50 cm^3 of gasoline are mixed with 50 cm^3 of water. What is the average density of the mixture?

Section 9.2 Pressure

5. || A 1.0-m-diameter vat of liquid, open to the air, is 2.0 m deep. The pressure at the bottom of the vat is 1.3 atm. What is the mass of the liquid in the vat?
6. || A 35-cm-tall, 5.0-cm-diameter cylindrical beaker is filled to its brim with water. What is the total downward force on the bottom of the beaker?
7. | The deepest point in the ocean is 11 km below sea level, deeper than Mt. Everest is tall. What is the pressure in atmospheres at this depth?
8. || a. What volume of water has the same mass as 8.0 m^3 of ethyl alcohol?
b. If this volume of water is in a cubic tank, what is the pressure at the bottom?
9. || A 50-cm-thick layer of oil floats on a 120-cm-thick layer of water. What is the pressure at the bottom of the water layer?
10. || The gauge pressure at the bottom of a cylinder of liquid is $p_g = 0.40$ atm. The liquid is poured into another cylinder with twice the radius of the first cylinder. What is the gauge pressure at the bottom of the second cylinder?
11. || A city uses a water tower to store water for times of high demand. When demand is light, water is pumped into the tower. When demand is heavy, water can flow from the tower without overwhelming the pumps. To provide water at a typical 350 kPa gauge pressure, how tall must the tower be?
12. ||| An ocean-going research submarine has a 20-cm-diameter window 8.0 cm thick. The manufacturer says the window can withstand forces up to 1.0×10^6 N. What is the submarine's maximum safe depth? The pressure inside the submarine is maintained at 1.0 atm.
13. ||| Glycerin is poured into an open U-shaped tube until the height in both sides is 20 cm. Ethyl alcohol is then poured into one arm until the height of the alcohol column is 20 cm. The two liquids do not mix. What is the difference in height between the top surface of the glycerin and the top surface of the alcohol?
14. || What is the minimum hose diameter of an ideal vacuum cleaner that could lift a 10 kg (22 lb) dog off the floor?

Section 9.3 Buoyancy

15. | A 6.00-cm-diameter sphere with a mass of 89.3 g is neutrally buoyant in a liquid. Identify the liquid.
16. || A 2.0 cm × 2.0 cm × 6.0 cm block floats in water with its long axis vertical. The length of the block above water is 2.0 cm. What is the block's mass density?
17. || Hippos spend much of
BIO their lives in water, but amazingly, they don't swim. They have, like manatees, very little body fat. The density of a hippo's body is approximately 1030 kg/m^3, so it sinks to the bottom of the freshwater lakes and rivers it frequents—and then it simply walks on the bottom. A 1500 kg hippo is completely submerged, standing on the bottom of a lake. What is the approximate value of the upward normal force on the hippo?

18. || Astronauts visiting a new planet find a lake filled with an unknown liquid. They have with them a plastic cube, 6.0 cm on each side, with a density of 840 kg/m^3. First they weigh the cube with a spring scale, measuring a weight of 21 N. Then they float the cube in the lake and find that two-thirds of the cube is submerged. At what depth in the lake will the pressure be twice the atmospheric pressure of 85 kPa?

19. || A sphere completely submerged in water is tethered to the bottom with a string. The tension in the string is one-third the weight of the sphere. What is the density of the sphere?

20. || What is the tension of the string in Figure P9.20?

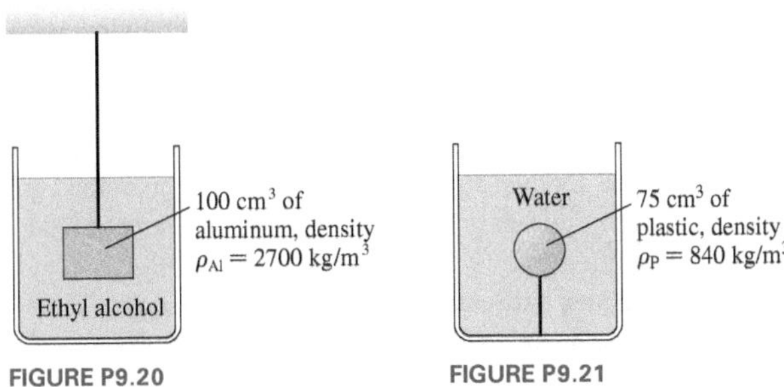

FIGURE P9.20

FIGURE P9.21

21. || What is the tension in the string in Figure P9.21?

22. || To determine the body fat of athletes, they are weighed first
BIO in air and then again while completely underwater, as discussed on page 275. One athlete weighed 690 N when weighed in air and 42 N when weighed underwater. What is that athlete's average density?

23. || You need to determine the density of a ceramic statue. If you suspend it from a spring scale, the scale reads 28.4 N. If you then lower the statue into a tub of water, so that it is completely submerged, the scale reads 17.0 N. What is the statue's density?

24. || Styrofoam has a density of 150 kg/m^3. What is the maximum mass that can hang without sinking from a 50-cm-diameter Styrofoam sphere in water? Assume the volume of the mass is negligible compared to that of the sphere.

25. || A 1.0-mm-diameter glass bead is dropped into a cylinder of oil. It falls 25 cm to the bottom in 60 s at a steady speed. The glass and oil have densities of 2500 kg/m^3 and 900 kg/m^3, respectively. What is the oil's viscosity?

Section 9.4 Surface Tension and Capillary Action

26. || Figure P9.26 shows a thin liquid film bounded on the right
INT side by a sliding wire that is attached to a spring with spring constant 0.50 N/m. The spring is stretched by 1.5 cm. What is the liquid's surface tension in mN/m?

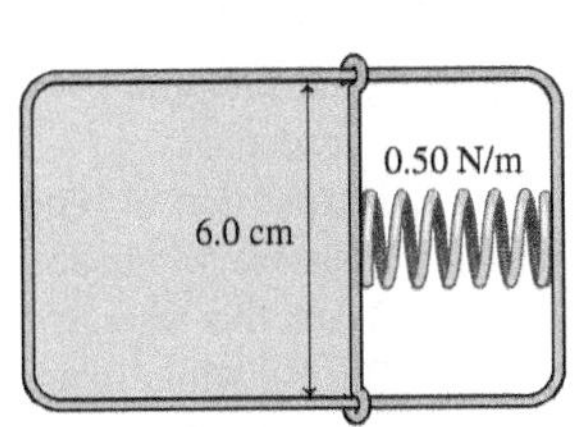

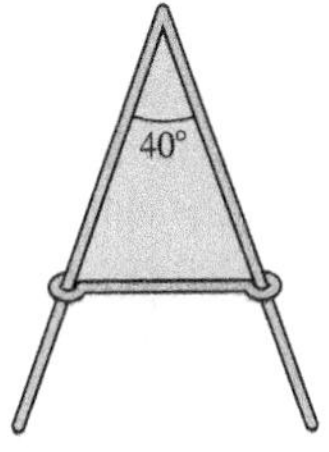

FIGURE P9.26

FIGURE P9.27

27. || The two wires in Figure P9.27 form an inverted V shape
INT with a 40° upper angle. A third wire, with a mass of 0.19 g, is arranged so that it can slide up and down the V while remaining horizontal. A film of soapy water with surface tension 35 mN/m fills the triangle defined by the wires. At equilibrium, how far is the horizontal wire below the top of the triangle?

28. || What is the pressure inside a 10-μm-diameter air bubble that is 10 m below the surface of a lake?

29. || A droplet of water sits on a leaf. The contact angle of the water with the leaf is 135°, and the circumference of the circular contact line is 12 mm. What is the magnitude of the net upward force exerted on the leaf by the droplet's surface tension?

30. || The left tube of the water-filled reservoir in Figure P9.30 has an inside diameter of 1.0 mm. The inside diameter of the right tube is 2.0 mm. What is the difference in the heights of the two water columns? Assume that the glass is very smooth and clean.

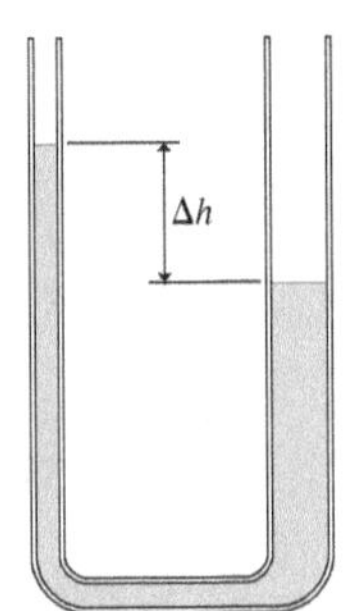

FIGURE P9.30

31. || *Proteinuria,* excess protein in the
BIO urine, is a symptom of many kidney diseases. A marker for proteinuria is a decrease in the surface tension of a urine sample over the course of a few hours, typically from 31 mN/m to 29 mN/m; the surface tension of normal urine does not change with time. The change in surface tension can be measured by monitoring the height of the liquid in a capillary tube. For a patient who has proteinuria, how high in mm will the liquid initially rise in a 0.80-mm-diameter capillary tube? How far will it fall over the course of a few hours? You can assume that the density of urine is the same as that of water.

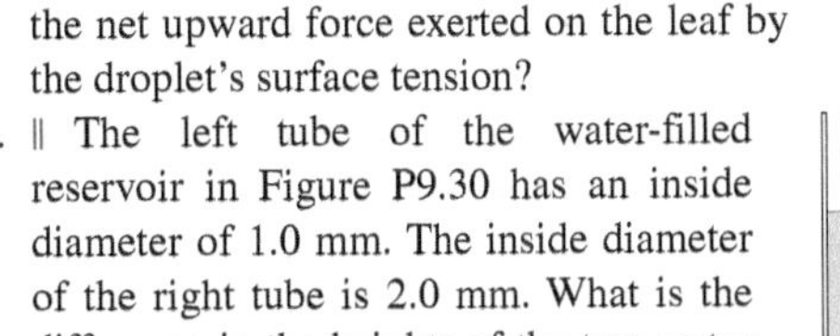

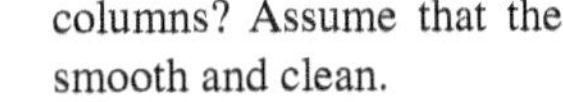

Section 9.5 Fluids in Motion

32. || Your low-flow showerhead is delivering water at 1.1×10^{-4} m^3/s, about 1.8 gallons per minute. If this is the only water being used in your house, how fast is the water moving through your house's water supply line, which has a diameter of 0.021 m (about ¾ of an inch)?

33. || Water flowing through a 2.0-cm-diameter pipe can fill a 300 L bathtub in 5.0 min. What is the speed of the water in the pipe?

34. || A 1.0-cm-diameter pipe widens to 2.0 cm, then narrows to 0.50 cm. Liquid flows through the first segment at a speed of 4.0 m/s.
    a. What are the speeds in the second and third segments?
    b. What is the volume flow rate through the pipe?

35. || Water flows through a 2.5-cm-diameter hose at 3.0 m/s. How long, in minutes, will it take to fill a 600 L child's wading pool?

Section 9.6 Ideal Fluid Dynamics

36. || A square tube 1.0 cm on a side gradually changes shape to become a circular tube 1.0 cm in diameter. Gasoline flows through the tube at 0.75 L/s.
    a. What are the flow speeds in the square tube and the circular tube?
    b. What is the pressure difference between the square tube and the circular tube?

37. ||| Glycerin flows through a tube that expands from a 1.00 cm^2 cross-section area at point 1 to a 4.00 cm^2 cross-section area farther downstream at point 2. The pressure difference between points 1 and 2 is 9.45 kPa. What are the speeds of the glycerin at points 1 and 2? Assume that the glycerin flows as an ideal fluid.

38. || A child's water pistol shoots a stream of water through a 1.0-mm-diameter nozzle at a speed of 5.0 m/s. Squeezing the trigger pressurizes the water reservoir inside the pistol. It is reasonable to assume that the water in the reservoir is at rest. Assume that the water is an ideal fluid.
 a. What is the volume flow rate in mL/s as the trigger is being squeezed?
 b. What is the gauge pressure inside the reservoir?

39. || The 3.0-cm-diameter water line in Figure P9.39 splits into two 1.0-cm-diameter pipes. All pipes are circular and at the same elevation. At point A, the water speed is 2.0 m/s and the gauge pressure is 50 kPa. What is the gauge pressure at point B?

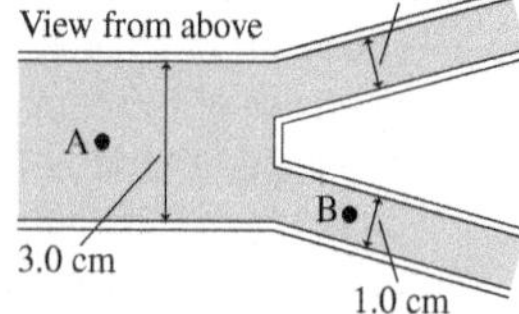

FIGURE P9.39

Section 9.7 Viscous Fluid Dynamics

40. ||| What pressure difference is required between the ends of a 2.0-m-long, 1.0-mm-diameter horizontal tube for 40°C water to flow through it at an average speed of 4.0 m/s?

41. || What is the average speed in mm/s through a 4.0-mm-diameter tube of (a) 20°C motor oil and (b) 100°C motor oil if the pressure gradient is 1500 Pa/m?

42. ||| The piping system in a dairy must deliver 1.5 L/min of 20°C milk through a 12-m-long, 1.0-cm-diameter pipe. What is the pressure difference in kPa between the ends of the pipe?

43. ||| An unknown liquid flows smoothly through a 6.0-mm-diameter horizontal tube where the pressure gradient is 600 Pa/m. Then the tube diameter gradually shrinks to 3.0 mm. What is the pressure gradient in this narrower portion of the tube?

44. || *Balloon angioplasty* is a technique to restore blood flow
BIO through coronary arteries that have been narrowed by the build-up of plaque. The surgeon threads a catheter into the coronary artery, then inflates a balloon that compresses the plaque against the wall of the artery. Suppose a diseased coronary artery has a diameter of 2.0 mm. To what diameter must it be expanded to increase the blood flow by a factor of 5.0 with no change in pressure?

45. ||| Figure P9.45 shows a syringe filled with 20°C olive oil. What is the gauge pressure in atm at point P, where the needle meets the wider body of the syringe? The pressure at the exit of the needle is 1.0 atm.

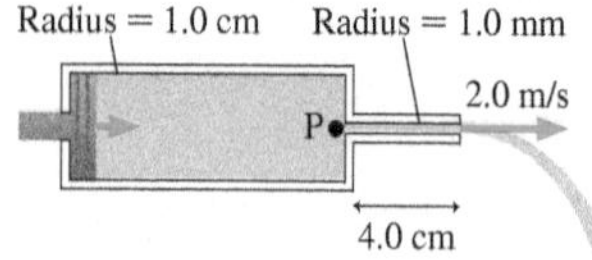

FIGURE P9.45

Section 9.8 The Circulatory System

46. | When you hold your hands at your sides, you may have noticed
BIO that the veins sometimes bulge—the height difference between your heart and your hands produces increased pressure in the veins. The same thing happens in the arteries. Estimate the distance that your hands are below your heart. If the average arterial pressure at your heart is a typical 100 mm Hg, what is the average arterial pressure in your hands when they are held at your side?

47. | The top of your head is about 30 cm above your heart. What
BIO is the blood pressure difference between your heart and the top of your head?

48. | To keep blood from pooling in their lower legs on plane trips,
BIO some people wear compression socks. These socks are sold by the pressure they apply; a typical rating is 20 mm Hg. Over what vertical distance can this pressure move the blood?

49. ||| Using the data in Figure 9.45 for the arterioles, and assuming a typical blood flow rate of 5.0 L per minute,
 a. What is the flow speed in the arterioles?
 b. What is the pressure difference across a 1.0 cm length of arteriole?

50. || Coronary arteries are responsible for supplying oxygenated
BIO blood to heart muscle. Most heart attacks are caused by the narrowing of these arteries due to arteriosclerosis, the deposition of plaque along the arterial walls. A common physiological response to this condition is an increase in blood pressure. A healthy coronary artery is 3.0 mm in diameter and 4.0 cm in length.
 a. Consider a diseased artery in which the artery diameter has been reduced to 2.5 mm. What is the ratio $Q_{\text{diseased}}/Q_{\text{healthy}}$ if the pressure gradient along the artery does not change?
 b. The body attempts to compensate for the reduced open area by increasing the blood pressure. By what factor would the pressure gradient along the diseased artery need to increase to achieve a volume flow rate equal to that in the healthy artery?

51. ||| Blood in a carotid artery carrying blood to the head is moving
BIO at 0.15 m/s when it reaches a section where plaque has narrowed the artery to 80% of its normal diameter. What pressure drop occurs when the blood reaches this narrow section?

General Problems

52. ||| The two 60-cm-diameter cylinders in Figure P9.52, closed at one end, open at the other, are joined to form a single cylinder, then the air inside is removed.
 a. How much force does the atmosphere exert on the flat end of each cylinder?
 b. Suppose one cylinder is bolted to a sturdy ceiling. How many 100 kg football players would need to hang from the lower cylinder to pull the two cylinders apart?

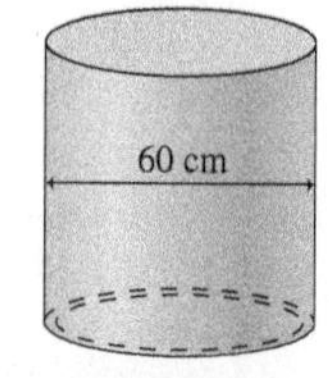

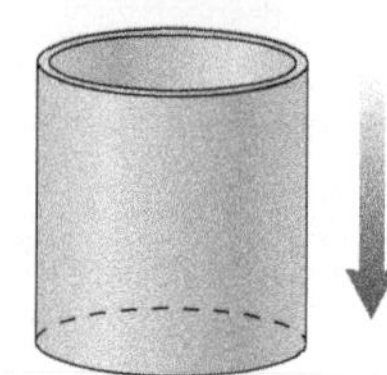

FIGURE P9.52

53. || *Postural hypotension* is the occur-
BIO rence of low systolic blood pressure when a person stands up too quickly from a reclining position. A brain blood pressure lower than 90 mm Hg can cause fainting or lightheadedness. In a healthy adult, the automatic constriction and expansion of blood vessels keep the brain blood pressure constant while posture is changing, but disease or aging can weaken this response. If the blood pressure in your brain is 118 mm Hg while lying down, what would it be when you stand up if this automatic response failed? Assume your brain is 35 cm from your heart and the density of blood is 1060 kg/m^3.

54. || A U-shaped tube, open to the air on both ends, contains mercury. Water is poured into the left arm until the water column is 10.0 cm deep. How far upward from its initial position does the mercury in the right arm rise?

55. || The average density of the body of a fish is 1080 kg/m^3. To BIO keep from sinking, a fish increases its volume by inflating an internal air bladder, known as a swim bladder, with air. By what percent must the fish increase its volume to be neutrally buoyant in fresh water? The density of air at 20°C is 1.19 kg/m^3.

56. || A person's percentage of body fat can be estimated by BIO weighing the person both in air and underwater, from which the average density ρ_{avg} can be calculated. A widely used equation that relates average body density to fat percentage is the *Siri equation*, % body fat $= (495/\rho_{avg}) - 450$, where ρ_{avg} is measured in g/cm^3. One person's weight in air is 947 N while their weight underwater is 44 N. What is their percentage of body fat?

57. ||| An aquarium of length L, width (front to back) W, and depth CALC D is filled to the top with liquid of density ρ.
   a. Find an expression for the force of the liquid on the bottom of the aquarium.
   b. Find an expression for the force of the liquid on the front window of the aquarium.
   c. Evaluate the forces for a 100-cm-long, 35-cm-wide, 40-cm-deep aquarium filled with water.

58. ||| It's possible to use the ideal-gas law to show that the density CALC of the earth's atmosphere decreases exponentially with height. That is, $\rho = \rho_0 \exp(-z/z_0)$, where z is the height above sea level, ρ_0 is the density at sea level (you can use the Table 9.1 value), and z_0 is called the *scale height* of the atmosphere.
   a. Determine the value of z_0.
   Hint: What is the weight of a column of air?
   b. What is the density of the air in Denver, at an elevation of 1600 m? What percent of sea-level density is this?

59. | Alligators float in ponds and lakes with their snouts and eyes BIO protruding above the surface. A 130 kg gator with a density of 950 kg/m^3 floats at the surface of a pond. What fraction of the gator's volume is above the water's surface?

60. || A 30-cm-tall, 4.0-cm-diameter plastic tube has a sealed bottom. 250 g of lead pellets are poured into the bottom of the tube, whose mass is 30 g, then the tube is lowered into a liquid. The tube floats with 5.0 cm extending above the surface. What is the density of the liquid?

61. ||| You learned in Chapter 7 that a person's maximum walking BIO INT speed is determined by the length of their legs and the strength of gravity. Now, consider walking underwater, where the buoyant force is a factor. Suppose a person's body, after they exhale the air in their lungs, has a density of 1050 kg/m^3. What is this individual's maximum walking speed on the bottom of a swimming pool if their legs are 0.90 m long?

62. ||| What is the terminal speed with which a 20-cm-diameter INT helium-filled balloon rises on a 20°C day?

63. ||| Is it possible to "fly" by holding on to enough helium-filled INT balloons? Consider a 20-cm-diameter helium balloon at 20°C.
   a. What is the net upward force on the balloon?
   b. How many such balloons would be needed to lift a 25 kg child?

64. ||| An aquatic organism needs to be neutrally buoyant to stay BIO at a constant depth. Fish accomplish this with an internal *swim bladder* they can fill with air that they take in from the water through their gills. One complication is that the pressure in the swim bladder matches that of the surrounding water, but the water pressure changes with depth. Because the volume of a gas is inversely proportional to pressure (as you may already know if you have studied the ideal-gas law), the volume of air in a fish's swim bladder decreases with depth unless the fish actively adds more air.
   a. Consider a 2.5 kg freshwater fish whose tissues have an average density of 1050 kg/m^3. To what volume in mL must the swim bladder be inflated for the fish to be neutrally buoyant at the surface?
   b. Suppose the fish descends to a depth of 20 m. If the fish does nothing, will its average density increase, decrease, or be unchanged?
   c. What volume of air would have to be added to the swim bladder for the fish to remain neutrally buoyant at a depth of 20 m?

65. ||| A 355 mL soda can is 6.2 cm in diameter and has a mass of 20 g. Such a soda can half full of water is floating upright in water. What length of the can is above the water level?

66. ||| Water is flowing smoothly at 25 m/s through a narrow pipe INT where the gauge pressure is 1.0 atm. The pipe then widens and the flow speed drops to 5.0 m/s. An air bubble is trapped in the water and moves along with it. If the bubble's diameter is 12 mm in the narrow section of the pipe, what is its diameter in the wider section? You'll need to recall that Boyle's law, which you learned in chemistry, says that $p_i V_i = p_f V_f$ for a gas at constant temperature. The surface-tension contribution to the pressure is negligible for bubbles of this size, so there's no difference in pressure between the inside and the outside of the bubble.

67. || Water striders are well known for being able to stand on and BIO walk across the still surface of a pond. Figure P9.67 shows a cross-section view of the strider's leg on the water surface; the leg extends into the page. There is no buoyant force because the leg is not submerged, but there is an upward force due to surface tension. A typical water strider has a mass of 10 mg, and it is supported by four 3.0-mm-long leg segments. Natural surfactants in pond water reduce the surface tension to 62 mN/m, somewhat less than for pure water.
   a. For a water strider at rest on the surface, what is the angle ϕ for the shallow dimple that the legs create?
   b. The maximum surface tension force occurs for a deep dimple with $\phi = 90°$. What is the maximum strider mass in mg that could be supported on these 3.0-mm-long legs? You'll find that a stationary strider has a significant margin of safety, but the water's ability to support larger forces is needed for locomotion when the strider pushes harder against the surface.

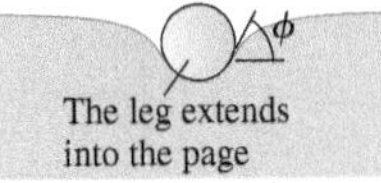

FIGURE P9.67

68. ||| One day when you come into physics lab you find several plastic hemispheres floating like boats in a tank of fresh water. Each lab group is challenged to determine the heaviest rock that can be placed in the bottom of a plastic boat without sinking it. You get one try. You begin by measuring one of the hemispheres, finding a mass of 21 g and a diameter of 8.0 cm. What is the mass of the heaviest rock that, in perfectly still water, won't sink the plastic boat?

69. ||| A clean glass tube has an inner diameter of 3.0 mm. Inside the tube, and concentric with it, is a 2.5-mm-diameter clean glass rod. This tube and rod combination is lowered vertically into a beaker of 20°C water. How high does the water rise into the space between the tube and the rod?

70. || A 2.0 mL syringe has an inner diameter of 6.0 mm, a needle inner diameter of 0.25 mm, and a plunger pad diameter (where you place your finger) of 1.2 cm. A nurse uses the syringe to inject medicine into a patient whose blood pressure is 140 / 100. Assume the liquid is an ideal fluid.
BIO
   a. What is the minimum force the nurse needs to apply to the syringe?
   b. The nurse empties the syringe in 2.0 s. What is the flow speed of the medicine through the needle?

71. ||| Water flowing out of a 16-mm-diameter faucet fills a 2.0 L bottle in 10 s. At what distance below the faucet has the water stream narrowed to 10 mm diameter?
INT

72. || Figure P9.72 shows the instantaneous flow rate in the aorta during one heartbeat for a patient who has normal cardiovascular health.
BIO
   a. The aorta has a diameter of about 2.5 cm. What is the maximum blood speed in the aorta?
   b. Estimate the total flow rate in L / min for a resting heart rate of 72.

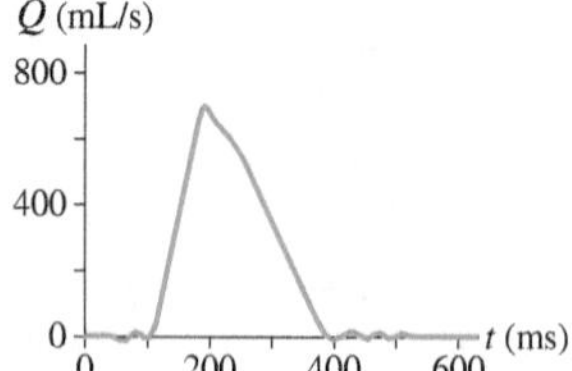

FIGURE P9.72

73. | A hurricane wind blows across a 6.0 m $\times$ 15.0 m flat roof at a speed of 130 km / h.
   a. Is the air pressure above the roof higher or lower than the pressure inside the house? Explain.
   b. What is the pressure difference?
   c. How much force is exerted on the roof? If the roof cannot withstand this much force, will it "blow in" or "blow out"?

74. ||| Air flows through the tube shown in Figure P9.74 at a rate of 1200 cm^3/s. Assume that air is an ideal fluid. What is the height h of mercury in the right side of the U-tube?

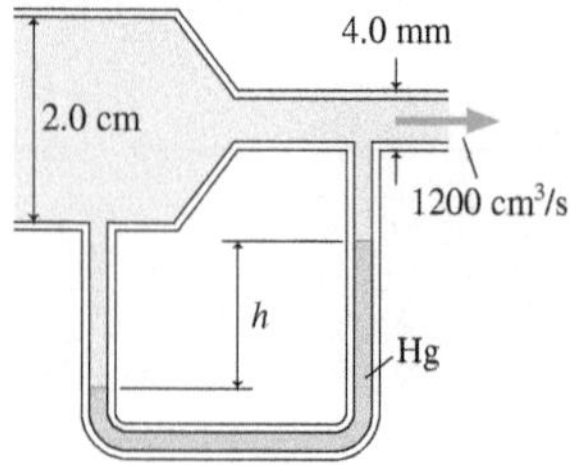

FIGURE P9.74

75. ||| The aorta, the main blood vessel that takes oxygenated blood out of the heart, has a diameter of about 2.5 cm. The aorta branches into several major arteries with a total cross-section area of about 20 cm^2.
BIO
   a. What is the volume flow rate of the blood passing through the aorta if the blood moves with an average speed of 100 cm/s?
   b. All the blood that flows through the aorta then branches into the major arteries. What is the average speed of blood in the major arteries?
   c. The blood flowing in the major arteries then branches into the capillaries. The average speed in the capillaries has been measured to be 0.070 cm/s. What is the total cross-section area of the capillaries?

76. || An electric power plant draws 3.0×10^6 L/min of cooling water from the ocean through two parallel, 3.0-m-diameter pipes. What is the water speed in each pipe?

77. |||| There are two carotid arteries that feed blood to the brain, one on each side of the neck and head. One patient's carotid arteries are each 11.2 cm long and have an inside diameter of 5.2 mm. Near the middle of the left artery, however, is a 2.0-cm-long *stenosis,* a section of the artery with a smaller diameter of 3.4 mm. For the same blood flow rate, what is the ratio of the pressure drop along the patient's left carotid artery to the drop along the right artery?
BIO

78. ||| Smoking tobacco is bad for your circulatory health. In an attempt to maintain the blood's capacity to deliver oxygen, the body increases its red blood cell production, and this increases the viscosity of the blood. In addition, nicotine from tobacco causes arteries to constrict.
BIO

For a nonsmoker, normal blood flow requires a pressure difference of 8.0 mm Hg between the two ends of an artery. Assume that smoking increases viscosity by 10% and reduces the arterial diameter to 90% of its previous value. For a smoker, what pressure difference is needed to maintain the same blood flow?

79. ||| The transpiration rate of a large, 50-m-tall oak tree can reach 20 L/h at midday as sap (viscosity 2.0×10^{-3} Pa · s, density 1040 kg/m^3) rises through the 100-μm-diameter xylem. Microscope measurements of the density of xylem tubes suggest that a large oak should have approximately 450,000 xylem. As a simple model, assume that each xylem tube extends uninterrupted from the base to the top of the tree.
BIO
   a. What is the volume flow rate in L / h and m^3/s through one xylem tube?
   b. What is the sap flow speed in m / h?
   c. What is the pressure difference in atm driving this flow?
   d. The total pressure difference between the base and the top of the tree is the pressure difference you calculated in part c plus the hydrostatic pressure difference. We've seen that physically this pressure difference is a *tension* due to cohesive forces pulling sap up the xylem, but plant physiologists prefer to say that the top of the tree has a negative pressure. If the sap pressure at ground level is 1.0 atm, what is the pressure at the top of the tree?

80. || 20°C water flows through a 2.0-m-long, 6.0-mm-diameter pipe. What is the maximum flow rate in L / min for which the flow is laminar?

81. ||| The 30-cm-long left coronary artery is 4.6 mm in diameter. Blood pressure drops by 3.0 mm Hg over this distance. What are (a) the average blood speed and (b) the volume flow rate in L / min through this artery?
BIO

MCAT-Style Passage Problems

Blood Pressure and Blood Flow

As blood goes from the left ventricle through the arteries and veins of the human body, both its speed and pressure change. The arteries and arterioles can either constrict, reducing the area, or dilate, increasing the area, in response to certain conditions. Both of these changes can affect blood flow and blood pressure. An artery can also develop a permanent narrow area (stenosis) or a permanent wide area (aneurysm). Both of these changes can have significant health consequences.

82. Suppose that in response to some stimulus a small blood vessel narrows to 90% of its original diameter. If there is no change in the pressure across the vessel, what is the ratio of the new volume flow rate to the original flow rate?
   A. 0.66 B. 0.73
   C. 0.81 D. 0.90
83. A patient has developed a narrowing of the coronary arteries that provide blood to the heart. If the rate of blood flow through the arteries is to stay the same,
   A. The length of the arteries must decrease as well.
   B. The speed of the flow must not change.
   C. The pressure difference across the arteries must stay the same, requiring a decrease in blood pressure.
   D. The pressure difference across the arteries must increase, requiring an increase in blood pressure.
84. A patient has developed an aneurysm in the aorta, a short section where the diameter is twice the normal diameter. In the aneurysm, the speed of the blood is _______ than in the section before the aneurysm, and the pressure is _______ than in the section before the aneurysm.
   A. Greater, greater B. Greater, Less
   C. Less, greater D. Less, less

Center of Buoyancy

You are familiar with the concept of center of gravity; it is the point of an object at which we assume the downward weight force acts. We can take a similar approach to the buoyant force; the **center of buoyancy** is the point of an object at which we assume the upward buoyant force acts. Frogs can adjust their position in the water by shifting their center of buoyancy, which they do by using their respiratory muscles to shift their lungs inside their torso.

85. A frog is floating in a pond with its head just breaking the surface and its body at a 20° angle below the horizontal. How does the magnitude of the buoyant force compare to the frog's weight?
   A. The buoyant force is greater than the weight.
   B. The buoyant force is equal to the weight.
   C. The buoyant force is less than than the weight.
86. For the frog's position to be stable, where must the frog's center of buoyancy be relative to its center of gravity?
   A. The center of buoyancy must be directly above the center of gravity.
   B. The center of buoyancy must coincide with the center of gravity.
   C. The center of buoyancy must be directly below the center of gravity.
87. The frog repositions its lungs to move its center of buoyancy toward the front of the body and its center of gravity toward the rear. How does this affect the angle at which the frog floats in the water?
   A. The angle increases.
   B. The angle is unchanged.
   C. The angle decreases.

PART | SUMMARY

Force and Motion

FUNDAMENTAL CONCEPTS

Forces cause objects to *change* their motion—that is, to accelerate.
Objects interact by exerting equal but opposite forces on each other.
The momentum of an isolated system is conserved.

GENERAL PRINCIPLES

Newton's first law	An object with no net force acting on it will remain at rest, or, if moving, will move in a straight line at constant speed.	
Newton's second law	$\vec{F}_{\text{net}} = m\vec{a}$	A net force on an object causes the object to accelerate.
Newton's third law	$\vec{F}_{\text{A on B}} = -\vec{F}_{\text{B on A}}$	For every action, there is an equal but opposite reaction.
Momentum principle	$\Delta p_x = J_x$	An impulse delivered to an object causes the object's momentum to change.
Conservation of momentum	$\vec{P}_{\text{f}} = \vec{P}_{\text{i}}$	The total momentum of an isolated system does not change.

MOTION WITH CONSTANT ACCELERATION

Model the object as a particle. Acceleration is in the direction of the net force and is constant.

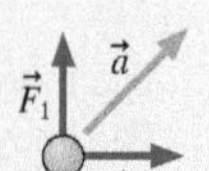

Newton's second law is

$$(F_{\text{net}})_x = \sum F_x = ma_x$$
$$(F_{\text{net}})_y = \sum F_y = ma_y$$

The kinematic equations of constant acceleration are

$$(v_x)_{\text{f}} = (v_x)_{\text{i}} + a_x\Delta t$$
$$x_{\text{f}} = x_{\text{i}} + (v_x)_{\text{i}}\Delta t + \frac{1}{2}a_x(\Delta t)^2$$
$$(v_x)_{\text{f}}^2 = (v_x)_{\text{i}}^2 + 2a_x\Delta x$$

a_x Horizontal line — The acceleration is constant.
v_x Straight line — The slope is a_x.
x Parabola

Uniform motion has $a_x = 0$. The position graph is a straight line with slope v_x.

Free fall is constant-acceleration motion with $a_y = -g$.

The sign of a_x indicates how $\vec{a}$ points, not whether the object is speeding up or slowing down.

- Speeding up: a_x and v_x have the same sign.
- Slowing down: a_x and v_x have opposite signs.

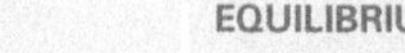

EQUILIBRIUM

An object in static equilibrium must have no net force acting on it and no net torque:

$$\sum F_x = 0 \quad \sum F_y = 0 \quad \sum \tau = 0$$

The center of gravity is the point at which gravity acts. The gravitational torque about a pivot point is $\tau = r_\perp Mg$.

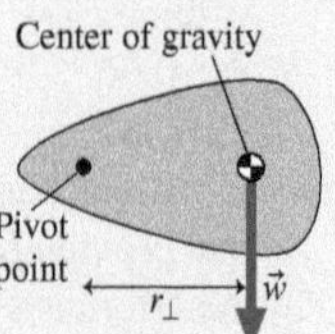

CIRCULAR AND ROTATIONAL MOTION

An object in uniform circular motion requires a net force toward the center to provide centripetal acceleration.

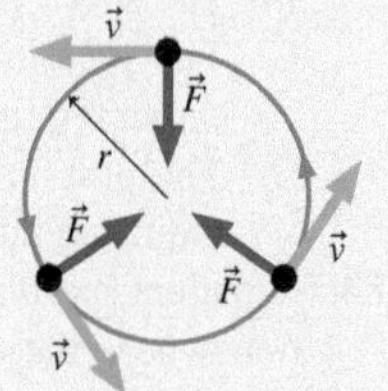

Newton's second law for circular motion is

$$\vec{F}_{\text{net}} = (mv^2/r, \text{ toward center of circle})$$

Speed, period, and frequency are related by $v = 2\pi fr = 2\pi r/T$.

The centripetal acceleration toward the center of the circle has magnitude $a = v^2/r$.

If a torque τ acts on an extended object with moment of inertia I, the object will undergo angular acceleration $\alpha = \tau/I$.

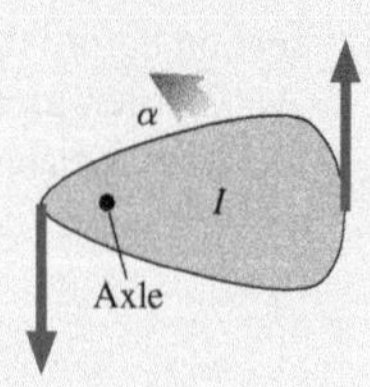

FORCE

Weight $\vec{w} = (mg, \text{down})$

Friction $\vec{f}_{\text{s}} = (0 \text{ to } \mu_{\text{s}}n, \text{ direction to prevent motion})$

$\vec{f}_{\text{k}} = (\mu_{\text{k}}n, \text{ opposite motion})$

Drag $D = \frac{1}{2}C_{\text{D}}\rho Av^2$ at high *Re*

$D = 6\pi\eta rv$ at low *Re*

Spring $(F_{\text{sp}})_x = -k\Delta x$

ELASTICITY

For a small deformation, stress $\sigma = F/A$ is related to strain $\varepsilon = \Delta L/L$ by $\sigma = Y\varepsilon$, where Y is Young's modulus. Tensile strength is the stress at which an object breaks.

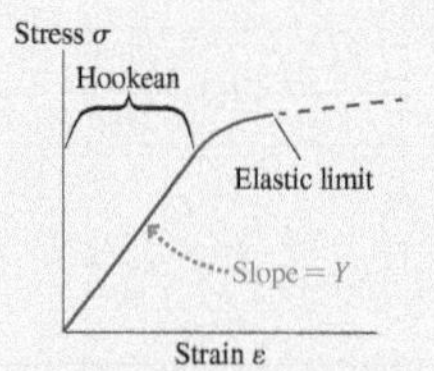

FLUIDS

The hydrostatic pressure at depth d in a liquid is $p = p_0 + \rho gd$, where p_0 is the pressure at the surface.

Archimedes' principle says that the upward buoyant force on an object is equal to the weight of the fluid displaced by the object.

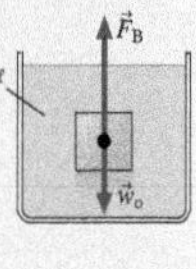

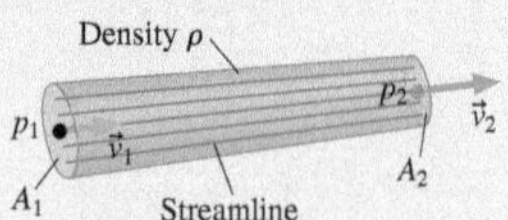

The equation of continuity for fluid flow is $Q = v_1A_1 = v_2A_2$. Bernoulli's equation for an ideal fluid is

$$p_1 + \tfrac{1}{2}\rho v_1{}^2 = p_2 + \tfrac{1}{2}\rho v_2{}^2$$

A viscous fluid obeys Poiseuille's equation

$$Q = v_{\text{avg}}A = \frac{\pi R^4}{8\eta}\frac{\Delta p}{L}$$

where $\Delta p/L$ is the pressure gradient.

TORQUE

Method 1: $\tau = rF_\perp$

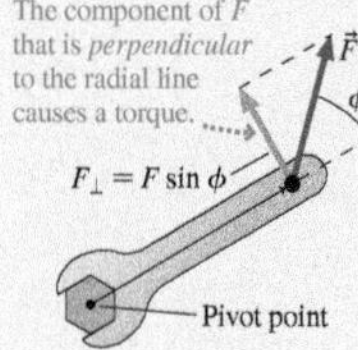

Method 2: $\tau = r_\perp F$

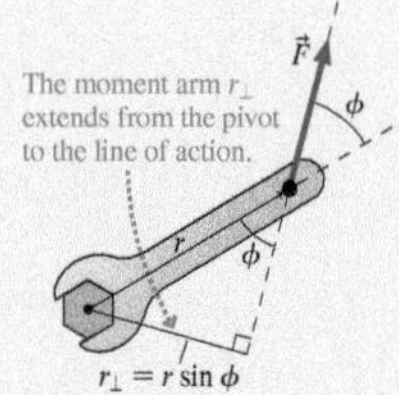

IMPULSE AND MOMENTUM

Momentum $\vec{p} = m\vec{v}$

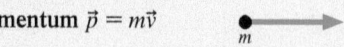

Impulse J_x = area under the force curve

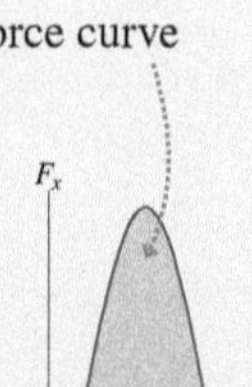

Momentum and impulse are related by the momentum principle $\Delta p_x = J_x$.

Momentum is conserved in an isolated system:

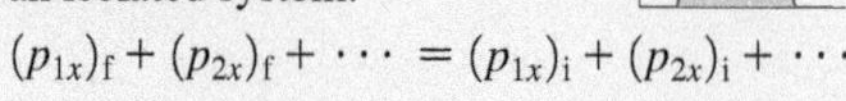

$$(p_{1x})_{\text{f}} + (p_{2x})_{\text{f}} + \cdots = (p_{1x})_{\text{i}} + (p_{2x})_{\text{i}} + \cdots$$

Dark Matter and the Structure of the Universe

The idea that the earth exerts a gravitational force on us is something we now accept without questioning. But when Isaac Newton developed this idea to show that the gravitational force also holds the moon in its orbit, it was a remarkable, ground-breaking insight. It changed the way that we look at the universe we live in.

Newton's laws of motion and gravity are tools that allow us to continue Newton's quest to better understand our place in the cosmos. But it sometimes seems that the more we learn, the more we realize how little we actually know and understand.

Here's an example. Advances in astronomy over the past 100 years have given us great insight into the structure of the universe. But everything our telescopes can see appears to be only a small fraction of what is out there. Approximately 80% of the mass in the universe is *dark matter*—matter that gives off no light or other radiation that we can detect. Everything that we have ever seen through a telescope is merely the tip of the cosmic iceberg.

What is this dark matter? Black holes? Neutrinos? Some form of exotic particle? We simply aren't sure. It could be any of these, or all of them—or something entirely different that no one has yet dreamed of. You might wonder how we know that such matter exists if no one has seen it. Even though we can't directly observe dark matter, we see its effects. And you now know enough physics to understand why.

Whatever dark matter is, it has mass, and so it has gravity. This picture of the Andromeda galaxy shows a typical spiral galaxy structure: a dense collection of stars in the center surrounded by a disk of stars and other matter. This is the shape of our own Milky Way galaxy.

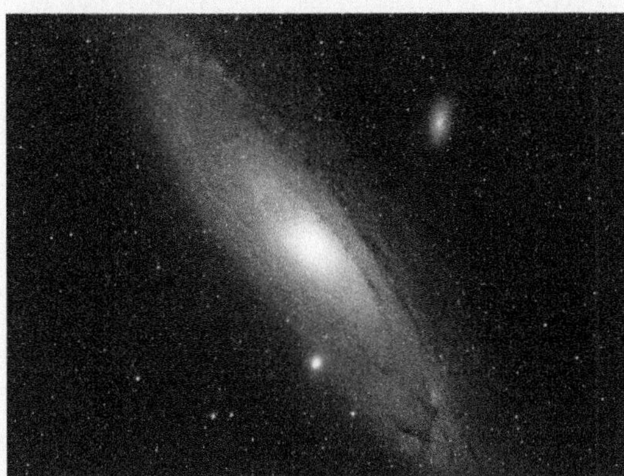

The spiral Andromeda galaxy.

This structure is reminiscent of the structure of the solar system: a dense mass (the sun) in the center surrounded by a disk of other matter (the planets, asteroids, and comets). The sun's gravity keeps the planets in their orbits, but the planets would fall into the sun if they were not in constant motion around it. The same is true of a spiral galaxy; everything in the galaxy orbits its center. Our solar system orbits the center of our galaxy with a period of about 200 million years.

The orbital speed of an object depends on the mass that pulls on it. If you analyze our sun's motion about the center of the Milky Way, or the motion of stars in the Andromeda galaxy about its center, you find that the orbits are much faster than they should be, based on how many stars we see. There must be some other mass present.

There's another problem with the orbital motion of stars around the center of their galaxies. We know that the orbital speeds of planets decrease with distance from the sun; Neptune orbits at a much slower speed than the earth. We might expect something similar for galaxies: Stars farther from the center should orbit at reduced speeds. But they don't. As we measure outward from the center of the galaxy, the orbital speed stays about the same—even as we get to the edge of the visible disk. There must be some other mass—the invisible dark matter—exerting a gravitational force on the stars. This dark matter, which far outweighs the matter we can see, seems to form a halo around the centers of galaxies, providing the gravitational force necessary to produce the observed rotation. Other observations of the motions of galaxies with respect to each other verify this basic idea.

On a cosmic scale, the picture is even stranger. The universe is currently expanding. The mutual gravitational attraction of all matter—regular and dark—in the universe should slow this expansion. But recent observations of the speeds of distant galaxies imply that the expansion of the universe is accelerating, so there must be yet another component to the universe, something that "pushes out." The best explanation at present is that the acceleration is caused by *dark energy*. The nature of dark matter isn't known, but the nature of dark energy is even more mysterious. If current theories hold, it's the most abundant stuff in the universe. And we don't know what it is.

This sort of mystery is what drives scientific investigation. It's what drove Newton to wonder about the connection between the fall of an apple and the motion of the moon, and what drove investigators to develop all of the techniques and theories you will learn about in the coming chapters.

The following questions are related to the passage "Dark Matter and the Structure of the Universe" on the previous page.

1. As noted in the passage, our solar system orbits the center of the Milky Way galaxy in about 200 million years. If there were no dark matter in our galaxy, this period would be
   A. Longer.
   B. The same.
   C. Shorter.
2. Saturn is approximately 10 times as far away from the sun as the earth. This means that its orbital acceleration is ________ that of the earth.
   A. 1/10
   B. 1/100
   C. 1/1000
   D. 1/10,000
3. Saturn is approximately 10 times as far away from the sun as the earth. If dark matter changed the orbital properties of the planets so that Saturn had the same orbital speed as the earth, Saturn's orbital acceleration would be ________ that of the earth.
   A. 1/10
   B. 1/100
   C. 1/1000
   D. 1/10,000
4. Which of the following might you expect to be an additional consequence of the fact that galaxies contain more mass than expected?
   A. The gravitational force between galaxies is greater than expected.
   B. Galaxies appear less bright than expected.
   C. Galaxies are farther away than expected.
   D. There are more galaxies than expected.

The following passages and associated questions are based on the material of Part I.

Animal Athletes BIO

Different animals have very different capacities for running. A horse can maintain a top speed of 20 m/s for a long distance but has a maximum acceleration of only 6.0 m/s^2, half what a good human sprinter can achieve with a block to push against. Greyhounds, dogs especially bred for feats of running, have a top speed of 17 m/s, but their acceleration is much greater than that of the horse. Greyhounds are particularly adept at turning corners at a run (Figure I.1).

FIGURE I.1

5. If a horse starts from rest and accelerates at the maximum value until reaching its top speed, how much time elapses, to the nearest second?
   A. 1 s B. 2 s
   C. 3 s D. 4 s
6. If a horse starts from rest and accelerates at the maximum value until reaching its top speed, how far does it run, to the nearest 10 m?
   A. 40 m B. 30 m
   C. 20 m D. 10 m
7. A greyhound on a racetrack turns a corner at a constant speed of 15 m/s with an acceleration of 7.1 m/s^2. What is the radius of the turn?
   A. 40 m B. 30 m
   C. 20 m D. 10 m
8. A human sprinter of mass 70 kg starts a run at the maximum possible acceleration, pushing backward against a block set in the track. What is the force of the sprinter's foot on the block?
   A. 1500 N B. 840 N
   C. 690 N D. 420 N
9. In the photograph of the greyhounds in Figure I.1, what is the direction of the net force on each dog?
   A. Up B. Down
   C. Left, toward the outside of the turn D. Right, toward the inside of the turn

Sticky Liquids BIO

At small scales, viscous drag becomes very important. To a paramecium (Figure I.2), a single-celled animal that can propel itself through water with fine hairs on its body, swimming through water feels like swimming through honey would to you. We can model a paramecium as a sphere of diameter 50 μm, with a mass of 6.5×10^{-11} kg.

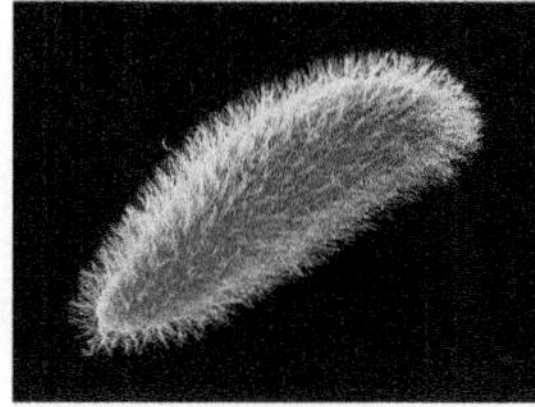

FIGURE I.2

10. A paramecium swimming at a constant speed of 0.25 mm/s ceases propelling itself and slows to a stop. At the instant it stops swimming, what is the magnitude of its acceleration?
    A. $0.2g$ B. $0.5g$
    C. $2g$ D. $5g$
11. If the acceleration of the paramecium in Problem 10 were to stay constant as it came to rest, approximately how far would it travel before stopping?
    A. 0.02 μm B. 0.2 μm
    C. 2 μm D. 20 μm
12. If the paramecium doubles its swimming speed, how does this change the drag force?
    A. The drag force decreases by a factor of 2.
    B. The drag force is unaffected.
    C. The drag force increases by a factor of 2.
    D. The drag force increases by a factor of 4.
13. You can test the viscosity of a liquid by dropping a steel sphere into it and measuring the speed at which it sinks. For viscous fluids, the sphere will rapidly reach a terminal speed. At this terminal speed, the net force on the sphere is
    A. Directed downward.
    B. Zero.
    C. Directed upward.

Pulling Out of a Dive BIO

Falcons are excellent fliers that can reach very high speeds by diving nearly straight down. To pull out of such a dive, a falcon extends its wings and flies through a circular arc that redirects its motion. The forces on the falcon that control its motion are its weight and an upward lift force—like an airplane—due to the air flowing over its wings. At the bottom of the arc, as in Figure I.3, a falcon can easily achieve an acceleration of 15 m/s^2.

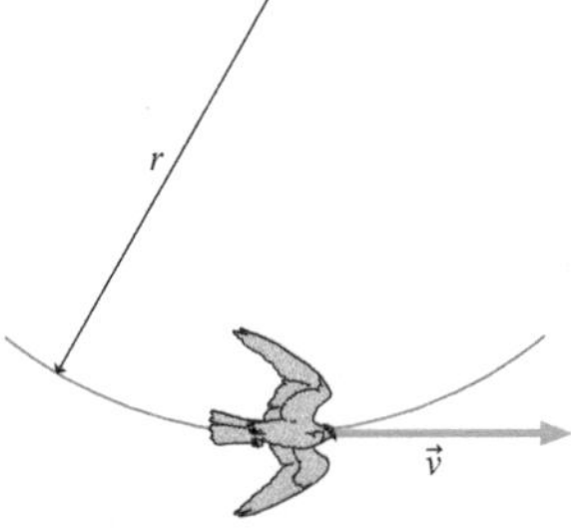

FIGURE I.3

14. At the bottom of the arc, as in Figure I.3, what is the direction of the net force on the falcon?
    A. To the left, opposite the motion
    B. To the right, in the direction of the motion
    C. Up
    D. Down
    E. The net force is zero.
15. Suppose the falcon weighs 8.0 N and is turning with an acceleration of 15 m/s^2 at the lowest point of the arc. What is the magnitude of the upward lift force at this instant?
    A. 8.0 N
    B. 12 N
    C. 16 N
    D. 20 N
16. A falcon starts from rest, does a free-fall dive from a height of 30 m, and then pulls out by flying in a circular arc of radius 50 m. Which segment of the motion has a higher acceleration?
    A. The free-fall dive
    B. The circular arc
    C. The two accelerations are equal.

Measuring Forces at Small Scales

Suppose you want to measure the force generated by a swimming nematode. The force is extremely small, and attaching a probe to this tiny creature is a challenge. But investigators have demonstrated a technique to measure this force using a long, thin micropipette. A small amount of suction in the micropipette allows it to clamp onto the tail of a nematode (or other small creature), and the force it applies to the tip of the micropipette causes the glass tube of the micropipette to deflect. The deflection is proportional to the applied force. Turning a measurement of deflection into a determination of force requires calibration; investigators used a microscope and a camera, while holding a small drop of water, to measure both the deflection of the tip and the size of the droplet. Sample data for a micropipette are presented in Figure I.4.

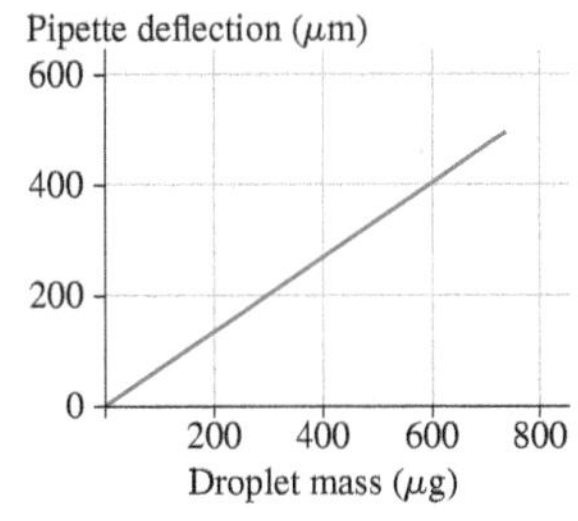

FIGURE I.4

17. What is the approximate spring constant of the micropipette of Figure I.4?
    A. 0.015 N/m
    B. 0.070 N/m
    C. 15 N/m
    D. 70 N/m
18. In a calibration experiment, an investigator measures the deflection caused by a drop of a certain size. Doubling the diameter of the drop increases the deflection by what factor?
    A. 2
    B. 4
    C. 8
    D. 16
19. A micropipette used for measuring swimming forces must have a very small spring constant; a typical value is 0.0050 N/m. A swimming nematode generates a maximum force of about 6 nN. How much does this force deflect the tip?
    A. 30 pm
    B. 1.2 nm
    C. 30 nm
    D. 1.2 μm

Teeing Off

A golf club has a lightweight flexible shaft with a heavy block of wood or metal (called the head of the club) at the end. A golfer making a long shot off the tee uses a driver, a club whose 300 g head is much more massive than the 46 g ball it will hit. The golfer swings the driver so that the club head is moving at 40 m/s just before it collides with the ball. The collision is so rapid that it can be treated as the collision of a moving 300 g mass (the club head) with a stationary 46 g mass (the ball); the shaft of the club and the golfer can be ignored. The collision takes 5.0 ms, and the ball leaves the tee with a speed of 63 m/s.

20. What is the change in momentum of the ball during the collision?
    A. 1.4 kg · m/s
    B. 1.8 kg · m/s
    C. 2.9 kg · m/s
    D. 5.1 kg · m/s
21. What is the speed of the club head immediately after the collision?
    A. 30 m/s
    B. 25 m/s
    C. 19 m/s
    D. 11 m/s
22. Is this a perfectly elastic collision?
    A. Yes
    B. No
    C. There is insufficient information to make this determination.

Additional Integrated Problems

23. You go to the playground and slide down the slide, a 3.0-m-long ramp at an angle of 40° with respect to horizontal. The pants that you've worn aren't very slippery; the coefficient of kinetic friction between your pants and the slide is $\mu_k = 0.45$. A friend gives you a very slight push to get you started. How long does it take you to reach the bottom of the slide?
24. If you stand on a scale at the equator, the scale will read slightly less than your true weight due to your circular motion with the rotation of the earth.
    a. Draw a free-body diagram to show why this is so.
    b. By how much is the scale reading reduced for a person with a true weight of 800 N?
25. Dolphins and other sea creatures can leap to great heights by swimming straight up and exiting the water at a high speed. A 210 kg dolphin leaps straight up to a height of 7.0 m. When the dolphin reenters the water, drag from the water brings it to a stop in 1.5 m. Assuming that the force of the water on the dolphin stays constant as it slows down,
    a. How much time does it take for the dolphin to come to rest?
    b. What is the force of the water on the dolphin as it is coming to rest?

PART II Energy and Thermodynamics

Humpback whales are warm-blooded mammals that spend their summers swimming among the icebergs of the Arctic and Antarctic Oceans, where water temperatures rarely exceed 0°C. In Part II you'll learn how a whale's metabolism transforms chemical energy into the mechanical energy of swimming and how marine mammals are adapted to minimize heat loss to their environment.

OVERVIEW

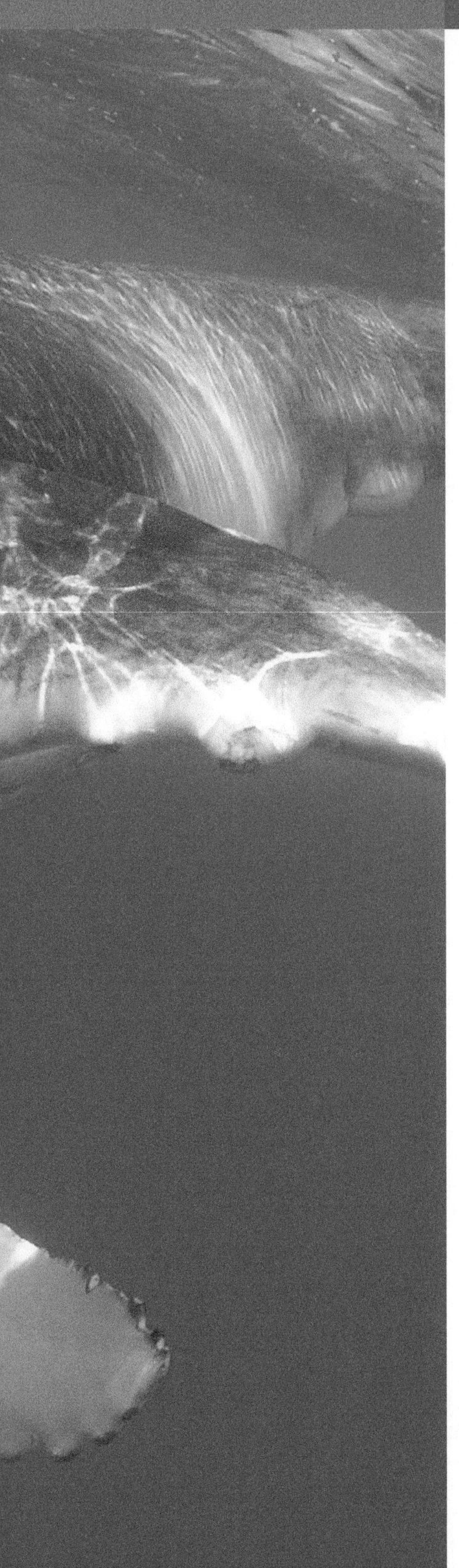

Energy and Life

Biological systems, from cells to ecosystems, survive by exchanging energy and matter with their environment. On a larger scale, energy transport on earth gives us weather and ocean currents. Energy exchange between the earth and the sun determines our climate and drives photosynthesis, the basis of life. Energy is essential to life; some scientists would go so far as to say that the law of conservation of energy is the most fundamental of all scientific laws. Energy will be *the* most important concept for us throughout the remainder of this text.

Part II will focus on what energy is and how energy is used in physics and in biology. We'll begin by looking at energy in simple systems, investigating how energy is transformed from one form into another and how energy is transferred between a system and its environment. This analysis will make frequent use of our earlier study of force and motion. Then we'll look closely at what *chemical energy* is and how systems ranging from organisms to jet engines transform chemical energy into a variety of other forms of energy, including kinetic energy of motion, sound and light energy, and thermal energy.

Thermodynamics is the science of energy in its broadest context, providing an understanding of how heat (*thermo*) and motion (*dynamics*) are related. We'll use thermodynamics to understand how chemical reactions power the body, how heat energy is used and then dissipated, and how matter is transported by diffusion and osmosis.

The Micro/Macro Connection

Our everyday world is described by parameters such as temperature and pressure. Macroscopic objects are characterized by properties such as density, specific heat, and thermal conductivity. But at the atomic level, all matter consists of atoms and molecules in a chaotic dance of random motion and collisions. For example, we'll find that a single air molecule in a room undergoes billions of collisions every second, moving only a few nanometers between impacts. Can this random microscopic motion be reconciled with the apparently steady and stable behavior of macroscopic systems?

This connection—what we'll call the *micro/macro connection*—provides an incredibly powerful basis for understanding *why* macroscopic systems behave as they do. We'll see:

- How pressure and temperature are related to molecular collisions.
- How molecular-level energy exchange allows a system to reach equilibrium.
- What actually happens in a chemical reaction.
- Why chemical reaction rates depend on temperature.
- What really goes on during diffusion and osmosis.

Random collisions disperse energy, spreading it to all the molecules, and we'll discover how this dispersal of energy can be measured with a quantity called *entropy*. Entropy's governing law, the second law of thermodynamics, is one of the most subtle but also most profound and far-reaching statements in science. We'll conclude Part II by learning how the second law of thermodynamics determines whether a process, such as a biochemical reaction, can occur spontaneously.

10 Work and Energy

These sled dogs do work by pulling the sled through a distance. The work becomes kinetic and thermal energy.

LOOKING AHEAD

Forms of Energy

These dolphins have a lot of *kinetic energy* as they leave the water. At their highest point, the energy will have been transformed into *potential energy*.

You'll meet several of the most important forms of energy: kinetic, potential, thermal, and chemical energy.

Work and Energy

When the archer lets go, the tension in the bowstring will do *work* on the arrow and increase the arrow's kinetic energy.

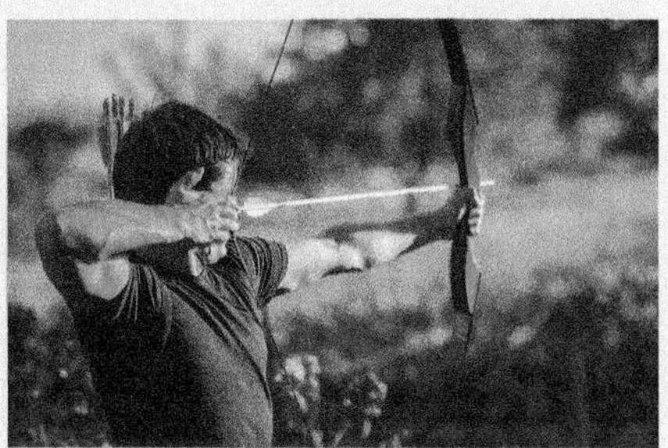

You'll learn how to calculate the work done by a force and how this work is related to the *change* in a system's kinetic energy.

A Model of Energy

A *basic energy model* provides a framework for thinking about how energy is transferred, transformed, and ultimately conserved.

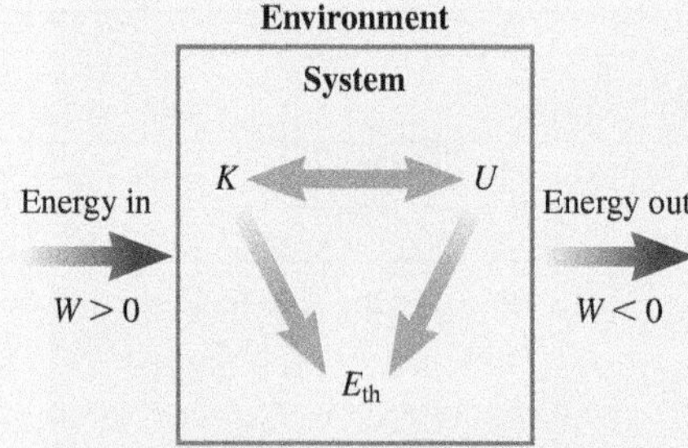

You'll discover that the relationship between work and energy is summarized by the *energy principle*.

GOAL To begin the study of how energy is transferred and transformed.

PHYSICS AND LIFE

Energy Is Essential for Life

Living organisms need energy to survive, grow, and reproduce. But not all of the energy obtained from the food you eat can be used to make you run faster or jump higher; much of it simply keeps your body warm. Why can't living creatures use energy more efficiently? What sets the limits on how energy is moved around and utilized? Our goal in the next five chapters is to develop a precise picture of how energy is transferred among different systems and changed from one form into another. We'll look at energy both from a macroscopic perspective of whole organisms and at the microscopic level of atoms and molecules. This detailed study of energy will allow us to understand many of the ways in which living organisms use energy, the underlying mechanisms of the biochemical reactions that fuel our cells, and how useful forms of energy are degraded into thermal energy.

Myosin, a molecular motor, does atomic-level work, powered by ATP, to contract a muscle. The contracting muscle does work by pushing or pulling on the environment.

10.1 Energy Overview

Energy. It's a word you hear all the time, and everyone has some sense of what "energy" means. Moving objects have energy; energy is the ability to make things happen; energy is associated with heat and with electricity; and we're constantly told to conserve energy. All living organisms must obtain and use energy, and energy flows help define ecosystems. Some scientists consider the *law of conservation of energy* to be the most important of all the laws of nature. But just what is energy?

Scientists across the disciplines agree that it is not easy to define in a general way just what energy is. Rather than starting with a formal definition, we're going to let the concept of energy expand slowly over the course of several chapters. Our goal is to understand the characteristics of energy, how energy is used, and how energy is transformed from one form into another. It's a complex story, so we'll take it step by step until all the pieces are in place.

Some important forms of energy

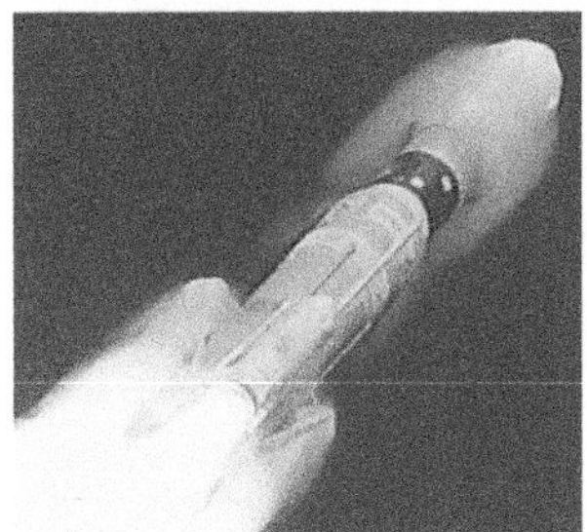

Kinetic energy K

Kinetic energy is the energy of motion. All moving objects have kinetic energy. The more massive an object or the faster it moves, the larger its kinetic energy.

Potential energy U

Potential energy is stored energy associated with an object's position. Gravitational potential energy depends on an object's height above the ground.

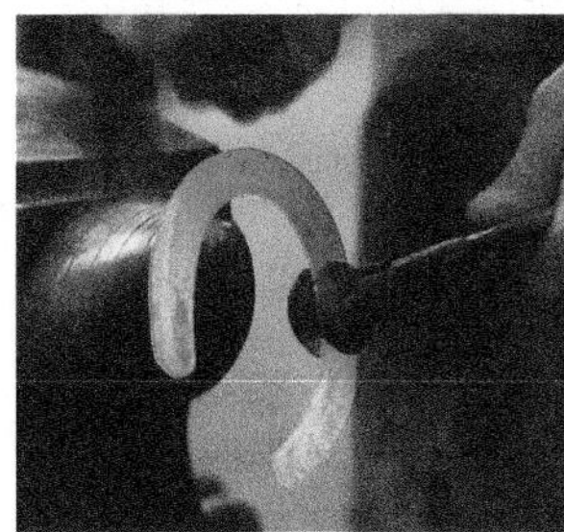

Thermal energy E_{th}

Thermal energy is associated with the random motions of the atoms that make up the system. A hot object has more thermal energy than a cold object.

Chemical energy E_{chem}

Chemical energy is potential energy associated with molecular bonds. It can be absorbed or released as the bonds are rearranged during a chemical reaction.

The Energy Principle

◀◀ SECTION 5.6 introduced *interaction diagrams* and the very important distinction between the **system,** those objects whose motion and interactions we wish to analyze, and the **environment,** objects external to the system but exerting forces on the system. The most important step in an energy analysis is to clearly define the system. Why? Because energy is not some disembodied, ethereal substance; it's the energy *of something.* Specifically, it's *the energy of a system.*

FIGURE 10.1 illustrates the idea pictorially. The system *has energy,* the **system energy,** which we'll designate E_{sys}. There are many kinds or forms of energy: kinetic energy K, potential energy U, thermal energy E_{th}, chemical energy E_{chem}, and so on. We'll introduce these one by one as we go along. The system energy is simply the sum of these:

$$E_{sys} = K + U + E_{th} + E_{chem} + \cdots \tag{10.1}$$

Within the system, energy can be *transformed without loss.* For example, chemical energy can be transformed into kinetic energy, which is then transformed into thermal energy. We can indicate this process symbolically as $E_{chem} \rightarrow K \rightarrow E_{th}$. As long as the system is not interacting with the environment, the total energy of the system is unchanged. You'll recognize this idea as an initial statement of the *law of conservation of energy.*

FIGURE 10.1 A system-environment perspective on energy.

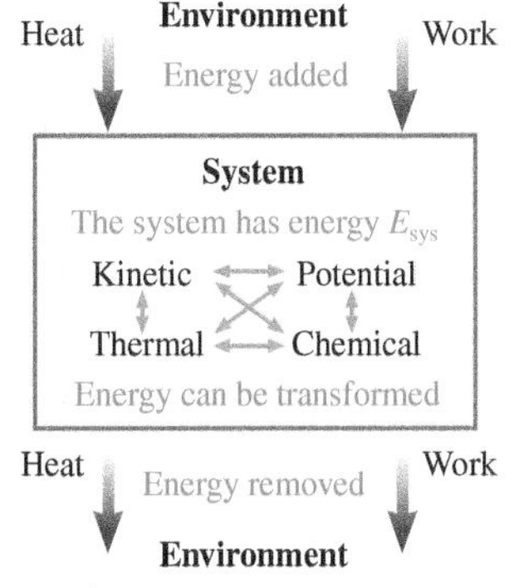

But systems often do interact with their environment. Those interactions *change* the energy of the system, either increasing it (energy added) or decreasing it (energy removed). We say that interactions with the environment *transfer* energy into or out of the system. Interestingly, there are only two ways to transfer energy. One is by mechanical means, using forces to push and pull on the system. A process that transfers energy to or from a system by mechanical means is called **work,** with the symbol W. We'll have a lot to say about work in this chapter. The second is by thermal means when the environment is hotter or colder than the system. A process that transfers energy to or from a system by thermal means is called **heat.** We'll defer a discussion of heat until Chapter 12, but we wanted to mention it now in order to gain an overview of what the energy story is all about.

NOTE ▸ *Work* is another example of physics giving a technical meaning to a common word in the English language. You might use "work" to mean physical effort ("working out") or what you do to earn a living. But in physics, "work" is the process of using forces to transfer energy. ◂

Some energy transfers ... **... and transformations**

Putting a shot
System: The shot
Transfer: $W \rightarrow K$
The athlete (the environment) does work pushing the shot to give it kinetic energy.

A falling diver
System: The diver and the earth
Transformation: $U \rightarrow K$
The diver is speeding up as gravitational potential energy is transformed into kinetic energy.

Pulling a slingshot
System: The slingshot
Transfer: $W \rightarrow U$
The boy (the environment) does work by stretching the rubber band to give it potential energy.

A sprinting pronghorn
System: The pronghorn
Transformation: $E_{\text{chem}} \rightarrow K$
Chemical energy released by ATP hydrolysis is being transformed into kinetic energy.

The key ideas are **energy transfer** between the environment and the system and **energy transformation** within the system. This is much like what happens with money. You may have several accounts at the bank—perhaps a checking account and a couple of savings accounts. You can move money back and forth between the accounts, thus transforming it without changing the total amount of money. Of course, you can also transfer money into or out of your accounts by making deposits or withdrawals. If we treat a withdrawal as a negative deposit—which is exactly what accountants do—simple accounting tells you that

$$\Delta(\text{balance}) = \text{net deposit}$$

That is, the change in your bank balance is simply the sum of all your deposits.

Energy accounting works the same way. Transformations of energy within the system move the energy around but don't change the total energy of the system. *Change* occurs only when there's a transfer of energy between the system and the environment. If we treat incoming energy as a positive transfer and outgoing energy as a negative transfer, and with work being the only energy-transfer process that we consider for now, we can write

$$\Delta E_{\text{sys}} = W \tag{10.2}$$

where W represents work that is done *on* the system *by* the environment. This very simple looking statement, which is just a statement of energy accounting, is called the **energy principle.** But don't let the simplicity fool you; this will turn out to be an incredibly powerful tool for analyzing physical situations and solving problems.

The Basic Energy Model

We'll complete our energy overview—a roadmap of the next two chapters—with the **basic energy model** illustrated in FIGURE 10.2. The key ideas of the basic energy model are as follows:

- Energy is a property of the system.
- The only forms of energy are kinetic energy, potential energy, and thermal energy. The only energy-transfer mechanism is work.
- Energy is *transformed* within the system without loss.
- Energy is *transferred* to and from the system by forces from the environment.
    - The forces do *work* on the system.
    - $W > 0$ for energy added.
    - $W < 0$ for energy removed.
- The energy principle is $\Delta E_{sys} = W$.
- The energy of an *isolated system*—one that doesn't interact with its environment—does not change. We say it is *conserved.*

FIGURE 10.2 The basic energy model.

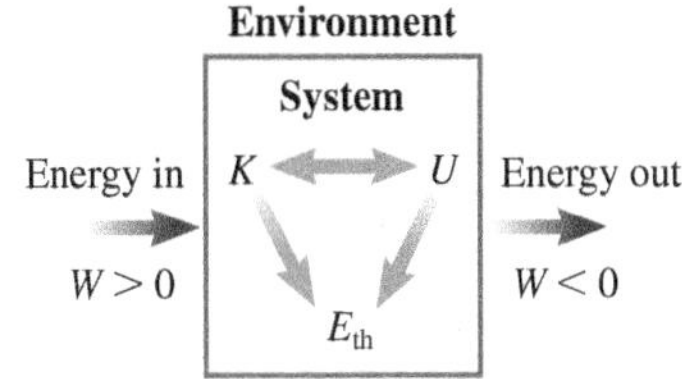

This is an excellent model for a mechanical process, but it's not complete. We'll expand the model in the next two chapters by adding chemical energy, another form of energy, and heat, another energy-transfer process. And this model, although basic, still has many complexities, so we'll be developing it piece by piece in this chapter and the next.

10.2 Work and Kinetic Energy

To begin, we'd like to understand how energy is transferred by forces. Let's look at a simple example—pulling a cart—in which an applied force increases an object's speed and thus its energy of motion. FIGURE 10.3 is a *before-and-after representation,* just as you learned to draw for momentum problems in Chapter 8. It shows a force $\vec{F}$ as it pulls a cart of mass m along a surface with no rolling friction. Let's define the system to be just the cart, which we'll model as a particle; force $\vec{F}$ is an external force from the environment. Initially the cart is rolling with velocity $(v_x)_i$; later, after having undergone displacement Δx, the cart's velocity is $(v_x)_f$.

FIGURE 10.3 The force increases the cart's kinetic energy. We say that the force *does work* on the cart.

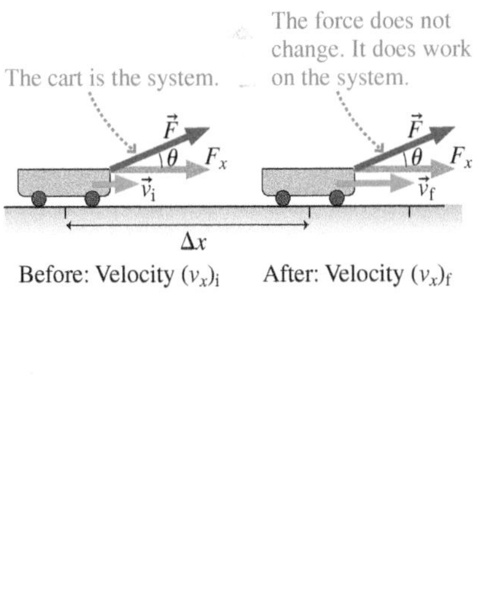

The cart is the system.
The force does not change. It does work on the system.
Before: Velocity $(v_x)_i$ After: Velocity $(v_x)_f$

The only force in the x-direction is the x-component of $\vec{F}$, so Newton's second law for the cart is

$$ma_x = (F_{net})_x = F_x = F\cos\theta \qquad (10.3)$$

Suppose we multiply both sides of this equation by the displacement Δx:

$$ma_x\Delta x = F_x\Delta x \qquad (10.4)$$

This may seem an odd thing to do, but recall that one of the kinematic equations for constant acceleration is

$$(v_x)_f^2 = (v_x)_i^2 + 2a_x\Delta x \qquad (10.5)$$

Solving Equation 10.5 for $a_x\Delta x$ and substituting the result into Equation 10.4 give

$$\tfrac{1}{2}m(v_x)_f^2 - \tfrac{1}{2}m(v_x)_i^2 = F_x\Delta x \qquad (10.6)$$

The sign of the velocity doesn't matter because the velocity is squared, so Equation 10.6 can be expressed more simply in terms of the initial and final *speeds:*

$$\tfrac{1}{2}mv_f^2 - \tfrac{1}{2}mv_i^2 = F_x\Delta x \qquad (10.7)$$

The left-hand side of Equation 10.7 is a final value minus an initial value; that is, the left-hand side appears to be the *change* in some quantity. With that in mind, let's define the **kinetic energy** of a particle to be

$$K = \tfrac{1}{2}mv^2 \qquad (10.8)$$

Kinetic energy of a particle of mass m with speed v

Kinetic energy is an energy of motion. It depends on the particle's mass and speed but not on its position.

With this definition, Equation 10.7 becomes

$$\Delta K = F_x \Delta x = F(\Delta x)\cos\theta \qquad (10.9)$$

where $F_x = F\cos\theta$ is the component of force in the direction of motion. The quantity $\Delta K = K_f - K_i$ is the *change* in kinetic energy as a force pushes or pulls a particle through displacement Δx. ΔK is positive if the particle speeds up (gains kinetic energy), negative if it slows down (loses kinetic energy). We started with a force pulling a cart, to have a concrete example, but Equation 10.9 is quite general and applies to any *constant* force that pushes or pulls a particle along a straight line.

NOTE ▸ By its definition, kinetic energy can *never* be negative. Finding a negative value for K while solving a problem is an indication that you've made a mistake somewhere. ◂

The unit of kinetic energy is mass multiplied by speed squared. In SI units, this is $\text{kg}\cdot\text{m}^2/\text{s}^2$. Because energy is so important, the unit of energy is given its own name, the **joule.** We define

$$1 \text{ joule} = 1 \text{ J} = 1 \text{ kg}\cdot\text{m}^2/\text{s}^2$$

All other forms of energy are also measured in joules.

To give you an idea about the size of a joule, consider a 0.5 kg mass (≈ 1 lb on earth) moving at 4 m/s (≈ 10 mph). Its kinetic energy is

$$K = \tfrac{1}{2}mv^2 = \tfrac{1}{2}(0.5\text{ kg})(4\text{ m/s})^2 = 4\text{ J}$$

This suggests that everyday objects moving at ordinary speeds have energies from a fraction of a joule up to, perhaps, a few thousand joules. A running person has $K \approx 1000$ J, while a high-speed truck might have $K \approx 10^6$ J.

NOTE ▸ You *must* have masses in kilograms and speeds in m/s in order to obtain energy in joules. ◂

STOP TO THINK 10.1 A 1000 kg car has a speed of 20 m/s. A 2000 kg truck has a speed of 10 m/s. Which has more kinetic energy?

A. The car. B. The truck. C. Their kinetic energies are the same.

Work

Now let's turn to the right-hand side of Equation 10.9. The quantity $F_x \Delta x = F(\Delta x)\cos\theta$ tells us *by how much* the kinetic energy changes due to the force. That is, this quantity is the energy transferred to or from the system by the force. We said earlier that a process that transfers energy to or from a system by mechanical means—by forces—is called work. So the right-hand side of Equation 10.9 must be *the work W done by force $\vec{F}$;* that is,

$$W = F_x \Delta x = F(\Delta x)\cos\theta \qquad (10.10)$$

Work done by a constant force acting at an angle θ to the displacement

Similarly, $F_y \Delta y$ is the work done during a vertical displacement Δy.

NOTE ▶ If you've previously studied physics, you may have been taught that "work is force times distance." This is *not* the general definition of work; it is true only for the special case in which the force is parallel to the displacement ($\theta = 0°$). ◀

Having identified the left side of Equation 10.9 with the changing kinetic energy of the system and the right side with the work done on the system, we can rewrite Equation 10.9 as

$$\Delta K = K_f - K_i = W \qquad (10.11)$$

Change in the system's kinetic energy; Final kinetic energy; Initial kinetic energy; Amount of work done by an external force

(Energy principle for a one-particle system)

This is our first version of the energy principle. Notice that it's a cause-and-effect statement: **The work done on a one-particle system causes the system's kinetic energy to change.**

The unit of work, that of force multiplied by distance, is the N·m. Recall that $1\text{ N} = 1\text{ kg}\cdot\text{m/s}^2$. Thus

$$1\text{ N}\cdot\text{m} = 1\ (\text{kg}\cdot\text{m/s}^2)\text{m} = 1\text{ kg}\cdot\text{m}^2/\text{s}^2 = 1\text{ J}$$

Thus the unit of work is really the unit of energy. This is consistent with the idea that work is a transfer of energy. Rather than use N·m, we will measure work in joules.

EXAMPLE 10.1 **Pulling a suitcase**

A strap inclined upward at a 45° angle pulls a suitcase 100 m through the airport. The tension in the strap is 20 N. How much work does the tension force do on the suitcase?

PREPARE We'll let the system consist of only the suitcase, which we model as a particle. The pulling force that does work is the tension force T. FIGURE 10.4 is a before-and-after pictorial representation.

SOLVE The motion is along the x-axis. We can use Equation 10.10 to find that the tension does work:

$$W = T(\Delta x)\cos\theta = (20\text{ N})(100\text{ m})\cos 45° = 1400\text{ J}$$

FIGURE 10.4 Pictorial representation of the suitcase.

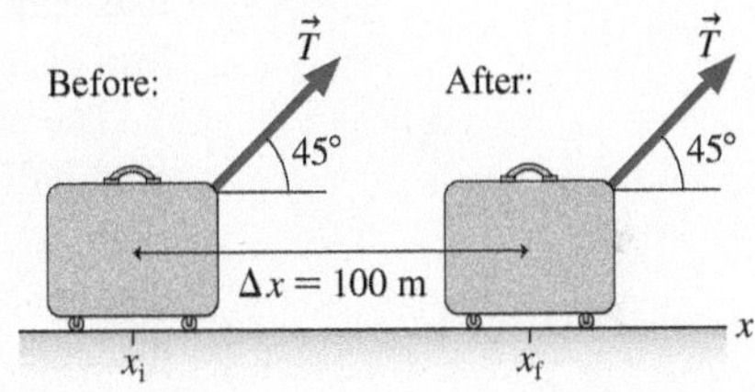

ASSESS Because a person pulls the strap, we would say informally that the person does 1400 J of work on the suitcase.

EXAMPLE 10.2 **Firing a cannonball**

A 5.0 kg cannonball is fired straight up at 35 m/s. What is its speed after rising 45 m?

PREPARE We'll let the system consist of only the cannonball, which we model as a particle. As is usual, we'll model the situation by assuming that air resistance is negligible. FIGURE 10.5 is a before-and-after pictorial representation. Notice two things: First, for problem solving we typically use numerical subscripts instead of the generic i and f. Second, our symbols are speeds, not velocities, so there's no x or y in the subscripts. Before-and-after representations are usually simpler than the pictorial representation used in dynamics problems, so you can include known information right on the diagram instead of making a Known table.

SOLVE Using work and energy is not the only way to solve this problem. You could solve it as a free-fall problem. Or you might

FIGURE 10.5 Before-and-after representation of the cannonball.

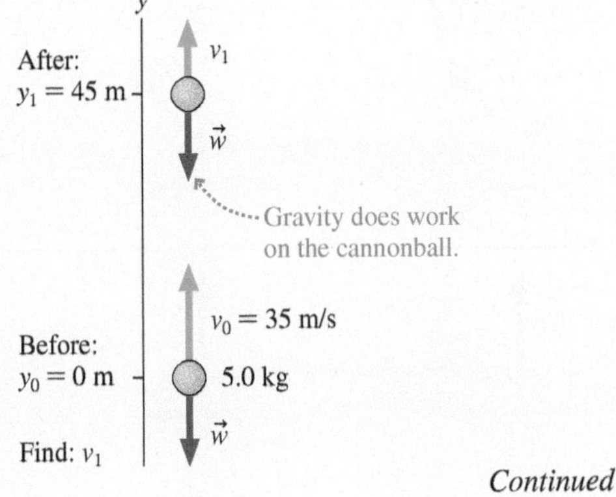

Continued

have previously learned to solve problems like this using potential energy, a topic we'll take up in the next chapter. But using work and energy emphasizes how these two key ideas are related, and it gives us a simple example of the *problem-solving process* before we get to more complex problems. The energy principle is $\Delta K = W$, where work is done by the weight force or, equivalently, by gravity. The cannonball is rising, so its displacement $\Delta y = y_1 - y_0$ is positive. But the force vector points down, with component $F_y = -mg$. Thus gravity does work

$$W = F_y\,\Delta y = -mg\,\Delta y = -(5.0\text{ kg})(9.80\text{ m/s}^2)(45\text{ m}) = -2210\text{ J}$$

as the cannonball rises 45 m. A negative work means that the system is losing energy, which is what we expect as the cannonball slows.

The cannonball's change of kinetic energy is $\Delta K = K_1 - K_0$. The initial kinetic energy is

$$K_0 = \tfrac{1}{2}mv_0^2 = \tfrac{1}{2}(5.0\text{ kg})(35\text{ m/s})^2 = 3060\text{ J}$$

Using the energy principle, we find the final kinetic energy to be $K_1 = K_0 + W = 3060\text{ J} - 2210\text{ J} = 850\text{ J}$. Thus

$$v_1 = \sqrt{\frac{2K_f}{m}} = \sqrt{\frac{2(850\text{ J})}{5.0\text{ kg}}} = 18\text{ m/s}$$

ASSESS 35 m/s ≈ 70 mph. A cannonball fired upward at that speed is going to go fairly high. To have lost half its speed at a height of 45 m ≈ 150 ft seems reasonable.

STOP TO THINK 10.2 A rock falls to the bottom of a deep canyon. Is the work done on the rock by gravity positive, negative, or zero?

Signs of Work

Work can be either positive or negative, but some care is needed to get the sign right when calculating work. The key is to remember that work is an energy transfer. If the force causes the particle to speed up, then the work done by that force is positive. Similarly, negative work means that the force is causing the object to slow and lose energy.

The sign of W is *not* determined by the direction the force vector points. That's only half the issue. The displacement Δx or Δy also has a sign, so you have to consider both the force direction *and* the displacement direction. Tactics Box 10.1 shows that the sign of W is determined entirely by the angle between the force and the displacement. And be sure to notice that there's no work at all $(W = 0)$ if the particle doesn't move!

TACTICS BOX 10.1 Calculating the work done by a constant force

Force and displacement	θ	Work W	Sign of W	Energy transfer
$\vec{F}$, Δx, $\vec{F}$, $\vec{v}_i$, $\vec{v}_f$	0°	$F(\Delta x)$	+	Energy is transferred into the system. The particle speeds up. K increases.
$\vec{F}$, θ, Δx, $\vec{F}$, $\vec{v}_i$, $\vec{v}_f$	< 90°	$F(\Delta x)\cos\theta$	+	
$\vec{F}$, θ, Δx, $\vec{F}$, $\vec{v}_i$, $\vec{v}_f$	90°	0	0	No energy is transferred. Speed and K are constant.

Force and displacement	θ	Work W	Sign of W	Energy transfer
$\vec{F}$, θ, Δx, $\vec{F}$, $\vec{v}_i$, $\vec{v}_f$	$> 90°$	$F(\Delta x)\cos\theta$	$-$	Energy is transferred out of the system. The particle slows down. K decreases.
θ, $\vec{F}$, Δx, $\vec{F}$, $\vec{v}_i$, $\vec{v}_f$	$180°$	$-F(\Delta x)$	$-$	

NOTE ▸ The sign of W depends on the angle between the force vector and the displacement vector, *not* on the coordinate axes. A force to the left does *positive* work if it pushes a particle to the left (the force and the displacement are in the same direction) even though the force component F_x is negative. Think about whether the force is trying to increase the particle's speed $(W > 0)$ or decrease the particle's speed $(W < 0)$. ◂

Extending the Model

Our initial model has been of a single particle acted on by a constant force parallel to the displacement. We can easily make some straightforward extensions of this model to slightly more complex—and interesting—situations:

- **Multiple forces:** If multiple forces act on a system, their works add. That is, $\Delta K = W_{tot}$, where the total work done is

$$W_{tot} = W_1 + W_2 + W_3 + \cdots \tag{10.12}$$

- **Multiparticle systems:** If a system has more than one particle, the system's energy is the total kinetic energy of all the particles:

$$E_{sys} = K_{tot} = K_1 + K_2 + K_3 + \cdots \tag{10.13}$$

K_{tot} is truly a *system* energy, not the energy of any one particle. How does K_{tot} change when work is done? You can see from its definition that ΔK_{tot} is the sum of all the individual kinetic-energy changes, and each of those changes is the work done on that particular particle. Thus

$$\Delta K_{tot} = W_{tot} \tag{10.14}$$

where now W_{tot} is the total work done on *all* the particles in the system.

STOP TO THINK 10.3 A crane uses a single cable to lower a steel girder into place. The girder moves with constant speed. The cable tension does work W_T and gravity does work W_G. Which statement is true?

A. W_T is positive and W_G is positive.
B. W_T is positive and W_G is negative.
C. W_T is negative and W_G is positive.
D. W_T is negative and W_G is negative.
E. W_T and W_G are both zero.

Zero-Work Situations

There are three common situations where *no* work is done. The most obvious is when the object doesn't move $(\Delta x = 0)$. If you were to hold a 200 lb weight over your head, you might break out in a sweat and your arms would tire. You might feel that you had done a lot of work, and, indeed, the motor proteins in your muscles have done work. However, you have done zero work *on the weight* because the weight was not displaced and thus you transferred no energy to it. **A force acting on a particle does no work unless the particle is displaced.**

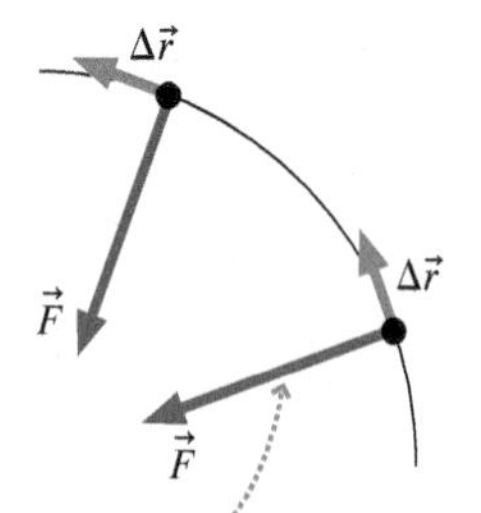

FIGURE 10.6 A perpendicular force does no work.

Second, FIGURE 10.6 shows a particle moving in uniform circular motion. As you learned in Chapter 7, uniform circular motion requires a force pointing toward the center of the circle. How much work does this force do?

Zero! Tactics Box 10.1 showed that a force perpendicular to the displacement does no work. Circular motion extends this idea to motion along a curve. A particle in uniform circular motion has a constant speed, so there's no change in kinetic energy and hence no energy is transferred into or out of the system. Thus **a force everywhere perpendicular to the motion does no work.**

Last, consider the roller skater in FIGURE 10.7 who straightens her arms and pushes off from a wall. She applies a force to the wall and thus, by Newton's third law, the wall applies a force $\vec{F}_{\text{W on S}}$ to her. How much work does force $\vec{F}_{\text{W on S}}$ do?

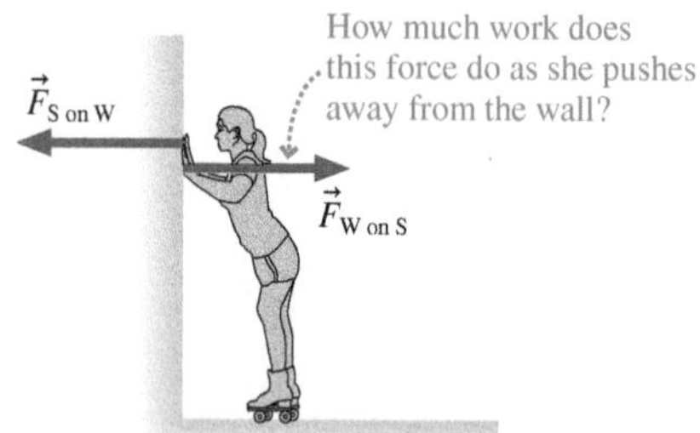

FIGURE 10.7 Does the wall do work on the skater?

Surprisingly, zero. The reason is subtle but worth discussing because it gives us insight into how energy is transferred and transformed. The skater differs from suitcases and cannonballs in two important ways. First, the skater, as she extends her arms, is a *deformable object.* We cannot use the particle model for a deformable object. Second, the skater has an *internal source of energy.* Because she's a living object, she has an internal store of chemical energy that is available through metabolic processes.

Although the skater's center of gravity is displaced, *the palms of her hands—where the force is exerted—are not.* The particles on which force $\vec{F}_{\text{W on S}}$ acts have no displacement, and we've just seen that there's no work without displacement. The force acts, but the force doesn't push any physical thing through a displacement. Hence no work is done.

But the skater indisputably gains kinetic energy. How? A system can gain kinetic energy without any work being done on the system *if* it can transform some other energy into kinetic energy. In this case, the skater transforms chemical energy into kinetic energy. The same is true if you jump straight up from the ground. The ground applies an upward force to your feet, but that force does no work because the point of application—the soles of your feet—has no displacement while you're jumping. Instead, your increased kinetic energy comes via a decrease in your body's chemical energy. A brick cannot jump or push off from a wall because it cannot deform and has no usable source of internal energy.

EXAMPLE 10.3 The speed of a skier

A 70 kg skier is gliding at 2.0 m/s when he starts down a very slippery 50-m-long, 10° slope. A headwind exerts a constant 25 N force parallel to the slope as he skis. What is his speed at the bottom?

PREPARE We'll model the skier as a particle and interpret "very slippery" to mean frictionless. We will use the energy principle to find his final speed. FIGURE 10.8 shows a pictorial representation.

SOLVE Three forces act on the skier. The normal force is perpendicular to the motion and thus does no work. The angle between the weight force and the displacement is smaller than 90°, so gravity does positive work speeding up the skier:

FIGURE 10.8 Pictorial representation of the skier.

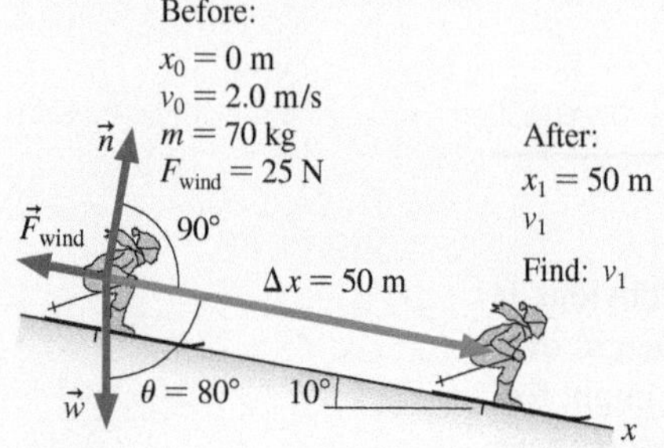

$$W_{\text{grav}} = mg(\Delta x)\cos\theta = (70\text{ kg})(9.8\text{ m/s}^2)(50\text{ m})\cos 80°$$
$$= 5960\text{ J}$$

The force of the wind is directly opposite the displacement, at an angle of 180°, so it does negative work slowing down the skier:

$$W_{\text{wind}} = -F_{\text{wind}}\Delta x = -(25\text{ N})(50\text{ m}) = -1250\text{ J}$$

The total work done on the skier is $W_{\text{tot}} = W_{\text{grav}} + W_{\text{wind}} = 4710$ J. Thus, from the energy principle, we find

$$\Delta K = \tfrac{1}{2}mv_1^2 - \tfrac{1}{2}mv_0^2 = W_{\text{tot}}$$

$$v_1 = \sqrt{v_0^2 + \frac{2W_{\text{tot}}}{m}} = \sqrt{(2.0\text{ m/s})^2 + \frac{2(4710\text{ J})}{70\text{ kg}}} = 12\text{ m/s}$$

ASSESS 12 m/s, or ≈25 mph, is a fairly typical speed for a skier, so our answer is reasonable.

NOTE ▶ While in the midst of the mathematics of calculating work, do not lose sight of what the energy principle is all about. It is a statement about *energy transfer:* The total work done on a particle causes the particle's kinetic energy to either increase or decrease. ◀

STOP TO THINK 10.4 A car accelerates away from a stop sign. Is the work done on the car positive, negative, or zero?

10.3 Work Done by a Force That Changes

You've learned how to calculate the work done on an object by a constant force, but what happens when the force changes as the object moves? For example, the force on an object attached to a spring changes as the spring stretches. Fortunately, it takes only a small modification to Equation 10.10 to handle a changing force.

Suppose a particle moves from an initial position x_i to a final position x_f while acted on by a force $\vec{F}$ whose magnitude or direction might change as the particle moves. For example, FIGURE 10.9 shows how F_x, the x-component of force that does the work, might change as the particle moves. We can divide the total displacement into a large number of very small displacements dx. According to Equation 10.10, the small amount of work dW done by the force during one such displacement is

$$dW = F_x\,dx$$

To find the total work done as the particle moves from x_i to x_f, we simply sum the work done during all the little displacements.

Calculus tells us that such a sum is an integral and that an integral can be found graphically as the area under a curve. Thus a general expression for the work done on a particle is

$$W = \int_{x_i}^{x_f} F_x\,dx = \text{area under the force graph} \qquad (10.15)$$

The integral simply adds up the work done during each step along the trajectory. If the force is constant, we can take it outside the integral and rediscover $W = F_x \Delta x$. But in general we must evaluate the integral either by actually carrying out the integration or, where possible, by evaluating the area under the curve.

FIGURE 10.9 The work done by a changing force can be calculated by adding the small amounts of work dW done during a large number of very small displacements dx.

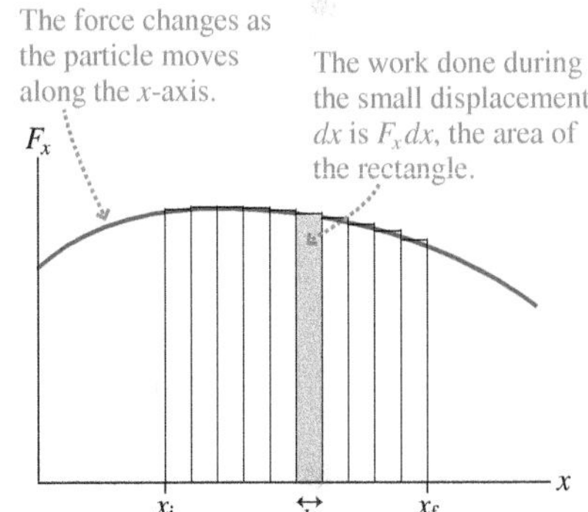
The force changes as the particle moves along the x-axis.
The work done during the small displacement dx is F_x dx, the area of the rectangle.
F_x
x_i
dx
x_f
x

EXAMPLE 10.4 Using work to find the speed of an accelerating car

A 1500 kg car is towed, starting from rest. FIGURE 10.10 shows the tension force in the tow rope as the car travels from $x = 0$ m to $x = 200$ m. What is the car's speed after being pulled 200 m?

PREPARE We'll let the system consist of only the car, which we model as a particle. We'll neglect rolling friction. Two vertical forces, the normal force and the weight force, are perpendicular to the motion and thus do no work.

FIGURE 10.10 Force graph for a car.

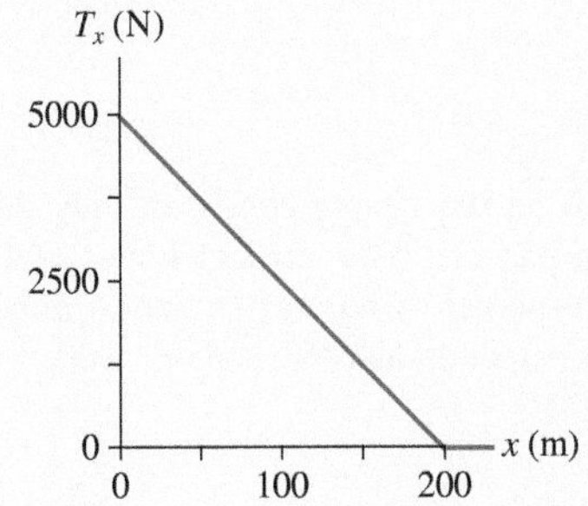
T_x (N)
5000
2500
0
0
100
200
x (m)

SOLVE We can solve this problem with the energy principle, $\Delta K = K_f - K_i = W$. Here W is the work done by the tension force, but the force is not constant so we have to use the full definition of work as an integral. In this case, we can do the integral graphically:

$$W = \int_{0\text{ m}}^{200\text{ m}} T_x\,dx$$

$$= \text{area under the force curve from 0 m to 200 m}$$

$$= \tfrac{1}{2}(5000\text{ N})(200\text{ m}) = 500{,}000\text{ J}$$

The initial kinetic energy is zero, so the final kinetic energy is simply the energy transferred to the system by the work of the tension: $K_f = W = 500{,}000$ J. Then, from the definition of kinetic energy,

$$v_f = \sqrt{\frac{2K_f}{m}} = \sqrt{\frac{2(500{,}000\text{ J})}{1500\text{ kg}}} = 26\text{ m/s}$$

ASSESS 26 m/s ≈ 55 mph is a reasonable final speed after being towed 200 m.

FIGURE 10.11 The spring does work on the object.

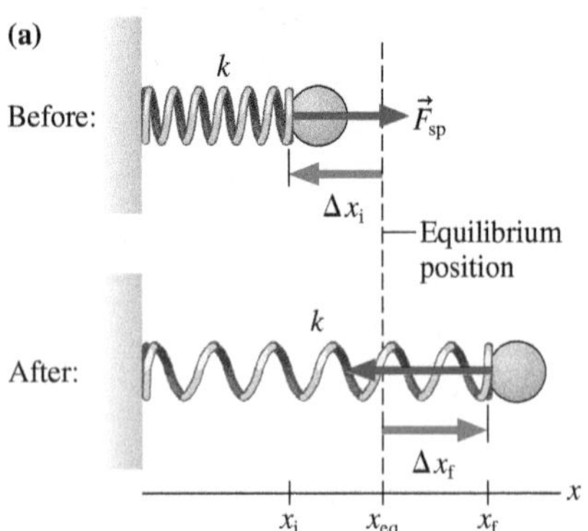

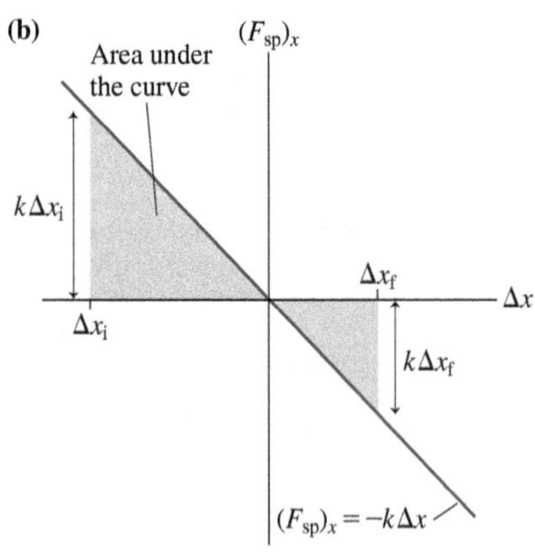

Work Done by Springs

FIGURE 10.11a shows a spring pushing an object from position x_i to position x_f. How much work does the spring do? The force exerted by an ideal spring was introduced in Chapter 5 as a *linear restoring force* given by Hooke's law: $(F_{sp})_x = -k(x - x_{eq}) = k\,\Delta x$, where x_{eq} is the equilibrium position of the end of the spring. This is a variable force, one that depends on the spring's extension, so we can either integrate or find the area under the force curve. We'll do both.

The spring's force graph is shown in FIGURE 10.11b. The graph has a negative slope, which indicates that a compressed spring ($\Delta x < 0$) gives a force to the right.

You can see that the area under the curve between Δx_i and Δx_f consists of two triangles. You may recall, from calculus, that an area *beneath* the x-axis is considered to be a negative area. Consequently, the area under the force curve is the area of the left triangle *minus* the area of the right triangle. The area of a triangle is $\frac{1}{2} \times$ base $\times$ height, so we can write

$$\begin{aligned} W_{sp} &= \text{area under the force graph} \\ &= \tfrac{1}{2}(\Delta x_i)(k\,\Delta x_i) - \tfrac{1}{2}(\Delta x_f)(k\,\Delta x_f) \\ &= \tfrac{1}{2}k(\Delta x_i)^2 - \tfrac{1}{2}k(\Delta x_f)^2 \end{aligned} \tag{10.16}$$

The displacements are squared, so it makes no difference whether the initial and final displacements are extensions or compressions.

Now we can use integration to calculate the work done by a spring. We know that $(F_{sp})_x = -k(x - x_{eq})$, so the work done by the spring as the object moves from x_i to x_f is

$$W_{sp} = \int_{x_i}^{x_f} (F_{sp})_x\,dx = -k\int_{x_i}^{x_f} (x - x_{eq})\,dx \tag{10.17}$$

This is most easily integrated with a change of variables to $u = x - x_{eq}$, in which case $dx = du$. Changing the integration variable also requires changing the limits of integration. When $x = x_i$, $u = x_i - x_{eq} = \Delta x_i$. Similarly, $u = \Delta x_f$ when $x = x_f$. Thus the change of variables leaves us with

$$W_{sp} = -k\int_{\Delta x_i}^{\Delta x_f} u\,du = -\tfrac{1}{2}ku^2\Big|_{\Delta x_i}^{\Delta x_f} = \tfrac{1}{2}k(\Delta x_i)^2 - \tfrac{1}{2}k(\Delta x_f)^2 \tag{10.18}$$

where we used $\int u\,du = \frac{1}{2}u^2$. Not surprisingly, our two methods arrive at the same expression for the work done by a spring:

$$W_{sp} = \tfrac{1}{2}k(\Delta x_i)^2 - \tfrac{1}{2}k(\Delta x_f)^2 \tag{10.19}$$

Work done by a spring

The work done by a spring is energy transferred to the object by the force of the spring. We can use this—and the energy principle—to solve problems that we were unable to solve with a direct application of Newton's laws.

EXAMPLE 10.5 Using the energy principle for a spring

The "pincube machine" was an ill-fated predecessor of the pinball machine. A 100 g cube is launched by pulling a spring back 12 cm and releasing it. The spring's spring constant is 65 N/m. What is the cube's launch speed as it leaves the spring? Assume that the surface is frictionless.

PREPARE We'll let the system consist of only the cube, which we model as a particle. Two vertical forces, the normal force and gravity, are perpendicular to the cube's displacement, and we've seen that perpendicular forces do no work. Only the spring force does work. FIGURE 10.12 is a before-and-after pictorial representation.

SOLVE We can solve this problem with the energy principle, $\Delta K = K_1 - K_0 = W$, where W is the work done by the spring. The initial displacement from equilibrium is $\Delta x_0 = -0.12$ m.

FIGURE 10.12 Pictorial representation of the pincube machine.

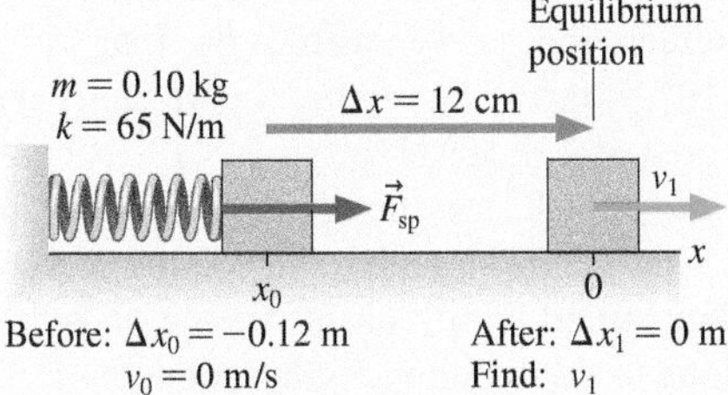

The cube will separate from the spring when the spring has expanded back to its equilibrium length, so the final displacement is $\Delta x_1 = 0$ m. From Equation 10.19, the spring does work

$$W = -\left(\tfrac{1}{2}k(\Delta x_1)^2 - \tfrac{1}{2}k(\Delta x_0)^2\right) = \tfrac{1}{2}(65\text{ N/m})(-0.12\text{ m})^2 - 0$$

$$= 0.468\text{ J}$$

The initial kinetic energy is zero, so the final kinetic energy is simply the energy transferred to the system by the work of the spring: $K_1 = W = 0.468$ J. Then, from the definition of kinetic energy,

$$v_1 = \sqrt{\frac{2K_1}{m}} = \sqrt{\frac{2(0.468\text{ J})}{0.10\text{ kg}}} = 3.1\text{ m/s}$$

ASSESS 3.1 m/s ≈ 6 mph seems a reasonable final speed for a small, spring-launched cube.

STOP TO THINK 10.5 A compressed spring pushes two equal-mass blocks apart, as shown. Is the work done by the spring positive, negative, or zero?

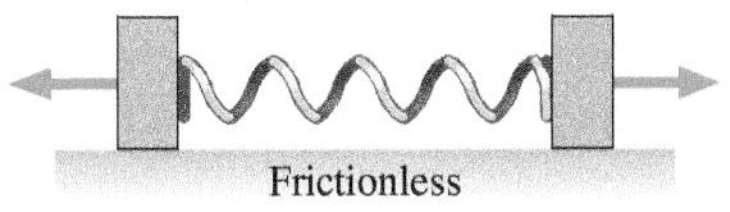

Work to Stretch Elastic Materials

The spring constant of a spring relates the spring's restoring force to its elongation. You learned in ◀◀ SECTION 6.7 that elastic materials, especially biomaterials, are characterized not by force and elongation but by *stress* and *strain.* Suppose a material has length L and cross-section area A. If a force F stretches this material by the amount ΔL, then the material has strain $\varepsilon = \Delta L/L$, the fractional change in length, and it is under stress $\sigma = F/A$, the force per unit area in N/m^2.

For example, FIGURE 10.13 is the stress-strain curve of spider silk that you saw in Chapter 6. It shows how much stress is needed to produce a given amount of strain. Stretching an elastic material takes work—whether it's a strand of spider silk or a tendon or an artery—because a force has to pull the end of the material through a distance. We want to know how the amount of work is related to the stress-strain curve.

FIGURE 10.13 The stress-strain curve for stretching spider silk.

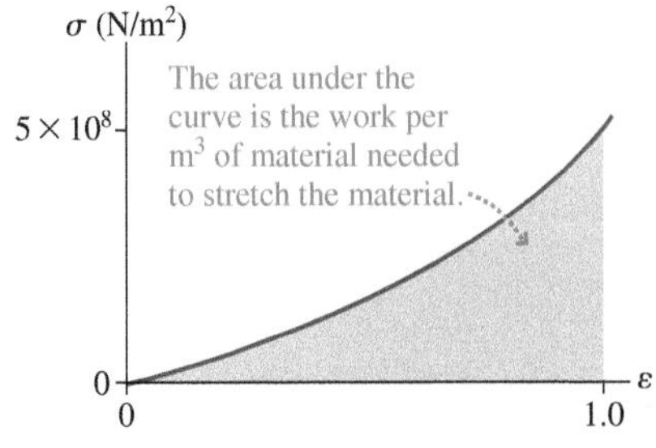

Suppose an elastic material is stretched by the small distance dL. The small amount of work needed to do this is $dW = F\,dL$. If we multiply the right-hand side of the equation by AL/AL, which equals one and doesn't change anything, we can write the small amount of work as

$$dW = \frac{AL}{AL}F\,dL = AL\frac{F}{A}\frac{dL}{L} = V(\sigma\, d\varepsilon) \tag{10.20}$$

where $V = AL$ is the volume of the material and $d\varepsilon$ is an infinitesimal change in the strain.

Integrating gives the work done, and dividing by V gives the work *per unit volume* of material; that is,

$$\frac{W}{V} = \text{work per unit volume in J/m}^3 = \int \sigma\, d\varepsilon$$
$$= \text{area under the stress-strain curve} \tag{10.21}$$

In other words, the area under the stress-strain curve is the work done per unit volume of material (i.e., per m³) to stretch the material. This area is highlighted in

Figure 10.13. You'll see in the next chapter how this work is related to the elastic potential energy stored in a stretched material.

NOTE ▶ For Equation 10.19, the system is the object attached to the spring; thus W is the work done on the object *by* the spring. In contrast, we've now taken the elastic material to be the system, so Equation 10.21 gives the work done *on* the material by the external force that stretches it. This is a subtle but important distinction. ◀

If the curve of Figure 10.13 were a straight line out to a stress of $5.0 \times 10^8\ \text{N/m}^2$ at a strain of 1.0, the area under the curve would be that of a triangle: $\frac{1}{2}(1.0)(5.0 \times 10^8\ \text{N/m}^2) = 2.5 \times 10^8\ \text{N/m}^2$. The actual curve falls beneath the straight line, so we'll estimate an area of $2 \times 10^8\ \text{N/m}^2 = 2 \times 10^8\ \text{N} \cdot \text{m/m}^3$. Notice how we changed the units. But $1\ \text{N} \cdot \text{m} = 1\ \text{J}$, so the area is $2 \times 10^8\ \text{J/m}^3$, which is the work done per m^3 of material. That seems large, but the volume of a strand of spider silk is very small. A 20-cm-long strand of silk has a volume of $\approx 5 \times 10^{-12}\ \text{m}^3$, so the actual work that must be done to double the length, producing a strain of 1.0, is $\approx 1 \times 10^{-3}\ \text{J} = 1\ \text{mJ}$.

You may recall, from Section 6.7, that a *linearly elastic material* is one for which $\sigma = Y\varepsilon$, where Y is Young's modulus for the material, and thus the stress-strain curve is a straight line with slope Y. Young's modulus, which is a property of the material, plays the same role for a linear stress-strain curve as the spring constant does for a force-elongation curve. We'll leave it as a homework problem to show that for a linearly elastic material,

$$\frac{W}{V} = \frac{1}{2}Y\varepsilon^2 \quad \text{(linearly elastic material)} \qquad (10.22)$$

This looks very much like the work done by a spring, which isn't surprising, but in this case the calculation gives the work done per unit volume to stretch or compress a linearly elastic material. To get the actual work, in joules, we need to multiply by the material's volume. You'll see in Chapter 11 that we can think of this as energy stored in the material.

FIGURE 10.14 Two perspectives of motion and energy.

(a)

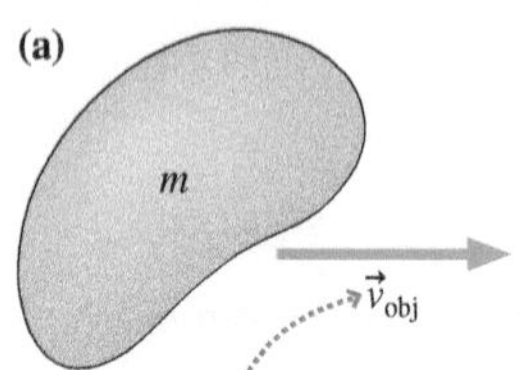

This is macroscopic motion of the object as a whole.

(b)

Hot object: Fast-moving atoms have lots of kinetic and elastic potential energy.

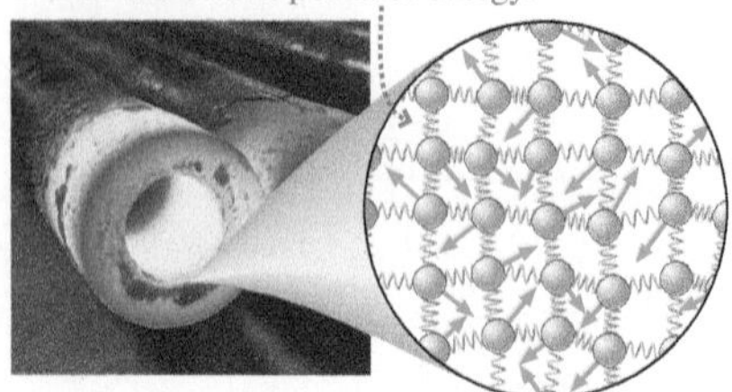

Cold object: Slow-moving atoms have little kinetic and elastic potential energy.

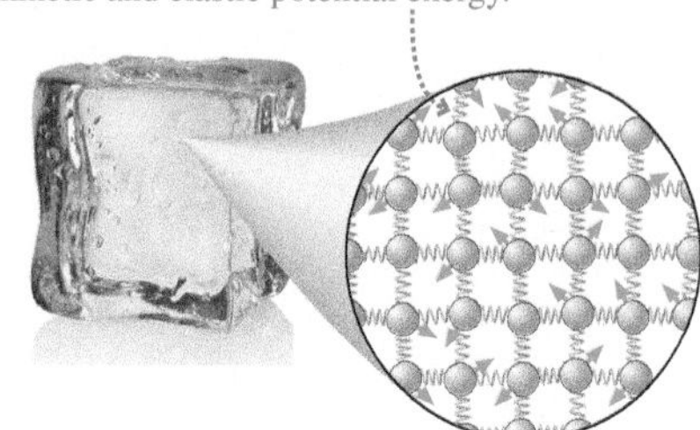

10.4 Dissipative Forces and Thermal Energy

Suppose you drag a heavy sofa across the floor at a steady speed. You are doing work, but the sofa is not gaining kinetic energy. And when you stop pulling, the sofa almost instantly stops moving. Where is the energy going that you're adding to the system? And what happens to the sofa's kinetic energy when you stop pulling?

You know that rubbing things together raises their temperature, in extreme cases making them hot enough to start a fire. As the sofa slides across the floor, friction causes the bottom of the sofa and the floor to get hotter. An increasing temperature is associated with increasing *thermal energy,* so in this situation the work done by pulling is increasing the system's thermal energy instead of its kinetic energy. Our goal in this section is to understand what thermal energy is and how it is related to *dissipative forces.*

Energy at the Microscopic Level

In FIGURE 10.14a you see an object of mass m moving as a whole with velocity v_{obj}. As a consequence of its motion, the object has *macroscopic* kinetic energy $K = \frac{1}{2}m(v_{obj})^2$.

NOTE ▶ You recognize the prefix *micro,* meaning "small." You may not be familiar with *macro,* which means "large." Everyday objects, which consist of vast numbers of particle-like atoms, are *macroscopic objects.* ◀

As you know, this macroscopic object is made up of atoms. But the atoms are not sitting quietly at rest. Instead, as FIGURE 10.14b shows, each of these atoms is randomly jiggling about and has kinetic energy. As the atoms move, they stretch and compress the spring-like bonds between them. We'll study potential energy in Chapter 11, but you'll recall from the energy overview at the beginning of this chapter that potential energy is *stored* energy. Stretched and compressed springs store energy, so the bonds have potential energy.

The energy of one atom is exceedingly small, but there are enormous numbers of atoms in a macroscopic object. As all these atoms undergo random motion, the combined microscopic kinetic and potential energy of the atoms—the energy of the jiggling atoms and stretching bonds—is called the **thermal energy** of the system, E_{th}. This energy is distinct from the macroscopic energy of the object as a whole. Thermal energy is hidden from view in our macrophysics perspective, but it is quite real. You will discover later, when we reach thermodynamics, that the thermal energy is related to the *temperature* of the system. Raising the temperature causes the atoms to move faster and the bonds to stretch more, giving the system more thermal energy.

NOTE ▶ The microscopic energy of atoms is *not* called "heat." As was mentioned earlier, heat is a *process,* similar to work, for transferring energy between the system and the environment. We'll have a lot more to say about heat in future chapters. ◀

With the inclusion of thermal energy, a system can have both macroscopic kinetic energy *and* thermal energy: $E_{sys} = K + E_{th}$. With this, the energy principle becomes

$$\Delta E_{sys} = \Delta K + \Delta E_{th} = W \tag{10.23}$$

where W is the work due to external forces acting on the object. Work done on the system might increase the system's kinetic energy, its thermal energy, or both. Or, in the absence of work, kinetic energy can be transformed into thermal energy as long as the total energy change is zero. Recognizing thermal energy greatly expands the range of problems we can analyze with the energy principle.

NOTE ▶ A particle has no internal structure and cannot have thermal energy. Macroscopic objects that have thermal energy must be modeled as *extended objects.* ◀

FIGURE 10.15 Motion with friction leads to thermal energy.

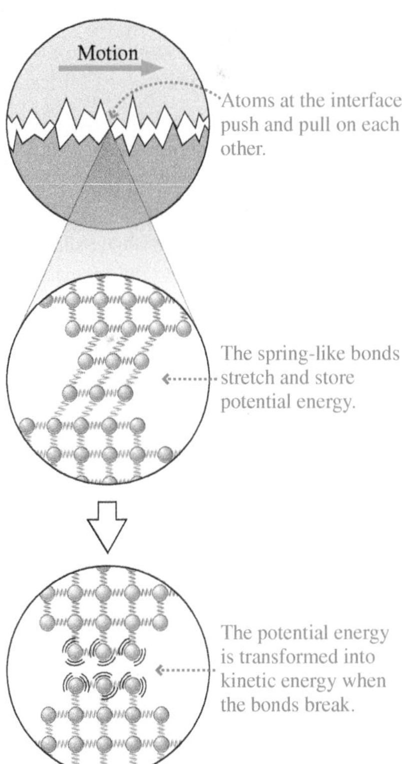

Motion
Atoms at the interface push and pull on each other.
The spring-like bonds stretch and store potential energy.
The potential energy is transformed into kinetic energy when the bonds break.

Dissipative Forces

Forces such as friction and drag cause the macroscopic kinetic energy of a system to be *dissipated* as thermal energy. Hence these are called **dissipative forces.** FIGURE 10.15 shows how microscopic interactions are responsible for transforming macroscopic kinetic energy into thermal energy when two objects slide against each other. Because friction causes *both* objects to get warmer, with increased thermal energy, **we must define the system to include both objects whose temperature changes.**

For example, FIGURE 10.16 shows a box being pulled at constant speed across a horizontal surface with friction. As you can imagine, both the surface and the box are getting warmer—increasing thermal energy—but the kinetic energy is not changing. If we define the system to be box + surface, then the increasing thermal energy of the system is entirely due to the work being done on the system by tension in the rope: $\Delta E_{th} = W_{tension}$.

The work done by tension in pulling the box through distance Δx is simply $W_{tension} = T\Delta x$; thus $\Delta E_{th} = T\Delta x$. Because the box is moving with constant velocity, and thus no net force, the tension force has to exactly balance the friction force: $T = f_k$. Consequently, the increase in thermal energy due to the dissipative force of friction is

$$\Delta E_{th} = f_k \Delta x \tag{10.24}$$

Notice that the increase in thermal energy is directly proportional to the total distance of sliding. **Dissipative forces always increase the thermal energy; they never decrease it.**

FIGURE 10.16 Work done by tension is dissipated as thermal energy.

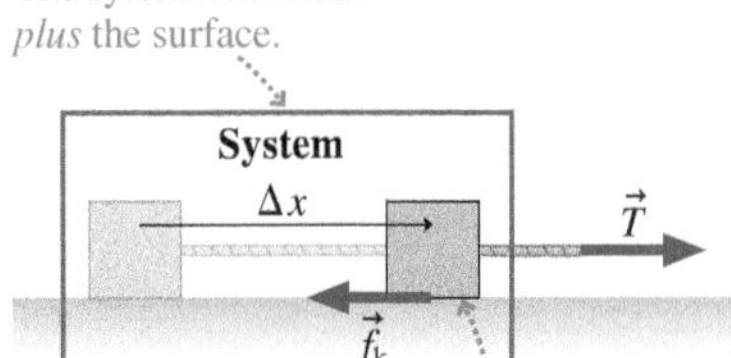

You might wonder why we didn't simply calculate the work done by friction. The rather subtle reason is that work is defined only for forces acting on a *particle.* There is work being done on individual atoms at the boundary as they are pulled this way and that, but we would need a detailed knowledge of atomic-level friction forces to calculate this work. The friction force $\vec{f}_k$ is an average force on the object as a whole; it is not a force on any particular particle, so we cannot use it to calculate work. Furthermore, increasing thermal energy is not an energy transfer—the definition of work—from the box to the surface or from the surface to the box; both the box *and* the surface are gaining thermal energy. **The techniques used to calculate the work done on a particle cannot be used to calculate the work done by dissipative forces.**

NOTE ▸ The considerations that led to Equation 10.24 allow us to calculate the total increase in thermal energy of the entire system, but we cannot determine what fraction of ΔE_{th} goes to the box and what fraction goes to the surface. ◂

EXAMPLE 10.6 Pushing a sofa

Suppose you use a 95 N horizontal force to push a 35 kg sofa, starting from rest, 2.0 m across a horizontal floor. The coefficient of kinetic friction between the sofa and the floor is 0.20. What is the increase in thermal energy? What is the sofa's final speed?

PREPARE We'll model the sofa as an extended object and let the system consist of both the sofa and the floor. Your push is an external force that does work on the system.

SOLVE The energy principle, Equation 10.23, is $\Delta K + \Delta E_{th} = W$. The friction force on an object sliding across a horizontal surface is $f_k = \mu_k n = \mu_k mg$, so the increase in thermal energy, given by Equation 10.24, is

$$\Delta E_{th} = f_k \Delta x = \mu_k mg \Delta x$$

$$= (0.20)(35 \text{ kg})(9.80 \text{ m/s}^2)(2.0 \text{ m}) = 140 \text{ J}$$

Your push, which is parallel to the displacement with $\theta = 0°$, does work $W_{ext} = F_{push} \Delta x = (95 \text{ N})(2.0 \text{ m}) = 190 \text{ J}$. Consequently, the change in the sofa's kinetic energy is $\Delta K = W - \Delta E_{th} = 190 \text{ J} - 140 \text{ J} = 50 \text{ J}$. The initial kinetic energy is zero, so $K_f = \Delta K = 50$ J. We can now use the definition of kinetic energy to find that the sofa's final speed is

$$v_f = \sqrt{\frac{2K_f}{m}} = \sqrt{\frac{2(50 \text{ J})}{35 \text{ kg}}} = 1.7 \text{ m/s}$$

ASSESS 1.7 m/s is approximately 3.5 mph, a fast walking speed. That's about the speed you would expect for the sofa after a hard shove, so our answer is believable. Most of your effort, though, went to increasing thermal energy instead of moving the sofa. That's not uncommon when dissipative forces are present. Notice that it's the thermal energy of the sofa *and* the floor that increases by 140 J. We cannot determine ΔE_{th} for the sofa (or floor) alone.

10.5 Power and Efficiency

Work is a transfer of energy between the environment and a system. In many situations we would like to know *how quickly* the energy is transferred. Does the force act quickly and transfer the energy very rapidly, or is it a slow and lazy transfer of energy? If a predator needs to reach a speed of 20 m/s, it makes a *big* difference whether its muscles have to do this in 2 s or 20 s.

The question How quickly? implies that we are talking about a *rate.* For example, the velocity of an object—how quickly it is moving—is the *rate of change* of position. So when we raise the issue of how quickly the energy is transferred, we are talking about the *rate of transfer* of energy. The rate at which energy is transferred or transformed is called the **power** P, and it is defined as

$$P = \frac{dE_{sys}}{dt} \tag{10.25}$$

The English unit of power is the *horsepower.* The conversion factor to watts is

1 horsepower = 1 hp = 746 W

Many common appliances, such as motors, are rated in hp.

The unit of power is the **watt,** which is defined as 1 watt = 1 W = 1 J/s. Common prefixes used with power are mW (milliwatts), kW (kilowatts), and MW (megawatts).

Equation 10.25 is the *instantaneous* power. In many cases, however, we know only the energy change ΔE_{sys} over a time interval Δt. Then the *average power* is $P_{avg} = \Delta E_{sys}/\Delta t$. For example, you pushed the sofa in Example 10.6 with a force of

95 N and, by doing so, transferred 190 J of energy to the system. You could calculate that it took 2.3 s to push the sofa, so on average energy was being transferred at the rate of 83 J/s. We would say that you had an *average power output* of 83 W.

The idea of power as a *rate* of energy transfer applies no matter what the form of energy. FIGURE 10.17 shows three examples that give you some idea about the power requirements of various machines. For now, we want to focus primarily on *work* as the source of energy transfer. Within this more limited scope, power is simply the **rate of doing work:** $P = dW/dt$. If a particle moves through a small displacement dx while acted on by force $\vec{F}$, the force does a small amount of work dW given by

$$dW = F_x\,dx = F\cos\theta\,dx$$

where θ is the angle between the force and the displacement. Dividing both sides by dt, to give a rate of change, yields

$$\frac{dW}{dt} = F\cos\theta\,\frac{dx}{dt}$$

But dx/dt is the velocity v, so we can write the power as

$$P = Fv\cos\theta \qquad (10.26)$$

These ideas will become clearer with some examples.

FIGURE 10.17 Examples of power.

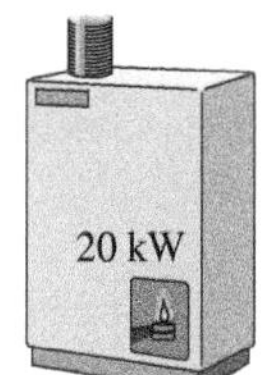

EXAMPLE 10.7 Pulling a sled

During training, a football player uses a harness to drag a 200 kg sled across the grass. The coefficient of kinetic friction between the sled and the grass is 0.40. What is the athlete's power output when dragging the sled at a steady 0.60 m/s?

PREPARE The force applied by the athlete is the harness tension pulling forward on the sled. We'll assume that the tension force is parallel to the ground, so the angle between the force and the sled's displacement is $\theta = 0°$. There's no net force on the sled, which is moving with constant velocity, so the tension force T balances the friction force f_k.

SOLVE The athlete's power output, through the harness tension, is

$$P = Tv\cos 0° = Tv = f_k v$$

The upward normal force and the downward weight force are the only vertical forces. There's no vertical acceleration, so in this case $n = w = mg$ and the kinetic friction is $f_k = \mu_k n = \mu_k mg$. Thus the athlete's power output is

$$P = \mu_k mgv = (0.40)(200\text{ kg})(9.8\text{ m/s}^2)(0.60\text{ m/s})$$
$$= 470\text{ W}$$

ASSESS 470W is roughly $\frac{2}{3}$ hp. As you'll see, this is typical of the power output of well-trained athletes, so the result is reasonable.

EXAMPLE 10.8 Power output of a car engine

A 1500 kg car has a front profile that is 1.6 m wide by 1.4 m high and a drag coefficient of 0.50. The coefficient of rolling friction is 0.02. What power must the engine provide to drive at a steady 30 m/s (≈65 mph) if 25% of the power is "lost" before it reaches the drive wheels?

PREPARE The net force on a car moving at a steady speed is zero, with the forward propulsion force $\vec{F}_{\text{ground on car}}$ being opposed by the drag and friction forces. At first this seems to be a problem of calculating the rate at which work is done by the propulsion force, but a moment's thought shows otherwise. The propulsion force, like the force of the wall on the skater in Figure 10.7, is a force that does no work. The reason is that the bottom of a car tire is at rest on the road surface—otherwise, the car would be skidding—so the particles on which the propulsion force acts have no displacement. In this case, the kinetic energy of motion comes not from work but from an *energy transformation*—the transformation of chemical energy in the fuel into kinetic energy. We need to use the more general definition of power as the rate of transfer of energy.

SOLVE A car will move at a steady speed only if energy from the engine is delivered at exactly the same *rate* at which energy is dissipated by friction and drag. The rate of transferring energy is power, so we can write this requirement as

$$P_{\text{total}} = P_{\text{car}} + P_{\text{drag}} + P_{\text{fric}} = 0$$

where P_{car} is the rate at which energy is delivered through the engine and drivetrain (the drive power) while P_{drag} and P_{fric}, which are both negative, are the rates at which energy is dissipated. We can find P_{car} by calculating P_{drag} and P_{fric}.

The drag force does the actual work of pushing as the car undergoes a displacement. Drag has magnitude $\frac{1}{2}C_D\rho Av^2$ and acts

Continued

opposite the displacement with $\theta = 180°$. We can use Equation 10.26 to write the power dissipation from drag as

$$P_{\text{drag}} = F_{\text{drag}} v \cos 180° = -\tfrac{1}{2} C_D \rho A v^3$$

The minus sign indicates that energy is being dissipated. A is the front cross-section area of the car, and the density of air at 20°C, from Chapter 5, is $1.2\ \text{kg/m}^3$. Thus

$$\begin{aligned} P_{\text{drag}} &= -\tfrac{1}{2}(0.50)(1.2\ \text{kg/m}^3)(1.6\ \text{m} \times 1.4\ \text{m})(30\ \text{m/s})^3 \\ &= -18{,}100\ \text{W} \end{aligned}$$

That is, drag dissipates energy at the rate of $\approx 18\ \text{kW} = 18\ \text{kJ/s}$.

We found in Section 10.4 that we can't directly calculate the work done by friction, but we showed that the increase in thermal energy due to kinetic friction is $\Delta E_{\text{th}} = f_k \Delta x$. The same argument would find $\Delta E_{\text{th}} = f_r \Delta x = \mu_r n \Delta x$ for rolling friction, so the *rate* at which thermal energy is generated in the tires and the road surface is

$$\frac{\Delta E_{\text{th}}}{\Delta t} = \frac{\mu_r n \Delta x}{\Delta t} = \mu_r n v$$

where $v = \Delta x / \Delta t$ is the car's speed. From the car's perspective, this is an energy *loss* at the rate

$$\begin{aligned} P_{\text{fric}} &= -\mu_r n v = -\mu_r m g v \\ &= -(0.02)(1500\ \text{kg})(9.8\ \text{m/s}^2)(30\ \text{m/s}) = -8800\ \text{W} \end{aligned}$$

We used $n = mg$ because there are only two vertical forces, weight and the normal force, and they must balance for horizontal motion. Drag is the larger energy-loss mechanism at this speed, but rolling friction cannot be neglected.

To balance these losses, the drive power of transforming chemical energy into kinetic energy has to be

$$P_{\text{car}} = -(P_{\text{drag}} + P_{\text{fric}}) = 27{,}000\ \text{W} = 36\ \text{hp}$$

where we used $1\ \text{hp} = 746\ \text{W}$ to convert the power to horsepower, the units of automotive power in the United States. This is the power needed to drive the car forward against the dissipative forces of drag and friction. The power output of the engine is larger because some energy is used to run the water pump, the power steering, and other accessories. In addition, some energy is lost to friction in the drivetrain between the engine and the wheels. If 25% of the power is lost (a typical value), which leads to $P_{\text{car}} = 0.75 P_{\text{engine}}$, the engine's power output is

$$P_{\text{engine}} = \frac{P_{\text{car}}}{0.75} = 36{,}000\ \text{W} = 48\ \text{hp}$$

ASSESS Automobile engines are typically rated at 150 hp or more. Much of that power is reserved for fast acceleration and climbing hills, so finding that only ≈ 50 hp is needed for steady highway driving makes sense. It is notable that the rate of energy dissipation from drag is proportional to v^3, so the power requirement for steady driving increases rapidly as the speed increases. Increasing your highway speed from 30 m/s (≈ 65 mph) to 36 m/s (≈ 80 mph) increases the magnitude of P_{drag} from 18 kW (24 hp) to 31 kW (42 hp), with a matching decrease in the car's miles-per-gallon fuel economy.

Specific Power

A 100 kg weightlifter can produce more output power than a 50 kg sprinter, which makes sense. It's worthwhile to consider what we'll call the **specific power**—the output power divided by the mass of the person (or animal, device, or machine) doing the transformation or the work:

$$\text{specific power} = \frac{\text{power of a transformation or a transfer}}{\text{mass of agent causing the transformation or transfer}}$$

Let's compute the specific power for the weightlifter and the sprinter:

$$\text{Sprinter:} \quad \text{specific power} = \frac{1000\ \text{W}}{50\ \text{kg}} = 20\ \text{W/kg}$$

$$\text{Weightlifter:} \quad \text{specific power} = \frac{2000\ \text{W}}{100\ \text{kg}} = 20\ \text{W/kg}$$

The numbers for the sprinter and the weightlifter are at the high end of what humans are capable of; these are numbers typical of world-class athletes. It's interesting to note that the specific power for both cases is the same. Humans in peak condition who are skilled at athletic pursuits are capable of short bursts of about 20 W/kg. Larger athletes can produce more output power, but the power per kilogram is about the same. However, humans can't sustain this level of power output; this number applies for only short bursts that use the large muscles of the body.

We can use the concept of specific power to compare animals that have very different masses. For example, smaller animals are generally capable of higher specific powers. A bushbaby, a 200 g primate that gets around by executing rapid leaps in the trees it calls home, is able to push off with its legs with sufficient force to accelerate to 6.7 m/s in 0.16 s, corresponding to a specific power of 140 W/kg, seven times that of an elite sprinter or weightlifter.

Of course, the notion of specific power can be applied to other systems as well. We can do a similar calculation for a passenger car, either starting from rest or climbing a hill, to find a specific power in the range of 90 W/kg. This is an interesting measure of a vehicle.

STOP TO THINK 10.6 Four students run up the stairs in the times shown. Rank in order, from largest to smallest, their power outputs P_A through P_D.

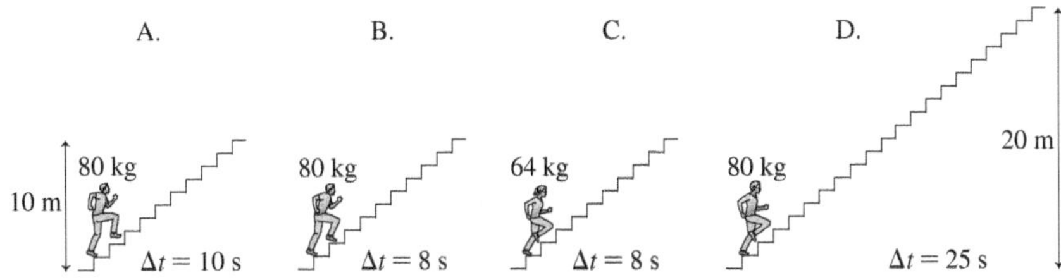

Human Power

Your muscles do work as they exert forces while pushing or pulling against external objects. Remember that no work is done if there's no displacement, so holding a position with tense muscles uses metabolic energy—we'll look at this issue in the next chapter—but no work is being done on the object held.

Exercise physiologists have gathered a large amount of data on the power output of humans engaged in various tasks. For example, simply walking at a modest 5 km/h ($\approx$3 mph) requires about 100 W of power because you're constantly lifting your center of gravity up and down and swinging your arms and legs. Running at a steady 15 km/h ($\approx$10 mph) boosts the required power to $\approx$300 W, but cycling at 15 km/h needs only 120 W, not much more than walking. Cycling is *more efficient* than walking or running.

The power of rowing Elite rowers generate a lot of power. One study found that rowers generate, on average, 450 W of mechanical power during a 5 min race. A measurement of their oxygen consumption showed that their metabolic power, the rate at which they were oxidizing fuel, was 1950 W. Thus their efficiency—the power generated versus the rate of fuel consumption—was 23%. This is typical of the efficiencies of many human activities.

Elite athletes are at the extreme edge of human performance. Cyclists have been especially well studied. An elite cyclist can generate a sustained 500 W of power ($\approx\frac{2}{3}$ hp) for ten minutes or more and can reach 1500 W ($\approx$2 hp) for short bursts. Elite sprinters have similar short-burst power output.

As a general rule, a reasonably fit adult can produce a specific power output of 3 W/kg of body mass for a sustained period of an hour. Well-trained amateur athletes are at the level of 5 W/kg, while elite athletes can reach 6 W/kg or more. An athlete with more mass can generate more power, but he or she also has to accelerate a larger mass; thus a larger total body mass does not necessarily lead to better performance. But having more *muscle mass* as a percentage of body mass is an advantage in performance because there's less "extra" mass that needs to be moved.

Humans generate mechanical power by "burning" calories—that is, by metabolizing the chemical energy stored in carbohydrates and proteins. The rate at which the fuels are oxidized in the body is called *metabolic power.* Your body is not 100% efficient at turning chemical energy into mechanical energy, so the metabolic power required for an activity is larger—often much larger—than the body's mechanical power output. Chapter 11 will take a closer look at the use of energy in the body.

Efficiency

Your body's cells use more than 100 J of chemical energy stored in food in order for your muscles to perform 100 J of work. Your car's engine burns more than 100 J of chemical energy stored in gasoline in order for the wheels to perform 100 J of work. Gravity may be able to do work directly on a falling ball, but organisms and machines are able to do work only by utilizing some kind of stored energy. And converting that stored energy into work is never 100% efficient.

The work done is the desired outcome—what we want to happen. Using stored energy is the price we have to pay. With that in mind, we can define the **efficiency** e of an organism or machine as

$$\text{efficiency } e = \frac{\text{what you get}}{\text{what you had to pay}} = \frac{\text{work done}}{\text{energy used}} \tag{10.27}$$

Efficiency can also be expressed in terms of power: the ratio of mechanical power generated to the *rate* at which stored energy is consumed.

Example 10.8 found that a car driving at highway speeds has a mechanical power output of ≈ 35 kW. We noted that the *mechanical efficiency* of the car is $\approx 75\%$, which means that 75% of the engine's power output reaches the drive wheels while 25% runs other parts of the car or is used to overcome friction. But the *automotive efficiency* of a car, the mechanical power divided by the rate at which gasoline energy is released by burning, is $\approx 25\%$. That is, the engine is burning fuel at ≈ 140 kW (≈ 190 hp) in order to produce the 35 kW needed at the wheels. The majority of the "missing energy" is the thermal energy of heating up the engine, the brakes, and the surrounding air.

In general, neither machines nor organisms are very efficient at converting stored energy into mechanical energy. Your body's efficiency at converting the chemical energy in food into work is similar to the efficiency of a car—about 25%. We'll look at the details in Chapter 11. A fossil-fuel power plant generates steam by burning coal or natural gas, then does work by spinning a turbine that generates electricity. The efficiency of a typical power plant is about 40%. A solar panel converts only about 20% of the incident solar energy into electric energy—but the sun's energy is free.

In some cases, poor efficiency is a consequence of poor design; we could, in principle, design a better process with fewer losses. But in many cases the energy losses are *fundamental limitations* due to the laws of physics, especially the law of thermodynamics. We'll keep looking at the idea of efficiency as we continue our study of energy and then thermodynamics.

CHAPTER 10 INTEGRATED EXAMPLE Powering a bicycle

A cyclist is riding over level ground by pedaling at a cadence of 80 rpm. As he pedals, his feet move in a 32-cm-diameter circle. The mechanics of cycling are complex, but we will use a simple model that assumes that only the downward-moving foot is applying power, that the pushing force of the foot on the pedal is straight down through the movement, and that the magnitude of the pushing force is constant during the half revolution that it is applied.

a. If the bicycle's mechanical efficiency is 95%, what force must the cyclist apply to the pedals to generate a power output of 250 W, roughly the power needed to ride at 20 mph?
b. If the stress on the cyclist's knees is too large, he can reduce the force by changing gears to ride at a faster cadence. What cadence would be required to reduce the knee stress by 25% without changing speed?

PREPARE The foot pushing on the pedal does work as the pedal rotates, but the work changes as the angle of the pedal changes and so we'll need to do an integration. FIGURE 10.18a shows the force when the pedal is at angle θ. To find the power, we'll need

FIGURE 10.18 The pushing force of the cyclist when the pedal is at angle θ.

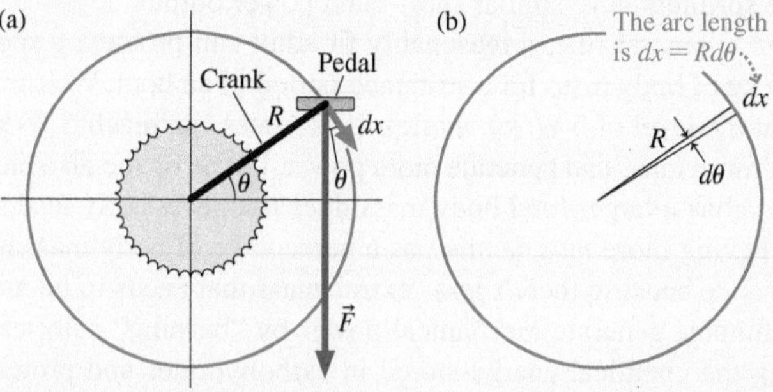

to know the time during which the work is done. The mechanical efficiency is the ratio P_{out}/P_{pedal}, where $P_{out} = 250$ W is the power that moves the bike against friction and air resistance while P_{pedal} is the power the rider generates by pushing the pedals. This ratio is less than one due to energy losses in the chain and gearing.

SOLVE Each foot of the cyclist exerts a downward force for half a cycle. At 80 rpm, the pedaling frequency is

$$f = 80 \text{ rev/min} \times \frac{1 \text{ min}}{60 \text{ s}} = 1.333 \text{ rev/s}$$

The period for one revolution is the inverse of the frequency: $T = 1/f = 0.75$ s. The force is applied for half of this, or $\Delta t = 0.375$ s.

a. You learned in this chapter that the work done by a constant force is $W = F(\Delta x)\cos\theta$, where θ is the angle between the force and the displacement. **FIGURE 10.18b** shows the cyclist's foot moving through a small displacement dx. From geometry, we see that the angle between the force and the displacement is the same as the angle θ of the pedal. Thus the small amount of work done as the pedal moves through dx is

$$dW = F\cos\theta\, dx$$

The work dW is zero at the top of the circle, where $\theta = -90°$ and $\cos\theta = 0$. This makes sense because at the top the downward force of the foot is perpendicular to the forward motion of the pedal. Halfway down, at $\theta = 0°$, the force and displacement are parallel, giving $dW = F\,dx$. During the half cycle when work is being done, these small amounts of work increase from zero to a maximum, then decrease back to zero. That is, the work dW changes from one increment to the next even though the force is constant. We can add up all these small amounts of work by integrating.

We need to integrate this expression for dW from the top to the bottom of the circle. To do so, we write the small amount of work dW not in terms of a small displacement dx but, instead, in terms of the small angle $d\theta$ through which the pedal has rotated. If dx is very small, which we're assuming, then dx is essentially a tiny piece of arc of the circle spanning a very small angle $d\theta$. Figure 10.18b reminds us that the relationship between arc length and angle, which you've seen in trigonometry and calculus, is $dx = R\,d\theta$. Thus the work done by the foot as the pedal rotates through a small angle $d\theta$ is

$$dW = FR\cos\theta\, d\theta$$

One subtle but important task in problems like this is establishing the initial and final angles. When the foot is at the top, starting down, does $\theta_i = 90°$ or $\theta_i = -90°$? The key is that the small increment of angle $d\theta$ needs to be positive. Because the pedal is rotating clockwise, $d\theta$ will be positive if θ is *increasing* as the stroke proceeds. So the pedal moves from $\theta_i = -90°$ at the top to $\theta_f = +90°$ at the bottom.

With that, we can find the total work done by integrating from the initial position of the foot at $\theta_i = -90°$ to the final position at $\theta_f = +90°$:

$$W = \int_{-90°}^{90°} FR\cos\theta\, d\theta = FR\sin\theta \Big|_{-90°}^{90°}$$
$$= FR(1-(-1)) = 2FR$$

FR is a constant that we took outside the integration; then we used $\int \cos\theta\, d\theta = \sin\theta$. We can now work backward, knowing the required power, to determine F.

The average pedaling power is the average *rate* at which work is done, or $P_{pedal} = W/\Delta t = 2FR/\Delta t$. We're given that $P_{out} = 250\text{ W} = 0.95P_{pedal}$. Thus

$$P_{pedal} = \frac{2FR}{\Delta t} = \frac{P_{out}}{0.95}$$

$$F = \frac{P_{out}\Delta t}{2(0.95)R} = \frac{(250\text{ W})(0.375\text{ s})}{(1.90)(0.16\text{ m})} = 310\text{ N}$$

The cyclist needs to press down with 310 N of force, alternating from one side to the other, to produce 250 W of output power.

b. The output power is proportional to the ratio $F/\Delta t$. To keep pedaling at the same speed requires the same output power, so this ratio cannot change. Thus the force F can be reduced by 25% only if Δt is reduced by 25%. The new Δt will be $\Delta t = 0.75 \times 0.375\text{ s} = 0.281$ s. The period for one revolution of the crank is twice this, or $T = 0.562$ s, so the new frequency is

$$f = \frac{1}{T} = \frac{1}{0.562\text{ s}} \times \frac{60\text{ s}}{1\text{ min}} = 107\text{ rpm}$$

ASSESS 310 N is about the weight of a 31 kg mass. A cyclist likely weighs 50 kg to 80 kg, so 310 N is roughly half the cyclist's weight. That seems reasonable for the effort needed to ride at a brisk 20 mph. We assumed that the force is constant and straight down. While the biomechanics of the leg don't allow this to be exactly true, it's a reasonably good approximation. We also assumed that only the downward-moving foot exerts power. This is true for a rider in sneakers, so our calculation is most accurate for that situation, not for a cyclist who is wearing cleats. Cleats allow the upward-moving foot to exert some power, thus reducing the load on the downward-moving leg, so our calculation is probably a bit too high for a cyclist with cleats. If you've ridden a multispeed bike, you know that shifting to a lower gear makes it easier to pedal, but you have to pedal faster to maintain the same speed. Studies have found that many recreational riders use too high a gear, pedaling at 60–70 rpm, and overstress their joints. The optimum cadence for most recreational riders is 80–90 rpm; elite cyclists often have cadences above 100 rpm.

SUMMARY

GOAL To begin the study of how energy is transferred and transformed.

GENERAL PRINCIPLES

The Basic Energy Model

- Energy is a property of the system.
- Energy is *transformed* within the system without loss.
- Energy is *transferred* to and from the system by forces that do work W.
- $W > 0$ for energy added.
- $W < 0$ for energy removed.

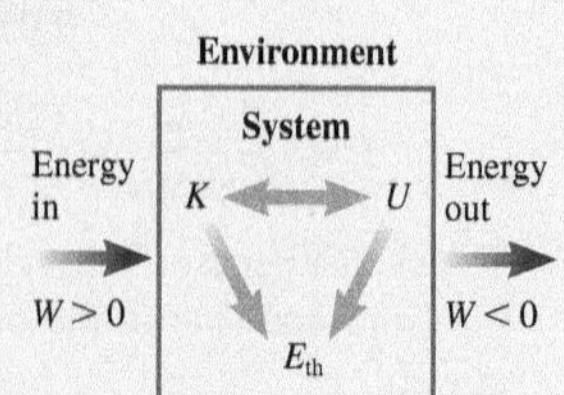

The Energy Principle

Doing work on a system changes the **system energy:**

$$\Delta E_{sys} = W$$

For systems of noninteracting objects with only kinetic and thermal energy, the system energy is $E_{sys} = K + E_{th}$. In this case the energy principle is

$$\Delta K + \Delta E_{th} = W_{tot}$$

where W_{tot} is the total work done on all the particles.

IMPORTANT CONCEPTS

Kinetic energy is an energy of motion: $K = \frac{1}{2}mv^2$

Potential energy is stored energy associated with an object's position.

Thermal energy is the microscopic energy of randomly moving atoms and stretched bonds.

Chemical energy is the energy associated with molecular bonds that is released during chemical reactions.

Dissipative forces such as friction and drag transform macroscopic energy into thermal energy. For friction:

$$\Delta E_{th} = f_k \Delta x$$

The **work** done by a force on a particle as it moves from x_i to x_f is

$$W = \int_{x_i}^{x_f} F_x\,dx = \text{area under the force curve}$$

The work done by a constant force is

$$W = F_x \Delta x = F(\Delta x)\cos\theta$$

where θ is the angle between the force and the displacement.

The work done by a spring is

$$W_{sp} = \tfrac{1}{2}k(\Delta x_i)^2 - \tfrac{1}{2}k(\Delta x_f)^2$$

where Δx is the displacement of the end of the spring from equilibrium.

APPLICATIONS

The sign of work

$\vec{F}$, Δx	$\theta = 0°$	$W = F\Delta x > 0$	$\Delta K > 0$
$\vec{F}$, θ, Δx	$\theta < 90°$	$W = F(\Delta x)\cos\theta > 0$	$\Delta K > 0$
$\vec{F}$, θ, Δx	$\theta = 90°$	$W = 0$	$\Delta K = 0$
$\vec{F}$, θ, Δx	$\theta > 90°$	$W = F(\Delta x)\cos\theta < 0$	$\Delta K < 0$
$\vec{F}$, θ, Δx	$\theta = 180°$	$W = -F\Delta x < 0$	$\Delta K < 0$

Power is the rate at which energy is transferred or transformed:

$$P = dE_{sys}/dt$$

For a particle with velocity $\vec{v}$, the power delivered to the particle by force $\vec{F}$ is

$$P = Fv\cos\theta$$

Zero work is done when

- A particle undergoes no displacement.
- The force is always perpendicular to a particle's displacement.
- The point on an extended object where a force acts undergoes no displacement.

LEARNING OBJECTIVES After studying this chapter, you should be able to:

- Explain the transfer and transformation of energy using the basic energy model. *Conceptual Questions 10.7, 10.12, 10.13*
- Calculate the kinetic energy of and the work done on an object. *Conceptual Questions 10.1, 10.4, 10.9; Problems 10.1, 10.9, 10.10, 10.12, 10.15*
- Calculate the work done by a force that changes. *Problems 10.17, 10.18, 10.21, 10.22, 10.24*
- Understand what thermal energy is and how to calculate the change in thermal energy. *Problems 10.26, 10.27, 10.29, 10.31, 10.33*
- Determine the power required to transfer or transform energy. *Conceptual Question 10.14; Problems 10.34, 10.36, 10.38, 10.42, 10.47*

STOP TO THINK ANSWERS

Stop to Think 10.1: A. Kinetic energy depends linearly on the mass but on the *square* of the velocity. A factor of 2 change in velocity is more significant than a factor of 2 change in mass.

Stop to Think 10.2: Positive. The force (gravity) and the displacement are in the same direction. The rock gains kinetic energy.

Stop to Think 10.3: C. The upward tension force is opposite the displacement, so it does negative work. The downward gravitational force is parallel to the displacement, so it does positive work.

Stop to Think 10.4: Zero. The road does exert a forward force on the car, but the point of application does not move because the tires are not skidding on the road. The car's increasing kinetic energy is a transformation of chemical energy to kinetic energy, not a transfer of energy from the road to the car.

Stop to Think 10.5: Positive. Both blocks gain kinetic energy, so work was done *on* the system.

Stop to Think 10.6: $P_B > P_A = P_C > P_D$. The work done is $mg\Delta y$, so the power output is $mg\Delta y/\Delta t$. Runner B does the same work as runner A but in less time. The ratio $m/\Delta t$ is the same for runners A and C. Runner D does twice the work of runner A but takes more than twice as long.

QUESTIONS

Conceptual Questions

1. If a particle's speed increases by a factor of 3, by what factor does its kinetic energy change?
2. Particle A has half the mass and eight times the kinetic energy of particle B. What is the speed ratio v_A/v_B?
3. An elevator held by a single cable is ascending but slowing down. Is the work done by tension positive, negative, or zero? What about the work done by gravity? Explain.
4. The rope in Figure Q10.4 pulls the box to the left across a rough surface. Is the work done by tension positive, negative, or zero? Explain.

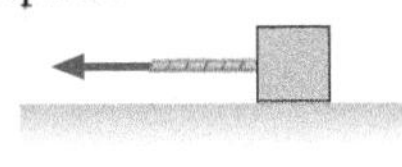

FIGURE Q10.4

5. A particle moving to the left is slowed by a force pushing to the right. Is the work done on the particle positive or negative? Or is there not enough information to tell? Explain.
6. A 0.2 kg plastic cart and a 20 kg lead cart both roll without friction on a horizontal surface. Equal forces are used to push both carts forward a distance of 1 m, starting from rest. After traveling 1 m, is the kinetic energy of the plastic cart greater than, less than, or equal to the kinetic energy of the lead cart? Explain.
7. A baseball pitcher can throw a baseball (mass 0.14 kg) much faster than a football quarterback can throw a football (mass 0.42 kg). Use energy concepts to explain why you would expect this to be true.
8. A particle moves in a vertical plane along the *closed* path seen in Figure Q10.8, and eventually returns to its starting point. Is the work done by gravity positive, negative, or zero? Explain.

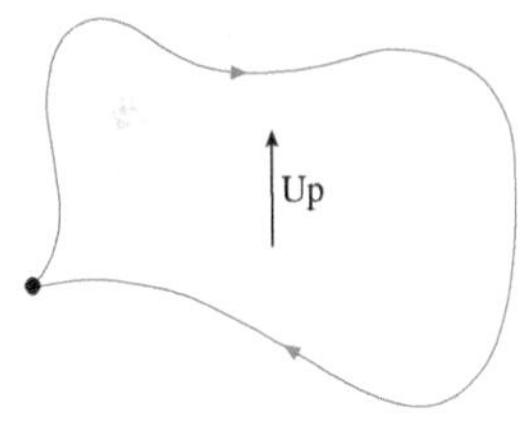

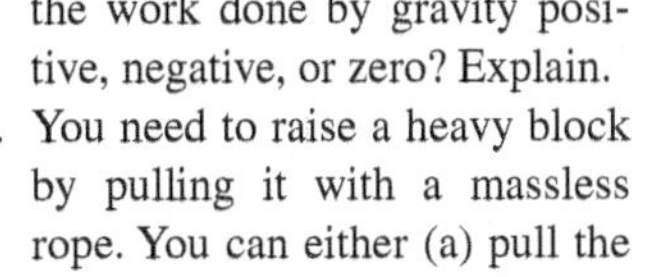

FIGURE Q10.8

9. You need to raise a heavy block by pulling it with a massless rope. You can either (a) pull the block straight up height h, or (b) pull it up a long, frictionless plane inclined at a 15° angle until its height has increased by h. Assume you will move the block at constant speed either way. Will you do more work in case a or case b? Or is the work the same in both cases? Explain.
10. A ball on a string travels once around a circle with a circumference of 2.0 m. The tension in the string is 5.0 N. How much work is done by tension?
11. A sprinter accelerates from rest. Is the work done on the sprinter by the ground positive, negative, or zero? Explain.
12. Give a specific example of a situation in which the energy transformation is (a) $W \rightarrow K$ and (b) $K \rightarrow W$.
13. Give a specific example of a situation in which the energy transformation is (a) $W \rightarrow E_{th}$ and (b) $K \rightarrow E_{th}$.
14. The motor of a crane uses power P to lift a steel beam. By what factor must the motor's power increase to lift the beam twice as high in half the time?

Problem difficulty is labeled as | (straightforward) to ||||| (challenging). Problems labeled INT integrate significant material from earlier chapters; BIO are of biological or medical interest; CALC require calculus to solve.

Multiple-Choice Questions

15. | At approximately what speed does a 1000 kg compact car have the same kinetic energy as a 20,000 kg truck going 20 km/h?
    A. 45 km/h B. 90 km/h
    C. 180 km/h D. 400 km/h
16. | You're pulling a 25 kg child in a 5 kg wagon. The handle is inclined upward at an angle of 40° above horizontal. How much work do you do by pulling the wagon 100 m at a constant speed with a force of 20 N?
    A. 600 J B. 1300 J C. 1500 J D. 2000 J
17. || A 20 g particle is moving to the left at 30 m/s. A force on the particle causes it to change direction and move to the right at 30 m/s. How much work does the force do on the particle?
    A. 0 J B. 1.2 J C. 9.0 J D. 18 J
18. | Skidding to a stop on your bicycle heats the tire and the road. Suppose an 80 kg cyclist, including the mass of the bicycle, who is riding at 8.0 m/s brakes hard and leaves 4.1-m-long skid marks. About how much thermal energy is deposited in the tire and the road surface? The coefficient of kinetic friction is 0.80.
    A. 650 J B. 1300 J C. 1600 J D. 2600 J
19. | A woman uses a pulley and a rope to raise a 20 kg weight to a height of 2 m. If it takes 4 s to do this, about how much power is she supplying?
    A. 100 W B. 200 W
    C. 300 W D. 400 W
20. || BIO A dog can provide sufficient power to pull a sled with a 60 N force at a steady 2.0 m/s. Suppose the dog is hitched to a different sled that requires 120 N to move at a constant speed. How fast can the dog pull this second sled?
    A. 0.50 m/s B. 1.0 m/s
    C. 1.5 m/s D. 2.0 m/s
21. || Most of the energy you expend in cycling is dissipated by the drag force. If you double your speed, you increase the drag force by a factor of 4. This increases the power to cycle at this greater speed by what factor?
    A. 2 B. 4 C. 8 D. 16
22. | BIO The graph of Figure Q10.22 shows the measured speed of a tree frog launching itself from a fixed perch. What does the graph tell you about the power provided by the frog's muscles during different phases of the motion?

FIGURE Q10.22

    A. The average power provided by the muscles is larger in the interval from 0 ms to 50 ms.
    B. The average power provided by the muscles is larger in the interval from 50 ms to 100 ms.
    C. The average power provided by the muscles is roughly constant throughout the time interval shown.

PROBLEMS

Section 10.2 Work and Kinetic Energy

1. | How fast would an 80 kg man need to run in order to have the same kinetic energy as an 8.0 g bullet fired at 400 m/s?
2. || A mother has four times the mass of her young son. Both are running with the same kinetic energy. What is the ratio v_{son}/v_{mother} of their speeds?
3. | A 60 kg runner in a sprint moves at 11 m/s. A 60 kg cheetah in a sprint moves at 33 m/s. By what factor does the kinetic energy of the cheetah exceed that of the human runner?
4. | A car is traveling at 10 m/s.
    a. How fast would the car need to go to double its kinetic energy?
    b. By what factor does the car's kinetic energy increase if its speed is doubled to 20 m/s?
5. || A typical meteor that hits the earth's upper atmosphere has a mass of only 2.5 g, about the same as a penny, but it is moving at an impressive 40 km/s. As the meteor slows, the resulting thermal energy makes a glowing streak across the sky, a shooting star. The small mass packs a surprising punch. At what speed would a 900 kg compact car need to move to have the same kinetic energy?

6. || A baseball thrown at 100 mph, about the limit of what a professional pitcher can do, slows to 93 mph on its way to the plate. What fraction of its initial kinetic energy is lost?
7. | A 2.0 kg book is lying on a 0.75-m-high table. You pick it up and place it on a bookshelf 2.25 m above the floor.
    a. How much work does gravity do on the book?
    b. How much work does your hand do on the book?
8. | A horizontal rope with 15 N tension drags a 25 kg box 2.0 m to the left across a horizontal surface. How much work is done by (a) tension and (b) gravity?
9. | A 25 kg box sliding to the left across a horizontal surface is brought to a halt in a distance of 35 cm by a horizontal rope pulling to the right with 15 N tension. How much work is done by (a) tension and (b) gravity?
10. ||| The two ropes seen in Figure P10.10 are used to lower a 255 kg piano 5.00 m from a second-story window to the ground. How much work is done by each of the three forces?

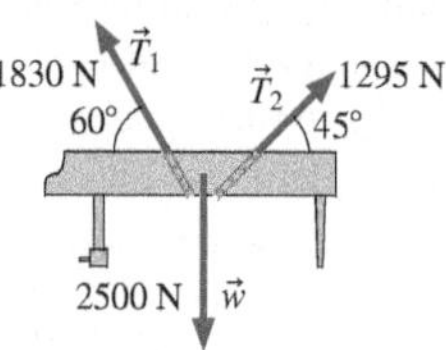

FIGURE P10.10

11. ||| The three ropes shown in the bird's-eye view of Figure P10.11 are used to drag a crate 3.0 m across the floor. How much work is done by each of the three forces?

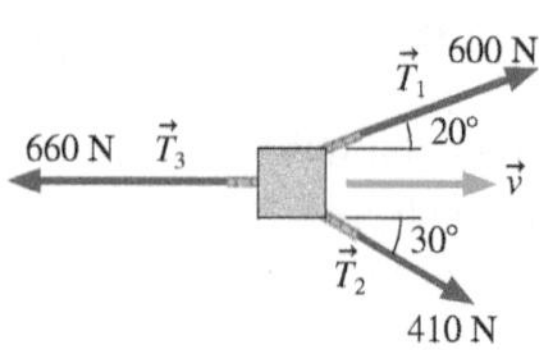

FIGURE P10.11

12. || BIO A typical muscle fiber is 2.0 cm long and has a cross-section area of 3.1×10^{-9} m^2. When the muscle fiber is stimulated, it pulls with a force of 1.2 mN. What is the work done by the muscle fiber as it contracts to a length of 1.6 cm?

13. || Figure P10.13 is the kinetic-energy graph for a 2.0 kg object moving along the x-axis. Determine the work done on the object during each of the four intervals AB, BC, CD, and DE.

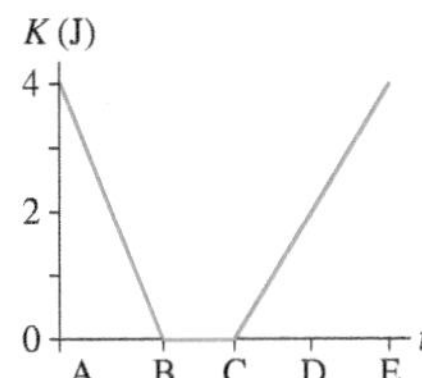

FIGURE P10.13

14. || You throw a 5.5 g coin straight down at 4.0 m/s from a 35-m-high bridge.
   a. How much work does gravity do as the coin falls to the water below?
   b. What is the coin's change in kinetic energy?
   c. What is the speed of the coin just as it hits the water?

15. || The cable of a crane is lifting a 750 kg girder. The girder increases its speed from 0.25 m/s to 0.75 m/s in a distance of 3.5 m.
   a. What is the girder's change in kinetic energy?
   b. How much work is done by gravity?
   c. How much work is done by tension?

Section 10.3 Work Done by a Force That Changes

16. || Figure P10.16 is the force graph for a particle moving along the x-axis. Determine the work done on the particle during each of the three intervals 0–1 m, 1–2 m, and 2–3 m.

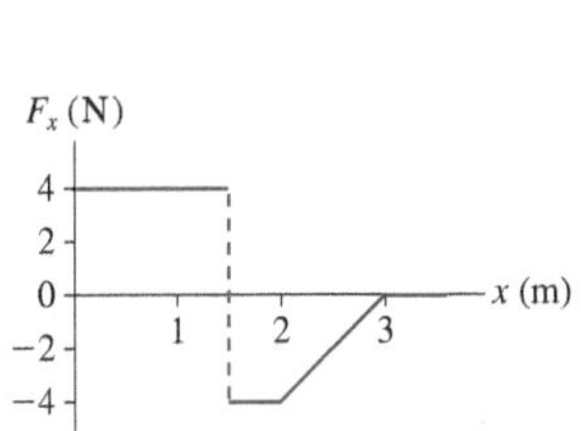

FIGURE P10.16

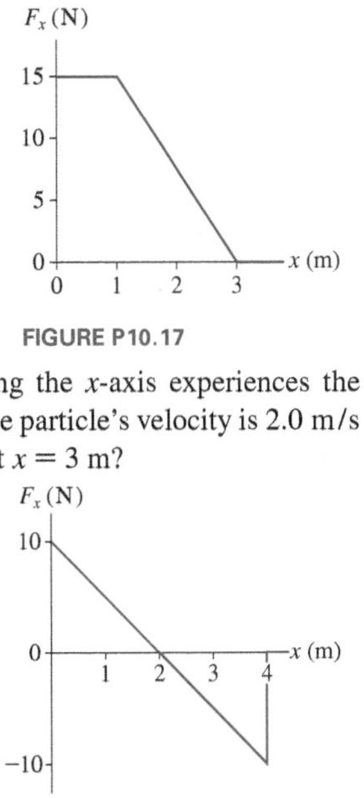

FIGURE P10.17

17. || A 500 g particle moving along the x-axis experiences the force shown in Figure P10.17. The particle's velocity is 2.0 m/s at $x = 0$ m. What is its velocity at $x = 3$ m?

18. || A 2.0 kg particle moving along the x-axis experiences the force shown in Figure P10.18. The particle's velocity is 4.0 m/s at $x = 0$ m. What is its velocity at $x = 2$ m and $x = 4$ m?

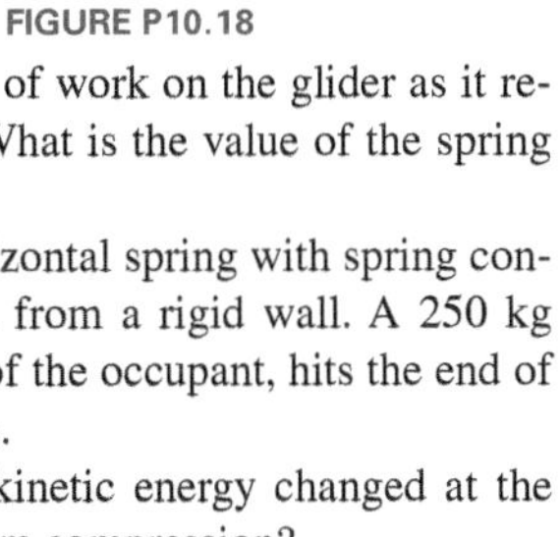

FIGURE P10.18

19. || A spring is stretched by 35 cm and attached to a 500 g air-track glider, which is then released. The spring does 0.25 J of work on the glider as it returns to its equilibrium length. What is the value of the spring constant?

20. || At the amusement park, a horizontal spring with spring constant 150 N/m extends outward from a rigid wall. A 250 kg bumper car, including the mass of the occupant, hits the end of the spring at a speed of 0.50 m/s.
   a. By how much has the car's kinetic energy changed at the instant the spring has maximum compression?
   b. How much work does the spring do on the car from the moment of impact until the spring has maximum compression?
   c. What is the maximum compression of the spring?

21. ||| A particle moving along the x-axis experiences the force
CALC $F_x = (12\ \text{N/m}^2)x^2$. How much work does the force do on the particle as the particle moves from the origin to $x = 2.0$ m?

22. |||| A 75 g particle moving to the right along the x-axis passes
CALC the origin at a speed of 2.0 m/s. As it continues to move, it experiences the force $F_x = (0.25\ \text{N})e^{-x/(2.0\ \text{m})}$. At what position on the x-axis has the particle's speed increased to 3.5 m/s?

23. || Figure P10.23 is a simplified stress-strain curve for a
INT metal wire that is stretched into the region of permanent deformation.
   a. What is the work done per unit volume to stretch the wire by 1.0% of its initial length?
   b. How much work must be done to stretch the length of a 1.0-mm-diameter, 1.0-m-long wire by 1.0%?

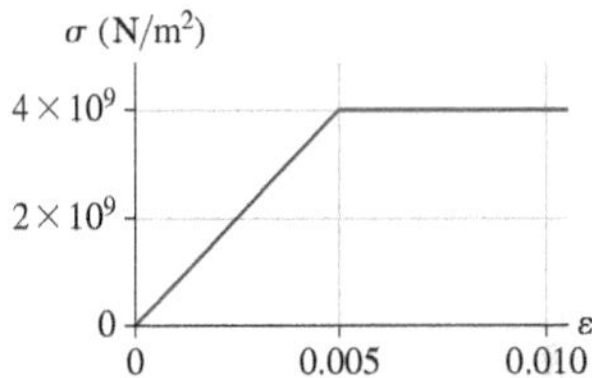

FIGURE P10.23

24. || Your biceps muscle is attached to your elbow by the distal
BIO biceps tendon. An average adult's tendon has a 6.0 mm by
INT 3.0 mm rectangular cross section and is 63 mm in length. Table 6.1 showed that a tendon has a Young's modulus of $1.5 \times 10^9\ \text{N/m}^2$. How much work must be done to stretch the length of the distal biceps tendon by 10%? Assume that the tendon is a linearly elastic material and that stretching does not alter its cross section.

Section 10.4 Dissipative Forces and Thermal Energy

25. | One mole (6.02×10^{23} atoms) of helium atoms in the gas phase has 3700 J of microscopic kinetic energy at room temperature. If we assume that all atoms move with the same speed, what is that speed? The mass of a helium atom is 6.68×10^{-27} kg.

26. | A 55 kg softball player slides into second base, generating 950 J of thermal energy in her legs and the ground. How fast was she running?

27. | Baseballs aren't especially bouncy. A baseball fired at 26 m/s at a solid wall rebounds at 14 m/s. What fraction of the initial kinetic energy is transformed into thermal energy?

28. || Mark pushes his broken car 150 m down the block to his friend's house. He has to exert a 110 N horizontal force to push the car at a constant speed. How much thermal energy is created in the tires and road during this short trip?

29. ||| When you skid to a stop on your bike, you can significantly
INT heat the small patch of tire that rubs against the road surface. Suppose a person skids to a stop by hitting the brake on his back tire, which supports half the 80 kg combined mass of the bike and rider, leaving a skid mark that is 40 cm long. Assume a coefficient of kinetic friction of 0.80. How much thermal energy is deposited in the tire and the road surface?

30. ||| A 900 N crate slides 12 m at constant speed down a ramp that is tilted 35° above horizontal.
   a. How much work does gravity do on the crate as it slides the 12 m?
   b. What is the change in thermal energy?

31. ‖ If you slide down a rope, it's possible to create enough thermal energy to burn your hands or your legs where they grip the rope. Suppose a 40 kg child slides down a rope at a playground, descending 2.0 m at a constant speed.
   a. How much work does gravity do on the child as he slides down the rope?
   b. What is the change in thermal energy?
32. ‖ A 1500 kg car traveling at an unsafe speed of 130 km/h skids to a halt, leaving 85-m-long skid marks. What is the coefficient of kinetic friction between the car's tires and the road?
33. ||| Some runners train with parachutes that trail behind them to provide a large drag force. These parachutes are designed to have a large drag coefficient. One model expands to a square 1.8 m on a side, with a drag coefficient of 1.4. A runner completes a 200 m run at 5.0 m/s with this chute trailing behind. How much thermal energy is added to the air by the drag force?
BIO INT

Section 10.5 Power and Efficiency

34. | a. How much work does an elevator motor do to lift a 1000 kg elevator a height of 100 m?
   b. How much power must the motor supply to do this in 50 s at constant speed?
35. ‖ a. How much work must you do to push a 10 kg block of steel across a steel table at a steady speed of 1.0 m/s for 3.0 s? The coefficient of kinetic friction for steel on steel is 0.60.
   b. What is your power output while doing so?
36. ‖ How much energy is consumed by (a) a 1.2 kW hair dryer used for 10 min and (b) a 10 W night light left on for 24 h?
37. ‖ At midday, solar energy strikes the earth with an intensity of about 1 kW/m^2. What is the area of a solar collector that could collect 150 MJ of energy in 1 h? This is roughly the energy content of 1 gallon of gasoline.
38. ||| A shooting star is actually the track of a meteor, typically a small chunk of debris from a comet that has entered the earth's atmosphere. As the drag force slows the meteor down, its kinetic energy is converted to thermal energy, leaving a glowing trail across the sky. A typical meteor has a surprisingly small mass, but what it lacks in size it makes up for in speed. Assume that a meteor has a mass of 1.5 g and is moving at an impressive 50 km/s, both typical values. What average power is generated if the meteor burns up in a typical time of 2.0 s? You can see how this tiny object can make a glowing trail that can be seen from hundreds of kilometers away.
39. ||| Suppose a 70 kg sprinter, starting from rest, runs the first 12 m in 3.0 s at constant acceleration, a reasonable model. BIO
   a. What is the magnitude of the horizontal ground force acting on the sprinter?
   b. What is the sprinter's power output at 1.0 s, 2.0 s, and 3.0 s?
40. ‖ A 70 kg human sprinter can accelerate from rest to 10 m/s in 3.0 s. During the same time interval, a 30 kg greyhound can accelerate from rest to 20 m/s. What is the average specific power for each of these athletes? BIO
41. ‖ A 75 kg major league pitcher can propel a 145 g baseball from 0 m/s to 45 m/s in 0.11 s. What are (a) the average power output and (b) the specific power output of the pitcher during this action? BIO
42. ‖ The human heart pumps the average adult's 6.0 L (6000 cm^3) of blood through the body every minute. The heart must do work to overcome friction forces that resist blood flow. The average adult blood pressure is 1.3×10^4 N/m^2. BIO
   a. How much work does the heart do to move the 6.0 L of blood completely through the body?
   b. What average power output must the heart have to do this task once a minute?

   Hint: When the heart contracts, it applies force to the blood. Pressure is force/area. Model the circulatory system as a single closed tube, with cross-section area A and volume $V = 6.0$ L, filled with blood to which the heart applies a force.
43. ‖ A 95 kg quarterback accelerates a 0.42 kg ball from rest to 24 m/s in 0.083 s. What is his average power output during the throw? BIO
44. ||| An elevator weighing 2500 N ascends at a constant speed of 8.0 m/s. How much power in hp must the motor supply to do this?
45. ‖ Humans can produce an output power as great as 20 W/kg during extreme exercise. Sloths are not so energetic. At its maximum speed, a 4.0 kg sloth can climb a height of 6.0 m in 2.0 min. BIO

   a. How much work does the sloth do against gravity to climb 6.0 m?
   b. What is the sloth's specific power output during this climb?
46. | A 10 cm × 10 cm photovoltaic cell delivers 1.9 W of electric power when illuminated by sunlight with an intensity of 1100 W/m^2. What is the efficiency of the cell?
47. ‖ A typical wind turbine extracts 40% of the kinetic energy of the wind that blows through the area swept by the blades. For a large turbine, 110,000 kg of air moves past the blades at 15 m/s every second. If the wind turbine extracts 40% of this kinetic energy, and if 90% of this energy is converted to electric energy, what is the power output of the generator?
48. ||| A 55 kg woman athlete can generate a sustained specific power output of 5.0 W/kg. One training exercise is to push a heavy sled across the grass at a steady speed, which takes a horizontal pushing force of 150 N. BIO
   a. What speed can she maintain for a sustained period of time?
   b. How much stored chemical energy will she "burn" in 10 minutes if her body is 25% efficient at converting chemical energy into mechanical energy?
   c. Calories are more familiar units than joules for thinking about the body's energy use. A useful conversion factor is 1 Cal = 4.2 kJ, where Cal is the abbreviation for a food calorie. How many calories does the woman burn during the 10 min workout?

General Problems

49. ‖ A 1000 kg elevator accelerates upward at 1.0 m/s^2 for 10 m, starting from rest.
   a. How much work does gravity do on the elevator?
   b. How much work does the tension in the elevator cable do on the elevator?
   c. What is the elevator's kinetic energy after traveling 10 m?
50. |||| Susan's 10 kg baby brother Paul sits on a mat. Susan pulls the mat across the floor using a rope that is angled 30° above the floor. The tension is a constant 30 N and the coefficient of friction is 0.20. Use work and energy to find Paul's speed after being pulled 3.0 m.
51. ‖ A 2.5 g particle moving along the x-axis has a velocity that changes with position as $v_x = (150 \text{ m/s}) \cos(x/(2.0 \text{ m}) \text{ rad})$. How much work is done on the particle as it moves from the origin to $x = 3.0$ m?

52. ||| A 50 kg ice skater is gliding along the ice, heading due north at 4.0 m/s. The ice has a small coefficient of static friction, to prevent the skater from slipping sideways, but $\mu_k = 0$. Suddenly, a wind from the northeast exerts a force of 4.0 N on the skater.
   a. Use work and energy to find the skater's speed after gliding 100 m in this wind.
   b. What is the minimum value of μ_s that allows her to continue moving straight north?

53. ||| Hooke's law describes an ideal spring. Many real springs are CALC better described by the restoring force $(F_{sp})_x = -k\,\Delta x - q(\Delta x)^3$, where q is a constant. Consider a spring with $k = 250$ N/m and $q = 800$ N/m^3.
   a. How much work must you do to compress this spring 15 cm? Note that, by Newton's third law, the work you do *on* the spring is the negative of the work done *by* the spring.
   b. By what percent has the cubic term increased the work over what would be needed to compress an ideal spring?
   Hint: Let the spring lie along the x-axis with the equilibrium position of the end of the spring at $x = 0$. Then $\Delta x = x$.

54. ||| The gravitational attraction between two objects with masses CALC m_1 and m_2, separated by distance x, is $F = Gm_1m_2/x^2$, where $G = 6.67 \times 10^{-11}$ N $\cdot$ m^2/kg^2 is the *gravitational constant.*
   a. How much work is done by gravity when the separation changes from x_1 to x_2? Assume $x_2 < x_1$.
   b. If one mass is much greater than the other, the larger mass stays essentially at rest while the smaller mass moves toward it. Suppose a 1.5×10^{13} kg comet is passing the orbit of Mars, heading straight for the sun at a speed of 3.5×10^4 m/s. What will its speed be when it crosses the orbit of Mercury? The mass of the sun is 2.0×10^{30} kg. Mercury and Mars orbit at distances of 5.8×10^{10} m and 2.3×10^{11} m from the sun, respectively.

55. ||| Water and other polar molecules are electric dipoles. An CALC electric dipole can be modeled as two equal but opposite electric charges, $+q$ and $-q$, separated by a small distance d. Suppose a dipole is located at the origin, aligned so that the $+q$ charge is at $x = +d/2$ and the $-q$ charge is at $x = -d/2$. If a positive charge Q is placed on the x-axis at distance $x \gg d$, it is repelled by the dipole with an electric force of magnitude $F = KqQd/x^3$, where K is a constant. How much work does the electric force do if charge Q moves along the x-axis from distance x_1 to distance x_2?

56. || A 50 g rock is placed in a slingshot and the rubber band is stretched. The magnitude of the force of the rubber band on the rock is shown by the graph in Figure P10.56. The rubber band is stretched 30 cm and then released. What is the speed of the rock?

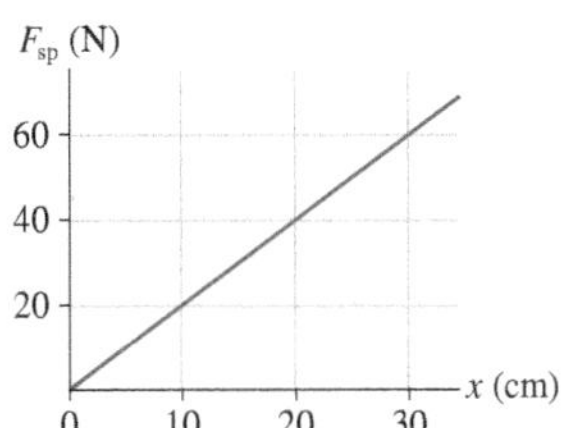

FIGURE P10.56

57. ||| A 45 kg gymnast is bouncing on a trampoline. Model the INT gymnast as a particle and the trampoline as a spring with a spring constant of 15,000 N/m. The spring is compressed by 30 cm when the gymnast is at her lowest point. We want to determine how high she goes above the trampoline.
   a. The trampoline pushes the gymnast 30 cm upward, until the trampoline surface is flat, at which point she is launched into the air. How much work does the trampoline do on the gymnast?
   b. How much work does gravity do on the gymnast as she moves 30 cm upward?
   c. With what speed is the gymnast launched upward from the trampoline?
   d. Use free-fall kinematics to determine the gymnast's maximum height above the trampoline.

58. ||| A horizontal spring with spring constant 250 N/m is compressed by 12 cm and then used to launch a 250 g box across the floor. The coefficient of kinetic friction between the box and the floor is 0.23. What is the box's launch speed?

59. || There's strong evidence that dinosaurs became extinct INT 65 million years ago when an approximately 2-km-diameter asteroid with a mass of 1×10^{13} kg hit the earth at a speed of about 40 km/s relative to the earth. What happened to the asteroid's energy? Was it transformed into the kinetic energy of a recoiling earth, into thermal energy, or into a combination of both? We can analyze the inelastic collision in a coordinate system in which the earth was initially at rest. This is an estimation problem because details of the asteroid are not known, so one-significant-figure answers are appropriate.
   a. What was the asteroid's initial kinetic energy?
   b. What was the earth's recoil speed? The mass of the earth is 6×10^{24} kg.
   c. What was the kinetic energy of the recoiling earth?
   d. How much energy, then, must have been dissipated as thermal energy? For comparison, each of the atomic bombs used in World War II released approximately 10^{14} J of energy. The explosive eruption of the Krakatoa volcano in 1883, which caused significant climate effects around the world for more than a year, released an estimated 10^{17} J.

60. ||| You are driving your 1500 kg car at 20 m/s down a hill with a 5.0° slope when a deer suddenly jumps out onto the roadway. You slam on your brakes, skidding to a stop. How far do you skid before stopping if the kinetic friction force between your tires and the road is 1.2×10^4 N? Solve this problem using work and energy.

61. | A 90 kg firefighter needs to climb the stairs of a 20-m-tall building while carrying a 40 kg backpack filled with gear. How much power does the firefighter need to reach the top in 55 s?

62. || A hydroelectric power plant uses spinning turbines to transform the kinetic energy of moving water into electric energy with 90% efficiency. That is, 90% of the kinetic energy becomes electric energy. A small hydroelectric plant at the base of a dam generates 50 MW of electric power when the falling water has a speed of 18 m/s. What is the water flow rate—kilograms of water per second—through the turbines?

63. ||| When you ride a bicycle at constant speed, nearly all the energy you expend goes into the work you do against the drag force of the air. Model a cyclist as having cross-section area 0.45 m^2 and, because the human body is not aerodynamically shaped, a drag coefficient of 0.90.
BIO
   a. What is the cyclist's power output while riding at a steady 7.3 m/s (16 mph)?
   b. *Metabolic power* is the rate at which your body "burns" fuel to power your activities. For many activities, your body is roughly 25% efficient at converting the chemical energy of food into mechanical energy. What is the cyclist's metabolic power while cycling at 7.3 m/s?
   c. The food calorie is equivalent to 4190 J. How many Calories does the cyclist burn while riding over level ground at 7.3 m/s for 1.0 h?

64. || A hedgehog flea, with a mass of 0.70 mg, can launch itself at a speed of 1.3 m/s. The energy for this jump comes primarily from two small *resilin pads,* one in each hind leg. Resilin, found in many insects, is an elastic protein with a Young's modulus of $1.8 \times 10^6\ N/m^2$. The flea, as it bends its legs, uses its muscles to slowly stretch the rubber-band-like resilin, doing work on the resilin; it then uses a latch to hold the resilin in its stretched state. You can think of the work done as stored energy. When the latch is released, the stored energy is rapidly transformed into the flea's kinetic energy, somewhat like a catapult. The *resilience* of resilin—the energy it can release when contracting relative to the work done to stretch it—is 95%.
BIO INT
   a. Each resilin pad can be approximated as having a box-like shape that is 85 μm long, 60 μm wide, and 30 μm thick. Research suggests that each pad is stretched enough to double its length. How much work must be done on a pad to double its length? Assume that resilin is a linearly elastic material and that, because the width and thickness contract, the pad's volume is not appreciably changed during stretching.
   b. With what speed will the flea be launched when the stored energy is released? (Be careful with unit conversions, especially of mass.) Your answer should be close to but somewhat less than the observed speed of 1.3 m/s, which tells us that stretched resilin is not the only source of energy. Researchers think that additional energy is stored in a bent cuticle that surrounds the resilin.

65. || A linearly elastic material is one for which the stress σ and strain ε are related by $\sigma = Y\varepsilon$, where Y is Young's modulus for the material. Demonstrate that the work done per unit volume to stretch a linearly elastic material is $W/V = \frac{1}{2}Y\varepsilon^2$.
INT

66. || A leaping Cuban tree frog can reach a launch speed of 3.0 m/s in a time of 100 ms.
BIO INT
   a. What is the specific power for this leap?
   b. If the frog jumped straight up, how high would it go?

67. |||| The power output of athletes is of both theoretical and practical interest in biomechanics. The velocity of a 70 kg elite sprinter who runs the 100 m dash in 10.0 s can be modeled as
BIO CALC

$$v_x = (11.7\ m/s)(1 - e^{-(0.689\ s^{-1})t})$$

where t is in s. The ground does no work on the sprinter because the point of contact, the sprinter's foot, does not move as the propulsion force $\vec{F}_{\text{ground on sprinter}}$ is exerted. Instead, the sprinter transforms chemical energy into kinetic energy. Even so, the sprinter's power output can be calculated as $P = Fv$ because he is, in effect, pushing the earth away from his body with force $\vec{F}_{\text{sprinter on ground}}$. Early in the race, before speeds have reached a point where drag becomes significant, $\vec{F}_{\text{ground on sprinter}}$ is the only horizontal force and thus $\vec{F}_{\text{ground on sprinter}} = m\vec{a}$. Determine the sprinter's power output at 0.5 s intervals from 0.0 s to 2.0 s. It's interesting that a sprinter's power peaks early in the race and then declines, although the decline is somewhat less than this model suggests because the power needed to push against drag becomes significant after about 2 s.

68. ||| Pregnancy dramatically affects the energetics of swimming for female dolphins. Dolphins have an approximately circular cross section and are very streamlined. A typical dolphin has a girth—the circumference at the widest part of the body—of 1.4 m and a drag coefficient of 0.090. Advanced pregnancy increases the girth to 1.7 m and, with a less streamlined shape, the drag coefficient to 0.22. Dolphins typically cruise at a speed of 3.4 m/s. What will be the cruising speed of a pregnant dolphin if she swims with the same power output as before becoming pregnant?
BIO INT

69. ||| Six dogs pull a two-person sled that has a total mass of 220 kg. The coefficient of kinetic friction between the sled and the snow is 0.080. The sled accelerates at 0.75 m/s^2 until it reaches a cruising speed of 12 km/h. What is the team's (a) maximum power output during the acceleration phase and (b) power output during the cruising phase?

MCAT-Style Passage Problems

Work and Power in Cycling

When you ride a bicycle at constant speed, almost all of the energy you expend goes into the work you do against the drag force of the air. In this problem, assume that *all* of the energy expended goes into working against drag. As we saw in Section 5.4, the drag force on an object is approximately proportional to the square of its speed with respect to the air. For this problem, assume that $F \propto v^2$ exactly and that the air is motionless with respect to the ground unless noted otherwise. Suppose a cyclist and her bicycle have a combined mass of 60 kg and she is cycling along at a speed of 5 m/s.

70. If the drag force on the cyclist is 10 N, how much energy does she use in cycling 1 km?
   A. 6 kJ
   B. 10 kJ
   C. 50 kJ
   D. 100 kJ

71. Under these conditions, how much power does she expend as she cycles?
   A. 10 W
   B. 50 W
   C. 100 W
   D. 200 W

72. If she doubles her speed to 10 m/s, how much energy does she use in cycling 1 km?
   A. 20 kJ
   B. 40 kJ
   C. 200 kJ
   D. 400 kJ

73. How much power does she expend when cycling at that speed?
   A. 100 W
   B. 200 W
   C. 400 W
   D. 1000 W

74. Upon reducing her speed back down to 5 m/s, she hits a headwind of 5 m/s. How much power is she expending now?
   A. 100 W
   B. 200 W
   C. 500 W
   D. 1000 W

11 Interactions and Potential Energy

The odd hopping gait of a kangaroo is actually quite efficient because the kangaroo can store a significant amount of energy in the stout tendons in its legs.

LOOKING AHEAD

Potential Energy

Potential energy is transformed into kinetic energy as this bungee jumper falls, then back into potential energy as the cord stretches.

Potential energy is the *interaction energy* within a system. You'll learn to calculate gravitational and elastic potential energy.

Conservation of Energy

This *energy diagram* is a graphical tool for seeing how kinetic and potential energy change in an isolated system.

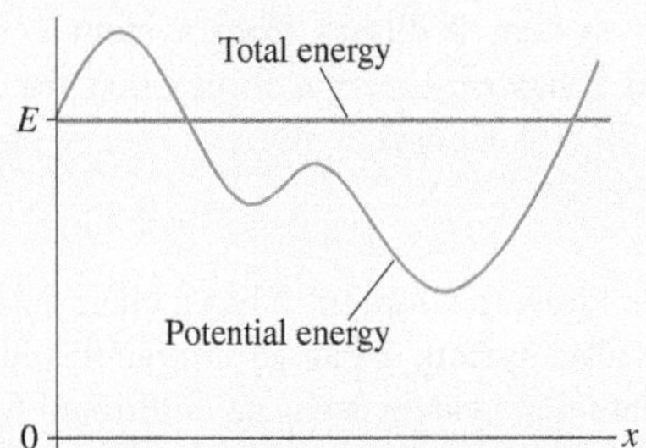

You'll see how the concept of *energy conservation* leads to an understanding of molecular bonds and chemical energy.

Energy Use in the Body

The energy to run, to move, and to stay warm comes from chemical energy stored in the molecules of the food we eat.

You'll learn to calculate how much energy your body requires to complete a range of tasks.

GOAL To develop a better understanding of energy and its application to living systems.

PHYSICS AND LIFE

Energy and the Body

How does the food we eat power our bodies? Why does some of the energy we get from food warm our bodies even on hot days when we don't need it? To answer questions like these, we're going to continue our investigation of energy by looking at how *interactions* between objects, from the atoms in molecules to the pull of earth's gravity on all living things, involve energy. We will examine how tendons and other biomaterials store energy and how we can understand chemical bonds in terms of energy. We'll also look at how efficient our body is at converting the chemical energy of food into the energy of motion and why biochemical metabolism converts carbohydrates from our food to ATP as an intermediate step instead of powering our cells directly from sugars.

Surprisingly, the heat of a candle flame is not the energy released by breaking molecular bonds but the energy released when new and stronger bonds are formed.

11.1 Potential Energy

If you press a ball against a spring and release it, the ball shoots forward. It certainly seems like the spring had a supply of stored energy that was transferred to the ball. Or imagine tossing the ball straight up. Where does its kinetic energy go as it slows? And from where does it acquire kinetic energy as it falls? There's again a sense that the energy is stored somewhere as the ball rises, then released as the ball falls. But is energy really stored? And if so, where? And how? Answering these questions is key to expanding our understanding of the basic energy model.

The key idea of Chapter 10 was that energy is a property of a *system* and that forces from the *environment*—external forces—can change the system's energy by doing work on the system. Chapter 10 focused on systems of one or more particles where all forces are external forces that originate in the environment. But that's not the only way to define the system. An alternative is to define the system with the forces inside. How does that affect the system's energy?

FIGURE 11.1 Two ways to think about a spring pushing objects apart.

(a)

Before:

After:

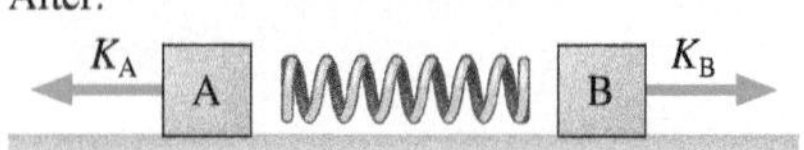

(b) External forces do work, *transferring* kinetic energy to A and B.

System 1

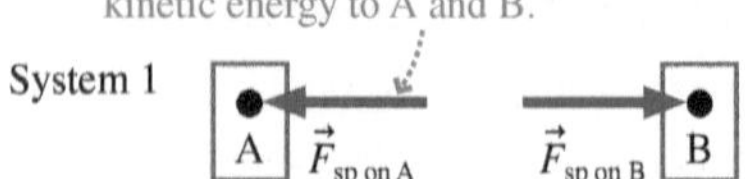

(c) The interaction energy (potential energy) is *transformed* to kinetic energy.

System 2

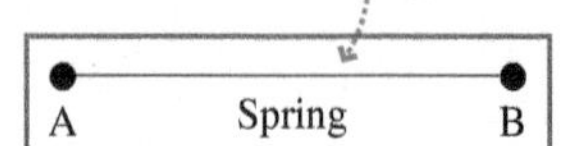

For example, suppose objects A and B are connected by a compressed spring. If we release them, they'll fly apart and gain kinetic energy. FIGURE 11.1a is a before-and-after representation and FIGURE 11.1b is an analysis, based on the discussion of Chapter 10, in which we've defined system 1 to consist of only the two objects. The spring forces $\vec{F}_{sp\ on\ A}$ and $\vec{F}_{sp\ on\ B}$ do work W_A on A and W_B on B. For system 1, the energy principle

$$\Delta E_{sys\ 1} = \Delta K_{tot} = \Delta K_A + \Delta K_B = W_A + W_B \tag{11.1}$$

tells us that the work done by the spring changes the system's kinetic energy.

But recall that forces are interactions. With that in mind, we can consider a different choice of system, system 2, in which the spring—the interaction—is inside the system. FIGURE 11.1c is an interaction diagram, like those we drew in Chapter 5, showing that A and B are interacting with each other via the spring. It's important to recognize that a system is an analysis tool, not a physical thing. We can define the system however we want; our choice does not change the behavior of physical objects. Objects A and B are oblivious to our choice of system, so ΔK_{tot} for system 2 is exactly the same as for system 1.

But system 2 differs from system 1 in one important way. Unlike system 1, system 2 has *no* external forces that transfer energy to or from the environment; hence $W = 0$. Consequently, the energy principle for system 2 is

$$\Delta E_{sys\ 2} = W = 0 \tag{11.2}$$

But we know that system 2 has a changing kinetic energy, so how can $\Delta E_{sys\ 2} = 0$?

Because system 2 has an interaction inside the system that system 1 lacks, let's postulate that system 2 has an additional form of energy associated with the interaction. That is, system 1 has $E_{sys\ 1} = K_{tot}$, because particles have only kinetic energy, but system 2 has $E_{sys\ 2} = K_{tot} + U$, where U, called **potential energy**, is the energy of the interaction.

If this is true, we can combine $\Delta E_{sys\ 2} = 0$, from Equation 11.2, with our knowledge of ΔK_{tot} from Equation 11.1 to write

$$\Delta E_{sys\ 2} = \Delta K_{tot} + \Delta U = (W_A + W_B) + \Delta U = 0 \tag{11.3}$$

That is, system 2 can have $\Delta E_{sys} = 0$ if it has a potential energy that changes by just the right amount to offset the change in kinetic energy:

$$\Delta U = -(W_A + W_B) = -W_{int} \tag{11.4}$$

where W_{int} is the total work done *inside the system* by the interaction forces.

Equation 11.3 tells us that the system's kinetic energy can increase ($\Delta K > 0$) if its potential energy decreases ($\Delta U < 0$) by the same amount. In effect, **the interaction stores energy inside the system** with the *potential* to be converted to kinetic energy (or, in other situations, to thermal energy)—hence the name *potential energy*. This idea will become more concrete as we start looking at specific examples.

NOTE ▶ Kinetic energy is the energy of an object. In contrast, potential energy is the energy of an interaction. You can say "The ball has kinetic energy" but not "The ball has potential energy." We'll look at the best way to describe potential energy when we get to specific examples. ◀

The power of water Hydroelectric power is all about the transfer and transformation of energy. The gravitational potential energy of water stored behind the dam is transformed into kinetic energy as water falls to the turbines at the base of the dam. There the water does work as it pushes on the turbine blades, transferring energy to the rotating turbines. The turbines, in turn, do work by spinning the coils of a generator, transforming mechanical energy into electric energy. That energy is transported over the grid to end users where some of it does useful work (e.g., turning electric motors) and the remainder is dissipated as thermal energy.

Systems Matter

In Figure 11.1, system 1 is a restricted system of just the particles, so system 1 has only kinetic energy. All the interaction forces are external forces that do work. Thus system 1 obeys

$$\Delta E_{\text{sys 1}} = \Delta K_{\text{tot}} = W_A + W_B$$

System 2 includes the interaction, so system 2 has both kinetic and potential energy. But the choice of the system boundary means that no work is done by external forces. So for system 2,

$$\Delta E_{\text{sys 2}} = \Delta K_{\text{tot}} + \Delta U = 0$$

Both mathematical statements are true because they refer to different systems. Notice that, for system 2, kinetic energy can be transformed into potential energy, or vice versa, but **the total energy of the system does not change, $\Delta E_{\text{sys 2}} = 0$,** because no external forces do work on it. This is in keeping with the idea, from the basic energy model, that the energy of an isolated system does not change.

The point to remember is that **any choice of system is acceptable, but you can't mix and match.** You can define the system so that you have to calculate work, or you can define the system where you use potential energy, but using both work *and* potential energy is incorrect because it double counts the contribution of the interaction. Thus **the most critical step in an energy analysis is to clearly define the system you're working with.**

Gravitational Potential Energy

We'll start our exploration of potential energy with **gravitational potential energy,** the interaction energy associated with the gravitational interaction between two masses. The symbol for gravitational potential energy is U_G.

FIGURE 11.2 shows a ball of mass m moving upward from an initial vertical position y_i to a final vertical position y_f. The earth exerts force $\vec{F}_{\text{E on B}}$ on the ball and, by Newton's third law, the ball exerts an equal-but-opposite force $\vec{F}_{\text{B on E}}$ on the earth.

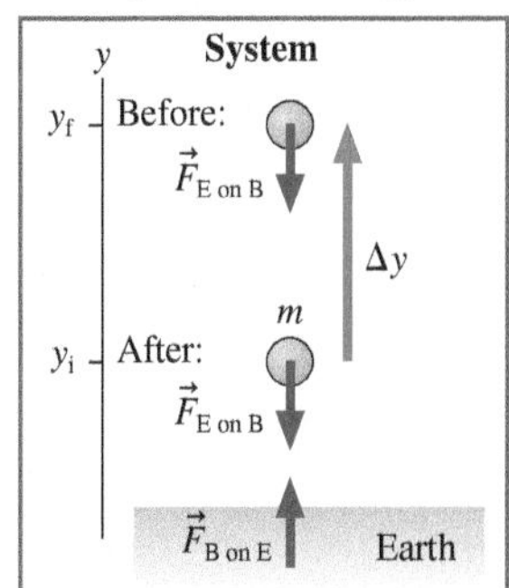

FIGURE 11.2 The ball + earth system has a gravitational potential energy.

We could define the system to consist of only the ball, in which case the force of gravity is an external force that does work on the ball, changing its kinetic energy. We did exactly this in Chapter 10. Now let's define the system to be ball + earth. This brings the interaction inside the system, so there are no external forces. Instead, we have an energy of interaction—the gravitational potential energy—described by Equation 11.4:

$$\Delta U_G = -W_{\text{int}} = -(W_B + W_E) \tag{11.5}$$

where W_B is the work gravity does on the ball and W_E is the work gravity does on the earth. The latter, practically speaking, is zero. $\vec{F}_{\text{E on B}}$ and $\vec{F}_{\text{B on E}}$ have equal magnitudes, by Newton's third law, but the earth's displacement is completely insignificant compared to the ball's displacement. Because work is a product of force and displacement, the work done on the earth is essentially zero and we can write

$$\Delta U_G = -W_B \tag{11.6}$$

You learned in Chapter 10 to compute the work of gravity on the ball: $W_B = w_y \Delta y = -mg\,\Delta y$, where $w_y = -mg$ is the y-component of the weight force. So if the ball changes its vertical position by Δy, the gravitational potential energy changes by

$$\Delta U_G = -W_B = mg\,\Delta y \tag{11.7}$$

Notice that increasing the ball's height ($\Delta y > 0$) increases the gravitational potential energy ($\Delta U_G > 0$), as we would expect.

Our energy analysis has given us an expression for ΔU_G, the *change* in potential energy, but not an expression for U_G itself. If we write out what the Δ in Equation 11.7 means—final value minus initial value—we have

$$U_{G\,f} - U_{G\,i} = mgy_f - mgy_i \tag{11.8}$$

Consequently, we define the gravitational potential energy to be

$$U_G = mgy \tag{11.9}$$

Gravitational potential energy for an object of mass m at height y

Notice that **gravitational potential energy is an energy of position.** It depends on the object's position but not on its speed. You should convince yourself that the units of mass times acceleration times position are joules, the unit of energy.

As a quick example, suppose a weightlifter lifts a 50 kg barbell from the floor to a height of 2.1 m as it's held over his head. The gravitational potential energy of the barbell + earth system increases by

$$\Delta U_G = mg\Delta y = (50\text{ kg})(9.8\text{ m/s}^2)(2.1\text{ m}) = 1000\text{ J}$$

STOP TO THINK 11.1 Rank in order, from largest to smallest, the gravitational potential energies of the ball + earth system when the ball is at positions A, B, C, and D.

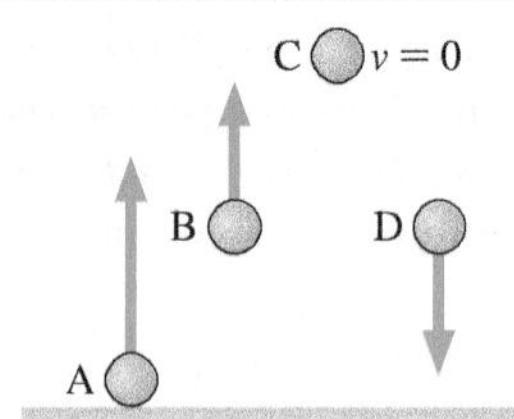

11.2 Conservation of Energy

We noticed that the total system energy does not change if no work is done by external forces. It's time to expand on this idea. One of the most powerful statements in physics is the **law of conservation of energy:**

Law of conservation of energy The total energy $E_{sys} = K + U + E_{th}$ of an isolated, nonreactive ($\Delta E_{chem} = 0$) system is a constant. The kinetic, potential, and thermal energy within the system can be transformed into each other, but their sum cannot change. Further, the *mechanical energy* $E_{mech} = K + U$ is conserved if the system is isolated, nonreactive, and nondissipative.

NOTE ▶ For now we're using the *basic energy model,* which considers only kinetic, potential, and thermal energies and in which work is the only process of energy transfer. We'll consider chemical energy and chemical reactions later in the chapter and in upcoming chapters. ◀

The key is that energy is conserved for an **isolated system,** a system that does not exchange energy with its environment either because it has no interactions with the environment or because those interactions do no work. **FIGURE 11.3** shows our basic energy model for an isolated system.

FIGURE 11.3 The basic energy model for an isolated system.

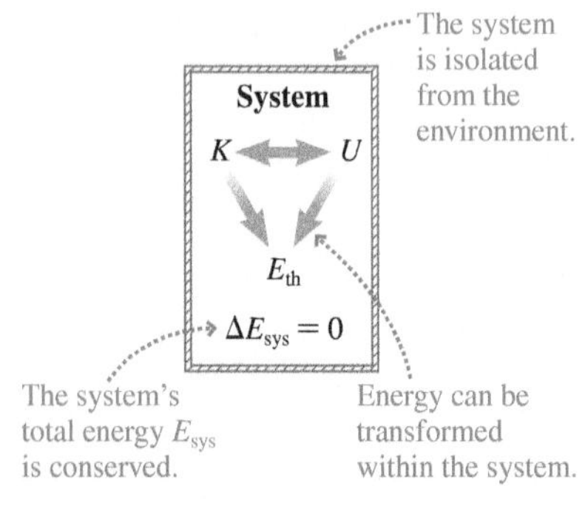

The total macroscopic kinetic and potential energy $K + U$ is often called the *mechanical energy* E_{mech}. Mechanical energy is transformed into thermal energy by any dissipative forces such as friction or drag. The real world always has dissipative forces, but we'll often find it useful to model processes as being nondissipative, in which case the mechanical energy is conserved.

It Depends on the System

As significant as the law of conservation of energy is, it's critical to notice that the law does *not* say "Energy is always conserved." The law of conservation of energy refers to the energy *of a system*—hence our emphasis on systems in Chapters 10 and 11. Energy is conserved for some choices of system, but not others. For example, energy is conserved for a ball sailing through the air (ignoring air resistance) if we define the system to be the ball + earth but not if we define the system to be only the ball.

In addition, the law of conservation of energy comes with an important qualification: Is the system isolated? Energy is certainly not conserved if an external force does work on the system. Thus the answer to the question "Is energy conserved?" is "It depends on the system."

A Strategy for Energy Problems

To say that energy is constant or conserved is to say that the *final* energy, after an interaction has occurred, equals the *initial* energy: $(E_{sys})_i = (E_{sys})_f$. This idea is the basis of a problem-solving strategy for energy problems:

PROBLEM-SOLVING STRATEGY 11.1 **Energy-conservation problems**

PREPARE Define the system so that there are no external forces or so that any external forces do no work on the system. If there's friction, bring both surfaces into the system. Draw a before-and-after pictorial representation. A free-body diagram may be needed to visualize forces.

SOLVE If the system is both isolated and nondissipative, then the mechanical energy is conserved:

$$K_i + U_i = K_f + U_f$$

where K is the total kinetic energy of all moving objects and U is the total potential energy of all interactions within the system. If there's friction, then

$$K_i + U_i = K_f + U_f + \Delta E_{th}$$

where the thermal energy increase due to friction is $\Delta E_{th} = f_k \Delta x$.

ASSESS Check that your result has correct units and significant figures, is reasonable, and answers the question.

STOP TO THINK 11.2 Two identical projectiles are fired with the same speed but at different angles. Neglect air resistance. At the elevation shown as a dashed line,

A. The speed of 1 is greater than the speed of 2.
B. The speed of 1 is the same as the speed of 2.
C. The speed of 1 is less than the speed of 2.

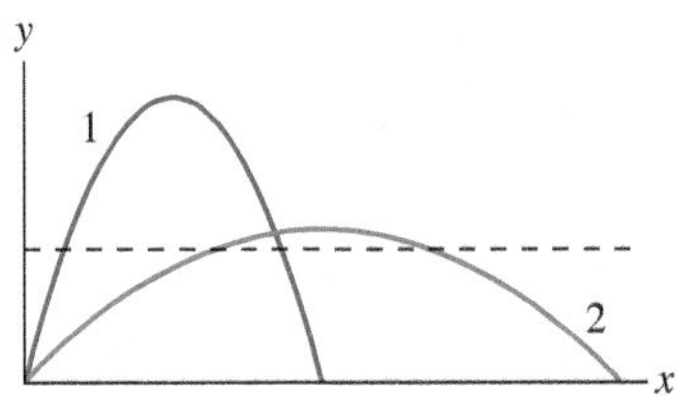

Energy Bar Charts

If there are no dissipative forces, then the conservation of mechanical energy

$$K_i + U_i = K_f + U_f \quad (11.10)$$

can be represented graphically with an **energy bar chart.** For example, FIGURE 11.4 is a bar chart showing how energy is transformed when a ball is tossed straight up. Kinetic energy is gradually transformed into potential energy as the ball rises, then potential energy is transformed into kinetic energy as it falls, but **the combined height of the bars does not change.** That is, the mechanical energy of the ball + earth system is conserved.

FIGURE 11.4 Energy bar charts for a ball tossed into the air.

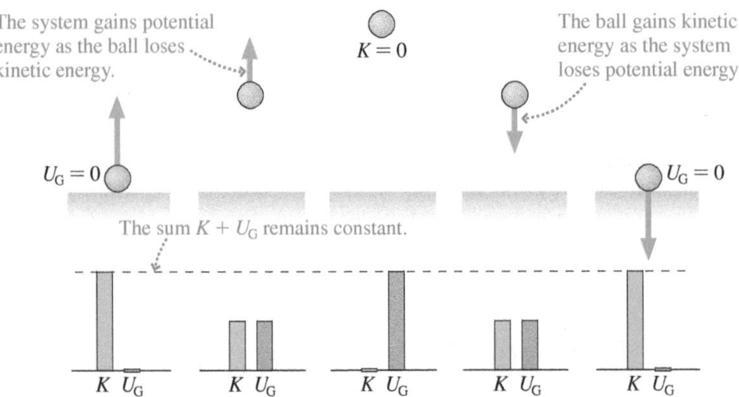

NOTE ▶ Most bar charts have no numbers. Their purpose is to think about the relative changes—what's increasing, what's decreasing, and what remains constant; there's no significance to how tall a bar is. ◀

The Zero of Potential Energy

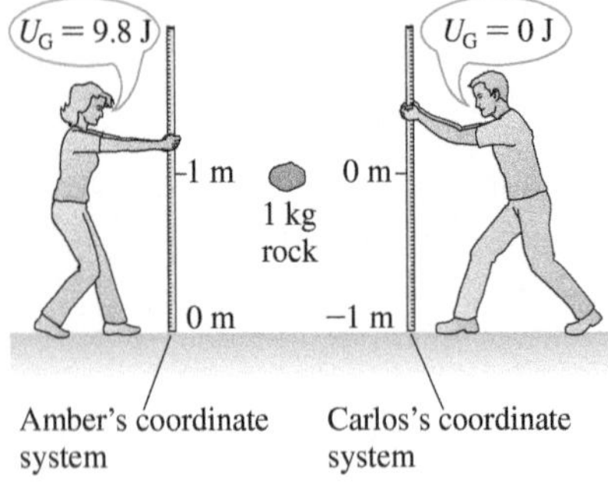

FIGURE 11.5 Amber and Carlos use different coordinate systems to determine the gravitational potential energy.

Our expression for the gravitational potential energy $U_G = mgy$ seems straightforward. But you might notice, on further reflection, that the value of U_G depends on where you choose to put the origin of your coordinate system. Consider FIGURE 11.5, where Amber and Carlos are attempting to determine the potential energy when a 1 kg rock is 1 m above the ground. Amber chooses to put the origin of her coordinate system on the ground, measures $y_{rock} = 1$ m, and quickly computes $U_G = mgy = 9.8$ J. Carlos, on the other hand, read Chapter 2 very carefully and recalls that it is entirely up to him where to locate the origin of his coordinate system. So he places his origin next to the rock, measures $y_{rock} = 0$ m, and declares that $U_G = mgy = 0$ J!

How can the potential energy have two different values? The source of this apparent difficulty comes from our interpretation of Equation 11.7. We found that the potential energy *changes* by $\Delta U_G = mg(y_f - y_i)$. Our claim that $U_G = mgy$ is consistent with this finding, but so also would be a claim that $U_G = mgy + C$, where C is any constant.

In other words, potential energy does not have a uniquely defined value. Adding or subtracting the same constant from all potential energies in a problem has no physical consequences because our analysis uses only ΔU_G, the *change* in the potential energy, never the actual value of U_G. In practice, we work with potential energies by setting a *reference point* or *reference level* where $U_G = 0$. This is the **zero of potential energy.** Where you place the reference point is entirely up to you; it makes no difference as long as every potential energy in the problem uses the same reference point. For gravitational potential energy, we choose the reference level by placing the origin of the y-axis at that point. Where $y = 0$, $U_G = 0$. In Figure 11.5, Amber has placed her zero of potential energy at the ground, whereas Carlos has set a reference level 1 m above the ground. Either is perfectly acceptable as long as Amber and Carlos use their reference levels consistently.

But what happens when the rock falls? When it gets to the ground, Amber measures $y = 0$ m and computes $U_G = 0$ J. No problem. But Carlos measures $y = -1$ m and thus computes $U_G = -9.8$ J. A negative potential energy may seem surprising, but it's not wrong; it simply means that the potential energy is less than at the reference point. The potential energy with the rock on the ground is certainly less than when the rock was 1 m above the ground, so for Carlos—with an elevated reference level—the potential energy is negative. The important point is that both Amber and Carlos agree that the gravitational potential energy *changes* by $\Delta U_G = -9.8$ J as the rock falls.

NOTE ▶ It may be tempting to refer to *the rock's potential energy,* but potential energy is not a property of an object. Potential energy is an interaction energy and thus is a property of the system. ◀

EXAMPLE 11.1 Dropping a watermelon

A 5.0 kg watermelon is dropped from a third-story balcony, 11 m above the street. Unfortunately, the water department forgot to replace the cover on a manhole, and the watermelon falls into a 3.0-m-deep hole. How fast is the watermelon going just before it hits bottom?

PREPARE Let the system consist of both the earth and the watermelon. Assume that air resistance is negligible. There are no external forces and no dissipation, so the system's mechanical energy is conserved.

FIGURE 11.6 shows both a before-and-after pictorial representation and an energy bar chart. Initially the system has gravitational potential energy but no kinetic energy. Potential energy is transformed into kinetic energy as the watermelon falls. Our choice of the y-axis origin has placed the zero of potential energy at ground level, so the potential energy is negative when the watermelon reaches the bottom of the hole. Even so, the combined height of the two bars has not changed.

FIGURE 11.6 Pictorial representation and energy bar chart of the watermelon + earth system.

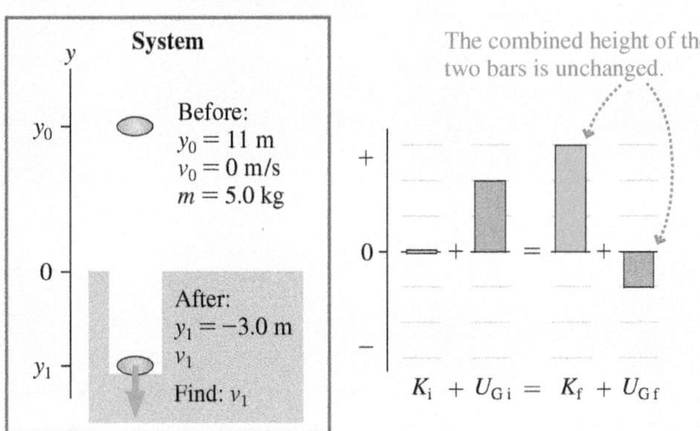

SOLVE The energy conservation statement for the watermelon + earth system is

$$K_i + U_{Gi} = 0 + mgy_0 = K_f + U_{Gf} = \tfrac{1}{2}mv_1^2 + mgy_1$$

Solving for the impact speed, we find

$$\begin{aligned} v_1 &= \sqrt{2g(y_0 - y_1)} \\ &= \sqrt{2(9.80\ \text{m/s}^2)(11.0\ \text{m} - (-3.0\ \text{m}))} \\ &= 17\ \text{m/s} \end{aligned}$$

ASSESS A speed of $17\ \text{m/s} \approx 35$ mph seems reasonable for the watermelon after falling ≈ 4 stories. In thinking about this problem, you might be concerned that, once below ground level, potential energy continues being transformed into kinetic energy even though the potential energy is "less than none." Keep in mind that the actual value of U is not relevant because we can place the zero of potential energy anywhere we wish, so a negative potential energy is just a number with no implication that it's "less than none." There's no "storehouse" of potential energy that might run dry. As long as the interaction acts, potential energy can continue being transformed into kinetic energy.

Digging Deeper into Gravitational Potential Energy

The concept of gravitational potential energy would be of little use if it applied only to vertical free fall. We can expand the idea. The object in FIGURE 11.7 is sliding down a curved, frictionless surface. Now there are two forces acting on it: weight and the normal force. Does this affect our analysis of gravitational potential energy?

No! First, notice that the normal force is always perpendicular to the direction of motion. You learned in Chapter 10 that a force that is always perpendicular to the displacement does no work, so we can ignore the normal force in energy calculations. In general, **forces that are always perpendicular to the direction of motion do not affect a system's energy.**

Second, the weight force is perpendicular to any horizontal motion, so no work is done on any horizontal component of the object's displacement. The work done by gravity, $-mg\Delta y$, depends on only the object's vertical displacement Δy. Thus **the change in gravitational potential energy depends on only an object's vertical displacement.** The shape of the trajectory—straight or curved—does not matter.

FIGURE 11.7 For motion on any frictionless surface, only the vertical displacement Δy affects the energy.

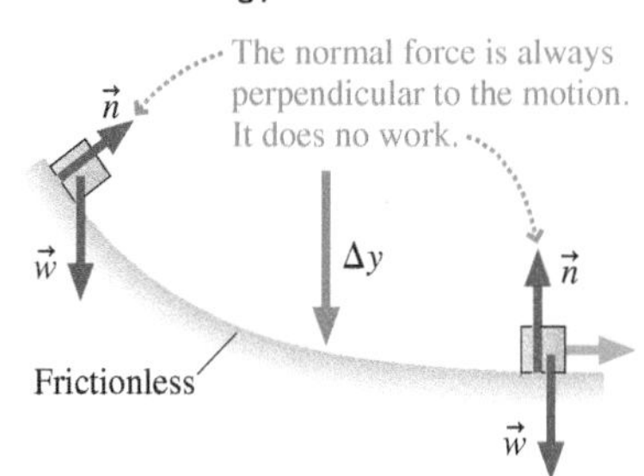

EXAMPLE 11.2 **The speed of a pendulum**

A pendulum is created by attaching one end of a 78-cm-long string to the ceiling and tying a 150 g steel ball to the other end. The ball is pulled back until the string is 60° from vertical, then released. What is the speed of the ball at its lowest point?

PREPARE We'll let the system consist of the earth and the ball. The tension force, like a normal force, is always perpendicular to the motion and does no work, so this is an isolated system with no friction. Its mechanical energy is conserved. It's important to recognize that the change in gravitational potential energy depends on only the ball's vertical displacement Δy, *not* the shape of its trajectory as it swings.

FIGURE 11.8 shows a before-and-after pictorial representation, where we've placed the zero of potential energy at the lowest point of the ball's swing. Notice how trigonometry is used to determine the ball's initial height. The energy bar chart shows that gravitational potential energy is transformed entirely into kinetic energy as the ball swings to the bottom of its arc.

SOLVE Conservation of mechanical energy is

$$K_i + U_{G\,i} = 0 + mgy_0 = K_f + U_{G\,f} = \tfrac{1}{2}mv_1^2 + 0$$

We've noted that the ball starts with zero kinetic energy and ends with zero potential energy. Thus the ball's speed at the bottom is

$$v_1 = \sqrt{2gy_0} = \sqrt{2(9.80\ \text{m/s}^2)(0.39\ \text{m})} = 2.8\ \text{m/s}$$

The speed is exactly the same as if the ball had simply fallen 0.39 m.

ASSESS To solve this problem directly from Newton's laws of motion requires advanced mathematics because of the complex way the net force changes with angle. But we can solve it in a few lines with an energy analysis!

FIGURE 11.8 Pictorial representation of a pendulum.

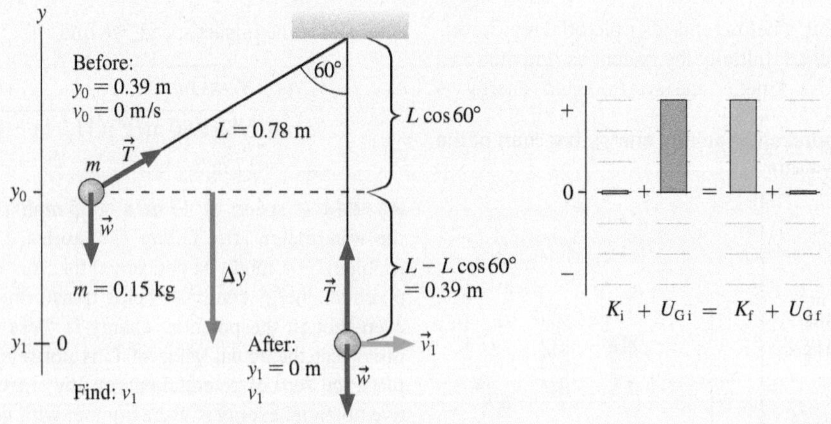

STOP TO THINK 11.3 A small child slides down the four frictionless slides A–D. Each has the same height. Rank in order, from largest to smallest, her speeds v_A to v_D at the bottom.

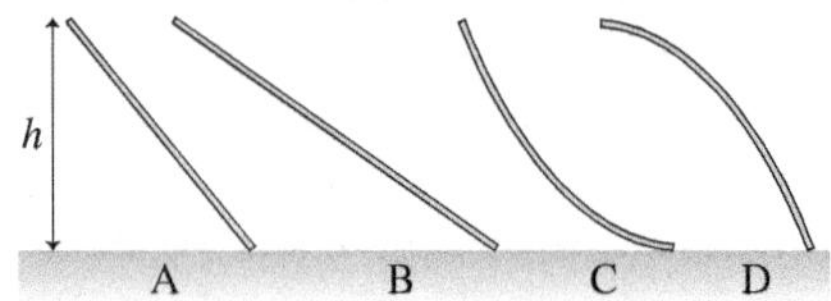

Motion with Gravity and Friction

What if there's friction? You learned in ◀◀ SECTION 10.4 that friction increases the thermal energy of the system—defined to include *both* objects—by $\Delta E_{th} = f_k \Delta x$. For an isolated system with both gravitational potential energy and friction, the energy principle becomes

$$\Delta K + \Delta U_G + \Delta E_{th} = 0 \tag{11.11}$$

or, equivalently, the energy conservation statement is

$$K_i + U_{G\,i} = K_f + U_{G\,f} + \Delta E_{th} \tag{11.12}$$

Mechanical energy $K + U_G$ is *not* conserved if there is friction. Because $\Delta E_{th} > 0$ (friction always makes surfaces hotter, never cooler), the final mechanical energy is less than the initial mechanical energy. That is, some fraction of the initial kinetic and potential energy is transformed into thermal energy during the motion. Friction causes objects to slow down, and motion ceases when all the mechanical energy has been transformed into thermal energy. Mechanical energy is conserved only when there are no dissipative forces and thus $\Delta E_{th} = 0$.

NOTE ▶ We can write the energy principle in terms of initial and final values of the kinetic energy and the potential energy, but *not* the thermal energy. Objects always have thermal energy—the atoms are constantly in motion—but we have no way to know how much. All we can calculate is the *change* in thermal energy. ◀

Although mechanical energy is not conserved, *the system's energy is*. Equation 11.11 tells us that the sum of kinetic, potential, and thermal energy—the energy of the system—does not change as the object moves on a surface with friction. The initial mechanical energy does not disappear; it's merely transformed into an equal amount of thermal energy.

EXAMPLE 11.3 Tobogganing penguins

Penguins are flightless birds that are well adapted to swimming but awkward on land. Walking through snow on short legs is energetically demanding, so penguins are often observed gliding across the snow on their stomachs, a mode of transport known as *tobogganing*. They use their feet and flippers for propulsion across level snow but simply slide down slopes. Researchers in Antarctica determined that a typical coefficient of kinetic friction between penguin feathers and snow is 0.13 by measuring the force needed to drag a recently deceased penguin across level snow. Suppose a 4.5 kg penguin tobogganing at 1.0 m/s reaches a 6.0-m-long, 8.0° slope and slides to the bottom. How much of the penguin's initial mechanical energy is transformed into thermal energy? What is its speed at the bottom?

PREPARE Let the system consist of the penguin, the earth, and the snow. Mechanical energy is not conserved because of friction, but the total energy is conserved because we've included both the penguin and the snow in the system. FIGURE 11.9 shows a pictorial representation. Gravitational potential energy depends on only the height above the reference level, and we've chosen to place the origin of the y-axis at the bottom of the slope. Trigonometry is then used to find the initial height y_0.

FIGURE 11.9 Pictorial representation of a penguin sliding down a slope.

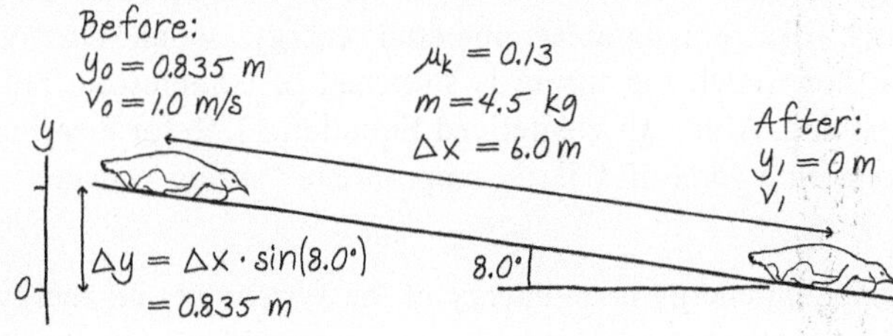

SOLVE The initial energy of the system, with the penguin at the top of the slope, is

$$(E_{sys})_i = K_i + U_{Gi} = \tfrac{1}{2}mv_0^2 + mgy_0$$

$$= \tfrac{1}{2}(4.5\text{ kg})(1.0\text{ m/s})^2 + (4.5\text{ kg})(9.8\text{ m/s}^2)(0.835\text{ m}) = 39.1\text{ J}$$

As the penguin slides down, the increase in thermal energy of the penguin and the snow is $\Delta E_{th} = f_k \Delta x$, with $\Delta x = 6.0$ m measured along the slope. Sliding friction is $f_k = \mu_k n$, and recall—from Chapter 5—that the normal force acting on an object on a slope is $n = mg\cos\theta$ (draw a free-body diagram if you're not sure). Thus

$$\Delta E_{th} = f_k \Delta x = (\mu_k mg\cos\theta)\Delta x$$

$$= (0.13)(4.5\text{ kg})(9.8\text{ m/s}^2)(\cos 8.0°)(6.0\text{ m}) = 34.1\text{ J}$$

We see that most (87%) of the initial energy is transformed into thermal energy.

With friction, the energy conservation statement is

$$K_i + U_{Gi} = K_f + U_{Gf} + \Delta E_{th}$$

The potential energy at the bottom of the slope is zero, so the final kinetic energy is

$$K_f = K_i + U_{Gi} - \Delta E_{th} = (E_{sys})_i - \Delta E_{th} = 5.0\text{ J}$$

Thus the penguin's speed at the bottom is

$$v_1 = \sqrt{\frac{2K_f}{m}} = \sqrt{\frac{2(5.0\text{ J})}{4.5\text{ kg}}} = 1.5\text{ m/s}$$

ASSESS Most of the initial energy is dissipated as thermal energy, so the penguin glides down the slope without effort but also without gaining much speed. A kinetic friction coefficient of 0.13 is not especially small—feathers have little ridges that probably help them grip the snow—and an 8° slope is not especially steep. Thus a fairly small increase in speed is not surprising.

STOP TO THINK 11.4 A skier glides down a gentle slope at constant speed. Let the system consist of the skier and the earth, including the snowy slope. What energy transformation is taking place?

A. $U_G \rightarrow K$
B. $U_G \rightarrow K + E_{th}$
C. $U_G \rightarrow E_{th}$
D. $K \rightarrow E_{th}$
E. No energy transformation is occurring.

11.3 Elastic Potential Energy

FIGURE 11.10 The block + spring + wall system has an elastic potential energy.

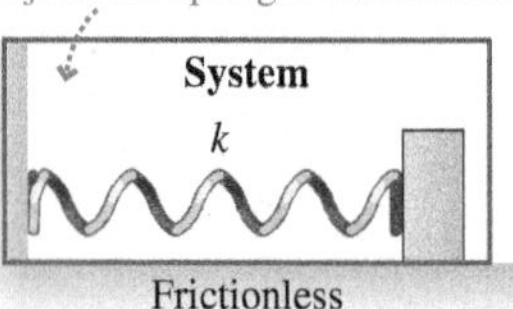
The system is the spring and the objects the spring is attached to.
System
k
Frictionless

Much of what you've just learned about gravitational potential energy carries over to the *elastic potential energy* of a spring. FIGURE 11.10 shows a spring exerting a force on a block while the block moves on a frictionless, horizontal surface. We'll define the system to be block + spring + wall. That is, the system is the spring and the objects that are interacting with the spring. The surface and the earth exert forces on the block—the normal force and the weight force—but those forces are always perpendicular to the displacement and do not transfer any energy to the system.

Let's model the spring as being massless, which is a reasonable model as long as the spring's mass is much less than the mass of the block. With this model, the spring has no kinetic energy but, instead, provides an *interaction* between the block and the wall. Because the interaction is inside the system, it has an interaction energy, the **elastic potential energy,** given by

$$\Delta U_{sp} = -W_{int} = -(W_B + W_W) \tag{11.13}$$

where W_B is the work the spring does on the block and W_W is the work done on the wall. But the wall is rigid and has no displacement, so $W_W = 0$ and thus $\Delta U_{sp} = -W_B$.

We calculated the work done by an ideal spring—one that has a linear restoring force for all displacement—in ◀◀ SECTION 10.3. If the block moves from an initial position x_i, where the spring's displacement is $\Delta x_i = x_i - x_{eq}$, to a final position x_f with displacement $\Delta x_f = x_f - x_{eq}$, the spring does work

$$W_B = \tfrac{1}{2}k(\Delta x_i)^2 - \tfrac{1}{2}k(\Delta x_f)^2 \tag{11.14}$$

With the minus sign of Equation 11.13, we have

$$\Delta U_{sp} = U_f - U_i = -W_B = \tfrac{1}{2}k(\Delta x_f)^2 - \tfrac{1}{2}k(\Delta x_i)^2 \tag{11.15}$$

Thus the elastic potential energy is

$$U_{sp} = \tfrac{1}{2}k(\Delta x)^2 \tag{11.16}$$

Elastic potential energy of a stretched or compressed spring with spring constant k

where Δx is the displacement of the spring from its equilibrium length. Elastic potential energy, like gravitational potential energy, is an *energy of position*. It depends on how much the spring is stretched or compressed, not on how fast the block is moving. Although we derived Equation 11.16 for a spring, it applies to *any* linear restoring force if k is the appropriate "spring constant" for that force.

NOTE ▶ Elastic potential energy is an energy of the *system,* not an energy of the spring. ◀

EXAMPLE 11.4 **An air-track glider compresses a spring**

In a laboratory experiment, your instructor challenges you to figure out how fast a 500 g air-track glider is traveling when it collides with a horizontal spring attached to the end of the track. He pushes the glider, and you notice that the spring compresses 2.7 cm before the glider rebounds. After discussing the situation with your lab partners, you decide to hang the spring on a hook and suspend the glider from the bottom end of the spring. This stretches the spring by 3.5 cm. Based on your measurements, how fast was the glider moving?

PREPARE Let the system consist of the track, the spring, and the glider. The spring is inside the system, so the elastic interaction will be treated as a potential energy. The weight force and the normal force of the track on the glider are perpendicular to the glider's displacement, so they do no work and do not enter into an energy analysis. An air track is essentially frictionless, and there are no other external forces. FIGURE 11.11 shows a before-and-after pictorial representation of the collision, an energy bar chart, and a free-body diagram of the suspended glider.

SOLVE The glider's kinetic energy is gradually transformed into elastic potential energy as it compresses the spring. The point of maximum compression—After in Figure 11.11—is a turning point in the motion. The velocity is instantaneously zero, the glider's kinetic energy is zero, and thus—as the bar chart shows—all the energy has been transformed into potential energy. The spring will expand and cause the glider to rebound, but that's not part of this problem. Conservation of mechanical energy gives

$$K_i + U_{sp\,i} = \tfrac{1}{2}mv_0^2 + 0 = K_f + U_{sp\,f} = 0 + \tfrac{1}{2}k(\Delta x_1)^2$$

where we utilized our knowledge that the initial elastic potential energy and the final kinetic energy, at the turning point, are zero. Solving for the glider's initial speed, we find

$$v_0 = \sqrt{\frac{k}{m}}\,\Delta x_1$$

It was at this point that you and your lab partners realized you needed to determine the spring constant k. One way to do so is to measure the stretch caused by a suspended mass. The hanging glider is in equilibrium with no net force, and the free-body diagram shows that the upward spring force exactly balances the downward weight force. From Hooke's law, the *magnitude* of the spring force is $F_{sp} = k|\Delta y|$. Thus Newton's first law for the suspended glider is

$$F_{sp} = k|\Delta y| = w = mg$$

from which the spring's spring constant is

$$k = \frac{mg}{|\Delta y|} = \frac{(0.50\text{ kg})(9.80\text{ m/s}^2)}{0.035\text{ m}} = 140\text{ N/m}$$

Knowing k, you can now find that the glider's speed was

$$v_0 = \sqrt{\frac{k}{m}}\,\Delta x_1 = \sqrt{\frac{140\text{ N/m}}{0.50\text{ kg}}}(0.027\text{ m}) = 0.45\text{ m/s}$$

ASSESS A speed of ≈ 0.5 m/s is typical for gliders on an air track.

FIGURE 11.11 Pictorial representation of the experiment.

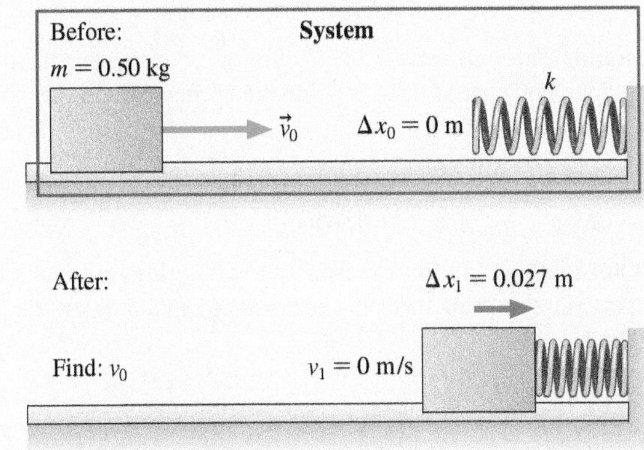

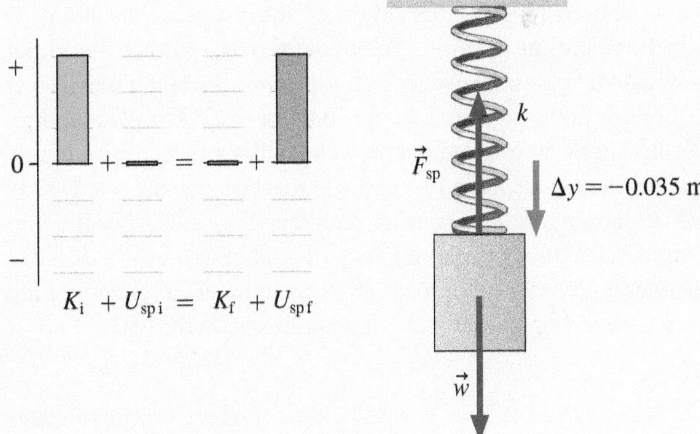

Including Additional Forms of Energy

Now that you see how the basic energy model works, we can extend these ideas to new situations. If a problem has both a spring *and* a vertical displacement, we define the system so that both the gravitational interaction and the elastic interaction are inside the system. Then we have both elastic *and* gravitational potential energy. That is,

$$U = U_G + U_{sp} \tag{11.17}$$

You have to be careful with the energy accounting because there are more ways that energy can be transformed, but nothing fundamental has changed by having two potential energies rather than one.

And we know how to include the increased thermal energy if there's friction. Thus for a system that has gravitational interactions, elastic interactions, and friction, but no external forces that do work, the energy principle for an isolated system is

$$\Delta E_{sys} = \Delta K + \Delta U_G + \Delta U_{sp} + \Delta E_{th} = 0 \tag{11.18}$$

or, in conservation form,

$$K_i + U_{G\,i} + U_{sp\,i} = K_f + U_{G\,f} + U_{sp\,f} + \Delta E_{th} \tag{11.19}$$

This is looking a bit more complex as we have more and more energies to keep track of, but the message of Equations 11.18 and 11.19 is both simple and profound: **For a system that has no other interactions with its environment, the total energy of the system does not change.** It can be transformed in many ways by the interactions, but the total does not change.

EXAMPLE 11.5 A vertical spring gun

A spring gun uses a compressed spring to fire a small ball. Instead of firing the gun horizontally, suppose you fire a spring gun straight up. When the spring is compressed by 4.0 cm, a 50 g ball reaches a height of 2.50 m above the end of the barrel. What is the spring constant of the spring?

PREPARE We'll let the system consist of the earth, the spring, and the ball, so there will be two potential energies, U_G and U_{sp}. We'll model the situation by assuming that friction and drag can be neglected, so this is an isolated system in which mechanical energy is conserved. **FIGURE 11.12** shows a pictorial representation. An important consideration is where to place the origin. We've chosen to place the origin of the y-axis at the point of launch, where the spring is fully compressed, so that the initial gravitational potential energy is zero. Hence the ball's position at its highest point is $y_1 = 2.50\text{ m} + 0.04\text{ m} = 2.54\text{ m}$. The energy bar chart tells an important story: The ball starts from rest and is again instantaneously at rest at the highest point, so $K_i = K_f = 0$. We've chosen a reference level such that $U_{G\,i} = 0$. After the gun is fired, when the spring is no longer compressed, $U_{sp\,f} = 0$. Thus the energy transfer is from entirely elastic potential energy at the start to entirely gravitational potential energy at the end.

SOLVE You might think we would need to find the ball's speed as it leaves the gun. That would be true if we were solving this problem with Newton's laws, but an energy analysis allows us to compare the system's pre-launch energy with its energy when the ball reaches its highest point. The ball certainly has kinetic energy at all points in between, but we don't need to know its value.

With both elastic and gravitational potential energies included, the energy principle is

$$K_i + U_{G\,i} + U_{sp\,i} = K_f + U_{G\,f} + U_{sp\,f}$$

$$0 + 0 + \tfrac{1}{2}k(\Delta y_0)^2 = 0 + mgy_1 + 0$$

Be careful! The ball travels to position y_1, but the end of the spring does not; that is, $\Delta y_1 = 0$ for the spring. Solving for the spring constant, we find

$$k = \frac{2mgy_1}{(\Delta y_0)^2} = \frac{2(0.050\text{ kg})(9.80\text{ m/s}^2)(2.54\text{ m})}{(-0.040\text{ m})^2} = 1600\text{ N/m}$$

ASSESS 1600 N/m is for a reasonably stiff spring, but that's to be expected for a spring that can shoot a 50 g ball 2.5 m, or ≈8 ft, straight up.

FIGURE 11.12 Pictorial representation of a spring gun.

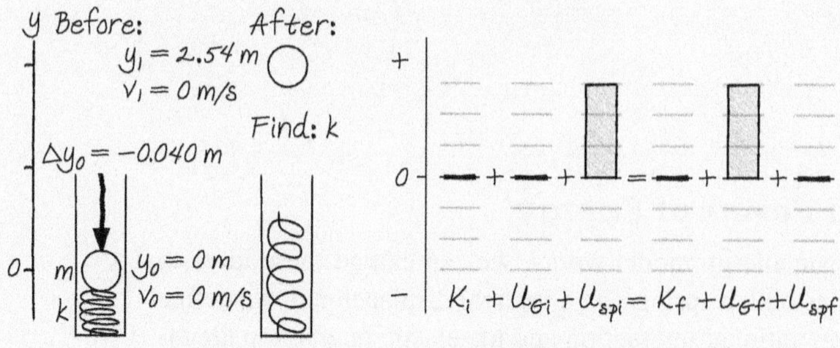

STOP TO THINK 11.5 A spring-loaded pop gun shoots a plastic ball with a speed of 4 m/s. If the spring is compressed twice as far, the ball's speed will be

A. 2 m/s B. 4 m/s
C. 8 m/s D. 16 m/s

Stored Energy in Biomaterials

The elastic potential energy of a spring is stored energy that can be transformed into kinetic energy or, if there's friction, into thermal energy. Similarly, biomaterials such as tendons, ligaments, and cartilage can store energy when they are stretched or compressed, and some or all of it can be released at a later time.

Biomaterials are characterized not by a spring constant but by Young's modulus and stress-strain curves. In ◀◀ SECTION 10.3 we calculated the work per unit volume—that is, the work per m^3 of material—needed to stretch or compress a material:

$$\frac{W}{V} = \text{work per unit volume in J/m}^3 = \int \sigma\, d\epsilon$$
$$= \text{area under the stress-strain curve} \quad (11.20)$$

where ε is the stress and σ is the strain in N/m^2.

You've now learned that doing work to stretch or compress a spring causes it to store elastic potential energy, and the same is true for any elastic material. In this case, though, the calculation of Equation 11.20 gives the stored energy per m^3 of material, which is an **energy density** with units of J/m^3. We'll use a lowercase u to represent energy density; that is, an object with volume V and stored energy U has energy density $u = U/V$. Thus the energy density of a stretched or compressed material is

$$u_{elas} = \int \sigma\, d\epsilon = \text{area under the stress-strain curve} \quad (11.21)$$

Let's look at an example.

EXAMPLE 11.6 Energy stored in a stretched tendon

The Achilles tendon is the largest tendon in the body. It stretches when you walk or run, and some of the stored energy helps you push back off the ground. The data points in FIGURE 11.13 are *in vivo* measurements of the stress generated in the Achilles tendon for different amounts of strain. This is not a linear stress-strain curve that can be characterized by a value of Young's modulus. However, data analysis finds that the stress-strain curve can be well modeled by the equation $\sigma = (4.2 \times 10^{10}\ \text{N/m}^2)\varepsilon^2$, which describes the parabola seen in the figure. Trained runners stretch their Achilles tendon to a maximum strain of about 0.075, or 7.5%. That's beyond the data shown in the graph, but we'll assume that the parabolic model is valid over the entire range. An Achilles tendon has a somewhat irregular shape, but a reasonable model is a flattened cylinder with a length of 160 mm and a cross-section area of 25 mm^2. Consider a 55 kg athlete running at 5.2 m/s (5 min miles). What is the maximum energy stored in her stretched Achilles tendon in J and as a fraction of her kinetic energy?

FIGURE 11.13 A measured stress-strain curve of an Achilles tendon.

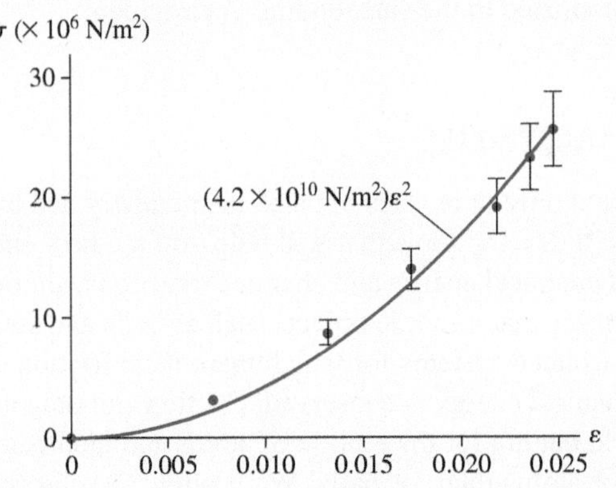

PREPARE The energy density of the stretched tendon is the area under the stress-strain curve. We'll need to carry out the integration rather than finding the area with geometry. For a 160-mm-long tendon, a strain of 0.075 results from an elongation of $\Delta L_{max} = 0.075 \times 160\ \text{mm} = 12\ \text{mm}$.

SOLVE The stored energy density of the tendon is given by Equation 11.21. The initial strain is zero. We can let ε_{max} be the

Continued

maximum strain that's reached, which occurs at the maximum elongation. Thus

$$u_{elas} = \int_0^{\varepsilon_{max}} \sigma d\varepsilon = (4.2 \times 10^{10}\ \text{N/m}^2)\int_0^{\varepsilon_{max}} \varepsilon^2 d\varepsilon$$

$$= (4.2 \times 10^{10}\ \text{N/m}^2)\left.\frac{\varepsilon^3}{3}\right|_0^{\varepsilon_{max}}$$

$$= (1.4 \times 10^{10}\ \text{J/m}^3)(\varepsilon_{max})^3$$

In the last step we used the fact that 1 N·m = 1 J to write N/m^2 as J/m^3, the correct units for energy density. For $\varepsilon_{max} = 0.075$ we find

$$u_{elas} = (1.4 \times 10^{10}\ \text{J/m}^3)(0.075)^3 = 5.9 \times 10^6\ \text{J/m}^3$$

Knowing the dimensions of the tendon, we can calculate its volume:

$$V = AL = (25\ \text{mm}^2)(160\ \text{mm}) \times \left(\frac{1\ \text{m}}{10^3\ \text{mm}}\right)^3 = 4.0 \times 10^{-6}\ \text{m}^3$$

Thus the energy stored at maximum stretch is

$$U_{elas} = u_{elas}V = (5.9 \times 10^6\ \text{J/m}^3)(4.0 \times 10^{-6}\ \text{m}^3) = 24\ \text{J}$$

The actual stored energy is a reasonable number of joules. The runner's kinetic energy is

$$K = \tfrac{1}{2}mv^2 = \tfrac{1}{2}(55\ \text{kg})(5.2\ \text{m/s})^2 = 740\ \text{J}$$

Consequently, the maximum stored energy, just before the athlete begins the next stride, is 3.2% of her running kinetic energy.

ASSESS Because a runner doesn't obviously slow down each time she plants a foot on the ground, it makes sense that the stored energy is not a large fraction of her running energy. But a runner's kinetic energy is more complex than a simple $\frac{1}{2}mv^2$. Her torso moves at a fairly steady speed, but each leg alternately slows down (with the foot on the ground) and speeds up (as the leg lifts). We don't know the changing kinetic energy of a leg—that would take a more complex analysis—but it's reasonable to think that 24 J of stored energy is a substantial fraction of the kinetic energy that a leg loses as it slows and needs to regain as the tendon expands during the next push-off.

An ideal spring is able to release 100% of its stored energy, but elastic biomaterials are not ideal. Chapter 6 introduced the idea of *resilience*—the fraction of stored energy that can be recovered. Tendons, which are mostly collagen, have a resilience of ≈90%, close to ideal behavior; however, many biomaterials are far from ideal.

FIGURE 11.14 A material's resilience is determined by its loading and unloading curves.

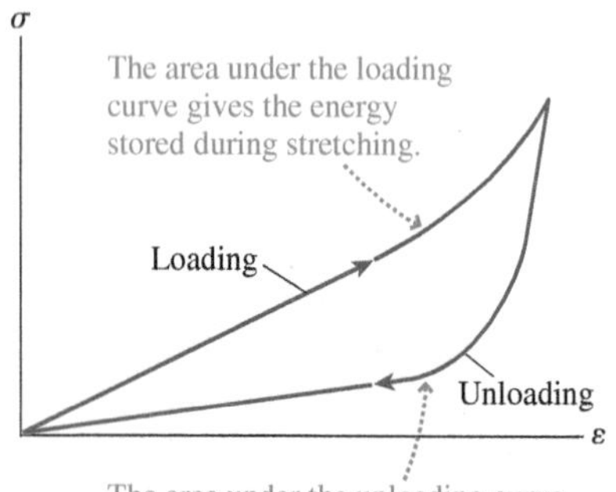
σ
The area under the loading curve gives the energy stored during stretching.
Loading
Unloading
ε
The area under the unloading curve gives the energy recovered as the strain is released.

FIGURE 11.14, which you saw earlier as Figure 6.42, shows the *loading and unloading curves* of a more typical biomaterial. The loading curve is the stress-strain curve as the material is stretched or compressed, and the area under this curve is the stored energy density. An ideal material would follow the same curve as the strain is released, but the unloading curve of a real material falls below the loading curve. The area under the unloading curve is the energy density of recovered mechanical energy (i.e., transformed back into kinetic energy) during relaxation. **The material's resilience is simply the ratio of the area under the unloading curve to the area under the loading curve.** Elastin, the stretchy material in skin and arteries, has a resilience of ≈75%, while spider silk's resilience is a low 35%.

What happens to the energy that's not returned to mechanical energy? The energy principle for an isolated material is $\Delta K + \Delta U + \Delta E_{th} = 0$. If less than 100% of the stored potential energy is transformed into kinetic energy ($\Delta K < \Delta U$), then the remainder must be transformed into thermal energy ($\Delta E_{th} > 0$). In other words, a repeated loading/unloading of a biomaterial generates thermal energy that heats up the material and has to be dissipated to the surrounding environment.

11.4 Energy Diagrams

Energy is a central concept not only in physics but also in chemistry and biology. In this section and the next, we'll develop tools that will help you see how energy concepts are used to understand chemical energy and chemical reactions, although we'll use them first to look at ordinary macroscopic objects such as balls and springs.

We're going to focus on isolated systems for which there is no friction or drag—that is, systems whose mechanical energy is conserved. Friction and drag are important in the macroscopic world but not for the motion of atoms and molecules, so this is actually a better model for atoms than for balls! We'll omit the subscript sys for now and write the conservation of mechanical energy as simply

$$E = K + U = \text{constant}$$

Kinetic energy K depends on an object's speed, but a system's potential energy U depends on the *positions* of the objects in the system. If a ball is tossed upward, the gravitational potential energy of the ball + earth system depends on the ball's height y. Similarly, the elastic potential energy of a compressed spring depends on the displacement Δx. Other potential energies also depend in some way on position. A graph showing a system's potential energy and total energy as a function of position is called an **energy diagram.** We'll spend some time learning about energy diagrams so that we can use them to think about bonds and chemical reactions in the next section.

FIGURE 11.15 is the energy diagram of a ball in free fall. This is a bit different from most graphs we've seen. It doesn't include time; instead, the horizontal axis is the vertical position y, and the vertical axis represents energy. The gravitational potential energy increases with the vertical position; the mathematical relationship is $U_G = mgy$, and a graph of mgy versus y is a straight line through the origin with slope mg. The resulting blue *potential-energy curve* is labeled PE. The horizontal brown line labeled TE is the *total energy line* showing the system's mechanical energy E. This line is always horizontal because the mechanical energy of an isolated system is conserved; it has the same value at every position.

FIGURE 11.15 The energy diagram of a ball in free fall.

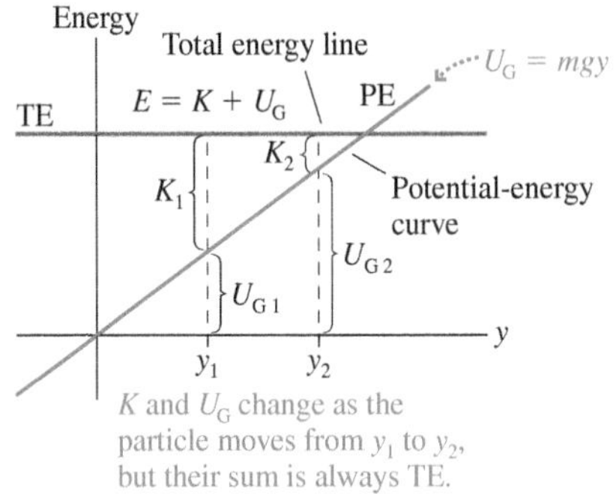

Suppose we consider a ball that is at a vertical position y_1 and is moving upward. When the ball is at height y_1, the distance from the axis up to the potential-energy curve is the potential energy U_{G1} at that position. Because $K_1 = E - U_{G1}$, the kinetic energy is represented graphically as the distance between the potential-energy curve and the total energy line. Later, as the ball continues to rise, it will be at height y_2. The energy diagram shows that the potential energy U_{G2} has increased while the ball's kinetic energy K_2 has decreased, as we know must be the case. Kinetic energy has been transformed into potential energy, but their sum has not changed.

NOTE ▶ In graphs like this, the potential-energy curve PE is determined by the physical properties of the system—for example, the mass or the spring constant. But the total energy line TE is under experimental control. If you change the *initial conditions,* such as throwing the ball upward with a different speed or compressing a spring by a different amount, the total energy line will appear at a different height. We can thus use an energy diagram to see how changing the initial conditions affects the subsequent motion. ◀

FIGURE 11.16 is the energy diagram of an object attached to a horizontal spring. In this case, we've placed the end of the unstretched spring at $x_{eq} = 0$. This makes the displacement Δx simply x and the elastic potential energy a simpler $U_{sp} = \frac{1}{2}kx^2$. The blue potential-energy curve is thus a parabola centered at $x = 0$, the equilibrium position of the end of the spring. The blue PE curve is determined by the spring constant; we can't change it. But we can set the brown TE line to any height we wish by stretching or compressing the spring to different lengths. The figure shows one possible TE line.

FIGURE 11.16 The energy diagram of an object on a horizontal spring.

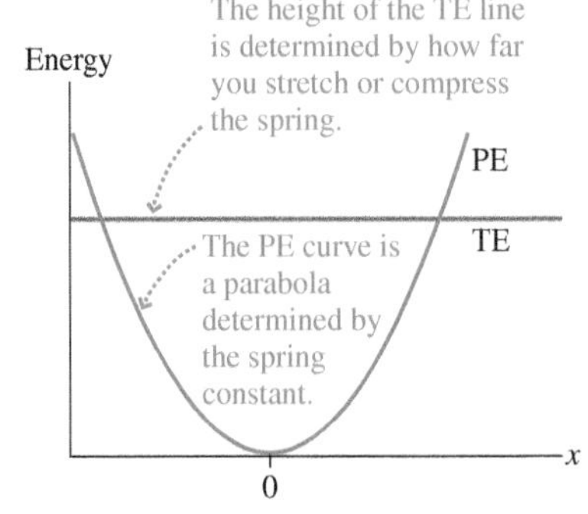

To see how to "read" an energy diagram, suppose you pull the object out to position x_R and release it from rest. FIGURE 11.17 (next page) shows a five-frame "movie" of the subsequent motion. Initially, in frame a, the energy is entirely potential—the energy of a stretched spring—so the TE line has been drawn to cross the PE line at $x_a = x_R$. This is the graphical statement that initially $E = U_{sp}$ and $K = 0$.

The restoring force pulls the object toward the origin. In frame b, where the object has reached x_b, the potential energy has decreased while the kinetic energy—the distance *above* the PE curve—has increased. The object continues to speed up until it reaches maximum speed at $x_c = 0$, where the PE curve is at a minimum and the distance above the PE curve is maximum. At position x_d, the object has started to slow down as it begins to transform kinetic energy back into elastic potential energy.

The object continues moving to the left until, in frame e, it reaches position x_L, where the total energy line crosses the potential-energy curve. This point, where

FIGURE 11.17 A five-frame movie of an object oscillating on a spring.

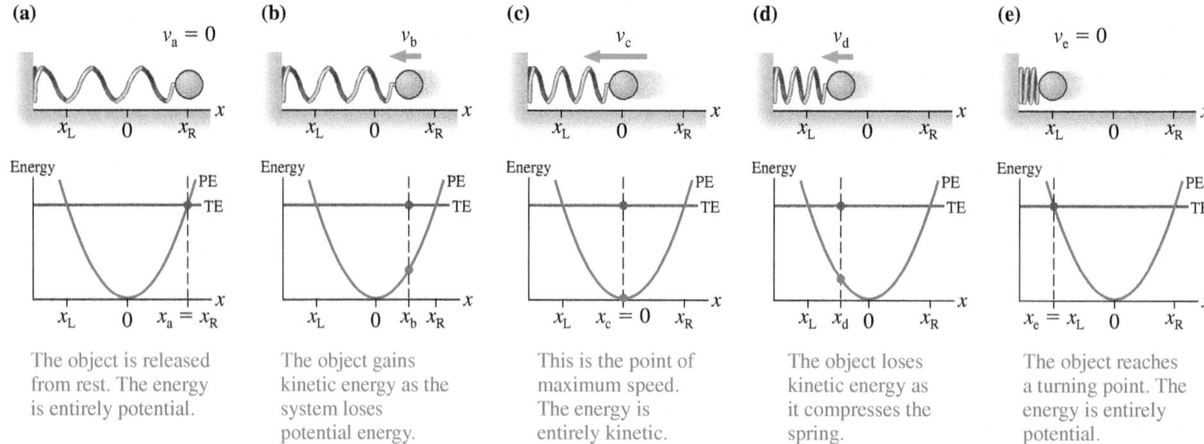

(a)
$v_a = 0$
Energy
PE
TE
x_L 0 $x_a = x_R$
The object is released from rest. The energy is entirely potential.
(b)
v_b
Energy
PE
TE
x_L 0 x_b x_R
The object gains kinetic energy as the system loses potential energy.
(c)
v_c
Energy
PE
TE
x_L $x_c = 0$ x_R
This is the point of maximum speed. The energy is entirely kinetic.
(d)
v_d
Energy
PE
TE
x_L x_d 0 x_R
The object loses kinetic energy as it compresses the spring.
(e)
$v_e = 0$
Energy
PE
TE
$x_e = x_L$ 0 x_R
The object reaches a turning point. The energy is entirely potential.

$K = 0$ and the energy is entirely potential, is a *turning point* where the object reverses direction. An object would need negative kinetic energy to be to the left of x_L, and that's not physically possible. You should be able to see, from the energy diagram, that the object will *oscillate* back and forth between positions x_L and x_R, having maximum kinetic energy (and thus maximum speed) each time it passes through $x = 0$.

Now, let's consider a different initial condition. Suppose you pull the object out to a greater initial distance. You've increased the potential energy in the system, and thus the total energy. The brown TE line is now at a greater height, and it will intersect the PE graph at two points that are farther from the equilibrium point. With this new initial condition, the object will oscillate back and forth between two points at a greater distance from equilibrium.

Interpreting Energy Diagrams

The lessons we learn from Figure 11.17 can be summarized in a Tactics Box.

TACTICS BOX 11.1 Interpreting an Energy Diagram

1. At any position, the distance from the axis to the PE curve is the system's potential energy. The distance from the PE curve to the TE line is its kinetic energy.
2. The object cannot be at a position where the PE curve is above the TE line.
3. A position where the TE line crosses the PE curve is a turning point where the object reverses direction.
4. If the TE line crosses the PE curve at two positions, the object will oscillate between those two positions. Its speed will be maximum at the position where the PE curve is a minimum.

CONCEPTUAL EXAMPLE 11.7 Interpreting an energy diagram

FIGURE 11.18 is a more general energy diagram. We don't know how this potential energy was created, but we can still use the energy diagram to understand how a particle will move. Suppose a particle begins at rest at the position x_1 shown in the figure and is then released. Describe its subsequent motion.

REASON We've added details to the graph and sketched out details of the motion in FIGURE 11.19. The particle is at rest at the starting point, so $K = 0$ and the total energy is equal to the potential energy. We can draw the horizontal TE line through this

FIGURE 11.18 A more general energy diagram.

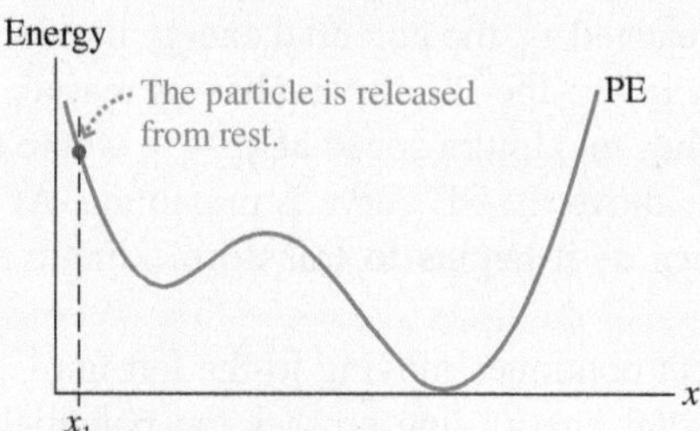

Energy
The particle is released from rest.
PE
x
x_1

FIGURE 11.19 The motion of the particle in the potential energy of Figure 11.18.

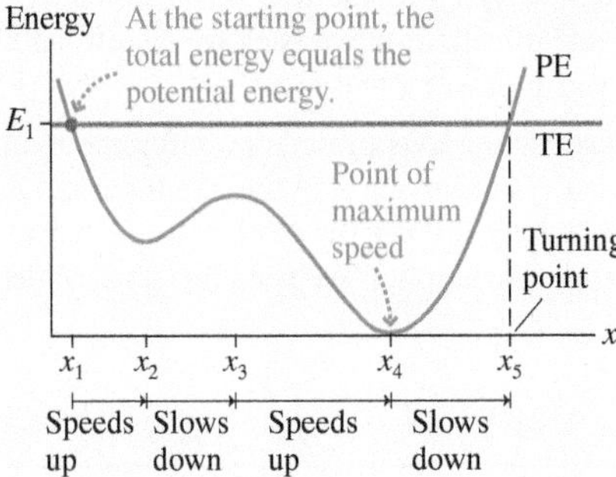

point. The PE curve tells us the system's potential energy at each position. The distance between the PE curve and the TE line is the particle's kinetic energy. The particle cannot move to the left because that would require a negative kinetic energy, so it begins moving to the right. The particle speeds up from x_1 to x_2 because U decreases and thus K must increase. It then slows down (but doesn't stop) from x_2 to x_3 as it goes over the "potential-energy hill." It speeds up after x_3 until it reaches maximum speed at x_4, where the PE curve is a minimum. The particle then steadily slows from x_4 to x_5 as kinetic energy is transformed into an increasing potential energy. Position x_5 is a turning point, a position where the TE line crosses the PE curve. The particle is instantaneously at rest and then reverses direction. Because the TE line crosses the PE curve at both x_1 and x_5 the particle will oscillate back and forth between these two points, speeding up and slowing down as described.

ASSESS Our results make sense. The particle is moving fastest where the PE line is lowest, and it turns around where the TE and PE lines cross, meaning $K = 0$ and the particle is at rest.

Equilibrium Positions

Positions x_2, x_3, and x_4 in Figure 11.19, where the potential energy has a local minimum or maximum, are special positions. Consider a system with the total energy E_2 shown in FIGURE 11.20. A particle can be at rest at x_2, with $K = 0$, but it cannot move away from x_2. In other words, a system with energy E_2 is in *equilibrium* at x_2. If you disturb the particle, giving it a small kinetic energy and a total energy just *slightly* larger than E_2, the particle will undergo a very small oscillation centered on x_2, like a marble in the bottom of a bowl. An equilibrium for which small disturbances cause small oscillations is called a point of **stable equilibrium.** You should recognize that *any* minimum in the PE curve is a point of stable equilibrium. Position x_4 is also a point of stable equilibrium, in this case for a particle with $E = 0$.

FIGURE 11.20 Points of stable and unstable equilibrium.

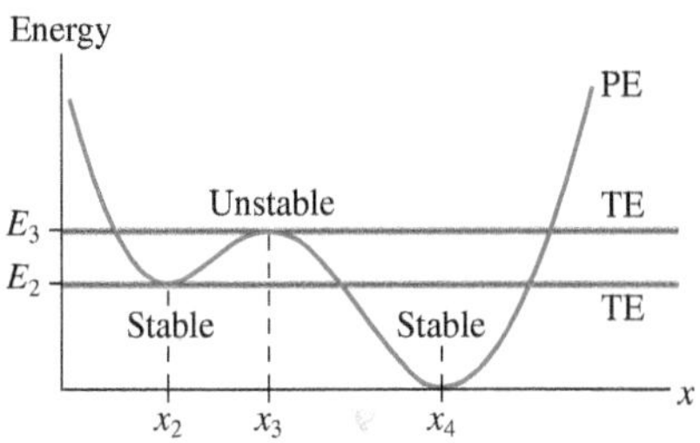

Figure 11.20 also shows a system with energy E_3 that is tangent to the PE curve at x_3. If a particle is placed *exactly* at x_3, it will stay there at rest $(K = 0)$. But if you disturb the particle at x_3, giving it an energy only slightly more than E_3, it will speed up as it moves away from x_3. This is like trying to balance a marble on top of a hill. The slightest displacement will cause the marble to roll down the hill. A point of equilibrium for which a small disturbance causes the particle to move away is called a point of **unstable equilibrium.** Any maximum in the PE curve, such as x_3, is a point of unstable equilibrium.

In the next section, we'll see how ideas about stable and unstable equilibrium help us understand molecular bonds and chemical reactions.

EXAMPLE 11.8 Finding maximum speed from an energy diagram

FIGURE 11.21 is the potential-energy curve for a system in which a 25 g particle is the only moving object. At $t = 0$ s, the particle is at $x = 1.0$ m and moving to the right with 1.0 J of kinetic energy. What is the particle's maximum speed? Where is its turning point?

PREPARE The graph shows the system's potential energy as a function of the particle's position. Initially, when the particle is at $x = 1.0$ m, the system has 3.0 J of potential energy. That, along with the particle's 1.0 J of kinetic energy, makes the total energy $E = K + U = 4.0$ J. Energy is conserved, so the system will continue to have 4.0 J of energy as the particle moves. The total energy line would be a horizontal line across the graph at a height of 4.0 J.

FIGURE 11.21 The potential-energy curve for Example 11.8.

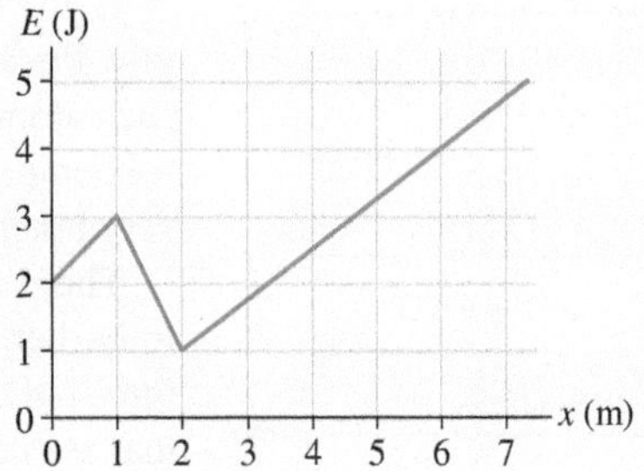

Continued

SOLVE As the particle moves to the right, the system's potential energy U decreases and the particle's kinetic energy K—the distance from the blue PE line up to the total energy line—increases. Kinetic energy will be maximum when potential energy is minimum, and you can see that U reaches a minimum $U_{min} = 1.0$ J at $x = 2.0$ m. At that point, the kinetic energy is $K_{max} = E - U_{min} = 3.0$ J. Kinetic energy is $K = \frac{1}{2}mv^2$, so the particle's maximum speed is

$$v_{max} = \sqrt{\frac{2K_{max}}{m}} = \sqrt{\frac{2(3.0\text{ J})}{0.025\text{ kg}}} = 15\text{ m/s}$$

If we extend the total energy line at 4.0 J to the right, it crosses the potential-energy curve at $x = 6.0$ m. Thus the turning point where instantaneously $v = 0$ and $K = 0$ is at $x = 6.0$ m. The particle will reverse direction, move back to the left, and disappear from our view as it passes $x = 0$ m.

ASSESS Conservation laws provide no information about time. We can use energy conservation to find the turning point, but we don't know how long it takes the particle to get there. Newton's laws are the appropriate tool if we need that level of detail.

STOP TO THINK 11.6 The figures below show potential-energy curves and total energy lines for four identical particles. Which particle has the highest maximum speed?

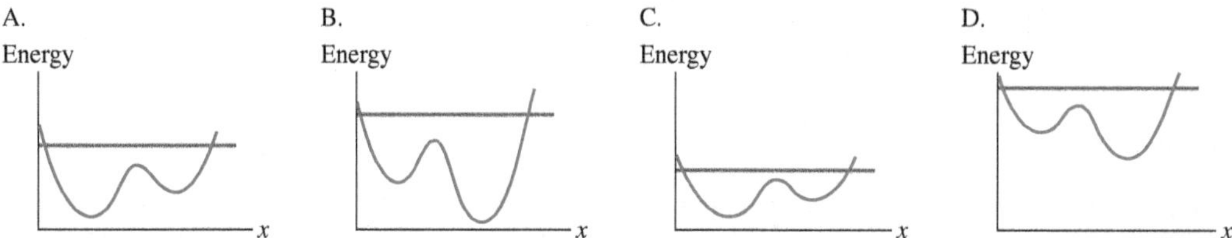

11.5 Molecular Bonds and Chemical Energy

With few exceptions, the materials of everyday life are made of atoms bound together into larger molecules. The *molecular bond* that holds two atoms together is an electric interaction between the atoms' negative electrons and positive nuclei. The electric force, like the gravitational force, is a force that can store energy. Fortunately, we don't need to know any details about electric potential energy—a topic we'll take up in Chapter 22—to understand the energy diagram of a molecular bond.

We've noted that molecular bonds are somewhat analogous to springs: The normal force when an object rests on a table arises from the compression of spring-like bonds, and thermal energy is due, in part, to spring-like vibrations of atoms around an equilibrium position. This suggests that the energy diagram of two atoms connected by a molecular bond should look similar to the Figure 11.16 energy diagram of an object attached to a spring.

FIGURE 11.22 The energy diagram and bond energy of the diatomic molecule HCl.

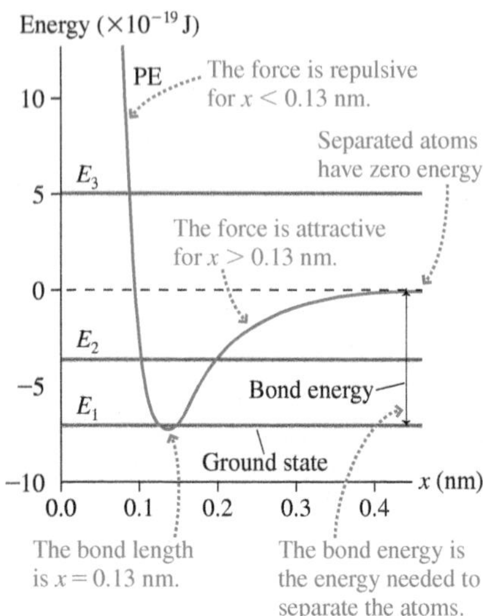

FIGURE 11.22 shows the experimentally determined energy diagram of the diatomic molecule HCl (hydrogen chloride). Distance x is the *atomic separation,* the distance between the hydrogen and chlorine atoms. Note the very small distances: 1 nm $= 10^{-9}$ m. The left side of the PE curve looks very much like the PE curve of a spring; the right side starts out similar to the PE curve of a spring and then levels off. You can interpret and understand this potential-energy diagram by using what you learned in Section 11.4.

If you try to push the atoms too close together, the potential energy rises very rapidly. Physically, this is an electric repulsion between the negative electrons orbiting each atom, but it's analogous to the increasingly strong repulsive force we get when we compress a spring. Thus the PE curve to the left of the equilibrium position looks very much like the PE curve of a spring.

There are also attractive forces between two atoms. These can be the attractive force between two oppositely charged ions, as is the case for HCl; the attractive forces of covalent bonds when electrons are shared; or even weak *polarization forces* that are related to the static electricity force by which a comb that has been brushed through your hair attracts small pieces of paper. For any of these, the attractive force resists if you try to pull the atoms apart—analogous to stretching a spring—and thus potential energy increases to the right.

The repulsive force gets stronger as you push the atoms closer together, but the attractive force gets *weaker* as you pull them farther apart. If you pull too hard, the bond breaks and the atoms come apart. Consequently, the PE curve becomes *less steep* as x increases, eventually leveling off when the atoms are so far apart that they cease interacting with each other. This difference between the attractive and repulsive forces explains the asymmetric PE curve in Figure 11.22.

Notice that there's a clear minimum in the HCl potential energy at an atomic separation of 0.13 nm. As you learned in the preceding section, a potential-energy minimum is a position of *stable equilibrium* where the repulsive and attractive forces are exactly balanced. Because x is the atomic separation, not the position of a specific atom, the minimum in the PE curve corresponds to an HCl molecule in which the H and Cl atoms are 0.13 nm apart and at rest. The atomic separation at the minimum of the PE curve is the **bond length** of the molecule. The PE curve of Figure 11.22 is specific to HCl, but any molecular bond has a similar looking PE curve. The bond length is the atomic separation at the minimum potential energy.

It turns out, for quantum physics reasons, that the atoms in a molecule cannot have $K = 0$ and thus cannot simply be at rest at the equilibrium separation. The *ground state* of the molecule, its lowest possible energy, is represented in Figure 11.22 by the energy line E_1 that is just barely above the minimum of the PE curve. This small amount of kinetic energy gives the ground state a tiny oscillation or *vibration* as the two atoms alternately move toward and away from each other.

Perhaps the most surprising feature of Figure 11.22 is that the ground-state energy E_1 is negative. What is the significance of a negative energy? Earlier, when we introduced gravitational potential energy, we noted that potential energy is not uniquely defined. We can pick any height we wish to be the zero of gravitational potential energy; our choice doesn't matter because it doesn't affect the *change* in potential energy. The same is true for the potential energy of a molecular bond: We can choose any atomic separation we wish to be the zero of potential energy.

The choice we make—and the choice made in most physics and chemistry textbooks—is that zero energy describes two atoms that are very far apart and at rest. The atoms are not moving and they're not interacting, and that seems a reasonable definition of zero energy. Thus you can see, in Figure 11.22, that the PE curve approaches zero as the atomic separation x gets large. Zero energy corresponds to "free" H and Cl atoms that are too far apart to feel any attraction toward each other.

With this choice, **a negative potential energy represents a bound system.** An HCl molecule with negative energy E_1 is a stable molecule with the two atoms bound to each other. An HCl molecule with the larger but still negative energy E_2 is also a bound molecule, although with this energy it has a much larger oscillation than a ground-state molecule, oscillating between turning points of $x \approx 0.10$ nm and $x \approx 0.19$ nm. We'll see a number of instances in future chapters where bound states are associated with negative energies.

If two atoms are bound together as a molecule, it would take energy to pry them apart. The energy needed to separate the two atoms in a ground-state molecule is called the **bond energy;** that is, the bond energy is the energy needed to break a molecular bond and separate the atoms. Figure 11.22 shows the bond energy as the vertical separation between the ground-state energy E_1 and the energy of the separated atoms. But the latter, by our choice, is zero, so the bond energy is $E_{bond} = |E_1|$, the absolute value of the negative ground-state energy. You can see from Figure 11.22 that the bond energy of HCl is approximately 7.5×10^{-19} J.

NOTE ▸ Physicists most often work directly with the energy per molecule or energy per bond in J. Chemists and biologists usually specify the total energy of a mole of a substance, using units of kJ/mol or kcal/mol. To find energies in kJ/mol, simply multiply the bond energy by Avogadro's number. For example, the 7.5×10^{-19} J bond energy of HCl becomes 450 kJ/mol. ◂

Suppose the molecule's energy is increased to $E_3 = 5 \times 10^{-19}$ J. For example, the molecule might reach this energy by absorbing a photon of light. Energy E_3 is *not* a bound system. With positive energy, the two atoms can move apart until the molecule dissociates into free atoms. In other words, we've broken the molecular bond. The breaking of molecular bonds by the absorption of light is called **photodissociation.** It's a common outcome of irradiating molecules with high-energy ultraviolet light. Other light-mediated reactions, from sun tanning to photosynthesis to vision, are very similar to photodissociation, but in these processes the higher-energy molecular state drives a conformational change in macromolecules rather than breaking bonds.

EXAMPLE 11.9 **Does the photon have enough energy?**

An energy diagram for molecular oxygen, O_2, is shown in FIGURE 11.23. A germicidal lamp for sterilizing equipment uses short-wavelength ultraviolet radiation at 185 nm. At this wavelength, each photon, or quantum, of ultraviolet light has 11×10^{-19} J of energy. If a molecule of O_2 at room temperature absorbs one photon of light from the lamp, does this provide enough energy to split the molecule? If so, what will be the kinetic energy of the atoms after they have separated?

PREPARE We can use the molecular energy diagram of Figure 11.23 to determine what happens. We'll assume that the ground-state energy of the molecule is very close to the bottom of the potential-energy curve.

FIGURE 11.23 The energy diagram for molecular oxygen, O_2.

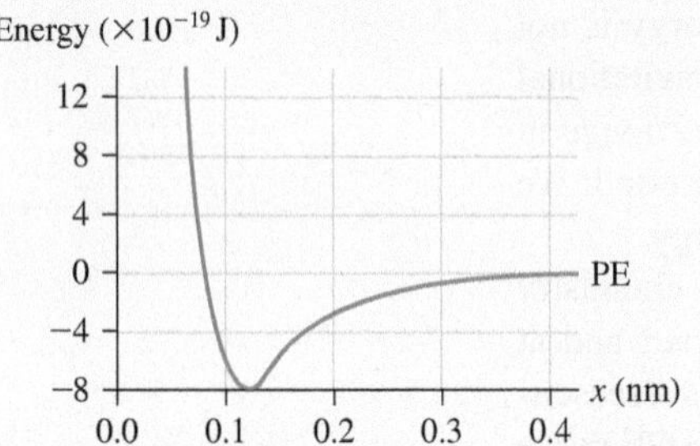

SOLVE The ground-state energy E_1, shown in FIGURE 11.24, is approximately -8×10^{-19} J. Absorbing a photon of energy 11×10^{-19} J will raise the molecule's energy to $E_2 \approx 3 \times 10^{-19}$ J. The molecule's energy is now greater than zero, so the two oxygen atoms are no longer bound and will separate. The bond has been broken. The residual kinetic energy of the two atoms is the height of the total energy line above zero, which is approximately 3×10^{-19} J.

ASSESS Germicidal lamps at this wavelength are known to produce O_3, ozone, a very reactive form of oxygen, so it's clear that the photons have enough energy to break apart the normally stable O_2 molecules. Therefore, our answer makes sense.

FIGURE 11.24 The molecular energy after absorbing a UV photon.

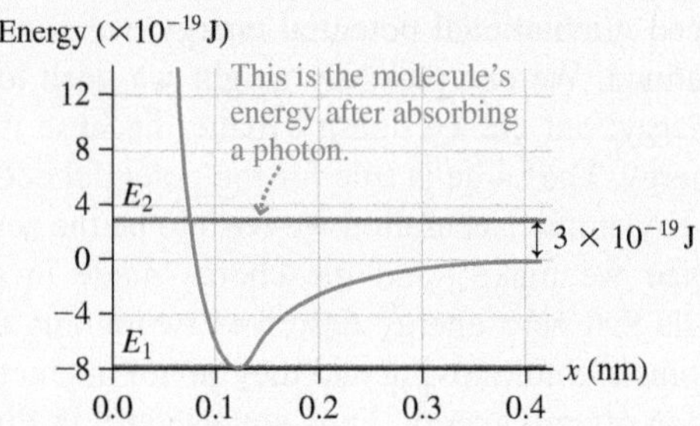

Does Breaking Bonds Release Energy?

Two widely held ideas are that chemical energy is stored in bonds and that energy is released when bonds are broken. Thus, for example, the biologically important molecule ATP (adenosine triphosphate) is often said to have "high-energy phosphate bonds" that release their energy when they are broken during the hydrolysis of ATP.

These ideas may be common, but they reflect an incomplete understanding of molecular bonds and of what happens during a chemical reaction. We saw in Figure 11.22 that the ground state of a molecule, the most stable configuration, has *less* energy than two free atoms and that, in fact, we have to *add* energy—the bond energy—to break a bond. That is, breaking bonds is *endothermic* (requiring energy), not *exothermic* (releasing energy). A bond, unlike a compressed spring, does not have stored energy just waiting to be released. A better analogy for a bond is two magnets stuck together that we can free only by doing work—an input of energy—to pry them apart.

What actually happens during a chemical reaction? The reactant molecules begin with thermal energy, so they are moving and vibrating. A reaction is initiated when two reactant molecules collide. The two molecules could just bounce apart in an elastic collision, and this often happens. Alternatively, some or all of their kinetic energy could be absorbed into a molecule, promoting it to a higher energy level. This is an energy transformation $K \rightarrow U$ at the molecular level. If the molecular energy gain exceeds E_{bond}, the bond will be broken and the atoms will fly apart. The essential idea is that **the energy to break bonds comes from a transformation of kinetic energy into potential energy during collisions between molecules.**

Exothermic reactions release energy. If the released energy doesn't come from breaking bonds, where does it comes from? From *forming new bonds,* specifically the bonds of the reaction products. A bound molecule has less energy than the free atoms, so atoms have to give up energy—release energy—to form a molecule. They do that in two ways. One is by transforming potential energy into the kinetic energy of the product molecules—again through collisions with other molecules—and we recognize those faster-moving molecules as the increased temperature in an exothermic reaction. The second is by the emission of light, both visible and infrared.

Let's summarize how energy flows in an exothermic chemical reaction. An input of energy is required to break the bonds of the reactants, an input that usually comes from the thermal energy of moving and colliding molecules. Then a larger amount of energy is released during the formation of new bonds in the products. A fast reaction generally indicates that *weakly* bound reactants, whose bonds are easily broken, are converted to much more tightly bound products.

Seen in this light, the hydrolysis of ATP is a good way to release energy because the terminal phosphate bond in ATP is quite weak, requiring little energy to break, and the inorganic phosphate produced upon the hydrolysis of ATP to ADP (adenosine diphosphate) is a tightly bound molecule. Relatively little input energy is needed to break the phosphate bond in ATP, and then a large amount of energy is released during the bond formation of the inorganic phosphate. In general, **energy is released when less tightly bound molecules undergo a chemical reaction or conformational change to become more tightly bound molecules.**

FIGURE 11.25 illustrates the processes and energy flows of a generic chemical reaction AB + CD → AC + BD in which the bonds of diatomic molecules AB and CD are broken and, subsequently, new molecules AC and BD are formed. We'll return to look at how the reaction depends on temperature during our upcoming study of thermodynamics.

FIGURE 11.25 A chemical reaction involves the breaking and re-forming of molecular bonds.

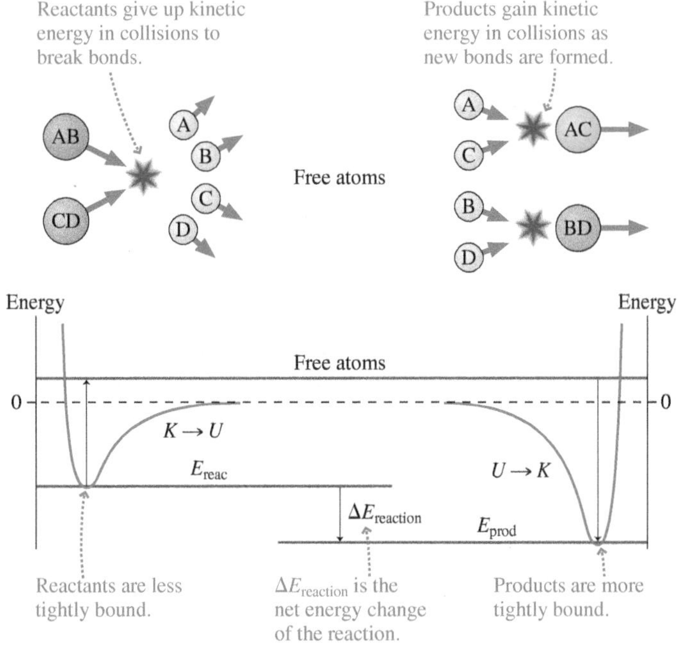

The reactant molecules in their ground states begin with energy E_{reac}. At the conclusion of the reaction, the product molecules have gone to their ground states and have energy E_{prod}. The energy difference $\Delta E_{reaction} = E_{prod} - E_{reac}$ is the *reaction energy*, the energy gained or released in *one* reaction due to the breaking and re-forming of bonds.

Any realistic chemical reaction consists of a vast number of these molecular interactions. The change in **chemical energy** is the total energy gained or released by all the reactions. If N reactions take place, then

$$\Delta E_{chem} = N\Delta E_{reaction} \tag{11.22}$$

An exothermic reaction, like the one illustrated in Figure 11.25, has $\Delta E_{chem} < 0$; that is, chemical energy is *released*. If the system is isolated, then the total energy of the system must be conserved. Hence the loss of chemical energy is accompanied by a matching *gain* of thermal energy, which we see as an increased temperature.

But notice that no energy had been stored anywhere. Energy is released in a chemical reaction simply by the molecular bonds changing their configuration to reach a more stable, more tightly bound state.

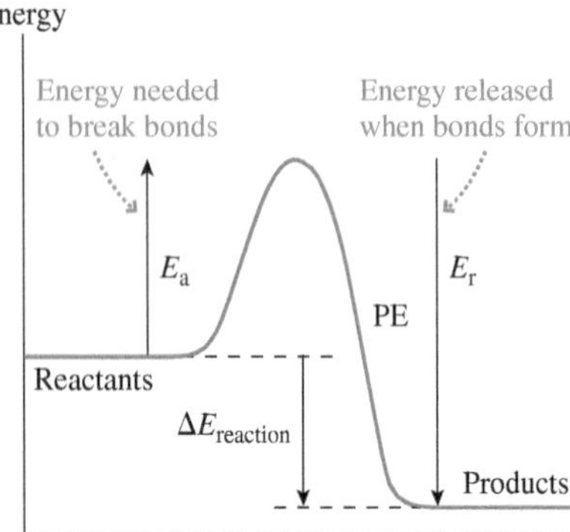

FIGURE 11.26 A reaction energy diagram.

Biochemical reactions are much more complex than a reaction between two diatomic molecules, but the key ideas do not change. FIGURE 11.26 is an energy diagram of a generic chemical reaction. It's very much like the energy diagrams we've been using, but with one important difference: The position coordinate of the horizontal axis has been replaced with an abstract *reaction coordinate*. The reaction coordinate is not a physical quantity that can be measured; instead, it shows in a general sense the progress of bond breaking and bond formation as a reaction moves from reactants, on the left, to products, on the right.

All reaction energy diagrams have a bump, or energy barrier, in the middle. It represents the energy required to break the bonds of the reactant molecules. For the reaction to take place, the reactants must increase their potential energy by the amount E_a, called the **activation energy.** Graphically, the activation energy is the height of the energy barrier above the initial potential energy of the reactants. Then, as the reaction goes to completion, energy E_r is released during the formation of new bonds.

As before, the reaction can proceed only if the reactants have sufficient thermal energy. If the temperature is too low, so that the reactant molecules are moving slowly, the potential energy gained during a collision is less than the activation energy; thus the bonds don't break and the reaction doesn't occur. Gasoline and oxygen don't react at room temperature, even though the combustion reaction is energetically favorable, because the activation energy is too large. To burn gasoline, you need to increase the thermal energy of a portion of the fuel with, say, the flame from a match or the spark of a spark plug. Energy released from the initial reactions can then trigger further reactions. Once gasoline is ignited, the flame spreads quickly.

An **exothermic reaction** of combustion or oxidation has $E_r > E_a$, releasing more energy than was required to initiate the reaction. This release of chemical energy causes the final temperature of the products to be higher than the initial temperature of the reactants. In contrast, an **endothermic reaction** releases less energy than was required to initiate it. Endothermic reactions can be driven by an external input of energy or, as we'll see later, by an appropriate change of *entropy*.

NOTE ▶ Chemists and biologists usually describe reactions in terms of what is called *free energy*. Free energy is a more appropriate description when some or all of the energy released in a reaction is used to do work rather than increasing the thermal energy. This is often the case in biology, where the energy released by a reaction does work by moving molecules around or changing the configuration of macromolecules instead of simply heating up the cell. We'll return to free energy in Chapter 14. ◀

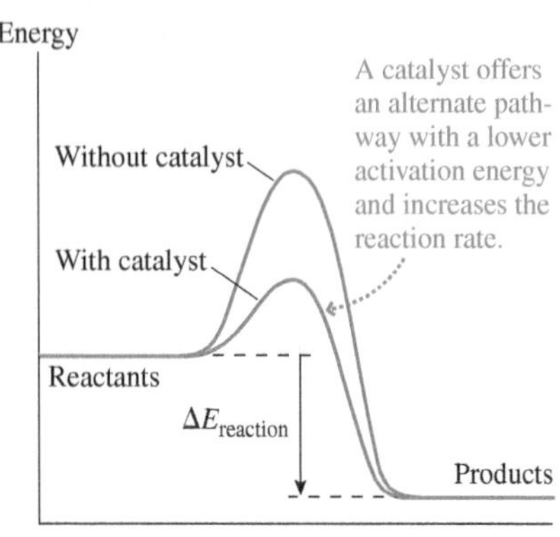

FIGURE 11.27 A reaction energy diagram for a chemical reaction with and without a catalyst.

Reaction Rates and Catalysts

The reaction energy diagram tells us nothing about the *rate of reaction*—how fast a reaction proceeds. A more detailed theory, which we'll look at during our study of thermodynamics, finds that the rate of reaction increases *exponentially* as the activation energy decreases. An exponential change means that even a small decrease in activation energy can produce a large increase in the rate of reaction.

The role of a catalyst is to provide an alternate reaction pathway with a lower activation energy, thus dramatically speeding up the rate of reaction. FIGURE 11.27 shows an exothermic reaction, one that is energetically favorable but where the energy barrier is so high that this reaction will not happen at room temperature because the reaction rate is essentially zero. In Figure 11.27, we see that a catalyst offers an alternate pathway whose activation energy is easily exceeded by room-temperature molecules. A catalyst can dramatically increase reaction rates.

Most of biochemistry is mediated by catalysts in the form of *enzymes*. Processes such as respiration, photosynthesis, and protein synthesis involve energetically favorable exothermic (also called *exergonic*) reactions, but the activation energy is so high that the reactants, on their own, would react barely, if at all, at normal temperatures. Enzymes catalyze these reactions, allowing them to proceed at a rate sufficient for cellular functions.

11.6 Connecting Potential Energy to Force

We introduced potential energy as an *interaction energy:* If interaction forces inside the system do work W_{int}, then the associated potential energy changes by $\Delta U = -W_{int}$. For example, we found the elastic potential energy of a spring by including the spring within the system and then calculating the work done by the spring. It's sometimes useful to reverse this procedure—that is, to use an already known potential energy to determine the interaction force.

Suppose an object has a very small displacement dx while being acted on by force $\vec{F}$. During this displacement, a small amount of work $dW_{int} = F_x dx$ is done by the interaction forces and the system's potential energy changes by

$$dU = -dW_{int} = -F_x dx \tag{11.23}$$

We can re-write Equation 11.23 as

$$F_x = -\frac{dU}{dx} = \text{the negative of the slope of the PE curve at position } x \tag{11.24}$$

The connection between potential energy and force

The quantity dU/dx is the derivative of the potential energy with respect to position, which, graphically, is the slope of the potential-energy curve in an energy diagram.

It's important not to overlook the minus sign. You can see that

- A positive slope corresponds to a negative force: to the left or downward.
- A negative slope corresponds to a positive force: to the right or upward.
- The steeper the slope, the larger the force.

Conceptually, Equation 11.24 tells us that the interaction force points along the direction in which potential energy decreases. You may have heard in a biology course that matter and systems tend to "move down energy gradients." Why? A *gradient* occurs if some quantity increases or decreases with distance through space. A region of space with a changing potential energy has a *potential-energy gradient,* and moving down the gradient means moving from higher potential energy to lower potential energy. That's the direction in which the force points! In many cases, the force—a pressure force or an electric force—has the ability to drive a flow of matter "down the gradient."

As an example, consider the elastic potential energy $U_{sp} = \frac{1}{2}kx^2$ for a horizontal spring with $x_{eq} = 0$ so that $\Delta x = x$. FIGURE 11.28a shows that the potential-energy curve is a parabola, with changing slope. If an object attached to the spring is at position x, the force on the object is

$$F_x = -\frac{dU_{sp}}{dx} = -\frac{d}{dx}\left(\tfrac{1}{2}kx^2\right) = -kx$$

This is just Hooke's law for an ideal spring, with the minus sign indicating that Hooke's law is a restoring force. FIGURE 11.28b is a graph of force versus x. At each

FIGURE 11.28 Elastic potential energy and force graphs.

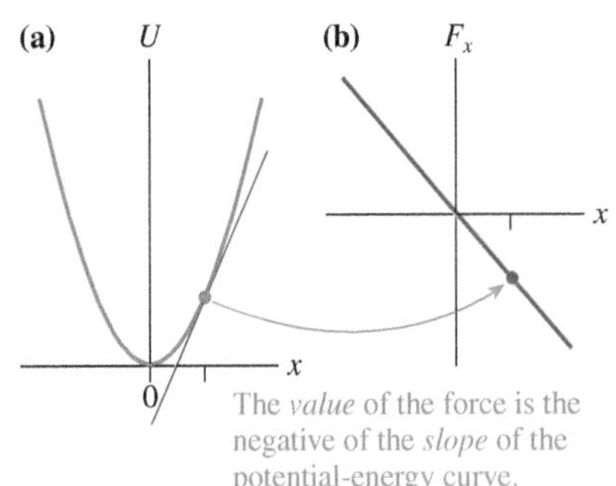
(a)
U
(b)
F_x
x
x
0
The *value* of the force is the negative of the *slope* of the potential-energy curve.

position x, the *value* of the force is equal to the negative of the *slope* of the PE curve.

We already knew Hooke's law, of course, so the point of this particular exercise was to illustrate the meaning of Equation 11.24. But if we had *not* known the force, we see that it's possible to find the force from the PE curve. An example will show how.

EXAMPLE 11.10 Find the force on ions

FIGURE 11.29 is a hypothetical potential-energy curve for ions inside a cell. Draw and interpret a graph of the force on the ions.

FIGURE 11.29 A potential-energy curve for ions inside a cell.

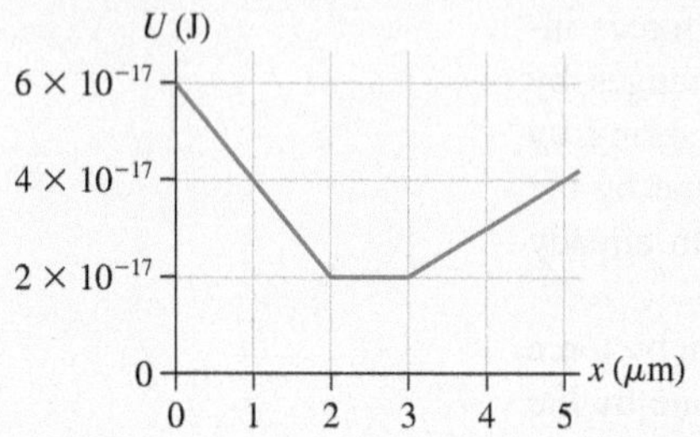

PREPARE The x-component of the force on the ions can be found as the *negative* of the slope of the PE curve, which we can determine graphically.

SOLVE The graph slopes downward—a negative slope—from $x = 0\ \mu\text{m}$ to $x = 2\ \mu\text{m}$. The slope is

$$\text{slope} = \frac{\Delta U}{\Delta x} = \frac{2.0 \times 10^{-17}\ \text{J} - 6.0 \times 10^{-17}\ \text{J}}{2.0 \times 10^{-6}\ \text{m}}$$

$$= -2.0 \times 10^{-11}\ \text{N} = -20\ \text{pN}$$

We used the fact that $1\ \text{J} = 1\ \text{N} \cdot \text{m}$ to see that $1\ \text{J/m} = 1\ \text{N}$, then converted to piconewtons (pN) as an appropriate scale for forces within the cell. Thus the force over the range $0\ \mu\text{m}$ to $2\ \mu\text{m}$ is $F_x = 20\ \text{pN}$.

The curve is horizontal with zero slope from $x = 2\ \mu\text{m}$ to $x = 3\ \mu\text{m}$, so the force in this region is zero. Finally, from $x = 3\ \mu\text{m}$ to $x = 5\ \mu\text{m}$, the force is

$$F_x = -\frac{\Delta U}{\Delta x} = -\frac{4.0 \times 10^{-17}\ \text{J} - 2.0 \times 10^{-17}\ \text{J}}{2.0 \times 10^{-6}\ \text{m}} = -10\ \text{pN}$$

A force curve with this information is shown in FIGURE 11.30. Our interpretation is aided by drawing force vectors beneath the curve. The positive force from $0\ \mu\text{m}$ to $2\ \mu\text{m}$ is a force to the right. The negative force from $3\ \mu\text{m}$ to $5\ \mu\text{m}$ is a smaller force to the left. In both cases, the force points down the potential-energy gradient toward the lower potential energy at the center of the cell, with the steeper gradient on the left giving a stronger force.

ASSESS We've seen other examples, such as Example 5.13 that found the spring constant of DNA, where the forces within cells are in the pN range, so the answer seems plausible. Real cellular potential energies are much more complex than the simple graph of Figure 11.29, but this type of reasoning still applies.

FIGURE 11.30 The corresponding force curve.

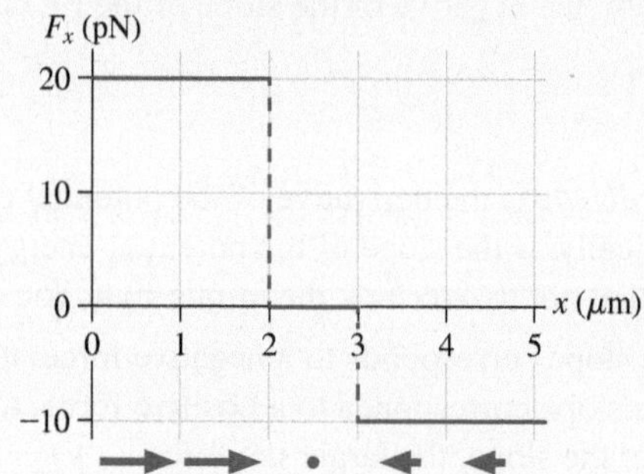

STOP TO THINK 11.7 A particle moves along the x-axis with the potential energy shown. The x-component of the force on the particle when it is at $x = 4$ m is

A. 4 N
B. 2 N
C. 1 N
D. −4 N
E. −2 N
F. −1 N

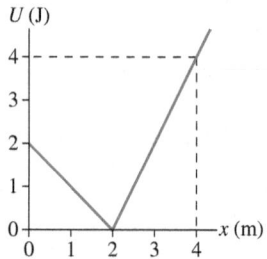

11.7 The Expanded Energy Model

We opened Chapter 10 by introducing the basic energy model and energy principle—basically a statement of energy accounting—but noted that we would need to develop many new ideas to make sense of energy. We've now explored kinetic energy, potential energy, work, thermal and chemical energy, and much more. It's time to return to the basic energy model and start pulling together the many ideas introduced in Chapters 10 and 11.

Our starting point was to consider a single particle or perhaps a group of noninteracting particles. A force acting on a particle changes its kinetic energy, and we captured this idea in our first statement of the energy principle: $W = \Delta K$, where W is the work done by the force. We noted that the work is zero if the particle has no displacement or if the force is always perpendicular to the displacement.

The next step was to recognize that real objects are not particles but consist of atoms and molecules. Those microscopic entities also have energy—thermal energy—and *dissipative force* such as friction can transform work or the macroscopic kinetic energy of the object as a whole into thermal energy. We can adjust our energy accounting to allow for this by restating the energy principle as

$$W = \Delta K + \Delta E_{th}$$

The idea is the same, but we've expanded the ways in which a system can have energy.

Continuing from there, we noted that many forces—gravity, spring forces, electric forces—are best thought of as interactions between two objects. Thus we distinguished between *interaction forces,* which we include in the system, and *external forces*, which are pushes and pulls from agents outside the system. The interaction forces do work, but we can account for the work they do in terms of a *potential energy* that changes by $\Delta U = -W_{int}$. We need to calculate the work only for any external forces, so the energy principle becomes

$$W = \Delta K + \Delta U + \Delta E_{th}$$

Finally, we introduced another way that systems can change energy: by changing the configurations of the molecules inside the system. The breaking and re-forming of molecular bonds changes what deep down is an electric potential energy that, for practical purposes, we call *chemical energy*. With chemical energy included as yet another way that energy can change, our final version—for now—of the energy principle is

$$W = \Delta E_{sys} = \Delta K + \Delta U + \Delta E_{th} + \Delta E_{chem} \tag{11.25}$$

The expanded energy principle

The qualification "for now" is because we've not yet included *heat* as a second energy-transfer mechanism. We'll turn to heat in the next chapter; then the energy principle will truly be complete and will become what is called the *first law of thermodynamics.*

Section 11.4 defined an *isolated system* as one that does not exchange energy with its environment. That is, an isolated system is one on which no work is done on the system by external forces: $W = 0$. Thus an important conclusion from Equation 11.25 is that **the total energy E_{sys} of an isolated system does not change.**

This idea—the law of conservation of energy—is one of the most powerful statements in physics. The kinetic, potential, thermal, and chemical energies inside an isolated system can be transformed into one another—and real-world processes often do so in very messy ways—but their sum does not change. If, in addition, there's no friction and no chemical reactions, then the mechanical energy $E_{mech} = K + U$ is conserved.

Biological systems are rarely isolated, but the law of conservation of energy still applies. Any change in the energy of a nonisolated system can be accounted for by the transfer of energy in or out via work or, as is more typical for biological systems, heat, the topic of Chapter 12. Energy can be transformed from one form into

FIGURE 11.31 An expanded energy model.

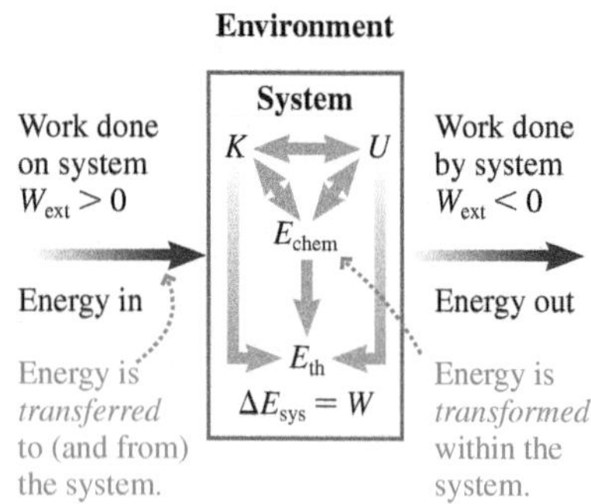

another, it can be transferred into or out of the system, but it can never be created or destroyed.

FIGURE 11.31 is a pictorial representation of Equation 11.25. It expands upon the basic energy model that we introduced at the beginning of Chapter 10 because it now includes chemical energy.

STOP TO THINK 11.8 A weight attached to a rope is released from rest. As the weight falls, picking up speed, the rope spins a generator that causes a lightbulb to glow. Define the system to be the weight and the earth. In this situation,

A. $U \rightarrow K + W$. E_{mech} is not conserved but E_{sys} is.
B. $U + W \rightarrow K$. Both E_{mech} and E_{sys} are conserved.
C. $U \rightarrow K + E_{th}$. E_{mech} is not conserved but E_{sys} is.
D. $U \rightarrow K + W$. Neither E_{mech} nor E_{sys} is conserved.
E. $W \rightarrow K + U$. E_{mech} is not conserved but E_{sys} is.

FIGURE 11.32 Energy of the body, considered as the system.

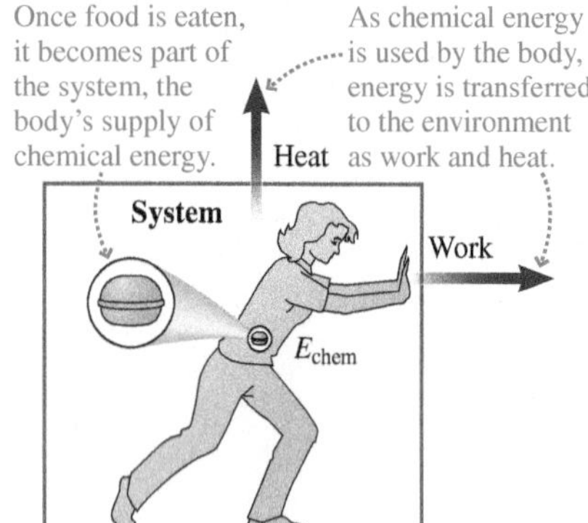

11.8 Energy in the Body

In this final section we will look at energy in the body—your *metabolism*. This will give us the opportunity to explore a number of energy transfers and transformations of practical importance. FIGURE 11.32 shows the body considered as a system for energy analysis. The chemical energy from the food you eat provides the necessary energy input for your body to function. Part of this chemical energy is transformed into thermal energy (your body stays warm) and kinetic energy (your body is in motion). The remainder is transferred to the environment as work you do against forces in the environment (gravity, drag, etc.) and as heat. We haven't yet said much about heat, but being aware of heat loss is part of the story of energy in the body.

Energy Units

Biologists, exercise physiologists, nutritionists, and athletic trainers are all interested in metabolism and energy use, but they use a variety of different units. The SI unit of energy is the joule; the kilojoule (kJ) is widely used in more modern discussions of metabolism. You'll often see the energy content of fuels, such as sugars or fats, given in kJ/g (kilojoules per gram of fuel).

The traditional energy unit of metabolism is the Calorie (Cal), but you need to be aware that there are two different "calories." The scientific calorie (cal), with a lowercase c, used in chemistry, was originally the energy needed to raise the temperature of 1 g of water by 1°C. While you often see this stated as "the definition," in reality the *thermochemical calorie* has been defined in terms of the joule since 1925: 1 cal = 4.184 J.

The energy content of food is often given in Calories (Cal), with an uppercase C, which is the *food calorie:* 1 Cal = 1000 cal = 1 kcal. However, many books and websites on diet and nutrition blur the distinction between food calories and scientific calories by using *calories* instead of the correct *Calories*. So you need to be cautious whenever calories are used.

TABLE 11.1 gives the relevant conversions to three significant figures.

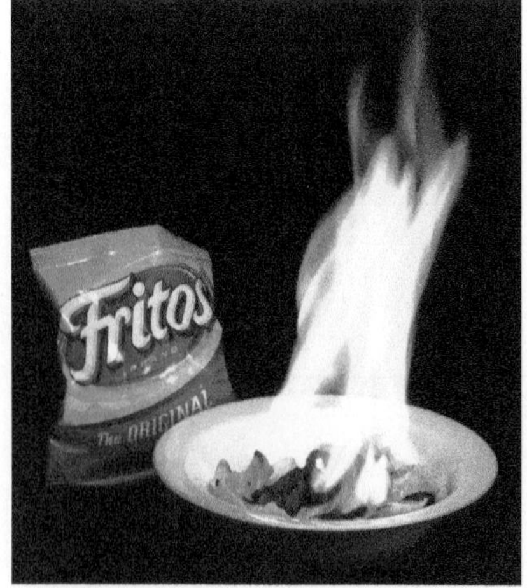

Counting calories Most dry foods burn quite well, as this photo of corn chips illustrates. You could set food on fire to measure its energy content, but this isn't really necessary. The chemical energies of the basic components of food (carbohydrates, proteins, fats) have been carefully measured—by burning—in a device called a *calorimeter.* Foods are analyzed to determine their composition, and their chemical energy can then be calculated.

Getting Energy from Food: Energy Inputs

When you walk up a flight of stairs, where does the energy come from to increase your body's potential energy? At some point, the energy came from the food you ate, but what were the intermediate steps? The chemical energy in food is made available to the cells in the body by a two-step process. First, the digestive system breaks down food into simpler molecules such as glucose, a simple sugar. Then these molecules are delivered via the bloodstream to cells in the body, where they are metabolized by reacting with oxygen via the exothermic reaction.

Glucose from the digestion of food combines with oxygen that is breathed in to produce . . .

. . . carbon dioxide, which is exhaled; water, which can be used by the body; and energy.

$$C_6H_{12}O_6 + 6O_2 \longrightarrow 6CO_2 + 6H_2O + 4.65 \times 10^{-18}\ \text{J} \tag{11.26}$$

Glucose Oxygen Carbon dioxide Water Energy

Although this is the overall reaction, it doesn't happen in one step because the net energy released—4.65×10^{-18} J per reaction or, after multiplying by Avogadro's number, 2800 kJ/mol—is too large for the body to use directly. Instead, glucose is oxidized through a complex sequence of reactions that use the released energy to produce a molecule called ATP. Cells in the body use this ATP to provide energy for all the processes of life: Muscle cells use it to contract, nerve cells use it to produce electrical signals.

TABLE 11.1 **Conversion factors for chemical energy**

1 cal	= 4.18 J
1 Cal	= 1000 cal = 1 kcal
	= 4180 J = 4.18 kJ
1 J	= 0.239 cal
1 kJ	= 1000 J = 239 cal
	= 0.239 Cal = 0.239 kcal

EXAMPLE 11.11 Computing basic biochemical efficiency

Power for cellular processes comes from the conversion of ATP to ADP in a process called ATP hydrolysis. The energy released in this reaction depends on the chemical environment, but within the cell $\Delta E_{\text{chem}} \approx 8.3 \times 10^{-20}$ J/reaction. The cell then uses chemical energy from the oxidation of glucose to convert ADP back to ATP. The metabolism of one glucose molecule typically produces 32 molecules of ATP. (The theoretical production is 38 molecules of ATP, but losses in the metabolic pathway reduce the actual production.) What is the overall efficiency of this energy conversion process?

PREPARE You learned in Chapter 10 that efficiency is the ratio of what you get to what you had to pay. You have to pay for the energy obtained by metabolizing one molecule of glucose, which Equation 11.26 shows is 4.65×10^{-18} J. What you get is 32 ATP molecules, each with the ability to provide 8.3×10^{-20} J of energy to the cellular processes.

SOLVE The efficiency is

$$e = \frac{\text{what you get}}{\text{what you had to pay}} = \frac{(32)(8.3 \times 10^{-20}\ \text{J})}{4.65 \times 10^{-18}\ \text{J}} = 0.57$$

ASSESS The efficiency of using the energy released from the oxidation of glucose to synthesize ATP, the body's "energy currency," is about 57%. The other 43% of the released energy increases the body's thermal energy—it helps maintain body temperature—and is ultimately released to the environment as heat.

If you were to use the density of glucose along with its energy release of 2800 kJ/mol, you would find that metabolizing 1 g of glucose (or any other carbohydrate) from food or body stores releases approximately 17 kJ of energy; 1 g of fat from food or body stores provides 37 kJ. TABLE 11.2 compares the energy contents of carbohydrates and other foods with other common sources of chemical energy. The energy available from foods is similar to that of burning wood but substantially less—on an equal-mass basis—than the energy released by burning gasoline.

TABLE 11.2 **Energy in fuels**

Fuel	Energy in 1 g of fuel (in kJ)
Hydrogen	121
Gasoline	44
Fat (in food)	37
Coal	27
Carbohydrates	17
Protein	17
Wood chips	15

EXAMPLE 11.12 How much energy does a soda provide?

A 12 oz can of soda contains 40 g (or a bit less than 1/4 cup) of sugar, a simple carbohydrate. What is the chemical energy in joules? How many Calories is this?

SOLVE From Table 11.2, 1 g of sugar contains 17 kJ of energy, so 40 g contains

$$40\ \text{g} \times \frac{17 \times 10^3\ \text{J}}{1\ \text{g}} = 68 \times 10^4\ \text{J} = 680\ \text{kJ}$$

Converting to Calories, we get

$$680\ \text{kJ} = (680\ \text{kJ})\frac{1\ \text{Cal}}{4.18\ \text{kJ}} = 160\ \text{Cal}$$

ASSESS 160 Calories is a typical value for the energy content of a 12 oz can of soda (check the nutrition label on one to see), so this result seems reasonable.

The first item on the nutrition label on packaged foods is Calories—a measure of the chemical energy in the food. (In other countries where SI units are standard, you will find the energy contents listed in kJ.) The energy contents of some common foods are given in TABLE 11.3.

TABLE 11.3 Energy content of foods

Food	Energy content in Cal	Energy content in kJ	Food	Energy content in Cal	Energy content in kJ
Carrot (large)	30	125	Slice of pizza	300	1260
Fried egg	100	420	Frozen burrito	350	1470
Apple (large)	125	525	Apple pie slice	400	1680
Beer (can)	150	630	Milkshake	650	2700
BBQ chicken wing	180	750	Fast-food meal	1350	5660
Latte (whole milk)	260	1090			

TABLE 11.4 Energy use at rest

Organ	Resting power (W) of 68 kg individual
Liver	26
Brain	19
Kidneys	11
Heart	7
Skeletal muscle	18
Remainder of body	19
Total	**100**

TABLE 11.5 Metabolic power use during activities

Activity	Metabolic power (W) of 68 kg individual
Typing	125
Household chores	250
Walking at 5 km/h	380
Cycling at 15 km/h	480
Swimming at 3 km/h	800
Running at 15 km/h	1150

Using Energy in the Body: Energy Outputs

You know that your body uses energy when you exercise. But even at rest, your body uses energy to build and repair tissue, digest food, and keep warm. TABLE 11.4 lists the amounts of power used by different tissues in the resting body. All told, your body uses approximately 100 W, or 100 J per second, when at rest, a quantity called your **basal metabolic rate.** All of this energy is ultimately converted to thermal energy and, because your body stays at a constant temperature, is dissipated as heat into the environment. This process has practical consequences. If you take physics in a class of 100 people in a lecture hall, the class, as a whole, dissipates 10,000 W into the room—the equivalent of seven electric space heaters running at full blast. In the winter, this extra heat is welcome; in the summer, the air conditioning system must work harder to keep the room cool.

Activity increases your metabolic rate as additional energy is needed by the muscles. The *rate* at which chemical energy is released through metabolic processes—in J/s—is called **metabolic power.** The metabolic power of an activity depends on an individual's size and level of fitness, but TABLE 11.5 lists some typical values. The metabolic power needed for typing is little more than the 100 W basal metabolic rate of resting. At the other extreme, running at 15 km/h (approximately 6.5 min miles) requires the body's metabolism to increase its output by more than a factor of 10.

NOTE ▸ Energy is not "used up" when you exercise. The chemical energy of glucose is transformed into thermal energy and into work you do against the environment, so energy moves from your body to your surroundings, but the total energy is unchanged. However, your stores of chemical energy are depleted, so your body's *available energy* is used up. ◂

EXAMPLE 11.13 How many miles per gallon do you get?

Automobile *fuel economy* is rated in miles per gallon (mpg). A similar rating for human-powered locomotion is kilometers per Calorie. How far can you travel per Cal of food energy? What is the fuel economy in kilometers per Calorie for the four types of locomotion listed in Table 11.5?

PREPARE Metabolic power in watts is the rate in J/s at which chemical energy is "burned" in metabolic reactions. Table 11.5 specifies a speed for each of the four types of locomotion. It's mostly a matter of unit conversions to change the power for a given speed to km/Cal.

SOLVE The quantity we're looking for is $\Delta x/|\Delta E_{\text{chem}}|$, the distance traveled per unit of energy used. ΔE_{chem} is negative because metabolic reactions release energy, but we need only the *quantity* of energy—hence the absolute value signs. Dividing both numerator and denominator by Δt, the time interval over which the motion occurs, gives

$$\text{distance per energy} = \frac{\Delta x/\Delta t}{|\Delta E_{\text{chem}}|/\Delta t} = \frac{v}{P_{\text{m}}}$$

where P_m is the metabolic power, the *rate* of using chemical energy. The values of speed (in km/h) and metabolic power (in W or J/s) in Table 11.5 will give an answer in km · h/J · s, so some unit conversions are needed:

$$\text{kilometers per Calorie} = \frac{v(\text{in km/h})}{P_m(\text{in J/s})} \times \frac{1\text{ h}}{3600\text{ s}} \times \frac{4184\text{ J}}{1\text{ Cal}}$$

$$= 1.16\frac{v(\text{in km/h})}{P_m(\text{in J/s})}\frac{\text{km}}{\text{Cal}}$$

The inverse of km/Cal—namely, Cal/km—is also interesting; it is the food energy needed to move 1 km at the given speed. Perhaps more useful in the United States is Cal/mi, which requires one more unit conversion. We can now use the data in Table 11.5 to find:

$$\text{Fuel economy} = \begin{cases} \text{Walking} & 0.015\text{ km/Cal} & 110\text{ Cal/mi} \\ \text{Cycling} & 0.036\text{ km/Cal} & 44\text{ Cal/mi} \\ \text{Swimming} & 0.004\text{ km/Cal} & 370\text{ Cal/mi} \\ \text{Running} & 0.015\text{ km/Cal} & 110\text{ Cal/mi} \end{cases}$$

ASSESS This calculation produces some interesting results. First, the fuel economy of walking and running are the same! Running at 15 km/h uses roughly three times more metabolic power than walking at 5 km/h, but it takes only one-third the time to cover the same distance. A good rule of thumb is that direct human locomotion, walking or running, burns ≈100 Cal per mile. You'll need to walk or run 6.5 mi to burn off a 650 Cal milkshake, although you can do it more quickly by running. Second, cycling is much more efficient—more than a factor of 2—than walking or running, even with the addition of 15 kg or so of extra mass that has to be moved. And swimming may be a good way to burn calories quickly, but it's horribly inefficient as a mode of human transport because our bodies aren't designed for it. In contrast, fish and marine mammals swim with excellent fuel economy.

You might wonder how metabolic power is measured. Your cells use the hydrolysis of ATP to power cellular processes, so your body must constantly replenish its stores of ATP by oxidizing carbohydrates, fats, and other fuels through reactions such as the one shown in Equation 11.26. These reactions all require oxygen, and laboratory experiments have established how the rate of chemical energy release—metabolic power—is related to the rate at which oxygen is consumed.

A breathing apparatus like the one shown in FIGURE 11.33 measures the rate at which a person's body is taking up oxygen while engaged in different activities, and the oxygen uptake rate is then converted to metabolic power. This includes all the body's basic processes plus whatever additional energy is needed to perform an activity.

FIGURE 11.33 The mask measures the oxygen used by the athlete.

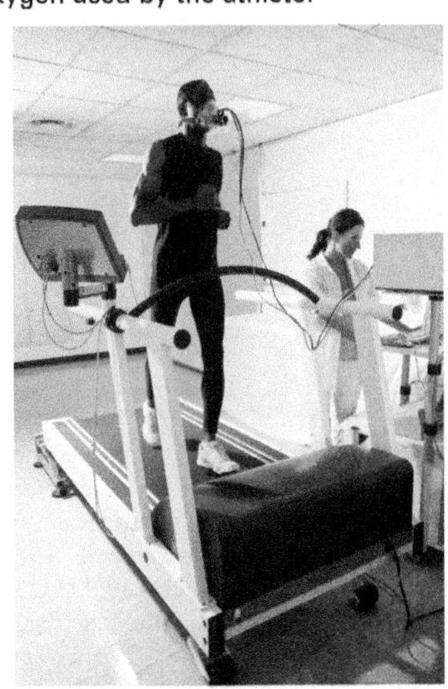

These are called VO2 measurements for "volume of oxygen." A so-called "VO2 max" measurement determines a person's rate of oxygen uptake when exercising at his or her maximum sustained effort. Results are usually given as mL of O_2 per minute per kilogram of body mass, or mL/kg · min. Larger bodies use oxygen more rapidly because there's more muscle mass, so dividing the oxygen uptake rate in mL/min by the person's mass gives a standardized value that can be compared to norms for that person's age and sex.

A young, healthy, but untrained adult male has a VO2 max of ≈35 mL/kg · min; for females the value is ≈30 mL/kg · min. These values decline with age. Elite athletes who compete in endurance sports can reach ≈80 mL/kg · min. Values as high as 240 mL/kg · min have been reported for Alaskan sled dogs.

The laboratory measurements referred to above find that an oxygen uptake rate of 1 mL/kg · min corresponds to a metabolic power of 0.34 W/kg. Thus a 75 kg elite cyclist with a VO2 max of 80 mL/kg · min would have a maximum metabolic power of ≈2000 W. That's substantially higher than the value for cycling in Table 11.5, but this is a professional cycling at 50 km/h rather than a weekend athlete cycling at 15 km/h. However, we noted in Chapter 10 that an elite cyclist can produce a sustained output power of ≈500 W. This is an apparent discrepancy that needs to be resolved.

Efficiency of the Human Body

Suppose you climb several flights of stairs at a constant speed, as shown in FIGURE 11.34. What is your body's efficiency for this process? In Chapter 10 we defined efficiency e as

$$e = \frac{\text{what you get}}{\text{what you had to pay}} \tag{11.27}$$

FIGURE 11.34 Climbing a set of stairs.

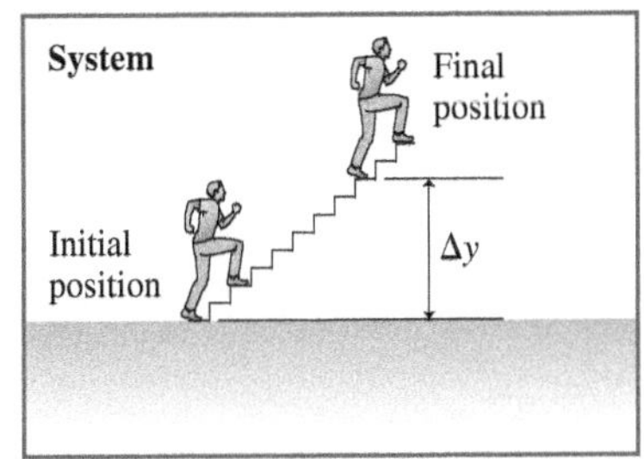

Your kinetic energy is not changing if you climb at a constant speed, so in terms of energy, what you get when climbing stairs is an increase in gravitational potential energy.

There's no external input of energy, and no work is done on you as there would be if you took the elevator. The normal force does push up on your foot, but your foot has no vertical displacement while it's on the step; this is one of the zero-work situations discussed in Chapter 10. Instead, the price of gaining potential energy is an expenditure of chemical energy. The energy principle with $\Delta K = 0$ for constant-speed motion is

$$\Delta U_G + \Delta E_{th} + \Delta E_{chem} = 0 \tag{11.28}$$

The change in chemical energy is negative because chemical energy is released in metabolic reactions; hence

$$|\Delta E_{chem}| = \Delta U_G + \Delta E_{th} \tag{11.29}$$

Chemical energy from your breakfast is used to increase your potential energy *and* your thermal energy: You get hotter while climbing stairs and eventually dissipate this energy as heat.

Suppose your mass is 71 kg and you climb three flights of stairs with $\Delta y = 10$ m. The potential-energy gain is

$$\Delta U_G = mg\,\Delta y = 7000 \text{ J}$$

Your metabolic energy expenditure can be determined by measuring your oxygen uptake as you climb. A climb of 10 m requires $|\Delta E_{chem}| \approx 28{,}000$ J. Thus your efficiency is

$$e = \frac{\text{what you get}}{\text{what you had to pay}} = \frac{\Delta U_G}{|\Delta E_{chem}|} \approx \frac{7000 \text{ J}}{28{,}000 \text{ J}} = 0.25 = 25\%$$

That is, only about 25% of the chemical energy available from food is useful at accomplishing a task; the other 75% simply increases your thermal energy and is then dissipated as heat.

Measurements find that nearly all human activities—walking, running, cycling, swimming—have an efficiency in the range of 20–30%. You learned in Example 11.11 that the efficiency of using glucose oxidation to produce ATP is $\approx$57%, so the body's efficiency for converting chemical energy to useful work would be only 57% if converting ATP energy into muscular effort were 100% efficient. In fact, the reactions by which ATP drives muscular contractions are only about 50% efficient and that reduces your overall efficiency to roughly 25%. Throughout this text, **we will use 25% as a typical value for the human body's efficiency.**

High heating costs? The daily energy use of mammals is much higher than that of reptiles, largely because mammals use energy to maintain a constant body temperature. A 40 kg timber wolf uses approximately 19,000 kJ during the course of a day. A Komodo dragon, a reptilian predator of the same size, uses only 2100 kJ.

We just saw that an elite cyclist might have a maximum metabolic power of 2000 W but his maximum sustained power output is only 500 W. This seemed to be a discrepancy, but it's not because these numbers represent two different things. The 500 W is what he gets, the muscle power *output* needed to sustain his motion in the face of friction and drag. The 2000 W is what he pays, the power *input* from burning fuel. Thus the cyclist's efficiency of transforming chemical energy into useful cycling energy is 25%, exactly as we would expect.

EXAMPLE 11.14 Determining the energy usage for a cyclist

A cyclist pedals for 20 min at a speed of 15 km/h. How much metabolic energy is required? How much energy is used for forward propulsion?

PREPARE A cyclist expends metabolic energy to overcome the resistive forces of drag and friction. Table 11.5 gives 480 W as the metabolic power used in cycling at 15 km/h. This power is used for 20 min = 1200 s at an assumed efficiency of 25%.

SOLVE We know the power and the time, so we can compute the energy needed by the body as follows:

$$|\Delta E_{chem}| = P\Delta t = (480 \text{ J/s})(1200 \text{ s}) = 580 \text{ kJ}$$

If we assume an efficiency of 25%, then only 25%, or 140 kJ, of this energy is used for forward propulsion. The remainder goes into thermal energy.

ASSESS This result is reasonable. 15 km/h isn't that speedy; most recreational cyclists could keep up this pace, so we don't expect the energy expenditure to be all that great. A glance at Table 11.3 shows that 580 kJ is slightly more than the energy available in a large apple.

Your body is really quite efficient. The typical 25% efficiency that we've used might not sound very large, but it's better than the efficiency of a typical automobile or that of other devices that use chemical energy as a power source. This efficiency is something to be aware of, though, if you have overindulged and plan to "work off" the calories. Based on the previous examples, the energy in a typical fast-food meal would power the cyclist for more than 3 hours!

EXAMPLE 11.15 **How high can you climb?**

How many flights of stairs could you climb on the energy contained in a 12 oz can of soda? During the climb, how much energy is transformed into thermal energy? Assume that your mass is 68 kg and that a flight of stairs has a vertical height of 2.7 m.

PREPARE What you get in this case is an increase in potential energy. This is the energy output. What you have to pay is the energy input, the energy in the soda. Then, if we assume a typical efficiency, 25% of this energy will go to the increase in potential energy. The rest will go to thermal energy.

SOLVE In Example 11.12, we determined that the energy content of a can of soda is 680 kJ. This is the chemical energy input. At 25% efficiency, the amount of chemical energy transformed into increased potential energy is

$$\Delta U_G = (0.25)(680 \times 10^3\ \text{J}) = 1.7 \times 10^5\ \text{J}$$

Because $\Delta U_G = mg\Delta y$, the height gained is

$$\Delta y = \frac{\Delta U_G}{mg} = \frac{1.7 \times 10^5\ \text{J}}{(68\ \text{kg})(9.8\ \text{m/s}^2)} = 255\ \text{m}$$

With each flight of stairs having a height of 2.7 m, the number of flights climbed is

$$\frac{255\ \text{m}}{2.7\ \text{m}} \approx 94\ \text{flights}$$

The remainder of the energy is transformed into thermal energy:

$$\Delta E_{th} = |\Delta E_{chem}| - \Delta U_G = 6.8 \times 10^5\ \text{J} - 1.7 \times 10^5\ \text{J}$$
$$= 5.1 \times 10^5\ \text{J} = 510\ \text{kJ}$$

ASSESS 94 flights is almost enough to get to the top of the Empire State Building! But this makes sense; there is a lot of energy in a can of soda. And if you climb several sets of stairs, you know that your body warms up, so the large increase in thermal energy makes sense as well.

Energy Storage

If the energy that the body gets from food is not used, it is stored. A small amount of energy needed for immediate use is stored as ATP. A larger amount of energy is stored as chemical energy of simple carbohydrates in muscle tissue and the liver. A healthy adult might store 400 g of these carbohydrates, which is a little more carbohydrate than is typically consumed in one day.

If the energy input from food continuously exceeds the energy outputs of the body, this energy will be stored in the form of fat under the skin and around the organs. From an energy point of view, gaining weight is simply explained!

EXAMPLE 11.16 **How far can you run?**

The body stores about 400 g of carbohydrates. Approximately how far could a 68 kg runner travel at 15 km/h on this stored energy?

PREPARE Table 11.2 gives an energy value of 17 kJ per g of carbohydrate. When oxidized, the 400 g of carbohydrates in the body release energy

$$|\Delta E_{chem}| = (400\ \text{g})(17 \times 10^3\ \text{J/g}) = 6.8 \times 10^6\ \text{J}$$

SOLVE Table 11.5 gives the power used in running at 15 km/h as 1150 W. The time that the stored chemical energy will last at this rate is

$$\Delta t = \frac{|\Delta E_{chem}|}{P} = \frac{6.8 \times 10^6\ \text{J}}{1150\ \text{W}} = 5.91 \times 10^3\ \text{s} = 1.64\ \text{h}$$

Continued

And the distance that can be covered during this time at 15 km/h is

$$\Delta x = v\Delta t = (15\text{ km/h})(1.64\text{ h}) = 25\text{ km}$$

to two significant figures.

ASSESS A marathon is longer than this—just over 42 km. Even with "carbo loading" before the event (eating high-carbohydrate meals), many marathon runners "hit the wall" before the end of the race as they reach the point where they have exhausted their store of carbohydrates. Given that this is a problem for long-distance runners, our answer makes sense. But if runners deplete their carbohydrate reserves, how can they finish the race? The body has other energy stores (in fats, for instance), but the rate that they can be drawn on is much lower.

Energy and Locomotion

Why does your body need energy to walk at a steady speed on level ground? Where does this energy go?

FIGURE 11.35 Human locomotion analysis.

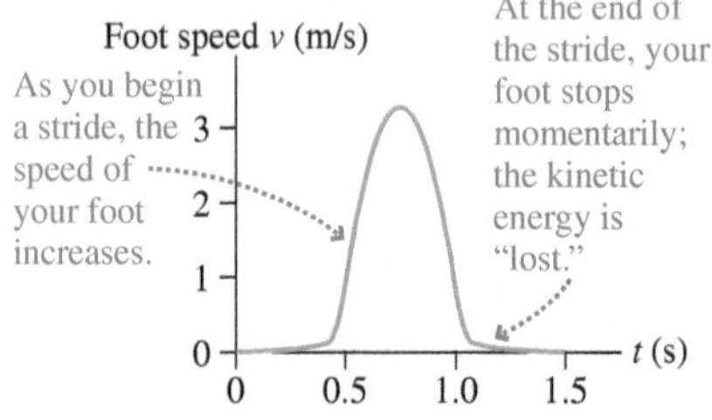
Foot speed v (m/s)
As you begin a stride, the speed of your foot increases.
At the end of the stride, your foot stops momentarily; the kinetic energy is "lost."
3
2
1
0
0
0.5
1.0
1.5
t (s)

We use energy to walk because of mechanical inefficiencies in our gait. Although your center of gravity moves with constant speed, each foot and leg must speed up and then slow down with every step. FIGURE 11.35 shows how the speed of one foot typically changes during each stride. The kinetic energy of your leg and foot increases, only to go to zero at the end of the stride. Chemical energy is used during the increase, then the kinetic energy is mostly transformed into thermal energy in your muscles and in your shoes. This thermal energy is lost; it can't be used for making more strides.

This inefficiency is a process limitation. It's possible to do better. Footwear can be designed to minimize the loss of kinetic energy to thermal energy. A spring in the sole of the shoe can store potential energy, which can be returned to kinetic energy during the next stride. Such a spring will make the collision with the ground more elastic. You saw earlier in this chapter that the tendons in the ankle store a certain amount of energy during a stride; very stout tendons in the legs of kangaroos store energy even more efficiently. Their peculiar hopping gait is quite efficient at high speeds.

◄ **Where do you wear the weights?** If you wear a backpack with a mass equal to 1% of your body mass, your energy expenditure for walking will increase by 1%. But if you wear ankle weights with a combined mass of 1% of your body mass, the increase in energy expenditure is 6%, because you must repeatedly accelerate this extra mass. If you want to "burn more fat," wear the weights on your ankles, not on your back! If you are a runner who wants to shave seconds off your time in the mile, you might try lighter shoes.

STOP TO THINK 11.9 A runner is moving at a constant speed on level ground. Chemical energy in the runner's body is being transformed into other forms of energy. Most of the chemical energy is transformed into

A. Kinetic energy.
B. Potential energy.
C. Thermal energy.

CHAPTER 11 INTEGRATED EXAMPLE Stopping a runaway truck

A truck's brakes can overheat and fail while descending mountain highways, leading to an extremely dangerous runaway truck. Some highways have *runaway-truck ramps* to safely bring out-of-control trucks to a stop. These uphill ramps are covered with a deep bed of gravel. The uphill slope and the large coefficient of rolling friction as the tires sink into the gravel bring the truck to a safe halt.

A 22,000 kg truck heading down a 3.5° slope at 20 m/s ($\approx$45 mph) suddenly has its brakes fail. Fortunately, there's a runaway-truck ramp 600 m ahead. The ramp slopes upward at an angle of 10°, and the coefficient of rolling friction between the truck's tires and the loose gravel is $\mu_r = 0.40$. Ignore air resistance and rolling friction as the truck rolls down the highway.

a. How far along the ramp does the truck travel before stopping?
b. By how much does the thermal energy of the truck and ramp increase as the truck stops?

PREPARE We'll use an energy conservation approach and follow the steps of Problem-Solving Strategy 11.1. We start by defining the system as the truck + ramp + earth. The change in thermal energy of the ramp and the truck will then be an internal transformation of kinetic into thermal energy.

FIGURE 11.36 shows a before-and-after pictorial representation. Notice that Δx is measured along the road, not horizontally. Because we're going to need to determine friction forces to calculate the increase in thermal energy, we've also drawn a free-body diagram for the truck as it moves up the ramp. One slight complication is that the y-axis of free-body diagrams is drawn perpendicular to the slope, whereas the calculation of gravitational potential energy needs a vertical y-axis to measure height. We've dealt with this by labeling the free-body diagram axis the y'-axis.

FIGURE 11.36 Pictorial representation of the runaway truck.

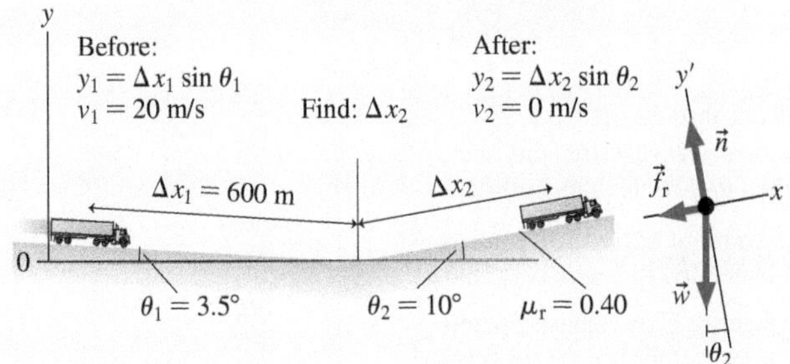

SOLVE

a. The energy equation for the motion of the truck, from the moment its brakes fail to when it finally stops, is

$$K_i + U_{G\,i} = K_f + U_{G\,f} + \Delta E_{th}$$

Because friction is present only for distance Δx_2 along the ramp, thermal energy will increase only as the truck moves up the ramp. This increase in thermal energy is $\Delta E_{th} = f_r\,\Delta x_2$. The conservation of energy equation then is

$$\tfrac{1}{2}mv_1^2 + mgy_1 = \tfrac{1}{2}mv_2^2 + mgy_2 + f_r\Delta x_2$$

From Figure 11.36 we have $y_1 = \Delta x_1 \sin\theta_1$, $y_2 = \Delta x_2 \sin\theta_2$, and $v_2 = 0$, so the equation becomes

$$\tfrac{1}{2}mv_1^2 + mg\Delta x_1 \sin\theta_1 = mg\Delta x_2 \sin\theta_2 + f_r\Delta x_2$$

To find $f_r = \mu_r n$ we need to find the normal force n. The free-body diagram shows that

$$\sum F_{y'} = n - mg\cos\theta_2 = a_{y'} = 0$$

from which $f_r = \mu_r n = \mu_r mg\cos\theta_2$. With this result for f_r, our conservation of energy equation is

$$\tfrac{1}{2}mv_1^2 + mg\Delta x_1 \sin\theta_1 = mg\Delta x_2 \sin\theta_2 + \mu_r mg\cos\theta_2\,\Delta x_2$$

which, after we divide both sides by mg, simplifies to

$$\frac{v_1^2}{2g} + \Delta x_1 \sin\theta_1 = \Delta x_2 \sin\theta_2 + \mu_r \cos\theta_2 \Delta x_2$$

Solving this for Δx_2 gives

$$\Delta x_2 = \frac{\dfrac{v_1^2}{2g} + \Delta x_1 \sin\theta_1}{\sin\theta_2 + \mu_r \cos\theta_2}$$

$$= \frac{\dfrac{(20\text{ m/s})^2}{2(9.8\text{ m/s}^2)} + (600\text{ m})(\sin 3.5°)}{\sin 10° + 0.40(\cos 10°)} = 100\text{ m}$$

b. We know that $\Delta E_{th} = f_r\Delta x_2 = (\mu_r mg\cos\theta_2)\Delta x_2$, so that

$$\Delta E_{th} = (0.40)(22{,}000\text{ kg})(9.8\text{ m/s}^2)(\cos 10°)(100\text{ m})$$
$$= 8.5 \times 10^6\text{ J}$$

ASSESS A stopping distance of 100 m seems reasonable and practical for the length of a runaway-truck ramp. However, the truck's kinetic energy, $K_1 = \frac{1}{2}mv_1^2 = 4.4 \times 10^6$ J, when the brakes fail is less than ΔE_{th}. That seems surprising, but the truck will continue to pick up speed as it rolls to the beginning of the ramp. Its kinetic energy as it enters the ramp will be larger than K_1, and most of that kinetic energy will be transformed into thermal energy. So the answer for ΔE_{th} seems reasonable.

SUMMARY

GOAL To develop a better understanding of energy and its application to living systems.

GENERAL PRINCIPLES

The Energy Principle Revisited

- Energy is *transformed* within the system.
- Energy is *transferred* to and from the system by work W.

The expanded energy principle is

$$W = \Delta E_{sys} = \Delta K + \Delta U + \Delta E_{th} + \Delta E_{chem}$$

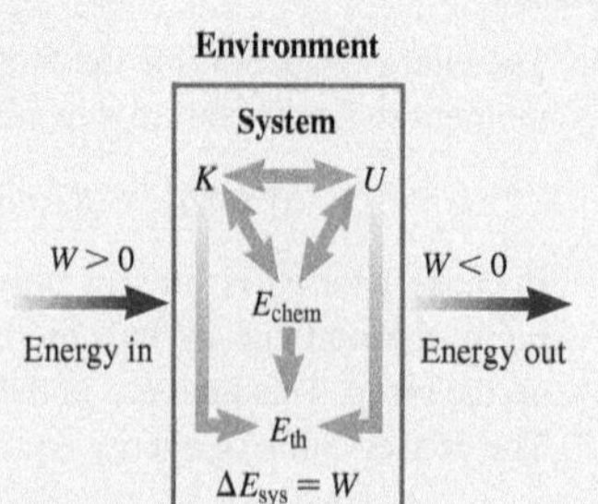

Solving Energy Problems

PREPARE Define the system. Draw a before-and-after pictorial representation and an energy bar chart.

SOLVE Use the before-and-after version of the energy principle:

$$K_i + U_i + W = K_f + U_f + \Delta E_{th} + \Delta E_{chem}$$

- $W = 0$ for an isolated system.
- $\Delta E_{th} = 0$ if there's no friction or drag.

ASSESS Is the result reasonable?

Conservation of Energy

- **Isolated system:** $W = 0$. The total system energy $E_{sys} = K + U + E_{th} + E_{chem}$ is conserved. $\Delta E_{sys} = 0$.
- **Isolated, nonreactive systems:** $W = 0$ and $\Delta E_{chem} = 0$. The energy $K + U + E_{th}$ is conserved.
- **Isolated, nonreactive, nondissipative system:** $W = 0$ and $\Delta E_{chem} = 0$ and $\Delta E_{th} = 0$. The mechanical energy $E_{mech} = K + U$ is conserved.

IMPORTANT CONCEPTS

Potential energy is the energy of interactions within the system.

- The work W_{int} done by interaction forces causes $\Delta U = -W_{int}$.
- Force $F_x = -dU/dx = -$(slope of the PE curve).
- Potential energy is an energy of the system, not of a specific object.

Gravitational potential energy: $U_G = mgy$

Elastic potential energy: $U_{sp} = \frac{1}{2}k(\Delta x)^2$

Energy diagrams show the potential-energy curve PE and the total mechanical energy line TE.

- From the axis to the curve is U. From the curve to the TE line is K.
- Turning points occur where the TE line crosses the PE curve.
- Minima and maxima in the PE curve are, respectively, positions of **stable** and **unstable equilibrium.**

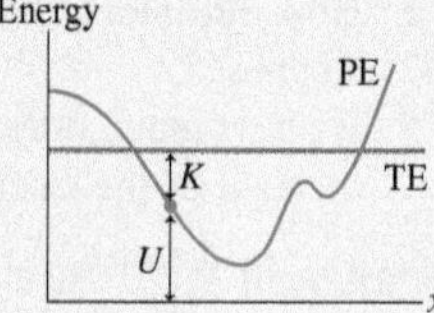

APPLICATIONS

Energy bar charts

An energy bar chart is a graphical representation of the energy principle.

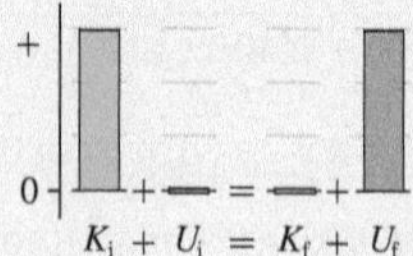

Energy in the body

Cells in the body metabolize chemical energy in food. The body's efficiency for most actions is $\approx 25\%$.

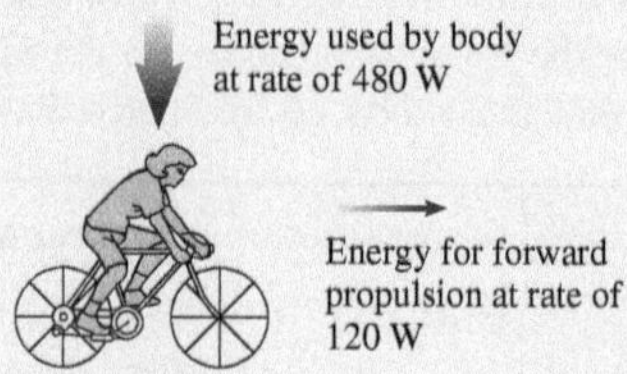

Molecular bonds

A molecular bond can be represented by a potential-energy curve.

- The **bond length** is the separation at the minimum of the PE curve.
- Negative potential energy indicates a bound state.
- The **bond energy** is the energy needed to separate a ground-state molecule into free atoms.

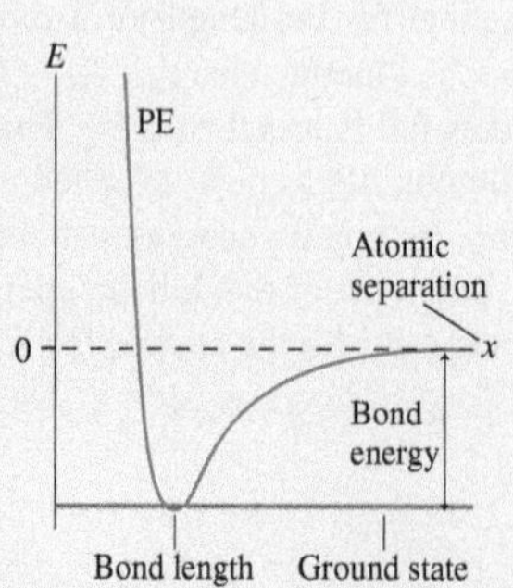

Chemical reactions

A chemical reaction can occur when two molecules collide.

- An input **activation energy** E_a is needed to *break* bonds.
- Energy E_r is released when new, tighter bonds are formed.
- The **chemical energy** released in N reactions is

$$\Delta E_{chem} = N\,\Delta E_{reaction}$$

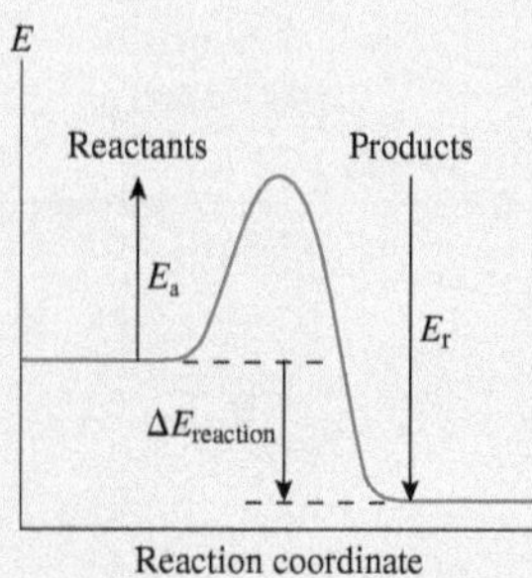

LEARNING OBJECTIVES After studying this chapter, you should be able to:

- Determine a system's potential energy using the work done in an interaction. *Conceptual Question 11.11; Problem 11.1*
- Compute and solve problems related to gravitational potential energy. *Conceptual Questions 11.7, 11.8; Problems 11.2, 11.4–11.7*
- Solve conservation of energy problems using the problem-solving strategy. *Conceptual Questions 11.2, 11.3; Problems 11.8, 11.9, 11.12–11.14*
- Calculate and solve problems related to elastic potential energy. *Conceptual Questions 11.9, 11.10; Problems 11.15, 11.21, 11.22, 11.24, 11.25*
- Draw and interpret energy diagrams. *Conceptual Question 11.12; Problems 11.27–11.31*
- Determine bond energy and chemical energy from molecular energy diagrams. *Problems 11.32–11.34*
- Calculate interaction forces from a potential-energy graph or function. *Conceptual Questions 11.13, 11.14; Problems 11.35–11.39*
- Apply the expanded energy principle to solve problems that involve a changing thermal energy. *Problems 11.40–11.42*
- Demonstrate how energy is used by the human body, and calculate the energy efficiency of different activities. *Conceptual Question 11.15; Problems 11.43–11.46, 11.49*

STOP TO THINK ANSWERS

Stop to Think 11.1: $(U_G)_C > (U_G)_B = (U_G)_D > (U_G)_A$. Gravitational potential energy depends only on height, not on speed.

Stop to Think 11.2: B. Potential energy depends only on the vertical displacement. At the elevation of the dashed line, both have gained the same gravitational potential energy, so both have lost the same kinetic energy.

Stop to Think 11.3: $v_A = v_B = v_C = v_D$. Her increase in kinetic energy depends only on the vertical distance through which she falls, not on the shape of the slide.

Stop to Think 11.4: C. Constant speed means no change of kinetic energy. But for motion on a slope, constant speed requires friction. All the gravitational potential energy is being transformed into thermal energy.

Stop to Think 11.5: C. U_{sp} depends on $(\Delta x)^2$. Doubling the compression increases U_{sp} by a factor of 4. All potential energy is transformed into kinetic energy, so K increases by a factor of 4. But K depends on v^2, so v increases by only a factor of 2.

Stop to Think 11.6: B. The kinetic energy is the difference between the total energy and the potential energy. This is highest at the bottom of the right well in B.

Stop to Think 11.7: E. Force is the negative of the slope of the potential-energy diagram. At $x = 4$ m the potential energy has risen by 4 J over a distance of 2 m, so the slope is 2 J/m = 2 N.

Stop to Think 11.8: D. The system is losing potential energy as the weight falls. It's gaining speed, so some U is tranformed into K. Energy is also being transferred out of the system to the environment via negative work done by the rope tension. The system is not isolated, so neither E_{mech} nor E_{sys} is conserved.

Stop to Think 11.9: C. As the body uses chemical energy from food, approximately 75% is transformed into thermal energy. Also, kinetic energy of motion of the legs and feet is transformed into thermal energy with each stride. Most of the chemical energy is transformed into thermal energy.

QUESTIONS

Conceptual Questions

1. Upon what basic quantity does kinetic energy depend? Upon what basic quantity does potential energy depend?
2. A diver leaps from a high platform, speeds up as she falls, and then slows to a stop in the water. Which objects should be included within the system in order to make an energy analysis as easy as possible?
3. A compressed spring launches a block up an incline. Which objects should be included within the system in order to make an energy analysis as easy as possible?
4. Give a specific example of a situation in which the energy transformation is $U \rightarrow K$.
5. Give a specific example of a situation in which the energy transformation is $U \rightarrow E_{th}$.
6. Give a specific example of a situation in which the energy transformation is $W \rightarrow U$.
7. A roller-coaster car rolls down a frictionless track, reaching speed v_0 at the bottom. If you want the car to go twice as fast at the bottom, by what factor must you increase the height of the track? Explain.
8. Sandy and Chris stand on the edge of a cliff and throw identical mass rocks at the same speed. Sandy throws her rock horizontally while Chris throws his upward at an angle of 45° to the horizontal. Are the rocks moving at the same speed when they hit the ground, or is one moving faster than the other? If one is moving faster, which one? Explain.
9. A spring is compressed 1.0 cm. How far must you compress a spring with twice the spring constant to store the same amount of energy?
10. A spring gun shoots out a plastic ball at speed v_0. The spring is then compressed twice the distance it was on the first shot. By what factor is the ball's speed increased? Explain.

Problem difficulty is labeled as | (straightforward) to ||||| (challenging). Problems labeled INT integrate significant material from earlier chapters; BIO are of biological or medical interest; CALC require calculus to solve.

11. A process occurs in which a system's potential energy increases while the environment does work on the system. Does the system's kinetic energy increase, decrease, or stay the same? Or is there not enough information to tell? Explain.
12. Figure Q11.12 is the energy bar chart for a firefighter sliding down a fire pole from the second floor to the ground. Let the system consist of the firefighter, the pole, and the earth. What are the bar heights of K_f and U_{Gf}?

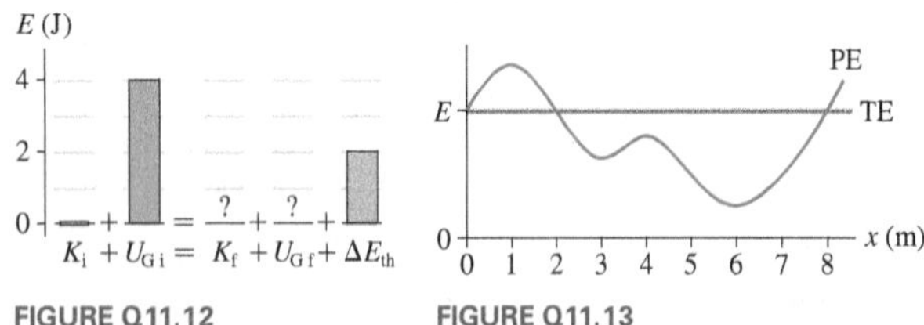

FIGURE Q11.12

FIGURE Q11.13

13. A particle with the potential energy shown in Figure Q11.13 is moving to the right at $x = 5$ m with total energy E.
    a. At what value or values of x is this particle's speed a maximum?
    b. Does this particle have a turning point or points in the range of x covered by the graph? If so, where?
    c. If E is changed appropriately, could the particle remain at rest at any point or points in the range of x covered by the graph? If so, where?
14. Figure Q11.14 shows a potential-energy diagram for a particle. The particle is at rest at point A and is then given a slight nudge to the right. Describe the subsequent motion.

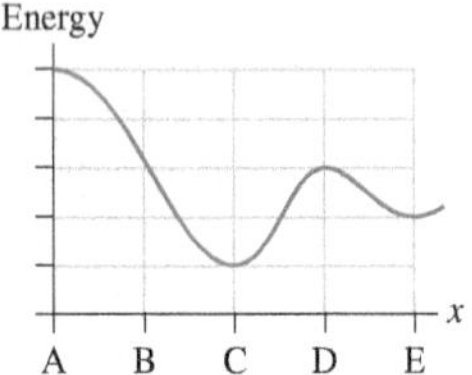

FIGURE Q11.14

15. If your running speed is three times your walking speed, then your metabolic power while running is three times larger than your metabolic power while walking. For the same expenditure of energy, is your running distance three times, the same as, or one-third your walking distance?

Multiple-Choice Questions

16. || A roller coaster starts from rest at its highest point and then descends on its (frictionless) track. Its speed is 30 m/s when it reaches ground level. What was its speed when its height was half that of its starting point?
    A. 11 m/s B. 15 m/s C. 21 m/s D. 25 m/s
17. | A hockey puck sliding along frictionless ice with speed v to the right collides with a horizontal spring and compresses it by 2.0 cm before coming to a momentary stop. What will be the spring's maximum compression if the same puck hits it at a speed of $2v$?
    A. 2.0 cm B. 2.8 cm C. 4.0 cm
    D. 5.6 cm E. 8.0 cm
18. || A block slides down a smooth ramp, starting from rest at a height h. When it reaches the bottom it's moving at speed v. It then continues to slide up a second smooth ramp. At what height is its speed equal to $v/2$?
    A. $h/4$ B. $h/2$ C. $3h/4$ D. $2h$
19. | A wrecking ball is suspended from a 5.0-m-long cable that makes a 30° angle with the vertical. The ball is released and swings down. What is the ball's speed at the lowest point?
    A. 7.7 m/s B. 4.4 m/s C. 3.6 m/s D. 3.1 m/s
20. | On a clear day, sunlight delivers approximately 1000 J each second to a 1 m^2 surface; smaller areas receive proportionally less—half the area receives half the energy. A 12% efficient solar cell, a square 15 cm on a side, is in bright sunlight. How much electric power does it produce?
    A. 0.9 W B. 1.8 W C. 2.7 W D. 3.6 W
21. || For locations that have no electric service, companies are designing bicycle-powered generators that are nearly 100% efficient that can be used to charge cell phones or power lights. If you are the person chosen to pedal and your friends need a total of 150 W of electric power, how much metabolic power will you use as you pedal?
    A. 150 W B. 300 W C. 450 W D. 600 W
22. || A person is walking on level ground at constant speed. What energy transformation is taking place?
    A. Chemical energy is being transformed to thermal energy.
    B. Chemical energy is being transformed to kinetic energy.
    C. Chemical energy is being transformed to kinetic energy and thermal energy.
    D. Chemical energy and thermal energy are being transformed to kinetic energy.
23. | A person walks 1 km, turns around, and runs back to where he started. Compare the energy used and the power during the two segments.
    A. The energy used and the power are the same for both.
    B. The energy used while walking is greater, the power while running is greater.
    C. The energy used while running is greater, the power while running is greater.
    D. The energy used is the same for both segments, the power while running is greater.
24. | An energy bar contains 26 g of carbohydrates and 5 g of fat. How many Calories will you obtain by eating the bar?
    A. 150 Cal B. 250 Cal
    C. 610 Cal D. 2600 Cal
25. | The graph in Figure Q11.25 shows the metabolic power used by a starling in level flight. If the starling needs to fly a certain distance, it will use the least metabolic energy if it flies at
    A. 6 m/s
    B. 10 m/s
    C. 14 m/s
    D. It will use about the same energy at each of these speeds.

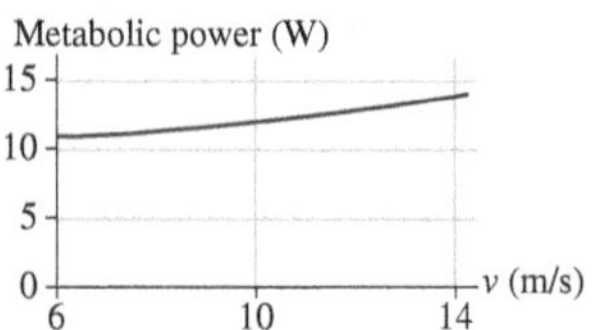

FIGURE Q11.25

PROBLEMS

Section 11.1 Potential Energy

1. || A system of two objects has $\Delta K_{tot} = 7$ J and $\Delta U_{int} = -5$ J.
   a. How much work is done by interaction forces?
   b. How much work is done by external forces?
2. | The lowest point in Death Valley is 85 m below sea level. The summit of nearby Mt. Whitney has an elevation of 4420 m. What is the change in potential energy when an energetic 65 kg hiker makes it from the floor of Death Valley to the top of Mt. Whitney?
3. | a. What is the kinetic energy of a 1500 kg car traveling at a speed of 30 m/s ($\approx$65 mph)?
   b. From what height would the car have to be dropped to have this same amount of kinetic energy just before impact?
4. | a. With what minimum speed must you toss a 100 g ball straight up to just touch the 10-m-high roof of the gymnasium if you release the ball 1.5 m above the ground? Solve this problem using energy.
   b. With what speed does the ball hit the ground?
5. | The world's fastest humans can reach speeds of about 11 m/s. In order to increase his gravitational potential energy by an amount equal to his kinetic energy at full speed, how high would such a sprinter need to climb?
6. || The maximum energy a bone can absorb without breaking is
BIO surprisingly small. Experimental data show that the leg bones of a healthy, 60 kg human can absorb about 200 J. From what maximum height could a 60 kg person jump and land rigidly upright on both feet without breaking his legs? Assume that all energy is absorbed by the leg bones in a rigid landing.
7. || In a hydroelectric dam, water falls 25 m and then spins a turbine to generate electricity.
   a. What is ΔU_G of 1.0 kg of water?
   b. Suppose the dam is 80% efficient at converting the water's potential energy to electrical energy. How many kilograms of water must pass through the turbines each second to generate 50 MW of electricity? This is a typical value for a small hydroelectric dam.

Section 11.2 Conservation of Energy

8. | What height does a frictionless playground slide need so that a 35 kg child reaches the bottom at a speed of 4.5 m/s?
9. | A 72 kg bike racer climbs a 1200-m-long section of road that has a slope of 4.3°. By how much does his gravitational potential energy change during this climb?
10. || A 1000 kg wrecking ball hangs from a 15-m-long cable. The ball is pulled back until the cable makes an angle of 25° with the vertical. By how much has the gravitational potential energy of the ball changed?
11. | A 55 kg skateboarder wants to just make it to the upper edge of a "quarter pipe," a track that is one-quarter of a circle with a radius of 3.0 m. What speed does she need at the bottom?
12. | What minimum speed does a 100 g puck need to make it to the top of a 3.0-m-long, 20° frictionless ramp?
13. || A 20 kg child is on a swing that hangs from 3.0-m-long chains. What is her maximum speed if she swings out to a 45° angle?
14. || A 1500 kg car traveling at 10 m/s suddenly runs out of gas while approaching the valley shown in Figure P11.14. The alert driver immediately puts the car in neutral so that it will roll. If there are no dissipative losses, what will be the car's speed as it coasts into the gas station on the other side of the valley?

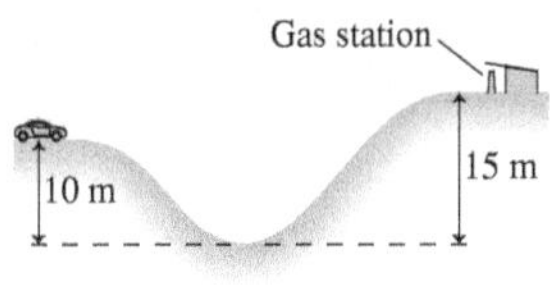

FIGURE P11.14

Section 11.3 Elastic Potential Energy

15. | How far must you stretch a spring with $k = 1000$ N/m to store 200 J of energy?
16. || How much energy can be stored in a spring with a spring constant of 500 N/m if its maximum possible stretch is 20 cm?
17. || A stretched spring stores 2.0 J of energy. How much energy will be stored if the spring is stretched three times as far?
18. | A student places her 500 g physics book on a frictionless table. She pushes the book against a spring, compressing the spring by 4.0 cm, then releases the book. What is the book's speed as it slides away? The spring constant is 1250 N/m.
19. | A block sliding along a horizontal frictionless surface with speed v collides with a spring and compresses it by 2.0 cm. What will be the compression if the same block collides with the spring at a speed of $2v$?
20. | A 10 kg runaway grocery cart runs into a spring with spring constant 250 N/m and compresses it by 60 cm. What was the speed of the cart just before it hit the spring?
21. || As a 15,000 kg jet plane lands on an aircraft carrier, its tail hook snags a cable to slow it down. The cable is attached to a spring with spring constant 11,000 N/m. If the spring stretches 81 m to stop the plane, what was the plane's landing speed?
22. || The elastic energy stored in your tendons can contribute up
BIO to 35% of your energy needs when running. Sports scientists find that (on average) the knee extensor tendons in sprinters stretch 41 mm while those of nonathletes stretch only 33 mm. The spring constant of the tendon is the same for both groups, 33 N/mm. What is the difference in maximum stored energy between the sprinters and the nonathletes?
23. || The spring in Figure P11.23a is compressed by Δx. It launches the block across a frictionless surface with speed v_0. The two springs in Figure P11.23b are identical to the spring of Figure P11.23a. They are compressed by the same Δx and used to launch the same block. What is the block's speed now?

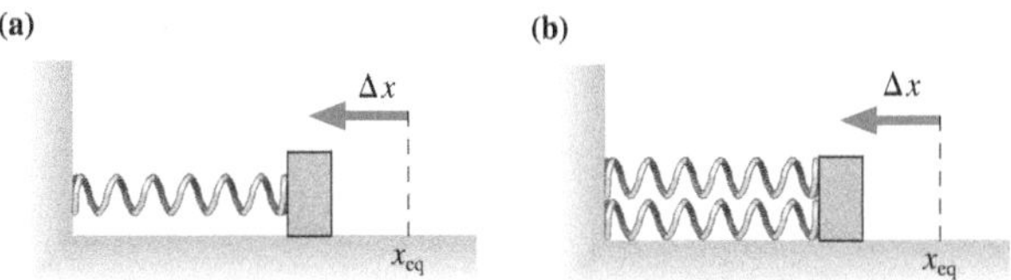

FIGURE P11.23

24. || Scallops use muscles to close their shells. Opening the shell BIO INT is another story—muscles can only pull, they can't push. Instead of muscles, the shell is opened by a spring, a pad of a very elastic biological material called abductin. When the shell closes, the pad compresses; a restoring force then pushes the shell back open. The energy to open the shell comes from the elastic energy that was stored when the shell was closed. Figure P11.24 shows smoothed data for the restoring force of an abductin pad versus the compression. When the shell closes, the pad compresses by 0.15 mm. How much elastic potential energy is stored?

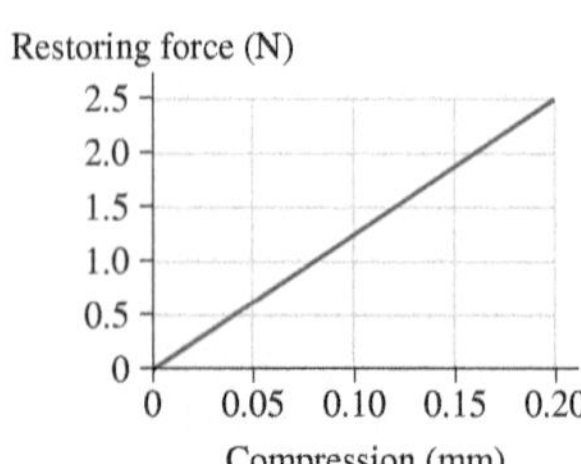

FIGURE P11.24

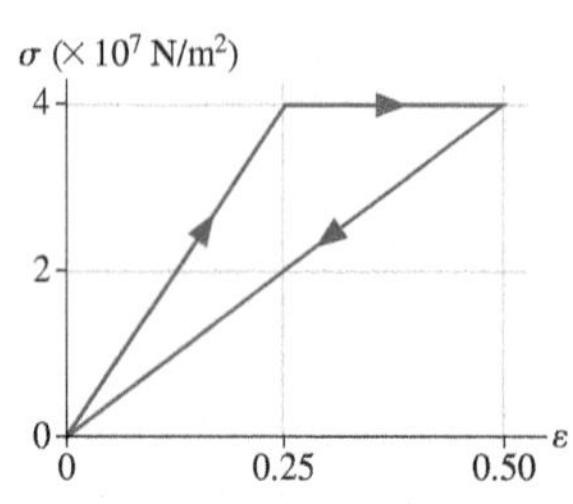

FIGURE P11.25

25. || Figure P11.25 is the stress-strain curve of a 2.0 mm × 2.0 mm × 12 mm tendon. BIO INT
a. How much elastic energy is stored in the tendon if its length is stretched by 50%?
b. What is the resilience of this tendon? Give your answer as a percent.

Section 11.4 Energy Diagrams

26. || Figure P11.26 is the potential-energy diagram for a 20 g particle that is released from rest at $x = 1.0$ m.
a. Will the particle move to the right or to the left?
b. What is the particle's maximum speed? At what position does it have this speed?
c. Where are the turning points of the motion?

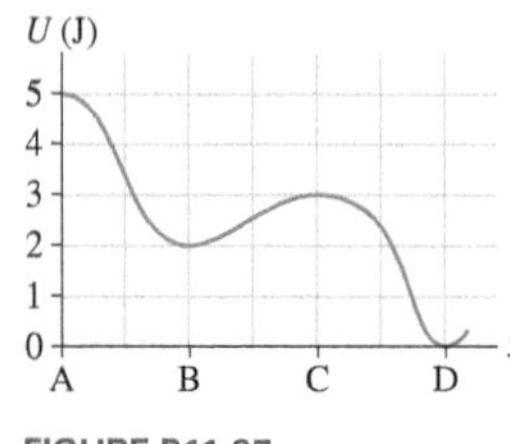

FIGURE P11.26

27. || Figure P11.27 is the potential-energy diagram for a 500 g particle that is released from rest at A. What are the particle's speeds at B, C, and D?

FIGURE P11.27

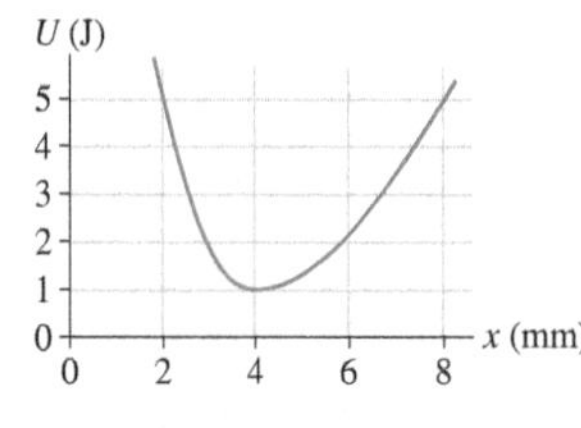

FIGURE P11.28

28. || In Figure P11.28, what is the maximum speed of a 2.0 g particle that oscillates between $x = 2.0$ mm and $x = 8.0$ mm?

29. | a. In Figure P11.29, what minimum speed does a 100 g particle need at point A to reach point B?
b. What minimum speed does a 100 g particle need at point B to reach point A?

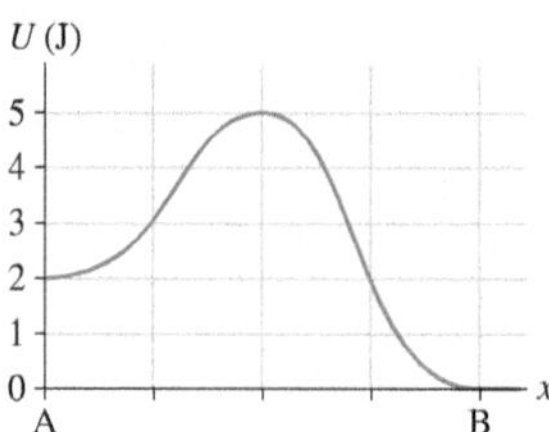

FIGURE P11.29

30. || Figure P11.30 shows the potential energy of a 500 g particle as it moves along the x-axis. Suppose the particle's mechanical energy is 12 J.
a. Where are the particle's turning points?
b. What is the particle's speed when it is at $x = 4.0$ m?
c. What is the particle's maximum speed? At what position or positions does this occur?
d. Suppose the particle's energy is lowered to 4.0 J. Can the particle ever be at $x = 2.0$ m? At $x = 4.0$ m?

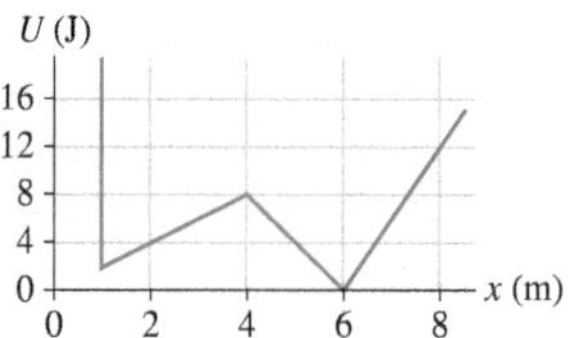

FIGURE P11.30

31. || In Figure P11.30, what is the maximum speed a 200 g particle could have at $x = 2.0$ m and never reach $x = 6.0$ m?

Section 11.5 Molecular Bonds and Chemical Energy

Hydrogen H_2 is a diatomic molecule composed of two hydrogen atoms. Figure P11.32 is an energy diagram for hydrogen as a function of the atomic separation x. Use the information to solve Problems 32 to 34.

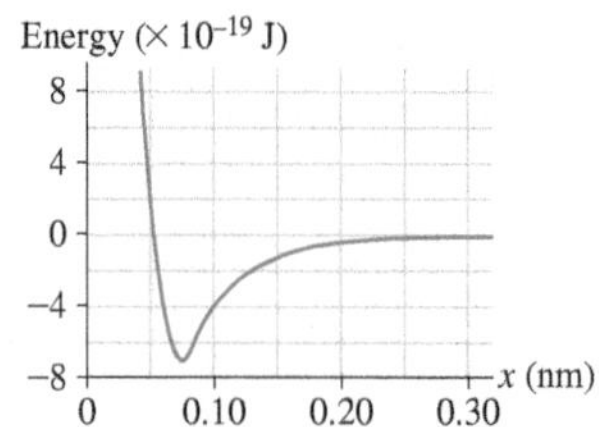

FIGURE P11.32

32. | What is the bond length of a hydrogen molecule?
33. || What is the minimum light-photon energy that can dissociate a hydrogen molecule?
34. ||| Suppose a hydrogen molecule in its ground state is dissociated by absorbing a photon of ultraviolet light, causing the two hydrogen atoms to fly apart. What photon energy will give each atom a speed of 17 km/s? The mass of a hydrogen atom is 1.7×10^{-27} kg.

Section 11.6 Connecting Potential Energy to Force

35. || A system in which only one particle can move has the potential energy shown in Figure P11.35. What is the x-component of the force on the particle at $x = 5$, 15, and 25 cm?

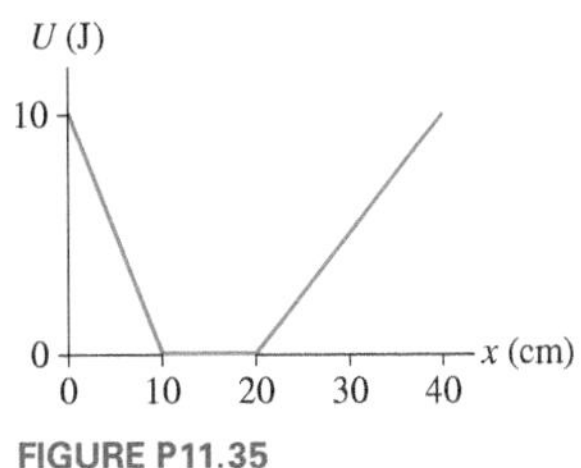

FIGURE P11.35

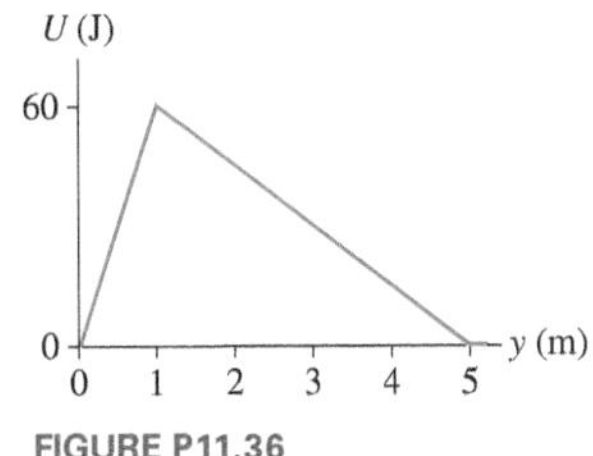

FIGURE P11.36

36. || A system in which only one particle can move has the potential energy shown in Figure P11.36. What is the y-component of the force on the particle at $y = 0.5$ m and 4 m?

37. || Figure P11.37 shows the potential energy of a system in which a particle moves along the x-axis. Draw a graph of the force F_x as a function of position x.

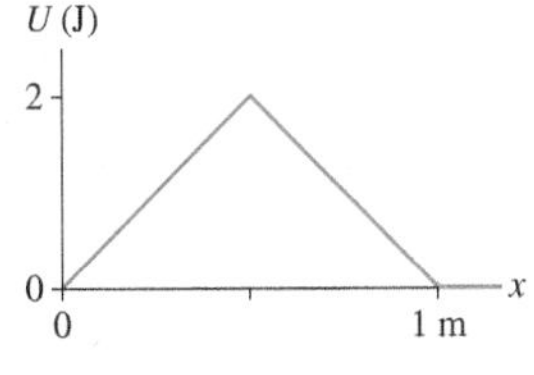

FIGURE P11.37

38. ||| A particle moving along the y-axis is in a system with potential energy $U = 4y^3$ J, where y is in m. What is the y-component of the force on the particle at $y = 0$ m, 1 m, and 2 m?
CALC

39. ||| A particle moving along the x-axis is in a system with potential energy $U = 10/x$ J, where x is in m. What is the x-component of the force on the particle at $x = 2$ m, 5 m, and 8 m?
CALC

Section 11.7 The Expanded Energy Model

40. | A system loses 400 J of potential energy. In the process, it does 400 J of work on the environment and the thermal energy increases by 100 J. Show this process on an energy bar chart.

41. | What is the final kinetic energy of the system for the process shown in Figure P11.41?

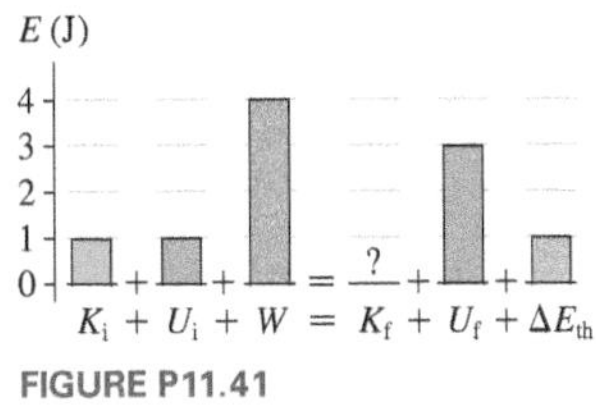

FIGURE P11.41

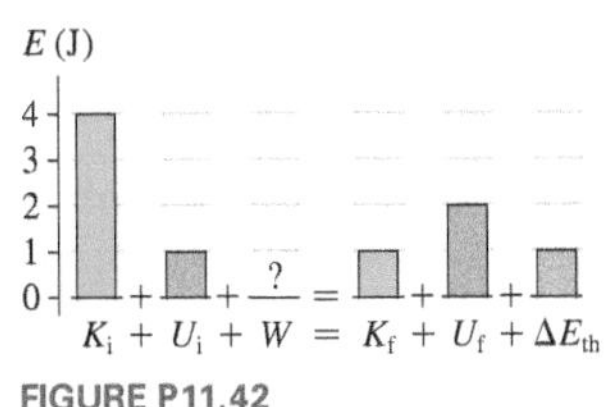

FIGURE P11.42

42. | How much work is done by the environment in the process shown in Figure P11.42? Is energy transferred from the environment to the system or from the system to the environment?

Section 11.8 Energy in the Body

43. | A sleeping 68 kg student has a metabolic power of 71 W. How many Calories does the student burn during an 8.0 hour sleep?
BIO

44. || In an average human, basic life processes require energy to be supplied at a steady rate of 100 W. What daily energy intake, in Calories, is required to maintain these basic processes?
BIO

45. || Jessie and Jaime complete a 5.0 km race. Each has a mass of 68 kg. Jessie runs the race at 15 km/h; Jaime walks it at 5 km/h. How much metabolic energy does each use to complete the course?
BIO

46. || An "energy bar" contains 22 g of carbohydrates. If the energy bar was his only fuel, how far could a 68 kg person walk at 5.0 km/h?
BIO

47. || Each time he does one pushup, Jose, who has a mass of 75 kg, raises his center of gravity by 25 cm. He completes an impressive 150 pushups in 5 minutes, exercising at a steady rate.
BIO
    a. If we assume that lowering his body has no energetic cost, what is his metabolic power during this workout?
    b. In fact, it costs Jose a certain amount of energy to lower his body—about half of what it costs to raise it. If you include this in your calculation, what is his metabolic power?

48. ||| Tessa and Jody, each of mass 68 kg, go out for some exercise together. Tessa runs at 15 km/h; Jody cycles alongside at the same speed. After 20 minutes, how much metabolic energy has each used?
BIO

49. || The basis of muscle action is the power stroke of the myosin protein pulling on an actin filament. It takes the energy of one molecule of ATP, 5.1×10^{-20} J, to produce a displacement of 10 nm against a force of 1.0 pN. What is the efficiency?
BIO INT

50. || A weightlifter curls a 30 kg bar, raising it each time a distance of 0.60 m. How many times must he repeat this exercise to burn off the energy in one slice of pizza?
BIO

General Problems

51. ||| A very slippery ice cube slides in a *vertical* plane around the inside of a smooth, 20-cm-diameter horizontal pipe. The ice cube's speed at the bottom of the circle is 3.0 m/s. What is the ice cube's speed at the top?

52. ||| You have been hired to design a spring-launched roller coaster that will carry two passengers per car. The car goes up a 10-m-high hill, then descends 15 m to the track's lowest point. You've determined that the spring can be compressed a maximum of 2.0 m and that a loaded car will have a maximum mass of 400 kg. For safety reasons, the spring constant should be 10% larger than the minimum needed for the car to just make it over the top.
    a. What spring constant should you specify?
    b. What is the maximum speed of a 350 kg car if the spring is compressed the full amount?

53. ||| Fleas have remarkable jumping ability. A 0.50 mg flea, jumping straight up, would reach a height of 40 cm if there were no air resistance. In reality, air resistance limits the height to 20 cm.
BIO
    a. What is the flea's kinetic energy as it leaves the ground?
    b. At its highest point, what fraction of the initial kinetic energy has been converted to potential energy?

54. | As you walk, your center of gravity moves up and down by approximately 1.5 cm but your mechanical energy remains nearly constant. As a result, you speed up and slow down during every step as potential energy is transformed into kinetic energy, and then back again. Suppose your speed is 1.2 m/s when your center of gravity it at its lowest point. What is your speed when your center of gravity reaches its highest point?
BIO

55. ||| You are driving your 1500 kg car at 20 m/s down a hill with a 5.0° slope when a deer suddenly jumps out onto the roadway. You slam on your brakes, skidding to a stop. How far do you skid before stopping if the kinetic friction force between your tires and the road is 1.2×10^4 N? Solve this problem using conservation of energy.

56. |||| A block of mass m slides down a frictionless track, then around the inside of a circular loop-the-loop of radius R. From what minimum height h must the block start to make it around without falling off? Give your answer as a multiple of R.

57. ||| A 1000 kg safe is 2.0 m above a heavy-duty spring when the rope holding the safe breaks. The safe hits the spring and compresses it 50 cm. What is the spring constant of the spring?

58. || Mosses don't spread by dispersing seeds; they disperse tiny
BIO spores. The spores are so small that they will stay aloft and move with the wind, but getting them to be windborne requires the moss to shoot the spores upward. Some species do this by using a spore-containing capsule that dries out and shrinks. The pressure of the air trapped inside the capsule increases until, at a certain point, the capsule pops, and a stream of spores is ejected upward at 3.6 m/s, reaching an ultimate height of 20 cm. What fraction of the initial kinetic energy is transformed into the final potential energy? What happens to the "lost" energy?

59. ||| When you stand on a trampoline, the surface depresses below
INT equilibrium, and the surface pushes up on you, as the data for a real trampoline in Figure P11.59 show. The linear variation of the force as a function of distance means that we can model the restoring force as that of a spring. A 72 kg gymnast jumps on the trampoline. At the lowest point of his motion, he is 0.80 m below equilibrium. If we assume that all of the energy stored in the trampoline goes into his motion, how high above this lowest point will he rise?

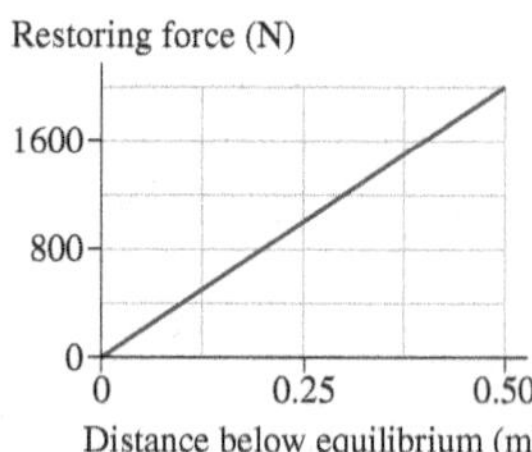

FIGURE P11.59

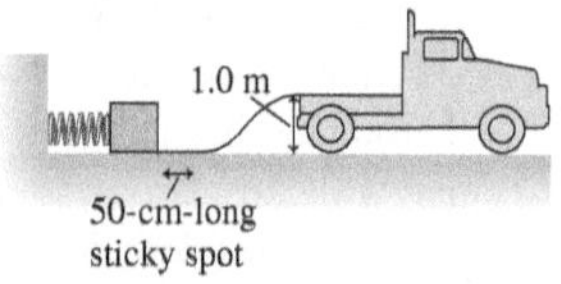

1.0 m
50-cm-long sticky spot

FIGURE P11.60

60. ||| A freight company uses a compressed spring to shoot 2.0 kg packages up a 1.0-m-high frictionless ramp into a truck, as Figure P11.60 shows. The spring constant is 500 N/m and the spring is compressed 30 cm.
a. What is the speed of the package when it reaches the truck?
b. A careless worker spills soda on the ramp. This creates a 50-cm-long sticky spot with a coefficient of kinetic friction 0.30. Will the next package make it into the truck?

61. || a. A 50 g ice cube can slide without friction up and down a 30° slope. The ice cube is pressed against a spring at the bottom of the slope, compressing the spring 10 cm. The spring constant is 25 N/m. When the ice cube is released, what total distance will it travel up the slope before reversing direction?
b. The ice cube is replaced by a 50 g plastic cube whose coefficient of kinetic friction is 0.20. How far will the plastic cube travel up the slope? Use work and energy.

62. |||| A material's stress-strain
INT curve for $0 \leq \varepsilon \leq 1$, shown
CALC in Figure P11.62, can be written as $\sigma = A\sin(\pi\varepsilon/2)$, where $A = 5.0 \times 10^8\ \text{N/m}^2$ and the angle of the sine function is in radians. What is the stored energy density if the material is stretched to a strain of (a) 0.50 and (b) 1.00? Recall that $\int \sin(ax)\,dx = -\cos(ax)/a$.

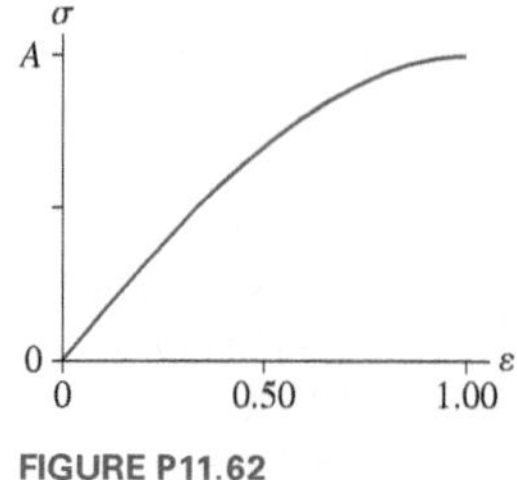

FIGURE P11.62

63. ||| In a physics lab experiment, a spring clamped to the table
INT shoots a 20 g ball horizontally. When the spring is compressed 20 cm, the ball travels horizontally 5.0 m and lands on the floor 1.5 m below the point at which it left the spring. What is the spring constant?

64. || In an amusement park water slide, people slide down an es-
INT sentially frictionless tube. The top of the slide is 3.0 m above the bottom where they exit the slide, moving horizontally, 1.2 m above a swimming pool. What horizontal distance do they travel from the exit point before hitting the water? Does the mass of the person make any difference?

65. || Two coupled boxcars are rolling along at 2.5 m/s when they
INT collide with and couple to a third, stationary boxcar.
a. What is the final speed of the three coupled boxcars?
b. What fraction of the cars' initial kinetic energy is transformed into thermal energy?

66. || A package of mass m is re-
INT leased from rest at a warehouse loading dock and slides down a 3.0-m-high frictionless chute to a waiting truck. Unfortunately, the truck driver went on a break without having removed the previous package, of mass $2m$, from the bottom of the chute as shown in Figure P11.66.

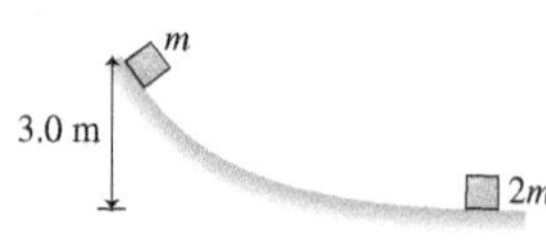

FIGURE P11.66

a. Suppose the packages stick together. What is their common speed after the collision?
b. Suppose the collision between the packages is perfectly elastic. To what height does the package of mass m rebound?

67. ||| A particle that can move along the x-axis is part of a system
CALC with potential energy

$$U(x) = \frac{A}{x^2} - \frac{B}{x}$$

where A and B are positive constants.
a. Where are the particle's equilibrium positions?
b. For each, is it a point of stable or unstable equilibrium?

68. ||| A 100 g particle experiences the one-dimensional force F_x shown in Figure P11.68.

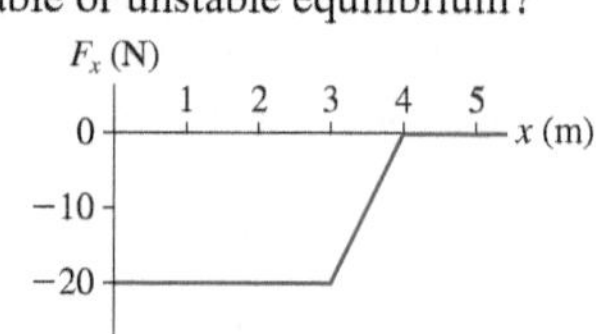

FIGURE P11.68

a. Let the zero of potential energy be at $x = 0$ m. What is the potential energy at $x = 1.0$, 2.0, 3.0, and 4.0 m?
Hint: Use the definition of potential energy and the geometric interpretation of work.
b. Suppose the particle is shot to the right from $x = 1.0$ m with a speed of 25 m/s. Where is its turning point?

69. ||| A particle that can move along the x-axis experiences an
CALC interaction force $F_x = (3x^2 - 5x)$ N, where x is in m. Find an expression for the system's potential energy.

70. ||| The periodontal ligament is the tissue between the tooth and
BIO the jaw that protects teeth from shock forces when biting. A
CALC vertical force F_z pushing on a tooth compresses the ligament by distance z. Measurements find that the relationship between force and compression is well modeled by the equation $F_z = (2.0 \times 10^{10}\ \text{N/m}^2)z^2$.
a. What is the ligament's compression in μm for a bite that exerts a 200 N force on a tooth?
b. How much work is done to compress the ligament this distance?
c. The resilience of the ligament tissue is 85%. How much thermal energy is created in the tissue during one such bite?

71. || The basal metabolism of a resting adult is approximately
BIO 100 W. Most of this is provided by the hydrolysis of ATP, which in resting muscle has an estimated energy release of 60 kJ/mol.
    a. On average, how many moles of ATP are hydrolyzed per hour in the body? This is an estimate, so an answer with one significant figure is appropriate.
    b. It's estimated that the human body contains 30 trillion (3×10^{13}) cells. On average, how many ATP molecules are hydrolyzed in each cell every hour?
72. || Biology textbooks often state that "the mitochondrion is the
BIO powerhouse of the cell." Let's check this assertion. In a study of rat livers, one mitochondrion hydrolyzed 4.2×10^5 molecules of ATP per second, with an energy release of 52 kJ/mol of ATP. A mitochondrion has a volume of 2.7×10^{-19} m^3 and a density of 1140 kg/m^3.
    a. What is the specific power—the power per kilogram—of a mitochondrion?
    b. An elite athlete has a sustained specific power output of about 6 W/kg. How many times larger than this is the specific power of a mitochondrion?
73. || A liter of gasoline has a chemical energy content of 34 MJ.
    a. If the fuel economy of a car is 25 mpg (miles per gallon), what is its fuel economy in Cal/mi?
    b. How many times larger is this than the fuel economy of a walking person?
74. ||| For how long would a 68 kg athlete have to swim at 3 km/h
BIO to use all the energy available in a typical fast-food meal?
75. ||| The label on a candy bar says 400 Calories. Assuming a typi-
BIO cal efficiency for energy use by the body, if a 60 kg person were to use the energy in this candy bar to climb stairs, how high could she go?
76. ||| In an extreme marathon, participants run a total of 100 km;
BIO world-class athletes maintain a pace of 15 km/h. How many 230 Calorie energy bars would be required to fuel such a run for a 68 kg athlete?
77. ||| Suppose your body was able to use the chemical energy in
BIO gasoline. How far could you pedal a bicycle at 15 km/h on the energy in 1 gal of gas? (1 gal of gas has a mass of 3.2 kg.)
78. ||| A 68 kg hiker walks at 5.0 km/h up a 7% slope. What is the
BIO necessary metabolic power? Hint: You can model her power needs as the sum of the power to walk on level ground plus the power needed to raise her body by the appropriate amount.

MCAT-Style Passage Problems

Tennis Ball Testing

A tennis ball bouncing on a hard surface compresses and then rebounds. The details of the rebound are specified in tennis regulations. Tennis balls, to be acceptable for tournament play, must have a mass of 57.5 g. When dropped from a height of 2.5 m onto a concrete surface, a ball must rebound to a height of 1.4 m. During impact, the ball compresses by 6.0 mm.

79. How fast is the ball moving when it hits the concrete surface? (Ignore air resistance.)
    A. 5 m/s B. 7 m/s C. 25 m/s D. 50 m/s
80. Model the ball as a spring. Estimate the spring constant by temporarily assuming that mechanical energy is conserved.
    A. 230 N/m B. 280 N/m
    C. 7.8×10^4 N/m D. 7.8×10^7 N/m
81. The ball's kinetic energy just after the bounce is less than just before the bounce. In what form does this lost energy end up?
    A. Elastic potential energy
    B. Gravitational potential energy
    C. Thermal energy
    D. Rotational kinetic energy
82. By approximately what percent does the kinetic energy decrease?
    A. 35% B. 45% C. 55% D. 65%

Kangaroo Locomotion

Kangaroos have very stout tendons in their legs that can be used to store energy. When a kangaroo lands on its feet, the tendons stretch, transforming kinetic energy of motion to elastic potential energy. Much of this energy can be transformed back into kinetic energy as the kangaroo takes another hop. The kangaroo's peculiar hopping gait is not very efficient at low speeds but is quite efficient at high speeds.

Figure P11.83 shows the energy cost of human and kangaroo locomotion as measured by oxygen uptake (in mL/s) per kg of body mass, allowing a direct comparison between the two species.

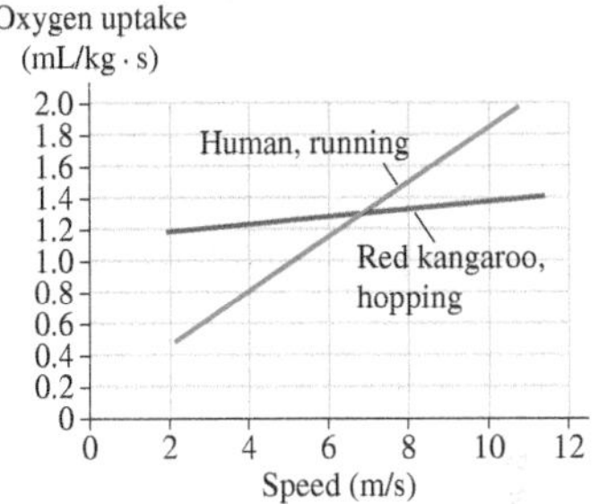

FIGURE P11.83

For humans, the energy used per second (i.e., power) is proportional to the speed. That is, the human curve nearly passes through the origin, so running twice as fast takes approximately twice as much power. For a hopping kangaroo, the graph of energy use has only a very small slope. In other words, the energy used per second changes very little with speed. Going faster requires very little additional power.

83. A person runs 1 km. How does the runner's speed affect the total energy needed to cover this distance?
    A. A faster speed requires less total energy.
    B. A faster speed requires more total energy.
    C. The total energy is about the same for a fast speed and a slow speed.
    D. The graph does not allow us to say how the runner's speed affects the energy needed.
84. A kangaroo hops 1 km. How does its speed affect the total energy needed to cover this distance?
    A. A faster speed requires less total energy.
    B. A faster speed requires more total energy.
    C. The total energy is about the same for a fast speed and a slow speed.
    D. The graph does not allow us to say how the kangaroo's speed affects the energy needed.
85. At approximately what speed would a human use half the power of an equal-mass kangaroo moving at the same speed?
    A. 3 m/s B. 4 m/s C. 5 m/s D. 6 m/s

12 Thermodynamics

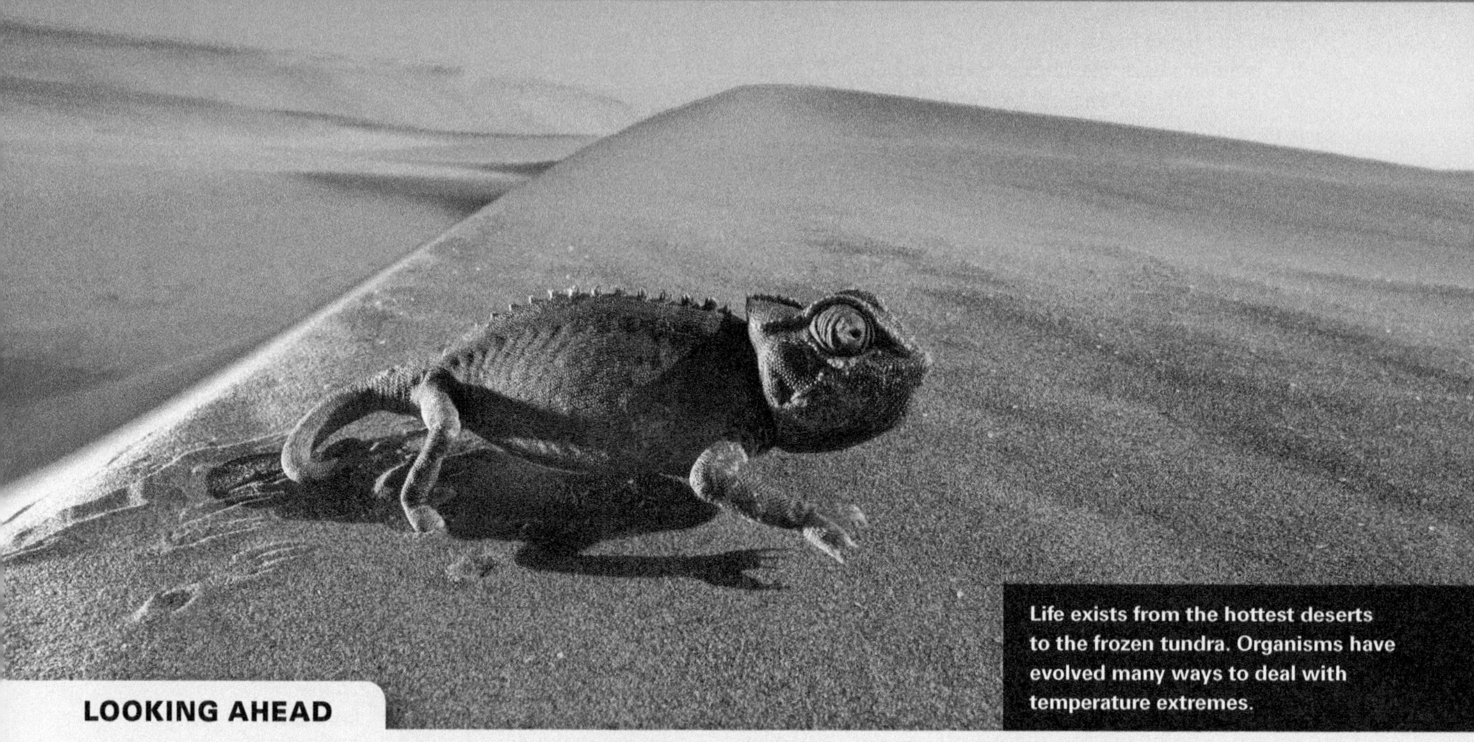

Life exists from the hottest deserts to the frozen tundra. Organisms have evolved many ways to deal with temperature extremes.

LOOKING AHEAD

Heat and Temperature

Adding ice cools your drink as heat is transferred from the warm liquid to the cold ice. Even more heat is used to melt the ice.

You'll learn how to calculate the heat flow that occurs when the temperature changes and during a *phase change* such as melting.

Heat Transfer

An elephant's large ears keep the animal cool as the many blood vessels radiate heat to the environment. In addition, flapping creates cooling air currents.

You'll learn about four different ways that heat is transferred: conduction, convection, radiation, and evaporation.

The Ideal Gas

Gases are all around us. We breathe the air, we generate carbon dioxide and water vapor, and we have many uses for compressed gases.

The *ideal gas* is an extremely important model system because we can study its properties in great detail.

GOAL To learn about heat, temperature, and the first law of thermodynamics.

PHYSICS AND LIFE

Organisms Have Many Ways to Deal with Temperature Extremes

Living organisms have adapted to environments ranging from very cold, where water is frozen, to very hot, where water is a liquid that evaporates rapidly. Liquid water is essential to all life as we know it. Organisms that live in freezing temperatures either have evolved physiological mechanisms to keep their bodies warmer than their environment or have developed ways to survive freezing with minimal damage. Desert plants and animals have adaptations to stay cool and minimize evaporative losses. This chapter and the next look at the physics that underlies these survival mechanisms. We'll examine the connection between heat and temperature, and we'll see how energy is involved in phase changes such as freezing and evaporation.

A penguin's short, dense feathers trap air that provides the excellent thermal insulation needed to survive in Antarctica.

12.1 Heat and the First Law of Thermodynamics

Thermodynamics arose hand in hand with the industrial revolution as the systematic study of converting heat into mechanical motion and work. Two centuries later, thermodynamics encompasses all the aspects of physics, chemistry, and biology that involve energy: its transfer, its transformation from one form into another, and its end use. You can think of **thermodynamics** as being the science of energy in its broadest context.

Thermodynamics is fundamental to the life sciences. It has a great deal to say about the energetics of biochemical reactions, the synthesis of macromolecules, the processes by which organisms exchange energy and matter with their environment, and the ability of organisms to develop and grow in the face of ever-increasing entropy. Our goal in this chapter, and in the related Chapters 13 and 14, is to help you understand the physical mechanisms that underlie many of the processes you will study in other courses.

Macroscopic and Microscopic

Thermodynamics began with the study of macroscopic systems: solid, liquids, and gases. A macroscopic system can be characterized by quantities such as volume, pressure, density, temperature, and thermal energy. The parameters used to describe a macroscopic system are known as **state variables** because, taken all together, they describe the **state** of the system—its condition at one instant of time. Macroscopic thermodynamics is the focus of this chapter.

Later, with the discovery of atoms and molecules, it was found that the observable characteristics of macroscopic systems have *microscopic* underpinnings. That is, state variables such as temperature and pressure can be understood in terms of the motions and collisions of molecules. You've already seen a hint of this in our discussion of thermal energy. The microscopic view of thermodynamics, which will deepen your understanding, will be fully developed in Chapter 13. Even so, we will refer to atoms and molecules in this chapter when doing so provides insight into the behavior of macroscopic systems.

Most macroscopic systems can exist as a solid, liquid, or gas—the three most common **phases** of matter. FIGURE 12.1 reminds you of some of the macroscopic and microscopic characteristics of solids, liquids, and gases that will be important to us.

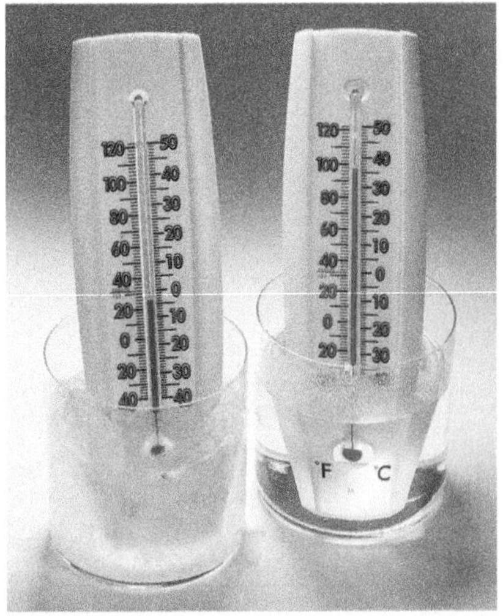

The temperature rises Thermal expansion of the liquid in the thermometer tube pushes it higher in the hot water than in the ice water. Measuring the height of the liquid is one way to measure temperature.

Temperature

Temperature, with the symbol T, is one of the most important state variables. We are all familiar with the idea of temperature, but what physical property of the system have you determined if you measure its temperature? Let's begin with the

FIGURE 12.1 The three phases of matter.

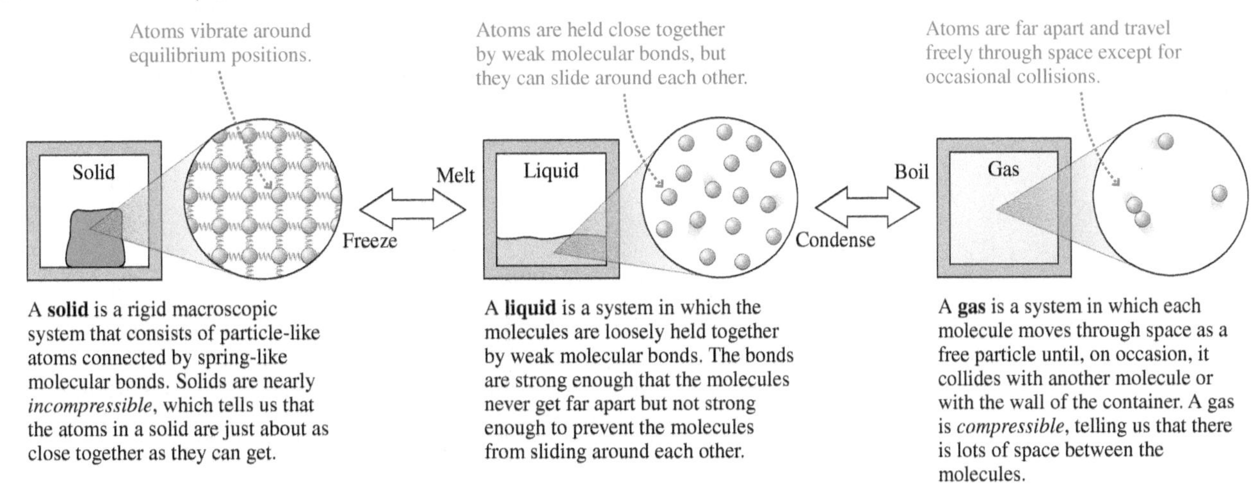

Atoms vibrate around equilibrium positions.
Atoms are held close together by weak molecular bonds, but they can slide around each other.
Atoms are far apart and travel freely through space except for occasional collisions.
Solid
Melt
Freeze
Liquid
Boil
Condense
Gas
A **solid** is a rigid macroscopic system that consists of particle-like atoms connected by spring-like molecular bonds. Solids are nearly *incompressible*, which tells us that the atoms in a solid are just about as close together as they can get.
A **liquid** is a system in which the molecules are loosely held together by weak molecular bonds. The bonds are strong enough that the molecules never get far apart but not strong enough to prevent the molecules from sliding around each other.
A **gas** is a system in which each molecule moves through space as a free particle until, on occasion, it collides with another molecule or with the wall of the container. A gas is *compressible*, telling us that there is lots of space between the molecules.

commonsense idea that temperature is a measure of how "hot" or "cold" a system is. As we develop these ideas, we'll find that temperature is related to a system's *thermal energy.*

In 1742, the Swedish scientist Anders Celsius sealed mercury into a small capillary tube and observed that it moved up and down the tube as the temperature changed. He selected two temperatures that anyone could reproduce, the freezing and boiling points of pure water, and labeled them 0 and 100. He then marked off the glass tube into 100 equal intervals between these two reference points. By doing so, he invented the temperature scale that we today call the *Celsius scale.* The units of the Celsius temperature scale are "degrees Celsius," which we abbreviate as °C. Note that the degree symbol ° is part of the unit, not part of the number.

FIGURE 12.2 The pressure in a constant-volume gas thermometer extrapolates to zero at $T_0 = -273°C$.

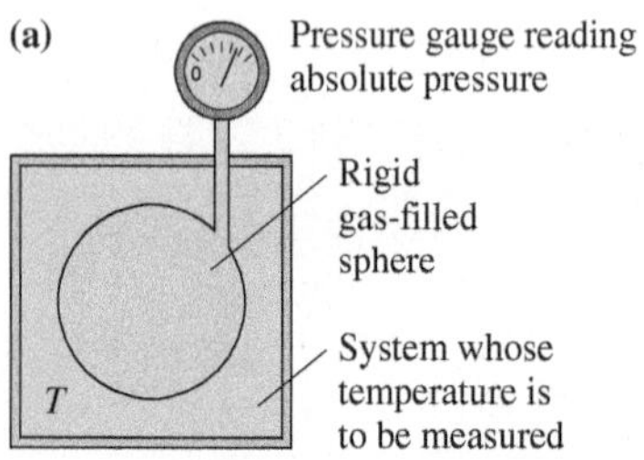

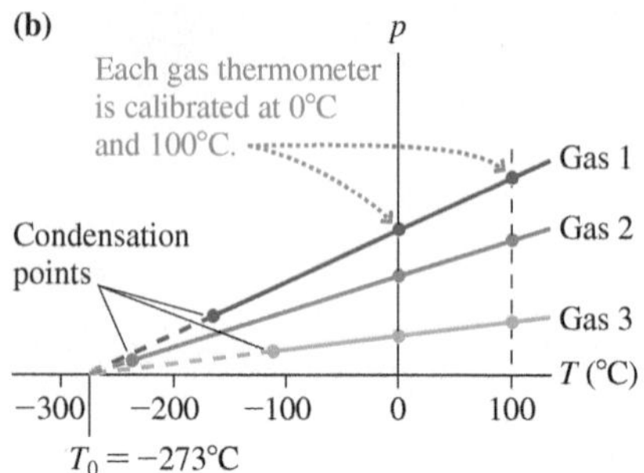

Any physical property that changes with temperature can be used as a thermometer. In practice, the most useful thermometers have a physical property that changes *linearly* with temperature. An important scientific thermometer is the constant-volume gas thermometer shown in FIGURE 12.2a. This thermometer depends on the fact that the *absolute* pressure (not the gauge pressure) of a gas in a sealed container increases linearly as the temperature increases. (You may know this idea from chemistry as *Gay-Lussac's law.*)

A gas thermometer is calibrated by recording the pressure at two reference temperatures, such as the boiling and freezing points of water. These two points are plotted on a pressure-versus-temperature graph and a straight line is drawn through them. The gas bulb is then brought into contact with the system whose temperature is to be measured. The pressure is measured, then the corresponding temperature is read off the graph.

FIGURE 12.2b shows the pressure-temperature relationship for three different gases. Notice two important things about this graph:

1. There is a *linear* relationship between temperature and pressure.
2. For all gases, the relationship extrapolates to *zero pressure* at the same temperature: $T_0 = -273°C$. No gas actually gets that cold without condensing into a liquid, although helium comes very close, but it is surprising that we get the same zero-pressure temperature for any gas and any starting pressure.

As we'll explore further in Chapter 13, the pressure in a gas is caused by collisions of the molecules with one another and with the walls of the container. A pressure of zero would mean that all motion, and thus all collisions, had ceased. If there were no atomic motion, the system's thermal energy would be zero. The temperature at which all motion would cease, and at which $E_{th} = 0$, is called **absolute zero.** Because temperature is related to thermal energy, absolute zero is the lowest temperature that has physical meaning. We see from the gas-thermometer data that $T_0 = -273°C$.

It is useful to have a temperature scale that has its zero point at absolute zero. Such a temperature scale is called an **absolute temperature scale.** The absolute temperature scale that has the same unit size as the Celsius scale is called the *Kelvin scale.* It is the SI scale of temperature. The units of the Kelvin scale are *kelvins,* abbreviated as K. The conversion between the Celsius scale and the Kelvin scale is

$$T_K = T_C + 273 \tag{12.1}$$

NOTE ▶ The units are simply "kelvins," *not* "degrees Kelvin." ◀

FIGURE 12.3 Temperatures measured with different scales.

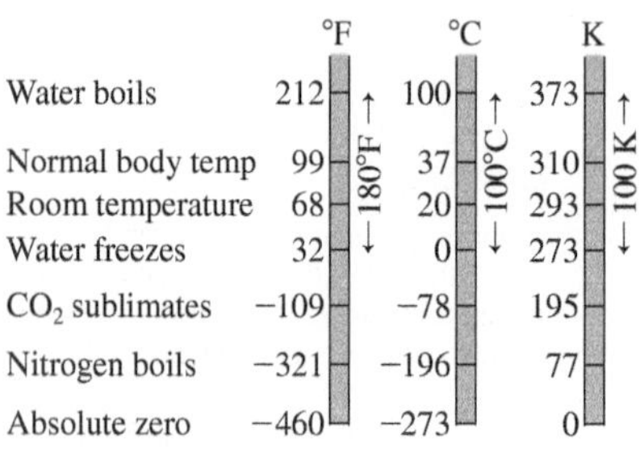

FIGURE 12.3 shows several temperatures measured on the Celsius and Kelvin scales and also on the Fahrenheit scale, which is more familiar to most people in the United States. On the Kelvin scale, absolute zero is 0 K, the freezing point of water is 273 K, and the boiling point of water is 373 K.

STOP TO THINK 12.1 The temperature of a glass of water increases from 20°C to 30°C. What is ΔT?

A. 10 K B. 283 K C. 293 K D. 303 K

Heat

Heat is a more elusive concept than work. We use the word "heat" very loosely in the English language, often to mean hot. We might say on a very hot day, "This heat is oppressive." Expressions like this date to a time when it was thought that heat was a *substance* with fluid-like properties.

Our concept of heat changed with the work of British physicist James Joule in the 1840s. Joule was the first to carry out careful experiments to learn what causes an object's temperature to change. Using experiments like those shown in FIGURE 12.4, Joule found that the temperature of a beaker of water can be raised by two entirely different means:

1. Heating it with a flame, or
2. Doing work on it (stirring it) with a rapidly spinning paddle wheel.

FIGURE 12.4 Joule's experiments to show the equivalence of heat and work.

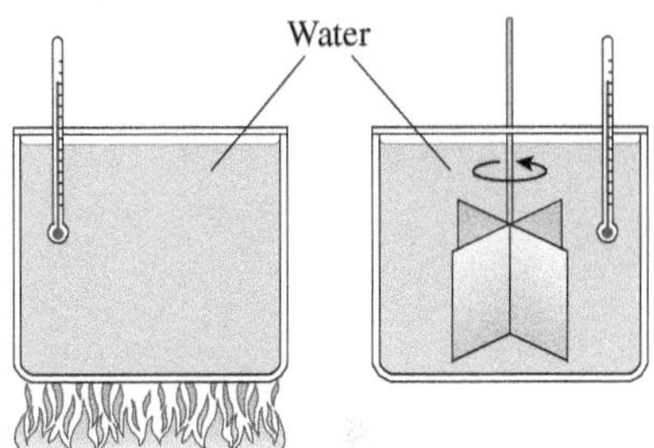

Water
The flame heats the water. The temperature increases.
The paddle does work on the water. The temperature increases.

The final state of the water is *exactly* the same in both cases. This implies that heat and work are essentially equivalent. In other words, heat is not a substance. Instead, like work, **heat is a way of transferring energy** to or from a system. Thus the SI unit of heat is the joule, the unit of energy, although cal, kcal, and Cal are often used in chemistry and when discussing energy use in the body.

To be specific, **heat** is energy transferred between a system and the environment as a consequence of a *temperature difference* between them. This happens when *faster* molecules in the hotter object collide with *slower* molecules in the cooler object. On average, these collisions cause the faster molecules to lose energy and the slower molecules to gain energy. The net result—we'll look at the details in Chapter 13—is that energy is transferred from the hotter object to the colder object.

When you place a pan of water on the stove, energy is transferred *from* the hotter flame *to* the cooler water. This transfer of energy is heat. If you place the water in a freezer, heat is the energy transferred from the warmer water to the colder air in the freezer. A system is in **thermal equilibrium** with its environment, or two systems are in thermal equilibrium with each other, if there is no temperature difference. No net energy transfer takes place between two systems that are in thermal equilibrium.

Heat, like work, is positive when energy is transferred *from* the environment *to* the system, increasing the system's energy. The symbol for heat we will use in this text is Q. (You may see a lowercase q in chemistry texts as well as a lowercase w for work.) FIGURE 12.5 shows how to interpret the sign of Q.

FIGURE 12.5 Interpreting the sign of heat.

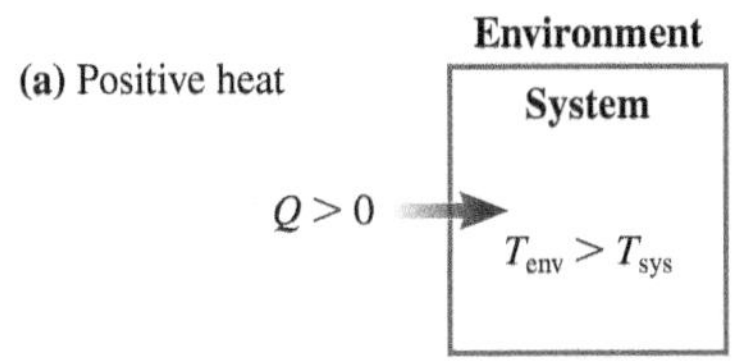

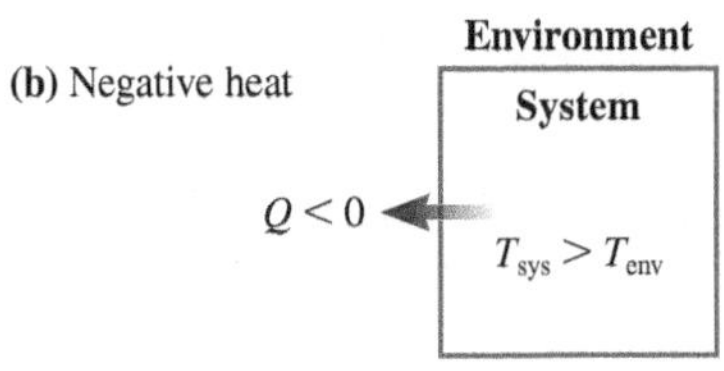

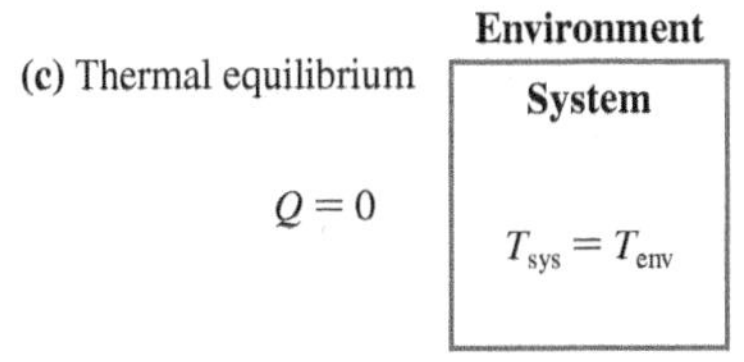

It's important to note that neither work nor heat is a state variable. State variables, such as pressure and temperature, are properties *of* the system, quantities that describe the system. A state variable changes as the system changes, so we can meaningfully talk about "the change in kinetic energy." In contrast, work and heat are *interactions*—transfers of energy—between the system and its environment; that is, **heat is not a property of the system.** Instead, heat is energy that moves from the system to the environment (or vice versa) during a thermal interaction. It is not meaningful to talk about a "change of heat" or a "change of work." Thus heat and work are shown as Q and W, never as ΔQ or ΔW.

It is important to distinguish between *heat* and *thermal energy*. These ideas are related, but the differences are crucial. In brief,

- Thermal energy is the energy *of the system* due to the motion of its atoms and molecules and the stretching and compressing of spring-like molecular bonds. It is a *form* of energy. Thermal energy is a state variable, and we will regularly consider how E_{th} changes during a process. A system has thermal energy even if it is isolated and not interacting with its environment.

- Heat is energy that is transferred *between* the system and the environment as they interact. You can think of heat as energy in transit. Heat is *not* a particular form of energy, nor is it a state variable. $Q = 0$ if a system does not interact thermally with its environment because no energy is transferred. A heat transfer may cause the system's thermal energy to increase or decrease, but that doesn't make heat and thermal energy the same.

Talking About Heat

Many of the phrases and terms we use to talk about heat, such as heat flow, heat loss, and heat capacity, are, strictly speaking, incorrect because heat is not a substance or a "thing." There's nothing to flow or get lost. These are linguistic relics of an obsolete, 18th-century theory in which heat was thought to be a fluid, called *caloric,* that flowed from hot objects to cold objects.

Our scientific understanding has changed, but, for the most part, our language for describing heat has not. Practicing scientists continue to use these terms, but with a tacit understanding that they are simply metaphors for a transfer of energy due to a temperature difference. We also will use this terminology, but with the caveat that the phrases should not be interpreted literally. **Heat is not a substance that can be moved around.**

STOP TO THINK 12.2 Which one or more of the following processes involves heat?

A. The brakes in your car get hot when you stop.
B. A test tube is held over a flame.
C. You place a bowl of water in the freezer.
D. You quickly push a piston into a cylinder of gas, increasing the temperature of the gas.
E. You place a cylinder of gas in hot water. The gas expands, causing a piston to rise and lift a weight. The temperature of the gas does not change.

The First Law of Thermodynamics

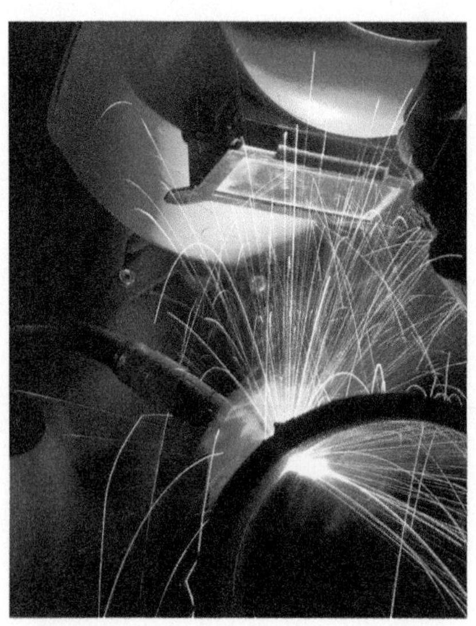

Heat is the energy transferred in a thermal interaction.

Heat was the missing piece that we needed to arrive at a completely general statement of the law of conservation of energy. If heat is included alongside work as a means of energy transfer, then the energy principle of Chapter 11 becomes

$$\Delta E_{\text{sys}} = \Delta K + \Delta U + \Delta E_{\text{th}} + \Delta E_{\text{chem}} = W + Q \tag{12.2}$$

Work and heat are simply two different ways of changing a system's energy.

In thermodynamics we are not interested in the macroscopic motion of the system as a whole. Moving macroscopic systems were important for many chapters, but now, as we investigate the thermal properties of a system, we would like the system as a whole to rest peacefully on the laboratory bench as we study it. Thus we will assume, throughout the remainder of Part II, that $\Delta K = \Delta U = 0$.

The two remaining types of energy, thermal energy and chemical energy, involve the microscopic motions and configurations of molecules. Together, they are what we call the **internal energy** of the system:

$$E_{\text{int}} = E_{\text{th}} + E_{\text{chem}} \tag{12.3}$$

With these assumptions and definitions, the energy principle becomes the **first law of thermodynamics:**

First law of thermodynamics For systems in which only the internal energy changes, the change in internal energy is equal to the energy transferred into or out of the system as work W, heat Q, or both:

$$\Delta E_{\text{int}} = W + Q \tag{12.4}$$

Chapter 10 introduced the basic energy model—*basic* because it included work but not heat and thermal energy but not chemical energy. The first law of thermodynamics is the basis for the more general energy model shown in FIGURE 12.6, a **thermodynamic energy model** in which work and heat are on an equal footing.

This more general energy model is useful across all the sciences, and you may have encountered some of these ideas in chemistry or biochemistry. Physicists, chemists, and biologists sometimes use different notation, and they focus on different examples and applications, but the fundamental ideas are the same. One of the goals of this text is to help you see how thermodynamics and kinetic theory, the topics of Chapters 12–14, give you a firmer grasp of what happens in chemical and biological systems. We'll begin by exploring the implications of the first law, especially the issue of what happens to an object or a system that is heated.

FIGURE 12.6 The thermodynamic energy model.

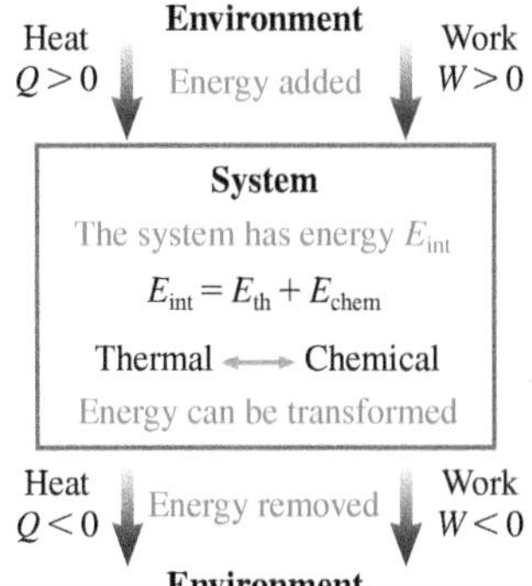

12.2 Thermal Expansion

Objects expand when heated. This **thermal expansion** is why the liquid rises in a thermometer and why pipes, highways, and bridges have expansion joints. FIGURE 12.7 shows an object of length L that changes by ΔL when the temperature is changed from T to $T + \Delta T$. The ratio $\Delta L/L$ is called the *fractional change* in length. For most solids, the fractional change in length is directly proportional to the temperature change ΔT as long as $\Delta L/L \ll 1$ (that is, if the length change ΔL is very small compared to the overall length L). We can write this proportionality as

$$\frac{\Delta L}{L} = \alpha \Delta T \qquad (12.5)$$

where α (Greek alpha) is the material's **coefficient of linear expansion.** Equation 12.5 characterizes both expansion ($\Delta L > 0$ if the temperature increases) and contraction ($\Delta L < 0$ if the temperature falls).

TABLE 12.1 gives the coefficients of linear expansion for a few common materials. Because the fractional change in length is dimensionless, α has units of °C^{-1}, read as "per degree Celsius." The units may also be written K^{-1}, because a temperature *change* ΔT is the same in °C and K, but practical measurements are made in °C, not K.

NOTE ▶ This chapter has several tables of data. Many end-of-chapter problems require you to find and use data from these tables. *Please use these values for end-of-chapter problems,* not values from an online search. There can be small differences in the values, depending on the source of the data, and using a slightly different value that you find online could result in your answer being marked incorrect. ◀

FIGURE 12.7 An object's length changes when the temperature changes.

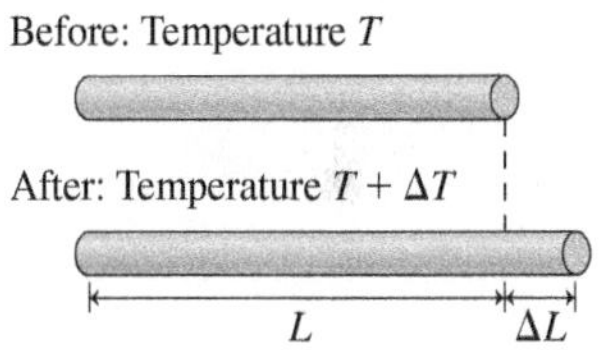

TABLE 12.1 **Coefficients of linear and volume expansion**

Material	α (°C^{-1})
Aluminum	2.3×10^{-5}
Copper	1.7×10^{-5}
Glass	2.7×10^{-5}
Ice	5.5×10^{-5}
Stainless steel	1.7×10^{-5}
Material	**β (°C^{-1})**
Acetic acid	1.1×10^{-3}
Ethyl alcohol	1.1×10^{-3}
Glycerin	5.0×10^{-4}
Mercury	1.8×10^{-4}

EXAMPLE 12.1 A thermal contraction

A cryogenic storage unit keeps tissue samples frozen at the liquid nitrogen temperature of −196°C. Suppose a 55-cm-tall stainless steel storage rack is lowered into the liquid. By how much does it contract from its room-temperature (20°C) length?

PREPARE The temperature change is large, but we'll assume $\Delta L/L \ll 1$ so that we can use Equation 12.5. From Table 12.1, the coefficient of thermal expansion of stainless steel is 1.7×10^{-5} °C^{-1}.

SOLVE The storage rack's temperature change is $\Delta T = T_f - T_i = -196°C - 20°C = -216°C$. The contraction is given by Equation 12.5:

$$\Delta L = \alpha L \Delta T = (1.7 \times 10^{-5}\,°C^{-1})(55\text{ cm})(-216°C)$$
$$= -0.20\text{ cm} = -2.0\text{ mm}$$

ASSESS 2 mm is only a small fraction of the storage rack's length, so our assumption that $\Delta L/L \ll 1$ is valid, but 2 mm is not a negligible distance. The designer of the storage unit has to take thermal contraction and expansion into consideration. The tissue samples themselves also change size with temperature, but tissue is a complex material that does not show this simple linear dependence on ΔT.

Volume expansion is treated the same way. If an object's volume changes by ΔV during a temperature change ΔT, the *fractional* change in volume is

$$\frac{\Delta V}{V} = \beta \Delta T \tag{12.6}$$

where β (Greek beta) is the material's **coefficient of volume expansion.**

Liquids are constrained by the shape of their container, so liquids are characterized by a volume-expansion coefficient but not by a linear-expansion coefficient. A few values are given in Table 12.1, where you can see that β, like α, has units of $°C^{-1}$. Thus, for example, 1000 mL of glycerin will expand by 10 mL if its temperature is increased by 20°C, a not insignificant change.

Special Properties of Water

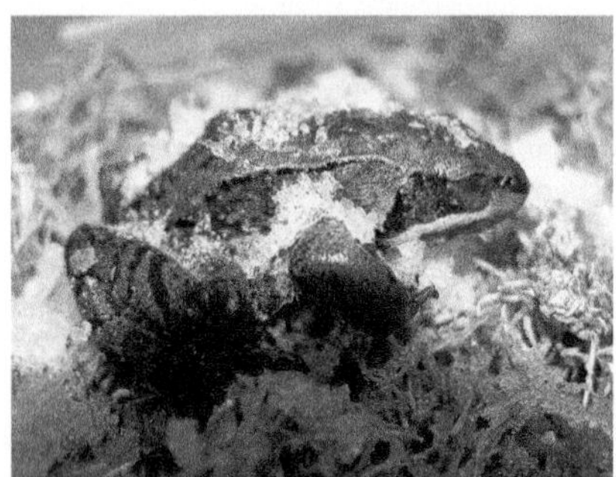

Frozen frogs It seems impossible, but common wood frogs survive the winter with much of their bodies frozen. When you dissolve substances in water, the freezing point lowers. Although the liquid water *between* cells in the frogs' bodies freezes, the water *inside* their cells remains liquid because of high concentrations of dissolved glucose. This prevents the cell damage that accompanies the freezing and subsequent thawing of tissues. When spring arrives, the frogs thaw and appear no worse for their winter freeze.

You may have noticed that Table 12.1 does not include water. Water, the most important molecule for life, differs from other liquids in many important ways. You expect a liquid's volume to decrease as you cool it toward the freezing point. Water does contract as it's cooled, but only until the temperature reaches 4°C. With further cooling, water actually *expands* and increases in volume. FIGURE 12.8a shows the volume per mole of water over a temperature range from 0°C to 10°C; you can see that the volume is minimum—maximum contraction—at 4°C. You can look up water's coefficient of volume expansion at a specific temperature if you need to do a calculation, but there's no single value for a table such as Table 12.1.

Water has another surprise when it freezes. All other common liquids contract upon freezing, so that the solid phase is denser than the liquid phase. In a mixture of solid and liquid, the denser solid sinks to the bottom and is covered by the liquid. In contrast, water *expands* upon freezing; solid water (ice) is less dense than liquid water, so ice floats on water. This increase of volume upon freezing is illustrated in FIGURE 12.8b. Notice that the vertical scale differs from Figure 12.8a; the volume increase upon freezing is nearly 10%, so the much smaller variation of the liquid's volume with temperature is not visible in Figure 12.8b.

The unusual properties of water arise because water is a *polar molecule*—its positive and negative charges are somewhat separated—that links to other water molecules through hydrogen bonds. In liquid water, hydrogen bonds are constantly being broken and re-formed; this occurs easily at higher temperatures where the water has more thermal energy, less readily as water is cooled. It turns out that a hydrogen bond forces two molecules to stay farther away from each other than would be the case for randomly moving molecules, so a water crystal with a full set of fixed hydrogen bonds—in other words, ice—takes up more volume than the randomly moving molecules of the liquid. Consequently, water expands upon freezing. Even before freezing, the molecules of liquid water start to form extended

FIGURE 12.8 Volume per mole of water near the freezing point.

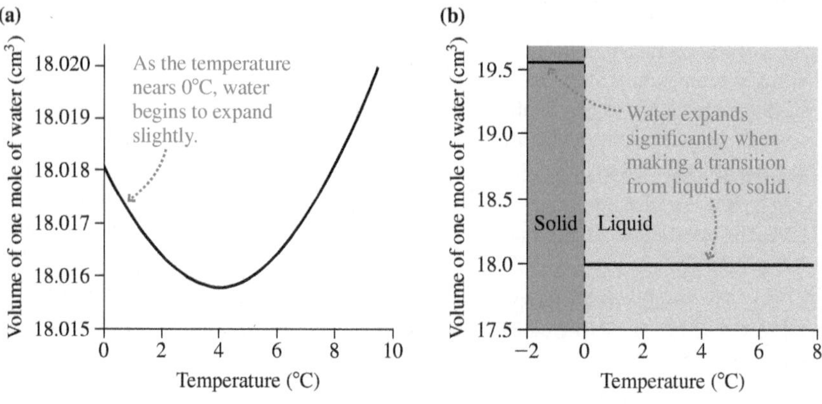

hydrogen-bonded clusters at temperatures below 4°C, gradually pushing the molecules apart and thus increasing the liquid's volume.

The expansion of water upon freezing has important consequences. The bursting of exposed water pipes during a freeze is well known to people who live in cold climates. Similarly, cells are susceptible to rupture if they are frozen, so special techniques are required to freeze samples without damage. (A few organisms have what is effectively antifreeze in their cells to prevent freezing.) When lakes begin to freeze, the ice floats on the surface rather than sinking to the bottom. The ice layer insulates the water below from the colder air above and prevents all except the shallowest lakes from freezing solid; this allows aquatic life to survive even the harshest winters.

STOP TO THINK 12.3 A steel plate has a 2.000-cm-diameter hole through it. If the plate is heated, does the diameter of the hole increase or decrease?

12.3 Specific Heat and Heat of Transformation

Suppose you start with a system in its solid phase and heat it at a slow but steady rate. **FIGURE 12.9** shows the cumulative energy Q that must be added as heat to increase the temperature to the value T. Initially the temperature increases as energy is added, but then the temperature remains at T_m for an extended period as energy continues to be added. Another warming is then followed by an extended period during which the system is heated but the temperature remains at T_b.

Two different things can happen as a system as heated. It is not surprising that heating one of the system's three phases—solid, liquid, or gas—increases its temperature. But we also have two temperatures at which heating causes a *phase change* rather than a temperature change. In this case, the added energy goes to breaking molecular bonds rather than increasing molecular speeds. The temperature remains constant as energy is added until all the bonds are broken and the system has completely transitioned to the next phase.

Melting and freezing are familiar changes between a substance's solid and liquid phases. The change occurs at a temperature T_m that is called either the **melting point** or the **freezing point,** depending on the direction of the change. A system at its melting point is in **phase equilibrium,** which means that any amount of solid can coexist with any amount of liquid. If we turn off the heating, the system will remain a mixture of solid and liquid at temperature T_m; there's no tendency of the system to move one way or the other. But the slightest shift in the temperature will move the system to being entirely solid or entirely liquid.

Similarly, there's a phase equilibrium between the liquid and gas phases at the **boiling point** T_b or, if the change is in the opposite direction, the **condensation point.**

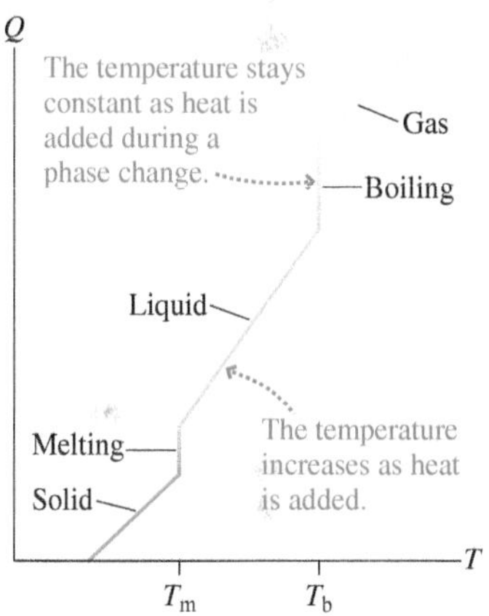

FIGURE 12.9 The energy needed to increase a system's temperature. The vertical segments correspond to phase changes.

Temperature Change

The most common result of heating an object is that it gets hotter—its temperature increases. If you do an experiment in which you transfer 4190 J of energy to 1 kg water, you will find that the temperature increases by 1 K. If you were fortunate enough to have 1 kg of gold, you would need to add only 129 J of energy to raise its temperature by 1 K.

The amount of energy that raises the temperature of 1 kg of a substance by 1 K is called the **specific heat** of that substance. The symbol for specific heat is c. Water has specific heat $c_{water} = 4190$ J/kg · K. The specific heat of gold is $c_{gold} = 129$ J/kg · K. Specific heat depends only on the material from which an object is made. **TABLE 12.2** provides some specific heats for common liquids and solids.

TABLE 12.2 Specific heats and (for pure substances) molar specific heats of solids and liquids

Substance	c (J/kg · K)	C (J/mol · K)
Solids		
Aluminum	900	24.3
Copper	385	24.4
Gold	129	25.4
Ice	2090	37.6
Mammalian body	3400	–
Stainless steel	502	–
Liquids		
Acetic acid	2040	123
Ethyl alcohol	2400	110
Mercury	140	28
Water	4190	75

We'll also have occasion to use the **molar specific heat** of a substance, denoted by an uppercase C, which is the energy that raises the temperature of 1 mol of the substance by 1 K. Table 12.2 includes molar specific heats.

If energy c is required to raise the temperature of 1 kg of a substance by 1 K, then energy Mc is needed to raise the temperature of mass M by 1 K and $(Mc)\Delta T$ is needed to raise the temperature of mass M by ΔT. In other words, the heat required for a temperature change ΔT is

$$Q = Mc\,\Delta T \qquad (12.7)$$

Heat needed to change the temperature of an object with mass M and specific heat c

Referring back to the graph of Figure 12.9, the slope of the graph is the derivative dQ/dT, and the initial constant slope tells us that dQ/dT is constant. We can understand this by considering Equation 12.7: The heat required for an infinitesimal temperature change dT would be $dQ = Mc\,dT$, so $dQ/dT = Mc$. That is, the slope of this graph during any phase is Mc, so a constant specific heat implies a constant slope. In fact, you can measure c from such a graph.

NOTE ▸ The different slopes indicate that the solid, liquid, and gas phases of a substance have different specific heats. ◂

Because $\Delta T = Q/Mc$, it takes more heat energy to change the temperature of a substance with a large specific heat than to change the temperature of a substance with a small specific heat. We can think of specific heat as measuring the *thermal inertia* of a substance. Metals, with small specific heats, warm up and cool down quickly. A piece of aluminum foil can be safely held within seconds of removing it from a hot oven. Water, with a very large specific heat, is slow to warm up and slow to cool down.

EXAMPLE 12.2 Running a fever

A 70 kg student catches the flu, and the body temperature increases from 37.0°C (98.6°F) to 39.0°C (102.2°F). How much energy is required to raise the body's temperature?

PREPARE Energy is supplied by the chemical reactions of the body's metabolism. These exothermic reactions transfer heat to the body. Normal metabolism provides enough heat to offset energy losses (radiation, evaporation, etc.) while maintaining a normal body temperature of 37°C. We need to calculate the additional energy needed to raise the body's temperature by 2.0°C, or 2.0 K. Table 12.2 gives the specific heat of a mammalian body as 3400 J/kg · K.

SOLVE The necessary heat is

$$Q = Mc\,\Delta T = (70\text{ kg})(3400\text{ J/kg}\cdot\text{K})(2.0\text{ K}) = 4.8 \times 10^5\text{ J}$$

ASSESS This appears to be a lot of energy, but a joule is actually a very small amount of energy. The required heat is only 110 Cal, approximately the energy gained by eating an apple.

The heat needed to bring about a temperature change ΔT of n moles of substance is calculated in the same way but uses the molar specific heat C instead of the specific heat c:

$$Q = nC\,\Delta T \qquad (12.8)$$

Molar specific heats of pure substances are listed in Table 12.2. Look at the three elemental solids. All have C very near 25 J/mol · K. If we were to expand the table, we would find that most elemental solids have $C \approx 25$ J/mol · K. This can't be a coincidence, but what is it telling us? This is a puzzle we will address in Chapter 13, where we will explore thermal energy at the atomic level.

Phase Changes

The vertical segments in the graph of Figure 12.9 show that energy is being transferred to the system but the temperature *isn't* changing. These are phase changes during which the thermal energy continues to increase but the temperature remains constant. **A phase change is characterized by an increase in thermal energy without a change of temperature.**

The amount of heat that causes 1 kg of a substance to undergo a phase change is called the **heat of transformation** of that substance. For example, laboratory experiments show that 3.33×10^5 J of heat are needed to melt 1 kg of ice at 0°C. The symbol for heat of transformation is L. The heat required for the entire system of mass M to undergo a phase change is

$$Q = ML \qquad \text{(phase change)} \qquad (12.9)$$

This is the height of a vertical segment of the graph in Figure 12.9.

Heat of transformation is a generic term that refers to any phase change. Two specific heats of transformation are the **heat of fusion** L_f, the heat of transformation between a solid and a liquid, and the **heat of vaporization** L_v, the heat of transformation between a liquid and a gas. The heat needed for these phase changes is

$$Q = \begin{cases} \pm ML_f & \text{melt/freeze} \\ \pm ML_v & \text{boil/condense} \end{cases} \qquad (12.10)$$

Liquid rock Lava—molten rock—undergoes a phase change when it contacts the much colder water. This is one way in which new islands are formed.

where the $\pm$ indicates that heat must be *added* to the system during melting or boiling but *removed* from the system during freezing or condensing. **You must explicitly include the minus sign when it is needed.**

TABLE 12.3 gives the heats of transformation of a few substances. Notice that the heat of vaporization is always much larger than the heat of fusion. We can understand this. Melting breaks just enough molecular bonds to allow the system to lose rigidity and flow. Even so, the molecules in a liquid remain close together and loosely bonded. Vaporization breaks all bonds completely and sends the molecules flying apart. This process requires a larger increase in the thermal energy and thus a larger quantity of heat.

TABLE 12.3 **Melting/boiling temperatures and heats of transformation**

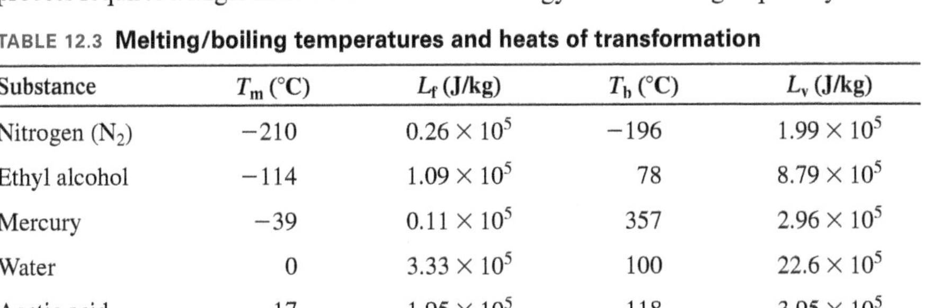

Substance	T_m (°C)	L_f (J/kg)	T_b (°C)	L_v (J/kg)
Nitrogen (N_2)	−210	0.26×10^5	−196	1.99×10^5
Ethyl alcohol	−114	1.09×10^5	78	8.79×10^5
Mercury	−39	0.11×10^5	357	2.96×10^5
Water	0	3.33×10^5	100	22.6×10^5
Acetic acid	17	1.95×10^5	118	3.95×10^5

NOTE ▶ An older term for heat of transformation, but one still in use, is *latent heat*. ◀

EXAMPLE 12.3 How much energy is needed to melt a popsicle?

A girl eats a 45 g frozen popsicle that was taken out of a −10°C freezer. How much energy does her body use to bring the popsicle up to body temperature? How does this compare to the amount of chemical energy she will gain by eating the popsicle?

PREPARE We will assume that the popsicle is pure water. There are three parts to the problem: The popsicle must be warmed to 0°C, the popsicle must melt, and then the resulting water must be warmed to body temperature. Normal body temperature is 37°C. The specific heats of ice and liquid water are given in Table 12.2; the heat of fusion of water is given in Table 12.3.

SOLVE The heat needed to warm the frozen water by $\Delta T = 10°\text{C} = 10$ K to the melting point is

$$Q_1 = Mc_{\text{ice}}\Delta T = (0.045\text{ kg})(2090\text{ J/kg}\cdot\text{K})(10\text{ K}) = 940\text{ J}$$

Note that we use the specific heat of water ice, not liquid water. Melting 45 g of ice requires heat

$$Q_2 = ML_f = (0.045\text{ kg})(3.33 \times 10^5\text{ J/kg}) = 15{,}000\text{ J}$$

The liquid water must now be warmed to body temperature; this requires heat

$$Q_3 = Mc_{\text{water}}\Delta T = (0.045\text{ kg})(4190\text{ J/kg}\cdot\text{K})(37\text{ K}) = 7000\text{ J}$$

The total energy is the sum of these three values: $Q_{\text{total}} = 23{,}000$ J.

A commercial popsicle has 40 Calories, which is about 170 kJ. Roughly 15% of the chemical energy in this frozen treat is used to bring it up to body temperature!

ASSESS More energy is needed to melt the ice than to warm the water, as we would expect.

FIGURE 12.10 Phase diagrams (not to scale) for water and carbon dioxide.

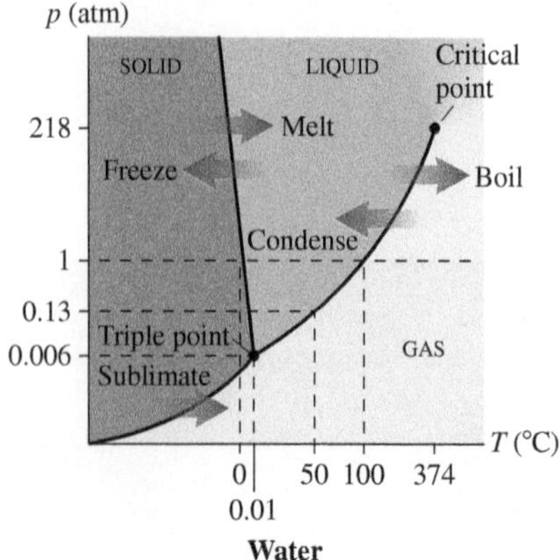

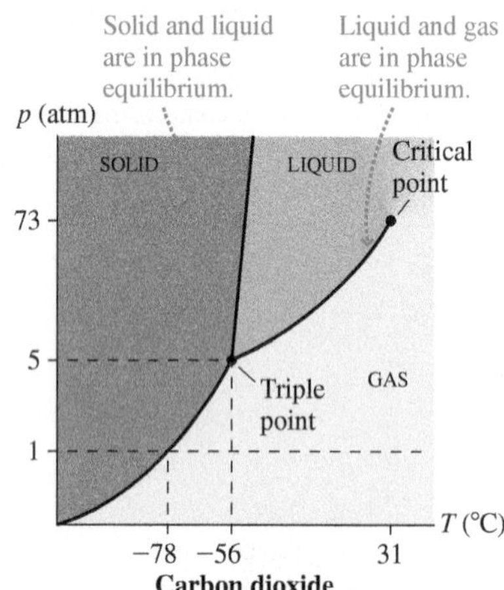

Phase Diagrams

A **phase diagram** is used to show how the phases and phase changes of a substance vary with both temperature and pressure. FIGURE 12.10 shows the phase diagrams for water and carbon dioxide. You can see that each diagram is divided into three regions corresponding to the solid, liquid, and gas phases. The boundary lines separating the regions indicate the phase transitions. The system is in phase equilibrium at a pressure-temperature point that falls on one of these lines.

Phase diagrams contain a great deal of information. Notice on the water phase diagram that the dashed line at $p = 1$ atm crosses the solid-liquid boundary at 0°C and the liquid-gas boundary at 100°C. These well-known melting and boiling point temperatures of water apply only at standard atmospheric pressure. In high-altitude Denver, where $p_{atmos} < 1$ atm, water melts at slightly above 0°C and boils at a temperature below 100°C. A *pressure cooker* works by allowing the pressure inside to exceed 1 atm. This raises the boiling point, so foods that are in boiling water are at a temperature above 100°C and cook faster.

Crossing the solid-liquid boundary corresponds to melting or freezing while crossing the liquid-gas boundary corresponds to boiling or condensing. But there's another possibility—crossing the solid-gas boundary. The phase change in which a solid becomes a gas is called **sublimation.** It is not an everyday experience with water, but you probably are familiar with the sublimation of dry ice. Dry ice is solid carbon dioxide. You can see on the carbon dioxide phase diagram that the dashed line at $p = 1$ atm crosses the solid-*gas* boundary, rather than the solid-liquid boundary, at $T = -78$°C. This is the *sublimation temperature* of dry ice.

Liquid carbon dioxide does exist, but only at pressures greater than 5 atm and temperatures greater than −56°C. A CO_2 fire extinguisher contains *liquid* carbon dioxide under high pressure. (You can hear the liquid slosh if you shake a CO_2 fire extinguisher.)

One important difference between the water and carbon dioxide phase diagrams is the slope of the solid-liquid boundary. For most substances, the solid phase is denser than the liquid phase and the liquid is denser than the gas. Pressurizing the substance compresses it and increases the density. If you start compressing CO_2 gas at room temperature, thus moving upward through the phase diagram along a vertical line, you'll first condense it to a liquid and eventually, if you keep compressing, change it into a solid.

As we've seen, water is a very unusual substance in that the density of ice is *less* than the density of liquid water. That is why ice floats. If you compress ice, making it denser, you eventually cause a phase transition in which the ice turns to liquid water! Consequently, the solid-liquid boundary for water slopes to the left.

Lowering the boiling point Food takes longer to cook at high altitudes because the boiling point of water is less than 100°C.

The liquid-gas boundary ends at a point called the **critical point.** Below the critical point, liquid and gas are clearly distinct and there is a phase change if you go from one to the other. But there is no clear distinction between liquid and gas at pressures or temperatures above the critical point. The system is a *fluid,* but it can be varied continuously between high density and low density without a phase change.

The final point of interest on the phase diagram is the **triple point** where the phase boundaries meet. Two phases are in phase equilibrium along the boundaries. The triple point is the *one* value of temperature and pressure for which all three phases can coexist in phase equilibrium. That is, any amounts of solid, liquid, and gas can happily coexist at the triple point. For water, the triple point occurs at $T_3 = 0.01$°C and $p_3 = 0.006$ atm.

STOP TO THINK 12.4 For which is there a sublimation temperature that is higher than a melting temperature?

A. Water B. Carbon dioxide C. Both D. Neither

Vapor Pressure and Evaporation

A sealed container of liquid is more than just the liquid; there are also molecules in the gas phase above the liquid. Molecules in the liquid have thermal energy, and some molecules at the surface are moving fast enough to break the weak intermolecular bonds and enter the gas phase. At the same time, a molecule in the gas phase may re-enter the liquid if it collides with the surface. At any temperature, not just the boiling point, there's a *phase equilibrium* of liquid and gas when the number of molecules that are leaving the liquid is balanced by the number that are returning to the liquid. At equilibrium, the pressure of the gas above the liquid is called the **vapor pressure.**

The vapor pressure is simply the pressure of the point on the liquid-gas boundary in the phase diagram that corresponds to the system's temperature. Referring back to Figure 12.10, you can see that the vapor pressure of 50°C water is 0.13 atm. That would be the pressure of the water vapor—its *partial pressure*—inside a 50°C sealed container that holds liquid water.

Vapor pressure increases with temperature until, when it equals the external pressure, the liquid boils. At sea level, the vapor pressure of water is 1 atm at 100°C. The vapor above a liquid can be a significant component of the gas in a sealed system even at temperatures far below the boiling temperature. For example, the vapor pressure of water at body temperature, 37°C, is 0.062 atm or 49 mm Hg—comparable to blood pressure.

A liquid *evaporates* when the gas pressure above the liquid is less than the vapor pressure. **Evaporation** is not an equilibrium situation because the number of molecules leaving the liquid is not balanced by an equal number returning; this creates a phase change of liquid to gas. Consequently, evaporation is a *cooling mechanism,* removing more thermal energy from the liquid than is returned to it.

The evaporation of water, both from sweat and in moisture you exhale, is one of the body's methods of exhausting the excess heat of metabolism to the environment, allowing you to maintain a steady body temperature. The heat needed to evaporate mass M of a liquid is $Q = ML_v$, but the heat of vaporization L_v at room or body temperature is a little larger than the boiling-point values given in Table 12.4 because the molecules have a little less thermal energy and thus need additional assistance to escape. For water,

$$L_v(15°\text{C}) = 24.7 \times 10^5 \text{ J/kg}$$

$$L_v(30°\text{C}) = 24.3 \times 10^5 \text{ J/kg}$$

Skin temperature is about 30°C, so you should use $L_V(30°\text{C})$ to calculate heat loss due to perspiration from the skin.

Freezing to stay warm Citrus crops cannot withstand freezing, and they are grown where freezing temperatures are rare. One way that farmers protect their crops from an occasional overnight freeze is counterintuitive—they spray the orchards all night with water that freezes on the fruit. Evaporation (and similarly melting) withdraws heat from an object to drive the phase change. Conversely, condensation and freezing *release* heat to an object as the molecules form tighter bonds. The released heat of fusion does not warm up the fruit, as you sometimes hear, but simply holds the temperature right at the freezing point. That is, the released heat keeps the fruit on the vertical freezing-melting segment of the graph in Figure 12.9 rather than dropping to the colder air temperature. In addition, any ice that does form provides a bit of insulation between the air and the fruit.

EXAMPLE 12.4 Computing heat loss by perspiration

The human body can produce 30 g of perspiration per minute (≈ 2 L per hour) during vigorous exercise. What is the rate of heat loss by evaporation for an athlete who is perspiring at this rate?

PREPARE Mathematically, a *rate* is a derivative with respect to time. For example, velocity, $v_x = dx/dt$, is the rate of change of position. Similarly, the rate of heat transfer is dQ/dt. A heat *loss* is a heat transfer with a negative value of dQ/dt, but a question that explicitly asks about "heat loss" rather than "heat transfer" can be answered with the absolute value of dQ/dt. The rate of energy transfer, joules per second, is a power, so the answer will be in watts. We'll assume that perspiration is pure water.

SOLVE The evaporation of 30 g of perspiration at a normal skin temperature of 30°C requires heat

$$Q = ML_v = (0.030 \text{ kg})(24.3 \times 10^5 \text{ J/kg}) = 7.3 \times 10^4 \text{ J}$$

Continued

This is the heat lost per minute; the rate of heat loss is

$$\left|\frac{dQ}{dt}\right| = \frac{7.3 \times 10^4 \text{ J}}{60 \text{ s}} = 1200 \text{ J/s} = 1200 \text{ W}$$

ASSESS Given the metabolic power required for different activities, as listed in Chapter 11, this is sufficient to keep the body cool even when exercising in hot weather—as long as the person drinks enough water to keep up this rate of perspiration.

Biological Phase Transitions

A protein molecule is a long chain of amino acids that, under physiological conditions, takes a complex three-dimensional shape that is essential to its function. This shape is called the *folded* configuration of the protein. If a protein is heated, it undergoes a transition from the folded configuration to an unfolded configuration. Because the folded configuration is considered the *native* state of a protein, the unfolding process is called *denaturation*. The denaturation of proteins is an important part of what happens when food is cooked.

FIGURE 12.11 shows the results of an experiment with a protein from the bacterium *Peptostreptococcus magnus,* called protein L, in which the temperature is increased from 40°C to 100°C. Initially the protein molecules are all in their native, folded configurations, so the unfolded fraction is essentially zero. The fraction of unfolded proteins increases rapidly over the temperature range 60–80°C, and by 90°C the proteins are completely denatured. If the process occurs slowly, and if the final temperature is not too high, the proteins will refold as the sample is cooled.

FIGURE 12.11 The *melting curve* of protein L. The data points are measured values of the fraction of the unfolded protein molecules; the solid line is a theoretical model of protein denaturation.

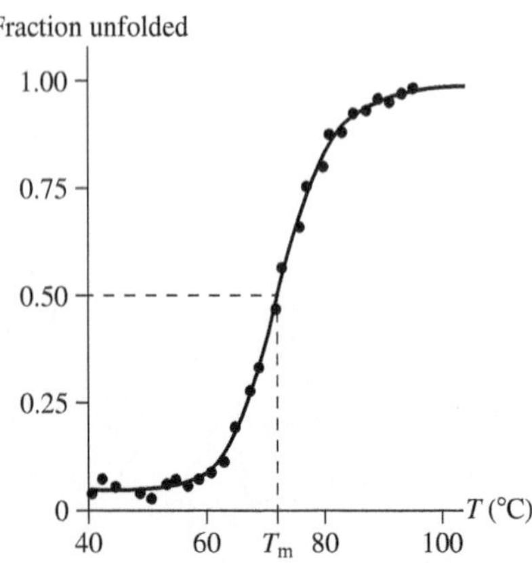

Because the process requires heating, is reversible, and happens at a specific temperature, this is just as much a *phase transition* as the melting of ice. Indeed, the process is often referred to as *protein melting,* although the term is used figuratively, not literally. One distinction between protein melting and water melting is that the phase transition occurs over a range of temperatures instead of at a sharp, well-defined temperature. The difference arises because water is a pure substance, meaning that all the bonds are identical, whereas a folded protein has many slightly different types of bonds that must break as it unfolds.

A macromolecule's melting temperature T_m is defined to be the temperature at which the unfolded fraction is 50%. From Figure 12.11, we see that the melting temperature of protein L under these experimental conditions is about 72°C.

Other biological macromolecules also undergo phase transitions. *DNA melting* occurs when heating causes the double-stranded structure to separate into single strands. The melting of DNA is an important step in the polymerase chain reaction (PCR) that is used to make copies of a DNA segment. The melting temperature depends on the ratio of adenine-thymine base pairs (A-T pairs) to guanine-cytosine base pairs (G-C pairs). A-T pairs have somewhat weaker hydrogen bonds, and a DNA strand that consists of only A-T pairs melts at $\approx$70°C. A strand consisting of only G-C pairs, which are somewhat more strongly bonded, melts at $\approx$110°C. The melting temperature of naturally occurring DNA is typically around 90°C.

12.4 Calorimetry

The laboratory practice of **calorimetry** determines heat transfers by measuring changes in temperature. In its simplest form, two systems at different initial temperatures are combined in an insulated container so that all heat flows between the two systems with none going to the environment. Calorimetry is used in chemistry and biochemistry to measure heats of reaction.

FIGURE 12.12 shows two systems thermally interacting with each other but isolated from everything else. Suppose they start at different temperatures T_1 and T_2. As you know from experience, heat will be transferred from the hotter to the colder system until they reach a common final temperature T_f. The systems will then be in thermal equilibrium and the temperature will not change further.

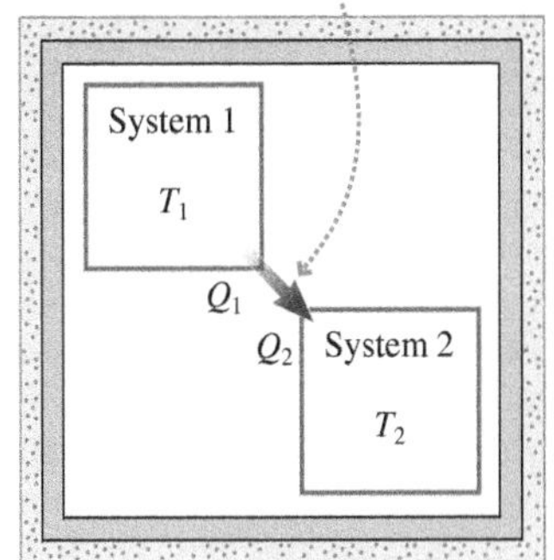
Heat is transferred from system 1 to system 2.
System 1
T_1
Q_1
Q_2
System 2
T_2
Energy conservation requires $Q_1 + Q_2 = 0$

FIGURE 12.12 Two systems interact thermally.

The insulation prevents any heat from being transferred to or from the environment, so energy conservation tells us that any energy leaving the hotter system must enter the colder system. That is, the systems *exchange* energy with no net loss or gain. The concept is straightforward, but to state the idea mathematically we need to be careful with signs.

Let Q_1 be the energy transferred to system 1 as heat. Similarly, Q_2 is the energy transferred to system 2. The fact that the systems are merely exchanging energy can be written $|Q_1| = |Q_2|$. The energy *lost* by the hotter system is the energy *gained* by the colder system, so Q_1 and Q_2 have opposite signs: $Q_1 = -Q_2$. No energy is exchanged with the environment, hence it makes more sense to write this relationship as

$$Q_{net} = Q_1 + Q_2 = 0 \qquad (12.11)$$

This idea is not limited to the interaction of only two systems. If three or more systems are combined in isolation from the rest of their environment, each at a different initial temperature, they will all come to a common final temperature that can be found from the relationship

$$Q_{net} = Q_1 + Q_2 + Q_3 + \cdots = 0 \qquad (12.12)$$

NOTE ▸ The signs are very important in calorimetry problems. ΔT is always $T_f - T_i$, so ΔT and Q are negative for any system whose temperature decreases. The proper sign of Q for any phase change must be supplied *by you,* depending on the direction of the phase change. ◂

PROBLEM-SOLVING STRATEGY 12.1 **Calorimetry problems**

PREPARE Model the systems as interacting with each other but isolated from the larger environment. List known information and identify what you need to find. Convert all quantities to SI units.

SOLVE The mathematical representation, a statement of energy conservation, is

$$Q_{net} = Q_1 + Q_2 + \cdots = 0$$

- For systems that undergo a temperature change, $Q = Mc(T_f - T_i)$. Be sure to have the temperatures T_i and T_f in the correct order.
- For systems that undergo a phase change, $Q = \pm ML$. Supply the correct sign by observing whether energy enters or leaves the system.
- Some systems may undergo a temperature change *and* a phase change. Treat the changes separately. The heat is $Q = Q_{\Delta T} + Q_{phase}$.

ASSESS Is the final temperature in the middle? T_f that is higher or lower than all initial temperatures is an indication that something is wrong, usually a sign error.

NOTE ▸ You may have learned to solve calorimetry problems in other courses by writing $Q_{gained} = Q_{lost}$. That is, by balancing heat gained with heat lost. That approach works in simple problems, but it has two drawbacks. First, you often have to "fudge" the signs to make them work. Second, and more serious, you can't extend this approach to a problem with three or more interacting systems. Using $Q_{net} = 0$ is much preferred. ◂

EXAMPLE 12.5 **Calorimetry with a phase change**

Your 500 mL soda is at 20°C, room temperature, so you add 100 g of ice from the −20°C freezer. Does all the ice melt? If so, what is the final temperature? If not, what fraction of the ice melts? Assume that you have a well-insulated cup.

PREPARE We have a thermal interaction between the soda, which is essentially water, and the ice. We need to distinguish between three possible outcomes. If all the ice melts, then $T_f > 0°C$. It's also possible that the soda will cool to 0°C before all the ice has melted, leaving the ice and liquid in equilibrium at 0°C. A third possibility is that the soda will freeze solid before the ice warms up to 0°C. That seems unlikely here, but there are situations, such as the pouring of molten metal out of furnaces, when all the liquid does solidify. We need to distinguish between these before knowing how to proceed.

SOLVE Let's first calculate the heat needed to melt all the ice and leave it as liquid water at 0°C. To do so, we must warm the ice to 0°C, then change it to water. The heat input for this two-stage process is

$$Q_{melt} = M_i c_i (20\text{ K}) + M_i L_f = 37{,}500\text{ J}$$

where L_f is the heat of fusion of water. It is used as a *positive* quantity because we must *add* heat to melt the ice. Next, let's calculate how much heat will leave the soda if it cools all the way to 0°C. The volume is $V = 500\text{ mL} = 5.00 \times 10^{-4}\text{ m}^3$ and thus the mass is $M_s = \rho V = 0.500\text{ kg}$. The heat loss is

$$Q_{cool} = M_s c_w(-20\ K) = -41{,}900\text{ J}$$

where $\Delta T = -20$ K because the temperature decreases. Because $|Q_{cool}| > Q_{melt}$, the soda has sufficient energy to melt all the ice. Hence the final state will be all liquid at $T_f > 0$. (Had we found $|Q_{cool}| < Q_{melt}$, then the final state would have been an ice-liquid mixture at 0°C.)

Energy conservation requires $Q_{ice} + Q_{soda} = 0$. The heat Q_{ice} consists of three terms: warming the ice to 0°C, melting the ice to water at 0°C, then warming the 0°C water to T_f. The mass will still be M_i in the last of these steps because it is the "ice system," but we need to use the specific heat of *liquid water.* Thus

$$Q_{ice} + Q_{soda} = [M_i c_i(20\text{ K}) + M_i L_f + M_i c_w(T_f - 0°\text{C})] + M_s c_w(T_f - 20°\text{C}) = 0$$

We've already done part of the calculation, allowing us to write

$$37{,}500\text{ J} + M_i c_w(T_f - 0°\text{C}) + M_s c_w(T_f - 20°\text{C}) = 0$$

Solving for T_f gives

$$T_f = \frac{20 M_s c_w - 37{,}500}{M_i c_w + M_s c_w} = 1.8°\text{C}$$

ASSESS As expected, the soda has been cooled to nearly the freezing point.

Measuring the Heat of Combustion

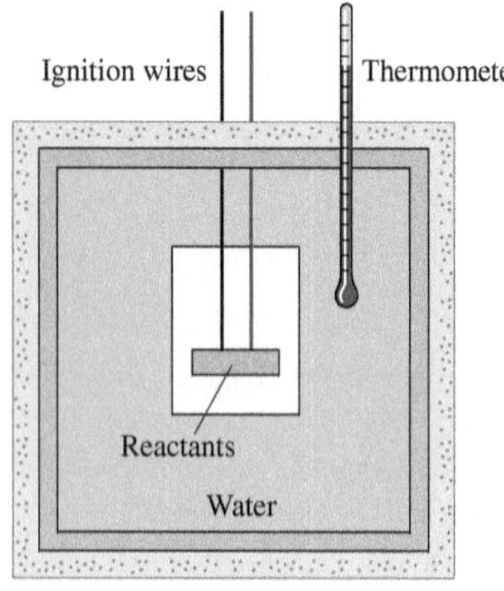

FIGURE 12.13 A calorimeter for measuring heats of reaction.

Measuring the energy change associated with a chemical or biochemical reaction, called the *heat of reaction,* is done in a device called a *calorimeter.* FIGURE 12.13 shows an example in which a sealed stainless steel container with a known amount of reactants is surrounded by a water bath and insulation. The reaction is initiated by a spark, and the heat Q released by the reaction increases the temperature of the container and the water. Equation 12.7 with two objects being heated to the same temperature is

$$Q = (M_{ss} c_{ss} + M_w c_w)\,\Delta T \tag{12.13}$$

In principle, knowing the masses and specific heats of the apparatus and measuring the temperature increase ΔT allow us to directly determine Q in joules.

The *heat of combustion* Q_{comb} is the heat evolved per mole of substance when it is completely burned in oxygen. Q_{comb} is negative for an exothermic reaction because heat *leaves* the reaction. If n moles of a substance are burned in the calorimeter, releasing a quantity of heat Q, the heat of combustion is

$$Q_{comb} = -\frac{Q}{n} \tag{12.14}$$

In practice, an experimenter must account for the influence of the wires, the thermometer, and less-than-ideal conditions. To deal with this, we can define an *effective heat capacity* $(Mc)_{eff}$ such that $Q = (Mc)_{eff}\Delta T$. The value of $(Mc)_{eff}$ is first established by calibrating the calorimeter with a reaction whose heat of combustion is well known, then used to measure an unknown heat of combustion.

EXAMPLE 12.6 **The heat of combustion of sucrose**

A calorimeter is calibrated by burning 0.500 g of benzoic acid (0.122 kg/mol), whose heat of combustion is known to be −3226 kJ/mol. This combustion causes a 5.35°C temperature rise. Then 0.500 g of sucrose ($C_{12}H_{22}O_{11}$, 0.342 kg/mol) is burned, and the temperature increase is measured to be 3.34°C. What is the heat of combustion of sucrose?

PREPARE The calibration data allow us to determine $(Mc)_{eff}$, so it's not necessary to know the details of the calorimeter.

SOLVE The heat delivered to the calorimeter by burning benzoic acid is

$$Q_{benzoic} = 0.500\text{ g} \times \frac{1\text{ kg}}{1000\text{ g}} \times \frac{1\text{ mol}}{0.122\text{ kg}} \times \frac{3226\text{ kJ}}{1\text{ mol}}$$

$$= 13.22\text{ kJ}$$

This causes a temperature rise of 5.35°C, so the effective heat capacity is

$$(Mc)_{eff} = \frac{Q_{benzoic}}{\Delta T} = \frac{13.22\text{ kJ}}{5.35°\text{C}} = 2.471\text{ kJ/°C}$$

Burning sucrose in the same apparatus increases the temperature by 3.34°C, so

$$Q_{sucrose} = (Mc)_{eff}\Delta T = (2.471\text{ kJ/°C})(3.34°\text{C}) = 8.253\text{ kJ}$$

We can use the molecular mass of 0.342 kg/mol to find that 0.500 g of sucrose is 1.46×10^{-3} mol; thus the heat of combustion is

$$(Q_{comb})_{sucrose} = -\frac{8.253\text{ kJ}}{1.462 \times 10^{-3}\text{ mol}} = -5650\text{ kJ/mol}$$

ASSESS Later in this chapter, in Section 12.8, we'll see how to relate this measurement to the enthalpy change of the reaction, which is how you would see it described in a chemistry textbook.

12.5 Heat Transfer

You feel warmer when the sun is shining on you, colder when sitting on a metal bench or when the wind is blowing, especially if your skin is wet. This is due to the transfer of heat. Although we've talked about heat a lot in this chapter, we haven't said much about *how* heat is transferred from a hotter object to a colder object. There are four basic mechanisms by which objects exchange heat with their surroundings. Evaporation was treated in Section 12.3; in this section, we will consider the other mechanisms.

Heat-transfer mechanisms

When two objects are in direct contact, such as the hot plate and the beaker, heat is transferred by *conduction*.

Air currents near a lighted candle rise, taking thermal energy with them in a process known as *convection*.

The lamp at the top shines on the lambs huddled below, warming them. The energy is transferred by *radiation*.

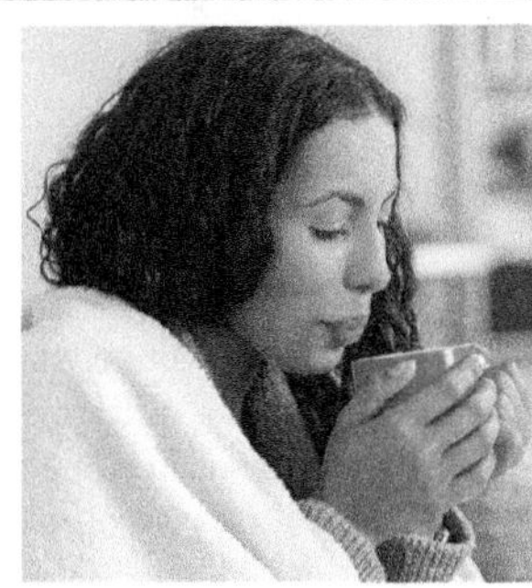

Blowing on a hot cup of tea or coffee cools it by *evaporation*.

Conduction

FIGURE 12.14 (next page) shows an object sandwiched between a hotter temperature T_H and a colder temperature T_C. The temperature *difference* causes heat to be transferred from the hot side to the cold side in a process known as **conduction.**

It is not surprising that more heat is transferred if the temperature difference ΔT is larger. A material with a larger cross section A transfers more heat, while a thicker material, increasing the distance L between the hot and cold sources, decreases the rate of heat transfer.

FIGURE 12.14 Conduction of heat.

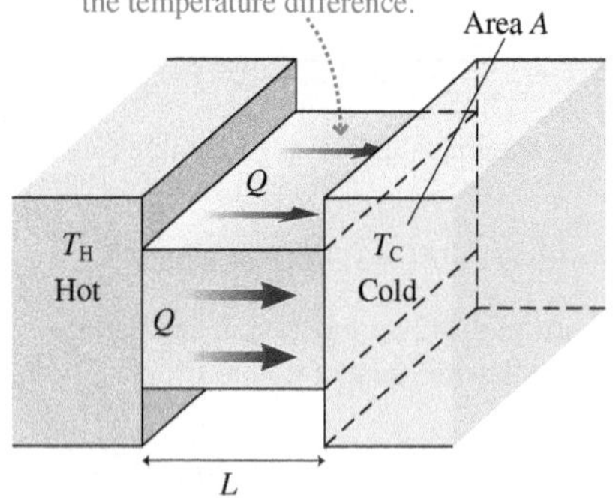

These observations about heat conduction can be summarized in a single formula. If a small amount of heat dQ is transferred in a small time interval dt, the *rate* of heat transfer is dQ/dt. For a material of cross-section area A and length L, spanning a temperature difference $\Delta T = T_H - T_C$, the rate of heat transfer is

$$\frac{dQ}{dt} = kA\frac{\Delta T}{L} \qquad (12.15)$$

The quantity k, which characterizes whether the material is a good conductor of heat or a poor conductor, is called the **thermal conductivity** of the material. Because the heat-transfer rate J/s is a *power,* measured in watts, the units of k are W/m · K. Values of k for common materials are given in TABLE 12.4; a material with a larger value of k is a better conductor of heat.

TABLE 12.4 **Thermal conductivities**

Material	k (W/m · K)
Diamond	2000
Copper	400
Aluminum	240
Stainless steel	14
Ice	1.7
Glass	0.80
Skin	0.50
Muscle	0.46
Fat	0.21
Styrofoam	0.035
Air (20°C, 1 atm)	0.023

NOTE ▸ The traditional symbol for thermal conductivity is k. We have used k to represent the spring constant and, unfortunately, it will be used again when we study waves. Whenever you see a k, be very alert to the context in which it is used. ◂

The quantity $\Delta T/L$, kelvins of temperature change per meter, is called the **temperature gradient.** A large temperature change over a small distance is a large temperature gradient. An important interpretation of the heat-conduction equation is that *a temperature gradient drives a thermal energy flow.* We met the idea that **gradients drive flows** in our discussion of fluid flows; it's an important concept that we'll see several more times.

Most good heat conductors are metals, which are also good conductors of electricity. One exception is diamond, in which the strong bonds among atoms that make diamond such a hard material lead to a rapid transfer of thermal energy. Biological materials, which are weakly bonded, tend to be poor heat conductors. Air is an especially poor conductor of heat because there are no bonds between adjacent molecules.

Some of our perceptions of hot and cold have more to do with thermal conductivity than with temperature. For example, a metal chair feels colder to your bare skin than a wooden chair not because it has a lower temperature—both are at room temperature—but because it has a much larger thermal conductivity that conducts heat away from your body at a much higher rate.

EXAMPLE 12.7 Warming the bench

Suppose you are sitting on a 10°C concrete bench. You are wearing thin clothing that provides negligible insulation. In this case, most of the insulation that protects your body's core (temperature 37°C) from the cold of the bench is provided by a 1.0-cm-thick layer of fat on the part of your body that touches the bench. (The thickness varies from person to person, but this is a reasonable average value.) A good estimate of the area of contact with the bench is 0.10 m^2. Given these details, what is your rate of heat loss by conduction?

PREPARE Heat is lost to the bench by conduction through the fat layer, whose thermal conductivity is given in Table 12.4. The skin is very thin compared to the fat layer, so we'll assume that the skin can be neglected. The temperature difference between your body's core and the bench is 27°C or 27 K. This temperature difference over a distance of 1.0 cm = 0.010 m is a temperature *gradient* of 2700 K/m.

SOLVE The rate of heat loss is given by Equation 12.15:

$$\frac{dQ}{dt} = kA\frac{\Delta T}{L} = (0.21\ \text{W/m}\cdot\text{K})(0.10\ \text{m}^2)(2700\ \text{K/m}) = 57\ \text{W}$$

ASSESS 57 W is more than half your body's resting power, which we learned in Chapter 11 is approximately 100 W. That's a significant loss, so your body will feel cold, a result that seems reasonable if you've ever sat on a cold bench for any length of time.

Convection

Air is a poor conductor of heat, but thermal energy is easily transferred through air, water, and other fluids because the air and water can flow. A pan of water on the stove is heated at the bottom. This heated water expands, becomes less dense than the water above it, and thus rises to the surface, while cooler, denser water sinks to take its place. The same thing happens to air. This transfer of thermal energy by the motion of a fluid—the well-known idea that "heat rises"—is called **convection.**

Convection is usually the main mechanism for heat transfer in fluids, both gases and liquids. On a small scale, convection mixes the pan of water that you heat on the stove; on a large scale, convection is responsible for making the wind blow and ocean currents circulate. Air is a very poor thermal conductor, but it is very effective at transferring energy by convection. To use air for thermal insulation, it is necessary to trap the air in small pockets to limit convection. And that's exactly what feathers, fur, double-paned windows, and fiberglass insulation do. Convection is much more rapid in water than in air, which is why people can die of hypothermia in 68°F (20°C) water but can live quite happily in 68°F air.

Because convection involves the often-turbulent motion of fluids, there is no simple equation for energy transfer by convection. It is often the dominant form of heat transfer, but our description must remain qualitative.

Warm water (colored) moves by convection.

Radiation

The sun *radiates* energy to earth through the vacuum of space. Similarly, you feel the warmth from the glowing red coals in a fireplace.

All objects emit energy in the form of **radiation,** electromagnetic waves—infrared and visible light waves—generated by oscillating electric charges in the atoms that form the object. These waves transfer energy from the object that emits the radiation to the object that absorbs it. Electromagnetic waves carry energy from the sun; this energy is absorbed when sunlight falls on your skin, warming you by increasing your thermal energy. Your skin also emits electromagnetic radiation, helping to keep your body cool by decreasing your thermal energy. Radiation is a significant part of the *energy balance* that keeps your body at the proper temperature.

You are familiar with radiation from objects hot enough to glow "red hot" or, at a high enough temperature, "white hot." The sun is simply a very hot ball of glowing gas, and the white light from an incandescent lightbulb is radiation emitted by a thin wire filament heated to a very high temperature by an electric current. Objects at lower temperatures also radiate, but at infrared wavelengths. You can't see this radiation (although you can sometimes feel it), but infrared-sensitive detectors can measure it and are used to make thermal images.

Global heat transfer This satellite image shows infrared radiation emitted by the ocean waters off the east coast of the United States. You can clearly see the warm waters of the Gulf Stream, a large-scale convection that transfers heat to northern latitudes.

The energy radiated by an object depends strongly on temperature. If a small amount of heat dQ is radiated during a small time interval dt by an object with surface area A and absolute temperature T, the *rate* of heat transfer is found to be

$$\frac{dQ}{dt} = e\sigma AT^4 \tag{12.16}$$

Because the rate of energy transfer is power (1 J/s = 1 W), dQ/dt is often called the *radiated power.* Notice the very strong fourth-power dependence on temperature. Doubling the absolute temperature of an object increases the radiated power by a factor of 16!

The parameter e in Equation 12.16 is the **emissivity** of the surface, a measure of how effectively it radiates. The value of e ranges from 0 to 1. σ is a constant, known as the Stefan-Boltzmann constant, with the value

$$\sigma = 5.67 \times 10^{-8}\ \text{W/m}^2 \cdot \text{K}^4$$

NOTE ▶ The temperature in Equation 12.16 *must* be in kelvins. ◀

Objects not only emit radiation, they also *absorb* radiation emitted by their surroundings. Suppose an object at temperature T is surrounded by an environment at temperature T_0. The *net* rate at which the object radiates heat—that is, radiation emitted minus radiation absorbed—is

$$\frac{dQ_{net}}{dt} = e\sigma A(T^4 - T_0^4) \tag{12.17}$$

This makes sense. An object should have no *net* radiation if it's in thermal equilibrium $(T = T_0)$ with its surroundings.

Notice that the emissivity e appears for absorption as well as emission; good emitters are also good absorbers. A perfect absorber $(e = 1)$, one absorbing all light and radiation impinging on it but reflecting none, would appear completely black. Thus a perfect absorber is sometimes called a **black body.** But a perfect absorber would also be a perfect emitter, so thermal radiation from an ideal emitter is called **black-body radiation.** It seems strange that black objects are perfect emitters, but think of black charcoal glowing bright red in a fire. At room temperature, it "glows" at infrared wavelengths.

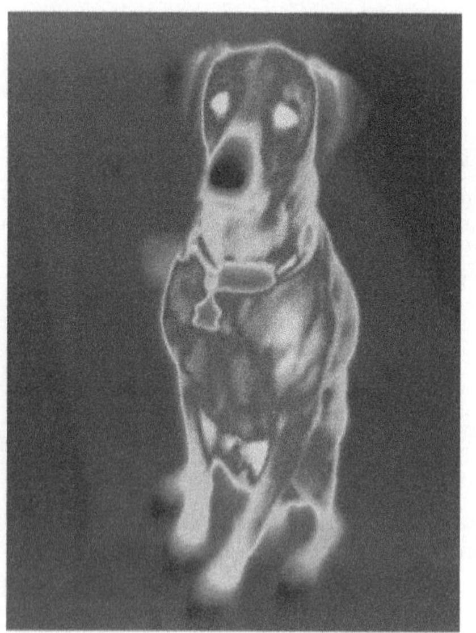

Thermal imaging The infrared radiation emitted by a dog is captured in this image. His cool nose and paws radiate much less energy than the rest of his body.

EXAMPLE 12.8 Determining energy loss by radiation for the body

A person with a skin temperature of 33°C is in a room at 24°C. What is the net rate of heat loss by radiation? The radiation is in the infrared portion of the electromagnetic spectrum, and it turns out that all human skin, regardless of its color at visible wavelengths, is an effective infrared radiator with emissivity 0.97.

PREPARE We know the temperature of the skin and the environment, so we will compute the net rate of transfer using Equation 12.17.

SOLVE The net radiation rate is

$$\frac{dQ}{dt} = e\sigma A(T^4 - T_0^4)$$
$$= (0.97)\left(5.67 \times 10^{-8} \frac{\text{W}}{\text{m}^2 \cdot \text{K}^4}\right)(1.8\ \text{m}^2)[(306\ \text{K})^4 - (297\ \text{K})^4] = 98\ \text{W}$$

ASSESS This is a reasonable value, roughly matching your resting metabolic power. When you are dressed (little convection) and sitting on wood or plastic (little conduction), radiation is your body's primary mechanism for dissipating the excess thermal energy of metabolism.

Climate and Global Warming

Thermal radiation plays a prominent role in climate and global warming. The earth as a whole is in thermal equilibrium. Consequently, it must radiate back into space exactly as much energy as it receives from the sun. The incoming radiation from the hot sun is mostly visible light. The earth's atmosphere is transparent to visible light, so this radiation reaches the surface and is absorbed. The cooler earth radiates infrared radiation, but the atmosphere is *not* completely transparent to infrared. Some components of the atmosphere, notably water vapor and carbon dioxide, are strong absorbers of infrared radiation. They hinder the emission of radiation and, rather like a blanket, keep the earth's surface warmer than it would be without these gases in the atmosphere.

The **greenhouse effect,** as it's called, is a natural part of the earth's climate. The earth would be much colder and mostly frozen were it not for naturally occurring carbon dioxide in the atmosphere. But carbon dioxide also results from the burning of fossil fuels, and human activities since the beginning of the industrial revolution have increased the atmospheric concentration of carbon dioxide by over 50%. This human contribution has amplified the greenhouse effect and is the primary cause of global warming.

STOP TO THINK 12.5 Suppose you are an astronaut in space, hard at work in your sealed spacesuit. The only way that you can transfer excess heat to the environment is by

A. Conduction. B. Convection. C. Radiation. D. Evaporation.

FIGURE 12.15 Heat loss from organisms.

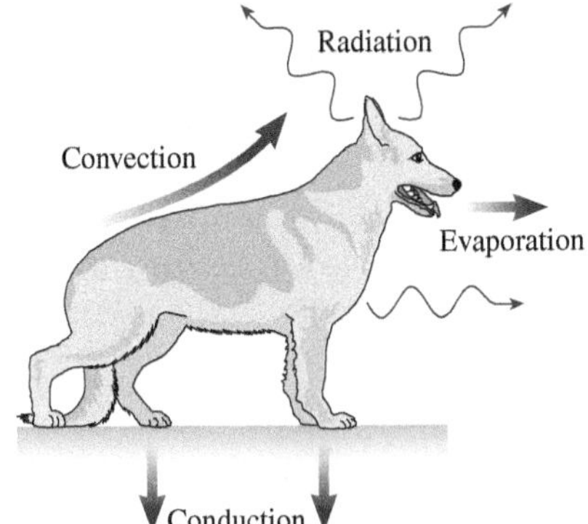
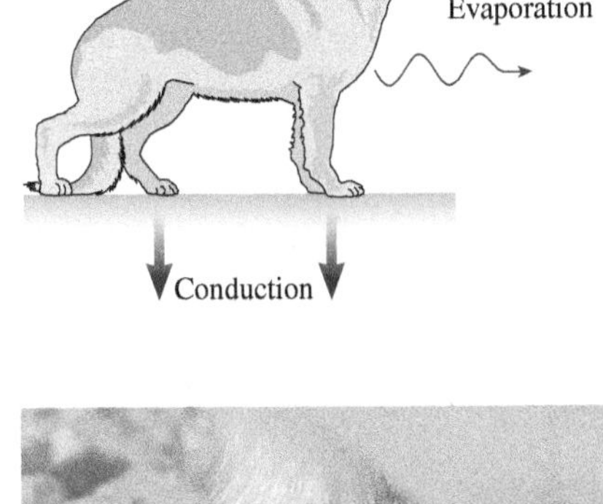

Thermoregulation

Mammals and birds are *endotherms,* species that can maintain a stable core temperature that differs, sometimes greatly, from the temperature of their environment. These organisms use a number of processes and mechanisms to regulate their core body temperature in the face of heat loss to (or occasionally heat gain from) their surroundings.

The primary mechanisms of heat loss from an organism are convection, radiation, and evaporation, as shown in FIGURE 12.15. Conduction plays a lesser role in moving heat from the organism to the environment, but conduction is the primary means by which heat flows from an animal's core to its surface.

Evaporation is an effective cooling mechanism during hot weather because of water's large heat of vaporization. Most animals achieve cooling by the evaporation of moisture from their lungs during respiration, perhaps supplemented by evaporation from the surface of a large, wet tongue. Humans are one of the few species for which evaporation of moisture from the body's surface—perspiration—is more important than evaporation through respiration.

Keeping warm rather than cool is the greater challenge for most animals because the ambient temperature is usually lower than the animal's core temperature. One way to do so is with fur and feathers. These are excellent insulators not because of the materials themselves but because fur and feathers trap a layer of air that provides insulation. Air by itself is *not* a good insulator because it is very efficient at transferring heat through convection. However, air becomes an effective insulator when it is trapped in many small pockets, too small—less than a few millimeters—for convection currents to be established. Fur and feathers do this. Arctic animals can maintain a core temperature near 37°C with only 2 or 3 cm of trapped air between their skin and outside temperatures of −40°C. You do the same when you wear a down-filled jacket.

An alternative to wearing insulation on the outside is to have insulation under the skin in the form of fat or blubber, which is a poor heat conductor. Many animals add extra layers of fat in the fall, both for insulation and for metabolic energy when other food is scarce during the winter. Marine mammals that live in cool or cold water depend on a thick layer of blubber, a vascularized fat tissue. In fact, the insulation of blubber is so good that whales have more problems with overheating in somewhat warmer waters than with staying warm in freezing waters.

Cold feet, warm heart A penguin standing on the cold Antarctic ice loses very little heat through its feet. Its thick skin with limited thermal conductivity helps, but more important, the penguin's feet are very cold, which minimizes heat loss by conduction. The penguin can have cold feet so close to the warmth of the body because of an adaptation called *countercurrent heat exchange*. Arteries carrying warm blood to the feet run next to veins carrying cold blood back from the feet. Heat goes from the arteries to the veins, cooling the blood going to the feet and warming the blood going to the body—so the feet stay cool while the body stays warm.

EXAMPLE 12.9 **The power of blubber**

An adult fin whale, the second largest of all whales, is typically 20 m in length with a mass of 55,000 kg. The whale's basal metabolic rate—not an easy thing to measure—is thought to be about 14 kW, 140 times larger than a human's 100 W, and its core temperature is 37°C. Fin whales visit the edge of the arctic ice sheet where the seawater temperature is −1°C. Develop a reasonable model to estimate the thickness of the layer of blubber that is needed to survive in these waters. The thermal conductivity of whale blubber is 0.25 W/m · K.

PREPARE First, we'll model the whale's shape as a uniform cylinder. 20 m would be an overestimate of the length because real whales taper toward the tail. If you imagine pushing the flukes and tail inward on a deformable whale until the whale is more of a constant-diameter cylinder, the resulting cylinder might be, somewhat arbitrarily, 17 m long. We're making an estimate, not a precise calculation, so calling it a 17-m-long cylinder is not unreasonable. Second, we'll assume that the whale's skin temperature is exactly that of the water, which minimizes convection losses to the water. If we ignore convection, then the heat loss is due to heat conduction between a 37°C interior and a −1°C exterior through thickness L of blubber. To maintain a constant body temperature, the whale's rate of heat loss via conduction

Continued

has to match its basal metabolic rate, the rate at which metabolic reactions release heat.

SOLVE The heat-conduction equation, Equation 12.15, is for heat flow along a straight line through a material with thickness L and cross-section area A. The same equation applies to heat flow from an inner sphere or cylinder to an outer sphere or cylinder if their separation L is small enough in comparison to the radius that there's no significant difference between the inner and outer surface area A. That's true in this case because a whale's blubber layer is small in comparison with the size of the whale. A whale is very close to being neutrally buoyant, so its average density must be very close to that of seawater, 1030 kg/m^3. The volume of a cylinder of length d is $\pi r^2 d$, so its mass is $M = (\pi r^2 d)\rho$. We know the mass, density, and length, so our cylindrical whale has radius

$$r = \sqrt{\frac{M}{\pi d \rho}} = \sqrt{\frac{55{,}000 \text{ kg}}{\pi (17 \text{ m})(1030 \text{ kg/m}^3)}} = 1.0 \text{ m}$$

The rate of heat loss by conduction is $dQ/dt = kA\Delta T/L$, where dQ/dt must match the basal metabolic rate of 14 kW and L is the blubber thickness that we want to find. The surface area of a cylinder consists of both the wall and two ends:

$$A = 2\pi r d + 2 \times \pi r^2 = 2\pi\left[(1.0 \text{ m})(17 \text{ m}) + (1.0 \text{ m})^2\right] = 113 \text{ m}^2$$

The temperature difference is 38°C, and we're given the blubber's thermal conductivity. Thus

$$L = \frac{kA\Delta T}{dQ/dt} = \frac{(0.25 \text{ W/m} \cdot \text{K})(113 \text{ m}^2)(38°\text{C})}{14{,}000 \text{ W}} = 0.077 \text{ m} \approx 8 \text{ cm}$$

If our model is valid, the thickness of a fin whale's blubber should be about 8 cm.

ASSESS The actual thickness of a fin whale's blubber averages about 12 cm, somewhat more around the central core and less near the extremities. Although our estimate is about 30% low, we can be pleased that such a simple model does fairly well. A real whale will certainly have convection losses, and that will require more insulation than we've calculated. But the fact that our simple estimate is close suggests that a whale's heat loss is primarily conduction to the surrounding water, with convection playing a smaller role. A simple model has given us insight into the thermoregulation of a whale.

FIGURE 12.16 A countercurrent heat exchanger carrying blood to an extremity at ≈0°C.

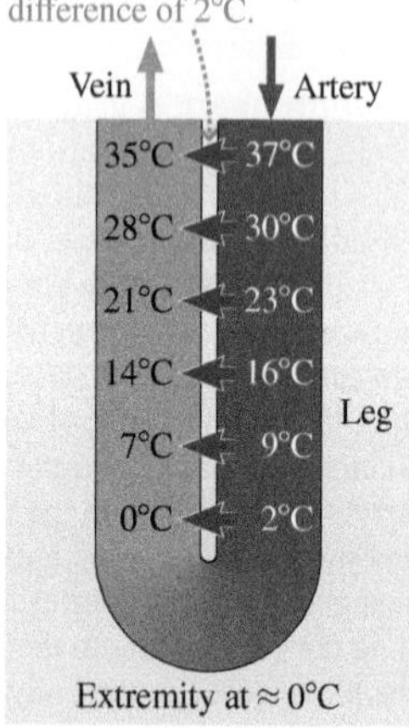

An animal's extremities, whether a wolf's footpads on bitterly cold snow or a waterfowl's feet in cold water, are more exposed and more difficult to insulate. The extremity itself may have adaptations that allow it to function at a temperature near freezing, but it has to be supplied with blood coming from the animal's warm core.

Remarkably, some animals have evolved *heat exchangers* in which arteries that carry warm blood to the extremity are in close thermal contact with veins that carry cold blood back toward the heart. As FIGURE 12.16 shows, simple heat conduction across a temperature gradient allows the blood to be chilled before it arrives at the extremity by transferring its heat to the cold returning blood. This avoids large heat losses from the extremity. The returning blood is then re-warmed almost to the core temperature. Engineers use the same process—a *countercurrent heat exchanger*—in many industrial processes, but nature designed it first.

12.6 The Ideal Gas: A Model System

You may have seen and used the ideal-gas law in other classes. The **ideal gas** is a model in which we treat molecules as hard spheres that do not interact except for occasional elastic collisions when two molecules come into contact and bounce apart. Many insights that we gain from studying ideal gases carry over to more complex systems.

Experiments show that the ideal-gas model is quite good for real gases if three conditions are met:

1. The density is low.
2. The temperature is well above the condensation point.
3. There are no chemical reactions; thus $E_{\text{int}} = E_{\text{th}}$.

If the density gets too high, or the temperature too low, then the attractive forces between the molecules begin to play an important role and our model, which ignores those attractive forces, fails. These are the forces that are responsible, under the right conditions, for the gas condensing into a liquid.

The Ideal-Gas Law

The **ideal-gas law** that you learned in chemistry is

$$pV = nRT \tag{12.18}$$

Ideal-gas law using the number of moles

where n is the number of moles of gas. **The ideal-gas law is a relationship among the four state variables—p, V, n, and T—that characterize a gas in thermal equilibrium.** The constant R is called the **universal gas constant.** Its experimentally determined value, in SI units, is

$$R = 8.31 \text{ J/mol} \cdot \text{K}$$

The units of R seem puzzling. The denominator mol • K is clear because R multiplies nT. But what about the joules? The left side of the ideal-gas law, pV, has units

$$\text{Pa} \cdot \text{m}^3 = \frac{\text{N}}{\text{m}^2} \cdot \text{m}^3 = \text{N} \cdot \text{m} = \text{joules}$$

Thus the product pV has units of joules.

NOTE ► You perhaps learned in chemistry to work gas problems using units of atmospheres and liters. To do so, you had a different numerical value of R expressed in those units. In physics, however, we usually work gas problems in SI units in which pressures are expressed in Pa, volumes in m^3, and temperatures in K. ◄

The surprising fact, and one worth commenting upon, is that the ideal-gas law, within its limits of validity, describes *all* gases with a single value of the constant R.

Quick Review: Molecules and Moles

Recall from chemistry that atoms of different elements have different masses. The mass of an atom is determined primarily by its most massive constituents: the protons and neutrons in its nucleus. The *sum* of the number of protons and neutrons is called the **atomic mass number** A:

$$A = \text{number of protons} + \text{number of neutrons}$$

A, which by definition is an integer, is written as a leading superscript on the atomic symbol. For example, the primary isotope of carbon, with six protons (which makes it carbon) and six neutrons, is ^{12}C. The radioactive isotope ^{14}C, used for the carbon dating of archeological finds, contains six protons and eight neutrons.

TABLE 12.5 lists the atomic mass numbers of some of the elements that we'll use in examples and homework problems. A complete periodic table of the elements, including atomic masses, is found in Appendix B.

The *atomic mass scale* is established by defining the mass of ^{12}C to be exactly 12 u, where u is the symbol for the **atomic mass unit.** Chemists often use the *Dalton* (Da), which is another name for the same unit; that is, 1 Da = 1 u. The conversion to SI units is

$$1 \text{ u} = 1 \text{ Da} = 1.66 \times 10^{-27} \text{ kg}$$

The numerical value of an atom's atomic mass in u is close to, but not exactly, its atomic mass number A. For our purposes, it will be sufficient to overlook the slight difference; thus **the integer atomic mass number is the value of the atomic mass in u.** That is, $m(^1\text{H}) = 1$ u, $m(^4\text{He}) = 4$ u, $m(^{16}\text{O}) = 16$ u, and so on. For molecules, the **molecular mass** is the sum of the atomic masses of the atoms that form

TABLE 12.5 Some atomic mass numbers

Element		A
^1H	Hydrogen	1
^4He	Helium	4
^{12}C	Carbon	12
^{14}N	Nitrogen	14
^{16}O	Oxygen	16
^{20}Ne	Neon	20
^{27}Al	Aluminum	27
^{40}Ar	Argon	40
^{207}Pb	Lead	207

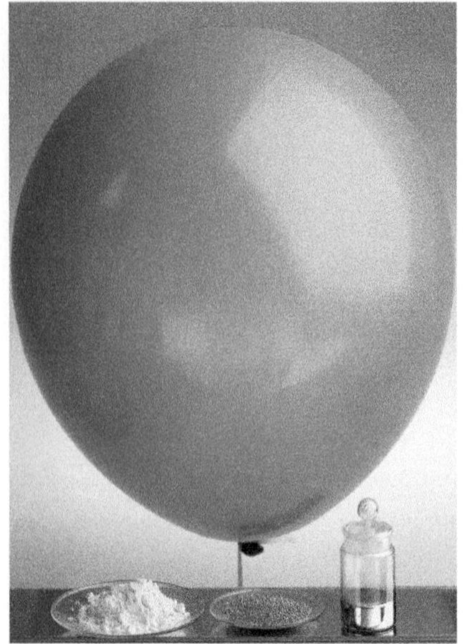
One mole of helium, sulfur, copper, and mercury.

the molecule. Thus the molecular mass of O_2, the constituent of oxygen gas, is $m(O_2) = 32$ u.

We often need to know the amount of *substance* in a macroscopic system. One way of doing so is to specify the system's mass M. Another is to measure the amount of substance in *moles*. By definition, 1 **mole** of matter is the amount of a substance that contains 6.02×10^{23} basic particles. The basic particle depends on the substance. Helium is a **monatomic gas,** which means that the basic particle is the helium atom. Thus 6.02×10^{23} helium atoms are 1 mol of helium. But oxygen gas is a **diatomic gas** because the basic particle is the two-atom diatomic molecule O_2. 1 mol of oxygen gas contains 6.02×10^{23} *molecules* of O_2 and thus $2 \times 6.02 \times 10^{23}$ oxygen atoms. TABLE 12.6 lists the monatomic and diatomic gases that we will use in examples and homework problems.

The number of basic particles per mole of substance, called **Avogadro's number** N_A, is $N_A = 6.02 \times 10^{23}\,\text{mol}^{-1}$. Despite its name, Avogadro's number is not simply "a number"; it has units. Because there are N_A particles per mole, the number of moles in a substance containing N basic particles is

$$n = \frac{N}{N_A} \tag{12.19}$$

The number N of atoms or molecules can be found from the total mass of the substance M and its molecular mass m, which must be expressed in the same units as M:

$$N = \frac{M}{m} \tag{12.20}$$

TABLE 12.6 **Monatomic and diatomic gases**

Monatomic		Diatomic	
He	Helium	H_2	Hydrogen
Ne	Neon	N_2	Nitrogen
Ar	Argon	O_2	Oxygen

Finally, the mass of 1 mol of substance is called the **molar mass** M_{mol}. The definition of the mole is such that a substance with atomic or molecular mass A u has a molar mass of almost exactly A g/mol. Thus the molar mass of ^{4}He, with $A = 4$, is 4.0 g/mol. Chemists usually work with g/mol, but in physics we specify molar masses in SI units of kg/mol. Thus our rule is that the molar mass is the atomic or molecular mass in u divided by 1000. The molar mass of ^{4}He is 0.0040 kg/mol, while that of O_2 is 0.032 kg/mol. For a system of mass M that consists of atoms or molecules with molar mass M_{mol}, the number of moles is

$$n = \frac{M}{M_{mol}} \tag{12.21}$$

EXAMPLE 12.10 Moles of oxygen

100 g of oxygen gas is how many moles of oxygen?

PREPARE Oxygen is a diatomic molecule. We can calculate the number of moles in two ways.

SOLVE First, let's determine the number of molecules in 100 g of oxygen. The diatomic oxygen molecule O_2 has molecular mass $m = 32$ u. Converting this to kg, we get the mass of one molecule:

$$m = 32\text{ u} \times \frac{1.66 \times 10^{-27}\text{ kg}}{1\text{ u}} = 5.31 \times 10^{-26}\text{ kg}$$

Thus the number of molecules in 100 g = 0.100 kg is

$$N = \frac{M}{m} = \frac{0.100\text{ kg}}{5.31 \times 10^{-26}\text{ kg}} = 1.88 \times 10^{24}$$

Knowing the number of molecules gives us the number of moles:

$$n = \frac{N}{N_A} = 3.13\text{ mol}$$

Alternatively, we can use Equation 12.21 to find

$$n = \frac{M}{M_{mol}} = \frac{0.100\text{ kg}}{0.032\text{ kg/mol}} = 3.13\text{ mol}$$

Using the Ideal-Gas Law

Let's look at some examples of using the ideal-gas law.

EXAMPLE 12.11 Calculating a gas pressure

100 g of pure oxygen gas is transferred into an evacuated 600 cm^3 container. What is the gas pressure at a temperature of 150°C?

PREPARE The gas can be treated as an ideal gas. Oxygen is a diatomic gas of O_2 molecules.

SOLVE From the ideal-gas law, the pressure is $p = nRT/V$. In Example 12.10 we calculated the number of moles in 100 g of O_2 and found $n = 3.13$ mol. Gas problems typically involve several conversions to get quantities into the proper units, and this example is no exception. The SI units of V and T are m^3 and K, respectively, thus

$$V = (600\ \text{cm}^3)\left(\frac{1\ \text{m}}{100\ \text{cm}}\right)^3 = 6.00 \times 10^{-4}\ \text{m}^3$$

$$T = (150 + 273)\ \text{K} = 423\ \text{K}$$

With this information, the pressure is

$$p = \frac{nRT}{V} = \frac{(3.13\ \text{mol})(8.31\ \text{J/mol} \cdot \text{K})(423\ \text{K})}{6.00 \times 10^{-4}\ \text{m}^3}$$

$$= 1.83 \times 10^7\ \text{Pa} = 181\ \text{atm}$$

NOTE ▸ Temperatures *must* be in kelvins when you use the ideal-gas law. Failure to convert °C to K is one of the most common errors in gas calculations. ◂

In this text we will consider only gases in sealed containers. The number of moles (and number of molecules) will not change during a problem. In that case,

$$\frac{pV}{T} = nR = \text{constant} \tag{12.22}$$

If the gas is initially in state i, characterized by the state variables p_i, V_i, and T_i, and at some later time in a final state f, the state variables for these two states are related by

$$\frac{p_f V_f}{T_f} = \frac{p_i V_i}{T_i} \tag{12.23}$$

Initial and final states for an ideal gas in a sealed container

This before-and-after relationship between the two states, reminiscent of a conservation law, will be valuable for many problems.

STOP TO THINK 12.6 Which system contains more atoms: 5 mol of helium ($A = 4$) or 1 mol of neon ($A = 20$)?

A. Helium. B. Neon. C. They have the same number of atoms.

EXAMPLE 12.12 Calculating a gas temperature

A cylinder of gas is at 0°C. A piston compresses the gas to half its original volume and three times its original pressure. What is the final gas temperature?

PREPARE We'll treat the gas as an ideal gas in a sealed container.

SOLVE The before-and-after relationship of Equation 12.23 can be written

$$T_2 = T_1 \frac{p_2}{p_1} \frac{V_2}{V_1}$$

In this problem, the compression of the gas results in $V_2/V_1 = \frac{1}{2}$ and $p_2/p_1 = 3$. The initial temperature is $T_1 = 0°\text{C} = 273$ K. With this information,

$$T_2 = 273\ \text{K} \times 3 \times \tfrac{1}{2} = 409\ \text{K} = 136°\text{C}$$

ASSESS We did not need to know actual values of the pressure and volume, just the *ratios* by which they change.

Chemists work mostly with moles, but in physics we will often want to refer to the number of molecules N in a gas rather than the number of moles n. This is an easy change to make. Because $n = N/N_A$, the ideal-gas law in terms of N is

$$pV = nRT = \frac{N}{N_A}RT = N\frac{R}{N_A}T \tag{12.24}$$

R/N_A, the ratio of two known constants, is known as the **Boltzmann constant** k_B:

$$k_B = \frac{R}{N_A} = 1.38 \times 10^{-23}\ \text{J/K}$$

The subscript B distinguishes the Boltzmann constant from a spring constant or other uses of the symbol k.

Ludwig Boltzmann was an Austrian physicist who did some of the pioneering work in statistical physics during the mid-19th century. The Boltzmann constant k_B can be thought of as the "gas constant per molecule," whereas R is the "gas constant per mole." With this definition, the ideal-gas law in terms of N is

$$pV = Nk_BT \tag{12.25}$$

Ideal-gas law using the number of molecules

Equations 12.18 and 12.25 are both the ideal-gas law, just expressed in terms of different state variables.

Finally, we will sometimes need to know the number of molecules per cubic meter. We call this quantity, given by N/V, the **number density.** It characterizes how densely the molecules are packed together. A rearrangement of Equation 12.25 gives the number density as

$$\frac{N}{V} = \frac{p}{k_BT} \tag{12.26}$$

This is a useful consequence of the ideal-gas law, but keep in mind that the pressure *must* be in SI units of pascals and the temperature *must* be in SI units of kelvins.

EXAMPLE 12.13 The distance between molecules

"Standard temperature and pressure," abbreviated **STP**, are $T = 0°\text{C}$ and $p = 1$ atm. Estimate the average distance between N gas molecules at STP.

PREPARE We'll consider the gas to be an ideal gas.

SOLVE We can start with a container of volume V that holds N gas molecules at STP. How do we estimate the distance between them? Imagine placing an imaginary sphere around each molecule, separating it from its neighbors. This divides the total volume V into N little spheres of volume v_i, where $i = 1$ to N. The spheres of two neighboring molecules touch each other, like a crate full of Ping-Pong balls of somewhat different sizes all touching their neighbors, so the distance between two molecules is the sum of the radii of their two spheres. Each of these spheres is somewhat different, but a reasonable *estimate* of the distance between molecules would be twice the *average* radius of a sphere.

The average volume of one of these little spheres is

$$v_{avg} = \frac{V}{N} = \frac{1}{N/V}$$

That is, the average volume per molecule (m^3 per molecule) is the inverse of the number density, the number of molecules per m^3. This is not the volume of the molecule itself, which is much smaller, but the average volume of space that each molecule can claim as its own. We can use Equation 12.26 to calculate the number density:

$$\frac{N}{V} = \frac{p}{k_BT} = \frac{1.01 \times 10^5\ \text{Pa}}{(1.38 \times 10^{-23}\ \text{J/K})(273\ \text{K})}$$
$$= 2.69 \times 10^{25}\ \text{molecules/m}^3$$

where we used the definition of STP in SI units. Thus the average volume per molecule is

$$v_{avg} = \frac{1}{N/V} = 3.72 \times 10^{-26}\ \text{m}^3$$

The volume of a sphere is $\frac{4}{3}\pi r^3$, so the average radius of a sphere is

$$r_{avg} = \left(\frac{3}{4\pi}v_{avg}\right)^{1/3} = 2.1 \times 10^{-9}\ \text{m} = 2.1\ \text{nm}$$

The average distance between two molecules, with their spheres touching, is twice r_{avg}. Thus

$$\text{average distance} = 2r_{avg} \approx 4 \text{ nm}$$

This is a simple estimate, so we've given the answer with only one significant figure.

ASSESS One of the assumptions of the ideal-gas model is that atoms or molecules are "far apart" in comparison to the sizes of atoms and molecules. Chemistry experiments find that small molecules, such as N_2 and O_2, are roughly 0.3 nm in diameter. For a gas at STP, we see that the average distance between molecules is more than 10 times the size of a molecule. Thus the ideal-gas model works very well for a gas at STP.

NOTE ▶ Don't confuse STP with the *standard state* of 1 atm and 25°C used in chemistry. ◀

STOP TO THINK 12.7 You have two containers of equal volume. One is full of helium gas. The other holds an equal mass of nitrogen gas. Both gases have the same pressure. How does the temperature of the helium compare to the temperature of the nitrogen?

A. $T_{helium} > T_{nitrogen}$ B. $T_{helium} = T_{nitrogen}$ C. $T_{helium} < T_{nitrogen}$

Changing the State of an Ideal Gas

The ideal-gas law is the connection between the state variables pressure, temperature, and volume. If the state variables change, as they would from heating or compressing the gas, the state of the gas changes. An *ideal-gas process* is the means by which the gas changes from one state to another.

We will have frequent occasion to look at ideal-gas processes in which one of the three variables—volume, pressure, or temperature—is held constant and does not change. For example, FIGURE 12.17a shows a gas being heated inside a container of constant, unchanging volume, often called a *rigid container*. Because **a gas completely fills its container,** the volume of the gas is the volume of the container. This heating is a *constant-volume process* in which the pressure and temperature change but the volume does not; that is, $V_2 = V_1$ and thus $p_2/T_2 = p_1/T_1$.

Other gas processes, especially chemical reactions, take place at a constant, unchanging pressure. A constant-pressure process is one for which $p_2 = p_1$ and thus $V_2/T_2 = V_1/T_1$.

An especially important process is an **isothermal process,** which takes place at a constant temperature. One possible isothermal process is illustrated in FIGURE 12.17b, where a piston is pushed in to compress a gas. If the piston is pushed *slowly,* then heat transfer through the walls of the cylinder keeps the gas at the same temperature as the surroundings. The pressure and volume change, but $T_2 = T_1$ and thus $p_1V_1 = p_2V_2$.

FIGURE 12.17 A (a) constant-volume and (b) constant-temperature process.

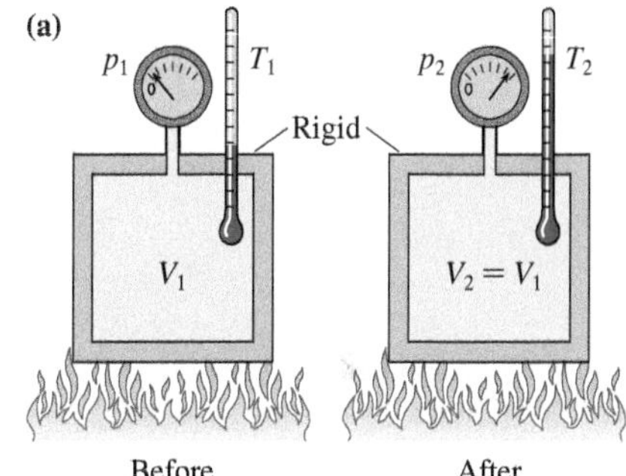

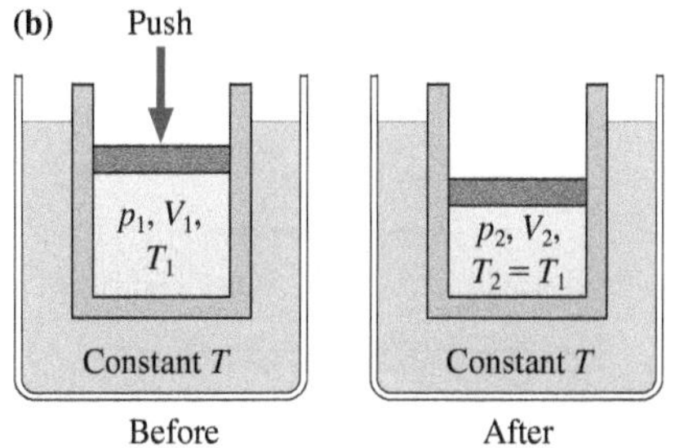

EXAMPLE 12.14 Compressing air in the lungs

An ocean snorkeler takes a deep breath at the surface, filling his lungs with 4.0 L of air. He then descends to a depth of 5.0 m. At this depth, what is the volume of air in the snorkeler's lungs?

PREPARE Air is quickly warmed to body temperature as it enters through the nose and mouth, and it remains at body temperature as the snorkeler dives, so we can consider this to be an isothermal process. At the surface, the pressure in the lungs is 1.00 atm. Because the body cannot sustain large pressure differences between inside and outside, the air pressure in the lungs rises—and the volume decreases—to match the surrounding water pressure as he descends.

SOLVE The ideal-gas law for an isothermal process is

$$V_2 = \frac{p_1}{p_2}V_1$$

Continued

We know $p_1 = 1.00$ atm $= 101{,}000$ Pa at the surface. We can find p_2 from the hydrostatic pressure equation, using the density of seawater:

$$p_2 = p_1 + \rho g d = 101{,}000 \text{ Pa} + (1030 \text{ kg/m}^3)(9.80 \text{ m/s}^2)(5.0 \text{ m})$$
$$= 151{,}000 \text{ Pa}$$

With this, the volume of the lungs at a depth of 5.0 m is

$$V_2 = \frac{101{,}000 \text{ Pa}}{151{,}000 \text{ Pa}} \times 4.0 \text{ L} = 2.7 \text{ L}$$

ASSESS The air in your lungs does compress—significantly—as you dive below the surface.

12.7 Thermodynamics of Ideal Gases

The ideal-gas law is called an *equation of state* because it is a relationship between the state variables that characterize a gas. But the ideal-gas law alone is insufficient for understanding ideal gases. We also need the first law of thermodynamics, $\Delta E_{\text{int}} = W + Q$, and we need to understand the processes by which work is done on or heat is transferred to a gas.

You learned in ◀◀ **SECTION 10.3** how to calculate work. If a *constant* force $\vec{F}$ pushes an object through a displacement Δx, the force does work $W = F_x \Delta x$. Work is positive if the force is in the direction of the displacement, negative if the force is opposite the displacement. More often than not, the force is not constant but, instead, changes as the object moves. In that case, we need to use calculus to add up the small amounts of work done over a large number of very tiny steps.

If the force $\vec{F}$ acts through a very small displacement dx, during which the force can be considered constant, it does the small amount of work $dW = F_x dx$. The total work done during a displacement from an initial position x_i to a final position x_f is then found by adding up—integrating—all these small amounts of work:

$$W = \int_{x_i}^{x_f} dW = \int_{x_i}^{x_f} F_x dx \quad (12.27)$$

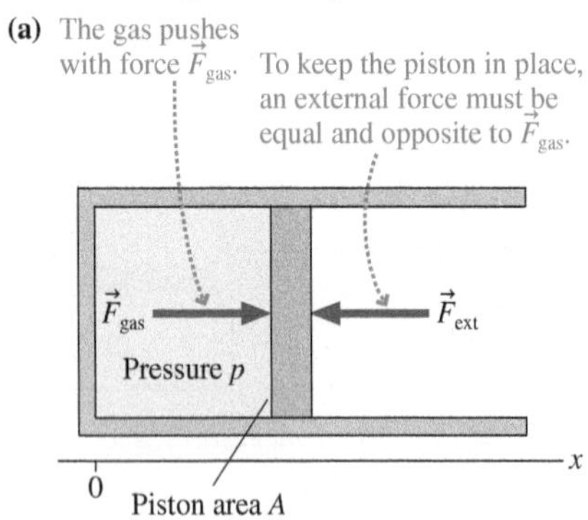
(a) The gas pushes with force $\vec{F}_{\text{gas}}$.
To keep the piston in place, an external force must be equal and opposite to $\vec{F}_{\text{gas}}$.
$\vec{F}_{\text{gas}}$
$\vec{F}_{\text{ext}}$
Pressure p
0
x
Piston area A

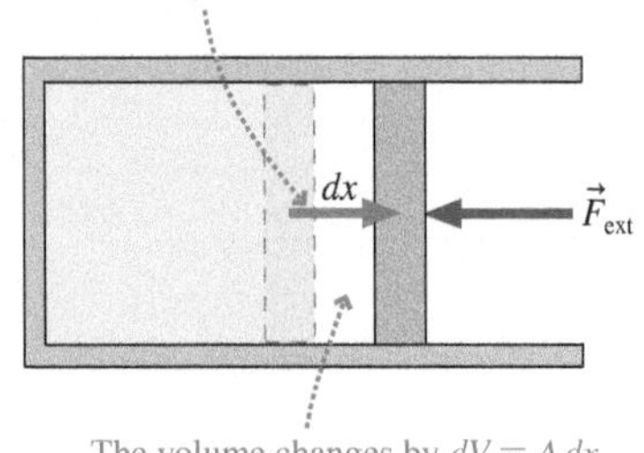
(b) As the piston moves dx, the external force does work $(F_{\text{ext}})_x dx$ on the gas.
dx
$\vec{F}_{\text{ext}}$
The volume changes by $dV = A\,dx$ as the piston moves dx.

FIGURE 12.18 The external force does work on the gas as the piston moves.

Let's apply this definition to calculate the work done when the volume of a gas changes. **FIGURE 12.18a** shows a gas of pressure p whose volume can be changed by a moving piston. The *system* is the gas in the cylinder, which exerts a force $\vec{F}_{\text{gas}}$ of magnitude pA on the left side of the piston. To prevent the pressure from blowing the piston out of the cylinder, there must be an equal but opposite external force $\vec{F}_{\text{ext}}$ pushing on the right side of the piston. Using the coordinate system of Figure 12.18a, we have

$$(F_{\text{ext}})_x = -(F_{\text{gas}})_x = -pA \quad (12.28)$$

Suppose the piston moves a small distance dx, as shown in **FIGURE 12.18b**. As it does so, the external force (i.e., the environment) does work

$$dW = (F_{\text{ext}})_x \, dx = -pA \, dx \quad (12.29)$$

If dx is positive (expanding gas), then dW is negative. This is to be expected because the force is opposite the displacement. dW is positive if the gas is slightly compressed $(dx < 0)$.

If the piston moves dx, the gas volume changes by $dV = A\,dx$. Thus the work is

$$dW = -p\,dV \quad (12.30)$$

If we allow the gas to change slowly from an initial volume V_i to a final volume V_f, the total work done by the environment—called **pressure-volume work**—is found by integrating Equation 12.30:

$$W = -\int_{V_i}^{V_f} p\,dV \quad (12.31)$$

Pressure-volume work on a gas

Equation 12.31 is a key result of thermodynamics. Although we used a cylinder to derive it, it turns out to be true for a container of any shape. Note that, in general, p is *not* a constant that can be brought outside the integral. You need to know how the pressure changes with volume before you can carry out the integration.

Work On or Work By?

Work is done *on* a gas when a force pushes a piston in to compress the gas. But consider the rapidly burning gasoline in the cylinder of your car's engine that pushes a piston out and drives the car forward. There it seems like work is being done *by* the expanding gas.

A little thought reveals that this is not an either-or situation because Newton's third law tells us that $\vec{F}_{\text{env on sys}} = -\vec{F}_{\text{sys on env}}$. When a force from the environment pushes a piston into a gas, the pressure of the gas—the system—pushes back on the piston with an equal but opposite force. *Both* forces do work, which we can call W_{env} for work done by the environment and W_{sys} for work done by the system, and they are related by $W_{\text{sys}} = -W_{\text{env}}$.

If the environment does positive work, adding energy to the system, we tend to say that work is done *on* the system. This is the case when we compress a gas. If energy flows from the system to the environment during a mechanical interaction, as when an expanding gas pushes a piston out, we tend to say that work is done *by* the system. The environment also does work, but in this case W_{env} is negative because $\vec{F}_{\text{env on sys}}$ is opposite the piston's displacement.

The important point is that W in the first law is explicitly W_{env}, the work done on the system by forces from the environment. If energy is transferred out of the system, we may *say* that "work is done by the system" but our equations require us to use W_{env}, which in this case is negative. The language that people use—work *on* or work *by*—reflects the direction of energy flow, but in solving problems we have to indicate this with a positive or negative value of W. In general:

- $W > 0$ when a gas is compressed. Volume decreases and energy is transferred from the environment to the gas (the system). We say that work is done *on* the gas.
- $W = 0$ in a constant-volume process. The pressure and temperature may change, but no work is done unless the volume changes.
- $W < 0$ when a gas expands. Volume increases and energy is transferred from the gas to the environment. We might say that work is done *by* the gas, but that's an indication that the W in the first law is negative.

NOTE ▸ Caution! Textbooks don't all agree about whether W in the first law is W_{env} or W_{sys}. For many, including this textbook, $W = W_{\text{env}}$ and the first law is written $\Delta E_{\text{int}} = Q + W$. But other textbooks prefer to look at things from the system's perspective and use $W = W_{\text{sys}}$. As a result, the first law is written as $\Delta E_{\text{int}} = Q - W$. Either approach is valid if used consistently, but as a reader you need to check to see which convention the author is following. ◂

Calculating the Work Done on a Gas

We've already noted that $W = 0$ in a constant-volume process. In a constant-pressure process, we can take p outside the integral of Equation 12.31 to find

$$W_{\text{P}} = -p\int_{V_{\text{i}}}^{V_{\text{f}}} dV = -p\Delta V \tag{12.32}$$

where the subscript P indicates that this is the work done as pressure is held constant. $\Delta V = V_{\text{f}} - V_{\text{i}}$ is the volume change. ΔV is *positive* if the gas expands ($V_{\text{f}} > V_{\text{i}}$), so W_{P} is negative. ΔV is *negative* if the gas is compressed ($V_{\text{f}} < V_{\text{i}}$), which makes W_{P} positive.

For a constant-temperature (isothermal) process, we need to know the pressure as a function of volume before we can integrate Equation 12.31. From the ideal-gas law, $p = nRT/V$. Thus the work on the gas as the volume changes from V_i to V_f is

$$W_T = -\int_{V_i}^{V_f} p\,dV = -\int_{V_i}^{V_f} \frac{nRT}{V}\,dV = -nRT\int_{V_i}^{V_f} \frac{dV}{V} \tag{12.33}$$

where the subscript T indicates that this is the work done as the temperature is held constant. The fact that T is a constant allowed us to take it outside the integral. We can use $\int dx/x = \ln x$ to complete the integration:

$$W_T = -nRT\int_{V_i}^{V_f} \frac{dV}{V} = -nRT \ln V\Big|_{V_i}^{V_f}$$

$$= -nRT(\ln V_f - \ln V_i) = -nRT\ln\left(\frac{V_f}{V_i}\right)$$

Because $nRT = p_iV_i = p_fV_f$ during a constant-temperature process, the work is

$$W_T = -nRT\ln\left(\frac{V_f}{V_i}\right) = -p_iV_i\ln\left(\frac{V_f}{V_i}\right) = -p_fV_f\ln\left(\frac{V_f}{V_i}\right) \tag{12.34}$$

Which version of Equation 12.34 is easiest to use will depend on the information you're given. The pressure, volume, and temperature *must* be in SI units.

EXAMPLE 12.15 The work of an isothermal compression

A cylinder contains 7.0 g of nitrogen gas. How much work must be done to compress the gas at a constant temperature of 80°C until the volume is halved?

PREPARE This is an isothermal ideal-gas process.

SOLVE Nitrogen gas is N_2, with molar mass $M_{mol} = 0.028$ kg/mol = 28 g/mol, so 7.0 g is 0.25 mol of gas. The temperature is $T = 353$ K. Although we don't know the actual volume, we do know that $V_f = \frac{1}{2}V_i$. The volume ratio is all we need to calculate the work:

$$W_T = -nRT\ln\left(\frac{V_f}{V_i}\right)$$

$$= -(0.25\text{ mol})(8.31\text{ J/mol}\cdot\text{K})(353\text{ K})\ln(1/2) = 510\text{ J}$$

ASSESS The work is positive because a force from the environment pushes the piston inward to compress the gas.

The Specific Heats of Gases

Specific heats were given in Table 12.2 for solids and liquids. Gases are harder to characterize because the heat required to cause a specified temperature change depends on the *process* by which the gas changes state. It is useful to define two different versions of the specific heat of gases, one for heating the gas at constant volume and one for heating the gas at constant pressure. We will define these as molar specific heats because we usually do gas calculations using moles instead of mass. The heat needed to change the temperature of n moles of gas by ΔT is

$$Q_V = nC_V\Delta T \quad \text{(temperature change at constant volume)}$$
$$Q_P = nC_P\Delta T \quad \text{(temperature change at constant pressure)} \tag{12.35}$$

where the subscripts V and P on Q show which state variable is being kept constant as the gas is heated. C_V is called the **molar specific heat at constant volume** and C_P is the **molar specific heat at constant pressure.** TABLE 12.7 gives the values of C_V and C_P for a few common monatomic and diatomic gases. The units are J/mol · K. Molar specific heats vary somewhat with temperature—we'll look at an example in the next chapter—but the values in the table are adequate for temperatures from 200 K to 800 K.

TABLE 12.7 **Molar specific heats of gases (J/mol · K) at $T = 0$°C**

Gas	C_P	C_V	$C_P - C_V$
Monatomic Gases			
He	20.8	12.5	8.3
Ne	20.8	12.5	8.3
Ar	20.8	12.5	8.3
Diatomic Gases			
H_2	28.7	20.4	8.3
N_2	29.1	20.8	8.3
O_2	29.2	20.9	8.3

NOTE ▶ C_V and C_P differ for gases because gases are compressible. Solids and liquids are virtually incompressible, so C_V and C_P are the same and Table 12.2 didn't need to make a distinction between them. ◀

EXAMPLE 12.16 Heating and cooling a gas

Three moles of O_2 gas are at 20.0°C. 600 J of heat are transferred to the gas at constant pressure, then 600 J are removed at constant volume. What is the final temperature?

PREPARE O_2 is a diatomic ideal gas. The gas is heated at constant pressure, then cooled at constant volume.

SOLVE The heat transferred during the constant-pressure process causes a temperature rise

$$\Delta T = T_2 - T_1 = \frac{Q_P}{nC_P} = \frac{600\ \text{J}}{(3.0\ \text{mol})(29.2\ \text{J/mol}\cdot\text{K})} = 6.8°\text{C}$$

where C_P for oxygen was taken from Table 12.7. The gas expands during this heating, which leaves the gas at temperature $T_2 = T_1 + \Delta T = 26.8°\text{C}$. The temperature then falls as heat is removed during the constant-volume process:

$$\Delta T = T_3 - T_2 = \frac{Q_V}{nC_V} = \frac{-600\ \text{J}}{(3.0\ \text{mol})(20.9\ \text{J/mol}\cdot\text{K})} = -9.6°\text{C}$$

We used a *negative* value for Q_V because heat is transferred from the gas to the environment. The final temperature of the gas is $T_3 = T_2 + \Delta T = 17.2°\text{C}$.

ASSESS The final temperature is lower than the initial temperature because $C_P > C_V$.

You may have noticed two curious features in Table 12.7. First, the molar specific heats of monatomic gases are *all alike.* And the molar specific heats of diatomic gases, while different from monatomic gases, are again *very nearly alike.* We saw a similar feature in Table 12.3 for the molar specific heats of elemental solids. Second, the *difference* $C_P - C_V = 8.3\ \text{J/mol}\cdot\text{K}$ is the same in every case. And, most puzzling of all, the value of $C_P - C_V$ appears to be equal to the universal gas constant R! Why should this be?

The relationship between C_V and C_P hinges on one crucial idea: ΔE_{int}, the change in the internal energy of an ideal gas, is the same for *any* two processes that have the same ΔT. First, there are no chemical reactions in an ideal gas, so the internal energy is entirely thermal energy E_{th}. Second, the thermal energy of an ideal gas is the sum of all the molecular kinetic energies, and the speeds of little hard spheres can't depend on anything other than temperature. (We'll verify this in Chapter 13.)

With that in mind, consider two ideal gases that have the same number of moles and start with the same conditions. Let one gas undergo a constant-volume process and the other a constant-pressure process that has *the same* ΔT. Because they have the same change in temperature, they also have the same change in thermal energy: $(E_{\text{th}})_V = (E_{\text{th}})_P$. If we apply the first law of thermodynamics to the constant-volume process, and recall that no work is done if the volume doesn't change, we find that

$$(\Delta E_{\text{th}})_V = W_V + Q_V = 0 + nC_V\Delta T = nC_V\Delta T \qquad (12.36)$$

Doing the same for the constant-pressure process, where work *is* done, we find

$$(\Delta E_{\text{th}})_P = W_P + Q_P = -p\Delta V + nC_P\Delta T \qquad (12.37)$$

Because $(E_{\text{th}})_V = (E_{\text{th}})_P$, we can equate the right-hand sides of Equations 12.36 and 12.37:

$$-p\Delta V + nC_P\Delta T = nC_V\Delta T \qquad (12.38)$$

We can then use the ideal-gas law $pV = nRT$ to write $p\Delta V = nR\Delta T$ for a constant-pressure process. Substituting this expression for $p\,\Delta V$ into Equation 12.38 gives

$$-nR\Delta T + nC_P\Delta T = nC_V\Delta T \qquad (12.39)$$

The $n\,\Delta T$ cancels, and we are left with

$$C_P = C_V + R \qquad (12.40)$$

This result, which applies to all ideal gases, is exactly what we see in the data of Table 12.7.

But that's not the only conclusion we can draw. Equation 12.36 found that $(\Delta E_{th})_V = nC_V\Delta T$ for a constant-volume process. But we had just noted that ΔE_{th} is the same for *all* gas processes that have the same ΔT. Consequently, this expression for ΔE_{th} is equally true for any other process. That is

$$\Delta E_{th} = nC_V\Delta T \qquad (12.41)$$

Thermal energy change of any ideal-gas process

The thermal energy of an ideal gas does not depend at all on the pressure, only on the temperature.

Compare this result to Equations 12.35. We first made a distinction between constant-volume and constant-pressure processes, but now we're saying that Equation 12.41 is true for any process. Are we contradicting ourselves? No, the difference lies in what you need to calculate.

- The change in thermal energy when the temperature changes by ΔT is the same for any process. That's Equation 12.41.
- The *heat* required to bring about the temperature change depends on what the process is. That's Equations 12.35. A constant-pressure process requires more heat than a constant-volume process that produces the same ΔT.

The reason for the difference is seen by writing the first law as $Q = \Delta E_{th} - W$. In a constant-volume process, where $W = 0$, *all* the heat input is used to increase the gas temperature. But in a constant-pressure process, some of the energy that enters the system as heat then leaves the system as work $(W < 0)$ done by the expanding gas. Thus more heat is needed to produce the same ΔT.

Adiabatic Processes

The first law of thermodynamics for an ideal gas is $\Delta E_{th} = W + Q$. A constant-temperature process has $\Delta E_{th} = 0$. A constant-volume process has $W = 0$. What about a process in which there is no heating: $Q = 0$? A process in which no heat is transferred is called an **adiabatic process.** FIGURE 12.19 compares an adiabatic process with constant-temperature and constant-volume processes.

FIGURE 12.19 The relationship of three important processes to the first law of thermodynamics.

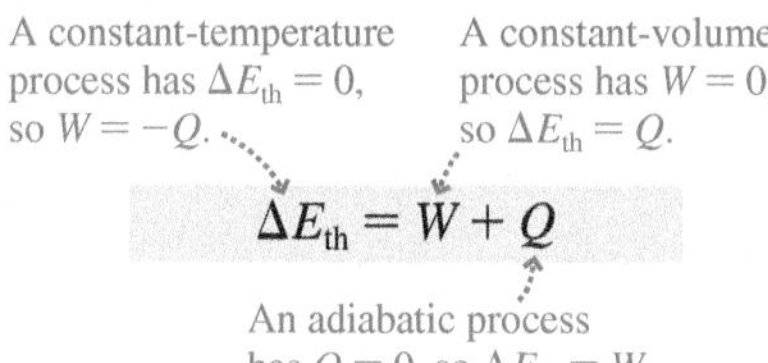
A constant-temperature process has $\Delta E_{th} = 0$, so $W = -Q$.
A constant-volume process has $W = 0$, so $\Delta E_{th} = Q$.
$\Delta E_{th} = W + Q$
An adiabatic process has $Q = 0$, so $\Delta E_{th} = W$.

In practice, there are two ways that an adiabatic process can come about. First, a gas cylinder can be completely surrounded by thermal insulation, such as thick pieces of Styrofoam. The environment can interact mechanically with the gas by pushing or pulling on the insulated piston, but there is no thermal interaction.

Second, the gas can be expanded or compressed very rapidly in what we call an *adiabatic expansion* or an *adiabatic compression.* In a rapid process there is essentially no time for heat to be transferred between the gas and the environment. We've already alluded to the idea that heat is transferred via atomic-level collisions. These collisions take time. If you stick one end of a copper rod into a flame, the other end will eventually get too hot to hold—but not instantly. Some amount of time is required for heat to be transferred from one end to the other. A process that takes place faster than the heat can be transferred is adiabatic.

CONCEPTUAL EXAMPLE 12.17 Squeezing air

A cylinder of air at room temperature has a tightly fitting but frictionless piston. Suppose you push the piston in quickly, compressing the gas. Does the temperature of the gas increase, decrease, or stay the same?

REASON From the ideal-gas law, $pV = nRT$, you might think that the increasing pressure means an increasing temperature. Or you might argue that an increasing value of p along with a decreasing value of V means that T doesn't change. It's true that p

increases and V decreases, but we don't have enough information to know whether these offset each other. The ideal-gas law alone cannot answer this question. But there is another law that the gas must obey: the first law of thermodynamics, $\Delta E_{th} = W + Q$. The fact that the piston is pushed in *quickly* suggests that this is an adiabatic process, one that happens too quickly for heat to be transferred between the gas and the environment. Thus $Q = 0$ and $\Delta E_{th} = W$. Pushing the piston *in* to compress the gas means that the environment does work *on* the system. That makes $W > 0$ and thus $\Delta E_{th} > 0$. The thermal energy of an ideal gas is proportional to its temperature, so an increased thermal energy means an increased temperature. Compressing the gas rapidly raises its temperature.

ASSESS If you've ever pumped up a bicycle tire with a hand pump, you might have noticed that the pump and the tire get warm. This is not from friction but because you're adiabatically compressing the air. Air released from a container of compressed air feels cold because it *is* colder. A rapid, adiabatic expansion lowers the temperature of a gas. Adiabatic compressions and expansions are examples of how work, rather than heat, can be used to change the temperature of a gas.

12.8 Enthalpy

Suppose, as shown in FIGURE 12.20, we want to create a system at temperature T with internal energy E_{int} starting with nothing more than atoms at rest. We first have to supply energy E_{int} to bring the atoms together to form molecules, then we set all the molecules in motion with speeds appropriate to temperature T. But that's not all: We also have to make space for the system, which has volume V, by pushing the air out of the way to make room for it. That takes more energy.

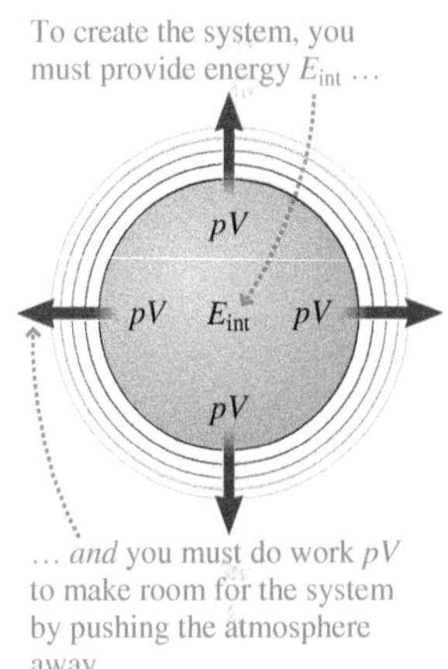
To create the system, you must provide energy E_{int} ...
pV
pV
E_{int}
pV
pV
... *and* you must do work pV to make room for the system by pushing the atmosphere away.

FIGURE 12.20 To create a system, we have to provide it with energy *and* make room for it.

You learned earlier that $W_{on\ sys} = -p\Delta V$ is the work done *by* the environment *on* the system in a constant-pressure process. The work we have to do to move the air out of the way is work *on* the environment, or $W_{on\ env} = +p\Delta V$. We need to create space for the entire volume of the system, so $\Delta V = V$ and the required work, an energy input, is $W_{on\ env} = pV$. Thus the total expenditure of energy to create the system *and* make space for it is $E_{int} + pV$.

This seems an odd idea, but it turns out to be very useful. Let's define the **enthalpy** H of a system as

$$H = E_{int} + pV \qquad (12.42)$$

Enthalpy of a system with internal energy E_{int}

Enthalpy, like internal energy and temperature and pressure, is a *state variable*. Physically, enthalpy is the energy needed to create a system *and* make space for it in an atmosphere at pressure p. The SI units of enthalpy are joules.

Consider a process that takes place at constant pressure. As the system changes from an initial state i to a final state f, the enthalpy changes by

$$\Delta H = \Delta E_{int} + \Delta(pV) = \Delta E_{int} + p\Delta V = \Delta E_{int} - W_P = Q_P \qquad (12.43)$$

We first took p outside the delta because we're assuming a constant pressure. Then, for such a process, $W_P = -p\Delta V$. Finally, we used the first law of thermodynamics, $\Delta E_{int} = W + Q$, to recognize that $\Delta E_{int} - W_P$ is the heat flow to the system—specifically the heat flow Q_P at constant pressure. Thus **the enthalpy change during a constant-pressure process equals the heat required for the process.**

This is interesting for two reasons. First, chemical reactions, especially biochemical reactions within organisms, usually take place in solution at a constant pressure and temperature dictated by the environment. Second, heat flow is measurable. If we use a calorimeter to measure the heat Q_P required by or released by a constant-pressure process—such as a chemical reaction or a phase change—then we're measuring the system's change in enthalpy.

For comparison, consider a constant-volume process with heat flow Q_V. No work is done on the system if there's no expansion or contraction ($W_V = 0$), so the first law for a constant-volume process is $\Delta E_{int} = Q_V$. In summary,

$$\begin{aligned} \Delta E_{int} &= Q_V && \text{constant-volume process} \\ \Delta H &= Q_P && \text{constant-pressure process} \end{aligned} \qquad (12.44)$$

It is easy to measure heat, so we can determine a system's change in energy or change in enthalpy by measuring the heat with a constant volume or with a constant pressure.

Phase Changes

Although a phase change can be initiated by a change in pressure—squeezing a liquid until it solidifies—most phase changes are caused by heating or cooling a system at constant pressure. Thus the heat $Q = \pm ML$ is the enthalpy change of the phase change: $\pm ML = Q_P = \Delta H$. To turn liquid water at 100°C into steam at 100°C, we have to increase the thermal energy of the molecules *and* push the air away to make room for the increased volume of the gas-phase water, which is exactly what an enthalpy change indicates.

Indeed, what physicists call a substance's *heat of fusion* L_f and *heat of vaporization* L_v, from Table 12.3, are usually referred to by chemists as *enthalpy of fusion* ΔH_{fus} and *enthalpy of vaporization* ΔH_{vap}. And while physicists use J/kg, you'll usually see units of kJ/mol in chemistry books. Table 12.3 showed that the heat of vaporization of water is $L_v = 22.6 \times 10^5$ J/kg. The molar mass of water is 0.018 kg/mol, so

$$\Delta H_{vap} = 22.6 \times 10^5 \frac{\text{J}}{\text{kg}} \times \frac{0.018 \text{ kg}}{1 \text{ mol}} \times \frac{1 \text{ kJ}}{1000 \text{ J}} = 40.7 \text{ kJ/mol}$$

This is exactly the value you will find in a chemistry book. The terminology and units differ, but the *ideas* are the same.

EXAMPLE 12.18 Energy and enthalpy of boiling water

The enthalpy of vaporization of water at 100°C is 40.7 kJ/mol. What are the change in enthalpy and the change in thermal energy when 2.00 mol of water (36 g) are boiled at 100°C?

PREPARE Enthalpy is $H = E_{int} + pV$. Boiling the water doesn't change its chemical energy, so in this case $\Delta E_{int} = \Delta E_{th}$. The tabulated enthalpy of vaporization is the enthalpy change *per mole*.

SOLVE The enthalpy change is simply

$$\Delta H = n\Delta H_{vap} = (2.00 \text{ mol})(40.7 \text{ kJ/mol}) = 81.4 \text{ kJ}$$

Boiling is a constant-pressure process in which the water volume expands from V_{liq} to V_{gas}. We can use the definition of enthalpy, $H = E_{int} + pV$, to write the change in thermal (internal) energy as

$$\Delta E_{th} = \Delta H - \Delta(pV) = \Delta H - p\Delta V = \Delta H - p(V_{gas} - V_{liq})$$

The volume of a gas is much larger than the volume of a condensed liquid, so to a very good approximation $p(V_{gas} - V_{liq}) \approx pV_{gas}$. We can use the ideal-gas law to write $pV_{gas} = nRT$, thus

$$\begin{aligned} \Delta E_{th} &\approx \Delta H - nRT \\ &= 81.4 \text{ kJ} - (2.00 \text{ mol})(8.31 \text{ J/mol} \cdot \text{K})(373 \text{ K}) \\ &= 75.2 \text{ kJ} \end{aligned}$$

ASSESS The heat needed to boil the water is 81.4 kJ. All the heat goes to an increase of enthalpy, but it doesn't all go to an increase of thermal energy. The thermal energy increases by 75.2 kJ while the remaining 6.2 kJ of the heat does the work of pushing the air away to make room for the water vapor. We could include the initial volume of the liquid for a more precise result, but it makes little difference.

Phase changes and chemical reactions that involve only solids and liquids have $\Delta E_{int} \approx \Delta H$ because there's very little change in volume. But ΔE_{th} and ΔH differ for any process that involves a gas.

NOTE ▶ Enthalpy is an energy-like quantity, with units of joules, but **enthalpy is not a conserved quantity.** Only energy itself is conserved. The water energy in Example 12.18 changes by $\Delta E_{int} = W + Q$, the energy exchanged between the system and the environment in the form of work and heat. This is a conservation statement. There's no equivalent statement for enthalpy. ◀

Chemical Reactions

If we burn natural gas to heat a beaker of solution or cook dinner, we care about the *heat* released by the reaction, not about any work that the reaction does to expand and push away the surrounding air. That is, we care more about the reaction's change in enthalpy than about its change in energy. Enthalpy is one of the most important thermodynamic tools for analyzing chemical reactions because one of the easiest things to measure for a reaction is the heat liberated or absorbed.

It is often said that "biochemical reactions take place at constant pressure and temperature," but this isn't strictly true; the temperature and pressure might change significantly while a reaction is under way. What *is* true is that the temperature and pressure return to their initial values—values set by the surrounding environment—after a reaction has gone to completion; that is, $p_f = p_i$ and $T_f = T_i$. Because quantities such as energy and enthalpy are state variables, any change in their values depends on only the initial and final states of the system, not what happens as the change is occurring. Thus ΔH and ΔE_{int} *for the entire reaction* are exactly the same as they would be for a reaction that occurs entirely at constant pressure and temperature.

Burning calories The caloric content in foods, as shown on labels, can be determined by quite literally burning the food—though in a calorimeter rather than out in the open as seen here. If food is burned at a constant pressure, the heat given off is the change in enthalpy. Food calorimeters used to be widely used, but today labels usually show a calculated value based on the amount of fat, carbohydrates, and protein in the food.

If we use a calorimeter to measure the heat Q_P of a reaction—the heat absorbed from or transferred to the environment—we are measuring what chemists call the *enthalpy of reaction* ΔH_{rxn}. A reaction that liberates heat—that is, one that transfers heat from the system of chemicals to the environment—is called **exothermic.** $Q_P < 0$ for heat flow from the system to the environment, according to our sign convention for energy transfers; thus $\Delta H_{rxn} = Q_P < 0$ for an exothermic reaction. Similarly, an **endothermic** reaction is one in which heat flows from the environment to the system; thus $Q_P > 0$ and $\Delta H_{rxn} = Q_P > 0$. The chemical "rules" for using an increase or decrease in enthalpy to judge whether a reaction is endothermic or exothermic are based on the physical idea that ΔH reveals both the quantity *and direction* of heat flow.

In many situations, especially in biochemistry, it might be difficult or impossible to isolate and measure the heat of one reaction in a complex sequence of reactions. Fortunately, doing so is not necessary. In chemistry, we use tabulated values of *enthalpies of formation* to calculate the value of ΔH_{rxn} for a given reaction. That is, we can calculate how much heat a reaction will liberate or absorb.

Our goal in a physics text is not to do reaction calculations but to understand their physical basis. Enthalpies of formation are similar to gravitational potential energy in that we need to set a *reference level* that we call zero energy or zero enthalpy. For gravitational potential energy, you can set any level you wish as $y = 0$; then the potential energy $U_G = mgy$ is the potential energy *relative to* that level. A negative value of U_G doesn't mean "less than zero energy" but "less energy than at the reference level."

In chemistry, enthalpies are measured relative to what is called the **standard state** of 25°C and 1 atm = 101 kPa (or sometimes 1 bar = 100 kPa). By convention, the enthalpy of an *element* in its most stable form at the conditions of the standard state is zero. For example, the stable form of oxygen at 25°C and 1 atm pressure is the diatomic gas O_2. Consequently, the enthalpy of gas-phase oxygen at 25°C and 1 atm pressure is zero.

The standard state is an arbitrary reference level but a practical one: It's easy to make laboratory measurements at the conditions of the standard state. For a substance that is not an element, the **standard enthalpy of formation** is the heat released or absorbed in creating 1 mol of that substance at the standard state from its basic elements.

We started this section by asking how much energy it would take to create a system starting from atoms at rest, and that led to the idea of enthalpy. This is almost exactly what a standard enthalpy of formation tells us, the difference being that we start with atoms at 25°C instead of 0 K. Consequently, much or all of the necessary thermal energy is present in the starting state. A standard enthalpy of formation is the additional thermal energy needed to create the system *plus* the change in chemical energy as the substance is created *plus* the all-important work needed to make space for the system.

Most tabulated enthalpies of formation are negative, telling us that the reactions to create a substance from its elements are exothermic. This is not surprising. Stable molecules are more stable than free atoms; otherwise, molecules would spontaneously dissociate. You learned in Chapter 11 that bound molecular systems have *negative energy* relative to free atoms. When free atoms react to form stable molecules, the chemical energy decreases ($\Delta E_{chem} < 0$) relative to that of the free atoms, and this negative change in chemical energy upon formation is the primary reason that stable compounds have negative enthalpies of formation. Said another way, we would have to heat a stable compound to dissociate it into its zero-enthalpy free atoms.

EXAMPLE 12.19 Heat of combustion of glucose

The metabolism of glucose in the body occurs in a complex multistep reaction sequence, but the net result is the combustion reaction

$$C_6H_{12}O_6(s) + 6O_2(g) \rightarrow 6CO_2(g) + 6H_2O(l)$$

where the letters in parentheses indicate the solid, liquid, or gaseous state. Tabulations of standard enthalpies of formation, usually given the symbol ΔH_f°, are -1273.3 kJ/mol for glucose, -393.5 kJ/mol for gaseous carbon dioxide, and -285.5 kJ/mol for liquid water. What is the heat of combustion of glucose?

PREPARE Let's imagine carrying out the reaction by using heat to completely dissociate the reactant glucose and oxygen molecules into their constituent elements at 25°C, then measuring the heat liberated as those atoms combine to form carbon dioxide and water at 25°C. No enthalpy of formation was given for $O_2(g)$ because this is the most stable form of oxygen at 25°C and thus, by convention, $\Delta H_f^\circ[O_2(g)] = 0$. Oxygen is starting at the reference level.

SOLVE The enthalpy of formation of glucose is the enthalpy change—the heat liberated—when glucose is formed from its constituent atoms. We can, in principle, reverse this and dissociate glucose by inputting heat $Q_{in} = -\Delta H_f^\circ[C_6H_{12}O_6(s)]$. Dissociating glucose is more complex than simply heating it, but those details are better explained in chemistry. We might have to go through several steps to dissociate glucose, but *energetically* the net heat required is just Q_{in}.

The formation of carbon dioxide and water is exothermic, and the heat released, $Q_{out} = 6\Delta H_f^\circ[CO_2(g)] + 6\Delta H_f^\circ[H_2O(l)]$, is negative because heat leaves the system. Thus the *net* heat of the reaction for 1 mol of glucose, which is what "heat of combustion" means, is

$$\begin{aligned}\Delta H_{rxn}[C_6H_{12}O_6(s)] &= Q_{net} = Q_{in} + Q_{out}\\ &= 6\Delta H_f^\circ[CO_2(g)] + 6\Delta H_f^\circ[H_2O(l)]\\ &\quad -\Delta H_f^\circ[C_6H_{12}O_6(s)] - \Delta H_f^\circ[O_2(g)]\\ &= 6(-393.5\text{ kJ/mol}) + 6(-285.5\text{ kJ/mol})\\ &\quad -(-1273.3\text{ kJ/mol}) + (0\text{ kJ/mol})\\ &= -2803\text{ kJ/mol}\end{aligned}$$

We included the enthalpy of formation of oxygen for completeness, although it does not contribute to the enthalpy of this reaction.

ASSESS $\Delta H_{rxn}[C_6H_{12}O_6(s)] = -2803$ kJ/mol is a value you can find in chemical tables and is the value we used in Example 11.11 to calculate the efficiency of converting the energy of glucose to the energy of ATP. This is an exothermic reaction.

If you have taken chemistry, you might have learned a different approach to doing a calculation like this. Our approach focuses on thinking through the problem stepwise, as physicists like to do.

STOP TO THINK 12.8 If a gas is heated at constant pressure, is the increase in enthalpy larger than, smaller than, or equal to the increase in thermal energy?

CHAPTER 12 INTEGRATED EXAMPLE Breathing in cold air

On a cold day, breathing costs your body energy that you use to warm the air coming into your lungs. You expend additional energy evaporating water that you exhale as water vapor. Your lungs hold several liters of air, but only a small portion of the air is exchanged with each breath. A person at rest typically takes 12 breaths per minute, with each breath drawing in 0.50 L of outside air that is warmed to body temperature (37°C). The air expands as it's warmed, and it also becomes saturated with water vapor; that is, enough water is evaporated from the surface of the lungs for the partial pressure of the water vapor in the lungs to equal the vapor pressure of water at 37°C, which is 0.062 atm.

Suppose the outside air temperature is a chilly −10°C. What volume of air is exhaled with each breath? What percentage of the body's 100 W basal metabolic rate goes to warming the air and evaporating the water? There is an exchange of oxygen for carbon dioxide, but the number of moles of gas in your lungs is very nearly constant. The heat of vaporization of water at 37°C is 24.1×10^5 J/kg.

PREPARE Your lungs draw in air by expanding their volume, not by having a lower pressure. There's very little pressure difference between your lungs and the air, so we can consider the air intake and warming to be a constant-pressure process at $p = 1 \text{ atm} = 1.01 \times 10^5$ Pa. The molar specific heat at constant pressure of air is that of nitrogen and oxygen, both 29 J/mol · K. We'll need to calculate the mass of water vapor in the lungs in order to determine how much heat is needed to vaporize it. Air is very dry at −10°C, so we'll assume that the air has no water vapor as it is breathed in. The ideal-gas law applies to any gas, so the changing composition is not important as long as the number of moles of gas doesn't change.

SOLVE The air is warmed from $T_i = -10°\text{C} = 263$ K to $T_f = 37°\text{C} = 310$ K, so $\Delta T = 47$ K. We can use the ideal-gas law to find that the 0.50 L of air breathed in at −10°C expands to

$$V_f = V_i \frac{p_i T_f}{p_f T_i} = (0.50 \text{ L}) \times 1 \times \frac{310 \text{ K}}{263 \text{ K}} = 0.59 \text{ L}$$

The expansion is nearly 20%—not insignificant. Here we did not need to change the units of volume because we were working with ratios, but we do need to use SI units to determine the number of moles of air that need to be heated. $1 \text{ m}^3 = 1000$ L, so $V_i = 5.0 \times 10^{-4}$ m^3. With this, we can compute the number of moles of cold air drawn in:

$$n_{air} = \frac{p_i V_i}{RT_i} = \frac{(1.01 \times 10^5 \text{ Pa})(5.0 \times 10^{-4} \text{ m}^3)}{(8.31 \text{ J/mol} \cdot \text{K})(263 \text{ K})} = 0.023 \text{ mol}$$

The heat required to warm this one breath of air at constant pressure is

$$Q_P = n_{air} C_P \Delta T = (0.023 \text{ mol})(29 \text{ J/mol} \cdot \text{K})(47 \text{ K}) = 31 \text{ J}$$

You might think that we should also include pressure-volume work in the energy balance, but in this case the work is essentially zero. Pushing in a piston to reduce the volume of a sealed container of gas requires doing work to compress the gas. However, the lungs are not sealed, so expanding and contracting the lungs simply pull air in and push air out with almost no expenditure of energy.

Now let's turn to the heat needed to evaporate water. The pressure of water vapor in the exhaled air is $0.062 \text{ atm} = 6260$ Pa. The lung will have expanded to 0.59 L at 37°C by the time all the water has vaporized, so the number of moles of water vapor is

$$n_{water} = \frac{p_{water} V_f}{RT_f} = \frac{(6260 \text{ Pa})(5.9 \times 10^{-4} \text{ m}^3)}{(8.31 \text{ J/mol} \cdot \text{K})(310 \text{ K})} = 1.4 \times 10^{-3} \text{ mol}$$

The molar mass of water is 0.018 kg/mol, so the mass of evaporated water is

$$M_{water} = nM_{mol} = (1.4 \times 10^{-3} \text{ mol})(0.018 \text{ kg/mol}) = 2.5 \times 10^{-5} \text{ kg}$$

or, equivalently, 25 mg. The heat required to vaporize the water is

$$Q_{vap} = ML_v = (2.5 \times 10^{-5} \text{ kg})(24.1 \times 10^5 \text{ J/kg}) = 60 \text{ J}$$

The total heat-energy demands per breath are $Q_P + Q_{vap} = 91$ J. At 12 breaths per second, each breath takes 5.0 s. Thus the heating power is

$$P = \frac{Q}{\Delta t} = \frac{91 \text{ J}}{5.0 \text{ s}} = 18 \text{ W}$$

This is 18% of the body's basal metabolic rate of approximately 100 W.

ASSESS About one-third of the energy demand on your body goes to warming the air, two-thirds to evaporating the water vapor that you exhale. The extra effort for warming is not large, but it is part of the additional energy you must "burn" on a cold day to stay warm. It seems reasonable that approximately 12 W of your basal metabolic rate of 100 W would be used to evaporate water because this is an important cooling mechanism on more weather-friendly days when dissipating heat is more important than staying warm.

SUMMARY

GOAL To learn about heat, temperature, and the first law of thermodynamics.

GENERAL PRINCIPLES

First Law of Thermodynamics

The first law is a general statement of energy conservation.

$$\Delta E_{int} = \Delta E_{th} + \Delta E_{chem} = W + Q$$

Work W and heat Q depend on the process by which the system is changed.

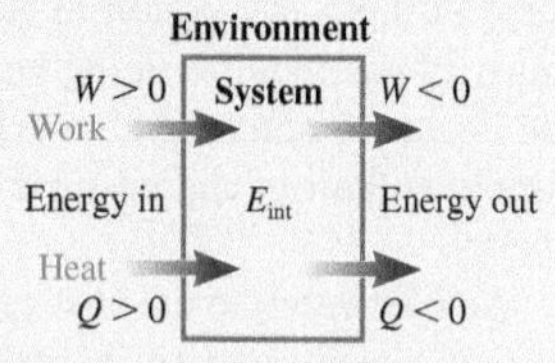

Energy

Internal energy E_{int} is the microscopic energy of moving molecules (thermal energy) and the chemical energy of molecular bonds. Internal energy is a state variable.

Work W is energy transferred in a mechanical interaction. Work is not a state variable.

Heat Q is energy transferred in a thermal interaction when there is a temperature difference. Heat is not a state variable.

IMPORTANT CONCEPTS

The **specific heat** c is the heat needed to raise the temperature of 1 kg of a substance by 1°C (1 K).

$$Q = Mc\Delta T$$

The **heat of transformation** L is the heat needed to cause 1 kg of a substance to undergo a phase change.

$$Q = \pm ML$$

The **molar specific heat** C is the heat needed to raise the temperature of 1 mol of a substance by 1°C (1 K).

$$Q = nC\Delta T$$

The molar specific heat of gases depends on the *process* by which the temperature is changed:

C_V = **molar specific heat at constant volume**

$C_P = C_V + R$ = **molar specific heat at constant pressure**

Heat is transferred by **conduction, convection, radiation,** and **evaporation.**

Conduction: $dQ/dt = kA(\Delta T/L)$

Radiation: $dQ/dt = e\sigma AT^4$

Phases of Matter

Solid	Liquid	Gas
Rigid	Loosely bound molecules	Free molecules
Incompressible	Incompressible	Compressible

Ideal-Gas Law

The **state variables** of an ideal gas are related by the ideal-gas law

$$pV = nRT \quad \text{or} \quad pV = Nk_BT$$

where $R = 8.31$ J/mol · K is the universal gas constant and $k_B = 1.38 \times 10^{-23}$ J/K is the Boltzmann constant. p, V, and T must be in SI units of Pa, m^3, and K.

For a gas in a sealed container, with constant n:

$$\frac{p_2V_2}{T_2} = \frac{p_1V_1}{T_1}$$

Work Done on an Ideal Gas

The work done on a gas, called **pressure-volume work,** is

$$W = -\int_{V_i}^{V_f} p\,dV = \begin{cases} -p\Delta V & \text{constant pressure} \\ -nRT\ln(V_f/V_i) & \text{constant temperature} \end{cases}$$

APPLICATIONS

Thermal expansion

For a temperature change ΔT,

$$\Delta L/L = \alpha\Delta T \qquad \Delta V/V = \beta\Delta T$$

Water has special properties, with a minimum volume at 4°C and with ice being less dense than liquid water.

Phase changes

The boundaries separating the regions of a **phase diagram** are lines of phase equilibrium. Any amounts of the two phases can coexist. A phase change absorbs or liberates heat, also called the enthalpy of fusion or enthalpy of vaporization.

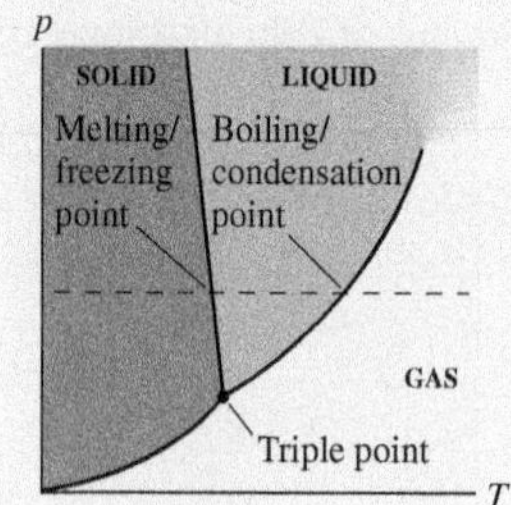

Solving calorimetry problems

When two or more systems interact thermally, they come to a common final temperature determined by

$$Q_{net} = Q_1 + Q_2 + \cdots = 0$$

Enthalpy

Enthalpy H is the energy needed to create a system and make space for it in an atmosphere at pressure p. For a constant-pressure process, such as a phase change or a chemical reaction,

$$\Delta H = \Delta E_{int} + p\Delta V = Q_P$$

LEARNING OBJECTIVES After studying this chapter, you should be able to:

- Distinguish between work, heat, and thermal energy. *Conceptual Questions 12.1, 12.6; Problem 12.4*
- Apply the first law of thermodynamics to situations where a system is heated. *Conceptual Question 12.13; Problems 12.5, 12.15*
- Calculate the thermal expansions of solids and liquids. *Conceptual Question 12.4; Problems 12.6–12.9*
- Determine the heat required to change an object's temperature or cause a phase change. *Conceptual Question 12.7; Problems 12.14–12.16, 12.21, 12.22*
- Solve calorimetry problems for solids, liquids, and gases. *Conceptual Question 12.5; Problems 12.24–12.28*
- Calculate the rate of heat transfer via conduction and radiation. *Problems 12.29–12.32, 12.35*
- Solve problems about ideal gases. *Conceptual Questions 12.11, 12.12; Problems 12.36–12.40*
- Calculate work and heat transfer for ideal-gas processes. *Conceptual Questions 12.14, 12.15; Problems 12.49–12.53*
- Interpret and use the enthalpy change of phase changes and chemical reactions. *Problems 12.56–12.60*

STOP TO THINK ANSWERS

Stop to Think 12.1: A. The step size on the Kelvin scale is the same as the step size on the Celsius scale. A *change* of 10°C is a *change* of 10 K.

Stop to Think 12.2: B, C, and E. The temperature rise in A and D is from doing work, not from heat. E involves heat, but there's no temperature rise because heat is used to do work.

Stop to Think 12.3: Increase. When an object undergoes thermal expansion, all dimensions increase by the same percentage.

Stop to Think 12.4: A. On the water phase diagram, you can see that for a pressure just slightly below the triple-point pressure, the solid/gas transition occurs at a higher temperature than does the solid/liquid transition at high pressures. This is not true for carbon dioxide.

Stop to Think 12.5: C. Conduction, convection, and evaporation require matter. Only radiation transfers energy through the vacuum of space.

Stop to Think 12.6: A. The number of atoms depends only on the number of moles, not the substance.

Stop to Think 12.7: C. $T = pV/nR$. Pressure and volume are the same, but n differs. The number of moles in mass M is $n = M/M_{mol}$. Helium, with the smaller molar mass, has a larger number of moles and thus a lower temperature.

Stop to Think 12.8: Larger than. For a gas, with $E_{int} = E_{th}$, the change in enthalpy in a constant-pressure process is $\Delta H = \Delta E_{th} + \Delta(pV) = \Delta E_{th} + p\,\Delta V$. $\Delta V > 0$ when the gas is heated and expands, so $\Delta H > \Delta E_{th}$.

QUESTIONS

Conceptual Questions

1. Do (a) temperature, (b) heat, and (c) thermal energy describe a property of a system, an interaction of the system with its environment, or both? Explain.
2. Rank in order, from highest to lowest, the temperatures $T_1 = 0$ K, $T_2 = 0$°C, and $T_3 = 0$°F.
3. The sample in an experiment is initially at 10°C. If the sample's temperature is doubled, what is the new temperature in °C?
4. You need to measure the height of a small tree. Your metal measuring tape has been sitting in the sun on a hot summer day while your coworker's measuring tape has been in an air-conditioned car. Will your measurement of the tree's height be greater than, less than, or the same as your coworker's measurement? Explain.
5. Materials A and B have equal densities, but A has a larger specific heat than B. You have 100 g cubes of each material. Cube A, initially at 0°C, is placed in good thermal contact with cube B, initially at 200°C. The cubes are inside a well-insulated container where they don't interact with their surroundings. Is their final temperature greater than, less than, or equal to 100°C? Explain.
6. Two blocks of copper, one with a mass of 1 kg and the other with a mass of 3 kg, are both at a temperature of 50°C. Which block, if either, has more thermal energy? If the blocks are placed in thermal contact, will the temperature of the 1 kg block increase, decrease, or stay the same?
7. If you are exposed to water vapor at 100°C, you are likely to experience a worse burn than if you are exposed to liquid water at 100°C. Why is water vapor more damaging than liquid water at the same temperature?
8. An aquanaut lives in an underwater apartment 100 m beneath the surface of the ocean. Compare the freezing and boiling points of water in the aquanaut's apartment to their values at the surface. Are they higher, lower, or the same? Explain.
9. a. A sample of water vapor in an enclosed cylinder has an initial pressure of 500 Pa at an initial temperature of −0.01°C. A piston squeezes the sample smaller and smaller, without limit. Describe what happens to the water as the squeezing progresses.
   b. Repeat part a if the initial temperature is 0.03°C warmer.

Problem difficulty is labeled as | (straightforward) to ||||| (challenging). Problems labeled INT integrate significant material from earlier chapters; BIO are of biological or medical interest; CALC require calculus to solve.

10. The cylinder in Figure Q12.10 is divided into two compartments by a frictionless piston that can slide back and forth. If the piston is in equilibrium, is the pressure on the left side greater than, less than, or equal to the pressure on the right? Explain.

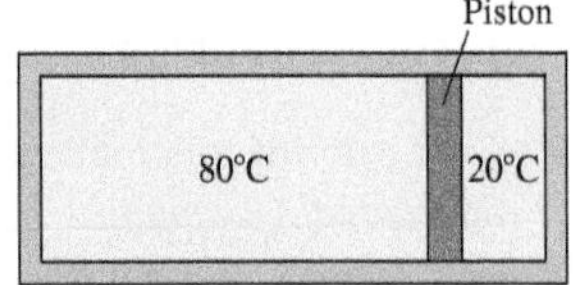

FIGURE Q12.10

11. A gas is in a sealed container. By what factor does the gas temperature change if:
    a. The volume and pressure are both doubled?
    b. The volume is halved and the pressure is tripled?
12. A gas is in a sealed container. By what factor does the gas pressure change if:
    a. The volume is doubled and the temperature is halved?
    b. The volume is halved and the temperature is tripled?
13. Two containers hold equal masses of nitrogen gas at equal temperatures. You supply 10 J of heat to container A while not allowing its volume to change, and you supply 10 J of heat to container B while not allowing its pressure to change. Afterward, is temperature T_A greater than, less than, or equal to T_B? Explain.
14. You need to raise the temperature of a gas by 10°C. To use the least amount of heat, should you heat the gas at constant pressure or at constant volume? Explain.
15. The gas cylinder in Figure Q12.15 is well insulated except for the bottom surface, which is in contact with a block of ice. The piston can slide without friction. The initial gas temperature is >0°C. During the process that occurs until the gas reaches equilibrium, are (i) ΔT, (ii) W, and (iii) Q greater than, less than, or equal to zero? Explain.

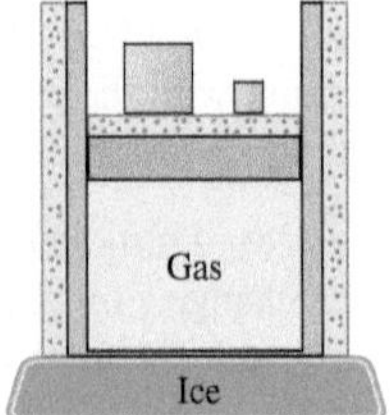

FIGURE Q12.15

Multiple-Choice Questions

16. | A 25-cm-long glass stirring rod at a laboratory temperature of 20°C is placed in an autoclave and heated to 150°C. By about how much does the rod expand?
    A. 0.3 mm B. 0.9 mm C. 3 mm D. 9 mm
17. | A cup of water is heated with a heating coil that delivers 100 W of heat. In one minute, the temperature of the water rises by 20°C. What is the mass of the water?
    A. 72 g B. 140 g
    C. 720 g D. 1.4 kg
18. | Three identical beakers each hold 1000 g of water at 20°C. 100 g of liquid water at 0°C is added to the first beaker, 100 g of ice at 0°C is added to the second beaker, and the third beaker gets 100 g of aluminum at 0°C. The contents of which container end up at the lowest final temperature?
    A. The first beaker.
    B. The second beaker.
    C. The third beaker.
    D. All end up at the same temperature.
19. || Steam at 100°C causes worse burns than liquid water at 100°C. This is because
    BIO
    A. The steam is hotter than the water.
    B. Heat is transferred to the skin as steam condenses.
    C. Steam has a higher specific heat than water.
    D. Evaporation of liquid water on the skin causes cooling.
20. || How much heat is needed to change 25 g of acetic acid at 20°C into acetic acid vapor at the boiling point?
    A. 4 kJ B. 8 kJ C. 12 kJ D. 15 kJ
21. | The number of atoms in a rigid container is increased by a factor of 2 while the temperature is held constant. The pressure
    A. Decreases by a factor of 4.
    B. Decreases by a factor of 2.
    C. Stays the same.
    D. Increases by a factor of 2.
    E. Increases by a factor of 4.
22. ||| A gas is compressed by an isothermal process that decreases its volume by a factor of 2. In this process, the pressure
    A. Does not change.
    B. Increases by a factor of less than 2.
    C. Increases by a factor of 2.
    D. Increases by a factor of more than 2.
23. | The thermal energy of a rigid container of helium gas is halved. What happens to the temperature, in kelvins?
    A. It decreases to one-fourth its initial value.
    B. It decreases to one-half its initial value.
    C. It stays the same.
    D. It increases to twice its initial value.
24. || 200 J of heat is added to two gases, each in a sealed container. Gas 1 is in a rigid container that does not change volume. Gas 2 expands as it is heated, pushing out a piston that lifts a small weight. Which gas has the greater increase in its thermal energy?
    A. Gas 1
    B. Gas 2
    C. Both gases have the same increase.
25. | A rigid container holds 4.0 g of nitrogen gas at 20°C. How much heat is needed to increase the gas temperature to 200°C?
    A. 15 J B. 530 J C. 750 J D. 1100 J

PROBLEMS

Section 12.1 Heat and the First Law of Thermodynamics

1. | The lowest and highest natural temperatures ever recorded on earth are −129°F in Antarctica and 134°F in Death Valley. What are these temperatures in °C and in K?
2. || At what temperature does the numerical value in °F match the numerical value in °C?
3. || A scientist creates a new temperature scale, the "Z scale." He decides to call the boiling point of nitrogen 0°Z and the melting point of iron 1000°Z.
    a. What is the boiling point of water on the Z scale?
    b. Convert 500°Z to degrees Celsius and to kelvins.

4. | 500 J of work are done on a system in a process that decreases the system's internal energy by 200 J. How much heat is transferred to or from the system?
5. | 10 J of heat are removed from a gas sample while it is being compressed by a piston that does 20 J of work. What is the change in the internal energy of the gas? Does the temperature of the gas increase or decrease?

Section 12.2 Thermal Expansion

6. | A 35-cm-long icicle hangs from the eave of a house on a day when the temperature is −2°C. By how many millimeters does the icicle shrink if a bitterly cold wind drops the temperature to −30°C?
7. | A surveyor has a stainless steel measuring tape that is calibrated to be 100.000 m long (i.e., accurate to ±1 mm) at 20°C. If she measures the distance between two stakes to be 65.175 m on a 3°C day, does she need to add or subtract a correction factor to get the true distance? How large, in mm, is the correction factor?
8. | Two students each build a piece of scientific equipment that uses a 655-mm-long rod. One student uses a glass rod, the other a copper rod. If the temperature increases by 5.0°C, how much more does the glass rod expand than the copper rod?
9. ||| The density of mercury is 13,600 kg/m^3 at 0°C. What is the density of mercury at 200°C?

Section 12.3 Specific Heat and Heat of Transformation

10. || How much heat must be added to a 6.0-cm-diameter copper sphere to raise its temperature from −50°C to 150°C? The density of copper is 8920 kg/m^3.
11. | a. 100 J of heat are transferred to 20 g of mercury. By how much does the temperature increase?
    b. How much heat is needed to raise the temperature of 20 g of water by the same amount?
12. || How much heat is needed to change 20 g of water at 20°C into water vapor at the boiling point?
13. || What is the maximum mass of ethyl alcohol you could boil with 1000 J of heat, starting from 20°C?
14. | A scientist whose scale is broken but who has a working 2.5 kW heating coil and a thermometer decides to improvise to determine the mass of a block of aluminum she has recently acquired. She heats the aluminum for 30 s and finds that its temperature increases from 20°C to 35°C. What is the mass of the aluminum?
15. || Two cars collide head-on while each is traveling at 80 km/h. Suppose all their kinetic energy is transformed into the thermal energy of the wrecks. What is the temperature increase of each car? You can assume that each car's specific heat is that of aluminum.
16. || An experiment measures the temperature of a 500 g substance while steadily supplying heat to it. Figure P12.16 shows the results of the experiment. What are the (a) specific heat of the solid phase, (b) specific heat of the liquid phase, (c) melting and boiling temperatures, and (d) heats of fusion and vaporization?

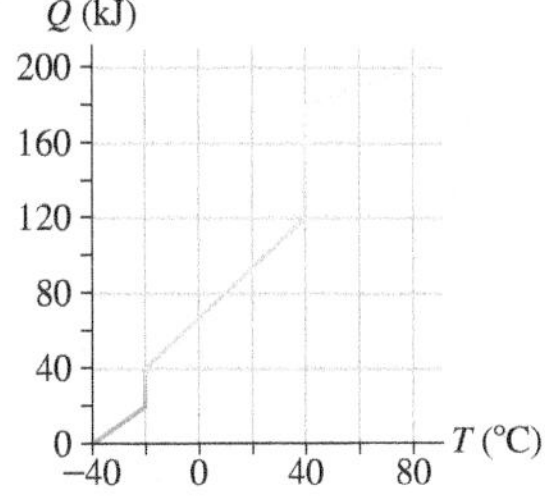

FIGURE P12.16

17. || A 5.0-m-diameter garden pond is 30 cm deep. Solar energy is incident on the pond at an average rate of 400 W/m^2. If the water absorbs all the solar energy and does not exchange energy with its surroundings, how many hours will it take to warm from 15°C to 25°C?
18. || A 150 L (≈40 gal) electric hot-water tank has a 5.0 kW heater. How many minutes will it take to raise the water temperature from 65°F to 140°F?
19. | Suppose that the 800 W of radiation in a microwave oven is absorbed by 250 g of water in a very lightweight cup. How long will it take for the water to warm up from 20°C to 80°C?
20. || Dogs keep themselves cool by panting, rapidly breathing air BIO in and out. Panting results in evaporation from moist tissues of the airway and lungs, which cools the animal. Measurements show that, on a 35°C day with a relative humidity of 50%, a 12 kg dog loses 1.0 g of water per minute if it is panting vigorously. What rate of heat loss, in watts, does this achieve?
21. || A pronghorn can run at a remarkable 18 m/s for up to 10 BIO minutes, almost triple the speed that an elite human runner can maintain. For a 32 kg pronghorn, this requires an astonishing 3.4 kW of metabolic power, which leads to a significant increase in body temperature. If the pronghorn had no way to exhaust heat to the environment, by how much would its body temperature increase during this run? (In fact, it will lose some heat, so the rise won't be this dramatic, but it will be quite noticeable, requiring adaptations that keep the pronghorn's brain cooler than its body in such circumstances.)
22. |||| Over the course of a day, 4.0 kg of water evaporates from the BIO leaves of a corn plant.
    a. How much energy is required to evaporate the water? (Assume that the temperature of the leaves is 30°C.)
    b. If the plant is active for 12 hours, how much power does this correspond to? You can think of this as the necessary power to drive transport in the plant.

Section 12.4 Calorimetry

23. || 30 g of copper pellets are removed from a 300°C oven and immediately dropped into 100 mL of water at 20°C in an insulated cup. What will the new water temperature be?
24. | A 750 g aluminum pan is removed from the stove and plunged into a sink filled with 10.0 L of water at 20.0°C. The water temperature quickly rises to 24.0°C. What was the initial temperature of the pan in °C?
25. || A 50.0 g thermometer is used to measure the temperature of 200 mL of water. The specific heat of the thermometer, which is mostly glass, is 750 J/kg·K, and it reads 20.0°C while lying on the table. After being completely immersed in the water, the thermometer's reading stabilizes at 71.2°C. What was the actual water temperature before it was measured?
26. ||| A 65 cm^3 block of copper is removed from an 800°C furnace and immediately dropped into 200 mL of 20°C water. What fraction of the water boils away?
27. || It's possible to boil water by adding hot rocks to it, a technique that has been used in many societies over time. If you heat a rock in the fire, you can easily get it to a temperature of 500°C. If you use granite or other similar stones, the specific heat is about 800 J/kg·K. If 5.0 kg of water at 10°C is in a leak-proof vessel, what minimum number of 1.0 kg stones must be added to bring the water to a boil?

28. || If a person has a dangerously high fever, submerging her in ice water is a bad idea, but an ice pack can help to quickly bring her body temperature down. How many grams of ice at 0°C will be melted in bringing down a 60 kg patient's fever from 40°C to 39°C?
BIO

Section 12.5 Heat Transfer

29. | You are boiling pasta and absentmindedly grab a copper stirring spoon rather than your wooden spoon. The copper spoon has a 20 mm × 1.5 mm rectangular cross section, and the distance from the boiling water to your 35°C hand is 18 cm. How long does it take the spoon to transfer 25 J of energy to your hand?

30. | The ends of a 19-cm-long, 2.0-cm-diameter rod are maintained at 0°C and 100°C by immersion in an ice-water bath and boiling water. Heat is conducted through the rod at 140 kJ per hour. Of what material is the rod made?

31. | Windows are a major source of heat loss from a house because glass has a higher thermal conductivity than the other materials typically used in construction and because insulation can be placed inside walls but not windows. A typical house has 30 m^2 of windows, and window glass is typically 4.0 mm thick.
    a. What is the rate of heat loss through windows on a chilly −5°C day from a typical house with single-pane windows if the interior temperature of the house is 20°C (68°F)?
    b. Double-pane windows are two panes of glass separated by a 3.0-mm-wide gap filled with dry air. The gap has to be narrow to prevent convection currents in the air. Air is such a better insulator than glass that, to a good approximation, the thermal conductivity of a double-pane window is simply that of the layer of air. What is the rate of heat loss on the −5°C day if the house has double-pane windows?

32. ||| Radiation from the head is a major source of heat loss from the human body. Model a head as a 20-cm-diameter, 20-cm-tall cylinder with a flat top. If the body's surface temperature is 35°C, what is the net rate of heat loss on a chilly 5°C day? All human skin is effectively black in the infrared where the radiation occurs, so use an emissivity of 0.97.
BIO

33. ||| Electronics and inhabitants of the International Space Station generate a significant amount of thermal energy that the station must get rid of. The only way that the station can exhaust thermal energy is by radiation, which it does using thin, 1.8-m-by-3.6-m panels that have a working temperature of 6.0°C and an emissivity of 0.85. How much power is radiated from each panel? Assume that the panels are in the shade so that the absorbed radiation will be negligible. The temperature of deep space, which the panels face, is close to 0 K.
    Hint: Don't forget that the panels have two sides!

34. || The glowing filament in a lamp is radiating energy at a rate of 60 W. At the filament's temperature of 1500°C, the emissivity is 0.23. What is the surface area of the filament?

35. | A 30 kg male emperor penguin under a clear sky in the Antarctic winter loses very little heat to the environment by convection; its feathers provide very good insulation. It does lose some heat through its feet to the ice, and some heat due to evaporation as it breathes; the combined power is about 12 W. The outside of the penguin's body is a chilly −22°C, but its surroundings are an even chillier −38°C. The penguin's surface area is 0.56 m^2, and its emissivity is 0.97.
BIO
    a. What is the net rate of energy loss by radiation?
    b. If the penguin has a 45 W basal metabolic rate, will it feel warm or cold under these circumstances?

Section 12.6 The Ideal Gas: A Model System

36. | 3.0 mol of gas at a temperature of −120°C fills a 2.0 L container. What is the gas pressure?

37. || A rigid container holds 2.0 mol of gas at a pressure of 1.0 atm and a temperature of 30°C.
    a. What is the container's volume?
    b. What is the pressure if the temperature is raised to 130°C?

38. | Commercial planes routinely fly at altitudes of 10 km, where the atmospheric pressure is less than 0.3 atm, but the pressure inside the cabin is maintained at 0.75 atm. Suppose you have an inflatable travel pillow that, once you reach cruising altitude, you inflate to a volume of 1.5 L and use to take a nap. You manage to sleep through the rest of the flight and awaken when the plane is about to land. What is the volume of the pillow after landing? Ignore any effect of the elasticity of the pillow's material.

39. || A gas at 100°C fills volume V_0. If the pressure is held constant, what is the volume if (a) the Celsius temperature is doubled and (b) the Kelvin temperature is doubled?

40. || The total lung capacity of a typical adult is 5.0 L. Approximately 20% of the air is oxygen. At sea level and at a body temperature of 37°C, how many oxygen molecules do the lungs contain at the end of a strong inhalation?
BIO

41. ||| A 20-cm-diameter cylinder that is 40 cm long contains 50 g of oxygen gas at 20°C.
    a. How many moles of oxygen are in the cylinder?
    b. How many oxygen molecules are in the cylinder?
    c. What is the number density of the oxygen?
    d. What is the reading of a pressure gauge attached to the tank?

42. || The solar corona is a very hot atmosphere surrounding the visible surface of the sun. X-ray emissions from the corona show that its temperature is about 2×10^6 K. The gas pressure in the corona is about 0.03 Pa. Estimate the number density of particles in the solar corona.

43. || A rigid, hollow sphere is submerged in boiling water in a room where the air pressure is 1.0 atm. The sphere has an open valve with its inlet just above the water level. After a long period of time has elapsed, the valve is closed. What will be the pressure inside the sphere if it is then placed in (a) a mixture of ice and water and (b) an insulated box filled with dry ice?

44. | A rigid container holds hydrogen gas at a pressure of 3.0 atm and a temperature of 20°C. What will the pressure be if the temperature is lowered to −20°C?

45. || A 24-cm-diameter vertical cylinder is sealed at the top by a frictionless 20 kg piston. The piston is 84 cm above the bottom when the gas temperature is 303°C. The air above the piston is at 1.00 atm pressure.
    a. What is the gas pressure inside the cylinder?
    b. What will the height of the piston be if the temperature is lowered to 15°C?

46. || 0.10 mol of argon gas is admitted to an evacuated 50 cm^3 container at 20°C. The gas then undergoes a constant-pressure heating to a temperature of 300°C. What is the final volume of the gas?

47. ||| A football is inflated in the locker room before the game. The air warms as it is pumped, so it enters the ball at a temperature of 27°C. The ball is inflated to a gauge pressure of 13 psi. The ball is used for play at 10°C. Once the ball cools, what is the pressure in the ball? Assume that atmospheric pressure is 14.7 psi.

48. || A weather balloon rises through the atmosphere, its volume expanding from 4.0 m^3 to 12 m^3 as the temperature drops from 20°C to −10°C. If the initial gas pressure inside the balloon is 1.0 atm, what is the final pressure?

Section 12.7 Thermodynamics of Ideal Gases

49. || A 2000 cm^3 container holds 0.10 mol of helium gas at 300°C. How much work must be done to compress the gas to 1000 cm^3 at (a) constant pressure and (b) constant temperature?
50. || 500 J of work must be done to compress a gas to half its initial volume at constant temperature. How much work must be done to compress the gas by a factor of 10, starting from its initial volume?
51. | A gas is compressed from 600 cm^3 to 200 cm^3 at a constant pressure of 400 kPa. At the same time, 100 J of heat is transferred out of the gas. What is the change in thermal energy of the gas during this process?
52. || A container holds 4.0 g of neon at a pressure of 5.0 atm and a temperature of 15°C. How much work must be done on the gas to reduce the volume by 50% in (a) a constant-pressure process and (b) a constant-temperature process?
53. | A container holds 1.0 g of argon at a pressure of 8.0 atm.
    a. How much heat is required to increase the temperature by 100°C at constant volume?
    b. How much will the temperature increase if this amount of heat is transferred to the gas at constant pressure?
54. || A container holds 1.0 g of oxygen at a pressure of 8.0 atm.
    a. How much heat is required to increase the temperature by 100°C at constant pressure?
    b. How much will the temperature increase if this amount of heat is transferred to the gas at constant volume?
55. ||| A cube 20 cm on each side contains 3.0 g of helium at 20°C. 1000 J of heat are transferred to this gas. What are (a) the final pressure if the process is at constant volume and (b) the final volume if the process is at constant pressure?

Section 12.8 Enthalpy

56. | Carbon tetrachloride (CCl_4) boils at 77°C. The measured enthalpy of vaporization is 29.8 kJ/mol. What is the heat of vaporization in SI units?
57. || The combustion of methane is the reaction $CH_4(g) + 2O_2(g) \rightarrow CO_2(g) + 2H_2O(l)$, where (g) indicates that all components are in their gas phase. The standard enthalpies of formation for methane, carbon dioxide, and water vapor are −74.6 kJ/mol, −393.5 kJ/mol, and −285.8 kJ/mol, respectively. $O_2(g)$ is the most stable form of oxygen at 25°C. What is the heat of combustion of methane?
58. ||| Dissolving sodium hydroxide (NaOH) in water is an exothermic process that increases the temperature. The reaction is $H_2O(l) + NaOH(s) \rightarrow H_2O(l) + NaOH(aq)$, where (aq) indicates an aqueous solution. In one experiment, dissolving 4.0 g of NaOH in 100 mL of 27.0°C water increased the temperature to 36.9°C. The standard enthalpies of formation for sodium hydroxide and liquid water are −425.9 kJ/mol and −285.8 kJ/mol, respectively. What is the standard enthalpy of formation of aqueous sodium hydroxide? The solution is dilute, so you can assume that the thermal properties of the solution are the same as those of pure water.
59. ||| The enthalpy of 36 g of water vapor increases by 1830 J when its temperature increases from 150°C to 175°C. Assume that water vapor is an ideal gas.
    a. What is the molar specific heat at constant pressure in SI units for water vapor?
    b. By how much did the water vapor's thermal energy increase?
60. |||| Nitroglycerin ($C_3H_5N_3O_9$) undergoes the spontaneous explosive reaction

$$4C_3H_5N_3O_9(l) \rightarrow 12CO_2(g) + 10H_2O(g) + 6N_2(g) + O_2(g)$$

In one measurement, exploding 10.0 g of nitroglycerin in a calorimeter surrounded by 2.00 L of water increased the water temperature from 15.0°C to 22.4°C. The standard enthalpies of formation for carbon dioxide and water vapor are −393.5 kJ/mol and −241.8 kJ/mol, respectively.
    a. What is the heat of combustion of nitroglycerin?
    b. What is the standard enthalpy of formation of nitroglycerin?

General Problems

61. || An aluminum ring with inner diameter 2.00 cm and outer diameter 3.00 cm needs to fit over a 2.00-cm-diameter stainless steel rod, but at 20°C the hole through the aluminum ring is 50 μm too small in diameter. To what temperature, in °C, must the rod and ring be heated so that the ring just barely slips over the rod?
62. |||| A 15°C, 2.0-cm-diameter aluminum bar just barely slips between two rigid steel walls 10.0 cm apart. If the bar is warmed to 25°C, how much force does it exert on each wall? Young's modulus for aluminum is 6.9×10^{10} N/m^2.
63. || Older railroad tracks in the U.S. are made of 12-m-long pieces of steel. When the tracks are laid, gaps are left between the sections to prevent buckling when the steel thermally expands. If a track is laid at 16°C, how large should the gaps be if the track is not to buckle when the temperature is as high as 50°C? The coefficient of linear expansion of soft steel is 1.1×10^{-5} °C^{-1}.
64. ||| At 15°C, a 3.00-cm-diameter glass cylinder is filled with glycerin to a height of 15.0 cm. If the temperature increases, the glycerin expands in volume *and* the diameter of the cylinder undergoes a linear expansion. What will be the height of glycerin at a temperature of 65°C?
65. || A 5.0 g ice cube at −20°C is in a rigid, sealed container from which all the air has been evacuated. How much heat is required to change this ice cube into steam at 200°C? Steam has $c_V = 1500$ J/kg · K and $c_P = 1960$ J/kg · K.
66. || When air is inhaled, it quickly becomes saturated with water
BIO vapor as it passes through the moist airways. Consequently, an adult human exhales about 25 mg of evaporated water with each breath. Evaporation—a phase change—requires heat, and the heat is removed from your body. Evaporation is much like boiling, only water's heat of vaporization at 35°C is a somewhat larger 24×10^5 J/kg because at lower temperatures more energy is required to break the molecular bonds. At 12 breaths/min, on a dry day when the inhaled air has almost no water content, what is the body's rate of energy loss (in J/s) due to exhaled water? (For comparison, the energy loss from radiation, usually the largest loss on a cool day, is about 100 J/s.)

67. ||| 512 g of an unknown metal at a temperature of 15°C is dropped into a 100 g aluminum container holding 325 g of water at 98°C. A short time later, the container of water and metal stabilizes at a new temperature of 78°C. Identify the metal.

68. || A lava flow is threatening to engulf a small town. A 400-m-wide, 35-cm-thick tongue of 1200°C lava is advancing at the rate of 1.0 m per minute. The mayor devises a plan to stop the lava in its tracks by flying in large quantities of 20°C water and dousing it. The lava has density 2500 kg/m^3, specific heat 1100 J/kg · K, melting temperature 800°C, and heat of fusion 4.0×10^5 J/kg. How many liters of water per minute, at a minimum, will be needed to save the town?

69. || Your 300 mL cup of coffee is too hot to drink when served at 90°C. What is the mass of an ice cube, taken from a −20°C freezer, that will cool your coffee to a pleasant 60°C?

70. || Muskoxen have a number of adaptations that allow them to survive in very frigid winter conditions, chief among them a thick coat that has very low thermal conductivity. Their heat loss to the environment is primarily via radiation, convection, and conduction to the ground through their hooves. In one study, which took place at a temperature of −32°C, thermal cameras found that the surface of the coat had a temperature of −28°C, only slightly warmer than the environment. Other measurements found that the top of the 2.5-cm-thick keratin layer (thermal conductivity 0.63 W/m · K) that makes up the bottom of the hooves was only 9°C warmer than the −32°C ground. Researchers estimated that convective losses were two-thirds of radiation losses. When not foraging, muskoxen minimize heat loss by standing in tight groups. Most of their heat loss to the environment comes when they are foraging.
BIO

   a. Assuming a surface area of 3.7 m^2, an emissivity of 0.98, and a total hoof area in contact with the ground of 100 cm^2, determine the rates of heat loss due to radiation, convection, and conduction.
   b. In the study, an adult muskox spent 350 minutes foraging, during which she consumed 35,000 kJ of food. What fraction of the energy intake was lost as heat during this time?

71. ||| On a cold day, you lose heat more quickly to the environment when the wind is blowing. For exposed skin, an empirical formula for the rate of heat loss to the environment in W is
BIO

$$\frac{dQ}{dt} = A(12.2 + 11.6\sqrt{v_{\text{wind}}} - 1.16v_{\text{wind}}) \times (33°\text{C} - T)$$

where A is the person's surface area in m^2, v_{wind} is the wind speed in m/s, and T is the air temperature in °C. 33°C is the assumed temperature of exposed skin, so the term $(33°\text{C} - T)$ is the temperature difference between the skin and the air. Suppose a 70 kg person with a surface area of 1.7 m^2 is outside in light clothing that provides little insulation. If the air temperature is 5.0°C (≈40°F) and a stiff wind of 15 m/s is blowing, estimate how long it will take the person's core temperature to drop from the usual 37°C to 35°C, at which time hypothermia will set in. Assume a resting metabolic rate of 100 W. The situation isn't quite as dire as this calculation suggests because the body directs blood flow from the skin to the core, so the $(33°\text{C} - T)$ term would be replaced by a smaller number; even so, hypothermia can set in quickly during exposure to wind and cold.

72. || You know that some animals make it through a long winter by hibernating, which reduces their body temperature and energy expenditure. Some smaller animals, like the 8.5 g Australian fairy wren, do this on a daily basis, allowing their body temperature to drop at night to reduce heat lost to the environment. On a 3°C night, a wren lets its body temperature drop from 42°C to 30°C. By what factor does this reduction in temperature reduce the energy lost by radiation?
BIO

73. || Much of Australia is hot and dry. This poses a problem for animals that are looking to keep cool without losing too much water to the environment. A team of investigators found that, on hotter days, koalas spend more time resting in acacia trees, which tend to have cooler trunks and branches than the eucalyptus trees in which the koalas feed. The koalas straddle cool branches, which compresses their fur. This reduces the thickness of the fur layer and, by squeezing out air, increases thermal conductivity. Does this behavior provide appreciable cooling? Suppose an 11 kg koala, with a 36°C body temperature, rests with 13% of its 0.40 m^2 surface area in contact with a 31°C tree branch. Its compressed fur is 2.0 mm thick and has thermal conductivity 0.057 W/m · K. What fraction of the koala's 12 W resting metabolic power could be transferred to the tree?
BIO

74. ||| The specific heat of most solids is nearly constant over a wide temperature range. Not so for diamond. Between 200 K and 600 K, the specific heat of diamond is reasonably well described by $c = 2.8T - 350$ J/kg·K, where T is in K. For gemstone diamonds, 1 carat = 200 mg. How much heat is needed to raise the temperature of a 3.5 carat diamond from −50°C to 250°C?
CALC

75. ||| An ideal gas undergoes a process in which the pressure changes linearly with the volume as $p = (4.0 \times 10^8\ \text{Pa/m}^3)V$. How much work must be done on the gas to compress it from 300 cm^3 to 200 cm^3?
CALC

76. || A 68 kg woman cycles at a constant 15 km/h. All of the metabolic energy that does not go to forward propulsion is converted to thermal energy in her body. If the only way her body has to keep cool is by evaporation, how many kilograms of water must she lose to perspiration each hour to keep her body temperature constant?
BIO INT

77. || A 5000 kg African elephant has a resting metabolic rate of 2500 W. On a hot day, the elephant's environment is likely to be nearly the same temperature as the animal itself, so cooling by radiation is not effective. The only plausible way to keep cool is by evaporation, and elephants spray water on their body to accomplish this. If this is the only possible means of cooling, how many kilograms of water per hour must be evaporated from an elephant's skin to keep it at a constant temperature?
BIO INT

78. ||| The 3.0-m-long pipe in Figure P12.78 is closed at the top end. It is slowly pushed straight down into the water until the top end of the pipe is level with the water's surface. What is the length L of the trapped volume of air?

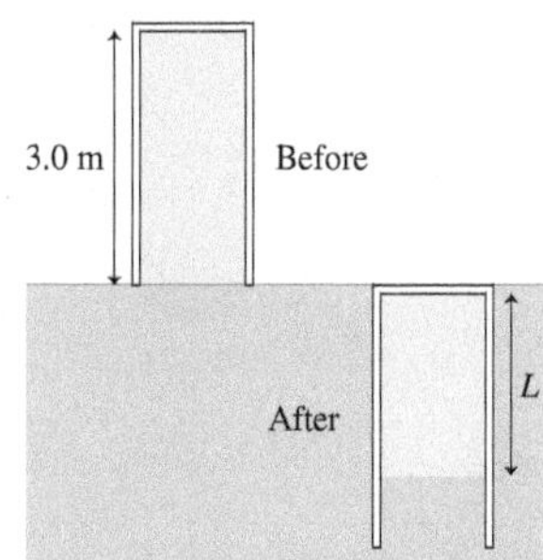

FIGURE P12.78

79. || On a cool morning, when the temperature is 15°C, you measure the pressure in your car tires to be 30 psi. After driving 20 mi on the freeway, the temperature of your tires is 45°C. What pressure will your tire gauge now show?

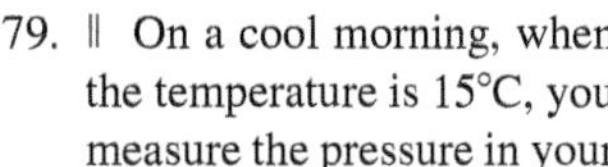

80. || 10,000 cm^3 of 200°C steam at a pressure of 20 atm is cooled until it condenses. What is the volume of the liquid water? Give your answer in cm^3.

81. || Suppose you take and hold a deep breath on a chilly day, inhaling 3.0 L of air at 0°C and 1 atm pressure. Assume that the air consists entirely of nitrogen and that the pressure of the gas stays constant as it warms to your core body temperature of 37°C.
BIO
   a. By how much does the volume of the air increase as it warms?
   b. How much heat is supplied by your body to warm the air?

82. || A 10-cm-diameter cylinder contains argon gas at 10 atm pressure and a temperature of 50°C. A piston can slide in and out of the cylinder. The cylinder's initial length is 20 cm. 2500 J of heat are transferred to the gas, causing the gas to expand at constant pressure. What are (a) the final temperature and (b) the final length of the cylinder?

83. ||| 5.0 g of nitrogen gas at 20°C and an initial pressure of 3.0 atm undergo a constant-pressure expansion until the volume has tripled.
   a. What are the gas volume and temperature after the expansion?
   b. How much heat is transferred to the gas to cause this expansion?

   The gas pressure is then decreased at constant volume until the original temperature is reached.
   c. What is the gas pressure after the decrease?
   d. How much heat is transferred from the gas as its pressure decreases?

MCAT-Style Passage Problems

Thermal Properties of the Oceans

Seasonal temperature changes in the ocean only affect the top layer of water, to a depth of 500 m or so. This "mixed" layer is thermally isolated from the cold, deep water below. The average temperature of this top layer of the world's oceans, which has area $3.6 \times 10^8\ km^2$, is approximately 17°C.

In addition to seasonal temperature changes, the oceans have experienced an overall warming trend over the last century that is expected to continue as the earth's climate changes. A warmer ocean means a larger volume of water; the oceans will rise. Suppose the average temperature of the top layer of the world's oceans were to increase from a temperature T_i to a temperature T_f. The area of the oceans will not change, as this is fixed by the size of the ocean basin, so any thermal expansion of the water will cause the water level to rise, as shown in Figure P12.84. The original volume is the product of the original depth and the surface area, $V_i = Ad_i$. The change in volume is given by $\Delta V = A\Delta d$.

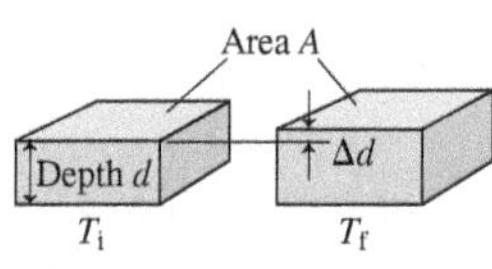

FIGURE P12.84

84. If the top 500 m of ocean water increased in temperature from 17°C to 18°C, what would be the resulting rise in ocean height? The coefficient of volume expansion of water is $1.7 \times 10^{-4}\ °C^{-1}$ at 17°C.
   A. 0.085 m
   B. 0.170 m
   C. 0.255 m
   D. 0.340 m

85. Approximately how much energy would be required to raise the temperature of the top layer of the oceans by 1°C? (1 m^3 of water has a mass of 1000 kg.)
   A. 1×10^{24} J
   B. 1×10^{21} J
   C. 1×10^{18} J
   D. 1×10^{15} J

86. An increase in temperature of the water may cause other changes. An increase in surface water temperature is likely to ________ the rate of evaporation.
   A. Increase
   B. Not affect
   C. Decrease

87. The ocean is mostly heated from the top, by light from the sun. The warmer surface water doesn't mix much with the colder deep ocean water. This lack of mixing can be ascribed to a lack of
   A. Conduction.
   B. Convection.
   C. Radiation.
   D. Evaporation.

13 Kinetic Theory

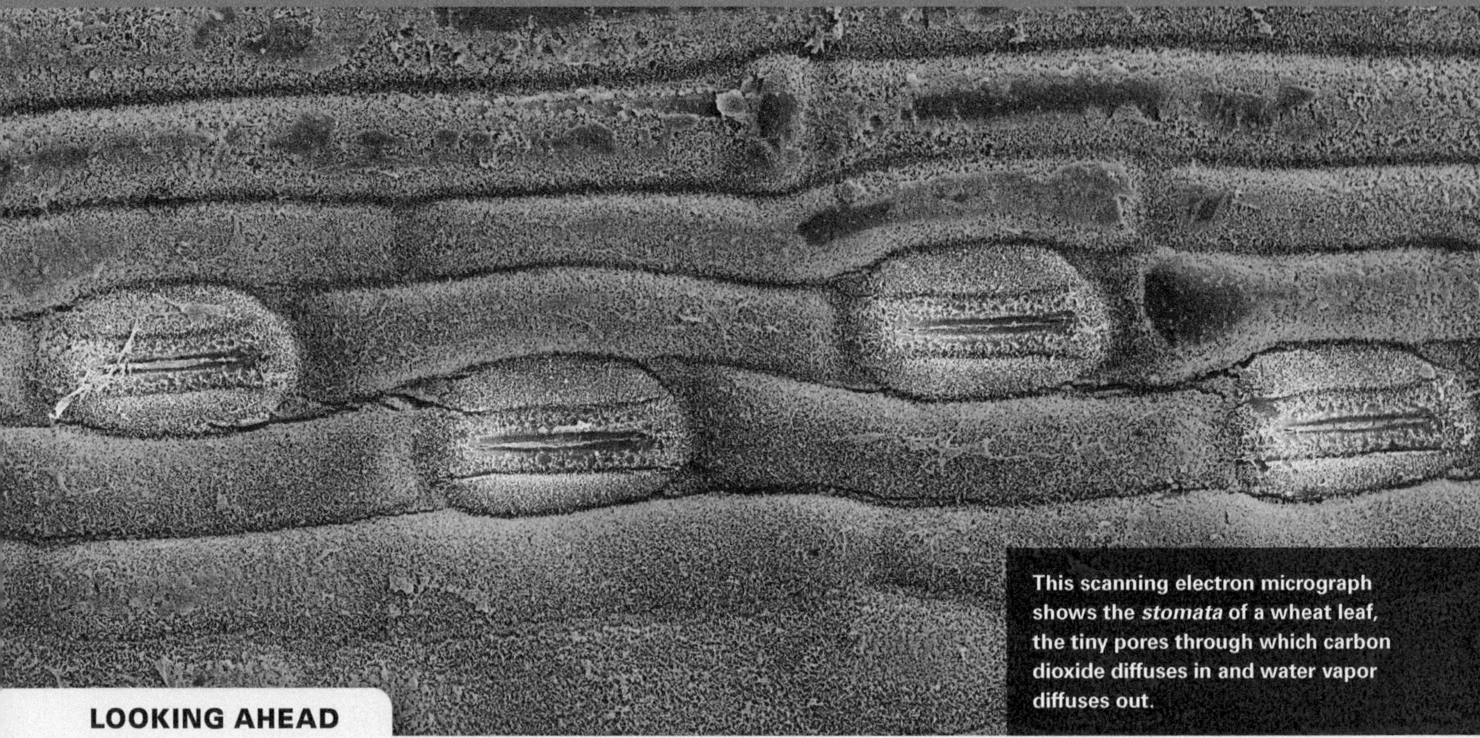

This scanning electron micrograph shows the *stomata* of a wheat leaf, the tiny pores through which carbon dioxide diffuses in and water vapor diffuses out.

LOOKING AHEAD

Kinetic Theory of Gases

The pressure in a diver's air tank is due to countless collisions between the air molecules inside and the tank's walls.

You'll learn how pressure, temperature, specific heat, and other state variables are related to the microscopic motions of molecules.

The Physics of Chemistry

Chemical reactions occur when the kinetic energy of colliding molecules is sufficient to break molecular bonds, allowing stronger bonds to form.

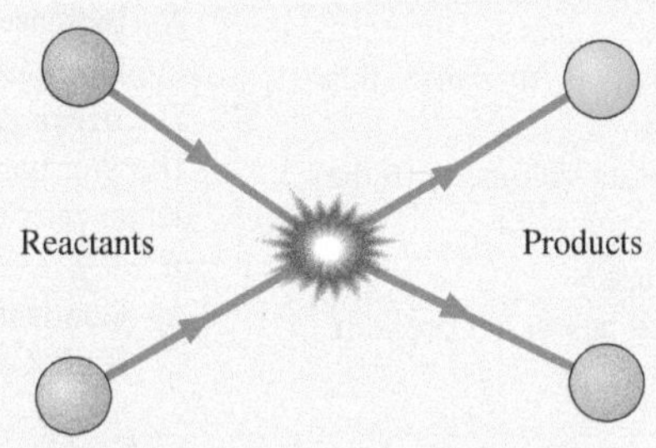

You'll see that the temperature dependence of chemical reaction rates can be explained by kinetic theory.

Diffusion

Oxygen moves from the air to your blood cells by diffusing through a thin tissue separating the alveoli of your lungs from the pulmonary capillaries.

You'll study how *diffusion* is able to transport molecules over small distances with no expenditure of energy.

GOAL To see how macroscopic properties of matter depend on the microscopic motions of atoms and molecules.

PHYSICS AND LIFE

Life Depends on the Motions of Molecules

In this chapter we will explore the microscopic picture of how energy and matter move and how life forms have developed structures to control this movement when individual cells or entire organisms need to maintain a different temperature or different concentration than their surroundings. *Diffusion,* the flow of molecules across a concentration gradient, is of critical importance to living organisms. This same microscopic picture will help us understand how chemical reactions occur, why reaction rates depend on temperature, and how catalysts speed up biochemical reactions that would otherwise be unfavorable. In addition, we'll see how this microscopic picture gives us a better intuitive understanding of pressure, heat transfer, and other macroscopic phenomena.

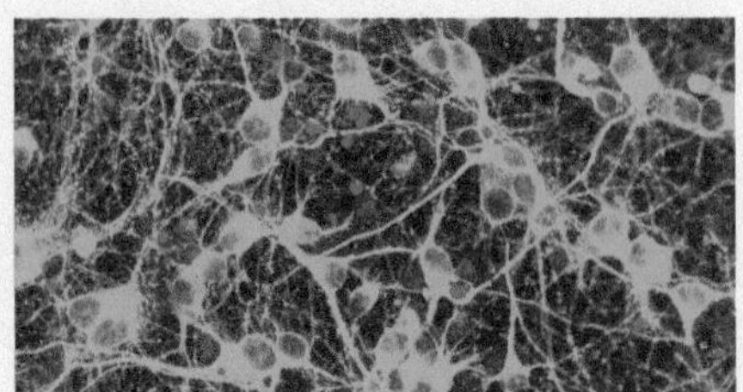

These hippocampus neurons are connected by synapses. Signals travel from neuron to neuron as neurotransmitter molecules diffuse across the synaptic cleft.

13.1 Connecting the Microscopic and the Macroscopic

Of all the scientific discoveries of the past 200 years, one of the most profound—some would argue *the* most profound—is that matter is made of atoms. All of chemistry flows from this discovery. Atoms join together as molecules, and smaller molecules join together as the macromolecules of biochemistry. But these molecules aren't just sitting there; they are in constant motion, and their collisions are responsible for everything from the pressure of a gas to the reactions that occur within cells.

Chapter 12 was about thermodynamics at the macroscopic level—the behavior of systems we can measure and manipulate in the laboratory. We've said little about atoms and molecules other than to note that thermal energy is the microscopic energy of moving molecules. Now we're going to shift gears and look at thermodynamics from a microscopic perspective. What are all those atoms and molecules doing? How do their motions and interactions influence the macroscopic physics?

The goal of this chapter is to understand how much of what we observe at the macroscopic level can be *explained* in terms of the underlying microscopic physics, what we call the *micro/macro connection.* In particular:

- Macroscopic state variables, such as pressure, temperature, and thermal energy, are directly related to the *average* behavior of moving and colliding molecules.
- The conduction of heat energy from one object to another is a direct result of molecular collisions.
- Diffusion, an essential element of biology, is also the result of molecular collisions.
- Reaction rates and the role of enzymes are governed by the *distribution* of molecular speeds.

A key idea underlying the micro/macro connection, one that we'll see throughout this chapter, is that **atomic-level collisions and interactions continuously redistribute a system's thermal energy among all the atoms and molecules.** Yet this chaotic behavior at the microscopic level leads to very steady, predictable behavior at the macroscopic level. Exploration of this micro/macro connection will give you increased insight and intuition about the myriad cellular-level processes that take place in the body.

FIGURE 13.1 An experiment to measure the speeds of molecules in a gas.

Only molecules with speed $L/\Delta t$ reach the detector.
Molecular beam
Velocity selector
Detector
Source of molecules
Axle
L
A vacuum inside prevents molecules from colliding.
The axle rotates once every Δt s.

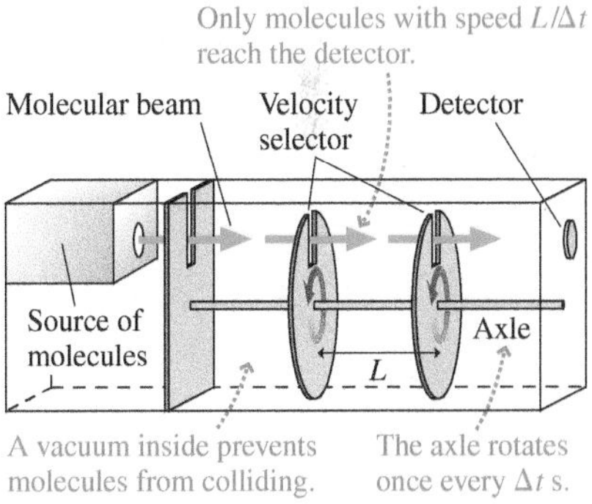

13.2 Molecular Speeds and Collisions

Let's begin by looking at a gas. Do all the molecules in the gas move with the same speed? Or is there a range of speeds? This is an important question that can be answered experimentally.

FIGURE 13.1 shows an experiment to measure the speeds of molecules in a gas. The two rotating disks form a *velocity selector.* Once every revolution, the slot in the first disk allows a small pulse of molecules to pass through. By the time these molecules reach the second disk, the slots have rotated. The molecules can pass through the second slot and be detected *only* if they have exactly the right speed $v = L/\Delta t$ to travel between the two disks during time interval Δt it takes the axle to complete one revolution. Molecules having any other speed are blocked by the second disk. By changing the rotation speed of the axle, this apparatus can measure how many molecules have each of many possible speeds.

FIGURE 13.2 shows the results for nitrogen gas (N_2) at $T = 20°C$. The data are presented in the form of a *histogram,* a bar chart in which the height of each bar tells how many (or, in this case, what percentage) of the molecules have a speed in the *range* of speeds shown below the bar. For example, 12% of the molecules have speeds in the range from 600 m/s to 700 m/s. All the bars sum to 100%, showing that this histogram describes *all* of the molecules leaving the source.

FIGURE 13.2 The distribution of molecular speeds in a sample of nitrogen gas.

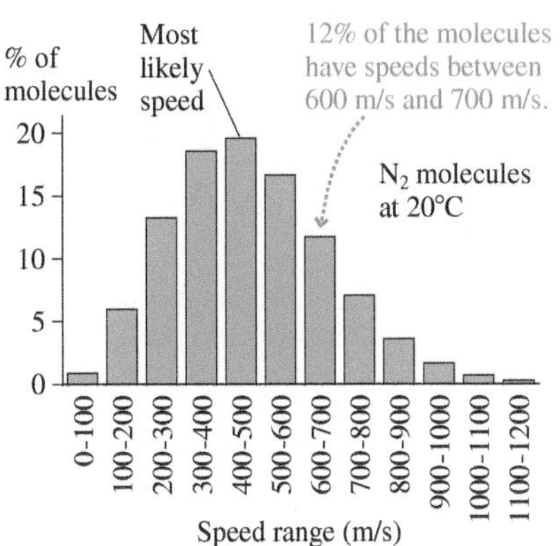

It turns out that the molecules have what is called a *distribution* of speeds, ranging from as low as ≈100 m/s to as high as ≈1200 m/s. But not all speeds are equally likely; there is a *most likely speed* of ≈450 m/s. This is really fast, ≈1000 mph! Changing the temperature or changing to a different gas changes the most likely speed, as you'll learn later in the chapter, but it does not change the *shape* of the distribution.

If you were to repeat the experiment, you would again find the most likely speed to be ≈450 m/s and that 12% of the molecules have speeds between 600 m/s and 700 m/s. This is an important lesson. Although a gas consists of a vast number of molecules, each moving randomly, *averages,* such as the average number of molecules in the speed range 600 to 700 m/s, have precise, predictable values. **The micro/macro connection is built on the idea that the macroscopic properties of a system, such as temperature or pressure, are related to the *average* behavior of the atoms and molecules.**

Mean Free Path

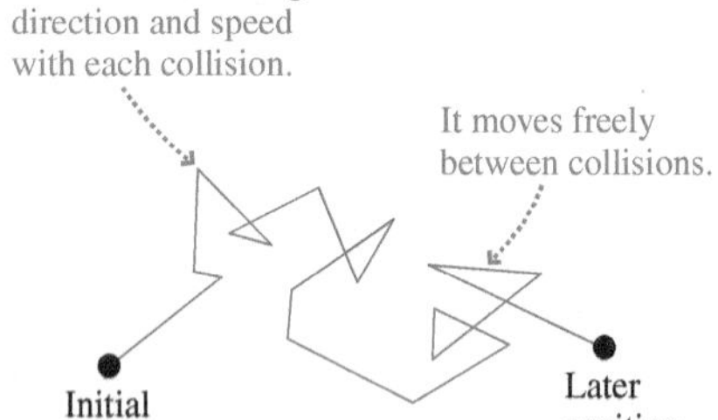

The molecule changes direction and speed with each collision.
It moves freely between collisions.
Initial position
Later position

FIGURE 13.3 A single molecule follows a zig-zag path through a gas.

A molecule in a gas does not travel in a straight line because it is constantly colliding with other molecules. Instead, as FIGURE 13.3 shows, a molecule follows a zig-zag path in which short linear segments vary in both length and direction because of the random nature of the collisions.

Rather than focusing on individual collisions, let's look at the *average* distance a molecule moves between collisions. If a molecule has N_{coll} collisions as it travels distance L, the average distance between collisions, which is called the **mean free path** λ (lowercase Greek lambda), is

$$\lambda = \frac{L}{N_{\text{coll}}} \tag{13.1}$$

FIGURE 13.4a shows two molecules approaching each other. We will assume that the molecules are hard spheres of radius r. In that case, the molecules will collide if the distance between their *centers* is less than $2r$. They will miss if the distance is greater than $2r$.

FIGURE 13.4 A "sample" molecule collides with "target" molecules.

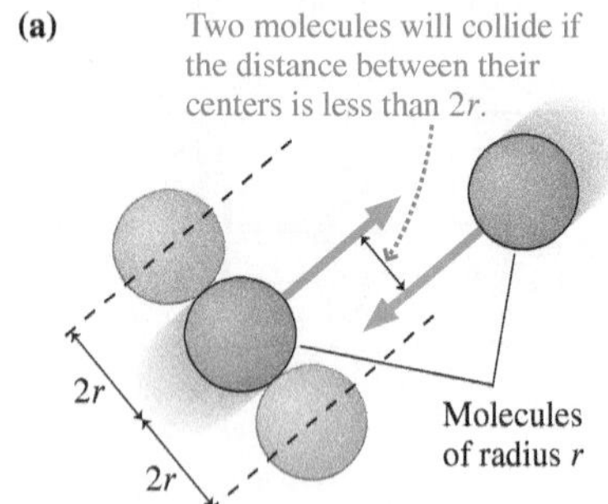

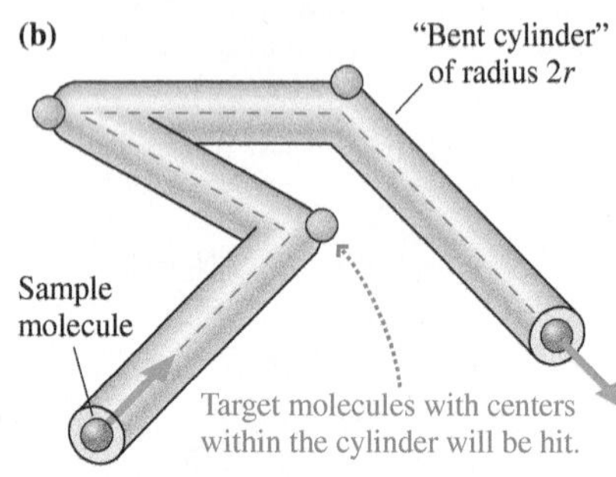

In FIGURE 13.4b we've drawn a cylinder of radius $2r$ centered on the trajectory of a "sample" molecule. The sample molecule collides with any "target" molecule whose center is located within the cylinder, causing the cylinder to bend at that point. Hence the number of collisions N_{coll} is equal to the number of molecules in a cylindrical volume of length L.

The volume of a cylinder is $V_{\text{cyl}} = AL = \pi(2r)^2L$. If the number density of the gas is N/V particles per m^3, then the number of collisions along a trajectory of length L is

$$N_{\text{coll}} = \frac{N}{V}V_{\text{cyl}} = \frac{N}{V}\pi(2r)^2L = 4\pi\frac{N}{V}r^2L \tag{13.2}$$

Thus the mean free path between collisions is

$$\lambda = \frac{L}{N_{\text{coll}}} = \frac{1}{4\pi(N/V)r^2}$$

We made a tacit assumption in this derivation that the target molecules are at rest. While the general idea behind our analysis is correct, a more detailed calculation with all the molecules moving introduces an extra factor of $\sqrt{2}$, giving

$$\lambda = \frac{1}{4\sqrt{2}\,\pi(N/V)r^2} \tag{13.3}$$

Mean free path for atoms or molecules of radius r

Laboratory measurements are necessary to determine atomic and molecular radii, but a reasonable rule of thumb is to assume that atoms in a monatomic gas have $r \approx 0.5 \times 10^{-10}$ m and small molecules have $r \approx 1.0 \times 10^{-10}$ m.

EXAMPLE 13.1 The mean free path at room temperature

What is the mean free path of a nitrogen molecule at 1.0 atm pressure and room temperature (20°C)?

PREPARE Nitrogen is a diatomic molecule, so $r \approx 1.0 \times 10^{-10}$ m. We'll need to use the ideal-gas law to determine the number density of the gas.

SOLVE We'll use the ideal-gas law in the form $pV = Nk_BT$:

$$\frac{N}{V} = \frac{p}{k_BT} = \frac{101{,}000 \text{ Pa}}{(1.38 \times 10^{-23} \text{ J/K})(293 \text{ K})} = 2.5 \times 10^{25} \text{ m}^{-3}$$

Thus the mean free path is

$$\lambda = \frac{1}{4\sqrt{2}\,\pi(N/V)r^2}$$

$$= \frac{1}{4\sqrt{2}\,\pi(2.5 \times 10^{25} \text{ m}^{-3})(1.0 \times 10^{-10} \text{ m})^2}$$

$$= 2.3 \times 10^{-7} \text{ m} = 230 \text{ nm}$$

ASSESS You learned in Example 12.13 that the average separation between gas molecules at STP is ≈ 4 nm. It seems that any given molecule can slip between its neighbors, which are spread out in three dimensions, and travel—on average—about 60 times the average spacing before it collides with another molecule.

13.3 The Kinetic Theory of Gases

Whether a tossed coin lands with heads or tails up is unpredictable; we call it a *random event.* While a single event cannot be predicted, we can predict with great certainty that on *average* half of a very large number of tosses will be heads, half tails. An individual molecule's motion, which we've seen is a sequence of random collisions, is completely unpredictable. Yet, just as with the coins, an average over a very large number of molecules washes out the randomness and allows us to make precise predictions about average behavior. **Kinetic theory** is the name for a theory that explains the physical properties of matter in terms of the motions of its constituent molecules. How many properties of an ideal gas can we understand simply by a suitable averaging over random collisions?

To begin, why does a gas have pressure? In Chapter 9, where pressure was introduced, we suggested that the pressure in a gas is due to collisions of the molecules with the walls of its container. The force due to one such collision may be unmeasurably tiny, but the steady rain of a vast number of molecules striking a wall each second exerts a measurable macroscopic force. The gas pressure is the force per unit area ($p = F/A$) resulting from these molecular collisions.

Our task in this section is to calculate the pressure by doing the appropriate averaging over molecular motions and collisions. This task can be divided into three main pieces:

1. Calculate the momentum change of a single molecule during a collision.
2. Find the force due to all collisions.
3. Introduce an appropriate average speed.

Molecular Collisions Exert a Force on a Wall

We'll begin with one molecule. FIGURE 13.5 shows a molecule with an x-component of velocity v_x having a perfectly elastic collision with a wall and rebounding with its velocity changed to $-v_x$. This one molecule has a momentum *change*

$$\Delta p_x = m(-v_x) - mv_x = -2mv_x \qquad (13.4)$$

Now suppose there is not one but N_{coll} collisions with the wall during a small interval of time Δt. Further, suppose that all the molecules have the same speed. (This latter assumption isn't really necessary, and we'll soon remove this constraint,

FIGURE 13.5 A molecule colliding with the wall exerts an impulse on it.

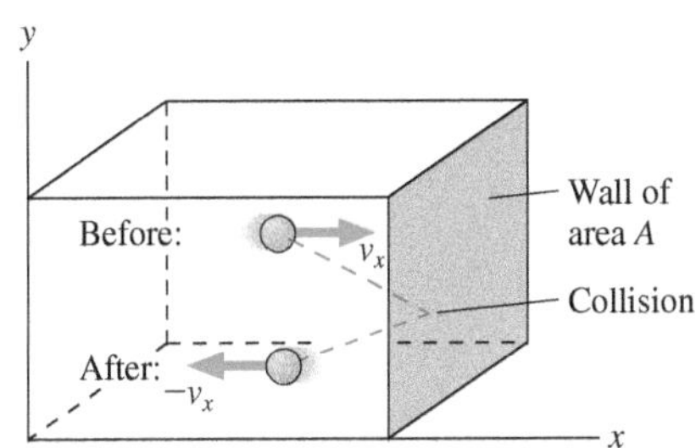

but it helps us focus on the physics without getting lost in the math.) Then the total momentum change of the gas during Δt is

$$\Delta P_x = N_{\text{coll}}\,\Delta p_x = -2N_{\text{coll}}mv_x \tag{13.5}$$

You learned in Chapter 8 that the momentum principle can be written as

$$\Delta P_x = (F_{\text{avg}})_x \Delta t \tag{13.6}$$

Thus the average force of the wall *on the gas* is

$$(F_{\text{on gas}})_x = \frac{\Delta P_x}{\Delta t} = -\frac{2N_{\text{coll}}mv_x}{\Delta t} \tag{13.7}$$

Equation 13.7 has a negative sign because, as we've set it up, the collision force of the wall on the gas molecules is to the left. But Newton's third law is $(F_{\text{on gas}})_x = -(F_{\text{on wall}})_x$, so the force *on the wall* due to these collisions is

$$(F_{\text{on wall}})_x = \frac{2N_{\text{coll}}mv_x}{\Delta t} \tag{13.8}$$

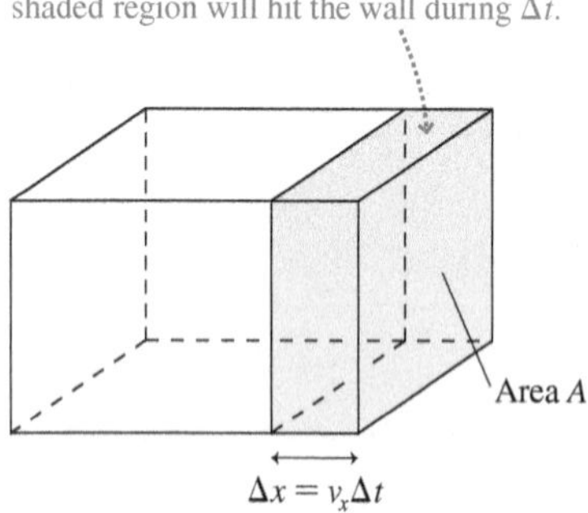
Only molecules moving to the right in the shaded region will hit the wall during Δt.
Area A
$\Delta x = v_x\Delta t$

FIGURE 13.6 Determining the rate of collisions.

We need to determine how many collisions occur during Δt. Assume that Δt is smaller than the mean time between collisions, so no collisions alter the molecular velocities during this interval. **FIGURE 13.6** has shaded a volume of the gas of length $\Delta x = v_x\,\Delta t$. *Every one* of the molecules in this shaded region *that is moving to the right* will reach and collide with the wall during Δt. Molecules outside this region will not reach the wall and will not collide.

The shaded region has volume $A\,\Delta x$, where A is the area of the wall. If the gas has number density N/V, the number of molecules in the shaded region is $(N/V)A\,\Delta x = (N/V)Av_x\,\Delta t$. But only half these molecules are moving to the right, so the number of collisions during Δt is

$$N_{\text{coll}} = \frac{1}{2}\frac{N}{V}Av_x\,\Delta t \tag{13.9}$$

Substituting Equation 13.9 into Equation 13.8, we see that Δt cancels and that the force of the molecules on the wall is

$$F_{\text{on wall}} = \frac{N}{V}mv_x^2A \tag{13.10}$$

Notice that this expression for $F_{\text{on wall}}$ does not depend on any details of the collisions.

A key aspect of the micro/macro connection is the recognition that state variables, such as pressure, are determined by appropriate *averages* of molecular properties. We can relax the assumption that all molecules have the same speed by replacing the squared velocity v_x^2 in Equation 13.10 with its average value. That is,

$$F_{\text{on wall}} = \frac{N}{V}m(v_x^2)_{\text{avg}}A \tag{13.11}$$

where $(v_x^2)_{\text{avg}}$ is the quantity v_x^2 averaged over all the molecules in the container.

The Root-Mean-Square Speed

We need to be careful when averaging velocities. The velocity component v_x has a sign. At any instant of time, half the molecules in a container move to the right and have positive v_x while the other half move to the left and have negative v_x. Thus the *average velocity* is $(v_x)_{\text{avg}} = 0$. If this weren't true, the entire container of gas would move away!

The speed of a molecule is $v = (v_x^2 + v_y^2 + v_z^2)^{1/2}$. Thus the average of the speed squared is

$$(v^2)_{\text{avg}} = (v_x^2 + v_y^2 + v_z^2)_{\text{avg}} = (v_x^2)_{\text{avg}} + (v_y^2)_{\text{avg}} + (v_z^2)_{\text{avg}} \tag{13.12}$$

The square root of $(v^2)_{avg}$ is called the **root-mean-square speed** v_{rms}:

$$v_{rms} = \sqrt{(v^2)_{avg}} \tag{13.13}$$

This is usually called the *rms speed.* You can remember its definition by noting that its name is the *opposite* of the sequence of operations: First you square all the speeds, then you average the squares (find the mean), then you take the square root. Because the square root "undoes" the square, v_{rms} must, in some sense, give an average speed.

NOTE ▶ We use the root-mean-square speed rather than the average speed because root-mean-square values tend to arise naturally in many scientific and statistical calculations. It turns out that v_{rms} differs from v_{avg} by less than 10%, so for practical purposes we can interpret v_{rms} as being essentially the average speed of a molecule in a gas. ◀

There's nothing special about the x-axis. The coordinate system is something that *we* impose on the problem, so *on average* it must be the case that

$$(v_x^2)_{avg} = (v_y^2)_{avg} = (v_z^2)_{avg} \tag{13.14}$$

Hence we can use Equation 13.12 and the definition of v_{rms} to write

$$v_{rms}{}^2 = (v_x^2)_{avg} + (v_y^2)_{avg} + (v_z^2)_{avg} = 3(v_x^2)_{avg} \tag{13.15}$$

Consequently, $(v_x^2)_{avg}$ is

$$(v_x^2)_{avg} = \tfrac{1}{3}v_{rms}{}^2 \tag{13.16}$$

Using this result in Equation 13.11 gives us the net force on the wall of the container:

$$F_{\text{on wall}} = \frac{1}{3}\frac{N}{V}mv_{rms}{}^2A \tag{13.17}$$

Thus the pressure on the wall of the container due to all the molecular collisions is

$$p = \frac{F_{\text{on wall}}}{A} = \frac{1}{3}\frac{N}{V}mv_{rms}{}^2 \tag{13.18}$$

We have met our goal. Equation 13.18 expresses the macroscopic pressure in terms of the microscopic physics. The pressure depends on the number density of molecules in the container and on how fast, on average, the molecules are moving.

EXAMPLE 13.2 The rms speed of helium atoms

A container holds helium at a pressure of 200 kPa and a temperature of 60°C. What is the rms speed of the helium atoms?

PREPARE The rms speed can be found from the pressure and the number density.

SOLVE Using the ideal-gas law gives us the number density:

$$\frac{N}{V} = \frac{p}{k_BT} = \frac{200{,}000\ \text{Pa}}{(1.38\times10^{-23}\ \text{J/K})(333\ \text{K})} = 4.35\times10^{25}\ \text{m}^{-3}$$

The mass of a helium atom is $m = 4\ \text{u} = 6.64\times10^{-27}$ kg. Thus

$$v_{rms} = \sqrt{\frac{3p}{(N/V)m}} = 1440\ \text{m/s}$$

ASSESS 1440 m/s is extremely fast, approximately 3000 mph. It's difficult to assess whether this is reasonable, although Figure 13.2 did show that the most likely speed of a nitrogen molecule at 20°C is ≈450 m/s, and it makes sense that helium, being lighter than nitrogen, moves even faster.

STOP TO THINK 13.1 The speed of every molecule in a gas is suddenly increased by a factor of 4. As a result, v_{rms} increases by a factor of

A. 2.
B. <4 but not necessarily 2.
C. 4.
D. >4 but not necessarily 16.
E. 16.
F. v_{rms} doesn't change.

Partial Pressure

Suppose a container holds two or more different gases that we label A, B, C, ... The molecules of each gas collide with the container walls independently of what the other gases are doing, so each gas exerts its own force. Forces add, so the total force on the container wall is $F_{total} = F_A + F_B + F_C + \cdots$. Thus the total pressure in the container, the total force divided by the wall area, is

$$p_{total} = p_A + p_B + p_C + \cdots \quad (13.19)$$

The pressure contributed by each gas is called the **partial pressure** of that gas, and the total pressure—what we measure with a pressure gauge—is the sum of all the partial pressures. You may have learned this in chemistry as *Dalton's law of partial pressures.*

For example, dry air consists of 78% N_2, 21% O_2, and 1% Ar (plus smaller amounts of trace gases), which means that the partial pressures of nitrogen, oxygen, and argon are 0.78 atm, 0.21 atm, and 0.01 atm, respectively. Each exerts pressure independently of the others, with the net result being a pressure of 1 atm = 101 kPa.

We can use the ideal-gas law to write

$$\begin{aligned} p_{total} &= \frac{n_A RT}{V} + \frac{n_B RT}{V} + \frac{n_C RT}{V} + \cdots \\ &= \left(\frac{n_A}{n} + \frac{n_B}{n} + \frac{n_B}{n} + \cdots\right)\frac{nRT}{V} = (x_A + x_B + x_C + \cdots)p_{total} \end{aligned} \quad (13.20)$$

where $n = n_A + n_B + n_C + \cdots$ is the total number of moles of gas in the container. The ratio $x_A = n_A/n$ is the *mole fraction* of gas A, a ratio that shows up frequently in chemistry and is usually denoted x. What we see is that the partial pressure of gas A

$$p_A = x_A p_{total} \quad (13.21)$$

is simply the total pressure multiplied by the mole fraction of A.

EXAMPLE 13.3 Partial pressure of carbon dioxide

Prior to the industrial revolution, the level of carbon dioxide in the earth's atmosphere was 270 parts per million (ppm). Today, as a result of burning fossil fuels, deforestation, and other causes, the level is approximately 410 ppm. What is the partial pressure of CO_2 at sea level in Pa and in mm Hg?

PREPARE 410 ppm is the mole fraction of CO_2 in the atmosphere.

SOLVE The mole fraction of CO_2 is

$$x_{CO_2} = 410 \text{ ppm} = 410 \times 10^{-6} = 4.1 \times 10^{-4}$$

The total pressure at sea level is 1 atm = 101 kPa, so

$$p_{CO_2} = x_{CO_2} p_{total} = (4.1 \times 10^{-4})(101{,}000 \text{ Pa}) = 41 \text{ Pa}$$

Converting to mm Hg, we have

$$p_{CO_2} = 41 \text{ Pa} \times \frac{760 \text{ mm Hg}}{101{,}000 \text{ Pa}} = 0.31 \text{ mm Hg}$$

ASSESS These numbers are a tiny fraction of the atmospheric pressure, which is consistent with the fact that carbon dioxide is a *trace gas;* its partial pressure makes little contribution to atmospheric pressure. Nonetheless, it plays a large role in the earth's climate because of its strong absorption of the infrared radiation that the earth radiates to space.

Proper lung function requires careful regulation of the partial pressures of oxygen, carbon dioxide, and water vapor. Their partial pressures in the lung's alveoli, where gas exchange with the blood occurs via diffusion, can be quite different from their partial pressures in ambient air. In particular, the body tries to maintain the partial pressure of carbon dioxide in the alveoli at ≈40 mm Hg, a factor of more than 100 larger than its 0.3 mm Hg partial pressure in the air. This is because CO_2 is not simply a waste product to be exhausted but also part of the bicarbonate buffer system that manages the blood's pH.

Hyperventilation—excessively rapid breathing—forces CO_2 from the lungs and lowers the alveolar partial pressure. This in turn lowers the CO_2 partial pressure in the blood, causing the blood pH to rise and creating symptoms that can be mistaken for a heart attack.

Vapor pressure, which we discussed in ◀◀ SECTION 12.3, is another example of a partial pressure. The evaporation of water into a closed container increases the number of molecular collisions with the container's wall, so the total pressure is that of any air *plus* that of the water vapor. If a bowl of water is sealed into a container of dry air with pressure $p_{\text{air}} = 100\text{ kPa}$ at a temperature where the water vapor pressure is $p_{\text{water}} = 10\text{ kPa}$, then enough water will evaporate to create a total pressure $p_{\text{total}} = 110\text{ kPa}$.

Temperature

An individual molecule of mass m and velocity v has translational kinetic energy

$$\epsilon = \tfrac{1}{2}mv^2 \tag{13.22}$$

We'll use ϵ (lowercase Greek epsilon) to distinguish the energy of a molecule from the system energy E. The *average* translational kinetic energy, averaged over all the molecules in a gas, is

$$\begin{aligned}\epsilon_{\text{avg}} &= \text{average translational kinetic energy of a molecule}\\ &= \tfrac{1}{2}m(v^2)_{\text{avg}} = \tfrac{1}{2}mv_{\text{rms}}{}^2\end{aligned} \tag{13.23}$$

We've included the word "translational" to distinguish ϵ from rotational kinetic energy, which we will consider later in this chapter. Notice how the rms speed arises naturally in this average.

We can write the gas pressure, Equation 13.18, in terms of the average translational kinetic energy as

$$p = \frac{2}{3}\frac{N}{V}\left(\tfrac{1}{2}mv_{\text{rms}}{}^2\right) = \frac{2}{3}\frac{N}{V}\epsilon_{\text{avg}} \tag{13.24}$$

The pressure is directly proportional to the average molecular translational kinetic energy. This makes sense. More-energetic molecules will hit the walls harder as they bounce and thus exert more force on the walls.

It's instructive to write Equation 13.24 as

$$pV = \tfrac{2}{3}N\epsilon_{\text{avg}} \tag{13.25}$$

We know, from the ideal-gas law, that

$$pV = Nk_{\text{B}}T \tag{13.26}$$

Comparing these two equations, we reach the significant conclusion that the average translational kinetic energy per molecule is

$$\epsilon_{\text{avg}} = \tfrac{3}{2}k_{\text{B}}T \tag{13.27}$$

Average translational kinetic energy at temperature T

where k_{B} is the Boltzmann constant and the temperature T is in kelvins. For example, the average translational kinetic energy of a molecule at room temperature (20°C) is

$$\epsilon_{\text{avg}} = \tfrac{3}{2}(1.38 \times 10^{-23}\text{ J/K})(293\text{ K}) = 6.1 \times 10^{-21}\text{ J}$$

NOTE ▶ A molecule's average translational kinetic energy depends *only* on the temperature, not on the molecule's mass. If two gases have the same temperature, their molecules have the same average translational kinetic energy. ◀

Equation 13.27 is especially satisfying because it finally gives real meaning to the concept of temperature. Writing it as

$$T = \frac{2}{3k_{\text{B}}}\epsilon_{\text{avg}} \tag{13.28}$$

FIGURE 13.7 The micro/macro connection for pressure and temperature.

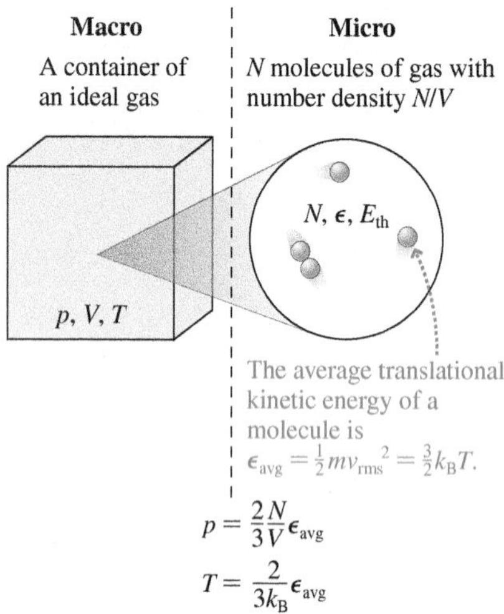
Macro
Micro
A container of an ideal gas
N molecules of gas with number density N/V
N, ϵ, E_th
p, V, T
The average translational kinetic energy of a molecule is
$\epsilon_{avg} = \frac{1}{2}mv_{rms}^2 = \frac{3}{2}k_BT$.

$p = \frac{2}{3}\frac{N}{V}\epsilon_{avg}$

$T = \frac{2}{3k_B}\epsilon_{avg}$

we can see that, for a gas, **this thing we call *temperature* measures the average translational kinetic energy.** A higher temperature corresponds to a larger value of ϵ_{avg} and thus to higher molecular speeds. This concept of temperature also gives meaning to *absolute zero* as the temperature at which $\epsilon_{avg} = 0$ and all molecular motion ceases. (Quantum effects at very low temperatures prevent the motions from actually stopping, but our classical theory predicts that they would.) FIGURE 13.7 summarizes what we've learned thus far about the micro/macro connection.

We can now justify our assumption that molecular collisions are perfectly elastic. Suppose they were not. If kinetic energy was lost in collisions, the average translational kinetic energy ϵ_{avg} of the gas would decrease and we would see a steadily decreasing temperature. But that doesn't happen. The temperature of an isolated system remains constant, indicating that ϵ_{avg} is not changing with time. Consequently, the collisions must be perfectly elastic.

EXAMPLE 13.4 Total microscopic kinetic energy

What is the total translational kinetic energy of the molecules in 1.0 mol of gas at STP?

PREPARE The total translational kinetic energy is the number of molecules multiplied by the average kinetic energy of a molecule.

SOLVE The average translational kinetic energy of each molecule is

$$\epsilon_{avg} = \frac{3}{2}k_BT = \frac{3}{2}(1.38 \times 10^{-23}\ \text{J/K})(273\ \text{K})$$
$$= 5.65 \times 10^{-21}\ \text{J}$$

1.0 mol of gas contains N_A molecules; hence the total kinetic energy is

$$K_{micro} = N_A\epsilon_{avg} = 3400\ \text{J}$$

ASSESS The energy of any one molecule is incredibly small. Nonetheless, a macroscopic system has substantial thermal energy because it consists of an incredibly large number of molecules.

By definition, $\epsilon_{avg} = \frac{1}{2}mv_{rms}^2$. Using the ideal-gas law, we found $\epsilon_{avg} = \frac{3}{2}k_BT$. By equating these expressions we find that the rms speed of molecules at temperature T is

$$v_{rms} = \sqrt{\frac{3k_BT}{m}} \qquad (13.29)$$

Root-mean-square speed of molecules at temperature T

The rms speed depends on the square root of the temperature and inversely on the square root of the molecular mass. For example, room-temperature nitrogen (molecular mass 28 u) has rms speed

$$v_{rms} = \sqrt{\frac{3(1.38 \times 10^{-23}\ \text{J/K})(293\ \text{K})}{28(1.66 \times 10^{-27}\ \text{kg})}} = 509\ \text{m/s}$$

This is in excellent agreement with the experimental results of Figure 13.2.

NOTE ► Recall that a gas always fills its container. At a lower temperature, the molecules move more slowly, but they continue to fill the container and to interact only via collisions. Only at the condensation temperature do molecules begin to attract each other and stick together. ◄

EXAMPLE 13.5 Mean time between collisions

Estimate the mean time between collisions for a nitrogen molecule at 1.0 atm pressure and room temperature (20°C).

PREPARE Because v_{rms} is essentially the average molecular speed, the *mean time between collisions* is simply the time needed to travel distance λ, the mean free path, at speed v_{rms}.

SOLVE We found $\lambda = 2.3 \times 10^{-7}$ m in Example 13.1 and $v_{rms} = 509$ m/s above. Thus the mean time between collisions is

$$(\Delta t)_{avg} = \frac{\lambda}{v_{rms}} = \frac{2.3 \times 10^{-7}\ \text{m}}{509\ \text{m/s}} = 4.5 \times 10^{-10}\ \text{s}$$

ASSESS The air molecules around us move very fast, they collide with their neighbors about two billion times every second, and they manage to move, on average, only about 230 nm between collisions.

STOP TO THINK 13.2 The speed of every molecule in a gas is suddenly increased by a factor of 4. As a result, the temperature T increases by a factor of

A. 2.
B. <4 but not necessarily 2.
C. 4.
D. >4 but not necessarily 16.
E. 16.
F. T doesn't change.

13.4 Thermal Energy and Specific Heat

In ◀◀ SECTION 10.4 we defined the thermal energy E_{th} of a system to be the sum of the microscopic kinetic and potential energies of the jiggling atoms and stretched molecular bonds that make up the system. A closer look will give us insight into thermal energy and, surprisingly, will allow us to predict the molar specific heat—a macroscopic state parameter—of gases and other substances. This shows how powerful the micro/macro connection is!

Monatomic Gases

FIGURE 13.8 shows a monatomic gas such as helium or neon. The atoms in an ideal gas have no molecular bonds with their neighbors; hence their thermal energy is entirely the kinetic energy of the moving atoms. Furthermore, the kinetic energy of a monatomic gas particle is entirely translational kinetic energy ϵ. Thus the thermal energy of a monatomic gas of N atoms is

$$E_{th} = \epsilon_1 + \epsilon_2 + \epsilon_3 + \cdots + \epsilon_N = N\epsilon_{avg} \tag{13.30}$$

where ϵ_i is the translational kinetic energy of atom i. We found that $\epsilon_{avg} = \frac{3}{2}k_B T$; hence the thermal energy is

$$E_{th} = \tfrac{3}{2}Nk_B T = \tfrac{3}{2}nRT \tag{13.31}$$

Thermal energy of a monatomic gas

where we used $N = nN_A$ and the definition of the Boltzmann constant, $k_B = R/N_A$.

We've noted for the last three chapters that thermal energy is associated with temperature. Now we have an explicit result for a monatomic gas: E_{th} is directly proportional to the temperature. Notice that E_{th} is independent of the atomic mass. **Any two monatomic gases will have the same thermal energy if they have the same temperature and the same number of atoms (or moles).**

If the temperature of a monatomic gas changes by ΔT, its thermal energy changes by

$$\Delta E_{th} = \tfrac{3}{2}nR\,\Delta T \tag{13.32}$$

In Chapter 12 we found that the change in thermal energy for *any* ideal-gas process is related to the molar specific heat at constant volume by

$$\Delta E_{th} = nC_V\,\Delta T \tag{13.33}$$

FIGURE 13.8 The atoms in a monatomic gas have only translational kinetic energy.

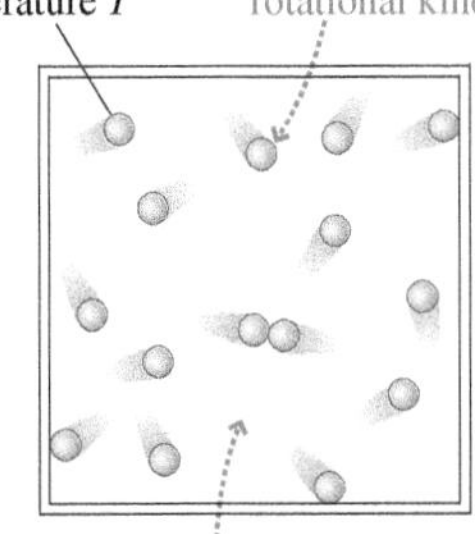
N atoms in a gas at temperature T
Atom i has translational kinetic energy ϵ_i but no potential energy or rotational kinetic energy.
The thermal energy of the gas is $E_{th} = \epsilon_1 + \epsilon_2 + \epsilon_3 + \cdots = N\epsilon_{avg}$.

Equation 13.32 is a microscopic result that we obtained by relating the temperature to the average translational kinetic energy of the atoms. Equation 13.33 is a macroscopic result that we arrived at from the first law of thermodynamics. We can make a micro/macro connection by combining these two equations. Doing so gives us a *prediction* for the molar specific heat:

$$C_V = \tfrac{3}{2}R = 12.5 \text{ J/mol} \cdot \text{K} \qquad (13.34)$$

Molar specific heat of a monatomic gas

This was exactly the value of C_V for all three monatomic gases in Table 12.7. There we had no explanation for the curious fact that helium, neon, and argon have identical molar specific heats. Now we do. The perfect agreement of theory and experiment is strong evidence that gases really do consist of moving, colliding molecules.

The Equipartition Theorem

The particles of a monatomic gas are atoms. Their energy consists exclusively of their translational kinetic energy. A particle's translational kinetic energy can be written

$$\epsilon = \tfrac{1}{2}mv^2 = \tfrac{1}{2}mv_x^2 + \tfrac{1}{2}mv_y^2 + \tfrac{1}{2}mv_z^2 = \epsilon_x + \epsilon_y + \epsilon_z \qquad (13.35)$$

where we have written separately the energy associated with translational motion along the three axes. Because each axis in space is independent, we can think of ϵ_x, ϵ_y, and ϵ_z as independent *modes* of storing energy within the system.

Other systems have additional modes of energy storage. For example,

- Two atoms joined by a spring-like molecular bond can vibrate back and forth. Both kinetic and potential energy are associated with this vibration.
- A diatomic molecule, in addition to translational kinetic energy, has rotational kinetic energy if it rotates end-over-end like a dumbbell.

We define the number of **degrees of freedom** as the number of distinct and independent modes of energy storage. A monatomic gas has three degrees of freedom. Systems that can vibrate or rotate have more degrees of freedom.

Collisions between molecules constantly move energy between one degree of freedom and another. For example, a collision could cause a diatomic molecule to rotate faster, increasing its rotational kinetic energy by decreasing its translational kinetic energy. The proof is beyond what we can do in this textbook, but it's possible to show that the net result of the vast number of collisions is to cause the thermal energy to, on average, be shared equally among all the degrees of freedom. This conclusion is known as the *equipartition theorem,* meaning that the energy is equally divided.

Equipartition theorem The thermal energy of a system of particles is equally divided among all the possible degrees of freedom. For a system of N particles at temperature T, the energy stored in each mode (each degree of freedom) is $\tfrac{1}{2}Nk_BT$ or, in terms of moles, $\tfrac{1}{2}nRT$.

A monatomic gas has three degrees of freedom and thus, as we found above, $E_{th} = \tfrac{3}{2}Nk_BT$.

FIGURE 13.9 A simple model of a solid.

Each atom has microscopic translational kinetic energy *and* microscopic potential energy along all three axes.

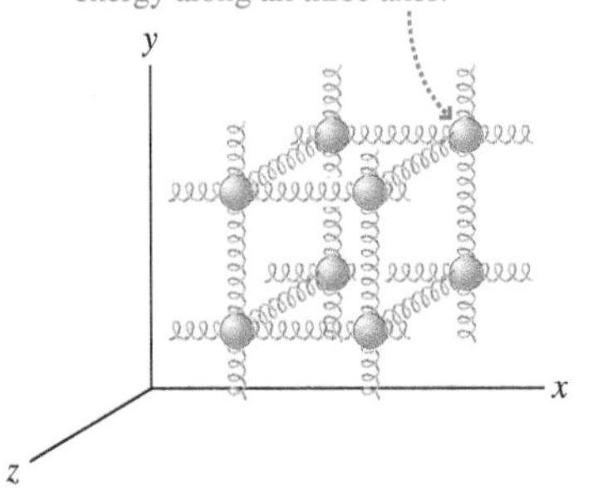

Solids

FIGURE 13.9 reminds you of our "bedspring model" of a solid with particle-like atoms connected by a lattice of spring-like molecular bonds. How many degrees of freedom does a solid have? Three degrees of freedom are associated with the kinetic energy, just as in a monatomic gas. In addition, the molecular bonds can be compressed or stretched independently along the x-, y-, and z-axes. Three additional degrees of freedom are associated with these three modes of potential energy. Altogether, a solid has six degrees of freedom.

The energy stored in each of these six degrees of freedom is $\frac{1}{2}Nk_BT$. The thermal energy of a solid is the total energy stored in all six modes, or

$$E_{th} = 3Nk_BT = 3nRT \quad (13.36)$$

Thermal energy of a solid

We can use this result to predict the molar specific heat of a solid. If the temperature changes by ΔT, then the thermal energy changes by

$$\Delta E_{th} = 3nR\,\Delta T \quad (13.37)$$

In Chapter 12 we defined the molar specific heat of a solid such that

$$\Delta E_{th} = nC\,\Delta T \quad (13.38)$$

By comparing Equations 13.37 and 13.38 we can predict that the molar specific heat of a solid is

$$C = 3R = 25.0 \text{ J/mol} \cdot \text{K} \quad \text{(solid)} \quad (13.39)$$

Not bad. The three elemental solids in Table 12.2 had molar specific heats clustered right around 25 J/mol • K. They ranged from 24.3 J/mol • K for aluminum to 25.4 J/mol • K for gold. There are two reasons the agreement between theory and experiment isn't quite as perfect as it was for monatomic gases. First, our simple bedspring model of a solid isn't quite as accurate as our model of a monatomic gas. Second, quantum effects are beginning to make their appearance. More on this shortly. Nonetheless, our ability to predict C to within a few percent from a simple model of a solid is further evidence for the atomic structure of matter.

FIGURE 13.10 A diatomic molecule can rotate or vibrate.

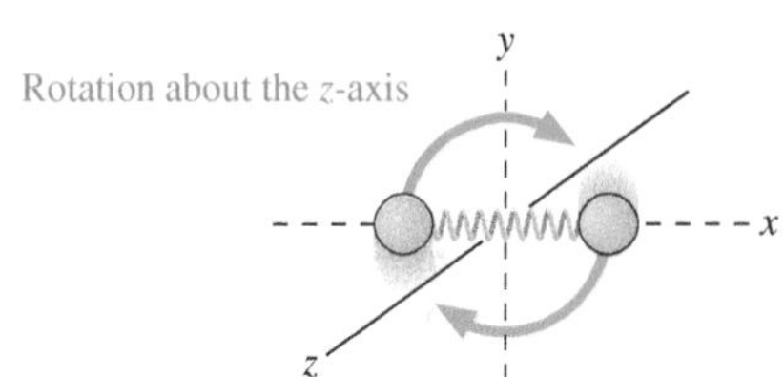

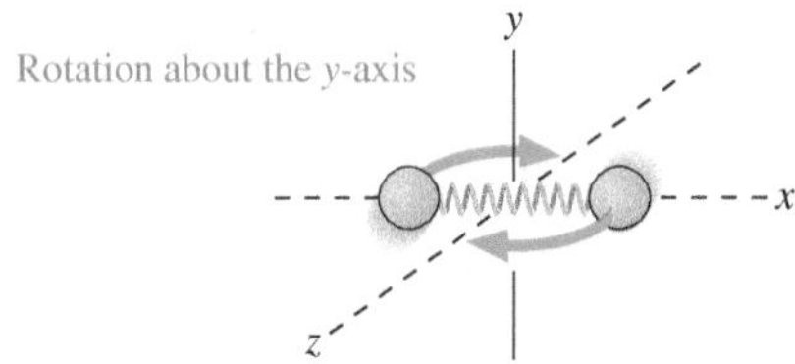

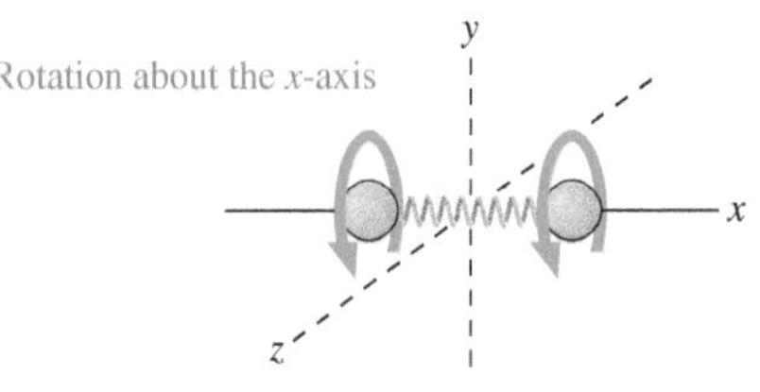

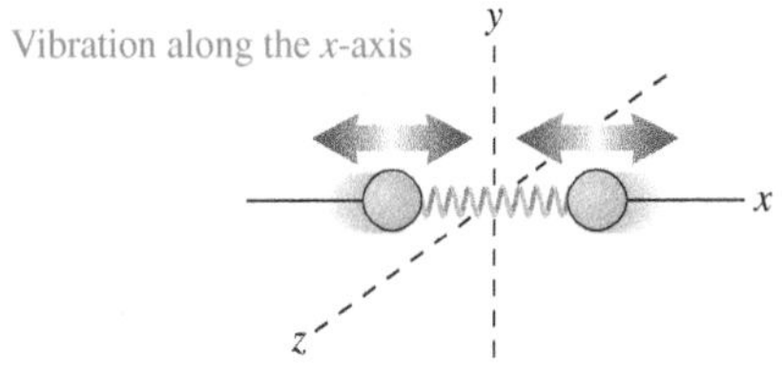

Diatomic Molecules

Diatomic molecules are a bigger challenge. How many degrees of freedom does a diatomic molecule have? **FIGURE 13.10** shows a diatomic molecule, such as molecular nitrogen N_2, oriented along the x-axis. Three degrees of freedom are associated with the molecule's translational kinetic energy. The molecule can have a dumbbell-like end-over-end rotation about either the y-axis or the z-axis. It can also rotate about its own axis. These are three rotational degrees of freedom. The two atoms can also vibrate back and forth, stretching and compressing the molecular bond. This vibrational motion has both kinetic and potential energy—thus two more degrees of freedom.

Altogether, then, a diatomic molecule has eight degrees of freedom, and we would expect the thermal energy of a gas of diatomic molecules to be $E_{th} = 4Nk_BT$. The analysis we followed for a monatomic gas would then lead to the prediction $C_V = 4R = 33.2$ J/mol • K. As compelling as this reasoning seems to be, this is *not* the experimental value of C_V that was reported for diatomic gases in Table 12.7. Instead, we found $C_V = 20.8$ J/mol • K.

Why should a theory that works so well for monatomic gases and solids fail so miserably for diatomic molecules? To see what's going on, notice that 20.8 J/mol • K $= \frac{5}{2}R$. A monatomic gas, with three degrees of freedom, has $C_V = \frac{3}{2}R$. A solid, with six degrees of freedom, has $C = 3R$. A diatomic gas would have $C_V = \frac{5}{2}R$ if it had five degrees of freedom, not eight.

This discrepancy was a major conundrum as kinetic theory developed in the late 19th century. Although it was not recognized as such at the time, we are here seeing our first evidence for the breakdown of classical Newtonian physics. Classically, a diatomic molecule has eight degrees of freedom. The equipartition theorem doesn't distinguish between them; all eight should have the same energy. But atoms and molecules are not classical particles. It took the development of quantum theory in the 1920s to accurately characterize the behavior of atoms and molecules. We don't yet have the tools needed to see why, but quantum effects prevent three of the modes—the two vibrational modes and the rotation of the molecule about its own axis—from being active at room temperature.

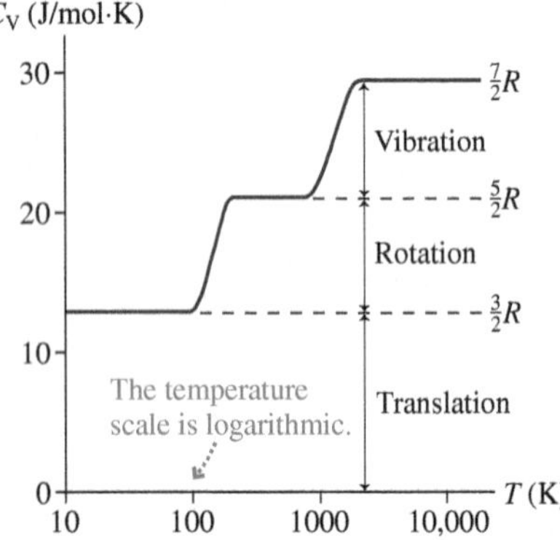

FIGURE 13.11 Hydrogen molar specific heat at constant volume as a function of temperature.

FIGURE 13.11 shows C_V as a function of temperature for hydrogen gas. C_V is right at $\frac{5}{2}R$ for temperatures from ≈ 200 K up to ≈ 800 K. But at very low temperatures C_V drops to the monatomic-gas value $\frac{3}{2}R$. The two rotational modes become "frozen out" and the nonrotating molecule has only translational kinetic energy. Quantum physics can explain this, but not classical Newtonian physics. You can also see that the two vibrational modes *do* become active at very high temperatures, where C_V rises to $\frac{7}{2}R$. Thus the real answer to What's wrong? is that Newtonian physics is not the right physics for describing atoms and molecules. We are somewhat fortunate that Newtonian physics is adequate to understand monatomic gases and solids, at least at room temperature.

Accepting the quantum result that a diatomic gas has only five degrees of freedom at commonly used temperatures (the translational degrees of freedom and the two end-over-end rotations), we find

$$E_{th} = \tfrac{5}{2}Nk_BT = \tfrac{5}{2}nRT$$
$$C_V = \tfrac{5}{2}R = 20.8 \text{ J/mol} \cdot \text{K} \qquad (13.40)$$

Thermal energy and molar specific heat of a diatomic gas

A diatomic gas has more thermal energy than a monatomic gas at the same temperature because the molecules have rotational as well as translational kinetic energy.

While the micro/macro connection firmly establishes the atomic structure of matter, it also heralds the need for a new theory of matter at the atomic level. That is a task we will take up in Part VI. For now, TABLE 13.1 summarizes what we have learned from kinetic theory about thermal energy and molar specific heats.

TABLE 13.1 **Kinetic theory predictions for the thermal energy and the molar specific heat**

System	Degrees of freedom	E_{th}	C_V
Monatomic gas	3	$\frac{3}{2}Nk_BT = \frac{3}{2}nRT$	$\frac{3}{2}R = 12.5$ J/mol · K
Diatomic gas	5	$\frac{5}{2}Nk_BT = \frac{5}{2}nRT$	$\frac{5}{2}R = 20.8$ J/mol · K
Elemental solid	6	$3Nk_BT = 3nRT$	$3R = 25.0$ J/mol · K

EXAMPLE 13.6 The rotational frequency of a molecule

The nitrogen molecule N_2 has a bond length of 0.12 nm. Estimate the rotational frequency of N_2 at 20°C.

PREPARE The molecule can be modeled as a rigid dumbbell of length $L = 0.12$ nm rotating about its center.

SOLVE Each atom is moving in uniform circular motion with angular frequency ω at radius $r = L/2$ from the center of the molecule. Recall that the speed of a particle in uniform circular motion is $v = \omega r = \omega L/2$. Thus, with two atoms, the rotational kinetic energy is

$$\epsilon_{rot} = \tfrac{1}{2}mv^2 + \tfrac{1}{2}mv^2 = m\left(\frac{\omega L}{2}\right)^2 = m\left(\frac{2\pi f L}{2}\right)^2$$
$$= \pi^2 mL^2 f^2$$

where we used $\omega = 2\pi f$ to relate the rotational frequency f to the angular frequency ω. From the equipartition theorem, the energy associated with this mode is $\frac{1}{2}Nk_BT$, so the *average* rotational kinetic energy per molecule is

$$(\epsilon_{rot})_{avg} = \tfrac{1}{2}k_BT$$

Equating these two expressions for ϵ_{rot} gives us

$$\pi^2 mL^2 f^2 = \tfrac{1}{2}k_BT$$

Thus the rotational frequency is

$$f = \sqrt{\frac{k_BT}{2\pi^2 mL^2}} = 7.8 \times 10^{11} \text{ rev/s}$$

We evaluated f at $T = 293$ K, using $m = 14$ u $= 2.34 \times 10^{-26}$ kg for each *atom*.

ASSESS This is a *very* high frequency, but these values are typical of molecular rotations.

STOP TO THINK 13.3 How many degrees of freedom does a bead on a rigid rod have?

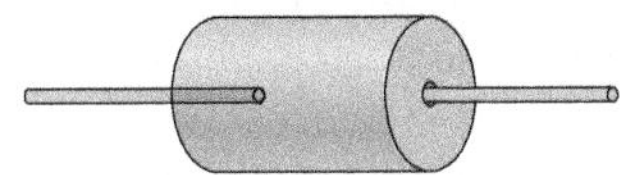

A. 1 B. 2 C. 3 D. 4 E. 5 F. 6

Heat Transfer and Thermal Equilibrium

What happens when two systems at different temperatures interact with each other? In Chapter 12 we said that "heat is transferred," but what does that mean? What is the *mechanism* by which energy is transferred from a hotter object to a colder object? An atomic-level view will improve our understanding of heat and thermal equilibrium.

FIGURE 13.12 shows a rigid, insulated container divided into two sections by a very thin membrane. The left side, which we'll call system 1, has N_1 atoms at an initial temperature T_{1i}. System 2 on the right has N_2 atoms at an initial temperature T_{2i}. The membrane is so thin that atoms can collide at the boundary as if the membrane were not there, yet it is a barrier that prevents atoms from moving from one side to the other. The situation is analogous, on an atomic scale, to basketballs colliding through a shower curtain.

FIGURE 13.12 Two gases can interact thermally through a very thin barrier.

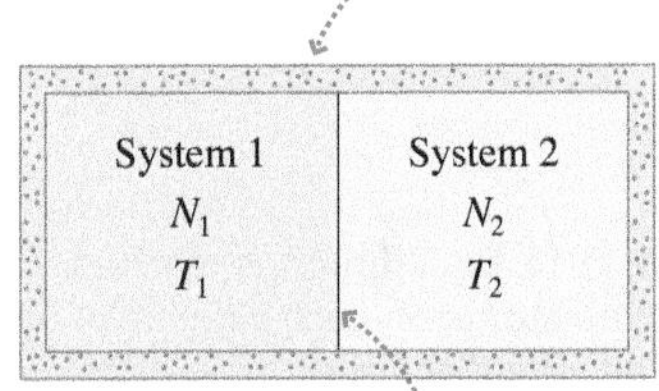
Insulation prevents heat from entering or leaving the container.
System 1
N_1
T_1
System 2
N_2
T_2
A thin barrier prevents atoms from moving from system 1 to 2 but still allows them to collide. The barrier is clamped in place and cannot move.

Suppose that system 1 is initially at a higher temperature: $T_{1i} > T_{2i}$. This is not an equilibrium situation. The temperatures will change with time until the systems eventually reach a common final temperature T_f. If you *watch* the gases as one warms and the other cools, you see nothing happening. This interaction is quite different from a mechanical interaction in which, for example, you might see a piston move from one side toward the other. The only way in which the gases can interact is via molecular collisions at the boundary. This is a *thermal interaction,* and our goal is to understand how thermal interactions bring the systems to thermal equilibrium.

System 1 and system 2 begin with thermal energies

$$\begin{aligned} E_{1i} &= \tfrac{3}{2}N_1 k_B T_{1i} = \tfrac{3}{2}n_1 R T_{1i} \\ E_{2i} &= \tfrac{3}{2}N_2 k_B T_{2i} = \tfrac{3}{2}n_2 R T_{2i} \end{aligned} \tag{13.41}$$

We've written the energies for monatomic gases; you could do the same calculation if one or both of the gases is diatomic by replacing the $\frac{3}{2}$ with $\frac{5}{2}$. Notice that we've omitted the subscript "th" to keep the notation manageable.

The total energy of the combined systems is $E_{tot} = E_{1i} + E_{2i}$. As systems 1 and 2 interact, their individual thermal energies E_1 and E_2 can change but their sum E_{tot} remains constant. The system will have reached thermal equilibrium when the individual thermal energies reach final values E_{1f} and E_{2f} that no longer change.

FIGURE 13.13 shows a fast atom and a slow atom approaching the barrier from opposite sides. They undergo a perfectly elastic collision at the barrier. Although no net energy is lost in a perfectly elastic collision, in most such collisions the more-energetic atom loses energy while the less-energetic atom gains energy. In other words, there's an energy *transfer* from the more-energetic atom's side to the less-energetic atom's side.

FIGURE 13.13 On average, collisions transfer energy from more-energetic atoms to less-energetic atoms.

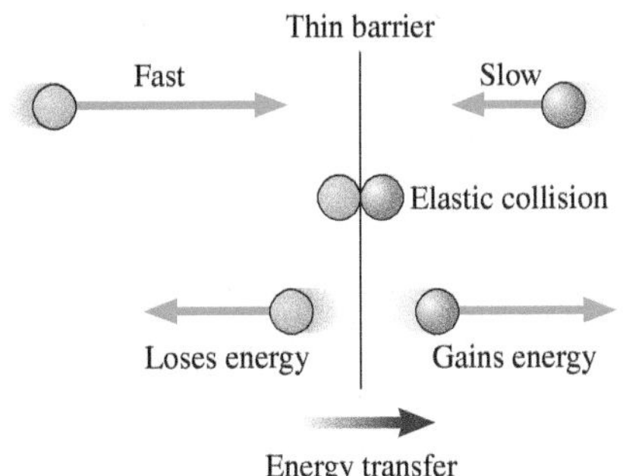
Thin barrier
Fast
Slow
Elastic collision
Loses energy
Gains energy
Energy transfer

The average translational kinetic energy per atom is directly proportional to the temperature: $\epsilon_{avg} = \frac{3}{2}k_B T$. Because $T_{1i} > T_{2i}$, the atoms in system 1 are, on average, more energetic than the atoms in system 2. Thus *on average* the collisions transfer energy from system 1 to system 2. Not in every collision: sometimes a fast atom in system 2 collides with a slow atom in system 1, transferring energy from 2 to 1. But the net energy transfer, from all collisions, is from the warmer system 1 to the cooler system 2. In other words, **heat is the energy transferred *via collisions* between the more-energetic (warmer) atoms on one side and the less-energetic (cooler) atoms on the other.**

How do the systems "know" when they've reached thermal equilibrium? Energy transfer continues until the atoms on both sides of the barrier have the *same average translational kinetic energy.* Once the average translational kinetic energies are the same, there is no tendency for energy to flow in either direction. This is the state of thermal equilibrium, so the condition for thermal equilibrium is $(\epsilon_1)_{avg} = (\epsilon_2)_{avg}$, where, as before, ϵ is the translational kinetic energy of an atom.

Because the average energies are directly proportional to the final temperatures, $\epsilon_{avg} = \frac{3}{2}k_B T_f$, thermal equilibrium is characterized by the macroscopic condition

$$T_{1f} = T_{2f} = T_f \tag{13.42}$$

Condition for thermal equilibrium

In other words, **two thermally interacting systems reach a common final temperature *because* they exchange energy via collisions until the atoms on each side have, on average, equal translational kinetic energies.** This is a very important idea.

Equation 13.42 can be used to determine the equilibrium thermal energies. Because these are monatomic gases, $E_{th} = N\epsilon_{avg}$. Thus the equilibrium condition $(\epsilon_1)_{avg} = (\epsilon_2)_{avg} = (\epsilon_{tot})_{avg}$ implies

$$\frac{E_{1f}}{N_1} = \frac{E_{2f}}{N_2} = \frac{E_{tot}}{N_1 + N_2} \tag{13.43}$$

from which we can conclude

$$E_{1f} = \frac{N_1}{N_1 + N_2}E_{tot} = \frac{n_1}{n_1 + n_2}E_{tot}$$
$$E_{2f} = \frac{N_2}{N_1 + N_2}E_{tot} = \frac{n_2}{n_1 + n_2}E_{tot} \tag{13.44}$$

where in the last step we used moles rather than molecules.

Notice that $E_{1f} + E_{2f} = E_{tot}$, verifying that energy has been conserved even while being redistributed between the systems.

No work is done on either system because the barrier has no macroscopic displacement, so the first law of thermodynamics is

$$Q_1 = \Delta E_1 = E_{1f} - E_{1i}$$
$$Q_2 = \Delta E_2 = E_{2f} - E_{2i} \tag{13.45}$$

As a homework problem you can show that $Q_1 = -Q_2$, as required by energy conservation. That is, the heat lost by one system is gained by the other. $|Q_1|$ is the quantity of heat that is transferred from the warmer gas to the cooler gas during the thermal interaction.

NOTE ▸ In general, the equilibrium thermal energies of the system are *not* equal. That is, $E_{1f} \neq E_{2f}$. They will be equal only if $N_1 = N_2$. Equilibrium is reached when the average translational kinetic energies in the two systems are equal—that is, when $(\epsilon_1)_{avg} = (\epsilon_2)_{avg}$, not when $E_{1f} = E_{2f}$. The distinction is important. FIGURE 13.14 summarizes these ideas. ◂

FIGURE 13.14 Equilibrium is reached when the atoms on each side have, on average, equal energies.

Collisions transfer energy from the warmer system to the cooler system as more-energetic atoms lose energy to less-energetic atoms.

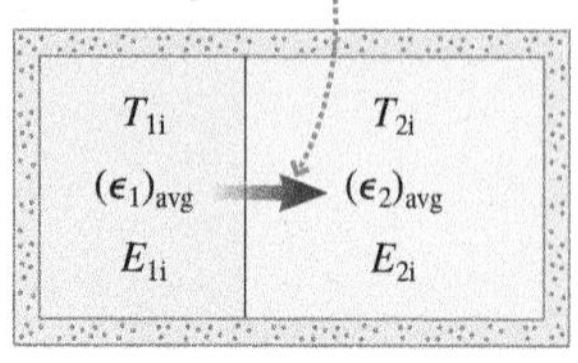

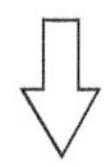

Thermal equilibrium occurs when the systems have the same average translational kinetic energy and thus the same temperature.

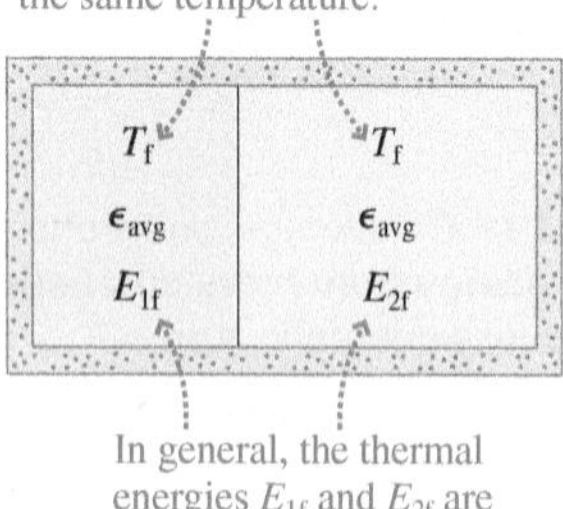

In general, the thermal energies E_{1f} and E_{2f} are *not* equal.

EXAMPLE 13.7 A thermal interaction

A sealed, insulated container has 2.0 g of helium at an initial temperature of 300 K on one side of a barrier and 10.0 g of argon at an initial temperature of 600 K on the other side.

a. How much heat energy is transferred, and in which direction?
b. What is the final temperature?

PREPARE The systems start with different temperatures, so they are not in thermal equilibrium. Energy will be transferred via collisions from the argon to the helium until both systems have the same average molecular energy.

SOLVE

a. Let the helium be system 1. Helium has molar mass $M_{mol} =$ 0.004 kg/mol, so $n_1 = M/M_{mol} = 0.50$ mol. Similarly, argon has $M_{mol} = 0.040$ kg/mol, so $n_2 = 0.25$ mol. The initial thermal energies of the two monatomic gases are

$$E_{1i} = \tfrac{3}{2}n_1RT_{1i} = 225R = 1870\text{ J}$$
$$E_{2i} = \tfrac{3}{2}n_2RT_{2i} = 225R = 1870\text{ J}$$

The systems start with *equal* thermal energies, but they are not in thermal equilibrium. The total energy is $E_{tot} = 3740$ J. In equilibrium, this energy is distributed between the two systems as

$$E_{1f} = \frac{n_1}{n_1 + n_2}E_{tot} = \frac{0.50}{0.75} \times 3740\text{ J} = 2493\text{ J}$$
$$E_{2f} = \frac{n_2}{n_1 + n_2}E_{tot} = \frac{0.25}{0.75} \times 3740\text{ J} = 1247\text{ J}$$

The heat entering or leaving each system is

$$Q_1 = Q_{He} = E_{1f} - E_{1i} = 623\text{ J}$$
$$Q_2 = Q_{Ar} = E_{2f} - E_{2i} = -623\text{ J}$$

The helium and the argon interact thermally via collisions at the boundary, causing 623 J of heat to be transferred from the warmer argon to the cooler helium.

b. These are constant-volume processes, thus $Q = nC_V\Delta T$. $C_V = \tfrac{3}{2}R$ for monatomic gases, so the temperature changes are

$$\Delta T_{He} = \frac{Q_{He}}{\tfrac{3}{2}nR} = \frac{623\text{ J}}{1.5(0.50\text{ mol})(8.31\text{ J/mol}\cdot\text{K})} = 100\text{ K}$$
$$\Delta T_{Ar} = \frac{Q_{Ar}}{\tfrac{3}{2}nR} = \frac{-623\text{ J}}{1.5(0.25\text{ mol})(8.31\text{ J/mol}\cdot\text{K})} = -200\text{ K}$$

Both gases reach the common final temperature $T_f = 400$ K.

ASSESS $E_{1f} = 2E_{2f}$ because there are twice as many atoms in system 1.

The main idea of this section is that two systems reach a common final temperature not by magic or by a prearranged agreement but simply from the energy exchange of vast numbers of molecular collisions. Real interacting systems, of course, are separated by walls rather than our unrealistic thin membrane. As the systems interact, the energy is first transferred via collisions from system 1 into the wall and subsequently, as the cooler molecules collide with a warm wall, into system 2. That is, the energy transfer is $E_1 \rightarrow E_{wall} \rightarrow E_2$. This is still heat because the energy transfer is occurring via molecular collisions rather than mechanical motion.

STOP TO THINK 13.4 Systems A and B are interacting thermally. At this instant of time,

A. $T_A > T_B$
B. $T_A = T_B$
C. $T_A < T_B$

A	B
$N = 1000$	$N = 2000$
$\epsilon_{avg} = 1.0 \times 10^{-20}$ J	$\epsilon_{avg} = 0.5 \times 10^{-20}$ J
$E_{th} = 1.0 \times 10^{-17}$ J	$E_{th} = 1.0 \times 10^{-17}$ J

13.5 k_BT and the Boltzmann Factor

A molecule undergoes billions of collisions every second. Each collision changes the molecule's speed and kinetic energy, but *on average* its translational kinetic energy at temperature T is

$$\epsilon_{avg} = \tfrac{3}{2}k_BT$$

where, you will recall, the Boltzmann constant $k_B = 1.38 \times 10^{-23}$ J/K is effectively the "gas constant per molecule." We found this result in our analysis of an ideal gas (Equation 13.27), and it also follows from the equipartition theorem.

The quantity k_BT is important, and it occurs throughout kinetic theory and statistical physics. **k_BT sets the scale of atomic-level energies** in the sense that k_BT is a typical molecular energy. The translational kinetic energy of a molecule is roughly k_BT. The kinetic and potential energies of vibrating molecules are roughly k_BT. It's true that molecules have a range or distribution of energies—that's what the histogram of molecular speeds in Figure 13.2 shows—but nearly all molecular energies are within a factor of 2 or 3 of k_BT; very few molecules have an energy that is either much less or much greater than k_BT. As you'll see, this has implications that range from the rates of chemical reactions to the physics underlying magnetic resonance imaging (MRI).

It is straightforward to calculate that $k_B T = 4.0 \times 10^{-21}$ J at room temperature. Thus a room-temperature molecule, whether it's H_2 or a macromolecule, has a kinetic energy of approximately 4×10^{-21} J. Because very light and very heavy molecules have roughly the same kinetic energies, a light atom or molecule has a much larger rms speed than a heavy macromolecule.

EXAMPLE 13.8 **How fast do water and protein molecules move?**

Just as gas molecules are constantly moving and colliding, so are water molecules and molecules in solution. Figure 13.2 showed that, at room temperature, a typical nitrogen molecule moves with a speed of ≈ 450 m/s. The molecular mass of water is 18 Da = 18 u. A fairly typical protein has a molecular mass of ≈ 100 kDa $= 10^5$ u. Estimate the speed of a water molecule and of a typical protein in the cell.

PREPARE All molecules, whether light or heavy, have the same average kinetic energy at the same temperature. The temperature inside a cell is slightly higher than room temperature, but we can overlook the difference for the purpose of making an estimate.

SOLVE We can solve this problem with ratio reasoning. On average,

$$K_{N_2} = \tfrac{1}{2} m_{N_2} (v_{N_2})^2 = K_{H_2O} = \tfrac{1}{2} m_{H_2O} (v_{H_2O})^2$$
$$= K_{protein} = \tfrac{1}{2} m_{protein} (v_{protein})^2$$

Thus

$$v_{H_2O} = \sqrt{\frac{m_{N_2}}{m_{H_2O}}} v_{N_2} = \sqrt{\frac{28\text{ u}}{18\text{ u}}} \times 450 \text{ m/s} \approx 560 \text{ m/s}$$

Similarly,

$$v_{protein} = \sqrt{\frac{m_{N_2}}{m_{protein}}} v_{N_2} = \sqrt{\frac{28\text{ u}}{100{,}000\text{ u}}} \times 450 \text{ m/s} \approx 8 \text{ m/s}$$

ASSESS These are *average* speeds, but the speed of any specific molecule won't differ from this by much. Proteins and other large macromolecules lumber through the cell at a few meters per second, while water molecules and other light molecules flit about at hundreds of meters per second. Even so, the *kinetic energy* of a macromolecule is essentially the same as that of a water molecule. Keep in mind, though, that these molecules move only a fraction of a nanometer between collisions, so their motion is jittery, not straight-line motion. We'll return to this idea later in the chapter when we discuss diffusion.

Populations

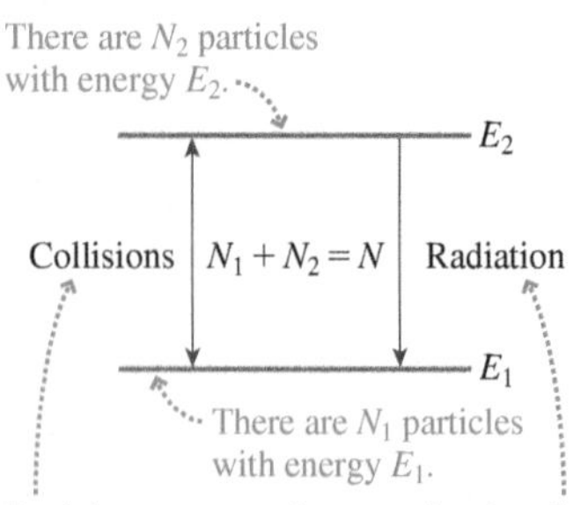
There are N_2 particles with energy E_2.
E_2
Collisions
$N_1 + N_2 = N$
Radiation
E_1
There are N_1 particles with energy E_1.
Particles can move from one level to the other due to radiation and collisions.

FIGURE 13.15 The populations in a two-level system are influenced by both radiation and collisions.

You learned in chemistry that energy is *quantized,* which means that atoms and molecules have discrete energy levels. Suppose we have a system of N atomic particles, each of which can be in one of only two energy levels, E_1 or E_2. FIGURE 13.15 shows that there are N_1 particles in the lower energy level and N_2 in the upper energy level, and these add up to the total number of particles: $N_1 + N_2 = N$. We say that N_1 and N_2 are the *populations* of the two energy levels.

A particle doesn't stay in one energy level. A quantum system can radiate electromagnetic waves in the form of discrete photons, so there's a tendency for the particles in the upper energy level to jump to the lower energy level by emitting a photon. At the same time, a collision between two particles could, if the particles have sufficient kinetic energy, move a particle from the lower energy level to the upper energy level. In other words, collisions and radiation are constantly causing particles to move from one energy level to the other.

Suppose the system of N particles is in equilibrium, which means that the populations N_1 and N_2 of the two energy levels are stable. What can we say about how many particles are in each level?

We'll use the notation $\delta E = E_2 - E_1$ to represent the *energy separation* between the two energy levels. First, let's suppose the temperature T is low enough to make $k_B T \ll \delta E$. We've seen that $k_B T$ is roughly the kinetic energy of each particle, so the maximum kinetic energy available in a collision is only a few $k_B T$. Because $\delta E \gg k_B T$, this is not enough energy to move a particle from energy level E_1 to energy level E_2. Only rarely can a particle gain enough energy in a collision to make it to the upper level. Further, any particle that does reach the upper level quickly drops back to the lower level, emitting a photon in the process. Thus we expect $N_1 \approx N$ and $N_2 \approx 0$. In other words, at low temperatures ($k_B T \ll \delta E$) we expect that essentially all of the population will be in the lower energy level, none in the upper energy level.

What about the opposite situation: a high temperature with $k_BT \gg \delta E$? Now every collision has more than enough energy to move a particle from E_1 to E_2, or to drive it from E_2 back to E_1, so we expect that collisions will be far more important than radiation, leading to $N_1 \approx N_2 \approx N/2$. The point—and it's an important point—is that **k_BT determines how a population of atomic particles is distributed among the various possible energy levels.**

A detailed analysis that is beyond the scope of this text finds that the *ratio* of the populations of the two levels is

$$\frac{N_2}{N_1} = e^{-\delta E/k_BT} \tag{13.46}$$

Populations in a two-level system at temperature T

The exponential term $e^{-\delta E/k_BT}$ is called the **Boltzmann factor.**

At low temperatures, where $k_BT \ll \delta E$, the term $\delta E/k_BT$ in the exponent of the Boltzmann factor is a large number: $\delta E/k_BT \gg 1$. But there's a minus sign in the exponent, and you'll recall that $e^{-x} \to 0$ as $x \to \infty$. Thus the Boltzmann factor predicts that $N_2 \approx 0$ and $N_1 \approx N$ at low temperatures, consistent with our reasoning above.

At high temperatures, $k_BT \gg \delta E$ makes $\delta E/k_BT \approx 0$. You will recall that $e^0 = 1$, so the Boltzmann factor is very close to one. Thus $N_2 \approx N_1 \approx N/2$ and at high temperatures the two levels are almost equally populated, exactly as we had foreseen.

By combining $N_1 + N_2 = N$ with $N_2 = N_1 e^{-\delta E/k_BT}$, we can solve for N_1 and N_2:

$$N_1 = \frac{1}{1 + e^{-\delta E/k_BT}}N \qquad N_2 = \frac{e^{-\delta E/k_BT}}{1 + e^{-\delta E/k_BT}}N \tag{13.47}$$

FIGURE 13.16 shows graphs of N_1 and N_2 as a function of k_BT. You can see that the low-temperature behavior and the high-temperature behavior are as predicted. The temperature at which $k_BT = \delta E$ finds $N_1 = 0.73N$ and $N_2 = 0.27N$, each about midway between its two extremes, so $k_BT = \delta E$ represents a balance between photon emissions that move particles to the lower level and collisions that move particles to the upper level.

FIGURE 13.16 Graphs of the populations of a two-level system as the temperature is increased.

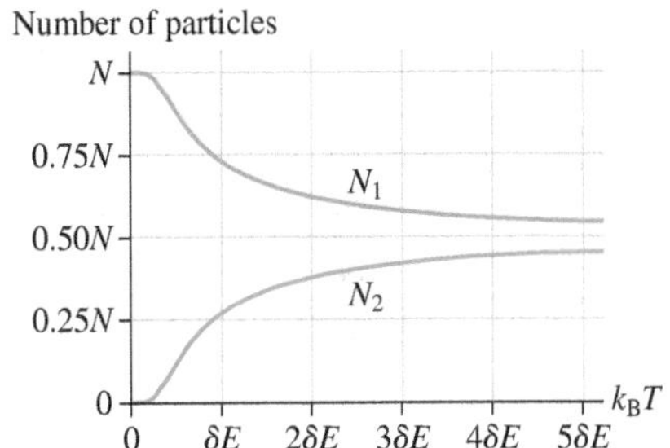

EXAMPLE 13.9 Proton populations in magnetic resonance imaging

Magnetic resonance imaging (MRI), which we'll study in Chapter 26, depends on the fact that a proton (e.g., the nucleus of a hydrogen atom) in a magnetic field has two possible energy levels: a lower-energy "spin aligned" level in which the proton's inherent spin is aligned with the magnetic field and a higher-energy "spin anti-aligned" level in which the proton's spin points opposite to the magnetic field. The "resonance" in MRI comes about by tuning high-frequency radio waves until they are at just the right frequency to match the energy difference δE between the two proton energy levels. At resonance, the radio waves cause the proton spins to flip back and forth between the two orientations. Detecting this spin flipping produces an MRI signal with strength proportional to $\Delta N/N$, where $\Delta N = N_1 - N_2$ is the population difference. In other words, detecting a signal depends on there being more energy-absorbing spins flipping from aligned to anti-aligned (E_1 to E_2) than energy-emitting spins flipping from anti-aligned to aligned (E_2 to E_1). A commercial MRI machine generates a magnetic field for which the energy difference between the spin states is $\delta E = 8.4 \times 10^{-26}$ J, corresponding to an MRI frequency of 128 MHz. What is $\Delta N/N$ at room temperature?

PREPARE The populations of the two states are given by Equation 13.47. We've noted that $k_BT = 4.0 \times 10^{-21}$ J at room temperature. This is a situation with $k_BT \gg \delta E$, so we expect to find $N_2 \approx N_1$ and thus $\Delta N/N \ll 1$.

SOLVE The Boltzmann factor is $e^{-\delta E/k_BT}$. For this MRI machine,

$$\frac{\delta E}{k_BT} = \frac{8.4 \times 10^{-26}\ \text{J}}{4.0 \times 10^{-21}\ \text{J}} = 2.1 \times 10^{-5}$$

Thus the two populations are

$$N_1 = \frac{1}{1 + e^{-\delta E/k_BT}}N = \frac{1}{1 + e^{-0.000021}}N = 0.500005N$$

$$N_2 = \frac{e^{-\delta E/k_BT}}{1 + e^{-\delta E/k_BT}}N = \frac{e^{-0.000021}}{1 + e^{-0.000021}}N = 0.499995N$$

Continued

The populations are almost identical, with a tiny fractional difference of only

$$\frac{\Delta N}{N} = \frac{N_1 - N_2}{N} = 0.500005 - 0.499995 = 10 \times 10^{-6}$$
$$= 10 \text{ parts per million}$$

ASSESS The energy difference between the two spin states is very small compared to $k_B T$. The Boltzmann factor is very close to one, and thermal interactions see to it that the spin aligned and anti-aligned populations are almost the same. Even so, the very slight excess of spins flipping from aligned to anti-aligned over those flipping from anti-aligned to aligned, a difference of only 10 parts per million, can be detected, and that is the basis for generating an MRI image. The signal would be stronger, allowing a more detailed MRI image, if $\delta E/k_B T$ were larger. The temperature of a patient cannot be significantly lowered, but the energy separation δE can be increased by using a stronger magnetic field. The magnetic field strengths used in MRI have increased over the years and now are very near the limits of magnet technology.

Real molecular systems almost always have many possible energy levels, not just two. Suppose a system has energy levels E_i numbered $i = 1, 2, 3, \ldots$. We can generalize Equation 13.46 by writing the population of energy level i as

$$N_i = \text{population of energy level } i = Ce^{-E_i/k_B T} \qquad (13.48)$$

where C is a constant. We again see the Boltzmann factor, but this time written directly in terms of the energy E_i rather than δE. Equation 13.46 follows from this more general statement if we calculate the ratio N_2/N_1.

FIGURE 13.17 shows the idea graphically in a system with 10 equally spaced energy levels (that is, $E_i = E_1, 2E_1, 3E_1, \ldots$ for $i = 1, 2, 3, \ldots$) at a temperature T such that $k_B T = 3E_1$. You can see that **the population of energy levels decreases exponentially with increasing energy.** Energy levels with $E_i < k_B T$ are well populated, energy levels with $E_i > k_B T$ have rapidly decreasing populations, and any energy levels with $E_i \gg k_B T$ have essentially no populations. Once again, $k_B T$ determines how a population of atomic particles is distributed among the various possible energy levels. Raising or lowering the temperature has a huge impact on which energy levels are populated.

FIGURE 13.17 Histogram showing the populations of 10 energy levels at a temperature T such that $k_B T = 3E_1$.

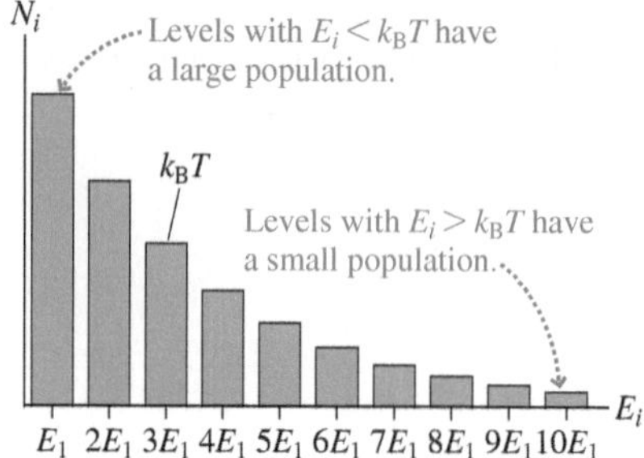

STOP TO THINK 13.5 In a two-level system with energy separation δE, the population ratio N_2/N_1 can be increased by (choose all that apply)

A. Increasing δE.
B. Decreasing δE.
C. Increasing T.
D. Decreasing T.
E. Increasing N.

The Maxwell-Boltzmann Speed Distribution

Our study of thermodynamics has been a classical description in terms of molecular speeds, not a quantum description in terms of energy levels. Rather than finding the probability that a molecule is in a particular energy level, we'd rather know the probability that a molecule has a particular speed. The process of translating from energy levels to speeds is the topic of an advanced course in physical chemistry or statistical physics, so here we will present the result without proof.

Energy levels are discrete. A particle is in energy level i or energy level j, but not in between. Thus it makes sense to talk about the populations N_i and N_j of these energy levels. But when it comes to molecular speeds, we can't specify a mathematically *exact* speed. Speed varies continuously, rather than discretely, and it would make no sense to talk about the number of molecules that have an exact value of speed.

However, we can talk about the number or the fraction of molecules that fall within a specified *range* of speeds. That's exactly what the histogram of nitrogen speeds in Figure 13.2 showed. There we saw, for example, that 12% of a group of room-temperature nitrogen molecules have a speed between 600 m/s and 700 m/s.

Now let's narrow the range to those molecules that have a speed between v and $v + dv$, where dv is very small but not zero. For example, we might look at molecules that have a speed between 600.00 m/s and 600.01 m/s, in which case $dv = 0.01$ m/s. Given a sample of molecules of mass m at temperature T, the fraction that have speeds in the range from v to $v + dv$ turns out to be

$$\text{fraction that have speeds between } v \text{ and } v + dv = P(v)\,dv \qquad (13.49)$$

where

$$P(v) = 4\pi\left(\frac{m}{2\pi k_B T}\right)^{3/2} v^2 e^{-mv^2/2k_B T} \qquad (13.50)$$

Equation 13.50 is called the **Maxwell-Boltzmann distribution** of speeds.

You can recognize the last term in the expression as the Boltzmann factor $e^{-E/k_B T}$ with $E = \frac{1}{2}mv^2$ for the molecular kinetic energy. Otherwise, it's hard to make much sense of Equation 13.50 without graphing it, which we do in FIGURE 13.18 for three different gases at 20°C. Please compare this to Figure 13.2 to see that the curve for nitrogen is a continuous version of the experimental histogram. The atoms of the heavy gas xenon move much more slowly than those of the light gas helium, but their average kinetic energy is the same.

The quantity $P(v)$ is called a *distribution function* because it tells us how the probability is *distributed* over a range of speeds. What we find is that the never-ending collisions see to it that most molecules have a speed within about a factor of 2 of $v_{\text{most prob}}$, the *most probable* speed that corresponds to the maximum of the curve. There's a distribution of speeds, but it's not a wide distribution. We'll leave it as a homework problem for you to prove that the peak occurs when

$$\epsilon_{\text{most prob}} = \tfrac{1}{2}m(v_{\text{most prob}})^2 = k_B T \qquad (13.51)$$

Thus the significance of $k_B T$ appears once again.

FIGURE 13.18 The Maxwell-Boltzmann distribution of molecular speeds at 20°C. The numerical values of $P(v)$ are not important.

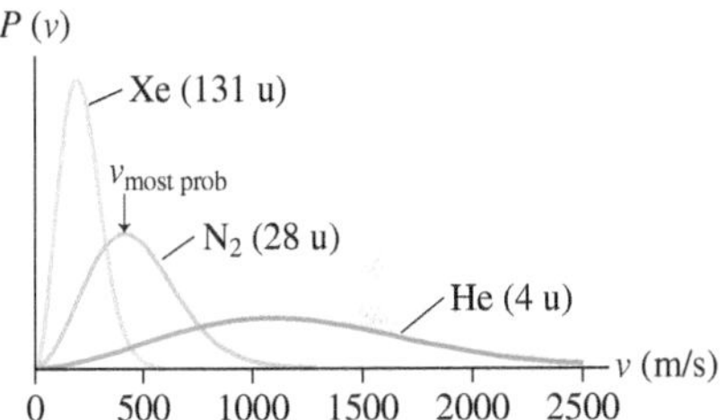

13.6 Reaction Kinetics and Catalysis

The hydrolysis of ATP to ADP and inorganic phosphate happens quickly in the cell, in the presence of the enzyme ATPase, but a solution of pure ATP is quite stable even though it is not in equilibrium. What's the reason for the difference? What determines whether a chemical reaction runs quickly or slowly? Our goal in this section is to see how the Boltzmann factor governs the rate of a reaction. The study of how reactions evolve with time at the molecular level is called *reaction kinetics*.

Consider a generic chemical reaction that goes to completion:

$$\text{reactants} \rightarrow \text{products} \qquad (13.52)$$

FIGURE 13.19 shows that if we start with all reactants, then over some period of time the number of reactant molecules decreases and the number of product molecules increases. Our interest is not the chemical details but the physics of what happens in a reaction and what determines the rate of a reaction.

◀◀ SECTION 11.5 introduced two important ideas about how a chemical reaction comes about:

- Molecules have thermal energy and are in motion. A reaction is initiated when two reactant molecules *collide* with each other.
- If the colliding molecules have sufficient kinetic energy, the collision may break the molecular bonds that hold the reactant molecules together and allow the formation of new bonds. The newly bonded molecules are the products of the reaction.

FIGURE 13.20 (next page) illustrates the idea; part a shows the collision and part b the molecular energy diagram. The energy required to break the bonds of the reactants is

FIGURE 13.19 Time dependence of a chemical reaction.

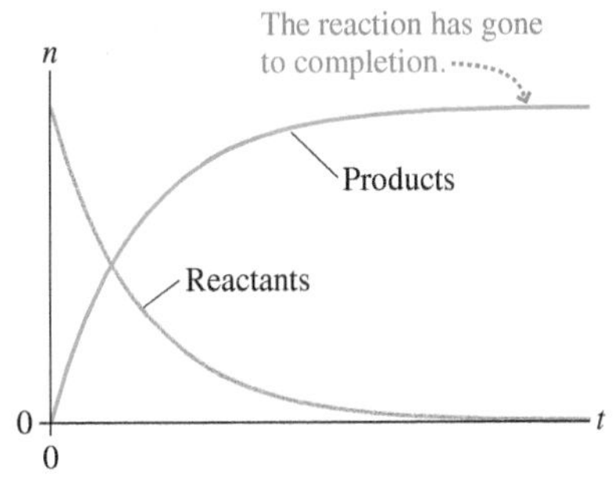

FIGURE 13.20 A reaction can take place if the collision energy of the reactants exceeds the activation energy E_a seen on the molecular energy diagram.

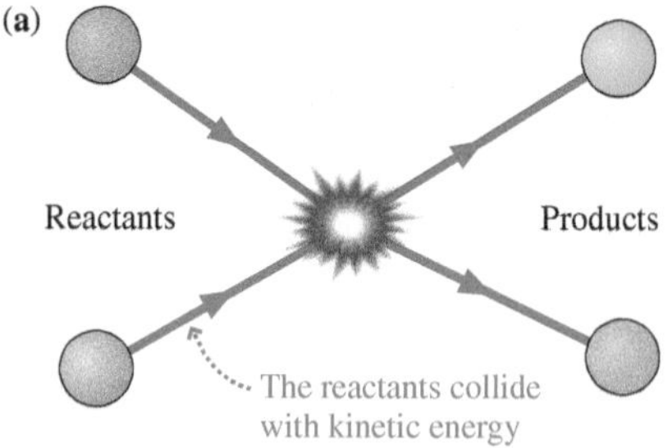
(a)
Reactants
Products
The reactants collide with kinetic energy

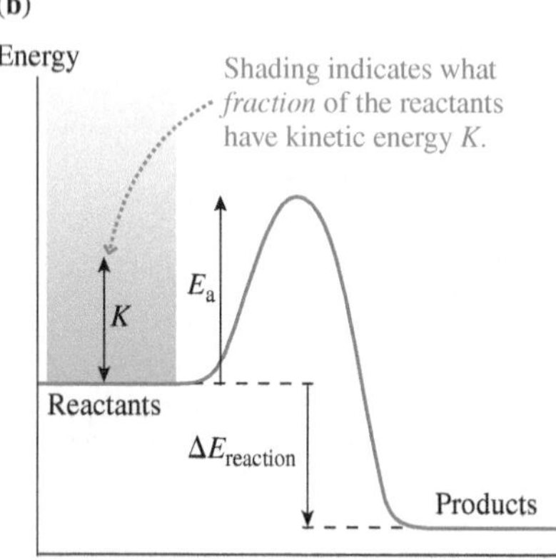
(b)
Energy
Shading indicates what fraction of the reactants have kinetic energy K.
K
Ea
Reactants
ΔEreaction
Products
Reaction coordinate

called the *activation energy* E_a, and you saw in Chapter 11 that the activation energy appears as a barrier on the energy diagram.

The Reaction Rate Constant

Reactions go faster at higher temperatures. We can use kinetic theory and the Boltzmann factor to predict the temperature dependence of a reaction rate. The overall reaction rate R—the rate in mol/s at which reactants are consumed and products are generated—depends on the concentrations of the reactant and product molecules; we'll leave those details to chemistry. The essential result is that, for any reaction, we can write

$$\text{reaction rate } R = k \times \text{concentration factors} \tag{13.53}$$

where all the physics of the reaction—how it depends on energy and temperature—are contained in the proportionality constant k, called the **rate constant** for the reaction.

NOTE ▶ Chemists invariably use the symbol k for the rate constant, although this is yet another use of the letter k. We also use the Boltzmann constant k_B in this chapter, so don't forget to include the subscript B on the latter. ◀

The energy available to the reactant molecules is their thermal energy of random motion. As you learned in the preceding section, the fraction of reactants that have kinetic energy K is proportional to the Boltzmann factor e^{-K/k_BT}. This fraction is shaded in Figure 13.20b. Most of the molecules collide with kinetic energy less than the activation energy, $K < E_a$, and energy conservation tells us that these collisions cannot initiate a reaction by breaking the bonds.

But you can see that, in this case, the shading extends higher than the activation-energy barrier. If molecules collide with a total kinetic energy that exceeds the activation-energy barrier, $K > E_a$, then there's a possibility that their kinetic energy can be used to break bonds. The rate at which the reaction proceeds is determined by the *fraction* of the collisions that have $K > E_a$, a fraction that is proportional to e^{-E_a/k_BT}.

The overall reaction rate depends on both the rate of molecular collisions, which is determined by the concentrations, *and* on the probability that a collision will break the bonds. We can capture this idea by writing the rate constant as

$$k = Ae^{-E_a/k_BT} \tag{13.54}$$

where A is a constant that depends on the specific reaction. Note that the rate "constant" is not really a constant because it depends on temperature. If E_a is too large or T too low, the shading in Figure 13.20b will not extend as high as the activation-energy barrier and the reaction, though energetically favorable, will not go.

Before looking at the significance of Equation 13.54, we need to switch from a physicist's perspective (where the activation energy is measured in joules) to a chemist's perspective (where the activation energy is given in J/mol or, more typically, kJ/mol). We can do so by multiplying both the numerator and denominator of the exponent in the Boltzmann factor by Avogadro's number N_A and recalling the relationship between the Boltzmann constant and Avogadro's number: $k_B = R/N_A$. Doing so gives

$$\frac{E_a \text{ (in J)}}{k_BT} \times \frac{N_A}{N_A} = \frac{E_a \text{ (in J/mol)}}{(N_Ak_B)T} = \frac{E_a \text{ (in J/mol)}}{RT}$$

Thus the chemist's version of Equation 13.54 is

$$k = Ae^{-E_a/RT} \tag{13.55}$$

Temperature dependence of a reaction rate

with the activation energy E_a in J/mol.

No fire without a spark Fuel, whether it's candle wax, gasoline, or gunpowder, doesn't spontaneously combust. Combustion begins with a flame or a spark that increases the temperature of just a small portion of the fuel to the *ignition point,* the temperature at which the highly exothermic combustion reaction can go quickly. The large kinetic energy of the product molecules is then transferred, by collisions, to nearby reactant molecules, giving them sufficient energy to react. Once initiated, the reaction is self-sustaining.

Equation 13.55, which gives the temperature dependence of reaction rates, is known as the **Arrhenius equation.** It was proposed in 1888 by Swedish chemist Svante Arrhenius, the scientist who first recognized the role that carbon dioxide plays in regulating earth's temperature and first suggested that burning fossil fuels could alter the earth's climate.

FIGURE 13.21 shows how the rate constant depends on temperature for a reaction that has an activation energy of 100 kJ/mol, a typical value, over a temperature range of 290 K (17°C) to 350 K (77°C). This is not an especially large temperature range, but the Arrhenius equation shows that $k_{350} \approx 1300k_{290}$, a remarkable increase in the reaction rate.

FIGURE 13.21 Temperature dependence of a rate constant for a reaction that has an activation energy of 100 kJ/mol.

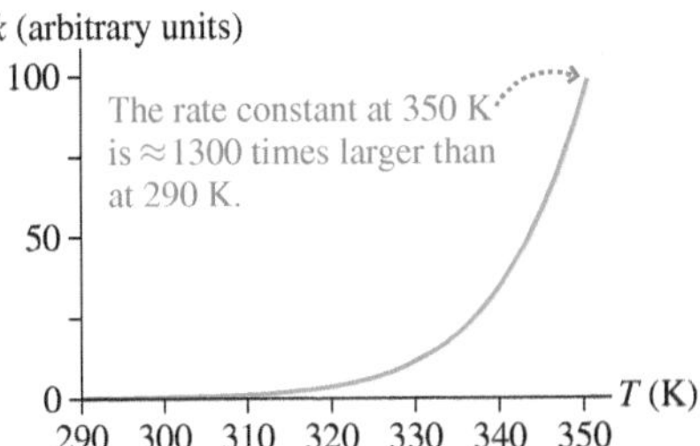

EXAMPLE 13.10 Determining a reaction rate constant

The reaction $CO_2(g) + H_2O(l) \rightarrow HCO_3^- + H^+$, in which carbon dioxide dissolves in water to produce the bicarbonate ion HCO_3^-, is exceptionally important for buffering pH. Laboratory measurements find that the rate constant for the uncatalyzed reaction is $0.0171\ M^{-1}s^{-1}$ at 20°C and a larger $0.106\ M^{-1}s^{-1}$ at 50°C. The unit M is the *molarity,* the concentration in mol/L. What is the activation energy for this reaction, and what is the rate constant at a body temperature of 37°C?

PREPARE The temperature dependence of the rate constant is described by the Arrhenius equation. The data points are at temperatures 293 K and 323 K. We can use natural logarithms to solve for the activation energy.

SOLVE The rate constant for this reaction is

$$k = Ae^{-E_a/RT}$$

We can isolate the activation energy by taking the natural logarithm of both sides and using the logarithm property $\ln(xy) = \ln x + \ln y$:

$$\ln k = \ln A + \ln(e^{-E_a/RT}) = \ln A - \frac{E_a}{RT}$$

There are two unknowns, A and E_a, but we have two data points. Subtracting $\ln(k_{293})$ from $\ln(k_{323})$ eliminates the $\ln A$ term:

$$\ln(k_{323}) - \ln(k_{293}) = \ln\left(\frac{k_{323}}{k_{293}}\right) = \left(-\frac{E_a}{R(323\text{ K})}\right) - \left(-\frac{E_a}{R(293\text{ K})}\right)$$

$$= \frac{E_a}{R}\left(\frac{1}{293\text{ K}} - \frac{1}{323\text{ K}}\right)$$

Solving this for the activation energy gives

$$E_a = R\ln\left(\frac{k_{323}}{k_{293}}\right) \times \left(\frac{1}{293\text{ K}} - \frac{1}{323\text{ K}}\right)^{-1} = 48\text{ kJ/mol}$$

We could now solve for the constant A, but it's easier to use ratios to calculate k_{310} at 37°C:

$$\frac{k_{310}}{k_{293}} = \frac{Ae^{-E_a/R(310\text{ K})}}{Ae^{-E_a/R(293\text{ K})}}$$

The A cancels, and we can then use $e^a/e^b = e^{a-b}$ to write

$$\frac{k_{310}}{k_{293}} = \exp\left(\frac{E_a}{R}\left(\frac{1}{293\text{ K}} - \frac{1}{310\text{ K}}\right)\right) = 2.95$$

Consequently, $k_{310} = 2.95k_{293} = 0.0504\ M^{-1}s^{-1}$.

ASSESS We previously noted that 100 kJ/mol is a typical activation energy, so 48 kJ/mol for this reaction is not unreasonable.

Catalysts

If you've ever poured hydrogen peroxide on a wound, you know that it fizzes and bubbles quite vigorously. Why?

Hydrogen peroxide decomposes to water and oxygen very slowly via the reaction $2H_2O_2 \rightarrow 2H_2O + O_2$ with rate constant $k \approx 10^{-8}\ s^{-1}$. At that rate, it would take more than a year for 50% of the hydrogen peroxide to decompose. But if we add the enzyme *catalase,* the reaction speeds up by a factor of nearly 10^{12} to $k \approx 6 \times 10^3\ s^{-1}$. Now half of the hydrogen peroxide decomposes in about 100 μs. The enzyme catalase happens to be present in almost all living tissue, and what you see when you pour hydrogen peroxide on a wound is oxygen bubbles from the rapidly catalyzed

► **It's not easy being green** Plant temperatures rise and fall with the temperature of the air, and that affects photosynthesis. The process of photosynthesis involves a series of catalyzed reactions, but, as in all reactions, the reaction rates change dramatically with temperature. For plants that use the C3 cycle, the most common photosynthetic cycle, experiments have found that the overall photosynthesis rate at 5°C is only about 20% of the rate at 30°C. One reason deciduous trees drop their leaves is that photosynthesis nearly ceases in winter, even on sunny days. Evergreen trees have adaptations that allow photosynthesis to continue throughout the winter, but only at a very low level.

decomposition of the hydrogen peroxide. (This is useful because the oxygen acts as a disinfectant.)

NOTE ▶ The units in which the rate constant k is written depend on the details of the reaction. The units were $M^{-1}s^{-1}$ in Example 13.10, but here they are s^{-1}. We don't need to be concerned about units because our use of the rate constant always involves *ratios* in which the units cancel. ◀

A **catalyst** is a material or molecule that speeds up a reaction rate without itself being consumed. Metal surfaces, particularly metals like platinum, are important catalysts in applications ranging from fuel cells to the *catalytic converter* that removes carbon monoxide from the exhaust of your car engine. Catalysts are essential to life because nearly all biochemical reactions are extremely slow in the absence of a catalyst. Most biological catalysts are proteins called *enzymes;* catalase, as its name suggests, is an exemplar.

An enzyme is a complex molecule with one or more *active sites* that binds to a very specific reactant molecule called, in biochemistry terminology, the *substrate.* ATP-powered processes can then alter the substrate or join it to another bound substrate molecule to form the *product.* The molecular details of how particular enzymes accomplish this are extraordinary and are an important topic in biochemistry. The essential *physical* idea is that the substrate-enzyme interactions allow alternative reaction paths that have lower activation energies than does the direct interaction of two substrate molecules. For example, the substrate-enzyme interaction might stress specific substrate bonds in such a way as to facilitate the breaking of those bonds.

FIGURE 13.22 A reaction energy diagram for a chemical reaction with and without a catalyst.

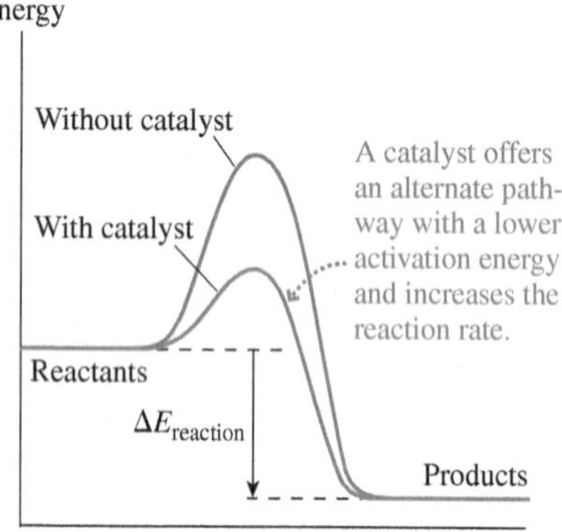
Energy
Without catalyst
With catalyst
A catalyst offers an alternate pathway with a lower activation energy and increases the reaction rate.
Reactants
$\Delta E_{reaction}$
Products
Reaction coordinate

Regardless of the details, the result of the substrate-enzyme interaction, shown in FIGURE 13.22, is to lower the energy barrier between reactants and products *without changing the reaction energy*. That is, the barrier is lowered but the overall result of the reaction, the energy released, is unchanged.

The effect of the catalyst on the reaction is given by the Arrhenius equation $k = Ae^{-E_a/RT}$. We've seen that a fairly small change in temperature can have a very large effect on the rate constant. Similarly, it doesn't take much lowering of the activation energy E_a to cause a large increase in the rate constant.

EXAMPLE 13.11 Measuring an activation energy

The hydrogen peroxide decomposition reaction $2H_2O_2 \rightarrow 2H_2O + O_2$ has a measured activation energy of 75.3 kJ/mol. At 300 K, the addition of the enzyme catalase changes the rate constant for this reaction from $1.0 \times 10^{-8}\ s^{-1}$ to $6.0 \times 10^{3}\ s^{-1}$. Assume that sufficient enzyme is added so that lack of the enzyme is not a limiting factor. What is the activation energy of the catalyzed reaction?

PREPARE The dependence of the rate constant on the activation energy is given by the Arrhenius equation $k = Ae^{-E_a/RT}$.

SOLVE The rate constant for the uncatalyzed reaction is

$$k = Ae^{-E_a/RT}$$

while that of the catalyzed reaction is

$$k_{cat} = Ae^{-E_{cat}/RT}$$

where we've represented the activation energy of the catalyzed reaction by E_{cat}. The constant A can be eliminated by using ratios:

$$\frac{k_{cat}}{k} = \frac{Ae^{-E_{cat}/RT}}{Ae^{-E_a/RT}} = e^{(E_a - E_{cat})/RT}$$

Taking the natural logarithm of both sides then gives

$$\ln\left(\frac{k_{cat}}{k}\right) = \frac{E_a - E_{cat}}{RT}$$

Finally, we can solve for the activation energy of the catalyzed reaction:

$$E_{cat} = E_a - RT\ln\left(\frac{k_{cat}}{k}\right)$$

$$= 75{,}300\ \text{J/mol} - (8.31\ \text{J/mol}\cdot\text{K})(300\ \text{K})\ln\left(\frac{6.0 \times 10^{3}\ s^{-1}}{1.0 \times 10^{-8}\ s^{-1}}\right)$$

$$= 7.7\ \text{kJ/mol}$$

ASSESS The activation energy of the catalyzed reaction is much less than that of the uncatalyzed reaction and only slightly higher than $RT = 2.4$ kJ/mol at room temperature, but that's not surprising for such a fast reaction.

Another benefit of catalyzed reactions is that they essentially run in one direction only. The enzyme molecule is not physically built to enable the reverse reaction, so the reverse-reaction rate of a catalyzed reaction is essentially zero. The hydrolysis of ATP to ADP and inorganic phosphate is a catalyzed reaction. Energy input from the metabolism of glucose does rebuild ATP from ADP and inorganic phosphate, but by using a completely different catalyzed reaction rather than by reversing the hydrolysis reaction.

STOP TO THINK 13.6 In the presence of a catalyst that reduces the activation energy by 50%, a reaction's rate constant will be unchanged if

A. The reaction energy is reduced by 50%.
B. The reaction energy is doubled.
C. The absolute temperature is reduced by 50%.
D. The absolute temperature is doubled.

13.7 Diffusion

In 1827, the botanist Robert Brown was observing pollen grains through a microscope. He noticed that pollen grains suspended in water were not stationary, as expected, but moved about continuously in a jittery, erratic fashion rather like that shown in FIGURE 13.23. This apparently random motion, today called **Brownian motion,** is caused by the ceaseless bombardment of the pollen grains by individual water molecules.

FIGURE 13.23 The random motion of a pollen grain undergoing Brownian motion.

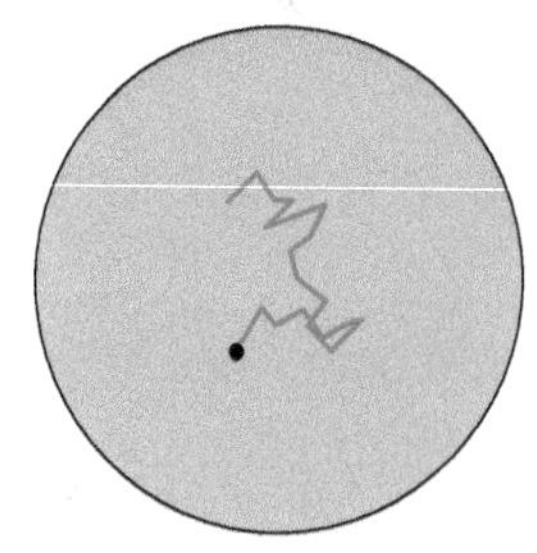

Brownian motion is an example of a **random walk.** Suppose you stood at the origin of an *xy*-coordinate system, turned through a randomly chosen angle, and then took a step with a randomly chosen length. Wherever you end up, you repeat the process again. And again. You would be engaging in a random walk, and your path would look something like the path of the particle in Figure 13.23.

We can't see or track individual molecules, but their motion—the motion of thermal energy—is also a random walk. At atmospheric pressure, a molecule in air can move only about 200 nm before it collides with another molecule. Each collision sends the molecule off in a new direction, and the trajectory of an individual molecule also looks similar to Figure 13.23—a sequence of straight lines of varying lengths connected by random turns.

FIGURE 13.24 Random walks cause a group of molecules to diffuse outward.

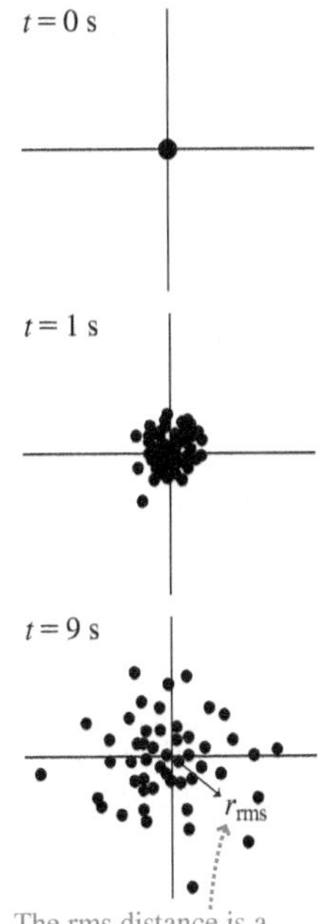
$t = 0$ s
$t = 1$ s
$t = 9$ s
r_{rms}
The rms distance is a measure of how far the molecules have diffused.

Imagine using a pipette to put a very tiny droplet of dye molecules in a beaker of very still water. The droplet will gradually disperse into the water. This net movement of molecules from a region of higher concentration (the initial droplet) to a region of lower concentration (the water) is called **diffusion.** Microscopically, diffusion is simply the random walks of the dye molecules as they collide both with each other and with the more numerous water molecules.

FIGURE 13.24 shows 50 molecules initially grouped so close together that we can't distinguish them. Then each begins a random walk, and their positions are shown at $t = 1$ s and $t = 9$ s. We can see that the molecules are diffusing outward. A random step is just as likely to go right as to go left, or to go up as to go down, so the *average position* of all the molecules is $(0, 0)$. However, the *average distance* from the origin is not zero. And it increases with time.

We modeled diffusion as a random walk with a fixed step size in ◀◀ SECTION 1.3; a review of that section is highly recommended. There we found that the root-mean-square distance r_{rms}, a rough measure of how far the molecules diffuse in time t, is

$$r_{rms} = \sqrt{6Dt} \tag{13.56}$$

where D is called the **diffusion coefficient.** Its value, with units of m^2/s, depends on the type of molecules that are diffusing, the type of molecules that they are diffusing

TABLE 13.2 Diffusion coefficients

Substance	D (m^2/s)
Diffusion in air, 1 atm, 25°C	
H_2O	2.5×10^{-5}
CO_2	1.9×10^{-5}
O_2	1.8×10^{-5}
Diffusion in water, 25°C	
O_2	2.3×10^{-9}
CO_2	2.0×10^{-9}
Urea	1.3×10^{-9}
Glucose	7.1×10^{-10}
Acetylcholine	4.0×10^{-10}
Hemoglobin	6.9×10^{-11}
Collagen	6.9×10^{-12}
DNA	1.3×10^{-13}

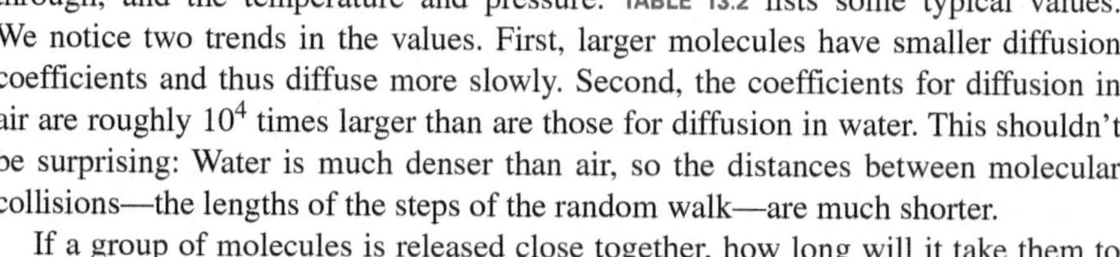

through, and the temperature and pressure. TABLE 13.2 lists some typical values. We notice two trends in the values. First, larger molecules have smaller diffusion coefficients and thus diffuse more slowly. Second, the coefficients for diffusion in air are roughly 10^4 times larger than are those for diffusion in water. This shouldn't be surprising: Water is much denser than air, so the distances between molecular collisions—the lengths of the steps of the random walk—are much shorter.

If a group of molecules is released close together, how long will it take them to diffuse a distance d? There's no definitive answer to this question because the diffusing molecules don't all arrive at once. However, we've seen that r_{rms} is about how far the molecules have diffused, so we can estimate a typical time needed to diffuse a distance d by setting $r_{rms} = d$ in Equation 13.56 and solving for time:

$$t = \frac{d^2}{6D} \tag{13.57}$$

Estimate of time to diffuse a distance d with diffusion coefficient D

The *diffusion time,* as it's often called, is proportional to the square of the distance, so an increase in the distance results in a much larger increase in time, as Example 13.12 illustrates.

EXAMPLE 13.12 Diffusion of oxygen through cytoplasm

A cell doesn't need a circulatory system, but your body does. Let's do a quick calculation to see why. A typical cell has a diameter of 10 μm. The smallest mammals in the world, shrews, are about 10 mm across. For small molecules, the diffusion coefficients in cytoplasm, the intercellular fluid, are about 25% of the coefficient for pure water because there are other structures and dissolved molecules that impede diffusion. Estimate the diffusion time for oxygen molecules through cytosplasm for these two distances.

PREPARE We will use Equation 13.57 to find the diffusion times for the two distances. Table 13.2 gives the diffusion coefficient of oxygen through water at 25°C as $D = 2.3 \times 10^{-9}\,\text{m}^2/\text{s}$, so for cytoplasm $D \approx 6 \times 10^{-10}\,\text{m}^2/\text{s}$ at 25°C. For the purposes of an estimate we'll assume that D has the same value at a body temperature of 37°C.

SOLVE For the two distances we're considering, we find

$$\text{For } d = 10\ \mu\text{m},\ t = \frac{(10 \times 10^{-6}\,\text{m}^2)^2}{6(6 \times 10^{-10}\,\text{m}^2/\text{s})} \approx 0.03\,\text{s}$$

$$\text{For } d = 10\ \text{mm},\ t = \frac{(10 \times 10^{-3}\,\text{m}^2)^2}{6(6 \times 10^{-10}\,\text{m}^2/\text{s})} \approx 30{,}000\,\text{s}$$

The time for oxygen to diffuse across the full diameter of a cell is about 30 ms; this is fast enough that there is no need for active transport. For a size typical of the smallest mammals, the diffusion time is much longer—about 8 hours! Clearly, mammals need a circulatory system to deliver oxygen to the cells of the body.

ASSESS Diffusion works for a cell, but not for a shrew. This result makes sense. We know that all but the smallest animals have circulatory systems. The diffusion time increases as the *square* of the distance, so it takes 1,000,000 times longer to diffuse 1000 times as far!

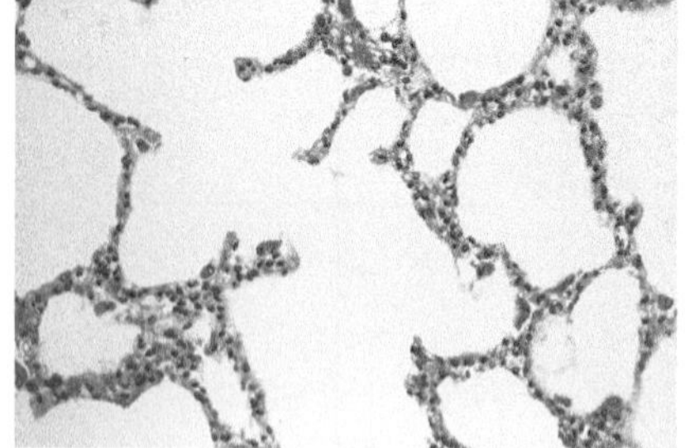

Gas exchange Gas exchange between your body and the air occurs by diffusion across a 2-μm-thick barrier (one epithelium lung cell and one endothelium capillary cell) that separates the alveoli of the lungs from hemoglobin in the capillaries. A higher concentration of oxygen in the alveoli drives diffusion into the capillaries, while a higher concentration of carbon dioxide in the blood drives diffusion into the alveoli. If you've been at high elevations and experienced difficulty breathing, the problem is not so much "lack of oxygen" (there are plenty of oxygen molecules in your lungs) but a reduced rate of oxygen diffusion into the capillaries due to a reduced concentration gradient.

Your circulatory system delivers oxygen and other nutrients to different parts of your body, but once the oxygen reaches the cells, diffusion can take over. The cellular transport of oxygen, nutrients, and smaller macromolecules is often by passive diffusion:

- Air in the alveoli of the lungs, the small gas-filled sacs through which gas exchange with the blood takes place, is separated from blood in the pulmonary capillaries by a membrane that is about 2 μm thick. Oxygen in the lungs rapidly diffuses through this membrane into the capillaries, while excess carbon dioxide in the capillaries diffuses into the lungs and is exhaled.

- Oxygen, glucose, and other molecules diffuse across the cell wall from capillaries into cells. Waste products, such as carbon dioxide or urea, diffuse from the cell into the capillaries for removal.
- A nerve impulse moves across a synapse when a neurotransmitter, such as acetylcholine, is released from the pre-synaptic membrane and diffuses across the synaptic cleft to the post-synaptic membrane. The transmission takes less than a microsecond because the synaptic cleft is so narrow.

Diffusion is very effective at transporting small molecules over distances of a few micrometers or less, but the body needs active transport systems, such as the circulatory system, to move molecules through distances that are more than a few cell diameters.

Large macromolecules move more slowly than small molecules at the same temperature, and the distances between collisions are shorter, so their diffusion coefficients are smaller. The diffusion coefficient for a collagen molecule in water is only 0.3% that of an oxygen molecule, and—although measurements are difficult—it is thought that the diffusion coefficients of large macromolecules in cytoplasm are roughly a factor of 10 less than in pure water. Instead of crossing a 10-mm-diameter cell in 30 ms, as oxygen does, a collagen molecule would need more than 20 s. Consequently, active transport is needed even at the cellular level to move large macromolecules. Cells use ATP-powered molecular motors running along microtubules to transport macromolecules faster than the diffusion limit.

Finally, we need to note that diffusion is *not* the reason that the odor from an opened bottle of ammonia or perfume spreads quickly throughout a room, even though this is often cited as an example of diffusion. It would take almost 10 h for molecules to diffuse even 2 m through air, as you can show in a homework problem. Instead, the odor is spread quickly by *convection currents* of moving air.

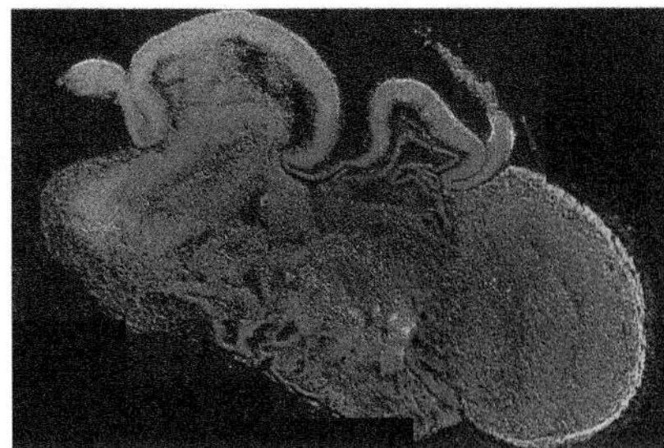

Brain organoids Pluripotent stem cells can be coaxed to grow into three-dimensional clumps of neuronal tissue called *brain organoids* or *cerebral organoids*. These neurons develop many of the connections seen in real brains, and it is hoped that they will be a powerful model for studying neurodegenerative diseases. Organoids are cultured in a high-oxygen environment, but even so, the size of an organoid is limited to about 3 mm because diffusion is the only mechanism to supply the cells with oxygen and nutrients.

Diffusion and Kinetic Theory

Our random-walk model from Section 1.3, in which each molecule takes steps of a fixed length d, found that the diffusion coefficient is $D = d^2/2\,\Delta t$, where Δt is the time between steps. But $d/\Delta T$ is a molecule's speed, so our model also predicts $D = \frac{1}{2}vd$. We used some simple estimates of v and d to estimate diffusion coefficients in water and air. Our estimates were in the right range, which is gratifying for such a simple model, but too large by about a factor of 5.

Two obvious deficiencies of such a simple model are that molecules don't all have the same speed and that the distances traveled between collisions, the step sizes, are not all the same. We've dealt with both issues in this chapter by introducing the *mean free path* λ as an average distance traveled between collisions and the *rms speed* v_{rms} as an average speed that arises naturally because of its close connection to the average kinetic energy. We'll skip the details of the analysis, but a more robust random-walk model based on kinetic theory finds that

$$D = \tfrac{1}{3}v_{\text{rms}}\lambda \qquad (13.58)$$

This looks much like our simpler version, but now speeds and step sizes are appropriate averages.

Earlier we found that $v_{\text{rms}} = \sqrt{3k_BT/m}$, where m is the molecular mass in kg and $\lambda = (4\sqrt{2}\pi(N/V)r^2)^{-1}$ for the mean free path of a spherical particle of radius r. For diffusion in a gas, we can use the ideal-gas law to write $N/V = p/k_BT$. Using these results in Equation 13.58 leads, after some algebra, to the prediction

$$D = \frac{(k_BT)^{3/2}}{4\pi\sqrt{6m}\,pr^2} \qquad (13.59)$$

for diffusion in a gas at temperature T and pressure p.

First notice that Equation 13.59 predicts that the diffusion coefficient scales with the absolute temperature as $T^{3/2}$. This is well confirmed by experiment. The

Wrinkly respiration The hellbender salamander, native to the eastern United States, can be over half a meter long, with a mass of 2 kg. Remarkably, this large amphibian doesn't have lungs or gills. Instead, it extracts oxygen from water by diffusion through its skin. The wrinkles on the sides of the salamander aren't just for looks; they increase the surface area of its skin so that diffusion can deliver sufficient oxygen.

diffusion coefficients in Table 13.2 are about 13% higher at 50°C than the reported values at 25°C.

Second, a molecule's mean free path depends on its radius, but we don't have a way to directly measure the radius of a molecule. Earlier we gave some rough guidelines, but they were only estimates. We can now turn this around and use a measured diffusion coefficient, a macroscopic laboratory measurement, to determine molecular radii. Thus diffusion is another important link in the micro/macro connection.

EXAMPLE 13.13 The size of an oxygen molecule

Table 13.2 gives the diffusion coefficient for oxygen molecules in 25°C air, at 1.0 atm pressure, as $1.8 \times 10^{-5}\ \text{m}^2/\text{s}$. Use this value to determine the radius of an oxygen molecule.

PREPARE 25°C is 298 K and 1.0 atm is 101,000 Pa.

SOLVE Solving Equation 13.59 for r gives

$$r = \left[\frac{(k_B T)^{3/2}}{4\pi\sqrt{6m}\,pD}\right]^{1/2}$$

$$= \left[\frac{((1.38 \times 10^{-23}\ \text{J/K})(298\ \text{K}))^{3/2}}{4\pi\sqrt{6(32 \times 1.66 \times 10^{-27}\ \text{kg})}(101{,}000\ \text{Pa})(1.8 \times 10^{-5}\ \text{m}^2/\text{s})}\right]^{1/2}$$

$$= 1.4 \times 10^{-10}\ \text{m} = 140\ \text{pm}$$

This is a straightforward computation, although one that requires care not to make calculation errors.

ASSESS The result is about 40% larger than the rough estimate that "diatomic molecules have $r \approx 1.0 \times 10^{-10}$ m" from earlier in the chapter, but this more precise value is confirmed by other experiments.

Fick's Law

It's useful to know how far and how fast molecules can diffuse, but many of the most important examples of diffusion in biology involve the *steady* diffusion of molecules across a membrane or some other permeable barrier from a higher concentration on one side to a lower concentration on the other. For example, oxygen diffuses from the lungs into the pulmonary capillaries while carbon dioxide diffuses from the capillaries into the lungs. Oxygen and nutrients diffuse through the cell wall into the cell while waste products diffuse in the opposite direction. Kidneys and hemodialysis both use the diffusion of urea and small proteins through a thin membrane to filter the blood. The diffusive transport of matter across a thin barrier from a higher to a lower concentration is an integral part of many physiological processes.

FIGURE 13.25 Diffusion of molecules across a permeable barrier separating a region of higher concentration from a region of lower concentration.

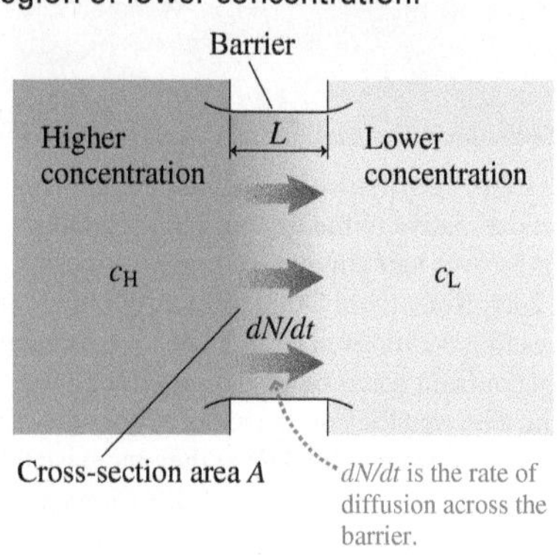

FIGURE 13.25 shows two regions that contain different concentrations of some substance: a higher concentration c_H and a lower concentration c_L. If N molecules occupy volume V, their **number concentration** is

$$c = \frac{N}{V} \tag{13.60}$$

Number concentration, equivalent to number density, is the number of molecules per cubic meter in m^{-3}.

NOTE ▸ Chemists and biologists often measure concentration in other units, such as moles per liter (the *molar concentration,* represented by an upper case C) or parts per million. The concepts of this section still apply, but diffusion coefficients have different units. Don't confuse this use of c with specific heat. ◂

The two regions in the figure are separated by a permeable barrier that has length or thickness L and cross-section area A. How rapidly do molecules diffuse across the barrier from higher to lower concentration? You met a similar situation in Chapter 12 where we looked at heat transfer between two objects at different temperatures. The quantity of interest in conductive heat transfer was dQ/dt, the rate (J/s) at which heat energy moves from a higher to a lower temperature. The analogous quantity for diffusive matter transfer is dN/dt, the rate (molecules/s) at which molecules move from a higher to a lower concentration. We'll skip the derivation, but a direct application of the random-walk model of Chapter 1 finds that the *diffusion rate* between two regions with a concentration difference $\Delta c = c_H - c_L$ is

$$\frac{dN}{dt} = DA\frac{\Delta c}{L} \tag{13.61}$$

Fick's law of diffusion along a concentration gradient

where D is the diffusion coefficient *inside* the separating barrier. Equation 13.61 is called **Fick's law of diffusion.**

The heat conduction equation is $dQ/dt = kA(\Delta T/L)$. You can see that Fick's law is exactly analogous, with the diffusion coefficient D playing the same role as the thermal conductivity constant k. More significant was the earlier observation that $\Delta T/L$ is a *temperature gradient,* and we interpreted the heat conduction equation as telling us that *a temperature gradient drives an energy flow*. Similarly, $\Delta c/L$ is a **concentration gradient,** the change of concentration per meter, and Fick's law tells us that *a concentration gradient drives a matter flow*. Biologists tend to say that molecules diffuse "down" a concentration gradient.

It may seem puzzling that diffusion can cause a net transport of matter. After all, if diffusion is a random walk, and if a molecule can move either left or right with equal probability, then it seems like, on average, nothing should happen. The key to understanding diffusive transport is that *net transport occurs if the concentration of molecules changes with position*. This is the case if there are conditions that maintain a higher concentration in one region and a lower concentration in another, such as your lungs bringing in air with a higher concentration of oxygen while your blood oxygen concentration is lower.

FIGURE 13.26 shows a cross section of the permeable barrier connecting the region of high concentration and the region of low concentration. Every molecule shown has an equal probability of moving distance d to the right or d to the left during the next time interval Δt, *but there are more molecules on the left side than on the right side.* Consequently, more molecules will cross the highlighted plane from left to right than from right to left. That is, diffusion *with a concentration difference* is responsible for a net flux of molecules from higher concentration to lower concentration.

FIGURE 13.26 Molecules undergoing random walks as the concentration decreases from left to right.

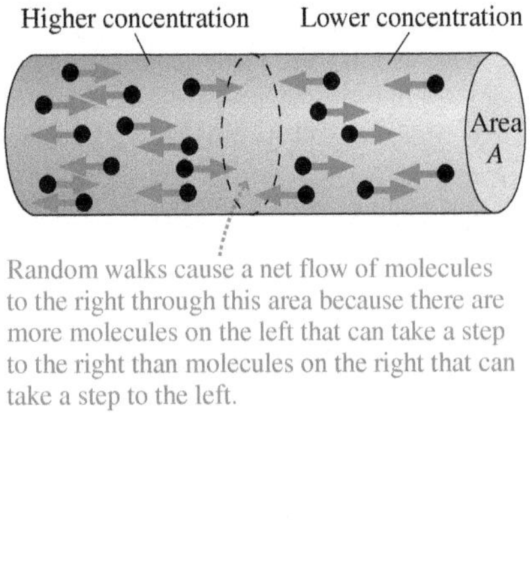

Higher concentration

Lower concentration

Area A

Random walks cause a net flow of molecules to the right through this area because there are more molecules on the left that can take a step to the right than molecules on the right that can take a step to the left.

An important lesson from the random-walk model is that the statement "a concentration gradient drives a matter flow" cannot be taken too literally. Our other example of gradient-driven matter flow was Poiseuille's equation in Chapter 9, where we found that a pressure gradient drives a fluid flow. A fluid flow is a *bulk flow* of matter driven by actual forces on the fluid. *All* the matter moves in one direction. In contrast, the diffusive flow of matter is simply the consequence of randomly moving molecules. Some molecules are diffusing "upstream" from lower to higher concentration, but overall more are diffusing from higher to lower concentration, and the *rate* of net flow of the diffusing molecules increases as the concentration gradient increases. There's no force "driving" the flow.

Finally, it's worth noting that diffusive transport is passive; it requires no expenditure of metabolic energy. Whenever a living system can take advantage of diffusion, it's free!

STOP TO THINK 13.7 A drop of blue dye is placed near the left end of a horizontal tube filled with very still water. Several hours later, the tube looks as shown in the figure. At the point in the tube marked with a dot,

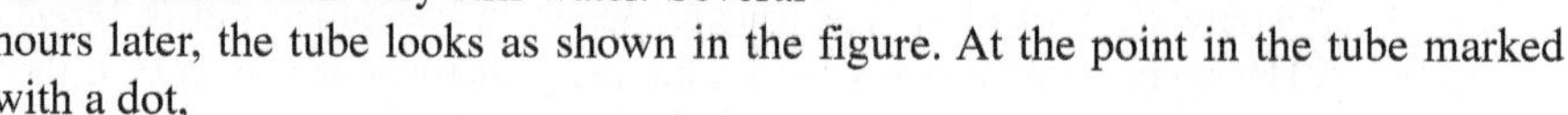

A. The dye molecules have stopped moving because the tube is in equilibrium.
B. All dye molecules are moving to the right.
C. All dye molecules are moving to the left.
D. Equal numbers of dye molecules are moving right and left.
E. Dye molecules are moving both right and left, but more are moving to the right.
F. Dye molecules are moving both right and left, but more are moving to the left.

EXAMPLE 13.14 Carbon dioxide diffusion in photosynthesis

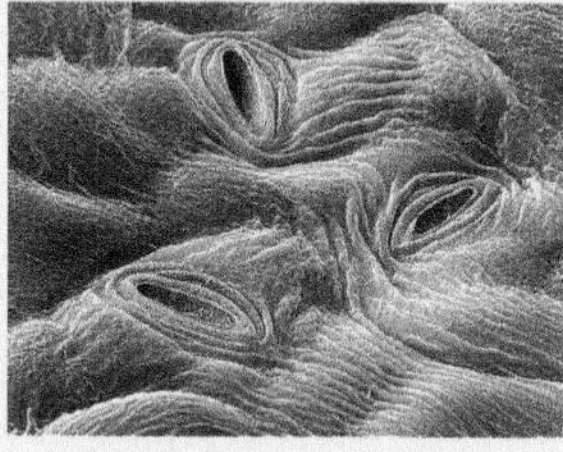

Carbon dioxide makes up about 0.040% of the atmosphere. Carbon dioxide enters a green leaf, where photosynthesis occurs, by diffusing through small pores called *stomata,* pores with a depth of about 10 μm. When the stomata are open, we can assume that their total area is 2.0% of the leaf area. Photosynthesis consumes CO_2, and the concentration within the leaf is about half the atmospheric value. Estimate, in both molecules/s and μmol/s, the rate of CO_2 uptake by a leaf with an area of $7.0 \times 10^{-4}\ \text{m}^2$.

PREPARE The atmosphere and the leaf are regions with different concentrations of CO_2 connected by pores through which matter transfer can occur via diffusion. Fick's law describes the rate of CO_2 uptake. We'll assume a temperature of 25°C.

SOLVE From Table 13.2, the diffusion coefficient of carbon dioxide through air at 25°C is $1.9 \times 10^{-5}\ \text{m}^2/\text{s}$. The depth of the stomata, 10×10^{-6} m, is the distance L between the air and the inside of the leaf through which CO_2 diffuses. The total cross-section area of all the stomata is 2.0% of the leaf area, or $A = 1.4 \times 10^{-5}\ \text{m}^2$.

We can use the ideal-gas law to determine the concentration of CO_2 in the atmosphere, which is the region of high concentration. Because CO_2 is 0.040% of the atmosphere, the partial pressure of CO_2 is 4.0×10^{-4} atm. Thus

$$c_\text{H} = \frac{N}{V} = \frac{p}{k_\text{B}T}$$
$$= \frac{(4.0 \times 10^{-4}\ \text{atm})(1.01 \times 10^5\ \text{Pa/atm})}{(1.38 \times 10^{-23}\ \text{J/K})(298\ \text{K})} = 9.8 \times 10^{21}\ \text{m}^{-3}$$

We're told that $c_\text{L} \approx \frac{1}{2}c_\text{H}$, so the concentration difference is $\Delta c \approx \frac{1}{2}c_\text{H}$.

The rate at which molecules diffuse from the air into the leaf can now be calculated with Fick's law:

$$\frac{dN}{dt} = DA\frac{\Delta c}{L}$$
$$= (1.9 \times 10^{-5}\ \text{m}^2/\text{s})(1.4 \times 10^{-5}\ \text{m}^2)\frac{\frac{1}{2}(9.8 \times 10^{21}\ \text{m}^{-3})}{10 \times 10^{-6}\ \text{m}}$$
$$= 1.3 \times 10^{17}\ \text{molecules/s}$$

We can convert to moles by dividing by Avogadro's number:

$$\frac{dN}{dt} = \frac{1.3 \times 10^{17}\ \text{molecules/s}}{6.02 \times 10^{23}\ \text{molecules/mol}}$$
$$= 2.2 \times 10^{-7}\ \text{mol/s} = 0.22\ \mu\text{mol/s}$$

ASSESS Measured rates of CO_2 uptake are about a third of what we've calculated, which indicates, not surprisingly, that our simple model of gas transfer in photosynthesis is a bit too simple. Nonetheless, our calculations give the correct order of magnitude and provide insight into how the physical process of diffusion plays a key role in the biological process of photosynthesis.

CHAPTER 13 INTEGRATED EXAMPLE Oxygen consumption

A person's metabolic rate can be determined by measuring their oxygen consumption, using the VO2 measurement described in Chapter 11. The typical resting VO2 for an adult male is 350 mL/min. How well can we predict this measured value using a model of diffusive uptake of O_2 in the lungs?

Diffusion happens across a 2.0-μm-thick tissue layer that separates air in the lung's alveoli from red blood cells in the pulmonary capillaries. The diffusion coefficient of O_2 through this tissue is reported to be $1.1 \times 10^{-11}\ \text{m}^2/\text{s}$. The total surface area of the alveoli is frequently said to be about 70 m^2, the size of a tennis court, but diffusion doesn't occur through this entire area. **FIGURE 13.27**, based on photomicrographs, shows that the capillaries form a mesh around each alveolus. Diffusion takes place only through the *contact area* where a capillary and an alveolus are in close contact. The pulmonary capillaries are 6.0 μm in diameter and are reported to contain a total volume of 100 mL of blood.

FIGURE 13.27 The pulmonary capillaries form a mesh around each alveolus in the lungs.

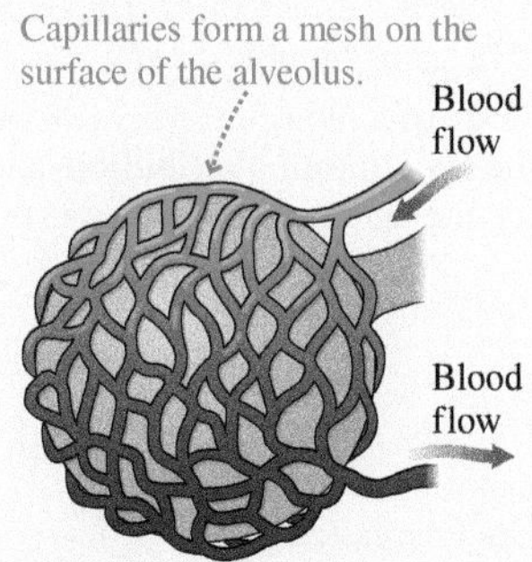
Capillaries form a mesh on the surface of the alveolus.
Blood flow
Blood flow

The partial pressure of O_2 in the air is 160 mm Hg (21% of the air), but the partial pressure of O_2 in the alveoli stays at a fairly steady 100 mm Hg throughout an entire respiratory cycle. Venous blood enters a pulmonary capillary with an O_2 partial pressure of 40 mm Hg. A red blood cell moves through its pulmonary capillary in 0.75 s, but the pressure has equilibrated with the alveoli at 100 mm Hg after 0.25 s. (These values can change to increase the oxygen uptake during exercise.) Use this information to model diffusion in the alveoli and to predict the rate of oxygen consumption.

PREPARE We can condense this information into a fairly simple model based on Fick's law. Oxygen-depleted blood first reaches the alveoli with an O_2 pressure of 40 mm Hg, but 0.25 s later the pressure has increased to 100 mm Hg. Instead of dealing with a time-dependent pressure, we'll assume that the blood becomes fully oxygenated in 0.25 s at its *average* O_2 partial pressure of 70 mm Hg. Thus we will use $\Delta p = 30$ mm Hg as the oxygen pressure difference between the alveoli and the blood for the 0.25 s needed to become fully oxygenated. We can use Fick's law to calculate the diffusion rate between the higher and lower concentrations. However, we're told that the blood stays in an alveolar capillary for 0.75 s even though it is fully oxygenated after 0.25 s, so there is an interval of 0.50 s during which $\Delta p = 0$ and no diffusion occurs. That is, no diffusion takes place over $\frac{2}{3}$ of the length of contact between a capillary and an alveolus. We'll account for this by using $A_{\text{eff}} = \frac{1}{3}A_{\text{contact}}$ as the *effective area* in Fick's law.

SOLVE Fick's law gives the oxygen uptake rate in molecules/s, which we can convert to mol/s or mol/min. For comparison, we need to express the measured oxygen consumption rate in the same units. An oxygen consumption of 350 mL/min refers to a 350 mL volume of oxygen *at the oxygen partial pressure in air* of 160 mm Hg. A mL is a cubic centimeter, so 350 mL is 350×10^{-6} m^3. The pressure of 160 mm Hg converts to 21 kPa. We can now use the ideal-gas law to find out how many moles of oxygen are in 350 mL:

$$n = \frac{pV}{RT} = \frac{(21{,}000 \text{ Pa})(350 \times 10^{-6} \text{ m}^3)}{(8.31 \text{ J/mol} \cdot \text{K})(310 \text{ K})}$$

$$= 2.9 \times 10^{-3} \text{ mol} = 2.9 \text{ mmol}$$

We used 310 K = 37°C as the temperature of the oxygen at the time of uptake by the body. Thus the VO2 measurement is equivalent to 2.9 mmol/min of oxygen consumption.

Fick's law of diffusion is $dN/dt = DA(\Delta c/L)$, where the concentration c is N/V, the number of molecules per m^3. We are given data about pressure differences, not concentration differences, so we need another application of the ideal-gas law to convert concentration to pressure: $c = N/V = p/k_B T$. Thus Fick's law can be written

$$\frac{dN}{dt} = \frac{DA}{k_B T}\frac{\Delta p}{L}$$

Our model uses $\Delta p = 30$ mm Hg $= 4000$ Pa as the average oxygen pressure difference between the alveoli and the blood. We're given the diffusion coefficient D, and we know that $L = 2.0 \times 10^{-6}$ m and $T = 310$ K. Because blood is taking up oxygen by diffusion for only $\frac{1}{3}$ of the time it spends in contact with the alveoli, the effective diffusing area is $A_{\text{eff}} = \frac{1}{3}A_{\text{contact}}$. The challenge is to determine the area of contact between the alveoli and capillaries.

Suppose we take length L of capillary and press it against a surface so that it's nearly flat. Seen from above, the capillary—a long, skinny rectangle—would cover area $A_{\text{contact}} = Ld$, where d is the width (the diameter) of the capillary, which we know to be 6.0 μm. The total *volume* of this capillary of length L is the volume of a cylinder: $V = \pi r^2 L = \pi(d/2)^2 L$. We're given that the total volume of the pulmonary capillaries is 100 mL $= 1.0 \times 10^{-4}$ m^3. If this were one long capillary, its length would be

$$L = \frac{V}{\pi(d/2)^2} = \frac{1.0 \times 10^{-4} \text{ m}^3}{\pi(3.0 \times 10^{-6} \text{ m})^2} = 3.5 \times 10^6 \text{ m}$$

L is more than 2000 miles! And while there are a vast number of pulmonary capillaries, not just one, L really is the *combined length* of all of them. When pressed against the surfaces of the alveoli, these capillaries cover an area

$$A_{\text{contact}} = Ld = (3.5 \times 10^6 \text{ m})(6.0 \times 10^{-6} \text{ m}) = 21 \text{ m}^2$$

This is the total area over which the capillaries are pressed closely against the alveoli, and it's a believable number. 21 m^2 is 30% of the total alveolar surface area of ≈ 70 m^2, very much in line with what we would expect from a mesh of capillaries around each alveolus.

We now know that $A_{\text{eff}} = \frac{1}{3}A_{\text{contact}} = 7.0$ m^2, so we can use Fick's law to calculate

$$\frac{dN}{dt} = \frac{(1.1 \times 10^{-11} \text{ m}^2)(7.0 \text{ m}^2)}{(1.38 \times 10^{-23} \text{ J/K})(310 \text{ K})} \times \frac{4000 \text{ Pa}}{2.0 \times 10^{-6} \text{ m}}$$

$$= 3.6 \times 10^{19} \frac{\text{molecules}}{\text{s}}$$

Converting to mmol/min, for comparison with the measured uptake rate, gives

$$\frac{dN}{dt} = 3.6 \times 10^{19} \frac{\text{molecules}}{\text{s}} \times \frac{60 \text{ s}}{1 \text{ min}} \times \frac{1 \text{ mol}}{6.02 \times 10^{23} \text{ molecules}}$$

$$= 3.6 \text{ mmol/min}$$

ASSESS A simple model of oxygen diffusion from the lungs into the blood has used a number of measured physiological quantities to predict that the diffusion rate—a person's oxygen consumption—should be 3.6 mmol/min. This is only 25% larger than the typical measured value for adult males of 2.9 mmol/min. The agreement is remarkably good for such a simple model. It confirms the importance of diffusion for biological processes and illustrates how randomly moving molecules have very real effects.

SUMMARY

GOAL To see how macroscopic properties of matter depend on the microscopic motions of atoms and molecules.

GENERAL PRINCIPLES

Kinetic Theory of Gases

Pressure in a gas is due to the forces of molecules colliding with the walls.

Temperature is a measure of the average kinetic energy of the molecules.

Molecules have a range of speeds given by the **Maxwell-Boltzmann distribution.**

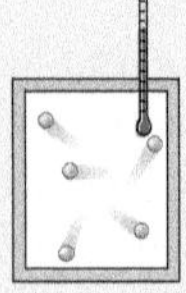

The Equipartition Theorem

Collisions distribute the thermal energy of the system equally among all possible **degrees of freedom.** The energy stored in each mode is $\frac{1}{2}Nk_BT$ or, in terms of moles, $\frac{1}{2}nRT$.

k_BT is a typical energy of a molecule. Very few molecules have energy $\gg k_BT$. The **Boltzmann distribution** says that the number of molecules having energy E_i decreases exponentially as

$$N_i \propto e^{-E_i/k_BT}$$

IMPORTANT CONCEPTS

The **thermal energy** of a system is the microscopic kinetic and potential energy of its molecules:

$$E_{th} = \text{translational kinetic energy} + \text{rotational kinetic energy} + \text{vibrational energy}$$

- **Monatomic gas** $E_{th} = \frac{3}{2}Nk_BT = \frac{3}{2}nRT$
- **Diatomic gas** $E_{th} = \frac{5}{2}Nk_BT = \frac{5}{2}nRT$
- **Elemental solid** $E_{th} = 3Nk_BT = 3nRT$

Heat is energy transferred via collisions from more-energetic molecules on one side to less-energetic molecules on the other. Equilibrium is reached when $(\epsilon_1)_{avg} = (\epsilon_2)_{avg}$, which implies $T_{1f} = T_{2f}$.

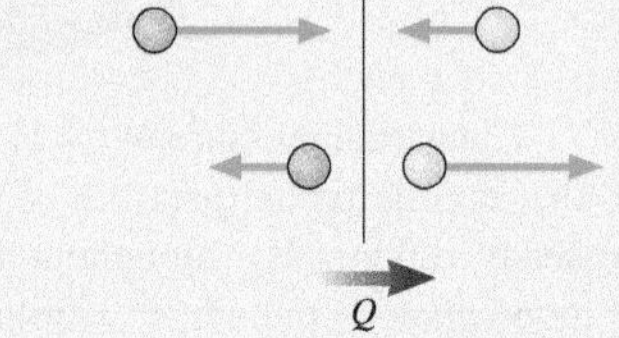

Rate Constants

Chemical reactions occur when molecules collide with sufficient energy. The **rate constant** of a chemical reaction varies with temperature according to the **Arrhenius equation** as

$$k = Ae^{-E_a/k_BT}$$

where E_a is the *activation energy* barrier.

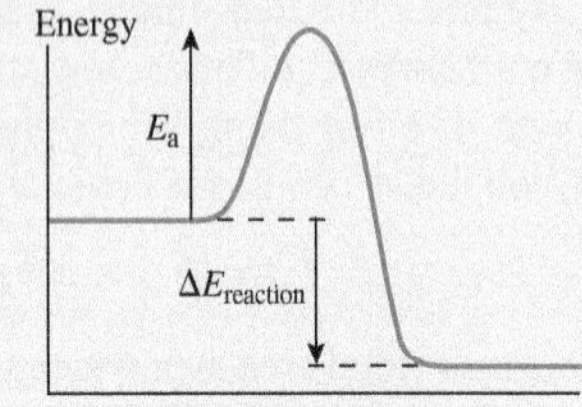

The Boltzmann factor gives the fraction of molecules with sufficient available collision energy to initiate a reaction.

Diffusion

The random motion of molecules causes diffusion across a permeable barrier from a higher concentration to a lower concentration. **Fick's law** of diffusion says that the rate of diffusive transport is

$$\frac{dN}{dt} = DA\frac{\Delta c}{L}$$

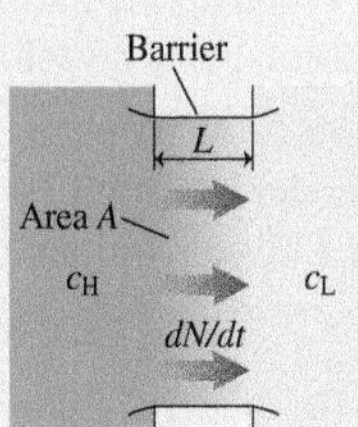

where $\Delta c/L$ is the **concentration gradient** and D the **diffusion coefficient.** The time needed for molecules to diffuse distance d is

$$t_{diff} = \frac{d^2}{6D}$$

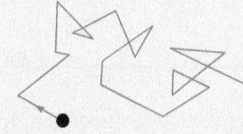

APPLICATIONS

The **mean free path** of a molecule of radius r through a gas with number density N/V is

$$\lambda = \frac{1}{4\sqrt{2}\pi(N/V)r^2}$$

Molar specific heats can be predicted from the thermal energy because $\Delta E_{th} = nC\Delta T$.

- **Monatomic gas** $C_V = \frac{3}{2}R$
- **Diatomic gas** $C_V = \frac{5}{2}R$
- **Elemental solid** $C = 3R$

A **catalyst** speeds up a chemical reaction by providing an alternative reaction pathway with a lower activation energy E_a. The rate constant depends exponentially on E_a, so a modest change in the activation energy can have an enormous effect on the reaction rate.

The **partial pressure** of a gas in a mixture is the pressure that gas would exert if it occupied the container by itself. The total pressure is the sum of the partial pressures of all the gases.

The **root-mean-square-speed** of molecules in a gas, which can be interpreted as a typical speed, is

$$v_{rms} = \sqrt{\frac{3k_BT}{m}}$$

Heavier molecules move more slowly than lighter molecules at the same temperature but, on average, have the same kinetic energy.

The populations in a two-level system are established by collisions and radiation. Their ratio is given by the **Boltzmann factor:**

$$\frac{N_2}{N_1} = e^{-(E_2-E_1)/k_BT} = e^{-\delta E/k_BT}$$

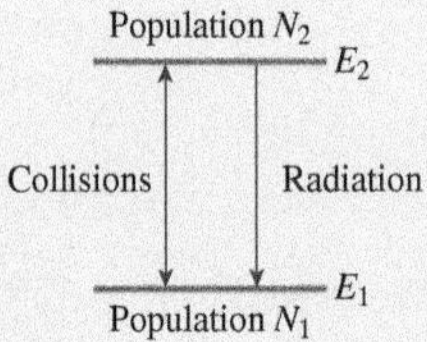

LEARNING OBJECTIVES After studying this chapter, you should be able to:

- Calculate molecular mean free paths. *Conceptual Questions 13.3, 13.4; Problems 13.3–13.6*
- Interpret and use root-mean-square speeds. *Conceptual Questions 13.6, 13.7; Problems 13.7–13.10, 13.12*
- Relate temperature to the average kinetic energy of molecules. *Conceptual Questions 13.8, 13.10; Problems 13.15–13.17, 13.19, 13.20*
- Interpret and use the equipartition theorem. *Conceptual Question 13.13; Problems 13.24–13.26, 13.28, 13.29*
- Solve problems utilizing the Boltzmann factor. *Conceptual Question 13.12; Problems 13.31–13.35*
- Perform calculations and solve problems using reaction kinetics. *Conceptual Question 13.14; Problems 13.36–13.40*
- Calculate diffusion rates and times. *Conceptual Question 13.15; Problems 13.41–13.45*

STOP TO THINK ANSWERS

Stop to Think 13.1: C. Each v^2 increases by a factor of 16 but, after averaging, v_{rms} takes the square root.

Stop to Think 13.2: E. Temperature is proportional to the average energy. The energy of a gas molecule is kinetic, proportional to v^2. The average energy, and thus T, increases by 4^2.

Stop to Think 13.3: B. The bead can slide along the wire (one degree of translational motion) and rotate around the wire (one degree of rotational motion).

Stop to Think 13.4: A. Temperature measures the average translational kinetic energy *per molecule,* not the thermal energy of the entire system.

Stop to Think 13.5: B and C. The ratio N_2/N_1 increases if collisions play a more important role in populating the upper level. This can be done by decreasing the energy required (decreasing δE) or increasing the thermal energy of the particles (increasing T).

Stop to Think 13.6: C. The rate constant depends on the ratio $E_a/k_B T$. The rate constant doesn't change if the temperature and the activation energy change by the same factor.

Stop to Think 13.7: E. The collision-induced random walks never stop, so molecules are always moving both to the right and to the left. But there are more molecules on the left than on the right, a higher concentration, so more molecules are moving to the right than to the left.

QUESTIONS

Conceptual Questions

1. Solids and liquids resist being compressed. They are not totally incompressible, but it takes large forces to compress them even slightly. If it is true that matter consists of atoms, what can you infer about the microscopic nature of solids and liquids from their incompressibility?
2. Gases, in contrast with solids and liquids, are very compressible. What can you infer from this observation about the microscopic nature of gases?
3. The density of air at STP is about $\frac{1}{1000}$ the density of water. What is the approximate ratio of the average distance between air molecules to the average distance between water molecules?
4. The mean free path of molecules in a gas is 200 nm.
   a. What will be the mean free path if the pressure is doubled while the temperature is held constant?
   b. What will be the mean free path if the absolute temperature is doubled while the pressure is held constant?
5. If the pressure of a gas is really due to the *random* collisions of molecules with the walls of the container, why do pressure gauges—even very sensitive ones—give perfectly steady readings? Shouldn't the gauge be continually jiggling and fluctuating? Explain.
6. If you double the typical speed of the molecules in a gas, by what factor does the pressure change? Give a simple explanation why the pressure changes by this factor.
7. Suppose you could suddenly increase the speed of every molecule in a gas by a factor of 2.
   a. Would the rms speed of the molecules increase by a factor of $2^{1/2}$, 2, or 2^2? Explain.
   b. Would the gas pressure increase by a factor of $2^{1/2}$, 2, or 2^2? Explain.
8. Suppose you could suddenly increase the speed of every molecule in a gas by a factor of 2.
   a. Would the temperature of the gas increase by a factor of $2^{1/2}$, 2, or 2^2? Explain.
   b. Would the molar specific heat at constant volume change? If so, by what factor? If not, why not?
9. Two gases have the same number of molecules per cubic meter (N/V) and the same rms speed. The molecules of gas 2 are more massive than the molecules of gas 1.
   a. Do the two gases have the same pressure? If not, which is larger?
   b. Do the two gases have the same temperature? If not, which is larger?

Problem difficulty is labeled as | (straightforward) to ||||| (challenging). Problems labeled INT integrate significant material from earlier chapters; BIO are of biological or medical interest; CALC require calculus to solve.

10. If the temperature T of an ideal gas doubles, by what factor does the average kinetic energy of the atoms change?
11. A bottle of helium gas and a bottle of argon gas contain equal numbers of atoms at the same temperature. Which bottle, if either, has the greater total thermal energy?
12. As a liquid evaporates, the remaining liquid is chilled to a temperature below room temperature. Why? Explain this cooling by reasoning about what is happening at the molecular scale.
13. The two containers of gas in Figure Q13.13 are in good thermal contact with each other but well insulated from the environment. They have been in contact for a long time and are in thermal equilibrium.

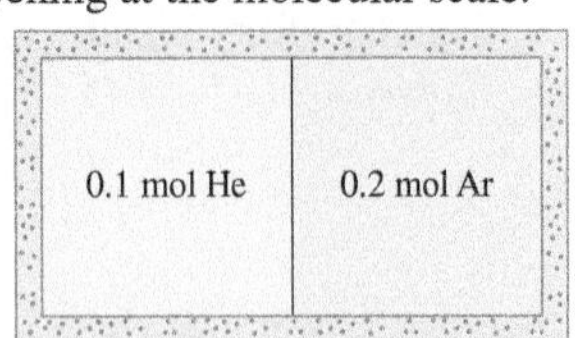

FIGURE Q13.13

    a. Is v_{rms} of helium greater than, less than, or equal to v_{rms} of argon? Explain.
    b. Does the helium have more thermal energy, less thermal energy, or the same amount of thermal energy as the argon? Explain.
14. Suppose a catalyst reduces the activation energy of a reaction by 50%. Does the reaction rate constant increase by a factor of 2, less than a factor of 2, or more than a factor of 2? Or is it impossible to say without knowing the temperature?
15. A gas diffuses 10 μm in 10 s. How far will it diffuse in 20 s?

Multiple-Choice Questions

16. | At what temperature does neon ($m = 20$ u) have an rms speed of 750 m/s?
    A. 0°C B. 180°C C. 450°C D. 1100°C
17. || What is the thermal energy of a 1.0 m × 1.0 m × 1.0 m box of nitrogen at a pressure of 3.0 atm?
    A. 380 kJ B. 450 kJ C. 760 kJ D. 910 kJ
18. || A reaction with an activation energy of 50 kJ/mol has a rate constant of $2.5 \times 10^{-9}\ s^{-1}$ at 10°C. What is the rate constant at 50°C?
    A. $1.8 \times 10^{-10}\ s^{-1}$ B. $2.8 \times 10^{-9}\ s^{-1}$
    C. $1.2 \times 10^{-8}\ s^{-1}$ D. $3.5 \times 10^{-8}\ s^{-1}$
19. || Oxygen diffuses across the membranes that separate the air in the alveoli in the lungs from the blood in the capillaries. If you double the thickness of the separating membrane, by what factor will the diffusion time increase?
BIO
    A. $\sqrt{2}$ B. 2 C. $2\sqrt{2}$ D. 4
20. | The diffusion coefficient of CO_2 in air is about 10^4 times larger than the diffusion coefficient of CO_2 in water. In the time it takes CO_2 to diffuse a distance d in water, how far will it diffuse in air?
    A. $d/10^4$ B. $d/10^2$ C. $10^2 d$ D. $10^4 d$
21. || On average, the partial pressure of oxygen is about 150 mm Hg in the alveoli and 50 mm Hg in the pulmonary capillaries. The partial pressure of carbon dioxide is about 45 mm Hg in the capillaries and 40 mm Hg in the alveoli. The use of oxygen in the body produces one CO_2 molecule for every O_2 molecule consumed. For the respiratory tissue of the lungs,
BIO
    A. The diffusion coefficient of CO_2 is about the same as the diffusion coefficient of O_2.
    B. The diffusion coefficient of CO_2 is about 3 times the diffusion coefficient of O_2.
    C. The diffusion coefficient of CO_2 is about 20 times the diffusion coefficient of O_2.
    D. The diffusion coefficient of O_2 is about 20 times the diffusion coefficient of CO_2.

PROBLEMS

Section 13.2 Molecular Speeds and Collisions

1. || The number density of an ideal gas at STP is called the *Loschmidt number.* Calculate the Loschmidt number.
2. ||| A 1.0 m × 1.0 m × 1.0 m cube of nitrogen gas is at 20°C and 1.0 atm. Using the histogram of Figure 13.2, estimate the number of molecules in the cube with a speed between 700 m/s and 1000 m/s.
3. ||| At what pressure will the mean free path in room-temperature (20°C) nitrogen be 1.0 m?
4. ||| Integrated circuits are manufactured in vacuum chambers in which the air pressure is 1.0×10^{-10} mm Hg. What are (a) the number density and (b) the mean free path of a molecule? Assume $T = 20°C$.
5. | The mean free path of a molecule in a gas is 300 nm. What will the mean free path be if the gas temperature is doubled at (a) constant volume and (b) constant pressure?
6. || A lottery machine uses blowing air to keep 2000 Ping-Pong balls bouncing around inside a 1.0 m × 1.0 m × 1.0 m box. The diameter of a Ping-Pong ball is 3.0 cm. What is the mean free path between collisions? Give your answer in cm.

Section 13.3 The Kinetic Theory of Gases

7. | Eleven molecules have speeds 15, 16, 17, . . . , 25 m/s. Calculate (a) v_{avg} and (b) v_{rms}.
8. | Figure P13.8 is a histogram showing the speeds of the molecules in a very small gas. What are (a) the most probable speed, (b) the average speed, and (c) the rms speed?

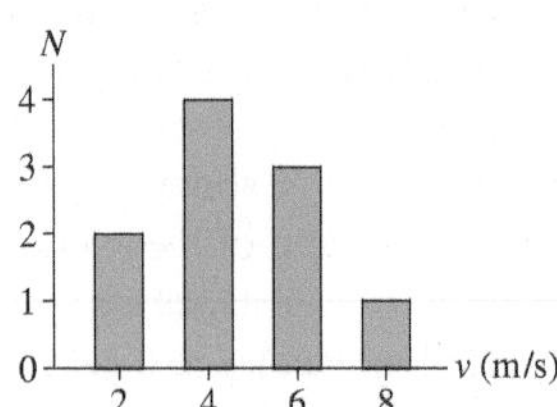

FIGURE P13.8

9. | A gas consists of a mixture of neon and argon. The rms speed of the neon atoms is 400 m/s. What is the rms speed of the argon atoms?
10. || What are the rms speeds of (a) argon atoms and (b) hydrogen molecules at 800°C?
11. ||| 1.5 m/s is a typical walking speed. At what temperature (in °C) would nitrogen molecules have an rms speed of 1.5 m/s?
12. | The atmosphere on Mars is composed almost entirely of carbon dioxide at a pressure of 610 Pa. A typical surface temperature is −63°C. What are (a) the rms speed of a molecule in Mars's atmosphere and (b) the mean free path of these molecules?

13. || By what factor does the rms speed of a molecule change if the temperature is increased from 10°C to 1000°C?
14. || Atoms can be "cooled" to incredibly low temperatures by letting them interact with a laser beam. Various novel quantum phenomena appear at these temperatures. What is the rms speed of cesium atoms that have been cooled to a temperature of 100 nK?
15. | An ideal gas is at 20°C. If we double the average kinetic energy of the gas atoms, what is the new temperature in °C?
16. || A typical helium balloon contains 1.1 g of helium gas. At 20°C room temperature, what is the total kinetic energy of the helium in the balloon?
17. || The number density in a container of neon gas is $5.00 \times 10^{25}\ m^{-3}$. The atoms are moving with an rms speed of 660 m/s. What are (a) the temperature and (b) the pressure inside the container?
18. || A cylinder contains gas at a pressure of 2.0 atm and a number density of $4.2 \times 10^{25}\ m^{-3}$. The rms speed of the atoms is 660 m/s. Identify the gas.
19. ||| During a physics experiment, helium gas is cooled to a temperature of 10 K at a pressure of 0.10 atm. What are (a) the mean free path in the gas, (b) the rms speed of the atoms, and (c) the average energy per atom?
20. | What are (a) the average kinetic energy and (b) the rms speed of a proton in the center of the sun, where the temperature is 2.0×10^7 K?
21. || The atmosphere of the sun consists mostly of hydrogen *atoms* (not molecules) at a temperature of 6000 K. What are (a) the average translational kinetic energy per atom and (b) the rms speed of the atoms?

Section 13.4 Thermal Energy and Specific Heat

22. || A 10 g sample of neon gas has 1700 J of thermal energy. Estimate the average speed of a neon atom.
23. || The rms speed of the atoms in a 2.0 g sample of helium gas is 700 m/s. What is the thermal energy of the gas?
24. || A 6.0 m × 8.0 m × 3.0 m room contains air at 20°C. What is the room's thermal energy?
25. || The thermal energy of 1.0 mol of a substance is increased by 1.0 J. What is the temperature change if the system is (a) a monatomic gas, (b) a diatomic gas, and (c) a solid?
26. ||| What is the thermal energy of 100 cm^3 of aluminum at 100° C? The density of aluminum is 2700 kg/m^3.
27. ||| A cylinder of nitrogen gas has a volume of 15,000 cm^3 and a pressure of 100 atm.
    a. What is the thermal energy of this gas at room temperature (20°C)?
    b. What is the mean free path in the gas?
    c. The valve is opened and the gas is allowed to expand slowly and isothermally until it reaches a pressure of 1.0 atm. What is the change in the thermal energy of the gas?
28. || A rigid container holds 0.20 g of hydrogen gas. How much heat is needed to change the temperature of the gas
    a. From 50 K to 100 K?
    b. From 250 K to 300 K?
    c. From 2250 K to 2300 K?
29. | 2.0 mol of monatomic gas A initially has 5000 J of thermal energy. It interacts with 3.0 mol of monatomic gas B, which initially has 8000 J of thermal energy.
    a. Which gas has the higher initial temperature?
    b. What is the final thermal energy of each gas?
30. ||| 4.0 mol of monatomic gas A interacts with 3.0 mol of monatomic gas B. Gas A initially has 9000 J of thermal energy, but in the process of coming to thermal equilibrium it transfers 1000 J of heat energy to gas B. How much thermal energy did gas B have initially?

Section 13.5 $k_B T$ and the Boltzmann Factor

31. || A two-level system has $N_1 = 6000$ and $N_2 = 4000$ when the temperature is 500 K.
    a. What is δE?
    b. At what temperature would 90% of the particles be in the lower energy level?
32. || A two-state system has population ratio $N_2/N_1 = 1/3$ at a temperature of 400 K. At what temperature is the ratio 2/3?
33. || A system has equally spaced energy levels $E_i = E_1, 2E_1, 3E_1, \ldots$ for $i = 1, 2, 3, \ldots,$ where $E_1 = 1.0 \times 10^{-20}$ J. What is the population ratio N_5/N_3 of the fifth and third energy levels at a temperature of (a) 300 K and (b) 3000 K?
34. || An experiment with protons uses a magnetic field for which the energy difference between the aligned and anti-aligned spin states is 8.4×10^{-26} J, the same as used for a medical MRI. At what temperature will the fractional population difference be $\Delta N/N = 0.010$, a 1% difference? Temperatures this low are routinely achieved in *low-temperature physics* experiments.
35. || A container at 300 K contains helium and neon atoms. What is the ratio of the fraction of helium atoms that have a speed between 999 m/s and 1001 m/s to the fraction of neon atoms in the same range of speeds?

Section 13.6 Reaction Kinetics and Catalysis

36. || If the temperature of a reaction is increased from 20°C to 50°C, by what factor does the rate constant increase if the activation energy is (a) 10 kJ/mol, (b) 50 kJ/mol, and (c) 90 kJ/mol?
37. || Ozone (O_3) in the upper atmospheric layer called the *stratosphere* protects life on the earth's surface by absorbing ultraviolet radiation from the sun. Ozone can be depleted by reacting with nitrogen oxides. One important reaction is $NO + O_3 \rightarrow NO_2 + O_2$, which has an activation energy of 13 kJ/mol. Laboratory measurements at 25°C have found that the rate constant is $2.2 \times 10^7\ M^{-1}s^{-1}$. What is the rate constant at a typical stratospheric temperature of −70°C?
38. ||| BIO Soft-boiling an egg causes egg albumin protein in the egg white to coagulate but leaves the yolk runny. It takes 6.0 min to soft-boil an egg at sea level in Los Angeles. In Denver, where water boils at 95°C, it takes 7.0 min. What is the activation energy for the coagulation of egg albumin protein?
39. || It is sometimes said that a reaction rate doubles for every 10°C increase in temperature. This statement is not generally true, but it can be a reasonable approximation for reactions that take place near room temperature with a "typical" activation energy. For what activation energy in kJ/mol would be this be exactly true for a temperature increase from 20°C to 30°C?
40. || BIO The reversible reaction $CO_2 + H_2O \rightleftharpoons HCO_3^- + H^+$ is especially important for stabilizing the pH of blood. This reaction is catalyzed by the enzyme carbonic anhydrase. Laboratory experiments find that the uncatalyzed reaction has an activation energy of 67 kJ/mol. The enzyme lowers the activation energy to 32 kJ/mol. By what factor does the enzyme speed up the reaction at a body temperature of 37°C?

Section 13.7 Diffusion

41. | Aromatic molecules like those in perfume have a diffusion coefficient in air of approximately 2×10^{-5} m^2/s. Estimate, to one significant figure, how many hours it takes perfume to diffuse 3 m, about 10 ft, in still air.

42. ||| A 3.0-cm-diameter, 0.50-mm-thick spherical plastic shell holds carbon dioxide at 2.0 atm pressure and 25°C. CO_2 molecules diffuse out of the shell into the surrounding air, where the carbon dioxide concentration is essentially zero. The diffusion coefficient of carbon dioxide in the plastic is 2.5×10^{-12} m^2/s.
    a. What is the diffusion rate in molecules/s of carbon dioxide out of the shell?
    b. If the rate from part a is maintained, how long in hours will it take for the carbon dioxide pressure to decrease to 1.0 atm? The actual rate slows with time as the concentration difference decreases, but assuming a constant rate gives a reasonable estimate of how long the shell will contain the carbon dioxide.

43. ||| A nerve impulse is propagated across a synapse when the
BIO neurotransmitter acetylcholine is released from the pre-synaptic membrane and diffuses across the 20-nm-wide synaptic cleft to the post-synaptic membrane.
    a. How long does it take a nerve signal to cross a synapse? You can assume that the synaptic fluid is essentially water.
    b. What is the effective speed of transmission, in m/s?

44. || The small capillaries in the lungs
BIO are in close contact with the alveoli. A red blood cell takes up oxygen during the 0.75 s that it squeezes through a capillary at the surface of an alveolus. What is the diffusion time for oxygen across the 2-μm-thick membrane separating air from blood? The diffusion coefficient for oxygen in tissue is 1.1×10^{-11} m^2/s.

45. || The partial pressure of oxygen in the lungs is 150 mm Hg.
BIO This corresponds to a concentration of 5.3×10^{24} molecules per m^3. In the oxygen-depleted blood entering the pulmonary capillaries, the concentration is 1.4×10^{24} molecules per m^3. The blood is separated from air in the alveoli of the lungs by a 2-μm-thick membrane. What is the rate of transfer of oxygen to the blood through the 5×10^{-9} m^2 surface area of one alveolus? Give your answer in both molecules/s and μmol/s. The diffusion coefficient for oxygen in tissue is 1.1×10^{-11} m^2/s.

General Problems

46. || What is the thermal energy of a 1.0 m $\times$ 1.0 m $\times$ 1.0 m box of helium at a pressure of 5.0 atm?

47. || A sealed vessel contains 4.0 g of nitrogen and 6.0 g of helium at 300 K. The total pressure in the vessel is 150 kPa. What are the partial pressures in kPa of each of the gases?

48. | The range of temperatures over which life exists is about
BIO -40°C to 70°C. We perceive this as a large range, but what does it imply about molecular speeds and energies?
    a. What is the ratio of the average translational kinetic energy per molecule at 70°C to that at -40°C?
    b. What is the ratio of the rms speed of a molecule at 70°C to that at -40°C?

49. || For a planet to have an atmosphere, gravity must be sufficient to keep the gas from escaping. The *escape speed* a particle needs to escape the earth's gravitational attraction is 1.1×10^4 m/s. The motion of projectiles never depends on mass, so this escape speed applies equally to rockets and to molecules in the earth's upper atmosphere.
    a. At what temperature does the rms speed of nitrogen molecules equal the escape speed?
    b. At what temperature does the rms speed of hydrogen molecules equal the escape speed?
    c. The temperature of the earth's thin upper atmosphere is approximately 1000°C due to heating from the sun's ultraviolet radiation. This is a high temperature, but much lower than the temperatures you calculated in parts a and b. However, a gas can slowly escape from a planet's atmosphere if just a small fraction of the molecules have speeds that exceed the escape speed. These are the molecules at the far right of the Maxwell-Boltzmann distribution. A rule of thumb based on the Maxwell-Boltzmann distribution is that, given sufficient time, a gas can escape if $v_{esc} < 6v_{rms}$. According to this rule, what is the largest molecular mass, in atomic mass units, that can escape the earth's upper atmosphere?
    d. Can N_2 escape the earth? What about H_2?

50. || The space between the stars isn't totally empty. In interstellar space, far from any stars, there is about one hydrogen atom per cm^3 at the very low temperature of 3 K. This is an estimation problem, so answers with one significant figure are appropriate.
    a. What is the mean free path of the hydrogen atoms in interstellar space?
    b. About how long in years does a typical hydrogen atom travel between encounters with neighboring atoms?

51. || Figure P13.51 shows the thermal energy of 0.14 mol of gas as a function of temperature. What is C_V for this gas?

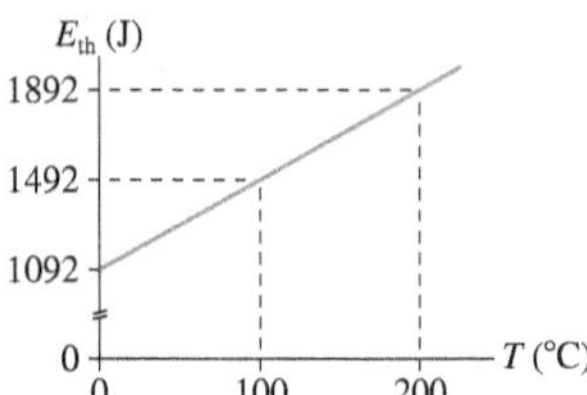

FIGURE P13.51

52. || The 2010 Nobel Prize in Physics was awarded for the discovery of graphene, a two-dimensional form of carbon in which the atoms form a two-dimensional crystal-lattice sheet only one atom thick. Predict the molar specific heat of graphene. Give your answer as a multiple of R.

53. ||| a. Find an expression in terms of k_BT for the most probable
CALC speed $v_{most\ prob}$ of the Maxwell-Boltzmann distribution.
    b. What is the translational kinetic energy $\epsilon_{most\ prob}$ of a molecule with the most probable speed?

54. ||| a. What is the rms speed of a gas of argon atoms at 300 K?
    b. What fraction of the argon atoms have speeds in the range from 200 m/s to 201 m/s, 400 m/s to 401 m/s, and 600 m/s to 601 m/s?

55. ||| Ion channels are active pores in a cell membrane that transport ions between the cell's interior and exterior. Although generally opened or closed by various cellular mechanisms, for inert cells studied in the laboratory the pores can open or close due to thermal effects alone. In other words, we can model an ion channel as a two-state system with open energy E_{open} and closed energy E_{closed}. In one study, performed at 37°C, the channels were found to be open 10% of the time and closed 90% of the time.
BIO
a. Which energy is higher: E_{open} or E_{closed}?
b. What is the energy difference between the two states?

56. ||| Suppose a molecule has three energy levels that are equally spaced in energy. In a collection of such molecules, 10,000 are found to be in the lowest energy level and 7000 in the next higher energy level. How many are in the highest energy state?

57. ||| At 300 K, 3.8×10^{15} molecules in a gas are found to be in energy level 1 while only 4.8×10^{6} molecules are in energy level 2. What is the energy separation between the two energy levels?

58. ||| In practice, activation energies are determined by measuring reaction rates at several different temperatures. If we take the natural logarithm of both sides of the Arrhenius equation, we find

$$\ln k = \ln A - \left(\frac{E_a}{k_B}\right)\frac{1}{T}$$

Thus a graph of $\ln k$ versus $1/T$ should be a straight line, and the slope of the trend line can be used to determine E_a. Figure P13.58 shows data from one set of experiments. What is the activation energy for this reaction?

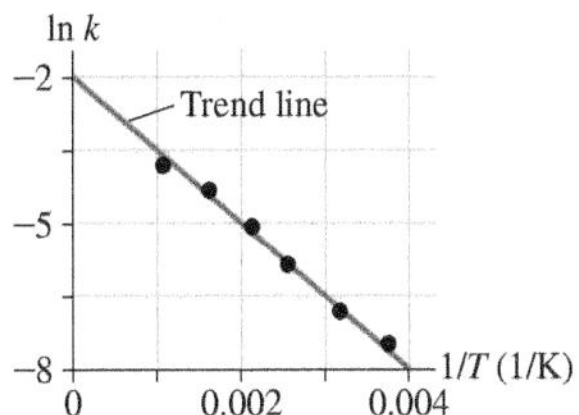

FIGURE P13.58

59. |||| A *first-order reaction* is one for which the concentration of a reactant A, which chemists write as [A], changes at the rate $d[A]/dt = -k[A]$, where k is the rate constant. That is, the rate of change of A depends on how much of A is present: A large concentration of A means that the concentration is changing rapidly. This is similar to population dynamics: The more individuals of a species there are, the more rapidly the population grows. In this case, because of the minus sign, the concentration *decreases* as the reactants are used up.
CALC

You can solve the rate equation to find how the concentration [A] changes with time. To do so, put all the terms involving [A] on one side of the equation and all the terms involving t on the other, then integrate. You'll have an integration constant that you can determine by setting the concentration to an initial $[A]_0$ at $t = 0$. The decomposition of hydrogen peroxide, the subject of Example 13.11, is a first-order reaction in which [A] is the concentration of hydrogen peroxide. The rate constant of the uncatalyzed reaction is $1.0 \times 10^{-8}\ s^{-1}$. How long in years must you wait for the concentration of a sample of hydrogen peroxide to decrease by 50%?

60. || Your cornea doesn't have blood vessels, so the living cells of the cornea must get their oxygen from other sources. Cells in the front of the cornea obtain their oxygen from the air. Wearing a contact lens interferes with this oxygen uptake, so contact lenses are designed to permit the diffusion of oxygen. The diffusion coefficient of one brand of soft contact lenses was measured to be $1.3 \times 10^{-13}\ m^2/s$. We can model the lens as a 14-mm-diameter disk with a thickness of 40 μm. The partial pressure of oxygen at the front of the lens is 20% of atmospheric pressure, and the partial pressure at the rear is 7.3 kPa. At 30°C how many oxygen molecules cross the lens in 1 h?
BIO

61. ||| A typical mammalian cell is a 15-μm-diameter sphere with an oxygen partial pressure of 40 mm Hg. It is separated from the higher oxygen concentration in blood by a capillary wall with a thickness of 1.0 μm. (The much thinner cell-wall thickness can be neglected.) The oxygen partial pressure of blood varies with distance along the capillary, but 75 mm Hg is a typical value. The diffusion coefficient of oxygen through the capillary wall is reported to be $1.1 \times 10^{-11}\ m^2/s$.
BIO
a. What is the oxygen diffusion time in ms from the capillary to the cell?
b. Suppose that 25% of the cell wall is in contact with a capillary, the remaining 75% being in contact with other cells. What is the diffusion rate in molecules/s of oxygen into the cell? Assume a body temperature of 37°C.

62. |||| A 3.0-cm-diameter, 0.50-mm-thick spherical plastic shell holds carbon dioxide at 2.0 atm pressure and 25°C. We would like to know how long it will take the CO_2 molecules to diffuse out of the shell. The diffusion coefficient of carbon dioxide in the plastic is $2.5 \times 10^{-12}\ m^2/s$. Suppose there are N CO_2 molecules inside the sphere, which has volume V. Because the CO_2 concentration in the air is essentially zero, the concentration difference Δc between the inside and outside of the sphere is $\Delta c = c_{out} - c_{in} = 0 - N/V = -N/V$. Thus Fick's law for the diffusion rate of CO_2 molecules is
CALC

$$\frac{dN}{dt} = -\left(\frac{DA}{LV}\right)N$$

where the minus sign indicates that the number of molecules inside the sphere is decreasing with time.

You can solve this equation to find how N changes with time. To do so, put all the terms involving N on one side of the equation and all the terms involving t on the other, then integrate. You'll have an integration constant that you can determine by setting the number of molecules to N_0 at $t = 0$. How long in hours must you wait for the carbon dioxide concentration to decrease to 1.0 atm?

MCAT-Style Passage Problems

The Dual Role of Eggshells

The outer shell of a bird's egg is composed of mostly calcium carbonate crystals. The shell must be sufficiently thick and strong to protect the embryo. At the same time, a developing embryo needs to "breathe" by exchanging oxygen and carbon dioxide with the atmosphere. To allow for this, an eggshell has microscopic pores a few μm across, such as those seen in the micrograph of Figure P13.63,

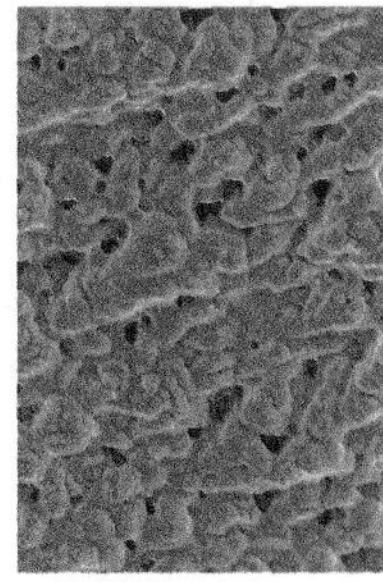

FIGURE P13.63

through which diffusion can occur. Avian researchers can improve their understanding of embryo development by using the size and distribution of pores to calculate the rates of diffusion of oxygen into and carbon dioxide out of the shell. A greater rate of gas exchange implies a greater rate of embryonic development.

63. Geese lay a clutch of many eggs over the course of several days, but all the eggs hatch at approximately the same time. Which of the following helps to explain this observation?
   A. Eggs laid earlier have a greater number of pores than eggs laid later.
   B. All eggs have approximately the same number of pores.
   C. Eggs laid later have a greater number of pores than eggs laid earlier.
64. As time goes on, the embryo inside the egg grows and consumes more oxygen. The rate of gas exchange must increase even though the size and distribution of pores in the eggshell don't change. This means that
   A. The concentration of oxygen inside the shell increases as the embryo develops.
   B. The concentration of oxygen inside the shell stays the same as the embryo develops.
   C. The concentration of oxygen inside the shell decreases as the embryo develops.
65. Cowbirds are brood parasites—they lay their eggs in the nests of other birds so that the host birds will take care of their young when the eggs hatch. A cowbird egg has a thicker shell with no more pores than the eggs of a red-winged blackbird, a common host, but cowbird eggs typically hatch a day or two in advance of red-winged blackbird eggs in the same nest. Which of the following helps to explain this observation?
   A. The individual pores in a cowbird egg are smaller than those in a red-winged blackbird egg.
   B. The individual pores in a cowbird egg are about the same size as those in a red-winged blackbird egg.
   C. The individual pores in a cowbird egg are larger than those in a red-winged blackbird egg.
   D. There's not enough information to compare the pore sizes in the two eggs.
66. The pores in the eggshell permit diffusion of not only oxygen and carbon dioxide but also water vapor. Consider what the relative concentrations of water vapor inside and outside the shell must be. Given this, we would expect that
   A. As the embryo develops, an egg will have a net loss of water to the environment.
   B. As the embryo develops, an egg will experience no net gain or loss of water to the environment.
   C. As the embryo develops, an egg will have a net gain of water from the environment.

14 Entropy and Free Energy

Saltwater fish lose water to the saltier sea via osmosis. To compensate, most saltwater fish drink seawater and then actively excrete salt through their gills.

LOOKING AHEAD

Microstates and Macrostates

This roll is just one of six *microstates* that correspond to the *macrostate* of "seven." The number of microstates is called the *multiplicity*.

The *entropy* of a physical system is defined in terms of the multiplicity of microstates that correspond to a physical macrostate.

Free Energy and Spontaneity

An instant cold pack spontaneously lowers its temperature to near freezing. How does it do so without violating the laws of physics?

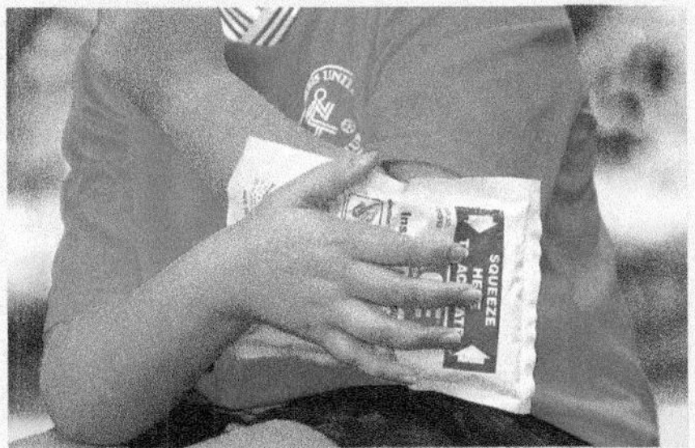

You'll see how the *Gibbs free energy* uses entropy and the second law of thermodynamics to predict spontaneity.

Osmosis

Water is removed along with urea when blood is filtered by your kidneys or in dialysis. Nearly all of the water is then reabsorbed via *osmosis*.

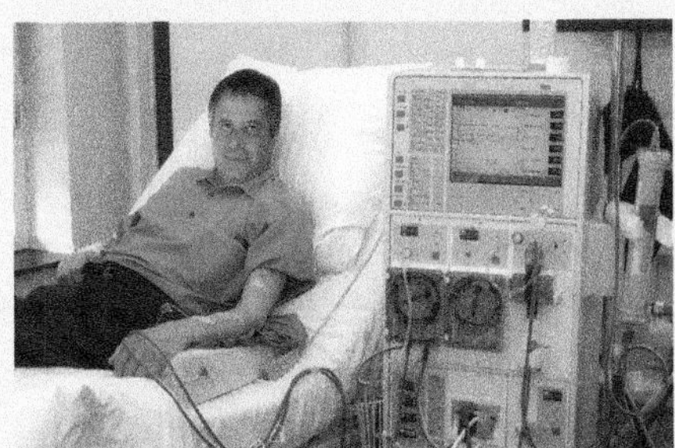

You'll learn that osmosis, along with the creation of osmostic pressure, is an entropically driven process.

GOAL To understand how entropy and the second law of thermodynamics apply to living and nonliving systems.

PHYSICS AND LIFE

Entropy and Life

Entropy, often associated with disorder and decay, might seem to be the antagonist of life. Yet one hallmark of life is that it generates astonishingly beautiful and functional structures. Not only does this seem to fly in the face of entropy, but it turns out that entropy plays a critical role in generating those structures by means of the hydrophobic effect, osmotic pressure, and diffusion, all of which are fundamentally entropic processes. Life proves to be ingenious at harnessing entropy as well as energy to its service. The quantity that helps us figure out how entropy combines with energy to drive molecular processes is called *free energy*. Entropy and free energy are subtle ideas, but they lead to deep insights into the most basic processes of life.

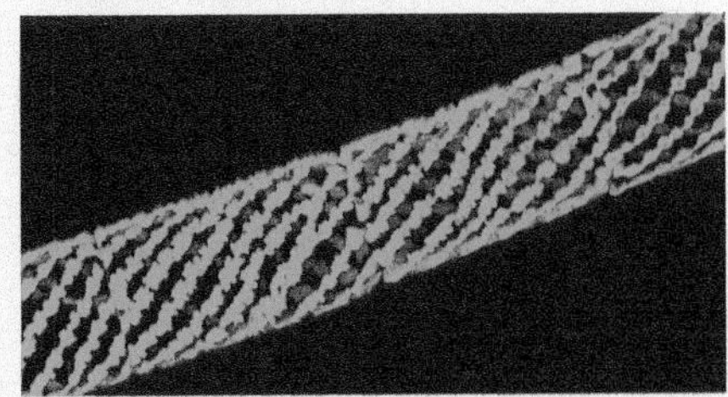

These *Spirogyra* algal chloroplasts use sunlight to make glucose. Energetically unfavorable steps in the cycle are driven "uphill" by using the *available energy* from ATP hydrolysis.

14.1 Reversible and Irreversible Processes

An aqueous solution of adenosine triphosphate (ATP), the "energy currency" of the cell, is quite stable. But if we add just a small amount of the enzyme ATPase, all the ATP is quickly converted to adenosine diphosphate (ADP) and inorganic phosphate. In contrast, a solution of ADP and inorganic phosphate never becomes ATP. The hydrolysis of ATP is not only *spontaneous,* it is also *irreversible.*

Irreversible processes abound. If you ignite a piece of paper—mostly cellulose—it burns to produce carbon dioxide, water vapor, and ash; a mixture of carbon dioxide, water vapor, and ash never produces cellulose. An ice cube left on the counter turns into a puddle of water, but a puddle of water on the counter never turns into an ice cube. Cream mixes into your coffee; they never separate. Gas expands into a vacuum, but all the molecules of a gas never spontaneously move to one side of their container to leave a vacuum on the other side. Why not?

We are so accustomed to the fact that most processes are irreversible—that they have a one-way direction—that it seems almost silly to ask Why? Nonetheless, it's the question we're going to focus on in this chapter, and it turns out to be a profound question with many implications ranging from how cellular processes work to the very existence of life.

If you think about it, none of the processes just mentioned would violate any of the laws of physics we have studied so far if they ran backward. Energy is conserved as heat flows from a warm counter to a melting ice cube, and energy could be conserved just as well by heat flowing from a puddle of water to the counter, lowering the water's temperature until it freezes. Yet that doesn't happen. There must be another law of physics at work that prevents some energy-conserving processes from occurring.

This new law—the *second law of thermodynamics*—is subtle, so we'll start with a qualitative overview before we dive into a quantitative analysis. We'll spend three sections understanding what the second law and its essential concept, *entropy,* are all about, then conclude our study of thermodynamics by looking at the essential roles that entropy and the second law play in biological processes. In particular, you will see that an idea you may have learned in chemistry—that a reaction is spontaneous if the change in the Gibbs free energy is negative—is actually just a restatement of the second law of thermodynamics in terms of quantities that can be measured in chemistry or biochemistry.

Thermal Energy Disperses

The main point of Chapter 13 was that macroscopic phenomena such as pressure, heat transfer, and chemical reactions are caused by collisions of molecules. FIGURE 14.1a shows two frames from a simple "movie" of a collision between two gas molecules. FIGURE 14.1b is the same movie played in reverse. This is also a possible molecular collision. That is, nothing in either movie looks wrong, and no measurements we might make would reveal that either of these two collisions violates any laws of physics. We can't label one of these movies as "real" and the other "impossible." In other words, **interactions at the molecular level are reversible.**

FIGURE 14.1 Two movies showing that molecular collisions are reversible.

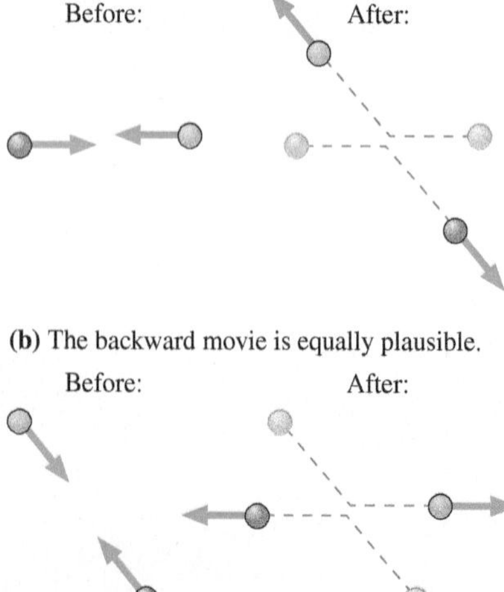

Now contrast this with movies of any of the processes we described earlier. A movie in which a puddle of water turns into an ice cube while a nearby thermometer shows a reading of 25°C is clearly running backward. It shows an impossible process. But what has been violated in the backward movie?

Processes such as phase changes, chemical reactions, and the mixing of molecules are molecular processes. **If molecular interactions are reversible, why are macroscopic processes—which consist of molecular interactions—irreversible?** If reversible collisions transfer thermal energy from hot to cold, why don't they ever transfer thermal energy from cold to hot? This is a more exact framing of the question we need to answer.

As we begin to consider this question, you may recall that the thermal energy of a macroscopic object is the total energy of all the moving atoms and molecules that make up the object. However, knowing the thermal energy tells us nothing about how the energy is distributed among the molecules. FIGURE 14.2a shows a container of gas in which one molecule is moving while all the others are at rest. The thermal energy of the gas is simply the kinetic energy of the one moving molecule. This is one possible way that the energy could be distributed, but this is not a stable, equilibrium situation. The moving molecule will collide with other molecules, giving them kinetic energy while the original fast molecule loses energy. Eventually, we expect that every atom in the gas will be moving, something like the situation shown in FIGURE 14.2b. The thermal energy is the same in both cases; it's just distributed differently.

FIGURE 14.2 Two different distributions of the thermal energy of a gas.

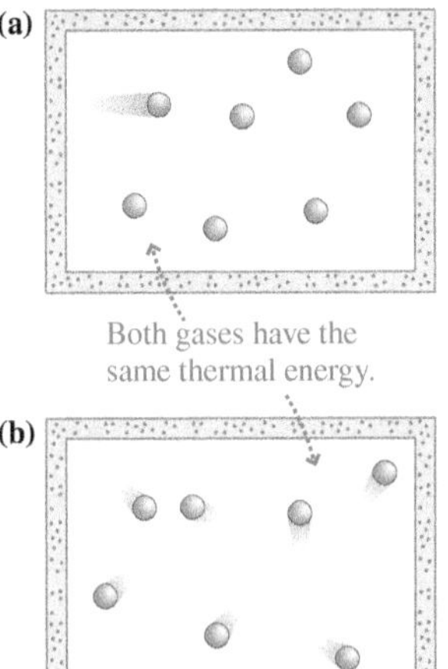

Let's think about what happened here. Initially, the system's thermal energy was concentrated in a single molecule. As time went on, collisions and interactions caused the system's energy to spread out, or disperse, until it was shared among all the constituents. And this spreading will not spontaneously reverse. Ordinary molecular collisions cause a concentrated energy to spread out, but we never see a system spontaneously concentrating dispersed energy into a single constituent. **The individual collisions may be reversible, but the dispersal of energy is an irreversible process.**

All thermal interactions involve the spreading of energy. Suppose a hot object and a cold object are brought into thermal contact, as in FIGURE 14.3. The thermal energy is initially concentrated in the hot object; it is not shared equally. As we've seen, collisions at the boundary transfer energy from the more-energetic molecules on the warmer side to the less-energetic molecules on the cooler side until the thermal energy is shared equally among the constituents of the combined system. After some time, thermal equilibrium is reached, which means that no additional sharing or spreading is possible.

FIGURE 14.3 A thermal interaction causes energy to spread until it is shared equally among all the molecules.

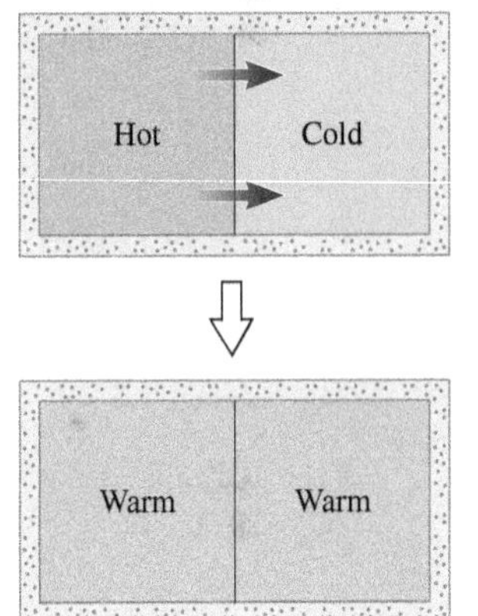

The *spreading* of thermal energy happens spontaneously—it's simply what happens when macroscopic systems interact. You can imagine the heat-transfer arrows of Figure 14.3 being reversed, making the cold side colder and the hot side hotter. But that would concentrate the thermal energy rather than dispersing it. The *concentration* of thermal energy does not happen spontaneously, and thus heat energy is never transferred spontaneously from a colder object to a hotter object.

Introducing Entropy

Scientists use the term **entropy** to quantify the spread or dispersal of thermal energy. A system in which energy is concentrated—not spread out—has low entropy. The initial situations in Figures 14.2 and 14.3 are states of low entropy. In each case, the system spontaneously evolves to a state in which the energy is more spread out. These are states of higher entropy; the entropy increases as thermal energy becomes more dispersed. Because equilibrium is a state in which thermal energy is maximally dispersed, or shared among all the constituents, **the entropy of a system is maximum when the system is in equilibrium.**

This observation about how thermal energy spreads and how macroscopic systems evolve irreversibly toward equilibrium is a new law of physics, the **second law of thermodynamics,** that is usually stated in terms of entropy:

Second law of thermodynamics The entropy of an isolated system never decreases. The entropy either increases until the system reaches equilibrium or, if the system began in equilibrium, stays the same. Entropy is maximum when the system is in equilibrium.

NOTE ► The qualifier "isolated" is crucial. We could reduce a system's entropy by reaching in from the outside, perhaps using tiny tweezers to give most of the thermal energy to a few fast molecules while freezing the others in place. Similarly, we can transfer heat from cold to hot by using a refrigerator. The second law is about what a system can or cannot do spontaneously, on its own, without outside intervention. ◄

The second law of thermodynamics is often stated in equivalent but more informal versions. One of these, and the one most relevant to our discussion, is

Second law, informal statement #1 When two systems at different temperatures interact, heat energy is transferred spontaneously from the hotter to the colder system, never from the colder to the hotter.

Thermodynamics is a science of energy, and we need two laws of energy to understand how macroscopic systems behave. We can restate the two laws of energy as follows:

- **First law of thermodynamics:** Energy is conserved. It can be transferred or transformed, but the total amount does not change.
- **Second law of thermodynamics:** Energy spreads out. An isolated system evolves until the thermal energy is maximally dispersed among all the system's constituents.

It is the second law that creates the "arrow of time" that we observe in irreversible processes. Stirring blends your coffee and cream; it never unmixes them. Systems evolve over time only in ways that continue to spread out thermal energy, thus increasing entropy, not in ways that concentrate thermal energy, which would decrease entropy.

Hence another statement of the second law is

Second law, informal statement #2 The time direction in which the entropy of an isolated macroscopic system increases is "the future."

Establishing the arrow of time is one of the most profound implications of the second law of thermodynamics.

Reasoning with Entropy

Heating a system increases its thermal energy. By increasing the thermal energy, we also increase the ability of the system to disperse thermal energy, so we increase the entropy. Similarly, removing heat energy from a system reduces its ability to disperse thermal energy, so the entropy is reduced.

FIGURE 14.4 The free expansion of a gas increases its entropy.

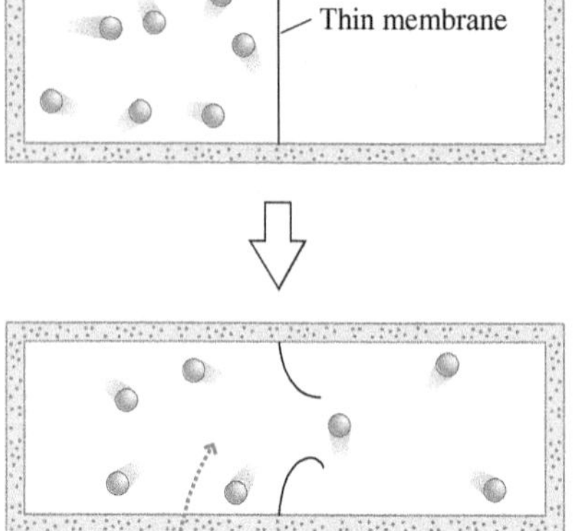

Thin membrane
The thermal energy hasn't changed, but the entropy has.

The entropy can also change when there is no change in thermal energy. FIGURE 14.4 shows an insulated container of a gas with the molecules kept on one side by a thin membrane. The container is an isolated system; no heat is transferred to or from the environment. If the membrane breaks, the gas expands to fill the container—what we call a *free expansion.* The volume has changed and the pressure will change. But if we think about energy, we can see that this expansion does not change the gas temperature. No work is done because the expanding gas isn't pushing against anything, and no heat is transferred from the environment, so the first law of thermodynamics tells us there's no change in thermal energy and hence no change in temperature. But the system's entropy has increased! The thermal energy may not have changed in value, but it's more spatially dispersed because each molecule has more room to move around. And a larger dispersal of energy corresponds to greater entropy. The fact that entropy increases makes the free expansion an irreversible process; the reverse process of all the gas molecules moving into the left half of the container never occurs spontaneously.

Something else that changes is the ability of the gas to do work. Instead of breaking the membrane, we could have allowed the compressed gas to do work by pushing on a piston. This would have transferred some of the thermal energy to the environment by doing mechanical work. After expanding freely, the gas has the same thermal energy but less ability to do work. In general, our ability to extract useful work from thermal energy depends not only on the quantity of thermal energy but also on the system's entropy. Increased entropy, with the energy more dispersed, means less ability to do work.

This is why biochemical reactions are analyzed using *free energy,* which, loosely speaking, is the amount of work that can be done by the system. Any increase in entropy decreases the work that can be done and thus decreases the free energy even if the total energy is unchanged. We'll return to this idea in Section 14.6.

CONCEPTUAL EXAMPLE 14.1 **Does entropy change when ice melts?**

A phase change, such as melting or freezing, occurs at a fixed temperature. What happens to the entropy of the water if a block of frozen water at 0°C becomes liquid water at 0°C?

REASON The entropy can change, even at the same temperature, if one of the phases of matter—solid or liquid in this case—is able to spread out the energy more. Ice is a solid, so water molecules are frozen into a lattice. Molecules jiggle around—that's what thermal energy is—but their energy is confined to small volumes around the lattice positions. In contrast, the molecules in liquid water can move freely throughout the volume of the liquid. This allows the thermal energy to be more spatially dispersed, just as it was for the free expansion of a gas. Thus the entropy increases as ice melts.

ASSESS This result makes sense. You know that you need to heat ice to melt it, so you expect this change to increase the entropy.

Is Entropy Disorder?

You may have learned elsewhere that entropy is a measure of *disorder,* or *chaos,* and that increasing entropy shows a tendency of an ordered system to become disordered. But consider the following:

- A beaker of water and cracked-ice pieces of various sizes appears disordered. After the ice has all melted, the beaker of water looks more orderly. But the beaker of water has a higher entropy, not a lower entropy.
- A container of oil and water that has been vigorously shaken looks quite disordered. But the oil and water soon separate to what appears to be a more orderly state. However, the separated oil and water actually have a higher entropy, not a lower entropy.
- The free expansion of a gas into a vacuum starts with completely random moving molecules and ends with completely random moving molecules at the same average speed. There doesn't seem to be any difference in how ordered or disordered these two states are, but the expanded gas has a higher entropy.

The difficulty is that "disorder" is not a technical term; it has no scientific definition. What seems disordered to one person may not to another. And, are we talking about macroscopic disorder, such as the cracked pieces of ice in a beaker of water, or microscopic disorder?

Disorder is a metaphor for entropy, and we don't want to imply that it's never useful. But, all in all, it's much more productive to begin to think of entropy as a measure of the spread or dispersal of energy.

STOP TO THINK 14.1 Which of the following processes does not involve a change in entropy?

A. An electric heater raises the temperature of a cup of water by 20°C.
B. A ball rolls up a ramp, decreasing in speed as it rolls higher.
C. A basketball is dropped from 2 m and bounces until it comes to rest.
D. The sun shines on a black surface and warms it.

14.2 Microstates, Multiplicity, and Entropy

The idea behind the second law of thermodynamics is that thermal energy spreads or disperses until, in equilibrium, the energy is maximally dispersed. How can this idea be quantified? We'll start with an analogy.

TABLE 14.1 The 16 possible microstates when tossing four coins

TTTT	HTTH
TTTH	HTHT
TTHT	HHTT
THTT	HHHT
HTTT	HHTH
TTHH	HTHH
THTH	THHH
THHT	HHHH

TABLE 14.2 Multiplicities of the five possible macrostates when tossing four coins

Macrostate	Multiplicity (Ω)
H = 0	1
H = 1	4
H = 2	6
H = 3	4
H = 4	1

Suppose you toss four coins simultaneously. TABLE 14.1 shows that there are 16 possible outcomes of heads and tails. Any one of these outcomes—say, HTHH—is called a **microstate,** so there are a total of 16 microstates.

Perhaps you care only about the *number* of heads, not about the details of how each coin lands. The number of heads ranges from 0 to 4, and each possibility is a **macrostate.** By examining Table 14.1 you can see that there is only one way (TTTT) the macrostate H = 0 can occur but six ways to have H = 2, with an equal number of heads and tails. The number of microstates associated with a particular macrostate is called the **multiplicity** of that macrostate, denoted by Ω (uppercase Greek omega). TABLE 14.2 shows the multiplicity of each of the five macrostates when tossing four coins.

If you assume that each coin is fair, with a 50% probability of giving a head or a tail, then each of the 16 microstates is equally likely to occur. That is, the chances of seeing TTTT are exactly the same—1 in 16—as seeing HTHT. But the probabilities of the macrostates are *not* equal. There's only one way to achieve the H = 0 macrostate, so its probability is 1/16. There are six ways to have H = 2, so its probability is 6/16. If you toss these four coins over and over, you will see the H = 2 macrostate six times as often as you see H = 0. **Some macrostates are more probable than others,** and that is the key to understanding entropy and the second law.

If you toss N coins, there are 2^N possible outcomes and thus 2^N microstates. There's always only one way to achieve the H = 0 macrostate—a toss that gives all tails—so the probability of macrostate H = 0 is $1/2^N$. You wouldn't be terribly surprised to see H = 0 when you toss only four coins because, on average, this macrostate occurs once in every 16 tosses. But if you toss 1000 coins, the odds of H = 0 (that is, the odds that all 1000 coins are tails) are reduced to $1/2^{1000} \approx 10^{-300}$, or once in every 10^{300} tosses. The age of the universe is only $\approx 10^{17}$ s, so it would be remotely unlikely to have seen H = 0 even if you had been tossing 1000 coins once a second since the Big Bang ≈14 billion years ago. With 1000 coins, it's safe to say that the macrostate H = 0 *never* occurs.

FIGURE 14.5 shows the macrostate probability graphically. The heights of the histogram bars for $N = 4$ in Figure 14.5a are taken from Table 14.2. If N is very large, Figure 14.5b shows that the only macrostates with any likelihood of occurring are clustered very near H = $N/2$, with equal numbers of heads and tails. It's not that *only* H = $N/2$ occurs, but the macrostates that do occur are clustered so close to H = $N/2$ as to be indistinguishable. That's not yet entirely true for $N = 1000$, but think of the situation with $N = 10^{22}$, roughly the number of molecules in a liter of gas at atmospheric pressure. For *very* large N, the only macrostates that will ever be seen are indistinguishable from H = $N/2$; a macrostate that differs from $N/2$ enough to be experimentally noticed, even with the best equipment, will never be seen.

FIGURE 14.5 The number of microstates that show H heads when four coins are tossed and when a very large number of coins are tossed.

(a) $N = 4$ coins

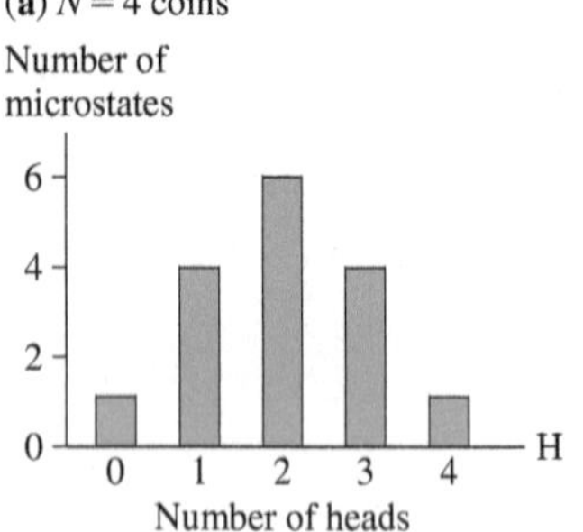

(b) N large

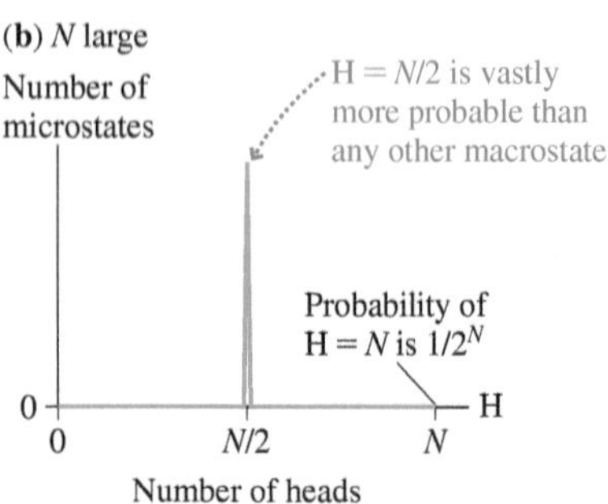

Coin flipping may seem to have nothing to do with thermodynamics, but consider a macroscopic system with pressure p, temperature T, thermal energy E_{th}, and so on. The values of the state variables $p, T, E_{th}, \ldots$ define a state of the system—in particular, a *macrostate.* Just as knowing that H = 2 tells us nothing about how each coin lands, knowing the state variables of a macroscopic system tells us nothing about what individual molecules are doing.

At the atomic level, each possible arrangement of the atoms or molecules in a system—their positions, their velocities, their rotations and vibrations—is a microstate. The number of microstates that give rise to a particular macrostate is the multiplicity Ω of that macrostate. There are an enormous number of ways the positions and velocities of the molecules can be shuffled around without changing the observable state variables, so the multiplicity is an unimaginably large number—on the order of $10^{10^{23}}$ for any real-world macrostate.

Once we recognize that a vast number of microstates (different atomic-level arrangements) correspond to each macrostate, we can reach important conclusions using four steps of logic:

1. Random collisions and atomic-level interactions, like coin flips, constantly change the microstate. Every macrostate has a vast number of microstates, each of which is equally likely to occur.
2. One macrostate is overwhelmingly more probable than the others, which means that its multiplicity Ω is vastly larger than the multiplicities of other possible macrostates. This idea is illustrated in Figure 14.5b by the fact that, for large N, almost all the microstates correspond to macrostates that are *extremely* close to $H = N/2$.
3. The most probable macrostate is the system's equilibrium state. Once an isolated system is in equilibrium, atomic-level interactions will never cause it to change because the probability of being in any other macrostate is essentially zero.
4. If an isolated system is not in equilibrium, atomic-level interactions will cause it to evolve from a less-probable macrostate to the most-probable macrostate. That is, a nonequilibrium system will evolve toward equilibrium—not because it "knows" where equilibrium is or has any "desire" to be in equilibrium but simply because **equilibrium is overwhelmingly the most probable macrostate to be in.**

The importance of these key ideas cannot be overstated. They rely on the fact that any macroscopic system has an enormous number of atoms or molecules. We will sometimes use examples that have relatively small numbers of particles to develop a feel for how the statistics work, but it is essential to keep the reality of large N in mind.

NOTE ▸ These conclusions apply to only an *isolated system* whose behavior is determined entirely by atomic-level collisions. ◂

Multiplicity of a Physical System

We began this section looking for a way to quantify the dispersal of energy. The multiplicity Ω does this; each microstate provides a different way to spread the energy, and maximal dispersion—that is, equilibrium—occurs for the macrostate that has the largest value of Ω. We need to find a way to compute the multiplicity Ω for a physical system such as a gas or a liquid.

An unlikely event Tossing all heads, while not impossible, is extremely unlikely, and the probability of doing so rapidly decreases as the number of coins increases.

To begin, consider a container divided into 1000 small bins that can each "store" a molecule; that is, there are 1000 distinct spatial locations for each molecule. Suppose the container holds one molecule. No macroscopic measurements can reveal which of the 1000 bins the molecule is in, so each of the 1000 possible locations is a microstate corresponding to the same one-molecule macrostate. The multiplicity of this macrostate is $\Omega = 1000$.

If there are two molecules, each placed randomly into a bin, the number of possible arrangements—each a microstate—is $1000 \times 1000 = 1000^2$. For three molecules it is $1000 \times 1000 \times 1000 = 1000^3$, and for N molecules 1000^N. We might insist on physical grounds that there can be no more than one molecule per bin—two molecules can't occupy the same point in space—but as long as N is much less than the number of available bins (a dilute system), then this constraint has little effect on the calculation and the multiplicity of a macrostate with N molecules is $\Omega = 1000^N$.

Now suppose we allow the container volume V to vary *without changing the bin size.* For example, doubling the volume would cause there to be 2000 bins, and then the multiplicity would be $\Omega = 2000^N$. Because *the number of bins is proportional to* V, the multiplicity of a system that consists of N molecules in volume V must be

$$\Omega = cV^N \tag{14.1}$$

where c is an unknown proportionality constant. (Don't confuse this use of c with the system's specific heat or concentration.) It might depend on the temperature or other state variables, but c does not depend on the system's volume; the dependence of Ω on V is now fully specified.

NOTE ▸ This analysis in terms of artificial bins is intended to be a plausibility argument, not a rigorous derivation. However, the complete analysis, which is beyond the scope of this textbook, reaches the same conclusion. ◂

STOP TO THINK 14.2 If the volume of a gas doubles at a constant temperature, the multiplicity increases by a *factor* of

A. 2. B. 2^N. C. V^N. D. $(2V)^N$.

FIGURE 14.6 The multiplicity of a gas increases as it undergoes a slow isothermal expansion.

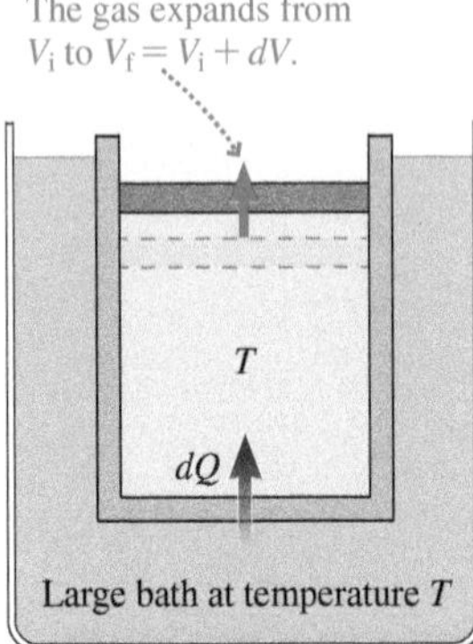
The gas expands from V_i to $V_f = V_i + dV$.
T
dQ
Large bath at temperature T

It's not immediately obvious how Equation 14.1 is useful, so let's consider a specific example. FIGURE 14.6 shows a container of gas with initial volume V_i that is undergoing a slow isothermal expansion (no change in temperature). A process that happens so slowly that it's always essentially in internal equilibrium is called a **quasi-static process**. You learned in Chapter 12 that an isothermal expansion requires a heat input, so we'll let the environment supply an infinitesimal amount of heat dQ to cause an infinitesimal volume expansion to $V_f = V_i + dV$.

For infinitesimal changes, the first law, $\Delta E_{int} = W + Q$, becomes $dE_{int} = dW + dQ$, where dW and dQ are infinitesimal amounts of energy transferred to or from the system in mechanical and thermal interactions. There's no change in the chemical energy, and the thermal energy of a gas does not change in an isothermal process, so $dE_{int} = 0$. We found in Chapter 12 that the pressure-volume work done on an ideal gas is $dW = -p\,dV$. Thus the isothermal expansion is characterized by

$$dQ = dE_{int} - dW = 0 - (-p\,dV) = \frac{Nk_BT}{V}dV \tag{14.2}$$

where, in the last step, we used the ideal-gas law for p. We can rearrange Equation 14.2 to show that the *fractional* change in volume during the isothermal heating is directly proportional to the heat dQ:

$$\frac{dV}{V} = \frac{dQ}{Nk_BT} \tag{14.3}$$

As the gas expands, its multiplicity—the number of microstates—increases from $\Omega_i = cV_i^N$ to $\Omega_f = cV_f^N = c(V_i + dV)^N$. We don't know the proportionality constant c, but we don't need it if we form a ratio in which c cancels:

$$\frac{\Omega_f}{\Omega_i} = \frac{(V_i + dV)^N}{V_i^N} = \left(1 + \frac{dV}{V}\right)^N \tag{14.4}$$

We dropped the subscript on V at the end because, with the volume changing only infinitesimally, there's essentially no difference between V_i and V_f.

The exponent N is extremely large. One way to deal with large exponents is to use logarithms. Suppose we take the natural logarithm of Equation 14.4 and use the logarithm properties $\ln(a/b) = \ln a - \ln b$ and $\ln a^n = n\ln a$ to write

$$\ln\left(\frac{\Omega_f}{\Omega_i}\right) = \ln\Omega_f - \ln\Omega_i = N\ln\left(1 + \frac{dV}{V}\right) \tag{14.5}$$

Now $dV/V \ll 1$ because it's an infinitesimal expansion, and a very useful approximation is $\ln(1 + a) \approx a$ when $a \ll 1$. You can use your calculator to check that, for example, $\ln(1.001) = 0.0009995 \approx 0.001$. (This approximation is the first term in the Taylor-series expansion of $\ln(1 + x)$, which you may have studied in calculus.) We can use this approximation and Equation 14.3 to write Equation 14.5 as

$$\ln\Omega_f - \ln\Omega_i = \Delta(\ln\Omega) = N\frac{dV}{V} = \frac{dQ}{k_BT} \tag{14.6}$$

Equation 14.6 is an important micro/macro connection because the measurable macroscopic quantities dQ and T allow us to calculate $\Delta(\ln\Omega) = \ln\Omega_f - \ln\Omega_i$, the *change* in the logarithm of the multiplicity.

Defining Entropy

We've seen that state variables, such as p and T, undergo changes that we denote as Δp and ΔT. Thus the quantity $\ln\Omega$, which changes by $\Delta(\ln\Omega)$ during the infinitesimal isothermal expansion, appears to be a state variable. Specifically, $\ln\Omega$ seems to be a state variable that measures the dispersal of energy in a system—exactly what we are looking for.

With that in mind, we define the entropy S of a macrostate that has multiplicity Ω as

$$S = k_B \ln\Omega \tag{14.7}$$

Entropy of a macrostate with multiplicity Ω

That is, **entropy is a state variable that measures the number of ways a macrostate can differ microscopically.** The Boltzmann constant is included in the definition for historical reasons. As a result, the units of entropy are the same as the units of the Boltzmann constant: J/K.

Equation 14.7 defines entropy in terms of the multiplicity of the macrostate, but in practice we rarely know the multiplicity Ω. On the other hand, we can easily measure heat transfer. For an infinitesimal heating, such as we've been considering, the quantity $\Delta(\ln\Omega) = \Delta S/k_B$ on the left-hand side of Equation 14.6 is better written in terms of an infinitesimal change in entropy as dS/k_B. Thus Equation 14.6 is

$$dS = \frac{dQ}{T} \tag{14.8}$$

Entropy change in a quasi-static process

Equation 14.8 is equivalent to Equation 14.7 but will turn out to be much more useful because it directly relates entropy to heat. Although we used the slow heating of a gas to arrive at Equation 14.8, it turns out to be true in general for any slow quasi-static process.

EXAMPLE 14.2 Calculating entropy

Earlier we suggested that the multiplicity of a real-world macrostate is on the order of $10^{10^{23}}$. What is the entropy of such a state?

PREPARE Entropy is defined in Equation 14.7.

SOLVE We can use the property $\ln a^b = b\ln a$ to write

$$S = k_B \ln\Omega = k_B \ln 10^{10^{23}} = 10^{23} k_B \ln 10 = 3.2\ \text{J/K}$$

ASSESS Despite the unimaginable size of the multiplicity, the entropy of this state is an ordinary-size number. That's because of the small size of the Boltzmann constant.

The Second Law Revisited

The second law of thermodynamics says that the entropy of an isolated system never decreases. It either increases until the system reaches thermal equilibrium or, if the system is already in equilibrium, remains constant. We had not yet defined entropy when we gave our initial statement of the second law. We've now defined the entropy S in terms of the multiplicity of microstates, and we've seen that a system's evolution toward equilibrium occurs because atomic-level interactions move the system from less-probable (lower multiplicity) to more-probable (higher multiplicity) macrostates. With this in mind, a more precise statement of the second law is

$$\Delta S_{\text{isolated system}} \geq 0 \tag{14.9}$$

The second law of thermodynamics for an isolated system

where the equality holds only for a system in thermal equilibrium.

STOP TO THINK 14.3 Two boxes are identical, and each contains 1,000,000 molecules. In box A, 600,000 molecules happen to be in the left half of the box while 400,000 are in the right half. In box B, 499,900 molecules are in the left half while 500,100 are in the right half. At this instant,

A. The entropy of box A is greater than the entropy of box B.
B. The entropy of box A is equal to the entropy of box B.
C. The entropy of box A is less than the entropy of box B.
D. There's not enough information to compare the entropies of A and B.

14.3 Using Entropy

Entropy is a powerful concept, but a subtle one. This section will look at some examples of how entropy is calculated and used.

Isothermal Processes

An isothermal process has a constant T, so we can immediately integrate Equation 14.8 to find that the entropy change in a slow isothermal process is

$$\Delta S_{\text{isothermal}} = \frac{Q}{T} \tag{14.10}$$

where Q is the heat absorbed or released by the process. Equation 14.10 is especially important for *phase changes,* which occur at a constant temperature.

EXAMPLE 14.3 **The entropy change of melting ice**

What is the entropy change of water when 100 g of 0°C ice are slowly melted to 0°C liquid water?

PREPARE This phase change is a slow isothermal process.

SOLVE The heat needed for the phase change is

$$Q = ML_{\text{f}} = (0.100\text{ kg})(3.33 \times 10^5\text{ J/kg}) = 3.33 \times 10^4\text{ J}$$

where the heat of fusion of ice is taken from Table 12.3. Thus the increase in entropy of the water is

$$\Delta S = \frac{Q}{T} = \frac{3.33 \times 10^4\text{ J}}{273\text{ K}} = 122\text{ J/K}$$

ASSESS Liquid water has a larger entropy than ice even at the same temperature because molecules in the liquid have more spatial mobility than do molecules in the solid and thus more ways to disperse the thermal energy.

EXAMPLE 14.4 **The multiplicity increase of melting ice**

By what factor does the multiplicity increase when 100 g of 0°C ice are melted to 0°C liquid water?

PREPARE Example 14.3 found that the entropy increase is $\Delta S = 122$ J/K. Equation 14.7 defines entropy in terms of the multiplicity. We can reverse this and use the increase in entropy to find the increase in multiplicity.

SOLVE The definition of entropy is $S = k_{\text{B}} \ln \Omega$. Thus the *change* in entropy is

$$\Delta S = S_{\text{f}} - S_{\text{i}} = k_{\text{B}} \ln \Omega_{\text{f}} - k_{\text{B}} \ln \Omega_{\text{i}} = k_{\text{B}} \ln\left(\frac{\Omega_{\text{f}}}{\Omega_{\text{i}}}\right)$$

We can use the identity $e^{\ln a} = a$ to solve for Ω_f/Ω_i. First we divide by k_B, then we make both sides of the equation a power of e:

$$e^{\ln(\Omega_f/\Omega_i)} = \frac{\Omega_f}{\Omega_i} = e^{\Delta S/k_B} = e^{(122\ \mathrm{J/K})/(1.38\times10^{-23}\ \mathrm{J/K})} = e^{8.8\times10^{24}}$$

This is hard to comprehend, but it might make a bit more sense as a power of 10. Because $\log(e) = 0.434$, using base-10 logarithms, we can write $e = 10^{0.434}$. You will recall that $(10^a)^b = 10^{ab}$. With this, we find that melting the ice increases the water's multiplicity by a factor of

$$\frac{\Omega_f}{\Omega_i} = e^{8.8\times10^{24}} = (10^{0.434})^{8.8\times10^{24}} = 10^{3.8\times10^{24}}$$

ASSESS It's almost impossible to understand how staggeringly large this number is—a consequence of the extremely large number of molecules in 100 g of water. $100 = 10^2$ is a 1 followed by 2 zeros. 10^{24}, roughly the number of molecules in the 100 g of water, is a 1 followed by 24 zeros. Thus $10^{3.8\times10^{24}}$ is a 1 followed by 3.8×10^{24} zeros. In print books, the zeros in a number are about 2 mm apart. Our answer, if printed, would stretch for nearly 800,000 light years, or about 8 times the diameter of our Milky Way galaxy. When we say that one macrostate is more probable than another because it has a higher multiplicity, we don't mean 10 times more probable or even a million times more probable but something like $10^{10^{24}}$ times more probable. It is truly the case that the less probable macrostate will *never* occur.

Entropy of Heating

Equation 14.8, $dS = dQ/T$, is for an infinitesimal heating at temperature T. We can find the entropy change for a finite heat transfer, for which T changes, by integrating.

You learned in Chapter 12 that the heat required to change the temperature of a solid or liquid of mass M with specific heat c is $Q = Mc\,\Delta T$. For an infinitesimal heating, $dQ = Mc\,dT$ and thus $dS = Mc\,dT/T$. We can integrate from the initial entropy S_i at initial temperature T_i to the final values S_f at T_f. If we assume that the specific heat remains constant—usually a good assumption if the temperature change is modest—then

$$\Delta S_{\text{heat}} = \int_{S_i}^{S_f} dS = Mc\int_{T_i}^{T_f}\frac{dT}{T} = Mc(\ln T_f - \ln T_i) = Mc\ln\left(\frac{T_f}{T_i}\right) \qquad (14.11)$$

Increasing entropy The volume of a gas is much larger than the volume of its liquid, so boiling increases the number of microstates—the number of ways that the molecules can be arranged to share the energy. A liquid-to-gas phase change is always accompanied by a large increase in entropy.

EXAMPLE 14.5 **Heating water**

Example 14.2 found that the entropy of 100 g of water increases by 122 J/K in changing from 0°C ice to 0°C liquid water. To what temperature does the water need to be heated to increase its entropy by an additional 122 J/K?

PREPARE We'll assume that the heating is carried out slowly.

SOLVE We can solve Equation 14.11 for T_f by first dividing through by Mc, then using $e^{\ln a} = a$:

$$T_f = T_i e^{\Delta S/Mc} = (273\ \mathrm{K})e^{(122\ \mathrm{J/K})/(0.100\ \mathrm{kg})(4190\ \mathrm{J/kg\cdot K})} = 365\ \mathrm{K} = 92°\mathrm{C}$$

ASSESS A substantial temperature increase is needed to match the entropy increase of melting. The water's entropy increases with temperature because faster molecules have more ways to disperse the kinetic energy.

Standard Molar Entropy

Calculations in chemistry often use the **standard molar entropy** of a substance, written $S°$, which is the *absolute* entropy per mole of the substance in J/mol · K at the standard state of 25°C (298 K) and 1 atm pressure. Standard molar entropies are based on the idea that the entropy of a pure crystalline substance at 0 K is zero. A crystal at 0 K has every atom in an exactly specified position with no motion. This is like a macrostate of all heads, with only one way to achieve it; thus the multiplicity is $\Omega = 1$ and the entropy is $S = k_B \ln \Omega = 0$. This statement—that the entropy of a crystal is zero at absolute zero—is often called the *third law of thermodynamics*.

The standard molar entropy is given by the integral of Equation 14.11 with an initial temperature of 0 K *plus* any entropy increases associated with phase changes:

$$S^\circ = \int_0^{298\text{ K}} \frac{C\,dT}{T} + \Delta S_{\text{phase changes}} \tag{14.12}$$

Equation 14.12 is simply adding all contributions to the entropy as the temperature is raised from 0 K to 298 K. The integral—this time using the molar specific heat C—adds up all the entropy changes caused by heating 1 mol of the substance. In addition, liquids and gases at 25°C have undergone one or more phase changes, so the entropy of each phase change has to be included.

An important difference from the above example of heating water, where we considered the specific heat to be constant, is that the molar specific heat is *not* constant over this large range of temperatures. In particular, the molar specific heat of a solid changes rapidly with temperature at very low temperatures. To determine S°, the molar specific heat C has to be measured in the laboratory from temperatures very near absolute zero up to 25°C; then the integral of Equation 14.12 is evaluated numerically. A great deal of effort over many decades has gone into determining the standard entropies that are tabulated in chemistry books. We will leave the details to chemistry and biochemistry, but understanding where those values come from does connect directly to our discussion of entropy.

The Entropy of an Ideal Gas

There's no change in chemical energy during an ideal-gas process, so we can use the first law of thermodynamics to help us calculate how the entropy of an ideal gas changes. For infinitesimal changes, the first law is $dE_{\text{th}} = dW + dQ$. Using dQ in $dS = dQ/T$ gives

$$dS = \frac{dQ}{T} = \frac{dE_{\text{th}}}{T} - \frac{dW}{T} \tag{14.13}$$

Chapter 12 gave explicit expressions for the thermal energy of and work done on an ideal gas. In differential form, the expressions are $dE_{\text{th}} = nC_{\text{V}}dT$ and $dW = -p\,dV = -(nRT/V)\,dV$. With these substitutions, Equation 14.13 becomes

$$dS = nC_{\text{V}}\frac{dT}{T} + nR\frac{dV}{V} \tag{14.14}$$

We can integrate from entropy S_{i} at initial temperature T_{i} and volume V_{i} to S_{f} at T_{f} and V_{f}. The integration is like the one we did above for the entropy associated with a temperature change, which leads to

$$\Delta S_{\text{gas}} = nC_{\text{V}}\ln\left(\frac{T_{\text{f}}}{T_{\text{i}}}\right) + nR\ln\left(\frac{V_{\text{f}}}{V_{\text{i}}}\right) \tag{14.15}$$

Entropy change of an ideal gas

NOTE ▶ An integration to find S itself (the absolute entropy) would have an unknown integration constant. We can't determine an absolute entropy, but we don't need to because changes in S are all we need to know. ◀

You learned in Chapter 12 that heating a gas at constant pressure requires more heat, for the same temperature change, than heating the gas at constant volume. The second term in Equation 14.15 vanishes for a constant-volume process because $\ln(1) = 0$. A constant-pressure process has $V_{\text{f}}/V_{\text{i}} = T_{\text{f}}/T_{\text{i}}$, so the two logarithms are the same. Using this along with $C_{\text{P}} = C_{\text{V}} + R$, which we proved in Chapter 12, leads to

$$\Delta S_{\text{gas}} = \begin{cases} nC_{\text{V}}\ln(T_{\text{f}}/T_{\text{i}}) & \text{constant-volume gas process} \\ nC_{\text{P}}\ln(T_{\text{f}}/T_{\text{i}}) & \text{constant-pressure gas process} \end{cases} \tag{14.16}$$

EXAMPLE 14.6 Heating a gas

2 mol of nitrogen gas are heated at constant pressure until the volume has doubled. What is the entropy change of the gas?

PREPARE A constant-pressure process has $T_f/T_i = V_f/V_i$. If the volume doubles, so does the absolute temperature; that is, $T_f/T_i = V_f/V_i = 2$.

SOLVE The entropy change for a doubling of the volume and temperature at constant pressure is

$$\Delta S = nC_P \ln\left(\frac{T_f}{T_i}\right) = nC_P \ln 2$$

The specific heat at constant pressure for nitrogen is given in Table 12.7 as 29.1 J/mol · K. Thus

$$\Delta S = nC_P \ln 2 = (2.0 \text{ mol})(29.1 \text{ J/mol} \cdot \text{K})\ln 2 = 40 \text{ J/K}$$

ASSESS Interestingly, the change in entropy does not depend on the initial temperature of the gas.

STOP TO THINK 14.4 Chapter 12 defined an adiabatic process as one with $Q = 0$. It follows that an adiabatic process has no change of entropy: $\Delta S = 0$. Suppose an ideal gas undergoes an adiabatic expansion to a larger volume. In this process,

A. The gas temperature decreases.
B. The gas temperature does not change.
C. The gas temperature increases.
D. There's not enough information to tell how the temperature changes.

Interactions

You've seen that we can calculate ΔS for various processes. In addition, entropy and the second law explain why interactions proceed in one direction but not the other. FIGURE 14.7 shows a system that consists of object C, at a colder temperature T_C, and object H, at the hotter temperature T_H. These objects can interact with each other, but the two objects form an *isolated system* that does not interact with the rest of the universe. Note that *this system is not in thermal equilibrium.*

FIGURE 14.7 An interaction of two objects at different temperatures.

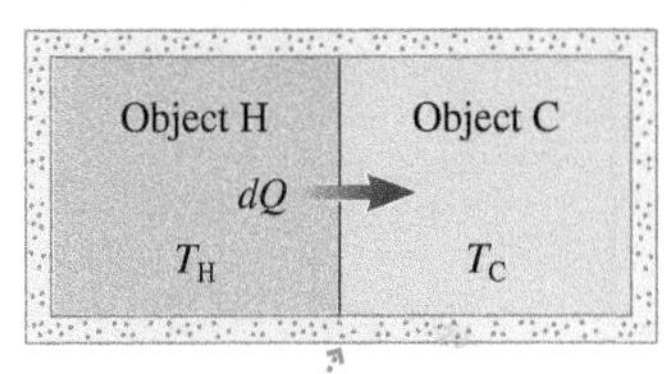

The system is the combination of C and H. What are the multiplicity and the entropy of this system? For each microstate in C there are Ω_H microstates in H, so the total number of microstates, the multiplicity of the system, is the product $\Omega_{tot} = \Omega_C \Omega_H$. For example, if C has 4 microstates while H has 6, each of the 4 microstates of C can be paired with any one of the 6 microstates of H to give $\Omega_{tot} = 6 + 6 + 6 + 6 = 4 \times 6 = 24$. Thus, using the definition of entropy, we find

$$S_{sys} = k_B \ln \Omega_{sys} = k_B \ln(\Omega_C \Omega_H) = k_B \ln \Omega_C + k_B \ln \Omega_H = S_C + S_H \quad (14.17)$$

We see that **the entropy of a combined system is the sum of the entropies of each part.** That is, entropy is additive, a useful property for a state variable.

Suppose a small amount of heat dQ is transferred from H to C. The heat leaves H and enters C, so $dQ_H = -dQ$ and $dQ_C = +dQ$. This small energy exchange causes the entropy of the total system to change by

$$dS_{sys} = dS_C + dS_H = \frac{dQ_C}{T_C} + \frac{dQ_H}{T_H} = \left(\frac{1}{T_C} - \frac{1}{T_H}\right) dQ \quad (14.18)$$

Notice that the expression in parentheses is positive because $T_C < T_H$.

If $dQ > 0$, which describes an energy transfer from hot to cold, then $dS_{sys} > 0$. The system's entropy increases, and it will continue to increase with every little energy transfer until $T_C = T_H$, at which time $dS_{sys} = 0$. This is the second law of thermodynamics! The total system is an isolated system, and initially it is *not* in equilibrium because the temperatures of C and H differ. The system's entropy increases as C and H interact—the ability to disperse energy is increasing—until $T_C = T_H$. At that point, **the entropy is maximized and the system is in thermal equilibrium.**

Notice that *the entropy of object H decreases* as heat is withdrawn from it. The second law does not say that entropy can never decrease, only that the entropy of an *isolated system* cannot decrease. H is not isolated, but its entropy decrease is more than made up for by the entropy increase of C, so the entropy of the combined system, which *is* isolated, increases.

What would happen if, instead, heat was transferred from cold to hot, making C colder and H hotter? A cold-to-hot heat transfer would have $dQ < 0$. Looking at Equation 14.18, you can see that the associated change in the system's entropy would be $dS_{sys} < 0$, taking the system to a macrostate with a lower multiplicity and a lower entropy. An energy transfer from cold to hot would not violate the first law of thermodynamics—energy would be conserved—but it would violate the second law of thermodynamics by requiring a decrease in the entropy of an isolated system. That's why a heat transfer from cold to hot never happens.

The key idea here is that a system initially *not* in thermal equilibrium evolves toward thermal equilibrium because objects within the system interact with each other by exchanging energy. The exchanges *conserve energy* but *increase entropy*. As a result, a system evolves toward thermal equilibrium by moving through a series of macrostates of ever-increasing entropy.

FIGURE 14.8 A system interacting with its environment increases the entropy of the universe.

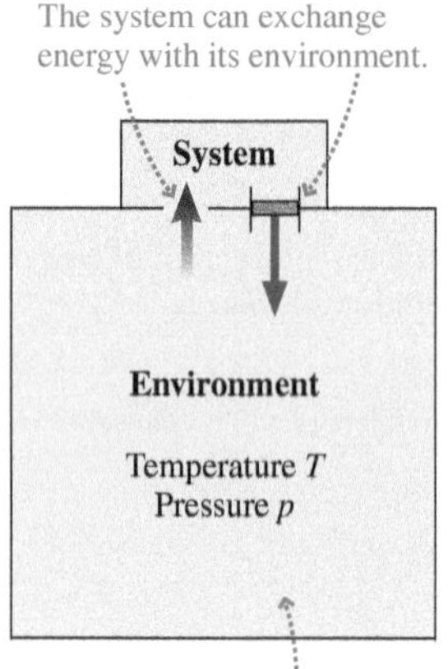
The system can exchange energy with its environment.
System
Environment
Temperature T
Pressure p
The system and its environment make up the entire universe.

We can make this idea more general. FIGURE 14.8 shows a small system interacting both thermally (heat is exchanged) and mechanically (work is done by a moving piston) with a much larger environment. This system is *not* isolated, so we can't apply the second law of thermodynamics to the system alone. In principle there's no limit to the size of the environment, so we can think of the system and its environment as forming the entire universe; that is, universe = system + environment. In most cases only the nearby environment is affected by the system, so "universe" is more figurative than literal, but there are situations where a system radiates heat into deep space and truly interacts with the entire universe.

A system may not be isolated, but the universe as a whole is the ultimate isolated system. Thus the second law of thermodynamics applied to the universe is $\Delta S_{universe} \geq 0$; that is, nothing can cause the entropy of the universe to decrease. Entropy is additive, so $S_{universe} = S_{sys} + S_{env}$ and we can write the second law for the universe as

$$\Delta S_{universe} = \Delta S_{sys} + \Delta S_{env} \geq 0 \qquad \text{with} \begin{cases} = & \text{for a reversible process} \\ > & \text{for an irreversible process} \end{cases} \tag{14.19}$$

Only idealized *reversible processes* have $\Delta S_{universe} = 0$; any realistic interaction between a system and its environment has $\Delta S_{universe} > 0$ and is *irreversible* because reversing the process, which would require $\Delta S_{universe} < 0$, would violate the second law.

For an isolated system, one that has no interactions with its environment, $\Delta S_{sys} \geq 0$ because the system is the only part of the universe that is changing. The entropy is unchanging only if the system is in equilibrium. Otherwise, the entropy will increase as the system evolves toward equilibrium. If the system is *not* isolated—and most are not—then the second law tells us that any interactions of the system with its environment must increase the entropy of the universe. It's possible for the system's entropy to decrease, but only if the environment's entropy increases by a larger amount.

NOTE ▸ Entropy, unlike energy, is *not* a conserved quantity. Just because the entropy of a system decreases doesn't mean that its entropy "went somewhere." It simply means that the system moved to a macrostate that has a lower multiplicity. Similarly, an increasing entropy doesn't mean that entropy was added from somewhere; it means only that the system changed to a macrostate that has a higher probability of occurrence. ◂

Analyzing an Irreversible Process

Our primary result for how the entropy changes, $dS = dQ/T$, is for a slow quasi-static process in which the system is always essentially in equilibrium. But most real-world processes don't meet these criteria. For example, in Section 14.1 we looked at the *free expansion* of a gas; FIGURE 14.9 reminds you of the situation. This irreversible process is fast, not quasi-static, and the gas is *not* in equilibrium during the expansion. We argued that the entropy of the gas must increase even though its thermal energy and temperature do not. Can we calculate the value of $\Delta S_{\text{free expansion}}$?

We can, but not by using Equation 14.8, $dS = dQ/T$, because it applies to only a slow quasi-static process. If we were to try, we would erroneously conclude that $\Delta S_{\text{free expansion}} = 0$ because no heat is transferred during a free expansion.

However, we know that entropy is a state variable. We also know that an irreversible process, such as a free expansion, *starts* from an equilibrium state (before the membrane is broken) and *ends* in an equilibrium state (after the gas fills the larger container). The value of a state variable depends on only the state of the system, not the process by which that state was reached. As a result, **the entropy change of an irreversible process is exactly the same as the entropy change of any slow quasi-static process that connects the initial and final states.** This is important because it allows us to find the entropy change of an irreversible process by calculating the entropy change due to a *reversible* process between the same two states.

For example, instead of breaking the membrane, we could let the gas fill the container in a quasi-static isothermal process in which a piston slowly moves outward while the addition of heat keeps the temperature from changing. This is a reversible process because we could always slowly push the piston back in, while removing heat, to restore the initial state. We can use Equation 14.15 for the entropy change of an ideal gas to write

$$\Delta S_{\text{slow expansion}} = nR\ln\left(\frac{V_f}{V_i}\right) \tag{14.20}$$

The first term in Equation 14.15 is zero because $\ln(T_f/T_i) = \ln(1) = 0$ for an isothermal expansion.

Because the initial and final states are the same, $\Delta S_{\text{free expansion}}$ for the irreversible free expansion is exactly the same as $\Delta S_{\text{slow expansion}}$: that is,

$$\Delta S_{\text{free expansion}} = \Delta S_{\text{slow expansion}} = nR\ln\left(\frac{V_f}{V_i}\right) \tag{14.21}$$

Thus a free expansion that doubles the volume of a gas increases its entropy by $\Delta S_{\text{sys}} = nR\ln 2$.

If $\Delta S_{\text{free expansion}} = \Delta S_{\text{slow expansion}}$, then what's the difference? Why is one a reversible process but not the other? The answer lies not in the gas but in the environment.

A gas undergoing a free expansion is an isolated system; nothing in the environment changes. A doubling of the volume is a process with $\Delta S_{\text{sys}} = \Delta S_{\text{free expansion}} = nR\ln 2$ and $\Delta S_{\text{env}} = 0$, so the entropy change of the universe is $\Delta S_{\text{universe}} = \Delta S_{\text{sys}} + \Delta S_{\text{env}} = nR\ln 2 > 0$. A free expansion increases the entropy of the universe. We can't reverse the free expansion because a reversal would require a decrease in the entropy of the universe, which the second law doesn't allow.

We can contrast this with doubling the volume in a slow quasi-static expansion. The system's entropy change is the same, $\Delta S_{\text{sys}} = nR\ln 2$, but now the system isn't the only thing changing. A slow isothermal expansion requires a heat input from the environment to keep the temperature constant, so the environment also has an entropy change. If Q_{env} is the heat flow relative to the environment and the exchange takes place at a constant temperature, then $\Delta S_{\text{env}} = Q_{\text{env}}/T$.

Any heat removed from the environment is added to the system, so $Q_{\text{env}} = -Q_{\text{sys}}$. An isothermal process has $\Delta E_{\text{th}} = 0$ and thus, from the first law of thermodynamics, $Q_{\text{sys}} = -W$. In Chapter 12 you learned that the work done during an isothermal process is $W = -nRT\ln(V_f/V_i)$. Thus $Q_{\text{env}} = -nRT\ln(V_f/V_i) = -nRT\ln 2$ for a doubling of the volume, and $\Delta S_{\text{env}} = Q_{\text{env}}/T = -nR\ln 2$, exactly opposite that of the system. As a result, the entropy change of the universe is $\Delta S_{\text{universe}} = \Delta S_{\text{sys}} + \Delta S_{\text{env}} = 0$.

Thus the difference between the free expansion and the slow isothermal expansion is not the entropy change of the gas, which is the same in both cases, but the

FIGURE 14.9 The free expansion of a gas.

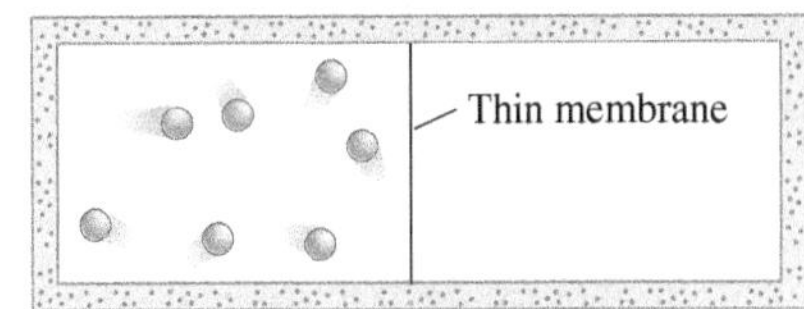

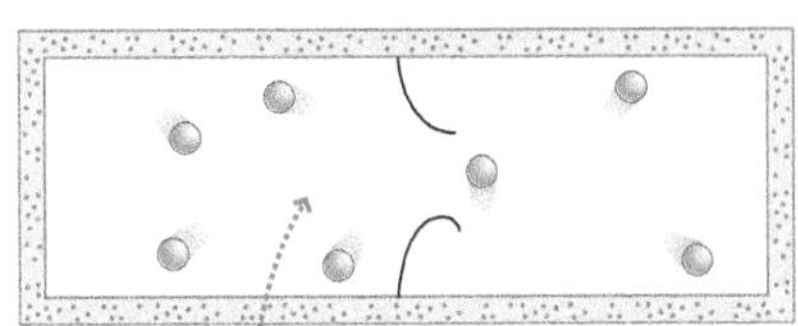

Neither the temperature nor the thermal energy of the gas changes, but its entropy increases because this is an irreversible process.

entropy change of the environment. The slow isothermal expansion is reversible because the entropy of the universe is unchanged, but we cannot reverse the free expansion without violating the second law of thermodynamics.

Sealed, but not isolated This glass container is a completely sealed system that holds living organisms, shrimp and algae. The reason this is possible is that the glass sphere and its contents, though sealed, are not an isolated system. Energy can be transferred in and out as light and heat. The organisms will quickly perish if the container is isolated from its environment by being wrapped in dark insulation.

Entropy and Life

The second law of thermodynamics is sometimes interpreted as saying that "things run down," that order becomes disorder. But living systems seem to violate this rule:

- Plants grow from simple seeds to complex, organized entities.
- Single-cell fertilized eggs grow into complex, organized adults.
- Over billions of years, life has evolved from simple unicellular organisms to the complex forms we see today.

Rather than running down, living systems seem to be ramping up and becoming more complex. How can this be?

The important qualification in the second law is that the inexorable increase in entropy applies to isolated systems, those that do not exchange energy or matter with their environment. Living organisms are neither isolated nor in equilibrium. All living organisms are maintained in a nonequilibrium state by a steady flow of energy passing through them, taking in low-entropy, high-quality energy in the form of food or sunlight and then producing thermal energy *that increases the entropy of the environment.* An organism can stay in a low-entropy state, or even decrease its entropy, because this flow of energy steadily *increases* the entropy of the environment. Life is possible without violating any laws of physics.

14.4 Spontaneity and Gibbs Free Energy

The second law of thermodynamics tells us that the entropy of the universe is always increasing. So what? This seems like a fairly arcane statement that, even if true, has little practical relevance. Our task for the remainder of this chapter is to show that, in fact, the second law is fundamental to cellular processes and to life itself. And you've been using the second law in your chemistry and biology classes, possibly without realizing it, every time you use *free energy.*

What does the second law really tell us? At its core, the second law is about microstates and macrostates. A system in thermal equilibrium with its environment is in the *most probable* macrostate, the macrostate that has by far the most microstates. Entropy is maximized and never changes: $\Delta S = 0$. As a result, *nothing ever happens!* Equilibrium is the most bland and boring state possible.

For any process—whether an expanding gas or a biochemical reaction—to *do* something useful, it has to start from a nonequilibrium state. A nonequilibrium state is *not* the most probable macrostate. However, the system may move toward equilibrium by undergoing one or more processes in which the system *does work* on the environment or *transfers heat* to the environment. This evolution is one of increasing multiplicity, increasing energy dispersal, and increasing entropy. Thus a process that does something interesting or useful increases the entropy of the universe: $\Delta S_{\text{universe}} > 0$.

Suppose a system has two macrostates that we label A and B. These might be, for example, the solid phase and the liquid phase of a substance or the reactants and products of a chemical reaction. There are two fundamental questions that, as we'll see, relate directly to the second law of thermodynamics:

- If the system is in A, can it *spontaneously*—without outside intervention—evolve to B? Will a solid A spontaneously melt to become liquid B? Can reactant molecules A undergo a reaction to form product molecules B?
- If the process $A \rightarrow B$ can happen, how much *useful energy* can be obtained from it? How far will the combustion of one gallon of gasoline move your car? How many molecules of ATP can be synthesized from the oxidation of one molecule of glucose?

It will turn out that these are really the same question, and we can reach the answer by looking at a state variable called the *Gibbs free energy*. We'll focus on the first question in this section, then the second in the next section.

Spontaneous Processes

What does it mean to say that a process is *spontaneous*? Our everyday use of the term suggests a hair-trigger reaction in which, for example, fuel suddenly bursts into flame. However, chemists and biochemists use the term "spontaneous process" in a different sense. It doesn't mean that the process happens quickly or that it is exothermic, merely that the process is energetically and entropically favorable. For example, the structure of carbon changing from diamond to graphite is classified as a spontaneous process, but the rate is so slow that, in practice, it never happens.

Looking back at Figure 14.8, we see a system interacting both thermally and mechanically with its environment. We'll make two assumptions:

- The environment is so large that its temperature T and pressure p are not affected by anything that happens inside the system.
- As the system changes from macrostate A into a different macrostate B, the system remains at *the same temperature and pressure as the environment.*

That is, the environment stabilizes the system's temperature and pressure. If something tries to lower the system's temperature, the environment will send enough heat to keep that from happening. If a process tries to lower the system's pressure, the environment will push in on the piston, doing work, to prevent that from happening.

With these assumptions, Figure 14.8 is a model of processes that take place at both *constant temperature* and *constant pressure.* These are the conditions of much of chemistry and essentially all of biochemistry. Reactions take place at constant pressure, usually 1 atm, and when the reaction is *completely* over, the temperature of the products is the same as the original temperature of the reactants.

Suppose this system is initially in a macrostate that we'll call A. Can the system with *no outside intervention* (other than being connected to its environment) evolve to a different macrostate B? Is the process A $\rightarrow$ B physically allowed, or is it like hoping that the cream and coffee will spontaneously separate?

The criteria for spontaneity are the two laws of thermodynamics. The process A $\rightarrow$ B will certainly not happen unless it conserves energy, so the first law of thermodynamics has to be satisfied. In addition, A $\rightarrow$ B can't be a process that will decrease the entropy of the universe because that would violate the second law of thermodynamics. So to say that a **spontaneous process** is one that is energetically and entropically allowed is to say that it obeys the first and second laws of thermodynamics.

As we've seen, the second law of thermodynamics for any real, irreversible process is

$$\Delta S_{\text{universe}} = \Delta S_{\text{sys}} + \Delta S_{\text{env}} > 0 \tag{14.22}$$

An energy-conserving process A $\rightarrow$ B is spontaneous if $\Delta S_{\text{universe}} > 0$. It is a process that can occur naturally, without outside intervention. A process for which $\Delta S_{\text{universe}} < 0$ cannot happen spontaneously even if energy is conserved, although it might happen with outside assistance.

The requirement that $\Delta S_{\text{universe}} > 0$ is a formal statement of spontaneity, but not a practical one because we need to know how the entropy of the universe is changing. It would be much more useful to have a criterion for judging the spontaneity of a process that depends on only knowing the properties of the system itself. Fortunately, our assumptions of constant temperature and pressure allow us to do so.

We've specified that the temperature of the environment remains constant, so we can use Equation 14.10 for isothermal processes to write the environment's entropy change as $\Delta S_{\text{env}} = Q_{\text{env}}/T_{\text{env}}$. Any heat gained (or lost) by the environment as it interacts with the system is lost (or gained) by the system, so $Q_{\text{env}} = -Q_{\text{sys}}$.

Furthermore, we're considering only processes for which the system's temperature T_{sys} matches the environment's temperature: $T_{env} = T_{sys}$. Consequently, the entropy change of the environment is $\Delta S_{env} = Q_{env}/T_{env} = -Q_{sys}/T_{sys}$. If we use this result for ΔS_{env} in Equation 14.22, the requirement for spontaneity becomes

$$\Delta S_{universe} = \Delta S_{sys} - \frac{Q_{sys}}{T_{sys}} > 0 \quad (14.23)$$

Now we can bring in the first law, energy conservation, by writing $Q_{sys} = (\Delta E_{int})_{sys} - W_{env}$, where the W in the first law is W_{env}, the work done *on* the system *by* the environment. For a constant-pressure process, which we're assuming, the environment does pressure-volume work $W_{env} = -p_{sys}\,\Delta V_{sys}$ as the volume of the system changes. (We calculated this work in Chapter 12.) For now, we'll assume that pressure-volume work is the only work done in the process, in which case $Q_{sys} = (\Delta E_{int})_{sys} + p_{sys}\,\Delta V_{sys}$.

Substituting this expression for Q_{sys} into Equation 14.23 gives

$$\Delta S_{universe} = \Delta S_{sys} - \frac{(\Delta E_{int})_{sys}}{T_{sys}} - \frac{p_{sys}\,\Delta V_{sys}}{T_{sys}} > 0 \quad (14.24)$$

We've been using subscripts to clearly indicate which parameters refer to the system and which to the environment. However, we've reached a point where *everything* is a system parameter, so we can now drop the subscripts. If we multiply through by $-T$, which changes the $>$ to $<$, we find

$$\Delta E_{int} + p\Delta V - T\Delta S = -T\Delta S_{universe} < 0 \quad (14.25)$$

We're considering only constant-pressure, constant-temperature processes, so we can bring the p and T inside the deltas:

$$\Delta E_{int} + p\Delta V - T\Delta S = \Delta E_{int} + \Delta(pV) - \Delta(TS) = \Delta(E_{int} + pV - TS) \quad (14.26)$$

Thus the Equation 14.25 criterion that a process can occur spontaneously is

$$\Delta(E_{int} + pV - TS) < 0 \quad (14.27)$$

With Equation 14.27 we've reached our goal of establishing a criterion for spontaneity that depends on only properties of the system. Still, it's not exactly clear what Equation 14.27 is telling us, so we need to look at some examples.

Gibbs Free Energy

Salt water Salt dissolves in water spontaneously because the entropy of widely dispersed, dissociated sodium and chlorine ions is much higher than the entropy of the same ions sitting in an orderly crystal. The entropy of the universe increases when salt dissolves. A different way of stating this is to say that the Gibbs free energy of the salt-water system decreases.

The quantity $E_{int} + pV - TS$, which depends on only measurable state variables, is called the **Gibbs free energy** G; that is,

$$G = E_{int} + pV - TS = H - TS \quad (14.28)$$

Gibbs free energy

where you'll recall that $H = E_{int} + pV$ is the system's enthalpy. Gibbs free energy is another state variable, such as S or T or E_{int}, that describes the state of the system. Its units are those of energy, J or kJ. You'll see in the next section why it's called "free" energy.

NOTE ▸ When biology textbooks refer to "free energy," they mean *Gibbs free energy*. There is another form of free energy, *Helmholtz free energy,* that is useful in engineering but rarely in biology. We'll usually use the full name to avoid any possible ambiguity. ◂

In terms of the Gibbs free energy, the Equation 14.27 criterion for spontaneity is

$$\Delta G < 0 \quad (14.29)$$

Criterion for spontaneity of a constant-pressure, constant-temperature process

The idea that spontaneous processes are characterized by $\Delta G < 0$, a decrease in the Gibbs free energy, is one you may have encountered in chemistry or biology, but you may not have appreciated *why* it's true. Looking back at Equation 14.25, we see that

$$\Delta S_{\text{universe}} = -\frac{\Delta G}{T} \tag{14.30}$$

Apart from a scale factor of $1/T$, the change in the Gibbs free energy is the *negative* of the change in the entropy of the universe. Thus testing whether the free energy decreases is equivalent to testing whether the entropy of the universe increases—which the second law requires of any allowed process. In other words, **the $\Delta G < 0$ requirement for a spontaneous process is simply an alternative statement of the second law of thermodynamics that any allowed process must satisfy, $\Delta S_{\text{universe}} > 0$.**

A process for which $\Delta G < 0$ can occur spontaneously because it increases the entropy of the universe. A process that would require $\Delta G > 0$ cannot occur spontaneously because it would violate the second law of thermodynamics. A "process" for which $\Delta G = 0$ is not really a process, simply a statement that the system and the environment are in thermal equilibrium and so nothing happens.

NOTE ► The spontaneity requirement $\Delta G < 0$ was derived under the assumption of constant temperature and pressure. ΔG can be calculated for *any* process because G is a state variable, but using it as a test of spontaneity is valid only for processes with constant temperature and pressure. ◄

We've been doing quite a few manipulations; now we need to work on understanding the significance of Equation 14.29. A lot of information is packed into this seemingly simply statement.

EXAMPLE 14.7 Spontaneity of a phase change

In Example 12.18 we calculated that the enthalpy change of boiling 2.00 mol of water at 100°C is $\Delta H = 81.4$ kJ. Table 12.3 gives the heat of vaporization of water as 2260 kJ/kg. Is the change of water from liquid to gas spontaneous at 100°C? What if the temperature is 100.1°C or 99.9°C?

PREPARE A process is spontaneous if $\Delta G < 0$, where $\Delta G = \Delta H - T\Delta S$ is the change in the Gibbs free energy. We'll consider 2.00 mol of water so that we can use the previously calculated ΔH.

SOLVE You learned in the preceding section that the entropy change of a phase change is $\Delta S = Q/T$. To boil 2.00 mol of water, with a molar mass of 0.018 kg/mol, requires heat

$$Q = ML_v = (0.036\text{ kg})(2260\text{ kJ/kg}) = 81.4\text{ kJ}$$

Thus the entropy increase at 373 K is $\Delta S = Q/T = 0.2182$ kJ/K and the change in the Gibbs free energy is

$$\Delta G = \Delta H - T\Delta S = 81.4\text{ kJ} - (373\text{ K})(0.2182\text{ kJ/K}) = 0$$

Finding $\Delta G = 0$ may seem surprising, but a little thought shows why it is true. First, because phase changes occur at constant pressure, as you learned in Chapter 12, $\Delta H = Q_P$. Thus the entropy increase is $\Delta S = Q_P/T = \Delta H/T$ and so $\Delta G = \Delta H - \Delta H = 0$. Second, $\Delta G = 0$ indicates a system in thermal equilibrium. But that's exactly right! Another lesson of Chapter 12 was that the liquid and vapor are in equilibrium with each other *at the transition temperature,* so any amount of liquid can coexist with any amount of vapor. There's no tendency of the system to move one way or the other, so *nothing happens.* That's what $\Delta G = 0$ implies.

A very small change in temperature has almost no effect on ΔH or ΔS. Consequently, any slight increase in T makes $\Delta G = \Delta H - T\Delta S < 0$. Now the liquid-to-gas transition is a spontaneous process and all the liquid boils. A slight decrease in T makes $\Delta G = \Delta H - T\Delta S > 0$. Now boiling is prohibited—it would violate the second law—but the reverse process of a gas-to-liquid transition has a negative ΔG and is spontaneous. At any temperature below 100°C, no matter how small the difference, the vapor spontaneously condenses to a liquid.

ASSESS Under carefully controlled conditions, liquid water can be *superheated* to above 100° and steam can be *supercooled* below 100°C. But these are very unstable states, like balancing a pencil on its tip, that rapidly and spontaneously boil or condense with even the slightest disturbance. Only at exactly 100°C (for $p = 1.00$ atm) can liquid and vapor be in equilibrium.

Balancing Enthalpy and Entropy

A process in which the free energy decreases, $\Delta G < 0$, is called an *exergonic process.* It is a spontaneous process. An *endergonic process* with $\Delta G > 0$ does not happen spontaneously but, as we'll explore later in the chapter, it can sometimes be driven with an external energy input.

NOTE ▶ Be careful not to confuse exergonic ($\Delta G < 0$) and endergonic ($\Delta G > 0$) processes with exothermic ($\Delta H < 0$) and endothermic ($\Delta H > 0$) processes that release or absorb heat. ◀

The minus sign in the definition of the Gibbs free energy tells us that the free-energy change $\Delta G = \Delta H - T\Delta S$ of a process is a balance between enthalpy and entropy. We can distinguish four cases:

1. An exothermic process ($\Delta H < 0$) that increases the system's entropy ($\Delta S > 0$) always has $\Delta G < 0$. It is exergonic and spontaneous. This describes a great many chemical reactions.
2. An endothermic process ($\Delta H > 0$) that decreases the system's entropy ($\Delta S < 0$) always has $\Delta G > 0$. It is endergonic and never happens spontaneously.
3. An exothermic process ($\Delta H < 0$) that *decreases* the system's entropy ($\Delta S < 0$) could be either exergonic (spontaneous) or endergonic. For example, a strongly exothermic reaction with a large negative value of ΔH can have $\Delta G < 0$ in spite of an entropy decrease; we say that such a process is *enthalpy dominated,* where the enthalpy change pushes the process to be exergonic. The combustion of hydrogen and oxygen, which we'll analyze in the Integrated Example at the end of the chapter, is in this category. In contrast, a weakly exothermic reaction may be endergonic and not happen spontaneously if it is *entropy dominated,* where a decreasing entropy makes $\Delta G > 0$.
4. An endothermic process ($\Delta H > 0$) that *increases* the system's entropy ($\Delta S > 0$) could be either exergonic (spontaneous) or endergonic. We'll look at the example of a cold pack that seems to magically reduce its temperature well below room temperature.

EXAMPLE 14.8 Chilling out with ammonium nitrate

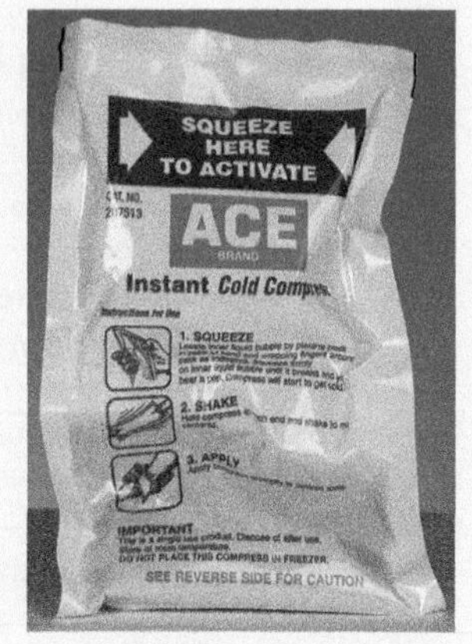

An instant *cold pack,* used for first aid and sports injuries, is a plastic bag with 60 g of the salt ammonium nitrate (NH_4NO_3) and an inner bag that contains 200 g of water. Squeezing the pack to break the inner bag causes the ammonium nitrate to dissolve in the water, quickly lowering the water temperature to near freezing. The dissociation of the ionic compound is $NH_4NO_3(s) \xrightarrow{H_2O} NH_4^+(aq) + NO_3^-(aq)$. Data on the molar masses, enthalpies of formation, and standard molar entropies of the salt and the ions are listed in the table:

	M_{mol} (kg/mol)	H_f° (kJ/mol)	S° (J/mol · K)
$NH_4NO_3(s)$	0.080	–365.6	151.1
$NH_4^+(aq)$	0.018	–133.3	113.4
$NO_3^-(aq)$	0.062	–206.9	146.4

What are (a) the lowest temperature of the water and (b) the change in entropy as the salt dissolves? (c) Explain how this can be a spontaneous process. The specific heat of liquid water was given in Chapter 12 as 4190 J/kg · K. Assume the process starts at the standard-state conditions of 25°C and 1 atm pressure.

PREPARE The table gives molar quantities as you would find them in a chemical reference. Note that tables usually give enthalpies of formation in kJ/mol but standard molar entropies in J/mol · K. The dissolution of a salt is not a chemical reaction (covalent bonds are not being broken and reformed) but a dissociation of the molecules. Although the temperature of the system (the cold pack) decreases temporarily, interactions with the environment will eventually return the temperature to its initial value. Thus the *complete* process is one of constant temperature and pressure, the conditions for which spontaneity can be judged by looking at the change in the Gibbs free energy.

SOLVE

a. The heat needed to change the temperature of mass M of water by ΔT is $Q_{water} = Mc\Delta T$, where c is the specific heat. This heat is provided by the dissolution of the ammonium nitrate. The heat released is $Q_{salt} = \Delta H$, the change in the salt's enthalpy. If heat is exchanged between the salt and the water with no losses, which we'll assume, then $Q_{water} = -Q_{salt} = -\Delta H$.

To calculate ΔH, we could imagine completely dissociating the reactants into their constituent atoms and then constructing the products. The result of doing so, as Example 12.19 showed, is that the change in enthalpy is simply the enthalpy of the products minus the enthalpy of the reactants. We see, from the table above, that 60 g of ammonium nitrate is 0.75 mol. It dissociates to 0.75 mol of each of the two ions. Thus the enthalpy change is

$$\Delta H = (0.75\text{ mol})(-133.3\text{ kJ/mol} - 206.9\text{ kJ/mol} - (-365.6\text{ kJ/mol}))$$
$$= 19.1\text{ kJ}$$

ΔH is positive, so this is an endothermic process in which the dissociating salt *absorbs* heat. Eventually, when the system returns to equilibrium, that heat will have come from the environment, but initially it comes from the water. The water has $Q_{\text{water}} = -\Delta H = -19.1$ kJ, so its temperature change is

$$\Delta T = \frac{Q_{\text{water}}}{Mc} = \frac{-19{,}100\text{ J}}{(0.20\text{ kg})(4190\text{ J/kg}\cdot\text{K})} = -23°\text{C}$$

The final temperature, ideally, is 2°C.

b. Endothermic reactions are usually not spontaneous. To understand why this one is, we need to consider the entropy. The entropy change, like the enthalpy change, is calculated as the total entropy of the products minus the total entropy of the reactants. In this case,

$$\Delta S = (0.75\text{ mol})(113.4\text{ J/mol}\cdot\text{K} + 146.4\text{ J/mol}\cdot\text{K} - 151.1\text{ J/mol}\cdot\text{K})$$
$$= 81.5\text{ J/K} = 0.0815\text{ kJ/K}$$

Entropy *increases* when a substance dissolves because the molecules in a low-entropy crystal become randomly spread throughout the water.

c. An enthalpy increase works against a process being spontaneous, while an entropy increase promotes spontaneity. Neither alone can predict what will happen. It's the Gibbs free energy that balances the two and predicts whether a process will be spontaneous. For this process,

$$\Delta G = \Delta H - T\Delta S = 19.1\text{ kJ} - (298\text{ K})(0.0815\text{ kJ/K})$$
$$= -5.2\text{ kJ}$$

$\Delta G < 0$ for this process, so it is a spontaneous process that increases the entropy of the universe. Notice, though, that the size of the entropic contribution pushing ΔG negative decreases with decreasing temperature. At a sufficiently low temperature, ΔG becomes positive and the process is no longer spontaneous. The limitation is not meaningful here because the water would be frozen and the salt would not dissolve, but there are many reactions in which spontaneity can be turned off or on by changing the temperature.

ASSESS A cold pack is an example of an *entropically driven process,* where an otherwise unfavorable change in enthalpy is more than offset by a large increase in entropy. Many biochemical processes, especially protein folding and unfolding, are controlled by the interplay between enthalpy and entropy. Our calculated low temperature of 2°C is quite close to the measured temperatures of cold packs, which are in the 3°–5°C range. In actual use, some of the heat absorbed by the dissociated molecules would come from the environment rather than the water; our model of the process as a perfect heat transfer with $Q_{\text{water}} = -Q_{\text{salt}}$ doesn't account for this.

14.5 Doing Useful Work

We've noted earlier, and you probably know from biology, that adenosine triphosphate (ATP) is the "energy currency" of the cell. What does this really mean? The hydrolysis of ATP is the reaction ATP + $H_2O \rightarrow$ ADP + P_i, where ADP is adenosine diphosphate and P_i represents a phosphate ion, called *inorganic phosphate.* This is an exergonic reaction with $\Delta G \approx -57$ kJ/mol in the cell. (The exact value depends on cellular conditions.)

What happens to the energy that is released in this reaction? Some of the released energy becomes heat that increases the cell's thermal energy and temperature. That's what happens when ATP is hydrolyzed in a test tube and, to some extent, in your body to maintain your core temperature. Some released energy might do pressure-volume work of causing the system to expand in volume. In the cell, however, much of the released energy is used for more practical and useful forms of work. For example, energy from ATP hydrolysis is used to:

- Transport macromolecules through the cell along microtubules.
- Push ions into and out of the cell through ion pumps in the cell membrane.
- Contract muscle cells.
- Drive unfavorable endergonic reactions that otherwise would not occur.

These cellular processes and many others require *work* to be done. A force is needed to move ions or molecules, and that force does work. Molecular forces do work as they change the shapes of enzymes so that reactions can occur. Collectively, the energy used by these processes is the **useful work** done by ATP hydrolysis.

We began the last section with two questions: How do we judge whether a process $A \rightarrow B$ is spontaneous, and, if so, how much useful energy can be obtained in the form of work? We've answered the first by finding that a process is spontaneous if the Gibbs free energy decreases: $\Delta G < 0$. We'll now turn our attention to the second. How much of the energy released in a reaction can be used to do useful work? Is there an upper limit to the amount of work that can be done? Interestingly, we'll see that ΔG also tells us how much useful energy can be obtained from an exergonic process.

Available Energy

A gallon of gasoline is a large store of chemical energy. You can tap into this store to drive your car by burning the gasoline in an engine. It takes work to propel a car, and the engine is designed to extract that work from the energy released in the combustion reaction. The useful work that an energy supply can do, if harnessed, is called its **available energy.** A gallon of gasoline has a large available energy.

FIGURE 14.10 Available energy can do useful work.

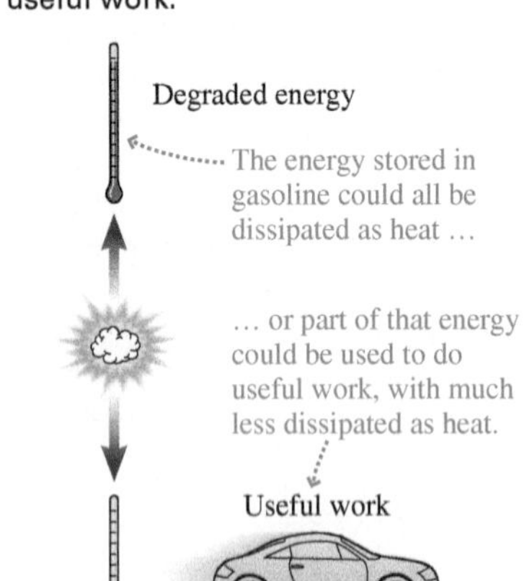

Degraded energy
The energy stored in gasoline could all be dissipated as heat ...
... or part of that energy could be used to do useful work, with much less dissipated as heat.
Useful work

Alternatively, you can simply burn the gasoline in air, ending up with some hot gases. The total amount of energy hasn't changed—energy has to be conserved—but it has been transformed from concentrated chemical energy into dispersed and less useful thermal energy. You can't power your car with the reaction products of carbon dioxide and water, so we would say that the reaction products have little or no available energy. This less useful thermal energy is often called *degraded energy.* FIGURE 14.10 helps you visualize these two alternatives.

What has changed when available energy becomes an equal quantity of degraded energy? Entropy! Useful energy is the energy of a low-entropy macrostate, a state in which energy is concentrated. If the proper conditions are established—such as a spark—a *spontaneous process* occurs in which useful work can be done. In contrast, the reaction products are a high-entropy macrostate in which the energy has been dispersed. The energy is still present, but it is no longer available to do useful work.

NOTE ▸ Discussions of energy in a societal context, whether that be fossil-fuel energy or solar energy, are really about low-entropy useful energy rather than the physics definition of energy. After all, why should we make efforts to "conserve energy" if energy is always conserved? The total amount of energy in all its forms may be conserved, but the low-entropy energy that we can put to work isn't. In this context, "conserving energy" means conserving useful, available energy. ◂

STOP TO THINK 14.5 Gases 1 and 2 have an equal number of moles of gas at the same temperature. Both can do work by pushing the piston out. Which is true?

A. Gases 1 and 2 have the same thermal energy and the same available energy.
B. Gases 1 and 2 have the same thermal energy, but gas 1 has more available energy.
C. Gases 1 and 2 have the same thermal energy, but gas 1 has less available energy.
D. Gases 1 and 2 have the same available energy, but gas 1 has more thermal energy.
E. Gases 1 and 2 have the same available energy, but gas 1 has less thermal energy.

Understanding Gibbs Free Energy

FIGURE 14.11 shows a system with internal energy E_{int}. How much of this energy can be removed or extracted from the system to do useful work? That is, what is the *available* energy?

Figure 14.11 is similar to the earlier Figure 12.20, where we associated enthalpy $H = E_{int} + pV$ with the energy needed to *create* a system with internal energy E_{int} *and* to move the atmosphere aside to make space for it. Suppose we reverse the process. We can extract energy E_{int} from the system by disassembling all the molecules. We can also extract energy pV as the pressure-volume work done when the atmosphere fills the vacated space. That will yield energy $E_{int} + pV$, the enthalpy, but not all of this energy is available to do work.

The reason is that the system has entropy. To not violate the second law of thermodynamics with a reduction of entropy, we need to move the system's entropy S to the environment. We can do this with a heat flow in the amount $Q = T\Delta S = TS$, where $\Delta S = S$ because we're transferring *all* of the system's entropy. The heat transfer TS is energy that is *not* available for doing work, so the maximum energy that we can obtain in the form of work is

$$\text{maximum work} = E_{int} + pV - TS \qquad (14.31)$$

The more entropy a system has—that is, the more its energy is dispersed and uniformly distributed—the less of its energy is available for doing work. That's exactly what we said in comparing a gallon of low-entropy gasoline to the high-entropy reaction products.

Interestingly, the quantity $E_{int} + pV - TS$ is none other than the Gibbs free energy G, which we defined in Equation 14.28. In other words, **the maximum energy that can be extracted from a system as useful work is the system's Gibbs free energy.** Now we have a physical interpretation of the Gibbs free energy.

Needless to say, destroying a system in order to extract its energy is hardly practical, so the magnitude of G is not especially useful. What we'll work with, instead, is ΔG, the amount by which the Gibbs free energy *changes* during a process, such as a chemical reaction. You'll see that ΔG is the maximum amount of energy that can be harnessed from the process for doing useful work. Thus the "free" in *free energy* refers to the amount of energy that a process makes available "for free" to do useful work.

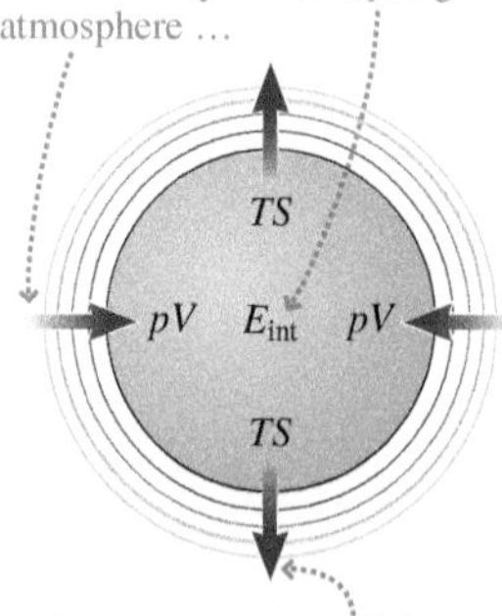
If the system is destroyed, you can get out energy E_{int} plus the work done by the collapsing atmosphere ...
TS
pV
E_{int}
pV
TS
... but you have to send the system's entropy to the environment with a heat flow TS.

FIGURE 14.11 To extract the maximum energy from a system, you must account for the system's entropy.

Maximum Useful Work

In our analysis of spontaneity in Section 14.4, we wrote the second law of thermodynamics as $\Delta S_{universe} = \Delta S - Q/T > 0$ and then used the first law to find $Q = \Delta E_{int} + p\Delta V$. Substituting Q into $\Delta S_{universe}$ led to the $\Delta G < 0$ criterion for spontaneity.

That analysis made the explicit assumption that pressure-volume work $W_{env} = -p\Delta V$ due to the system's volume change is the only work being done. Now let's suppose that, in addition, other forms of work—various kinds of cellular work—are being done. In that case, $W_{env} = -p\Delta V + W_{other}$, where W_{other} is, collectively, all nonexpansion forms of work. With additional work being done, the first law becomes $Q = \Delta E_{int} + p\Delta V - W_{other}$.

The work W that appears in the first law, including W_{other}, is work done by the environment. However, the "useful work" that we've been discussing is work done by the system; we'll call it W_{useful}. We noted in ◀◀ SECTION 12.7 that W_{sys}, the work done by the system, and W_{env}, the work done by the environment, are both valid perspectives that simply have opposite signs: $W_{sys} = -W_{env}$. If we shift to the system's perspective, $W_{useful} = -W_{other}$ and the heat transferred to the system becomes $Q = \Delta E_{int} + p\Delta V + W_{useful}$.

NOTE ▶ Pressure-volume work is useful work in machines; it's the work that propels your car when expanding hot gases push on the pistons. But pressure-volume work is not important in cellular processes. That's why we've defined "useful work" as any work *other than* pressure-volume work. ◀

We'll skip the details, which are almost identical to the analysis in the preceding section, but substituting this new expression for Q into $\Delta S_{\text{universe}} = \Delta S - Q/T \geq 0$ leads to

$$W_{\text{useful}} \leq -\Delta G \tag{14.32}$$

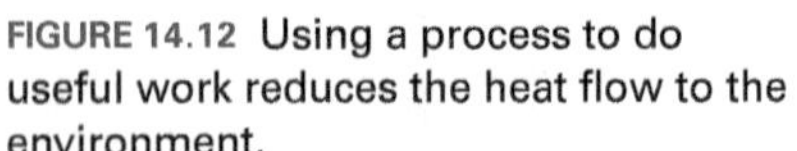

FIGURE 14.12 Using a process to do useful work reduces the heat flow to the environment.

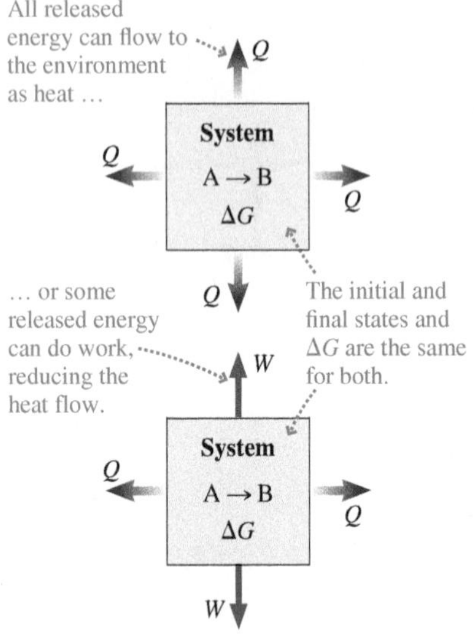

All released energy can flow to the environment as heat ...
Q
System
A → B
ΔG
Q
Q
Q
... or some released energy can do work, reducing the heat flow.
The initial and final states and ΔG are the same for both.
W
Q
System
A → B
ΔG
Q
W

We know that ΔG is negative for an exergonic process, so $-\Delta G$ is positive. Equation 14.32 tells us that an exergonic process can do useful work, but not an unlimited amount; the upper limit is $-\Delta G = |\Delta G|$. That is, **the maximum useful work that can be done in an exergonic process is $W_{\text{max}} = |\Delta G|$.**

We can summarize what we've learned:

- A process A → B is spontaneous if $\Delta G < 0$ for the process.
- *If* process A → B is spontaneous, then the maximum useful work that can be done by the process is $W_{\text{max}} = |\Delta G|$.

To understand this result, it's important to recognize that **the Gibbs free energy G is *not* a conserved quantity.** G has units of energy, but only the actual energy E is conserved. Unlike true energy, the Gibbs free energy cannot be transferred or transformed; it's simply a number that tells us how much useful work can be done.

What *is* conserved is energy, so, as FIGURE 14.12 illustrates, any useful work comes not by reducing the Gibbs free energy but by reducing the heat flow to the environment: More work means less heat. And less heat flowing to the environment means less entropy increase of the environment. The maximum work that can be done, $W_{\text{max}} = |\Delta G|$, occurs when enough heat has been diverted to work that the entropy change of the universe has been reduced to $\Delta S_{\text{universe}} = 0$. That is, the maximum useful work that a spontaneous process can do is limited by how much heat can be diverted to work without violating the second law of thermodynamics.

EXAMPLE 14.9 Work and free energy in a reaction

An exothermic reaction at constant temperature and pressure has an enthalpy decrease of 400 kJ and an entropy decrease of 1.25 kJ/K.

a. For what temperatures is this reaction spontaneous? Assume that ΔH and ΔS are independent of temperature.
b. What is the maximum useful work that can be done at a temperature of 20°C?
c. How much heat flows to the environment with and without the maximum useful work being done?

PREPARE The Gibbs free energy change is $\Delta G = \Delta H - T\Delta S$. This is a reaction with $\Delta S < 0$, a decrease in entropy, so ΔG will be negative for some temperatures (a spontaneous reaction) but not others.

SOLVE

a. We're given $\Delta H = -400$ kJ and $\Delta S = -1.25$ kJ/K. The reaction is spontaneous if

$$\Delta G = \Delta H - T\Delta S = -400 \text{ kJ} + (1.25 \text{ kJ/K})T < 0$$

This will be true if

$$T < \frac{400 \text{ kJ}}{1.25 \text{ kJ/K}} = 320 \text{ K} = 47°\text{C}$$

The reaction proceeds spontaneously at temperatures lower than 47°C.

b. The maximum useful work that can be done is $W_{\text{max}} = |\Delta G|$. At 20°C = 293 K, the free-energy change is

$$\Delta G = \Delta H - T\Delta S$$
$$= -400 \text{ kJ} - (293 \text{ K})(-1.25 \text{ kJ/K}) = -34 \text{ kJ}$$

Thus the maximum useful work that can be done—the available energy—is

$$W_{\text{max}} = |\Delta G| = 34 \text{ kJ}$$

c. The heat transfer to the system in this reaction is

$$Q_{\text{sys}} = \Delta E_{\text{int}} + p\,\Delta V + W_{\text{useful}} = \Delta H + W_{\text{useful}}$$

If only pressure-volume work is done, not useful or cellular work, then

$$Q_{\text{sys}} = \Delta H = -400 \text{ kJ}$$

Q_{sys} is negative, which indicates that 400 kJ of heat flow from the system to the environment, as expected for an exothermic reaction. If the maximum useful work is done, then

$$Q_{\text{sys}} = \Delta H + W_{\text{useful}} = -400 \text{ kJ} + 34 \text{ kJ} = -366 \text{ kJ}$$

With useful work being done, the heat flow to the environment is reduced to 366 kJ.

ASSESS This exothermic reaction releases 400 kJ of heat. You might think that, if you were clever enough, you could figure out a way to use all of that heat to do useful work. But it turns out that the laws of thermodynamics allow only a small amount of that heat—at most 34 kJ—to be diverted to work. Any work that is done comes not by reducing the free energy but by reducing the heat flow to the environment.

STOP TO THINK 14.6 A chemical reaction is harnessed to perform 100 J of useful work. Which of the following is true?

A. $\Delta G \geq 100$ J
B. ΔG is between 0 J and 100 J
C. ΔG is between -100 J and 0 J
D. $\Delta G \leq -100$ J

14.6 Using Gibbs Free Energy

Let's look at two more uses of the Gibbs free energy that are relevant to biochemistry: equilibrium and coupled reactions.

Equilibrium

A process A $\rightarrow$ B that lowers the Gibbs free energy will happen spontaneously. Then what? If there's another process B $\rightarrow$ C that further lowers the Gibbs free energy, that process will also occur. Processes will continue to occur spontaneously until there is no accessible macrostate that has a lower value of G. With no further change of macrostate, the system will be in equilibrium. In other words, **a system goes down the free-energy hill until it reaches the minimum free energy, at which point it is in equilibrium.** The energy available for useful work is the magnitude of the free-energy fall.

As an example of these ideas, consider the isomerization of butane. Butane (C_4H_{10}) is a gas at room temperature but a liquid under modest pressure, where it's used as the fuel in cigarette lighters and torches. Butane has two *isomers,* different stable structures of the molecule: normal butane (n-butane), in which the four carbons form a chain, and isobutane (i-butane), in which three carbons attach to a central carbon. Collisions cause the two isomers to change back and forth in the reversible reaction n-butane $\rightleftharpoons$ i-butane, a process called *isomerization.* Under standard conditions of 25°C and 1 atm, a sample of pure n-butane spontaneously undergoes isomerization until it reaches an equilibrium that is 23% n-butane and 77% i-butane. We can say that the process n-butane $\rightarrow$ i-butane goes 77% to completion.

FIGURE 14.13 shows graphically what's going on. The horizontal axis is the "extent of reaction," going from 0 on the left (all n-butane) to 1 on the right (all i-butane). The midpoint, 0.50, is a mixture of 50% each. The vertical axis is the Gibbs free energy. A sample of pure n-butane starts at the left edge. A move to the right edge, a complete change of n-butane to i-butane, would have $\Delta G = -2.3$ kJ/mol. That would be a spontaneous process, so why don't we end up with *all* i-butane?

FIGURE 14.13 The Gibbs free energy for the isomerization of butane.

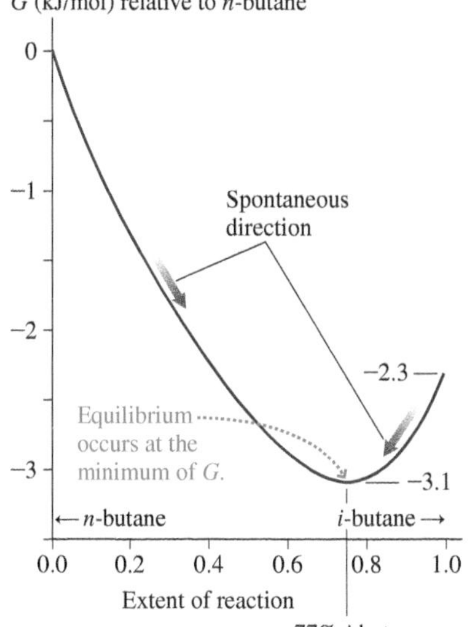

The reason is that a sample of pure i-butane would not be in equilibrium. You can see that the Gibbs free energy is not a linear decline from left to right but, instead, has a minimum. A sample of pure i-butane could lower its free energy—in a spontaneous process—by moving the end point of the reaction to the left. Equilibrium is the state in which there are no processes that will further reduce G. **Equilibrium is the macrostate with the minimum Gibbs free energy.** Thus the system, no matter what mixture of isomers it starts with, goes downhill in free energy until it reaches the equilibrium mixture of 77% i-butane and 23% n-butane.

To understand the significance of minimizing the Gibbs free energy, recall that $\Delta S_{\text{universe}} = -\Delta G/T$; that is, the change in G tells us directly how a process affects the entropy of the universe. Saying that a reaction continues until it reaches the minimum possible value of G is equivalent to saying that the reaction continues until the entropy of the universe is maximized. But that's the second law of thermodynamics! It's more practical to analyze processes by using free energy rather than entropy because G depends on only the system and not the environment, but keep in mind that the various conditions about spontaneity or equilibrium are really just restatements of the second law of thermodynamics.

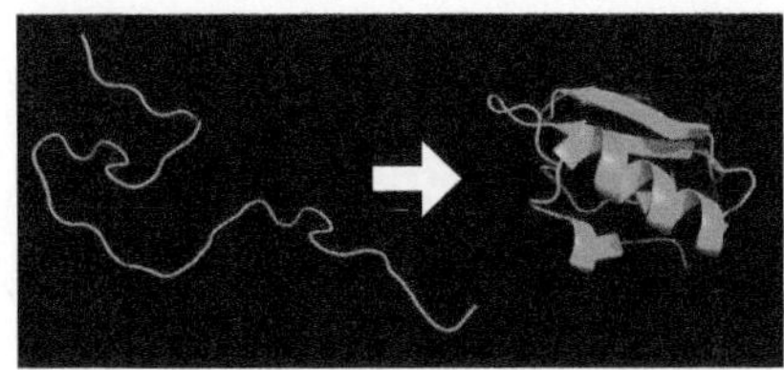

Protein folding Proteins are synthesized as long linear chains of amino acids, but the functioning of a protein depends on it having the correct three-dimensional structure. There are a vast number of ways that a protein can fold, and each fold—driven by thermal energy—either slightly increases or slightly decreases the molecule's free energy. The process of reaching the optimum folded configuration is sometimes described in terms of a *free-energy landscape* in which each individual fold is seen as a step across a landscape that goes either uphill or downhill in free energy. Any uphill steps are quickly reversed by thermal fluctuations, but a downhill step clears the way for subsequent downhill steps as the protein structure evolves toward the deepest valley in the free-energy landscape. Through trial and error—accepting downhill steps but rejecting uphill steps—the amino acid chain quickly folds itself into a shape that minimizes its free energy.

But why is there a minimum in the free-energy curve? Why doesn't G steadily decrease from the left side of Figure 14.13 to the right side? The answer, once again, is entropy. In the next section you'll discover that the entropy of two gases is higher when they're mixed together than if they are kept separated. That is, there's an additional *entropy of mixing* when the two isomers are both present that is absent when either isomer is alone. Because entropy enters the free energy as $-TS$, with a minus sign, this additional entropy of mixing pushes the free-energy curve down in the middle but not at the ends. This downward push creates a minimum in the curve at some point along the horizontal axis.

No reaction goes 100% to completion, although for many reactions, such as combustion reactions with large ΔG, the minimum in the free-energy curve is so far to the right that it's reasonable to treat the reaction as complete. In contrast, many biochemical reactions, like the butane isomerization, have an equilibrium point at which both reactants and products are present.

Coupled Reactions

A primary role of ATP in the cell is to use the energy of ATP hydrolysis to do useful work in the form of driving unfavorable endergonic reactions ($\Delta G > 0$) that otherwise would not occur. When energy from an exergonic reaction drives an endergonic reaction, we say that they are **coupled reactions.** Coupled reactions are essential to the biochemistry of the cell.

For example, proteins are synthesized when amino acids are linked together. The bond between two amino acids is a *peptide bond,* and the reaction that forms a peptide bond is endergonic with, typically, $\Delta G \approx +10\ \text{kJ/mol}$. Amino acids do not, on their own, link up to form proteins. However, "spending" some of the ≈ 57 kJ/mol of available energy from the hydrolysis of an ATP molecule can move the reaction *uphill* on the free-energy curve. The mechanism of how this is done with enzymes is a topic for biochemistry, but the energetics are something we can understand from our analysis of how free energy is used.

An example we can study graphically is the oxidation of glucose, the primary fuel of cells. The direct burning of glucose would provide a large and sudden release of energy that is of little value to a cell. Instead, the cellular oxidation of glucose is broken into many small enzyme-mediated steps that are grouped into three phases. We'll consider only the first phase, called *glycolysis,* which requires 10 steps to turn one molecule of glucose into two molecules of the compound *pyruvate.* FIGURE 14.14a shows the required free energy, relative to the initial glucose, for each of the nine intermediate steps. You can see that steps $0 \rightarrow 1$ and $2 \rightarrow 3$ are endergonic, requiring an increase in the Gibbs free energy. Overall, the sequence of reactions is highly favorable, with $\Delta G = -320$ kJ/mol, but the two uphill steps are highly unfavorable, and so this metabolic sequence cannot occur on its own.

FIGURE 14.14 The energetics of glycolysis (a) without and (b) with coupled reactions. The vertical scales differ.

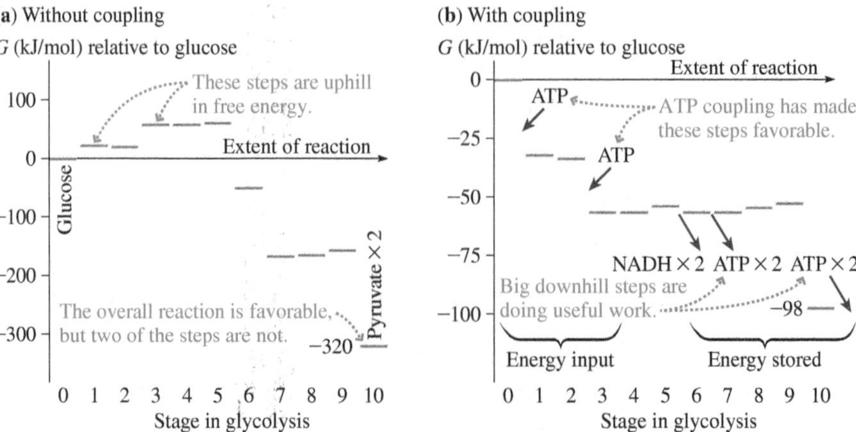

But the situation is entirely different when several steps are coupled to other reactions, as shown in FIGURE 14.14b. (Notice that the vertical scale differs from the scale in Figure 14.14a.) For example, the unfavorable step $0 \rightarrow 1$ has $\Delta G = +23$ kJ/mol, but coupling this reaction to the $\Delta G = -57$ kJ/mol from hydrolyzing one molecule of ATP makes it favorable with a net $\Delta G = -34$ kJ/mol. The unfavorable step $2 \rightarrow 3$ is also coupled to ATP.

Some of the steps, such as $4 \rightarrow 5$, are very slightly endergonic with $\Delta G \approx +2$ kJ/mol. These can proceed without outside intervention because $k_B T$, roughly the thermal energy of molecules, is 2.6 kJ/mol at body temperature. That is, thermal energy can drive endergonic reactions that have $\Delta G < RT$.

Steps $5 \rightarrow 6$, $6 \rightarrow 7$, and $9 \rightarrow 10$ are also coupled reactions, but in the opposite sense. Each of these steps in the oxidation of glucose has a fairly large downward step in free energy that, by itself, would create a fairly large amount of "wasted" thermal energy. Instead, these steps are coupled to the endergonic reactions by which ATP is rebuilt from ADP (and, similarly, another energy-storage molecule called NADH is assembled). That is, reaction energy that otherwise would have been released as heat is, instead, used to drive an uphill reaction.

Overall, this phase of the glucose metabolic pathway has two coupled reactions in which energy from ATP hydrolysis drives unfavorable endergonic processes, then three coupled reactions in which energy from the oxidation of glucose drives the endergonic processes of synthesizing ATP and NADH.

EXAMPLE 14.10 Efficiency of glycolysis

What is the overall efficiency of glycolysis for storing the energy of glucose for use elsewhere in the cell?

PREPARE Figure 14.14 shows the reaction sequence of glycolysis. The 10 uncoupled reactions have an overall $\Delta G = -320$ kJ/mol. This is the available energy of the reactions. The coupled reactions, with $\Delta G = -98$ kJ/mol, first hydrolyze two molecules of ATP, then build two molecules of NADH and four molecules of ATP. The net formation of energy-storage molecules is two NADH and two ATP.

SOLVE The net energy that has been extracted as useful work and ultimately stored in the four energy-storage molecules is $320 \text{ kJ/mol} - 98 \text{ kJ/mol} = 222 \text{ kJ/mol}$. This is energy that otherwise would have been "wasted" as thermal energy heating up the cell. As a fraction of the available energy,

$$\text{efficiency} = \frac{\text{stored energy}}{\text{available energy}} = \frac{222 \text{ kJ/mol}}{320 \text{ kJ/mol}} = 0.69 = 69\%$$

ASSESS Simply oxidizing glucose, whether quickly in one step or slowly in several, releases the energy as heat but doesn't store any of that energy for other uses. The coupled reactions of glycolysis divert 69% of that heat into stored energy that later, through more coupled reactions of ATP and NADH, can be used to build macromolecules.

STOP TO THINK 14.7 A reaction is exothermic ($\Delta H < 0$). Which of the following (perhaps more than one) are true?

A. The reaction is always spontaneous.
B. The reaction is never spontaneous.
C. The reaction is spontaneous if $\Delta S > 0$.
D. The reaction is spontaneous if $\Delta S < 0$ and $T > |\Delta H|/|\Delta S|$.
E. The reaction is spontaneous if $\Delta S < 0$ and $T < |\Delta H|/|\Delta S|$.

14.7 Mixing and Osmosis

You learned in Chapter 13 that diffusion is a *passive* transport mechanism, which means that molecules can move from one point to another with no expenditure of metabolic energy. Osmosis, which most often occurs in solutions, is another passive transport mechanism that is utilized in many biological processes. **Osmosis** is the flow of a solvent from a region of lower concentration of the *solute* (the dissolved substance) to a region of higher concentration, leading to the idea of *osmotic pressure.* We'll begin our exploration of osmosis with the mixing of two gases, which is easier to understand, and then see how the lessons learned carry over to solutions.

Entropy of Mixing

FIGURE 14.15 Two gases separated by a semipermeable membrane that lets blue molecules pass but not amber molecules.

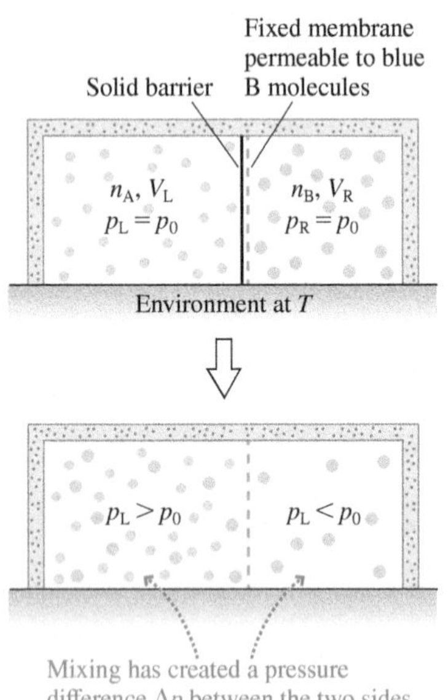

FIGURE 14.15 shows a container divided by a barrier into two compartments with n_A moles of gas A (for *amber*) on the left side (volume V_L) and n_B moles of gas B (for *blue*) on the right (volume V_R). Both sides have the same pressure p_0, and the container is in thermal contact with an environment at temperature T that keeps the gas temperatures constant.

In addition to the barrier, the container is divided by a **semipermeable membrane,** a material with small pores that let some molecules pass but block others. For example, a cell membrane is a semipermeable membrane that allows water and other small molecules to pass but blocks ions and macromolecules. The blue membrane, which is fixed in place, is permeable to blue B molecules but not to amber A molecules.

NOTE ▸ A semipermeable membrane is not a one-way passage. Molecules are either completely blocked or allowed to pass freely in both directions as if the membrane were not there. ◂

What happens when the solid barrier is removed? The pressures on both sides are equal, so you might expect nothing to happen. Instead, the B molecules, which can pass through the membrane, spread out to fill the container, while the membrane keeps the A molecules confined to the left side. This flow of B molecules across the barrier creates a *pressure difference* Δp between the two sides.

You will recall that the total pressure is the sum of the *partial pressures* of all gases in a mixture. Initially, the pressure on the left (all A molecules) was $p_L = (p_A)_i = p_0$ and the pressure on the right (all B molecules) was $p_R = (p_B)_i = p_0$. The pressure of gas A hasn't changed: $(p_A)_f = p_0$. However, gas B has spread out and now has the same lower partial pressure $(p_B)_f < p_0$ in *both* compartments. Thus the final pressures on the two sides are

$$\begin{aligned} p_L &= (p_A)_f + (p_B)_f = p_0 + (p_B)_f > p_0 \\ p_R &= (p_B)_f < p_0 \end{aligned} \tag{14.33}$$

We see that the mixing of the gases has created a pressure difference $\Delta p = p_L - p_R = p_0$ across the semipermeable membrane. In fact, since the A molecules in the left compartment obey $p_0 V_L = n_A RT$, we can write

$$\Delta p = \frac{n_A RT}{V_L} = C_A RT \tag{14.34}$$

where $C_A = n_A/V_L$ is the *molar concentration* (mol/m^3) of A in volume V_L. As you'll see, this is exactly the osmotic pressure of a solution.

Something else changed when the barrier was removed: the entropy of the system. Gas A didn't change, but gas B had a free expansion into the left side of the container. We earlier found that the entropy change for the free expansion of an ideal gas is

$$\Delta S_{mix} = n_B R \ln\left(\frac{V_f}{V_i}\right) = n_B R \ln\left(\frac{V_R + V_L}{V_R}\right) > 0 \tag{14.35}$$

We've not changed the number of molecules or their thermal energy, but we have *mixed* them together. This mixing, just like coffee and cream, is an irreversible process, so the entropy must have increased. In this context, ΔS is called the *entropy of mixing*.

Osmosis and Osmotic Pressure

A standard demonstration of osmosis, illustrated in FIGURE 14.16, uses a tube closed at one end by a semipermeable membrane that lets water molecules pass but not molecules of a solute dissolved in the water. Initially, before the solute is added, the water level inside the tube is the same as the level in the beaker. Dissolving a solute in the water inside the tube causes water to flow in through the semipermeable membrane; solvent flows from the side of lower (zero, in this case) solute concentration to the side of higher solute concentration. At equilibrium, the solution in the tube stands at height h above the water level in the beaker. Can we predict how high the solution will rise?

An analysis of Figure 14.16 is tricky because the hydrostatic pressure changes with depth. Let's start with a simpler experiment where the pressure on each side is the same throughout the volume. This will allow us to determine how osmotic pressure depends on the solute concentration, and we can then use our findings to predict the height in Figure 14.16.

FIGURE 14.16 Water molecules flow through the semipermeable membrane into the tube, where the solute concentration is higher, thus raising the water level above the level in the beaker.

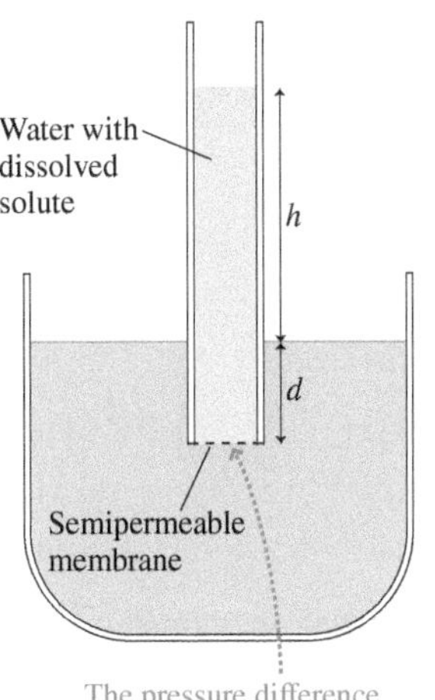
Water with dissolved solute
h
d
Semipermeable membrane
The pressure difference across the membrane is the osmotic pressure.

FIGURE 14.17 A pure solvent and a solution are separated by a semipermeable membrane that allows only water molecules to pass. Equilibrium requires a larger force on the left side.

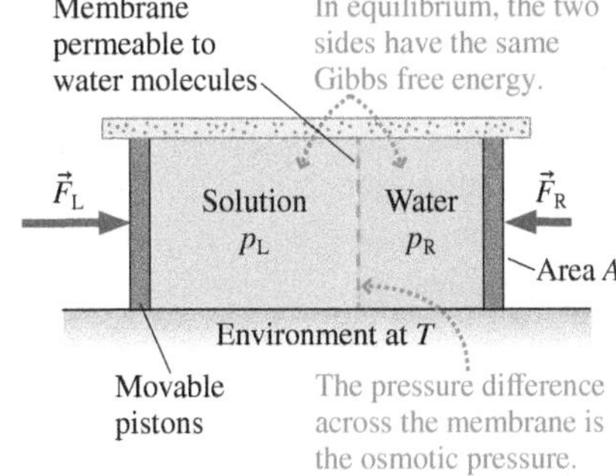
Membrane permeable to water molecules
In equilibrium, the two sides have the same Gibbs free energy.
$\vec{F}_L$
Solution
p_L
Water
p_R
$\vec{F}_R$
Area A
Environment at T
Movable pistons
The pressure difference across the membrane is the osmotic pressure.

FIGURE 14.17 shows this experiment—a two-compartment cylinder filled with a solvent, typically water, that we call B. The right compartment (volume V_R) contains *only* solvent, while the left (volume V_L) holds a solution in which a solute, which we call A, is dissolved in the solvent. Its molar concentration is $C_A = n_A/V_L$. The cylinder is divided by a semipermeable membrane with pores that let solvent molecules pass but not solute molecules. Notice that the situation in Figure 14.17 is very similar to that of Figure 14.15 after the gases have been mixed on the left side.

The left and right ends of the cylinder are pistons that external forces push on to create pressures $p_L = F_L/A$ in the left compartment and $p_R = F_R/A$ in the right. If the external forces differ, as shown in the figure, that difference must be balanced internally by a pressure difference $\Delta p = p_L - p_R$ across the membrane. In the context of osmosis, this pressure difference across the membrane is called the **osmotic pressure;** it is represented by the symbol Π (uppercase Greek pi); that is, $\Pi = p_L - p_R$.

Our goal is to find how the osmotic pressure depends on the solute concentration. The tool for doing so is the Gibbs free energy. We won't go through all the details, but we can outline the reasoning. Suppose the Gibbs free energy of the solution on the left is less than the Gibbs free energy of the pure solvent on the right: $G_L < G_R$. In that case, the free energy of the system would decrease if solvent spontaneously flowed across the membrane from right to left. Similarly, solvent would spontaneously flow from left to right if $G_R < G_L$. Thus the equilibrium situation with no flow is characterized by $G_R = G_L$; there's no change in free energy across the membrane.

NOTE ▶ The notation is a bit awkward, but we need to distinguish between the left and right compartments (subscripts L and R), and we also need to distinguish between solute and solvent molecules (subscripts A and B). ◀

We'll make two assumptions that are quite reasonable in many circumstances. First, the solution is a *dilute solution* in which the solute concentration is much lower than the water concentration. Second, this is what chemists call an *ideal solution.* The details need not concern us, but an ideal solution is a model of a solution in which the solvent-solvent, solute-solute, and solute-solvent molecular interactions are all assumed to have the same strength. We won't prove it, but it follows from this assumption that the solute-solvent mixture can be characterized mathematically in much the same way as a mixture of two different ideal gases.

If we were to write expressions for the Gibbs free energy and then apply these assumptions to the equilibrium condition $G_R = G_L$, we would find that the osmotic pressure—the pressure difference across the membrane—is

$$\Pi = \frac{T\Delta S_{mix}}{V_L} \tag{14.36}$$

Equation 14.36 tells us that **osmosis is an entropy-driven process.** The entropy increase of adding the solute—the entropy of mixing—is responsible for the osmotic pressure.

For gases, the entropy of mixing is $\Delta S_{mix} = n_B T \ln((V_R + V_L)/V_R)$. Because ideal solutions act much like ideal gases, it turns out that the entropy of mixing for a dilute ideal solution is the same except that the ratio of volumes is replaced by a ratio of moles:

$$\begin{aligned}\Delta S_{mix} &= n_B R \ln\left(\frac{n_A + n_B}{n_B}\right) = -n_B R \ln\left(\frac{n_B}{n_A + n_B}\right) \\ &= -n_{tot} x_B R \ln x_B\end{aligned} \tag{14.37}$$

where $n_{tot} = n_A + n_B$ is the total amount of substance and $x_B = n_B/n_{tot}$ is the mole fraction of solvent B.

We're assuming a dilute solution, with $x_B \approx 1$ for the solvent and $x_A \ll 1$ for the solute, so we can write $\Delta S_{mix} = -n_{tot} R \ln x_B$. We can't, however, use $x_B = 1$ in the logarithm because $\ln(1) = 0$ and we wouldn't have any entropy of mixing at all. We can, though, write $x_B = 1 - x_A$, which gives

$$\Delta S_{mix} = -n_{tot} R \ln(1 - x_A) \tag{14.38}$$

Earlier in this chapter we used the approximation $\ln(1 + x) \approx x$ when $x \ll 1$ to simplify Equation 14.5. For a dilute solution, the solute mole fraction is $x_A \ll 1$. With the minus sign, the approximation becomes $\ln(1 - x_A) \approx -x_A$. If we use this approximation in Equation 14.38, we find that the entropy of mixing for a dilute ideal solution is

$$\Delta S_{mix} = x_A n_{tot} R = n_A R \tag{14.39}$$

where, in the last step, we used the definition of the mole fraction to write $x_A n_{tot} = n_A$, the number of moles of solute. We've made several approximations to arrive at Equation 14.39, but they are reasonable for a great many solutions.

Having determined the entropy of mixing, we can now return to Equation 14.36 to find that the osmotic pressure is

$$\Pi = \frac{T\Delta S_{mix}}{V_L} = \frac{n_A RT}{V_L} = C_{solute} RT \tag{14.40}$$

Osmotic pressure for a solute with molar concentration C

where $C_{solute} = n_A/V_L$ is the molar concentration of the solute. This is the same pressure difference that we found for the mixing of gases at the beginning of this section.

NOTE ▸ Don't confuse this use of the symbol C with our earlier use of C to represent the molar specific heat or with the use of a lower case c to represent the number concentration N/V. There are more concepts in science than there are letters in the alphabet, so many symbols have multiple possible meanings. As always, you need to be alert to the context in which a symbol is used. ◂

Equation 14.40 assumes that the membrane separated a solution from a pure solvent. More generally, the membrane may separate two solutions that have different solute concentrations. In that case, the osmotic pressure—the pressure difference across the membrane—is

$$\Pi = \Delta C_{solute} RT \tag{14.41}$$

where ΔC_{solute} is the difference in concentrations.

Equation 14.40 is called *van't Hoff's law,* after Dutch chemist Jacobus van't Hoff, winner of the first Nobel Prize in Chemistry in 1901, who discovered this relationship experimentally. Van't Hoff's law, like Hooke's law and other "laws" that you'll meet, is not truly a law of physics but simply a model that works well in many, but not all, circumstances.

EXAMPLE 14.11 Calculating osmotic pressure

What is the molarity of a 25°C solution that has an osmotic pressure of 1.0 atm?

PREPARE Osmotic pressure is given by Equation 14.40 with the solute concentration C_{solute} in mol/m^3. The molarity of a solution is the moles of solute per *liter* of solution, so we'll need to use $1\ \text{L} = 10^3\ \text{cm}^3 = 10^{-3}\ \text{m}^3$ for a unit conversion. The temperature is 298 K.

SOLVE 1.0 atm is 1.01×10^5 Pa. The solute concentration that gives an osmotic pressure of 1.0 atm is

$$C_{solute} = \frac{\Pi}{RT} = \frac{1.01 \times 10^5\ \text{Pa}}{(8.31\ \text{J/mol} \cdot \text{K})(298\ \text{K})} = 41\ \text{mol/m}^3$$

The unit conversion to mol/L gives

$$C_{solute} = 41\ \text{mol/m}^3 \times \frac{10^{-3}\ \text{m}^3}{1\ \text{L}} = 0.041\ \text{mol/L} = 0.041\ \text{M}$$

ASSESS A concentration of 0.041 M is, indeed, a dilute solution. Even so, the solution's entropy of mixing is sufficient to establish an osmotic pressure of 1.0 atm, the same as a 760-mm-tall column of mercury.

Figure 14.17 shows an equilibrium situation, with no flow of solvent. What you see, from the piston forces, is that **osmotic flow is halted by an external pressure equal to the osmotic pressure.** That is, osmotic pressure does not *cause* the flow of solvent across the membrane; it is the pressure that *stops* any flow of solvent. A slight reduction in force F_L on the left side of Figure 14.17 would upset the equilibrium; both pistons would move to the left as solvent flowed from right to left through the membrane. This would reduce the volume on the right, increase the volume on the left, and reduce the solute concentration C until a new equilibrium was established at a somewhat lower osmotic pressure.

Conversely, an increase in force F_L would cause both pistons to move to the right, with solvent flowing from left to right, from the solution side to the pure solvent side. That is, the solvent can be squeezed out of the solution by an external force on the left that applies a pressure larger than the osmotic pressure, giving more pure solvent and leaving behind a solution with a higher concentration of solute. Using an external force to remove solvent from a solution by pressing it through a semipermeable membrane is called **reverse osmosis.** It is used industrially in some arid regions of the world to obtain fresh water from seawater. The process works technically, but the very high pressures needed make the process very energy intensive and thus very costly (the details will be left as a homework problem).

Returning to the vertical tube of Figure 14.16, we can now determine the height h to which the liquid rises above the level in the beaker. Inside the tube, the membrane is at total depth $d + h$ below the surface, so the hydrostatic pressure just above the membrane is $p_{tube} = p_{atm} + \rho g(d + h)$. In the pure water in the beaker, the water just below the membrane is at depth d and has hydrostatic pressure $p_{water} = p_{atm} + \rho g d$. The difference between these is the pressure difference across the membrane, which is the osmotic pressure Π; that is,

$$\begin{aligned}\Pi = C_{solute}RT &= p_{tube} - p_{water} \\ &= [p_{atm} + \rho g(d + h)] - [p_{atm} + \rho g d] = \rho g h\end{aligned} \tag{14.42}$$

We are assuming that the density of the solution is the same as the density of the solvent, which is a reasonable approximation for a dilute solution. Thus the height of the liquid is

$$h = \frac{C_{solute}RT}{\rho g} \tag{14.43}$$

EXAMPLE 14.12 Water rise due to a sugar solution

To what height does a dilute 0.020 M sucrose solution rise above a beaker of pure water at 25°C?

PREPARE Molarity is moles per liter, so we'll need to use $1\ \text{L} = 10^3\ \text{cm}^3 = 10^{-3}\ \text{m}^3$ for the unit conversion. The temperature is 298 K, and the density of water is $1000\ \text{kg/m}^3$.

SOLVE We first need to convert the concentration to SI units of mol/m^3:

$$C_{solute} = 0.020\ \text{M} = 0.020\ \text{mol/L} \times \frac{1\ \text{L}}{10^{-3}\ \text{m}^3} = 20\ \text{mol/m}^3$$

Osmotic flow causes this concentration of solution to rise to a height

$$h = \frac{C_{solute}RT}{\rho g} = \frac{(20\ \text{mol/m}^3)(8.31\ \text{J/mol}\cdot\text{K})(298\ \text{K})}{(1000\ \text{kg/m}^3)(9.8\ \text{m/s}^2)} = 5.0\ \text{m}$$

ASSESS A height of 5 m seems astounding, and practical limitations (e.g, real membranes aren't perfect) might prevent such a rise. However, osmosis is the primary mechanism by which roots, whose cell sap is a solution of various salts, bring pure water up from the soil into the xylem of trees, where mechanisms we discussed in Chapter 9 then lift it upward to the crown. Large trees need an osmotic lift of several meters.

Osmosis in Biology

Water that you drink is absorbed into your body primarily from the large intestine. How? The large intestine is also where salt is absorbed. Dissolved salts are ions, which don't pass through cell membranes by either diffusion or osmosis. Instead, the cells in the wall of the large intestine use the active transport of ATP-powered ion channels to move Na^+ and K^+ ions from the intestine into the body. (We'll look at the physics of ion channels when we get to our study of electricity.)

Removing salt makes the water in the large intestine a less concentrated solution than the intracellular fluid of the intestinal wall. The membrane of intestinal wall cells is a semipermeable membrane that readily allows water molecules to pass, so water is moved *by osmosis* from the less-concentrated solution in the intestine into the more-concentrated solution of the intestinal wall cells, from where it is transported to the rest of the body.

Sometimes things don't go quite right, and the removal of too much water can cause constipation. Many over-the-counter remedies for constipation are a soluble polymer, such as polyethylene glycol, that you drink. These are large molecules that cannot pass through cell membranes, so they travel through the digestive system into the large intestine. This causes the liquid in the large intestine to have a high concentration of dissolved polymer molecules, so osmosis reverses direction and pulls water from the lower-concentration intestinal wall cells into the large intestine.

For much the same reason, drinking seawater—should you happen to be stranded at sea—will kill you rather than keep you alive. Although your cellular fluids are basically salt water, the salts are less concentrated than in seawater. If you drink seawater, there will be an osmotic flow of water *out* of your cells into the more concentrated salt water in your intestines. That is, drinking seawater further dehydrates rather than rehydrates.

These are just of few of the many situations in which osmosis plays an important transport role in biology. Liquids in the body and in plants are all solutions of one type or another, and cell walls are semipermeable membranes that allow water and other small molecules to pass but not ions or large macromolecules. Consequently, osmosis occurs wherever there's a concentration difference on opposite sides of a membrane. Osmosis takes place throughout the digestive system, in the kidneys, and, as mentioned in Example 14.12, in the uptake of water by plant roots.

Osmosis plays a critical role in maintaining water balance in the cell. As FIGURE 14.18 shows, a cell surrounded by extracellular fluid that has a lower concentration of solutes—a *hypotonic solution*—will gain water by osmosis and swell. This is dangerous for animal cells because too much inflow will cause the cell to rupture (the biological term is to *lyse*) and die.

FIGURE 14.18 The water balance of a cell depends on whether the extracellular fluid is hypotonic, hypertonic, or isotonic.

Hypotonic: Solute concentration is higher inside the cell

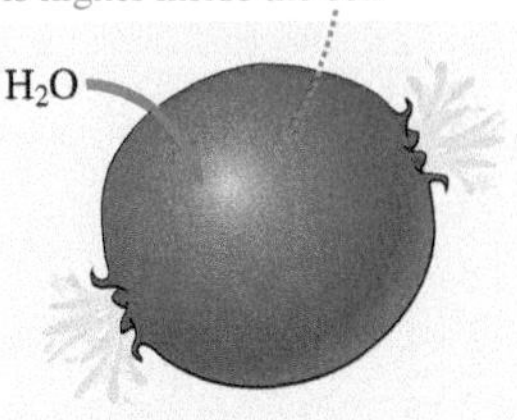

Isotonic: Solute concentrations inside and outside are equal

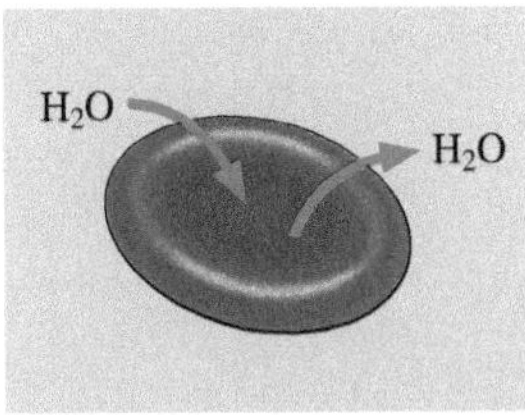

Hypertonic: Solute concentration is higher outside the cell

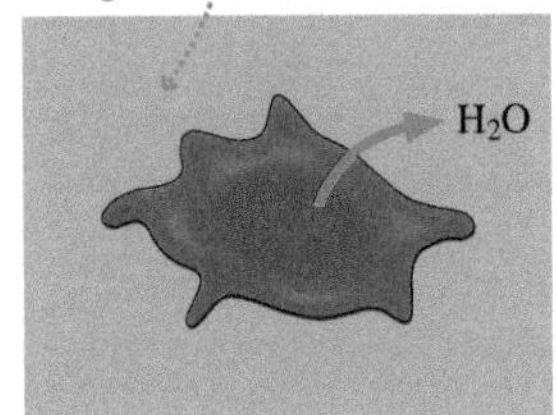

For plant cells, being surrounded by lower-concentration fluid is normal. The extracellular fluid in plants is a relatively low-concentration solution, hypotonic, so osmosis pulls water into the cells. But plant cells, unlike animal cells, have a rigid cell wall that can withstand the increased pressure. Water inflow continues, and a cell swells until the internal pressure equals the osmotic pressure, which prevents further flow. This makes normal plant cells *turgid* and gives even nonwoody plant stems stiffness, allowing them to stay erect.

The opposite condition—a *hypertonic solution*—occurs when the extracellular fluid has a higher concentration of solutes than the cell. Then osmosis draws water from the cell, causing the cell to shrivel. A plant wilts if it is not watered because the extracellular fluid becomes more concentrated and then osmosis begins to pull water out of the cells.

An *isotonic solution* is one that matches the solute concentration of the cell. Water and nutrients can then pass back and forth by diffusion, but there's no osmotic flow. This is the normal state of animal cells, and the body has many mechanisms for maintaining isotonicity. When fluids are medically administered, whether liquids or plasma, it is critical to ensure that they do not disrupt the body's tonicity.

Misconceptions About Osmosis

It's not easy to understand entropically driven processes, such as osmosis, so there's a desire to look for "simple" explanations. Unfortunately, many of the common explanations of osmosis are incorrect. Two, in particular, are worth noting.

One purported explanation is that osmosis occurs because of an attraction between solute molecules and water molecules. This attraction allows the solute molecules to "pull" water molecules through the semipermeable membrane. Solute molecules generally do attract water molecules, but water molecules also attract water molecules; it's the presence of attractive forces that distinguishes a liquid from a gas and holds the liquid together. These short-range forces are *not* pulling water molecules through the membrane.

A more commonly seen explanation is that osmosis is a form of diffusion in which water moves down the concentration gradient from a higher water concentration on the pure-solvent side to a lower water concentration on the solution side. You might read that the water on the solution side has been "diluted" by the solute. While this may sound plausible, it's based on an erroneous picture of molecules as solid spheres.

If you have a large box completely filled with blue balls (solvent molecules), you can add amber balls (solute molecules) to the box only by removing an equal number of blue balls. In this case, increasing the concentration of amber balls forces a decrease in the concentration of blue balls. But molecules don't work this way! Water molecules are connected in a rather loose framework by hydrogen bonds, and it's quite possible for small solute molecules to slip between the water molecules *without changing the water concentration.* In fact, the electric charge of ions in a salt solution can disrupt the hydrogen bonding in such a way as to bring the water molecules closer together, actually *increasing* the density and hence the concentration of water. In this case, the osmotic flow is not down a concentration gradient but *up* a concentration gradient. The tacit assumption that an increase in solute concentration causes a decrease in water concentration is simply wrong. Osmosis is not diffusion from a higher water concentration to a lower water concentration.

Instead, osmosis occurs because a flow of solvent molecules from a region of lower solute concentration to a region of higher solute concentration increases entropy, due to the entropy of mixing. Any process that increases entropy happens spontaneously. That, ultimately, is the message of this chapter.

Oil and Water: The Hydrophobic Effect

Sugar (a solid), hydrogen peroxide (a liquid), and oxygen (a gas) all dissolve in water to create well-mixed solutions. But oil and water don't mix. Why not? What is different? A common, but incorrect, explanation is that oil and water separate because the hydrogen bonds between two water molecules are strong whereas the interaction between a polar water molecule and a nonpolar oil molecule is weak. This may sound plausible, but calculations of the enthalpies of these interactions don't support this idea. Rather, entropy is the key player: Like osmosis, the separation of oil and water—an example of the biologically essential phenomenon called the *hydrophobic effect*—is an entropically driven process.

Whether oil and water mix depends on whether $\Delta G = \Delta H - T\Delta S$ is negative (mixing is spontaneous) or positive (the reverse, separation, is spontaneous). It turns out that $\Delta H \approx 0$, being slightly endothermic (favoring separation) for some hydrocarbons but slightly exothermic (favoring mixing) for others. Either way, the fact that ΔH is very small means that the sign of ΔG is determined primarily by the change in entropy. Thus far we've looked at examples where the entropy of mixing is positive ($\Delta S > 0$), which makes mixing a spontaneous process ($\Delta G < 0$).

Ice—water in its solid state—is a hydrogen-bonded crystal with low entropy. The crystal structure cannot be maintained above 0°C, but liquid water maintains a constantly shifting network of hydrogen bonds. These hydrogen bonds account for the unusual properties of water, such as contracting upon melting and having an extremely large specific heat.

The introduction of a nonpolar hydrocarbon molecule into water disrupts this network because a nonpolar molecule cannot form hydrogen bonds. Detailed theoretical and experimental studies have found that water molecules rearrange themselves to form a hydrogen-bonded, cage-like structure around each hydrocarbon molecule; biochemists call it a *clathrate cage.* This cage has less flexibility than free water molecules (i.e., it is somewhat more like a crystal) and is thus, at the molecular level, a more ordered structure.

As a result, the cage-like water structure around a hydrocarbon molecule has *less* entropy than free water. Thus, counterintuitively, the introduction of nonpolar hydrocarbon molecules into water usually causes a *reduction* of entropy ($\Delta S < 0$; for oil and water the entropy of mixing is negative) and a corresponding increase in free energy ($\Delta G > 0$). The equilibrium state, which minimizes free energy and maximizes entropy, is one of separated oil and water rather than mixed oil and water.

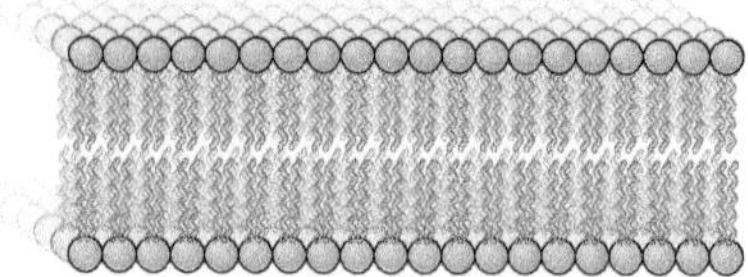

Cell membranes Phospholipids are molecules that have a polar head and two hydrocarbon tails. Phospholipids in water spontaneously organize themselves into two-layered structures, called *lipid bilayers,* with the polar heads (which can form hydrogen bonds with water molecules) on the surface and the nonpolar hydrocarbons coalesced on the inside, away from the water. Spontaneous organization would seem to defy the second law of thermodynamics, but this is another example of the hydrophobic effect. What appears to be an organized, low-entropy configuration actually has a *higher* entropy—it's the equilibrium state—than phospholipids uniformly mixed throughout the water. The formation of a lipid bilayer—the cell membrane—is an entropically driven process.

This entropically driven separation is called the **hydrophobic effect.** That is a poor term because it suggests—incorrectly—that the hydrocarbon molecules are actively repelled by water molecules. Instead, the random motions of molecules favor configurations in which oil molecules coalesce because those are the macrostates that have the largest multiplicity and the highest entropy. The equilibrium state of oil floating on water, which visually appears to be ordered, is actually more disordered (higher entropy) at the molecular level than a uniform oil-water mixture.

The temperature in $\Delta G = \Delta H - T\Delta S$ plays an important role in the hydrophobic effect. Generally one expects solubility to increase with increasing temperature, but hydrocarbons become *less* soluble with increasing temperature. Because the entropy of mixing is negative ($\Delta S < 0$), a higher temperature makes ΔG even more positive and further reduces the chances of a hydrocarbon molecule being in solution. Conversely, ΔG could become negative at sufficiently low temperatures if the mixing is slightly exothermic ($\Delta H < 0$). A stabilized hydrocarbon-water mixture at low temperatures is called a *clathrate*. Methane clathrates occur naturally in deep-ocean sediments where the high pressure and low temperature create an ice-like solid of methane dissolved in water. These structures are unstable, and there is some concern that a warming of the deep ocean due to climate change could cause the clathrates to release their methane, which is a more potent greenhouse gas than carbon dioxide.

STOP TO THINK 14.8 The two circular, frictionless pistons in the figure, each with cross-section area A, are yoked to slide back and forth together. The semipermeable membrane is attached to the walls of the cylinder and cannot move. Which is true of mass m?

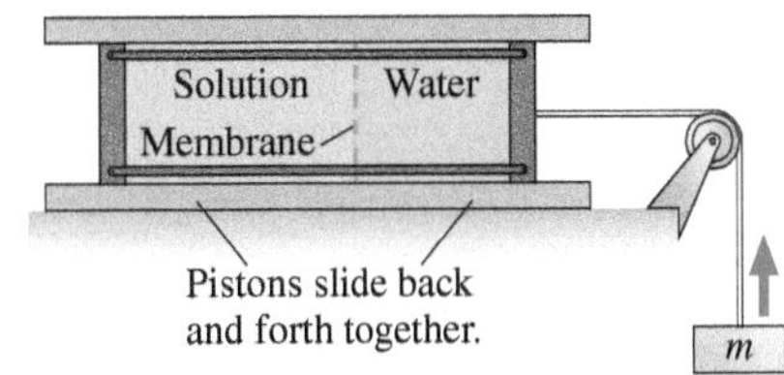

A. Mass m rises if $\Pi A > mg$.
B. Mass m rises if $\Pi A = mg$.
C. Mass m rises if $\Pi A < mg$.
D. Mass m always rises.
E. Mass m never rises.

CHAPTER 14 INTEGRATED EXAMPLE Making water

One option for an alternative to fossil-fuel vehicles would be a vehicle that burns hydrogen, whose only combustion product is water. Hydrogen doesn't occur naturally, but it could be generated by solar-powered electrolysis of seawater, and there is ongoing research to produce hydrogen with algae. The direct combustion of hydrogen is impractical for various reasons, so proposals for a hydrogen economy usually envision using a catalyst to carry out the reaction in a fuel cell that would run an electric motor. That changes the details, but this problem illustrates the basic concept.

Hydrogen and oxygen combust to form water via the reaction $2H_2(g) + O_2(g) \rightarrow 2H_2O(l)$. We will assume that the reaction is carried out with constant temperature and pressure at the standard state of 25°C and 1 atm pressure, so the final state of the water is liquid. Data on the molar masses, enthalpies of formation, and standard molar entropies are listed here:

	M_{mol} (kg/mol)	H_f° (kJ/mol)	S° (J/mol · K)
$H_2(g)$	0.002	0	130.7
$O_2(g)$	0.032	0	205.2
$H_2O(l)$	0.018	−285.8	70.0

Let's consider the reaction of 4.0 g of hydrogen with 32 g of oxygen to form 36 g of water. What are ΔG, ΔH, ΔS, and ΔE_{int} for this reaction? What is the maximum useful work that can be done? What are ΔS_{env} and $\Delta S_{universe}$ if no useful work is done and if the maximum useful work is done?

PREPARE The Gibbs free energy is most often applied to chemical reactions. We'll leave more complex reactions to your chemistry and biochemistry courses, but it's worth using physical principles to examine the energy and entropy changes in a simple reaction. The table gives molar quantities as they are listed in a chemical reference. The reaction equation shows that 2 mol of H_2 react with 1 mol of O_2 to produce 2 mol of H_2O.

SOLVE The Gibbs free-energy change is $\Delta G = \Delta H - T\Delta S$, so we'll start with enthalpy and entropy. Example 12.19 showed that ΔH is the enthalpy of the products minus the enthalpy of the reactants:

$$\begin{aligned}\Delta H &= H_f^\circ(\text{products}) - H_f^\circ(\text{reactants}) \\ &= (2.0\text{ mol})(-285.8\text{ kJ/mol}) - (2.0\text{ mol})(0\text{ kJ/mol}) - \\ &\quad (1.0\text{ mol})(0\text{ kJ/mol}) \\ &= -572\text{ kJ}\end{aligned}$$

The change in entropy is calculated in a similar fashion:

$$\begin{aligned}\Delta S &= S^\circ(\text{products}) - S^\circ(\text{reactants}) \\ &= (2.0\text{ mol})(70\text{ J/mol}\cdot\text{K}) - (2.0\text{ mol})(130.7\text{ J/mol}\cdot\text{K}) - \\ &\quad (1.0\text{ mol})(205.2\text{ J/mol}\cdot\text{K}) \\ &= -330\text{ J/K} = -0.330\text{ kJ/K}\end{aligned}$$

Thus the change in the Gibbs free energy is

$$\begin{aligned}\Delta G &= \Delta H - T\Delta S = -572\text{ kJ} - (298\text{ K})(-0.330\text{ kJ/K}) \\ &= -474\text{ kJ}\end{aligned}$$

$\Delta G < 0$, so this reaction does occur spontaneously.

Enthalpy is defined as $H = E_{int} + pV$, so at constant pressure the change in energy is $\Delta E_{int} = \Delta H - p\Delta V$. The volume change is $\Delta V = V_{products} - V_{reactants}$. The product is liquid whereas the reactants are gases, so, to a very good approximation, $\Delta V = -V_{reactants}$. The reactants have $n_{tot} = 3.0$ mol of gas, and we can use the ideal-gas law for $pV_{reactants}$ to find

$$\begin{aligned}\Delta E_{int} &= \Delta H + pV_{reactants} = \Delta H + n_{tot}RT \\ &= (-572\text{ kJ}) + (3.0\text{ mol})(8.31\text{ J/mol}\cdot\text{K})(298\text{ K}) \\ &= -565\text{ kJ}\end{aligned}$$

If no useful work is done—that is, no work other than pressure-volume work—then the heat released in the reaction

(Continued)

is $Q_{sys} = \Delta H = -572$ kJ. The heat is negative because this is an exothermic reaction, with heat leaving the system. This heat flows to the environment as $Q_{env} = -Q_{sys} = 572$ kJ and increases the environment's entropy by

$$\Delta S_{env} = \frac{Q_{env}}{T} = \frac{572 \text{ kJ}}{298 \text{ K}} = 1.92 \text{ kJ/K} = 1920 \text{ J/K}$$

The total entropy change of the universe is thus

$$\Delta S_{universe} = \Delta S_{env} + \Delta S = 1920 \text{ J/K} - 330 \text{ J/K} = 1590 \text{ J/K}$$

If, instead, the reaction takes place under circumstances where we can harness some of this energy to do useful work, the maximum useful work is $W_{useful} = |\Delta G| = 474$ kJ. In this case, the heat released in the reaction is reduced to

$$Q_{sys} = \Delta H + W_{useful} = -572 \text{ kJ} + 474 \text{ kJ} = -98 \text{ kJ}$$

Thus $Q_{env} = -Q_{sys} = 98$ kJ and the environment's entropy increase is

$$\Delta S_{env} = \frac{Q_{env}}{T} = \frac{98 \text{ kJ}}{298 \text{ K}} = 0.330 \text{ kJ/K} = 330 \text{ J/K}$$

With maximum useful work being done, the entropy change of the universe is

$$\Delta S_{universe} = \Delta S_{env} + \Delta S = 330 \text{ J/K} - 330 \text{ J/K} = 0 \text{ J/K}$$

ASSESS $\Delta H < 0$ indicates that this is an exothermic reaction; it releases heat. Interestingly $\Delta S < 0$, so the system *loses* entropy as a result of the reaction. The reason is that gases, which are dispersed, high-entropy systems, are changed to a lower-entropy liquid. Nonetheless, we still find a negative ΔG, which tells us that the entropy of the environment increases enough for the reaction to proceed spontaneously. Energy, the one conserved quantity, changes by $\Delta E_{int} = -564$ kJ. The large negative value is mostly a change in chemical energy as weaker bonds in the gases are replaced by much stronger bonds in the water molecules. However, the magnitude of ΔE_{int} is less than the magnitude of ΔH, the heat released in the reaction. The additional released energy comes from the negative pressure-volume work done as the gases collapse in volume to a liquid.

Of the 565 kJ reduction in energy, the amount of available energy—the energy that can do useful work—is 474 kJ. That is, at most 84% of the initial chemical energy can be used for useful work. This is a high percentage, but the reason it isn't 100% has to do with entropy. If no useful work is done, then the positive entropy change of the environment is much larger than the entropy reduction of the system. But, because energy is conserved, doing work comes at the expense of heat flow to the environment. Less heat flow means less entropy increase of the environment and thus less entropy increase of the universe. Doing 474 kJ of work reduces the heat flow to the point where $\Delta S_{universe} = 0$. Extracting any more work from the reaction would require $\Delta S_{universe} < 0$, which is impossible. The limit set by the second law of thermodynamics is that—at best—only 474 kJ of the 565 kJ reduction in energy can be harnessed for useful work, while at least 91 kJ becomes degraded energy that cannot be put to use.

This has been a long example, but it's well worth studying in detail because it illustrates how the seemingly arcane laws of physics have very real consequences for chemical reactions.

SUMMARY

GOAL To understand how entropy and the second law of thermodynamics apply to living and nonliving systems.

GENERAL PRINCIPLES

Second Law of Thermodynamics

The entropy of the universe never decreases:

$$\Delta S_{\text{universe}} \geq 0 \text{ with } \begin{cases} = \text{ for a reversible process} \\ > \text{ for an irreversible process} \end{cases}$$

- The entropy of an isolated system increases as it evolves toward equilibrium.
- Equilibrium is the state of maximum entropy.
- A process in which a system interacts with its environment is allowed only if the total entropy $S_{\text{sys}} + S_{\text{env}}$ increases.

Gibbs Free Energy

The **Gibbs free energy** of a system is $G = H - TS$.

- At constant temperature and pressure, a process $A \rightarrow B$ can occur spontaneously if

$$\Delta G = \Delta H - T\Delta S < 0$$

  This is an alternative statement of the second law.
- The maximum useful work that can be done by the process is

$$W_{\text{max}} = |\Delta G|$$

- An *exergonic process* has $\Delta G < 0$.
  An *endergonic process* has $\Delta G > 0$.

IMPORTANT CONCEPTS

A set of values of the variables $p, T, S, \ldots$ of a system defines a **macrostate.** Each possible arrangement of the positions and velocities of the molecules is a **microstate.** The number of microstates corresponding to a given macrostate is the **multiplicity** Ω of the state.

The most probable macrostate, with the largest Ω, is the equilibrium state. A nonequilibrium state evolves toward equilibrium because *equilibrium is overwhelmingly the most probable macrostate to be in.*

Entropy is a state variable that measures the spread or dispersal of thermal energy. It is defined as

$$S = k_B \ln \Omega$$

where Ω is the multiplicity, the number of microstates.

A slow, quasi-static process that transfers heat dQ to a system increases the system's entropy by

$$dS = dQ/T$$

Available energy

A system that is initially not in equilibrium can do work on the environment as it comes to equilibrium. The *amount* of work that can be done is limited by the second-law requirement that $\Delta S_{\text{universe}} \geq 0$. The maximum work that can be done, called the *available energy,* is

$$W_{\text{max}} = |\Delta G|$$

Useful work is done at the expense of heat energy that would otherwise flow to the environment.

Osmosis

Osmosis is the flow of a solvent from a region of lower solute concentration to a region of higher concentration. It is an entropically driven process, not diffusion.

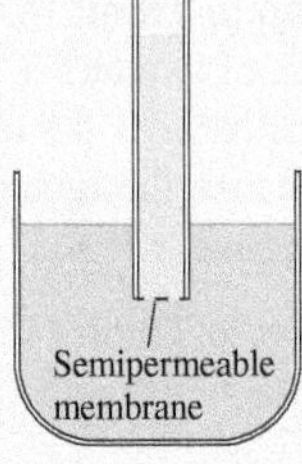

The **osmotic pressure** is

$$\Pi = C_{\text{solute}} RT$$

where C_{solute} is the solute concentration in mol/m^3.

APPLICATIONS

Entropy changes

- Phase change: $\Delta S = Q/T$
- Heating: $\Delta S = Mc\ln(T_f/T_i)$
- Ideal gas: $\Delta S = nC_V\ln(T_f/T_i) + nR\ln(V_f/V_i)$

Spontaneity

A process or reaction can occur spontaneously if $\Delta G = \Delta H - T\Delta S < 0$.

- An exothermic process ($\Delta H < 0$) that increases entropy ($\Delta S > 0$) is exergonic and always spontaneous.
- An endothermic process ($\Delta H > 0$) that decreases entropy ($\Delta S < 0$) is endergonic and never spontaneous.
- An exothermic process ($\Delta H < 0$) that decreases entropy ($\Delta S < 0$) could be exergonic or endergonic. If exergonic and spontaneous, it is an *enthalpy-dominated process.*
- An endothermic process ($\Delta H > 0$) that increases entropy ($\Delta S > 0$) could be exergonic or endergonic. If exergonic and spontaneous, it is an *entropy-dominated process.*

Equilibrium

An exergonic process or reaction proceeds by moving down the free-energy curve until it reaches a minimum value of G. The minimum of G is the equilibrium state.

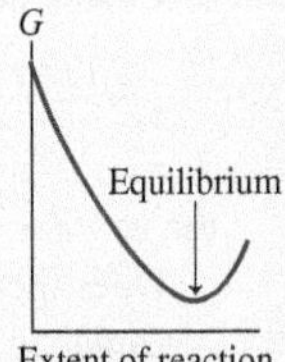

Coupled reactions

An endergonic reaction would need to move up the free-energy curve. The second law forbids this for an isolated system, but an endergonic reaction becomes allowed if it is *coupled* to an exergonic reaction. This happens in the cell when an endergonic reaction is coupled, via enzymes, to the strongly exergonic hydrolysis of ATP or other energy-storage molecules.

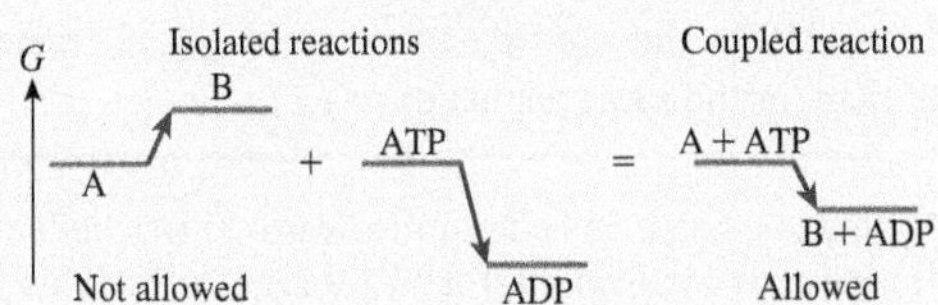

LEARNING OBJECTIVES After studying this chapter, you should be able to:

- Distinguish between reversible and irreversible processes. *Multiple-Choice Question 14.12*
- Interpret and use the second law of thermodynamics. *Conceptual Questions 14.2, 14.5, 14.6; Problems 14.14, 14.18*
- Determine the multiplicity of a simple system. *Problems 14.1–14.4, 14.16*
- Calculate the change in entropy for different processes. *Conceptual Questions 14.3, 14.7; Problems 14.10–14.13, 14.15*
- Determine the change in the Gibbs free energy and use it to judge whether a process is spontaneous. *Conceptual Questions 14.1, 14.8; Problems 14.21–14.25*
- Calculate the maximum useful work that can be done in a process. *Problems 14.27–14.30*
- Determine the equilibrium conditions of a system. *Conceptual Question 14.9; Problems 14.31, 14.32*
- Calculate the osmotic pressure of a solution. *Conceptual Question 14.10; Problems 14.33–14.35, 14.37, 14.38*

STOP TO THINK ANSWERS

Stop to Think 14.1: B. In this case, kinetic energy is transformed into potential energy; there is no entropy change. In the other cases, energy is transformed into thermal energy, meaning entropy increases.

Stop to Think 14.2: B. The multiplicity increases from cV^N to $c(2V)^N = c2^NV^N$. This is an increase by a factor of 2^N.

Stop to Think 14.3: C. Box B, with nearly equal numbers of molecules on both sides, is like a coin toss with nearly equal numbers of heads and tails. This is a very likely outcome with a high multiplicity. Box A is a very unlikely situation with a low multiplicity. Thus $S_A < S_B$.

Stop to Think 14.4: A. The entropy change of an ideal gas is the sum of two terms. In an expansion, with $V_f > V_i$, the volume term is positive. For ΔS to be zero, the temperature term must be negative. Because the temperature ratio appears in a logarithm, this term is negative if $T_f < T_i$.

Stop to Think 14.5: B. The gases have the same number of moles at the same temperature, so they have the same thermal energy. Gas 1 is more compressed, and it can do more work by exerting a larger force through a greater distance.

Stop to Think 14.6: D. The reaction is spontaneous, so $\Delta G < 0$. The work done is $W_{useful} \leq -\Delta G$, so the absolute value of ΔG must be at least 100 J.

Stop to Think 14.7: C and E. A reaction is spontaneous if $\Delta G = \Delta H - T\Delta S < 0$. This exothermic reaction has $\Delta H < 0$, so $\Delta S > 0$ makes $\Delta G < 0$. $\Delta G < 0$ is also possible with $\Delta S < 0$, depending on the temperature. Rearranging $\Delta H - T\Delta S < 0$ gives $T\Delta S > \Delta H$. With ΔH and ΔS both negative, multiplying both sides by -1 to get absolute values changes $>$ to $<$, so $T < |\Delta H|/|\Delta S|$.

Stop to Think 14.8: A. The system is in equilibrium if $\Pi A = mg$ because the excess pressure Π in the solution pushes the pistons to the left with exactly the same force as the hanging mass m pulls them to the right. But if $\Pi A > mg$, the osmotic pressure pulls the mass up as the pistons move left to draw more water into the solution.

QUESTIONS

Conceptual Questions

1. A reaction happens in a rigid, sealed, insulated container. Let the interior of the container be the system.
   a. Does the energy of the system increase, decrease, or stay the same?
   b. Does the Gibbs free energy of the system increase, decrease, or stay the same?
2. Every cubic meter of air has a rather significant amount of thermal energy. A scientist claims to have invented a car that can power itself by extracting thermal energy from the air. Is that possible? Why or why not?
3. Gas 1 undergoes a free expansion from volume V to volume $2V$, then later another free expansion from volume $2V$ to volume $3V$. Gas 2 undergoes a free expansion from volume V directly to volume $3V$. Is the entropy increase of gas 2 larger than, smaller than, or the same as that of gas 1?
4. An ideal gas with pressure p_0 and volume V_0 undergoes three processes. First it is heated so that its pressure, temperature, and volume all increase. Then it is cooled at constant volume until its pressure returns to p_0. Finally it is cooled at constant pressure until its volume returns to V_0. Is the net change in the entropy of the gas positive, negative, or zero? Or do you not have enough information to tell?
5. A system that can interact with its environment undergoes a natural, spontaneous process.
   a. Is ΔS_{sys} of the system positive, negative, or zero? Or do you not have enough information to tell?
   b. Is ΔS_{env} of the environment positive, negative, or zero? Or do you not have enough information to tell?
   c. Is $\Delta S_{universe}$ of the universe positive, negative, or zero? Or do you not have enough information to tell?

Problem difficulty is labeled as | (straightforward) to ||||| (challenging). Problems labeled INT integrate significant material from earlier chapters; BIO are of biological or medical interest; CALC require calculus to solve.

6. A hot metal sphere is dropped into a beaker of cold liquid inside an insulated container. The metal and the liquid quickly reach a common final temperature. Let the metal and the beaker of liquid be the system. In this process,
   a. Does the energy of the system increase, decrease, or stay the same?
   b. Does the entropy of the system increase, decrease, or stay the same?
7. Copper blocks 1 and 2 have the same mass. Block 1 is heated from 300 K to 350 K. Block 2 is heated from 400 K to 450 K.
   a. Is the heat Q_1 required to heat block 1 larger than, smaller than, or the same as the heat Q_2 needed to heat block 2?
   b. Is the entropy increase ΔS_1 of block 1 larger than, smaller than, or the same as the entropy increase ΔS_2 of block 2?
8. A reaction is spontaneous at higher temperatures but not at lower temperatures. Is the reaction exothermic or endothermic? Explain.
9. Reactions 1 and 2 occur at the same temperature with equal numbers of moles of reactants. Both reactions go to completion. Reaction 1 has $\Delta G = -20$ kJ; reaction 2 has $\Delta G = -40$ kJ. Does reaction 1 go to completion faster or slower than reaction 2? Or do you not have enough information to tell?
10. When red blood cells are placed in an unknown liquid, the cells soon shrivel. Is the salt concentration in the liquid higher or lower than the salt concentration inside the cells? Explain.

Multiple-Choice Questions

11. | Which of the following does not involve an increase of entropy of the specified system?
   A. Wax melting (the system is the wax)
   B. Sugar dissolving in water (the system is sugar + water)
   C. Water condensing on the side of a cold glass (the system is the water)
   D. Dry ice sublimating (the system is dry ice + vapor)
12. | A system is in thermal contact with its environment, and together they form an isolated system that undergoes an irreversible process. Consider these situations:
   I. The entropy of the system increases by 10 J/K; the entropy of the environment decreases by 10 J/K.
   II. The entropy of the system increases by 10 J/K; the entropy of the environment decreases by 5 J/K.
   III. The entropy of the system decreases by 5 J/K; the entropy of the environment increases by 10 J/K.
   IV. The entropy of the system increases by 5 J/K; the entropy of the environment decreases by 10 J/K.

   Which of these situations are physically possible?
   A. I only B. IV only
   C. II and III D. I, II, and IV
   E. I, II, and III
13. | 1 L of water is heated by 10°C. The largest entropy increase will occur if the initial temperature is
   A. 0°C
   B. 50°C
   C. 90°C
   D. The entropy increase is the same for all three.
14. || The standard entropies of formation for $N_2(g)$, $O_2(g)$, and NO(g) are 192 J/mol · K, 205 J/mol · K, and 211 J/mol · K, respectively. What is the entropy change if 1 mol of nitrogen and 1 mol of oxygen undergo the reaction $N_2(g) + O_2(g) \rightarrow 2NO(g)$?
   A. 819 J/K B. 608 J/K C. 25 J/K
   D. −25 J/K E. −186 J/K
15. | Carbon dioxide crosses the cell membrane by
   BIO
   A. Osmosis.
   B. Diffusion.
   C. Both osmosis and diffusion.
   D. Active transport.
16. | Osmosis is driven by
   A. Entropy.
   B. Diffusion.
   C. Temperature differences.
   D. Pressure differences.
17. || Suppose a reaction has $\Delta H = -30$ kJ/mol and $\Delta S = -60$ J/mol · K. For what temperatures is the reaction spontaneous?
   A. Below 500 K B. Above 500 K
   C. Above 0.5 K D. All temperatures
18. || A chemical reaction run at 60°C has $\Delta H = -40$ kJ and $\Delta S = 30$ J/K. What is the maximum useful work this reaction can do?
   A. 30 kJ
   B. 42 kJ
   C. 50 kJ
   D. This reaction can't do useful work.
19. | An endothermic reaction is spontaneous
   A. Always.
   B. Never.
   C. If there's a sufficient increase of entropy.
   D. If there's a sufficient decrease of entropy.

PROBLEMS

Section 14.2 Microstate, Multiplicity, and Entropy

1. || A tetrahedron has four faces numbered 1 to 4. If you roll a tetrahedron, the value of the roll is the number on the bottom face. Suppose you roll two tetrahedrons. Each possible outcome is a microstate. Let the macrostate of the two tetrahedrons be the sum of the two faces.
   a. How many macrostates are there?
   b. What is the multiplicity of the most likely macrostate?
2. || Suppose you roll two six-sided dice. Let the macrostate S of the two dice be the sum of their top faces. What are the multiplicities of the states (a) S = 2 and (b) S = 7?
3. || A box that contains five particles, labeled 1 to 5, is divided into equal halves. The particles are equally likely to be anywhere in the box. The macrostate of the box is defined by the number of particles N_L that are in the left half of the box.
   a. How many macrostates are there?
   b. What is the multiplicity of the macrostate $N_L = 0$?
   c. What is the multiplicity of the macrostate $N_L = 1$?
4. || What is the multiplicity of a state that has entropy 1.0 J/K? Give your answer as a power of 10.
5. || A container has 1000 cells that can hold molecules.
   a. What is the entropy as a multiple of k_B if the container holds one molecule, two molecules, and three molecules?
   b. How does entropy scale with the number of molecules N? That is, entropy is proportional to what power of N?

6. ||| A system at a temperature of 300 K has an entropy of 100 J/K. It undergoes an isothermal process that increases the entropy to 101 J/K.
   a. How much heat is required for this process?
   b. By what *factor* does the system's multiplicity increase? Give your answer as a power of 10.
7. || A gas confined to a container of volume V has 5.0×10^{22} molecules. If the volume of the container is doubled while the temperature remains constant, by how much does the entropy of the gas increase?

Section 14.3 Using Entropy

8. | What is the entropy change when 50 g of nitrogen gas at the boiling-point temperature condense to liquid nitrogen?
9. ||| What is the entropy change upon heating 200 g of ice at $-50°C$ to steam at 100°C?
10. | Freon-12 (CCl_2F_2) was once used as the coolant in refrigerators and air conditioners. Its boiling point is $-29.8°C$ and its heat of vaporization is 20.2 kJ/mol. What is the change in entropy as 2.0 mol of Freon-12 is transformed from liquid into vapor?
11. | Suppose you take a deep breath on a cold day, bringing in 2.0 L of $-10°C$ air at a pressure of 1.0 atm. The air, once in your lungs, expands at roughly constant pressure as it warms up to 37°C. What is the air's change in entropy? BIO
12. || 3.0 mol of argon gas are sealed in a container at 1.0 atm. The container is then heated until the pressure triples. What is the change in entropy of the gas?
13. || 1.2 mol of a monatomic ideal gas are at a temperature of 300 K and a pressure of 1.0 atm. What is the entropy change in a process that brings the gas to 350 K and 1.3 atm?
14. || A system consists of a 1.0 kg block of copper at 20°C and 1.5 kg block of copper at 300°C. The two blocks, isolated from the environment, are placed in contact until thermal equilibrium is reached. What is the change in entropy of the system? The specific heat of copper is 385 J/kg · K. INT
15. || A container is divided into two equal volumes by a very thin membrane. It is held at a fixed temperature by immersion in a water bath. The left half holds 1.0 mol of gas, and the right half holds 2.0 mol of the same gas. The membrane is then punctured so that gas can flow freely between the two sides. What is the change in the entropy of the system after the gas has come to equilibrium?
16. ||| What is the ratio of the multiplicity of water vapor at 100°C to that of liquid water at 100°C for 1.0 g of water at a pressure of 1.0 atm? Give your answer as a power of 10.
17. || The specific heat of liquid water is 4190 J/kg · K.
   a. What is the molar specific heat of liquid water in J/mol · K? Give your answer as a multiple of R, rounded to the nearest half integer.
   b. Equal moles of liquid water and helium gas are heated at constant pressure from the same initial temperature to the same final temperature. By what factor is the entropy increase of the water larger than the entropy increase of the helium?
18. || A 1.0 kg block of ice with a temperature of $-10°C$ is placed on a large stone slab with a temperature of $+10°C$. The stone slab is so large that its temperature does not change. The ice and the slab are isolated from the rest of the universe. What are (a) ΔS_{ice}, (b) ΔS_{stone}, and (c) ΔS_{tot} as the system comes to equilibrium?
19. || A container with 2.0 mol of oxygen gas at 300 K is placed on a large stone slab with a temperature of 600 K. The stone slab is so large that its temperature does not change. The gas is heated at constant pressure until it reaches equilibrium. What are (a) ΔS_{gas}, (b) ΔS_{stone}, and (c) ΔS_{tot} as the system comes to equilibrium? Assume that the entropy change of the container itself is negligible.
20. || 8.0 g of helium gas undergo an irreversible free expansion from a 2.0 L container at 20°C into an 8.0 L insulated container at 100°C. What is the change in entropy?

Section 14.4 Spontaneity and Gibbs Free Energy

21. | Hot object A and cold object B are allowed to interact with each other but are isolated from the rest of the universe. As they interact, are the following positive, negative, or zero: (a) ΔH_{sys}, (b) ΔS_{sys}, (c) ΔG_{sys}?
22. || An important atmospheric chemistry process in the production of smog is the reaction of gaseous nitric oxide (NO) with oxygen to form nitrogen dioxide (NO_2) via the reaction $2NO(g) + O_2(g) \rightarrow 2NO_2(g)$. In this reaction, the enthalpy change is -117 kJ/mol and the entropy change is -153 J/mol · K. Below what temperature in °C is this reaction spontaneous? Assume that the enthalpy and entropy changes are independent of temperature.
23. || The commercial production of fertilizer depends on the Haber-Bosch process for producing ammonia (NH_3). The reaction is $N_2(g) + 3H_2(g) \rightarrow 2NH_3(g)$, with $\Delta H° = -92$ kJ/mol at the standard state of 298 K and 1 atm. The standard molar entropies of nitrogen, hydrogen, and ammonia are 192 J/mol · K, 131 J/mol · K, and 193 J/mol · K, respectively. What is the highest temperature for which this reaction is spontaneous at 1 atm pressure? Assume that ΔH and ΔS are independent of temperature. In fact, the process is too slow to be practical under these conditions; the discovery by Fritz Haber that makes this an important industrial process is the use of a catalyst and very high pressures, typically 250 atm.
24. || The decomposition of nitrogen pentoxide in the reaction $2N_2O_5(s) \rightarrow 4NO_2(g) + O_2(g)$ is endothermic with $\Delta H = +110$ kJ/mol. Even so, this reaction is spontaneous for temperatures above 230 K because the change of a solid into a gas is a large increase of entropy.
   a. What is ΔS in J/mol · K for this reaction?
   b. What is ΔG in kJ/mol at 20°C?
25. || The standard enthalpy of formation and standard molar entropy of carbon in its diamond form are 1.9 kJ/mol and 2.38 J/mol · K, respectively. For carbon in its graphite form the values are 0 kJ/mol and 5.74 J/mol · K.
   a. What is ΔG for the reaction C(s, diamond) $\rightarrow$ C(s, graphite) in which diamond changes to graphite, the material of pencil lead, at 25°C?
   b. Is this a spontaneous process?
26. || The rusting of iron is the oxidation reaction $4Fe + 3O_2 \rightarrow 2Fe_2O_3$. The enthalpies of formation are zero for each of the reactants, which are in their standard states, and -824 kJ/mol for the iron oxide. The standard molar entropies of iron, oxygen, and iron oxide are 27 J/mol · K, 205 J/mol · K, and 87 J/mol · K, respectively. What is ΔG in kJ/mol for this reaction at 25°C?

Section 14.5 Doing Useful Work

27. || A spontaneous exothermic reaction at a constant temperature and pressure of 50°C and 2.0 atm has an enthalpy decrease of 350 kJ. This reaction performs 150 kJ of useful work, which is 50% of the maximum work that could be done. What is the reaction's change of entropy?

28. || Propane (C_3H_8) is a widely used fuel. The heat of combustion of propane is –2220 kJ/mol, and the entropy change of the combustion reaction is –375 J/K · mol. A gas grill for cooking uses a cylinder with 20 lb (9.1 kg) of propane. If the combustion of propane is used to do useful work rather than heating food, what is the maximum useful work that could be done by one 20 lb propane cylinder?

29. | The decomposition of calcium carbonate, $CaCO_3(s) \rightarrow CaO(s) + CO_2(g)$, has $\Delta G = 130$ J/mol at room temperature and does not occur spontaneously. The reaction does become spontaneous at very high temperatures, so it can be driven by being coupled to a combustion reaction that raises the temperature. For example, the combustion of methane, $CH_4 + 2O_2 \rightarrow CO_2 + 2H_2O$, has $\Delta G = -818$ kJ/mol. How many moles of calcium carbonate can be decomposed by the burning of 1.0 mol of methane?

30. ||| The combustion of methane is the reaction $CH_4(g) + 2O_2(g) \rightarrow CO_2(g) + 2H_2O(l)$. Assume that the reaction is carried out at the standard state of 25°C and 1 atm pressure. Data on enthalpies of formation and standard molar entropies of the reactants and products are given in the table. If 32 g of methane are burned, what is the magnitude of the heat flow to the environment if (a) no useful work is done and (b) maximum useful work is done?

	H_f° (kJ/mol)	S° (J/mol · K)
$CH_4(g)$	–74.6	186.3
$O_2(g)$	0	205.2
$CO_2(g)$	–393.5	213.8
$H_2O(l)$	–285.8	70.0

Section 14.6 Using Gibbs Free Energy

31. | Figure P14.31 shows how the Gibbs free energy changes with the extent of the reaction $C_4H_8 \rightleftharpoons C_4H_6 + H_2$, known as the dehydrogenation of 1-butene, at a temperature of 780 K. What is ΔG as the reaction proceeds to equilibrium if the starting point is (a) 2.0 mol of C_4H_8, (b) 2.0 mol each of C_4H_6 and H_2, and (c) 1.0 mol each of C_4H_8, C_4H_6, and H_2?

FIGURE P14.31

32. | BIO Alanine and glycine are two of the amino acids that are used to form proteins. The reaction to connect an alanine molecule to a glycine molecule, to form a peptide bond, has $\Delta G = 29$ kJ/mol at 37°C.
    a. Do these two amino acids bond spontaneously?
    b. In the body, this bonding reaction is coupled to the hydrolysis of one ATP molecule, a reaction that has $\Delta G = -57$ kJ/mol. What is ΔG_{tot} for the coupled reaction?

Section 14.7 Mixing and Osmosis

33. || BIO The osmotic pressure in tree roots can reach 19 atm. What molarity of a tree sap solution would cause this pressure on a day when the root temperature is 15°C?

34. || What is the osmotic pressure in atm of a solution made by dissolving 25 g of sucrose ($C_{12}H_{22}O_{11}$, molar mass 0.342 mol/kg) in enough water to make 250 mL of solution at 50°C?

35. || Fresh water can be squeezed out of seawater, through a semipermeable membrane, by using a piston to raise the pressure on the seawater side of the membrane to a value larger than the osmotic pressure. To achieve a sufficient flow of fresh water, the operating pressure is typically twice the osmotic pressure. The membrane has to be able to withstand this pressure difference and also not become clogged too rapidly with any sediment or microorganisms that are too small to be filtered out of the seawater. The density of seawater is 1030 kg/m^3, and a typical salinity is 3.5% NaCl by weight. Note that 1 mol of NaCl dissociates to 2 mol of solute particles in solution. Assume that the temperature is 20°C. What operating pressure in atm is needed by a reverse osmosis system?

36. || Where freshwater rivers meet the ocean, it's possible to generate electricity with a system that separates the fresh water and salt water with a semipermeable membrane that allows water molecules to pass but not dissolved ions. The resulting osmotic pressure can be used to drive water through a turbine. The density of seawater is 1030 kg/m^3, and its salinity is 3.5% NaCl by weight. Note that 1 mol of NaCl dissociates to 2 mol of solute particles in solution. Assume that the temperatures are 20°C. What pressure difference is possible?

37. || BIO When a dehydrated patient is given an intravenous saline solution, this solution must be matched to the osmotic pressure of red blood cells, which is approximately 8.0 atm. To make such a solution, how many grams of NaCl should be added to 1.0 L of water? Note that 1 mol of NaCl dissociates to 2 mol of solute particles in solution.

38. || BIO If cells are placed in a 150 mol/m^3 solution of sodium chloride (NaCl) at 37°C, there is no osmotic pressure difference across the cell membrane. What will be the pressure difference across the cell membrane if the cells are placed in pure water at 20°C? Note that 1 mol of NaCl dissociates to 2 mol of solute particles in solution.

39. ||| BIO INT Figure P14.39 shows two compartments separated by a semipermeable membrane that is permeable to the solvent water but not to the solute. The concentration of the solute is 140 mM on the right side of the membrane and 60 mM on the left side.

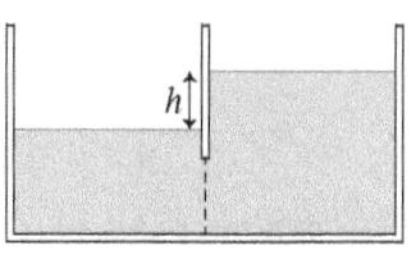

FIGURE P14.39

    a. What is the height difference h between the two water surfaces at a temperature of 20°C?
    b. Suppose that the membrane is suddenly made permeable to the solute, as can happen in cells when voltage-gated ion channels open, and that the diffusion constant for the solute through the membrane is 5.0×10^{-10} m^2/s. What is the initial rate of flow of the solute in molecules/s through 10 μm^2 of a 7.0-nm-thick membrane? These values are comparable to the surface area and thickness of a cell membrane.

General Problems

40. || Suppose you toss 100 coins.
   a. What is the number of microstates?
   b. What is the number of macrostates?
   c. What is the multiplicity of the H = 1 macrostate, which has 1 head and 99 tails?
   d. What is the entropy of the H = 1 macrostate as a multiple of k_B?

41. || Suppose systems A and B each have 25 bins, and each bin can store one unit of energy. If a system has one unit of energy, its multiplicity is $\Omega = 25$ because the unit of energy can go into any of the 25 bins. If there are two units of energy, the second to arrive has only 24 choices of bins and so the multiplicity is $\Omega = 25 \times 24$. With three units of energy, the multiplicity is $\Omega = 25 \times 24 \times 23$, and so on. The entropy of a state is $S = k_B \ln \Omega$, but if we measure entropy in multiples of k_B, which we'll do in this problem, the entropy is simply $S = \ln \Omega$.
   a. Suppose that A and B between them have 10 units of energy to share. What are S_A, S_B, and S_{tot} if $E_A = 8$ and $E_B = 2$? Calculate entropies to two decimal places for this problem.
   b. What are ΔS_A, ΔS_B, and ΔS_{tot} if an interaction between A and B reduces E_A to 7? Is this process spontaneous?
   c. What is the maximum value of S_{tot}? What are the values of E_A and E_B that correspond to this equilibrium state?

42. | Your body temperature is a fairly constant 37°C. Suppose
BIO that you are at rest in a room with an air temperature of 20°C and that your metabolic rate is 125 W. By how much does your metabolism increase the entropy of the room over 1 hour?

43. || A container holds 2.0 mol of argon gas at a constant pressure of 1.0 atm. The initial temperature is 300 K. Suppose the gas is slowly heated until its entropy has changed by 12 J/K. What is the final temperature of the argon?

44. || Ideal-gas processes are often shown on a *pV diagram* that graphs pressure versus volume. Figure P14.44 is the *pV* diagram for 2.0 mol of a monatomic gas that undergo a process taking them from initial state i to final state f. What are (a) the change in temperature and (b) the change in entropy?

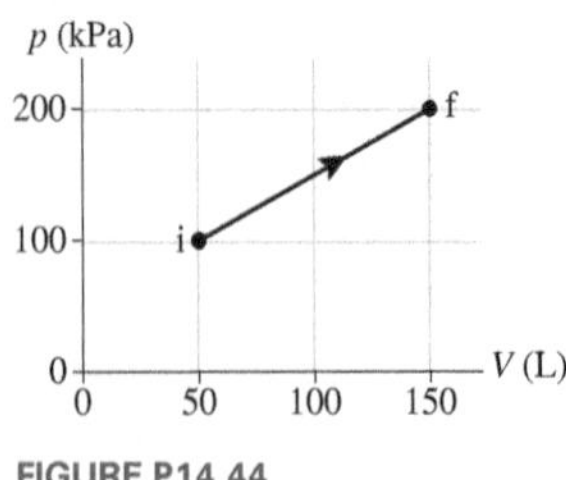

FIGURE P14.44

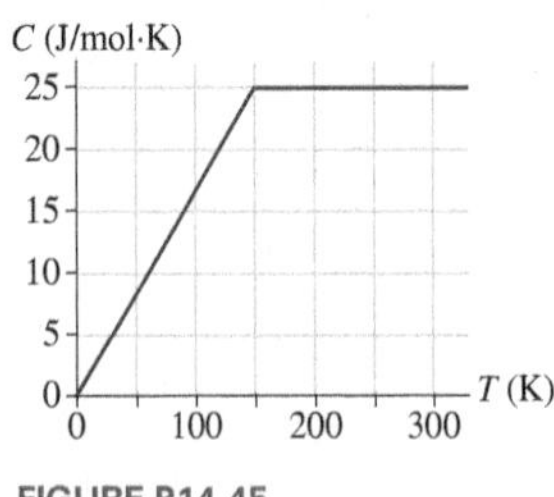

FIGURE P14.45

45. || Figure P14.45 shows a simplified graph of the molar specific
CALC heat of silver. What is the standard molar entropy of silver?

46. || At extremely low temperatures, the specific heats of metals
CALC are proportional to the absolute temperature T. For example, the specific heat of copper is $c = (0.011\ \text{J/kg} \cdot \text{K}^2)T$ for temperatures below 2 K. What is the entropy change of 150 g of copper as it is heated from 0.5 K to 2.0 K?

47. || At extremely low temperatures, the specific heats of metals
CALC are proportional to the absolute temperature T. For example, the molar specific heat of platinum is $C = (6.5\ \text{mJ/mol} \cdot \text{K}^2)T$ for temperatures below 2 K. What is the molar entropy of platinum at 1.0 K?

48. ||| The molar heat capacity of diamond changes with tempera-
CALC ture, but for temperatures lower than 500 K it is reasonably well modeled as being proportional to T^3; that is, $C = aT^3$, where a is a constant. At the standard state, 25°C, the measured specific heat of diamond is 520 J/kg · K.
   a. What is the molar specific heat in J/mol · K of diamond at 25°C?
   b. Use this model to predict the standard molar entropy of diamond. The accepted value in reference tables is 2.4 J/mol · K, which indicates that this model is reasonably good but not perfect.

49. || Two containers, one with a volume of 1.0 L and the other with a volume of 2.0 L, are separated by a closed valve. The 1.0 L container holds 2.0 mol of helium gas; the 2.0 L container holds 1.0 mol of neon gas. If the valve is opened, what is ΔS_{mix}, the entropy of mixing?

50. ||| An instant *hot pack* is a plastic bag with 15 g of the salt calcium chloride ($CaCl_2$) and an inner bag filled with 200 g of water. Squeezing the pack to break the inner bag causes the calcium chloride to dissolve in the water, which quickly increases the water temperature. The dissociation of the ionic compound is $CaCl_2(s) \xrightarrow{H_2O} Ca^{2+}(aq) + 2Cl^-(aq)$. Assume the reaction begins at 25°C and 1 atm pressure. Calcium chloride has a molar mass of 0.111 kg/mol. The specific heat of the salt solution is 3800 J/kg · K. Data on the enthalpies of formation and standard molar entropies of the salt and the ions are listed here:

	H_f° (kJ/mol)	S° (J/mol · K)
$CaCl_2(s)$	–365.6	151.1
$Ca^{2+}(aq)$	–133.3	113.4
$Cl^-(aq)$	–206.9	146.4

   a. What temperature will the water reach?
   b. What is the change in the Gibbs free energy?

51. || Consider the formation of ammonium chloride (NH_4Cl) in its solid form from the gaseous reactants nitrogen, hydrogen, and chlorine according to the reaction $\frac{1}{2}N_2(g) + 2H_2(g) + \frac{1}{2}Cl_2(g) \rightarrow NH_4Cl(s)$. Data on the enthalpies of formation and standard molar entropies for each substance are listed here:

	H_f° (kJ/mol)	S° (J/mol · K)
$N_2(g)$	0	191.6
$H_2(g)$	0	130.7
$Cl_2(g)$	0	223.1
$NH_4Cl(s)$	–314.6	94.9

   a. What is the change in the Gibbs free energy during this reaction at the standard state of 25°C and 1 atm pressure?
   b. Can the reaction occur spontaneously under these conditions?

52. ||| The sap that comes from sugar maple trees is about 2% sucrose ($C_{12}H_{22}O_{11}$) and 98% water. The sap is then concentrated to create the thick maple syrup you pour on pancakes. In the traditional method, boiling concentrated the sap. Today, many producers start by using reverse osmosis. The kit for a small producer runs at a pressure of 480 kPa. With 25°C sap entering at 2.0% sucrose, what is the maximum concentration of the resulting concentrate?

53. ||| The yeast cells used in beer fermentation excrete ethanol BIO (ethyl alcohol), C_2H_6O, as a waste product. The yeast cells find themselves in a hypertonic ethanol-water solution as the ethanol concentration increases. The resulting osmotic pressure across the cell membrane creates what's called *ethanol stress,* causing the yeast metabolism to shut down. This stress limits the ethanol concentration that can be produced by simple fermentation. The ethanol concentration of beer is typically 5.0% by volume. What is the osmotic pressure in atm across the cell membrane at this concentration at a fermentation temperature of 20°C? The ethanol concentration inside a yeast cell is essentially zero because ethanol is actively removed from the cell as it is produced.

54. ||| A solution that contains 30 g of protein per liter has an os- BIO motic pressure of 9.5 mm Hg at 25°C. What is the molecular mass of the protein in kDa?

55. || Creatures that live in fresh water have blood and body fluids BIO that are saltier than the water in which they swim. This leads to water entering the body by osmosis that must be excreted in the urine, which is therefore less salty than the blood. A freshwater crayfish has 0.42 mol of dissolved salts per liter of blood, compared to 0.12 mol of dissolved salts per liter of urine. What is the osmotic pressure difference at 25°C?

MCAT-Style Passage Problems

Leaky Capillaries

The walls of the capillaries in your circulatory system are thin enough to permit the diffusion of gases and nutrients. The walls are also somewhat leaky; a significant movement of fluid into and out of the capillaries is driven by differences in pressure. Two competing processes are at work. First, the blood pressure inside the capillaries is higher than the pressure in the surrounding tissue; this pressure difference drives fluid out of the capillaries. Second, large molecules in the blood that cannot pass through the walls of the capillaries (the most important of which is a family of proteins called *albumin*) lead to an osmotic pressure that drives fluid back into the capillaries. If we consider a single capillary, illustrated in Figure P14.56, a general pattern emerges. Osmotic pressure is constant along the length of the capillary, but blood pressure decreases with distance along the direction of flow. The blood pressure exceeds the osmotic pressure at the start of the capillary, so there is a net motion of fluid out of the capillary. At the end of the capillary, the osmotic pressure is higher than the reduced blood pressure, so there is a net motion of fluid into the capillary. Under normal circumstances the volume of fluid leaving the capillary is nearly balanced by the fluid entering.

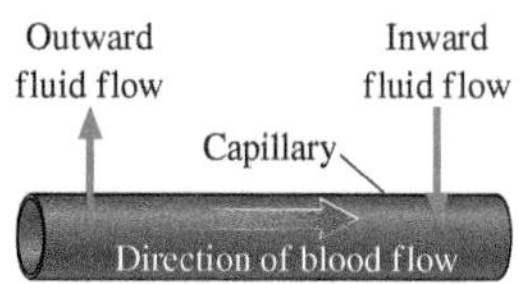

FIGURE P14.56

56. If the concentration of albumin in a capillary decreases,
    A. The outward flow will exceed the inward flow.
    B. The outward and inward flows will remain balanced.
    C. The outward flow will be less than the inward flow.

57. If a capillary narrows but the pressure at the start of the capillary stays the same,
    A. The outward flow will exceed the inward flow.
    B. The outward and inward flows will remain balanced.
    C. The outward flow will be less than the inward flow.

58. If the viscosity of the blood increases,
    A. The outward flow will exceed the inward flow.
    B. The outward and inward flows will remain balanced.
    C. The outward flow will be less than the inward flow.

PART II SUMMARY

Energy and Thermodynamics

FUNDAMENTAL CONCEPTS

Energy is conserved, but it can be transformed or transferred in mechanical (work) or thermal (heat) interactions.

Macroscopic properties of matter can be understood in terms of microscopic atomic-level collisions and interactions.

Macroscopic processes are irreversible; the entropy of the universe always increases.

GENERAL PRINCIPLES

Energy principle	$\Delta E_{sys} = \Delta K + \Delta U + \Delta E_{th} + \Delta E_{chem} = W + Q$
Conservation of energy	For an isolated system, the total energy $E_{sys} = K + U + E_{int}$ is conserved. For an isolated, nondissipative system, the mechanical energy $E_{mech} = K + U$ is conserved.
First law of thermodynamics	$\Delta E_{int} = W + Q$ Energy is conserved.
Second law of thermodynamics	$\Delta S_{universe} \geq 0$ Energy is dispersed.
Spontaneity	A process will occur spontaneously only if $\Delta G_{sys} < 0$.

SYSTEM AND ENVIRONMENT

Energy is a property of the system.

Energy is transformed within the system or transferred between the system and its environment.

Work is a mechanical transfer of energy by forces.

Heat is a thermal transfer of energy when there's a temperature difference.

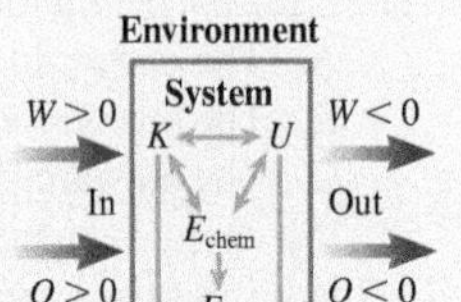

PHASES OF MATTER

Solid: Rigid, nearly incompressible.

Liquid: Molecules are loosely bound but flow. Nearly incompressible.

Gas: Noninteracting particles are freely moving. Highly compressible.

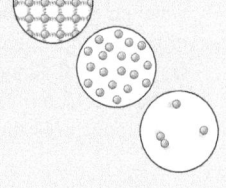

- Thermal expansion: $\Delta L/L = \alpha\,\Delta T$
- Heating: $Q = Mc\,\Delta T$
- Phase change: $Q = \pm ML$

ENERGY AND MATTER FLOWS

Systems exchange energy and matter.

- Thermal conductivity: $dQ/dt = kA(\Delta T/L)$
- Radiation: $dQ/dt = e\sigma AT^4$
- Diffusion: $dN/dt = DA(\Delta c/L)$
- Osmotic pressure: $\Pi = C_{solute}RT$

IDEAL GASES

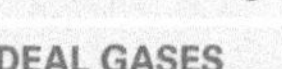

The ideal-gas law is $pV = nRT$ or $pV = Nk_BT$

- Constant temperature: $\Delta E_{th} = 0,\ W = -nRT\ln(V_f/V_i)$
- Constant pressure: $Q_P = nC_P\Delta T \quad W = -p\Delta V$
- Constant volume: $Q_V = nC_V\Delta T \quad W = 0$
- Molar specific heats: $C_P = C_V + R$

ENTROPY

Entropy measures the multiplicity of a macrostate: $S = k_B \ln \Omega$

- Constant temperature: $\Delta S = Q/T$
- Heating: $\Delta S = Mc\ln(T_f/T_i)$
- Ideal gas: $\Delta S = nC_V\ln(T_f/T_i) + nR\ln(V_f/V_i)$

ENTHALPY AND GIBBS FREE ENERGY

Enthalpy: $H = E_{int} + pV$ Gibbs free energy: $G = H - TS$

At constant pressure, $\Delta H = Q_P$ is the heat released in a reaction.

At constant pressure and temperature, $\Delta G = \Delta H - T\Delta S$.

A spontaneous process $\Delta G < 0$ can be enthalpy or entropy driven.

ENERGY

- Kinetic energy: $K = \frac{1}{2}mv^2$
- Gravitational potential energy: $U_G = mgy$
- Elastic potential energy: $U_{sp} = \frac{1}{2}k(\Delta x)^2$
- Thermal energy of atomic motion: E_{th}
- Chemical energy of molecular bonds: E_{chem}

MOLECULAR BONDS

Bound states have negative energies.

Bond energy is the energy needed to separate a ground-state molecule into free atoms.

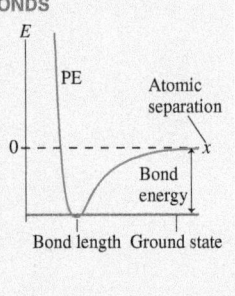

PE, Atomic separation, Bond energy, Bond length, Ground state

ENERGY DIAGRAMS

Energy diagrams are used to visualize speeds and turning points. Minima and maxima are points of stable and unstable equilibrium.

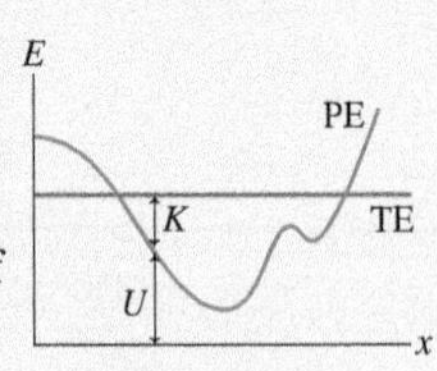

CHEMICAL REACTIONS

Reactions occur when molecules collide with enough energy to break bonds. Energy is released when new, tighter bonds are formed.

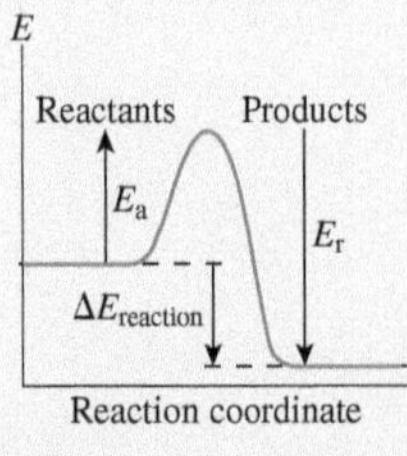

The Arrhenius equation for how a reaction rate changes with temperature is

$$k = Ae^{-E_a/k_BT}$$

WORK

Work is done by a force on a particle:

$$W = \int_{x_i}^{x_f} F_x\,dx$$

- Constant force: $W = F_x\Delta x = F(\Delta x)\cos\theta$
- Spring: $W_{sp} = \frac{1}{2}k(\Delta x_i)^2 - \frac{1}{2}k(\Delta x_f)^2$

KINETIC THEORY

Pressure is due to collisions of molecules with each other and the walls.

- Mean free path: $\lambda = 1/(4\sqrt{2}\pi(N/V)r^2)$
- rms speed: $v_{rms} = \sqrt{3k_BT/m}$

Equipartition theorem: The energy stored in each mode is $\frac{1}{2}Nk_BT = \frac{1}{2}nRT$.

Boltzmann factor: The number of molecules having energy E_i decreases exponentially as

$$N_i \propto e^{-E_i/k_BT}$$

Entropy and the Living World

Your body is a complex collection of systems, a pocket of order in a disorderly world. This might seem to violate the law that entropy always increases (How could such an orderly structure come about in a world where entropy is always on the rise?) but, as we have seen, it does not.

Over the course of a day, you take in about 8 million joules of high-quality chemical energy in the food you eat. From the start of one day to the start of the next, your temperature stays about the same, and your kinetic energy and potential energy don't change. This means that you must exhaust 8 million joules of thermal energy into the environment, which results in an overall increase in entropy. It seems that the process of living increases the entropy of the universe. This is true of other creatures as well; the process of living increases the entropy of the universe.

We've spoken of increasing entropy as the spreading of energy. As time goes on, heat flows from hot to cold, smoothing out temperature differences. Life can help this process along. The photo of a landscape, with a visible light image on the left and a thermal image on the right, shows that where there is mixed forest, the temperature is approximately constant. The thermal energy has spread out until it is nearly uniform. This isn't the case where the trees have been cut for roads and structures; there are hot spots and cold spots. Looking at the picture through the lens of entropy, we see that the entropy is higher, the energy is more spread out, where the ecosystem is most healthy. In fact, the variation of temperature is a measure of the health of an ecosystem. If the entropy is high, with the thermal energy evenly distributed, the ecosystem is thriving, with living creatures filling available niches. Life serves the cause of entropy increase; the more vibrant the ecosystem, the greater the rate of increase.

Though it may seem paradoxical, the spreading of energy can be aided by the existence of ordered systems. On a sunny summer day, puffy cumulus clouds form in a regular pattern. The clouds represent regions of rising air; the clear sky between represents regions of falling air. The orderly pattern of clouds and clear sky leads to a more rapid transfer of energy from the warm surface of the earth to the upper, cooler layers of the atmosphere, reducing temperature differences. The orderly, regular pattern of clouds thus enhances the rate of entropy increase.

Cumulus clouds on a summer afternoon.

Such patterns also arise in living systems. An aerial view of a salt marsh, like the one shown, illustrates the complex pattern of water channels that develops over time. As tides rise and fall, water flows in and out of the marsh. But the flow doesn't carve simple, straight channels; instead, we see a complex network of sinuous, branching channels with no raised area far from water. This pattern is produced by a complex interaction of water, soil, and vegetation; saltmarshes in arid areas with limited vegetation develop much less complex patterns. The living plants in the saltmarsh help their environment develop in a very orderly way, and this orderly pattern leads to a greater mass of plants and the animals that depend on them. The teeming life in a saltmarsh, like all life, increases entropy, and the increased density of life means a greater rate of entropy increase.

Now, let's consider the largest ecosystem we know, the earth. Over time, the earth's average temperature changes only slowly. This implies that the magnitude of the thermal energy radiated by the earth into space must equal the incoming energy from the sun; the energy of the earth system is conserved. But this change from incoming solar radiation to outgoing thermal energy involves an increase in entropy. We know that, at smaller scales, healthier ecosystems produce greater rates of entropy increase. Some researchers speculate that we could use the rate of entropy increase produced by the earth system as a measure of the health of the biosphere, and that this could be an important measure to assess the earth's changing climate.

A thermal image of a landscape.

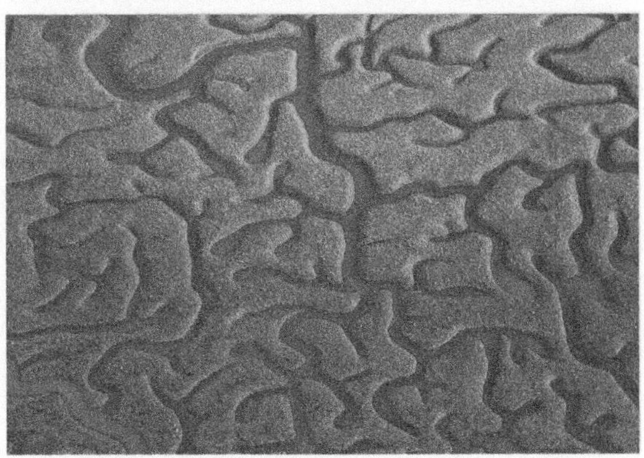

Sinuous channels in a saltmarsh.

PART II PROBLEMS

The following questions are related to the passage "Entropy and the Living World" on the previous page.

1. Living creatures
   A. Reduce the entropy of their environment.
   B. Have no effect on the entropy of their environment.
   C. Increase the entropy of their environment.
2. Over the course of a day you transfer heat to the environment. This reduces the entropy of your body, but it increases the entropy of the environment. Your skin is at a higher temperature than that of the environment. The change in entropy of your body is thus ____ the change in entropy of the environment.
   A. Greater than
   B. The same as
   C. Less than
3. Plants convert some of the energy of sunlight to chemical energy via photosynthesis; the balance goes to increasing thermal energy. If a plant undergoing photosynthesis stays at a constant temperature,
   A. The entropy of the environment increases.
   B. The entropy of the environment stays the same.
   C. The entropy of the environment decreases.
4. The entropy changes of lake systems have been extensively studied; doing so is a good way to monitor the health of a lake. Suppose an investigator measures the daily entropy change in a lake environment as the lake experiences a large algae bloom. Such blooms deplete the oxygen in the lake and cause a die-off of many of the species in the lake. It's likely that the investigator will find
   A. An increase in the daily entropy change, and then a decrease.
   B. A constant daily entropy change.
   C. A decrease in the daily entropy change, and then an increase.
5. Some ecosystems have very uniform vegetation; other ecosystems have alternating patterns of vegetation and open water or soil. Which of the following statements is true?
   A. The ecosystem with uniform vegetation has higher entropy and is healthier.
   B. The ecosystem with patchy vegetation has lower entropy and is healthier.
   C. Both ecosystems could be healthy; different vegetation patterns can lead to a maximum entropy increase in different circumstances.

The following passages and associated questions are based on the material of Part II.

Keeping Your Cool BIO

A 68 kg cyclist is pedaling down the road at 15 km/h, using a total metabolic power of 480 W. A certain fraction of this energy is used to move the bicycle forward, but the balance ends up as thermal energy in his body, which he must get rid of to keep cool. On a very warm day, conduction, convection, and radiation transfer little energy, and so he does this by perspiring, with the evaporation of water taking away the excess thermal energy.

6. If the cyclist reaches his 15 km/h cruising speed by rolling down a hill, what is the approximate height of the hill?
   A. 22 m B. 11 m C. 2 m D. 1 m
7. As he cycles at a constant speed on level ground, at what rate is chemical energy being converted to thermal energy in his body, assuming a typical efficiency of 25% for the conversion of chemical energy to the mechanical energy of motion?
   A. 480 W B. 360 W C. 240 W D. 120 W
8. To keep from overheating, the cyclist must get rid of the excess thermal energy generated in his body. If he cycles at this rate for 2 hours, how many liters of water must he perspire, to the nearest 0.1 liter?
   A. 0.4 L B. 0.9 L C. 1.1 L D. 1.4 L
9. Being able to exhaust this thermal energy is very important. If he isn't able to get rid of any of the excess heat, by how much will the temperature of his body increase in 10 minutes of riding, to the nearest 0.1°C?
   A. 0.3°C B. 0.6°C C. 0.9°C D. 1.2°C

Weather Balloons

The data used to generate weather forecasts are gathered by hundreds of weather balloons launched from sites throughout the world. A typical balloon is made of latex and filled with hydrogen. A packet of sensing instruments (called a *radiosonde*) transmits information back to earth as the balloon rises into the atmosphere. At the beginning of its flight, the average density of the weather balloon package (total mass of the balloon plus cargo divided by their volume) is less than the density of the surrounding air, so the balloon rises. As it does, the density of the surrounding air decreases, as shown in Figure II.1. The balloon will rise to the point at which the buoyant force of the air exactly balances its weight. This would not be very high if the balloon couldn't expand. However, the latex envelope of the balloon is very thin and very stretchy, so the balloon can, and does, expand, allowing the volume to increase by a factor of 100 or more. The expanding balloon displaces an ever-larger volume of the lower-density air, keeping the buoyant force greater than the weight force until the balloon rises to an altitude of 40 km or more.

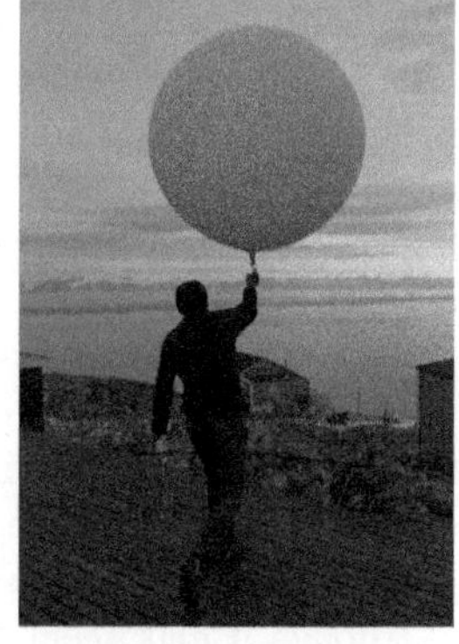

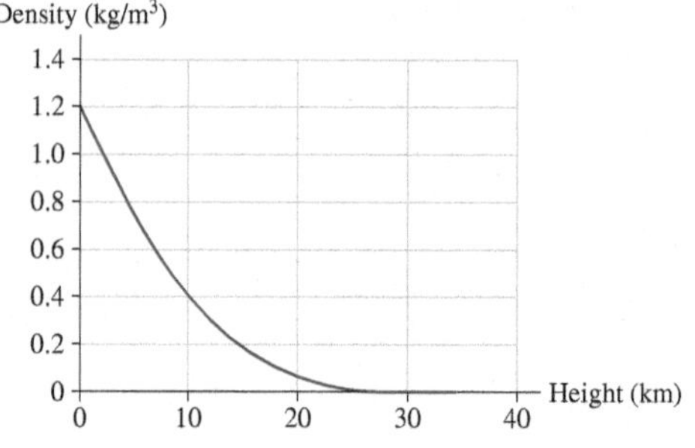

FIGURE II.1

10. A balloon launched from sea level has a volume of approximately 4 m^3. What is the approximate buoyant force on the balloon?
A. 50 N B. 40 N C. 20 N D. 10 N
11. A balloon launched from sea level with a volume of 4 m^3 will have a volume of about 12 m^3 on reaching an altitude of 10 km. What is the approximate buoyant force now?
A. 50 N B. 40 N C. 20 N D. 10 N
12. The balloon expands as it rises, keeping the pressures inside and outside the balloon approximately equal. If the balloon rises slowly, heat transfers will keep the temperature inside the same as the outside air temperature. A balloon with a volume of 4.0 m^3 is launched at sea level, where the atmospheric pressure is 100 kPa and the temperature is 15°C. It then rises slowly to a height of 5500 m, where the pressure is 50 kPa and the temperature is –20°C. What is the volume of the balloon at this altitude?
A. 5.0 m^3 B. 6.0 m^3 C. 7.0 m^3 D. 8.0 m^3
13. If the balloon rises quickly, so that no heat transfer is possible, the temperature inside the balloon will drop as the gas expands. If a 4.0 m^3 balloon is launched at a pressure of 100 kPa and rapidly rises to a point where the pressure is 50 kPa, the volume of the balloon will be
A. Greater than 8.0 m^3.
B. 8.0 m^3.
C. Less than 8.0 m^3.
14. At the end of the flight, the radiosonde is dropped and falls to earth by parachute. Suppose the parachute achieves its terminal speed at a height of 30 km. As it descends into the atmosphere, how does the terminal speed change?
A. It increases.
B. It stays the same.
C. It decreases.

Testing Tennis Balls

Tennis balls are tested by being dropped from a height of 2.5 m onto a concrete floor. The 57 g ball hits the ground, compresses, then rebounds. A ball will be accepted for play if it rebounds to a height of about 1.4 m; it will be rejected if the bounce height is much more or much less than this.

15. Consider the sequence of energy transformations in the bounce. When the dropped ball is motionless on the floor, compressed, and ready to rebound, most of the energy is in the form of
A. Kinetic energy.
B. Gravitational potential energy.
C. Thermal energy.
D. Elastic potential energy.
16. If a ball is "soft," it will spend more time in contact with the floor and won't rebound as high as it is supposed to. The force on the floor of the "soft" ball is ____ the force on the floor of a "normal" ball.
A. Greater than
B. The same as
C. Less than
17. Suppose a ball is dropped from 2.5 m and rebounds to 1.4 m. What is the increase in thermal energy of the ball + earth system?
A. 0 J B. 0.6 J C. 1.2 J D. 2.4 J

Additional Integrated Problems

18. Football players measure their acceleration by seeing how fast they can sprint 40 yards (37 m). A zippy player can, from a standing start, run 40 yards in 4.1 s, reaching a top speed of about 11 m/s. For an 80 kg player, what is the average power output for this sprint?
A. 300 W B. 600 W C. 900 W D. 1200 W
19. The unit of horsepower was defined by considering the power output of a typical horse. Working-horse guidelines in the 1900s called for them to pull with a force equal to 10% of their body weight at a speed of 3.0 mph. For a typical working horse of 1200 lb, what power does this represent in W and in hp?
20. A swift blow with the hand can break a pine board. As the hand hits the board, the kinetic energy of the hand is transformed into elastic potential energy of the bending board; if the board bends far enough, it breaks. Applying a force to the center of a particular pine board deflects the center of the board by a distance that increases in proportion to the force. Ultimately the board breaks at an applied force of 800 N and a deflection of 1.2 cm.
a. To break the board with a blow from the hand, how fast must the hand be moving? Use 0.50 kg for the mass of the hand.
b. If the hand is moving this fast and comes to rest in a distance of 1.2 cm, what is the average force on the hand?
21. As you know, the pressure and density of the atmosphere decrease as you go up in altitude. It's found that the pressure decreases exponentially with height z above sea level as $p = p_0 e^{-z/H}$, where p_0 is the sea-level pressure. The quantity H, called the *scale height* of the atmosphere, is the height at which the pressure has decreased to $1/e = 0.37$ of its sea-level value. Similarly, the density decreases as $\rho = \rho_0 e^{-z/H}$, where ρ_0 is the density at sea level. There is another way to interpret the scale height H: If the atmosphere had a constant density ρ_0, all of the atmosphere would be contained in a uniform slab of thickness H. The total mass of the atmosphere is thus $\rho_0 AH$, where A is the area of the earth's surface. (The thickness of the atmosphere is negligible compared to the radius of the earth, so we can ignore the curvature of the earth's surface.)
a. The average mass of a molecule in the earth's atmosphere is 29 u, and the average temperature of the earth's atmosphere is 288 K. Use the Boltzmann factor to determine the scale height of the earth's atmosphere.
b. Demonstrate the second interpretation of the scale height of the atmosphere by integrating the density of the atmosphere from 0 to ∞. If you multiply this result by A, you will get the mass of the atmosphere.
c. Based on your calculations, what is the total mass of the earth's atmosphere in kg?

PART III Oscillations and Waves

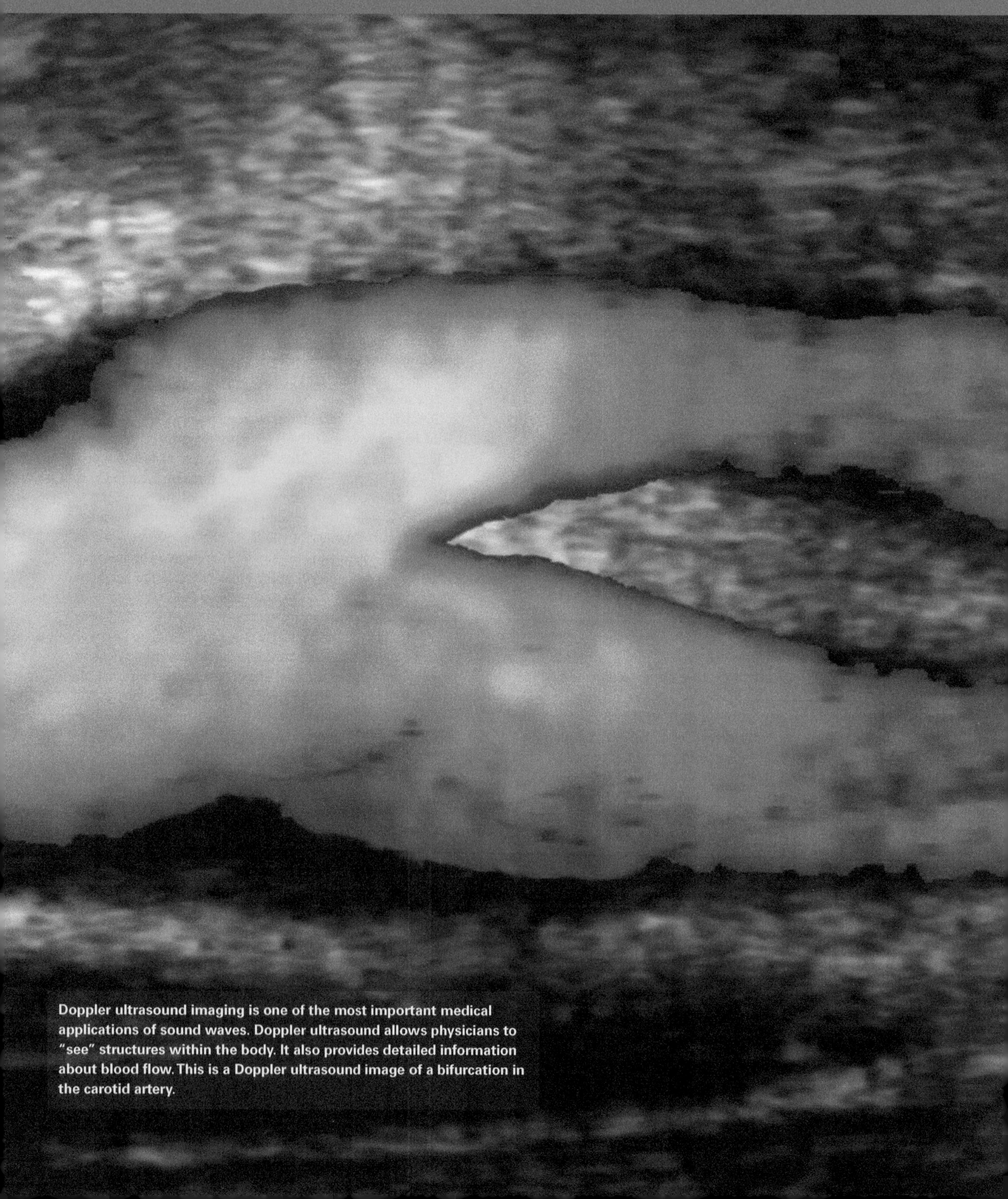

Doppler ultrasound imaging is one of the most important medical applications of sound waves. Doppler ultrasound allows physicians to "see" structures within the body. It also provides detailed information about blood flow. This is a Doppler ultrasound image of a bifurcation in the carotid artery.

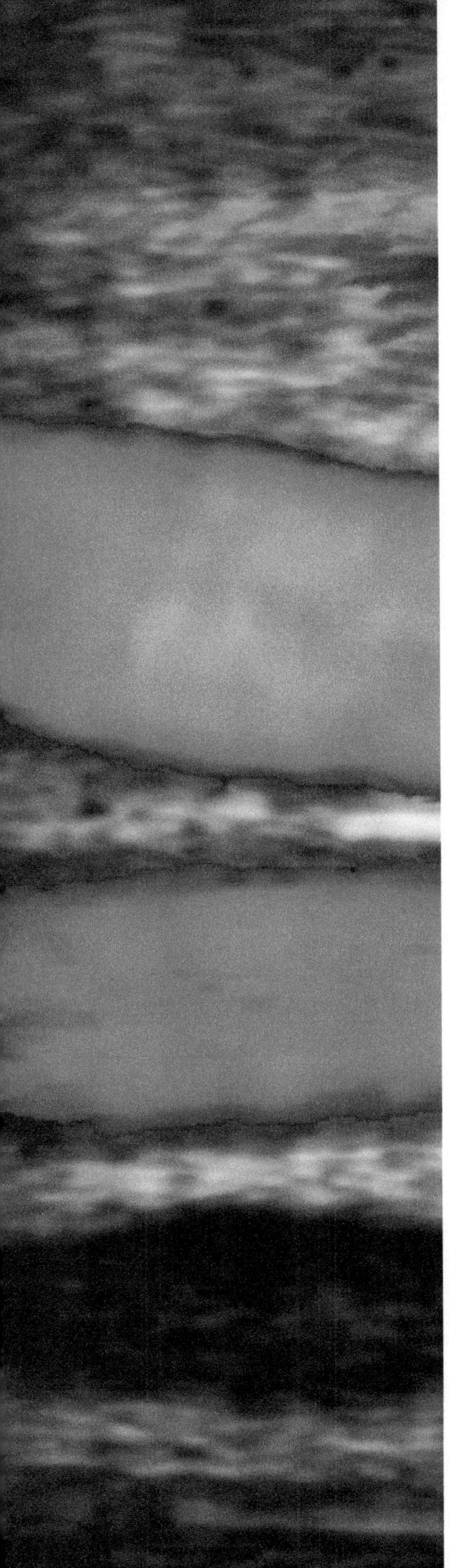

OVERVIEW

Motion That Repeats

Up to this point in the text, we have generally considered processes that have a clear starting and ending point, such as a car accelerating from rest to a final speed, or a solid being heated from an initial to a final temperature. In Part III, we begin to consider processes that are *periodic*—they repeat. A child on a swing, a boat bobbing on the water, and even the repetitive bass beat of a rock song are *oscillations* that happen over and over without a starting or ending point. The *period*, the time for one cycle of the motion, will be a key parameter for us to consider as we look at oscillatory motion.

Our first goal will be to develop the language and tools needed to describe oscillations, ranging from the swinging of the bob of a pendulum clock to the bouncing of a car on its springs. Once we understand oscillations, we will extend our analysis to consider oscillations that travel—*waves*.

The Wave Model

We've had great success modeling the motion of complex objects as the motion of one or more particles. We were even able to explain the macroscopic properties of matter, such as pressure and temperature, in terms of the motion of the atomic particles that comprise all matter.

Now it's time to explore another way of looking at nature, the *wave model*. Familiar examples of waves include

- Ripples on a pond.
- The vibrations of a guitar string.
- The swaying ground of an earthquake.
- The ultrasound echoes of medical imaging.
- The colors of a rainbow.

Despite the great diversity of types and sources of waves, there is a single, elegant physical theory that can describe them all. Our exploration of wave phenomena will call upon water waves, sound waves, and light waves for examples, but our goal will be to emphasize the unity and coherence of the ideas that are common to *all* types of waves. As was the case with the particle model, we will use the wave model to explain a wide range of phenomena.

When Waves Collide

The collision of two particles is a dramatic event. Energy and momentum are transferred as the two particles head off in different directions. Something much gentler happens when two waves come together—the two waves pass through each other unchanged. Where they overlap, we get a *superposition* of the two waves. We will finish our discussion of waves by analyzing the standing waves that result from the superposition of two waves traveling in opposite directions. The physics of standing waves will allow us to understand how your vocal tract can produce such a wide range of sounds, and how your ears are able to analyze them.

15 Oscillations

An electrocardiogram provides information about the mechanical oscillations of the heart.

LOOKING AHEAD

Simple Harmonic Motion

A vibrating guitar string oscillates back and forth in *simple harmonic motion.* The frequency is the same for both large and small amplitudes.

You'll see why an oscillating mass on a spring is a model for all oscillations, and you'll learn how to calculate the frequency and the period.

Pendulums

A swinging pendulum, another example of simple harmonic motion, is one of the oldest technologies for keeping time.

A freely swinging pendulum eventually runs down. You'll see how we can characterize a *damped oscillation* as an exponential decay.

Resonance

You get a large response when you make a system oscillate at its natural frequency. We call this response a *resonance*.

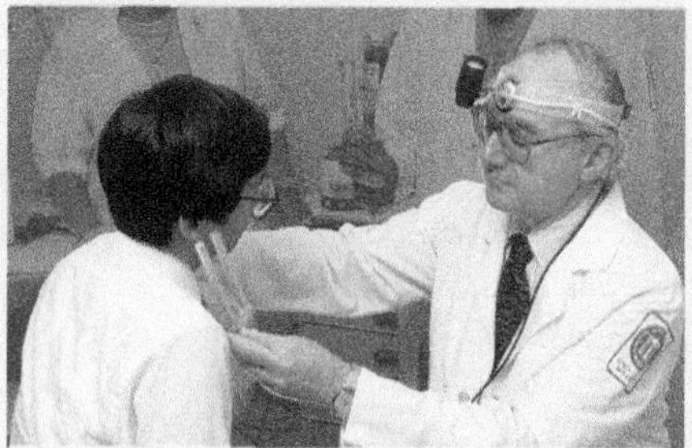

You'll learn how the resonance of a membrane in your inner ear lets you determine the pitch of a musical note.

GOAL To learn about systems that oscillate with simple harmonic motion.

PHYSICS AND LIFE

Many of the Motions of an Organism Are Repetitive

Life is always in motion. Only some of that motion is used by creatures to get from place to place; a great deal of it is repetitive internal motion that plays an essential part in a multicellular organism's survival and perception of its surroundings. Your heart beats, your eardrums vibrate, your arms and legs swing like pendulums, the membranes of your cells stretch and release like ripples on the surface of water, and your body has a circadian rhythm. In addition, every molecular bond in your body acts like a vibrating spring that connects the atoms on either end, vibrations that are an essential part of the ambient thermal energy of the cell. In short, it's impossible to overstate the importance of vibrations, also called *oscillations,* in helping us understand the many different ways that living systems move.

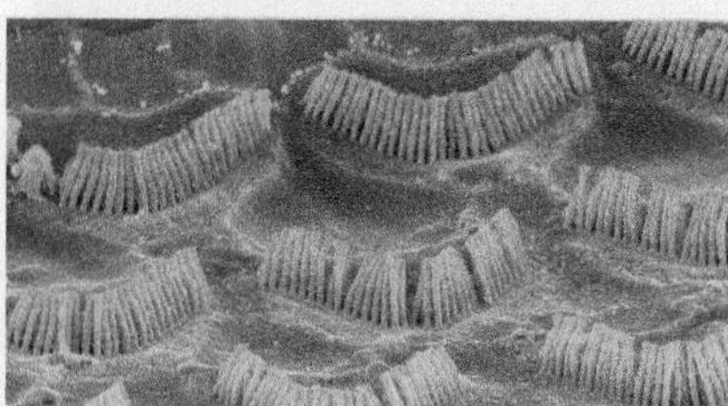

Microscopic hair cells in your inner ear, called *stereocilia,* oscillate in response to sound waves. Their oscillations trigger auditory nerve signals to your brain.

15.1 Simple Harmonic Motion

A marble rolling back and forth in the bottom of a bowl is undergoing **oscillatory motion,** one of the most common and most important types of motion. Oscillations are all around you—from vibrating guitar strings and vibrating eardrums to electric oscillations in your phone and neural oscillations in your brain. In addition, oscillations are the source of *waves,* our subject in the next two chapters.

FIGURE 15.1 shows position graphs for two different oscillating systems. The shape of the graph depends on the details of the oscillator, but all oscillators have two things in common:

1. The oscillations take place around an *equilibrium position.*
2. The motion is *periodic,* repeating at regular intervals of time.

The time to complete one full cycle, or one oscillation, is called the **period** of the oscillation. Period is represented by the symbol T. Be careful! It's easy to confuse this use of the symbol T with the tension in a string. The context in which a symbol is used alerts you to its meaning.

A system can oscillate in many ways, but the most fundamental oscillation is the smooth *sinusoidal* oscillation (i.e., like a sine or cosine) of Figure 15.1. This sinusoidal oscillation is called **simple harmonic motion,** abbreviated SHM. We'll focus on SHM because it is the basis for understanding all types of oscillations.

The prototype of simple harmonic motion is a mass oscillating on a spring. FIGURE 15.2 shows an air-track glider attached to a spring. If the glider is pulled out a few centimeters and released, it oscillates back and forth around its equilibrium position. The graph shows data from an actual air-track measurement in which the glider's position was recorded 20 times per second. This is a position graph that has been rotated 90° from its usual orientation to match the motion of the glider. You can see that it's a sinusoidal oscillation—simple harmonic motion.

As Figures 15.1 and 15.2 show, an oscillator moves back and forth between $x = -A$ and $x = +A$, where A, the **amplitude** of the motion, is the maximum displacement from equilibrium. Notice that the amplitude is the distance from the *axis* to a maximum or minimum, *not* the distance from the minimum to the maximum. That is, $x_{max} = +A$ and $x_{min} = -A$.

Period and amplitude are two important characteristics of oscillatory motion. A third is the **frequency,** f, which is the number of cycles or oscillations completed per second. If one cycle takes T seconds, the oscillator can complete $1/T$ cycles each second. That is, period and frequency are inverses of each other:

$$f = \frac{1}{T} \quad \text{or} \quad T = \frac{1}{f} \tag{15.1}$$

The units of frequency are **hertz,** abbreviated Hz, named in honor of the German physicist Heinrich Hertz, who produced the first artificially generated radio waves in 1887. By definition,

$$1 \text{ Hz} = 1 \text{ cycle per second} = 1 \text{ s}^{-1}$$

We will often deal with very rapid oscillations and make use of the units shown in TABLE 15.1. For example, ultrasound waves oscillating at 10 MHz have an oscillation period $T = 1/(10 \times 10^6 \text{ Hz}) = 10^{-7} \text{ s} = 100 \text{ ns}$.

NOTE ► Uppercase and lowercase letters *are* important. 1 MHz is 1 megahertz = 10^6 Hz, but 1 mHz is 1 millihertz = 10^{-3} Hz! ◄

FIGURE 15.1 Examples of oscillatory motion.

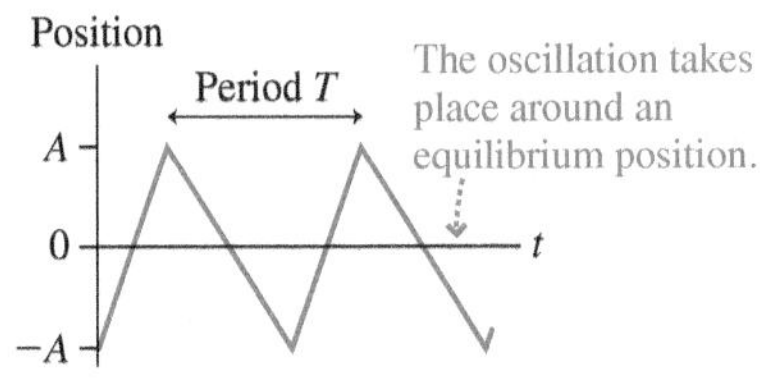

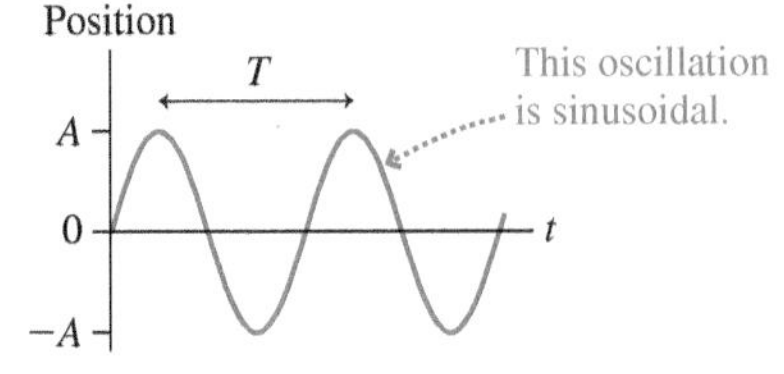

FIGURE 15.2 A prototype simple-harmonic-motion experiment.

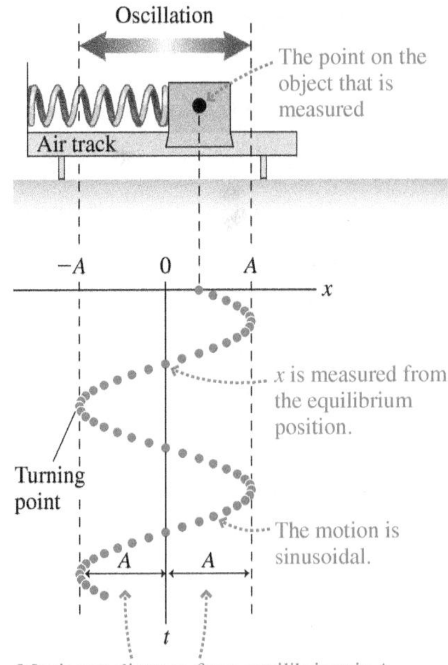

TABLE 15.1 **Units of frequency**

Frequency	Period
10^3 Hz = 1 kilohertz = 1 kHz	1 ms
10^6 Hz = 1 megahertz = 1 MHz	1 μs
10^9 Hz = 1 gigahertz = 1 GHz	1 ns

Examples of simple harmonic motion

Oscillating system		Related real-world example	
Mass on a spring		**Vibrations in the ear**	
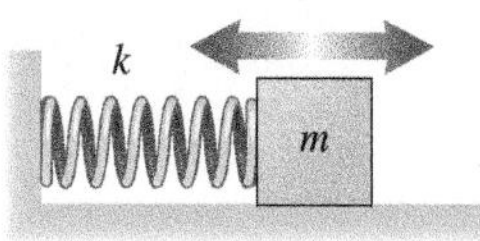	The mass oscillates back and forth due to the restoring force of the spring. The period depends on the mass and the stiffness of the spring.	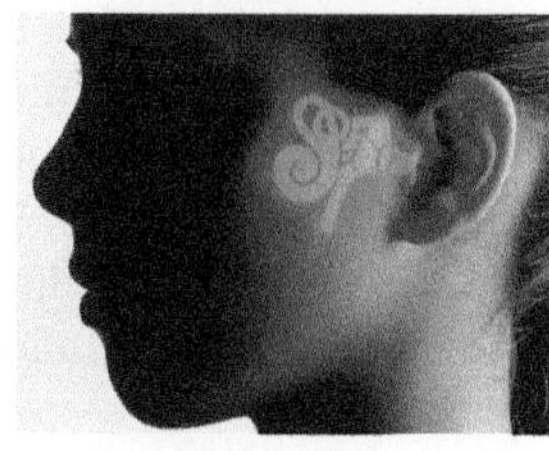	Sound waves entering the ear cause the oscillation of a membrane in the cochlea, part of the inner ear. The vibration can be modeled as a mass on a spring. The period of oscillation of a segment of the membrane depends on mass (the thickness of the membrane) and stiffness (the rigidity of the membrane).
Pendulum		**Motion of legs while walking**	
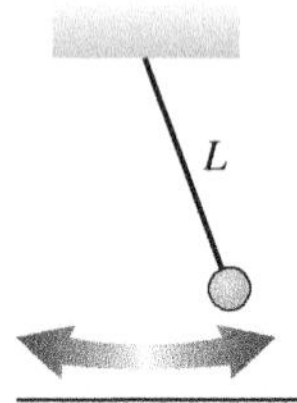	The mass oscillates back and forth due to the restoring gravitational force. The period depends on the length of the pendulum and the free-fall acceleration g.		The motion of a walking animal's legs can be modeled as pendulum motion. The rate at which the legs swing depends on the length of the legs and the free-fall acceleration g.

Kinematics of Simple Harmonic Motion

We'll start by *describing* simple harmonic motion mathematically—that is, with kinematics. Then in Section 15.4 we'll take up the dynamics of how forces *cause* simple harmonic motion.

What does it mean to say that SHM is a sinusoidal oscillation? We've used the trigonometry of right triangles to find the components of vectors, but now we need the more general idea that sine and cosine are *oscillatory functions.* FIGURE 15.3 reminds you that $\cos\theta$ is a function of the angle θ, a function that oscillates from $+1$ when $\theta = 0°$ to -1 at $\theta = 180°$ and back to $+1$ when $\theta = 360°$. That is, $\cos\theta$ goes through one complete cycle as the angle θ increases from 0° to 360°, the angle of a full circle. The function $\sin\theta$ also oscillates through one cycle, starting from 0, as θ goes from 0° to 360°. Both have an amplitude of 1.

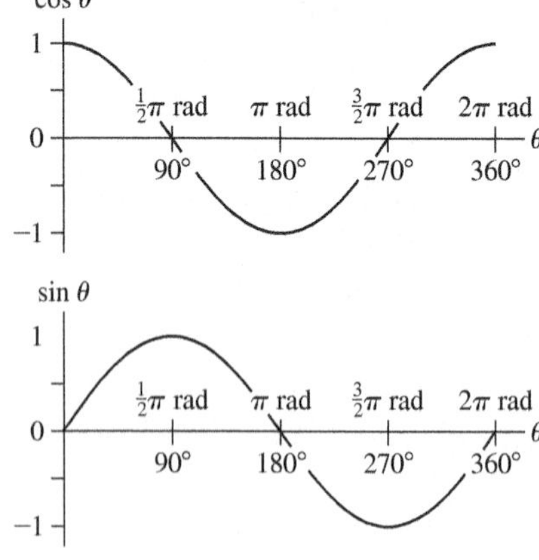

FIGURE 15.3 Sine and cosine are oscillatory functions with amplitude 1 and period 360° or 2π rad.

The equations we'll use to describe oscillations work better when the angle is specified not in degrees but in *radians,* abbreviated rad, where 2π rad $= 360°$. That is, the angle of a full circle is 2π rad. In terms of radians, $\cos(0 \text{ rad}) = +1$, $\cos(\pi \text{ rad}) = -1$, and $\cos(2\pi \text{ rad}) = +1$.

FIGURE 15.4 shows a SHM position graph—such as the one generated by the air-track glider—drawn in its normal orientation with a horizontal time axis. For the moment we'll assume that the oscillator starts at maximum displacement $(x = +A)$ at $t = 0$. Also shown is the oscillator's velocity graph. Recall from Chapter 3 that, at any instant, velocity is the slope of the position graph. In particular, you can see that

- The instantaneous velocity is zero at the instants when $x = \pm A$ because the slope of the position graph is zero. These are the *turning points* in the motion.
- The position graph has maximum slope as the oscillator passes through the equilibrium position at $x = 0$, so these are points of maximum speed. When $x = 0$ with a positive slope—maximum speed to the right—the instantaneous velocity is $v_x = +v_{max}$, where v_{max} is the amplitude of the velocity curve. Similarly, $v_x = -v_{max}$ when $x = 0$ with a negative slope—maximum speed to the left.

You can see that the position graph in Figure 15.4, with a maximum at $t = 0$, shows two cycles of a cosine function with amplitude A and period T. We can describe the oscillator's position with the equation

$$x(t) = A\cos\left(\frac{2\pi t}{T}\right) \tag{15.2}$$

where the notation $x(t)$ indicates that x is a *function* of time t.

Let's make sure we understand why Equation 15.2 describes the position graph. At $t = 0$, Equation 15.2 is $x(t) = A\cos(0 \text{ rad}) = A$ because $\cos(0 \text{ rad}) = +1$. This agrees with the graph. At $t = T$, Equation 15.2 is $x(t) = A\cos(2\pi \text{ rad}) = A$ because $\cos(2\pi \text{ rad}) = +1$, also agreeing with the graph and showing that one cycle of the oscillation has been completed in one period. Notice that x passes through zero at $t = \frac{1}{4}T$ and at $t = \frac{3}{4}T$ because $\cos\left(\frac{1}{2}\pi \text{ rad}\right) = 0$ and $\cos\left(\frac{3}{2}\pi \text{ rad}\right) = 0$. Similarly, the oscillator reaches its most negative displacement $x = -A$ at $t = \frac{1}{2}T$ when $\cos(\pi \text{ rad}) = -1$.

NOTE ▶ Be sure to set your calculator to radian mode before doing calculations that involve oscillations or, in upcoming chapters, waves. ◀

In practice we often describe oscillations in terms of frequency rather than period. Because the oscillation frequency is $f = 1/T$, we can write the oscillator's position as

$$x(t) = A\cos(2\pi f t) \tag{15.3}$$

Equations 15.2 and 15.3 do *not* express different ideas; they are simply alternative ways of writing the time-dependent position of an oscillator.

Just as the position graph is a cosine function, you can see that the velocity graph in Figure 15.4 is an "upside-down" sine function. That is, $v_x = 0$ at $t = 0$, like a sine function, but then the value of v_x goes down (negative) instead of up (positive). Thus we can write the velocity function as

$$v_x(t) = -v_{max}\sin\left(\frac{2\pi t}{T}\right) = -v_{max}\sin(2\pi f t) \tag{15.4}$$

where the minus sign inverts the graph. This function is zero at $t = 0$ and again at $t = T$.

NOTE ▶ v_{max} is the maximum *speed* and thus is a *positive* number. ◀

We deduced Equation 15.4 from the experimental results, but we could equally well find it from the position function of Equation 15.2. After all, velocity is the time derivative of position. TABLE 15.2 reminds you of the derivatives of the sine and cosine functions. Using the derivative of the position function, we find

$$v_x(t) = \frac{dx}{dt} = -\frac{2\pi A}{T}\sin\left(\frac{2\pi t}{T}\right) = -2\pi f A\sin(2\pi f t) \tag{15.5}$$

Comparing Equation 15.5, the mathematical definition of velocity, to Equation 15.4, the empirical description, we see that the maximum speed of an oscillation is

$$v_{max} = \frac{2\pi A}{T} = 2\pi f A \tag{15.6}$$

Maximum speed of an oscillation

Not surprisingly, the object has a greater maximum speed if we stretch the spring farther and give the oscillation a larger amplitude.

FIGURE 15.4 Position and velocity graphs for simple harmonic motion.

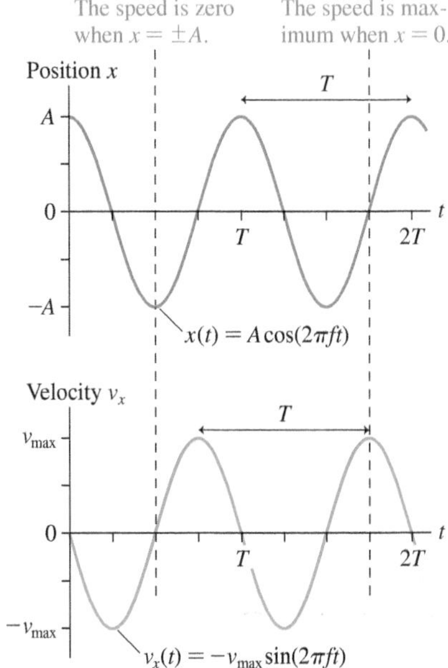

The beat of a heart An electrocardiogram (EKG) is an oscillating electric signal that provides information about the mechanical oscillation of the heart. Once available only at medical clinics, patients can now monitor their EKG on a personal device.

TABLE 15.2 **Derivatives of sine and cosine functions**

$\frac{d}{dt}(a\sin(bt + c)) = +ab\cos(bt + c)$
$\frac{d}{dt}(a\cos(bt + c)) = -ab\sin(bt + c)$

EXAMPLE 15.1 **A system in simple harmonic motion**

An air-track glider is attached to a spring, pulled 20.0 cm to the right, and released at $t = 0$ s. It makes 15 oscillations in 10.0 s.

a. What is the period of oscillation?
b. What is the object's maximum speed?
c. What are the position and velocity at $t = 0.800$ s?

PREPARE An object oscillating on a spring is in SHM.

SOLVE

a. The oscillation frequency is

$$f = \frac{15 \text{ oscillations}}{10.0 \text{ s}} = 1.50 \text{ oscillations/s} = 1.50 \text{ Hz}$$

Thus the period is $T = 1/f = 0.667$ s.

b. The oscillation amplitude is $A = 0.200$ m. Thus

$$v_{max} = \frac{2\pi A}{T} = \frac{2\pi(0.200 \text{ m})}{0.667 \text{ s}} = 1.88 \text{ m/s}$$

c. The object starts at $x = +A$ at $t = 0$ s. This is exactly the oscillation described by Equations 15.2 and 15.4. The position at $t = 0.800$ s is

$$x = A\cos\left(\frac{2\pi t}{T}\right) = (0.200 \text{ m})\cos\left(\frac{2\pi(0.800 \text{ s})}{0.667 \text{ s}}\right)$$
$$= (0.200 \text{ m})\cos(7.54 \text{ rad}) = 0.0625 \text{ m} = 6.25 \text{ cm}$$

The velocity at this instant of time is

$$v_x = -v_{max}\sin\left(\frac{2\pi t}{T}\right) = -(1.88 \text{ m/s})\sin\left(\frac{2\pi(0.800 \text{ s})}{0.667 \text{ s}}\right)$$
$$= -(1.88 \text{ m/s})\sin(7.54 \text{ rad}) = -1.79 \text{ m/s} = -179 \text{ cm/s}$$

ASSESS $t = 0.800$ is slightly more than one period, so the glider will have just begun a second oscillation. Its position should be less than A, and it should have a negative velocity because it will be moving to the left. Our calculations agree with these expectations. Notice the use of radians in the calculations.

EXAMPLE 15.2 **Lynxes and hares**

The populations of species in an ecosystem change with time. In most cases, the time dependence is very complex. However, it is well known—from field observations and theoretical modeling—that in very simple ecosystems, ecosystems dominated by one predator species and one prey species, the populations can undergo SHM-like oscillations. A classic example is the interaction between lynxes and hares in arctic ecosystems. In one study area, near Hudson Bay, Canada, the lynx population oscillates with a period of 10 years. The lynx population grows while hares, their prey, are abundant. Eventually the population of hares is depleted and the lynx population also declines. Then, with less pressure, the hare population recovers, and the lynx population begins to grow again. In this study area, the lynx population varies between a low of about 10 and a high of about 70. How many years after the lynx population peaks does it take to decline to 55?

PREPARE We've noted that SHM is important because it is the model for many kinds of oscillations. In this case, we'll model the lynx population N as undergoing SHM around an equilibrium population. We can write this as $N = N_{equil} + \Delta N$, where the oscillation—the changing population—is

$$\Delta N = A\cos\left(\frac{2\pi t}{T}\right)$$

The equilibrium population, midway between the low of 10 and the high of 70, is $N_{equil} = 40$. Thus the amplitude of the oscillation that causes the population to swing between a low of 10 and a high of 70 is $A = 30$. The period is $T = 10$ yr. The lynx population will be $N = 55$ when $\Delta N = \frac{1}{2}A = 15$.

SOLVE A graph of ΔN looks like the position graph in Figure 15.4. The population is a maximum $N = 70$ at $t = 0$ when $\Delta N = A$. It declines to $N = N_{equil} = 40$ at $t = \frac{1}{4}T$, one-quarter of a period, when $\Delta N = 0$. You might expect the oscillation to take $\frac{1}{8}T$ to reach $\frac{1}{2}A$, but that is not the case because the SHM graph is not linear between $\Delta N = A$ and $\Delta N = 0$. We need to use the oscillation equation to solve for the time $t_{1/2}$ at which $\Delta N = \frac{1}{2}A$. First we write the oscillation equation as

$$\Delta N = \tfrac{1}{2}A = A\cos\left(\frac{2\pi t_{1/2}}{T}\right)$$

We can solve this for $t_{1/2}$ by using the inverse-cosine function (also called arccos):

$$\frac{2\pi t_{1/2}}{T} = \cos^{-1}\left(\frac{\frac{1}{2}A}{A}\right) = \cos^{-1}\left(\tfrac{1}{2}\right) = \tfrac{1}{3}\pi \text{ rad}$$

You might remember from trigonometry that $\cos^{-1}\left(\frac{1}{2}\right) = 60° = \frac{1}{3}\pi$ rad and $\sin^{-1}\left(\frac{1}{2}\right) = 30° = \frac{1}{6}\pi$ rad. If not, you can use your calculator (set to radian mode!) to find $\cos^{-1}\left(\frac{1}{2}\right) = 1.047$ rad, which is $\frac{1}{3}\pi$ rad. We can now complete the calculation to find

$$t_{1/2} = \frac{T}{2\pi}\frac{\pi}{3} = \tfrac{1}{6}T = 1.7 \text{ yr}$$

ASSESS The population change is slow at the beginning, so it takes longer for the population to fall from $\Delta N = A$ to $\Delta N = \frac{1}{2}A$ than it does to fall from $\Delta N = \frac{1}{2}A$ to $\Delta N = 0$. Notice that the answer is simply a fraction of the period T; it is independent of the amplitude A.

STOP TO THINK 15.1 An object moves with simple harmonic motion. If the amplitude and the period are both doubled, the object's maximum speed is

A. Quadrupled. B. Doubled. C. Unchanged.
D. Halved. E. Quartered.

15.2 SHM and Circular Motion

The graphs of Figure 15.4 and the position function $x(t) = A\cos(2\pi ft)$ are for an oscillation in which the object just happened to be at $x_0 = A$ at $t = 0$. But you will recall that $t = 0$ is an arbitrary choice, the instant of time when you or someone else starts a stopwatch. What if you had started the stopwatch when the object was at $x_0 = -A$, or when the object was somewhere in the middle of an oscillation? In other words, what if the oscillator had different *initial conditions*? The position graph would still show an oscillation, but neither Figure 15.4 nor $x(t) = A\cos(2\pi ft)$ would describe the motion correctly.

To learn how to describe the oscillation for other initial conditions it will help to turn to a topic you studied in Chapter 7—uniform circular motion. There's a very close connection between simple harmonic motion and circular motion.

Imagine you have a turntable with a small ball glued to the edge. FIGURE 15.5a shows how to make a "shadow movie" of the ball by projecting a light past the ball and onto a screen. The ball's shadow oscillates back and forth as the turntable rotates. This is certainly periodic motion, with the same period as the turntable, but is it simple harmonic motion?

To find out, you could attach a block to a spring directly below the shadow, as shown in FIGURE 15.5b. If you did so, and if you adjusted the turntable to have the same period as the spring, you would find that the shadow's motion exactly matches the simple harmonic motion of the block. **Uniform circular motion projected onto one dimension is simple harmonic motion.**

FIGURE 15.5 A projection of the circular motion of a rotating ball matches the simple harmonic motion of an object on a spring.

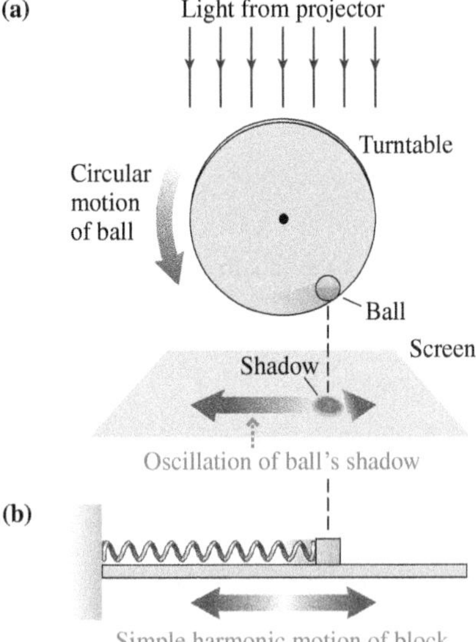

To understand this, consider the particle in FIGURE 15.6. It is in uniform circular motion, moving *counterclockwise* at constant speed in a circle with radius A. We can locate the particle on the circle by the angle ϕ measured counterclockwise (ccw) from the x-axis. Projecting the ball's shadow onto a screen in Figure 15.5 is equivalent to observing just the x-component of the particle's motion. Figure 15.6 shows that the x-component, when the particle is at angle ϕ, is

$$x = A\cos\phi \tag{15.7}$$

The angle ϕ increases with time as the particle moves around the circle. For linear motion with constant velocity v_x, the position of a particle that starts from $x_0 = 0$ at $t = 0$ is $x = v_x t$. By analogy, the *angle* of a particle that starts from $\phi_0 = 0$ at $t = 0$ can be written as

$$\phi = \omega t \tag{15.8}$$

where ω (Greek omega, *not* w) is the object's *angular velocity* as it moves around the circle. Angular velocity, measured in rad/s, is simply the rate at which the angle ϕ is increasing in Figure 15.6. Using this in Equation 15.7, we see that the shadow's position, as it oscillates back and forth, is given by

$$x(t) = A\cos\omega t \tag{15.9}$$

Because the ball's shadow matches the motion of the block oscillating on the spring, Equation 15.9 is another way of writing the position of an object in simple harmonic motion. However, when used to describe SHM, the quantity ω is called the **angular frequency** rather than the angular velocity. The angular frequency of an oscillator has the same numerical value, in rad/s, as the angular velocity of the corresponding particle in circular motion.

FIGURE 15.6 A particle in uniform circular motion with radius A and angular velocity ω.

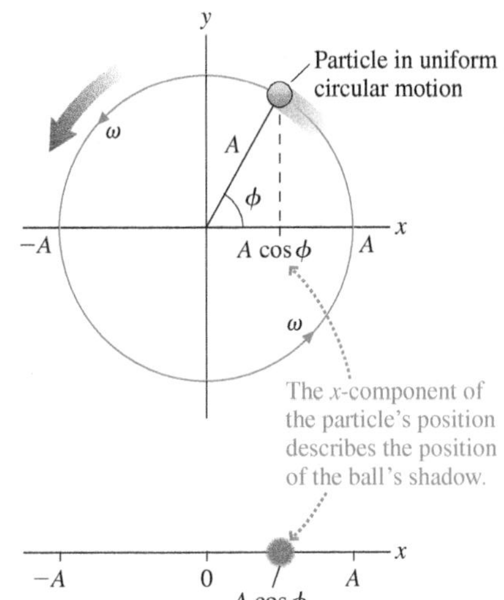

Comparing Equation 15.9 to Equations 15.2 and 15.3, our earlier expressions for $x(t)$, we can see that the angular frequency ω is related to the actual oscillator frequency f and the period T by

$$\omega\,(\text{in rad/s}) = 2\pi f(\text{in Hz}) = \frac{2\pi}{T(\text{in s})} \tag{15.10}$$

This makes sense. Angle ϕ in Figure 15.6 increases by 2π rad for each cycle of the SHM oscillation, so multiplying f in cycles/s by 2π rad/cycle gives ω in rad/s. For example, the air-track glider in Example 15.1 that oscillates with a frequency of 1.5 Hz has angular frequency $\omega = 2\pi f = 3\pi$ rad/s.

NOTE ▶ Be careful! Both f and ω are frequencies, but they have different units and are not interchangeable. Frequency f is the true oscillation frequency, in cycles per second, and it's always measured in Hz. Angular frequency ω is useful for the mathematical description of oscillations and waves, and it's always measured in rad/s. ◀

Initial Conditions: The Phase Constant

Now we're ready to consider the issue of other initial conditions. The particle in Figure 15.6 started at $\phi_0 = 0$. This was equivalent to a horizontal oscillator starting at the far right edge, $x_0 = A$. FIGURE 15.7 shows a more general situation in which the initial angle ϕ_0 can have any value. The angle at a later time t is then

FIGURE 15.7 A particle in uniform circular motion with initial angle ϕ_0.

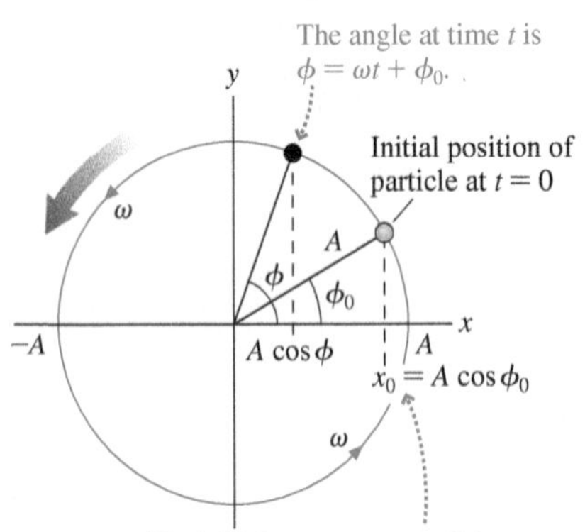

The angle at time t is $\phi = \omega t + \phi_0$.
Initial position of particle at $t = 0$
y
ω
A
ϕ
ϕ_0
x
$-A$
$A\cos\phi$
A
$x_0 = A\cos\phi_0$
ω
The initial x-component of the particle's position can be anywhere between $-A$ and A, depending on ϕ_0.

$$\phi = \omega t + \phi_0 \tag{15.11}$$

In this case, the particle's projection onto the x-axis at time t is

$$x(t) = A\cos(\omega t + \phi_0) \tag{15.12}$$

If Equation 15.12 describes the particle's projection, then it must also be the position of an oscillator in simple harmonic motion. The oscillator's velocity v_x is found by taking the derivative dx/dt. The resulting equations,

$$\begin{aligned} x(t) &= A\cos(\omega t + \phi_0) \\ v_x(t) &= -\omega A\sin(\omega t + \phi_0) = -v_{max}\sin(\omega t + \phi_0) \end{aligned} \tag{15.13}$$

Kinematic equations of SHM

are the two primary kinematic equations of simple harmonic motion. Here we see that

$$v_{max} = \omega A \tag{15.14}$$

is another expression for the maximum speed.

The quantity $\phi = \omega t + \phi_0$, which steadily increases with time, is called the **phase** of the oscillation. The phase will be especially important in our study of waves. It is simply the *angle* of the circular-motion particle whose shadow matches the oscillator. The constant ϕ_0 is called the **phase constant.** It is determined by the *initial conditions* of the oscillator.

SHM in your microwave The next time you are warming a cup of water in a microwave oven, try this: As the turntable rotates, moving the cup in a circle, stand in front of the oven with your eyes level with the cup and watch it, paying attention to the side-to-side motion. You'll see something like the turntable demonstration. The cup's apparent motion is the horizontal component of the turntable's circular motion—simple harmonic motion!

To see what the phase constant means, we can set $t = 0$ in Equations 15.13:

$$\begin{aligned} x_0 &= A\cos\phi_0 \\ (v_x)_0 &= -\omega A\sin\phi_0 \end{aligned} \tag{15.15}$$

The position x_0 and velocity $(v_x)_0$ at $t = 0$ are the initial conditions. **Different values of the phase constant correspond to different starting points on the circle and thus to different initial conditions.**

The cosine function of Figure 15.4 and the equation $x(t) = A\cos\omega t$ are for an oscillation with $\phi_0 = 0$ rad. You can see from Equations 15.13 that $\phi_0 = 0$ rad implies $x_0 = A$ and $(v_x)_0 = 0$. That is, the particle starts from rest at the point of maximum displacement.

FIGURE 15.8 illustrates these ideas by looking at three values of the phase constant: $\phi_0 = \pi/3$ rad (60°), $-\pi/3$ rad (−60°), and π rad (180°). Notice that $\phi_0 = \pi/3$ rad and $\phi_0 = -\pi/3$ rad have the same starting position, $x_0 = \frac{1}{2}A$, because of the fact that $\cos(-\theta) = \cos(\theta)$. But these are *not* the same initial conditions. In one case the oscillator starts at $\frac{1}{2}A$ while moving to the left, in the other case it starts at $\frac{1}{2}A$ while moving to the right. You can distinguish between the two by visualizing the motion.

All values of the phase constant ϕ_0 between 0 and π rad correspond to a particle in the upper half of the circle and *moving to the left.* Thus v_{0x} is negative. All values of the phase constant ϕ_0 between π and 2π rad (or, as they are usually stated, between $-\pi$ and 0 rad) have the particle in the lower half of the circle and *moving to the right.* Thus v_{0x} is positive. If you're told that the oscillator is at $x = \frac{1}{2}A$ and moving to the right at $t = 0$, then the phase constant must be $\phi_0 = -\pi/3$ rad, not $+\pi/3$ rad.

FIGURE 15.8 Different initial conditions are described by different values of the phase constant.

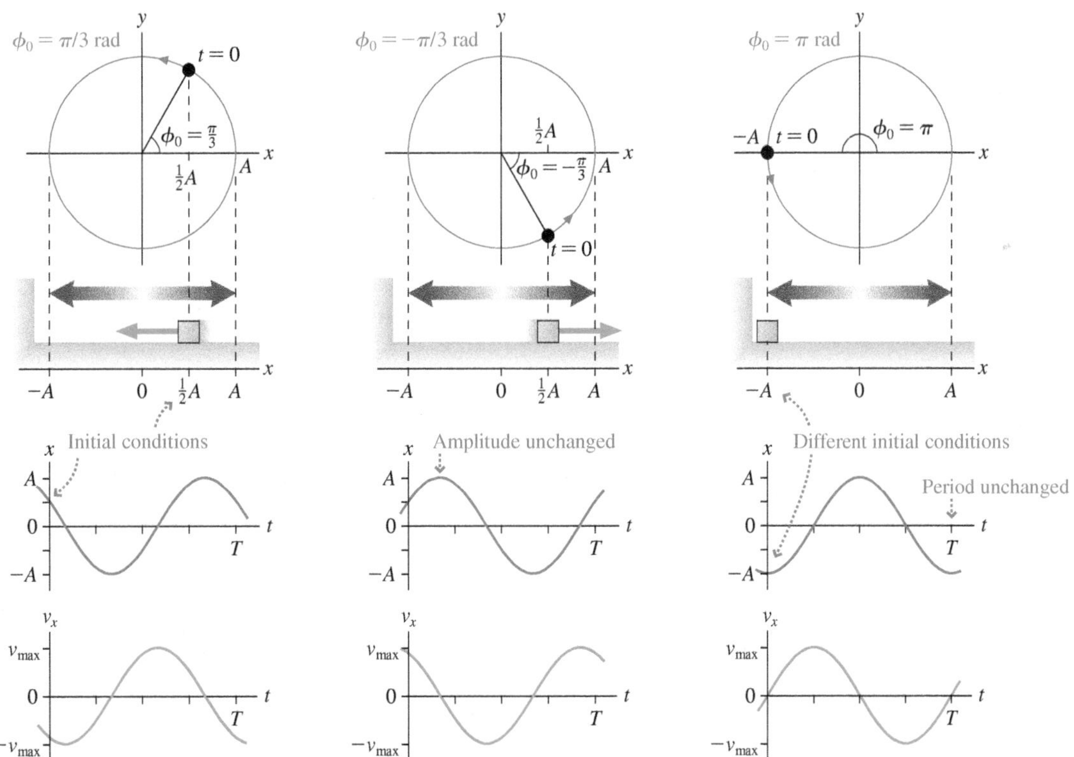

EXAMPLE 15.3 Using the initial conditions

An object on a spring oscillates with a period of 0.80 s and an amplitude of 10 cm. At $t = 0$ s, it is 5.0 cm to the left of equilibrium and moving to the left. What are its position and direction of motion at $t = 2.0$ s?

PREPARE An object oscillating on a spring is in simple harmonic motion. This oscillation does not start at maximum displacement, so we need to use the initial conditions to find the phase constant.

Continued

SOLVE We can find the phase constant ϕ_0 from the initial condition $x_0 = -5.0$ cm $= A\cos\phi_0$. This condition gives

$$\phi_0 = \cos^{-1}\left(\frac{x_0}{A}\right) = \cos^{-1}\left(-\frac{1}{2}\right) = \pm\tfrac{2}{3}\pi \text{ rad} = \pm 120°$$

Because the oscillator is moving to the *left* at $t = 0$, it is in the upper half of the circular-motion diagram and must have a phase constant between 0 and π rad. Thus ϕ_0 is $\frac{2}{3}\pi$ rad. The angular frequency is

$$\omega = \frac{2\pi}{T} = \frac{2\pi}{0.80 \text{ s}} = 7.85 \text{ rad/s}$$

Thus the object's position at time $t = 2.0$ s is

$$\begin{aligned} x(t) &= A\cos(\omega t + \phi_0) \\ &= (10 \text{ cm})\cos\left((7.85 \text{ rad/s})(2.0 \text{ s}) + \tfrac{2}{3}\pi\right) \\ &= (10 \text{ cm})\cos(17.8 \text{ rad}) = 5.0 \text{ cm} \end{aligned}$$

The object is now 5.0 cm to the right of equilibrium. But which way is it moving? There are two ways to find out. The direct way is to calculate the velocity at $t = 2.0$ s:

$$v_x = -\omega A\sin(\omega t + \phi_0) = +68 \text{ cm/s}$$

The velocity is positive, so the motion is to the right. Alternatively, we could note that the phase at $t = 2.0$ s is $\phi = 17.8$ rad. Dividing by π, you can see that

$$\phi = 17.8 \text{ rad} = 5.67\pi \text{ rad} = (4\pi + 1.67\pi) \text{ rad}$$

The 4π rad represents two complete revolutions. The "extra" phase of 1.67π rad falls between π and 2π rad, so the particle in the circular-motion diagram is in the lower half of the circle and moving to the right.

ASSESS Using two different ways to determine the direction of motion gives us confidence in the result.

NOTE ▶ The equation $\phi_0 = \cos^{-1}(x_0/A)$ has multiple solutions. Your calculator returns a single value, an angle between 0 rad and π rad, but the negative of this angle is also a solution. As Example 15.3 demonstrates, you must use additional information to choose between them. ◀

STOP TO THINK 15.2 The figure shows four oscillators at $t = 0$. Which one has the phase constant $\phi_0 = \pi/4$ rad?

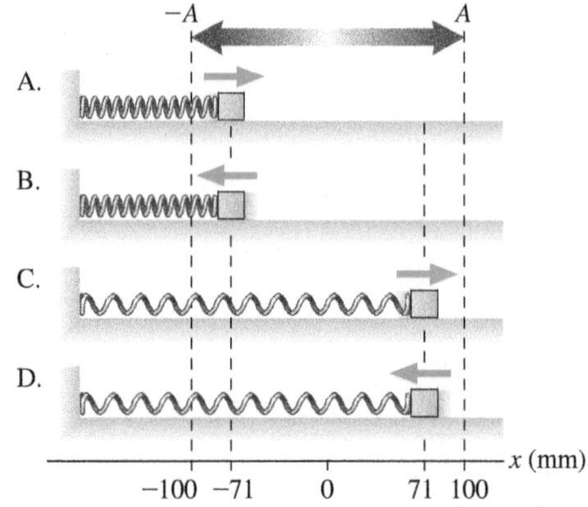

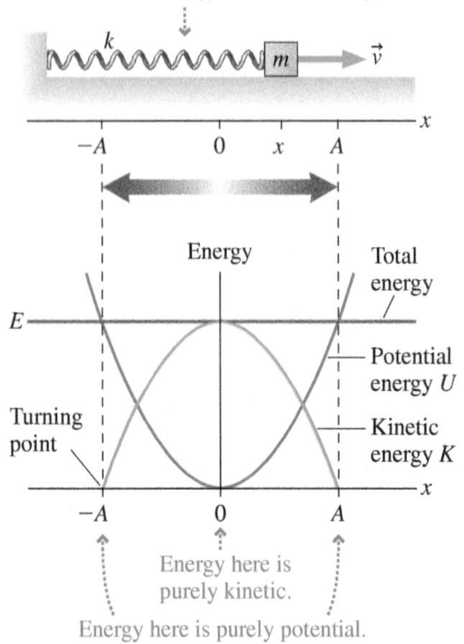

FIGURE 15.9 Energy transformations during SHM.

15.3 Energy in SHM

We've begun to develop the mathematical language of simple harmonic motion, but thus far we haven't included any physics. We've made no mention of the mass of the object or the spring constant of the spring. An energy analysis, using the tools of Chapter 11, is a good starting place.

FIGURE 15.9 shows an object oscillating on a spring, our prototype of simple harmonic motion. Now we'll specify that the object has mass m, the spring has spring constant k, and the motion takes place on a frictionless surface. You learned in Chapter 11 that the elastic potential energy when the object is at position x is $U_{sp} = \frac{1}{2}k(\Delta x)^2$, where $\Delta x = x - x_{eq}$ is the object's displacement from the equilibrium position x_{eq}. In this chapter we'll always use a coordinate system in which $x_{eq} = 0$, making $\Delta x = x$. We won't need to use gravitational potential energy because the object's height doesn't change, so we can omit the subscript sp and write the elastic potential energy as

$$U = \tfrac{1}{2}kx^2 \qquad (15.16)$$

Thus the mechanical energy of an object oscillating on a spring is

$$E = K + U = \tfrac{1}{2}mv^2 + \tfrac{1}{2}kx^2 \tag{15.17}$$

The lower portion of Figure 15.9 is an energy diagram, showing the parabolic potential-energy curve $U = \frac{1}{2}kx^2$ and the kinetic energy $K = E - U$. Recall that a particle oscillates between the *turning points* where the total energy line E crosses the potential-energy curve. The left turning point is at $x = -A$, and the right turning point is at $x = +A$. To go beyond these points would require a negative kinetic energy, which is physically impossible.

You can see that **the particle has purely potential energy at $x = \pm A$ and purely kinetic energy as it passes through the equilibrium point at $x = 0$.** At maximum displacement, with $x = \pm A$ and $v = 0$, the energy is

$$E(\text{at } x = \pm A) = U = \tfrac{1}{2}kA^2 \tag{15.18}$$

At $x = 0$, where $v = \pm v_{max}$, the energy is

$$E(\text{at } x = 0) = K = \tfrac{1}{2}m(v_{max})^2 \tag{15.19}$$

The system's mechanical energy is conserved because the surface is frictionless and there are no external forces, so the energy at maximum displacement and the energy at maximum speed, Equations 15.18 and 15.19, must be equal. That is

$$\tfrac{1}{2}m(v_{max})^2 = \tfrac{1}{2}kA^2 \tag{15.20}$$

Thus the maximum speed is related to the amplitude by

$$v_{max} = \sqrt{\frac{k}{m}}A \tag{15.21}$$

This is a relationship based on the physics of the situation.

Earlier, using kinematics, we found that

$$v_{max} = \frac{2\pi A}{T} = 2\pi f A = \omega A \tag{15.22}$$

Comparing Equations 15.21 and 15.22, we see that the frequency and period of an oscillating spring are determined by the spring constant k and the object's mass m:

$$\omega = \sqrt{\frac{k}{m}} \qquad f = \frac{1}{2\pi}\sqrt{\frac{k}{m}} \qquad T = 2\pi\sqrt{\frac{m}{k}} \tag{15.23}$$

Frequency and period of a mass on a spring

These three expressions are really only one equation. They say the same thing, but each expresses it in slightly different terms.

Equations 15.23 tell us that the period and frequency are related to the object's mass m and the spring constant k. It is perhaps surprising, but **the period and frequency do not depend on the amplitude A.** A small oscillation and a large oscillation have the same period.

EXAMPLE 15.4 Weighing DNA molecules

It has recently become possible to "weigh" individual DNA molecules by measuring the influence of their mass on a nanoscale oscillator. FIGURE 15.10 shows a thin rectangular cantilever etched out of silicon. The cantilever has a mass of 3.7×10^{-16} kg. If pulled down and released, the end of the cantilever vibrates with simple harmonic motion, moving up and down like a diving board after a jump. When the end of the cantilever is bathed with DNA molecules whose ends have been modified to bind to the cantilever's surface, one or more molecules may attach to the end of the cantilever. The addition of their mass causes a very slight—but measurable—decrease in the oscillation frequency.

FIGURE 15.10 A nanoscale cantilever.

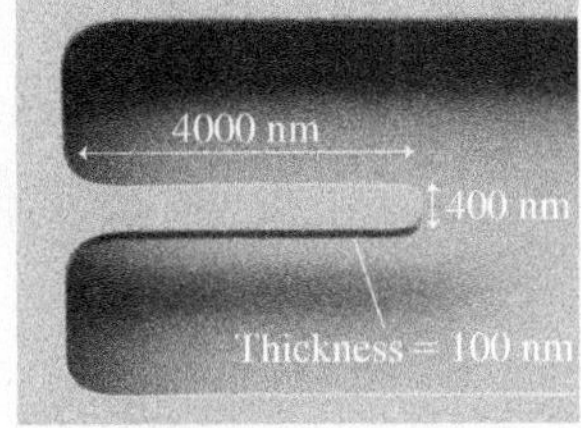

A vibrating cantilever of mass M can be modeled as a simple block of mass $\frac{1}{3}M$ attached to a spring. (The factor of $\frac{1}{3}$ arises

Continued

from the moment of inertia of a bar pivoted at one end: $I = \frac{1}{3}ML^2$.) Neither the mass nor the spring constant can be determined very accurately—perhaps to only two significant figures—but the oscillation frequency can be measured with very high precision simply by counting the oscillations. In one experiment, the cantilever was initially vibrating at exactly 12 MHz. Attachment of a DNA molecule caused the frequency to decrease by 50 Hz. What was the mass of the DNA molecule?

PREPARE According to Equations 15.23, the oscillation frequency of a mass on a spring depends on the mass and the spring constant. We will model the cantilever as a block of mass $m = \frac{1}{3}M = 1.2 \times 10^{-16}$ kg oscillating on a spring with spring constant k. Attaching a DNA molecule does not change the spring constant, but the oscillation frequency decreases from $f_0 = 12{,}000{,}000$ Hz to $f_1 = 11{,}999{,}950$ Hz.

SOLVE We can solve Equations 15.23 for the spring constant to get $k = m(2\pi f)^2$. The spring constant doesn't change with the addition of mass, so we have

$$k = m(2\pi f_0)^2 = (m + m_{DNA})(2\pi f_1)^2$$

The 2π terms cancel, and we can rearrange this equation to give

$$\frac{m + m_{DNA}}{m} = 1 + \frac{m_{DNA}}{m} = \left(\frac{f_0}{f_1}\right)^2 = \left(\frac{12{,}000{,}000\text{ Hz}}{11{,}999{,}950\text{ Hz}}\right)^2$$
$$= 1.0000083$$

Subtracting 1 from both sides gives

$$\frac{m_{DNA}}{m} = 0.0000083$$

and thus

$$m_{DNA} = 0.0000083m = (0.0000083)(1.2 \times 10^{-16}\text{ kg})$$
$$= 1.0 \times 10^{-21}\text{ kg} = 1.0 \times 10^{-18}\text{ g}$$

ASSESS By using 1 Da = 1 u = 1.7×10^{-24} g, we can quickly show that the mass is $\approx$600 kDa, which is a reasonable mass for a DNA molecule. It's a remarkable technical achievement to be able to measure a mass this small. With further improvements in sensitivity, scientists will be able to determine the number of base pairs in a strand of DNA simply by weighing it!

Conservation of Energy

Because energy is conserved, we can combine Equations 15.17, 15.18, and 15.19 to write

$$E = \tfrac{1}{2}mv^2 + \tfrac{1}{2}kx^2 = \tfrac{1}{2}kA^2 = \tfrac{1}{2}m(v_{max})^2 \quad (15.24)$$

This is the mechanical energy at any point in the oscillation.

At maximum displacement, the energy is entirely potential energy.

When passing through equilibrium, the energy is entirely kinetic energy.

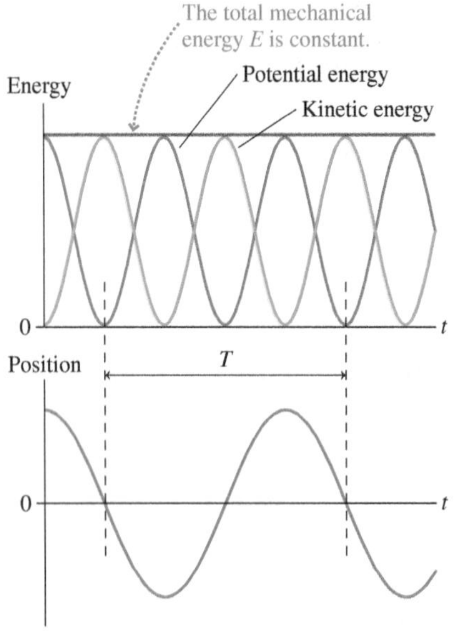

FIGURE 15.11 Kinetic energy, potential energy, and the total mechanical energy for simple harmonic motion.

Any pair of these expressions may be useful, depending on the known information. For example, you can use the amplitude A to find the speed at any point x by combining the first and second expressions for E. The speed v at position x is

$$v = \sqrt{\frac{k}{m}(A^2 - x^2)} = \omega\sqrt{A^2 - x^2} \quad (15.25)$$

FIGURE 15.11 shows graphically how the kinetic and potential energy change with time. They both oscillate but remain *positive* because x and v are squared. Energy is continuously being transformed back and forth between the kinetic energy of the moving block and the stored potential energy of the spring, but their sum remains constant. Notice that K and U both oscillate *twice* each period; make sure you understand why.

EXAMPLE 15.5 Using conservation of energy

A 500 g block on a spring is pulled a distance of 20 cm and released. The subsequent oscillations are measured to have a period of 0.80 s.

a. At what position or positions is the block's speed 1.0 m/s?
b. What is the spring constant?

PREPARE The motion is SHM. Energy is conserved.

SOLVE

a. The block starts from the point of maximum displacement, where $E = U = \frac{1}{2}kA^2$. At a later time, when the position is x and the speed is v, energy conservation requires

$$\tfrac{1}{2}mv^2 + \tfrac{1}{2}kx^2 = \tfrac{1}{2}kA^2$$

Solving for x, we find

$$x = \sqrt{A^2 - \frac{mv^2}{k}} = \sqrt{A^2 - \left(\frac{v}{\omega}\right)^2}$$

where we used $k/m = \omega^2$ from Equations 15.23. The angular frequency is easily found from the period: $\omega = 2\pi/T = 7.85$ rad/s. Thus

$$x = \sqrt{(0.20\text{ m})^2 - \left(\frac{1.0\text{ m/s}}{7.85\text{ rad/s}}\right)^2} = \pm 0.15\text{ m} = \pm 15\text{ cm}$$

There are two positions because the block has this speed on either side of equilibrium.

b. Although part a did not require that we know the spring constant, it is straightforward to find from Equations 15.23:

$$T = 2\pi\sqrt{\frac{m}{k}}$$

$$k = \frac{4\pi^2 m}{T^2} = \frac{4\pi^2(0.50\text{ kg})}{(0.80\text{ s})^2} = 31\text{ N/m}$$

ASSESS 31 N/m is a fairly small spring constant, indicating a "soft" spring. That's consistent with the observation that the block oscillates fairly slowly.

STOP TO THINK 15.3 In the figure to the right, the four springs have been compressed from their equilibrium position at $x = 0$ cm. When released, the attached mass will start to oscillate. Rank in order, from highest to lowest, the maximum speeds $(v_{max})_A$ to $(v_{max})_D$ of the masses.

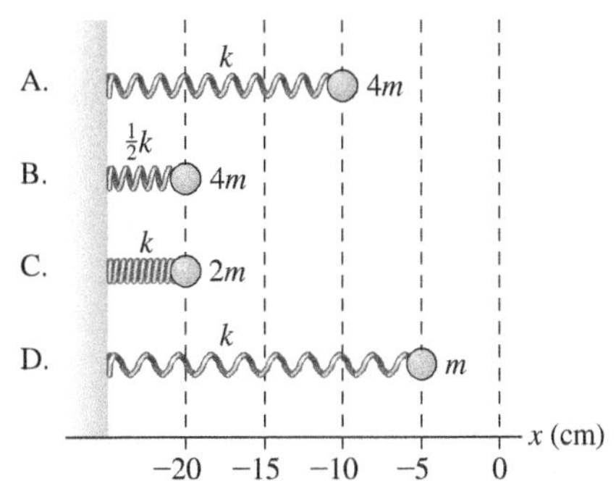

Molecular Vibrations

FIGURE 15.12 is the energy diagram of the diatomic molecule HCl, a diagram you saw in Chapter 11. There you learned that a minimum in the potential-energy curve is a point of stable equilibrium and that a system with mechanical energy E oscillates around equilibrium with turning points where the energy line crosses the potential-energy curve. Although our analysis in Chapter 11 predicted oscillations, we did not then have the tools to determine the oscillation frequency. Now we do.

The horizontal axis in Figure 15.12 is the atomic separation between the hydrogen and the chlorine atoms. The minimum at 0.135 nm is the equilibrium separation or, equivalently, the *bond length*. It occurs at energy $E = -7.4 \times 10^{-19}$ J. Recall that the zero-energy reference level for molecules corresponds to separated, noninteracting atoms, so a negative energy—less energy than two free atoms—represents a bound state. It would take 7.4×10^{-19} J to break the bond and separate the two atoms.

The energy diagram is asymmetrical, but the lowest portion of the potential-energy curve, nearest the equilibrium point, looks very much like the parabolic SHM potential-energy curve of a mass on a spring. Consequently, we can model *small oscillations* around equilibrium as SHM. The two vertical dashed lines in Figure 15.12 show the edges of an oscillation that has an amplitude of 0.020 nm—that is, an atomic separation that oscillates between 0.115 nm and 0.155 nm. From the graph, we see that an energy line that crosses the potential-energy curve at these two turning points has $E = -6.6 \times 10^{-19}$ J.

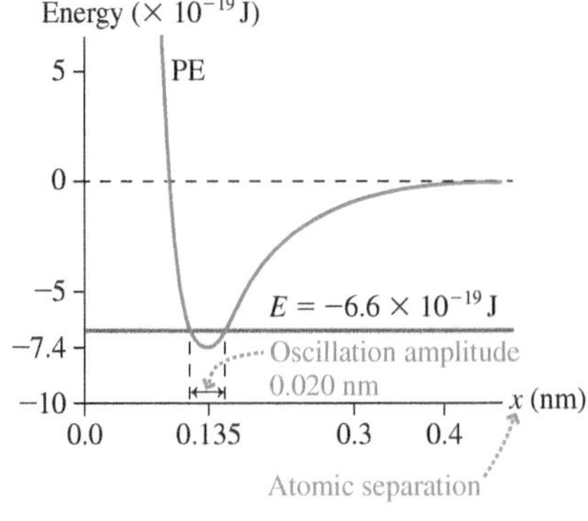

FIGURE 15.12 The energy diagram of the molecule HCl and an energy that corresponds to an oscillation amplitude of 0.020 nm.

NOTE ▶ The SHM oscillation frequency does not depend on the amplitude. There's nothing special about our choice of a 0.020 nm amplitude; any amplitude will predict the same frequency as long as it's a *small* amplitude where the potential-energy curve is parabolic. ◀

We need two more ideas. First, the zero of energy is a reference level of our choice. If we shift the zero of energy to the minimum of the potential-energy curve, we can write the potential energy as $U = \frac{1}{2}kx^2$, where x is the displacement from equilibrium. The energy line at -6.6×10^{-19} J is 0.8×10^{-19} J above the minimum, so shifting the zero of energy means that an oscillation with amplitude $A = 0.020$ nm has $E = 0.8 \times 10^{-19}$ J.

Second, the actual oscillation is somewhat complex because both atoms move in and out. However, the lighter hydrogen atom moves significantly more than the much heavier chlorine atom, so we'll make the simplifying assumption that the hydrogen atom oscillates back and forth on the spring-like molecular bond while the chlorine atom, like an immovable wall, remains at rest.

At the turning points, $x = \pm A$, the energy is entirely potential energy: $E = \frac{1}{2}kA^2$. We can solve this for the spring constant of the molecular bond:

$$k = \frac{2E}{A^2} = \frac{2(0.8 \times 10^{-19}\ \text{J})}{(0.020 \times 10^{-9}\ \text{m})^2} = 400\ \text{N/m}$$

It is interesting that the spring constant of the bond is similar in magnitude to that of a spring you might use in the lab or find in a piece of equipment.

The piece of information that we lacked in Chapter 11 was Equations 15.23, the frequency of a mass oscillating on a spring. The mass is that of a hydrogen atom, $m = 1\ \text{u} = 1.66 \times 10^{-27}$ kg. Thus the molecule's vibration frequency is

$$f = \frac{1}{2\pi}\sqrt{\frac{k}{m}} = \frac{1}{2\pi}\sqrt{\frac{400\ \text{N/m}}{1.66 \times 10^{-27}\ \text{kg}}} = 8 \times 10^{13}\ \text{Hz}$$

Only one significant figure is justified due to the limited accuracy with which we can read the graph, but our calculation is in excellent agreement with the measured HCl vibration frequency of 9.0×10^{13} Hz.

This may seem like a very high frequency, but it is typical of the frequencies with which atoms vibrate back and forth along molecular bonds. You'll see in the next chapter that an electromagnetic wave with this frequency is in the infrared portion of the spectrum. And, indeed, the infrared absorption spectra of molecules, used to identify molecules and characterize bonds, are caused by the excitation of these molecular vibrations.

Not only is simple harmonic motion an excellent model for the vibrations of molecules, but also it is an excellent model for anything else that has elastic potential energy and can vibrate if displaced from equilibrium. Living organisms are full of such oscillators—from cell membranes and the cytoskeleton to the walls of blood vessels and the systems in the ear that facilitate hearing.

15.4 Linear Restoring Forces

Our analysis thus far has been based on the experimental observation that the oscillation of a spring "looks" sinusoidal. It's time to look at force and acceleration and to see that Newton's second law *predicts* sinusoidal motion.

A motion diagram will help us visualize the object's acceleration. FIGURE 15.13 shows one cycle of the motion, separating motion to the left and motion to the right to make the diagram clear. As you can see, the object's velocity is large as it

FIGURE 15.13 Motion diagram of simple harmonic motion. The left and right motions are separated vertically for clarity but really occur along the same line.

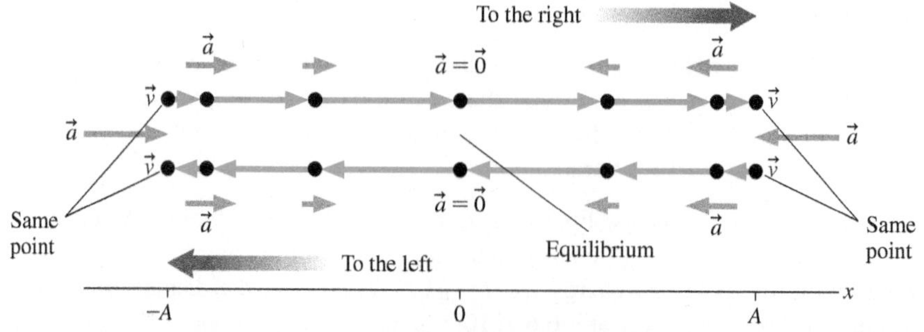

passes through the equilibrium point at $x = 0$, but $\vec{v}$ is *not changing* at that point. Acceleration measures the *change* of the velocity; hence $\vec{a} = \vec{0}$ at $x = 0$.

In contrast, the velocity is changing rapidly at the turning points. At the right turning point, $\vec{v}$ changes from a right-pointing vector to a left-pointing vector. Thus the acceleration $\vec{a}$ at the right turning point is large and *to the left.* In one-dimensional motion, the acceleration component a_x has a large *negative* value at the right turning point. Similarly, the acceleration $\vec{a}$ at the left turning point is large and *to the right.* Consequently, a_x has a large positive value at the left turning point.

Our motion-diagram analysis suggests that the acceleration a_x is most positive when the displacement is most negative, most negative when the displacement is a maximum, and zero when $x = 0$. This is confirmed by taking the derivative of the velocity:

$$a_x = \frac{dv_x}{dt} = \frac{d}{dt}(-\omega A \sin \omega t) = -\omega^2 A \cos \omega t \qquad (15.26)$$

then graphing it.

FIGURE 15.14 shows the position graph that we started with in Figure 15.4 and the corresponding acceleration graph. Comparing the two, you can see that the acceleration graph looks like an upside-down position graph. In fact, because $x = A \cos \omega t$, Equation 15.26 for the acceleration can be written

$$a_x = -\omega^2 x \qquad (15.27)$$

That is, **the acceleration is proportional to the negative of the displacement.** The acceleration is, indeed, most positive when the displacement is most negative and is most negative when the displacement is most positive. The maximum acceleration is

$$a_{max} = \omega^2 A \qquad (15.28)$$

FIGURE 15.14 Position and acceleration graphs for an oscillating spring. We've chosen $\phi_0 = 0$.

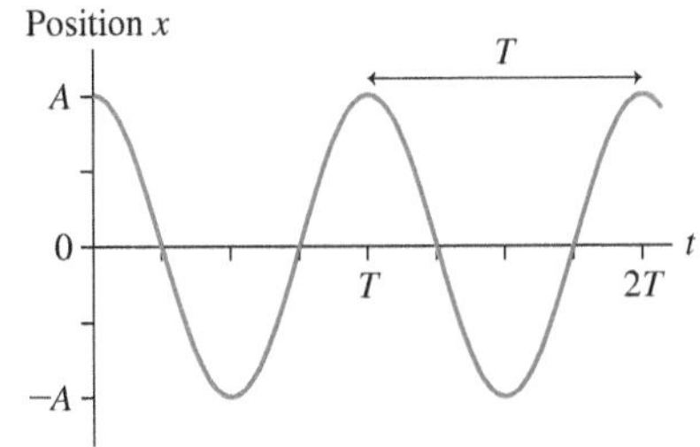

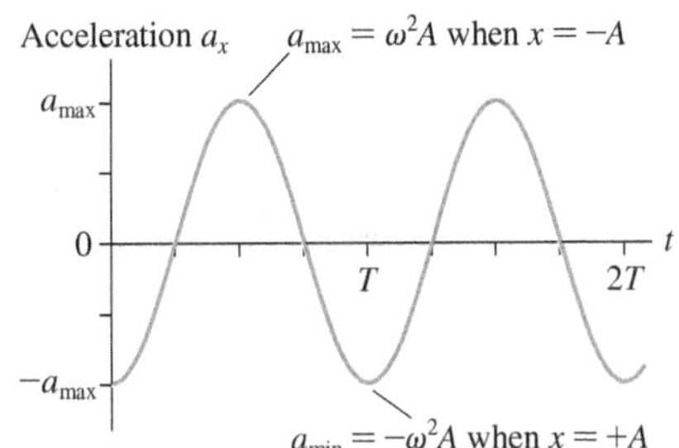

EXAMPLE 15.6 Measuring the sway of a tall building

The John Hancock Center in Chicago is 100 stories tall. Strong winds can cause the building to sway back and forth. On some windy days, the top oscillates with an amplitude of 40 cm ($\approx$16 in) and a period of 7.7 s. What is the maximum acceleration at the top of the building?

PREPARE We'll model the sideways swaying of the top of the building as SHM.

SOLVE The frequency and angular frequency can be computed from the period:

$$f = \frac{1}{T} = \frac{1}{7.7 \text{ s}} = 0.13 \text{ Hz}$$

$$\omega = 2\pi f = 0.82 \text{ rad/s}$$

The amplitude of the motion is $A = 0.40$ m. Thus the maximum acceleration is

$$a_{max} = \omega^2 A = (0.82 \text{ rad/s})^2 (0.40 \text{ m}) = 0.27 \text{ m/s}^2$$

In terms of the free-fall acceleration, $a_{max} = 0.027g$.

ASSESS The acceleration is quite small, even though the amplitude is not insignificant, because the period is so large. That's important; otherwise, people wouldn't be able to work in tall buildings.

Applying Newton's Laws

Recall that the acceleration is related to the net force by Newton's second law. Consider again our prototype mass on a spring, shown in FIGURE 15.15. This is the simplest possible oscillation, with no distractions due to friction or weight forces. We will assume the spring itself to be massless.

You learned in Chapter 5 that the spring force is given by Hooke's law:

$$(F_{sp})_x = -k\,\Delta x \qquad (15.29)$$

The minus sign in Hooke's law tells us that the spring force is a *restoring force,* a force that always points back toward the equilibrium position. In particular, Hooke's law is what we call a **linear restoring force** because the force is directly (i.e., linearly) proportional to the displacement from equilibrium. There are many examples of linear restoring forces, and what we're going to show for a mass on a spring—that it oscillates in SHM—applies to all linear restoring forces. That's why we've

FIGURE 15.15 The prototype of simple harmonic motion: a mass oscillating on a horizontal spring without friction.

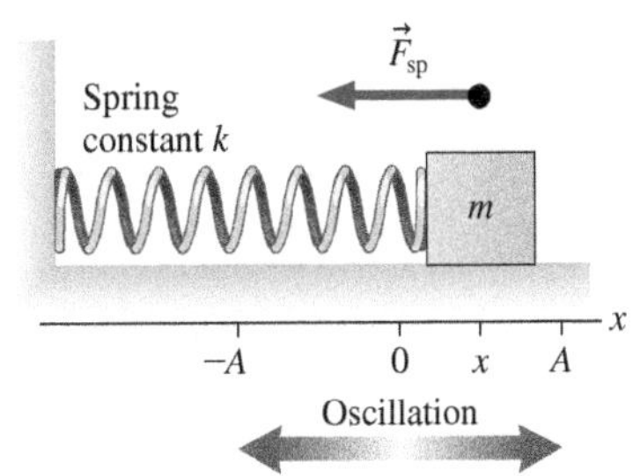

emphasized that a mass on a spring is a model for all oscillations; we can apply what we learn here to any situation with a linear restoring force.

If we place the origin of the coordinate system at the equilibrium position, as we've done throughout this chapter, then $\Delta x = x$ and Hooke's law is simply $(F_{sp})_x = -kx$. Thus Newton's second law for the object attached to the spring is

$$(F_{net})_x = (F_{sp})_x = -kx = ma_x \tag{15.30}$$

Equation 15.30 is easily rearranged to read

$$a_x = -\frac{k}{m}x \tag{15.31}$$

You can see that Equation 15.31 is identical to Equation 15.27 if the system oscillates with angular frequency $\omega = \sqrt{k/m}$. We previously found this expression for ω from an energy analysis. Our experimental observation that the acceleration is proportional to the *negative* of the displacement is exactly what Hooke's law would lead us to expect. That's the good news.

The bad news is that a_x is not a constant. As the object's position changes, so does the acceleration. Nearly all of our kinematic tools have been based on constant acceleration. We can't use those tools to analyze oscillations, so we must go back to the very definition of acceleration:

$$a_x = \frac{dv_x}{dt} = \frac{d^2x}{dt^2}$$

Acceleration is the second derivative of position with respect to time. If we use this definition in Equation 15.31, it becomes

$$\frac{d^2x}{dt^2} = -\frac{k}{m}x \tag{15.32}$$

Equation 15.32 is the *equation of motion* for a mass on a spring. We would like to solve this equation to find $x(t)$, the mass's position as a function of time. Unlike some other equations we've seen, however, Equation 15.32 cannot be solved simply by integrating both sides.

In algebra, if you don't know how to solve an equation such as $x \ln x = 3$, you can *test* a proposed solution by seeing whether the two sides of the equation agree. We can do that with Equation 15.32. We have experimental reasons to believe that the position of an oscillating object on a spring is given by the equation

$$x(t) = A\cos(\omega t + \phi_0) \tag{15.33}$$

Can we show that this is a solution to Equation 15.32?

To find out, we need the second derivative of $x(t)$. That is straightforward. Starting with Equation 15.33, we have

$$\begin{aligned} \frac{dx}{dt} &= -\omega A \sin(\omega t + \phi_0) \\ \frac{d^2x}{dt^2} &= -\omega^2 A \cos(\omega t + \phi_0) \end{aligned} \tag{15.34}$$

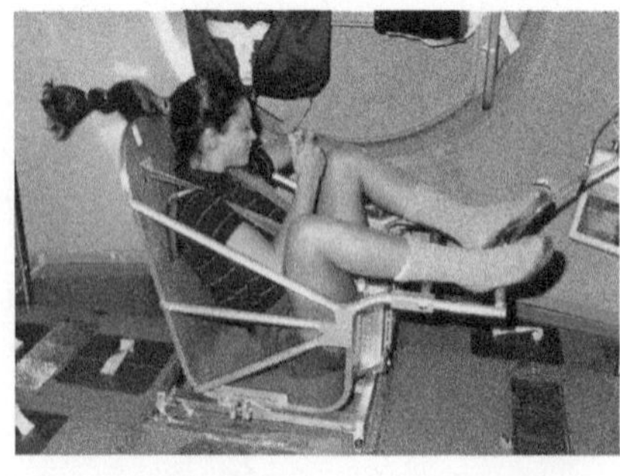

Measuring mass in space Astronauts on extended space flights monitor their mass to track the effects of weightlessness on their bodies. But because they are weightless, they can't just hop on a scale! Instead, they use an ingenious device in which an astronaut sitting on a platform oscillates back and forth due to the restoring force of a spring. The astronaut is the moving mass in a mass-spring system, so by measuring the period of her motion, she can determine her mass.

If we now substitute Equation 15.33 and the second of Equations 15.34 into Equation 15.32, we find

$$-\omega^2 A\cos(\omega t + \phi_0) = -\frac{k}{m}A\cos(\omega t + \phi_0) \tag{15.35}$$

Equation 15.35 is not automatically true, but it *will* be true, which means that Equation 15.33 is the correct solution if $\omega^2 = k/m$. There do not seem to be any restrictions on the two constants A and ϕ_0—they are determined by the initial conditions. So we have shown that the solution to the equation of motion for a mass oscillating on a spring is

$$x(t) = A\cos(\omega t + \phi_0) \tag{15.36}$$

where the angular frequency

$$\omega = 2\pi f = \sqrt{\frac{k}{m}} \tag{15.37}$$

is determined by the mass and the spring constant.

NOTE ► Once again we see that the oscillation frequency is independent of the amplitude A. ◄

Equations 15.36 and 15.37 seem somewhat anticlimactic because we've been using these results for the last several pages. But keep in mind that we had been *assuming* $x = A\cos\omega t$ simply because the experimental observations "looked" like a cosine function. We've now justified that assumption by showing that Equation 15.36 really is the solution to Newton's second law for a mass on a spring. **The *theory* of oscillation, based on Hooke's law for a spring and Newton's second law, is in good agreement with the experimental observations.**

EXAMPLE 15.7 **Analyzing an oscillator**

A 200 g block oscillates on a spring at 2.0 Hz. At $t = 0$ s, the block is 5.0 cm from equilibrium and is moving at 30 cm/s. What are (a) the amplitude of the motion, (b) the maximum speed of the block, and (c) the maximum force on the block?

PREPARE The block is in simple harmonic motion.

SOLVE

a. We can start by using $f = (1/2\pi)\sqrt{k/m}$ to find the spring constant:

$$k = m(2\pi f)^2 = (0.20\text{ kg})(4\pi\text{ rad/s})^2 = 32\text{ N/m}$$

The quantity $2\pi f$ is the angular frequency ω, which is why the units of 4π are given as rad/s. Knowing the spring constant, we can use conservation of energy $E = \frac{1}{2}mv^2 + \frac{1}{2}kx^2 = \frac{1}{2}kA^2$ to determine the amplitude. We first need to convert all quantities to SI units:

$$A = \sqrt{\frac{m}{k}v^2 + x^2} = \sqrt{\frac{0.20\text{ kg}}{32\text{ N/m}}(0.30\text{ m/s})^2 + (0.050\text{ m})^2}$$
$$= 0.055\text{ m} = 5.5\text{ cm}$$

b. The block's maximum speed is

$$v_{max} = \omega A = (2\pi f)A = (4\pi\text{ rad/s})(0.055\text{ m})$$
$$= 0.69\text{ m/s} = 69\text{ cm/s}$$

c. The maximum acceleration is $a_{max} = \omega^2 A$, so the maximum force is

$$F_{max} = ma_{max} = m\omega^2 A = m(2\pi f)^2 A$$
$$= (0.20\text{ kg})(4\pi\text{ rad/s})^2(0.055\text{ m}) = 1.7\text{ N}$$

ASSESS Neither the block's mass nor the oscillation frequency is large, so it is reasonable to find a modest maximum speed and maximum force.

STOP TO THINK 15.4 This is the position graph of a mass on a spring. What can you say about the velocity and the force at the instant indicated by the dashed line?

A. Velocity positive; force to the right.
B. Velocity negative; force to the right.
C. Velocity zero; force to the right.
D. Velocity positive; force to the left.
E. Velocity negative; force to the left.
F. Velocity zero; force to the left.
G. Velocity and force both zero.

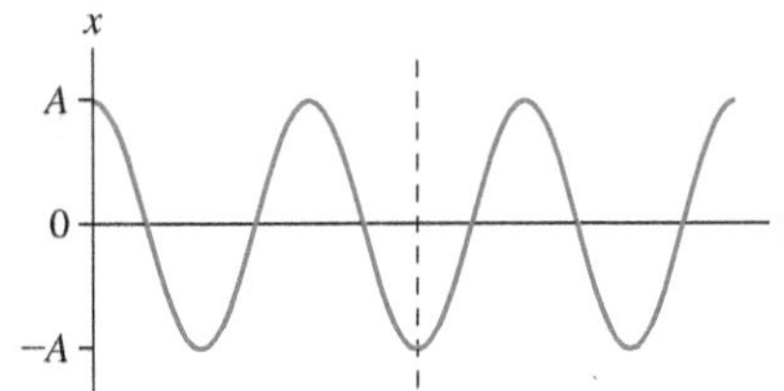

Vertical Oscillations

We have focused our analysis on a horizontally oscillating spring. But the typical demonstration you'll see in class is a mass bobbing up and down on a spring hung vertically from a support. Do vertical oscillations have the same behavior as horizontal oscillations, or does gravity change the motion?

FIGURE 15.16 Gravity stretches the spring.

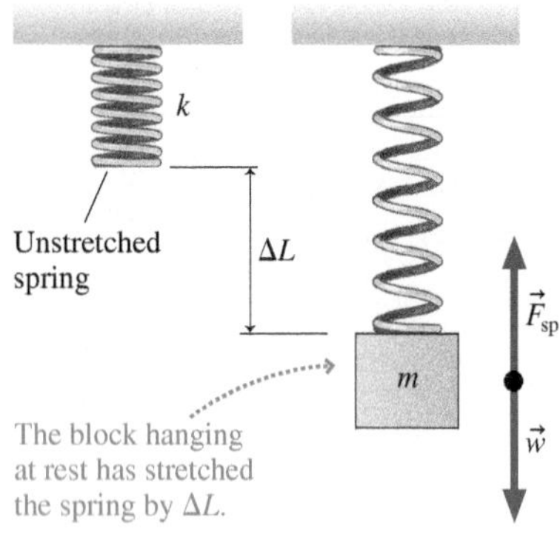

FIGURE 15.16 shows a block of mass m hanging from a spring of spring constant k. An important fact to notice is that the equilibrium position of the block is *not* where the spring is at its unstretched length. At the equilibrium position of the block, where it hangs motionless, the spring has stretched by ΔL.

Finding ΔL is an equilibrium problem in which the upward spring force balances the downward weight force on the block. The y-component of the spring force is given by Hooke's law:

$$(F_{sp})_y = -k\,\Delta y = +k\,\Delta L \tag{15.38}$$

Equation 15.38 makes a distinction between ΔL, which is simply a *distance* and is a positive number, and the displacement Δy. The block is displaced downward, so $\Delta y = -\Delta L$. Newton's second law with $a_y = 0$ is

$$(F_{net})_y = (F_{sp})_y + w_y = k\,\Delta L - mg = 0 \tag{15.39}$$

from which we can find

$$\Delta L = \frac{mg}{k} \tag{15.40}$$

This is the distance the spring stretches when the block is attached to it.

FIGURE 15.17 The block oscillates around the equilibrium position.

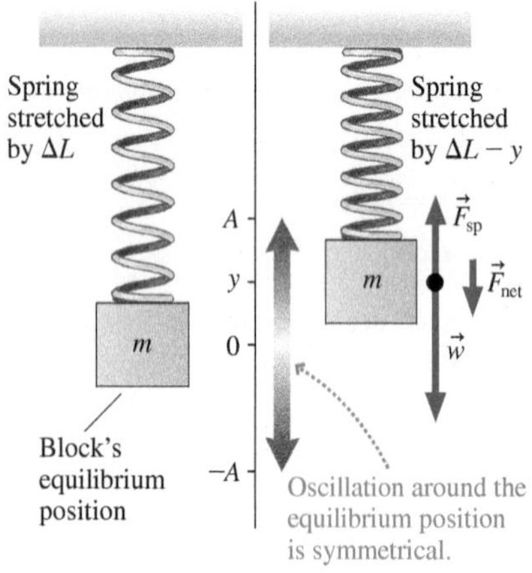

Let the block oscillate around this equilibrium position, as shown in FIGURE 15.17. We've now placed the origin of the y-axis at the block's equilibrium position in order to be consistent with our analyses of oscillations throughout this chapter. If the block moves upward, as the figure shows, the spring gets shorter compared to its equilibrium length, but the spring is still *stretched* compared to its unstretched length in Figure 15.16. When the block is at position y, the spring is stretched by an amount $\Delta L - y$ and hence exerts an *upward* spring force $F_{sp} = k(\Delta L - y)$. The net force on the block at this point is

$$(F_{net})_y = (F_{sp})_y + w_y = k(\Delta L - y) - mg = (k\,\Delta L - mg) - ky \tag{15.41}$$

But $k\,\Delta L - mg$ is zero, from Equation 15.39, so the net force on the block is simply

$$(F_{net})_y = -ky \tag{15.42}$$

Equation 15.42 for vertical oscillations is *exactly* the same as Equation 15.30 for horizontal oscillations, where we found $(F_{net})_x = -kx$. That is, the restoring force for vertical oscillations is identical to the linear restoring force for horizontal oscillations. The role of gravity is to determine where the equilibrium position is, but it doesn't affect the oscillatory motion around the equilibrium position. Thus **the vertical oscillations of a mass on a spring are the same simple harmonic motion as those of a block on a horizontal spring.**

EXAMPLE 15.8 Bungee oscillations

An 83 kg student hangs from a bungee cord with spring constant 270 N/m. The student is pulled down to a point where the cord is 5.0 m longer than its unstretched length, then released. Where is the student, and what is his velocity 2.0 s later?

PREPARE A bungee cord can be modeled as a spring. Vertical oscillations on the bungee cord are SHM around the equilibrium position. FIGURE 15.18 shows the situation.

FIGURE 15.18 A student on a bungee cord oscillates about the equilibrium position.

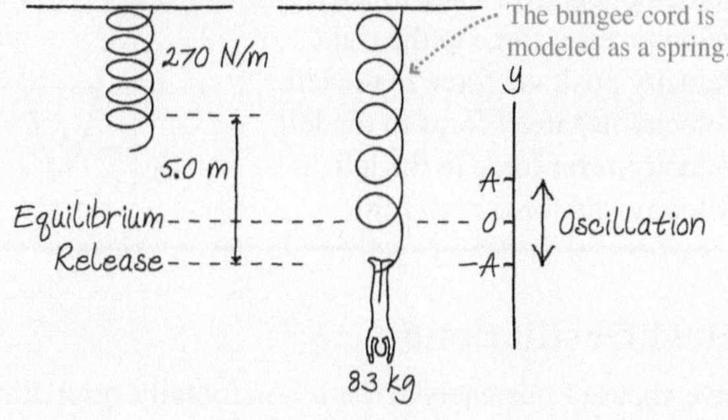

SOLVE Although the cord is stretched by 5.0 m when the student is released, this is *not* the amplitude of the oscillation. Oscillations occur around the equilibrium position, so we have to begin by finding the equilibrium point where the student hangs motionless. The cord stretch at equilibrium is given by Equation 15.40:

$$\Delta L = \frac{mg}{k} = 3.0 \text{ m}$$

Stretching the cord 5.0 m pulls the student 2.0 m below the equilibrium point, so $A = 2.0$ m. That is, the student oscillates with amplitude $A = 2.0$ m about a point 3.0 m beneath the bungee cord's original end point. The student's position as a function of time, as measured from the equilibrium position, is

$$y(t) = (2.0\text{ m})\cos(\omega t + \phi_0)$$

where $\omega = \sqrt{k/m} = 1.80$ rad/s.

The initial condition

$$y_0 = A\cos\phi_0 = -A$$

requires the phase constant to be $\phi_0 = \pi$ rad. At $t = 2.0$ s the student's position and velocity are

$$y = (2.0\text{ m})\cos\big((1.80\text{ rad/s})(2.0\text{ s}) + \pi\text{ rad}\big) = 1.8\text{ m}$$

$$v_y = -\omega A\sin(\omega t + \phi_0) = -1.6\text{ m/s}$$

The student is 1.8 m *above* the equilibrium position, or 1.2 m *below* the original end of the cord. Because his velocity is negative, he's passed through the highest point and is heading down.

ASSESS The bungee cord is always stretched—from a 1.0 m stretch when the student is at his highest point to a 5.0 m stretch at the lowest point. Even so, the motion is SHM around the equilibrium point where the student would hang at rest.

15.5 The Pendulum

Another common oscillator is a *pendulum*. FIGURE 15.19a shows an object of mass m that is attached to a string of length L and free to swing back and forth. The pendulum swings in a circle of radius $r = L$, and its position at any instant of time is best described by the *arc length s*, analogous to the position x of a horizontal oscillator. The pendulum is in equilibrium at $s = 0$ when it hangs straight down. The arc length and the corresponding angle θ are positive when the pendulum is to the right of center (a counterclockwise angle), negative when it is to the left (a clockwise angle).

The pendulum is certainly an oscillator, but is its motion simple harmonic motion? To find out, we need to see whether the restoring force is a linear restoring force. The free-body diagram of FIGURE 15.19b shows that two forces act on the pendulum: the string tension $\vec{T}$ and the weight $\vec{w}$. The restoring force is the component of the net force that is tangent to the arc in what we can call the s-direction; force components perpendicular to the arc affect the string tension but don't play a role in returning the mass to equilibrium. You can see that the tension is always perpendicular to the arc, so the restoring force is the tangential component of the weight:

$$F_s = w_s = -w\sin\theta = -mg\sin\theta \tag{15.43}$$

where, as in Hooke's law, the minus sign shows that this is a restoring force.

FIGURE 15.20 reminds you how angles, arc lengths, and distances are related. You can see that the adjacent and far sides of the right triangle are $L\cos\theta$ and $L\sin\theta$, respectively. And, by definition, angle θ in radians is the ratio of the arc length to the radius: $\theta = s/L$. Consequently, $\sin\theta = \sin(s/L)$ and thus the restoring force of Equation 15.43 is

$$F_s = -mg\sin(s/L) \tag{15.44}$$

The restoring force on a pendulum is not directly proportional to s and thus is *not* a linear restoring force. In general, the oscillations of a pendulum are *not* SHM. It turns out that the oscillation frequency of a pendulum decreases as the amplitude increases. This differs from SHM, where the frequency is the same for all amplitudes.

Suppose, however, that we limit the pendulum's motion to *small angles*. If the angle θ in Figure 15.20 is very small, then the adjacent side of the right triangle $L\cos\theta$ is very nearly the same as the string length L (that is, $L\cos\theta \approx L$), which implies that $\cos\theta \approx 1$. Similarly, the far side $L\sin\theta$ is almost exactly the same as the arc length $s = L\theta$, so $\sin\theta \approx \theta$. That is, if θ is very small, two good approximations are

$$\sin\theta \approx \theta$$
$$\cos\theta \approx 1 \tag{15.45}$$

Small-angle approximations for $\theta \ll 1$ rad

FIGURE 15.19 Pendulum motion

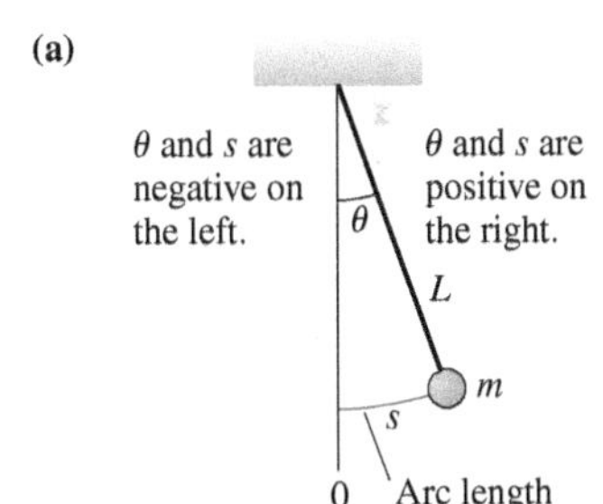

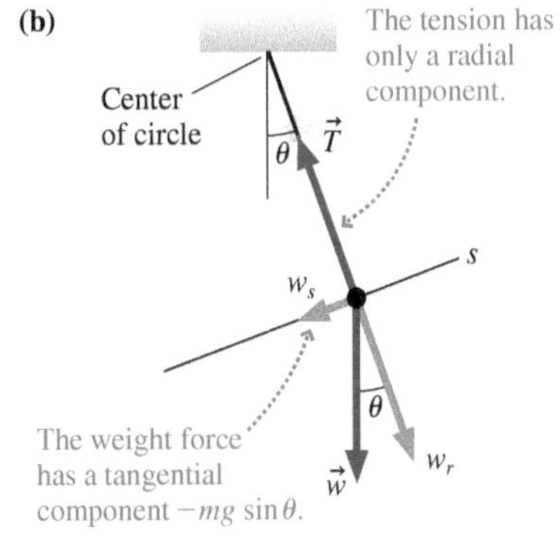

FIGURE 15.20 The geometry of pendulum oscillations.

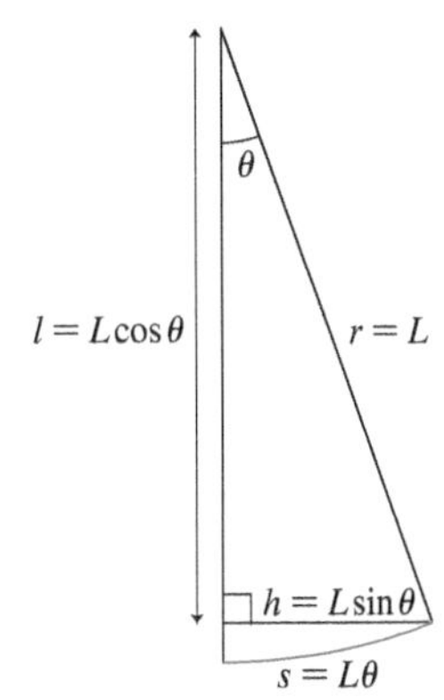

The two expressions in Equations 15.45, **which apply only when angle θ is in radians,** are called the **small-angle approximations.** We will have several uses for them throughout this text.

Equations 15.45 are formally justified in calculus, but even without a proof you can use your calculator to convince yourself that they're true. Consider, for example, $\theta = 0.1$ rad, which is approximately 6°. You can calculate

$$\sin(0.1 \text{ rad}) = 0.0998 \approx 0.1$$

$$\cos(0.1 \text{ rad}) = 0.9950 \approx 1$$

That is, the approximations differ from the true values by 0.5% or less. In practice, it is safe to use the small-angle approximations up to angles of about 10°. **Unless stated otherwise, you can assume that all pendulums in this text oscillate with maximum angles smaller than 10°.**

For a pendulum whose oscillation angles are always $\theta < 10°$, the $\sin\theta$ in Equation 15.43 can be replaced by $\sin\theta \approx \theta = s/L$. In that case, the restoring force is

$$F_s \approx -mg\theta = -mg\frac{s}{L} = -\left(\frac{mg}{L}\right)s \qquad (15.46)$$

For small-angle oscillations, the restoring force on a pendulum *is* a linear restoring force, directly proportional to the displacement s, and so **small-angle oscillations of a pendulum are simple harmonic motion.**

Furthermore, we can immediately deduce the frequency and period. If we compare Equation 15.46 to the earlier restoring force of a spring, $F_x = -kx$, we see that for a pendulum the quantity mg/L plays the role of the spring constant k. We simply need to substitute mg/L for k in the earlier equations for frequency and period to see that

$$\omega = \sqrt{\frac{g}{L}} \qquad f = \frac{1}{2\pi}\sqrt{\frac{g}{L}} \qquad T = 2\pi\sqrt{\frac{L}{g}} \qquad (15.47)$$

Frequency and period of a simple pendulum

A mass on a string forms what physicists call a *simple pendulum.* It is interesting and surprising that **the frequency and period are independent of the mass.** They depend on only the length of the pendulum and the strength of gravity. Our earlier equations for the position, velocity, and phase constant of an object oscillating on a spring apply equally well to a pendulum.

EXAMPLE 15.9 Designing a grandfather clock

A grandfather clock is designed so that one swing of the pendulum, from one side to the other, takes 1.00 s. What is the length of the pendulum? What is the maximum speed of the pendulum bob if the pendulum swings to an angle of 8.0° on either side?

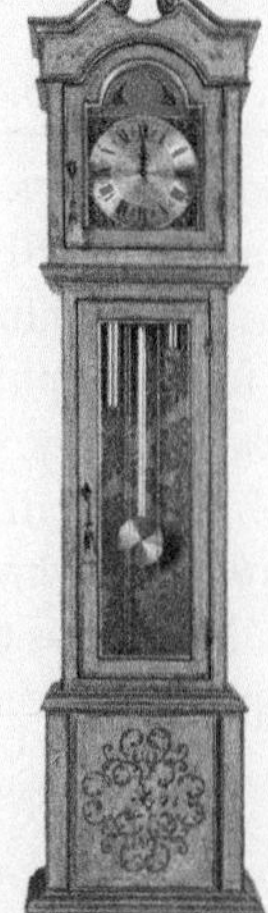

PREPARE The pendulum of a grandfather clock supports the pendulum bob with a metal rod or metal wires rather than a string. However, the swinging mass is much heavier than the rod, so we can neglect the mass of the rod and model the grandfather clock pendulum as a simple pendulum. The maximum angle is smaller than 10°, so the motion is SHM.

SOLVE The period of the pendulum is two 1.00 s swings, or $T = 2.00$ s. From Equations 15.47, we see that the period, which depends on only the pendulum length, is

$$T = 2\pi\sqrt{\frac{L}{g}}$$

Solving for L, we find

$$L = g\left(\frac{T}{2\pi}\right)^2 = (9.80 \text{ m/s}^2)\left(\frac{2.00 \text{ s}}{2\pi}\right)^2$$

$$= 0.993 \text{ m} = 99.3 \text{ cm}$$

The mass's maximum speed, which occurs at the bottom of the arc, is $v_{max} = \omega A$. The angular frequency is $\omega = 2\pi/T = 3.14$ rad/s. The amplitude A is the maximum arc length s_{max}, which we can find from the maximum angle of the swing as $s_{max} = L\theta_{max}$. However, this relationship requires the angle to be in radians, so we need a conversion:

$$\theta_{max} = 8.0° \times \frac{2\pi \text{ rad}}{360°} = 0.140 \text{ rad}$$

With this value of θ_{max} we find

$$A = s_{max} = L\theta_{max} = (0.993 \text{ m})(0.140 \text{ rad}) = 0.139 \text{ m}$$

$$v_{max} = \omega A = (3.14 \text{ rad/s})(0.139 \text{ m}) = 0.437 \text{ m/s}$$

ASSESS A pendulum clock with a "tick" or "tock" each second requires a long pendulum—just about 1 m in length. This is consistent with grandfather clocks that you've seen.

STOP TO THINK 15.5 One person swings on a swing and finds that the period is 3.0 s. A second person of equal mass joins him. With two people swinging, the period is

A. 6.0 s
B. >3.0 s but not necessarily 6.0 s
C. 3.0 s
D. <3.0 s but not necessarily 1.5 s
E. 1.5 s
F. Can't tell without knowing the length

The Physical Pendulum

A mass on a string is a simple pendulum. But you can make a pendulum from any solid object that swings back and forth on a pivot under the influence of gravity. This is called a *physical pendulum.*

FIGURE 15.21 shows a physical pendulum of mass M for which the distance between the pivot and the center of gravity is l. The moment arm of the weight force acting at the center of gravity is $d = l\sin\theta$, so the gravitational torque, which you learned to calculate in Chapter 6, is

$$\tau = -Mgd = -Mgl\sin\theta$$

The torque is negative because, for positive θ, it's causing a clockwise rotation. If we restrict the angle to being small ($\theta < 10°$), as we did for the simple pendulum, we can use the small-angle approximation to write

$$\tau = -Mgl\theta \tag{15.48}$$

Gravity exerts a linear restoring torque on the pendulum—that is, the torque is directly proportional to the angular displacement θ—so we expect the physical pendulum to undergo SHM.

A full analysis of the physical pendulum requires the rotational version of Newton's second law. The result is that a physical pendulum swings in SHM with angular frequency

$$\omega = 2\pi f = \sqrt{\frac{Mgl}{I}} \tag{15.49}$$

where I is the object's moment of inertia about the pivot point. It appears that the frequency depends on the mass of the pendulum, but recall that the moment of inertia is directly proportional to M. Thus M cancels and the frequency of a physical pendulum, like that of a simple pendulum, is independent of mass. The frequency does, however, depend on how the mass is *distributed,* which is captured by the moment of inertia.

FIGURE 15.21 A physical pendulum.

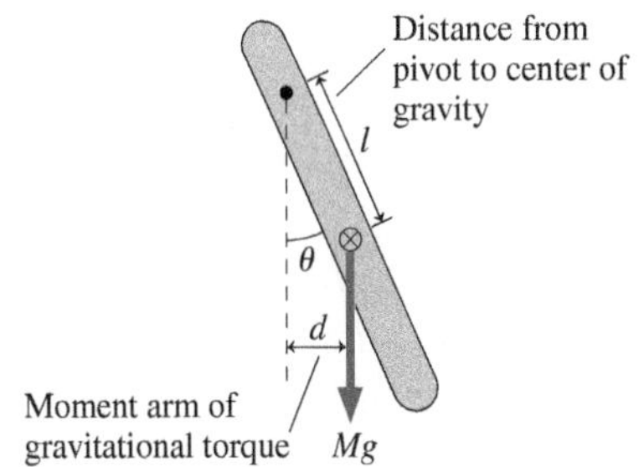

How do you hold your arms? You maintain your balance when walking or running by moving your arms back and forth opposite the motion of your legs. At a normal walking pace, your arms are extended and naturally swing at the same period as your legs. When you run, your gait is more rapid. To decrease the period of the pendulum motion of your arms to match, you bend them at the elbows, shortening their effective length and increasing the natural frequency of oscillation.

EXAMPLE 15.10 **A swinging leg as a pendulum**

A student in a biomechanics lab measures the length of her leg, from hip to heel, to be 0.90 m. What is the frequency of the pendulum motion of the student's leg? What is the period?

PREPARE We can model a human leg reasonably well as a rod of uniform cross section, pivoted at one end (the hip) to form a physical pendulum. For small-angle oscillations it will undergo SHM. The center of gravity of a uniform leg is at the midpoint, so $l = L/2$.

SOLVE Recall from Chapter 7 that the moment of inertia of a rod pivoted about one end is $I = \frac{1}{3}ML^2$. Consequently, the pendulum frequency is

$$f = \frac{1}{2\pi}\sqrt{\frac{Mgl}{I}} = \frac{1}{2\pi}\sqrt{\frac{Mg(L/2)}{ML^2/3}} = \frac{1}{2\pi}\sqrt{\frac{3g}{2L}} = 0.64\ \text{Hz}$$

The corresponding period is $T = 1/f = 1.6$ s. Notice that we didn't need to know the mass.

ASSESS As you walk, your legs do swing as physical pendulums as you bring them forward. The frequency is fixed by the length of your legs and their distribution of mass; it doesn't depend on amplitude. Consequently, you don't increase your walking speed by taking more rapid steps—changing the frequency is difficult. You simply take longer strides, changing the amplitude but not the frequency.

STOP TO THINK 15.6 A pendulum clock is made with a metal rod. It keeps perfect time at a temperature of 20°C. At a higher temperature, the metal rod lengthens. How will this change the clock's timekeeping?

A. The clock will run fast; the dial will be ahead of the actual time.
B. The clock will keep perfect time.
C. The clock will run slow; the dial will be behind the actual time.

15.6 Damped Oscillations

Heavy damping The shock absorbers in cars and trucks are heavily damped springs. The vehicle's vertical motion, after hitting a rock or a pothole, is a damped oscillation.

A pendulum left to itself gradually slows down and stops. The sound of a ringing bell gradually dies away. All real oscillators do run down—some very slowly but others quite quickly—as friction or other dissipative forces transform their mechanical energy into the thermal energy of the oscillator and its environment. An oscillation that runs down and stops is called a **damped oscillation.**

There are many possible reasons for the dissipation of energy, such as air resistance, friction, and internal forces within a metal spring as it flexes. The forces involved in dissipation are complex, but a simple *linear drag* model gives a reasonably accurate description of most damped oscillations. That is, we'll assume a drag force that depends linearly on the velocity as

$$\vec{D} = -b\vec{v} \tag{15.50}$$

where the minus sign is the mathematical statement that the force is always opposite in direction to the velocity in order to slow the object.

The **damping constant** b depends on the shape of the object *and* on the viscosity of the medium in which the particle moves; we're not going to be concerned with predicting a value for b. The units of b need to be such that they will give units of force when multiplied by units of velocity. As you can confirm, these units are kg/s. A value $b = 0$ kg/s corresponds to the limiting case of no resistance, in which case the mechanical energy is conserved.

FIGURE 15.22 An oscillating mass in the presence of a drag force.

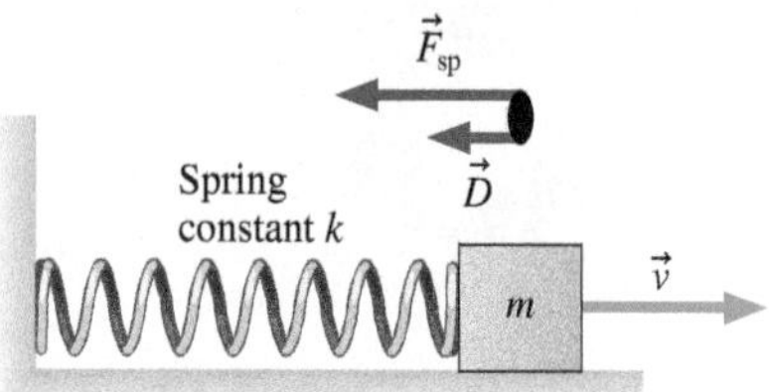

FIGURE 15.22 shows a mass oscillating on a spring in the presence of a drag force. With the drag included, Newton's second law is

$$(F_{net})_x = (F_{sp})_x + D_x = -kx - bv_x = ma_x \tag{15.51}$$

Using $a_x = d^2x/dt^2$, we can write Equation 15.51 as

$$\frac{d^2x}{dt^2} = -\frac{k}{m}x - \frac{b}{m}v_x \tag{15.52}$$

Equation 15.52 is the equation of motion of a damped oscillator. If you compare it to Equation 15.32, the equation for oscillatory motion without friction, you'll see that it differs by the inclusion of the term involving v_x.

We will simply assert without proof that the position function $x(t)$ that satisfies Equation 15.52 is

$$x(t) = Ae^{-bt/2m}\cos(\omega t + \phi_0) \tag{15.53}$$

where the angular frequency is given by

$$\omega = \sqrt{\frac{k}{m} - \frac{b^2}{4m^2}} = \sqrt{\omega_0^2 - \frac{b^2}{4m^2}} \tag{15.54}$$

Here $\omega_0 = \sqrt{k/m}$ is the angular frequency of an undamped oscillator ($b = 0$). The constant e is the base of natural logarithms, so $e^{-bt/2m}$ is an *exponential function.* Because $e^0 = 1$, Equation 15.53 reduces to our previous $x(t) = A\cos(\omega t + \phi_0)$ when $b = 0$. This makes sense and gives us confidence in Equation 15.53.

Lightly Damped Oscillators

Equation 15.54 shows that, in general, damping lowers the oscillation frequency. However, we'll limit our analysis to situations in which the damping constant b is small enough to make $b/2m \ll \omega_0$, in which case $\omega = \omega_0$ is a good approximation. This is called a *lightly damped oscillator,* one that oscillates many times before stopping. Light damping barely affects the oscillation frequency.

Let's define the **time constant** τ (Greek tau) to be

$$\tau = \frac{2m}{b} \tag{15.55}$$

Because b has units of kg/s, τ has units of seconds; it is an interval of time. With this, Equation 15.53 for a lightly damped oscillator is

$$x(t) = Ae^{-t/\tau}\cos(\omega_0 t + \phi_0) \tag{15.56}$$

FIGURE 15.23 is a graph of the position $x(t)$ for a lightly damped oscillator, as given by Equation 15.53. To keep things simple, we've assumed that the phase constant is zero. We see that **the position of a lightly damped oscillator decays exponentially with time constant τ.**

You encountered *exponential decay* in Chapter 3, but a quick review will be helpful. An exponential function is a function of the form

$$u = Ae^{-t/\tau} = A\exp(-t/\tau) \tag{15.57}$$

where $e = 2.71828\ldots$ is the base of natural logarithms in the same way that 10 is the base of ordinary logarithms. FIGURE 15.24, a graph of u versus t, starts from $u = A$ at $t = 0$, because $e^0 = 1$, and then asymptotically approaches zero for very large values of t.

No matter what u represents, when $t = \tau$, $u = e^{-1}A = 0.37A$; that is, **the quantity decays to 37% of its initial value when one time constant has elapsed.** When $t = 2\tau$, $u = e^{-2}A = 0.13A$; the quantity has decayed to 13% of its initial value after two time constants have elapsed. These properties are true for any quantity that is undergoing exponential decay.

For a lightly damped oscillator, the term $Ae^{-t/\tau}$, which is shown by the dashed line in Figure 15.23, acts as a slowly decreasing amplitude:

$$x_{max}(t) = Ae^{-t/\tau} \tag{15.58}$$

where A is the *initial* amplitude, at $t = 0$. The oscillation keeps bumping up against this line, slowly dying out with time.

A slowly changing curve that provides a border to a rapid oscillation is called the **envelope** of the oscillations. In this case, the oscillations have an *exponentially*

FIGURE 15.23 Position graph for a lightly damped oscillator.

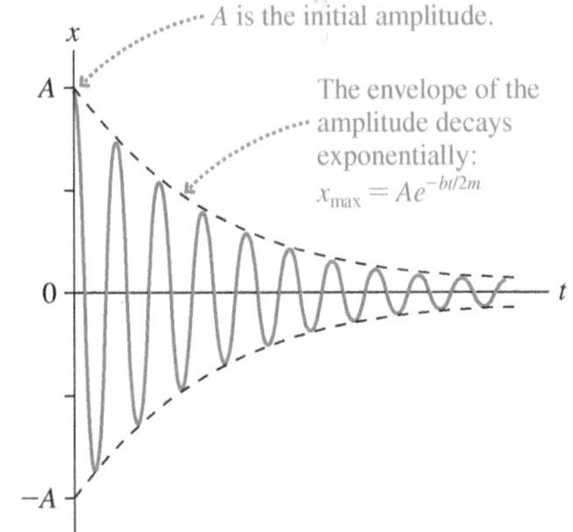

FIGURE 15.24 Exponential decay.

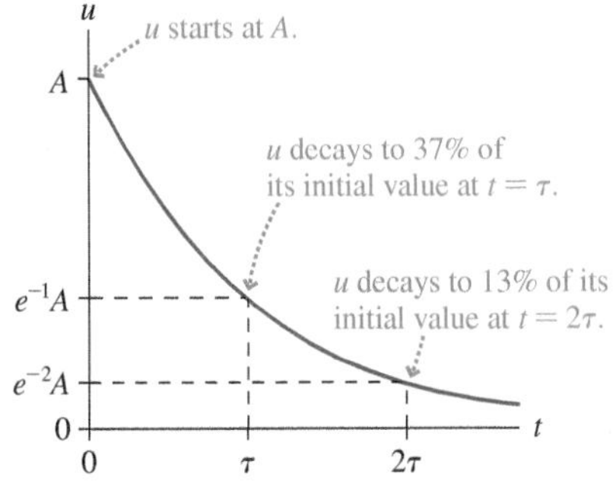

FIGURE 15.25 Oscillation envelopes for a mass of 1.0 kg with several values of b.

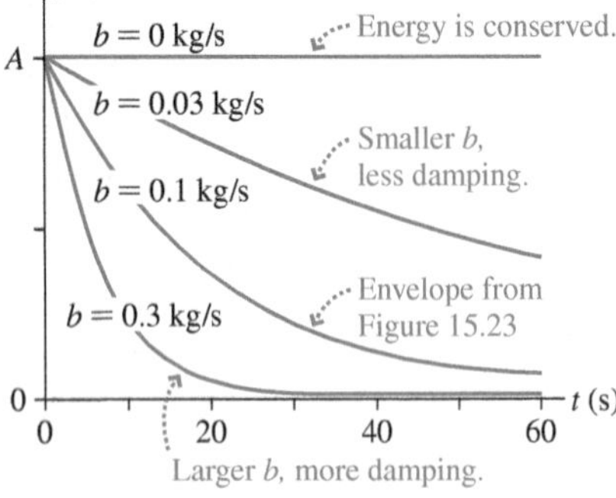

decaying envelope. Make sure you study Figure 15.23 long enough to see how both the oscillations and the decaying amplitude are related to Equation 15.53.

Changing the amount of damping, by changing the value of b, affects how quickly the oscillations decay. FIGURE 15.25 shows just the envelope $x_{max}(t)$ for several oscillators that are identical except for the value of the damping constant b. (You need to imagine a rapid oscillation within each envelope, as in Figure 15.23.) Increasing b causes the oscillations to damp more quickly, while decreasing b makes them last longer.

For practical purposes, we can speak of the time constant as the *lifetime* of the oscillation—about how long it lasts. Mathematically, there is never a time when the oscillation is "over." The decay approaches zero asymptotically, but it never gets there in any finite time. The best we can do is define a characteristic time when the motion is "almost over," and that is what the time constant τ does.

The mechanical energy of a damped oscillator is *not* conserved because of the drag force. We previously found the energy of an undamped oscillator to be $E = \frac{1}{2}kA^2$. This is still valid for a lightly damped oscillator if we replace A with the slowly decaying amplitude x_{max}. Thus

$$E(t) = \tfrac{1}{2}k(x_{max})^2 = \tfrac{1}{2}k(Ae^{-t/\tau})^2 = E_0 e^{-2t/\tau} \quad (15.59)$$

where

$$E_0 = \tfrac{1}{2}kA^2 \quad (15.60)$$

is the initial mechanical energy at $t = 0$. In other words, **the mechanical energy of a lightly damped oscillator also decays exponentially,** but with a faster decay due to the additional factor of 2 in the exponent.

EXAMPLE 15.11 **A damped pendulum**

A 500 g mass swings on a 60-cm-long string as a pendulum. The amplitude is observed to decay to half its initial value after 35 oscillations.

a. What is the time constant for this oscillator?
b. At what time will the *energy* have decayed to half its initial value?

PREPARE The motion is a lightly damped oscillation.

SOLVE

a. The initial amplitude at $t = 0$ is $x_{max} = A$. After 35 oscillations the amplitude is $x_{max} = \frac{1}{2}A$. The period of the pendulum is

$$T = 2\pi\sqrt{\frac{L}{g}} = 2\pi\sqrt{\frac{0.60\text{ m}}{9.8\text{ m/s}^2}} = 1.55\text{ s}$$

so 35 oscillations have occurred at $t = 54.2$ s.

The amplitude of oscillation at time t is given by Equation 15.57: $x_{max}(t) = Ae^{-t/\tau}$. In this case,

$$\tfrac{1}{2}A = Ae^{-(54.2\text{ s})/\tau}$$

Notice that we do not need to know A itself because it cancels out. To solve for τ, we take the natural logarithm of both sides of the equation:

$$\ln\left(\tfrac{1}{2}\right) = -\ln 2 = \ln e^{-(54.2\text{ s})/\tau} = -\frac{54.2\text{ s}}{\tau}$$

This is easily rearranged to give

$$\tau = \frac{54.2\text{ s}}{\ln 2} = 78\text{ s}$$

If desired, we could now determine the damping constant to be $b = 2m/\tau = 0.013$ kg/s.

b. The energy at time t is given by

$$E(t) = E_0 e^{-2t/\tau}$$

The time at which an exponential decay is reduced to $\frac{1}{2}E_0$, half its initial value, has a special name. It is called the **half-life** and given the symbol $t_{1/2}$. The concept of the half-life is widely used in applications such as radioactive decay. To relate $t_{1/2}$ to τ, we first write

$$E(\text{at } t = t_{1/2}) = \tfrac{1}{2}E_0 = E_0 e^{-2t_{1/2}/\tau}$$

The E_0 cancels, giving

$$\tfrac{1}{2} = e^{-2t_{1/2}/\tau}$$

Again, we take the natural logarithm of both sides:

$$\ln\left(\tfrac{1}{2}\right) = -\ln 2 = \ln e^{-2t_{1/2}/\tau} = -2t_{1/2}/\tau$$

Finally, we solve for $t_{1/2}$:

$$t_{1/2} = \tfrac{1}{2}\tau \ln 2 = \tfrac{1}{2}(78\text{ s}) \ln 2 = 27\text{ s}$$

The result is that half of the energy has been dissipated after 27 s.

ASSESS The oscillator loses energy faster than it loses amplitude. This is what we should expect because the energy depends on the *square* of the amplitude.

STOP TO THINK 15.7 Rank in order, from largest to smallest, the time constants τ_A to τ_D of the decays shown in the figure. All the graphs have the same scale.

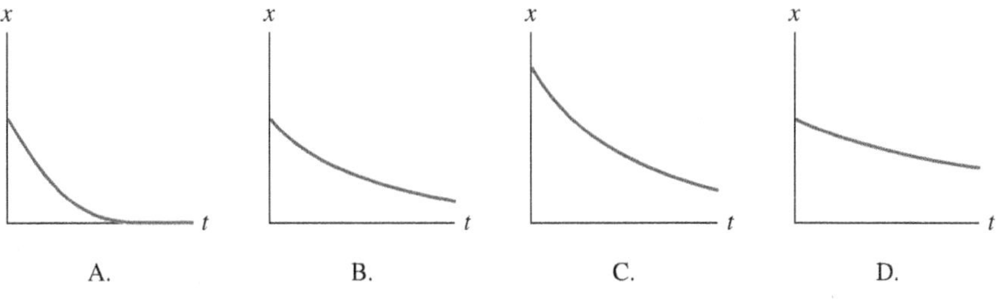

15.7 Driven Oscillations and Resonance

Thus far we have focused on the free oscillations of an isolated system. Some initial disturbance displaces the system from equilibrium, and it then oscillates freely until its energy is dissipated. These are very important situations, but they do not exhaust the possibilities. Another important situation is an oscillator that is subjected to a periodic external force. Its motion is called a **driven oscillation.**

A simple example of a driven oscillation is pushing a child on a swing, where your push is a periodic external force applied to the swing. A more complex example is a car driving over a series of equally spaced bumps. Each bump causes a periodic upward force on the car's shock absorbers, which are big, heavily damped springs. The electromagnetic coil on the back of a loudspeaker cone provides a periodic magnetic force to drive the cone back and forth, causing it to send out sound waves.

Consider an oscillating system that, when left to itself, oscillates at a frequency f_0. We will call this the **natural frequency** of the oscillator. The natural frequency for a mass on a spring is $\sqrt{k/m}/2\pi$, but it might be given by some other expression for another type of oscillator. Regardless of the expression, f_0 is simply the frequency of the system if it is displaced from equilibrium and released.

Suppose that this system is subjected to a *periodic* external force of frequency f_{ext}. This frequency, which is called the **driving frequency,** is completely independent of the oscillator's natural frequency f_0. Somebody or something in the environment selects the frequency f_{ext} of the external force, causing the force to push on the system f_{ext} times every second.

Although it is possible to solve Newton's second law with an external driving force, we will be content to look at a graphical representation of the solution. The most important result is that the oscillation amplitude depends very sensitively on the frequency f_{ext} of the driving force. The response to the driving frequency is shown in FIGURE 15.26 for a system with $m = 1.0$ kg, a natural frequency $f_0 = 2.0$ Hz, and a damping constant $b = 0.20$ kg/s. This graph of amplitude versus driving frequency, called the **response curve,** occurs in many different applications.

When the driving frequency is substantially different from the oscillator's natural frequency, at the right and left edges of Figure 15.26, the system oscillates but the amplitude is very small. The system simply does not respond well to a driving frequency that differs much from f_0. As the driving frequency gets closer and closer to the natural frequency, the amplitude of the oscillation rises dramatically. After all, f_0 is the frequency at which the system "wants" to oscillate, so it is quite happy to respond to a driving frequency near f_0. Hence the amplitude reaches a maximum when the driving frequency exactly matches the system's natural frequency: $f_{ext} = f_0$.

The amplitude can become exceedingly large when the frequencies match, especially if the damping constant is very small. FIGURE 15.27 shows the same oscillator with three different values of the damping constant. There's very little response if the damping constant is increased to 0.80 kg/s, but the amplitude for $f_{ext} = f_0$ becomes very large when the damping constant is reduced to 0.08 kg/s. This large-amplitude

FIGURE 15.26 The response curve of a driven oscillator at frequencies near its natural frequency of 2.0 Hz.

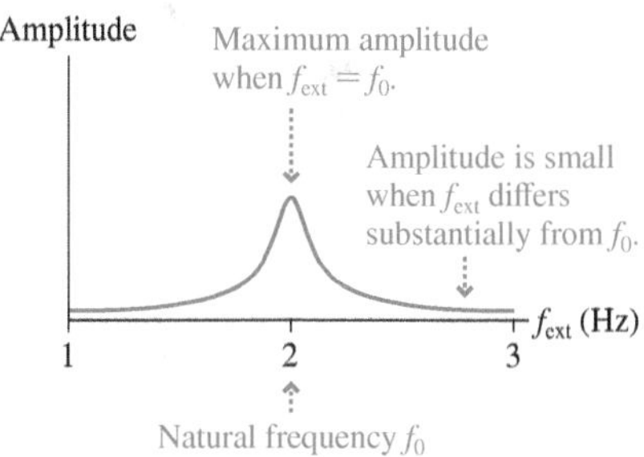

FIGURE 15.27 The resonance amplitude becomes higher and narrower as the damping constant decreases.

Amplitude
$f_0 = 2.0$ Hz
A lightly damped system has a very tall and very narrow response curve.
$b = 0.08$ kg/s
$b = 0.20$ kg/s
$b = 0.80$ kg/s
A heavily damped system has little response.
f_{ext} (Hz)
1
2
3

Myth or reality? Can a trained singer really shatter a crystal goblet by matching the goblet's natural oscillation frequency? Yes, but the sound must be *extremely* loud. Most demonstrations of goblet shattering use amplified sound, but there are a few verified instances of a singer doing so unassisted. Note that this image is an artist's impression, not an actual photo.

response to a driving force whose frequency matches the natural frequency of the system is a phenomenon called **resonance.** The condition for resonance is

$$f_{ext} = f_0 \tag{15.61}$$

Within the context of driven oscillations, the natural frequency f_0 is often called the **resonance frequency.**

An important feature of Figure 15.27 is how the amplitude and width of the resonance depend on the damping constant. A heavily damped system responds fairly little, even at resonance, but it responds to a wide range of driving frequencies. Very lightly damped systems can reach exceptionally high amplitudes, but notice that the range of frequencies to which the system responds becomes narrower and narrower as b decreases.

This allows us to understand why a singer's voice might be able to shatter crystal goblets but not inexpensive, everyday glasses. An inexpensive glass gives a "thud" when tapped, but a fine crystal goblet "rings" for several seconds. In physics terms, the goblet has a much longer time constant than the glass. That, in turn, implies that the goblet is very lightly damped while the ordinary glass is heavily damped (because the internal forces within the glass are not those of a high-quality crystal structure).

The singer causes a sound wave to impinge on the goblet, exerting a small driving force at the frequency of the note she is singing. If the singer's frequency matches the natural frequency of the goblet—resonance! Even at resonance, only a lightly damped goblet, like the top curve in Figure 15.27, can reach amplitudes large enough to shatter. The restriction, though, is that its natural frequency has to be matched very precisely. And the sound has to be *very* loud!

Resonance and Hearing

Resonance in a system means that certain frequencies produce a large response and others do not. The phenomenon of resonance is responsible for the frequency discrimination of the ear.

As we will see in the next chapter, sound is a vibration in air. FIGURE 15.28 provides an overview of the structures by which sound waves that enter the ear produce vibrations in the *cochlea,* the coiled, fluid-filled, sound-sensing organ of the inner ear.

FIGURE 15.28 The structures of the ear.

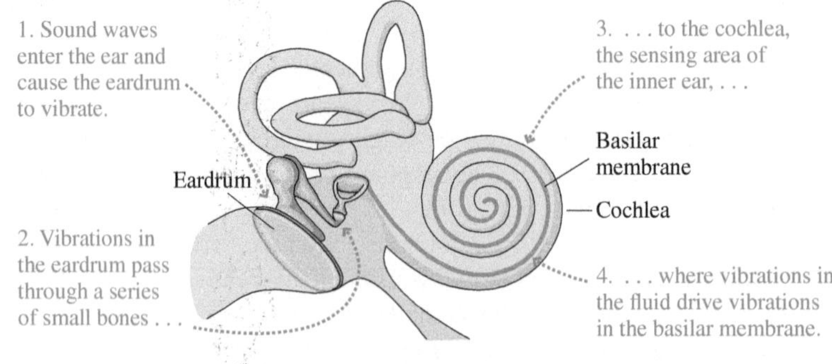
1. Sound waves enter the ear and cause the eardrum to vibrate.
Eardrum
2. Vibrations in the eardrum pass through a series of small bones . . .
3. . . . to the cochlea, the sensing area of the inner ear, . . .
Basilar membrane
Cochlea
4. . . . where vibrations in the fluid drive vibrations in the basilar membrane.

FIGURE 15.29 A simplified model of the cochlea.

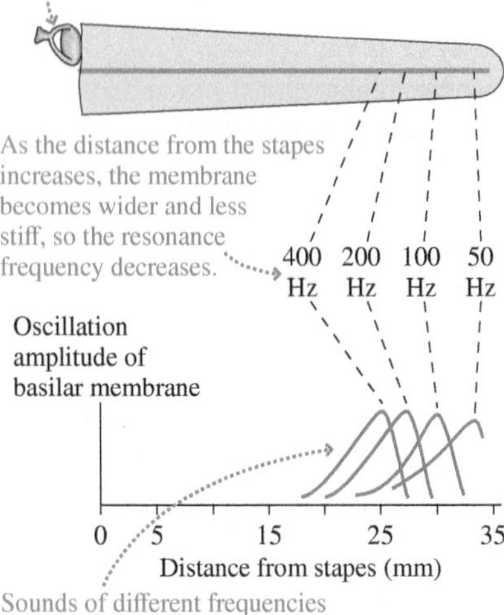
The stapes, the last of the small bones, transfers vibrations into fluid in the cochlea.
As the distance from the stapes increases, the membrane becomes wider and less stiff, so the resonance frequency decreases.
400 Hz
200 Hz
100 Hz
50 Hz
Oscillation amplitude of basilar membrane
0
5
15
25
35
Distance from stapes (mm)
Sounds of different frequencies cause different responses.

FIGURE 15.29 shows a very simplified model of the cochlea. As a sound wave travels down the cochlea, it causes a large-amplitude vibration of the basilar membrane at the point where the membrane's natural oscillation frequency matches the sound frequency—a resonance. Lower-frequency sound causes a response farther from the stapes. Sensitive hair cells on the membrane sense the vibration and send nerve signals to your brain. The fact that different frequencies produce maximal response at different positions allows your brain to very accurately determine frequency because a small shift in frequency causes a detectable change in the position of the maximal response. Most people, whether or not they have had musical training, can listen to two notes and easily determine which is at a higher pitch.

CHAPTER 15 INTEGRATED EXAMPLE Springboard diving

Flexible diving boards that are designed for large deflections are called *springboards*. A diver bobbing up and down on the end of the board can be modeled as a mass oscillating on a spring. Suppose that a light springboard deflects by 15 cm when a 65 kg diver stands at the end. He jumps and lands on the end of the board, depressing it by 25 cm; then he moves up and down at the end of the board.

a. Initially, what is the maximum speed of his up-and-down motion?
b. What is the damping constant of the board if the oscillation amplitude decreases by 50% after three cycles?
c. The diver can increase the amplitude of his oscillations by "pumping" the board with his feet. What pumping frequency is most effective?
d. At what oscillation amplitude will the diver lose contact with the board and become briefly airborne at one point in the cycle?

PREPARE We'll model the diver as a mass on a lightly damped, vertical spring. We're told that the board is "light," so we assume that the board's mass can be neglected. Vertical oscillations occur around the equilibrium position to which the board is depressed by the diver's weight.

SOLVE

a. We need the board's spring constant, which we can find by knowing that the board is depressed 15 cm when the diver is standing at rest. At this point, the upward spring force $k\Delta y$ balances the diver's weight force mg; thus

$$k = \frac{mg}{\Delta y} = \frac{(65 \text{ kg})(9.8 \text{ m/s}^2)}{0.15 \text{ m}} = 4250 \text{ N/m}$$

With k determined, we can find the frequency and period of oscillation:

$$\omega = \sqrt{\frac{k}{m}} = \sqrt{\frac{4250 \text{ N/m}}{65 \text{ kg}}} = 8.09 \text{ rad/s}$$

$$f = \frac{\omega}{2\pi} = 1.29 \text{ Hz}$$

$$T = \frac{1}{f} = 0.777 \text{ s}$$

We'll keep three significant figures to avoid round-off errors. Depressing the board to 25 cm is 10 cm past the equilibrium point, so the board and diver initially oscillate with amplitude $A = 10$ cm about equilibrium. Under these conditions, the maximum speed, as the board passes through equilibrium, is

$$v_{\max} = \omega A = (8.09 \text{ rad/s})(0.10 \text{ m}) = 0.81 \text{ m/s}$$

However, the maximum speed will decrease with each cycle as the oscillation amplitude decays.

b. We can determine the damping constant b from the time constant τ. The amplitude of a lightly damped oscillator decreases as $x_{\max}(t) = Ae^{-t/\tau}$, and we know that the amplitude has decreased to $x_{\max} = \frac{1}{2}A$ at $t = 3T = 2.33$ s. Thus

$$\tfrac{1}{2}A = Ae^{-3T/\tau}$$

Canceling the A and taking the logarithm of both sides of this equation gives

$$\ln\left(\tfrac{1}{2}\right) = -\ln 2 = \ln(e^{-3T/\tau}) = -\frac{3T}{\tau}$$

$$\tau = \frac{3T}{\ln 2} = \frac{2.33 \text{ s}}{\ln 2} = 3.36 \text{ s}$$

The time constant is defined as $\tau = 2m/b$, so the board's damping constant is

$$b = \frac{2m}{\tau} = \frac{2(65 \text{ kg})}{3.36 \text{ s}} = 39 \text{ kg/s}$$

c. The diver provides a periodic external force when he pumps with his feet. He has the largest effect by pumping at the resonance frequency, which is the natural oscillation frequency $f_0 = 1.3$ Hz.

d. The diver is simply standing on the board, rather than attached to the board, so the upward "spring force" on the diver is just the normal force $\vec{n}$ of the board pushing against his feet. The board can only push, not pull, so the normal force could reach zero but it cannot be negative. A normal force of zero would mean that, at that instant, the board is not pushing against the diver at all, so for that instant the diver will lose contact with the board and, very briefly, be airborne. Thus we're looking for the oscillation amplitude that makes $n = 0$ at some point in the cycle.

The normal force appears in Newton's laws. The only downward force is the diver's weight, so Newton's second law for the diver is

$$(F_{\text{net}})_y = ma_y = n - w = n - mg$$

If we set $n = 0$, we find that the acceleration at which the diver loses contact is $a_y = -g$. That makes sense; if there's no normal force, then the diver is acted on only by gravity and is thus, very briefly, in free fall. The acceleration of an object in SHM was found to be

$$a_y = -\omega^2 A \cos \omega t$$

The acceleration will be at its most negative when $\cos \omega t = 1$, which is the instant when $y = A$ and the diver is at the highest point in the oscillation. Reaching an acceleration of $a_y = -g$ for just an instant requires $-g = -\omega^2 A$. This will occur when the diver's oscillations reach an amplitude of

$$A = \frac{g}{\omega^2} = \frac{9.80 \text{ m/s}^2}{(8.09 \text{ rad/s})^2} = 0.15 \text{ m} = 15 \text{ cm}$$

Because the equilibrium position is a deflection of 15 cm, we find the interesting result that the diver will lose contact with the board at the instant the board becomes perfectly straight, with no deflection.

ASSESS If you've been on a springboard or watched divers, you may have noticed that, indeed, springboards oscillate with a period of approximately 1 s and the oscillations die out within a few seconds. Our results seem reasonable.

SUMMARY

GOAL To learn about systems that oscillate with simple harmonic motion.

GENERAL PRINCIPLES

Frequency and Period

Simple harmonic motion (SHM) occurs when a **linear restoring force** acts to return a system to the equilibrium position. The frequency and period depend on the details of the oscillator but not on the amplitude.

Mass on spring

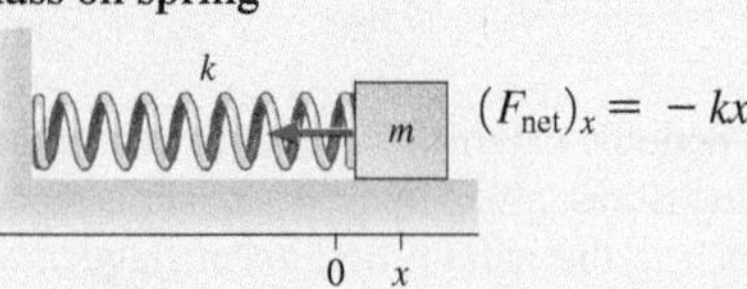

$(F_{net})_x = -kx$

The frequency and period of a mass on a spring depend on the mass and the spring constant: They are the same for horizontal and vertical systems.

$\omega = 2\pi f = \sqrt{\dfrac{k}{m}} \quad T = 2\pi\sqrt{\dfrac{m}{k}}$

Pendulum

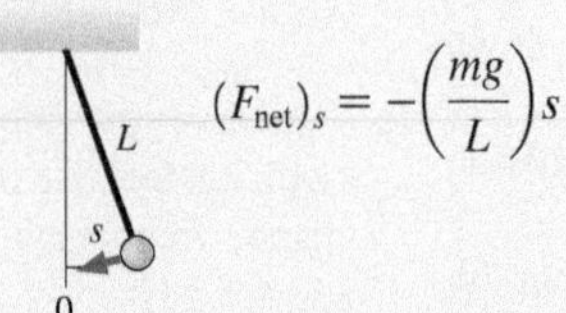

$(F_{net})_s = -\left(\dfrac{mg}{L}\right)s$

The frequency and period of a pendulum depend on the length and the free-fall acceleration. They do not depend on the mass.

$\omega = 2\pi f = \sqrt{\dfrac{g}{L}} \quad T = 2\pi\sqrt{\dfrac{L}{g}}$

Energy

If there is no friction or dissipation, kinetic and potential energies are alternately transformed into each other in SHM, with the sum of the two conserved.

$$E = \tfrac{1}{2}mv^2 + \tfrac{1}{2}kx^2 = \tfrac{1}{2}m(v_{max})^2 = \tfrac{1}{2}kA^2$$

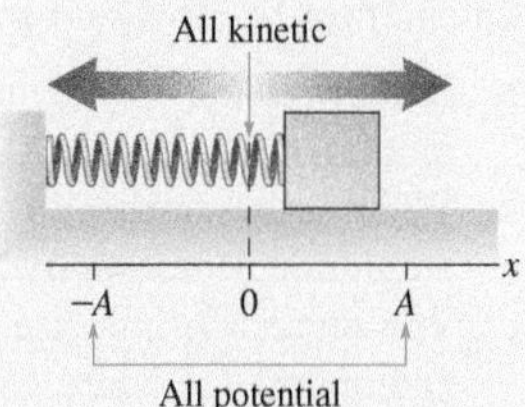

Small-amplitude molecular vibrations can be understood by modeling the bottom part of the potential-energy curve as a SHM potential energy.

IMPORTANT CONCEPTS

Simple harmonic motion (SHM)

Simple harmonic motion is a sinusoidal oscillation with frequency f and period $T = 1/f$. The **angular frequency** is $\omega = 2\pi f$. All systems that undergo SHM can be described by the same equations.

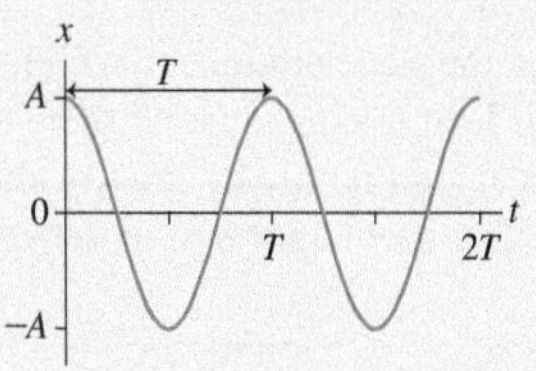

$x(t) = A\cos(\omega t + \phi_0)$

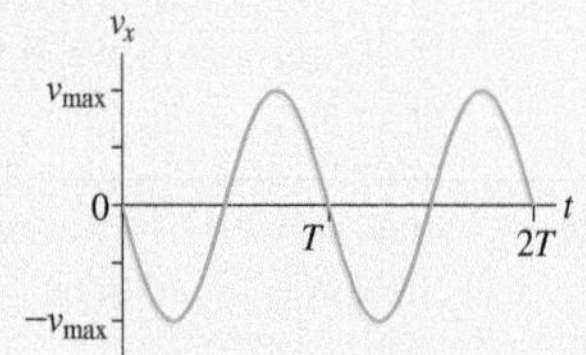

$v_x(t) = -v_{max}\sin(\omega t + \phi_0)$

$v_{max} = \omega A = 2\pi f A$

The maximum acceleration is $a_{max} = \omega^2 A$.

SHM is the projection onto the x-axis of uniform circular motion. The quantity

$$\phi = \omega t + \phi_0$$

is the **phase** of the motion.

The **phase constant** ϕ_0 is determined by the *initial conditions:*

$$x_0 = A\cos\phi_0$$

$$(v_x)_0 = -\omega A\sin\phi_0$$

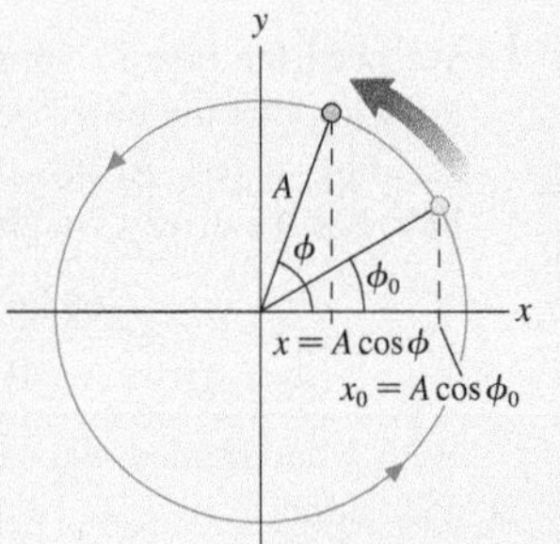

APPLICATIONS

Damping

Simple harmonic motion with damping (due to drag) decreases in amplitude over time. The **time constant** τ determines how quickly the amplitude decays.

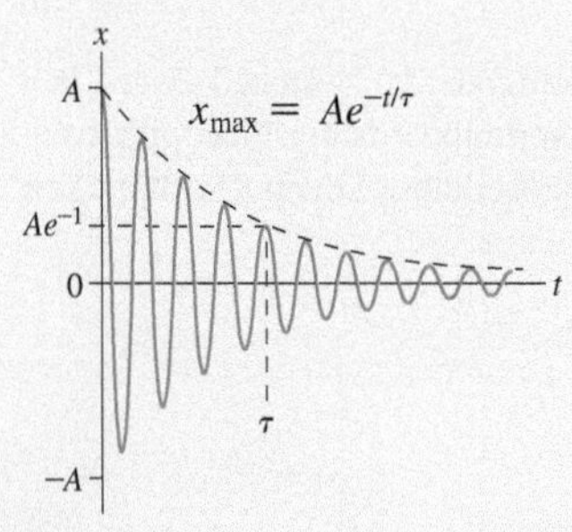

Resonance

A system that oscillates has a **natural frequency** of oscillation f_0. **Resonance** occurs if the system is driven with a frequency f_{ext} that matches this natural frequency. This may produce a large amplitude of oscillation.

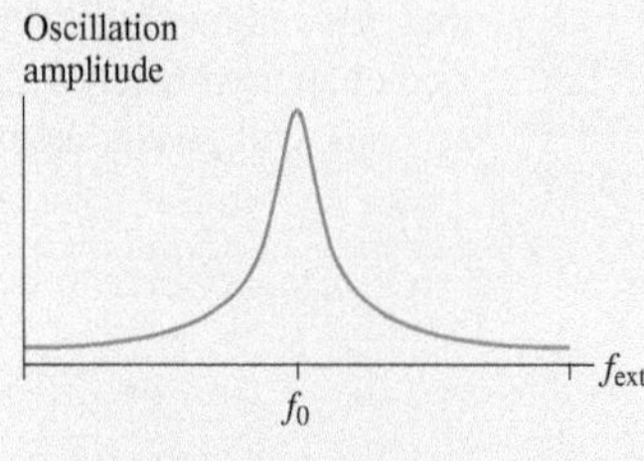

Physical pendulum

A **physical pendulum** is a pendulum with mass distributed along its length. The frequency depends on the position of the center of gravity and the moment of inertia.

The motion of legs during walking can be described using a physical pendulum model.

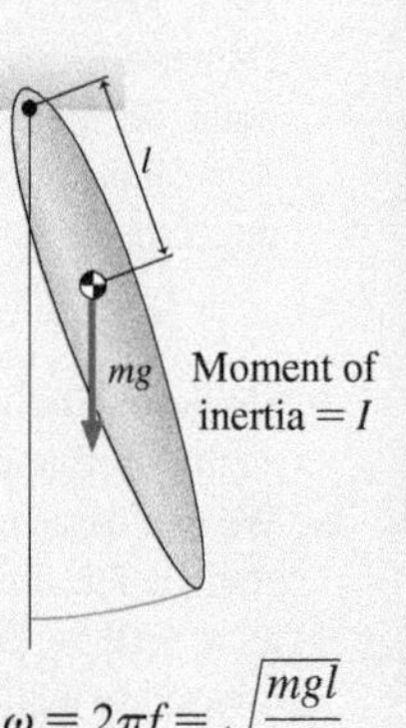

$\omega = 2\pi f = \sqrt{\dfrac{mgl}{I}}$

LEARNING OBJECTIVES After studying this chapter, you should be able to:

- Determine the period, frequency, angular frequency, and amplitude of simple harmonic motion. *Conceptual Question 15.1; Problems 15.1–15.5*
- Solve problems using both graphical and mathematical representations of SHM. *Conceptual Question 15.2; Problems 15.6, 15.9, 15.12, 15.13*
- Determine the phase constant of SHM. *Conceptual Question 15.3; Problems 15.11, 15.14–15.16*
- Apply energy conservation to SHM. *Conceptual Questions 15.7, 15.10; Problems 15.17, 15.19–15.23*
- Calculate the force acting on an object in SHM. *Conceptual Questions 15.4, 15.9; Problems 15.24, 15.25, 15.27, 15.28, 15.31*
- Solve problems that involve simple and physical pendulums. *Conceptual Questions 15.5, 15.15; Problems 15.34–15.37, 15.41*
- Analyze damped harmonic motion. *Conceptual Question 15.16; Problems 15.42–15.44, 15.46, 15.47*
- Recognize and apply the concept of resonance. *Conceptual Question 15.17; Problems 15.48–15.51*

STOP TO THINK ANSWERS

Stop to Think 15.1: C. $v_{max} = 2\pi A/T$. Doubling A and T leaves v_{max} unchanged.

Stop to Think 15.2: D. Think of circular motion. At 45°, the particle is in the first quadrant (positive x) and moving to the left (negative v_x).

Stop to Think 15.3: C > B > A = D. Energy conservation $\frac{1}{2}kA^2 = \frac{1}{2}m(v_{max})^2$ gives $v_{max} = \sqrt{k/m}\,A$. k or m has to be increased or decreased by a factor of 4 to have the same effect as increasing or decreasing A by a factor of 2.

Stop to Think 15.4: C. $v_x = 0$ because the slope of the position graph is zero. The negative value of x shows that the particle is left of the equilibrium position, so the restoring force is to the right.

Stop to Think 15.5: C. The period of a pendulum does not depend on its mass.

Stop to Think 15.6: C. The increase in length will cause the frequency to decrease and thus the period will increase. The time between ticks will increase, so the clock will run slow.

Stop to Think 15.7: $\tau_D > \tau_B = \tau_C > \tau_A$. The time constant is the time to decay to 37% of the initial height. The time constant is independent of the initial height.

QUESTIONS

Conceptual Questions

1. A person's heart rate is given in beats per minute. Is this a period or a frequency?
2. Figure Q15.2 shows the position graph of a particle in SHM.
   a. At what time or times is the particle moving right at maximum speed?
   b. At what time or times is the particle moving left at maximum speed?
   c. At what time or times is the particle instantaneously at rest?

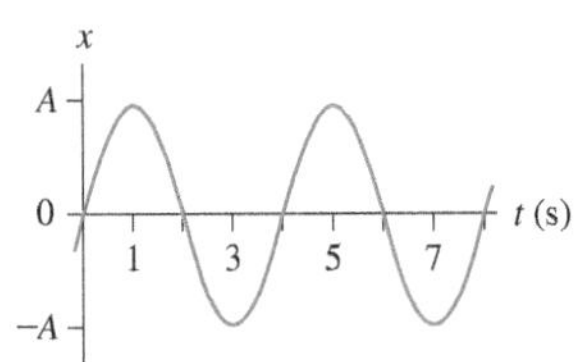

FIGURE Q15.2

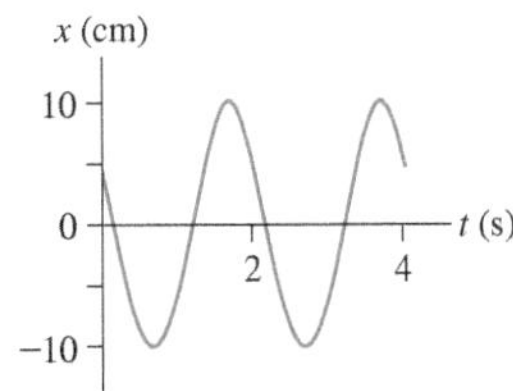

FIGURE Q15.3

3. Figure Q15.3 shows a position graph for a particle in SHM. What are (a) the amplitude A, (b) the angular frequency ω, and (c) the phase constant ϕ_0?
4. A block oscillating on a spring has period $T = 2$ s. What is the period if:
   a. The block's mass is doubled? Explain. Note that you do not know the value of either m or k, so do *not* assume any particular values for them. The required analysis involves thinking about ratios.
   b. The value of the spring constant is quadrupled?
   c. The oscillation amplitude is doubled while m and k are unchanged?
5. A pendulum on Planet X, where the value of g is unknown, oscillates with a period $T = 2$ s. What is the period of this pendulum if:
   a. Its mass is doubled? Explain. Note that you do not know the value of m, L, or g, so do not assume any specific values. The required analysis involves thinking about ratios.
   b. Its length is doubled?
   c. Its oscillation amplitude is doubled?
6. Equation 15.24 states that $\frac{1}{2}kA^2 = \frac{1}{2}m(v_{max})^2$. What does this mean? Write a couple of sentences explaining how to interpret this equation.
7. A block oscillating on a spring has an amplitude of 20 cm. What will the amplitude be if the total energy is doubled? Explain.

Problem difficulty is labeled as | (straightforward) to ||||| (challenging). Problems labeled INT integrate significant material from earlier chapters; BIO are of biological or medical interest; CALC require calculus to solve.

8. A block oscillating on a spring has a maximum speed of 20 cm/s. What will the block's maximum speed be if the total energy is doubled? Explain.
9. A mass, a spring, and a string are taken to the moon.
   a. If the mass oscillates vertically on the spring, is its period greater than, less than, or the same as its period on earth?
   b. If the mass swings at the end of the string as a pendulum, is its period greater than, less than, or the same as its period on earth?
10. Figure Q15.10 shows the potential-energy diagram and the total energy line of a particle oscillating on a spring.
   a. What is the spring's equilibrium length?
   b. Where are the turning points of the motion? Explain.
   c. What is the particle's maximum kinetic energy?
   d. What will be the turning points if the particle's total energy is doubled?

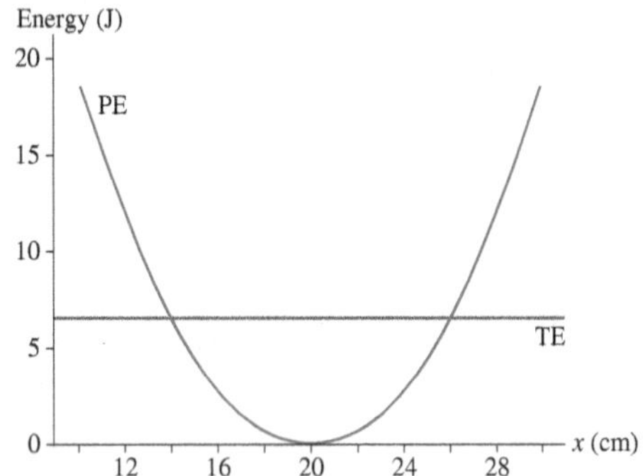

FIGURE Q15.10

11. Denver is at a higher elevation than Miami; the free-fall acceleration is slightly less at this higher elevation. If a pendulum clock keeps perfect time in Miami, will it run fast or slow in Denver? Explain.
12. Two identical pendulums are pulled back from their equilibrium positions. One is pulled back by 5° and the other by 10°. They are released simultaneously. Which pendulum gets to the bottom of its swing first?
13. BIO Sprinters push off from the ball of their foot, then bend their knee to bring their foot up close to the body as they swing their leg forward for the next stride. Why is this an effective strategy for running fast?

14. BIO Gibbons, small apes of Southeast Asia, move through the trees by swinging from successive handholds. To increase their speed, gibbons may bring their legs close to their bodies. How does this help them move more quickly?
15. A solid rod of mass M and length L swings as a pendulum from a pivot at one end of the rod. Is the rod's period of oscillation larger than, smaller than, or the same as the period of a simple pendulum of mass M that swings on a string of length L? Explain.
16. Suppose the damping constant b of an oscillator increases.
   a. Is the medium more resistive or less resistive?
   b. Do the oscillations damp out more quickly or less quickly?
   c. Is the time constant τ increased or decreased?
17. What is the difference between the driving frequency and the natural frequency of an oscillator?

Multiple-Choice Questions

18. | A spring has an unstretched length of 20 cm. A 100 g mass hanging from the spring stretches it to an equilibrium length of 30 cm.
   a. Suppose the mass is pulled down to where the spring's length is 40 cm. When it is released, it begins to oscillate. What is the amplitude of the oscillation?
      A. 5.0 cm
      B. 10 cm
      C. 20 cm
      D. 40 cm
   b. For the data given above, what is the frequency of the oscillation?
      A. 0.10 Hz
      B. 0.62 Hz
      C. 1.6 Hz
      D. 10 Hz
   c. Suppose this experiment were done on the moon, where the free-fall acceleration is approximately 1/6 of that on the earth. How would this change the frequency of the oscillation?
      A. The frequency would decrease.
      B. The frequency would increase.
      C. The frequency would stay the same.
19. || Figure Q15.19 represents the motion of a mass on a spring.
   a. What is the period of this oscillation?
      A. 12 s B. 24 s C. 36 s D. 48 s E. 50 s

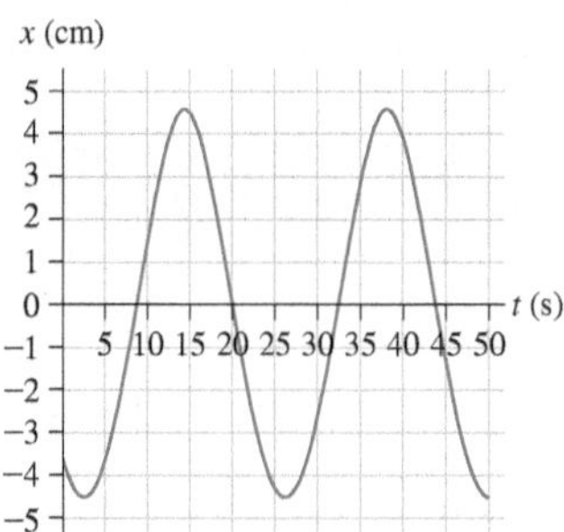

FIGURE Q15.19

   b. What is the amplitude of the oscillation?
      A. 1.0 cm
      B. 2.5 cm
      C. 4.5 cm
      D. 5.0 cm
      E. 9.0 cm
   c. What is the position of the mass at time $t = 30$ s?
      A. −4.5 cm
      B. −2.5 cm
      C. 0.0 cm
      D. 4.5 cm
      E. 30 cm
   d. When is the first time the velocity of the mass is zero?
      A. 0 s B. 2 s C. 8 s D. 10 s E. 13 s
   e. At which of these times does the kinetic energy have its maximum value?
      A. 0 s B. 8 s C. 13 s D. 26 s E. 30 s

20. | A ball of mass m oscillates on a spring with spring constant $k = 200$ N/m. The ball's position is $x = (0.350 \text{ m})\cos(15.0t)$, with t measured in seconds.
 a. What is the amplitude of the ball's motion?
  A. 0.175 m B. 0.350 m
  C. 0.700 m D. 7.50 m
  E. 15.0 m
 b. What is the frequency of the ball's motion?
  A. 0.35 Hz B. 2.39 Hz
  C. 5.44 Hz D. 6.28 Hz
  E. 15.0 Hz
 c. What is the value of the mass m?
  A. 0.45 kg B. 0.89 kg
  C. 1.54 kg D. 3.76 kg
  E. 6.33 kg
 d. What is the total mechanical energy of the oscillator?
  A. 1.65 J B. 3.28 J
  C. 6.73 J D. 10.1 J
  E. 12.2 J
 e. What is the ball's maximum speed?
  A. 0.35 m/s B. 1.76 m/s
  C. 2.60 m/s D. 3.88 m/s
  E. 5.25 m/s

21. | A 0.20 kg mass on a horizontal spring is pulled back 2.0 cm and released. If, instead, a 0.40 kg mass were used in this same experiment, the total energy of the system would
 a. Double.
 b. Remain the same.
 c. Be half as large.

22. | A 0.20 kg mass on a horizontal spring is pulled back a certain distance and released. The maximum speed of the mass is measured to be 0.28 m/s. If, instead, a 0.40 kg mass were used in this same experiment, the maximum speed would be
 a. 0.14 m/s
 b. 0.20 m/s
 c. 0.28 m/s
 d. 0.40 m/s
 e. 0.56 m/s

23. | A heavy brass ball is used to make a pendulum with a period of 5.5 s. How long is the cable that connects the pendulum ball to the ceiling?
 a. 4.7 m b. 6.2 m
 c. 7.5 m d. 8.7 m

PROBLEMS

Section 15.1 Simple Harmonic Motion

1. | When a guitar string plays the note "A," the string vibrates at 440 Hz. What is the period of the vibration?

2. | In taking your pulse, you count 75 heartbeats in 1 min. What
BIO are the period (in s) and frequency (in Hz) of your heart's oscillations?

3. | An air-track glider attached to a spring oscillates between the 10 cm mark and the 60 cm mark on the track. The glider completes 10 oscillations in 33 s. What are the (a) period, (b) frequency, (c) angular frequency, (d) amplitude, and (e) maximum speed of the glider?

4. || An air-track glider is attached to a spring. The glider is pulled to the right and released from rest at $t = 0$ s. It then oscillates with a period of 2.0 s and a maximum speed of 40 cm/s.
 a. What is the amplitude of the oscillation?
 b. What is the glider's position at $t = 0.25$ s?

5. ||| An object in SHM oscillates with a period of 4.0 s and an amplitude of 10 cm. How long does the object take to move from $x = 0.0$ cm to $x = 6.0$ cm?

6. | What are the (a) amplitude and (b) frequency of the oscillation shown in Figure P15.6?

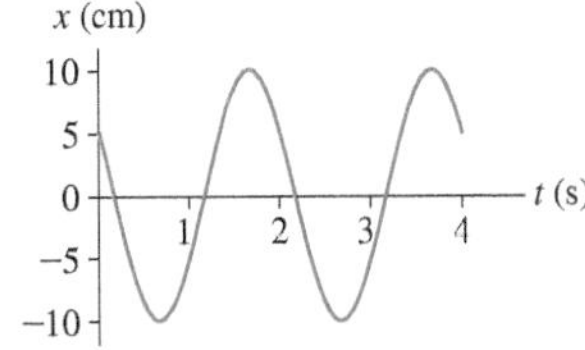

FIGURE P15.6

7. | We can model the motion of a bumblebee's wing as simple
BIO harmonic motion. A bee beats its wings 250 times per second, and the wing tip moves at a maximum speed of 2.5 m/s. What is the amplitude of the wing tip's motion?

Section 15.2 SHM and Circular Motion

8. | What are the (a) amplitude, (b) frequency, and (c) maximum speed of the oscillation shown in Figure P15.8?

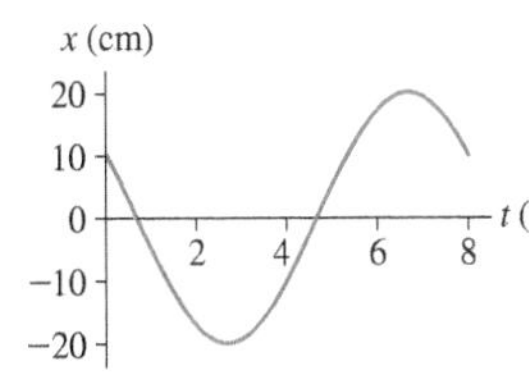

FIGURE P15.8

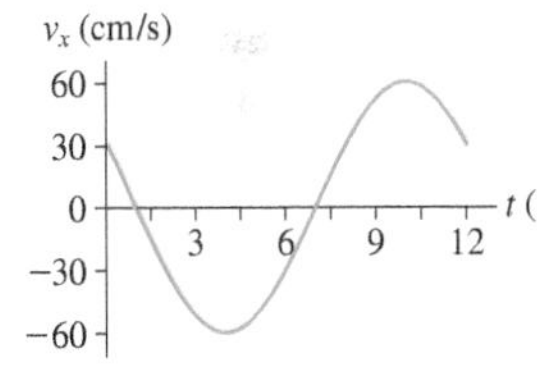

FIGURE P15.9

9. || What are (a) the maximum speed, (b) the frequency, and (c) the amplitude of the oscillation shown in Figure P15.9?

10. || What are the (a) amplitude, (b) frequency, and (c) phase constant of the oscillation shown in Figure P15.10?

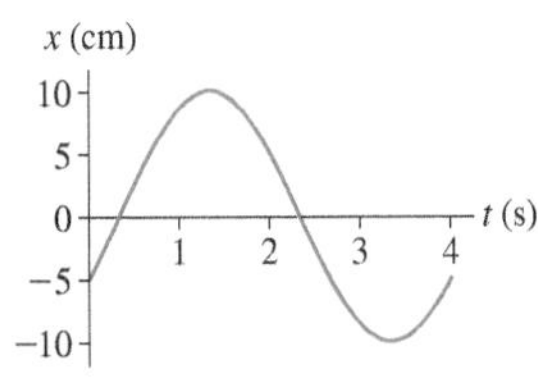

FIGURE P15.10

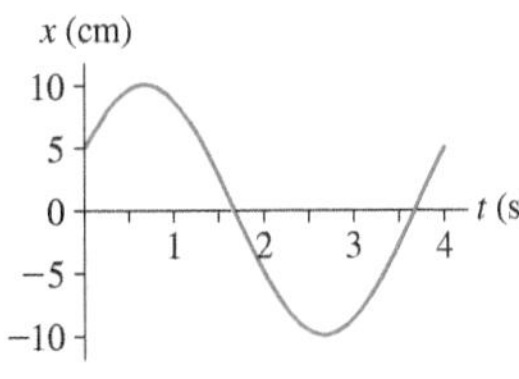

FIGURE P15.11

11. || Figure P15.11 is the position graph of a particle in simple harmonic motion.
 a. What is the phase constant?
 b. What is the velocity at $t = 0$ s?
 c. What is v_{max}?

12. || An object in simple harmonic motion has an amplitude of 4.0 cm, a frequency of 2.0 Hz, and a phase constant of $2\pi/3$ rad. Draw a position graph showing two cycles of the motion.
13. || An object in simple harmonic motion has an amplitude of 8.0 cm, a frequency of 0.25 Hz, and a phase constant of $-\pi/2$ rad. Draw a position graph showing two cycles of the motion.
14. | An object in simple harmonic motion has amplitude 4.0 cm and frequency 4.0 Hz, and at $t = 0$ s it passes through the equilibrium point while moving to the right. Write the function $x(t)$ that describes the object's position.
15. | An object in simple harmonic motion has amplitude 8.0 cm and frequency 0.50 Hz. At $t = 0$ s it has its most negative position. Write the function $x(t)$ that describes the object's position.
16. || An air-track glider attached to a spring oscillates with a period of 1.5 s. At $t = 0$ s the glider is 5.00 cm left of the equilibrium position and moving to the right at 36.3 cm/s.
    a. What is the phase constant?
    b. What is the phase at $t = 0$ s, 0.5 s, 1.0 s, and 1.5 s?

Section 15.3 Energy in SHM

17. || The position of a 50 g oscillating mass is given by $x(t) = (2.0\text{ cm})\cos(10t - \pi/4)$, where t is in s. Determine:
    a. The amplitude.
    b. The period.
    c. The spring constant.
    d. The phase constant.
    e. The initial conditions.
    f. The maximum speed.
    g. The total energy.
    h. The velocity at $t = 0.40$ s.
18. | A block attached to a spring with unknown spring constant oscillates with a period of 2.0 s. What is the period if
    a. The mass is doubled?
    b. The mass is halved?
    c. The amplitude is doubled?
    d. The spring constant is doubled?

    Parts a to d are independent questions, each referring to the initial situation.
19. || A 200 g air-track glider is attached to a spring. The glider is pushed in 10 cm and released. A student with a stopwatch finds that 10 oscillations take 12.0 s. What is the spring constant?
20. || A 200 g mass attached to a horizontal spring oscillates at a frequency of 2.0 Hz. At $t = 0$ s, the mass is at $x = 5.0$ cm and has $v_x = -30$ cm/s. Determine:
    a. The period.
    b. The angular frequency.
    c. The amplitude.
    d. The phase constant.
    e. The maximum speed.
    f. The maximum acceleration.
    g. The total energy.
    h. The position at $t = 0.40$ s.
21. || A 1.0 kg block is attached to a spring with spring constant 16 N/m. While the block is sitting at rest, a student hits it with a hammer and almost instantaneously gives it a speed of 40 cm/s. What are
    a. The amplitude of the subsequent oscillations?
    b. The block's speed at the point where $x = \frac{1}{2}A$?
22. || The motion of a particle is given by $x(t) = (25\text{ cm})\cos(10t)$, where t is in s. What is the first time at which the kinetic energy is twice the potential energy?
23. || Figure P15.23 is a kinetic-energy graph of a block oscillating on a *very* long horizontal spring. What is the spring constant?

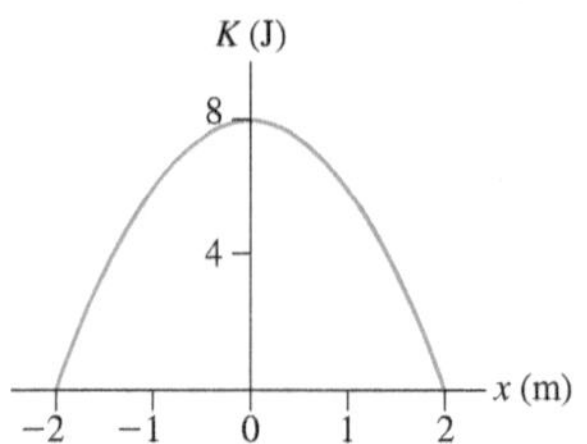

FIGURE P15.23

Section 15.4 Linear Restoring Forces

24. || The acceleration of an oscillator undergoing simple harmonic motion is described by the equation $a_x(t) = -(18.0\text{ m/s}^2)\cos(30t)$, where the time t is measured in seconds. What is the amplitude of this oscillator?
25. || Some passengers on an ocean cruise may suffer from motion sickness as the ship rocks back and forth on the waves. At one end of the ship, passengers experience a vertical motion of amplitude 1.5 m with a period of 15 s.
    a. What is the maximum acceleration of the passengers during this motion?
    b. What fraction is this of g?
26. || The New England Merchants Bank Building in Boston is 152 m high. On windy days it sways with a frequency of 0.17 Hz, and the acceleration of the top of the building can reach 2.0% of the free-fall acceleration, enough to cause discomfort for occupants. What is the total distance, side to side, that the top of the building moves during such an oscillation?
27. || In a loudspeaker, an electromagnetic coil rapidly drives a paper cone back and forth, sending out sound waves. If the cone of a loudspeaker moves sinusoidally at 1.2 kHz with an amplitude of 3.5 μm, what are the cone's maximum speed and acceleration?
28. || A 75 g block oscillates horizontally with an amplitude of 12 cm. At what frequency does the maximum force on the block equal the block's weight?
29. | A spring is hanging from the ceiling. Attaching a 500 g physics book to the spring causes it to stretch 20 cm in order to come to equilibrium.
    a. What is the spring constant?
    b. From equilibrium, the book is pulled down 10 cm and released. What is the period of oscillation?
    c. What is the book's maximum speed?
30. || A spring with spring constant 15 N/m hangs from the ceiling. A ball is attached to the spring and allowed to come to rest. It is then pulled down 6.0 cm and released. If the ball makes 30 oscillations in 20 s, what are its (a) mass and (b) maximum speed?
31. || A spring is hung from the ceiling. When a block is attached to its end, it stretches 2.0 cm before reaching its new equilibrium length. The block is then pulled down slightly and released. What is the frequency of oscillation?
32. || BIO The end of a nanoscale cantilever used for weighing DNA molecules (see Example 15.4) oscillates at 8.8 MHz with an amplitude of 5.0 nm. What are (a) the maximum speed and (b) the maximum acceleration of the end of the cantilever?

33. || Hummingbirds may seem BIO fragile, but their wings are capable of sustaining very large forces and accelerations. Figure P15.33 shows data for the vertical position of the wing tip of a rufous hummingbird. What is the maximum acceleration of the wing tip in m/s^2 and as a multiple of g?

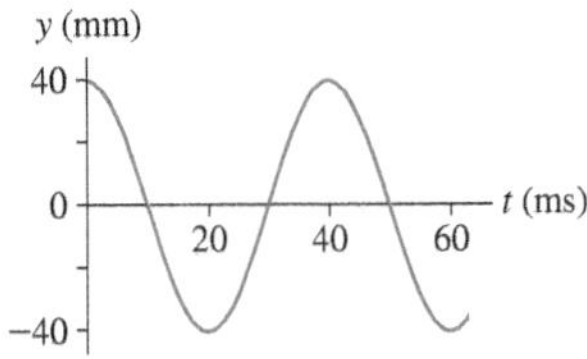

FIGURE P15.33

Section 15.5 The Pendulum

34. || A 200 g ball is tied to a string. It is pulled to an angle of 8.0° and released to swing as a pendulum. A student with a stopwatch finds that 10 oscillations take 12 s. How long is the string?
35. | A mass on a string of unknown length oscillates as a pendulum with a period of 4.0 s. What is the period if
    a. The mass is doubled?
    b. The string length is doubled?
    c. The string length is halved?
    d. The amplitude is doubled?

    Parts a to d are independent questions, each referring to the initial situation.
36. | The free-fall acceleration on the moon is 1.62 m/s ^{2}. What is the length of a pendulum whose period on the moon matches the period of a 2.00-m-long pendulum on the earth?
37. | Astronauts on the first trip to Mars take along a pendulum that has a period on earth of 1.50 s. The period on Mars turns out to be 2.45 s. Use this data to calculate the Martian free-fall acceleration.
38. || In a science museum, you may have seen a Foucault pendulum, which is used to demonstrate the rotation of the earth. In one museum's pendulum, the 110 kg bob swings from a 15.8-m-long cable with an amplitude of 5.0°.
    a. What is the period of this pendulum?
    b. What is the bob's maximum speed?
    c. What is the pendulum's maximum kinetic energy?
    d. When the bob is at its maximum displacement, how much higher is it than when it is at its equilibrium position?
39. ||| A uniform steel bar swings from a pivot at one end with a period of 1.2 s. How long is the bar?
40. || Interestingly, there have been several studies using cadavers BIO to determine the moment of inertia of human body parts by letting them swing as a pendulum about a joint. In one study, the center of gravity of a 5.0 kg lower leg was found to be 18 cm from the knee. When pivoted at the knee and allowed to swing, the oscillation frequency was 1.6 Hz. What was the moment of inertia of the lower leg?
41. || An elephant's legs have BIO a reasonably uniform cross section from top to bottom, and they are quite long, pivoting high on the animal's body. When an elephant moves at a walk, it uses very little energy to bring its legs forward, simply allowing them to swing like pendulums. For fluid walking motion, this time should be half the time for a complete stride; as soon as the right leg finishes swinging forward, the elephant plants the right foot and begins swinging the left leg forward.

    a. An elephant has legs that stretch 2.3 m from its shoulders to the ground. How much time is required for one leg to swing forward after completing a stride?
    b. What would you predict for this elephant's stride frequency? That is, how many steps per minute will the elephant take?

Section 15.6 Damped Oscillations

42. || The amplitude of an oscillator decreases to 36.8% of its initial value in 10.0 s. What is the value of the time constant?
43. || A damped pendulum has a period of 0.66 s and a time constant of 4.1 s. How many oscillations will this pendulum make before its amplitude has decreased to 20% of its initial amplitude?
44. ||| In a science museum, a 110 kg brass pendulum bob swings at the end of a 15.0-m-long wire. The pendulum is started at exactly 8:00 A.M. every morning by pulling it 1.5 m to the side and releasing it. Because of its compact shape and smooth surface, the pendulum's damping constant is only 0.010 kg/s. At exactly 12:00 noon, how many oscillations will the pendulum have completed and what is its amplitude?
45. || A physics department has a Foucault pendulum, a long-period pendulum suspended from the ceiling. The pendulum has an electric circuit that keeps it oscillating with a constant amplitude. When the circuit is turned off, the oscillation amplitude decreases by 50% in 22 minutes. What is the pendulum's time constant? How much additional time elapses before the amplitude decreases to 25% of its initial value?
46. || A small earthquake starts a lamppost vibrating back and forth. The amplitude of the vibration of the top of the lamppost is 6.5 cm at the moment the quake stops, and 8.0 s later it is 1.8 cm.
    a. What is the time constant for the damping of the oscillation?
    b. What was the amplitude of the oscillation 4.0 s after the quake stopped?
47. ||| The common field cricket makes its characteristic loud BIO chirping sound using a specialized vibrating structure in its wings. The motion of this structure—and the sound intensity that it produces—can be modeled as a damped oscillation. The sound intensity of such a cricket is shown in Figure P15.47.
    a. What is the frequency of the oscillations?
    b. What is the time constant for the decay of these oscillations?

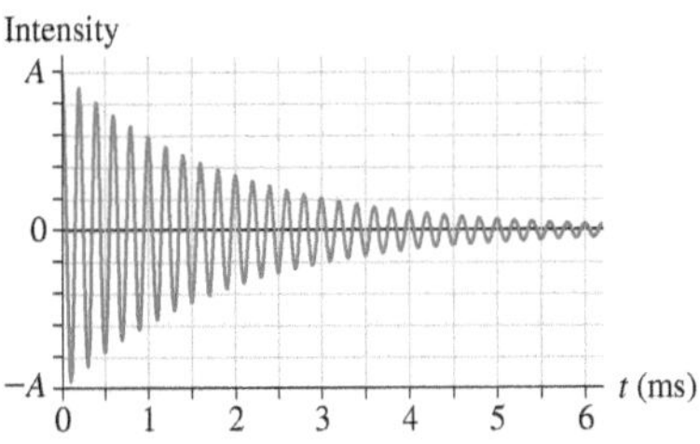

FIGURE P15.47

Section 15.7 Driven Oscillations and Resonance

48. || A 25 kg child sits on a 2.0-m-long rope swing. You are going to give the child a small, brief push at regular intervals. If you want to increase the amplitude of her motion as quickly as possible, how much time should you wait between pushes?

49. || A 2.0 g spider is dangling at the end of a silk thread. You can make the spider bounce up and down on the thread by tapping lightly on his feet with a pencil. You soon discover that you can give the spider the largest amplitude on his little bungee cord if you tap exactly once every second. What is the spring constant of the silk thread?

50. | Vision is blurred if the head is vibrated at 29 Hz because the BIO vibrations are resonant with the natural frequency of the eyeball in its socket. If the mass of the eyeball is 7.5 g, a typical value, what is the effective spring constant of the musculature that holds the eyeball in the socket?

51. || Your car rides on springs, so it has a natural frequency of oscillation. Figure P15.51 shows data for the amplitude of vertical motion of a car that is bounced up and down at different frequencies. The car is driven at 20 mph over a washboard road with bumps spaced 10 feet apart; the resulting ride is quite bouncy. Should the driver speed up or slow down for a smoother ride?

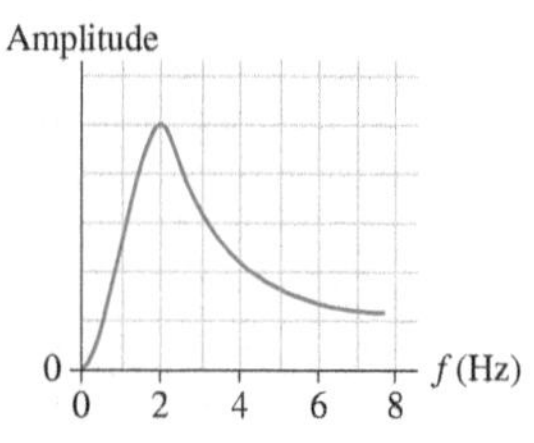

FIGURE P15.51

General Problems

52. | In one study of hummingbird wingbeats, the tip of a 5.0-cm-BIO long wing moved up and down in simple harmonic motion through a total distance of 2.5 cm at a frequency of 40 Hz. What were (a) the maximum speed and (b) the maximum acceleration of the wing tip?

53. || If you wear a backpack, the motion of your body drives the back-BIO INT pack up and down. Researchers are exploring the idea of using this motion to power a generator to produce electricity. Figure P15.53 shows the motion of a 29 kg backpack on a person who is walking at 1.5 m/s. In one test, it required only 19 W of additional metabolic power for a person to generate an energy output of 12 W.
a. What is the maximum kinetic energy of the backpack?
b. What maximum force is applied to the backpack? According to Newton's third law, this is the force that the backpack could exert on the generator system.
c. What is the efficiency, as a percent, for the conversion of metabolic power to electric power?

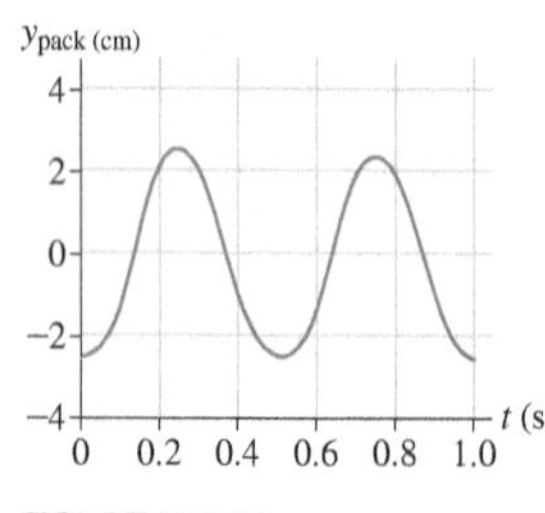

FIGURE P15.53

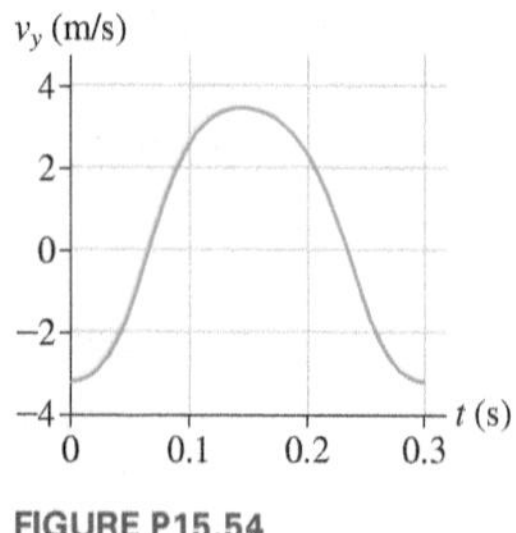

FIGURE P15.54

54. || Wild sheep have very elastic tendons in their legs that provide BIO an energy return as they trot. Figure P15.54 shows a velocity graph for the vertical motion of the 81 kg torso of a ram while it is trotting. As you can see, the motion is approximately SHM.
a. At what approximate time does the ram's torso reach its lowest point? Its highest point?
b. What is the total vertical displacement of the ram, from the lowest point to the highest point?
c. If the ram's leg system is modeled as a single spring, what is its spring constant?

55. ||| In yeast cells, a close BIO coupling of the reactions that metabolize glucose causes the concentrations of the reaction products to oscillate. The process is known as a *glycolytic oscillation*. Figure P15.55 shows smoothed data from an experiment that measured the concentrations of two energy-storage molecules that are produced from glucose: ATP (adenosine triphosphate) and NADH (nicotinamide adenine dinucleotide). The ATP concentration was measured directly in the yeast cells, and its concentration is shown in mM. NADH was monitored using a fluorescence technique that gives the relative concentration but not an absolute value.
a. What is the glycolytic oscillation frequency in mHz?
b. Relative to ATP, what is the phase constant for the NADH oscillation?
c. For ATP, what is the maximum rate in mM/min at which the concentration changes?

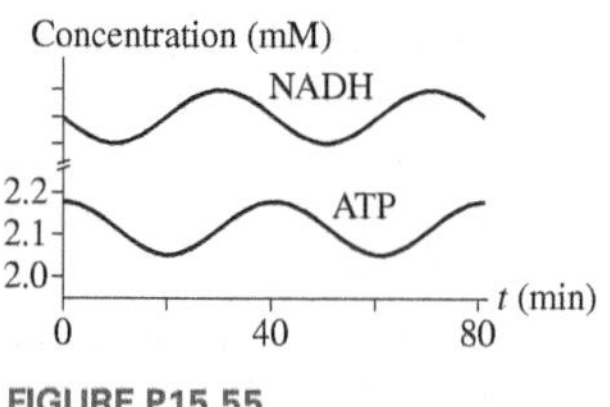

FIGURE P15.55

56. || a. When the displacement of a mass on a spring is $\frac{1}{2}A$, what fraction of the energy is kinetic energy and what fraction is potential energy?
b. At what displacement, as a fraction of A, is the energy half kinetic and half potential?

57. ||| A 0.300 kg oscillator has a speed of 95.4 cm/s when its displacement is 3.00 cm and 71.4 cm/s when its displacement is 6.00 cm. What is the oscillator's maximum speed?

58. || An ultrasonic transducer, of the type used in medical ultra-BIO sound imaging, is a very thin disk ($m = 0.10$ g) driven back and forth in SHM at 1.0 MHz by an electromagnetic coil.
a. The maximum restoring force that can be applied to the disk without breaking it is 40,000 N. What is the maximum oscillation amplitude that won't rupture the disk?
b. What is the disk's maximum speed at this amplitude?

59. || Astronauts in space cannot BIO weigh themselves by standing on a bathroom scale. Instead, they determine their mass by oscillating on a large spring. Suppose an astronaut attaches one end of a large spring to her belt and the other end to a hook on the wall of the space capsule. A fellow astronaut then pulls her away from the wall and releases her. The spring's length as a function of time is shown in Figure P15.59.
a. What is her mass if the spring constant is 240 N/m?
b. What is her speed when the spring's length is 1.2 m?

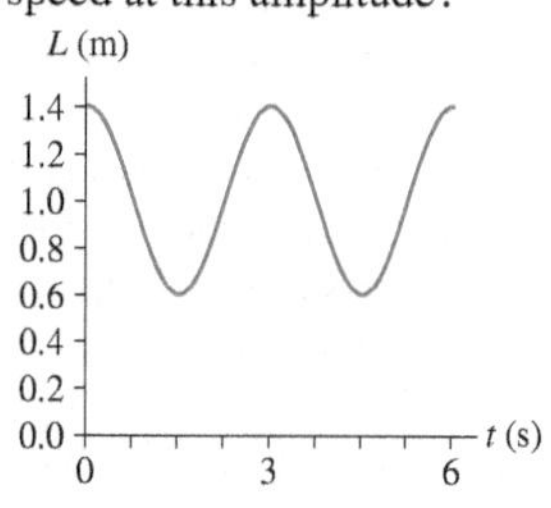

FIGURE P15.59

60. || Figure P15.60 is the energy diagram of the oxygen molecule O_2. The vibration of an O_2 molecule consists of both oxygen atoms oscillating in and out with equal amplitudes. A more advanced analysis of the oscillations finds that two equal masses m oscillating at the ends of a spring with spring constant k are equivalent—have the same frequency and energy—to a single mass $m/2$ oscillating at one end of the spring while the other end is fixed in place. What is the vibration frequency of an oxygen molecule for small oscillations of the bond length? The atomic mass of oxygen is 16 u.

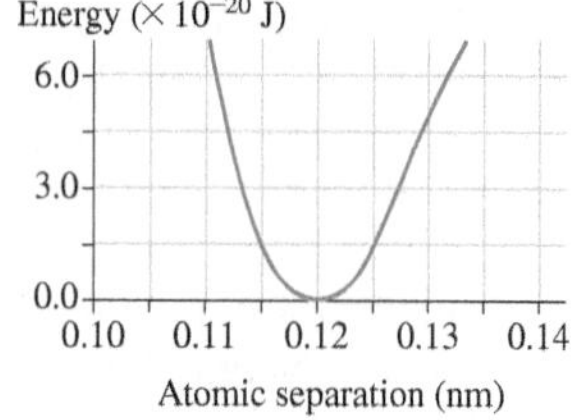

FIGURE P15.60

61. ||| A block attached to a horizontal spring is pulled back a certain distance from equilibrium, then released from rest at $t = 0$ s. If the frequency of the block is 0.72 Hz, what is the earliest time after the block is released that its kinetic energy is exactly one-half of its potential energy?

62. ||| When Kayla stands on her trampoline, it sags by 0.21 m. Now she starts bouncing. How much time elapses between the instant when she first lands on the trampoline's surface and when she passes the same point on the way up?

63. ||| A 200 g block hangs from a spring with spring constant 10 N/m. At $t = 0$ s the block is 20 cm below the equilibrium point and moving upward with a speed of 100 cm/s. What are the block's
   a. Oscillation frequency?
   b. Distance from equilibrium when the speed is 50 cm/s?
   c. Distance from equilibrium at $t = 1.0$ s?

64. ||| A block hangs in equilibrium from a vertical spring. When a second identical block is added, the original block sags by 5.0 cm. What is the oscillation frequency of the two-block system?

65. |||| A mass hanging from a spring oscillates with a period of
INT 0.35 s. Suppose the mass and spring are swung in a horizontal circle, with the free end of the spring at the pivot. What rotation frequency, in rpm, will cause the spring's length to stretch by 15%?

66. ||| A compact car has a mass of 1200 kg. Assume that the car
INT has one spring on each wheel, that the springs are identical, and that the mass is equally distributed over the four springs.
   a. What is the spring constant of each spring if the empty car bounces up and down 2.0 times each second?
   b. What will be the car's oscillation frequency while carrying four 70 kg passengers?

67. || A 1.00 kg block is attached to a horizontal spring with spring
INT constant 2500 N/m. The block is at rest on a frictionless surface. A 10 g bullet is fired into the block, in the face opposite the spring, and sticks. What was the bullet's speed if the subsequent oscillations have an amplitude of 10.0 cm?

68. || A 500 g air-track glider attached to a spring with spring constant 10 N/m is sitting at rest on a frictionless air track. A 250 g glider is pushed toward it from the far end of the track at a speed of 120 cm/s. It collides with and sticks to the 500 g glider. What are the amplitude and period of the subsequent oscillations?

69. |||| A circular cylinder has a diameter of 2.0 cm and a mass of
INT 10 g. It floats in water with its long axis perpendicular to the water's surface. It is pushed down into the water by a small distance and released; it then bobs up and down. What is the oscillation frequency?

70. || It is said that Galileo discovered a basic principle of the pendulum—that the period is independent of the amplitude—by using his pulse to time the period of swinging lamps in the cathedral as they swayed in the breeze. Suppose that one oscillation of a swinging lamp takes 5.5 s.
   a. How long is the lamp chain?
   b. What maximum speed does the lamp have if its maximum angle from vertical is 3.0°?

71. || Orangutans can move by *brachiation,* swinging like a pen-
BIO dulum beneath successive handholds. If an orangutan has arms that are 0.90 m long and repeatedly swings to a 20° angle, taking one swing after another, estimate its speed of forward motion in m/s. While this is somewhat beyond the range of validity of the small-angle approximation, the standard results for a pendulum are adequate for making an estimate.

72. ||| The pendulum shown in Figure P15.72 is
INT pulled to a 10° angle on the left side and released.
   a. What is the period of this pendulum?
   b. What is the pendulum's maximum angle on the right side?

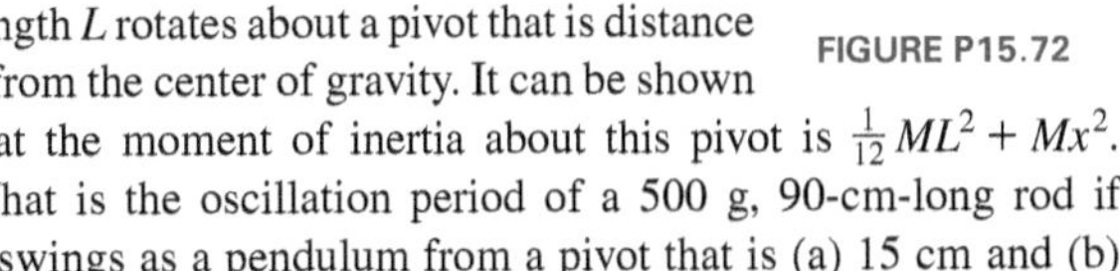

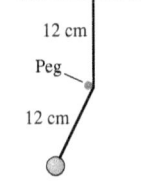

FIGURE P15.72

73. || Suppose a uniform rod of mass M and
INT length L rotates about a pivot that is distance x from the center of gravity. It can be shown that the moment of inertia about this pivot is $\frac{1}{12}ML^2 + Mx^2$. What is the oscillation period of a 500 g, 90-cm-long rod if it swings as a pendulum from a pivot that is (a) 15 cm and (b) 30 cm from the center of gravity?

74. || On your first trip to Planet X you happen to take along a 200 g mass, a 40-cm-long spring, a meter stick, and a stopwatch. You're curious about the free-fall acceleration on Planet X, where ordinary tasks seem easier than on earth, but you can't find this information in your Visitor's Guide. One night you suspend the spring from the ceiling in your room and hang the mass from it. You find that the mass stretches the spring by 31.2 cm. You then pull the mass down 10.0 cm and release it. With the stopwatch you find that 10 oscillations take 14.5 s. Based on this information, what is g?

75. || The typical American man has a leg length of 0.85 m and
BIO walks at a speed of 1.4 m/s. A giraffe's legs are 1.8 m long. At what speed do you expect a giraffe to walk? **Hint:** An animal's speed is proportional to the length of its legs times the frequency of its strides.

76. |||| A jellyfish can propel itself with jets of water pushed out
BIO of its bell, a flexible structure on top of its body. The elastic bell and the water it contains function as a mass-spring system, greatly increasing efficiency. Normally, the jellyfish emits one jet right after the other, but we can get some insight into the jet system by looking at a single jet thrust. Figure P15.76 shows a graph of the motion of one point in the wall of the bell for such a single jet; this is the pattern of a damped oscillation. The spring constant for the bell can be estimated to be 1.2 N/m.
   a. What is the period for the oscillation?
   b. Estimate the effective mass participating in the oscillation. This is the mass of the bell itself plus the mass of the water.
   c. Consider the peaks of positive displacement in the graph. By what factor does the amplitude decrease over one period? Given this, what is the time constant for the damping?

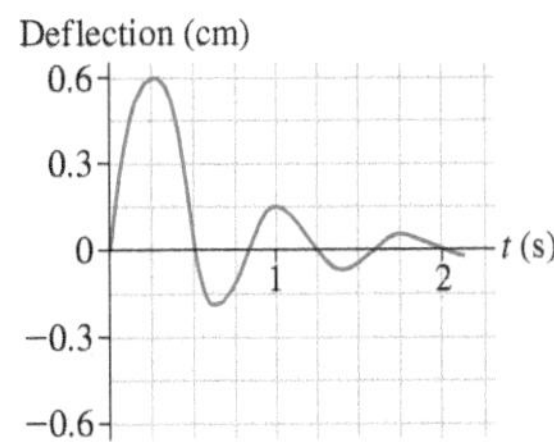

FIGURE P15.76

77. ||| While seated on a tall bench, extend your lower leg a small BIO amount and then let it swing freely about your knee joint, with no muscular engagement. It will oscillate as a damped pendulum. Figure P15.77 is a graph of the lower leg angle versus time in such an experiment. Estimate (a) the period and (b) the time constant for this oscillation.

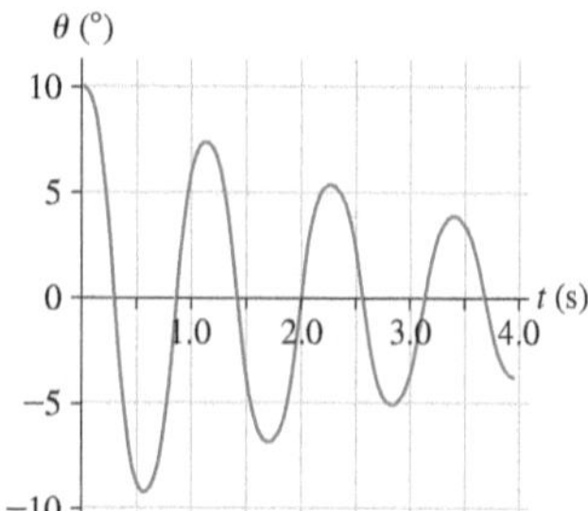

FIGURE P15.77

78. || An oscillator with a mass of 500 g and a period of 0.50 s has an amplitude that decreases by 2.0% during each complete oscillation. If the initial amplitude is 10 cm, what will be the amplitude after 25 oscillations?

79. ||| A 250 g air-track glider is attached to a spring with spring constant 4.0 N/m. The damping constant due to air resistance is 0.015 kg/s. The glider is pulled out 20 cm from equilibrium and released. How many oscillations will it make during the time in which the amplitude decays to e^{-1} of its initial value?

MCAT-Style Passage Problems

The Spring in Your Step

In Chapter 11, we saw that a runner's Achilles tendon will stretch like a spring and then rebound, storing and returning energy during a step. We can model this as the simple harmonic motion of a mass-spring system. When the foot rolls forward, the tendon spring begins to stretch as the weight moves to the ball of the foot, transforming kinetic energy into elastic potential energy. This is the first phase of an oscillation. The spring then rebounds, converting potential energy to kinetic energy as the foot lifts off the ground. The oscillation is fast: Sprinters running a short race keep each foot in contact with the ground for about 0.10 second, and some of that time corresponds to the heel strike and subsequent rolling forward of the foot.

80. We can make a static measurement to deduce the spring constant to use in the model. If a 61 kg woman stands on a low wall with her full weight on the ball of one foot and the heel free to move, the stretch of the Achilles tendon will cause her center of gravity to lower by about 2.5 mm. What is the spring constant?
A. 1.2×10^4 N/m B. 2.4×10^4 N/m
C. 1.2×10^5 N/m D. 2.4×10^5 N/m

81. If, during a stride, the stretch causes her center of gravity to lower by 10 mm, what is the stored energy?
A. 3.0 J B. 6.0 J
C. 9.0 J D. 12 J

82. If we imagine a full cycle of the oscillation, with the woman bouncing up and down and the tendon providing the restoring force, what will her oscillation period be?
A. 0.10 s B. 0.15 s
C. 0.20 s D. 0.25 s

83. Given what you have calculated for the period of the full oscillation in this model, what is the landing-to-liftoff time for the stretch and rebound of the sprinter's foot?
A. 0.050 s B. 0.10 s
C. 0.15 s D. 0.20 s

Web Spiders and Oscillations

All spiders have special organs that make them exquisitely sensitive to vibrations. Web spiders detect vibrations of their web to determine what has landed in their web, and where.

In fact, spiders carefully adjust the tension of strands to "tune" their web. Suppose an insect lands and is trapped in a web. The silk of the web serves as the spring in a mass-spring system while the body of the insect is the mass. The frequency of oscillation depends on the restoring force of the web and the mass of the insect. Spiders respond more quickly to larger—and therefore more valuable—prey, which they can distinguish by the web's oscillation frequency.

Suppose a 12 mg fly lands in the center of a horizontal spider's web, causing the web to sag by 3.0 mm.

84. Assuming that the web acts like a spring, what is the spring constant of the web?
A. 0.039 N/m B. 0.39 N/m
C. 3.9 N/m D. 39 N/m

85. Modeling the motion of the fly on the web as a mass on a spring, at what frequency will the web vibrate when the fly hits it?
A. 0.91 Hz B. 2.9 Hz
C. 9.1 Hz D. 29 Hz

86. If the web were vertical rather than horizontal, how would the frequency of oscillation be affected?
A. The frequency would be higher.
B. The frequency would be lower.
C. The frequency would be the same.

87. Spiders are more sensitive to oscillations at higher frequencies. For example, a low-frequency oscillation at 1 Hz can be detected for amplitudes down to 0.1 mm, but a high-frequency oscillation at 1 kHz can be detected for amplitudes as small as 0.1 μm. For these low- and high-frequency oscillations, we can say that
A. The maximum acceleration of the low-frequency oscillation is greater.
B. The maximum acceleration of the high-frequency oscillation is greater.
C. The maximum accelerations of the two oscillations are approximately equal.

16 Traveling Waves and Sound

The surfer is "catching a wave." At the same time, he is seeing light waves and hearing sound waves.

LOOKING AHEAD

Wave Motion

This woman is sending a *disturbance* through a *medium*—the rope—at a well-defined *wave speed.* These are characteristics of a *traveling wave.*

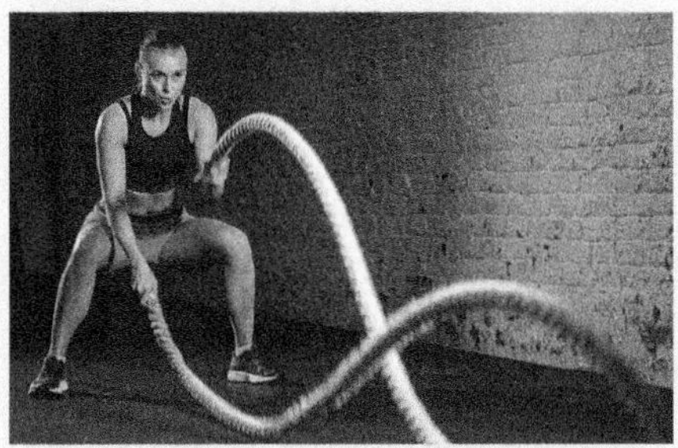

You'll learn about the wave model and how the wave speed is determined by the elastic properties of the medium.

Sound and Light

Two of the most important waves are *sound,* a longitudinal wave, and *light,* a transverse wave.

The loudness of sound is measured using decibels.

The colors of light correspond to different wavelengths.

You'll see how the physics of the ear determines the range of frequencies that humans and other animals can hear.

The Doppler Effect

The pitch of an ambulance siren drops as it races past you. This is an example of an important wave phenomenon called the *Doppler effect.*

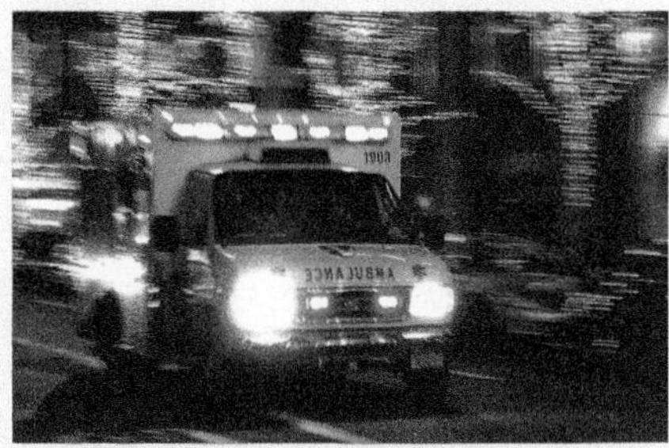

You'll learn how the shift in frequency depends on the motion of the source and the listener.

GOAL To learn the basic properties of traveling waves.

PHYSICS AND LIFE

Sound Waves Can Examine and Image the Inside of the Body

Humans and animals use sound waves to communicate and to perceive their environment. Sound waves are also medically important in the form of ultrasound imaging of the interior of the body. Ultrasound is less expensive than MRI and, unlike x rays or CT scans, does not expose the patient to ionizing radiation. In addition to imaging, ultrasound can measure the speed of blood flow through arteries by utilizing the Doppler effect. Expert use of ultrasound technology and the correct reading of ultrasound images require an understanding of the physics of sound and of how sound waves travel through different media.

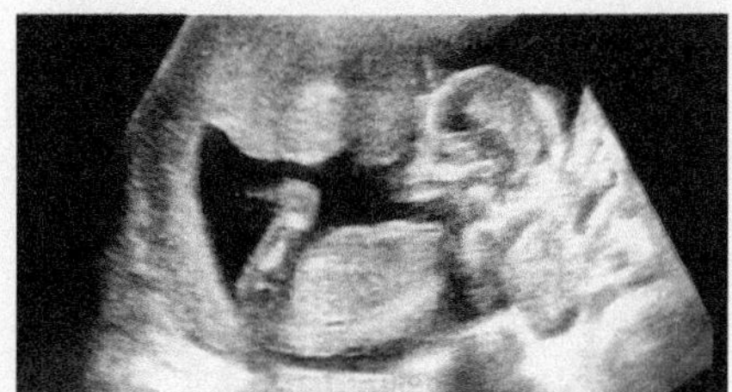

Ultrasound imaging is one of the most important applications of waves, providing an essential tool for obstetrics, cardiology, and other fields of medicine.

16.1 An Introduction to Waves

From sound and light to ocean waves and seismic waves, we're surrounded by waves. Understanding hearing, ultrasound, or the medical use of lasers requires a knowledge of waves. With this chapter we shift our focus from the particle model to a new **wave model** that emphasizes those aspects of wave behavior common to all waves.

The wave model is built around the idea of a **traveling wave,** which is an organized disturbance traveling with a well-defined wave speed. We'll begin our study of traveling waves by looking at two distinct wave motions.

Two types of wave motion

A transverse wave

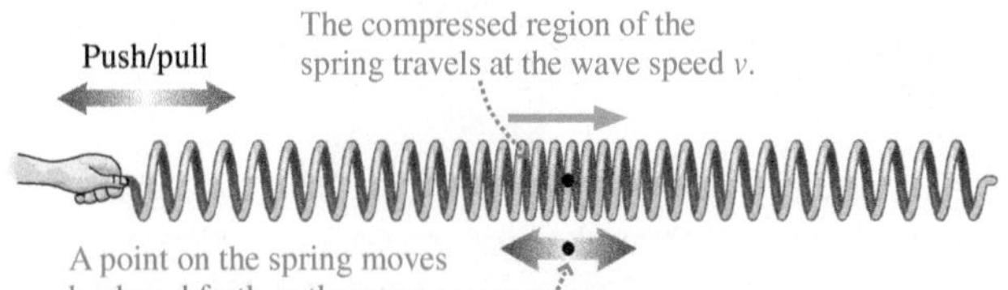
The bump in the rope travels at the wave speed v.
Up/down
A point on the rope moves up and down as the wave passes.

For mechanical waves, a **transverse wave** is a wave in which the particles in the medium move *perpendicular* to the direction in which the wave travels. Shaking the end of a stretched string creates a wave that travels along the string in a horizontal direction while the particles that make up the string oscillate vertically. Electromagnetic waves are also transverse waves in which the electromagnetic field oscillates perpendicular to the direction of travel.

A longitudinal wave

Push/pull

The compressed region of the spring travels at the wave speed v.

A point on the spring moves back and forth as the wave passes.

In a **longitudinal wave,** the particles in the medium move *parallel* to the direction in which the wave travels. Quickly moving the end of a spring back and forth sends a wave—in the form of a compressed region—down the spring. The particles that make up the spring oscillate horizontally as the wave passes. Sound waves in gases and liquids are the most well known longitudinal waves.

NOTE ▸ Transverse and longitudinal waves are two *models* of waves. Many waves, such as light and sound, are well described by these models. Some waves, such as water waves, have both transverse and longitudinal characteristics. Their analysis is more complex, so we'll focus on waves that can be modeled as purely transverse or purely longitudinal. ◂

We can also classify waves on the basis of what is "waving":

1. **Mechanical waves** travel only within a material *medium,* such as air or water. Two familiar mechanical waves are sound waves and water waves.
2. **Electromagnetic waves,** from radio waves to visible light to x rays, are a self-sustaining oscillation of the *electromagnetic field.* Electromagnetic waves require no material medium and can travel through a vacuum.

FIGURE 16.1 Ripples on a pond are a traveling wave.

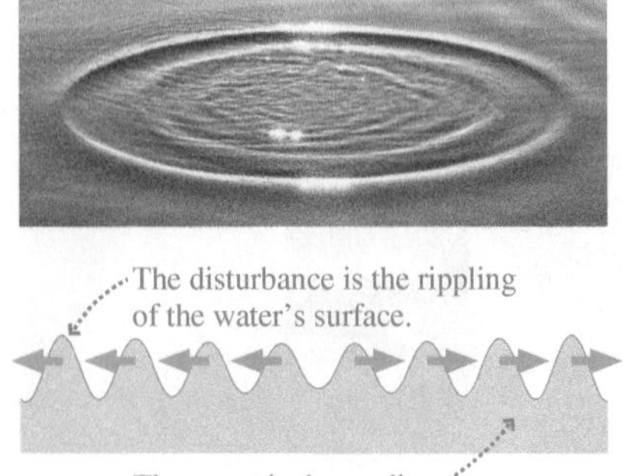
The disturbance is the rippling of the water's surface.
The water is the medium.

The **medium** of a mechanical wave is the substance through or along which the wave moves. For example, the medium of a water wave is the water, the medium of a sound wave is the air, and the medium of a wave on a stretched string is the string. A medium must be *elastic.* That is, a restoring force of some sort brings the medium back to equilibrium after it has been displaced or disturbed. The tension in a stretched string pulls the string back straight after you pluck it. Gravity restores the level surface of a lake after the wave generated by a boat has passed by.

As a wave passes through a medium, the molecules of the medium—we'll simply call them the particles of the medium—are displaced from equilibrium. This is a *disturbance* of the medium. The water ripples of FIGURE 16.1 are a disturbance of the water's surface. A pulse traveling down a string is a disturbance, as are the wake of a boat and the sonic boom created by a jet traveling faster than the speed of sound. **The disturbance of a wave is an *organized* motion of the particles in the medium,** in contrast to the *random* molecular motions of thermal energy.

NOTE ▶ The disturbance propagates through the medium, but the medium as a whole does not move! The ripples on the pond (the disturbance) move outward from the splash of the rock, but there is no outward flow of water from the splash. **A wave transfers energy, but it does not transfer any material or substance outward from the source.** ◀

Wave Speed

A wave disturbance is created by a *source*. The source of a wave might be a rock thrown into water, your hand plucking a stretched string, or an oscillating loudspeaker cone pushing on the air. Once created, the disturbance travels outward through the medium at the **wave speed** v. This is the speed with which a ripple moves across the water or a pulse travels down a string.

A mass oscillating on a spring was our prototype for simple harmonic motion. Similarly, our prototype for waves will be a disturbance traveling along a stretched string that is under tension T_s. The lessons we learn from waves on strings will apply to other types of waves.

NOTE ▶ The subscript s on the symbol T_s for the string's tension distinguishes it from the symbol T for the oscillation period. ◀

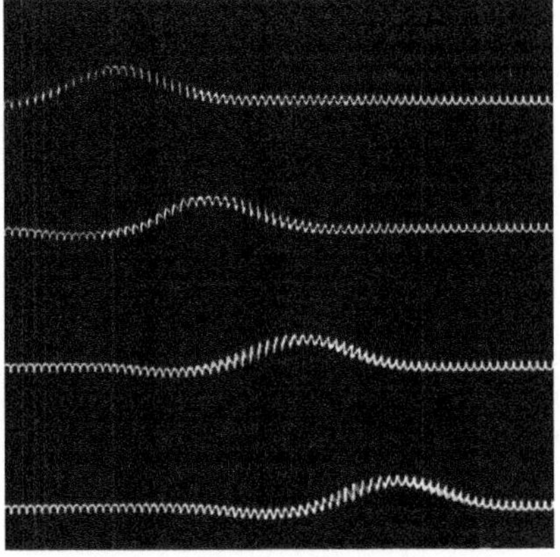
This sequence of photographs shows a wave pulse traveling along a spring.

If you stretch a string and pluck one end, what determines how quickly the pulse travels down the string? As you can imagine, pulling the string tighter—increasing the tension—makes the pulse go faster. It also seems likely that replacing a lightweight string with a heavy rope under the same tension would slow the pulse because more mass would have to accelerate up and down as the pulse passes. But the determining factor is not the mass itself—doubling the string length doubles the mass without changing the wave speed—but the mass-to-length ratio m/L. This quantity is the same no matter the length of the string.

An object's density ρ is its mass-to-volume ratio. Analogously, the mass-to-length ratio is called the **linear density,** represented by the symbol μ (Greek mu). That is, a string of length L and mass m has linear density

$$\mu = \frac{m}{L} \tag{16.1}$$

The SI unit of linear density is kg/m, although it is common to see values given in g/m. If two strings are made of the same material, the fatter string has a larger value of μ than the skinnier string. Similarly, a steel wire has a larger value of μ than a plastic string of the same diameter. We will assume that strings are *uniform,* which means that the linear density is the same everywhere along the string.

It can be shown (with a lengthy derivation) that the wave speed on a stretched string is

$$v_{\text{string}} = \sqrt{\frac{T_s}{\mu}} \tag{16.2}$$

Wave speed on a stretched string

Equation 16.2 is the wave *speed,* not the wave velocity, so v_{string} always has a positive value. Every point on a wave travels with this speed. You can increase the wave speed either by *increasing* the string's tension (make it tighter) or by *decreasing* the string's linear density (make it skinnier).

The wave speed on a string is a property of the string—its tension and linear density. In general, **the wave speed is a property *of the medium.*** The wave speed depends on the restoring forces within the medium but not at all on the shape or size of the pulse, how the pulse was generated, or how far it has traveled.

EXAMPLE 16.1 **How quickly does the spider sense its lunch?**

All spiders are very sensitive to vibrations. An orb spider will sit at the center of its large, circular web and monitor radial threads for vibrations created when an insect lands. Assume that these threads are made of silk with a linear density of 0.010 g/m under a tension of 0.15 N, both typical values. If an insect lands in the web 30 cm from the spider, how long will it take for the spider to find out?

PREPARE When the insect hits the web, a wave pulse will be transmitted along the silk fibers. The speed of the wave depends on the properties of the silk. In SI units, the linear density of the silk is 1.0×10^{-5} kg/m.

SOLVE First, we determine the speed of the wave:

$$v_{\text{string}} = \sqrt{\frac{T_s}{\mu}} = \sqrt{\frac{0.15\text{ N}}{1.0 \times 10^{-5}\text{ kg/m}}} = 120\text{ m/s}$$

The time for the wave to travel a distance $d = 30$ cm to reach the spider is

$$\Delta t = \frac{d}{v_{\text{string}}} = \frac{0.30\text{ m}}{120\text{ m/s}} = 2.5\text{ ms}$$

ASSESS Spider webs are made of very light strings under significant tension, so the wave speed is quite high and we expect a short travel time—important for the spider to quickly respond to prey caught in the web. Our answer makes sense.

STOP TO THINK 16.1 Which of the following actions would make a pulse travel faster along a stretched string? More than one answer may be correct. If so, give all that are correct.

A. Move your hand up and down more quickly as you generate the pulse.
B. Move your hand up and down a larger distance as you generate the pulse.
C. Use a heavier string of the same length, under the same tension.
D. Use a lighter string of the same length, under the same tension.
E. Stretch the string tighter to increase the tension.
F. Loosen the string to decrease the tension.
G. Put more force into the wave.

FIGURE 16.2 A snapshot graph of a wave pulse shows how it looks at one instant of time.

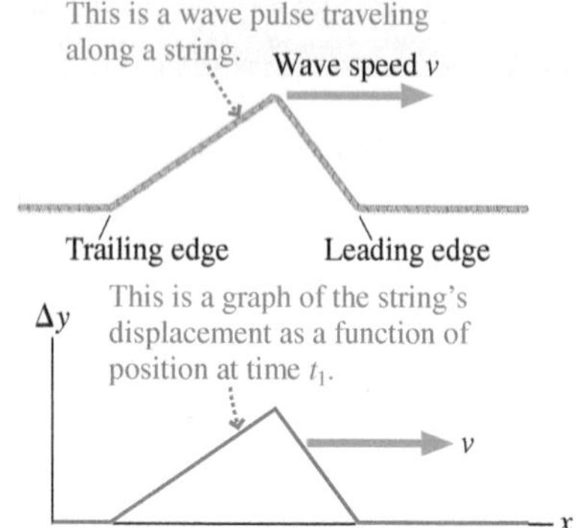

FIGURE 16.3 A sequence of snapshot graphs shows the wave in motion.

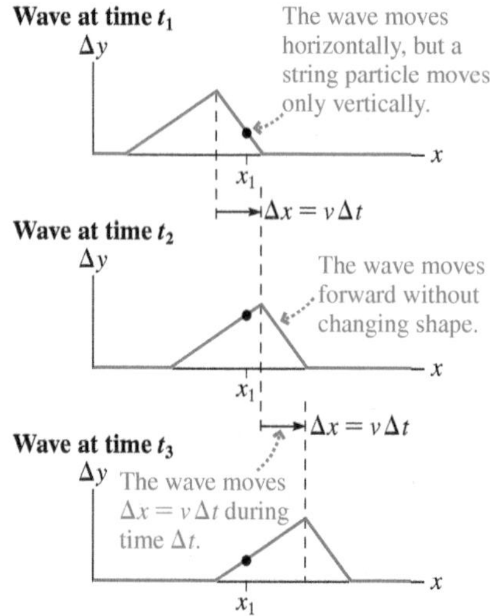

16.2 Visualizing Wave Motion

To understand waves we must deal with functions of *two* variables. Until now, we have been concerned with quantities that depend only on time, such as $x(t)$ or $v(t)$. Functions of the one variable t are appropriate for a particle because a particle is only in one place at a time, but a wave is not localized. It is spread out through space at each instant of time. To describe a wave mathematically requires a function that specifies not only an instant of time (when) but also a point in space (where).

Rather than leaping into mathematics, we will start by thinking about waves graphically. Consider the wave pulse shown moving along a stretched string in FIGURE 16.2. (We will consider somewhat artificial triangular and square-shaped pulses in this section to make clear where the edges of the pulse are.) Graphs of waves show the **displacement** from equilibrium of the particles in the medium. For a wave on a string, the displacement Δy at position x is the sideways displacement of the string at that position. The graph of Figure 16.2 shows the string's displacement Δy at a particular instant of time t_1 at every position x along the string. We'll call this a **snapshot graph** because it's much like a snapshot you might make with a camera whose shutter is opened briefly at t_1. For a wave on a string, a snapshot graph is literally a picture of the wave at this instant.

FIGURE 16.3 shows a sequence of snapshot graphs as the wave of Figure 16.2 continues to move. These are like successive frames from a video. Notice that the wave pulse moves forward distance $\Delta x = v\,\Delta t$ during the time interval Δt. That is, the wave moves with constant speed.

A snapshot graph tells only half the story. It tells us *where* the wave is and how it varies with position, but only at one instant of time. It gives us no information about how the wave *changes* with time. As a different way of portraying the wave, suppose we follow the dot marked on the string in Figure 16.3 and produce a graph showing how the displacement of this dot changes with time. The result, shown in FIGURE 16.4, is called a **history graph.** It tells the history of that particular point in the medium.

NOTE ▸ A velocity vector is shown on a snapshot graph but not on a history graph. The wave pulse travels along the axis, moving either right or left, but it doesn't travel through time. ◂

You might think we have made a mistake; the graph of Figure 16.4 is reversed compared to Figure 16.3. It is not a mistake, but it requires careful thought to see why. As the wave moves toward the dot, the steep **leading edge** causes the dot to rise quickly. On the displacement graph, *earlier* times (smaller values of t) are to the *left* and later times (larger t) to the right. Thus the leading edge of the wave is on the *left* side of the Figure 16.4 history graph. As you move to the right on Figure 16.4 you see the slowly falling **trailing edge** of the wave as it moves past the dot at later times.

The snapshot graph of Figure 16.2 and the history graph of Figure 16.4 portray complementary information. The snapshot graph tells us how things look throughout all of space, but at only one instant of time. The history graph tells us how things look at all times, but at only one position in space. We need them both to have the full story of the wave. An alternative representation of the wave is the series of graphs in FIGURE 16.5, where we can get a clearer sense of the wave moving forward. But graphs like these are essentially impossible to draw by hand, so it is necessary to move back and forth between snapshot graphs and history graphs.

FIGURE 16.4 A history graph for the dot on the string in Figure 16.3 shows how the displacement at this point changes with time.

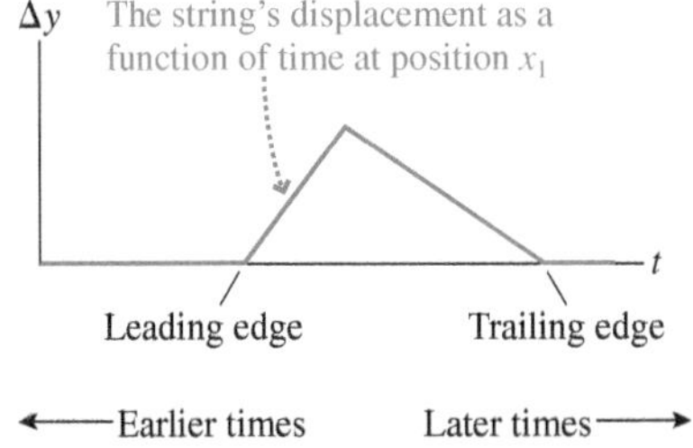

The wave You've probably seen or participated in "the wave" at a sporting event. The wave moves around the stadium, but the people (the medium) simply undergo small displacements from their equilibrium positions.

FIGURE 16.5 An alternative look at a traveling wave.

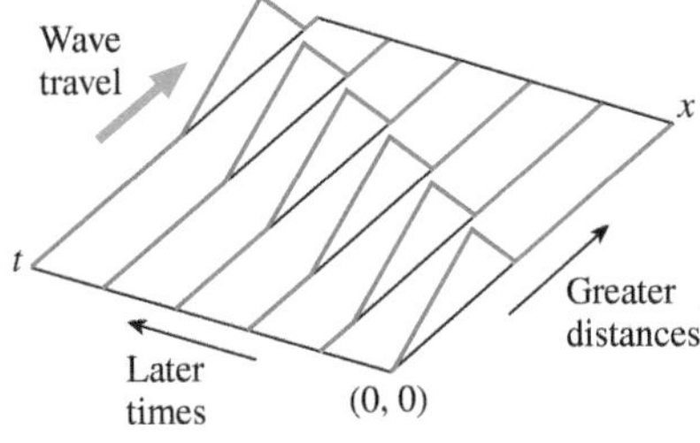

CONCEPTUAL EXAMPLE 16.2 **Finding a history graph from a snapshot graph**

FIGURE 16.6 is a snapshot graph at $t = 0$ s of a wave moving to the right at a speed of 2.0 m/s. Draw a history graph for the position $x = 8.0$ m.

FIGURE 16.6 A snapshot graph at $t = 0$ s.

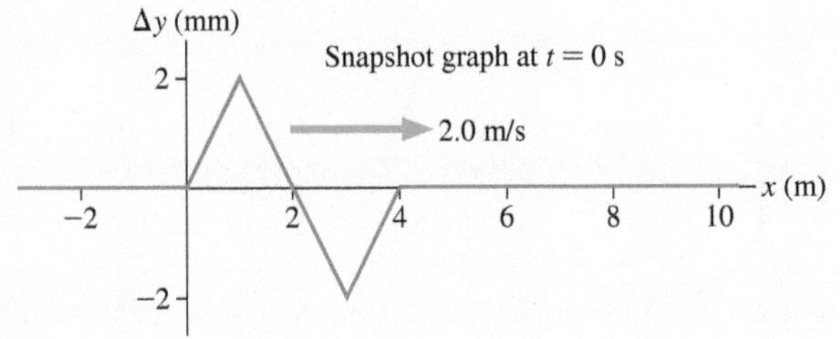

REASON This is a wave traveling at constant speed. The pulse moves 2.0 m to the right every second. The snapshot graph of Figure 16.6 shows the wave at all points on the x-axis at $t = 0$ s. You can see that nothing is happening at $x = 8.0$ m at this instant of time because the wave has not yet reached $x = 8.0$ m. In fact, at $t = 0$ s the leading edge of the wave is still 4.0 m away from $x = 8.0$ m. Because the wave is traveling at 2.0 m/s, it will take 2.0 s for the leading edge to reach $x = 8.0$ m. Thus the history graph for $x = 8.0$ m will be zero until $t = 2.0$ s. The first part of the wave causes a *downward* displacement of the medium, so immediately after $t = 2.0$ s the displacement at $x = 8.0$ m will be negative. The negative portion of the wave pulse is 2.0 m wide and takes 1.0 s to pass $x = 8.0$ m, so the midpoint of the pulse reaches $x = 8.0$ m at $t = 3.0$ s. The positive portion takes another 1.0 s to go past, so the trailing edge of the pulse arrives at

Continued

$t = 4.0$ s. You could also note that the trailing edge was initially 8.0 m away from $x = 8.0$ m and needed 4.0 s to travel that distance at 2.0 m/s. The displacement at $x = 8.0$ m returns to zero at $t = 4.0$ s and remains zero for all later times. This information is all portrayed on the history graph of FIGURE 16.7.

ASSESS Figures 16.6 and 16.7 use complementary information to describe the same wave. The first shows the pulse moving through space, the second how the medium changes with time at one point in space.

FIGURE 16.7 The corresponding history graph at $x = 8.0$ m.

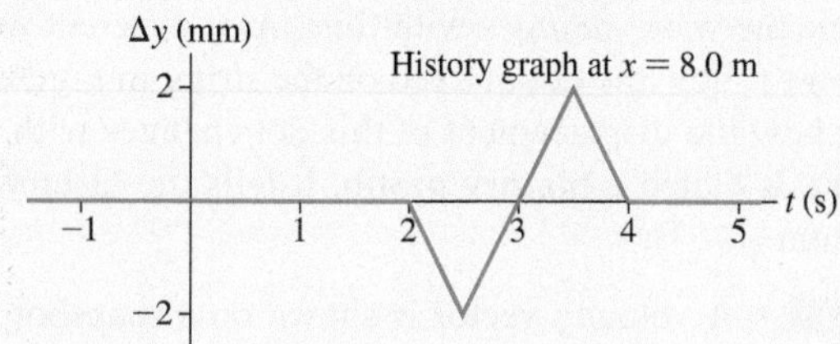

STOP TO THINK 16.2 The graph at the right is the history graph at $x = 4.0$ m of a wave traveling to the right at a speed of 2.0 m/s. Which is the history graph of this wave at $x = 0$ m?

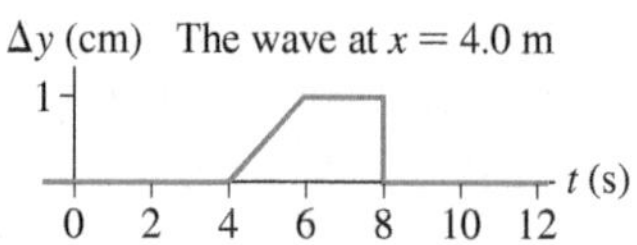

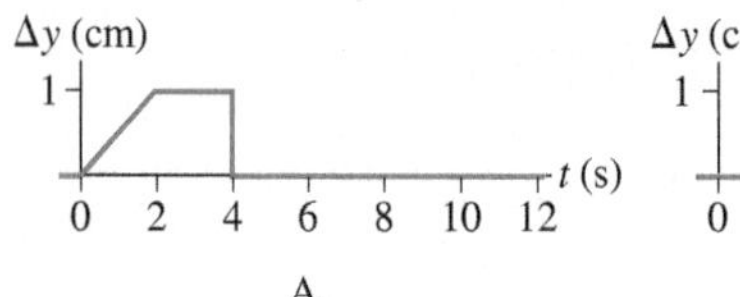

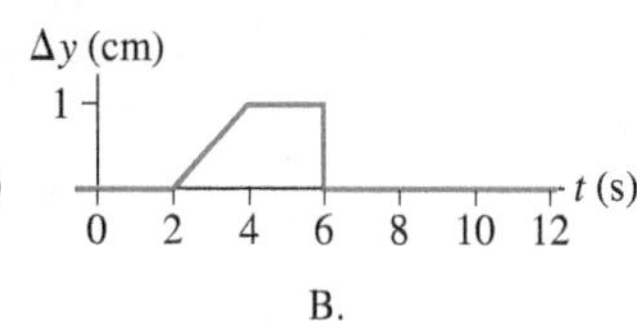

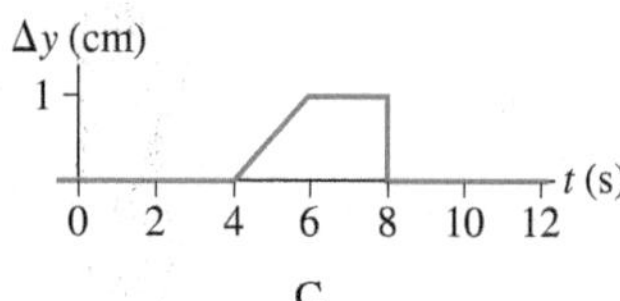

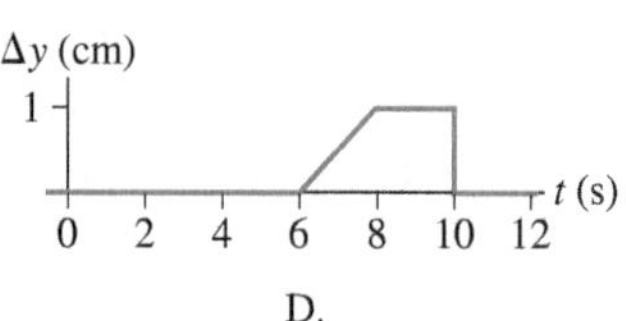

Longitudinal Waves

For a wave on a string, a transverse wave, the snapshot graph is literally a picture of the wave. Not so for a longitudinal wave, where the particles in the medium are displaced parallel to the direction in which the wave is traveling. Thus the displacement is Δx rather than Δy, and a snapshot graph is a graph of Δx versus x.

FIGURE 16.8a is a snapshot graph of a longitudinal wave, such as a sound wave. It's purposefully drawn to have the same shape as the string wave in Conceptual Example 16.2. Without practice, it's not clear what this graph tells us about the particles in the medium.

To help you find out, FIGURE 16.8b provides a tool for visualizing longitudinal waves. In the second row, we've used information from the graph to displace the particles in the medium to the right or to the left of their equilibrium positions. For example, the

FIGURE 16.8 Visualizing a longitudinal wave.

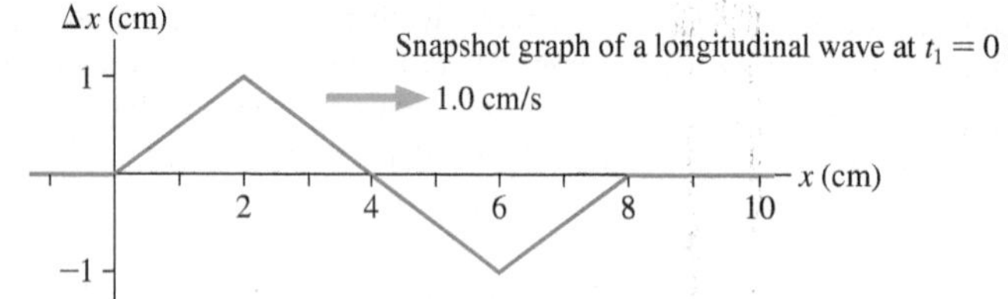

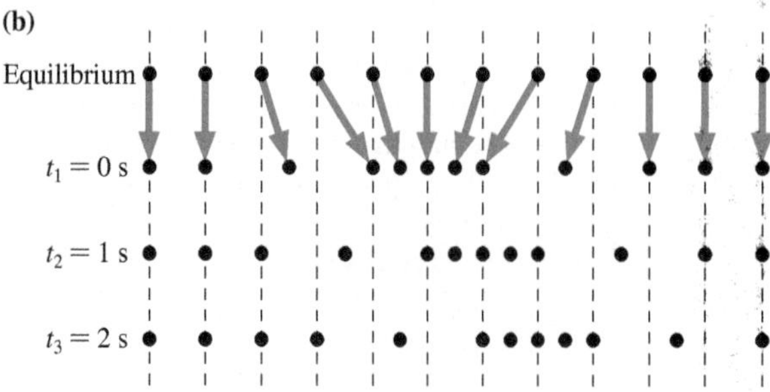

particle at $x = 1.0$ cm has been displaced 0.5 cm to the right because the snapshot graph shows $\Delta x = 0.5$ cm at $x = 1.0$ cm. We now have a picture of the longitudinal wave pulse at $t_1 = 0$ s. You can see that the medium is compressed to higher density at the center of the pulse and, to compensate, expanded to lower density at the leading and trailing edges. Two more lines show the medium at $t_2 = 1$ s and $t_3 = 2$ s so that you can see the wave propagating through the medium at 1.0 cm/s.

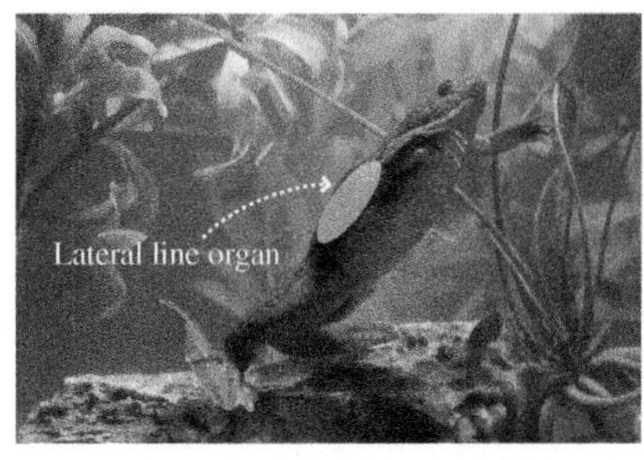

Sensing water waves The African clawed frog has a simple hunting strategy: It sits and "listens" for prey animals. It detects water waves, not sound waves, using an array of sensors called the *lateral line organ* on each side of its body. Each sensor, called a *neuromast,* converts the slight motion of microscopic hairs to nerve impulses. An incoming wave reaches different parts of the organ at different times, and this allows the frog to determine where the waves come from.

The Displacement

A traveling wave causes the particles of the medium to be displaced from their equilibrium positions. Because one of our goals is to develop a mathematical representation to describe all types of waves, we'll use the generic symbol D to stand for the *displacement* of a wave of any type. But what do we mean by a "particle" in the medium? And what about electromagnetic waves, for which there is no medium?

For a string, you can think of very small segments of the string as being the particles of the medium. D is then the perpendicular displacement Δy of a point on the string. For a sound wave, D is the longitudinal displacement Δx of a small volume of fluid. For any other mechanical wave, D is the appropriate displacement. Even electromagnetic waves can be described within the same mathematical representation if D is interpreted as an *electromagnetic field strength,* a topic we'll return to in Part V.

Because the displacement of a particle in the medium depends both on *where* the particle is (position x) and on *when* you observe it (time t), D must be a function of the two variables x and t. That is,

$$D(x, t) = \text{the displacement at time } t \text{ of a particle at position } x$$

The values of *both* variables—where and when—must be specified before you can evaluate the displacement D.

EXAMPLE 16.3 Tsunami!

A *tsunami* (sometimes mistakenly called a *tidal wave*) is an ocean wave, but it differs in many respects from normal ocean waves lapping at the shore. A tsunami is created by an underwater earthquake or landslide, and it can be modeled, as shown in FIGURE 16.9, as a single wave pulse traveling outward from the source at speed v. A tsunami traveling across the open ocean is *very* wide ($w \approx 200$ km ≈ 120 mi) and not at all tall ($h \approx 1$ m). A tsunami passes beneath ships at sea as a minor swell that is not even noticed. Because the ocean depth d is much smaller than the width of the tsunami pulse, a tsunami is classified as a *shallow-water wave.* It can be shown that the speed of a shallow-water wave is $v = \sqrt{gd}$. A tsunami slows as it approaches land because the ocean depth decreases. Cars on a highway get more bunched together when they encounter an obstacle that forces them to slow down. Similarly, the width of the tsunami pulse gets smaller as it slows; it can be shown that the width is directly proportional to the speed: $w \propto v$.

a. The average depth of the open ocean is 4000 m. What is the average speed of a tsunami in the open ocean? How long does it take a 200-km-wide tsunami to pass beneath a ship?
b. What are the speed and width of this tsunami as it enters a harbor with a depth of 10 m?
c. What is the height of the tsunami in the harbor? How long does it take to move onshore?

FIGURE 16.9 A tsunami can be modeled as a wave pulse.

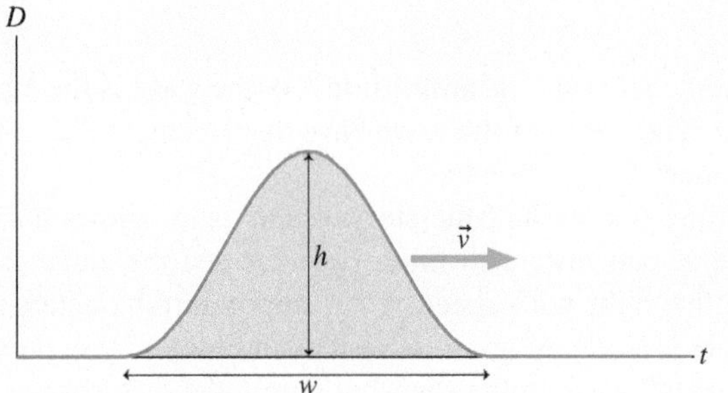

PREPARE We'll model the tsunami as a wave pulse traveling along a line. Its motion is like that of a wave along a stretched string, but its speed changes as the ocean depth changes.

SOLVE

a. The tsunami's speed as it travels through the 4000-m-deep ocean is

$$v_{\text{ocean}} = \sqrt{gd} = \sqrt{(9.8 \text{ m/s}^2)(4000 \text{ m})} = 200 \text{ m/s} \approx 450 \text{ mph}$$

This is nearly the speed of a jet flying over the ocean! At this speed, a 200-km-wide pulse passes beneath a ship in a time of

$$\Delta t = \frac{w_{\text{ocean}}}{v_{\text{ocean}}} = \frac{200{,}000 \text{ m}}{200 \text{ m/s}} = 1000 \text{ s} \approx 16 \text{ min}$$

A 1-m-high swell that takes 8 min to rise and another 8 min to fall will certainly not be noticed.

Continued

b. As the ocean depth decreases to 10 m, the tsunami slows to a speed of

$$v_{harbor} = \sqrt{(9.8 \text{ m/s}^2)(10 \text{ m})} = 10 \text{ m/s} \approx 20 \text{ mph}$$

This is more typical of the speed of ordinary ocean waves approaching the shore. The pulse width is proportional to the speed, so we can use ratios to write

$$\frac{w_{harbor}}{v_{harbor}} = \frac{w_{ocean}}{v_{ocean}}$$

Thus the width of the tsunami in the harbor, as it reaches the shore, is

$$w_{harbor} = \frac{v_{harbor}}{v_{ocean}} \times w_{ocean} = \frac{10 \text{ m/s}}{200 \text{ m/s}} \times 200 \text{ km} = 10 \text{ km}$$

This is still a very wide pulse of water, although much smaller than the 200 km width in the open ocean.

c. Water is nearly incompressible. As the width of the pulse narrows, the pulse has to get taller to account for the volume of water in the pulse. In our one-dimensional model, the area of the pulse, which is proportional to width × height, must remain constant. Thus

$$w_{harbor} \times h_{harbor} = w_{ocean} \times h_{ocean}$$

We can calculate that the height of the tsunami in the harbor is

$$h_{harbor} = \frac{w_{ocean}}{w_{harbor}} \times h_{ocean} = \frac{200 \text{ km}}{10 \text{ km}} \times 1 \text{ m} = 20 \text{ m} \approx 60 \text{ ft}$$

The time it takes this pulse of water to come ashore is

$$\Delta t = \frac{w_{harbor}}{v_{harbor}} = \frac{10{,}000 \text{ m}}{10 \text{ m/s}} = 1000 \text{ s} \approx 16 \text{ min}$$

Because the pulse narrows as it slows, the time it takes to pass is the same in the harbor as at sea.

ASSESS Our results are quite consistent with tsunamis that have been observed after major undersea earthquakes. Unlike movie versions, a tsunami is not a large breaking wave. It is more like a humongous tidal surge in which the water level just keeps rising and rising for ≈8 min until it is the height of a four or five story building. The ocean may push several miles inland during this time, causing great destruction and loss of life. Then the water recedes, rushing outward for another ≈8 min. Tsunamis often have secondary pulses that continue to come and go every 15 to 20 minutes, but these are usually much smaller than the initial pulse.

16.3 Sinusoidal Waves

In Chapter 15 we studied objects, such as springs and pendulums, that oscillate sinusoidally with simple harmonic motion (SHM). An object in SHM oscillates in a small, confined region of space. However, a SHM oscillator that is connected to an elastic medium can generate a **sinusoidal wave** that travels outward through the medium. For example, a loudspeaker cone that oscillates in SHM produces a sinusoidal sound wave. The sinusoidal electromagnetic waves broadcast by cell towers and FM radio stations are generated by electrons oscillating back and forth in the antenna wire with SHM. **The frequency f of the wave is the frequency of the oscillating source.**

FIGURE 16.10 A sinusoidal wave moving along the x-axis.

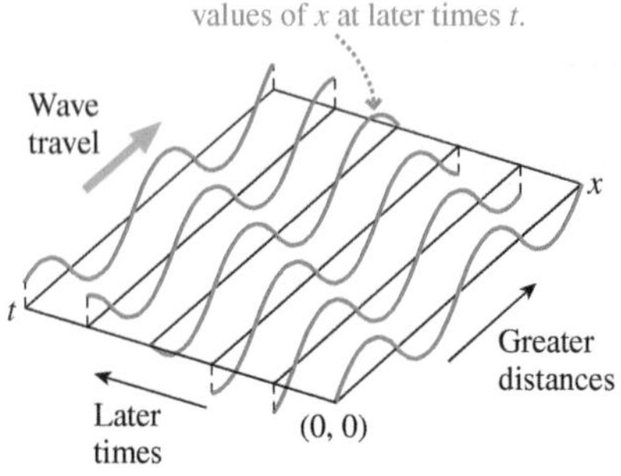
The wave crests move with steady speed toward larger values of x at later times t.
Wave travel
x
t
Greater distances
Later times
(0, 0)

FIGURE 16.10 shows a sinusoidal wave moving along the x-axis. To understand how this wave propagates, let's look at history and snapshot graphs. FIGURE 16.11a is a history graph, showing the displacement of the medium at one point in space. Each particle in the medium undergoes simple harmonic motion with frequency f, so this graph of SHM is identical to the graphs you learned to work with in Chapter 15. The *period* of the wave, shown on the graph, is the time interval for one cycle of the motion. The period is related to the wave frequency f by

$$T = \frac{1}{f} \tag{16.3}$$

exactly as in simple harmonic motion. The **amplitude** A of the wave is the maximum value of the displacement. The crests of the wave have displacement $D_{crest} = A$ and the troughs have displacement $D_{trough} = -A$.

Displacement versus time is only half the story. FIGURE 16.11b shows a snapshot graph for the same wave at one instant in time. Here we see the wave stretched out in space, moving to the right with speed v. An important characteristic of a sinusoidal wave is that it is *periodic in space* as well as in time. As you move from left to right along the "frozen" wave in the snapshot graph, the disturbance repeats itself over and over. The distance spanned by one cycle of the motion is called

FIGURE 16.11 History and snapshot graphs for a sinusoidal wave.

(a) A history graph shows displacement versus time at one point in space.

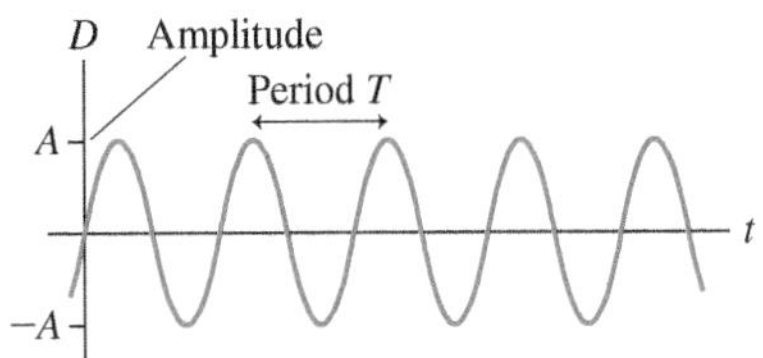

(b) A snapshot graph shows displacement versus position at one instant of time.

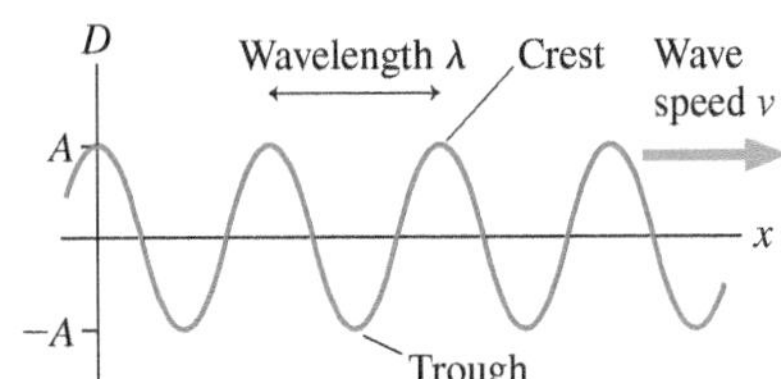

the **wavelength** of the wave. Wavelength is symbolized by λ (Greek lambda) and, because it is a length, it is measured in meters. The wavelength is shown in Figure 16.11b as the distance between two crests, but it could equally well be the distance between two troughs.

NOTE ▸ Wavelength is the spatial analog of period. The period T is the *time* in which the disturbance at a single point in space repeats itself. The wavelength λ is the *distance* in which the disturbance at one instant of time repeats itself. ◂

The Fundamental Relationship for Sinusoidal Waves

There is an important relationship between the wavelength and the period of a wave. FIGURE 16.12 shows this relationship through five snapshot graphs of a sinusoidal wave at time increments of one-quarter of the period T. One full period has elapsed between the first graph and the last, which you can see by observing the motion at a fixed point on the x-axis. Each point in the medium has undergone exactly one complete oscillation.

The critical observation is that the wave crest marked by an arrow has moved one full wavelength between the first graph and the last. That is, **each crest of a sinusoidal wave travels forward a distance of exactly one wavelength λ during a time interval of exactly one period T.** Because speed is distance divided by time, the wave speed must be

$$v = \frac{\text{distance}}{\text{time}} = \frac{\lambda}{T} \tag{16.4}$$

Because $f = 1/T$, it is customary to write Equation 16.4 in the form

$$v = \lambda f \tag{16.5}$$

The fundamental relationship for periodic waves

Although Equation 16.5 has no special name, it is *the* fundamental relationship for periodic waves. When using it, keep in mind the *physical* meaning that a wave moves forward a distance of one wavelength during a time interval of one period.

NOTE ▸ Wavelength and period are defined only for *periodic* waves, so Equations 16.4 and 16.5 apply only to periodic waves. A wave pulse has a wave speed, but it doesn't have a wavelength or a period. Hence Equations 16.4 and 16.5 cannot be applied to wave pulses. ◂

Because the wave speed is a property of the medium while the wave frequency is a property of the oscillating source, it is often useful to write Equation 16.5 as

$$\lambda = \frac{v}{f} = \frac{\text{property of the medium}}{\text{property of the source}} \tag{16.6}$$

The wavelength is a *consequence* of a wave of frequency f traveling through a medium in which the wave speed is v.

FIGURE 16.12 A series of snapshot graphs at time increments of one-quarter of the period T.

STOP TO THINK 16.3 What is the frequency of this traveling wave?

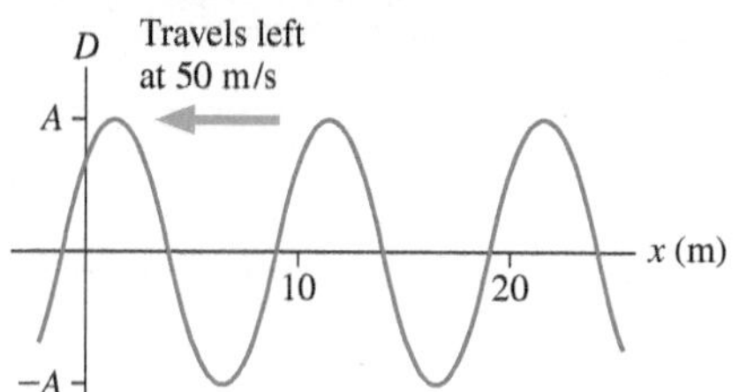

A. 0.1 Hz
B. 0.2 Hz
C. 2 Hz
D. 5 Hz
E. 10 Hz
F. 500 Hz

The Mathematics of Sinusoidal Waves

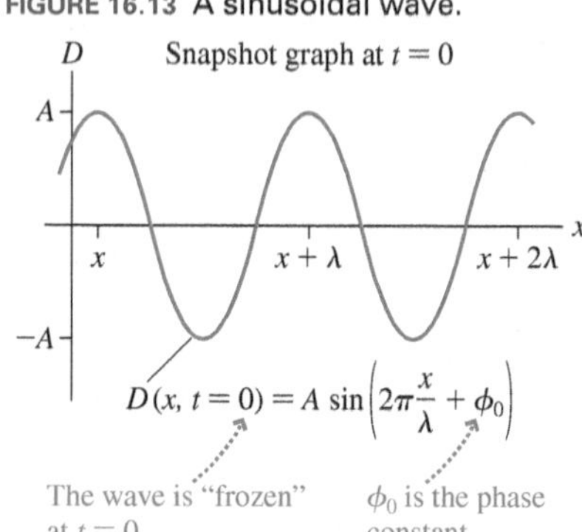

FIGURE 16.13 A sinusoidal wave.

FIGURE 16.13 shows a snapshot graph at $t = 0$ of a sinusoidal wave. The sinusoidal function that describes the displacement of this wave is

$$D(x, t = 0) = A \sin\left(2\pi \frac{x}{\lambda} + \phi_0\right) \tag{16.7}$$

where the notation $D(x, t = 0)$ means that we've frozen the time at $t = 0$ to make the displacement a function of only x. The term ϕ_0 is a *phase constant* that characterizes the initial conditions. (We'll return to the phase constant momentarily.)

The function of Equation 16.7 is periodic with period λ. We can see this by writing

$$D(x + \lambda) = A \sin\left(2\pi \frac{(x + \lambda)}{\lambda} + \phi_0\right) = A \sin\left(2\pi \frac{x}{\lambda} + \phi_0 + 2\pi \text{ rad}\right)$$

$$= A \sin\left(2\pi \frac{x}{\lambda} + \phi_0\right) = D(x)$$

where we used the fact that $\sin(a + 2\pi \text{ rad}) = \sin a$. In other words, the disturbance created by the wave at $x + \lambda$ is exactly the same as the disturbance at x.

The next step is to set the wave in motion. We can do this by replacing x in Equation 16.7 with $x - vt$. To see why this works, recall that the wave moves distance vt during time t. In other words, the wave's displacement at position x at time t must be the same as the displacement at position $x - vt$ at the earlier time $t = 0$. Mathematically, this idea can be captured by writing

$$D(x, t) = D(x - vt, t = 0) \tag{16.8}$$

Make sure you understand how this statement describes a wave moving in the positive x-direction at speed v.

This is what we were looking for. $D(x, t)$ is the general function describing the traveling wave. It's found by taking the function that describes the wave at $t = 0$—the function of Equation 16.7—and replacing x with $x - vt$. Thus the displacement equation of a sinusoidal wave traveling in the positive x-direction at speed v is

$$D(x, t) = A \sin\left(2\pi \frac{x - vt}{\lambda} + \phi_0\right) = A \sin\left(2\pi\left(\frac{x}{\lambda} - \frac{t}{T}\right) + \phi_0\right) \tag{16.9}$$

In the last step we used $v = \lambda f = \lambda/T$ to write $v/\lambda = 1/T$. The function of Equation 16.9 is not only periodic in space with period λ, it is also periodic in time with period T. That is, $D(x, t + T) = D(x, t)$.

It will be useful to introduce two new quantities. First, recall from simple harmonic motion the *angular frequency*

$$\omega = 2\pi f = \frac{2\pi}{T} \tag{16.10}$$

The units of ω are rad/s, although many textbooks use simply s^{-1}.

You can see that ω is 2π times the reciprocal of the period in time. This suggests that we define an analogous quantity, called the **wave number** k, that is 2π times the reciprocal of the period in space:

$$k = \frac{2\pi}{\lambda} \quad (16.11)$$

The units of k are rad/m, although some textbooks use simply m^{-1}.

NOTE ▶ The wave number k is *not* a spring constant, even though it uses the same symbol. This is a most unfortunate use of symbols, but every major textbook and professional tradition uses the same symbol k for these two very different meanings, so we have little choice but to follow along. ◀

We can use the fundamental relationship $v = \lambda f$ to find an analogous relationship between ω and k:

$$v = \lambda f = \frac{2\pi}{k}\frac{\omega}{2\pi} = \frac{\omega}{k} \quad (16.12)$$

which is usually written

$$\omega = vk \quad (16.13)$$

Equation 16.13 contains no new information. It is a variation of Equation 16.5, but one that is convenient when working with k and ω.

If we use the definitions of Equations 16.10 and 16.11, Equation 16.9 for the displacement can be written

$$D(x, t) = A\sin(kx - \omega t + \phi_0) \quad (16.14)$$

Sinusoidal wave traveling in the positive x-direction

A sinusoidal wave traveling in the negative x-direction is $A\sin(kx + \omega t + \phi_0)$. Equation 16.14 is graphed versus x and t in FIGURE 16.14.

You learned in ◀◀ SECTION 15.2 that the initial conditions of an oscillator can be characterized by a phase constant. The same is true for a sinusoidal wave. At $(x, t) = (0\text{ m}, 0\text{ s})$ Equation 16.14 becomes

$$D(0\text{ m}, 0\text{ s}) = A\sin\phi_0 \quad (16.15)$$

Different values of the phase constant ϕ_0 describe different initial conditions for the wave.

FIGURE 16.14 Interpreting the equation of a sinusoidal traveling wave. The phase constant has been set to zero.

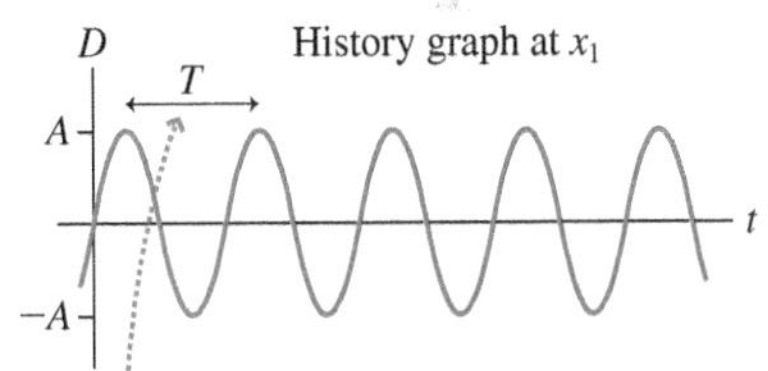

If x is fixed, $D(x_1, t) = A\sin(kx_1 - \omega t + \phi_0)$ gives a sinusoidal history graph at one point in space, x_1. It repeats every T s.

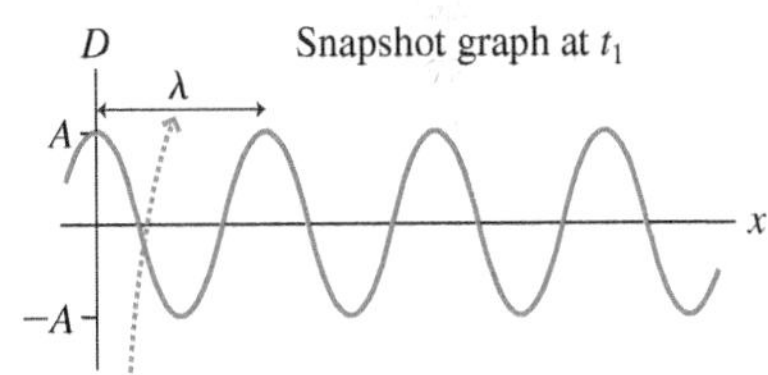

If t is fixed, $D(x, t_1) = A\sin(kx - \omega t_1 + \phi_0)$ gives a sinusoidal snapshot graph at one instant of time, t_1. It repeats every λ m.

EXAMPLE 16.4 **Analyzing a sinusoidal wave**

A sinusoidal wave with an amplitude of 1.00 cm and a frequency of 100 Hz travels at 200 m/s in the positive x-direction. At $t = 0$ s, the point $x = 1.00$ m is on a crest of the wave.

a. Determine the values of A, v, λ, k, f, ω, T, and ϕ_0 for this wave.
b. Write the equation for the wave's displacement as it travels.
c. Draw a snapshot graph of the wave at $t = 0$ s.

PREPARE The snapshot graph will be sinusoidal, but we must do some numerical analysis before we know how to draw it.

SOLVE

a. There are several numerical values associated with a sinusoidal traveling wave, but they are not all independent. From the problem statement itself we learn that

$$A = 1.00\text{ cm} \qquad v = 200\text{ m/s} \qquad f = 100\text{ Hz}$$

We can then find:

$$\lambda = v/f = 2.00\text{ m}$$
$$k = 2\pi/\lambda = \pi\text{ rad/m or } 3.14\text{ rad/m}$$
$$\omega = 2\pi f = 628\text{ rad/s}$$
$$T = 1/f = 0.0100\text{ s} = 10.0\text{ ms}$$

The phase constant ϕ_0 is determined by the initial conditions. We know that a wave crest, with displacement $D = A$, is passing $x_0 = 1.00$ m at $t_0 = 0$ s. Equation 16.14 at x_0 and t_0 is

$$D(x_0, t_0) = A = A\sin[k(1.00\text{ m}) + \phi_0]$$

This equation is true only if $\sin[k(1.00\text{ m}) + \phi_0] = 1$, which requires

$$k(1.00\text{ m}) + \phi_0 = \frac{\pi}{2}\text{ rad}$$

Continued

Solving for the phase constant gives

$$\phi_0 = \frac{\pi}{2}\text{ rad} - (\pi\text{ rad/m})(1.00\text{ m}) = -\frac{\pi}{2}\text{ rad}$$

b. With the information gleaned from part a, the wave's displacement is

$$D(x, t) = (1.00\text{ cm}) \times \sin[(3.14\text{ rad/m})x - (628\text{ rad/s})t - \pi/2\text{ rad}]$$

Notice that we included units with A, k, ω, and ϕ_0.

c. We know that $x = 1.00$ m is a wave crest at $t = 0$ s and that the wavelength is $\lambda = 2.00$ m. FIGURE 16.15 is a snapshot graph that portrays this information.

FIGURE 16.15 A snapshot graph at $t = 0$ s of the sinusoidal wave of Example 16.4.

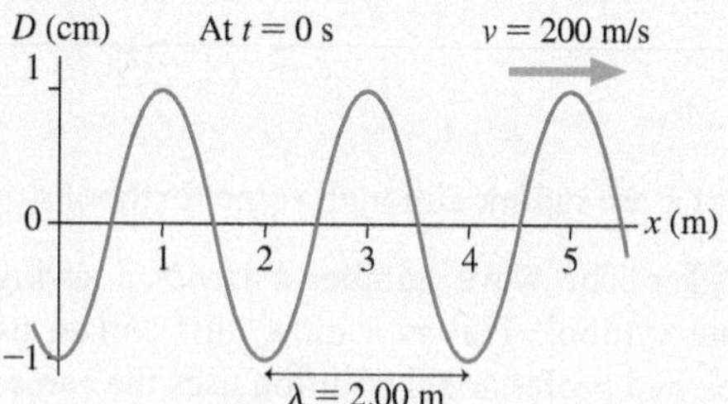

ASSESS We expect to find a wave trough at the origin at $t = 0$ s because the origin is $\lambda/2$ away from the crest at $x = 1.00$ m. The graph matches our expectation.

The Velocity of a Particle in the Medium

As a sinusoidal wave travels along the x-axis with speed v, the particles of the medium oscillate back and forth in SHM. For a transverse wave, such as a wave on a string, the oscillation is in the y-direction. For a longitudinal sound wave, the particles oscillate in the x-direction, parallel to the propagation. We can use the displacement equation, Equation 16.14, to find the velocity of a particle in the medium.

At time t, the displacement of a particle in the medium at position x is

$$D(x, t) = A\sin(kx - \omega t + \phi_0) \qquad (16.16)$$

The velocity of the particle—**which is not the same as the velocity of the wave along the string**—is the time derivative of Equation 16.16:

$$v_{\text{particle}} = \frac{dD}{dt} = -\omega A\cos(kx - \omega t + \phi_0) \qquad (16.17)$$

FIGURE 16.16 A snapshot graph of a wave on a string with vectors showing the velocity *of the string* at various points.

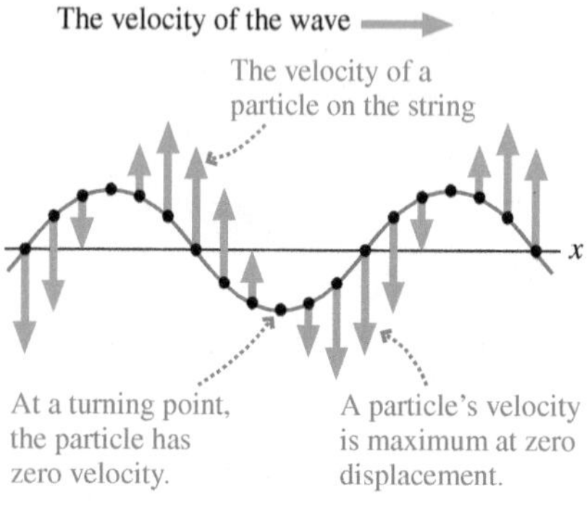

The velocity of the wave

The velocity of a particle on the string

x

At a turning point, the particle has zero velocity.

A particle's velocity is maximum at zero displacement.

Thus the maximum speed of particles in the medium is $v_{max} = \omega A$. This is the same result we found for simple harmonic motion because the motion of the medium *is* simple harmonic motion. FIGURE 16.16 shows velocity vectors *of the particles* at different points along a string as a sinusoidal wave moves from left to right.

NOTE ▶ Creating a wave of larger amplitude increases the speed of particles in the medium, but it does *not* change the speed of the wave *through* the medium. ◀

EXAMPLE 16.5 Generating a sinusoidal wave

A very long string with $\mu = 2.0$ g/m is stretched along the x-axis with a tension of 5.0 N. At $x = 0$ m it is tied to a 100 Hz simple harmonic oscillator that vibrates perpendicular to the string with an amplitude of 2.0 mm. The oscillator is at its maximum positive displacement at $t = 0$ s.

a. Write the displacement equation for the traveling wave on the string.

b. At $t = 5.0$ ms, what are the string's displacement and vertical speed at a point 2.7 m from the oscillator?

PREPARE The oscillator generates a sinusoidal traveling wave on a string. The displacement of the wave has to match the displacement of the oscillator at $x = 0$ m.

SOLVE

a. The equation for the string's displacement is

$$D(x, t) = A\sin(kx - \omega t + \phi_0)$$

with A, k, ω, and ϕ_0 to be determined. The wave amplitude is the same as the amplitude of the oscillator that generates the wave, so $A = 2.0$ mm. The oscillator has its maximum displacement $y_{osc} = A = 2.0$ mm at $t = 0$ s, thus

$$D(0\text{ m}, 0\text{ s}) = A\sin(\phi_0) = A$$

This requires the phase constant to be $\phi_0 = \pi/2$ rad. The wave's frequency is $f = 100$ Hz, the frequency of the source;

therefore the angular frequency is $\omega = 2\pi f = 200\pi$ rad/s. We still need $k = 2\pi/\lambda$, but we do not know the wavelength. However, we have enough information to determine the wave speed, and we can then use either $\lambda = v/f$ or $k = \omega/v$. The speed is

$$v_{\text{string}} = \sqrt{\frac{T_s}{\mu}} = \sqrt{\frac{5.0\text{ N}}{0.0020\text{ kg/m}}} = 50\text{ m/s}$$

Using v_{string}, we find $\lambda = 0.50$ m and $k = 2\pi/\lambda = 4\pi$ rad/m. Thus the wave's displacement equation is

$$D(x,t) = (2.0\text{ mm}) \times \sin[(4\pi\text{ rad/m})x - (200\pi\text{ rad/s})t + \pi/2\text{ rad}]$$

b. The wave's displacement at $t = 5.0$ ms $= 0.0050$ s and $x = 2.7$ m (calculator set to radians!) is

$$\begin{aligned} D(2.7\text{ m}, 0.0050\text{ s}) &= (2.0\text{ mm})\sin[(4\pi\text{ rad/m})(2.7\text{ m}) - (200\pi\text{ rad/s})(0.0050\text{ s}) + \pi/2\text{ rad}] \\ &= 1.6\text{ mm} \end{aligned}$$

The vertical velocity of a piece of the string, Equation 16.17, is $v_{\text{particle}} = -\omega A\cos(kx - \omega t + \phi_0)$. Thus

$$\begin{aligned} v_{\text{particle}}(2.7\text{ m}, 0.0050\text{ s}) &= -(200\pi\text{ rad/s})(2.0\text{ mm}) \times \cos[(4\pi\text{ rad/m})(2.7\text{ m}) - (200\pi\text{ rad/s})(0.0050\text{ s}) + \pi/2\text{ rad}] \\ &= -740\text{ mm/s} \end{aligned}$$

At this instant, this piece of the string is displaced upward by 1.6 mm and it is moving downward at 0.74 m/s. The vertical speeds of particles in the string are much less than the 50 m/s wave speed along the string.

ASSESS During one cycle, of period T, a piece of the string moves through a total vertical distance of $4A$. Thus the *average* vertical speed is $v = 4A/T$, which is 800 mm/s for an amplitude of 2.0 mm and a period of 0.01 s. The vertical speed changes during a cycle, being sometimes higher than average and sometimes lower, but the fact that our calculated value of v_{particle} is close to the average speed gives us confidence in the result.

16.4 Sound and Light

Although there are many kinds of waves in nature, two are especially significant for us as humans. These are sound waves and light waves, the basis of hearing and seeing.

Sound Waves

We usually think of sound waves traveling in air, but sound can travel through any gas, through liquids, and even through solids. FIGURE 16.17 shows a loudspeaker cone vibrating back and forth in a fluid such as air or water. Each time the cone moves forward, it collides with the molecules and pushes them closer together. A half cycle later, as the cone moves backward, the fluid has room to expand and the density decreases a little. These regions of higher and lower density (and thus higher and lower pressure) are called **compressions** and **rarefactions.**

This periodic sequence of compressions and rarefactions travels outward from the loudspeaker as a longitudinal sound wave. When the wave reaches your ear, the oscillating pressure causes your eardrum to vibrate. These vibrations are transferred into your inner ear and perceived as sound.

The speed of sound waves depends on the compressibility of the medium. As TABLE 16.1 shows, the speed is faster in liquids and solids (relatively incompressible) than in gases (highly compressible).

Earlier, in ◀◀ SECTION 13.3, we found that the typical speed of an atom or molecule of mass m in a gas is the root-mean-square speed

$$v_{\text{rms}} = \sqrt{\frac{3k_B T}{m}} \tag{16.18}$$

where k_B is the Boltzmann constant and T is the absolute temperature in kelvin.

It seems likely that the speed of sound in a gas is related to the speed with which the molecules move. A thorough analysis finds that the sound speed is slightly less than the rms speed, but it has the same dependence on temperature and mass:

$$v_{\text{sound}} = \sqrt{\frac{\gamma k_B T}{m}} = \sqrt{\frac{\gamma RT}{M}} \tag{16.19}$$

Speed of sound in a gas

FIGURE 16.17 A sound wave is a sequence of compressions and rarefactions.

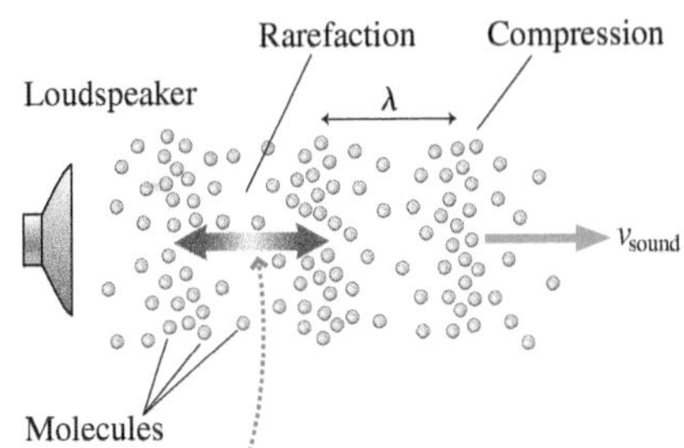
Rarefaction
Compression
Loudspeaker
λ
v_{sound}
Molecules
Individual molecules oscillate back and forth. As they do so, the compressions propagate forward at speed v_{sound}. Compressions are regions of higher pressure, so a sound wave is a pressure wave.

TABLE 16.1 **The speed of sound**

Medium	Speed (m/s)
Air (0°C)	331
Air (20°C)	343
Helium (0°C)	970
Ethyl alcohol	1170
Fat	1450
Water	1480
Soft tissue	1540
Bone	4000
Granite	6000
Aluminum	6420

Distance to a lightning strike Sound travels approximately 1 km in 3 s, or 1 mi in 5 s. When you see a lightning flash, start counting seconds. When you hear the thunder, stop counting. Divide the result by 3, and you will have the approximate distance to the lightning strike in kilometers; divide by 5 and you will have the approximate distance in miles.

In the second version, R is the gas constant and M is the molar mass in kg/mol. The constant γ (Greek gamma) is the ratio C_P/C_V of molar specific heat at constant pressure to molar specific heat at constant volume. For ideal gases, $\gamma = 5/3 = 1.67$ for monatomic gases, such as He, and $\gamma = 7/5 = 1.40$ for diatomic gases, such as N_2 and O_2.

For sound waves in gases, we see that:

- The speed of sound increases with temperature.
- The speed of sound decreases as the molecular mass of the gas increases.
- The speed of sound does not depend on the pressure of the gas.

For example, air—a mixture of approximately 80% N_2 and 20% O_2—has an effective molecular mass of 0.029 kg/mol. At room temperature (20°C), the speed of sound in air is

$$v_{\text{sound in air}} = \sqrt{\frac{(1.40)(8.31\ \text{J/mol}\cdot\text{K})(293\ \text{K})}{(0.029\ \text{kg/mol})}} = 343\ \text{m/s}$$

Use this value when solving problems unless you're given a different value of the temperature.

A speed of 343 m/s is high, but not extraordinarily so. A distance as small as 100 m is enough to notice a slight delay between when you see something, such as a person hammering a nail, and when you hear it. The time required for sound to travel 1 km is $\Delta t = (1000\ \text{m})/(343\ \text{m/s}) \approx 3$ s. You may have learned to estimate the distance to a bolt of lightning by timing the number of seconds between when you see the flash and when you hear the thunder. Because sound takes 3 s to travel 1 km, the time divided by 3 gives the distance in kilometers. Or, in English units, the time divided by 5 gives the distance in miles.

TABLE 16.2 Range of hearing for animals

Animal	Range of hearing (Hz)
Chicken	120–1,200
Owl	200–12,000
Human	20–20,000
Dog	40–45,000
Mouse	1000–80,000
Bat	2000–100,000
Dolphin	75–150,000

Your ears are able to detect sound waves with frequencies between roughly 20 Hz and 20,000 Hz, or 20 kHz. You can use the fundamental relationship $v_{\text{sound}} = \lambda f$ to calculate that a 20 Hz sound wave has a 17-m-long wavelength, while the wavelength of a 20 kHz note is a mere 17 mm. Low frequencies are perceived as "low-pitch" bass notes, while high frequencies are heard as "high-pitch" treble notes. Your high-frequency range of hearing can deteriorate with age (10 kHz is the average upper limit at age 65) or as a result of exposure to very loud sounds.

Many animals have much better high-frequency hearing than humans. TABLE 16.2 shows that dogs, well known for responding to high-pitched dog whistles that we can't hear, can sense sound waves up to about 45 kHz, and the hearing of dolphins and other cetaceans extends above 100 kHz. It is interesting that the hearing range of birds is significantly narrower than that of humans.

FIGURE 16.18 Ultrasound imaging is based on echolocation as a narrow beam of ultrasound is swept across the patient.

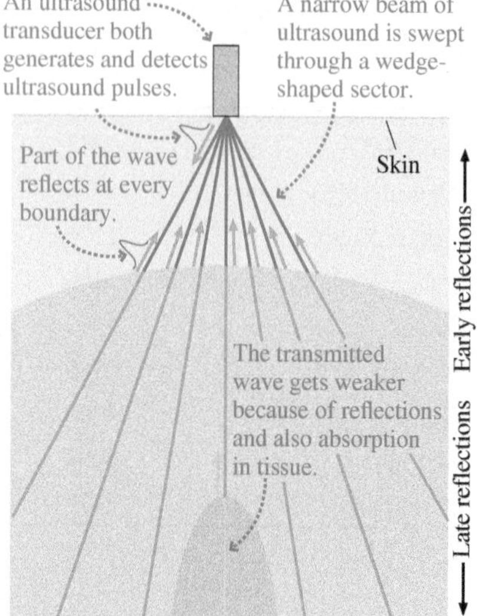

Ultrasound Imaging

Bats, which catch insects at night, use **echolocation** to find their prey in the dark. They emit loud chirps at **ultrasound** frequencies—higher than humans can hear—and then listen for echoes. Medical ultrasound imaging works in much the same way.

Sound-wave frequencies above 20 kHz are called *ultrasonic*. Ultrasound can be produced from 20 kHz to many hundreds of MHz, but medical ultrasound devices operate mostly in the range from 2 MHz to 20 MHz. Table 16.1 shows that the speed of sound in soft tissue is about 1540 m/s, and we can use $v = \lambda f$ to calculate that the wavelengths of these sound waves in the body decrease from approximately 0.8 mm at 2 MHz to 0.08 mm (or 80 μm) at 20 MHz. As we'll see, wavelengths this small are needed for imaging.

FIGURE 16.18 shows the setup for ultrasound imaging. A **transducer,** which is both the generator and the detector of ultrasound, generates a short pulse of ultrasound at $t = 0$. The sound waves encounter numerous boundaries as they pass through muscle, fat, organs, and bone. At each boundary, a portion of the wave's energy is reflected while the remainder continues to move forward. If a boundary is at distance d from the transducer, the round-trip distance of a reflected wave—an echo—is $2d$.

The waves travel through the body at speed v_{sound}, so the echo is detected at time $t = 2d/v_{sound}$ after the burst is emitted. A wave pulse typically encounters several boundaries, producing several echoes, and a careful timing of when each echo is detected is then used to determine the distance to each.

To produce an image, a narrow ultrasound beam is scanned from one side to the other, which produces a wedge-shaped sector of ultrasound entering the patient. At each angle, the timing of the reflections is used to produce one vertical line of the image, going from earlier reflections to the later reflections. Thus the top of an ultrasound image is the anterior (skin) side of objects closest to the probe, and the bottom is the posterior side of objects farthest from the probe. That is, the plane of the image is a slice of increasing depth *into* the body, not, like a photograph, a plane parallel to the viewer. Current technology can acquire a full frame in less than $\frac{1}{20}$ s, so it's possible to produce 20 frames/s ultrasound videos that easily show a beating heart.

The strength of a reflection depends on the *difference* in sound speed between the tissues on opposite sides of the boundary. Table 16.1 shows that, other than bone, there's not a great difference in sound speed between the soft tissues, fat, and water that make up much of the body, so ultrasound reflections tend to be weak.

One exception is a boundary between tissue and air, which has a much lower sound speed. In this case, the large difference in speeds causes ultrasound to be almost entirely reflected. The gel that's placed between the transducer and a patient's skin is designed to have a sound speed almost exactly the same as skin; this avoids large reflections at the air-skin boundary, and the beam enters the body with very little initial reflection.

You'll learn in the discussion of optical instruments in Chapter 20 that the *resolution* of an image—the size of the smallest feature that can be resolved—is *at best* one wavelength. This is a property of waves that applies to both ultrasound (imaging with sound waves) and microscopy (imaging with light waves). In principle, a 2 MHz ultrasound image could resolve features as small as 0.8 mm, while 20 MHz could—just barely—detect features only 80 μm wide, not much larger than the diameter of a hair.

Practical issues with the technology prevent the resolution from being quite this good, but ultrasound at the higher frequencies does easily see features that are smaller than 0.5 mm. For example, FIGURE 16.19 is a high-resolution ultrasound image of the carotid artery. The scale on the right is in cm, so the tick marks are 2 mm apart. The resolution appears to be about 0.25 mm. The color, which indicates the speed of blood flow, is obtained from the Doppler effect, a topic we'll study later in this chapter.

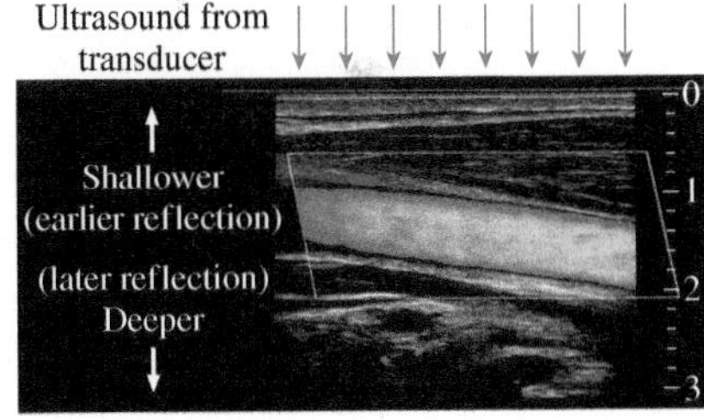

FIGURE 16.19 An ultrasound image of the carotid artery with a resolution of about 0.25 mm. The scale is in cm.

Electromagnetic Waves

A light wave is an *electromagnetic wave,* a self-sustaining oscillation of the electromagnetic field. Other electromagnetic waves, such as radio waves, microwaves, and ultraviolet light, have the same physical characteristics as light waves even though we cannot sense them with our eyes. It is easy to demonstrate that light will pass unaffected through a container from which all the air has been removed, and light reaches us from distant stars through the vacuum of interstellar space. Such observations raise interesting but difficult questions. If light can travel through a region in which there is no matter, then what is the *medium* of a light wave? What is it that is waving?

It took scientists over 50 years, most of the 19th century, to answer this question. We will examine the answers in more detail in Chapter 27 after we introduce the ideas of electric and magnetic fields. For now we can say that light waves are a "self-sustaining oscillation of the electromagnetic field." That is, the displacement D is an electric or magnetic field. Being self-sustaining means that electromagnetic waves require *no material medium* in order to travel; hence electromagnetic waves are not mechanical waves. Fortunately, we can learn about the wave properties of light without having to understand electromagnetic fields.

It was predicted theoretically in the 1860s, and has been subsequently confirmed, that all electromagnetic waves travel through vacuum with the same speed, called the *speed of light.* The value of the speed of light is

$$v_{\text{light}} = c = 3.00 \times 10^8 \text{ m/s}$$

where the special symbol c is used to designate the speed of light. (This is the c in Einstein's famous formula $E = mc^2$.) Now *this* is really moving—about one million times faster than the speed of sound in air!

White light passing through a prism is spread out into a band of colors called the *visible spectrum.*

The wavelengths of light are extremely small. You will learn in Chapter 18 how these wavelengths are determined, but for now we will note that visible light is an electromagnetic wave with a wavelength (in air) in the range of roughly 400 nm $(400 \times 10^{-9} \text{ m})$ to 700 nm $(700 \times 10^{-9} \text{ m})$. Each wavelength is perceived as a different color, with the longer wavelengths seen as orange or red light and the shorter wavelengths seen as blue or violet light. A prism is able to spread the different wavelengths apart, from which we learn that "white light" is all the colors, or wavelengths, combined. The spread of colors seen with a prism, or seen in a rainbow, is called the *visible spectrum.*

If the wavelengths of light are unbelievably small, the oscillation frequencies are unbelievably large. The frequency for a 600 nm wavelength of light (orange) is

$$f = \frac{v}{\lambda} = \frac{3.00 \times 10^8 \text{ m/s}}{600 \times 10^{-9} \text{ m}} = 5.00 \times 10^{14} \text{ Hz}$$

The frequencies of light waves are roughly a factor of a trillion (10^{12}) higher than sound frequencies.

Electromagnetic waves exist at many frequencies other than the rather limited range that our eyes detect. One of the major technological advances of the 20th century was learning to generate and detect electromagnetic waves at many frequencies, ranging from low-frequency radio waves to the extraordinarily high frequencies of x rays. FIGURE 16.20 shows that the visible spectrum is a small slice of the much broader **electromagnetic spectrum**.

FIGURE 16.20 The electromagnetic spectrum from 10^6 Hz to 10^{18} Hz.

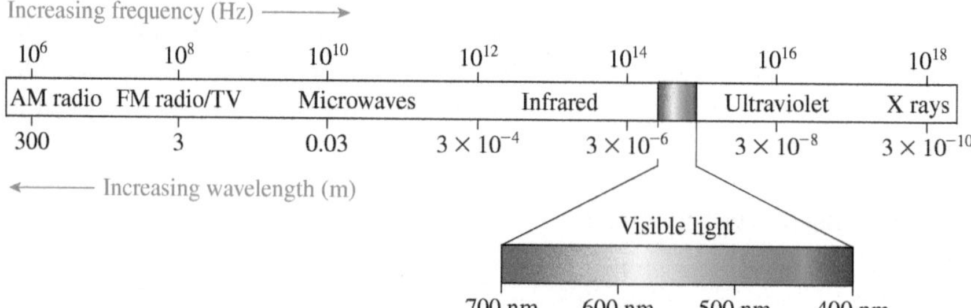

EXAMPLE 16.6 Traveling at the speed of light

A spacecraft exploring Jupiter transmits data to the earth as a radio wave with a frequency of 200 MHz. What is the wavelength of the electromagnetic wave, and how long does it take the signal to travel 800 million kilometers from Jupiter to the earth?

PREPARE Radio waves are sinusoidal electromagnetic waves traveling with speed c.

SOLVE We can use the fundamental relationship for periodic waves to calculate the wavelength:

$$\lambda = \frac{c}{f} = \frac{3.00 \times 10^8 \text{ m/s}}{2.00 \times 10^8 \text{ Hz}} = 1.5 \text{ m}$$

The time needed to travel $800 \times 10^6 \text{ km} = 8.0 \times 10^{11}$ m is

$$\Delta t = \frac{\Delta x}{c} = \frac{8.0 \times 10^{11} \text{ m}}{3.00 \times 10^8 \text{ m/s}} = 2700 \text{ s} = 45 \text{ min}$$

ASSESS If you've ever watched documentaries about planetary exploration, you've probably heard that it takes a long time to communicate with distant spacecraft. Our calculation confirms this.

The Index of Refraction

Light waves travel with speed c in a vacuum, but they slow down as they pass through transparent materials such as water or glass or even, to a very slight extent, air. The slowdown is a consequence of interactions between the electromagnetic field of the wave and the electrons in the material. The speed of light in a material is characterized by the material's **index of refraction** n, defined as

$$n = \frac{\text{speed of light in a vacuum}}{\text{speed of light in the material}} = \frac{c}{v} \qquad (16.20)$$

The index of refraction of a material is always greater than 1 because $v < c$. A vacuum has $n = 1$ exactly. TABLE 16.3 shows the index of refraction for several materials. You can see that liquids and solids have larger indices of refraction than gases.

TABLE 16.3 Typical indices of refraction

Material	Index of refraction
Vacuum	1 exactly
Air	1.0003
Water	1.33
Glass	1.50
Diamond	2.42

NOTE ▶ An accurate value for the index of refraction of air is relevant only in very precise measurements. We will assume $n_{air} = 1.00$ in this text. ◀

If the speed of a light wave changes as it enters into a transparent material, such as glass, what happens to the light's frequency and wavelength? Because $v = \lambda f$, either λ or f or both have to change when v changes.

As an analogy, think of a sound wave in the air as it impinges on the surface of a pool of water. As the air oscillates back and forth, it periodically pushes on the surface of the water. These pushes generate the compressions of the sound wave that continues on into the water. Because each push of the air causes one compression of the water, the frequency of the sound wave in the water must be *exactly the same* as the frequency of the sound wave in the air. In other words, **the frequency of a wave is the frequency of the source. It does not change as the wave moves from one medium to another.**

The same is true for electromagnetic waves; the frequency does not change as the wave moves from one material to another.

FIGURE 16.21 shows a light wave passing through a transparent material with index of refraction n. As the wave travels through vacuum it has wavelength λ_{vac} and frequency f_{vac} such that $\lambda_{vac} f_{vac} = c$. In the material, $\lambda_{mat} f_{mat} = v = c/n$. The frequency does not change as the wave enters ($f_{mat} = f_{vac}$), so the wavelength must. The wavelength in the material is

$$\lambda_{mat} = \frac{v}{f_{mat}} = \frac{c}{n f_{mat}} = \frac{c}{n f_{vac}} = \frac{\lambda_{vac}}{n} \qquad (16.21)$$

The wavelength in the transparent material is less than the wavelength in vacuum. This makes sense. Suppose a marching band is marching at one step per second at a speed of 1 m/s. Suddenly they slow their speed to $\frac{1}{2}$ m/s but maintain their march at one step per second. The only way to go slower while marching at the same pace is to take *smaller steps.* When a light wave enters a material, the only way it can go slower while oscillating at the same frequency is to have a *smaller wavelength.*

FIGURE 16.21 Light passing through a transparent material with index of refraction n.

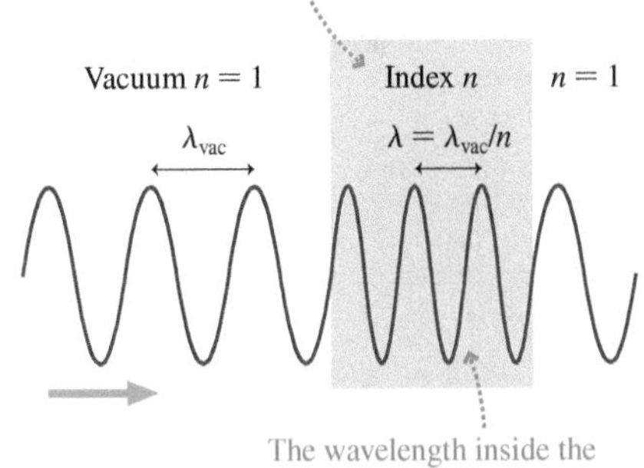
A transparent material in which light travels slower, at speed $v = c/n$
Vacuum $n = 1$
Index n
$n = 1$
λ_{vac}
$\lambda = \lambda_{vac}/n$
The wavelength inside the material decreases, but the frequency doesn't change.

EXAMPLE 16.7 Light traveling through glass

Orange light with a wavelength of 600 nm is incident upon a 1.00-mm-thick glass microscope slide.

a. What is the light speed in the glass?
b. How many wavelengths of the light are inside the slide?

PREPARE The speed and wavelength of light depend on the index of refraction.

SOLVE

a. From Table 16.3 we see that the index of refraction of glass is $n_{glass} = 1.50$. Thus the speed of light in glass is

$$v_{glass} = \frac{c}{n_{glass}} = \frac{3.00 \times 10^8 \text{ m/s}}{1.50} = 2.00 \times 10^8 \text{ m/s}$$

b. The wavelength inside the glass is

$$\lambda_{glass} = \frac{\lambda_{vac}}{n_{glass}} = \frac{600 \text{ nm}}{1.50} = 400 \text{ nm} = 4.00 \times 10^{-7} \text{ m}$$

N wavelengths span a distance $d = N\lambda$, so the number of wavelengths in $d = 1.00$ mm is

$$N = \frac{d}{\lambda} = \frac{1.00 \times 10^{-3} \text{ m}}{4.00 \times 10^{-7} \text{ m}} = 2500$$

ASSESS The fact that 2500 wavelengths fit within 1 mm shows how small the wavelengths of light are.

STOP TO THINK 16.4 A light wave travels from left to right through three transparent materials of equal thickness. Rank in order, from largest to smallest, the indices of refraction n_1, n_2, and n_3.

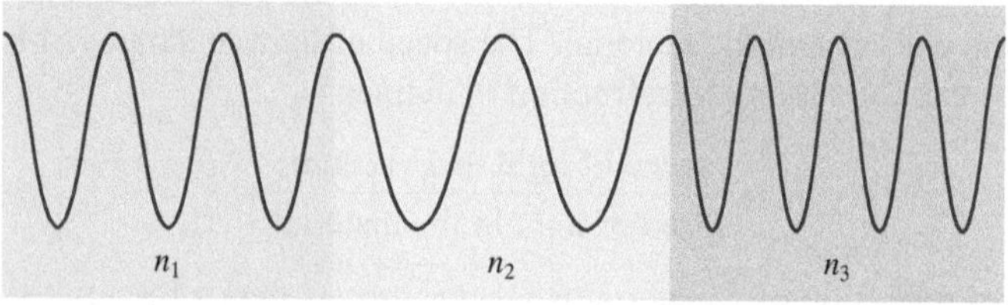

16.5 Circular and Spherical Waves

FIGURE 16.22 The wave fronts of a circular or spherical wave.

(a) Wave fronts are the crests of the wave. They are spaced one wavelength apart.

v v Source λ λ λ v v

The circular wave fronts move outward from the source at speed v.

(b)

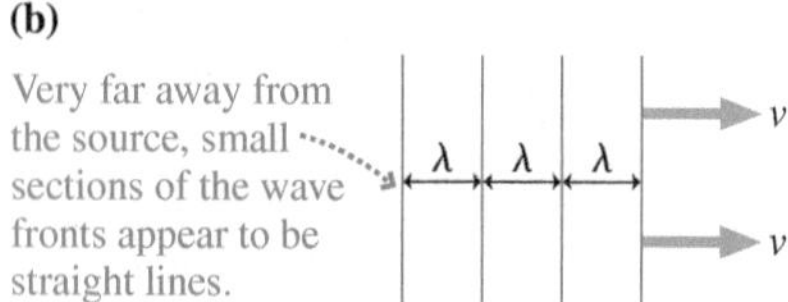

FIGURE 16.23 A plane wave.

Very far from the source, small segments of spherical wave fronts appear to be planes. The wave is cresting at every point in these planes.

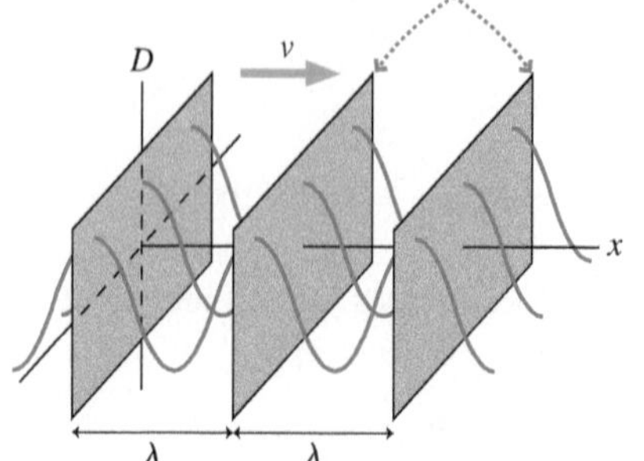

Suppose you were to take a photograph of ripples spreading on a pond. If you mark the location of the *crests* on the photo, your picture would look like FIGURE 16.22a. The lines that locate the crests are called **wave fronts,** and they are spaced precisely one wavelength apart. The diagram shows only a single instant of time, but you can imagine a movie in which you would see the wave fronts moving outward from the source at speed v. A wave like this is called a **circular wave.** It is a two-dimensional wave that spreads across a surface.

Although the wave fronts are circles, you would hardly notice the curvature if you observed a small section of the wave front very, very far away from the source. That is, the wave fronts would appear to be parallel lines, still spaced one wavelength apart and traveling at speed v. A good example is an ocean wave reaching a beach. Ocean waves are generated by storms and wind far out at sea, hundreds or thousands of miles away. By the time they reach the beach where you are working on your tan, the crests appear to be straight lines. An aerial view of the ocean would show a wave diagram like FIGURE 16.22b.

Many waves of interest, such as sound waves or light waves, move in three dimensions. For example, loudspeakers and lightbulbs emit **spherical waves.** That is, the crests of the wave form a series of concentric spherical shells separated by the wavelength λ. In essence, the waves are three-dimensional ripples. It will still be useful to draw wave-front diagrams such as Figure 16.22, but now the circles are slices through the spherical shells locating the wave crests.

If you observe a spherical wave very, very far from its source, the small piece of the wave front that you can see is a little patch on the surface of a very large sphere. If the radius of the sphere is sufficiently large, you will not notice the curvature and this little patch of the wave front appears to be a plane. FIGURE 16.23 illustrates the idea of a **plane wave.**

To visualize a plane wave, imagine standing on the x-axis facing a sound wave as it comes toward you from a very distant loudspeaker. Sound is a longitudinal wave, so the particles of medium oscillate toward you and away from you. If you were to locate all of the particles that, at one instant of time, were at their maximum displacement toward you, they would all be located in a plane perpendicular to the travel direction. This is one of the wave fronts in Figure 16.23, and all the particles in this plane are doing exactly the same thing at that instant of time. This plane is moving toward you at speed v. There is another plane one wavelength behind it where the molecules are also at maximum displacement, yet another two wavelengths behind the first, and so on.

Because a plane wave's displacement depends on x but not on y or z, the displacement function $D(x, t)$ describes a plane wave just as readily as it does a one-dimensional wave. Once you specify a value for x, the displacement is the same at every point in the yz-plane that slices the x-axis at that value (i.e., one of the planes shown in Figure 16.23).

NOTE ▶ There are no perfect plane waves in nature, but many waves of practical interest can be *modeled* as plane waves. It's a very useful idealization. ◀

We can describe a circular wave or a spherical wave by changing the mathematical description from $D(x, t)$ to $D(r, t)$, where r is the radial distance measured outward from the source. Then the displacement of the medium will be the same at every point on a spherical surface. In particular, a sinusoidal spherical wave with wave number k and angular frequency ω is written

$$D(r, t) = A(r)\sin(kr - \omega t + \phi_0) \qquad (16.22)$$

Other than the change of x to r, the only difference is that the amplitude is now a function of r. A one-dimensional wave propagates with no change in the wave amplitude. But circular and spherical waves spread out to fill larger and larger volumes of space. To conserve energy, an issue we'll look at later in the chapter, the wave's amplitude has to decrease with increasing distance r. This is why sound and light decrease in intensity as you get farther from the source.

Phase and Phase Difference

◀◀ SECTION 15.2 introduced the concept of *phase* for an oscillator in simple harmonic motion. The **phase** of a sinusoidal wave, denoted ϕ, is the quantity $(kx - \omega t + \phi_0)$. Phase will be an important concept in Chapter 17, where we will explore the consequences of adding various waves together. For now, we can note that the wave fronts seen in Figures 16.22 and 16.23 are "surfaces of constant phase." To see this, write the displacement as simply $D(x, t) = A\sin\phi$. Because each point on a wave front has the same displacement, the phase must be the same at every point.

It will be useful to know the *phase difference* $\Delta\phi$ between two different points on a wave. FIGURE 16.24 shows two points on a sinusoidal wave at time t. The phase difference between these points is

$$\begin{aligned}\Delta\phi = \phi_2 - \phi_1 &= (kx_2 - \omega t + \phi_0) - (kx_1 - \omega t + \phi_0) \\ &= k(x_2 - x_1) = k\,\Delta x = 2\pi\frac{\Delta x}{\lambda}\end{aligned} \qquad (16.23)$$

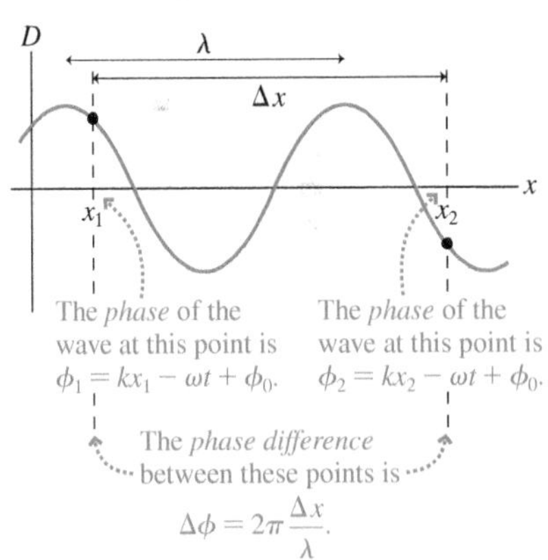
D
λ
Δx
x
x_1
x_2
The *phase* of the wave at this point is $\phi_1 = kx_1 - \omega t + \phi_0$.
The *phase* of the wave at this point is $\phi_2 = kx_2 - \omega t + \phi_0$.
The *phase difference* between these points is $\Delta\phi = 2\pi\frac{\Delta x}{\lambda}$.

FIGURE 16.24 The phase difference between two points on a wave.

That is, **the phase difference between two points on a wave depends on only the ratio of their separation Δx to the wavelength λ.** For example, two points on a wave separated by $\Delta x = \frac{1}{2}\lambda$ have a phase difference $\Delta\phi = \pi$ rad.

An important consequence of Equation 16.23 is that **the phase difference between two adjacent wave fronts is $\Delta\phi = 2\pi$ rad.** This follows from the fact that two adjacent wave fronts are separated by $\Delta x = \lambda$. This is an important idea. Moving from one crest of the wave to the next corresponds to changing the *distance* by λ and changing the *phase* by 2π rad.

EXAMPLE 16.8 The phase difference between two points on a sound wave

A 100 Hz sound wave travels with a wave speed of 343 m/s.

a. What is the phase difference between two points 60.0 cm apart along the direction the wave is traveling?
b. How far apart are two points whose phase differs by 90°?

PREPARE We'll treat the wave as a plane wave traveling in the positive x-direction.

SOLVE

a. The phase difference between two points is

$$\Delta\phi = 2\pi\frac{\Delta x}{\lambda}$$

In this case, $\Delta x = 60.0\text{ cm} = 0.600\text{ m}$. The wavelength is

$$\lambda = \frac{v}{f} = \frac{343\text{ m/s}}{100\text{ Hz}} = 3.43\text{ m}$$

and thus

$$\Delta\phi = 2\pi\frac{0.600\text{ m}}{3.43\text{ m}} = 0.350\pi\text{ rad} = 63.0°$$

b. A phase difference $\Delta\phi = 90°$ is $\pi/2$ rad. This will be the phase difference between two points when $\Delta x/\lambda = \frac{1}{4}$, or when $\Delta x = \lambda/4$. Here, with $\lambda = 3.43$ m, $\Delta x = 85.8$ cm.

ASSESS The phase difference increases as Δx increases, so we expect the answer to part b to be larger than 60 cm.

STOP TO THINK 16.5 What is the phase difference between the crest of a wave and the adjacent trough?

A. -2π rad
B. 0 rad
C. $\pi/4$ rad
D. $\pi/2$ rad
E. π rad
F. 3π rad

16.6 Power, Intensity, and Decibels

A traveling wave transfers energy from one point to another. The sound wave from a loudspeaker sets your eardrum into motion. Light waves from the sun warm the earth. The *power* of a wave is the rate, in joules per second, at which the wave transfers energy. As you learned in Chapter 10, power is measured in watts. A loudspeaker might emit 2 W of power, meaning that energy in the form of sound waves is radiated at the rate of 2 joules per second.

A focused light, like that of a projector, is more *intense* than the diffuse light that goes in all directions. Similarly, a loudspeaker that beams its sound forward into a small area produces a louder sound in that area than a speaker of equal power that radiates the sound in all directions. Quantities such as brightness and loudness depend not only on the rate of energy transfer, or power, but also on the *area* that receives that power.

FIGURE 16.25 Plane waves of power P impinge on area a with intensity $I = P/a$.

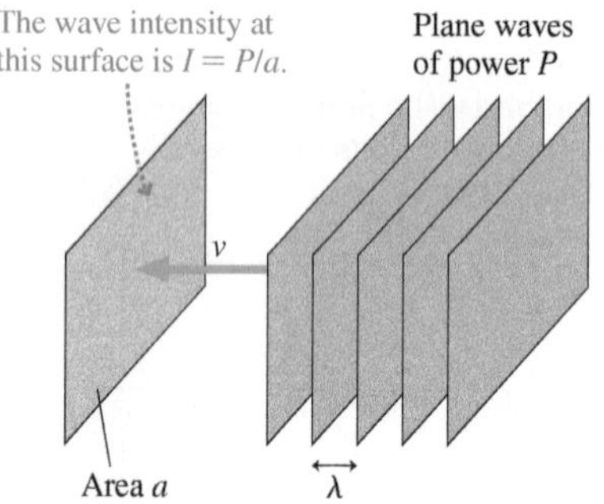

FIGURE 16.25 shows a wave impinging on a surface of area a that is perpendicular to the direction in which the wave is traveling. (We use a lowercase a for area to avoid confusion with the uppercase A used for amplitude.) This might be a real, physical surface, such as your eardrum or a photovoltaic cell, but it could equally well be a mathematical surface in space that the wave passes right through. If the wave has power P, we define the **intensity** I of the wave to be

$$I = \frac{P}{a} = \text{power-to-area ratio} \tag{16.24}$$

The SI units of intensity are W/m^2. Because intensity is a power-to-area ratio, a wave focused into a small area will have a larger intensity than a wave of equal power that is spread out over a large area.

EXAMPLE 16.9 The intensity of a laser beam

A typical red laser pointer emits 1.0 mW of light power into a 1.0-mm-diameter laser beam. What is the intensity of the laser beam?

PREPARE We'll model the laser beam as a plane wave.

SOLVE The light waves of the laser beam pass through a mathematical surface that is a circle of diameter 1.0 mm. The intensity of the laser beam is

$$I = \frac{P}{a} = \frac{P}{\pi r^2} = \frac{0.0010\text{ W}}{\pi(0.00050\text{ m})^2} = 1300\text{ W/m}^2$$

ASSESS This is roughly the intensity of sunlight at noon on a summer day. The difference between the sun and a small laser is not their intensities, which are about the same, but their powers. The laser has a small power of 1 mW. It can produce a very intense wave only because the area through which the wave passes is very small. The sun, by contrast, radiates a total power $P_{sun} \approx 4 \times 10^{26}$ W. This immense power is spread through *all* of space, producing an intensity of 1400 W/m^2 at a distance of 1.5×10^{11} m, the radius of the earth's orbit.

If a source of spherical waves radiates uniformly in all directions, then, as FIGURE 16.26 shows, the power at distance r is spread uniformly over the surface of a sphere of radius r. The surface area of a sphere is $a = 4\pi r^2$, so the intensity of a uniform spherical wave is

$$I = \frac{P_{source}}{4\pi r^2} \tag{16.25}$$

The inverse-square dependence of r is really just a statement of energy conservation. The source emits energy at the rate P joules per second. The energy is spread over a larger and larger area as the wave moves outward. Consequently, the energy *per unit area* must decrease in proportion to the surface area of a sphere.

If the intensity at distance r_1 is $I_1 = P_{source}/4\pi r_1^2$ and the intensity at r_2 is $I_2 = P_{source}/4\pi r_2^2$, then you can see that the intensity *ratio* is

$$\frac{I_1}{I_2} = \frac{r_2^2}{r_1^2} \tag{16.26}$$

You can use Equation 16.26 to compare the intensities at two distances from a source without needing to know the power of the source.

NOTE ▶ Wave intensities are strongly affected by reflections and absorption. Equations 16.25 and 16.26 apply to situations such as the light from a star or the sound from a firework exploding high in the air. Indoor sound does *not* obey a simple inverse-square law because of the many reflecting surfaces. ◀

FIGURE 16.26 A source emitting uniform spherical waves.

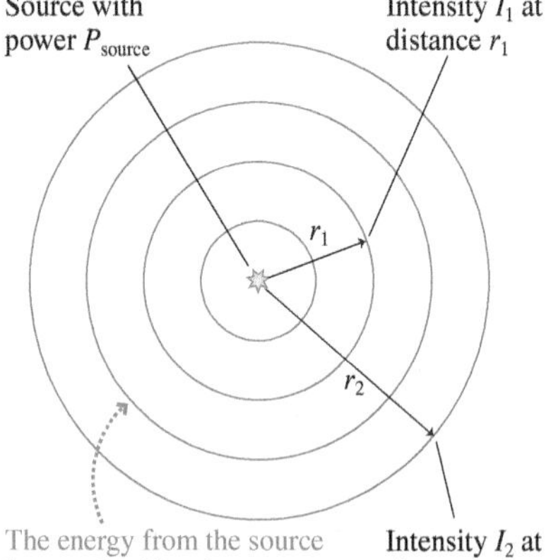

Decibels

The loudness of sound is a perception, not a physical quantity, so we can't truly measure loudness. However, psychoacoustic experiments have found that the perception of loudness is generally proportional not to a sound's intensity, as you might expect, but to the logarithm of the sound's intensity.

Humans can perceive an extremely wide range of intensities—from a *threshold of hearing* at $\approx 1 \times 10^{-12}\ \text{W/m}^2$ to a *threshold of pain* at $\approx 10\ \text{W/m}^2$. That is, intensities relevant to hearing span 13 orders of magnitude (i.e., a factor of 10^{13}). However, the logarithm of the intensity $\log_{10}(I)$ ranges only from –12 to 1, a much more manageable set of values.

We could imagine a "loudness scale" that goes from –12 to 1, but it's more logical to have a scale that starts from zero at the threshold of hearing. To do this, acousticians define the **sound intensity level** β (Greek beta), which is expressed in **decibels** (dB), as

$$\beta = (10\ \text{dB})\log_{10}\left(\frac{I}{I_0}\right) \tag{16.27}$$

Sound intensity level in decibels for a sound of intensity I

where $I_0 = 1 \times 10^{-12}\ \text{W/m}^2$ is approximately the threshold of hearing. Notice that β is computed as a base-10 logarithm, not a natural logarithm.

The decibel is named after Alexander Graham Bell, inventor of the telephone. Sound intensity level is actually dimensionless because it's formed from the ratio of two intensities, so decibels are just a *name* to remind us that we're dealing with an intensity *level* rather than a true intensity.

Right at the threshold of hearing, where $I = I_0$, the sound intensity level is

$$\beta = (10\ \text{dB})\ \log_{10}\left(\frac{I_0}{I_0}\right) = (10\ \text{dB})\ \log_{10}(1) = 0\ \text{dB}$$

Note that 0 dB doesn't mean no sound; it means that, for most people, no sound is heard. Dogs have more sensitive hearing than humans, and most dogs can easily perceive a 0 dB sound. The sound intensity level at the pain threshold is

$$\beta = (10\ \text{dB})\ \log_{10}\left(\frac{10\ \text{W/m}^2}{10^{-12}\ \text{W/m}^2}\right) = (10\ \text{dB})\ \log_{10}(10^{13}) = 130\ \text{dB}$$

The loudest animal in the world The blue whale is the largest animal in the world, up to 30 m (about 100 ft) long, weighing 150,000 kg or more. It is also the loudest, with a sound intensity level measured at 180 dB at a distance of 1 m

The major point to notice is that the sound intensity level increases by 10 dB each time the actual intensity increases by a *factor* of 10. For example, the sound intensity level increases from 70 dB to 80 dB when the sound intensity increases from 10^{-5} W/m^2 to 10^{-4} W/m^2. Perception experiments find that sound is perceived as "twice as loud" when the intensity increases by a factor of 10. In terms of decibels, we can say that the perceived loudness of a sound doubles with each increase in the sound intensity level by 10 dB.

TABLE 16.4 gives the sound intensity levels for a number of sounds. Although 130 dB is the threshold of pain, quieter sounds can damage your hearing. A fairly short exposure to 120 dB can cause damage to the hair cells in the ear, but lengthy exposure to sound intensity levels of over 85 dB can produce damage as well.

TABLE 16.4 Sound intensity levels of common sounds

Sound	β (dB)
Threshold of hearing	0
Person breathing, at 3 m	10
A whisper, at 1 m	20
Quiet room	30
Outdoors, no traffic	40
Quiet restaurant	50
Normal conversation, at 1 m	60
Busy traffic	70
Vacuum cleaner, for user	80
Niagara Falls, at viewpoint	90
Snowblower, at 2 m	100
Stereo, at maximum volume	110
Rock concert	120
Threshold of pain	130

EXAMPLE 16.10 Blender noise

The blender making a smoothie produces a sound intensity level of 83 dB. What is the intensity of the sound? What will the sound intensity level be if a second blender is turned on?

PREPARE Working with decibels requires working with logarithms. You will recall that if $a = \log_{10}(b)$, then $b = 10^a$.

SOLVE We can solve Equation 16.27 for the sound intensity, finding $I = I_0 \times 10^{\beta/10\text{ dB}}$. In this case,

$$I = (1.0 \times 10^{-12}\text{ W/m}^2) \times 10^{8.3} = 2.0 \times 10^{-4}\text{ W/m}^2$$

A second blender doubles the sound power and thus raises the intensity to $I = 4.0 \times 10^{-4}$ W/m^2. The new sound intensity level is

$$\beta = (10\text{ dB})\log_{10}\left(\frac{4.0 \times 10^{-4}\text{ W/m}^2}{1.0 \times 10^{-12}\text{ W/m}^2}\right) = 86\text{ dB}$$

ASSESS In general, doubling the actual sound intensity increases the decibel level by 3 dB.

EXAMPLE 16.11 Finding the loudness of a shout

A person shouting at the top of his lungs emits about 1.0 W of energy as sound waves. What is the sound intensity level 1.0 m from such a person?

PREPARE We'll assume that the shouting person emits a spherical sound wave.

SOLVE At a distance of 1.0 m, the sound intensity is

$$I = \frac{P}{4\pi r^2} = \frac{1.0\text{ W}}{4\pi(1.0\text{ m})^2} = 0.080\text{ W/m}^2$$

Thus the sound intensity level is

$$\beta = (10\text{ dB})\log_{10}\left(\frac{0.080\text{ W/m}^2}{1.0 \times 10^{-12}\text{ W/m}^2}\right) = 110\text{ dB}$$

ASSESS This is quite loud (compare with values in Table 16.4), as you might expect.

STOP TO THINK 16.6 Four trumpet players are playing the same note. If three of them suddenly stop, the sound intensity level decreases by

A. 40 dB B. 12 dB C. 6 dB D. 4 dB

Hearing

Sound is a longitudinal wave, so sound waves cause the eardrum—the *tympanic membrane*—to vibrate in and out at the wave frequency. The eardrum is attached to three tiny bones, the *ossicles*, that transmit the vibrations to the *cochlea*, the sound-sensing organ. The bones act as levers to magnify the amplitude of the vibrations.

You learned in Chapter 15 that sounds of different frequencies produce responses at different positions along the cochlea's *basilar membrane*—a resonance response—and that subsequent oscillations of microscopic hairs on the basilar membrane trigger nerve signals to the brain.

This complex electromechanical system does not respond the same at all frequencies. The physical size of the cochlea doesn't allow the necessary resonances at lower frequencies, so hearing sensitivity decreases with decreasing frequency. The inertia of the tympanic membrane and the ossicles limits their ability to respond at very high frequencies.

The graph of FIGURE 16.27 shows the threshold of hearing over the frequency range from 20 Hz to 20,000 Hz. Notice that the horizontal axis is a logarithmic scale. The vertical scale, because it displays decibels, is also a logarithmic scale. Hence this is a log-log graph, like ones you saw in Chapter 1.

FIGURE 16.27 A graph showing how the threshold of hearing varies with frequency.

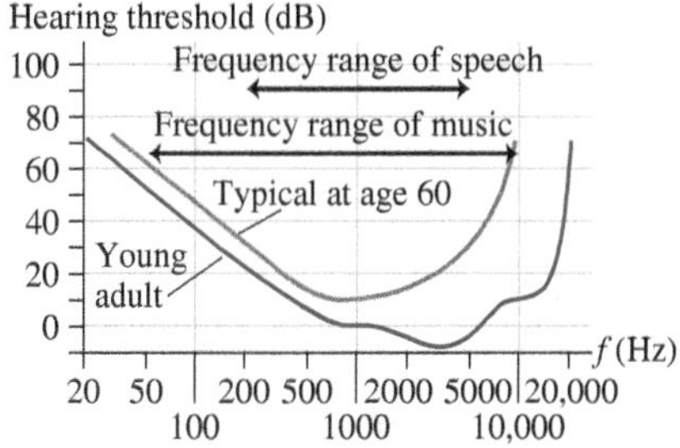

The quantity I_0, which appears in the equation for the sound intensity level as the "threshold of hearing," is actually the threshold only at frequencies near 1000 Hz. 1000 Hz is the frequency of a flute note in the middle of its range or of one of the highest notes you can play on an acoustic guitar, so it's perceived as a fairly high pitch. (You'll see in Chapter 17 that the notes played by most musical instruments are a few hundred Hz.)

The hearing sensitivity decreases at frequencies below 1000 Hz, so by the time you get down to 100 Hz, a fairly low bass note, the threshold of hearing is about 35 dB. That means that a 100 Hz note has to be $10^{3.5}$ or ≈ 3000 times more intense than a 1000 Hz note for you to barely perceive it in a quiet environment. Similarly, the threshold for hearing rises rapidly for frequencies above about 12,000 Hz, which means that your ear's sensitivity is rapidly decreasing.

The hearing response between 1000 Hz and 10,000 Hz is complex, but that complexity is due to the physics of the ear. You can see that maximum sensitivity is reached at a frequency of about 3500 Hz, where the threshold of hearing is about -8 dB. The minus sign indicates that the threshold intensity at this frequency is *less* than $I_0 = 10^{-12}\ \text{W/m}^2$. You'll learn in Chapter 17 that your ear canal—the outer ear—has a *standing-wave resonance* at ≈ 3500 Hz due to its length. This standing-wave resonance acts as an amplifier to boost the oscillating air pressure on your eardrum, and this amplification increases your hearing sensitivity at frequencies near 3500 Hz. There's another standing-wave resonance at a frequency three times higher, $\approx 10{,}500$ Hz. That one doesn't produce a dip in the hearing threshold curve, but it is strong enough to cause the curve to flatten out near 10,000 Hz before rising sharply at higher frequencies.

There are many reasons a person's hearing sensitivity may decrease, which is seen as an *increase* in the threshold of hearing: exposure to loud sounds, disease, and age. The upper curve in Figure 16.27 is a typical hearing threshold curve at age 60. There's some hearing loss at all frequencies, with the threshold increasing by ≈ 10 dB at frequencies of 1000 Hz or less, but the loss is especially pronounced at higher frequencies. The upper limit for hearing is now around 10,000 Hz. These losses—both a decrease in overall sensitivity and a loss of high-frequency hearing—continue with increasing age, which is why many senior citizens need hearing aids.

When does hearing loss begin to matter? We'll look at speech and music more closely in Chapter 17, but the sound energy involved in speech is concentrated between roughly 200 Hz and 5000 Hz. Despite the loss of high-frequency hearing, a typical 60-year-old has no difficulty understanding speech. Music, however, spans a larger range of frequencies, mostly 50 Hz to 10,000 Hz. A typical 60-year-old experiences some decreased ability to hear music at the lowest and especially the highest frequencies.

16.7 The Doppler Effect

Our final topic for this chapter is an interesting effect that occurs when you are in motion relative to a wave source. It is called the *Doppler effect.* You've likely noticed that the pitch of an ambulance's siren drops as it goes past you. Why?

FIGURE 16.28a shows a source of sound waves moving away from Pablo and toward Nancy at a steady speed v_s. The subscript s indicates that this is the speed of the source, not the speed of the waves. The source is emitting sound waves of frequency f_0 as it travels. The figure is a motion diagram showing the position of the source at times $t = 0, T, 2T$, and $3T$, where $T = 1/f_0$ is the period of the waves.

Nancy measures the frequency of the wave emitted by the *approaching source* to be f_+. At the same time, Pablo measures the frequency of the wave emitted by the *receding source* to be f_-. Our task is to relate f_+ and f_- to the source frequency f_0 and speed v_s.

After a wave crest leaves the source, its motion is governed by the properties of the medium. That is, the motion of the source cannot affect a wave that has already been emitted. Thus each circular wave front in FIGURE 16.28b is centered on the point from which it was emitted. The wave crest from point 3 was emitted just as this figure was made, but it hasn't yet had time to travel any distance.

The wave crests are bunched up in the direction the source is moving, stretched out behind it. The distance between one crest and the next is one wavelength, so the wavelength λ_+ Nancy measures is *less* than the wavelength $\lambda_0 = v/f_0$ that would be emitted if the source were at rest. Similarly, λ_- behind the source is larger than λ_0.

These crests move through the medium at the wave speed v. Consequently, the frequency $f_+ = v/\lambda_+$ detected by the observer whom the source is approaching is *higher* than the frequency f_0 emitted by the source. Similarly, $f_- = v/\lambda_-$ detected behind the source is *lower* than frequency f_0. This change of frequency when a source moves relative to an observer is called the **Doppler effect**.

The distance labeled d in Figure 16.28b is the difference between how far the wave has moved and how far the source has moved at time $t = 3T$. These distances are

$$\begin{aligned}\Delta x_{\text{wave}} &= vt = 3vT \\ \Delta x_{\text{source}} &= v_s t = 3v_s T\end{aligned} \tag{16.28}$$

The distance d spans three wavelengths; thus the wavelength of the wave emitted by an approaching source is

$$\lambda_+ = \frac{d}{3} = \frac{\Delta x_{\text{wave}} - \Delta x_{\text{source}}}{3} = \frac{3vT - 3v_s T}{3} = (v - v_s)T \tag{16.29}$$

FIGURE 16.28 A motion diagram showing the wave fronts emitted by a source as it moves to the right at speed v_s.

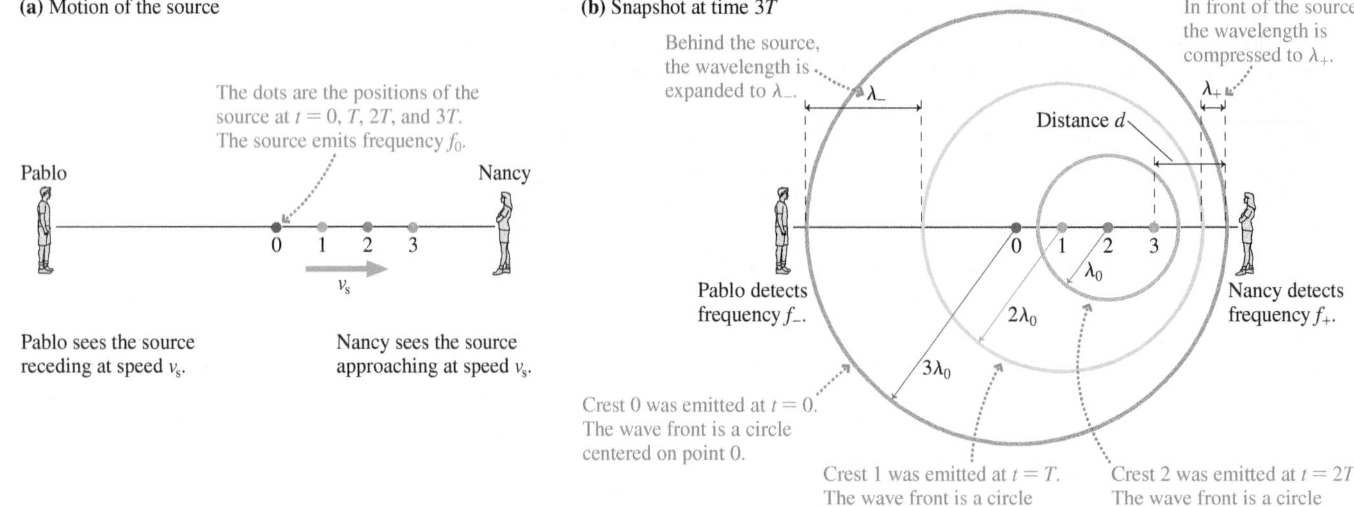
(a) Motion of the source
The dots are the positions of the source at t = 0, T, 2T, and 3T. The source emits frequency f0.
Pablo
Nancy
0 1 2 3
vs
Pablo sees the source receding at speed vs.
Nancy sees the source approaching at speed vs.
(b) Snapshot at time 3T
Behind the source, the wavelength is expanded to λ−.
λ−
In front of the source, the wavelength is compressed to λ+.
λ+
Distance d
0 1 2 3
λ0
2λ0
3λ0
Pablo detects frequency f−.
Nancy detects frequency f+.
Crest 0 was emitted at t = 0. The wave front is a circle centered on point 0.
Crest 1 was emitted at t = T. The wave front is a circle centered on point 1.
Crest 2 was emitted at t = 2T. The wave front is a circle centered on point 2.

You can see that our arbitrary choice of three periods was not relevant because the 3 cancels. The frequency detected in Nancy's direction is

$$f_+ = \frac{v}{\lambda_+} = \frac{v}{(v - v_s)T} = \frac{v}{(v - v_s)} f_0 \qquad (16.30)$$

where $f_0 = 1/T$ is the frequency of the source and is the frequency you would detect if the source were at rest. We say that the wave Nancy detects has been *Doppler shifted.*

We'll find it convenient to write the detected frequency as

$$f_+ = \frac{f_0}{1 - v_s/v} \quad \text{(Doppler effect for an approaching source)}$$

$$f_- = \frac{f_0}{1 + v_s/v} \quad \text{(Doppler effect for a receding source)} \qquad (16.31)$$

Doppler effect for a source moving at speed v_s

Storm trackers Doppler weather radar uses the Doppler shift of reflected radar signals to measure wind speeds and thus better gauge the severity of a storm.

Proof of the second equation, for the frequency f_- of a receding source, is similar. You can see that $f_+ > f_0$ in front of the source, because the denominator is less than 1, and $f_- < f_0$ behind the source.

EXAMPLE 16.12 **How fast are the police traveling?**

A police siren has a frequency of 550 Hz as the police car approaches you, 450 Hz after it has passed you and is receding. How fast are the police traveling? The temperature is 20°C.

PREPARE The siren's frequency is altered by the Doppler effect. The frequency is f_+ as the car approaches and f_- as it moves away.

SOLVE To find v_s, we rewrite Equations 16.31 as

$$f_0 = (1 + v_s/v)f_-$$
$$f_0 = (1 - v_s/v)f_+$$

We subtract the second equation from the first, giving

$$0 = f_- - f_+ + \frac{v_s}{v}(f_- + f_+)$$

This is easily solved to give

$$v_s = \frac{f_+ - f_-}{f_+ + f_-} v = \frac{100 \text{ Hz}}{1000 \text{ Hz}} \times 343 \text{ m/s} = 34.3 \text{ m/s}$$

ASSESS If you now solve for the siren frequency when at rest, you will find $f_0 = 495$ Hz. Surprisingly, the at-rest frequency is not halfway between f_- and f_+.

A Stationary Source and a Moving Observer

Suppose the police car in Example 16.12 is at rest while you drive toward it at 34.3 m/s. You might think that this is equivalent to having the police car move toward you at 34.3 m/s, but it isn't. Mechanical waves move through a medium, and the Doppler effect depends not just on how the source and the observer move with respect to each other but also on how they move with respect to the medium. We'll omit the proof, but it's not hard to show that the frequencies heard by an observer moving at speed v_o relative to a stationary source emitting frequency f_0 are

$$f_+ = (1 + v_o/v)f_0 \quad \text{(observer approaching a source)}$$

$$f_- = (1 - v_o/v)f_0 \quad \text{(observer receding from a source)} \qquad (16.32)$$

Doppler effect for an observer moving at speed v_o

A quick calculation shows that the frequency of the police siren as you approach it at 34.3 m/s is 545 Hz, not the 550 Hz you heard as it approached you at 34.3 m/s.

STOP TO THINK 16.7 Amy and Zack are both listening to the source of sound waves that is moving to the right. Compare the frequencies each hears.

A. $f_{Amy} > f_{Zack}$
B. $f_{Amy} = f_{Zack}$
C. $f_{Amy} < f_{Zack}$

10 m/s 10 m/s 10 m/s
Amy f_0 Zack

Doppler Ultrasound

Doppler imaging is based on the idea that the distance to a reflecting object can be determined by measuring the time between when a pulse of ultrasound is emitted and when the echo is detected. But more information is available than simply distance; the sound waves reflected from a moving object are *Doppler shifted,* and measuring the frequency of the echo, in addition to its arrival time, allows us to determine the object's speed toward or away from the ultrasound transducer.

FIGURE 16.29 Doppler ultrasound is based on the Doppler effect that occurs when sound waves reflect from a moving object.

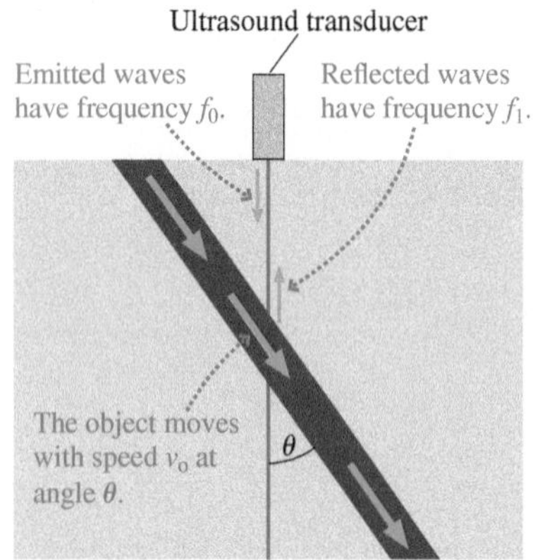
Ultrasound transducer
Emitted waves have frequency f_0.
Reflected waves have frequency f_1.
The object moves with speed v_o at angle θ.
θ

FIGURE 16.29 shows ultrasound at frequency f_0 being aimed at an object, such as flowing blood cells, that is moving at speed v_o at an angle θ relative to the ultrasound beam. This situation differs from the two situations we discussed earlier: a moving source and a moving observer. Now the stationary transducer is both the source *and* the observer while the object is moving.

In this case, the frequency f_1 of the returning echo has been Doppler shifted *twice.* To see this, let's first calculate the frequency of the sound wave as experienced by the moving object. For this part, the object is an observer receding from a stationary source, the transducer. The object's speed along the direction of wave travel is $v_o \cos\theta$, so the frequency seen by the moving object is given in Equations 16.32:

$$f_o = f_- = (1 - v_o \cos\theta / v) f_0 \qquad (16.33)$$

where v is the speed of sound through the medium.

The object reflects this wave back toward the transducer. Reflection doesn't change the frequency, so f_o is the frequency of the reflected wave as it leaves the moving object. From the transducer's perspective, the moving object is a receding source emitting waves of frequency f_o, so we need to use Equations 16.31 to deal with this second Doppler shift. The transducer finds that the returning echo has frequency

$$f_1 = f_- = \frac{f_o}{(1 + v_o \cos\theta / v)} = \frac{(1 - v_o \cos\theta / v)}{(1 + v_o \cos\theta / v)} f_0 \qquad (16.34)$$

Equation 16.34 is exact, but we can simplify it for medical applications because any moving object of medical interest, such as blood cells or the wall of the heart, has a speed *much* less than the speed of sound: $v_o \ll v$. The necessary approximation, which we won't prove, is

$$\frac{1 - x}{1 + x} \approx 1 - 2x \text{ if } x \ll 1 \qquad (16.35)$$

You can convince yourself that this works by checking it for $x = 0.001$:

$$\frac{1 - 0.001}{1 + 0.001} = \frac{0.999}{1.001} = 0.998002 \approx 0.998 = 1 - 2(0.001)$$

We can apply Equation 16.35 to Equation 16.34 by letting $v_o \cos\theta / v = x$:

$$f_1 = \frac{(1 - v_o \cos\theta / v)}{(1 + v_o \cos\theta / v)} f_0 \approx \left(1 - 2\frac{v_o \cos\theta}{v}\right) f_0 = f_0 - \frac{2 v_o \cos\theta}{v} f_0 \qquad (16.36)$$

Thus the *frequency shift* of the reflected waves, $\Delta f = f_1 - f_0$, which is what actually gets measured, is

$$\Delta f = -\frac{2v_o \cos\theta}{v} f_0 \qquad (16.37)$$

Frequency shift of waves reflected from an object moving with speed v_o at angle θ

An object moving away from the transducer has $\theta < 90^\circ$ and $\cos\theta > 0$, so $\Delta f < 0$, which indicates that the returning echo's frequency has been Doppler shifted *downward* to a lower frequency. This is the situation shown in Figure 16.29. An object moving toward the transducer is traveling at angle $\theta > 90^\circ$ and thus $\cos\theta < 0$. Its echo is shifted *upward* in frequency to have $\Delta f > 0$. Notice that $\Delta f = 0$ (no frequency shift) if the object is at rest ($v_o = 0$) or if it is moving perpendicular to the ultrasound beam ($\cos\theta = 0$).

Doppler images that include velocity information—usually color coded—are called *Doppler ultrasound.* As an example, FIGURE 16.30 is a Doppler ultrasound imaging of blood flow through the heart. It has been color coded so that red shows flow away from the transducer and blue flow toward the transducer. The scale at the right shows that the flow speeds are a few tens of centimeters per second. Images such as this convey a tremendous amount of information to a trained practitioner.

FIGURE 16.30 A Doppler ultrasound image of blood flow in the heart. Red indicates flow away from the transducer; blue is flow toward the transducer.

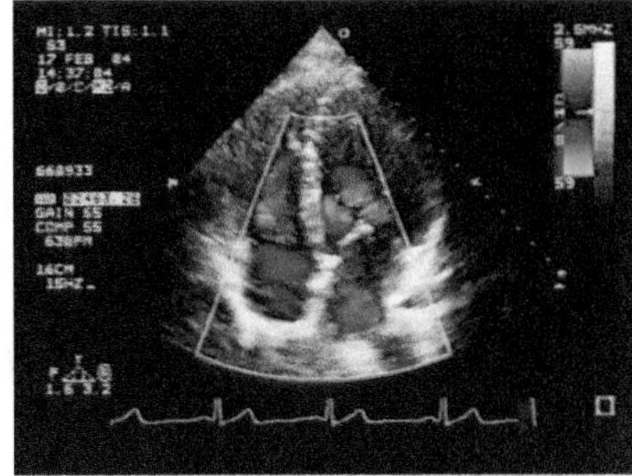

CHAPTER 16 INTEGRATED EXAMPLE Diagnosing aortic stenosis

Aortic stenosis, a heart disease, occurs when there's a narrowing—a stenosis—in or around the aortic valve between the heart's left ventricle and the ascending aorta artery (the major artery carrying oxygenated blood to the upper body and the head). The resulting restriction of blood flow can be a life-threatening condition. A diagnosis is made on the basis of the pressure difference across the valve, between the left ventricle and the aorta. A direct measurement is difficult, but an indirect measurement can be made using Doppler ultrasound. A trained physician can use images like the one in Figure 16.30 to determine the speed of blood flow in the aorta as it is ejected from the heart. For one patient, the physician uses 5.00 MHz ultrasound and places the transducer on a patient's upper chest at a location where, from earlier imaging, she knows that the ultrasound beam crosses the artery at a 30° angle. She finds that the maximum ultrasound frequency shift is 10.1 kHz. The average diameter of the left ventricle is 5.5 cm and that of the aorta is 2.8 cm. For this patient, what is the pressure difference across the aortic valve, and is it large enough (>25 mm Hg) to make a diagnosis of aortic stenosis?

PREPARE The Doppler echo is frequency shifted because the target, flowing blood, is moving. The shift to a higher frequency ($\Delta f > 0$) indicates that blood is flowing toward the transducer. We can refer to the geometry of Figure 16.29 to see that the angle θ is 150° rather than 30°. The flow speed in the aorta changes throughout one heartbeat cycle, but, by measuring the maximum frequency shift, the physician has focused on the highest flow speed that occurs as the heart contracts and ejects blood into the aorta.

We can use Bernoulli's equation to relate the changing speed of a fluid to a pressure difference. The blood is moving in both the ventricle and the aorta, but only the aortic speed is measured. However, we can use the equation of continuity for a fluid to relate the flow speed in the aorta to the flow speed in the ventricle. The ventricle has a complex shape, but we'll model it as a 5.5-cm-diameter cylinder. The density of blood is 1050 kg/m^3.

SOLVE We first need to find the blood flow speeds v_a and v_v in the aorta and the ventricle. We can solve Equation 16.37 to determine $v_a = v_o$ from the measured frequency shift. Doing so, with a soft-tissue sound speed of 1540 m/s, gives

$$v_a = v_o = -\frac{v}{2\cos\theta}\frac{\Delta f}{f_0} = -\frac{1540 \text{ m/s}}{2\cos 150^\circ} \times \frac{10.1 \times 10^3 \text{ Hz}}{5.00 \times 10^6 \text{ Hz}}$$
$$= 1.8 \text{ m/s}$$

In Chapter 9 we found that the flow speeds at different points in a tube with a changing cross section are related by the *equation of continuity*. In this situation, where blood flows through the ventricle and then the aorta, with cross-section areas A_v and A_a, the equation of continuity is $v_v A_v = v_a A_a$. The equation of continuity is the mathematical statement that the blood volume flow rate remains constant. We now have enough information to find the flow speed through the ventricle:

$$v_v = \frac{A_a}{A_v} v_a = \frac{\pi r_a^2}{\pi r_v^2} v_a = \left(\frac{2.8 \text{ cm}}{5.5 \text{ cm}}\right)^2 \times 1.8 \text{ m/s} = 0.47 \text{ m/s}$$

Not surprisingly, the flow speed through the ventricle is much slower than through the aorta. In fact, a speed of 47 cm/s is very consistent with the cardiac Doppler ultrasound in Figure 16.30.

Knowing the flow speeds on both sides of the valve, we can now use Bernoulli's equation to find the pressure difference across the valve:

$$p_v + \tfrac{1}{2}\rho v_v^2 = p_a + \tfrac{1}{2}\rho v_a^2$$
$$\Delta p = p_v - p_a = \tfrac{1}{2}\rho(v_a^2 - v_v^2)$$
$$= \tfrac{1}{2}(1050 \text{ kg/m}^3)((1.8 \text{ m/s})^2 - (0.47 \text{ m/s})^2)$$
$$= 1590 \text{ Pa} \times \frac{760 \text{ mm Hg}}{101{,}000 \text{ Pa}} = 12 \text{ mm Hg}$$

ASSESS The pressure difference across the aortic valve is within the normal range; this patient does not have aortic stenosis.

SUMMARY

GOAL To learn the basic properties of traveling waves.

GENERAL PRINCIPLES

The Wave Model

The wave model is based on the idea of a **traveling wave,** which is an organized disturbance traveling at a well-defined **wave speed** v.

- In a **transverse wave,** the particles of the medium move *perpendicular* to the direction in which the wave travels.

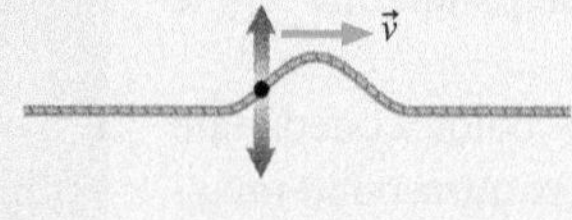

- In a **longitudinal wave,** the particles of the medium move *parallel* to the direction in which the wave travels.

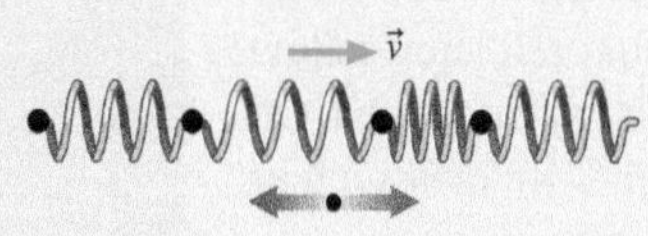

A wave transfers energy, but there is no material or substance transferred.

Mechanical waves require a material **medium.** The speed of the wave is a property of the medium, not the wave. The speed does not depend on the size or shape of the wave.

- For a wave on a string, the string is the medium.

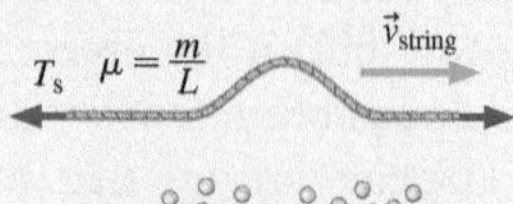

- A sound wave is a wave of compressions and rarefactions of a medium such as air.

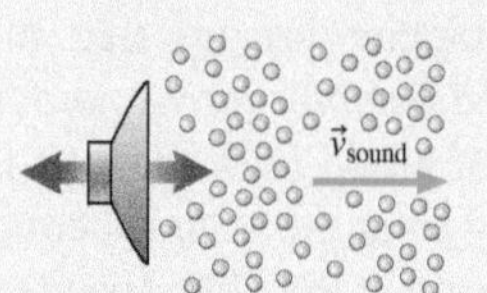

Electromagnetic waves are waves of the electromagnetic field. They do not require a medium. All electromagnetic waves travel at the same speed in a vacuum, $c = 3.00 \times 10^8$ m/s.

IMPORTANT CONCEPTS

The **displacement** D of a wave is a function of both position (where) and time (when).

- A **snapshot graph** shows the wave's displacement as a function of position at a single instant of time.
- A **history graph** shows the wave's displacement as a function of time at a single point in space.

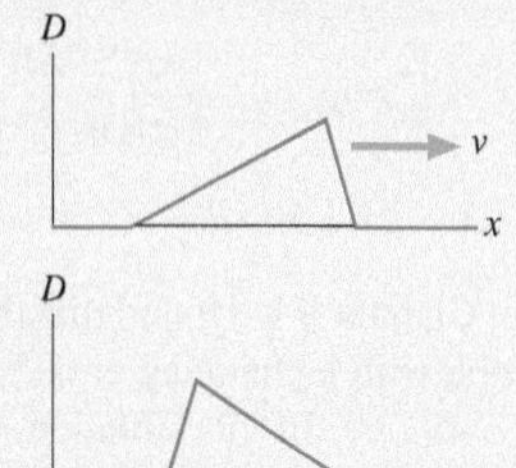

For a transverse wave on a string, the snapshot graph is a picture of the wave. The displacement of a longitudinal wave is parallel to the motion; thus the snapshot graph of a longitudinal sound wave is *not* a picture of the wave.

Sinusoidal waves are periodic in both time (period T) and space (wavelength λ):

$$D(x, t) = A\sin[2\pi(x/\lambda - t/T) + \phi_0]$$
$$= A\sin(kx - \omega t + \phi_0)$$

where A is the **amplitude,** $k = 2\pi/\lambda$ is the **wave number,** $\omega = 2\pi f = 2\pi/T$ is the angular frequency, and ϕ_0 is the phase constant that describes initial conditions.

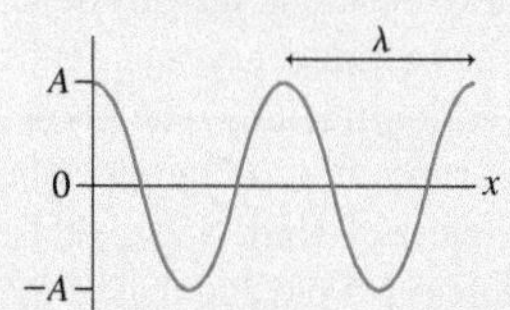

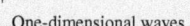

One-dimensional waves

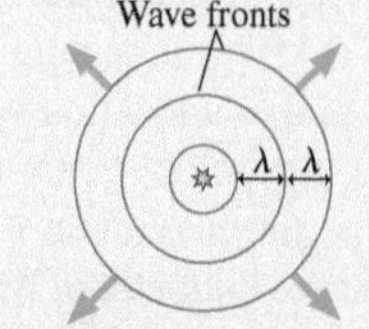

Two- and three-dimensional waves

The *fundamental relationship* for any sinusoidal wave is $v = \lambda f$.

APPLICATIONS

- **String** (transverse): $v_{string} = \sqrt{T_s/\mu}$
- **Sound** (longitudinal): $v_{sound} = \sqrt{\gamma RT/M}$ in a gas
- **Light** (transverse): $v = c/n$, where $c = 3.00 \times 10^8$ m/s is the speed of light in a vacuum and n is the material's **index of refraction.**

The wave **intensity** is the power-to-area ratio: $I = P/a$

For a circular or spherical wave: $I = P_{source}/4\pi r^2$

The **sound intensity level** is

$$\beta = (10\text{ dB})\log_{10}(I/1.0 \times 10^{-12}\text{ W/m}^2)$$

$I_0 = 1.0 \times 10^{-12}$ W/m² is the *threshold of hearing.*

The **Doppler effect** occurs when a wave source and detector are moving with respect to each other: The frequency detected differs from the frequency f_0 emitted.

Approaching source

$$f_+ = \frac{f_0}{1 - v_s/v}$$

Observer approaching a source

$$f_+ = (1 + v_o/v)f_0$$

Receding source

$$f_- = \frac{f_0}{1 + v_s/v}$$

Observer receding from a source

$$f_- = (1 - v_o/v)f_0$$

Reflected waves

$$\Delta f = \frac{2v_o\cos\theta}{v}f_0$$ for an object with speed v_o at angle θ

LEARNING OBJECTIVES After studying this chapter, you should be able to:

- Calculate the speed of waves on strings and the speed of sound in gases. *Conceptual Questions 16.1, 16.2; Problems 16.1–16.4, 16.19*
- Draw and interpret snapshot graphs and history graphs of wave pulses. *Conceptual Question 16.3; Problems 16.5–16.9*
- Solve problems using the mathematical representation of sinusoidal waves. *Conceptual Question 16.5; Problems 16.11–16.15*
- Solve problems that involve sound waves and light waves. *Conceptual Questions 16.4, 16.6; Problems 16.17, 16.18, 16.22, 16.23, 16.25*
- Draw and interpret wave-front diagrams for waves in two and three dimensions. *Conceptual Question 16.7; Problems 16.30–16.33*
- Calculate and interpret sound intensity levels. *Conceptual Questions 16.8, 16.10; Problems 16.35, 16.36, 16.38, 16.42, 16.47*
- Understand and use the Doppler effect. *Conceptual Questions 16.11, 16.12; Problems 16.48, 16.49, 16.53–16.55*

STOP TO THINK ANSWERS

Stop to Think 16.1: D and E. The wave speed depends on properties of the medium, not on how you generate the wave. For a string, $v = \sqrt{T_s/\mu}$. Increasing the tension or decreasing the linear density (lighter string) will increase the wave speed.

Stop to Think 16.2: B. The wave is traveling to the right at 2.0 m/s, so each point on the wave passes $x = 0$ m, the point of interest, 2.0 s before reaching $x = 4.0$ m. The graph has the same shape, but everything happens 2.0 s earlier.

Stop to Think 16.3: D. The wavelength—the distance between two crests—is seen to be 10 m. The frequency is $f = v/\lambda = (50 \text{ m/s})/(10 \text{ m}) = 5$ Hz.

Stop to Think 16.4: $n_3 > n_1 > n_2$. $\lambda = \lambda_{vac}/n$, so a shorter wavelength corresponds to a larger index of refraction.

Stop to Think 16.5: E. A crest and an adjacent trough are separated by $\lambda/2$. This is a phase difference of π rad.

Stop to Think 16.6: C. Any factor-of-2 change in intensity changes the sound intensity level by 3 dB. One trumpet is $\frac{1}{4}$ the original number, so the intensity has decreased by two factors of 2.

Stop to Think 16.7: C. Zack hears a higher frequency as he and the source approach. Amy is moving with the source, so $f_{Amy} = f_0$.

QUESTIONS

Conceptual Questions

1. The three wave pulses in Figure Q16.1 travel along the same stretched string. Rank in order, from largest to smallest, their wave speeds v_A, v_B, and v_C. Explain.

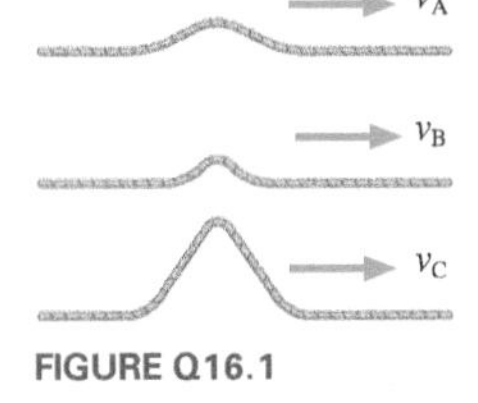

FIGURE Q16.1

2. A wave pulse travels along a string at a speed of 200 cm/s. What will be the speed if:
   a. The string's tension is doubled?
   b. The string's mass is quadrupled (but its length is unchanged)?
   c. The string's length is quadrupled (but its mass is unchanged)?
   d. The string's mass and length are both quadrupled?

   Note that parts a–d are independent and refer to changes made to the original string.
3. Figure Q16.3 shows a snapshot graph *and* a history graph for a wave pulse on a stretched string. They describe the same wave from two perspectives.
   a. In which direction is the wave traveling? Explain.
   b. What is the speed of this wave?

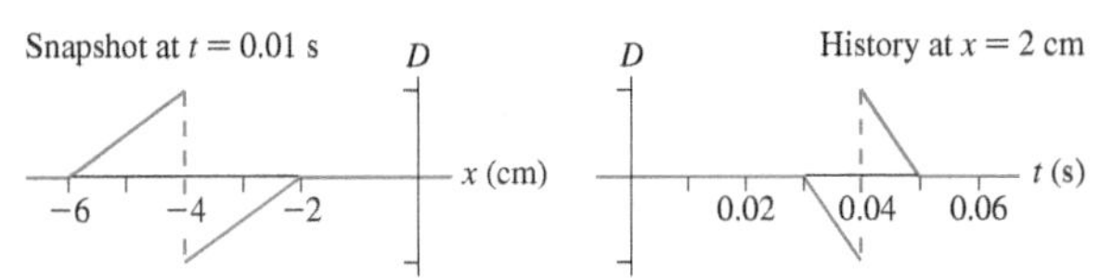

FIGURE Q16.3

4. When in air, a waterproof speaker emits sound waves that have a frequency of 1000 Hz. When the speaker is lowered into water, does the frequency of the sound increase, decrease, or remain the same? Does the wavelength of the sound increase, decrease, or remain the same?
5. Rank in order, from largest to smallest, the wavelengths λ_a, λ_b, and λ_c for sound waves having frequencies $f_a = 100$ Hz, $f_b = 1000$ Hz, and $f_c = 10{,}000$ Hz. Explain.
6. Ultrasound measuring tools can be used to quickly measure the size of rooms. These instruments work by sending out a pulse of ultrasonic sound, then measuring the time it takes for this pulse to reflect off a wall and return to the device. If you used such a device on a particularly hot day, would you measure the length of a given room to be longer or shorter than on a cool day?

Problem difficulty is labeled as | (straightforward) to ||||| (challenging). Problems labeled INT integrate significant material from earlier chapters; BIO are of biological or medical interest; CALC require calculus to solve.

7. Figure Q16.7 shows the wave fronts of a circular wave. What is the phase difference between (a) points A and B, (b) points C and D, and (c) points E and F?

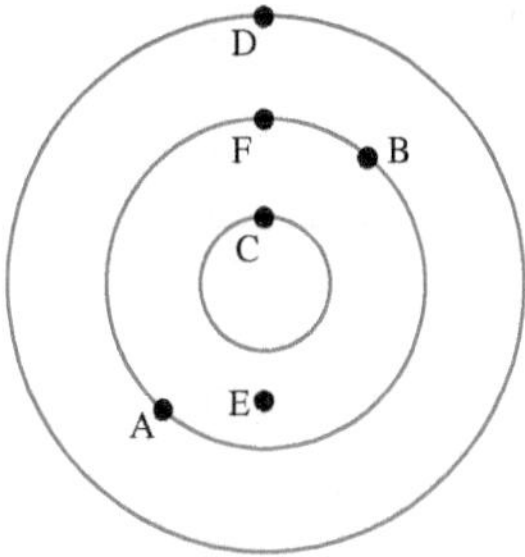

FIGURE Q16.7

8. Sound wave A delivers 2 J of energy in 2 s. Sound wave B delivers 10 J of energy in 5 s. Sound wave C delivers 2 mJ of energy in 1 ms. Rank in order, from largest to smallest, the sound powers P_A, P_B, and P_C of these three sound waves. Explain.
9. A laser beam has intensity I_0.
   a. What is the intensity, in terms of I_0, if a lens focuses the laser beam to 1/10 its initial diameter?
   b. What is the intensity, in terms of I_0, if a lens defocuses the laser beam to 10 times its initial diameter?
10. One physics professor talking produces a sound intensity level of 52 dB. It's a frightening idea, but what would be the sound intensity level of 100 physics professors talking simultaneously?
11. Some bat species have auditory systems that work best over a narrow range of frequencies. To account for this, the bats adjust the sound frequencies they emit so that the returning, Doppler-shifted sound pulse is in the correct frequency range. As a bat increases its forward speed, should it increase or decrease the frequency of the emitted pulses to compensate?
12. You are standing at $x = 0$ m, listening to seven identical sound sources described by Figure Q16.12. At $t = 0$ s, all seven are at $x = 343$ m and moving as shown below. The sound from all seven will reach your ear at $t = 1$ s. Rank in order, from highest to lowest, the seven frequencies f_1 to f_7 that you hear at $t = 1$ s. Explain.

1 → 50 m/s, speeding up
2 → 50 m/s, steady speed
3 → 50 m/s, slowing down
4 At rest
50 m/s, speeding up ← 5
50 m/s, steady speed ← 6
50 m/s, slowing down ← 7

FIGURE Q16.12

Multiple-Choice Questions

13. || An AM radio listener is located 5.0 km from the radio antenna. If he is listening to the radio at the frequency of 660 kHz, how many wavelengths fit in the distance from the antenna to his house?
   A. 11 B. 22 C. 33 D. 132
14. || The probe used in a medical ultrasound examination emits
BIO sound waves in air that have a wavelength of 0.12 mm. What is the wavelength of the sound waves in the patient?
   A. 0.027 mm B. 0.12 mm
   C. 0.26 mm D. 0.54 mm
15. | Ultrasound can be used to deliver energy to tissues for ther-
BIO apy. It can penetrate tissue to a depth approximately 200 times its wavelength. What is the approximate depth of penetration of ultrasound at a frequency of 5.0 MHz?
   A. 0.29 mm B. 1.4 cm
   C. 6.2 cm D. 17 cm
16. | A sinusoidal wave traveling on a string has a period of 0.20 s, a wavelength of 32 cm, and an amplitude of 3 cm. The speed of this wave is
   A. 0.60 cm/s. B. 6.4 cm/s.
   C. 15 cm/s. D. 160 cm/s.
17. || A scientist measures the speed of sound in a monatomic gas to be 449 m/s at 20°C. What is the gas?
   A. Hydrogen B. Helium
   C. Neon D. Argon
18. || Two strings of different linear density are joined together and pulled taut. A sinusoidal wave on these strings is traveling to the right, as shown in Figure Q16.18. When the wave goes across the boundary from string 1 to string 2, the frequency is unchanged. What happens to the velocity?
   A. The velocity increases.
   B. The velocity stays the same.
   C. The velocity decreases.

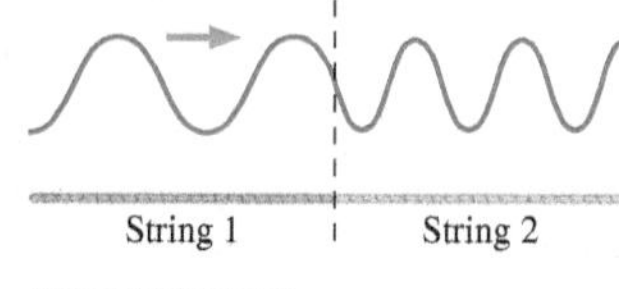

FIGURE Q16.18

19. || You stand at $x = 0$ m, listening to a sound that is emitted at frequency f_s. Figure Q16.19 shows the frequency you hear during a four-second interval. Which of the following describes the motion of the sound source?
   A. It moves from left to right and passes you at $t = 2$ s.
   B. It moves from right to left and passes you at $t = 2$ s.
   C. It moves toward you but doesn't reach you. It then reverses direction at $t = 2$ s.
   D. It moves away from you until $t = 2$ s. It then reverses direction and moves toward you but doesn't reach you.

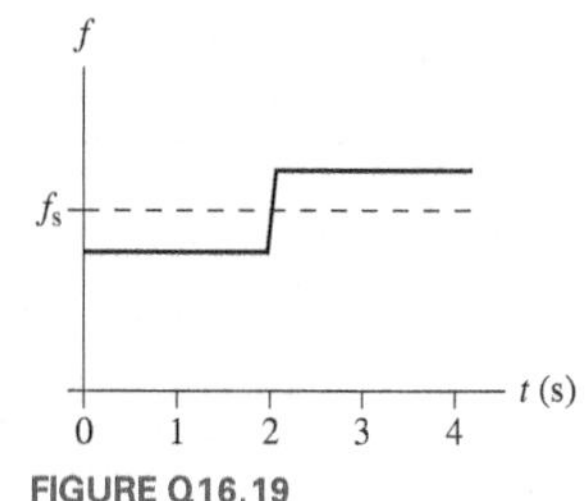

FIGURE Q16.19

PROBLEMS

Section 16.1 An Introduction to Waves

1. || The wave speed on a string under tension is 200 m/s. What is the speed if the tension is halved?
2. | The wave speed on a string is 150 m/s when the tension is 75 N. What tension will give a speed of 180 m/s?
3. || A 25 g string is under 20 N of tension. A pulse travels the length of the string in 50 ms. How long is the string?
4. | String 1 in Figure P16.4 has linear density 2.0 g/m and string 2 has linear density 4.0 g/m. A student sends pulses in both directions by quickly pulling up on the knot, then releasing it. What should the string lengths L_1 and L_2 be if the pulses are to reach the ends of the strings simultaneously?

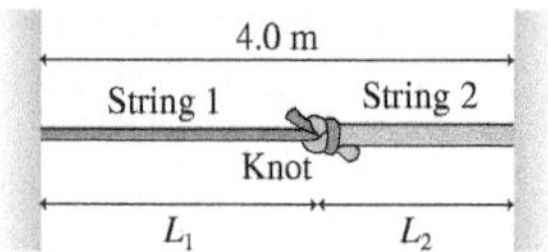

FIGURE P16.4

Section 16.2 Visualizing Wave Motion

5. ‖ Draw the history graph $D(x = 4.0\text{ m}, t)$ at $x = 4.0$ m for the wave shown in Figure P16.5.

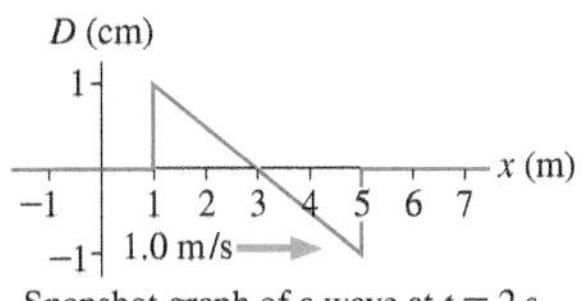

FIGURE P16.5

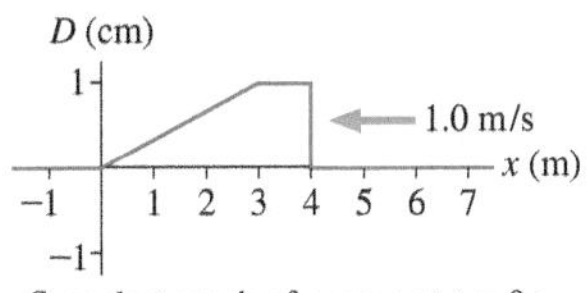

FIGURE P16.6

6. ‖ Draw the history graph $D(x = 0\text{ m}, t)$ at $x = 0$ m for the wave shown in Figure P16.6.
7. ‖ Draw the snapshot graph $D(x, t = 0\text{ s})$ at $t = 0$ s for the wave shown in Figure P16.7.

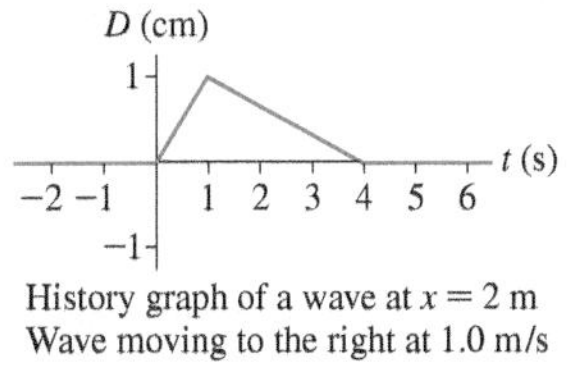

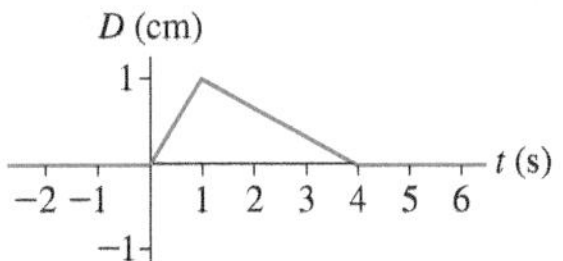

FIGURE P16.7

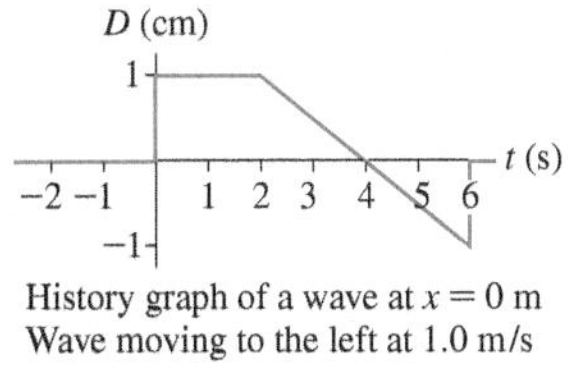

D (cm)

History graph of a wave at $x = 0$ m
Wave moving to the left at 1.0 m/s

FIGURE P16.8

8. ‖ Draw the snapshot graph $D(x, t = 1.0\text{ s})$ at $t = 1.0$ s for the wave shown in Figure P16.8.
9. ‖ Figure P16.9 is a picture at $t = 0$ s of the particles in a medium as a longitudinal wave is passing through. The equilibrium spacing between the particles is 1.0 cm. Draw the snapshot graph $D(x, t = 0\text{ s})$ of this wave at $t = 0$ s.

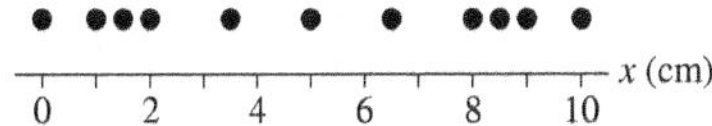

FIGURE P16.9

10. ‖ Figure P16.10 is the snapshot graph at $t = 0$ s of a *longitudinal* wave. Draw the corresponding picture of the particle positions, as was done in Figure 16.8b. Let the equilibrium spacing between the particles be 1.0 cm.

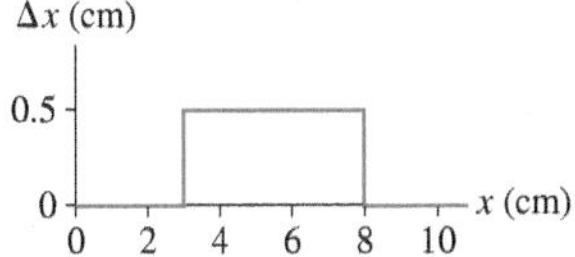

FIGURE P16.10

Section 16.3 Sinusoidal Waves

11. | A wave has angular frequency 30 rad/s and wavelength 2.0 m. What are its (a) wave number and (b) wave speed?
12. | A wave travels with speed 200 m/s. Its wave number is 1.5 rad/m. What are its (a) wavelength and (b) frequency?
13. | The displacement of a wave traveling in the negative y-direction is $D(y, t) = (5.2\text{ cm})\sin(5.5y + 72t)$, where y is in m and t is in s. What are the (a) frequency, (b) wavelength, and (c) speed of this wave?
14. ‖ What are the amplitude, frequency, and wavelength of the wave in Figure P16.14?

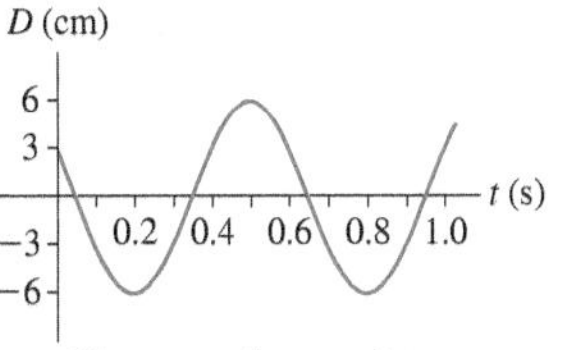

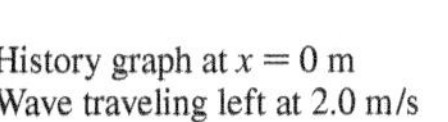

History graph at $x = 0$ m
Wave traveling left at 2.0 m/s

FIGURE P16.14

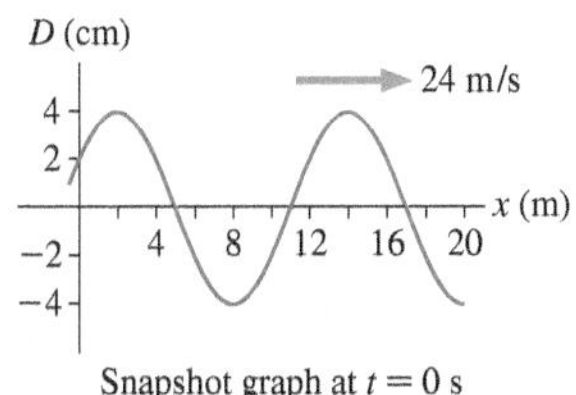

FIGURE P16.15

15. ‖ What are the amplitude, wavelength, frequency, and phase constant of the traveling wave in Figure P16.15?
16. ‖ A string that is under 50.0 N of tension has linear density 5.0 g/m. A sinusoidal wave with amplitude 3.0 cm and wavelength 2.0 m travels along the string. What is the maximum speed of a particle on the string?

Section 16.4 Sound and Light

17. ‖ The back wall of an auditorium is 26.0 m from the stage. If you are seated in the middle row, how much time elapses between a sound from the stage reaching your ear directly and the same sound reaching your ear after reflecting from the back wall? Assume that the temperature is 20°C.
18. | A medical ultrasound imaging system sends out a steady BIO stream of very short pulses. To simplify analysis, the reflection of one pulse should be received before the next is transmitted. If the system is being used to create an image of tissue 12 cm below the skin, what is the minimum time between pulses? Give your answer in μs.
19. | Air, a mixture of nitrogen and oxygen, has an effective molar mass of 0.029 kg/mol. What is the speed of sound in the stratosphere, 20 km above the earth's surface, where the temperature is –80°C?
20. ‖ a. What is the wavelength of a 2.0 MHz ultrasound wave traveling through bone?
    b. What frequency of electromagnetic wave would have the same wavelength as the ultrasound wave of part a?
21. | a. What is the frequency of an electromagnetic wave with a wavelength of 20 cm?
    b. What would be the wavelength of a sound wave in water with the same frequency as the electromagnetic wave of part a?
22. | a. What is the frequency of blue light that has a wavelength of 450 nm?
    b. What is the frequency of red light that has a wavelength of 650 nm?
    c. What is the index of refraction of a material in which the red-light wavelength is 450 nm?
23. | a. An FM radio station broadcasts at a frequency of 101.3 MHz. What is the wavelength?
    b. What is the frequency of a sound source that produces the same wavelength in 20°C air?
24. | a. Telephone signals are often transmitted over long distances by microwaves. What is the frequency of microwave radiation with a wavelength of 3.0 cm?
    b. Microwave signals are beamed between two mountaintops 50 km apart. How long does it take a signal to travel from one mountaintop to the other?

25. | A hammer taps on the end of a 4.00-m-long metal bar at room temperature. A microphone at the other end of the bar picks up two pulses of sound, one that travels through the metal and one that travels through the air. The pulses are separated in time by 9.00 ms. What is the speed of sound in this metal?

26. | Cell phone conversations are transmitted by high-frequency radio waves. Suppose the signal has wavelength 35 cm while traveling through air. What are the (a) frequency and (b) wavelength as the signal travels through 3-mm-thick window glass into your room?

27. || a. How long does it take light to travel through a 3.0-mm-thick piece of window glass?
  b. Through what thickness of water could light travel in the same amount of time?

28. | A light wave has a 670 nm wavelength in air. Its wavelength in a transparent solid is 420 nm.
  a. What is the speed of light in this solid?
  b. What is the light's frequency in the solid?

29. | A 440 Hz sound wave in 20°C air propagates into the water of a swimming pool. What are the wave's (a) frequency and (b) wavelength in the water?

Section 16.5 Circular and Spherical Waves

30. | A circular wave travels outward from the origin. At one instant of time, the phase at $r_1 = 20$ cm is 0 rad and the phase at $r_2 = 80$ cm is 3π rad. What is the wavelength of the wave?

31. || A spherical wave with a wavelength of 2.0 m is emitted from the origin. At one instant of time, the phase at $r = 4.0$ m is π rad. At that instant, what is the phase at $r = 3.5$ m and at $r = 4.5$ m?

32. || A loudspeaker at the origin emits a 120 Hz tone on a day when the speed of sound is 340 m/s. The phase difference between two points on the x-axis is 5.5 rad. What is the distance between these two points?

33. || A sound source is located somewhere along the x-axis. Experiments show that the same wave front simultaneously reaches listeners at $x = -7.0$ m and $x = +3.0$ m.
  a. What is the x-coordinate of the source?
  b. A third listener is positioned along the positive y-axis. What is her y-coordinate if the same wave front reaches her at the same instant it does the first two listeners?

Section 16.6 Power, Intensity, and Decibels

34. || Sound is detected when a sound wave causes the eardrum
BIO to vibrate. Typically, the diameter of the eardrum is about 8.4 mm in humans. When someone speaks to you in a normal tone of voice, the sound intensity at your ear is approximately 1.0×10^{-6} W/m^2. How much energy is delivered to your eardrum each second?

35. ||| At a rock concert, the sound intensity 1.0 m in front of the
BIO bank of loudspeakers is 0.10 W/m^2. A fan is 30 m from the loudspeakers. Her eardrums have a diameter of 8.4 mm. How much sound energy reaches each eardrum in 1.0 s? Assume the sound is broadcast equally in all directions.

36. || The intensity of electromagnetic waves from the sun is 1.4 kW/m^2 just above the earth's atmosphere. Eighty percent of this reaches the surface at noon on a clear summer day. Suppose you model your back as a 30 cm $\times$ 50 cm rectangle. How many joules of solar energy fall on your back as you work on your tan for 60 min?

37. ||| A large solar panel on a spacecraft in earth orbit produces 1.0 kW of power when the panel is turned toward the sun. What power would the solar cell produce if the spacecraft were in orbit around Saturn, 9.5 times as far from the sun?

38. ||| LASIK eye surgery uses pulses of laser light to shave off tis-
BIO sue from the cornea, reshaping it. A typical LASIK laser emits a 1.0-mm-diameter laser beam with a wavelength of 193 nm. Each laser pulse lasts 15 ns and contains 1.0 mJ of light energy.
  a. What is the power of one laser pulse?
  b. During the very brief time of the pulse, what is the intensity of the light wave?

39. || Using a dish-shaped mirror, a *solar cooker* concentrates the sun's energy onto a pot for cooking. A cooker with a 1.5-m-diameter dish focuses the sun's energy onto a pot with a diameter of 25 cm. Given that the intensity of sunlight is about 1000 W/m^2, what is the intensity at the base of the pot?

40. || During takeoff, the sound intensity level of a jet engine is 140 dB at a distance of 30 m. What is the sound intensity level at a distance of 1.0 km?

41. | What are the sound intensity levels for sound waves of intensity (a) 3.0×10^{-6} W/m^2 and (b) 3.0×10^{-2} W/m^2?

42. || A loudspeaker on a tall pole broadcasts sound waves equally in all directions. What is the speaker's power output if the sound intensity level is 90 dB at a distance of 20 m?

43. || The sound intensity level 5.0 m from a large power saw is 100 dB. At what distance will the sound be a more tolerable 80 dB?

44. || The sound intensity from a jack hammer breaking concrete is 2.0 W/m^2 at a distance of 2.0 m from the point of impact. This is sufficiently loud to cause permanent hearing damage if the operator doesn't wear ear protection. What are (a) the sound intensity and (b) the sound intensity level for a person watching from 50 m away?

45. || The African cicada is the world's loudest insect, producing a
BIO sound intensity level of 107 dB at a distance of 0.50 m. What is the intensity of its sound (in W/m^2) as heard by someone standing 3.0 m away?

46. || Your ears are sensitive to differences in pitch, but they are not
BIO very sensitive to differences in intensity. You are not capable of detecting a difference in sound intensity level of less than 1 dB. By what factor does the sound intensity increase if the sound intensity level increases from 60 dB to 61 dB?

47. || 30 seconds of exposure to 115 dB sound can damage your
BIO hearing, but a much quieter 94 dB may begin to cause damage after 1 hour of continuous exposure. You are going to an outdoor concert, and you'll be standing near a speaker that emits 50 W of acoustic power as a spherical wave. What minimum distance should you be from the speaker to keep the sound intensity level below 94 dB?

Section 16.7 The Doppler Effect

48. | A friend of yours is loudly singing a single note at 400 Hz while racing toward you at 25.0 m/s on a day when the speed of sound is 340 m/s.
  a. What frequency do you hear?
  b. What frequency does your friend hear if you suddenly start singing at 400 Hz?

49. | An opera singer in a convertible sings a note at 600 Hz while cruising down the highway at 90 km/h. What is the frequency heard by
   a. A person standing beside the road in front of the car?
   b. A person on the ground behind the car?

50. || A bat locates insects by emitting ultrasonic "chirps" and then
BIO listening for echoes from the bugs. Suppose a bat chirp has a frequency of 25 kHz. How fast would the bat have to fly, and in what direction, for you to just barely be able to hear the chirp at 20 kHz?

51. || An osprey's call is a distinct whistle at 2200 Hz. An osprey
BIO calls while diving at you, to drive you away from her nest. You hear the call at 2300 Hz. How fast is the osprey approaching?

52. || A whistle you use to call your hunting dog has a frequency of 21 kHz, but your dog is ignoring it. You suspect the whistle may not be working, but you can't hear sounds above 18 kHz. To test it, you ask a friend to blow the whistle, then you hop on your bicycle. In which direction should you ride (toward or away from your friend) and at what minimum speed to know if the whistle is working?

53. || An echocardiogram uses 4.4 MHz ultrasound to measure
BIO blood flow in the aorta. The blood is moving directly away from the probe at 1.4 m/s. What is the frequency shift of the reflected ultrasound?

54. || A train whistle is heard at 300 Hz as the train approaches town. The train cuts its speed in half as it nears the station, and the sound of the whistle is then 290 Hz. What is the speed of the train before and after slowing down?

55. || A Doppler unit that measures blood flow emits 5.0 MHz ultra-
BIO sound. The ultrasound waves reflect from blood flowing through an artery that makes a 45° angle with the ultrasound beam. The technician measures an upward frequency shift of 4.6 kHz in the returning sound waves. What is the speed of the blood flow?

General Problems

56. || Figure P16.56 is a history graph at $x = 0$ m of a wave traveling in the positive x-direction at 4.0 m/s.
   a. What is the wavelength?
   b. What is the phase constant of the wave?
   c. Write the displacement equation for this wave.

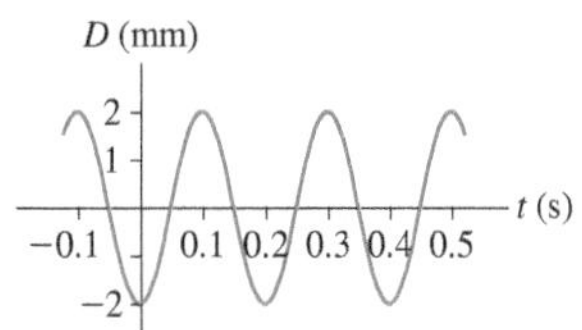

FIGURE P16.56

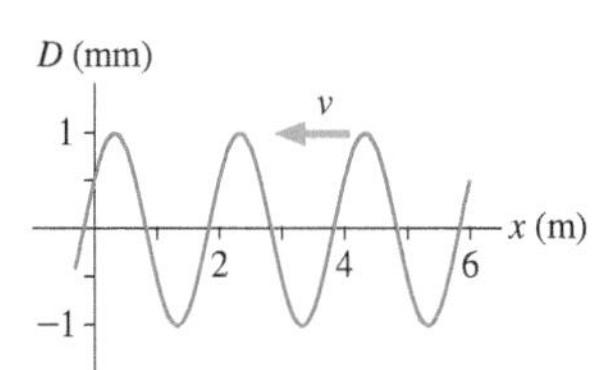

FIGURE P16.57

57. || Figure P16.57 is a snapshot graph at $t = 0$ s of a 5.0 Hz wave traveling to the left.
   a. What is the wave speed?
   b. What is the phase constant of the wave?
   c. Write the displacement equation for this wave.

58. ||| A female orb spider has a mass of 0.50 g. She is suspended
BIO from a tree branch by a 1.1 m length of 0.0020-mm-diameter
INT silk. Spider silk has a density of 1300 kg/m^3. If you tap the branch and send a vibration down the thread, how long does it take to reach the spider?

59. ||| Figure P16.59 shows two
INT masses hanging from a steel wire. The mass of the wire is 60.0 g. A wave pulse travels along the wire from point 1 to point 2 in 24.0 ms. What is mass m?

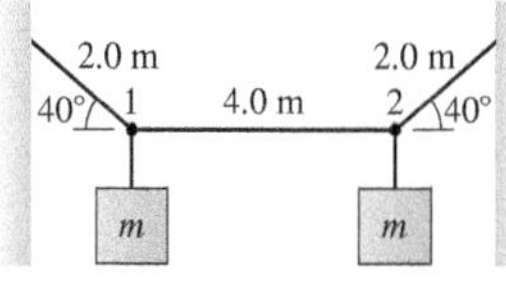

FIGURE P16.59

60. ||| The G string on a guitar is a 0.46-mm-diameter steel string
INT with a linear density of 1.3 g/m. When the string is properly tuned to 196 Hz, the wave speed on the string is 250 m/s. Tuning is done by turning the tuning screw, which slowly tightens—and stretches—the string. By how many mm does a 75-cm-long G string stretch when it's first tuned? Young's modulus for steel is 2.0×10^{11} N/m^2.

61. || Oil explorers set off explosives to make loud sounds, then listen for the echoes from underground oil deposits. Geologists suspect that there is oil under 500-m-deep Lake Physics. It's known that Lake Physics is carved out of a granite basin. Explorers detect a weak echo 0.94 s after exploding dynamite at the lake surface. If it's really oil, how deep will they have to drill into the granite to reach it?

62. ||| A coyote can locate a sound source with good accuracy by
BIO comparing the arrival times of a sound wave at its two ears. Suppose a coyote is listening to a bird whistling at 1000 Hz. The bird is 3.0 m away, directly in front of the coyote's right ear. The coyote's ears are 15 cm apart.
   a. What is the difference in the arrival times of the sound at the left ear and the right ear?
   b. What is the ratio of this time difference to the period of the sound wave?

   Hint: You are looking for the difference between two numbers that are nearly the same. What does this near equality imply about the necessary precision during intermediate stages of the calculation?

63. | A technician is using 5.0 MHz ultrasound to image a
BIO patient's gallbladder. She sees two echoes, the first 78 μs after the ultrasound pulse is emitted and a second 136 μs after the pulse. What is the diameter of the patient's gallbladder?

64. || A helium-neon laser beam has a wavelength in air of 633 nm. It takes 1.38 ns for the light to travel through 30 cm of an unknown liquid. What is the wavelength of the laser beam in the liquid?

65. | A simple linear (rod-shaped) antenna is most efficient at sending and receiving signals when its length is one-quarter of the wavelength of the signal. Cell phone antennas are built into the case, so the antenna length has to be less than the length of the case. To one significant figure, the antenna length is 10 cm ($\approx$4 in). Estimate, to one significant figure, the frequency in MHz of the cell phone signal.

66. || Earthquakes are essentially sound waves—called seismic waves—traveling through the earth. Because the earth is solid, it can support both longitudinal and transverse seismic waves. The speed of longitudinal waves, called P waves, is 8000 m/s. Transverse waves, called S waves, travel at a slower 4500 m/s. A seismograph records the two waves from a distant earthquake. If the S wave arrives 2.0 min after the P wave, how far away was the earthquake? You can assume that the waves travel in straight lines, although actual seismic waves follow more complex routes.

67. || Helium (density 0.18 kg/m^3 at 0°C and 1 atm pressure) remains a gas until the extraordinarily low temperature of 4.2 K. What is the speed of sound in helium at 5 K?

68. || A wave on a string is described by $D(x, t) = (3.0 \text{ cm}) \times \sin[2\pi(x/(2.4 \text{ m}) + t/(0.20 \text{ s}) + 1)]$, where x is in m and t is in s.
    a. In what direction is this wave traveling?
    b. What are the wave speed, the frequency, and the wave number?
    c. At $t = 0.50$ s, what is the displacement of the string at $x = 0.20$ m?
69. || A wave on a string is described by $D(x, t) = (2.00 \text{ cm}) \times \sin[(12.57 \text{ rad/m})x - (638 \text{ rad/s})t]$, where x is in m and t in s. The linear density of the string is 5.00 g/m. What are
    a. The string tension?
    b. The maximum displacement of a point on the string?
    c. The maximum speed of a point on the string?
70. || Figure P16.70 shows a snapshot graph of a wave traveling to the right along a string at 45 m/s. At this instant, what is the velocity of points 1, 2, and 3 on the string?

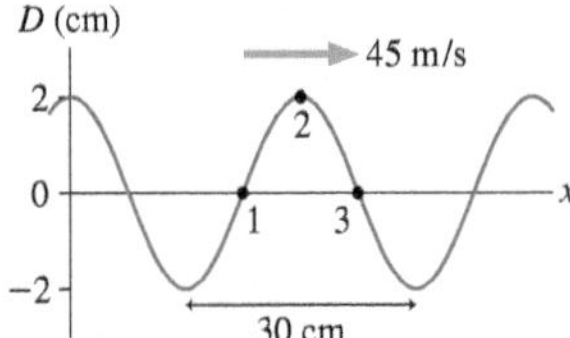

FIGURE P16.70

71. || A sun-like star is barely visible to naked-eye observers on earth when it is a distance of 7.0 light years, or 6.6×10^{16} m, away. The sun emits a power of 3.8×10^{26} W. Using this information, at what distance would a candle that emits a power of 0.20 W just be visible?
72. || Solar cells convert the energy of incoming light to electric energy; a good quality cell operates at an efficiency of 15%. Each person in the United States uses energy (for lighting, heating, transportation, etc.) at an average rate of 11 kW. Although sunlight varies with season and time of day, solar energy falls on the United States at an average intensity of 200 W/m^2. Assuming you live in an average location, what total solar-cell area would you need to provide all of your energy needs with energy from the sun?
73. || An AM radio station broadcasts with a power of 25 kW at a frequency of 920 kHz. Estimate the intensity of the radio wave at a point 10 km from the broadcast antenna.
74. || The sound intensity 50 m from a wailing tornado siren is 0.10 W/m^2.
    a. What is the intensity at 1000 m?
    b. The weakest intensity likely to be heard over background noise is $\approx 1\ \mu$W/m^2. Estimate the maximum distance at which the siren can be heard.
75. ||| A harvest mouse can detect sounds below the threshold of BIO human hearing, as quiet as −10 dB. Suppose you are sitting in a field on a very quiet day while a harvest mouse sits nearby. A very gentle breeze causes a leaf 1.5 m from your head to rustle, generating a faint sound right at the limit of your ability to hear it. The sound of the rustling leaf is also right at the threshold of hearing of the harvest mouse. How far is the harvest mouse from the leaf?
76. |||| A loudspeaker, mounted on a tall pole, is engineered to emit 75% of its sound energy into the forward hemisphere, 25% toward the back. You measure an 85 dB sound intensity level when standing 3.5 m in front of and 2.5 m below the speaker. What is the speaker's power output?
77. ||| A physics professor demonstrates the Doppler effect by tying INT a 600 Hz sound generator to a 1.0-m-long rope and whirling it around her head in a horizontal circle at 100 rpm. What are the highest and lowest frequencies heard by a student in the classroom?
78. || On a 20°C night, a bat hovering in the air emits an ultrasonic BIO chirp that has a frequency of 45 kHz. It hears an echo 55 ms later.
    a. How far away is the object that reflected the ultrasound?
    b. Suppose the bat's timing uncertainty is ±1 ms; that is, what the bat perceives as a 55 ms delay could actually be anywhere in the range 54–56 ms. What is the uncertainty in the distance to the object? Give your answer in cm.
    c. Suppose the object is an insect flying straight away from the bat. What is the insect's speed if the ultrasonic echo is shifted down in frequency by 750 Hz?
79. || Cardiologists use Doppler ultrasound to determine the BIO flow speed of blood through the heart. A typical system uses 3.5 MHz ultrasound and can detect a frequency change as small as 0.1 kHz. What is the smallest flow speed measurable with this device? Give your answer in cm/s.
80. || A 1000 Hz sound wave is at the threshold of hearing for a BIO young adult. By approximately what factor must the intensity of the sound be increased to be barely perceptible to a typical 60-year-old?
81. || A loudspeaker in an extremely quiet room emits a 1.0 kHz BIO tone and a 10 kHz tone with equal intensities. A young adult can barely perceive the 1.0 kHz tone when she is 4.0 m from the loudspeaker. What is the maximum distance from the speaker at which she can barely hear the 10 kHz tone? Assume that the speaker broadcasts equally in all directions.
82. |||| A loudspeaker on a pole is radiating 100 W of sound energy CALC in all directions. You are walking directly toward the speaker at 0.80 m/s. When you are 20 m away, what are (a) the sound intensity level and (b) the rate (dB/s) at which the sound intensity level is increasing? **Hint:** Use the chain rule and the relationship $\log_{10} x = \ln x / \ln 10$.
83. |||| A water wave is a *shallow-water wave* if the water depth d CALC is less than $\approx \lambda/10$. It is shown in hydrodynamics that the speed of a shallow-water wave is $v = \sqrt{gd}$, so waves slow down as they move into shallower water. Ocean waves, with wavelengths of typically 100 m, are shallow-water waves when the water depth is less than ≈ 10 m. Consider a beach where the depth increases linearly with distance from the shore until reaching a depth of 5.0 m at a distance of 100 m. How long does it take a wave to move the last 100 m to the shore? Assume that the waves are so small that they don't break before reaching the shore.

MCAT-Style Passage Problems

Echolocation

As discussed in the chapter, many species of bats find flying insects by emitting pulses of ultrasound and listening for the reflections. This technique is called *echolocation*. Bats possess several adaptations that allow them to echolocate very effectively.

84. Although we can't hear them, the ultrasonic pulses are very loud. In order not to be deafened by the sound they emit, bats can temporarily turn off their hearing. Muscles in the ear cause the bones in their middle ear to separate slightly, so that they don't transmit vibrations to the inner ear. After an ultrasound pulse ends, a bat can hear an echo from an object a minimum of 1 m away. Approximately how much time after a pulse is emitted is the bat ready to hear its echo?
    A. 0.5 ms B. 1 ms C. 3 ms D. 6 ms

85. Bats are sensitive to very small changes in frequency of the reflected waves. What information does this allow them to determine about their prey?
 A. Size B. Speed
 C. Distance D. Species
86. Some bats have specially shaped noses that focus ultrasound echolocation pulses in the forward direction. Why is this useful?
 A. Increasing intensity reduces the time delay for a reflected pulse.
 B. The energy of the pulse is concentrated in a smaller area, so the intensity is larger; reflected pulses will have a larger intensity as well.
 C. Increasing intensity allows the bat to use a lower frequency and still have the same spatial resolution.
87. Some bats utilize a sound pulse with a rapidly decreasing frequency. A decreasing-frequency pulse has
 A. Decreasing wavelength. B. Decreasing speed.
 C. Increasing wavelength. D. Increasing speed.

17 Superposition and Standing Waves

The notes played by this guitar are due to standing waves on the vibrating strings.

LOOKING AHEAD

Standing Waves

This *standing-wave* oscillation of a string is actually due to two traveling waves moving in opposite directions.

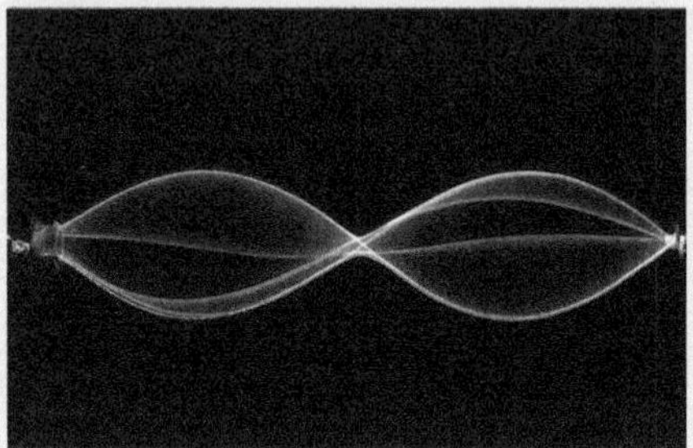

You'll learn to recognize the patterns of standing waves on strings and standing sound waves in columns of air.

Acoustics

The note played by a musical instrument, whether a trumpet or a piccolo, depends on the length of the instrument.

You'll find that the frequencies of the notes and the character of the instrument's sound are due to standing waves.

Interference

Where these two water waves overlap, they produce an *interference pattern* having distinct regions of maximum and minimum intensity.

You'll discover that interference is due to the *superposition*—the point-by-point addition—of two waves that have the same frequency.

GOAL To use the principle of superposition to understand the phenomena of standing waves and interference.

PHYSICS AND LIFE

The Physics of Sound Determines the Sounds of Speech

Humans and other animals communicate primarily with sound. You learned in Chapter 15 that we are able to hear different pitches or frequencies because of resonances in the cochlea of our ears. Somewhat similarly, we are able to produce a wide variety of different sounds—the sounds of speech—by changing the standing-wave resonances of our vocal tract. The frequencies of these resonances depend on the lengths and shapes of the throat, tongue, and lips in the same way that the note played by a wind instrument depends on the instrument's length and shape. It is our ability to manipulate the length and shape of our vocal tract, altering its acoustical properties, that allows us to produce the tremendous variety of sounds required for speech. A knowledge of these characteristics of sound waves will lay the groundwork for understanding the interference of light waves, which plays an essential role in vision and in optical instruments.

The sounds of human speech are made of many complex and rapidly changing waveforms that are generated by your vocal cords and then modified by your vocal tract.

17.1 The Principle of Superposition

We began our study of waves in Chapter 16 by looking at individual waves that travel in one direction—along an axis or radially outward from a source. Focusing on isolated waves allowed us to determine the physical properties and the mathematical description of waves. Now we turn our attention to the more realistic situation in which multiple waves pass through each other. This can occur when waves have more than one source or, of particular significance, when one wave reflects back on itself to produce a *standing wave*. The physics of simultaneous waves is relevant to topics ranging from speech and hearing to the resolving power of a microscope.

To begin, FIGURE 17.1a shows two baseball players, Anthony and Bill, at batting practice. Unfortunately, someone has turned the pitching machines so that pitching machine A throws baseballs toward Bill while machine B throws toward Anthony. If two baseballs are launched at the same time, and with the same speed, they collide at the crossing point. Two particles cannot occupy the same point of space at the same time.

FIGURE 17.1 Unlike particles, two waves can pass directly through each other.

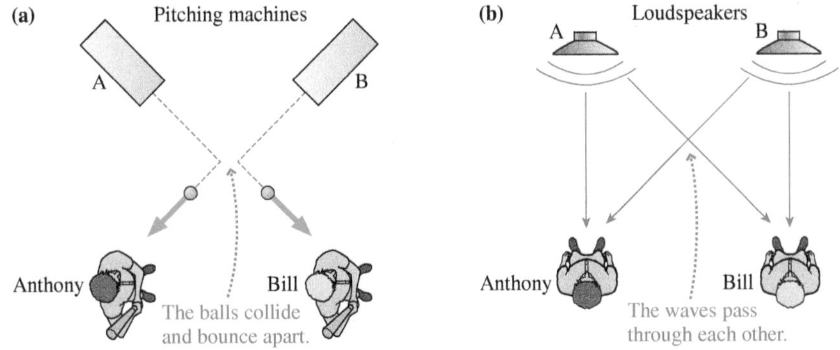

But waves, unlike particles, can pass directly through each other. In FIGURE 17.1b Anthony and Bill are listening to the stereo system in the locker room after practice. Because both hear the music quite well, the sound wave that travels from loudspeaker A toward Bill must pass through the wave traveling from loudspeaker B toward Anthony.

What happens to the medium at a point where two waves are present simultaneously? If wave 1 displaces a particle in the medium by D_1 and wave 2 *simultaneously* displaces it by D_2, the net displacement of the particle is simply $D_1 + D_2$. This is a very important idea because it tells us how to combine waves. It is known as the *principle of superposition.*

Principle of superposition When two or more waves are *simultaneously* present at a single point in space, the displacement of the medium at that point is the sum of the displacements due to each individual wave.

Mathematically, the net displacement of a particle in the medium is

$$D_{\text{net}} = D_1 + D_2 + \cdots = \sum_i D_i \tag{17.1}$$

where D_i is the displacement that would be caused by wave i alone. We will make the simplifying assumption that the displacements of the individual waves are along the same line so that we can add displacements as scalars rather than vectors.

To use the principle of superposition you must know the displacement caused by each wave if traveling alone. Then you go through the medium *point by point* and add the displacements due to each wave *at that point* to find the net displacement at that point.

To illustrate, FIGURE 17.2 shows snapshot graphs taken 1 s apart of two waves traveling at the same speed (1 m/s) in opposite directions. The principle of superposition

FIGURE 17.2 The superposition of two waves as they pass through each other.

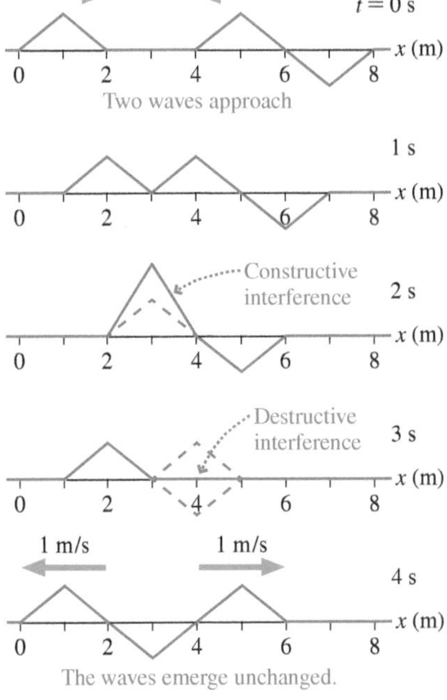

comes into play wherever the waves overlap. The solid line is the sum *at each point* of the two displacements at that point. This is the displacement that you would actually observe as the two waves pass through each other.

Notice how two overlapping positive displacements at $t = 2$ s add to give a displacement twice that of the individual waves. This is called *constructive interference.* Similarly, *destructive interference* is occurring at $t = 3$ s where positive and negative displacements add to give a superposition with zero displacement. We will defer the main discussion until later in this chapter, but you can already see that *interference is a consequence of superposition.*

STOP TO THINK 17.1 Two pulses on a string approach each other at speeds of 1 m/s. What is the shape of the string at $t = 6$ s?

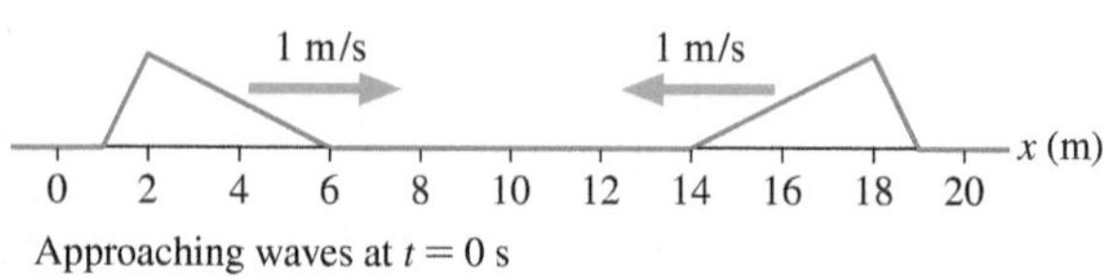

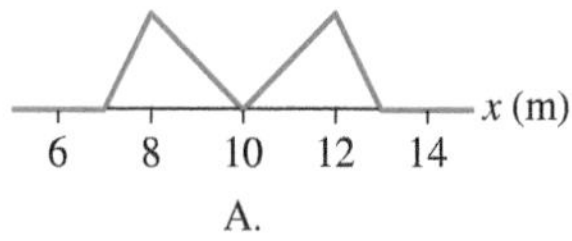

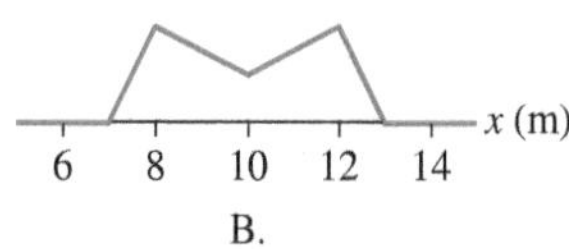

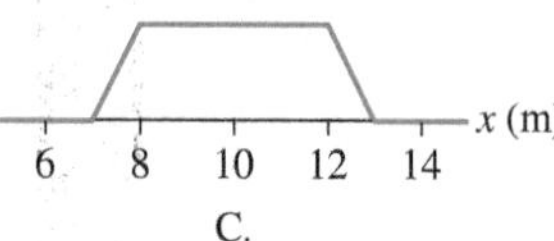

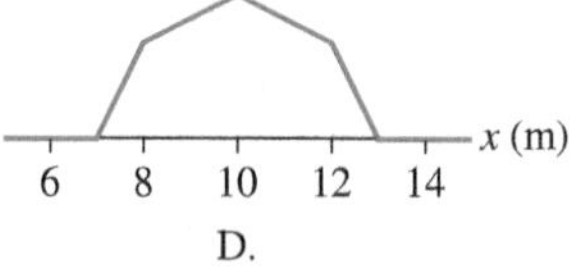

17.2 Standing Waves

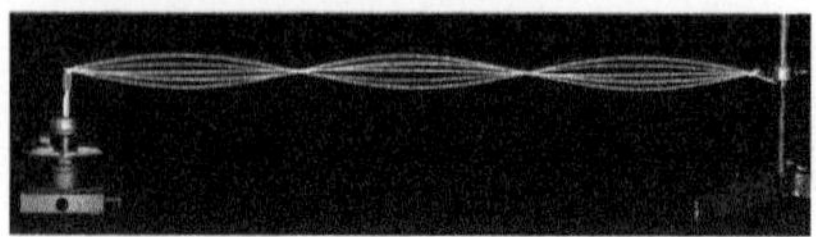
FIGURE 17.3 A vibrating string is an example of a standing wave.

If you pluck a guitar string or a stretched rubber band, you create a wave like the one shown in FIGURE 17.3. A wave like this, in which the particles of the medium oscillate up and down but the wave itself does *not* travel, is called a **standing wave.** Notice that the string is clamped at both ends, unable to move, so any disturbance on the string is "trapped" between two boundaries. This may appear to be a new kind of wave, but actually it is the superposition of two traveling waves moving in opposite directions.

Standing Waves Are Created by Superposition

To help you understand this, FIGURE 17.4a shows two traveling waves with the same frequency and amplitude moving in opposite directions on a string. FIGURE 17.4b then shows nine snapshot graphs at intervals of $\frac{1}{8}T$. This is a complex figure, but you can see that the red and green waves in Figure 17.4b are traveling waves by following the dots attached to crests of the waves.

As these two waves pass through each other, the principle of superposition tells us that the net displacement of the string at any point on the axis—what we'll actually observe—is found by adding the red displacement and the green displacement at that point. This has to be done point by point along the string. The blue wave in Figure 17.4b is the result.

Notice that the red and green waves exactly overlap at times $t = \frac{1}{8}T$ and $t = \frac{5}{8}T$. Crests align with crests, and troughs with troughs, and at every point on the axis $D_{green} = D_{red}$. At these instants, $D_{blue} = D_{red} + D_{green} = 2D_{red}$ *at every point* on the axis. The result, as you can see, is that the blue wave has twice the amplitude of the red or green wave.

At time $t = \frac{3}{8}T$ and again at $t = \frac{7}{8}T$, the crests of the red wave align with the troughs of the green wave. *At every point* $D_{green} = -D_{red}$ and thus $D_{blue} = D_{red} + D_{green} = 0$. At these instants, the blue wave has no displacement at any point.

FIGURE 17.4c shows only the blue wave, the superposition of the red and green waves. This is the string disturbance we would observe. The blue dot shows that the blue wave is moving neither right nor left. Instead, the crests and troughs "stand in place" as the string oscillates—hence, the name *standing wave.* Note that a standing wave is formed only if the oppositely directed traveling waves have the same frequency.

FIGURE 17.4 The superposition of two sinusoidal waves traveling in opposite directions.

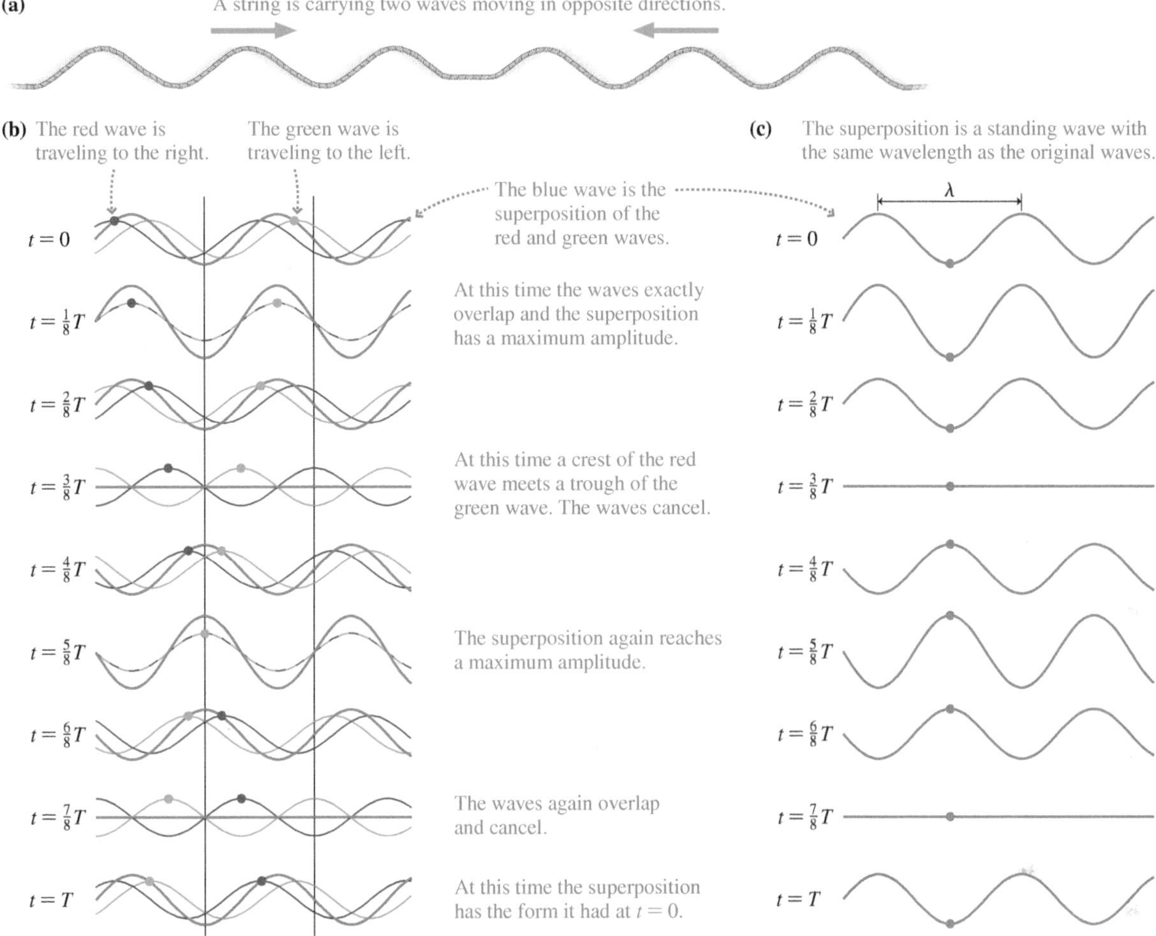
(a)
A string is carrying two waves moving in opposite directions.
(b) The red wave is traveling to the right.
The green wave is traveling to the left.
The blue wave is the superposition of the red and green waves.
(c) The superposition is a standing wave with the same wavelength as the original waves.
λ
t = 0
t = 1/8 T
At this time the waves exactly overlap and the superposition has a maximum amplitude.
t = 2/8 T
t = 3/8 T
At this time a crest of the red wave meets a trough of the green wave. The waves cancel.
t = 4/8 T
t = 5/8 T
The superposition again reaches a maximum amplitude.
t = 6/8 T
t = 7/8 T
The waves again overlap and cancel.
t = T
At this time the superposition has the form it had at t = 0.
Antinode
Node

Nodes and Antinodes

Finally, FIGURE 17.5a has collapsed the nine graphs of Figure 17.4c into a single graphical representation of a standing wave. Compare this to the Figure 17.3 photograph of a vibrating string. A striking feature of a standing-wave pattern is the existence of **nodes,** points with zero amplitude that *never move*! **The nodes are spaced $\lambda/2$ apart.** Halfway between the nodes are the points where the particles in the medium oscillate with maximum displacement. These points of maximum amplitude are called **antinodes,** and you can see that they are also spaced $\lambda/2$ apart.

It seems surprising and counterintuitive that some particles in the medium have no motion at all. To understand this, look closely at the two traveling waves in Figure 17.4a. You will see that the nodes occur at points where at *every instant* of time the displacements of the two traveling waves have equal magnitudes but *opposite signs.* That is, nodes are points of destructive interference where the net displacement is always zero. In contrast, antinodes are points of constructive interference where two displacements of the same sign always add to give a net displacement larger than that of the individual waves.

In Chapter 15 we saw that the energy of SHM is proportional to the square of the amplitude. Analogously, the *intensity* of a wave is proportional to the square of the amplitude at that point: $I \propto A^2$. You can see in FIGURE 17.5b that maximum intensity occurs at the antinodes and that the intensity is zero at the nodes. If this is a sound wave, the loudness is maximum at the antinodes and zero at the nodes. A standing light wave is bright at the antinodes, dark at the nodes. The key idea is

FIGURE 17.5 The intensity of a standing wave is maximum at the antinodes, zero at the nodes.

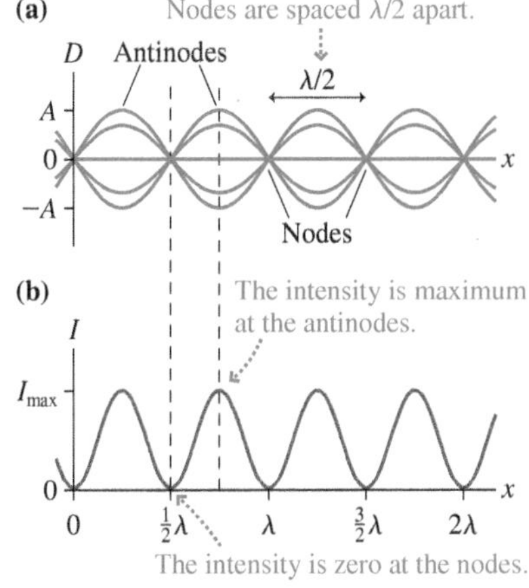
(a)
Nodes are spaced λ/2 apart.
D
Antinodes
λ/2
A
0
−A
x
Nodes
(b)
The intensity is maximum at the antinodes.
I
I_{max}
0
x
0
½λ
λ
3/2λ
2λ
The intensity is zero at the nodes.

Standing waves on a bridge This photograph shows the Tacoma Narrows suspension bridge on the day in 1940 when it experienced a wind-induced standing-wave oscillation that led to its collapse. The red line shows the original line of the deck of the bridge. You can clearly see the large amplitude of the oscillation and the node at the center of the span.

that **the intensity is maximum at points of constructive interference and zero at points of destructive interference.**

The Mathematics of Standing Waves

A sinusoidal wave traveling to the right along the x-axis with angular frequency $\omega = 2\pi f$, wave number $k = 2\pi/\lambda$, and amplitude a is

$$D_R = a\sin(kx - \omega t) \qquad (17.2)$$

An equivalent wave traveling to the left is

$$D_L = a\sin(kx + \omega t) \qquad (17.3)$$

We previously used the symbol A for the wave amplitude, but here we will use a lowercase a to represent the amplitude of each individual wave and reserve A for the amplitude of the net wave. For now, we'll assume that the phase constants are zero.

According to the principle of superposition, the net displacement of the medium at position x when both waves are present is the sum of D_R and D_L:

$$D(x, t) = D_R + D_L = a\sin(kx - \omega t) + a\sin(kx + \omega t) \qquad (17.4)$$

We can simplify Equation 17.4 by using the trigonometric identity

$$\sin(\alpha \pm \beta) = \sin\alpha\cos\beta \pm \cos\alpha\sin\beta$$

Doing so gives

$$\begin{aligned} D(x, t) &= a(\sin kx\cos\omega t - \cos kx\sin\omega t) + a(\sin kx\cos\omega t + \cos kx\sin\omega t) \\ &= (2a\sin kx)\cos\omega t \end{aligned} \qquad (17.5)$$

It is useful to write Equation 17.5 as

$$D(x, t) = A(x)\cos\omega t \qquad (17.6)$$

where $A(x)$ is defined as

$$A(x) = 2a\sin kx \qquad (17.7)$$

$A(x)$ is the amplitude of the standing-wave oscillation at position x on the axis. The amplitude reaches a maximum value $A_{max} = 2a$ at points where $\sin kx = 1$.

The displacement $D(x, t)$ given by Equation 17.6 is neither a function of $x - vt$ nor a function of $x + vt$; hence it is *not* a traveling wave. Instead, the $\cos\omega t$ term in Equation 17.6 describes a medium in which a particle at position x oscillates in simple harmonic motion with amplitude $A(x)$ and frequency $f = \omega/2\pi$.

FIGURE 17.6 graphs Equation 17.6 at several different instants of time. Notice that the graphs are identical to those of Figure 17.5a, showing us that Equation 17.6 is the mathematical description of a standing wave.

FIGURE 17.6 The net displacement resulting from two counter-propagating sinusoidal waves.

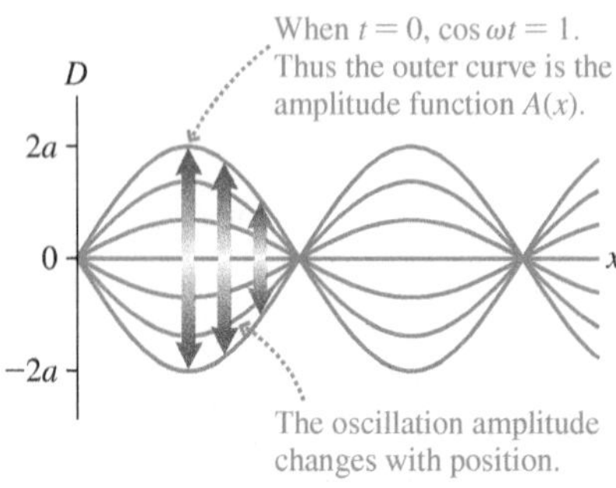
When $t = 0$, $\cos\omega t = 1$. Thus the outer curve is the amplitude function $A(x)$.
D
2a
0
x
−2a
The oscillation amplitude changes with position.

The nodes of the standing wave are the points at which the amplitude is zero. They are located at positions x for which

$$A(x) = 2a\sin kx = 0 \qquad (17.8)$$

The sine function is zero if the angle is an integer multiple of π rad, so Equation 17.8 is satisfied if

$$kx_m = \frac{2\pi x_m}{\lambda} = m\pi \qquad m = 0, 1, 2, 3, \ldots \qquad (17.9)$$

Thus the position x_m of the mth node is

$$x_m = m\frac{\lambda}{2} \qquad m = 0, 1, 2, 3, \ldots \qquad (17.10)$$

You can see that the spacing between two adjacent nodes is $\lambda/2$, in agreement with Figure 17.5b. The nodes are *not* spaced by λ, as you might have expected.

EXAMPLE 17.1 Node spacing

A very long string has a linear density of 5.0 g/m and is stretched with a tension of 8.0 N. 100 Hz waves with amplitudes of 2.0 mm are generated at the ends of the string.

a. What is the node spacing along the resulting standing wave?
b. What is the maximum displacement of the string?

PREPARE Two counter-propagating waves of equal frequency create a standing wave. The standing wave looks like Figure 17.5a.

SOLVE

a. The speed of the waves on the string is

$$v = \sqrt{\frac{T_s}{\mu}} = \sqrt{\frac{8.0\text{ N}}{0.0050\text{ kg/m}}} = 40\text{ m/s}$$

and the wavelength is

$$\lambda = \frac{v}{f} = \frac{40\text{ m/s}}{100\text{ Hz}} = 0.40\text{ m} = 40\text{ cm}$$

Thus the spacing between adjacent nodes is $\lambda/2 = 20$ cm.

b. The maximum displacement is $A_{max} = 2a = 4.0$ mm.

17.3 Standing Waves on a String

Wiggling both ends of a very long string is not a practical way to generate standing waves. Instead, as in the photograph in Figure 17.3, standing waves are usually seen on a string that is fixed at both ends. To understand why this condition causes standing waves, we need to examine what happens when a traveling wave encounters a discontinuity.

FIGURE 17.7a shows a *discontinuity* between a string with a larger linear density and one with a smaller linear density. The tension is the same in both strings, so the wave speed is slower on the left, faster on the right. Whenever a wave encounters a discontinuity, some of the wave's energy is *transmitted* forward and some is *reflected.*

You encountered this idea in Chapter 16 when we found that ultrasound imaging depends on the sound wave being partially reflected and partially transmitted at a boundary. Light waves exhibit an analogous behavior when they encounter a piece of glass. Most of the light wave's energy is transmitted through the glass, which is why glass is transparent, but a small amount of energy is reflected. That is how you see your reflection dimly in a storefront window.

In FIGURE 17.7b, an incident wave encounters a discontinuity at which the wave speed decreases. In this case, the reflected pulse is *inverted.* A positive displacement of the incident wave becomes a negative displacement of the reflected wave. Because $\sin(\phi + \pi) = -\sin\phi$, we say that the reflected wave has a *phase change of π upon reflection.* This aspect of reflection will be important later when we look at the interference of light waves.

The wave in FIGURE 17.7c reflects from a *boundary.* This is like Figure 17.7b in the limit that the string on the right becomes infinitely massive. Thus the reflection in Figure 17.7c looks like that of Figure 17.7b with one exception: Because there is no transmitted wave, *all* the wave's energy is reflected. Hence **the amplitude of a wave reflected from a boundary is unchanged.**

FIGURE 17.7 A wave reflects when it encounters a discontinuity or a boundary.

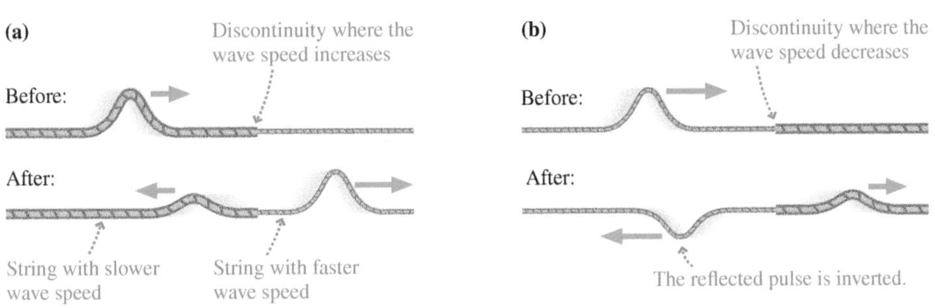

Creating Standing Waves

FIGURE 17.8 Reflections at the two boundaries cause a standing wave on the string.

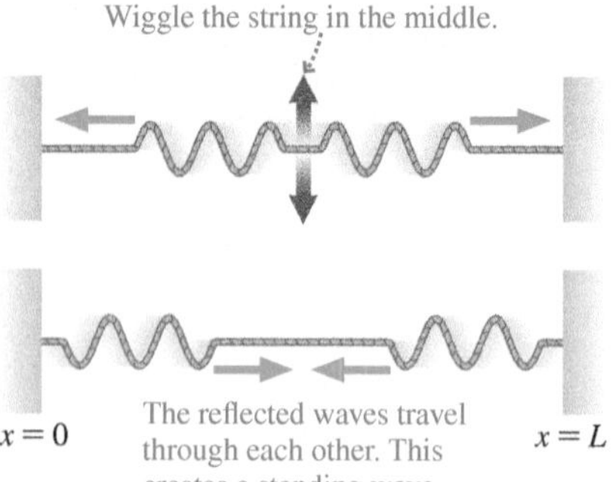
Wiggle the string in the middle.
$x = 0$
The reflected waves travel through each other. This creates a standing wave.
$x = L$

FIGURE 17.8 shows a string of length L tied at $x = 0$ and $x = L$. If you wiggle the string in the middle, sinusoidal waves travel outward in both directions and soon reach the boundaries. Because the speed of a reflected wave does not change, **the wavelength and frequency of a reflected sinusoidal wave are unchanged.** Consequently, reflections at the ends of the string cause two waves of *equal amplitude and wavelength* to travel in opposite directions along the string. As we've just seen, these are the conditions that cause a standing wave!

To connect the mathematical analysis of standing waves in Section 17.2 with the physical reality of a string tied down at the ends, we need to impose *boundary conditions*. A **boundary condition** is a mathematical statement of any constraint that *must* be obeyed at the boundary or edge of a medium. Because the string is tied down at the ends, the displacements at $x = 0$ and $x = L$ must be zero at all times. Thus the standing-wave boundary conditions are $D(x = 0, t) = 0$ and $D(x = L, t) = 0$. Stated another way, we require nodes at both ends of the string.

We found that the displacement of a standing wave is $D(x, t) = (2a \sin kx)\cos \omega t$. This equation already satisfies the boundary condition $D(x = 0, t) = 0$. That is, the origin has already been located at a node. The second boundary condition, at $x = L$, requires $D(x = L, t) = 0$. This condition will be met at all times if

$$2a \sin kL = 0 \qquad \text{(boundary condition at } x = L) \tag{17.11}$$

Equation 17.11 will be true if $\sin kL = 0$, which in turn requires

$$kL = \frac{2\pi L}{\lambda} = m\pi \qquad m = 1, 2, 3, 4, \ldots \tag{17.12}$$

kL must be an integer multiple of π, but $m = 0$ is excluded because neither k nor L is zero.

For a string of fixed length L, the only quantity in Equation 17.12 that can vary is λ. That is, the boundary condition is satisfied only if the wavelength has one of the values

$$\lambda_m = \frac{2L}{m} \qquad m = 1, 2, 3, 4, \ldots \tag{17.13}$$

Wavelengths of standing waves on a string of length L

A standing wave can exist on the string *only* if its wavelength is one of the values given by Equation 17.13. The mth possible wavelength $\lambda_m = 2L/m$ is just the right size so that its mth node is located at the end of the string (at $x = L$).

NOTE ▶ Other wavelengths, which would be perfectly acceptable wavelengths for a traveling wave, cannot exist as a *standing* wave of length L because they cannot meet the boundary conditions requiring a node at each end of the string. ◀

If standing waves are possible only for certain wavelengths, then only a few specific oscillation frequencies are allowed. Because $\lambda f = v$ for a sinusoidal wave, the oscillation frequency corresponding to wavelength λ_m is

$$f_m = \frac{v}{\lambda_m} = \frac{v}{2L/m} = m\frac{v}{2L} \qquad m = 1, 2, 3, 4, \ldots \tag{17.14}$$

Frequencies of standing waves on a string of length L

The lowest allowed frequency

$$f_1 = \frac{v}{2L} \qquad \text{(fundamental frequency)} \tag{17.15}$$

which corresponds to wavelength $\lambda_1 = 2L$, is called the **fundamental frequency** of the string. The allowed frequencies can be written in terms of the fundamental frequency as

$$f_m = mf_1 \qquad m = 1, 2, 3, 4, \ldots \tag{17.16}$$

The allowed standing-wave frequencies are all integer multiples of the fundamental frequency. If, for example, a vibrating string's fundamental frequency is 50 Hz, then the allowed higher-frequency standing waves have frequencies 100 Hz, 150 Hz, 200 Hz, 250 Hz, and so on. The higher-frequency standing waves are called **harmonics,** with the $m = 2$ wave at frequency f_2 called the *second harmonic,* the $m = 3$ wave called the *third harmonic,* and so on.

NOTE ▸ The v in Equation 17.14 is the speed of *traveling* waves on the string, as determined by the string's tension and linear density. It may seem as if a standing wave has no wave speed, but keep in mind that the standing-wave pattern is the superposition of two traveling waves. ◂

FIGURE 17.9 graphs the first four possible standing waves on a string of fixed length L. These possible standing waves are called the **modes** of the string, or sometimes the *normal modes.* Each mode, numbered by the integer m, has a unique wavelength and frequency. Keep in mind that these drawings simply show the *envelope,* or outer edge, of the oscillations. The string is continuously oscillating at all positions between these edges, as we showed in more detail in Figure 17.5a.

There are three things to note about the modes of a string.

1. m is the number of *antinodes* on the standing wave, not the number of nodes. You can tell a string's mode of oscillation by counting the number of antinodes.
2. The *fundamental mode,* with $m = 1$, has $\lambda_1 = 2L$, not $\lambda_1 = L$. Only half of a wavelength is contained between the boundaries, a direct consequence of the fact that the spacing between nodes is $\lambda/2$.
3. The frequencies of the normal modes form a series: $f_1, 2f_1, 3f_1, 4f_1, \ldots$. The fundamental frequency f_1 can be found as the *difference* between the frequencies of any two adjacent modes. That is, $f_1 = \Delta f = f_{m+1} - f_m$.

FIGURE 17.9 The first four modes for standing waves on a string of length L.

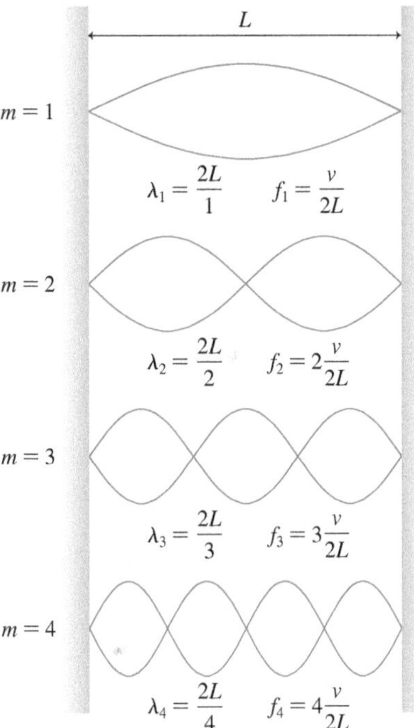

EXAMPLE 17.2 Measuring g

Standing-wave frequencies can be measured very accurately. Consequently, standing waves are often used in experiments to make accurate measurements of other quantities. One such experiment, shown in FIGURE 17.10, uses standing waves to measure the free-fall acceleration g. A heavy mass is suspended from a 1.65-m-long, 5.85 g steel wire; then an oscillating magnetic field (because steel is magnetic) is used to excite the $m = 3$ standing wave on the wire. Measuring the frequency for different masses yields the data given in the table. Analyze these data to determine the local value of g.

FIGURE 17.10 An experiment to measure g.

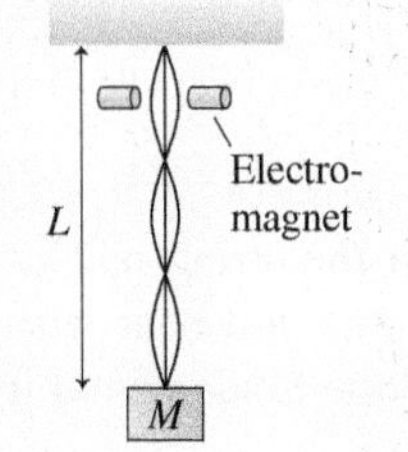

Mass (kg)	f_3 (Hz)
2.00	68
4.00	97
6.00	117
8.00	135
10.00	152

PREPARE The hanging mass creates tension in the wire. This establishes the wave speed along the wire and thus the frequencies of standing waves. We have data for several masses, so our analysis will be based on plotting the data and using a trend line.

Masses of a few kg might stretch the wire a mm or so, but that doesn't change the length L until the third decimal place. The mass of the wire itself is insignificant in comparison to that of the hanging mass. We'll be justified in determining g to three significant figures.

SOLVE The frequency of the third harmonic is

$$f_3 = \frac{3}{2}\frac{v}{L}$$

The wave speed on the wire is

$$v = \sqrt{\frac{T_s}{\mu}} = \sqrt{\frac{Mg}{m/L}} = \sqrt{\frac{MgL}{m}}$$

where Mg is the weight of the hanging mass, and thus the tension in the wire, while m is the mass of the wire. Combining these two equations, we have

$$f_3 = \frac{3}{2}\sqrt{\frac{Mg}{mL}} = \frac{3}{2}\sqrt{\frac{g}{mL}}\sqrt{M}$$

Continued

Squaring both sides gives

$$f_3^2 = \frac{9g}{4mL}M$$

In other words, a graph of the square of the standing-wave frequency versus mass M should be a straight line passing through the origin with slope $9g/4mL$. We can use the experimental slope to determine g.

FIGURE 17.11 is a graph of f_3^2 versus M. The slope of the trend line is 2289 kg^{-1} s^{-2}, from which we find

$$g = \text{slope} \times \frac{4mL}{9}$$

$$= 2289\ \text{kg}^{-1}\,\text{s}^{-2} \times \frac{4(0.00585\ \text{kg})(1.65\ \text{m})}{9} = 9.82\ \text{m/s}^2$$

ASSESS The fact that the graph is linear and passes through the origin confirms our analysis. This is an important reason it's better to have multiple data points rather than using only one mass.

FIGURE 17.11 Graph of the data.

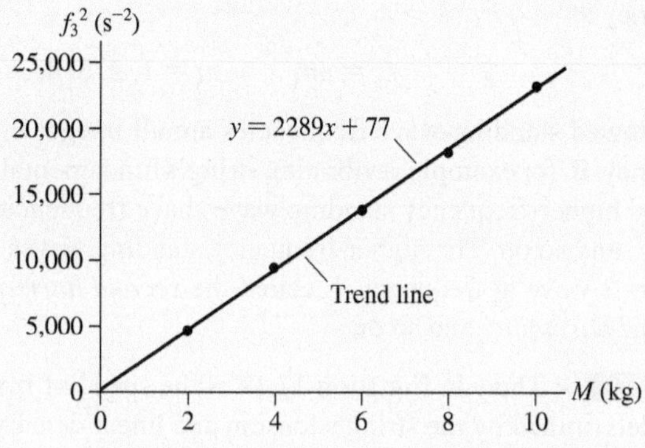

STOP TO THINK 17.2 A standing wave on a string vibrates as shown at the right. Suppose the string tension is quadrupled while the frequency and the length of the string are held constant. Which standing-wave pattern is produced?

Original standing wave

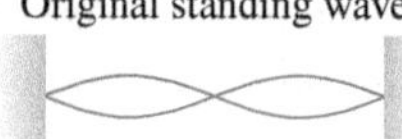

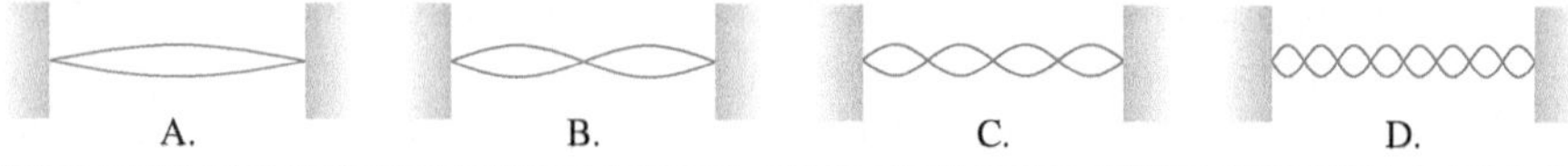

Stringed Instruments

An important application of standing waves is to musical instruments. Instruments such as the guitar, the piano, and the violin have strings fixed at the ends and tightened to create tension. A disturbance generated on the string by plucking, striking, or bowing it creates a standing wave on the string.

The fundamental frequency of a vibrating string is

$$f_1 = \frac{v}{2L} = \frac{1}{2L}\sqrt{\frac{T_s}{\mu}}$$

where T_s is the tension in the string and μ is its linear density. The fundamental frequency is the musical note you hear when the string is sounded. Increasing the tension in the string raises the fundamental frequency, which is how stringed instruments are tuned.

NOTE ▶ v is the wave speed *on the string,* not the speed of sound in air. ◀

For the guitar or the violin, the strings are all the same length and under approximately the same tension. Were that not the case, the neck of the instrument would tend to twist toward the side of higher tension. The strings have different frequencies because they differ in linear density: The lower-pitched strings are thick, while the higher-pitched strings are thin. This difference changes the frequency by changing the wave speed. *Small* adjustments are then made in the tension to bring each string to the exact desired frequency. Once the instrument is tuned, you play it by using your fingertips to alter the effective length of the string. As you shorten the string's length, the frequency and pitch go up.

A piano covers a much wider range of frequencies than a guitar or violin. This range cannot be produced by changing only the linear densities of the strings. The high end would have strings too thin to use without breaking, and the low end would have solid rods rather than flexible wires! So a piano is tuned through a combination of changing the linear density *and* the length of the strings. The bass note strings are not only thicker, they are also longer.

Vibrating strings The strings on a harp vibrate as standing waves. Their frequencies determine the notes that you hear.

EXAMPLE 17.3 Setting the tension in a guitar string

The fifth string on a guitar plays the musical note A, at a frequency of 110 Hz. On a typical guitar, this string is stretched between two fixed points 0.640 m apart, and this length of string has a mass of 2.86 g. What is the tension in the string?

PREPARE Strings sound at their fundamental frequency, so 110 Hz is f_1.

SOLVE The linear density of the string is

$$\mu = \frac{m}{L} = \frac{2.86 \times 10^{-3}\ \text{kg}}{0.640\ \text{m}} = 4.47 \times 10^{-3}\ \text{kg/m}$$

We can rearrange Equation 17.15 for the fundamental frequency to solve for the tension in terms of the other variables:

$$T_s = (2Lf_1)^2\mu = [2(0.640\ \text{m})(110\ \text{Hz})]^2(4.47 \times 10^{-3}\ \text{kg/m})$$
$$= 88.6\ \text{N}$$

ASSESS If you have ever strummed a guitar, you know that the string tension is quite large, so this result seems reasonable. If each of the guitar's six strings has approximately the same tension, the total force on the neck of the guitar is a bit more than 500 N.

17.4 Standing Sound Waves

A standing wave can be created on a string when a disturbance is trapped between two boundaries. A gas inside a long, narrow pipe or tube is also a wave medium confined between two boundaries. In this case, it's possible to create standing sound waves inside the tube. Sound waves are longitudinal waves; these are somewhat trickier than string waves because a graph—showing displacement *parallel* to the tube—is not a picture of the wave.

To illustrate the ideas, FIGURE 17.12 is a series of three graphs and pictures that show the $m = 2$ standing wave inside a column of air closed at both ends. We call this a *closed-closed tube.* The air at the closed ends cannot oscillate because the air molecules are pressed up against the wall, unable to move; hence **a closed end of a**

FIGURE 17.12 The $m = 2$ standing sound wave in a closed-closed tube of air.

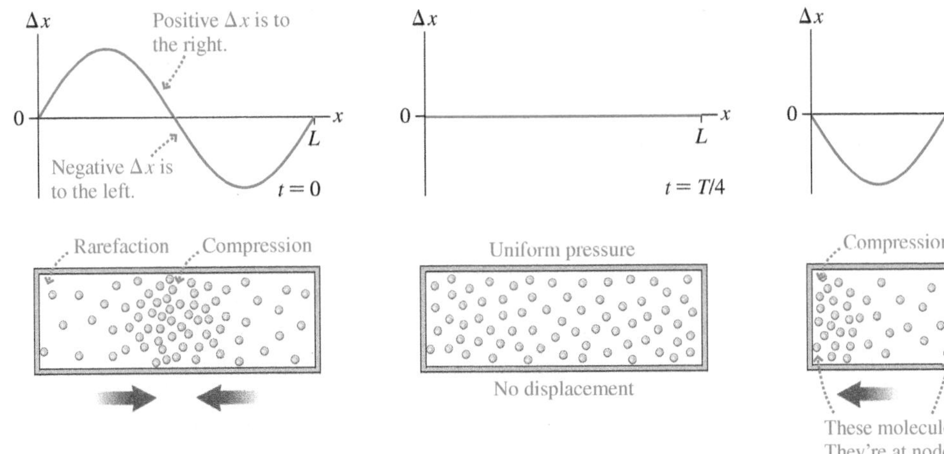

column of air must be a displacement node. Thus the boundary conditions—nodes at the ends—are the same as for a standing wave on a string.

Although the graph looks familiar, it is now a graph of *longitudinal* displacement. At $t = 0$, positive displacements in the left half and negative displacements in the right half cause all the air molecules to converge at the center of the tube. A half cycle later, at $t = T/2$, the molecules have rushed to the ends of the tube. Try to visualize the air molecules sloshing back and forth this way.

FIGURE 17.13 combines these illustrations into a single picture showing where the molecules are oscillating (antinodes) and where they're not (nodes). A graph of the displacement Δx looks just like the $m = 2$ graph of a standing wave on a string. Because the boundary conditions are the same, the possible wavelengths and frequencies of standing waves in a closed-closed tube are the same as for a string of the same length.

FIGURE 17.13 The $m = 2$ longitudinal standing wave can be represented as a displacement wave or as a pressure wave.

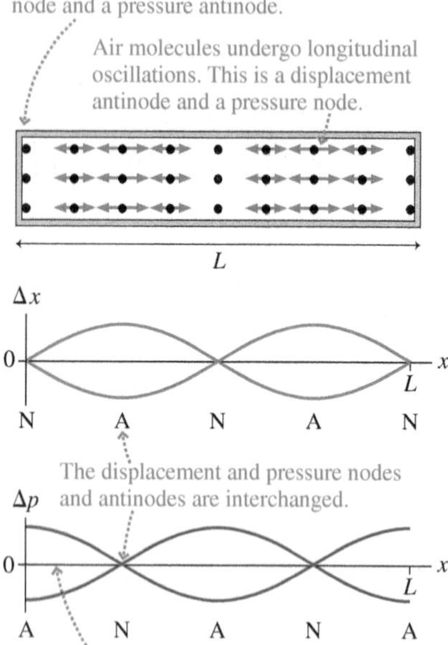

The closed end is a displacement node and a pressure antinode.
Air molecules undergo longitudinal oscillations. This is a displacement antinode and a pressure node.
L
Δx
0
x
L
N A N A N
The displacement and pressure nodes and antinodes are interchanged.
Δp
0
x
L
A N A N A
The pressure is oscillating around atmospheric pressure p_{atmos}.

It is often useful to think of sound as a *pressure wave* rather than a displacement wave. As Figure 17.12 shows, the molecules converge at the center of the tube at $t = 0$, which increases the density—a *compression* in the terminology of Chapter 16—and thus also increases the pressure. At $t = T/2$, molecules moving to the ends of the tube reduce the density and pressure at the center—a *rarefaction*—while increasing them at the ends. So while the center of the tube and the ends are nodes of displacement—air molecules at those points never move—they are *antinodes* of pressure where pressure undergoes its largest oscillations.

In general, **the nodes and antinodes of the pressure standing wave are interchanged with those of the displacement standing wave.** You can see this in the bottom graph in Figure 17.13. This graph shows not p itself but Δp, the variation above or below atmospheric pressure. The ends of the tubes are pressure antinodes as the gas molecules are alternately pushed against the ends and then pulled away. *Pressure nodes* are points where the pressure never varies from atmospheric pressure.

EXAMPLE 17.4 Singing in the shower

A shower stall is 2.45 m (8 ft) tall. For what frequencies less than 500 Hz are there standing sound waves in the shower stall?

PREPARE The shower stall, to a first approximation, is a column of air 2.45 m long. It is closed at the ends by the ceiling and floor, and the displacement standing wave will have nodes at those points. Assume a 20°C speed of sound.

SOLVE The fundamental frequency for a standing sound wave in this air column is

$$f_1 = \frac{v}{2L} = \frac{343 \text{ m/s}}{2(2.45 \text{ m})} = 70 \text{ Hz}$$

The possible standing-wave frequencies are integer multiples of the fundamental frequency. These are 70 Hz, 140 Hz, 210 Hz, 280 Hz, 350 Hz, 420 Hz, and 490 Hz.

ASSESS The many possible standing waves in a shower cause the sound to *resonate,* which helps explain why some people like to sing in the shower. Our approximation of the shower stall as a one-dimensional tube is actually a bit too simplistic. A three-dimensional analysis would find additional modes, making the "sound spectrum" even richer.

Tubes with Openings

Air columns closed at both ends are of limited interest unless, as in Example 17.4, you are inside the column. Columns of air that *emit* sound are open at one or both ends. Many musical instruments fit this description. For example, a flute is a tube of air open at both ends. The flutist blows across one end to create a standing wave inside the tube, and a note of that frequency is emitted from both ends of the flute. (The blown end of a flute is open on the side, rather than across the tube. That is necessary for practical reasons of how flutes are played, but from a physics perspective this is the "end" of the tube because it opens the tube to the atmosphere.) A trumpet, however, is open at the bell end but is *closed* by the player's lips at the other end.

You saw earlier that a wave is partially transmitted and partially reflected at a discontinuity. When a sound wave traveling through a tube of air reaches an open end, some of the wave's energy is transmitted out of the tube to become the sound that you hear and some portion of the wave is reflected back into the tube. These reflections, analogous to the reflection of a string wave from a boundary, allow standing sound waves to exist in a tube of air that is open at one or both ends.

Not surprisingly, the *boundary condition* at the open end of a column of air is not the same as the boundary condition at a closed end. The air pressure at the open end of a tube is constrained to match the atmospheric pressure of the surrounding air. Consequently, the open end of a tube must be a pressure node, a point where the pressure doesn't oscillate. Because pressure nodes and antinodes are interchanged with those of the displacement wave, **an open end of an air column is required to be a displacement antinode.** (A careful analysis shows that the antinode is actually just outside the open end, but for our purposes we'll assume the antinode is exactly at the open end.)

FIGURE 17.14 shows displacement and pressure graphs of the first three standing-wave modes of a tube closed at both ends (a *closed-closed tube*), a tube open at both ends (an *open-open tube*), and a tube open at one end but closed at the other (an *open-closed tube*), all with the same length L. Notice the pressure and displacement boundary conditions. The standing wave in the open-open tube looks like the wave in the closed-closed tube except that the positions of the nodes and antinodes are interchanged. In both cases there are m half-wavelength segments between the ends; thus the wavelengths and frequencies of an open-open tube and a closed-closed tube are the same as those of a string tied at both ends:

$$\begin{aligned} \lambda_m &= \frac{2L}{m} \\ f_m &= m\frac{v}{2L} = mf_1 \end{aligned} \qquad m = 1, 2, 3, 4, \ldots \qquad (17.17)$$

Wavelengths and frequencies of standing waves in open-open and closed-closed tubes

Fiery nodes In this apparatus, a loudspeaker at one end of the metal tube emits a sinusoidal wave. The wave reflects from the other end, which is closed, to make a counter-propagating wave and set up a standing sound wave in the tube. The tube is filled with propane gas that exits through small holes on top. The burning propane allows us to easily discern the nodes and the antinodes of the standing sound wave.

FIGURE 17.14 The first three standing sound wave modes in columns of air with different boundary conditions.

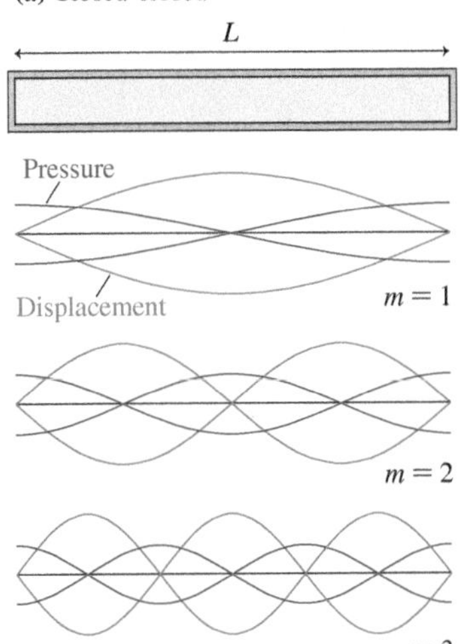

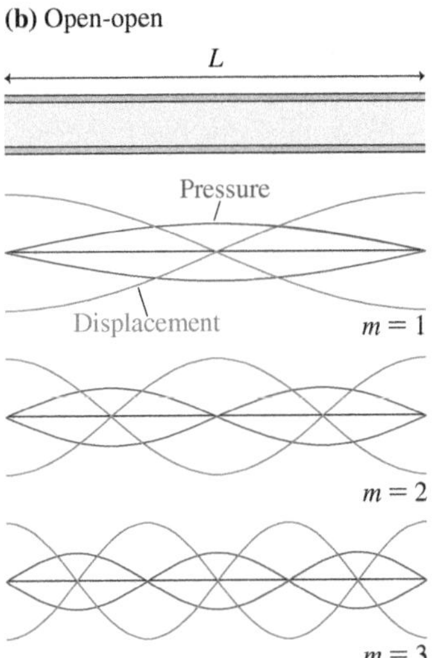

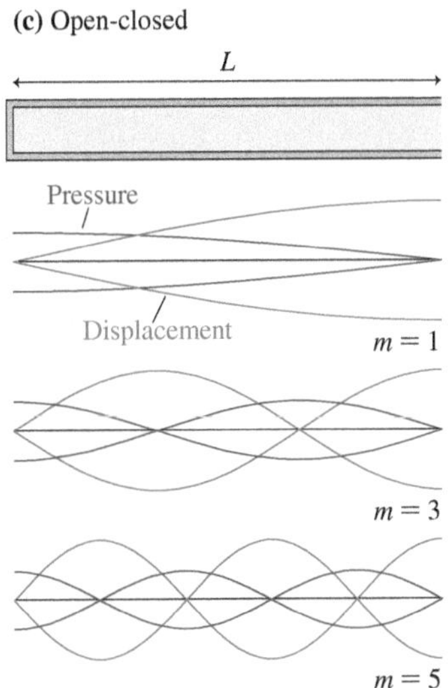

The open-closed tube is different. The fundamental mode has only one-quarter of a wavelength in a tube of length L; hence the $m = 1$ wavelength is $\lambda_1 = 4L$. This is twice the λ_1 wavelength of an open-open or a closed-closed tube. Consequently, **the fundamental frequency of an open-closed tube is half that of an open-open or a closed-closed tube of the same length.** In general:

$$\lambda_m = \frac{4L}{m} \qquad f_m = m\frac{v}{4L} = mf_1 \qquad m = 1, 3, 5, 7, \ldots \tag{17.18}$$

Wavelengths and frequencies of standing waves in open-closed tubes

Notice that m in Equation 17.18 takes on only *odd* values.

EXAMPLE 17.5 Resonances of the ear canal

The eardrum, which transmits sound vibrations to the sensory organs of the inner ear, lies at the end of the ear canal. As FIGURE 17.15 shows, the ear canal in adults is about 2.5 cm in length. What frequency standing waves can occur in the ear canal that are within the range of human hearing? The speed of sound in the warm air of the ear canal is 350 m/s.

PREPARE The ear canal is open to the air at one end, closed by the eardrum at the other. We can model it as an open-closed tube. The standing waves will be those of Figure 17.14c.

FIGURE 17.15 The anatomy of the ear.

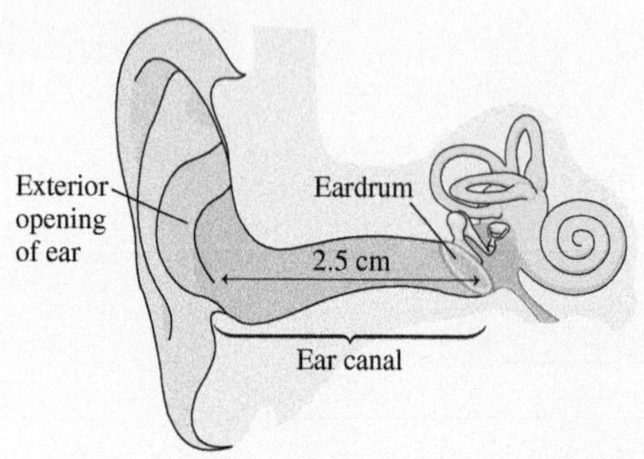

SOLVE The lowest standing-wave frequency is the fundamental frequency for a 2.5-cm-long open-closed tube:

$$f_1 = \frac{v}{4L} = \frac{350 \text{ m/s}}{4(0.025 \text{ m})} = 3500 \text{ Hz}$$

Standing waves also occur at the harmonics, but an open-closed tube has only odd harmonics. These are

$$f_3 = 3f_1 = 10{,}500 \text{ Hz}$$
$$f_5 = 5f_1 = 17{,}500 \text{ Hz}$$

Higher harmonics are beyond the range of human hearing.

ASSESS The ear canal is short, so we expected the standing-wave frequencies to be relatively high. The air in your ear canal responds readily to sounds at these frequencies—what we call a *resonance* of the ear canal—and transmits these sounds to the eardrum. Consequently, as we noted in Section 16.6, your ear actually is slightly more sensitive to sounds with frequencies around 3500 Hz and 10,500 Hz than to sounds at nearby frequencies.

STOP TO THINK 17.3 An open-open tube of air supports standing waves at frequencies of 300 Hz and 400 Hz and at no frequencies between these two. The second harmonic of this tube has frequency

A. 100 Hz B. 200 Hz C. 400 Hz D. 600 Hz E. 800 Hz

Wind Instruments

With a wind instrument, such as a flute or a trumpet, blowing into the mouthpiece creates a standing sound wave inside a tube of air. The player changes the notes by using her fingers to cover holes or open valves, changing the length of the tube and thus its frequency. The fact that the holes are on the side makes very little difference; the first open hole becomes an antinode because the air is free to oscillate in and out of the opening.

A wind instrument's frequency depends on the speed of sound *inside* the instrument. But the speed of sound depends on the temperature of the air. When a wind player first blows into the instrument, the air inside starts to rise in temperature. This increases the sound speed, which in turn raises the instrument's frequency for each note until the air temperature reaches a steady state. Consequently, wind players must "warm up" before tuning their instrument.

Many wind instruments have a "buzzer" at one end of the tube, such as a vibrating reed on a saxophone or vibrating lips on a trombone. Buzzers generate a continuous range of frequencies rather than single notes, which is why they sound like a "squawk" if you play on just the mouthpiece without the rest of the instrument. When a buzzer is connected to the body of the instrument, most of those frequencies cause no response of the air molecules. But the frequency from the buzzer that matches the fundamental frequency of the instrument causes the buildup of a large-amplitude response at just that frequency—a standing-wave resonance. This is the energy input that generates and sustains the musical note.

EXAMPLE 17.6 **Flutes and clarinets**

A clarinet is 66.0 cm long. A flute is nearly the same length, with 63.6 cm between the hole the player blows across and the end of the flute. What are the frequencies of the lowest note and the next higher harmonic on a flute and on a clarinet? The speed of sound in warm air is 350 m/s.

PREPARE The flute is an open-open tube, open at the end as well as at the hole the player blows across. A clarinet is an open-closed tube because the player's lips and the reed seal the tube at the upper end.

SOLVE The lowest frequency is the fundamental frequency. For the flute, an open-open tube, this is

$$f_1 = \frac{v}{2L} = \frac{350 \text{ m/s}}{2(0.636 \text{ m})} = 275 \text{ Hz}$$

The clarinet, an open-closed tube, has

$$f_1 = \frac{v}{4L} = \frac{350 \text{ m/s}}{4(0.660 \text{ m})} = 133 \text{ Hz}$$

The next higher harmonic on the flute's open-open tube is $m = 2$ with frequency $f_2 = 2f_1 = 550$ Hz. An open-closed tube has only odd harmonics, so the next higher harmonic of the clarinet is $f_3 = 3f_1 = 399$ Hz.

ASSESS The clarinet plays a much lower note than the flute—musically, about an octave lower—because it is an open-closed tube. It's worth noting that neither of our fundamental frequencies is exactly correct because our open-open and open-closed tube models are a bit too simplified to adequately describe a real instrument. However, both calculated frequencies are close because our models do capture the essence of the physics.

A vibrating string plays the musical note corresponding to the fundamental frequency f_1, so stringed instruments must use several strings to obtain a reasonable range of notes. In contrast, wind instruments can sound at the second or third harmonic of the tube of air (f_2 or f_3). These higher frequencies are sounded by *overblowing* (flutes, brass instruments) or with keys that open small holes to encourage the formation of an antinode at that point (clarinets, saxophones). The controlled use of these higher harmonics gives wind instruments a wide range of notes.

17.5 The Physics of Speech

When you hear a note played on a guitar, it sounds very different from the same note played on a trumpet or a saxophone. You can clearly distinguish a singer's "oo" vowel sound from her "ee" vowel sound even when they are sung at the same pitch. Clearly, there is more to your brain's perception of sound than pitch alone.

The Frequency Spectrum

To this point, we have pictured sound waves as sinusoidal waves, with a well-defined frequency. In fact, most of the sounds that you hear are not pure sinusoidal waves. Most sounds are a mix, or superposition, of different frequencies. For example, we have seen how certain standing-wave modes are possible on a stretched string. When you pluck a string on a guitar, you generally don't excite just one standing-wave mode—you simultaneously excite many different modes.

If you play the note "middle C" on a guitar, the fundamental frequency is 262 Hz. There will be a standing wave at this frequency, but there will also be standing waves at the frequencies 524 Hz, 786 Hz, 1048 Hz, . . ., all the higher harmonics predicted by Equation 17.16.

FIGURE 17.16 The frequency spectrum and a graph of the sound wave of a guitar playing a note with fundamental frequency 262 Hz.

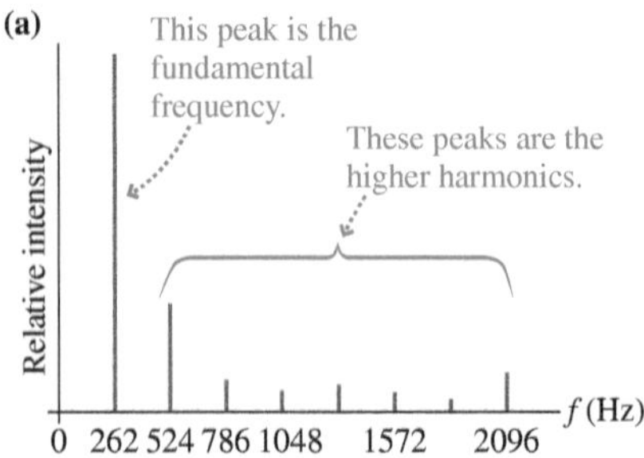
(a)
This peak is the fundamental frequency.
These peaks are the higher harmonics.
Relative intensity
f (Hz)
0 262 524 786 1048 1572 2096

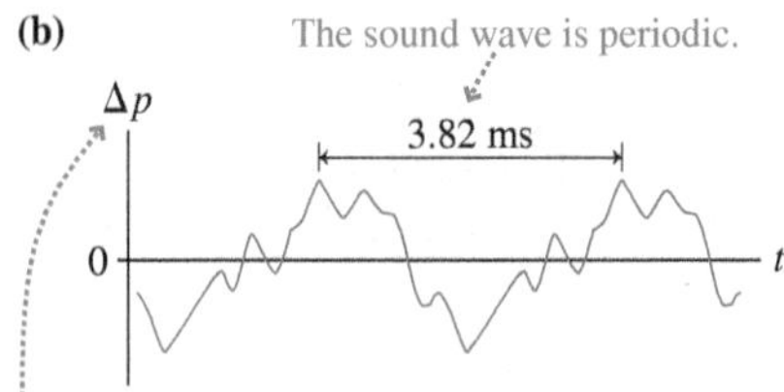
(b)
The sound wave is periodic.
Δp
3.82 ms
0
t
The vertical axis is the change in pressure from atmospheric pressure due to the sound wave.

FIGURE 17.16a is a bar chart showing all the frequencies present in the sound of the vibrating guitar string. The height of each bar shows the relative intensity of that harmonic. A bar chart showing the relative intensities of the different frequencies is called the **frequency spectrum** of the sound.

When your brain interprets the mix of frequencies from the guitar in Figure 17.16a, it identifies the fundamental frequency as the *pitch.* 262 Hz corresponds to middle C, so you will identify the pitch as middle C, even though the sound consists of many different frequencies. Your brain uses the higher harmonics to determine the **tone quality,** which is also called the *timbre.* The tone quality—and therefore the higher harmonics—is what makes a middle C played on a guitar sound quite different from a middle C played on a trumpet. The frequency spectrum of a trumpet would show a very different pattern of the relative intensities of the higher harmonics.

The sound wave produced by a guitar playing middle C is shown in FIGURE 17.16b. The sound wave is periodic, with a period of 3.82 ms that corresponds to the 262 Hz fundamental frequency. **The higher harmonics don't change the period of the sound wave; they change only its shape.** The sound wave of a trumpet playing middle C would also have a 3.82 ms period, but its shape would be entirely different.

Vowels and Formants

Try this: Keep your voice at the same pitch and say the "ee" sound, as in "beet," then the "oo" sound, as in "boot," and then the "ah" sound, as in "father." Pay attention to how you reshape your mouth as you move back and forth between the three sounds. The vowel sounds are at the same pitch, but they sound quite different. The difference in sound arises from the difference in the higher harmonics, just as for musical instruments. As you speak, you adjust the properties of your vocal tract to produce different mixes of harmonics that make the "ee," "oo," "ah," and other vowel sounds.

Speech begins with the vibration of your vocal cords, stretched bands of tissue in your throat. The vibration is similar to that of a wave on a stretched string. In ordinary speech, the average fundamental frequency for adult males and females is about 150 Hz and 225 Hz, respectively, but you can change the vibration frequency by changing the tension of your vocal cords. That's how you make your voice higher or lower as you sing.

FIGURE 17.17 The frequency spectrum from the vocal cords.

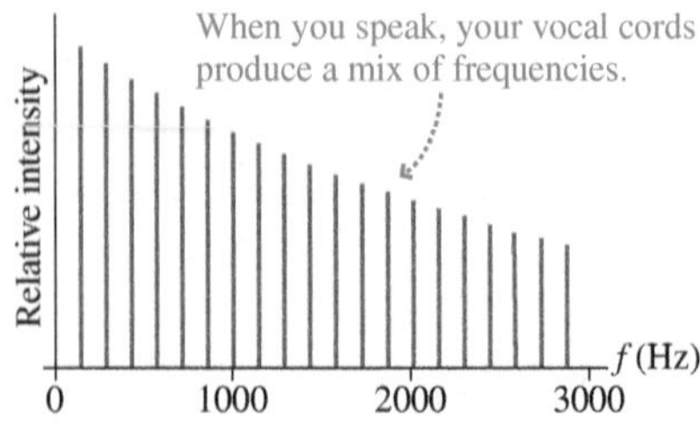
When you speak, your vocal cords produce a mix of frequencies.
Relative intensity
f (Hz)
0
1000
2000
3000

Your vocal cords produce a mix of different frequencies as they vibrate—the fundamental frequency and a rich mixture of higher harmonics. If you put a microphone in your throat and measured the sound waves right at your vocal cords, the frequency spectrum would appear as in FIGURE 17.17. You can see that your vocal cords produce many harmonics.

There is more to the story, though. Before reaching the opening of your mouth, sound from your vocal cords must pass through your *vocal tract,* which includes your throat, mouth, and nose. Each of these is a hollow cavity that, like a tube, allows only certain frequencies—those that can set up standing waves—to pass through. These frequencies are enhanced as they pass through the vocal tract, while frequencies that fail to set up standing waves are suppressed.

For example, FIGURE 17.18a shows the frequency spectrum of the vowel sound "ee." Each bar in this bar chart is one of the frequencies produced by your vocal cords—the frequencies of Figure 17.17—but now the frequencies have been filtered by the vocal tract. You can see a band of enhanced frequencies near 400 Hz and another band of enhanced frequencies near 2200 Hz. These are the frequencies that resonate in the vocal tract as you say "ee." However, the frequencies from 800 Hz to 1800 Hz have been strongly suppressed because they are not allowed standing-wave frequencies of the vocal tract cavities.

Each band of enhanced frequencies is called a **formant.** A formant is not a specific frequency but a *range* of frequencies, usually a few hundred Hz, that are enhanced by the standing-wave resonances of the vocal tract. Starting with the lowest frequencies, they are called the first formant, the second formant, and—if present—the third formant.

As you speak, the motions of your jaw, tongue, and lips change the physical size and shape of the cavities in the vocal tract and thus alter the formants, the vocal-cord frequencies that are enhanced. You switch from the vowel "ee" to the vowel "ah" by opening your mouth wider and lowering your tongue. You can see in FIGURE 17.18b that this physical modification of your vocal tract raises the frequency of the first formant, significantly lowers the frequency of the second formant, and creates a third formant. The "ee" and "ah" vowel sounds start with the same vocal-cord vibrations but filter those vibrations differently to produce very different sounds.

A very simple model of the vocal tract is a tube that is closed at one end by the larynx but open at the mouth. For adult males, the average length of the vocal tract is about 17 cm, and the allowed frequencies of this open-closed tube are approximately 500 Hz, 1500 Hz, 2500 Hz, and so on. This simple model makes a reasonably good prediction of the formants of an "ah" sound, where the formants are centered at about 600 Hz, 1200 Hz, and 2600 Hz. However, the formants for the "ee" sound and for most other vowel sounds differ significantly, which tells us that the vocal tract is too complex to model as a simple open-closed tube.

FIGURE 17.18 The frequency spectrum of spoken vowel sounds after passing through the vocal tract.

(a) Vowel sound "ee"

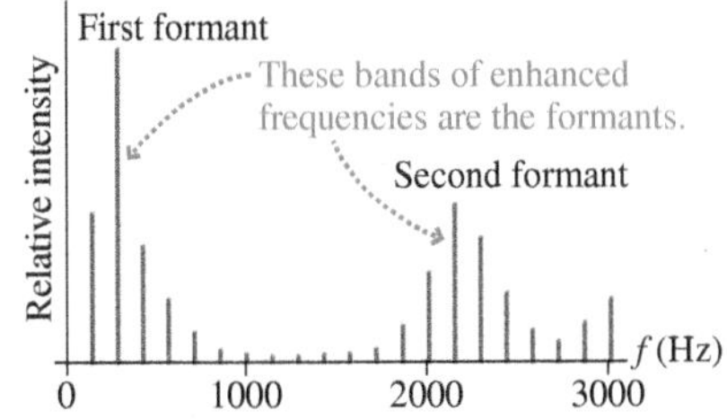

(b) Vowel sound "ah"

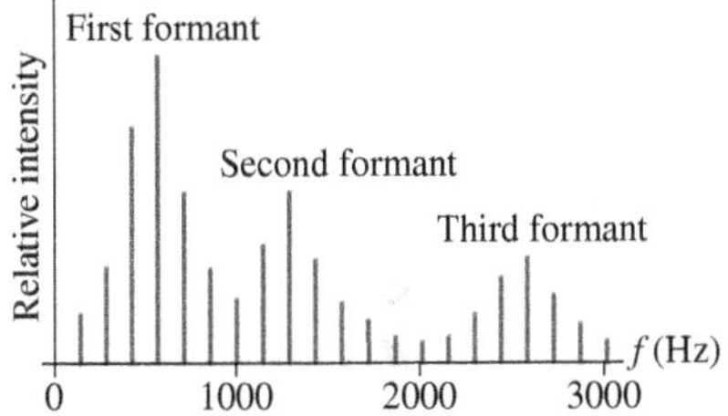

CONCEPTUAL EXAMPLE 17.7 High-frequency hearing loss

As you age, your hearing sensitivity will decrease. For most people, the loss of sensitivity is greater for higher frequencies. The loss of sensitivity at high frequencies may make it difficult to understand what others are saying. Why is this?

REASON It is the high-frequency components of speech that allow us to distinguish different vowel sounds. A decrease in sensitivity to these higher frequencies makes it more difficult to make such distinctions.

ASSESS This result makes sense. In Figure 17.18a, the lowest frequency is less than 200 Hz, but the second formant is over 2000 Hz. There's a big difference between what we hear as the pitch of someone's voice and the frequencies we use to interpret speech.

STOP TO THINK 17.4 If you speak at a certain pitch, then hold your nose and continue speaking at the same pitch, your voice sounds very different. This is because

A. The fundamental frequency of your vocal cords has changed.
B. The frequencies of the harmonics of your vocal cords have changed.
C. The pattern of resonance frequencies of your vocal tract has changed, thus changing the formants.

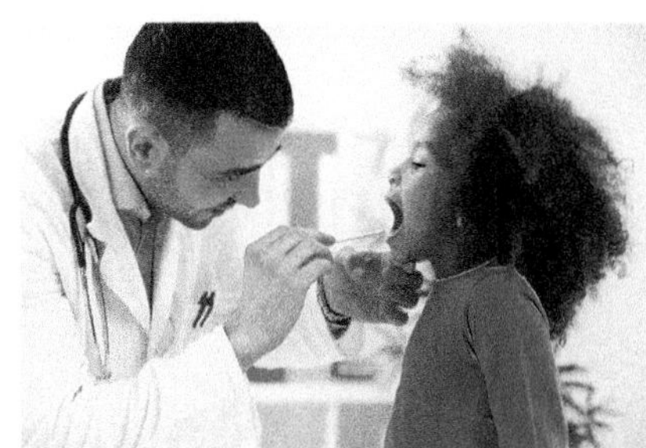

Saying "ah" Why, during a throat exam, does a doctor ask you to say "ah"? This particular vowel sound is formed by opening the mouth and the back of the throat wide—giving a clear view of the tissues of the throat.

17.6 Interference Along a Line

You may have used noise-canceling headphones. You've certainly used optical devices—from cameras and microscopes to eyeglasses—for which the lenses have "light-canceling" antireflection coatings. These technologies (and others) are applications of the phenomenon of **interference** that arises from the superposition of two equal-frequency waves traveling in the *same* direction. Although most of the important examples of interference occur when waves spread out in two or three dimensions, we'll lay the groundwork by first looking at the simpler case of the interference of two waves traveling along the same line.

FIGURE 17.19 Two overlapped sound waves travel along the x-axis.

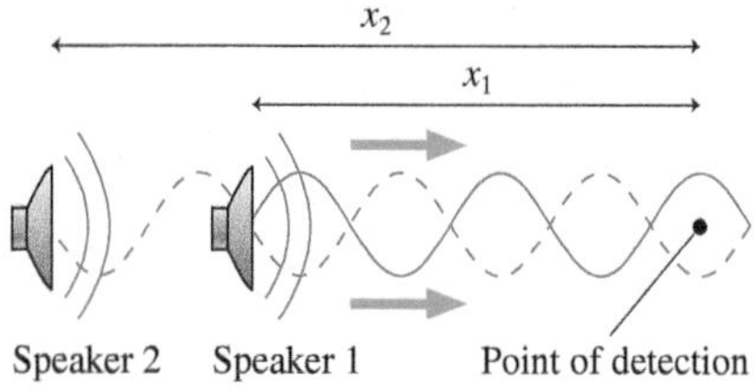

FIGURE 17.20 Interference of two waves traveling along the x-axis.

(a) Maximum constructive interference

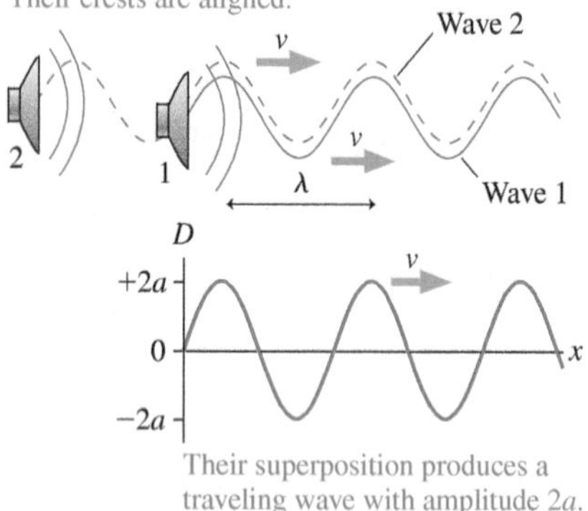

(b) Perfect destructive interference

These two waves are out of phase. The crests of one wave are aligned with the troughs of the other.

Wave 2

2 1 λ Wave 1

D $+2a$ 0 x $-2a$

Their superposition produces a wave with zero amplitude.

Interference can occur for any type of wave (Chapter 18 will explore the interference of light waves), but we'll use sound waves to illustrate the ideas. FIGURE 17.19 shows two loudspeakers emitting sound waves that have the *same frequency and wavelength;* interference is always between waves that have the same wavelength. The wave from loudspeaker 2 joins with the wave from loudspeaker 1 so that two overlapped sound waves are traveling to the right along the x-axis.

Suppose you place a microphone or your ear on the x-axis at a point that is distance x_1 from speaker 1 and x_2 from speaker 2. You would not hear or detect two distinct waves but, instead, their superposition. Hence our goal in this section is to see what happens when two waves traveling in the same direction are superimposed.

Let's examine interference graphically before getting into the mathematics. FIGURE 17.20 shows two important situations. In part a, the crests of the two individual sound waves are aligned as they travel along the x-axis. Waves that are aligned crest to crest and trough to trough are said to be **in phase.** They march along in step. In part b, the crests of one wave align with the troughs of the other; the waves are out of step. Two waves aligned crest to trough are said to be *180° out of phase* or, more generally, just **out of phase.**

NOTE ▶ Textbook illustrations like Figure 17.20 can be misleading because they are frozen in time. You need to imagine how this situation unfolds as the waves travel to the right. These are *not* standing waves with nodes and antinodes. ◀

The two in-phase waves of Figure 17.20a have the same displacement at every point along the x-axis: $D_1(x) = D_2(x)$. The principle of superposition tells us to add the displacements *at each point,* so these waves combine to give a wave whose net displacement at each point is twice that of each individual wave. The superposition of two waves to create a traveling wave with a *larger* amplitude, and thus a greater intensity, is called **constructive interference.** When the waves are exactly in phase, so that the amplitude of the superposition is $A = 2a$, we have *maximum constructive interference.* The individual waves reinforce each other to create a larger wave.

In contrast, the out-of-phase waves of Figure 17.20b have $D_1(x) = -D_2(x)$; that is, at every point on the axis, the displacement of one wave is always opposite the displacement of the other. The superposition of the two waves, found by adding their displacements, is $D_{net} = D_1 + D_2 = 0$. In other words, the waves *cancel* each other to give no wave at all! The superposition of two waves to create a traveling wave with a *smaller* amplitude, and thus a weaker intensity, is called **destructive interference.** When the waves exactly cancel, so that the amplitude of the superposition is zero, we have *perfect destructive interference.*

NOTE ▶ Perfect destructive interference occurs only if the two waves have exactly equal amplitudes and are exactly 180° out of phase. A 180° phase difference always produces *maximum destructive interference,* but the cancellation isn't perfect if the two waves differ in amplitude. Similarly, two equal-amplitude waves interfere destructively but without perfect cancellation if their phase difference is close to but not exactly 180°. ◀

Interference with Identical Sources

What determines whether interference is constructive, destructive, or somewhere in between? We'll begin our exploration of interference by considering the wave sources to be two loudspeakers that oscillate back and forth in phase with each other. We call these **identical sources.**

FIGURE 17.21a shows identical sources (i.e., the speakers are doing the same thing at the same time) that are separated by exactly one wavelength: $\Delta x = \lambda$. $\Delta x = x_2 - x_1$ is the extra distance traveled by a wave from speaker 2; it's called the **path-length difference.** Because waves travel one wavelength in one period, a wave emitted by speaker 1 is in phase with the wave that left speaker 2 exactly one period earlier. As a result, the two waves are exactly in phase and interfere constructively to produce a wave of amplitude $A = 2a$. The situation is the same if the speakers are separated by exactly two wavelengths, exactly three wavelengths, and so on. In other words, **two identical sources produce maximum constructive interference when the path-length difference is an integer number of wavelengths.**

FIGURE 17.21b differs by showing identical sources that are separated by exactly one-half wavelength: $\Delta x = \frac{1}{2}\lambda$. Now the two waves are 180° out of phase and, when added, cancel to produce perfect destructive interference. As you can imagine, the result is the same if the speakers are separated by one and a half wavelengths, two and a half wavelengths, and so on. Thus **two identical sources produce maximum destructive interference when the path-length difference is a half-integer number of wavelengths.**

We can summarize the conditions for maximum constructive or destructive interference as

$$\begin{aligned} &\text{maximum constructive:} \quad \Delta x = m\lambda \\ &\text{maximum destructive:} \quad \Delta x = \left(m + \tfrac{1}{2}\right)\lambda \end{aligned} \qquad m = 0, 1, 2, 3, \ldots \quad (17.19)$$

Interference conditions for identical sources

FIGURE 17.21 Constructive and destructive interference for two identical sources.

(a) Separation is an integer number of wavelengths.

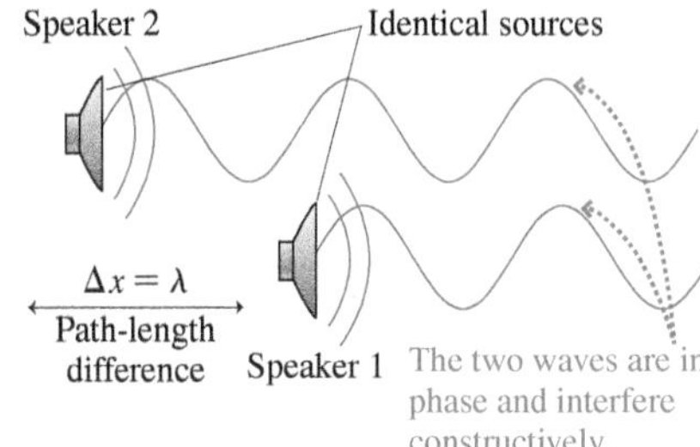

(b) Separation is a half-integer number of wavelengths.

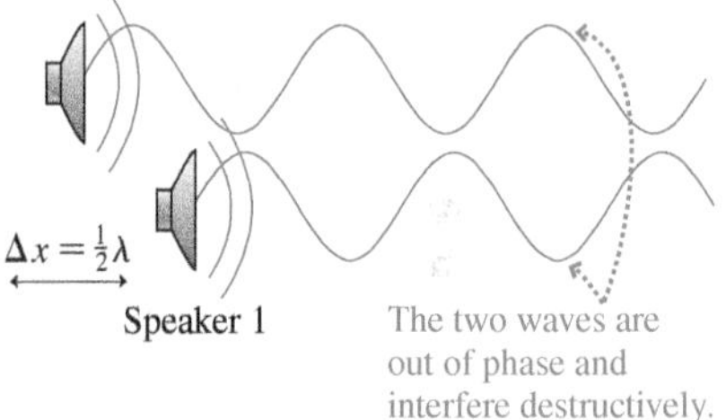

EXAMPLE 17.8 Interference with a changing frequency

Mei stands directly in front of two loudspeakers that are in line with each other. The closer speaker is 5.00 m away; the farther 6.00 m away. The speakers are driven by the same amplifier, and Mei hears a maximum sound intensity when the frequency is 688 Hz. Then the frequency of the source is slowly increased until, at some point, the sound intensity drops to zero. What is the frequency when this cancellation occurs? The speed of sound in the room is 344 m/s.

PREPARE The two loudspeakers are driven by the same amplifier, so they are identical sources oscillating back and forth in phase with each other. Hearing a maximum intensity at 688 Hz implies that this is a situation of constructive interference between the two sound waves. Because the interference is maximum constructive interference, the separation between the speakers (i.e., the path-length difference) satisfies $\Delta x = m\lambda_0$, where m is an integer. Increasing the frequency shortens the wavelength, and the interference will become destructive when $\Delta x = (m + \frac{1}{2})\lambda_1$.

SOLVE The initial wavelength of the sound wave is

$$\lambda_0 = \frac{v_{sound}}{f_0} = \frac{344 \text{ m/s}}{688 \text{ Hz}} = 0.500 \text{ m}$$

The path-length difference is $\Delta x = 1.00$ m, thus $m = \Delta x/\lambda = 2$. That is, speaker 2 is exactly two wavelengths behind speaker 1. Destructive interference at a higher frequency will be observed when the wavelength decreases to the point at which $\Delta x = 2.5\lambda_1$. This occurs at wavelength

$$\lambda_1 = \frac{1.00 \text{ m}}{2.5} = 0.400 \text{ m}$$

which corresponds to a frequency of

$$f_1 = \frac{v_{sound}}{\lambda_1} = \frac{344 \text{ m/s}}{0.400 \text{ m}} = 860 \text{ Hz}$$

ASSESS 860 Hz is an increase of 172 Hz, one-fourth of the original frequency. That makes sense. Originally, two cycles of the wave "fit" into the 1 m path-length difference; now 2.5 cycles "fit," an increase of one-fourth of the original.

The Phase Difference

The interference conditions of Equations 17.19 apply only when the two sources are identical. What if they're not? The path-length difference is still important, but we also have to account for the inherent phase difference between the sources.

We'll assume that the two waves have the same wavelength λ (and wave number $k = 2\pi/\lambda$), the same frequency f (and angular frequency $\omega = 2\pi f$), and the same amplitude a, and that they travel to the right along the x-axis. In that case, we can write the displacements of the two waves as

$$\begin{aligned} D_1(x_1, t) &= a\sin(kx_1 - \omega t + \phi_{10}) = a\sin\phi_1 \\ D_2(x_2, t) &= a\sin(kx_2 - \omega t + \phi_{20}) = a\sin\phi_2 \end{aligned} \tag{17.20}$$

where ϕ_1 and ϕ_2 are the *phases* of the waves, a concept we introduced in Chapter 16. Distances x_1 and x_2 are measured from the loudspeakers, as shown in Figure 17.19.

You learned in Chapter 16 that the phase constant ϕ_0 is determined by what the *sources* are doing at time $t = 0$. As examples, **FIGURE 17.22** shows snapshot graphs at $t = 0$ of waves emitted by three sources that have phase constants 0 rad, $\pi/2$ rad, and π rad. That is, these are graphs of Equations 17.20 at $t = 0$. Figure 17.22a shows zero displacement at $x = 0$, the location of the speaker, so initially the speaker cone is centered. But, as it oscillates, is it moving forward or backward? You can see that there's a displacement peak at $x = \lambda/4$. The wave is traveling to the right, so this peak must have left the speaker a quarter-cycle ago, at $t = -T/4$, and the loudspeaker cone is now moving backward. In other words, the phase constant $\phi_0 = 0$ rad describes a loudspeaker for which the cone is at its center position and moving backward at $t = 0$.

FIGURE 17.22 Waves from three sources having phase constants $\phi_0 = 0$ rad, $\phi_0 = \pi/2$ rad, and $\phi_0 = \pi$ rad.

(a) Snapshot graph at $t = 0$ for $\phi_0 = 0$ rad

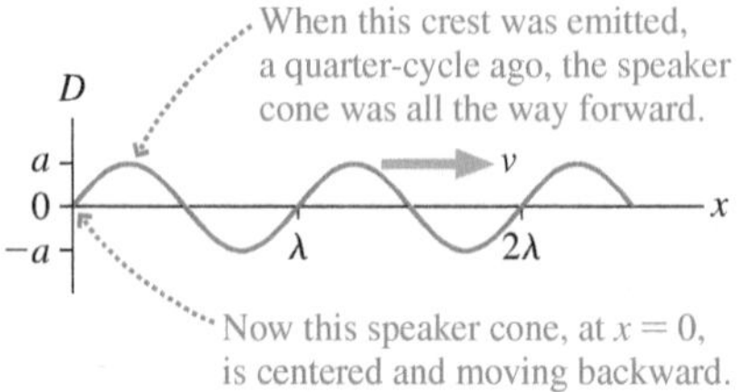
When this crest was emitted, a quarter-cycle ago, the speaker cone was all the way forward.
D
a
0
−a
v
λ
2λ
x
Now this speaker cone, at x = 0, is centered and moving backward.

(b) Snapshot graph at $t = 0$ for $\phi_0 = \pi/2$ rad

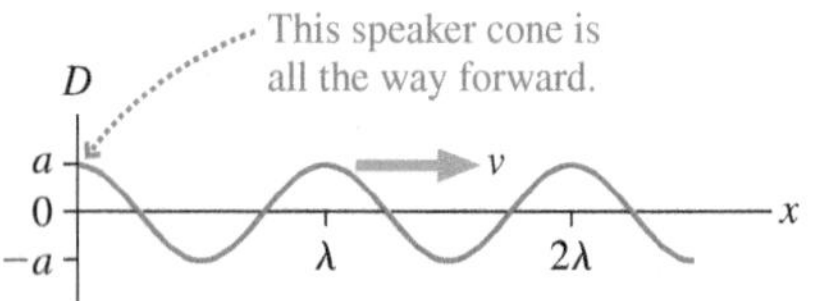
This speaker cone is all the way forward.
D
a
0
−a
v
λ
2λ
x

(c) Snapshot graph at $t = 0$ for $\phi_0 = \pi$ rad

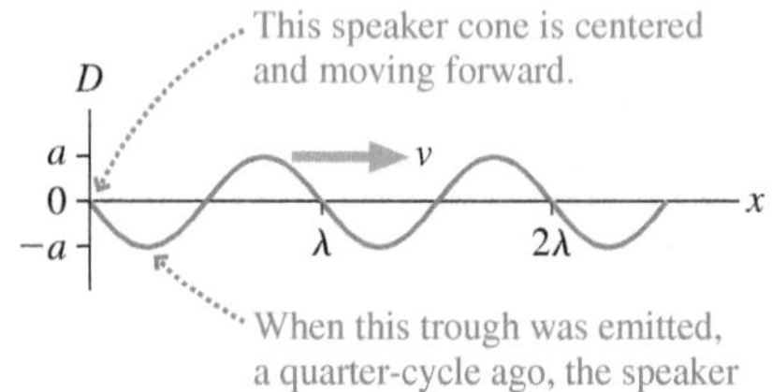
This speaker cone is centered and moving forward.
D
a
0
−a
v
λ
2λ
x
When this trough was emitted, a quarter-cycle ago, the speaker cone was all the way back.

By applying similar reasoning, you can see from Figures 17.22b and 17.22c that the phase constant $\phi_0 = \pi/2$ rad describes a loudspeaker cone that is all the way forward at $t = 0$ while $\phi_0 = \pi$ rad is a speaker cone that is centered and moving forward. Each possible initial condition of what the source is doing at $t = 0$ is described by a different value of the phase constant. Looking back to Figure 17.20b, you can see that loudspeaker 1 has phase constant $\phi_{10} = 0$ rad and loudspeaker 2 has $\phi_{20} = \pi$ rad.

We're going to focus on the phases of the two waves, which are

$$\begin{aligned} \phi_1 &= kx_1 - \omega t + \phi_{10} \\ \phi_2 &= kx_2 - \omega t + \phi_{20} \end{aligned} \tag{17.21}$$

The difference between the two phases ϕ_1 and ϕ_2, called the **phase difference** $\Delta\phi$, is

$$\begin{aligned} \Delta\phi = \phi_2 - \phi_1 &= (kx_2 - \omega t + \phi_{20}) - (kx_1 - \omega t + \phi_{10}) \\ &= k(x_2 - x_1) + (\phi_{20} - \phi_{10}) \\ &= 2\pi\frac{\Delta x}{\lambda} + \Delta\phi_0 \end{aligned} \tag{17.22}$$

You can see that there are two contributions to the phase difference. $\Delta x = x_2 - x_1$, the distance between the two sources, is the path-length difference. It is the extra distance traveled by wave 2 on the way to the point where the two waves are combined. $\Delta\phi_0 = \phi_{20} - \phi_{10}$ is the *inherent phase difference* between the sources.

The condition of being in phase, where crests are aligned with crests and troughs with troughs, is $\Delta\phi = 0, 2\pi, 4\pi$, or any integer multiple of 2π rad. Thus the condition for maximum constructive interference is

$$\Delta\phi = 2\pi\frac{\Delta x}{\lambda} + \Delta\phi_0 = m \cdot 2\pi \text{ rad} \qquad m = 0, 1, 2, 3, \ldots \tag{17.23}$$

Condition for maximum constructive interference

Maximum destructive interference, where the crests of one wave are aligned with the troughs of the other, occurs when two waves are *out of phase*, meaning that $\Delta\phi = \pi, 3\pi, 5\pi$, or any odd multiple of π rad. Thus the condition for maximum destructive interference is

$$\Delta\phi = 2\pi\frac{\Delta x}{\lambda} + \Delta\phi_0 = \left(m + \tfrac{1}{2}\right) \cdot 2\pi \text{ rad} \qquad m = 0, 1, 2, 3, \ldots \tag{17.24}$$

Condition for maximum destructive interference

Equations 17.23 and 17.24 are identical to Equations 17.19 for identical sources ($\Delta\phi_0 = 0$), but these more general conditions are needed if the sources themselves are not in phase.

For example, two waves can be out of phase because the sources are located at different positions, because the sources themselves are out of phase, or because of a combination of these two. FIGURE 17.23 illustrates these ideas by showing three different ways in which two waves interfere destructively. Each of these three arrangements creates waves with $\Delta\phi = \pi$ rad.

FIGURE 17.23 Destructive interference three ways.

(a) The sources are out of phase.

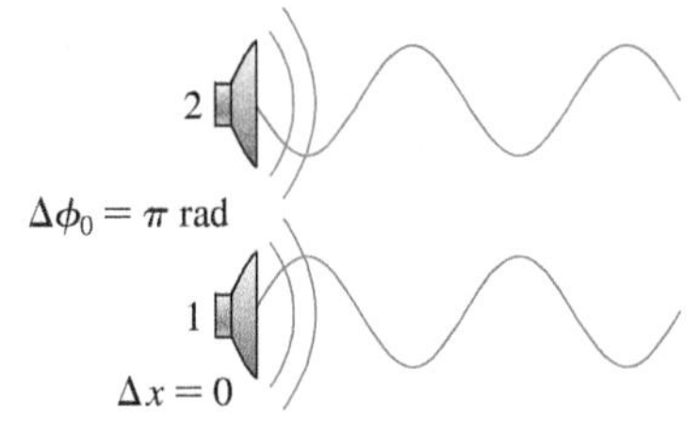

(b) Identical sources are separated by half a wavelength.

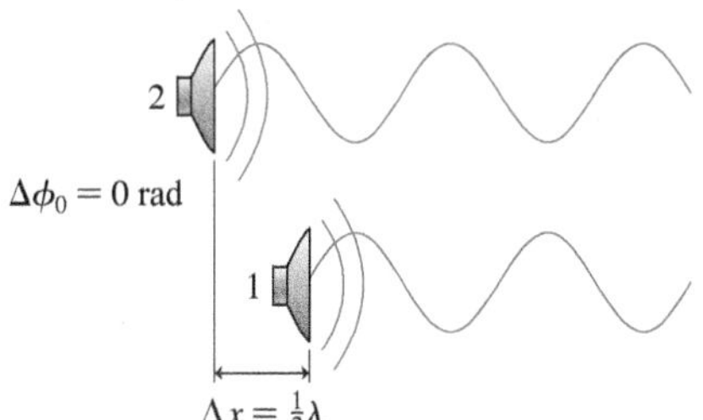

(c) The sources are both separated and partially out of phase.

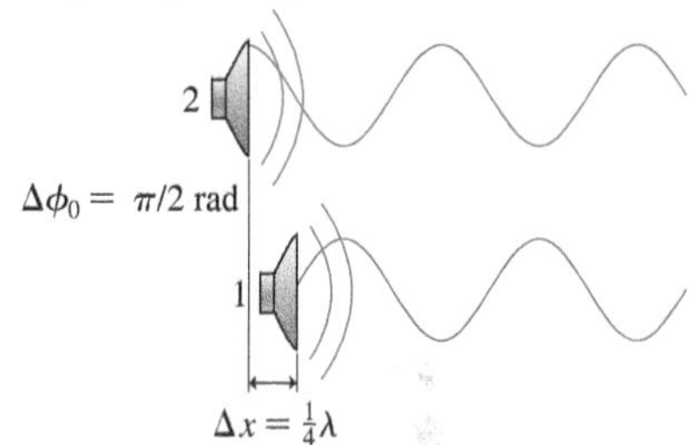

NOTE ▶ Don't confuse the phase difference of the waves ($\Delta\phi$) with the phase difference of the sources ($\Delta\phi_0$). It is $\Delta\phi$, the phase difference of the waves, that governs interference. ◀

EXAMPLE 17.9 Interference between two sound waves

You are standing in front of two side-by-side loudspeakers playing sounds of the same frequency. Initially there is almost no sound at all. Then one of the speakers is moved slowly away from you. The sound intensity increases as the separation between the speakers increases, reaching a maximum when the speakers are 0.75 m apart. Then, as the speaker continues to move, the intensity starts to decrease. What is the distance between the speakers when the sound intensity is again a minimum?

PREPARE The changing sound intensity is due to the interference of two overlapped sound waves. Moving one speaker relative to the other changes the phase difference between the waves.

SOLVE A minimum sound intensity implies that the two sound waves are interfering destructively. Initially the loudspeakers are side by side, so the situation is as shown in Figure 17.23a with $\Delta x = 0$ and $\Delta\phi_0 = \pi$ rad. That is, the speakers themselves are out of phase. Moving one of the speakers does not change $\Delta\phi_0$, but it does change the path-length difference Δx and thus increases the overall phase difference $\Delta\phi$. Constructive interference, causing maximum intensity, is reached when

$$\Delta\phi = 2\pi\frac{\Delta x}{\lambda} + \Delta\phi_0 = 2\pi\frac{\Delta x}{\lambda} + \pi = 2\pi \text{ rad}$$

where we used $m = 1$ because this is the first separation giving constructive interference. The speaker separation at which this occurs is $\Delta x = \lambda/2$. This is the situation shown in FIGURE 17.24.

Because $\Delta x = 0.75$ m is $\lambda/2$, the sound's wavelength is $\lambda = 1.50$ m. The next point of destructive interference, with $m = 1$, occurs when

$$\Delta\phi = 2\pi\frac{\Delta x}{\lambda} + \Delta\phi_0 = 2\pi\frac{\Delta x}{\lambda} + \pi = 3\pi \text{ rad}$$

Thus the distance between the speakers when the sound intensity is again a minimum is

$$\Delta x = \lambda = 1.50 \text{ m}$$

ASSESS A separation of λ gives constructive interference for two *identical* speakers ($\Delta\phi_0 = 0$). Here the phase difference of π rad between the speakers (one is pushing forward as the other pulls back) gives destructive interference at this separation.

FIGURE 17.24 Two out-of-phase sources generate waves that are in phase if the sources are one half-wavelength apart.

The sources are out of phase, $\Delta\phi_0 = \pi$ rad.

2

1

$\Delta x = \frac{1}{2}\lambda$

The sources are separated by half a wavelength.

As a result, the waves are in phase.

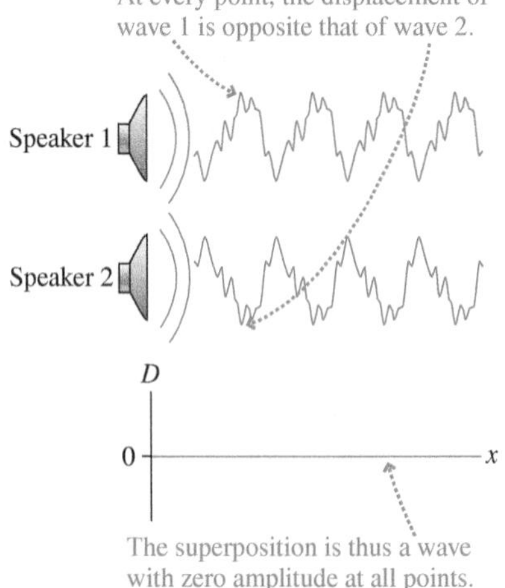
At every point, the displacement of wave 1 is opposite that of wave 2.
Speaker 1
Speaker 2
D
0
x
The superposition is thus a wave with zero amplitude at all points.

FIGURE 17.25 Perfect destructive interference can occur for any waveform if the two waves are exact opposites of each other.

An interesting and important application of interference is to noise-cancelling headphones that use what is called *active noise reduction*. Although we've analyzed only sinusoidal waves, FIGURE 17.25 shows that perfect destructive interference can be achieved for any waveform as long as speaker 2 produces a wave whose displacement, at every instant, is exactly opposite the displacement of the wave from speaker 1.

A headphone with active noise reduction uses a microphone on the outside to measure the ambient sound of the environment. A circuit inside the headphones then produces an inverted version of the microphone signal and sends it to the headphone speakers. As two sound waves converge on your ear, the ambient sound and the sound from the speaker, your ear detects and responds to the *superposition* of these waves. Ideally, the superposition would be complete silence, an exact cancellation of the noise by perfect destructive interference. There are many reasons the destructive interference is not perfect, though, not the least of which is that the ambient sound arrives from all directions and not along a line. Independent tests find that noise-canceling headphones can reduce the noise level by 25 dB to 30 dB. That's significant in a noisy environment, reducing the 85 dB sound intensity level inside an airplane cabin to the 60 dB of normal conversation, but it is far from eliminating noise altogether.

STOP TO THINK 17.5 Two loudspeakers emit waves with $\lambda = 2.0$ m. Speaker 2 is 1.0 m in front of speaker 1. What, if anything, can be done to cause constructive interference between the two waves?

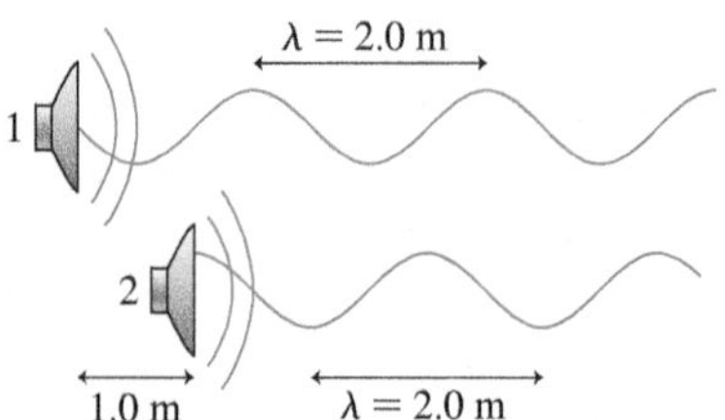
$\lambda = 2.0$ m
1
2
1.0 m
$\lambda = 2.0$ m

A. Move speaker 1 forward (to the right) 1.0 m.
B. Move speaker 1 forward (to the right) 0.5 m.
C. Move speaker 1 backward (to the left) 0.5 m.
D. Move speaker 1 backward (to the left) 1.0 m.
E. Nothing. The situation shown already causes constructive interference.
F. Constructive interference is not possible for any placement of the speakers.

The Mathematics of Interference

Let's look more closely at the superposition of two waves. As two waves of equal amplitude and frequency travel together along the x-axis, the net displacement of the medium is

$$\begin{aligned} D = D_1 + D_2 &= a\sin(kx_1 - \omega t + \phi_{10}) + a\sin(kx_2 - \omega t + \phi_{20}) \\ &= a\sin\phi_1 + a\sin\phi_2 \end{aligned} \tag{17.25}$$

where the phases ϕ_1 and ϕ_2 were defined in Equations 17.21.

A useful trigonometric identity is

$$\sin\alpha + \sin\beta = 2\cos\left[\tfrac{1}{2}(\alpha - \beta)\right]\sin\left[\tfrac{1}{2}(\alpha + \beta)\right] \tag{17.26}$$

This identity is certainly not obvious, although it is easily proven by working backward from the right side. We can use this identity to write the net displacement of Equation 17.25 as

$$D = \left[2a\cos\left(\frac{\Delta\phi}{2}\right)\right]\sin\left[kx_{avg} - \omega t + (\phi_0)_{avg}\right] \tag{17.27}$$

where $\Delta\phi = \phi_2 - \phi_1$ is the phase difference between the two waves, exactly as in Equation 17.22. $x_{avg} = (x_1 + x_2)/2$ is the average distance to the two sources and $(\phi_0)_{avg} = (\phi_{10} + \phi_{20})/2$ is the average phase constant of the sources.

The sine term has the form of a traveling wave, so an observer would see a sinusoidal wave moving along the x-axis with the *same* wavelength and frequency as the original waves.

But how *big* is this wave compared to the two original waves? They each had amplitude a, but the amplitude of their superposition is

$$A = \left|2a\cos\left(\frac{\Delta\phi}{2}\right)\right| \tag{17.28}$$

where we have used an absolute value sign because amplitudes must be positive. Depending upon the phase difference of the two waves, the amplitude of their superposition can be anywhere from zero (perfect destructive interference) to $2a$ (maximum constructive interference).

The amplitude has its maximum value $A = 2a$ if $\cos(\Delta\phi/2) = \pm 1$. This occurs when

$$\Delta\phi = m \cdot 2\pi \qquad \text{(maximum amplitude } A = 2a\text{)} \tag{17.29}$$

where m is an integer. Similarly, the amplitude is zero if $\cos(\Delta\phi/2) = 0$, which occurs when

$$\Delta\phi = \left(m + \tfrac{1}{2}\right) \cdot 2\pi \qquad \text{(minimum amplitude } A = 0\text{)} \tag{17.30}$$

Equations 17.29 and 17.30 are identical to the conditions of Equations 17.23 and 17.24 for constructive and destructive interference. We initially found these conditions by considering the alignment of the crests and troughs. Now we have confirmed them with an algebraic addition of the waves.

It is entirely possible, of course, that the two waves are neither exactly in phase nor exactly out of phase. Equation 17.28 allows us to calculate the amplitude of the superposition for any value of the phase difference. As an example, FIGURE 17.26 shows the calculated interference of two waves that differ in phase by 40°, by 90°, and by 160°.

FIGURE 17.26 The interference of two waves for three different values of the phase difference.

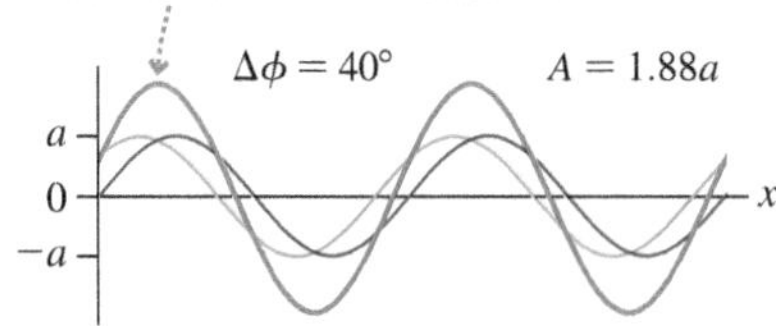

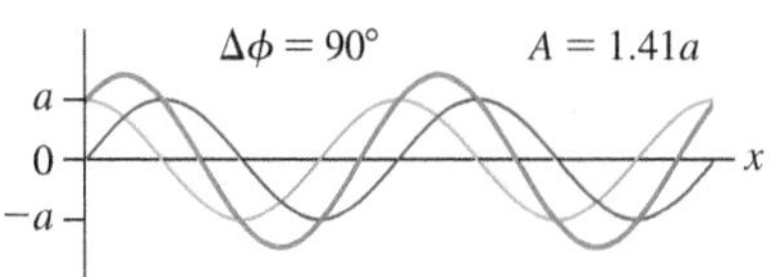

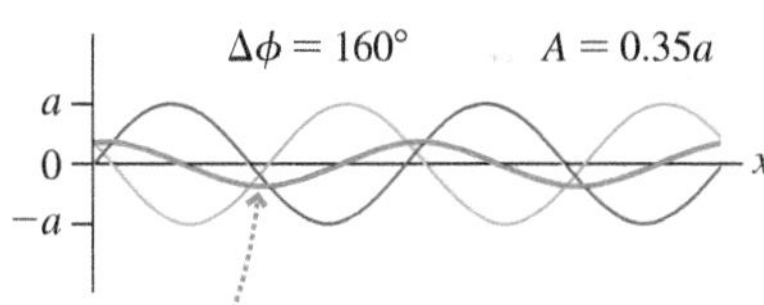

EXAMPLE 17.10 More interference of sound waves

Two loudspeakers emit 500 Hz sound waves with an amplitude of 0.10 mm. Speaker 2 is 1.00 m behind speaker 1, and the phase difference between the speakers is 90°. What is the amplitude of the sound wave at a point 2.00 m in front of speaker 1?

PREPARE The amplitude is determined by the interference of the two waves. Assume that the speed of sound has a room-temperature (20°C) value of 343 m/s.

SOLVE The amplitude of the sound wave is

$$A = |2a\cos(\Delta\phi/2)|$$

where $a = 0.10$ mm and the phase difference between the waves is

$$\Delta\phi = \phi_2 - \phi_1 = 2\pi\frac{\Delta x}{\lambda} + \Delta\phi_0$$

The sound's wavelength is

$$\lambda = \frac{v}{f} = \frac{343 \text{ m/s}}{500 \text{ Hz}} = 0.686 \text{ m}$$

Distances $x_1 = 2.00$ m and $x_2 = 3.00$ m are measured from the speakers, so the path-length difference is $\Delta x = 1.00$ m. We're given that the inherent phase difference between the speakers is $\Delta\phi_0 = \pi/2$ rad. Thus the phase difference at the observation point is

$$\Delta\phi = 2\pi\frac{\Delta x}{\lambda} + \Delta\phi_0 = 2\pi\frac{1.00 \text{ m}}{0.686 \text{ m}} + \frac{\pi}{2} \text{ rad} = 10.73 \text{ rad}$$

and the amplitude of the wave at this point is

$$A = \left|2a\cos\left(\frac{\Delta\phi}{2}\right)\right| = \left|(0.200 \text{ mm})\cos\left(\frac{10.73}{2}\right)\right| = 0.121 \text{ mm}$$

ASSESS The interference is constructive because $A > a$, but less than maximum constructive interference.

17.7 Interference in Two and Three Dimensions

Ripples on a lake move in two dimensions. The glow from a lightbulb spreads outward as a spherical wave. A circular or spherical wave, illustrated in FIGURE 17.27 (next page), can be written

$$D(r, t) = a\sin(kr - \omega t + \phi_0) \tag{17.31}$$

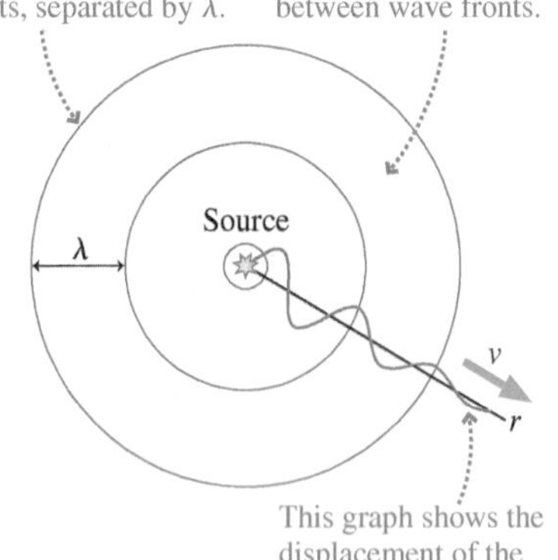
The wave fronts are crests, separated by λ.
Troughs are halfway between wave fronts.
Source
λ
v
r
This graph shows the displacement of the medium.

FIGURE 17.27 A circular or spherical wave.

where r is the distance measured outward from the source. Equation 17.31 is our familiar wave equation with the one-dimensional coordinate x replaced by a more general radial coordinate r. Recall that the wave fronts represent the *crests* of the wave and are spaced by the wavelength λ.

What happens when two circular or spherical waves overlap? For example, imagine two paddles oscillating up and down on the surface of a pond. We will assume that the two paddles oscillate with the same frequency and amplitude and that they are in phase. FIGURE 17.28 shows the wave fronts of the two waves. The ripples overlap as they travel, and, as was the case in one dimension, this causes interference. An important difference, though, is that amplitude decreases with distance as waves spread out in two or three dimensions—a consequence of energy conservation—so the two overlapped waves generally do *not* have equal amplitudes. Consequently, destructive interference rarely produces perfect cancellation.

Maximum constructive interference occurs where two crests align or two troughs align. Several locations of constructive interference are marked in Figure 17.28. Intersecting wave fronts are points where two crests are aligned. It's a bit harder to visualize, but two troughs are aligned when a midpoint between two wave fronts is overlapped with another midpoint between two wave fronts. Maximum, but usually not perfect, destructive interference occurs where the crest of one wave aligns with a trough of the other wave. Several points of destructive interference are also indicated in Figure 17.28.

FIGURE 17.28 The overlapping ripple patterns of two sources. Several points of constructive and destructive interference are noted.

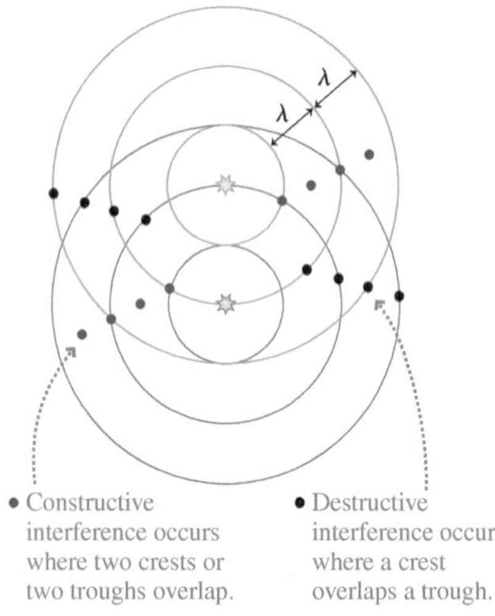
Two in-phase sources emit circular or spherical waves.
λ
λ
• Constructive interference occurs where two crests or two troughs overlap.
• Destructive interference occurs where a crest overlaps a trough.

A picture on a page is static, but **the wave fronts are in motion.** Try to imagine the wave fronts of Figure 17.28 expanding outward as new circular rings are born at the sources. The waves will move forward half a wavelength during half a period, causing the crests in Figure 17.28 to be replaced by troughs while the troughs become crests.

The important point to recognize is that **the motion of the waves does not affect the points of constructive and destructive interference.** Points in the figure where two crests overlap will become points where two troughs overlap, but this overlap is still constructive interference with the superposition oscillating at twice the amplitude of the individual waves. Similarly, points in the figure where a crest and a trough overlap will become a point where a trough and a crest overlap—still destructive interference.

The mathematical description of interference in two or three dimensions is very similar to that of one-dimensional interference. The net displacement of a particle in the medium is

$$D = D_1 + D_2 = a_1 \sin(kr_1 - \omega t + \phi_{10}) + a_2 \sin(kr_2 - \omega t + \phi_{20}) \qquad (17.32)$$

The only differences between Equation 17.32 and the earlier one-dimensional Equation 17.25 are that the linear coordinates have been changed to radial coordinates and we've allowed the amplitudes to differ. These changes do not affect the phase difference, which, with x replaced by r, is now

$$\Delta\phi = 2\pi \frac{\Delta r}{\lambda} + \Delta\phi_0 \qquad (17.33)$$

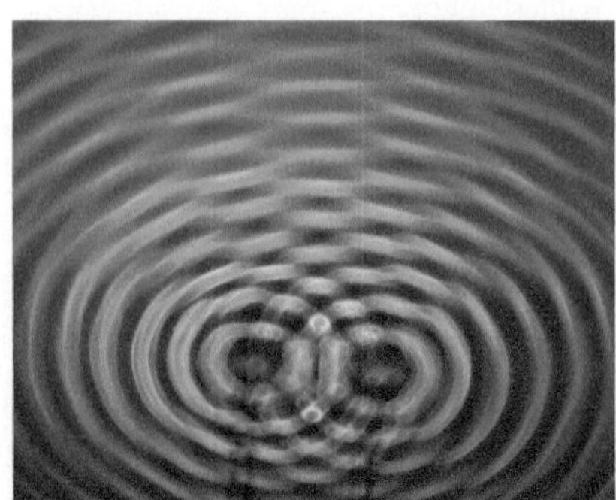

Interference in two dimensions Two overlapping water waves create an interference pattern.

The term $2\pi(\Delta r/\lambda)$ is the phase difference that arises when the waves travel different distances from the sources to the point at which they combine. Δr itself is the *path-length difference.* As before, $\Delta\phi_0$ is any inherent phase difference of the sources themselves.

The conditions for maximum constructive and destructive interference have not changed—crests aligning with crests and crests aligning with troughs—so the earlier Equations 17.23 and 17.24 still apply if x is replaced by the radial distance r:

$$\text{maximum constructive:} \quad \Delta\phi = 2\pi \frac{\Delta r}{\lambda} + \Delta\phi_0 = m \cdot 2\pi$$
$$\text{maximum destructive:} \quad \Delta\phi = 2\pi \frac{\Delta r}{\lambda} + \Delta\phi_0 = \left(m + \tfrac{1}{2}\right) \cdot 2\pi \qquad m = 0, 1, 2, 3, \ldots \qquad (17.34)$$

Identical Sources

For two identical sources (i.e., sources that oscillate in phase with $\Delta\phi_0 = 0$), the conditions for constructive and destructive interference are simple:

$$\begin{aligned} \text{constructive:} &\quad \Delta r = m\lambda \\ \text{destructive:} &\quad \Delta r = \left(m + \frac{1}{2}\right)\lambda \end{aligned} \qquad (17.35)$$

The waves from two identical sources interfere constructively at points where the path-length difference is an integer number of wavelengths because, for these values of Δr, crests are aligned with crests and troughs with troughs. **The waves interfere destructively at points where the path-length difference is a half-integer number of wavelengths** because, for these values of Δr, crests are aligned with troughs. These two statements are the essence of interference.

Wave fronts are spaced exactly one wavelength apart; hence we can measure the distances r_1 and r_2 simply by counting the rings in the wave-front pattern. In FIGURE 17.29, which is based on Figure 17.28, point A is distance $r_1 = 3\lambda$ from the first source and $r_2 = 2\lambda$ from the second. The path-length difference is $\Delta r_A = 1\lambda$, the condition for the maximum constructive interference of identical sources. Point B has $\Delta r_B = \frac{1}{2}\lambda$, so it is a point of maximum destructive interference.

NOTE ▶ Interference is determined by Δr, the path-length *difference*, rather than by r_1 or r_2. ◀

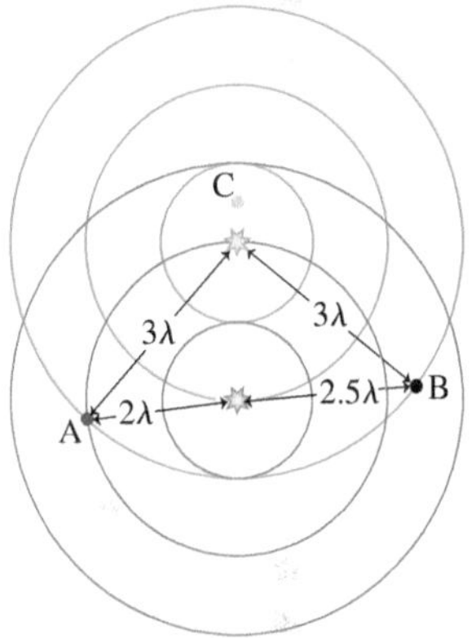

• At A, $\Delta r_A = \lambda$, so this is a point of constructive interference.
C
3λ
3λ
2λ
2.5λ
A
B
• At B, $\Delta r_B = \frac{1}{2}\lambda$, so this is a point of destructive interference.

FIGURE 17.29 For identical sources, the path-length difference Δr determines whether the interference at a particular point is constructive or destructive.

STOP TO THINK 17.6 The interference at point C in Figure 17.29 is

A. Maximum constructive.
B. Constructive, but less than maximum.
C. Maximum destructive.
D. Destructive, but less than maximum.
E. There is no interference at point C.

We can now locate the points of maximum constructive interference by drawing a line through *all* the points at which $\Delta r = 0$, another line through all the points at which $\Delta r = \lambda$, and so on. These lines, shown in red in FIGURE 17.30, are called **antinodal lines.** They are analogous to the antinodes of a standing wave, hence the

FIGURE 17.30 The points of constructive and destructive interference fall along antinodal and nodal lines.

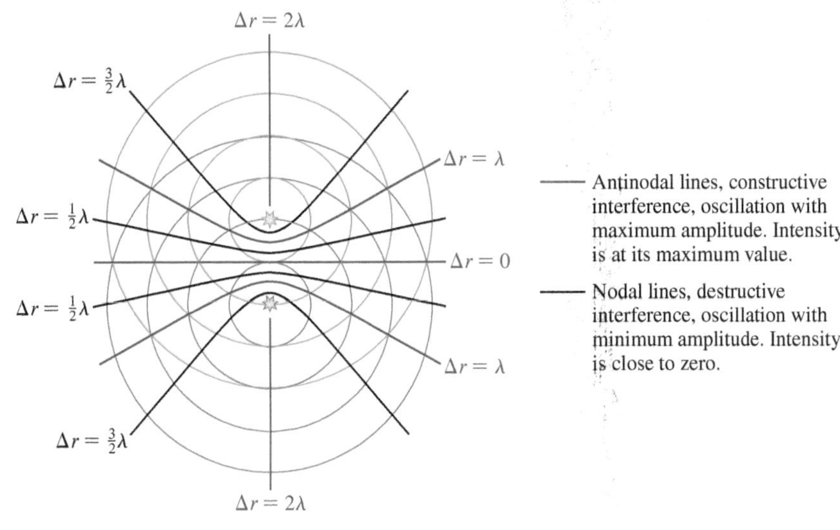

$\Delta r = 2\lambda$
$\Delta r = \frac{3}{2}\lambda$
$\Delta r = \lambda$
$\Delta r = \frac{1}{2}\lambda$
$\Delta r = 0$
$\Delta r = \frac{1}{2}\lambda$
$\Delta r = \lambda$
$\Delta r = \frac{3}{2}\lambda$
$\Delta r = 2\lambda$
Antinodal lines, constructive interference, oscillation with maximum amplitude. Intensity is at its maximum value.
Nodal lines, destructive interference, oscillation with minimum amplitude. Intensity is close to zero.

name. An antinode is a *point* of maximum constructive interference; for circular waves, oscillation at maximum amplitude occurs along a continuous *line.* Similarly, destructive interference occurs along lines called **nodal lines.** The amplitude is a minimum along a nodal line, usually close to zero, just as it is at a node in a standing-wave pattern.

Figure 17.30 will be essential for understanding the interference of light waves in Chapter 18, so it's well worth making sure that you understand what it is showing.

A Problem-Solving Strategy for Interference Problems

The information in this section is the basis of a strategy for solving interference problems. This strategy applies equally well to interference in one dimension if you use Δx instead of Δr. We've used sound waves to develop our understanding of interference, but we'll apply this strategy to the interference of light waves in the next chapter.

PROBLEM-SOLVING STRATEGY 17.1 **Interference of two waves**

PREPARE Model the waves as linear, circular, or spherical. Draw a picture showing the sources of the waves and the point where the waves interfere. Give relevant dimensions. Identify the distances r_1 and r_2 from the sources to the point. Note any phase difference $\Delta\phi_0$ between the two sources.

SOLVE The interference depends on the path-length difference $\Delta r = r_2 - r_1$ and the source phase difference $\Delta\phi_0$.

$$\text{Constructive:} \quad \Delta\phi = 2\pi\frac{\Delta r}{\lambda} + \Delta\phi_0 = m \cdot 2\pi$$

$$m = 0, 1, 2, \ldots$$

$$\text{Destructive:} \quad \Delta\phi = 2\pi\frac{\Delta r}{\lambda} + \Delta\phi_0 = \left(m + \tfrac{1}{2}\right) \cdot 2\pi$$

For identical sources $(\Delta\phi_0 = 0)$, the interference is maximum constructive if $\Delta r = m\lambda$, maximum destructive if $\Delta r = (m + \frac{1}{2})\lambda$.

ASSESS Check that your result has correct units and significant figures, is reasonable, and answers the question.

EXAMPLE 17.11 **Two-dimensional interference between two loudspeakers**

Two loudspeakers in a plane are 2.0 m apart and in phase with each other. Both emit 700 Hz sound waves into a room where the speed of sound is 341 m/s. A listener stands 5.0 m in front of the loudspeakers and 2.0 m to one side of the center. Is the interference at this point maximum constructive, maximum destructive, or in between? How will the situation differ if the loudspeakers are out of phase?

PREPARE The two speakers are sources of in-phase, spherical waves. The overlap of these waves causes interference. FIGURE 17.31 shows the loudspeakers and defines the distances r_1 and r_2 to the point of observation. The figure includes dimensions and notes that $\Delta\phi_0 = 0$ rad.

SOLVE It's not r_1 and r_2 that matter, but the *difference* Δr between them. From the geometry of the figure we can calculate that

$$r_1 = \sqrt{(5.0\text{ m})^2 + (1.0\text{ m})^2} = 5.10\text{ m}$$

$$r_2 = \sqrt{(5.0\text{ m})^2 + (3.0\text{ m})^2} = 5.83\text{ m}$$

FIGURE 17.31 Pictorial representation of the interference between two loudspeakers.

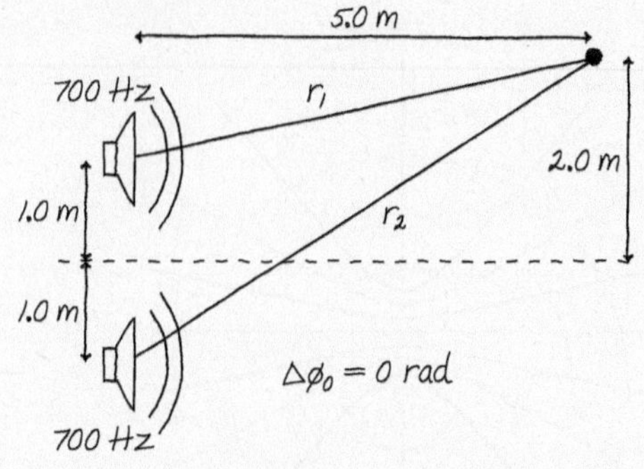

Thus the path-length difference is $\Delta r = r_2 - r_1 = 0.73$ m. The wavelength of the sound waves is

$$\lambda = \frac{v}{f} = \frac{341 \text{ m/s}}{700 \text{ Hz}} = 0.487 \text{ m}$$

In terms of wavelengths, the path-length difference is $\Delta r/\lambda = 1.50$, or

$$\Delta r = \tfrac{3}{2}\lambda$$

Because the sources are in phase ($\Delta\phi_0 = 0$), this is the condition for *destructive* interference. If the sources were out of phase ($\Delta\phi_0 = \pi$ rad), then the phase difference of the waves at the listener would be

$$\Delta\phi = 2\pi\frac{\Delta r}{\lambda} + \Delta\phi_0 = 2\pi\left(\frac{3}{2}\right) + \pi \text{ rad} = 4\pi \text{ rad}$$

This is an integer multiple of 2π rad, so in this case the interference would be *constructive.*

ASSESS Both the path-length difference *and* any inherent phase difference of the sources must be considered when evaluating interference.

STOP TO THINK 17.7 These two loudspeakers are in phase. They emit equal-amplitude sound waves with a wavelength of 1.0 m. At the point indicated, is the interference maximum constructive, maximum destructive, or something in between?

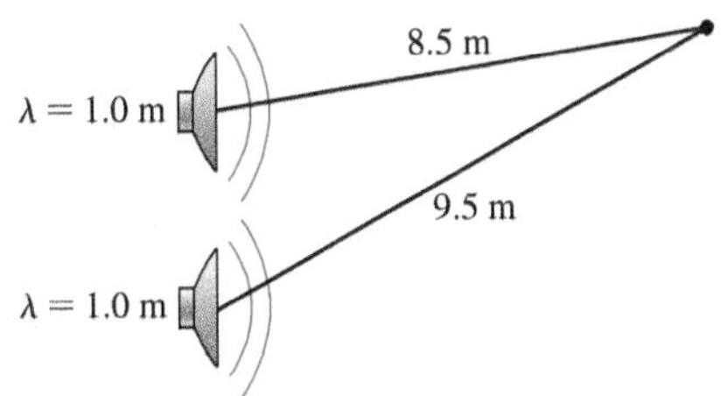

17.8 Beats

Thus far we have looked at the superposition of waves that have the same wavelength and frequency. Now we'll consider two sinusoidal waves traveling toward your ear, as shown in FIGURE 17.32, with the same amplitude but slightly different frequencies. The wave shown as a red curve has a slightly longer wavelength and slightly smaller frequency than the green one. This slight difference causes the superposition of the waves—the blue curve—to alternate between constructive and destructive interference.

FIGURE 17.32 Beats are created by the superposition of two waves that have slightly different frequencies.

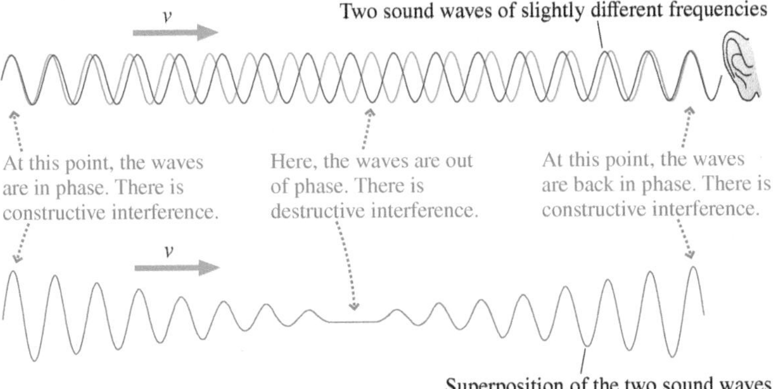

If the frequency difference between two sounds is large, such as a high note and a low note, we will hear two distinct tones. But if the frequency difference is small, only one or two hertz, then we will hear a single tone whose volume, like the blue curve, alternates loud, soft, loud, soft, and so on, making a distinctive sound pattern known as **beats.**

To analyze beats, let's consider two sinusoidal waves traveling along the x-axis with angular frequencies $\omega_1 = 2\pi f_1$ and $\omega_2 = 2\pi f_2$. The two waves are

$$\begin{aligned} D_1 &= a\sin(k_1x - \omega_1 t + \phi_{10}) \\ D_2 &= a\sin(k_2x - \omega_2 t + \phi_{20}) \end{aligned} \tag{17.36}$$

where the subscripts 1 and 2 indicate that the frequencies, wave numbers, and phase constants of the two waves may be different.

To simplify the analysis, let's make several assumptions:

1. The two waves have the same amplitude a,
2. A detector, such as your ear, is located at the origin $(x = 0)$,
3. The two sources are in phase $(\phi_{10} = \phi_{20})$, and
4. The source phases happen to be $\phi_{10} = \phi_{20} = \pi$ rad.

None of these assumptions is essential to the outcome. All could be otherwise and we would still come to the same conclusion, but the mathematics would be far messier. Making these assumptions allows us to emphasize the physics with the least amount of mathematics.

With these assumptions, the two waves as they reach the detector at $x = 0$ are

$$\begin{aligned} D_1 &= a\sin(-\omega_1 t + \pi) = a\sin\omega_1 t \\ D_2 &= a\sin(-\omega_2 t + \pi) = a\sin\omega_2 t \end{aligned} \tag{17.37}$$

where we've used the trigonometric identity $\sin(\pi - \theta) = \sin\theta$. The principle of superposition tells us that the *net* displacement of the medium at the detector is the sum of the displacements of the individual waves. Thus

$$D = D_1 + D_2 = a(\sin\omega_1 t + \sin\omega_2 t) \tag{17.38}$$

Earlier, for interference, we used the trigonometric identity

$$\sin\alpha + \sin\beta = 2\cos\left[\tfrac{1}{2}(\alpha - \beta)\right]\sin\left[\tfrac{1}{2}(\alpha + \beta)\right]$$

We can use this identity again to write Equation 17.38 as

$$\begin{aligned} D &= 2a\cos\left[\tfrac{1}{2}(\omega_1 - \omega_2)t\right]\sin\left[\tfrac{1}{2}(\omega_1 + \omega_2)t\right] \\ &= \left[2a\cos(\omega_{mod}t)\right]\sin(\omega_{avg}t) \end{aligned} \tag{17.39}$$

where $\omega_{avg} = \frac{1}{2}(\omega_1 + \omega_2)$ is the *average* angular frequency and $\omega_{mod} = \frac{1}{2}|\omega_1 - \omega_2|$ is called the *modulation frequency*. We've used the absolute value because the modulation depends only on the frequency *difference* between the sources, not on which has the larger frequency.

We are interested in the situation when the two frequencies are very nearly equal: $\omega_1 \approx \omega_2$. In that case, ω_{avg} hardly differs from either ω_1 or ω_2 while ω_{mod} is very near to—but not exactly—zero. When ω_{mod} is very small, the term $\cos(\omega_{mod}t)$ oscillates *very* slowly. We have grouped it with the $2a$ term because, together, they provide a slowly changing "amplitude" for the rapid oscillation at frequency ω_{avg}.

FIGURE 17.33 is a history graph of the wave at the detector $(x = 0)$. It shows the oscillation of the air against your eardrum at frequency $f_{avg} = \omega_{avg}/2\pi = \frac{1}{2}(f_1 + f_2)$. This oscillation determines the note you hear; it differs little from the two notes at frequencies f_1 and f_2. We are especially interested in the time-dependent amplitude, shown as a dashed line, that is given by the term $2a\cos(\omega_{mod}t)$. This periodically varying amplitude is called a **modulation** of the wave, which is where ω_{mod} gets its name.

As the amplitude rises and falls, the sound alternates as loud, soft, loud, soft, and so on. But that is exactly what you hear when you listen to beats! The alternating loud and soft sounds arise from the two waves being alternately in phase and out of phase, causing constructive and then destructive interference.

Imagine two people walking side by side at just slightly different paces. Initially both of their right feet hit the ground together, but after a while they get out of step.

FIGURE 17.33 Beats are caused by the superposition of two waves of nearly identical frequency.

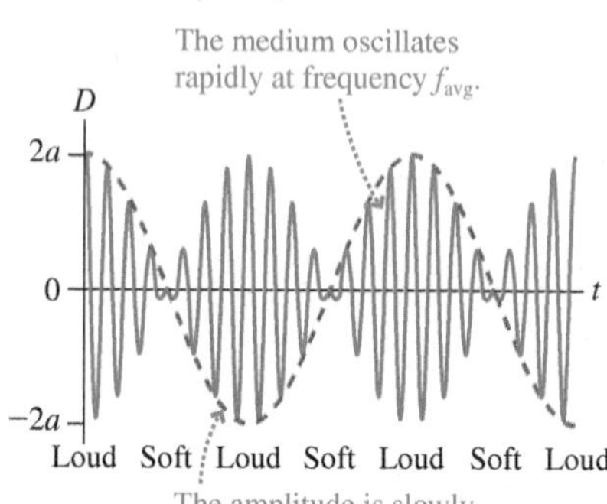
The medium oscillates rapidly at frequency f_{avg}.
D
2a
0
t
−2a
Loud Soft Loud Soft Loud Soft Loud
The amplitude is slowly modulated as $2a\cos(\omega_{mod}t)$.

A little bit later they are back in step and the process alternates. The sound waves are doing the same. Initially the crests of each wave, of amplitude a, arrive together at your ear and the net displacement is doubled to $2a$. But after a while the two waves, being of slightly different frequency, get out of step and a crest of one arrives with a trough of the other. When this happens, the two waves cancel each other to give a net displacement of zero. This process alternates over and over, loud and soft.

Notice, in Figure 17.33, that the sound intensity rises and falls *twice* during one cycle of the modulation envelope. Each "loud-soft-loud" is one beat, so the **beat frequency** f_{beat}, which is the number of beats per second, is *twice* the modulation frequency $f_{\text{mod}} = \omega_{\text{mod}}/2\pi$. From the above definition of ω_{mod}, the beat frequency is

$$f_{\text{beat}} = 2f_{\text{mod}} = 2\frac{\omega_{\text{mod}}}{2\pi} = 2 \cdot \frac{1}{2}\left|\frac{\omega_1}{2\pi} - \frac{\omega_2}{2\pi}\right| = |f_1 - f_2| \qquad (17.40)$$

The beat frequency is simply the *difference* between the two individual frequencies.

EXAMPLE 17.12 **Hearing bats with beats**

The little brown bat is a common species in North America. It emits echolocation pulses at a frequency of 40 kHz, well above the range of human hearing. To allow researchers to "hear" these bats, the bat detector shown in FIGURE 17.34 combines the bat's sound wave at frequency f_1 with a wave of frequency f_2 from a tunable oscillator. The resulting beat frequency is then amplified and sent to a loudspeaker. To what frequency should the tunable oscillator be set to produce an audible beat frequency of 1 kHz?

FIGURE 17.34 The operation of a bat detector.

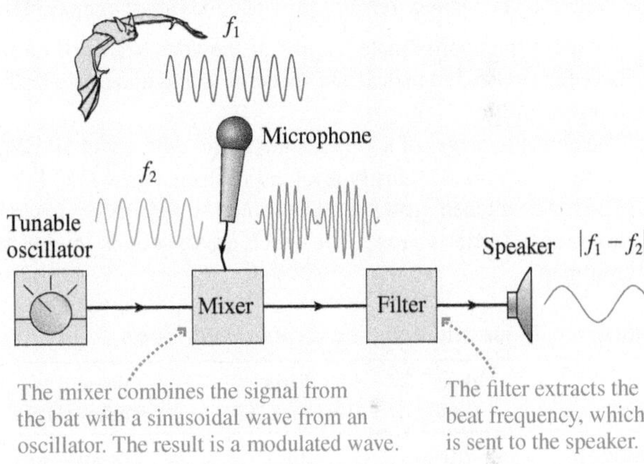
f_1
Microphone
f_2
Tunable oscillator
Speaker $|f_1 - f_2|$
Mixer
Filter
The mixer combines the signal from the bat with a sinusoidal wave from an oscillator. The result is a modulated wave.
The filter extracts the beat frequency, which is sent to the speaker.

PREPARE Beats occur when the two frequencies to be superimposed are close together.

SOLVE Combining two waves with different frequencies gives a beat frequency

$$f_{\text{beat}} = |f_1 - f_2|$$

A beat frequency will be generated at 1 kHz if the oscillator frequency and the bat frequency *differ* by 1 kHz. An oscillator frequency of either 41 kHz or 39 kHz will work nicely.

ASSESS The frequency is close to 40 kHz, as expected. The electronic circuitry of radios, televisions, and cell phones makes a similar use of *mixers* to generate difference frequencies.

Beats aren't limited to sound waves. FIGURE 17.35 shows a graphical example of beats. Two "fences" of slightly different frequencies are superimposed on each other. The difference in the two frequencies is two lines per inch. You can confirm, with a ruler, that the figure has two "beats" per inch, in agreement with Equation 17.40.

Beats are important in many other situations. For example, you have probably seen movies where rotating wheels seem to turn slowly backward. Why is this? Suppose the movie camera is shooting at 30 frames per second but the wheel is rotating 32 times per second. The combination of the two produces a "beat" of 2 Hz, meaning that the wheel appears to rotate only twice per second. The same is true if the wheel is rotating 28 times per second, but in this case, where the wheel frequency slightly lags the camera frequency, it appears to rotate *backward* twice per second!

FIGURE 17.35 A graphical example of beats.

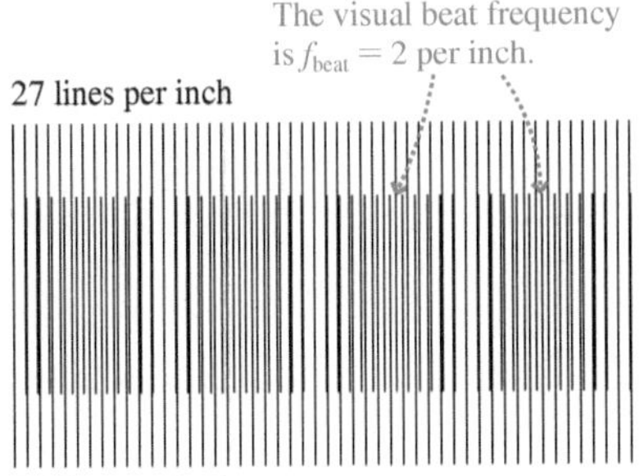
The visual beat frequency is $f_{\text{beat}} = 2$ per inch.
27 lines per inch
25 lines per inch

STOP TO THINK 17.8 You hear three beats per second when two sound tones are generated. The frequency of one tone is 610 Hz. The frequency of the other is

A. 604 Hz.
B. 607 Hz.
C. 613 Hz.
D. 616 Hz.
E. Either A or D.
F. Either B or C.

CHAPTER 17 INTEGRATED EXAMPLE The size of a dog determines the sound of its growl

A dog's vocalizations, like those of a human, result from an interplay between oscillations of the vocal cords and the standing-wave resonances of the vocal tract. A dog's vocal tract is simpler than that of a human, however, and is well modeled as a tube closed at the larynx and open at the lips.

All dogs growl at a low pitch because the fundamental frequency of their vocal cords is low. But the growls of large and small dogs sound different because they have different formants as determined by the length of their vocal tracts. Data on a small dog and a large dog are given in TABLE 17.1. Most of the energy is in the first formant, so your perception of a Doberman's growl is at frequencies near 350 Hz, while a Terrier's growl sounds at around 650 Hz.

a. What are the approximate lengths of the vocal tracts of these two dogs? Use 350 m/s as the speed of sound at a dog's body temperature.
b. The acoustic power of a growl is not large, only about 50 μW. What is the sound intensity level 2.0 m from a growling dog?
c. The lower-pitched growl of a Doberman sounds more menacing than a Terrier's growl, but which growl sounds *louder* to a human?

TABLE 17.1 **Mass and acoustic data for two dogs**

Breed	Mass (kg)	First formant (Hz)	Second formant (Hz)
Terrier	7.0	650	1950
Doberman	38	350	1050

PREPARE We'll model the canine vocal tract as an open-closed tube, as shown in FIGURE 17.36. Because an open-closed tube has only odd-numbered harmonics, the first two formants correspond to standing-wave resonances with $m = 1$ and $m = 3$.

FIGURE 17.36 A model of the canine vocal tract.

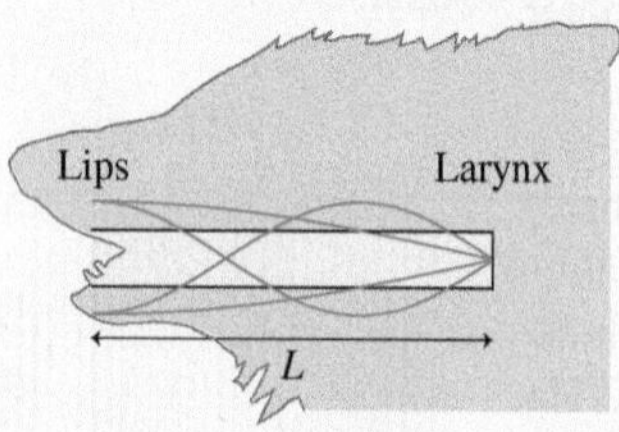

SOLVE

a. The allowed frequencies of standing waves in an open-closed tube of length L are $f_m = mv_{\text{sound}}/4L$, where $m = 1, 3, 5, \ldots$ is an odd integer. The two formants correspond to the first two standing waves, with $m = 1$ and $m = 3$, so our simple model predicts that the frequency of the second formant should be three times that of the first formant. This prediction is confirmed by the data in Table 17.1. We can use the frequency of the first formant to determine the lengths of the vocal tracts:

$$L = \frac{1 \times v_{\text{sound}}}{4f_1} = \begin{cases} \dfrac{350 \text{ m/s}}{4(650 \text{ Hz})} = 13 \text{ cm} & \text{Terrier} \\[2ex] \dfrac{350 \text{ m/s}}{4(350 \text{ Hz})} = 25 \text{ cm} & \text{Doberman} \end{cases}$$

The vocal tract is longer in the Doberman, which we expect given such a large difference in their masses and sizes.

b. You learned in Chapter 16 that the sound intensity level is $\beta = (10 \text{ dB})\log_{10}(I/I_0)$, where $I_0 = 1.0 \times 10^{-12} \text{ W/m}^2$ is the threshold of hearing at frequencies near 1000 Hz. If a dog growls with power $P = 50 \times 10^{-6}$ W, the sound wave intensity at distance $r = 2.0$ m is

$$I = \frac{P}{4\pi r^2} = \frac{50 \times 10^{-6} \text{ W}}{4\pi(2.0 \text{ m})^2} = 1.0 \times 10^{-6} \text{ W/m}^2$$

Thus the sound intensity level at 2.0 m is

$$\beta = (10 \text{ dB})\log_{10}\left(\frac{1.0 \times 10^{-6} \text{ W/m}^2}{1.0 \times 10^{-12} \text{ W/m}^2}\right) = 60 \text{ dB}$$

60 dB is roughly the sound level of normal conversation, and that seems about right for the loudness of a dog growl.

c. The graph of Figure 16.27 (in Chapter 16) showed the threshold of hearing as a function of frequency. For frequencies below about 1000 Hz, the threshold rises as the frequency decreases. Consequently, a lower-frequency sound with a higher threshold is perceived as less loud than a higher-frequency sound of equal intensity. If both dogs growl with the same acoustic power, the higher-frequency Terrier growl at frequencies around 650 Hz will sound louder than the lower-frequency Doberman growl at frequencies around 350 Hz.

How much louder? The curve in Figure 16.27 showed that the hearing threshold is about 12 dB at 350 Hz and 4 dB at 650 Hz. A 60 dB growl is 48 dB above the threshold at the Doberman's 350 Hz growl but 56 dB above the threshold at the Terrier's 650 Hz. Thus a human perceives the Terrier's growl as about 8 dB louder than the Doberman's growl.

ASSESS Our discussion of scaling in Chapter 1 noted that an animal's body mass M scales as $M \propto L^3$, where L is a characteristic length. Equivalently, we could say that the length of some feature of an animal scales with the animal's mass as $L \propto M^{1/3}$. We know the masses of the two dogs, so scaling arguments would suggest that the lengths of their vocal tracts should be in the ratio

$$\frac{L_{\text{Doberman}}}{L_{\text{Terrier}}} = \left(\frac{M_{\text{Doberman}}}{M_{\text{Terrier}}}\right)^{1/3} = \left(\frac{38 \text{ kg}}{7.0 \text{ kg}}\right)^{1/3} = 1.8$$

The lengths that we determined from the dog's acoustical properties have a ratio of 1.9, in almost perfect agreement. The sound of a dog is determined by its size.

SUMMARY

GOAL To use the principle of superposition to understand the phenomena of standing waves and interference.

GENERAL PRINCIPLES

Principle of Superposition

The displacement of a medium when more than one wave is present is the sum at each point of the displacements due to each individual wave.

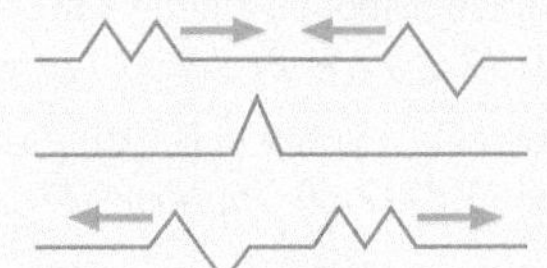

IMPORTANT CONCEPTS

Standing Waves

Standing waves are due to the superposition of two waves traveling in opposite directions.

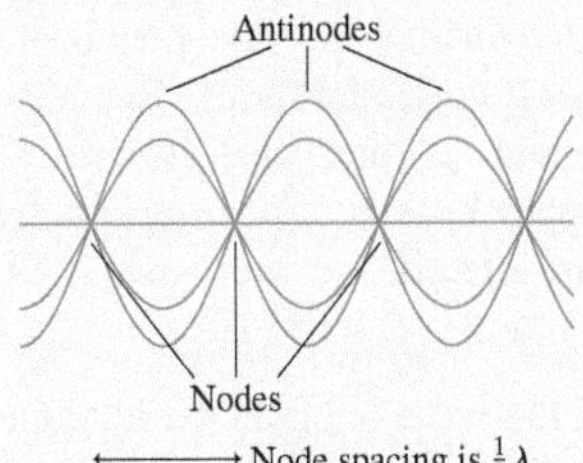

The **nodes** of a standing wave, where the particles never move, are spaced half a wavelength apart. The allowed standing waves are the **modes** of the system.

A standing wave on a string must have a node at each end. A standing wave in a tube of air has a displacement node at a closed end and a displacement antinode at an open end. The allowed frequencies, called **harmonics,** are integer multiples of the **fundamental frequency** f_1.

String

$$f_m = m\frac{v}{2L} = mf_1$$

$$\lambda_m = \frac{2L}{m}$$

$m = 1, 2, 3, \ldots$

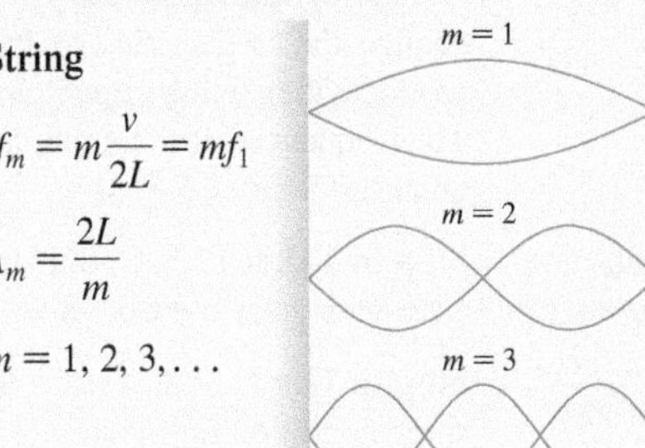

Closed-closed tube

$$f_m = m\frac{v}{2L}$$

$m = 1, 2, 3, \ldots$

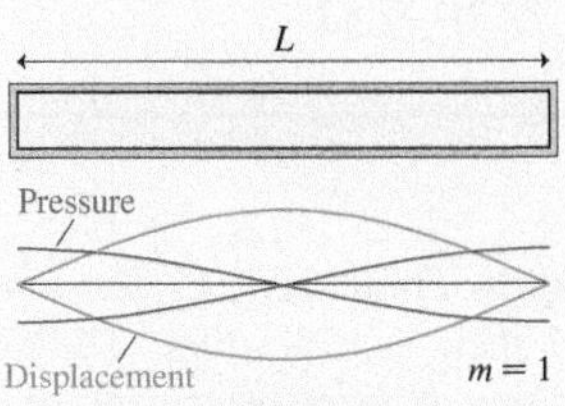

Open-open tube

$$f_m = m\frac{v}{2L}$$

$m = 1, 2, 3, \ldots$

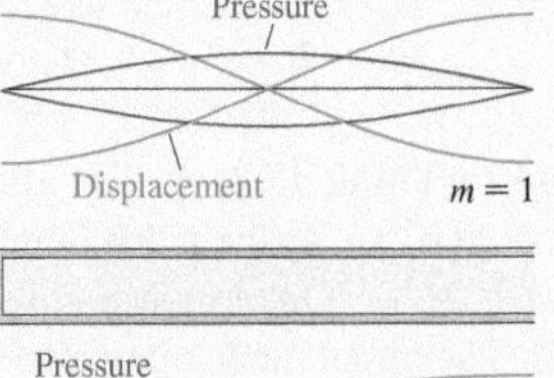

Open-closed tube

$$f_m = m\frac{v}{4L}$$

$m = 1, 3, 5, \ldots$

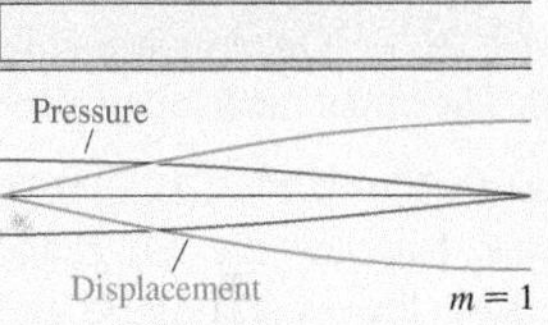

Interference

Maximum constructive interference occurs when the waves are **in phase:** Crests are aligned with crests and troughs with troughs. The waves add to give an amplitude twice that of the individual waves.

Maximum destructive interference occurs when the waves are **out of phase:** Crests are aligned with troughs. The waves cancel to give zero amplitude.

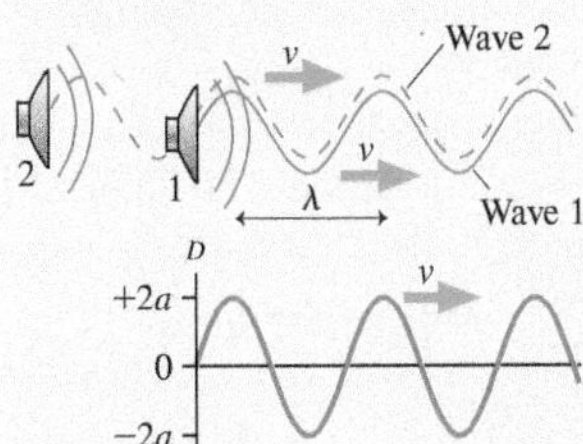

D, +2a, 0, −2a, x, Wave 2, Wave 1, λ

Interference in two or three dimensions occurs along **nodal lines** and **antinodal lines.**

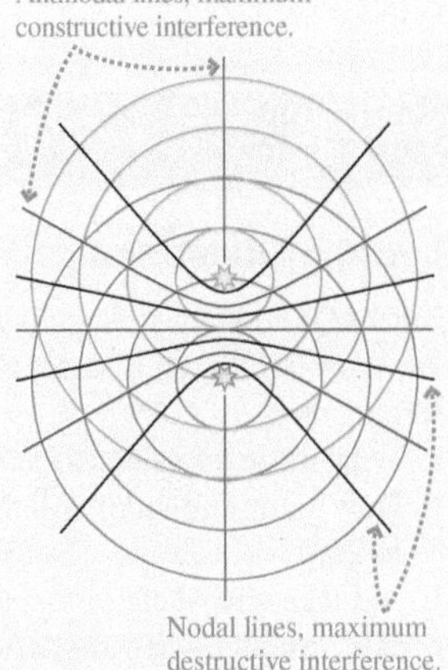

The interference conditions are

$$\text{constructive:}\quad \Delta\phi = 2\pi\frac{\Delta x}{\lambda} + \Delta\phi_0 = m\cdot 2\pi$$

$m = 0, 1, 2, 3, \ldots$

$$\text{destructive:}\quad \Delta\phi = 2\pi\frac{\Delta x}{\lambda} + \Delta\phi_0 = (m + \tfrac{1}{2})\cdot 2\pi$$

where Δx is the path-length difference, and $\Delta\phi_0$ is any inherent phase difference between the sources. For identical (in phase) sources:

constructive: $\Delta x = m\lambda$ destructive: $\Delta x = (m + \frac{1}{2})\lambda$

Beats (loud-soft modulations of intensity) occur when two waves of slightly different frequency are superimposed.

The **beat frequency** is

$$f_{\text{beat}} = |f_1 - f_2|$$

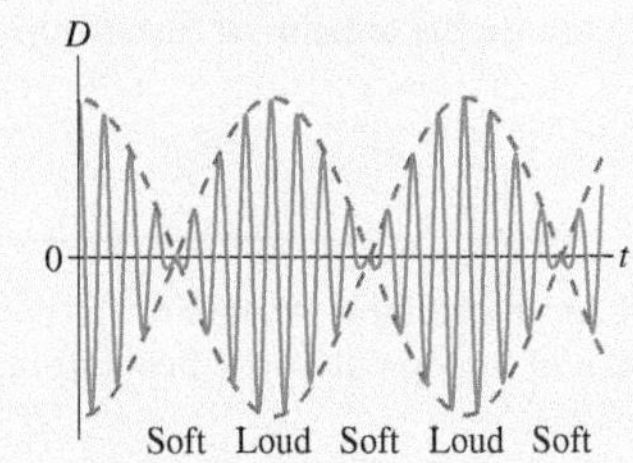

LEARNING OBJECTIVES After studying this chapter, you should be able to:

- Understand and use the principle of superposition. *Problems 17.1–17.4*
- Locate the positions of the nodes and antinodes of standing waves. *Conceptual Questions 17.1, 17.5; Problems 17.5, 17.8, 17.27*
- Solve problems about standing waves on strings. *Conceptual Questions 17.3, 17.9; Problems 17.6, 17.7, 17.10, 17.12, 17.13*
- Solve problems relating to standing sound waves. *Conceptual Questions 17.6, 17.10; Problems 17.18, 17.19, 17.24, 17.29, 17.30*
- Apply the concepts of standing waves to the production of human speech. *Conceptual Question 17.4; Problems 17.31–17.33*
- Determine whether the interference of two waves is constructive or destructive. *Conceptual Question 17.11; Problems 17.34, 17.35, 17.37, 17.40, 17.41*
- Locate nodal and antinodal lines in the interference patterns of two- and three-dimensional waves. *Conceptual Question 17.12; Problems 17.42–17.45*
- Calculate the beat frequency of two sound waves. *Conceptual Question 17.13; Problems 17.46–17.50*

STOP TO THINK ANSWERS

Stop to Think 17.1: C. The figure shows the two waves at $t = 6$ s and their superposition. The superposition is the *point-by-point* addition of the displacements of the two individual waves.

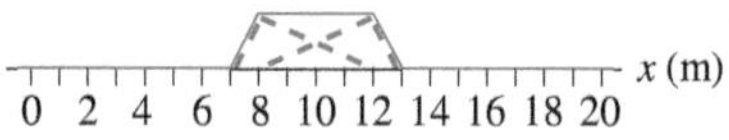

Stop to Think 17.2: A. The allowed standing-wave frequencies are $f_m = m(v/2L)$, so the mode number of a standing wave of frequency f is $m = 2Lf/v$. Quadrupling T_s increases the wave speed v by a factor of 2. The initial mode number was 2, so the new mode number is 1.

Stop to Think 17.3: B. 300 Hz and 400 Hz are allowed standing waves, but they are not f_1 and f_2 because $400\text{ Hz} \neq 2 \times 300\text{ Hz}$. Because there's a 100 Hz difference between them, these must be $f_3 = 3 \times 100$ Hz and $f_4 = 4 \times 100$ Hz, with a fundamental frequency $f_1 = 100$ Hz. Thus the second harmonic is $f_2 = 2 \times 100\text{ Hz} = 200\text{ Hz}$.

Stop to Think 17.4: C. Holding your nose does not affect the way your vocal cords vibrate, so it won't change the fundamental frequency or the harmonics of the vocal-cord vibrations. But it will change the frequencies of the formants of the vocal system: The nasal cavities will be open-closed tubes instead of open-open tubes. The frequencies that are amplified are altered, and your voice sounds quite different.

Stop to Think 17.5: C. Shifting the top wave 0.5 m to the left aligns crest with crest and trough with trough.

Stop to Think 17.6: A. $r_1 = 0.5\lambda$ and $r_2 = 2.5\lambda$, so $\Delta r = 2.0\lambda$. This is the condition for maximum constructive interference.

Stop to Think 17.7: Maximum constructive. The path-length difference is $\Delta r = 1.0\text{ m} = \lambda$. For identical sources, interference is constructive when Δr is an integer multiple of λ.

Stop to Think 17.8: F. The beat frequency is the difference between the two frequencies.

QUESTIONS

Conceptual Questions

1. Figure Q17.1 shows a standing wave oscillating on a string at frequency f_0.
   a. What mode (m-value) is this?
   b. How many antinodes will there be if the frequency is doubled to $2f_0$?

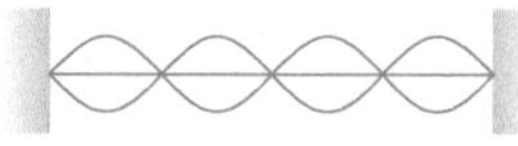

FIGURE Q17.1

2. If you take snapshots of a standing wave on a string, there are certain instants when the string is totally flat. What has happened to the energy of the wave at those instants?
3. A guitarist finds that the pitch of one of her strings is slightly flat—the frequency is a bit too low. Should she increase or decrease the tension of the string? Explain.
4. BIO Certain illnesses inflame your vocal cords, causing them to absorb water and swell. How does this affect the pitch of your voice? Explain.
5. Figure Q17.5 shows the displacement of a standing sound wave in a 32-cm-long horizontal tube of air open at both ends.
   a. What mode (m-value) is this?
   b. Are the air molecules moving horizontally or vertically? Explain.
   c. At what distances from the left end of the tube do the molecules oscillate with maximum amplitude?
   d. At what distances from the left end of the tube does the air pressure oscillate with maximum amplitude?

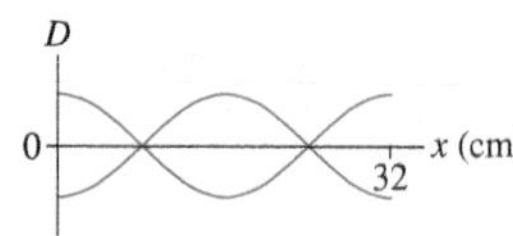

FIGURE Q17.5

Problem difficulty is labeled as | (straightforward) to ||||| (challenging). Problems labeled INT integrate significant material from earlier chapters; BIO are of biological or medical interest; CALC require calculus to solve.

6. An organ pipe is tuned to exactly 384 Hz when the room temperature is 20°C. If the room temperature later increases to 22°C, does the pipe's frequency increase, decrease, or stay the same? Explain.
7. A flute filled with helium will, until the helium escapes, play notes at a much higher pitch than normal. Why?
8. If you pour liquid into a tall, narrow glass, you may hear sound with a steadily rising pitch. What is the source of the sound? And why does the pitch rise as the glass fills?
9. In music, two notes are said to be an *octave* apart when one note is exactly twice the frequency of the other. Suppose you have a guitar string playing frequency f_0. To increase the frequency by an octave, to $2f_0$, by what factor would you have to (a) increase the tension or (b) decrease the length?
10. Figure Q17.10 shows frequency spectra of the same note played on a flute (modeled as an open-open tube) and on a clarinet (a closed-open tube). Which figure corresponds to the flute, and which to the clarinet? Explain.

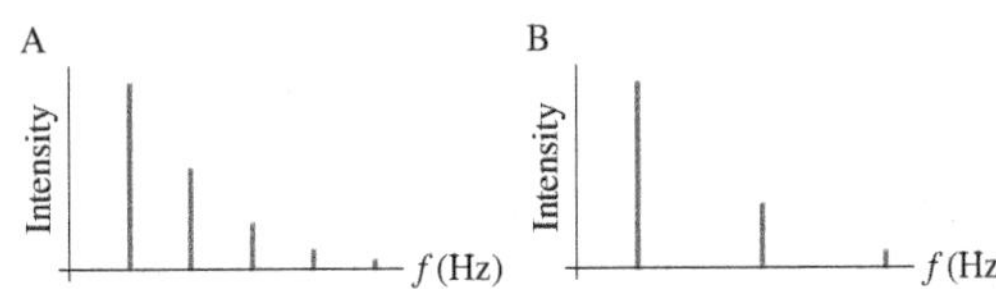

FIGURE Q17.10

11. Figure Q17.11 is a snapshot graph of two plane waves passing through a region of space. Each wave has a 2.0 mm amplitude and the same wavelength. What is the net displacement of the medium at points A, B, and C?

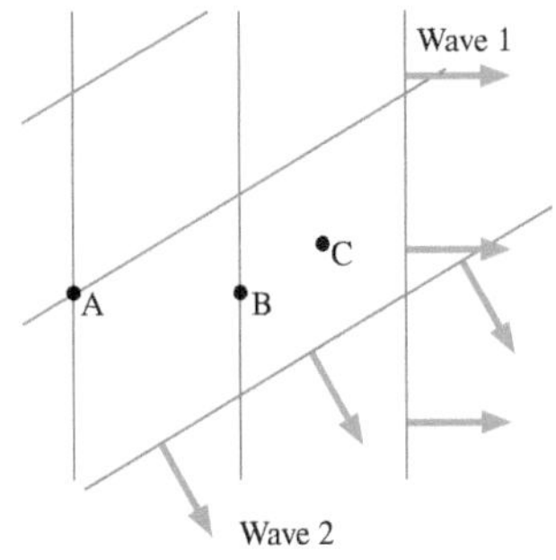

FIGURE Q17.11

FIGURE Q17.12

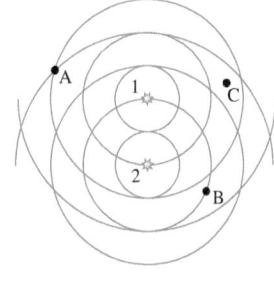

12. Figure Q17.12 shows the circular waves emitted by two in-phase sources. Are A, B, and C points of maximum constructive interference, maximum destructive interference, or in between?
13. A trumpet player hears 5 beats per second when she plays a note and simultaneously sounds a 440 Hz tuning fork. After pulling her tuning valve out to slightly increase the length of her trumpet, she hears 3 beats per second against the tuning fork. Was her initial frequency 435 Hz or 445 Hz? Explain.

Multiple-Choice Questions

Questions 14 through 16 refer to the snapshot graph of Figure Q17.14.

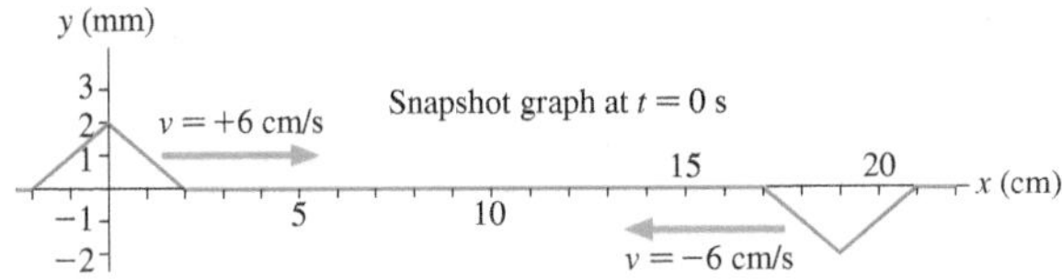

FIGURE Q17.14

14. | At $t = 1$ s, what is the displacement y of the string at $x = 7$ cm?
A. −1.0 mm B. 0 mm C. 0.5 mm
D. 1.0 mm E. 2.0 mm
15. | At $x = 3$ cm, what is the earliest time that y will equal 2 mm?
A. 0.5 s B. 0.7 s C. 1.0 s
D. 1.5 s E. 2.5 s
16. | At $t = 1.5$ s, what is the value of y at $x = 10$ cm?
A. −2.0 mm B. −1.0 mm C. −0.5 mm
D. 0 mm E. 1.0 mm
17. | A small wave pulse and a large wave pulse approach each other on a string; the large pulse is moving to the right. Some time after the pulses have met and passed each other, which of the following statements is correct? (More than one answer may be correct.) Explain.
A. The large pulse continues unchanged, moving to the right.
B. The large pulse continues moving to the right but is smaller in amplitude.
C. The small pulse is reflected and moves off to the right with its original amplitude.
D. The small pulse is reflected and moves off to the right with a smaller amplitude.
E. The two pulses combine into a single pulse moving to the right.
18. | In a tube, standing-wave modes are found at 200 Hz and 400 Hz. The tube could *not* be
A. Open-closed.
B. Open-open.
C. Closed-closed.
D. It could be any of these.
19. | A student in her physics lab measures the standing-wave modes of a tube. The lowest frequency that makes a resonance is 20 Hz. As the frequency is increased, the next resonance is at 60 Hz. What will be the next resonance after this?
A. 80 Hz
B. 100 Hz
C. 120 Hz
D. 180 Hz
20. | Resonances of the ear canal lead to increased sensitivity of
BIO hearing, as we've seen. Dogs have a much longer ear canal—5.2 cm—than humans. What are the two lowest frequencies at which dogs have an increase in sensitivity? The speed of sound in the warm air of the ear is 350 m/s.
A. 1700 Hz, 3400 Hz
B. 1700 Hz, 5100 Hz
C. 3400 Hz, 6800 Hz
D. 3400 Hz, 10,200 Hz
21. | The frequency of the lowest standing-wave mode on a 1.0-m-long string is 20 Hz. What is the wave speed on the string?
A. 10 m/s B. 20 m/s C. 30 m/s D. 40 m/s
22. | Suppose you pluck a string on a guitar and it produces the note A at a frequency of 440 Hz. Now you press your finger down on the string against one of the frets, making this point the new end of the string. The newly shortened string has 4/5 the length of the full string. When you pluck the string, its frequency will be
A. 350 Hz B. 440 Hz C. 490 Hz D. 550 Hz

PROBLEMS

Section 17.1 The Principle of Superposition

Section 17.2 Standing Waves

1. || Figure P17.1 is a snapshot graph at $t = 0$ s of two waves approaching each other at 1.0 m/s. List the values of the displacement at position $x = 5.0$ m at 1 s intervals from $t = 0$ s to $t = 6$ s.

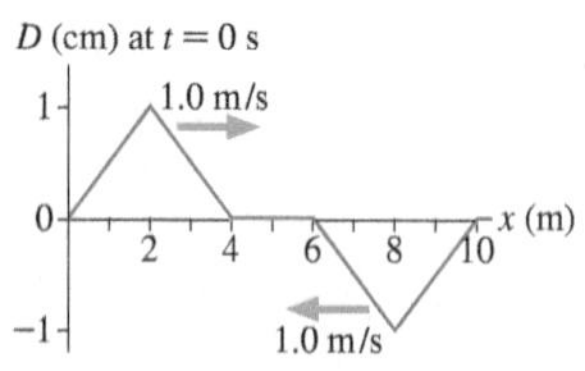

FIGURE P17.1

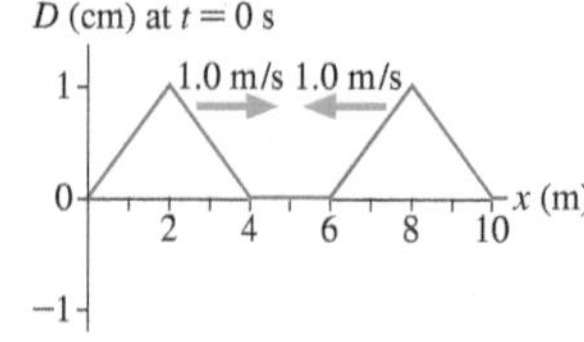

FIGURE P17.2

2. || Figure P17.2 is a snapshot graph at $t = 0$ s of two waves approaching each other at 1.0 m/s. Draw six snapshot graphs, stacked vertically, showing the string at 1 s intervals from $t = 1$ s to $t = 6$ s.

3. || Figure P17.3a is a snapshot graph at $t = 0$ s of two waves approaching each other at 1.0 m/s. At what time was the snapshot graph in Figure P17.3b taken?

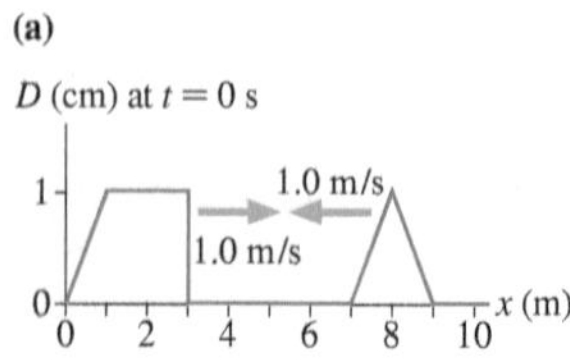

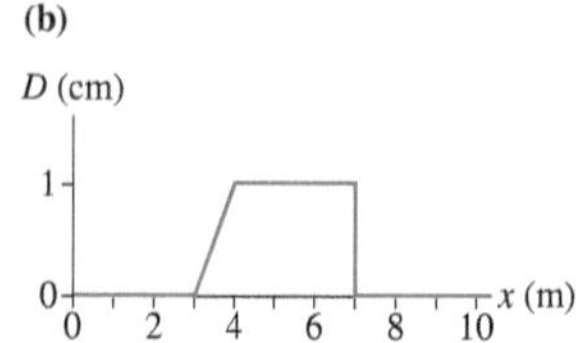

FIGURE P17.3

4. || Figure P17.4 is a snapshot graph at $t = 0$ s of two waves moving to the right at 1.0 m/s. The string is fixed at $x = 8.0$ m. Draw four snapshot graphs, stacked vertically, showing the string at $t = 2$, 4, 6, and 8 s.

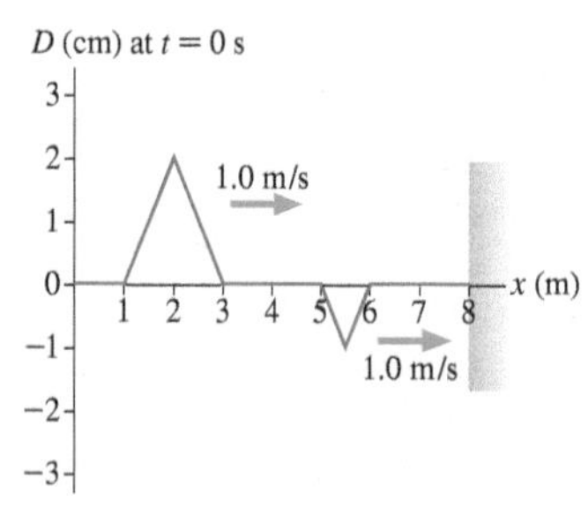

FIGURE P17.4

5. || Ocean waves of wavelength 26 m are moving directly toward a concrete barrier wall at 4.4 m/s. The waves reflect from the wall, and the incoming and reflected waves overlap to make a standing wave with an antinode at the wall. (Such waves are a common occurrence in certain places.) A kayaker is bobbing up and down with the water at the first antinode out from the wall. How far from the wall is she? What is the period of her up-and-down motion?

Section 17.3 Standing Waves on a String

6. || Figure P17.6 shows a standing wave oscillating at 100 Hz on a string. What is the wave speed?

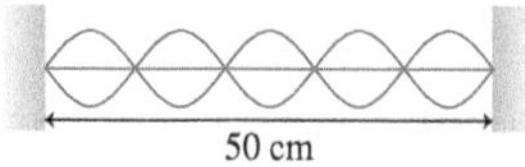

FIGURE P17.6

7. || Figure P17.7 shows a standing wave on a 2.0-m-long string that has been fixed at both ends and tightened until the wave speed is 40 m/s. What is the frequency?

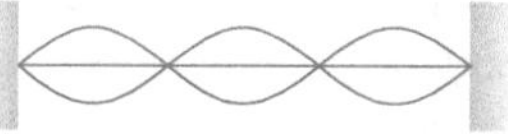

FIGURE P17.7

8. | Figure P17.8 shows a standing wave on a string that is oscillating at 100 Hz.

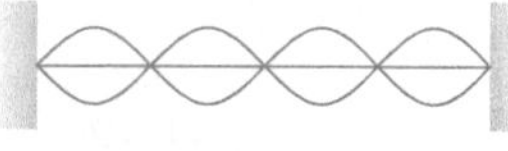

FIGURE P17.8

   a. How many antinodes will there be if the frequency is increased to 200 Hz?
   b. If the tension is increased by a factor of 4, at what frequency will the string continue to oscillate as a standing wave that looks like the one in the figure?

9. | a. What are the three longest wavelengths for standing waves on a 240-cm-long string that is fixed at both ends?
   b. If the frequency of the second-longest wavelength is 50 Hz, what is the frequency of the third-longest wavelength?

10. | Standing waves on a 1.0-m-long string that is fixed at both ends are seen at successive frequencies of 36 Hz and 48 Hz.
    a. What are the fundamental frequency and the wave speed?
    b. Draw the standing-wave pattern when the string oscillates at 48 Hz.

11. || The two highest-pitch strings on a violin are tuned to 440 Hz (the A string) and 659 Hz (the E string). What is the ratio of the mass of the A string to that of the E string? Violin strings are all the same length and under essentially the same tension.

12. || The lowest note on a five-string bass guitar has a frequency of 31 Hz. The vibrating length of string is 89 cm long.
    a. What is the wave speed on this string?
    b. What tension is needed to tune this string if the string's linear mass density is 59 g/m?

13. ||| The G string on a guitar is 59 cm long and has a fundamental frequency of 196.0 Hz. A guitarist can play different notes by pushing the string against various frets, which changes the string's length. The first fret from the neck gives the musical note A-flat (207.7 Hz); the second fret gives A (220.0 Hz). What is the distance between the first and second frets?

14. || The lowest note on a grand piano has a frequency of 27.5 Hz. The entire string is 2.00 m long and has a mass of 400 g. The vibrating section of the string is 1.90 m long. What tension is needed to tune this string properly?

15. || A violin string is 30 cm long. It sounds the musical note A (440 Hz) when played without fingering. How far from the end of the string should you place your finger to play the note C (523 Hz)?

16. ||| INT A steel wire is used to stretch the spring of Figure P17.16. An oscillating magnetic field drives the steel wire back and forth. A standing wave with three antinodes is created when the spring is stretched 8.0 cm. What stretch of the spring produces a standing wave with two antinodes?

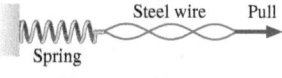

Steel wire Pull Spring

FIGURE P17.16

Section 17.4 Standing Sound Waves

17. | The lowest frequency in the audible range is 20 Hz. What are the lengths of (a) the shortest open-open tube and (b) the shortest open-closed tube needed to produce this frequency?

18. | What are the three longest wavelengths for standing sound waves in a 121-cm-long tube that is (a) open at both ends and (b) open at one end, closed at the other?

19. | Figure P17.19 shows a standing sound wave in an 80-cm-long tube. The tube is filled with an unknown gas. What is the speed of sound in this gas?

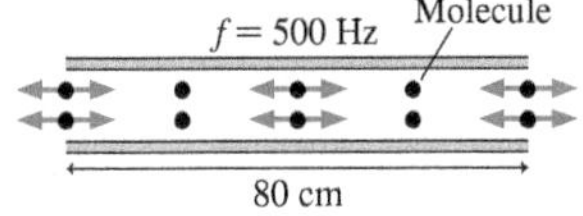

FIGURE P17.19

20. | The fundamental frequency of an open-open tube is 1500 Hz when the tube is filled with 0°C helium. What is its frequency when filled with 0°C air?

21. | The world's longest organ pipe, in the Boardwalk Hall Auditorium in Atlantic City, is 64 feet long. What is the fundamental frequency of this open-open pipe?

22. | An organ pipe is made to play a low note at 27.5 Hz, the same as the lowest note on a piano. Assuming a sound speed of 343 m/s, what length open-open pipe is needed? What length open-closed pipe would suffice?

23. | *Parasaurolophus* was a dinosaur whose distinguishing feature was a hollow crest on the head. The 1.5-m-long hollow tube in the crest had connections to the nose and throat, leading some investigators to hypothesize that the tube was a resonant chamber for vocalization. If you model the tube as an open-closed system, what are the first three resonant frequencies? Assume a speed of sound of 350 m/s.
BIO

24. | A bass clarinet can be modeled as a 120-cm-long open-closed tube. A bass clarinet player starts playing in a 20°C room, but soon the air inside the clarinet warms to where the speed of sound is 352 m/s. Does the fundamental frequency increase or decrease? By how much?

25. || To study the physical basis of underwater hearing in frogs, scientists used a vertical tube filled with 20°C water to a depth of 1.4 m. A microphone at the bottom of the tube was used to create standing sound waves in the water column. Frogs were lowered to different depths where the standing waves created large or small pressure variations. Because the microphone creates the sound, the bottom of the tube is a pressure antinode; the water's surface, fixed at atmospheric pressure, is a node.
BIO
    a. What is the fundamental frequency of this water-filled tube?
    b. A frog sits on a platform located 0.28 m from the bottom. What is the lowest frequency that would result in a sound node at this point?

26. | You know that you sound better when you sing in the shower. This has to do with the amplification of frequencies that correspond to the standing-wave resonances of the shower enclosure. Suppose a 85 cm × 95 cm × 210 cm shower stall is completely enclosed when the door is closed. Standing sound waves can be set up along any axis of the enclosure. What are the two lowest frequencies that correspond to resonances on each axis of the shower? These frequencies will be especially amplified. Assume a sound speed of 350 m/s.

27. || When a sound wave travels directly toward a hard wall, the incoming and reflected waves can combine to produce a standing wave. There is a pressure antinode right at the wall, just as at the end of a closed tube, so the sound near the wall is loud. You are standing beside a brick wall listening to a 50 Hz tone from a distant loudspeaker. How far from the wall must you move to find the first quiet spot? Assume a sound speed of 340 m/s.

28. | A longitudinal standing wave can be created in a long, thin aluminum rod by stroking the rod with very dry fingers. This is often done as a physics demonstration, creating a high-pitched, very annoying whine. From a wave perspective, the standing wave is equivalent to a sound standing wave in an open-open tube. As Figure P17.28 shows, both ends of the rod are antinodes. What is the fundamental frequency of a 2.0-m-long aluminum rod?

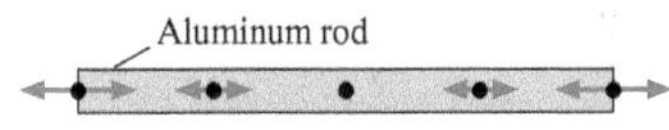

FIGURE P17.28

29. || An open-open organ pipe is 78.0 cm long. An open-closed pipe has a fundamental frequency equal to the third harmonic of the open-open pipe. How long is the open-closed pipe?

30. || The 40-cm-long tube of Figure P17.30 has a 40-cm-long insert that can be pulled in and out. A vibrating tuning fork is held next to the tube. As the insert is slowly pulled out, the sound from the tuning fork creates standing waves in the tube when the total length L is 42.5 cm, 56.7 cm, and 70.9 cm. What is the frequency of the tuning fork? Assume $v_{sound} = 343$ m/s.

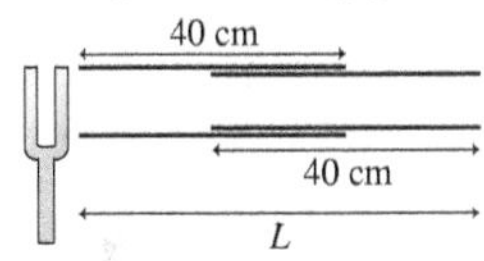

FIGURE P17.30

Section 17.5 The Physics of Speech

31. || The simplest possible model of the vocal tract is an open-closed tube, with the closed end being your vocal cords and the open end your lips.
BIO
    a. Estimate the frequency of the first formant of the "ee" sound in Figure 17.18a and then determine the length of an open-closed tube for which this is the fundamental frequency. Assume a sound speed of 350 m/s.
    b. Does your result seem reasonable? To answer, use a ruler to make a rough measurement of the length of your own vocal tract.
    c. Is an open-closed tube an acceptable model of the vocal tract when making the vowel sound "ee"?

32. || The first and second formants when you make an "ee" vowel sound are approximately 270 Hz and 2300 Hz. The speed of sound in your vocal tract is approximately 350 m/s. If you breathe a mix of oxygen and helium (as deep-sea divers often do), the speed increases to 750 m/s. With this mix of gases, what are the frequencies of the two formants?
BIO

33. || Figure P17.33 shows the two lowest resonances recorded in BIO the vocal tract of the eastern towhee, a small songbird.
   a. Is this bird's vocal tract better modeled as an open-open tube or an open-closed tube?
   b. Estimate the length of the towhee's vocal tract. Assume a sound speed of 350 m/s.

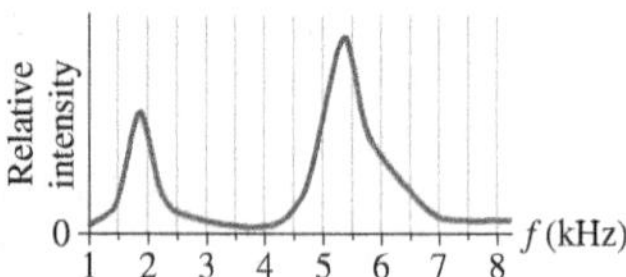

FIGURE P17.33

Section 17.6 Interference Along a Line

34. | Two loudspeakers emit sound waves along the x-axis. The sound has maximum intensity when the speakers are 20 cm apart. The sound intensity decreases as the distance between the speakers is increased, reaching zero at a separation of 60 cm.
   a. What is the wavelength of the sound?
   b. If the distance between the speakers continues to increase, at what separation will the sound intensity again be a maximum?
35. || Two loudspeakers in a 20°C room emit 686 Hz sound waves along the x-axis.
   a. If the speakers are in phase, what is the smallest distance between the speakers for which the interference of the sound waves to the right of the speakers is maximum destructive?
   b. If the speakers are out of phase, what is the smallest distance between the speakers for which the interference of the sound waves is maximum constructive?
36. ||| Two loudspeakers, 1.0 m apart, emit sound waves with the same frequency along the positive x-axis. Victor, standing on the axis to the right of the speakers, hears no sound. As the frequency is slowly tripled, Victor hears the sound go through the sequence loud-soft-loud-soft-loud before becoming quiet again. What was the original sound frequency?
37. || In noisy factory environments, it's possible to use a loudspeaker to cancel persistent low-frequency machine noise at the position of one worker. The details of practical systems are complex, but we can present a simple example that gives you the idea. Suppose a machine 5.0 m away from a worker emits a persistent 85 Hz hum. This hum can be canceled by installing a loudspeaker between the worker and the machine that exactly duplicates the machine's hum. How far from the worker should the speaker be placed? Assume a sound speed of 340 m/s.
38. ||| Two in-phase loudspeakers emit identical 1000 Hz sound waves along the x-axis. What distance should one speaker be placed behind the other for the sound on the x-axis to have an amplitude 1.5 times that of each speaker alone?
39. ||| Two loudspeakers emit sound waves of the same frequency along the x-axis. The amplitude of each wave is a. The sound intensity is minimum when speaker 2 is 10 cm behind speaker 1. The intensity increases as speaker 2 is moved forward and first reaches maximum, with amplitude $2a$, when it is 30 cm in front of speaker 1. What is
   a. The wavelength of the sound?
   b. The phase difference between the two loudspeakers?
   c. The amplitude of the sound (as a multiple of a) if the speakers are placed side by side?

Section 17.7 Interference in Two and Three Dimensions

40. || Figure P17.40 shows the circular wave fronts emitted by two wave sources.
   a. Are these sources in phase or out of phase? Explain.
   b. Make a table with rows labeled P, Q, and R and columns labeled r_1, r_2, Δr, and C/D. Fill in the table for points P, Q, and R, giving the distances as multiples of λ and indicating, with a C or a D, whether the interference at that point is constructive or destructive.

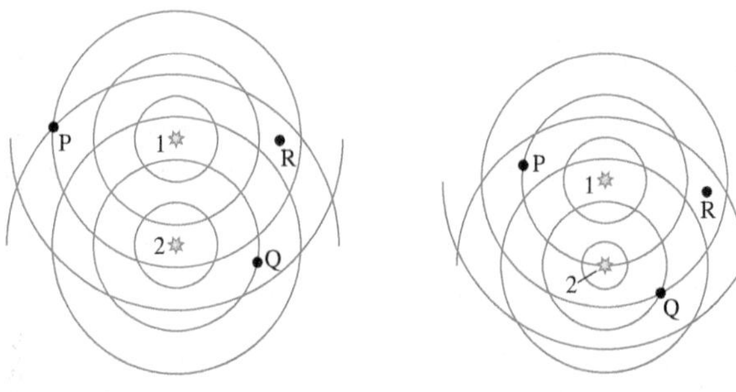

FIGURE P17.40　　FIGURE P17.41

41. ||| Figure P17.41 shows the circular wave fronts emitted by two wave sources.
   a. Are these sources in phase or out of phase? Explain.
   b. Make a table with rows labeled P, Q, and R and columns labeled r_1, r_2, Δr, and C/D. Fill in the table for points P, Q, and R, giving the distances as multiples of λ and indicating, with a C or a D, whether the interference at that point is constructive or destructive.
42. || Two in-phase speakers 2.0 m apart in a plane are emitting 1800 Hz sound waves into a room where the speed of sound is 340 m/s. Is the point 4.0 m in front of one of the speakers, perpendicular to the plane of the speakers, a point of maximum constructive interference, maximum destructive interference, or something in between?
43. || Two identical loudspeakers 2.0 m apart are emitting sound waves into a room where the speed of sound is 340 m/s. Abby is standing 5.0 m in front of one of the speakers, perpendicular to the line joining the speakers, and hears a maximum in the intensity of the sound. What is the lowest possible frequency of sound for which this is possible?
44. ||| Two in-phase loudspeakers, which emit sound in all directions, are sitting side by side. One of them is moved sideways by 3.0 m, then forward by 4.0 m. Afterward, constructive interference is observed $\frac{1}{4}$ and $\frac{3}{4}$ of the distance between the speakers along the line that joins them. What is the maximum possible wavelength of the sound waves?
45. || Two out-of-phase radio antennas at $x = \pm 300$ m on the x-axis are emitting 3.0 MHz radio waves. Is the point $(x, y) = (300 \text{ m}, 800 \text{ m})$ a point of maximum constructive interference, maximum destructive interference, or something in between?

Section 17.8 Beats

46. | Two strings are adjusted to vibrate at exactly 200 Hz. Then the tension in one string is increased slightly. Afterward, three beats per second are heard when the strings vibrate at the same time. What is the new frequency of the string that was tightened?

47. | A flute player hears four beats per second when she compares her note to a 523 Hz tuning fork (the note C). She can match the frequency of the tuning fork by pulling out the "tuning joint" to lengthen her flute slightly. What was her initial frequency?

48. | Traditional Indonesian music uses an ensemble called a *gamelan* that is based on tuned percussion instruments somewhat like gongs. In Bali, the gongs are often grouped in pairs that are slightly out of tune with each other. When both are played at once, the beat frequency lends a distinctive vibrating quality to the music. Suppose a pair of gongs are tuned to produce notes at 151 Hz and 155 Hz. How many beats are heard if the gongs are struck together and both ring for 2.5 s?

49. || Two microwave signals of nearly equal wavelengths can generate a beat frequency if both are directed onto the same microwave detector. In an experiment, the beat frequency is 100 MHz. One microwave generator is set to emit microwaves with a wavelength of 1.250 cm. If the second generator emits the longer wavelength, what is that wavelength?

50. ||| A student waiting at a stoplight notices that her turn signal, which has a period of 0.85 s, makes one blink exactly in sync with the turn signal of the car in front of her. The blinker of the car ahead then starts to get ahead, but 17 s later the two are exactly in sync again. What is the period of the blinker of the other car?

51. |||| Figure P17.51 shows the superposition of two sound waves. What are their frequencies?

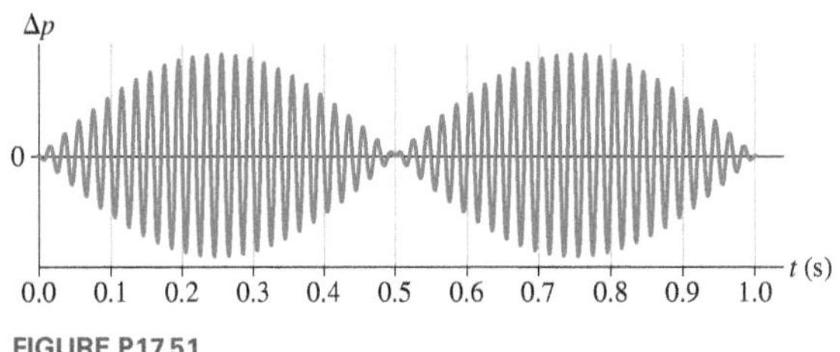

FIGURE P17.51

52. ||| Two loudspeakers emit 400 Hz notes. One speaker sits on the ground. The other speaker is in the back of a pickup truck. You hear eight beats per second as the truck drives away from you. What is the truck's speed?
INT

General Problems

53. ||| An 80-cm-long steel string with a linear density of 1.0 g/m is under 200 N tension. It is plucked and vibrates at its fundamental frequency. What is the wavelength of the sound wave that reaches your ear in a 20°C room?
INT

54. || A string vibrates at its third-harmonic frequency. The amplitude at a point 30 cm from one end is half the maximum amplitude. How long is the string?

55. || Tendons are, essentially, elastic cords stretched between two fixed ends. As such, they can support standing waves. A woman has a 20-cm-long Achilles tendon—connecting the heel to a muscle in the calf—with a cross-section area of 90 mm^2. The density of tendon tissue is 1100 kg/m^3. For a reasonable tension of 500 N, what will be the fundamental frequency of her Achilles tendon?
BIO

56. ||| Biologists think that some spiders "tune" strands of their web to give enhanced response at frequencies corresponding to those at which desirable prey might struggle. Orb spider web silk has a typical diameter of 20 μm, and spider silk has a density of 1300 kg/m^3. To have a fundamental frequency at 100 Hz, to what tension must a spider adjust a 12-cm-long strand of silk?
BIO

57. || A particularly beautiful note reaching your ear from a rare Stradivarius violin has a wavelength of 39.1 cm. The room is slightly warm, so the speed of sound in air is 344 m/s. If the string's linear density is 0.600 g/m and the tension is 150 N, how long is the vibrating section of the violin string?

58. |||| A guitar player can change the frequency of a string by "bending" it—pushing it along a fret that is perpendicular to its length. This stretches the string, increasing its tension and its frequency. The G string on a guitar is 64 cm long and has a tension of 74 N. The guitarist pushes this string down against a fret located at the center of the string, which gives it a frequency of 392 Hz. He then bends the string, pushing with a force of 4.0 N so that it moves 8.0 mm along the fret. What is the new frequency?
INT

59. || Lake Erie is prone to remarkable *seiches*—standing waves that slosh water back and forth in the lake basin from the west end at Toledo to the east end at Buffalo. Figure P17.59 shows smoothed data for the displacement from normal water levels along the lake at the high point of one particular seiche. 3 hours later the water was at normal levels throughout the basin; 6 hours later the water was high in Toledo and low in Buffalo.
 a. What is the wavelength of this standing wave?
 b. What is the frequency?
 c. What is the wave speed?

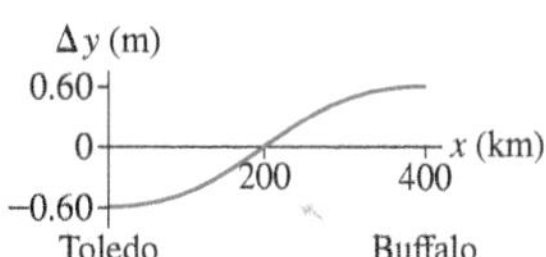

FIGURE P17.59

60. || Microwaves are electromagnetic waves. Microwaves confined between two parallel metal plates can form standing waves that are exactly analogous to standing waves on a string—that is, a node at each end and a node spacing of $\lambda/2$. The designers of microwave ovens try to avoid the conditions for standing waves because antinodes and nodes are, respectively, hot spots and cold spots for cooking food. A microwave oven that is 30.5 cm wide would be an especially poor design because it will support a standing wave with six nodes: two at the edges and four in between. What is the frequency in MHz of the microwave oven?
INT

61. |||| A 75 g bungee cord has an equilibrium length of 1.20 m. The cord is stretched to a length of 1.80 m, then vibrated at 20 Hz. This produces a standing wave with two antinodes. What is the spring constant of the bungee cord?
INT

62. || What is the fundamental frequency of the steel wire in Figure P17.62?
INT

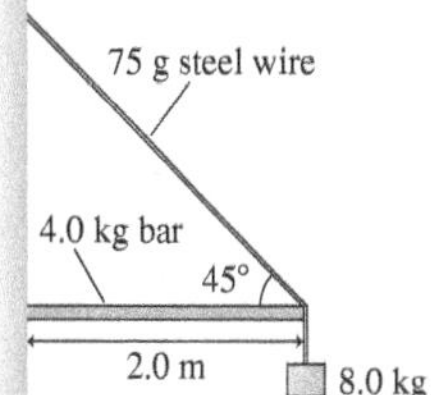

FIGURE P17.62

63. In 1866, the German scientist Adolph Kundt developed a technique for accurately measuring the speed of sound in various gases. A long glass tube, known today as a Kundt's tube, has a vibrating piston at one end and is closed at the other. Very finely ground particles of cork are sprinkled in the bottom of the tube before the piston is inserted. As the vibrating piston is slowly moved forward, there are a few positions that cause the cork particles to collect in small, regularly spaced piles along the bottom. Figure P17.63 shows an experiment in which the tube is filled with pure oxygen and the piston is driven at 400 Hz. What is the speed of sound in oxygen?

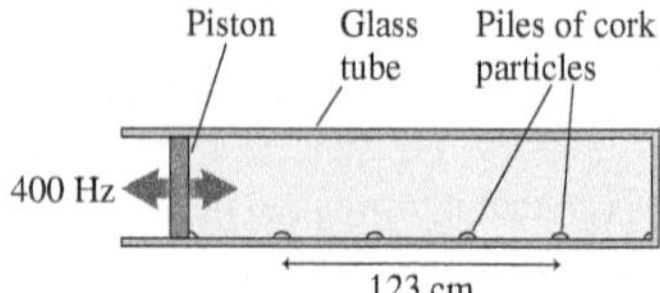

FIGURE P17.63

64. Male Bornean tree-hole frogs, like other frogs, call to attract
BIO females for mating. Investigators suspected that these frogs adjust the frequency of their calls to match the resonance frequency of the hole they inhabit, thus enhancing the volume of their calls. To test this idea, investigators placed a frog in a vertical plastic tube partially filled with water but open at the top. As the water depth varied, the frequency of the frog's call did, indeed, change to match the fundamental frequency of the tube. In one test, the frog's call dropped from 925 Hz to 810 Hz. What lengths of the air column do these frequencies correspond to? Assume a sound speed of 350 m/s.

65. American alligators increase their rate of bellowing dur-
BIO ing mating season, and part of the bellow is to advertise the size of the gator. The fundamental frequency of the bellow, which is determined by the tissues that vibrate, doesn't vary with size. However, the frequencies of the formants, which are determined by the length of the vocal tract, do vary. Lower-frequency formants mean a larger gator. Investigators found a nearly perfect correlation between the size of alligators and the *difference* in frequency between the first and second formants—the first and second resonances of the vocal tract. For a gator with head length 40 cm, this difference was 300 Hz. What length vocal tract is indicated by this measurement? Model the vocal tract as a simple tube closed at the glottis (the source of the vibrations) and open at the nostrils. Alligators are cold-blooded but live in warm environments, so assume a sound speed of 350 m/s.

66. Some caterpillars, when under
BIO attack by birds, emit high-pitched whistles by sharply forcing air through openings in their thorax. Figure P17.66 is an experimental graph of intensity versus frequency for a caterpillar whistle.

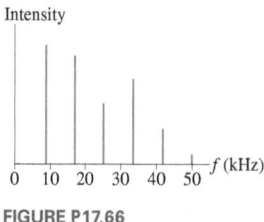

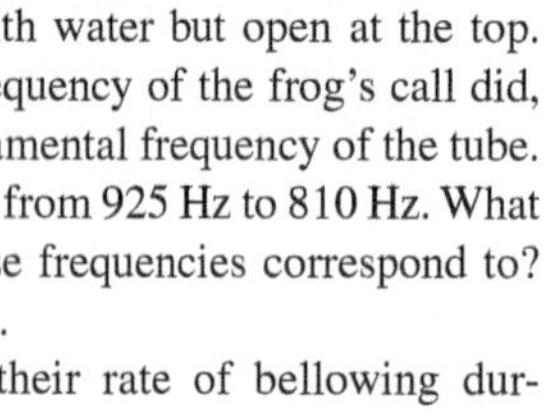

FIGURE P17.66

a. Is the structure that produces the whistle an open-open or open-closed tube? Explain. Assume a sound speed of 340 m/s.
b. What is the length of the tube?

67. A 1.0-m-tall vertical tube is filled with 20°C water. A tuning fork vibrating at 580 Hz is held just over the top of the tube as the water is slowly drained from the bottom. At what water heights, measured from the bottom of the tube, will there be a standing wave in the tube above the water?

68. An old mining tunnel disappears into a hillside. You would like to know how long the tunnel is, but it's too dangerous to go inside. Recalling your recent physics class, you decide to try setting up standing-wave resonances inside the tunnel. Using your subsonic amplifier and loudspeaker, you find resonances at 4.5 Hz and 6.3 Hz, and at no frequencies between these. It's rather chilly inside the tunnel, so you estimate the sound speed to be 335 m/s. Based on your measurements, how far is it to the end of the tunnel?

69. A water wave is called a *deep-water wave* if the water's depth is more than one-quarter of the wavelength. Unlike the waves we've considered in this chapter, the speed of a deep-water wave depends on its wavelength:

$$v = \sqrt{\frac{g\lambda}{2\pi}}$$

Longer wavelengths travel faster. Let's apply this to standing waves. Consider a diving pool that is 5.0 m deep and 10.0 m wide. Standing water waves can set up across the width of the pool. Because water sloshes up and down at the sides of the pool, the boundary conditions require antinodes at $x = 0$ and $x = L$. Thus a standing water wave resembles a standing sound wave in an open-open tube.

a. What are the wavelengths of the first three standing-wave modes for water in the pool? Do they satisfy the condition for being deep-water waves?
b. What are the wave speeds for each of these waves?

70. When mass M is tied to the bottom of a long, thin wire suspended from the ceiling, the wire's second-harmonic frequency is 200 Hz. Adding an additional 1.0 kg to the hanging mass increases the second-harmonic frequency to 245 Hz. What is M?

71. You are standing 2.5 m directly in front of one of the two loudspeakers shown in Figure P17.71. They are 3.0 m apart and both are playing a 686 Hz tone in phase. As you begin to walk directly away from the speaker, at what distances from the speaker do you hear a *minimum* sound intensity? The room temperature is 20°C.

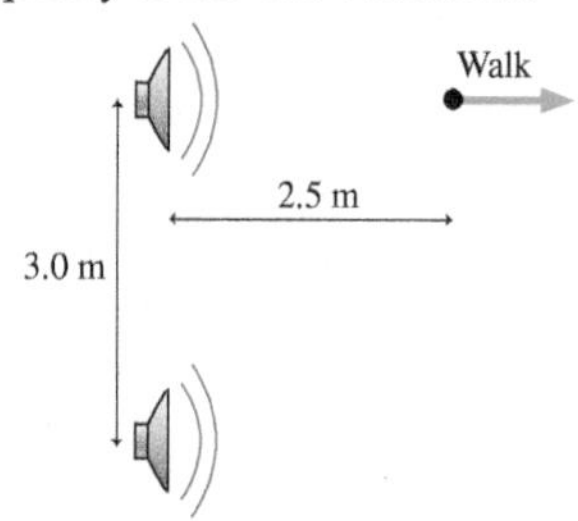

FIGURE P17.71

72. Two loudspeakers in a plane, 5.0 m apart, are playing the same frequency. If you stand 12.0 m in front of the plane of the speakers, centered between them, you hear a sound of maximum intensity. As you walk parallel to the plane of the speakers, staying 12.0 m in front of them, you first hear a minimum of sound intensity when you are directly in front of one of the speakers. What is the frequency of the sound? Assume a sound speed of 340 m/s.

73. Piano tuners tune pianos by listening to the beats between the *harmonics* of two different strings. When properly tuned, the note A should have a frequency of 440 Hz and the note E should be at 659 Hz.

a. What is the frequency difference between the third harmonic of the A and the second harmonic of the E?
b. A tuner first tunes the A string very precisely by matching it to a 440 Hz tuning fork. She then strikes the A and E strings simultaneously and listens for beats between the harmonics. What beat frequency indicates that the E string is properly tuned?
c. The tuner starts with the tension in the E string a little low, then tightens it. What is the frequency of the E string when she hears four beats per second?

74. | | A flutist assembles her flute in a room where the speed of sound is 342 m/s. When she plays the note A, it is in perfect tune with a 440 Hz tuning fork. After a few minutes, the air inside her flute has warmed to where the speed of sound is 346 m/s.
    a. How many beats per second will she hear if she now plays the note A as the tuning fork is sounded?
    b. How far does she need to extend the "tuning joint" of her flute to be in tune with the tuning fork?

75. | | A Doppler blood flowmeter emits ultrasound at a frequency of 5.0 MHz. What is the beat frequency between the emitted waves and the waves reflected from blood cells moving away from the emitter at 0.15 m/s?
BIO INT

76. | An ultrasound unit is being used to measure a patient's heartbeat by combining the emitted 2.0 MHz signal with the sound waves reflected from the moving tissue of one point on the heart. The beat frequency between the two signals has a maximum value of 520 Hz. What is the maximum speed of the heart tissue?
BIO INT

MCAT-Style Passage Problems

Harmonics and Harmony

You know that certain musical notes sound good together—harmonious—whereas others do not. This harmony is related to the various harmonics of the notes.

The musical notes C (262 Hz) and G (392 Hz) make a pleasant sound when played together; we call this consonance. As Figure P17.77 shows, the harmonics of the two notes are either far from each other or very close to each other (within a few Hz). This is the key to consonance: harmonics that are spaced either far apart or very close. The close harmonics have a beat frequency of a few Hz that is perceived as pleasant. If the harmonics of two notes are close but not too close, the rather high beat frequency between the two is quite unpleasant. This is what we hear as dissonance. Exactly how much a difference is maximally dissonant is a matter of opinion, but harmonic separations of 30 or 40 Hz seem to be quite unpleasant for most people.

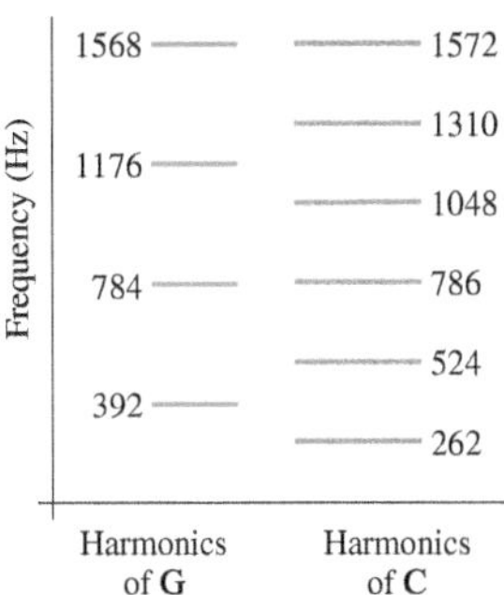

FIGURE P17.77

77. What is the beat frequency between the second harmonic of G and the third harmonic of C?
    A. 1 Hz
    B. 2 Hz
    C. 4 Hz
    D. 6 Hz

78. Would a G-flat (frequency 370 Hz) and a C played together be consonant or dissonant?
    A. Consonant
    B. Dissonant

79. An organ pipe open at both ends is tuned so that its fundamental frequency is a G. How long is the pipe?
    A. 43 cm
    B. 87 cm
    C. 130 cm
    D. 173 cm

80. If the C were played on an organ pipe that was open at one end and closed at the other, which of the harmonic frequencies in Figure P17.77 would be present?
    A. All of the harmonics in the figure would be present.
    B. 262, 786, and 1310 Hz
    C. 524, 1048, and 1572 Hz
    D. 262, 524, and 1048 Hz

PART III SUMMARY

Oscillations and Waves

FUNDAMENTAL CONCEPTS

All simple harmonic motion (SHM) can be described with the same mathematical equations.

Unlike particles, which are localized and discrete, waves are diffuse and spread out.

Two waves can pass through each other, causing standing waves and interference.

GENERAL PRINCIPLES

Linear restoring force An object oscillates about an equilibrium position in simple harmonic motion when it is subjected to a linear restoring force $F_x = -kx$.

Conservation of energy For SHM: $E = \frac{1}{2}mv^2 + \frac{1}{2}kx^2 = \frac{1}{2}m(v_{max})^2 = \frac{1}{2}kA^2$ For waves: $I = P/4\pi r^2$

Fundamental relationship for sinusoidal waves $v = \lambda f = \omega/k$

Principle of superposition The net displacement at each point is the sum of the displacements due to each wave.

SIMPLE HARMONIC MOTION

An object with a linear restoring force can undergo SHM. This is sinusoidal motion with

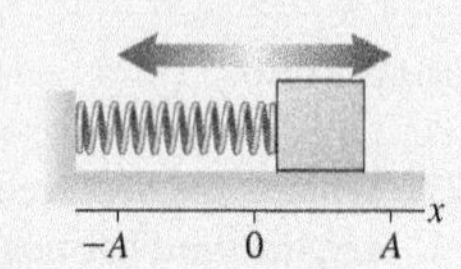

$$x = A\cos(\omega t + \phi_0)$$

$$v_x = -v_{max}\sin(\omega t + \phi_0) \qquad v_{max} = \omega A$$

Mechanical energy is conserved if there is no friction.

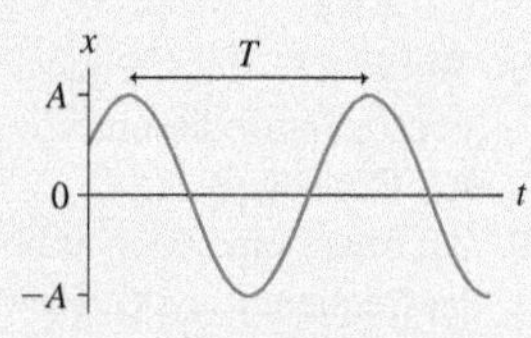

With friction, the oscillations decay. A simple model of a damped oscillator predicts that the oscillations decay exponentially with time.

$$x_{max} = Ae^{-t/\tau}$$

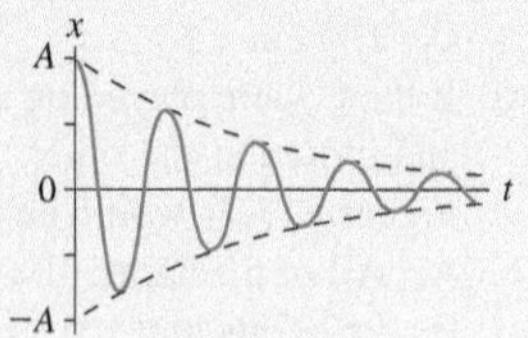

Resonance is a large-amplitude response when an oscillator is driven at its natural frequency.

WAVES

A wave is a disturbance that travels.

- *Mechanical waves* travel through a medium.
- *Electromagnetic waves* travel through a vacuum.
- Waves can be transverse or longitudinal.

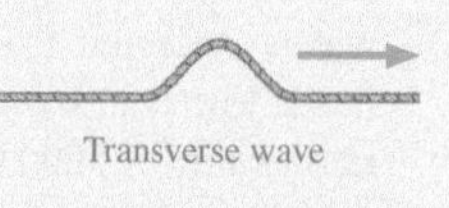
Transverse wave

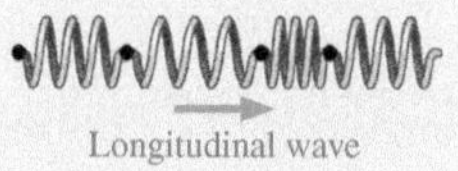
Longitudinal wave

Waves are functions of both time and position.

Wave speed is a property of the medium.

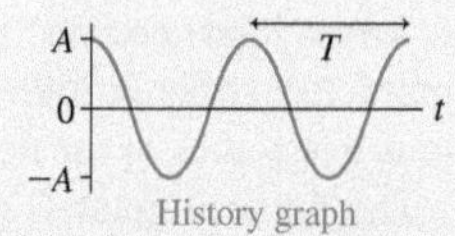

History graph

Sinusoidal waves are periodic in both time (period T) and space (wavelength λ). They obey $v = \lambda f$.

Waves obey the principle of superposition.

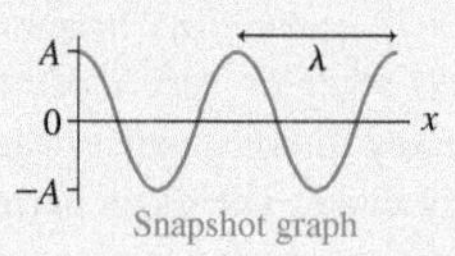

Snapshot graph

OSCILLATION PERIOD

Frequency $f = 1/T$

Angular frequency $\omega = 2\pi f = 2\pi/T$

Spring $T = 2\pi\sqrt{m/k}$

Pendulum $T = 2\pi\sqrt{L/g}$

Wave $f = v/\lambda$

SINUSOIDAL WAVES

Displacement is a function of x and t:

$$D = (x, t) = A\sin(kx \mp \omega t + \phi_0)$$

$-\omega t$ for motion to the right

$+\omega t$ for motion to the left

The *wave number* is $k = 2\pi/\lambda$.

WAVE SPEED

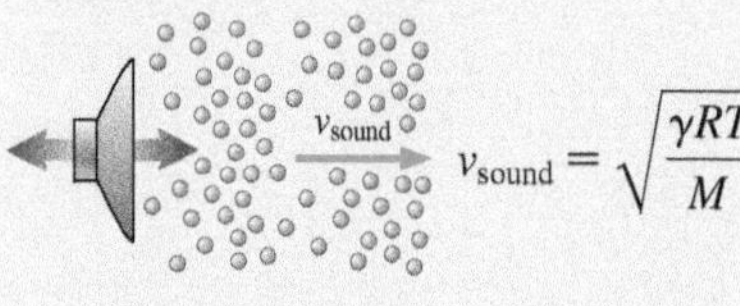

$$v_{sound} = \sqrt{\frac{\gamma RT}{M}}$$

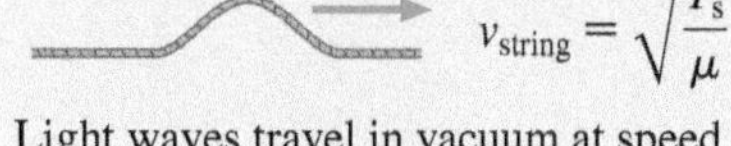

$$v_{string} = \sqrt{\frac{T_s}{\mu}}$$

Light waves travel in vacuum at speed

$$v_{light} = c = 3.00 \times 10^8 \text{ m/s}$$

PHASE

The quantity $\phi = \omega t + \phi_0$ is called the phase of simple harmonic motion.

The quantity $\phi = kx - \omega t + \phi_0$ is the phase of a sinusoidal wave.

The phase constant ϕ_0 is determined by the initial conditions.

DOPPLER EFFECT

Frequency shifts when a sound source moves relative to an observer:

Approaching $(-)$ or receding $(+)$ source:

$$f = f_0/(1 \mp v_{source}/v_{sound})$$

Observer approaching $(+)$ or receding $(-)$:

$$f = (1 \pm v_{observer}/v_{sound})f_0$$

Frequency shift upon reflection:

$$\Delta f = -(2v_{object}\cos\theta/v_{sound})f_0$$

SOUND INTENSITY LEVEL

$$\beta = (10 \text{ dB})\log_{10}(I/I_0)$$

where $I_0 = 10^{-12}$ W/m^2 is the threshold of hearing at frequencies near 1000 Hz.

BEATS

The superposition of waves with different frequencies causes a beat frequency

$$f_{beat} = |f_1 - f_2|$$

STANDING WAVES

Standing waves are due to the superposition of two waves traveling in opposite directions.

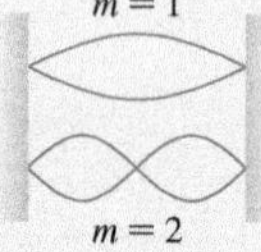

Points that never move are called nodes. Node spacing is $\lambda/2$.

Strings, open-open tubes, closed-closed tubes:

$$f_m = mf_1 \qquad m = 1, 2, 3, 4, \ldots$$

$$f_1 = v/2L = \text{ fundamental frequency}$$

Open-closed tubes have only odd harmonics:

$$f_m = mf_1 \qquad m = 1, 3, 5, 7, \ldots$$

$$f_1 = v/4L = \text{ fundamental frequency}$$

INTERFERENCE

Constructive interference if waves in phase:

Crests are aligned with crests

$\Delta\phi = m \cdot 2\pi$ rad

$\Delta r = m\lambda$ for identical sources

Amplitude $A = 2a$

Destructive interference if waves out of phase:

Crests are aligned with troughs

$\Delta\phi = (m + \frac{1}{2}) \cdot 2\pi$ rad

$\Delta r = (m + \frac{1}{2})\lambda$ for identical sources

Amplitude $A = 0$

Waves in the Earth and the Ocean

In December 2004, a large earthquake off the coast of Indonesia produced a devastating water wave, called a *tsunami,* that caused tremendous destruction thousands of miles away from the earthquake's epicenter. The tsunami was a dramatic illustration of the energy carried by waves.

It was also a call to action. Many of the communities hardest hit by the tsunami were struck hours after the waves were generated, long after seismic waves from the earthquake that passed through the earth had been detected at distant recording stations, long after the possibility of a tsunami was first discussed. With better detection and more accurate models of how a tsunami is formed and how a tsunami propagates, the affected communities could have received advance warning. The study of physics may seem an abstract undertaking with few practical applications, but on this day a better scientific understanding of these waves could have averted tragedy.

Let's use our knowledge of waves to explore the properties of a tsunami. In Chapter 16, we saw that a vigorous shake of one end of a rope causes a pulse to travel along it, carrying energy as it goes. The earthquake that produced the Indian Ocean tsunami of 2004 caused a sudden upward displacement of the seafloor that produced a corresponding rise in the surface of the ocean. This was the *disturbance* that produced the tsunami, very much like a quick shake on the end of a rope. The resulting wave propagated through the ocean, as we see in the figure.

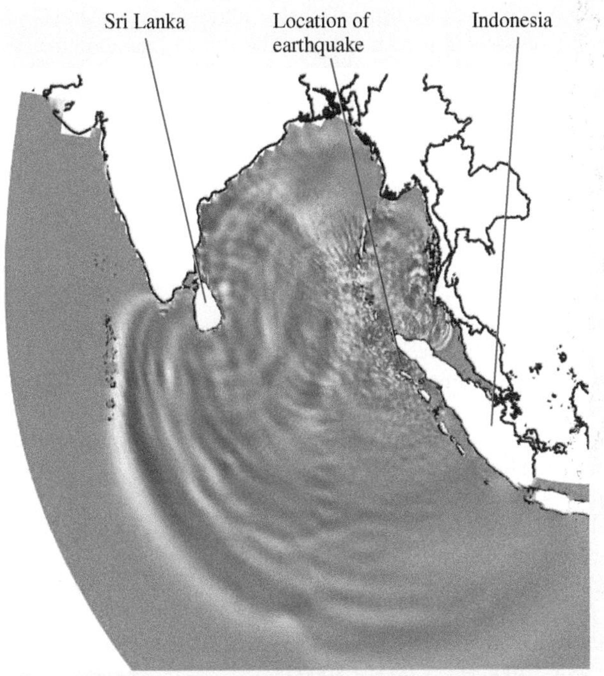

One frame from a computer simulation of the Indian Ocean tsunami three hours after the earthquake that produced it. The disturbance propagating outward from the earthquake is clearly seen, as are wave reflections from the island of Sri Lanka.

This simulation of the tsunami looks much like the ripples that spread when you drop a pebble into a pond. But there is a big difference—the scale. The fact that you can see the individual waves on this diagram that spans 5000 km is quite revealing. To show up so clearly, the individual wave pulses must be very wide—up to hundreds of kilometers from front to back.

A tsunami is actually a "shallow-water wave," even in the deep ocean, because the depth of the ocean is much less than the width of the wave. Consequently, a tsunami travels differently than normal ocean waves. In Chapter 16 we learned that wave speeds are fixed by the properties of the medium. That is true for normal ocean waves, but the great width of the wave causes a tsunami to "feel the bottom." Its wave speed is determined by the depth of the ocean: The greater the depth, the greater the speed. In the deep ocean, a tsunami travels at hundreds of kilometers per hour, much faster than a typical ocean wave. Near shore, as the ocean depth decreases, so does the speed of the wave.

The height of the tsunami in the open ocean was about half a meter. Why should such a small wave—one that ships didn't even notice as it passed—be so fearsome? Again, it's the *width* of the wave that matters. Because a tsunami is the wave motion of a considerable mass of water, great energy is involved. As the front of a tsunami wave nears shore, its speed decreases, and the back of the wave moves faster than the front. Consequently, the width decreases. The water begins to pile up, and the wave dramatically increases in height.

The Indian Ocean tsunami had a height of up to 15 m when it reached shore, with a width of up to several kilometers. This tremendous mass of water was still moving at high speed, giving it a great deal of energy. A tsunami reaching the shore isn't like a typical wave that breaks and crashes. It is a kilometers-wide wall of water that moves onto the shore and just keeps on coming. In many places, the water reached 2 km inland.

The impact of the Indian Ocean tsunami was devastating, but it was the first tsunami for which scientists were able to use satellites and ocean sensors to make planet-wide measurements. An analysis of the data has helped us better understand the physics of these ocean waves. We won't be able to stop future tsunamis, but with a better knowledge of how they are formed and how they travel, we will be better able to warn people to get out of their way.

The following questions are related to the passage "Waves in the Earth and the Ocean" on the previous page.

1. Rank from fastest to slowest the following waves according to their speed of propagation:
   A. An earthquake wave B. A tsunami
   C. A sound wave in air D. A light wave
2. The increase in height as a tsunami approaches shore is due to
   A. The increase in frequency as the wave approaches shore.
   B. The increase in speed as the wave approaches shore.
   C. The decrease in speed as the wave approaches shore.
   D. The constructive interference with the wave reflected from shore.
3. In the middle of the Indian Ocean, the tsunami referred to in the passage was a train of pulses approximating a sinusoidal wave with speed 200 m/s and wavelength 150 km. What was the approximate period of these pulses?
   A. 1 min B. 3 min C. 5 min D. 15 min
4. If a train of pulses moves into shallower water as it approaches a shore,
   A. The wavelength increases.
   B. The wavelength stays the same.
   C. The wavelength decreases.
5. The tsunami described in the passage produced a very erratic pattern of damage, with some areas seeing very large waves and nearby areas seeing only small waves. Which of the following is a possible explanation?
   A. Certain areas saw the wave from the primary source, others only the reflected waves.
   B. The superposition of waves from the primary source and reflected waves produced regions of constructive and destructive interference.
   C. A tsunami is a standing wave, and certain locations were at nodal positions, others at antinodal positions.

The following passages and associated questions are based on the material of Part III.

Deep-Water Waves

Water waves are called *deep-water waves* when the depth of the water is much greater than the wavelength of the wave. The speed of deep-water waves depends on the wavelength as follows:

$$v = \sqrt{\frac{g\lambda}{2\pi}}$$

Suppose you are on a ship at rest in the ocean, observing the crests of a passing sinusoidal wave. You estimate that the crests are 75 m apart.

6. Approximately how much time elapses between one crest reaching your ship and the next?
   A. 3 s B. 5 s C. 7 s D. 12 s
7. The captain starts the engines and sails directly opposite the motion of the waves at 4.5 m/s. Now how much time elapses between one crest reaching your ship and the next?
   A. 3 s B. 5 s C. 7 s D. 12 s
8. In the deep ocean, a longer-wavelength wave travels faster than a shorter-wavelength wave. Thus, a higher-frequency wave travels ____ a lower-frequency wave.
   A. Faster than
   B. At the same speed as
   C. Slower than

Attenuation of Ultrasound BIO

Ultrasound is absorbed in the body; this complicates the use of ultrasound to image tissues. The intensity of a beam of ultrasound decreases by a factor of 2 after traveling a distance of 40 wavelengths. Each additional travel of 40 wavelengths results in a decrease by another factor of 2.

9. A beam of 1.0 MHz ultrasound begins with an intensity of 1000 W/m^2. After traveling 12 cm through tissue with no significant reflection, the intensity is about
   A. 750 W/m^2 B. 500 W/m^2
   C. 250 W/m^2 D. 125 W/m^2
10. A physician is making an image with ultrasound of initial intensity 1000 W/m^2. When the frequency is set to 1.0 MHz, the intensity drops to 500 W/m^2 at a certain depth in the patient's body. What will be the intensity at this depth if the physician changes the frequency to 2.0 MHz?
    A. 750 W/m^2 B. 500 W/m^2
    C. 250 W/m^2 D. 125 W/m^2
11. A physician is using Doppler ultrasound to measure the motion of a patient's heart. The device measures the beat frequency between the emitted and the reflected waves. Increasing the frequency of the ultrasound will
    A. Increase the beat frequency.
    B. Not affect the beat frequency.
    C. Decrease the beat frequency.

The Bellbird BIO

The male bellbird, a small Central American rainforest bird (typical mass 0.25 kg), uses extremely loud calls to attract mates, the loudest known call of any bird. One type of call can have a sound intensity level of 125 dB at a distance of 1.0 m, with a fundamental frequency of 4000 Hz. The bellbird's vocal tract is relatively simple; we can model it as an open-closed tube.

12. What are the three lowest frequencies in the bellbird's call?
    A. 4000 Hz, 6000 Hz, 8000 Hz
    B. 4000 Hz, 8000 Hz, 12,000 Hz
    C. 4000 Hz, 8000 Hz, 16,000 Hz
    D. 4000 Hz, 12,000 Hz, 20,000 Hz
13. What is the approximate length of the bellbird's vocal tract?
    A. 2 cm B. 4 cm C. 6 cm D. 8 cm
14. If a call is 125 dB at a distance of 1.0 m, approximately how loud will it be at 6.0 m, if we assume that the sound spreads out uniformly in all directions?
    A. 120 dB B. 115 dB C. 110 dB D. 105 dB

Measuring the Speed of Sound

A student investigator is measuring the speed of sound by looking at the time for a brief, sinusoidal pulse from a loudspeaker to travel down a tube, reflect from the closed end, and reach a microphone. The apparatus is shown in Figure III.1a; typical data recorded by the microphone are graphed in Figure III.1b. The first pulse is the sound directly from the loudspeaker; the second pulse is the reflection from the closed end. A portion of the returning wave reflects from the open end of the tube and makes another round trip before being detected by the microphone; this is the third pulse seen in the data.

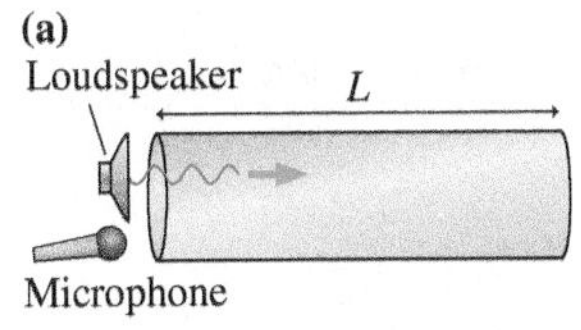

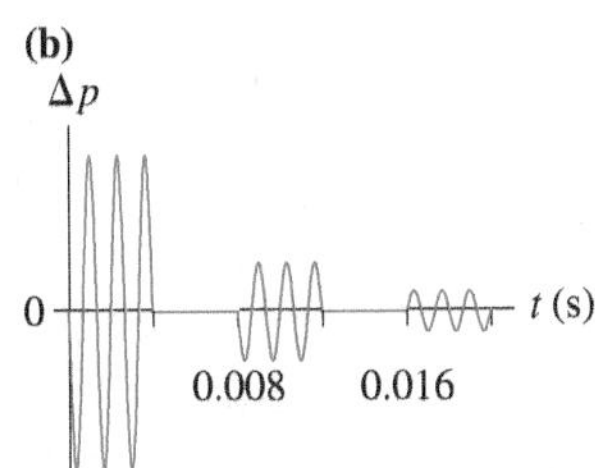

FIGURE III.1

15. What was the approximate frequency of the sound wave used in this experiment?
    A. 250 Hz B. 500 Hz
    C. 750 Hz D. 1000 Hz
16. What can you say about the reflection of sound waves at the ends of a tube?
    A. Sound waves are inverted when reflected from both open and closed tube ends.
    B. Sound waves are inverted when reflected from a closed end, not inverted when reflected from an open end.
    C. Sound waves are inverted when reflected from an open end, not inverted when reflected from a closed end.
    D. Sound waves are not inverted when reflected from either open or closed tube ends.
17. What was the approximate length of the tube?
    A. 0.35 m B. 0.70 m C. 1.4 m D. 2.8 m
18. An alternative technique to determine sound speed is to measure the frequency of a standing wave in the tube. What is the wavelength of the lowest resonance of this tube?
    A. $L/2$ B. L C. $2L$ D. $4L$

In the Swing

A rope swing is hung from a tree right at the edge of a small creek. The rope is 5.0 m long; the creek is 3.0 m wide.

19. You sit on the swing, and your friend gives you a gentle push so that you swing out over the creek. How long will it be until you swing back to where you started?
    A. 4.5 s B. 3.4 s C. 2.2 s D. 1.1 s
20. Now you switch places with your friend, who has twice your mass. You give your friend a gentle push so that he swings out over the creek. How long will it be until he swings back to where he started?
    A. 4.5 s B. 3.4 s
    C. 2.2 s D. 1.1 s
21. Your friend now pushes you over and over, so that you swing higher and higher. At some point you are swinging all the way across the creek—at the top point of your arc you are right above the opposite side. How fast are you moving when you get back to the lowest point of your arc?
    A. 6.3 m/s B. 5.4 m/s
    C. 4.4 m/s D. 3.1 m/s

Additional Integrated Problems

22. BIO The jumping gait of the kangaroo is efficient because energy is stored in the stretch of stout tendons in the legs; the kangaroo literally bounces with each stride. We can model the bouncing of a kangaroo as the bouncing of a mass on a spring. A 70 kg kangaroo hits the ground, the tendons stretch to a maximum length, and the rebound causes the kangaroo to leave the ground approximately 0.10 s after its feet first touch.
    a. Modeling this as the motion of a mass on a spring, what is the period of the motion?
    b. Given the kangaroo mass and the period you've calculated, what is the spring constant?
    c. If the kangaroo speeds up, it must bounce higher and farther with each stride, and so must store more energy in each bounce. How does this affect the time and the amplitude of each bounce?

23. BIO A brand of earplugs reduces the sound intensity level by 27 dB. By what factor do these earplugs reduce the acoustic intensity?
24. BIO Sperm whales, just like bats, use echolocation to find prey. A sperm whale's vocal system creates a single sharp click, but the emitted sound consists of several equally spaced clicks of decreasing intensity. Researchers use the time interval between the clicks to estimate the size of the whale that created them. Explain how this might be done.
    Hint: The head of a sperm whale is complex, with air pockets at either end.

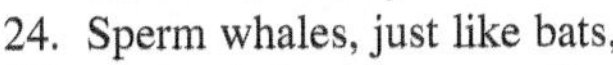

PART
IV Optics

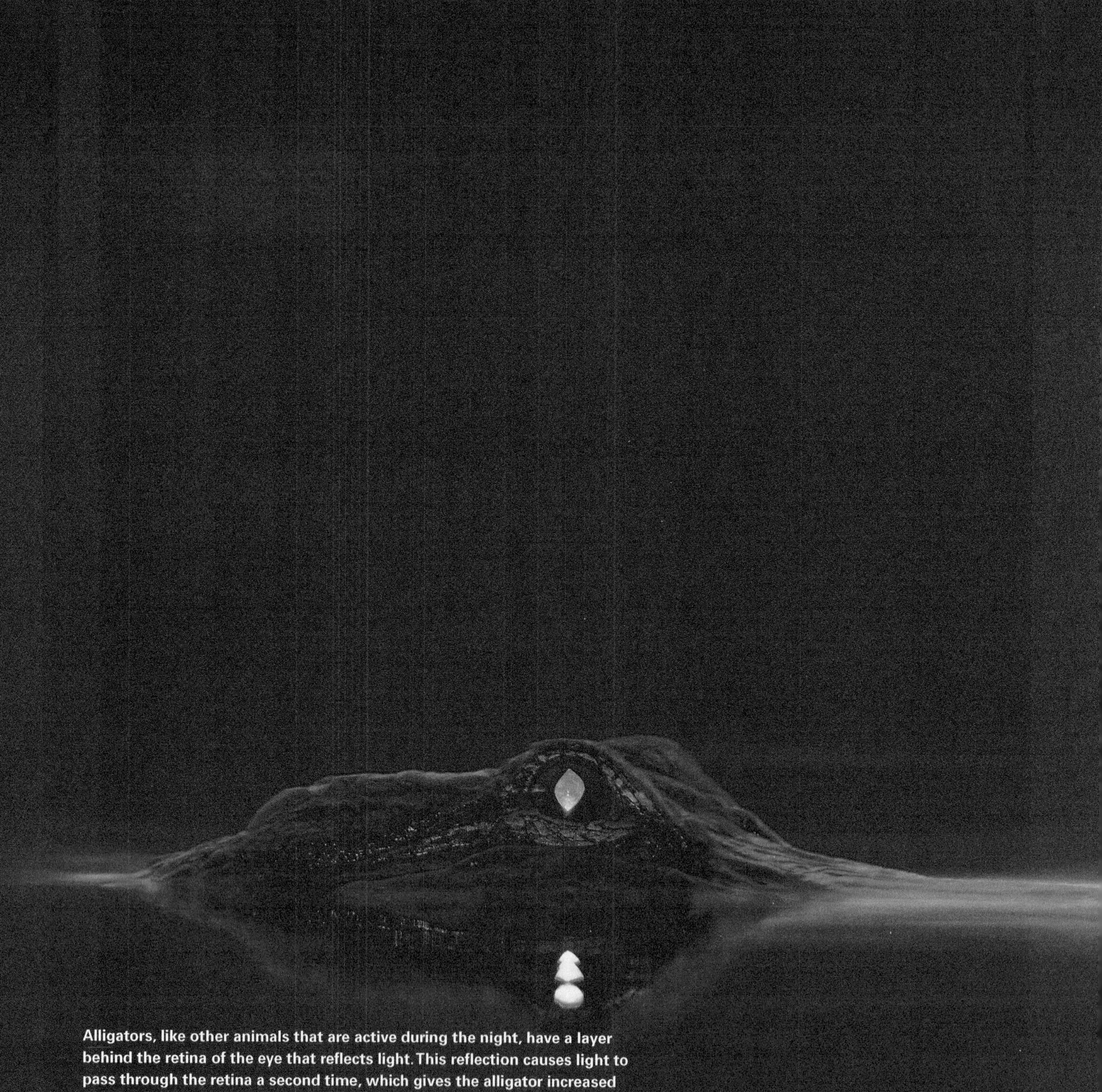

Alligators, like other animals that are active during the night, have a layer behind the retina of the eye that reflects light. This reflection causes light to pass through the retina a second time, which gives the alligator increased vision sensitivity in low-light conditions. If you take a flash photo of a gator at night, this layer reflects light back toward you, as this image clearly shows. In the coming section, we'll explore the visual systems of humans and other animals and demonstrate how the right combination of transparent tissues at the back of the eye can lead to a strong reflection—the "eyeshine" that you've seen in dogs, cats, and other animals.

OVERVIEW

The Wave Model of Light

Isaac Newton is best known for his studies of mechanics and the three laws that bear his name, but he also did important early work on optics. He was the first person to carefully study how a prism breaks white light into colors. Newton was a strong proponent of the "corpuscle" theory of light, arguing that light consists of a stream of particles.

That's one possible model of light. But, as you will see, the beautiful colors of a peacock's feathers and the shimmery rainbow of a soap bubble both require us to model light as a wave, not a particle. The wave theory we developed in Part III will be put to good use in Part IV as we begin our investigation of light and optics with an analysis of the *wave model* of light.

The Ray Model of Light

Newton was correct in his observation that light seems to travel in straight lines, something we wouldn't expect a wave to do. Consequently, our investigations of how light works will be greatly aided by another model of light, the *ray model,* in which light travels in straight lines, bounces from mirrors, and is bent by lenses.

The ray model will be an excellent tool for analyzing many of the practical applications of optics. When you look in a mirror, you see an *image* of yourself that appears to be behind the mirror. We will use the ray model of light to determine just how it is that mirrors and lenses form images. At the same time, we will need to reconcile the wave and ray models, learning how they are related to each other and when it is appropriate to use each.

Working with Light

The nature of light is quite subtle and elusive. In Parts V and VI, we will turn to the question of just what light is. As we will see, light has both wave-like *and* particle-like aspects. For now, however, we will set this question aside and work with the wave and ray models to develop a practical understanding of light. This will lead us, in Chapter 20, to an analysis of some common optical instruments. We will explore how a camera captures images and how microscopes work.

Ultimately, the fact that you are reading this text is due to the optics of the first optical instrument you ever used, your eye! We will investigate the optics of the eye, learn how the cornea and lens bend light to create an image on your retina, and see how glasses or contact lenses can be used to correct the image should it be out of focus.

18 Wave Optics

The vivid colors of this peacock feather are not due to pigments. Instead, the colors are caused by the interference of light waves as they interact with microscopic structures in the feathers.

LOOKING AHEAD

The Wave Model of Light

The colors reflected from this soap film can be understood using the *wave model of light*, the subject of this chapter.

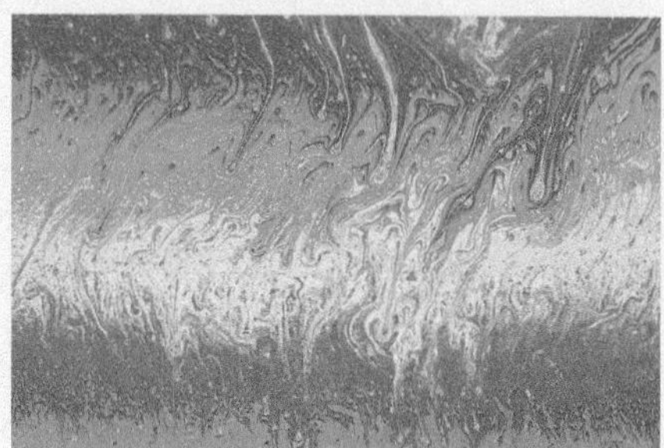

You studied the *interference* of sound waves in Chapter 17. Now you'll learn that light waves also exhibit interference.

Interference

Spectroscopy is an important analytical tool used in chemistry and biochemistry for identifying substances by the wavelengths of the light they emit.

You'll see how a spectrum is created when light waves interfere after passing through a *diffraction grating.*

Diffraction

The light waves passing around this lightbulb filament create a ripple pattern of intensity known as *diffraction*.

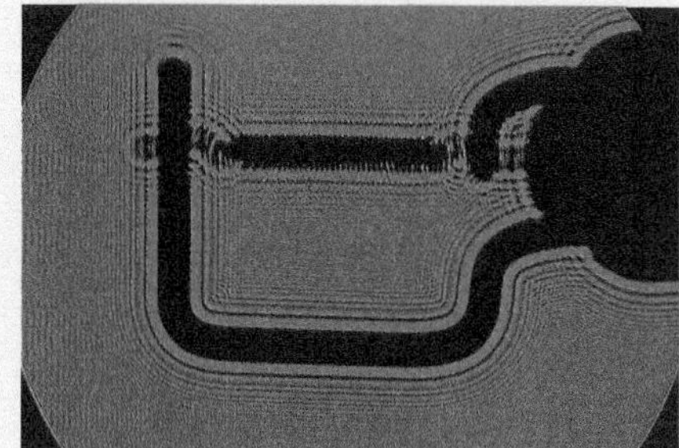

You'll discover that diffraction, a characteristic of waves, occurs when light passes around or through very small objects and apertures.

GOAL To apply the wave model of light to the phenomena of interference and diffraction.

PHYSICS AND LIFE

Light Waves Allow Us To Explore the Microcosm and the Macrocosm

We rely on light to sense and study the world around us. Many of the most spectacular colors we see in the natural world arise not from pigments or dyes but from the interference of light due to microscopic structures that are found in shells and feathers. Light is also an important tool, and scientists have developed a wide range of instruments and experimental techniques that use light of many wavelengths—from infrared to x rays—to detect biomolecules and determine their structure. An understanding of the wave nature of light enables scientists and health professionals to use instruments such as spectrometers and microscopes effectively and to interpret the data they provide.

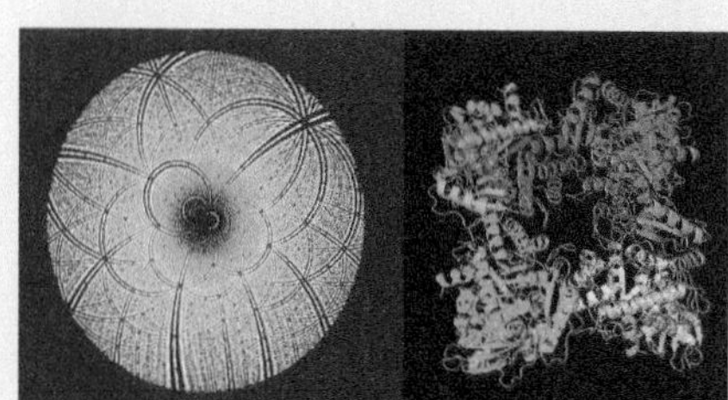

X-ray diffraction is used to determine the structure of biomolecules. This is the diffraction pattern and structure of the enzyme rubisco that is used in photosynthesis.

18.1 Models of Light

Vision is one of our most acute senses. Much of what we know about our world, and nearly all of what we know about the universe, comes to us in the form of light. The study of light is called **optics.** But what is light?

The first Greek scientists did not make a distinction between light and vision. Gradually, there arose a view that light is some kind of physical entity that exists whether or not someone is looking. Even then, it took 300 years—from the early 17th century to the early 20th century—to understand what light is.

One of the difficulties is that light is the chameleon of the physical world. Under some circumstances, light acts like a collection of particles traveling in straight lines. But if the circumstances change, light shows the same kinds of wave-like behavior as sound waves or water waves. Change the circumstances yet again and light exhibits behavior that is neither wave-like nor particle-like but has characteristics of both.

Rather than an all-encompassing "theory of light," it is better to develop three *models of light,* as shown in the following table. Each model successfully explains the behavior of light within a certain range of physical situations. We begin with the wave model in this chapter. We'll explore the ray model in Chapter 19 and the photon model in Chapter 28.

Three models of light

The Wave Model	The Ray Model	The Photon Model
The wave model of light is responsible for the well-known "fact" that light is a *wave.* Indeed, under many circumstances light exhibits the same behavior as sound or water waves. Lasers are best described by the wave model of light. Some aspects of the wave model were introduced in Chapters 16 and 17, and it is the primary focus of this chapter.	An equally well-known "fact" is that light travels in straight lines. These straight-line paths are called *light rays.* The properties of prisms, mirrors, and lenses are best understood in terms of light rays. Ray optics is the subject of Chapter 19.	The photon model of light focuses on the fact that light consists of *photons* that have both wave-like and particle-like properties. Each photon has a fixed amount of energy that is determined by its frequency. Molecules absorb and emit light in quantities determined by the photon energy. We'll examine the photon model in Part VI.

NOTE ▸ Chapter 16 introduced the idea that light is an electromagnetic wave. However, the science of optics was developed before the nature of light was understood. This chapter requires an understanding of waves, from Chapters 16 and 17, but it does not require an understanding of electromagnetic fields. You can study Part IV either before or after your study of electricity and magnetism in Part V. ◂

Is Light a Wave?

FIGURE 18.1 shows sunlight passing through openings and making sharp-edged shadows as it falls on the ground. This is behavior you would expect if light consists of rays that travel in straight lines. Indeed, the presence of sharp-edged shadows is one reason Newton and other scientists of the 17th and 18th centuries postulated that light consists of very small, light particles that travel in straight lines.

FIGURE 18.1 Light passing through an opening makes sharp-edged shadows.

How would we know if light is actually a wave? What characteristics distinguish waves from particles? The most important, which you learned about in Chapter 17, is that waves obey the principle of superposition. Consequently, waves—but not particles—undergo interference. Let's use water waves to illustrate some properties that are common to all waves. The photographs in FIGURE 18.2 (next page) look down on the surface of a water tank. A plane wave enters from the top of the picture and then passes through an opening in a barrier. In Figure 18.2a, where the opening is very narrow—comparable to the wavelength of the water waves—the wave *spreads out* to fill the space behind the opening. This spreading of the wave is called **diffraction.** We'll

FIGURE 18.2 A water wave passing through narrow and wide openings in a barrier.

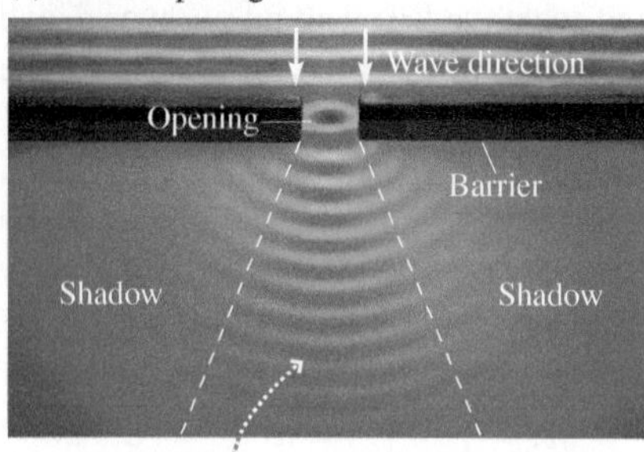

The wave spreads out behind a narrow opening whose width is comparable to the wavelength.

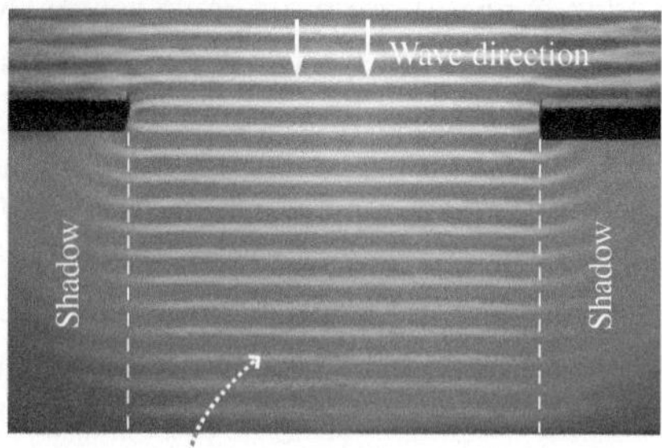

The wave moves nearly straight ahead behind a wide opening whose width is much larger than the wavelength.

study diffraction later in this chapter and see that it is a form of wave interference. Thus diffraction is a sure sign that whatever is passing through the opening is a wave.

Figure 18.2b shows a very different result; there's almost no spreading when the opening is very wide, much larger than the wavelength. In this case, the wave continues to move straight ahead with a well-defined boundary between the wave and its "shadow" where there is no wave. The demarcation of the shadow is not perfect in Figure 18.2 because the width of the opening (about 15 wavelengths) is not enormously larger than the wavelength. But if the width of the opening were thousands of wavelengths, as is typical for light passing through openings, we would see the sharp-edged shadows of Figure 18.1.

FIGURE 18.3 Light, just like water waves, spreads out behind an opening in a barrier if the opening is sufficiently narrow.

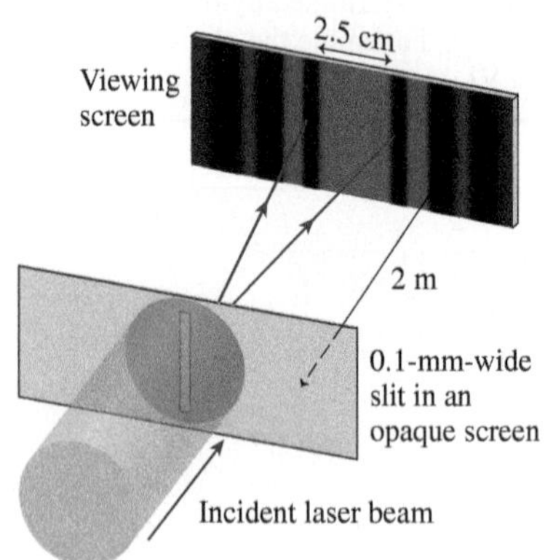

The scientists of the 17th and 18th centuries might have reached a different conclusion about the nature of light if they had been able to perform the experiment illustrated in FIGURE 18.3. Here light of a single wavelength (or color) passes through a narrow slit that is only 0.1 mm wide, about twice the width of a human hair. The figure shows how the light appears on a viewing screen 2 m behind the slit. If light consists of particles traveling in straight lines, we should see a narrow strip of light, about 0.1 mm wide, with dark shadows on both sides. Instead, we see a band of light about 2.5 cm wide, much wider than the opening, with fainter patches of light extending even farther.

If we compare Figure 18.3 to Figure 18.2a, we can see that *the light wave spreads out* behind the opening. It is exhibiting diffraction, the sure sign of waviness. However, the wavelength of light implied by this experiment (roughly 500 nm, as you'll see later in this chapter) is incredibly tiny. Scientists of earlier centuries could only observe light passing through openings that were very much larger than a wavelength, openings that cast sharp shadows with no evidence of spreading.

The experimental evidence leads to two important ideas:

- The *wave model of light,* the subject of this chapter, is the appropriate model when light interacts with openings or objects that are comparable in size to the wavelength. Diffraction, interference, and other wave phenomena are important.
- The *ray model of light,* the subject of Chapter 19, is the appropriate model when light interacts with openings or objects that are much larger than the wavelength. In that case, light appears to travel in straight lines and casts sharp shadows.

For visible light, we'll use the wave model when openings or objects are smaller than about 1 mm. The wave model will help us understand how light behaves when it passes through very small openings and how it interacts with the microscopic structures of feathers and scales. The ray model is a better model when openings or objects are larger than 1 mm. The lens of your eye and the lenses of a microscope are larger than 1 mm, so we'll use the ray model to analyze them.

18.2 Thin-Film Interference

We begin our examination of wave optics with an optical analog of the sound-wave interference you studied in Chapter 17. You learned in ◀◀ SECTION 17.6 that two equal-wavelength sound waves traveling in the same direction exhibit interference—constructive or destructive, depending on whether the waves are in phase or out of phase. We looked at *noise-canceling headphones* as an example of destructive interference.

Light waves can also interfere in this way. Two out-of-phase light waves can interfere destructively to cancel each other; this is exactly what happens when an *antireflection coating* is applied to the lenses of eyeglasses, microscopes, and other optical equipment. An optical coating is a very thin film of transparent material, and antireflection coatings are an example of what we call **thin-film interference.** We'll see that thin-film interference is responsible for the colors of soap bubbles and the iridescence of feathers and scales.

FIGURE 18.4 Two reflections are visible in the window, one from each surface.

Interference of Reflected Light Waves

FIGURE 18.4 shows that there are *two* reflections when light travels through a sheet of glass, one from the front surface and one from the back surface. These are weak reflections, typically only a few percent of the intensity of the incident wave, but even weak reflections can degrade the performance of optical instruments. In general, a light wave is partially reflected from *any* boundary between two transparent materials that have different indices of refraction. We're going to look at the interference between the two reflected waves when the distance between the two reflecting surfaces is very small, comparable to the wavelength of the light.

NOTE ▶ Recall from Chapter 16 that a material's index of refraction is $n = c/v$, where c is the speed of light in vacuum and v is the speed in the material. A slower speed causes the wave crests to bunch together, so the wavelength in the material is reduced to $\lambda_{mat} = \lambda_{vac}/n$. ◀

An important aspect of wave reflections was shown for strings in Figure 17.7 of the last chapter: If a wave moves from a string with a higher wave speed to a string with a lower wave speed, the reflected wave is *inverted* with respect to the incoming wave. It is not inverted if the wave moves from a string with a lower wave speed to a string with a higher wave speed.

The same thing happens for light waves. When a light wave moves from a material with a higher light speed (lower index of refraction) to a material with a lower light speed (higher index of refraction), the reflected wave is inverted. This inversion of the wave, corresponding to a *phase shift* of π rad, is equivalent to adding an extra half-wavelength $\lambda/2$ to the distance the wave travels. You can see this in FIGURE 18.5, where a reflected wave with a phase shift is compared to a reflection without a phase shift. In summary, we can say that **a light wave undergoes a phase shift of π rad if it reflects from a boundary at which the index of refraction increases.** There's no phase shift at a boundary where the index of refraction decreases.

FIGURE 18.6 (next page) shows a light wave approaching a thin, transparent film with thickness t and index of refraction n coated onto a piece of glass. We'll assume that the wave is incident at right angles to the surface. Most of the light is transmitted into the film, but a small fraction reflects from the air-film boundary. This is a boundary where the index of refraction increases, so the first reflected wave has a phase shift of π rad. The transmitted wave continues through the film; then a small fraction reflects from the second (film-glass) boundary. The two reflected waves, which have exactly the same wavelength, travel back out into the air where they overlap and interfere. As we learned in Chapter 17, the two reflected waves will interfere constructively to cause a *strong reflection* if they are *in phase* (i.e., if their crests overlap). If the two reflected waves are *out of phase,* with the crests of one

FIGURE 18.5 Reflected waves with and without a phase shift.

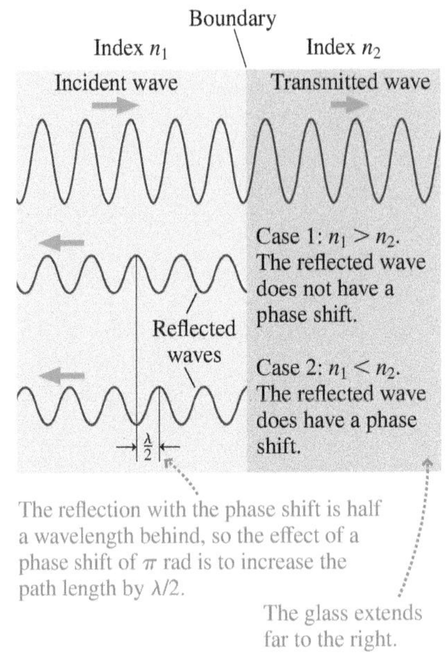
Boundary
Index n_1
Index n_2
Incident wave
Transmitted wave
Case 1: $n_1 > n_2$. The reflected wave does not have a phase shift.
Reflected waves
Case 2: $n_1 < n_2$. The reflected wave does have a phase shift.
$\frac{\lambda}{2}$
The reflection with the phase shift is half a wavelength behind, so the effect of a phase shift of π rad is to increase the path length by $\lambda/2$.
The glass extends far to the right.

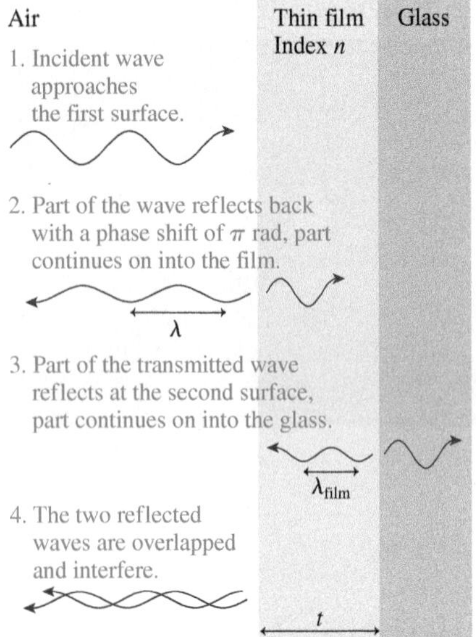

Air
Thin film
Index n
Glass
1. Incident wave approaches the first surface.
2. Part of the wave reflects back with a phase shift of π rad, part continues on into the film.
λ
3. Part of the transmitted wave reflects at the second surface, part continues on into the glass.
λfilm
4. The two reflected waves are overlapped and interfere.
t

FIGURE 18.6 In thin-film interference, two reflections, one from the film and one from the glass, overlap and interfere.

wave overlapping the troughs of the other, they will interfere destructively to cause a *weak reflection* or, if their amplitudes are equal, *no reflection* at all.

We found the interference of two sound waves to be constructive if their path-length difference is $\Delta x = m\lambda$ and destructive if $\Delta x = \left(m + \frac{1}{2}\right)\lambda$, where m is an integer. The same idea holds true for reflected light waves, for which the path-length difference is the extra distance traveled by the wave that reflects from the second surface. Because this wave travels twice through a film of thickness t, the path-length difference is $\Delta x = 2t$.

In addition, we have to account for any phase shifts due to reflections. We saw in Figure 18.5 that the phase shift when a light wave reflects from a boundary at which the index of refraction increases is equivalent to adding an extra half-wavelength to the distance traveled. This leads to two situations:

1. If *neither* or *both* waves have a phase shift due to reflection, the net addition to the path-length difference is zero. The *effective path-length difference* is $\Delta x_{eff} = 2t$.
2. If only *one* wave has a phase shift due to reflection, the effective path-length difference is increased by one half-wavelength to $\Delta x_{eff} = 2t + \frac{1}{2}\lambda$.

The interference of the two reflected waves is then constructive if $\Delta x_{eff} = m\lambda_{film}$ and destructive if $\Delta x_{eff} = \left(m + \frac{1}{2}\right)\lambda_{film}$. Why λ_{film}? Because the extra distance is traveled inside the film, so we need to compare Δx_{eff} to the wavelength in the film. The film's index of refraction is n, so the wavelength in the film is $\lambda_{film} = \lambda/n$, where λ is the wavelength of the light in vacuum or air.

With this information, we can write the conditions for constructive and destructive interference of the light waves reflected by a thin film:

$$2t = m\frac{\lambda}{n} \qquad m = 1, 2, 3, \ldots \tag{18.1}$$

Condition for constructive interference with either 0 or 2 reflective phase shifts
Condition for destructive interference with only 1 reflective phase shift

$$2t = \left(m + \tfrac{1}{2}\right)\frac{\lambda}{n} \qquad m = 0, 1, 2, \ldots \tag{18.2}$$

Condition for destructive interference with either 0 or 2 reflective phase shifts
Condition for constructive interference with only 1 reflective phase shift

NOTE ► Equations 18.1 and 18.2 give the film thicknesses that yield constructive or destructive interference. At other thicknesses, the waves will interfere neither fully constructively nor fully destructively, and the reflected intensity will fall somewhere between these two extremes. ◄

These conditions are the basis of a procedure to analyze thin-film interference.

TACTICS BOX 18.1 Analyzing thin-film interference

Follow the light wave as it passes through the film. The wave reflecting from the second boundary travels an extra distance $2t$.

❶ Note the indices of refraction of the material in which the wave is traveling before it reaches the film, the film itself, and the material beyond the film. The first and third may be the same. There's a reflective phase shift at any boundary where the index of refraction increases.

❷ If *neither* or *both* reflected waves undergo a phase shift, the phase shifts cancel and the effective path-length difference is $\Delta x_{\text{eff}} = 2t$. Use Equation 18.1 for constructive interference and 18.2 for destructive interference.

❸ If *only one* wave undergoes a phase shift, the effective path-length difference is $\Delta x_{\text{eff}} = 2t + \frac{1}{2}\lambda$. Use Equation 18.1 for destructive interference and 18.2 for constructive interference.

EXAMPLE 18.1 Designing an antireflection coating

To keep unwanted light from reflecting from the surface of eyeglasses or other lenses, which also maximizes the light being transmitted, a thin film of magnesium fluoride, which has an index of refraction $n = 1.38$, is coated onto the plastic lens ($n = 1.55$). What is the thinnest film that will minimize reflection for $\lambda = 550$ nm, the middle of the visible-light spectrum?

The glasses on the top have an antireflection coating on them. Those on the bottom do not.

PREPARE The film coats the plastic lens, so there is a reflection from the front surface of the film, where it meets the air, and the rear surface of the film, where it meets the plastic. We want to find the thickness for which there is destructive interference between these two reflections, minimizing the overall reflection.

We follow the steps of Tactics Box 18.1. At the first surface the index of refraction increases from that of air ($n = 1.00$) to that of the film ($n = 1.38$), so there is a reflective phase shift. The index of refraction also increases at the rear surface from that of the film ($n = 1.38$) to that of the plastic ($n = 1.55$). With two phase shifts, Tactics Box 18.1 tells us that we should use Equation 18.2 for destructive interference.

SOLVE We can solve Equation 18.2 for the thickness t that causes destructive interference:

$$t = \frac{\lambda}{2n}\left(m + \tfrac{1}{2}\right)$$

The thinnest film is the one for which $m = 0$, which gives

$$t = \frac{550\text{ nm}}{2(1.38)} \times \frac{1}{2} = 100\text{ nm}$$

ASSESS Interference effects occur when path-length differences are comparable to a wavelength, so our answer of 100 nm seems reasonable. Note that the reflected light will be perfectly eliminated—light cancellation—only if the two reflected waves have equal amplitudes and only at this one wavelength. In practice, antireflection coatings substantially reduce the reflected light intensity over much of the range of visible wavelengths (400–700 nm). However, increasing reflections at the ends of the visible spectrum ($\lambda \approx 400$ nm and $\lambda \approx 700$ nm), where the condition for destructive interference is no longer met, give a reddish-purple tinge to the coated lenses on optical equipment.

A thin film can *reduce* reflection with destructive interference, but it can also *enhance* reflection with constructive interference. This is important in biological systems. You have certainly seen the reflected light from the eyes of a cat or dog at night. This "eyeshine" is the enhanced constructive interference of reflections from the two sides of microscopic crystals in a thin tissue at the back of the eye called the *tapetum lucidum* (Latin for "bright carpet"). The tapetum is a common structure in the eyes of animals that must see in low light. Light that passes through the retina is reflected by the tapetum back through the cells of the retina, giving these cells a second chance to detect the light.

Eyeshine A *tapetum lucidum* is common in the eyes of animals that are nocturnal or that, like fish, live in dimly lit environments. Differences in the thickness of the tapetum can cause constructive interference in the reflected light at yellow, green, or even blue wavelengths.

EXAMPLE 18.2 **Eyeshine**

Sharks and related fish have a very well-developed tapetum. FIGURE 18.7 shows a camera flash reflected from a shark's eye back toward the camera. This reflected light is much brighter than the diffuse reflection from the body of the shark.

FIGURE 18.7 Eyeshine in a flash photo of a shark.

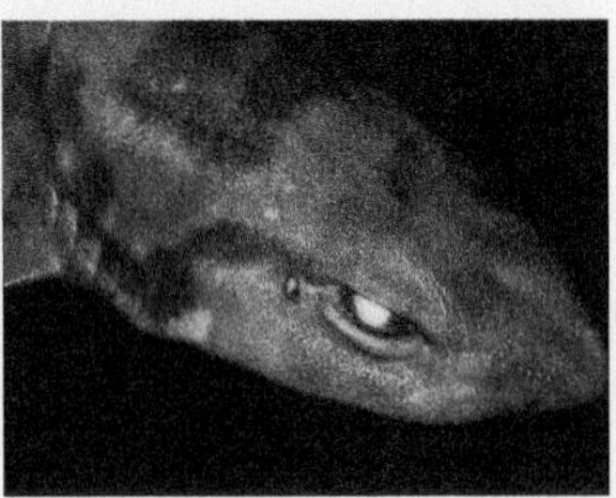

FIGURE 18.8 shows the structure responsible for the enhanced reflection. It consists of two layers of nearly transparent cells (whose index of refraction is essentially that of water) with a stack of guanine crystals sandwiched between them. Light is reflected from the interface at both sides of the stack of crystals, and the reflection is especially strong if there is constructive interference between these two reflections. If the layer of crystals is 80 nm thick, for what visible-light wavelengths is there constructive interference?

FIGURE 18.8 Structure of the tapetum in a shark.

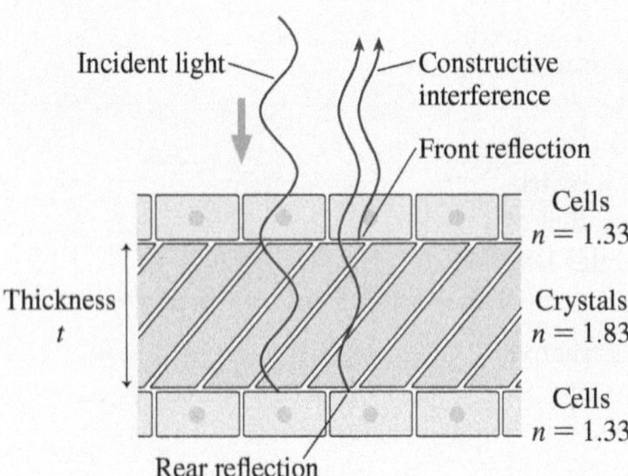

PREPARE We are looking for constructive interference between two reflections, so we'll use the steps of Tactics Box 18.1. The film in this case is the layer of crystals. We are looking for a visible-light wavelength, so we'll look for a wavelength or wavelengths between 400 nm and 700 nm.

For the front reflection in Figure 18.8, the index increases at the boundary, so there is a reflective phase shift. For the rear reflection, the index decreases, so there is no reflective phase shift. We are thus looking for constructive interference with 1 reflective phase shift, so we can use Equation 18.2.

SOLVE We rearrange Equation 18.2 to solve for the wavelength, using $n = 1.83$, the index of refraction of the crystals:

$$\lambda = \frac{2tn}{(m + \frac{1}{2})} = \frac{2(80 \times 10^{-9}\ \text{m})(1.83)}{(m + \frac{1}{2})}$$

The first possible solution is for $m = 0$:

$$\lambda(m = 0) = 590\ \text{nm}$$

For $m > 1$, we get wavelengths beyond the visible-light spectrum, so this is our final result. The reflection will be enhanced for wavelengths around 590 nm, which corresponds to yellow light.

ASSESS In Figure 18.7, the reflection looks distinctly yellow, which is consistent with our calculation.

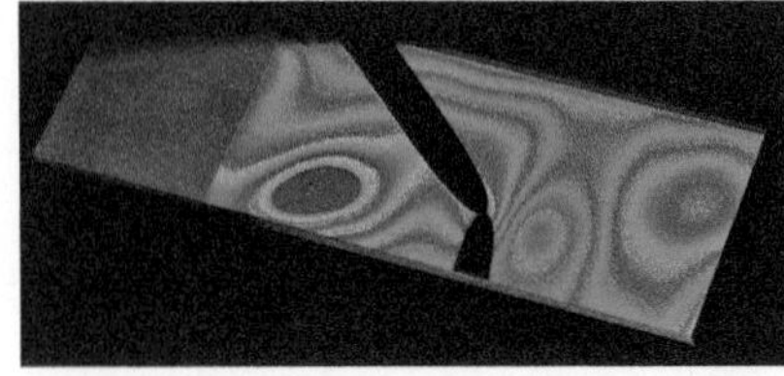

FIGURE 18.9 Light and dark patches caused by thin-film interference due to the air layer between two microscope slides.

A thin layer of air sandwiched between two glass surfaces also exhibits thin-film interference due to the waves that reflect off both interior air-glass boundaries. FIGURE 18.9 shows two microscope slides being pressed together. The light and dark patches occur because the slides are not exactly flat and they touch each other only at a few points. Everywhere else there is a thin layer of air between them. At some points, the air layer's thickness is such as to give constructive interference (light patches), while at other places its thickness gives destructive interference (dark patches).

Another feature of thin-film interference is also illustrated by Figure 18.9. The patches aren't just light and dark; they are also colored. The slides are illuminated with white light, which, as we've seen, spans a range of wavelengths. For certain thicknesses of the air film, some wavelengths will experience destructive interference, and other wavelengths will experience constructive interference. This gives the reflection different colors at different locations. The materials responsible for the reflection—glass and air—are both transparent, but these colorless materials can produce bright colors.

Structural Color from Thin Films

The colors of a soap bubble.

You've certainly noticed the bright colors in soap bubbles. When illuminated by white light from the sun, the reflections from the film that defines the bubble, the interfaces between the soapy water and the air, result in constructive and destructive interference that suppresses some wavelengths and enhances others. The watery film is colorless, but interference can produce bright colors, with different thicknesses leading to different colors. This is an everyday example of **structural color,** color that depends not on pigment, but on an object's structure.

CONCEPTUAL EXAMPLE 18.3 What color is the film?

FIGURE 18.10 shows a soap film in a metal ring that is held vertically. Explain why the colors shift as you move down the film, and why the film at the top reflects no light at all.

REASON The pull of gravity causes the thickness of the film to increase from top to bottom. The color we see corresponds to the wavelength that best satisfies the condition for constructive interference in a thin film. The wavelength that does so changes as the film becomes thicker, cycling through the visible spectrum several times for different values of the integer m.

FIGURE 18.10 A soap film in a metal ring.

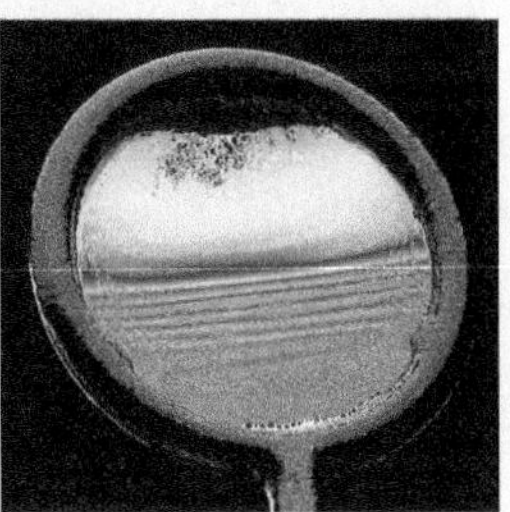

The very top of the film is extremely thin, thinner than the wavelength of light. When the film is very thin, there is almost no path-length difference between the waves reflected from the front and the back of the film. However, the wave reflected from the front undergoes a reflective phase shift and is out of phase with the wave reflected from the back. The two waves thus always interfere destructively, no matter what their wavelength. Thus all reflection is suppressed and the film appears dark.

ASSESS The absence of reflected light at the top of the film may seem surprising, but our explanation makes sense given what we know about thin films and reflection, which gives us confidence in our reasoning.

The brilliant interference colors produced by structures are generally more intense than those produced by pigments. Some of the most brightly colored animals and plants have little or no pigment. The brilliant colors of peacock feathers shown in the photo at the start of the chapter result from the interaction of light with structures that are themselves quite dull in color. The mother of pearl, or *nacre,* layer inside an abalone shell is made largely of transparent materials, but the interaction of light with complex structures produces the bright, striking colors shown in FIGURE 18.11a. A micrograph of a cross section of the thin nacre layer in FIGURE 18.11b shows stacks of approximately 500-nm-thick plates that produce strong reflections for certain wavelengths and suppress reflections for others. Slight variations in thickness and spacing produce dramatic color variations across the surface.

FIGURE 18.11 Brilliant colors in the abalone shell and the microscopic structures responsible for them.

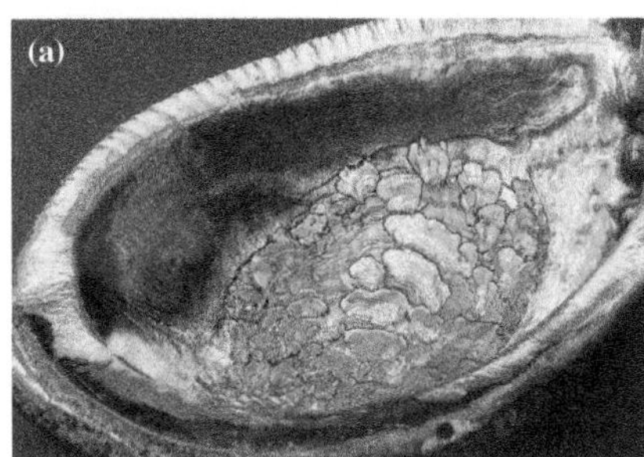

FIGURE 18.12 Changing color with angle.

The glass tile in FIGURE 18.12 has a transparent thin-film layer sandwiched between glass layers. There is no pigment in the tile; it is made of transparent materials. The color comes from thin-film interference, just as for the air layer between the microscope slides in Figure 18.9. Notice that different locations on the tile appear different colors. What is the source of this variation? For this photo, a diffuse white light source was above the tile. At the top edge of the tile, the reflected angle is large, so the light has traveled through the thin film at an oblique angle, leading to a large value of Δx_{eff}. This leads to destructive interference for the long wavelengths of red light. Eliminating the red light leaves blue and green, so the reflection appears blue-green at the top of the tile. At the bottom edge of the tile, the reflected angle is smaller, and so is Δx_{eff}. This leads to destructive interference for the shorter

FIGURE 18.13 An iridescent beetle.

wavelengths of green light. Eliminating the green light leaves blue and red, so the reflection appears magenta.

This change in color with angle is a hallmark of structural color, as is the metallic, shiny nature of the reflection. Such colors are often called **iridescent,** which means "displaying a play of lustrous colors like those of the rainbow." Hummingbird feathers are iridescent as a result of thin-film interference in air-layer structures within the feathers. As a result, the color of a hummingbird's head and neck changes as the incident angle of sunlight changes. Iridescence is especially common in the hard shells of beetles, such as the one in FIGURE 18.13, and in other insects. Their color, like that of the abalone, is due to stacked layers of thin, transparent materials. It turns out that blue pigments are fairly rare in nature, so any blue—such as the head of the beetle—is almost certainly due to structural color.

STOP TO THINK 18.1 Reflections from a thin layer of air between two glass plates cause constructive interference for a particular wavelength of light λ. By how much must the thickness of this layer be increased for the interference to be destructive?

A. $\lambda/8$ B. $\lambda/4$ C. $\lambda/2$ D. λ

18.3 Double-Slit Interference

In ◀◀ SECTION 17.7 you learned that waves of equal wavelength emitted from two sources—such as the sound waves emitted from two loudspeakers—can overlap and *interfere*. Constructive interference when the path-length difference is $\Delta r = m\lambda$ leads to a large amplitude; destructive interference when the path-length difference is $\Delta r = \left(m + \frac{1}{2}\right)\lambda$ produces a small or even zero amplitude. Interference is inherently a wave phenomenon, so if light acts like a wave we should be able to observe the interference of light waves.

We found in Section 17.7 that the conditions for maximum constructive interference are met everywhere along lines that are called *antinodal lines*. For sound waves, the sound intensity (loudness) is a maximum everywhere along an antinodal line. For light waves, the intensity (brightness) is a maximum wherever an antinodal line intersects a viewing screen.

Similarly, maximum destructive interference occurs along *nodal lines*—analogous to the *nodes* of a one-dimensional standing wave. For light waves, a viewing screen is dark where it intersects a nodal line. ◀◀ FIGURE 17.30 illustrates these ideas; it is well worth a quick review.

To observe interference, we need *two* light sources whose waves can overlap and interfere. FIGURE 18.14a shows an experiment in which a laser beam is aimed at an opaque screen containing two long, narrow slits that are very close together. This pair of slits is called a **double slit,** and in a typical experiment they are ≈ 0.1 mm wide and spaced ≈ 0.5 mm apart. We will assume that the laser beam illuminates both slits equally and that any light passing through the slits strikes a viewing screen. Such a double-slit experiment was first performed by Thomas Young in 1801, using sunlight instead of a laser beam.

What should we expect to see on the viewing screen? FIGURE 18.14b is a view from above the experiment, looking down on the top ends of the slits and the top edge of the viewing screen. Because the slits are very narrow, **light spreads out behind each slit** as it did in Figure 18.3, and these two spreading waves overlap in the region between the slits and the screen.

The screen is bright with constructive interference at points where an antinodal line intersects the screen. These points occur where the path-length difference Δr for the two waves, from leaving the slits to meeting on the screen, is an integer number of wavelengths. The screen is dark where a nodal line intersects the screen. These are

FIGURE 18.14 A double-slit interference experiment.

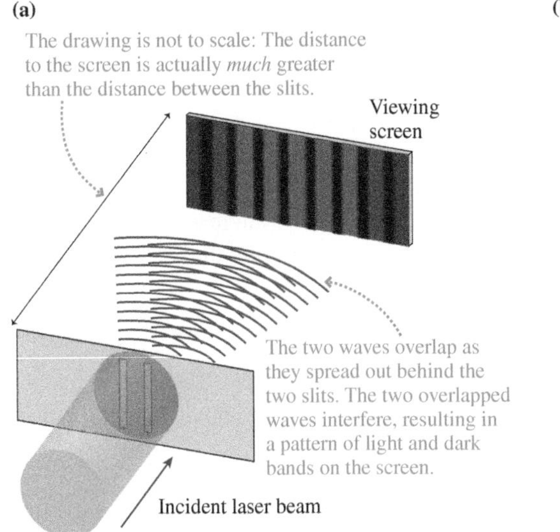
(a)
The drawing is not to scale: The distance to the screen is actually *much* greater than the distance between the slits.
Viewing screen
The two waves overlap as they spread out behind the two slits. The two overlapped waves interfere, resulting in a pattern of light and dark bands on the screen.
Incident laser beam

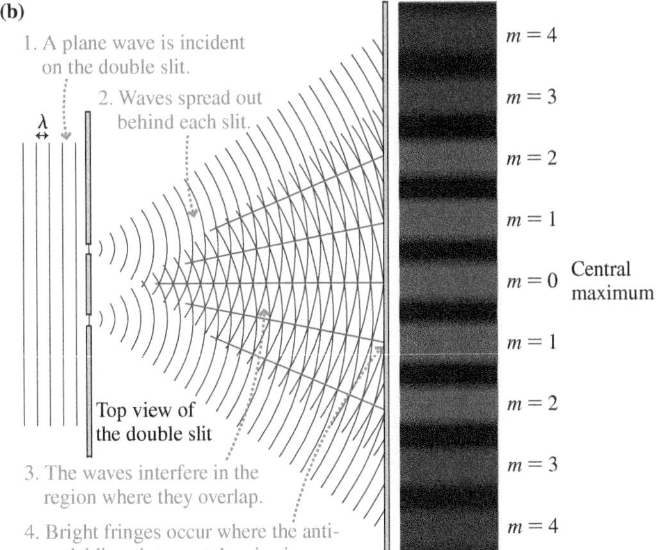
(b)
1. A plane wave is incident on the double slit.
λ
2. Waves spread out behind each slit.
Top view of the double slit
3. The waves interfere in the region where they overlap.
4. Bright fringes occur where the anti-nodal lines intersect the viewing screen.
m = 4
m = 3
m = 2
m = 1
m = 0
Central maximum
m = 1
m = 2
m = 3
m = 4

points of destructive interference where the path-length difference is a half-integer number of wavelengths.

The image in Figure 18.14b shows how the screen looks. As we move along the screen, the path-length difference Δr alternates between being an integer number m of wavelengths and a half-integer number $m + \frac{1}{2}$ of wavelengths, leading to a series of alternating bright and dark bands of light called **interference fringes.** The bright fringes are numbered by the integer $m = 1, 2, 3, \ldots$, going outward from the center. The brightest fringe, at the midpoint of the viewing screen, has $m = 0$ and is called the **central maximum.**

STOP TO THINK 18.2 In Figure 18.14b, suppose that for some point P on the screen $r_1 = 5{,}002{,}248.5\lambda$ and $r_2 = 5{,}002{,}251.5\lambda$, where λ is the wavelength of the light. The interference at point P is

A. Constructive.
B. Destructive.
C. Something in between.

Analyzing Double-Slit Interference

Figure 18.14b showed qualitatively that interference is produced behind a double slit by the overlap of the light waves spreading out behind each opening. Now let's analyze the experiment more carefully. FIGURE 18.15 shows the geometry of a double-slit experiment in which the spacing between the two slits is d and the distance to the viewing screen is L. **We will assume that L is *very* much larger than d.**

Our goal is to determine if the interference at a particular point on the screen is constructive, destructive, or something in between. As we've just noted, constructive interference between two waves from identical sources occurs at points for which the path-length difference $\Delta r = r_2 - r_1$ is an integer number of wavelengths, which we can write as

$$\Delta r = m\lambda \qquad m = 0, 1, 2, 3, \ldots \tag{18.3}$$

Thus the interference at a particular point is constructive, producing a bright fringe, if $\Delta r = m\lambda$ at that point.

FIGURE 18.15 Geometry of the double-slit experiment.

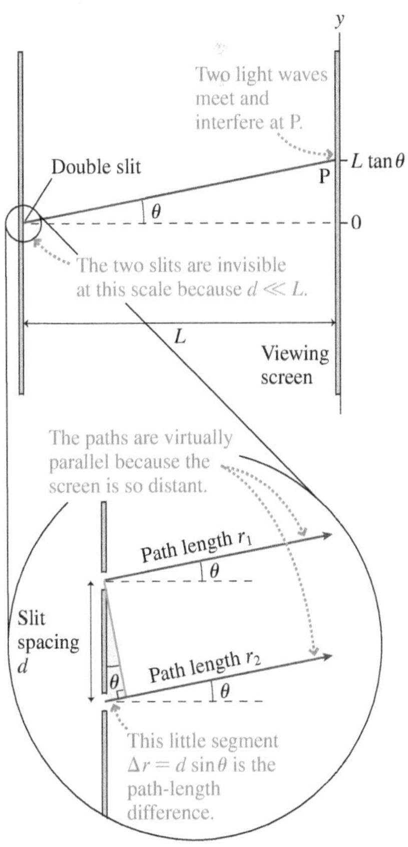
y
Two light waves meet and interfere at P.
Double slit
P
L tan θ
θ
0
The two slits are invisible at this scale because d ≪ L.
L
Viewing screen
The paths are virtually parallel because the screen is so distant.
Path length r₁
θ
Slit spacing d
θ
Path length r₂
θ
This little segment Δr = d sin θ is the path-length difference.

Figure 18.15 shows a point P on the screen that is a distance r_1 from one slit and r_2 from the other. We can specify point P either by its distance y from the center of the viewing screen or by the angle θ shown in Figure 18.15a; angle θ and distance y are related by

$$y = L \tan \theta \tag{18.4}$$

Because the screen is very far away compared to the spacing between the slits, the two paths to point P are virtually parallel, and the small triangle that is shaded blue in the enlargement of Figure 18.15 is a right triangle whose angle is also θ. The path-length difference—the extra distance traveled by the wave from the lower slit—is the short side of this triangle, so

$$\Delta r = d \sin \theta \tag{18.5}$$

If we use this result for the path-length difference in Equation 18.3, we find that the screen is bright at point P due to constructive interference if the angle θ_m is such that

$$\Delta r = d \sin \theta_m = m\lambda \qquad m = 0, 1, 2, 3, \ldots \tag{18.6}$$

We have added the subscript m to denote that θ_m is the angle of the mth bright region—an interference fringe—starting with $m = 0$ at the center.

The center of the viewing screen at $y = 0$ is equally distant from both slits, so $\Delta r = 0$. This point of constructive interference, with $m = 0$, is the bright fringe identified as the central maximum in Figure 18.14b. The path-length difference increases as we move away from the center of the screen, and the $m = 1$ fringes occur at the positions where $\Delta r = 1\lambda$. That is, one wave has traveled exactly one wavelength farther than the other. In general, **the mth bright fringe occurs where one wave has traveled m wavelengths farther than the other and thus $\Delta r = m\lambda$.**

In practice, the angle θ in a double-slit experiment is always very small (<1°). In ◂◂ SECTION 15.5 you learned the *small-angle approximation* $\sin \theta \approx \theta$ when $\theta \ll 1$ rad, which is the case here. If we apply this approximation to Equation 18.6, we find that the bright fringes occur at angles

$$\theta_m = m\frac{\lambda}{d} \qquad m = 0,\ 1,\ 2,\ 3, \ldots \tag{18.7}$$

Angles of the bright fringes for double-slit interference with slit spacing d

We can see that the double-slit interference depends on the ratio λ/d. It's not hard to measure a deflection of 0.1°, so interference with a laser-light source can be seen for slit spacings up to $d \approx 1000\lambda$. Interference is not seen, and the ray model of light becomes more appropriate, when light passes through slits that have a separation very large compared to the wavelength.

In practice, it's much easier to measure distances than angles. The positions on the screen of the bright fringes, measured from the center of the screen, are found by substituting Equation 18.7 into Equation 18.6. Recall that $\tan \theta = \sin \theta / \cos \theta$. The small-angle approximation is $\sin \theta \approx \theta$ and $\cos \theta \approx 1$. Consequently, $\tan \theta \approx \theta$ if $\theta \ll 1$ rad. With this approximation, we find that the mth bright fringe occurs at position

$$y_m = \frac{m\lambda L}{d} \qquad m = 0, 1, 2, 3, \ldots \tag{18.8}$$

Positions of bright fringes for double-slit interference at screen distance L

The interference pattern is symmetrical, so there is an mth bright fringe at the same distance on both sides of the center. You can see this in Figure 18.14b.

EXAMPLE 18.4 How far do the waves travel?

Light from a helium-neon laser ($\lambda = 633$ nm) illuminates two slits spaced 0.40 mm apart. A viewing screen is 2.0 m behind the slits. A bright fringe is observed at a point 9.5 mm from the center of the screen. What is the fringe number m, and how much farther does the wave from one slit travel to this point than the wave from the other slit?

PREPARE A bright fringe is observed when one wave has traveled an integer number of wavelengths farther than the other. Thus we know that Δr must be $m\lambda$, where m is an integer.

SOLVE Solving Equation 18.8 for m gives

$$m = \frac{y_m d}{\lambda L} = \frac{(9.5 \times 10^{-3}\text{ m})(0.40 \times 10^{-3}\text{ m})}{(633 \times 10^{-9}\text{ m})(2.0\text{ m})} = 3$$

Then the extra distance traveled by one wave compared to the other is

$$\Delta r = m\lambda = 3(633 \times 10^{-9}\text{ m}) = 1.9 \times 10^{-6}\text{ m}$$

ASSESS We expect that the path-length differences in two-slit interference are generally very small, just a few wavelengths of light, so the result is reasonable.

Equation 18.8 predicts that **the interference pattern is a series of equally spaced bright lines** on the screen, exactly as shown in Figure 18.14b. How do we know the fringes are equally spaced? The **fringe spacing** between fringe m and fringe $m + 1$ is

$$\Delta y = y_{m+1} - y_m = \frac{(m+1)\lambda L}{d} - \frac{m\lambda L}{d}$$

which simplifies to

$$\Delta y = \frac{\lambda L}{d} \qquad (18.9)$$

Spacing between any two adjacent bright fringes

Because Δy is independent of m, *any* two adjacent bright fringes have the same spacing.

The dark fringes are bands of destructive interference. These occur where the path-length difference of the waves is a half-integer number of wavelengths:

$$\Delta r = (m + \tfrac{1}{2})\lambda \qquad m = 0, 1, 2, 3, \ldots \qquad (18.10)$$

We can use Equation 18.6 for Δr and the small-angle approximation to find that the dark fringes are located at positions

$$y'_m = (m + \tfrac{1}{2})\frac{\lambda L}{d} \qquad m = 0, 1, 2, 3, \ldots \qquad (18.11)$$

Positions of dark fringes for double-slit interference

We have used y'_m, with a prime, to distinguish the location of the mth minimum from the mth maximum at y_m. You can see from Equation 18.11 that **the dark fringes are located exactly halfway between the bright fringes.**

FIGURE 18.16 is a graph of the double-slit intensity versus y. Notice the unusual orientation of the graph, with the intensity increasing toward the left so that the y-axis can match the experimental layout. You can see that the intensity oscillates between dark fringes, where the intensity is zero, and equally spaced bright fringes of maximum intensity. The maxima occur at positions where $y_m = m\lambda L/d$.

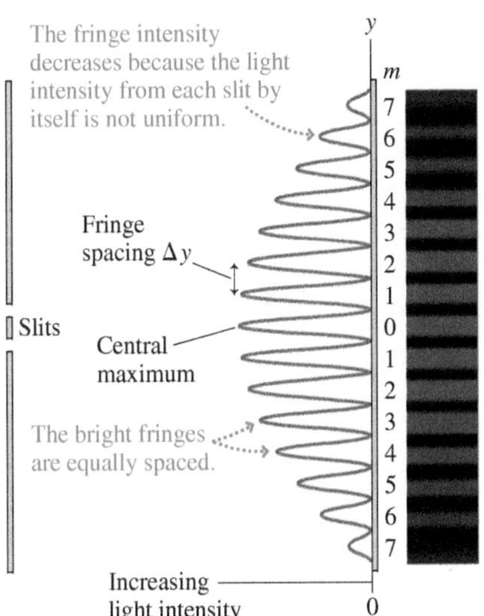

FIGURE 18.16 Intensity of the interference fringes in the double-slit experiment.

EXAMPLE 18.5 Measuring the wavelength of light

A double-slit interference pattern is observed on a screen 1.0 m behind two slits spaced 0.30 mm apart. From the center of one particular fringe to the center of the ninth bright fringe from this one is 1.6 cm. What is the wavelength of the light?

PREPARE It is not always obvious which fringe is the central maximum. Slight imperfections in the slits can make the interference fringe pattern less than ideal. However, we do not need to identify the $m = 0$ fringe because we can make use of the fact, expressed

Continued

in Equation 18.9, that the fringe spacing Δy is uniform. The interference pattern looks like the photograph of Figure 18.16.

SOLVE The fringe spacing is

$$\Delta y = \frac{1.6 \text{ cm}}{9} = 1.78 \times 10^{-3} \text{ m}$$

Using this fringe spacing in Equation 18.9, we find that the wavelength is

$$\lambda = \frac{d}{L}\Delta y = \frac{3.0 \times 10^{-4} \text{ m}}{1.0 \text{ m}}(1.78 \times 10^{-3} \text{ m})$$
$$= 5.3 \times 10^{-7} \text{ m} = 530 \text{ nm}$$

It is customary to express the wavelengths of visible light in nanometers. Be sure to do this as you solve problems.

ASSESS We've noted that visible light spans the wavelength range 400–700 nm, so finding a wavelength in this range is reasonable. The double-slit experiment not only demonstrates that light is a wave but also provides a way of determining the wavelength of visible light using measurements that we can make with a ruler.

STOP TO THINK 18.3 Light of wavelength λ_1 illuminates a double slit, and interference fringes are observed on a screen behind the slits. When the wavelength is changed to λ_2, the fringes get closer together. Is λ_2 larger or smaller than λ_1?

18.4 The Diffraction Grating

Suppose we were to replace the double slit with an opaque screen that has N closely spaced slits. When illuminated from one side, each of these slits becomes the source of a light wave that diffracts, or spreads out, behind the slit. Such a multi-slit device is called a **diffraction grating.** The light intensity pattern on a screen behind a diffraction grating is due to the interference of N overlapped waves.

FIGURE 18.17 Top view of a diffraction grating with $N = 10$ slits.

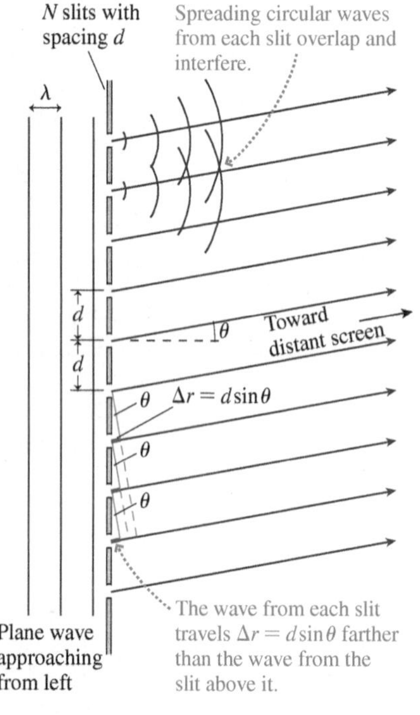
N slits with spacing d
Spreading circular waves from each slit overlap and interfere.
λ
d
d
θ
Toward distant screen
θ Δr = d sin θ
θ
θ
The wave from each slit travels Δr = d sin θ farther than the wave from the slit above it.
Plane wave approaching from left

FIGURE 18.17 shows a diffraction grating in which N slits are equally spaced a distance d apart. This is a top view of the grating, as we look down on the experiment, and the slits extend above and below the page. Only 10 slits are shown here, but a practical grating will have hundreds or even thousands of slits. Suppose a plane wave of wavelength λ approaches from the left. The crest of a plane wave arrives *simultaneously* at each of the slits, causing the wave emerging from each slit to be *in phase* with the wave emerging from every other slit—that is, all the emerging waves crest and trough simultaneously. Each of these emerging waves spreads out, just like the light wave in Figure 18.3, and after a short distance they all overlap with each other and interfere.

NOTE ▶ The terms "interference" and "diffraction" have historical roots that precede our modern understanding of wave optics, and their use can be confusing. Physically, both arise from the superposition of overlapped waves. *Interference* usually describes a superposition of waves that come from distinct sources—as in double-slit interference. *Diffraction* usually describes how the superposition of different portions of a wave front causes light to bend or spread after encountering an obstacle, such as a narrow slit. Diffraction is important for a diffraction grating because the light does spread out behind each slit, but the intensity pattern seen behind a diffraction grating is due to the interference of the light waves coming from each of the slits. It would make more sense to call it an *interference grating*; don't let the name confuse you. ◀

We want to know how the interference pattern will appear on a screen behind the grating. The light wave at the screen is the superposition of N waves, from N slits, as they spread and overlap. As we did with the double slit, we'll assume that the distance L to the screen is very large in comparison with the slit spacing d; hence the path followed by the light from one slit to a point on the screen is *very nearly* parallel to the path followed by the light from neighboring slits. You can see in Figure 18.17 that the wave from one slit travels distance $\Delta r = d \sin\theta$ farther than the wave from the slit above it and $\Delta r = d \sin\theta$ less than the wave below it. This is the same reasoning we used in Figure 18.15 to analyze the double-slit experiment.

Figure 18.17 was a magnified view of the slits. FIGURE 18.18 steps back to where we can see the viewing screen. Notice that most of the screen is dark—destructive interference. However, if the angle θ is such that $\Delta r = d \sin\theta = m\lambda$, where m is an integer, then the light wave arriving at the screen from one slit will travel *exactly m* wavelengths more or less than light from the two slits next to it, so these waves will be *exactly in phase* with each other. But each of those waves is in phase with waves from the slits next to them, and so on until we reach the end of the grating. In other words, **N light waves, from N different slits, will *all* be in phase with each other and will interfere constructively when they arrive at a point on the screen at angle θ_m such that**

$$d \sin\theta_m = m\lambda \qquad m = 0, 1, 2, 3, \ldots \qquad (18.12)$$

Angles of bright fringes for a diffraction grating with slit spacing d

FIGURE 18.18 Interference for a grating with five slits.

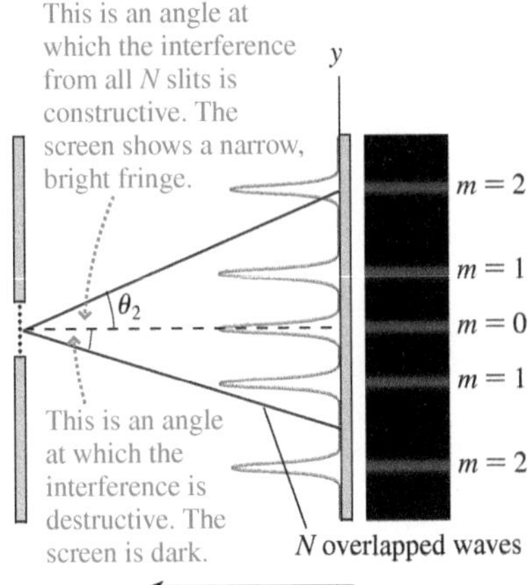
This is an angle at which the interference from all N slits is constructive. The screen shows a narrow, bright fringe.
y
$m = 2$
$m = 1$
θ_2
$m = 0$
$m = 1$
This is an angle at which the interference is destructive. The screen is dark.
$m = 2$
N overlapped waves
Increasing light intensity

Equation 18.12 for the constructive interference of N slits is the same as Equation 18.6 for the constructive interference of two slits. This makes sense: The condition for constructive interference is met for pairs of slits across the grating. In the case of a diffraction grating, though, the angles for the bright fringes are not necessarily small, so we can't use the small-angle approximation.

The pattern on the screen will have bright constructive-interference fringes at the values of θ_m given by Equation 18.12. When this happens, we say that the light is "diffracted at angle θ_m." The geometry for this pattern is the same as it was for the double-slit case, so we can again use Equation 18.4 to specify the positions of the bright fringes. We can't use the small-angle approximation, so the position y_m of the mth maximum is given as

$$y_m = L \tan\theta_m \qquad (18.13)$$

Positions of bright fringes for a diffraction grating at screen distance L

The integer m is called the **order** of the diffraction. Because d is usually very small, it is customary to characterize a grating by the number of *lines per millimeter.* Here "line" is synonymous with "slit," so the number of lines per millimeter is simply the inverse of the slit spacing d in millimeters. Equation 18.12 requires a value of d in meters, so you'll need to make this conversion as well.

In Equation 18.12, larger values of the wavelength λ imply larger diffraction angles θ_m. When white light, a mix of all colors, shines through a diffraction grating, different colors diffract at different angles—the colors of light are spread into a spectrum, a rainbow.

EXAMPLE 18.6 Exploring diffraction from a feather

White light passing through a feather is spread into a rainbow diffraction pattern by the feather's barbules, which are small, evenly spaced structures visible in the micrograph on the right in FIGURE 18.19. If the barbules are spaced at 50 per mm, a typical value, what is the first-order diffraction angle for blue light of 450 nm and red light of 650 nm?

PREPARE The barbules are evenly spaced, with open spaces between them, so they function as a diffraction grating. We will find the diffraction angles using Equation 18.12.

FIGURE 18.19 The rainbow-like diffraction of the light seen through a feather is due to diffraction from the barbules.

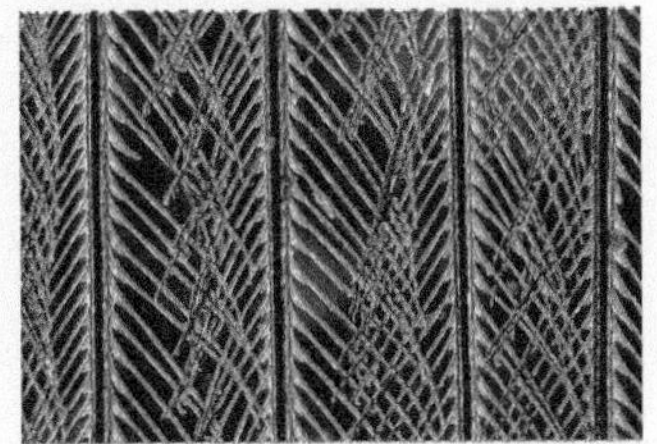

Continued

SOLVE We are told that there are 50 barbules per mm. In meters, the grating spacing d is

$$d = \frac{1}{50}\text{ mm} = 0.020\text{ mm} = 2.0 \times 10^{-5}\text{ m}$$

For the two colors noted, the first-order diffraction angle is

$$\sin\theta_1(\text{blue}) = \frac{\lambda}{d} = \frac{450 \times 10^{-9}\text{ m}}{2.0 \times 10^{-5}\text{ m}} = 0.0225$$

$$\theta_1(\text{blue}) = 1.3°$$

$$\sin\theta_1(\text{red}) = \frac{\lambda}{d} = \frac{650 \times 10^{-9}\text{ m}}{2.0 \times 10^{-5}\text{ m}} = 0.0325$$

$$\theta_1(\text{red}) = 1.9°$$

ASSESS In the photo of the diffraction pattern from the feather barbules, the rainbow pattern has a small spread, so a small angle seems reasonable.

There is an important difference between the intensity pattern of double-slit interference shown in Figure 18.16 and the intensity pattern of a multiple-slit diffraction grating shown in Figure 18.18: The bright fringes of a diffraction grating are *much* narrower. In general, as the number of slits N increases, the bright fringes get narrower and brighter. It can be shown that the intensities of the bright fringes of a grating with N slits are

$$I_{\text{max}} = N^2 I_1 \tag{18.14}$$

where I_1 is the intensity of the wave from a single slit.

This trend is shown in **FIGURE 18.20** for gratings with two slits (double-slit interference), 10 slits, and 50 slits. For 50 slits, which is not a very large number, the intensities of the fringes are $50^2 = 2500$ times that of the wave arriving from a single slit. For a practical diffraction grating, which may have thousands of closely spaced slits, the interference pattern consists of a small number of *very* bright and *very* narrow fringes while most of the screen remains dark.

FIGURE 18.20 The intensity on the screen due to three diffraction gratings. Notice that the intensity axes have different scales.

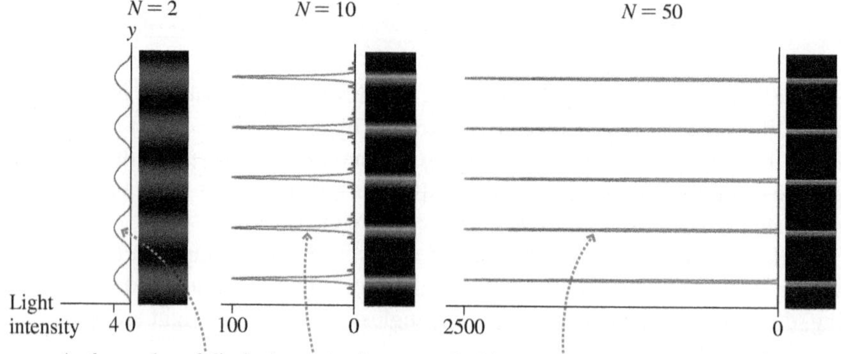

Spectroscopy

The wavelengths of light that are emitted from or absorbed by a sample are the sample's *spectrum,* and the science of detecting and measuring those wavelengths is called **spectroscopy.** Spectroscopy is an important scientific tool in biology, chemistry, and medicine because the absorption or emission spectrum provides a molecular fingerprint of the sample. This is the case because, as we'll discuss in Chapter 29, the absorption and emission of light correspond to a change in the energy state of the molecule—whether electron energy states or vibrational states of the bonds. Measurements of spectra allow very sensitive detection of small quantities of molecules, such as drugs in a urine sample or chlorophyll in a public water supply. Astronomers similarly use spectroscopy to identify the elements present in stars and to search for oxygen in the atmospheres of exoplanets as evidence of life.

Because their bright fringes are so distinct, diffraction gratings are an ideal tool for spectroscopy. Suppose the light incident on a grating consists of two slightly different wavelengths. According to Equation 18.12, each wavelength will diffract at a slightly different angle and, if N is sufficiently large, we'll see two distinct fringes on the screen. FIGURE 18.21 illustrates this idea.

FIGURE 18.21 A diffraction grating separates two different wavelengths of light.

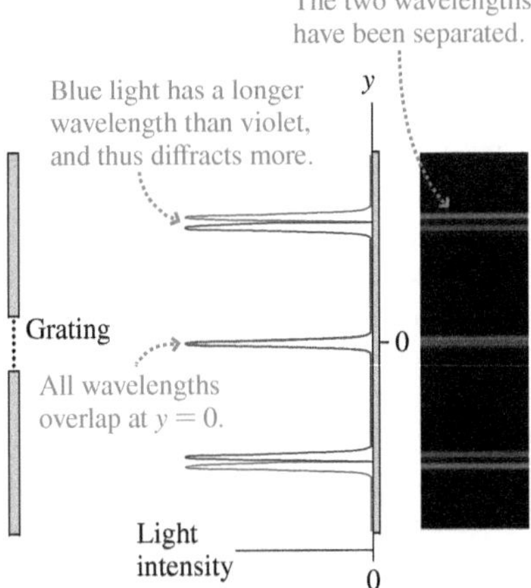

EXAMPLE 18.7 Measuring wavelengths emitted by sodium atoms

Light from a sodium lamp passes through a diffraction grating that has 1000 slits per millimeter. The interference pattern is viewed on a screen 1.000 m behind the grating. Two bright yellow fringes are visible 72.88 cm and 73.00 cm from the central maximum. What are the wavelengths of these two fringes?

PREPARE This situation is similar to that in Figure 18.21. Light passes through the diffraction grating, leading to two sets of fringes that correspond to the two wavelengths. The two fringes are very close together, so we expect the wavelengths to be only slightly different. No other yellow fringes are mentioned, so we assume these two fringes correspond to the first-order diffraction ($m = 1$).

SOLVE The distance y_m of a bright fringe from the central maximum is related to the diffraction angle by $y_m = L \tan \theta_m$. Thus the diffraction angles of these two fringes are

$$\theta_1 = \tan^{-1}\left(\frac{y_1}{L}\right) = \begin{cases} 36.085^\circ & \text{fringe at 72.88 cm} \\ 36.129^\circ & \text{fringe at 73.00 cm} \end{cases}$$

These angles must satisfy the interference condition $d \sin \theta_1 = 1 \cdot \lambda$, so the wavelengths are

$$\lambda = d \sin \theta_1$$

What is d? If a 1 mm length of the grating has 1000 slits, then the spacing from one slit to the next must be 1/1000 mm, or $d = 1.00 \times 10^{-6}$ m. Thus the wavelengths creating the two bright fringes are

$$\lambda = d \sin \theta_1 = \begin{cases} 589.0 \text{ nm} & \text{fringe at 72.88 cm} \\ 589.6 \text{ nm} & \text{fringe at 73.00 cm} \end{cases}$$

ASSESS In Chapter 16 you learned that yellow light has a wavelength of about 600 nm, so our answer is reasonable. The two wavelengths are nearly equal, which is not surprising given that the two observed fringes are very close together. Notice that we had data accurate to four significant figures, and all four are necessary to distinguish the two wavelengths. This pair of wavelengths, called the *sodium doublet,* is used to establish the presence of sodium in a sample.

FIGURE 18.22 A reflection grating.

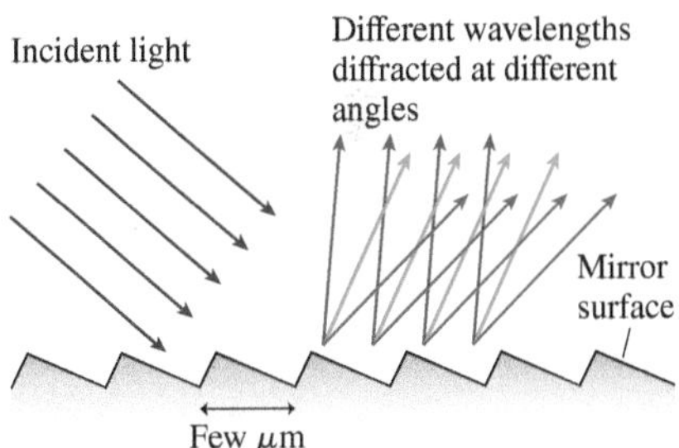

A reflection grating can be made by cutting parallel grooves in a mirror surface.

Reflection Gratings

We have analyzed what is called a *transmission grating,* with many parallel slits. We also want to consider *reflection gratings.* The simplest reflection grating, shown in FIGURE 18.22, is a mirror with hundreds or thousands of narrow, parallel grooves cut into the surface. The grooves divide the surface into many parallel reflective stripes, each of which, when illuminated, becomes the source of a spreading wave. Thus an incident light wave is divided into N overlapped waves. The interference pattern is exactly the same as the interference pattern of light transmitted through N parallel slits, and so **Equation 18.13 applies to reflection gratings as well as to transmission gratings.**

The rainbow of colors seen on the surface of a DVD is an everyday display of this phenomenon. The surface of a DVD is smooth plastic with a mirror-like reflective coating. As shown in FIGURE 18.23, billions of microscopic holes, each about 320 nm in diameter, are "burned" into the surface with a laser. The presence or

FIGURE 18.23 A DVD's colors are caused by diffraction.

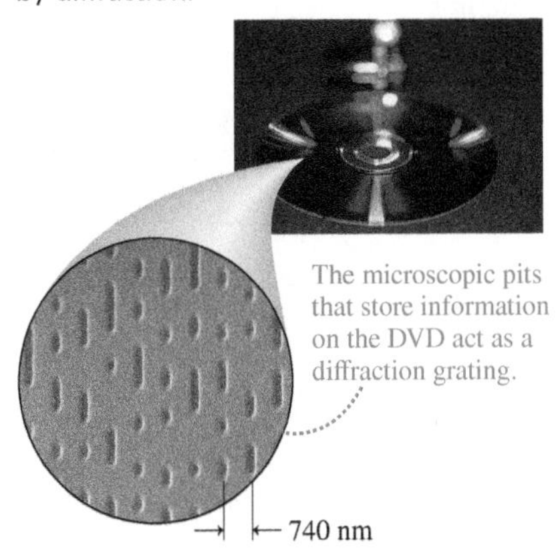

absence of a hole at a particular location on the disc is interpreted as the 0 or 1 of digitally encoded information. But from an optical perspective, the array of holes in a shiny surface is a two-dimensional version of the reflection grating shown in Figure 18.22.

Nature has many examples of reflection gratings as well, as the next example shows.

EXAMPLE 18.8 **Beetle colors**

The hard outer coating of many beetles has microscopic structures that produce amazing colors. In some beetles, parallel ridges in the cuticle function as a diffraction grating—a reflection grating—that spreads white light into the rainbow colors in the reflection from the beetle in FIGURE 18.24. An analysis of the grating spacing can provide valuable taxonomic information.

FIGURE 18.24 This beetle's rainbow of colors is the interference pattern due to the ridges in the beetle's cuticle.

A researcher illuminated a specimen of beetle with a beam of yellow light with $\lambda = 570$ nm. The resulting diffraction produced clear maxima at 13° and 25°. What is the spacing of the ridges that produces this diffraction effect?

PREPARE The ridges are working as a reflection grating. The spacing between the ridges is the grating spacing. We can assume that the angles correspond to the first two orders, so $\theta_1 = 13°$ and $\theta_2 = 25°$. We can do a separate calculation of the grating spacing d for each of the angles.

SOLVE We rearrange Equation 18.12 to solve for the spacing d for the two angles given:

$$m = 1: d = \frac{\lambda}{\sin\theta_1} = \frac{570 \times 10^{-9}\text{ m}}{\sin(13°)} = 2.5\ \mu\text{m}$$

$$m = 2: d = 2\frac{\lambda}{\sin\theta_2} = 2\frac{570 \times 10^{-9}\text{ m}}{\sin(25°)} = 2.7\ \mu\text{m}$$

The two orders of diffraction are produced by the same grating, so they should give the same spacing. The details given in the problem are real numbers from a research paper; the fact that the results do not agree exactly means that there was some uncertainty in the measurement. Our best estimate of the grating spacing is the average of the two results, 2.6 μm.

ASSESS The grating spacing is comparable to other gratings we've seen, and our two results are a reasonable match to each other, so we can have confidence in our results.

STOP TO THINK 18.4 White light passes through a diffraction grating and forms rainbow patterns on a screen behind the grating. For each rainbow,

A. The red side is on the right, the violet side on the left.
B. The red side is on the left, the violet side on the right.
C. The red side is closest to the center of the screen, the violet side is farthest from the center.
D. The red side is farthest from the center of the screen, the violet side is closest to the center.

18.5 Single-Slit Diffraction

We opened this chapter with a photograph (Figure 18.2) of a water wave passing through a hole in a barrier, then spreading out on the other side. We then saw an image (Figure 18.3) showing that light also spreads out—it *diffracts*—after it passes through a very narrow slit. We're now ready to look at the details of this type of diffraction.

FIGURE 18.25 A single-slit diffraction experiment.

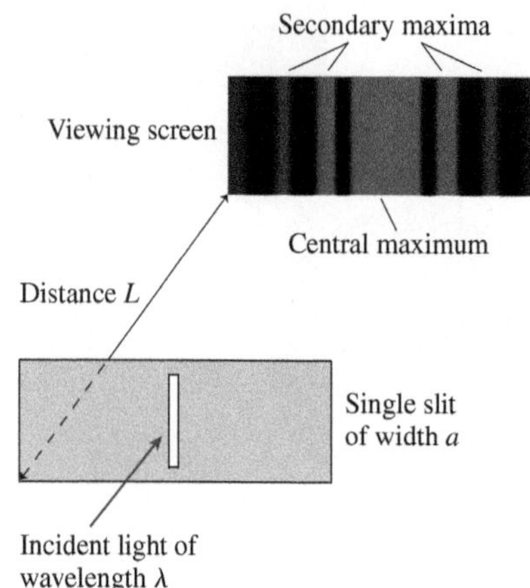

FIGURE 18.25 again shows the experimental arrangement for observing the diffraction of light through a narrow slit of width a. Diffraction through a tall, narrow slit of width a is known as **single-slit diffraction.** A viewing screen is placed a distance L behind the slit, and we will assume that $L \gg a$. The light pattern on the viewing screen consists of a *central maximum* flanked by a series of weaker **secondary maxima** and dark fringes. Notice that the central maximum is significantly broader than the secondary maxima. It is also significantly brighter than the secondary maxima, although that is hard to tell here because this photograph has been overexposed to make the secondary maxima show up better.

Huygens' Principle

Our analysis of the superposition of waves from distinct sources, such as two loudspeakers or the two slits in a double-slit experiment, has tacitly assumed that the sources are *point sources,* with no measurable extent. To understand single-slit diffraction, we need to think about the propagation of an *extended* wave front. This problem was first considered by the Dutch scientist Christiaan Huygens, a contemporary of Newton.

In ◀◀ SECTION 16.5 you learned how wave fronts—the "crests" of a wave—propagate outward from a source as circular waves or spherical waves. Huygens developed a geometrical model to visualize how *any* wave, such as a wave passing through a narrow slit, propagates. **Huygens' principle** has two parts:

1. Each point on a wave front is the source of a spherical *wavelet* that spreads out at the wave speed.
2. At a later time, the shape of the wave front is the curve that is tangent to all the wavelets.

FIGURE 18.26 illustrates Huygens' principle for a plane wave and a spherical wave. As you can see, the curve tangent to the wavelets of a plane wave is a plane that has propagated to the right. The curve tangent to the wavelets of a spherical wave is a larger sphere.

FIGURE 18.26 Huygens' principle applied to the propagation of plane waves and spherical waves.

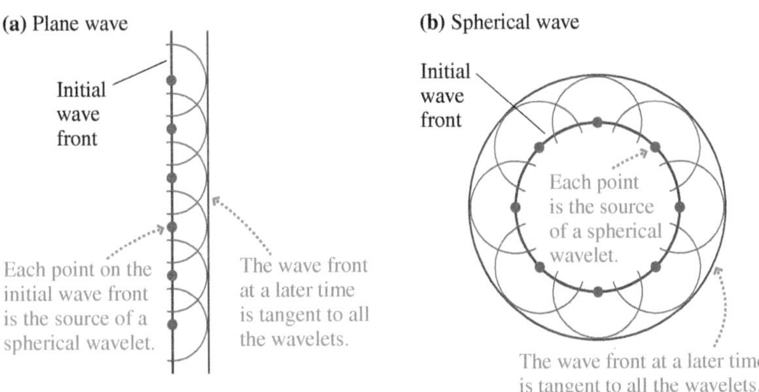
(a) Plane wave
Initial wave front
Each point on the initial wave front is the source of a spherical wavelet.
The wave front at a later time is tangent to all the wavelets.
(b) Spherical wave
Initial wave front
Each point is the source of a spherical wavelet.
The wave front at a later time is tangent to all the wavelets.

Analyzing Single-Slit Diffraction

FIGURE 18.27a shows a wave front passing through a narrow slit of width a. According to Huygens' principle, each point on the wave front can be thought of as the source of a spherical wavelet. These wavelets overlap and interfere, producing the diffraction pattern seen on the viewing screen.

FIGURE 18.27 Each point on the wave front is a source of spherical wavelets. The superposition of these wavelets produces the diffraction pattern on the screen.

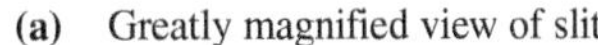
(a) Greatly magnified view of slit

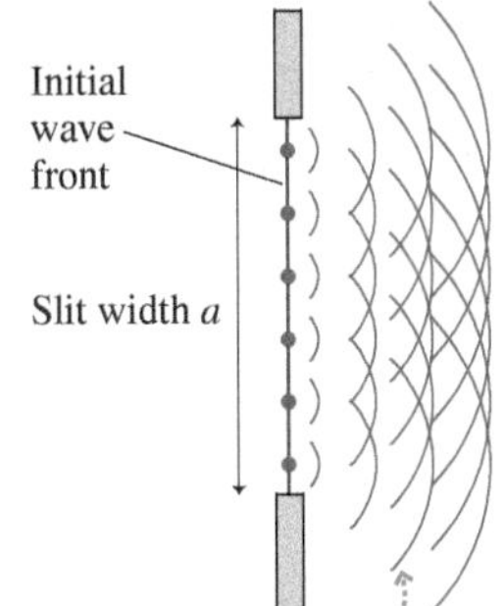
Initial wave front
Slit width a
The wavelets from each point on the initial wave front overlap and interfere, creating a diffraction pattern on the screen.

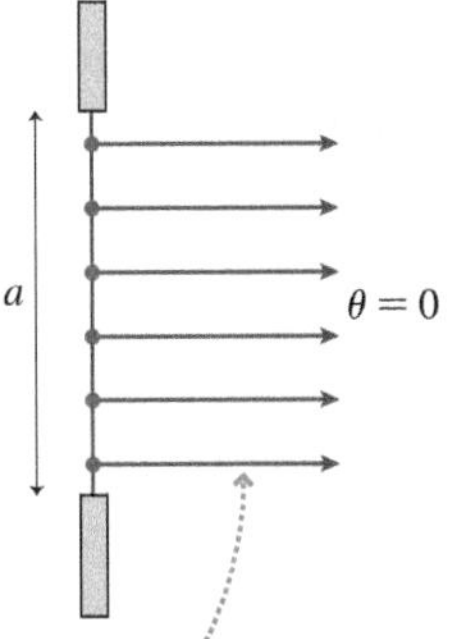
(b)
a
θ = 0
The wavelets going straight forward all travel the same distance to the screen. Thus they arrive in phase and interfere constructively to produce the central maximum.

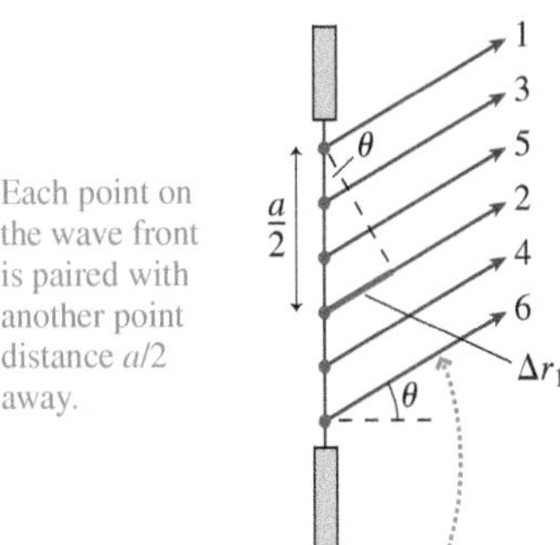
(c)
Each point on the wave front is paired with another point distance a/2 away.
a/2
θ
1
3
5
2
4
6
Δr12
These wavelets all meet on the screen at angle θ. Wavelet 2 travels distance Δr12 = (a/2) sin θ farther than wavelet 1.

Diffraction at the beach Ocean waves can be seen diffracting between two islands that form the entrance to a lagoon.

FIGURE 18.27b shows the paths of several wavelets as they travel straight ahead to the central point on the screen. (The screen is *very* far to the right in this magnified view of the slit.) The paths to the screen are very nearly parallel to each other; thus all the wavelets travel the same distance and arrive at the screen *in phase* with each other. The *constructive interference* between these wavelets produces the central maximum of the diffraction pattern at $\theta = 0$.

The situation is different at points away from the center of the screen. Wavelets 1 and 2 in FIGURE 18.27c start from points that are distance $a/2$ apart. If the angle is such that Δr_{12}, the extra distance traveled by wavelet 2, happens to be $\lambda/2$, then wavelets 1 and 2 arrive out of phase and interfere destructively. But if Δr_{12} is $\lambda/2$, then the difference Δr_{34} between paths 3 and 4 and the difference Δr_{56} between paths 5 and 6 are also $\lambda/2$. Those pairs of wavelets also interfere destructively. The superposition of all the wavelets produces perfect destructive interference.

Figure 18.27c happens to show six wavelets, but our conclusion is valid for any number of wavelets. The key idea is that **every point on the wave front can be paired with another point that is distance *a*/2 away.** If the path-length difference is $\lambda/2$, the wavelets that originate at these two points will arrive at the screen out of phase and interfere destructively. When we sum the displacements of all N wavelets, they will—pair by pair—add to zero. The viewing screen at this position will be dark. This is the main idea of the analysis, one worth thinking about carefully.

You can see from Figure 18.27c that $\Delta r_{12} = (a/2)\sin\theta$. This path-length difference will be $\lambda/2$, the condition for destructive interference, if

$$\Delta r_{12} = \frac{a}{2}\sin\theta_1 = \frac{\lambda}{2} \tag{18.15}$$

or, equivalently, $\sin\theta_1 = \lambda/a$.

We can extend this idea to find other angles of perfect destructive interference. Suppose each wavelet is paired with another wavelet from a point $a/4$ away. If Δr between these wavelets is $\lambda/2$, then all N wavelets will again cancel in pairs to give complete destructive interference. The angle θ_2 at which this occurs is found by replacing $a/2$ in Equation 18.15 with $a/4$, leading to the condition $a\sin\theta_2 = 2\lambda$. This process can be continued, and we find that the general condition for complete destructive interference is

$$a\sin\theta_p = p\lambda \qquad p = 1, 2, 3, \ldots \tag{18.16}$$

For any situation of single-slit diffraction you are likely to encounter, $\theta_p \ll 1$ rad and thus we can use the small-angle approximation $\sin\theta_p \cong \theta_p$ (with θ_p in radians) to simplify Equation 18.16:

$$\theta_p = p\frac{\lambda}{a} \qquad p = 1, 2, 3, \ldots \tag{18.17}$$

Angles of the dark fringes for single-slit diffraction with slit width a

FIGURE 18.28 A graph of the intensity of a single-slit diffraction pattern.

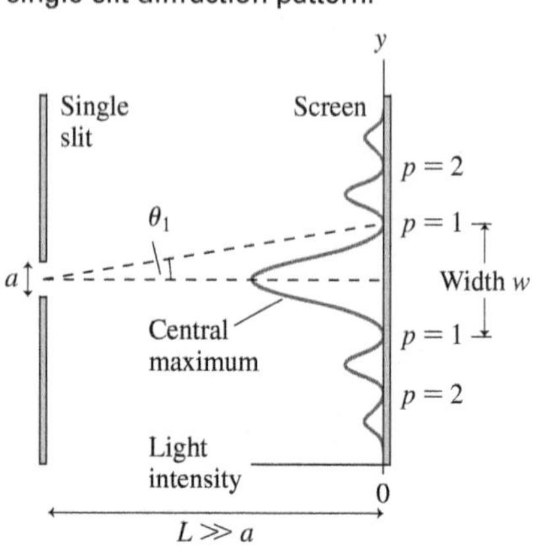

Equations 18.16 and 18.17 let us determine the angles *in radians* to the dark minima in the diffraction pattern of a single slit. Notice that $p = 0$ is explicitly *excluded.* $p = 0$ corresponds to the straight-ahead position at $\theta = 0$, but you saw in Figures 18.25 and 18.27b that $\theta = 0$ is the central *maximum,* not a minimum.

NOTE ▶ Equation 18.16 is *mathematically* the same as the condition for the *m*th *maximum* of the double-slit interference pattern. But the physical meaning here is quite different. Equation 18.16 locates the *minima* (dark fringes) of the single-slit diffraction pattern. There is not a simple equation for angles of the maxima. ◀

A graph of the single-slit intensity pattern is shown in FIGURE 18.28. You can see the bright central maximum at $\theta = 0$, the weaker secondary maxima, and the dark points of destructive interference at the angles given by Equation 18.16. Compare this graph to the photograph of Figure 18.25 and make sure you see the agreement between the two.

The Width of a Single-Slit Diffraction Pattern

It's easier to measure *positions* on the screen rather than angles. The position of the pth dark fringe, at angle θ_p, is $y_p = L \tan \theta_p$, where L is the distance from the slit to the viewing screen. Using Equation 18.17 for θ_p and the small-angle approximation $\tan \theta_p \approx \theta_p$, we find that the dark fringes in the single-slit diffraction pattern are located at

$$y_p = \frac{p\lambda L}{a} \qquad p = 1, 2, 3, \ldots \tag{18.18}$$

Positions of dark fringes for single-slit diffraction with screen distance L

Again, $p = 0$ is explicitly excluded because the midpoint on the viewing screen is the central maximum, not a dark fringe.

A single-slit diffraction pattern is dominated by the central maximum, which is much brighter than the secondary maxima. The width w of the central maximum, shown in Figure 18.28, is defined as the distance between the two $p = 1$ minima on either side of the central maximum. Because the pattern is symmetrical, the width is simply $w = 2y_1$. This is

$$w = \frac{2\lambda L}{a} \tag{18.19}$$

Width of the central maximum for single-slit diffraction

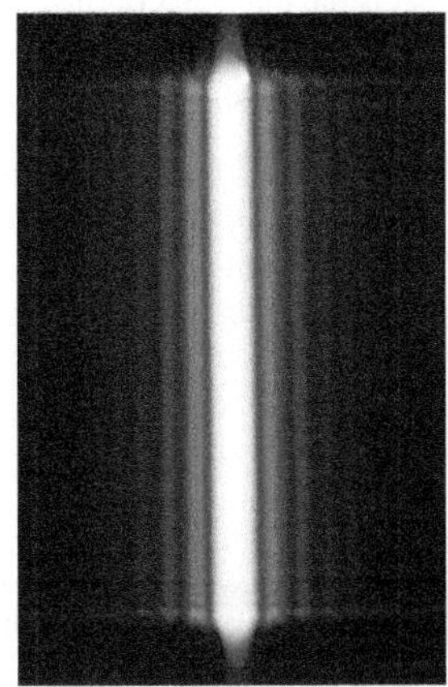

The central maximum is wider and brighter The central maximum of this single-slit diffraction pattern appears white because the photo is overexposed. The width of the central maximum is clear.

An important implication of Equation 18.19 is that a narrower slit (smaller a) causes a *wider* diffraction pattern. **The smaller the opening a wave squeezes through, the *more* it spreads out on the other side.**

EXAMPLE 18.9 Finding the width of a slit

Light from a helium-neon laser ($\lambda = 633$ nm) passes through a narrow slit and is seen on a screen 2.0 m behind the slit. The first minimum in the diffraction pattern is 1.2 cm from the middle of the central maximum. How wide is the slit?

PREPARE The first minimum in a single-slit diffraction pattern corresponds to $p = 1$. The position of this minimum is given as $y_1 = 1.2$ cm. We can then use Equation 18.18 to find the slit width a.

SOLVE Equation 18.18 gives

$$a = \frac{p\lambda L}{y_p} = \frac{(1)(633 \times 10^{-9}\text{ m})(2.0\text{ m})}{0.012\text{ m}}$$
$$= 1.1 \times 10^{-4}\text{ m} = 0.11\text{ mm}$$

ASSESS This value is typical of the slit widths used to observe single-slit diffraction.

FIGURE 18.29 The diffraction pattern of laser light behind a thin vertical wire. The exposure was chosen to image the weaker secondary maxima, so the center is overexposed and appears white.

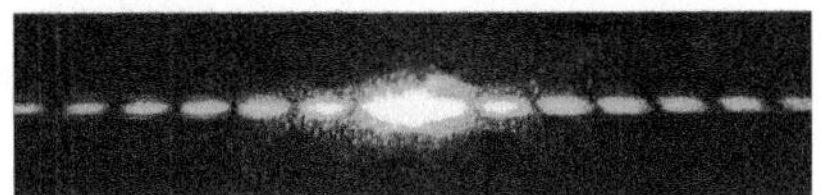

If a laser beam shines on a thin wire that blocks the beam itself, the pattern that appears on a screen behind the wire shows up as in FIGURE 18.29. You'll note that this image appears the same as the single-slit diffraction pattern. One of the surprising facts about diffraction, known as *Babinet's principle,* is that the diffraction pattern from an opaque body is the same as that from a hole or slit of the same size and shape. That is, the diffraction pattern for light bending around a 100-μm-diameter wire is the same as the diffraction pattern for light passing through a 100-μm-wide slit, and all of the equations that work for slits will work just as well for wires. This can be very practical for measuring the size of small objects; it may be easier to measure distances on the diffraction pattern than to measure physical dimensions on the scale where diffraction is important.

EXAMPLE 18.10 Using diffraction to measure the width of a hair

A student is doing a lab experiment in which she uses diffraction to measure the width of a shaft of her hair. She shines a 530 nm green laser pointer on a single hair, which produces a diffraction pattern on a screen 1.2 m away. The width of the central maximum of the pattern is 14 mm.

a. What is the thickness of the hair?
b. If she chose a wider shaft of hair, how would this change the width of the central maximum?

PREPARE The diffraction pattern produced by a hair of width a will be the same as that from a slit of width a. We are given the width of the central maximum, so we can use Equation 18.19 to analyze the situation, with a as the width of the hair instead of the width of a slit.

SOLVE

a. We rearrange Equation 18.19 to solve for a, the width of the hair:

$$a = \frac{2\lambda L}{w} = \frac{2(530 \times 10^{-9}\ \text{m})(1.2\ \text{m})}{14 \times 10^{-3}\ \text{m}} = 91\ \mu\text{m}$$

b. For a single slit, we know that a wider slit gives a narrower pattern. The same is true for hair, so a wider shaft of hair would produce a *narrower* central maximum.

ASSESS This is a fairly typical value for the thickness of a hair, so our result is reasonable.

STOP TO THINK 18.5 The figure shows two single-slit diffraction patterns. The distance between the slit and the viewing screen is the same in both cases. Which of the following could be true?

A. The slits are the same for both; $\lambda_1 > \lambda_2$
B. The slits are the same for both; $\lambda_2 > \lambda_1$
C. The wavelengths are the same for both; $a_1 > a_2$
D. The wavelengths are the same for both; $a_2 > a_1$

18.6 Circular-Aperture Diffraction

Diffraction occurs if a wave passes through an opening of any shape. A common situation of practical importance is diffraction of a wave by a **circular aperture.**

Consider some examples. A loudspeaker cone generates sound by the rapid oscillation of a diaphragm, but the sound wave must pass through the circular aperture defined by the outer edge of the speaker cone before it travels into the room beyond. This is diffraction by a circular aperture. Telescopes and microscopes are the reverse. Light waves from outside need to enter the instrument. To do so, they must pass through a circular lens. In fact, the performance limit of optical instruments is determined by the diffraction of the circular openings through which the waves must pass. This is an issue we'll look at more closely in Chapter 20.

FIGURE 18.30 shows a circular aperture of diameter D. Light waves that pass through this aperture spread out to generate a *circular* diffraction pattern. You should compare this to Figure 18.25 for a single slit to note the similarities and differences. The diffraction pattern still has a *central maximum,* now circular, and it is surrounded by a series of secondary bright fringes. Most of the intensity is contained within the central maximum.

FIGURE 18.30 The diffraction of light by a circular opening.

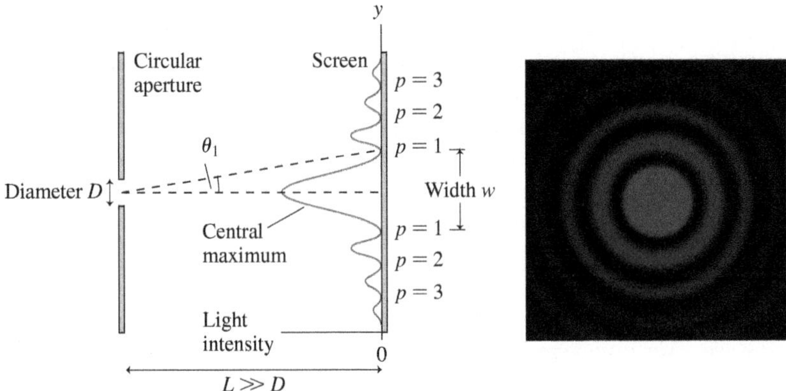

Angle θ_1 locates the first minimum in the intensity, where there is perfect destructive interference. A mathematical analysis of circular diffraction finds that

$$\theta_1 = \frac{1.22\lambda}{D} \tag{18.20}$$

where D is the *diameter* of the circular opening. Equation 18.20 has assumed the small-angle approximation, which is valid for all practical cases we'll consider. Equation 18.20 gives the angle *in radians.*

The diameter of the central maximum on a screen a distance L from the aperture is twice the distance from the center of the pattern to the first minimum; this is

$$w = 2y_1 = 2L \tan \theta_1 \tag{18.21}$$

Assuming the small-angle approximation, we can replace $\tan \theta_1$ with the value of θ_1 from Equation 18.20 to find

$$w = \frac{2.44\lambda L}{D} \tag{18.22}$$

Diameter of the central maximum for diffraction through a circular aperture of diameter D

The diameter of the diffraction pattern increases with distance L, showing that light spreads out behind a circular aperture, but it decreases if the size D of the aperture is increased.

EXAMPLE 18.11 Finding the right viewing distance

Light from a helium-neon laser ($\lambda = 633$ nm) passes through a 0.50-mm-diameter hole. How far away should a viewing screen be placed to observe a diffraction pattern whose central maximum is 3.0 mm in diameter?

PREPARE The size of the central maximum is determined by the wavelength and the aperture size, which are given, and the distance to the screen, which we will determine.

SOLVE We can rearrange Equation 18.22 to solve for L, the distance to the screen:

$$L = \frac{wD}{2.44\lambda} = \frac{(3.0 \times 10^{-3}\ \text{m})(5.0 \times 10^{-4}\ \text{m})}{2.44(633 \times 10^{-9}\ \text{m})} = 0.97\ \text{m}$$

ASSESS The hole is relatively large compared to the wavelength of light, so the spreading will be relatively small. To see a central maximum of the noted size, we expect a large screen distance.

18.7 X Rays and X-Ray Diffraction

In 1895, the German physicist Wilhelm Röntgen was studying how electrons travel through a vacuum. He sealed an electron-producing cathode and a metal target electrode into a vacuum tube. A high voltage pulled electrons from the cathode and accelerated them to very high speed before they struck the target electrode. One day, by chance, Röntgen left a sealed envelope containing film near the vacuum tube. He was later surprised to discover that the film had been exposed even though it had never been removed from the envelope. Some sort of penetrating radiation from the tube had exposed the film.

Röntgen had no idea what was coming from the tube, so he called them x rays, using the algebraic symbol x meaning "unknown." X rays were unlike anything, particle or wave, ever discovered before. Röntgen was not successful at reflecting the rays or at focusing them with a lens. He showed that they travel in straight lines, like particles, but they also pass right through most solid materials with very little absorption, something no known particle could do. The experiments of Röntgen and others led scientists to suspect that these mysterious rays were electromagnetic waves with very short wavelengths.

X-Ray Images

FIGURE 18.31 Creating an x-ray image.

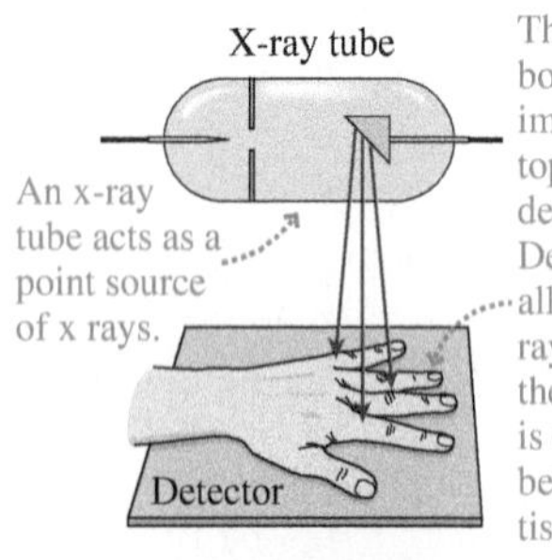
X-ray tube
An x-ray tube acts as a point source of x rays.
The part of the body to be imaged is on top of an x-ray detector.
Dense tissues allow few x rays to pass; the detector is not exposed below these tissues.
Detector

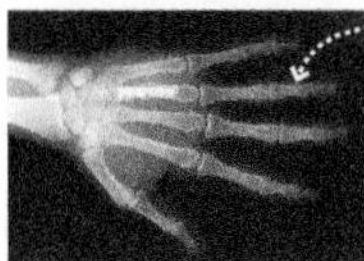
The image is a negative. Brighter areas show where bones and dense tissues have blocked the x rays.

X rays are penetrating, and Röntgen immediately realized that x rays could be used to create an image of the interior of the body. One of Röntgen's first images showed the bones in his wife's hand, dramatically demonstrating the medical potential of these newly discovered rays. Substances with high atomic numbers, such as lead or the minerals in bone, are effective at stopping the rays; materials with low atomic numbers, such as the water and organic compounds of soft tissues in the body, diminish them only slightly. As illustrated in FIGURE 18.31, an x-ray image is essentially a shadow of the bones and dense components of the body; where these tissues stop the x rays, the detector is not exposed.

X-Ray Diffraction

At about the same time, scientists were deducing that the size of an atom is ≈ 0.1 nm, and it was suggested that solids might consist of atoms arranged in a regular crystalline *lattice.* In 1912, the German scientist Max von Laue noted that if x rays are waves, then x rays passing through a crystal ought to undergo diffraction from the "three-dimensional grating" of the crystal in much the same way that visible light diffracts from a diffraction grating. Such x-ray diffraction by crystals was soon confirmed experimentally, and measurements confirmed that x rays are indeed electromagnetic waves with wavelengths in the range 0.01 nm to 10 nm—a much shorter wavelength than visible light.

FIGURE 18.32 X rays incident on a simple cubic lattice crystal.

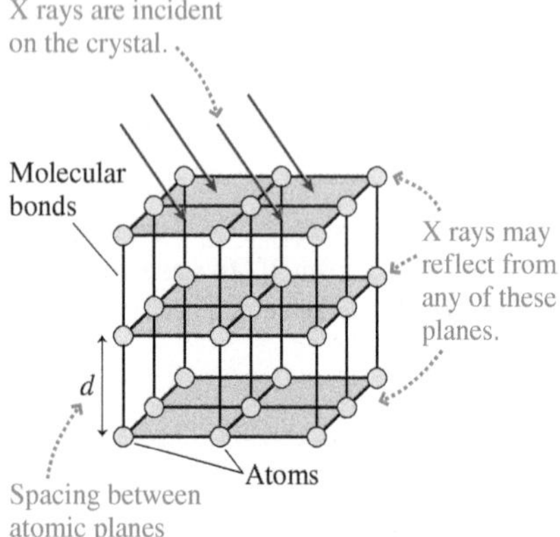
X rays are incident on the crystal.
Molecular bonds
X rays may reflect from any of these planes.
d
Atoms
Spacing between atomic planes

To understand x-ray diffraction, we begin by looking at the arrangement of atoms in a solid. FIGURE 18.32 shows x rays striking a crystal with a *simple cubic lattice.* This is a very straightforward arrangement, with the atoms in planes with spacing d between them.

FIGURE 18.33a shows a side view of the x rays striking one plane of atoms in the crystal, with the x rays incident at angle θ. (Angles of incidence, as we'll emphasize more in Chapter 19, are measured from a line perpendicular to the plane.) Most of the x rays are transmitted through the plane, but a small fraction of the wave is reflected, much like the weak reflection of light from a sheet of glass. The reflected wave obeys the law of reflection—the angle of reflection equals the angle of incidence—and the figure has been drawn accordingly.

As we saw in Figure 18.32, a solid has not one single plane of atoms but many parallel planes. As x rays pass through a solid, a small fraction of the wave reflects from each of the parallel planes of atoms shown in FIGURE 18.33b. The *net* reflection from the solid is the *superposition* of the waves reflected by each atomic plane. For most angles of incidence, the reflected waves are out of phase and their superposition is very nearly zero. However, as in thin-film interference, there are a few specific angles of incidence for which the reflected waves are in phase. For these angles of incidence, the reflected waves interfere constructively to produce a strong reflection. This strong x-ray reflection at a few specific angles of incidence is called **x-ray diffraction.**

You can see from Figure 18.33b that the wave reflecting from any particular plane travels an extra distance $\Delta r = 2d\cos\theta$ before combining with the reflection from the plane immediately above it, where d is the spacing between the atomic planes. If Δr is an integer number of wavelengths, then these two waves will be in phase when they recombine. But if the reflections from two neighboring planes are in phase, then *all* the reflections from *all* the planes are in phase and will interfere constructively to produce a strong reflection. Consequently, x rays will reflect from the crystal when the angle of incidence θ_m satisfies the **Bragg condition**:

$$\Delta r = 2d\cos\theta_m = m\lambda \qquad m = 1, 2, 3, \ldots \tag{18.23}$$

The Bragg condition for constructive interference of x rays reflected from a solid

FIGURE 18.33 X-ray reflections from parallel atomic planes.

(a) X rays are transmitted and reflected at one plane of atoms.

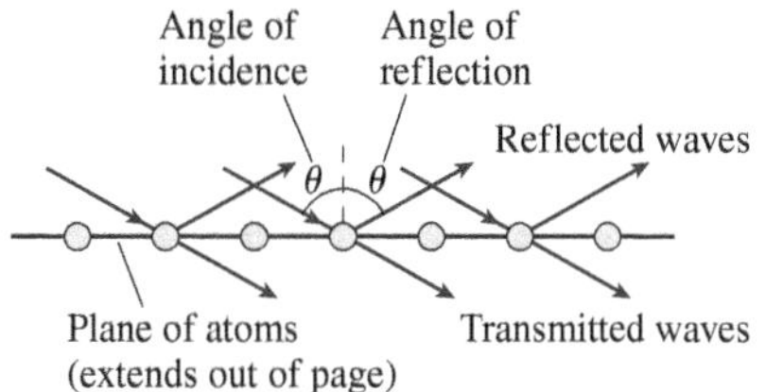

(b) The reflections from parallel planes interfere.

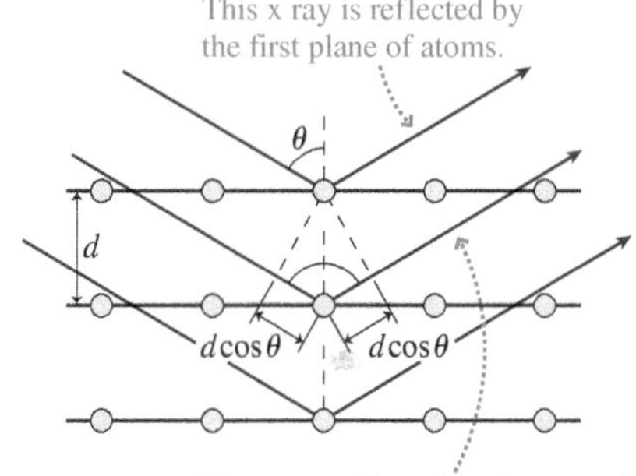

Equation 18.23 is similar to the equation for the angles of the bright fringes of a diffraction grating. In both cases, we get constructive interference at only a few well-defined angles.

NOTE ▶ You may see the Bragg condition written elsewhere in terms of $\sin\theta$ rather than $\cos\theta$ because traditional x-ray diffraction studies measured the angle θ from the surface rather than from the perpendicular. We've chosen to measure θ from the perpendicular for consistency with the way angles are measured in ray optics, as you'll see in Chapter 19. The two versions of the Bragg condition are equivalent if the angle is properly interpreted. ◀

EXAMPLE 18.12 Analyzing x-ray diffraction

X rays with a wavelength of 0.105 nm are diffracted by a crystal with a simple cubic lattice. Diffraction maxima are observed at angles 31.6° and 55.4° and at no angles between these two. What is the spacing between the atomic planes that cause this diffraction?

PREPARE The angles must satisfy the Bragg condition. We don't know the values of m, but we know that they are two consecutive integers. We'll assume that the two maxima occur at m and $m+1$. In Equation 18.23 θ_m *decreases* as m increases, so 31.6° corresponds to the larger value of m. We will assume that 55.4° corresponds to m and 31.6° to $m+1$.

SOLVE The values of d and λ are the same for both diffractions, so we can use the Bragg condition to find

$$\frac{m+1}{m} = \frac{\cos 31.6^\circ}{\cos 55.4^\circ} = 1.50 = \frac{3}{2}$$

Thus 55.4° is the second-order diffraction and 31.6° is the third-order diffraction. With this information we can use the Bragg condition with the second-order diffraction data to find

$$d = \frac{2\lambda}{2\cos\theta_2} = \frac{0.105\text{ nm}}{\cos 55.4^\circ} = 0.185\text{ nm}$$

ASSESS We learned above that the size of atoms is ≈ 0.1 nm, so this is a reasonable value for the atomic spacing in a crystal.

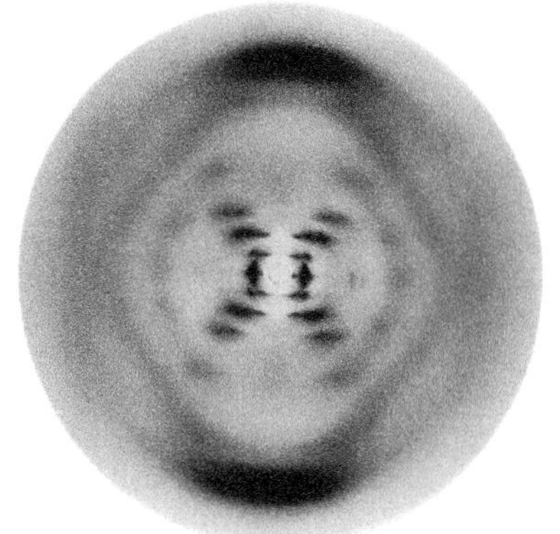

X marks the spot Rosalind Franklin, an English chemist and x-ray crystallographer, obtained this x-ray diffraction pattern for DNA in 1953. The cross of dark bands in the center of the diffraction pattern reveals something about the arrangement of atoms in the DNA molecule—that the molecule has the structure of a helix. This x-ray diffraction image was a key piece of information in the effort to unravel the structure of the DNA molecule.

Example 18.12 illustrates how x-ray diffraction data contain information about the structure of the crystal—in this case, the spacing between atomic planes. However, the example is significantly oversimplified by considering just one set of reflecting atomic planes. A real crystal has many such sets of planes, and they have many different orientations in three dimensions. Consequently, a real x-ray diffraction pattern is a complex pattern of interference maxima at many points on the inside surface of a sphere that surrounds the sample. Sophisticated mathematical procedures have been developed to infer the three-dimensional structure of a crystal by working backward from its diffraction pattern. X-ray diffraction was used in the first half of the 20th century to determine the crystal structures of many metals, minerals, and salts that form crystals.

Determining the Structure of Macromolecules

X-ray diffraction is a powerful tool for elucidating atomic structure, but it works only for crystals. The structures of biological macromolecules govern their function, and determining those structures has played a central role in the progress of biomedical research. However, growing crystals of macromolecules that are regular enough to produce high-quality diffraction patterns is a major experimental challenge. The first x-ray diffraction patterns of protein crystals were acquired in the 1930s, but it wasn't until 1958 that a complete protein structure—that of *myoglobin*—was determined. Since then, the structures of thousands of proteins, nucleic acids, and other macromolecules have been determined from their x-ray diffraction patterns.

FIGURE 18.34 illustrates the challenge of working with macromolecules. Figure 18.34a shows a crystal of a quite simple five-atom molecule that has a tetrahedral structure. The blue, green, and red lines indicate three possible sets of reflecting atomic planes of just the blue atoms. Each of these planes contributes constructive interference at several different angles when the Bragg condition is satisfied for the incident x rays. As you can imagine, there are many more atomic planes made up of green and gold atoms. Even a simple molecule like this has dozens of interference maxima in its x-ray diffraction pattern. The x-ray diffraction pattern of a macromolecule, such as the one seen in Figure 18.34b, has thousands of interference maxima, many of which are not individually resolved but instead blended together to form arcs and swirls of intensity maxima. Powerful computers are needed to deduce structural information from these data, but doing so is now a routine part of the scientific discipline of *x-ray crystallography*.

FIGURE 18.34 The x-ray diffraction pattern of a large molecule is very complex because of the large number of reflecting atomic planes.

(a) Three sets of reflecting planes for the blue atoms.

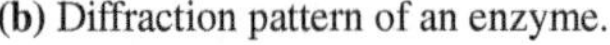

(b) Diffraction pattern of an enzyme.

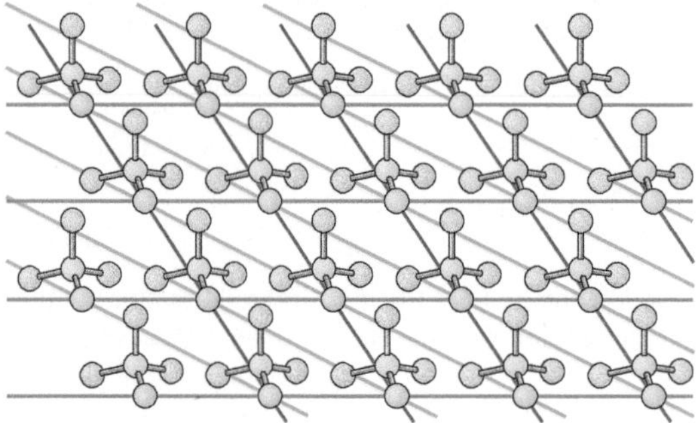

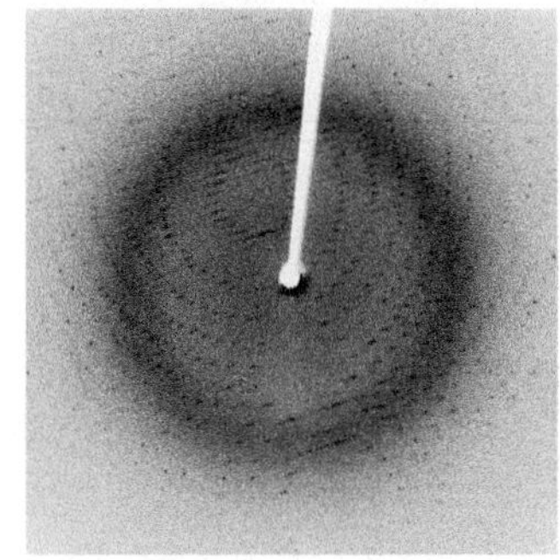

STOP TO THINK 18.6 The first-order diffraction of x rays from two crystals with simple cubic structure is measured. The first-order diffraction from crystal A occurs at an angle of 20°. The first-order diffraction of the same x rays from crystal B occurs at 30°. Which crystal has the larger atomic spacing?

CHAPTER 18 INTEGRATED EXAMPLE How good is your eye?

The human eye is a remarkable optical instrument. A simple model of the eye, shown in FIGURE 18.35, is a 25-mm-diameter water-filled sphere with a circular aperture (the pupil) of diameter D on one side and a detector (the retina) on the other. A lens just behind the pupil focuses light onto the retina by changing the wave fronts to converging spherical waves, a process we'll look at in Chapter 19. The pupil diameter varies from about 2 mm in very bright light to about 8 mm when the eye is fully dark adapted. Diameters larger than 1 mm suggest that we should use the ray model of light to analyze the optics of the eye, and we will do so in Chapter 20. However, the fact that light is a wave means that the performance of an optical system is ultimately limited by the very small amount of diffraction, usually unnoticeable, that occurs when light passes through the large openings of the pupils and lenses. In other words, light is not focused to a perfect point, as the ray model would predict, but instead makes a small circular-aperture diffraction pattern of width w on the retina. Equation 18.22 for the diameter of the central maximum assumes that the viewing-screen distance L is much larger than the aperture diameter D. That's not true in the eye, where the retina is close by, but it turns out that the presence of the focusing lens allows the equation to be valid if for L we use the lens-to-retina distance of about 18 mm.

The retina has two types of light-detecting cells: rods and cones. Sharpest vision occurs for light focused at the *fovea* of the eye, an area about 1 mm in diameter that consists almost entirely of cone cells. Cone cells are about 1.5 μm in diameter and are spaced about 2 μm apart. As a simple model, we can assume that the cones in the fovea form a square lattice of 2 μm × 2 μm detectors. (This is fairly close to the size of the pixels in the detector of a digital camera.) Physiological and neurological studies have found that about five adjacent cones need to be stimulated for the brain to receive a reliable visual signal.

a. If the eye is otherwise optically perfect, what is the minimum diameter on the retina of a focused spot of light if the pupil diameter is 2 mm, 5 mm, and 8 mm?

b. In reality, imperfections in the eye's focusing ability are a more serious issue than diffraction at larger pupil diameters. (We'll look at some of these imperfections in Chapter 20.) For most people, maximum acuity is achieved at a pupil diameter of about 3 mm. Estimate the number of cone cells that are stimulated by a focused spot of light when the pupil diameter is 3 mm.

PREPARE Diffraction of light by the circular aperture of the pupil causes light that otherwise would be focused to a perfect point to instead create a diffraction pattern that has a central maximum with diameter $w = 2.44\lambda_{eye}L/D$. For L we can use the lens-to-retina distance of 18 mm. The light diffracts *inside* the eye, where its wavelength is $\lambda_{eye} = \lambda_{air}/n$. We'll use $\lambda_{air} = 550$ nm, the midpoint of the visible spectrum, as a typical wavelength. We can model the eye as being filled with water, so we'll use the index of refraction of water: $n = 1.33$.

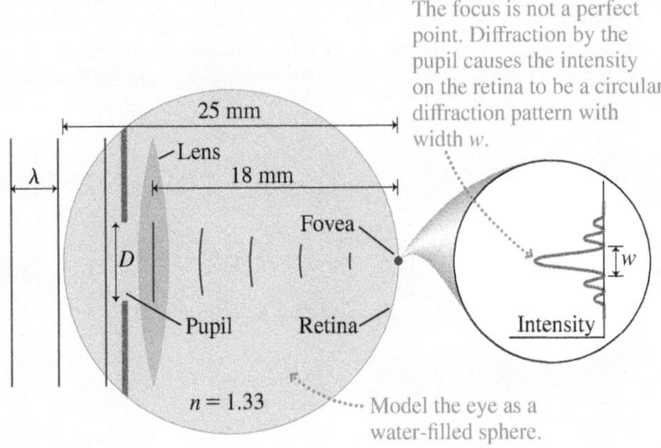
The focus is not a perfect point. Diffraction by the pupil causes the intensity on the retina to be a circular diffraction pattern with width w.
25 mm
Lens
λ
18 mm
Fovea
D
w
Pupil
Retina
Intensity
$n = 1.33$
Model the eye as a water-filled sphere.

FIGURE 18.35 A simple model of the human eye.

SOLVE

a. The wavelength of the light as it travels inside the eye is

$$\lambda_{eye} = \frac{\lambda_{air}}{n} = \frac{550 \text{ nm}}{1.33} = 410 \text{ nm}$$

We can use this value to calculate the diameter of the pupil's circular diffraction pattern on the retina. For a 2-mm-diameter pupil,

$$w = \frac{2.44\lambda_{eye}L}{D} = \frac{2.44(410 \times 10^{-9} \text{ m})(0.018 \text{ m})}{0.002 \text{ m}}$$

$$= 9.0 \times 10^{-6} \text{ m} \times \frac{1\ \mu\text{m}}{10^{-6} \text{ m}} = 9\ \mu\text{m}$$

In other words, light that would be focused to a perfect point according to the ray model of straight-line travel is actually spread into a 9-μm-diameter circle on the retina because of diffraction by the pupil. It's a very small circle of light, but larger than the size of a cone cell. Repeating the calculation for pupils of 5 mm and 8 mm gives

$$w = \begin{cases} 9\ \mu\text{m} & D = 2 \text{ mm} \\ 4\ \mu\text{m} & D = 5 \text{ mm} \\ 2\ \mu\text{m} & D = 8 \text{ mm} \end{cases}$$

b. For diffraction, the spreading of a wave *decreases* as the aperture size *increases*. It seems as if your eye would be sharpest, with the smallest focal spot, when fully dilated to a diameter of 8 mm, and that would be true if your eye were otherwise a perfect optical instrument. But certain imperfections in the eye get worse as the pupil diameter expands to more than about 3 mm, and the focusing degradation they cause is larger than any improvement in sharpness due to decreased diffraction. As a result, maximum visual sharpness occurs for a pupil diameter of about 3 mm, a size that gives a diffraction spot diameter of 6 μm.

The number of cone cells stimulated by this small spot of light is the area A_{spot} of the spot (using a radius of 3 μm) divided by the area occupied by a cone cell. We're modeling the cone cells as being on a square 2 μm × 2 μm lattice, so each cone occupies an area $A_{cone} = (2\ \mu\text{m})^2$. Thus the number of stimulated cones is

$$N_{cones} = \frac{A_{spot}}{A_{cone}} = \frac{\pi(3\ \mu\text{m})^2}{(2\ \mu\text{m})^2} = 7$$

ASSESS Our model of the eye predicts that, at maximum sharpness, light that would ideally be focused to a point on the retina is spread out by diffraction to cover and stimulate about seven cone cells. This answer is in good agreement with physiological and neurological studies showing that about five adjacent cone cells need to be stimulated in order for a reliable signal to be sent to the brain. A smaller-diameter pupil would increase the spot size due to increased diffraction, while a larger-diameter pupil would increase the spot size due to imperfections in the eye. The optics of the eye are complex, but a simple model has given us insight into the close connection between the eye's physical design and its physiological function.

SUMMARY

GOAL To apply the wave model of light to the phenomena of interference and diffraction.

GENERAL PRINCIPLES

The Wave Model

The wave model considers light to be a wave with wavelength λ propagating through space.

- Diffraction and interference are important.
- The wave model is appropriate when visible light interacts with objects smaller than about 1 mm.

IMPORTANT CONCEPTS

Huygens' principle says that each point on a wave front is the source of a spherical wavelet. The wave front at a later time is tangent to all the wavelets.

Diffraction is the spreading of a wave after it passes through an opening.

Constructive and destructive interference are due to the overlap of two or more waves as they spread behind openings.

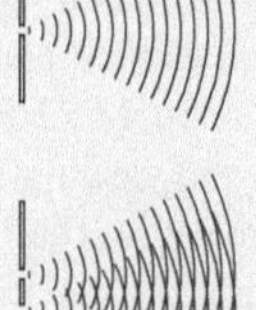

APPLICATIONS

Interference from multiple slits

Waves overlap as they spread out behind slits. Bright fringes are seen on the viewing screen at positions where the path-length difference Δr between successive slits is equal to $m\lambda$, where m is an integer.

Double slit with spacing d:

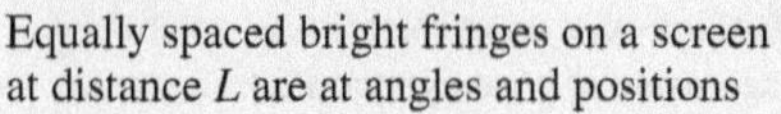

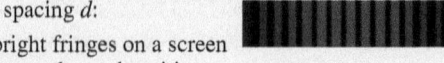

Equally spaced bright fringes on a screen at distance L are at angles and positions

$$\theta_m = m\frac{\lambda}{d} \quad y_m = \frac{m\lambda L}{d} \quad m = 0, 1, 2, \ldots$$

The fringe spacing is $\Delta y = \lambda L/d$.

Diffraction grating with slit spacing d:

Bright, narrow fringes at distance L are at angles and positions

$$d\sin\theta_m = m\lambda \quad y_m = L\tan\theta_m \quad m = 0, 1, 2, \ldots$$

Diffraction

Diffraction of light through a **single slit** of width a has a bright central maximum at distance L of width

$$w = \frac{2\lambda L}{a}$$

It is flanked by weaker secondary maxima. Dark fringes between the secondary maxima are located at angles and positions

$$\theta_p = p\frac{\lambda}{a} \quad y_p = \frac{p\lambda L}{a} \quad p = 1, 2, 3, \ldots$$

Diffraction of light through a **circular aperture** of diameter D has a bright central maximum at distance L of width

$$w = \frac{2.44\lambda L}{D}$$

For an aperture of any shape, a smaller opening causes a greater spreading of the light wave behind the opening.

Interference by reflection

Interference occurs between the waves reflected from the two surfaces of a thin film with index of refraction n. A wave that reflects from a surface at which the index of refraction increases has a phase shift of π rad. For a film of thickness t:

Interference	0 or 2 phase shifts	1 phase shift
Constructive	$2t = m\frac{\lambda}{n}$	$2t = (m + \frac{1}{2})\frac{\lambda}{n}$
Destructive	$2t = (m + \frac{1}{2})\frac{\lambda}{n}$	$2t = m\frac{\lambda}{n}$

X-ray diffraction

X rays with wavelength λ undergo constructive interference to produce strong reflections from atomic planes spaced by distance d when the angle of incidence satisfies the Bragg condition

$$2d\cos\theta_m = m\lambda \quad m = 1, 2, 3, \ldots$$

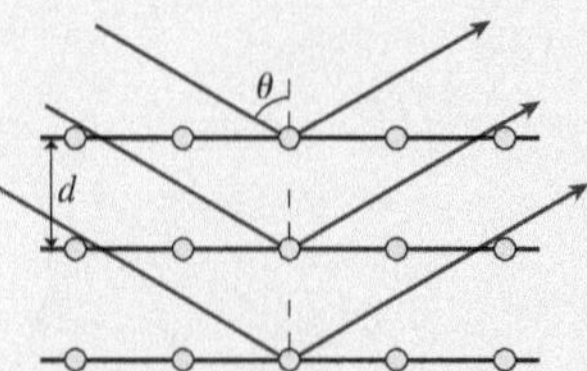

LEARNING OBJECTIVES After studying this chapter, you should be able to:

- Determine whether the interference of the reflections from a thin, transparent film is constructive or destructive. *Conceptual Questions 18.1, 18.2; Problems 18.1, 18.2, 18.5, 18.6, 18.8*
- Calculate the interference pattern of a double slit. *Conceptual Questions 18.3, 18.6; Problems 18.11–18.13, 18.15, 18.16*
- Interpret and apply the interference pattern of a diffraction grating. *Conceptual Questions 18.4, 18.11; Problems 18.18–18.21, 18.25*
- Calculate the diffraction pattern of a single slit. *Conceptual Questions 18.5, 18.8; Problems 18.27–18.30, 18.32*
- Determine the width of the diffraction pattern of a circular aperture. *Conceptual Questions 18.9, 18.10; Problems 18.34–18.37, 18.40*
- Solve problems about x-ray diffraction. *Conceptual Question 18.12; Problems 18.41–18.45*

STOP TO THINK ANSWERS

Stop to Think 18.1: B. An extra path difference of $\lambda/2$ must be added to change from constructive to destructive interference. In thin-film interference, one wave passes *twice* through the film. To increase the path length by $\lambda/2$, the thickness needs to be increased by only one-half this, or $\lambda/4$.

Stop to Think 18.2: A. The type of interference observed depends on the path-length *difference* $\Delta r = r_2 - r_1$, not on the values of r_1 and r_2 alone. Here, $\Delta r = 3\lambda$, so the path-length difference is an integer number of wavelengths, giving constructive interference.

Stop to Think 18.3: Smaller. The fringe spacing Δy is directly proportional to the wavelength λ.

Stop to Think 18.4: D. Longer wavelengths have larger diffraction angles. Red light has a longer wavelength than violet light, so red light is diffracted farther from the center.

Stop to Think 18.5: B or C. The width of the central maximum, which is proportional to λ/a, has increased. This could occur either because the wavelength has increased or because the slit width has decreased.

Stop to Think 18.6: B. The Bragg condition $2d\cos\theta_1 = \lambda$ tells us that larger values of d go with larger values of θ_1.

QUESTIONS

Conceptual Questions

1. Figure Q18.1 shows a light wave incident on and passing through a thin soap film. Reflections from the front and back surfaces of the film create smaller waves (not shown in the figure) that travel to the left of the film, where they interfere. Is the interference constructive, destructive, or something in between? Explain.

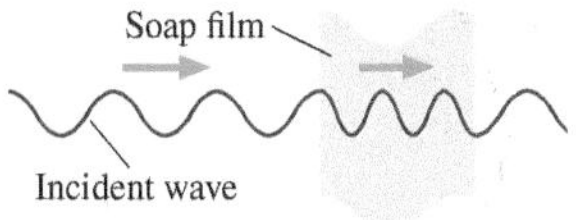

Soap film
Incident wave

FIGURE Q18.1

2. Antireflection coatings for glass usually have an index of refraction that is less than that of glass. Explain how this permits a thinner coating.
3. Figure Q18.3 shows the viewing screen in a double-slit experiment with monochromatic light. Fringe C is the central maximum.
   a. What will happen to the fringe spacing if the wavelength of the light is decreased?
   b. What will happen to the fringe spacing if the spacing between the slits is decreased?
   c. What will happen to the fringe spacing if the distance to the screen is decreased?
   d. Suppose the wavelength of the light is 500 nm. How much farther is it from the dot on the screen in the center of fringe E to the left slit than it is from the dot to the right slit?

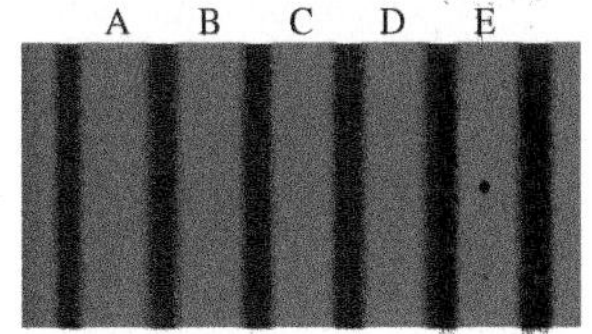

FIGURE Q18.3

4. Figure Q18.3 is the interference pattern seen on a viewing screen behind 2 slits. Suppose the 2 slits were replaced by 20 slits having the same spacing d between adjacent slits.
   a. Would the number of fringes on the screen increase, decrease, or stay the same?
   b. Would the fringe spacing increase, decrease, or stay the same?
   c. Would the width of each fringe increase, decrease, or stay the same?
   d. Would the brightness of each fringe increase, decrease, or stay the same?
5. Figure Q18.5 shows light waves passing through two closely spaced, narrow slits. The graph shows the intensity of light on a screen behind the slits. Reproduce these graph axes, including the zero and the tick marks locating the double-slit fringes, then draw a graph to show how the light-intensity pattern will appear if the right slit is blocked, allowing light to go through only the left slit. Explain your reasoning.

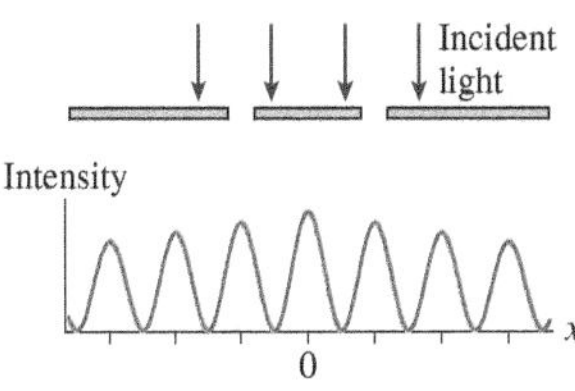

FIGURE Q18.5

Problem difficulty is labeled as | (straightforward) to ||||| (challenging). Problems labeled INT integrate significant material from earlier chapters; BIO are of biological or medical interest; CALC require calculus to solve.

6. In a double-slit interference experiment, which of the following actions (perhaps more than one) would cause the fringe spacing to increase? (a) Increasing the wavelength of the light. (b) Increasing the slit spacing. (c) Increasing the distance to the viewing screen. (d) Submerging the entire experiment in water.
7. A double-slit interference experiment shows fringes on a screen. The entire experiment is then immersed in water. Do the fringes on the screen get closer together, farther apart, remain the same, or disappear entirely? Explain.
8. Figure Q18.8 shows the light intensity on a viewing screen behind a single slit of width a. The light's wavelength is λ. Is $\lambda < a$, $\lambda = a$, $\lambda > a$, or is it not possible to tell? Explain.

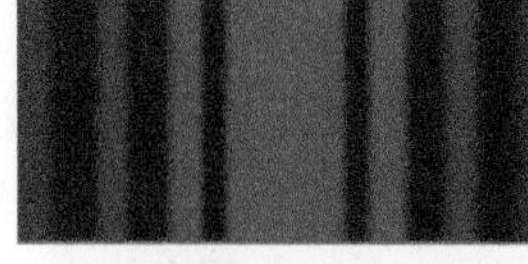
FIGURE Q18.8

9. Figure Q18.9 shows the light intensity on a viewing screen behind a circular aperture. What happens to the width of the central maximum if
   a. The wavelength is increased?
   b. The diameter of the aperture is increased?
   c. How will the screen appear if the aperture diameter is less than the light wavelength?

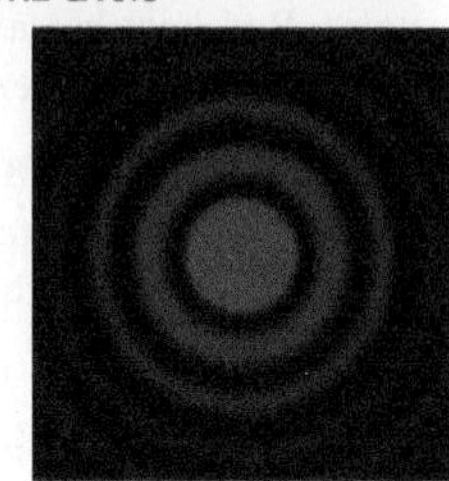
FIGURE Q18.9

10. a. Green light shines through a 100-mm-diameter hole and is observed on a screen. If the hole diameter is increased by 20%, does the circular spot of light on the screen decrease in diameter, increase in diameter, or stay the same? Explain.
    b. Green light shines through a 100-μm-diameter hole and is observed on a screen. If the hole diameter is increased by 20%, does the circular spot of light on the screen decrease in diameter, increase in diameter, or stay the same? Explain.
11. If you look at the light spectrum reflected from the surface of a DVD compared to the spectrum reflected from the surface of a audio CD, you'll see that the rainbows from the DVD are more spread out, with greater angular separation. What does this tell you about the relative track spacing on these two types of discs?
12. A beam of x rays that have wavelength λ impinges on a solid surface at a 30° angle above the surface. These x rays produce a strong reflection. Suppose the wavelength is slightly decreased. To continue to produce a strong reflection, does the angle of the x-ray beam above the surface need to be increased, decreased, or maintained at 30°?

Multiple-Choice Questions

13. | Light of wavelength 500 nm in air enters a glass block with index of refraction $n = 1.5$. When the light enters the block, which of the following properties of the light will not change?
    A. The speed of the light
    B. The frequency of the light
    C. The wavelength of the light
14. ||| Reflected light from a thin film of oil gives constructive interference for light with a wavelength inside the film of λ_{film}. By how much would the film thickness need to be increased to give destructive interference?
    A. $2\lambda_{film}$ B. λ_{film} C. $\lambda_{film}/2$ D. $\lambda_{film}/4$
15. | In a double-slit experiment, the fringe spacing on a screen 100 cm behind the slits is 4 mm. What will the fringe spacing be if the screen distance is doubled to 200 cm?
    A. 2 mm B. 4 mm C. 8 mm D. 16 mm
16. | Light passes through a diffraction grating with a slit spacing of 0.001 mm. A viewing screen is 100 cm behind the grating. If the light is blue, with a wavelength of 450 nm, at about what distance from the center of the interference pattern will the first-order maximum appear?
    A. 5 cm B. 25 cm
    C. 50 cm D. 100 cm
17. || Blue light of wavelength 450 nm passes through a diffraction grating with a slit spacing of 0.001 mm and makes an interference pattern on the wall. How many bright fringes will be seen?
    A. 1 B. 3 C. 5 D. 7
18. | Yellow light of wavelength 590 nm passes through a diffraction grating and makes an interference pattern on a screen 80 cm away. The first bright fringes are 1.9 cm from the central maximum. How many lines per mm does this grating have?
    A. 20 B. 40 C. 80 D. 200
19. | Light passes through a 10-μm-wide slit and is viewed on a screen 1 m behind the slit. If the width of the slit is narrowed, the band of light on the screen will
    A. Become narrower.
    B. Become wider.
    C. Stay about the same.
20. | A laser beam with a wavelength of 530 nm passes through a narrow slit and illuminates a screen 2.0 m behind the slit. A diffraction pattern is observed in which the central maximum has a width of 2.0 cm. Approximately how wide is the slit?
    A. 50 μm B. 100 μm
    C. 200 μm D. 1000 μm
21. || A laser shines through a pinhole, and it makes a diffraction pattern on a screen behind. The area of the circular hole is reduced by a factor of 2. What can you say about the new intensity?
    A. It is less than 1/2 the original intensity.
    B. It is 1/2 the original intensity.
    C. Is is the same as the original intensity.
    D. It is greater than the original intensity.
22. | You want to estimate the diameter of a very small circular pinhole that you've made in a piece of aluminum foil. To do so, you shine a red laser pointer ($\lambda = 632$ nm) at the hole and observe the diffraction pattern on a screen 3.5 m behind the foil. You measure the width of the central maximum to be 15 mm. What is the diameter of the hole?
    A. 0.18 mm B. 0.29 mm
    C. 0.36 mm D. 1.1 mm

PROBLEMS

Section 18.1 Models of Light

Section 18.2 Thin-Film Interference

1. || What is the thinnest film of MgF_2 ($n = 1.38$) on glass that produces a strong reflection for orange light with a wavelength of 600 nm?
2. || A very thin oil film ($n = 1.25$) floats on water ($n = 1.33$). What is the thinnest film that produces a strong reflection for green light with a wavelength of 500 nm?
3. || A thin film of MgF_2 ($n = 1.38$) coats a piece of glass. Constructive interference is observed for the reflection of light with wavelengths of 500 nm and 625 nm. What is the thinnest film for which this can occur?
4. || Solar cells are given antireflection coatings to maximize their efficiency. Consider a silicon solar cell ($n = 3.50$) coated with a layer of silicon dioxide ($n = 1.45$). What is the minimum coating thickness that will minimize the reflection at the wavelength of 700 nm, where solar cells are most efficient?
5. || Sunglasses often have an antireflection coating applied to the BIO *inner* surface of the lens to prevent the wearer from seeing a reflection of his or her eye.
   a. A 90-nm-thick coating is applied to the lens. What must be the coating's index of refraction to be most effective at 480 nm? Assume that the coating's index of refraction is less than that of the lens.
   b. If the index of refraction of the coating is 1.38, what thickness should the coating be so as to be most effective at 480 nm? The thinnest possible coating is best.
6. || The blue-ringed octopus BIO reveals the bright blue rings that give it its name as a warning display. The rings have a stack of reflectin (a protein used for structural color in many cephalopods) plates with index of refraction $n = 1.59$ separated by cells with index $n = 1.37$. The plates have thickness 62 nm. What is the longest wavelength in air of light that reflects strongly from the reflecting plates?

7. || Fossilization does not preserve pigments, but it does preserve BIO structure. Specimens from a rare cache of 50-million-year-old beetle fossils still show the microscopic layers that produced structural colors in the living creatures, and we can deduce the colors from an understanding of thin-film interference. One fossil had 80-nm-thick plates of fossilized chitin (modern samples have index of refraction $n = 1.56$) embedded in fossilized tissue (for which we can assume $n = 1.33$). What is the longest wavelength for which there is constructive interference for reflections from opposite sides of the chitin layers? If this wavelength is enhanced by reflection, what color was the beetle?
8. | A soap bubble is essentially a thin film of water surrounded by air. The colors you see in soap bubbles are produced by interference. What visible wavelengths of light are strongly reflected from a 390-nm-thick soap bubble? What color would such a soap bubble appear to be?
9. || Pigeons and doves have iri- BIO descent feathers around their necks. The color is produced by a 330-nm-thick layer of keratin ($n = 1.56$) with air on both sides that is found around the edge of the feather barbules. What wavelength or wavelengths of visible light are strongly reflected by this structure?

Section 18.3 Double-Slit Interference

10. || Light of 630 nm wavelength illuminates two slits that are 0.25 mm apart. Figure P18.10 shows the intensity pattern seen on a screen behind the slits. What is the distance to the screen?

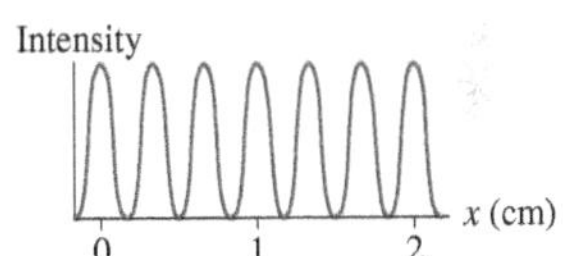

FIGURE P18.10

11. | Two narrow slits 50 μm apart are illuminated with light of wavelength 500 nm. The light shines on a screen 1.2 m distant. What is the angle in degrees of the $m = 2$ bright fringe? How far is this fringe from the center of the pattern?
12. || Light from a sodium lamp ($\lambda = 589$ nm) illuminates two narrow slits. The fringe spacing on a screen 150 cm behind the slits is 4.0 mm. What is the spacing (in mm) between the two slits?
13. || Two narrow slits are illuminated by light of wavelength λ. The slits are spaced 20 wavelengths apart. What is the angle, in radians, between the central maximum and the $m = 1$ bright fringe?
14. || Figure P18.14 shows the fringes observed in a double-slit interference experiment when the two slits are illuminated by white light. The central maximum is white because all of the colors overlap. This is not true for the other fringes. The $m = 1$ fringe clearly shows bands of color, with red appearing farther from the center of the pattern, and blue closer. If the slits that create this pattern are 20 μm apart and are located 0.85 m from the screen, what are the $m = 1$ distances from the central maximum for red (700 nm) and violet (400 nm) light?

FIGURE P18.14

15. || A double-slit experiment is performed with light of wavelength 600 nm. The bright interference fringes are spaced 1.8 mm apart on the viewing screen. What will the fringe spacing be if the light is changed to a wavelength of 400 nm?
16. || Light illuminating a pair of slits contains two wavelengths, 500 nm and an unknown wavelength. The 10th bright fringe of the unknown wavelength overlaps the 9th bright fringe of the 500 nm light. What is the unknown wavelength?

17. || Two narrow slits are 0.12 mm apart. Light of wavelength 550 nm illuminates the slits, causing an interference pattern on a screen 1.0 m away. Light from each slit travels to the $m = 1$ maximum on the right side of the central maximum. How much farther did the light from the left slit travel than the light from the right slit?

Section 18.4 The Diffraction Grating

18. || A diffraction grating with 750 slits/mm is illuminated by light that gives a first-order diffraction angle of 34.0°. What is the wavelength of the light?
19. || A commercial diffraction grating has 500 lines per mm. When a student shines a 530 nm laser through this grating, how many bright spots are seen on a screen behind the grating?
20. || A 1.0-cm-wide diffraction grating has 1000 slits. It is illuminated by light of wavelength 550 nm. What are the angles of the first two diffraction orders?
21. || Light of wavelength 620 nm illuminates a diffraction grating. The second-order maximum is at angle 39.5°. How many lines per millimeter does this grating have?
22. ||| A physics instructor wants to project a spectrum of visible-light colors from 400 nm to 700 nm as part of a classroom demonstration. She shines a beam of white light through a diffraction grating that has 500 lines per mm, projecting a pattern on a screen 2.4 m behind the grating.
    a. How wide is the spectrum that corresponds to $m = 1$?
    b. How much distance separates the end of the $m = 1$ spectrum and the start of the $m = 2$ spectrum?
23. || A diffraction grating with 600 lines/mm is illuminated with light of wavelength 500 nm. A very wide viewing screen is 2.0 m behind the grating.
    a. What is the distance between the two $m = 1$ fringes?
    b. How many bright fringes can be seen on the screen?
24. || A diffraction grating is illuminated simultaneously with red light of wavelength 660 nm and light of an unknown wavelength. The fifth-order maximum of the unknown wavelength exactly overlaps the third-order maximum of the red light. What is the unknown wavelength?
25. || BIO Figure P18.25 shows a microscopic view of muscle tissue. In the figure, structures called *sarcomeres* are bordered by ridges. The regular pattern of the ridges means that the muscle can operate as a reflection grating, and a measurement of the resulting diffraction pattern can give a measurement of the length of the sarcomeres, which is the distance between the ridges. This has been used to measure the sarcomere length during exercise, as the sarcomeres lengthen and shorten. In one case, an investigator shined a 632 nm laser on exposed muscle tissue in a patient's forearm as he moved his wrist back and forth through a 100° angle, contracting and then stretching the muscle. The resulting diffraction pattern was observed on a screen 2.4 cm from the muscle. The investigator measured the distance between the two $m = 2$ fringes on either side of the central maximum. This distance varied from 1.9 cm to 2.7 cm. What were the minimum and maximum values of the sarcomere length?

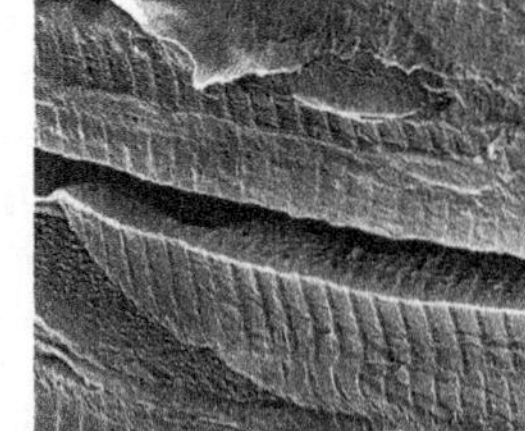
FIGURE P18.25

26. || BIO Glass catfish lack pigmentation and are largely transparent. If light passes through or reflects from one of these fish, regularly spaced striations in their muscles act as a diffraction grating. The reflection of white light can produce rainbow stripes along the fish. As the fish swims, moving its tail from side to side, muscle contractions and relaxations cause the striations to alternately shorten and lengthen. Investigators measured the changing striation width by passing a laser beam (wavelength 632 nm) through a swimming catfish, from one side to the other, and observing the diffraction pattern on a screen 30.0 cm on the other side of the fish. They found that the distance between the two first-order ($m = 1$) diffraction spots of the laser beam oscillated between 15.2 cm and 15.8 cm as the fish swam. What were the minimum and maximum striation widths in μm of the swimming fish? The diffraction occurred within the body of the fish, which has an index of refraction that is essentially that of water.

Section 18.5 Single-Slit Diffraction

27. || Figure P18.27 shows the light intensity on a screen behind a single slit. The wavelength of the light is 500 nm and the screen is 1.0 m behind the slit. What is the width (in mm) of the slit?

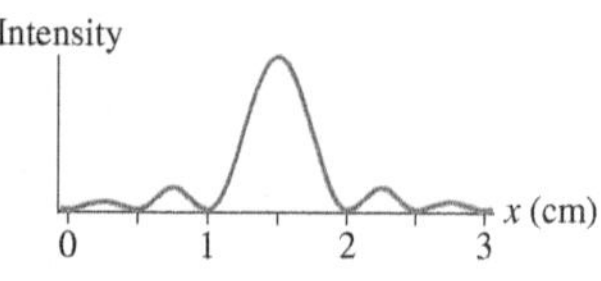

FIGURE P18.27

28. || A helium-neon laser ($\lambda = 633$ nm) illuminates a single slit and is observed on a screen 1.50 m behind the slit. The distance between the first and second minima in the diffraction pattern is 4.75 mm. What is the width (in mm) of the slit?
29. || For a demonstration, a professor uses a razor blade to cut a thin slit in a piece of aluminum foil. When she shines a laser pointer ($\lambda = 680$ nm) through the slit onto a screen 5.5 m away, a diffraction pattern appears. The bright band in the center of the pattern is 8.0 cm wide. What is the width of the slit?
30. || A 0.50-mm-wide slit is illuminated by light of wavelength 500 nm. What is the width of the central maximum on a screen 2.0 m behind the slit?
31. || For what slit-width-to-wavelength ratio does the first minimum of a single-slit diffraction pattern appear at (a) 30°, (b) 60°, and (c) 90°?
32. || An early practical application of diffraction measured the diameters of wool fibers. One instrument used a collimated beam of light with a wavelength of 560 nm to produce a diffraction pattern on a screen 30 cm behind a stretched fiber. In one measurement, the width of the central maximum was 21 mm. What was the fiber diameter in μm?
33. || Quality control systems have been developed to remotely measure the diameter of wires using diffraction. A wire with a stated diameter of 170 μm blocks the beam of a 633 nm laser, producing a diffraction pattern on a screen 50.0 cm distant. The width of the central maximum is measured to be 3.77 mm. The wire should have a diameter within 1% of the stated value. Does this wire pass the test?

Section 18.6 Circular-Aperture Diffraction

34. || A 0.50-mm-diameter hole is illuminated by light of wavelength 500 nm. What is the width of the central maximum on a screen 2.0 m behind the slit?

35. || Infrared light of wavelength 2.5 μm illuminates a 0.20-mm-diameter hole. What is the angle of the first dark fringe in radians? In degrees?

36. || Light from a helium-neon laser ($\lambda = 633$ nm) passes through a circular aperture and is observed on a screen 4.0 m behind the aperture. The width of the central maximum is 2.5 cm. What is the diameter (in mm) of the hole?

37. ||| You want to photograph a circular diffraction pattern whose central maximum has a diameter of 1.0 cm. You have a helium-neon laser ($\lambda = 633$ nm) and a 0.12-mm-diameter pinhole. How far behind the pinhole should you place the viewing screen?

38. || Your artist friend is designing an exhibit inspired by circular-aperture diffraction. A pinhole in a red zone will be illuminated with a red laser beam of wavelength 670 nm, while a pinhole in a violet zone is going to be illuminated with a violet laser beam of wavelength 410 nm. She wants all the diffraction patterns seen on a distant screen to have the same size. For this to work, what must be the ratio of the red pinhole's diameter to that of the violet pinhole?

39. || Investigators measure the size of fog droplets using the diffraction of light. A camera records the diffraction pattern on a screen as the droplets pass in front of a laser, and a measurement of the size of the central maximum gives the droplet size. In one test, a 690 nm laser creates a pattern on a screen 30 cm from the droplets. If the central maximum of the pattern is 0.28 cm in diameter, how large is the droplet?

40. ||| Diffraction can be used to provide a quick test of the size BIO of red blood cells. Blood is smeared onto a slide, and a laser shines through the slide. The size of the cells is very consistent, so the multiple diffraction patterns overlap and produce an overall pattern that is similar to what a single cell would produce. Ideally, the diameter of a red blood cell should be between 7.5 and 8.0 μm. If a laser beam with a wavelength of 633 nm passes through a slide and produces a pattern on a screen 24.0 cm distant, what range of diameters of the central maximum should be expected? Values outside this range might indicate a health concern and warrant further study.

Section 18.7 X Rays and X-Ray Diffraction

41. | X rays with a wavelength of 0.085 nm diffract from a crystal in which the spacing between atomic planes is 0.18 nm. How many diffraction orders are observed?

42. ||| X rays diffract from a crystal in which the spacing between atomic planes is 0.175 nm. The second-order diffraction occurs at 45.0°. What is the angle of the first-order diffraction?

43. ||| X rays with a wavelength of 0.20 nm undergo first-order diffraction from a crystal at a 54° angle of incidence. At what angle does first-order diffraction occur for x rays with a wavelength of 0.15 nm?

44. || The spacing between atomic planes in a crystal is 115 pm. If x rays with a wavelength of 90.0 pm are diffracted by this crystal, what are the angles of (a) first-order and (b) second-order diffraction?

45. || A crystal sample of bacteriorhodopsin, a light-sensitive pro- BIO tein found in halobacteria that responds to light energy, has crystal planes separated by 0.20 nm. If a beam of x rays with a wavelength of 0.11 nm illuminates a sample, what angles will give diffraction maxima?

General Problems

46. || A sheet of glass is coated with a 500-nm-thick layer of oil ($n = 1.42$).
    a. For what *visible* wavelengths of light do the reflected waves interfere constructively?
    b. For what *visible* wavelengths of light do the reflected waves interfere destructively?
    c. What is the color of reflected light? What is the color of transmitted light?

47. ||| Scientists are testing a new thin-film coating that has an index of refraction that is less than that of glass. They deposit a 280-nm-thick layer on glass and then measure the reflectivities for different wavelengths of incident light. The reflectivity is a maximum when the wavelength is 400 nm, and it steadily decreases as the wavelength is increased until it first reaches a minimum when the wavelength is 533 nm. What is the coating's index of refraction?

48. ||| Scientists are testing a transparent material whose index of refraction for visible light varies with wavelength as $n = 30.0\ \text{nm}^{1/2}/\lambda^{1/2}$, where λ is in nm. If a 295-nm-thick coating is placed on glass ($n = 1.50$) for what visible wavelengths will the reflected light have maximum constructive interference?

49. |||| A laboratory dish, 20 cm in diameter, is half filled with water. INT One at a time, 0.50 μL drops of oil from a micropipette are dropped onto the surface of the water, where they spread out into a uniform thin film. After the first drop is added, the intensity of 600 nm light reflected from the surface is very low. As more drops are added, the reflected intensity increases, then decreases again to a minimum after a total of 13 drops have been added. What is the index of refraction of the oil?

50. || Figure P18.50 shows the light intensity on a screen behind a double slit. The slit spacing is 0.20 mm and the wavelength of the light is 600 nm. What is the distance from the slits to the screen?

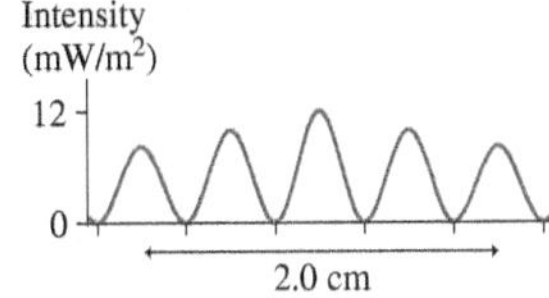

FIGURE P18.50

51. || Figure P18.50 shows the light intensity on a screen behind a double slit. The slit spacing is 0.20 mm and the screen is 2.0 m behind the slits. What is the wavelength of the light?

52. || Figure P18.52 shows the light intensity on a screen 2.5 m behind a double slit. The wavelength of the light is 532 nm. What is the spacing between the slits?

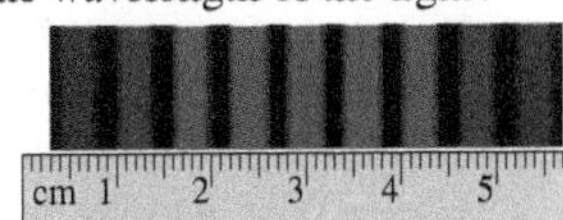

FIGURE P18.52

53. || In a double-slit experiment, the slit separation is 200 times the wavelength of the light. What is the angular separation (in degrees) between two adjacent bright fringes?

54. || A double-slit interference pattern is created by two narrow slits spaced 0.25 mm apart. The distance between the first and the fifth minimum on a screen 60 cm behind the slits is 5.5 mm. What is the wavelength (in nm) of the light used in this experiment?

55. || The two most prominent wavelengths in the light emitted by a hydrogen discharge lamp are 656 nm (red) and 486 nm (blue). Light from a hydrogen lamp illuminates a diffraction grating with 500 lines/mm, and the light is observed on a screen 1.50 m behind the grating. What is the distance between the first-order red and blue fringes?

56. ||| White light (400–700 nm) is incident on a 600 line/mm diffraction grating. What is the width of the first-order rainbow on a screen 2.0 m behind the grating?

57. |||| A small spectrometer for chemical analysis uses a diffraction grating with 800 slits/mm that is set 25.0 mm in front of a detector like those used in digital cameras. The detector can barely resolve two bright fringes that are 30 μm apart; two fringes that are less than 30 μm apart on the detector blend into a single, unresolved fringe. The *resolution* of a spectrometer is defined to be the wavelength separation $\Delta\lambda$ between two distinct wavelengths that can barely be distinguished. What is the first-order resolution of this spectrometer at a wavelength of 600 nm? That is, what is the separation in nm between two barely distinguished wavelengths if one of them is 600.0 nm?

58. | Figure P18.58 shows the interference pattern on a screen 1.0 m behind an 800 line/mm diffraction grating. What is the wavelength of the light?

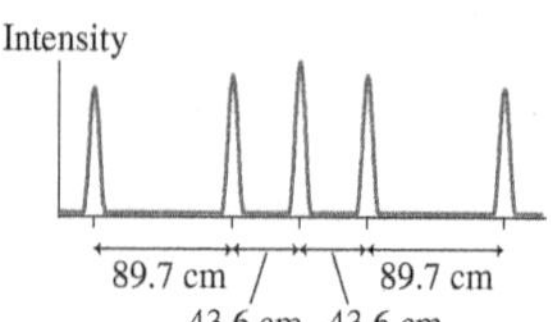

FIGURE P18.58

59. || Figure P18.58 shows the interference pattern on a screen 1.0 m behind a diffraction grating. The wavelength of the light is 600 nm. How many lines per millimeter does the grating have?

60. || The shiny surface of a DVD is imprinted with millions of tiny pits, arranged in a pattern of thousands of essentially concentric circles that act like a reflection grating when light shines on them. You decide to determine the distance between those circles by aiming a laser pointer (with $\lambda = 680$ nm) perpendicular to the disc and measuring the diffraction pattern reflected onto a screen 1.5 m from the disc. The central bright spot you expected to see is blocked by the laser pointer itself. You do find two other bright spots separated by 1.4 m, one on either side of the missing central spot. The rest of the pattern is apparently diffracted at angles too great to show on your screen. What is the distance between the circles on the DVD's surface?

61. || BIO The wings of some beetles have closely spaced parallel lines of melanin, causing the wing to act as a reflection grating. Suppose sunlight shines straight onto a beetle wing. If the melanin lines on the wing are spaced 2.0 μm apart, what is the first-order diffraction angle for green light ($\lambda = 550$ nm)?

62. || BIO The bright colors on the head and neck of a male Anna's hummingbird are structural colors due not to pigments but to constructive thin-film interference from layered structures in the feathers. The electron micrograph in Figure P18.62 shows layers of air trapped in a matrix of keratin and melanin. What is the minimum-thickness air layer that gives a strong reflection at the experimentally observed wavelength of 670 nm?

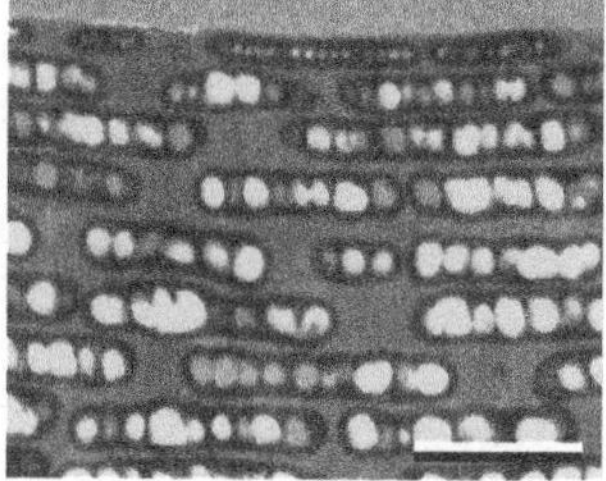

FIGURE P18.62

63. |||| Light emitted by element X passes through a diffraction grating that has 1200 slits/mm. The interference pattern is observed on a screen 75.0 cm behind the grating. First-order maxima are observed at distances of 56.2 cm, 65.9 cm, and 93.5 cm from the central maximum. What are the wavelengths of light emitted by element X?

64. ||| Light of a single wavelength is incident on a diffraction grating with 500 slits/mm. Several bright fringes are observed on a screen behind the grating, including one at 45.7° and one next to it at 72.6°. What is the wavelength of the light?

65. || INT Because sound is a wave, it's possible to make a diffraction grating for sound from a large board of sound-absorbing material with several parallel slits cut for sound to go through. When 10 kHz sound waves pass through such a grating, listeners 10 m from the grating report "loud spots" 1.4 m on both sides of center. What is the spacing between the slits? Use 340 m/s for the speed of sound.

66. || A chemist identifies compounds by identifying bright lines in their spectra. She does so by heating the compounds until they glow, sending the light through a diffraction grating, and measuring the positions of first-order spectral lines on a detector 15.0 cm behind the grating. Unfortunately, she has lost the card that gives the specifications of the grating. Fortunately, she has a known compound that she can use to calibrate the grating. She heats the known compound, which emits light at a wavelength of 461 nm, and observes a spectral line 9.95 cm from the center of the diffraction pattern. What are the wavelengths emitted by compounds A and B that have spectral lines detected at positions 8.55 cm and 12.15 cm, respectively?

67. || A 600 line/mm diffraction grating is in an empty aquarium tank. The index of refraction of the glass walls is $n_{glass} = 1.50$. A helium-neon laser ($\lambda = 633$ nm) is outside the aquarium. The laser beam passes through the glass wall and illuminates the diffraction grating.

a. What is the first-order diffraction angle of the laser beam?

b. What is the first-order diffraction angle of the laser beam after the aquarium is filled with water ($n_{water} = 1.33$)?

68. || You need to use your cell phone, which broadcasts an 830 MHz signal, but you're in an alley between two massive, radio-wave-absorbing buildings that have only a 15 m space between them. What is the angular width, in degrees, of the electromagnetic wave after it emerges from between the buildings?

69. || In a single-slit experiment, the slit width is 200 times the wavelength of the light. What is the width (in mm) of the central maximum on a screen 2.0 m behind the slit?

70. || Light from a helium-neon laser ($\lambda = 633$ nm) is incident on a single slit. What is the largest slit width for which there are no minima in the diffraction pattern?

71. || Figure P18.71 shows the light intensity on a screen 2.5 m behind an aperture. The aperture is illuminated with light of wavelength 600 nm.
   a. Is the aperture a single slit or a double slit? Explain.
   b. If the aperture is a single slit, what is its width? If it is a double slit, what is the spacing between the slits?

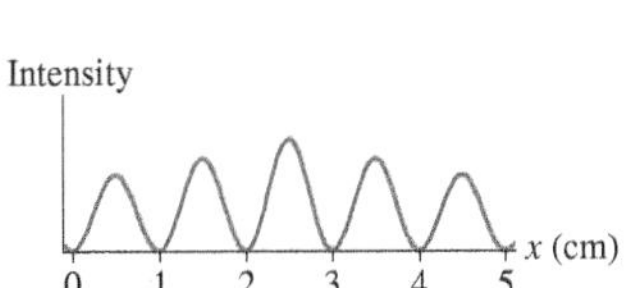

FIGURE P18.71

FIGURE P18.72

72. ||| Occasionally, when the full moon is covered by a very thin cloud layer, you may see a *lunar corona*, a bright area wider than the moon with a reddish fringe. A lunar corona is produced when moonlight passes through a thin layer of water droplets of very uniform size. Each droplet diffracts the light, and the diffraction patterns from individual droplets overlap to produce a pattern similar to that of a single drop. The bright area of the lunar corona is the central maximum of the diffraction pattern, and the red fringe occurs because longer wavelengths are diffracted at larger angles. The photograph in Figure P18.72 shows a lunar corona in which the outer, reddish edge of the central maximum has a diameter about 10 times that of the moon. The angular diameter of a full moon is almost exactly 0.50°. What is the diameter in μm of the droplets that cause this lunar corona? Use 650 nm for the wavelength of red light.

73. ||| One day, after pulling down your window shade, you notice that sunlight is passing through a pinhole in the shade and making a small patch of light on the far wall. Having recently studied optics in your physics class, you're not too surprised to see that the patch of light seems to be a circular diffraction pattern. It appears that the central maximum is about 3 cm across, and you estimate that the distance from the window shade to the wall is about 3 m. Knowing that the average wavelength of sunlight is about 500 nm, estimate the diameter of the pinhole.

74. || A radar for tracking aircraft broadcasts a 12 GHz microwave
INT beam from a 2.0-m-diameter circular radar antenna. From a wave perspective, the antenna is a circular aperture through which the microwaves diffract.
   a. What is the diameter of the radar beam at a distance of 30 km?
   b. If the antenna emits 100 kW of power, what is the average microwave intensity at 30 km?

75. || X rays with a wavelength of 0.0700 nm diffract from a crystal. Two adjacent angles of x-ray diffraction are 45.6° and 21.0°. What is the distance in nm between the atomic planes responsible for the diffraction?

MCAT-Style Passage Problems

The Blue Morpho Butterfly BIO

The brilliant blue color of a blue morpho butterfly is, like the colors of peacock feathers, due to interference. Figure P18.76a shows an easy way to demonstrate this: If a drop of the clear solvent acetone is placed on the wing of a blue morpho butterfly, the color changes from a brilliant blue to an equally brilliant green—returning to blue once the acetone evaporates. There would be no change if the color were due to pigment.

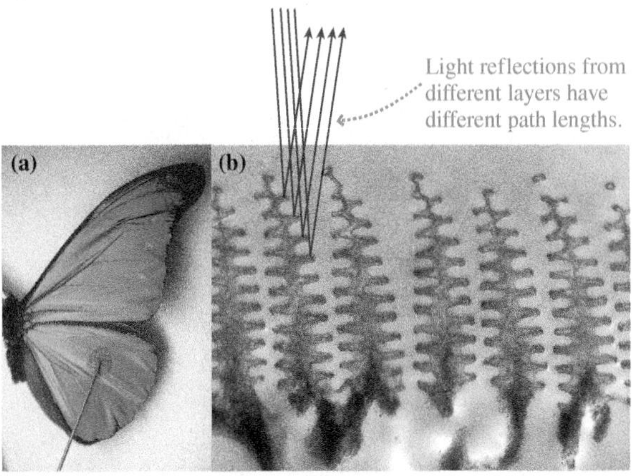

FIGURE P18.76

A cross section of a scale from the wing of a blue morpho butterfly reveals the source of the butterfly's color. As Figure P18.76b shows, the scales are covered with structures that look like small Christmas trees. Light striking the wings reflects from different layers of these structures, and the differing path lengths cause the reflected light to interfere constructively or destructively, depending on the wavelength. For light at normal incidence, blue light experiences constructive interference while other colors undergo destructive interference and cancel. Acetone fills the spaces in the scales with a fluid of index of refraction $n = 1.38$; this changes the conditions for constructive interference and results in a change in color.

76. The coloring of the blue morpho butterfly is protective. As the butterfly flaps its wings, the angle at which light strikes the wings changes. This causes the butterfly's color to change and makes it difficult for a predator to follow. This color change is because
   A. A diffraction pattern appears only at certain angles.
   B. The index of refraction of the wing tissues changes as the wing flexes.
   C. The motion of the wings causes a Doppler shift in the reflected light.
   D. As the angle changes, the differences in paths among light reflected from different surfaces change, resulting in constructive interference for a different color.

77. The change in color when acetone is placed on the wing is due to the difference between the indices of refraction of acetone and air. Consider light of some particular color. In acetone,
   A. The frequency of the light is less than in air.
   B. The frequency of the light is greater than in air.
   C. The wavelength of the light is less than in air.
   D. The wavelength of the light is greater than in air.

78. The scales on the butterfly wings are actually made of a transparent material with index of refraction 1.56. Light reflects from the surface of the scales because
   A. The scales' index of refraction is different from that of air.
   B. The scales' index of refraction is similar to that of glass.
   C. The scales' density is different from that of air.
   D. Different colors of light have different wavelengths.

19 Ray Optics

The colors of a rainbow are created when light rays from the sun refract through and reflect from water droplets in the air.

LOOKING AHEAD

Reflection

Light rays can *reflect* from a surface. The still surface of the water allows us to see both the willet and its mirror image.

You'll learn that the *law of reflection* can be used to understand how images are formed by mirrors and other flat, reflective surfaces.

Refraction

The pencil is not really broken. Instead, light rays from the pencil have been bent as they cross the boundary between the air and the plastic block.

This bending of light rays at a boundary is called *refraction.* You'll learn that the angles of refraction are given by *Snell's law.*

Lenses and Mirrors

Refraction and reflection at curved surfaces allow lenses and mirrors to form *images.* Sometimes, as for this ladybug, the image is magnified.

You'll learn both graphical and mathematical techniques for locating and characterizing the images made by lenses and mirrors.

GOAL To apply the ray model of light to the phenomena of reflection, refraction, and image formation.

PHYSICS AND LIFE

Seeing the World

Eyes form images that allow creatures to perceive and respond to their surroundings. Myriad types of eyes have evolved to function in a stunning variety of biological niches and to address a wide range of biological needs. Some eyes respond to very dim light, others to very bright light. Some eyes function underwater, and yet others respond to the polarization of light waves. The image-forming optics of almost all eyes are well described by the ray model. The ray model also explains how light travels through optical fibers to allow physicians to view images from inside a patient, making possible minimally invasive diagnostics and laparoscopic surgery. This chapter explores the basic principles by which light is focused and guided. Then in Chapter 20 we'll apply this knowledge to eyes and to optical instruments that correct or extend what eyes can do.

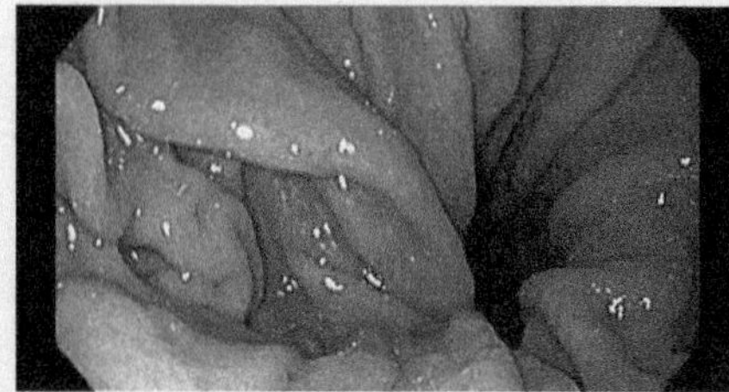

An *endoscope* uses ray optics to peer inside the body. Here we see a hiatal hernia in which the stomach is protruding into the thorax through a hole in the diaphragm.

19.1 The Ray Model of Light

A flashlight makes a beam of light through the night's darkness, sunbeams stream into a darkened room through a small hole in the shade, and laser beams are even more well defined. Our everyday experience that light travels in straight lines is the basis of the ray model of light.

The ray model is an oversimplification of reality, but nonetheless is very useful within its range of validity. As we saw in Chapter 18, diffraction and other wave aspects of light are important only for apertures and objects comparable in size to the wavelength of light. Because the wavelength is so small, typically 0.5 μm, the wave nature of light is not apparent when light interacts with ordinary-sized objects. The ray model of light, which ignores diffraction, is a more appropriate model as long as any apertures through which the light passes (lenses, mirrors, holes, and the like) are larger than about 1 mm.

To begin, let us define a **light ray** as a line in the direction along which light energy is flowing. A light ray is an abstract idea, not a physical entity or a "thing." Any narrow beam of light, such as the laser beam in FIGURE 19.1, is actually a bundle of many parallel light rays. You can think of a single light ray as the limiting case of a laser beam whose diameter approaches zero. Laser beams are good approximations of light rays, certainly adequate for demonstrating ray behavior, but any real laser beam is a bundle of many parallel rays.

FIGURE 19.1 A laser beam is a bundle of parallel light rays.

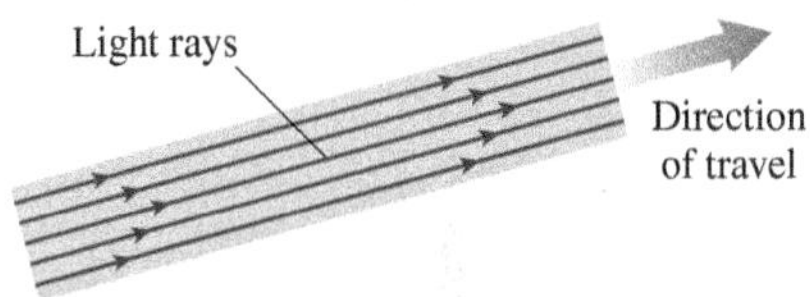

The following table outlines five basic ideas of the ray model of light.

The ray model of light

Light rays are straight lines.

Light travels through a vacuum or a transparent material in straight lines called light rays. The speed of light in a material is $v = c/n$, where n is the index of refraction of the material.

Light rays can cross.

Light rays do not interact with each other. Two rays can cross without either being affected in any way.

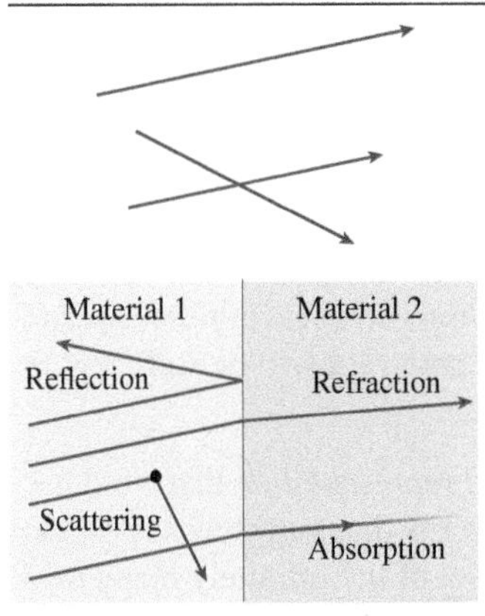

A light ray goes on forever unless it interacts with matter.

A light ray continues forever unless it has an interaction with matter that causes the ray to change direction or to be absorbed. Light interacts with matter in four different ways:

- At an interface between two materials, light can be *reflected, refracted,* or both.
- Within a material, light can be either *scattered* or *absorbed.*

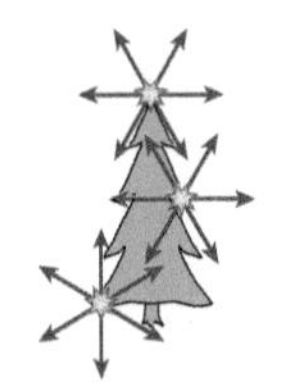

An object is a source of light rays.

An **object** is a source of light rays. Rays originate from *every* point on the object, and each point sends rays in *all* directions. Objects may be self-luminous—they create light rays—or they may be reflective objects that reflect only rays that originate elsewhere.

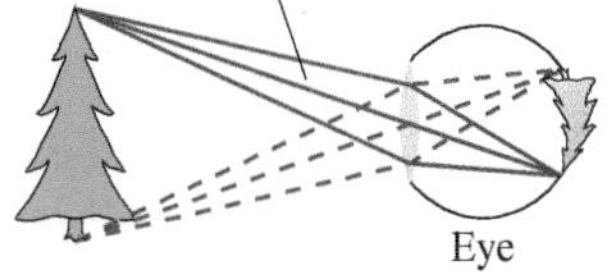

The eye sees by focusing a bundle of rays.

The eye sees an object when *diverging* bundles of rays from each point on the object enter the pupil and are focused to an image on the retina. Imaging is discussed later in the chapter, and the eye will be treated in much greater detail in Chapter 20.

Light Sources

FIGURE 19.2 (next page) illustrates the idea that sources of light—what we call *objects*—can be either *self-luminous,* such as the sun, flames, and lightbulbs, or *reflective.* Most objects are reflective. A tree, unless it is on fire, is seen or photographed by virtue of

FIGURE 19.2 Self-luminous and reflective objects.

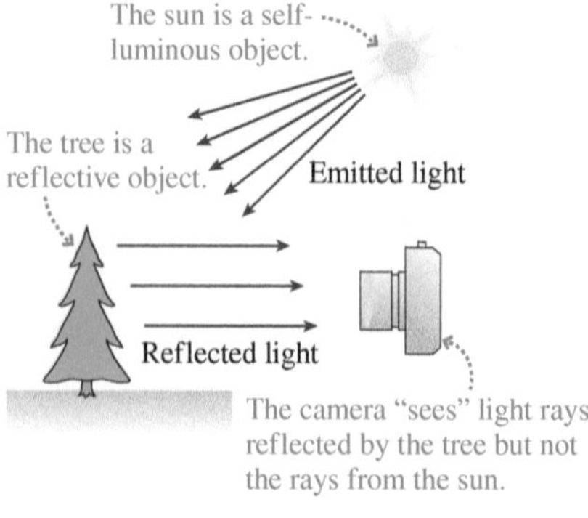

The sun is a self-luminous object.
The tree is a reflective object.
Emitted light
Reflected light
The camera "sees" light rays reflected by the tree but not the rays from the sun.

reflected sunlight or reflected skylight. People, houses, and a page in a printed book reflect light from self-luminous sources. In this chapter we are concerned not with how the light originates but with how it behaves after leaving the object.

Light rays from an object are emitted in all directions, but you are not *aware* of light rays unless they enter the pupil of your eye. Consequently, most light rays go completely unnoticed. For example, light rays travel from the sun to the tree in Figure 19.2, but you're not aware of these unless the tree reflects some of them into your eye.

Or consider a laser beam. You've probably noticed that it's almost impossible to see a laser beam from a laser pointer as it crosses the room unless there's dust in the air. The dust scatters a few of the light rays toward your eye, but in the absence of dust you would be completely unaware of a very powerful light beam traveling past you. **Light rays exist independently of whether you are seeing them.**

FIGURE 19.3 shows two idealized sets of light rays. The diverging rays from a **point source** are emitted in all directions. It is useful to think of each point on an object as a point source of light rays. A **parallel bundle** of rays could be a laser beam. Alternatively it could represent a *distant object,* an object such as a star so far away that the rays arriving at the observer are essentially parallel to each other.

FIGURE 19.3 Point sources and parallel bundles represent idealized objects.

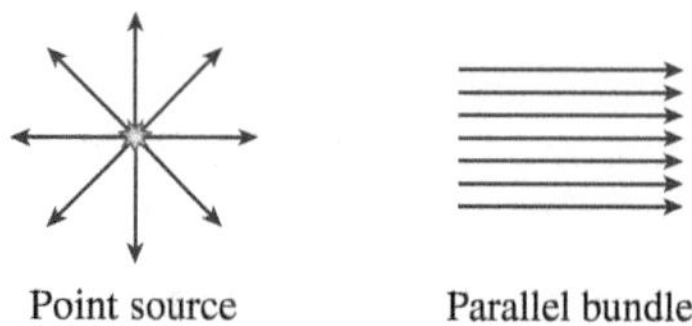

Ray Diagrams

FIGURE 19.4 A ray diagram simplifies the situation by showing only a few rays.

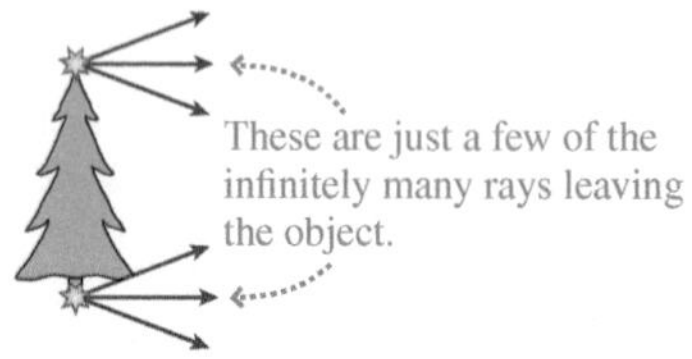

Rays originate from *every* point on an object and travel outward in *all* directions, but a diagram trying to show all these rays would be hopelessly messy and confusing. To simplify the picture, we usually use a **ray diagram** showing only a few rays. For example, FIGURE 19.4 is a ray diagram showing only a few rays leaving the top and bottom points of the object and traveling to the right. These rays will be sufficient to show us how the object is imaged by lenses or mirrors.

NOTE ▸ Ray diagrams are the basis for a *pictorial representation* that we'll use throughout this chapter. Be careful not to think that a ray diagram shows all of the rays. The rays shown on the diagram are just a subset of the infinitely many rays leaving the object. ◂

Apertures

FIGURE 19.5 A camera obscura.

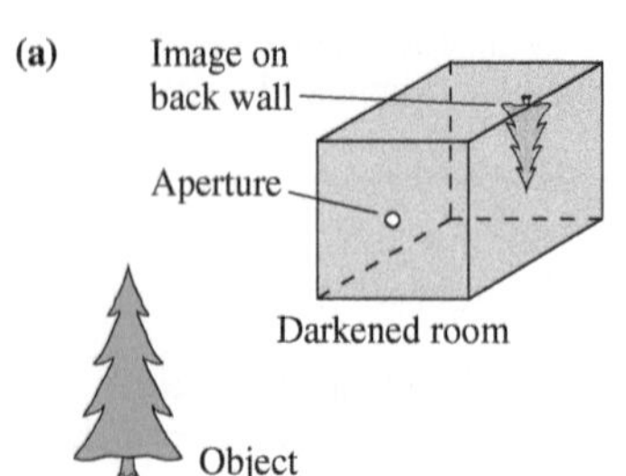

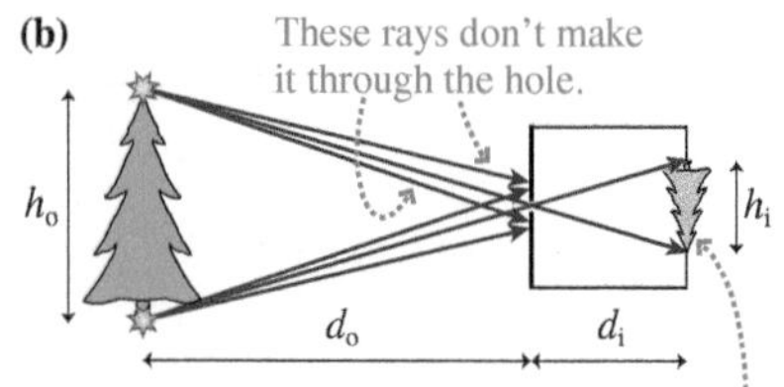

It has been known since ancient times that light passing through a small hole in an opaque barrier casts a dim but full-color image of the outside world onto a wall, as shown in FIGURE 19.5a. Such a device is known as a *camera obscura,* Latin for "dark room." A *pinhole camera* is a miniature version of a camera obscura. But there is one problem with viewing the image: It's upside down!

A hole through which light passes is called an **aperture**. FIGURE 19.5b uses the ray model of light passing through a small aperture to explain how the camera obscura works. Each point on an object emits light rays in all directions, but only a very few of these rays pass through the aperture and reach the back wall. As the figure illustrates, the geometry of the rays causes the image to be upside down.

Actually, as you may have realized, each *point* on the object illuminates a small but extended *patch* on the wall. This is because the nonzero size of the aperture—needed for the image to be bright enough to see—allows several rays from each point on the object to pass through at slightly different angles. As a result, the image is slightly blurred and out of focus. (Diffraction also becomes an issue if the hole gets too small.) We'll later discover how a modern camera, with a lens, improves on the camera obscura.

You can see from the similar triangles in Figure 19.5b that the object and image heights are related by

$$\frac{h_i}{h_o} = \frac{d_i}{d_o} \tag{19.1}$$

where d_o is the distance to the object and d_i is the depth of the camera obscura. Any realistic camera obscura has $d_i < d_o$; thus the image is smaller than the object.

We can apply the ray model to more complex apertures, such as the L-shaped aperture in **FIGURE 19.6**. The pattern of light on the screen is found by tracing all the straight-line paths—the ray trajectories—that start from the point source and pass through the aperture. We will see an enlarged L on the screen, with a sharp boundary between the image and the dark shadow.

FIGURE 19.6 Light through an aperture.

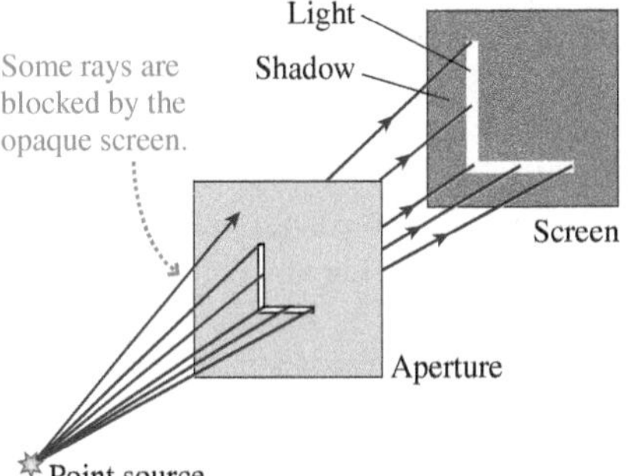

STOP TO THINK 19.1 A long, thin lightbulb illuminates a vertical aperture. Which pattern of light do you see on a viewing screen behind the aperture?

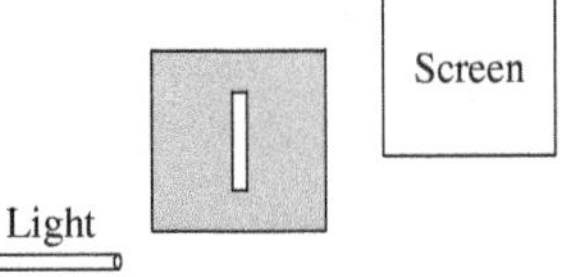

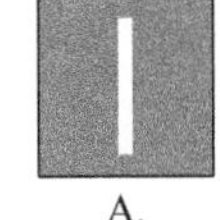
A.

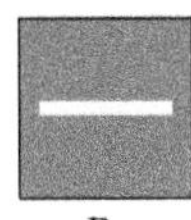
B.

C.

D.

Pinhole vision The eye of the chambered nautilus has no lens; it is a pinhole camera. A nautilus can acquire a general sense of nearby objects and possible predators, but its visual acuity is very low.

19.2 Reflection

As you know, light is reflected from smooth metal surfaces, like the silvered backs of mirrors. You see your reflection in the bathroom mirror first thing every morning as well as reflections in your car's rearview mirror as you drive to school. Reflection from a smooth, shiny surface is called **specular reflection,** from *speculum,* Latin for "mirror."

FIGURE 19.7a shows a bundle of parallel light rays reflecting from a mirror-like surface. You can see that the incident and reflected rays are both in a plane that is normal, or perpendicular, to the reflective surface. A three-dimensional perspective accurately shows the relationship between the light rays and the surface, but figures such as this are hard to draw by hand. Instead, it is customary to represent reflection with the simpler pictorial representation of **FIGURE 19.7b**. In this figure:

- The incident and reflected rays are in the plane of the page. The reflective surface extends into and out of the page.
- A *single* light ray represents the entire bundle of parallel rays. This is oversimplified, but it keeps the figure and the analysis clear.

The angle θ_i between the incident ray and a line perpendicular to the surface—the *normal* to the surface—is called the **angle of incidence.** Similarly, the **angle of reflection** θ_r is the angle between the reflected ray and the normal to the surface. The **law of reflection,** easily demonstrated with simple experiments, states:

1. The incident ray and the reflected ray are both in the same plane, which is perpendicular to the surface, and
2. The angle of reflection equals the angle of incidence: $\theta_r = \theta_i$.

NOTE ► Optics calculations *always* use the angle measured from the normal, not the angle between the ray and the surface. ◄

FIGURE 19.7 Specular reflection of light.

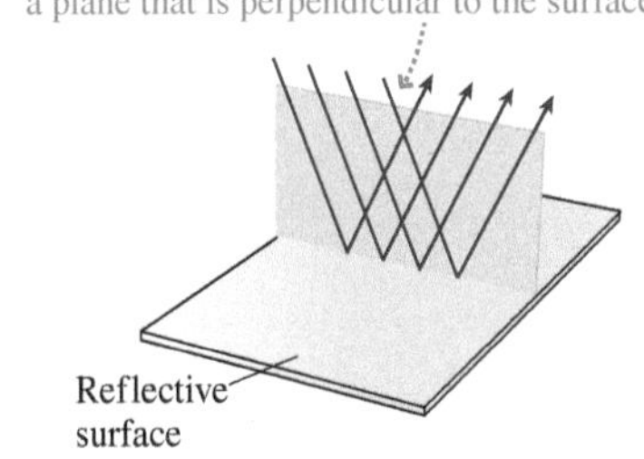

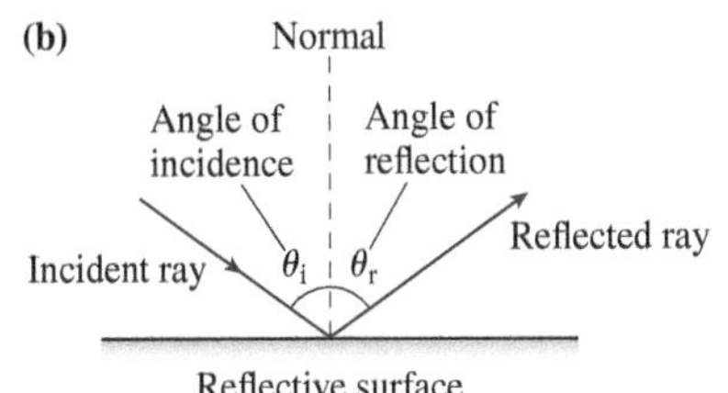

EXAMPLE 19.1 Light reflecting from a mirror

A full-length mirror on a closet door is 2.0 m tall. The bottom touches the floor. A bare lightbulb hangs 1.0 m from the closet door, 0.5 m above the top of the mirror. How long is the streak of reflected light across the floor?

PREPARE We will treat the lightbulb as a point source, then trace rays of light from the bulb and use the law of reflection to see where they strike the floor. The range of the rays that strike the floor will be the length of the streak. FIGURE 19.8 is a pictorial representation of the light rays. We need to consider only the two rays that strike the edges of the mirror. All other reflected rays will fall between these two.

FIGURE 19.8 Pictorial representation of light rays reflecting from a mirror.

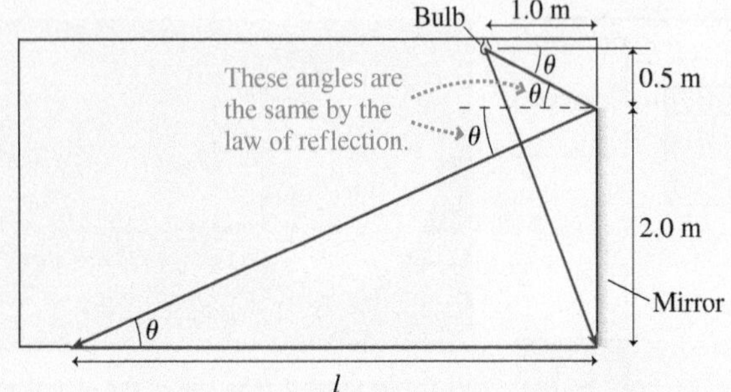

SOLVE The ray that strikes the bottom of the mirror reflects from it and hits the floor just where the mirror meets the floor. For the top ray, Figure 19.8 has used the law of reflection to set the angle of reflection equal to the angle of incidence; we call both θ. We can use geometry to see that θ is also the angle of the incident ray below horizontal and the angle that the reflected ray makes with the floor. From the small triangle at the upper right,

$$\theta = \tan^{-1}\left(\frac{0.5\text{ m}}{1.0\text{ m}}\right) = 26.6°$$

But we also have $\tan\theta = (2.0\text{ m})/l$, or

$$l = \frac{2.0\text{ m}}{\tan\theta} = \frac{2.0\text{ m}}{\tan 26.6°} = 4.0\text{ m}$$

The lower ray struck right at the mirror's base, so the total length of the reflected streak is 4.0 m.

ASSESS The pictorial representation was drawn more or less to scale, and the length of the streak on the floor seems about twice as long as the mirror, so we can have confidence in the results of our calculation.

Diffuse Reflection

Most objects are seen by virtue of their reflected light. On a microscopic scale, the surface of most objects is rough or irregular, with texture features larger than the wavelength of light. The law of reflection $\theta_r = \theta_i$ is still obeyed at each point, but the irregularities of the surface cause the reflected rays to leave in many random directions. This situation, shown in FIGURE 19.9, is called **diffuse reflection.** It is how you see the wall, your hand, and so on. Diffuse reflection is much more common than the mirror-like specular reflection.

FIGURE 19.9 Diffuse reflection from an irregular surface.

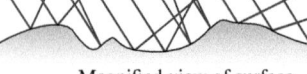

Each ray obeys the law of reflection at that point, but the irregular surface causes the reflected rays to leave in many random directions.

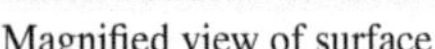

Magnified view of surface

The Plane Mirror

When you look at yourself in a mirror, how does the reflection of light lead to your seeing an image of yourself? FIGURE 19.10a shows rays from point source P reflecting from a flat mirror, called a **plane mirror.** Consider the ray shown in FIGURE 19.10b. This ray reflects with $\theta_r = \theta_i$ and then travels along a line that appears to have passed through point P′ behind the mirror. Because our argument applies to any incoming ray, *all* reflected rays appear to be coming from point P′, as FIGURE 19.10c shows. We call P′, the point from which the reflected rays diverge and from where they *appear* to originate, the **virtual image** of P. The image is "virtual" in the sense that no rays actually leave P′, which is in darkness behind the mirror. But as far as your eye is concerned, the light rays act exactly *as if* the light really originated at P′. So while you may say "I see P in the mirror," what you are actually seeing is the virtual image of P.

FIGURE 19.10 The light rays reflecting from a plane mirror.

(a)

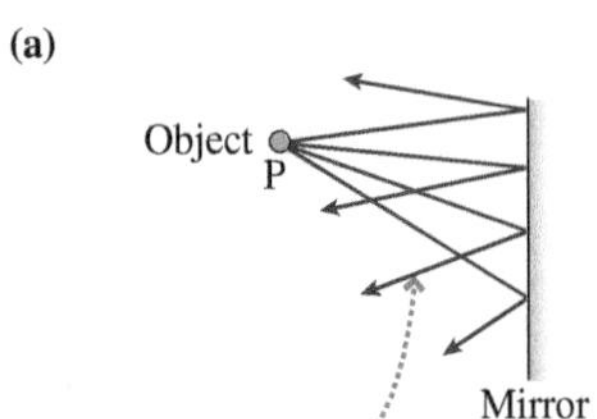

Rays from P reflect from the mirror. Each ray obeys the law of reflection.

(b)

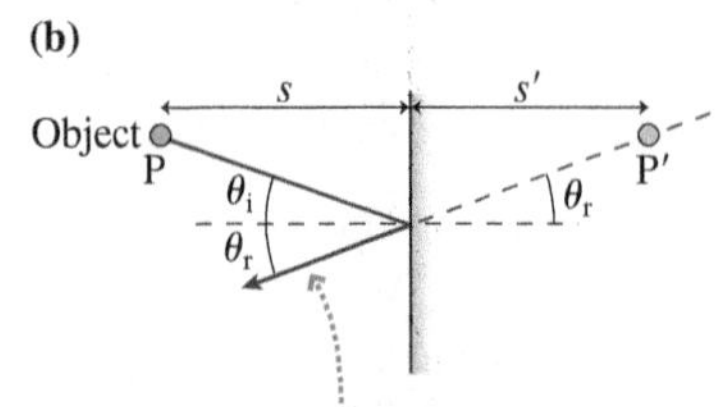

This reflected ray appears to have been traveling along a line that passed through point P′.

(c)

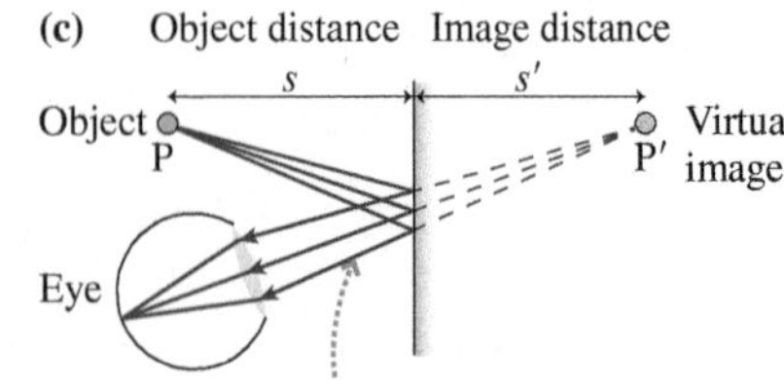

The reflected rays *all* diverge from P′, which appears to be the source of the reflected rays. Your eye collects the bundle of diverging rays and "sees" the light coming from P′.

In Figure 19.10b, simple geometry dictates that P′ is the same distance behind the mirror as P is in front of the mirror. That is, the **image distance** s' is equal to the **object distance** s:

$$s' = s \quad \text{(plane mirror)} \tag{19.2}$$

NOTE ▶ In ray optics we use primes to distinguish images from objects. Thus P′ is an image point and s' is an image distance. ◀

For an extended object, such as the one in FIGURE 19.11, each point on the object has a corresponding image point an equal distance on the opposite side of the mirror. The eye captures and focuses diverging bundles of rays from each point of the image in order to see the full image in the mirror. Two facts are worth noting:

- Rays from each point on the object spread out in all directions and strike *every point* on the mirror. Only a very few of these rays enter your eye, but the other rays are real and might be seen by other observers.
- Rays from points P and Q enter your eye after reflecting from *different* areas of the mirror. This is why you can't always see the full image of an object in a very small mirror.

FIGURE 19.11 Each point on an extended object has a corresponding image point an equal distance on the opposite side of the mirror.

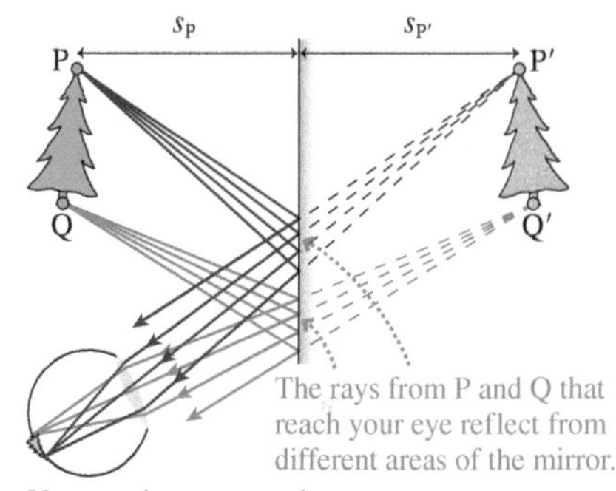

EXAMPLE 19.2 **How high is the mirror?**

If your height is h, what is the shortest mirror on the wall in which you can see your full image?

PREPARE We'll use the ray model of light. If the mirror is tall enough, there will be a path for rays from all parts of your body to reach your eyes, and you will be able to see your full image. FIGURE 19.12 is a pictorial representation of the light rays. We need to consider only the two rays that leave the top of your head and your feet and reflect into your eye.

FIGURE 19.12 Pictorial representation of light rays from your head and feet reflecting into your eye.

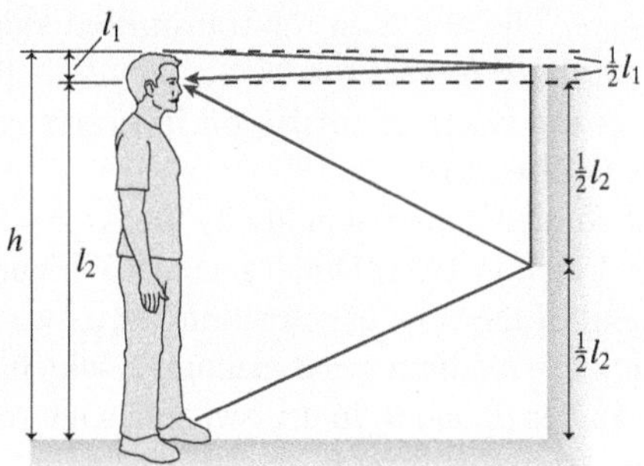

SOLVE Let the distance from your eyes to the top of your head be l_1 and the distance to your feet be l_2. Your height is $h = l_1 + l_2$. A light ray from the top of your head that reflects from the mirror at $\theta_r = \theta_i$ and enters your eye must, by congruent triangles, strike the mirror a distance $\frac{1}{2}l_1$ above your eyes. Similarly, a ray from your foot to your eye strikes the mirror a distance $\frac{1}{2}l_2$ below your eyes. The distance between these two points on the mirror is $\frac{1}{2}l_1 + \frac{1}{2}l_2 = \frac{1}{2}h$. A ray from anywhere else on your body will reach your eye if it strikes the mirror between these two points. Thus the shortest mirror in which you can see your full reflection is $\frac{1}{2}h$.

ASSESS At no point in our derivation did we use the distance from you to the mirror. This makes sense. Getting closer to the mirror doesn't allow you to see more of your image—it just gets you closer to the image. If you doubt this result, try it!

Seeing yourself The dancers stand in front of the mirror. You can see their images in the mirror. Each image is the same distance behind the mirror as the dancer is in front of it.

STOP TO THINK 19.2 An object is placed in front of a mirror. The observer is positioned as shown. Which of the points, A, B, or C, best indicates where the observer would perceive the image to be located?

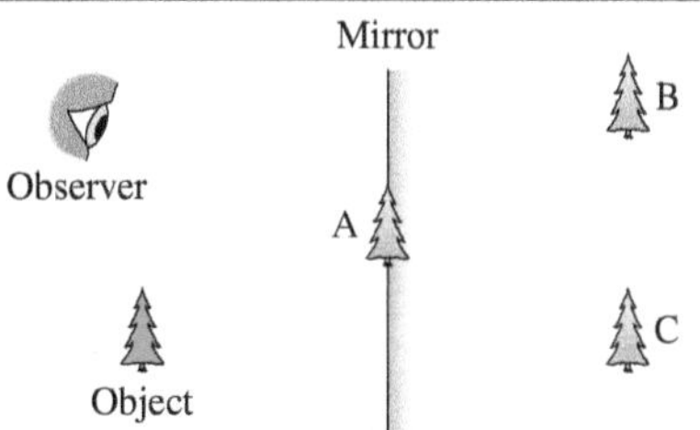

19.3 Refraction

FIGURE 19.13 A light beam refracts twice in passing through a glass prism.

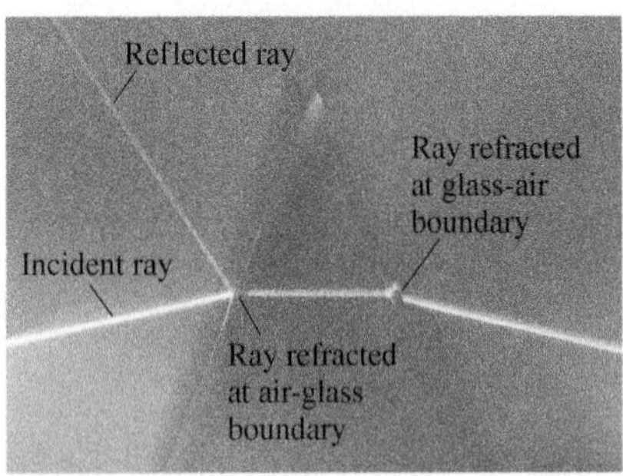

FIGURE 19.13 shows light passing through a glass prism. The light goes from the air into the glass, then back into the air. Two things happen when a light ray crosses the boundary between the air and the glass:

- Part of the light *reflects* from the boundary, obeying the law of reflection. This is how you see reflections from pools of water or storefront windows, even though water and glass are transparent.
- Part of the light continues into the glass. It is *transmitted* rather than reflected, but the transmitted ray changes direction as it crosses the boundary. The transmission of light from one medium to another, but with a change in direction, is called **refraction.**

NOTE ▸ The transparent material through which light travels is called the medium (plural *media*). ◂

In Figure 19.13, the ray changes direction as the light enters the glass, and also as it leaves. This is the central concept we'll deal with when we consider refraction, the change in direction that occurs when light crosses from one medium into another. Reflection from the boundary between transparent media is usually weak. Our goal in this section is to understand refraction, so we will usually ignore the weak reflection and focus on the transmitted light.

FIGURE 19.14 Refraction of light rays.

(a) Refraction of parallel and point-source rays

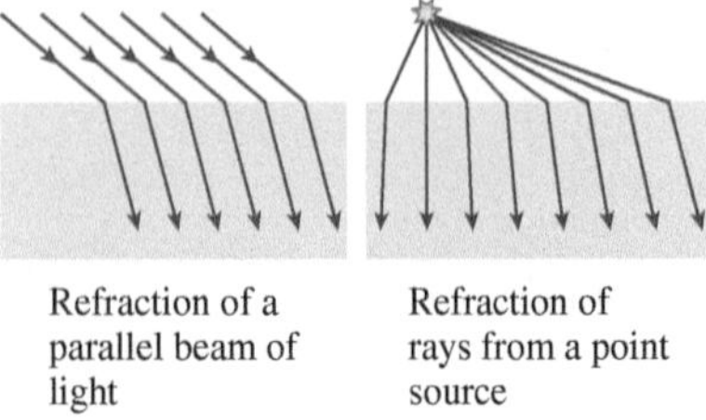

Refraction of a parallel beam of light

Refraction of rays from a point source

(b) Refraction from lower-index medium to higher-index medium

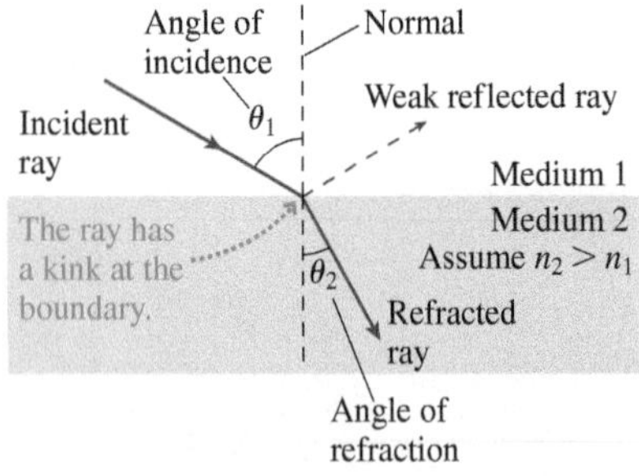

(c) Refraction from higher-index medium to lower-index medium

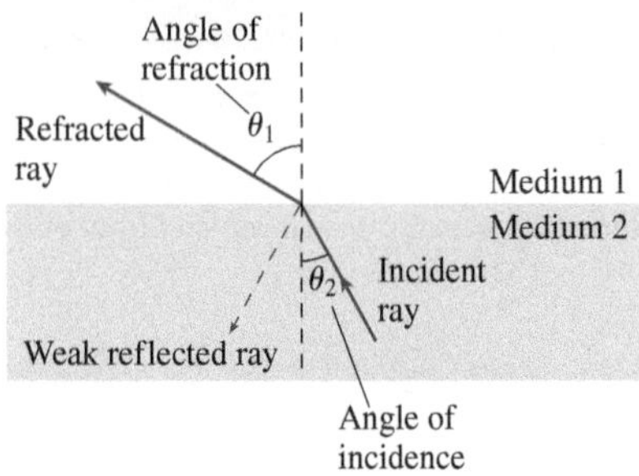

FIGURE 19.14a shows the refraction of light rays from a parallel beam of light, such as a laser beam, and rays from a point source. These pictures remind us that an infinite number of rays are incident on the boundary, but our analysis will be simplified if we focus on a single light ray. FIGURE 19.14b is a ray diagram showing the refraction of a single ray at a boundary between medium 1 and medium 2. Let the angle between the ray and the normal be θ_1 in medium 1 and θ_2 in medium 2. Just as for reflection, the angle between the incident ray and the normal is the *angle of incidence.* The angle on the transmitted side, *measured from the normal,* is called the **angle of refraction.** Notice that θ_1 is the angle of incidence in Figure 19.14b but is the angle of refraction in FIGURE 19.14c, where the ray is traveling in the opposite direction.

Refraction was first studied experimentally by the Arab scientist Ibn al-Haitham in about the year 1000. Later, in 1621, Dutch scientist Willebrord Snellius proposed a mathematical statement of the "law of refraction" or, as we know it today, **Snell's law.** If a ray refracts between medium 1 and medium 2, which have indices of refraction n_1 and n_2, the ray angles θ_1 and θ_2 in the two media are related by

$$n_1 \sin\theta_1 = n_2 \sin\theta_2 \tag{19.3}$$

Snell's law for refraction between two media

Notice that Snell's law does not mention which is the incident angle and which the refracted angle.

TABLE 19.1 lists the indices of refraction for several media. The n in the law of refraction, Equation 19.3, is the same index of refraction n we studied in ◂◂ SECTION 16.4. There we found that the index of refraction determines the speed of a light wave in a medium according to $v = c/n$. Huygens' principle can be used to show that Snell's law is a *consequence* of the change in the speed of light as it moves across a boundary between media.

Examples of Refraction

Look back at Figure 19.14. As the ray in Figure 19.14b moves from medium 1 to medium 2, where $n_2 > n_1$, it bends closer to the normal. In Figure 19.14c, where the ray moves from medium 2 to medium 1, it bends away from the normal. This is a general conclusion that follows from Snell's law:

- When a ray is transmitted into a material with a higher index of refraction, it bends to make a smaller angle with the normal.
- When a ray is transmitted into a material with a lower index of refraction, it bends to make a larger angle with the normal.

This rule becomes a central idea in a procedure for analyzing refraction problems.

TACTICS BOX 19.1 Analyzing refraction

1. **Draw a ray diagram.** Represent the light beam with one ray.
2. **Draw a line normal (perpendicular) to the boundary.** Do this at each point where the ray intersects a boundary.
3. **Show the ray bending in the correct direction.** The angle is larger on the side with the smaller index of refraction. This is the qualitative application of Snell's law.
4. **Label angles of incidence and refraction.** Measure all angles from the normal.
5. **Use Snell's law.** Calculate the unknown angle or unknown index of refraction.

TABLE 19.1 Indices of refraction

Medium	n
Vacuum	1 exactly
Air (actual)	1.0003
Air (accepted)*	1.00
Water	1.33
Ethyl alcohol	1.36
Oil (typical)	1.46
Glass (typical)	1.50
Polystyrene plastic	1.59
Cubic zirconia	2.18
Diamond	2.42

*Use this value in problems.

Erasing boundaries Light reflects and refracts at the boundary between two transparent media, but only if there is a difference in their indices of refraction. You can make nearly invisible repairs to glass objects by using glue with an index of refraction that matches the index of the glass being repaired. Then light hits the boundary between the glue and the glass and continues without reflecting or refracting, so there is no way to tell that the boundary is there. Similarly, an immersion objective lens on a microscope fills the space between the cover slip and the lens with a fluid whose index of refraction is almost exactly that of glass.

EXAMPLE 19.3 Measuring the index of refraction

FIGURE 19.15 shows a laser beam deflected by a 30°–60°–90° prism. What is the prism's index of refraction?

FIGURE 19.15 A prism deflects a laser beam.

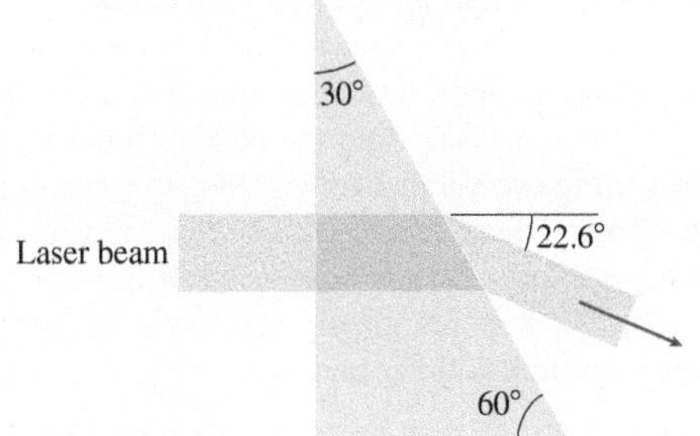

FIGURE 19.16 Pictorial representation of a laser beam passing through the prism.

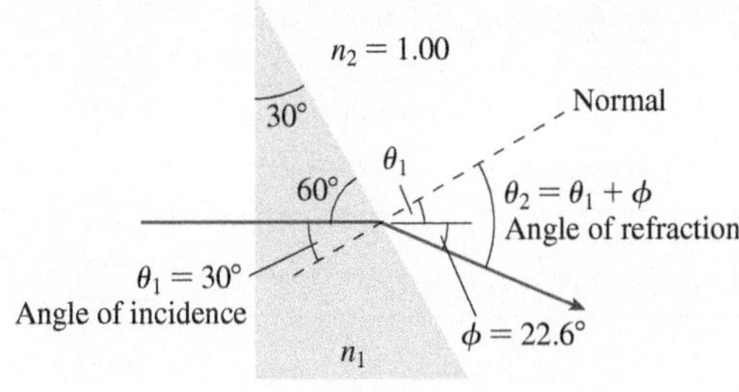

PREPARE We will represent the laser beam with a single ray and use the ray model of light. FIGURE 19.16 uses the steps of Tactics Box 19.1 to draw a ray diagram. The ray is incident perpendicular to the front face of the prism ($\theta_i = 0°$); thus it is transmitted through the first boundary without deflection. At the second boundary it is especially important to *draw the normal to the surface* at the point of incidence and to *measure angles from the normal.*

SOLVE From the geometry of the triangle we find that the laser's angle of incidence on the hypotenuse of the prism is $\theta_1 = 30°$, the same as the apex angle of the prism. The ray exits the prism at angle θ_2 such that the deflection is $\phi = \theta_2 - \theta_1 = 22.6°$. Thus $\theta_2 = 52.6°$. Knowing both angles and $n_2 = 1.00$ for air, we use Snell's law to find n_1:

$$n_1 = \frac{n_2 \sin\theta_2}{\sin\theta_1} = \frac{1.00 \sin 52.6°}{\sin 30°} = 1.59$$

ASSESS Referring to the indices of refraction in Table 19.1, we see that the prism could be made of polystyrene plastic.

EXAMPLE 19.4 Refraction at the cornea

The human eye of an adult is approximately spherical with a radius of 12 mm. The cornea—the front portion of the eye that has a thickness of about 0.5 mm—protrudes slightly, and its radius of curvature is a somewhat smaller 7.0 mm. The cornea's index of refraction is 1.38. As FIGURE 19.17 shows, a light ray parallel to the axis of the eye refracts at the front surface of the cornea and bends toward the axis. There is almost no refraction at the internal boundary between the cornea and the aqueous humor, the eye's fluid, because they have almost identical indices of refraction. At what distance behind the cornea does a light ray cross the axis if its initial height above the axis is (a) 2.0 mm and (b) 4.0 mm?

PREPARE There's essentially no refraction at the internal boundary between the cornea and the aqueous humor, so we can model the refraction as being due to a single spherical surface at the boundary between the air and the cornea. Figure 19.17 is a pictorial representation that shows an incident ray parallel to the axis. Let's call the height of the ray h, the radius of the spherical cornea surface R, and the distance at which the refracted ray crosses the axis d. The dashed radial line drawn from the cornea's center of curvature to the point of incidence is the normal to the boundary, and that defines the angle of incidence θ_1. The figure defines three distances, x_1, x_2, and x_3, that we'll use in the solution. The distance we seek is $d = x_1 + x_3$.

SOLVE The angle of incidence θ_1 is the same as the angle between the axis and the dashed radial line. Given ray height h and radius R, which are the opposite side and hypotenuse of the right triangle with angle θ_1, we see that

$$\sin\theta_1 = \frac{h}{R}$$

FIGURE 19.17 Pictorial representation for refraction of light at the cornea.

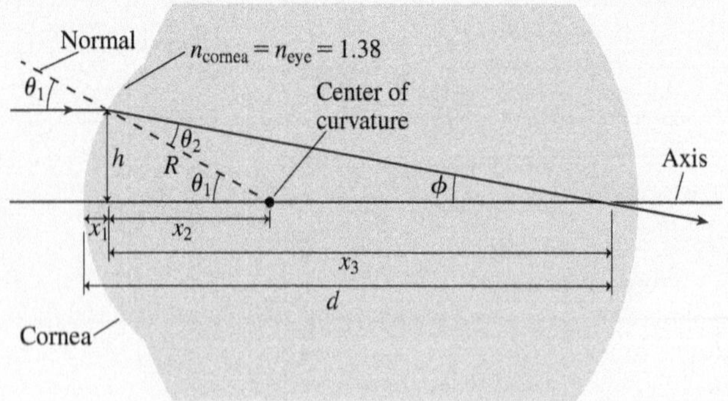

We can use this result directly in Snell's law, $n_1 \sin\theta_1 = n_2 \sin\theta_2$, with $n_1 = n_{air} = 1$ and $n_2 = n_{cornea}$, to write that the sine of the angle of refraction is

$$\sin\theta_2 = \frac{\sin\theta_1}{n_{cornea}} = \frac{h}{Rn_{cornea}}$$

Now it's a geometry problem to determine the distance d. First we use the Pythagorean theorem to write $x_2 = \sqrt{R^2 - h^2}$. Then, because $x_1 + x_2 = R$, we find

$$x_1 = R - \sqrt{R^2 - h^2}$$

The ray from the refraction point to the axis is the hypotenuse of a right triangle with angle ϕ. h and x_3 are the opposite and adjacent sides and their ratio is the tangent of the angle, so

$$x_3 = \frac{h}{\tan\phi}$$

The three angles of a triangle add to 180°, which allows us to write $\phi + \theta_2 + (180° - \theta_1) = 180°$. Thus

$$\phi = \theta_1 - \theta_2$$

Once x_1 and x_3 are computed, the distance at which the ray crosses the axis is $d = x_1 + x_3$. It's helpful to show each step of the calculation in a table for the two cases of (a) $h = 2.0$ mm and (b) $h = 4.0$ mm.

h	θ_1	θ_2	ϕ	x_1	x_3	d
2.0 mm	16.6°	11.9°	4.7°	0.3 mm	24.3 mm	25 mm
4.0 mm	34.8°	24.5°	10.3°	1.3 mm	22.0 mm	23 mm

If the eye had no lens and was sufficiently long, the ray would cross the axis 25 mm behind the cornea if it enters at a height of 2.0 mm and at an almost identical 23 mm behind the cornea if it enters at a height of 4.0 mm.

ASSESS We see that the cornea is *focusing* the light by bringing all the incident light rays to (almost) a single point on the axis. In fact, as we'll examine more closely in Chapter 20, most of the eye's focusing ability is due to the cornea, not the lens. The distance at which the light rays cross the axis, about 24 mm, is almost exactly the distance from the cornea to the retina. Light rays that enter the eye parallel to the axis, as we're considering here, come from a distance object. The lens of your eye is relaxed when you view a distant object. A relaxed lens fine-tunes the focus a bit but does not greatly add to the focusing already achieved by refraction at the cornea.

Total Internal Reflection

What would have happened in Example 19.3 if the prism angle had been 45° rather than 30°? The light rays would approach the rear surface of the prism at an angle of incidence $\theta_1 = 45°$. When we try to calculate the angle of refraction at which the ray emerges into the air, we find

$$\sin\theta_2 = \frac{n_1}{n_2}\sin\theta_1 = \frac{1.59}{1.00}\sin 45° = 1.12$$

$$\theta_2 = \sin^{-1}(1.12) = ???$$

Angle θ_2 does not exist because the sine of an angle can't be greater than 1. The ray is unable to refract through the boundary. Instead, 100% of the light *reflects* from the boundary back into the prism. This process is called **total internal reflection,** often abbreviated TIR. That it really happens is illustrated in FIGURE 19.18. Here, three light beams traveling through water are incident on the air-water boundary at increasing angles of incidence. The two beams with the smallest angles of incidence refract out of the water, but the beam with the largest angle of incidence undergoes total internal reflection at the water's surface.

FIGURE 19.18 One of the three beams of light undergoes total internal reflection.

FIGURE 19.19 shows several rays leaving a point source in a medium with index of refraction n_1. The medium on the other side of the boundary has $n_2 < n_1$. As we've seen, crossing a boundary into a material with a lower index of refraction causes the ray to bend away from the normal. Two things happen as angle θ_1 increases. First, the refraction angle θ_2 approaches 90°. Second, the fraction of the light energy that is transmitted decreases while the fraction reflected increases.

A **critical angle** θ_c is reached when $\theta_2 = 90°$. Snell's law becomes $n_1 \sin\theta_c = n_2 \sin 90°$, or

$$\theta_c = \sin^{-1}\left(\frac{n_2}{n_1}\right) \tag{19.4}$$

Critical angle of incidence for total internal reflection

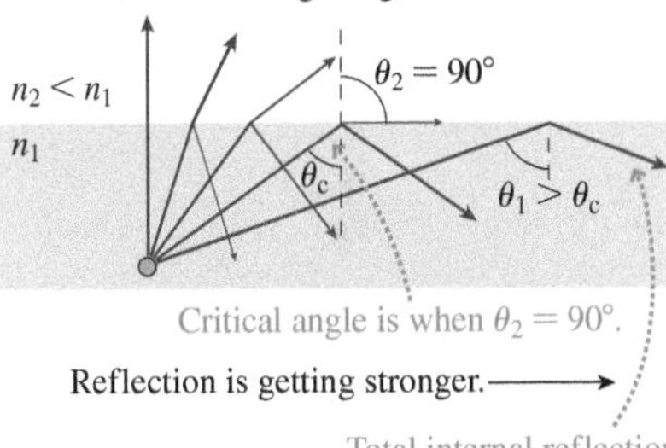
Angle of incidence is increasing.
Transmission is getting weaker.
$\theta_2 = 90°$
$n_2 < n_1$
n_1
θ_c
$\theta_1 > \theta_c$
Critical angle is when $\theta_2 = 90°$.
Reflection is getting stronger.
Total internal reflection occurs when $\theta_1 > \theta_c$.

FIGURE 19.19 Refraction and reflection of rays as the angle of incidence increases.

The refracted light vanishes at the critical angle and the reflection becomes 100% for any angle $\theta_1 > \theta_c$. The critical angle is well defined because of our assumption that $n_2 < n_1$. **There is no critical angle and no total internal reflection if $n_2 > n_1$.**

We can compute the critical angle in a typical piece of glass at the glass-air boundary as

$$\theta_{c\ \text{glass}} = \sin^{-1}\left(\frac{1.00}{1.50}\right) = 42°$$

The fact that the critical angle is smaller than 45° has important applications. For example, FIGURE 19.20 shows a pair of binoculars. The lenses are much farther apart than your eyes, so the light rays need to be brought together before exiting the eyepieces. Rather than using mirrors, binoculars use a pair of prisms on each side. Thus the light undergoes two TIRs and emerges from the eyepiece. (The actual prism arrangement in binoculars is a bit more complex, but this illustrates the basic idea.)

FIGURE 19.20 Binoculars and other optical instruments make use of total internal reflection.

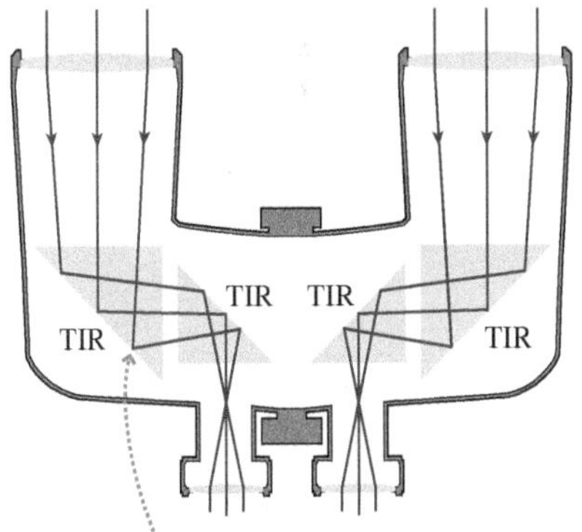
TIR
TIR
TIR
TIR
Angles of incidence exceed the critical angle.

EXAMPLE 19.5 **Where can you see the sea star?**

A woman lies on a paddleboard floating on the still surface of a tropical lagoon, scanning the water below. A sea star rests on the sandy bottom of the lagoon, 3.0 m below the surface. What is the diameter of the circle on the surface of the water inside which the woman can see the sea star?

PREPARE If we can see something, light rays from the object must be reaching our eyes. We can model the sea star as a point source of light rays and consider the path of the rays to see if they make it from the water into the air where they can be seen.

FIGURE 19.21 is a pictorial representation showing several rays. Rays that strike the surface at an angle greater than the critical angle undergo total internal reflection back down into the water

FIGURE 19.21 Pictorial representation of the rays of light from the sea star.

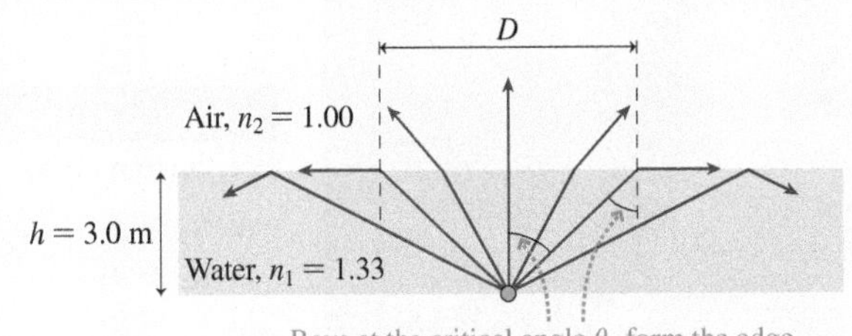
D
Air, $n_2 = 1.00$
$h = 3.0$ m
Water, $n_1 = 1.33$
Rays at the critical angle θ_c form the edge of the circle of light seen from above.

Continued

where they can't be seen by the woman on the paddleboard. The diameter of the circle in which the woman is able to see the sea star is the distance D between the two points at which rays strike the surface at the critical angle.

SOLVE From trigonometry, the circle diameter is $D = 2h \tan \theta_c$, where h is the depth of the water. The critical angle for the water-air boundary is $\theta_c = \sin^{-1}(1.00/1.33) = 48.8°$. The diameter of the circle is thus

$$D = 2(3.0\text{ m}) \tan(48.8°) = 6.9\text{ m}$$

ASSESS This result seems reasonable if you've ever had the experience of looking for objects below the surface of the water. You can see objects directly below you, but you can't see objects off to the side.

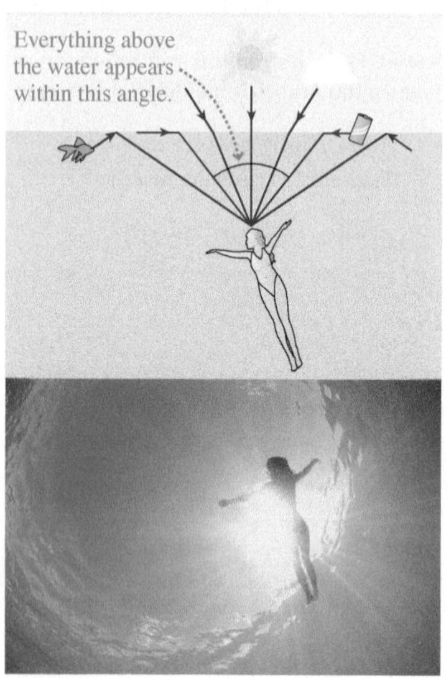

Everything above the water appears within this angle.

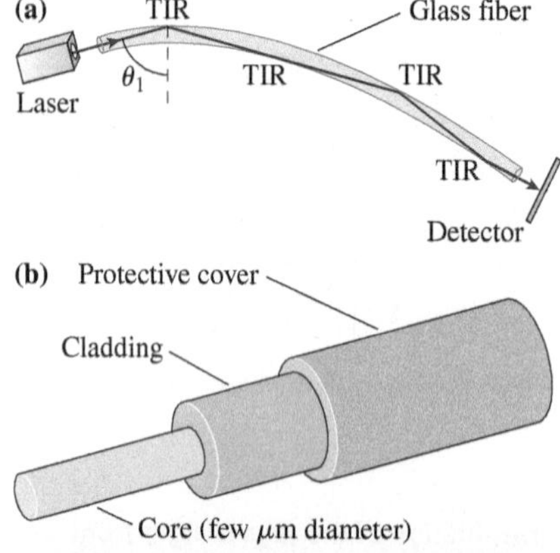

Snell's window We can understand what a diver sees when she's underwater by reversing the direction of all the rays in Figure 19.21. The drawing shows that she can see the sun overhead and clouds at larger angles. She can even see objects sitting at the waterline—but they appear at the edge of a circle as she looks up. Anything outside of this circle is a reflection of something in the water. The photo shows what she sees: a bright circle from the sky above—Snell's window—surrounded by the dark reflection of the water below.

Fiber Optics

The most important modern application of total internal reflection is the transmission of light through optical fibers. FIGURE 19.22a shows a laser beam shining into the end of a long, narrow-diameter glass fiber. The light rays pass easily from the air into the glass, but they then strike the inside wall of the fiber at an angle of incidence θ_1 approaching 90°. This is much larger than the critical angle, so the laser beam undergoes TIR and remains inside the glass. The laser beam continues to "bounce" its way down the fiber as if the light were inside a pipe. Indeed, optical fibers are sometimes called "light pipes." The rays have an angle of incidence *smaller* than the critical angle ($\theta_1 \approx 0$) when they finally reach the flat end of the fiber; thus they refract out without difficulty and can be detected.

While a simple glass fiber can transmit light, dependence on a glass-air boundary is not sufficiently reliable for commercial use. Any small scratch on the side of the fiber alters the rays' angle of incidence and allows leakage of light. FIGURE 19.22b shows the construction of a practical optical fiber. A small-diameter glass *core* is surrounded by a layer of glass *cladding*. The glasses used for the core and the cladding have $n_{\text{core}} > n_{\text{cladding}}$. Thus, light undergoes TIR at the core-cladding boundary and remains confined within the core. This boundary is not exposed to the environment and hence retains its integrity even under adverse conditions.

Optical fibers have important applications in medical diagnosis and treatment. Thousands of small fibers can be fused together to make an *endoscope,* a flexible bundle capable of transmitting high-resolution images along its length. During *endoscopic surgery,* surgeons perform operations by using an endoscope inserted through a small incision. The endoscope allows the surgeon to observe the procedure, which is performed with instruments inserted through another incision. The recovery time for such surgery is usually much shorter than for conventional operations that require a full incision. Laparoscopy, colonoscopy, and arthroscopy are different names for the endoscopic procedures used in different fields of medicine.

FIGURE 19.22 Light rays are confined within an optical fiber by TIR.

(a) TIR, Glass fiber, θ_1, TIR, TIR, Laser, TIR, Detector

(b) Protective cover, Cladding, Core (few μm diameter)

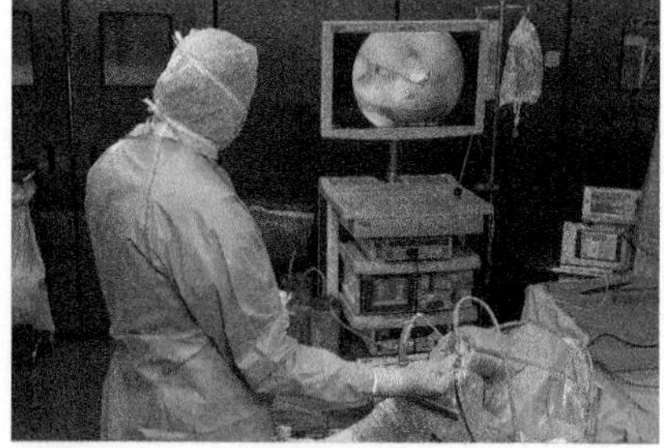

Arthroscopic surgery The surgeon uses a monitor to view the endoscope image while performing arthroscopic surgery on a knee.

STOP TO THINK 19.3 A light ray travels from medium 1 to medium 3 as shown. For these media,

A. $n_3 > n_1$
B. $n_3 = n_1$
C. $n_3 < n_1$
D. We can't compare n_1 to n_3 without knowing n_2.

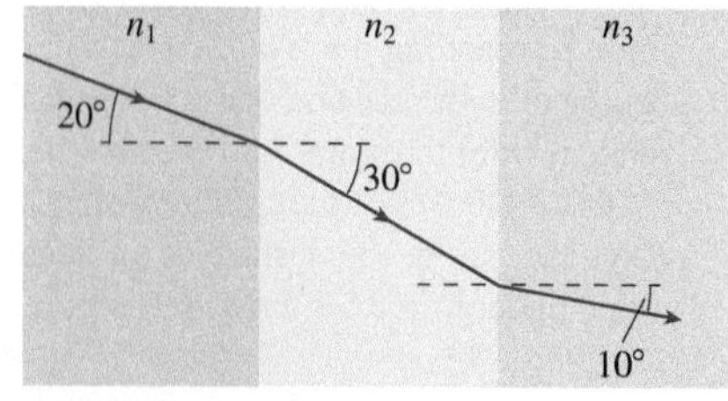

19.4 Image Formation by Refraction

FIGURE 19.23a shows a photograph of a ruler as seen through the front of an aquarium tank. The part of the ruler below the waterline appears *closer* than the part that is above water. FIGURE 19.23b shows why this is so. Rays that leave point P on the ruler refract away from the normal at the water-air boundary. (The thin glass wall of the aquarium has little effect on the refraction of the rays and can be ignored.) To your eye, outside the aquarium, these rays appear to diverge not from the object at point P, but instead from point P′ that is *closer* to the boundary. The same argument holds for every point on the ruler, so that **the ruler appears closer than it really is because of refraction of light at the boundary.**

We found that the rays reflected from a mirror diverge from a point that is not the object point. We called that point a *virtual image.* Similarly, if rays from an object point P refract at a boundary between two media such that the rays then diverge from a point P′ and *appear* to come from P′, we call P′ a virtual image of point P. The virtual image of the ruler is what you see.

Let's examine this image formation a bit more carefully. FIGURE 19.24 shows a boundary between two transparent media that have indices of refraction n_1 and n_2. Point P, a source of light rays, is the object. Point P′, from which the rays *appear* to diverge, is the virtual image of P. The figure assumes $n_1 > n_2$, but this assumption isn't necessary. Distance s, measured from the boundary, is the object distance. Our goal is to determine the image distance s'.

The line through the object and perpendicular to the boundary is called the **optical axis**. Consider a ray that leaves the object at angle θ_1 with respect to the optical axis. θ_1 is also the angle of incidence at the boundary, where the ray refracts into the second medium at angle θ_2. By tracing the refracted ray backward, we can see that θ_2 is also the angle between the refracted ray and the optical axis at point P′.

The distance l is common to both the incident and the refracted rays, and we can see that $l = s \tan\theta_1 = s' \tan\theta_2$. However, **it is customary in optics for virtual image distances to be negative.** (The reason will be clear when we discuss image formation by lenses.) Hence we insert a minus sign and find that

$$s' = -\frac{\tan\theta_1}{\tan\theta_2} s \tag{19.5}$$

Snell's law relates the sines of angles θ_1 and θ_2; that is,

$$\frac{\sin\theta_1}{\sin\theta_2} = \frac{n_2}{n_1} \tag{19.6}$$

In practice, the angle between any of these rays and the optical axis is very small because the pupil of your eye is very much smaller than the distance between the object and your eye. (The angles in the figure have been greatly exaggerated.) The small-angle approximation $\sin\theta \approx \tan\theta \approx \theta$, where θ is in radians, is therefore applicable. Consequently,

$$\frac{\tan\theta_1}{\tan\theta_2} \approx \frac{\sin\theta_1}{\sin\theta_2} = \frac{n_2}{n_1} \tag{19.7}$$

Using this result in Equation 19.5, we find that the image distance is

$$s' = -\frac{n_2}{n_1} s \tag{19.8}$$

The index of refraction of the medium that the object is in (n_1). The index of refraction of the medium that the observer is in (n_2).

NOTE ▸ The fact that the result for s' is independent of θ_1 implies that *all* rays appear to diverge from the same point P′. This property of the diverging rays is essential in order to have a well-defined image. ◂

This section has given us a first look at image formation via refraction. We will extend this idea to image formation with lenses in the next section.

FIGURE 19.23 Refraction causes an object in an aquarium to appear closer than it really is.

(a) A ruler in an aquarium

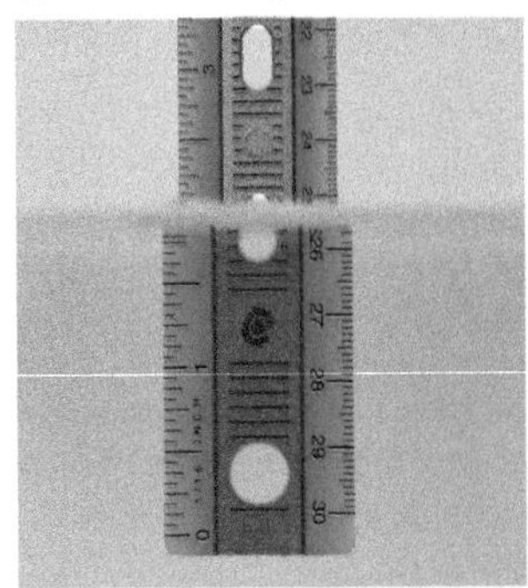

(b) Finding the image of the ruler

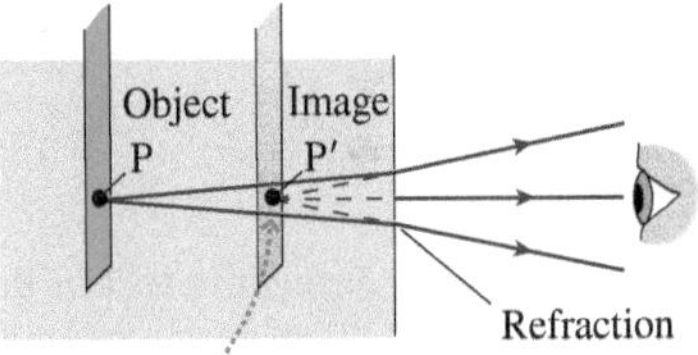

FIGURE 19.24 Finding the virtual image P′ of an object at P.

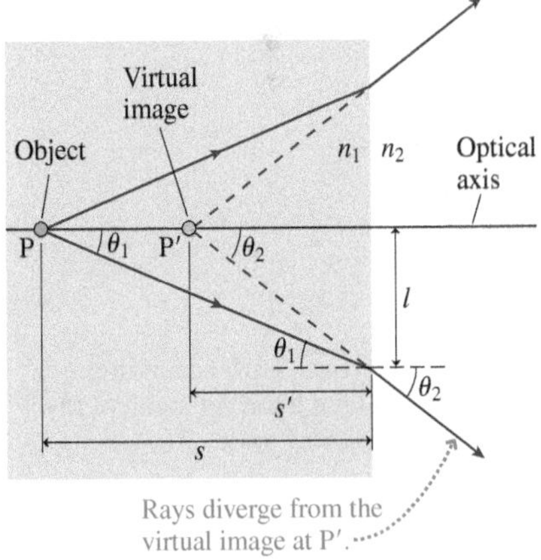

EXAMPLE 19.6 **Where are your feet?**

Aaliyah is walking in a shallow, clear bay, in still water just over her knees. When she looks down at her feet in the sand, she notes that they appear closer to her than usual. Normally, when she tips her head forward to look at her feet, her feet are 1.60 m from her eyes. How far away do her feet appear when they are under 0.60 m of water?

PREPARE Aaliyah's feet are the object, the source of light rays heading toward the surface. The index of refraction of water is higher than the index of refraction of air. Consequently, as Figure 19.24 shows, the rays refract away from the normal and appear to be coming from a point—the virtual image—that is closer to the surface. Aaliyah focuses on the virtual image of her feet because that is the point the rays appear to be coming from as they reach her eyes, so her feet seem closer than normal. The distances are shown in FIGURE 19.25.

FIGURE 19.25 Pictorial representation of a woman looking at her feet under the water.

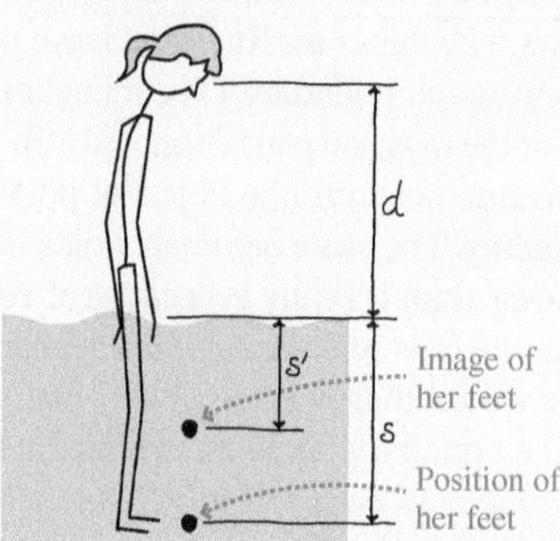

SOLVE The object distance s is the distance of her feet from the water's surface: $s = 0.60$ m. The light rays start in water and end in air, so $n_1 = n_{water} = 1.33$ and $n_2 = n_{air} = 1.00$. We can use Equation 19.8 to find the image distance:

$$s' = -\frac{n_2}{n_1}s = -\frac{1.00}{1.33}(0.60 \text{ m}) = -0.45 \text{ m}$$

The fact that s' is negative tells us that the image is virtual, on the water side of the boundary; that is, the image is 0.45 m below the water surface. The actual distance from Aaliyah's eyes to her feet is 1.60 cm, so having her feet submerged by 0.60 m means that the distance from her eyes to the water is $d = 1.00$ m. Consequently, the distance from her eyes to the image of her feet is

$$d + |s'| = 1.00 \text{ m} + 0.45 \text{ m} = 1.45 \text{ m}$$

ASSESS If you have ever looked down at your feet while standing in the water, you have observed this effect. Your feet really do seem too close!

19.5 Thin Lenses: Ray Tracing

A pinhole camera forms an image on a screen, but the image is faint and not well focused. The ability to create a bright, well-focused image is vastly improved by using a lens. A **lens** is a transparent object that uses refraction of light rays at *curved* surfaces to form an image. In this section we want to establish a pictorial method of understanding image formation. This method is called *ray tracing*.

FIGURE 19.26 Converging and diverging lenses.

(a) Converging lenses, which are thicker in the center than at the edges, refract parallel rays toward the optical axis.

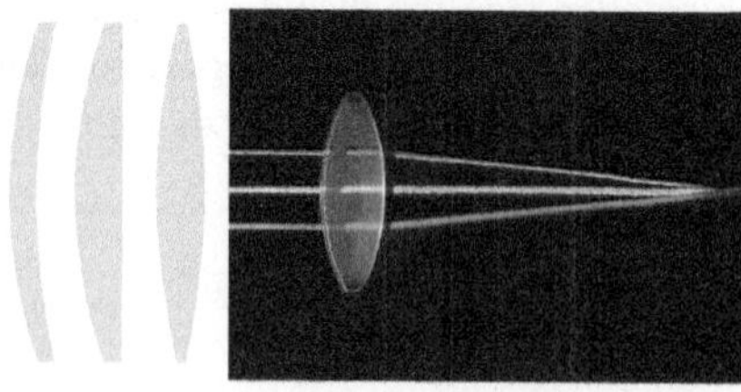

(b) Diverging lenses, which are thinner in the center than at the edges, refract parallel rays away from the optical axis.

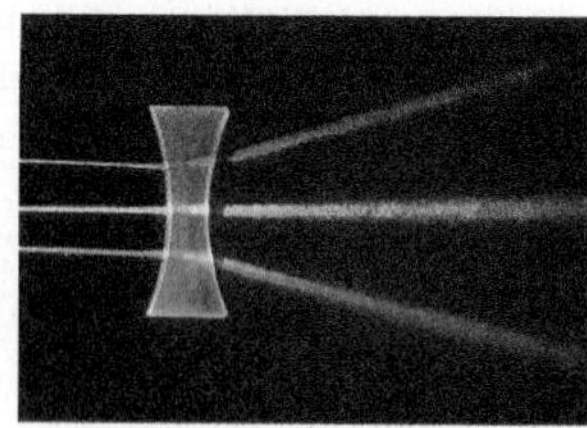

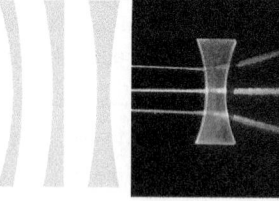

FIGURE 19.27 Both surfaces of this converging lens bend an incident ray toward the optical axis.

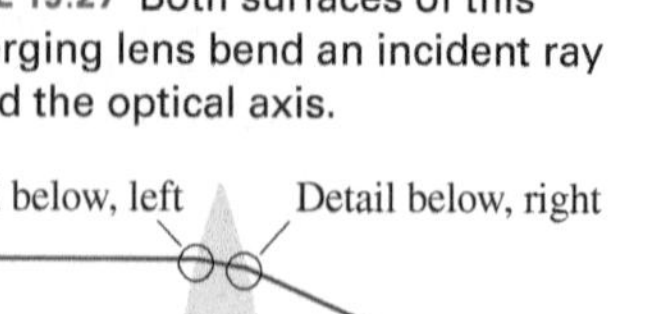

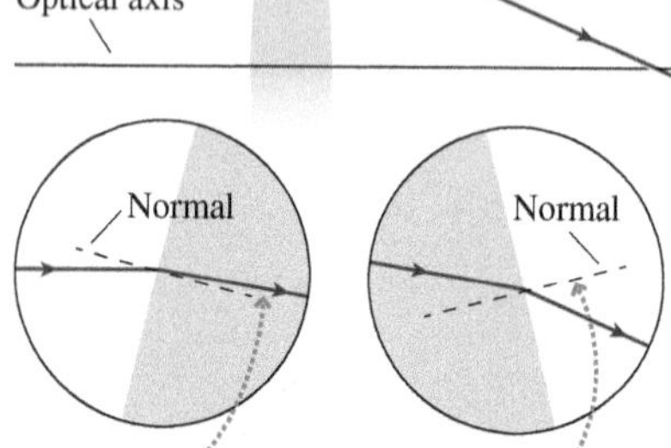

As the ray enters the lens, it bends toward the normal and the optical axis.

As the ray leaves the lens, it bends away from the normal but still toward the optical axis.

FIGURE 19.26 shows parallel light rays entering two different lenses. The lens in Figure 19.26a, called a **converging lens,** causes the rays to refract *toward* the optical axis. FIGURE 19.27 shows how this works. An incoming ray refracts toward the optical axis at both the first, air-to-glass boundary *and* the second, glass-to-air boundary. FIGURE 19.28a (next page) shows that *all* incoming rays initially parallel to the optical axis converge at the *same* point, the **focal point** of the lens. The distance of the focal point from the lens is called the **focal length** f of the lens.

If the parallel rays approached the lens from the right side instead, they would focus to a point on the left side of the lens, indicating that there is a focal point on *each* side of the lens. Both focal points are the same distance f from the lens. The focal point on the side from which the light is incident is the *near focal point*; the focal point on the other side is the *far focal point.*

The lens in Figure 19.26b, called a **diverging lens,** causes the rays to refract *away* from the axis. A diverging lens also has two focal points, although these are not as obvious as for a converging lens. FIGURE 19.28b clarifies the situation. A backward projection of the diverging rays shows that they all *appear* to have started from the same point. This is the near focal point of a diverging lens, and its distance from the lens is the focal length of the lens. For both types of lenses, **the focal length is the distance from the lens to the point at which rays parallel to the optical axis converge or from which they appear to diverge.**

NOTE ▶ The focal length f is a property *of the lens,* independent of how the lens is used. The focal length characterizes a lens in much the same way that a mass m characterizes an object or a spring constant k characterizes a spring. ◀

FIGURE 19.28 The focal point and focal length of converging and diverging lenses.

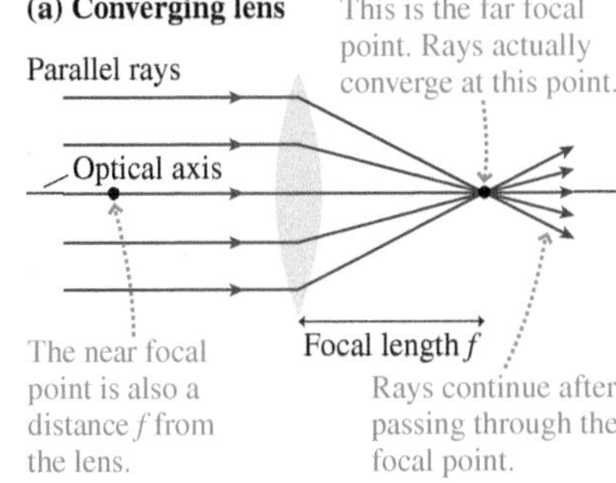

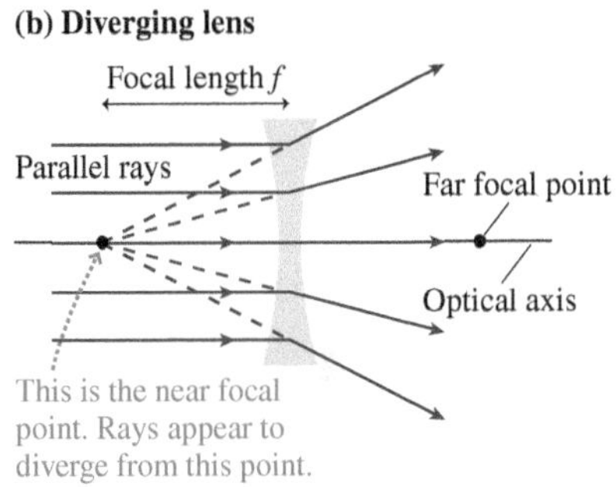

Converging Lenses

These basic observations about lenses are enough for us to understand image formation by a **thin lens,** an idealized lens whose thickness is zero and that lies entirely in a plane called the **lens plane.** Within this *thin-lens approximation,* **all refraction occurs as the rays cross the lens plane, and all distances are measured from the lens plane.** Fortunately, the thin-lens approximation is quite good for most practical applications of lenses.

NOTE ▶ We'll *draw* lenses as if they have a thickness, because it will serve as a visual reminder of the lens type, but our analysis will not depend on the details of the drawing. ◀

FIGURE 19.29 shows three important sets of light rays passing through a thin, converging lens. Part (a) is familiar from Figure 19.28a. If the direction of each of the rays in Figure 19.29a is reversed, Snell's law tells us that each ray will exactly retrace its path and emerge from the lens parallel to the optical axis. This leads to Figure 19.29b, which is the "mirror image" of part (a).

Figure 19.29c shows three rays passing through the *center* of the lens. At the center, the two sides of a lens are very nearly parallel to each other. Light rays do not bend when they pass through a thin piece of window glass that has parallel sides, and similarly rays pass through the center of a thin lens without being bent.

These three situations form the basis for ray tracing.

FIGURE 19.29 Three important sets of rays passing through a thin, converging lens.

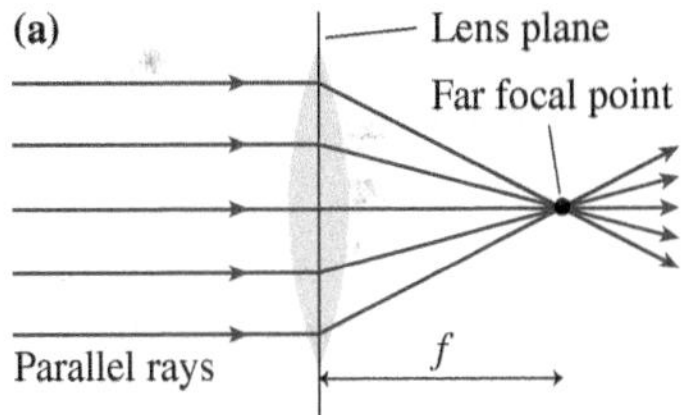

Any ray initially parallel to the optical axis will refract through the focal point on the far side of the lens.

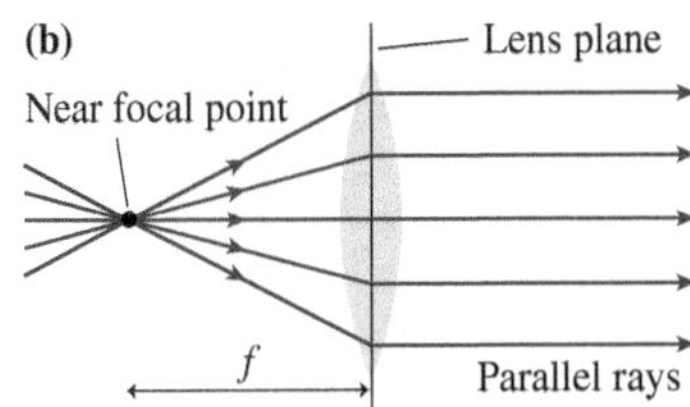

Any ray passing through the near focal point emerges from the lens parallel to the optical axis.

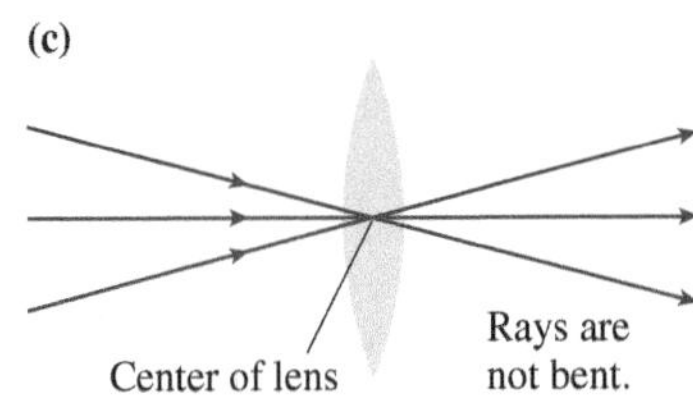

Any ray directed at the center of the lens passes through in a straight line.

Real Images

FIGURE 19.30 (next page) shows a lens and an object whose distance s from the lens is larger than the focal length. The figure illustrates the principles of ray tracing that we'll use to determine the location of an image formed by the lens. For now, we'll focus on one point on the object, labeled P. Rays leave this point in all directions, but we've shown only a few, and we'll further restrict our attention to three *special rays,* labeled in the figure, that have particular properties.

- The first special ray is parallel to the optical axis before the lens, so it goes through the far focal point.
- The second special ray goes through the center of the lens and so is not bent.
- The third special ray goes through the near focal point, so it exits the lens parallel to the optical axis.

These three rays converge at point P′ on the far side of the lens.

We've singled out three special rays, but *all* of the rays from point P on the object that pass through the lens are refracted by the lens so as to converge at point P′ on the opposite side of the lens, at a distance s' from the lens. When rays diverge from point P on the object and interact with a lens such that the refracted rays *converge* at point P′, then we call P′ a **real image** of point P. Contrast this with our prior definition of

FIGURE 19.30 Rays from an object point P are refracted by the lens and converge to a real image at point P′.

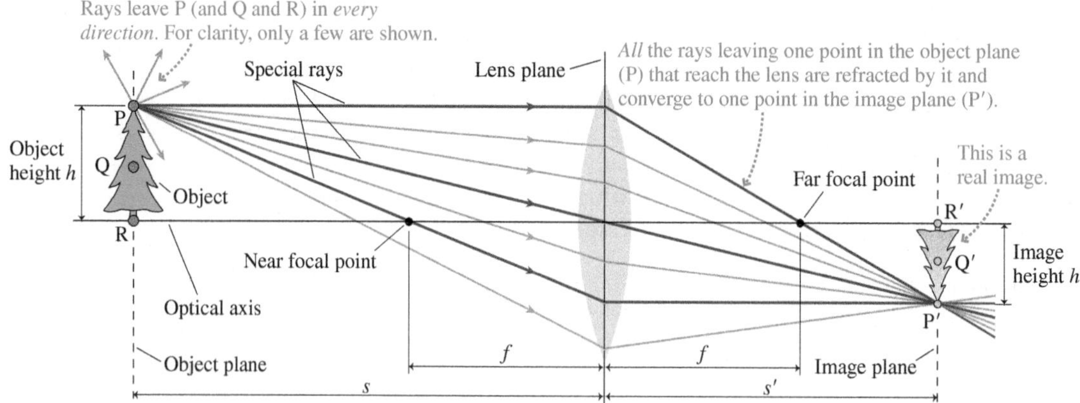

a *virtual image* as a point from which rays appear to *diverge,* but through which no rays actually pass.

FIGURE 19.31 The lamp's image is upside down.

In Figure 19.30, all the rays from point Q on the object converge at point Q′, and all the rays from point R converge at point R′. Ultimately, each point on the object gives rise to a corresponding point on the image. All object points that are in the same plane, the **object plane,** converge to image points in the **image plane.** Points Q and R in the object plane of Figure 19.30 have image points Q′ and R′ in the same plane as point P′. Once we locate *one* point in the image plane, such as point P′, we know that the full image lies in the same plane.

There are two additional observations to make about Figure 19.30. First, as also seen in FIGURE 19.31, the image is upside down with respect to the object. This is called an **inverted image,** and it is a standard characteristic of real-image formation. (The real image formed on the retina of your eye is inverted, but your brain is wired to "flip" the image and perceive it as right-side-up.) Second, rays from point P fill the entire lens surface, so that all portions of the lens contribute to the image. A larger lens will "collect" more rays and thus make a brighter image.

FIGURE 19.32 is a close-up view of the rays and images very near the image plane. The rays don't stop at P′ unless we place a screen in the image plane. When we do so, we see a sharp, well-focused image on the screen. If a screen is placed in front of or behind the image plane, an image is produced on the screen, but it's blurry and out of focus.

FIGURE 19.32 A close-up look at the rays and images near the image plane.

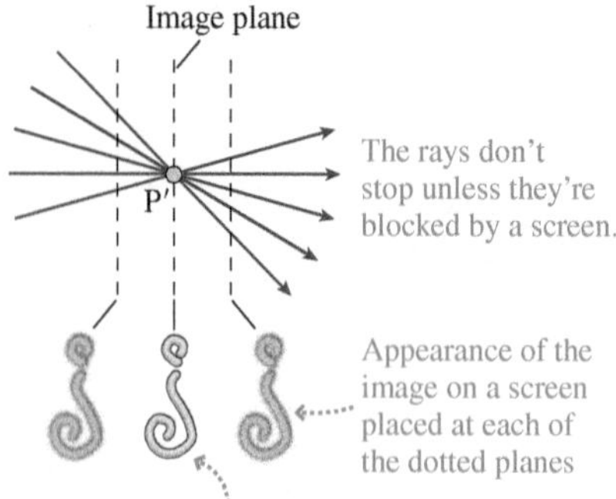

NOTE ▶ Our ability to see a real image on a screen sets real images apart from virtual images. But keep in mind that we need not *see* a real image in order to *have* an image. A real image exists at a point in space where the rays converge even if there's no viewing screen in the image plane. ◀

Figure 19.30 highlights the three "special rays" that are based on the three situations of Figure 19.29. Notice that these three rays alone are sufficient to locate the image point P′. The procedure known as *ray tracing* consists of locating the imaging by the use of these three rays.

TACTICS BOX 19.2 Ray tracing for a converging lens

1. **Draw an optical axis.** Use a straightedge or a ruler! Establish an appropriate scale.
2. **Center the lens on the axis.** Draw the lens plane perpendicular to the axis. Mark and label the focal points at distance f on either side.
3. **Represent the object with an upright arrow at distance s.** It's usually best to place the base of the arrow on the axis and to draw the arrow smaller than the radius of the lens.

❹ **Draw the three "special rays" from the tip of the arrow.** Use a straightedge or a ruler. The rays refract at the lens plane, *not* at the surfaces of the lens.
a. A ray initially parallel to the axis refracts through the far focal point.
b. A ray that enters the lens along a line passing through the near focal point emerges parallel to the axis.
c. A ray through the center of the lens does not bend.

❺ **Extend the rays until they converge.** The rays converge at the image point. Draw the rest of the image in the image plane. If the base of the object is on the axis, then the base of the image will also be on the axis.

❻ **Measure the image distance s'.** Also, if needed, measure the image height relative to the object height.

EXAMPLE 19.7 **Finding the image of a flower**

A 4.0-cm-diameter flower is 200 cm from the 50-cm-focal-length lens of a camera. How far should the plane of the camera's light detector be placed behind the lens to record a well-focused image? What is the diameter of the image on the detector?

PREPARE We will use ray tracing, following the steps of Tactics Box 19.2, to find the location of the image.

SOLVE FIGURE 19.33 shows the ray-tracing diagram and the steps of Tactics Box 19.2. The image has been drawn in the plane where the three special rays converge. You can see *from the drawing,* where tick marks are shown every 25 cm, that the image distance is $s' \approx 65$ cm. This is where the detector needs to be placed to record a focused image. The heights of the object and image are labeled h and h'. These are, respectively, the diameter of the flower and the diameter of the flower's image. The ray through the center of the lens is a straight line; thus the object and image both subtend the same angle θ. From similar triangles,

$$\frac{h'}{s'} = \frac{h}{s}$$

Solving for h' gives

$$h' = h\frac{s'}{s} = (4.0\text{ cm})\frac{65\text{ cm}}{200\text{ cm}} = 1.3\text{ cm}$$

The flower's image has a diameter of 1.3 cm.

ASSESS The ray-tracing process we have outlined has a built-in check. The first two rays converge at a point; the third ray should converge at the same point. If it doesn't, there is an error in our work. In this case, all three rays converge at the same point, so we can have confidence in our result.

FIGURE 19.33 Ray-tracing diagram for the image of a flower.

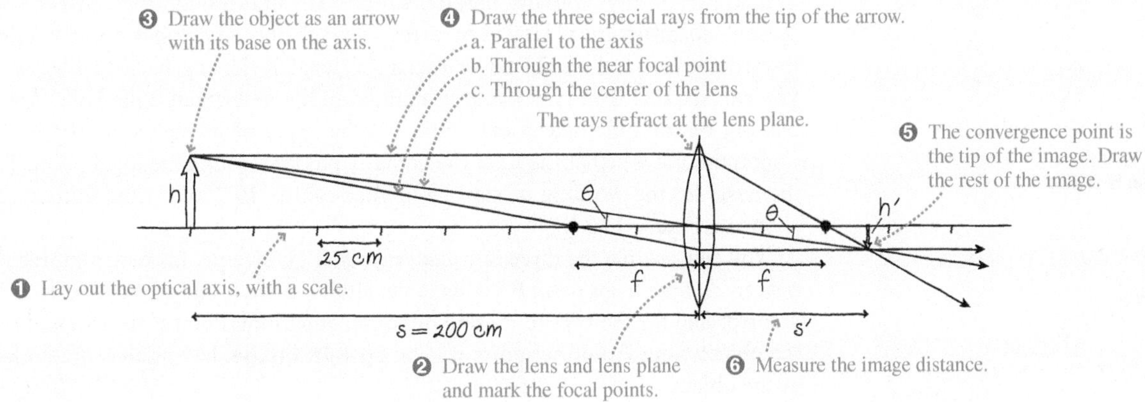
❸ Draw the object as an arrow with its base on the axis.
❹ Draw the three special rays from the tip of the arrow.
a. Parallel to the axis
b. Through the near focal point
c. Through the center of the lens
The rays refract at the lens plane.
❺ The convergence point is the tip of the image. Draw the rest of the image.
h
θ
θ
h′
25 cm
❶ Lay out the optical axis, with a scale.
f
f
s = 200 cm
s′
❷ Draw the lens and lens plane and mark the focal points.
❻ Measure the image distance.

STOP TO THINK 19.4 A lens produces a sharply focused, inverted image on a screen. What will you see on the screen if the lens is removed?

A. The image will be inverted and blurry.
B. The image will be upright and sharp.
C. The image will be upright and blurry.
D. The image will be much dimmer but otherwise unchanged.
E. There will be no image at all.

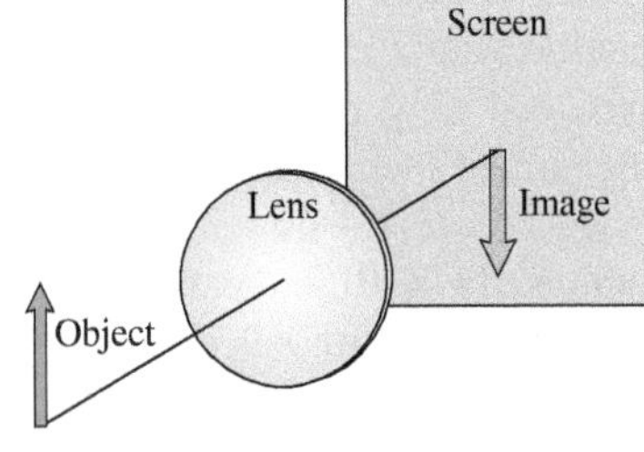
Screen
Lens
Image
Object

Water lens Small droplets of water bead up on a feather. Each droplet is a converging lens that makes a virtual image of the feather underneath, providing a clearer view of the feather's structure.

Lateral Magnification

The image can be either larger or smaller than the object, depending on the location and focal length of the lens. But there's more to a description of the image than just its size. We also want to know its *orientation* relative to the object. That is, is the image upright (right-side-up) or inverted (upside-down)? It is customary to combine size and orientation information into a single number. The **lateral magnification** m is defined as

$$m = -\frac{s'}{s} \tag{19.9}$$

You just saw in Example 19.7 that the image-to-object height ratio is $h'/h = s'/s$. Consequently, we interpret the lateral magnification m as follows:

1. A positive value of m indicates that the image is upright relative to the object. A negative value of m indicates that the image is inverted relative to the object.
2. The absolute value of m gives the size ratio of the image and object: $h'/h = |m|$.

The lateral magnification in Example 19.7 would be $m = -0.33$, indicating that the image is inverted and 33% the size of the object.

NOTE ▶ The image-to-object height ratio is called *lateral* magnification to distinguish it from angular magnification, which we'll introduce in the next chapter. In practice, m is simply called "magnification" when there's no chance of confusion. Although we usually think that "to magnify" means "to make larger," in optics the absolute value of the magnification can be either > 1 (the image is larger than the object) or < 1 (the image is smaller than the object). ◀

FIGURE 19.34 Rays from an object at distance $s < f$ are refracted by the lens and diverge to form a virtual image.

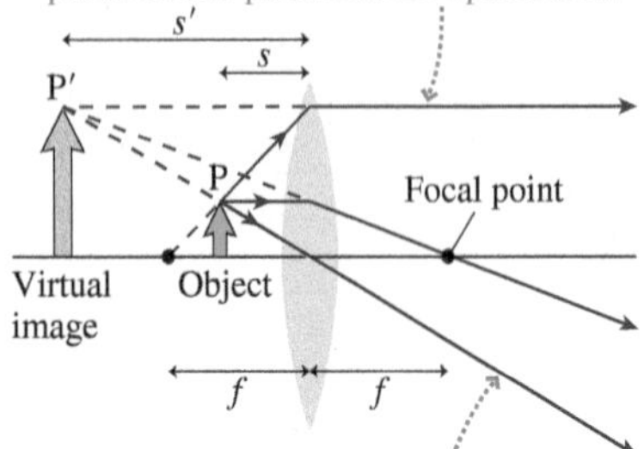
A ray *along a line* through the near focal point refracts parallel to the optical axis.
s'
s
P'
P
Focal point
Virtual image
Object
f
f
The refracted rays are diverging. They appear to come from point P'.

FIGURE 19.35 A converging lens is a magnifying glass when the object distance is less than f.

(a)

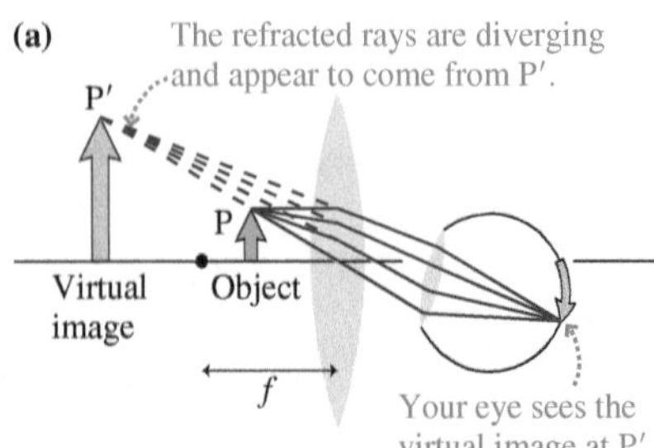
The refracted rays are diverging and appear to come from P'.
P'
P
Virtual image
Object
f
Your eye sees the virtual image at P'.

(b)

Virtual Images

The previous section considered a converging lens with the object at distance $s > f$. That is, the object was outside the focal point. What if the object is inside the focal point, at distance $s < f$? FIGURE 19.34 shows just this situation, and we can use ray tracing to analyze it.

The special rays initially parallel to the axis and through the center of the lens present no difficulties. However, a ray through the near focal point would travel toward the left and would never reach the lens! Referring back to Figure 19.29b, you can see that the rays emerging parallel to the axis entered the lens *along a line* passing through the near focal point. It's the angle of incidence on the lens that is important, not whether the light ray actually passes through the focal point. This was the basis for the wording of step 4b in Tactics Box 19.2 and is the third special ray shown in Figure 19.34.

You can see that the three refracted rays don't converge. Instead, all three rays appear to *diverge* from point P′. This is the situation we found for rays reflecting from a mirror and for the rays refracting out of an aquarium. Point P′ is a *virtual image* of the object point P. Furthermore, it is an **upright image,** having the same orientation as the object.

The refracted rays, which are all to the right of the lens, *appear* to come from P′, but none of the rays were ever at that point. No image would appear on a screen placed in the image plane at P′. So what good is a virtual image?

Your eye collects and focuses bundles of diverging rays; thus, as FIGURE 19.35a shows, you can see a virtual image by looking *through* the lens. This is exactly what you do with a magnifying glass, producing a scene like the one in FIGURE 19.35b. In fact, you view a virtual image anytime you look *through* the eyepiece of an optical instrument such as a microscope or binoculars.

As before, **the image distance s' for a virtual image is defined to be a *negative number* ($s' < 0$),** indicating that the image is on the opposite side of the lens from a real image. With this choice of sign, the definition of magnification, $m = -s'/s$, is still valid. A virtual image with negative s' has $m > 0$, thus the image is upright. This agrees with the rays in Figure 19.34 and the photograph of Figure 19.35b.

NOTE ▶ A lens thicker in the middle than at the edges is classified as a converging lens. The light rays from an object *can* converge to form a real image after passing through such a lens, but only if the object distance is larger than the focal length of the lens: $s > f$. If $s < f$, the rays diverge to produce a virtual image. ◀

EXAMPLE 19.8 **Magnifying a flower**

To see a flower better, you hold a 6.0-cm-focal-length magnifying glass 4.0 cm from the flower. What is the magnification?

PREPARE The flower is in the object plane. We use ray tracing to locate the image. Once the image distance is known, we can use Equation 19.9 to find the magnification.

SOLVE FIGURE 19.36 shows the ray-tracing diagram. The three special rays diverge from the lens, but we can use a straightedge to extend the rays backward to the point from which they diverge. This point, the image point, is seen to be 12 cm to the left of the lens. Because this is a virtual image, the image distance is $s' = -12$ cm. From Equation 19.9 the magnification is

$$m = -\frac{s'}{s} = -\frac{-12\text{ cm}}{4.0\text{ cm}} = 3.0$$

The image is three times as large as the object and, because m is positive, upright.

FIGURE 19.36 Ray-tracing diagram for a magnifying glass.

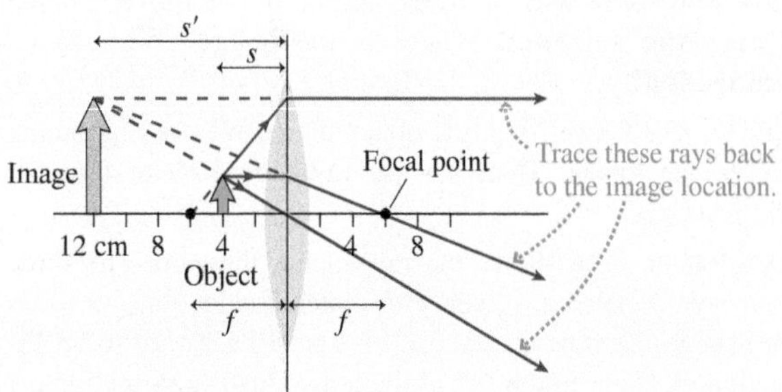

ASSESS The three special rays diverge from a single point, and the image orientation and size in our ray-tracing diagram match the computed magnification—the image is upright and larger than the object—so we have confidence in our results.

Diverging Lenses

As Figure 19.26b showed, a *diverging lens* is one that is thinner at its center than at its edges. FIGURE 19.37 shows three important sets of rays passing through a diverging lens. These are based on Figures 19.26b and 19.28b, where you saw that rays initially parallel to the axis diverge after passing through a diverging lens.

FIGURE 19.37 Three important sets of rays passing through a thin, diverging lens.

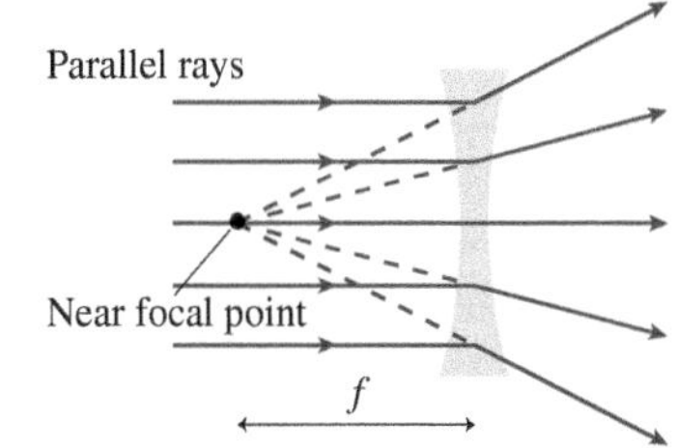

Any ray initially parallel to the optical axis diverges along a line through the near focal point.

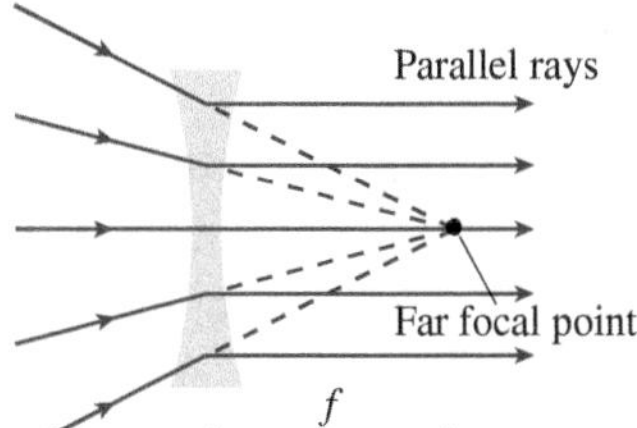

Any ray directed along a line toward the far focal point emerges from the lens parallel to the optical axis.

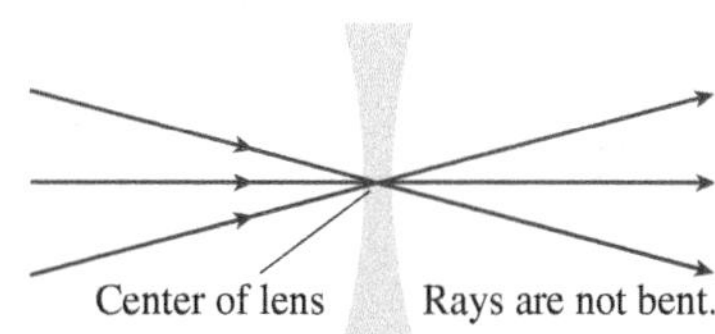

Any ray directed at the center of the lens passes through in a straight line.

Ray tracing follows the steps of Tactics Box 19.2 for a converging lens *except* that two of the three special rays in step 4 are different.

TACTICS BOX 19.3 **Ray tracing for a diverging lens**

❶–❸ **Follow steps 1 through 3 of Tactics Box 19.2.**

❹ **Draw the three "special rays" from the tip of the arrow.** Use a straightedge or a ruler. The rays refract at the lens plane.

a. A ray parallel to the axis diverges along a line through the near focal point.
b. A ray along a line toward the far focal point emerges parallel to the axis.
c. A ray through the center of the lens does not bend.

⑤ **Trace the diverging rays backward.** The point from which they are diverging is the image point, which is always a virtual image.

⑥ **Measure the image distance s'**, which, because the image is virtual, will be a negative number. Also, if needed, measure the image height relative to the object height.

EXAMPLE 19.9 **Demagnifying a flower**

A diverging lens with a focal length of 50 cm is placed 100 cm from a flower. Where is the image? What is its magnification?

PREPARE The flower is in the object plane. We use ray tracing to locate the image. Then we use Equation 19.9 to find the magnification.

SOLVE FIGURE 19.38 shows the ray-tracing diagram. The three special rays (labeled a, b, and c to match Tactics Box 19.3) do not converge. However, they can be traced backward to an intersection ≈ 33 cm to the left of the lens. This is a virtual image because the rays appear to diverge from this point, so the image distance is $s' = -33$ cm. The magnification is

$$m = -\frac{s'}{s} = -\frac{-33\text{ cm}}{100\text{ cm}} = 0.33$$

The image, which can be seen by looking *through* the lens, is one-third the size of the object and upright.

FIGURE 19.38 Ray-tracing diagram for a diverging lens.

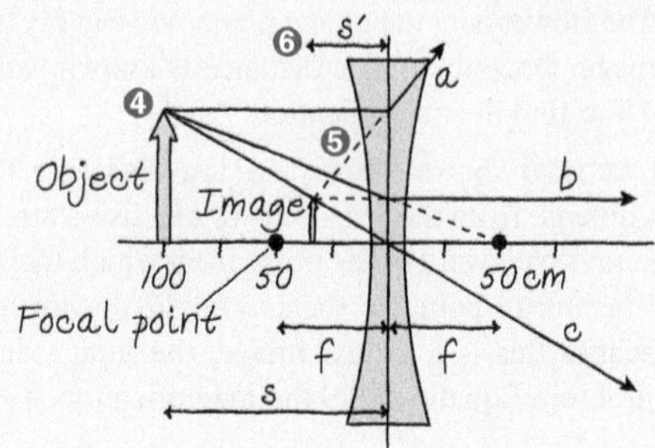

ASSESS The three special rays appear to diverge from a single point, and the image orientation and size in our ray-tracing diagram match the computed magnification—the image is upright and smaller than the object—so we have confidence in our results.

Diverging lenses *always* make virtual images and, for this reason, are rarely used alone. However, they have important applications when used in combination with converging lenses, including the converging lens of your eye. For example, microscope objectives and eyepieces often incorporate diverging lenses. And, as we'll see in the next chapter, your eyeglass lenses are diverging lenses if you are nearsighted.

STOP TO THINK 19.5 An object and lens are positioned to form a well-focused image on a viewing screen. Then a piece of cardboard is lowered just in front of the lens to cover the top half of the lens. What happens to the image on the screen?

A. Nothing
B. The upper half of the image vanishes.
C. The lower half of the image vanishes.
D. The image becomes fuzzy and out of focus.
E. The image becomes dimmer but remains in focus.

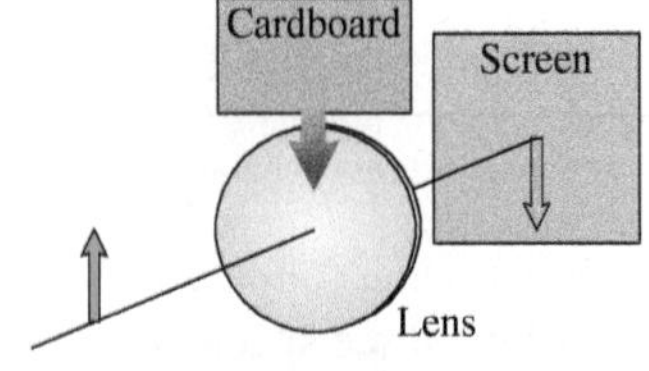

19.6 The Thin-Lens Equation

Ray tracing is an important tool, but it doesn't provide detailed information about the image. For more precise work, we would like a mathematical expression that relates the three fundamental quantities of an optical system: the focal length f of the lens, the object distance s, and the image distance s'. We can find such an expression by considering the converging lens in FIGURE 19.39. Two of the special rays are shown.

FIGURE 19.39 Deriving the thin-lens equation.

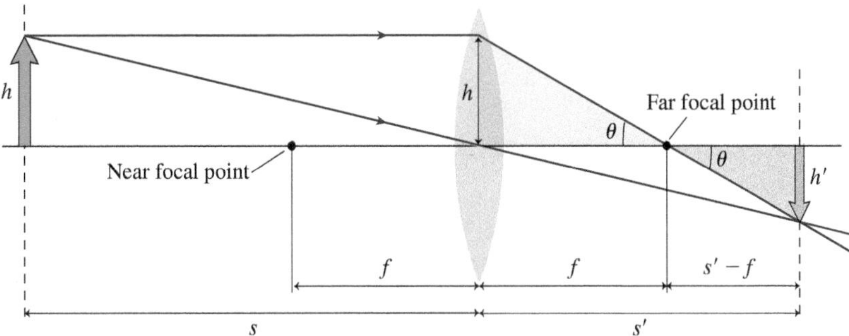

Consider the two right triangles highlighted in green and pink. Because they both have one 90° angle and a second angle θ that is the same for both, the two triangles are *similar.* This means that they have the same shape, although their sizes may be different. For similar triangles, the ratios of any two similar sides are the same. Thus we have

$$\frac{h'}{h} = \frac{s' - f}{f} \tag{19.10}$$

Further, we found in Example 19.7 that

$$\frac{h'}{h} = \frac{s'}{s} \tag{19.11}$$

Combining Equations 19.10 and 19.11 gives

$$\frac{s'}{s} = \frac{s' - f}{f}$$

We then divide both sides by s':

$$\frac{1}{s} = \frac{s' - f}{s'f} = \frac{1}{f} - \frac{1}{s'}$$

which we can write as

$$\frac{1}{s} + \frac{1}{s'} = \frac{1}{f} \tag{19.12}$$

Thin-lens equation relating object and image distances to focal length

This equation, the **thin-lens equation,** relates the locations of the object and image to the focal length of the lens.

We derived the thin-lens equation for a converging lens that produces a real image. However, the results are valid for any lens as long as all quantities are given the appropriate signs. TABLE 19.2 shows the sign convention used with the thin-lens equation. We had already seen that a virtual image has a negative image distance. In addition, the thin-lens equation requires that we assign a negative focal length to a diverging lens.

TABLE 19.2 **Sign convention for thin lenses**

	Positive	Negative
f	Converging lens, thicker in center	Diverging lens, thinner in center
s'	Real image, opposite side from object	Virtual image, same side as object

It's worth checking that the thin-lens equation describes what we already know about lenses. In Figure 19.29a, we saw that rays initially parallel to the optical axis are focused at the focal point of a converging lens. Initially parallel rays come from an object extremely far away, with $s \to \infty$. Because $1/\infty = 0$, the thin-lens equation tells us that the image distance is $s' = f$, as we expected. Or suppose an object is located right at the focal point, with $s = f$. Then, according to Equation 19.12, $1/s' = 1/f - 1/s = 0$. This implies that the image distance is infinitely far away $(s' = \infty)$, so the rays leave the lens parallel to the optical axis. Indeed, this is what Figure 19.29b showed. Now, it's true that no real object or image can be at infinity. But if either the object or image is more than several focal lengths from the lens

($s \gg f$ or $s' \gg f$), then it's an excellent approximation to consider the distance to be infinite, the rays to be parallel to the axis, and the reciprocal ($1/s$ or $1/s'$) in the thin-lens equation to be zero.

EXAMPLE 19.10 **Focusing a microscope**

The objective lens of a microscope is a planoconvex (flat on one side) lens that has a short focal length of 7.6 mm. The focusing knob on the microscope moves the object toward or away from the objective lens. The microscope is designed so that the object will be seen in focus through the eyepiece when the objective lens forms a real image of the object 160 mm behind the objective lens. At what object distance is the microscope focused? What is the lateral magnification of the objective lens?

PREPARE We can use the thin-lens equation to solve for the object distance. Even so, it is a good idea to start with a ray diagram to help make sense of the situation. **FIGURE 19.40** is a ray diagram of an object being imaged by a microscope objective. Real microscope objectives are complex optical systems built with several internal lenses, but we'll model the objective as a single lens.

FIGURE 19.40 Ray diagram of a microscope objective.

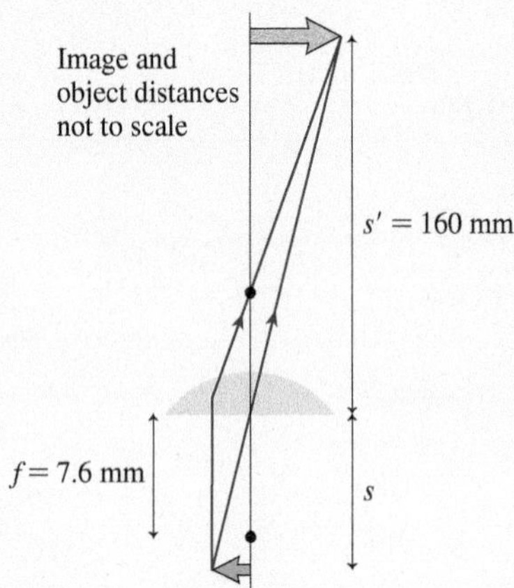

SOLVE We're given the objective's focal length $f = 7.6$ mm. For proper focusing, we need the image to fall at distance $s' = 160$ mm. This is a real image, so the image distance is positive. We can use the thin-lens equation to find that the *inverse* of the object distance is

$$\frac{1}{s} = \frac{1}{f} - \frac{1}{s'} = \frac{1}{7.6 \text{ mm}} - \frac{1}{160 \text{ mm}} = 0.125 \text{ mm}^{-1}$$

Notice that the units of $1/s$ are mm^{-1}. Now we invert this result to find that the required object distance is

$$s = \frac{1}{0.125 \text{ mm}^{-1}} = 8.0 \text{ mm}$$

Finally, we calculate that the lateral magnification of the objective lens is

$$m = -\frac{s'}{s} = -\frac{160 \text{ mm}}{8.0 \text{ mm}} = -20$$

The image is inverted with respect to the object and is 20 times larger in diameter.

ASSESS The object is very close to the objective lens, which we know is true for microscopes. We also know that microscopes typically have an overall magnification—due to the objective and the eyepiece combined—of several hundred, so a first-stage magnification of 20 seems reasonable. Notice that we can use the thin-lens equation without converting distances to meters as long as all the distances have the same units.

EXAMPLE 19.11 **Analyzing a magnifying lens**

An entomologist uses a magnifying lens that sits 2.0 cm above a specimen. The magnification is +4.0. What is the focal length of the lens?

PREPARE A magnifying lens is a converging lens. You look *through* the lens to view an upright virtual object, so the object distance is less than the focal length ($s < f$). It's always a good idea to start an optics problem with a ray diagram even when, as in this example, we will use the thin-lens equation to obtain a precise result. In this case, the ray diagram shows that the lens will form a magnified, virtual image of the specimen on the same side of the lens as the object. The diagram is shown in **FIGURE 19.41**.

SOLVE The magnification is related to the object and image distances as $m = -s'/s$; thus

$$s' = -ms = -(4.0)(2.0 \text{ cm}) = -8.0 \text{ cm}$$

We can use s and s' in the thin-lens equation to find the focal length:

$$\frac{1}{f} = \frac{1}{s} + \frac{1}{s'} = \frac{1}{2.0 \text{ cm}} + \frac{1}{-8.0 \text{ cm}} = 0.375 \text{ cm}^{-1}$$

Thus

$$f = \frac{1}{0.375 \text{ cm}^{-1}} = 2.7 \text{ cm}$$

FIGURE 19.41 Ray-tracing diagram of a magnifying lens.

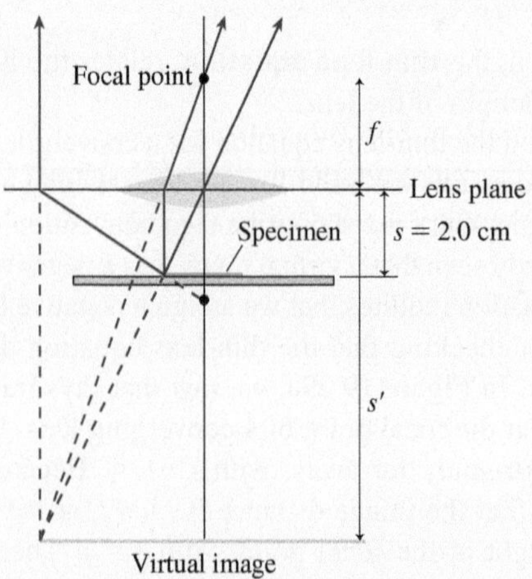

ASSESS $f > 2$ cm, as expected because the object has to be inside the focal point.

EXAMPLE 19.12 Estimating the focal length of a diverging lens

Alejandra is nearsighted, and the lenses that correct her vision are diverging lenses. To estimate the focal length of the lenses, she holds one lens 24 cm above a coin on the table, looks at the coin through the lens, and estimates that the diameter of the coin she sees is two-thirds the actual diameter of the coin. What is the focal length of her lenses?

PREPARE Alejandra sees a virtual image of the coin. You learned in Example 19.9 that a diverging lens *demagnifies,* so the coin she sees is smaller than the actual coin. Its magnification $m = 2/3$ is less than 1. We can use this and the known object distance $s = 24$ cm to find the focal length, which we expect to be negative because the lens is a diverging lens. We start with the ray diagram of FIGURE 19.42.

FIGURE 19.42 Ray-tracing diagram for the image in a diverging lens.

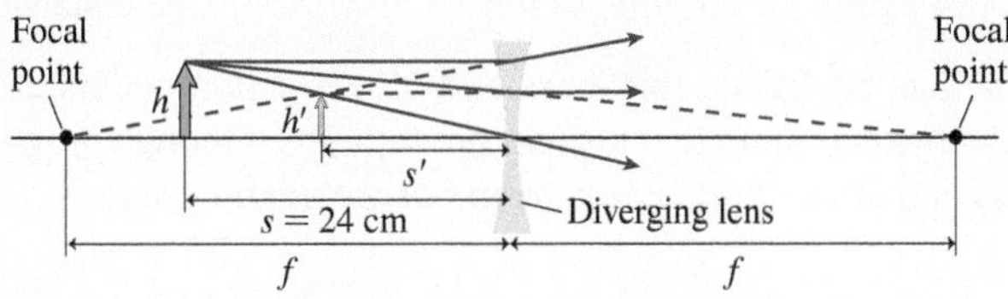

SOLVE We know the object distance and the magnification, so we rearrange Equation 19.9 to find the image distance:

$$s' = -ms = -\left(\frac{2}{3}\right)(24\text{ cm}) = -16\text{ cm}$$

We expected s' to be negative because the image is a virtual image. We now use the thin-lens equation to find the focal length:

$$f = \frac{1}{\left(\dfrac{1}{s} + \dfrac{1}{s'}\right)} = \frac{1}{\left(\dfrac{1}{24\text{ cm}} + \dfrac{1}{-16\text{ cm}}\right)} = -48\text{ cm}$$

ASSESS The focal length is negative, as we expect. In addition, as we'll see in the next chapter, this is a reasonable value for the focal length of the corrective lens of a person who is moderately nearsighted.

STOP TO THINK 19.6 A lens forms a real image of a lightbulb, but the image of the bulb on a viewing screen is a little blurry because the screen is slightly in front of the image plane. To focus the image on the screen, should you move the lens slightly toward the bulb or slightly away from the bulb?

19.7 Image Formation with Spherical Mirrors

Curved mirrors—such as those used in security and rearview mirrors—can be used to form images, and their images can be analyzed with ray diagrams similar to those used with lenses. We'll consider only the important case of **spherical mirrors,** whose surface is a section of a sphere.

Concave Mirrors

FIGURE 19.43 shows a **concave mirror,** a mirror in which the edges curve *toward* the light source. Rays parallel to the optical axis reflect from the surface of the mirror so as to pass through a single point on the optical axis. This is the focal point of the mirror. The focal length is the distance from the mirror surface to the focal point. A concave mirror is analogous to a converging lens, but it has only one focal point.

FIGURE 19.43 The focal point and focal length of a concave mirror.

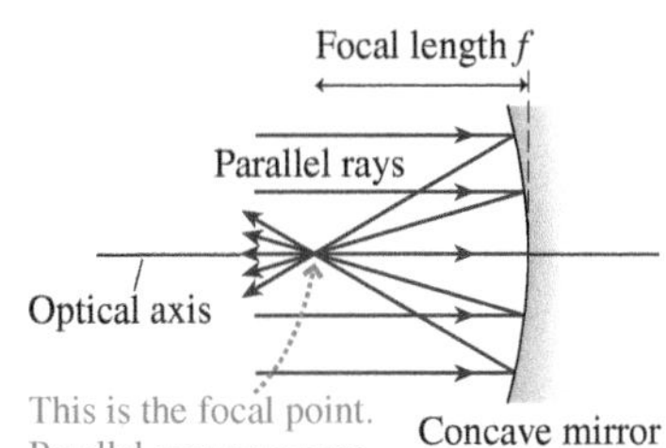

Let's begin by considering the case where the object's distance s from the mirror is greater than the focal length ($s > f$), as shown in FIGURE 19.44 (next page). We see that the image is *real* (and inverted) because rays from the object point P converge at the image point P′. Although an infinite number of rays from P all meet at P′, each ray obeying the law of reflection, you can see that three "special rays" are enough to determine the position and size of the image:

- A ray parallel to the axis reflects through the focal point.
- A ray through the focal point reflects parallel to the axis.
- A ray striking the center of the mirror reflects at an equal angle on the opposite side of the axis.

FIGURE 19.44 A real image formed by a concave mirror.

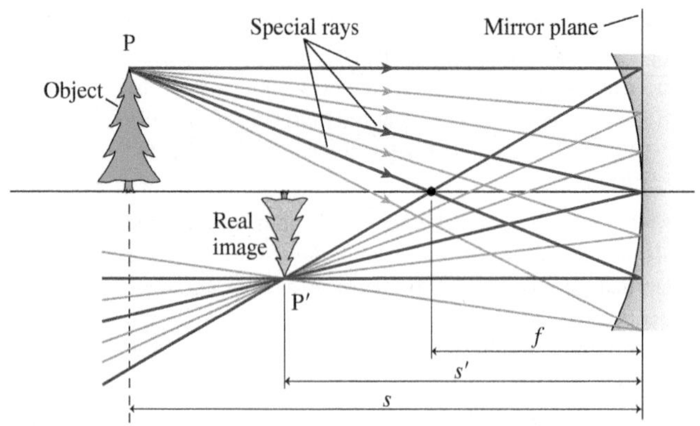

FIGURE 19.45 The focal point and focal length of a convex mirror.

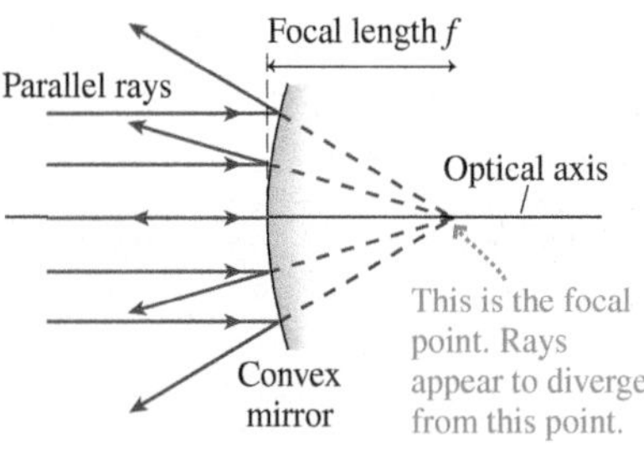

FIGURE 19.46 A city skyline is reflected in this polished sphere.

These three rays also locate the image if $s < f$, but in that case the image is *virtual* and behind the mirror. Once again, virtual images have a *negative* image distance s'.

NOTE ▸ A thin lens has negligible thickness and all refraction occurs at the lens plane. Similarly, we will assume that mirrors are thin (even though drawings may show a thickness) and thus *all reflection occurs at the mirror plane.* ◂

Convex Mirrors

FIGURE 19.45 shows parallel light rays approaching a mirror in which the edges curve *away from* the light source. This is called a **convex mirror.** In this case, the reflected rays appear to come from a point behind the mirror. This is the focal point for a convex mirror.

A common example of a convex mirror is a silvered ball, such as a tree ornament. You may have noticed that if you look at your reflection in such a ball, your image appears right-side-up but is quite small. As another example, FIGURE 19.46 shows a city skyline reflected in a polished metal sphere. Let's use ray tracing to understand why the skyscrapers all appear to be so small.

FIGURE 19.47 shows an object in front of a convex mirror. In this case, the reflected rays—each obeying the law of reflection—create an upright image of reduced height behind the mirror. We see that the image is virtual because no rays actually converge at the image point P′. Instead, diverging rays *appear* to come from this point. Once again, three special rays are enough to find the image.

FIGURE 19.47 A virtual image formed by a convex mirror.

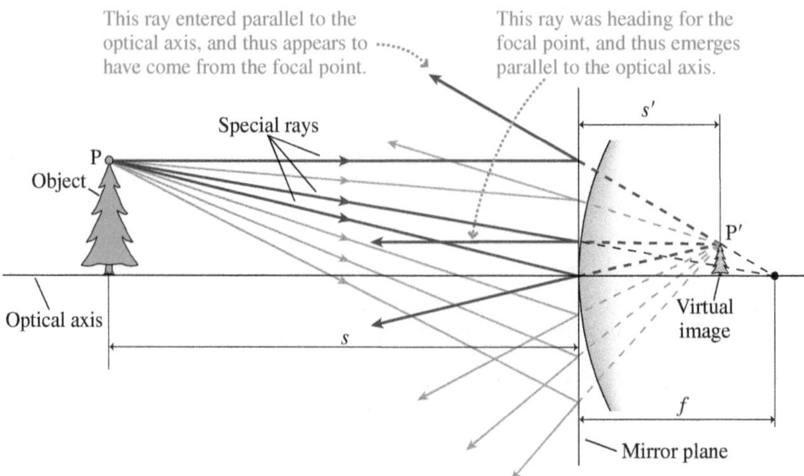

Convex mirrors are used for a variety of safety and monitoring applications, such as passenger-side rearview mirrors and the round mirrors used in stores to keep an eye on the customers. When an object is reflected in a convex mirror, the image appears smaller than the object itself. Because the image is, in a sense, a miniature version of the object, you can *see much more of it* within the edges of the mirror than you could with an equal-sized flat mirror.

TACTICS BOX 19.4 Ray tracing for a spherical mirror

1. **Draw an optical axis.** Use a straightedge or a ruler! Establish a scale.
2. **Center the mirror on the axis.** Mark and label the focal point at distance f from the mirror's surface.
3. **Represent the object with an upright arrow at distance s.** It's usually best to place the base of the arrow on the axis and to draw the arrow about half the radius of the mirror.
4. **Draw the three "special rays" from the tip of the arrow.** All reflections occur at the mirror plane.
   a. A ray parallel to the axis reflects through (concave) or away from (convex) the focal point.
   b. An incoming ray passing through (concave) or heading toward (convex) the focal point reflects parallel to the axis.
   c. A ray that strikes the center of the mirror reflects at an equal angle on the opposite side of the optical axis.
5. **Extend the rays forward or backward until they converge.** This is the image point. Draw the rest of the image in the image plane. If the base of the object is on the axis, then the base of the image will also be on the axis.
6. **Measure the image distance s'.** Also, if needed, measure the image height relative to the object height.

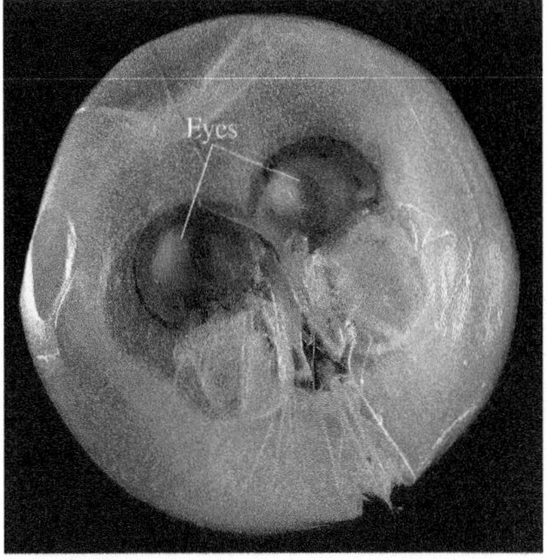
Eyes

Look into my eyes The eyes of most animals use lenses to focus an image. *Gigantocypris,* a deep-sea crustacean about 3 cm in diameter, is unusual in that it uses two converging mirrors to focus light onto its retina. Because it lives at depths where no sunlight penetrates, *Gigantocypris* is thought to use its mirror eyes to hunt bioluminescent animals.

EXAMPLE 19.13 Analyzing a concave mirror

A 3.0-cm-high object is located 60 cm from a concave mirror. The mirror's focal length is 40 cm. Use ray tracing to find the position and height of the image.

PREPARE We will use the ray-tracing steps of Tactics Box 19.4, as shown in **FIGURE 19.48**.

SOLVE We can use a ruler to find that the image position is $s' \approx 120$ cm in front of the mirror and its height is $h' \approx 6$ cm.

ASSESS The image is a *real* image because light rays converge at the image point.

FIGURE 19.48 Ray-tracing diagram for a concave mirror.

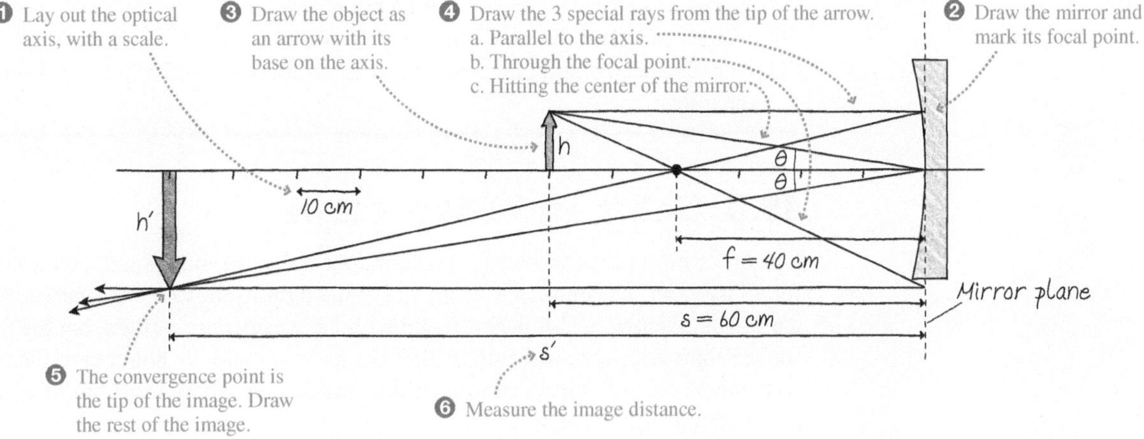

1 Lay out the optical axis, with a scale.
3 Draw the object as an arrow with its base on the axis.
4 Draw the 3 special rays from the tip of the arrow.
a. Parallel to the axis.
b. Through the focal point.
c. Hitting the center of the mirror.
2 Draw the mirror and mark its focal point.
h
θ
θ
h′
10 cm
f = 40 cm
Mirror plane
s = 60 cm
s′
5 The convergence point is the tip of the image. Draw the rest of the image.
6 Measure the image distance.

The Mirror Equation

The thin-lens equation assumes lenses have negligible thickness (so a single refraction occurs in the lens plane) and the rays are nearly parallel to the optical axis (paraxial rays). If we make the same assumptions about spherical mirrors—the mirror has negligible thickness and so paraxial rays reflect at the mirror plane—then the object and image distances are related exactly as they were for thin lenses:

$$\frac{1}{s}+\frac{1}{s'}=\frac{1}{f} \tag{19.13}$$

TABLE 19.3 Sign convention for spherical mirrors

	Positive	Negative
f	Concave toward the object	Convex toward the object
s'	Real image, same side as object	Virtual image, opposite side from object

TABLE 19.3 shows the sign convention used with spherical mirrors. A concave mirror (analogous to a converging lens) has a positive focal length while a convex mirror (analogous to a diverging lens) has a negative focal length. The lateral magnification of a spherical mirror is computed exactly as for a lens:

$$m=-\frac{s'}{s} \tag{19.14}$$

EXAMPLE 19.14 Analyzing a concave mirror

A 3.0-cm-high object is located 20 cm in front of a concave mirror that has a focal length of 40 cm. Determine the position, orientation, and height of the image.

PREPARE We'll model the mirror as a thin mirror. The sign convention of Table 19.3 tells us that the focal length is positive. The three special rays in FIGURE 19.49 show that the image is a magnified, virtual image behind the mirror.

FIGURE 19.49 Pictorial representation of Example 19.14.

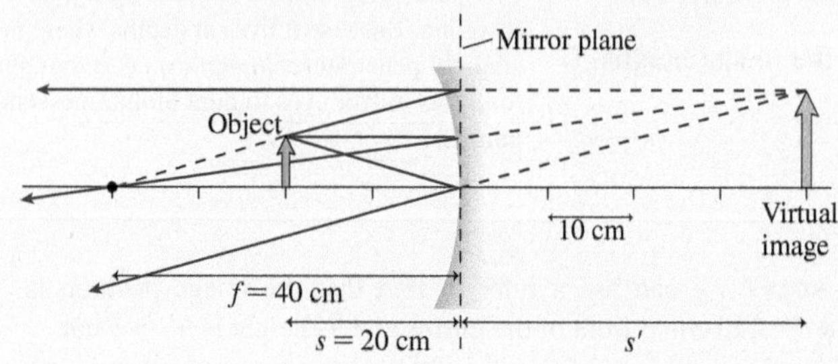

SOLVE The thin-mirror equation is

$$\frac{1}{20\text{ cm}}+\frac{1}{s'}=\frac{1}{40\text{ cm}}$$

This is easily solved to give $s'=-40$ cm, in agreement with the ray tracing. The negative sign tells us this is a virtual image behind the mirror. The magnification is

$$m=-\frac{-40\text{ cm}}{20\text{ cm}}=+2.0$$

Consequently, the image is 6.0 cm tall and upright.

ASSESS This is a virtual image because light rays diverge from the image point. You could see this enlarged image by standing behind the object and looking into the mirror. In fact, this is how magnifying cosmetic mirrors work.

STOP TO THINK 19.7 A concave mirror of focal length f forms an image of the moon. Where is the image located?

A. At the mirror's surface
B. Almost exactly a distance f behind the mirror
C. Almost exactly a distance f in front of the mirror
D. At a distance behind the mirror equal to the distance of the moon in front of the mirror

19.8 Color and Dispersion

One of the most obvious visual aspects of light is the phenomenon of color. Yet color, for all its vivid sensation, is not inherent in the light itself. Color is a *perception,* not a physical quantity. Color is associated with the wavelength of light, but the fact that we see light with a wavelength of 650 nm as "red" tells us how our visual system responds to electromagnetic waves of this wavelength. There is no "redness" associated with the light wave itself.

Most of the results of optics do not depend on color; a microscope works the same with red light and blue light. Even so, indices of refraction are slightly wavelength dependent, which will be important in the next chapter where we consider the resolution of optical instruments. And color in nature is an interesting subject, one worthy of a short digression.

Color

It has been known since antiquity that irregularly shaped glass and crystals cause sunlight to be broken into various colors. A common idea was that the glass or crystal somehow altered the properties of the light by *adding* color to the light. Newton suggested a different explanation. He first passed a sunbeam through a prism, producing the familiar rainbow of light. We say that the prism *disperses* the light. Newton's novel idea, shown in FIGURE 19.50a, was to use a second prism, inverted with respect to the first, to "reassemble" the colors. He found that the light emerging from the second prism was a beam of pure, white light.

But the emerging light beam is white only if *all* the rays are allowed to move between the two prisms. Blocking some of the rays with small obstacles, as in FIGURE 19.50b, causes the emerging light beam to have color. This suggests that color is associated with the light itself, not with anything that the prism is doing to the light. Newton tested this idea by inserting a small aperture between the prisms to pass only the rays of a particular color, such as green. If the prism alters the properties of light, then the second prism should change the green light to other colors. Instead, the light emerging from the second prism is unchanged from the green light entering the prism.

These and similar experiments show that

1. What we perceive as white light is a mixture of all colors. White light can be dispersed into its various colors and, equally important, mixing all the colors produces white light.
2. The index of refraction of a transparent material differs slightly for different colors of light. Glass has a slightly larger index of refraction for violet light than for green light or red light. Consequently, different colors of light refract at slightly different angles. A prism does not alter the light or add anything to the light; it simply causes the different colors that are inherent in white light to follow slightly different trajectories.

FIGURE 19.50 Newton used prisms to study color.

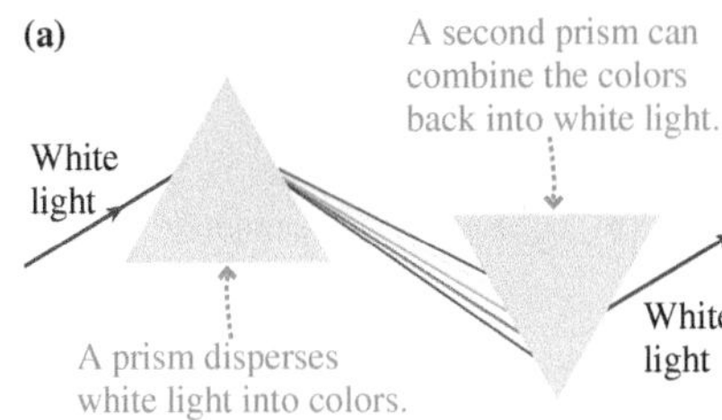

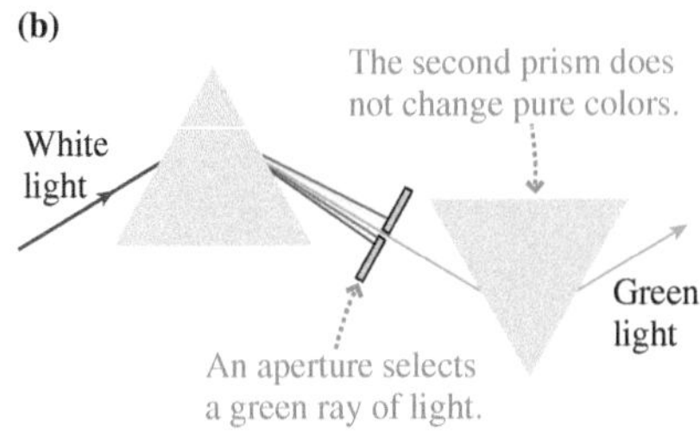

TABLE 19.4 A brief summary of the visible spectrum of light

Color	Approximate wavelength
Deepest red	700 nm
Red	650 nm
Green	550 nm
Blue	450 nm
Deepest violet	400 nm

Dispersion

It was Thomas Young, with his two-slit interference experiment, who showed that different colors are associated with light of different wavelengths. The longest wavelengths are perceived as red light and the shortest as violet light. TABLE 19.4 is a brief summary of the *visible spectrum* of light. Visible-light wavelengths are used so frequently that it is well worth committing this short table to memory.

The slight variation of index of refraction with wavelength is known as **dispersion**. FIGURE 19.51 shows the *dispersion curves* of two common glasses. Notice that **n is *larger* when the wavelength is *shorter*,** thus violet light refracts more than red light.

FIGURE 19.51 Dispersion curves show how the index of refraction varies with wavelength.

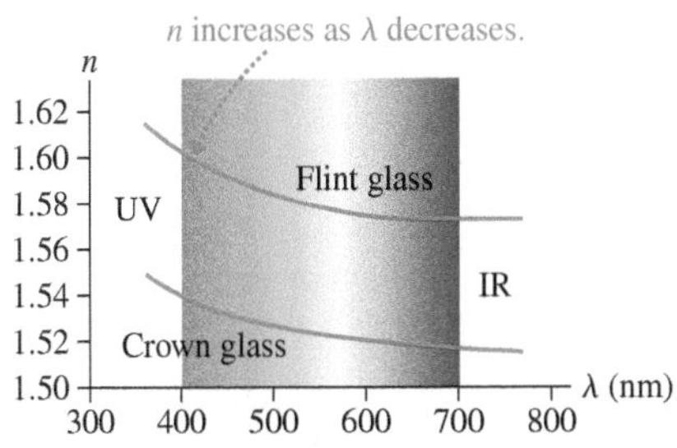

EXAMPLE 19.15 Dispersing light with a prism

Example 19.3 found that a ray incident on a 30° prism is deflected by 22.6° if the prism's index of refraction is 1.59. Suppose this is the index of refraction of deep violet light and deep red light has an index of refraction of 1.54.

a. What is the deflection angle for deep red light?
b. If a beam of white light is dispersed by this prism, how wide is the rainbow spectrum on a screen 2.0 m away?

PREPARE Figure 19.16 showed the geometry. A ray of any wavelength is incident on the hypotenuse of the prism at $\theta_1 = 30°$.

SOLVE

a. If $n_1 = 1.54$ for deep red light, the refraction angle is

$$\theta_2 = \sin^{-1}\left(\frac{n_1 \sin\theta_1}{n_2}\right) = \sin^{-1}\left(\frac{1.54 \sin 30°}{1.00}\right) = 50.4°$$

Continued

Example 19.3 showed that the deflection angle is $\phi = \theta_2 - \theta_1$, so deep red light is deflected by $\phi_{red} = 20.4°$. This angle is slightly smaller than the previously observed $\phi_{violet} = 22.6°$.

b. The entire spectrum is spread between $\phi_{red} = 20.4°$ and $\phi_{violet} = 22.6°$. The angular spread is

$$\delta = \phi_{violet} - \phi_{red} = 2.2° = 0.038 \text{ rad}$$

At distance r, the spectrum spans an arc length

$$s = r\delta = (2.0 \text{ m})(0.038 \text{ rad}) = 0.076 \text{ m} = 7.6 \text{ cm}$$

ASSESS The angle is so small that there's no appreciable difference between arc length and a straight line. The spectrum will be 7.6 cm wide at a distance of 2.0 m.

Rainbows

One of the most interesting sources of color in nature is the rainbow. The details get somewhat complicated, but FIGURE 19.52a shows that the basic cause of the rainbow is a combination of refraction, reflection, and dispersion.

Figure 19.52a might lead you to think that the top edge of a rainbow is violet. In fact, the top edge is red, and violet is on the bottom. The rays leaving the drop in Figure 19.52a are spreading apart, so they can't all reach your eye. As FIGURE 19.52b shows, a ray of red light reaching your eye comes from a drop *higher* in the sky than a ray of violet light. In other words, the colors you see in a rainbow refract toward your eye from different raindrops, not from the same drop. You have to look higher in the sky to see the red light than to see the violet light.

FIGURE 19.52 Light seen in a rainbow has undergone refraction + reflection + refraction in a raindrop.

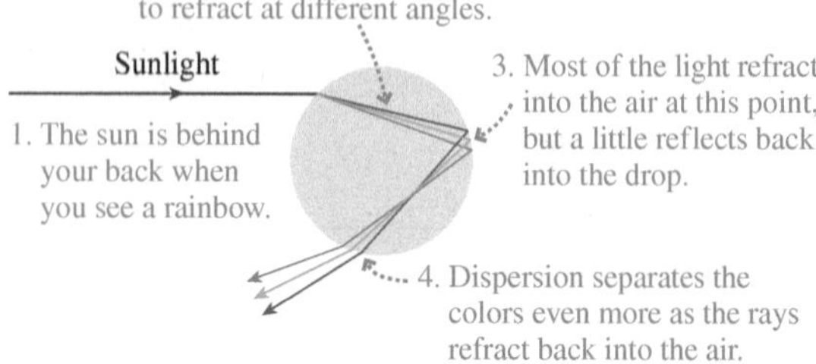
(a)
2. Dispersion causes different colors to refract at different angles.
Sunlight
3. Most of the light refracts into the air at this point, but a little reflects back into the drop.
1. The sun is behind your back when you see a rainbow.
4. Dispersion separates the colors even more as the rays refract back into the air.
(b)
Sunlight
42.5°
40.8°
Eye
Red light is refracted predominantly at 42.5°. The red light reaching your eye comes from drops higher in the sky.
Violet light is refracted predominantly at 40.8°. The violet light reaching your eye comes from drops lower in the sky.
You see a rainbow with red on the top, violet on the bottom.

Colored Filters and Colored Objects

White light passing through a piece of green glass emerges as green light. A possible explanation would be that the green glass *adds* "greenness" to the white light, but Newton found otherwise. Green glass is green because it *absorbs* any light that is "not green." We can think of a piece of colored glass or plastic as a *filter* that removes all wavelengths except a chosen few.

FIGURE 19.53 No light at all passes through both a green and a red filter.

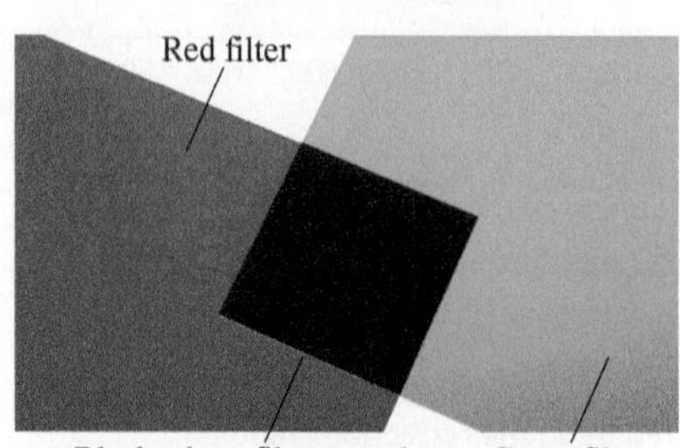
Red filter
Black where filters overlap
Green filter

If a green filter and a red filter are overlapped, as in FIGURE 19.53, *no* light gets through. The green filter transmits only green light, which is then absorbed by the red filter because it is "not red."

This behavior is true not just for glass filters, which transmit light, but for *pigments* that absorb light of some wavelengths but *reflect* light at other wavelengths. For example, red paint contains pigments reflecting light at wavelengths near 650 nm while absorbing all other wavelengths. Pigments in paints, inks, and natural objects are responsible for most of the color we observe in the world, from the red of lipstick to the blue of a blueberry.

As an example, FIGURE 19.54 shows the absorption curve of *chlorophyll.* Chlorophyll is essential for photosynthesis in green plants. The chemical reactions of photosynthesis are able to use red light and blue/violet light, thus chlorophyll absorbs red light and blue/violet light from sunlight and puts it to use. But green and yellow light are not absorbed. Instead, these wavelengths are mostly *reflected* to give the object a greenish-yellow color. When you look at the green leaves on a tree, you're seeing the light that was reflected because it *wasn't* needed for photosynthesis.

FIGURE 19.54 The absorption curve of chlorophyll.

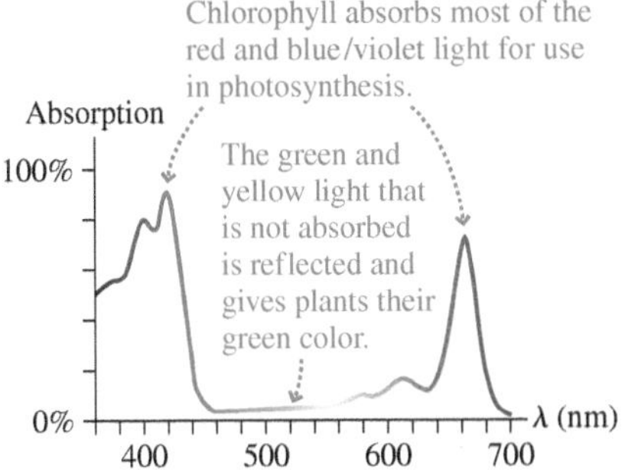
Chlorophyll absorbs most of the red and blue/violet light for use in photosynthesis.
Absorption
100%
The green and yellow light that is not absorbed is reflected and gives plants their green color.
0%
400
500
600
700
λ (nm)

Light Scattering: Blue Skies and Red Sunsets

In the ray model of Section 19.1 we noted that light within a medium can be scattered or absorbed. As we've now seen, the absorption of light can be wavelength dependent and can create color in objects. What are the effects of scattering?

Light can scatter from small particles that are suspended in a medium. If the particles are large compared to the wavelengths of light—even though they may be microscopic and not readily visible to the naked eye—the light essentially reflects off the particles. The law of reflection doesn't depend on wavelength, so all colors are scattered equally. White light scattered from many small particles makes the medium appear cloudy and white. Two well-known examples are clouds, where micrometer-size water droplets scatter the light, and milk, which is a colloidal suspension of microscopic droplets of fats and proteins.

A more interesting aspect of scattering occurs at the atomic level. The atoms and molecules of a transparent medium are much smaller than the wavelengths of light, so they can't scatter light simply by reflection. Instead, the oscillating electric field of the light wave interacts with the electrons in each atom in such a way that the light is scattered. This atomic-level scattering is called **Rayleigh scattering.**

Red sky at night Sunsets are red because all the blue light has scattered as the sunlight passes through the atmosphere.

Unlike the scattering by small particles, Rayleigh scattering from atoms and molecules *does* depend on the wavelength. A detailed analysis shows that the intensity of scattered light depends inversely on the fourth power of the wavelength: $I_{\text{scattered}} \propto \lambda^{-4}$. This wavelength dependence explains why the sky is blue and sunsets are red.

As sunlight travels through the atmosphere, the λ^{-4} dependence of Rayleigh scattering causes the shorter wavelengths to be preferentially scattered. If we take 650 nm as a typical wavelength for red light and 450 nm for blue light, the intensity of scattered blue light relative to scattered red light is

$$\frac{I_{\text{blue}}}{I_{\text{red}}} = \left(\frac{650}{450}\right)^4 \approx 4$$

Four times more blue light is scattered toward us than red light and thus, as FIGURE 19.55 shows, the sky appears blue.

Because of the earth's curvature, sunlight has to travel much farther through the atmosphere when we see it at sunrise or sunset than it does during the midday hours. In fact, the path length through the atmosphere at sunset is so long that essentially all the short wavelengths have been lost due to Rayleigh scattering. Only the longer wavelengths remain—orange and red—and they make the colors of the sunset.

FIGURE 19.55 Rayleigh scattering by molecules in the air gives the sky and sunsets their color.

At midday the scattered light is mostly blue because molecules preferentially scatter shorter wavelengths.

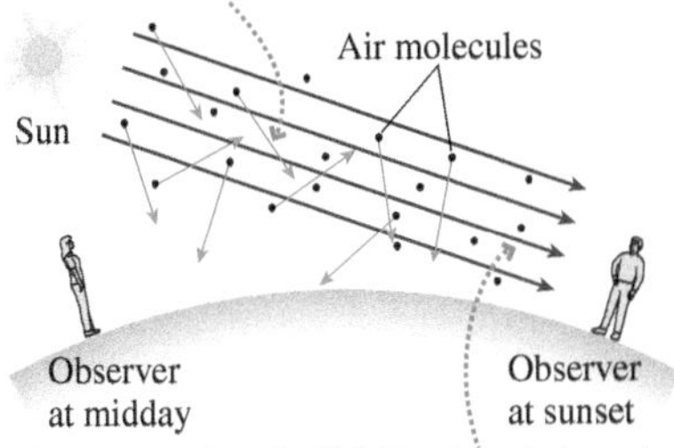
Air molecules
Sun
Observer at midday
Observer at sunset

At sunset, when the light has traveled much farther through the atmosphere, the light is mostly red because the shorter wavelengths have been lost to scattering.

STOP TO THINK 19.8 A swatch of pure red fabric is viewed through a green filter. The fabric appears

A. Red B. Green C. Yellow D. Black E. White

CHAPTER 19 INTEGRATED EXAMPLE **Optical fiber imaging**

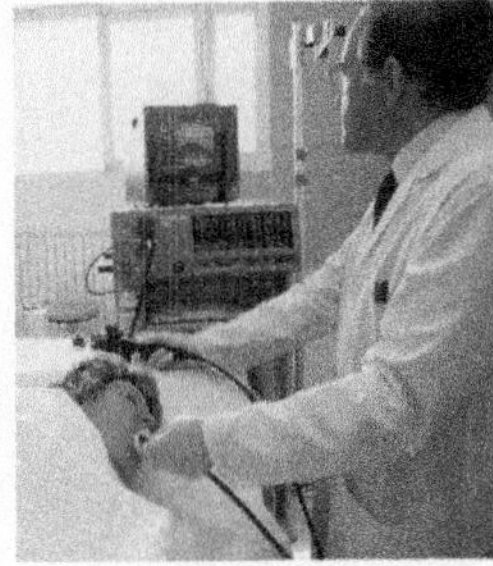

An *endoscope* is a thin bundle of optical fibers that can be inserted through a bodily opening or small incision to view the interior of the body. As FIGURE 19.56 (next page) shows, an *objective* lens forms a real image on the entrance face of the fiber bundle. Individual fibers, using total internal reflection (TIR), transport the light to the exit face, where it emerges. The physician (or, more likely, a camera) observes the object by viewing the exit face through an *eyepiece* lens.

Consider an endoscope having a 3.0-mm-diameter objective lens with a focal length of 1.1 mm. These are typical values. The indices of refraction of the core and the cladding of the optical fibers are 1.62 and 1.50, respectively. To give maximum brightness, the objective lens is positioned so that, for an on-axis object, rays passing through the outer edge of the lens have the maximum angle of incidence for undergoing TIR

Continued

FIGURE 19.56 An endoscope.

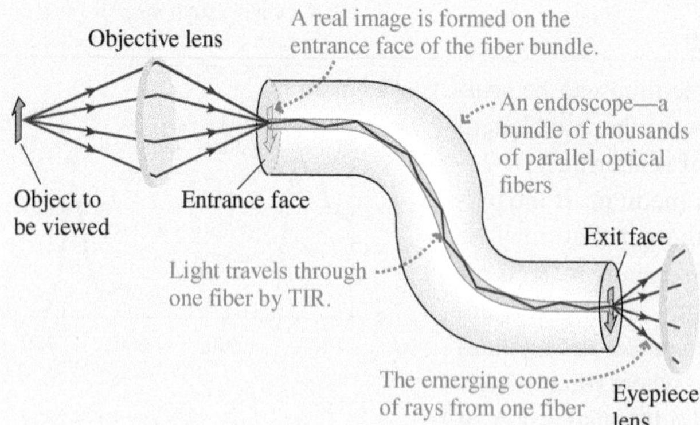

in the fiber. Suppose a physician wishes to view a 1.5-mm-diameter lesion.

a. How far from the lesion must the objective lens be positioned?
b. What is the diameter of the image of the lesion on the entrance face of the fiber bundle?

PREPARE We'll represent the object as an on-axis point source and use the ray model of light. FIGURE 19.57 shows the real image being focused on the entrance face of the endoscope. Inside the fiber, rays that strike the cladding at an angle of incidence greater than the critical angle θ_c undergo TIR and stay in the fiber; rays are lost if their angle of incidence is less than θ_c. We found in Section 19.3 that the critical angle at a boundary between indices of refraction n_1 and n_2 is $\theta_c = \sin^{-1}(n_2/n_1)$.

For maximum brightness, the lens is positioned so that a ray passing through the outer edge refracts into the fiber at the maximum angle of incidence θ_{max} for which TIR is possible. A smaller-diameter lens would sacrifice light-gathering power, whereas the outer rays from a larger-diameter lens would impinge on the core-cladding boundary at less than θ_c and would not undergo TIR.

FIGURE 19.57 Magnified view of the entrance of an optical fiber.

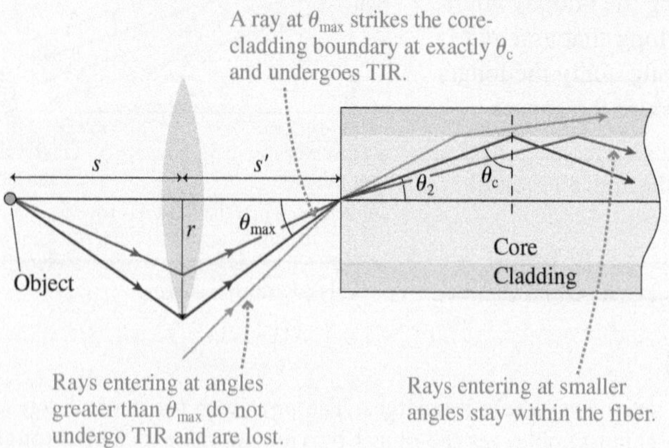

SOLVE We know the focal length of the lens. We can use the geometry of the ray at the critical angle to find the distance between the lens and the fiber, which is the image distance s', then use the thin-lens equation to find the object distance s. The critical angle for TIR inside the fiber is

$$\theta_c = \sin^{-1}\left(\frac{n_{cladding}}{n_{core}}\right) = \sin^{-1}\left(\frac{1.50}{1.62}\right) = 67.8°$$

A ray incident on the core-cladding boundary at exactly the critical angle must have entered the fiber, at the entrance face, at angle $\theta_2 = 90° - \theta_c = 22.2°$. For optimum lens placement, this ray passed through the outer edge of the lens and was incident on the entrance face at angle θ_{max}. Snell's law at the entrance face is

$$n_{air} \sin\theta_{max} = 1.00 \sin\theta_{max} = n_{core} \sin\theta_2$$

and thus

$$\theta_{max} = \sin^{-1}(1.62 \sin 22.2°) = 37.7°$$

We know the lens radius, $r = 1.5$ mm, so the distance of the lens from the fiber—the image distance s'—is

$$s' = \frac{r}{\tan\theta_{max}} = \frac{1.5 \text{ mm}}{\tan(37.7°)} = 1.9 \text{ mm}$$

a. Now we can use the thin-lens equation to calculate the object distance to the lesion:

$$\frac{1}{s} = \frac{1}{f} - \frac{1}{s'} = \frac{1}{1.1 \text{ mm}} - \frac{1}{1.9 \text{ mm}}$$
$$s = 2.6 \text{ mm}$$

The physician, viewing the exit face of the fiber bundle, will see a focused image when the objective lens is 2.6 mm from the lesion she wishes to view.

b. The entrance face of the fiber bundle is like a viewing screen on which an image of the lesion is focused. The size of the image is determined by the magnification:

$$m = -\frac{s'}{s} = -\frac{1.9 \text{ mm}}{2.6 \text{ mm}} = -0.73$$

That is, the image on the entrance face is inverted relative to the object and is slightly demagnified to 73% the size of the object. Thus the diameter of the image of the 1.5-mm-diameter lesion is

$$h' = |m|h = (0.73)(1.5 \text{ mm}) = 1.1 \text{ mm}$$

ASSESS The object and image distances are both greater than the focal length of the objective lens, which is correct for forming a real image.

SUMMARY

GOAL To apply the ray model of light to the phenomena of reflection, refraction, and image formation.

GENERAL PRINCIPLES

Reflection

Law of reflection: $\theta_r = \theta_i$

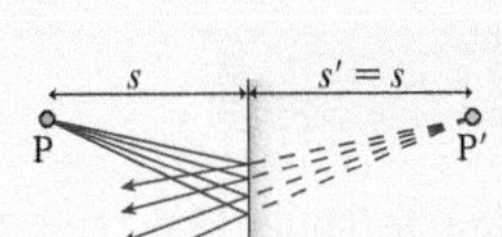

Reflection can be specular (mirror-like) or diffuse (from rough surfaces).

Plane mirrors: A virtual image is formed at P′ with $s' = s$, where s is the object distance and s' is the image distance.

Refraction

Snell's law of refraction:

$$n_1 \sin\theta_1 = n_2 \sin\theta_2$$

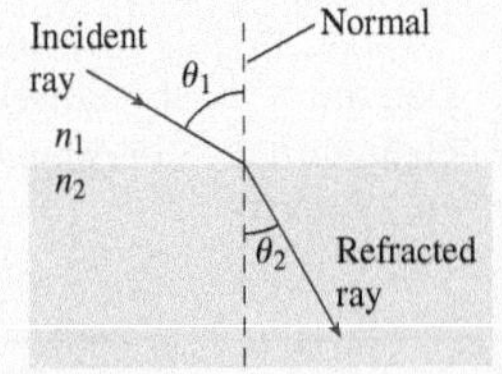

The **index of refraction** is $n = c/v$. The ray is closer to the normal on the side that has the larger index of refraction.

If $n_2 < n_1$, **total internal reflection** (TIR) occurs when the angle of incidence θ_1 is greater than $\theta_c = \sin^{-1}(n_2/n_1)$.

IMPORTANT CONCEPTS

The ray model of light

The ray model is used when apertures (lenses, mirrors, prisms, etc.) are larger than about 1 mm.

Light travels through a medium along straight lines called **light rays.**

A light ray continues forever unless an interaction with matter causes it to reflect, refract, scatter, or be absorbed.

Light rays come from **objects**, which can be self-luminous or reflective. Each point on the object sends rays in all directions.

Ray diagrams use only a few select rays to represent all the rays emitted by an object.

The eye sees an object (or an image) when diverging rays are collected by the pupil and focused on the retina.

Image formation

If rays diverge from P and interact with a lens or mirror so that the refracted rays *converge* at P′, then P′ is a **real image** of P. Rays actually pass through a real image.

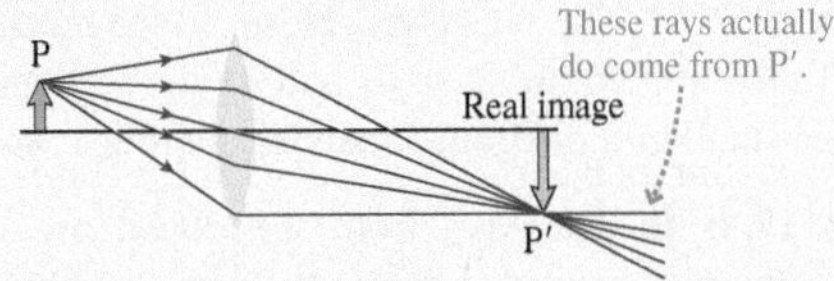

If rays diverge from P and, after interacting with a lens or mirror, *appear* to diverge from P′ without actually passing through P′, then P′ is a **virtual image** of P.

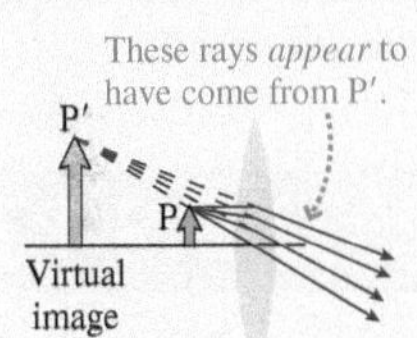

APPLICATIONS

Ray tracing for lenses

Three special rays in three basic situations:

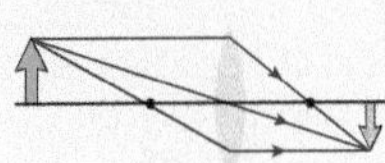

Converging lens
Real image

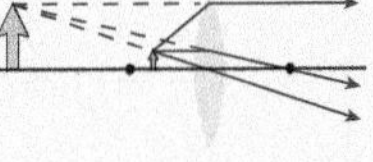

Converging lens
Virtual image

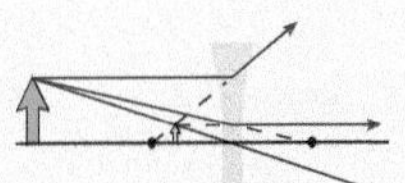

Diverging lens
Virtual image

Ray tracing for mirrors

Three special rays in three basic situations:

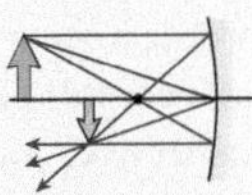

Concave mirror
Real image

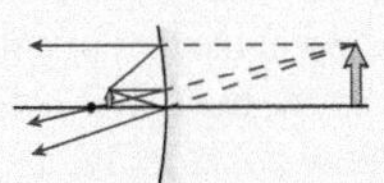

Concave mirror
Virtual image

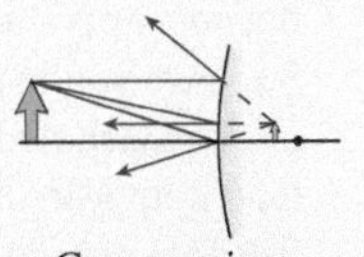

Convex mirror
Virtual image

The thin-lens equation

For a lens or curved mirror, the object distance s, the image distance s', and the focal length f are related by the thin-lens equation:

$$\frac{1}{s} + \frac{1}{s'} = \frac{1}{f}$$

The **lateral magnification** of a lens or mirror is $m = -s'/s$.

Sign conventions are used with the thin-lens equation:

Quantity	Positive	Negative
f	Converging lens or concave mirror	Diverging lens or convex mirror
s	Always	Not treated here
s'	Real image. For a lens, opposite side from object. For a mirror, same side as object.	Virtual image. For a lens, same side as object. For a mirror, opposite side from object.
m	Upright image	Inverted image

LEARNING OBJECTIVES After studying this chapter, you should be able to:

- Draw ray diagrams. *Conceptual Questions 19.1, 19.2; Problems 19.2–19.4*
- Solve problems of reflection. *Conceptual Question 19.3; Problems 19.5–19.8, 19.11*
- Calculate refraction angles and total internal reflection. *Conceptual Question 19.4; Problems 19.12–19.14, 19.19, 19.21*
- Determine the locations of images caused by refraction at a plane surface. *Conceptual Questions 19.5, 19.7; Problems 19.23–19.27*
- Determine the locations of real and virtual images of lenses using ray tracing. *Conceptual Questions 19.8, 19.10; Problems 19.28–19.32*
- Solve optics problems with the thin-lens equation. *Conceptual Question 19.11; Problems 19.33–19.36, 19.40*
- Determine the locations of real and virtual images of spherical mirrors. *Conceptual Questions 19.12, 19.13; Problems 19.42–19.45, 19.47*
- Determine the effects of dispersion on refraction. *Problems 19.48–19.50*

STOP TO THINK ANSWERS

Stop to Think 19.1: C. The light spreads vertically as it goes through the vertical aperture. The light spreads horizontally due to different points on the horizontal lightbulb.

Stop to Think 19.2: C. The image due to a plane mirror is always located on the opposite side of the mirror as the object, along a line passing through the object and perpendicular to the mirror, and the same distance from the mirror as the object. The position of the observer has no bearing on the position of the image.

Stop to Think 19.3: A. The ray travels closer to the normal in both media 1 and 3 than in medium 2, so n_1 and n_3 are both greater than n_2. The angle is smaller in medium 3 than in medium 1, so $n_3 > n_1$.

Stop to Think 19.4: E. The rays from the object are diverging. Without a lens, the rays cannot converge to form any kind of image on the screen.

Stop to Think 19.5: E. From Figure 19.30, the image at P′ is created by many rays from P that go through *all* parts of the lens. If the upper half of the lens is obscured, all the rays that go through the lower half still focus at P′. The image remains sharp but, because of the fewer rays reaching it, becomes dimmer.

Stop to Think 19.6: Away from. You need to decrease s' to bring the image plane onto the screen. s' is decreased by increasing s.

Stop to Think 19.7: C. A concave mirror forms a real image in front of the mirror. Because the object distance is $s \approx \infty$, the image distance is $s' \approx f$.

Stop to Think 19.8: D. The fabric is "pure red," so it reflects only red light. A green filter lets through only green light, so it blocks all of the reflected light. No light from the fabric can pass through the filter, so it appears black.

QUESTIONS

Conceptual Questions

1. On the earth, you can see the ground in someone's shadow; on the moon, you can't—the shadow is deep black. Explain the difference.
2. Suppose you have two pinhole cameras. The first has a small round hole in the front. The second is identical except it has a square hole of the same area as the round hole in the first camera. Would the pictures taken by these two cameras, under the same conditions, be different in any obvious way? Explain.

FIGURE Q19.1

3. You are looking at the image of a pencil in a mirror, as shown in Figure Q19.3.
   a. What happens to the image if the top half of the mirror, down to the midpoint, is covered with a piece of cardboard? Explain.
   b. What happens to the image if the bottom half of the mirror is covered with a piece of cardboard? Explain.

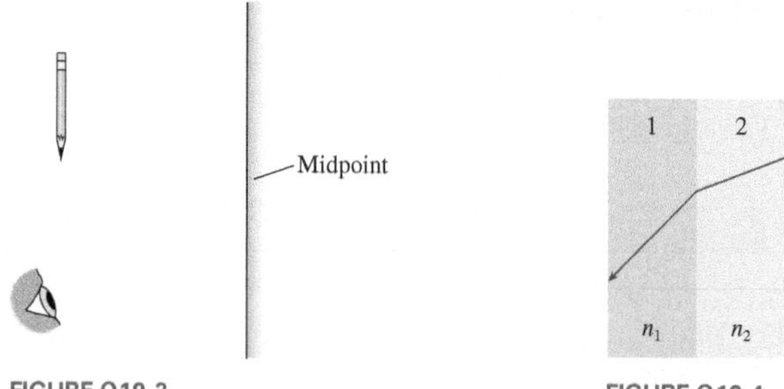

FIGURE Q19.3

FIGURE Q19.4

4. A light beam passing from medium 2 to medium 1 is refracted as shown in Figure Q19.4. Is n_1 larger than n_2, is n_1 smaller than n_2, or is there not enough information to tell? Explain.
5. Suppose you're standing at the end of a dock, and you see a fish in the water. As you know, the fish is actually at a greater depth than it appears. But now consider the fish's perspective. As the fish looks at you, does your head appear to be at a greater or lesser height above the surface of the dock than it actually is?

Problem difficulty is labeled as | (straightforward) to ||||| (challenging). Problems labeled INT integrate significant material from earlier chapters; BIO are of biological or medical interest; CALC require calculus to solve.

6. A fish in an aquarium with flat sides looks out at a hungry cat. To the fish, does the distance to the cat appear to be less than the actual distance, the same as the actual distance, or more than the actual distance? Explain.
7. You are looking straight into the front of an aquarium. You see a fish off to your right. Is the fish actually in the direction you're looking, farther to the right, or farther to the left? Explain.
8. The object and lens in Figure Q19.8 are positioned to form a well-focused, inverted image on a viewing screen. Then a piece of cardboard is lowered just in front of the lens to cover the top half of the lens. Describe what you see on the screen when the cardboard is in place.

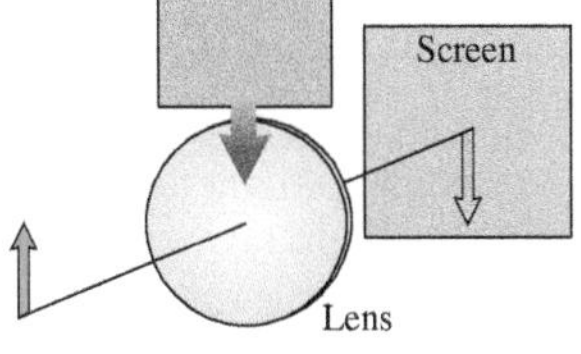

FIGURE Q19.8

9. a. Consider *one* point on an object near a lens. What is the minimum number of rays needed to locate its image point?
   b. For each point on the object, how many rays from this point actually strike the lens and refract to the image point?
10. A converging lens creates the image shown in Figure Q19.10. Is the object distance less than the focal length f, between f and $2f$, or greater than $2f$? Explain.

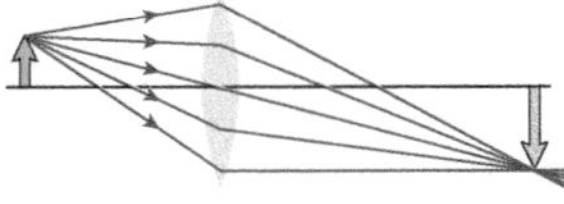
FIGURE Q19.10

11. A converging lens and a converging mirror have the same focal length in air. Which one has a longer focal length if they are used underwater?
12. A concave mirror brings the sun's rays to a focus in front of the mirror. Suppose the mirror is submerged in a swimming pool but still pointed up at the sun. Will the sun's rays be focused nearer to, farther from, or at the same distance from the mirror? Explain.
13. When you look at your reflection in the bowl of a spoon, it is upside down. Why?
14. A red card is illuminated by red light. What color will the card appear? What if it's illuminated by blue light?

Multiple-Choice Questions

Questions 15 through 17 are concerned with the situation sketched in Figure Q19.15, in which a beam of light in the air encounters a transparent block with index of refraction $n = 1.53$. Some of the light is reflected and some is refracted.

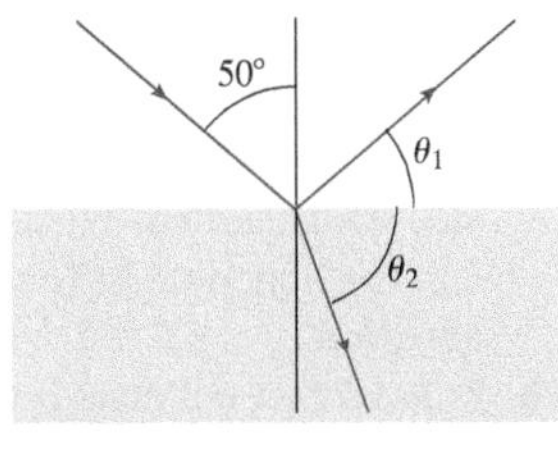

FIGURE Q19.15

15. | What is θ_1?
    A. 40° B. 45° C. 50° D. 90°
16. | What is θ_2?
    A. 20°
    B. 30°
    C. 50°
    D. 60°
17. | Is there an angle of incidence between 0° and 90° such that all of the light will be reflected?
    A. Yes, at an angle greater than 50°
    B. Yes, at an angle less than 50°
    C. No
18. | A 2.0-m-tall man is 5.0 m from the converging lens of a camera. His image appears on a detector that is 50 mm behind the lens. How tall is his image on the detector?
    A. 10 mm B. 20 mm C. 25 mm D. 50 mm
19. || You are 2.4 m from a plane mirror, and you would like to take a picture of yourself in the mirror. You need to manually adjust the focus of the camera by dialing in the distance to what you are photographing. What distance do you dial in?
    A. 1.2 m B. 2.4 m C. 3.6 m D. 4.8 m
20. | As shown in Figure Q19.20, an object is placed in front of a convex mirror. At what position is the image located?

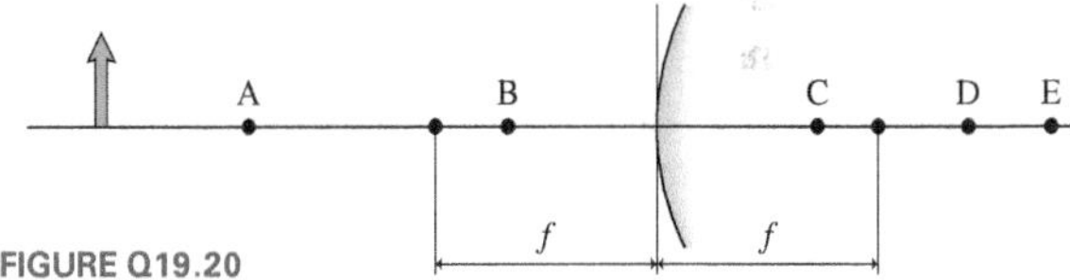

FIGURE Q19.20

21. | A virtual image of an object can be formed by
    A. A converging lens
    B. A diverging lens
    C. A plane mirror
    D. A concave mirror
    E. Any of the above
22. | An object is 40 cm from a converging lens with a focal length of 30 cm. A real image is formed on the other side of the lens, 120 cm from the lens. What is the magnification?
    A. −2.0 B. −3.0 C. −4.0
    D. −1.33 E. −0.33
23. | The lens in Figure Q19.23 is used to produce a real image of a candle flame. What is the focal length of the lens?
    A. 9.0 cm B. 12 cm
    C. 24 cm D. 36 cm
    E. 48 cm

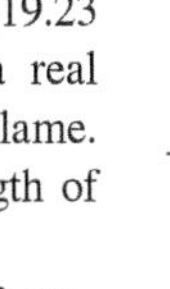

FIGURE Q19.23

24. | You look at yourself in a convex mirror. Your image is
    A. Upright.
    B. Inverted.
    C. It's impossible to tell without knowing how far you are from the mirror and its focal length.

PROBLEMS

Section 19.1 The Ray Model of Light

1. || a. How long (in ns) does it take light to travel 1.0 m in vacuum?
   b. What distance does light travel in water, glass, and cubic zirconia during the time that it travels 1.0 m in vacuum?
2. | A 5.0-ft-tall girl stands on level ground. The sun is 25° above the horizon. How long is her shadow?
3. || A point source of light illuminates an aperture 2.00 m away. A 12.0-cm-wide bright patch of light appears on a screen 1.00 m behind the aperture. How wide is the aperture?
4. | A student has built a 15-cm-long pinhole camera for a science fair project. She wants to photograph her 180-cm-tall friend and have the image on the film be 5.0 cm high. How far should the front of the camera be from her friend?

Section 19.2 Reflection

5. | The mirror in Figure P19.5 deflects a horizontal laser beam by 60°. What is the angle ϕ?

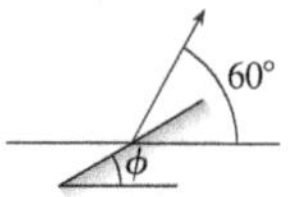

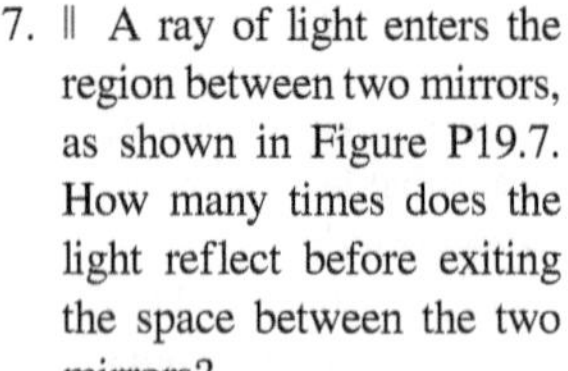

FIGURE P19.5

6. | It is 165 cm from your eyes to your toes. You're standing 200 cm in front of a tall mirror. How far is it from your eyes to the image of your toes?
7. || A ray of light enters the region between two mirrors, as shown in Figure P19.7. How many times does the light reflect before exiting the space between the two mirrors?

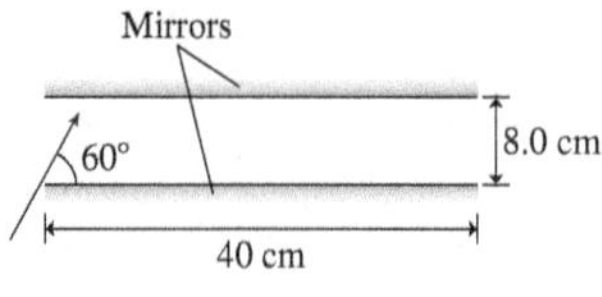

FIGURE P19.7

8. || You are standing 1.5 m from a mirror, and you want to use a classic camera to take a photo of yourself. This camera requires you to select the distance of whatever you focus on. What distance do you choose?
9. || Starting 3.5 m from a tall wall mirror, Suzanne walks toward INT the mirror at 1.5 m/s for 2.0 s. How far is Suzanne from her image in the mirror after 2.0 s?
10. ||| The lightbulb in Figure P19.10 is 50 cm from a mirror. It INT emits 1.5 W of visible light. A small barrier blocks the direct rays of light from the bulb from reaching a sensor 70 cm to the right, but not the reflected rays. What is the light intensity at the sensor?

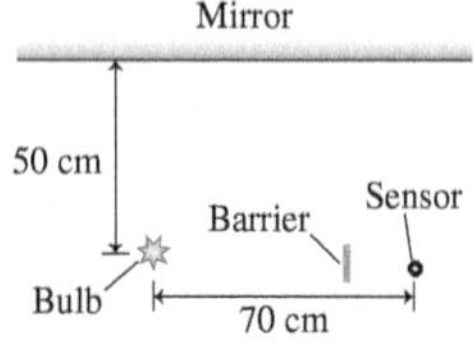

FIGURE P19.10

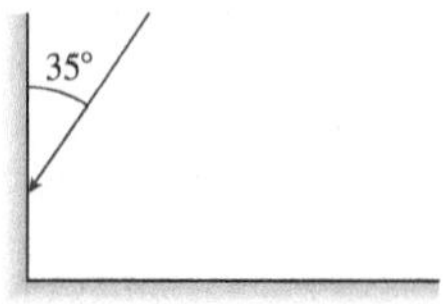

FIGURE P19.11

11. ||| A ray of light impinges on a mirror as shown in Figure P19.11. A second mirror is fastened at 90° to the first. At what angle above horizontal does the ray emerge after reflecting from both mirrors?

Section 19.3 Refraction

12. | An underwater diver sees the sun 50° above horizontal. How high is the sun above the horizon to a fisherman in a boat above the diver?
13. || A laser beam in air is incident on a liquid at an angle of 37° with respect to the normal. The laser beam's angle in the liquid is 26°. What is the liquid's index of refraction?
14. | A 1.0-cm-thick layer of water stands on a horizontal slab of glass. A light ray in the air is incident on the water 60° from the normal. After entering the glass, what is the ray's angle from the normal?
15. || Figure P19.15 shows a ray of light entering an equilateral prism, with all sides and angles equal to each other. The ray traverses the prism parallel to the bottom and emerges at the same angle at which it entered. If $\theta = 40°$, what is the index of refraction of the prism?

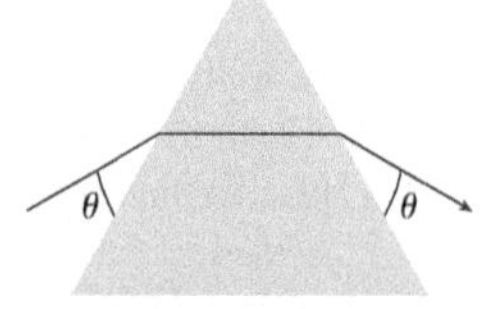

FIGURE P19.15

16. || A 4.0-m-wide swimming pool is filled to the top. The bottom of the pool becomes completely shaded in the afternoon when the sun is 20° above the horizon. How deep is the pool?
17. || You are on a diving trip. Deep below the surface, you look up at the surface of the water. Right at sunset, at what angle from the vertical do you see the sun?
18. || A ray of light traveling through air encounters a 1.2-cm-thick sheet of glass at a 35° angle of incidence. How far does the light ray travel in the glass before emerging on the far side?
19. | A thin glass rod is submerged in oil. What is the critical angle for light traveling inside the rod?
20. || A light ray travels inside a horizontal plate of glass, striking its upper surface at an angle of incidence of 60°. This ray is totally internally reflected at the glass-air boundary. A liquid is then poured on top of the glass. What is the largest index of refraction that the liquid could have such that the ray is still totally internally reflected?
21. || A 1.0-mm-diameter glass BIO optical fiber is inserted into the artery of a patient. The index of refraction of the surrounding blood is 1.37. Light from inside the artery is incident on the flat face of the fiber, as shown in Figure P19.21. What is the maximum angle θ_0 for which the light ray will undergo TIR inside the fiber?

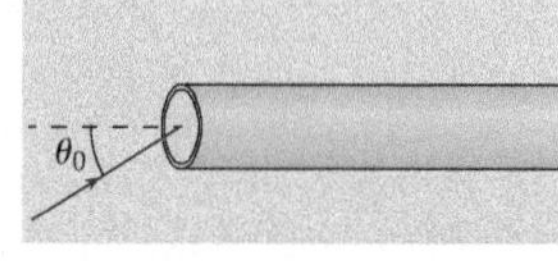

FIGURE P19.21

22. || The glass core of an optical fiber has an index of refraction 1.60. The index of refraction of the cladding is 1.48. What is the maximum angle a light ray can make with the wall of the core if it is to remain inside the fiber?

Section 19.4 Image Formation by Refraction

23. | A biologist keeps a specimen of her favorite beetle embedded in a cube of polystyrene plastic. The hapless bug appears to be 2.0 cm within the plastic. What is the beetle's actual distance beneath the surface?
24. || The composition of the ancient atmosphere can be determined by analyzing bubbles of air trapped in amber, which is fossilized tree resin. An air bubble appears to be 7.2 mm below the flat surface of a piece of amber, which has index of refraction 1.54. How long a needle is required to reach the bubble?
25. || A fish in a flat-sided aquarium sees a can of fish food on the counter. To the fish's eye, the can looks to be 30 cm outside the aquarium. What is the actual distance between the can and the aquarium? (You can ignore the thin glass wall of the aquarium.)
26. || A 1.75-m-tall diver is standing completely submerged on the bottom of a swimming pool, in 3.00 m of water. You are sitting on the end of the diving board, almost directly over her. How tall does the diver appear to be?
27. || To a fish in an aquarium, the 4.00-mm-thick walls appear to be only 3.50 mm thick. What is the index of refraction of the walls?

Section 19.5 Thin Lenses: Ray Tracing

28. | An object is 30 cm in front of a converging lens with a focal length of 10 cm. Use ray tracing to determine the location of the image. Is the image upright or inverted? Is it real or virtual?
29. | An object is 30 cm in front of a converging lens with a focal length of 15 cm. Use ray tracing to determine the location of the image. Is the image upright or inverted? Is it real or virtual?

30. | An object is 6.0 cm in front of a converging lens with a focal length of 10 cm. Use ray tracing to determine the location of the image. Is the image upright or inverted? Is it real or virtual?

31. || An object is 20 cm in front of a diverging lens with a focal length of −10 cm. Use ray tracing to determine the location of the image. Is the image upright or inverted? Is it real or virtual?

32. || An object is 15 cm in front of a diverging lens with a focal length of −10 cm. Use ray tracing to determine the location of the image. Is the image upright or inverted? Is it real or virtual?

Section 19.6 The Thin-Lens Equation

33. | A 2.0-cm-tall object is 40 cm in front of a converging lens that has a 20 cm focal length. Determine the position, orientation, and height of the image.

34. || A 1.0-cm-tall object is 10 cm in front of a converging lens that has a 30 cm focal length. Determine the position, orientation, and height of the image.

35. | A 2.0-cm-tall object is 15 cm in front of a diverging lens that has a −20 cm focal length. Determine the position, orientation, and height of the image.

36. | A 1.0-cm-tall object is 60 cm in front of a diverging lens that has a −30 cm focal length. Determine the position, orientation, and height of the image.

37. | A 1.0-cm-tall candle flame is 60 cm from a lens with a focal length of 20 cm. What are the image distance and the height of the flame's image?

38. || You are using a converging lens to look at a splinter in your finger. The lens has a 9.0 cm focal length, and you place the splinter 6.0 cm from the lens. How far from the lens is the image? What is the magnification?

39. | At what distance from a converging lens with a focal length of 20 cm should an object be placed so that its image is the same distance from the lens as the object?

40. || A toy insect viewer for kids consists of a plastic container with a lens in the lid. The lid is 12 cm from the bottom of the container. The lens produces a magnified image of an insect on the bottom of the container. If the magnification is 4.0, what is the focal length of the lens?

41. || The sun is 1.5×10^{11} m from earth; its diameter is
INT 1.4×10^{9} m. A hiker who wishes to start a fire uses a 4.0-cm-diameter lens with a focal length of 10 cm to focus the sun's rays onto a piece of wood.
    a. How far should the lens be from the piece of wood to get the best-focused image of the sun?
    b. What is the diameter of the sun's image?
    c. On a clear day, the intensity of midday sunlight is about 1000 W/m^2. What is the light intensity at the focus of the lens? Assume that the lens reflects 10% of the sun's light and transmits 90%; these are typical values for a lens that does not have an antireflection coating.

Section 19.7 Image Formation with Spherical Mirrors

42. | A 3.0-cm-tall object is 15 cm in front of a concave mirror that has a 25 cm focal length. Determine the position, orientation, and height of the image.

43. | A 3.0-cm-tall object is 45 cm in front of a concave mirror that has a 25 cm focal length. Determine the position, orientation, and height of the image.

44. | A 3.0-cm-tall object is 15 cm in front of a convex mirror that has a −25 cm focal length. Determine the position, orientation, and height of the image.

45. || A 3.0-cm-tall object is 45 cm in front of a convex mirror that has a −25 cm focal length. Determine the position, orientation, and height of the image.

46. || The illumination lights in an operating room use a concave mirror to focus an image of a bright lamp onto the surgical site. One such light has a mirror with a focal length of 15 cm. How far should the lamp be from the mirror so that the light is focused at a distance of 120 cm from the mirror?

47. || Consider a typical convex passenger-side mirror with a focal length of −80 cm. A 1.5-m-tall cyclist on a bicycle is 25 m from the mirror. You are 1.0 m from the mirror, and suppose, for simplicity, that the mirror, you, and the cyclist all lie along a line.
    a. How far are you from the image of the cyclist?
    b. What is the image height?

Section 19.8 Color and Dispersion

48. || A sheet of glass has $n_{red} = 1.52$ and $n_{violet} = 1.55$. A narrow beam of white light is incident on the glass at 30°. What is the angular spread (i.e., $\theta_{red} - \theta_{violet}$) of the light inside the glass?

49. ||| A ray of white light strikes the surface of a 4.0-cm-thick slab of flint glass as shown in Figure P19.49. As the ray enters the glass, it is dispersed into its constituent colors. Use information from Figure 19.51 to estimate how far apart the rays of deepest red and deepest violet light are when they reach the bottom surface.

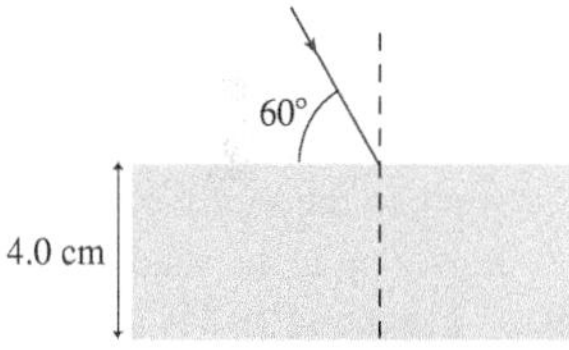

FIGURE P19.49

50. || Infrared telescopes, which use special infrared detectors, are able to peer farther into star-forming regions of the galaxy because infrared light is not scattered as strongly as is visible light by the tenuous clouds of hydrogen gas from which new stars are created. For what wavelength of light is the scattering only 1% that of light with a visible wavelength of 500 nm?

General Problems

51. | A chambered nautilus has a simple eye: a 10-mm-diameter
BIO hollow sphere with a 1-mm-diameter opening on one side and a region of light-sensitive tissue on the other. About how tall is the image of a piece of 1-m-tall seaweed that is 10 m away?

52. || At what angle ϕ should the laser beam in Figure P19.52 be aimed at the mirrored ceiling in order to hit the midpoint of the far wall?

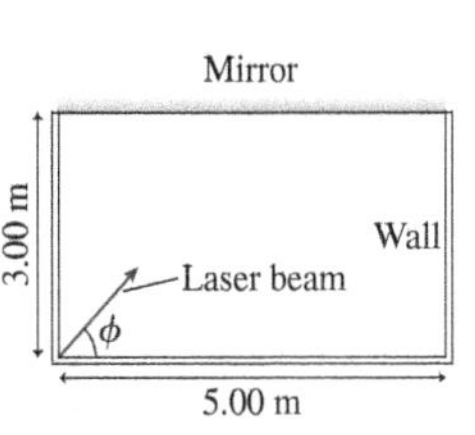

FIGURE P19.52

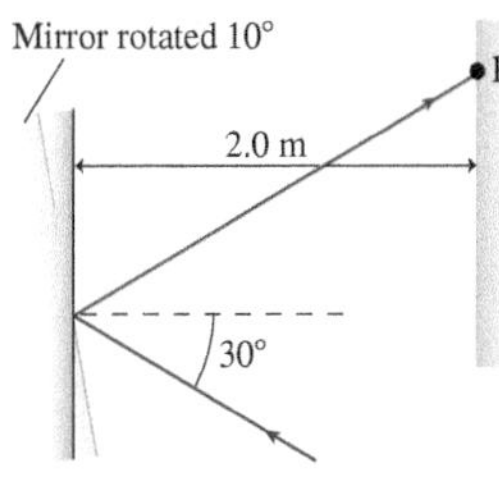

FIGURE P19.53

53. |||| A laser beam is incident on a mirror at an angle of 30°, as shown in Figure P19.53. It reflects off the mirror and strikes a wall 2.0 m away at point P. By what distance does the laser spot on the wall move if the mirror is rotated by 10°?

54. ||| The place you get your hair cut has two nearly parallel mirrors 5.0 m apart. As you sit in the chair, your head is 2.0 m from the nearer mirror. Looking toward this mirror, you first see your face and then, farther away, the back of your head. (The mirrors need to be slightly nonparallel for you to be able to see the back of your head, but you can treat them as parallel in this problem.) How far away does the back of your head appear to be? Neglect the thickness of your head.

55. || What is the angle of incidence in air of a light ray whose angle of refraction in glass is half the angle of incidence?

56. || Figure P19.56 shows a light ray incident on a glass cylinder. What is the angle α of the ray after it has entered the cylinder?

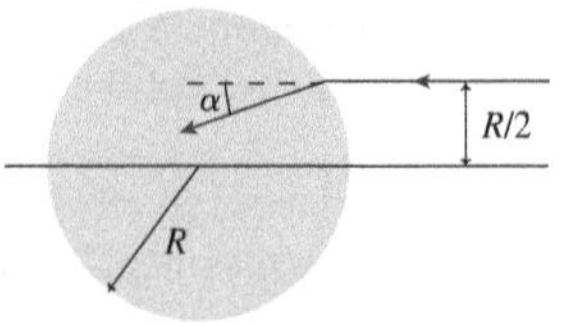

FIGURE P19.56

57. ||| It's nighttime, and you've dropped your goggles into a swimming pool that is 3.0 m deep. If you hold a laser pointer 1.0 m directly above the edge of the pool, you can illuminate the goggles if the laser beam enters the water 2.0 m from the edge. How far are the goggles from the edge of the pool?

58. ||| The sun is 60° above the horizon. Rays from the sun strike the still surface of a pond and cast a shadow of a stick that is stuck in the sandy bottom of the pond. If the stick is 10 cm tall, how long is the shadow?

59. || Figure P19.59 shows a meter stick lying on the bottom of a 100-cm-long tank with its zero mark against the left edge. You look into the tank at a 30° angle, with your line of sight just grazing the upper left edge of the tank. What mark do you see on the meter stick if the tank is (a) empty, (b) half full of water, and (c) completely full of water?

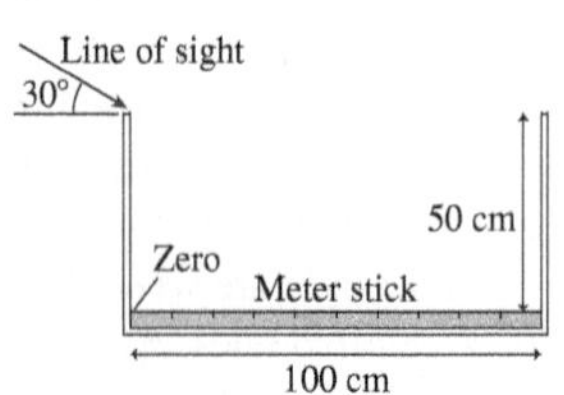

FIGURE P19.59

60. || The 80-cm-tall, 65-cm-wide tank shown in Figure P19.60 is completely filled with water. The tank has marks every 10 cm along one wall, and the 0 cm mark is barely submerged. As you stand beside the opposite wall, your eye is level with the top of the water.
   a. Can you see the marks from the top of the tank (the 0 cm mark) going down, or from the bottom of the tank (the 80 cm mark) coming up? Explain.
   b. Which is the lowest or highest mark, depending on your answer to part a, that you can see?

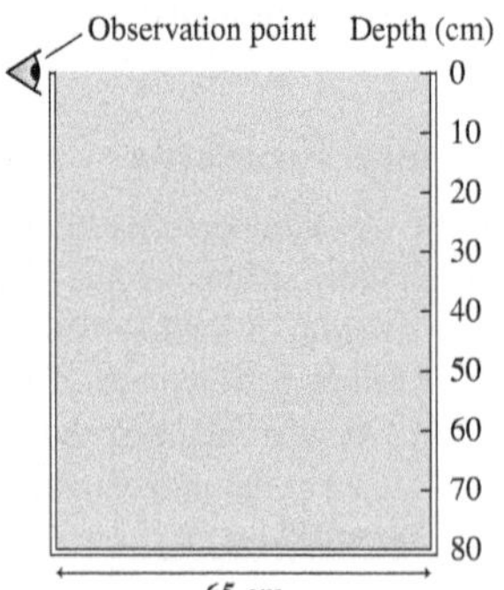

FIGURE P19.60

61. ||| A 1.0-cm-thick layer of water stands on a horizontal slab of glass. Light from within the glass is incident on the glass-water boundary. What is the maximum angle of incidence for which a light ray can emerge into the air above the water?

62. || What is the exit angle θ from the glass prism in Figure P19.62?

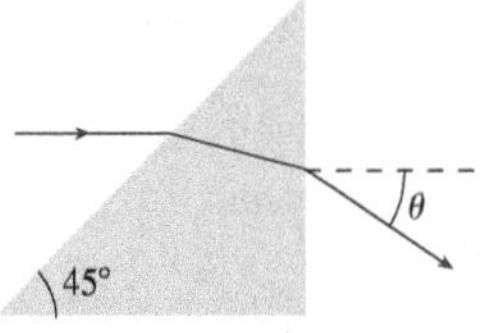

FIGURE P19.62

63. ||| What is the smallest angle θ_1 for which a laser beam will undergo total internal reflection on the hypotenuse of the glass prism in Figure P19.63?

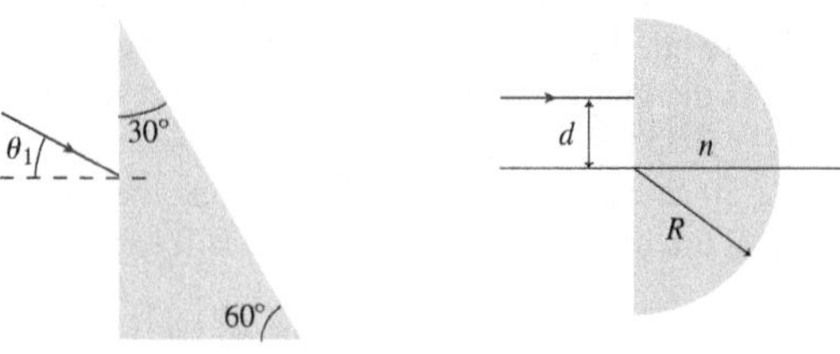

FIGURE P19.63 FIGURE P19.64

64. || Figure P19.64 is a transparent hemisphere with radius R and index of refraction n. What is the maximum distance d for which a light ray parallel to the axis refracts out through the curved surface?

65. || The glass core of an optical fiber has index of refraction 1.60. The index of refraction of the cladding is 1.48. What is the maximum angle between a light ray and the wall of the core if the ray is to remain inside the core?

66. || A microscope is focused on an amoeba. When a 0.15-mm-
BIO thick cover glass ($n = 1.50$) is placed over the amoeba, by how far must the microscope objective be moved to bring the organism back into focus? Must it be raised or lowered?

67. || You need to use a 24-cm-focal-length lens to produce an inverted image twice the height of an object. At what distance from the object should the lens be placed?

68. ||| An LCD (liquid crystal display) projector works by illuminating a small LCD display—not unlike a small computer screen—with a very bright light and using a lens to form an image of the display on the viewing screen. A small LCD projector has a 4.00-cm-tall LCD display, and the lens needs to create a 2.00-m-tall image. The screen is 5.00 m from the LCD display.
   a. How far should the lens be from the LCD display? Assume it is a thin lens.
   b. What focal length does the lens need?

69. |||| You can focus on an ob-
BIO ject about 25 cm (10 in) in front of you, but not closer. A cat can focus on an object at roughly half that distance, and a mouse can see objects only a few centimeters away. Why the difference? The optics of the eye are complex (see Chapter 20), but we can gain some insight by looking at the focusing of light by a solid, transparent sphere. A sphere is *not* a thin lens, but Figure P19.69 shows that the two refracting surfaces of a sphere have an effective focal length, measured from the center, at which parallel light rays converge. An application of Snell's law at each spherical surface finds that

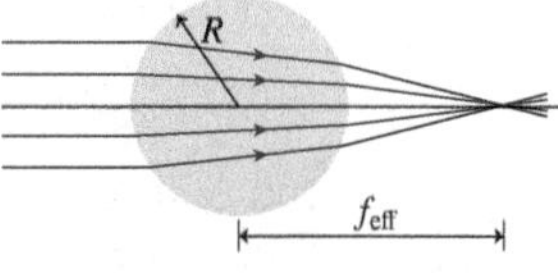

FIGURE P19.69

$$f_{eff} = \frac{nR}{2(n-1)}$$

   a. Find an algebraic expression for the distance l between the object and the image of a sphere when the object and image distances are equal.
   b. Eyes are not refracting spheres, but, like a sphere, the effective focal length of the eye is directly proportional to the radius of the eyeball. And the radius of the eyeball scales with the mass M of the animal; experimental data for vertebrates show that $R \propto M^{0.22}$. The average masses of humans, cats, and mice are 65 kg, 4.5 kg, and 20 g, respectively, and the human eye has a 12 mm radius. Evaluate l for three water-filled spheres that have radii equal to the expected eyeball radii of humans, cats, and mice.

70. || Luis is nearsighted. To correct his vision, he wears a BIO diverging eyeglass lens with a focal length of -0.50 m. When wearing glasses, Luis looks not at an object but at the virtual image of the object because that is the point from which diverging rays enter his eye. Suppose Luis, while wearing his glasses, looks at a vertical 12-cm-tall pencil that is 2.0 m in front of his glasses.
   a. How far from his glasses is the image of the pencil?
   b. What is the height of the image?
   c. Your answer to part b might seem to suggest that Luis sees everything as being very tiny. However, the apparent size of an object (or a virtual image) is determined not by its height but by the *angle* it spans. In the absence of other visual cues, a nearby short object is perceived as being the same size as a distant tall object if they span the same angle at your eye. From the position of the lens, what angle is spanned by the actual pencil 2.0 m away that Luis sees without his glasses? And what angle is spanned by the virtual image of the pencil that he sees when wearing his glasses?
71. ||| A 2.0-cm-tall candle flame is 2.0 m from a wall. You happen to have a lens with a focal length of 32 cm. How many places can you put the lens to form a well-focused image of the candle flame on the wall? For each, how far is the lens from the flame, and what is the height of the image?
72. ||| A 2.0-cm-diameter spider is 2.0 m from a wall. Determine the focal length and position (measured from the wall) of a lens that will make a half-size image of the spider on the wall.
73. ||| Figure P19.73 shows a meter stick held lengthwise along the optical axis of a concave mirror. How long is the image of the meter stick?

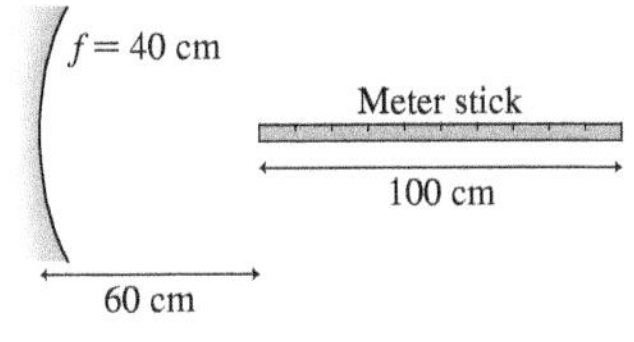

FIGURE P19.73

74. || A dentist uses a curved BIO mirror to view the back side of teeth in the upper jaw. Suppose she wants an upright image with a magnification of 1.5 when the mirror is 1.2 cm from a tooth. Should she use a convex or a concave mirror? What focal length should it have?
75. ||| A *keratometer* is an optical device used to measure the BIO radius of curvature of the eye's cornea—its entrance surface. This measurement is especially important when fitting contact lenses, which must match the cornea's curvature. Most light incident on the eye is transmitted into the eye, but some light reflects from the cornea, which, due to its curvature, acts like a convex mirror. A useful property of a convex mirror is that its focal length is related to its radius of curvature R by $f = -R/2$. The keratometer places a small, illuminated ring of known diameter 7.5 cm in front of the eye. The optometrist, using an eyepiece, looks through the center of this ring and sees a small virtual image of the ring that appears to be behind the cornea. The optometrist uses a scale inside the eyepiece to measure the diameter of the image and calculate its magnification. Suppose the optometrist finds that the magnification for one patient is 0.049. What is the radius of curvature of her cornea?
76. ||| White light is incident onto a 30° prism at the 40° angle shown in Figure P19.76. Violet light emerges perpendicular to the rear face of the prism. The index of refraction of violet light in this glass is 2.0% larger than the index of refraction of red light. At what angle ϕ does red light emerge from the rear face?

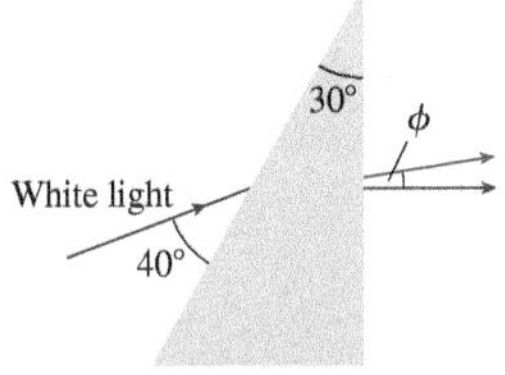

FIGURE P19.76

MCAT-Style Passage Problems

Mirages

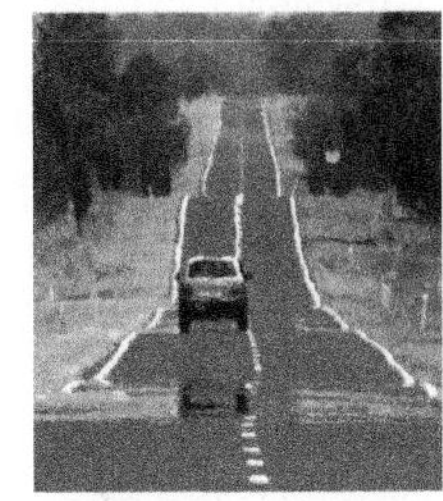

There is an interesting optical effect you have likely noticed while driving along a flat stretch of road on a sunny day. A small, distant dip in the road appears to be filled with water. You may even see the reflection of an oncoming car. But, as you get closer, you find no puddle of water after all; the shimmering surface vanishes, and you see nothing but empty road. It was only a *mirage,* the name for this phenomenon.

The mirage is due to the different index of refraction of hot and cool air. The actual bending of the light rays that produces the mirage is subtle, but we can make a simple model as follows. When air is heated, its density decreases and so does its index of refraction. Consequently, a pocket of hot air in a dip in a road has a lower index of refraction than the cooler air above it. Incident light rays with large angles of incidence (that is, nearly parallel to the road, as shown in Figure P19.77) experience total internal reflection. The mirage that you see is due to this reflection. As you get nearer, the angle goes below the critical angle and there is no more total internal reflection; the "water" disappears!

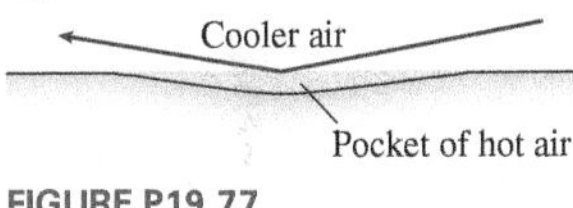

FIGURE P19.77

77. The pocket of hot air appears to be a pool of water because
   A. Light reflects at the boundary between the hot and cool air.
   B. Its density is close to that of water.
   C. Light refracts at the boundary between the hot and cool air.
   D. The hot air emits blue light that is the same color as the daytime sky.
78. Which of these changes would allow you to get closer to the mirage before it vanishes?
   A. Making the pocket of hot air nearer in temperature to the air above it
   B. Looking for the mirage on a windy day, which mixes the air layers
   C. Increasing the difference in temperature between the pocket of hot air and the air above it
   D. Looking at it from a greater height above the ground
79. If you could clearly see the image of an object that was reflected by a mirage, the image would appear
   A. Magnified.
   B. With up and down reversed.
   C. Farther away than the object.
   D. With right and left reversed.

20 Optical Instruments

Owls and other raptors have remarkable visual acuity, much better than humans.

LOOKING AHEAD

The Camera

A camera is an essential tool of wildlife biologists and wildlife photographers. A telephoto lens allows them to observe from a safe distance.

You'll see how the knowledge of lenses you gained in Chapter 19 is applied to real optical instruments.

The Human Eye

Our most important optical instruments are our own eyes, which use the cornea and the lens to focus an image onto the light-sensitive retina.

You'll learn how near- and farsightedness can be corrected with eyeglasses or contact lenses.

The Microsope

This pathologist is examining cells under a microscope, looking for cancer cells or other abnormalities.

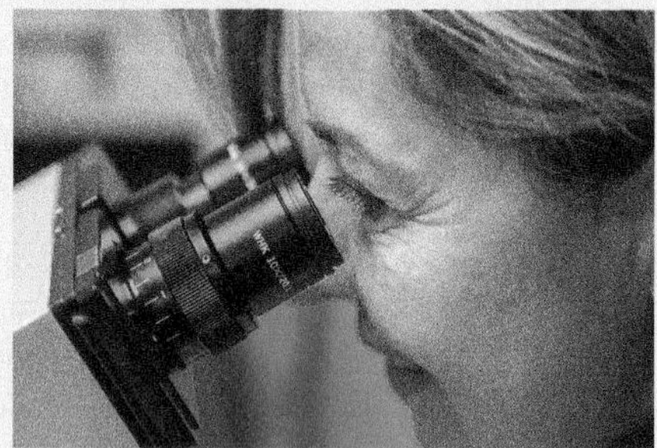

You'll learn how microscopes work and how several different microscopy techniques are applied to biology and medicine.

GOAL To understand how optical instruments work and how diffraction ultimately limits their performance.

PHYSICS AND LIFE

Optical Instruments Magnify and Extend Our Vision

Human vision is astonishingly flexible and effective. Even so, optical instruments can optimize our vision with eyeglasses or contact lenses and extend our vision to the cellular scale with microscopes. There many different microscopy techniques, and each new development in microscopy brings scientific advances as scientists explore their newfound ability to image structures that couldn't previously be seen. But microscopy has its limits. Light is a wave, and the diffraction of light sets the ultimate limit on the smallest features that can be resolved with a microscope. Understanding the basic principles of how optical instruments work—and their limitations—will help you choose the most suitable imaging strategy for a given scientific problem.

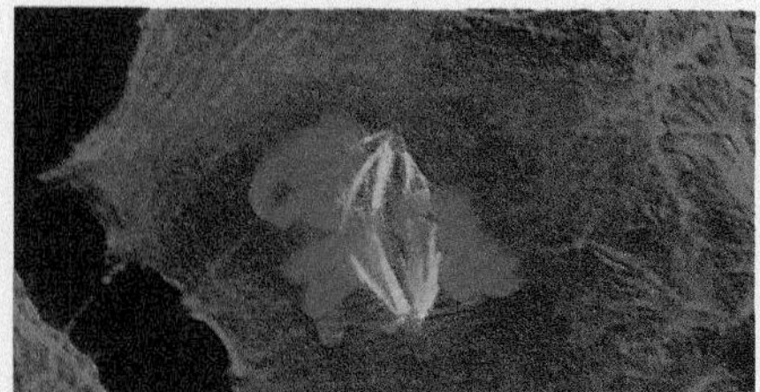

The microtubles (green) and chromosomes (blue) are indistinct in this micrograph of cell division. This is a limitation of optical microscopy due to the diffraction of light.

20.1 Lenses in Combination

If you need eyeglasses or contact lenses to correct your vision, the corrective lens works together with the cornea and lens of your eye to create a focused image on your retina. Other optical instruments, such as cameras and microscopes, also use two or more lenses in combination. The reason, as we'll see, is to improve the image quality.

The analysis of a multi-lens system requires only one new rule: **The image of the first lens acts as the object for the second lens.** To see why this is so, FIGURE 20.1 shows a simple microscope consisting of an *objective* lens and an *eyepiece* lens. For each lens, the figure highlights the three special rays you learned to use in Chapter 19:

- A ray parallel to the optical axis refracts through the focal point.
- A ray through the focal point refracts parallel to the optical axis.
- A ray through the center of the lens is undeviated.

FIGURE 20.1 Ray-tracing diagram of a simple microscope.

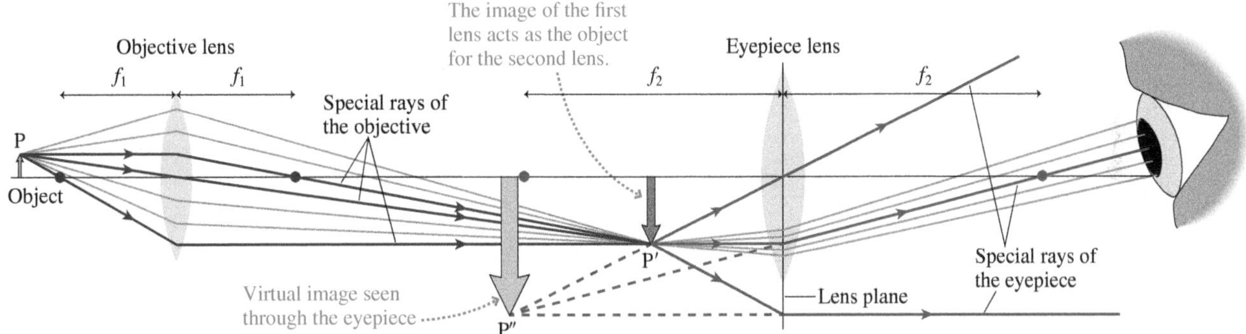

The rays passing through the objective converge to a real image at P′, **but they don't stop there.** Instead, light rays *diverge* from P′ as they approach the second lens. **As far as the eyepiece is concerned, the rays are coming from P′, and thus P′ acts as the object for the second lens.** The three special rays passing through the objective lens are sufficient to locate the image P′, but these rays are generally *not* the special rays for the second lens. However, other rays converging at P′ leave at the correct angles to be the special rays for the eyepiece. That is, a new set of special rays is drawn from P′ to the second lens and used to find the final image point P″.

NOTE ▶ One ray seems to "miss" the eyepiece lens, but this isn't a problem. All rays passing through the lens converge to (or diverge from) a single point, and the purpose of the special rays is to locate that point. To do so, we can let the special rays refract as they cross the *lens plane,* regardless of whether the physical lens really extends that far. ◀

EXAMPLE 20.1 A magnifying lens

A magnifying "lens" (used in the singular) is usually a combination of two or more lenses. The combination has better optical properties than a single lens because, as you'll learn later in the chapter, the combination reduces or eliminates *aberrations* that impair a lens's ability to form a sharp image. Consider a magnifying lens in which light passes first through a diverging lens with $f_1 = -10$ mm and then through a converging lens with $f_2 = 20$ mm, with the two lenses spaced 12 mm apart. Suppose we use this combination lens to view a small bug by looking through the converging lens when the bug is 20 mm from the diverging lens. Where is the bug's image, and what is its lateral magnification?

PREPARE Each lens is a thin lens. The image of the first lens is the object for the second lens. FIGURE 20.2 (next page) shows a pictorial representation and a few of the rays. We see that the image is an upright, virtual image to the left of the diverging lens.

SOLVE For the first lens, the bug is at object distance $s_1 = 20$ mm. Its image distance is found with the thin-lens equation:

$$\frac{1}{s_1'} = \frac{1}{f_1} - \frac{1}{s_1} = \frac{1}{-10\text{ mm}} - \frac{1}{20\text{ mm}} = -0.150\text{ mm}^{-1}$$

$$s_1' = -6.67\text{ mm}$$

This is a virtual image 6.7 mm to the left of the first lens. The magnification of the first lens is

$$m_1 = -\frac{s_1'}{s_1} = -\frac{-6.67\text{ mm}}{20\text{ mm}} = +0.33$$

Continued

The virtual image of the diverging lens is the object for the converging lens. The object distance—the distance from the virtual image to the second lens—is $s_2 = 6.7\text{ mm} + 12\text{ mm} = 18.7\text{ mm}$. Another application of the thin-lens equation yields

$$\frac{1}{s_2'} = \frac{1}{f_2} - \frac{1}{s_2} = \frac{1}{20\text{ mm}} - \frac{1}{18.7\text{ mm}} = -0.00348\text{ mm}^{-1}$$

$$s_2' = -290\text{ mm}$$

This is a virtual image 290 mm to the left of the converging lens with magnification

$$m_2 = -\frac{s_2'}{s_2} = -\frac{-290\text{ mm}}{18.7\text{ mm}} = +15$$

The second lens magnifies the image of the first lens, so **the total magnification is the product of the individual magnifications:**

$$m_{\text{total}} = m_1 m_2 = +5.0$$

Thus there is an upright image 290 mm in front of the converging lens with a total lateral magnification of 5.0. You would see this image by looking at the bug through the converging lens.

ASSESS Handheld magnifiers used by naturalists, geologists, and jewelers are slightly more complex: a three-element combination with diverging lenses on both sides of a central converging lens. The second diverging lens improves the optical quality, but a three-element magnifier functions very much like the two-element lens analyzed in this example.

FIGURE 20.2 Pictorial representation of a combination lens.

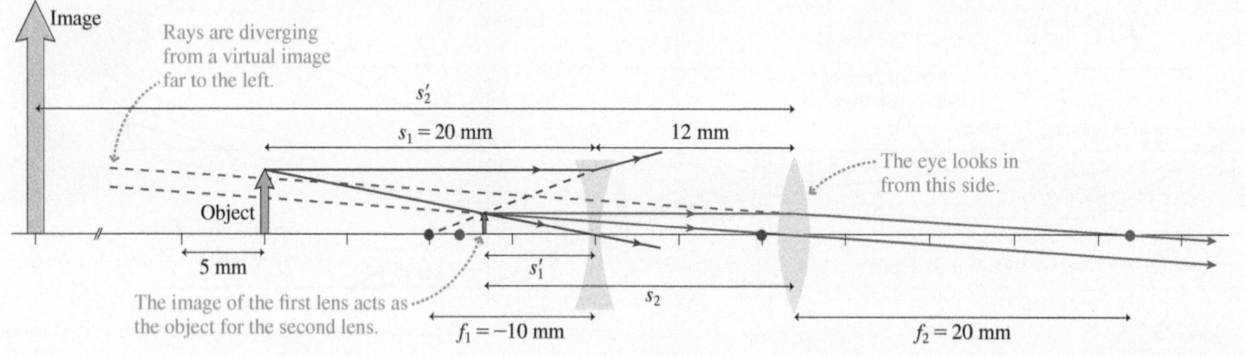

A combination lens, such as a magnifier, a camera lens, or the eyepiece lens on a microscope, is characterized by an **effective focal length** that describes how it produces images if we think of it as a single lens. For example, a multicomponent camera lens might be specified as a "35 mm lens," which means that the combination has an effective focal length of 35 mm.

STOP TO THINK 20.1 The second lens in this optical instrument

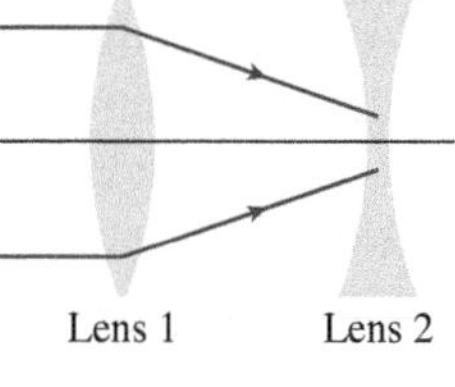

A. Causes the light rays to focus closer than they would with the first lens acting alone.
B. Causes the light rays to focus farther away than they would with the first lens acting alone.
C. Inverts the image but does not change where the light rays focus.
D. Prevents the light rays from reaching a focus.

20.2 The Camera

FIGURE 20.3 A camera.

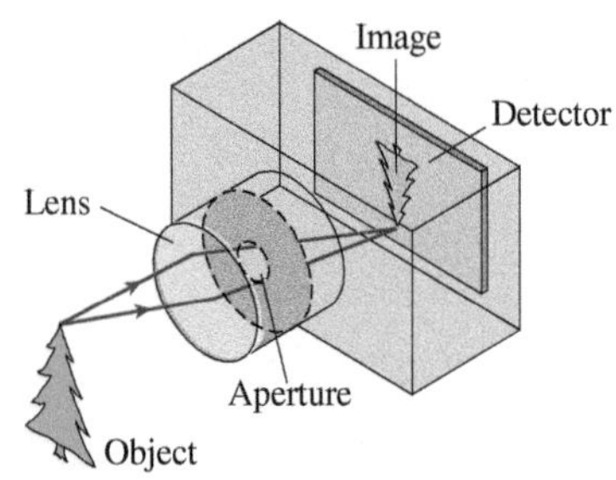

A **camera,** shown in FIGURE 20.3, takes a picture by using a lens to focus a real, inverted image on a light-sensitive detector. The camera lens is always a combination of two or more individual lenses. A high-quality camera lens can use as many as eight individual lenses; however, the simple two-lens model of FIGURE 20.4 illustrates the main ideas. This combination of a converging and a diverging lens corrects some of the defects inherent in a single lens, as we'll discuss later in the chapter.

NOTE ▸ The camera in your smartphone works exactly the same way as the cameras described in this section, only with very small lenses and with automatic exposure adjustments. ◂

FIGURE 20.4 A simple camera lens is a combination of a converging lens and a diverging lens.

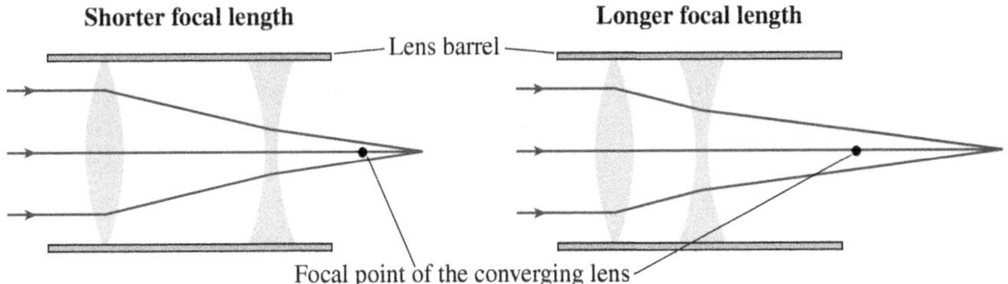

A *zoom lens* varies the effective focal length by changing the spacing between the converging and diverging lenses. If an object is more than about 10 focal lengths from the lens, which is typically no more than a couple of meters, then the condition $s \gg f_{eff}$ (and thus $1/s \ll 1/f_{eff}$) leads to the conclusion $s' \approx f_{eff}$. In other words, objects more than about 10 focal lengths away are essentially at infinity, so parallel rays are focused one focal length behind the lens.

For such an object, the lateral magnification of the image is

$$m = -\frac{s'}{s} \approx \frac{f}{s} \tag{20.1}$$

Equation 20.1 tells us that **the size of the image is directly proportional to the effective focal length of the lens.** A lens that can be varied from $f_{eff\,min} = 6$ mm to $f_{eff\,max} = 18$ mm gives magnifications spanning a factor of 3; that is why you see it specified as a 3× zoom lens.

EXAMPLE 20.2 Focusing a camera

Your digital camera lens, with an effective focal length of 10.0 mm, is focused on a flower 20.0 cm away. You then turn to take a picture of a distant landscape. How far, and in which direction, must the lens move to bring the landscape into focus?

PREPARE We'll model the camera's combination lens as a single thin lens with $f = 10.0$ mm. Distances are measured from an *effective lens plane,* but we don't need to know where that plane is to find the *change* in the focus.

SOLVE The flower is at object distance $s = 20.0$ cm $= 200$ mm. When the camera is focused, the image distance between the effective lens plane and the detector is found by solving the thin-lens equation $1/s + 1/s' = 1/f$ to give

$$s' = \left(\frac{1}{f} - \frac{1}{s}\right)^{-1} = \left(\frac{1}{10.0\text{ mm}} - \frac{1}{200\text{ mm}}\right)^{-1} = 10.5\text{ mm}$$

The distant landscape is effectively at object distance $s = \infty$, so its image distance is $s' = f = 10.0$ mm. To refocus as you shift scenes, the lens must move 0.5 mm closer to the detector.

ASSESS The required motion of the lens is very small, about the diameter of the lead used in a mechanical pencil.

Controlling the Exposure

The camera also must control the amount of light reaching the detector. Too little light results in photos that are *underexposed;* too much light gives *overexposed* pictures. Both the shutter and the lens diameter help control the exposure.

The *shutter* is "opened" for a selected amount of time as the image is recorded. Older cameras used a spring-loaded mechanical shutter that literally opened and closed; digital cameras electronically control the amount of time the detector is active. Either way, the exposure—the amount of light captured by the detector—is directly proportional to the time the shutter is open. Typical exposure times range from 1/1000 s or less for a sunny scene to 1/30 s or more for dimly lit or indoor scenes. The exposure time is generally referred to as the *shutter speed.*

The amount of light passing through the lens is controlled by an adjustable aperture, also called an *iris* because it functions much like the iris of your eye. The aperture sets the effective diameter D of the lens. The full area of the lens is used when the aperture is fully open, but a *stopped-down* aperture allows light to pass through only the central portion of the lens.

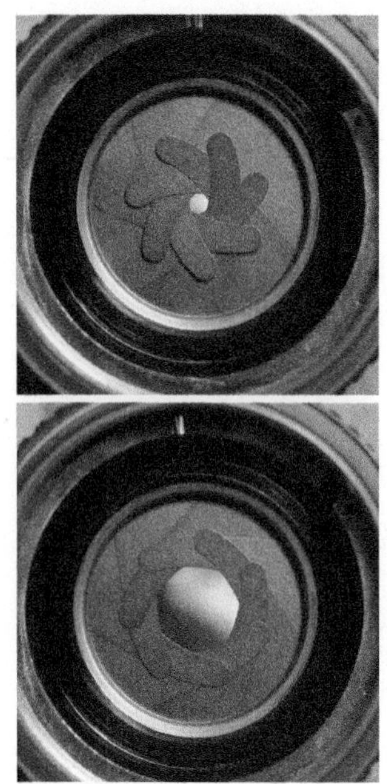

Letting in the light An iris can change the effective diameter of a lens and thus the amount of light reaching the detector.

The light intensity on the detector is directly proportional to the area of the aperture; a lens aperture with twice as much area will collect and focus twice as many light rays from the object to make an image twice as bright. The aperture area is proportional to the square of its diameter, so the intensity I is proportional to D^2. The light intensity—power per square meter—is also *inversely* proportional to the area of the image. That is, the light reaching the detector is more intense if the rays collected from the object are focused into a small area than if they are spread out over a large area. The lateral size of the image is proportional to the focal length of the lens, as we saw in Equation 20.1, so the *area* of the image is proportional to f^2 and thus I is proportional to $1/f^2$. Altogether, $I \propto D^2/f^2$.

By long tradition, the light-gathering ability of a lens is specified by its **f-number,** defined as

$$\text{f-number} = \frac{f}{D} \tag{20.2}$$

The f-number of a lens may be written either as *f*/4.0, to mean that the f-number is 4.0, or as F4.0. The instruction manuals with some digital cameras call this the *aperture value* rather than the f-number. A digital camera in fully automatic mode does not display shutter speed or f-number, but that information is displayed if you set your camera to any of the other modes. For example, the display 1/125 F5.6 means that your camera is going to achieve the correct exposure by adjusting the diameter of the lens aperture to give $f/D = 5.6$ and by opening the shutter for 1/125 s. If your lens's effective focal length is 10 mm, the diameter of the lens aperture will be

$$D = \frac{f}{\text{f-number}} = \frac{10\text{ mm}}{5.6} = 1.8\text{ mm}$$

Fundamentals of photography Focal length and f-number information is stamped on a camera lens. This camera lens is labeled 5.8–34.8 mm 1:2.8–4.8. The first numbers are the range of focal lengths. They span a factor of 6, so this is a 6× zoom lens. The second numbers show that the minimum f-number ranges from F2.8 (for the 5.8 mm focal length) to F4.8 (for the 34.8 mm focal length).

NOTE ▶ The f in f-number is not the focal length f; it's just a name. And the / in *f*/4 does not mean division; it's just a notation. These both derive from the long history of photography. ◀

Because the aperture diameter is in the denominator of the f-number, a *larger-diameter* aperture, which gathers more light and makes a brighter image, has a *smaller* f-number. The light intensity on the detector is related to the lens's f-number by

$$I \propto \frac{D^2}{f^2} = \frac{1}{(\text{f-number})^2} \tag{20.3}$$

Historically, a lens's f-numbers could be adjusted in the sequence 2.0, 2.8, 4.0, 5.6, 8.0, 11, 16. Each differs from its neighbor by a factor of $\sqrt{2}$, so changing the lens by one "f stop" changed the light intensity by a factor of 2. A modern digital camera is able to adjust the f-number continuously.

The exposure, the total light reaching the detector while the shutter is open, depends on the product $I\,\Delta t_{\text{shutter}}$. A small f-number (large aperture diameter D) and short $\Delta t_{\text{shutter}}$ can produce the same exposure as a larger f-number (smaller aperture) and a longer $\Delta t_{\text{shutter}}$. It might not make any difference for taking a picture of a distant mountain, but action photography needs very short shutter times to "freeze" the action. Thus action photography requires a large-diameter lens with a small f-number.

EXAMPLE 20.3 Capturing the action

Before a race, a photographer finds that she can make a perfectly exposed photo of the track while using a shutter speed of 1/250 s and a lens setting of F8.0. To freeze the sprinters as they go past, she plans to use a shutter speed of 1/1000 s. To what f-number must she set her lens?

PREPARE The exposure depends on $I\Delta t_{\text{shutter}}$, and the light intensity depends inversely on the square of the f-number.

SOLVE Changing the shutter speed from 1/250 s to 1/1000 s will reduce the light reaching the detector by a factor of 4. To compensate, she needs to let 4 times as much light through the lens. Because $I \propto 1/(\text{f-number})^2$, the intensity will increase by a factor of 4 if she *decreases* the f-number by a factor of 2. Thus the correct lens setting is F4.0.

The Detector

For traditional cameras, the light-sensitive detector was film. Today's digital cameras use a rectangular array of many millions of small light detectors called **pixels**. When light hits one of these pixels, it generates an electric charge proportional to the light intensity. Thus an image is recorded on the detector in terms of little packets of charge. After the image has been recorded, the charges are read out, the signal levels are digitized, and the picture is stored in the digital memory of the camera.

FIGURE 20.5a shows a detector "chip" and, schematically, the magnified appearance of the pixels on its surface. To record color information, different pixels are covered by red, green, or blue filters. A pixel covered by a green filter, for instance, records only the intensity of the green light hitting it. Later, the camera's microprocessor interpolates nearby colors to give each pixel an overall true color. The pixels are so small—a few micrometers across—that the picture looks "smooth" even after some enlargement. Even so, as you can see in FIGURE 20.5b, sufficient magnification reveals the individual pixels.

FIGURE 20.5 The detector used in a digital camera.

(a)

(b)

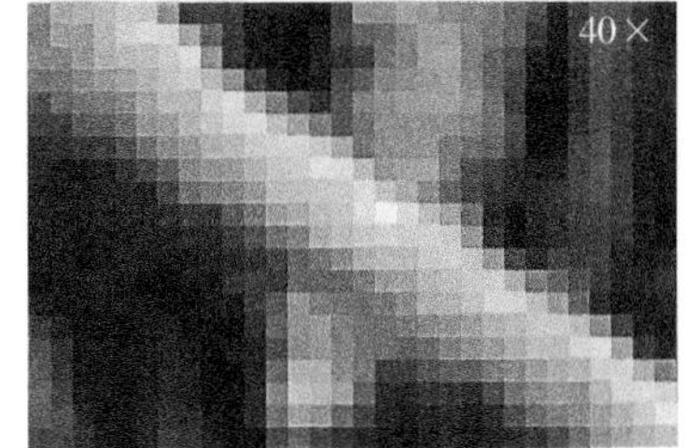

STOP TO THINK 20.2 A photographer has adjusted his camera for a correct exposure with a short-focal-length lens. He then decides to zoom in by increasing the focal length. To maintain a correct exposure without changing the shutter speed, the diameter of the lens aperture should

A. Be increased. B. Be decreased. C. Stay the same.

20.3 The Human Eye

The human eye is a marvelous and intricate organ that functions very much like a camera. Like a camera, the eye has refracting surfaces that focus incoming light rays, an adjustable iris to control the light intensity, and a light-sensitive detector.

FIGURE 20.6 shows the basic structure of the eye. It is roughly spherical, about 2.4 cm in diameter. The transparent *cornea,* which is somewhat more sharply curved, and the *lens* are the eye's refractive elements. The eye is filled with a clear, jellylike fluid called the *aqueous humor* (in front of the lens) and the *vitreous humor* (behind the lens). The indices of refraction of the aqueous and vitreous humors are 1.34, only slightly different from the 1.33 of pure water. The lens, although not uniform, has an average index of 1.40. The *pupil,* a variable-diameter aperture in the *iris,* automatically opens and closes to control the light intensity. A fully dark-adapted eye can open to ≈ 8 mm, and the pupil closes down to ≈ 1.5 mm in bright sun.

FIGURE 20.6 The human eye.

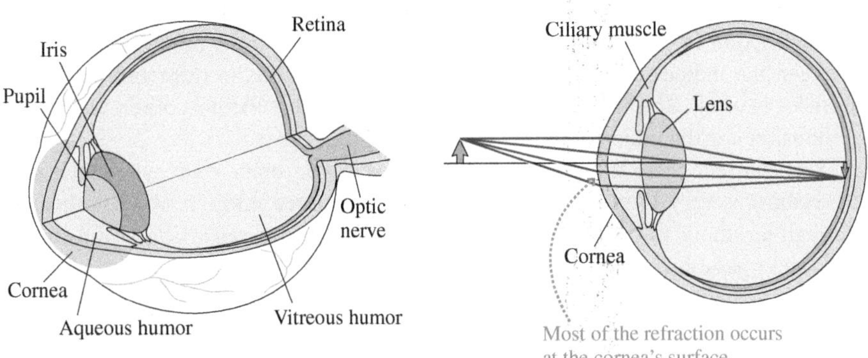

The eye's detector, the *retina,* consists of specialized light-sensitive cells called *rods* and *cones.* The rods, sensitive mostly to light and dark, are most important in very dim lighting. Color vision, which requires somewhat more light, is due to the

FIGURE 20.7 Wavelength sensitivity of the three types of cones in the human retina.

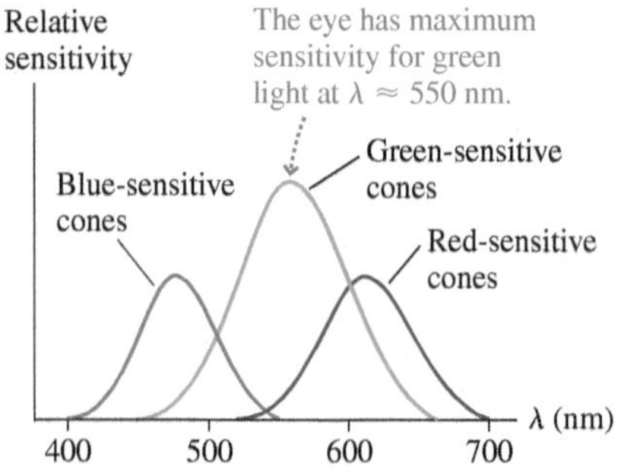

cones, of which there are three types. FIGURE 20.7 shows the wavelength responses of the cones. They have overlapping ranges, especially the red- and green-sensitive cones, so two or even all three cones respond to light of any particular wavelength. The relative response of the different cones is interpreted by your brain as light of a particular color. Color is a *perception,* a response of our sensory and nervous systems, not something inherent in the light itself.

Different sensory cells, which use different pigments, have evolved in other animals. Many, or even most, birds and reptiles have ultraviolet-sensitive cones that humans lack, allowing them to see light in the 320–400 nm range. The red-sensitive cones in fish have a response that is shifted toward longer wavelengths by ≈ 50 nm, compared to our red-sensitive cones, which provides vision to ≈ 750 nm.

Refractive Power

The eye and corrective lenses are characterized not by their focal lengths but by their *refractive powers*. The **refractive power** P of a lens or curved surface is the inverse of its focal length:

$$P = \frac{1}{f} \tag{20.4}$$

Refractive power of a lens with focal length f

A lens with a higher refractive power (a shorter focal length) causes light rays to refract through a larger angle and to converge more quickly. The SI unit of refractive power is the **diopter,** abbreviated D, which is defined as $1\text{ D} = 1\text{ m}^{-1}$. Thus a lens with focal length $f = 50\text{ cm} = 0.50\text{ m}$ has power $P = 2.0\text{ D}$.

FIGURE 20.8 The power of two lenses.

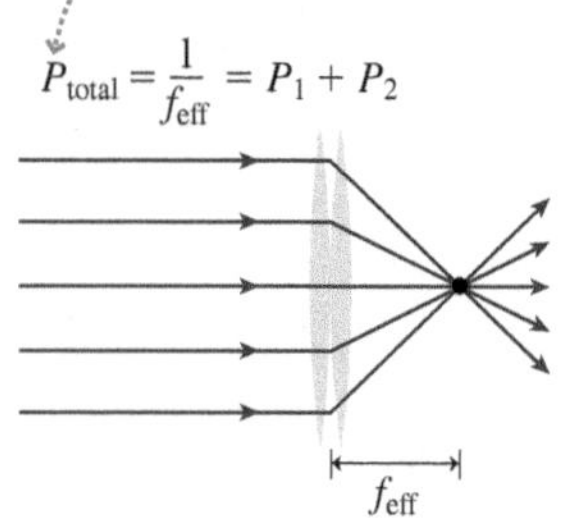

If two lenses are placed in contact with each other, as in FIGURE 20.8, the total refractive power—the inverse of the effective focal length of the lens combination—is the sum of the two individual powers:

$$P_{total} = \frac{1}{f_{eff}} = P_1 + P_2 \tag{20.5}$$

This useful relationship—that the refractive powers add for two lenses in contact—is an important reason refractive power is used to characterize corrective lenses. Note that Equation 20.5 does *not* apply if the lenses are separated and not in contact.

Focusing and Accommodation

The eye, like a camera, focuses light rays to an inverted image on the retina. Perhaps surprisingly, most of the refractive power of the eye is due to the cornea, not the lens. The cornea is a sharply curved, spherical surface, and the rather large difference between the index of refraction of air and that of the aqueous humor causes a significant refraction of light rays at the cornea. In contrast, there is much less difference between the indices of the lens and its surrounding fluid, so refraction at the lens surfaces is weak. The lens fine-tunes the eye's focus, but the air-cornea boundary is responsible for the majority of the refraction.

You can recognize the power of the cornea if you open your eyes underwater. Everything is very blurry! When light enters the cornea through water, rather than through air, there's almost no difference in the indices of refraction at the surface. Light rays pass through the cornea with almost no refraction, so what little focusing ability you have while underwater is due to the lens alone.

A camera focuses by moving the lens. The eye focuses by changing the focal length of the lens, a feat it accomplishes by using the *ciliary muscles* to change the curvature of the lens surface. The ciliary muscles are relaxed when you look at a distant scene. Thus the lens surface is relatively flat and the lens has its longest focal length. As you shift your gaze to a nearby object, the ciliary muscles contract and cause the lens to bulge. This process, called **accommodation,** decreases the lens's focal length.

The farthest distance at which a relaxed eye can focus is called the eye's **far point** (FP). The far point of a normal eye is infinity; that is, the relaxed eye can focus on objects extremely far away. The closest distance at which an eye can focus, using maximum accommodation, is the eye's **near point** (NP). (Objects can be *seen* closer than the near point, but they're not sharply focused on the retina.) A near point of 25 cm—typical of young adults—is considered to be normal vision. Both situations are shown in FIGURE 20.9.

FIGURE 20.9 Normal vision of far and near objects.

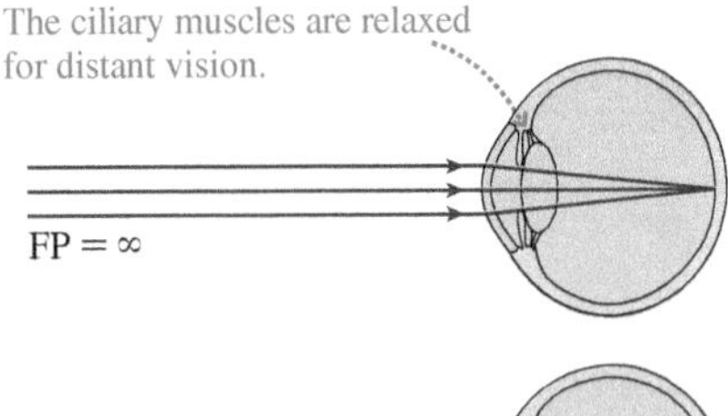
The ciliary muscles are relaxed for distant vision.
FP = ∞

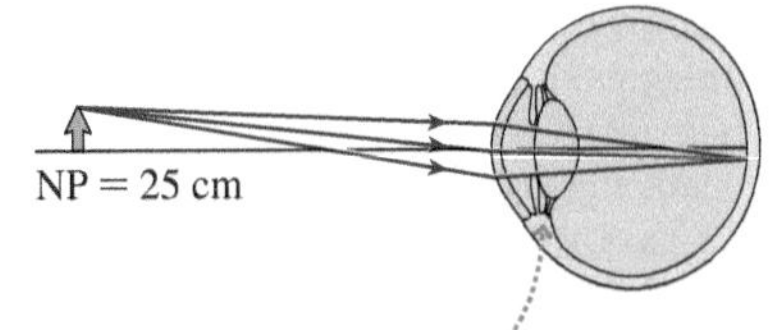
NP = 25 cm
The ciliary muscles are contracted for near vision, causing the lens to curve more.

◀◀ EXAMPLE 19.4 analyzed refraction at the cornea and found that, if the lens were not there, parallel light rays would be focused about 25 mm behind the cornea. In other words, the cornea has a focal length of about 25 mm and thus a refractive power of about 40 D.

Individuals vary, but measurements find that the total refractive power of a relaxed eye is about 60 D. A refractive power of 60 D implies that the effective focal length of a relaxed eye—the combined effect of the cornea and lens—is

$$f_{\text{relax}} = \frac{1}{P} = 0.017\text{ m} = 17\text{ mm} \tag{20.6}$$

That is, the combination of the cornea and the lens is equivalent to a single lens with a 17 mm focal length that is placed 17 mm in front of the retina. This single-lens optical model of the eye is a useful tool for thinking about vision defects and how they can be corrected.

Accommodation increases the power of the lens, but by how much does the total refractive power of the eye change? To answer this question, we start with a relaxed eye focused on an object at its far point. We'll call the object distance $s = \text{FP}$. In our single-lens model of the eye, the lens produces a focused image at distance $s' = d_{\text{ret}}$, the distance between the lens and the retina. The eye's refractive power is a minimum P_{min} when the lens is fully relaxed, and we can use $P = 1/f$ to write the thin-lens equation for a relaxed eye as

$$P_{\text{min}} = \frac{1}{f_{\text{relax}}} = \frac{1}{s} + \frac{1}{s'} = \frac{1}{\text{FP}} + \frac{1}{d_{\text{ret}}} \tag{20.7}$$

We expect $f_{\text{relax}} \approx 17$ mm and $P_{\text{min}} \approx 60$ D if the eye has a normal far point at infinity.

The eye's refractive power increases to a maximum P_{max} as the eye uses maximum accommodation to focus on an object at its near point, where the object distance is $s = \text{NP}$. In this case, the thin-lens equation is

$$P_{\text{max}} = \frac{1}{f_{\text{accom}}} = \frac{1}{s} + \frac{1}{s'} = \frac{1}{\text{NP}} + \frac{1}{d_{\text{ret}}} \tag{20.8}$$

Distance d_{ret} is a fixed property of an individual's eye that cannot change. We can eliminate d_{ret} by subtracting Equation 20.7 from Equation 20.8:

$$\Delta P_{\text{accom}} = P_{\text{max}} - P_{\text{min}} = \frac{1}{\text{NP}} - \frac{1}{\text{FP}} \tag{20.9}$$

Equation 20.9 is the *total accommodation* of the eye as the lens goes from fully relaxed to fully tensed. $\Delta P_{\text{accom}} \approx 4$ D for a young adult who has a normal near point of 25 cm and a normal far point of infinity. In other words, the eye changes from a refractive power of about 60 D when relaxed to about 64 D with maximum accommodation; this is a fairly small range.

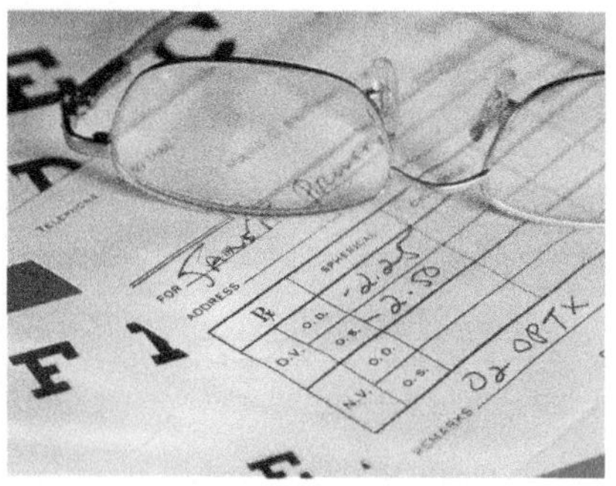

Vision correction The optometrist's prescription is −2.25 D for the right eye (top) and −2.50 D for the left (bottom), the minus sign indicating that these are diverging lenses. The optometrist doesn't write the D because the lens maker already knows that prescriptions are in diopters. Most people's eyes are not exactly the same, so each eye usually gets a different lens.

Refractive Errors of the Eye and Their Correction

The near point of normal vision is considered to be 25 cm, but the near point of any individual changes with age. The near point of young children can be as little as 10 cm. The "normal" 25 cm near point is characteristic of young adults, but the near point of most individuals begins to move outward by age 40 or 45 and can reach 200 cm by age 60. This loss of accommodation, which arises because the lens loses flexibility, is called **presbyopia,** meaning "elder eyes." Even if their vision is otherwise normal, individuals with presbyopia need reading glasses to bring their near point back to 25 or 30 cm, a comfortable distance for reading.

Presbyopia is known as a *refractive error* of the eye. **Hyperopia,** or *farsightedness,* is a somewhat similar refractive error. A person who is farsighted can see faraway objects, but the eye's near point is larger than 25 cm, often much larger, so it cannot focus on nearby objects. The cause of hyperopia is an eyeball that is too short for the refractive power of the cornea and lens. As FIGURES 20.10a and b show, no amount of accommodation allows the eye to focus on an object 25 cm away, the normal near point.

With hyperopia, the eye needs assistance to focus the rays from a near object onto the closer-than-normal retina. This assistance is obtained by *adding* refractive power with the converging lens shown in FIGURE 20.10c. An eye with more refracting power (i.e., a shorter effective focal length) can focus on closer objects, bringing the near point closer. Presbyopia, the loss of accommodation with age, is corrected in the same way.

FIGURE 20.10 Hyperopia.

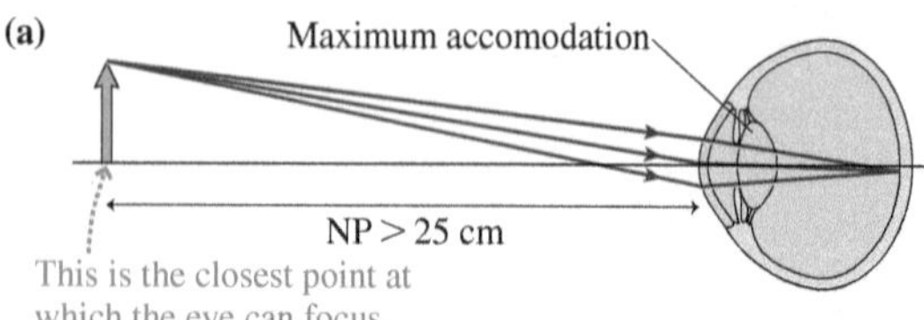
(a)
Maximum accomodation
NP > 25 cm
This is the closest point at which the eye can focus.

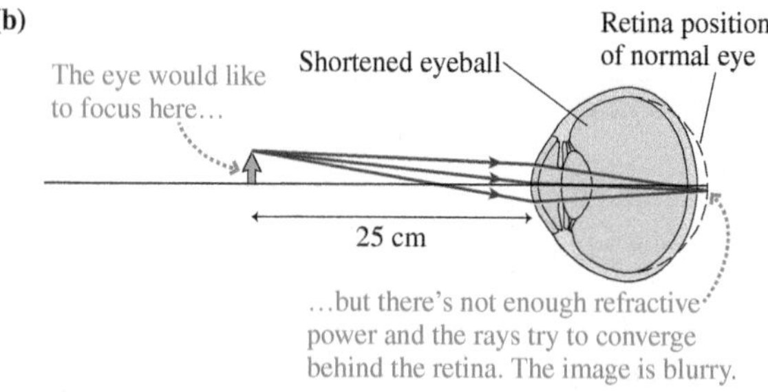
(b)
The eye would like to focus here...
Shortened eyeball
Retina position of normal eye
25 cm
...but there's not enough refractive power and the rays try to converge behind the retina. The image is blurry.

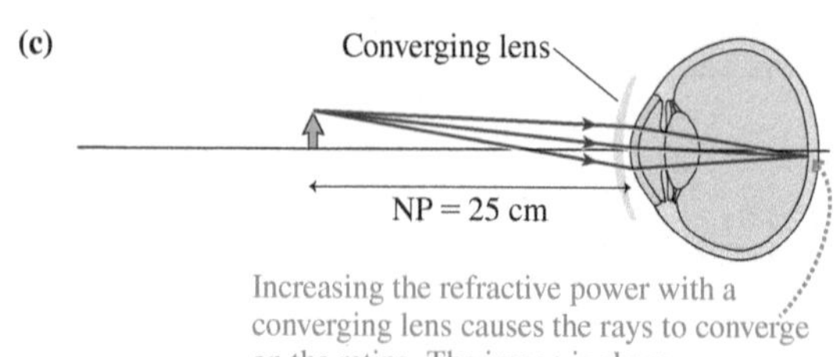
(c)
Converging lens
NP = 25 cm
Increasing the refractive power with a converging lens causes the rays to converge on the retina. The image is sharp.

FIGURE 20.11 Myopia.

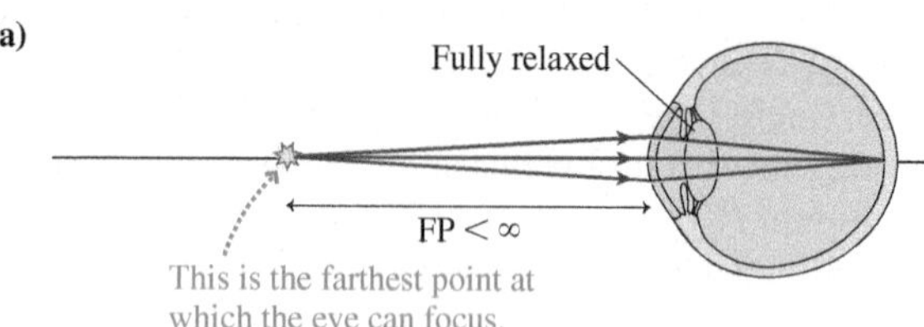
(a)
Fully relaxed
FP < ∞
This is the farthest point at which the eye can focus.

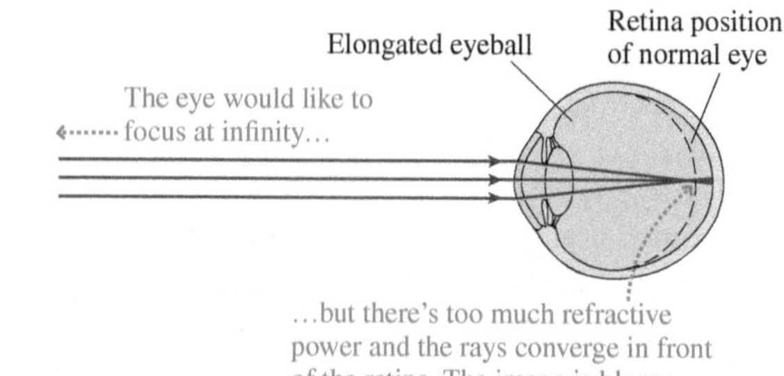
(b)
Elongated eyeball
Retina position of normal eye
The eye would like to focus at infinity...
...but there's too much refractive power and the rays converge in front of the retina. The image is blurry.

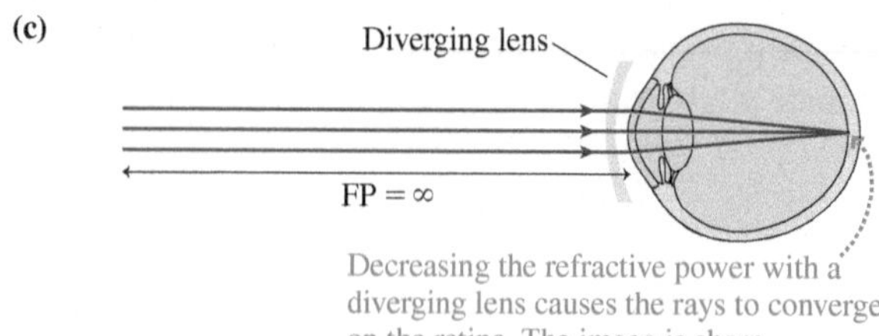
(c)
Diverging lens
FP = ∞
Decreasing the refractive power with a diverging lens causes the rays to converge on the retina. The image is sharp.

NOTE ► Figures 20.10 and 20.11 show the corrective lenses as they are actually shaped—called *meniscus lenses*—rather than with our usual lens shape. Nonetheless, the lens in Figure 20.10c is a converging lens because it's thicker in the center than at the edges. The lens in Figure 20.11c is a diverging lens because it's thicker at the edges than in the center. ◄

EXAMPLE 20.4 Correcting hyperopia

Sanjay has hyperopia. The near point of his left eye is 150 cm. What prescription lens will allow him to read a book at a normal distance of 25 cm?

PREPARE Sanjay's eye is at maximum refractive power when he is focused on an object at his near-point distance of 150 cm; he lacks sufficient power to focus on anything closer. Additional refractive power is needed to bring Sanjay's near point in to a normal 25 cm. If we ignore the small space between a corrective lens and the eye, treating them as if they are in contact, then refractive powers add: The total refractive power is the sum of the eye's refractive power and the corrective lens's refractive power.

SOLVE The refractive power of Sanjay's eye when focused on an object at his near point is

$$P_{\text{max}} = \frac{1}{\text{NP}} + \frac{1}{d_{\text{ret}}} = \frac{1}{1.50\text{ m}} + \frac{1}{d_{\text{ret}}} = 0.67\text{ D} + \frac{1}{d_{\text{ret}}}$$

The refractive power that he needs to focus on an object 25 cm away is

$$P_{\text{needed}} = \frac{1}{0.25\text{ m}} + \frac{1}{d_{\text{ret}}} = 4.0\text{ D} + \frac{1}{d_{\text{ret}}}$$

The difference between these powers is the additional refractive power that a corrective lens needs to provide:

$$P_{\text{lens}} = P_{\text{needed}} - P_{\text{max}} = 3.33\text{ D}$$

The optometrist will write the prescription as +3.33.

ASSESS Hyperopia and presbyopia are always corrected with a converging lens (which has positive power) that supplements the eye's refractive power, thus allowing the individual to focus more closely.

A person who is *nearsighted* can clearly see nearby objects when the eye is relaxed (and extremely close objects by using accommodation), but no amount of relaxation allows the eye to focus on distant objects. Nearsightedness—called **myopia**—is caused by an eyeball that is too long. As FIGURE 20.11a (previous page) shows, rays from a distant object come to a focus in front of the retina and have begun to diverge by the time they reach the retina. The eye's far point, shown in FIGURE 20.11b, is less than infinity.

A myopic eye has too much refractive power when relaxed, which causes light rays from a distant object to converge too quickly. The solution, shown in FIGURE 20.11c, is to reduce the eye's refractive power with a diverging lens that has negative refractive power. A reduced refractive power reduces the convergence of the light rays and moves the focus back to the retina.

EXAMPLE 20.5 Correcting myopia

Martina has myopia. The far point of her left eye is 200 cm. What prescription lens will allow her to see distant objects clearly?

PREPARE Martina's eye is at minimum refractive power when she is focused on an object at her far-point distance of 200 cm. Even this minimum power is too strong to focus on anything farther away. The refractive power needs to be reduced so that Martina's far point moves out to infinity. If we ignore the small space between a corrective lens and the eye, treating them as if they are in contact, then the total refractive power is the sum of the eye's refractive power and the corrective lens's refractive power.

SOLVE The refractive power of Martina's eye when focused on an object at her far point is

$$P_{\text{min}} = \frac{1}{\text{FP}} + \frac{1}{d_{\text{ret}}} = \frac{1}{2.00\text{ m}} + \frac{1}{d_{\text{ret}}} = 0.50\text{ D} + \frac{1}{d_{\text{ret}}}$$

The refractive power that she needs to focus on a very distant object, one effectively at infinity, is

$$P_{\text{needed}} = \frac{1}{\infty\text{ m}} + \frac{1}{d_{\text{ret}}} = \frac{1}{d_{\text{ret}}}$$

The difference between these powers is the additional refractive power that a corrective lens needs to provide:

$$P_{\text{lens}} = P_{\text{needed}} - P_{\text{min}} = -0.50\text{ D}$$

The optometrist will write the prescription as −0.50.

ASSESS Myopia is always corrected with a diverging lens (which has negative power) that weakens the eye's refractive power, allowing the individual to focus on more distant objects.

Other Vision Corrections

We have assumed that the cornea and lens have smooth, spherical surfaces with the same curvature at every point. For many people, however, especially as they age, the refracting surfaces become distorted and irregular. The simplest form of distortion, shown in FIGURE 20.12 (next page), occurs if a spherical refracting surface is squeezed along one axis while expanding along a perpendicular axis.

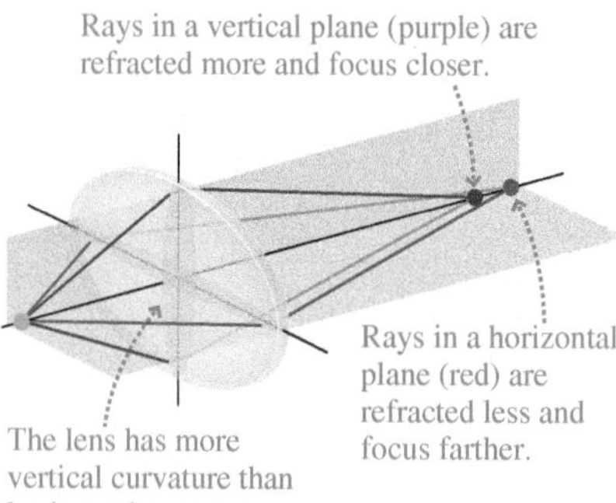

Rays in a vertical plane (purple) are refracted more and focus closer.
Rays in a horizontal plane (red) are refracted less and focus farther.
The lens has more vertical curvature than horizontal curvature.

FIGURE 20.12 Front view of two refracting surfaces.

The lens in Figure 20.12 has been squeezed vertically so that the surface bulges outward more—it has more curvature—in a vertical plane than it does in a horizontal plane. The two purple rays travel in a vertical plane, and the two red rays are in a horizontal plane. An ideal lens would focus both sets of rays at the same point. But the purple rays pass through a section of the lens that, because it's more curved, has more refractive power. As a result, the purple rays converge at a closer image distance than do the red rays. In other words, vertical and horizontal rays from an object aren't focused to the same point, so the image is blurry. This refractive error due to a distortion in the shapes of the refracting surfaces—mostly the cornea but sometimes the lens—is called **astigmatism.**

Astigmatism is corrected with a compensating distortion in the shape of the eyeglass lens. Opticians call this a *cylindrical correction* because light that passes through a transparent cylinder is focused in one plane but not the other. Thus a prescription for eyeglasses or contact lenses looks something like this:

EYE	SPH	CYL	AXIS
OD	+2.50	+0.50	100
OS	+2.75	+0.75	120

OD and OS are Latin-based abbreviations for your right and left eyes. SPH, for *spherical,* is the primary prescription in diopters; CYL, for *cylindrical,* is the additional power needed along one axis to correct for astigmatism; and AXIS is the angle in degrees of that axis from horizontal. In this case, because refractive powers add, a 2.50 D converging lens (for the right eye) will be given enough additional curvature to increase its power to 3.00 D along an axis at 100°.

Sometimes, especially for younger people, a single corrective lens is sufficient to provide good vision from a near point at 25 cm to a far point at infinity. At other times, especially for older people for whom presbyopia is significant, near vision and distant vision need different corrections. The solution is to wear *bifocals,* which have different prescriptions for the top half (for distant vision) and bottom half (for reading or close work).

For some individuals, an alternative to wearing glasses or contact lenses is to have their cornea reshaped. We've seen that most of the eye's focusing is caused by refraction at the cornea, so a careful reshaping can dramatically improve vision. The most common form of surgical vision correction is called LASIK, an acronym for *laser-assisted in situ keratomileusis,* in which the physician carefully removes cornea tissue with pulses from an ultraviolet laser until the appropriate shape is achieved. The computer software that guides the LASIK process is simply a detailed application of Snell's law to a carefully mapped shape of the cornea.

STOP TO THINK 20.3 You need to improvise a magnifying glass to read some very tiny print. Should you borrow the eyeglasses from your hyperopic friend or from your myopic friend?

A. The hyperopic friend
B. The myopic friend
C. Either will do.
D. Neither will work.

20.4 Magnifiers and Microscopes

A camera allows us to capture images of events that take place too quickly for our unaided eye to resolve. Another use of optical systems is to magnify—to see objects smaller or closer together than our eye can see.

The easiest way to magnify an object requires no extra optics at all; simply get closer! The closer you get, the bigger the object appears. Obviously the actual size of the object is unchanged as you approach it, so what exactly is getting "bigger"?

Consider the green arrow in FIGURE 20.13a. We can determine the size of its image on the retina by tracing the rays that are undeviated as they pass through the center of the lens. (Here we're modeling the eye's optical system as one thin lens.) If we get closer to the arrow, now shown as red, we find the arrow makes a larger image on the retina. Our brain interprets the larger image as a larger-appearing object. The object's actual size doesn't change, but its *apparent size* gets larger as it gets closer.

Technically, we say that closer objects look larger because they subtend a larger angle θ, called the **angular size** of the object. The red arrow has a larger angular size than the green arrow, $\theta_2 > \theta_1$, so the red arrow looks larger and we can see more detail. But you can't keep increasing an object's angular size by moving closer because you can't focus on the object if it's closer than your near point, which we'll take to be a normal 25 cm. FIGURE 20.13b defines the angular size θ_{NP} of an object at your near point. If the object's height is h and if we assume the small-angle approximation $\tan\theta \approx \theta$, the maximum angular size viewable by your unaided eye is

$$\theta_{NP} = \frac{h}{25 \text{ cm}} \tag{20.10}$$

FIGURE 20.13 Angular size.

(a) Same object at two different distances

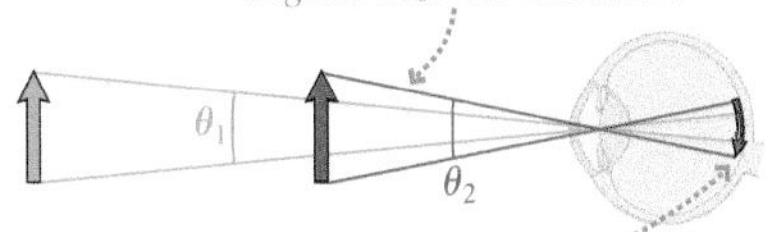
As the object gets closer, the angle it subtends becomes larger. Its *angular size* has increased.
θ_1
θ_2
Further, the size of the image on the retina gets larger. The object's *apparent size* has increased.

(b)

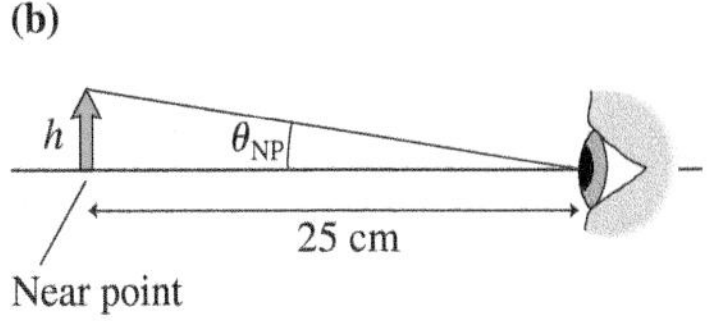
h
θ_{NP}
25 cm
Near point

Suppose we view the same object, of height h, through the single converging lens in FIGURE 20.14. If the object's distance from the lens is less than the lens's focal length, we'll see an enlarged, upright image. Used in this way, the lens is called a **magnifier** or *magnifying glass*. The eye sees the virtual image subtending angle θ, and it can focus on this virtual image as long as the image distance is more than 25 cm. Within the small-angle approximation, the image subtends angle $\theta = h/s$. In practice, we usually want the image to be at distance $s' \approx \infty$ so that we can view it with a relaxed eye as a "distant object." This will be true if the object is very near the focal point: $s \approx f$. In this case, the image subtends angle

$$\theta = \frac{h}{s} \approx \frac{h}{f} \tag{20.11}$$

FIGURE 20.14 The magnifier.

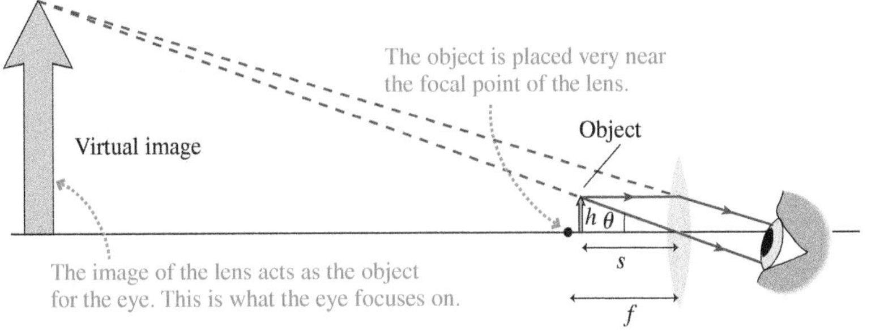
The object is placed very near the focal point of the lens.
Object
Virtual image
h
θ
s
f
The image of the lens acts as the object for the eye. This is what the eye focuses on.

Let's define the **angular magnification** M as

$$M = \frac{\theta}{\theta_{NP}} \tag{20.12}$$

Angular magnification is the increase in the *apparent size* of the object that you achieve by using a magnifying lens rather than simply holding the object at your near point. Substituting from Equations 20.10 and 20.11, we find the angular magnification of a magnifying glass is

$$M = \frac{25 \text{ cm}}{f} \tag{20.13}$$

Movie magic? Even if the more distant of two equally sized objects *appears* smaller, we don't usually believe it actually *is* smaller because there are abundant visual clues that tell our brain that it's farther away. If those clues are removed, however, the brain readily accepts the illusion that the farther object is smaller. The technique of *forced perspective* is a special effect used in movies to give this illusion. Here, Camelia, who is actually closer to the camera, looks like a giant compared to Kevin. The lower photo shows how the trick was done.

The angular magnification depends on the focal length of the lens but not on the size of the object. Although it would appear we could increase angular magnification without limit by using lenses with shorter and shorter focal lengths, the inherent limitations of lenses we discuss later in the chapter limit the angular magnification of a simple lens to about 4×. Slightly more complex magnifiers with two or three lenses reach 20×, but beyond that one would use a microscope.

NOTE ▸ Don't confuse angular magnification M with lateral magnification m. Lateral magnification compares the height of an object to the height of its image. The lateral magnification of a magnifying glass is $\approx \infty$ because the virtual image is at $s' \approx \infty$, but that doesn't make the object seem infinitely big. Its apparent size is determined by the angle subtended on your retina, and that angle remains finite. Thus angular magnification tells us how much bigger things appear. ◂

CONCEPTUAL EXAMPLE 20.6 The angular size of a magnified image

An object is placed right at the focal point of a magnifier. How does the apparent size of the image depend on where the *eye* is placed relative to the lens?

REASON When the object is precisely at the focal point of the lens, Equation 20.11 holds exactly. The angular size $\theta = h/f$ is *independent* of the position of the eye. Thus the object's apparent size is independent of the eye's position as well. FIGURE 20.15 shows a calculator at the focal point of a magnifier. The apparent size of the COS button is the same whether the camera taking the picture is close to or far from the lens.

FIGURE 20.15 Looking through a magnifier with the object at its focal point.

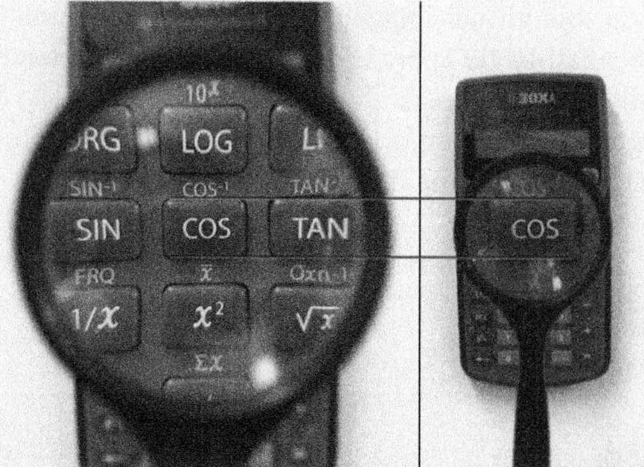

ASSESS When the object is at the magnifier's focus, the virtual image is at infinity. The situation is similar to observing any "infinitely" distant object, such as the moon. If you walk closer to or farther from the moon, its apparent size doesn't change at all. The same holds for a virtual *image* at infinity: Its apparent size is independent of the point from which you observe it.

STOP TO THINK 20.4 You are using a handheld magnifier to observe the tip of a flower stamen. The stamen tip is just slightly beyond the focal point of the lens. What do you see as you look through the lens?

A. An upright image, smaller than the object.
B. An upright image, larger than the object.
C. An inverted image, smaller than the object.
D. An inverted image, larger than the object.
E. No image.

The Microscope

Microscopes are ubiquitous in biology and medicine, and high-quality microscopic images are of profound importance to life scientists. Every advance in microscopy has led to new scientific discoveries as scientists image structures that couldn't previously be seen. You may have used a compound microscope, like the one shown in FIGURE 20.16, in a laboratory course. More sophisticated microscopes record the image directly on a digital camera. We'll analyze a basic microscope in this section and then look at some of the details in Section 20.6.

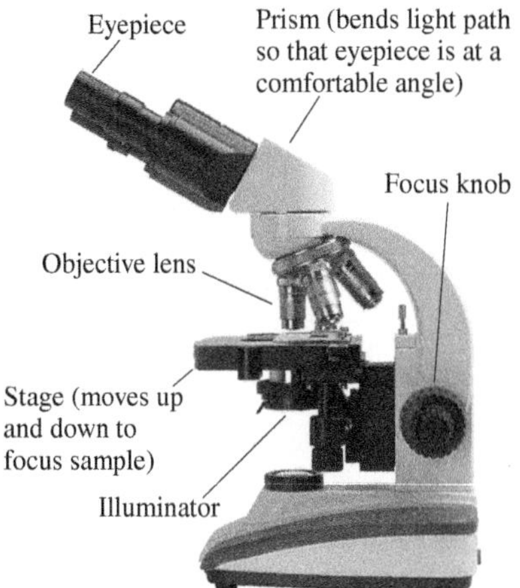

FIGURE 20.16 A compound microscope.

NOTE ▶ Many microscopes, like the one in Figure 20.16, use a prism to bend the light path to a comfortable viewing angle. This doesn't change the imaging properties, so we'll consider a simplified model in which the light travels along a straight tube. ◀

FIGURE 20.17 is a two-lens model of a microscope. The object is located just slightly outside the focal point of the **objective,** a converging lens with a short focal length. The objective lens creates a magnified real image inside the microscope tube. That real image is then observed and further magnified by the **eyepiece,** which functions as a magnifier to produce a distant virtual image that is viewed by a relaxed eye. The overall magnification of a microscope, an *angular* magnification M, is the product of the lateral magnification of the objective and the angular magnification of the eyepiece:

$$M = m_{\text{obj}} M_{\text{eye}} \tag{20.14}$$

The angular magnification of the eyepiece is given by Equation 20.13: $M_{\text{eye}} = (25\text{ cm})/f_{\text{eye}}$. The most common microscope eyepiece is a 10× magnifier, but 5× and 20× eyepieces are sometimes used.

FIGURE 20.17 The optics of a microscope.

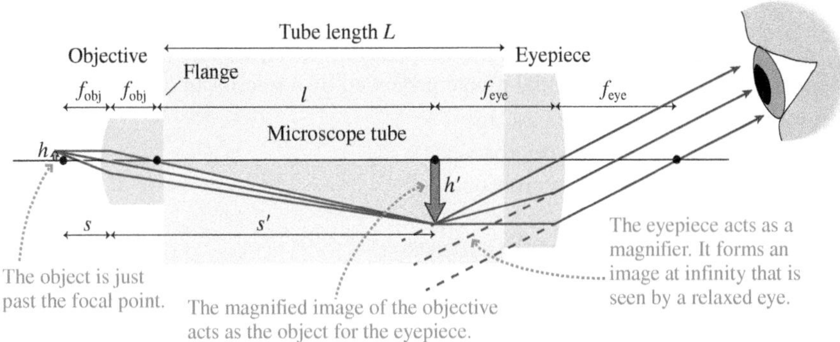

We can find an approximate expression for the magnification of the objective by considering the two shaded triangles in Figure 20.17. The triangle on the left has height h, the object height, and width f_{obj}. The triangle on the right has height h', the image height, and width l, the distance between the focal point and the image. These are similar triangles because they have all the same angles, so $h'/h = l/f_{\text{obj}}$. But h'/h, the ratio of the image height to the object height, is $|m|$, the absolute value of the lateral magnification. We need to include a negative sign to show that the image is inverted, so the lateral magnification of the objective lens is

$$m_{\text{obj}} = -\frac{h'}{h} = -\frac{l}{f_{\text{obj}}} \tag{20.15}$$

Unfortunately, there's no easy way to determine l, and it varies from one objective lens to another.

One thing that doesn't change, though, is the physical length of the microscope tube measured between the flanges at the ends where the objective and eyepiece are screwed in. This is called the *tube length L*. Most biological microscopes have a standard tube length of $L = 160$ mm. It turns out that length l in Equation 20.15 differs from the tube length L by only a few percent; thus $m_{\text{obj}} \approx -L/f_{\text{obj}}$ is a good approximation for the magnification of the objective. We can use this to see that a 20× objective has focal length $f_{obj} \approx (160 \text{ mm})/m_{obj} = 8 \text{ mm}$. If we use this approximate expression for m_{obj}, we can write the microscope's overall magnification as

$$M \approx -\frac{L}{f_{obj}} \frac{25 \text{ cm}}{f_{eye}} \tag{20.16}$$

with the eyepiece focal length in cm.

In practice, the true magnifications of the objective (without the minus sign) and the eyepiece are stamped on the barrels, and the microscope's overall magnification is the product of the two. A set of objectives on a rotating turret might include 10×, 20×, 40×, and 100×. When combined with a 10× eyepiece, the microscope's angular magnification ranges from 100× to 1000×.

EXAMPLE 20.7 Viewing blood cells

A pathologist inspects a sample of 7-μm-diameter human blood cells under a microscope. She selects a 40× objective and a 10× eyepiece. What size object, viewed from 25 cm, has the same apparent size as a blood cell seen through the microscope?

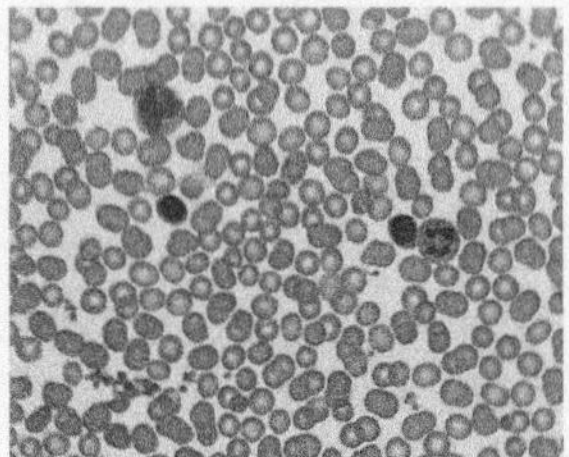

PREPARE Angular magnification compares the magnified angular size to the angular size seen at the near-point distance of 25 cm.

SOLVE The microscope's angular magnification is $M = -(40) \times (10) = -400$. The magnified cells will have the same apparent size as an object $400 \times 7\ \mu\text{m} \approx 3 \text{ mm}$ in diameter seen from a distance of 25 cm.

ASSESS 3 mm is about the size of a capital O in a print textbook, so a blood cell seen through the microscope will have about the same apparent size as an O seen from a comfortable reading distance.

STOP TO THINK 20.5 The final image produced by a microscope is

A. A real image inside the microscope.
B. A virtual image inside the microscope.
C. A real image outside the microscope.
D. A virtual image outside the microscope.
E. Either B or D, depending on the eyepiece.

20.5 The Resolution of Optical Instruments

Suppose you want to study the *E. coli* bacterium. It's quite small, about 2 μm long and 0.5 μm wide. You might imagine that you could pair a 150× objective (the highest magnification available) with a 25× eyepiece to get a total magnification of 3750. At that magnification, the *E. coli* would appear about 8 mm across—about the size of Lincoln's head on a penny—with much fine detail revealed. But if you tried this, you'd be disappointed. Although you would see the general shape of a bacterium, you wouldn't be able to make out any details. All real optical instruments are limited in the details they can observe. Some limits are practical: Lenses are never perfect; they suffer from *aberrations*. But even a perfect lens has a fundamental limit to the smallest details that can be seen. As we'll see, this limit is set by the diffraction of light, and so is intimately related to the wave nature of light itself. Together, lens aberrations and diffraction set a limit on an optical system's *resolution*—its ability to make out the fine details of an object.

Aberrations

We saw in Chapter 19 that any lens has dispersion. That is, its index of refraction varies slightly with wavelength. Because the index of refraction for violet light is larger than for red light, a lens's focal length is shorter for violet light than for red light. Consequently, different colors of light come to a focus at slightly different distances from the lens. If red light is sharply focused on a viewing screen, then blue and violet wavelengths are not well focused. This imaging error, illustrated in FIGURE 20.18a, is called **chromatic aberration.**

Our analysis of thin lenses was based on paraxial rays traveling nearly parallel to the optical axis. A more exact analysis, taking all the rays into account, finds that rays incident on the outer edges of a spherical surface are not focused at exactly the same point as rays incident near the center. This imaging error, shown in FIGURE 20.18b, is called **spherical aberration.** Spherical aberration, which causes the image to be slightly blurred, gets worse as the lens diameter increases.

Fortunately, the chromatic and spherical aberrations of a converging lens and a diverging lens are in opposite directions. When a converging lens and a diverging lens made of glasses with different dispersions are used in combination, their aberrations tend to cancel. A combination lens, such as the one in FIGURE 20.18c, can produce a much sharper focus than a single lens with the equivalent focal length. This is the main reason that optical instruments use combination lenses rather than single lenses. Converging and diverging lenses of matched curvatures are often cemented together to form what is called an *achromatic doublet.*

FIGURE 20.18 Chromatic aberration and spherical aberration prevent simple lenses from forming perfect images.

(a) Chromatic aberration

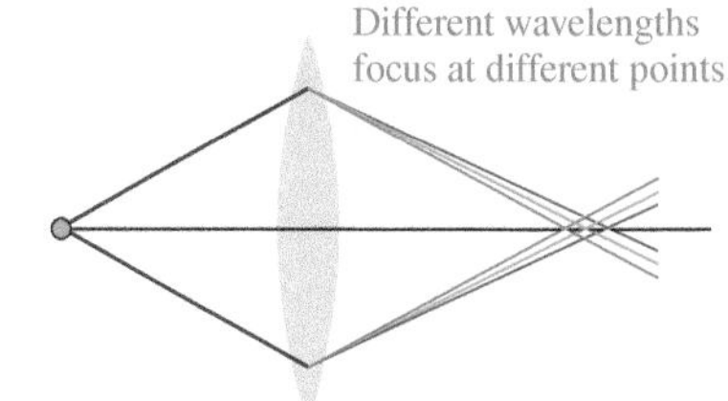

(b) Spherical aberration

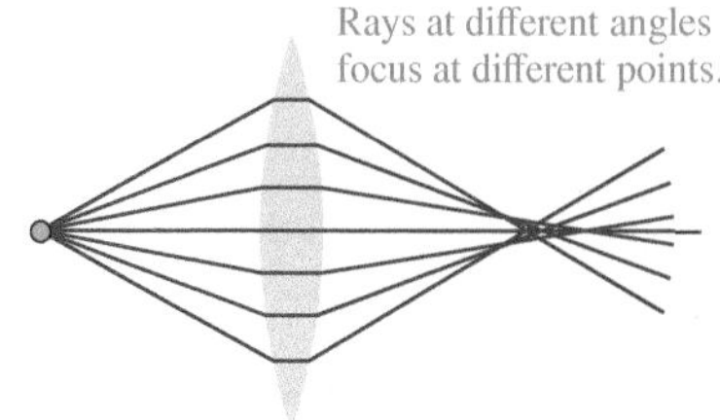

Rays at different angles focus at different points.

(c) Correcting aberrations

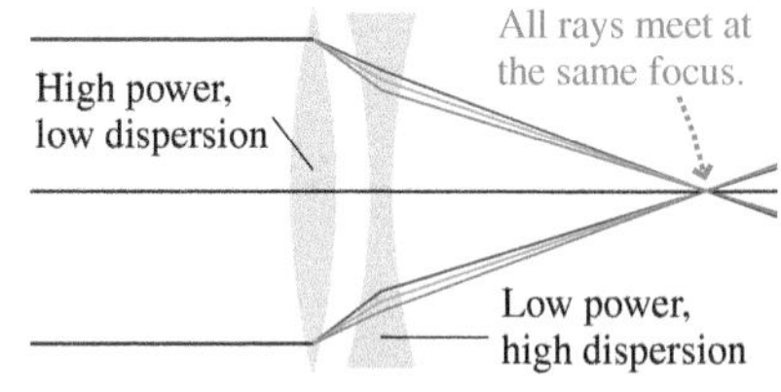

All rays meet at the same focus.
High power, low dispersion
Low power, high dispersion

Diffraction Limits Resolution

According to the ray model of light, a perfect lens (one with no aberrations) should be able to form a perfect image. But the ray model of light, though a very good model for lenses, is not an absolutely correct description of light. If we look closely, the wave aspects of light haven't entirely disappeared. In fact, the performance of optical equipment is limited by the diffraction of light.

FIGURE 20.19a shows a plane wave, with parallel light rays, being focused by a lens of diameter D. According to the ray model of light, a perfect lens would focus parallel rays to a perfect point. Notice, though, that only a piece of each wave front passes *through* the lens and gets focused. In effect, **the lens itself acts as a circular aperture** in an opaque barrier, allowing through only a portion of each wave front. Consequently, **the lens diffracts the light wave.** The diffraction is usually very small because D is usually much greater than the wavelength of the light; nonetheless, this small amount of diffraction is the limiting factor in how well the lens can focus the light.

FIGURE 20.19b separates the diffraction from the focusing by modeling the lens as an actual aperture of diameter D followed by an "ideal" diffractionless lens. You learned in Chapter 18 that a circular aperture produces a diffraction pattern with a bright central maximum surrounded by dimmer fringes. A converging lens brings this diffraction pattern to a focus in the image plane, as shown in FIGURE 20.19c. As a result, a perfect lens focuses parallel light rays not to a perfect point of light, as we expected, but to a small, circular diffraction pattern.

FIGURE 20.19 A lens both focuses and diffracts the light passing through.

(a) A lens acts as a circular aperture.

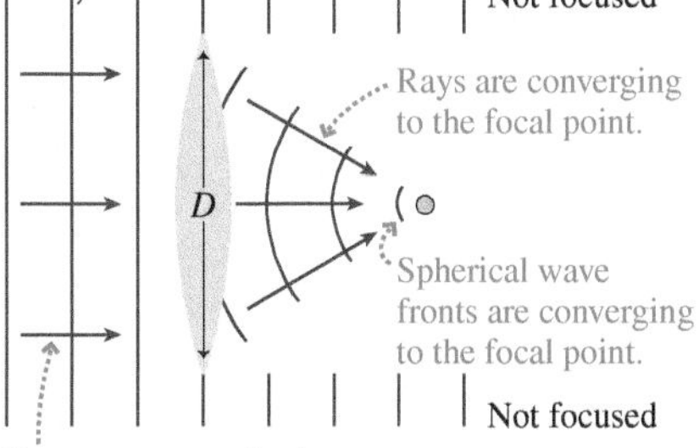

Wave fronts
Not focused
Rays are converging to the focal point.
D
Spherical wave fronts are converging to the focal point.
Not focused
Rays are perpendicular to the wave fronts.

(b) The aperture and focusing effects can be separated.

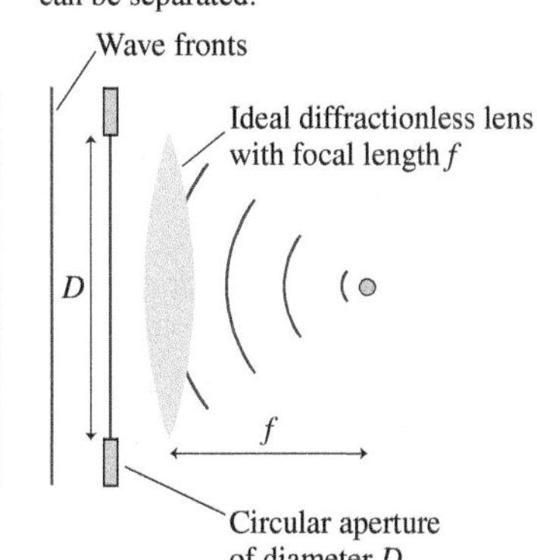

(c) The lens focuses the diffraction pattern in the focal plane.

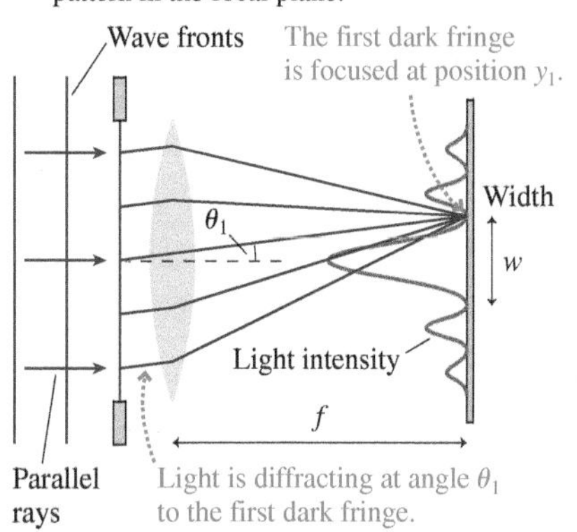

The angle to the first minimum of a circular diffraction pattern is $\theta_1 = 1.22\lambda/D$. The ray that passes through the center of a lens is not bent, so Figure 20.19c uses this ray to show that the position of the dark fringe is $y_1 = f\tan\theta_1 \approx f\theta_1$. Thus the width of the central maximum in the focal plane is

$$w_{min} \approx 2f\theta_1 = \frac{2.44\lambda f}{D} \tag{20.17}$$

This is the **minimum spot size** to which a lens can focus light.

Lenses are often limited by aberrations, so not all lenses can focus parallel light rays to a spot this small. A well-crafted lens, for which Equation 20.17 is the minimum spot size, is called a *diffraction-limited lens.* No optical design can overcome the spreading of light due to diffraction, and it is because of this spreading that the image point has a minimum spot size. The image of an actual object, rather than of parallel rays, becomes a mosaic of overlapping diffraction patterns, so even the most perfect lens inevitably forms an image that is slightly fuzzy.

Your eye is close to being diffraction limited when the pupil diameter is smaller than 3 mm. The ◀◀ CHAPTER 18 INTEGRATED EXAMPLE calculated that the diffraction of light by a 3-mm-diameter pupil turns what would be a perfect focus into a 6-μm-diameter spot on the retina, a spot large enough to stimulate several cone cells. The spot size increases with a smaller diameter pupil due to increased diffraction, and it increases with a larger diameter pupil due to increased aberrations, primarily spherical aberration. Thus the human eye has its best focus and maximum acuity with a pupil diameter of about 3 mm.

For various reasons, it is difficult to produce a diffraction-limited lens having a focal length that is less than half its diameter; that is, $f > 0.5D$. This implies that **the smallest diameter to which you can focus a spot of light, no matter how hard you try, is $w_{min} \approx \lambda$.** This is a fundamental limit on the performance of optical equipment. In the next section, we will look at how this affects microscopy.

EXAMPLE 20.8 Seeing stars

A photographer takes a picture of the night sky with a high-quality camera. The camera lens has a focal length of 75 mm and an aperture diameter of 18 mm. What is the diameter on the detector of the image of a star? How does this compare to the detector's pixel size of 8 μm? Assume that the earth's atmosphere is not a limiting factor.

PREPARE Stars are so far away that they appear as points in space. An ideal diffractionless lens would focus their light to a perfect point. Diffraction prevents this.

SOLVE Equation 20.17 for the minimum spot size due to diffraction is

$$w_{min} = \frac{2.44\lambda f}{D}$$

The camera's focal length and aperture diameter appear as the ratio f/D. Stars emit white light, and the longest wavelengths spread the most and determine the spot size. If we use $\lambda = 700$ nm as the approximate upper limit of visible wavelengths, we find

$$w_{min} = 2.44\frac{(700 \times 10^{-9}\ \text{m})(75\ \text{mm})}{18\ \text{mm}} \approx 7\ \mu\text{m}$$

This is approximately the pixel size, which is by design.

ASSESS A high-quality camera lens is close to being diffraction limited. A smaller aperture would increase the spot size and illuminate several pixels, which would blur the image. A larger aperture would let in more light, which might be good for night-sky photography, but then aberrations become a more limiting factor. A camera lens, like your eye, has an optimum aperture for maximum acuity. In reality, the atmospheric turbulence that causes the twinkling of stars prevents ground-based photography from being this good, but space-based telescopes do take pictures that are diffraction limited.

Resolution

Suppose you point a telescope at two nearby stars in a galaxy far, far away. If you use the best possible detector, will you be able to distinguish separate images for the two stars, or will they blur into a single blob of light? A similar question could be asked of a microscope. Can two microscopic objects, very close together, be distinguished if sufficient magnification is used? Or is there some size limit at which their images will blur together and never be separated? These are important questions about the *resolution* of optical instruments.

Because of diffraction, the image of a distant star or a distant headlight is not a point but a circular diffraction pattern. Our question, then, really is: How close together can two diffraction patterns be before you can no longer distinguish them? One of the major scientists of the 19th century, Lord Rayleigh, studied this problem and suggested a reasonable rule that today is called **Rayleigh's criterion.**

FIGURE 20.20 shows two distant point sources being imaged by a lens of diameter D. The angular separation between the objects, as seen from the lens, is ϕ. Rayleigh's criterion states that

- The two objects are resolvable if $\phi > \theta_{min}$, where $\theta_{min} = \theta_1 = 1.22\lambda/D$ is the angle of the first dark fringe in the circular diffraction pattern.
- The two objects are not resolvable if $\phi < \theta_{min}$ because their diffraction patterns are too overlapped.
- The two objects are marginally resolvable if $\phi = \theta_{min}$. The central maximum of one image falls exactly on top of the first dark fringe of the other image. This is the situation shown in the figure.

FIGURE 20.20 Two images that are marginally resolved.

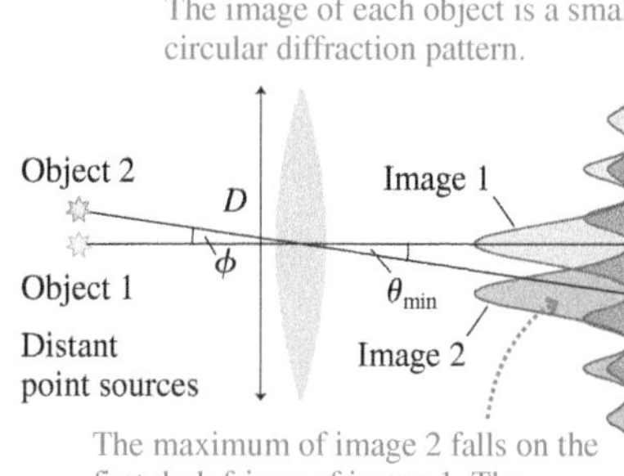

FIGURE 20.21 shows enlarged photographs of the images of two point sources. The images are circular diffraction patterns, not points. The two images are close but distinct where the objects are separated by $\phi > \theta_{min}$. Two objects really were recorded in the photo at the bottom, but their separation is $\phi < \theta_{min}$ and their images have blended together. In the middle photo, with $\phi = \theta_{min}$, you can see that the two images are just barely resolved.

The angle

$$\theta_{min} = \frac{1.22\lambda}{D} \qquad (20.18)$$

Angular resolution of a lens

is called the **angular resolution** of a lens. The angular resolution of a lens, such as a camera lens or the input lens (or mirror) of a telescope, depends on the diameter of the lens and the wavelength of the light; magnification is not a factor. Two images will remain overlapped and unresolved no matter what the magnification if their angular separation is less than θ_{min}. For visible light, where λ is pretty much fixed, the only parameter over which an astronomer has any control is the diameter of the lens or mirror of the telescope. The urge to build ever-larger telescopes is motivated, in part, by a desire to improve the angular resolution. (Another motivation is to increase the light-gathering power so as to see objects farther away.)

The performance of a microscope is also limited by the diffraction of light passing through the objective lens. Just as light cannot be focused to a spot smaller than about a wavelength, the most perfect microscope cannot resolve the features of objects separated by less than about one wavelength, or roughly 500 nm. This limitation is not simply a matter of needing a better design or more precise components; it is a fundamental limit set by the wave nature of the light with which we see. We'll look at microscope resolution more carefully in the next section.

FIGURE 20.21 Enlarged photographs of the images of two point sources.

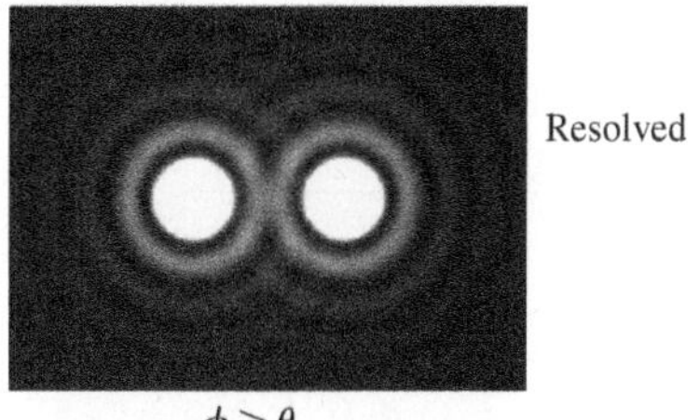

$\phi > \theta_{min}$

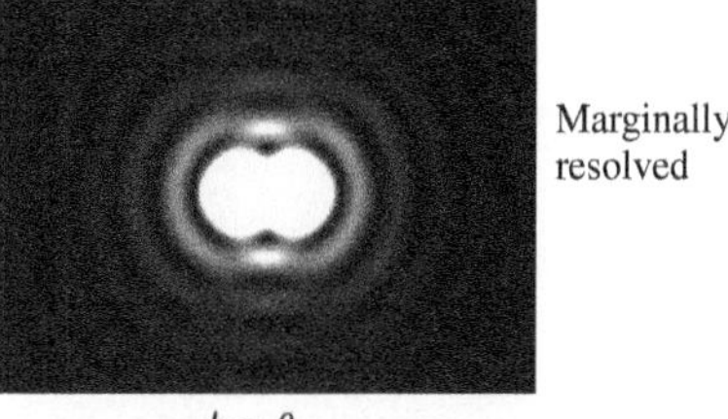

$\phi = \theta_{min}$

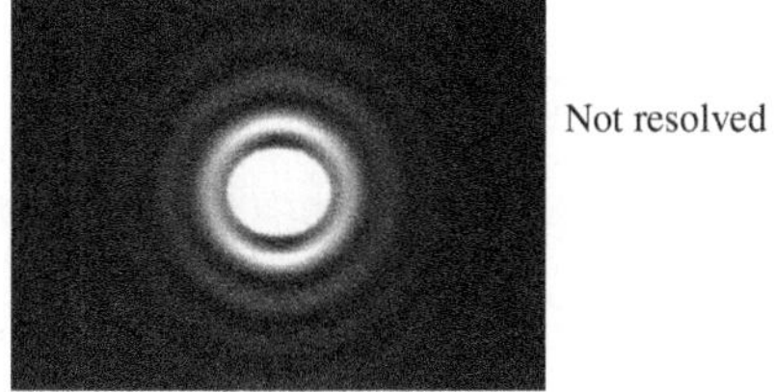

$\phi < \theta_{min}$

STOP TO THINK 20.6 Four diffraction-limited lenses focus plane waves of light with the same wavelength λ. Rank in order, from largest to smallest, the spot sizes w_1 to w_4.

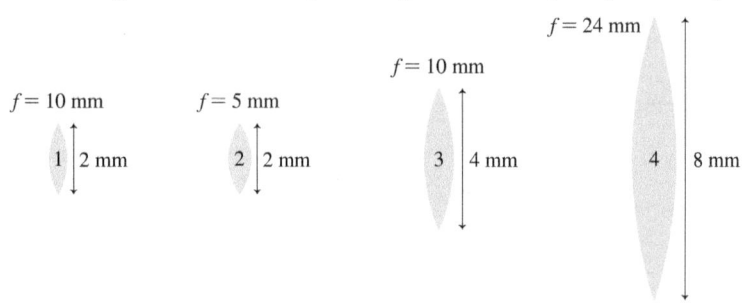

$f = 10$ mm, 1, 2 mm $f = 5$ mm, 2, 2 mm $f = 10$ mm, 3, 4 mm $f = 24$ mm, 4, 8 mm

20.6 Microscopy

Microscopes have been one of the most important tools in biology and medicine since Antonie van Leeuwenhoek developed the first modern microscope in the 1670s and become the founder of microbiology. We conclude this chapter by using what you've learned about optics to take a closer look at some important applications of microscopy to the life sciences.

The Resolution of a Microscope

FIGURE 20.22 The resolution of a microscope objective.

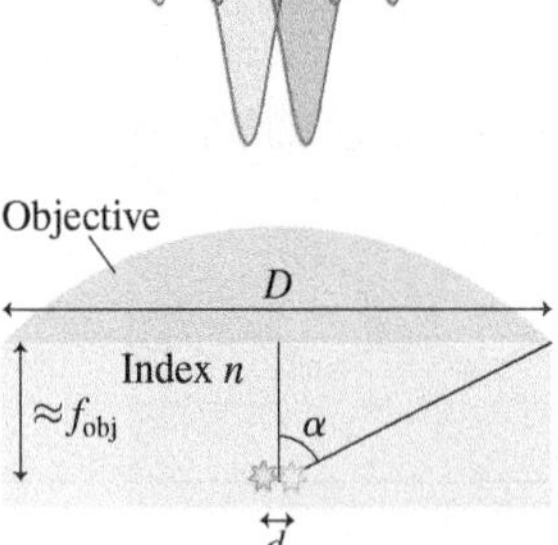
The images are circular diffraction patterns. The objects cannot be resolved if the diffraction patterns are too overlapped.
Objective
D
Index n
≈f_obj
α
d

A microscope differs from a camera or a telescope because it magnifies objects that are very close to the lens, not very far away. Even so, the wave nature of light sets a limit on the resolution of a microscope. FIGURE 20.22 shows a microscope objective of diameter D viewing two point-like objects separated by distance d. (The first lens in a multielement objective is usually a planoconvex lens with the flat side facing the specimen.) A microscope objective is designed to focus on objects at a distance only slightly farther than the focal length, so the object-to-lens distance is $\approx f_{obj}$.

The medium between the objects and the lens—often an oil, for reasons that we'll discover—has index of refraction n. Because of diffraction by the lens, the real image inside the microscope tube will not be two points but, instead, two circular diffraction patterns. The two objects cannot be resolved if their diffraction patterns are too overlapped.

Rayleigh's criterion says that the two objects are marginally resolvable if the central maximum of one image falls on top of the first dark fringe of the other image. This is the situation shown in the figure. The analysis that led to Equation 20.18 for the angular resolution of a camera or a telescope was based on an assumption of *plane waves* being incident on the diffracting aperture (see Figure 20.19c). That's not the case for a microscope because the objects are so close to the lens. A full analysis—beyond the scope of this text—finds that two point-like objects are marginally resolved when the separation between them is

$$d_{min} = \frac{0.61\lambda}{n \sin\alpha} \tag{20.19}$$

where the angle α, shown in Figure 20.22, is the *half-angle* of the lens as seen by the object. The numerator 0.61λ comes from $\theta_1 = 1.22\lambda/D$ for circular diffraction and shows that diffraction is the ultimate limit of a microscope's resolution. Angle α is the interior angle of a right triangle with adjacent and opposite sides $\approx f_{obj}$ and $D/2$, where D is the lens diameter, so the angle can be found from

$$\tan\alpha \approx \frac{D/2}{f_{obj}} \tag{20.20}$$

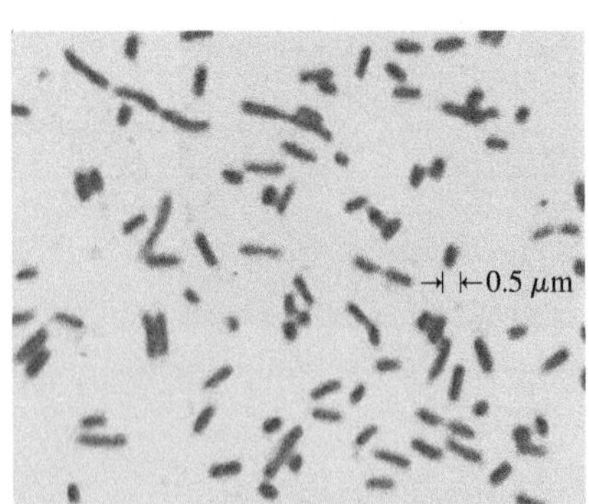
0.5 μm

At the limit This micrograph of the bacillus *E. coli* shows individual bacteria cells but no structural detail. The width of a bacterium is about 500 nm = 0.5 μm, just about the average wavelength of visible light, so the individual cells can be resolved by using an objective with the largest possible value of NA. Structures within the cells are smaller than 0.1 μm and thus cannot be resolved by a visible-light microscope. Additional magnification will enlarge the images of the cells but will not show additional detail.

The quantity in the denominator of Equation 20.19 is called the **numerical aperture** of the objective, abbreviated NA:

$$\text{NA} = n \sin\alpha \tag{20.21}$$

In general, the value of NA increases as the magnification of the objective increases because the focal length decreases and the objects get closer and closer to the lens, which increases the angle α. A low-power objective might have NA = 0.1, but a 100X oil-immersion objective (discussed later) with $n_{oil} \approx 1.5$ has $\alpha \approx 60°$ and NA $\approx$ 1.3.

The minimum resolvable distance of a microscope, called the **resolution,** is thus

$$R = d_{min} = \frac{0.61\lambda}{\text{NA}} \tag{20.22}$$

Resolution of a microscope

The smaller the resolution, the better the microscope is at seeing details. Resolution is improved by using an objective that has a larger value of NA, but NA $\approx$ 1.45 is about the upper limit that is technically possible. Thus the smallest achievable resolution is

$$R_{\text{min}} \approx 0.4\lambda \tag{20.23}$$

In other words, **the minimum resolution of a microscope, and thus the size of the smallest observable detail, is just slightly less than half the wavelength of light.** This is a *fundamental limit* set by the wave nature of light. For white-light viewing, where the limit is set by the longest visible wavelengths of about 700 nm, the best possible resolution is about 350 nm = 0.3 μm. This could be pushed to 0.25 μm or a bit less if the object is viewed using only blue light at the short-wavelength edge of the visible spectrum.

NOTE ▶ Don't confuse the roles of resolution and magnification of a microscope. Magnification makes the image larger, and initially that allows us to see more detail. But magnification doesn't improve the resolution. If we've magnified to the point that diffraction becomes an issue and the image begins to appear fuzzy, then further magnification will produce a larger fuzzy image but no additional detail. ◀

The numerical aperture of a microscope objective is about more than simply resolution; it also describes the objective's light-gathering power and thus the brightness of an image. FIGURE 20.23 shows a realistic situation in which a specimen is under a thin cover glass. The left side shows the path of the extreme ray if the space between the cover glass and the objective is filled with a liquid (usually oil) whose index of refraction n is almost identical to that of the cover glass. An objective designed to use this thin layer of oil is called an *oil-immersion objective*.

First, notice that an objective with a larger angle α, and thus a larger value of NA, will collect more of the light that is leaving the object. This is an important practical matter so that the user is able to view a specimen without needing an overly bright source of illumination. Second, introducing oil eliminates refraction at the top surface of the cover glass because there's no change in the index of refraction. This has two advantages:

- Angle α_{oil} is larger than angle α_{glass}, so the oil allows more light to be collected and produces a brighter image.
- Refraction at a glass-air boundary introduces chromatic and spherical aberration, which lowers the contrast and quality of the image. These aberrations are avoided at the glass-oil boundary because there's no refraction.

In addition, oil increases the NA by a factor of $n_{\text{oil}} \approx 1.5$, which improves the resolution. For these reasons, biological microscopes commonly use an oil-immersion objective to achieve the optimum resolution at high magnifications. Metallurgical and geological microscopes that view samples without a cover glass are usually "dry," which means they do not use oil.

Magnification
Tube length
Cover glass thickness range
Cover glass adjustment gauge
Numerical aperture
Correction collar
Nikon
Plan Apo
60x / 0.95
DIC M
∞/0.11-0.23 WD 0.15

The anatomy of a microscope objective The barrel of a microscope objective provides a wealth of information about the lens. The two most important properties of an objective, its magnification and its numerical aperture, are prominently displayed. The abbreviations above the magnification, Plan Apo in this case, provide details of how well the lens corrects for aberrations. Many biological observations are made through a cover glass, so the objective states the thickness range of the cover glasses at which it is designed to work and a correction collar to set the exact thickness of the cover being used. Finally, an objective is designed to work with a specific tube length, and that value is given at the lower left. Older or less expensive microscopes usually have a tube length of 160 mm, but newer research microscopes, called *infinity corrected,* use a somewhat different optical design that is equivalent to having an infinitely long tube. They display the symbol ∞.

FIGURE 20.23 A closer look at a microscope objective.

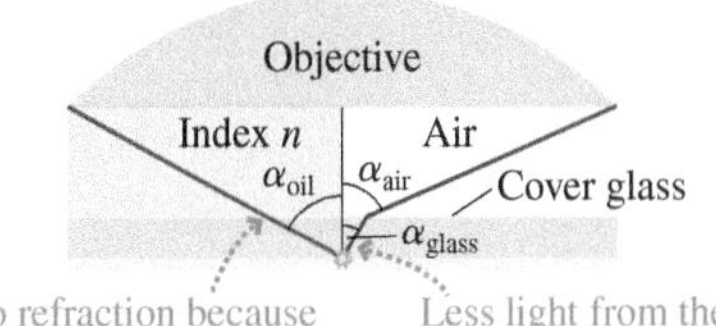

EXAMPLE 20.9 The resolving power of a microscope

A 40× objective has a diameter of 6.8 mm. The objective is designed to work in air with a tube length of 160 mm. Estimate the resolution of the lens at a typical visible-light wavelength of 550 nm.

PREPARE Resolution depends on the numerical aperture of the lens, which we need to determine. We can do that after we use the magnification to calculate the focal length.

SOLVE A microscope objective focuses at an object distance that is only slightly greater than the focal length; that is, $s \approx f_{\text{obj}}$. This approximation is shown in Figure 20.21, where it led to Equation 20.20:

$$\tan\alpha \approx \frac{D/2}{f_{\text{obj}}}$$

We found that a good approximation for the magnification of an objective is $m_{\text{obj}} \approx -L/f_{\text{obj}}$, where L is the tube length. We're given that the tube length of this microscope is 160 mm, so

Continued

$$f_{obj} \approx -\frac{L}{m_{obj}} = -\frac{160\text{ mm}}{(-40)} = 4.0\text{ mm}$$

We can use the focal length to calculate that the half-angle is

$$\alpha \approx \tan^{-1}\left(\frac{D/2}{f_{obj}}\right) = \tan^{-1}\left(\frac{3.4\text{ mm}}{4.0\text{ mm}}\right) = 40°$$

Thus the numerical aperture of the lens, with $n = 1$ for air, is

$$\text{NA} = n\sin\alpha = \sin(40.4°) = 0.65$$

Finally, we use Equation 20.22 to calculate the resolution:

$$R = \frac{0.61\lambda}{\text{NA}} = \frac{(0.61)(550\text{ nm})}{0.65} = 520\text{ nm} \approx 0.5\ \mu\text{m}$$

ASSESS The resolution is approximately the wavelength of the light, which we expected.

Digital Microscopes

Traditional microscopes used film cameras to photograph images directly through the eyepiece. Digital cameras have significantly altered the way in which microscopes are used by providing a real-time display of the image on a monitor.

FIGURE 20.24 shows a digital camera attached to a microscope in a way that allows the user to display the image or view the image through the eyepiece. (Less expensive digital microscopes provide only the camera image and don't allow direct viewing.) Unlike the cameras we use to photograph scenes around us, a microscope camera has no lens! Instead, the microscope objective acts as the camera lens, focusing a real image—the same image we view through the eyepiece—directly onto the camera's light-sensitive detector. The piece of optical equipment that makes this possible is a *beam splitter* inserted into the microscope between the objective and the eyepiece. A beam splitter is a flat piece of glass that uses specially designed thin-film coatings to transmit half the light and reflect half the light, thus "splitting" a beam of light into two components.

FIGURE 20.24 Using a digital camera with a microscope.

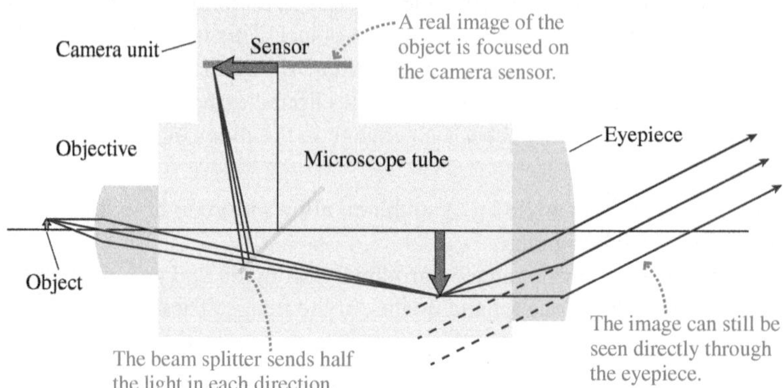
Camera unit
Sensor
A real image of the object is focused on the camera sensor.
Objective
Microscope tube
Eyepiece
Object
The beam splitter sends half the light in each direction.
The image can still be seen directly through the eyepiece.

A 10× eyepiece is perfect for visually observing the image of the objective with additional magnification. In contrast, the small pixel size of a camera's detector allows the camera to resolve the smallest details of the image without the need for any additional magnification.

EXAMPLE 20.10 Resolution of a digital camera

The digital camera attached to a microscope is a 2.8 megapixel camera that uses a square detector with 4.5 μm × 4.5 μm pixels. What are (a) the size of the camera's field of view and (b) the camera's resolution when the microscope uses a 75× objective?

PREPARE The camera's detector records the magnified real image of the objective. The detector is an $N \times N$ square array of small, light-sensitive pixels, where N is the number of pixels along each edge. The total number of pixels is N^2.

SOLVE The detector has 2.8×10^6 pixels, which is the square of the number of pixels along each edge. The pixel's width is 4.5 μm, so the edge length L of the detector is $4.5N$ μm. We can calculate that

$$N = \sqrt{2.8 \times 10^6} = 1673$$

$$L = (1673)(4.5\ \mu\text{m}) = 7530\ \mu\text{m} \approx 7.5\text{ mm}$$

That is, the physical size of the detector is 7.5 mm × 7.5 mm.

a. An image that is 7.5 mm tall or 7.5 mm wide, filling the detector, is the magnified image of an object whose height or width is $7.5/m_{obj}$ mm. The objective's magnification is 75, so an object whose height or width is 0.10 mm = 100 μm has an image the size of the detector. Or, equivalently, we say that the camera's field of view is 100 μm × 100 μm. Anything within this square is imaged on the detector.

b. Both directions on the detector—vertical or horizontal—have 1673 pixels to record the image of a 100-μm-long object. Thus each pixel records a segment of length 100/1673 μm = 0.06 μm. That is, the camera is capable of recording an image with a resolution of 0.06 μm.

ASSESS A fairly typical cell has a diameter of about 10 μm, so a 100 μm × 100 μm field of view allows the user to view specimens with a few dozen cells. Micrographs often show a few dozen cells, so our calculated field of view seems reasonable. However, the calculated pixel resolution of 0.06 μm shows that the camera is not a limiting factor. The resolution of the microscope itself is roughly an order of magnitude larger at ≈0.6 μm. Diffraction will cause light from each point on the object to spread over roughly 10 pixels in each direction. Cameras used for portraits or landscapes vie to have the most pixels, but there's no advantage to increasing the number of pixels in a microscope camera because the resolution of the objective, not the pixel size, is the ultimate limiting factor.

Fluorescence Microscopy

An important variation on traditional microscopy, especially to monitor cell physiology, is **fluorescence microscopy.** Fluorescence microscopy is based on the fact that some organic molecules, known as **fluorophores,** absorb light at one wavelength and then emit light at a longer wavelength. This process, which we'll examine more closely in Chapter 29, is called *fluorescence.*

Fluorescence was discovered in the 19th century when it was found that some mineral crystals glow with visible light when they are irradiated with shorter-wavelength ultraviolet light. It was subsequently learned that many organic molecules have similar properties. FIGURE 20.25 shows the *absorption spectrum* and the *emission spectrum* for green fluorescent protein (GFP), a protein from the jellyfish *Aequorea victoria* that is one of the most important fluorophores because it can be used to signal which genes are active. The absorption spectrum shows that the protein absorbs light over a range of wavelengths from roughly 400 nm to 500 nm (violet and blue). After absorbing short-wavelength light, the protein emits light over a range of longer wavelengths, typically 500 nm to 550 nm (green). A sample of GFP glows bright green while being irradiated with violet light.

FIGURE 20.25 Absorption and emission spectra of green fluorescent protein.

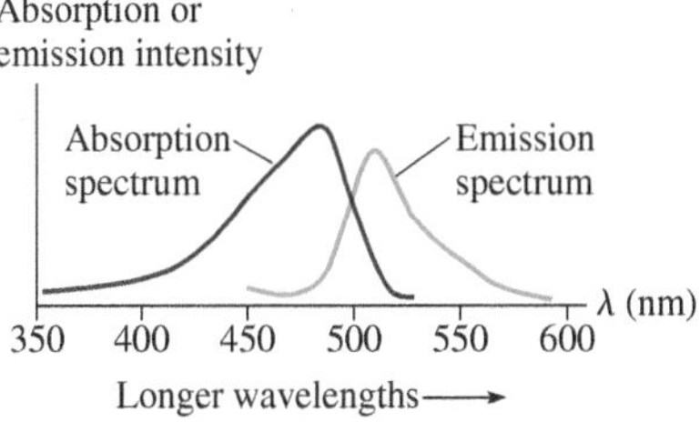

Fluorescence microscopy images only the fluorophore molecules in a sample by illuminating the sample with short-wavelength light and then, with filters, observing only the longer-wavelength fluorescence. FIGURE 20.26 illustrates how this is done. Fluorescence light is usually dim, so a very bright illumination source is needed. That is sometimes a laser, but most often the light source is a mercury lamp. A filter

FIGURE 20.26 A fluorescence microscope.

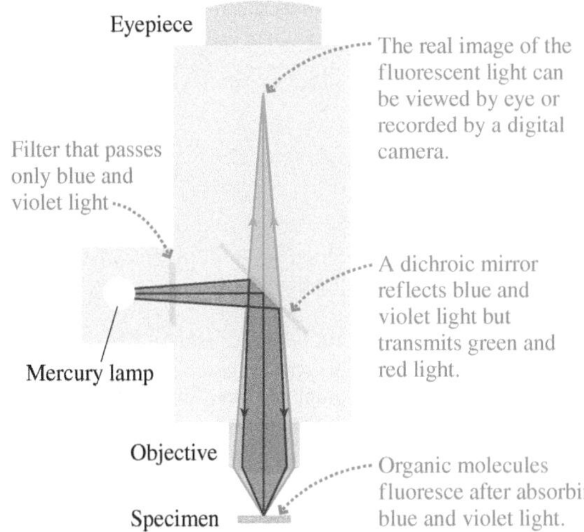

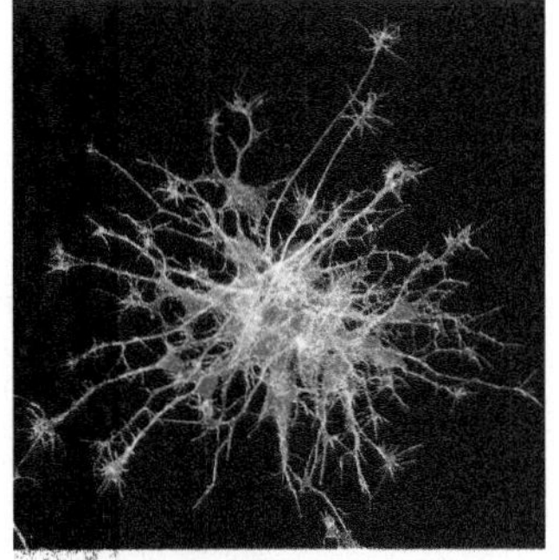

Fluorescent microscope image of neurons derived from stem cells

passes only the light with shorter wavelengths, violet and blue, to excite the fluorescence, while blocking light with longer wavelengths.

A special feature of a fluorescence microscope is a *dichroic mirror,* a piece of glass that transmits some wavelengths of light but reflects others. A dichroic mirror that transmits wavelengths longer than 500 nm but reflects shorter wavelengths is transparent to red and green wavelengths but acts as a mirror for blue wavelengths. The dichroic mirror used in a fluorescence microscope reflects short-wavelength violet and blue light but passes longer wavelengths in the green and red parts of the spectrum.

The optical design of Figure 20.26 uses a dichroic mirror to illuminate the sample from above with light focused through the objective. Fluorescent light is emitted in all directions by the sample. Some of that light is collected by the objective and directed upward through the dichroic mirror—which, as we've seen, acts as a filter to pass the longer wavelengths while reflecting the shorter wavelengths—to form an image for the viewer or a digital camera.

Some specimens are prepared with several different fluorophores that emit light of different wavelengths, so a fluorescence micrograph, such as the one in Figure 20.26, can be quite colorful. Fluorescence micrographs have a distinctive dark background where no longer-wavelength light is emitted.

Confocal Laser Scanning Microscopy

One notable disadvantage of conventional microscopy is a lack of contrast due both to most cell structures being nearly transparent and to issues with *depth of field.* Depth of field refers to the fact that only a very thin slice of the sample—the object plane—is at the proper distance from the objective to be well focused on the detector or in the image plane viewed through the eyepiece. Portions of the sample above and below the object plane are illuminated by the source light and contribute to the light collected by the objective, but they are not well focused. Thus the image has out-of-focus light from above and below the object plane superimposed on the focused portion of the image that we wish to view.

A variation of fluorescence microscopy sidesteps these difficulties and, in addition, allows for three-dimensional imaging. FIGURE 20.27 shows a microscope in which a blue or ultraviolet laser beam reflects downward from a dichroic mirror, much like in fluorescence microscopy, and is focused through the objective onto the sample. The laser beam is focused to an extremely small and intense spot, roughly one wavelength ($\approx 0.5\ \mu m$) in diameter. Portions of the sample just above or below the focus point are illuminated by less intense light while, unlike in fluorescence microscopy, the rest of the sample is not illuminated at all.

Fluorescent light from the sample—the green rays in Figure 20.27—is collected by the objective and, because it has a longer wavelength, passes upward through the dichroic mirror. But instead of a real image being focused on a detector, a second lens focuses

FIGURE 20.27 A confocal laser scanning microscope.

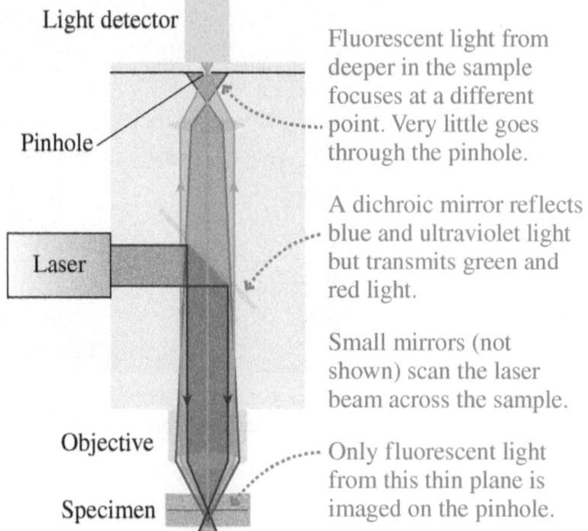

Light detector
Fluorescent light from deeper in the sample focuses at a different point. Very little goes through the pinhole.
Pinhole
A dichroic mirror reflects blue and ultraviolet light but transmits green and red light.
Laser
Small mirrors (not shown) scan the laser beam across the sample.
Objective
Only fluorescent light from this thin plane is imaged on the pinhole.
Specimen

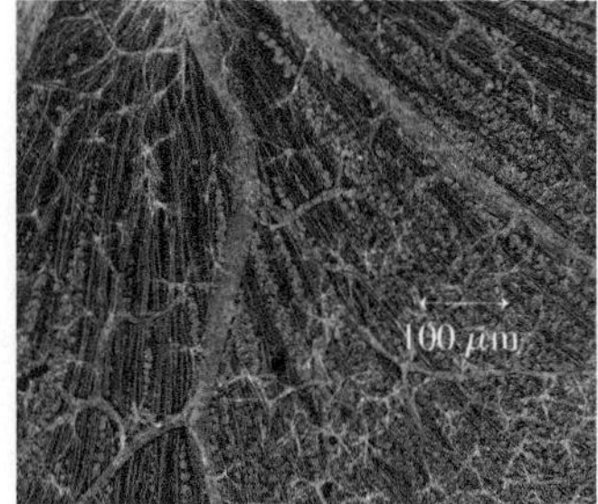

100 μm

A confocal laser scanning micrograph of blood veins and nerve cells in the retina

light emitted from the laser spot onto a *pinhole,* a tiny hole in a thin metal sheet. The fluorescent light intensity from such a small spot is weak, so the light that passes through the pinhole strikes a sensitive detector that amplifies extremely weak light signals.

The advantage of this arrangement is that **only fluorescence from the wavelength-size spot at the laser focus is detected.** Portions of the sample above and below this spot are illuminated by the laser and fluoresce, although less intensely because the illumination is less intense. But light from those portions of the sample—the red rays—is focused in front of or behind the pinhole, so very little light passes through the pinhole to reach the detector. This design has a very thin and very well-defined depth of field because it rejects light that comes from shallower or deeper layers of the sample. It is called a *confocal design* because only light from the one point of interest is focused on the pinhole,

But the advantage also appears to be a disadvantage: The microscope obtains an excellent signal from that one point in the sample, but it images *only* that one point. To deal with this, small motorized mirrors (not shown in Figure 20.27) *scan* the laser beam back and forth across the sample until the entire sample has been covered. Computer software then generates an image on the basis of the detected intensity as the laser is aimed at each point on the sample. The complete system is called a **confocal laser scanning microscope.**

The extremely well-defined depth of field of a confocal microscope—typically about 1 μm—allows the acquisition of three-dimensional images. This is done by recording an image, moving the stage up 1 μm, recording another image, moving the stage up another 1 μm, recording another image, and so on until the entire depth of the specimen has been sampled. Computer software then pieces these images together in many different ways to provide three-dimensional information.

Optical Tweezers

Optical tweezers, a relatively new tool, use beams of laser light to hold and manipulate microscopic objects that range from living cells to biological molecules. We can use Snell's law to understand how optical tweezers work.

FIGURE 20.28a shows a small transparent sphere in an intense but uniform laser beam. This might be a cell or, as is often the case, a glass or polystyrene bead about 1 μm in diameter. Light rays that pass through the sphere are refracted twice. Light rays carry not only energy but also, as we'll discuss in Chapter 27, momentum. Refraction changes the light's momentum vector, so the light exerts an impulsive force on the sphere. A full analysis finds that the forces from the two rays shown in Figure 20.28a act in the directions shown, imparting a net forward force on the sphere. The laser beam gently (the forces are very tiny even in an intense laser beam) pushes the sphere forward.

The situation differs if the laser beam is not uniform, as in FIGURE 20.28b. In this case, there's a net force in the direction of the more intense light. That is, **light forces due to refraction push the sphere toward regions of higher light intensity.**

Suppose, as in FIGURE 20.28c, that a microscope objective is used to focus a laser beam to a very small spot, roughly 1 μm in diameter. The light intensity is highest at the center of the focus; it decreases away from the center in both the radial and longitudinal directions. Consequently, light forces push a small bead into the center of the focus and hold it there. This is called an **optical trap.** One tweezers-like use of an optical trap is to hold and manipulate individual cells. A cell can be transported from one point to another by using mirrors to slowly move the point at which the laser beam is focused.

The force on a sphere in an optical trap is very small, only a few piconewtons. But it is a *restoring force*—it always points toward the center of the trap—and it increases with distance. As a result, we can model a sphere in an optical trap as a bead attached to a spring. Once the trap has been calibrated, so that its spring constant is known, it can be used to measure the very small forces exerted by biological molecules and molecular motors.

FIGURE 20.29 shows a biological molecule with one end attached to a glass slide and the other to a bead that is held in an optical trap. The optical trap is at the focal point

FIGURE 20.28 Light forces push a transparent bead toward regions of higher light intensity. The bead is trapped at the focus of a laser beam.

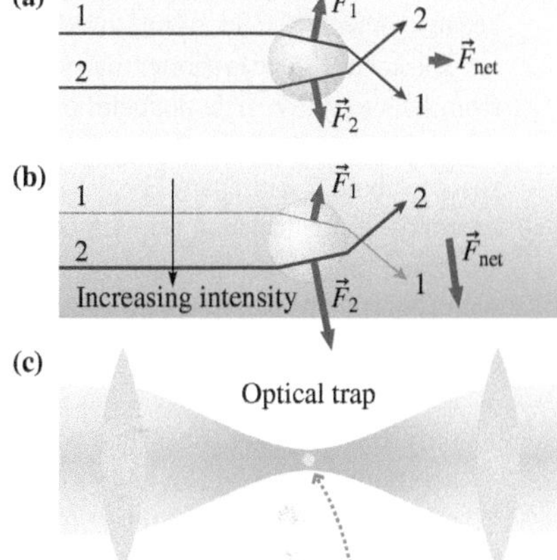
(a)
1
2
$\vec{F}_1$
$\vec{F}_2$
$\vec{F}_{net}$
(b)
Increasing intensity
(c)
Optical trap
The bead is pulled to and held at the laser focus where the intensity is maximum. Diffraction prevents the light from being focused to a perfect point.

FIGURE 20.29 Optical tweezers are used to measure forces exerted by biological molecules.

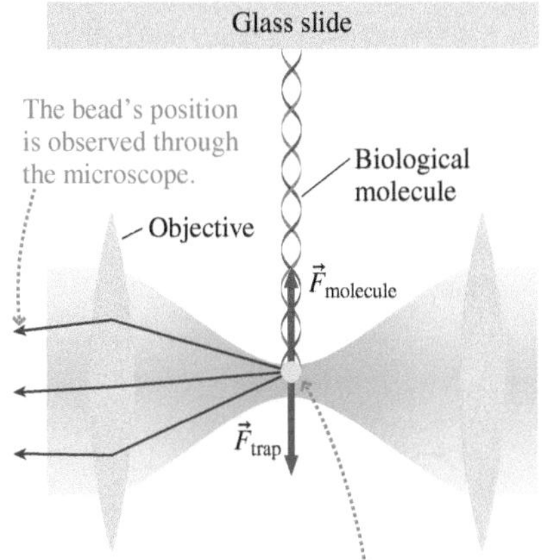
Glass slide
The bead's position is observed through the microscope.
Biological molecule
Objective
$\vec{F}_{molecule}$
$\vec{F}_{trap}$
The molecule pulls the bead to one side of the trap. The molecular force is measured by using the observed displacement of the bead and the optical trap's spring constant.

of a microscope's objective, and the microscope (not shown) is used to observe the position of the bead. If the optical trap is moved sideways, which is done with mirrors, the molecule tugs on the bead and pulls it away from the center of the trap. The measured displacement of the bead and the trap spring constant are then used to determine the strength of the molecular force. As examples, optical tweezers are used to measure the force needed to stretch DNA and the force exerted by a molecular motor.

CHAPTER 20 INTEGRATED EXAMPLE Determining the visual acuity of a kestrel

Like most birds of prey, the American kestrel has excellent eyesight. The smallest angular separation between two objects that an eye can resolve is called its *visual acuity*. A smaller visual acuity means better eyesight because objects that are closer together can be resolved.

The kestrel's eye is shorter than yours, with a distance of 0.90 cm from lens to retina. The diameter of its pupil is about the same as yours, 3.0 mm. Its retina has photoreceptors that are more closely spaced, about 2 μm apart. For the purposes of this problem, we'll model the optical system very simply, with a single lens projecting an image onto the retina, and we'll assume that the eye is filled with fluid that has an index of refraction the same as that of water.

a. Laboratory measurements indicate that the kestrel can just resolve two small objects that have an angular separation of only 0.013°. How does this result compare with the visual acuity predicted by Rayleigh's criterion? Take the wavelength of light in air to be 550 nm.
b. What is the distance on the retina between the images of two small objects that can just be resolved? How does this distance compare to the 2 μm distance between two photoreceptors?
c. The kestrel's excellent visual acuity allows it to detect small motions. If a single object moves laterally to produce two subsequent images that have the noted angular separation, the kestrel can, in principle, detect it. Suppose the kestrel is hunting on the ground directly below its 18-m-high perch in a tree. How far would a mouse need to move for the kestrel to notice this motion?

PREPARE We are interested in the kestrel's ability to resolve two closely spaced objects. FIGURE 20.30 shows the situation. θ_{min} is the smallest angle that the kestrel can resolve, s and s' are the object and image distances, d is the distance between the two objects, and d' is the distance between the two images.

FIGURE 20.30 The kestrel's eye observing two closely spaced objects.

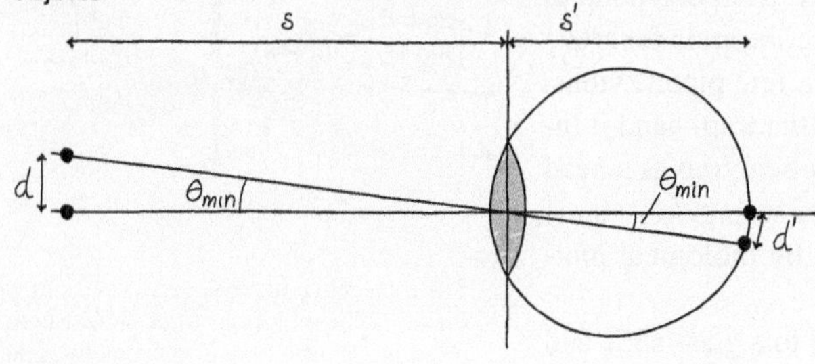

Inside the eye, the wavelength of light is shorter than its wavelength in air because of the index of refraction of the liquid within the eye. Thus for part (a) we will use $\lambda = (550\text{ nm})/1.33 = 410\text{ nm}$ in Rayleigh's criterion. For part (b), Figure 20.30 shows that distance d' on the retina is actually the arc of a circle. Thus the distance is $d' = \theta_{min}s'$, where θ_{min} is in radians. Similarly, for part (c), the distance d between the two small objects subtending an angle θ_{min} is simply $d = \theta_{min}s$.

SOLVE

a. For a perfect lens, with no aberrations, the smallest resolvable angular separation between two objects is given by Rayleigh's criterion as

$$\theta_{min} = \frac{1.22\lambda}{D} = \frac{(1.22)(410 \times 10^{-9}\text{ m})}{3.0 \times 10^{-3}\text{ m}} = 1.7 \times 10^{-4}\text{ rad}$$

Recalling that there are 360° in 2π rad, this angle is

$$\theta_{min} = (1.7 \times 10^{-4}\text{ rad}) \times \frac{360°}{2\pi\text{ rad}} = 0.0097° \simeq 0.01°$$

This is only slightly less than the experimentally observed resolution of 0.013°. Kestrels are able to resolve images at separations near the Rayleigh limit, a testament to the excellence of their vision.

b. The measured angle of 0.013° corresponds to 2.3×10^{-4} rad. Thus the distance between the images of the two small objects is

$$d' = \theta_{min}s' = (2.3 \times 10^{-4}\text{ rad})(9.0\text{ mm}) = 2.1 \times 10^{-3}\text{ mm} = 2.1\ \mu\text{m}$$

This is just about the same as the photoreceptor distance. This is about as good as it gets; the light from the two objects falls on adjacent photoreceptors. Better resolution wouldn't be possible without packing photoreceptors more tightly in the fovea.

c. The distance d is computed similarly to d', with an object distance $s = 18$ m:

$$d = \theta_{min}s = (2.3 \times 10^{-4}\text{ rad})(18\text{ m}) = 4.1\text{ mm}$$

This distance is very small—about $\frac{1}{6}$ inch—and the kestrel can, in principle, spot this motion from a perch 60 feet above the ground. You might not notice motion at the limits of your resolution, but a raptor's optical system is well adapted to detect changes in the visual field corresponding to such small movements.

ASSESS We know that raptors have excellent vision, so the fact that our calculations show that their vision works at almost the theoretical limits gives us confidence in our results.

SUMMARY

GOAL To understand how optical instruments work and how diffraction ultimately limits their performance.

IMPORTANT CONCEPTS

Lens Combinations

The image of the first lens acts as the object for the second lens.

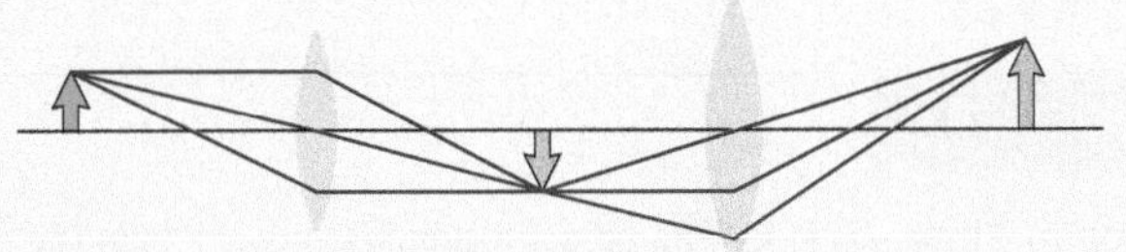

Lens **refractive power:** $P = \frac{1}{f}$ diopters, $1 \text{ D} = 1 \text{ m}^{-1}$

Resolution

The **angular resolution** of a lens of diameter D is

$$\theta_{min} = 1.22\lambda/D$$

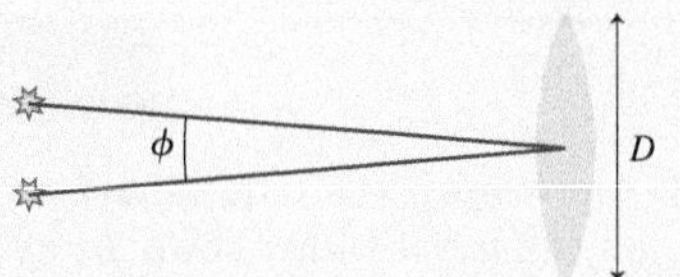

Rayleigh's criterion states that two objects separated by an angle ϕ are marginally resolvable if $\phi = \theta_{min}$.

APPLICATIONS

Cameras

A camera lens forms a real, inverted image on a detector. A lens with focal length f and aperture diameter D has **f-number**

$$\text{f-number} = \frac{f}{D}$$

The light intensity on the detector is

$$I \propto \frac{1}{(\text{f-number})^2}$$

Vision

Refraction at the cornea is responsible for most of the focusing. The lens provides fine-tuning by changing its shape (**accommodation**).

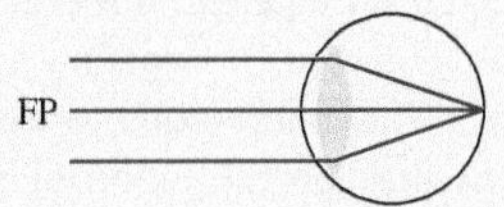

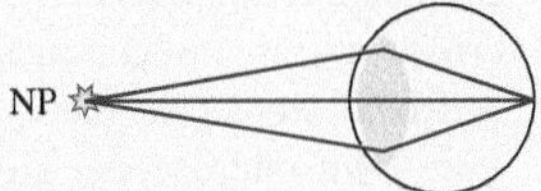

In normal vision, the eye can focus from a far point (FP) at ∞ (relaxed eye) to a near point (NP) at ≈ 25 cm (maximum accommodation).

- **Hyperopia** (farsightedness) is corrected with a converging lens.
- **Myopia** (nearsightedness) is corrected with a diverging lens.

Focusing and spatial resolution

The minimum spot size to which a lens of focal length f and diameter D can focus light is limited by diffraction to

$$w_{min} = \frac{2.44\lambda f}{D}$$

With the best lenses that can be manufactured, $w_{min} \approx \lambda$.

Magnifiers

For relaxed eye viewing, the **angular magnification** is

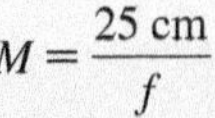

$$M = \frac{25 \text{ cm}}{f}$$

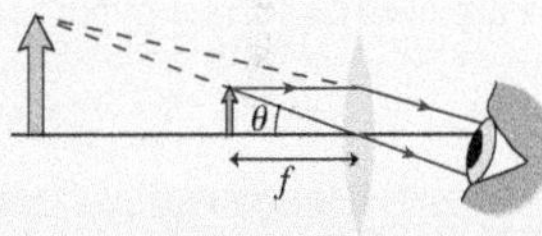

The eye views an upright, virtual image of the object. Angular magnification, not lateral magnification, determines how big the image appears.

Microscopes

The object is very close to the focal point of the objective.

The angular magnification is $M = m_{obj} M_{eye} \approx -\frac{L}{f_{obj}} \frac{25 \text{ cm}}{f_{eye}}$.

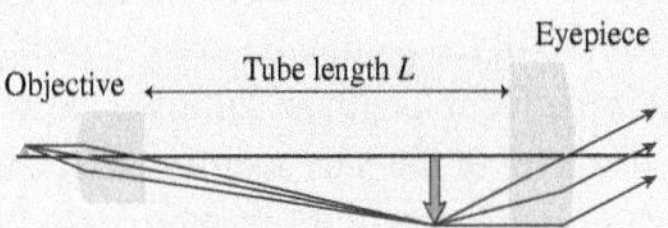

The eyepiece acts as a magnifier to view the real image formed by the objective lens.

The **resolution** of a microscope—the minimum separation of two points that can be distinguished—is

$$R = \frac{0.61\lambda}{\text{NA}}$$

The **numerical aperture** is

$$\text{NA} = n \sin\alpha$$

where α is the half-angle of the objective as seen from the object and n is the index of refraction of the medium between the object and the lens. The numerical aperture affects both the resolution and the amount of light collected by the objective.

LEARNING OBJECTIVES After studying this chapter, you should be able to:

- Determine the image position and magnification of two lenses used in combination. *Problems 20.1–20.4, 20.15*
- Explain how a camera works. *Conceptual Questions 20.1, 20.3; Problems 20.5, 20.6, 20.9, 20.11, 20.12*
- Analyze the optics of the human eye and specify appropriate corrective lenses. *Conceptual Questions 20.5, 20.6; Problems 20.16, 20.17, 20.22–20.24*
- Calculate the angular magnification of a magnifying lens and a microscope. *Conceptual Questions 20.8, 20.9; Problems 20.30, 20.33, 20.34, 20.36, 20.37*
- Calculate the resolution limits of an optical system. *Conceptual Questions 20.10, 20.12; Problems 20.41–20.45*
- Analyze microscopes of different designs. *Problems 20.46–20.49, 20.51*

STOP TO THINK ANSWERS

Stop to Think 20.1: B. A diverging lens refracts rays away from the optical axis, so the rays will travel farther down the axis before converging.

Stop to Think 20.2: A. Because the shutter speed doesn't change, the f-number must remain unchanged. The f-number is f/D, so increasing f requires increasing D.

Stop to Think 20.3: A. A magnifier is a converging lens. Converging lenses are used to correct hyperopia.

Stop to Think 20.4: E. The object distance is greater than the focal length of the lens, so the lens forms a real image of the stamen tip on the same side of the lens as your eye. There are no diverging rays coming from anywhere for your eye to collect and focus on.

Stop to Think 20.5: D. The objective forms a real image, but the image of the eyepiece—a magnifier—is a virtual image at infinity that a relaxed eye can view.

Stop to Think 20.6: $w_1 > w_4 > w_2 = w_3$. The spot size is proportional to f/D.

QUESTIONS

Conceptual Questions

1. A photographer focuses his camera on his subject. The subject then moves closer to the camera. To refocus, should the lens be moved closer to or farther from the detector? Explain.
2. A photographer realizes that with the lens she is currently using, she can't fit the entire landscape she is trying to photograph into her picture. Should she switch to a lens with a longer or shorter focal length? Explain.
3. Suppose a camera's exposure is correct when the lens has a focal length of 8.0 mm. Will the picture be overexposed, underexposed, or still correct if the focal length is "zoomed" to 16.0 mm without changing the diameter of the lens aperture? Explain.
4. A camera has a circular aperture immediately behind the lens. Reducing the aperture diameter to half its initial value will
   A. Make the image blurry.
   B. Cut off the outer half of the image and leave the inner half unchanged.
   C. Make the image less bright.
   D. All the above.
   Explain your choice.
5. Suppose you wanted special glasses designed to let you see underwater without a face mask. Should the glasses use a converging or diverging lens? Explain.
6. Everyone has a *blind spot* in each eye where the optic nerve exits the eye and there are no light-sensitive cells. To locate the blind spot of your right eye, focus on the X in Figure Q20.6 from a distance of approximately 30 cm (1 ft), place your index finger on the X, and close your left eye. Then slowly move your finger first to the right, then to the left, while following it with your right eye. You will see the X in your peripheral vision, except at a certain point the X will disappear and then reappear. Is your right eye's blind spot on the right or left side of the retina? Explain.

FIGURE Q20.6

7. When you swim underwater, your vision is compromised, as we've seen. The same is true of water-dwelling animals that venture up into the air.
   a. When you open your eyes underwater, are you nearsighted or farsighted?
   b. When a dolphin pokes its head out of the water, is it nearsighted or farsighted?
8. A student makes a microscope using an objective lens and an eyepiece. If she moves the lenses closer together, does the microscope's magnification increase or decrease? Explain.
9. A friend lends you the eyepiece of his microscope to use on your own microscope. He claims that the resolution of your microscope will be halved, allowing you to see more detail, because his eyepiece has twice the diameter and twice the magnification as yours. Is his claim valid? Explain.
10. A diffraction-limited lens can focus light to a 2-μm-diameter spot on a screen. Do each of the following actions make the spot diameter larger, make it smaller, or leave it unchanged?
    a. Decreasing the wavelength of the light.
    b. Decreasing the lens diameter.
    c. Decreasing the lens focal length.
    d. Decreasing the lens-to-screen distance.
11. To focus parallel light rays to the smallest possible spot, should you use a lens with a small f-number or a large f-number? Explain.
12. An astronomer is trying to observe two distant stars. The stars are marginally resolved when she looks at them through a filter that passes green light with a wavelength near 550 nm. Which of the following actions would improve the resolution? Assume that the resolution is not limited by the atmosphere.
    A. Changing the filter to a different wavelength. If so, should she use a shorter or a longer wavelength?
    B. Using a telescope with an objective lens of the same diameter but a different focal length. If so, should she select a shorter or a longer focal length?

C. Using a telescope with an objective lens of the same focal length but a different diameter. If so, should she select a larger or a smaller diameter?

D. Using an eyepiece with a different magnification. If so, should she select an eyepiece with more or less magnification?

Multiple-Choice Questions

13. | You are using a 50-mm-focal-length lens to photograph a tree. If you change to a 100-mm-focal-length lens and refocus, the image height on the detector changes by a factor of
A. $\frac{1}{4}$ B. $\frac{1}{2}$ C. 2 D. 4
E. The image height does not change.

14. | A camera takes a properly exposed photo when the f-number is F8.0 and the shutter speed is 1/125 s. What f-number should be used if the shutter speed is changed to 1/500 s to freeze the motion of a sprinter?
A. F4.0 B. F5.6 C. F11 D. F16
E. The f-number does not need to be changed.

15. | A nearsighted person has a near point of 20 cm and a far point of 40 cm. When he is wearing glasses to correct his distant vision, what is his near point?
BIO
A. 10 cm B. 20 cm C. 40 cm D. 1.0 m

16. | A nearsighted person has a near point of 20 cm and a far point of 40 cm. What refractive power lens is necessary to correct this person's vision to allow her to see distant objects?
BIO
A. −5.0 D B. −2.5 D C. +2.5 D D. +5.0 D

17. | A 60-year-old man has a near point of 100 cm. What refractive power reading glasses would he need to focus on a newspaper held at a comfortable distance of 40 cm?
BIO
A. −2.5 D B. −1.5 D C. +1.5 D D. +2.5 D

18. || A cataract is a clouding or opacity that develops in the eye's lens. In extreme cases, the lens of the eye may need to be removed. This would have the effect of leaving a person
BIO
A. Nearsighted.
B. Farsighted.
C. Neither nearsighted nor farsighted.

19. | What is the refractive power of a lens that has a 5× angular magnification?
A. 5 D B. 10 D C. 20 D D. 50 D

20. | A microscope has a tube length of 20 cm. What combination of objective and eyepiece focal lengths will give an overall magnification of approximately 100?
A. 15 mm, 3 cm B. 10 mm, 4 cm
C. 10 mm, 5 cm D. 30 mm, 8 cm

21. | All the photographs taken by a camera show objects with a slightly red edge. This is because the camera lens has
A. Astigmatism B. Distortion
C. Spherical aberration D. Chromatic aberration

22. || An oil-immersion microscope objective has NA = 1.00 when used with oil that has index of refraction 1.50. The diameter of the objective lens is 3.0 mm. About how far is the specimen from the objective lens?
A. 1.2 mm B. 1.5 mm C. 1.7 mm D. 2.0 mm

PROBLEMS

Section 20.1 Lenses in Combination

1. || Two converging lenses with focal lengths of 40 cm and 20 cm are 10 cm apart. A 2.0-cm-tall object is 15 cm in front of the 40-cm-focal-length lens.
   a. Use ray tracing to find the position and height of the final image. Do this accurately using a ruler or paper with a grid, then make measurements on your diagram.
   b. Calculate the image position (relative to the second lens) and its height. Compare with your ray-tracing answers in part a.
2. || A converging lens with a focal length of 40 cm and a diverging lens with a focal length of −40 cm are 160 cm apart. A 2.0-cm-tall object is 60 cm in front of the converging lens.
   a. Use ray tracing to find the position and height of the final image. Do this accurately using a ruler or paper with a grid, then make measurements on your diagram.
   b. Calculate the image position (relative to the second lens) and its height. Compare with your ray-tracing answers in part a.
3. || A 2.0-cm-tall object is 20 cm to the left of a lens with a focal length of 10 cm. A second lens with a focal length of 15 cm is 30 cm to the right of the first lens.
   a. Use ray tracing to find the position and height of the final image. Do this accurately using a ruler or paper with a grid, then make measurements on your diagram.
   b. Calculate the image position (relative to the second lens) and its height. Compare with your ray-tracing answers in part a.
4. || A 2.0-cm-tall object is 20 cm to the left of a lens with a focal length of 10 cm. A second lens with a focal length of 5 cm is 30 cm to the right of the first lens.
   a. Use ray tracing to find the position and height of the final image. Do this accurately using a ruler or paper with a grid, then make measurements on your diagram.
   b. Calculate the image position (relative to the second lens) and its height. Compare with your ray-tracing answers in part a.

Section 20.2 The Camera

5. || A 2.0-m-tall man is 10 m in front of a camera that has a 15-mm-focal-length lens. How tall is his image on the detector?
6. || A photographer uses his camera, whose lens has a 50 mm focal length, to focus on an object 2.0 m away. He then wants to take a picture of an object that is 40 cm away. How far, and in which direction, must the lens move to focus on this second object?
7. || Turning the barrel of a 50-mm-focal-length lens on a manual-focus camera moves the lens closer to or farther from the sensor to focus on objects at different distances. The lens has a stated range of focus from 0.45 m to infinity. How far does the lens move between these two extremes?

Problem difficulty is labeled as | (straightforward) to ||||| (challenging). Problems labeled INT integrate significant material from earlier chapters; BIO are of biological or medical interest; CALC require calculus to solve.

8. || An older camera has a lens with a focal length of 50 mm and uses 36-mm-wide film. Using this camera, a photographer takes a picture of the Golden Gate Bridge that completely spans the width of the film. Now she wants to take a picture of the bridge using her digital camera with its 12-mm-wide detector. What focal length should this camera's lens have for the image of the bridge to cover the entire detector?
9. || A camera takes a properly exposed photo with a 3.0-mm-diameter aperture and a shutter speed of 1/125 s. What is the appropriate aperture diameter for a 1/500 s shutter speed?
10. || What is the f-number of a lens with a 35 mm focal length and a 7.0-mm-diameter aperture?
11. || What is the aperture diameter of a 12-mm-focal-length lens set to F4.0?
12. || A camera takes a properly exposed photo at F5.6 and 1/125 s. What shutter speed should be used if the lens is changed to F4.0?

Section 20.3 The Human Eye

13. | A lens with $f = +15$ cm is in contact with a lens with $f = -20$ cm. What is the effective focal length of the combination?
14. | Two converging lenses with focal lengths of 20 cm and 24 cm are combined to make a single lens. What is the effective focal length of the combination?
15. || A +2.0 D lens is being used to make an image of a distant object on a screen. The image is 2.4 cm tall. A second +2.0 D lens is added to the first, and the lens combination is moved to refocus the image. How tall is the new image?
16. || Ramon has contact lenses with the prescription +2.0 D.
BIO a. What eye condition does Ramon have?
b. What is his near point without the lenses?
17. || Ellen wears eyeglasses with the prescription −1.0 D.
BIO a. What eye condition does Ellen have?
b. What is her far point without the glasses?
18. || What is the f-number of a relaxed eye with the pupil fully di-
BIO lated to 8.0 mm? Model the eye as a single lens 1.7 cm in front of the retina.
19. | If the retina is 17 mm from the lens in the eye, how large is
BIO the image on the retina of a person of height 1.5 m standing 8.0 m away?
20. || At a distance of 6 meters, a person with normal vision is able
BIO to clearly read letters 1.0 cm high. Approximately how tall are the images of the letters on the retina? Assume that the retina is 17 mm from the lens.
21. || The rod and cone cells in the central part of the retina—the
BIO fovea—are packed closer together, giving a more detailed view. This area of increased rod and cone density has a diameter of about 1.5 mm. When you read a book, you want the image of the text you are reading to fall on the fovea. If you hold a book 30 cm from your eyes, how wide is the spot on the page whose image just fills the fovea? Assume that the retina is 17 mm from the lens.
22. || A farsighted person has a near point of 50 cm. What strength
BIO lens, in diopters, is needed to bring his near point to 25 cm?
23. || Rachel has good distant vision but has a touch of presby-
BIO opia. Her near point is 0.60 m. When she wears +2.0 D reading glasses, what is her near point? Her far point?
24. | A nearsighted woman has a far point of 300 cm. What kind
BIO of lens, converging or diverging, should be prescribed for her to see distant objects more clearly? What refractive power should the lens have?
25. || A nearsighted person has a near point of 12 cm and a far
BIO point of 40 cm. What power corrective lens is needed for her to have clear distant vision? With this corrective lens in place, what is her new near point?
26. ||| Martin has severe myopia, with a far point of only 17 cm. He
BIO wants to get glasses that he'll wear while using his computer, whose screen is 65 cm away. What refractive power is needed for Martin to view the screen with his eyes relaxed?
27. ||| Mary's glasses have +4.0 D converging lenses. This gives
BIO her a near point of 20 cm. What is the location of her near point when she is not wearing her glasses?
28. || Rank the following people from the most nearsighted to the
BIO most farsighted:
- Kareem has a prescription of +2.0 D.
- Carol needs diverging lenses with a focal length of −0.33 m.
- Maria wears converging lenses with a focal length of 1.00 m.
- Janet has a prescription of +2.5 D.
- Warren's prescription is −3.2 D.

29. ||| With −5.0 D corrective lenses, Juliana's distant vision is
BIO quite sharp. She has a pair of −3.5 D computer glasses that puts her computer screen right at her far point. How far away is her computer screen?

Section 20.4 Magnifiers and Microscopes

30. || The diameter of a penny is 19 mm. A full moon subtends an angle of 0.50° in the sky. How far from your eye must a penny be held so that it has the same apparent size as the moon?
31. || A jeweler is wearing a 20 D magnifying lens directly in front of his eye. If his near point is a typical 25 cm, how close can he hold a gem that he is inspecting?
32. || Oliver has had a stamp collection since he was a boy. In those days, holding a stamp 10 cm from his eye gave him a clear image. Now, his near point has receded to 90 cm, so he holds a magnifying lens directly in front of his eye to let him bring stamps closer. To the nearest diopter, what power lens enables him to focus on a stamp 10 cm away?
33. ||| A magnifier has a magnification of 5×. How far from the lens should an object be held so that its image is seen at the near-point distance of 25 cm? Assume that your eye is immediately behind the lens.
34. ||| Anna holds a 3× magnifier directly in front of her eye to get a close look at a 19-mm-diameter penny. What is the closest possible distance that she can hold the coin to have it appear in focus? At this distance, how large is the image of the coin on her retina? (Assume a typical 25 cm near point and a distance of 17 mm between the lens and the retina.)
35. || Chromatic aberration can be significantly reduced by using an *achromatic doublet,* a converging and a diverging lens combination with the lenses in contact. You need to make a 10× achromatic eyepiece in which the converging lens has a focal length of 2.0 cm. What focal length diverging lens should you use?

36. || You are using a microscope with a 10× eyepiece. What is the approximate focal length of an objective lens will give a total magnification of 200×? Assume a tube length of 160 mm.

37. || A microscope has a 20 cm tube length. Approximately what focal-length objective will give total magnification 500× when used with an eyepiece having a focal length of 5.0 cm?

38. || A biological microscope with a 160 mm tube length has an 8.0-mm-focal-length objective. Approximately what focal-length eyepiece should be used to achieve a total magnification of 100×?

39. || A 6.0-mm-diameter microscope objective has a focal length of 9.0 mm. What object distance gives a lateral magnification of −40?

40. || A forensic scientist examines a hair with a microscope that has a 15× objective and a 5× eyepiece. The magnified hair has the same apparent size as a 2.0-cm-wide ribbon seen from a distance of 1.0 m. What is the diameter of the hair?

Section 20.5 The Resolution of Optical Instruments

41. || A scientist needs to focus a helium-neon laser beam ($\lambda = 633$ nm) to a 10-μm-diameter spot 8.0 cm behind a lens.
   a. What focal-length lens should she use?
   b. What minimum diameter must the lens have?

42. || Two lightbulbs are 1.0 m apart. From what distance can these lightbulbs be marginally resolved by a 4.0-cm-diameter objective lens? Assume that the lens is diffraction limited and $\lambda = 600$ nm.

43. ||| Once dark adapted, the pupil of your eye is approximately
BIO 7 mm in diameter. The headlights of an oncoming car are 120 cm apart. If the lens of your eye is diffraction limited, at what distance are the two headlights marginally resolved? Assume a wavelength of 600 nm and that the index of refraction inside the eye is 1.33. (Your eye is not really good enough to resolve headlights at this distance, due both to aberrations in the lens and to the size of the receptors in your retina, but it comes reasonably close.)

44. || The Hubble Space Telescope is a diffraction-limited reflecting telescope with a 2.4-m-diameter mirror. The angular resolution of a mirror is the same as the resolution of a lens of the same diameter because both are limited by circular-aperture diffraction.
   a. What is the angular resolution of the Hubble Space Telescope? Use 600 nm for the wavelength of light.
   b. What is the distance in meters between two objects on the moon, 380,000 km away, that can be marginally resolved by the Hubble Space Telescope?

45. || Camera makers sometimes use the pixel count as an indicator of quality, and each year new cameras appear with more pixels. But having more pixels improves picture quality only if diffraction is not a limiting factor. If the minimum spot size of the camera lens—assuming the lens is otherwise perfect—is larger than a pixel, adding more pixels will not improve the image. Diffraction is an issue in smartphone cameras because of the small lenses. One smartphone camera has a 2.2-mm-diameter lens with a focal length of 4.8 mm. The detector has 21 MP (megapixels) with a distance of 1.2 μm between pixels.
   a. What is the minimum spot size of the lens? Use 600 nm for the wavelength of light.
   b. Would increasing the number of pixels in the detector improve the image quality?

Section 20.6 Microscopy

46. | For technical reasons, the maximum half-angle of a microscope objective lens is about 75°. For this half-angle, what are the numerical apertures of (a) an objective lens that is used in air and (b) an oil-immersion objective that uses oil with an index of refraction of 1.50?

47. || A 50× oil-immersion objective is marked NA 1.2. It uses oil that has an index of refraction of 1.50. What is the lens diameter?

48. || A 1.0-cm-diameter microscope objective has a focal length of 2.8 mm. It is used with light of wavelength of 550 nm.
   a. What is the objective's resolution if used in air?
   b. What is the resolution of the objective if it is used in an oil-immersion microscope with $n_{oil} = 1.45$?

49. ||| A microscope with an objective of focal length 1.6 mm is used to inspect the tiny features of a computer chip. It is desired to resolve two objects only 400 nm apart. What diameter objective is needed if the microscope is used in air with light of wavelength 550 nm?

50. || The magnification of a digital microscope is $m_{total} = m_{obj}m_{digital}$, where m_{obj} is the magnification of the objective lens and $m_{digital}$ is the *digital magnification.* If the sensor in the digital camera has width L_{sensor} and if the image is displayed on a monitor that has width $L_{monitor}$, then $m_{digital} = L_{monitor}/L_{sensor}$. This is the factor by which the image is scaled up in going from the sensor to the display. The sensor used in many microscope cameras has 1920 × 1080 pixels, with each pixel being a 3.0 μm × 3.0 μm square. What is the magnification if the user selects a 50× objective and displays the image on a 30-cm-wide monitor?

51. | Figure P20.51 shows idealized absorption and emission spectra of a fluorophore. To use this fluorophore to obtain images with a fluorescence microscope, what range of wavelengths should be passed by (a) the filter in front of the mercury lamp and (b) the dichroic mirror?

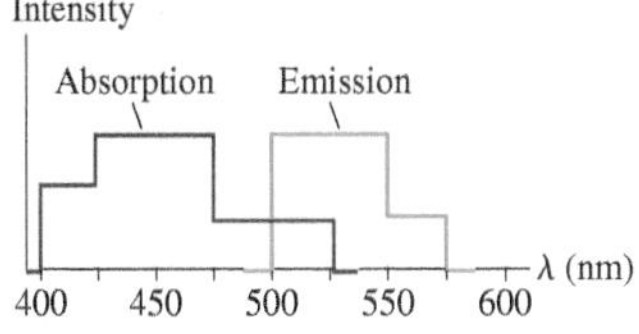

FIGURE P20.51

General Problems

52. ||| A 1.0-cm-tall object is 110 cm from a screen. A diverging lens with focal length −20 cm is 20 cm in front of the object. What are the focal length and distance from the screen of a second lens that will produce a well-focused, 2.0-cm-tall image on the screen?

53. ||| A 15-cm-focal-length converging lens is 20 cm to the right of a 7.0-cm-focal-length converging lens. A 1.0-cm-tall object is distance L to the left of the first lens.
   a. For what value of L is the final image of this two-lens system halfway between the two lenses?
   b. What are the height and orientation of the final image?

54. || The rays leaving the two-component optical system of Figure P20.54 (next page) produce two distinct images of the 1.0-cm-tall object. What are the position (relative to the lens), orientation, and height of each image?

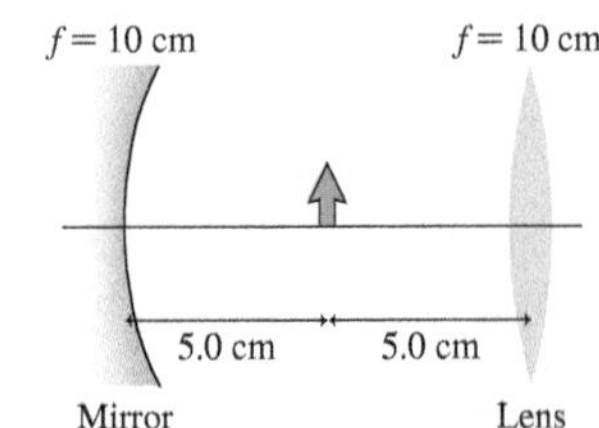

FIGURE P20.54

55. ||| Rays that leave the object in Figure P20.55 traveling to the right eventually leave the system traveling to the left. What are the position, height, and orientation of the final image? Give the position as a distance to the right or left of the lens.

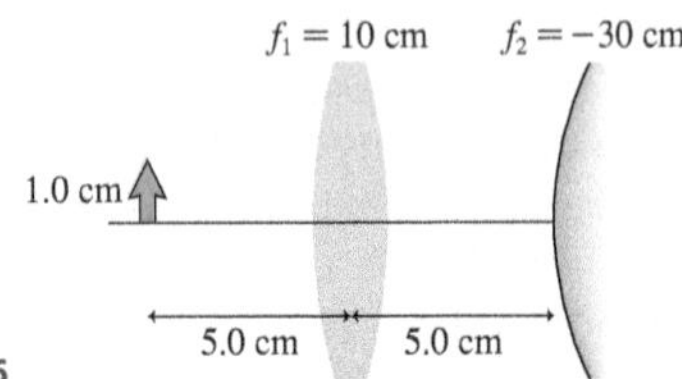

FIGURE P20.55

56. || Yang can focus on objects 150 cm away with a relaxed eye. BIO With full accommodation, she can focus on objects 20 cm away. After her eyesight is corrected for distance vision, what will her near point be while wearing her glasses?

57. || Susan is quite nearsighted; without her glasses, her far point BIO is 35 cm and her near point is 15 cm. Her glasses allow her to view distant objects with her eye relaxed. With her glasses on, what is the closest object on which she can focus?

58. |||| Frank is nearsighted and his glasses require a prescription BIO of −1.5 D. One day he can't find his glasses, but he does find an older pair with a prescription of −1.0 D. What is the most distant object that Frank can focus on while wearing this older pair of glasses?

59. || You need two eyes for depth BIO perception—to determine the distance to an object that you are looking at—but a chameleon needs only one. This is handy because each eye can track independently. A chameleon adjusts the lens to bring an object into focus and uses the required degree of accommodation to determine distance, which it does very precisely for catching insects with its long tongue. This works for the chameleon because its eye focuses at very close distances, but it wouldn't work for you. Let's see why.

a. Suppose a chameleon focuses on an insect 30 cm away and then shifts its focus to a different insect only 3.0 cm away, roughly at its near point. A single-lens model of a chameleon eye places the retina 1.3 cm behind the lens. What is the chameleon eye's change in accommodation—that is, the change in its eye's total refractive power?
b. Suppose you focus on a piece of pie 300 cm away and then shift your focus to a piece of food on your fork only 30 cm away, roughly at your near point. This is a factor-of-10 change in distance, the same as for the chameleon. A single-lens model of the human eye places the retina 1.7 cm behind the lens. What is your eye's change in accommodation—that is, the change in your eye's total refractive power?

60. ||| Your *visual acuity* is expressed by saying that you have 20/20 BIO vision, which is considered normal, or perhaps something like 20/40. Acuity of 20/40 means that you can barely read the text at a distance of 20 ft that a person with normal 20/20 vision can read at a distance of 40 ft. Or, equivalently, you need to be 10 ft from an eye chart to read the letters that a person with 20/20 vision can read from 20 ft. We've noted that visual acuity is limited by aberrations for pupil diameters larger than about 3 mm, but visual acuity becomes diffraction limited for smaller diameters. Let's see how this applies to reading an eye chart.
a. If your eye is diffraction limited, what is its angular resolution θ_{min} in degrees at a bright-light diameter of 3.0 mm? Use 650 nm as the wavelength of light.
b. The O on the 20/20 line of a standard eye chart is 8.8 mm tall and the central white area is 5.2 mm tall. To recognize it as an O, you need to distinguish the top line from the bottom line. What is the angle θ_O between the bottom of the top line and the top of the bottom line when the O is viewed from a distance of 20 ft?
c. What is the ratio θ_O/θ_{min}?
Your answer to part c shows that the O can be resolved by an ideal eye but not by a large factor. Even minor aberrations, which exist in a 3-mm-diameter pupil, reduce acuity in an otherwise normal eye. In addition, an imperfect response of the retina's cones and of the brain's perceptual system prevents your actual visual acuity from being quite as good as a theoretical calculation.

61. || A 10× microscope objective designed to work in air has a *working distance* (the distance from the lens to the object) of 7.0 mm and a numerical aperture of 0.25. What is the diameter of the lens?

62. ||| Modern microscopes are more likely to use a camera than human viewing. One way to do so—different from the method described in Section 20.6—is to replace the eyepiece with a *photo-ocular* that focuses the image of the objective to a real image on the sensor of a digital camera. A typical sensor is 22.5 mm wide and consists of 5625 4.0-μm-wide pixels. Suppose a microscopist pairs a 40× objective with a 2.5× photo-ocular.
a. What is the field of view? That is, what width on the microscope stage, in mm, fills the sensor?
b. The photo of a cell is 120 pixels in diameter. What is the cell's actual diameter, in μm?

63. |||| Your task in physics laboratory is to make a microscope from two lenses. One lens has a focal length of 2.0 cm, the other 1.0 cm. You plan to use the more powerful lens as the objective, and you want the eyepiece to be 16 cm from the objective.
a. For viewing with a relaxed eye, how far should the sample be from the objective lens?
b. What is the magnification of your microscope?

64. || Lasers are used in surgery to cut tissue and cauterize blood BIO INT vessels. One commonly used laser is an infrared laser that emits light with a wavelength of 1.38 μm. The laser beam is delivered through an optical fiber and focused by a 1.0-mm-diameter lens at the end of the fiber. The focal length of the lens is 3.0 cm.
a. What is the spot size of the focused laser beam if the lens is diffraction limited?
b. Cutting tissue requires a light intensity of 50 kW/cm^2 at the focal point. What is the necessary laser power?

65. ||| The resolution of a digital camera is limited by two factors: INT diffraction by the lens, a limit of any optical system, and the fact that the sensor is divided into discrete pixels. Consider a typical point-and-shoot camera that has a 20-mm-focal-length lens and a sensor with 2.5-μm-wide pixels.
   a. First, assume an ideal, diffractionless lens. At a distance of 100 m, what is the smallest distance, in cm, between two point sources of light that the camera can barely resolve? In answering this question, consider what has to happen on the sensor to show two image points rather than one. You can use $s' = f$ because $s \gg f$.
   b. You can achieve the pixel-limited resolution of part a only if the diffraction width of each image point is no greater than 1 pixel in diameter. For what lens diameter is the minimum spot size equal to the width of a pixel? Use 600 nm for the wavelength of light.
   c. What is the f-number of the lens for the diameter you found in part b? Your answer is a quite realistic value of the f-number at which a camera transitions from being pixel limited to being diffraction limited. For f-numbers smaller than this (larger-diameter apertures), the resolution is limited by the pixel size and does not change as you change the aperture. For f-numbers larger than this (smaller-diameter apertures), the resolution is limited by diffraction, and it gets worse as you "stop down" to smaller apertures.

66. || The Hubble Space Telescope is a diffraction-limited reflecting telescope with a 2.4-m-diameter mirror. The angular resolution of a mirror is the same as the resolution of a lens of the same diameter because both are limited by circular-aperture diffraction. Suppose the telescope is used to photograph stars at the center of our galaxy, 30,000 light years away, using red light with a wavelength of 650 nm. A light year is the distance that light travels in 1 year at the speed of light.
   a. What is the distance in km between two stars that are marginally resolved?
   b. For comparison, what is this distance as a multiple of the distance of the earth from the sun, 1.5×10^8 km?

67. ||| It's important when using a digital microscope to match the objective lens to the size and number of pixels of the camera sensor. Having too much magnification or too many pixels smears without improving resolution; having too little magnification or too few pixels means that information is lost. The optimum objective is one for which the lens resolution—the spacing between two marginally resolved objects—is magnified to match the sensor resolution—the minimum pixel spacing between two marginally resolved signals. Mathematically, the optimum objective is one for which $R_{sensor} = m_{obj}R_{obj}$. An ideal sensor would be able to resolve two signals that are two pixels apart, allowing for one "off" pixel between two "on" pixels. In practice, the way that sensors are designed to deal with color means that two signals have to be four pixels apart to be resolved. The sensor used on many microscope cameras has 1920×1080 pixels, with each pixel being a $3.0\ \mu\text{m} \times 3.0\ \mu\text{m}$ square.
   a. Which objective is better matched to this sensor: a 10× objective with NA = 0.30 or a 40× objective with NA = 0.75? Use 600 nm for the wavelength of light.
   b. For the better objective, what is the image size on the sensor and on a 30-cm-wide monitor of a 15-μm-diameter cell?

68. ||| How big should the pinhole be in a confocal laser scanning microscope? Light diffracts as it passes through the objective lens, and a perfect point source creates a circular-aperture diffraction pattern in the objective's image plane with width $w = 2m_{obj}R$, where R is the resolution of the objective. (The factor of 2 appears because, as Figure 20.19 showed, two objects are marginally resolved when the spacing between their images is half the width of the diffraction pattern.) A confocal laser scanning microscope uses a second lens to focus the light, but the objective lens determines the resolution and the plane of the pinhole is the objective's image plane. The pinhole diameter should be chosen to match the diffraction spot size. A smaller pinhole sacrifices light and produces a weaker signal; a larger pinhole accepts light from more than one point in the specimen and degrades the resolution. What pinhole diameter, rounded to the nearest 10 μm, should be used with (a) a 10× objective with NA = 0.30 and (b) a 100× objective with NA = 1.40? Use 660 nm as the wavelength of the fluorescent light.

MCAT-Style Passage Problems

Surgical Vision Correction

The optics of your visual system have a total refractive power of about +60 D—about +20 D from the lens in your eye and +40 D from the curved shape of your cornea. Surgical procedures to correct vision generally do not work on the lens; they work to reshape the cornea. In the most common procedure, a laser is used to remove tissue from the center of the cornea, reducing its curvature. This change in shape can correct certain kinds of vision problems.

69. Flattening the cornea would be a good solution for someone who was
   A. Nearsighted.
   B. Farsighted.
   C. Either nearsighted or farsighted.

70. Suppose a woman has a far point of 50 cm. How much should the refractive power of her cornea be changed to correct her vision?
   A. −2.0 D
   B. −1.0 D
   C. +1.0 D
   D. +2.0 D

71. The length of your eye decreases slightly as you age, making the lens a bit closer to the retina. Suppose a man had his vision surgically corrected at age 30. At age 70, once his eyes had decreased slightly in length, he would be
   A. Nearsighted.
   B. Farsighted
   C. Neither nearsighted nor farsighted.

PART IV SUMMARY

Optics

FUNDAMENTAL CONCEPTS Two important models of light are the wave model and the ray model:

Light waves
- Are electromagnetic waves.
- Spread out after passing through openings.
- Interfere.

Light rays
- Travel in straight lines.
- Do not interact or interfere.
- Can form images.

GENERAL PRINCIPLES

Principle of superposition Constructive interference occurs where wave crests overlap.

Law of reflection $\theta_r = \theta_i$

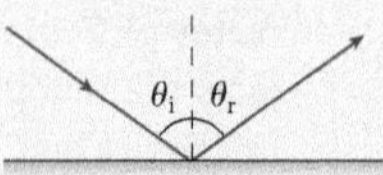

Law of refraction (Snell's law)
$n_1 \sin\theta_1 = n_2 \sin\theta_2$

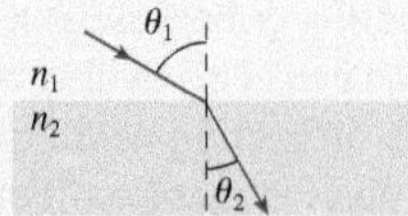

Rayleigh's criterion Two objects can be resolved by a lens of diameter D if their angular separation exceeds $1.22\lambda/D$.

WAVE MODEL

Light is an electromagnetic wave.
- Light travels through vacuum at speed c.
- Wavelength and frequency are related by $\lambda f = c$.

Light exhibits diffraction and interference.
- Light spreads out after passing through an opening.
- Equal-wavelength light waves interfere.
- Interference depends on the path-length difference.

The wave model is usually appropriate for openings smaller than about 1 mm.

RAY MODEL

Light rays travel in straight lines.
- The speed is $v = c/n$, where n is the index of refraction.

Light rays travel forever unless they interact with matter.
- Reflection and refraction
- Scattering and absorption

An object is a source of rays.
- Rays originate at every point.
- The eye sees by focusing a diverging bundle of rays.

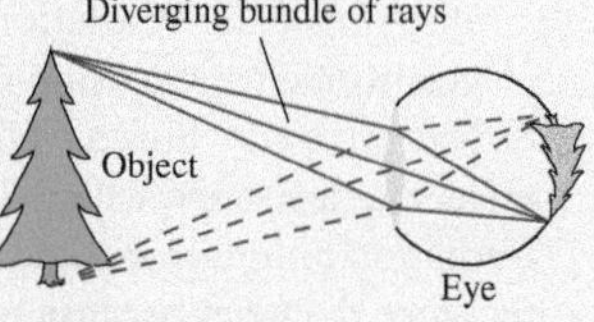

The ray model is usually appropriate for openings larger than about 1 mm.

SINGLE-SLIT DIFFRACTION

Dark fringes in a single-slit diffraction pattern are at $\theta_p = p\lambda/a \quad p = 1, 2, \ldots$

Slit width a

At distance L, the central maximum of the diffraction pattern has width

$$w = 2\lambda L/a$$

CIRCULAR-APERTURE DIFFRACTION

The first dark fringe in a circular-aperture diffraction pattern is at angle $\theta_1 = 1.22\lambda/D$.

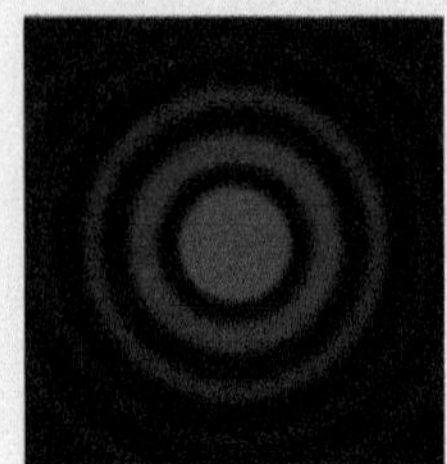

The width of the central maximum is

$$w = 2.44\lambda L/D$$

where D is the hole diameter.

DOUBLE-SLIT INTERFERENCE

Bright equally spaced fringes are located at

$$y_m = mL\lambda/d \qquad m = 0, 1, 2, \ldots$$

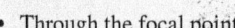

Slit spacing d

DIFFRACTION GRATING

Very narrow bright fringes are at

$$d \sin\theta_m = m\lambda \qquad y_m = L \tan\theta_m$$

RAY TRACING

For lenses and mirrors, three special rays locate the image:
- Parallel to the axis
- Through the focal point
- Through the center

IMAGES

If rays converge at P′, then P′ is a real image and s' is positive.
If rays diverge from P′, then P′ is a virtual image and s' is negative.

THIN LENSES AND MIRRORS

The thin-lens and thin-mirror equation for focal length f is

$$\frac{1}{s} + \frac{1}{s'} = \frac{1}{f}$$

Lateral magnification is $m = -s'/s$.

MAGNIFIERS

For relaxed-eye viewing, the angular magnification is

$$M = \frac{25\text{ cm}}{f}$$

VISION

Hyperopia occurs when the eye's near point is too far away. It is corrected with a converging lens.
Myopia occurs when the eye's far point is too close. It is corrected with a diverging lens.

MICROSCOPES

The angular magnification is

$$M = m_{obj}M_{eye} \approx -\frac{L}{f_{obj}}\frac{25\text{ cm}}{f_{eye}}$$

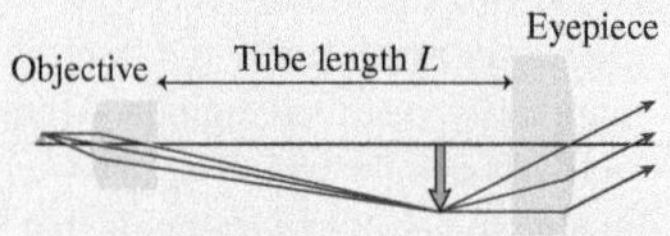

The numerical aperture is $\text{NA} = n \sin\alpha$, where α is the half-angle of the objective.

RESOLUTION

Diffraction limits optical instruments.
- The smallest spot to which light can be focused is $w_{min} = 2.44\lambda f/D$.
- Two objects can be resolved if their angular separation exceeds $1.22\lambda/D$.
- The resolution of a microscope is

$$R = \frac{0.61\lambda}{\text{NA}}$$

ONE STEP BEYOND

Phase-Contrast Microscopy

One of the difficulties of imaging cells and other biological samples is that they are transparent, or nearly so. Suppose we use a regular microscope to view a slide with cells. Very little light is absorbed as it passes through the cells, so there is very little contrast between a cell and its surroundings. We can stain the cells to highlight certain elements and increase the contrast that way, but staining doesn't work if we want to make images of living cells.

There is little change in intensity when light traverses a boundary between two transparent media—say, between a cell and the nutrient bath in which it is submerged. But any difference in the index of refraction causes the light rays to bend and also leads to *phase differences.* Phase-contrast microscopy uses these differences to produce striking images of living cells and similar systems.

Let's start with a related example. The photograph of a burning candle was produced with a technique that emphasizes slight variations in the index of refraction between the cool room air and the hot gases rising from the flame. We can't see these gases directly, but the difference in the indices of refraction generates an optical pattern that is there if we know how to look for it. Recall that light travels through a medium with speed $v = c/n$. Light waves traveling through the hot gas, which has a lower index of refraction, get slightly ahead of waves traveling through the cooler air. This offset in their wave fronts is a *phase difference,* and the goal is to turn that small phase difference into a visual contrast that we can see. This is what is done in phase-contrast microscopy.

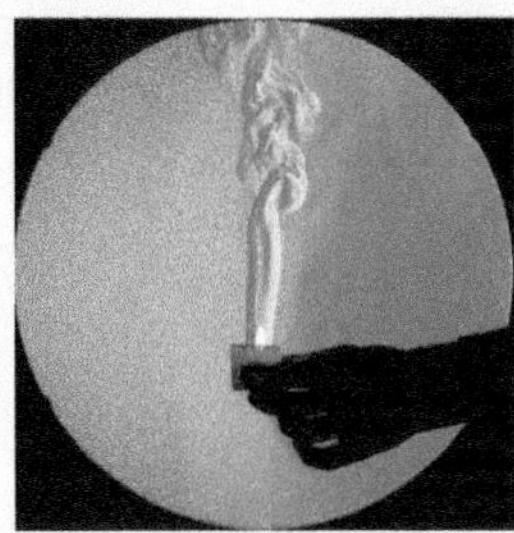

A shadowgraph of a burning candle brings out small difference in the indices of the refraction of the gases.

Consider a microscope that is used to image cells with an index of refraction of 1.360 while they are submerged in a nutrient bath with an index of refraction of 1.335. The small difference in the indices of refraction causes a slight delay in the light that passes through the cells compared to the light that passes through the surrounding liquid. We describe the delay mathematically as a phase difference between the light that passes through the cells and the light that passes through the nutrient bath. The phase difference is about 90°, corresponding to a path-length difference of $\lambda/4$, for light that passes through a cell that has a typical thickness of 5 μm. A phase-contrast microscope turns this difference in phase into a difference in brightness in the final image. The micrographs of a paramecium illustrate the great increase in contrast and detail that a phase-contrast microscope provides.

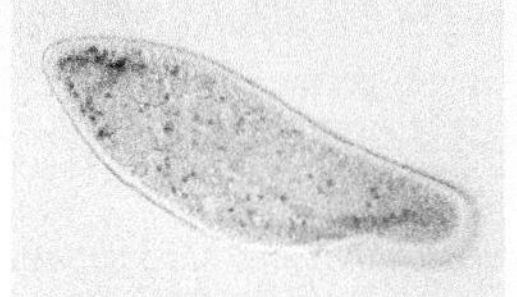
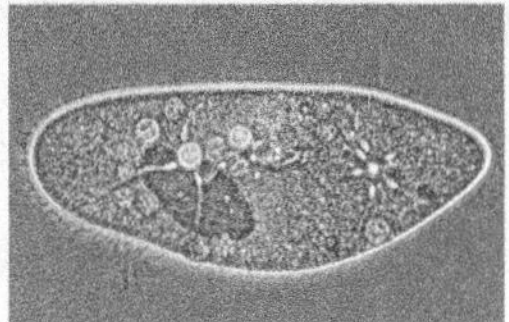

A paramecium imaged using regular light microscopy (left) and phase-contrast microscopy (right).

The figure below shows the basic idea of how a phase-contrast microscope works. Light rays from the source pass through an annulus, a circular aperture in an opaque screen that blocks all but a thin circle of light. Light passing through the annulus is focused onto the specimen, which we'll assume is a nutrient solution that contains cells. Some light passes through the solution without encountering a cell and is not deflected; this light—called the direct light—is colored purple. Light that passes through a cell, however, is scattered in many different directions and also, as we've seen, shifted in phase by about 90° relative to the direct light because of the slight additional time that is needed to travel through the higher index of refraction. The scattered light is colored pink.

The two sets of rays are collected by the objective lens and then pass through what is called a phase plate. The phase plate is another annulus, but this time made of a thin piece of transparent glass. The direct light passes through a ring of thinner glass, while the scattered light passes through slightly thicker glass. It takes a bit longer for the scattered light to pass through the thicker glass, and this creates another phase shift. The phase plate is designed to shift the phase of the scattered light by 90° relative to the direct light.

The scattered light from the cells has now undergone two 90° phase shifts relative to the direct light: one shift in the specimen and one in the phase plate. Consequently, the scattered light and the direct light are 180° out of phase when they meet at the image plane, and this is the condition for destructive interference. Because of interference, the structures in the cell that scattered the light appear very dark in the image, much darker than they would when viewed by a regular microscope. In other words, the image has much more contrast than the image of a regular microscope.

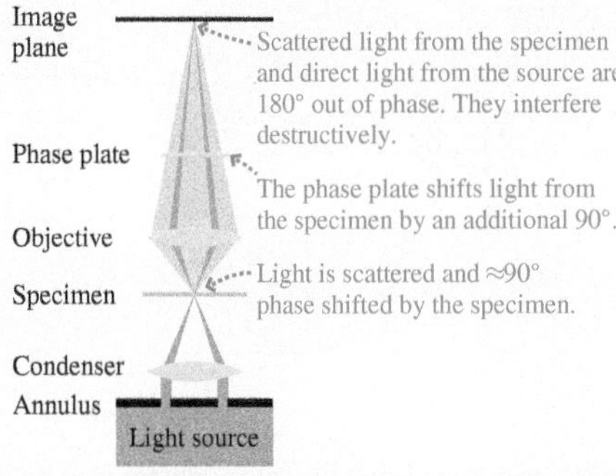
Image plane
Scattered light from the specimen and direct light from the source are 180° out of phase. They interfere destructively.
Phase plate
The phase plate shifts light from the specimen by an additional 90°.
Objective
Light is scattered and ≈90° phase shifted by the specimen.
Specimen
Condenser
Annulus
Light source

A simplified phase-contrast microscope.

The advent of phase-contrast microscopy allowed biologists to observe living cells in real time in unprecedented detail, and the development of the technique won Frits Zernike the Nobel Prize in Physics in 1953. Today it is a common imaging technique in labs, along with other variants of this technique in which deflected and undeflected rays are treated differently.

The following problems are related to the passage "Phase-Contrast Microscopy" on the previous page.

1. The basic problem that makes it difficult to create an image of living cells is that
   A. The cells are always on the move and hard to see.
   B. The cells are nearly opaque and pass little light.
   C. The cells are nearly transparent and create little intensity variation in transmitted light.
2. In the diagram of the phase-contrast microscope in the passage, the specimen is ________ of the condenser lens.
   A. in front of the focal point
   B. at the focal point
   C. behind the focal point
3. For the numbers given for the index of refraction of the cells and the nutrient bath,
   A. The light that passes through the cells travels slightly faster than the light that passes through the nutrient bath.
   B. The light that passes through the cells travels slightly slower than the light that passes through the nutrient bath.
   C. The light that passes through the cells travels at the same speed as the light that passes through the nutrient bath.
4. In the passage, light that traversed cells of thickness 5 μm had a phase shift of 90° relative to light that passed through the nutrient bath surrounding the cells. Which of these changes would *not* increase the phase shift?
   A. An increase in the index of refraction of the cells.
   B. An increase in the index of refraction of the nutrient bath surrounding the cells.
   C. An increase in the thickness of the cells.
5. The 90° phase shift of light that traverses the cells is equivalent to a path-length difference of $\lambda/4$ relative to light that passes through the nutrient bath surrounding the cells. What additional path-length difference must be added by the phase plate to give complete destructive interference at the image plane?
   A. $\lambda/4$
   B. $\lambda/2$
   C. $3\lambda/4$

The following passages and associated questions are based on the material of Part IV.

Horse Sense BIO

The ciliary muscles in a horse's eye can make only small changes to the shape of the lens, so a horse can't change the shape of the lens to focus on objects at different distances as humans do. Instead, a horse relies on the fact that its eyes aren't spherical. As Figure IV.1 shows, different points at the back of the eye are at somewhat different distances from the front of the eye. We say that the eye has a "ramped retina"; images that form on the top of the retina are farther from the cornea and lens than those that form at lower positions. The horse uses this ramped retina to focus on objects at different distances, tipping its head so that light from an object forms an image at a vertical location on the retina that is at the correct distance for sharp focus.

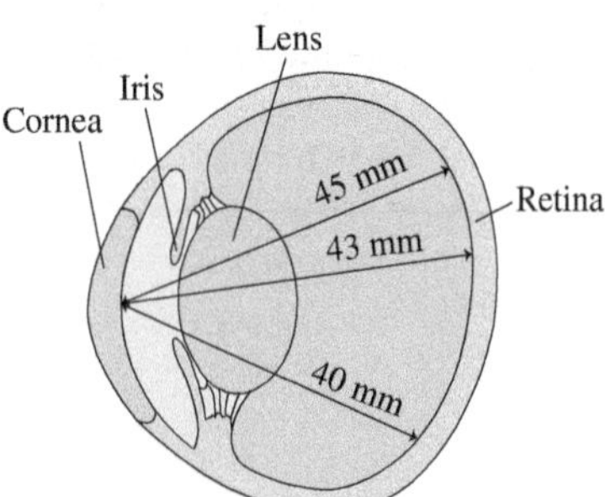

FIGURE IV.1

6. In a horse's eye, the image of a close object will be in focus
   A. At the top of the retina.
   B. At the bottom of the retina.
7. In a horse's eye, the image of a distant object will be in focus
   A. At the top of the retina.
   B. At the bottom of the retina.
8. A horse is looking straight ahead at a person who is standing quite close. The image of the person spans much of the vertical extent of the retina. What can we say about the image on the retina?
   A. The person's head is in focus; the feet are out of focus.
   B. The person's feet are in focus; the head is out of focus.
   C. The person's head and feet are both in focus.
   D. The person's head and feet are both out of focus.
9. Certain medical conditions can change the shape of a horse's eyeball; these changes can affect vision. If the lens and cornea are not changed but all of the distances in Figure IV.1 are increased slightly, then the horse will be
   A. Nearsighted.
   B. Farsighted.
   C. Unable to focus clearly at any distance.

Mirror Eyes BIO

Most animals—humans included—have eyes that use lenses to form images. The eyes of scallops are different. A typical scallop eye forms images largely by reflection from a mirror-like surface at the back of the eye. Figure IV.2 shows the important features of a typical scallop eye. The lens causes very little redirection of incoming light rays; it is the spherical surface in the back of the eye that brings rays of light to a focus on the cells of the retina. (For simplicity, we've shown no refraction by the lens, although the lens does cause some refraction that seems to help to make the image sharper by correcting for the spherical aberration introduced by the mirror.) The reflection is due to thin-film interference from the front and back faces of 80-nm-thick transparent crystals of guanine, index $n = 1.83$, that are embedded in cytoplasm with index $n = 1.34$. The individual eyes are quite small. A typical scallop has 40 to 60 eyes, each with a 450-μm-diameter pupil and a reflecting surface at the back of the eye with a focal length of only 200 μm.

The unusual imaging system of the scallop eye makes it very sensitive to light. The ratio of the focal length to the aperture of an optical system is known as the f-number. A smaller f-number implies greater light sensitivity. A telescope optimized for light gathering might have an f-number of 4; the scallop's f-number of less than 0.5 means that its small eyes work well in very dim light.

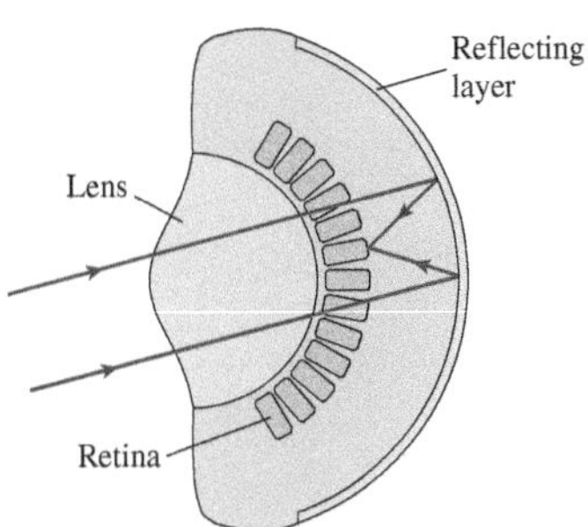

FIGURE IV.2

10. There is very little refraction when the light enters the lens of the scallop eye. This means that
    A. The index of refraction of the lens is much greater than that of water.
    B. The index of refraction of the lens is approximately equal to that of water.
    C. The index of refraction of the lens is much less than that of water.
11. Consider the reflection of light from the front and back faces of the guanine crystals. What is the number of reflective phase changes?
    A. 0
    B. 1
    C. 2
12. What is the longest wavelength for which there is a strong reflection from the guanine crystals, to the nearest 100 nm?
    A. 400 nm
    B. 500 nm
    C. 600 nm
    D. 700 nm
13. The image formed on the scallop's retina is
    A. Real and inverted.
    B. Real and upright.
    C. Virtual and inverted.
    D. Virtual and upright.
14. For the typical pupil diameter of 3.0 mm, what is the approximate f-number for a human eye?
    A. 2 B. 4 C. 6 D. 8

Pupil Size BIO

The lens of your eye changes to focus on objects at different distances; your eye's pupil changes to adapt to different light levels. This causes changes in your vision as well. Suppose you are in a dark movie theater, with your pupils fully dilated to 8.0 mm in diameter. Now you emerge into the bright daylight, and your pupils quickly contract to 2.0 mm in diameter.

15. When your pupil contracts, by what factor does this change decrease the total power of the light entering your eye?
    A. 2
    B. 4
    C. 8
    D. 16
16. When your pupil contracts, how does this affect the spherical aberration of your eye's optical system?
    A. The spherical aberration is increased.
    B. The spherical aberration does not change.
    C. The spherical aberration decreases.
17. When your pupil contracts, how does this affect the angular resolution of your eye?
    A. The resolution increases.
    B. The resolution does not change.
    C. The resolution decreases.

Additional Integrated Problems

18. BIO The pupil of your eye is smaller in bright light than in dim light. Explain how this makes images seen in bright light appear sharper than images seen in dim light.
19. BIO People with good vision can make out an 8.8-mm-tall letter on an eye chart at a distance of 6.1 m. Approximately how large is the image of the letter on the retina? Assume that the distance from the lens to the retina is 24 mm.
20. A photographer uses a lens with $f = 50$ mm to form an image of a distant object on the detector in a digital camera. The image is 1.2 mm high, and the intensity of light on the detector is 2.5 W/m^2. She then switches to a lens with $f = 300$ mm that is the same diameter as the first lens. What are the height of the image and the intensity now?
21. Sound and other waves undergo diffraction just as light does. Suppose a loudspeaker in a 20°C room is emitting a steady tone of 1200 Hz. A 1.0-m-wide doorway in front of the speaker diffracts the sound wave. A person on the other side walks parallel to the wall in which the door is set, staying 12 m from the wall. When he is directly in front of the doorway, he can hear the sound clearly and loudly. As he continues walking, the sound intensity decreases. How far must he walk from the point where he was directly in front of the door until he reaches the first quiet spot?

PART

V

Electricity and Magnetism

To make a quick escape, a squid ejects water from its mantle, causing it to rocket backward. The nerve impulse from the brain that causes the mantle to contract is carried along an especially large-diameter nerve fiber, the *squid giant axon*. Easy experimental access to such a large nerve fiber has made it a model for studying how electrical signals are transmitted through the nervous system, a topic that we'll study in Part V.

OVERVIEW

Charges, Currents, and Fields

The early Greeks discovered that a piece of amber that has been rubbed briskly can attract feathers or small pieces of straw. They also found that certain stones from the region they called *Magnesia* can pick up pieces of iron. These first experiences with the forces of electricity and magnetism began a chain of investigations that has led to today's high-speed computers, lasers, fiber-optic communications, and magnetic resonance imaging, as well as mundane modern-day miracles such as the lightbulb.

The development of a successful electromagnetic theory, which occupied the leading physicists of Europe for most of the 19th century, led to sweeping revolutions in both science and technology. The complete formulation of the theory of the electromagnetic field has been called by no less than Einstein "the most important event in physics since Newton's time."

The basic phenomena of electricity and magnetism are not as familiar to most people as are those of mechanics. We will deal with this lack of experience by placing a large emphasis on these basic phenomena. We will begin where the Greeks did—by looking at the forces between objects that have been briskly rubbed, exploring the concept of *electric charge*. We will first make systematic observations of how charges behave, and this will lead us to consider the forces between charges and how charges behave in different materials. *Electric current,* whether the motion of electrons through a lightbulb or the passage of sodium ions through an ion channel, is simply a controlled motion of charges through conducting materials. One of our goals will be to understand how charges move through electric circuits.

When we turn to magnetic behavior, we will again start where the Greeks did, noting how magnets stick to some metals. Magnets also affect compass needles. And, as we will see, an electric current can affect a compass needle in exactly the same way as a magnet. This observation shows the close connection between electricity and magnetism, which leads us to the phenomenon of *electromagnetic waves.*

Our theory of electricity and magnetism will introduce the entirely new concept of a *field.* Electricity and magnetism are about the long-range interactions of charges, both static charges and moving charges, and the field concept will help us understand how these interactions take place.

Microscopic Models

The field theory provides a macroscopic perspective on the phenomena of electricity and magnetism, but we can also take a microscopic view. At the microscopic level, we want to know what charges are, how they are related to atoms and molecules, and how they move through various kinds of materials. Electromagnetic waves are composed of electric and magnetic fields. The interaction of electromagnetic waves with matter can be analyzed in terms of the interactions of these fields with the charges in matter. When you heat food in a microwave oven, you are using the interactions of electric and magnetic fields with charges in a very fundamental way.

21 Electric Forces and Fields

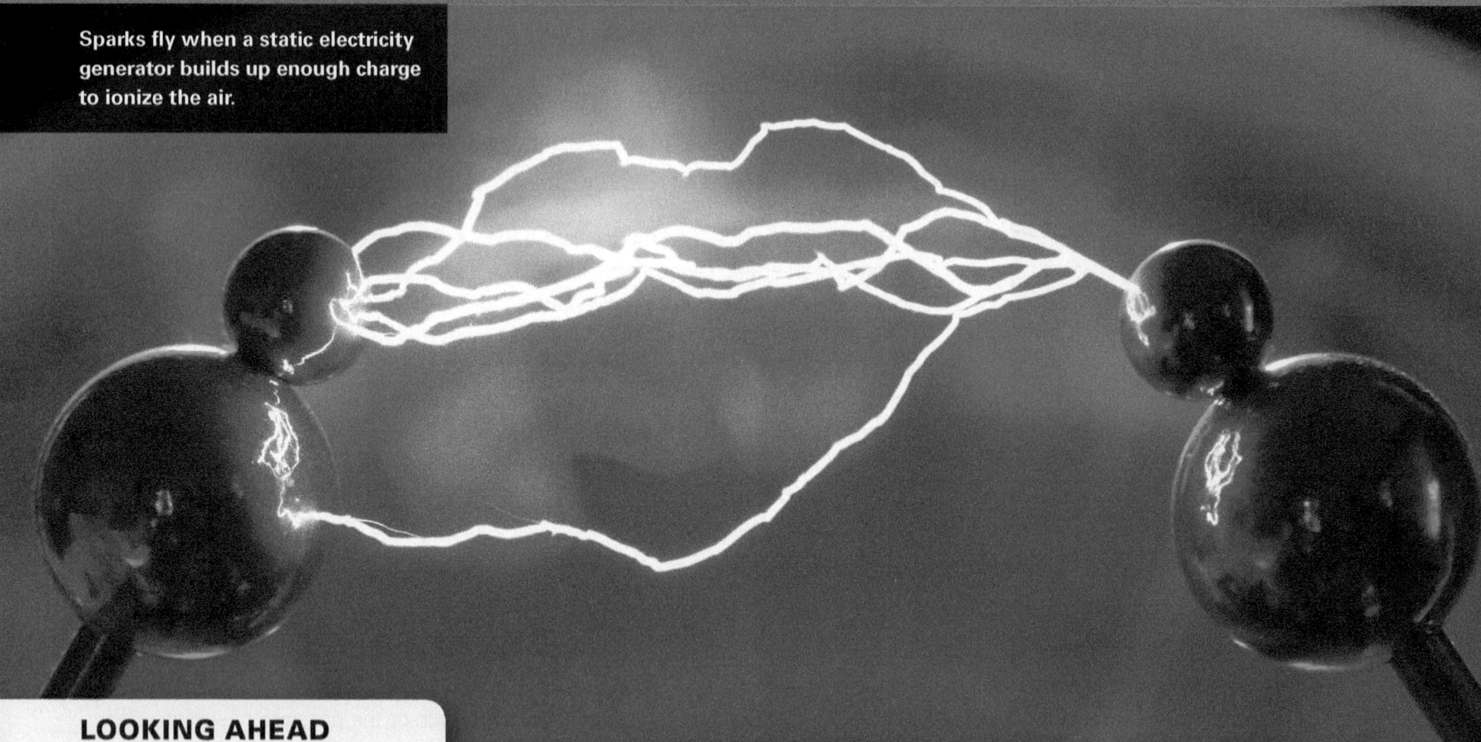

Sparks fly when a static electricity generator builds up enough charge to ionize the air.

LOOKING AHEAD

Electric Charge

A comb passed through your hair attracts a stream of water. The *charge model* of electricity explains this and other electric phenomena.

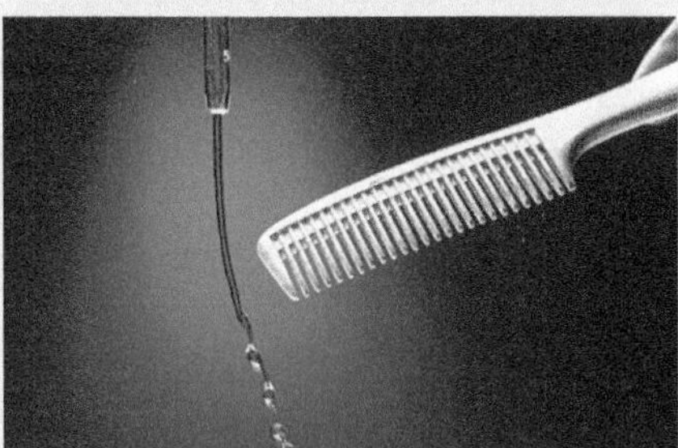

You'll use the charge model to explain attractive and repulsive forces, the nature of insulators and conductors, and much more.

Electric Forces

The two strands of a DNA molecule are held together by hydrogen bonds. These are electric forces between electric dipoles.

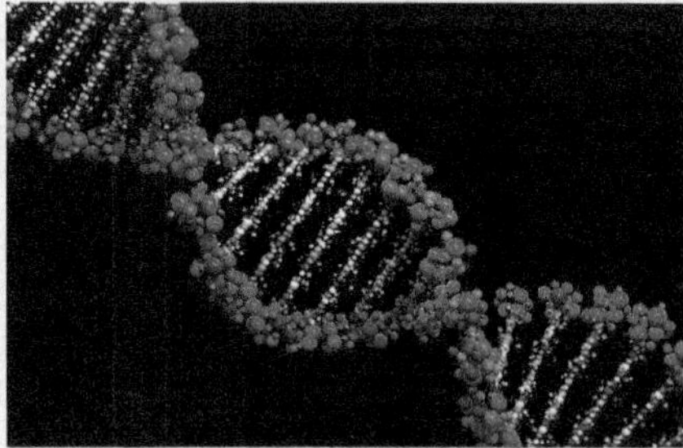

You'll discover that *Coulomb's law* for the force between two charged particles depends inversely on the square of the distance between them.

Electric Fields

Charges create *electric fields.* Electric fields in storms can be strong enough to ionize the air, causing lightning.

You'll learn how to calculate the electric fields of point charges, electric dipoles, and other important arrangements of charges.

GOAL To understand basic electric phenomena in terms of charges, forces, and electric fields.

PHYSICS AND LIFE

Electricity Is Essential for Life

Life would not exist without electricity. Life depends on water, and water has unique electrical properties that make biochemistry possible. The arrangement of electric charges on both sides of the cell membrane creates a membrane potential that is essential to a large number of biochemical and molecular biological processes. For example, nerve impulses and muscle contraction involve carefully timed changes in the charge state of the membrane. Finally, electrically based technologies ranging from gel electrophoresis to electrocardiograms are essential to modern biology and medicine. This is the first of several chapters that will explore the importance of electricity to the life sciences.

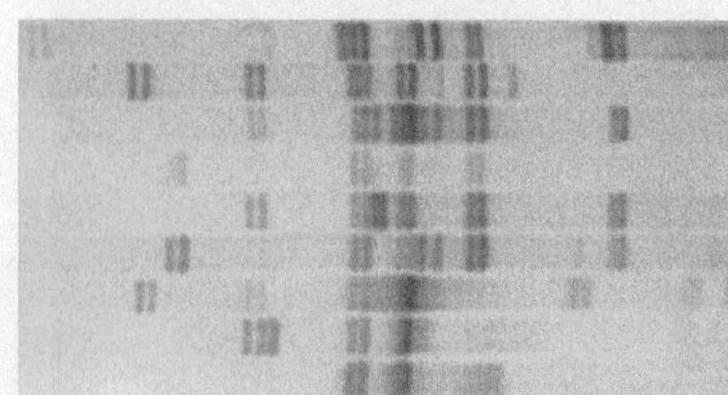

The technique of *gel electrophoresis* uses an electric field to separate fragments of DNA that have different lengths.

21.1 The Charge Model

You can receive a mildly unpleasant shock and produce a little spark if you touch a metal doorknob after scuffing your shoes across a carpet. A plastic comb that you've run through your hair will pick up bits of paper and other small objects. In both of these cases, two objects are *rubbed* together. Why should rubbing an object cause forces and sparks? What kind of forces are these? These are the questions with which we begin our study of electricity.

Electric forces and interactions are especially important in biology and medicine, from hydrogen bonds and membrane potentials to ion channels and nerve impulses. If you have taken chemistry, you're already familiar with some ideas about electricity—for example, that matter is made up of negatively charged electrons and positively charged nuclei, and that like charges repel and opposite charges attract. We will build on that knowledge, but we're going to start with some simple observations designed to help you develop your intuition about electric forces. The theory of electricity was well established long before the electron was discovered, and the basic concepts of electricity make no mention of atoms or electrons.

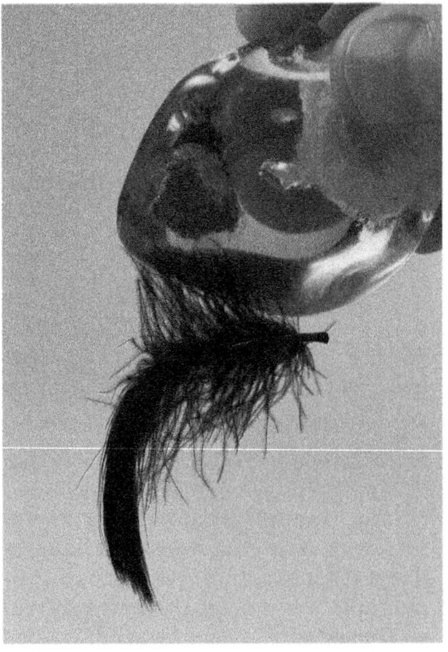

It's electric The ancient Greeks observed electric forces when they found that amber, a fossilized tree resin, attracts bits of feather or straw after being rubbed with a cloth. The Greek word for amber, *elektron,* is the source of our words "electric," "electricity," and—of course—"electron."

Experimenting with Charges

Let's imagine working in a modest laboratory where we can make observations of electric phenomena. The major tools in the lab are:

- A number of plastic and glass rods, each several inches long. These can be held in your hand or suspended by threads from a support.
- Pieces of wool and silk.
- Small metal spheres, an inch or two in diameter, on wood stands.

We will manipulate and use these tools with the goal of developing a theory to explain the phenomena we see. The experiments and observations described below are very much like those of early investigators of electric phenomena.

Discovering electricity I

Experiment 1

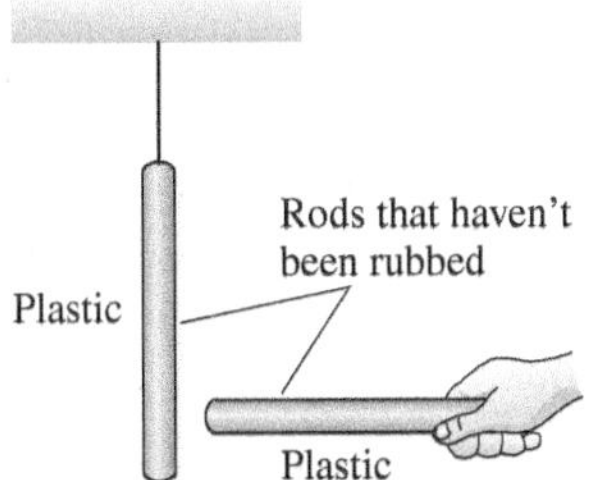

Take a plastic rod that has been undisturbed for a long period of time and hang it by a thread. Pick up another undisturbed plastic rod and bring it close to the hanging rod. Nothing happens to either rod.

Interpretation: There are no special electrical properties to these undisturbed rods. We say that they are **neutral.**

Experiment 2

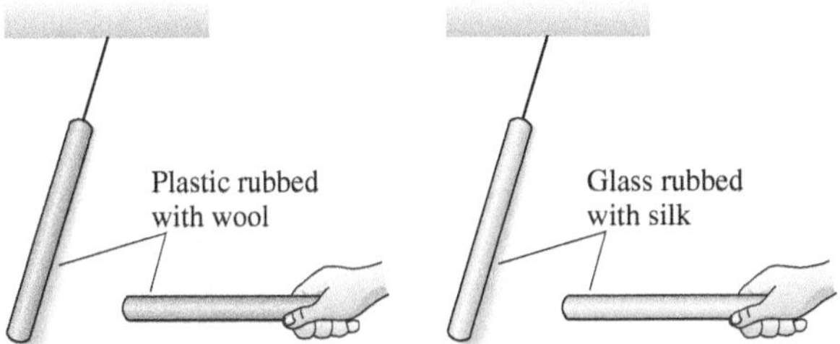

Vigorously rub both the hanging plastic rod and the handheld plastic rod with wool. Now the hanging rod *moves away* from the handheld rod when you bring the two close together. Rubbing two glass rods with silk produces the same result: The two rods repel each other.

Interpretation: Rubbing a rod somehow changes its properties so that forces now act between two such rods. We call this process of rubbing **charging** and say that a rubbed rod is charged, or that it has acquired a **charge.**

Experiment 2 shows that there is a *long-range repulsive force* (i.e., a force requiring no contact) between two identical objects that have been charged in the *same* way, such as two plastic rods both rubbed with wool or two glass rods rubbed with silk. The force between charged objects is called the **electric force.** We have seen a long-range force before, gravity, but the gravitational force is always attractive. This is the first time we've observed a repulsive long-range force. However, the electric force is not always repulsive, as the next experiment shows.

Discovering electricity II

Experiment 3

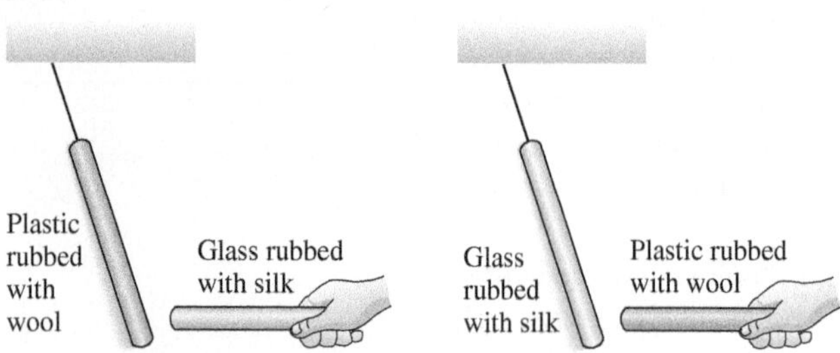

Bring a glass rod that has been rubbed with silk close to a hanging plastic rod that has been rubbed with wool. These two rods *attract* each other.

Interpretation: We can explain this experiment as well as Experiment 2 by assuming that there are two *different* kinds of charge. By convention, a glass rod that has been rubbed with silk has a **positive** charge and a plastic rod that has been rubbed with wool has a **negative** charge. The experiments show that **like charges** (positive/positive or negative/negative) exert repulsive forces on each other while **opposite charges** (positive/negative) exert attractive forces on each other.

Experiment 4

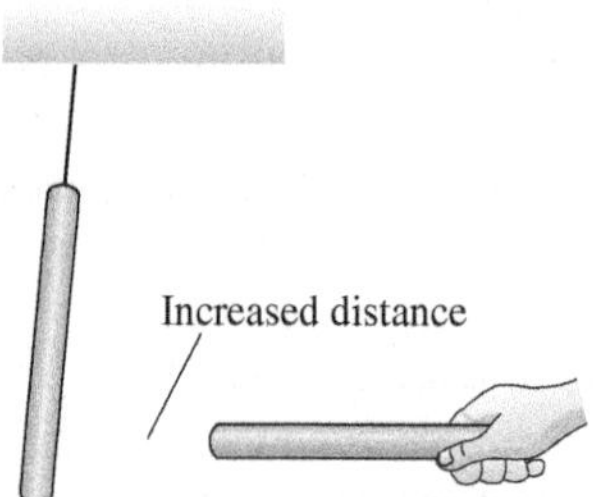

- If the two rods are held farther from each other, the force between them decreases.
- The strength of the force is greater for rods that have been rubbed more vigorously.

Interpretation: Like the gravitational force, the electric force decreases with the distance between the charged objects. And, the greater the charge on the two objects, the greater the force between them.

Additional experiments confirm that there are two and only two kinds of charge: positive and negative. An object's charge can be determined by seeing how it interacts with objects known to be positive and negative. For example, rubbing a balloon against your hair charges the balloon, and then it sticks to your clothes. You would find that the balloon attracts a charged glass rod that has been rubbed with silk and repels a plastic rod that has been rubbed with wool, so the balloon's charge is negative.

Visualizing Charge

FIGURE 21.1 Visualizing charge.

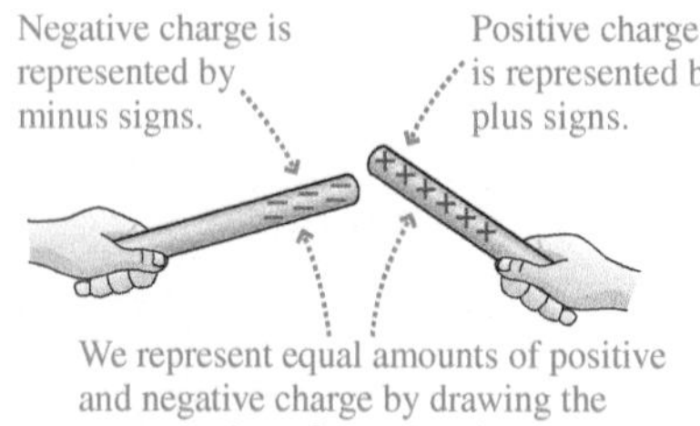

Negative charge is represented by minus signs.
Positive charge is represented by plus signs.
We represent equal amounts of positive and negative charge by drawing the same number of + and − signs.
More charge is represented by more + or − signs.

Diagrams are going to be an important tool for understanding and explaining charges and the forces between charged objects. FIGURE 21.1 shows how to draw a *charge diagram* that shows the distribution of charge on an object. It's important to realize that the + and − signs drawn in Figure 21.1 do not represent "individual" charges. At this point, we are thinking of charge only as something that can be acquired by an object by rubbing, so in charge diagrams the + and − signs represent the charge only in a general way.

We can gain an important insight into the nature of charge by investigating what happens when we bring together a plastic rod and the wool used to charge it, as the following experiment shows.

Discovering electricity III

Experiment 5

Start with a neutral, uncharged hanging plastic rod and a piece of wool. Rub the plastic rod with the wool, then hold the wool close to the rod. The rod is *attracted* to the wool.

Interpretation: From Experiment 3 we know that the plastic rod has a negative charge. Because the wool attracts the rod, the wool must have a *positive* charge.

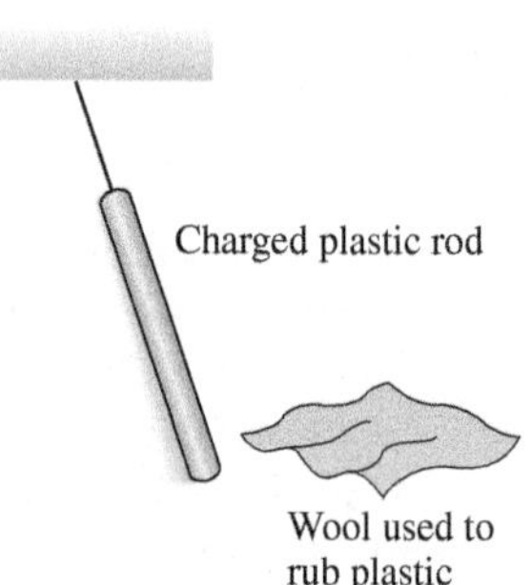

Experiment 5 shows that when a plastic rod is rubbed by wool, not only does the plastic rod acquire a negative charge, but the wool used to rub it acquires a positive charge. This observation can be explained if we postulate that a neutral object is not one that has no charge at all; rather, a neutral object contains **equal amounts of positive and negative charge.** Just as in ordinary addition, where $2 + (-2) = 0$, equal amounts of opposite charge "cancel," leaving no overall or *net* charge.

Rubbing two objects together somehow *separates* the positive and negative charges, leaving one with an excess of positive charge and the other with an excess of negative charge. FIGURE 21.2 shows how this works when a plastic rod is rubbed with wool, leaving the wool positively charged and the rod negatively charged. Notice that a charge diagram shows only the *excess* charge.

FIGURE 21.2 How a plastic rod and wool acquire charge during the rubbing process.

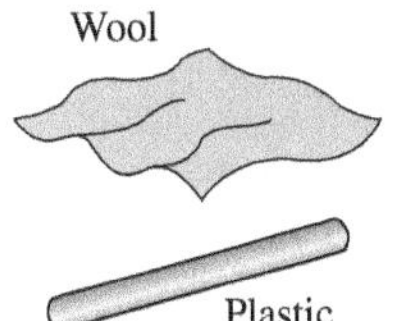

1. The rod and wool are initially neutral.

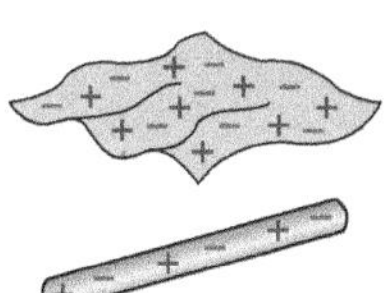
2. "Neutral" means that an object has equal amounts of positive and negative charge.

3. Now we rub the plastic and wool together. Rubbing separates the positive and negative charges.

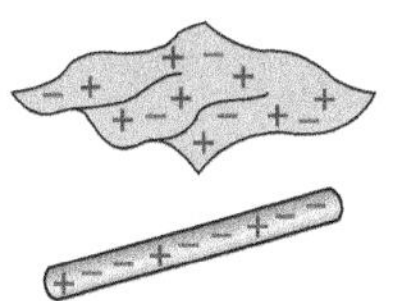
4. This leaves the rod with extra negative charge. The wool is left with more positive charge than negative.

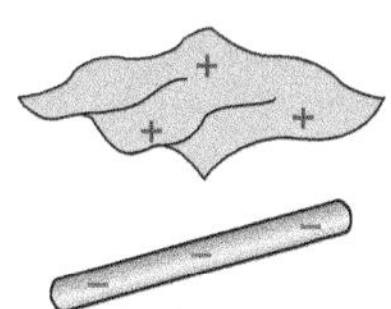
5. In a charge diagram, we draw only the *excess* charge.

There is another crucial fact about charge implicit in Figure 21.2: Nowhere in the rubbing process was charge either created or destroyed. Charge was merely transferred from one place to another. It turns out that this fact is a fundamental law of nature, the **law of conservation of charge.** Charge can be transferred or moved around, but the net charge cannot change.

The results of our experiments, and our interpretation of them in terms of positive and negative charge, can be summarized in the following **charge model.**

Charge model, part I The basic postulates of our model are:

- Frictional forces, such as rubbing, add something called *charge* to an object or remove it from the object. The process itself is called *charging.* More vigorous rubbing produces a larger quantity of charge.
- There are two kinds of charge, positive and negative.
- Two objects with *like charge* (positive/positive or negative/negative) exert repulsive forces on each other. Two objects with *opposite charge* (positive/negative) exert attractive forces on each other. We call these *electric forces.*
- The force between two charged objects is a long-range force. The magnitude of the force increases as the quantity of charge increases and decreases as the distance between the charges increases.
- *Neutral* objects have *equal amounts* of positive and negative charge. The rubbing process charges the objects by *transferring* charge from one to the other. The objects acquire equal but opposite charges.
- Charge is conserved: It cannot be created or destroyed.

Dust in the wind Collisions between sand grains and dust particles can transfer charges between them, usually leaving the smaller dust particles negative and the sand grains positive. These charges are separated by winds that loft the dust particles while leaving the sand grains closer to the surface. This electrifies the dust cloud, which in turn can cause a dramatic increase in the number of particles leaving the surface. When you see images of an intense sandstorm, know that it might be more than the wind that's lifting up the grains—it could also be the electric forces between them.

Insulators and Conductors

Experiments 2, 3, and 5 involved a transfer of charge from one object to another. Let's do some more experiments with charge to look at how charge *moves* on different materials.

Discovering electricity IV

Experiment 6

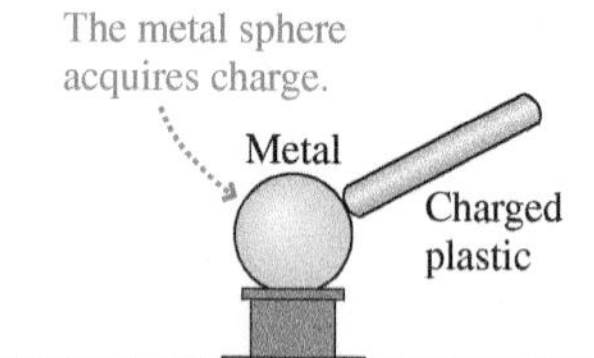

Charge a plastic rod by rubbing it with wool. Touch a neutral metal sphere with the rubbed area of the rod. The metal sphere then repels a charged, hanging plastic rod. The metal sphere appears to have acquired a charge of the same sign as the plastic rod.

Experiment 7

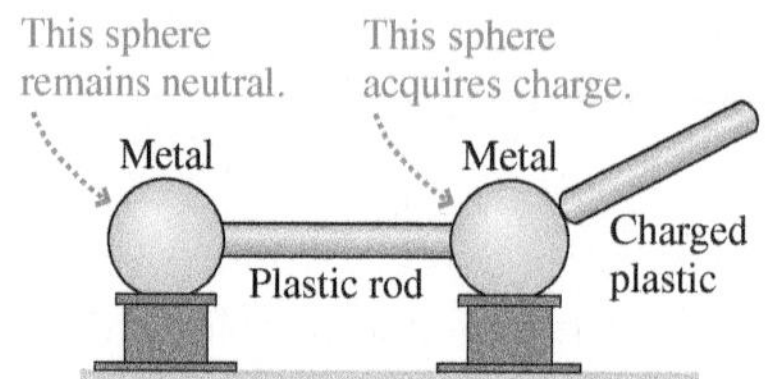

Place two metal spheres close together with a plastic rod connecting them. Charge a second plastic rod, by rubbing, and touch it to one of the metal spheres. Afterward, the metal sphere that was touched repels a charged, hanging plastic rod. The other metal sphere does not.

Experiment 8

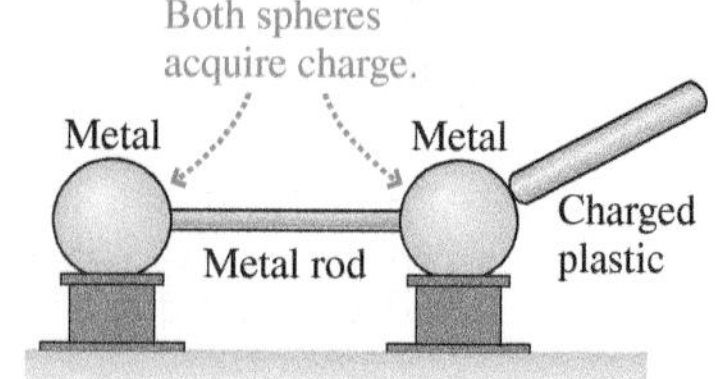

Repeat Experiment 7 with a metal rod connecting the two metal spheres. Touch one metal sphere with a charged plastic rod. Afterward, *both* metal spheres repel a charged, hanging plastic rod.

Our final set of experiments has shown that charge can be transferred from one object to another only when the objects *touch*. Contact is required. Removing charge from an object, which you can do by touching it, is called **discharging.**

In Experiments 7 and 8, charge is transferred from the charged rod to the metal sphere as the two are touched together. In Experiment 7, the other sphere remains neutral, indicating that no charge moved along the plastic rod connecting the two spheres. In Experiment 8, by contrast, the other sphere is found to be charged; evidently charge has moved along the metal rod that connects the spheres, transferring some charge from the first sphere to the second. We define **conductors** as those materials through or along which charge easily moves, and **insulators** as those materials on or in which charges remain immobile. Glass and plastic are insulators; metal and salt water are conductors.

This new information allows us to add more postulates to our charge model:

Charge model, part II

- There are two types of materials. Conductors are materials through or along which charge easily moves. Insulators are materials on or in which charges are immobile.
- Charge can be transferred from one object to another by contact.

NOTE ▶ Both insulators and conductors can be charged. They differ in the ability of charge to *move.* ◀

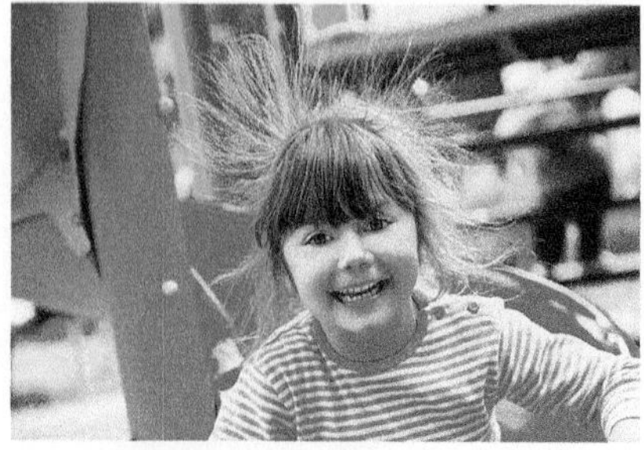

A bad hair day A dry day, a plastic slide, and a child with clothes of the right fabric lead to a lovely demonstration of electric charges and forces. The child becomes charged as her clothes rub against the plastic slide. The body is a good conductor, so the charges spread across her body and her hair. The resulting repulsion produces a dramatic result!

CONCEPTUAL EXAMPLE 21.1 Transferring charge

In Experiment 8, touching a metal sphere with a charged plastic rod caused a second metal sphere, connected by a metal rod to the first, to become charged with the same type of charge as the rod. Use the postulates of the charge model to construct a charge diagram for the process.

REASON We need the following ideas from the charge model:

- **Charge is transferred upon contact.** The plastic rod was charged by rubbing with wool, giving it a negative charge. The charge doesn't move around on the rod, an insulator, but some of the charge is transferred to the metal upon contact.
- **Metal is a conductor.** Once in the metal, which is a conductor, the charges are free to move around.
- **Like charges repel.** Because like charges repel, these negative charges quickly move as far apart as they possibly can. Some move through the connecting metal rod to the second sphere. Consequently, the second sphere acquires a net negative charge. The repulsive forces drive the negative charges as far apart as they can possibly get, causing them to end up on the *surfaces* of the conductors.

The charge diagram in FIGURE 21.3 illustrates these three steps.

ASSESS The two spheres share the total charge that was transferred by contact, but the net charge doesn't change.

FIGURE 21.3 A charge diagram for Experiment 8.

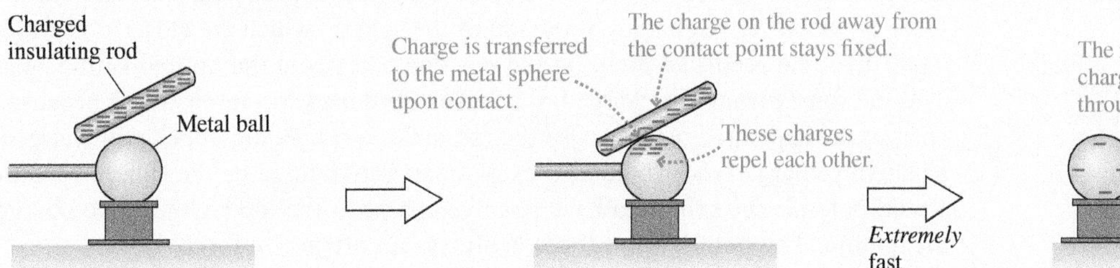
Charged insulating rod
Metal ball
Charge is transferred to the metal sphere upon contact.
The charge on the rod away from the contact point stays fixed.
These charges repel each other.
Extremely fast
The repulsive forces cause the charge to spread out by moving through the conducting metal rod.

In Conceptual Example 21.1, charge placed on the conductor rapidly distributes itself over the conductor's surface. This movement of charge is *extremely* fast. Other than this very brief interval during which the charges are adjusting, the charges on an isolated conductor are in static equilibrium with the charges at rest. This condition is called **electrostatic equilibrium.**

CONCEPTUAL EXAMPLE 21.2 Drawing a charge diagram for an electroscope

Many electricity demonstrations are carried out with the help of an *electroscope* like the one shown in FIGURE 21.4. Touching the sphere at the top of an electroscope with a charged plastic rod causes the leaves to fly apart and remain hanging at an angle. Use charge diagrams to explain why.

REASON We will use the charge model and our understanding of insulators and conductors to make a series of charge diagrams in FIGURE 21.5 that show the charging of the electroscope.

ASSESS The charges move around, but, because charge is conserved, the total number of negative charges doesn't change from picture to picture.

FIGURE 21.4 A charged electroscope.

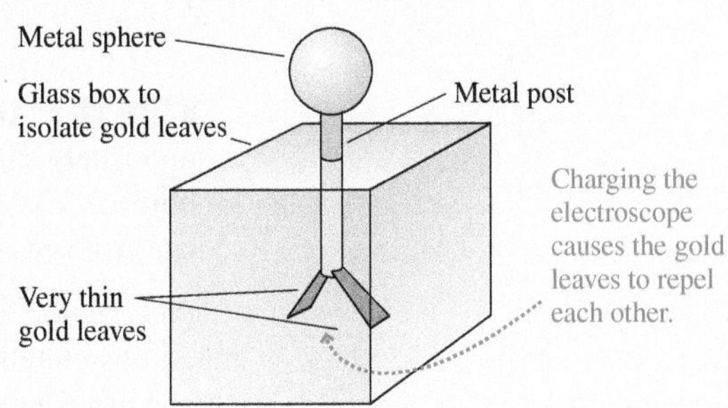
Metal sphere
Glass box to isolate gold leaves
Metal post
Charging the electroscope causes the gold leaves to repel each other.
Very thin gold leaves

FIGURE 21.5 Charging an electroscope.

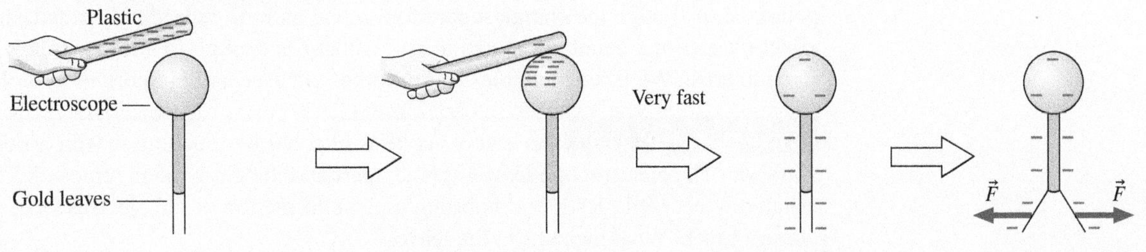
Plastic
Electroscope
Gold leaves
Very fast
$\vec{F}$
$\vec{F}$
1. Negative charges are transferred from the rod to the metal sphere upon contact.
2. Metal is a conductor. Therefore charge spreads (very rapidly) throughout the entire electroscope. The leaves become negatively charged.
3. Like charges repel. The negatively charged leaves exert repulsive forces on each other, causing them to spread apart.

Picking up pollen Rubbing a rod with a cloth gives the rod an electric charge. In a similar fashion, the rapid motion of a bee's wings through the air gives the bee a small positive electric charge. As small pieces of paper are attracted to a charged rod, so are tiny grains of pollen attracted to the charged bee, making the bee a much more effective pollinator.

Polarization

At the beginning of this chapter you saw a picture of a small feather being picked up by a piece of amber that had been rubbed with fur. The amber was charged by rubbing, but the feather was *neutral.* How can our charge model explain the attractive force that a charged object exerts on a neutral object?

Although a feather is an insulator, it's easiest to understand this phenomenon by first considering how a neutral *conductor* is attracted to a charged object. FIGURE 21.6 shows how this works. Because the charged rod doesn't touch the sphere, no charge is added to or removed from the sphere. Instead, the rod attracts some of the sphere's negative charge to the side of the sphere near the rod. This leaves a deficit of negative charge on the opposite side of the sphere, so that side is now positively charged. This slight *separation* of the positive and negative charge in a neutral object when a charged object is brought near is called **charge polarization.** We say that the object is *polarized.*

Figure 21.6 also shows that there's a net attractive force between the charged rod and the neutral sphere. The electric force weakens with distance, so the attractive force on the negative charges at the top of the sphere, which are closer to the rod, is larger than the repulsive force on the positive charges at the bottom of the sphere. This net force toward the charged rod, called a **polarization force,** arises because the charges are separated, *not* because the rod and the sphere are oppositely charged. A negatively charged rod would also exert an attractive force on the sphere because it would polarize the sphere to have positive charge at the top and negative charge at the bottom. **The polarization force is always an attractive force.**

FIGURE 21.6 The polarization force is due to charge separation. The net charge remains zero.

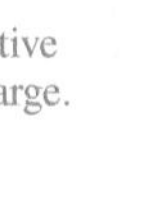
The neutral sphere contains equal amounts of positive and negative charge.

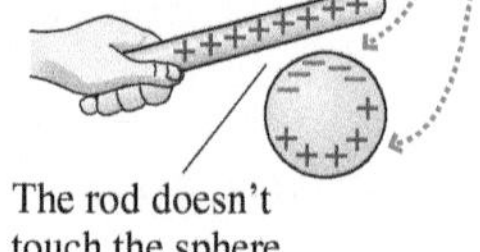
Negative charge is attracted to the positive rod. This leaves behind positive charge on the other side of the sphere.
The rod doesn't touch the sphere.

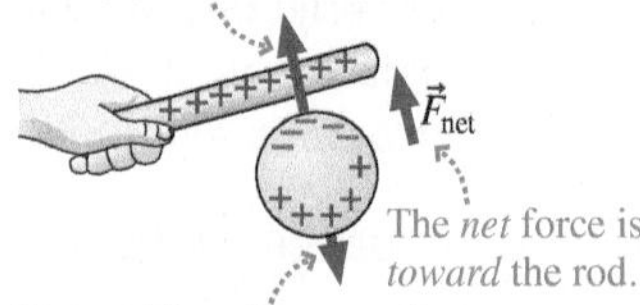
The negative charge on the sphere is close to the rod, so it is strongly attracted to the rod.
$\vec{F}_{net}$
The *net* force is *toward* the rod.
The positive charge on the sphere is far from the rod, so it is weakly repelled by the rod.

NOTE ▶ An attractive electric force is *not* an indication that the objects are oppositely charged. The force could be a polarization force that arises if one object is charged but the other is neutral. The only sure indicator that an object is charged is a repulsive force from a like charge. ◀

Polarization explains why forces arise between a charged object and a neutral metal object along which charge can freely move. But a feather is an insulator, and charge can't move through an insulator. Nevertheless, the attractive force between a charged object and a neutral insulator is also a polarization force. As you'll learn in the next section, the charge in each *atom* that makes up an insulator can be slightly polarized. Although the charge separation in one atom is exceedingly small, the net effect over all the countless atoms in an insulator is enough to allow a polarization force to arise. We'll see that polarization forces are especially important in biology.

STOP TO THINK 21.1 An electroscope is charged by touching it with a positive glass rod. The electroscope leaves spread apart and the glass rod is removed. Then a negatively charged plastic rod is brought close to the top of the electroscope, but it doesn't touch. What happens to the leaves?

A. The leaves move closer together.
B. The leaves spread farther apart.
C. The leaves do not change their position.

21.2 A Microscopic Model of Charge

We have been speaking about giving objects positive or negative charge without explaining what is happening at an atomic level. You already know that the basic constituents of atoms—the nucleus and the electrons surrounding it—are charged. In this section we will connect our observations of the previous section with our understanding of the atomic nature of matter.

Our current model of the atom is that it is made up of a very small and dense positively charged *nucleus* that contains positively charged *protons* as well as neutral particles called *neutrons.* The nucleus is surrounded by much-less-massive negatively charged *electrons* that form an **electron cloud,** as illustrated in FIGURE 21.7. The atom is held together by the attractive electric force between the positive nucleus and the negative electrons.

Experiments show that **charge, like mass, is an inherent property of electrons and protons.** It's no more possible to have an electron without charge than it is to have an electron without mass.

FIGURE 21.7 Our modern view of the atom.

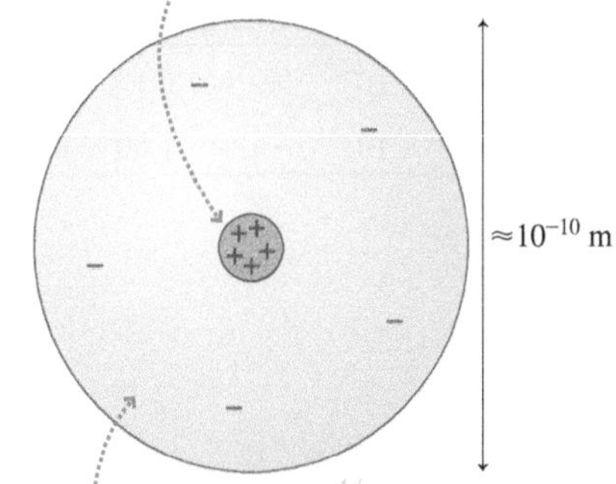

The nucleus, exaggerated in size for clarity, contains positive protons.
≈10^-10 m
The electron cloud is negatively charged.

An Atomic View of Charging

Electrons and protons are the basic charges in ordinary matter. **There are no other sources of charge.** Consequently, the observations we made in Section 21.1 need to be explained in terms of electrons and protons.

Experimentally, electrons and protons are found to have charges of opposite sign but *exactly* equal magnitude. Thus, because charge is due to electrons and protons, **an object is charged if it has an unequal number of electrons and protons.** An object with a negative charge has more electrons than protons; an object with a positive charge has more protons than electrons. Most macroscopic objects have an *equal number* of protons and electrons. Such an object has no *net* charge; we say it is *electrically neutral.*

In practice, objects acquire a positive charge not by gaining protons but by losing electrons. Protons are *extremely* tightly bound within the nucleus and cannot be added to or removed from atoms. Electrons, however, are bound much more loosely than the protons and can be removed with little effort.

The process of removing an electron from the electron cloud of an atom is called **ionization.** An atom that is missing an electron is called a *positive ion,* referred to in chemistry as a *cation.* Some atoms can accommodate an *extra* electron and thus become a *negative ion* or *anion.* FIGURE 21.8 shows positive and negative ions.

FIGURE 21.8 Positive and negative ions.

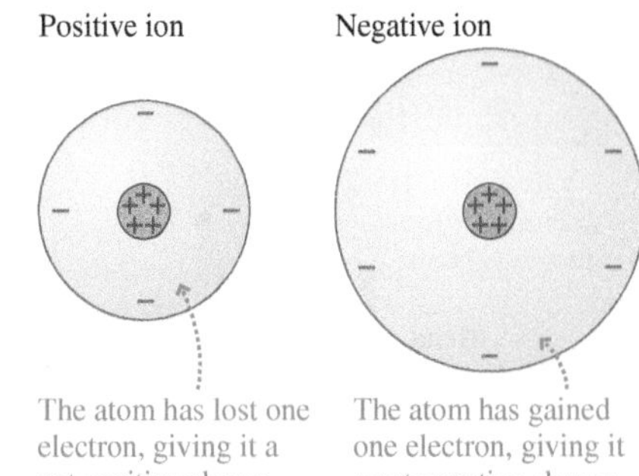

Positive ion
Negative ion
The atom has lost one electron, giving it a net positive charge.
The atom has gained one electron, giving it a net negative charge.

The charging processes we observed in Section 21.1 involved rubbing and friction. The forces of friction often cause molecular bonds at the surface to break as two materials slide past each other. Molecules are electrically neutral, but FIGURE 21.9 shows that *molecular ions* can be created when one of the bonds in a large molecule is broken. If the positive molecular ions remain on one material and the negative ions on the other, one of the objects being rubbed ends up with a net positive charge and the other with a net negative charge. This is how a plastic rod is charged by rubbing with wool or a comb is charged by passing through your hair.

Molecules in solution become charged by chemical rather than mechanical processes. Salts such as sodium chloride (NaCl) dissociate into oppositely charged ions (Na^+ and Cl^- in the case of NaCl) because it is energetically and entropically favorable to do so. Similar but more complex dissociations cause DNA and proteins to become charged—the basis for *electrophoresis,* as we'll discuss later in the chapter.

FIGURE 21.9 Charging by friction may result from molecular ions produced as bonds are broken.

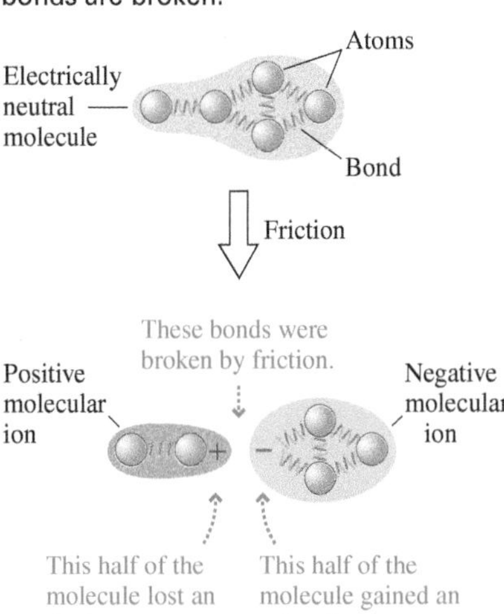

These bonds were broken by friction.

This half of the molecule lost an electron.

Units of Charge

Charge is represented by the symbol q (or sometimes Q). The SI unit of charge is the **coulomb** (C), named for French scientist Charles Coulomb, one of many scientists investigating electricity in the late 18th century.

Protons and electrons have the same amount of charge, but of opposite signs. We use the symbol e for the **fundamental charge,** the magnitude of the charge of an electron or a proton. The fundamental charge e has been measured to have the value

$$e = 1.60 \times 10^{-19}\ \text{C}$$

This is a very small amount of charge. Stated another way, 1 C is the net charge of roughly 6×10^{18} protons. TABLE 21.1 lists the masses and charges of protons and electrons.

TABLE 21.1 **Protons and electrons**

Particle	Mass (kg)	Charge (C)
Proton	1.67×10^{-27}	$+e = 1.60 \times 10^{-19}$
Electron	9.11×10^{-31}	$-e = -1.60 \times 10^{-19}$

NOTE ▶ The amount of charge produced by rubbing plastic or glass rods is typically in the range 1 nC (10^{-9} C) to 100 nC (10^{-7} C). This corresponds to an excess or deficit of 10^{10} to 10^{12} electrons. These may seem like large numbers, but for macroscopic objects they represent an excess or deficit of only perhaps 1 electron in 10^{13}. ◀

The fact that charge is associated with electrons and protons explains why charge is conserved. Because electrons and protons are neither created nor destroyed in ordinary processes, their associated charge is conserved as well.

Insulators and Conductors

We've seen that there are two classes of materials, insulators and conductors, that have very different electric properties. FIGURE 21.10 looks inside an insulator and two kinds of conductors. The electrons in the insulator are all tightly bound to the positive nuclei and so are not free to move around. Charging an insulator by friction leaves patches of molecular ions on the surface, but these patches are immobile.

FIGURE 21.10 A microscopic look at an insulator and two conductors.

Insulator

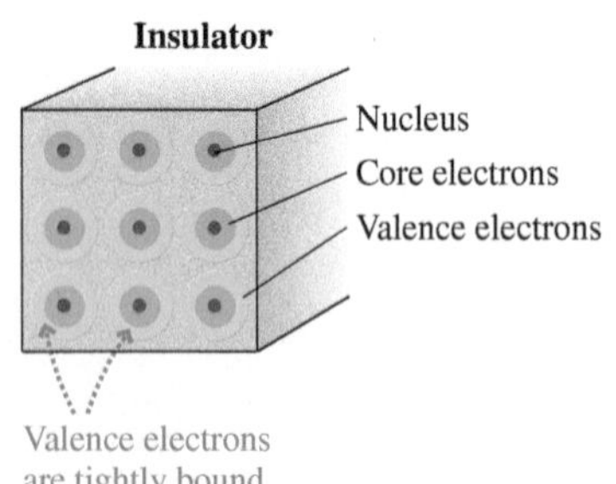
Nucleus
Core electrons
Valence electrons
Valence electrons
are tightly bound.

Metal

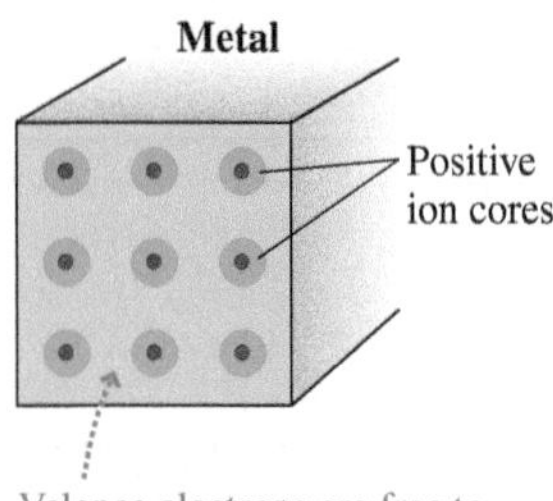
Positive
ion cores
Valence electrons are free to
move throughout the metal.

Ionic Solution

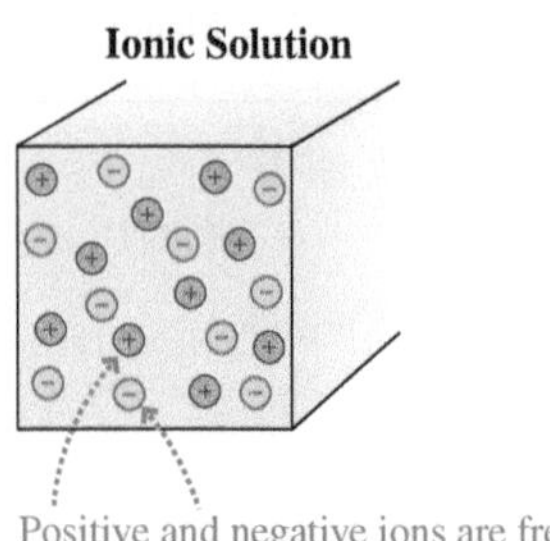
Positive and negative ions are free
to move through the solution.

In metals, the outer atomic electrons (called the *valence electrons* in chemistry) are only weakly bound to the nuclei. As the atoms come together to form a solid, these outer electrons become detached from their parent nuclei and are free to wander about through the entire solid. The solid *as a whole* remains electrically neutral, because we have not added or removed any electrons, but the electrons are now rather like a negatively charged gas or liquid—what physicists like to call a **sea of electrons**—permeating an array of positively charged **ion cores.**

The electrons in a metal are highly mobile. They can quickly and easily move through the metal in response to electric forces. The motion of charges through a conductor is what we'll later call a *current,* and the charges that physically move are called the **charge carriers.** The charge carriers in metals are electrons. Although the valence electrons are mobile, they are still attracted to and weakly bound to the ion cores; they cannot leave the metal.

Metals aren't the only conductors. Ionic solutions, such as salt water and the electrolytes in batteries, are also good conductors. The charge carriers in an ionic solution are the ions, not electrons. Both the positive and negative ions move—but in opposite directions—in response to an electric force. The human body is filled with salt water and is a good conductor except for the skin, which is an insulator as long as it stays dry.

The Electric Dipole

In the last section we saw that an insulator, such as a feather, is attracted to a charged object. An atomic-level explanation shows why. Consider what happens when we bring a positive charge near an atom. As FIGURE 21.11 shows, the charge *polarizes the atom.* The electron cloud doesn't move far, because the force from the positive nucleus pulls it back, but the center of positive charge and the center of negative charge are now slightly separated.

Two opposite charges with a slight separation between them form what is called an **electric dipole.** (The actual distortion from a perfect sphere is minuscule, nothing like the distortion shown in the figure.) The attractive force on the dipole's near end *slightly* exceeds the repulsive force on its far end because the near end is closer to the

FIGURE 21.11 A neutral atom is polarized by and attracted toward an external charge.

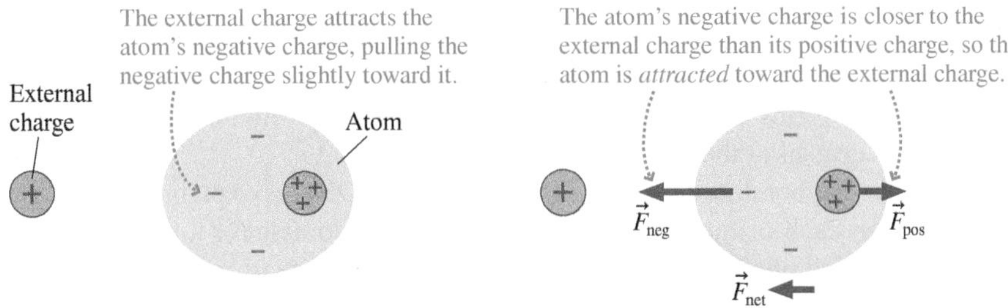

external charge. The net force, an *attractive* force between the charge and the atom, is another example of a polarization force.

An insulator has no sea of electrons to shift if an external charge is brought close. Instead, as FIGURE 21.12 shows, all the individual atoms inside the insulator become polarized. The polarization force acting *on each atom* produces a net polarization force toward the external charge. This solves the puzzle. A charged object picks up a feather by

- Polarizing the atoms in the feather,
- Then exerting an attractive polarization force on each atom.

FIGURE 21.12 The atoms in an insulator are polarized by an external charge.

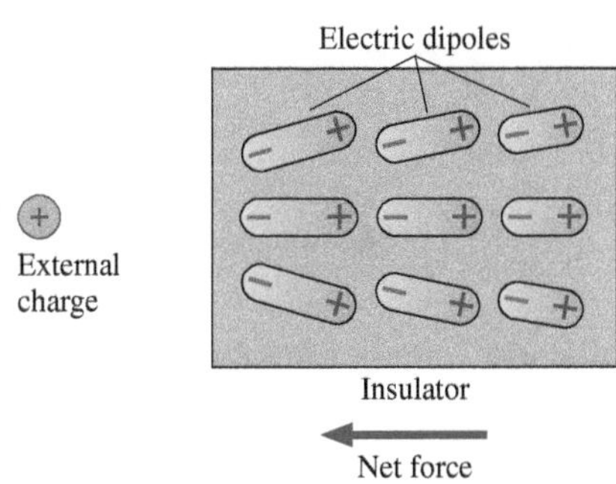

Polar Molecules and Hydrogen Bonding

Polarization is not limited to atoms. You may have encountered the idea of *polar molecules,* such as HCl and NH_3, in chemistry. Polar molecules have an asymmetry in their charge distribution that makes them *permanent electric dipoles.* An important example is the water molecule. Bonding between the hydrogen and oxygen atoms results in an unequal sharing of charge that, as shown in FIGURE 21.13a, leaves the hydrogen atoms with a small positive charge and the oxygen atom with a small negative charge.

One of the photographs on the opening page of this chapter shows a charged comb deflecting a stream of water. The reason is not that the water is charged but that water's permanent electric dipole allows the charged comb to exert an exceptionally large polarization force on the water.

Water molecules also interact directly with each other through their dipole moments. When two water molecules are close, the attractive electric force between the positive hydrogen atom of one molecule and the negative oxygen atom of the second molecule can form a weak bond, called a **hydrogen bond,** as illustrated in FIGURE 21.13b. These weak bonds create a certain "stickiness" between water molecules that is responsible for many of water's special properties, including its expansion on freezing, the wide range of temperatures over which it is liquid, and its high heat of vaporization.

FIGURE 21.13 Hydrogen bonding between water molecules.

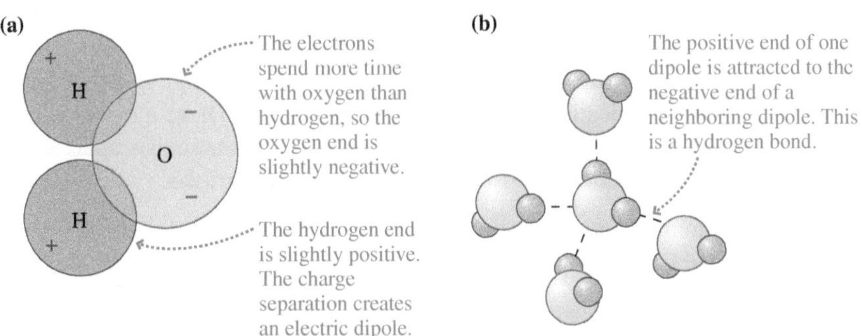

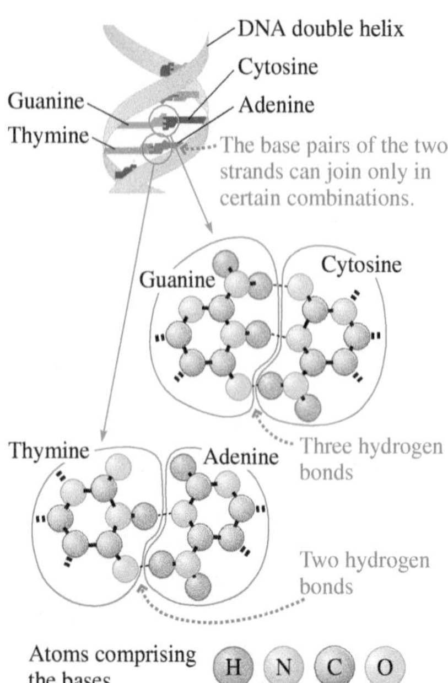

FIGURE 21.14 Hydrogen bonds in DNA base pairs.

Hydrogen bonds are essential to biochemistry. For example, information in DNA is coded in the *nucleotides,* the four molecules guanine, thymine, adenine, and cytosine. The nucleotides on one strand of the DNA helix form hydrogen bonds with the nucleotides on the opposite strand. The nucleotides bond only in certain pairs: Cytosine always forms a bond with guanine, adenine with thymine. This specific base pairing is crucial to the structure and function of DNA.

The preferential bonding of nucleotide base pairs in DNA is explained by hydrogen bonding. In each of the nucleotides, the hydrogen atoms have a small positive charge, oxygen and nitrogen a small negative charge. The positive hydrogen atoms on one nucleotide attract the negative oxygen or nitrogen atoms on another. As the detail in FIGURE 21.14 shows, the geometry of the nucleotides allows cytosine to form a hydrogen bond only with guanine, adenine only with thymine.

STOP TO THINK 21.2 Rank in order, from most positive to most negative, the charges q_A to q_E of these five systems.

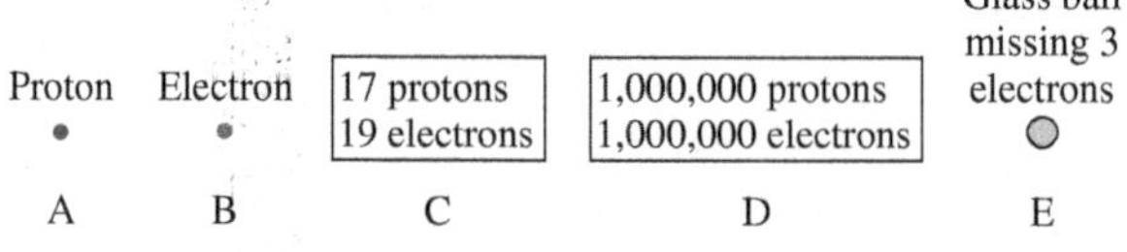

21.3 Coulomb's Law

The last two sections established a *model* of charges and electric forces. This model is very good at explaining electric phenomena and providing a general understanding of electricity. Now we need to make the model quantitative. Experiment 4 in Section 21.1 found that the electric force increases for objects with more charge and decreases as charged objects are moved farther apart. The force law that describes this behavior is known as *Coulomb's law.*

Coulomb's law

Magnitude: If two charged particles that have charges q_1 and q_2 are a distance r apart, the particles exert forces on each other of magnitude

$$F_{1 \text{ on } 2} = F_{2 \text{ on } 1} = \frac{K|q_1||q_2|}{r^2} \tag{21.1}$$

where the charges are in coulombs (C), and $K = 8.99 \times 10^9 \text{ N} \cdot \text{m}^2/\text{C}^2$ is called the **electrostatic constant.** These forces are an interaction pair, equal in magnitude and opposite in direction. It is customary to round K to $9.0 \times 10^9 \text{ N} \cdot \text{m}^2/\text{C}^2$ for all but extremely precise calculations, and we will do so.

Direction: The forces are directed along the line joining the two particles. The forces are *repulsive* for two like charges and *attractive* for two opposite charges.

We sometimes speak of the "force between charge q_1 and charge q_2," but keep in mind that we are really dealing with charged *objects* that also have a mass, a size, and other properties. Charge is not some disembodied entity that exists apart from matter. Coulomb's law describes the force between charged *particles.*

NOTE ▶ Coulomb's law applies only to **point charges.** A point charge is an idealized material object with charge and mass but with no size or extension. For practical purposes, two charged objects can be modeled as point charges if they are much smaller than the separation between them. ◀

Coulomb's law, like Newton's law of gravity in ◀◀ SECTION 5.7, is an *inverse-square law*; that is, the force depends inversely on the square of the distance between the charges. The force is only one-fourth as large if the separation is doubled. There's one key difference between the electric force and the gravitational force, however. Mass m is always positive, and the gravitational force is always attractive. But charge q can be either positive or negative, so the force can be attractive or repulsive. Consequently, the absolute value signs in Equation 21.1 are especially important. The first part of Coulomb's law gives only the *magnitude* of the force, which is always positive. The direction must be determined from the second part of the law. FIGURE 21.15 shows the forces between different combinations of positive and negative charges.

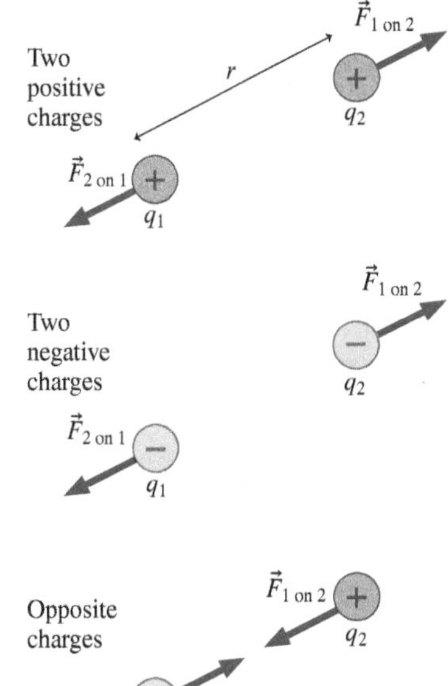

FIGURE 21.15 Attractive and repulsive forces between charged particles.

Using Coulomb's Law

Coulomb's law is a force law, and forces are vectors. **Electric forces, like other forces, can be superimposed.** If multiple charges 1, 2, 3, ... are present, the *net* electric force on charge j due to all other charges is therefore the sum of all the individual forces due to each charge; that is,

$$\vec{F}_{\text{net}} = \vec{F}_{1\text{ on }j} + \vec{F}_{2\text{ on }j} + \vec{F}_{3\text{ on }j} + \cdots \tag{21.2}$$

where each of the forces $\vec{F}_{i\text{ on }j}$ is given by Equation 21.1.

These ideas are the basis of a strategy for using Coulomb's law to solve electrostatic force problems.

PROBLEM-SOLVING STRATEGY 21.1 Electric forces and Coulomb's law

We can use Coulomb's law to find the electric force on a charged particle due to one or more other charged particles.

PREPARE Identify point charges or objects that can be modeled as point charges. Think about the forces involved, their likely directions and magnitudes. Draw a pictorial representation in which you establish a coordinate system, show the positions of the charges, show the force vectors on the charges, and define distances and angles.

SOLVE The magnitude of the force between point charges is given by Coulomb's law:

$$F_{1\text{ on }2} = F_{2\text{ on }1} = \frac{K|q_1||q_2|}{r^2}$$

- Show the directions of the forces—repulsive for like charges, attractive for opposite charges—on the pictorial representation.
- Draw the lengths of your force vectors according to Coulomb's law: A particle with a greater charge, or one that is closer, will lead to a longer force vector.
- When possible, do graphical vector addition on the pictorial representation. While not exact, it tells you the type of answer you should expect.
- Write each force vector in terms of its x- and y-components, then add the components to find the net force. Use the pictorial representation to determine which components are positive and which are negative.

ASSESS Check that your result has the correct units, is reasonable, and answers the question.

EXAMPLE 21.3 Adding electric forces in one dimension

Two +10 nC charged particles are 2.0 cm apart on the x-axis. What is the net force on a +1.0 nC charge midway between them? What is the net force on the +1.0 nC charge if the charged particle on the right is replaced by a −10 nC charged particle?

PREPARE We will proceed using the steps of Problem-Solving Strategy 21.1 We will model the charged particles as point charges. The pictorial representation of FIGURE 21.16 establishes a coordinate system and shows the forces $\vec{F}_{1\text{ on }3}$ and $\vec{F}_{2\text{ on }3}$. The +1.0 nC charge is repelled by positive charges and attracted to a negative charge. Figure 21.16a shows a +10 nC charged particle on the right; Figure 21.16b shows a −10 nC charged particle.

SOLVE Electric forces are vectors, and the net force on q_3 is the *vector* sum $\vec{F}_{\text{net}} = \vec{F}_{1\text{ on }3} + \vec{F}_{2\text{ on }3}$. Charged particles q_1 and q_2 each exert a repulsive force on q_3, but these forces are equal in magnitude and opposite in direction. Consequently, $\vec{F}_{\text{net}} = \vec{0}$. The situation changes if q_2 is negative, as in Figure 21.16b. In this case, the two forces are equal in magnitude but in the *same* direction, so $\vec{F}_{\text{net}} = 2\vec{F}_{1\text{ on }3}$. The magnitude of the force is given by Coulomb's law. The force due to q_1 is

$$F_{1\text{ on }3} = \frac{K|q_1||q_3|}{r_{13}^2}$$

$$= \frac{(9.0 \times 10^9\ \text{N}\cdot\text{m}^2/\text{C}^2)(10 \times 10^{-9}\ \text{C})(1.0 \times 10^{-9}\ \text{C})}{(0.010\ \text{m})^2}$$

$$= 9.0 \times 10^{-4}\ \text{N}$$

FIGURE 21.16 A pictorial representation of the forces for the two cases.

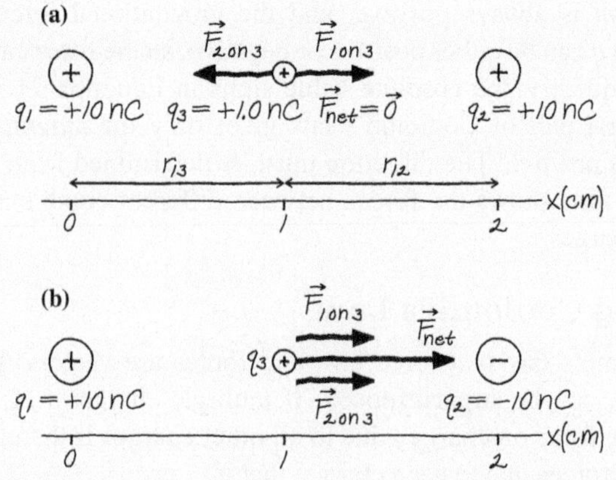

There is an equal force due to q_2, so the net force on the 1.0 nC charge is $\vec{F}_{\text{net}} = (1.8 \times 10^{-3}\ \text{N}$, to the right).

ASSESS This example illustrates the important idea that electric forces are *vectors*. An important part of assessing our answer is to see if it is "reasonable." In the second case, the net force on the charge is approximately 1 mN. Generally, charges of a few nC separated by a few cm experience forces in the range from a fraction of a mN to several mN. With this guideline, the answer appears to be reasonable.

EXAMPLE 21.4 Comparing electric and gravitational forces

A small plastic sphere is charged to −10 nC. It is held 1.0 cm above a small glass bead at rest on a table. The bead has a mass of 15 mg and a charge of +10 nC. Will the glass bead "leap up" to the plastic sphere?

PREPARE We'll model the bead and the sphere as point charges and measure distances between their centers. There is an attractive electric force from the sphere that will tend to pull the bead upward. If this force is greater than the downward weight force, then the bead will leap upward. The pictorial representation in FIGURE 21.17 establishes a y-axis, identifies the plastic sphere as q_1 and the glass bead as q_2, and shows a free-body diagram. The glass bead will rise if $F_{1\text{ on }2} > w$; if $F_{1\text{ on }2} < w$, the bead will remain at rest on the table, which then exerts a normal force $\vec{n}$ on the bead.

SOLVE Using the values provided, we have

$$F_{1\text{ on }2} = \frac{K|q_1||q_2|}{r^2} = 9.0 \times 10^{-3}\ \text{N}$$

$$w = m_2 g = 1.5 \times 10^{-4}\ \text{N}$$

$F_{1\text{ on }2}$ exceeds the bead's weight by a factor of 60, so the glass bead will leap upward.

ASSESS The values used in this example are realistic for spheres ≈2 mm in diameter. In general, as this example shows, electric forces are *much* larger than weight forces. You can neglect gravity when working electric-force problems.

FIGURE 21.17 A pictorial representation showing the charges and forces.

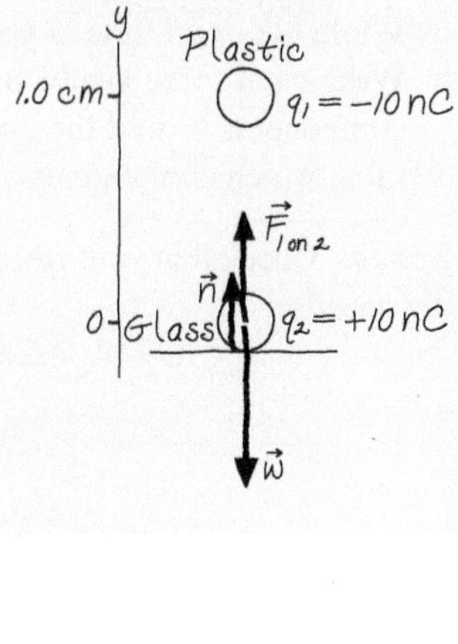

EXAMPLE 21.5 Three charges

Three charged particles with charges $q_1 = -50$ nC, $q_2 = +50$ nC, and $q_3 = +30$ nC are placed on the corners of the 5.0 cm × 10.0 cm rectangle shown in FIGURE 21.18. What is the net force on charge q_3 due to the other two charges?

FIGURE 21.18 The three charges.

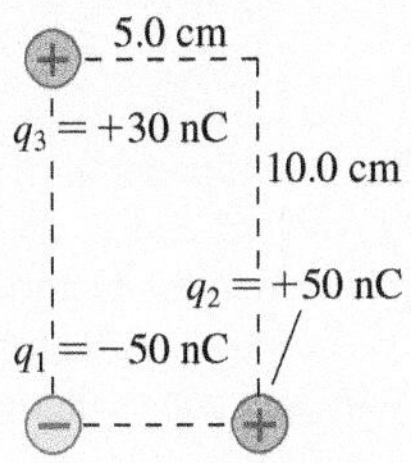

PREPARE We can model the charged particles as point charges. The pictorial representation of FIGURE 21.19 establishes a coordinate system. q_1 and q_3 are opposite charges, so force vector $\vec{F}_{1 \text{ on } 3}$ is an attractive force toward q_1. q_2 and q_3 are like charges, so force vector $\vec{F}_{2 \text{ on } 3}$ is a repulsive force away from q_2. q_1 and q_2 have equal magnitudes, but $\vec{F}_{2 \text{ on } 3}$ has been drawn shorter than $\vec{F}_{1 \text{ on } 3}$ because q_2 is farther from q_3. Vector addition has been used to draw the net force vector $\vec{F}_3$ and to define its angle ϕ.

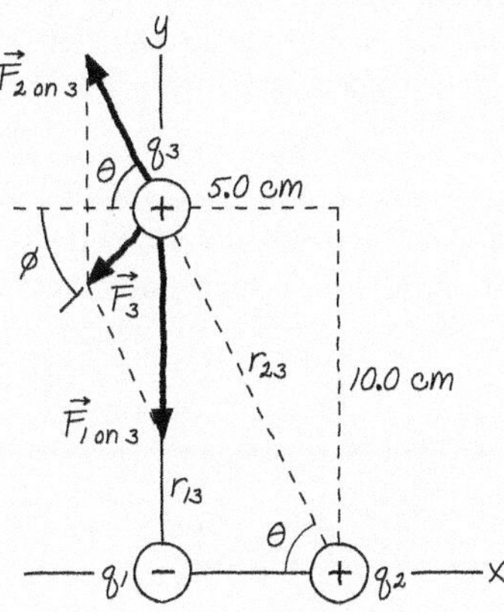

FIGURE 21.19 A pictorial representation of the charges and forces.

SOLVE The electric force, like all forces, is a vector. It has been many chapters since we made extensive use of vectors, so let's undertake a brief review. Recall that vectors can be decomposed into *components*. For example, vector $\vec{A}$ in FIGURE 21.20a has components

$$A_x = A\cos\theta$$

$$A_y = A\sin\theta$$

where A is the *magnitude* of the vector. If we know the components, we can use the Pythagorean theorem to find the magnitude of $\vec{A}$:

$$A = |\vec{A}| = \sqrt{A_x^2 + A_y^2}$$

FIGURE 21.20 Using vectors.

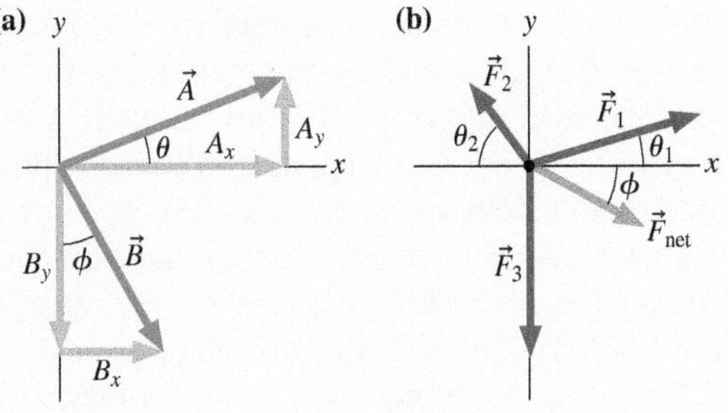

The angle used to label a vector's direction is our choice; we use an angle that's convenient. Vector $\vec{A}$ is measured from the $+x$-axis, but vector $\vec{B}$ is measured from the $-y$-axis. Thus

$$B_x = B\sin\phi$$

$$B_y = -B\cos\phi$$

The negative sign of B_y comes from *looking at the drawing*. Getting the signs right is essential, but there's no automatic formula for doing so; we have to pay attention to the direction in which a vector points. If we know the components, we can use trigonometry to find the direction of $\vec{B}$:

$$\phi = \tan^{-1}\left|\frac{B_x}{B_y}\right|$$

Only the magnitudes of the components are needed to find the angle, not their signs. However, we again need to *look at the drawing* to see if we need the arctangent of the x-over-y components or the y-over-x components.

Forces are added using *vector addition*. If three electric forces act on a charge, as shown in FIGURE 21.20b, the net force is $\vec{F}_{net} = \vec{F}_1 + \vec{F}_2 + \vec{F}_3$. We could add graphically with the parallelogram method of Chapter 4, but most often we use components:

$$(F_{net})_x = F_{1x} + F_{2x} + F_{3x} = F_1\cos\theta_1 - F_2\cos\theta_2$$

$$(F_{net})_y = F_{1y} + F_{2y} + F_{3y} = F_1\sin\theta_1 + F_2\sin\theta_2 - F_3$$

Thus we use this procedure:

- Determine individual force vectors.
- Decompose the vectors into x- and y-components, paying attention to directions to get the signs right.
- Add the components to find the components of the net force or the net field.
- Use the Pythagorean theorem and an arctangent to find the magnitude of the vector and a direction, such as angle ϕ.

The question asks for the force on q_3, so our answer will be the *vector* sum $\vec{F}_3 = \vec{F}_{1 \text{ on } 3} + \vec{F}_{2 \text{ on } 3}$. We need to determine the components of $\vec{F}_{1 \text{ on } 3}$ and $\vec{F}_{2 \text{ on } 3}$. The magnitude of force $\vec{F}_{1 \text{ on } 3}$ can be found using Coulomb's law:

$$F_{1 \text{ on } 3} = \frac{K|q_1||q_3|}{r_{13}^2}$$

$$= \frac{(9.0 \times 10^9\ \text{N}\cdot\text{m}^2/\text{C}^2)(50 \times 10^{-9}\ \text{C})(30 \times 10^{-9}\ \text{C})}{(0.100\ \text{m})^2}$$

$$= 1.35 \times 10^{-3}\ \text{N}$$

where we used $r_{13} = 10.0$ cm $= 0.100$ m.

The pictorial representation in Figure 21.19 shows that $\vec{F}_{1 \text{ on } 3}$ points downward, in the negative y-direction, so

$$(F_{1 \text{ on } 3})_x = 0\ \text{N}$$

$$(F_{1 \text{ on } 3})_y = -1.35 \times 10^{-3}\ \text{N}$$

To calculate $\vec{F}_{2 \text{ on } 3}$ we first need the distance r_{23} between the charges:

$$r_{23} = \sqrt{(5.0\ \text{cm})^2 + (10.0\ \text{cm})^2} = 11.2\ \text{cm}$$

Continued

The magnitude of $\vec{F}_{2\text{ on }3}$ is thus

$$F_{2\text{ on }3} = \frac{K|q_2||q_3|}{r_{23}^2}$$

$$= \frac{(9.0 \times 10^9\ \text{N}\cdot\text{m}^2/\text{C}^2)(50 \times 10^{-9}\ \text{C})(30 \times 10^{-9}\ \text{C})}{(0.112\ \text{m})^2}$$

$$= 1.08 \times 10^{-3}\ \text{N}$$

We need the angle θ, which is defined in Figure 21.19, before we can calculate the components of $\vec{F}_{2\text{ on }3}$. We can use trigonometry to find

$$\theta = \tan^{-1}\left(\frac{10.0\ \text{cm}}{5.0\ \text{cm}}\right) = \tan^{-1}(2.0) = 63.4°$$

The pictorial representation shows that the x-component of $\vec{F}_{2\text{ on }3}$ is negative and the y-component is positive. Thus

$$(F_{2\text{ on }3})_x = -F_{2\text{ on }3}\cos\theta = -4.83 \times 10^{-4}\ \text{N}$$

$$(F_{2\text{ on }3})_y = F_{2\text{ on }3}\sin\theta = 9.66 \times 10^{-4}\ \text{N}$$

We can now add $\vec{F}_{1\text{ on }3}$ and $\vec{F}_{2\text{ on }3}$:

$$(F_3)_x = (F_{1\text{ on }3})_x + (F_{2\text{ on }3})_x = -4.83 \times 10^{-4}\ \text{N}$$

$$(F_3)_y = (F_{1\text{ on }3})_y + (F_{2\text{ on }3})_y = -3.84 \times 10^{-4}\ \text{N}$$

Both components are negative, which tells us that $\vec{F}_3$ points downward and to the left. This agrees with the graphical vector addition in the pictorial representation.

Two more calculations give the magnitude and direction of $\vec{F}_3$. With angle ϕ as defined in Figure 21.19, these are

$$F_3 = \sqrt{F_{3x}^2 + F_{3y}^2} = 6.2 \times 10^{-4}\ \text{N}$$

$$\phi = \tan^{-1}\left|\frac{F_{3y}}{F_{3x}}\right| = 38°$$

Thus $\vec{F}_3 = (6.2 \times 10^{-4}\ \text{N}, 38°$ below the negative x-axis).

ASSESS The forces are not large, but they are typical of electrostatic forces. Even so, you'll soon see that these forces can produce very large accelerations because the masses of the charged objects are usually very small.

STOP TO THINK 21.3 Charged spheres A and B exert repulsive forces on each other. $q_A = 4q_B$. Which statement is true?

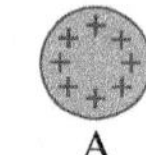
A

B

A. $F_{A\text{ on }B} > F_{B\text{ on }A}$ B. $F_{A\text{ on }B} = F_{B\text{ on }A}$ C. $F_{A\text{ on }B} < F_{B\text{ on }A}$

FIGURE 21.21 Visualizing the electric field.

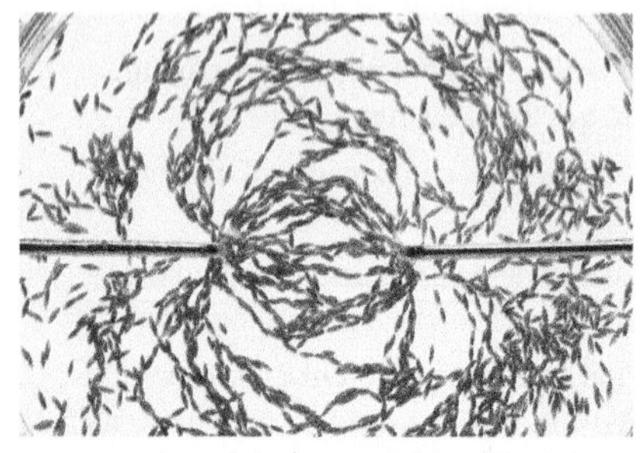

FIGURE 21.22 The force and field models for the interaction between two particles.

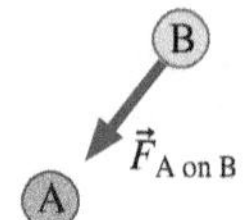
B
$\vec{F}_{A\text{ on }B}$
A
In the force model, A exerts a force directly on B.

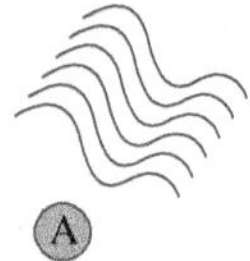
A
In the field model, A alters the space around it. (The wavy lines are poetic license. We'll soon learn a better representation.)

B
$\vec{F}_{\text{field on }B}$
Particle B then responds to the altered space. The altered space is the agent that exerts the force on B.

21.4 The Electric Field

The electric force, like the gravitational force, is a long-range interaction; no contact is required for one charged particle to exert a force on another. But this raises some troubling issues. Suppose a negatively charged particle is attracted to a positively charged particle. How does the negative charge "know" that the positive charge is there? Coulomb's law tells us how to calculate the magnitude and direction of the force, but it doesn't tell us how the force is transmitted through empty space from one charge to the other. To answer this question, we will introduce the *field model,* first suggested in the early 19th century by Michael Faraday, a British investigator of electricity and magnetism.

FIGURE 21.21 shows a photograph of the surface of a shallow pan of oil with tiny grass seeds floating on it. When charged wires, one positive and one negative, touch the surface of the oil, the grass seeds line up to form a regular pattern. The pattern suggests that some kind of electric influence from the charges *fills the space* around the charges. Perhaps the grass seeds are reacting to this influence, creating the pattern that we see. This alteration of the space around the charges could be the *mechanism* by which the long-range Coulomb's law force is exerted.

This is the essence of the field model. Consider an attractive force between particle A and particle B. FIGURE 21.22 shows the difference between the force model, which we have been using, and the field model. **In the field model, A alters the space around it, and the alteration of space is the agent that exerts a force on B.** This alteration of space is what we call a **field.** In other words, B responds to the field of A rather than directly to A. This is a *local* interaction, rather like a contact force.

The field model applies to many branches of physics. The space around a charge is altered to create the **electric field.** The alteration of the space around a mass is called the **gravitational field.** The alteration of the space around a magnet is called the **magnetic field,** which we will consider in Chapter 26.

The field idea was not taken seriously at first; it seemed more like a cartoon. But the importance of the field concept grew throughout the 19th century, and the discovery that light is an electromagnetic wave established the fact that fields are real.

NOTE ▶ The concept of a field is in sharp contrast to the concept of a particle. A particle exists at one *point* in space. Newton's laws tell us how the particle moves from point to point along a trajectory. A field exists simultaneously at *all* points in space. ◀

A platypus bill The platypus uses its unusual bill to forage on the bottoms of creek and ponds. Its bill is sensitive to the electric fields created by heartbeats and muscle action, helping the platypus find its dinner. Many fish and insects have an electric sense that helps them navigate and find prey. The electric field is very real to these creatures.

The Field Model

We begin our investigation of electric fields by postulating a field model that describes how charges interact:

1. A group of charges, which we call the **source charges,** alter the space around them by creating an *electric field* $\vec{E}$.
2. If another charge q is then placed in this electric field, it experiences a force $\vec{F}$ exerted *by the field.* That is, the source charges exert a force on q through the electric field that they create.

Suppose the positive charge q in FIGURE 21.23a experiences an electric force $\vec{F}_{\text{on } q}$ due to other charges. The strength and direction of this force vary as q is moved from point to point in space. We define the electric field $\vec{E}$ at the point (x, y, z) as

$$\vec{E} \text{ at } (x, y, z) = \frac{\vec{F}_{\text{on } q} \text{ at } (x, y, z)}{q} \tag{21.3}$$

Electric field at a point defined by the force on charge q

FIGURE 21.23 Charge q is a probe of the electric field.

(a)

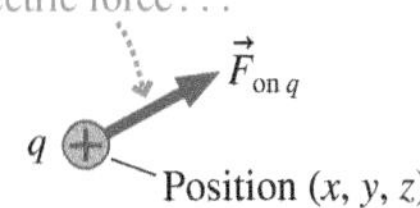
If the probe charge feels
an electric force . . .
$\vec{F}_{\text{on } q}$
q
Position (x, y, z)

(b)

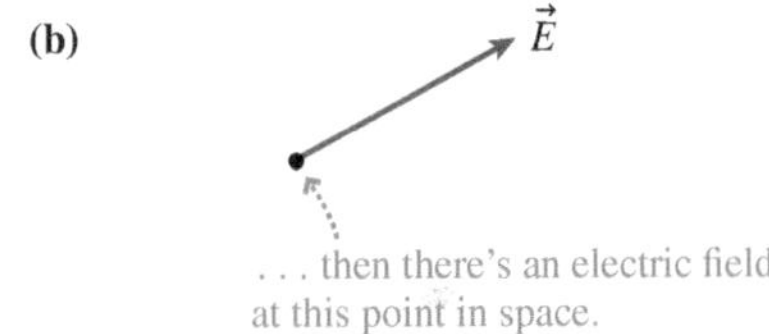
$\vec{E}$
. . . then there's an electric field
at this point in space.

We're *defining* the electric field as a force-to-charge ratio; hence the units of the electric field are newtons per coulomb, or N/C. The magnitude E of the electric field is called the **electric field strength.**

You can think of charge q as a *probe* to determine whether an electric field is present. If charge q experiences an electric force at a point in space, then, as FIGURE 21.23b shows, there is an electric field at that point. We can imagine "mapping out" the electric field by moving the charge q throughout all space.

TABLE 21.2 lists some typical electric field strengths. Fields in air—the main topic of this chapter—are generally between 1 N/C and 10^6 N/C. The earth has an electric field of about 100 N/C that is maintained by thunderstorms. A field of 3×10^6 N/C in air will cause a spark to jump and discharge the objects that are creating the field. The field inside a cell membrane is a factor of 10 larger than this, which suggests that electric interactions are important in the cell. We'll see numerous examples in the coming chapters.

TABLE 21.2 **Typical electric field strengths**

Field	Field strength (N/C)
Inside a current-carrying wire	10^{-2}
Earth's field, near the earth's surface	10^2
Near objects charged by rubbing	10^3 to 10^6
Needed to cause a spark in air	3×10^6
Inside a cell membrane	10^7
Inside an atom	10^{11}

The basic idea of the field model is that **the field is the agent that exerts an electric force on a particle with charge q.** Notice three important things about the field:

1. The electric field, a vector, exists at every point in space. Electric field diagrams show a sample of the vectors, but there is an electric field vector at every point whether one is shown or not.
2. If the probe charge q is positive, the electric field vector points in the same direction as the force on the charge; if negative, the electric field vector points opposite the force.
3. Because q appears in Equation 21.3, you might think that the electric field depends on the magnitude of the charge used to probe the field. It doesn't. We know from Coulomb's law that the force $\vec{F}_{\text{on } q}$ is proportional to q. Thus the electric field defined in Equation 21.3 is *independent* of the charge q that probes the field. The electric field depends on only the source charges that create the field.

FIGURE 21.24 Charge q' is used to probe the electric field of point charge q.

(a) What is the electric field of q at this point?

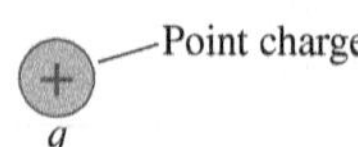

(b)

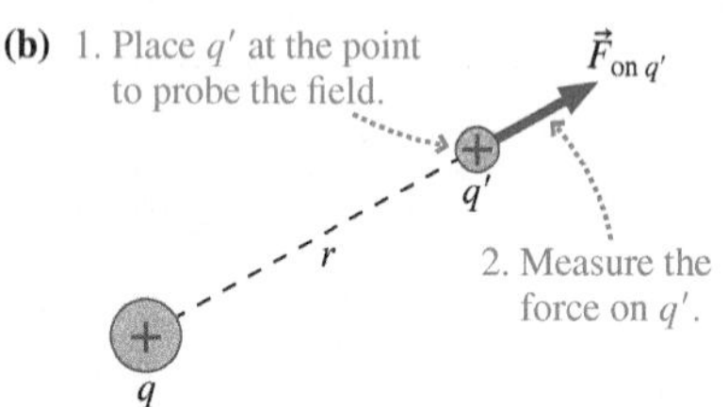

(c)

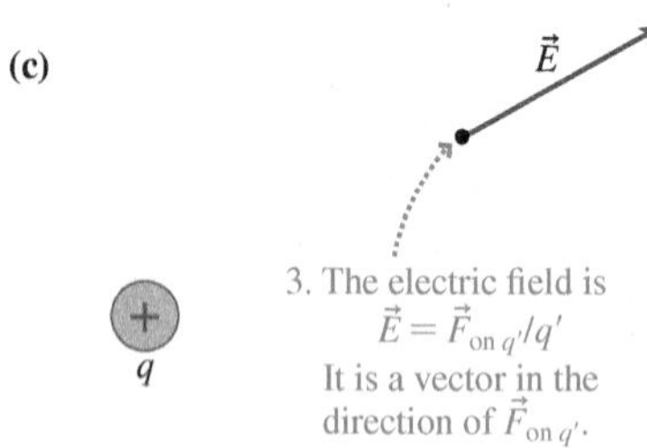

FIGURE 21.25 The electric field of positive and negative point charges.

(a)

(b)

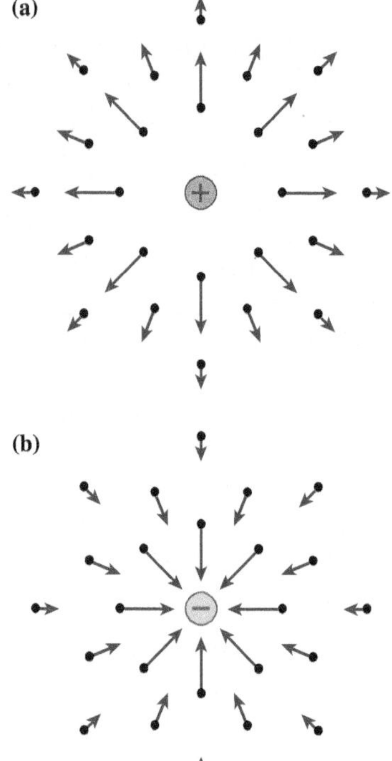

The Electric Field of a Point Charge

FIGURE 21.24a shows a point source charge q that creates an electric field at all points in space. We can use a second charge, shown as q' in FIGURE 21.24b, as a probe of the electric field created by charge q.

For the moment, assume both charges are positive. The force on q', which is repulsive and points directly away from q, is given by Coulomb's law:

$$\vec{F}_{\text{on } q'} = \left(\frac{Kqq'}{r^2}, \text{ away from } q\right) \tag{21.4}$$

Equation 21.3 defines the electric field in terms of the force on the probe charge as $\vec{E} = \vec{F}_{\text{on } q'}/q'$, so for a positive charge q,

$$\vec{E} = \left(\frac{Kq}{r^2}, \text{ away from } q\right) \tag{21.5}$$

The electric field is shown in FIGURE 21.24c.

If q is negative, the magnitude of the force on the probe charge is the same as in Equation 21.5, but the direction is toward q. Thus a general expression for the field is

$$\vec{E} = \left(\frac{K|q|}{r^2}, \begin{cases} \text{away} & q \text{ positive} \\ \text{toward} & q \text{ negative} \end{cases}\right) \tag{21.6}$$

Electric field of point charge q at a distance r from the charge

NOTE ▶ The expression for the electric field is similar to Coulomb's law. To distinguish the two, remember that Coulomb's law has the product of two charges in the numerator. It describes the force between *two* charges. The electric field has a single charge in the numerator. It is the field of a *single* charge. ◀

It's surprising, but we will find that Coulomb's law is not explicitly used in much of the theory of electricity. Most of our future analysis and calculations will be in terms of electric fields and potentials. It turns out that many equations are simpler if we rewrite Equation 21.6 with a different constant. Let's define the **permittivity constant** ϵ_0 (pronounced "epsilon zero" or "epsilon naught") as

$$\epsilon_0 = \frac{1}{4\pi K} = 8.85 \times 10^{-12}\ \text{C}^2/\text{N}\cdot\text{m}^2$$

Then we can rewrite Equation 21.6 as

$$\vec{E} = \left(\frac{1}{4\pi\epsilon_0}\frac{|q|}{r^2}, \begin{cases} \text{away} & q \text{ positive} \\ \text{toward} & q \text{ negative} \end{cases}\right) \tag{21.7}$$

Both expressions give the same electric field. And, in practice, we'll often use the value $9.0 \times 10^9\ \text{N}\cdot\text{m}^2/\text{C}^2$ for $1/4\pi\epsilon_0$.

By drawing electric field vectors at a number of points around a positive point charge, we can construct an **electric field diagram** such as the one shown in FIGURE 21.25a. Notice that the field vectors all point straight away from charge q. We can draw a field diagram for a negative point charge in a similar fashion, as in FIGURE 21.25b. In this case, the field vectors point toward the charge, as this would be the direction of the force on a positive probe charge.

In the coming sections, as we use electric field diagrams, keep these ideas in mind:

- The diagram is just a representative sample of electric field vectors. The field exists at all the other points. A well-drawn diagram gives a good indication of what the field would be like at a neighboring point.
- The arrow indicates the direction and the strength of the electric field *at the point to which it is attached*—at the point where the *tail* of the vector is placed. The length of any vector is significant only relative to the lengths of other vectors.

- Although we have to draw a vector across the page, from one point to another, an electric field vector does not "stretch" from one point to another. Each vector represents the electric field at *one point* in space.

EXAMPLE 21.6 **Calculating the electric field**

A -1.0 nC charged particle is located at the origin. Points 1, 2, and 3 have (x, y) coordinates (1 cm, 0 cm), (0 cm, 1 cm), and (1 cm, 1 cm), respectively. Determine the electric field $\vec{E}$ at these points, then show the vectors on an electric field diagram.

PREPARE The electric field—the field of a negative point charge—points straight *toward* the origin. It will be weaker at (1 cm, 1 cm), which is farther from the charge.

SOLVE The charge is $q = -1.0\text{ nC} = -1.0 \times 10^{-9}\text{ C}$. The distances to points 1 and 2 are equal: $r_1 = r_2 = 1.0\text{ cm} = 0.010\text{ m}$. The electric field strength at these two points is

$$E_1 = E_2 = \frac{K|q|}{r_1^2} = \frac{1}{4\pi\epsilon_0}\frac{|q|}{r_1^2}$$

$$= \frac{(9.0 \times 10^9\text{ N}\cdot\text{m}^2/\text{C}^2)(1.0 \times 10^{-9}\text{ C})}{(0.010\text{ m})^2} = 90{,}000\text{ N/C}$$

Point 1 is on the positive x-axis, so $\vec{E}_1$ points to the left toward the charge. Point 2 is on the positive y-axis, so $\vec{E}_2$ points down. Thus the electric fields at the first two points are

$$\vec{E}_1 = (90{,}000\text{ N/C, left})$$

$$\vec{E}_2 = (90{,}000\text{ N/C, down})$$

Point 3 is at distance $r_3 = \sqrt{2} \times 1\text{ cm} = 0.0141\text{ m}$. The field strength at this distance is

$$E_3 = \frac{1}{4\pi\epsilon_0}\frac{|q|}{r_3^2}$$

$$= \frac{(9.0 \times 10^9\text{ N}\cdot\text{m}^2/\text{C}^2)(1.0 \times 10^{-9}\text{ C})}{(0.0141\text{ m})^2} = 45{,}000\text{ N/C}$$

Point 3 is at a 45° angle above the x-axis. The field points toward the origin, so

$$\vec{E}_3 = (45{,}000\text{ N/C, left and down at 45°})$$

These vectors are shown on the electric field diagram of **FIGURE 21.26**.

FIGURE 21.26 The electric field diagram of a -1.0 nC charged particle.

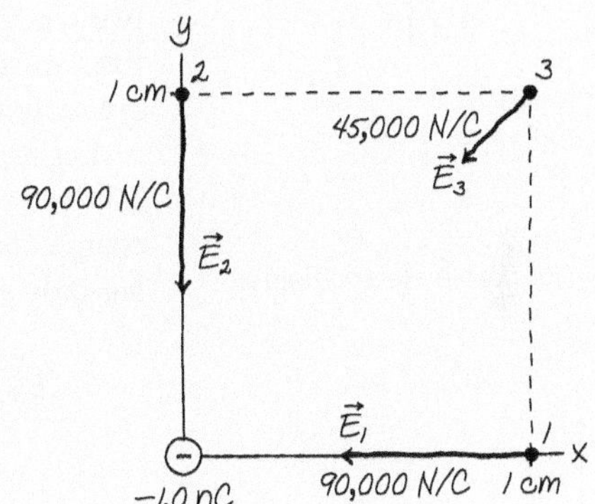

STOP TO THINK 21.4 Rank in order, from largest to smallest, the electric field strengths E_1 to E_4 at points 1 to 4.

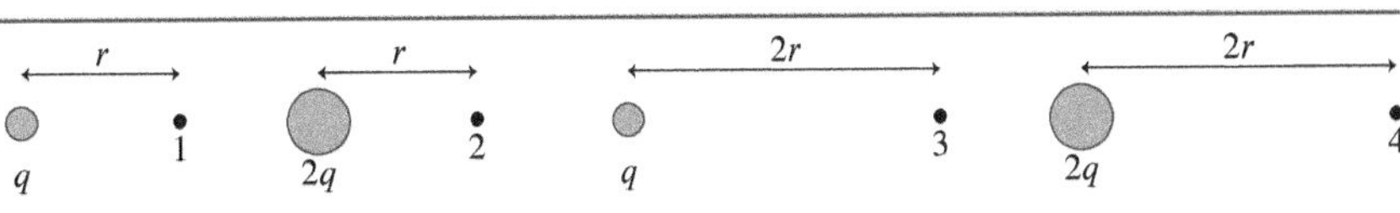

21.5 The Electric Field of Multiple Charges

What happens if there is more than one charge? The electric field was defined as $\vec{E} = \vec{F}_{\text{on } q}/q$, where $\vec{F}_{\text{on } q}$ is the electric force on charge q. Forces add as vectors, so the net force on q due to a group of point charges is the vector sum

$$\vec{F}_{\text{on } q} = \vec{F}_{1 \text{ on } q} + \vec{F}_{2 \text{ on } q} + \cdots$$

Consequently, the net electric field due to a group of point charges is

$$\vec{E}_{\text{net}} = \frac{\vec{F}_{\text{on } q}}{q} = \frac{\vec{F}_{1 \text{ on } q}}{q} + \frac{\vec{F}_{2 \text{ on } q}}{q} + \cdots = \vec{E}_1 + \vec{E}_2 + \cdots = \sum_i \vec{E}_i \qquad (21.8)$$

where $\vec{E}_i$ is the field from point charge i. That is, **the net electric field is the *vector sum* of the electric fields due to each charge.** In other words, electric fields obey the *principle of superposition.*

The Electric Field of a Dipole

FIGURE 21.27 Permanent and induced electric dipoles.

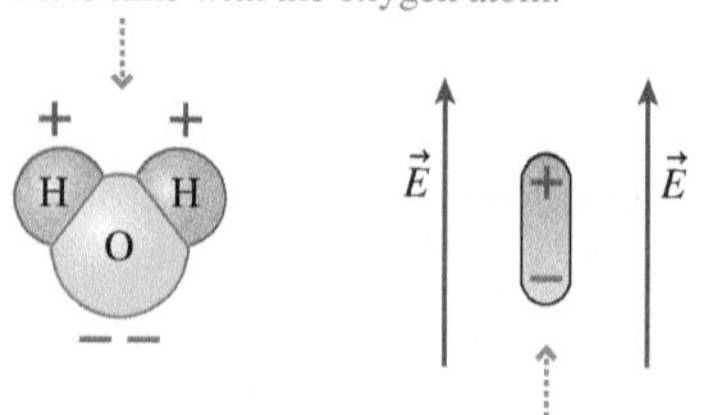
A water molecule is a *permanent* dipole because the negative electrons spend more time with the oxygen atom.
H
H
O
$\vec{E}$
$\vec{E}$
This dipole is *induced*, or stretched, by the electric field acting on the + and − charges.

Two equal but opposite charges separated by a small distance form an *electric dipole.* FIGURE 21.27 shows two examples. In a *permanent electric dipole,* such as the water molecule, the oppositely charged particles maintain a small permanent separation. We can also create an electric dipole, as you learned in Section 21.2, by polarizing a neutral atom with an external electric field. This is an *induced electric dipole.*

We can represent an electric dipole, whether permanent or induced, by two opposite charges $\pm q$ separated by the small distance s, as shown in FIGURE 21.28. The dipole has zero net charge, but it *does* have an electric field. Consider a point on the positive y-axis. This point is slightly closer to $+q$ than to $-q$, so the fields of the two charges do not cancel. You can see in the figure that $\vec{E}_{\text{dipole}}$ points in the positive y-direction. Similarly, vector addition shows that $\vec{E}_{\text{dipole}}$ points in the negative y-direction at points along the x-axis.

Let's calculate the electric field of a dipole at a point on the axis of the dipole. This is the y-axis in Figure 21.28. The point is distance $r_+ = y - s/2$ from the positive charge and $r_- = y + s/2$ from the negative charge. The net electric field at this point has only a y-component, and the sum of the fields of the two point charges gives

FIGURE 21.28 The dipole electric field at two points.

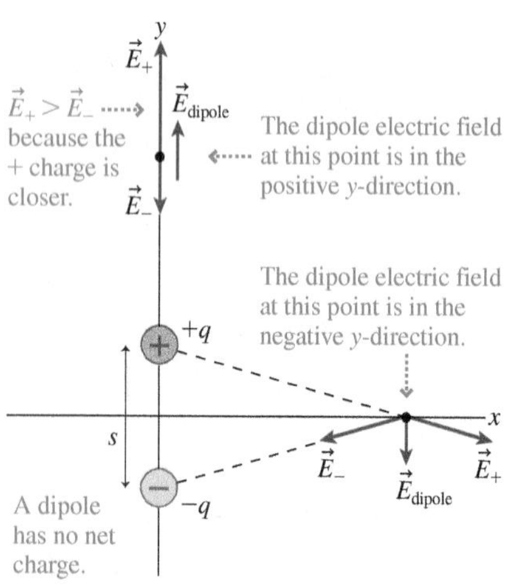
y
$\vec{E}_+$
$\vec{E}_+ > \vec{E}_-$ because the + charge is closer.
$\vec{E}_{\text{dipole}}$
The dipole electric field at this point is in the positive y-direction.
$\vec{E}_-$
The dipole electric field at this point is in the negative y-direction.
$+q$
x
s
$\vec{E}_-$
$\vec{E}_{\text{dipole}}$
$\vec{E}_+$
A dipole has no net charge.
$-q$

$$(E_{\text{dipole}})_y = (E_+)_y + (E_-)_y = \frac{1}{4\pi\epsilon_0}\frac{q}{(y-\frac{1}{2}s)^2} + \frac{1}{4\pi\epsilon_0}\frac{(-q)}{(y+\frac{1}{2}s)^2}$$
$$= \frac{q}{4\pi\epsilon_0}\left[\frac{1}{(y-\frac{1}{2}s)^2} - \frac{1}{(y+\frac{1}{2}s)^2}\right] \tag{21.9}$$

If we combine the two terms over a common denominator (and omit some of the algebraic steps), we find

$$(E_{\text{dipole}})_y = \frac{q}{4\pi\epsilon_0}\left[\frac{2ys}{(y-\frac{1}{2}s)^2(y+\frac{1}{2}s)^2}\right] \tag{21.10}$$

In practice, we almost always observe the electric field of a dipole at distances $y \gg s$—that is, for distances much larger than the charge separation. In such cases, the denominator can be approximated $(y-\frac{1}{2}s)^2(y+\frac{1}{2}s)^2 \approx y^4$. With this approximation, Equation 21.10 becomes

$$(E_{\text{dipole}})_y \approx \frac{1}{4\pi\epsilon_0}\frac{2qs}{y^3} \tag{21.11}$$

FIGURE 21.29 The dipole moment.

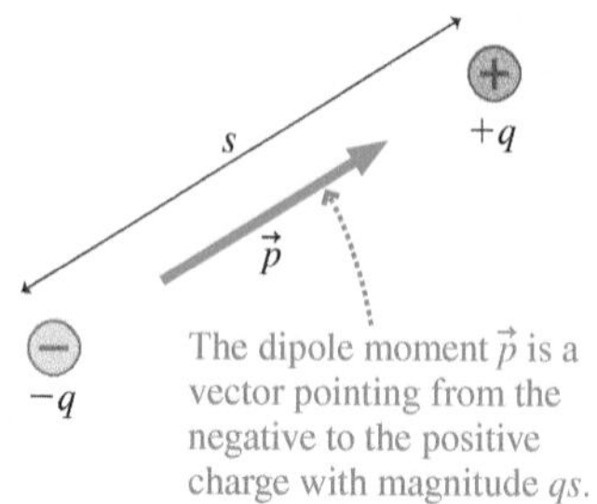
s
$+q$
$\vec{p}$
$-q$
The dipole moment $\vec{p}$ is a vector pointing from the negative to the positive charge with magnitude qs.

It is useful to define the **dipole moment** $\vec{p}$, shown in FIGURE 21.29, as the vector

$$\vec{p} = (qs, \text{ from the negative to the positive charge}) \tag{21.12}$$

The direction of $\vec{p}$ identifies the orientation of the dipole, and the dipole-moment magnitude $p = qs$ determines the electric field strength. The SI units of the dipole moment are C • m.

NOTE ► Some chemistry and biochemistry texts use the opposite convention in which the dipole moment points from positive to negative. ◄

We can use the dipole moment to write a succinct expression for the electric field at a point on the axis of a dipole:

$$\vec{E}_{\text{dipole}} \approx \frac{1}{4\pi\epsilon_0}\frac{2\vec{p}}{r^3} \tag{21.13}$$

Electric field on the axis of a dipole

where r is the distance measured from the *center* of the dipole. We've switched from y to r because we've now specified that Equation 21.13 is valid only along the axis of the dipole. Notice that the electric field along the axis points in the direction of the dipole moment $\vec{p}$.

A homework problem will let you calculate the electric field in the plane that bisects the dipole. This is the field shown on the x-axis in Figure 21.28. The field, for $r \gg s$, is

$$\vec{E}_{\text{dipole}} \approx -\frac{1}{4\pi\epsilon_0}\frac{\vec{p}}{r^3} \tag{21.14}$$

Electric field in the bisecting plane of a dipole

This field is *opposite* to $\vec{p}$, and it is only half the strength of the on-axis field at the same distance.

NOTE ▶ Do these inverse-cube equations violate Coulomb's law? Not at all. Coulomb's law describes the force between two *point charges,* and from Coulomb's law we found that the electric field of a *point charge* varies with the inverse square of the distance. But a dipole is not a point charge. The field of a dipole decreases more rapidly than that of a point charge, which is to be expected because the dipole is, after all, electrically neutral. ◀

EXAMPLE 21.7 The electric field of a water molecule

The water molecule H_2O has a permanent dipole moment of magnitude $6.2 \times 10^{-30}\ \text{C}\cdot\text{m}$. What is the electric field strength 1.0 nm from a water molecule at a point on the dipole's axis?

PREPARE The size of a molecule is ≈ 0.1 nm. Thus $r \gg s$, and we can use Equation 21.13 for the on-axis electric field of the molecule's dipole moment.

SOLVE The on-axis electric field strength at $r = 1.0$ nm is

$$E \approx \frac{1}{4\pi\epsilon_0}\frac{2p}{r^3} = (9.0\times10^9\ \text{N}\cdot\text{m}^2/\text{C}^2)\frac{2(6.2\times10^{-30}\ \text{C}\cdot\text{m})}{(1.0\times10^{-9}\ \text{m})^3}$$

$$= 1.1\times10^8\ \text{N/C}$$

ASSESS By referring to Table 21.2 you can see that the field strength is "strong" compared to our everyday experience with charged objects but "weak" compared to the electric field inside the atoms themselves. Even so, a container of water does not generate a macroscopic electric field like that of a charged rod. This is due both to the rapid inverse-cube decrease of field strength with distance and to the fact that water molecules are randomly oriented so that their dipole electric fields tend to cancel.

STOP TO THINK 21.5 At the dot, the electric field points

A. Left.
B. Right.
C. Up.
D. Down.
E. The electric field is zero.

Electric Field Lines

We can't see the electric field. Consequently, we need pictorial tools to help us visualize it in a region of space. One method, introduced in the previous section, is to picture the electric field by drawing electric field vectors at various points in space.

Another common way to picture the field is to draw **electric field lines.** As FIGURE 21.30 shows,

- Electric field lines are *continuous* curves tangent to the electric field vectors.
- Closely spaced field lines indicate a greater field strength; widely spaced field lines indicate a smaller field strength.

FIGURE 21.30 Electric field lines.

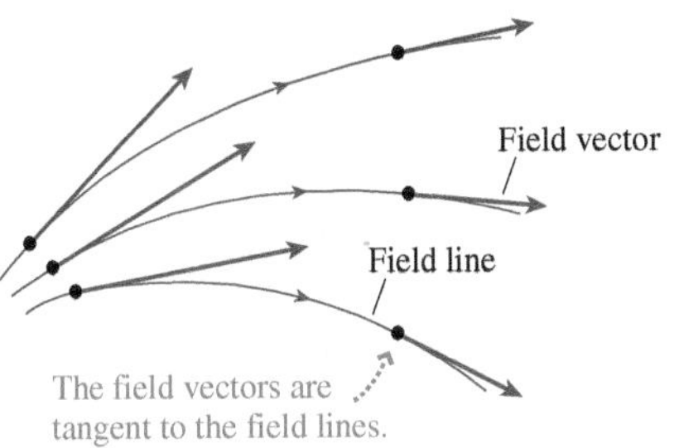
Field vector
Field line
The field vectors are tangent to the field lines.

- Electric field lines start on positive charges and end on negative charges.
- Electric field lines never cross.

FIGURE 21.31 shows three electric fields represented by electric field lines. Notice that the electric field of a dipole points in the direction of $\vec{p}$ (right to left) on both sides of the dipole, but points opposite to $\vec{p}$ in the bisecting plane.

FIGURE 21.31 The electric field lines of (a) a positive point charge, (b) a negative point charge, and (c) a dipole.

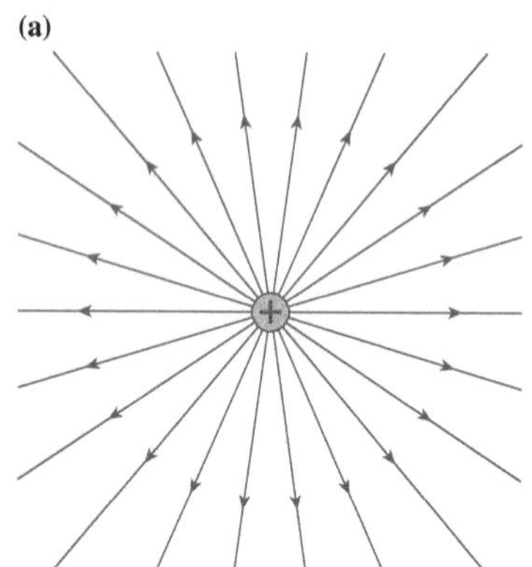
(a)

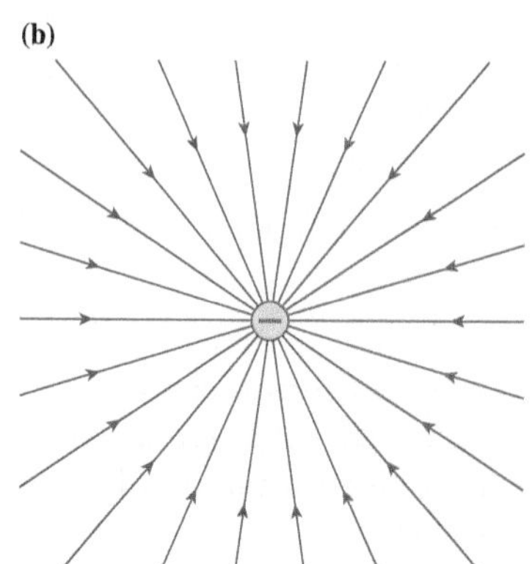
(b)

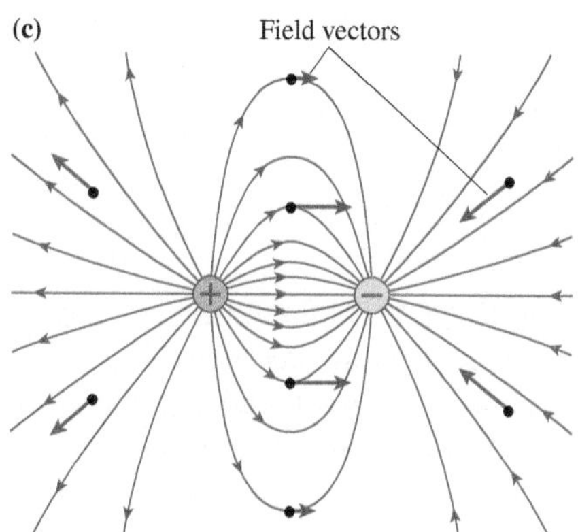
(c)
Field vectors

Neither field-vector diagrams nor field-line diagrams are perfect pictorial representations of an electric field. The field-vector diagrams are somewhat harder to draw, and they show the field at only a few points, but they do clearly indicate the direction and strength of the electric field at those points. Field-line diagrams perhaps look more elegant, and they're sometimes easier to sketch, but there's no formula for knowing where to draw the lines. We'll use both field-vector diagrams and field-line diagrams, depending on the circumstances.

The Electric Field of the Heart

A cell membrane is an insulator that encloses a conducting fluid and is surrounded by conducting fluid. While resting, the membrane is *polarized* with positive charges on the outside of the cell, negative charges on the inside. When a nerve or a muscle cell is stimulated, the polarity of the membrane switches; we say that the cell *depolarizes*. Later, when the charge balance is restored, we say that the cell *repolarizes*.

All nerve and muscle cells generate an electrical signal when depolarization occurs, but the largest electrical signal in the body comes from the heart. The rhythmic beating of the heart is produced by a highly coordinated wave of depolarization that sweeps across the tissue of the heart. We'll see in Chapter 22 how this depolarization generates an electric dipole moment that changes in strength and orientation during each beat of the heart.

That the heart creates a dipole field should not be too surprising. The heart is electrically neutral, so it cannot create a field like a point charge. But an electric dipole has zero net charge; its field is due to charge *separation,* exactly what happens on cell membranes. Thus the coordinated action of cell membranes during a heartbeat can produce a macroscopic dipole field. Similarly, dipole interactions are important throughout biology because electrically neutral molecules are often polarized.

The electric dipole of the heart generates a dipole electric field that extends throughout the torso, as shown in FIGURE 21.32. Measurement of the heart's electric field—an *electrocardiogram*—can be used to diagnose the operation of the heart. The field from the beating heart of a creature that lives in the water extends into the water around it. A platypus, a shark, or any of the other predators with an electric sense can use this field to locate prey.

FIGURE 21.32 The beating heart generates a dipole electric field.

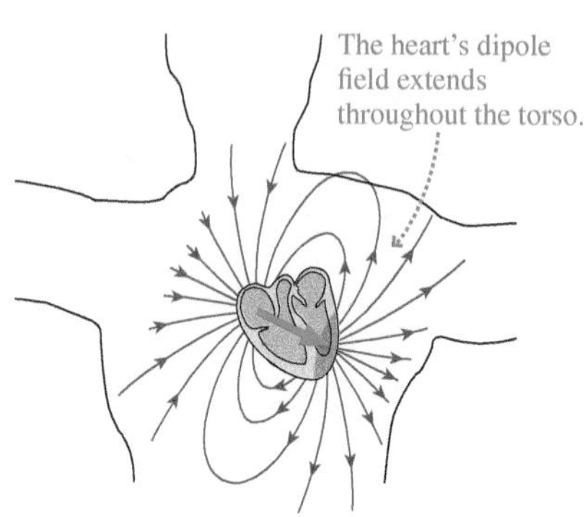
The heart's dipole field extends throughout the torso.

STOP TO THINK 21.6 Which of the following is the correct representation of the electric field created by two positive charges?

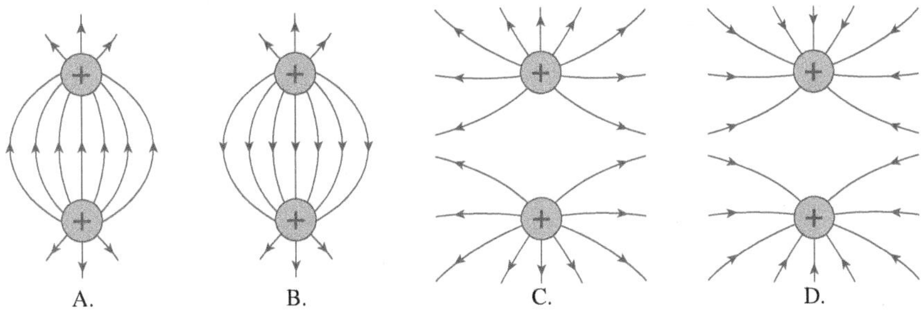
A.
B.
C.
D.

A Charged Sphere

Thus far we've considered only discrete charges. Suppose a sphere of radius R has total charge Q that is spread out uniformly on its surface or within its interior. (We will use the symbol Q for the total charge of an extended object, reserving q for point charges.)

What is the electric field of this charged sphere? A fairly complex calculation leads to the not surprising result that the field outside the sphere is exactly the same as that of a point charge Q located at the center of the sphere:

$$\vec{E} = \left(\frac{1}{4\pi\epsilon_0} \frac{|Q|}{r^2}, \begin{cases} \text{away} & Q \text{ positive} \\ \text{toward} & Q \text{ negative} \end{cases} \right) \text{ for } r > R \tag{21.15}$$

Electric field of a charged sphere with radius R and charge Q

FIGURE 21.33 The electric field of a sphere of positive charge.

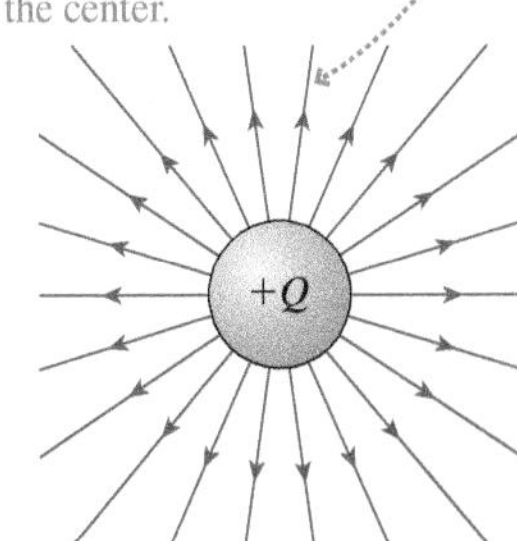
The electric field outside a sphere or spherical shell is the same as the field of a point charge Q at the center.
+Q

FIGURE 21.33 shows the electric field of a sphere of positive charge. The field of a negative sphere points inward.

The Parallel-Plate Capacitor

FIGURE 21.34 shows two metal plates, called **electrodes,** one with charge $+Q$ and the other with $-Q$, placed face-to-face a distance d apart. This arrangement of two electrodes, charged equally but oppositely, is called a **parallel-plate capacitor.** Capacitors have many applications, from heart defibrillators to electric circuits. In addition, the parallel-plate capacitor is an excellent model of the layers of charge on the inner and outer surfaces of the membrane of every living cell. In Chapter 25 we'll use a capacitor model of the membrane to understand the firing of neurons. Our goal in this section is to find the electric field between the capacitor plates between the capacitor.

FIGURE 21.34 A parallel-plate capacitor.

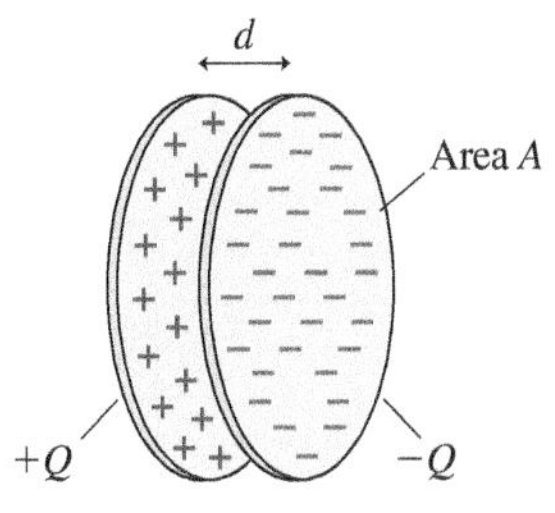
d
Area A
+Q
−Q

NOTE ▸ The *net* charge of a capacitor is zero. Capacitors are charged by transferring electrons from one plate to the other. The plate that gains N electrons has charge $-Q = N(-e)$. The plate that loses electrons has charge $+Q$. The symbol Q that we will use in equations refers to the *magnitude* of the charge on each plate. ◂

FIGURE 21.35 is an enlarged view of the capacitor plates, seen from the side. We will assume that the plate separation d is much less than their diameter, so the "broken off" look at the top and bottom indicates that the plates extend far above and below what we see in the figure. Notice that all of the charge is on the *inner* surfaces of the two plates. This is because opposite charges attract and because these are conductors within which charge can move.

The electric field of a parallel-plate capacitor is the vector sum of the electric fields of each and every charge on the electrodes. We can divide this sum into two

FIGURE 21.35 The electric fields inside and outside a parallel-plate capacitor.

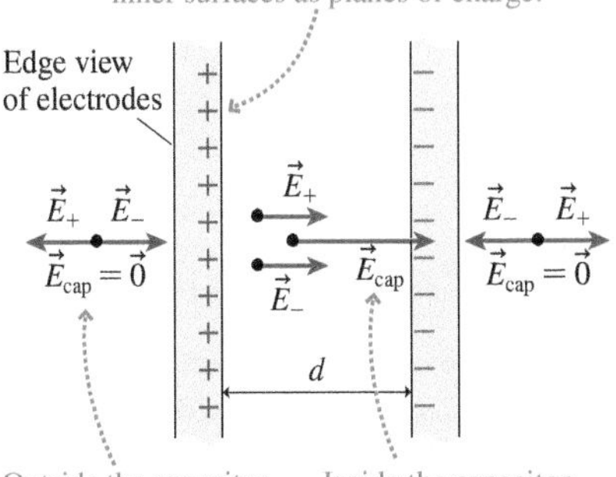
The capacitor's charge resides on the inner surfaces as planes of charge.
Edge view of electrodes
$\vec{E}_+$ $\vec{E}_-$
$\vec{E}_{cap} = \vec{0}$
$\vec{E}_+$
$\vec{E}_-$
$\vec{E}_{cap}$
$\vec{E}_-$ $\vec{E}_+$
$\vec{E}_{cap} = \vec{0}$
d
Outside the capacitor, $\vec{E}_+$ and $\vec{E}_-$ are opposite, so the net field is zero.
Inside the capacitor, $\vec{E}_+$ and $\vec{E}_-$ are parallel, so the net field is large.

parts: the field $\vec{E}_+$ due to all the positive charge on one electrode and the field $\vec{E}_-$ due to all the negative charge on the opposite electrode. The net field of the capacitor is then $\vec{E}_{cap} = \vec{E}_+ + \vec{E}_-$.

Figure 21.35 shows the fields $\vec{E}_+$ and $\vec{E}_-$ *perpendicular to the electrodes.* To see why this must be true, consider a point in Figure 21.35 to the right of the positive electrode. The field $\vec{E}_+$ is the vector sum of the fields of all the positive charges above this point and all the positive charges below it. Each of these individual fields has a horizontal component and a vertical component. All of the horizontal components point to the right; these add to give the net field a horizontal component to the right. But the vertical components from charges above this point are opposite to the vertical components from charges below it. The vertical components cancel, so the net field does not have a vertical component. Thus $\vec{E}_+$ points *straight away from* the positive electrode. Similarly, $\vec{E}_-$ points *straight toward* the negative electrode. Figure 21.35 shows this for points inside and outside the capacitor.

NOTE ▸ You might think the right capacitor plate would somehow "block" the electric field created by the positive plate and prevent the presence of an $\vec{E}_+$ field to the right of the capacitor. To see that it doesn't, consider an analogous situation with gravity. The strength of gravity above a table is the same as its strength below it. Just as the table doesn't block the earth's gravitational field, intervening matter or charges do not alter or block an object's electric field. ◂

FIGURE 21.36 Ideal versus real capacitors.

(a) Ideal capacitor—edge view

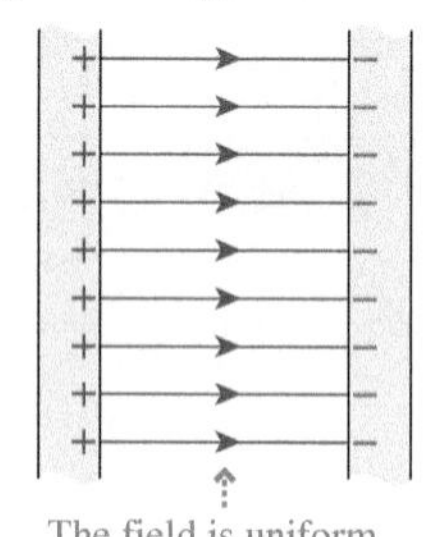

(b) Real capacitor—edge view

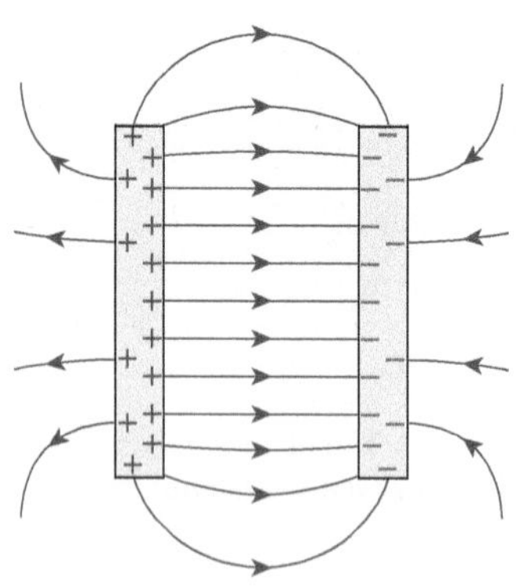

Outside the capacitor, $\vec{E}_+$ and $\vec{E}_-$ point in opposite directions. A more detailed calculation finds that these fields have equal magnitudes, so $\vec{E}_{cap} = \vec{E}_+ + \vec{E}_- = \vec{0}$. There's *no* electric field outside an ideal parallel-plate capacitor, one with electrodes that are infinite in extent. Inside the capacitor, between the plates, $\vec{E}_+$ and $\vec{E}_-$ both point from the positive to the negative plate. A detailed analysis finds that the electric field of an ideal parallel-plate capacitor with plate area A is

$$\vec{E}_{capacitor} = \begin{cases} \left(\dfrac{Q}{\epsilon_0 A}, \text{ from positive to negative}\right) & \text{inside} \\ \vec{0} & \text{outside} \end{cases} \qquad (21.16)$$

Electric field of a parallel-plate capacitor with plate area A and charge Q

FIGURE 21.36a shows the electric field—this time using field lines—of an ideal parallel-plate capacitor. Now, it's true that no real capacitor is infinite in extent, but the ideal parallel-plate capacitor is a very good approximation for all but the most precise calculations as long as the electrode separation d is much smaller than the electrodes' size. FIGURE 21.36b shows that the interior field of a real capacitor is virtually identical to that of an ideal capacitor but that the exterior field isn't quite zero. This weak field outside the capacitor is called the **fringe field.** We will keep things simple by always assuming the plates are very close together and using Equation 21.26 for the field inside a parallel-plate capacitor.

NOTE ▸ The shape of the electrodes—circular or square or any other shape—is not relevant as long as the electrodes are very close together. ◂

EXAMPLE 21.8 **The electric field between the plates of a capacitor**

Two 1.0 cm × 2.0 cm rectangular electrodes are 1.0 mm apart. What charge must be placed on each electrode to create a uniform electric field of strength 2.0×10^6 N/C? How many electrons must be moved from one electrode to the other to accomplish this?

PREPARE The electrodes can be modeled as an ideal parallel-plate capacitor because the spacing between them is much smaller than their lateral dimensions.

SOLVE The electric field strength between the plates of the capacitor is $E = Q/\epsilon_0 A$. Thus the charge to produce a field of strength E is

$$Q = (8.85 \times 10^{-12}\ \text{C}^2/\text{N} \cdot \text{m}^2)(2.0 \times 10^{-4}\ \text{m}^2)(2.0 \times 10^6\ \text{N/C})$$
$$= 3.5 \times 10^{-9}\ \text{C} = 3.5\ \text{nC}$$

The positive plate must be charged to $+3.5$ nC and the negative plate to -3.5 nC. In practice, the plates are charged by using a *battery* to move electrons from one plate to the other. The number of electrons in 3.5 nC is

$$N = \frac{Q}{e} = \frac{3.5 \times 10^{-9}\ \text{C}}{1.60 \times 10^{-19}\ \text{C/electron}} = 2.2 \times 10^{10}\ \text{electrons}$$

Thus 2.2×10^{10} electrons are moved from one electrode to the other. Note that the capacitor *as a whole* has no net charge.

ASSESS The plate spacing does not enter the result. As long as the spacing is much smaller than the plate dimensions, as is true in this example, the field is independent of the spacing.

Uniform Electric Fields

FIGURE 21.37 shows an electric field that is the *same*—in strength and direction—at every point in a region of space. This is called a **uniform electric field.** A uniform electric field is analogous to the uniform gravitational field near the surface of the earth.

The easiest way to produce a uniform electric field is with a parallel-plate capacitor, as you can see in Figure 21.36a. Indeed, our interest in capacitors is due in large measure to the fact that the electric field is uniform. Many electric field problems refer to a uniform electric field. Such problems carry an implicit assumption that the action is taking place between the plates of a parallel-plate capacitor.

FIGURE 21.37 A uniform electric field.

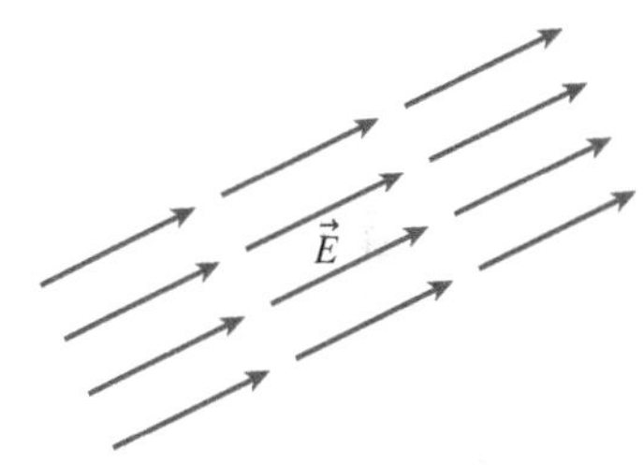
$\vec{E}$

STOP TO THINK 21.7 Rank in order, from largest to smallest, the electric field strengths E_1 to E_5 at points 1 to 5 in this parallel-plate capacitor.

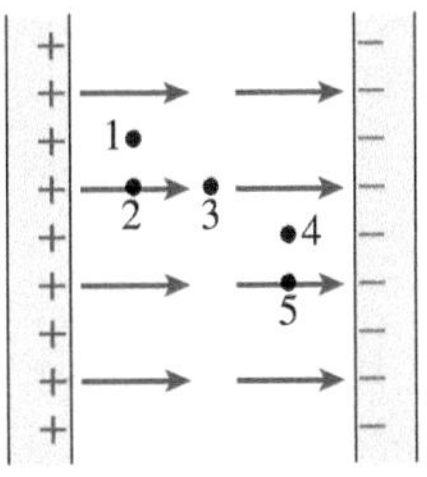
1
2
3
4
5

Electric Fields Around and Within Conductors

Recall that a conductor is in electrostatic equilibrium if none of its charges are moving. We can draw some important conclusions about the charges on and the electric fields of a conductor in electrostatic equilibrium.

First, the electric field inside the conductor must be zero. To see why, suppose there was an internal electric field within the conductor. Electric fields exert forces on charges, so an internal electric field would exert forces on the charges in the conductor. Because charges in a conductor are free to move, these forces would cause the charges to move. But that would violate the assumption that all the charges are at rest. Thus we're forced to conclude that **the electric field is zero at all points inside a conductor in electrostatic equilibrium.**

Second, **any excess charge on the conductor must lie at its surface,** as shown in FIGURE 21.38a. Any charge in the interior of the conductor would create an electric field there, in violation of our conclusion that the field inside is zero. Physically, excess charge ends up on the surface because the repulsive forces between like charges cause them to move as far apart as possible without leaving the conductor.

Third, FIGURE 21.38b shows that **the electric field right at the surface of a charged conductor is perpendicular to the surface.** To see that this is so, suppose $\vec{E}$ had a component tangent to the surface. This component of $\vec{E}$ would exert a force on

FIGURE 21.38 The electric field inside and outside a charged conductor.

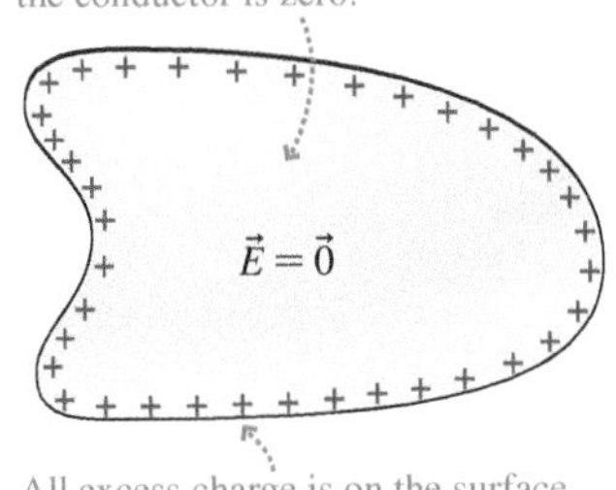
(a) The electric field inside the conductor is zero.
$\vec{E} = \vec{0}$
All excess charge is on the surface.

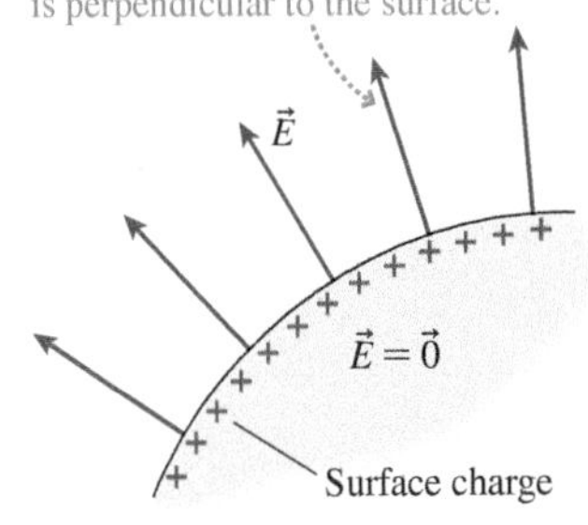
(b) The electric field at the surface is perpendicular to the surface.
$\vec{E}$
$\vec{E} = \vec{0}$
Surface charge

charges at the surface and cause them to move along the surface, thus violating the assumption that all charges are at rest. The only exterior electric field consistent with electrostatic equilibrium is one that is perpendicular to the surface.

21.6 The Motion of a Charged Particle in an Electric Field

Our motivation for introducing the concept of the electric field was to understand the long-range electric interaction of charges. Some charges, the *source charges,* create an electric field. Other charges then respond to that electric field. The first five sections of this chapter have focused on the electric field of the source charges. Now we turn our attention to the second half of the interaction.

FIGURE 21.39 shows a particle of charge q and mass m at a point where an electric field $\vec{E}$ has been produced by *other* charges, the source charges. The electric field exerts a force

$$\vec{F}_{\text{on } q} = q\vec{E}$$

on the charged particle. Notice that the force on a negatively charged particle is *opposite* in direction to the electric field vector. Signs are important!

FIGURE 21.39 The electric field exerts a force on a charged particle.

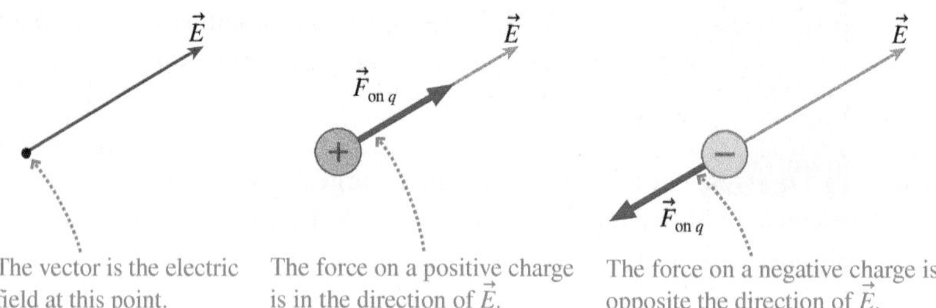
$\vec{E}$
$\vec{E}$
$\vec{F}_{\text{on } q}$
$\vec{E}$
$\vec{F}_{\text{on } q}$
The vector is the electric field at this point.
The force on a positive charge is in the direction of $\vec{E}$.
The force on a negative charge is opposite the direction of $\vec{E}$.

If $\vec{F}_{\text{on } q}$ is the only force acting on q, it causes the charged particle to accelerate with

$$\vec{a} = \frac{\vec{F}_{\text{on } q}}{m} = \frac{q}{m}\vec{E} \tag{21.17}$$

This acceleration is the *response* of the charged particle to the source charges that created the electric field. The ratio q/m is especially important for the dynamics of charged-particle motion. It is called the **charge-to-mass ratio.** Two *equal* charges, say a proton and a Na^+ ion, will experience *equal* forces $\vec{F} = q\vec{E}$ if placed at the same point in an electric field, but their accelerations will be *different* because they have different masses and thus different charge-to-mass ratios. Two particles with different charges and masses *but* with the same charge-to-mass ratio will undergo the same acceleration and follow the same trajectory.

Motion in a Uniform Field

The motion of a charged particle in a *uniform* electric field is especially important for its basic simplicity and because of its many valuable applications. A uniform field is *constant* at all points—constant in both magnitude and direction—within the region of space where the charged particle is moving. It follows, from Equation 21.17, that **a charged particle in a uniform electric field will move with constant acceleration.** The magnitude of the acceleration is

$$a = \frac{qE}{m} = \text{constant} \tag{21.18}$$

where E is the electric field strength, and the direction of $\vec{a}$ is parallel or antiparallel to $\vec{E}$, depending on the sign of q.

Identifying the motion of a charged particle in a uniform field as being one of constant acceleration brings into play all the kinematic machinery that we developed in Chapters 3 and 4 for constant-acceleration motion.

EXAMPLE 21.9 An electron moving between capacitor plates

Two 6.0-cm-diameter electrodes are spaced 5.0 mm apart. They are charged by transferring 1.0×10^{11} electrons from one electrode to the other. An electron is released from rest just above the surface of the negative electrode. How long does it take the electron to cross to the positive electrode? What is its speed as it collides with the positive electrode? Assume the space between the electrodes is a vacuum.

PREPARE The electrodes form a parallel-plate capacitor. The charges *on* the electrodes cannot escape, but any additional charges *between* the capacitor plates will be accelerated by the electric field. The electric field inside a parallel-plate capacitor is a uniform field, so the electron will have constant acceleration.

FIGURE 21.40 shows an edge view of the capacitor and the electron. The force on the negative electron is *opposite* the electric field, so the electron is repelled by the negative electrode as it accelerates across the gap of width d.

FIGURE 21.40 An electron accelerates across a capacitor (plate separation exaggerated).

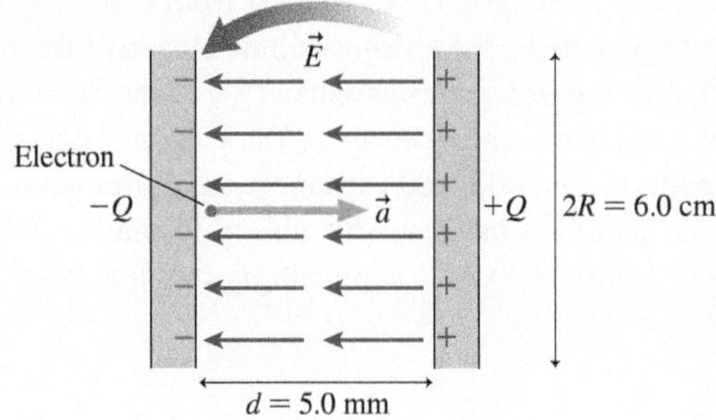
The capacitor was charged by transferring 10^11 electrons from the right electrode to the left electrode.
E
Electron
−Q
a
+Q
2R = 6.0 cm
d = 5.0 mm

SOLVE The electrodes are not point charges, so we cannot use Coulomb's law to find the force on the electron. Instead, we must analyze the electron's motion in terms of the electric field between the capacitor plates. The field is the agent that exerts the force on the electron, causing it to accelerate. The charge on the capacitor is $Q = Ne$, where N is the number of electrons transferred from one electrode to the other. The electric field strength inside a parallel-plate capacitor with charge Q is

$$E = \frac{Q}{\epsilon_0 A} = \frac{Ne}{\epsilon_0 \pi R^2} = 639{,}000 \text{ N/C}$$

The electron's acceleration in this field is

$$a = \frac{eE}{m} = 1.1 \times 10^{17} \text{ m/s}^2$$

where we used the electron mass $m = 9.11 \times 10^{-31}$ kg from Table 21.1. This is an enormous acceleration compared to accelerations we're familiar with for macroscopic objects. We can use one-dimensional kinematics, with $x_i = 0$ and $(v_i)_x = 0$, to find the time required for the electron to cross the capacitor:

$$x_f = d = \tfrac{1}{2}a(\Delta t)^2$$

$$\Delta t = \sqrt{\frac{2d}{a}} = 3.0 \times 10^{-10} \text{ s} = 0.30 \text{ ns}$$

The electron's speed as it reaches the positive electrode is

$$v = a\,\Delta t = 3.3 \times 10^7 \text{ m/s}$$

ASSESS We used e rather than $-e$ to find the acceleration because we already knew the direction; we needed only the magnitude. The electron's speed, after traveling a mere 5 mm, is approximately 10% the speed of light.

Parallel electrodes such as those in Example 21.9 are often used to accelerate charged particles. If the positive plate has a small hole in the center, a *beam* of electrons will pass through the hole and emerge with a speed of 3.3×10^7 m/s. This is the basic idea of the *electron gun* used until recently in *cathode-ray tube* (CRT) devices such as televisions and computer display terminals. (A negatively charged electrode is called a *cathode,* so the physicists who first learned to produce electron beams in the late 19th century called them *cathode rays.*)

Motion with Drag

A charged object that moves through a medium, such as a charged macromolecule in a solution, experiences a drag force. The motion is almost always at low Reynolds number, and you learned in ◀◀ SECTION 5.4 that the drag force is

$$\vec{D} = (bv, \text{ direction opposite the motion}) \tag{21.19}$$

where the drag constant b depends on the object's size and the viscosity of the medium.

An object that is pushed by an applied force $F_{applied}$ reaches terminal speed v_{term} when the drag force is equal in magnitude to the applied force: $bv_{term} = F_{applied}$. Thus a charged object that moves through a viscous medium in response to an electric force of magnitude $F_{applied} = qE$ will accelerate until it reaches terminal speed

$$v_{term} = \frac{qE}{b} \tag{21.20}$$

and will then move with constant speed.

One of the most important applications of this principle is *gel electrophoresis,* a powerful experimental technique used in molecular biology to sort and identify fragments of biomolecules. A sample of a biomolecule such as DNA is prepared by being placed in solution and then cut into fragments by enzymes. In solution, these fragments have a negative charge that is proportional to the length L of the fragment: $q \propto L$.

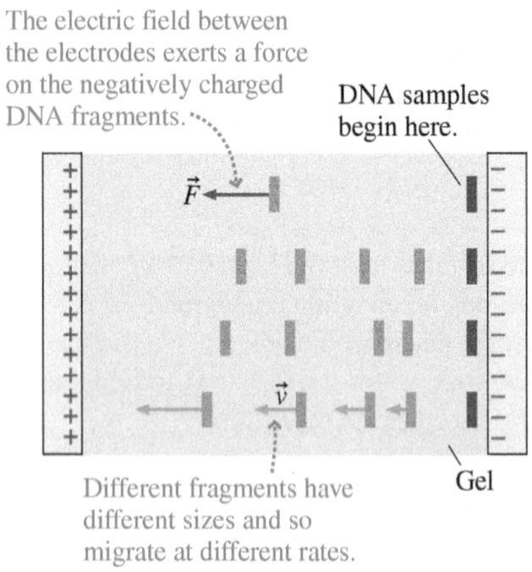

A drop of the solution is then placed into a well at one end of a tray of *gel* like the one seen in FIGURE 21.41. The gel is a semisolid gelatin material with billions of microscopic pores. Electrodes at opposite ends of the tray create an electric field in the gel of typically 500 N/C. This field pushes the negatively charged DNA fragments toward the positive electrode while the gel exerts a drag force. The fragments quickly reach a terminal speed given by Equation 21.20 and then migrate across the tray at constant speed.

FIGURE 21.41 Gel electrophoresis creates a "genetic fingerprint" by separating DNA fragments of different lengths.

The electric field between the electrodes exerts a force on the negatively charged DNA fragments.
DNA samples begin here.
$\vec{F}$
$\vec{v}$
Different fragments have different sizes and so migrate at different rates.
Gel

Longer fragments experience a larger electric force because they are more highly charged than shorter fragments, but longer fragments also experience a larger drag force. The key to understanding gel electrophoresis is that the drag coefficient b increases with length *more rapidly* than q increases. If q increases with the fragment length but b increases faster, then the ratio q/b *decreases* as the fragment length increases. Consequently, longer fragments have a slower terminal speed and shorter segments have a faster terminal speed.

After some time, typically 30 minutes, the fragments will have sorted themselves by size and will create distinct lines in the gel. Lines farthest from the negative electrode are the shortest fragments that moved the fastest; lines closer to the negative electrode are the more slowly moving, longer fragments. Two identical samples of DNA produce the same set of fragments when cut up by the enzymes and thus create the same electrophoresis pattern. The odds are extremely small that an unrelated DNA sample would produce an identical pattern. This ability to match DNA samples makes gel electrophoresis a critical tool in applications ranging from genetic research to forensics.

STOP TO THINK 21.8 An electron is placed at the position marked by the dot. The force on the electron is

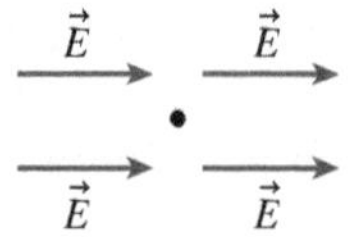

A. Zero.
B. To the right.
C. To the left.
D. There's not enough information to tell.

$\vec{E}$ $\vec{E}$
$\vec{E}$ $\vec{E}$

21.7 The Torque on a Dipole in an Electric Field

Let us conclude this chapter by returning to one of the more striking puzzles we faced when making the observations at the beginning of the chapter. There you found that charged objects of *either* sign exert forces on neutral objects, such as when a comb used to brush your hair picks up pieces of paper. Our qualitative understanding of the *polarization force* was that it required two steps:

- The charge polarizes the neutral object, creating an induced electric dipole.
- The charge then exerts an attractive force on the near end of the dipole that is slightly stronger than the repulsive force on the far end.

We are now in a position to make that understanding more quantitative.

Dipoles in a Uniform Field

FIGURE 21.42a shows an electric dipole in a *uniform* external electric field $\vec{E}$ that has been created by source charges that are not shown. That is, $\vec{E}$ is *not* the field of the dipole but, instead, is a field to which the dipole is responding. In this case, because the field is uniform, the dipole is presumably inside an unseen parallel-plate capacitor.

The net force on the dipole is the sum of the forces on the two charges forming the dipole. Because the charges $\pm q$ are equal in magnitude but opposite in sign, the two forces $\vec{F}_+ = +q\vec{E}$ and $\vec{F}_- = -q\vec{E}$ are also equal but opposite. Thus the net force on the dipole is

$$\vec{F}_{net} = \vec{F}_+ + \vec{F}_- = \vec{0} \qquad (21.21)$$

There is no net force on a dipole in a uniform electric field.

There may be no net force, but the electric field *does* affect the dipole. Because the two forces in Figure 21.42a are in opposite directions but not aligned with each other, the electric field exerts a *torque* on the dipole and causes the dipole to *rotate.*

The torque rotates the dipole until it is aligned with the electric field, as shown in FIGURE 21.42b. In this position, the dipole experiences not only no net force but also no torque. Thus Figure 21.42b represents the *equilibrium position* for a dipole in a uniform electric field. Notice that the positive end of the dipole is in the direction in which $\vec{E}$ points.

FIGURE 21.43 shows a sample of permanent dipoles, such as water molecules, in an external electric field. All the dipoles rotate until they are aligned with the electric field. This is the mechanism by which the sample becomes *polarized.* Once the dipoles are aligned, there is an excess of positive charge at one end of the sample and an excess of negative charge at the other end. The excess charges at the ends of the sample are the basis of the polarization forces we discussed in Section 21.1.

It's not hard to calculate the torque. Recall from Chapter 6 that the magnitude of a torque is the product of the force and the moment arm. FIGURE 21.44 shows that there are two forces of the same magnitude ($F_+ = F_- = qE$), each with the same moment arm ($d = \frac{1}{2}s\sin\theta$). Thus the torque on the dipole is

$$\tau = 2 \times dF_+ = 2(\tfrac{1}{2}s\sin\theta)(qE) = pE\sin\theta \qquad (21.22)$$

where $p = qs$ was our definition of the dipole moment. The torque is zero when the dipole is aligned with the field, making $\theta = 0$.

FIGURE 21.42 A dipole in a uniform electric field.

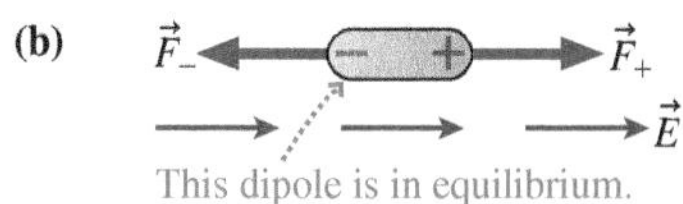

FIGURE 21.43 A sample of permanent dipoles is *polarized* in an electric field.

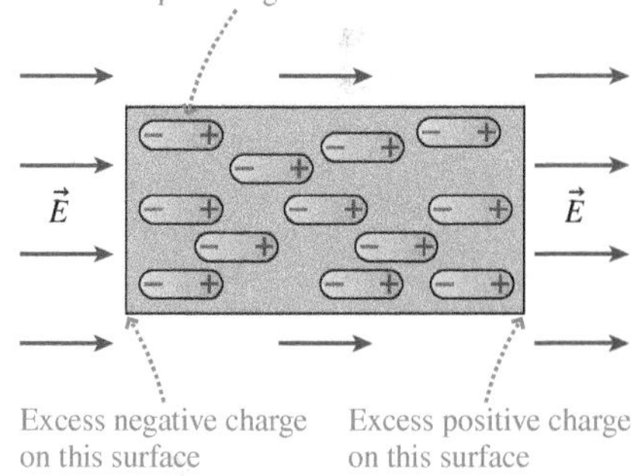

FIGURE 21.44 The torque on a dipole.

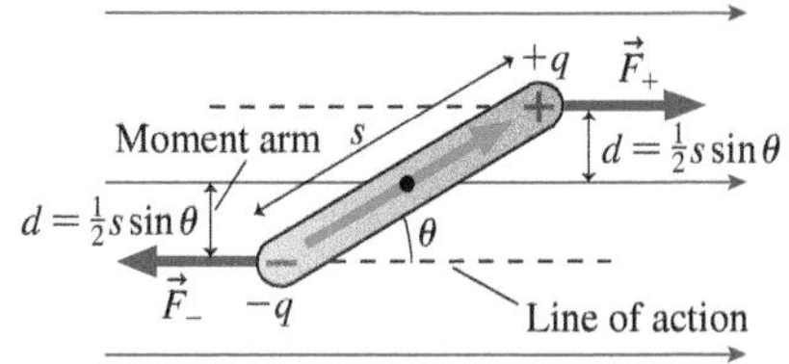

EXAMPLE 21.10 **The angular acceleration of a dipole dumbbell**

Two 1.0 g balls are connected by a 2.0-cm-long insulating rod of negligible mass. One ball has a charge of +10 nC, the other a charge of −10 nC. The rod is held in a 1.0×10^4 N/C uniform electric field at an angle of 30° with respect to the field, then released. What is its initial angular acceleration?

PREPARE The two oppositely charged balls form an electric dipole. The electric field exerts a torque on the dipole, causing an angular acceleration. FIGURE 21.45 shows the dipole in the electric field.

FIGURE 21.45 The dipole of Example 21.10.

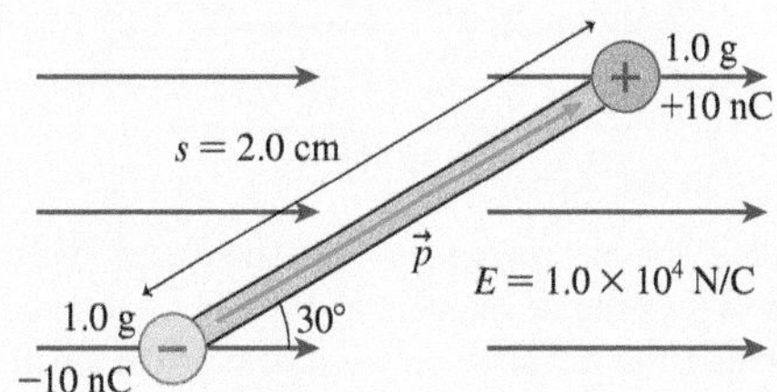

Continued

SOLVE The dipole moment is $p = qs = (1.0 \times 10^{-8}\ \text{C}) \times (0.020\ \text{m}) = 2.0 \times 10^{-10}\ \text{C} \cdot \text{m}$. The torque exerted on the dipole moment by the electric field is

$$\tau = pE\sin\theta = (2.0 \times 10^{-10}\ \text{C}\cdot\text{m})(1.0 \times 10^{4}\ \text{N/C})\sin 30^\circ$$
$$= 1.0 \times 10^{-6}\ \text{N}\cdot\text{m}$$

You learned in Chapter 7 that a torque causes an angular acceleration $\alpha = \tau/I$, where I is the moment of inertia. The dipole rotates about its center of mass, which is at the center of the rod, so the moment of inertia is

$$I = m_1 r_1^2 + m_2 r_2^2 = 2m(\tfrac{1}{2}s)^2 = \tfrac{1}{2}ms^2 = 2.0 \times 10^{-7}\ \text{kg}\cdot\text{m}^2$$

Thus the rod's angular acceleration is

$$\alpha = \frac{\tau}{I} = \frac{1.0 \times 10^{-6}\ \text{N}\cdot\text{m}}{2.0 \times 10^{-7}\ \text{kg}\cdot\text{m}^2} = 5.0\ \text{rad/s}^2$$

ASSESS This value of α is the initial angular acceleration, when the rod is first released. The torque and the angular acceleration will decrease as the rod rotates toward alignment with $\vec{E}$.

Dipoles in a Nonuniform Field

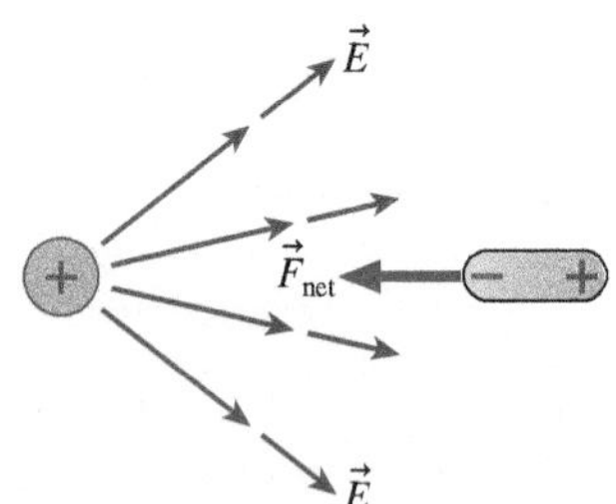

FIGURE 21.46 An aligned dipole is drawn toward a point charge.

Suppose that a dipole is placed in a nonuniform electric field, one in which the field strength changes with position. For example, FIGURE 21.46 shows a dipole in the nonuniform field of a point charge. The first response of the dipole is to rotate until it is aligned with the field at the dipole's position, with the dipole's positive end pointing in the same direction as the field. Now, however, there is a *slight difference* between the forces acting on the two ends of the dipole. This difference occurs because the electric field, which depends on the distance from the point charge, is stronger at the end of the dipole nearest the charge. This causes a net force to be exerted on the dipole.

Which way does the force point? Once the dipole is aligned, the leftward attractive force on its negative end is slightly stronger than the rightward repulsive force on its positive end. This causes a net force *toward* the point charge.

In fact, for any nonuniform electric field, **the net force on a dipole is toward the direction of the strongest field.** Because any finite-size charged object, such as a charged rod or a charged disk, has a field strength that increases as you get closer to the object, we can conclude that **a dipole will experience a net force toward any charged object.**

EXAMPLE 21.11 The force on a water molecule

The water molecule H_2O has a permanent dipole moment of magnitude $6.2 \times 10^{-30}\ \text{C}\cdot\text{m}$. A water molecule is located 10 nm from a Na^+ ion in a saltwater solution. What force does the ion exert on the water molecule?

PREPARE FIGURE 21.47 shows the ion and the dipole. The forces are an interaction pair.

FIGURE 21.47 The interaction between an ion and a permanent dipole.

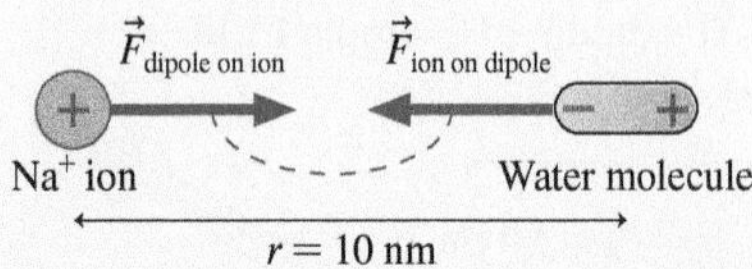

SOLVE A Na^+ ion has charge $q = +e$. The electric field of the ion aligns the water's dipole moment and exerts a net force on it. We could calculate the net force on the dipole as the small difference between the attractive force on its negative end and the repulsive force on its positive end. Alternatively, we know from Newton's third law that the force $\vec{F}_{\text{dipole on ion}}$ has the same magnitude as the force $\vec{F}_{\text{ion on dipole}}$ that we are seeking. We calculated the on-axis field of a dipole in Section 21.5. An ion of charge $q = e$ will experience a force of magnitude $F = qE_{\text{dipole}} = eE_{\text{dipole}}$ when placed in that field. The dipole's electric field, which we found in Equation 21.13, is

$$E_{\text{dipole}} = \frac{1}{4\pi\epsilon_0}\frac{2p}{r^3}$$

The force on the ion at distance $r = 1.0 \times 10^{-8}$ m is

$$F_{\text{dipole on ion}} = eE_{\text{dipole}} = \frac{1}{4\pi\epsilon_0}\frac{2ep}{r^3} = 1.8 \times 10^{-14}\ \text{N}$$

Thus the force on the water molecule is $F_{\text{ion on dipole}} = 1.8 \times 10^{-14}$ N.

ASSESS While 1.8×10^{-14} N may seem like a very small force, it is $\approx 10^{11}$ times larger than the size of the earth's gravitational force on these atomic particles. Forces such as these cause water molecules to cluster around any ions that are in solution. This clustering plays an important role in the microscopic physics of solutions studied in chemistry and biochemistry.

CHAPTER 21 INTEGRATED EXAMPLE Sorting cells

Flow cytometry, illustrated in FIGURE 21.48, is a technique to sort cells by type. A saline solution that contains individual cells is forced through a narrow nozzle that breaks the stream into small, 100-μm-diameter droplets traveling at 2.0 m/s. The cell concentration is adjusted so that, on average, 50% of the droplets hold one cell and 50% are vacant; very few droplets have more than one cell. The droplets pass through a laser-beam probe, and the presence or absence of fluorescence (detection optics not shown) is used to quickly determine whether a cell is one of two types, which we call A and B. The droplets then pass through a device called a *charging ring* that gives each droplet a charge q according to the rule

$$q = \begin{cases} +1.0 \times 10^{-13}\ \text{C} & \text{type A} \\ 0 & \text{empty droplet} \\ -1.0 \times 10^{-13}\ \text{C} & \text{type B} \end{cases}$$

The droplets then travel 2.0 cm through a horizontal, uniform electric field, created by a pair of electrodes, that deflects the charged droplets 3.0 cm sideways into collection tubes that are 10 cm below the electrodes. The lower part of the figure is not to scale.

a. How many cells are sorted per second if the centers of the droplets are 200 μm apart?
b. What is the electric field strength between the electrodes?

FIGURE 21.48 Flow cytometry is used to separate and sort cells.

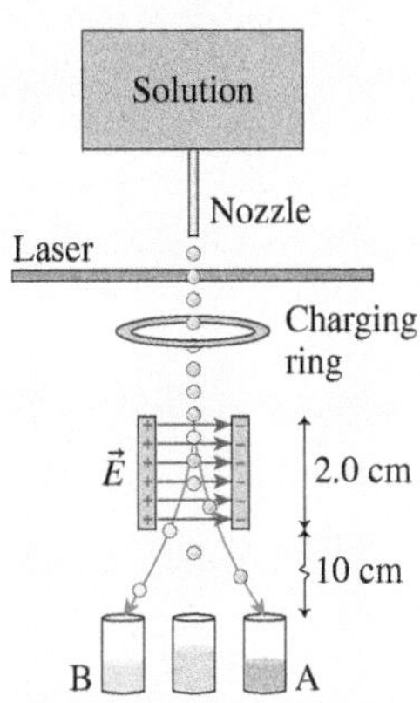

PREPARE We can model the droplets as point charges. The droplets move through the apparatus in a small fraction of a second, so we can ignore their free-fall acceleration and assume that the vertical motion is at constant velocity. The pictorial representation in FIGURE 21.49 shows that a droplet's trajectory is curved while it is between the electrodes, due to a horizontal acceleration, and then straight at angle θ after it leaves the electrodes.

SOLVE

a. The number of droplets that pass through the charging ring per second is the inverse of the time interval between droplet arrivals. For example, a droplet arriving every $\frac{1}{10}$ s means that 10 drops arrive per second. We're told that the droplets are 200 μm apart and traveling at 2.0 m/s, so the time interval between droplet arrivals is the time a droplet needs to move 200 μm:

$$\Delta t_{\text{arr}} = \frac{200 \times 10^{-6}\ \text{m}}{2.0\ \text{m/s}} = 1.0 \times 10^{-4}\ \text{s}$$

FIGURE 21.49 Pictorial representation of the trajectory of a charged droplet.

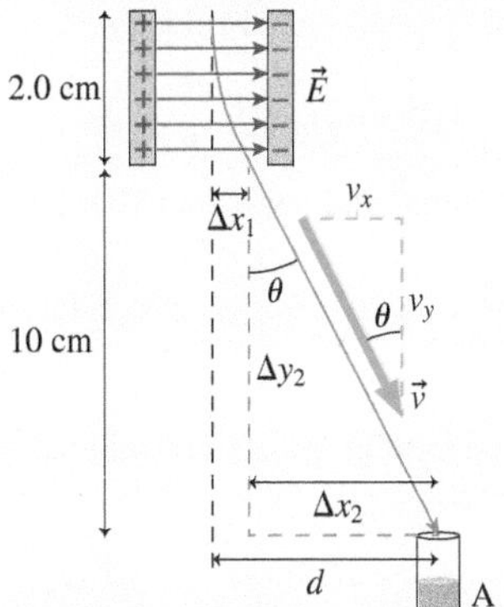

Thus the rate R of droplet arrivals is

$$R = \frac{1}{\Delta t_{\text{arr}}} = 1.0 \times 10^4\ \text{s}^{-1}$$

On average, half of the drops hold a cell. Consequently, cells are sorted at the rate of 5000 per s.

b. We can analyze the motion of a positive droplet holding a type A cell; the negative droplets have an identical motion in the opposite direction. The sideways deflection d of a droplet is the sum $d = \Delta x_1 + \Delta x_2$ of its deflections while in the electric field and after leaving the electric field. The uniform electric field causes the droplet to accelerate horizontally with $a_x = qE/m$ for the time Δt_{elec} that it takes to pass through the electrodes. We know from constant-acceleration kinematics that the deflection during this time is

$$\Delta x_1 = \tfrac{1}{2} a_x (\Delta t_{\text{elec}})^2 = \frac{qE}{2m} (\Delta t_{\text{elec}})^2$$

We need to work backward from the observed deflection to determine E, but we can calculate the other quantities. We're given that $q = 1.0 \times 10^{-13}$ C. We can use $m = \rho V$ to calculate the droplet's mass, where ρ and V are its density and volume. The solution is saline, but for a two-significant-figure calculation we can use the density of fresh water: $\rho = 1000\ \text{kg/m}^3$. The volume is that of a sphere with a radius of 50 μm, so

$$m = \rho V = \rho(\tfrac{4}{3}\pi r^3) = \tfrac{4}{3}\pi(1000\ \text{kg/m}^3)(50 \times 10^{-6}\ \text{m})^3$$
$$= 5.2 \times 10^{-10}\ \text{kg}$$

The droplet travels the 2.0 cm length of the electrodes at a constant 2.0 m/s, so the time spent in the electric field is

$$\Delta t_{\text{elec}} = \frac{0.020\ \text{m}}{2.0\ \text{m/s}} = 0.010\ \text{s}$$

After leaving the field, the droplet travels in a straight line at angle θ. The two right triangles in Figure 21.49—one with sides Δx_2 and Δy_2, the other with sides v_x and v_y—are similar triangles. Thus

$$\frac{\Delta x_2}{|\Delta y_2|} = \frac{v_x}{|v_y|}$$

Continued

where we use absolute value signs because Δy_2 and v_y are negative. We're assuming that there's no vertical acceleration, so $|v_y| = 2.0$ m/s. The constant horizontal acceleration while in the field gives the droplet a horizontal velocity of

$$v_x = a_x \Delta t_{\text{elec}} = \frac{qE}{m}\Delta t_{\text{elec}}$$

Thus the droplet's horizontal deflection after leaving the electric field is

$$\Delta x_2 = \frac{v_x}{|v_y|}|\Delta y_2| = \frac{qE\,\Delta t_{\text{elec}}}{m}\frac{|\Delta y_2|}{|v_y|}$$

Finally, we can combine the two individual deflections to write the total deflection d as

$$d = \Delta x_1 + \Delta x_2 = \frac{qE}{2m}(\Delta t_{\text{elec}})^2 + \frac{qE\,\Delta t_{\text{elec}}}{m}\frac{|\Delta y_2|}{|v_y|}$$

$$= \frac{qE\,\Delta t_{\text{elec}}}{m}\left(\frac{\Delta t_{\text{elec}}}{2} + \frac{|\Delta y_2|}{|v_y|}\right)$$

We know everything in this expression except E, so we can solve for the electric field strength:

$$E = \frac{md}{q\,\Delta t_{\text{elec}}\left(\dfrac{\Delta t_{\text{elec}}}{2} + \dfrac{|\Delta y_2|}{|v_y|}\right)}$$

$$= \frac{(5.2 \times 10^{-10}\text{ kg})(0.030\text{ m})}{(1.0 \times 10^{-13}\text{ C})(0.010\text{ s})\left(\dfrac{0.010\text{ s}}{2} + \dfrac{0.10\text{ m}}{2.0\text{ m/s}}\right)}$$

$$= 2.8 \times 10^5\text{ N/C}$$

Notice that we kept everything in terms of symbols until the very last step. Several expressions along the way included the field strength E, but we couldn't evaluate those expressions because E is what we were trying to find. An early replacement of symbols with numbers would have made this solution more difficult and increased the likelihood of errors.

ASSESS A 2.8×10^5 N/C electric field is strong, but that's expected because it has to accelerate macroscopic droplets rather than atomic particles. Even so, the field strength is only about 10% of the field strength that would cause a spark in air, so there's a large margin of safety. We'll see in Chapter 22 that a field of this strength can be easily created by applying 2800 volts across an electrode spacing of 1 cm. Flow cytometry is used when cells need to be sorted with great accuracy, but the collection rate is too slow for a bulk separation of cells.

SUMMARY

GOAL To understand basic electric phenomena in terms of charges, forces, and electric fields.

GENERAL PRINCIPLES

Coulomb's Law

The forces between two charged particles q_1 and q_2 separated by distance r are

$$F_{1\text{ on }2} = F_{2\text{ on }1} = \frac{K|q_1||q_2|}{r^2}$$

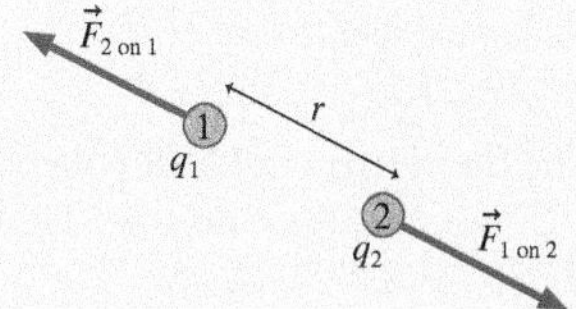

where the **electrostatic constant** is $K = 9.0 \times 10^9\ \text{N} \cdot \text{m}^2/\text{C}^2$.

- The forces are an interaction pair directed along the line joining the particles.
- The force is attractive for two opposite charges, repulsive for two like charges.
- The net force on a charged object is the vector sum of the forces due to all other charges.
- The SI unit of charge is the **coulomb** (C).

IMPORTANT CONCEPTS

The Charge Model

There are two kinds of **charge,** positive and negative.

- Fundamental charges are protons with charge $+e$ and electrons with charge $-e$, where $e = 1.60 \times 10^{-19}$ C.
- An object with equal positive and negative charge is **neutral.**

There are two kinds of materials, **insulators** and **conductors.**

- Charge remains fixed in or on an insulator.
- Charge moves easily through or along conductors.
- Charge is transferred by contact between objects.

Charged objects attract neutral objects.

- Charge polarizes atoms, creating electric dipoles.
- Charge polarizes conductors by shifting the sea of electrons.
- The **polarization force** is always an attractive force.

The Field Model

Charges interact with each other via the **electric field** $\vec{E}$.

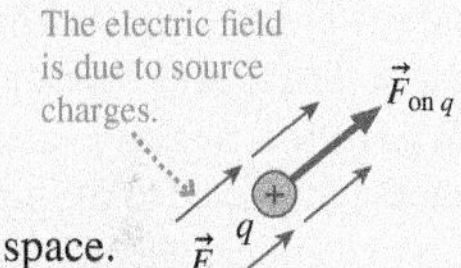

- The electric field exists at all points in space.
- The electric field vector shows the field at only one point, the point at the tail of the vector.
- The field is the agent that exerts a force. The force on charge q is

$$\vec{F}_{\text{on }q} = q\vec{E}$$

A **field diagram** shows electric field vectors at several points.

Electric field lines

- Go from positive to negative charges.
- Are parallel to the field vectors.
- Are closer together where the field is stronger, farther apart where the field is weaker.

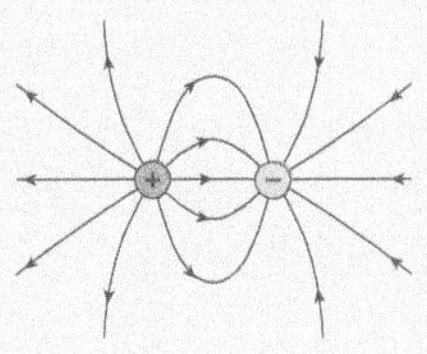

APPLICATIONS

Important electric fields

The electric field of a **point charge** or of a sphere of charge is

$$\vec{E} = \left(\frac{1}{4\pi\epsilon_0}\frac{|q|}{r^2}, \begin{cases}\text{away} & q \text{ positive}\\ \text{toward} & q \text{ negative}\end{cases}\right)$$

where $\epsilon_0 = 1/4\pi K = 8.85 \times 10^{-12}\ \text{C}^2/\text{N} \cdot \text{m}^2$ is the **permittivity constant.**

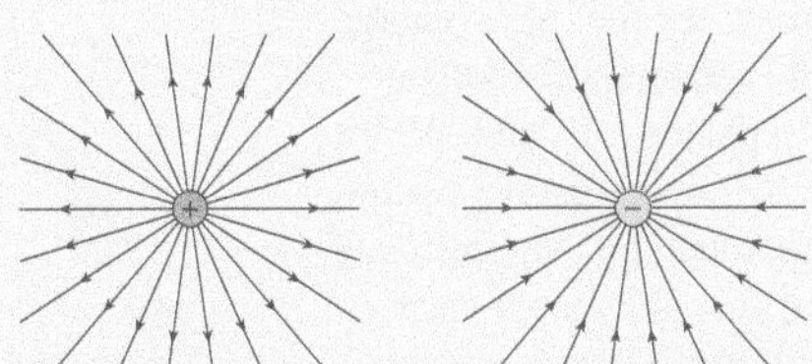

The electric field between the plates of a **parallel-plate capacitor** with plate area A and charge $\pm Q$ is

$$\vec{E} = \left(\frac{Q}{\epsilon_0 A}, \text{from positive to negative}\right)$$

This is a **uniform electric field.**

The electric dipole

Positive and negative charges $\pm q$ separated by distance s form an **electric dipole.** The **dipole moment** is

$$\vec{p} = (qs, \text{from negative to positive})$$

An electric field exerts a torque on a dipole:

$$\tau = pE\sin\theta$$

In a nonuniform electric field, a dipole has a net force in the direction of increasing field strength.

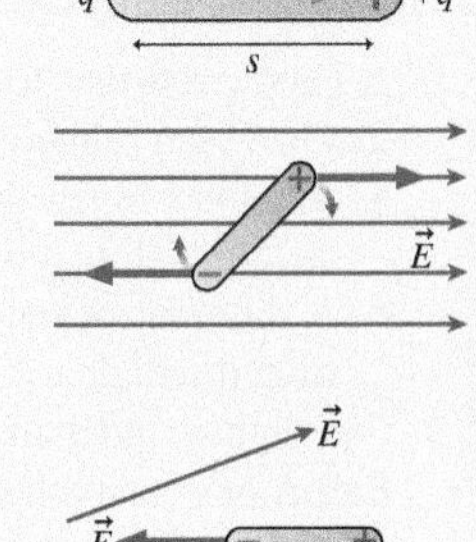

Conductors in electrostatic equilibrium

For a conductor in electrostatic equilibrium:

- Any excess charge is on the surface.
- The electric field inside the conductor is zero.
- The exterior electric field is perpendicular to the surface.

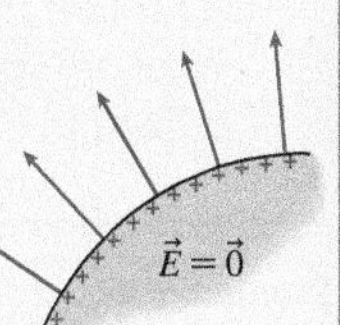

LEARNING OBJECTIVES After studying this chapter, you should be able to:

- Analyze basic electric phenomena with the charge model. *Conceptual Questions 21.2, 21.5; Problem 21.8*
- Use the atomic model of charge, conductors, insulators, and dipoles. *Conceptual Question 21.3; Problems 21.1, 21.3, 21.5, 21.7, 21.9*
- Calculate the force between charged particles using Coulomb's law. *Conceptual Questions 21.8, 21.9; Problems 21.11, 21.14, 21.18, 21.19, 21.21*
- Determine the electric fields of point charges and electric dipoles. *Conceptual Questions 21.10, 21.12; Problems 21.22, 21.24, 21.28, 21.33, 21.34*
- Solve problems involving parallel-plate capacitors and uniform electric fields. *Conceptual Question 21.15; Problems 21.40–21.43*
- Solve problems involving charged particles moving in electric fields. *Conceptual Question 21.14; Problems 21.45–21.49*
- Calculate the torque on a dipole in an electric field. *Problems 21.50–21.53*

STOP TO THINK ANSWERS

Stop to Think 21.1: A. The electroscope is originally given a positive charge. The charge spreads out, and the leaves repel each other. When a rod with a negative charge is brought near, some of the positive charge is attracted to the top of the electroscope, away from the leaves. There is less charge on the leaves, and so they move closer together.

Stop to Think 21.2: $q_E(+3e) > q_A(+1e) > q_D(0) > q_B(-1e) > q_C(-2e)$.

Stop to Think 21.3: B. The two forces are an interaction pair, opposite in direction but *equal* in magnitude.

Stop to Think 21.4: $E_2 > E_1 > E_4 > E_3$. The field is proportional to the charge, and inversely proportional to the square of the distance.

Stop to Think 21.5: C. From symmetry, the fields of the positive charges cancel. The net field is that of the negative charge, which is toward the charge.

Stop to Think 21.6: C. Electric field lines *start* on positive charges. Very near to each of the positive charges, the field lines should look like the field lines of a single positive charge.

Stop to Think 21.7: $E_1 = E_2 = E_3 = E_4 = E_5$. The field strength inside a capacitor is the same at all points. The electric field exists at all points whether or not a vector is shown at that point.

Stop to Think 21.8: C. There's an electric field at *all* points, whether an $\vec{E}$ vector is shown or not. The electric field at the dot is to the right. But an electron is a negative charge, so the force of the electric field on the electron is to the left.

QUESTIONS

Conceptual Questions

1. A plastic rod that has been rubbed with wool and a glass rod that has been rubbed with silk hang by threads.
   a. An object repels the plastic rod. Can you predict what it will do to the glass rod? If so, what? If not, why not?
   b. A different object attracts the plastic rod. Can you predict what it will do to the glass rod? If so, what? If not, why not?
2. Four lightweight balls A, B, C, and D are suspended by threads. Ball A has been touched by a plastic rod that was rubbed with wool. When the balls are brought close together, without touching, the following observations are made:
   - Balls B, C, and D are attracted to ball A.
   - Balls B and D have no effect on each other.
   - Ball B is attracted to ball C.

   What are the charge states (positive, negative, or neutral) of balls A, B, C, and D? Explain.
3. As shown in Figure Q21.3, metal sphere A has 4 units of negative charge and identical metal sphere B has 2 units of positive charge. The two spheres are brought into contact. What is the final charge state of each sphere? Explain.

FIGURE Q21.3

4. Metal spheres A and B in Figure Q21.4 are initially neutral and are touching. A positively charged rod is brought near A, but not touching. Is A now positive, negative, or neutral? Use both charge diagrams and words to explain.

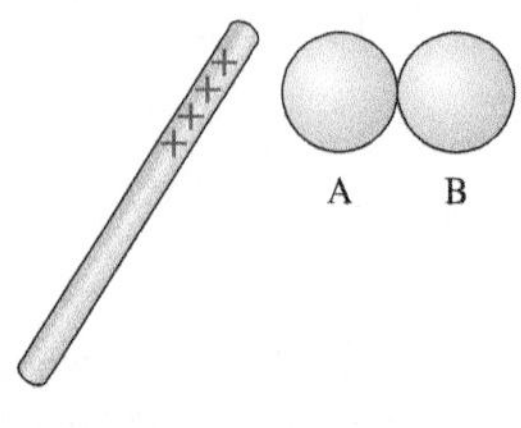

FIGURE Q21.4

Problem difficulty is labeled as | (straightforward) to ||||| (challenging). Problems labeled INT integrate significant material from earlier chapters; BIO are of biological or medical interest; CALC require calculus to solve.

5. A metal rod A and a metal sphere B, on insulating stands, touch each other as shown in Figure Q21.5. They are originally neutral. A positively charged rod is brought near (but not touching) the far end of A. While the charged rod is still close, A and B are separated. The charged rod is then withdrawn. Is the sphere then positively charged, negatively charged, or neutral? Explain.

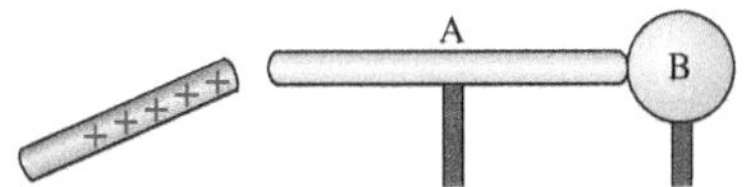

FIGURE Q21.5

6. A plastic balloon that has been rubbed with wool will stick to a wall. Can you conclude that the wall is charged? If so, where does the charge come from? If not, why does the balloon stick?
7. A hummingbird gains a significant electric charge while flying. This has consequences: When a charged bird approaches a flower, the stamens of the flower bend toward the bird, even though the stamens are uncharged. Explain how this happens.

8. Charged particle B in Figure Q21.8 has four times the charge of particle A. Reproduce the figure and then draw force vectors to show $\vec{F}_{A \text{ on } B}$ and $\vec{F}_{B \text{ on } A}$.

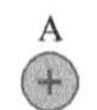

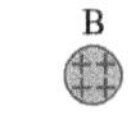

FIGURE Q21.8

9. Two charged particles are separated by 10 cm. Suppose the charge on each particle is doubled. By what factor does the electric force between the particles change?
10. Reproduce Figure Q21.10 on your paper. For each part, draw a dot or dots on the figure to show any position or positions (other than infinity) where $\vec{E} = \vec{0}$.

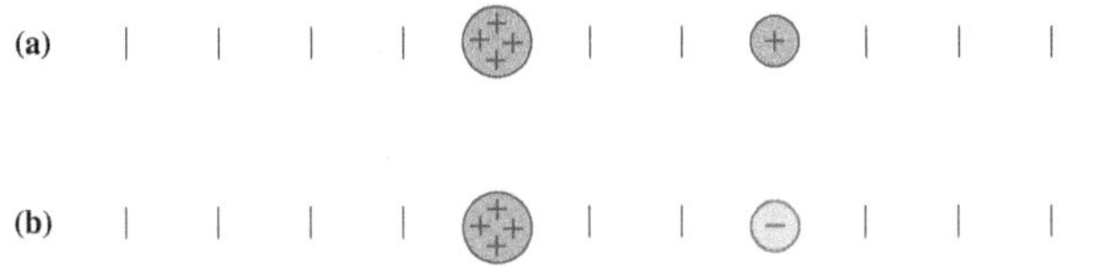

FIGURE Q21.10

11. Each part of Figure Q21.11 shows two points near two charges. Compare the electric field strengths E_1 and E_2 at these two points. Is $E_1 > E_2$, $E_1 = E_2$, or $E_1 < E_2$?

FIGURE Q21.11

12. Rank in order, from largest to smallest, the electric field strengths E_1 to E_4 at points 1 to 4 in Figure Q21.12. Explain.

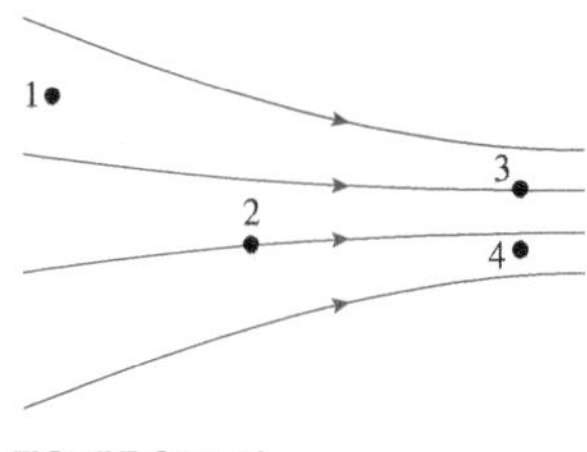

FIGURE Q21.12

13. Rank in order, from largest to smallest, the electric field strengths E_1 to E_5 at the five points in Figure Q21.13. Explain.
14. A proton and an electron are released from rest in the center of a capacitor.
    a. Is the force ratio F_p/F_e greater than 1, less than 1, or equal to 1? Explain.
    b. Is the acceleration ratio a_p/a_e greater than 1, less than 1, or equal to 1? Explain.

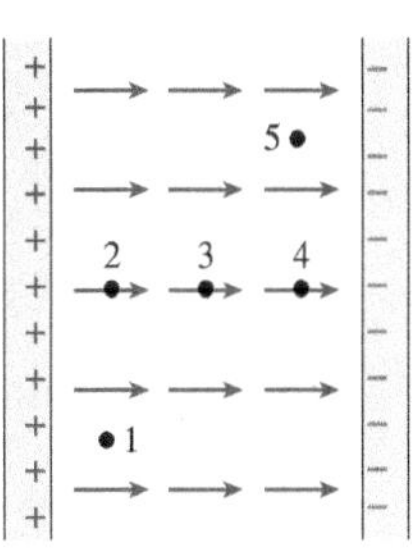

FIGURE Q21.13

15. Three charges are placed at the corners of the triangle in Figure Q21.15. The ++ charge has twice the quantity of charge of the two − charges; the net charge is zero. Is the triangle in equilibrium? If so, explain why. If not, draw the equilibrium orientation.

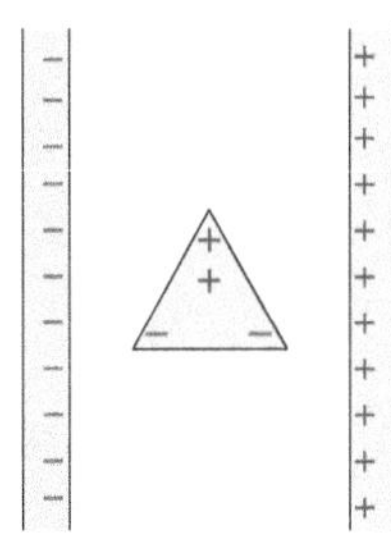

FIGURE Q21.15

Multiple-Choice Questions

16. | Two lightweight, electrically neutral conducting balls hang from threads. Choose the diagram in Figure Q21.16 that shows how the balls hang after:
    a. Both are touched by a negatively charged rod.
    b. Ball 1 is touched by a negatively charged rod and ball 2 is touched by a positively charged rod.
    c. Both are touched by a negatively charged rod but ball 2 picks up more charge than ball 1.
    d. Only ball 1 is touched by a negatively charged rod.

    Note that parts a through d are independent; these are not actions taken in sequence.

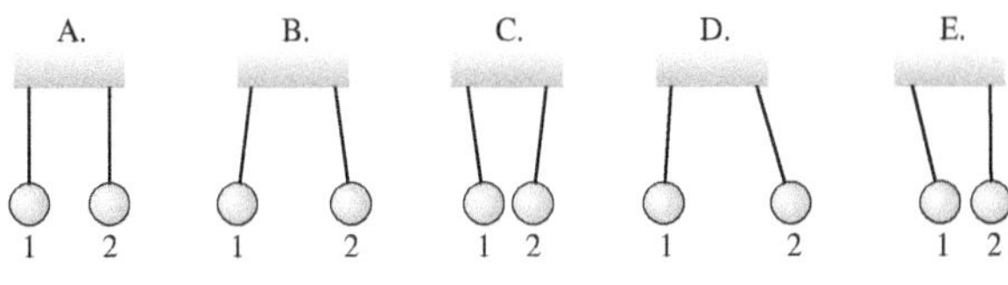

FIGURE Q21.16

17. | All the charges in Figure Q21.17 have the same magnitude. In which case does the electric field at the dot have the largest magnitude?

FIGURE Q21.17

18. | A glass bead charged to +3.5 nC exerts an 8.0×10^{-4} N repulsive electric force on a plastic bead 2.9 cm away. What is the charge on the plastic bead?
    A. +2.1 nC
    B. +7.4 nC
    C. +21 nC
    D. +740 nC

19. | A +7.5 nC point charge and a −2.0 nC point charge are 3.0 cm apart. What is the electric field strength at the midpoint between the two charges?
A. 3.3×10^3 N/C
B. 5.7×10^3 N/C
C. 2.2×10^5 N/C
D. 3.8×10^5 N/C

20. || Three point charges are arranged as shown in Figure Q21.20. Which arrow best represents the direction of the electric field vector at the position of the dot?

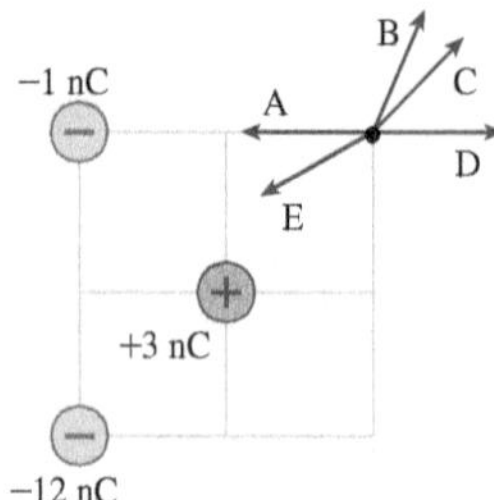

FIGURE Q21.20

21. | A positive charge is brought near to a dipole, as shown in Figure Q21.21. If the dipole is free to move, it
A. Rotates clockwise and moves to the right.
B. Rotates clockwise and moves to the left.
C. Rotates counterclockwise and moves to the right.
D. Rotates counterclockwise and moves to the left.
E. Does not rotate or move.

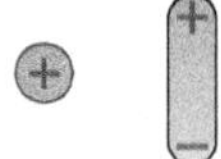

FIGURE Q21.21

Questions 22 through 25 concern an electric dipole in an electric field. Clinical studies suggest that an oscillating electric field can disrupt rapidly dividing tumor cells. Dividing cells form a structure called the *mitotic spindle* that is composed primarily of *microtubules.* Microtubules have a large electric dipole moment, and it is hypothesized that the response of the microtubules to an oscillating electric field interferes with cell division. Figure Q21.22 shows an electric dipole in a nonuniform electric field.

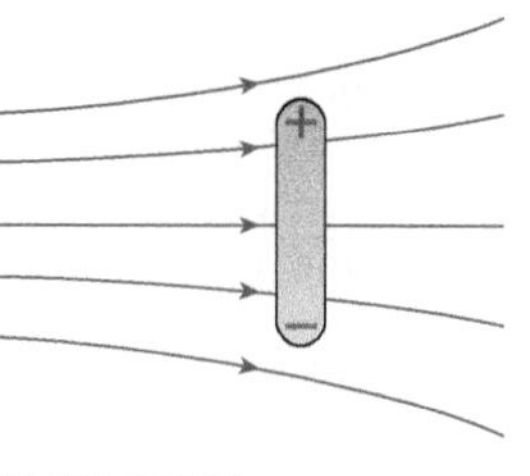

FIGURE Q21.22

22. | The strength of the field in Figure Q21.22 varies. The strength of the field is greatest
A. On the left side of the figure.
B. On the right side of the figure.
C. On one of the field lines.
D. In between the field lines.

23. | At the instant shown in Figure Q21.22, there is a net torque that will rotate the dipole
A. Clockwise.
B. Counterclockwise.

24. | The dipole will rotate to a position in which the net torque is zero. Once in this position,
A. There is a net force to the right.
B. There is a net force to the left.
C. There is no net force.

25. | Suppose the electric field of Figure Q21.22 oscillates at a low frequency, with the field direction reversing twice per cycle. The dipole will
A. Remain stationary.
B. Be driven steadily to the right.
C. Be driven steadily to the left.
D. Flip 180° twice per cycle.

PROBLEMS

Section 21.1 The Charge Model

Section 21.2 A Microscopic Model of Charge

1. || A glass rod is charged to +8.0 nC by rubbing.
a. Have electrons been removed from the rod or protons added?
b. How many electrons have been removed or protons added?
2. || A plastic rod is charged to −12 nC by rubbing.
a. Have electrons been added to the rod or protons removed?
b. How many electrons have been added or protons removed?
3. | When a honeybee flies through the air, it develops a charge
BIO of +17 pC. How many electrons did it lose in the process of acquiring this charge?
4. | A housefly walking across a surface may develop a sig-
BIO nificant electric charge through a process similar to frictional charging. Suppose a fly acquires a charge of +52 pC. How many electrons does it lose to the surface it is walking across?
5. || A plastic rod that has been charged to −15 nC touches a metal sphere. Afterward, the rod's charge is −10 nC.
a. What kind of charged particle was transferred between the rod and the sphere, and in which direction? That is, did it move from the rod to the sphere or from the sphere to the rod?
b. How many charged particles were transferred?
6. || A glass rod that has been charged to +12 nC touches a metal sphere. Afterward, the rod's charge is +8.0 nC.
a. What kind of charged particle was transferred between the rod and the sphere, and in which direction? That is, did it move from the rod to the sphere or from the sphere to the rod?
b. How many charged particles were transferred?
7. || What is the total charge of all the electrons in 1.0 L of liquid
INT water?
8. || If two identical conducting spheres are in contact, any excess charge will be evenly distributed between the two. Three identical metal spheres are labeled A, B, and C. Initially, A has charge q, B has charge $-q/2$, and C is uncharged. What is the final charge on each sphere if C is touched to B, removed, and then touched to A?
9. || Saline solution for medical use contains 9.0 g of sodium
BIO chloride (NaCl) dissolved in 1.0 L of water. What is the charge concentration in C/m^3 due to the positive charge carriers?
10. || Each atom in a piece of solid aluminum contributes three va-
INT lence electrons to the sea of electrons. The density and atomic mass of aluminum are 2700 kg/m^3 and 27 u. What is the charge carrier density of aluminum in electrons/m^3?

Section 21.3 Coulomb's Law

11. | Two 1.0 kg masses are 1.0 m apart (center to center) on a frictionless table. Each has $+10\ \mu C$ of charge.
   a. What is the magnitude of the electric force on one of the masses?
   b. What is the initial acceleration of this mass if it is released and allowed to move?
12. || Two small plastic spheres each have a mass of 2.0 g and a charge of -50.0 nC. They are placed 2.0 cm apart (center to center).
   a. What is the magnitude of the electric force on each sphere?
   b. By what factor is the electric force on a sphere larger than its weight?
13. || Falling raindrops frequently develop an electric charge. Does
INT this create noticeable forces between the droplets? Suppose two 1.8 mg drops each have a charge of +25 pC; these are typical values. The centers of the droplets are at the same height and 4.0 mm apart.
   a. What is the magnitude of the electric force of one drop on the other?
   b. What is the magnitude of the horizontal acceleration of one of the drops?
14. || Solid sodium chloride (NaCl) is an ionic crystal in which forces between positive Na^+ ions and negative Cl^- ions hold the crystal together. Two adjacent ions are 0.28 nm apart.
   a. What is the magnitude of the attractive force between a Na^+ ion and a Cl^- ion?
   b. Each Na^+ ion in the crystal has six Cl^- ion *nearest neighbors,* two on each side of it in the x-, y-, and z-directions. What is the magnitude of the net force on a Na^+ ion due to its nearest neighbors?
15. ||| A small metal bead, labeled A, has a charge of 25 nC. It is touched to metal bead B, initially neutral, so that the two beads share the 25 nC charge, but not necessarily equally. When the two beads are then placed 5.0 cm apart, the force between them is 5.4×10^{-4} N. What are the charges q_A and q_B on the beads?
16. ||| A small glass bead has been charged to +20 nC. A tiny ball 1.0 cm above the bead feels a 0.018 N downward electric force. What is the charge on the ball?
17. ||| A small metal sphere has a mass of 0.15 g and a charge of -23.0 nC. It is 10.0 cm directly above an identical sphere that has the same charge. This lower sphere is fixed and cannot move. If the upper sphere is released, it will begin to fall. What is the magnitude of its initial acceleration?
18. || In Figure P21.18, charge q_2 experiences no net electric force. What is q_1?

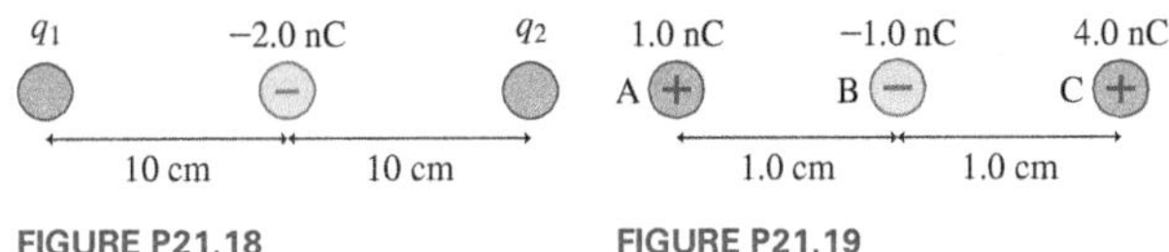

FIGURE P21.18

FIGURE P21.19

19. || What are the magnitude and direction of the electric force on on (a) charge A and (b) charge C in Figure P21.19? Specify the direction as "to the right" or "to the left."
20. || A 2.0 g plastic bead charged to -4.0 nC and a 4.0 g glass bead
INT charged to +8.0 nC are 2.0 cm apart and free to move. What are the accelerations of (a) the plastic bead and (b) the glass bead?
21. || Two positive point charges q and $4q$ are at $x = 0$ and $x = L$, respectively, and free to move. A third charge is placed so that the entire three-charge system is in static equilibrium. What are the magnitude, sign, and x-coordinate of the third charge?

Section 21.4 The Electric Field

22. | A 15 nC charged particle experiences an electric force (0.030 N, 25° above horizontal). What is the electric field at the particle's position?
23. | The electric field 2.0 cm from a small charged object is (1600 N/C, 20° above horizontal). What is the electric field 4.0 cm in the same direction from the object?
24. | What are the strength and direction of the electric field 1.0 mm from (a) a proton and (b) an electron?
25. || What magnitude charge creates a 1.0 N/C electric field at a point 1.0 m away?
26. || What are the strength and direction of the electric field 2.0 cm from a small glass bead that has been charged to +6.0 nC?
27. || A platypus foraging for prey can detect an electric field as
BIO small as 0.002 N/C. To give an idea of the sensitivity of the platypus's electric sense, how far from a +10 nC point charge does the field have this magnitude?
28. || A bumblebee can sense electric fields as the fields bend
BIO hairs on its body. Bumblebees have been conclusively shown to detect an electric field of 60 N/C. Could a bumblebee use this sense to detect the presence of another nearby bumblebee? Suppose a bumblebee has a charge of 24 pC. How far away could another bumblebee detect its presence?
29. | A 0.10 g honeybee acquires a charge of +23 pC while flying.
BIO a. The earth's electric field near the surface is typically (100 N/C, downward). What is the ratio of the electric force on the bee to the bee's weight?
   b. What electric field (strength and direction) would allow the bee to hang suspended in the air?
30. || What are the strength and direction of an electric field that will balance the weight of a 1.0 g plastic sphere that has been charged to -3.0 nC?
31. || A +12 nC charged particle is located at the origin. What are the electric fields at the (x, y) positions (a) (5.0 cm, 5.0 cm), (b) (0.0 cm, 5.0 cm), and (c) (3.0 cm, 4.0 cm)? Specify the direction as an angle above the $+x$-axis or the $-x$-axis.
32. || The electric field 2.0 cm from a small charged object points away from the object with a strength of 270,000 N/C. What is the object's charge?

Section 21.5 The Electric Field of Multiple Charges

33. || What are the strength and direction of the electric field at the position indicated by the dot in Figure P21.33? Specify the direction as an angle above or below horizontal.

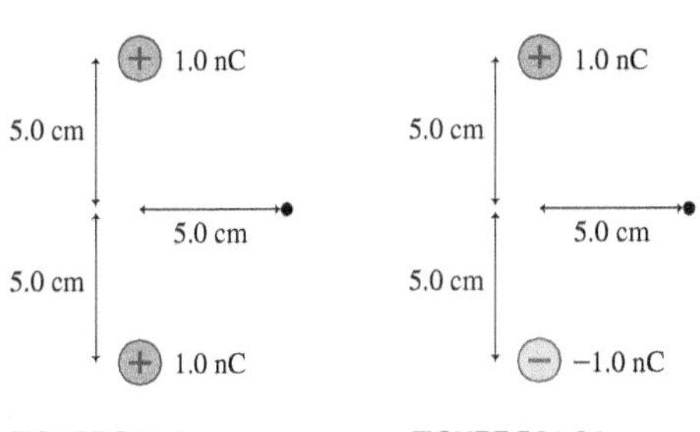

FIGURE P21.33

FIGURE P21.34

34. || What are the strength and direction of the electric field at the position indicated by the dot in Figure P21.34? Specify the direction as an angle above or below horizontal.

35. || What are the strength and direction of the electric field at the position indicated by the dot in Figure P21.35? Specify the direction as an angle above or below horizontal.

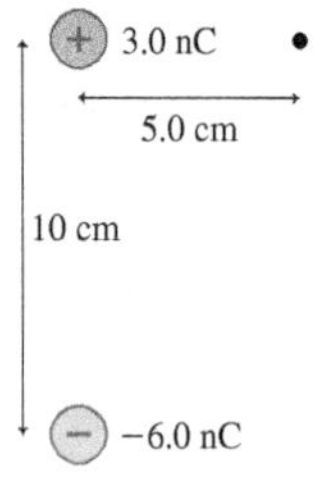

FIGURE P21.35

36. || An electric dipole is formed from two charges, $\pm q$, spaced 1.0 cm apart. The dipole is at the origin, oriented along the y-axis. The electric field strength at the point $(x, y) = (0\text{ cm}, 10\text{ cm})$ is 360 N/C.
   a. What is the charge q? Give your answer in nC.
   b. What is the electric field strength at the point $(x, y) = (10\text{ cm}, 0\text{ cm})$?

37. || An electric dipole is formed from ± 1.0 nC charges spaced 2.0 mm apart. The dipole is at the origin, oriented along the x-axis. What is the electric field strength at the points (a) $(x, y) = (10\text{ cm}, 0\text{ cm})$ and (b) $(x, y) = (0\text{ cm}, 10\text{ cm})$?

38. ||| An *electret* is similar to a magnet, but rather than being permanently magnetized, it has a permanent electric dipole moment. Suppose a small electret with electric dipole moment 1.0×10^{-7} C·m is 25 cm from a small ball charged to +25 nC, with the ball on the axis of the electric dipole. What is the magnitude of the electric force on the ball?

39. ||| The electric field strength 25 cm from the center of a uniformly charged, hollow metal sphere is 12,000 N/C. The sphere is 8.0 cm in diameter, and all the charge is on the surface. What is the magnitude of the surface charge density in nC/cm^2?

40. || A parallel-plate capacitor is constructed of two square plates, size $L \times L$, separated by distance d. The plates are given charge $\pm Q$. Let's consider how the electric field changes if one of these variables is changed while the others are held constant. What is the ratio E_f/E_i of the final electric field strength E_f to the initial electric field strength E_i if:
   a. Q is doubled?
   b. L is doubled?
   c. d is doubled?

41. || A parallel-plate capacitor is formed from two 4.0 cm × 4.0 cm electrodes spaced 2.0 mm apart. The electric field strength inside the capacitor is 1.0×10^6 N/C. What is the charge (in nC) on each electrode?

42. || Two circular disks spaced 0.50 mm apart form a parallel-plate capacitor. Transferring 3.0×10^9 electrons from one disk to the other causes the electric field strength to be 2.0×10^5 N/C. What are the diameters of the disks?

43. ||| Air "breaks down" when the electric field strength reaches 3.0×10^6 N/C, causing a spark. A parallel-plate capacitor is made from two 4.0 cm × 4.0 cm electrodes. How many electrons must be transferred from one electrode to the other to create a spark between the electrodes?

Section 21.6 The Motion of a Charged Particle in an Electric Field

44. || A 0.10 g plastic bead is charged by the addition of 1.0×10^{10} excess electrons. What electric field $\vec{E}$ (strength and direction) will cause the bead to hang suspended in the air?
INT

45. || A protein molecule in an electrophoresis gel has a negative charge. The exact charge depends on the pH of the solution, but 30 excess electrons is typical. What is the magnitude of the electric force on a protein with this charge in a 1500 N/C electric field?
BIO

46. || Two parallel plates 1.0 cm apart are equally and oppositely charged. An electron is released from rest at the surface of the negative plate and simultaneously a proton is released from rest at the surface of the positive plate. How far from the negative plate is the point at which the electron and proton pass each other?
INT

47. || Two 2.0-cm-diameter disks face each other, 1.0 mm apart. They are charged to ± 10 nC.
INT
   a. What is the electric field strength between the disks?
   b. A proton is shot from the negative disk toward the positive disk. What launch speed must the proton have to just barely reach the positive disk?

48. ||| An electron traveling parallel to a uniform electric field increases its speed from 2.0×10^7 m/s to 4.0×10^7 m/s over a distance of 1.2 cm. What is the electric field strength?
INT

49. ||| An electron is fired through a small hole in the positive plate of a parallel-plate capacitor at a speed of 1.0×10^7 m/s. The capacitor plates, which are in a vacuum chamber, are 2.0-cm-diameter disks spaced 3.0 mm apart. The electron travels 2.0 mm before being turned back. What is the capacitor's charge?
INT

Section 21.7 The Torque on a Dipole in an Electric Field

50. || The permanent electric dipole moment of the water molecule (H_2O) is 6.2×10^{-30} C·m. What is the maximum possible torque on a water molecule in a 5.0×10^8 N/C electric field?

51. || An ammonia molecule (NH_3) has a permanent electric dipole moment 5.0×10^{-30} C·m. A proton is 2.0 nm from the molecule in the plane that bisects the dipole. What is the torque on the molecule?

52. || Molecules of carbon monoxide are permanent electric dipoles due to unequal sharing of electrons between the carbon and oxygen atoms. Figure P21.52 shows the distance and charges.
INT

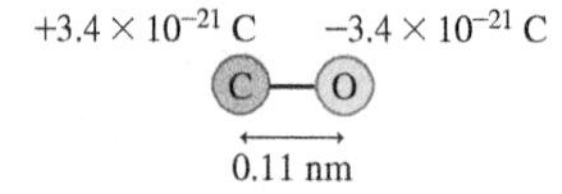

FIGURE P21.52

Suppose a carbon monoxide molecule with a horizontal axis is in a vertical electric field of strength 15,000 N/C.
   a. What is the magnitude of the net force on the molecule?
   b. What is the magnitude of the torque on the molecule?

53. || The display used in one kind of e-book reader consists of millions of tiny spheres that float in a thin fluid layer between two conducting transparent plates. Each sphere, which is black on one side and white on the other, is an electric dipole that can be modeled as opposite charges of magnitude 3.5×10^{-15} C separated by a distance of 100 μm. An electric field between the plates rotates the spheres so that either the white or black side is toward the reader. What is the maximum possible torque on a sphere if the electric field between the plates is 4.0×10^5 N/C?
INT

General Problems

54. || A 2.0-mm-diameter copper ball is charged to +50 nC. What fraction of its electrons have been removed? The density of copper is 8900 kg/m^3.
INT

55. || Two equally charged, 1.00 g spheres are placed with 2.00 cm between their centers. When released, each begins to accelerate at 225 m/s^2. What is the magnitude of the charge on each sphere?
INT

56. || Fungal spores develop a measurable charge when launched. BIO Mycologists are unsure whether the repulsive force between spores aids in their dispersal. A typical spore is nearly spherical, with a diameter of 4 μm and a density equal to that of water. Each spore loses approximately 200 electrons during launch.
   a. Estimate to one significant figure the electric force between two spores whose centers are 10 μm apart.
   b. At a 10 μm separation, what is the ratio of the electric force to the spore's weight?
   c. What will the ratio become when the spore separation has increased to 100 μm?

57. || A massless spring is attached to a support at one end and has INT a 2.0 μC charge glued to the other end. A $-4.0\ \mu$C charge is slowly brought near. The spring has stretched 1.2 cm when the charges are 2.6 cm apart. What is the spring constant of the spring?

58. || What is the force on the 1.0 nC charge in Figure P21.58? Give your answer as a magnitude and a direction. Specify the direction as an angle counterclockwise from the positive x-axis.

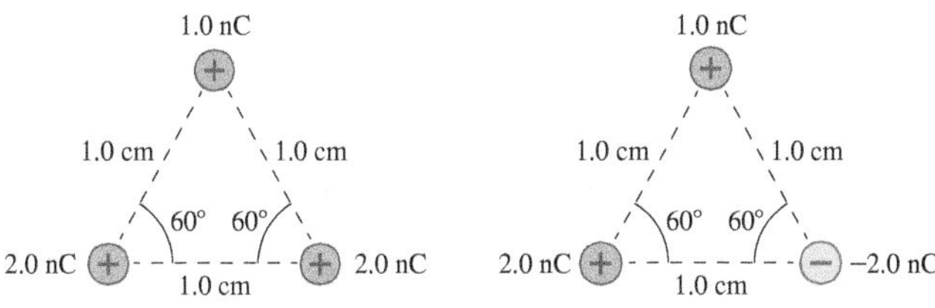

FIGURE P21.58

FIGURE P21.59

59. || What is the force on the 1.0 nC charge in Figure P21.59? Give your answer as a magnitude and a direction. Specify the direction as an angle counterclockwise from the positive x-axis.

60. | What is the magnitude of the force on the 1.0 nC charge in the middle of Figure P21.60 due to the four other charges?

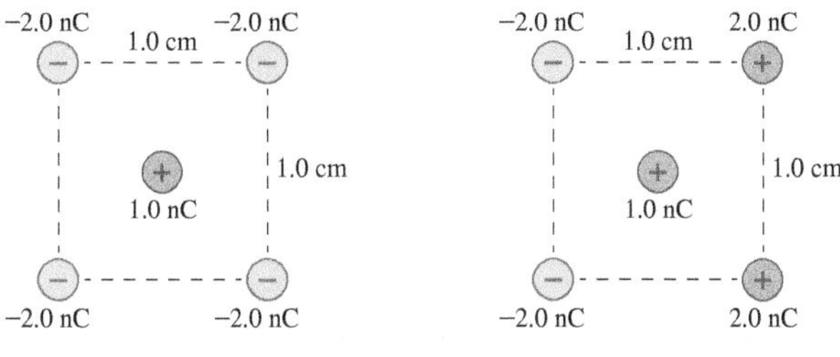

FIGURE P21.60

FIGURE P21.61

61. || What are the magnitude and direction of the force on the 1.0 nC charge in the middle of Figure P21.61 due to the four other charges? Specify the direction as an angle counterclockwise from the positive x-axis.

62. ||| A 2.0 g metal cube and a 4.0 g metal cube are 6.0 cm apart, INT measured between their centers, on a horizontal surface. For both, the coefficient of static friction is 0.65. Both cubes, initially neutral, are charged at a rate of 7.0 nC/s. How long after charging begins does one cube begin to slide away? Which cube moves first?

63. ||| Space explorers discover an 8.7×10^{17} kg asteroid that hap- INT pens to have a positive charge of 4400 C. They would like to place their 3.3×10^{5} kg spaceship in orbit around the asteroid. Interestingly, the solar wind has given their spaceship a charge of -1.2 C. What speed must their spaceship have to achieve a 7500-km-diameter circular orbit?

64. ||| A 2.0-mm-diameter glass sphere has a charge of $+1.0$ nC. INT What speed does an electron need to orbit the sphere 1.0 mm above the surface?

65. ||| You have a lightweight spring whose unstretched length is INT 4.0 cm. First, you attach one end of the spring to the ceiling and hang a 1.0 g mass from it. This stretches the spring to a length of 5.0 cm. You then attach two small plastic beads to the opposite ends of the spring, lay the spring on a frictionless table, and give each plastic bead the same charge. This stretches the spring to a length of 4.5 cm. What is the magnitude of the charge (in nC) on each bead?

66. || You sometimes create a spark when you touch a doorknob INT after shuffling your feet on a carpet. Why? The air always has a few free electrons that have been kicked out of atoms by cosmic rays. If an electric field is present, a free electron is accelerated until it collides with an air molecule. Most such collisions are elastic, so the electron collides, accelerates, collides, accelerates, and so on, gradually gaining speed. But if the electron's kinetic energy just before a collision is 2.0×10^{-18} J or more, it has sufficient energy to kick an electron out of the molecule it hits. Where there was one free electron, now there are two! Each of these can then accelerate, hit a molecule, and kick out another electron. Then there will be four free electrons. In other words, as Figure P21.66 shows, a sufficiently strong electric field causes a "chain reaction" of electron production. This is called a *breakdown* of the air. The current of moving electrons is what gives you the shock, and a spark is generated when the electrons recombine with the positive ions and give off excess energy as a burst of light.
   a. The average distance between ionizing collisions is 2.0 μm. (The electron's mean free path is less than this, but most collisions are elastic collisions in which the electron bounces with no loss of energy.) What acceleration must an electron have to gain 2.0×10^{-18} J of kinetic energy in this distance?
   b. What force must act on an electron to give it the acceleration found in part a?
   c. What strength electric field will exert this much force on an electron? This is the *breakdown field strength.* **Note:** The measured breakdown field strength is a little less than your calculated value because our model of the process is a bit too simple. Even so, your calculated value is close.
   d. Suppose a free electron in air is 1.0 cm away from a point charge. What minimum charge is needed to cause a breakdown and create a spark as the electron moves toward the point charge?

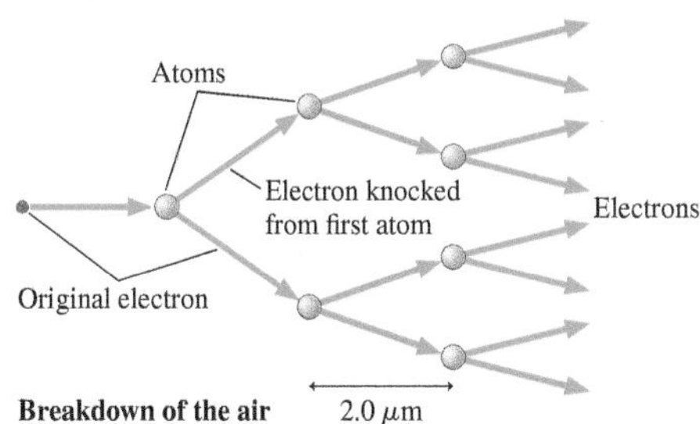

FIGURE P21.66

67. ||| What are the strength and direction of the electric field at the position indicated by the dot in Figure P21.67? Give your answer as a magnitude and angle measured cw or ccw (specify which) from the positive x-axis.

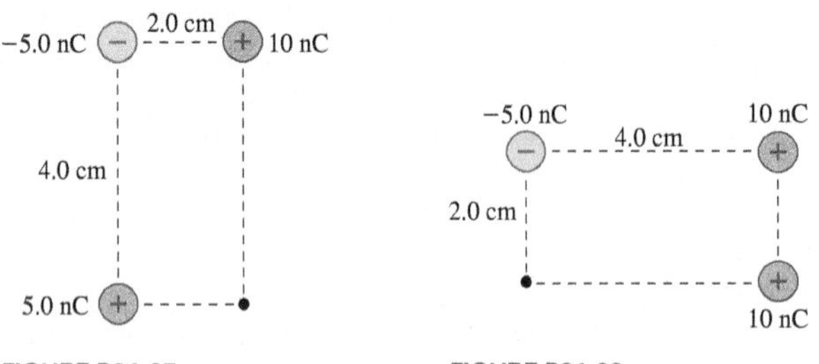

FIGURE P21.67 FIGURE P21.68

68. ||| What are the strength and direction of the electric field at the position indicated by the dot in Figure P21.68? Give your answer as a magnitude and angle measured cw or ccw (specify which) from the positive x-axis.

69. || *Ballooning* is a process by which some spiders travel through the air by releasing long strands of silk that catch a breeze. Under certain conditions, electric forces can provide much or even all of the upward force during liftoff. The earth has an electric field that averages 120 N/C pointing downward. Silk acquires a negative charge as it emerges from the spider's spinneret. (The spider's body stays neutral by discharging any positive charge to its surroundings.) Suppose a 0.20 mg spider deploys a long strand of silk with a total charge of -25 nC. If the spider lets go of a leaf, what is its initial upward acceleration while its speed is slow enough for drag to be neglected?
BIO INT

70. || Biotin, a B vitamin, bonds tightly to the bacterial protein streptavidin. Researchers measured the strength of the biotin-streptavidin bond by attaching a strand of DNA to a biotin molecule that was bonded to a tethered streptavidin molecule. DNA in solution is charged, with a charge of $-2e$ per base pair. An applied electric field exerts a force on the DNA, and researchers increased the field strength until the biotin-streptavidin bond broke. A 4.2×10^4 N/C field broke the bond when the attached DNA strand was 12 base pairs; a longer DNA strand with 48 base pairs required a 9.1×10^3 N/C field. Given these data, estimate the force necessary to break the biotin-streptavidin bond.
BIO

71. ||| As shown in Figure P21.71, a 5.0 nC charge sits at $x = 0$ in a uniform 4500 N/C electric field directed to the right. At what point along the x-axis would (a) a proton and (b) an electron experience no net force?

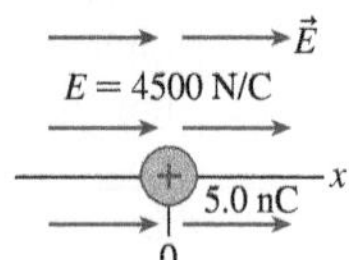

FIGURE P21.71

72. || The electron gun in a television tube uses a uniform electric field to accelerate electrons from rest to 5.0×10^7 m/s in a distance of 1.2 cm. What is the electric field strength?
INT

73. ||| An electron is launched at a 45° angle and a speed of 5.0×10^6 m/s from the positive plate of the parallel-plate capacitor shown in Figure P21.73. The electron lands 4.0 cm away.
INT

a. What is the electric field strength inside the capacitor?
b. What is the smallest possible spacing between the plates?

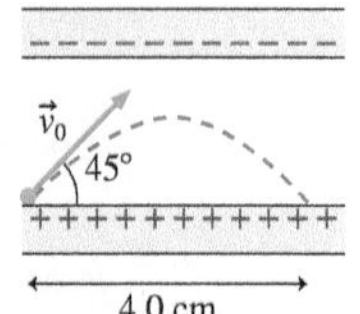

FIGURE P21.73

74. |||| The two parallel plates in Figure P21.74 are 2.0 cm apart and the electric field strength between them is 1.0×10^4 N/C. An electron is launched at a 45° angle from the positive plate. What is the maximum initial speed v_0 the electron can have without hitting the negative plate?
INT

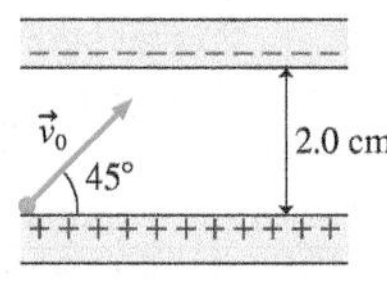

FIGURE P21.74

75. |||| The ozone molecule O_3 has a permanent dipole moment of 1.8×10^{-30} C·m. Although the molecule is very slightly bent—which is why it has a dipole moment—it can be modeled as a uniform rod of length 2.5×10^{-10} m with the dipole moment perpendicular to the axis of the rod. Suppose an ozone molecule is in a 5000 N/C uniform electric field. In equilibrium, the dipole moment is aligned with the electric field. But if the molecule is rotated by a *small* angle and released, it will oscillate back and forth in simple harmonic motion. What is the frequency f of oscillation?
INT

76. || We can model the electric dipole of the heart as ± 10 pC charges separated by 1.0 cm. The electric field of the heart is not an electric field in vacuum or in the air, which we have assumed in this chapter, but is an electric field in the body, which is mostly water. A material substance weakens an electric field because the molecules of the substance become polarized, and we'll see in Chapter 23 that $\vec{E}_{\text{substance}} = \vec{E}_{\text{vacuum}}/\kappa$, where κ (Greek kappa) is the *dielectric constant* of the substance. The dielectric constant of water is 80. What is the heart's electric field strength at a point in the body 20 cm from the center of the heart on the axis of the dipole?
BIO

77. ||| An electric field $\vec{E} = (100{,}000$ N/C, right) causes the 5.0 g ball in Figure P21.77 to hang at a 20° angle. What is the charge on the ball?
INT

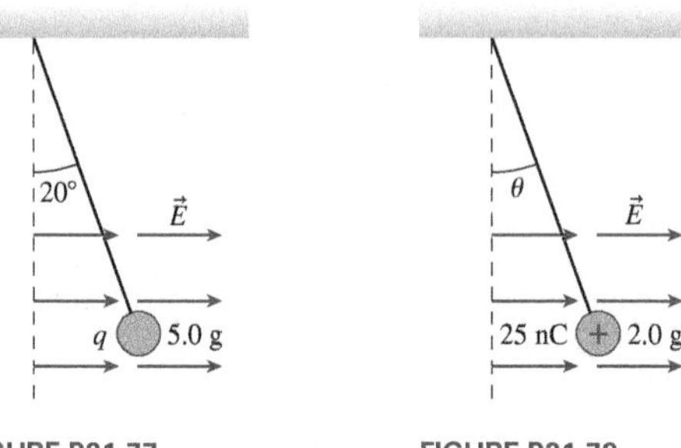

FIGURE P21.77 FIGURE P21.78

78. ||| An electric field $\vec{E} = (200{,}000$ N/C, right) causes the 2.0 g ball in Figure P21.78 to hang at an angle. What is θ?
INT

79. |||| Two 3.0 g spheres on 1.0-m-long threads repel each other after being equally charged, as shown in Figure P21.79. What is the charge q?
INT

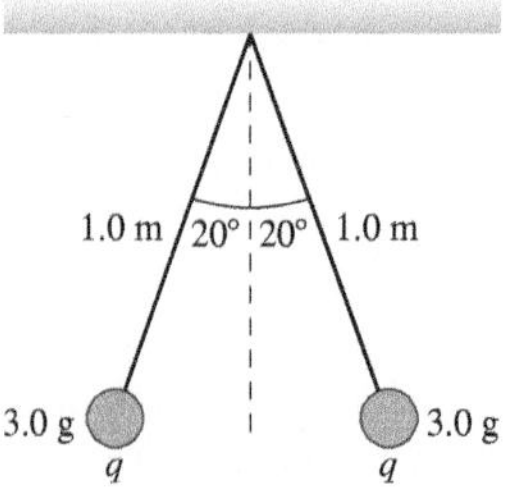

FIGURE P21.79

MCAT-Style Passage Problems

Flow Cytometry BIO

Flow cytometry, illustrated in Figure P21.80, is a technique used to sort cells by type. The cells are placed in a conducting saline solution which is then forced from a nozzle. The stream breaks up into small droplets, each containing one cell. A metal collar surrounds the stream right at the point where the droplets separate from the stream. Charging the collar polarizes the conducting liquid, causing the droplets to become charged as they break off from the stream. A laser beam probes the solution just upstream from the charging collar, looking for the presence of certain types of cells. All droplets containing one particular type of cell are given the same charge by the charging collar. Droplets with other desired types of cells receive a different charge, and droplets with no desired cell receive no charge. The charged droplets then pass between two parallel charged electrodes where they receive a horizontal force that directs them into different collection tubes, depending on their charge.

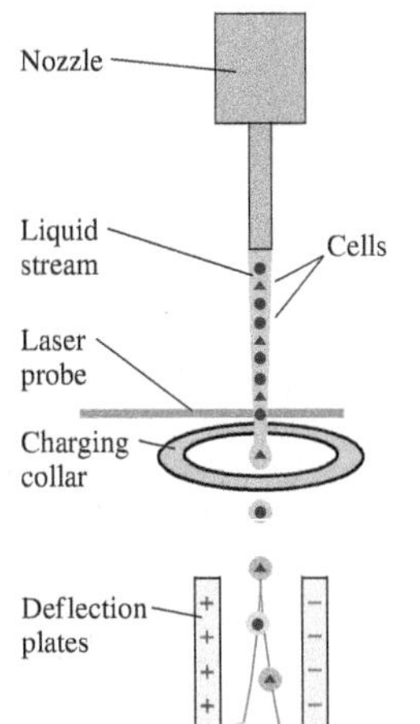

FIGURE P21.80

80. If the charging collar has a positive charge, the net charge on a droplet separating from the stream will be
   A. Positive.
   B. Negative.
   C. Neutral.
   D. The charge will depend on the type of cell.
81. Which of the following describes the charges on the droplets that end up in the five tubes, moving from left to right?
   A. $+2q, +q, 0, -q, -2q$
   B. $+q, +2q, 0, -2q, -q$
   C. $-q, -2q, 0, +2q, +q$
   D. $-2q, -q, 0, +q, +2q$
82. The droplets are conductors, and a droplet becomes polarized while it is in the region between the deflecting plates. Suppose a neutral droplet passes between the plates. The droplet's dipole moment will point
   A. Up.
   B. Down.
   C. Left.
   D. Right.
83. Another way to sort the droplets would be to give each droplet the same charge, then vary the electric field between the deflection plates. For the apparatus as sketched, this technique will not work because
   A. Several droplets are between the plates at one time, and they would all feel the same force.
   B. The cells in the solution have net charges that would affect the droplet charge.
   C. A droplet with a net charge would always experience a net force between the plates.
   D. The droplets would all repel each other, and this force would dominate the deflecting force.

22 Electric Potential

A ray cruising over the ocean floor uses special sensory organs to detect the electric potential of possible prey hiding among the rocks.

LOOKING AHEAD

Electric Potential

One of the best diagnostics of epilepsy is an *electroencephalogram,* which measures electric potentials generated by activity in the brain.

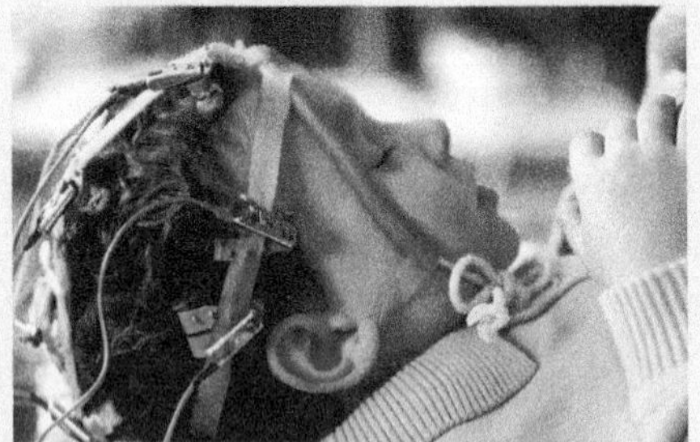

You'll learn to calculate the electric potential of charges and the electric potential energy of interacting charges.

Sources of Potential

Solar panels and wind turbines, much like batteries, generate a *voltage*—an electric potential difference between two points.

You'll discover that an electric potential difference is created when positive and negative charges are separated.

Potential and Field

There is a close connection between the *electric potential* and the electric field. They are two different perspectives of the same source charges.

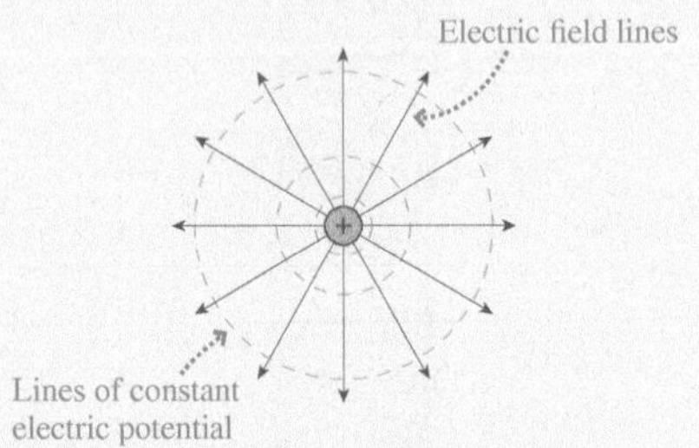

You'll learn how to move back and forth between the electric field and the electric potential.

GOAL To understand and use the electric potential and electric potential energy.

PHYSICS AND LIFE

Every Heartbeat Has Potential

The membrane of every cell in your body, and of any other eukaryotic cell, has a layer of positive charge on one side and a layer of negative charge on the other. These charge layers produce an electric potential difference across the membrane that is known as the *membrane potential.* One of the most important manifestations of the membrane potential is the electrocardiogram (EKG), an essential and widespread medical diagnostic. The EKG's revelatory power comes from the remarkable fact that the heart has an electric dipole moment that is created as the membrane potentials of heart cells change during the contractions of a heartbeat. An EKG is recorded by using electrodes to measure voltages—differences in the electric potential—on the surface of the body, which makes it quick, safe, and painless. These fluctuating voltages provide a wealth of information about the behavior and performance of the heart.

Special organs in an electric eel can create a head-to-tail potential difference of more than 600 volts.

22.1 Electric Potential Energy

Every living cell has an *electric potential difference* across its membrane—the *membrane potential*—that plays a critical role in the functioning of the cell, including but not limited to muscle contraction and nerve signaling. An *electrocardiogram* (EKG) provides crucial diagnostic information about the heart by measuring electric potential differences created by the heart's electric dipole moment. Electric potentials applied to specific areas of the brain can control epileptic seizures and the tremors of Parkinson's disease.

The *electric potential* is one of the most important concepts in physics, with applications ranging from electric power generation to the transmission of nerve impulses. However, the concept of the electric potential is subtle. In Chapter 21 we introduced the electric field after first looking at electric forces. This chapter follows a similar approach: We'll begin with an investigation of electric energy and that will lead us to the electric potential.

It's been many chapters since we dealt much with work and energy, but those ideas are now essential to our story. You will recall that mechanical energy $E_{mech} = K + U$ is conserved for an isolated system—a system with no external interactions and no dissipative forces. That is,

$$\Delta E_{mech} = \Delta K + \Delta U = 0 \tag{22.1}$$

We need to be careful with notation because we are now using E to represent the electric field strength. To avoid confusion, we will represent mechanical energy either as the sum $K + U$ or as E_{mech}, with an explicit subscript.

NOTE ► Recall that for any X, the *change* in X is $\Delta X = X_{final} - X_{initial}$. ◄

A key idea of Chapters 10 and 11 was that energy is the energy *of a system,* and clearly defining the system is crucial. Kinetic energy $K = \frac{1}{2}mv^2$ is a particle's *energy of motion*. For a multiparticle system, K is the sum of the kinetic energies of each particle in the system.

Potential energy U is the *interaction energy* of the system. Suppose the particles of the system move from some initial set of positions i to final positions f. As the particles move, the interaction pairs of forces between the particles do work and the system's potential energy changes. In ◄◄ SECTION 11.1 we defined the *change* in potential energy to be

$$\Delta U = -W_{interaction}(\mathrm{i} \rightarrow \mathrm{f}) \tag{22.2}$$

where the notation means the work done by the interaction forces as the configuration changes from i to f. This rather abstract definition will make more sense when we see specific applications.

Recall that *work* is done when a force acts on a particle as it is being displaced. In ◄◄ SECTION 10.2 you learned that a *constant* force $\vec{F}$ does work

$$W = F\,\Delta r \cos\theta \tag{22.3}$$

as the particle undergoes displacement $\Delta\vec{r}$, where θ is the angle between the force and the displacement. FIGURE 22.1 reminds you of the three special cases $\theta = 0°$, $90°$, and $180°$.

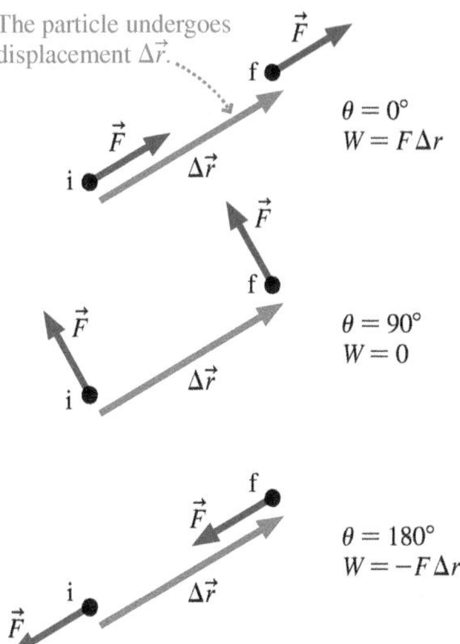

FIGURE 22.1 The work done by a constant force.

If the force is *not* constant, we can calculate the work by dividing the path into many small segments of length dx, finding the work done in each segment, and then summing (i.e., integrating) from the start of the path to the end. The work done in one such segment is $dW = F(x)\cos\theta\,dx$, where $F(x)$ indicates that the force is a function of position x. Thus the work done on the particle as it moves from x_i to x_f is

$$W = \int_{x_i}^{x_f} F(x)\cos\theta\,dx \tag{22.4}$$

A Gravitational Analogy

Gravity, like electricity, is a long-range force. Much as we defined the electric field $\vec{E} = \vec{F}_{\text{on } q}/q$, we can also define a *gravitational field*—the agent that exerts gravitational forces on masses—as $\vec{F}_{\text{on } m}/m$. But $\vec{F}_{\text{on } m} = m\vec{g}$ near the earth's surface; thus the familiar $\vec{g} = (9.80\ \text{N/kg, down})$ is really the gravitational field! Notice how we've written the units of $\vec{g}$ as N/kg—analogous to N/C for the electric field—but you can easily show that $\text{N/kg} = \text{m/s}^2$. The gravitational field near the earth's surface is a *uniform* field in the downward direction.

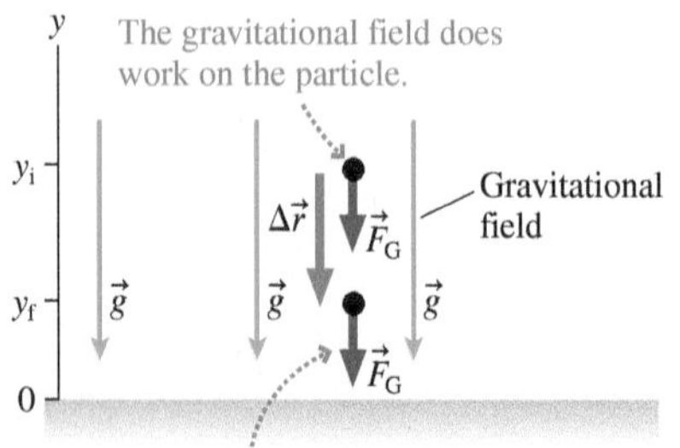
y
The gravitational field does work on the particle.
y_i
y_f
0
$\Delta\vec{r}$
$\vec{F}_G$
Gravitational field
$\vec{g}$
$\vec{F}_G$
The net force on the particle is down. It gains kinetic energy (i.e., speeds up) as it loses potential energy.

FIGURE 22.2 Potential energy is transformed into kinetic energy as a particle moves in a gravitational field.

FIGURE 22.2 shows a particle of mass m falling in the gravitational field. The gravitational force is in the same direction as the particle's displacement, so the gravitational field does a *positive* amount of work on the particle. The gravitational force is constant, hence the work done by the gravitational field is

$$W_G = F_G \Delta r \cos 0° = mg|y_f - y_i| = mgy_i - mgy_f \tag{22.5}$$

We have to be careful with signs because Δr, the magnitude of the displacement vector, must be a positive number.

Now we can see how the definition of ΔU in Equation 22.2 makes sense. The *change* in gravitational potential energy is

$$\Delta U_G = U_f - U_i = -W_G(i \rightarrow f) = mgy_f - mgy_i \tag{22.6}$$

Comparing the initial and final terms on the two sides of the equation, we see that the gravitational potential energy near the earth is the familiar quantity

$$U_G = mgy \tag{22.7}$$

We could add a constant to U_G, but we've made the customary choice that $U_G = 0$ at $y = 0$.

The Potential Energy of a Charge in a Uniform Electric Field

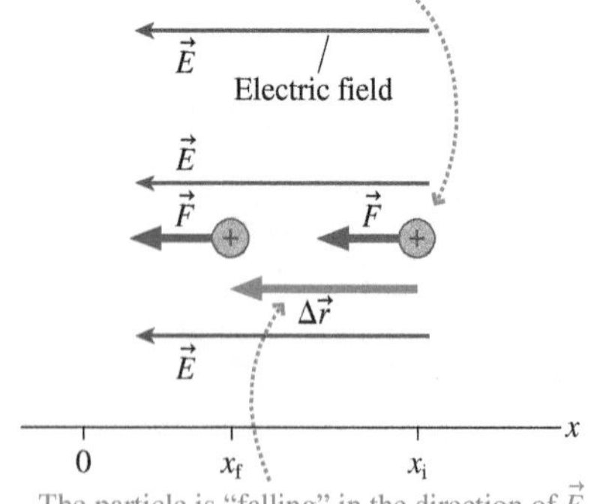
The electric field does work on the particle.
$\vec{E}$
Electric field
$\vec{E}$
$\vec{F}$
$\vec{F}$
$\Delta\vec{r}$
$\vec{E}$
x
0
x_f
x_i
The particle is "falling" in the direction of $\vec{E}$.

FIGURE 22.3 The electric field does work on the charged particle.

FIGURE 22.3 shows a charged particle q in a uniform electric field $\vec{E}$. The situation looks very much like Figure 22.2 for a particle in a uniform gravitational field with one exception: The gravitational field $\vec{g}$ always points down, but an electric field can point in any direction. To deal with this, we choose an x-axis that always points *opposite* the electric field direction in the same way that an upward y-axis points opposite the gravitational field direction.

A positive charge in the field experiences a force in the direction of the field. It speeds up and gains kinetic energy as it "falls downward" in the electric field. Is the charge losing potential energy as it gains kinetic energy? Indeed it is, and the calculation is just like the calculation of gravitational potential energy. The electric field exerts a *constant* force $F = qE$ on the charge in the direction of motion; thus the work done on the charge by the electric field is

$$W_{\text{elec}} = F \Delta r \cos 0° = qE|x_f - x_i| = qEx_i - qEx_f \tag{22.8}$$

where we again have to be careful with the signs because $x_f < x_i$.

The work done by the electric field causes the *electric* potential energy to change by

$$\Delta U_{\text{elec}} = U_f - U_i = -W_{\text{elec}}(i \rightarrow f) = qEx_f - qEx_i \tag{22.9}$$

Comparing the initial and final terms on the two sides of the equation, we see that the **electric potential energy** of charge q in a uniform electric field is

$$U_{\text{elec}} = qEx \tag{22.10}$$

Potential energy of a charge in a uniform electric field

We could add a constant to U_{elec}, but we've made the choice that $U_{elec} = 0$ at $x = 0$. Equation 22.10 was derived with the assumption that q is positive, but it is valid for either sign of q.

NOTE ▶ Although Equation 22.10 is sometimes called "the potential energy of charge q," it is really the interaction energy of the charge with the source charges that create the electric field. ◀

FIGURE 22.4 shows positive and negative charged particles moving in the uniform electric field between the plates of a parallel-plate capacitor with plate separation d. For a positive charge, U_{elec} decreases and K increases as the charge moves toward $x = 0$ at the negative plate. Thus a positive charge is going "downhill" if it moves in the direction of the electric field. A positive charge moving opposite the field direction is going "uphill," slowing as it transforms kinetic energy into electric potential energy.

FIGURE 22.4 A charged particle exchanges kinetic and potential energy as it moves in an electric field.

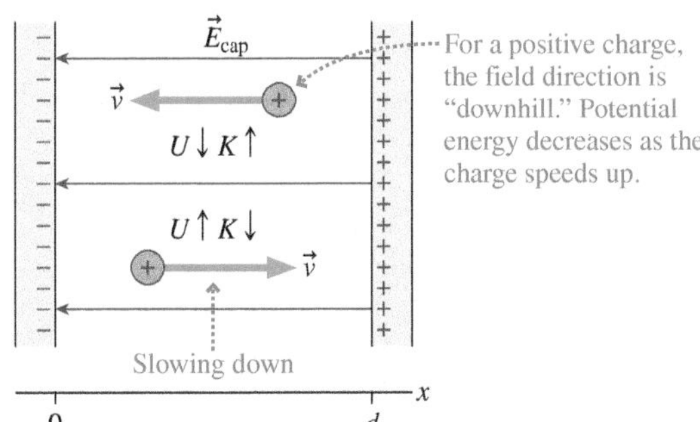

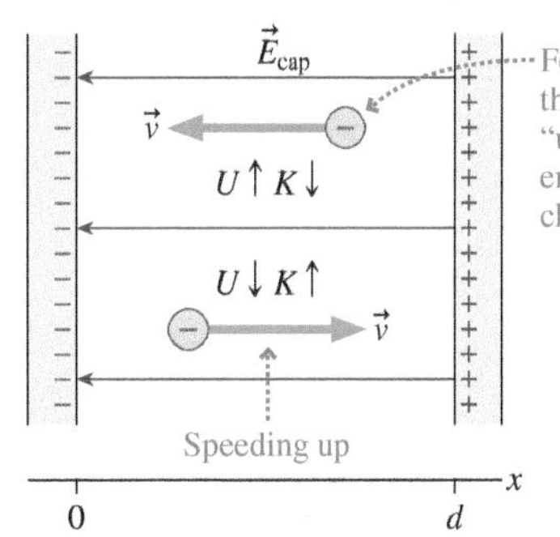

According to Equation 22.10, a negative charged particle has *negative* potential energy. You learned in Chapter 11 that there's nothing wrong with negative potential energy—it's simply less than the potential energy at some arbitrarily chosen reference location. The more important point, from Equation 22.10, is that the potential energy *increases* (becomes less negative) as a negative charge moves toward the negative plate. A negative charge moving in the field direction is going "uphill," transforming kinetic energy into electric potential energy as it slows.

FIGURE 22.5 is an *energy diagram* for a positively charged particle in an electric field. Recall that an energy diagram is a graphical representation of how kinetic and potential energies are transformed as a particle moves. For positive q, the electric potential energy given by Equation 22.10 increases linearly from 0 at the negative plate, which we've placed at $x = 0$, to qEd at the positive plate. The total mechanical energy—which is under your control—is constant. If $E_{mech} < qEd$, as shown here, a positively charged particle projected from the negative plate will gradually slow (transforming kinetic energy into potential energy) until it reaches a *turning point* where $U_{elec} = E_{mech}$. But if you project the particle with greater speed, such that $E_{mech} > qEd$, it will cross the gap to collide with the positive plate.

FIGURE 22.5 The energy diagram for a positively charged particle in a uniform electric field.

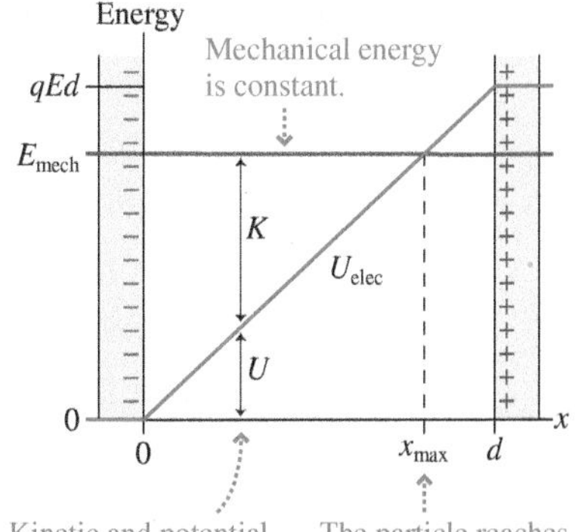

EXAMPLE 22.1 Conservation of energy

A 2.0 cm × 2.0 cm parallel-plate capacitor with a 2.0 mm spacing is charged to ±1.0 nC. First a proton and then an electron are released from rest midway between the capacitor plates. Assume the motion takes place in a vacuum.

a. What is each particle's energy?
b. What is each particle's speed as it reaches the plate?

PREPARE We'll define the system to be the charged particle and the charged electrodes. This is an isolated system with no dissipation, so the mechanical energy is conserved.

FIGURE 22.6 (next page) is a before-and-after pictorial representation, as you learned to draw in Part II. Each particle is

Continued

FIGURE 22.6 A proton and an electron in a capacitor.

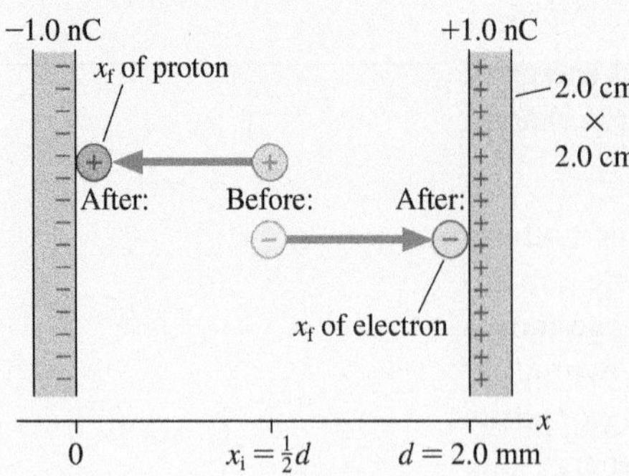

released from rest ($K = 0$) and moves "downhill" toward lower potential energy. Thus the proton moves toward the negative plate, the electron toward the positive plate. Equation 22.10 for the electric potential energy requires the x-axis to point opposite the field direction, so we've chosen an x-axis that goes from $x = 0$ at the negative plate to $x = d$ at the positive plate.

SOLVE

a. Both charged particles have $x_i = \frac{1}{2}d$, where $d = 2.0\,\text{mm}$ is the plate separation. The proton ($q = e$) begins with only potential energy, so its mechanical energy is

$$E_{\text{mech p}} = K_i + U_i = 0 + eEx_i = \tfrac{1}{2}eEd$$

Similarly, the electron ($q = -e$) has

$$E_{\text{mech e}} = K_i + U_i = 0 - eEx_i = -\tfrac{1}{2}eEd$$

The electric field inside the parallel-plate capacitor, from Chapter 21, is found to be

$$E = \frac{Q}{\epsilon_0 A} = 2.82 \times 10^5 \text{ N/C}$$

Thus the particles' energies can be calculated to be

$$E_{\text{mech p}} = 4.5 \times 10^{-17} \text{ J and } E_{\text{mech e}} = -4.5 \times 10^{-17} \text{ J}$$

Notice that the electron's mechanical energy is negative.

b. Conservation of mechanical energy requires $K_f + U_f = K_i + U_i = E_{\text{mech}}$. The proton collides with the negative plate, so $U_f = 0$, and the final kinetic energy is $K_f = \frac{1}{2}m_p v_f^2 = E_{\text{mech p}}$. Thus the proton's impact speed is

$$(v_f)_p = \sqrt{\frac{2E_{\text{mech p}}}{m_p}} = 2.3 \times 10^5 \text{ m/s}$$

Similarly, the electron collides with the positive plate at $x_f = d$, where $U_f = qEd = -eEd = 2E_{\text{mech e}}$. Thus energy conservation for the electron is

$$K_f = \tfrac{1}{2}m_e v_f^2 = E_{\text{mech e}} - U_f = E_{\text{mech e}} - 2E_{\text{mech e}} = -E_{\text{mech e}}$$

We found the electron's mechanical energy to be negative, so K_f is positive. The electron reaches the positive plate with speed

$$(v_f)_e = \sqrt{\frac{-2E_{\text{mech e}}}{m_e}} = 1.0 \times 10^7 \text{ m/s}$$

ASSESS Both particles have mechanical energy with the same magnitude, but the electron has a much greater final speed due to its much smaller mass.

STOP TO THINK 22.1 A plastic rod is negatively charged. The figure shows an end view of the rod. A positively charged particle moves in a circular arc around the glass rod. Is the work done on the charged particle by the rod's electric field positive, negative, or zero?

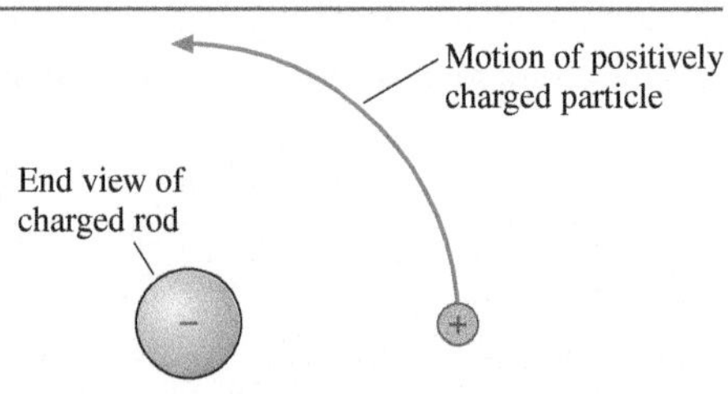

FIGURE 22.7 The interaction between two point charges.

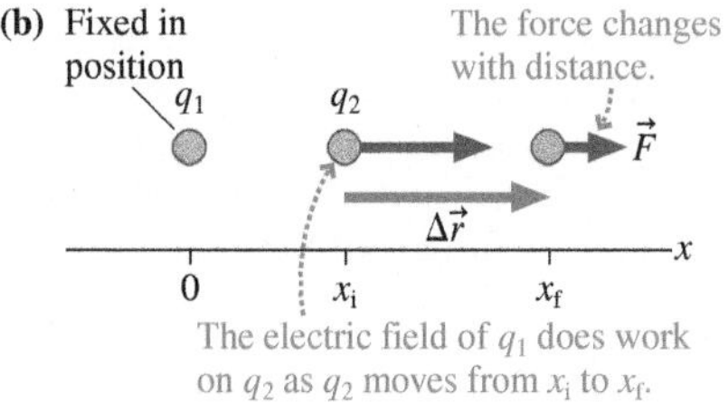

The Potential Energy of Two Point Charges

FIGURE 22.7a shows two charges q_1 and q_2, which we will assume to be like charges. These two charges interact, and the energy of their interaction can be found by calculating the work done by the electric field of q_1 on q_2 as q_2 moves from position x_i to position x_f. We'll assume that q_1 has been glued down and is unable to move, as shown in FIGURE 22.7b.

The force on q_2 is entirely in the direction of motion, so $\cos\theta = 1$. Thus the work done by the electric field as q_2 moves from x_i to x_f is

$$W_{\text{elec}} = \int_{x_i}^{x_f} F_{1\text{ on }2}\,dx = \int_{x_i}^{x_f} \frac{Kq_1q_2}{x^2}\,dx = Kq_1q_2 \frac{-1}{x}\bigg|_{x_i}^{x_f} = -\frac{Kq_1q_2}{x_f} + \frac{Kq_1q_2}{x_i} \quad (22.11)$$

The potential energy of the two charges is related to the work done by

$$\Delta U_{\text{elec}} = U_{\text{f}} - U_{\text{i}} = -W_{\text{elec}}(\text{i} \rightarrow \text{f}) = \frac{Kq_1q_2}{x_{\text{f}}} - \frac{Kq_1q_2}{x_{\text{i}}} \tag{22.12}$$

By comparing the left and right sides of the equation we see that the potential energy of the two-point-charge system is

$$U_{\text{elec}} = \frac{Kq_1q_2}{x} \tag{22.13}$$

We chose to integrate along the x-axis for convenience, but all that matters is the *distance* r between the charges. Thus a general expression for the electric potential energy is

$$U_{\text{elec}} = \frac{Kq_1q_2}{r} = \frac{1}{4\pi\epsilon_0}\frac{q_1q_2}{r} \tag{22.14}$$

Potential energy of two point charges separated by distance r

This is explicitly the energy *of the system,* not the energy of just q_1 or q_2.

NOTE ▶ The electric potential energy of two point charges looks *almost* the same as the force between the charges. One important difference is the r in the denominator of the potential energy compared to the r^2 in Coulomb's law. Another is that Equation 22.14 does *not* have absolute values of the charges, which means that U_{elec} can be negative. ◀

Three important points need to be noted:

- The potential energy of two charged particles is zero only when they are infinitely far apart. This makes sense because two charged particles cease interacting only when they are infinitely far apart.
- We derived Equation 22.14 for two like charges, but it is equally valid for two opposite charges. The potential energy of two like charges is *positive* and of two opposite charges is *negative.*
- Because the electric field outside a *sphere of charge* is the same as that of a point charge at the center, Equation 22.14 is also the electric potential energy of two charged spheres. Distance r is the distance between their centers.

FIGURE 22.8 The potential-energy diagrams for two like charges and two opposite charges.

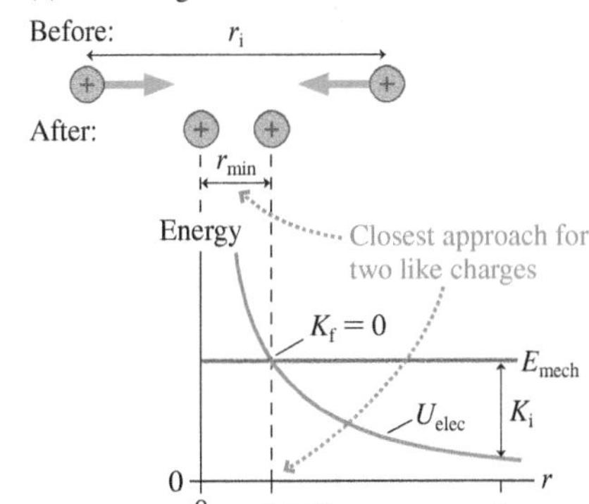

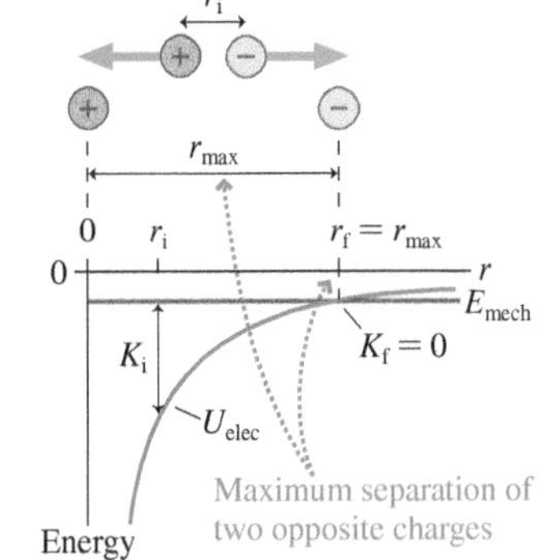

Charged-Particle Interactions

FIGURE 22.8a shows the potential-energy curve—a hyperbola—for two like charges as a function of the distance r between them. Also shown is the total energy line for two charged particles approaching each other with equal but opposite momenta.

You can see that the total energy line crosses the potential-energy curve at r_{min}. This is a turning point. The two charges gradually slow down, because of the repulsive force between them, until the distance separating them is r_{min}. At this point, the kinetic energy is zero and both particles are instantaneously at rest. Both then reverse direction and move apart, speeding up as they go. r_{min} is the *distance of closest approach.*

Two opposite charges are a little trickier because of the negative energies. Negative total energies seem troubling at first, but they characterize *bound systems.* **FIGURE 22.8b** shows two oppositely charged particles moving apart from each other with equal but opposite momenta. If $E_{\text{mech}} < 0$, as shown, then their total energy line crosses the potential-energy curve at r_{max}. That is, the particles slow down, lose kinetic energy, reverse directions at *maximum separation* r_{max}, and then "fall" back together. They cannot escape from each other. Although moving in three dimensions rather than one, the electron and proton of a hydrogen atom are a bound system, and their mechanical energy is negative.

Two oppositely charged particles *can* escape from each other if $E_{mech} > 0$. The threshold condition for escape is $E_{mech} = 0$, which will allow the particles to reach infinite separation $(U \rightarrow 0)$ at infinitesimally slow speed $(K \rightarrow 0)$. The initial speed that gives $E_{mech} = 0$ is called the *escape speed*.

NOTE ▶ Real particles can't be infinitely far apart, but $U_{elec} = 0$ is an excellent approximation when particles are so far apart that they have no meaningful interaction. Two charged particles for which $U_{elec} \approx 0$ are sometimes described as "far apart" or "far away." ◀

EXAMPLE 22.2 Proton radiation therapy

Proton radiation therapy destroys cancer cells by irradiating them with a beam of high-energy protons. Most of the proton energy disrupts bonds and damages DNA in tumor cells, but a small fraction of the protons come very close to the nucleus of an atom. Suppose a proton is aimed directly at the 5.4-fm-diameter nucleus of a ^{12}C atom, where 1 fm = 1 femtometer = 10^{-15} m. What initial speed must the proton have to just reach the surface of the nucleus? The proton mass is $m_p = 1.67 \times 10^{-27}$ kg.

PREPARE Carbon has atomic number 6, so a ^{12}C nucleus has six protons and charge $q_{carbon} = +6e$. The interaction between the proton with $q_{proton} = +e$ and the nucleus is an interaction between two charged particles. The atom's electrons are much less massive than the proton and the nucleus, and they are spread out over a distance much greater than the nuclear diameter, so we will assume that the proton's interactions with electrons can be ignored. The proton starts from "far away," so we will assume that $U_i = 0$.

FIGURE 22.9 is a before-and-after pictorial representation. To "just reach" the surface of the nucleus means that the proton comes to rest $(v_f = 0)$ as it reaches $r_f = R = 2.7 \times 10^{-15}$ m, the *radius* of the nucleus.

FIGURE 22.9 A proton approaching a nucleus.

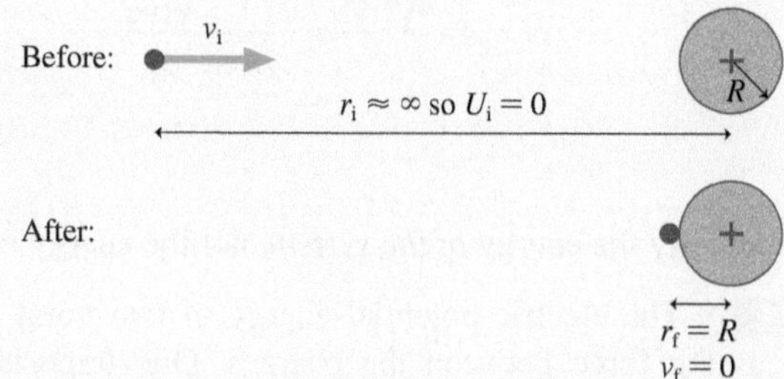

SOLVE Conservation of energy $K_f + U_f = K_i + U_i$ is

$$0 + \frac{Kq_{proton}q_{carbon}}{r_f} = \frac{6Ke^2}{R} = \tfrac{1}{2}m_p v_i^2 + 0$$

We can solve for the proton's initial speed:

$$v_i = \sqrt{\frac{12Ke^2}{m_p R}} = 2.5 \times 10^7 \text{ m/s}$$

ASSESS This is a very high speed, nearly 10% of the speed of light, but it is typical of proton speeds used in proton radiation therapy.

EXAMPLE 22.3 Escape speed

An interaction between two elementary particles causes an electron and a positron (a positive electron) to be shot out back to back with equal speeds. What minimum speed must each have when they are 100 fm apart in order to escape each other?

PREPARE Energy is conserved. The particles end "far apart," which we interpret as sufficiently far to make $U_f \approx 0$. FIGURE 22.10 shows the before-and-after pictorial representation. The minimum speed to escape is the speed that allows the particles to reach $r_f = \infty$ with $v_f = 0$.

FIGURE 22.10 An electron and a positron flying apart.

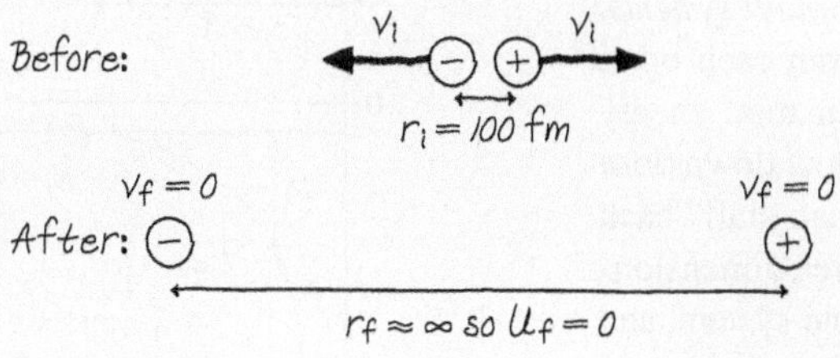

Before: v_i v_i
$r_i = 100$ fm
After: $v_f = 0$ $v_f = 0$
$r_f \approx \infty$ so $U_f = 0$

SOLVE U_{elec} is the potential energy of the electron + positron system. Similarly, the kinetic energy K is the *total* kinetic energy of the system. The electron and the positron, with equal masses and equal speeds, have equal kinetic energies. Conservation of energy $K_f + U_f = K_i + U_i$ is

$$0 + 0 + 0 = \tfrac{1}{2}mv_i^2 + \tfrac{1}{2}mv_i^2 + \frac{Kq_e q_p}{r_i} = mv_i^2 - \frac{Ke^2}{r_i}$$

Using $r_i = 100 \text{ fm} = 1.0 \times 10^{-13}$ m, we can calculate the minimum initial speed to be

$$v_i = \sqrt{\frac{Ke^2}{mr_i}} = 5.0 \times 10^7 \text{ m/s}$$

ASSESS v_i is a little more than 10% the speed of light. That's not too surprising because the attractive force between the particles is extremely strong when they are only 100 fm apart.

Multiple Point Charges

If more than two charges are present, their potential energy—the interaction energy of the group of charges—is the sum of the potential energies due to all pairs of charges. If the charges are numbered 1, 2, 3, . . . , then the potential energy is

$$U_{\text{elec}} = \sum_{\text{all pairs}} \frac{Kq_i q_j}{r_{ij}} \tag{22.15}$$

where r_{ij} is the distance between q_i and q_j. Be sure to include each pair only once!

EXAMPLE 22.4 The electric potential energy of a phosphate anion

Phosphate, which is critical to the functioning of both ATP and DNA, is one of the most important molecular groups in biochemistry. The stable form of phosphate in solution is the phosphate anion (a negative ion) PO_4^{3-}. This is a tetrahedral molecule, as shown in FIGURE 22.11, with a central P^+ ion surrounded by four O^- ions. The P-O bond lengths are 0.16 nm, which corresponds to an O-O distance—the edge length of the tetrahedron—of 0.26 nm. What is the electric potential energy of a phosphate anion?

PREPARE The phosphate anion is an electric interaction of a group of five point charges. The central P^+ interacts with each of the O^-; these four attractive interactions are equivalent. In addition, each O^- has a repulsive interaction with three other O^-. If we number the oxygen atoms 1 to 4, we find that there are six *pairs* of interactions: 1-2, 1-3, 1-4, 2-3, 2-4, and 3-4. All pairs have the same O-O distance, so the six interactions are equivalent.

FIGURE 22.11 The phosphate anion.

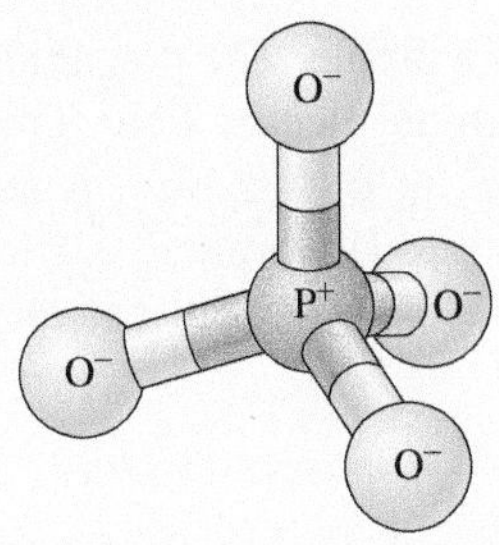

SOLVE The anion's electric potential energy is

$$U_{\text{elec}} = \sum_{\text{all pairs}} \frac{Kq_i q_i}{r_{ij}} = 4\frac{Kq_P q_O}{r_{\text{P-O}}} + 6\frac{Kq_O q_O}{r_{\text{O-O}}} = -\frac{4Ke^2}{r_{\text{P-O}}} + \frac{6Ke^2}{r_{\text{O-O}}}$$

$$= Ke^2\left(\frac{6}{r_{\text{O-O}}} - \frac{4}{r_{\text{P-O}}}\right)$$

$$= -4.4 \times 10^{-19}\ \text{J}$$

ASSESS A negative electric potential energy indicates that the phosphate ion is a bound system—a stable molecule—which we know is true. We would need to supply 4.4×10^{-19} J per molecule, or 265 kJ/mol, to disassemble a phosphate ion into its constituent atoms. It's also interesting to note that $U_{\text{elec}} \approx 100k_BT$ at room temperature, which indicates that thermal collisions with other molecules do not have enough energy to dissociate a phosphate anion.

STOP TO THINK 22.2 Rank in order, from largest to smallest, the potential energies U_A to U_D of these four pairs of charges. Each + symbol represents the same amount of charge.

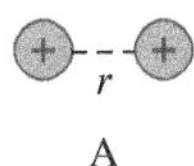

A

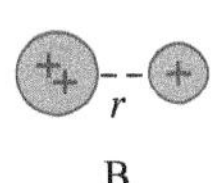

B

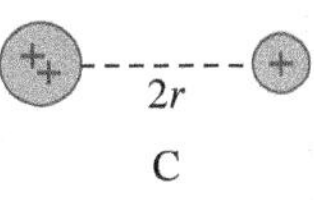

C

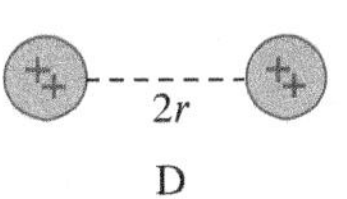

D

The Potential Energy of a Dipole

The electric dipole has been our model for understanding how charged objects interact with neutral objects. In Chapter 21 we found that an electric field exerts a *torque* on a dipole. We can complete the picture by calculating the potential energy of an electric dipole in a uniform electric field.

FIGURE 22.12 shows a dipole in an electric field $\vec{E}$. Recall that the dipole moment $\vec{p}$ for two charges separated by distance s is a vector that points from $-q$ to q with magnitude $p = qs$. The forces $\vec{F}_+$ and $\vec{F}_-$ exert a torque on the dipole, but now we're interested in calculating the *work* done by these forces as the dipole rotates from angle ϕ_i to angle ϕ_f.

FIGURE 22.12 The electric field does work as a dipole rotates.

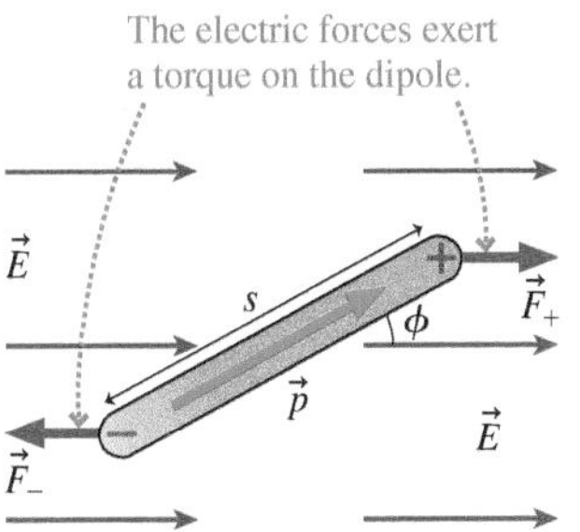

When a force component F_x acts through a small displacement dx, the force does work $dW = F_x\,dx$. If we exploit the rotational-linear motion analogy from Chapter 7, where torque τ is the analog of force and angular displacement $\Delta\phi$ is the analog of linear displacement, then a torque acting through a small angular displacement $d\phi$ does work $dW = \tau\,d\phi$. From Chapter 21, the torque on the dipole in Figure 22.12 is $\tau = -pE\sin\phi$, where the minus sign is due to the torque trying to cause a clockwise rotation. Thus the work done by the electric field on the dipole as it rotates through the small angle $d\phi$ is

$$dW_{\text{elec}} = -pE\sin\phi\,d\phi \tag{22.16}$$

The total work done by the electric field as the dipole turns from ϕ_i to ϕ_f is

$$W_{\text{elec}} = -pE\int_{\phi_i}^{\phi_f}\sin\phi\,d\phi = pE\cos\phi_f - pE\cos\phi_i \tag{22.17}$$

The potential energy associated with the work done on the dipole is

$$\Delta U_{\text{elec}} = U_f - U_i = -W_{\text{elec}}(i \rightarrow f) = -pE\cos\phi_f + pE\cos\phi_i \tag{22.18}$$

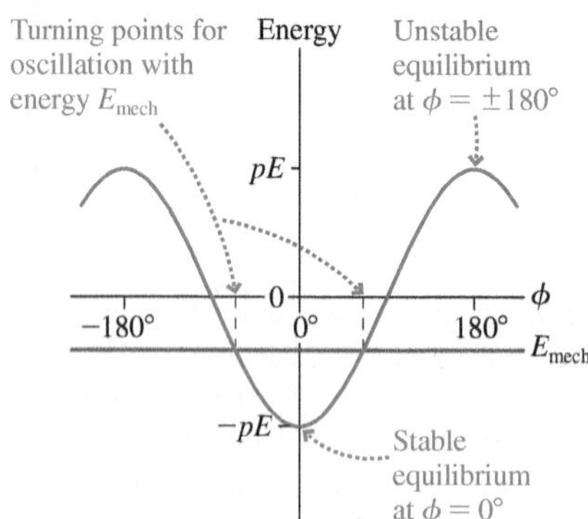

FIGURE 22.13 The energy of a dipole in an electric field.

By comparing the left and right sides of Equation 22.18, we see that the potential energy of an electric dipole $\vec{p}$ in a uniform electric field $\vec{E}$ is

$$U_{\text{elec}} = -pE\cos\phi \tag{22.19}$$

Potential energy of a dipole in an electric field

FIGURE 22.13 shows the energy diagram of a dipole. The potential energy is minimum at $\phi = 0°$ where the dipole is aligned with the electric field. This is a point of stable equilibrium. A dipole exactly opposite $\vec{E}$, at $\phi = \pm180°$, is at a point of unstable equilibrium. Any disturbance will cause it to flip around. A frictionless dipole with mechanical energy E_{mech} will oscillate back and forth between turning points on either side of $\phi = 0°$.

EXAMPLE 22.5 Rotating a molecule

The water molecule is a permanent electric dipole with dipole moment 6.2×10^{-30} C·m. A water molecule is aligned in an electric field with field strength 1.0×10^7 N/C. How much energy is needed to rotate the molecule 90°?

PREPARE The molecule is at the point of minimum energy. It won't spontaneously rotate 90°. However, an external force that supplies energy, such as a collision with another molecule, can cause the water molecule to rotate.

SOLVE The molecule starts at $\phi_i = 0°$ and ends at $\phi_f = 90°$. The increase in potential energy is

$$\Delta U_{\text{elec}} = U_f - U_i = -pE\cos 90° - (-pE\cos 0°)$$
$$= pE = 6.2 \times 10^{-23}\text{ J}$$

This is the energy needed to rotate the molecule 90°.

ASSESS ΔU_{elec} is significantly less than $k_B T$ at room temperature. Thus collisions with other molecules can easily supply the energy to rotate the water molecules and keep them from staying aligned with the electric field.

22.2 The Electric Potential

We introduced the concept of the *electric field* in Chapter 21 to provide an intermediary through which two charges exert forces on each other. Charge q_1 somehow alters the space around it by creating an electric field $\vec{E}_1$. Charge q_2 then responds to the field, experiencing force $\vec{F} = q_2\vec{E}_1$.

In defining the electric field, we separated the charges that are the *source* of the field from the charge *in* the field. The force on charge q is related to the electric field of the source charges by

force on q by sources = [charge q] × [alteration of space by the source charges]

Let's try a similar procedure for the potential energy. The electric potential energy is due to the interaction of charge q with other charges, so let's write

potential energy of q + sources
= [charge q] × [*potential* for interaction with the source charges]

FIGURE 22.14 shows this idea schematically.

In analogy with the electric field, we will define the **electric potential** V (or, for brevity, just *the potential*) as

$$V = \frac{U_{q\,+\,\text{sources}}}{q} \tag{22.20}$$

Charge q is used as a probe to determine the electric potential, but the value of V is *independent of q*. **The electric potential, like the electric field, is a property of the source charges.** And, like the electric field, the electric potential fills the space around the source charges. It is there whether or not another charge is there to experience it.

In practice, we're usually more interested in knowing the potential energy if a charge q happens to be at a point in space where the electric potential of the source charges is V. Turning Equation 22.20 around, we see that the electric potential energy is

$$U_{\text{elec}} = U_{q\,+\,\text{sources}} = qV \tag{22.21}$$

Once the potential has been determined, it's very easy to find the potential energy.

The unit of electric potential is the joule per coulomb, which is called the **volt** V:

$$1 \text{ volt} = 1 \text{ V} = 1 \text{ J/C}$$

This unit is named for Alessandro Volta, who invented the electric battery in the year 1800. Microvolts (μV), millivolts (mV), and kilovolts (kV) are commonly used units.

NOTE ► Once again, commonly used symbols are in conflict. The symbol V is widely used to represent *volume,* and now we're introducing the same symbol to represent *potential.* To make matters more confusing, V is the abbreviation for *volts.* In printed text, V for potential is italicized and V for volts is not, but you can't make such a distinction in handwritten work. This can be confusing, but these are the commonly accepted symbols. It's important to be alert to the *context* in which a symbol is used. ◄

If you've studied cell biology, you're already familiar with the concept of electric potential in the form of the cell's membrane potential. To say that a resting nerve cell has a membrane potential of –70 mV means that the electric potential inside the cell is 70 mV less than the potential outside, which is taken to be 0 V. Our goal in this chapter is to help you understand why the cell has an electric potential and how to interpret that information.

FIGURE 22.14 Source charges alter the space around them by creating an electric potential.

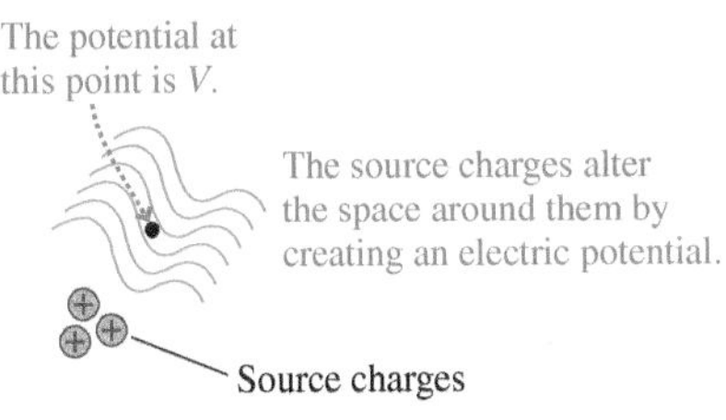
The potential at this point is V.
The source charges alter the space around them by creating an electric potential.
Source charges

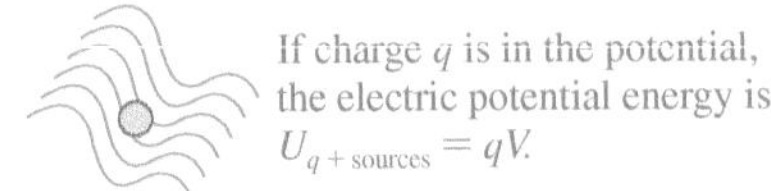
If charge q is in the potential, the electric potential energy is $U_{q\,+\,\text{sources}} = qV$.

AAA
Alkaline
MN2400
1.5 V
LR03
MAR 2018

Finding potential This battery is labeled 1.5 volts. As we'll soon see, a battery is a source of electric potential.

Using the Electric Potential

The electric potential is an abstract idea, and it will take some practice to see just what it means and how it is useful. We'll use multiple representations—words, pictures, graphs, and analogies—to explain and describe the electric potential.

NOTE ► It is unfortunate that the terms *potential* and *potential energy* are so much alike. Despite the similar names, they are very different concepts and are not interchangeable. TABLE 22.1 will help you to distinguish between the two. ◄

To begin, we'll consider charged particles that move in a vacuum without collisions. If we know the electric potential in a region of space we can determine whether a charged particle speeds up or slows down as it moves through that region.

TABLE 22.1 **Distinguishing electric potential and potential energy**

The *electric potential* is a property of the source charges and, as you'll soon see, is related to the electric field. The electric potential is present whether or not a charged particle is there to experience it. Potential is measured in J/C, or V.

The *electric potential energy* is the interaction energy of a charged particle with the source charges. Potential energy is measured in J.

FIGURE 22.15 illustrates this idea. Here a group of source charges, which remains hidden offstage, has created an electric potential V that increases from left to right. A charged particle q, which for now we'll assume to be positive, has electric potential energy $U_{elec} = qV$. If the particle moves to the right, its potential energy increases and so, by energy conservation, its kinetic energy must decrease. **A positive charge slows down as it moves into a region of higher electric potential.**

FIGURE 22.15 In the absence of other applied forces, a charged particle speeds up or slows down as it moves through a potential difference.

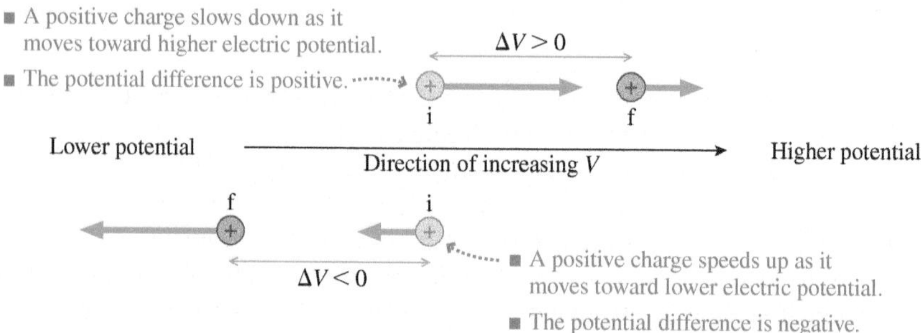

It is customary to say that the particle moves through a **potential difference** $\Delta V = V_f - V_i$. The potential difference between two points is often called the **voltage.** The particle that travels to the right moves through a positive potential difference ($\Delta V > 0$ because $V_f > V_i$), so we can say that a positively charged particle slows down as it moves through a positive potential difference.

The particle that moves to the left in Figure 22.15 travels in the direction of decreasing electric potential—through a negative potential difference—and loses potential energy. It speeds up as it transforms potential energy into kinetic energy. A negatively charged particle would slow down because its potential energy qV would increase as V decreases. TABLE 22.2 summarizes these ideas.

TABLE 22.2 Charged particles moving through an electric potential in a vacuum

	Electric potential	
	Increasing ($\Delta V > 0$)	Decreasing ($\Delta V < 0$)
+ charge	Slows down	Speeds up
− charge	Speeds up	Slows down

Because $U_{elec} = qV$, we can write the conservation of mechanical energy statement $K_f + U_f = K_i + U_i$ for a charged particle in a vacuum as

$$K_f + qV_f = K_i + qV_i \tag{22.22}$$

Equation 22.22 is the basis for solving problems about the motion of charged particles through a potential difference.

EXAMPLE 22.6 A proton moves through a potential difference

A proton with a speed of 2.0×10^5 m/s enters a region of space in which there is an electric potential. What is the proton's speed after it moves through a potential difference of 100 V? What will be the final speed if the proton is replaced by an electron?

PREPARE The system is the charge plus the unseen source charges that create the potential. We assume that the proton moves in a vacuum. This is an isolated system, so mechanical energy is conserved. FIGURE 22.16 is a before-and-after pictorial representation of a charged particle moving through a potential difference. A positive charge *slows down* as it moves into a region of higher potential ($K \rightarrow U$). A negative charge *speeds up* ($U \rightarrow K$).

FIGURE 22.16 A charged particle moving through a potential difference.

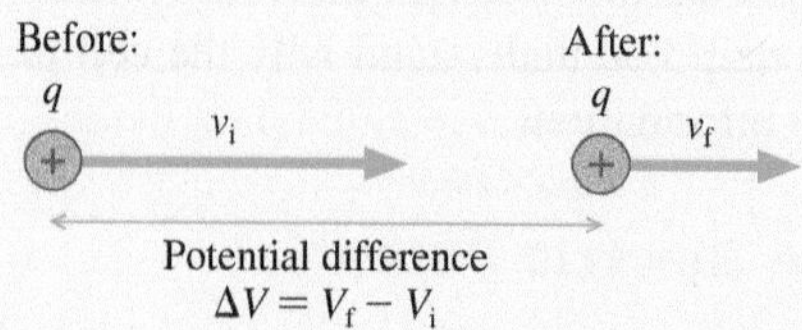

SOLVE The potential energy of charge q is $U = qV$. Conservation of energy, now expressed in terms of the electric potential V, is $K_f + qV_f = K_i + qV_i$, or

$$K_f = K_i - q\Delta V$$

where $\Delta V = V_f - V_i$ is the potential difference through which the particle moves. In terms of the speeds, energy conservation is

$$\tfrac{1}{2}mv_f^2 = \tfrac{1}{2}mv_i^2 - q\Delta V$$

We can solve this for the final speed:

$$v_f = \sqrt{v_i^2 - \frac{2q}{m}\Delta V}$$

For a proton, with $q = e$, the final speed is

$$(v_f)_p = \sqrt{(2.0 \times 10^5 \text{ m/s})^2 - \frac{2(1.60 \times 10^{-19} \text{ C})(100 \text{ V})}{1.67 \times 10^{-27} \text{ kg}}}$$
$$= 1.4 \times 10^5 \text{ m/s}$$

An electron, though, with $q = -e$ and a different mass, reaches speed $(v_f)_e = 5.9 \times 10^6$ m/s.

ASSESS The proton slowed down and the electron sped up, as we expected. Note that the electric potential *already existed* in space due to other charges that are not explicitly seen in the problem. The electron and proton have nothing to do with creating the potential. Instead, they *respond* to the potential by having potential energy $U = qV$.

The expanded energy principle of Chapter 11, which includes thermal energy but assumes no change in chemical energy, was

$$W = \Delta K + \Delta U + \Delta E_{th} \qquad (22.23)$$

where W is the work done by external forces. There are situations, such as ions moving at steady speed through inert gases, where electric potential energy is transformed not into kinetic energy but, via collisions, into thermal energy: $\Delta E_{th} = -\Delta U = -q\Delta V$.

A more important application of these ideas is the transport of ions across the potential difference of the cell membrane. In this case, work must be done—powered by the hydrolysis of ATP—to "lift" an ion to a higher electric potential: $W = \Delta U = q\Delta V$. We'll look at an example later in the chapter, but a quick calculation finds that the energy needed to transport an ion with $q = +e$ out of the cell, typically with $\Delta V = +70$ mV, is $\approx 1 \times 10^{-20}$ J/ion = 6 kJ/mol.

STOP TO THINK 22.3 A proton is released from rest in a vacuum at point B, where the potential is 0 V. Afterward, the proton

A. Remains at rest at B.
B. Moves toward A with a steady speed.
C. Moves toward A with an increasing speed.
D. Moves toward C with a steady speed.
E. Moves toward C with an increasing speed.

−100 V 0 V +100 V
A• B• C•

22.3 Calculating the Electric Potential

Earlier in the chapter we found an expression for the electric potential energy of a charged particle between the plates of a parallel-plate capacitor. Now we'll investigate the electric potential. **FIGURE 22.17** shows two parallel electrodes separated by distance d. As a specific example, we'll let $d = 3.00$ mm and the charge-to-area ratio $Q/A = 4.42 \times 10^{-9}$ C/m². The electric field inside the capacitor, as you learned in Chapter 21, is

$$\vec{E} = \left(\frac{Q}{\epsilon_0 A}, \text{ from positive toward negative}\right)$$
$$= (500 \text{ N/C, from right to left}) \qquad (22.24)$$

This electric field is due to the *source charges* on the capacitor plates.

In Section 22.1, we found that the electric potential energy of a charge q in the uniform electric field of a parallel-plate capacitor is

$$U_{elec} = U_{q+\text{sources}} = qEx \qquad (22.25)$$

U_{elec} is the energy of q interacting with the source charges on the capacitor plates.

FIGURE 22.17 A parallel-plate capacitor.

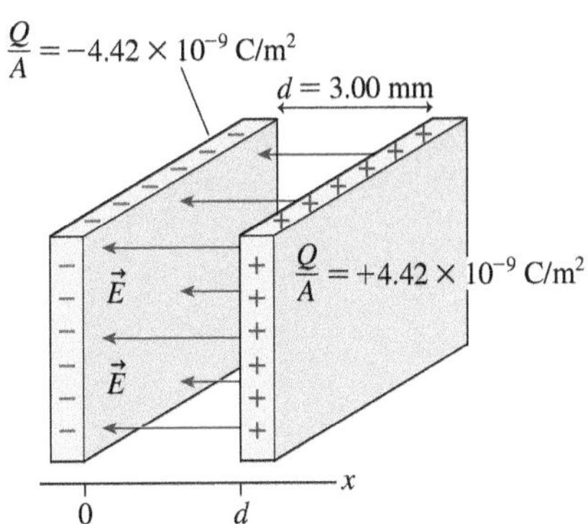

Our new view of the interaction is to separate the role of charge q from the role of the source charges by defining the electric potential $V = U_{q + \text{sources}}/q$. Thus the electric potential inside a parallel-plate capacitor is

$$V_{\text{cap}} = Ex \tag{22.26}$$

where **x is the distance from the negative electrode**, which we've taken to be the reference point where $V = 0$ V. The electric potential, like the electric field, exists at *all points* inside the capacitor. The electric potential is created by the source charges on the capacitor plates and exists whether or not charge q is inside the capacitor.

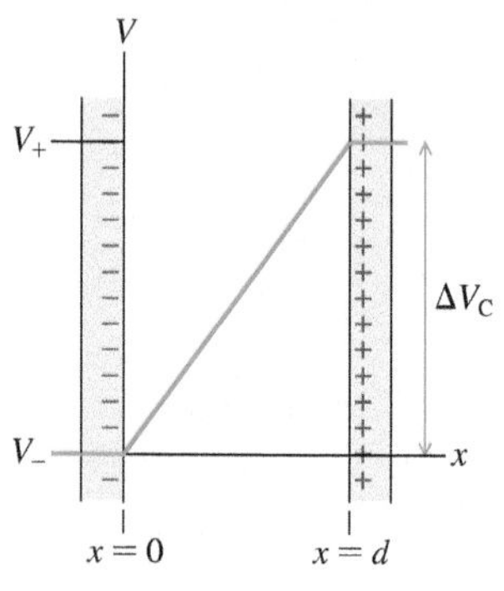

FIGURE 22.18 The electric potential of a parallel-plate capacitor increases linearly from the negative to the positive plate.

FIGURE 22.18 illustrates the important point that the electric potential increases linearly from the negative plate, where $V_- = 0$, to the positive plate at $x = d$, where $V_+ = Ed$. Let's define the *potential difference* ΔV_C between the two capacitor plates to be

$$\Delta V_C = V_+ - V_- = Ed \tag{22.27}$$

In our specific example, $\Delta V_C = (500 \text{ N/C})(0.0030 \text{ m}) = 1.5$ V. The units work out because $1.5 \text{ (N} \cdot \text{m)/C} = 1.5 \text{ J/C} = 1.5$ V.

NOTE ▸ People who work with circuits would call ΔV_C "the voltage across the capacitor" or simply "the capacitor voltage." ◂

Equation 22.27 has an interesting implication. Thus far, we've determined the electric field inside a capacitor by specifying the charge Q on the plates. Alternatively, we could specify the capacitor voltage ΔV_C (i.e., the potential difference between the capacitor plates) and then determine the electric field strength as

$$E = \frac{\Delta V_C}{d} \tag{22.28}$$

In fact, this is how E is determined in practical applications because it's easy to measure ΔV_C with a voltmeter but difficult, in practice, to know the value of Q.

Equation 22.28 implies that the units of electric field are volts per meter, or V/m. We have been using electric field units of newtons per coulomb. In fact, as you can show as a homework problem, these units are equivalent to each other. That is,

$$1 \text{ N/C} = 1 \text{ V/m}$$

NOTE ▸ Volts per meter are the electric field units used by scientists in practice. We will now adopt them as our standard electric field units. ◂

Returning to the electric potential, we can substitute $E = \Delta V_C/d$ into Equation 22.26 for V_{cap}. Thus the electric potential inside a capacitor is

$$V_{\text{cap}} = \frac{x}{d}\Delta V_C \tag{22.29}$$

Electric potential inside a parallel-plate capacitor

The potential increases linearly from $V_- = 0$ V at the negative plate ($x = 0$) to $V_+ = \Delta V_C$ at the positive plate ($x = d$).

Visualizing Electric Potential

Let's explore the electric potential inside the capacitor by looking at several different, but related, ways that the potential can be represented graphically.

Graphical representations of the electric potential between the plates of a capacitor

A graph of potential versus x. You can see the potential increasing from 0.0 V at the negative plate to 1.5 V at the positive plate.

A three-dimensional view showing **equipotential surfaces.** These are mathematical surfaces, not physical surfaces, with the same value of V at every point. The equipotential surfaces of a capacitor are planes parallel to the capacitor plates. The capacitor plates are also equipotential surfaces.

A two-dimensional **contour map.** The capacitor plates and the equipotential surfaces are seen edge-on, so you need to imagine them extending above and below the plane of the page.

A three-dimensional **elevation graph.** The potential is graphed vertically versus the x-coordinate on one axis and a generalized "yz-coordinate" on the other axis. Viewing the right face of the elevation graph gives you the potential graph.

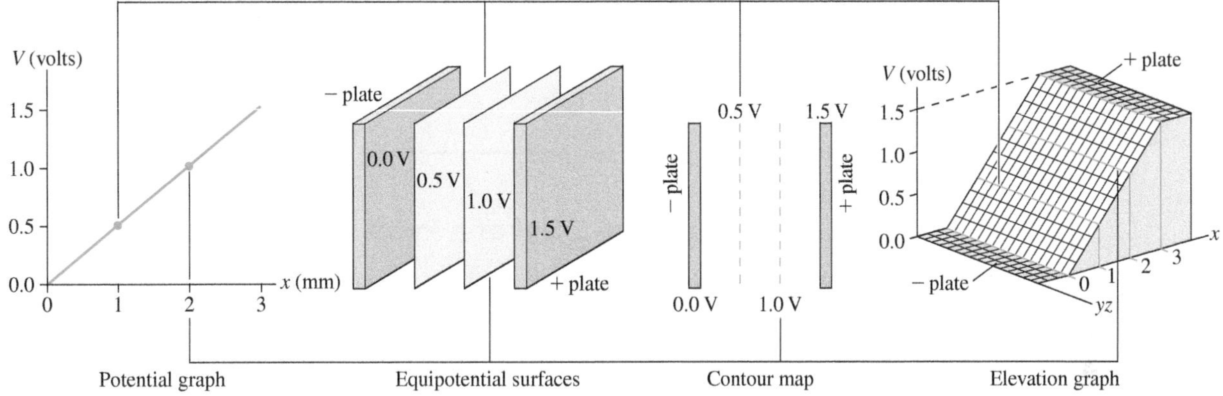

These four graphical representations show the same information from different perspectives, and the connecting lines help you see how they are related. If you think of the elevation graph as a "mountain," then the lines on the contour map are like the lines of a topographic map.

The potential graph and the contour map are the two representations most widely used in practice because they are easy to draw. Their limitation is that they are trying to convey three-dimensional information in a two-dimensional presentation. When you see graphs or contour maps, you need to imagine the three-dimensional equipotential surfaces or the three-dimensional elevation graph.

There's nothing special about showing equipotential surfaces or contour lines every 0.5 V. We chose these intervals because they were convenient. As an alternative, FIGURE 22.19 shows how the contour map looks if the contour lines are spaced every 0.3 V. Contour lines and equipotential surfaces are *imaginary* lines and surfaces drawn to help us visualize how the potential changes in space. Drawing the map more than one way reinforces the idea that there is an electric potential at *every* point inside the capacitor, not just at the points where we happened to draw a contour line or an equipotential surface.

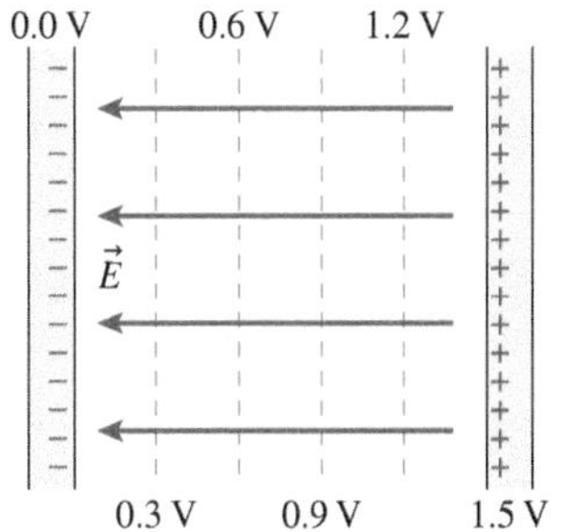

FIGURE 22.19 Equipotentials and electric field vectors inside a parallel-plate capacitor.

NOTE ► A particle moving on an equipotential surface has constant electric potential energy. If there are no external forces, then the particle's speed and kinetic energy are also constant. ◄

Figure 22.19 also shows the electric field vectors. Notice that

- The electric field vectors are perpendicular to the equipotential surfaces.
- The electric field points in the direction of decreasing potential. In other words, the electric field points "downhill" on a graph or map of the electric potential.

In Section 22.5 we'll take a more in-depth look at the connection between the electric field and the electric potential. There you will find that these observations are always true; they are not unique to the parallel-plate capacitor.

FIGURE 22.20 Using a battery to charge a capacitor to a precise value of ΔV_C.

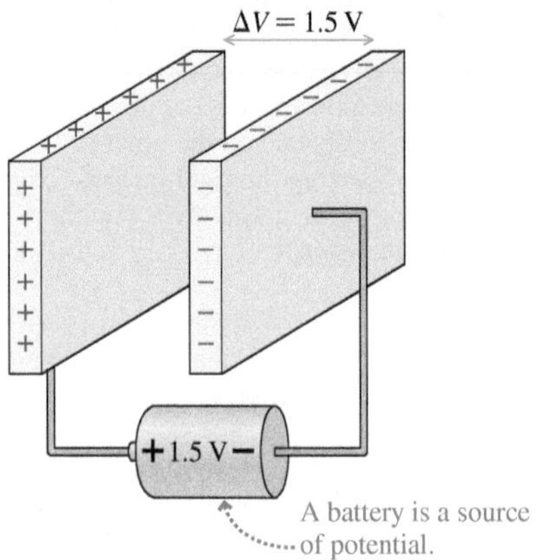
$\Delta V = 1.5$ V
+1.5 V−
A battery is a source of potential.

Finally, you might wonder how we can arrange a capacitor to have a charge-to-area ratio of precisely 4.42×10^{-9} C/m^2. Simple! As FIGURE 22.20 shows, we use wires to attach 3.00-mm-spaced capacitor plates to a 1.5 V battery. This is another topic that we'll explore later, but it's worth noting now that **a battery is a source of potential.** That's why batteries are labeled in volts, and it's a major reason we need to thoroughly understand the concept of potential.

STOP TO THINK 22.4 Rank in order, from largest to smallest, the potentials V_1 to V_5 at the points 1 to 5.

The Electric Potential of a Point Charge

FIGURE 22.21 Measuring the electric potential of charge q.

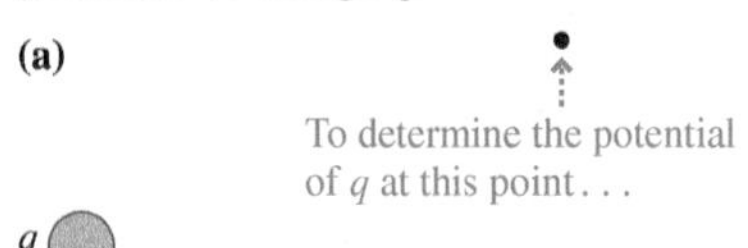
(a)
To determine the potential of q at this point . . .
q

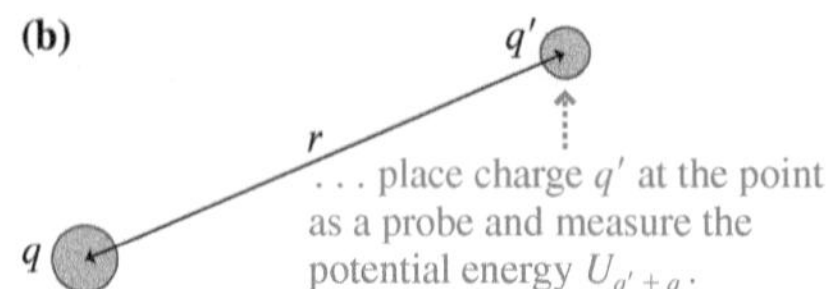
(b)
q'
r
q
. . . place charge q' at the point as a probe and measure the potential energy $U_{q'+q}$.

Another important electric potential is that of a point charge. FIGURE 22.21a shows charge q and a point in space at which we would like to know the electric potential. To do so, as shown in FIGURE 22.21b, we let a second charge q' probe the electric potential of q. The potential energy of the two point charges is

$$U_{q'+q} = \frac{1}{4\pi\epsilon_0}\frac{qq'}{r} \tag{22.30}$$

Thus, by definition, the electric potential of charge q is

$$V_{\text{point}} = \frac{U_{q'+q}}{q'} = \frac{1}{4\pi\epsilon_0}\frac{q}{r} \tag{22.31}$$

Electric potential of a point charge q

The potential of Equation 22.31 extends through all of space, but it weakens with distance as $1/r$. This expression for V assumes that we have chosen $V = 0$ V to be at $r = \infty$. This is the most logical choice for a point charge because the influence of charge q ends at infinity.

The expression for the electric potential of charge q is similar to that for the electric field of charge q. The difference most quickly seen is that V_{point} depends on $1/r$ whereas $\vec{E}$ depends on $1/r^2$. But it is also important to notice that **the potential is a scalar** whereas the field is a vector.

For example, the electric potential 1.0 cm from a +1.0 nC charge is

$$V_{1\text{ cm}} = \frac{1}{4\pi\epsilon_0}\frac{q}{r} = (9.0 \times 10^9\ \text{N}\cdot\text{m}^2/\text{C}^2)\,\frac{1.0 \times 10^{-9}\ \text{C}}{0.010\ \text{m}} = 900\ \text{V}$$

1 nC is typical of the electrostatic charge produced by rubbing, and you can see that such a charge creates a fairly large potential nearby. Why are we not shocked and injured when working with the "high voltages" of such charges? The sensation of being shocked is a result of current, not potential. Some high-potential sources simply do not have the ability to generate much current.

Visualizing the Potential of a Point Charge

FIGURE 22.22 shows four graphical representations of the electric potential of a point charge. These match the four representations of the electric potential inside a capacitor, and a comparison of the two is worthwhile. This figure assumes that q is positive; you may want to think about how the representations would change if q were negative.

FIGURE 22.22 Four graphical representations of the electric potential of a positive point charge.

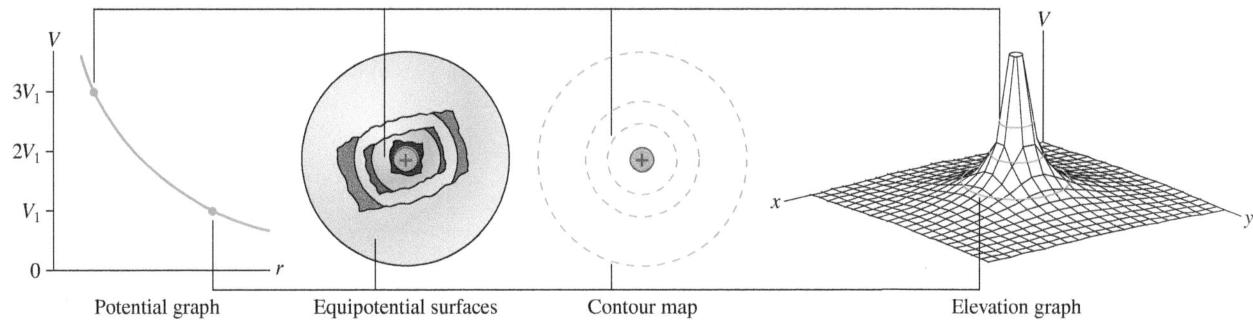

The Electric Potential of a Charged Sphere

In practice, you are more likely to work with a charged sphere, of radius R and total charge Q, than with a point charge. Outside a uniformly charged sphere, the electric potential is identical to that of a point charge Q at the center. That is,

$$V_{\text{sphere}} = \frac{1}{4\pi\epsilon_0}\frac{Q}{r} \tag{22.32}$$

We can cast this result in a more useful form. It is customary to speak of charging an electrode, such as a sphere, "to" a certain potential, as in "Bob charged the sphere to a potential of 3000 volts." This potential, which we will call V_0, is the potential right on the surface of the sphere. We can see from Equation 22.32 that

$$V_0 = V(\text{at } r = R) = \frac{Q}{4\pi\epsilon_0 R} \tag{22.33}$$

Consequently, a sphere of radius R that is charged to potential V_0 has total charge

$$Q = 4\pi\epsilon_0 R V_0 \tag{22.34}$$

If we substitute this expression for Q into Equation 22.32, we can write the potential outside a sphere that is charged to potential V_0 as

$$V_{\text{sphere}} = \frac{R}{r}V_0 \quad \text{(sphere charged to potential } V_0\text{)} \tag{22.35}$$

Equation 22.35 tells us that the potential of a sphere is V_0 on the surface and decreases inversely with the distance. The potential at $r = 3R$ is $\frac{1}{3}V_0$.

Breakdown A *plasma ball* consists of a small metal ball charged to a potential of about 2000 V inside a hollow glass sphere. The electric field of the high-voltage ball is sufficient to cause a gas breakdown at this pressure, creating "lightning bolts" between the ball and the glass sphere.

EXAMPLE 22.7 A proton and a charged sphere

A proton is released from rest at the surface of a 1.0-cm-diameter sphere that has been charged to +1000 V.

a. What is the charge of the sphere?
b. What is the proton's speed at 1.0 cm from the sphere?

PREPARE We'll assume that the motion takes place in a vacuum so that mechanical energy is conserved. The potential outside the charged sphere is the same as the potential of a point charge at the center. FIGURE 22.23 shows the situation.

FIGURE 22.23 A sphere and a proton.

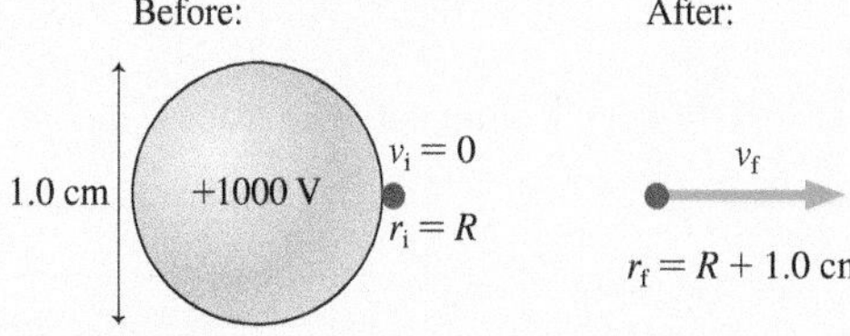

SOLVE

a. The charge of the sphere is

$$Q = 4\pi\epsilon_0 R V_0 = 0.56 \times 10^{-9}\text{ C} = 0.56\text{ nC}$$

b. A sphere charged to $V_0 = +1000$ V is positively charged. The proton will be repelled by this charge and move away from the sphere. The conservation of energy equation $K_f + eV_f = K_i + eV_i$, with Equation 22.35 for the potential of a sphere, is

$$\tfrac{1}{2}mv_f^2 + \frac{eR}{r_f}V_0 = \tfrac{1}{2}mv_i^2 + \frac{eR}{r_i}V_0$$

Continued

The proton starts from the surface of the sphere, $r_i = R$, with $v_i = 0$. When the proton is 1.0 cm from the *surface* of the sphere, it has $r_f = 1.0\text{ cm} + R = 1.5\text{ cm}$. Using these, we can solve for v_f:

$$v_f = \sqrt{\frac{2eV_0}{m}\left(1 - \frac{R}{r_f}\right)} = 3.6 \times 10^5 \text{ m/s}$$

ASSESS This example illustrates how the ideas of electric potential and potential energy work together, yet they are *not* the same thing.

The Electric Potential of Many Charges

Suppose there are many source charges $q_1, q_2, \ldots$. The electric potential V at a point in space is the sum of the potentials due to each charge:

$$V = \sum_i \frac{1}{4\pi\epsilon_0}\frac{q_i}{r_i} \tag{22.36}$$

where r_i is the distance from charge q_i to the point in space where the potential is being calculated. In other words, **the electric potential, like the electric field, obeys the principle of superposition.**

The electric dipole—two charges $\pm q$ separated by distance s—is an especially important multicharge system. ◀◀ SECTION 21.5 defined the electric dipole moment

$$\vec{p} = (qs, \text{ from the negative to the positive charge}) \tag{22.37}$$

and found expressions for the electric field on the axis of the dipole and in the bisecting plane of the dipole at distances $r \gg s$. In a homework problem you can show that the electric potential at distance r from a dipole is

$$V_{\text{dipole}} = \frac{p\cos\theta}{4\pi\epsilon_0 r^2} \tag{22.38}$$

FIGURE 22.24 The electric potential of an electric dipole depends on both distance and the angle from the dipole axis.

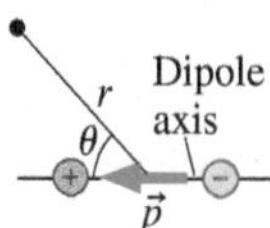

where θ, shown in FIGURE 22.24, is the angle between the dipole moment vector and the direction to the point. The electric potential of a dipole depends on both distance *and* direction. The potential in the bisecting plane, where $\theta = 90°$, is zero.

The contour map and elevation graph in FIGURE 22.25 show that the potential of an electric dipole is the sum of the potentials of the positive and negative charges. Potentials such as these have many practical applications. For example, electrical activity within the body can be monitored by measuring equipotential lines on the skin. FIGURE 22.25c shows that the equipotentials near the heart are a slightly distorted but recognizable electric dipole. We'll explore this idea in Section 22.7, where we look at the electrocardiogram.

FIGURE 22.25 The electric potential of an electric dipole.

(a) Contour map

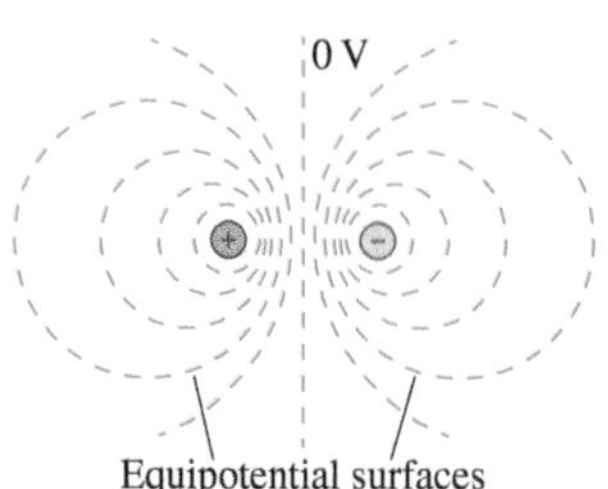

(b) Elevation graph

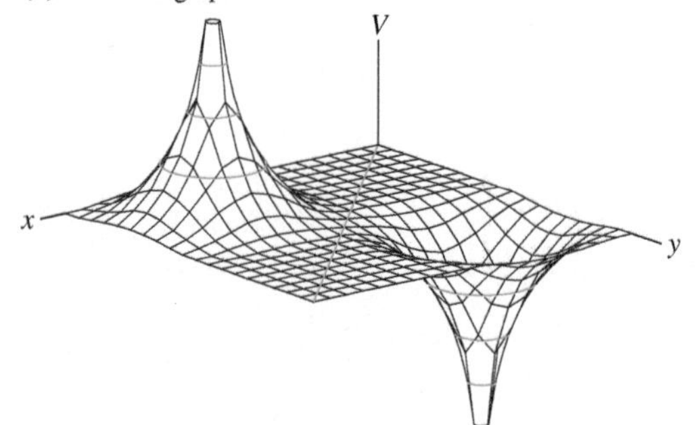

(c) Equipotentials near the heart

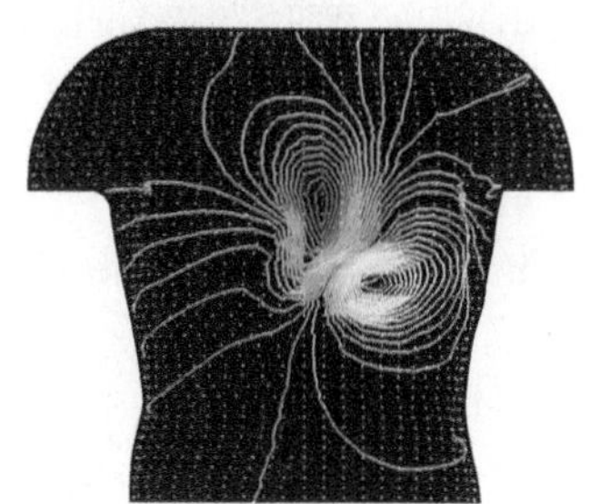

EXAMPLE 22.8 The potential of two charges

What is the electric potential at the point indicated in FIGURE 22.26?

FIGURE 22.26 Finding the potential of two charges.

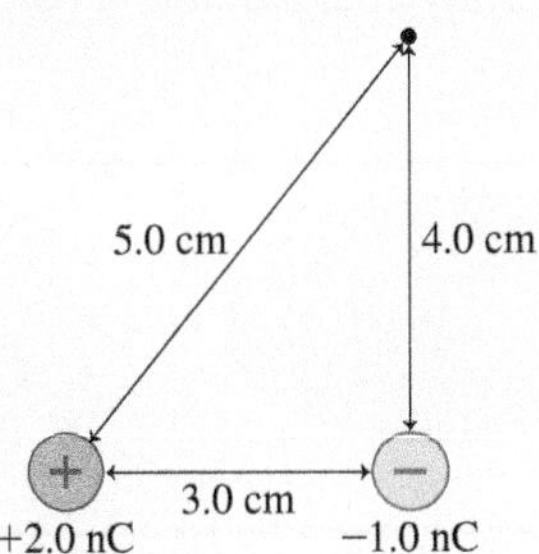
5.0 cm
4.0 cm
3.0 cm
+2.0 nC
−1.0 nC

PREPARE The potential is the sum of the potentials due to each charge.

SOLVE The potential at the indicated point is

$$V = \frac{1}{4\pi\epsilon_0}\frac{q_1}{r_1} + \frac{1}{4\pi\epsilon_0}\frac{q_2}{r_2}$$

$$= (9.0 \times 10^9\ \text{N}\cdot\text{m}^2/\text{C}^2)\left(\frac{2.0 \times 10^{-9}\ \text{C}}{0.050\ \text{m}} + \frac{-1.0 \times 10^{-9}\ \text{C}}{0.040\ \text{m}}\right)$$

$$= 135\ \text{V}$$

ASSESS The potential is a *scalar,* so we found the net potential by adding two numbers. We don't need any angles or components to calculate the potential.

STOP TO THINK 22.5 Rank in order, from largest to smallest, the potential differences ΔV_{12}, ΔV_{13}, and ΔV_{23} between points 1 and 2, points 1 and 3, and points 2 and 3.

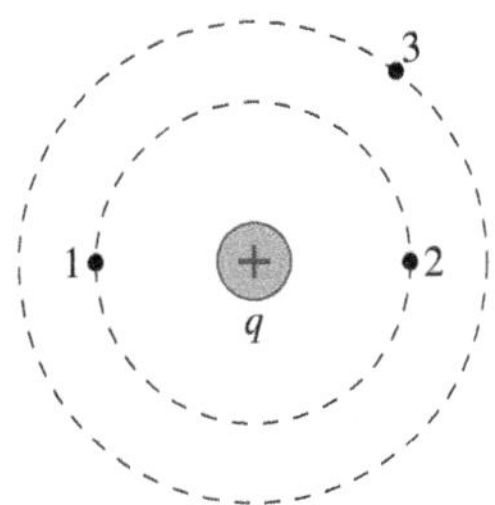
3
1
2
q

22.4 The Potential of a Continuous Distribution of Charge

Ordinary objects—tables, chairs, beakers of water—seem to our senses to be continuous distributions of matter; there is no obvious evidence for atoms, even though we have good reasons to believe that we would find atoms if we subdivided the matter sufficiently far. Thus it is easier, for many practical purposes, to consider matter to be continuous and to talk about the *density* of matter. Density—the number of kilograms of matter per cubic meter—allows us to describe the distribution of matter *as if* the matter were continuous rather than atomic.

Much the same situation occurs with charge. If a charged object contains a large number of excess electrons—for example, 10^{12} extra electrons on a metal rod—it is not practical to track every electron. It makes more sense to consider the charge to be *continuous* and to describe how it is distributed over the object.

FIGURE 22.27a shows an object of length L, such as a plastic rod or a metal wire, with charge Q spread uniformly along it. The **linear charge density** λ is defined to be

$$\lambda = \frac{Q}{L} \tag{22.39}$$

Linear charge density, which has units of C/m, is the amount of charge *per meter* of length. The linear charge density of a 20-cm-long wire with 40 nC of charge is 2.0 nC/cm or 2.0×10^{-7} C/m.

NOTE ▶ The linear charge density λ is analogous to the linear mass density μ that you used in Chapter 16 to find the speed of a wave on a string. ◀

We'll also be interested in charged surfaces. FIGURE 22.27b shows a two-dimensional distribution of charge across a surface of area A. We define the **surface charge density** η (lowercase Greek eta) to be

$$\eta = \frac{Q}{A} \tag{22.40}$$

FIGURE 22.27 One-dimensional and two-dimensional continuous charge distributions.

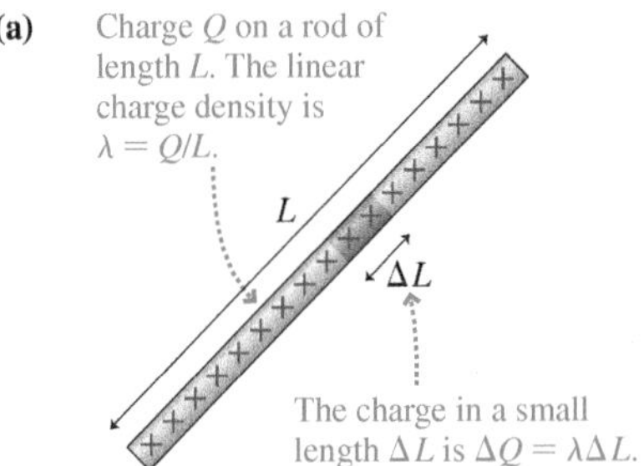
(a)
Charge Q on a rod of length L. The linear charge density is λ = Q/L.
L
ΔL
The charge in a small length ΔL is ΔQ = λΔL.

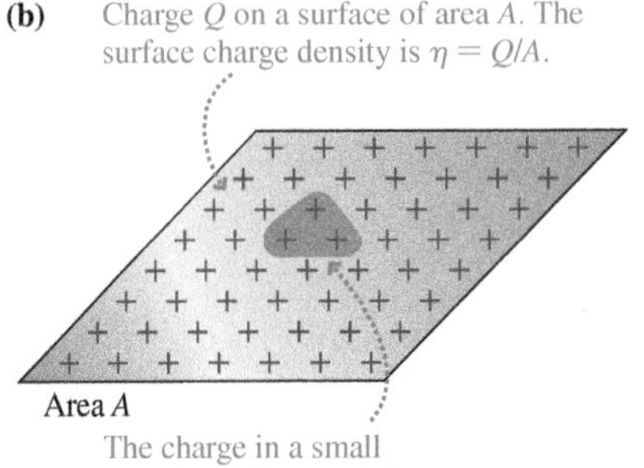
(b)
Charge Q on a surface of area A. The surface charge density is η = Q/A.
Area A
The charge in a small area ΔA is ΔQ = ηΔA.

Surface charge density, with units of C/m^2, is the amount of charge *per square meter.* A 1.0 mm $\times$ 1.0 mm square on a surface with $\eta = 2.0 \times 10^{-4}$ C/m^2 contains 2.0×10^{-10} C or 0.20 nC of charge.

Figure 22.27 and the definitions of Equations 22.39 and 22.40 assume that the object is **uniformly charged,** meaning that the charges are evenly spread over the object. We will assume objects are uniformly charged unless noted otherwise.

STOP TO THINK 22.6 A piece of plastic is uniformly charged with surface charge density η_A. The plastic is then broken into a large piece with surface charge density η_B and a small piece with surface charge density η_C. Rank in order, from largest to smallest, the surface charge densities η_A to η_C.

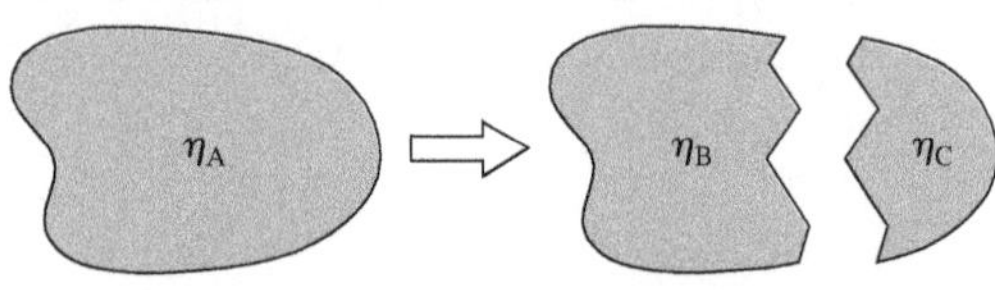

η_A
η_B
η_C

Integration Is Summation

Calculating the electric potential of a continuous charge distribution usually requires setting up and evaluating integrals—a skill you've been learning in calculus. It is common to think that "an integral is the area under a curve," an idea we used in our study of kinematics.

But integration is much more than a tool for finding areas. More generally, **integration is summation.** That is, an integral is a sophisticated way to add an infinite number of infinitesimally small pieces. The area under a curve happens to be a special case in which you're summing the small areas $y(x)dx$ of an infinite number of tall, narrow boxes, but the idea of integration as summation has many other applications.

Suppose, for example, that a charged object is divided into a large number of small pieces numbered $i = 1, 2, 3, \ldots, N$ having small quantities of charge $\Delta Q_1, \Delta Q_2, \Delta Q_3, \ldots, \Delta Q_N$. Figure 22.27 showed small pieces of charge for a charged rod and a charged sheet, but the object could have any shape. The total charge on the object is found by *summing* all the small charges:

$$Q = \sum_{i=1}^{N} \Delta Q_i \tag{22.41}$$

If we now let $\Delta Q_i \rightarrow 0$ and $N \rightarrow \infty$, then we *define* the integral:

$$Q = \lim_{\Delta Q \rightarrow 0} \sum_{i=1}^{N} \Delta Q_i = \int_{\text{object}} dQ \tag{22.42}$$

That is, integration is the summing of an infinite number of infinitesimally small pieces of charge. This use of integration has nothing to do with the area under a curve.

Although Equation 22.42 is a formal statement of "add up all the little pieces," it's not yet an expression that can actually be integrated with the tools of calculus. Integrals are carried out over coordinates, such as dx or dy, and we also need coordinates to specify what is meant by "integrate over the object." This is where densities come in.

FIGURE 22.28 Setting up an integral to calculate the charge on a rod.

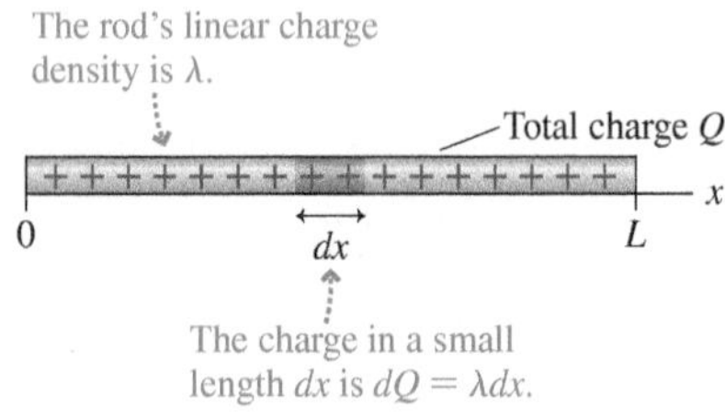

The rod's linear charge density is λ.
Total charge Q
x
0
dx
L
The charge in a small length dx is $dQ = \lambda dx$.

Suppose we want to find the total charge of a thin, charged rod of length L. First, we establish an x-axis with the origin at one end of the rod, as shown in FIGURE 22.28. Then we divide the rod into lots of tiny segments of length dx. Each of these little segments has a charge dQ, and the total charge on the rod is the sum of all the dQ values—that's what Equation 22.42 is saying. Now the critical step: The rod has some linear charge density λ. Consequently, the charge of a small segment of the rod is $dQ = \lambda\,dx$. **Densities are the link between quantities and coordinates.** Finally, "integrate over the rod" means to integrate from $x = 0$ to $x = L$. Thus the total charge on the rod is

$$Q = \int_{\text{rod}} dQ = \int_0^L \lambda\,dx \qquad (22.43)$$

Now we have an expression that can be integrated. If λ is constant, as it is for a uniformly charged rod, we can take it outside the integral to find $Q = \lambda L$. But we could also use Equation 22.43 for a nonuniformly charged rod where λ is some function of x.

This discussion reveals two key ideas that will be needed for calculating electric potentials:

- Integration is the tool for summing a vast number of small pieces.
- Density is the connection between quantities and coordinates.

A Problem-Solving Strategy

Our goal is to find the electric potential of a continuous distribution of charge, such as a charged rod or a charged disk. We have two tools to work with:

- The electric potential of a point charge, and
- Equation 22.36, which tells us that potentials add.

These are the basis for a problem-solving strategy:

PROBLEM-SOLVING STRATEGY 22.1 **The electric potential of a continuous distribution of charge**

PREPARE Model the charge distribution as a simple shape. For the pictorial representation:

- Draw a picture, establish a coordinate system, and identify the point P at which you want to calculate the electric potential.
- Divide the total charge Q into small pieces of charge ΔQ, using shapes for which you *already know* how to determine V. This division is often, but not always, into point charges.
- Identify distances that need to be calculated.

SOLVE The mathematical representation is $V = \sum V_i$.

- Use superposition to form an algebraic expression for the potential at P. Let the (x, y, z) coordinates of the point remain as variables.
- Replace the small charge ΔQ with an equivalent expression involving a *charge density* and a *coordinate,* such as dx. **This is the critical step in making the transition from a sum to an integral** because you need a coordinate to serve as the integration variable.
- All distances must be expressed in terms of the coordinates.
- Let the sum become an integral. The integration limits will depend on the coordinate system you have chosen.

ASSESS Check that your result is consistent with any limits for which you know what the potential should be.

EXAMPLE 22.9 The potential of a ring of charge

A thin, uniformly charged ring of radius R has total charge Q. Find the potential at distance z on the axis of the ring.

PREPARE Because the ring is thin, we'll assume the charge lies along a circle of radius R. FIGURE 22.29 illustrates the problem-solving strategy. We've chosen a coordinate system in which the ring lies in the xy-plane and point P is on the z-axis. We've then divided the ring into N small segments of charge ΔQ, each of which can be modeled as a point charge. The distance r_i between segment i and point P is

$$r_i = \sqrt{R^2 + z^2}$$

Note that r_i is the same for every charge segment.

FIGURE 22.29 Finding the potential of a ring of charge.

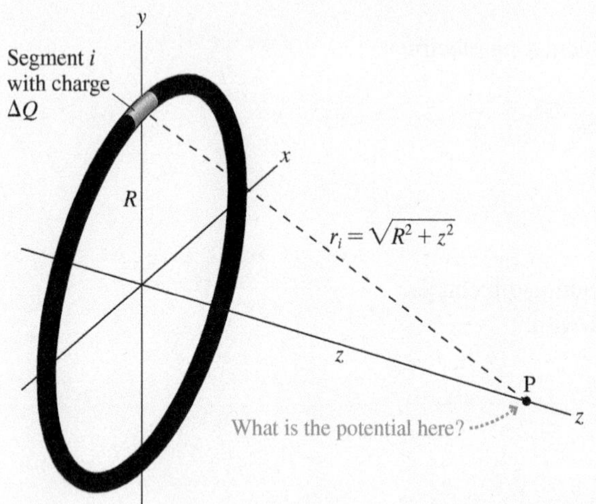

SOLVE The potential V at P is the sum of the potentials due to each segment of charge:

$$V = \sum_{i=1}^{N} V_i = \sum_{i=1}^{N} \frac{1}{4\pi\epsilon_0}\frac{\Delta Q}{r_i} = \frac{1}{4\pi\epsilon_0}\frac{1}{\sqrt{R^2+z^2}}\sum_{i=1}^{N}\Delta Q$$

We were able to bring all terms involving z to the front because z is a constant as far as the summation is concerned. Surprisingly, we don't need to convert the sum to an integral to complete this calculation. The sum of all the ΔQ charge segments around the ring is simply the ring's total charge, $\Sigma(\Delta Q) = Q$; hence the electric potential on the axis of a charged ring is

$$V_{\text{ring on axis}} = \frac{1}{4\pi\epsilon_0}\frac{Q}{\sqrt{R^2+z^2}}$$

ASSESS From far away, the ring appears as a point charge Q in the distance. Thus we expect the potential of the ring to be that of a point charge when $z \gg R$. You can see that $V_{\text{ring}} \approx Q/4\pi\epsilon_0 z$ when $z \gg R$, which is, indeed, the potential of a point charge Q.

EXAMPLE 22.10 The potential of a charged disk

A thin plastic disk of radius R is uniformly coated with charge until it receives total charge Q.

a. What is the potential at distance z along the axis of the disk?
b. What is the potential energy if an electron is 1.00 cm from a 3.50-cm-diameter disk that has been charged to + 5.00 nC?

PREPARE We can model the disk as a uniformly charged disk of zero thickness, radius R, and charge Q. The disk has uniform surface charge density $\eta = Q/A = Q/\pi R^2$. We can take advantage of now knowing the on-axis potential of a ring of charge.

We'll orient the disk in the xy-plane, as shown in FIGURE 22.30, with point P at distance z. Then we'll divide the disk into *rings* of equal width Δr. Ring i has radius r_i and charge ΔQ_i.

FIGURE 22.30 Finding the potential of a disk of charge.

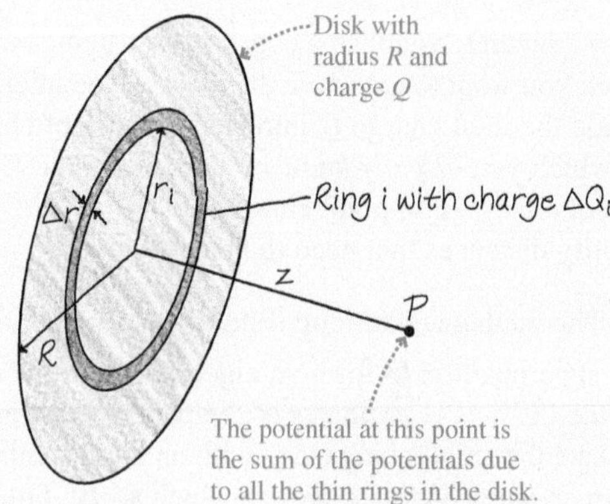

SOLVE

a. We can use the result of Example 22.9 to write the potential at distance z of ring i as

$$V_i = \frac{1}{4\pi\epsilon_0}\frac{\Delta Q_i}{\sqrt{r_i^2+z^2}}$$

The potential at P due to all the rings is the sum

$$V = \sum_i V_i = \frac{1}{4\pi\epsilon_0}\sum_{i=1}^{N}\frac{\Delta Q_i}{\sqrt{r_i^2+z^2}}$$

The critical step is to relate ΔQ_i to a coordinate. Because we now have a surface, rather than a line, the charge in ring i is $\Delta Q_i = \eta\,\Delta A_i$, where ΔA_i is the area of ring i. We can find ΔA_i, as you've learned to do in calculus, by "unrolling" the ring to form a narrow rectangle of length $2\pi r_i$ and width Δr. Thus the area of ring i is $\Delta A_i = 2\pi r_i\,\Delta r$ and the charge is

$$\Delta Q_i = \eta\,\Delta A_i = \frac{Q}{\pi R^2}2\pi r_i\,\Delta r = \frac{2Q}{R^2}r_i\,\Delta r$$

With this substitution, the potential at P is

$$V = \frac{1}{4\pi\epsilon_0}\sum_{i=1}^{N}\frac{2Q}{R^2}\frac{r_i\,\Delta r_i}{\sqrt{r_i^2+z^2}} \rightarrow \frac{Q}{2\pi\epsilon_0 R^2}\int_0^R \frac{r\,dr}{\sqrt{r^2+z^2}}$$

where, in the last step, we let $N \to \infty$ and the sum become an integral. This integral can be found in Appendix A, but it's not hard to evaluate with a change of variables. Let $u = r^2 + z^2$, in which case $r\,dr = \frac{1}{2}du$. Changing variables requires that we also change the integration limits. You can see that $u = z^2$ when $r = 0$, and $u = R^2 + z^2$ when $r = R$. With these changes, the on-axis potential of a charged disk is

$$V_{\text{disk on axis}} = \frac{Q}{2\pi\epsilon_0 R^2}\int_{z^2}^{R^2+z^2}\frac{\frac{1}{2}du}{u^{1/2}} = \frac{Q}{2\pi\epsilon_0 R^2}u^{1/2}\Big|_{z^2}^{R^2+z^2}$$

$$= \frac{Q}{2\pi\epsilon_0 R^2}\left(\sqrt{R^2+z^2} - z\right)$$

b. To calculate the potential energy, we first need to determine the potential of the disk at $z = 0.0100$ m. Using $R = 0.0175$ m and $Q = 5.00$ nC, you can calculate $V = 2980$ V. The electron's charge is $q = -e = -1.60 \times 10^{-19}$ C, so the potential energy with an electron at $z = 1.00$ cm is $U = qV = -4.77 \times 10^{-16}$ J.

ASSESS We earlier found that the potential 1 cm from a 1 nC point charge is 900 V. Thus a potential of 2980 V at the same distance from a 5 nC disk seems reasonable.

22.5 Sources of Electric Potential

We've now studied many properties of the electric potential, but we've not said much about how an electric potential is created. Simply put, **an electric potential difference is created by separating positive and negative charge.** Shuffling your feet on the carpet transfers electrons from the carpet to you, creating a potential difference between you and a doorknob that causes a spark and a shock as you touch it. Charging a capacitor by moving electrons from one plate to the other creates a potential difference across the capacitor. FIGURE 22.31 illustrates the idea.

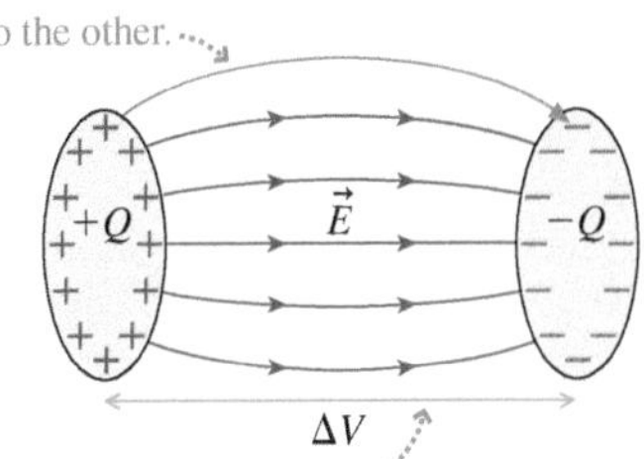

FIGURE 22.31 A charge separation creates a potential difference.

Now electric forces try to bring positive and negative charges together, so **a nonelectrical process is needed to separate charge.** As an example, the **Van de Graaff generator** shown in FIGURE 22.32a separates charges mechanically. A moving plastic or leather belt is charged, then the charge is mechanically transported via the conveyor belt to the spherical electrode at the top of the insulating column. The charging of the belt could be done by friction, but in practice a *corona discharge* created by the strong electric field at the tip of a needle is more efficient and reliable.

FIGURE 22.32 A Van de Graaff generator.

(a)

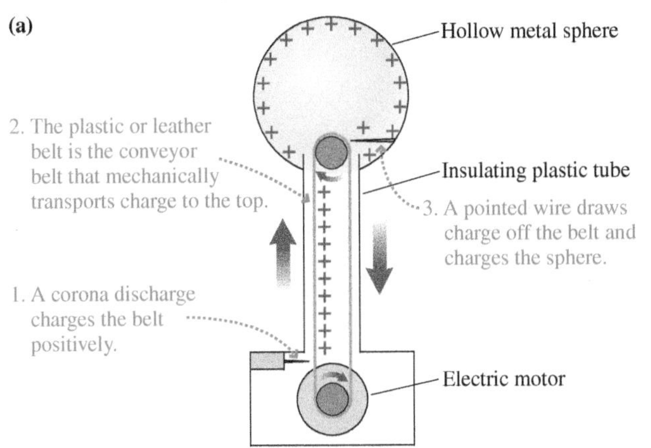

Hollow metal sphere
2. The plastic or leather belt is the conveyor belt that mechanically transports charge to the top.
Insulating plastic tube
3. A pointed wire draws charge off the belt and charges the sphere.
1. A corona discharge charges the belt positively.
Electric motor

(b)

A Van de Graaff generator has two noteworthy features:

- Charge is *mechanically* transported from the negative side to the positive side. This charge separation creates a potential difference between the spherical electrode and its surroundings.
- The electric field of the spherical electrode exerts a downward force on the positive charges moving up the belt. Consequently, *work must be done* to "lift" the positive charges. The work is done by the electric motor that runs the belt.

A classroom-demonstration Van de Graaff generator like the one shown in FIGURE 22.32b creates a potential difference of several hundred thousand volts between the upper sphere and its surroundings. The maximum potential is reached when the electric field near the sphere becomes large enough to cause a breakdown of the air. This produces a spark and temporarily discharges the sphere. A large Van de Graaff generator surrounded by vacuum can reach a potential of 20 MV or more. These generators are used to accelerate protons for nuclear physics experiments.

Batteries and emf

The most common source of electric potential is a **battery,** which uses *chemical reactions* to separate charge. A battery consists of chemicals, called *electrolytes,* sandwiched between two electrodes made of different metals. Chemical reactions in the electrolytes transport ions (i.e., charges) from one electrode to the other. This chemical process pulls positive and negative charges apart, creating a potential difference between the terminals of the battery. When the chemicals are used up, the reactions cease and the battery is dead.

FIGURE 22.33 Modeling a battery as a charge escalator.

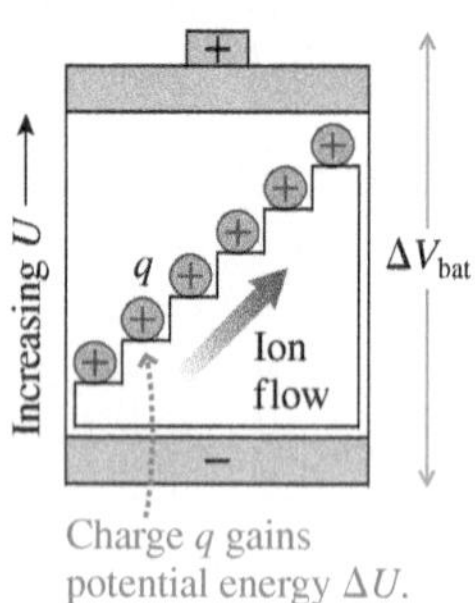

We can sidestep the chemistry details by introducing the **charge escalator model** of a battery shown in FIGURE 22.33:

- The charge escalator "lifts" positive charges from the negative terminal to the positive terminal. This requires *work,* with the energy being supplied by the chemical reactions.
- The work done *per charge* is called the **emf** of the battery: $\mathcal{E} = W_{chem}/q$.
- The charge separation creates a potential difference ΔV_{bat} between the terminals. An *ideal battery* has $\Delta V_{bat} = \mathcal{E}$.

Emf is pronounced as the sequence of letters e-m-f. The symbol for emf is $\mathcal{E}$, a script E, and the units of emf are joules per coulomb, or volts. The rating of a battery, such as 1.5 V or 9 V, is the battery's emf.

The key idea is that **emf is work,** specifically the work done *per charge* to pull positive and negative charges apart. This work can be done by mechanical forces, chemical reactions, or—as you'll see later—magnetic forces. *Because* work is done, charges gain potential energy and their separation creates a potential difference ΔV_{bat} between the positive and negative terminals of the battery. This is called the **terminal voltage.**

In an **ideal battery,** which has no internal energy losses, the work W_{chem} done to move charge q from the negative to the positive terminal goes entirely to increasing the potential energy of the charge, and so $\Delta V_{bat} = \mathcal{E}$. In practice, the terminal voltage is slightly less than the emf when current flows through a battery. We'll return to batteries when we begin to look at current and circuits in Chapter 24.

FIGURE 22.34 A cell's membrane potential is due to a charge separation created by ion pumps that actively transport ions through the membrane.

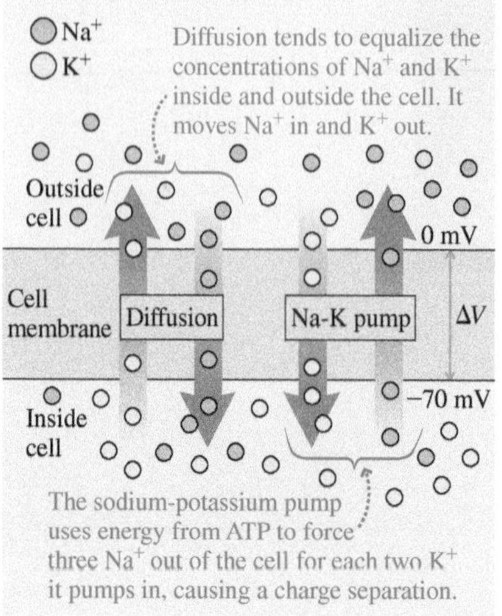

The Membrane Potential

The membrane that surrounds a eukaryotic cell—a lipid bilayer—is an insulator that separates conducting ionic fluids on the inside and outside of the cell. FIGURE 22.34 shows that ions are able to pass through the membrane by both passive (diffusion) and active (ion pumps) processes. Diffusion tends to equalize charge concentrations, but *sodium-potassium exchange pumps* continuously pump Na^+ ions out of the cell and K^+ ions into the cell.

One cycle of the pump pushes three Na^+ ions out of the cell and brings two K^+ ions in. The net result is the transfer of one ion with charge $+e$ out of the cell. In other words, **the ion exchange pumps create a charge separation in which the outside of the cell is positive while the inside is negative.** The resulting potential difference across the cell membrane is called the **membrane potential**.

The membrane potential is expressed as the potential inside the cell relative to the exterior: $\Delta V_{mem} = V_{in} - V_{out}$. We can think of the exterior as being the zero-volt reference point. The inside is more negative, so the membrane potential is negative. Typical values are -40 mV to -90 mV, where mV indicates *millivolts*. We'll look more closely at the membrane potential in Chapter 23 and then study its role in nerve conduction in Chapter 25.

EXAMPLE 22.11 Pumping ions

The membrane potential of a muscle cell is $\approx -90\text{ mV}$. How much work does a sodium-potassium exchange pump in a muscle cell do during one cycle? How does this compare to the $\approx 60\text{ kJ/mol}$ of energy released by the hydrolysis of ATP? Assume that work is done only to move ions across the membrane and that no work is required for conformational changes in the membrane proteins that form the pumps.

PREPARE One cycle of the pump moves three Na^+ ions out of the cell and brings two K^+ ions in. The inward-bound K^+ ions move through a potential difference $\Delta V_K = V_{in} - V_{out} = \Delta V_{mem}$; they experience a potential decrease. The outward-bound Na^+ ions move through a potential difference $\Delta V_{Na} = V_{out} - V_{in} = -\Delta V_{mem}$; they experience a potential increase. Each ion has charge $q = +e$.

SOLVE The electric potential energy changes not from a transformation of kinetic energy but because work is done by external forces to move the ions from rest on one side of the membrane to rest on the other side ($\Delta K \approx 0$). The work that must be done is

$$\begin{aligned} W &= \Delta U_{elec} = q_K \Delta V_K + q_{Na} \Delta V_{Na} = 2e\,\Delta V_{mem} - 3e\,\Delta V_{mem} \\ &= -e\,\Delta V_{mem} = -(1.6 \times 10^{-19}\text{ C})(-0.090\text{ V}) \\ &= 1.4 \times 10^{-20}\text{ J} \end{aligned}$$

The ion exchange pumps, like many cellular processes, are powered by the hydrolysis of ATP. The hydrolysis of one molecule of ATP releases energy

$$\Delta E_{ATP} \approx 60{,}000\text{ J/mol} \times \frac{1\text{ mol}}{6.02 \times 10^{23}\text{ molecules}} = 1.0 \times 10^{-19}\text{ J}$$

ASSESS The energy needs of the ion exchange pump are well matched to the energy available from ATP. In principle, one molecule of ATP could power seven cycles of the pump. However, the transformation of ATP energy to work is not 100% efficient—some energy is transformed into thermal energy—and our simple model does not consider the energy required for conformational change of the proteins or for any energy losses during the processes. In fact, each cycle of the pump uses exactly one ATP hydrolysis. Running the ion exchange pumps is a substantial fraction—sometimes more than half—of the total energy use of a cell.

22.6 Connecting Potential and Field

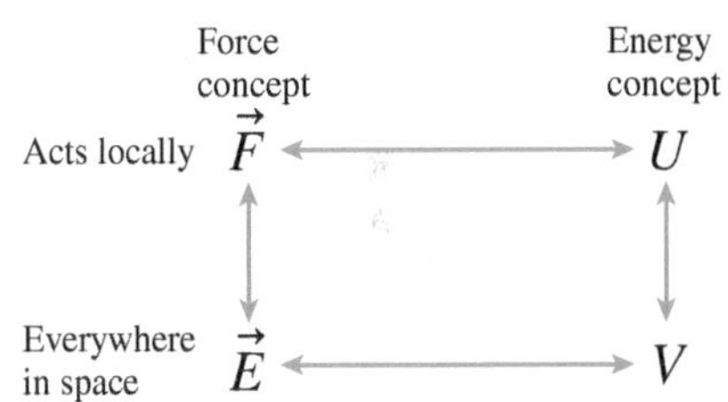

FIGURE 22.35 The four key ideas.

FIGURE 22.35 shows the four key ideas of force, field, potential energy, and potential. The electric field and the electric potential were based on force and potential energy. We know, from Chapters 10 and 11, that force and potential energy are closely related. We can now establish a similar relationship between the electric field and the electric potential. **The electric potential and electric field are not two distinct entities but, instead, simply two different ways to describe how source charges alter the space around them.**

If this is true, we should be able to determine the electric field from the electric potential in a manner similar to how we used potential energy to find force in Chapter 11. To see how, FIGURE 22.36 shows two points i and f separated by a very small distance Δx, so small that the electric field is essentially constant over this very short distance. The work done by the electric field as a charge q moves through this small distance is $W_{elec} = F_x \Delta x = qE_x \Delta x$. Consequently, the potential difference between these two points is

$$\Delta V = \frac{\Delta U_{q + \text{sources}}}{q} = \frac{-W_{elec}}{q} = -E_x \Delta x \tag{22.44}$$

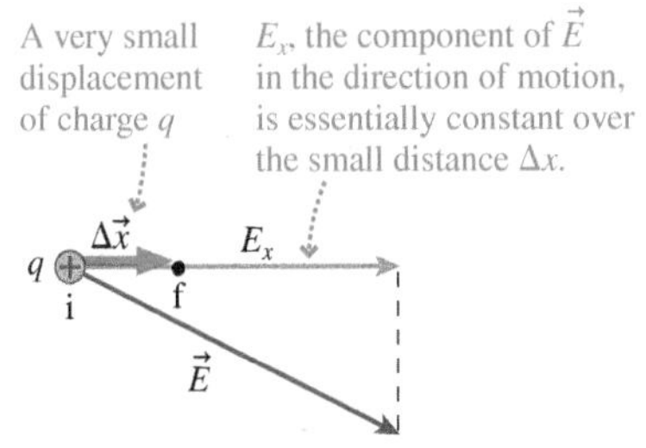

FIGURE 22.36 The electric field does work on charge q.

A very small displacement of charge q

E_x, the component of $\vec{E}$ in the direction of motion, is essentially constant over the small distance Δx.

In terms of the potential, the component of the electric field in the x-direction is $E_x = -\Delta V/\Delta x$. In the limit $\Delta x \to 0$,

$$E_x = -\frac{dV}{dx} \tag{22.45}$$

Equation 22.45 tells us that the strength of the x-component of the electric field is the rate at which the electric potential changes with distance along the x-axis; that is, a rapidly changing potential corresponds to a strong electric field. We can apply the same reasoning to the y- or z-component of the electric field.

Equation 22.45 is most useful if we can use symmetry to select a coordinate axis that is parallel to $\vec{E}$ and along which the perpendicular component of $\vec{E}$ is known to be zero. For example, suppose we knew the potential of a point charge to be $V = q/4\pi\epsilon_0 r$ but didn't remember the electric field. Symmetry requires that the field point straight outward from the charge, with only a radial component E_r. If we choose the x-axis to be in the radial direction, parallel to $\vec{E}$, we can use Equation 22.45 to find

$$E_r = -\frac{dV}{dr} = -\frac{d}{dr}\left(\frac{q}{4\pi\epsilon_0 r}\right) = \frac{1}{4\pi\epsilon_0}\frac{q}{r^2} \quad (22.46)$$

This is, indeed, the well-known electric field of a point charge.

Equation 22.45 is especially useful for a continuous distribution of charge because calculating V, which is a scalar, is usually much easier than calculating the vector $\vec{E}$ directly from the charge. Once V is known, $\vec{E}$ is found simply by taking a derivative.

EXAMPLE 22.12 The electric field of a ring of charge

In Example 22.9, we found the on-axis potential of a ring of radius R and charge Q to be

$$V_{\text{ring}} = \frac{1}{4\pi\epsilon_0}\frac{Q}{\sqrt{z^2 + R^2}}$$

Find the on-axis electric field of a ring of charge.

PREPARE Symmetry requires the electric field along the z-axis to point straight outward from the ring with only a z-component E_z.

SOLVE The electric field at position z is

$$E_z = -\frac{dV}{dz} = -\frac{d}{dz}\left(\frac{1}{4\pi\epsilon_0}\frac{Q}{\sqrt{z^2 + R^2}}\right)$$

$$= \frac{1}{4\pi\epsilon_0}\frac{zQ}{(z^2 + R^2)^{3/2}}$$

ASSESS From a very far distance $z \gg R$ the ring would appear to be a point charge. If we ignore the R in the denominator, the answer reduces to $Q/4\pi\epsilon_0 z^2$, which is the z-component of the electric field of a point charge at the origin. Our answer has the correct *limiting behavior,* which gives us confidence in the result.

A geometric interpretation of Equation 22.45 is that the electric field is the negative of the *slope* of the V-versus-x graph. This interpretation should be familiar. You learned in Chapter 11 that the force on a particle is the negative of the slope of the potential-energy graph: $F = -dU/dx$. In fact, Equation 22.45 is simply $F = -dU/dx$ with both sides divided by q to yield E and V. This geometric interpretation is an important step in developing an understanding of potential.

EXAMPLE 22.13 Finding E from the slope of V

FIGURE 22.37 is a graph of the electric potential in a region of space where $\vec{E}$ is parallel to the x-axis. Draw a graph of E_x versus x.

FIGURE 22.37 Graph of V versus position x.

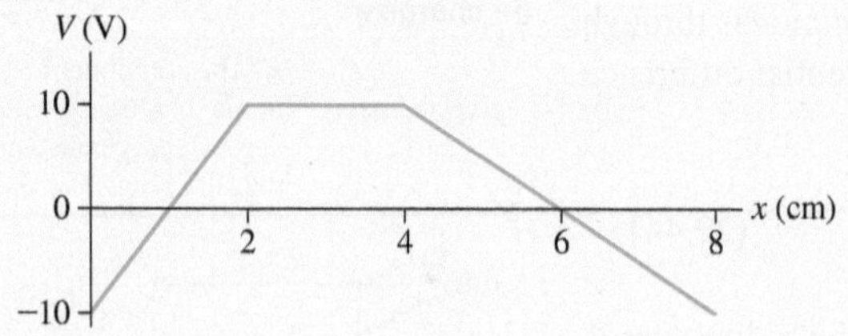

PREPARE The electric field is the *negative* of the slope of the potential graph.

SOLVE There are three regions of different slope:

$$0 < x < 2\text{ cm} \quad \begin{cases} \Delta V/\Delta x = (20\text{ V})/(0.020\text{ m}) = 1000\text{ V/m} \\ E_x = -1000\text{ V/m} \end{cases}$$

$$2 < x < 4\text{ cm} \quad \begin{cases} \Delta V/\Delta x = 0\text{ V/m} \\ E_x = 0\text{ V/m} \end{cases}$$

$$4 < x < 8\text{ cm} \quad \begin{cases} \Delta V/\Delta x = (-20\text{ V})/(0.040\text{ m}) = -500\text{ V/m} \\ E_x = 500\text{ V/m} \end{cases}$$

The results are shown in FIGURE 22.38.

FIGURE 22.38 Graph of E_x versus position x.

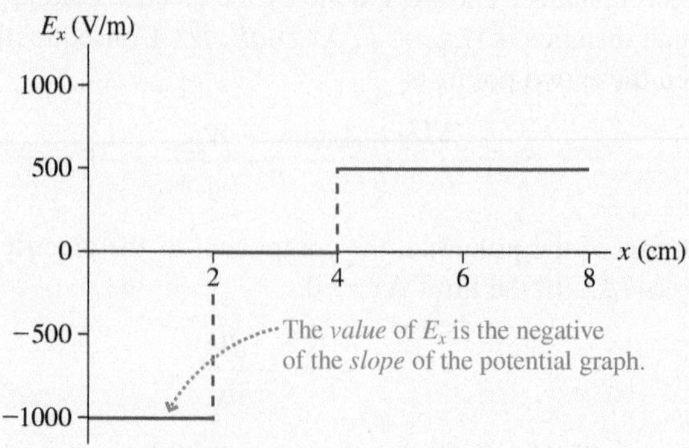

ASSESS The electric field $\vec{E}$ points to the left (E_x is negative) for $0 < x < 2$ cm and to the right (E_x is positive) for $4 < x < 8$ cm. Notice that **the electric field is zero in a region of space where the potential is not changing.**

STOP TO THINK 22.7 Which potential graph describes the electric field at the left?

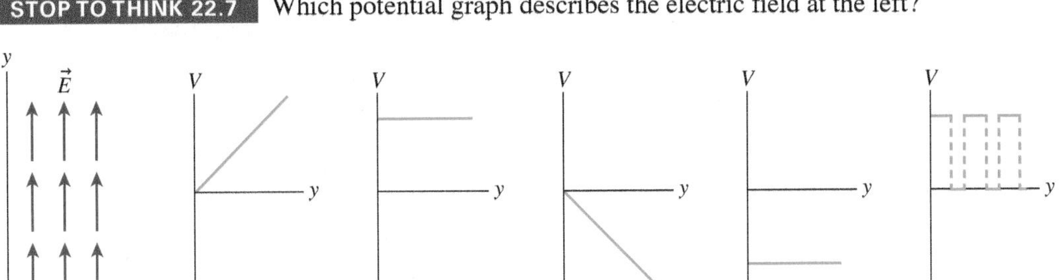
y
$\vec{E}$
x
V
y
A.
B.
C.
D.
E.

The Geometry of Potential and Field

FIGURE 22.39 shows two equipotential surfaces—surfaces on which the potential is constant—with V_+ positive relative to V_-. What can we say about the electric field at point P? Consider two paths that leave P. By definition, there is no potential difference along a path that follows the equipotential surface. The electric field is the rate at which the electric potential changes with distance. The potential doesn't change along an equipotential, so the electric field cannot have a component tangent to an equipotential. It would appear that the electric field at P must be *perpendicular* to the equipotential surfaces. This is the direction in which the potential changes most rapidly.

FIGURE 22.39 The electric field at P is related to the shape of the equipotential surfaces.

Direction of decreasing potential

There is no potential difference along a path that follows an equipotential surface.

V_-

V_+

P

Equipotential surfaces

d

The potential difference changes most rapidly along a path that is perpendicular to the equipotentials.

The perpendicular path in Figure 22.39 is pointing "uphill" from a lower potential toward a higher potential. If we let this be the x-axis, then Equation 22.45 is

$$E_x = -\frac{dV}{dx} \approx -\frac{\Delta V}{\Delta x} = -\frac{V_+ - V_-}{d} \qquad (22.47)$$

That is, we can estimate the field strength by knowing the potential difference between two equipotential surfaces that are distance d apart. Equipotential surfaces that are closer together represent a stronger electric field.

But what about the minus sign in Equation 22.47? We chose the x-axis to point "uphill," so the minus sign tells us that the electric field points "downhill." In other words, **the electric field $\vec{E}$ is perpendicular to the equipotential surfaces and points "downhill" in the direction of decreasing potential.** We can use Equation 22.47 to estimate the field strength.

These important ideas are summarized in **FIGURE 22.40**.

FIGURE 22.40 The geometry of the potential and the field.

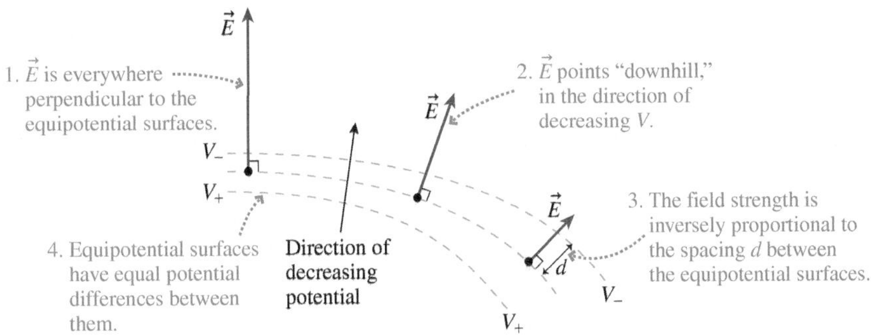
$\vec{E}$
1. $\vec{E}$ is everywhere perpendicular to the equipotential surfaces.
2. $\vec{E}$ points "downhill," in the direction of decreasing V.
V_-
V_+
3. The field strength is inversely proportional to the spacing d between the equipotential surfaces.
d
4. Equipotential surfaces have equal potential differences between them.
Direction of decreasing potential

EXAMPLE 22.14 Finding the electric field from the equipotential surfaces

In FIGURE 22.41 a 1 cm × 1 cm grid is superimposed on a contour map of the potential. Estimate the strength and direction of the electric field at points 1, 2, and 3. Show your results graphically by drawing the electric field vectors on the contour map.

FIGURE 22.41 Equipotential lines.

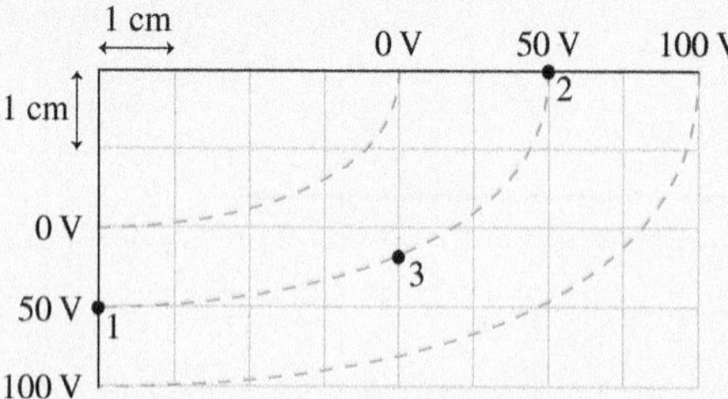

PREPARE The electric field is perpendicular to the equipotential lines, points "downhill," and depends on the slope of the potential hill. The potential is highest on the bottom and the right. An elevation graph of the potential would look like the lower-right quarter of a bowl or a football stadium.

SOLVE Some unseen source charges have created an electric field and potential. We do not need to see the source charges to relate the field to the potential. Because $E \approx -\Delta V/d$, the electric field is stronger where the equipotential lines are closer together and weaker where they are farther apart. If Figure 22.41 were a topographic map, we would interpret the closely spaced contour lines at the bottom of the figure as a steep slope.

FIGURE 22.42 shows how measurements of d from the grid are combined with values of ΔV to determine $\vec{E}$. Point 3 requires an estimate of the spacing between the 0 V and the 100 V surfaces. Notice that we're using the 0 V and 100 V equipotential surfaces to determine $\vec{E}$ at a point on the 50 V equipotential.

FIGURE 22.42 The electric field at points 1 to 3.

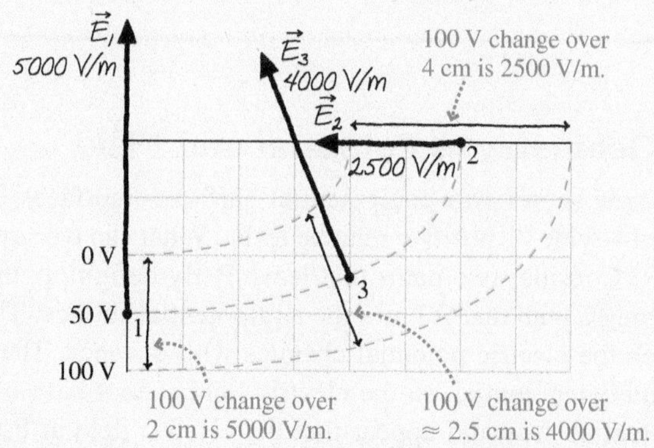

ASSESS The *directions* of $\vec{E}$ are found by drawing downhill vectors perpendicular to the equipotentials. The distances between the equipotential surfaces are needed to determine the field strengths.

Kirchhoff's Loop Law

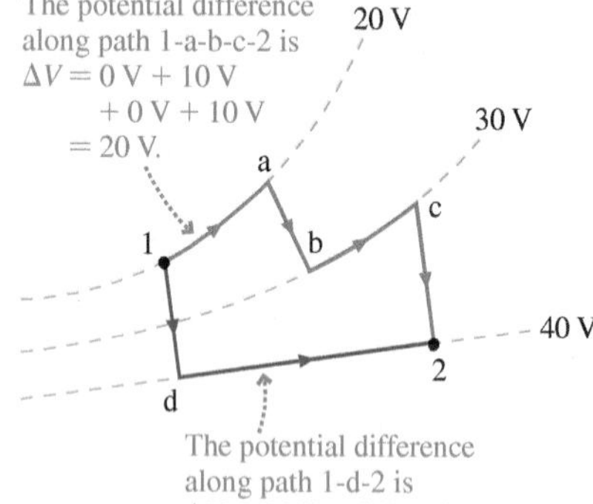

FIGURE 22.43 The potential difference between points 1 and 2 is the same along either path.

FIGURE 22.43 shows two points, 1 and 2, in a region of electric field and potential. The potential difference between points 1 and 2, $\Delta V = 20$ V, is the same whether we travel along the blue path, the red path, or any other path that connects the two points. This must be true in order for the idea of an equipotential surface to make sense.

Now consider the path 1–a–b–c–2–d–1 that ends where it started. What is the potential difference "around" this closed path? The potential increases by 20 V in moving from 1 to 2, but then decreases by 20 V in moving from 2 back to 1. Thus $\Delta V = 0$ V around the closed path.

The numbers are specific to this example, but the idea applies to any loop (i.e., a closed path) through an electric field. The situation is analogous to hiking on the side of a mountain. You may walk uphill during parts of your hike and downhill during other parts, but if you return to your starting point your *net* change of elevation is zero. So for any path that starts and ends at the same point, we can conclude that

$$\Delta V_{\text{loop}} = \sum_i (\Delta V)_i = 0 \tag{22.48}$$

Stated in words, **the sum of all the potential differences encountered while moving around a loop or closed path is zero.** This statement is known as **Kirchhoff's loop law.**

Kirchhoff's loop law is a statement of energy conservation because a charge that moves around a loop and returns to its starting point has $\Delta U = q\,\Delta V = 0$. Kirchhoff's loop law and a second Kirchhoff's law you'll meet in Chapter 24 will turn out to be the two fundamental principles of circuit analysis.

A Conductor in Electrostatic Equilibrium

In ◀◀ SECTION 21.6 you learned three important ideas about conductors in electrostatic equilibrium:

1. Any excess charge is on the surface.
2. The electric field inside is zero.
3. The exterior electric field is perpendicular to the surface.

Now we can add a fourth idea:

4. The entire conductor is at the same potential, and thus the surface is an equipotential surface.

To see why, FIGURE 22.44 shows two points separated by distance d inside a conductor. If there is a potential difference ΔV between the points, then there will be an electric field with strength $E \approx \Delta V/d$. But we already know that $E = 0$, so it must be the case that $\Delta V = 0$; the two points are at the same potential.

FIGURE 22.44 All points inside a conductor in electrostatic equilibrium are at the same potential.

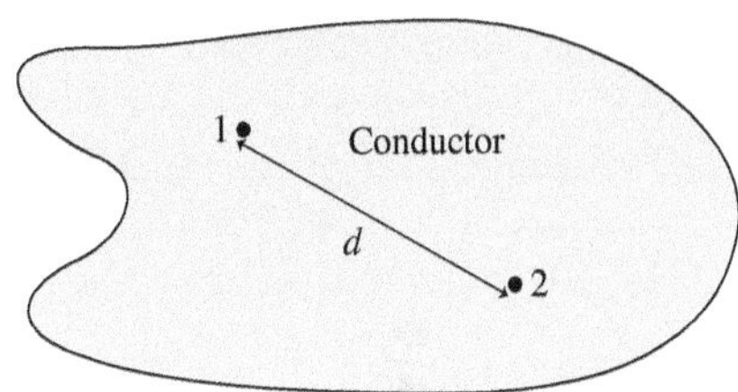

If a conductor is in electrostatic equilibrium, with all the charges at rest, then the *entire conductor* is at the same potential. If we charge a metal electrode, the entire electrode is at a single potential and the surface of the electrode is an equipotential surface. The fact that the exterior electric field is perpendicular to the surface is a special case of our conclusion that electric fields are always perpendicular to equipotential surfaces.

NOTE ▶ Two electrodes connected by a conducting wire form a single conductor. They exchange charge as needed to reach a common potential. ◀

STOP TO THINK 22.8 Three charged metal spheres of different radii are connected by a thin metal wire. The potential and electric field at the surface of each sphere are V and E. Which of the following is true?

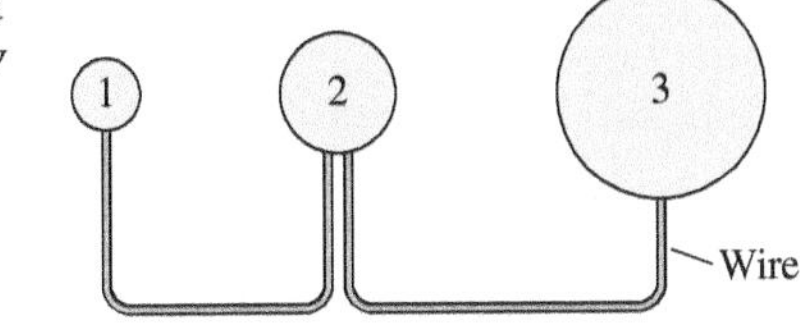

A. $V_1 = V_2 = V_3$ and $E_1 = E_2 = E_3$
B. $V_1 = V_2 = V_3$ and $E_1 > E_2 > E_3$
C. $V_1 > V_2 > V_3$ and $E_1 = E_2 = E_3$
D. $V_1 > V_2 > V_3$ and $E_1 > E_2 > E_3$
E. $V_3 > V_2 > V_1$ and $E_3 = E_2 = E_1$
F. $V_3 > V_2 > V_1$ and $E_3 > E_2 > E_1$

22.7 The Electrocardiogram

We noted in ◀◀ SECTION 21.5 that the heart generates an electric dipole moment as it beats. We're now in a position to look in more detail at the electric properties of the heart and to understand how an *electrocardiogram* measures potential differences created by the heart's dipole.

We'll begin at the level of the cell. We've seen that the ion exchange pumps in a cell wall separate charge and create a potential difference—the membrane potential—across the cell wall. For heart muscle cells, called *cardiac myocytes,* the resting membrane potential (when the muscle is not contracting) is about -90 mV. FIGURE 22.45a (next page) shows a myocyte in its resting state. The cell and its surroundings are electrically neutral, but the charge separation across the cell membrane creates small outward-pointing (from negative to positive) electric dipole moments that cross the cell membrane. The net dipole moment is zero because the small dipoles on opposite sides of the cell cancel, but because of the separation of charge, the myocyte is said to be *polarized.*

A muscle cell differs from most other cells in the body because it can *contract.* A myocyte contraction begins with a change in the membrane potential. The sodium-potassium exchange pumps of a resting myocyte have pushed most Na^+ ions out of the cell. When a myocyte is stimulated—by a nerve impulse or, as we'll see, by ions from a neighboring myocyte—ion channels in the cell wall open to let Na^+ and Ca^{2+} ions flow into the cell in a process called *depolarization.* This surge of

FIGURE 22.45 Depolarization of a heart muscle cell.

(a) Polarized cardiac myocyte

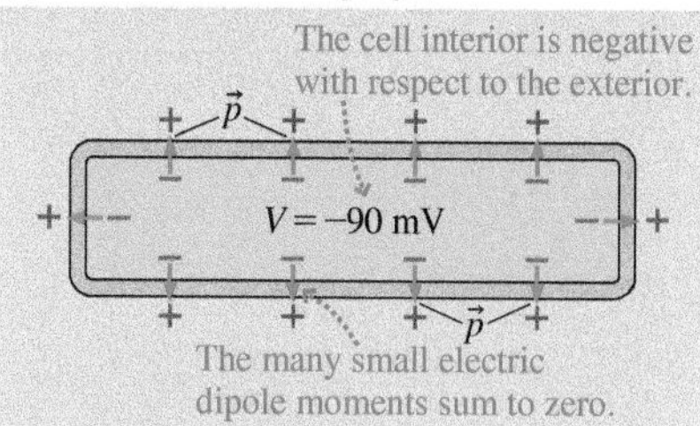
The cell interior is negative with respect to the exterior.
$\vec{p}$
$V = -90$ mV
The many small electric dipole moments sum to zero.

(b) Depolarizing myocyte

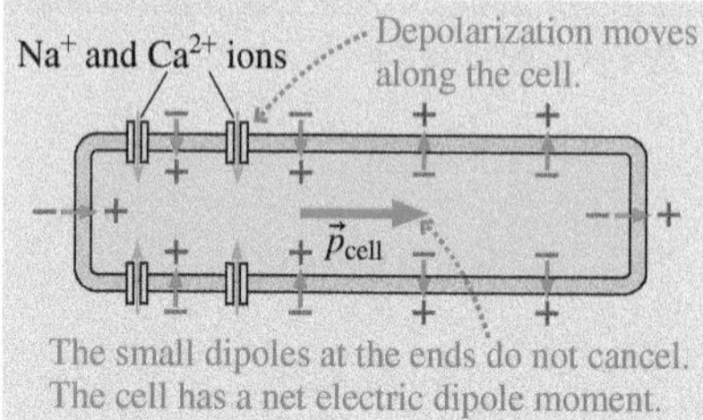
Na$^+$ and Ca^{2+} ions
Depolarization moves along the cell.
$\vec{p}_{cell}$
The small dipoles at the ends do not cancel. The cell has a net electric dipole moment.

(c) Depolarized myocyte

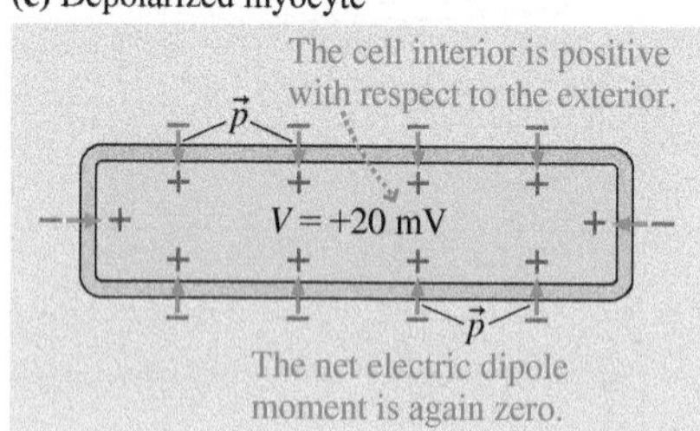
The cell interior is positive with respect to the exterior.
$\vec{p}$
$V = +20$ mV
The net electric dipole moment is again zero.

positive charge rapidly increases the membrane potential to about +20 mV, which means that the inside of the cell is now more positive than the exterior. (The Ca^{2+} ions initiate a series of steps that cause the myocyte to contract, but that's a different story.) The final stage, shown in FIGURE 22.45c, is a *depolarized* myocyte with no net dipole. After about 250 ms, the ion pumps restore the resting membrane potential of −90 mV; this restoration (not shown) is called *repolarization*.

Backing up a step to FIGURE 22.45b, we see that depolarization of a cardiac myocyte does not happen throughout the entire cell at once. Instead, the depolarization starts at one end of the myocyte and propagates down the cell. The dipole moments at the left end of this myocyte have reversed, but the dipole moments at the right end have not yet changed. These no longer cancel, so **a depolarizing cardiac myocyte has a net electric dipole moment $\vec{p}_{cell}$**. This net dipole moment lasts for only the brief interval, about 0.1 ms, that it takes the cell to depolarize. The dipole moment returns to zero when the cell is fully depolarized.

Heart cells are not isolated; instead, cardiac myocytes are connected end-to-end by small channels, called *gap junctions*. Ions from a depolarizing cell flow into the next cell, and their arrival triggers the next cell to depolarize. The net result, illustrated in FIGURE 22.46a, is that depolarization propagates like a wave along a chain of myocytes as each cell triggers the next. The electric dipole moment, which is present only in the cell that is depolarizing, moves down the chain of myocytes.

Heart tissue consists of bundles of parallel myocytes. As FIGURE 22.46b shows, the propagation of depolarization occurs simultaneously along a large number of parallel chains. The small dipole moments $\vec{p}_{cell}$ of each cell add to produce a large electric dipole moment $\vec{p}_{heart}$ at the moving boundary between polarized and depolarized myocytes. This moving dipole moment, with a magnitude of about 10^{-13} C · m, creates the potential differences measured during an electrocardiogram.

FIGURE 22.46 A depolarizing cardiac myocyte triggers its neighbor to depolarize, creating a net dipole moment that moves through the heart during a heartbeat.

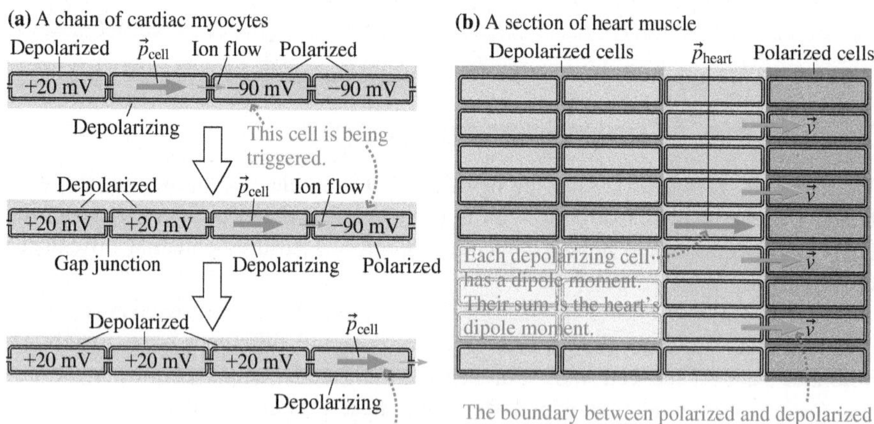
(a) A chain of cardiac myocytes
Depolarized
$\vec{p}_{cell}$
Ion flow
Polarized
+20 mV
−90 mV
−90 mV
Depolarizing
This cell is being triggered.
Depolarized
$\vec{p}_{cell}$
Ion flow
+20 mV
+20 mV
−90 mV
Gap junction
Depolarizing
Polarized
Depolarized
$\vec{p}_{cell}$
+20 mV
+20 mV
+20 mV
Depolarizing
This dipole moment propagates down the chain of myocytes.
(b) A section of heart muscle
Depolarized cells
$\vec{p}_{heart}$
Polarized cells
$\vec{v}$
Each depolarizing cell has a dipole moment. Their sum is the heart's dipole moment.
The boundary between polarized and depolarized myocytes sweeps across the heart with velocity $\vec{v}$ as the muscles contract.

Initially, all muscle cells in the heart are in their polarized state. A heartbeat begins with an electrical signal from the heart's *sinoatrial node*—the heart's pacemaker—in the right atrium. This sets off a wave of depolarization that travels downward through the heart, causing first the atria and then the ventricles to contract. FIGURE 22.47a shows the boundary between polarized and depolarized cells at an instant fairly early in the heartbeat. The heart's dipole moment $\vec{p}_{heart}$, which is located on this boundary and moves with the boundary, generates a dipole electric field and a dipole electric potential throughout a person's body. FIGURE 22.47b shows the equipotential surfaces of

FIGURE 22.47 A contracting heart generates the electric field of an electric dipole.

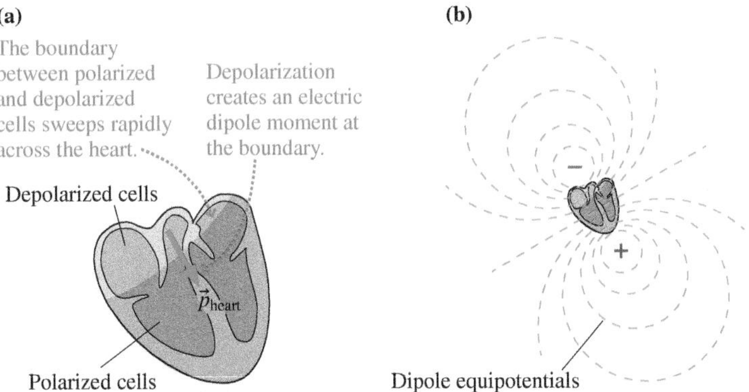
(a)
The boundary between polarized and depolarized cells sweeps rapidly across the heart.
Depolarization creates an electric dipole moment at the boundary.
Depolarized cells
$\vec{p}_{heart}$
Polarized cells
(b)
−
+
Dipole equipotentials

the dipole at this instant of time. These match the equipotential surfaces of a dipole shown earlier in Figure 22.25.

The heart's electric dipole moment can be determined by measuring the potential difference between two points in the electric field. In practice, several pairs of points are needed. A nurse or technician uses between 3 and 10 electrodes, as shown in FIGURE 22.48, depending on the level of detail that a physician wishes to see. A chart that shows how the potential differences change with time is known as an **electrocardiogram,** abbreviated either ECG or, from its German origin, EKG. An ECG shows how the heart's electric dipole changes throughout a heartbeat.

FIGURE 22.48 Between 3 and 10 electrodes are used to record an electrocardiogram.

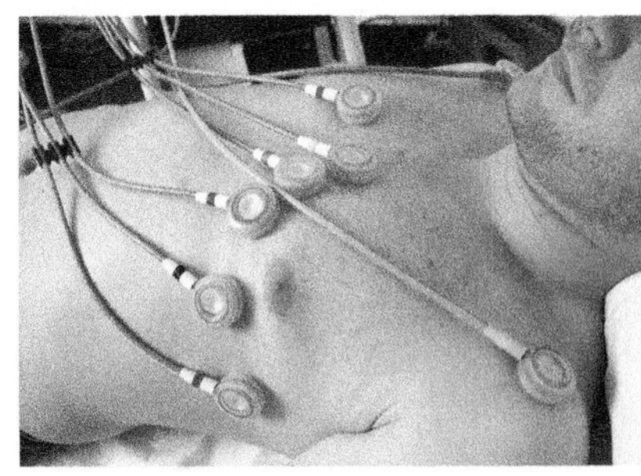

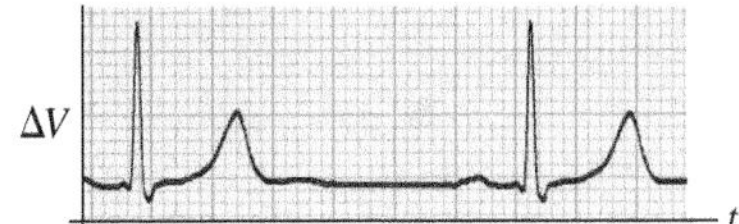
ΔV
t

Records of the potential differences between various pairs of electrodes allow the doctor to analyze the heart's condition.

FIGURE 22.49 shows two points—perhaps the locations of two electrodes—in the electric field of a dipole that, at this instant, is pointing down and to the right. The sign of the potential difference $\Delta V = V_2 - V_1$ can be determined by observing where the points are on the dipole's equipotential surfaces. In this case, point 2 is on a positive equipotential surface while point 1 has a negative potential; thus ΔV is positive.

To find a quantitative relationship between a measured potential difference and the heart's dipole moment, we need to return to Equation 22.38 for the electric potential of a dipole—with one important change. Equation 22.38 assumes that the dipole is either in a vacuum or perhaps in air, which has almost no effect on electric fields and potentials. In contrast, the heart's electric field extends through the body, which is much denser than air and is mostly water.

Electric fields and potentials in a material substance are the topic of Chapter 23, but we'll go ahead and use one of the results: In a material substance, the electric potential of charges and dipoles is reduced to $V_{substance} = V_{vacuum}/\kappa$, where κ (Greek kappa) is the *dielectric constant* of the substance. The dielectric constant of water is 80. Thus Equation 22.38 must be modified to

$$V_{heart} = \frac{p_{heart}\cos\theta}{4\pi\epsilon_0\kappa r^2} \qquad (22.49)$$

where θ is the angle away from the dipole moment vector—the direction that $\vec{p}_{heart}$ is pointing. Different pairs of electrodes, even if all electrodes are the same distance from the heart, measure different potential differences because the direction to each electrode is at a different angle.

FIGURE 22.50 (next page) shows a simplified model of the electrocardiogram signal recorded from the two electrodes on a patient's shoulders, electrodes that you can see in Figure 22.48. The heart's dipole moment changes in both strength and direction as the wave of depolarization moves through the entire heart during one heartbeat. Each point on the ECG graph in Figure 22.50d corresponds to a particular magnitude and orientation of $\vec{p}_{heart}$. By observing several simultaneously recorded potential differences along different lines, a physician can get a clear picture of how the heart's electric activity changes throughout a heartbeat.

FIGURE 22.49 The potential difference between two points is determined by where the points are on the dipole's equipotential surfaces.

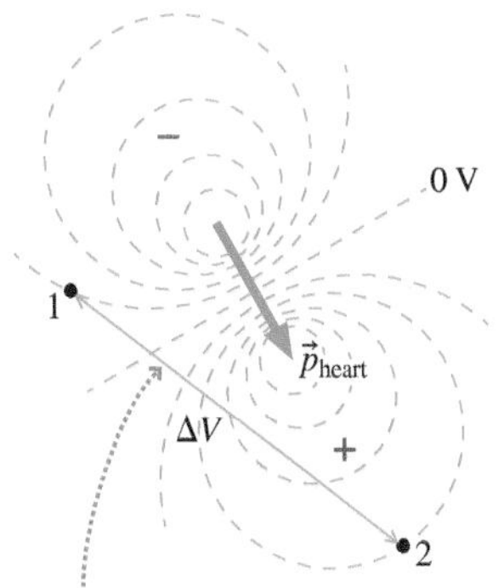
−
0 V
1
$\vec{p}_{heart}$
ΔV
+
2

The sign of the potential difference ΔV can be determined by observing which equipotential surfaces the points lie on.

FIGURE 22.50 The potential difference $\Delta V = V_2 - V_1$ between two points changes during a heartbeat as the dipole moment moves and rotates.

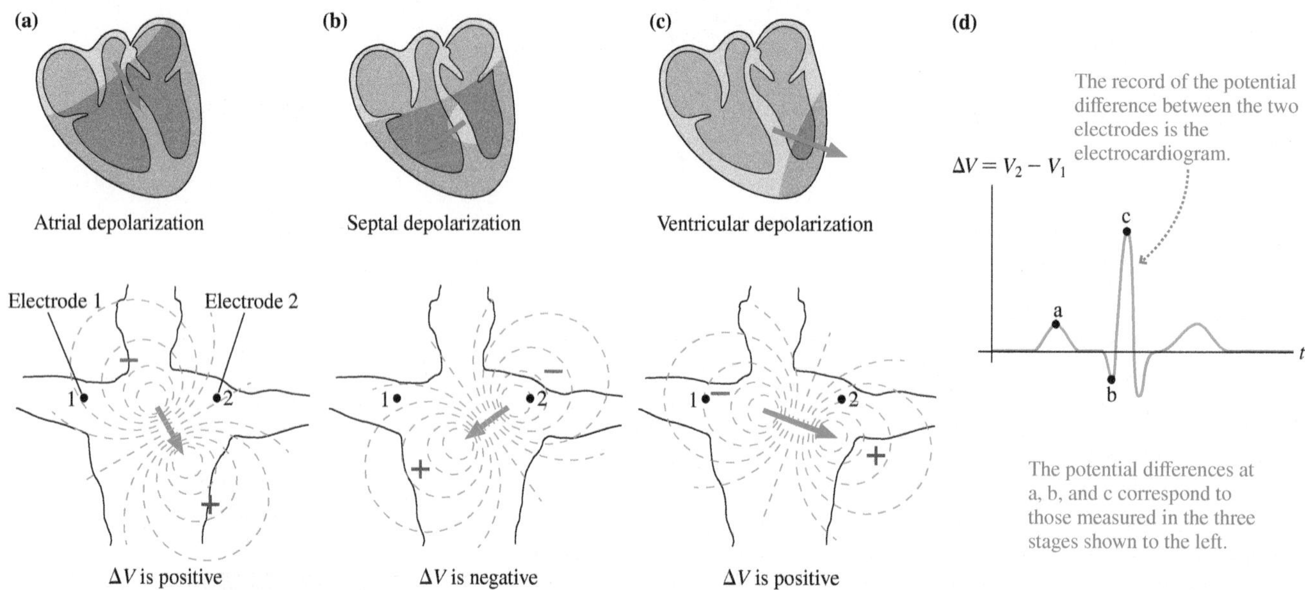

CONCEPTUAL EXAMPLE 22.15 **Determining the potential difference between EKG electrodes**

Technicians are recording an ECG for a patient using the three electrodes shown in FIGURE 22.51. At this instant, the dipole moment of the patient's heart is directed to the right as shown. Which potential difference has the largest positive value: $V_3 - V_1$, $V_2 - V_1$, or $V_2 - V_3$? Which has the smallest?

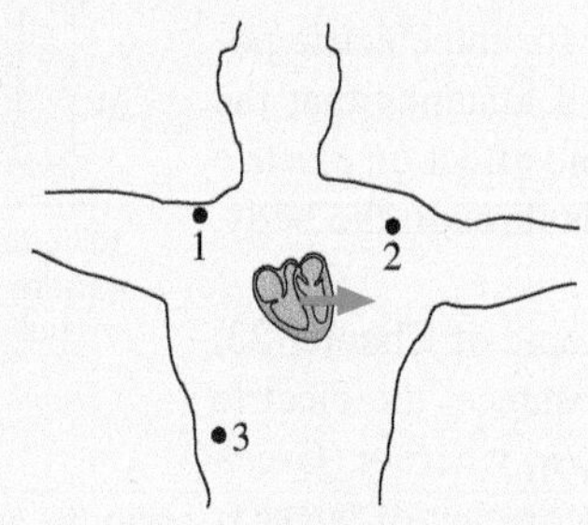

FIGURE 22.51 The dipole moment of a patient's heart during an ECG.

REASON Imagine the equipotential surfaces of a dipole overlaid onto the figure. Points 1 and 3 both have a negative potential, and their potential difference $V_3 - V_1$ is not large. V_2 is positive, and both $V_2 - V_1$ and $V_2 - V_3$ are fairly large. However, point 3 is less negative than point 1 because it lies on an equipotential surface farther from the negative charge. Thus $V_2 - V_1$ has the largest positive value and $V_3 - V_1$ the smallest.

ASSESS If we envision an overlay of the equipotential lines from the heart's dipole, we can see that this result makes sense.

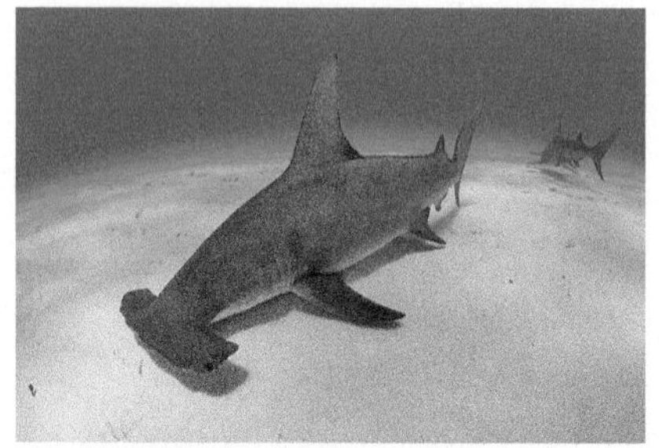

Looking for lunch A scalloped hammerhead shark uses its electric sense to find prey.

The beating of the heart—and the action of other muscles in the body—creates electric fields and potentials. For animals such as humans that live surrounded by air, these potentials can be detected at the surface of the body. For animals that live in the water, the fields and potentials can be detected in the water at some distance from the animal. Many fish, and at least a few water-dwelling mammals, are able to detect the electric fields and potentials produced by the prey they seek. Sharks and rays have the greatest sensitivity. Sharks can detect electric fields as small as 1×10^{-7} V/m, 1000 times smaller than the field from your heart. If you are swimming near a shark, the shark knows you are there. But the shark isn't really interested in you; it's looking for smaller, easier prey animals, which it can sense even when they are buried under the sand. The wide heads of hammerhead sharks and the wide bodies of rays, like the one in the photo that opened the chapter, enable them to scan a generous swath of ocean floor as they forage.

CHAPTER 22 INTEGRATED EXAMPLE Proton fusion in the sun

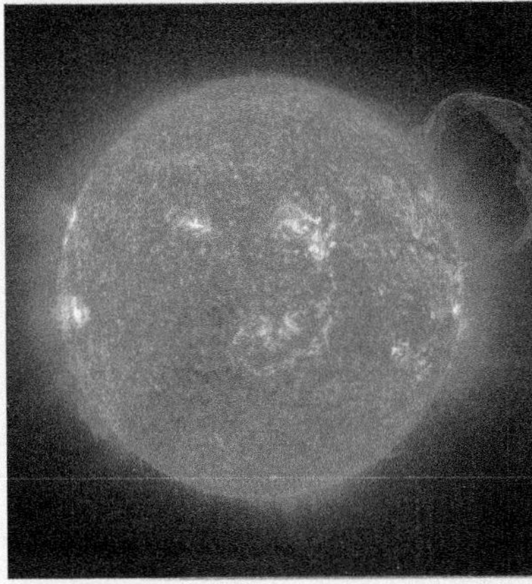

Life on earth depends on energy from the sun. But what has powered the sun for billions of years? The sun's energy comes from nuclear reactions that fuse lighter nuclei into heavier ones, releasing energy in the process. The solar fusion process begins when two protons (the nuclei of hydrogen atoms) merge to produce a *deuterium* nucleus. Deuterium is the "heavy" isotope of hydrogen, with a nucleus consisting of a proton *and* a neutron. To become deuterium, one of the protons that fused has to turn into a neutron. The nuclear-physics process by which this occurs—and which releases the energy—will be studied in Chapter 30. Our interest for now lies not with the nuclear physics but with the conditions that allow fusion to occur.

Before two protons can fuse, they must come into contact. However, the energy required to bring two protons into contact is considerable because the electric potential energy of the two protons increases rapidly as they approach each other. Fusion occurs in the core of the sun because the ultra-high temperature there gives the protons the kinetic energy they need to come together.

a. Two protons approach each other head-on, each with the same speed v_0. What value of v_0 is required for the protons to just touch each other? A proton can be modeled as a charged sphere of diameter 1.6×10^{-15} m with charge e and a mass of 1.7×10^{-27} kg.

b. What does the temperature of the sun's core need to be so that the rms speed v_{rms} of protons is equal to v_0?

PREPARE We will use conservation of energy to solve this problem. **FIGURE 22.52** shows a before-and-after pictorial representation. Both protons are initially moving with speeds $v_i = v_0$, so both contribute to the initial kinetic energy. We assume that they start out so far apart that $U_i = 0$. To "just touch" means that they've instantaneously come to rest ($K_f = 0$) at the point where the distance between their centers is equal to the diameter of a proton. Their energy at that point is the potential energy of two charges. We can use the energy-conservation equation to find the speed v_0 required to achieve contact. Then we can use the results of Chapter 13 to find the temperature at which v_0 is the rms speed of the protons.

FIGURE 22.52 Pictorial representation of two protons coming into contact.

SOLVE

a. Conservation of energy requires $K_f + U_f = K_i + U_i$. We've noted that $U_i = 0$, that $K_f = 0$, and that both protons contribute equally to K_i. The electric potential of a charged sphere is identical to that of a point charge Q at the center, so the electric potential energy of the two protons when separated by distance r_f is the potential energy of two point charges. Thus we have

$$0 + \frac{Ke^2}{r_f} = 0 + \tfrac{1}{2}mv_0^2 + \tfrac{1}{2}mv_0^2$$

Solving the energy equation for the initial speed v_0 gives

$$v_0 = \sqrt{\frac{Ke^2}{mr_f}} = \sqrt{\frac{(9.0 \times 10^9\ \text{N}\cdot\text{m}^2/\text{C}^2)(1.6 \times 10^{-19}\ \text{C})^2}{(1.7 \times 10^{-27}\ \text{kg})(1.6 \times 10^{-15}\ \text{m})}}$$
$$= 9.3 \times 10^6\ \text{m/s}$$

b. In Chapter 13, we found that the rms speed of particles in a gas at temperature T is

$$v_{rms} = \sqrt{\frac{3k_BT}{m}}$$

where k_B is the Boltzmann constant. It may seem strange to think of protons as a gas, but in the center of the sun, where all atoms are ionized into electrons and nuclei, the protons do act like a gas. The temperature at which $v_{rms} = v_0$ is

$$T = \frac{mv_0^2}{3k_B} = \frac{(1.7 \times 10^{-27}\ \text{kg})(9.3 \times 10^6\ \text{m/s})^2}{3(1.38 \times 10^{-23}\ \text{J/K})}$$
$$= 3.6 \times 10^9\ \text{K}$$

ASSESS An extraordinarily high temperature—over 3 billion kelvin—is required to give an average solar proton a speed of 9.29×10^6 m/s. In fact, the core temperature of the sun is "only" about 14 million kelvin, a factor of ≈ 200 less than we calculated. Protons can fuse at this lower temperature both because there are always a few protons moving much faster than average and because protons can reach each other even if their speeds are too low by the quantum-mechanical process of *tunneling,* which you'll learn about in Chapter 28. The core's relatively "low" temperature means that most protons bounce around in the sun for several billion years before fusing!

SUMMARY

GOAL To understand and use the electric potential and electric potential energy.

GENERAL PRINCIPLES

Electric Potential

The **electric potential** V, like the electric field, is created by source charges.

- Point charge: $V_{point} = \dfrac{1}{4\pi\epsilon_0}\dfrac{q}{r}$
- Sphere of charge: $V_{sphere} = \dfrac{1}{4\pi\epsilon_0}\dfrac{Q}{r}$
- Parallel-plate capacitor: $V_{cap} = \dfrac{x}{d}\Delta V_C$
  Field strength $E = \Delta V_C/d$
  Distance x is measured from the negative plate.

For multiple point charges:

$$V = V_1 + V_2 + V_3 + \cdots$$

For a continuous distribution of charge:
Divide the charge into point-like ΔQ.
Find the potential due to each ΔQ and add them.
Use a *charge density* to replace ΔQ with a coordinate Δx or ΔA.
Sum by integrating.

For a conductor in electrostatic equilibrium, the entire conductor is at the same potential. The surface is an equipotential surface.

$K_f + qV_f = K_i + qV_i$ for a charge moving in an electric potential.

Units

Electric potential: 1 V = 1 J/C
Electric field: 1 V/m = 1 N/C

Electric Potential Energy

A charge q placed in an electric potential V has **electric potential energy** (an interaction energy) $U_{elec} = qV$.

- Two point charges: $U_{elec} = \dfrac{Kq_1q_2}{r} = \dfrac{1}{4\pi\epsilon_0}\dfrac{q_1q_2}{r}$
- Charge in a uniform field: $U_{elec} = qEx$
- Electric dipole in a field: $U_{elec} = -pE\cos\theta$

Mechanical energy $K + U_{elec}$ is conserved for charged particles moving in a vacuum.

Connecting field and potential

The electric potential and the electric field are connected because they are two different perspectives on how source charges alter the space around them. They are related by

$$E_x = -\frac{dV}{dx}$$

= the negative of the slope of a graph of the potential

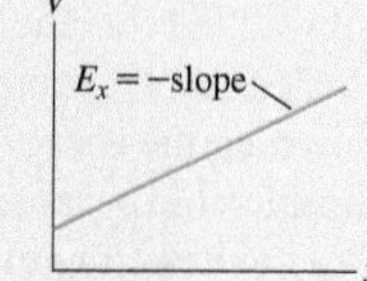
V
$E_x = -$slope
x

Kirchhoff's loop law

The sum of all potential differences around a closed path is zero:

$$\sum_i (\Delta V)_i = 0$$

Kirchhoff's loop law is a statement of energy conservation.

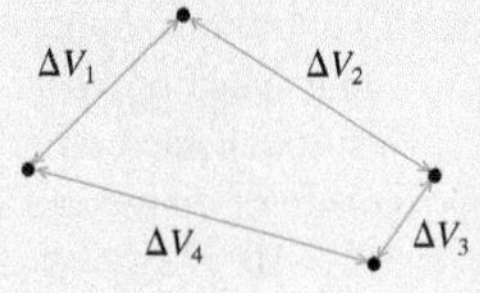
ΔV_1
ΔV_2
ΔV_3
ΔV_4

IMPORTANT CONCEPTS

Sources of potential

Potential differences are created by a *separation of charge*.

A **battery** uses chemical reactions to separate charge by moving charge from the negative terminal to the positive terminal:

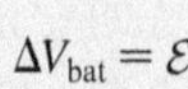

$$\Delta V_{bat} = \mathcal{E}$$

where the **emf** $\mathcal{E}$ is the work per charge done by the charge escalator.

The geometry of potential and field

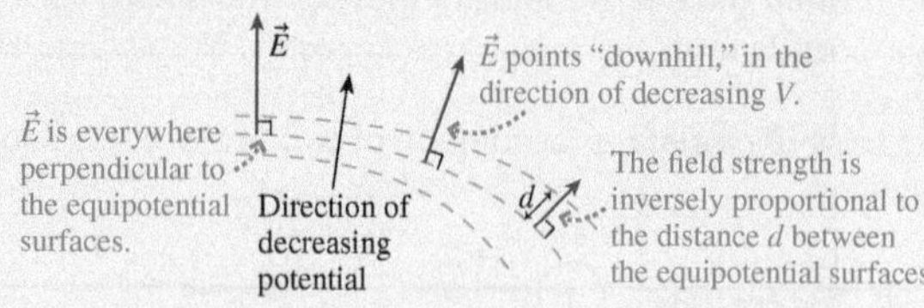
$\vec{E}$
$\vec{E}$ points "downhill," in the direction of decreasing V.
$\vec{E}$ is everywhere perpendicular to the equipotential surfaces.
Direction of decreasing potential
The field strength is inversely proportional to the distance d between the equipotential surfaces.

APPLICATIONS

Graphical representations of the potential

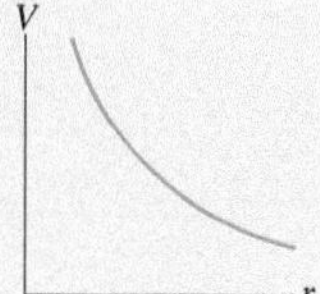
V
r

Potential graph

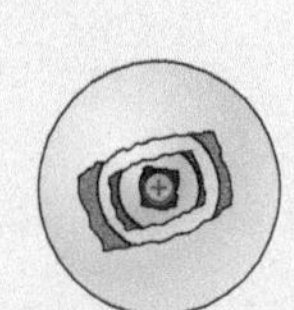

Equipotential surfaces

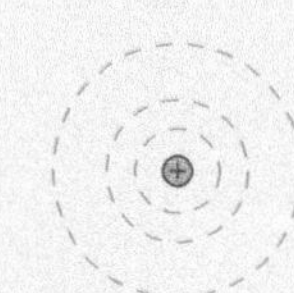

Contour map

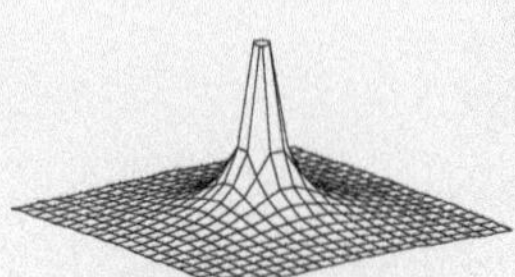

Elevation graph

LEARNING OBJECTIVES After studying this chapter, you should be able to:

- Solve problems using the electric potential and energy conservation. *Conceptual Questions 22.3, 22.4; Problems 22.2–22.4, 22.8, 22.10*
- Apply the concept of electric potential. *Conceptual Question 22.13; Problems 22.11, 22.12, 22.15, 22.17, 22.20*
- Calculate the electric potentials of charges and groups of charges. *Conceptual Questions 22.9, 22.10; Problems 22.22, 22.23, 22.26–22.28*
- Determine the electric potential of a continuous distribution of charge. *Conceptual Questions 22.7, 22.8; Problems 22.34–22.36*
- Solve problems involving batteries and emf. *Problems 22.37–22.40*
- Calculate the electric field from the potential. *Conceptual Questions 22.11, 22.14; Problems 22.41, 22.42, 22.44, 22.46, 22.50*
- Analyze the electric field of the heart. *Problems 22.52, 22.53, 22.55*

STOP TO THINK ANSWERS

Stop to Think 22.1: Zero. The motion is always perpendicular to the electric force.

Stop to Think 22.2: $U_B = U_D > U_A = U_C$. The potential energy depends inversely on *r*. The effects of doubling the charge and doubling the distance cancel each other.

Stop to Think 22.3: C. The proton gains speed by losing potential energy. It loses potential energy by moving in the direction of decreasing electric potential.

Stop to Think 22.4: $V_1 = V_2 > V_3 > V_4 = V_5$. The potential decreases steadily from the positive to the negative plate. It depends only on the distance from the positive plate.

Stop to Think 22.5: $\Delta V_{13} = \Delta V_{23} > \Delta V_{12}$. The potential depends only on the *distance* from the charge, not the direction. $\Delta V_{12} = 0$ because these points are at the same distance.

Stop to Think 22.6: $\eta_C = \eta_B = \eta_A$. All pieces of a uniformly charged surface have the same surface charge density.

Stop to Think 22.7: C. E_y is the negative of the slope of the *V*-versus-*y* graph. E_y is positive because $\vec{E}$ points up, so the graph has a negative slope. E_y has constant magnitude, so the slope has a constant value.

Stop to Think 22.8: B. Because of the connecting wire, the three spheres form a single conductor in electrostatic equilibrium. Thus all points are at the same potential. The electric field of a sphere is related to the sphere's potential by $E = V/R$, so a smaller-radius sphere has a larger *E*.

QUESTIONS

Conceptual Questions

1. a. Charge q_1 is distance r from a positive point charge Q. Charge $q_2 = q_1/3$ is distance $2r$ from Q. What is the ratio U_1/U_2 of their potential energies due to their interactions with Q?
   b. Charge q_1 is distance x from the negative plate of a parallel-plate capacitor. Charge $q_2 = q_1/3$ is distance $2x$ from the negative plate. What is the ratio U_1/U_2 of their potential energies?
2. Figure Q22.2 shows the potential energy of a proton $(q = +e)$ and a lead nucleus $(q = +82e)$. The horizontal scale is in units of *femtometers*, where 1 fm $= 10^{-15}$ m.
   a. A proton is fired toward a lead nucleus from very far away. How much initial kinetic energy does the proton need to reach a turning point 10 fm from the nucleus? Explain.
   b. How much kinetic energy does the proton of part a have when it is 20 fm from the nucleus and moving toward it, before the collision?

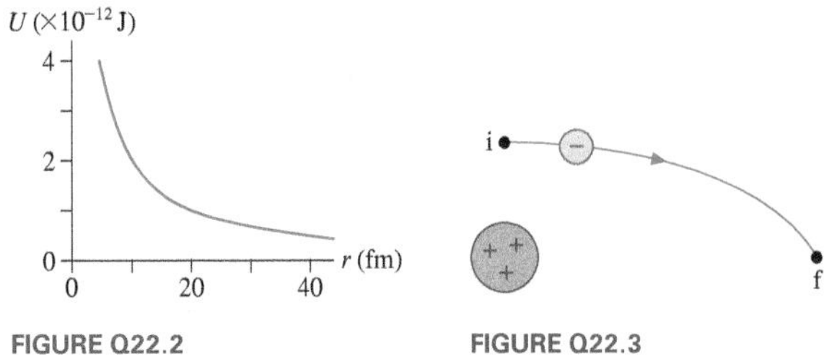

FIGURE Q22.2

FIGURE Q22.3

3. An electron moves along the trajectory of Figure Q22.3 from i to f.
   a. Does the electric potential energy increase, decrease, or stay the same? Explain.
   b. Is the electron's speed at f greater than, less than, or equal to its speed at i? Explain.

Problem difficulty is labeled as | (straightforward) to ||||| (challenging). Problems labeled INT integrate significant material from earlier chapters; BIO are of biological or medical interest; CALC require calculus to solve.

4. Two protons are launched with the same speed from point 1 inside the parallel-plate capacitor of Figure Q22.4. Points 2 and 3 are the same distance from the negative plate.
   a. Is $\Delta U_{1\rightarrow 2}$, the change in potential energy along the path $1 \rightarrow 2$, larger than, smaller than, or equal to $\Delta U_{1\rightarrow 3}$?
   b. Is the proton's speed v_2 at point 2 larger than, smaller than, or equal to v_3? Explain.

FIGURE Q22.4

5. Rank in order, from most positive to most negative, the potential energies U_A to U_F of the six electric dipoles in the uniform electric field of Figure Q22.5. Explain.

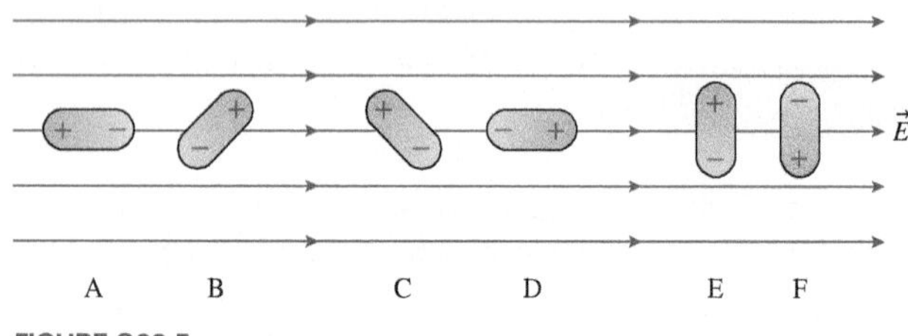

FIGURE Q22.5

6. A positively charged dust particle passes between the two oppositely charged plates of an electrostatic precipitator. There is an electric force directed toward the negative plate, but the drag force causes the particle to drift at a constant speed. What energy transformation takes place during this motion?
7. A capacitor with plates separated by distance d is charged to a potential difference ΔV_C. All wires and batteries are disconnected, then the two plates are pulled apart (with insulated handles) to a new separation of distance $2d$.
   a. Does the capacitor charge Q change as the separation increases? If so, by what factor? If not, why not?
   b. Does the electric field strength E change as the separation increases? If so, by what factor? If not, why not?
   c. Does the potential difference ΔV_C change as the separation increases? If so, by what factor? If not, why not?
8. Rank in order, from largest to smallest, the electric potentials V_1 to V_5 at points 1 to 5 in Figure Q22.8. Explain.

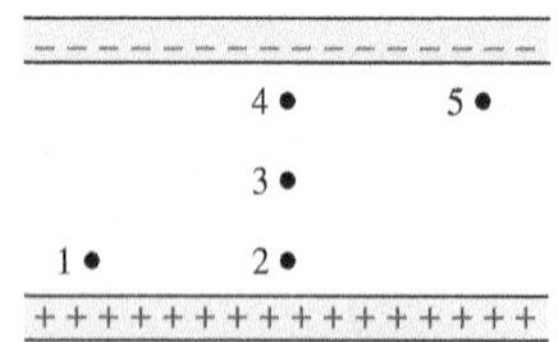

FIGURE Q22.8

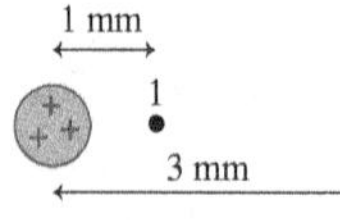

FIGURE Q22.9

9. Figure Q22.9 shows two points near a positive point charge.
   a. What is the ratio V_2/V_1 of the electric potentials? Explain.
   b. What is the ratio E_2/E_1 of the electric field strengths?
10. Reproduce Figure Q22.10 on your paper. Then draw a dot (or dots) on the figure to show the position (or positions) at which the electric potential is zero.

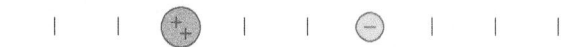
FIGURE Q22.10

11. Estimate the electric fields $\vec{E}_1$ and $\vec{E}_2$ at points 1 and 2 in Figure Q22.11. Don't forget that $\vec{E}$ is a vector.

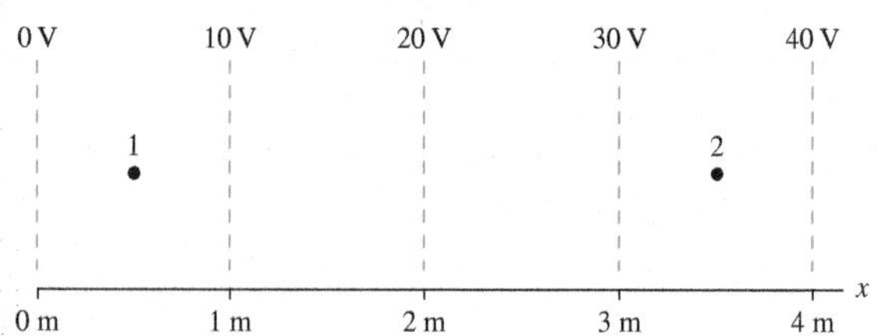

FIGURE Q22.11

12. Estimate the electric fields $\vec{E}_1$ and $\vec{E}_2$ at points 1 and 2 in Figure Q22.12. Don't forget that $\vec{E}$ is a vector.

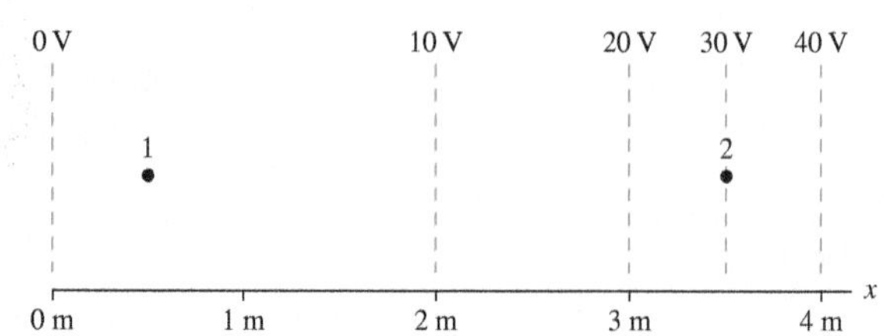

FIGURE Q22.12

13. An electron is released from rest at $x = 2$ m in the potential shown in Figure Q22.13. Does it move? If so, to the left or to the right? Explain.

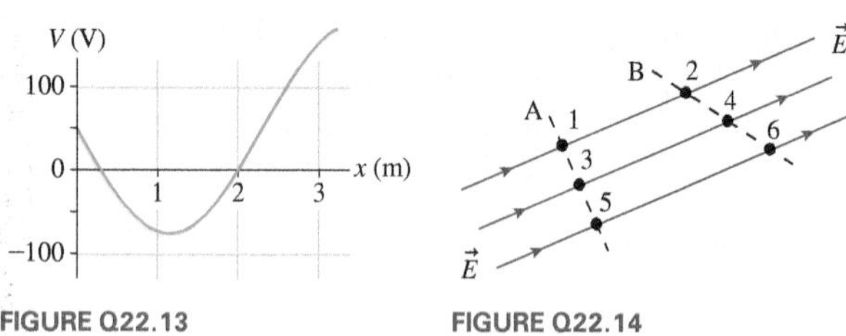

FIGURE Q22.13 FIGURE Q22.14

14. Figure Q22.14 shows an electric field diagram. Dashed lines A and B are two surfaces in space, not physical objects.
   a. Is the electric potential at point 1 higher than, lower than, or equal to the electric potential at point 2? Explain.
   b. Rank in order, from largest to smallest, the magnitudes of the potential differences ΔV_{12}, ΔV_{34}, and ΔV_{56}.
   c. Is surface A an equipotential surface? What about surface B? Explain why or why not.
15. The two metal spheres in Figure Q22.15 are connected by a metal wire with a switch in the middle. Initially the switch is open. Sphere 1, with the larger radius, is given a positive charge. Sphere 2, with the smaller radius, is neutral. Then the switch is closed. Afterward, sphere 1 has charge Q_1, is at potential V_1, and the electric field strength at its surface is E_1. The values for sphere 2 are Q_2, V_2, and E_2.
   a. Is V_1 larger than, smaller than, or equal to V_2? Explain.
   b. Is Q_1 larger than, smaller than, or equal to Q_2? Explain.
   c. Is E_1 larger than, smaller than, or equal to E_2? Explain.

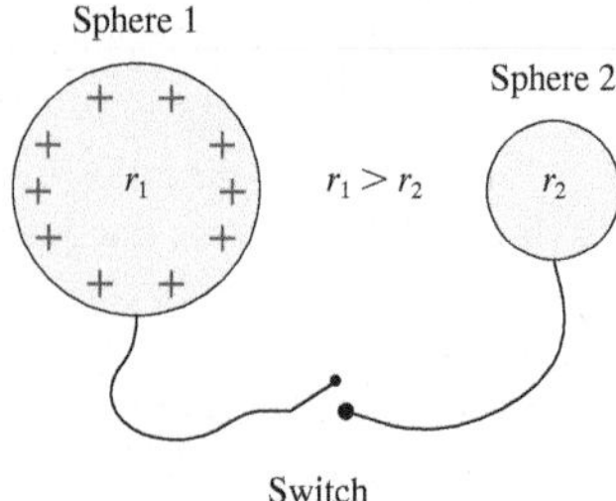

FIGURE Q22.15

Multiple-Choice Questions

16. | A 1.0 nC positive point charge is located at point A in Figure Q22.16. The electric potential at point B is

A. 9.0 V B. 9.0 sin 30° V
C. 9.0 cos 30° V D. 9.0 tan 30° V

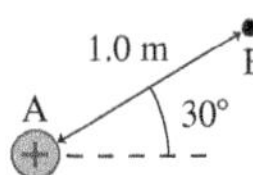

FIGURE Q22.16

17. | A bug zapper consists of two metal plates connected to a high-voltage power supply. The voltage between the plates is set to give an electric field slightly less than 1×10^6 V/m. When a bug flies between the two plates, it increases the field enough to initiate a spark that incinerates the bug. If a bug zapper has a 4000 V power supply, what is the approximate separation between the plates?

A. 0.05 cm B. 0.5 cm
C. 5 cm D. 50 cm

18. || An atom of helium and one of argon are singly ionized—one electron is removed from each. The two ions are then accelerated from rest by the electric field between two plates with a potential difference of 150 V. After accelerating from one plate to the other,

A. The helium ion has more kinetic energy.
B. The argon ion has more kinetic energy.
C. Both ions have the same kinetic energy.
D. There is not enough information to say which ion has more kinetic energy.

19. || The dipole moment of the heart is shown
BIO at a particular instant in Figure Q22.19. Which of the following potential differences will have the largest positive value?

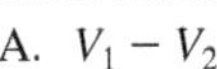

A. $V_1 - V_2$
C. $V_2 - V_3$.

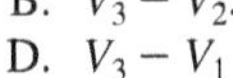

B. $V_3 - V_2$.
D. $V_3 - V_1$

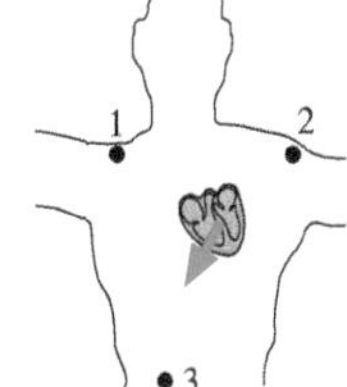

FIGURE Q22.19

Questions 20 through 22 concern Figure Q22.20, which shows the configuration of a gel electrophoresis experiment. Charged plates establish an electric field on a region of gel between the plates. A solution of DNA fragments, which acquire a negative charge in solution, is analyzed by using an electric field to drive the fragments through the gel. A drag force opposes the motion, causing the fragments to move at a constant speed. Larger fragments move more slowly, so the fragments separate by size to give a "fingerprint" of the sample.

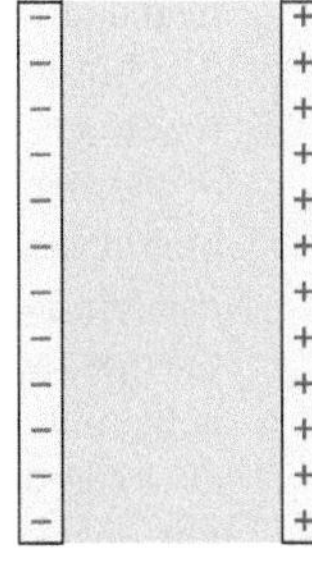

FIGURE Q22.20

20. | In the system shown in Figure Q22.20, the electric field is directed to the ________; the DNA fragments move to the ________.

A. right, right B. right, left
C. left, right D. left, left

21. | The fragments move in the direction of ____________ electric potential and ____________ electric potential energy.

A. higher, higher B. higher, lower
C. lower, higher D. lower, lower

22. || What is the energy transformation as the fragments move through the gel?

A. Electric potential energy to kinetic energy
B. Electric potential energy to thermal energy
C. Electric potential energy to kinetic energy and thermal energy
D. Kinetic energy to thermal energy

Questions 23 through 26 refer to Figure Q22.23, which shows equipotential lines in a region of space. The equipotential lines are spaced by the same difference in potential, and several of the potentials are given.

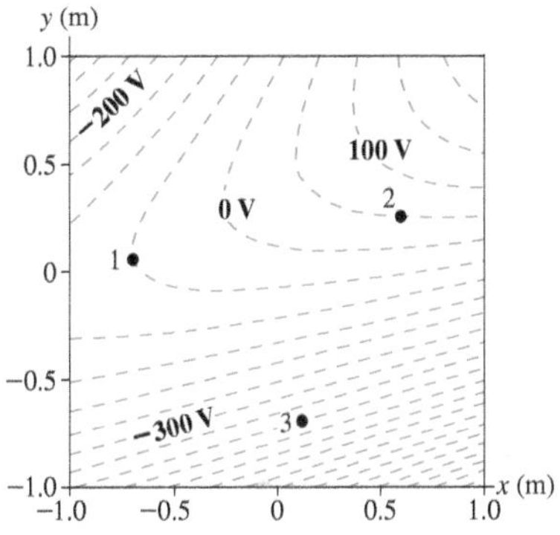

FIGURE Q22.23

23. | What is the potential at point 3?

A. −400 V B. −350 V
C. −100 V D. 350 V
E. 400 V

24. | At which point is the electric field the strongest?

A. Point 1 B. Point 2 C. Point 3

25. | What is the approximate magnitude of the electric field at point 3?

A. 100 V/m B. 300 V/m C. 800 V/m
D. 1500 V/m E. 3000 V/m

26. | The direction of the electric field at point 2 is closest to which direction?

A. Right B. Up
C. Left D. Down

PROBLEMS

Section 22.1 Electric Potential Energy

1. || The electric field strength is 20,000 N/C inside a parallel-plate capacitor with a 1.0 mm spacing. An electron is released from rest at the negative plate. What is the electron's speed when it reaches the positive plate?

2. || The electric field strength is 50,000 N/C inside a parallel-plate capacitor with a 2.0 mm spacing. A proton is released from rest at the positive plate. What is the proton's speed when it reaches the negative plate?

3. || A proton is released from rest at the positive plate of a parallel-plate capacitor. It crosses the capacitor and reaches the negative plate with a speed of 50,000 m/s. The experiment is repeated with a He^+ ion (charge e, mass 4 u). What is the ion's speed at the negative plate?
4. | Two small, equally charged positive spheres are 2.5 cm apart, measured between their centers. Their electric potential energy is 750 nJ. What is the charge in nC on each sphere?
5. || The graph in Figure P22.5 shows the electric potential energy as a function of separation for two point charges. If one charge is +0.44 nC, what is the other charge?

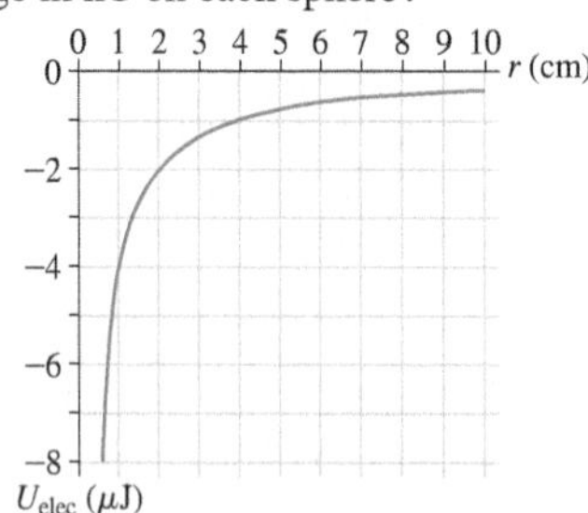

FIGURE P22.5

6. || What is the electric potential energy of the group of charges in Figure P22.6?

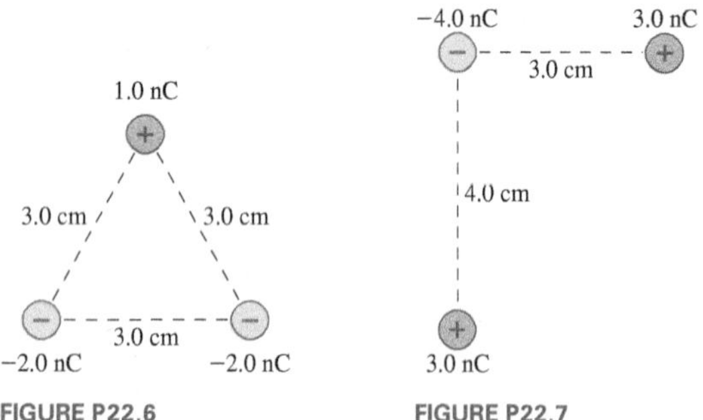

FIGURE P22.6

FIGURE P22.7

7. || What is the electric potential energy of the group of charges in Figure P22.7?
8. || Two positive point charges are 5.0 cm apart. If the electric potential energy is 72 μJ, what is the magnitude of the force between the two charges?
9. || A water molecule perpendicular to an electric field has 1.0×10^{-21} J more potential energy than a water molecule aligned with the field. The dipole moment of a water molecule is 6.2×10^{-30} C · m. What is the strength of the electric field?
10. || Figure P22.10 shows the potential energy of an electric dipole. Consider a dipole that oscillates between ±60°.
    a. What is the dipole's mechanical energy?
    b. What is the dipole's kinetic energy when it is aligned with the electric field?

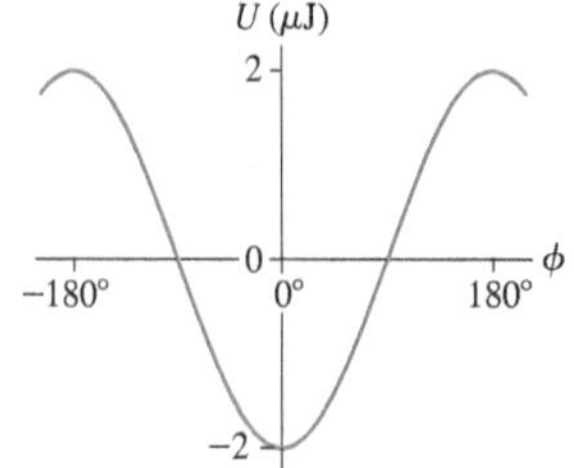

FIGURE P22.10

Section 22.2 The Electric Potential

11. | How much work does the electric field do on a sodium ion (Na^+) as the ion moves through a −1.5 V potential difference created by a 1.5 V battery?
12. | At one point in space, the electric potential energy of a 15 nC charge is 45 μJ.
    a. What is the electric potential at this point?
    b. If a 25 nC charge were placed at this point, what would its electric potential energy be?
13. || What are the speeds of (a) a proton that is accelerated from rest through a potential difference of −1000 V and (b) an electron that is accelerated from rest through a potential difference of 1000 V?
14. || What potential difference is needed to accelerate a He^+ ion (charge $+e$, mass 4 u) from rest to a speed of 2.0×10^6 m/s?
15. || BIO A patient's tumor is being treated with proton-beam therapy. The protons are accelerated through a potential difference of 50 MV. What is the speed of the protons?
16. || BIO A resting muscle cell has a 70 mV potential difference across the 7-nm-thick cell membrane, with the inside of the cell negative with respect to the outside. Ions such as Na^+ cross this potential difference through *ion channels* embedded in the membrane. The ions are *not* in vacuum—they interact with the biomolecules of the cell—but suppose they were. If a Na^+ ion starts from rest outside the cell, what will its speed be when it reaches the inside of the cell if it travels in vacuum? This is an estimate, so give your answer to one significant figure.
17. | A proton with an initial speed of 800,000 m/s is brought to rest by an electric field.
    a. Did the proton move into a region of higher potential or lower potential?
    b. What was the potential difference that stopped the proton?
18. || An electron with an initial speed of 500,000 m/s is brought to rest by an electric field.
    a. Did the electron move into a region of higher potential or lower potential?
    b. What was the potential difference that stopped the electron?
19. || BIO Bacteria can be identified by using a *time-of-flight mass spectrometer* to measure their chemical composition. First, a very short laser pulse vaporizes and ionizes a bacterial sample. The positive ions are accelerated, in vacuum, through a −15 kV potential difference, and then they travel at constant speed through a 1.5-m-long *drift tube* to a detector that records their arrival times. An ion's time of flight depends on its mass, so a record of the arrival times can be used to determine the masses of the biomolecules that were released from the bacteria. Each type of bacteria has a unique set of proteins with different masses, so the mass spectrum is a fingerprint for identifying bacteria. What is the mass in kDa of an ionized protein that is detected 51 μs after the laser pulse? You can assume that the protein is singly ionized ($q = +e$), which is mostly true in practice. You can also neglect the time needed to accelerate through the potential difference because it is very small compared to the drift time.
20. || BIO The scanning electron microscope image of a bacterium in Figure P22.20 was produced using a beam of electrons accelerated through a 30 kV potential difference. What is the speed of the electrons?

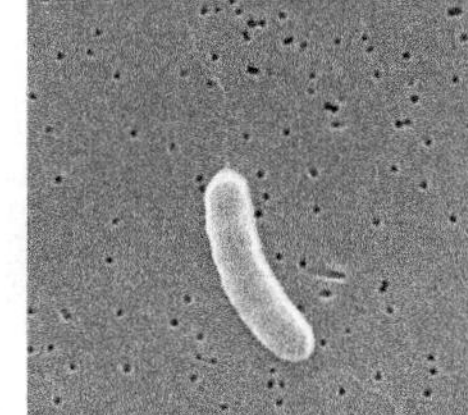
FIGURE P22.20

Section 22.3 Calculating the Electric Potential

21. | Show that 1 V/m = 1 N/C.

22. || a. What is the potential of an ordinary AA or AAA battery? (If you're not sure, find one and look at the label.)
    b. An AA battery is connected to a parallel-plate capacitor having 4.0 cm × 4.0 cm plates spaced 1.0 mm apart. How much charge does the battery supply to each plate?

23. || A 3.0-cm-diameter parallel-plate capacitor has a 2.0 mm spacing. The electric field strength inside the capacitor is 1.0×10^5 V/m.
    a. What is the potential difference across the capacitor?
    b. How much charge is on each plate?

24. || Two 2.0-cm-diameter disks spaced 2.0 mm apart form a parallel-plate capacitor. The electric field between the disks is 5.0×10^5 V/m.
    a. What is the voltage across the capacitor?
    b. An electron is launched from the negative plate. It strikes the positive plate at a speed of 2.0×10^7 m/s. What was the electron's speed as it left the negative plate?

25. || In Figure P22.25, a proton is fired with a speed of 200,000 m/s from the midpoint of the capacitor toward the positive plate.
    a. Show that this is insufficient speed to reach the positive plate.
    b. What is the proton's speed as it collides with the negative plate?

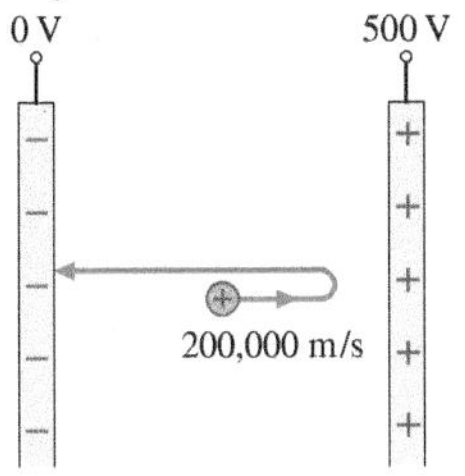

FIGURE P22.25

26. | The electric potential at a point that is halfway between two identical charged particles is 300 V. What is the potential at a point that is 25% of the way from one particle to the other?

27. | a. What is the electric potential at points A, B, and C in Figure P22.27?
    b. What are the potential differences $\Delta V_{AB} = V_B - V_A$ and $\Delta V_{CB} = V_B - V_C$?

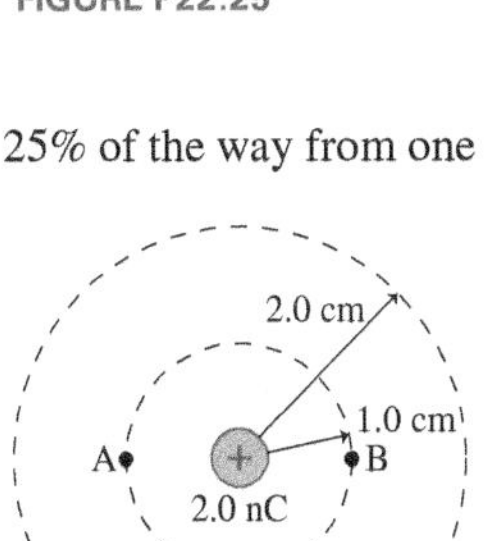

FIGURE P22.27

28. | What is the electric potential at the point indicated with the dot in Figure P22.28?

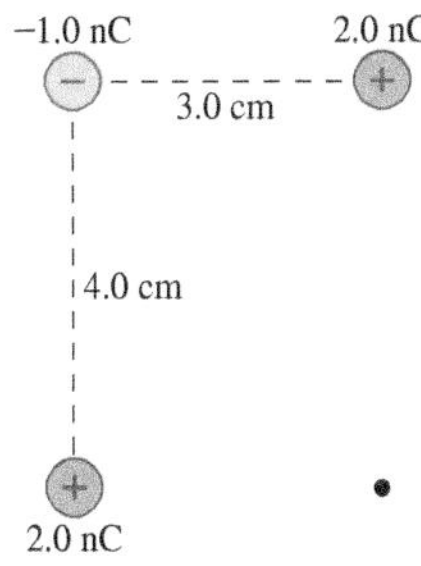

FIGURE P22.28

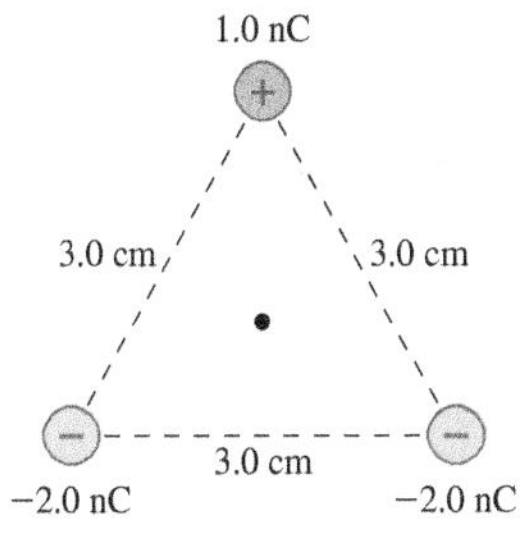

FIGURE P22.29

29. | What is the electric potential at the point indicated with the dot in Figure P22.29?

30. || A 1.0-mm-diameter ball bearing has 2.0×10^9 excess electrons. What is the ball bearing's potential?

31. || A flying hummingbird picks up charge as it moves through BIO the air. This creates a potential near the bird. What is the "voltage of a hummingbird"? Assume that the bird acquires a charge of +200 pC, a typical value, and model the bird as a sphere of radius 3 cm.

32. || Raindrops acquire an electric charge as they fall. Suppose a 2.5-mm-diameter drop has a charge of +15 pC, fairly typical values. What is the potential at the surface of the raindrop?

33. || An electric dipole is made of ±12 nC charges separated by 1.0 mm. What is the electric potential 25 cm from the dipole at angles of (a) 0°, (b) 45°, and (c) 90° from the direction of the dipole moment vector?

Section 22.4 The Potential of a Continuous Distribution of Charge

34. || The two halves of the rod in Figure P22.34 are uniformly charged to $\pm Q$. What is the electric potential at the point indicated by the dot?

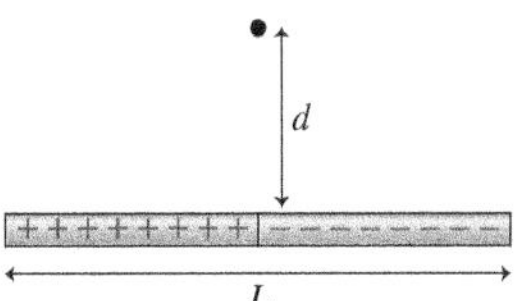

FIGURE P22.34

35. || A 2.0-cm-diameter metal sphere is charged to 1000 V. What is the sphere's surface charge density?

36. || Two 5.0-cm-diameter rings are facing each other 5.0 cm apart. Each is charged to +5.0 nC. What is the electric potential at the center of one of the rings?

Section 22.5 Sources of Electric Potential

37. | In 1 second, chemical reactions move 0.60 C of charge from the negative terminal to the positive terminal of a 1.5 V AA battery.
    a. How much work is done by the reactions?
    b. What is the power output of the reactions?

38. | In a typical mammalian cell, the net transport by the BIO sodium-potassium exchange pump across the 70 mV membrane potential is 500 singly charged ions per second. How much work does the pump do each second?

39. | How much charge does a 9.0 V battery transfer from the negative to the positive terminal while doing 27 J of work?

40. | Light from the sun allows a solar cell to move electrons from the positive to the negative terminal, doing 2.4×10^{-19} J of work per electron. What is the emf of this solar cell?

Section 22.6 Connecting Potential and Field

41. || a. In Figure P22.41, which point, A or B, has a higher electric potential?
    b. What is the potential difference between A and B?

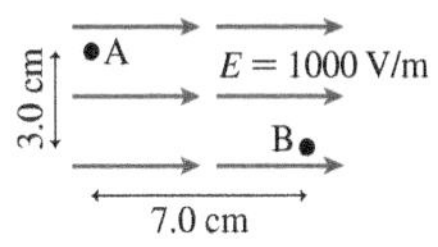

FIGURE P22.41

42. | Students in an introductory physics lab are producing a region of uniform electric field by applying a voltage to two 20-cm-diameter aluminum plates separated by 2.5 cm. They connect the two plates to the two terminals of a high-voltage power supply and then gradually turn up the voltage. In such a situation, sparks start to fly when the field exceeds 3×10^6 V/m. What is the highest voltage the students can use?

43. | Guiana dolphins are one of the few mammals able to detect
BIO electric fields. In a test of sensitivity, a dolphin was exposed to the variable electric field from a pair of charged electrodes. The magnitude of the electric field near the sensory organs was measured by detecting the potential difference between two measurement electrodes located 1.0 cm apart along the field lines. The dolphin could reliably detect a field that produced a potential difference of 0.50 mV between these two electrodes. What is the corresponding electric field strength?

44. | What are the magnitude and direction of the electric field at the dot in Figure P22.44?

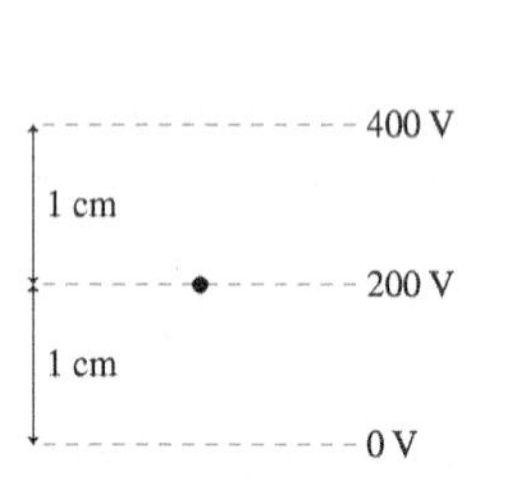

FIGURE P22.44

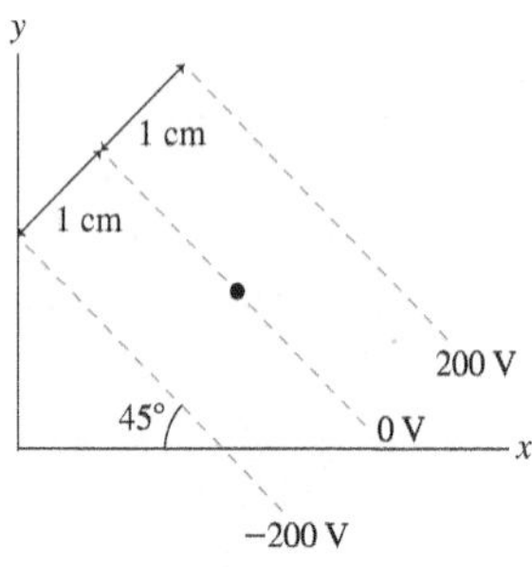

FIGURE P22.45

45. | What are the magnitude and direction of the electric field at the dot in Figure P22.45?

46. | Figure P22.46 shows a graph of V versus x in a region of space. The potential is independent of y and z. What is E_x at (a) $x = -2$ m, (b) $x = 0$ m, and (c) $x = 2$ m?

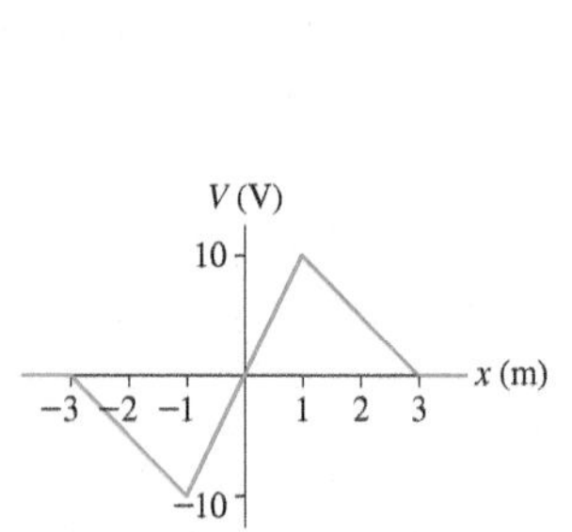

FIGURE P22.46

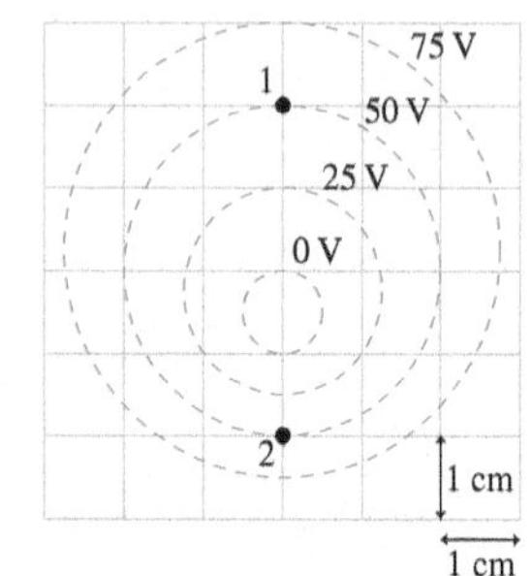

FIGURE P22.47

47. | Determine the magnitude and direction of the electric field at points 1 and 2 in Figure P22.47.

48. || Figure P22.48 is a graph of V versus x. Draw the corresponding graph of E_x versus x.

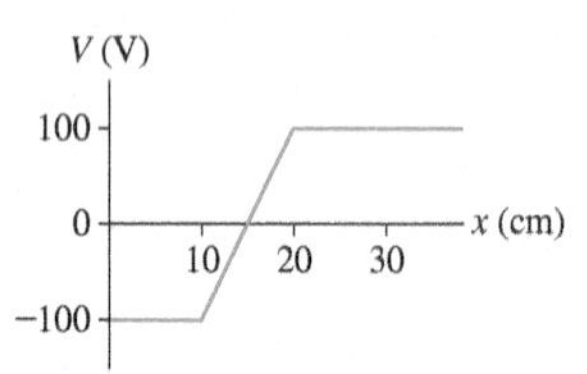

FIGURE P22.48

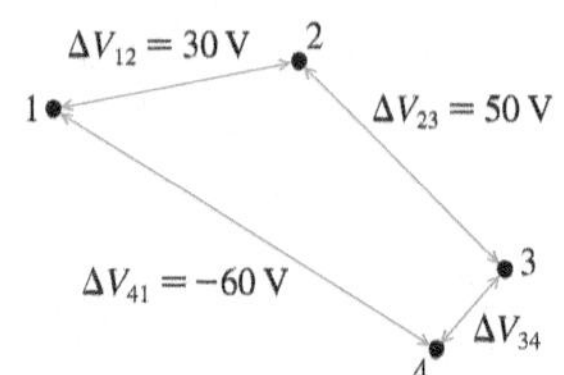

FIGURE P22.49

49. | What is the potential difference ΔV_{34} in Figure P22.49?

50. || The electric potential along the x-axis is $V = 100x^2$ V, where
CALC x is in meters. What is E_x at (a) $x = 0$ m and (b) $x = 1$ m?

51. || The electric potential along the x-axis is $V = 100e^{-2x}$ V, where
CALC x is in meters. What is E_x at (a) $x = 1.0$ m and (b) $x = 2.0$ m?

Section 22.7 The Electrocardiogram

52. | One standard location for a pair
BIO of electrodes during an ECG is shown in Figure P22.52. The potential difference $\Delta V = V_2 - V_1$ is recorded. For each of the three instants a, b, and c during the heart's cycle shown in Figure 22.50, will ΔV be positive, negative, or approximately zero? Explain.

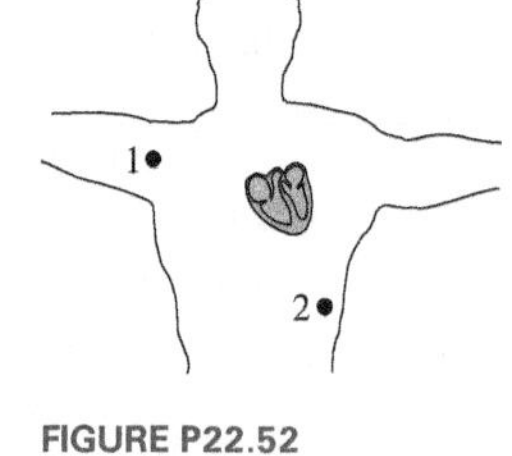

FIGURE P22.52

53. | Three electrodes, 1–3, are at-
BIO tached to a patient as shown in Figure P22.53. During ventricular depolarization (see Figure 22.50), across which pair of electrodes is the magnitude of the potential difference likely to be the smallest? Explain.

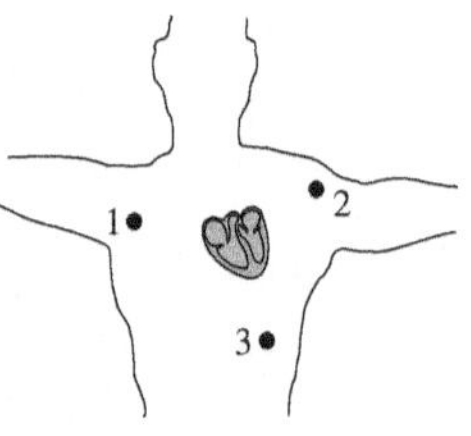

FIGURE P22.53

54. || The eye has a modest dipole moment, with the cornea positive
BIO and the back of the eye negative. This dipole moment rotates with the eyeball. If electrodes are placed on either side of the eye, as in Figure P22.54a, the potential difference between the electrodes can be used to measure the orientation of the eye. A record of this potential is known as an electrooculogram; it's similar in principle to the electrocardiogram.

a. Figure P22.54b is a top view showing the eyeball pointed straight ahead. For this case, is the potential difference $V_2 - V_1$ positive, negative, or zero?
b. Figure P22.54c is a top view showing the eyeball pointed to the right. For this case, is the potential difference $V_2 - V_1$ positive, negative, or zero?

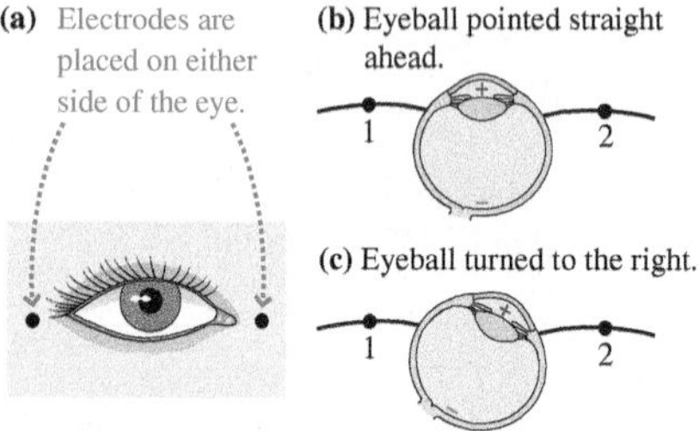

FIGURE P22.54

55. || In experimental tests, sharks
BIO have shown the ability to locate dipole electrodes (simulating the dipole fields of the heartbeats of prey animals) buried under the sand. In a test with young bonnethead sharks, sharks that detected the presence of a dipole usually swam toward the center of the dipole by following equipotential lines. Figure P22.55 shows a dipole electrode and three initial positions of a bonnethead shark. For each initial position, sketch the most likely path of the shark toward the center of the dipole.

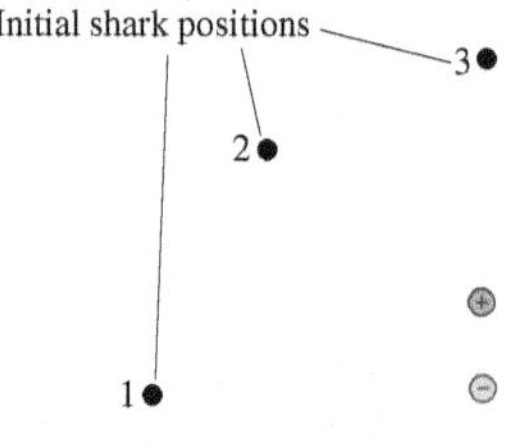

FIGURE P22.55

General Problems

56. ||| Two point charges 2.0 cm apart have an electric potential energy $-180\ \mu J$. The total charge is 30 nC. What are the two charges?
57. ||| A +3.0 nC charge is at $x = 0$ cm and a -1.0 nC charge is at $x = 4$ cm. At what point or points on the x-axis is the electric potential zero?
58. | A proton's speed as it passes point 1 is 50,000 m/s. It follows the trajectory shown in Figure P22.58. What is the proton's speed at point 2?

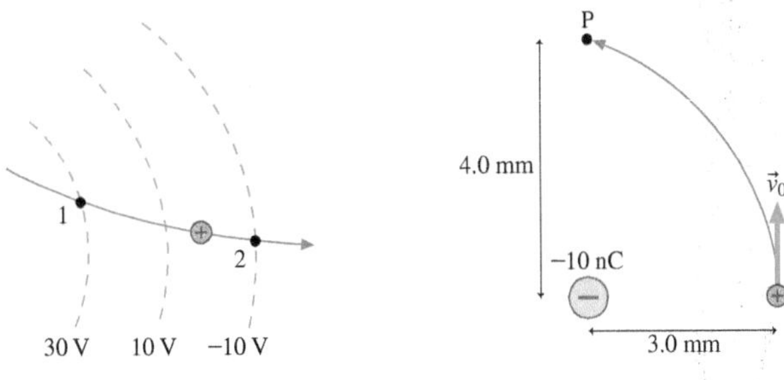

FIGURE P22.58

FIGURE P22.59

59. ||| A proton follows the path shown in Figure P22.59. Its initial speed is $v_0 = 1.9 \times 10^6$ m/s. What is the proton's speed as it passes through point P?
60. || Recent advances in technology have allowed researchers to **BIO** use nanoparticles to make the first measurements of electric fields inside cells. The field strengths are surprisingly large. In one type of cell, the electric field strength just outside the nucleus is approximately 50 kV/m. We can model the nucleus as an 8-μm-diameter sphere and assume that positive charges are uniformly distributed on its surface.
    a. What is the charge on the nucleus?
    b. If the charge is due to ions with charge $+e$, how many excess ions are on the surface of the nucleus?
    c. What is the electric potential at the surface of the nucleus?
61. ||| Three electrons form an equilateral triangle 1.0 nm on each side. A proton is at the center of the triangle. What is the potential energy of this group of charges?
62. || Physicists often use a different unit of energy, the *electron volt,* when dealing with energies at the atomic level. One electron volt, abbreviated eV, is defined as the amount of kinetic energy gained by an electron upon accelerating through a 1.0 V potential difference.
    a. What is 1.0 electron volt in joules?
    b. What is the speed of a proton with 5000 eV of kinetic energy?
63. || The 4000 V equipotential surface is 10.0 cm farther from a positively charged particle than the 5000 V equipotential surface. What is the charge on the particle?
64. ||| A -3.0 nC charge is on the x-axis at $x = -9$ cm and a +4.0 nC charge is on the x-axis at $x = 16$ cm. At what point or points on the y-axis is the electric potential zero?
65. ||| The four 1.0 g spheres shown in Figure P22.65 are released simultaneously and allowed to move away from each other. What is the speed of each sphere when they are very far apart?

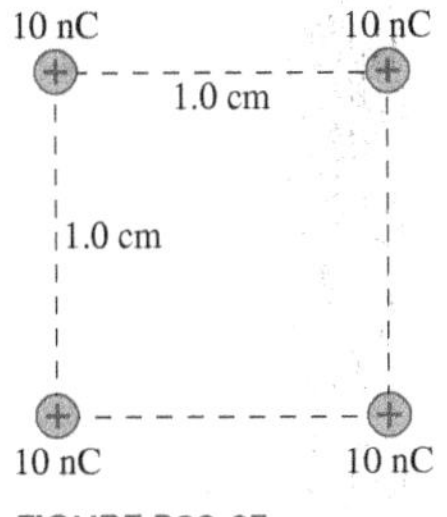

FIGURE P22.65

66. || Living cells "pump" singly ionized sodium ions, Na^+, from **BIO** the inside of the cell to the outside across the membrane potential $\Delta V_{membrane} = V_{in} - V_{out} = -70$ mV. It is called *pumping* because work must be done to move a positive ion from the negative inside of the cell to the positive outside, and it must go on continuously because sodium ions "leak" back through the cell wall by diffusion.
    a. How much work must be done to move one sodium ion from the inside of the cell to the outside?
    b. At rest, the human body uses energy at the rate of approximately 100 W to maintain basic metabolic functions. It has been estimated that 20% of this energy is used to operate the sodium pumps of the body. Estimate—to one significant figure—the number of sodium ions pumped per second.
67. ||| An arrangement of source charges produces the electric **INT** potential $V = 5000x^2$ along the x-axis, where V is in volts and x is in meters. What is the maximum speed of a 1.0 g, 10 nC charged particle that moves in this potential with turning points at ± 8.0 cm?
68. |||| An electric dipole has dipole moment p. If $r \gg s$, where s is the separation between the charges, show that the electric potential of the dipole can be written

$$V = \frac{1}{4\pi\epsilon_0}\frac{p\cos\theta}{r^2}$$

where r is the distance from the center of the dipole and θ is the angle from the dipole axis. You'll need to make two approximations: First, $r \gg s$ allows you to ignore s^2 when it occurs in combination with the squares of other distances. Second, $(1+x)^n \approx 1 + nx$ if $x \ll 1$.
69. || What is the escape speed of an electron launched from the surface of a 1.0-cm-diameter glass sphere that has been charged to 10 nC?
70. || A Van de Graaff generator is a device for generating a large electric potential by building up charge on a hollow metal sphere. A typical classroom-demonstration model has a diameter of 30 cm.
    a. How much charge is needed on the sphere for its potential to be 500,000 V?
    b. What is the electric field strength just outside the surface of the sphere when it is charged to 500,000 V?
71. ||| In the early 1900s, Robert Millikan used small charged **INT** droplets of oil, suspended in an electric field, to make the first quantitative measurements of the electron's charge. A 0.70-μm-diameter droplet of oil, having a charge of $+e$, is suspended in midair between two horizontal plates of a parallel-plate capacitor. The upward electric force on the droplet is exactly balanced by the downward force of gravity. The oil has a density of 860 kg/m^3, and the capacitor plates are 5.0 mm apart. What must the potential difference between the plates be to hold the droplet in equilibrium?
72. ||| A parallel-plate capacitor is charged to 5000 V. A proton is fired into the center of the capacitor at a speed of 3.0×10^5 m/s, as shown in Figure P22.72. The proton is deflected while inside the capacitor, and the plates are long enough that the proton will hit one of them before emerging from the far side of the capacitor. What is the impact speed of the proton?

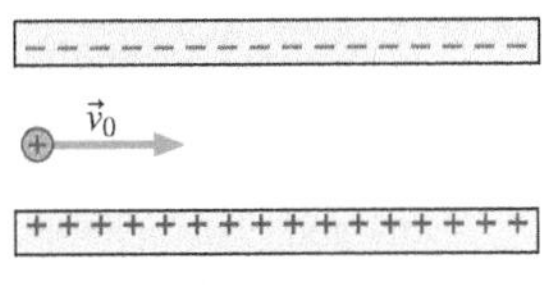

FIGURE P22.72

73. ||| Two 2.0-cm-diameter disks spaced 2.0 mm apart form a parallel-plate capacitor. The electric field between the disks is 5.0×10^5 V/m.
   a. What is the voltage across the capacitor?
   b. How much charge is on each disk?
   c. An electron is launched from the negative plate. It strikes the positive plate at a speed of 2.0×10^7 m/s. What was the electron's speed as it left the negative plate?

74. || In *proton-beam therapy,* a high-energy beam of protons is
BIO fired at a tumor. The protons come to rest in the tumor, depositing their kinetic energy and breaking apart the tumor's DNA, thus killing its cells. For one patient, it is desired that 0.10 J of proton energy be deposited in a tumor. To create the proton beam, the protons are accelerated from rest through a 10 MV potential difference. What is the total charge of the protons that must be fired at the tumor to deposit the required energy?

75. || Researchers measured the strength of the electric field pro-
BIO duced by an electric eel by placing electrodes inside the body of a dead prey fish. In one test, a potential difference of 5.6 V was measured between two electrodes that were 1.0 cm apart.
   a. What was the approximate electric field generated by the eel?
   b. The fish's body was 8.0 cm across. If the electric field was reasonably uniform, approximately what was the potential difference across the body of the fish?

76. |||| a. Find the potential at
CALC distance x on the axis of the charged rod shown in Figure P22.76.
   b. Use the result of part a to find the electric field at distance x on the axis of a rod.

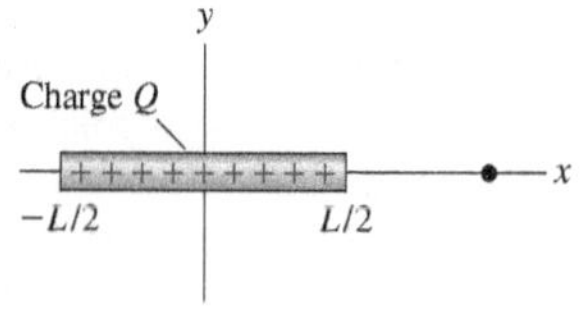

FIGURE P22.76

77. ||| The electric potential in a region of space is
CALC $V = (150x^2 - 200y^2)$ V, where x and y are in meters. What are the strength and direction of the electric field at $(x, y) = (2.0 \text{ m}, 2.0 \text{ m})$? Give the direction as an angle cw or ccw (specify which) from the positive x-axis.

78. | Figure P22.78 shows a series of equipotential curves.
   a. Is the electric field strength at point A larger than, smaller than, or equal to the field strength at point B? Explain.
   b. Is the electric field strength at point C larger than, smaller than, or equal to the field strength at point D? Explain.
   c. Determine the electric field $\vec{E}$ at point D. Express your answer as a magnitude and direction.

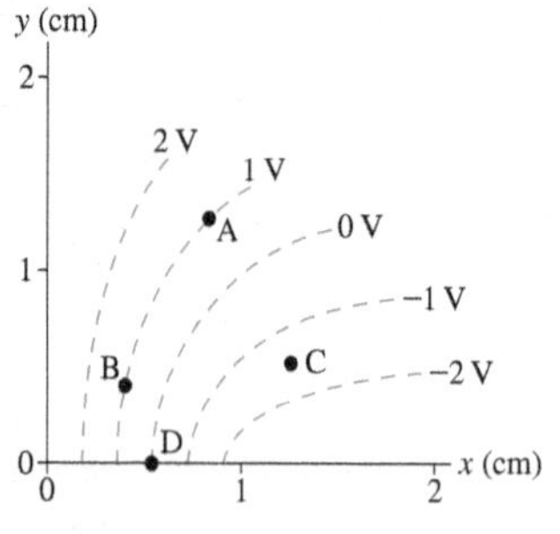

FIGURE P22.78

79. ||| Metal sphere 1 has a positive charge of 6.0 nC. Metal sphere 2, which is twice the diameter of sphere 1, is initially uncharged. The spheres are then connected together by a long, thin metal wire. What are the final charges on each sphere?

80. || ECG electrodes are placed on a patient in the locations shown
BIO in Figure P22.80. At one instant, when the heart's electric dipole moment points straight down, the potential difference $\Delta V = V_2 - V_1$ is 1.8 mV. What is the magnitude of the heart's electric dipole moment?

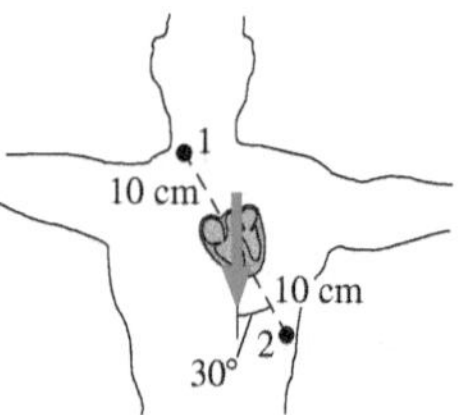

FIGURE P22.80

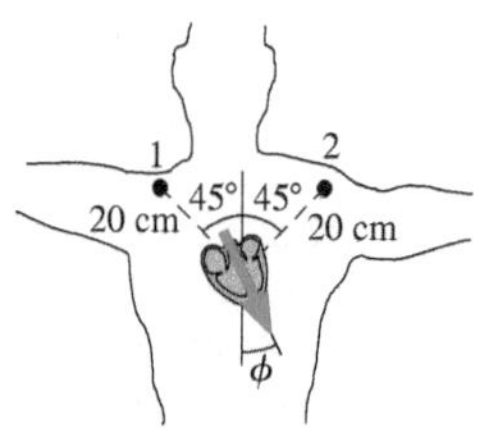

FIGURE P22.81

81. ||| ECG electrodes are placed on a patient in the locations
BIO shown in Figure P22.81. In this patient, the magnitude of the heart's electric dipole moment is 1.1×10^{-13} C · m. At one instant of the heartbeat, the potential difference $\Delta V = V_2 - V_1$ is 0.18 mV. At that instant, what angle ϕ does the electric dipole moment make with a vertical axis?

MCAT-Style Passage Problems

Mass Spectroscopy

Mass spectrometers are important tools in chemistry and biochemistry for identifying the molecules in a sample. Every molecule has a mass, often unique to that molecule. A mass spectrometer ionizes the molecules and then uses electric and/or magnetic fields to sort the molecules by mass. A graph showing the number of molecules that have each molecular mass is called a *mass spectrum.* One type of mass spectrometer, shown in Figure P22.82, is a *time-of-flight mass spectrometer.* A very short, intense pulse of laser light ionizes a sample of molecules very close to an electrode that is held at a potential of +18 kV. All the newly formed ions simultaneously begin to accelerate across a 1 cm gap toward a *grid*—a loose mesh of metal wire—that is at 0 V. Most ions pass through the grid and enter a 1-m-long *drift tube* in which the electric field is zero. Ions are detected when they reach a detector at the far end of the drift tube, and their time of flight since the laser pulse is recorded.

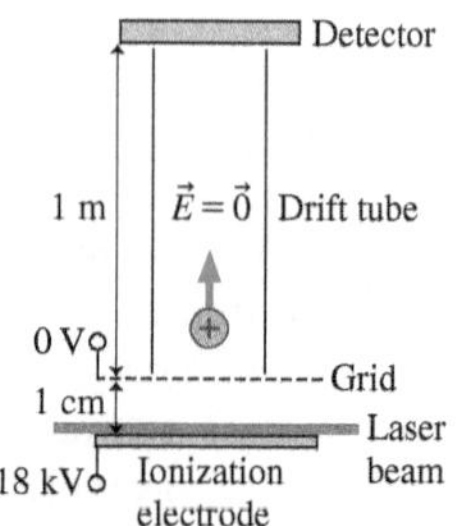

FIGURE P22.82

82. Which ions have the shortest time of flight and arrive at the detector first?
   A. Ions with the smallest mass.
   B. Ions with the largest mass.
   C. Ions of all masses arrive at the same time.

83. A carbon dioxide ion CO_2^+ has mass 44 u, where 1 u = 1 atomic mass unit = 1.66×10^{-27} kg. What is the approximate speed with which a carbon dioxide ion travels through the drift tube?
   A. 9×10^3 m/s
   B. 9×10^4 m/s
   C. 1×10^5 m/s
   D. 3×10^5 m/s

84. An ion would accelerate through the same potential difference if the ionization electrode were at 0 V and the grid at −18 kV. Why is this never done?
   A. Because it would detect negative ions instead of positive ions.
   B. Because placing a high voltage on the grid would be dangerous.
   C. Because it would create an electric field inside the drift tube.
   D. Because the ions would have negative potential energy.

23 Biological Applications of Electric Fields and Potentials

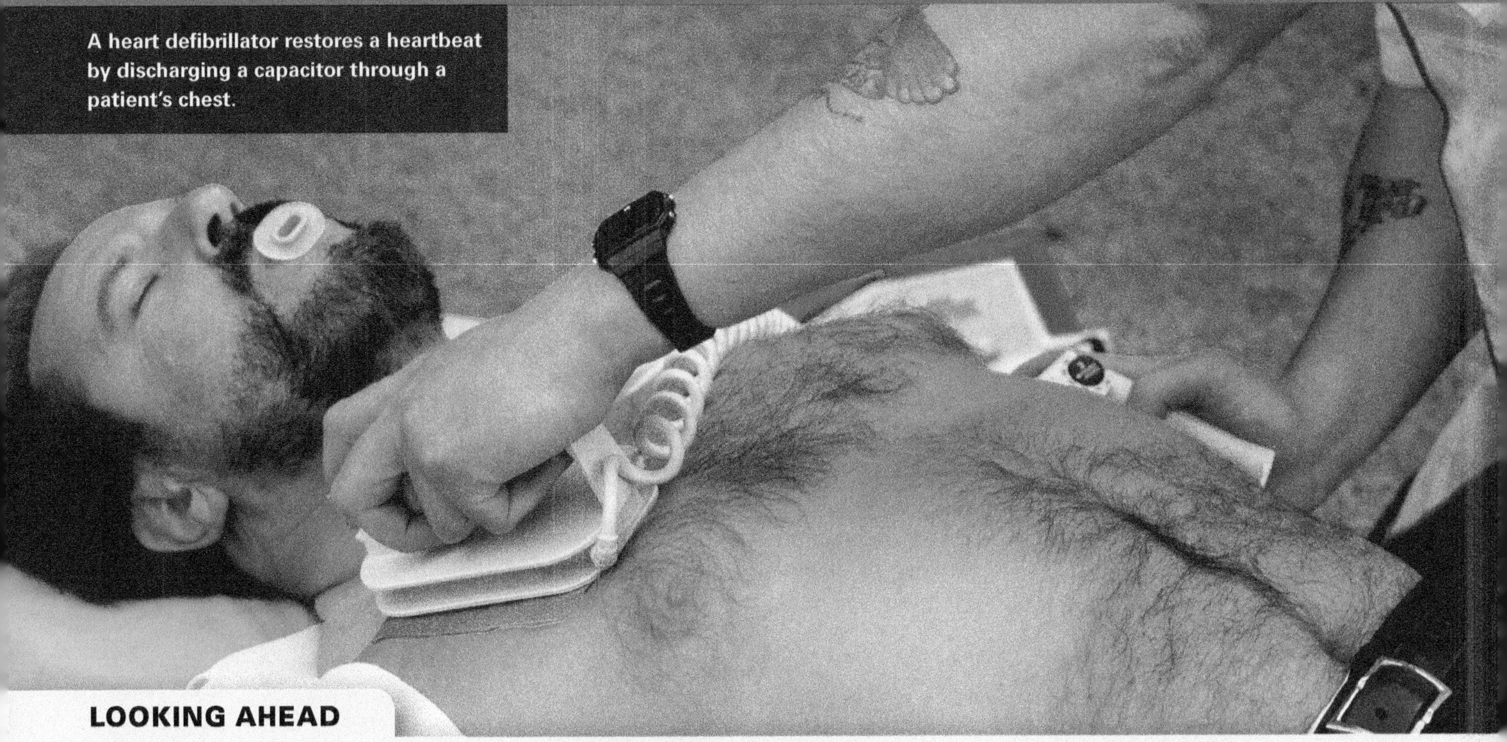

A heart defibrillator restores a heartbeat by discharging a capacitor through a patient's chest.

LOOKING AHEAD

Capacitors

A cardiac pacemaker depends on *capacitors*—devices that store charge and energy—to control abnormal heart rhythms.

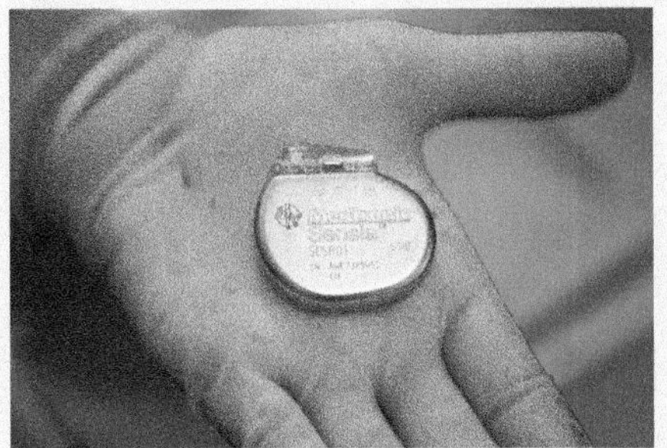

You'll learn how capacitors store charge and energy and how to calculate the equivalent capacitance of a combination of capacitors.

Electrostatics in Salt Water

Gel electrophoresis, an essential tool of modern biology, is able to sort fragments of protein and DNA because macromolecules are charged.

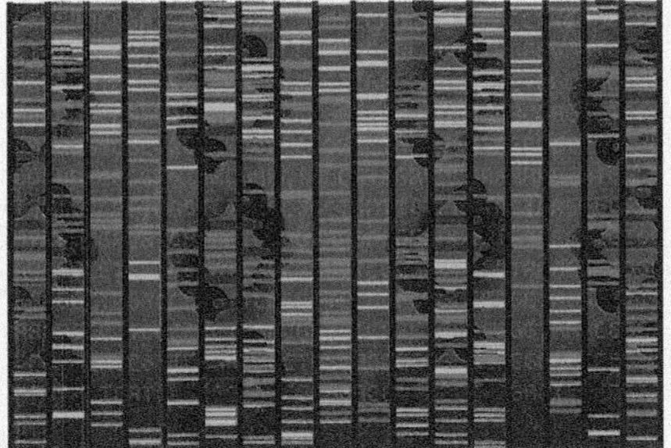

You'll learn why biological molecules are charged and how electrostatic interactions are reduced in salt water.

Membrane Potential

This researcher is studying how the flow of ions into and out of cells controls the potential difference across the cell membrane.

You'll learn how to calculate membrane potentials and the rate at which ions move through ion channels in the cell wall.

GOAL To study some of the applications of electric fields and potentials.

PHYSICS AND LIFE

Electric Interactions in the Cell Can Be Highly Specific

The lock-and-key interactions between an enzyme and a substrate are essential to biochemistry. These highly selective interactions are fundamentally electrical: The two molecules bind to each other in a very specific orientation by means of matching patterns of opposite charges on their surfaces. But given the strength of electric interactions, why don't all charged molecules in a cell interact electrically with each other even if they don't match in detail? Because dissolved ions in biological fluids cluster around charged molecules, screening the molecule's charge and turning the long-range electric force into a short-range force. This reduction in the range of the electric force allows charged molecules to diffuse freely throughout the cell until, by random motion, they come into close contact with a partner molecule that has just the right pattern of surface charges.

An atomic force microscope image shows ion channels and other proteins protruding above a cell membrane.

23.1 Capacitance and Capacitors

FIGURE 23.1 Two oppositely charged electrodes form a capacitor.

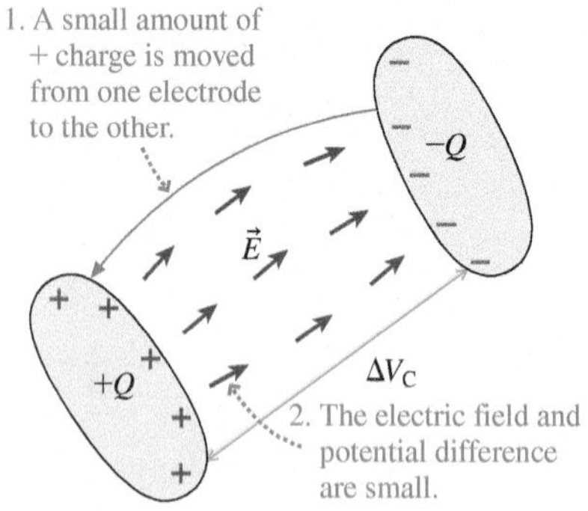
1. A small amount of + charge is moved from one electrode to the other.
−Q
$\vec{E}$
+Q
ΔV_C
2. The electric field and potential difference are small.

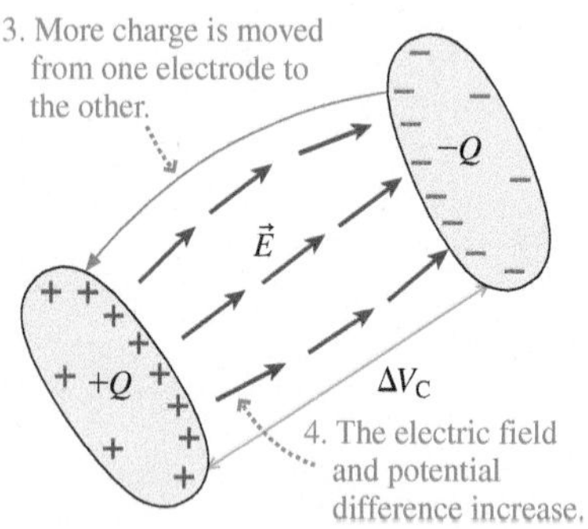
3. More charge is moved from one electrode to the other.
−Q
$\vec{E}$
+Q
ΔV_C
4. The electric field and potential difference increase.

Chapters 21 and 22 introduced the electric field and the electric potential. In this chapter, we will use electric fields and potentials to understand both practical devices and physiological processes. We will begin with *capacitors,* which are used to store charge and energy, and end with a detailed look at the potential difference that is generated across a cell membrane as ions in solution attempt to minimize their energy while maximizing their entropy.

In ◀◀ SECTION 22.5 we found that a potential difference is caused by the separation of charge. One common method of creating a charge separation, shown in FIGURE 23.1, is to move charge Q from one initially uncharged conductor to a second initially uncharged conductor. This results in charge $+Q$ on one conductor and $-Q$ on the other. Two conductors with equal but opposite charge form a **capacitor.** The two conductors that make up a capacitor are its *electrodes,* or *plates.* We've already looked at some of the properties of parallel-plate capacitors, but capacitors can come in any shape. All that's needed is to have two electrodes separated by a region of space or an insulating material.

As Figure 23.1 shows, the electric field strength E and the potential difference ΔV_C increase as the charge on each electrode increases. Doubling the charge moved from one electrode to the other doubles the field strength between the electrodes and thus doubles the potential difference. Stated another way, **the charge stored on a capacitor is directly proportional to the potential difference between its electrodes.** We can write the relationship between charge and potential difference as

$$Q = C\,\Delta V_C \tag{23.1}$$

Charge on a capacitor with potential difference ΔV_C

The constant of proportionality C between Q and ΔV_C is called the **capacitance** of the capacitor. Capacitance depends on the shape, size, and spacing of the two electrodes. A capacitor with a large capacitance holds more charge for a given potential difference than one with a small capacitance.

The SI unit of capacitance is the **farad.** One farad is defined as

$$1 \text{ farad} = 1 \text{ F} = 1 \text{ coulomb/volt} = 1 \text{ C/V}$$

One farad is an enormous amount of capacitance. The capacitors used in electric circuits typically range from picofareds (pF) to microfarads (μF).

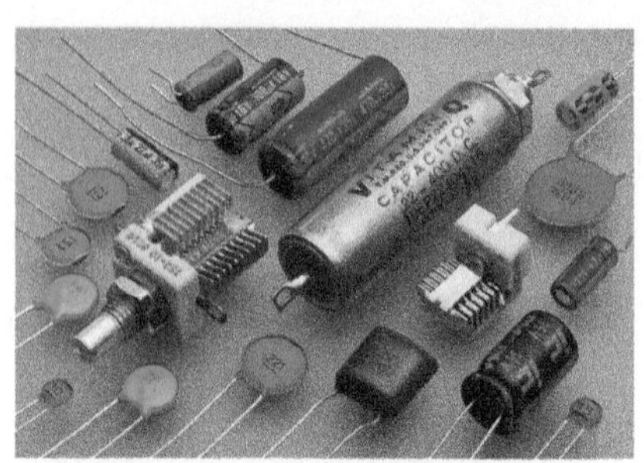

Capacitors are practical devices
Capacitors are important elements in electric circuits. They come in a wide variety of sizes and shapes.

Charging a Capacitor

To "charge" a capacitor, we need to move charge from one electrode to the other. The simplest way to do this is to use a source of potential difference such as a battery, as shown in FIGURE 23.2. We learned earlier that a battery uses its internal chemistry to maintain a fixed potential difference between its terminals. If we connect a capacitor to a battery, charge flows from the negative electrode of the capacitor, through the battery, and onto the positive electrode. This flow of charge continues until the potential difference between the capacitor's electrodes is the same as the fixed potential difference of the battery. If the battery is then removed, the capacitor remains charged with a potential equal to that of the battery that charged it because there's no conducting path for charge on the positive electrode to move back to the negative electrode. Thus **a capacitor can be used to store charge.**

FIGURE 23.2 Charging a capacitor using a battery.

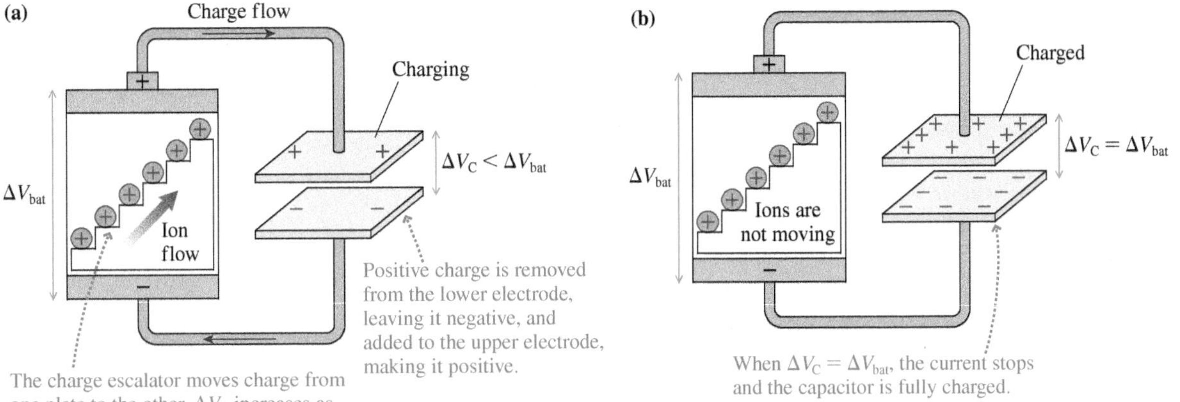
(a)
Charge flow
Charging
$\Delta V_C < \Delta V_{bat}$
ΔV_{bat}
Ion flow
Positive charge is removed from the lower electrode, leaving it negative, and added to the upper electrode, making it positive.
The charge escalator moves charge from one plate to the other. ΔV_C increases as the charge separation increases.
(b)
Charged
$\Delta V_C = \Delta V_{bat}$
ΔV_{bat}
Ions are not moving
When $\Delta V_C = \Delta V_{bat}$, the current stops and the capacitor is fully charged.

(c)

If the battery is removed, the capacitor remains charged, with ΔV_C still equal to the battery voltage.

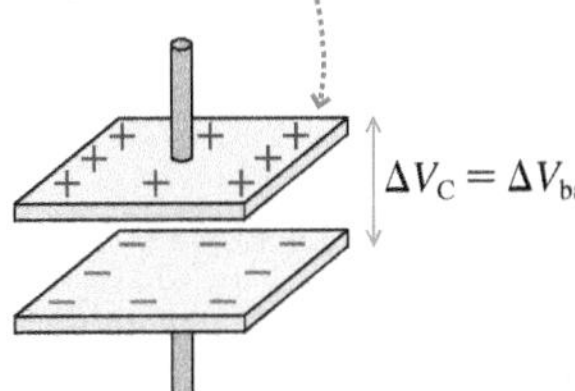
$\Delta V_C = \Delta V_{bat}$

EXAMPLE 23.1 **Charging a capacitor**

A 1.3 μF capacitor is connected to a 1.5 V battery. What is the charge on the capacitor?

PREPARE Charge flows through the battery from one capacitor electrode to the other until the potential difference ΔV_C between the electrodes equals that of the battery, or 1.5 V.

SOLVE The charge on the capacitor is given by Equation 23.1:

$$Q = C\,\Delta V_C = (1.3 \times 10^{-6}\ \text{F})(1.5\ \text{V}) = 2.0 \times 10^{-6}\ \text{C}$$

ASSESS The battery is only 1.5 V, but this charge is roughly 1000 times larger than the charge obtained by rubbing or friction.

The Parallel-Plate Capacitor

As we've seen, the *parallel-plate capacitor* is important because it creates a uniform electric field between its flat electrodes. In ◀◀ SECTION 21.5, we found that the electric field of a parallel-plate capacitor is

$$\vec{E} = \left(\frac{Q}{\epsilon_0 A}, \text{ from positive to negative}\right)$$

where A is the surface area of the electrodes and Q is the charge on the capacitor. We can use this result to relate the capacitance of a parallel-plate capacitor to the geometry of its plates.

Capacitor switches The keys on computer keyboards are capacitor switches. Pressing the key pushes two capacitor plates closer together, increasing their capacitance. A larger capacitor can hold more charge, so a momentary current carries charge from the battery (or power supply) to the capacitor. This current is sensed, and the keystroke is then recorded.

In Chapter 22, we found that the electric field strength of a parallel-plate capacitor is related to the potential difference ΔV_C and the plate spacing d by

$$E = \frac{\Delta V_C}{d}$$

Combining these two results, we see that

$$\frac{Q}{\epsilon_0 A} = \frac{\Delta V_C}{d}$$

or, equivalently,

$$Q = \frac{\epsilon_0 A}{d}\Delta V_C \tag{23.2}$$

If we compare Equation 23.2 to Equation 23.1, the definition of capacitance, we see that the capacitance of the parallel-plate capacitor is

$$C = \frac{\epsilon_0 A}{d} \tag{23.3}$$

Capacitance of a parallel-plate capacitor with plate area A and separation d

NOTE ► From Equation 23.3 we can see that the units of ϵ_0 can be written as F/m. These units are useful when working with capacitors. ◄

EXAMPLE 23.2 Making a parallel-plate capacitor

Suppose you want to make a 1.0 μF capacitor using square plates separated by 1.0 mm. How large do the plates need to be?

PREPARE The capacitance depends on the area and the spacing. The area of the plates is the square of the edge length L: $A = L^2$.

SOLVE Rearranging Equation 23.3, we solve for the area of the plates and thus the edge length:

$$A = L^2 = \frac{dC}{\epsilon_0}$$

$$L = \left(\frac{(0.0010 \text{ m})(1.0 \times 10^{-6} \text{ F})}{(8.85 \times 10^{-12} \text{ F/m})}\right)^{1/2} = 11 \text{ m}$$

ASSESS The plates would need to be enormous, so this is not a feasible way to make a capacitor. We'll look at how capacitors are actually made later in the chapter.

FIGURE 23.3 The charge escalator does work on charge dq as the capacitor is being charged.

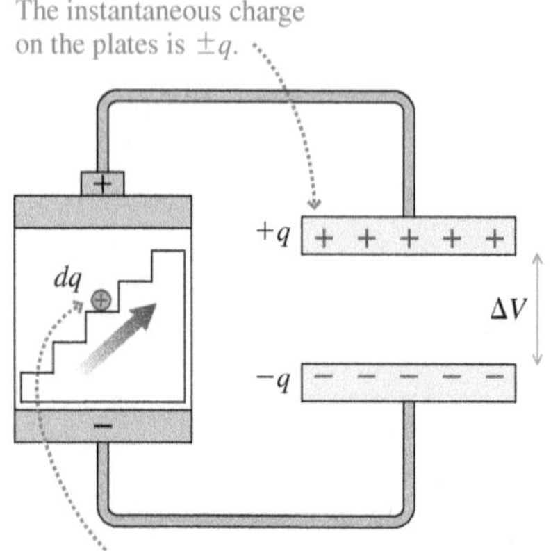
The instantaneous charge on the plates is $\pm q$.
+
$+q$
+ + + + +
dq
ΔV
$-q$
− − − − −
−
The charge escalator does work $dq\,\Delta V$ to move charge dq from the negative plate to the positive plate.

STOP TO THINK 23.1 If the potential difference across a capacitor is doubled, its capacitance

A. Doubles. B. Halves. C. Remains the same.

The Energy Stored in a Capacitor

Capacitors are important elements in electric circuits because of their ability to store not only charge but also energy. FIGURE 23.3 shows a capacitor being charged. The instantaneous value of the charge on the two plates is $\pm q$, and this charge separation has established a potential difference $\Delta V = q/C$ between the two electrodes.

An additional charge dq is in the process of being transferred from the negative to the positive electrode. The battery's charge escalator must do work to lift charge dq "uphill" to a higher potential. Consequently, the potential energy of the dq + capacitor system increases by

$$dU = dq\,\Delta V = \frac{q\,dq}{C} \tag{23.4}$$

as dq moves onto the positive electrode.

NOTE ▶ Energy must be conserved. This increase in the capacitor's potential energy is provided by the battery. ◀

The total energy transferred from the battery to the capacitor is found by integrating Equation 23.4 from the start of charging, when $q = 0$, until the end, when $q = Q$. Thus we find that the energy stored in a charged capacitor is

$$U_C = \frac{1}{C}\int_0^Q q\,dq = \frac{Q^2}{2C} \tag{23.5}$$

In practice, it is often easier to write the stored energy in terms of the capacitor's potential difference $\Delta V_C = Q/C$. This is

$$U_C = \frac{Q^2}{2C} = \tfrac{1}{2}C(\Delta V_C)^2 \tag{23.6}$$

The potential energy stored in a capacitor depends on the *square* of the potential difference across it. This result is reminiscent of the potential energy $U = \frac{1}{2}k(\Delta x)^2$ stored in a spring, and a charged capacitor really is analogous to a stretched spring. A stretched spring holds the energy until we release it, then that potential energy is transformed into kinetic energy. Likewise, a charged capacitor holds energy until we discharge it. Then the potential energy is transformed into the kinetic energy of moving charges (the current).

EXAMPLE 23.3 **Storing energy in a capacitor**

How much energy is stored in a 220 μF camera-flash capacitor that has been charged to 330 V? What is the average power dissipation if this capacitor is discharged in 1.0 ms?

PREPARE We can use Equation 23.6 to calculate the energy stored. Power is the *rate* at which the energy is dissipated.

SOLVE The energy stored in the charged capacitor is

$$U_C = \tfrac{1}{2}C(\Delta V_C)^2 = \tfrac{1}{2}(220 \times 10^{-6}\ \text{F})(330\ \text{V})^2 = 12\ \text{J}$$

If this energy is released in 1.0 ms, the average power dissipation is

$$P = \frac{\Delta E}{\Delta t} = \frac{12\ \text{J}}{1.0 \times 10^{-3}\ \text{s}} = 12{,}000\ \text{W}$$

ASSESS The stored energy is equivalent to raising a 1 kg mass 1.2 m. This is a rather large amount of energy, which you can see by imagining the damage a 1 kg mass could do after falling 1.2 m. When this energy is released very quickly, which is possible in an electric circuit, it provides an *enormous* amount of power.

Capacitors Can Release Energy Rapidly

The usefulness of a capacitor stems from the fact that it can be charged slowly (the charging rate is usually limited by the battery's ability to transfer charge) but then can release the energy very quickly. A mechanical analogy would be using a crank to slowly stretch the spring of a catapult, then quickly releasing the energy to launch a massive rock.

The capacitor described in Example 23.3 is typical of the capacitors used to power a camera flash. The camera batteries charge a capacitor, then the energy stored in the capacitor is quickly discharged into a *flashlamp.* The charging process in a camera takes several seconds, which is why you can't fire a camera flash twice in quick succession.

An important medical application of capacitors is the *defibrillator.* A heart attack or a serious injury can cause the heart to enter a state known as *fibrillation* in which

Taking a picture in a flash When you take a flash picture, the flash is fired using electric potential energy stored in a capacitor. Batteries are unable to deliver the required energy rapidly enough, but capacitors can discharge all their energy in only microseconds. A battery is used to slowly charge up the capacitor, which then rapidly discharges through the flash lamp.

the heart muscles twitch randomly and cannot pump blood. A strong electric shock through the chest completely stops the heart, giving the cells that control the heart's rhythm a chance to restore the proper heartbeat. A defibrillator, like the one shown in the chapter-opening photo, has a large capacitor that can store up to 360 J of energy. This energy is released in about 2 ms through two "paddles" pressed against the patient's chest. It takes several seconds to charge the capacitor, which is why, on television medical shows, you hear an emergency room doctor or nurse shout "Charging!"

The Energy in the Electric Field

We can "see" the potential energy of a stretched spring in the tension of the coils. If a charged capacitor is analogous to a stretched spring, where is the stored energy? It's in the electric field!

FIGURE 23.4 shows a parallel-plate capacitor in which the plates have area A and are separated by distance d. The potential difference across the capacitor is related to the electric field inside the capacitor by $\Delta V_C = Ed$. The capacitance, which we found in Equation 23.3, is $C = \epsilon_0 A/d$. Substituting these into Equation 23.6, we find that the energy stored in the capacitor is

$$U_C = \tfrac{1}{2}C(\Delta V_C)^2 = \frac{\epsilon_0 A}{2d}(Ed)^2 = \frac{\epsilon_0}{2}(Ad)E^2 \tag{23.7}$$

FIGURE 23.4 A capacitor's energy is stored in the electric field.

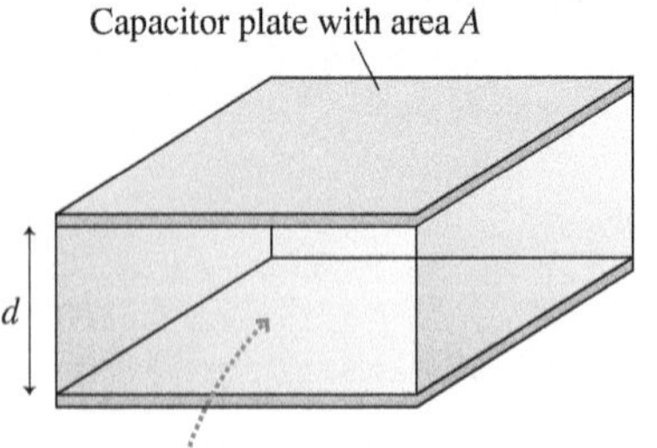
Capacitor plate with area A
d
The capacitor's energy is stored in the electric field in volume Ad between the plates.

The quantity Ad is the volume *inside* the capacitor, the region in which the capacitor's electric field exists. (Recall that an ideal capacitor has $\vec{E} = \vec{0}$ everywhere except between the plates.) Although we talk about "the energy stored in the capacitor," Equation 23.7 suggests that, strictly speaking, **the energy is stored in the capacitor's electric field.**

Because Ad is the volume in which the energy is stored, we can define an **energy density** u_E of the electric field:

$$u_E = \frac{\text{energy stored}}{\text{volume in which it is stored}} = \frac{U_C}{Ad} = \frac{\epsilon_0}{2}E^2 \tag{23.8}$$

The energy density has units J/m^3. We've derived Equation 23.8 for a parallel-plate capacitor, but it turns out to be the correct expression for any electric field.

From this perspective, charging a capacitor stores energy in the capacitor's electric field as the field grows in strength. Later, when the capacitor is discharged, the energy is released as the field collapses.

We first introduced the electric field as a way to visualize how a long-range force operates. But if the field can store energy, the field must be real, not merely a pictorial device. We'll explore this idea further in Chapter 27, where we'll find that the energy transported by a light wave—the very real energy of warm sunshine—is the energy of electric and magnetic fields.

23.2 Combinations of Capacitors

Two or more capacitors are often joined together. FIGURE 23.5 illustrates two basic combinations: **parallel capacitors** and **series capacitors.** Notice that a capacitor, no matter what its actual geometric shape, is represented in *circuit diagrams* by two parallel lines.

NOTE ► The terms "parallel capacitors" and "parallel-plate capacitor" do not describe the same thing. The former term describes how two or more capacitors are connected to each other, the latter describes how a particular capacitor is constructed. ◄

As we'll show, parallel or series capacitors (or, as is sometimes said, capacitors "in parallel" or "in series") can be represented by a single **equivalent capacitance.** We'll demonstrate this first with the two parallel capacitors C_1 and C_2 of FIGURE 23.6a.

FIGURE 23.5 Parallel and series capacitors.

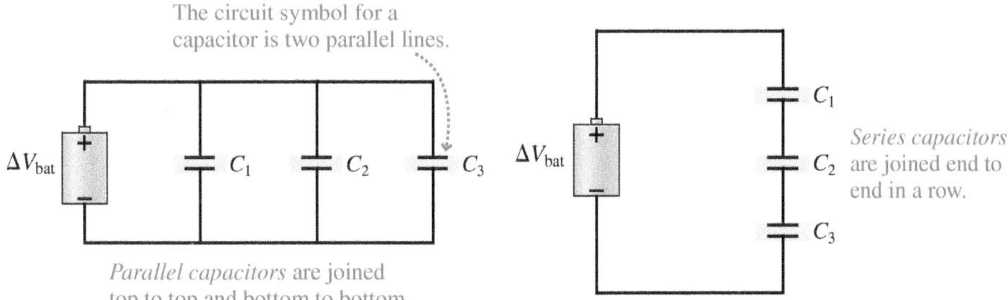

Because the two top electrodes are connected by a conducting wire, they form a single conductor in electrostatic equilibrium. Thus the two top electrodes are at the same potential. Similarly, the two connected bottom electrodes are at the same potential. Consequently, two (or more) capacitors in parallel each have the *same* potential difference ΔV_C between the two electrodes.

The charges on the two capacitors are $Q_1 = C_1\,\Delta V_C$ and $Q_2 = C_2\,\Delta V_C$. Altogether, the battery's charge escalator moved total charge $Q = Q_1 + Q_2$ from the negative electrodes to the positive electrodes. Suppose, as in FIGURE 23.6b, we replaced the two capacitors with a single capacitor having charge $Q = Q_1 + Q_2$ and potential difference ΔV_C. This capacitor is equivalent to the original two in the sense that the battery can't tell the difference. In either case, the battery has to establish the same potential difference and move the same amount of charge.

By definition, the capacitance of this equivalent capacitor is

$$C_{eq} = \frac{Q}{\Delta V_C} = \frac{Q_1 + Q_2}{\Delta V_C} = \frac{Q_1}{\Delta V_C} + \frac{Q_2}{\Delta V_C} = C_1 + C_2 \tag{23.9}$$

This analysis hinges on the fact that **parallel capacitors each have the same potential difference ΔV_C.** We could easily extend this analysis to more than two capacitors. If capacitors $C_1, C_2, C_3, \ldots$ are in parallel, their equivalent capacitance is

$$C_{eq} = C_1 + C_2 + C_3 + \cdots \tag{23.10}$$

Equivalent capacitance of parallel capacitors

Neither the battery nor any other part of a circuit can tell if the parallel capacitors are replaced by a single capacitor having capacitance C_{eq}.

Now consider the two series capacitors in FIGURE 23.7a. The center section, consisting of the bottom plate of C_1, the top plate of C_2, and the connecting wire, is electrically isolated. The battery cannot remove charge from or add charge to this section. If it starts out with no net charge, it must end up with no net charge. As a consequence, the two capacitors in series have equal charges $\pm Q$. The battery transfers Q from the bottom of C_2 to the top of C_1. This transfer polarizes the center section, as shown, but it still has $Q_{net} = 0$.

The potential differences across the two capacitors are $\Delta V_1 = Q/C_1$ and $\Delta V_2 = Q/C_2$. The total potential difference across both capacitors is $\Delta V_C = \Delta V_1 + \Delta V_2$. Suppose, as in FIGURE 23.7b, we replaced the two capacitors with a single capacitor having charge Q and potential difference $\Delta V_C = \Delta V_1 + \Delta V_2$. This capacitor is equivalent to the original two because the battery has to establish the same potential difference and move the same amount of charge in either case.

By definition, the capacitance of this equivalent capacitor is $C_{eq} = Q/\Delta V_C$. The inverse of the equivalent capacitance is thus

$$\frac{1}{C_{eq}} = \frac{\Delta V_C}{Q} = \frac{\Delta V_1 + \Delta V_2}{Q} = \frac{\Delta V_1}{Q} + \frac{\Delta V_2}{Q} = \frac{1}{C_1} + \frac{1}{C_2} \tag{23.11}$$

FIGURE 23.6 Replacing two parallel capacitors with an equivalent capacitor.

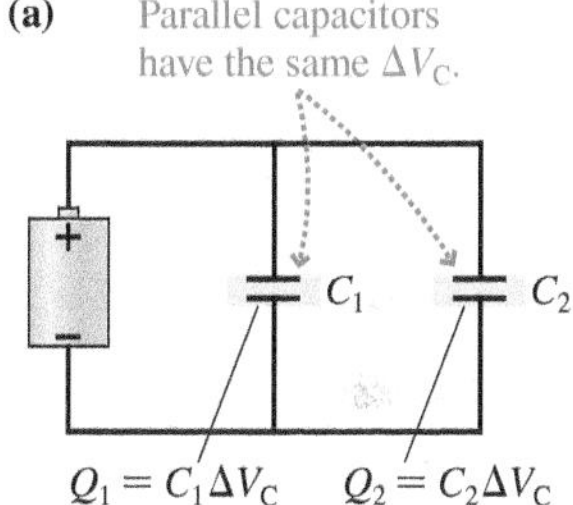

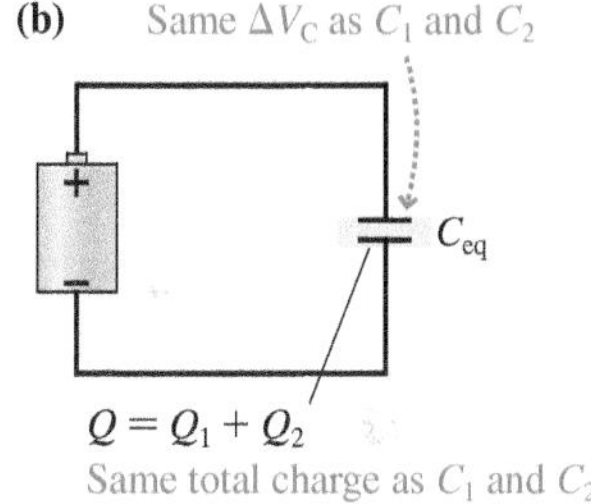

FIGURE 23.7 Replacing two series capacitors with an equivalent capacitor.

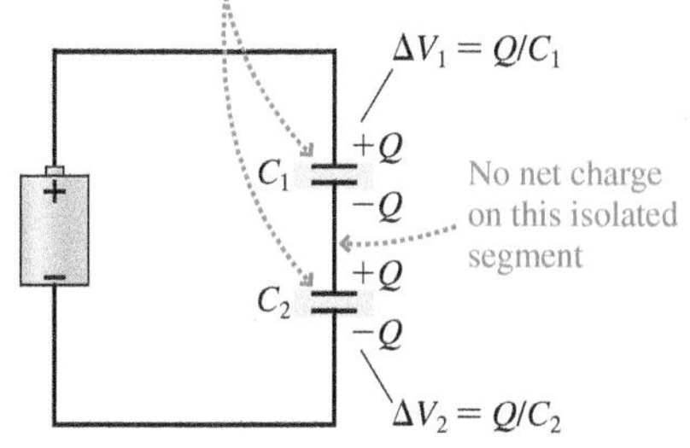

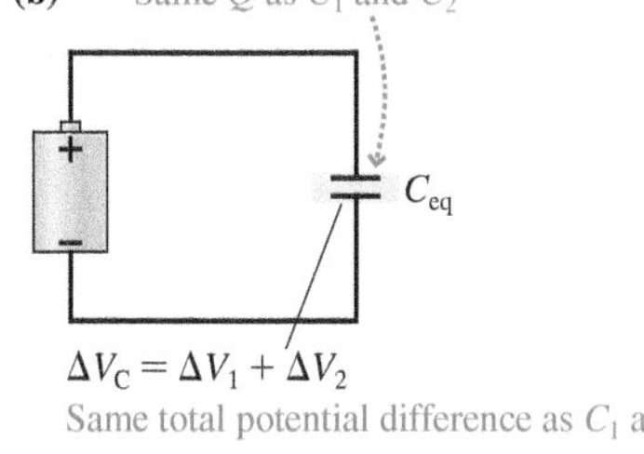

This analysis hinges on the fact that **series capacitors each have the same charge Q.** We could easily extend this analysis to more than two capacitors. If capacitors $C_1, C_2, C_3, \ldots$ are in series, their equivalent capacitance is

$$C_{eq} = \left(\frac{1}{C_1} + \frac{1}{C_2} + \frac{1}{C_3} + \cdots\right)^{-1} \quad (23.12)$$

Equivalent capacitance of series capacitors

NOTE ▶ After you add the inverses, be sure to invert the sum! ◀

Let's summarize the key facts before looking at a numerical example:

- Parallel capacitors all have the same potential difference ΔV_C. Series capacitors all have the same amount of charge $\pm Q$.
- The equivalent capacitance of a parallel combination of capacitors is *larger* than any single capacitor in the group. The equivalent capacitance of a series combination of capacitors is *smaller* than any single capacitor in the group.

EXAMPLE 23.4 A capacitor circuit

Find the charge on and the potential difference across each of the three capacitors in FIGURE 23.8.

FIGURE 23.8 A capacitor circuit.

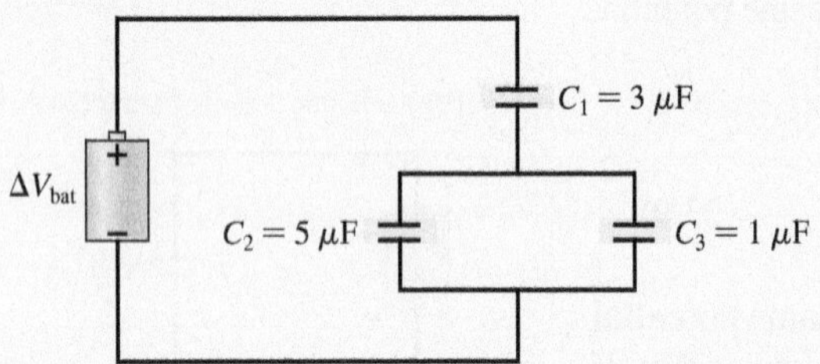

PREPARE We will use the results for parallel and series capacitors.

SOLVE The three capacitors are neither in parallel nor in series, but we can break them down into smaller groups that are. A useful method of *circuit analysis* is first to combine elements until reaching a single equivalent element, then to reverse the process and calculate values for each element. FIGURE 23.9 shows the analysis of this circuit. Notice that we redraw the circuit after every step. The equivalent capacitance of the 3 μF and 6 μF capacitors in series is found from

$$C_{eq} = \left(\frac{1}{3\ \mu F} + \frac{1}{6\ \mu F}\right)^{-1} = \left(\frac{2}{6} + \frac{1}{6}\right)^{-1} \mu F = 2\ \mu F$$

Once we get to the single equivalent capacitance, we find that $\Delta V_C = \Delta V_{bat} = 12$ V and $Q = C\,\Delta V_C = 24\ \mu C$. Now we can reverse direction. Capacitors in series all have the same charge, so the charge on C_1 and on C_{2+3} is $\pm 24\ \mu C$. This is enough to determine that $\Delta V_1 = 8$ V and $\Delta V_{2+3} = 4$ V. Capacitors in parallel all have the same potential difference, so $\Delta V_2 = \Delta V_3 = 4$ V. This is enough to find that $Q_2 = 20\ \mu C$ and $Q_3 = 4\ \mu C$. The charge on and the potential difference across each of the three capacitors are shown in the final step of Figure 23.9.

ASSESS Notice that we had two important checks of internal consistency. $\Delta V_1 + \Delta V_{2+3} = 8\text{ V} + 4\text{ V}$ add up to the 12 V we had found for the 2 μF equivalent capacitor. Then $Q_2 + Q_3 = 20\ \mu C + 4\ \mu C$ add up to the 24 μC we had found for the 6 μF equivalent capacitor. We'll do much more circuit analysis of this type in Chapter 25, but it's worth noting now that circuit analysis becomes nearly foolproof *if* you make use of these checks of internal consistency.

FIGURE 23.9 Analyzing the capacitor circuit.

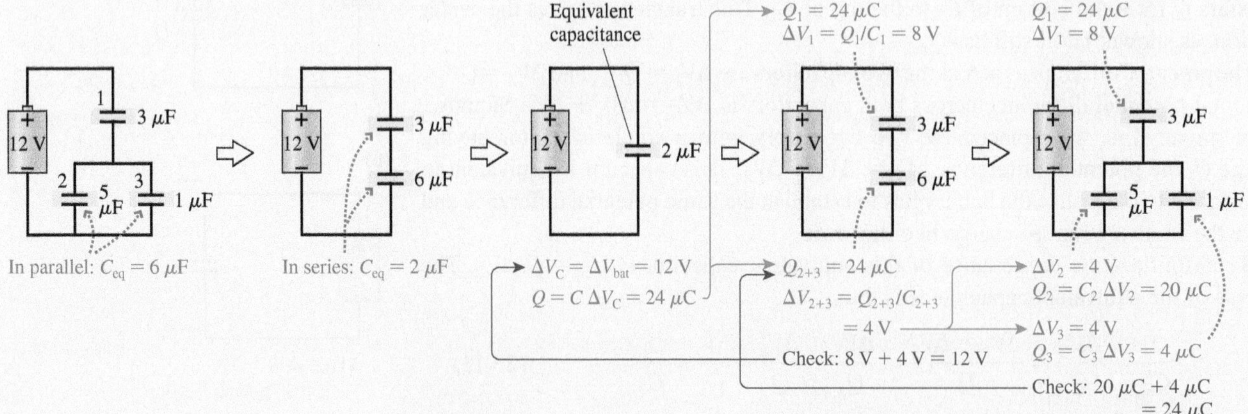

STOP TO THINK 23.2 Rank in order, from largest to smallest, the equivalent capacitance $(C_{eq})_A$ to $(C_{eq})_D$ of circuits A to D.

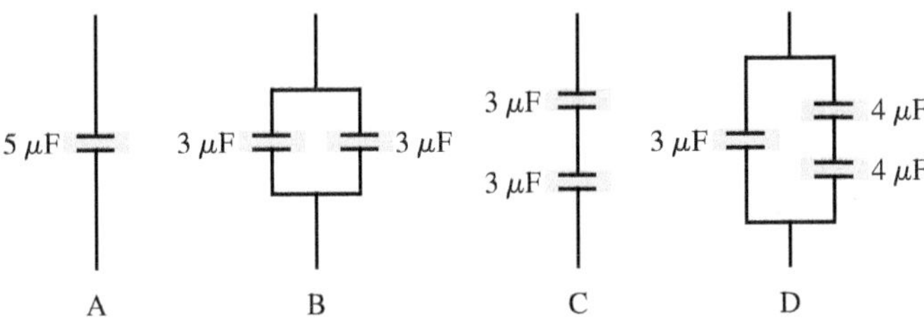

23.3 Dielectrics

FIGURE 23.10a shows a parallel-plate capacitor with the plates separated by vacuum, the perfect insulator. Suppose the capacitor is charged to voltage $(\Delta V_C)_0$, then *disconnected from the battery*. The charge on the plates will be $\pm Q_0$, where $Q_0 = C_0(\Delta V_C)_0$. We'll use a subscript 0 in this section to refer to a vacuum-insulated capacitor.

FIGURE 23.10 Vacuum-insulated and dielectric-filled capacitors.

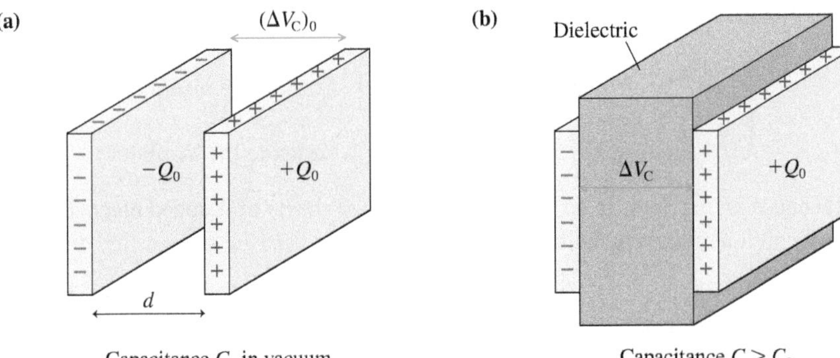

Now suppose, as in **FIGURE 23.10b**, an insulating material, such as oil or glass or plastic, is slipped between the capacitor plates. We'll assume for now that the insulator is of thickness d and completely fills the space. An insulator in an electric field is called a **dielectric,** for reasons that will soon become clear, so we call this a *dielectric-filled capacitor.* How does a dielectric-filled capacitor differ from the vacuum-insulated capacitor?

The charge on the capacitor plates does not change. The insulator doesn't allow charge to move through it, and the capacitor has been disconnected from the battery, so no charge can be added to or removed from either plate. That is, $Q = Q_0$. Nonetheless, measurements of the capacitor voltage with a voltmeter would find that the voltage has decreased: $\Delta V_C < (\Delta V_C)_0$. Consequently, based on the definition of capacitance, the capacitance has increased:

$$C = \frac{Q}{\Delta V_C} > \frac{Q_0}{(\Delta V_C)_0} = C_0$$

Example 23.2 found that the plate size needed to make a 1 μF capacitor is unreasonably large. It appears that we can get more capacitance *with the same plates* by filling the capacitor with an insulator.

We can utilize two tools you learned in Chapter 21, superposition and polarization, to understand the properties of dielectric-filled capacitors. Figure 21.43 showed how an insulating material becomes *polarized* in an external electric field. **FIGURE 23.11a** reproduces the basic ideas from that earlier figure. The electric dipoles

FIGURE 23.11 An insulator in an external electric field.

(a) The insulator is polarized.

Excess positive charge on this surface

Excess negative charge on this surface

$\vec{E}_0$

$\vec{E}_0$

(b) The polarized insulator—a dielectric—can be represented as two sheets of surface charge. This surface charge creates an electric field inside the insulator.

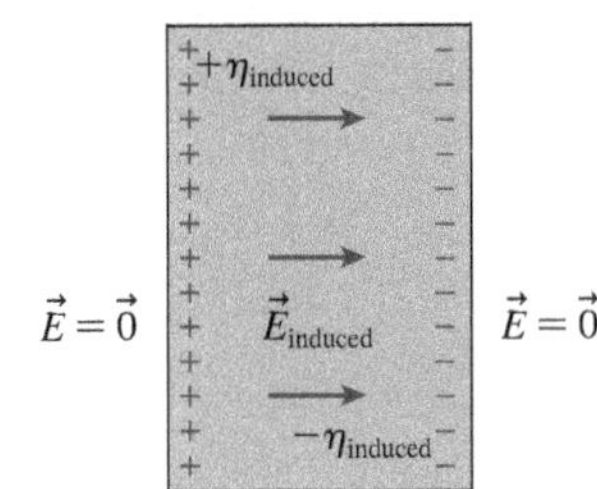

in Figure 23.11a could be permanent dipoles, such as water molecules, or simply induced dipoles due to a slight charge separation in the atoms. However the dipoles originate, their alignment in the electric field—the *polarization* of the material—causes there to be an excess positive charge on one surface, an excess negative charge on the other. This is called *induced charge.* The insulator as a whole is still neutral, but the external electric field separates positive and negative charge.

FIGURE 23.11b represents the polarized insulator as two sheets of induced charge on the surfaces. These two sheets of charge create an electric field—a situation we analyzed in Chapter 21. In essence, the two sheets of induced charge are equivalent to the two charged plates of a parallel-plate capacitor, and we know that a parallel-plate capacitor has a uniform electric field of strength $Q/\epsilon_0 A$, where A is the surface area of the electrodes. We see that the field between the plates of a capacitor depends not on the total charge Q but on the charge-to-area ratio Q/A—that is, on how much the charge is spread out. In ◀◀ SECTION 22.4 we defined the charge-to-area ratio as the *surface charge density* η (Greek eta):

$$\eta = \frac{Q}{A} \tag{23.13}$$

The units of surface charge density are C/m^2. Thus the electric field strength inside a parallel-plate capacitor is $E = \eta/\epsilon_0$.

Let's call the surface density of the induced charge $\eta_{induced}$. The electric field created by this charge, an **induced electric field** (the field due to the insulator responding to the external electric field), is

$$\vec{E}_{induced} = \begin{cases} \left(\dfrac{\eta_{induced}}{\epsilon_0}, \text{ from positive to negative}\right) & \text{inside the insulator} \\ \vec{0} & \text{outside the insulator} \end{cases} \tag{23.14}$$

It is because an insulator in an electric field has *two* sheets of induced *electric* charge that we call it a *dielectric,* with the prefix *di,* meaning *two,* the same as in "diatomic" and "dipole."

Inserting a Dielectric into a Capacitor

FIGURE 23.12 shows what happens when we insert a dielectric into a capacitor. The capacitor plates have their own surface charge density $\eta_0 = Q_0/A$. This creates the electric field $\vec{E}_0 = (\eta_0/\epsilon_0$, from positive to negative) into which the dielectric is placed. The dielectric responds with induced surface charge density $\eta_{induced}$ and the induced electric field $\vec{E}_{induced}$. Notice that $\vec{E}_{induced}$ points *opposite* to $\vec{E}_0$. By the

FIGURE 23.12 The consequences of filling a capacitor with a dielectric.

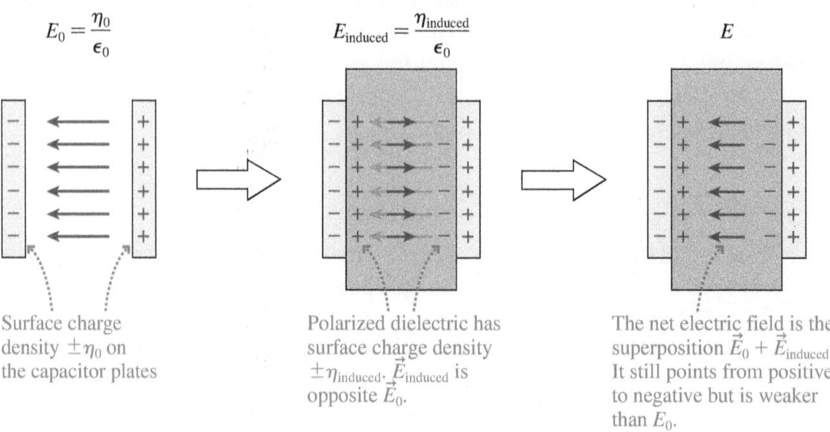
$E_0 = \frac{\eta_0}{\epsilon_0}$
$E_{induced} = \frac{\eta_{induced}}{\epsilon_0}$
E
Surface charge density $\pm\eta_0$ on the capacitor plates
Polarized dielectric has surface charge density $\pm\eta_{induced}$. $\vec{E}_{induced}$ is opposite $\vec{E}_0$.
The net electric field is the superposition $\vec{E}_0 + \vec{E}_{induced}$. It still points from positive to negative but is weaker than $\vec{E}_0$.

principle of superposition, another important lesson from Chapter 21, the net electric field between the capacitor plates is the *vector* sum of these two fields:

$$\vec{E} = \vec{E}_0 + \vec{E}_{\text{induced}} = (E_0 - E_{\text{induced}}, \text{ from positive to negative}) \qquad (23.15)$$

The presence of the dielectric weakens the electric field, from E_0 to $E_0 - E_{\text{induced}}$, but the field still points from the positive capacitor plate to the negative capacitor plate. The field is weakened because the induced surface charge in the dielectric acts to counter the electric field of the capacitor plates.

Let's define the **dielectric constant** κ (Greek *kappa*) as

$$\kappa = \frac{E_0}{E} \qquad (23.16)$$

Equivalently, the field strength inside a dielectric in an external field E_0 is $E = E_0/\kappa$. The dielectric constant is the factor by which a dielectric *weakens* the electric field produced by the source charges, so $\kappa \geq 1$. You can see from the definition that κ is a pure number with no units.

NOTE ▶ Chemists and biochemists often use the symbol D for the dielectric constant. Physicists use κ to avoid possible confusion with the diffusion coefficient. ◀

The dielectric constant, like density or specific heat, is a property of a material. Easily polarized materials have larger dielectric constants than materials not easily polarized. Vacuum has $\kappa = 1$ exactly, and low-pressure gases have $\kappa \approx 1$. (Air has $\kappa_{\text{air}} = 1.00$ to three significant figures, so we won't worry about the very slight effect air has on capacitors.) TABLE 23.1 lists the dielectric constants for different materials.

TABLE 23.1 **Properties of dielectrics**

Material	Dielectric constant κ	Dielectric strength E_{max} (10^6 V/m)
Vacuum	1	—
Air (1 atm)	1.0006	3
Teflon	2.1	60
Polystyrene plastic	2.6	24
Mylar	3.1	7
Olive oil	3.1	16
Pyrex glass	4.7	14
Cell membrane	5.0	200
Ethanol	24	—
Pure water (20°C)	80	—
Titanium dioxide	110	6
Strontium titanate	300	8

The electric field inside the capacitor, although weakened, is still uniform. Consequently, the potential difference across the capacitor is

$$\Delta V_C = Ed = \frac{E_0}{\kappa}d = \frac{(\Delta V_C)_0}{\kappa} \qquad (23.17)$$

where $(\Delta V_C)_0 = E_0 d$ was the voltage of the vacuum-insulated capacitor. The presence of a dielectric reduces the capacitor voltage, the observation with which we started this section. Now we see why; it is due to the polarization of the material. Further, the new capacitance is

$$C = \frac{Q}{\Delta V_C} = \frac{Q_0}{(\Delta V_C)_0/\kappa} = \kappa\frac{Q_0}{(\Delta V_C)_0} = \kappa C_0 \qquad (23.18)$$

Filling a capacitor with a dielectric increases the capacitance by a factor equal to the dielectric constant. This ranges from virtually no increase for an air-filled capacitor to a capacitance 300 times larger if the capacitor is filled with strontium titanate.

We'll leave it as a homework problem to show that the induced surface charge density is

$$\eta_{\text{induced}} = \eta_0\left(1 - \frac{1}{\kappa}\right) \qquad (23.19)$$

η_{induced} ranges from nearly zero when $\kappa \approx 1$ to $\approx \eta_0$ when $\kappa \gg 1$.

NOTE ▶ We assumed that the capacitor was disconnected from the battery after being charged, so Q couldn't change. An alternative is to insert a dielectric while the capacitor remains attached to the battery. In that case, it will be ΔV_C, fixed at the battery voltage, that can't change. As a result, additional charge will flow from the battery until $Q = \kappa Q_0$. In both cases, the capacitance increases to $C = \kappa C_0$. ◀

EXAMPLE 23.5 A water-filled capacitor

A 5.0 nF parallel-plate capacitor is charged to 160 V. It is then disconnected from the battery and immersed in distilled water. What are (a) the capacitance and voltage of the water-filled capacitor and (b) the energy stored in the capacitor before and after its immersion?

PREPARE Pure distilled water is a good insulator. (The conductivity of tap water is due to dissolved ions.) Thus the immersed capacitor has a dielectric between the electrodes.

SOLVE

a. From Table 23.1, the dielectric constant of water is $\kappa = 80$. The presence of the dielectric increases the capacitance to

$$C = \kappa C_0 = 80 \times 5.0\ \text{nF} = 400\ \text{nF}$$

At the same time, the voltage decreases to

$$\Delta V_C = \frac{(\Delta V_C)_0}{\kappa} = \frac{160\ \text{V}}{80} = 2.0\ \text{V}$$

b. The presence of a dielectric does not alter the derivation leading to Equation 23.6 for the energy stored in a capacitor. Right after being disconnected from the battery, the stored energy was

$$(U_C)_0 = \tfrac{1}{2}C_0(\Delta V_C)_0^2 = \tfrac{1}{2}(5.0 \times 10^{-9}\ \text{F})(160\ \text{V})^2 = 6.4 \times 10^{-5}\ \text{J}$$

After being immersed, the stored energy is

$$U_C = \tfrac{1}{2}C(\Delta V_C)^2 = \tfrac{1}{2}(400 \times 10^{-9}\ \text{F})(2.0\ \text{V})^2 = 8.0 \times 10^{-7}\ \text{J}$$

ASSESS Water, with its large dielectric constant, has a *big* effect on the capacitor. But where did the energy go? We learned in Chapter 21 that a dipole is drawn into a region of stronger electric field. The electric field inside the capacitor is much stronger than just outside the capacitor, so the polarized dielectric—the water—is actually *pulled* into the capacitor. The "lost" energy is the work the capacitor's electric field did pulling in the water.

EXAMPLE 23.6 Energy density of a defibrillator

A defibrillator unit contains a 150 μF capacitor that is charged to 2100 V. The capacitor plates are separated by a 0.050-mm-thick insulator with dielectric constant 120.

a. What is the area of the capacitor plates?
b. What are the stored energy and the energy density in the electric field when the capacitor is charged?

PREPARE We can model the defibrillator as a parallel-plate capacitor with a dielectric.

SOLVE

a. The capacitance of a parallel-plate capacitor in a vacuum is $C_0 = \epsilon_0 A/d$. A dielectric increases the capacitance by the factor κ, to $C = \kappa C_0$, so the area of the capacitor plates is

$$A = \frac{Cd}{\kappa\epsilon_0} = \frac{(150 \times 10^{-6}\ \text{F})(5.0 \times 10^{-5}\ \text{m})}{120(8.85 \times 10^{-12}\ \text{C}^2/\text{N}\cdot\text{m}^2)} = 7.1\ \text{m}^2$$

Although the surface area is very large, Figure 23.13 shows how very large sheets of very thin metal can be rolled up into capacitors that you hold in your hand.

b. The energy stored in the capacitor is

$$U_C = \tfrac{1}{2}C(\Delta V_C)^2 = \tfrac{1}{2}(150 \times 10^{-6}\ \text{F})(2100\ \text{V})^2 = 330\ \text{J}$$

Because the dielectric has increased C by a factor of κ, the energy density of Equation 23.8 is increased by a factor of κ to $u_E = \tfrac{1}{2}\kappa\epsilon_0 E^2$. The electric field strength in the capacitor is

$$E = \frac{\Delta V_C}{d} = \frac{2100\ \text{V}}{5.0 \times 10^{-5}\ \text{m}} = 4.2 \times 10^7\ \text{V/m}$$

Consequently, the energy density is

$$u_E = \tfrac{1}{2}(120)(8.85 \times 10^{-12}\ \text{C}^2/\text{N m}^2)(4.2 \times 10^7\ \text{V/m})^2$$
$$= 9.4 \times 10^5\ \text{J/m}^3$$

ASSESS 330 J is a substantial amount of energy—equivalent to that of a 1 kg mass traveling at 25 m/s. And it can be delivered very quickly as the capacitor is discharged through the patient's chest.

FIGURE 23.13 A practical capacitor.

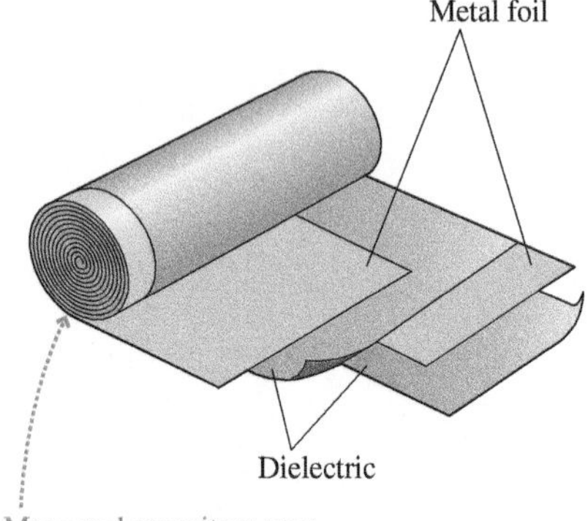

Metal foil
Dielectric
Many real capacitors are a rolled-up sandwich of metal foils and thin, insulating dielectrics.

Solid or liquid dielectrics allow a set of electrodes to have more capacitance than they would if filled with air. Not surprisingly, as FIGURE 23.13 shows, this is important in the production of practical capacitors. In addition, dielectrics allow capacitors to be charged to higher voltages. All materials have a maximum electric field they can sustain without *breakdown*—the production of a spark. The breakdown electric field of air, as we've noted previously, is about 3×10^6 V/m. In general, a material's maximum sustainable electric field is called its **dielectric strength.** Table 23.1 includes dielectric strengths for air and the solid dielectrics. (The breakdown of water is extremely sensitive to ions and impurities in the water, so water doesn't have a well-defined dielectric strength.)

Many materials have dielectric strengths much larger than air. Teflon, for example, has a dielectric strength 20 times that of air. Consequently, a Teflon-filled capacitor can be safely charged to a voltage 20 times larger than an air-filled capacitor with the same plate separation. An air-filled capacitor with a plate separation of 0.2 mm can be charged only to 600 V, but a capacitor with a 0.2-mm-thick Teflon sheet could be charged to 12,000 V.

A Point Charge in a Dielectric

A dielectric in the uniform electric field of a parallel-plate capacitor reduces the field strength and the electric potential by a factor of κ, the insulator's dielectric constant, compared to the field and potential of the same source charges in vacuum. This is a general conclusion that applies in any situation.

FIGURE 23.14 shows a point charge q embedded in a dielectric, such as an ion in a liquid. (The figure shows a positive charge, but our conclusions apply to a charge of either sign.) Whether the molecules of the insulator are polarizable or, as in water, permanent electric dipoles, the overall result is the same: The electric field of the embedded charge *polarizes* the dielectric, either by creating induced dipoles or by aligning permanent dipoles. The polarization creates an excess of negative charge close to the embedded charge and an equal excess of positive charge at the edges of the dielectric. The induced charge has an induced electric field opposite in direction to that of the point charge. The net electric field is the sum of the induced field and the field of a bare point charge. Just like the field of a capacitor, the point-charge electric field is reduced by a factor of κ to

FIGURE 23.14 A point charge in a dielectric.

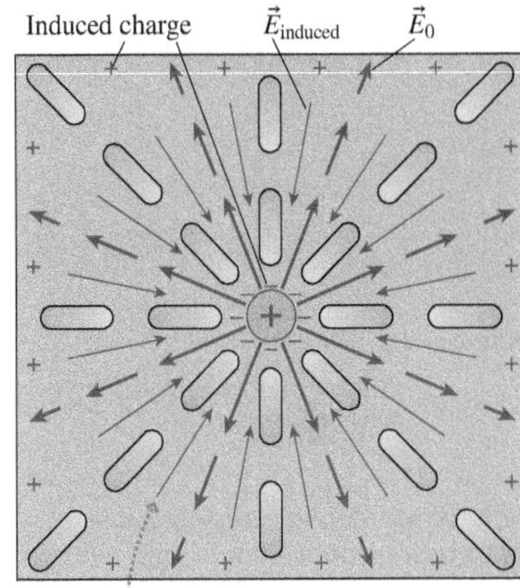
Induced charge
$\vec{E}_{\text{induced}}$
$\vec{E}_0$
The field of the induced charge is opposite to and weakens the field of the point charge.

$$\vec{E}_{\text{point}} = \left(\frac{1}{4\pi\kappa\epsilon_0}\frac{|q|}{r^2}, \begin{cases}\text{away} & q \text{ positive} \\ \text{toward} & q \text{ negative}\end{cases}\right) \tag{23.20}$$

The potential of a point charge (or a charged sphere) is reduced by the same factor:

$$V_{\text{point}} = \frac{1}{4\pi\kappa\epsilon_0}\frac{q}{r} \tag{23.21}$$

NOTE ► Some textbooks write the field strength as $E = |q|/4\pi\epsilon r^2$, where $\epsilon = \kappa\epsilon_0$ is the *permittivity* of the material. In that case, the permittivity constant ϵ_0 is called the *vacuum permittivity*. ◄

EXAMPLE 23.7 Comparing electric fields in oil and glass

A charged macromolecule on the surface of a Pyrex glass slide is covered with oil. We'll assume the oil has the same properties as olive oil. At what distance into the slide is the electric field strength the same as the field strength 75 μm into the oil?

PREPARE We can model the molecule as a point charge. Equation 23.20 gives the reduced electric field strength of a point charge in an insulator. The dielectric constants of Pyrex glass and olive oil are given in Table 23.1 as 4.7 and 3.1.

SOLVE The field strengths in the glass and oil are equal if

$$E_{\text{slide}} = \frac{1}{4\pi\kappa_{\text{slide}}\epsilon_0}\frac{|q|}{(r_{\text{slide}})^2} = E_{\text{oil}} = \frac{1}{4\pi\kappa_{\text{oil}}\epsilon_0}\frac{|q|}{(r_{\text{oil}})^2}$$

We don't need to know the amount of charge because it cancels. The distance into the slide at which the field strength matches the field strength at a distance of 75 μm into the oil is

$$r_{\text{slide}} = \sqrt{\frac{\kappa_{\text{oil}}}{\kappa_{\text{slide}}}}\,r_{\text{oil}} = \sqrt{\frac{3.1}{4.7}} \times (75\ \mu\text{m}) = 61\ \mu\text{m}$$

ASSESS Pyrex glass has a larger dielectric constant than oil, so the field strength decreases more quickly with distance into the glass. But there's not a great difference between the two dielectric constants, so we don't expect the distance into the slide to be significantly less than 75 μm. Our answer seems reasonable.

Much of chemistry and essentially all of biochemistry takes place in water, a *polar liquid* in which the molecules have a permanent electric dipole moment. In this case, the alignment of the molecular dipoles causes a physical rearrangement of the molecules that has important consequences for chemistry in a solution.

FIGURE 23.15 (next page) shows what happens when a positive ion is in water. First, the water molecules close to the ion experience a torque that aligns their dipole moments with the electric field of the ion; then the electrostatic force pulls them in very close to the ion. As a result, a weakly bound shell of water molecules, called

FIGURE 23.15 Water molecules form a hydration shell around a dissolved ion.

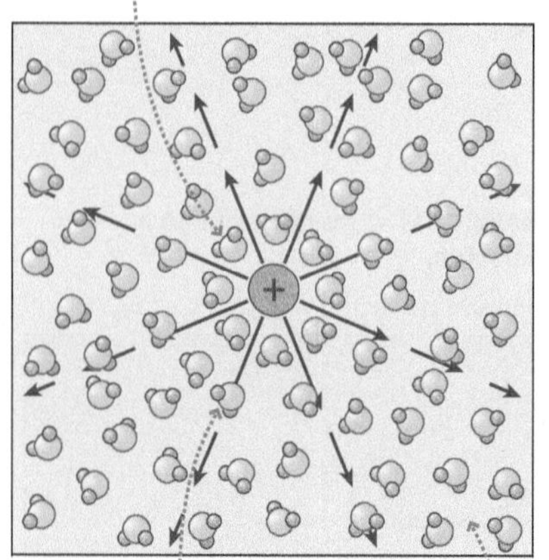
Water molecules nearest the ion are weakly bound to it by electrostatic attraction to form a hydration shell.
+
Water molecules farther away are somewhat aligned with the field...
...but distant water molecules are randomly oriented.

FIGURE 23.16 Mobile charge carriers—electrons in a metal, ions in salt water—shield a charged ball.

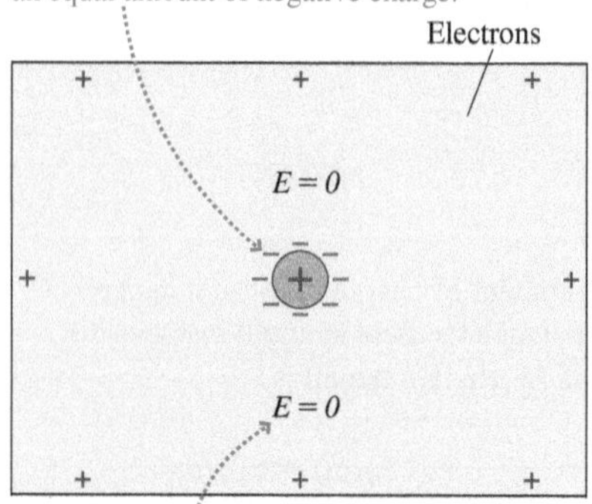
(a) Metal
The charge is neutralized by attracting an equal amount of negative charge.
Electrons
E = 0
E = 0
The electric field is zero.

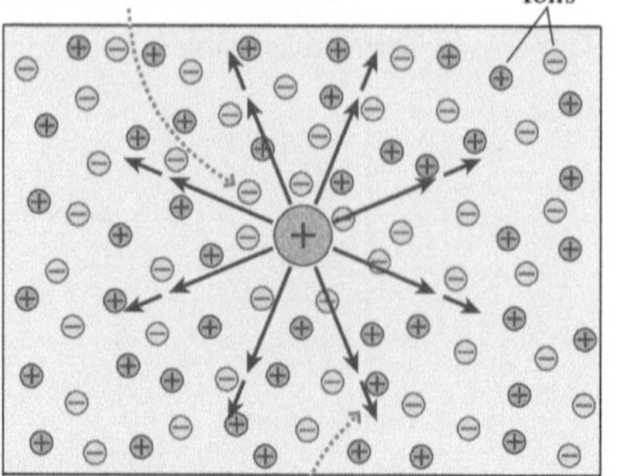
(b) Salt water
The charge attracts negative ions but is not neutralized.
Ions
The electric field strength decreases rapidly.

a **hydration shell,** forms around the ion. An ion moving through water drags the hydration shell with it, effectively increasing the ion's size and mass and thus slowing its speed. Figure 23.15 shows the hydration shell around a positive ion; the hydration shell of a negative ion is the same except that the water molecules are rotated 180°.

Hydration shells are especially important in biochemistry because most macromolecules are charged. The hydration shell around a charged protein is essential to its three-dimensional structure and proper functioning.

STOP TO THINK 23.3 A parallel-plate capacitor is charged to $\pm Q$, then disconnected from the battery. An insulator with dielectric constant κ_1 is inserted into the top half of the capacitor. An insulator with a larger dielectric constant κ_2 is inserted into the bottom half. Afterward, is the potential difference ΔV_{12} larger than, smaller than, or equal to the potential difference ΔV_{34}?

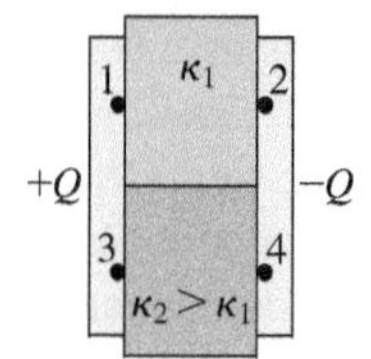
1
2
3
4
κ1
κ2 > κ1
+Q
−Q

23.4 Electrostatics in Salt Water

Most biologically important macromolecules, from proteins to DNA, are negatively charged (we'll see why later in this section), and electric forces are an important aspect of the lock-and-key mechanism by which macromolecules fit together to carry out their functions. We've seen that the polarizability of a medium weakens electric fields and lowers electric potentials. This is especially true for the fields and potentials of charges in water because the permanent electric dipole moments of water molecules give water an especially large dielectric constant.

Charged molecules in the body face an additional complication: The intracellular and extracellular fluids are not pure water but are *salt water.* And salt water, with its many dissolved ions, is a conductor. How does the presence of small, mobile ions affect electric fields and potentials?

To get an idea, suppose we could implant a small ball of positive charge $+Q$ inside a metal, such as copper or aluminum, where the electrons are free to move. We've seen that the electric field has to be zero inside a conductor in electrostatic equilibrium. To achieve this, enough electrons to have charge $-Q$ move almost instantly to the surface of the positive ball. This is shown in **FIGURE 23.16a**. The electrons *shield* the charge, effectively neutralizing it. The electric field inside the metal is then zero because there's no net charge to create a field. (The fact that electrons with charge $-Q$ moved to the ball leaves a net charge of $+Q$ on the *surface* of the metal, consistent with our earlier observation that any excess charge on a conductor resides on the surface.)

The same thing happens to a ball of positive charge $+Q$ in salt water, but the electric field of the ball is not completely neutralized because salt water is not as good a conductor as a metal and because ions have thermal energy that keeps them moving about. **FIGURE 23.16b** shows that the ball of charge attracts negative ions and repels positive ions. The negative ions do partially shield the ball of charge, but they don't completely neutralize it. Consequently, the ball's electric field does extend into the salt water, but its strength decreases much more rapidly than it would in pure water.

The ionic shielding of charged macromolecules plays a critical role in the biochemistry of the cell. We'll begin this section by looking at an idealized case of a charged, flat electrode in salt water, a model that can be applied to one side of a charged membrane. Then we'll see how the lessons learned apply to the more realistic situation of an object that can be modeled as a charged sphere.

A Charged Electrode in Salt Water

FIGURE 23.17a is an edge-on view of a flat, negatively charged electrode. We can think of it as the negative plate of a parallel-plate capacitor with the unseen positive plate outside the frame of the figure. You learned in ◀◀ SECTION 22.3 that the electric potential of a parallel-plate capacitor is $V = Ex$, where x is the distance from the negative electrode. We saw in Section 23.3 that the electric field strength of a dielectric-filled parallel-plate capacitor is $E = \eta/\kappa\epsilon_0$, where $\eta = Q/A$ is the magnitude of the surface charge density on the electrodes. In the absence of ions, the potential to the right of the electrode is

$$V = \frac{\eta}{\kappa\epsilon_0}x \tag{23.22}$$

and the electric field, which points toward the negative plate, is a constant

$$E_x = -\frac{dV}{dx} = -\frac{\eta}{\kappa\epsilon_0} \tag{23.23}$$

FIGURE 23.17 A negative electrode in salt water attracts positive ions and repels negative ions.

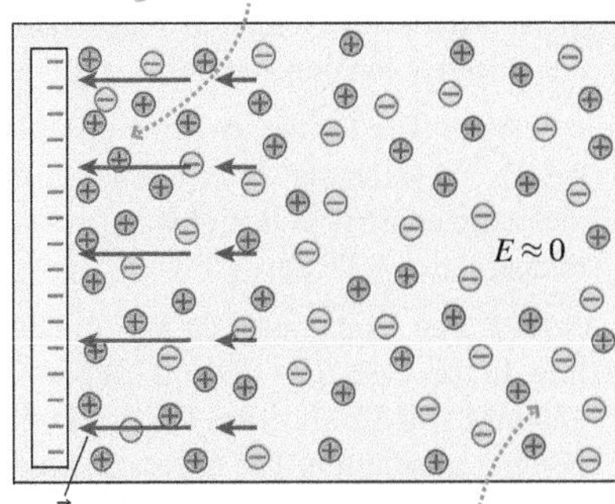

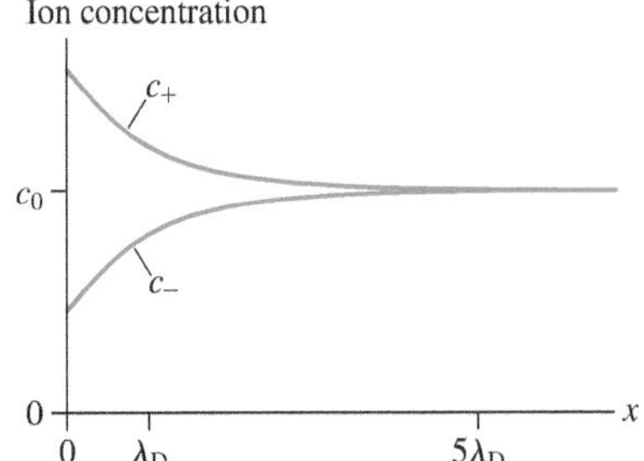

Suppose we now add salts, such as NaCl and KCl, that dissolve into positive and negative ions. For simplicity, we'll assume that all ions are monovalent with charge $\pm e$. Positive and negative ions are equal in number, which makes the solution electrically neutral, and we'll use the symbol c_0 to represent the uniform concentration of either positive or negative ions in the absence of any charged electrodes. However, electrodes or charged objects may cause the concentration c_+ of positive ions at one point in the solution to differ from the concentration c_- of negative ions. We expect $c_+ = c_- = c_0$ at points that are far from any electrode.

NOTE ▸ These concentrations have SI units of m^{-3}—that is, the number of ions per cubic meter. Concentrations in chemistry and biochemistry are usually given in terms of *molarity* M, the number of dissolved moles per liter of solution. Many problems require unit conversions. ◂

In salt water, the negative electrode attracts positive ions and repels negative ions, which leads to a situation like that seen in Figure 23.17a. We see an excess of positive charge and a deficit of negative charge near the electrode, but—unlike the case of a metal conductor—the charge on the electrode has not been completely neutralized and there is an electric field near the electrode. FIGURE 23.17b shows the result: The positive and negative ion concentrations c_+ and c_- differ near the electrode, but the difference decreases with increasing distance from the electrode.

Figure 23.17b has been drawn to suggest that the concentrations change exponentially with distance. Indeed, a quantitative analysis (beyond the scope of this text) finds that the positive and negative ion concentrations c_+ and c_- decay exponentially to the equilibrium concentration c_0, much like the decaying amplitude of a damped oscillator. For a damped oscillator, the *characteristic time* during which the amplitude decays to $e^{-1} \approx 37\%$ of its initial value is the *time constant.* For charged objects in a conducting fluid, the *characteristic distance* over which ion concentrations and, as we'll see, fields and potentials decay to e^{-1} of their initial values is called the **Debye length** λ_D. The Debye length is roughly the distance that electric interactions extend from a charged object. At points far from a charged object—by which we mean $x >> \lambda_D$—the electric field is zero and the ion concentrations have returned to c_0. In effect, **a charged object in salt water has no electric influence beyond a few Debye lengths.**

A detailed analysis of a charged electrode in salt water finds that the field strengths and concentrations change exponentially with the Debye length

$$\lambda_D = \sqrt{\frac{\kappa\epsilon_0 k_B T}{2e^2 c_0}} \tag{23.24}$$

where, you'll recall from Chapter 12, $k_B = 1.38 \times 10^{-23}$ J/K is the Boltzmann constant. The presence of the term k_BT tells us that the equilibrium ion concentrations are a balance between electric forces and thermal interactions.

EXAMPLE 23.8 **The Debye length in a cell**

What is the Debye length in a mammalian cell, where the positive and negative ion concentrations are typically 150 mM?

PREPARE The Debye length is given by Equation 23.24, but we need to do some unit conversions before we can calculate it. The dielectric constant is that of water, $\kappa = 80$. We can assume a body temperature of 37°C.

SOLVE The temperature in k_BT is always an absolute temperature. In this case, $T = (37 + 273)\text{ K} = 310\text{ K}$. A concentration of 150 mM = 0.15 M means 0.15 mol of positive and negative ions per liter of solution. The necessary conversion is

$$c_0 = \frac{0.15\text{ mol}}{1.0\text{ L}} \times \frac{6.02 \times 10^{23}\text{ ions}}{1\text{ mol}} \times \frac{1000\text{ L}}{1\text{ m}^3} = 9.0 \times 10^{25}\text{ ions/m}^3$$

Now we can calculate the Debye length:

$$\lambda_D = \sqrt{\frac{\kappa\epsilon_0 k_B T}{2e^2 c_0}}$$

$$= \sqrt{\frac{(80)(8.85 \times 10^{-12}\text{ C}^2/\text{N}\cdot\text{m}^2)(1.38 \times 10^{-23}\text{ J/K})(310\text{ K})}{2(1.60 \times 10^{-19}\text{ C})^2(9.0 \times 10^{25}\text{ m}^{-3})}}$$

$$= 8.1 \times 10^{-10}\text{ m} \approx 0.8\text{ nm}$$

ASSESS The Debye length in the cell is a little less than 1 nm. That's larger than atoms and small ions but smaller than many macromolecules. We'll look at some of the biological implications later in this section.

A complete analysis of the negatively charged electrode finds the electric potential, the electric field, and the ion concentrations. It is convenient to write the ion concentrations as $\Delta c_+ = c_+ - c_0$ and $\Delta c_- = c_- - c_0$, the amount by which the concentrations *differ* from the equilibrium value. The results for a weakly charged electrode (η is small) are

$$V = \frac{\eta\lambda_D}{\kappa\epsilon_0}(1 - e^{-x/\lambda_D})$$

$$E_x = -\frac{dV}{dx} = -\frac{\eta}{\kappa\epsilon_0}e^{-x/\lambda_D} \qquad (23.25)$$

$$\Delta c_\pm = \pm\frac{\eta}{2e\lambda_D}e^{-x/\lambda_D}$$

Both the electric field and the ion concentration differences decay exponentially with distance from the electrode, as Figure 23.17b shows, reaching e^{-1} of their values at the electrode's surface at $x = \lambda_D$. As a practical matter, E_x and both Δc are essentially zero by $x \approx 5\lambda_D$, which is ≈ 4 nm in a mammalian cell. A charged object in salt water has no influence—it's effectively invisible—beyond a distance of about 4 nm because of screening from the dissolved ions.

For $x \ll \lambda_D$, where $e^{-x/\lambda_D} \approx 1$, the electric field is $E_x \approx -\eta/\kappa\epsilon_0$. This is Equation 23.23 for the electric field *without* dissolved ions. In other words, the electric field of Equations 23.25 reduces to a familiar result at points so close to the electrode that the ions cannot provide any screening.

EXAMPLE 23.9 **Ion concentrations at an electrode**

Large macromolecules in solution have a surface charge density of about 10^{-2} C/m², a value we'll justify later in this section. A typical ion concentration in a mammalian cell is 150 mM. What are the positive and negative ion concentrations at the surface of a flat, negative electrode that has a surface charge density of 1.0×10^{-2} C/m²?

PREPARE Example 23.8 found the Debye length to be 0.81 nm and the equilibrium concentration c_0 to be 9.0×10^{25} m^{-3}.

SOLVE The surface of the electrode is $x = 0$, where $e^{-x/\lambda_D} = 1$. Thus the ion concentration differences are

$$\Delta c_\pm = \pm\frac{\eta}{2e\lambda_D} = \pm\frac{1.0 \times 10^{-2}\text{ C/m}^2}{2(1.60 \times 10^{-19}\text{ C})(8.1 \times 10^{-10}\text{ m})}$$

$$= \pm 3.9 \times 10^{25}\text{ m}^{-3}$$

Positive ions are attracted to the negative electrode, so their concentration increases to

$$c_+ = \Delta c_+ + c_0 = 12.9 \times 10^{25}\ \text{m}^{-3}$$

Negative ions are repelled, and their concentration decreases to

$$c_- = \Delta c_- + c_0 = 5.1 \times 10^{25}\ \text{m}^{-3}$$

ASSESS The *average* concentration is the equilibrium $9.0 \times 10^{25}\ \text{m}^{-3}$, but the composition is shifted in favor of positive ions. Both concentrations return to the equilibrium value at distances from the electrode greater than about 4 nm. A graph of the concentrations would look like Figure 23.17b.

Point Charges and Charged Spheres in Salt Water

It should come as no surprise that the field and potential of a point charge in salt water exhibit the same exponential decay with distance. If the ion concentration is low, a full analysis finds

$$\begin{aligned} V_{\text{point}} &= \frac{1}{4\pi\kappa\epsilon_0}\frac{q}{r}e^{-r/\lambda_D} \\ E_{\text{point}} &= \frac{1}{4\pi\kappa\epsilon_0}\frac{|q|}{r^2}e^{-r/\lambda_D} \end{aligned} \qquad (23.26)$$

For a charged sphere of radius R, the potential and the concentration differences are

$$\begin{aligned} V_{\text{sphere}} &= \frac{1}{4\pi\kappa\epsilon_0}\frac{Q}{r}\frac{e^{-(r-R)/\lambda_D}}{1+R/\lambda_D} \\ \Delta c &= \pm\frac{1}{8\pi e\lambda_D^2}\frac{|Q|}{r}\frac{e^{-(r-R)/\lambda_D}}{1+R/\lambda_D} \end{aligned} \qquad (23.27)$$

In this case, the exponential decay begins at the surface of the sphere.

EXAMPLE 23.10 Potential of hemoglobin

The hemoglobin protein can be modeled as a 5.0-nm-diameter sphere. Electrophoresis measurements find that, at a pH of approximately 8, a hemoglobin molecule in solution has a charge of $-10e$.

a. What is the surface charge density of hemoglobin if all the charge is on the surface?
b. What is the potential 1.5 nm from the surface of a hemoglobin molecule in air, in pure water, and in a mammalian cell?

PREPARE We can model the hemoglobin as a charged sphere with $Q = -10e = -1.6 \times 10^{-18}$ C. Example 23.8 found that the Debye length in a mammalian cell is 0.81 nm.

SOLVE

a. The surface of a sphere has area $A = 4\pi R^2$. If all the charge is on the surface, the surface charge density is

$$\eta = \frac{Q}{A} = \frac{1.6 \times 10^{-18}\ \text{C}}{4\pi(2.5 \times 10^{-9}\ \text{m})^2} = 0.020\ \text{C/m}^2$$

b. The potential of a charged sphere depends on the total charge Q. A point that is 1.5 nm from the surface is at distance $r = 1.5\ \text{nm} + R = 4.0\ \text{nm}$ from the center. The potentials at this distance are

$$\text{air:} \quad V_{\text{sphere}} = \frac{1}{4\pi\epsilon_0}\frac{Q}{r} = -3.6\ \text{V}$$

$$\text{pure water:} \quad V_{\text{sphere}} = \frac{1}{4\pi\kappa\epsilon_0}\frac{Q}{r} = -45\ \text{mV}$$

$$\text{cell:} \quad V_{\text{sphere}} = \frac{1}{4\pi\kappa\epsilon_0}\frac{Q}{r}\frac{e^{-(r-R)/\lambda_D}}{1+R/\lambda_D} = -1.7\ \text{mV}$$

ASSESS The electric influence of a charge is substantially less in the saltwater environment of a cell than in air. Both the dielectric constant of water and the screening from dissolved ions must be included when calculating electrostatic effects in the cell. Also, the calculation of this example finds that the surface charge density of hemoglobin is $2 \times 10^{-2}\ \text{C/m}^2$ *if* all the charge is on the surface. Some of the excess charge is likely to be buried deeper within the molecule, so the actual surface charge density is probably less than we calculated, compatible with the value of $1.0 \times 10^{-2}\ \text{C/m}^2$ in Example 23.9.

STOP TO THINK 23.4 The Debye length in a saltwater solution is 2.0 nm. If the ion concentration is doubled, the Debye length will be

A. 1.0 nm B. 1.4 nm C. 2.8 nm D. 4.0 nm

Macromolecules in Salt Water

We've noted several times that macromolecules in the cell are usually charged. This has two practical benefits. First, the tendency of macromolecules to form hydrogen bonds makes them rather sticky. That's a useful feature when a molecule needs to bend and fold into a proper shape, but it could cause all the macromolecules in a cell to stick together as useless sludge. The repulsive electrostatic force between two negative macromolecules, even if short range in salt water, keeps them far enough apart to prevent this.

Second, many biochemical reactions depend on the lock-and-key model in which an enzyme and a substrate fit together in a well-defined configuration. The specificity of these reactions depends on electric forces between a very specific pattern of surface charges on the enzyme and the substrate. The screening of electric forces in salt water allows molecules to diffuse freely until just the right contact is found; then the reduced short-range interaction holds the molecules together.

But *why* are macromolecules charged? And why negatively? You may recall from chemistry that the water molecules in pure water dissociate via the reaction $H_2O \rightleftharpoons H^+ + OH^-$. The fact that pure water has a pH of 7 indicates that the equilibrium concentration of H^+ ions (at 25°C) is 10^{-7} M. Physically, the dissociation of water depends on the fact that water is a polar molecule; as molecules randomly move and collide, the fluctuating forces between the positive and negative ends of the electric dipoles are occasionally large enough to break bonds.

Similarly, large molecules in solution are acted on by the polar water molecules, and they usually dissociate into a large negative molecular ion and one or more hydrogen ions (H^+). For example, the structural backbone of DNA is made of alternating sugar and phosphate groups. Inorganic phosphoric acid, H_3PO_4, dissociates as $H_3PO_4 \rightarrow PO_4^- + H^+ + H + H$. The two neutral hydrogen atoms are incorporated into other bonds, the H^+ ion is drawn into the solvent, and a negatively charged phosphate group PO_4^- becomes part of the DNA structure. There is one phosphate group per base pair on each of the two strands of DNA, so the charge of a DNA molecule is $-2e$ per base pair.

The final charge state of a macromolecule is a balance between energy and entropy. A macromolecule is in a low-entropy state in which the atoms are confined to specific locations. Entropy would increase if a macromolecule completely dissociated, allowing the atoms to wander freely throughout the entire cell. But systems also want to minimize energy, and macromolecules do so by bringing atoms together into bound states.

Shedding a few H^+ ions provides an entropy gain with very little energy penalty—because random collisions supply the energy—so this is a favorable process. But it takes more and more energy to pull a positive ion away from the molecule as it becomes increasingly negatively charged, and there comes a point where the energy price is not offset by the entropy increase.

This reasoning may sound familiar; it is exactly the reasoning we used in Chapter 14 to understand why chemical equilibrium is the set of conditions that minimizes the Gibbs free energy. The Gibbs free energy G is essentially energy minus entropy. A process that increases entropy, such as dissociation, decreases G *if* the process doesn't require too large an increase in energy. A process that reduces energy, such as bonding, decreases G *if* the process doesn't require too large a decrease in entropy. Equilibrium—the state in which G is a minimum—is an energy-entropy balance. We'll leave the details to biochemistry, but it's not surprising that the final charge state of a macromolecule in salt water is the state that minimizes the Gibbs free energy.

FIGURE 23.18 A macromolecule in salt water is surrounded by a cocoon-like layer of counterions.

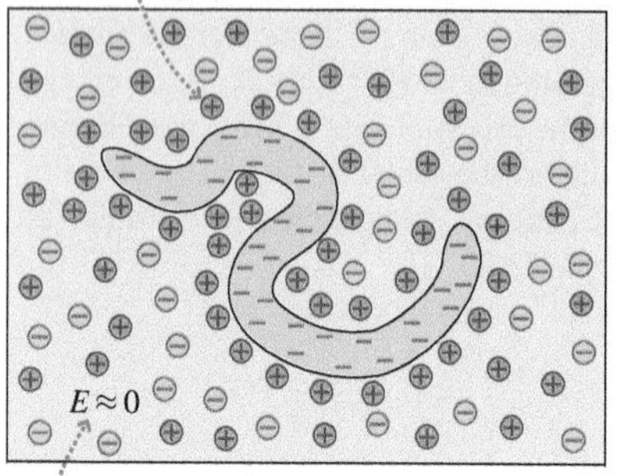

A negative macromolecule is surrounded by a layer of positive counterions of thickness $\approx \lambda_D$.
$E \approx 0$
The counterions shield the molecule so that the electric field nearby is zero.

FIGURE 23.18 shows a macromolecule in salt water. It is negatively charged, but the ions it shed—called *counterions*—haven't gone far; they form a cocoon-like layer around the molecule. The thickness of the counterion layer is approximately the Debye length λ_D, which we found to be slightly less than 1 nm under typical cell conditions. The Debye length is less than the size of many macromolecules, so the counterion layer is relatively thin compared to the size of the molecule. Even so, the

counterions shield the macromolecule so that its electric field and potential do not extend more than a few Debye lengths. Two macromolecules separated by more than about 4 nm simply don't interact even though both are charged objects. Biochemical processes involving macromolecules cannot begin until random motion brings the surfaces of the molecules extremely close together.

STOP TO THINK 23.5 A parallel-plate capacitor is submerged in salt water while it is connected to a 9 V battery. The capacitor plate spacing is much larger than the Debye length. At the point shown in the figure,

A. The electric field points to the right.
B. The electric field points to the left.
C. The electric field is approximately zero.

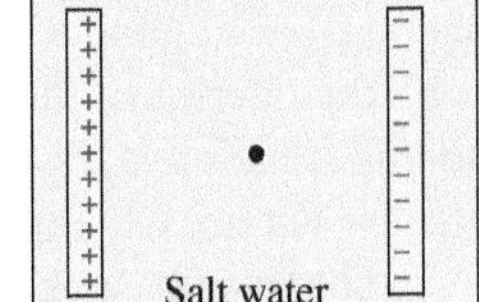

23.5 The Membrane Potential of a Cell

The environment of a cell is salt water, a conducting fluid with dissolved ions. Much of the physiology and biochemistry of the cell depend on electric forces and the motions of ions. ◀◀ SECTION 22.5 introduced the *membrane potential,* a permanent potential difference across the thin cell wall. Muscle contractions and nerve impulses occur when a changing membrane potential causes ions to flow into or out of the cell. In this section, we'll look at how and why a cell has a membrane potential, its *resting potential.* Later, after we develop the concept of an electric circuit in Chapter 25, we'll investigate the connection between the membrane potential and nerve impulses.

We begin with a simple model of the cell and the cell membrane in which there is only one dissolved salt, potassium chloride (KCl). We make this choice because the potassium ion K^+ is the most abundant positive ion inside a cell in its resting state. FIGURE 23.19a shows an impermeable membrane separating the interior and exterior of a cell. Measured concentrations of the potassium ion in skeletal muscle cells are $[K^+]_{in} = 150$ mM and $[K^+]_{out} = 4$ mM, which are very different. The negative Cl^- ions have the same concentrations, which make both sides electrically neutral. Both sides of the impermeable membrane are in equilibrium, and there is no potential difference across the membrane.

Real cell membranes are not impermeable. Suppose we open small, selective pores or channels in the membrane that allow K^+ ions to pass through, in both directions, but do not allow Cl^- ions to pass. Physiologically, *ion channels* are tube-like proteins embedded in the cell wall; their three-dimensional structure allows only one type of ion to pass through the tube. The type of channels we will consider first, called *leak channels,* are always open. You may have also encountered *voltage-gated ion channels* that can switch between being open and closed; we'll consider those later.

Opening the leak channels produces the situation of FIGURE 23.19b in which a *concentration gradient* between two connected containers allows K^+ ions to diffuse down the gradient. Recall that particles move in both directions, due to random collisions, but the net result is an outward flow of ions from the higher concentration inside the cell to the lower concentration outside.

A flow of *neutral* molecules would continue until a new equilibrium is established with equal concentrations on both sides. But potassium ions are not neutral. The outward flow of positive ions causes the exterior side of the membrane to become positively charged while the interior, which loses positive charge, becomes negatively charged. This charge separation establishes an electric field across the cell membrane, pointing inward, that exerts an opposing force on K^+ ions passing through the ion channel.

FIGURE 23.19 A cell membrane with selective ion channels leads to an equilibrium in which there is a potential difference across the membrane.

(a) Equilibrium

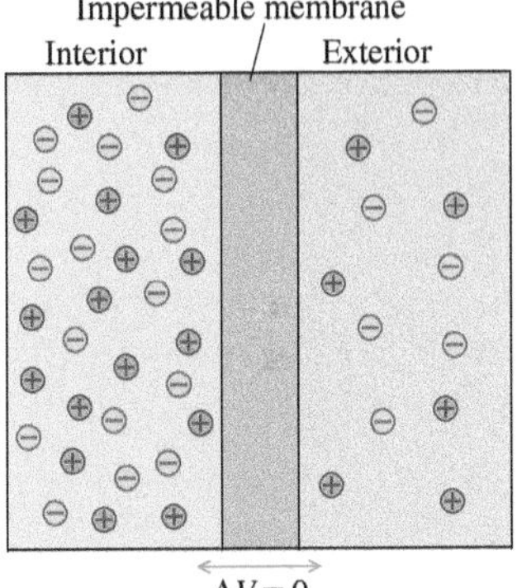

(b) Diffusion

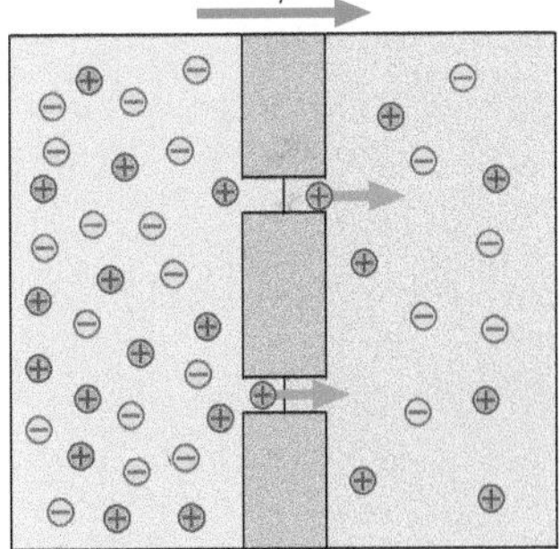

(c) New equilibrium

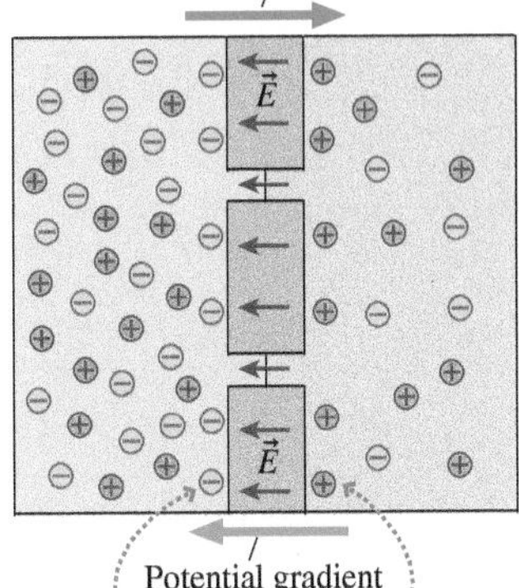

The charge separation and the electric field increase as more ions pass through, so the opposing electric force grows until, as FIGURE 23.19c shows, an equilibrium is reached in which the outward flow of ions due to the concentration gradient is exactly balanced by an inward flow due to the electric field. The separated charges—which reside on the surfaces of the membrane—create an electric potential, and you'll recall that an electric field points "downhill" in the direction of decreasing electric potential. In other words, at equilibrium there are both a concentration difference $\Delta c = c_{in} - c_{out}$ *and* a potential difference $\Delta V = V_{in} - V_{out}$ across the cell membrane, with the interior of the cell negative relative to the exterior. The key factor that leads to this situation is the *selective* permeability of the membrane, which allows K^+ ions to pass but not Cl^- ions.

◀◀ SECTION 13.5 looked at how equilibrium is established by random collisions in a two-level system in which particles could have either a smaller energy E_1 or a larger energy E_2. We found that the equilibrium population ratio is the *Boltzmann factor*

$$\frac{N_2}{N_1} = e^{-\delta E/k_B T} \tag{23.28}$$

where $\delta E = E_2 - E_1$. (Caution! The E in Equation 23.28 is energy, not electric field strength.)

The two sides of the cell membrane form a two-level system in which ions can have either potential energy $U_{in} = eV_{in}$ or potential energy $U_{out} = eV_{out}$. The reasoning of Chapter 13 finds that the equilibrium concentration ratio is

$$\frac{c_{out}}{c_{in}} = e^{-\delta U/k_B T} = e^{-(U_{out}-U_{in})/k_B T} = e^{e(V_{in}-V_{out})/k_B T} = e^{e\Delta V/k_B T} \tag{23.29}$$

We can solve this for the potential difference by taking the natural logarithm of both sides:

$$\Delta V = V_{Nernst} = \frac{k_B T}{e}\ln\left(\frac{c_{out}}{c_{in}}\right) = (27\text{ mV})\ln\left(\frac{c_{out}}{c_{in}}\right) \tag{23.30}$$

Nernst potential

where we evaluated $k_B T/e$ at the human body temperature of 37°C.

The equilibrium potential difference given by Equation 23.30 is called the **Nernst potential.** It is the potential difference across a membrane that exactly balances diffusion and brings the net flow of ions to zero. We can use the measured potassium ion concentrations $[K^+]_{in} = 150$ mM and $[K^+]_{out} = 4$ mM in skeletal muscle cells to calculate $V^{K}_{Nernst} = -98$ mV.

NOTE ▶ Concentrations are typically given as molarities rather than in SI units of m^{-3}. However, you can calculate *ratios* without converting molarity to m^{-3} because the conversion factors cancel. ◀

The actual potential difference across a cell membrane, the *membrane potential* V_{mem}, is easily measured with a voltmeter by inserting a microelectrode (typically 1 μm in diameter) into the cell. Measurements find $V_{mem} \approx -90$ mV for skeletal muscle cells. The calculated Nernst potential for K^+ is close to V_{mem}, which tells us that this simple model has captured the essential idea of why cells have a membrane potential. However, the fact that $V_{mem} \neq V^{K}_{Nernst}$ means that the potassium ions are *not* in equilibrium. We will consider a slightly more complex model later in this section.

The Capacitor Model of a Cell Membrane

Figure 23.19c showed a cell membrane that has equal but opposite charges on the two sides. This looks very much like a charged parallel-plate capacitor. Indeed, we can model the cell membrane as a capacitor, as shown in FIGURE 23.20. A cell wall is

FIGURE 23.20 The cell membrane can be modeled as a capacitor.

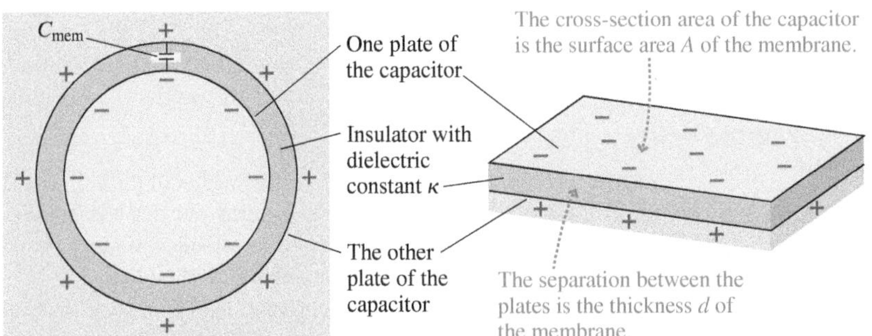

not a parallel-plate capacitor, but the parallel-plate capacitor is a reasonable model as long as the membrane thickness is much less than the radius of the cell. (Think about how a thin spherical shell can be flattened—like an orange peel—into two irregularly shaped but parallel surfaces after making several slices through it.) The capacitance of a parallel-plate capacitor in vacuum is $C_0 = \epsilon_0 A/d$, where A is the area of the electrode and d is the plate separation.

For a spherical cell,

$$C_{\text{mem}} \approx \frac{\kappa\epsilon_0 A}{d} \tag{23.31}$$

where κ is the membrane's dielectric constant (listed as 5.0 in Table 23.1), $A = 4\pi R^2$ is the surface area, and d is the thickness of the membrane. Cell membranes are too thin to observe with a light microscope (they are thinner than the optical resolution limit we discussed in Chapter 20), but electron microscope images show that $d \approx 7$ nm.

The charge separation across the membrane creates an inward-pointing electric field within the membrane that has field strength $E = |\Delta V|/d$. A potential difference of 90 mV across a membrane that is a mere 7 nm thick corresponds to an electric field strength of 1.3×10^7 V/m = 13 MV/m. This is a very strong field. It exceeds the breakdown field of 3 MV/m that would cause a spark in air, but it is a sustainable field within the membrane's lipid bilayer. This strong field exists *only* within the membrane. The external field, as we found in our study of capacitors in Chapter 21, is very close to zero, and any residual field extends only a few Debye lengths—a few nm—from the cell wall.

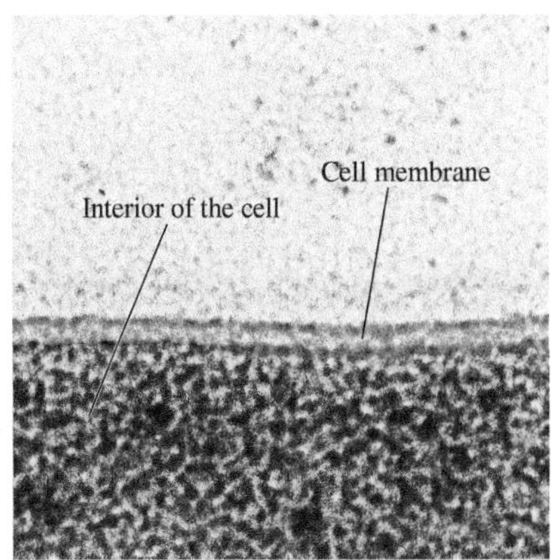

A thin line The cell membrane is an insulating layer that separates the conducting liquids inside and outside a cell. This false-color electron microscope image shows that the membrane is extremely thin, only about 7 nm wide.

EXAMPLE 23.11 The surface charge on a cell

A skeletal muscle cell, a fairly large cell, can be modeled as a 16-μm-diameter sphere.

a. How many K^+ ions have to diffuse out of the cell to establish the membrane potential of −90 mV?

b. What fraction is this of the number of K^+ ions inside the cell at a concentration of 150 mM?

PREPARE We can model the cell as a capacitor like the one shown in Figure 23.20. The charge on a capacitor is $Q = C\Delta V_C$, so we'll need to calculate the capacitance of the membrane. The membrane's thickness is 7 nm and its dielectric constant is 5.0.

SOLVE

a. A 16-μm-diameter sphere has surface area $A = 4\pi R^2 = 800\ \mu\text{m}^2$. Thus the membrane capacitance is

$$C_{\text{mem}} = \frac{\kappa\epsilon_0 A}{d}$$
$$= \frac{(5.0)(8.85 \times 10^{-12}\ \text{F/m})(800\ \mu\text{m}^2 \times 10^{-12}\ \text{m}^2/\mu\text{m}^2)}{7 \times 10^{-9}\ \text{m}}$$
$$= 5.1 \times 10^{-12}\ \text{F} = 5.1\ \text{pF}$$

A 5.1 pF capacitor charged to 90 mV has charge

$$Q = C_{\text{mem}}\Delta V_C = (5.1 \times 10^{-12}\ \text{F})(0.090\ \text{V}) = 5 \times 10^{-13}\ \text{C}$$

Only one significant figure is justified because the membrane thickness is known to one significant figure. This is a very small amount of charge. The number of K^+ surface ions needed to establish this charge is

$$N_{\text{surface}} = \frac{Q}{e} = 3 \times 10^6$$

Continued

In other words, 3×10^6 K^+ ions have to diffuse out of the cell and gather on the outer surface, leaving a surplus of 3×10^6 Cl^- ions inside, to establish the -90 mV membrane potential.

b. The number of K^+ ions inside the cell is $N_{inside} = c_{in}V$. The cell volume is $V = \frac{4}{3}\pi R^3 = 2100\ \mu m^3$. Here we do need to convert the concentration from molarity to SI units:

$$c = 0.15 \text{ mol/L} \times \frac{6.02 \times 10^{23} \text{ ions}}{1 \text{ mol}} \times \frac{1000 \text{ L}}{1 \text{ m}^{-3}}$$
$$= 9.0 \times 10^{25} \text{ ions/m}^3$$

Thus the number of K^+ ions inside the cell is

$$N_{inside} = c_{in}V = (9.0 \times 10^{25} \text{ ions/m}^3)\left(2100\ \mu\text{m}^3 \times \left(\frac{10^{-6} \text{ m}}{1\ \mu\text{m}}\right)^3\right)$$
$$= 1.9 \times 10^{11} \text{ ions}$$

The number of ions that diffuse out as a fraction of the number of ions available is

$$\frac{N_{surface}}{N_{inside}} = \frac{3 \times 10^6}{1.9 \times 10^{11}} \approx 2 \times 10^{-5}$$

ASSESS You might think that ions diffusing out of the cell would lower the ion concentration inside. It turns out that the number of ions needed on the exterior surface to establish the membrane potential is a tiny fraction of the number of available ions inside. The very slight decrease in concentration is not measurable, and the Nernst potential of potassium does not change due to this diffusion. Note that actual muscle cells are elongated, not spherical. Even so, a simple model of the cell as a sphere is adequate for doing one-significant-figure estimates. A cylindrical-cell model would be bit more complex, but it would not change the results.

A Steady-State Model

K^+ and Cl^- are not the only dissolved ions; Na^+, Ca^{2+}, and negatively charged macromolecules also play important roles. The measured membrane potential $V_{mem} = -90$ mV is established not by the K^+ ions alone, as our simple model would predict, but through the competing influence of all these ions.

The Nernst potential for K^+ ions, $V_{Nernst}^{K} = -98$ mV, is the membrane potential that stops the outward diffusion of K^+ ions and creates an equilibrium situation. The actual membrane potential of $V_{mem} = -90$ mV is a bit less, so the membrane electric field is not quite strong enough to prevent K^+ ions from flowing out through the leak channels. That is, K^+ ions are *not* in equilibrium but are steadily lost from the cell. This situation is not sustainable, so something must act to counter this outward flow.

We'll expand our model to include Na^+ ions but, in the interest of not getting too complex, we'll ignore other ions. The cell membrane has selective Na^+ ion leak channels that allow sodium ions, like potassium ions, to diffuse across the membrane. The measured sodium ion concentrations in skeletal muscle cells are $[Na^+]_{in} = 15$ mM and $[Na^+]_{out} = 150$ mM. These are similar to the potassium ion concentrations, but reversed. The sodium ion concentration is higher outside the cell, so the direction of diffusion—down the concentration gradient—is into the cell.

We can use Equation 23.30 to calculate that the Nernst potential for sodium is $V_{Nernst}^{Na} = +62$ mV. That is, equilibrium for Na^+ ions requires the interior of the cell to be positive, which creates an outward-pointing electric field to balance the inward diffusion. The fact that $V_{mem} = -90$ mV means that the Na^+ ion concentrations are very far from equilibrium. This is due in large part to there being far fewer leak channels for Na^+ than for K^+. The membrane's inward-pointing electric field actually *increases* the inward diffusion of Na^+; even so, there are not enough Na^+ leak channels to bring its concentrations close to equilibrium.

The ion concentrations in the cell may not be in equilibrium, but they are in what we call a *steady state.* This means that they are not changing with time despite the outward flow of K^+ ions and the inward flow of Na^+ ions. A system can be kept in a nonequilibrium steady state only if there's some outside intervention. For example, a swimming pool with a leak can be kept at a steady level if water is pumped in at exactly the same rate that it leaks out.

For cells, the outside intervention is the presence in the cell membrane of **sodium-potassium exchange pumps,** also called *ion pumps,* that use energy from ATP to actively pump Na^+ ions out of the cell and K^+ ions into the cell, thus countering the diffusive flow of these ions. FIGURE 23.21 illustrates the idea. One cycle of the pump, which is powered by the hydrolysis of one ATP molecule, ejects three Na^+ ions while pulling in two K^+ ions. In both cases, ions are moved *up* the concentration

FIGURE 23.21 A better model of the cell has both leak channels and ion pumps.

Each cycle of the sodium-potassium exchange pump moves three sodium ions out of the cell and two potassium ions in.

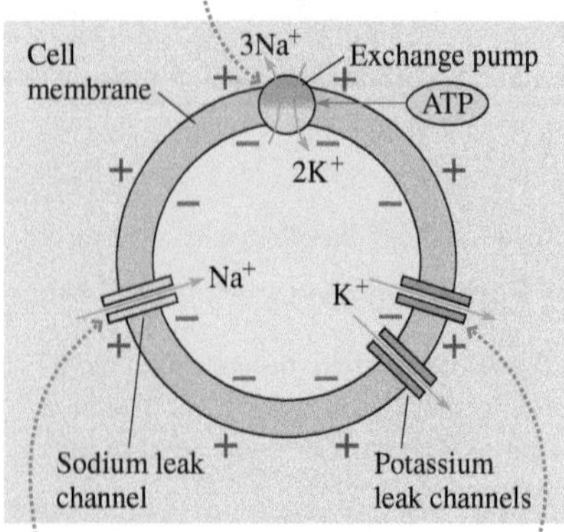

The sodium Nernst potential is higher than the membrane potential. Sodium ions diffuse into the cell.

The potassium Nernst potential is lower than the membrane potential. Potassium ions diffuse out of the cell.

gradient, toward a higher concentration. This cannot happen spontaneously, so an energy expenditure is required. Ion flow through the exchange pumps balances the outward diffusion of K^+ ions and the inward diffusion of Na^+ ions, which keeps the cell's ion concentrations from changing.

EXAMPLE 23.12 Potassium leakage

Ion channels and ion pumps are complex proteins embedded in the cell wall. Careful cell studies find that sodium-potassium exchange pumps are embedded in skeletal muscle cell walls at a density of about 1000 pumps/μm^2. Potassium leak channels that are open continuously have a lower density of about 4 channels/μm^2. The sodium-potassium exchange pumps operate at a speed of about 400 cycles/s.

a. In a steady state, at what rate do K^+ ions enter the cell through the sodium-potassium exchange pumps?

b. At what rate do K^+ ions diffuse out through each leak channel?

PREPARE We can model the cell as a 16-μm-diameter sphere, as we did in Example 23.11. We calculated in that example that the cell has surface area $A = 4\pi R^2 = 800\ \mu\text{m}^2$. The requirement for steady-state ion concentrations is that ions diffuse out of the cell at the same rate they are pumped in.

SOLVE

a. The number of ion pumps is

$$N_{\text{pumps}} = (1000\ \text{pumps}/\mu\text{m}^2)(800\ \mu\text{m}^2) = 800{,}000$$

Each pump completes 400 cycles/s, so the total pumping rate is

$$R = (8 \times 10^5\ \text{pumps})(400\ \text{cycles/s per pump}) \approx 3 \times 10^8\ \text{cycles/s}$$

Each cycle brings in two K^+ ions, so K^+ ions are pumped in at a rate of 6×10^8 ions/s.

b. The number of ion channels is

$$N_{\text{channels}} = (4\ \text{channels}/\mu\text{m}^2)(800\ \mu\text{m}^2) \approx 3000$$

A steady state requires an outward flow of 6×10^8 ions/s, so the rate of diffusion through each channel is

$$\frac{6 \times 10^8\ \text{ions/s}}{3000\ \text{channels}} \approx 2 \times 10^5\ \text{ions/s per channel}$$

ASSESS Multiplying the ion flow rate of 2×10^5 ions/s by e gives a charge flow rate of 3×10^{-14} C/s. In Chapter 24, we'll define *current* in amps (A) as the charge flow rate in C/s. Thus each leak channel carries a K^+ ion current of $3 \times 10^{-14}\ \text{A} = 0.03$ pA, where $1\ \text{pA} = 10^{-12}\ \text{A} = 1$ picoamp. Our calculation is in reasonable agreement with experimental determinations of the current through the ion leak channels.

STOP TO THINK 23.6 The Nernst potential of Ca^{2+}ions is $+137$ mV. Which side of the cell membrane has the higher concentration of Ca^{2+}ions: the interior or the exterior?

CHAPTER 23 INTEGRATED EXAMPLE Depolarization of cardiac myocytes

Our discussion of the electrocardiogram in ◀◀ SECTION 22.7 noted that a cardiac myocyte, a heart muscle cell, has a *resting* membrane potential of -90 mV when it is not contracting. However, stimulus from a neighboring cell causes a group of sodium ion channels to open for a brief interval. These channels allow a surge of Na^+ ions to enter the cell and raise the membrane potential to $+20$ mV. (We'll look more closely at these *voltage-gated channels* that open only briefly and then close again when we study nerve impulses in Chapter 25.) This reversal of the membrane potential is the *depolarization* of the cell that initiates a muscle contraction. Investigators are able to measure the ion flow through a single ion channel with the *patch clamp technique* illustrated in FIGURE 23.22. A micropipette makes a seal on a patch of cell membrane that includes one sodium channel. The micropipette is filled with salt water, a conductor, and a very sensitive meter measures the flow of Na^+ions from the salt water into the cell when the channel opens. One experiment with cardiac myocytes found an average current of $3.0\ \text{pA} = 3.0 \times 10^{-12}$ C/s for 2.0 ms, followed by the channel closing. How many voltage-gated sodium channels are needed to produce the observed rise in the membrane potential?

FIGURE 23.22 The patch clamp technique for measuring the current through a single sodium channel.

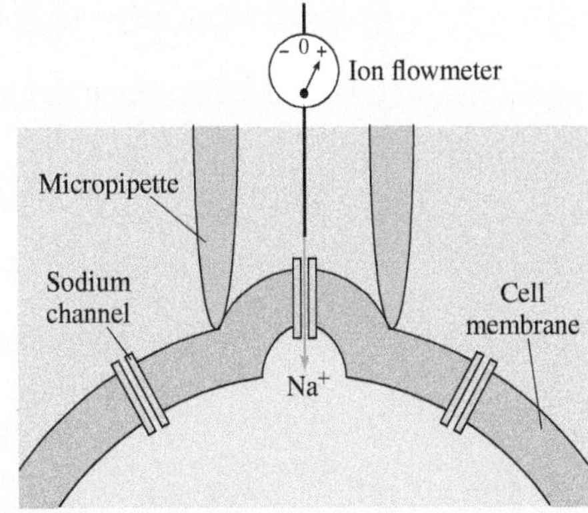

PREPARE We can model the cell as a 16-μm-diameter sphere. We calculated the membrane capacitance in Example 23.11 and found $C_{\text{mem}} = 5.1$ pF. We can use the capacitance and the observed change in the membrane potential to calculate the number

Continued

of Na^+ ions that enter the cell in 2.0 ms. The patch clamp data will be used to determine the number of ions that enter per channel. We can combine these two numbers to calculate the number of channels.

SOLVE The resting membrane potential is −90 mV, so the surface charge on the membrane is

$$Q_{\text{surface}} = C_{\text{mem}} \Delta V_C = (5.1 \times 10^{-12}\ \text{F})(0.090\ \text{V}) = 4.6 \times 10^{-13}\ \text{C}$$

The cell is negative with respect to its surroundings, so the initial charge on the inside surface of the membrane is $Q_i = -4.6 \times 10^{-13}\ \text{C}$. Opening the sodium channels changes the membrane potential to +20 mV. The surface charge is then

$$Q_{\text{surface}} = C_{\text{mem}} \Delta V_C = (5.1 \times 10^{-12}\ \text{F})(0.020\ \text{V}) = 1.0 \times 10^{-13}\ \text{C}$$

with a final $Q_f = +1.0 \times 10^{-13}$ C on the inside surface. The charge on the inside surface changes by $\Delta Q = Q_f - Q_i = 5.6 \times 10^{-13}\ \text{C}$. This charge is due to the Na^+ ions that entered the cell. The total number of entering ions is

$$N_{\text{total}} = \frac{Q}{e} = \frac{5.6 \times 10^{-13}\ \text{C}}{1.6 \times 10^{-19}\ \text{C}} = 3.5 \times 10^6$$

Turning now to the patch clamp data, we know that the flow through a single channel is 3.0×10^{-12} C/s for 2.0 ms. The amount of charge that flows into the cell through this one channel is

$$Q_{\text{channel}} = (3.0 \times 10^{-12}\ \text{C/s})(0.0020\ \text{s}) = 6.0 \times 10^{-15}\ \text{C}$$

Thus the number of ions that pass through this single channel is

$$N_{\text{single}} = \frac{Q}{e} = \frac{6.0 \times 10^{-15}\ \text{C}}{1.6 \times 10^{-19}\ \text{C}} = 3.8 \times 10^4$$

The total number of ions N_{total} is the number N_{single} of ions entering per channel multiplied by the number of channels N_{channel}. Thus

$$N_{\text{channel}} = \frac{N_{\text{total}}}{N_{\text{single}}} = 92$$

ASSESS The number of Na^+ ions that must enter the cell to switch the membrane potential is similar to the result in Example 23.11 for the number of K^+ ions needed to establish the resting membrane potential. It is a very small number compared to the number of ions already inside the cell, so we would not see any measurable change in the ion concentrations. These voltage-gated sodium channels carry a much larger current than the always-open leak channels: 3 pA compared to the 0.03 pA current we calculated for a potassium leak channel in Example 23.12. Thus it is not surprising that only a few channels are needed or that they can have a large effect on the cell in a very short time.

SUMMARY

GOAL To study some of the applications of electric fields and potentials.

APPLICATIONS

Capacitors

The **capacitance** of two conductors charged to $\pm Q$ is

$$C = \frac{Q}{\Delta V_C}$$

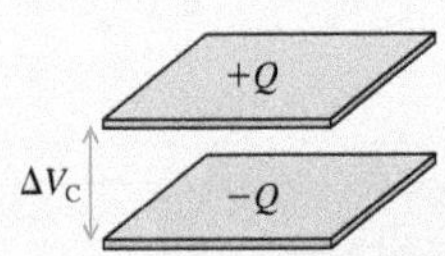

A **parallel-plate capacitor** has

$$C = \frac{\epsilon_0 A}{d}$$

Filling the space between the capacitor plates with a dielectric that has dielectric constant κ increases the capacitance to $C = \kappa C_0$.

The energy stored in a capacitor is $U_C = \frac{1}{2}C(\Delta V_C)$.

This energy is stored in the electric field at density $u_E = \frac{1}{2}\kappa\epsilon_0 E^2$.

Dielectrics

An insulator in an electric field, called a **dielectric,** is polarized by the field. Induced charges on the surfaces create an opposing *induced electric field* that weakens electric fields and potentials by a factor of κ, the **dielectric constant.**

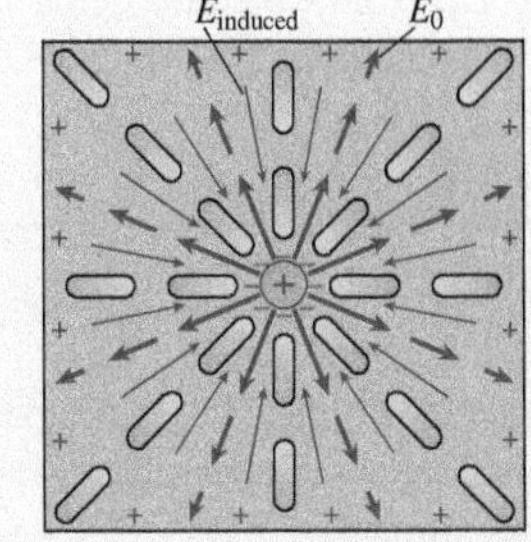

Point charge in a dielectric:

$$\vec{E}_{\text{point}} = \left(\frac{1}{4\pi\kappa\epsilon_0}\frac{|q|}{r^2}, \begin{cases}\text{away} & q \text{ positive}\\ \text{toward} & q \text{ negative}\end{cases}\right) \quad V_{\text{point}} = \frac{1}{4\pi\kappa\epsilon_0}\frac{q}{r}$$

Combinations of Capacitors

Series capacitors

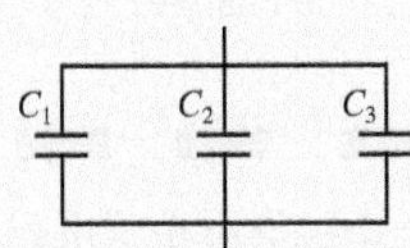

$$C_{\text{eq}} = \left(\frac{1}{C_1} + \frac{1}{C_2} + \frac{1}{C_3} + \cdots\right)^{-1}$$

Parallel capacitors

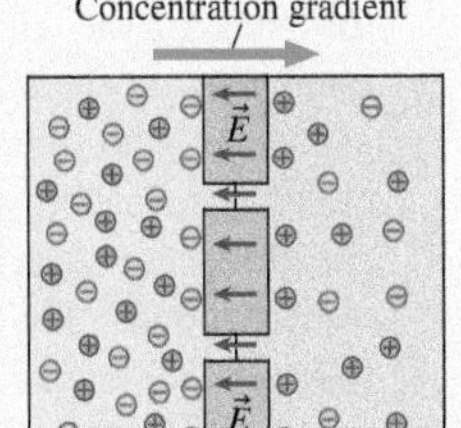

$$C_{\text{eq}} = C_1 + C_2 + C_3 + \cdots$$

Debye Length

An electrode or charged molecule in salt water is partially neutralized and shielded by oppositely charged ions that are attracted to it. Electric fields and potentials decrease exponentially with distance over a characteristic length called the **Debye length** λ_D:

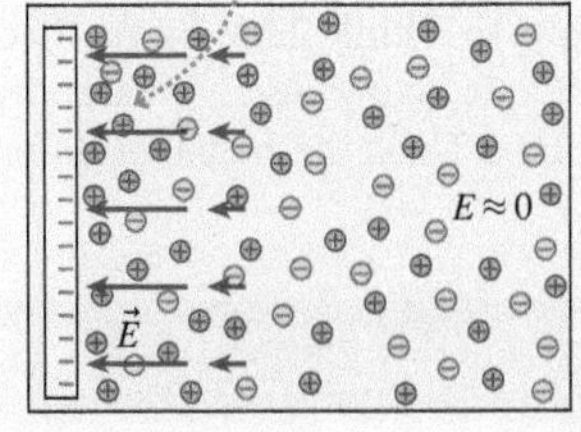

$$\lambda_D = \sqrt{\frac{\kappa\epsilon_0 k_B T}{2e^2 c_0}}$$

Point charge in salt water:

$$E_{\text{point}} = \frac{1}{4\pi\kappa\epsilon_0}\frac{|q|}{r^2}e^{-r/\lambda_D} \quad V_{\text{point}} = \frac{1}{4\pi\kappa\epsilon_0}\frac{q}{r}e^{-r/\lambda_D}$$

The Membrane Potential

Selective ion leak channels in the cell membrane allow some ions to diffuse down the concentration gradient. The resulting charge separation creates an opposing potential gradient. For one type of ion, equilibrium is reached when the membrane potential equals the **Nernst potential**:

$$V_{\text{Nernst}} = \frac{k_B T}{e}\ln\left(\frac{c_{\text{out}}}{c_{\text{in}}}\right)$$

$$= (27 \text{ mV})\ln\left(\frac{c_{\text{out}}}{c_{\text{in}}}\right)$$

Concentration gradient

Potential gradient

A cell membrane can be modeled as a capacitor with capacitance

$$C_{\text{mem}} = \kappa\epsilon_0 A/d$$

in which a nonequilibrium steady state is maintained by sodium-potassium exchange pumps that actively transport K^+ ions into and Na^+ ions out of the cell to balance diffusive leaks through the walls.

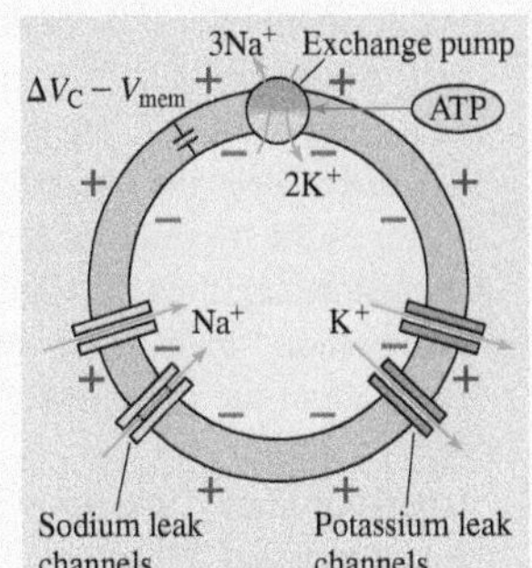

LEARNING OBJECTIVES After studying this chapter, you should be able to:

- Solve problems about capacitors and capacitance. *Conceptual Questions 23.4, 23.7; Problems 23.1, 23.2, 23.6, 23.9, 23.14*
- Calculate the equivalent capacitance of a combination of capacitors. *Conceptual Question 23.5; Problems 23.15–23.18, 23.22*
- Determine electric fields and potentials in dielectrics. *Conceptual Questions 23.6, 23.8; Problems 23.23–23.27*
- Calculate electric fields and potentials in salt water. *Conceptual Question 23.10; Problems 23.30–23.34*
- Solve problems about the cell's membrane potential. *Conceptual Questions 23.11, 23.12; Problems 23.35–23.38*

STOP TO THINK ANSWERS

Stop to Think 23.1: C. Capacitance is a property of the shape and position of the electrodes. It does not depend on the potential difference or charge.

Stop to Think 23.2: $(C_{eq})_B > (C_{eq})_A = (C_{eq})_D > (C_{eq})_C$. $(C_{eq})_B = 3\ \mu F + 3\ \mu F = 6\ \mu F$. The equivalent capacitance of series capacitors is less than any capacitor in the group, so $(C_{eq})_C < 3\ \mu F$. Only D requires any real calculation. The two 4 μF capacitors are in series and are equivalent to a single 2 μF capacitor. The 2 μF equivalent capacitor is in parallel with 3 μF, so $(C_{eq})_D = 5\ \mu F$.

Stop to Think 23.3: Same. Each electrode is a conductor in electrostatic equilibrium, so each electrode is at the same potential at all points. Thus, the potential difference is the same between any two points on opposite electrodes. However, the charge on each electrode will shift so that there is a larger surface charge density next to the dielectric that has the larger dielectric constant.

Stop to Think 23.4: B. The Debye length is inversely proportional to the square root of the concentration c_0. Doubling the concentration reduces the Debye length by a factor of $\sqrt{2}$.

Stop to Think 23.5: C. The point is at a distance from both plates much larger than the Debye length. Both plates are fully shielded by ions in the salt water, so the electric field is zero.

Stop to Think 23.6: Exterior. The Nernst potential is positive if $\ln(c_{out}/c_{in}) > 0$. This is true if $c_{out} > c_{in}$.

QUESTIONS

Conceptual Questions

1. Two separated electrodes are charged to $\pm Q$. If the amount of charge is doubled, is the capacitance halved, doubled, or unchanged? Explain.
2. Parallel-plate capacitor B is identical to parallel-plate capacitor A except that it is scaled up in size by a factor of 2, which doubles the width, height, and plate separation of capacitor A. What is the capacitance ratio C_B/C_A?
3. The surface area of a plant's roots is a good measure of the health of the plant: A healthy plant has a large root surface area. An investigator can measure the root area indirectly by measuring the capacitance of the plant. One electrode is the soil, the other is the tissue inside the plant. The membrane that separates the two has a reasonably constant thickness. If an investigator measures the capacitance of one plant and then finds that a second plant has twice the capacitance, what does that imply about the root area of the second plant compared to the first?
4. A parallel-plate capacitor with plate separation d is connected to a battery that has a potential difference ΔV_{bat}. Without breaking any of the connections, insulating handles are used to increase the plate separation to $2d$.
   a. Does the potential difference ΔV_C change as the separation increases? If so, by what factor? If not, why not?
   b. Does the capacitance change? If so, by what factor? If not, why not?
   c. Does the capacitor charge Q change? If so, by what factor? If not, why not?
5. Do three identical capacitors in series have more, less, or the same equivalent capacitance as the same three capacitors in parallel? Explain.
6. The touch screen on your phone has a grid of transparent electrodes etched onto the back of the glass plate. Each pair of electrodes is a capacitor, and the phone's circuitry measures all of the capacitances several times each second. The electric field of each capacitor—unlike that of an ideal parallel-plate capacitor—is not confined to the space between the electrodes. Instead, part of each capacitor's electric field extends above the glass. When your finger touches the screen, it changes the capacitance of the pair of electrodes directly beneath your finger. The change is detected by the circuitry, which then knows where your finger is. Does your finger increase or decrease the capacitance of that pair of electrodes? Explain.

Problem difficulty is labeled as | (straightforward) to ||||| (challenging). Problems labeled INT integrate significant material from earlier chapters; BIO are of biological or medical interest; CALC require calculus to solve.

7. Rank in order, from largest to smallest, the potential differences $(\Delta V_C)_1$ to $(\Delta V_C)_4$ of the four capacitors in Figure Q23.7. Explain.

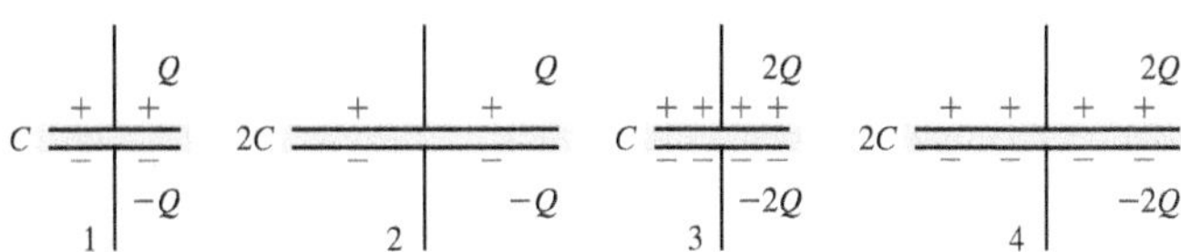

FIGURE Q23.7

8. Parallel-plate capacitors A and B are filled with the same dielectric and have plates of the same size. Capacitor B has twice the plate separation and thus twice the dielectric thickness as capacitor A. What is the capacitance ratio C_B/C_A?
9. A negative ion in pure water is attracted to a positive electrode. Will the attractive force be stronger, weaker, or unchanged if the water is replaced by ethanol? Explain.
10. The Debye length in a solution is 2.0 nm. What will the Debye length be if the ion concentration is doubled?
11. A cell's Nernst potential for K^+ ions is -60 mV. At one instant of time, the cell's membrane potential is -75 mV. At this instant, is the net flow of K^+ ions out of the cell or into the cell? Or is there no flow? Explain.
12. The concentration of an ion is the same both inside and outside a cell. Is the ion's Nernst potential positive, negative, or zero?

Multiple-Choice Questions

13. | A flower—a large conducting mass on top of a narrow stem—and the ground together act like a capacitor. The flower is one electrode, the earth is the other. A typical value of the capacitance is 0.80 pF. The electric field of the earth induces a charge on the flower and an opposite charge on the ground below. If a flower carries a charge of magnitude 40 pC, what is the approximate potential difference between the flower and the ground below?
A. 10 V B. 30 V C. 50 V D. 80 V
14. | A capacitor is connected to a 1.5 V battery. If the battery voltage is increased to 3 V,
A. The energy stored in the capacitor increases by a factor of $\sqrt{2}$.
B. The energy stored in the capacitor increases by a factor of 2.
C. The energy stored in the capacitor increases by a factor of $2\sqrt{2}$.
D. The energy stored in the capacitor increases by a factor of 4.
15. || A parallel-plate capacitor is charged by a battery. Then, with the battery still connected, an insulator with dielectric constant 2 is inserted between the plates. As a result,
A. The charge on the plates increases by a factor of 4.
B. The charge on the plates increases by a factor of 2.
C. The charge on the plates does not change.
D. The charge on the plates decreases by a factor of 2.
16. | The electric potential at distance r from a positive point charge has the largest value if the charge is in
A. Air.
B. Pure water.
C. Salt water.
D. The potential at distance r is the same in air, pure water, and salt water.
17. | As a general rule, mammalian cell membranes have a capaci-
BIO tance of about 1 μF per cm² of surface area, which can be used to estimate the surface area of a cell. The membrane capacitance of one particular cell is found to be 2.5 pF. What is the approximate membrane area?
A. 2.5 μm^2 B. 25 μm^2
C. 250 μm^2 D. 2500 μm^2
18. || The volume of the cell nucleus is perhaps 10% of the cell it
BIO is part of. The nucleus of a cell is separated from the rest of the cell by a nuclear membrane that is about twice the thickness of the cell membrane. If the capacitance of the cell membrane is 6 pF, approximately what is the capacitance of the nuclear membrane? Assume that the cell and the nucleus are both smooth spheres and that both membranes have the same dielectric constant.
A. 0.3 pF B. 0.6 pF C. 30 pF D. 60 pF
19. || Much of the experimental research on the propagation of
BIO nerve signals has focused on an especially large axon found in the squid. The axon's Nernst potential for K^+ ions is -115 mV at a temperature of 15°C. The ion concentration outside the axon is 2.0 mM. What is the concentration inside the axon?
A. 0.02 mM
B. 140 mM
C. 200 mM
D. 8000 mM
20. | The Nernst potential for sodium in a mammalian nerve cell is
BIO $+62$ mV. The resting membrane potential is -70 mV, but the membrane potential temporarily increases to $+40$ mV when the nerve fires. While the membrane potential is $+40$ mV,
A. Sodium ions are diffusing into the cell.
B. Sodium ions are diffusing out of the cell.
C. There is no net diffusion of sodium ions.

PROBLEMS

Section 23.1 Capacitance and Capacitors

1. || What is the capacitance of the two metal spheres shown in Figure P23.1?

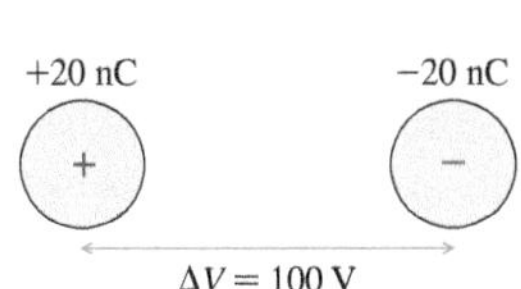

FIGURE P23.1

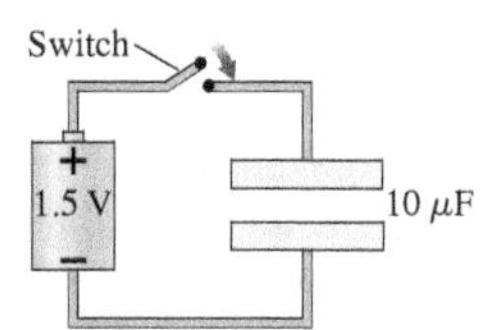

FIGURE P23.2

2. | Initially, the switch in Figure P23.2 is open and the capacitor is uncharged. How much charge flows through the switch after the switch is closed?
3. | A switch that connects a battery to a 10 μF capacitor is closed. Several seconds later you find that the capacitor plates are charged to ± 30 μC. What is the battery voltage?
4. || Two 2.0 cm × 2.0 cm square aluminum electrodes, spaced 0.50 mm apart, are connected to a 100 V battery.
a. What is the capacitance?
b. What is the charge on the positive electrode?

5. | Touching *Mimosa pudica,* the "sensitive plant," causes its
BIO leaflets to fold inward and droop. This response can also be triggered electrically. In one experiment, a 47 μF capacitor was first charged from a power supply. When a switch was flipped, the charged capacitor was disconnected from the power supply and connected to electrodes that had been inserted into the base of a leaf. It was found that the leaflets folded if the power supply voltage was at least 1.3 V. What minimum amount of charge must be delivered to the plant to trigger a response?

6. || When a hummingbird visits a flower, its wings rub against
BIO the flower and leaves, and this can result in a noticeable charge on the bird. There is an opposite charge in the earth. We can consider the hummingbird and the earth to be the two electrodes of a capacitor. The capacitance for one species has been estimated to be 1 pF. Measurements find that a bird's charge is typically +300 pC. To one significant figure, what is the potential difference between the hummingbird and the earth?

7. || The earth is negatively charged, while the *ionosphere*—the highest layer of the earth's atmosphere starting at about 100 km—is positively charged. We can think of the earth and the ionosphere as the electrodes of a large capacitor. Rocket experiments find that the potential difference between the earth and the ionosphere is typically 350 kV.
   a. The average surface charge density of the earth is measured to be $-1.0\ \text{nC/m}^2$. What is the total charge of the earth? The earth's radius is 6.37×10^6 m.
   b. What is the capacitance of the earth + ionosphere system?

8. ||| An uncharged capacitor is connected to the terminals of a 3.0 V battery, and 6.0 μC flows to the positive plate. The 3.0 V battery is then disconnected and replaced with a 5.0 V battery, with the positive and negative terminals connected in the same manner as before. How much additional charge flows to the positive plate?

9. || To what potential should you charge a 1.0 μF capacitor to store 1.0 J of energy?

10. | Capacitor 2 has half the capacitance and twice the potential difference as capacitor 1. What is the ratio $(U_C)_1/(U_C)_2$?

11. || The ability of a battery to deliver charge, and thus power, decreases with temperature. The same is not true of capacitors. For sure starts in cold weather, a truck has a 500 F capacitor alongside a battery. The capacitor is charged to the full 13.8 V of the truck's battery. How much energy does the capacitor store? How does the energy density of the 9.0 kg capacitor system compare to the 130,000 J/kg of the truck's battery?

12. || The 90 μF capacitor in a defibrillator unit supplies an average
BIO of 6500 W of power to the chest of the patient during a discharge
INT lasting 5.0 ms. To what voltage is the capacitor charged?

13. || A 2.0-cm-diameter parallel-plate capacitor with a spacing of 0.50 mm is charged to 200 V. What are (a) the total energy stored in the electric field and (b) the energy density?

14. || 50 pJ of energy is stored in a 2.0 cm × 2.0 cm × 2.0 cm region of uniform electric field. What is the electric field strength?

Section 23.2 Combinations of Capacitors

15. | A 6.0 μF capacitor, a 10 μF capacitor, and a 16 μF capacitor are connected in parallel. What is their equivalent capacitance?

16. | A 6.0 μF capacitor, a 10 μF capacitor, and a 16 μF capacitor are connected in series. What is their equivalent capacitance?

17. | You need a capacitance of 50 μF, but you don't have a 50 μF capacitor. You do have a 30 μF capacitor. What additional capacitor do you need to produce a total capacitance of 50 μF? Should you join the two capacitors in parallel or in series?

18. | You need a capacitance of 50 μF, but you don't happen to have a 50 μF capacitor. You do have a 75 μF capacitor. What additional capacitor do you need to produce a total capacitance of 50 μF? Should you join the two capacitors in parallel or in series?

19. || What is the equivalent capacitance of the three capacitors in Figure P23.19?

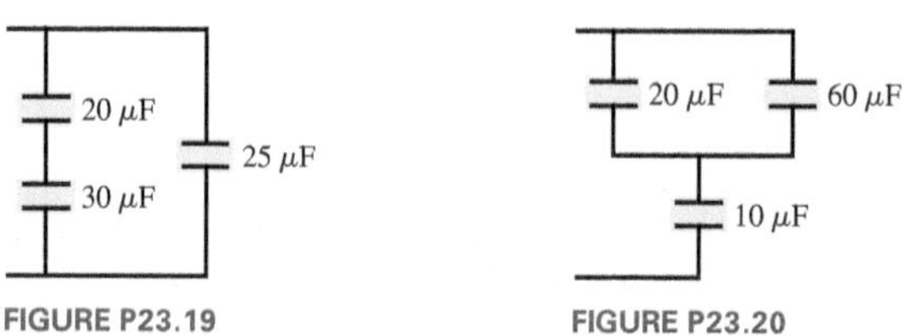

FIGURE P23.19 FIGURE P23.20

20. | What is the equivalent capacitance of the three capacitors in Figure P23.20?

21. ||| For the circuit of Figure P23.21,
   a. What is the equivalent capacitance?
   b. How much charge flows through the battery as the capacitors are being charged?

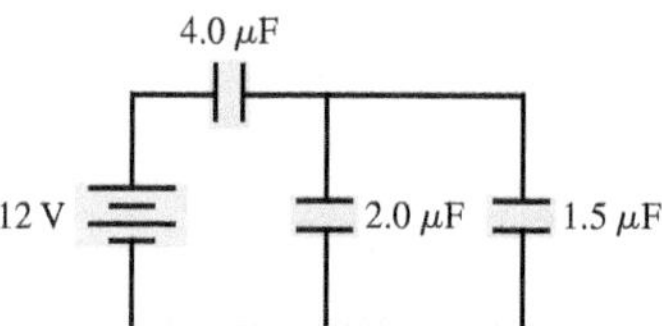

FIGURE P23.21

22. ||| For the circuit of Figure P23.22,
   a. What is the equivalent capacitance?
   b. What is the charge on each capacitor?

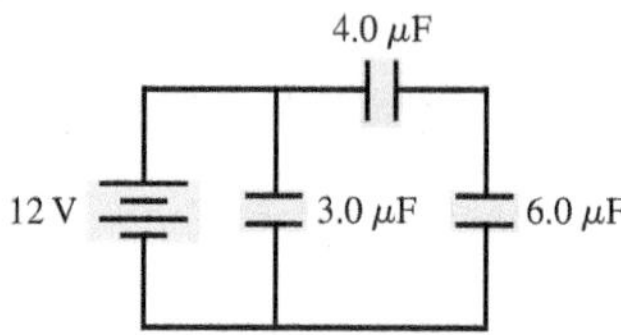

FIGURE P23.22

Section 23.3 Dielectrics

23. || Investigators are exploring ways to treat milk for longer shelf
BIO life by using pulsed electric fields to destroy bacterial contamination. One system uses 8.0-cm-diameter circular plates separated by 0.95 cm. The space between the plates is filled with milk, which has a dielectric constant the same as that of water. The plates are briefly charged to 30,000 V. What is the capacitance of the system, and how much charge is on each plate when they are fully charged?

24. || A 1.2 nF parallel-plate capacitor has an air gap between its plates. Its capacitance increases by 3.0 nF when the gap is filled by a dielectric. What is the dielectric constant of that dielectric?

25. ||| A 25 pF parallel-plate capacitor with an air gap between the plates is connected to a 100 V battery. A Teflon slab is then inserted between the plates and completely fills the gap. What is the change in the charge on the positive plate when the Teflon is inserted?

26. ||| A parallel-plate capacitor is connected to a battery and stores 4.4 nC of charge. Then, while the battery remains connected, a sheet of Teflon is inserted between the plates.
   a. Does the capacitor's charge increase or decrease?
   b. By how much does the charge change?
27. ||| A parallel-plate capacitor is charged by a 12.0 V battery, then the battery is removed.
   a. What is the potential difference between the plates after the battery is disconnected?
   b. What is the potential difference between the plates after a sheet of Teflon is inserted between them?
28. || The ionic bond between the sodium and chlorine atoms in NaCl can be modeled simply. We assume that the sodium atom transfers one electron to the chlorine atom, creating Na^+ and Cl^- ions. The ions attract each other because of their opposite charges, but they don't crash together because the system—like the earth and the moon—rotates about its center of gravity. For an inverse-square force, such as the electric force between point charges, it can be shown that the system's total mechanical energy is half the potential energy: $E_{mech} = \frac{1}{2}U$. We can approximate the molecule's dissociation energy as $|E_{mech}|$ when the ions are separated by the equilibrium bond length—0.24 nm for NaCl—because that is the energy that would be needed to pull the ions apart. This model is far from perfect, but it illustrates an important point: It is much easier to dissociate ionic bonds in water. According to this model, what is the dissociation energy in J and in kJ/mol for NaCl (a) in vacuum and (b) in pure water?
29. || The Debye length is a characteristic length for electrostatic interactions in salt water. There are similar length scales for electrostatic interactions in dielectrics. One example is the *Bjerrum length,* the distance at which electrostatic energies are approximately the same as thermal energies. We've seen that the average kinetic energy of atoms in an ideal gas is $\frac{3}{2}k_BT$. The Bjerrum length is the distance between two ions at which the interaction energy between the ions is equal to $\frac{3}{2}k_BT$. Physically, random thermal motion is more important than electrostatic interactions for ions separated by more than a few Bjerrum lengths. What are the Bjerrum lengths for two singly charged ions in (a) air, (b) the lipid bilayer of a cell membrane, and (c) pure water at 20°C?

Section 23.4 Electrostatics in Salt Water

30. || Pure water, with a pH of 7, has no dissolved salts, but there is a small dissociation of water molecules into H^+ and OH^- ions. You may recall that pH is defined as $-\log_{10}([H^+])$, so the concentrations of H^+ and OH^- ions in pure water are 10^{-7} M. What is the Debye length in pure water at 20°C?
31. || The phosphate groups that form the backbone of DNA molecules are 0.34 nm apart, and each has charge $-e$. Example 23.8 found that the Debye length in a mammalian cell is 0.8 nm.
BIO
   a. What is the electric force between two adjacent phosphate groups?
   b. By what factor is the force reduced from what it would be in vacuum?
32. || Two Na^+ ions are 3.0 nm apart. At 20°C, what is the magnitude of the electric force between the ions if they are in (a) vacuum, (b) pure water, and (c) seawater in which the positive and negative ions each have a concentration of 550 mM?
33. || The electric field strength 1.0 nm from a negatively charged electrode in 20°C salt water is half the field strength at that distance in pure water.
   a. What is the Debye length in the salt water?
   b. What is the ion concentration c_0?
34. || a. What is the Debye length in a 25°C solution with ion concentration 100 mM?
   b. At what distance from a point charge is the electric field strength 10% of the field strength at this distance in pure water?

Section 23.5 The Membrane Potential of a Cell

35. || A cell membrane can be modeled as a capacitor.
BIO
   a. How much energy is stored in a 16-μm-diameter spherical cell that has a membrane potential of -90 mV?
   b. For comparison, how many ATP molecules need to undergo hydrolysis to generate this much energy? In a cell, the energy released from the hydrolysis of ATP is approximately 60 kJ/mol.
36. || The chlorine ion Cl^- has an intracellular concentration of 10 mM and an extracellular concentration of 110 mM. Because Cl^- is a negative ion, the sign of the Nernst potential is opposite that of Equation 23.30. What is the Nernst potential of Cl^-?
BIO
37. || The calcium ion Ca^{2+} has charge $+2e$, so Equation 23.30 for the Nernst potential must have e replaced by $2e$. The Nernst potential of Ca^{2+} is $+135$ mV. If the Ca^{2+} ion concentration outside the cell is 2.0 mM, a typical value, what is its concentration inside the cell?
BIO
38. || Bacteria have a membrane potential, although the mechanisms of how it is maintained differ from those of mammalian cells. The transport of protons across the cell membrane is an important part of the mechanism, so a bacterium's membrane potential varies with the proton concentration—the pH—of the surrounding fluid. One strain of bacteria was found to have a membrane potential of -120 mV at a pH of 7.5. A bacterium can be modeled as a 1.5-μm-diameter sphere. How many positive ions are needed on the exterior surface to establish this membrane potential? (There are an equal number of negative ions on the interior surface.) Assume that the membrane properties are the same as those of mammalian cells.
BIO

General Problems

39. || Two 2.0 cm × 2.0 cm metal electrodes are spaced 1.0 mm apart and connected by wires to the terminals of a 9.0 V battery.
   a. What are the charge on each electrode and the potential difference between them?
   The wires are disconnected, and insulated handles are used to pull the plates apart to a new spacing of 2.0 mm.
   b. What are the charge on each electrode and the potential difference between them?
40. || Two 2.0 cm × 2.0 cm metal electrodes are spaced 1.0 mm apart and connected by wires to the terminals of a 9.0 V battery.
   a. What are the charge on each electrode and the potential difference between them?
   While the plates are still connected to the battery, insulated handles are used to pull them apart to a new spacing of 2.0 mm.
   b. What are the charge on each electrode and the potential difference between them?
41. ||| A science-fair radio uses a homemade capacitor made of two 35 cm × 35 cm sheets of aluminum foil separated by a 0.25-mm-thick sheet of paper. What is its capacitance?

42. ||| A capacitor consists of two 6.0-cm-diameter circular plates
INT separated by 1.0 mm. The plates are charged to 150 V, then the battery is removed.
   a. How much energy is stored in the capacitor?
   b. How much work must be done to pull the plates apart to where the distance between them is 2.0 mm?

43. | The highest magnetic fields in the world are generated when
INT large arrays, or "banks," of capacitors are discharged through the copper coils of an electromagnet. At the National High Magnetic Field Laboratory, the total capacitance of the capacitor bank is 32 mF. These capacitors can be charged to 16 kV.
   a. What is the energy stored in the capacitor bank when it is fully charged?
   b. When discharged, the entire energy from this bank flows through the magnet coil in 10 ms. What is the average power delivered to the coils during this time?

44. ||| The dielectric in a capacitor serves two purposes. It increases the capacitance, compared to an otherwise identical capacitor with an air gap, and it increases the maximum potential difference the capacitor can support. If the electric field in a material is sufficiently strong, the material will suddenly become able to conduct, creating a spark. The critical field strength, at which breakdown occurs, is 3.0 MV/m for air, but 60 MV/m for Teflon.
   a. A parallel-plate capacitor consists of two square plates, 15 cm on a side, spaced 0.50 mm apart with only air between them. What is the maximum energy that can be stored by the capacitor?
   b. What is the maximum energy that can be stored if the plates are separated by a 0.50-mm-thick Teflon sheet?

45. ||| The flash unit in a camera uses a special circuit to "step up"
INT the 3.0 V from the batteries to 300 V, which charges a capacitor. The capacitor is then discharged through a flashlamp. The discharge takes 10 μs, and the average power dissipated in the flashlamp is 10^5 W. What is the capacitance of the capacitor?

46. || A capacitor being charged has a *current* carrying charge to and away from the plates. In the next chapter we will define current to be dQ/dt, the rate of charge flow. What is the current in C/s to a 10 μF capacitor whose voltage is increasing at the rate of 2.0 V/s?

47. || The current that charges a capacitor transfers energy that is
CALC stored in the capacitor's electric field. Consider a 2.0 μF capacitor, initially uncharged, that is storing energy at a constant 200 W rate. What is the capacitor voltage 2.0 μs after charging begins?

48. || What are the charge on and the potential difference across each capacitor in Figure P23.48?

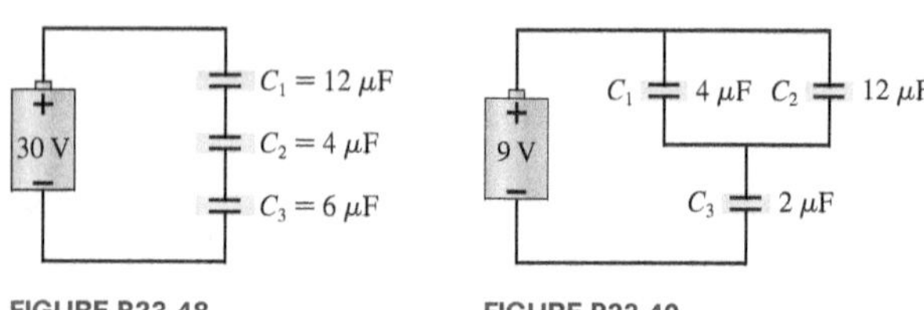

FIGURE P23.48 FIGURE P23.49

49. || What are the charge on and the potential difference across each capacitor in Figure P23.49?

50. || Initially, the switch in Figure P23.50 is in position a and capacitors C_2 and C_3 are uncharged. Then the switch is flipped to position b. Afterward, what are the charge on and the potential difference across each capacitor?

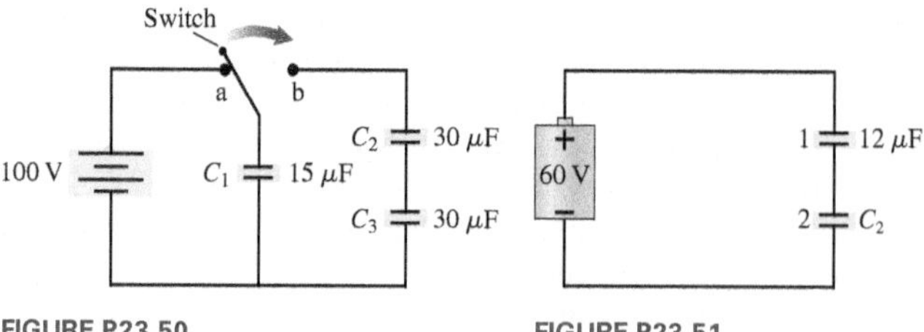

FIGURE P23.50 FIGURE P23.51

51. || A battery with an emf of 60 V is connected to the two capacitors shown in Figure P23.51. Afterward, the charge on capacitor 2 is 450 μC. What is the capacitance of capacitor 2?

52. || Capacitors $C_1 = 10\ \mu$F and $C_2 = 20\ \mu$F are each charged to 10 V, then disconnected from the battery without changing the charge on the capacitor plates. The two capacitors are then connected in parallel, with the positive plate of C_1 connected to the negative plate of C_2 and vice versa. Afterward, what are the charge on and the potential difference across each capacitor?

53. || An isolated 5.0 μF parallel-plate capacitor has 4.0 mC of
INT charge. An external force changes the distance between the electrodes until the capacitance is 2.0 μF. How much work is done by the external force?

54. | What is the dielectric constant of an insulator in a parallel-plate capacitor for which the induced surface charge density on the insulator is 50% of the surface charge density on the plates?

55. ||| Two 5.0-cm-diameter metal disks separated by a 0.50-mm-thick piece of Pyrex glass are charged to a potential difference of 1000 V. What are (a) the surface charge density on the disks and (b) the surface charge density on the glass?

56. || A virus can be modeled as a 50-nm-diameter sphere. Viruses,
BIO like biomolecules, are charged. A typical virus has a charge of $-300e$. Consider a virus inside a 37°C mammalian cell where the ion concentration is 150 mM.
   a. What are the electric potential, the positive ion concentration, and the negative ion concentration at the surface of the virus?
   b. What are the electric potential, the positive ion concentration, and the negative ion concentration 3.0 nm from the surface of the virus?

57. ||| One cycle of a sodium-potassium exchange pump, pow-
BIO ered by the hydrolysis of one molecule of ATP, moves three
INT Na^+ ions out of the cell and two K^+ ions into the cell. Each ion moves through a changing electric potential *and* a changing concentration. The change in the Gibbs free energy is $\Delta G = \Delta G_{elec} + \Delta G_{conc}$. The electric contribution is simply $\Delta G_{elec} = \Delta U_{elec}$. The concentration contribution is due to a change in entropy. This process is similar to osmosis, and a similar expression applies: $\Delta G_{conc} = N k_B T \ln(c_f/c_i)$, where N is the number of ions moved and c_i and c_f are the initial and final concentrations. Consider a muscle cell that has a membrane potential of -90 mV. The sodium ion concentrations are $[Na^+]_{in} = 15$ mM and $[Na^+]_{out} = 150$ mM. The potassium ion concentrations are $[K^+]_{in} = 150$ mM and $[K^+]_{out} = 4$ mM. Assume a temperature of 37°C.
   a. What is ΔG for one cycle of the sodium-potassium exchange pump?
   b. In a cell, the energy released from the hydrolysis of ATP is approximately 60 kJ/mol. What is the energy efficiency of the sodium-potassium exchange pump?

58. ||| A 4.0-nm-diameter protein is in a 0.05 M KCl solution at
BIO 25°C. The protein has 9 positive and 20 negative charges. Model the protein as a sphere with a uniform surface charge density. What is the electric potential of the protein (a) at the surface and (b) 2.0 nm from the surface?

59. ||| The electric field strength at the surface of a flat negative electrode in 20°C salt water is 20 kV/m. The field strength is 10 kV/m at a distance of 2.0 nm from the surface.
   a. What is the electrode's surface charge density in nC/cm^2?
   b. What is the molarity of the salt solution? Assume the salt is NaCl.

60. || In a mammalian cell, by how many mV does the Nernst po-
BIO tential of an ion increase if the external ion concentration is doubled?

MCAT-Style Passage Problems

A Lightning Strike

Friction-like collisions between ice particles cause a storm cloud to build up a large amount of negative charge near the base of the cloud. The charges dwell in *charge centers,* regions of concentrated charge. Suppose a cloud has -25 C in a 1.0-km-diameter spherical charge center located 10 km above the ground, as sketched in Figure P23.61. The negative charge center attracts a similar amount of positive charge that is spread on the ground below the cloud.

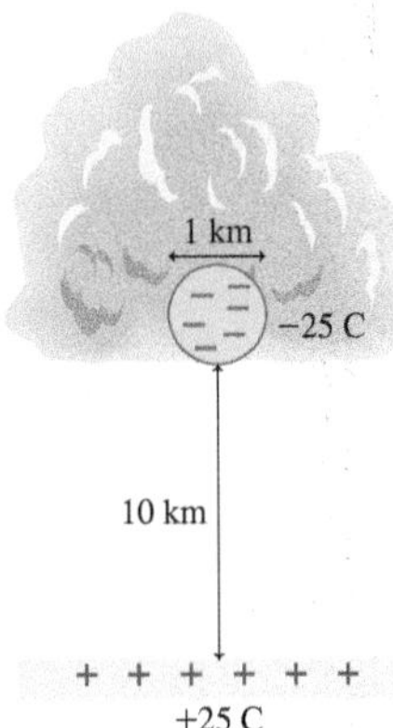

FIGURE P23.61

The charge center and the ground function as a charged capacitor, with a potential difference of approximately 4×10^8 V. The large electric field between these two "electrodes" may ionize the air, leading to a conducting path between the cloud and the ground. Charges will flow along this conducting path, causing a discharge of the capacitor—a lightning strike.

61. What is the approximate magnitude of the electric field between the charge center and the ground?
   A. 4×10^4 V/m
   B. 4×10^5 V/m
   C. 4×10^6 V/m
   D. 4×10^7 V/m

62. What is the approximate capacitance of the charge center + ground system?
   A. 6×10^{-8} F
   B. 2×10^7 F
   C. 4×10^6 F
   D. 8×10^6 F

63. If 12.5 C of charge is transferred from the cloud to the ground in a lightning strike, what fraction of the stored energy is dissipated?
   A. 12%
   B. 25%
   C. 50%
   D. 75%

64. If the cloud transfers all of its charge to the ground via several rapid lightning flashes lasting a total of 1 s, what is the average power?
   A. 1 GW
   B. 2 GW
   C. 5 GW
   D. 10 GW

24 Current and Resistance

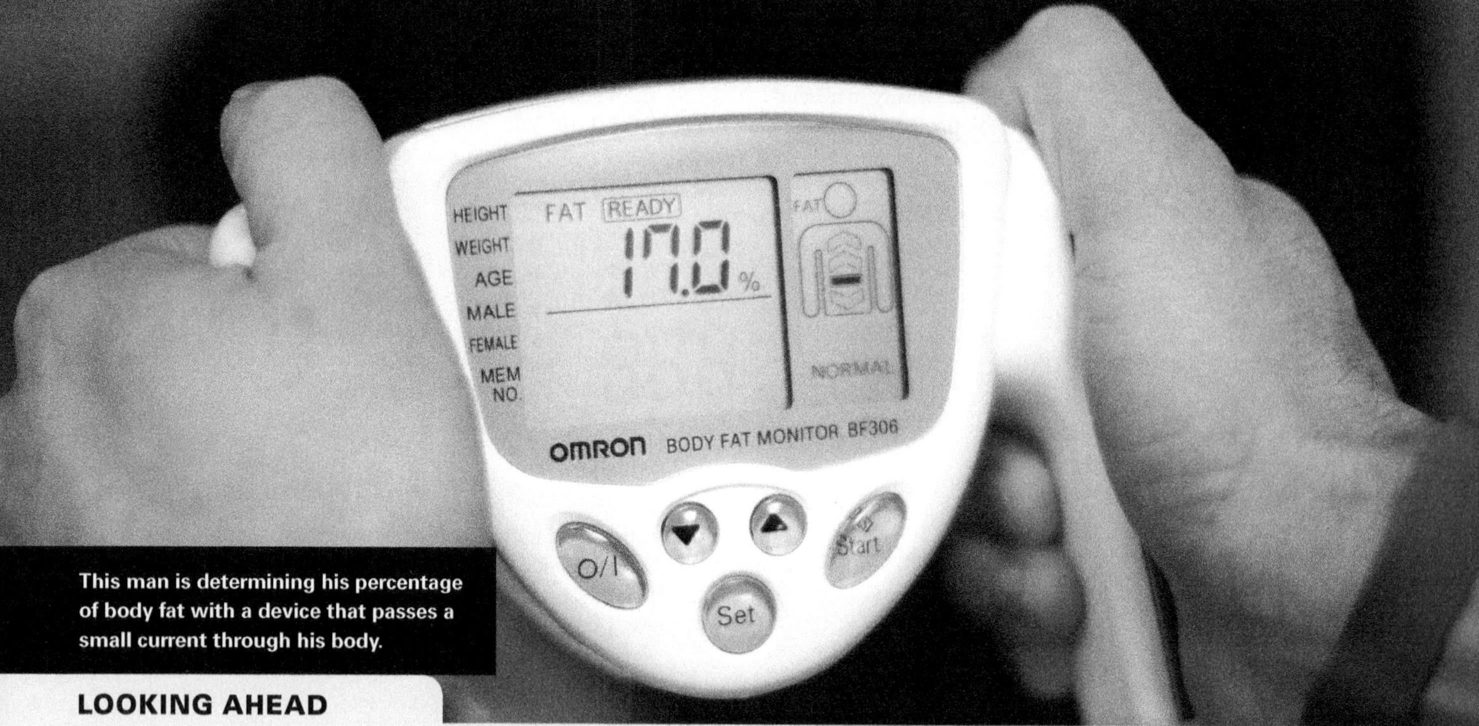

This man is determining his percentage of body fat with a device that passes a small current through his body.

LOOKING AHEAD

Current

A large electric *current* is recharging these electric cars. A charging station that provides more current allows the driver to recharge in less time.

You'll learn that current is the rate of charge flow through a conductor and that currents must obey the law of conservation of charge.

Resistance

A current passing through a flashlight bulb makes it glow. The size of the current is determined by the voltage of the battery and the *resistance* of the bulb.

You'll learn to use *Ohm's law* to calculate the current passing through resistors and to analyze circuits.

Electric Power

A hair dryer converts electric energy to thermal energy. The power consumption of electrical devices is measured in watts.

You'll learn how to calculate the power dissipation of resistors in both DC and AC circuits.

GOAL To learn how and why charge moves through a conductor as a current.

PHYSICS AND LIFE

Some Cells Respond to an Electric Current

Nerve cells fire and muscle fibers contract in response to an electric current—the flow of charges such as sodium and potassium ions. Electrophysiologists and neurobiologists probe the detailed workings of these electrically excitable cells by measuring and manipulating these currents. The current that passes through a single ion channel in the cell membrane, as small as it is, can be measured with a technique called a *patch clamp*. Electric current also plays an important role in the gel electrophoresis of proteins and nucleic acids; as the charged molecular fragments move through the gel, so do ions in the surrounding buffer. If not controlled, the ion current can heat the apparatus and even melt the gel. We'll develop the basic ideas of electric current in this chapter, then analyze the electrical workings of excitable cells in more detail in the next.

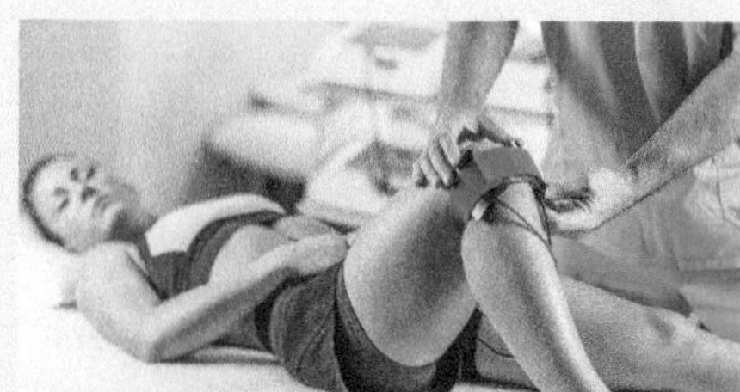

Electric stimulation therapy sends pulses of current through muscles, causing them to rhythmically contract.

24.1 A Model of Current

The previous three chapters focused on *electrostatics,* the interactions of charges at rest. These interactions are fundamental, but many of the most significant functions of electricity involve *current,* the motion of charges. Nerve signals and muscle contractions both rely on ion currents flowing across the cell membrane when ion channels open. A substantial current of ions also passes through the saltwater buffer solution during a gel electrophoresis experiment. Devices such as your phone and laptop have electric circuits that function by controlling the currents that pass through a myriad of small circuit elements. Magnetic fields, such as those in an MRI magnet, are created by currents. We'll explore circuits and magnetic fields in Chapters 25 and 26. First, we need to establish what a current is and how charge moves through a conductor.

Let's start our exploration of current with a very simple experiment. FIGURE 24.1a shows a charged parallel-plate capacitor. If we connect the two capacitor plates to each other with a metal wire, as shown in FIGURE 24.1b, the plates quickly become neutral. We say that the capacitor has been *discharged.*

The wire is a conductor, a material through which charge easily moves. Apparently the excess charge on one capacitor plate is able to move through the wire to the other plate, neutralizing both plates. The motion of charges through a material is called a **current,** so the capacitor is discharged by a current in the connecting wire.

If we observe the capacitor discharge more closely, we see other effects. As FIGURE 24.2 shows, the connecting wire gets warmer. If the wire is very thin in places, such as the thin filament in a lightbulb, the wire gets hot enough to glow. The current-carrying wire also deflects a compass needle, a phenomenon we'll study in Chapter 26. For now, we will use "makes the wire warmer" and "deflects a compass needle" as indicators that a current is present in a wire. In particular, we can use the brightness of a lightbulb to tell us the magnitude of the current; **more current means a brighter bulb.**

FIGURE 24.1 A capacitor is discharged by a metal wire.

(a)

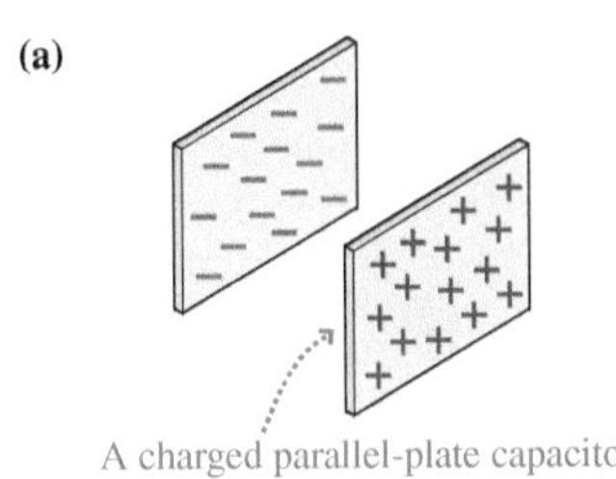
A charged parallel-plate capacitor

(b)

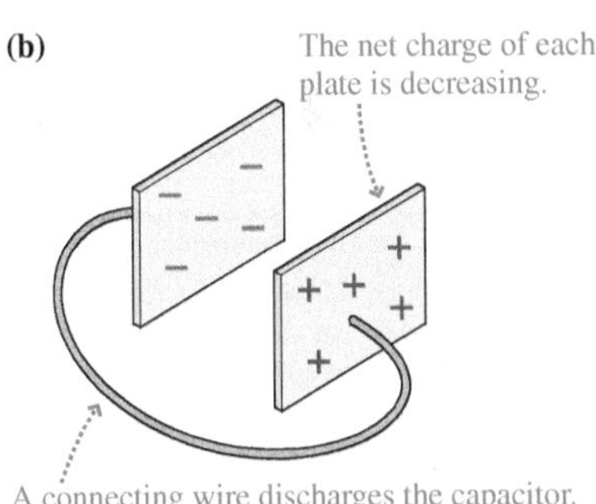
The net charge of each plate is decreasing.
A connecting wire discharges the capacitor.

FIGURE 24.2 Indicators of a current.

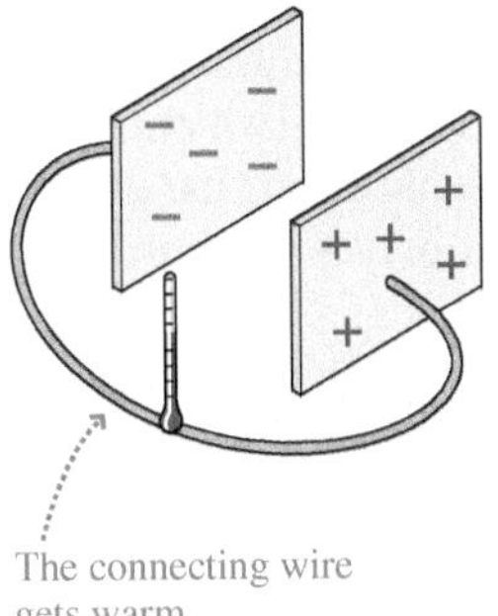
The connecting wire gets warm.

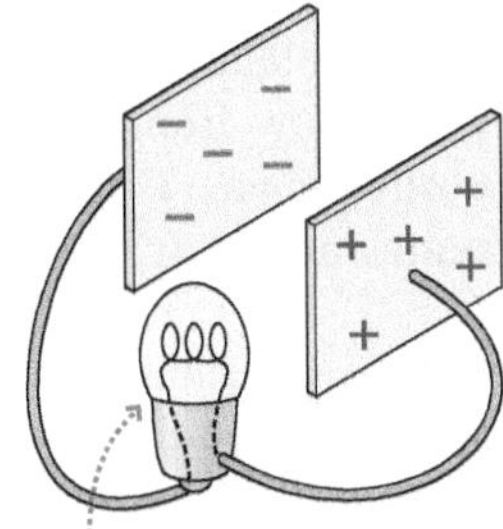
A lightbulb glows. The lightbulb filament is part of the connecting wire.

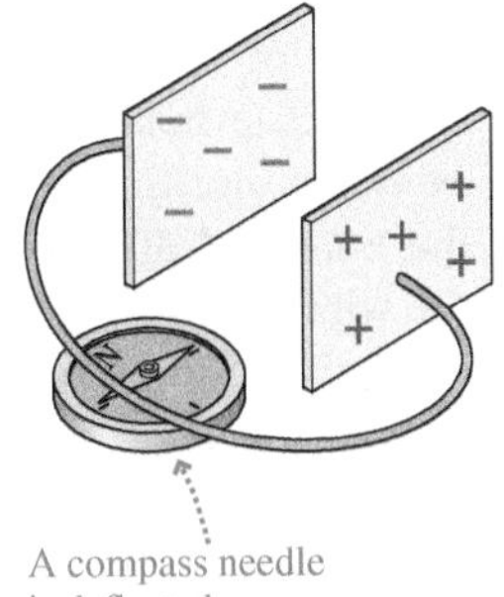
A compass needle is deflected.

The charges that move are called *charge carriers.* In a metal, the charge carriers are electrons. As FIGURE 24.3 shows, it is the motion of the *conduction electrons,* which are free to move around, that forms a current—a flow of charge—in the metal. An *insulator* does not have such free charges and cannot carry a current. Although electrons are the charge carriers in metals, other materials may have different charge carriers. In solutions, such as seawater, blood, and intercellular fluids, the charge carriers are ions, both positive and negative.

FIGURE 24.3 Conduction electrons in a metal.

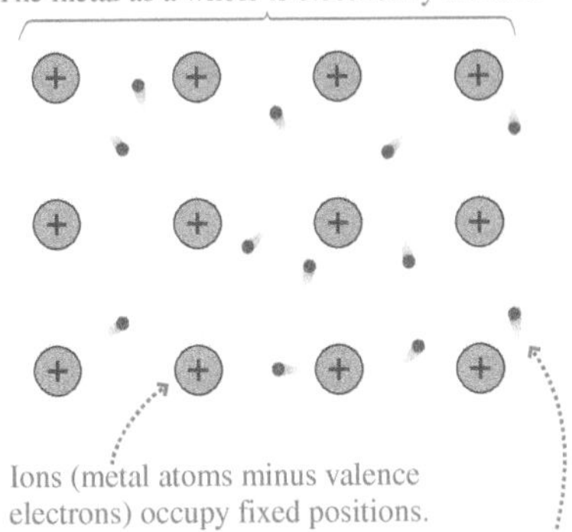
The metal as a whole is electrically neutral.
Ions (metal atoms minus valence electrons) occupy fixed positions.
The conduction electrons are bound to the solid as a whole, not any particular atom. They are free to move around.

Creating a Current

Suppose you want to slide a book across the table to your friend. You give it a quick push to start it moving, but it begins slowing down because of friction as soon as you take your hand off of it. The book's kinetic energy is transformed into thermal energy, leaving the book and the table slightly warmer. The only way to keep the book moving at a *constant* speed is to continue pushing it.

Something similar happens in a conductor. In the absence of a driving force, the conduction electrons, like the molecules in a liquid, undergo random thermal motions, but there is no *net* motion. If we push on the electrons, our push creates a fluid-like current of electrons moving through the conductor. But the electrons aren't moving in vacuum. Collisions between the electrons and the atoms of the metal slow the electrons down, which transforms their kinetic energy into the thermal energy of the metal and makes the metal warmer. The net motion of the electrons ends quickly *unless we keep pushing.*

FIGURE 24.4 The motion of an electron in a conductor.

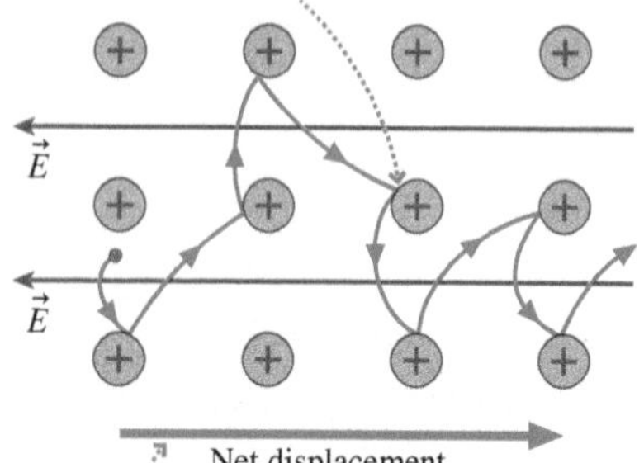
Each collision "resets" the motion of the electron. It then accelerates until the next collision.
$\vec{E}$
$\vec{E}$
Net displacement
A net displacement in the direction opposite to $\vec{E}$ is superimposed on the random thermal motion.

How do we push on electrons? With an electric field! FIGURE 24.4 shows a uniform electric field inside a conductor. The negative electrons experience a force of magnitude $F = eE$ in the direction opposite the field, and they accelerate in that direction. There's no acceleration perpendicular to the field, so each electron follows a parabolic trajectory like that of a projectile.

However, an electron doesn't travel far before it collides with an atom. Each collision "resets" the electron's motion, but the curvature of the trajectories causes the electrons to gradually drift in the direction opposite the field. The motion is similar to that of a ball in a pinball machine that has a downward tilt. The electrons move with a steady *drift speed* as long as they are pushed by the electric field, continuously transforming kinetic energy into thermal energy, but the drift ceases if we turn off the electric field that is pushing them along.

How can there be an electric field inside a conductor? One important conclusion of ◀◀ SECTION 21.5 was that $\vec{E} = \vec{0}$ inside a conductor that is in electrostatic equilibrium. However, a conductor in which the electrons have a steady drift speed is *not* in electrostatic equilibrium. **A current is a steady motion of charges sustained by an electric field inside the conductor.** And, as we'll see, an energy source—such as a battery—is needed to maintain the electric field.

FIGURE 24.5 Creating a current in a wire.

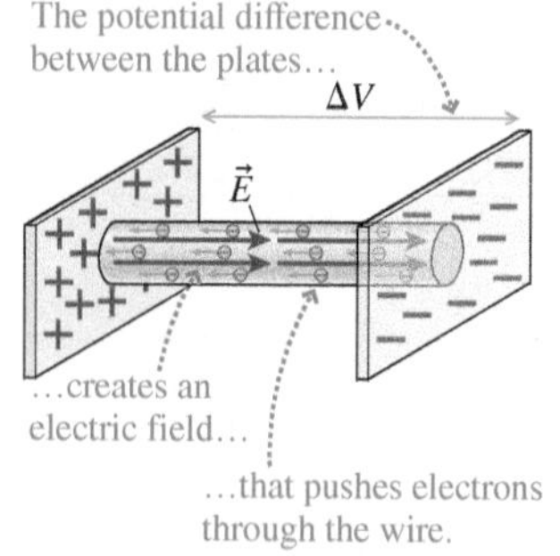
The potential difference between the plates...
ΔV
$\vec{E}$
...creates an electric field...
...that pushes electrons through the wire.

FIGURE 24.5 shows how a wire connected between the plates of a charged capacitor causes it to discharge. The initial separation of charges creates a potential difference between the two plates. We saw in ◀◀ SECTION 22.6 that wherever there's a potential difference, an electric field points from higher potential toward lower potential. Connecting a wire between the plates establishes an electric field in the wire, and this electric field causes electrons to flow from the negative plate (which has an excess of electrons) toward the positive plate. **The charge separation that creates the potential difference also creates an electric field that drives the current in the wire.**

As the current continues, and the charges flow, the plates discharge and the potential difference decreases. Eventually, the plates will be completely discharged, which means no more potential difference, no more field—and no more current. Finally, $\vec{E} = \vec{0}$ inside the conducting wire, and we have equilibrium.

24.2 Defining Current

The charge carriers in metals are electrons. Currents were known and used for decades before the electron was discovered, however. Early researchers established that charge moves through wires, but they had no knowledge of atomic particles and no way to discern what the charge carriers are. As a result, they arbitrarily defined current as the flow of *positive* charge, which moves in the direction of the electric field. Macroscopic measurements cannot distinguish between positive charges moving in the direction of the field and negative charges moving opposite to the field; the net result is the same. From now on, we'll think of current as the flow of positive charge.

NOTE ▶ The identity of the charge carriers *is* important for a microscopic analysis of processes. For example, understanding ion channels and ion exchange pumps in cellular membranes depends very much on knowing which ions are moving in which directions. But a macroscopic measurement of the current in an ion channel reveals only the net motion of the charges, nothing about what the charges are. ◀

FIGURE 24.6 shows a wire in which an electric field $\vec{E}$ pushes positive charge in the direction of the field. Because the coulomb is the SI unit of charge, and because currents are charges in motion, we define current I as the *rate,* in coulombs per second, at which charge moves through a wire:

$$I = \frac{dq}{dt} \quad (24.1)$$

Current

FIGURE 24.6 The current I is in the direction of the electric field.

The current I is considered to be the motion of positive charges in the electric field.

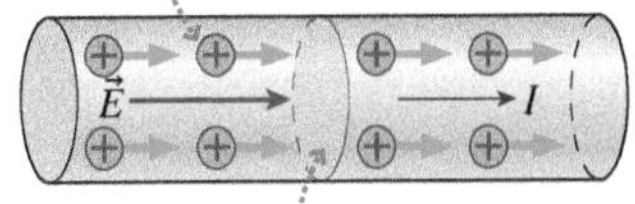

Current is the rate in C/s that charge moves through a cross section of the wire.

For a steady current in a wire, which will be our primary focus, the amount of charge that passes through any cross section of the wire during a time interval Δt is

$$Q = I\Delta t \quad (24.2)$$

The SI unit for current is the coulomb per second, which is called the **ampere** A:

$$1 \text{ ampere} = 1 \text{ A} = 1 \text{ coulomb per second} = 1 \text{ C/s}$$

The *amp* is an informal shortening. Household currents are typically a few amps. The currents in electronic devices are much smaller—typically measured in milliamps ($1 \text{ mA} = 10^{-3}$ A) or microamps ($1 \mu\text{A} = 10^{-6}$ A). Ion-channel currents that establish the cell membrane potential are no greater than a few picoamps ($1 \text{ pA} = 10^{-12}$ A).

EXAMPLE 24.1 **Discharging a laptop**

A laptop battery can supply a 1.0 A current for 6.0 hours. How much charge leaves the battery?

PREPARE Equation 24.2 gives the charge in terms of the current and the time interval.

SOLVE The total charge that leaves the battery and then passes through the computer circuits is

$$Q = I\Delta t = (1.0 \text{ A})\left(6.0 \text{ h} \times \frac{60 \text{ min}}{1 \text{ h}} \times \frac{60 \text{ s}}{1 \text{ min}}\right) = 2.2 \times 10^4 \text{ C}$$

ASSESS That's a lot of charge! Objects that are charged by rubbing or friction acquire a charge of a few nC. The charge that can be delivered by a modern battery is on an entirely different level.

STOP TO THINK 24.1 Suppose positive charges flow across a cell membrane from the outside to the inside of a cell. At the same time, an equal number of negative charges flow at the same rate from the inside to the outside. The total current is

A. Zero.
B. The rate of flow of the positive charges.
C. The rate of flow of the negative charges.
D. Twice the rate of flow of the positive charges.

Charge Conservation and Current

FIGURE 24.7 How does the brightness of bulb A compare to that of bulb B?

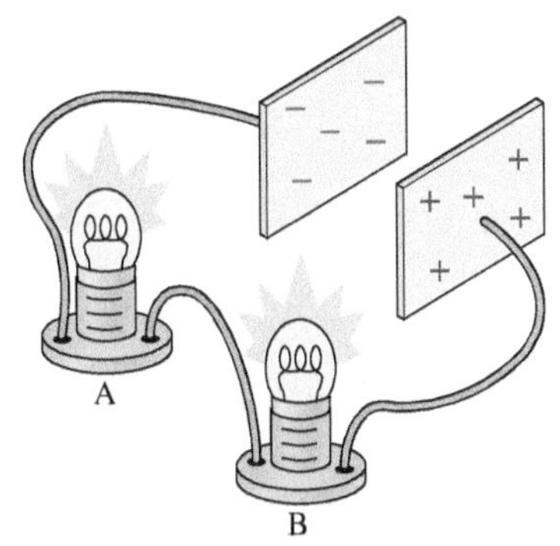

FIGURE 24.7 shows two identical lightbulbs in the wire connecting two charged capacitor plates. Both bulbs glow as the capacitor is discharged. How do you think the brightness of bulb A compares to that of bulb B? Is one brighter than the other? Or are they equally bright? Think about this before going on.

You might have predicted that B is brighter than A because the current I, which carries positive charges from plus to minus, reaches B first. In order to be glowing, B must use up some of the current, leaving less for A. Or perhaps you realized that the actual charge carriers are electrons, moving from minus to plus. The conventional current I may be mathematically equivalent, but physically it's the negative electrons

FIGURE 24.8 A current dissipates energy, but the flow is unchanged.

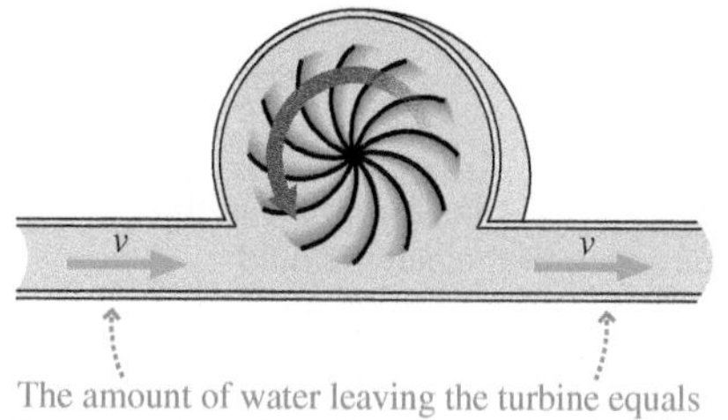

The amount of water leaving the turbine equals the amount entering; the number of electrons leaving the bulb equals the number entering.

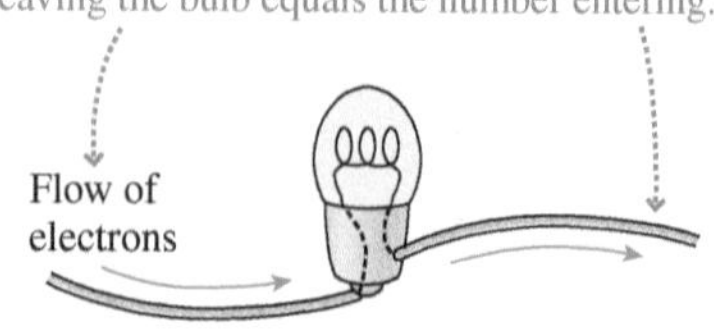

rather than positive charge that actually move. Because the electrons get to A first, you might have predicted that A is brighter than B.

In fact, both bulbs are equally bright. This is an important observation, one that demands an explanation. After all, "something" gets used up to make the bulb glow, so why don't we observe a decrease in the current? Current is the amount of charge moving through the wire per second. Electrons, the charge carriers, are charged particles. The lightbulb can't destroy electrons without violating both the law of conservation of mass and the law of conservation of charge. Thus the amount of charge (i.e., the *number* of electrons) cannot be changed by a lightbulb. Further, the lightbulb cannot store electrons. If it did, the bulb would become increasingly negatively charged until repulsive forces stopped the incoming flow of electrons and the light went out.

Consider the fluid analogy shown in FIGURE 24.8. Suppose the water flows into one end at a rate of 2.0 kg/s. Is it possible that the water, after turning a paddle wheel, flows out the other end at a rate of only 1.5 kg/s? That is, does turning the paddle wheel cause the water current to decrease?

We can't destroy water molecules any more than we can destroy electrons, we can't increase the density of water by pushing the molecules closer together, and there's nowhere to store extra water inside the pipe. Each drop of water entering the left end pushes a drop out the right end; hence water flows out at exactly the same rate it flows in.

The same is true for electrons in a wire. **The rate of electrons leaving a lightbulb (or any other device) is exactly the same as the rate of electrons entering the lightbulb. The current does not change.** A lightbulb doesn't "use up" current, but it *does*—like the paddlewheel in the fluid analogy—use energy. The kinetic energy of the electrons is dissipated by their collisions with the ions in the metal (atomic-level friction), making the wire hotter until, in the case of the lightbulb filament, it glows.

There are many issues that we'll need to look at before we can say that we understand how currents work, and we'll take them one at a time. For now, we draw a first important conclusion: **Due to conservation of charge, the current must be the same at all points in an individual current-carrying wire.**

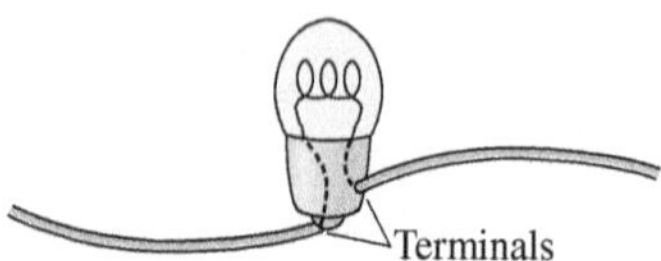

Let there be light A lightbulb—whether an incandescent bulb with a glowing filament or an LED bulb in your phone or flashlight—has *two* terminals. Current passes *through* the bulb, entering one terminal and exiting the other. For a screw-in bulb, the metal casing on the side of the bulb is one terminal, and the tip—separated by an insulator—is the other.

FIGURE 24.9a summarizes the situation in a single wire. But what about FIGURE 24.9b, where one wire splits into two and two wires merge into one? A point where a wire branches is called a **junction.** The presence of a junction doesn't change our basic reasoning. We cannot create or destroy electrons in the wire, and neither can we store them in the junction. The rate at which electrons flow into one *or many* wires must be exactly balanced by the rate at which they flow out of others. For a *junction,* the law of conservation of charge requires that

FIGURE 24.9 The sum of the currents into a junction must equal the sum of the currents leaving the junction.

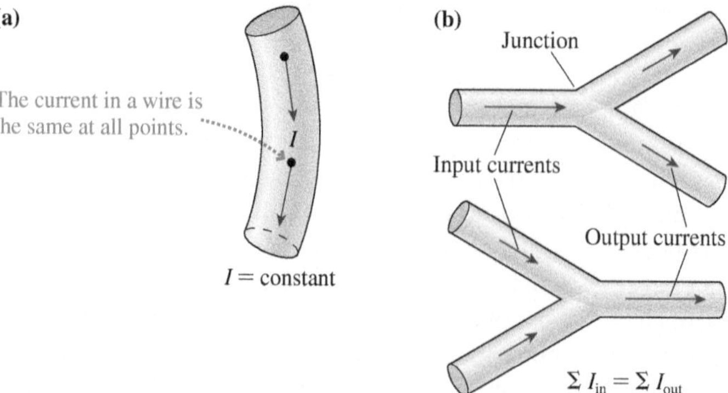

$$\sum I_{in} = \sum I_{out} \tag{24.3}$$

Kirchhoff's junction law

where, as usual, the Σ symbol means summation.

This basic conservation statement—that the sum of the currents into a junction equals the sum of the currents leaving—is called **Kirchhoff's junction law.** The junction law, together with *Kirchhoff's loop law* that you met in Chapter 22, will play an important role in circuit analysis in the next chapter.

Current in Solutions

Our definition of current also applies to the flow of dissolved ions in a solution, but the details differ. FIGURE 24.10 shows a solution—sometimes called an *electrolyte*—that contains both positive and negative ions. A positive ion in an electric field experiences a force toward the negative electrode and, while undergoing many collisions, slowly migrates in that direction. Similarly, a negative ion drifts toward the positive electrode. Both ions are charge carriers, and both contribute to the current.

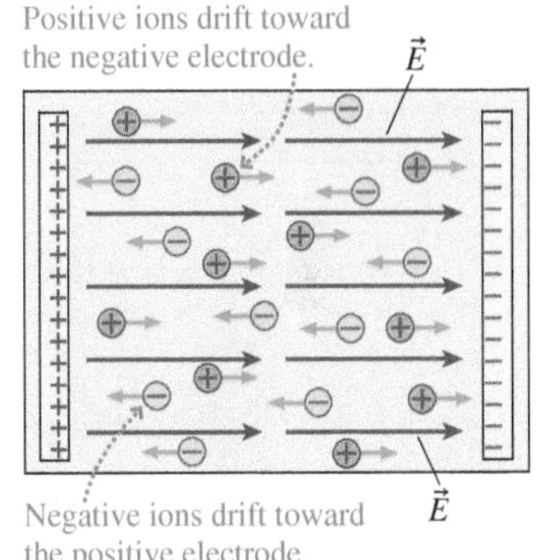
Positive ions drift toward the negative electrode.
$\vec{E}$
Negative ions drift toward the positive electrode.
$\vec{E}$

FIGURE 24.10 Both positive and negative ions are charge carriers in a solution.

Suppose that N_+ positive ions, each with charge q_+, arrive at the negative electrode during a time interval Δt. During the same interval, N_- negative ions, each with charge q_-, arrive at the positive electrode. The current through the solution is

$$I = q_+ \frac{N_+}{\Delta t} - q_- \frac{N_-}{\Delta t} \tag{24.4}$$

This is the net rate of charge flow. The minus sign appears because q_- is negative. The charge is $q_+ = +e$ for ions such as Na^+ and K^+, but Ca^{2+} has $q_+ = +2e$. The common negative ion Cl^- has $q_- = -e$, but the sulfate ion SO_4^{2-} has $q_- = -2e$.

NOTE ► It might seem like the positive ion and negative ion currents would cancel each other. But because q_- is negative, the negative sign in Equation 24.4 means that the current of negative ions *adds* to the current of positive ions. ◄

EXAMPLE 24.2 Silver-plating jewelry

Electroplating is a technique that deposits a thin metal coating on a conductor. A soluble compound of the metal is dissolved in water to create positive and negative ions, and then a battery creates a current between two electrodes. Positive metal ions migrate to and slowly coat the negative electrode. For silver-plating, which places a coating of silver on a cheaper metal, silver cyanide (AgCN) is dissolved to form Ag^+ and CN^- ions. The recommended current is 10 mA per cm^2 of electrode surface area. How many silver atoms are deposited on a 5.0 cm^2 metal piece of jewelry in 1.0 h?

PREPARE The piece of jewelry is the negative electrode. The solution stays electrically neutral, so we can assume that CN^- ions arrive at the positive electrode at the same rate that Ag^+ ions arrive at the negative electrode. Coating a 5.0 cm^2 piece requires a 50 mA current.

SOLVE Equal arrival rates of the positive and negative ions means that $N_- = N_+$. The ions have charges $q_+ = +e$ and $q_- = -e$. Consequently, the current through the solution is

$$I = q_+ \frac{N_+}{\Delta t} - q_- \frac{N_-}{\Delta t} = e\frac{N_+}{\Delta t} - (-e)\frac{N_+}{\Delta t} = 2e\frac{N_+}{\Delta t}$$

In 1 h = 3600 s, the number of Ag^+ ions that arrive at the negative piece of jewelry is

$$N_+ = \frac{I}{2e}\Delta t = \frac{0.050\text{ A}}{2(1.60 \times 10^{-19}\text{ C})}(3600\text{ s}) = 5.6 \times 10^{20}$$

ASSESS The atomic mass of silver is 108 g/mol. 5.6×10^{20} atoms is $\approx 10^{-3}$ mol, and 10^{-3} mol of silver has a mass of ≈ 0.1 g. Very little silver is needed to form a thin coating on an object, so our answer seems reasonable.

STOP TO THINK 24.2 What are the magnitude and the direction of the current in the fifth wire?

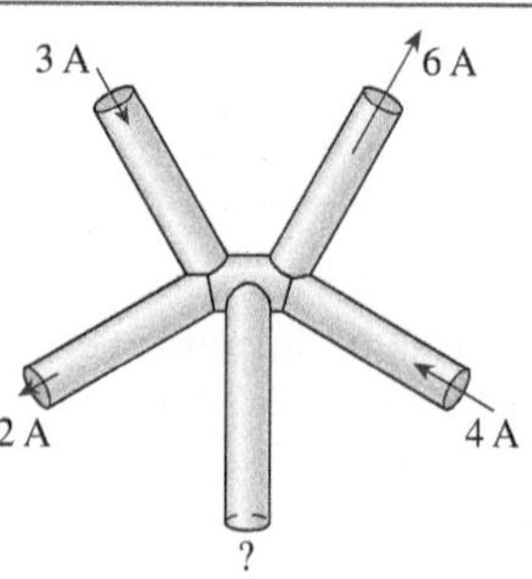

3 A
6 A
2 A
4 A
?

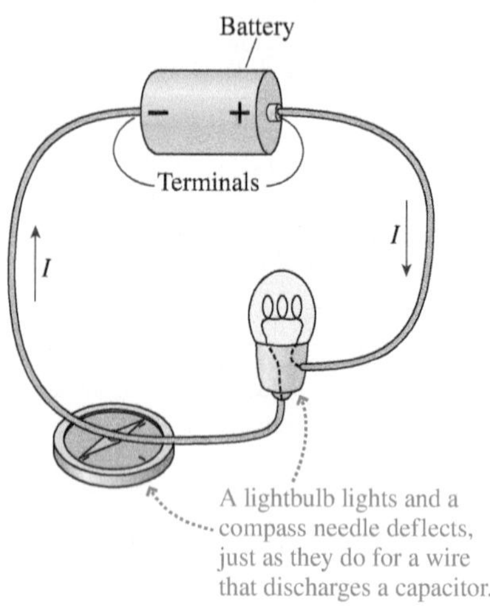

Battery
Terminals
I
I
A lightbulb lights and a compass needle deflects, just as they do for a wire that discharges a capacitor.

FIGURE 24.11 There is a current in a wire connecting the terminals of a battery.

24.3 Batteries and emf

There are practical devices, such as a camera flash, that use the charge on a capacitor to create a current. But a camera flash gives a single, bright flash of light; the capacitor discharges and the current ceases. If you want a light to illuminate your way along a dark path, you need a *continuous* source of light such as a flashlight. Continuous light requires the current to be continuous as well.

FIGURE 24.11 shows a wire connecting the two terminals of a battery, much like the wire that connected the capacitor plates in Figure 24.1. Just like that wire, the wire connecting the battery terminals gets warm, deflects a compass needle, and makes a lightbulb inserted into it glow brightly, which indicates that charges are flowing through the wire from one battery terminal to the other. Evidently the current in a wire is the same whether it is supplied by a capacitor or a battery. Everything you've learned so far about current applies equally well to the current supplied by a battery, with one important difference—the duration of the current.

The wire connecting the battery terminals *continues* to deflect the compass needle and *continues* to light the lightbulb. A capacitor quickly runs out of excess charge, but a battery can keep the charges in motion. In other words, a battery can *sustain* a current in a wire.

FIGURE 24.12 shows the charge escalator model of a battery that we introduced in ◀◀ SECTION 22.5. Charges are removed from the lower-potential negative terminal and "lifted" to the higher-potential positive terminal. By separating charges, the charge escalator establishes the potential difference ΔV_{bat} between the battery terminals. The potential difference established by a device, such as a battery, that can actively separate charges is called its emf $\mathcal{E}$; that is, $\Delta V_{bat} = \mathcal{E}$.

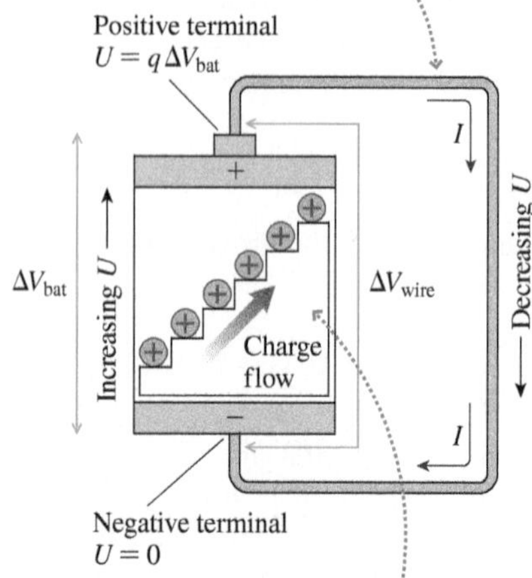

FIGURE 24.12 The charge escalator model of a battery.

An isolated battery has a potential difference but no current because the charges have nowhere to go. Connecting a wire across the terminals of the battery allows charges from the positive terminal—the charges that have gained potential energy on the charge escalator—to "fall downhill" through the wire until they reach the negative terminal. From there they can be lifted up to start the loop all over again. This flow of charge around a continuous loop—a current—is called a **complete circuit.**

Because the ends of the wire are connected to the battery terminals, the potential difference between the two ends of the wire is equal to the potential difference of the battery:

$$\Delta V_{wire} = \Delta V_{bat} \tag{24.5}$$

A potential difference between the ends of the wire creates an electric field in the wire that drives the current in the wire, but the battery's emf is the *cause* of the wire's potential difference.

The battery's charge escalator is doing work to separate the charges, and that requires energy. A battery consists of chemicals, called *electrolytes,* sandwiched between two electrodes made of different metals. The energy needed to separate the charges comes from chemical reactions between the electrolytes and the electrodes; these reactions move positive ions to one terminal and negative ions to the other.

In other words, it is chemical reactions rather than a mechanical conveyer belt that transport charge from one terminal to the other. A battery is "dead" when the supply of chemicals is exhausted.

A charge gains electric potential energy $\Delta U_{\text{elec}} = q\,\Delta V_{\text{bat}}$ as it moves from the negative terminal to the positive terminal. The charge gains kinetic energy as it moves into the connecting wire, because of the wire's electric field, but it soon undergoes a collision and transforms its kinetic energy into thermal energy of the wire. This acceleration-and-collision process is repeated over and over as the charge moves through the wire. Thus the energy transformations of a circuit are

$$E_{\text{chem}} \rightarrow U_{\text{elec}} \rightarrow K \rightarrow E_{\text{th}} \tag{24.6}$$

The battery's chemical energy powers the circuit, but the current slowly transforms that energy into thermal energy until the battery is depleted.

The *rating* or *voltage* of a battery, such as 1.5 V, is the battery's emf. It is determined by the specific chemical reaction used in the battery. An alkaline battery has an emf of 1.5 V. A rechargeable lithium ion battery has an emf of 3.7 V. Larger ΔV_{bat} can be created by placing several "cells" in a row, much like going from the first to the fourth floor of a building by taking three separate elevators. For example, a laptop battery uses either three or four 3.7 V lithium ion cells, which gives a battery a rating of 11.1 V or 14.8 V.

Devices other than batteries can generate an emf. Rooftop solar panels use the sun's energy to separate charge and generate an emf. A power plant generates an emf from the energy of expanding steam or falling water. We'll use batteries as the emf in most of our examples, but our conclusions apply to any source of emf.

A shocking predator? The torpedo ray captures and eats fish by paralyzing them with electricity. Special cells in the body of the ray called *electrocytes* produce an emf of a bit more than 0.10 V for a short time when stimulated. Such a small emf will not produce a large effect, but the torpedo ray has organs that contain clusters of hundreds of these electrocytes connected in series. The total emf can be 50 V or more, enough to immobilize nearby prey.

CONCEPTUAL EXAMPLE 24.3 Potential difference for batteries in series

Three batteries are connected one after the other as shown in FIGURE 24.13; we say they are connected in *series*. What's the total potential difference?

REASON Batteries have a distinct polarity: a + terminal of higher potential and a − terminal of lower potential. When they are connected as shown, we can think of this as three charge escalators, one after the other. Each one lifts charges to a higher potential. Because each battery raises the potential by 1.5 V, the total potential difference of the three batteries in series is 4.5 V.

ASSESS Common AA and AAA batteries are 1.5 V batteries. Many consumer electronics, such as digital cameras, use two or four of these batteries. Wires inside the device connect the batteries in series to produce a total 3.0 V or 6.0 V potential difference.

FIGURE 24.13 Three batteries in series.

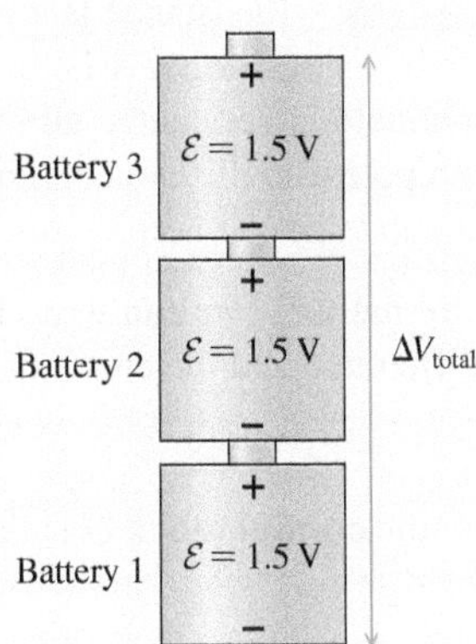

STOP TO THINK 24.3 A battery produces a current in a wire. As the current continues, which of the following quantities (perhaps more than one) decreases?

A. The positive charge in the battery
B. The emf of the battery
C. The chemical energy in the battery

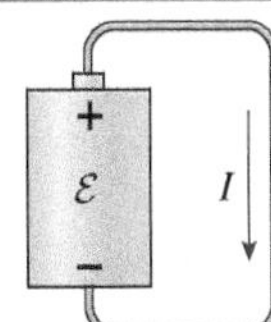

24.4 Resistance and Conductance

FIGURE 24.14 The current through a conductor is related to the potential difference ΔV.

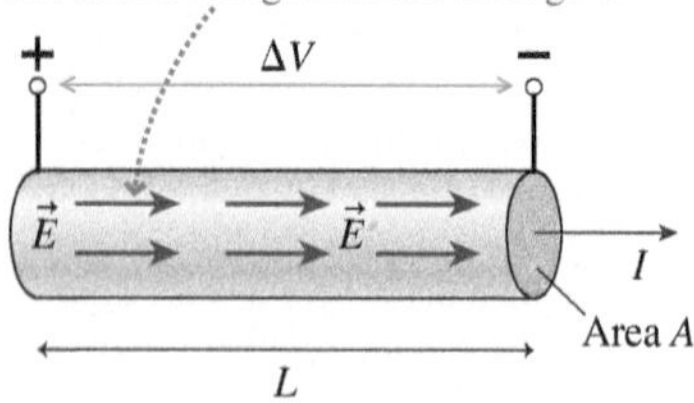
The potential difference creates an electric field inside the conductor and causes charges to flow through it.
+
ΔV
−
E
E
I
Area A
L

FIGURE 24.14 shows a conductor that has length L and cross-section area A. We've shown the conductor as a wire, but it could be a conductor of any size and shape. A potential difference ΔV has been established between the ends of the conductor—perhaps it is connected to a battery, making $\Delta V = \Delta V_{bat}$. The potential difference creates an electric field in the conductor that drives current I through the wire.

What determines the magnitude of the current? How is current I related to the potential difference ΔV? It's easy to demonstrate in the laboratory that I is directly proportional to ΔV; doubling the potential difference doubles the current. But the size of the current also depends on the properties of the conductor. Making the conductor longer or thinner reduces the current. And less current passes through an iron wire than through a copper wire of the same dimensions.

For any particular conductor, we can define a quantity called the **resistance** R that is a measure of how hard it is to push charges through the conductor. The larger the resistance, the smaller the current. Experiments show that the size of the current depends on both the potential difference (due to the battery) *and* the resistance (a property of the wire):

$$I = \frac{\Delta V}{R} \tag{24.7}$$

FIGURE 24.15 A graph of current versus potential difference is a straight line. The slope can be used to determine the conductor's resistance and conductance.

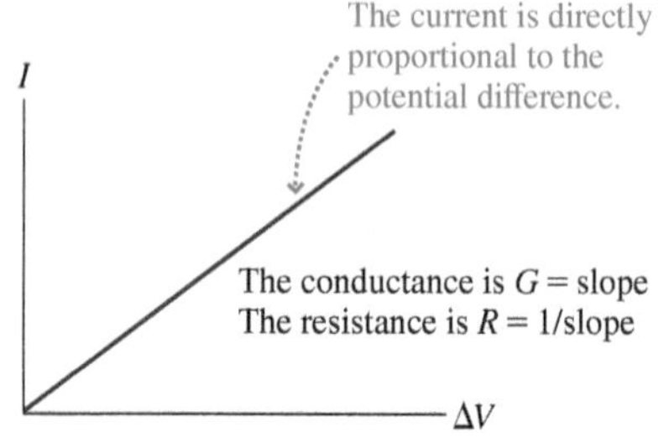
The current is directly proportional to the potential difference.
I
The conductance is G = slope
The resistance is R = 1/slope
ΔV

FIGURE 24.15 shows that a graph of current versus potential difference is a straight line with slope $1/R$.

The SI unit of resistance is the **ohm,** abbreviated Ω (uppercase Greek omega), which is defined as

$$1 \text{ ohm} = 1\ \Omega = 1 \text{ V/A}$$

The ohm is the basic unit of resistance, but kilohms ($1\text{ k}\Omega = 10^3\ \Omega$) and megohms ($1\text{ M}\Omega = 10^6\ \Omega$) are widely used.

NOTE ▸ Resistance is a property of a specific wire or conductor. The conductor has a resistance whether or not it is connected to a potential difference. A conductor's resistance allows us to calculate the current that flows in response to a potential difference in much the same way that an object's mass allows us to calculate how it accelerates in response to a force. ◂

Alternatively, we can write Equation 24.7 for the relationship between the potential difference and the current as

$$I = G\Delta V \tag{24.8}$$

where the **conductance** G is the inverse of the resistance:

$$G = \frac{1}{R} \tag{24.9}$$

The conductance is the slope of the graph in Figure 24.15.

The unit of conductance is the **siemens,** abbreviated S, where $1\text{ S} = 1\ \Omega^{-1} = 1\text{ A/V}$. Physicists and engineers typically analyze charge flow in terms of resistance, but biological circuits—such as ion channels—are often described in terms of their conductance. For the same potential difference, a conductor with a larger conductance carries a larger current than a conductor with a smaller conductance.

EXAMPLE 24.4 **The resistance and conductance of an arm**

Your arm, which is filled with conducting liquids, is a conductor. A potential difference of 1.0 V between microelectrodes inserted into the wrist and shoulder (to avoid the high resistance of the skin) creates a 3.2 mA current down the length of the arm—barely enough to be noticed. What are the resistance and conductance of the arm?

PREPARE Resistance relates current to the applied potential difference.

SOLVE We can use Equation 24.7 to compute the resistance:

$$R = \frac{\Delta V}{I} = \frac{1.0\ \text{V}}{0.0032\ \text{A}} = 310\ \Omega$$

Thus the conductance is

$$G = \frac{1}{R} = \frac{1}{310\ \Omega} = 3.2 \times 10^{-3}\ \text{S} = 3.2\ \text{mS}$$

ASSESS We'll use these values in Chapter 25 to look at electrical safety.

Cause and Effect

Distinguishing cause and effect is vitally important for understanding how a circuit works. FIGURE 24.16 illustrates the logic. A critical idea is that a battery is a source of constant potential difference, *not* a source of constant current. The battery's emf creates the potential difference, but the current that leaves a battery and passes through a circuit is determined jointly by the emf of the battery and the resistance of the circuit.

FIGURE 24.16 A cause-and-effect model of how a current is established in a circuit.

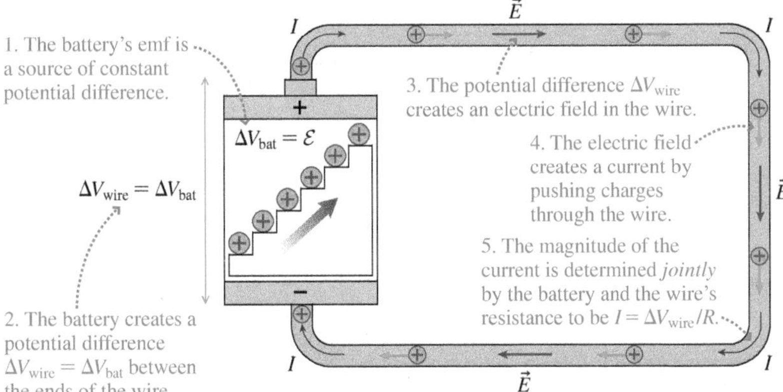
1. The battery's emf is a source of constant potential difference.
$\Delta V_{wire} = \Delta V_{bat}$
2. The battery creates a potential difference $\Delta V_{wire} = \Delta V_{bat}$ between the ends of the wire.
$\Delta V_{bat} = \mathcal{E}$
+
−
I
$\vec{E}$
3. The potential difference ΔV_{wire} creates an electric field in the wire.
4. The electric field creates a current by pushing charges through the wire.
5. The magnitude of the current is determined *jointly* by the battery and the wire's resistance to be $I = \Delta V_{wire}/R$.

Resistivity and Conductivity

An object's mass depends on its size and the density of the material from which it's made. Density is a property of the material; all pieces of copper, regardless of their size, have the same density. Similarly, an object's resistance and conductance depend on its size and the material from which it's made. It is useful to separate these and to have a way to characterize the electric properties of the material. We can do that with the material's **resistivity** ρ. Resistivity is a property of the material; all pieces of copper, regardless of their size, have the same resistivity. A good conductor has low resistivity; a poor conductor has high resistivity. The units of resistivity are $\Omega \cdot \text{m}$.

Alternatively, we can characterize a material by its **conductivity** σ (Greek sigma), which is the inverse of resistivity:

$$\sigma = \frac{1}{\rho} \tag{24.10}$$

A good conductor has a high conductivity. The units of conductivity are S/m. TABLE 24.1 gives the resistivities and conductivities of some common materials.

Metals are generally good conductors, with high conductivities and low resistivities, but some metals are better conductors than others. The conductivity of nichrome, an alloy used to make heater wires, is roughly a factor of 100 lower than that of copper. Pure water is an insulator, but seawater is a moderately good conductor (but much less so than metals) because the salts dissolve into ions that can carry current. The ion concentration in cytosol is lower than in seawater, so cytosol has a lower conductivity. Fat and muscle are relatively poor conductors.

TABLE 24.1 **Resistivity and conductivity of materials**

Material	Resistivity ($\Omega \cdot \text{m}$)	Conductivity (S/m)
Aluminum	2.8×10^{-8}	3.5×10^{7}
Copper	1.7×10^{-8}	6.0×10^{7}
Iron	9.7×10^{-8}	1.0×10^{7}
Nichrome	1.5×10^{-6}	6.7×10^{5}
Carbon	3.5×10^{-5}	2.9×10^{4}
Seawater	0.22	4.5
Cytosol	0.50	2.0
Blood	1.6	0.63
Muscle	8.0	0.13
Fat	25	4.0×10^{-2}
Cell membrane	4.0×10^{7}	2.5×10^{-8}

Figure 24.14 showed a conductor of length L and cross-section area A. Its resistance and conductance depend on its electric properties—its resistivity and conductivity—and on its dimensions:

$$R = \frac{\rho L}{A} \quad \text{and} \quad G = \frac{\sigma A}{L} \tag{24.11}$$

Resistance and conductance of a conductor

Keep in mind that resistance and conductance are not independent; each is the inverse of the other.

EXAMPLE 24.5 The resistivity of a leaf

Measuring the resistivity of the leaves of a corn plant is a good way to assess the plant's stress and overall health. To determine resistivity, the current is measured when a voltage is applied between two electrodes placed 20 cm apart on a leaf that is 5.0 cm wide and 0.25 mm thick. FIGURE 24.17 is a graph of current versus potential difference that shows several measurements and a trend line. What is the resistivity of the plant tissue?

FIGURE 24.17 A graph of current versus potential difference for a corn leaf.

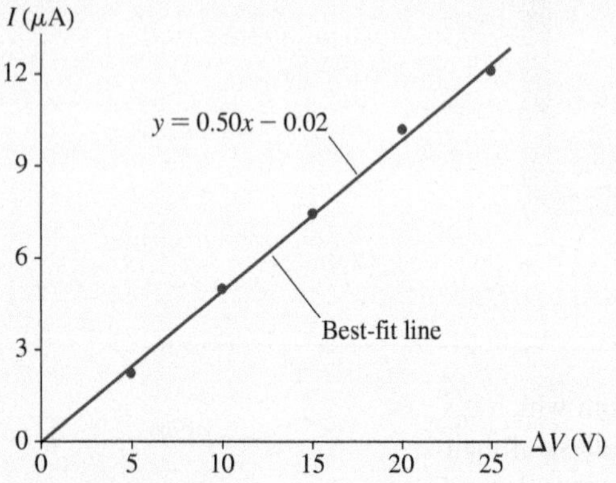

PREPARE We can model the portion of the leaf between the electrodes as a bar of length $L = 0.20$ m with a rectangular cross-section area $A = (0.050 \text{ m})(2.5 \times 10^{-4} \text{ m}) = 1.3 \times 10^{-5} \text{ m}^2$. The resistance R is the inverse of the slope of the trend line. We can use Equation 24.11 to find the resistivity from the resistance and the dimensions.

SOLVE Spreadsheets and graphing software give numerical values for the slope and intercept of the trend line, but not units. In this case, the graph plots current in μA versus potential difference in V. Thus the slope is $0.50\ \mu\text{A/V} = 0.50 \times 10^{-6}$ A/V. The leaf's resistance is the inverse of the slope:

$$R = \frac{1}{\text{slope}} = \frac{1}{0.50 \times 10^{-6} \text{ A/V}} = 2.0 \times 10^{6}\ \Omega$$

We can now use Equation 24.11 to calculate the resistivity:

$$\rho = \frac{AR}{L} = \frac{(1.3 \times 10^{-5} \text{ m}^2)(2.0 \times 10^{6}\ \Omega)}{0.20 \text{ m}} = 130\ \Omega \cdot \text{m}$$

The leaf's conductivity is $\sigma = 1/\rho = 7.7$ mS/m.

ASSESS The leaf's resistivity is fairly high, about a factor of 5 higher than the resistivity of fat. That seems reasonable. Plants have thick, fibrous cell walls, so we don't expect leaf tissue to be an especially good conductor.

EXAMPLE 24.6 Gel electrophoresis

In a gel electrophoresis apparatus, which is used to separate fragments of DNA and proteins, a 3.0-mm-thick gel is formed in the bottom of a 15 cm × 20 cm tray, then covered by a 3.0-mm-deep buffer. The buffer is mixed to have a conductivity of 60 mS/m. What is the current if a 150 V potential difference—a typical value—is applied between the electrodes at the ends of the tray? Charges move primarily through the buffer that covers the gel.

PREPARE We can model the apparatus as having a 3.0-mm-thick conducting solution on top of a 3.0-mm-thick insulator. The buffer is a conductor with a 20 cm length in the direction of the current and a 3.0 mm × 15 cm rectangular cross section. Its conductance can be calculated from Equation 24.11.

SOLVE The conductance of the buffer is

$$G = \frac{\sigma A}{L} = \frac{(0.060 \text{ S/m})(0.0030 \text{ m} \times 0.15 \text{ m})}{0.20 \text{ m}} = 1.4 \times 10^{-4} \text{ S}$$

We can now use Equation 24.8 to calculate the current:

$$I = G\Delta V = (1.4 \times 10^{-4} \text{ S})(150 \text{ V}) = 0.021 \text{ A} = 21 \text{ mA}$$

ASSESS The parameters used in gel electrophoresis vary widely, depending on the purpose of the separation, but these values of voltage and current are typical. It's important to keep the current fairly small because, as we'll see, power dissipation is proportional to I^2 and thus increases rapidly with increasing current. Too much power dissipation will melt the fragile gel.

CONCEPTUAL EXAMPLE 24.7 **Testing drinking water**

A house gets its drinking water from a well that has an intermittent problem with salinity. Before the water is pumped into the house, it passes between two electrodes in the circuit shown in FIGURE 24.18. The current passing through the water is measured with a meter. Which corresponds to increased salinity—an increased current or a decreased current?

REASON Increased salinity causes the water's resistivity to decrease. This decrease causes a decrease in resistance between the electrodes. Current is inversely proportional to resistance, so this leads to an increase in current.

ASSESS Increasing salinity means more ions in solution and thus more charge carriers, so an increase in current is expected. Electrical systems similar to this can therefore provide a quick check of water purity.

FIGURE 24.18 A water-testing circuit.

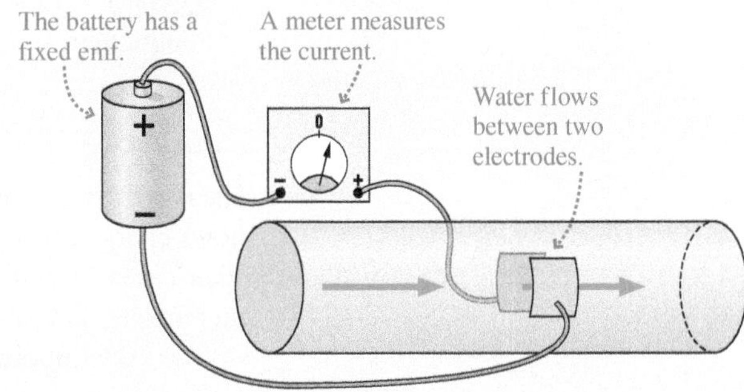
The battery has a fixed emf.
A meter measures the current.
Water flows between two electrodes.
+
−
0

Different tissues in the body have different resistivities, as we saw in Table 24.1. For example, fat has a higher resistivity than muscle. Consequently, a routine test to estimate the percentage of fat in a person's body is based on a measurement of the body's resistance, as illustrated in the photo at the start of the chapter. A higher resistance of the body means a higher proportion of fat.

More careful measurements of resistance can provide more detailed diagnostic information. Passing a small, safe current between pairs of electrodes on opposite sides of a person's torso permits the resistance of the intervening tissue to be measured, a technique known as *electrical resistance tomography* or *electrical impedance tomography*. ("Impedance" is a more general term for resistance that is used in AC circuits where the emf oscillates between positive and negative.) FIGURE 24.19 shows an image of a patient's torso generated from measurements of resistance between many pairs of electrodes. The image shows the change in resistance between two subsequent measurements; decreasing resistance shows in red, increasing resistance in blue. This image was created during the resting phase of the heart, when blood was leaving the lungs and entering the heart. Blood is a better conductor than the tissues of the heart and lungs, so the motion of blood decreased the resistance of the heart and increased that of the lungs. This patient was healthy, but in a patient with circulatory problems any deviation from normal blood flow would lead to abnormal patterns of resistance that would be revealed in such an image.

FIGURE 24.19 An electrical resistance map showing the cross section of a healthy patient's torso. Blue indicates an increased resistance; red a decreased resistance.

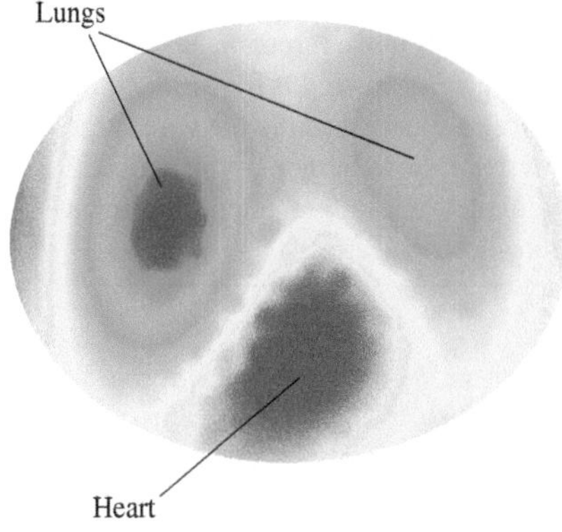
Lungs
Heart

STOP TO THINK 24.4 A wire connected between the terminals of a battery carries a current. The wire is removed and stretched, decreasing its cross-section area and increasing its length. When the wire is reconnected to the battery, the new current is

A. Larger than the original current.
B. The same as the original current.
C. Smaller than the original current.

24.5 Ohm's Law and Resistor Circuits

The relationship between the potential difference across a conductor and the current passing through it that we saw in the previous section was discovered by the German physicist Georg Ohm and is known as **Ohm's law**:

$$I = \frac{\Delta V}{R} = G\Delta V \qquad (24.12)$$

Ohm's law for a conductor of resistance R and conductance G

If we know that a wire of resistance R carries a current I, we can compute the potential difference between the ends of the wire as $\Delta V = IR$.

NOTE ► Ohm's law applies only to resistive devices such as wires or other conductors. It does *not* apply to devices such as batteries or capacitors. ◄

Resistors

The word "resistance" may have negative connotations—who needs something that slows charges and robs energy? In some cases resistance *is* undesirable. But in many other cases, circuit elements are designed to have a certain resistance for very practical reasons. We call these circuit elements **resistors.** There are a few basic types that will be very important as we start to look at electric circuits in detail.

Examples of resistors

Heating elements

As charges move through a resistive wire, their electric energy is transformed into thermal energy, heating the wire. Wires in a toaster, a stove burner, or the rear window defroster of a car are practical examples of this electric heating.

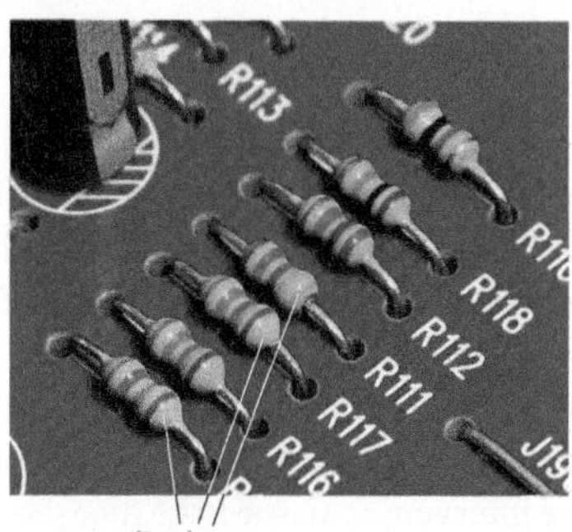
R113
R118
R112
R111
R117
R116
J19
Resistors

Circuit elements

Inside many electronic devices is a circuit board with many small cylinders. These cylinders are resistors that help control currents and voltages in the circuit. The colored bands on the resistors indicate their resistance values.

Light-sensitive resistor

Sensor elements

A resistor whose resistance changes in response to changing circumstances can be used as a sensor. The resistance of this night-light sensor changes when daylight strikes it. A circuit detects this change and turns off the light during the day.

FIGURE 24.20 The potential along a wire-resistor-wire combination.

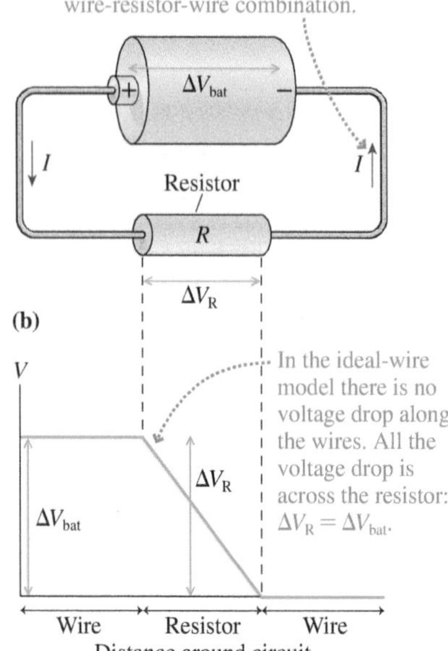
(a) The current is constant along the wire-resistor-wire combination.
ΔV_{bat}
I
I
Resistor
R
ΔV_R
(b)
V
In the ideal-wire model there is no voltage drop along the wires. All the voltage drop is across the resistor: $\Delta V_R = \Delta V_{bat}$.
ΔV_R
ΔV_{bat}
Wire
Resistor
Wire
Distance around circuit

We can identify three important classes of resistive materials:

1. *Wires* are metals with very small resistances $(R \ll 1\ \Omega)$. An **ideal wire** has $R = 0\ \Omega$; hence the potential difference between the ends of an ideal wire is $\Delta V = 0$ V *even if there is a current in it.* We will usually adopt the *ideal-wire model* of assuming that any connecting wires in a circuit are ideal.
2. *Resistors* are poor conductors with resistances usually in the range 10^1 to $10^6\ \Omega$. They are used to control the current in a circuit. Most resistors in a circuit have a specified value of R, such as 500 Ω.
3. *Insulators* are materials such as glass, plastic, or air. An **ideal insulator** has $R = \infty\ \Omega$; hence there is no current in an insulator even if there is a potential difference across it $(I = \Delta V/R = 0\text{ A})$. This is why insulators can be used to hold apart two conductors at different potentials. All practical insulators have $R \gg 10^9\ \Omega$ and can be treated, for our purposes, as ideal.

FIGURE 24.20a shows a resistor connected to a battery with current-carrying wires. The potential difference across the resistor, which we represent as ΔV_R, is given by Ohm's law: $\Delta V_R = IR$. This is often called the voltage *across* the resistor. There are no junctions in this circuit; hence the current I through the resistor is the same as the current in each wire. Because the wire's resistance is *much* less than that of the resistor, $R_{wire} \ll R$, the potential difference $\Delta V_{wire} = IR_{wire}$ between the ends of each wire is *much* less than the potential difference $\Delta V_R = IR$ across the resistor.

If we assume ideal wires with $R_{wire} = 0\ \Omega$, then $\Delta V_{wire} = 0$ V and *all* the voltage drop occurs across the resistor. In this **ideal-wire model,** shown in FIGURE 24.20b, the wires are equipotentials, and the segments of the voltage graph corresponding to the wires are horizontal. As we begin circuit analysis in the next chapter, we will assume that all wires are ideal unless stated otherwise. Thus our analysis will be focused on the resistors.

EXAMPLE 24.8 A battery and a resistor

What resistor would have a 15 mA current if connected across the terminals of a 9.0 V battery?

PREPARE We will assume the resistor is connected to the battery with ideal wires.

SOLVE Connecting the resistor to the battery with ideal wires makes $\Delta V_R = \Delta V_{bat} = 9.0$ V. From Ohm's law, the resistance that gives a 15 mA current is

$$R = \frac{\Delta V_R}{I} = \frac{9.0\ \text{V}}{0.015\ \text{A}} = 600\ \Omega$$

ASSESS Currents of a few mA and resistances of a few hundred ohms are quite typical of real circuits.

EXAMPLE 24.9 Analyzing a single-resistor circuit

FIGURE 24.21 shows a circuit in which a 15 Ω resistor is connected to the terminals of a 1.5 V battery.

a. Draw a graph showing the potential as a function of the distance traveled clockwise through the circuit, starting from $V = 0$ V at the negative terminal of the battery.
b. What is the current in the circuit?

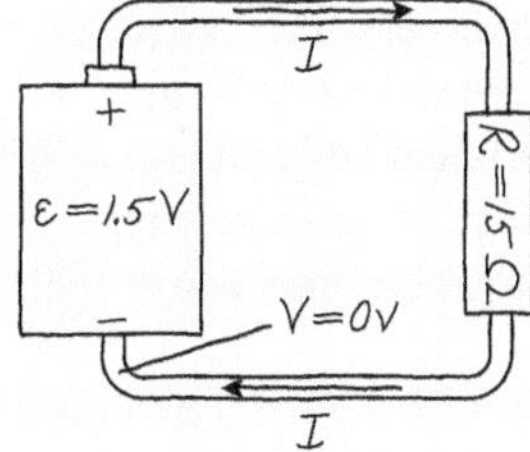

FIGURE 24.21 A single-resistor circuit.

PREPARE Kirchhoff's loop law, introduced in Chapter 22, says that there's no net change in potential around a closed path. The potential changes as we travel around a circuit, but at the completion of a loop the potential must return to its original value. We'll model the wires as being ideal, so there is no voltage drop along the wires; their potential is a constant. All changes in potential occur in the battery and the resistor.

SOLVE

a. Traveling clockwise around the circuit, starting from the negative terminal of the battery, we first gain potential as we go up the charge escalator. The potential at the top of the battery is 1.5 V more positive than at the bottom. The wires are ideal, so there's no change in potential as we move along the wire from the positive terminal of the battery to the resistor or from the resistor back to the battery's negative terminal. The potential returns to its original value after one complete loop, so the potential must *decrease* through the resistor by exactly the same 1.5 V that it increased in the battery. In other words, $\Delta V_R = \mathcal{E} = 1.5$ V. The change in potential as a function of the distance x around the circuit is graphed in FIGURE 24.22.

FIGURE 24.22 Potential-versus-position graph.

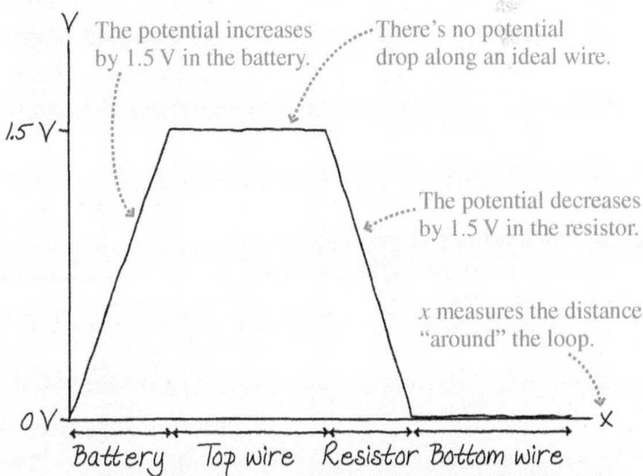

b. We can use Ohm's law and the potential difference across the resistor to compute the current through the resistor:

$$I = \frac{\Delta V_R}{R} = \frac{1.5\ \text{V}}{15\ \Omega} = 0.10\ \text{A}$$

There are no junctions in the circuit, so conservation of charge—Kirchhoff's junction law—tells us that the current has to be the same at every point. In other words, the battery's charge escalator lifts charge at the rate of 0.10 C/s, and charge flows through the wires and the resistor at the rate of 0.10 C/s.

ASSESS Fractions of an amp are typical of battery-powered devices.

EXAMPLE 24.10 **Using a thermistor**

A *thermistor* is a circuit element with a resistance that varies with temperature in a well-defined way. Measuring the current through a thermistor allows us to calculate its resistance and then determine the temperature. Thermistors are used in applications ranging from digital thermometers to temperature sensors in a car engine. One thermistor's resistance as a function of temperature is given by

$$R = (1000\ \Omega)e^{\beta(1/T - 1/T_0)}$$

where $T_0 = 25°\text{C} = 298\ \text{K}$ is a reference temperature, T is the temperature in K, and β is a calibration constant. The thermistor is first calibrated by measuring that its resistance is 2200 Ω at 0°C; then it is connected to a 1.50 V battery. What is the thermistor's temperature when the current through it is 2.70 mA?

PREPARE We can use Ohm's law to determine the thermistor's resistance, then calculate the temperature. But we first have to determine the calibration constant. We'll assume that the connecting wires are ideal, so the potential difference across the thermistor is the battery voltage of 1.5 V.

SOLVE The calibration found $R = 2200\ \Omega$ at $T = 0°\text{C} = 273\ \text{K}$. We can solve the resistance equation for β:

$$\beta = \left(\frac{1}{1/273\ \text{K} - 1/298\ \text{K}}\right)\ln\left(\frac{2200\ \Omega}{1000\ \Omega}\right) = 2570\ \text{K}$$

We can do a similar calculation to solve for the *inverse* of the temperature as a function of resistance:

$$\frac{1}{T} = \frac{1}{\beta}\ln\left(\frac{R}{1000\ \Omega}\right) + \frac{1}{T_0}$$

In this case, the current is $I = 2.70 \times 10^{-3}$ A when the potential difference is $\Delta V_R = 1.50$ V. We can use Ohm's law to calculate that the resistance at this temperature is

$$R = \frac{\Delta V_R}{I} = \frac{1.50\ \text{V}}{2.70 \times 10^{-3}\ \text{A}} = 555\ \Omega$$

Now that we know the thermistor's resistance, we can calculate the inverse of the temperature:

$$\frac{1}{T} = \frac{1}{(2570\ \text{K})}\ln\left(\frac{555\ \Omega}{1000\ \Omega}\right) + \frac{1}{298\ \text{K}} = 3.126 \times 10^{-3}\ \text{K}^{-1}$$

Then we invert the result to find that the temperature is

$$T = \frac{1}{3.126 \times 10^{-3}\ \text{K}^{-1}} = 320\ \text{K} = 47°\text{C}$$

ASSESS These values are quite typical of commercially used thermistors. We see that relatively modest changes in temperature lead to fairly large changes in the thermistor's resistance and current. This makes a thermistor a sensitive indicator of temperature.

STOP TO THINK 24.5 A wire connects the positive and negative terminals of a battery. Two identical wires connect the positive and negative terminals of an identical battery. Rank in order, from largest to smallest, the currents I_1 to I_4 at points 1 to 4.

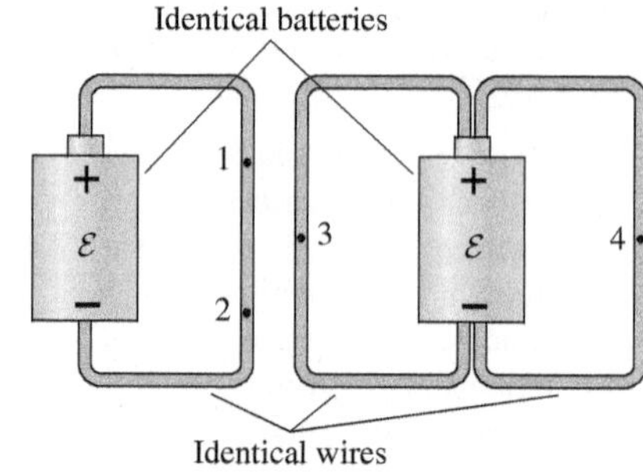
Identical batteries
1
2
3
4
ℰ
ℰ
Identical wires

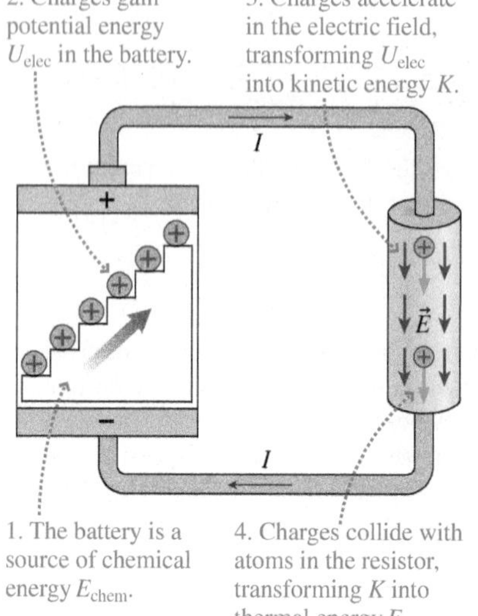
2. Charges gain potential energy U_{elec} in the battery.
3. Charges accelerate in the electric field, transforming U_{elec} into kinetic energy K.
I
I
E
1. The battery is a source of chemical energy E_{chem}.
4. Charges collide with atoms in the resistor, transforming K into thermal energy E_{th}.

FIGURE 24.23 The energy transformations that cause energy from the battery to be dissipated in the resistor.

24.6 Energy and Power

Flipping the switch on a flashlight connects the battery to a lightbulb, which then begins to glow. The bulb is radiating energy. Where does this energy come from?

It comes from the battery. A battery supplies not only a potential difference but also energy. The resistor circuit of **FIGURE 24.23** shows the energy transformations that occur in a resistor circuit. The battery's charge escalator is an energy-transfer process that transforms the chemical energy stored in the battery into the electric potential energy U_{elec} of the charges. The charges then accelerate in the electric field of the wire and resistor, which transforms potential energy into kinetic energy.

The charges frequently collide with atoms in the resistor, however, which transforms their kinetic energy into the thermal energy of atoms that vibrate around their equilibrium positions. The entire sequence of energy transfers and transformations is

$$E_{chem} \rightarrow U_{elec} \rightarrow K \rightarrow E_{th} \qquad (24.13)$$

The net result is that the battery's chemical energy is transformed into the thermal energy of the resistor, raising its temperature.

We need to consider both the *rate* at which the battery supplies energy and the rate at which the resistor dissipates energy. You learned in ◀◀ SECTION 10.5 that the rate at which energy is transformed is *power*, measured in joules per second or *watts*. If energy ΔU_{elec} is energy transferred by the battery to charge q, then the rate at which energy is transferred from the battery to the moving charges is

$$P_{\text{bat}} = \text{rate of energy transfer} = \frac{dU_{\text{elec}}}{dt} = \frac{dq}{dt}\mathcal{E} \qquad (24.14)$$

But dq/dt, the rate at which charge moves through the battery, is the current I. Hence the power supplied by a battery or any source of emf is

$$P_{\text{emf}} = I\mathcal{E} \qquad (24.15)$$

Power delivered by a source of emf

$I\mathcal{E}$ has units of J/s, or W.

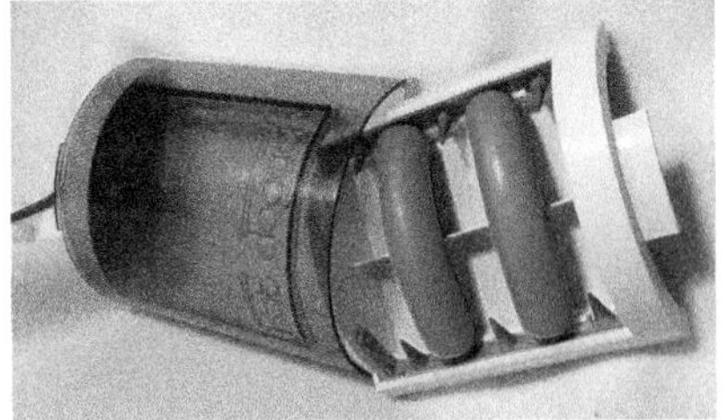

Hot dog resistors Before microwave ovens were common, there were devices that used a decidedly lower-tech approach to cook hot dogs. Prongs connected the hot dog to the 120 V of household electricity, making it the resistor in a circuit. The current through the hot dog dissipated energy as thermal energy, cooking the hot dog in about 2 minutes.

EXAMPLE 24.11 **Power delivered by a car battery**

A car battery has $\mathcal{E} = 12$ V. The battery current is 320 A while the car's starter motor is running. What power does the battery supply?

PREPARE The power is the product of the emf of the battery and the current delivered by the battery.

SOLVE The power supplied by the battery is

$$P_{\text{bat}} = I\mathcal{E} = (320\text{ A})(12\text{ V}) = 3.8\text{ kW}$$

ASSESS This is a lot of power, but this amount makes sense because turning over a car's engine is hard work. Car batteries are designed to reliably provide such intense bursts of power for starting the engine.

Now consider the resistor. The resistor has a potential difference ΔV_{R} between its ends, so a charge q that moves all the way through the resistor *loses* potential energy $\Delta U_{\text{elec}} = -q\Delta V_{\text{R}}$. That energy is all transformed into thermal energy, so the resistor *gains* thermal energy $\Delta E_{\text{th}} = q\Delta V_{\text{R}}$. The *rate* at which energy is transferred from the current to the resistor is

$$P_{\text{R}} = \frac{dE_{\text{th}}}{dt} = \frac{dq}{dt}\Delta V_{\text{R}} = I\Delta V_{\text{R}} \qquad (24.16)$$

We say that this power—so many joules per second—is *dissipated* by the resistor as charge flows through it.

Our analysis of the single-resistor circuit in Example 24.9 found that $\Delta V_{\text{R}} = \mathcal{E}$. That is, the potential difference across the resistor is exactly the emf supplied by the battery. Because the current is the same in the battery and the resistor, a comparison of Equations 24.15 and 24.16 shows that

$$P_{\text{R}} = P_{\text{bat}} \qquad (24.17)$$

The power dissipated in the resistor is exactly equal to the power supplied by the battery. The *rate* at which the battery supplies energy is exactly equal to the *rate* at which the resistor dissipates energy. This is, of course, exactly what we would have expected from energy conservation.

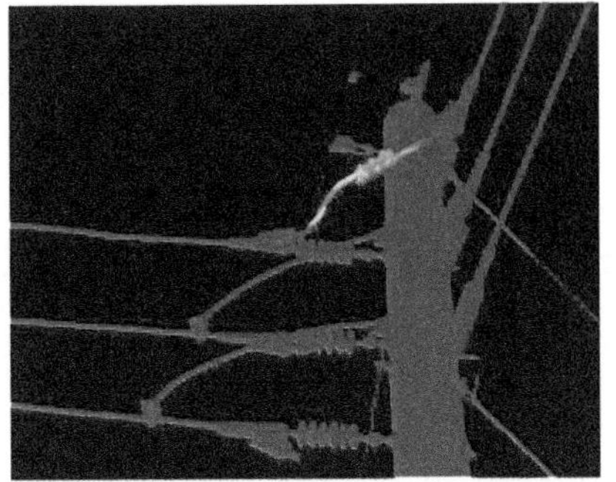

Hot wire This thermal camera image of power lines shows a hot spot where a large current passing through a bad connection between two wires—a connection with too much resistance—is causing the connection to overheat.

Most household appliances, such as a 100 W lightbulb or a 1500 W hair dryer, have a power rating. These appliances are intended for use at a standard household voltage of 120 V, and their rating is the power they will dissipate if operated with a potential difference of 120 V. Their power consumption will differ from the rating if they are operated at any other potential difference—for instance, if you use a lightbulb with a dimmer switch.

EXAMPLE 24.12 **Finding the current in a lightbulb**

The metabolic power of a person at rest is about 100 W. That is, the person dissipates energy at a rate of about 100 J/s, the same as that of a 100 W incandescent lightbulb. How much current is "drawn" by a 100 W lightbulb connected to a 120 V outlet?

PREPARE We will model the lightbulb as a resistor. Equation 24.15 relates the current to the power dissipated in the bulb and the outlet's emf.

SOLVE Because the lightbulb is operating as intended, at 120 V, it dissipates 100 W of power. We can rearrange Equation 24.16 to find

$$I = \frac{P_R}{\Delta V_R} = \frac{100\ \text{W}}{120\ \text{V}} = 0.83\ \text{A}$$

ASSESS We've said that we expect currents on the order of 1 A for lightbulbs and other household items, so our result seems reasonable.

A resistor obeys Ohm's law: $I = \Delta V_R/R$. This gives us two alternative ways of writing the power dissipated by a resistor. We can either substitute IR for ΔV_R or substitute $\Delta V_R/R$ for I. Thus

$$P_R = I\,\Delta V_R = I^2 R = \frac{(\Delta V_R)^2}{R} \qquad (24.18)$$

Power dissipated by resistance R with current I and potential difference ΔV_R

It is worth writing the different forms of this equation to illustrate that the power is proportional to the *square* of both the current and the potential difference.

EXAMPLE 24.13 **Determining the voltage of a stereo**

Most stereo speakers are designed to have a resistance of 8.0 Ω. If an 8.0 Ω speaker is connected to a stereo amplifier with a rating of 100 W, what is the maximum possible potential difference the amplifier can apply to the speaker?

PREPARE We will model the speaker as a resistor. The rating of an amplifier is the *maximum* power it can deliver. Most of the time it delivers far less, but the maximum might be needed for brief, intense sounds. The maximum potential difference will occur when the amplifier is providing the maximum power, so we will make our computation with this figure.

SOLVE The maximum potential difference occurs when the power is at a maximum. At the maximum power of 100 W,

$$P_R = 100\ \text{W} = \frac{(\Delta V_R)^2}{R} = \frac{(\Delta V_R)^2}{8.0\ \Omega}$$

$$\Delta V_R = \sqrt{(8.0\ \Omega)(100\ \text{W})} = 28\ \text{V}$$

This is the maximum potential difference the amplifier might provide.

ASSESS As a check on our result, we note that the resistance of the speaker is less than that of a lightbulb, so a smaller potential difference can provide 100 W of power.

STOP TO THINK 24.6 Rank in order, from largest to smallest, the powers P_A to P_D dissipated in resistors A to D.

A.	B.	C.	D.
ΔV	$2\Delta V$	ΔV	ΔV
R	R	$2R$	$\frac{1}{2}R$

24.7 Alternating Current

A battery creates a constant emf. In a battery-powered flashlight, the bulb carries a constant current and glows with a steady light. The electricity distributed to homes in your neighborhood is different. The picture on the right is a long-exposure photo of a string of LED minilights swung through the air. Each bulb appears as a series of dashes because each bulb in the string flashes on and off 60 times each second. This isn't a special property of the bulbs, but of the electricity that runs them. Household electricity does not have a constant emf; it has a sinusoidal variation that causes the light output of the bulbs to vary. The bulbs light when the emf is positive but not when the emf is negative. The resulting flicker is too rapid to notice under normal circumstances.

LED minilights flash on and off 60 times a second.

The electricity that is delivered to homes and offices is produced by a *generator.* We'll discuss generators in Chapter 26, but they work by rotating a coil of wire in a magnetic field. The steady rotation forces the charges in the wire to flow first in one direction and then, a half cycle later, in the other—an **alternating current,** abbreviated as AC. (If the emf is constant and the current is always in the same direction, as in a battery, we call the electricity *direct current,* abbreviated as DC.) The electricity from power outlets in your house is *AC electricity,* with an emf oscillating at a frequency of 60 Hz. Audio, radio, television, computer, and telecommunication equipment also make extensive use of AC circuits, with frequencies ranging from approximately 10^2 Hz in audio circuits to approximately 10^9 Hz in cell phones.

The instantaneous emf of an AC voltage source, shown graphically in FIGURE 24.24, can be written as

$$\mathcal{E} = \mathcal{E}_0 \cos(2\pi f t) = \mathcal{E}_0 \cos\left(\frac{2\pi t}{T}\right) \qquad (24.19)$$

emf of an AC voltage source

FIGURE 24.24 The emf of an AC voltage source.

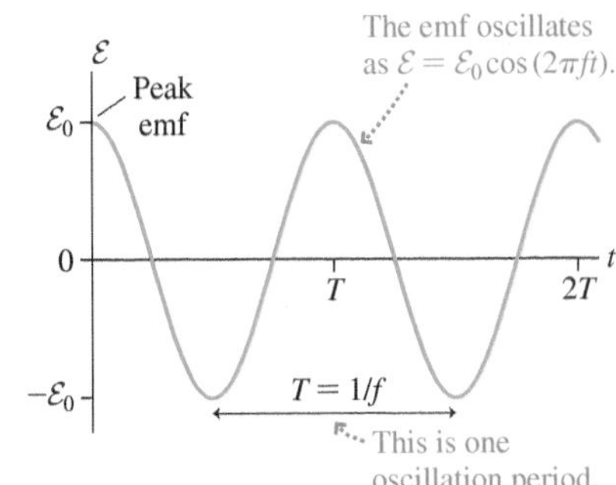
The emf oscillates as $\mathcal{E} = \mathcal{E}_0 \cos(2\pi f t)$.
$\mathcal{E}$
Peak emf
$\mathcal{E}_0$
0
$-\mathcal{E}_0$
T
$2T$
t
$T = 1/f$
This is one oscillation period.

where $\mathcal{E}_0$ is the peak, or maximum, emf; T is the period of oscillation (in s); and $f = 1/T$ is the oscillation frequency (in cycles per second, or Hz).

FIGURE 24.25 shows a simple circuit with a resistor R connected across an AC emf that oscillates at frequency f. The symbol —⊙— represents an AC generator; you can think of it as a battery with an oscillating potential difference. If the connecting wires are ideal, which we'll continue to assume, then the potential difference across the resistor is always equal to the emf. At an instant when the emf is $\mathcal{E}$, the resistor voltage is $v_R = \mathcal{E}$. We will use a lowercase v to represent an *instantaneous voltage* that is constantly changing. Thus the resistor voltage is also an oscillating voltage:

$$v_R = V_R \cos(2\pi f t) \qquad (24.20)$$

where $V_R = \mathcal{E}_0$ is the peak or maximum resistor voltage, the amplitude of the sinusoidally varying voltage.

FIGURE 24.25 An AC resistor circuit.

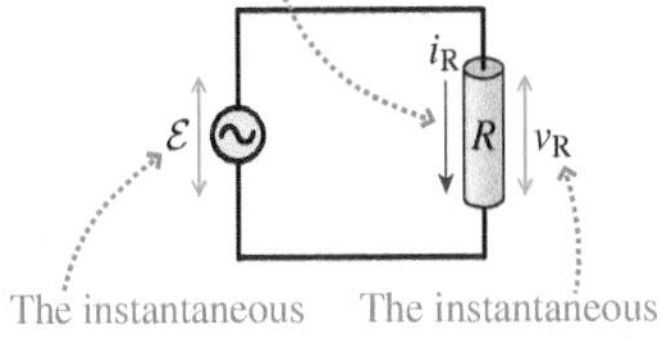
This is the instantaneous current direction when $\mathcal{E} > 0$. A half cycle later, the current will be in the opposite direction.
i_R
$\mathcal{E}$
R
v_R
The instantaneous emf is $\mathcal{E}$.
The instantaneous resistor voltage is v_R.

We can use Ohm's law to find the instantaneous current through the resistor:

$$i_R = \frac{v_R}{R} = \frac{V_R \cos(2\pi ft)}{R} = I_R \cos(2\pi ft) \tag{24.21}$$

where $I_R = V_R/R$ is the peak current through the resistor.

NOTE ▸ It is important to understand the distinction between instantaneous and peak quantities. The instantaneous current i_R, for example, is a quantity that is always changing with time according to Equation 24.21. However, the peak current I_R is a fixed number, the maximum value that the instantaneous current reaches. The instantaneous current oscillates between $+I_R$ and $-I_R$. ◂

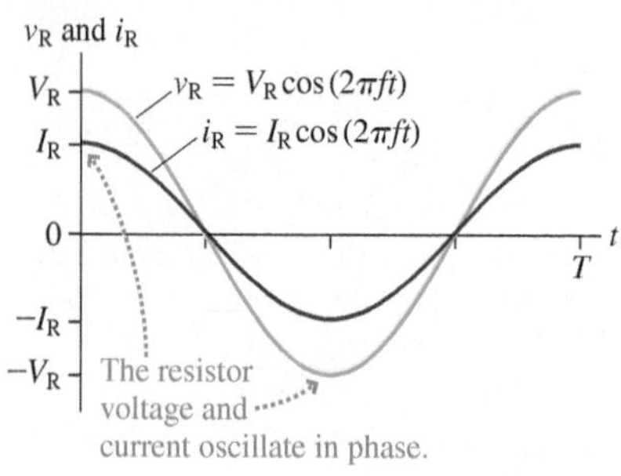

FIGURE 24.26 Graph of the current through and voltage across a resistor.

The resistor's instantaneous current and voltage both oscillate as $\cos(2\pi ft)$. FIGURE 24.26 shows the voltage and the current simultaneously on the same graph. The fact that the peak current I_R is drawn as being less than V_R has no significance. Current and voltage are measured in different units, so in a graph like this you can't compare the value of one to the value of the other. Showing the two different quantities on a single graph—a tactic that can be misleading if you're not careful—simply illustrates that they oscillate *in phase:* **The current is at its maximum value when the voltage is at its maximum, and the current is at its minimum value when the voltage is at its minimum.**

AC Power in Resistors

You learned in the previous section that the power dissipated by a resistor is $P = I\,\Delta V_R = I^2R$. In an AC circuit, the resistor current i_R and voltage v_R are constantly changing, as we saw in Figure 24.26, so the instantaneous power loss $p = i_R v_R = i_R^{\,2} R$ (note the lowercase p) is constantly changing as well. We can use Equations 24.20 and 24.21 to write this instantaneous power as

$$p = i_R^{\,2} R = [\,I_R \cos(2\pi ft)\,]^2 R = I_R^{\,2} R[\,\cos(2\pi ft)\,]^2 \tag{24.22}$$

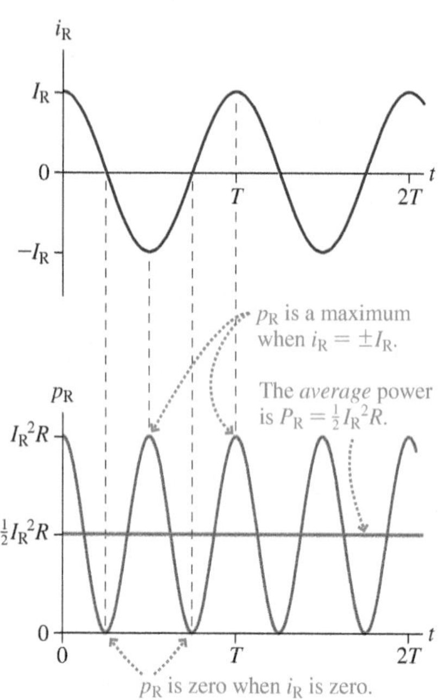

FIGURE 24.27 The instantaneous power loss in a resistor.

FIGURE 24.27 shows the instantaneous power graphically. You can see that, because the cosine is squared, the power oscillates *twice* during every cycle of the emf: The energy dissipation peaks both when $i_R = I_R$ and when $i_R = -I_R$. The energy dissipation doesn't depend on the current's direction through the resistor.

The current through a resistor reverses direction 120 times per second (twice per cycle), so the power reaches a maximum 120 times a second. In most cases, it makes more sense to pay attention to the *average power* than the instantaneous power. Figure 24.27 shows that the **average power** P_R is

$$P_R = \frac{1}{2} I_R^{\,2} R \tag{24.23}$$

We could do a similar analysis for voltage to show that the power is $P_R = \frac{1}{2}V_R^{\,2}/R$.

It is useful to write Equation 24.23 as

$$P_R = \left(\frac{I_R}{\sqrt{2}}\right)^2 R = (I_{rms})^2 R \tag{24.24}$$

where the quantity

$$I_{rms} = \frac{I_R}{\sqrt{2}} \tag{24.25}$$

is called the **root-mean-square current,** or rms current, I_{rms}.

You met the idea of an rms quantity in Chapter 1, where we looked at the rms distance x_{rms} traveled in a random walk, and again in Chapter 13, where we used the rms speed v_{rms} of molecules in a gas. An rms quantity is the square root of the average, or mean, of the quantity squared. We won't prove it, but for any sinusoidal oscillation, the rms value is the peak value divided by $\sqrt{2}$. Thus the rms resistor voltage is $V_{rms} = V_R/\sqrt{2}$.

Using the rms current and voltage, we can write the average power dissipated by a resistor in an AC circuit as

$$P_R = I_{rms}V_{rms} = (I_{rms})^2R = \frac{(V_{rms})^2}{R} \quad (24.26)$$

Average power loss in a resistor

These expressions are the same as the Equation 24.18 expressions for power loss in a DC circuit, with I_{rms} replacing I and V_{rms} replacing ΔV_R. The average power loss for a resistor in an AC circuit with $I_{rms} = 1$ A is the same as in a DC circuit with $I = 1$ A.

We can also define an rms value of the emf: $\mathcal{E}_{rms} = \mathcal{E}/\sqrt{2}$. It turns out that the voltage measured by an AC voltmeter is the rms voltage, and that is how an AC source is characterized. When we say that the wall outlet voltage is 120 V, as it is in the United States, we're saying that $\mathcal{E}_{rms} = 120$ V. The peak voltage at a wall outlet is higher by a factor of $\sqrt{2}$, so $\mathcal{E}_0 = 170$ V. The average power supplied by the emf is

$$P_{emf} = I_{rms}\mathcal{E}_{rms} \quad (24.27)$$

Average power delivered by an AC emf

This is the same as Equation 24.15 for the power delivered by a battery.

A better bulb The "120 V" on this LED lightbulb is its operating rms voltage. The "10 W" is its average power dissipation at this voltage. This efficient LED bulb provides the same light output as a 60 W traditional incandescent bulb.

EXAMPLE 24.14 The resistance and current of a toaster

The hot wire in a toaster dissipates 580 W when plugged into a 120 V outlet.

a. What is the wire's resistance?
b. What are the rms and peak currents through the wire?

PREPARE The 120 V outlet voltage is an rms value.

SOLVE

a. We can rearrange Equation 24.26 to find the resistance from the rms voltage and the average power:

$$R = \frac{(V_{rms})^2}{P_R} = \frac{(120\text{ V})^2}{580\text{ W}} = 25\ \Omega$$

b. A second rearrangement of Equation 24.26 allows us to find the current in terms of the power and the resistance, both of which are known:

$$I_{rms} = \sqrt{\frac{P_R}{R}} = \sqrt{\frac{580\text{ W}}{25\ \Omega}} = 4.8\text{ A}$$

From Equation 24.25, the peak current is

$$I_R = \sqrt{2}\,I_{rms} = \sqrt{2}\,(4.8\text{ A}) = 6.8\text{ A}$$

ASSESS We can do a quick check on our work by calculating the power for the rated voltage and computed current:

$$P_R = I_{rms}V_{rms} = (4.8\text{ A})(120\text{ V}) = 580\text{ W}$$

This agrees with the value in the problem statement, giving us confidence in our solution.

Kilowatt Hours

The product of watts and seconds is joules, the SI unit of energy. However, your local electric company prefers to use a different unit, called *kilowatt hours,* to measure the energy you use each month.

A device in your home that consumes P kW of electricity for Δt hours has used $P\Delta t$ kilowatt hours of energy, abbreviated kWh. For example, suppose you run a 1500 W electric water heater for 10 hours. The energy used is $(1.5\text{ kW})(10\text{ h}) = 15\text{ kWh}$.

Despite the rather unusual name, a kilowatt hour is a unit of energy because it is a power multiplied by a time. The conversion between kWh and J is

$$1.00\text{ kWh} = (1.00 \times 10^3\text{ W})(3600\text{ s}) = 3.60 \times 10^6\text{ J}$$

Feel the power Your home's electric meter records the kilowatt hours of electricity you use.

Your monthly electric bill specifies the number of kilowatt hours you used last month. This is the amount of energy that the electric company delivered to you that you transformed into light and thermal energy inside your home.

EXAMPLE 24.15 **Computing the cost of electric energy**

A typical electric space heater draws an rms current of 12.5 A on its highest setting. If electricity costs 15¢ per kilowatt hour (an approximate national average), how much does it cost to run the heater for 2 hours?

PREPARE We can find the energy in kWh as the product of the heater's power in kW and the time it runs in hours.

SOLVE The power dissipated by the heater is

$$P_R = I_{rms} V_{rms} = (12.5\ \text{A})(120\ \text{V}) = 1500\ \text{W} = 1.5\ \text{kW}$$

In 2 hours, the energy used is $(1.5\ \text{kW})(2.0\ \text{h}) = 3.0\ \text{kWh}$. At 15¢ per kWh, the cost is 45¢.

STOP TO THINK 24.7 An AC current with a peak value of 1.0 A passes through bulb A. A DC current of 1.0 A passes through an identical bulb B. Which bulb is brighter?

A. Bulb A
B. Bulb B
C. Both bulbs are equally bright.

CHAPTER 24 INTEGRATED EXAMPLE Electrical measurements of body composition

The man in the photo at the start of the chapter is gripping a device that measures his body fat by passing a small current through his body. The exact details of how the device works are beyond the scope of this chapter, but the basic principle is quite straightforward: The device applies a small potential difference, measures the resulting current, and then computes the resistance of the upper arm. This resistance depends sensitively on the percentage of body fat in the upper arm, and the percentage of body fat in the upper arm is a good predictor of the percentage of fat in the body overall.

FIGURE 24.28 models the upper arm as part muscle and part fat, showing the resistivity of each. Nonconductive elements, such as skin and bone, are ignored. This is obviously not a picture of the actual structure of the arm, but grouping all the fat tissue together and all the muscle tissue together predicts the arm's electrical character quite well.

FIGURE 24.28 A simple model of the resistance of the upper arm.

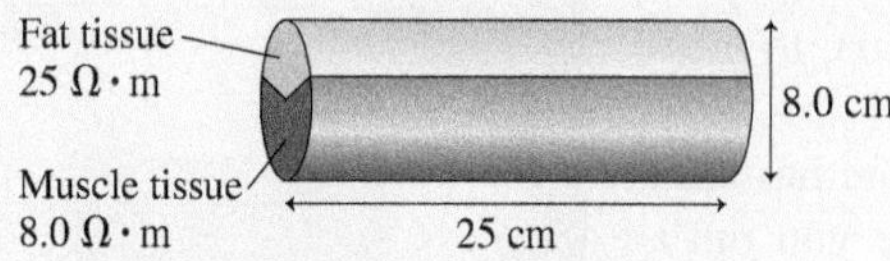

a. An experimental subject's upper arm, with the dimensions shown in the figure, is 40% fat and 60% muscle. A potential difference of 0.60 V is applied between the elbow and the shoulder. What current is measured?

b. A 0.60 V potential difference applied to the upper arm of a second subject with an arm of the same dimensions creates a current of 1.35 mA. What are the percentages of muscle and fat in this person's upper arm?

PREPARE FIGURE 24.29 shows how we can model the upper arm as two resistors that are connected together at the ends. We will use the ideal-wire model in which there's no loss of potential along the wires. Consequently, the potential difference across each of the two resistors is the full 0.60 V of the battery. The current splits at the junction between the two resistors, and Kirchhoff's junction law tells us that

$$I_{total} = I_{muscle} + I_{fat}$$

FIGURE 24.29 A circuit for passing current through the upper arm.

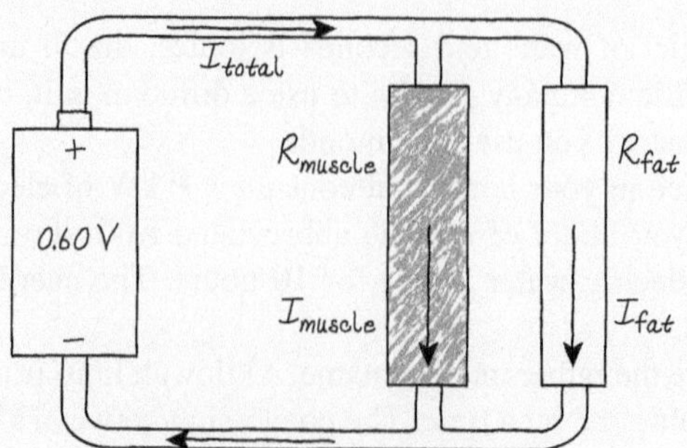

SOLVE

a. An object's resistance is related to its geometry and the resistivity of the material by $R = \rho L/A$. The cross-section area of the whole arm is $A_{\text{arm}} = \pi r^2 = \pi(0.040\text{ m})^2 = 0.00503\text{ m}^2$; the area of each segment is this number multiplied by the appropriate fraction. Thus the resistances of the muscle (60% of the area) and fat (40% of the area) segments are

$$R_{\text{muscle}} = \frac{\rho_{\text{muscle}}L}{A_{\text{muscle}}} = \frac{(8.0\ \Omega\cdot\text{m})(0.25\text{ m})}{(0.60)(0.00503\text{ m}^2)} = 663\ \Omega$$

$$R_{\text{fat}} = \frac{\rho_{\text{fat}}L}{A_{\text{fat}}} = \frac{(25\ \Omega\cdot\text{m})(0.25\text{ m})}{(0.40)(0.00503\text{ m}^2)} = 3110\ \Omega$$

We know that the potential difference across each resistor is 0.60 V. We can use Ohm's law, $I = \Delta V_R/R$, to find that the current in each resistor is

$$I_{\text{muscle}} = \frac{0.60\text{ V}}{663\ \Omega} = 0.904\text{ mA}$$

$$I_{\text{fat}} = \frac{0.60\text{ V}}{3110\ \Omega} = 0.193\text{ mA}$$

The total current measured by the device is

$$I_{\text{total}} = 0.904\text{ mA} + 0.193\text{ mA} = 1.1\text{ mA}$$

b. If we know the current, we can determine the amount of muscle and fat. Let the fraction of muscle tissue be x; the fraction of fat tissue is then $(1 - x)$. We repeat the steps of part a with these expressions in place:

$$R_{\text{muscle}} = \frac{\rho_{\text{muscle}}L}{A} = \frac{(8.0\ \Omega\cdot\text{m})(0.25\text{ m})}{(x)(0.00503\text{ m}^2)} = \frac{398\ \Omega}{x}$$

$$R_{\text{fat}} = \frac{\rho_{\text{fat}}L}{A} = \frac{(25\ \Omega\cdot\text{m})(0.25\text{ m})}{(1-x)(0.00503\text{ m}^2)} = \frac{1240\ \Omega}{1-x}$$

Thus the currents are

$$I_{\text{muscle}} = \frac{0.60\text{ V}}{398\ \Omega}x = 1.51(x)\text{ mA}$$

$$I_{\text{fat}} = \frac{0.60\text{ V}}{1240\ \Omega}(1-x) = 0.48(1-x)\text{ mA}$$

The sum of these currents is the total current:

$$I_{\text{total}} = 1.35\text{ mA} = 1.51(x)\text{ mA} + 0.48(1-x)\text{ mA}$$

Rearranging the terms on the right side gives

$$1.35\text{ mA} = (0.48 + 1.03x)\text{ mA}$$

Finally, we can solve for x:

$$x = 0.84$$

This man therefore has 0.84% muscle and 0.16% fatty tissue in the upper arm.

ASSESS The percentages seem reasonable for a healthy adult. A real measurement of body fat requires a more detailed model of the human body because the current passes through the length of each arm and across the chest, but the principles are the same.

SUMMARY

GOAL To learn how and why charge moves through a conductor as a current.

GENERAL PRINCIPLES

Current

Current is the motion of charges through a conductor when pushed by an electric field. The charge carriers in a metal are electrons; in an ionic fluid, they are positive and negative ions. Either way, current is defined as the rate in C/s that positive charge flows through a conductor: $I = dq/dt$.

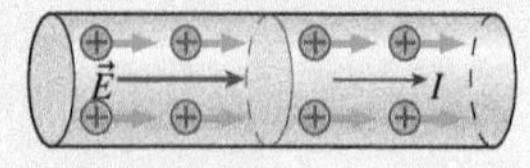

Conservation of Charge

The current is the same at any two points in a wire.

At a junction,

$$\sum I_{in} = \sum I_{out}$$

This is **Kirchhoff's junction law.**

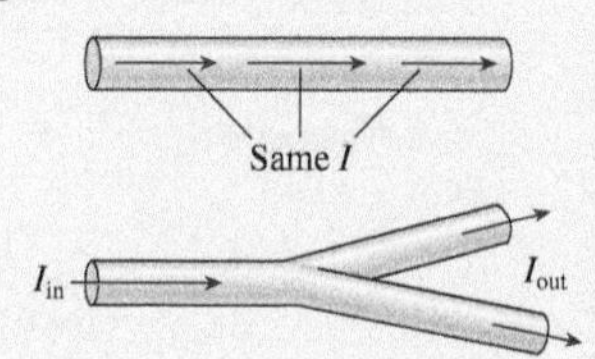

IMPORTANT CONCEPTS

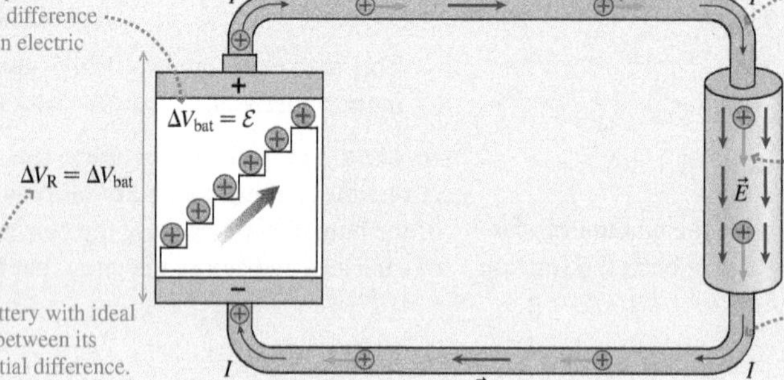

1. A battery uses chemical energy to drive a *charge escalator*. Charge separation in the battery creates a potential difference called the emf $\mathcal{E}$. Charges gain electric potential energy $q\mathcal{E}$.
2. A resistor connected to the battery with ideal wires has a potential difference between its ends equal to the battery's potential difference.
3. An electric field creates a current by pushing charges through the wire.
4. Conservation of charge dictates that the current is the same at all points in the wire.
5. Collisions of moving charges with atoms create resistance that impedes the flow. Collisions transform the kinetic energy of charges into thermal energy of the resistor.
6. The magnitude of the current is determined *jointly* by the battery and the resistance to be $I = \Delta V_R/R$.
$\Delta V_{bat} = \mathcal{E}$
$\Delta V_R = \Delta V_{bat}$
ΔV_R
$\vec{E}$
I

APPLICATIONS

Resistivity, resistance, and Ohm's law

Resistivity ρ and **conductivity** $\sigma = 1/\rho$ are properties of a material.

- Good conductors have low resistivity, high conductivity.
- Poor conductors have high resistivity, low conductivity.

The **resistance** R and **conductance** $G = 1/R$ of a conductor depend on its material properties and its dimensions:

$$R = \frac{\rho L}{A} \qquad G = \frac{\sigma A}{L}$$

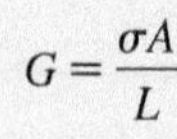

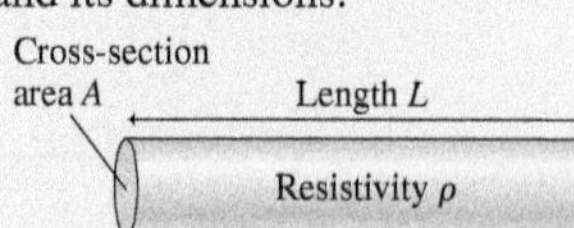

Ohm's law describes the relationship between the potential difference across and the current in a resistor:

$$I = \frac{\Delta V_R}{R} = G\Delta V_R$$

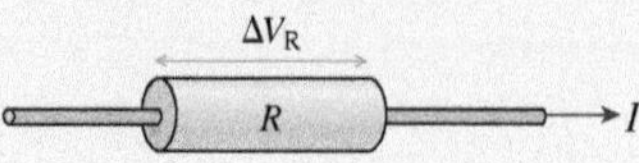

Energy and power

The energy used by a circuit is supplied by the emf of the battery through a series of energy transformations:

$$E_{chem} \rightarrow U_{elec} \rightarrow K \rightarrow E_{th}$$

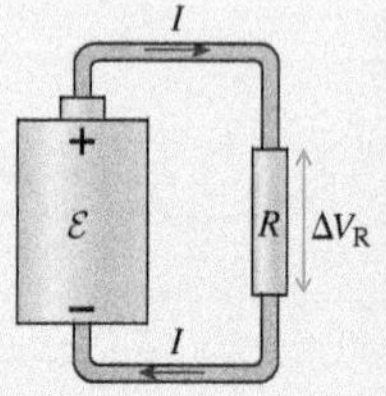

The battery *supplies* energy at the rate

$$P_{bat} = I\mathcal{E}$$

The resistor *dissipates* energy at the rate

$$P_R = I\Delta V_R = I^2R = \frac{(\Delta V_R)^2}{R}$$

In an AC circuit, the average power loss in a resistor is

$$P_R = I_{rms}V_{rms} = (I_{rms})^2R = \frac{(V_{rms})^2}{R}$$

where $I_{rms} = I_R/\sqrt{2}$ and $V_{rms} = V_R/\sqrt{2}$ are the rms current and voltage. The current and voltage oscillate with peak or maximum values I_R and V_R.

LEARNING OBJECTIVES After studying this chapter, you should be able to:

- Explain the motion of charged particles through a conductor. *Conceptual Questions 24.1, 24.2; Problems 24.11, 24.12, 24.34, 24.35*
- Calculate current as the rate of flow of charge. *Conceptual Question 24.3; Problems 24.1, 24.2, 24.5, 24.7, 24.10*
- Interpret and apply the emf of a battery. *Conceptual Question 24.6; Problems 24.13–24.17*
- Calculate the resistance of a wire from the resistivity of the material. *Conceptual Questions 24.4, 24.5; Problems 24.19, 24.21, 24.23, 24.25, 24.29*
- Apply Ohm's law to simple circuits with one resistor. *Conceptual Questions 24.8, 24.9; Problems 24.31–24.33, 24.36*
- Calculate the power dissipated by a resistor in DC circuits. *Conceptual Questions 24.10, 24.11; Problems 24.37, 24.39–24.41, 24.43*
- Calculate the power dissipated by a resistor in AC circuits. *Conceptual Question 24.13; Problems 24.45, 24.47–24.50*

STOP TO THINK ANSWERS

Stop to Think 24.1: D. Both the positive and negative ions contribute equally to the current. A negative ion moving from inside to outside is the same *net* motion of charge as a positive ion moving from outside to inside.

Stop to Think 24.2: 1 A into the junction. The total current entering the junction must equal the total current leaving the junction.

Stop to Think 24.3: C. Charge flows out of one terminal of the battery but back into the other; the amount of charge in the battery does not change. The emf is determined by the chemical reactions in the battery and is constant. But the chemical energy in the battery steadily decreases as the battery converts it to the potential energy of charges.

Stop to Think 24.4: C. Stretching the wire decreases the area and increases the length. Both of these changes increase the resistance of the wire. When the wire is reconnected to the battery, the resistance is greater but the potential difference is the same as in the original case, so the current will be smaller.

Stop to Think 24.5: $I_1 = I_2 = I_3 = I_4$. Conservation of charge requires $I_1 = I_2$. The current in each wire is $I = \Delta V_{wire}/R$. All the wires have the same resistance because they are identical, and they all have the same potential difference because each is connected directly to the battery, which is a *source of potential.*

Stop to Think 24.6: $P_B > P_D > P_A > P_C$. The power dissipated by a resistor is $P_R = (\Delta V_R)^2/R$. Increasing R decreases P_R; increasing ΔV_R increases P_R. But changing the potential has a larger effect because P_R depends on the square of ΔV_R.

Stop to Think 24.7: B. The power in the AC circuit is proportional to the square of the rms current, or to $(I_{rms})^2 = (I_R/\sqrt{2})^2 = \frac{1}{2}I_R^2 = \frac{1}{2}(1\text{ A})^2$. The power in the DC circuit is proportional to the square of the DC current, or to $I_R^2 = (1\text{ A})^2$. Thus the power in the DC circuit is twice that in the AC circuit, so bulb B is brighter.

QUESTIONS

Conceptual Questions

1. Both batteries in Figure Q24.1 are ideal and identical, and all lightbulbs are the same. Rank in order, from brightest to least bright, the brightness of bulbs A to C. Explain.

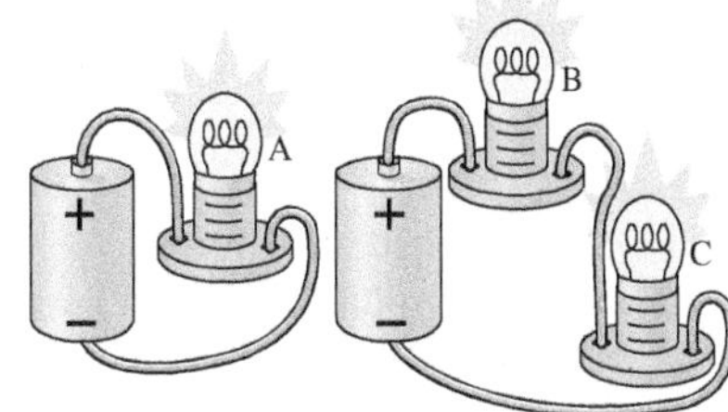

FIGURE Q24.1

2. Both batteries in Figure Q24.2 are ideal and identical, and all lightbulbs are the same. Rank in order, from brightest to least bright, the brightness of bulbs A to C. Explain.

FIGURE Q24.2

3. A wire carries a 4 A current. What is the current in a second wire that delivers twice as much charge in half the time?
4. The wire in Figure Q24.4 consists of two segments of different diameters but made from the same metal. The current in segment 1 is I_1. Compare the currents in the two segments. That is, is I_2 greater than, less than, or equal to I_1? Explain.

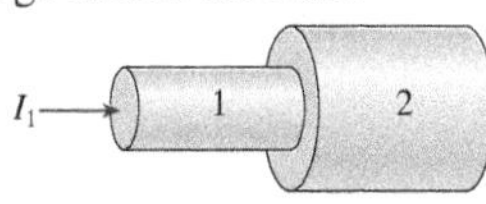

FIGURE Q24.4

Problem difficulty is labeled as | (straightforward) to ||||| (challenging). Problems labeled INT integrate significant material from earlier chapters; BIO are of biological or medical interest; CALC require calculus to solve.

5. The wires in Figure Q24.5 are all made of the same material. Rank in order, from largest to smallest, the resistances R_A to R_E of these wires. Explain.

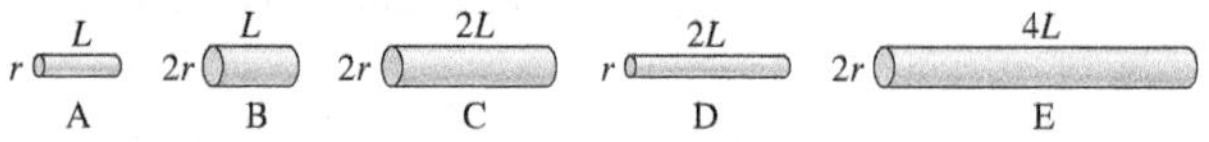

FIGURE Q24.5

6. Which, if any, of these statements are true? (More than one may be true.) Explain. Assume the batteries are ideal.
   a. A battery supplies the energy to a circuit.
   b. A battery is a source of potential difference; the potential difference between the terminals of the battery is always the same.
   c. A battery is a source of current; the current leaving the battery is always the same.
7. What is the current in the wire in Figure Q24.7?

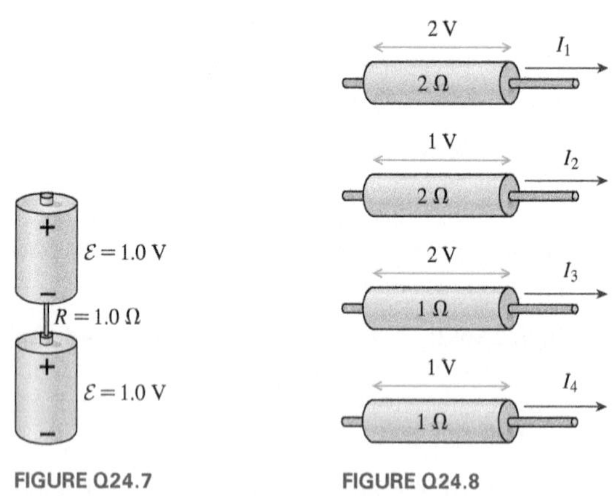

FIGURE Q24.7 FIGURE Q24.8

8. Rank in order, from largest to smallest, the currents I_1 to I_4 through the four resistors in Figure Q24.8. Explain.
9. The two circuits in Figure Q24.9 use identical batteries and wires made of the same material and of equal diameters. Rank in order, from largest to smallest, the currents I_1 to I_7 at points 1 to 7. Explain.

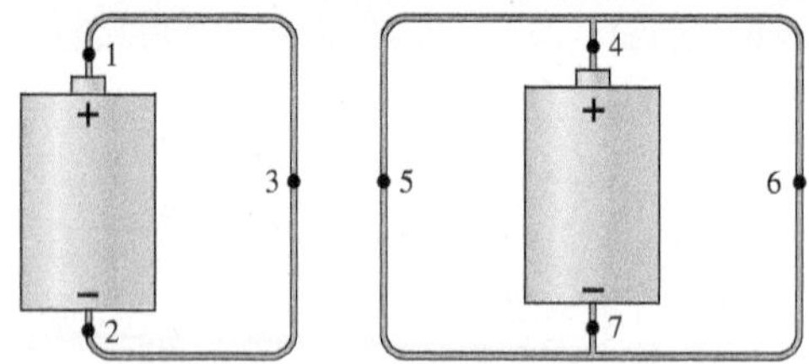

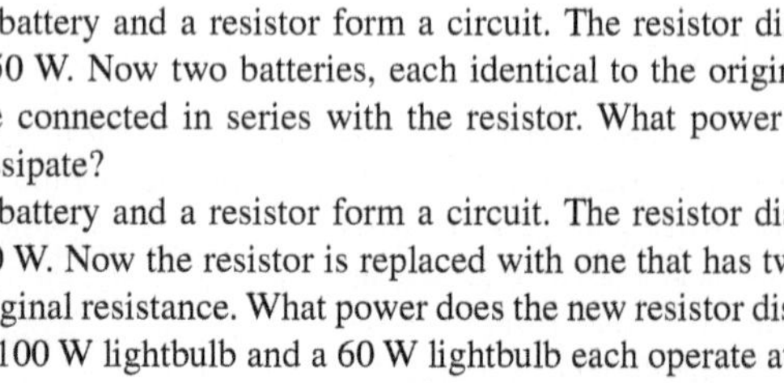

FIGURE Q24.9

10. A battery and a resistor form a circuit. The resistor dissipates 0.50 W. Now two batteries, each identical to the original one, are connected in series with the resistor. What power does it dissipate?
11. A battery and a resistor form a circuit. The resistor dissipates 4.0 W. Now the resistor is replaced with one that has twice the original resistance. What power does the new resistor dissipate?
12. A 100 W lightbulb and a 60 W lightbulb each operate at a voltage of 120 V. Which bulb carries more current?
13. Identical resistors are connected to separate 12 V AC sources. One source operates at 60 Hz, the other at 120 Hz. In which circuit, if either, does the resistor dissipate the greater average power?

Multiple-Choice Questions

14. | The potential difference across a length of wire is increased. Which of the following does *not* increase as well?
    A. The electric field in the wire
    B. The power dissipated in the wire
    C. The resistance of the wire
    D. The current in the wire
15. | A copper wire is stretched so that its length increases and its diameter decreases. What can you say about the wire?
    A. Its resistance decreases, but its resistivity stays the same.
    B. Its resistivity decreases, but its resistance stays the same.
    C. Its resistance increases, but its resistivity stays the same.
    D. Its resistivity increases, but its resistance stays the same.
16. | The resistance of an 8.3-m-long, 0.50-mm-diameter wire is 4.1 Ω. From what material is this wire likely made?
    A. Copper
    B. Aluminum
    C. Iron
    D. Nichrome
17. | Figure Q24.17 shows a side view of a wire of varying circular cross section. Rank in order the currents flowing in the three sections.

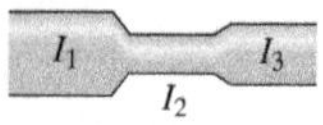

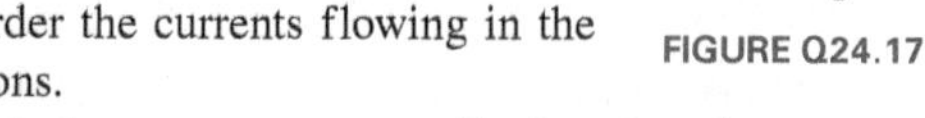

FIGURE Q24.17

    A. $I_1 > I_2 > I_3$ C. $I_1 = I_2 = I_3$
    B. $I_2 > I_3 > I_1$ D. $I_1 > I_3 > I_2$
18. | Wires 1 and 2 are made of the same material. Wire 2 has twice the length and twice the diameter of wire 1. What is the ratio R_2/R_1 of the resistances of the two wires?
    A. $\frac{1}{4}$ B. $\frac{1}{2}$ C. 2 D. 4
19. | The two segments of wire in Figure Q24.19 have the same length and the same diameter but different conductivities. Their ratio is $\sigma_2/\sigma_1 = 2$. Current I passes through the wire. What is the ratio $\Delta V_2/\Delta V_1$ of the potential difference across segment 2 to the potential difference across segment 1?

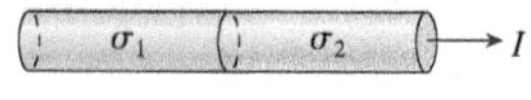

FIGURE Q24.19

    A. $\frac{1}{2}$ B. 1 C. 2
20. || A person gains weight by adding fat—and therefore adding girth—to his body and his limbs, with the amount of muscle remaining constant. How will this affect the electrical resistance of his limbs?
    BIO
    A. The resistance will increase.
    B. The resistance will stay the same.
    C. The resistance will decrease.
21. | Lightbulbs are typically rated by their power dissipation when operated at a given voltage. Which of the following lightbulbs has the largest current through it when operated at the voltage for which it's rated?
    A. 0.8 W, 1.5 V B. 6 W, 3 V
    C. 4 W, 4.5 V D. 8 W, 6 V
22. || Lightbulbs are typically rated by their power dissipation when operated at a given voltage. Which of the following lightbulbs has the largest resistance when operated at the voltage for which it's rated?
    A. 0.8 W, 1.5 V B. 6 W, 3 V
    C. 4 W, 4.5 V D. 8 W, 6 V
23. | If a 1.5 V battery stores 5.0 kJ of energy (a reasonable value for an inexpensive C cell), for how many minutes could it sustain a current of 1.2 A?
    A. 2.7 B. 6.9 C. 9.0 D. 46

PROBLEMS

Section 24.1 A Model of Current

Section 24.2 Defining Current

1. | The current in an electric hair dryer is 10.0 A. How many electrons flow through the hair dryer in 5.0 min?

2. | When a nerve cell fires, charge is transferred across the cell membrane to change the cell's potential from negative to positive. For a typical nerve cell, 9.0 pC of charge flows in a time of 0.50 ms. What is the average current through the cell membrane?
BIO

3. | A wire carries a 15 μA current. How many electrons pass a given point on the wire in 1.0 s?

4. || In a typical lightning strike, 2.5 C flows from cloud to ground in 0.20 ms. What is the current during the strike?

5. | Electroplating uses electrolysis to coat one metal with another. In a copper-plating bath, copper ions with a charge of $+2e$ move through the electrolyte from the copper anode to the cathode—the metal object to be plated. If the positive ion current through the system is 1.2 A, how many copper ions reach the cathode each second?

6. | Electroconvulsive therapy is a last-line treatment for certain mental disorders. In this treatment, an electric current is passed directly through the brain, inducing seizures. The total charge that passes through the brain is called the *dose.* In a typical session, a dose of 0.093 C is applied, using a pulse of current that lasts 1.0 ms. What is the current during this pulse?
BIO

7. || In an ionic solution, 5.0×10^{15} positive ions with charge $+2e$ pass to the right each second while 6.0×10^{15} negative ions with charge $-e$ pass to the left. What are the magnitude and direction of current in the solution?

8. || In proton beam therapy, a beam of high-energy protons is used to deliver radiation to a tumor, killing its cancerous cells. In one session, the radiologist calls for a dose of 4.9×10^8 protons to be delivered at a beam current of 120 nA. How long should the beam be turned on to deliver this dose?
BIO

9. ||| When a current passes through the body, it is mainly carried by positively charged sodium ions (Na^+) with charge $+e$ and negatively charged chloride ions (Cl^-) with charge $-e$. In a given electric field, these two kinds of ions do not move at the same speed; the chloride ions move 50% faster than the sodium ions. If a current of 150 μA is passed through the leg of a patient, how many chloride ions pass a cross section of the leg per second?
BIO

10. | A car battery is rated at 90 A · h, meaning that it can supply a 90 A current for 1 h before being completely discharged. If you leave your headlights on until the battery is completely dead, how much charge leaves the battery?

11. | What are the values of currents I_B and I_C in Figure P24.11? The directions of the currents are as noted.

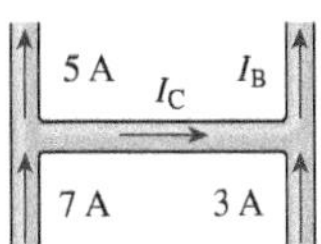

FIGURE P24.11

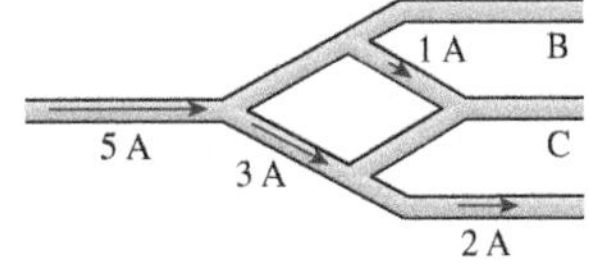

FIGURE P24.12

12. | The currents through several segments of a wire object are shown in Figure P24.12. What are the magnitudes and directions of the currents I_B and I_C in segments B and C?

Section 24.3 Batteries and emf

13. | An electric catfish can generate a significant potential difference using stacks of special cells called *electrocytes.* Each electrocyte develops a potential difference of 110 mV. How many cells must be connected in series to give the 350 V a large catfish can produce?
BIO

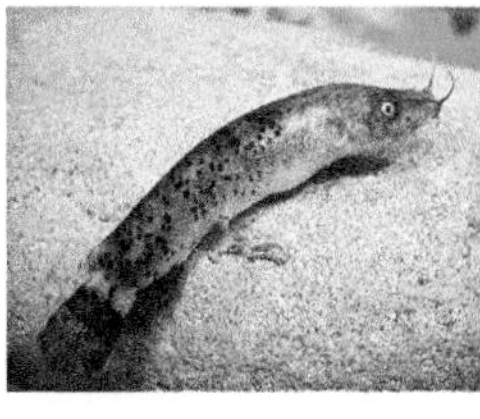

14. | How much electric potential energy does 1.0 μC of charge gain as it moves from the negative terminal to the positive terminal of a 1.5 V battery?

15. | What is the emf of a battery that increases the electric potential energy of 0.050 C of charge by 0.60 J as it moves it from the negative to the positive terminal?

16. || A 9.0 V battery supplies a 2.5 mA current to a circuit for 5.0 h.
    a. How much charge has been transferred from the negative to the positive terminal?
    b. How much electric potential energy has been gained by the charges that passed through the battery?

17. | A laptop battery has an emf of 11.1 V. The laptop uses 0.70 A while running.
    a. How much charge moves through the battery each second?
    b. By how much does the electric potential energy of this charge increase as it moves through the battery?

Section 24.4 Resistance and Conductance

18. | A wire with resistance R is connected to the terminals of a 6.0 V battery. What is the potential difference ΔV_{ends} between the ends of the wire and the current I through it if the wire has the following resistances? (a) 1.0 Ω (b) 2.0 Ω (c) 3.0 Ω.

19. || A motorcyclist is making an electric vest that, when connected to the motorcycle's 12 V battery, will warm her on cold rides. She is using 0.25-mm-diameter copper wire, and she wants a current of 4.0 A in the wire. What length wire must she use?

20. | Salt water is characterized by its *salinity* S, defined as grams of salt per kilogram of solution. The conductivity of water is directly proportional to its salinity: $\sigma = (0.135\ \text{S} \cdot \text{kg/m} \cdot \text{g})S$. What are the salinities of (a) seawater (conductivity 4.5 S/m), (b) human tears (conductivity 1.2 S/m), and (c) typical tap water (conductivity 50 mS/m)?

21. | a. What is the conductance of a 1.0-mm-diameter, 10-cm-long blood vessel filled with blood?
BIO
    b. What is the current in μA if a 10 V potential difference is applied across the ends of this vessel?

22. || A 15-cm-long nichrome wire is connected across the terminals of a 1.5 V battery. If the current in the wire is 2.0 A, what is the wire's diameter?

23. || The femoral artery is the large artery that carries blood to the leg. A person's femoral artery has an inner diameter of 1.0 cm. What is the resistance of a 20-cm-long column of blood in this artery?
BIO

24. || A 3.0 V potential difference is applied between the ends of a 0.80-mm-diameter, 50-cm-long nichrome wire. What is the current in the wire?

25. || Nerve impulses are carried along axons, which are long, thin projections of neuron cells. We can model an axon as a 10-μm-diameter tube of cytosol. What is the resistance of a 1.0-mm-long axon?
BIO

26. || A thermistor is connected to a 1.2 V battery. The thermistor's resistance at 20°C is 12 kΩ. The current through the thermistor increases by 30 μA as the temperature warms to 30°C. What is the thermistor's resistance at this higher temperature?

27. || The high resistivity of dry skin, about $5 \times 10^5\ \Omega \cdot$ m, combined with the 1.5 mm thickness of the skin on your palm can limit the flow of current deeper into tissues of the body. Suppose a worker accidentally places his palm against an electrified panel. The palm of an adult is approximately a 9 cm × 9 cm square. What is the approximate resistance of the worker's palm?
BIO

28. ||| A rear window defroster consists of a long, flat wire bonded to the inside surface of the window. When current passes through the wire, it heats up and melts ice and snow on the window. For one window the wire has a total length of 12.2 m, a width of 1.8 mm, and a thickness of 0.11 mm. The wire is connected to the car's 12.0 V battery and draws 7.5 A. What is the resistivity of the wire material?

29. ||| The resistance of a very fine aluminum wire with a 10 μm × 10 μm square cross section is 1000 Ω. A 1000 Ω resistor is made by wrapping this wire in a spiral around a 3.0-mm-diameter glass core. How many turns of wire are needed?

30. || A circuit calls for a 0.50-mm-diameter copper wire to be stretched between two points. You don't have any copper wire, but you do have aluminum wire in a wide variety of diameters. What diameter aluminum wire will provide the same resistance?

Section 24.5 Ohm's Law and Resistor Circuits

31. | Figure P24.31 shows a current-versus-potential-difference graph for a cylinder. What is the cylinder's resistance?

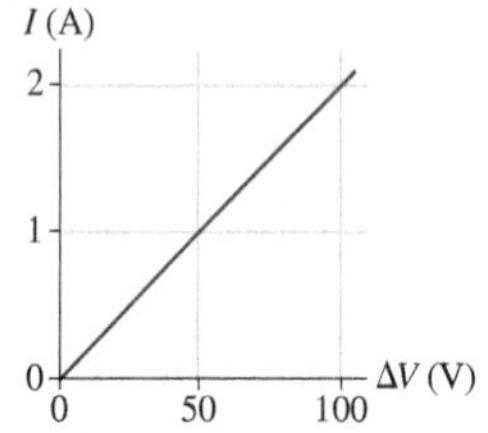

FIGURE P24.31

32. | Pencil "lead" is actually carbon. What is the current if a 9.0 V potential difference is applied between the ends of a 0.70-mm-diameter, 6.0-cm-long lead from a mechanical pencil?

33. || Household wiring often uses 2.0-mm-diameter copper wires. The wires can get rather long as they snake through the walls from the fuse box to the farthest corners of your house. What is the potential difference across a 20-m-long, 2.0-mm-diameter copper wire carrying an 8.0 A current?

34. | The electric field inside a 30-cm-long copper wire is 0.010 V/m. What is the potential difference between the ends of the wire?

35. || A copper wire is 1.0 mm in diameter and carries a current of 20 A. What is the electric field strength inside this wire?

36. || Two identical lightbulbs are connected in series to a single 9.0 V battery.
   a. Sketch the circuit.
   b. Sketch a graph showing the potential as a function of distance through the circuit, starting with $V = 0$ V at the negative terminal of the battery.

Section 24.6 Energy and Power

37. | What is the resistance of a 1500 W (120 V) hair dryer? What is the current in the hair dryer when it is used?

38. | On a sunny day, a rooftop solar panel delivers 150 W of power to the house at an emf of 17 V. How much current flows through the panel?

39. || A 1.5 V D-size battery stores 54 kJ of energy. What steady current could this battery provide for 20 h before being depleted?

40. || A 70 W electric blanket runs at 18 V.
   a. What is the resistance of the wire in the blanket?
   b. How much current does the wire carry?

41. || An electric eel develops a potential difference of 450 V, driving a current of 0.80 A for a 1.0 ms pulse. For this pulse, find (a) the power, (b) the total energy, and (c) the total charge that flows.
BIO

42. || When running on its 11.1 V battery, a laptop computer uses 8.3 W. The computer can run on battery power for 9.0 h before the battery is depleted.
INT
   a. What is the current delivered by the battery to the computer?
   b. How much energy, in joules, is this battery capable of supplying?
   c. How high off the ground could a 75 kg person be raised using the energy from this battery?

43. | A typical American family uses 1000 kWh of electricity a month.
   a. What is the average current in the 120 V power line to the house?
   b. On average, what is the resistance of a household?

44. | A waterbed heater uses 450 W of power. It is on 25% of the time, off 75%. What is the annual cost of electricity at a billing rate of \$0.15/kWh?

Section 24.7 Alternating Current

45. | A 200 Ω resistor is connected to an AC source with $\mathcal{E}_0 =$ 10 V. What is the peak current through the resistor if the emf frequency is (a) 100 Hz? (b) 100 kHz?

46. || A resistor dissipates 2.00 W when the rms voltage of the emf is 10.0 V. At what rms voltage will the resistor dissipate 10.0 W?

47. || A toaster oven is rated at 1600 W for operation at 120 V, 60 Hz.
   a. What is the resistance of the oven heater element?
   b. What is the peak current through it?
   c. What is the peak power dissipated by the oven?

48. | Figure P24.48 shows voltage and current graphs for a resistor.
   a. What is the value of the resistance R?
   b. What is the emf frequency f?

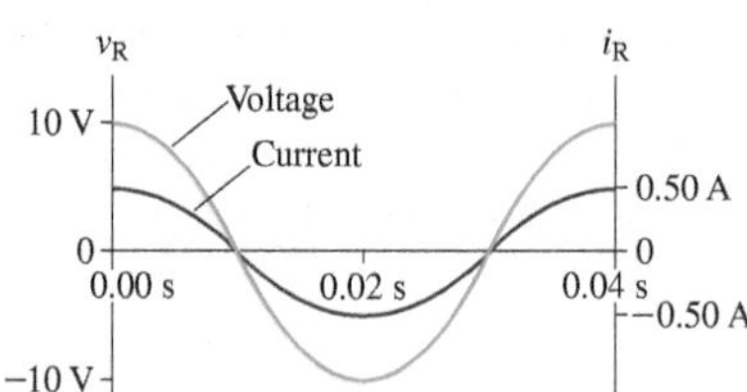

FIGURE P24.48

49. ||| The instantaneous power dissipated by a 20 Ω resistor connected to an AC source is shown in Figure P24.49.
   a. What is the rms voltage across the resistor?
   b. What is the peak current through the resistor?
   c. What is the frequency of the AC source?

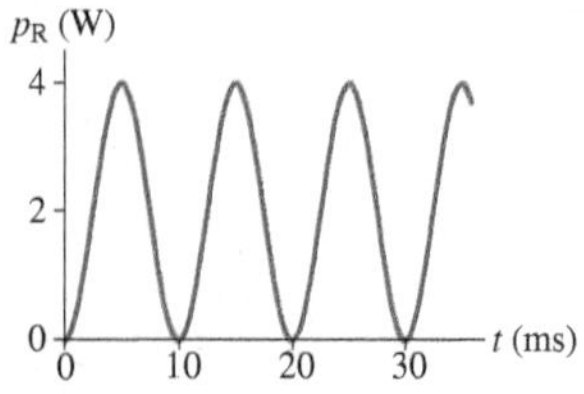

FIGURE P24.49

50. || A generator produces 40 MW of power and sends it to town at an rms voltage of 75 kV. What is the rms current in the transmission lines?

51. | Soles of boots that are designed to protect workers from electric shock are rated to pass a maximum rms current of 1.0 mA when connected across an 18,000 V AC source. What is the minimum allowed resistance of the sole?

General Problems

52. | A 3.0 V battery powers a flashlight bulb that has a resistance of 6.0 Ω. How much charge moves through the battery in 10 min?

53. | The total charge a battery can supply is rated in mA · h, the product of the current (in mA) and the time (in h) that the battery can provide this current. A battery rated at 1000 mA · h can supply a current of 1000 mA for 1.0 h, 500 mA current for 2.0 h, and so on. A typical AA rechargeable battery has a voltage of 1.2 V and a rating of 1800 mA · h. For how long could this battery drive current through a long, thin wire of resistance 22 Ω?

54. ||| The total amount of charge that has entered a wire at time t is CALC given by the expression $Q = (20\text{ C})(1 - e^{-t/(2.0\text{ s})})$, where t is in seconds and $t \geq 0$.
   a. Find an expression for the current in the wire at time t.
   b. What is the maximum value of the current?
   c. Graph I versus t for the interval $0 \leq t \leq 10$ s.

55. ||| The current in a wire at time t is given by the expression CALC $I = (2.0\text{ A})e^{-t/(2.0\,\mu\text{s})}$, where t is in microseconds and $t \geq 0$.
   a. Find an expression for the total amount of charge (in coulombs) that has entered the wire at time t. The initial conditions are $Q = 0$ C at $t = 0\,\mu$s.
   b. Graph Q versus t for the interval $0 \leq t \leq 10\,\mu$s.

56. || A 1.5 V flashlight battery is connected to a wire with a resistance of 3.0 Ω. Figure P24.56 shows the battery's potential difference as a function of time. What is the total charge lifted by the charge escalator?

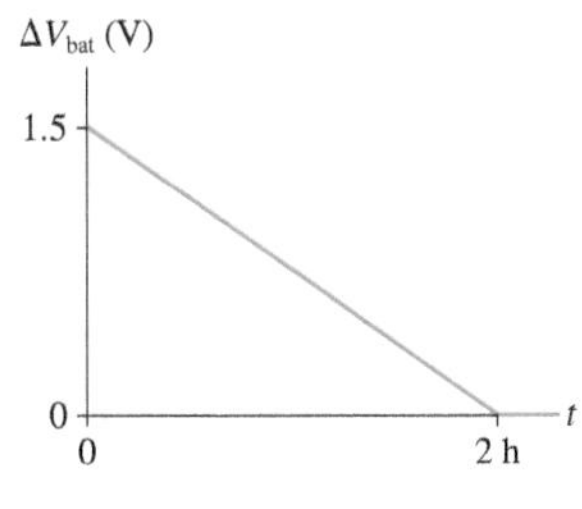

FIGURE P24.56

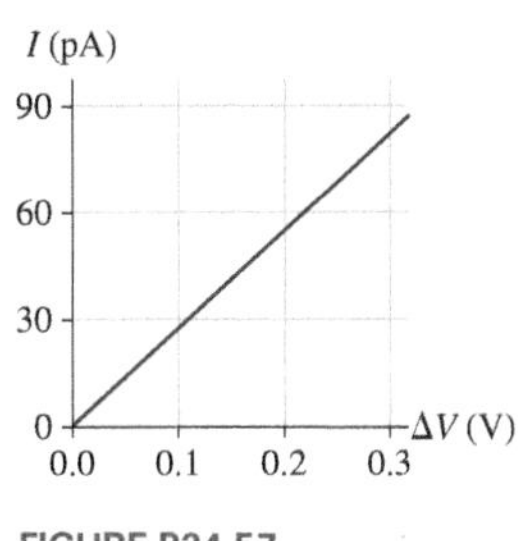

FIGURE P24.57

57. | Certain bacteria create microbial nanowires that they use for BIO electron transport in respiration. These channels have much lower resistivity than the typical fluids in a cell. Figure P24.57 shows current versus voltage data for a microbial nanowire of length 2.0 μm and diameter 1.8 nm. What is the resistivity of the nanowire?

58. ||| The current supplied by a battery slowly decreases as the battery CALC runs down. Suppose that the current as a function of time is $I = (0.75\text{ A})e^{-t/(6\text{ h})}$. What is the total number of electrons transported from the positive electrode to the negative electrode by the charge escalator from the time the battery is first used until it is completely dead?

59. || A simple model of a potassium ion leak channel in the cell BIO membrane is a 0.3-nm-diameter, 7-nm-long cylinder filled with cytosol. What is the approximate conductance of the channel? (This model is too simple, but your answer will have the correct order of magnitude.)

60. || Example 23.12 in Chapter 23 found that the ion flow rate BIO through an individual potassium ion leak channel in a skeletal muscle cell is about 2×10^5 ions/s. This flow rate occurs because the Nernst potential for K^+ ($V^{K}_{Nernst} = -98$ mV) is not equal to the membrane potential ($V_{mem} = -90$ mV), so the ion concentrations are not in equilibrium. The current through the channel is well modeled as $I = G(V_{mem} - V^{K}_{Nernst})$. In other words, the ion flow is driven by the difference between the Nernst potential and the membrane potential. What is the conductance in pS of a potassium leak channel?

61. | Measurements have shown that a voltage-gated potassium BIO ion channel in the cell membrane of an axon carries a 1.8 pA current during the 1.0 ms time interval that the channel is open. How many ions pass through?

62. || When a voltage-gated sodium ion channel opens in a cell BIO membrane, Na^+ ions flow through at the rate of 1.0×10^8 ions/s.
   a. What is the current through the channel?
   b. What is the power dissipation in the channel if the membrane potential is −70 mV?

63. || Air isn't a perfect electric insulator, but it has a very high INT resistivity. Dry air has a resistivity of 3.0×10^{13} Ω · m. A capacitor has square plates 10 cm on a side separated by 1.2 mm of dry air. If the capacitor is charged to 250 V, what fraction of the charge will flow across the air gap in 1 minute? Make the approximation that the potential difference doesn't change as the charge flows.

64. || Variations in the resistivity of blood can give valuable clues BIO to changes in the blood's viscosity and other properties. The resistivity is measured by applying a small potential difference and measuring the current. Suppose a medical device attaches electrodes into a 1.5-mm-diameter vein at two points 5.0 cm apart. What is the blood resistivity if a 9.0 V potential difference causes a 230 μA current through the blood in the vein?

65. || The resistivity of blood is related to its *hematocrit,* the BIO volume fraction of red blood cells in the blood. A commonly used equation relating the hematocrit h to the blood resistivity ρ (in Ω · m) is $\rho = 1.32/(1 - h) - 0.79$. In one experiment, blood filled a graduated cylinder with an inner diameter of 0.90 cm. The resistance of the blood between the 1.0 cm and 2.0 cm marks of the cylinder was measured to be 234 Ω. What was the hematocrit for this blood?

66. || The quality of water can be inferred from its conductivity. A test cell has 1.0 cm × 1.0 cm square electrodes separated by 2.0 cm. For ultrapure water, a 250 nA current is measured when a 9.0 V potential difference is applied between the electrodes. What is the conductivity in μS/m of ultrapure water?

67. ||| In recent years, investigators have explored alterations in BIO brain activity due to the passage of a small current through the brain via electrodes placed on the scalp, with intriguing results related to memory and cognition. In one study, electrodes on the scalp created a 1.0 V/m electric field in the brain tissue and caused a 1.0 mA current to pass through a 2.5×10^{-3} m² cross-section area of the brain. What resistivity does this imply for the brain tissue?

68. | The average resistivity of the human body (apart from surface BIO resistance of the skin) is about 5.0 Ω · m. The conducting path between the right and left hands can be approximated as a cylinder 1.6 m long and 0.10 m in diameter. The skin resistance can be made negligible by soaking the hands in salt water.
   a. What is the resistance between the hands if the skin resistance is negligible?
   b. If skin resistance is negligible, what potential difference between the hands is needed for a lethal shock current of 100 mA? Your result shows that even small potential differences can produce dangerous currents when skin is damp.

69. || You've made the finals of the Science Olympics! As one of your tasks, you're given 1.0 g of copper and asked to make a wire, using all the metal, with a resistance of 1.0 Ω. Copper has a density of 8900 kg/m³. What length and diameter will you choose for your wire?

70. || You've decided to protect your house by placing a 5.0-m-tall iron lightning rod next to the house. The top is sharpened to a point and the bottom is in good contact with the ground. From your research, you've learned that lightning bolts can carry up to 50 kA of current and last up to 50 μs.
   a. How much charge is delivered by a lightning bolt with these parameters?
   b. You don't want the potential difference between the top and bottom of the lightning rod to exceed 100 V. What minimum diameter must the rod have?

71. ||| A 0.60-mm-diameter wire made from an alloy (a combination CALC of different metals) has a conductivity that decreases linearly with distance from the center of the wire: $\sigma(r) = \sigma_0 - cr$, with $\sigma_0 = 5.0 \times 10^7$ S/m and $c = 1.2 \times 10^{11}$ S/m^2. What is the resistance of a 4.0 m length of this wire?

72. ||| A long, round wire has resistance *R*. What will the wire's resistance be if you stretch it to twice its initial length?

73. || The 120 V electric heater in a coffee maker has a resistance INT of 15 Ω. How long will it take for this heater to raise 5 cups (1100 g) of water from 20°C to the ideal brewing temperature of 90°C?

74. || The hot dog cooker described in the chapter heats hot dogs INT by connecting them to 120 V household electricity. A typical hot dog has a mass of 60 g and a resistance of 150 Ω. How long will it take for the cooker to raise the temperature of the hot dog from 20°C to 80°C? The specific heat of a hot dog is approximately 2500 J/kg · K.

75. || Laptop batteries are rated in W · h, the product of the power (in W) that the battery can provide and the time (in h) that it can provide this power. For instance, a 50 W · h battery can provide 50 W for 1.0 h, 25 W for 2.0 h, and so on. A 11.1 V laptop battery is rated at 76 W · h. What is the total charge the battery can provide to the laptop before the battery is depleted?

76. || The 400 V battery of a Tesla Model S electric car stores 3.0×10^8 J of energy. At 65 mph, the car can drive 250 mi before the battery is depleted. At this speed, what is the current delivered by the battery?

77. || An electric eel develops a 450 V potential difference be- BIO tween its head and tail. The eel can stun a fish or other prey by using this potential difference to drive a 0.80 A current pulse for 1.0 ms. What are (a) the energy delivered by this pulse and (b) the total charge that flows?

78. || A refrigerator has a 1000 W compressor, but the compressor runs only 20% of the time.
   a. If electricity costs \$0.10/kWh, what is the monthly (30 day) cost of running the refrigerator?
   b. A more energy-efficient refrigerator with an 800 W compressor costs \$100 more. If you buy the more expensive refrigerator, how many months will it take to recover your additional cost?

79. || It seems hard to justify spending \$4.00 for an LED lightbulb when an ordinary incandescent bulb costs \$0.50. To see if it makes sense, compare a 60 W incandescent bulb that lasts 1000 hours to a 10 W LED bulb that lasts 20,000 hours. (These are typical values.) Both bulbs produce the same amount of visible light and are interchangeable. If electricity costs \$0.15/kWh, about the national average, what is the total cost—purchase price plus energy cost—to obtain 20,000 hours of light from each type of bulb? This is called the *life-cycle cost*.

MCAT-Style Passage Problems

Measuring Blood Hematocrit

The blood cells that circulate throughout your body are held in suspension in *plasma,* a yellowish fluid that is mostly water with dissolved proteins and salts. The fraction of whole blood that is made up of red blood cells is known as the *hematocrit.* Measurements of hematocrit are a useful tool in diagnosing certain conditions. Hematocrit can be measured by separating the different components of blood in a centrifuge, but it is possible to make a more rapid determination by measuring the resistivity of blood. Plasma is a good conductor; blood cells, which are enclosed by an insulating cell membrane, are not. For this reason, a measurement of the resistivity provides a good estimate of the fraction of the blood made up of blood cells—the hematocrit. Blood is placed in a cell with two electrodes, a small voltage is applied, and the current passing through the blood is measured. The blood's resistance depends on its resistivity and on a *cell constant c* that is determined from the dimensions of the test cell. The resistance and resistivity are related by $R = c\rho$. Figure P24.80 shows the variation of current with hematocrit for blood in a cell with $c = 1950$ m^{-1} and an applied voltage of 50 mV.

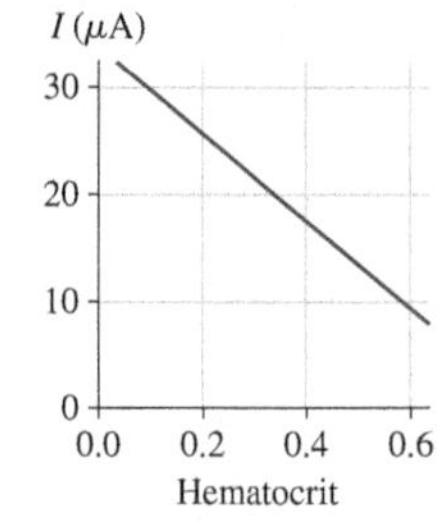

FIGURE P24.80

80. What happens to the cell constant if the distance between the electrodes is increased without changing their size?
   A. The cell constant increases.
   B. The cell constant does not change.
   C. The cell constant decreases.

81. How does the blood's resistivity change if the hematocrit increases?
   A. The resistivity increases.
   B. The resistivity does not change.
   C. The resistivity decreases.

82. What is the approximate resistance of the blood in the test cell for a hematocrit of 0.40?
   A. 1 kΩ B. 2 kΩ C. 3 kΩ D. 4 kΩ

83. What is the approximate resistivity of the blood for a hematocrit of 0.40?
   A. 1.0 $\Omega \cdot$ m
   B. 1.5 $\Omega \cdot$ m
   C. 2.0 $\Omega \cdot$ m
   D. 2.5 $\Omega \cdot$ m

25 Circuits

This South African wild dog is wearing a tracking collar, an electric circuit that sends radio signals to wildlife biologists.

LOOKING AHEAD

Analyzing Circuits

Practical circuits consist of many different elements—resistors, capacitors, batteries, and more—that are connected together.

You'll learn how to analyze circuits by breaking them down into simpler pieces and using Kirchhoff's circuit laws.

Series and Parallel Circuits

Holiday lights are wired in series; if you remove one bulb, an entire string goes out. Car headlights are wired in parallel to operate separately.

You'll learn how to calculate the *equivalent resistance* of a group of resistors that are in series or in parallel.

Electricity in the Body

The body's nervous system works by transmitting electrical signals along *axons,* the long nerve fibers shown here.

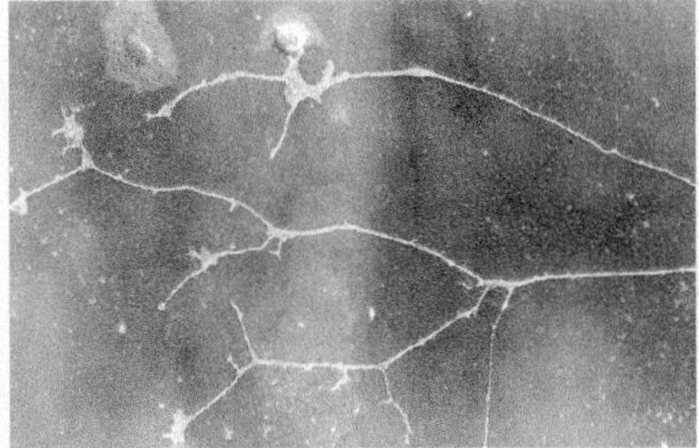

You'll learn to understand nerve impulses in terms of a model based on the resistance, capacitance, and potential of nerve cells.

GOAL To understand the fundamental physical principles that govern electric circuits.

PHYSICS AND LIFE

Nerves Are Electric

What are the physical mechanisms that send nerve signals racing from neuron to neuron in your brain and throughout your body? What determines the speed of a nerve signal? Why are some nerves covered with a thick insulating layer—the myelin sheath—but others are not? Why do some squid have a nerve axon—the *giant axon*—that can be a millimeter in diameter? The transmission of nerve signals is an electrical process that, because it involves the flow of ions, is closely related to the cell's membrane potential. It turns out that we can understand the propagation of nerve impulses by modeling a nerve axon as an electric circuit that uses batteries, resistors, and capacitors—a model that won Alan Hodgkin and Andrew Huxley a Nobel Prize. Some medical conditions, such as multiple sclerosis, arise when this circuitry is disrupted. In this chapter, we will develop the tools to understand the *action potential,* the electrical signal by which nerves fire and muscles contract.

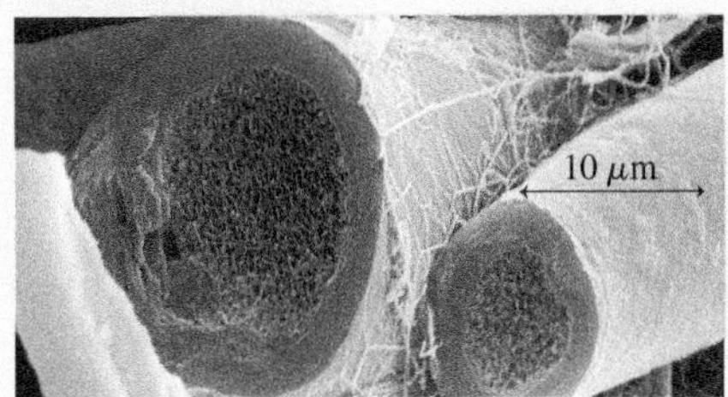

A scanning electron micrograph shows two axons covered with myelin sheaths (pink) that increase the speed of nerve impulse signals.

25.1 Circuit Elements and Diagrams

In Chapter 24, we analyzed the simplest possible electric circuit: a resistor connected across a battery. The electrical activity in firing neurons and contracting muscle cells involves the flow of charge through a much more complex circuit, one with multiple emfs, resistors, and capacitors. This greater complexity makes possible the complex electrical activity in these cells. Our goal for this chapter is to analyze a model circuit for a neuron. We must begin, however, by developing the tools we'll need to work with circuits that have more than one resistor and circuits that combine resistors and capacitors. One especially important tool is a pictorial representation of a circuit.

FIGURE 25.1 An electric circuit.

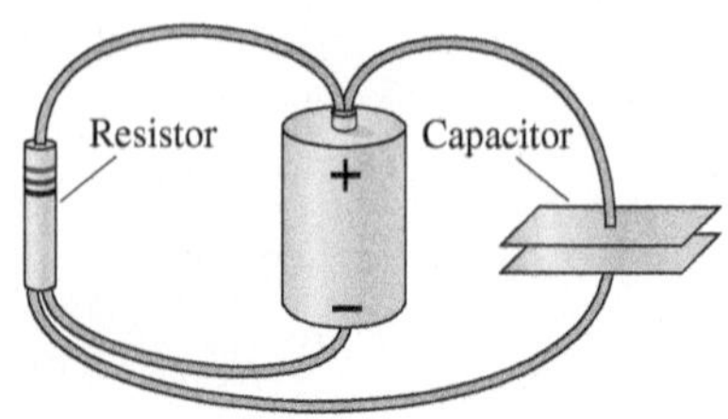

FIGURE 25.1 shows an electric circuit in which a resistor and a capacitor are connected by wires to a battery. To understand the operation of this circuit, we do not need to know whether the wires are bent or straight, or whether the battery is to the right or to the left of the resistor. The literal picture of Figure 25.1 provides many irrelevant details. It is customary when describing or analyzing circuits to use a more abstract picture called a **circuit diagram**. A circuit diagram is a *logical* picture of what is connected to what. The actual circuit, once it is built, may *look* quite different from the circuit diagram, but it will have the same logic and connections.

In a circuit diagram we replace pictures of the circuit elements with symbols. FIGURE 25.2 shows the basic symbols that we will need.

FIGURE 25.2 A library of basic symbols used for electric circuit drawings.

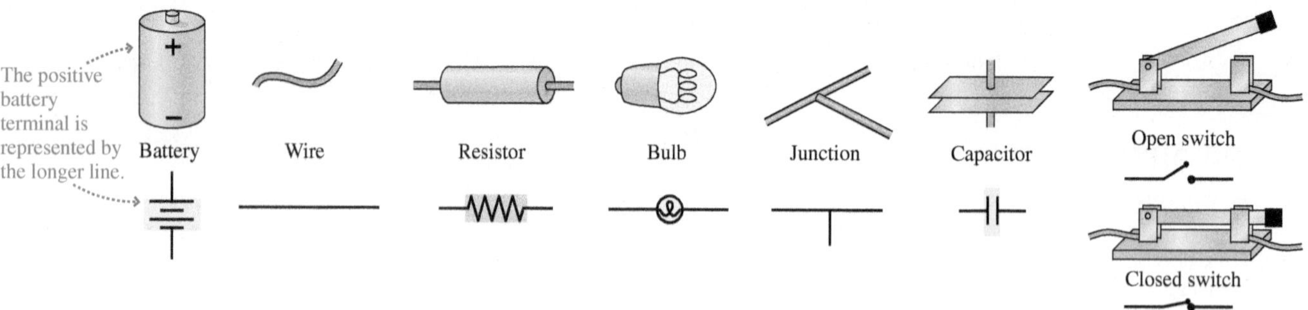

FIGURE 25.3 A circuit diagram for the circuit of Figure 25.1.

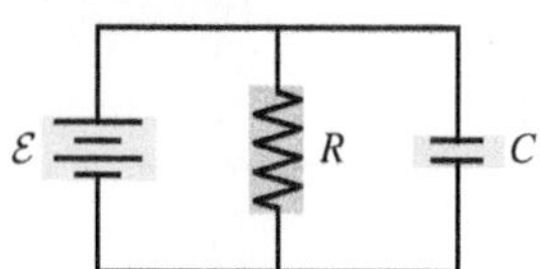

FIGURE 25.3 is a circuit diagram of the circuit shown in Figure 25.1. Notice how circuit elements are labeled. The battery's emf $\mathcal{E}$ is shown beside the battery, and the resistance R of the resistor and capacitance C of the capacitor are written beside them. We would include numerical values for $\mathcal{E}$, R, and C if they are known. The wires, which in reality may bend and curve, are shown as straight-line connections between the circuit elements. The positive potential of the battery is toward the top of the diagram; in general, we try to put higher potentials toward the top. You should get into the habit of drawing your own circuit diagrams in a similar fashion.

STOP TO THINK 25.1 Which of these diagrams represent the same circuit?

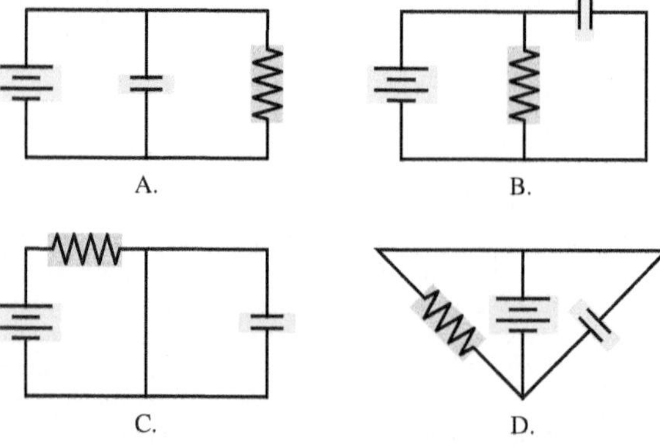

25.2 Using Kirchhoff's Laws

Once we have a diagram for a circuit, we can analyze it. Our tools and techniques for analyzing circuits will be based on the physical principles of potential differences and currents.

You learned in ◀◀ SECTION 24.2 that, as a result of charge conservation, the total current into a junction must equal the total current leaving the junction, as in FIGURE 25.4. This result was called *Kirchhoff's junction law,* which we wrote as

$$\sum I_{\text{in}} = \sum I_{\text{out}} \tag{25.1}$$

Kirchhoff's junction law

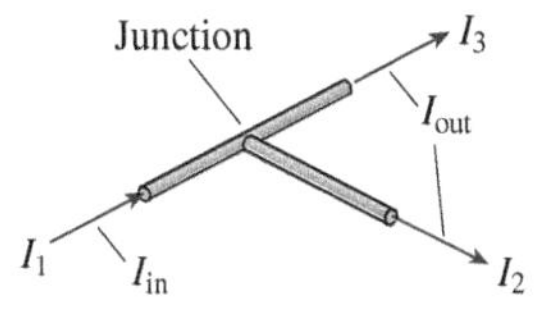

Junction law: $I_1 = I_2 + I_3$

FIGURE 25.4 Kirchhoff's junction law.

Kirchhoff's junction law isn't a new law of nature. It's an application of a law we already know: the conservation of charge.

We can also apply the law of conservation of energy. We found in ◀◀ SECTION 22.6 that all the potential differences around a closed loop must sum to zero because a charge going around the loop has no overall change in its electric potential energy. This was called *Kirchhoff's loop law.* FIGURE 25.5 shows how we apply the loop law to a circuit. If we start at the lower left corner and travel clockwise, the potential increases through the battery but then decreases through the resistors until it ends up at its starting value. Kirchhoff's loop law is written

$$\Delta V_{\text{loop}} = \sum_i \Delta V_i = 0 \tag{25.2}$$

Kirchhoff's loop law

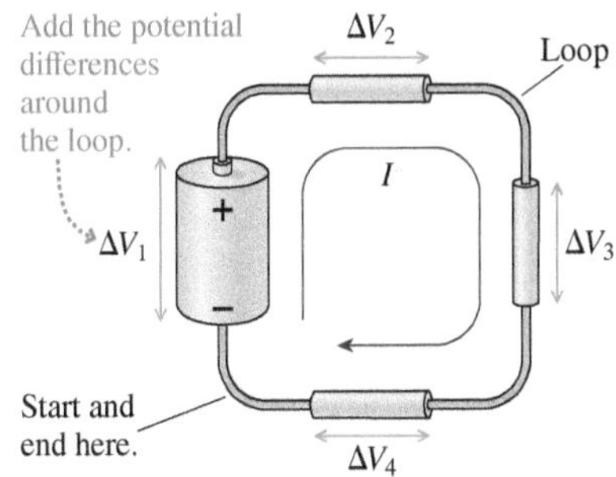

Loop law: $\Delta V_1 + \Delta V_2 + \Delta V_3 + \Delta V_4 = 0$

FIGURE 25.5 Kirchhoff's loop law.

In Equation 25.2, ΔV_i is the potential difference across the ith component in the loop. To apply the loop law, we need to explicitly identify which potential differences are positive and which are negative.

TACTICS BOX 25.1 Using Kirchhoff's loop law

❶ **Draw a circuit diagram.** Label all known and unknown quantities.

❷ **Assign a direction to the current.** Draw and label a current arrow I to show your choice.

- If you can determine the actual current direction, choose that direction.
- If you don't know the actual current direction, make an arbitrary choice. If the value that you calculate is positive, you've guessed correctly. If the value is negative, then the actual direction is opposite your choice.

❸ **"Travel" around the loop.** Start at any point in the circuit, then go all the way around the loop in the direction you assigned to the current in step 2. As you go through each circuit element, ΔV is interpreted to mean $\Delta V = V_{\text{downstream}} - V_{\text{upstream}}$.

- For a battery with current in the negative-to-positive direction:

$$\Delta V_{\text{bat}} = +\mathcal{E}$$

- For a battery with current in the positive-to-negative direction:

$$\Delta V_{\text{bat}} = -\mathcal{E}$$

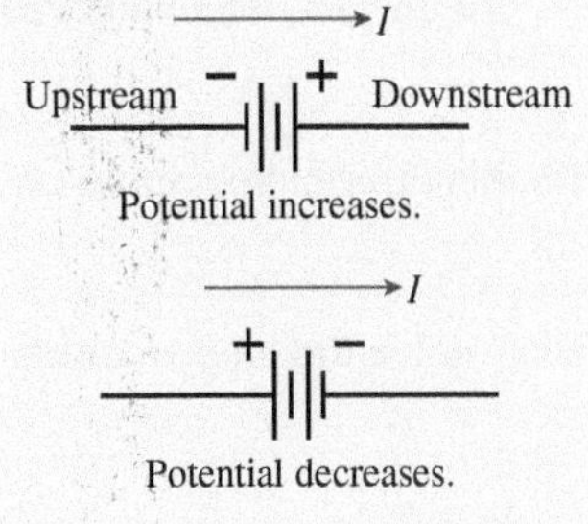

Continued

- For a resistor: $\Delta V_R = -IR$

❹ **Apply the loop law:** $\sum \Delta V_i = 0$

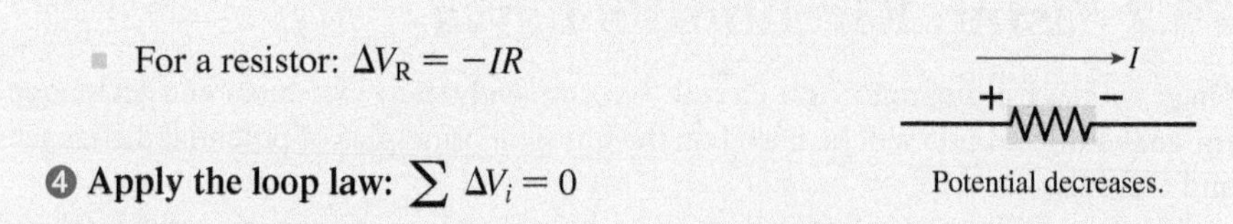
I
+
−
Potential decreases.

We usually think of the current in a battery as flowing in the negative-to-positive direction, as it certainly does in a simple circuit with one battery and one resistor. But in circuits that have more than one battery, the current can go through a battery in the "wrong," positive-to-negative direction when it is forced to do so by other, higher-voltage batteries.

Although ΔV_{bat} can be positive or negative for a battery, ΔV_R for a resistor is always negative because the potential in a resistor *decreases* along the direction of the current—charge flows "downhill," as we saw in ◀◀ SECTION 22.6. Because the potential across a resistor always decreases, we often speak of the *voltage drop* across the resistor.

NOTE ▸ The equation for ΔV_R in Tactics Box 25.1 seems to be the opposite of Ohm's law, but Ohm's law was concerned only with the *magnitude* of the potential difference. Kirchhoff's law requires us to recognize that the electric potential inside a resistor *decreases* in the direction of the current. ◂

The Basic Circuit

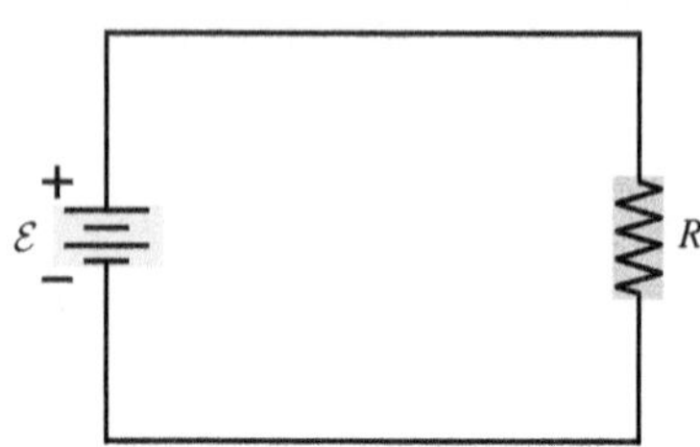
+
ℰ
−
R

FIGURE 25.6 The basic circuit of a resistor connected to a battery.

The most basic electric circuit is a single resistor connected to the two terminals of a battery, as in FIGURE 25.6. We considered this circuit in ◀◀ SECTION 24.5, but let's now apply Kirchhoff's laws to its analysis.

This basic circuit has no junctions, so the current is the same in all parts of the circuit. Kirchhoff's junction law is not needed. Kirchhoff's loop law is the tool we need to analyze this circuit.

FIGURE 25.7 shows the first three steps of Tactics Box 25.1. Notice that we're assuming the ideal-wire model in which there are no potential differences along the connecting wires. The fourth step is to apply Kirchhoff's loop law, $\sum \Delta V_i = 0$:

$$\Delta V_{\text{loop}} = \sum_i \Delta V_i = \Delta V_{\text{bat}} + \Delta V_R = 0 \tag{25.3}$$

FIGURE 25.7 Analysis of the basic circuit using Kirchhoff's loop law.

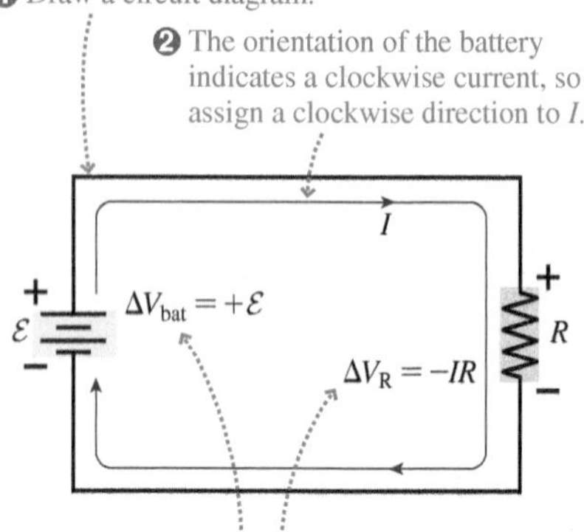
❶ Draw a circuit diagram.
❷ The orientation of the battery indicates a clockwise current, so assign a clockwise direction to I.
I
+
ℰ
−
ΔV_bat = +ℰ
ΔV_R = −IR
+
R
−
❸ Determine ΔV for each circuit element.

Let's look at each of the two terms in Equation 25.3:

1. The potential *increases* as we travel through the battery on our clockwise journey around the loop, as we see in the conventions in Tactics Box 25.1. We enter the negative terminal and, farther downstream, exit the positive terminal after having gained potential $\mathcal{E}$. Thus
$$\Delta V_{\text{bat}} = +\mathcal{E}$$
2. The *magnitude* of the potential difference across the resistor is $\Delta V = IR$, but Ohm's law does not tell us whether this should be positive or negative—and the difference is crucial. The potential of a resistor *decreases* in the direction of the current, which we've indicated with the + and − signs in Figure 25.7. Thus
$$\Delta V_R = -IR$$

With this information about ΔV_{bat} and ΔV_R, the loop equation becomes

$$\mathcal{E} - IR = 0 \tag{25.4}$$

We can solve the loop equation to find that the current in the circuit is

$$I = \frac{\mathcal{E}}{R} \tag{25.5}$$

This is exactly the result we saw in Chapter 24. Notice again that the current in the circuit depends on the size of the resistance. The emf of a battery is a fixed quantity; the current that the battery delivers depends jointly on the emf and the resistance.

EXAMPLE 25.1 Analyzing a circuit with two batteries

What is the current in the circuit of FIGURE 25.8? What is the potential difference across each resistor?

FIGURE 25.8 The circuit with two batteries.

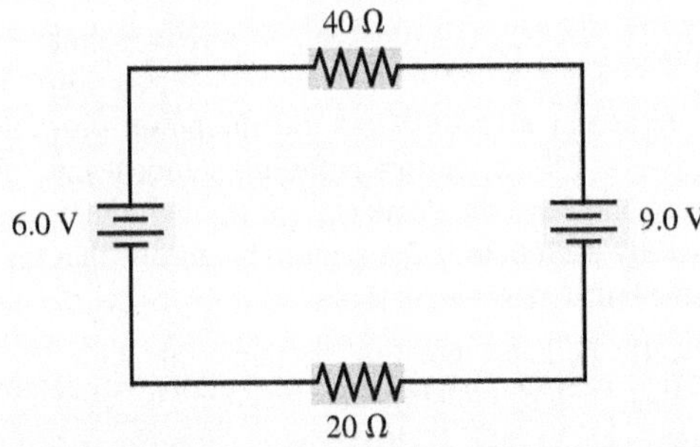

FIGURE 25.9 Analyzing the circuit.

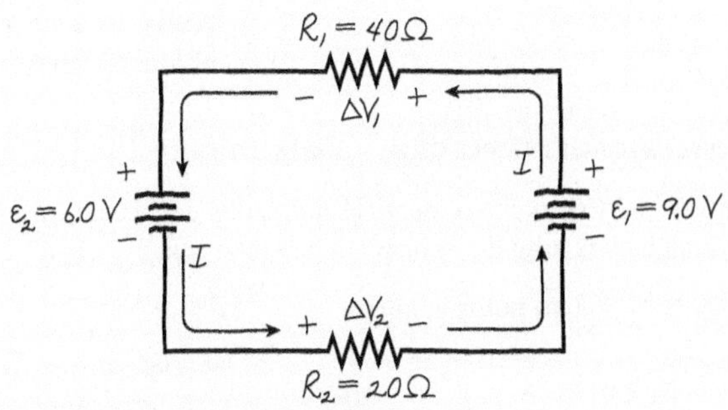

PREPARE We will first use Kirchhoff's loop law, outlined in Tactics Box 25.1, to find the current in the circuit. Then we will use Ohm's law to find the potential differences across the resistors.

Which direction should we assign to the current in this circuit with its *two* batteries? With two batteries, we anticipate that the current direction is determined by the larger 9.0 V battery—that is, counterclockwise. That is our choice. Remember, though, that if we incorrectly choose the current direction to be clockwise, we will still get the correct magnitude for the current, but it will be *negative*—telling us that the actual direction of the current is counterclockwise.

We have redrawn the circuit in FIGURE 25.9, showing the direction of the current and the direction of the potential difference for each circuit element.

SOLVE Kirchhoff's loop law tells us to add the potential differences as we travel around the circuit in the direction of the current. Let's do this starting at the negative terminal of the 9.0 V battery:

$$\sum_i \Delta V_i = +9.0\text{ V} - I\,(40\ \Omega) - 6.0\text{ V} - I\,(20\ \Omega) = 0$$

The 6.0 V battery has $\Delta V_{\text{bat}} = -\mathcal{E}$, in accord with Tactics Box 25.1, because the potential decreases as we travel through this battery in the positive-to-negative direction. We can solve this equation for the current:

$$I = \frac{3.0\text{ V}}{60\ \Omega} = 0.050\text{ A} = 50\text{ mA}$$

Now that the current is known, we can use Ohm's law, $\Delta V = IR$, to find the magnitude of the potential difference across each resistor. For the 40 Ω resistor,

$$\Delta V_1 = (0.050\text{ A})(40\ \Omega) = 2.0\text{ V}$$

and for the 20 Ω resistor,

$$\Delta V_2 = (0.050\text{ A})(20\ \Omega) = 1.0\text{ V}$$

ASSESS We can assess many circuit problems by verifying that our calculated values really do satisfy Kirchhoff's laws. In this case, the potential increases by 9.0 V in the first battery, then decreases by a calculated 2.0 V in the first resistor, by 6.0 V in the second battery, and by a calculated 1.0 V in the second resistor. The total decrease is 9.0 V, so the charge returns to its starting potential, a good check on our calculations.

STOP TO THINK 25.2 What is the potential difference across resistor R?

A. −3.0 V
B. −4.0 V
C. −5.0 V
D. −6.0 V
E. −10 V

$I = 1.0$ A
6 Ω
10 V
R

25.3 Series and Parallel Circuits

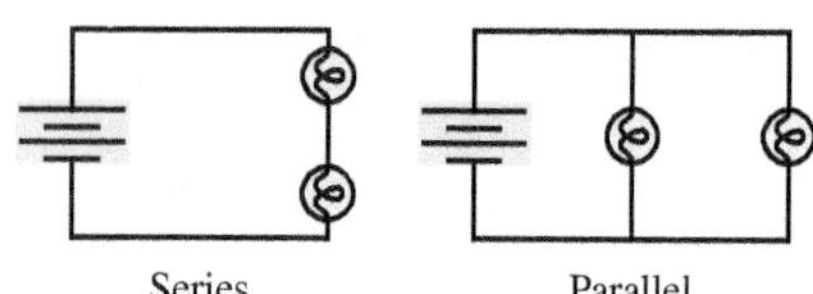

FIGURE 25.10 Series and parallel circuits.

Example 25.1 involved a circuit with multiple elements—two batteries and two resistors. As we introduce more circuit elements, we have possibilities for different types of connections. Suppose you use a single battery to light two lightbulbs. There are two possible ways that you can connect the circuit, as shown in FIGURE 25.10. These *series* and *parallel* circuits have very different properties. We will consider these two cases in turn.

We say two bulbs or resistors are connected in **series** if they are connected directly to each other with no junction in between. All series circuits share certain characteristics.

CONCEPTUAL EXAMPLE 25.2 Determining the brightness of bulbs in series

FIGURE 25.11 shows two identical lightbulbs connected in series. Which bulb is brighter: A or B? Or are they equally bright?

FIGURE 25.11 Two bulbs in series.

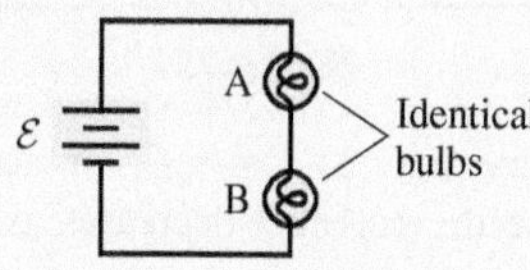

REASON Current is conserved, and there are no junctions in the circuit. Thus, as FIGURE 25.12 shows, the current is the same at all points.

FIGURE 25.12 The current in the series circuit.

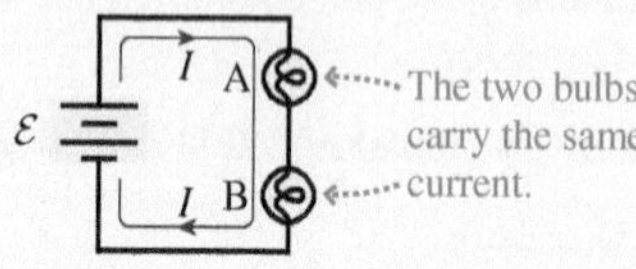

We learned in ◀◀ SECTION 24.6 that the power dissipated by a resistor is $P = I^2R$. If the two bulbs are identical (i.e., the same resistance) and have the same current through them, the power dissipated by each bulb is the same. This means that the brightness of the bulbs must be the same.

ASSESS It's perhaps tempting to think that bulb A will be brighter than bulb B, thinking that something is "used up" before the current gets to bulb B. It is true that *energy* is being transformed in each bulb, but charge must be conserved and so the current is the same through both. We can extend this logic to a special case: If one bulb burns out, so that current can no longer pass through it, the second bulb will go dark as well. If one bulb can no longer carry a current, neither can the other.

Series Resistors

FIGURE 25.13 Replacing two series resistors with an equivalent resistor.

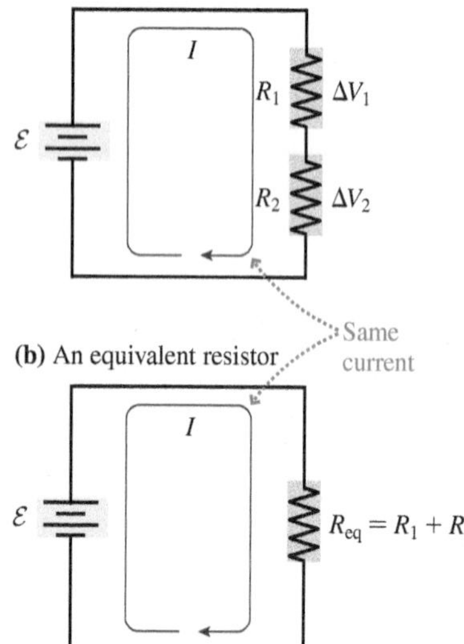

FIGURE 25.13a shows two resistors in series connected to a battery. Because there are no junctions, the current I must be the same in both resistors.

We can use Kirchhoff's loop law to look at the potential differences. Starting at the battery's negative terminal and following the current clockwise around the circuit, we find

$$\sum_i \Delta V_i = \mathcal{E} + \Delta V_1 + \Delta V_2 = 0 \qquad (25.6)$$

The voltage drops across the two resistors, in the direction of the current, are $\Delta V_1 = -IR_1$ and $\Delta V_2 = -IR_2$, so we can use Equation 25.6 to find the current in the circuit:

$$\mathcal{E} = -\Delta V_1 - \Delta V_2 = IR_1 + IR_2$$

$$I = \frac{\mathcal{E}}{R_1 + R_2} \qquad (25.7)$$

Suppose, as in FIGURE 25.13b, we replace the two resistors with a single *equivalent* resistor having the value $R_{eq} = R_1 + R_2$. The total potential difference across this resistor is still $\mathcal{E}$ because the potential difference is established by the battery. Further, the current in this single-resistor circuit is

$$I = \frac{\mathcal{E}}{R_{eq}} = \frac{\mathcal{E}}{R_1 + R_2}$$

which is the same as in the original two-resistor circuit. In other words, this single resistor is equivalent to the two series resistors in the sense that the circuit's current and potential difference are the same in both cases. Nothing anywhere else in the circuit would differ if we took out resistors R_1 and R_2 and replaced them with resistor R_{eq}.

We can extend this analysis to a case with more resistors. If we have resistors in series, their **equivalent resistance** is the sum of their individual resistances:

$$R_{eq} = R_1 + R_2 + \cdots + R_N \qquad (25.8)$$

Equivalent resistance of N series resistors

The current and the power output of the battery will be unchanged if the N series resistors are replaced by the single resistor R_{eq}.

EXAMPLE 25.3 Analyzing a series resistor circuit

What is the current in the circuit of FIGURE 25.14?

FIGURE 25.14 A series resistor circuit.

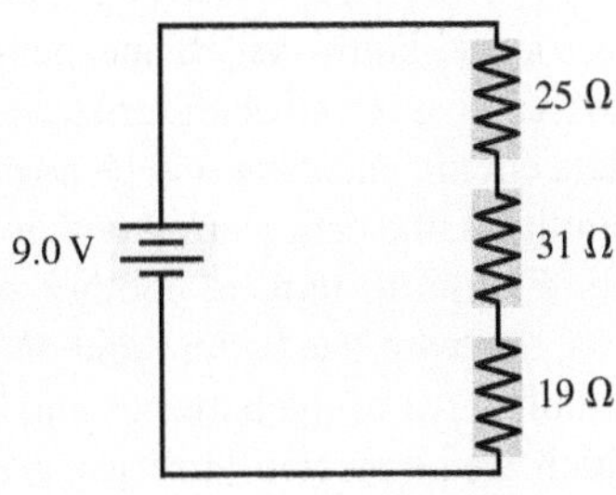

PREPARE The three resistors are in series, so we can replace them with a single equivalent resistor, then calculate the current using Ohm's law. The circuit with its equivalent resistor is shown in FIGURE 25.15.

SOLVE The equivalent resistance is calculated using Equation 25.8:

$$R_{eq} = 25\ \Omega + 31\ \Omega + 19\ \Omega = 75\ \Omega$$

FIGURE 25.15 Analyzing a circuit with series resistors.

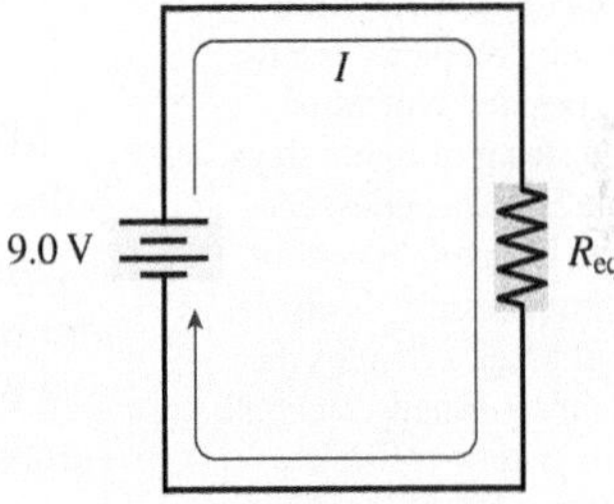

The current in the equivalent circuit of Figure 25.15 is

$$I = \frac{\mathcal{E}}{R_{eq}} = \frac{9.0\ \text{V}}{75\ \Omega} = 0.12\ \text{A}$$

This is also the current in the original circuit.

ASSESS The current in the circuit is the same whether there are three resistors or a single equivalent resistor. The equivalent resistance is the sum of the individual resistance values, and so it is always greater than any of the individual values. This is a good check on your work.

EXAMPLE 25.4 Finding the potential difference of a string of minilights

A string of holiday minilights consists of 50 bulbs wired in series. What is the potential difference across each bulb when the string is plugged into a 120 V outlet?

PREPARE The 50 bulbs are in series, so the current in the circuit, and hence through each bulb, will be the same if we replace the 50 bulbs with their equivalent resistance. Once we know the current, we will find the potential difference across one bulb using Ohm's law. FIGURE 25.16 shows the circuit, which has 50 bulbs in series. We assume that each bulb has resistance R.

FIGURE 25.16 50 bulbs connected in series.

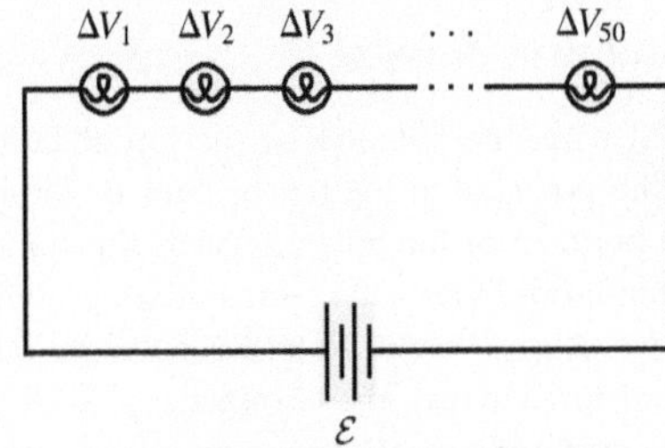

Continued

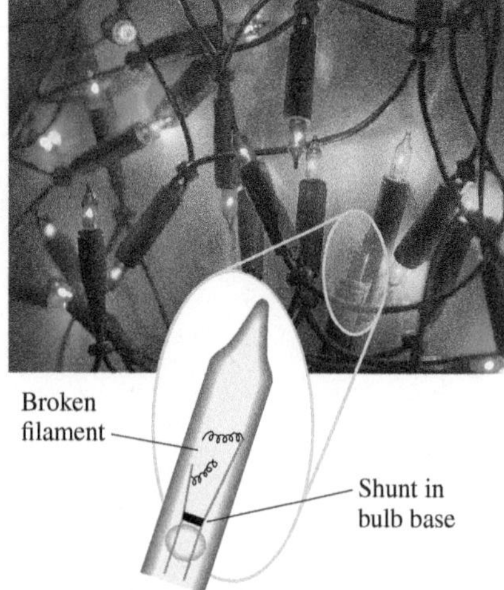

A seasonal puzzle Holiday minilights are connected in series. This is easy to verify: When you remove one bulb from a string of lights, the circuit is not complete, and the entire string of lights goes out. But when one bulb *burns* out, the string of lights stays lit. How is this possible? An incandescent minilight fails when the filament burns out. The bulb has a *shunt* that normally is an insulator, but a filament break activates the shunt to carry the current. A failed LED bulb becomes a *short circuit,* a zero-resistance connecting path that keeps the current flowing.

SOLVE The equivalent resistance of the 50 bulbs, each with resistance R, is

$$R_{eq} = R_1 + R_2 + \cdots + R_{50} = R + R + \cdots + R = 50R$$

Applying Ohm's law to a circuit with only this single equivalent resistor R_{eq}, we find that the current is

$$I = \frac{\mathcal{E}}{R_{eq}} = \frac{\mathcal{E}}{50R}$$

This current I through the equivalent resistor is the same as the current through each series resistor R. Thus we apply Ohm's law to find the potential drop across one bulb of resistance R:

$$\Delta V = IR = \left(\frac{\mathcal{E}}{50R}\right)R = \frac{\mathcal{E}}{50} = \frac{120\text{ V}}{50} = 2.4\text{ V}$$

ASSESS This result seems reasonable. The potential difference is "shared" by the bulbs in the circuit. Since the potential difference is shared among 50 bulbs, the potential difference across each bulb will be quite small.

Minilights are wired in series because the bulbs can be inexpensive low-voltage bulbs or LED bulbs. But there is a drawback that is true of all series circuits: If one bulb is removed, there is no longer a complete circuit, and there will be no current. Indeed, if you remove a bulb from a string of minilights, the entire string will go dark.

Let's use our knowledge of series circuits to look at another lightbulb puzzle. FIGURE 25.17 shows two different circuits, one with one battery and one lightbulb and a second with one battery and two lightbulbs. All of the batteries and bulbs are identical. You now know that B and C, which are connected in series, are equally bright, but how does the brightness of B compare to that of A?

FIGURE 25.17 How does the brightness of bulb B compare to that of bulb A?

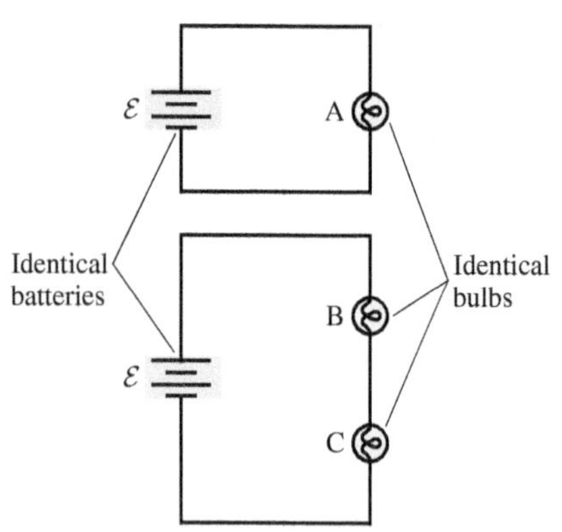

Suppose the resistance of each identical lightbulb is R. In the first circuit, the battery drives current $I_A = \mathcal{E}/R$ through bulb A. In the second circuit, bulbs B and C are in series, with an equivalent resistance $R_{eq} = R_A + R_B = 2R$, but the battery has the same emf $\mathcal{E}$. Thus the current through bulbs B and C is $I_{B,C} = \mathcal{E}/R_{eq} = \mathcal{E}/2R = \frac{1}{2}I_A$. Bulb B has only half the current of bulb A, so B is dimmer.

Many people predict that A and B should be equally bright. It's the same battery, so shouldn't it provide the same current to both circuits? No—recall that **a battery is a source of potential difference, *not* a source of current.** In other words, the battery's emf is the same no matter how the battery is used. When you buy a 1.5 V battery you're buying a device that provides a specified amount of potential difference, not a specified amount of current. The battery does provide the current to the circuit, but the *amount* of current depends on the resistance. Your 1.5 V battery causes 1 A to pass through a 1.5 Ω resistor but only 0.1 A to pass through a 15 Ω resistor. **The amount of current depends jointly on the battery's emf *and* the resistance of the circuit attached to the battery.**

Parallel Resistors

In the next example, we consider the second way of connecting two bulbs in a circuit. The two bulbs in FIGURE 25.18 are connected at *both* ends. We say that they are connected in **parallel.**

CONCEPTUAL EXAMPLE 25.5 Comparing the brightness of bulbs in parallel

Which lightbulb in the circuit of Figure 25.18 is brighter: A or B? Or are they equally bright?

REASON Both ends of the two lightbulbs are connected together by wires. Because there's no potential difference along an ideal wire, the potential at the top of bulb A must be the same as the potential at the top of bulb B. Similarly, the potentials at the bottoms of the bulbs must be the same. This means that the potential *difference* ΔV across the two bulbs must be the same, as we see in FIGURE 25.19. Because the bulbs are identical (i.e., equal resistances), the currents $I = \Delta V/R$ through the two bulbs are equal and thus the bulbs are equally bright.

FIGURE 25.18 Two bulbs in parallel.

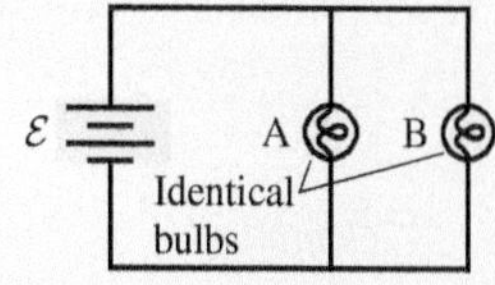

FIGURE 25.19 The potential differences of the bulbs.

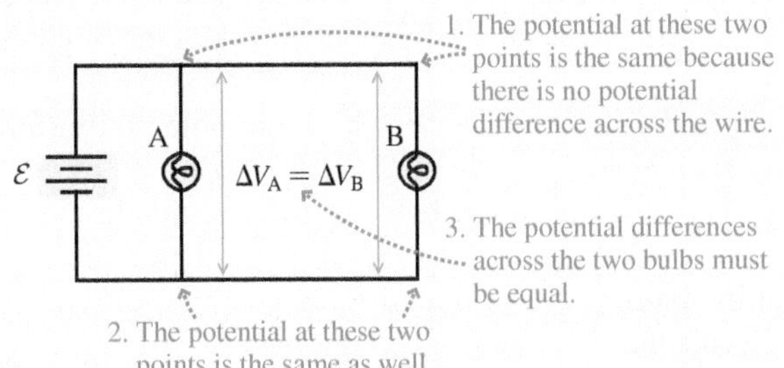
1. The potential at these two points is the same because there is no potential difference across the wire.
A
B
$\mathcal{E}$
$\Delta V_A = \Delta V_B$
3. The potential differences across the two bulbs must be equal.
2. The potential at these two points is the same as well.

ASSESS One might think that A would be brighter than B because current takes the "shortest route." But current is determined by potential difference, and two bulbs connected in parallel have the same potential difference.

Let's look at parallel circuits in more detail. The circuit of FIGURE 25.20a has a battery and two resistors connected in parallel. If we assume ideal wires, the potential differences across the two resistors are equal. In fact, the potential difference across each resistor is equal to the emf of the battery, because both resistors are connected directly to the battery with ideal wires; that is, $\Delta V_1 = \Delta V_2 = \mathcal{E}$.

Now we apply Kirchhoff's junction law. The current I_{bat} from the battery splits into currents I_1 and I_2 at the top junction in FIGURE 25.20b. According to the junction law,

$$I_{bat} = I_1 + I_2 \quad (25.9)$$

We can apply Ohm's law to each resistor to find that the battery current is

$$I_{bat} = \frac{\Delta V_1}{R_1} + \frac{\Delta V_2}{R_2} = \frac{\mathcal{E}}{R_1} + \frac{\mathcal{E}}{R_2} = \mathcal{E}\left(\frac{1}{R_1} + \frac{1}{R_2}\right) \quad (25.10)$$

Can we replace a group of parallel resistors with a single equivalent resistor as we did for series resistors? To be equivalent, the potential difference across the equivalent resistor must be $\Delta V = \mathcal{E}$, the same as for the two resistors it replaces. Further, the current through the equivalent resistor must be the same as it was through the parallel resistors, so that $I = I_{bat}$. A resistor with this current and potential difference must have resistance

$$R_{eq} = \frac{\Delta V}{I} = \frac{\mathcal{E}}{I_{bat}} = \left(\frac{1}{R_1} + \frac{1}{R_2}\right)^{-1} \quad (25.11)$$

where we used Equation 25.10 for I_{bat}. This is the *equivalent resistance,* so a single resistor R_{eq} acts exactly the same as the two resistors R_1 and R_2 as shown in FIGURE 25.20c.

We can extend this analysis to the case of N resistors in parallel. For this circuit, the equivalent resistance is the inverse of the sum of the inverses of the individual resistances:

$$R_{eq} = \left(\frac{1}{R_1} + \frac{1}{R_2} + \cdots + \frac{1}{R_N}\right)^{-1} \quad (25.12)$$

Equivalent resistance of N parallel resistors

The current and the power output of the battery will be unchanged if the N parallel resistors are replaced by the single resistor R_{eq}.

NOTE ▸ When you use Equation 25.12, don't forget to take the inverse of the sum that you compute. ◂

In Figure 25.20 each of the resistors is subject to the full potential difference of the battery. If one resistor were removed, the conditions of the second resistor would not change. This is an important property of parallel circuits.

FIGURE 25.20 Replacing two parallel resistors with an equivalent resistor.

(a) Two resistors in parallel

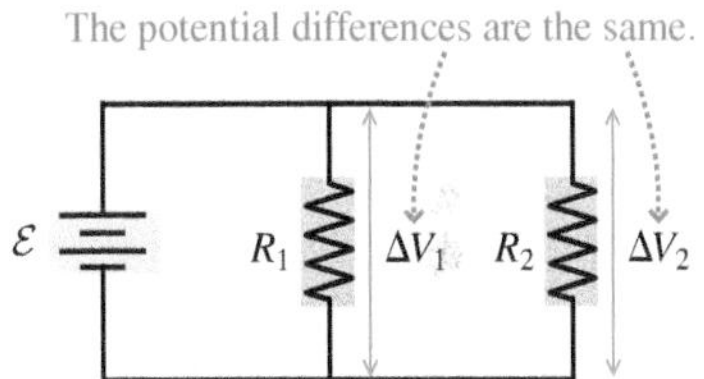
The potential differences are the same.
$\mathcal{E}$
R_1
ΔV_1
R_2
ΔV_2

(b) Applying the junction law

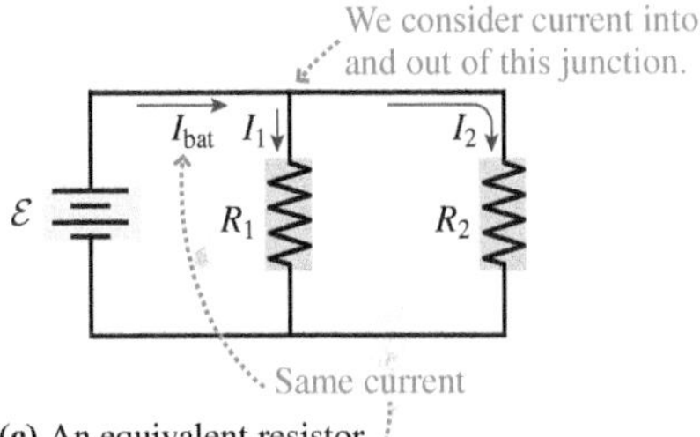
We consider current into and out of this junction.
I_{bat}
I_1
I_2
$\mathcal{E}$
R_1
R_2
Same current

(c) An equivalent resistor

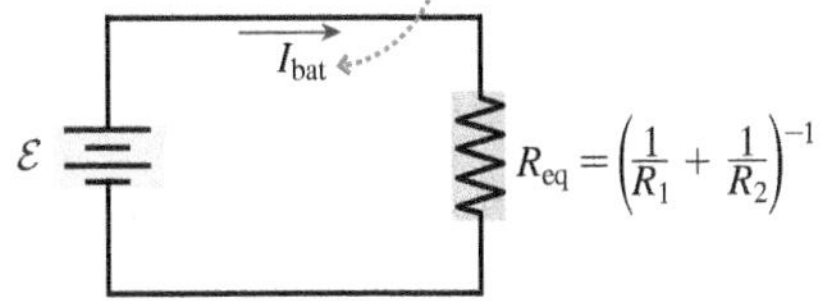
I_{bat}
$\mathcal{E}$
$R_{eq} = \left(\frac{1}{R_1} + \frac{1}{R_2}\right)^{-1}$

Parallel circuits for safety You have certainly seen cars with only one headlight lit. This tells us that automobile headlights are connected in parallel: The currents in the two bulbs are independent, so the loss of one bulb doesn't affect the other. The parallel wiring is very important so that the failure of one headlight will not leave the car without illumination.

FIGURE 25.21 How does the brightness of bulb B compare to that of bulb A?

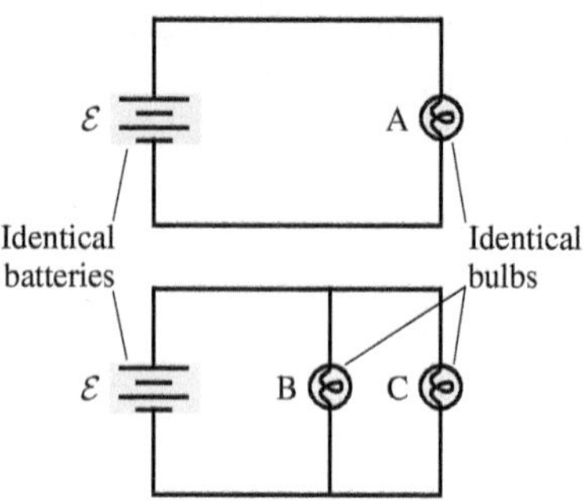

Now, let's look at a final lightbulb puzzle. FIGURE 25.21 shows two different circuits: one with one battery and one lightbulb and a second with one battery and two lightbulbs. As before, the batteries and the bulbs are identical. You know that B and C, which are connected in parallel, are equally bright, but how does the brightness of B compare to that of A?

Each of the bulbs A, B, and C is connected to the same potential difference, that of the battery, so they each have the *same* brightness. Though all of the bulbs have the same brightness, there is a difference between the circuits. In the second circuit, the battery must power two lightbulbs, and so it must provide twice as much current. Recall that the battery is a source of fixed potential difference; the current depends on the circuit that is connected to the battery. Adding a second lightbulb doesn't change the potential difference, but it does increase the current from the battery.

EXAMPLE 25.6 Current in a parallel resistor circuit

The three resistors of FIGURE 25.22 are connected to a 12 V battery. What current is provided by the battery?

FIGURE 25.22 A parallel resistor circuit.

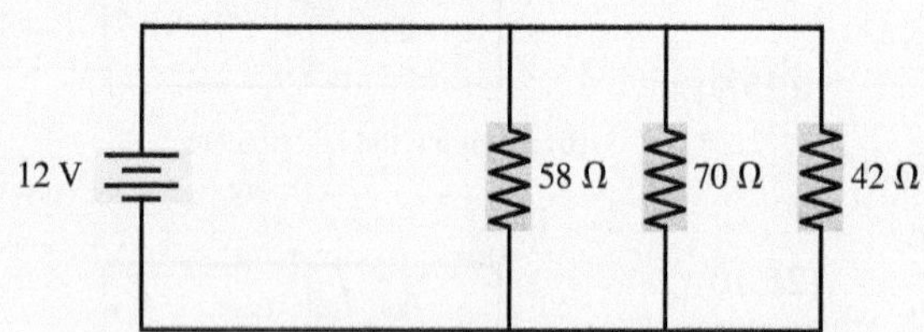

PREPARE The three resistors are in parallel, so we can reduce them to a single equivalent resistor. Then we will use Ohm's law to find the current. The circuit with its equivalent resistance is shown in FIGURE 25.23.

FIGURE 25.23 Analyzing a circuit with parallel resistors.

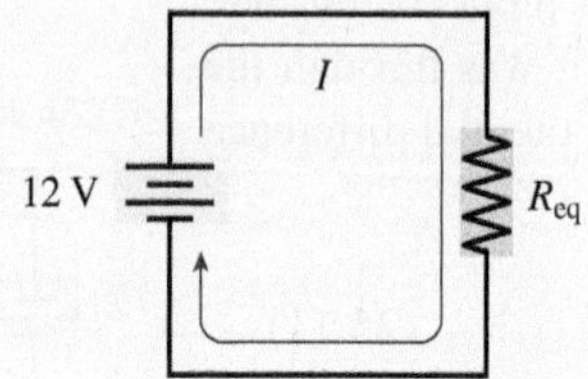

SOLVE We use Equation 25.12 to calculate the equivalent resistance:

$$R_{eq} = \left(\frac{1}{58\ \Omega} + \frac{1}{70\ \Omega} + \frac{1}{42\ \Omega}\right)^{-1} = 18.1\ \Omega$$

Once we know the equivalent resistance, we use Ohm's law to calculate the current leaving the battery:

$$I = \frac{\mathcal{E}}{R_{eq}} = \frac{12\ \text{V}}{18.1\ \Omega} = 0.66\ \text{A}$$

Because the battery can't tell the difference between the original three resistors and this single equivalent resistor, the battery in Figure 25.22 provides a current of 0.66 A to the circuit.

ASSESS As we'll see, the equivalent resistance of a group of parallel resistors is less than the resistance of any of the resistors in the group. 18 Ω is less than any of the individual values, a good check on our work.

The value of the total resistance in this example may seem surprising. The equivalent of a parallel combination of 58 Ω, 70 Ω, and 42 Ω is only 18 Ω. Shouldn't more resistors imply more resistance? The answer is yes for resistors in series, but not for resistors in parallel. Even though a resistor is an obstacle to the flow of charge, parallel resistors provide more pathways for charge to get through. Consequently, **the equivalent resistance of several resistors in parallel is always *less* than any single resistor in the group.** As an analogy, think about driving in heavy traffic. If there is an alternate route or an extra lane for cars to travel, more cars will be able to "flow."

This finding—that the equivalent resistance of several resistors in parallel is less than any single resistor in the group—is easier to understand if we think about not the resistance but its reciprocal, the *conductance* $G = 1/R$. We found in Chapter 24 that Ohm's law in terms of conductance rather than resistance is

$$I = G\Delta V \tag{25.13}$$

Let's re-analyze the parallel circuit of Figure 25.20 with conductances $G_1 = 1/R_1$ and $G_2 = 1/R_2$. We know that the potential differences are equal: $\Delta V_1 = \Delta V_2 = \mathcal{E}$. Thus Equation 25.9, Kirchhoff's junction law $I_{bat} = I_1 + I_2$, becomes

$$I_{bat} = I_1 + I_2 = G_1 \Delta V_1 + G_2 \Delta V_2 = (G_1 + G_2)\mathcal{E} \tag{25.14}$$

The battery would supply the same current if we replaced the two conductances with a single equivalent conductance $G_{eq} = G_1 + G_2$. In other words, the total conductance of two branches in parallel is the sum of the individual conductances, which makes intuitive sense.

Extending this to N parallel resistors that have conductances $G_1, G_2, \ldots, G_N$, we find that the equivalent conductance is

$$G_{eq} = G_1 + G_2 + \cdots + G_N \tag{25.15}$$

Resistances add for resistors in series because each further obstructs the flow of charge, but conductances add for resistors in parallel because each is an additional conducting pathway. If each G in Equation 25.15 is replaced with the appropriate $1/R$, then Equation 25.15 becomes Equation 25.12 for the equivalent resistance of N parallel resistors.

STOP TO THINK 25.3 Rank in order, from brightest to dimmest, the identical bulbs A to D.

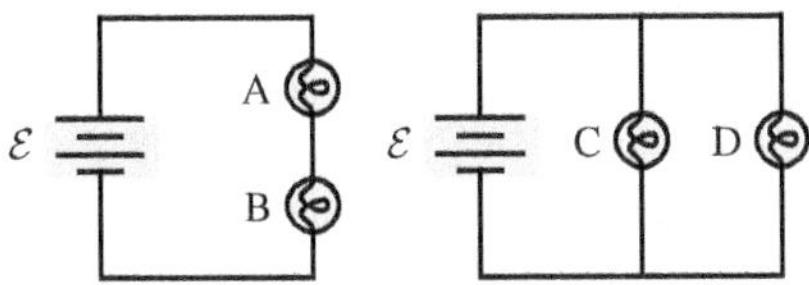

25.4 Measuring Voltage and Current

When you use a meter to measure the voltage or the current in a circuit, how do you connect the meter? The connection depends on the quantity you wish to measure.

A device that measures the current in a circuit element is called an **ammeter.** Because charge flows *through* circuit elements, an ammeter must be placed *in series* with the circuit element whose current is to be measured so that all the current that passes through the circuit element also passes through the meter.

FIGURE 25.24a shows a simple one-resistor circuit with a fixed emf $\mathcal{E} = 1.5$ V and an unknown resistance R. To determine the resistance, we must use an ammeter to measure the current in the circuit. We insert the ammeter in the circuit as shown in FIGURE 25.24b. We have to *break the connection* between the battery and the resistor in order to insert the ammeter. The resistor and the ammeter now have the same current because they are in series, so the reading of the ammeter is the current through the resistor.

Because the ammeter is in series with resistor R, the total resistance seen by the battery is $R_{eq} = R + R_{meter}$. In order to *measure* the current without *changing* the current, the ammeter's resistance must be much less than R. Thus **the resistance of an ideal ammeter is zero.** Real ammeters come quite close to this ideal.

The ammeter in Figure 25.24b reads 0.60 A, meaning that the current in the ammeter—and in the resistor—is $I = 0.60$ A. If the ammeter is ideal, which we will assume, then there is no potential difference across the ammeter ($\Delta V = IR = 0$ if $R = 0\ \Omega$) and thus the potential difference across the resistor is $\Delta V = \mathcal{E}$. The resistance can then be calculated as

$$R = \frac{\mathcal{E}}{I} = \frac{1.5\text{ V}}{0.60\text{ A}} = 2.5\ \Omega$$

FIGURE 25.24 An ammeter measures the current in a circuit.

(a)

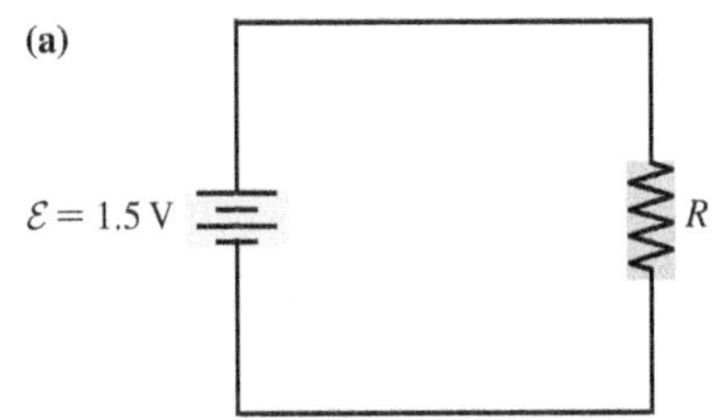

(b)

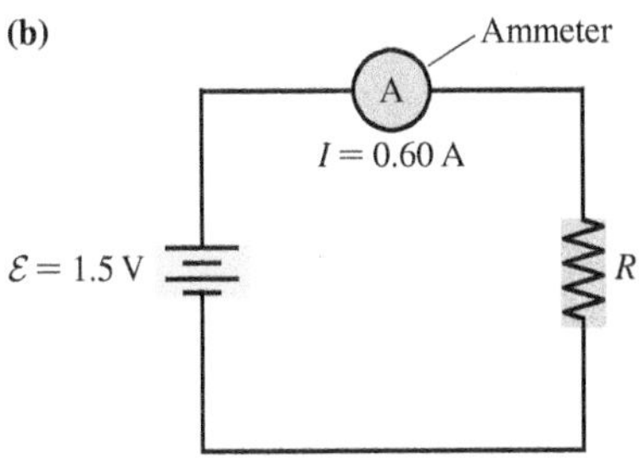

FIGURE 25.25 A voltmeter measures the potential difference across a circuit element.

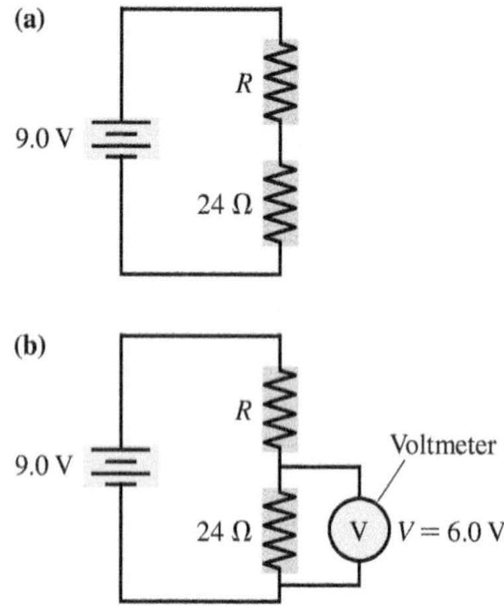

We use a **voltmeter** to measure potential differences in a circuit. Because a potential difference is measured *across* a circuit element, from one side to the other, a voltmeter is placed in *parallel* with the circuit element whose potential difference is to be measured. We want to *measure* the voltage without *changing* the voltage—without affecting the circuit. Because the voltmeter is in parallel with the resistor, the voltmeter's resistance must be very large so that it draws very little current. **An ideal voltmeter has infinite resistance.** Real voltmeters come quite close to this ideal.

FIGURE 25.25a shows a simple circuit in which a 24 Ω resistor is connected in series with an unknown resistance, with the pair of resistors connected to a 9.0 V battery. To determine the unknown resistance, we use a voltmeter to measure the potential difference across the known resistor, as shown in FIGURE 25.25b. The voltmeter is connected in parallel with the resistor; using a voltmeter does *not* require that we break the connections. The resistor and the voltmeter have the same potential difference because they are in parallel, so the reading of the voltmeter is the voltage across the resistor.

The voltmeter in Figure 25.25b tells us that the potential difference across the 24 Ω resistor is 6.0 V, so the current through the resistor is

$$I = \frac{\Delta V}{R} = \frac{6.0\text{ V}}{24\ \Omega} = 0.25\text{ A} \tag{25.16}$$

The two resistors are in series, so this is also the current in unknown resistor R. We can use Kirchhoff's loop law and the voltmeter reading to find the potential difference across the unknown resistor:

$$\sum_i \Delta V_i = 9.0\text{ V} + \Delta V_R - 6.0\text{ V} = 0 \tag{25.17}$$

from which we find $\Delta V_R = -3.0$ V. The minus sign tells us that the potential decreases across the resistor. We can use the magnitude $\Delta V = 3.0$ V and Ohm's law to calculate

$$R = \frac{\Delta V}{I} = \frac{3.0\text{ V}}{0.25\text{ A}} = 12\ \Omega \tag{25.18}$$

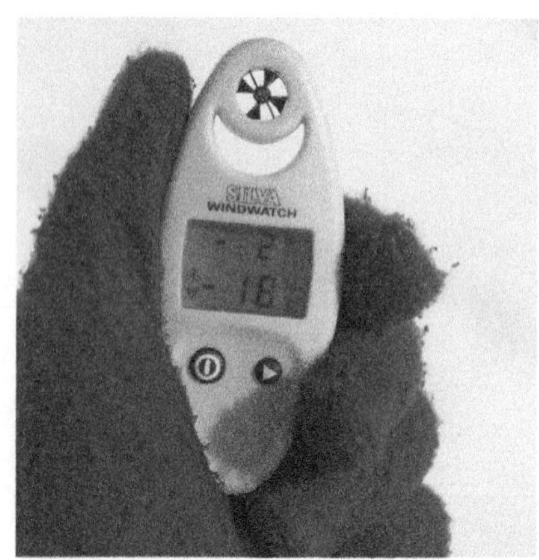

A circuit for all seasons This device displays wind speed and temperature, but these are computed from basic measurements of voltage and current. The wind turns a propeller attached to a generator; a rapid spin means a high voltage. A circuit in the device contains a *thermistor,* whose resistance varies with temperature; low temperatures mean high resistance and thus a small current.

STOP TO THINK 25.4 Which is the right way to connect the meters to measure the potential difference across and the current through the resistor?

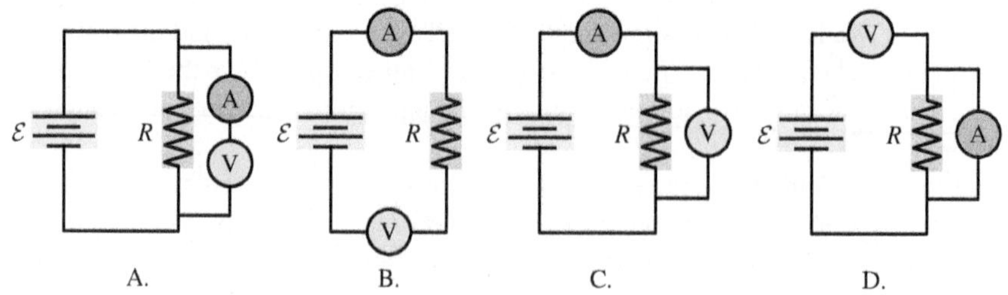

25.5 More Complex Circuits

In this section, we will consider circuits that involve both series and parallel resistors. Combinations of resistors can often be reduced to a single equivalent resistance through a step-by-step application of the series and parallel rules.

EXAMPLE 25.7 Combining resistors

What is the equivalent resistance of the group of resistors shown in FIGURE 25.26?

PREPARE We will analyze this circuit by looking for pairs of resistors that are in series or parallel, and then replacing them with their equivalent resistors. We will continue this process until the entire network of resistors is reduced to a single equivalent resistor.

FIGURE 25.26 A resistor circuit.

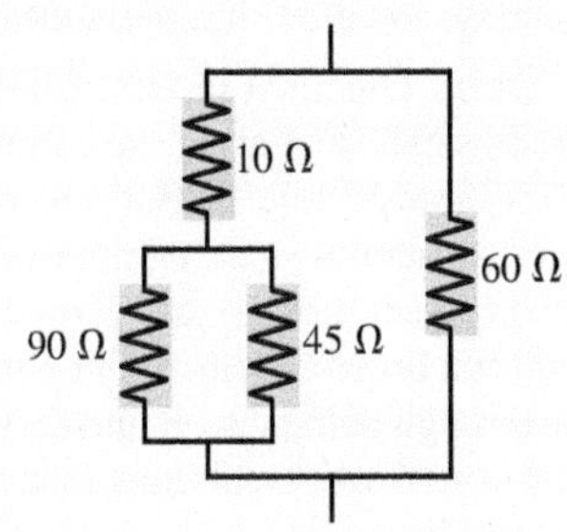

SOLVE The process of simplifying the circuit is shown in FIGURE 25.27. Note that the 10 Ω and 60 Ω resistors are *not* in parallel. They are connected at their top ends but not at their bottom ends. Resistors must be connected at *both* ends to be in parallel. Similarly, the 10 Ω and 45 Ω resistors are *not* in series because of the junction between them.

ASSESS The last step in the process is to reduce a combination of parallel resistors. The resistance of parallel resistors is always less than the smallest of the individual resistance values, so our final result must be less than 40 Ω. This is a good check on the result.

FIGURE 25.27 A combination of resistors is reduced to a single equivalent resistor.

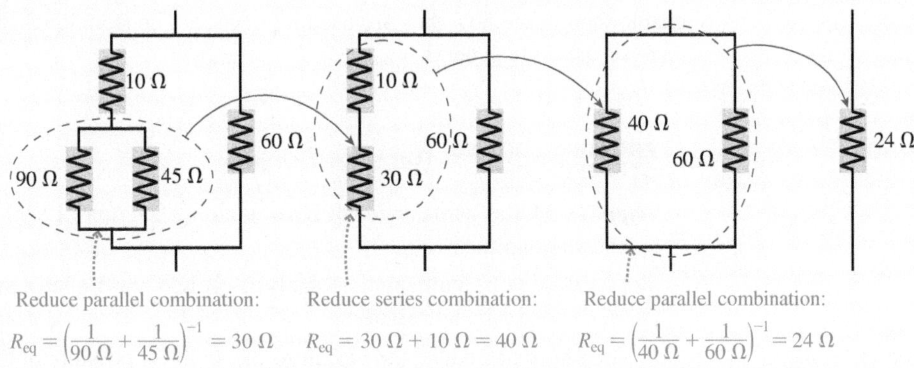

Two special cases (worth remembering for reducing circuits) are the equivalent resistances of two identical resistors $R_1 = R_2 = R$ in series and in parallel:

Two identical resistors in series: $R_{eq} = 2R$

Two identical resistors in parallel: $R_{eq} = \frac{R}{2}$

EXAMPLE 25.8 How does the brightness of bulbs change when a switch is closed?

Initially the switch in FIGURE 25.28 is open so that no current can flow through it. Bulbs A and B are equally bright, and bulb C is not glowing. What happens to the brightness of A and B when the switch is closed, connecting the wires on each side of the switch? And how does the brightness of C then compare to that of A and B?

FIGURE 25.28 A lightbulb circuit.

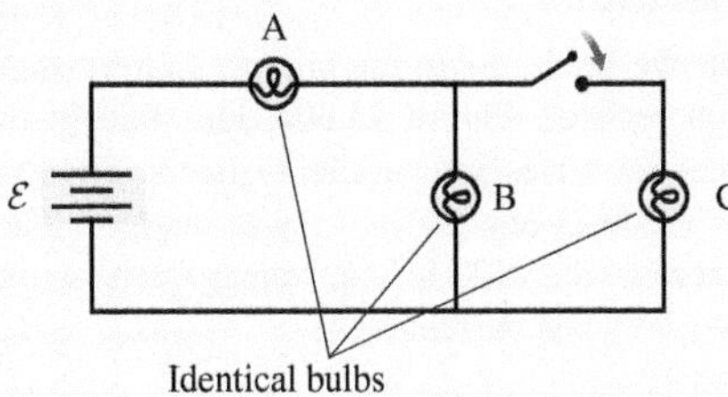

PREPARE We assume the bulbs are identical, with resistance *R*. We can then use the equivalent-resistors rules for identical resistors.

SOLVE Initially, before the switch is closed, bulbs A and B are in series; bulb C is not part of the circuit. A and B are identical resistors in series, so their equivalent resistance is $2R$ and the current from the battery is

$$I_{\text{before}} = \frac{\mathcal{E}}{R_{eq}} = \frac{\mathcal{E}}{2R} = \frac{1}{2}\frac{\mathcal{E}}{R}$$

This is the initial current in bulbs A and B, so they are equally bright.

Closing the switch places bulbs B and C in parallel with each other. The equivalent resistance of the two identical resistors in parallel is $R_{B,C} = R/2$. This equivalent resistance of B and C is

Continued

in series with bulb A; hence the total resistance of the circuit is $R_{eq} = R + \frac{1}{2}R = \frac{3}{2}R$, and the current leaving the battery is

$$I_{after} = \frac{\mathcal{E}}{R_{eq}} = \frac{\mathcal{E}}{3R/2} = \frac{2}{3}\frac{\mathcal{E}}{R} > I_{before}$$

Closing the switch *decreases* the total circuit resistance and thus *increases* the current leaving the battery.

All the current from the battery passes through bulb A, so A *increases* in brightness when the switch is closed. The current I_{after} then splits at the junction. Bulbs B and C have equal resistance, so the current divides equally. The current in B is $\frac{1}{3}(\mathcal{E}/R)$, which is *less* than I_{before}. Thus B *decreases* in brightness when the switch is closed. With the switch closed, bulbs B and C are in parallel, so bulb C has the same brightness as bulb B.

ASSESS Our final results make sense. Initially, bulbs A and B are in series, and all of the current that goes through bulb A also goes through bulb B. But when we add bulb C, the current has another option—it can go through bulb C. This will increase the total current, and all that current must go through bulb A, so we expect a brighter bulb A. But now the current through bulb A can go through both bulbs B and C. The current splits, so we'd expect that bulb B will be dimmer than before.

We can use the information in this chapter to analyze more complex but more realistic circuits. This will give us a chance to bring together the many ideas of this chapter and to see how they are used in practice. The techniques that we use for this analysis are quite general.

PROBLEM-SOLVING STRATEGY 25.1 Resistor Circuits

PREPARE Model wires as ideal. Draw a circuit diagram. Label all known and unknown quantities.

SOLVE Base your mathematical analysis on Kirchhoff's laws and on the rules for series and parallel resistors:

- Step by step, reduce the circuit to the smallest possible number of equivalent resistors.
- Determine the current through and potential difference across the equivalent resistors.
- Rebuild the circuit, using the facts that the current is the same through resistors in series and the potential difference is the same across parallel resistors.

ASSESS Use two important checks as you rebuild the circuit.

- Verify that the sum of the potential differences across series resistors matches ΔV for the equivalent resistor.
- Verify that the sum of the currents through parallel resistors matches I for the equivalent resistor.

EXAMPLE 25.9 Analyzing a complex circuit

Find the current through and the potential difference across each of the four resistors in the circuit shown in FIGURE 25.29.

FIGURE 25.29 A multiple-resistor circuit.

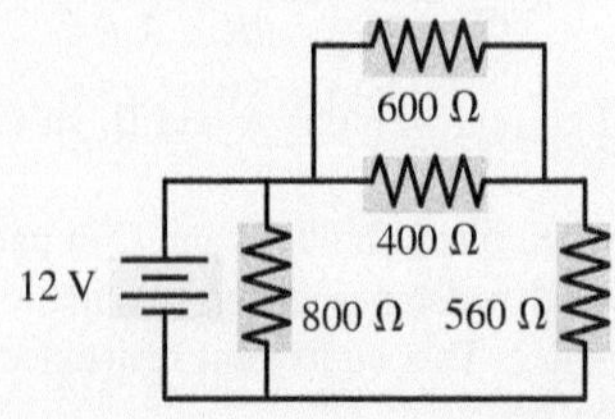

PREPARE We will analyze this complicated multi-resistor circuit by using the steps of Problem-Solving Strategy 25.1. FIGURE 25.30 shows the circuit diagram. We'll keep redrawing the diagram as we analyze the circuit.

SOLVE First, we break down the circuit, step-by-step, into one with a single resistor. Figure 25.30a does this in three steps, using the rules for series and parallel resistors. The final battery-and-resistor circuit is one that is easy to analyze. The potential difference across the 400 Ω equivalent resistor is $\Delta V_{400} = \Delta V_{bat} = \mathcal{E} = 12$ V. The current is

$$I = \frac{\mathcal{E}}{R} = \frac{12\text{ V}}{400\ \Omega} = 0.030\text{ A} = 30\text{ mA}$$

FIGURE 25.30 The step-by-step circuit analysis.

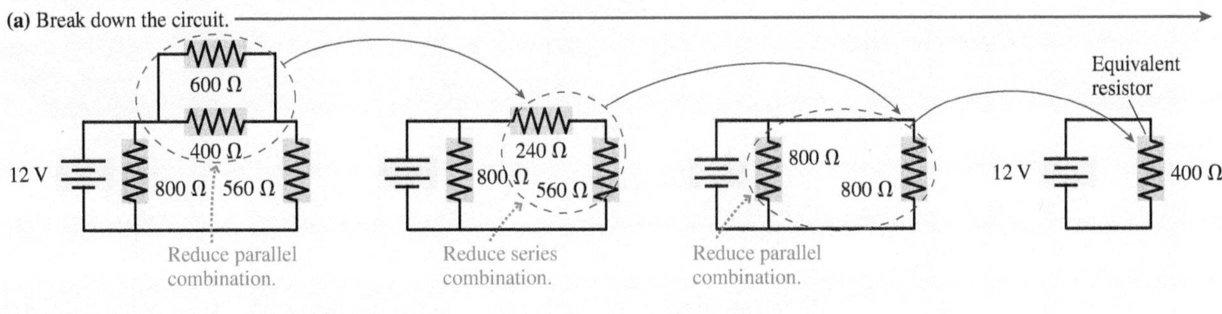

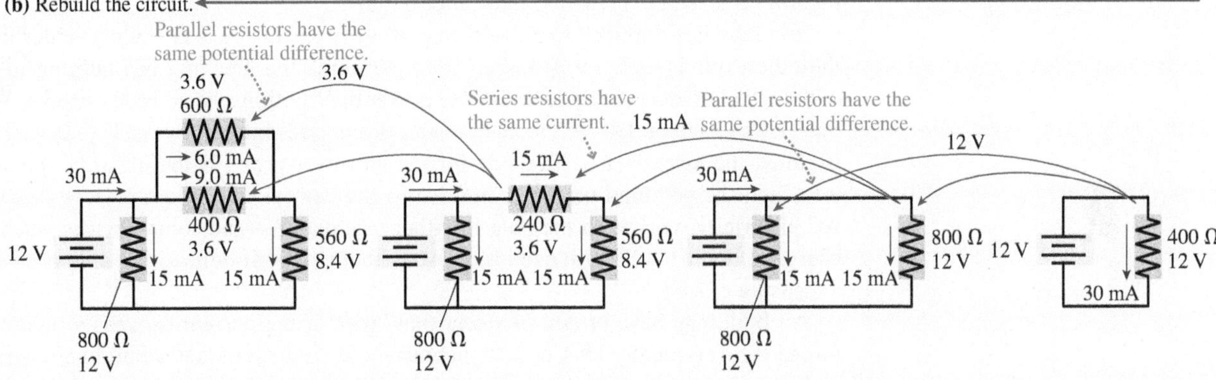

Second, we rebuild the circuit, step-by-step, finding the currents and potential differences at each step. Figure 25.30b repeats the steps of Figure 25.30a exactly, but in reverse order. The 400 Ω resistor came from two 800 Ω resistors in parallel. Because $\Delta V_{400} = 12$ V, it must be true that each $\Delta V_{800} = 12$ V. The current through each 800 Ω resistor is then $I = \Delta V/R = 15$ mA. A check on our work is to note that 15 mA + 15 mA = 30 mA.

The rightmost 800 Ω resistor was formed by combining 240 Ω and 560 Ω in series. Because $I_{800} = 15$ mA, it must be true that $I_{240} = I_{560} = 15$ mA. The potential difference across each is $\Delta V = IR$, so $\Delta V_{240} = 3.6$ V and $\Delta V_{560} = 8.4$ V. Here the check on our work is to note that 3.6 V + 8.4 V = 12 V = ΔV_{800}, so the potential differences add as they should.

Finally, the 240 Ω resistor came from 600 Ω and 400 Ω in parallel, so they each have the same 3.6 V potential difference as their 240 Ω equivalent. The currents are $I_{600} = 6.0$ mA and $I_{400} = 9.0$ mA. Note that 6.0 mA + 9.0 mA = 15 mA, which is a third check on our work. We now know all currents and potential differences.

ASSESS We *checked our work* at each step of the rebuilding process by verifying that currents summed properly at junctions and that potential differences summed properly along a series of resistances. This "check as you go" procedure is extremely important. It provides you, the problem solver, with a built-in error finder that will immediately inform you if you've made a mistake.

STOP TO THINK 25.5 Rank in order, from brightest to dimmest, the identical bulbs A to D.

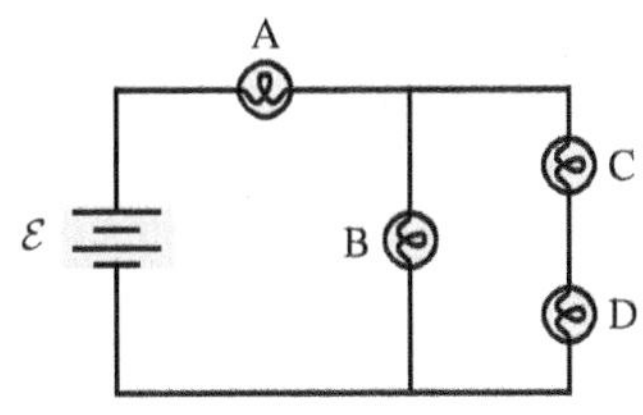

25.6 Electric Safety

FIGURE 25.31 (next page) is a circuit diagram for the electric outlets in a home or laboratory. The 120 V AC electricity is provided by the power company and distributed to outlets through wires in the walls. Electric outlets are wired in parallel

FIGURE 25.31 The 120 V AC wiring in a building.

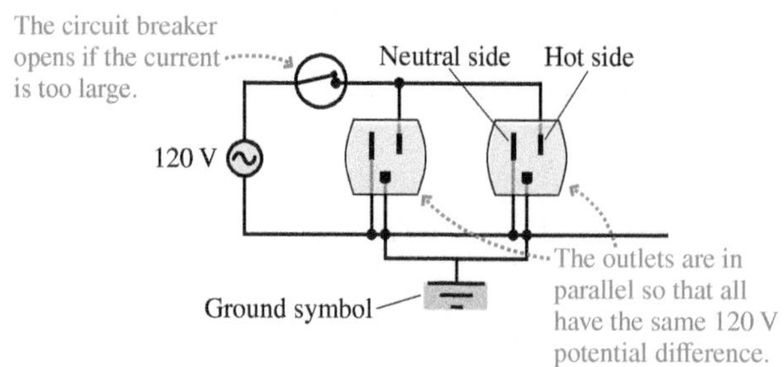

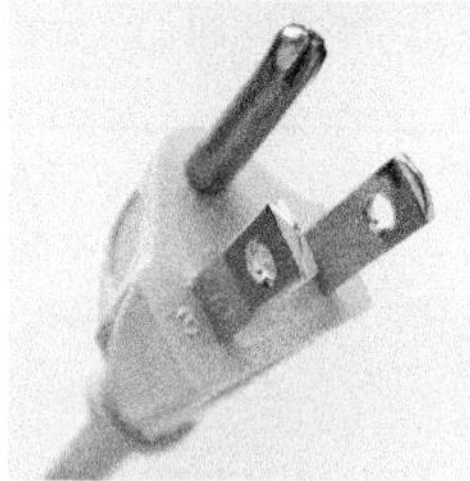
A three-prong plug

with the oscillating emf so that every outlet has the same 120 V potential difference regardless of what is plugged into other outlets.

You may have noticed that one prong of most power plugs is slightly wider than the other; this is called a *polarized plug,* although the meaning has nothing to do with the polarization of dielectrics. The power distribution system is designed so that one side of each outlet—the side the wider prong goes into—stays at $V \approx 0$ V. This is called the *neutral side.* The side for the narrower prong is the *hot side,* the side on which the potential oscillates relative to the neutral side. Most devices that use AC electricity work with the plug in either orientation—and some devices, such as chargers, don't have a polarized plug—but having a well-defined hot side provides a measure of safety.

All buildings have *circuit breakers* that "trip" if the current exceeds a predetermined value: typically 15 A or 20 A in household circuits. In fact, a building's wiring is divided into many independent 120 V circuits, each with its own circuit breaker; the circuit-breaker panel for a house usually has a dozen or more circuit breakers. A circuit breaker can be reset, but it will quickly trip again if the problem that causes too much current isn't fixed.

All modern electric outlets have a third hole that is matched to the round pin on a three-prong plug. This is a *ground connection,* and anything connected to this pin is said to be *grounded.* Figure 25.31 shows the circuit symbol for a ground connection. Ground connections are wired quite literally to a metal stake that is driven deep into the ground just outside the building.

FIGURE 25.32 A GFI outlet.

All laboratory equipment and many household appliances have a three-prong plug. The chassis of the device is usually metal (although it might be covered with decorative plastic so that the metal isn't obvious), and the metal chassis is grounded. This is an important safety precaution. If a malfunction causes a wire at 120 V to come into contact with the chassis, a surge of current to ground will immediately trip the circuit breaker or, for some laboratory equipment, blow an internal fuse. Without the ground connection, a malfunction could place the chassis of the device at 120 V, and anyone who touches the chassis could be electrocuted.

FIGURE 25.32 shows a special type of outlet, called a *ground-fault interrupter* (GFI) outlet, that is used in bathrooms and kitchens where electricity is in close proximity to water. A GFI outlet has a built-in sensor circuit that compares the currents in the hot side and the neutral side. Normally, all the current coming in from the hot side passes back out through the neutral side, so the two currents are equal. If they are not, current from the hot side is finding an alternative pathway—possibly through a person. This is a *ground fault,* and the GFI disconnects the circuit. A GFI outlet responds faster than a circuit breaker—in as little as $\frac{1}{30}$ s—and responds to current disruptions that are dangerous but that might not trip a circuit breaker.

Biological Response to Electric Current

It is current, not voltage, that produces physiological effects and damage when it passes through the body because it mimics nerve impulses and causes muscles—including heart muscles—to involuntarily contract. In general, higher voltages are more dangerous than lower voltages because they tend to produce larger currents, although some high-voltage sources, such as a Van de Graff generator, are not able to produce much current and are not dangerous.

TABLE 25.1 lists approximate values of current that produce difference physiological effects. Less AC current than DC current is required to produce the same response. Currents through the chest cavity—such as arm to arm—are especially dangerous because they can interfere with respiration and heartbeat. An AC current larger than 100 mA can cause fibrillation of the heart, in which it beats in a rapid, chaotic fashion.

To calculate likely currents through the body, FIGURE 25.33 models the body as several connected resistors. (These are average values; there is significant individual variation.) The interior of the body is largely salt water, and its resistance is fairly low. But current must pass through the skin to enter the body. Dry skin has a fairly high resistance, 100 kΩ or more, but the resistance of wet skin can be 1 kΩ or less.

TABLE 25.1 Physiological effects of current on the human body

Physiological effect	AC current (mA rms)	DC current (mA)
Threshold of sensation	1	3
Painful shock	12	40
Respiratory paralysis	25	75
Heart fibrillation, likely fatal	>100	>500

FIGURE 25.33 A resistance model of the human body.

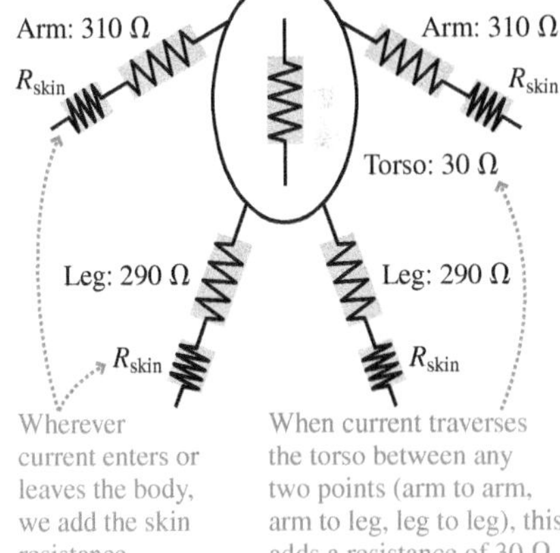

When current traverses the torso between any two points (arm to arm, arm to leg, leg to leg), this adds a resistance of 30 Ω.

EXAMPLE 25.10 Is the worker in danger?

A worker in a plant grabs a bare wire that he does not know is connected to a 480 V (rms) AC supply. His other hand is holding a grounded metal railing. The skin resistance of each of his hands, in full contact with a conductor, is 2200 Ω. He will receive a shock. Is it large enough to be dangerous?

PREPARE We will use the resistance model of the body shown in Figure 25.33. Here, the current path goes through the skin of one hand, up one arm, across the torso, down the other arm, and through the skin of the other hand.

SOLVE We can draw a circuit model for this situation as in FIGURE 25.34a; the worker's body completes a circuit between two points at a potential difference of 480 V. The current will depend on this potential difference and the resistance of his body, including the resistance of the skin.

FIGURE 25.34 A circuit model for the worker.

(a)

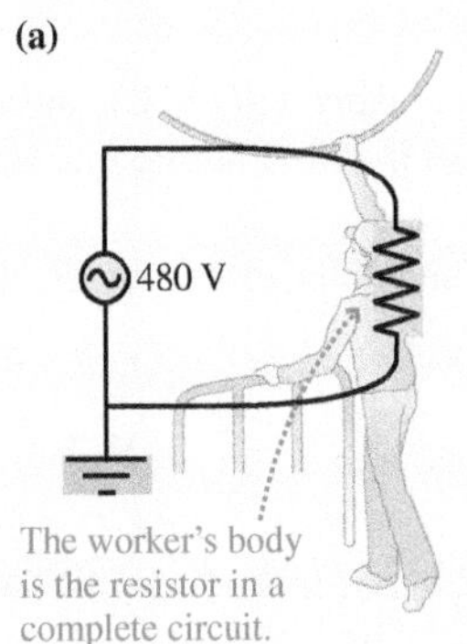

480 V
The worker's body is the resistor in a complete circuit.

(b) We add the resistances along the current path to find the equivalent resistance of the body.

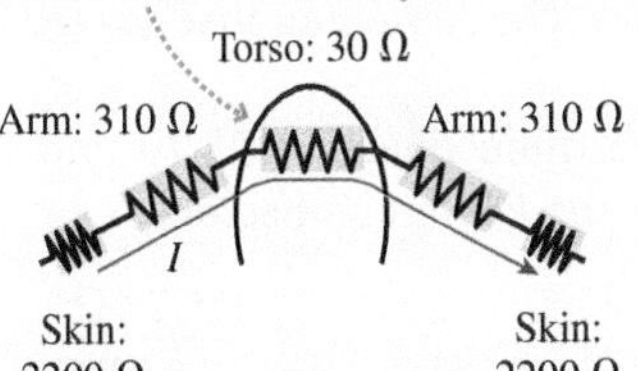

$$R_{eq} = 2200\ \Omega + 310\ \Omega + 30\ \Omega + 310\ \Omega + 2200\ \Omega = 5050\ \Omega$$

The individual resistances that determine the current are shown in FIGURE 25.34b. The equivalent resistance of the series combination is 5050 Ω, so the AC current through his body is

$$I_{rms} = \frac{\Delta V_{rms}}{R_{eq}} = \frac{480\ \text{V}}{5050\ \Omega} = 95\ \text{mA}$$

From Table 25.1 we see that this is a very dangerous, possibly fatal, current.

ASSESS The voltage is high and the resistance relatively low, so it's no surprise to find a dangerous level of current.

25.7 *RC* Circuits

FIGURE 25.35 Discharging a capacitor.

(a) Before the switch closes

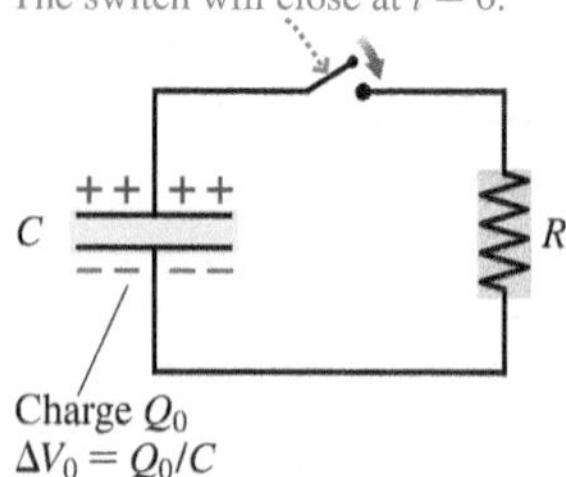

(b) Immediately after the switch closes

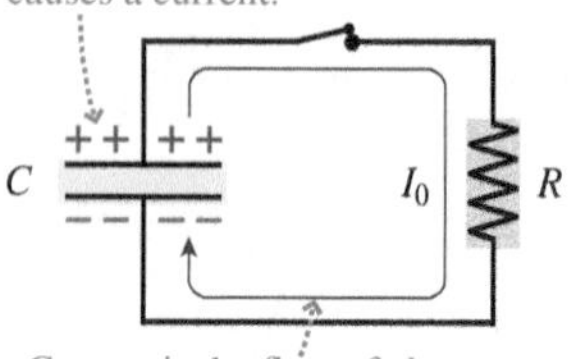

(c) At a later time

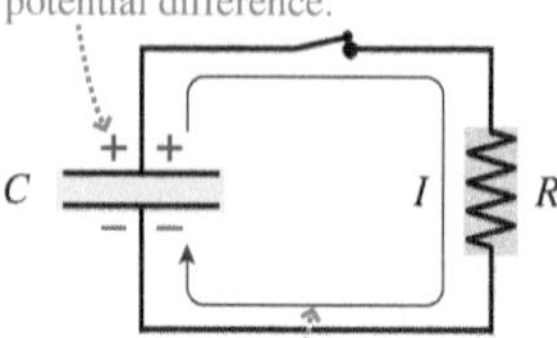

The current has reduced the charge on the capacitor. This reduces the potential difference.
C
I
R
The reduced potential difference leads to a reduced current.

A resistor circuit has a steady current. By adding a capacitor and a switch, we can make a circuit in which the current varies with time as the capacitor charges and discharges. Circuits with resistors and capacitors are called ***RC* circuits.** *RC* circuits are at the heart of timekeeping circuits in applications ranging from the intermittent windshield wipers on your car to computers and other digital electronics.

FIGURE 25.35a shows a charged capacitor, a switch, and a resistor. The capacitor has charge Q_0 and potential difference $\Delta V_0 = Q_0/C$. There is no current, so the potential difference across the resistor is zero. Then, at $t = 0$, the switch closes and the capacitor begins to discharge through the resistor. FIGURE 25.35b shows the circuit immediately after the switch closes as a current begins to discharge the capacitor.

How long does the capacitor take to discharge? How does the current through the resistor vary as a function of time? To answer these questions, FIGURE 25.35c shows the circuit at a later instant of time when the capacitor is partially discharged.

Kirchhoff's loop law is valid for any circuit, not just circuits with batteries. The loop law applied to the circuit of Figure 25.35c, going around the loop clockwise, is

$$\Delta V_C + \Delta V_R = \frac{Q}{C} - IR = 0 \tag{25.19}$$

Q and I in this equation are the *instantaneous* values of the capacitor charge and the resistor current as they vary with time.

The current I is the rate at which charge flows through the resistor: $I = dq/dt$. But the charge flowing through the resistor is charge that was *removed* from the capacitor. That is, an infinitesimal charge dq flows through the resistor when the capacitor charge *decreases* by dQ. Thus $dq = -dQ$, and the resistor current is related to the instantaneous capacitor charge by

$$I = -\frac{dQ}{dt} \tag{25.20}$$

Equation 25.20 tells us that I is positive when Q is decreasing $(dQ < 0)$, as we would expect. The reasoning that has led to Equation 25.20 is rather subtle but very important.

If we substitute Equation 25.20 into Equation 25.19 and then divide by R, the loop law for the *RC* circuit becomes

$$\frac{dQ}{dt} + \frac{Q}{RC} = 0 \tag{25.21}$$

We can solve Equation 25.21 for Q by direct integration. First, we rearrange Equation 25.21 to get all the charge terms on one side of the equation:

$$\frac{dQ}{Q} = -\frac{1}{RC}dt$$

The product RC is a constant for any particular circuit.

The capacitor charge was Q_0 at $t = 0$ when the switch was closed. We want to integrate from these starting conditions to charge Q at a later time t. That is,

$$\int_{Q_0}^{Q} \frac{dQ}{Q} = -\frac{1}{RC}\int_0^t dt \tag{25.22}$$

Both are well-known integrals, giving

$$\ln Q\Big|_{Q_0}^{Q} = \ln Q - \ln Q_0 = \ln\left(\frac{Q}{Q_0}\right) = -\frac{t}{RC}$$

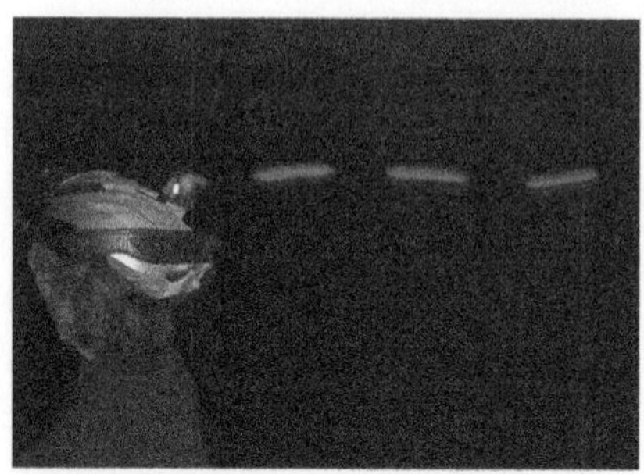

A bike flasher The rear flasher on a bike helmet flashes on and off. The timing is controlled by an *RC* circuit.

We can solve for the capacitor charge Q by taking the exponential of both sides, then multiplying by Q_0. Doing so gives

$$Q = Q_0 e^{-t/RC} \tag{25.23}$$

Notice that $Q = Q_0$ at $t = 0$, as expected.

It's not obvious, but by using the definitions of resistance and capacitance, we can show that the quantity RC has units of seconds. Consequently, we define the **time constant** τ to be

$$\tau = RC \tag{25.24}$$

We can then write Equation 25.23 as

$$Q = Q_0 e^{-t/\tau} \tag{25.25}$$

The capacitor voltage, $\Delta V = Q/C$, also decays exponentially as

$$\Delta V_C = \Delta V_0 e^{-t/\tau} \tag{25.26}$$

Voltage of a capacitor during discharge

where $\Delta V_0 = Q_0/C$.

The meaning of Equation 25.26 is easier to understand if we portray it graphically. **FIGURE 25.36a** shows the capacitor voltage as a function of time. The voltage decays exponentially, starting from ΔV_0 at $t = 0$ and asymptotically approaching zero as $t \to \infty$. The time constant τ is the time at which the voltage has decreased to e^{-1} (about 37%) of its initial value. At time $t = 2\tau$, the voltage has decreased to e^{-2} (about 13%) of its initial value. This is exactly the exponential decay that we saw previously for a damped harmonic oscillator.

NOTE ▶ The *shape* of the graph of ΔV_C is always the same, regardless of the specific value of the time constant τ. ◀

We find the resistor current by using Equation 25.20:

$$I = -\frac{dQ}{dt} = \frac{Q_0}{\tau}e^{-t/\tau} = \frac{Q_0}{RC}e^{-t/\tau} = \frac{\Delta V_0}{R}e^{-t/\tau} = I_0 e^{-t/\tau} \tag{25.27}$$

where $I_0 = \Delta V_0/R$ is the initial current, immediately after the switch closes. **FIGURE 25.36b** is a graph of the resistor current versus t. You can see that the current undergoes the same exponential decay, with the same time constant, as the capacitor voltage.

FIGURE 25.36 The decay curves of the capacitor voltage and the resistor current.

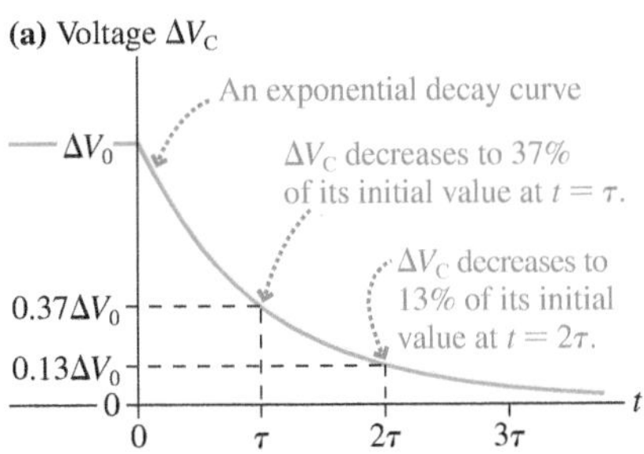

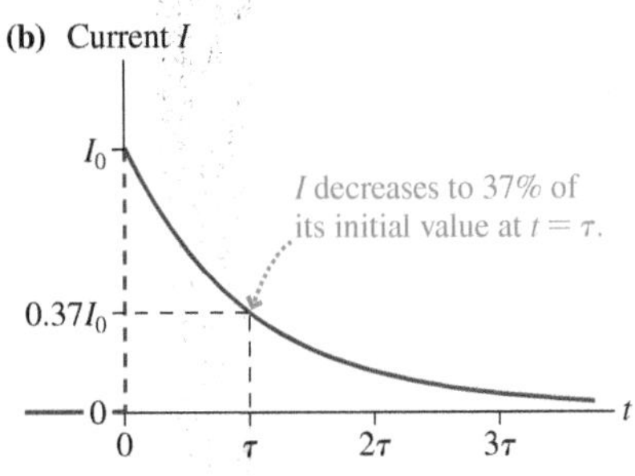

NOTE ▶ There's no specific time at which the capacitor has been completely discharged, because Q and ΔV_C approach zero asymptotically, but the charge and voltage have dropped to less than 1% of their initial values at $t = 5\tau$. Thus 5τ is a reasonable answer to the question How long does it take to discharge a capacitor? ◀

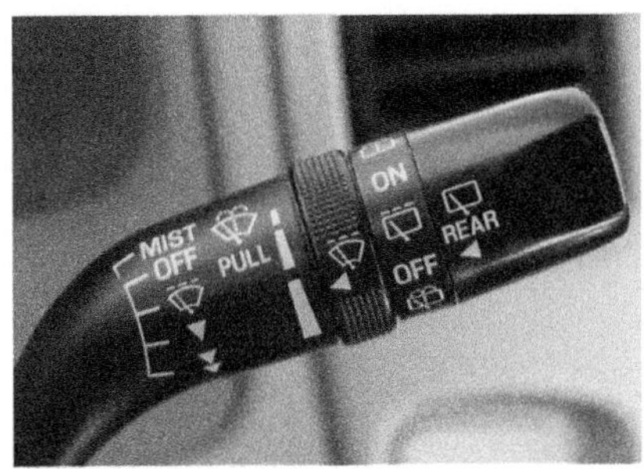

A rainy-day *RC* circuit When you adjust the dial to control the delay of the intermittent windshield wipers in your car, you are adjusting a variable resistor in an *RC* circuit that triggers the wipers. Increasing the resistance increases the time constant and thus produces a longer delay between swipes of the blades. A light mist calls for a long time constant and thus a large resistance.

Charging a Capacitor

FIGURE 25.37a shows a circuit that charges a capacitor. After the switch is closed, the battery's charge escalator moves charge from the bottom electrode of the capacitor to the top electrode. The resistor, by limiting the current, slows the process but doesn't stop it. The capacitor charges until $\Delta V_C = \mathcal{E}$; then the charging current ceases. The full charge of the capacitor is $Q_0 = C(\Delta V_C)_{max} = C\mathcal{E}$.

The analysis is much like that of discharging a capacitor. The capacitor voltage and the circuit current at time t are

$$\Delta V_C = \mathcal{E}(1 - e^{-t/\tau})$$
$$I = I_0 e^{-t/\tau} \qquad (25.28)$$

where $I_0 = \mathcal{E}/R$ and, again, $\tau = RC$. The capacitor charge's "upside-down decay" to $\mathcal{E}$ is shown graphically in FIGURE 25.37b.

FIGURE 25.37 Charging a capacitor.

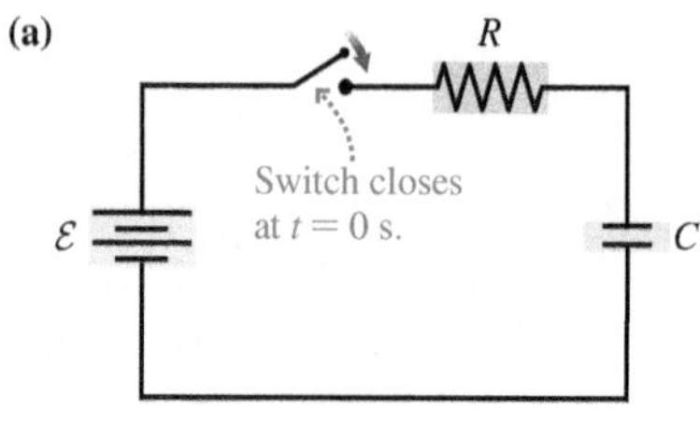

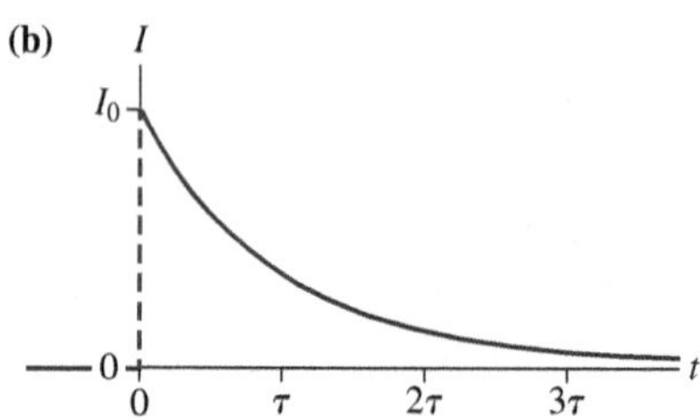

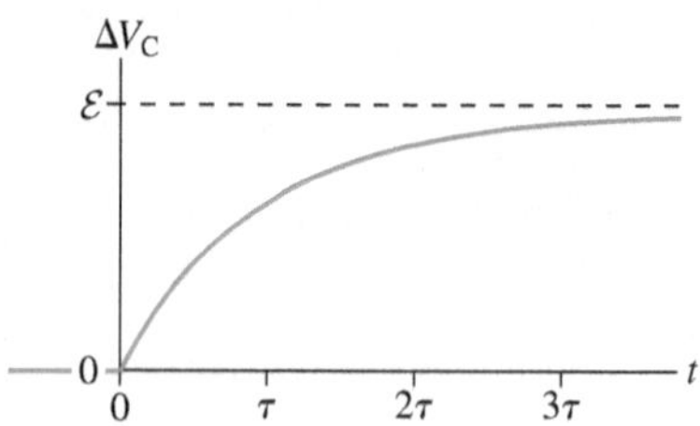

EXAMPLE 25.11 Measuring membrane capacitance

One method used to determine the capacitance of a cell membrane is to measure the time constant when the membrane is the capacitor in an RC circuit. In one experiment in which a microelectrode was inserted into a section of a rat hippocampal neuron, the current through the microelectrode was measured when a 5.0 mV potential difference was suddenly applied across the membrane. FIGURE 25.38a is a simple model of the experiment in which the membrane is represented as a capacitor, the resistor is the resistance of the microelectrode, and applying the potential difference is equivalent to closing a switch at $t = 0$ s. FIGURE 25.38b is a smoothed and slightly simplified current graph; actual curves in experiments like these are very noisy, so a large number of curves are averaged to reduce the noise. The investigators used microscopy to determine that the surface area of the membrane was 200 μm^2. Every cell has a somewhat different capacitance, depending on its surface area A, so electrophysiologists characterize cells by their *specific membrane capacitance* $c_{mem} = C_{mem}/A$, which should be the same for all cells of the same type. Specific membrane capacitance is usually given in units of $\mu F/cm^2$. What are (a) the specific membrane capacitance and (b) the membrane thickness of the hippocampal neuron?

FIGURE 25.38 Measuring the membrane capacitance.

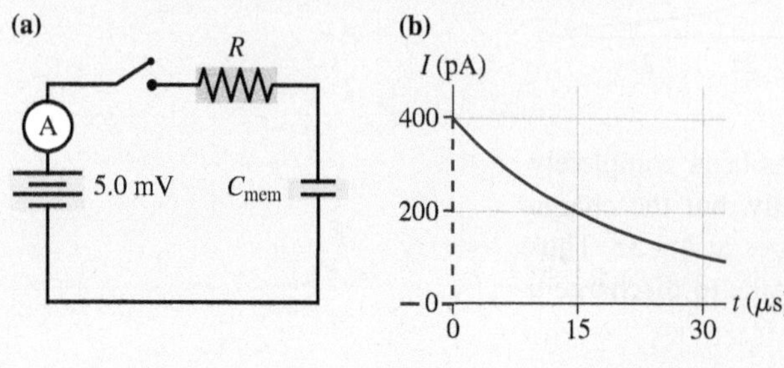

PREPARE The simplified model of Figure 25.38a is a capacitor charging circuit identical to Figure 25.37a. The observed current, for which we can use Equation 25.28, matches the graph of Figure 25.37b. First we'll measure the initial current I_0 and the time constant τ from the experimental data. Then we will use $I_0 = \mathcal{E}/R$ with $\mathcal{E} = 5.0$ mV to determine R and $\tau = RC_{mem}$ to determine C_{mem}. In Chapter 23, we found that the membrane capacitance is $C_{mem} \approx \kappa\epsilon_0 A/d$, and we can use this expression to determine the membrane thickness. The dielectric constant of the cell membrane was given as $\kappa = 5.0$ in Table 23.1.

SOLVE

a. We see from Figure 25.38b that the initial current, just after the switch is closed, is 400 pA. Thus the circuit resistance—an experimental resistance, not a property of the membrane—is

$$R = \frac{\mathcal{E}}{I} = \frac{5.0 \times 10^{-3}\ \text{V}}{400 \times 10^{-12}\ \text{A}} = 1.25 \times 10^{7}\ \Omega = 12.5\ \text{M}\Omega$$

Rather than trying to determine the time at which the current has decayed to e^{-1} of its initial value, it's easier to measure the *half-life* $t_{1/2}$ at which the current has decayed to 50% of its initial value. We see from Figure 25.38b that $t_{1/2} = 15\ \mu\text{s}$. Equation 25.28 allows us to relate $t_{1/2}$ to the time constant τ. By definition, $I = \frac{1}{2}I_0$ at $t = t_{1/2}$, so Equation 25.28 is

$$\tfrac{1}{2}I_0 = I_0 e^{-t_{1/2}/\tau}$$

The I_0 cancels. If we then take the natural logarithm of both sides, we get

$$\ln\left(\frac{1}{2}\right) = -\ln 2 = \ln\left(e^{-t_{1/2}/\tau}\right) = -\frac{t_{1/2}}{\tau}$$

$$\tau = \frac{t_{1/2}}{\ln 2} = \frac{t_{1/2}}{0.693}$$

We found $t_{1/2} = 15\ \mu\text{s}$, so the time constant is $\tau = 22\ \mu\text{s}$.

Thus the membrane capacitance is

$$C_{\text{mem}} = \frac{\tau}{R} = \frac{22 \times 10^{-6}\ \text{s}}{12.5 \times 10^{6}\ \Omega} = 1.8 \times 10^{-12}\ \text{F} = 1.8\ \text{pF}$$

The specific membrane capacitance is found by dividing by the surface area:

$$c_{\text{mem}} = \frac{C_{\text{mem}}}{A} = \frac{1.8 \times 10^{-12}\ \text{F} \times (1\ \mu\text{F}/10^{-6}\ \text{F})}{200\ \mu\text{m}^2 \times (10^{-8}\ \text{cm}^2/1\ \mu\text{m}^2)}$$

$$= 0.90\ \mu\text{F/cm}^2$$

b. We can use the Chapter 23 expression for the membrane capacitance to calculate the membrane thickness:

$$d = \frac{\kappa\epsilon_0 A}{C_{\text{mem}}} = \frac{(5.0)(8.85 \times 10^{-12}\ \text{C}^2/\text{N}\cdot\text{m}^2)(200 \times 10^{-12}\ \text{m}^2)}{1.8 \times 10^{-12}\ \text{F}}$$

$$= 4.9 \times 10^{-9}\ \text{m} = 4.9\ \text{nm}$$

ASSESS Physiology texts state that the specific membrane capacitance of a cell is approximately 1 μF/cm^2. Our result is consistent with that value but, for this specific type of cell, somewhat more accurate. The calculated membrane thickness is consistent with values of 5 nm to 7 nm that we've used previously.

The time constant τ in an RC circuit can be used to control the behavior of a circuit. For example, a bike flasher uses an RC circuit that alternately charges and discharges, over and over, as a switch opens and closes. A separate circuit turns the light on when the capacitor voltage exceeds some threshold voltage and turns the light off when the capacitor voltage goes below this threshold. The time constant of the RC circuit determines how long the capacitor voltage stays above the threshold and thus sets the length of the flashes. More complex RC circuits provide timing in computers and other digital electronics. As we will see in the next section, we can also use RC circuits to model the transmission of nerve impulses, and the time constant will be a key factor in determining the speed at which signals can be propagated in the nervous system.

STOP TO THINK 25.6 The time constant for the discharge of this capacitor is

A. 5 s B. 4 s
C. 2 s D. 1 s
E. The capacitor doesn't discharge because the resistors cancel each other.

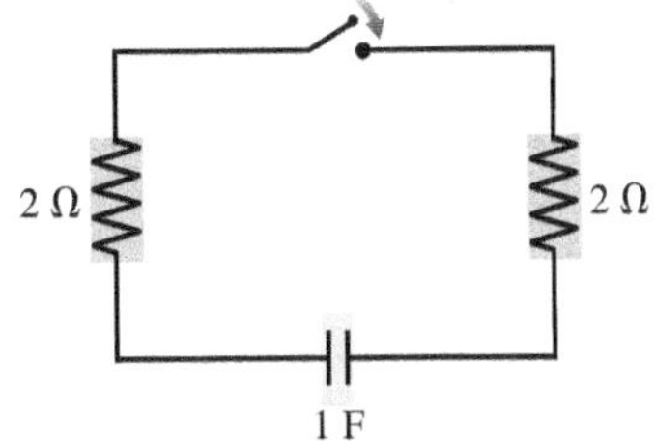

25.8 Electricity in the Nervous System

In the late 1700s, the Italian scientist Galvani discovered that animal tissue has an electrical nature. He found that a frog's leg would twitch when stimulated with electricity, even when no longer attached to the frog. Further investigations revealed that electrical signals can animate muscle cells, and that a small potential applied to the *axon* of a nerve cell can produce a signal that propagates down its length.

Our goal in this section will be to understand the nature of electrical signals in the nervous system. When your brain orders your hand to move, how does the signal get from your brain to your hand? Answering this question will use our knowledge of fields, potential, resistance, capacitance, and circuits, all of the knowledge and techniques that we have learned so far in Part V.

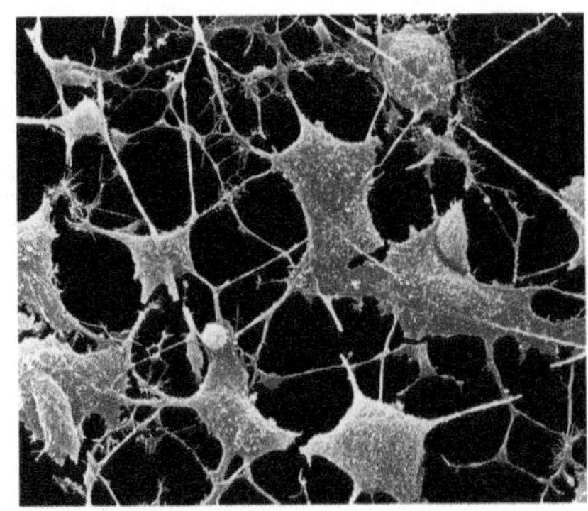

A nervous connection The axons connecting these nerve cells are clearly visible in the micrograph.

The Electrical Nature of Nerve Cells

Nerve signals are transmitted by cells called **neurons.** The transmission of a signal from the brain to a muscle is the function of a *motor neuron,* whose structure is sketched in FIGURE 25.39. The transmission takes place along the *axon* of the neuron, a long fiber that connects the cell body to a muscle fiber. The diameter of a human motor-neuron axon is typically 5 μm. Lengths vary greatly—from a few mm to $\approx$1 m for the sciatic nerve, the human body's longest nerve that runs from the lower back to a foot. The neuron shown in the figure has a myelin sheath around the axon, whose purpose we will discuss later, but not all axons do.

FIGURE 25.39 A motor neuron.

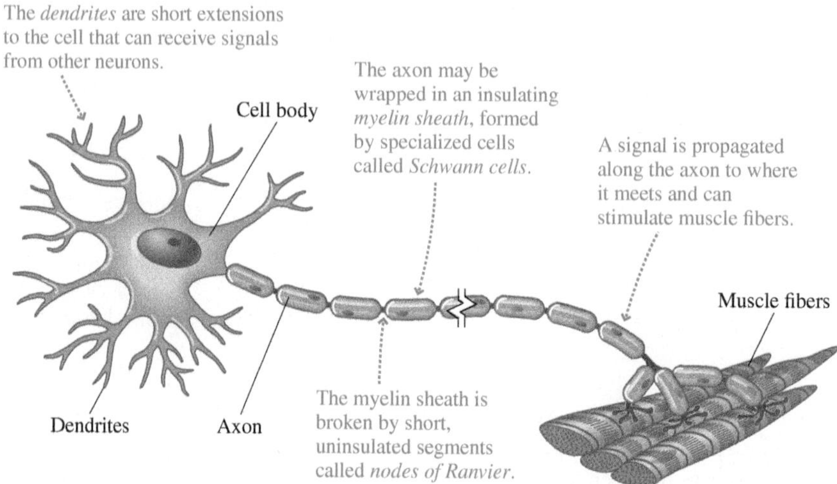

The *dendrites* are short extensions to the cell that can receive signals from other neurons.
Cell body
The axon may be wrapped in an insulating *myelin sheath,* formed by specialized cells called *Schwann cells.*
A signal is propagated along the axon to where it meets and can stimulate muscle fibers.
Muscle fibers
Dendrites
Axon
The myelin sheath is broken by short, uninsulated segments called *nodes of Ranvier.*

FIGURE 25.40 Ion flow through a cell membrane.

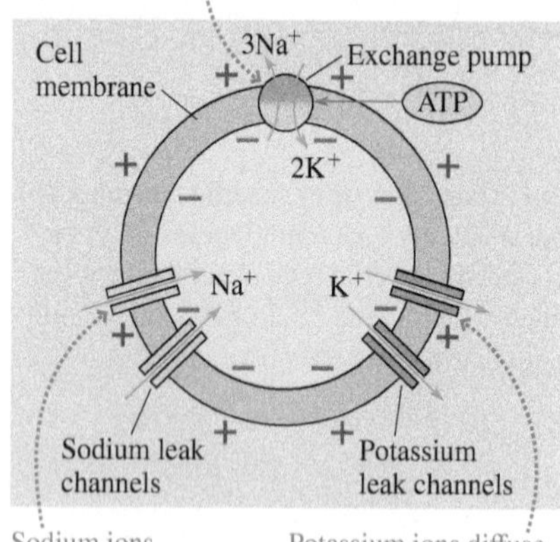

We start with a neuron in its resting state. We found in ◀◀ SECTION 23.5 that the inside of a cell is more negative than the outside. FIGURE 25.40 reminds us of the key ideas:

- Sodium and potassium *leak channels* in the cell membrane allow Na^+ and K^+ ions to diffuse down their concentration gradients. K^+ has a higher concentration inside the cell, so K^+ ions diffuse outward. Na^+ has a higher concentration outside the cell, so Na^+ ions diffuse inward.
- The *sodium-potassium exchange pumps* actively pump Na^+ ions out of the cell and K^+ ions into the cell, countering the diffusive motion.
- These two mechanisms establish a steady state in which $V_{in} < V_{out}$. The potential difference is called the *membrane potential.*

FIGURE 25.41 The resting potential of a neuron.

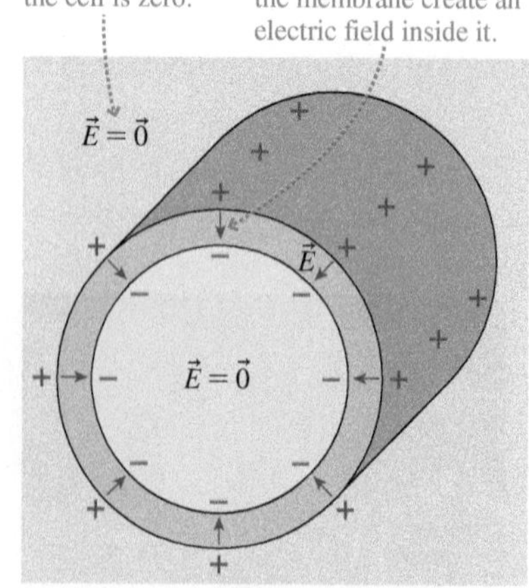

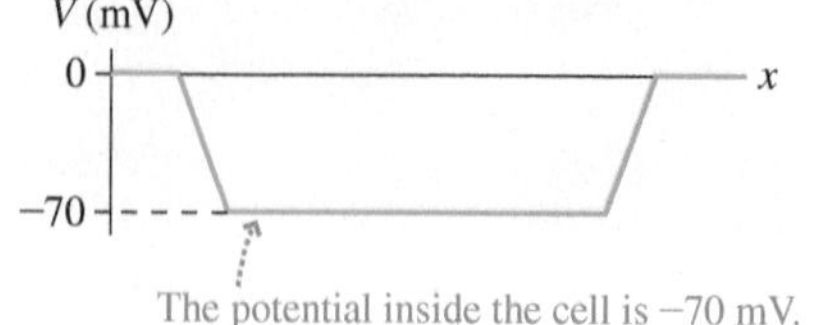

In Chapter 23 we focused on muscle cells for which the membrane potential is $V_{mem} = V_{in} - V_{out} \approx -90$ mV. The same mechanisms operate in nerve cells, but slight differences in the details make $V_{mem} \approx -70$ mV. This is the *resting potential* when the neuron is not firing. FIGURE 25.41, a cross section through an axon, shows how charge accumulates on the inside and outside surfaces of the cell. The 70 mV potential difference across a 7-nm-thick membrane creates an electric field inside the membrane that has field strength $E = \Delta V/d \approx 1 \times 10^7$ V/m. This is a very strong field.

The equal-but-opposite charges on the two sides of the cell membrane look like the charged plates of a parallel-plate capacitor, and we learned in Section 23.5 that a membrane with surface area A has capacitance

$$C_{mem} \approx \frac{\kappa \epsilon_0 A}{d} \tag{25.29}$$

where $\kappa \approx 5$ is the membrane's dielectric constant and $d \approx 7$ nm is its thickness. You can calculate that the capacitance of a 1-mm-long segment of a 5-μm-diameter axon has $C \approx 100$ pF.

A cell membrane also has an equivalent resistance. The membrane is not a uniform structure but, instead, consists of many low-conductivity pores—the leak channels—embedded in an insulating lipid bilayer. The ions passing through the pores are a current moving through a potential difference, and the equivalent resistance of the membrane is given by Ohm's law: $R_{\text{mem}} = \Delta V/I = |V_{\text{mem}}|/I$.

EXAMPLE 25.12 Finding the resistance and resistivity of a cell membrane

Researchers use microelectrodes to measure the ion currents into and out of cells. When a neuron is in its resting state, the current into a 1-mm-long segment of its 5-μm-diameter axon is $\approx$3 nA.

a. What are the equivalent resistance and the resistivity of the axon's membrane?

b. Measuring individual ion leak channels is difficult, but it's thought that the current through each channel is $\approx$0.1 pA. How many leak channels are in this 1-mm-long segment? What is their density in channels per μm^2?

PREPARE The membrane of an axon, as seen in Figure 25.41, is a thin-walled cylindrical shell that has radius R, length L, and thickness d. We can imagine unrolling the cylinder to obtain a thin, flat sheet of membrane of width $2\pi R$ (the cylinder's circumference), length L, and thickness d. V_{mem} is the potential difference across this sheet, and current I passes through from outside the cell to inside. We know from Chapter 24 that the resistance of a conductor of thickness d and cross-section area A is $R = \rho d/A$, where ρ is the resistivity.

SOLVE

a. We can use Ohm's law to calculate the equivalent resistance of the membrane:

$$R_{\text{mem}} = \frac{|V_{\text{mem}}|}{I} = \frac{0.070\ \text{V}}{3\times10^{-9}\ \text{A}} \approx 2\times10^{7}\ \Omega = 20\ \text{M}\Omega$$

The data are not precisely known, so one-significant-figure answers are appropriate. The membrane's resistivity is

$$\rho = \frac{AR_{\text{mem}}}{d} = \frac{(2\pi RL)R_{\text{mem}}}{d}$$
$$= \frac{2\pi(2.5\times10^{-6}\ \text{m})(0.001\ \text{m})(2\times10^{7}\ \Omega)}{7\times10^{-9}\ \text{m}}$$
$$= 4\times10^{7}\ \Omega\cdot\text{m}$$

b. The total current that passes into the cell through N leak channels is $I_{\text{total}} = NI_{\text{chan}}$. Thus, the number of channels is

$$N = \frac{I_{\text{total}}}{I_{\text{chan}}} \approx \frac{3\times10^{-9}\ \text{A}}{1\times10^{-13}\ \text{A}} = 3\times10^{4}$$

These 30,000 channels are spread out over the surface area $A = 2\pi RL$ of this segment of the axon, so the surface density of channels is

$$\frac{N}{A} = \frac{N}{2\pi RL} \approx \frac{3\times10^{4}\ \text{channels}}{2\pi(2.5\times10^{-6}\ \text{m})(0.001\ \text{m})}$$
$$= 2\times10^{12}\ \text{channels/m}^2 \times \left(\frac{10^{-6}\ \text{m}}{1\ \mu\text{m}}\right)^2 = 2\ \text{channels}/\mu\text{m}^2$$

ASSESS The equivalent resistance and resistivity are very high, but that is to be expected. The lipid bilayer cell membrane is an insulator, and it is only the relatively small number of ion leak channels that allow any conduction at all. Expressing the channel density as channels/μm^2 allows us to develop some intuition by thinking on a cellular scale. We see that the density of channels is fairly low at the cellular scale, but that is consistent with the membrane resistivity being very high.

A cell membrane has a resistance, a capacitance, and an emf, so we can model the membrane as the RC circuit seen in FIGURE 25.42. The membrane, like any RC circuit, has a time constant. We just calculated the resistance and capacitance of a 1-mm-long segment of an axon. We can use these numbers to compute the axon's time constant:

$$\tau = RC \approx (2\times10^{7}\ \Omega)(1\times10^{-10}\ \text{F}) = 2\times10^{-3}\ \text{s} = 2\ \text{ms}$$

Indeed, if researchers artificially change the membrane potential by a few mV (large enough to measure but not enough to trigger a neuron response), the potential decays back to its resting value with a time constant of $\approx$2 ms.

It seems as if the time constant should depend on the length and radius of the axon. However, the capacitance is $C = \kappa\epsilon_0 A/d$ and the resistance is $R = \rho d/A$. Consequently,

$$\tau = RC = \frac{\rho d}{A}\frac{\kappa\epsilon_0 A}{d} = \rho\kappa\epsilon_0 \approx 2\ \text{ms} \qquad (25.30)$$

In other words, the membrane's time constant depends on only its inherent electrical properties—its resistivity and dielectric constant—*not* on its dimensions.

FIGURE 25.42 The cell membrane can be modeled as an RC circuit.

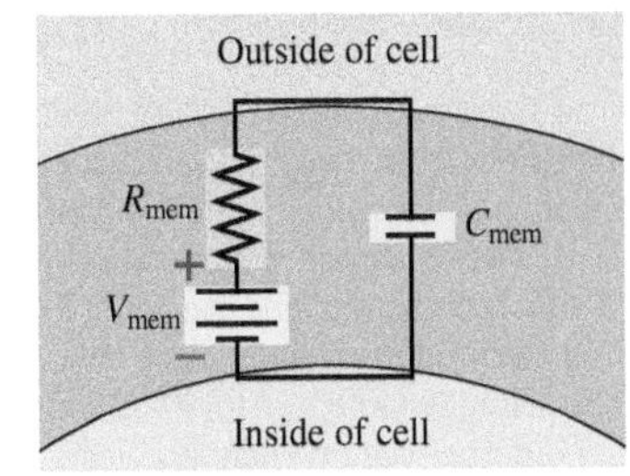

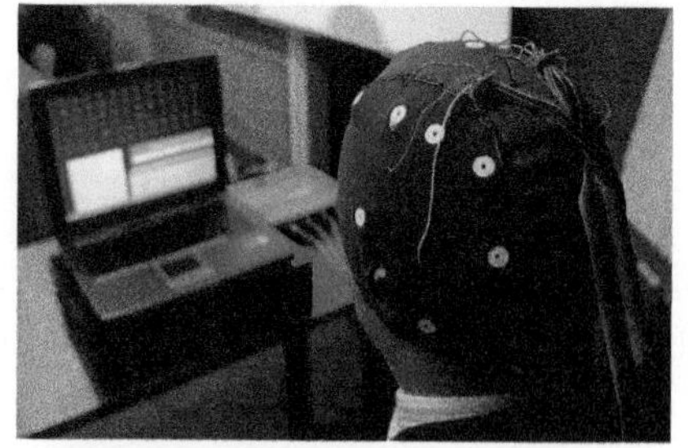

Touchless typing Different thought processes lead to different patterns of action potentials among the many neurons of the brain. The electrical activity of the cells and the motion of ions through the conducting fluid surrounding them lead to measurable differences in potential between points on the scalp. You can't use these potential differences to "read someone's mind," but it is possible to program a computer to recognize patterns and perform actions when they are detected. This man is using his thoughts—and the resulting pattern of electric potentials—to select and enter letters.

The Action Potential

A neuron at rest has $V_{mem} \approx -70$ mV; however, the membrane potential can change drastically in response to a stimulus. A neuron can be stimulated by neurotransmitter chemicals at a *synaptic junction,* where one neuron connects to the next, or by a changing electric potential, which is why Galvani saw the frog's leg twitch. Whatever the stimulus, the membrane potential undergoes a rapid change—a sudden rise and subsequent fall—that is called the **action potential.** The action potential is a spike in the membrane potential that corresponds to the "firing" of a nerve cell.

We need one more idea to understand the action potential: voltage-gated ion channels. The cell membrane's *leak channels* are always open, allowing Na^+ and K^+ ions to diffuse from higher to lower concentrations. In addition, the membrane has larger ion channels that are normally closed but that can open suddenly in response to a stimulus. The usual stimulus is a changing membrane potential, so these channels are called **voltage-gated ion channels.** There are two primary types in neurons:

- Voltage-gated sodium channels open if the membrane potential rises to ≈ -55 mV. This is called the *threshold voltage.* The channels close when the membrane potential reaches $\approx +40$ mV.
- Voltage-gated potassium channels open if the membrane potential rises to $\approx +40$ mV and then close when the membrane potential drops to ≈ -80 mV.

Voltage-gated ion channels carry currents of ≈ 5 pA, roughly a factor of 50 larger than the leak channels.

The three phases of the action potential are outlined in the following table:

The action potential

Depolarization

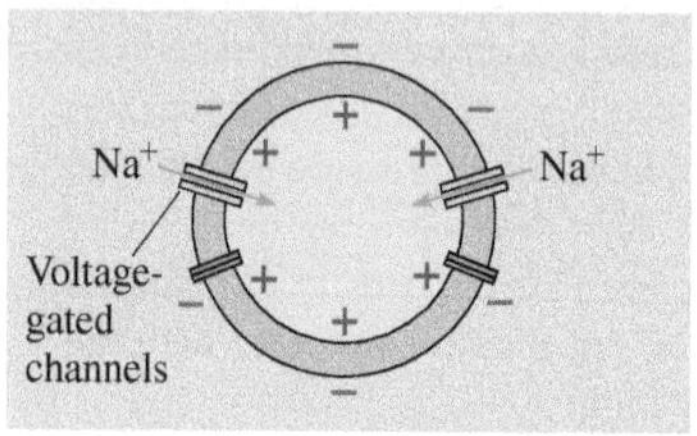

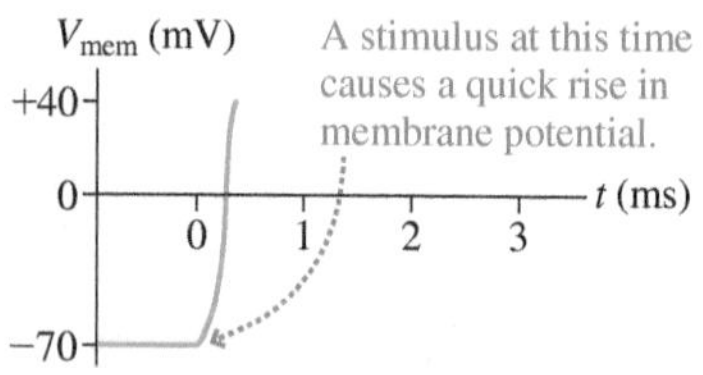

A cell *depolarizes* when a stimulus causes the opening of the voltage-gated sodium channels. The concentration of sodium ions is much higher outside the cell, so positive sodium ions flow into the cell, rapidly raising its potential to 40 mV, at which point the sodium channels close.

Repolarization

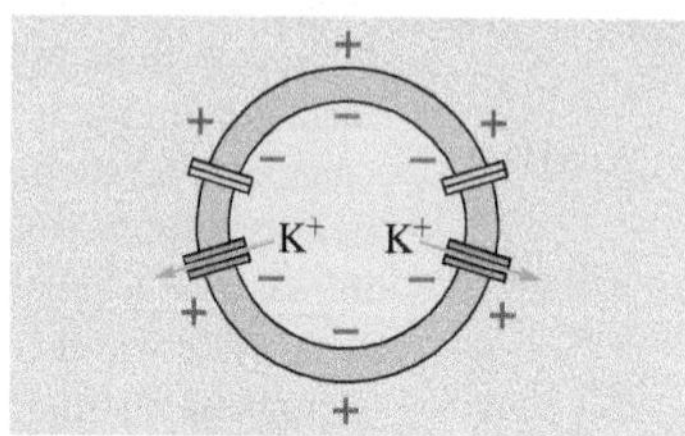

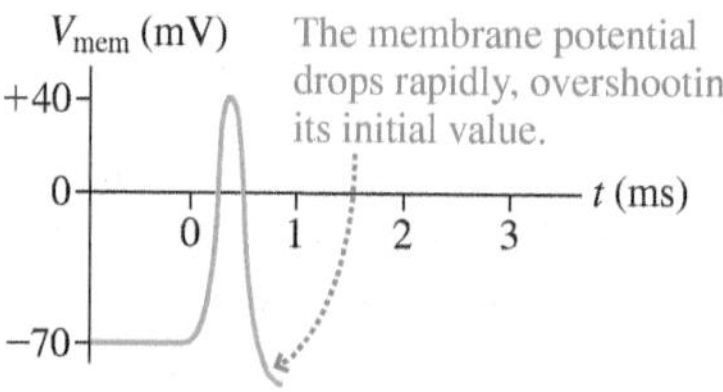

The cell *repolarizes* as the potassium channels open. The higher potassium concentration inside the cell drives these ions out of the cell. The potassium channels close when the membrane potential reaches about -80 mV, slightly *less* than the resting potential.

Reestablishing resting potential

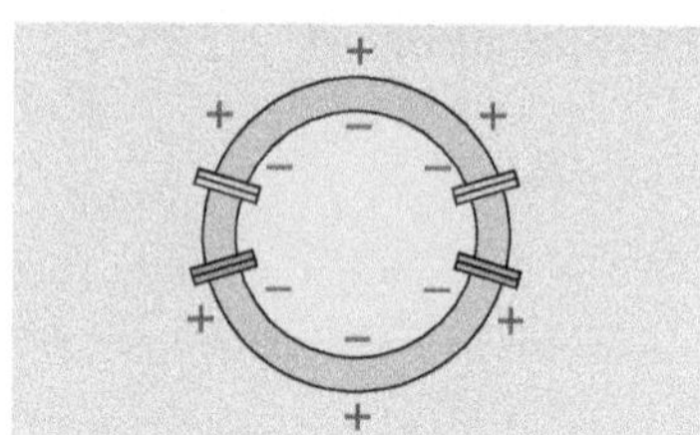

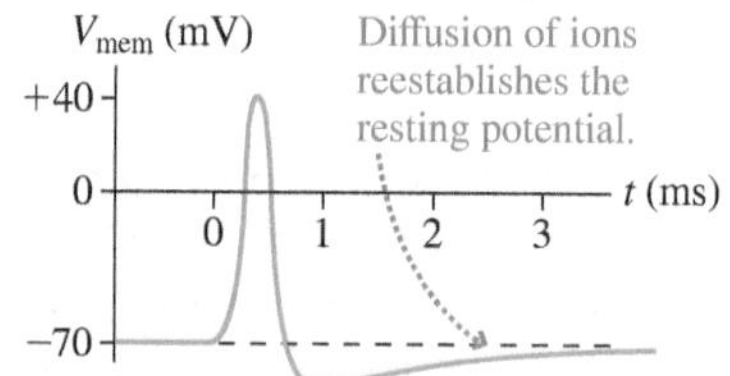

The reestablishment of the resting potential after the sodium and potassium channels close is a slower decay with the membrane time constant of ≈ 2 ms.

After the action potential is complete, there is a brief resting period, after which the cell is ready to be triggered again. The action potential is driven by ionic conduction through sodium and potassium channels, so the potential changes are quite rapid. The time for the potential to rise and then to fall is much less than the 2 ms time constant of the membrane.

This discussion has focused on nerve cells, but muscle cells undergo a similar cycle of depolarization and repolarization. The resulting potential changes are responsible for the signal that is measured by an electrocardiogram, which we learned about in ◀◀ SECTION 22.7.

The Propagation of Nerve Impulses

Let's return to the question posed at the start of the section: How is a signal transmitted from the brain to a muscle in the hand? The axon is long enough that different points on its membrane may have different potentials. When one point on the axon's membrane is stimulated, the membrane will depolarize at this point. The resulting action potential may trigger depolarization in adjacent parts of the membrane. Stimulating the axon's membrane at one point can trigger a *wave* of action potential—a nerve impulse—that travels along the axon. When this impulse reaches a muscle cell, the muscle cell depolarizes and produces a mechanical response.

Let's look at this process in more detail. We will start with a simple model of an axon in FIGURE 25.43a. The voltage-gated sodium channels are normally closed, but if the potential at some point is raised by ≈ 15 mV, from the resting potential of -70 mV to ≈ -55 mV, the sodium channels suddenly open, sodium ions rush into the cell, and an action potential is triggered. This is the key idea: **A small increase in the potential difference across the membrane causes the sodium channels to open, triggering a large action-potential response.**

This process begins at the cell body, in response to signals the neuron receives at its dendrites. If the cell's membrane potential goes up by ≈ 15 mV, an action potential is initiated in the cell body. As the membrane potential quickly rises to a peak of $+40$ mV, it causes the potential on the nearest section of the axon—where the axon attaches to the cell body—to rise by 15 mV. This triggers an action potential in this first section of the axon. The action potential in the first section of the axon triggers an action potential in the next section of the axon, which triggers an action potential in the next section, and so on down the axon until reaching the end.

As FIGURE 25.43b shows, this process causes a wave of action potential to propagate down the axon. The signal moves relatively slowly, at only about 1 m/s, because channels must open at each point on the membrane and ions must diffuse through, which takes time. If all nerve signals traveled at this speed, a signal telling your hand to move would take about 1 s to travel from your brain to your hand. Clearly, at least some neurons in the nervous system must transmit signals at a higher speed than this!

One way to make the signals travel more quickly is to increase an axon's diameter. The *giant axon* in the squid, which has long been a model for the study of nerve impulses, triggers a rapid escape response when the squid is threatened. This axon may have a diameter of 1 mm, a thousand times that of a typical human axon, providing for the necessary rapid signal transmission. But your nervous system consists of 300 billion neurons, and the axons can't all be 1 mm in diameter—there simply isn't enough space in your body. In your nervous system, higher neuron signal speed is achieved in a totally different manner.

Increasing Nerve-Impulse Speed by Insulation

The axons of motor neurons and most other neurons in your body can transmit signals at very high speeds because they are insulated with a myelin sheath. Look back at the structure of a motor neuron in Figure 25.39. Schwann cells wrap the axon with myelin, insulating it electrically and chemically, with breaks at the nodes of Ranvier. The ion channels are concentrated in these nodes because this is the only place where the extracellular fluid is in contact with the cell membrane. In an insulated axon, a signal propagates by jumping from one node to the next. This process is called *saltatory conduction,* from the Latin *saltare,* "to leap."

FIGURE 25.43 Propagation of a nerve impulse.

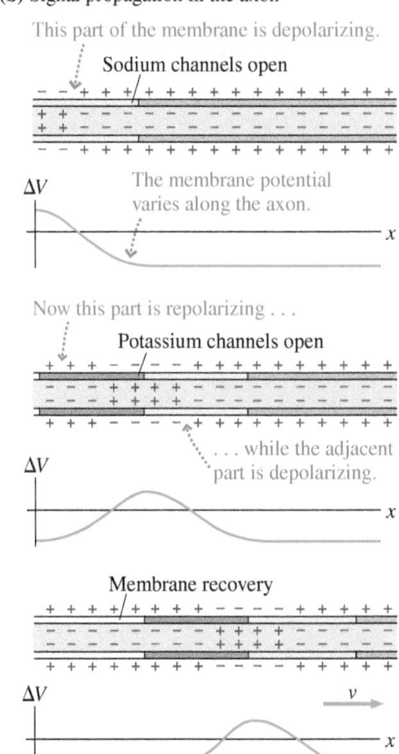
(a) A model of a neuron
Cell body
Axon
A close-up view of the axon shows that the cell membrane has the usual ion channels.
(b) Signal propagation in the axon
This part of the membrane is depolarizing.
Sodium channels open
ΔV
The membrane potential varies along the axon.
x
Now this part is repolarizing . . .
Potassium channels open
. . . while the adjacent part is depolarizing.
ΔV
x
Membrane recovery
ΔV
v
x
A changing potential at one point triggers the membrane to the right, leading to a wave of action potential that moves along the axon.

The myelin seals the axon so that, unlike the situation of Figure 25.43, there are neither leak channels nor voltage-gated channels between the nodes. FIGURE 25.44a is an electrical model for saltatory conduction in which the action potential, induced by the voltage-gated ion channels, occurs only at the nodes. Opening these ion channels to make the membrane potential positive at a node is like closing a switch to a battery.

Rather than triggering an action potential in a neighboring ion channel—because there isn't one—the action potential at one node acts as an emf to drive a current down the length of the axon toward the next node. This current charges the membrane capacitance and, as it does so, increases the potential "downstream" at the next node. An action potential at the next node is triggered when the potential increases by ≈15 mV. In this way, as FIGURE 25.44b shows, the triggering of action potentials jumps from node to node along an axon.

FIGURE 25.44 A circuit model of nerve-impulse propagation along myelinated axons.

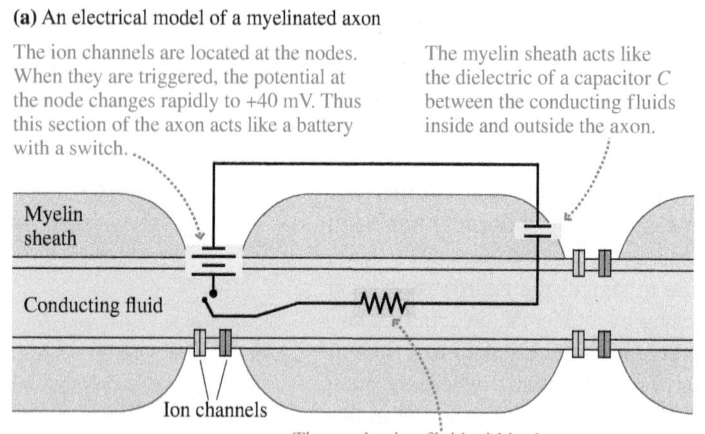

(a) An electrical model of a myelinated axon
The ion channels are located at the nodes. When they are triggered, the potential at the node changes rapidly to +40 mV. Thus this section of the axon acts like a battery with a switch.
The myelin sheath acts like the dielectric of a capacitor C between the conducting fluids inside and outside the axon.
Myelin sheath
Conducting fluid
Ion channels
The conducting fluid within the axon acts like a resistor R. The action potential emf drives a current through this resistor to charge the membrane capacitor.
(b) Impulse propagation in the myelinated axon
The ion channels at this node are triggered, generating an action potential.
Ion flow down the axon begins to charge the next segment.
Myelin sheath
Nodes
Axon
Once the potential reaches a threshold value, an action potential is triggered at the next node.
The process continues, with the signal triggering each node in sequence.

Saltatory conduction along myelinated axons is significantly faster than the propagation along unmyelinated axons that was illustrated in Figure 25.43. There are three primary reasons:

- Myelin insulates the axon. Ions that enter the cell during the action potential at one node cannot easily leak out of the axon and thus are available to flow as a current toward the next node.
- Capacitance is inversely proportional to the separation between the electrodes. The myelin sheath increases the separation between the conducting fluids inside and outside the axon, thus decreasing the membrane capacitance. In turn, a decreased capacitance allows the membrane potential at the next node to rise more quickly toward the threshold for triggering an action potential as charge flows down the axon toward it.
- Opening or closing a voltage-gated ion channel takes time. The speed with which depolarization propagates along an unmyelinated axon is determined in large part by the inherent response time of the ion channels. Saltatory conduction avoids many of these delays by using many fewer ion channels.

The speed of propagation along myelinated axons is 50 m/s to 100 m/s, a factor of 50 or more faster than propagation along unmyelinated axons. At this speed, your brain can send a signal to your hand in $\approx \frac{1}{50}$ s.

STOP TO THINK 25.7 In the axon model of Figure 25.44, if the thickness of the myelin sheath were increased, the propagation speed of nerve impulses would

A. Increase. B. Decrease. C. Remain the same.

CHAPTER 25 INTEGRATED EXAMPLE Measuring the moisture content of soil

The moisture content of soil is given in terms of its *volumetric fraction,* the ratio of the volume of water in the soil to the volume of the soil itself. Measuring this directly is quite time-consuming, so soil scientists use other means to reliably measure soil moisture. Water has a very large dielectric constant, so the dielectric constant of soil increases as its moisture content increases. FIGURE 25.45 shows data for the dielectric constant of soil versus the volumetric fraction; the increase in dielectric constant with soil moisture is quite clear. This strong dependence of dielectric constant on soil moisture allows a very sensitive—and simple—electrical test of soil moisture.

FIGURE 25.45 Variation of the dielectric constant with soil moisture.

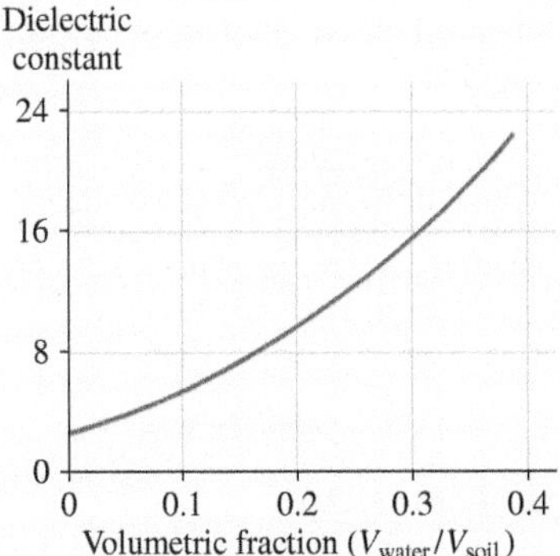
Dielectric constant
24
16
8
0
0
0.1
0.2
0.3
0.4
Volumetric fraction (V_{water}/V_{soil})

A soil moisture meter has a probe with two separated electrodes. When the probe is inserted into the soil, the electrodes form a capacitor whose capacitance depends on the dielectric constant of the soil between them. A circuit charges the capacitor probe to 3.0 V, then discharges it through a resistor. The decay time depends on the capacitance—and thus the soil moisture—so a measurement of the time for the capacitor to discharge allows a determination of the amount of moisture in the soil.

In air, the probe's capacitor takes 15 μs to discharge from 3.0 V to 1.0 V. In one particular test, when the probe was inserted into the ground, this discharge required 150 μs. What was the approximate volumetric fraction of water for the soil in this test?

PREPARE FIGURE 25.46 is a sketch of the measurement circuit of the soil moisture meter. The capacitor is first charged, then connected across a resistor to form an *RC* circuit. The decay of the capacitor voltage is governed by the time constant for the circuit. The time constant depends on the resistance and on the electrode capacitance, which depends on the dielectric constant of the soil between the electrodes.

SOLVE The capacitance of the probe in air is that of a parallel-plate capacitor: $C_{air} = \epsilon_0 A/d$. The capacitance of the probe in soil differs only by the additional factor of the dielectric constant of the material between the plates—in this case, the soil:

$$C_{soil} = \kappa_{soil}\left(\frac{\epsilon_0 A}{d}\right) = \kappa_{soil} C_{air}$$

FIGURE 25.46 The measurement circuit of the soil moisture meter.

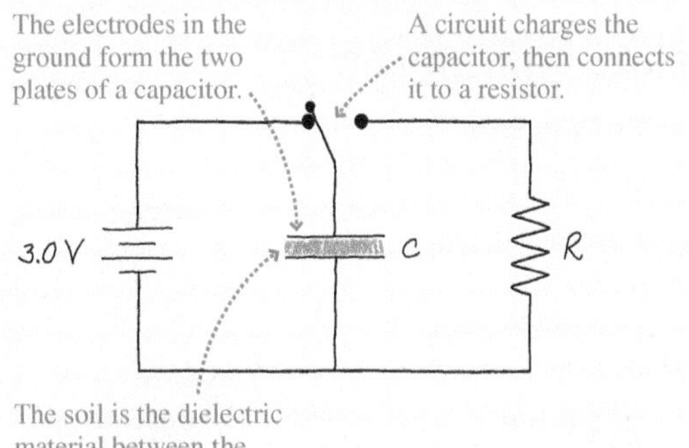
The electrodes in the ground form the two plates of a capacitor.
A circuit charges the capacitor, then connects it to a resistor.
3.0 V
C
R
The soil is the dielectric material between the plates of the capacitor.

The ratio of the capacitance values gives the dielectric constant:

$$\kappa_{soil} = \frac{C_{soil}}{C_{air}}$$

Once we know the dielectric constant, we can determine the volumetric fraction of water from the graph.

We are given the times for the decay in air and in soil, but not the capacitance or the resistance of the probe. However, we don't actually need the capacitance, only the *ratio* of the capacitances in air and in soil. Equation 25.26 gives the voltage decay of an *RC* circuit: $\Delta V_C = \Delta V_0 e^{-t/RC}$. In air, the decay is

$$1.0 \text{ V} = (3.0 \text{ V})e^{-(15\times10^{-6}\text{ s})/RC_{air}}$$

In soil, the decay is

$$1.0 \text{ V} = (3.0 \text{ V})e^{-(150\times10^{-6}\text{ s})/RC_{soil}}$$

Because the starting and ending points for the decay are the same, the exponents of the two expressions must be equal:

$$\frac{15\times10^{-6}\text{ s}}{RC_{air}} = \frac{150\times10^{-6}\text{ s}}{RC_{soil}}$$

We can solve this for the ratio of the capacitances in soil and air:

$$\frac{C_{soil}}{C_{air}} = \frac{150\times10^{-6}\text{ s}}{15\times10^{-6}\text{ s}} = 10$$

We saw previously that this ratio is the dielectric constant, so $\kappa_{soil} = 10$. We then use the graph of Figure 25.45 to determine that this dielectric constant corresponds to a volumetric water fraction of approximately 0.20.

ASSESS The decay times in air and in soil differ by a factor of 10, so the capacitance in the soil is much larger than that in air. This implies a large dielectric constant, meaning that there is a lot of water in the soil. A volumetric fraction of 0.20 means that 20% of the soil's volume is water (that is, 1.0 cm^3 of soil contains 0.20 cm^3 of water)—which is quite a bit, so our result seems reasonable.

SUMMARY

GOAL To understand the fundamental physical principles that govern electric circuits.

GENERAL PRINCIPLES

Kirchhoff's loop law

For a closed loop:

- Assign a direction to the current.
- Add potential differences around the loop:

$$\sum_i \Delta V_i = 0$$

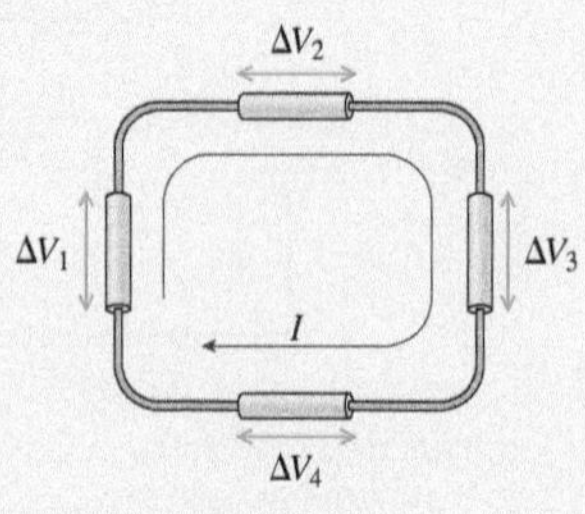

Kirchhoff's junction law

For a junction:

$$\sum I_{in} = \sum I_{out}$$

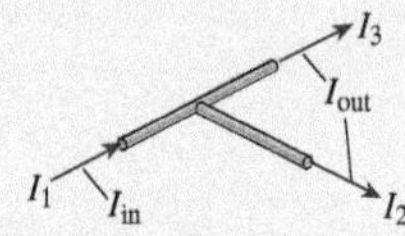

Analyzing Circuits

PREPARE Draw a circuit diagram.

SOLVE Use Kirchhoff's laws.

Break the circuit down:

- Reduce the circuit to the smallest possible number of equivalent resistors.
- Find the current and potential difference.

Rebuild the circuit:

- Find the current and potential difference for each resistor.

ASSESS Verify that

- The sum of the potential differences across series resistors equals the potential difference across the equivalent resistor.
- The sum of the currents through parallel resistors equals the current through the equivalent resistor.

IMPORTANT CONCEPTS

RC circuits

The discharge of a capacitor through a resistor is an exponential decay:

$$\Delta V_C = \Delta V_0 e^{-t/\tau}$$

$$I = I_0 e^{-t/\tau}$$

The **time constant** for the decay is

$$\tau = RC$$

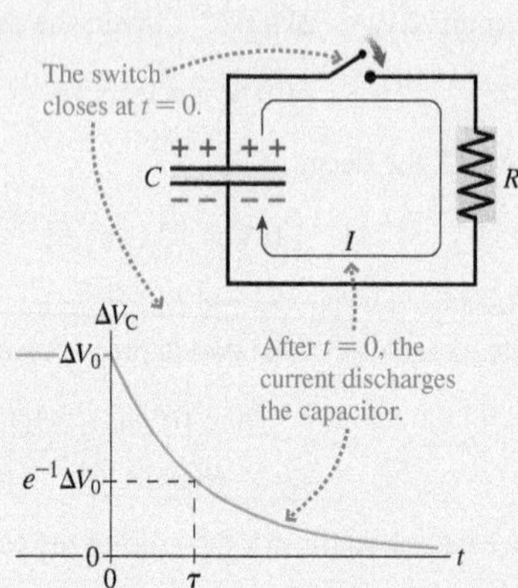

Series and parallel resistors

The current is the same through each resistor in a series. **Series resistors** can be reduced to an **equivalent resistance**:

$$R_{eq} = R_1 + R_2 + R_3 + \cdots$$

The potential difference is the same across each resistor in parallel. **Parallel resistors** can be reduced to an equivalent resistance:

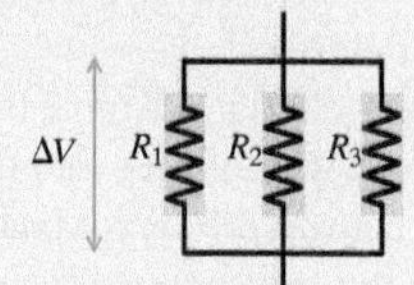

$$R_{eq} = \left(\frac{1}{R_1} + \frac{1}{R_2} + \frac{1}{R_3} + \cdots\right)^{-1}$$

$$G_{eq} = G_1 + G_2 + G_3 + \cdots$$

APPLICATIONS

Signs of ΔV for Kirchhoff's loop law

$\mathcal{E}$, Travel →, $\Delta V_{bat} = +\mathcal{E}$ $\mathcal{E}$, Travel →, $\Delta V_{bat} = -\mathcal{E}$ R, I →, $\Delta V_{res} = -IR$

Voltmeters and ammeters

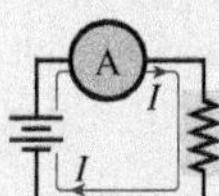

An ammeter measures the current that passes *through* the meter.

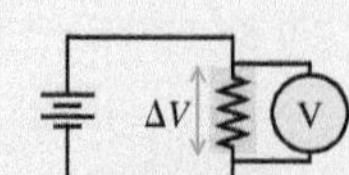

A voltmeter measures the potential difference *across* the meter.

Electricity in the nervous system

Cells in the nervous system maintain a negative potential inside the cell membrane. When triggered, the membrane depolarizes and generates an **action potential.**

An action potential travels as a wave along the axon of a neuron. More rapid saltatory conduction can be achieved by insulating the axon with myelin, which causes the action potential to jump from node to node.

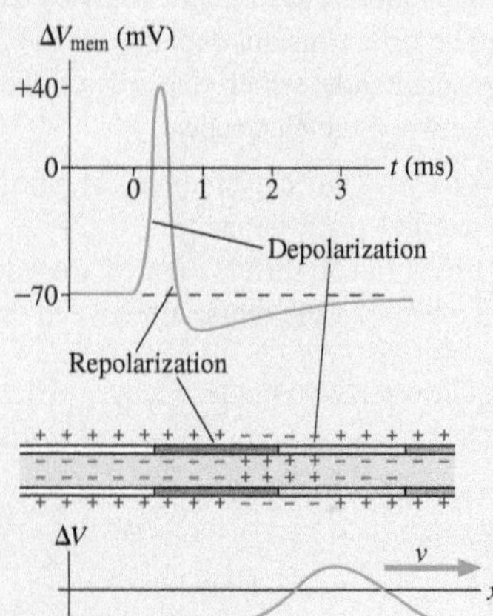

LEARNING OBJECTIVES After studying this chapter, you should be able to:

- Draw a circuit diagram. *Problems 25.1, 25.2, 25.7*
- Determine the currents and voltages in simple circuits. *Conceptual Questions 25.1, 25.2; Problems 25.3–25.6*
- Calculate the equivalent resistance of resistors in series and in parallel. *Conceptual Questions 25.3, 25.4; Problems 25.9, 25.12–25.14, 25.16*
- Measure current and voltage in a circuit using ammeters and voltmeters. *Conceptual Questions 25.9, 25.10; Problem 25.27*
- Analyze more complex circuits by repeated application of Kirchhoff's laws. *Conceptual Questions 25.5, 25.6; Problems 25.18–25.21, 25.25*
- Calculate the current that passes through the human body in an electrical accident. *Problems 25.30–25.33*
- Calculate the growth and decay of current and potential in an *RC* circuit. *Conceptual Questions 25.11, 25.12; Problems 25.34–25.36, 25.40, 25.41*
- Solve problems about action potentials and nerve impulses. *Conceptual Question 25.15; Problems 25.42–25.45*

STOP TO THINK ANSWERS

Stop to Think 25.1: A, B, and D. These three are the same circuit because the logic of the connections is the same. In each case, there is a junction that connects one side of each circuit element and a second junction that connects the other side. In C, the functioning of the circuit is changed by the extra wire connecting the two sides of the capacitor.

Stop to Think 25.2: B. The potential difference in crossing the battery is its emf, or 10 V. The potential drop across the 6 Ω resistor is $\Delta V = -IR = -6.0$ V. To make the sum of the potential differences zero, as Kirchhoff's loop law requires, the potential difference across resistor R must be -4.0 V.

Stop to Think 25.3: C = D > A = B. The two bulbs in series are of equal brightness, as are the two bulbs in parallel. But the two bulbs in series have a larger resistance than a single bulb, so there will be less current through the bulbs in series than the bulbs in parallel.

Stop to Think 25.4: C. The voltmeter must be connected in parallel with the resistor, and the ammeter in series.

Stop to Think 25.5: A > B > C = D. All the current from the battery goes through A, so it is brightest. The current divides at the junction, but not equally. Because B is in parallel with C + D, but has half the resistance of the two bulbs together, twice as much current travels through B as through C + D. So B is dimmer than A but brighter than C and D. C and D are equally bright because they are in series.

Stop to Think 25.6: B. The two 2 Ω resistors are in series and equivalent to a 4 Ω resistor. Thus $\tau = RC = 4$ s.

Stop to Think 25.7: A. A thicker sheath would space the "plates" of the capacitor—the inner and outer surfaces of the sheath—farther apart, thereby reducing the capacitance C_{mem}. The RC time constant would decrease, so the membrane potential at the next node would increase more rapidly to the threshold for triggering an action potential.

QUESTIONS

Conceptual Questions

1. The tip of a flashlight bulb is touching the top of the 3 V battery in Figure Q25.1. Does the bulb light? Why or why not?

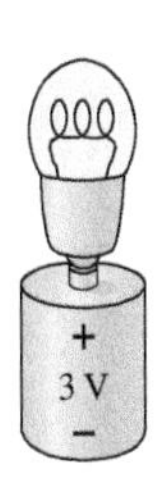

FIGURE Q25.1

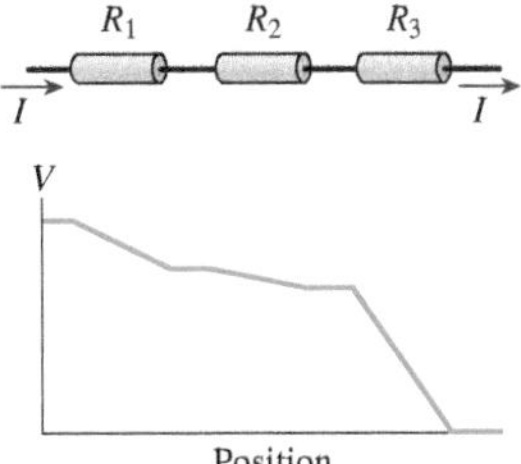

FIGURE Q25.2

2. Current I flows into three resistors connected together one after the other as shown in Figure Q25.2. The accompanying graph shows the value of the potential as a function of position. Rank in order, from largest to smallest, the three resistances R_1, R_2, and R_3. Explain.
3. The two lightbulbs in Figure Q25.3 are glowing. What happens to the brightness of bulb B if bulb A is removed from the circuit?

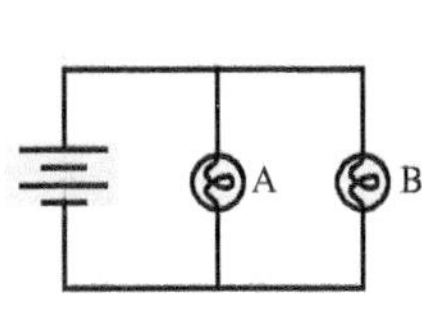

FIGURE Q25.3

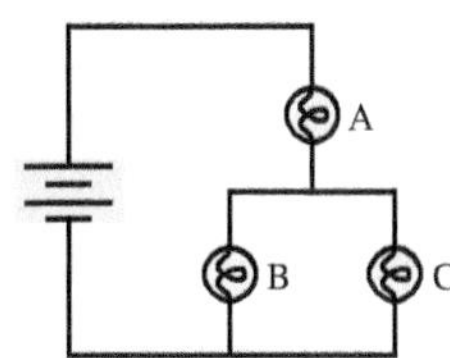

FIGURE Q25.4

4. a. The three bulbs in Figure Q25.4 are identical. All are glowing. Rank the bulbs from brightest to dimmest.
   b. Suppose bulb C is removed from the circuit. What will happen to the brightness of bulbs A and B? Explain.

Problem difficulty is labeled as | (straightforward) to ||||| (challenging). Problems labeled INT integrate significant material from earlier chapters; BIO are of biological or medical interest; CALC require calculus to solve.

5. In the circuit shown in Figure Q25.5, bulbs A and B are glowing. Then the switch is closed. What happens to each bulb? Does it get brighter, stay the same, get dimmer, or go out? Explain.

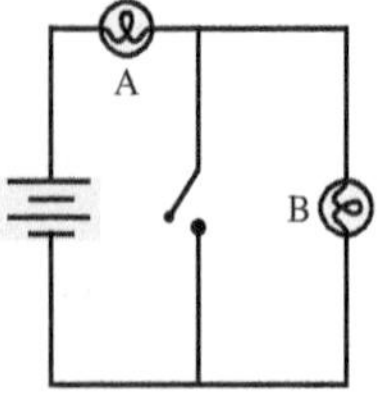

FIGURE Q25.5

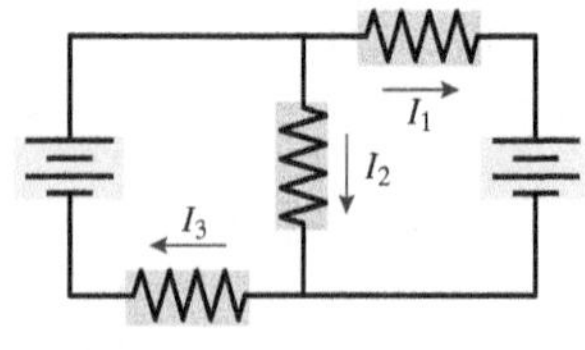

FIGURE Q25.6

6. The currents through three resistors in a circuit are shown. Which of the following statements is true? Explain.
A. $I_1 = I_2 + I_3$ B. $I_2 = I_1 + I_3$ C. $I_3 = I_1 + I_2$
7. The circuit in Figure Q25.7 has two resistors, with $R_1 > R_2$. Which resistor dissipates the larger amount of power? Explain.

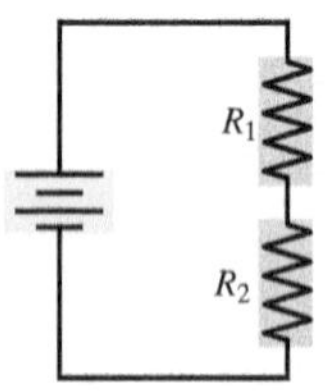

FIGURE Q25.7

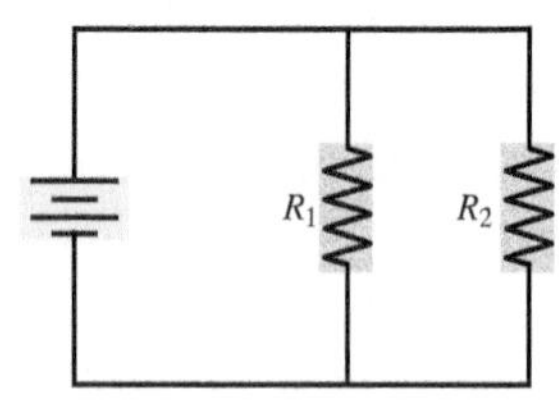

FIGURE Q25.8

8. The circuit in Figure Q25.8 has a battery and two resistors, with $R_1 > R_2$. Which resistor dissipates the larger amount of power? Explain.
9. A voltmeter is (incorrectly) inserted into a circuit as shown in Figure Q25.9.
a. What does the voltmeter read?
b. How would you change the circuit to correctly connect the voltmeter to measure the potential difference across the resistor?

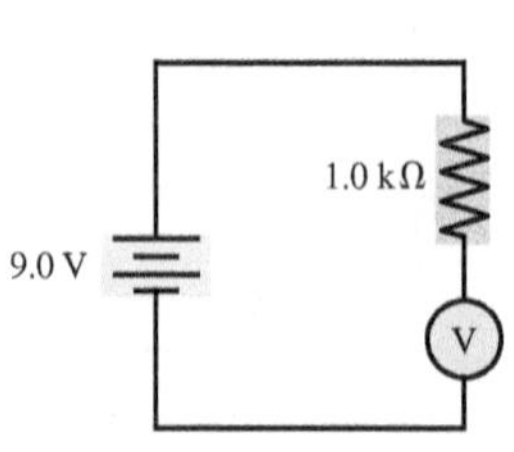

FIGURE Q25.9

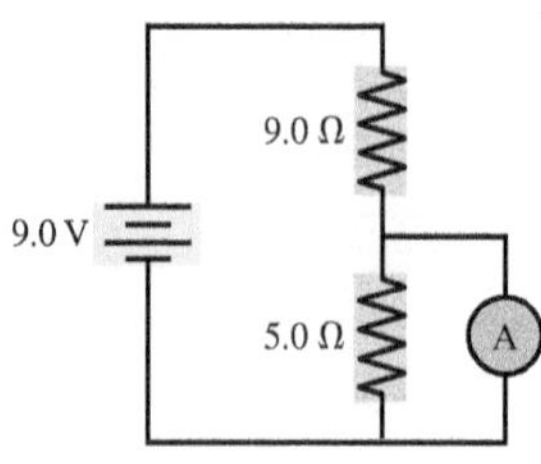

FIGURE Q25.10

10. An ammeter is (incorrectly) inserted into a circuit as shown in Figure Q25.10.
a. What does the ammeter read?
b. How would you change the circuit to correctly connect the ammeter to measure the current through the 5.0 Ω resistor?

11. Figure Q25.11 shows a circuit consisting of a battery, a switch, two identical lightbulbs, and a capacitor that is initially uncharged.
a. *Immediately* after the switch is closed, are either or both bulbs glowing? Explain.
b. If both bulbs are glowing, which is brighter? Or are they equally bright? Explain.
c. For any bulb (A or B or both) that lights up immediately after the switch is closed, does its brightness increase with time, decrease with time, or remain unchanged? Explain.

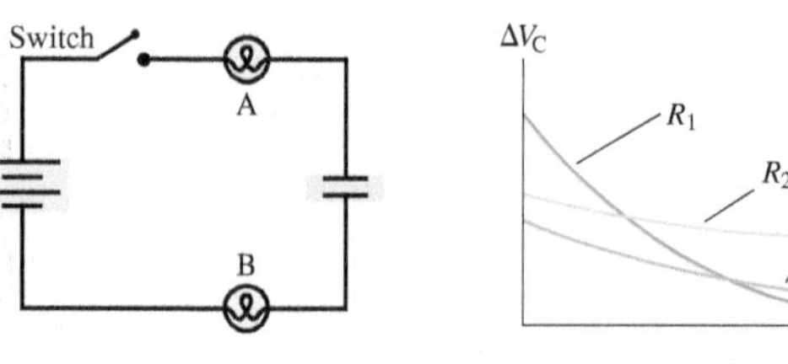

FIGURE Q25.11

FIGURE Q25.12

12. Figure Q25.12 shows the voltage as a function of time across a capacitor as it is discharged (separately) through three different resistors. Rank in order, from largest to smallest, the values of the resistances R_1 to R_3.
13. A charged capacitor could be connected to two identical resistors in either of the two ways shown in Figure Q25.13. Which configuration will discharge the capacitor in the shortest time once the switch is closed? Explain.

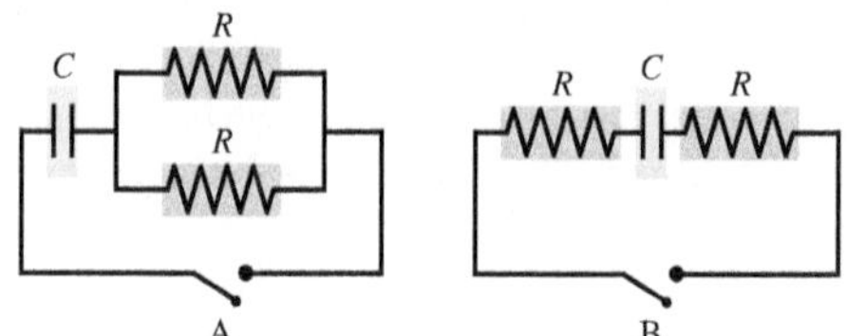

FIGURE Q25.13

14. Two students measure the time constant of an *RC* circuit. The first student charges the capacitor using a 12 V battery, then lets the capacitor discharge the resistor. The second student repeats the experiment using a 5 V battery. Which student, if either, measures the longer time constant? Explain.
15. If the membrane thickness of an unmyelinated axon were doubled, would the membrane time constant be doubled, halved, or unchanged? Explain.

Multiple-Choice Questions

16. | What is the current in the circuit of Figure Q25.16?
A. 1.0 A
B. 1.7 A
C. 2.5 A
D. 4.2 A
17. | Which resistor in Figure Q25.16 dissipates the most power?
A. The 4.0 Ω resistor.
B. The 6.0 Ω resistor.
C. Both dissipate the same power.

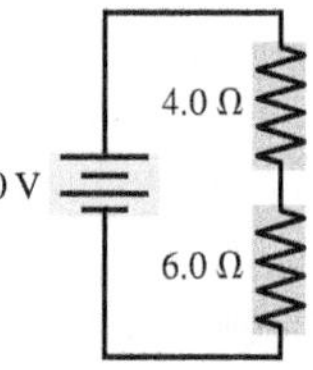

FIGURE Q25.16

18. || Normally, household lightbulbs are connected in parallel to a power supply. Suppose a 40 W and a 60 W lightbulb are, instead, connected in series, as shown in Figure Q25.18. Which bulb is brighter?
   A. The 60 W bulb.
   B. The 40 W bulb.
   C. The bulbs are equally bright.

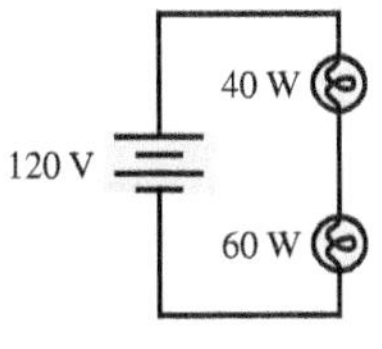

FIGURE Q25.18

19. ||| A metal wire of resistance R is cut into two pieces of equal length. The two pieces are connected together side by side. What is the resistance of the two connected wires?
   A. $R/4$ B. $R/2$ C. R D. $2R$ E. $4R$
20. || In Figure Q25.20, the current I_{bat} in the battery is equal to
   A. $\frac{1}{2}I_1$ B. $\frac{2}{3}I_1$ C. I_1 D. $\frac{3}{2}I_1$ E. $2I_1$

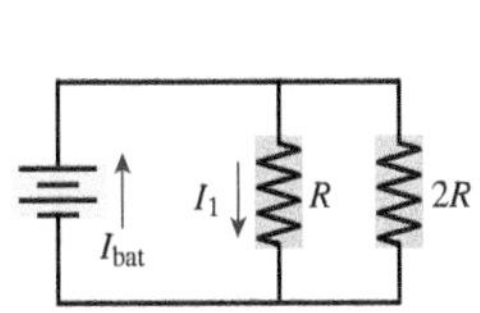

FIGURE Q25.20

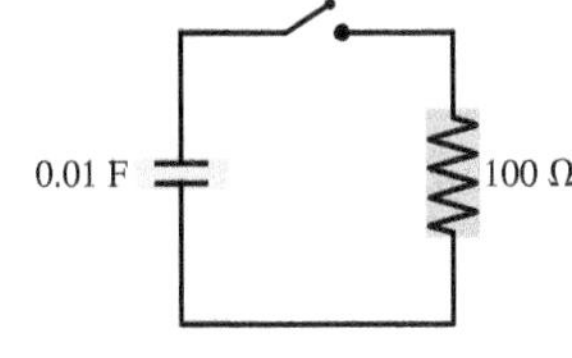

FIGURE Q25.21

21. | With the switch open, the capacitor in Figure Q25.21 has an initial charge of 1.0 C. The switch is then closed. 2.0 s later, the charge on the capacitor is
   A. 0.5 C B. $e^{-1}(1.0\text{ C}) = 0.37\text{ C}$
   C. 0.25 C D. $e^{-2}(1.0\text{ C}) = 0.14\text{ C}$

22. | Figure Q25.22 shows the capacitor voltage as a capacitor is discharged through a 4.0 kΩ resistor. Approximately what is the capacitance?
   A. 1 μF
   B. 3 μF
   C. 5 μF
   D. 7 μF

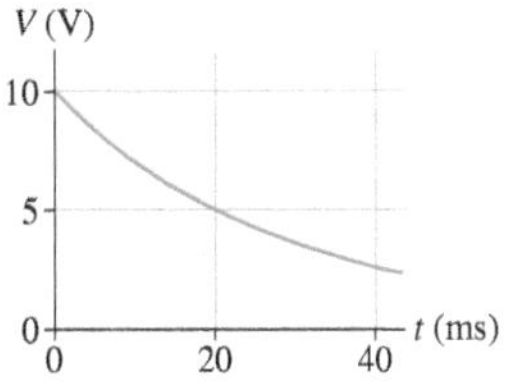

FIGURE Q25.22

23. | A fisherman has caught a torpedo ray in a net. As he picks it up, this electric fish applies a 50 V potential difference between the fisherman's hands. His hands are wet with salt water, so their skin resistance is a low 900 Ω. Approximately what current passes through his body?
   A. 1 mA B. 5 mA C. 10 mA D. 20 mA
24. | If a cell's membrane thickness doubles but the cell stays the same size, how do the resistance and the capacitance of the cell membrane change?
   BIO
   A. The resistance and the capacitance increase.
   B. The resistance increases, the capacitance decreases.
   C. The resistance decreases, the capacitance increases.
   D. The resistance and the capacitance decrease.
25. || If a spherical cell's diameter is reduced by 50% without changing the membrane thickness, how do the resistance and capacitance of the cell membrane change?
   BIO
   A. The resistance and the capacitance increase.
   B. The resistance increases, the capacitance decreases.
   C. The resistance decreases, the capacitance increases.
   D. The resistance and the capacitance decrease.

PROBLEMS

Section 25.1 Circuit Elements and Diagrams

1. | Draw a circuit diagram for the circuit of Figure P25.1.

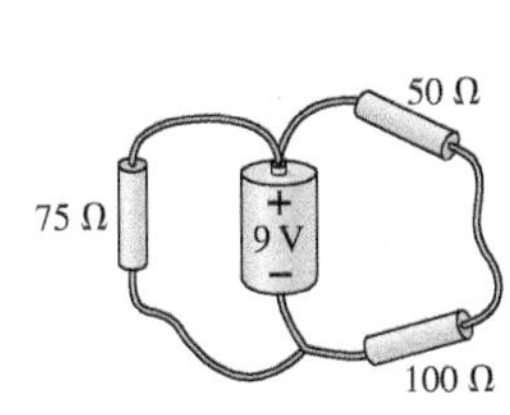

FIGURE P25.1

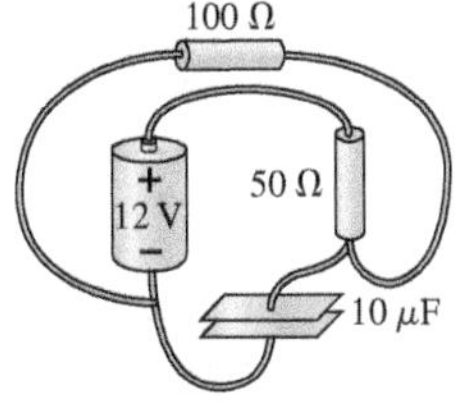

FIGURE P25.2

2. | Draw a circuit diagram for the circuit of Figure P25.2.

Section 25.2 Using Kirchhoff's Laws

3. | In Figure P25.3, what is the current in the wire above the junction? Does charge flow toward or away from the junction?

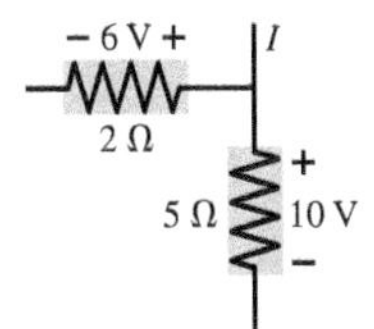

FIGURE P25.3

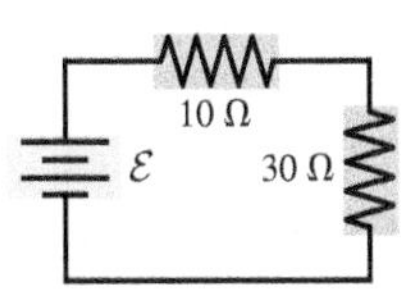

FIGURE P25.4

4. | The current through the 30 Ω resistor in Figure P25.4 is measured to be 0.25 A. What is the emf $\mathcal{E}$ of the battery?

5. | a. What is the potential difference across each resistor in Figure P25.5?
   b. Draw a graph of the potential as a function of the distance traveled through the circuit, traveling clockwise from $V = 0$ V at the lower left corner. See Figure P25.7 for an example of such a graph.

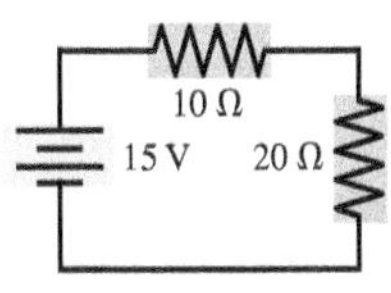

FIGURE P25.5

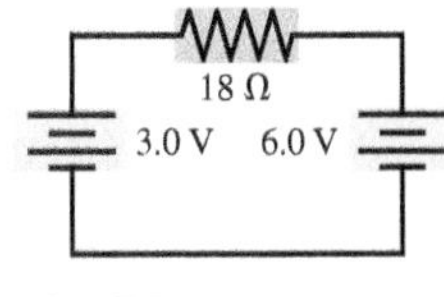

FIGURE P25.6

6. || a. What are the magnitude and direction of the current in the 18 Ω resistor in Figure P25.6?
   b. Draw a graph of the potential as a function of the distance traveled through the circuit, traveling clockwise from $V = 0$ V at the lower left corner. See Figure P25.7 for an example of such a graph.
7. | The current in a circuit with only one battery is 2.0 A. Figure P25.7 shows how the potential changes when going around the circuit in the clockwise direction, starting from the lower left corner. Draw the circuit diagram.

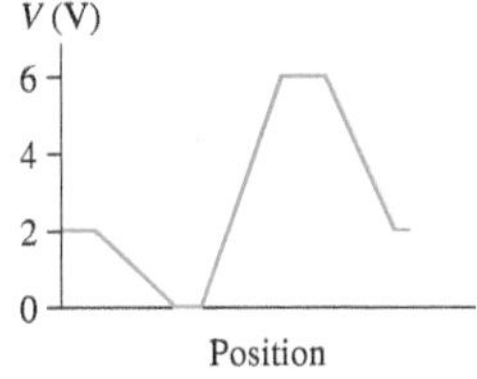

FIGURE P25.7

Section 25.3 Series and Parallel Circuits

8. | What is the value of resistor R in Figure P25.8?

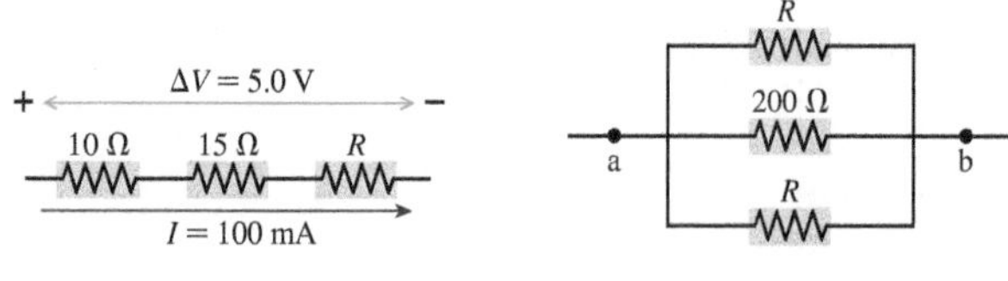

FIGURE P25.8

FIGURE P25.9

9. | Two of the three resistors in Figure P25.9 are unknown but equal. The total resistance between points a and b is 75 Ω. What is the value of R?
10. | Three resistors in parallel have an equivalent resistance of 5.0 Ω. Two of the resistors have resistances of 10 Ω and 30 Ω. What is the resistance of the third resistor?
11. | Three identical resistors have an equivalent resistance of 30 Ω when connected in parallel. What is their equivalent resistance when connected in series?
12. | You have a collection of 1.0 kΩ resistors. How can you connect four of them to produce an equivalent resistance of 0.25 kΩ?
13. | You have a collection of six 1.0 kΩ resistors. What is the smallest resistance you can make by combining them?
14. || When two resistors are wired in series with a 12 V battery, the current through the battery is 0.30 A. When they are wired in parallel with the same battery, the current is 1.6 A. What are the values of the two resistors?
15. ||| What is the value of resistor R in Figure P25.15?

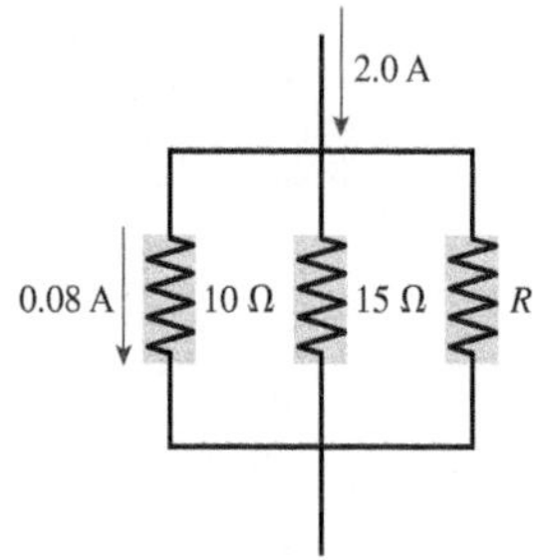

FIGURE P25.15

16. || What is the equivalent resistance between points a and b in Figure P25.16?

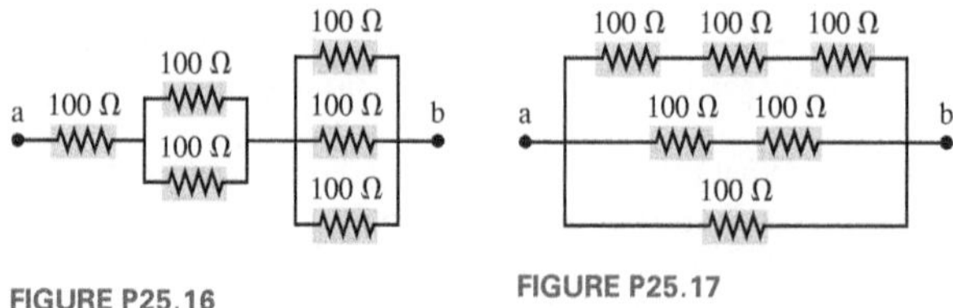

FIGURE P25.16

FIGURE P25.17

17. || What is the equivalent resistance between points a and b in Figure P25.17?

Section 25.4 Measuring Voltage and Current

Section 25.5 More Complex Circuits

18. || What is the equivalent resistance between points a and b in Figure P25.18?

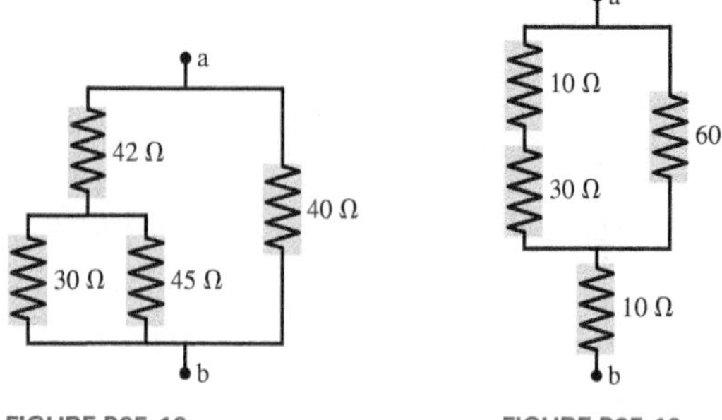

FIGURE P25.18

FIGURE P25.19

19. || What is the equivalent resistance between points a and b in Figure P25.19?
20. || Two batteries supply current to the circuit in Figure P25.20. The figure shows the potential difference across two of the resistors and the value of the third resistor. What current is supplied by the batteries?

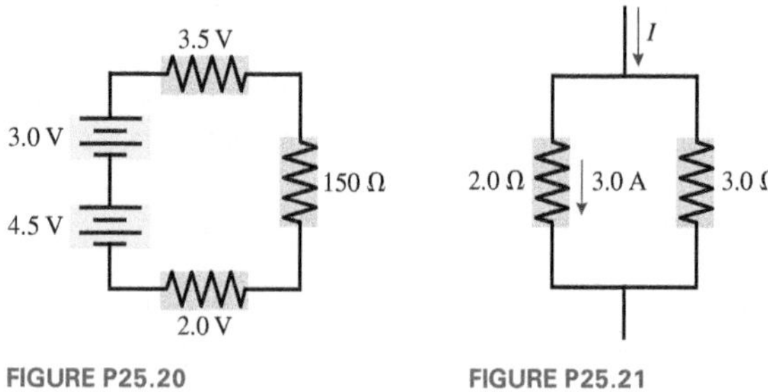

FIGURE P25.20

FIGURE P25.21

21. | Part of a circuit is shown in Figure P25.21.
    a. What is the current through the 3.0 Ω resistor?
    b. What is the value of the current I?
22. || What are the resistances R and the emf of the battery in Figure P25.22?

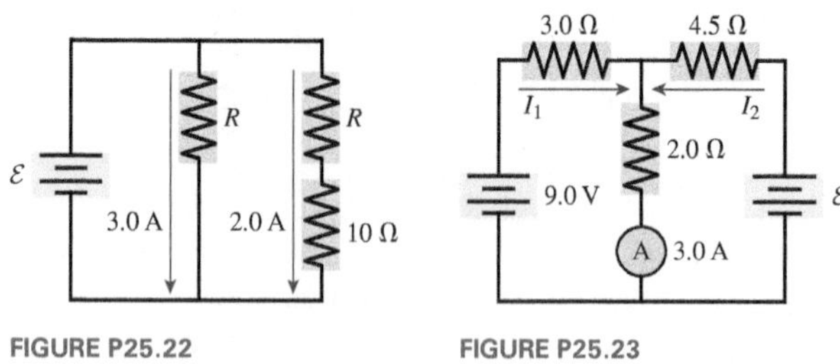

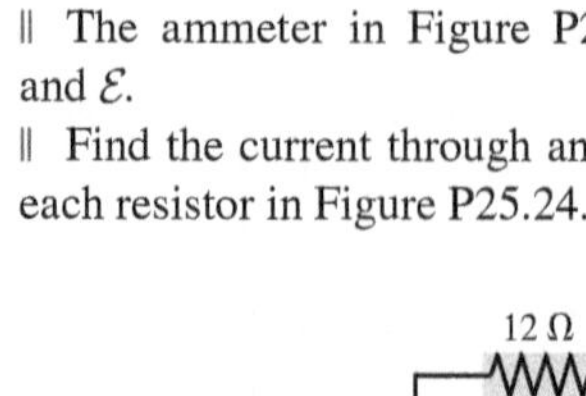

FIGURE P25.22

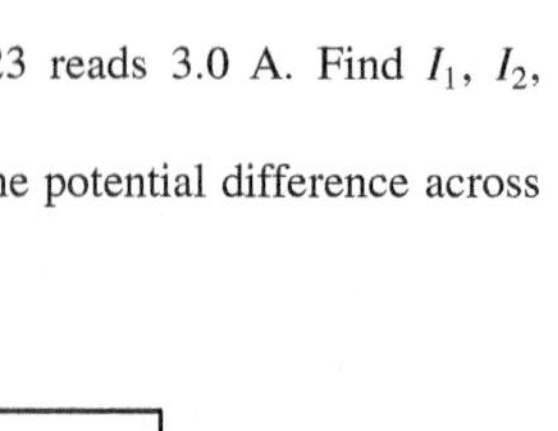

FIGURE P25.23

23. || The ammeter in Figure P25.23 reads 3.0 A. Find I_1, I_2, and $\mathcal{E}$.
24. || Find the current through and the potential difference across each resistor in Figure P25.24.

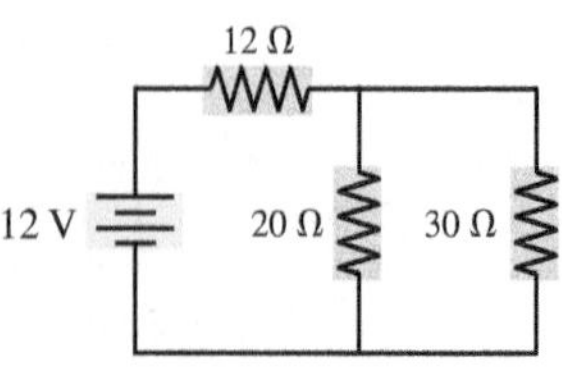

FIGURE P25.24

25. ||| Find the current through and the potential difference across each resistor in Figure P25.25.

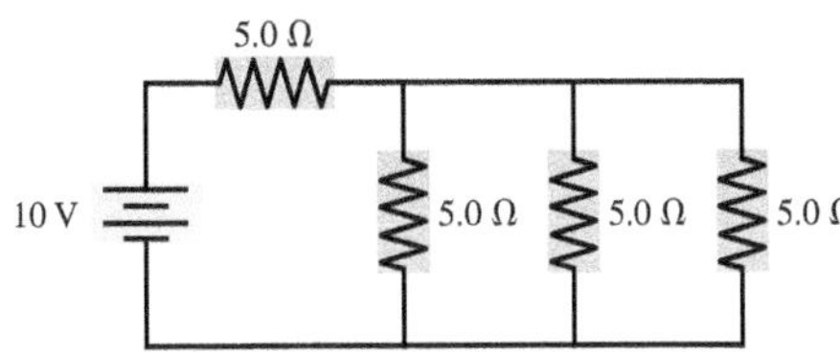

FIGURE P25.25

26. || For the circuit shown in Figure P25.26, find the current through and the potential difference across each resistor. Place your results in a table for ease of reading.

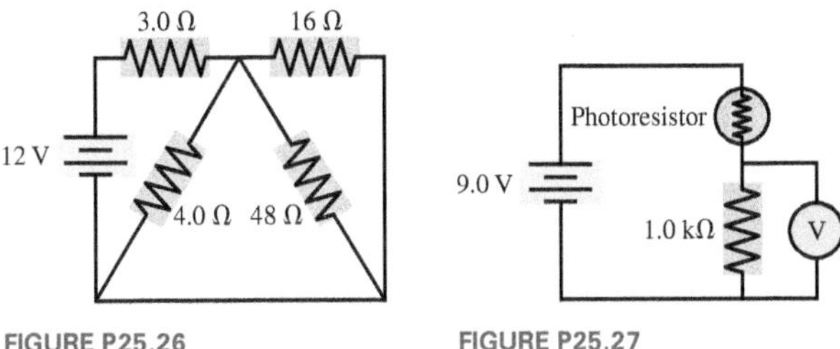

FIGURE P25.26

FIGURE P25.27

27. || A photoresistor, whose resistance decreases with light intensity, is connected in the circuit of Figure P25.27. On a sunny day, the photoresistor has a resistance of 0.56 kΩ. On a cloudy day, the resistance rises to 4.0 kΩ. At night, the resistance is 20 kΩ.
   a. What does the voltmeter read for each of these conditions?
   b. Does the voltmeter reading increase or decrease as the light intensity increases?

28. ||| The two unknown resistors in Figure P25.28 have the same resistance R. When the switch is closed, the current through the battery increases by 50%. What is R?

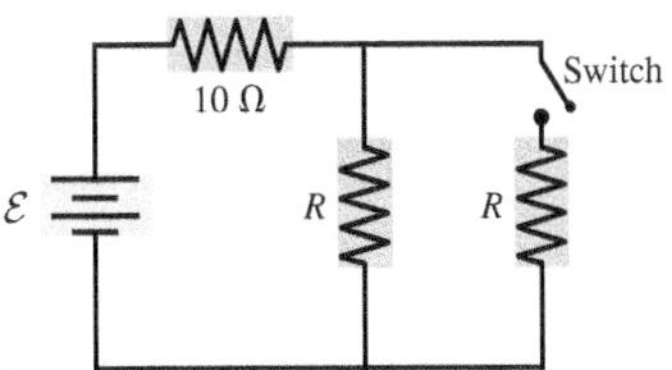

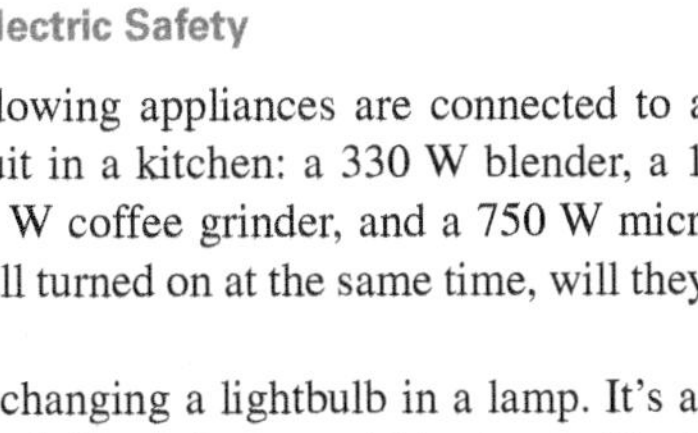

FIGURE P25.28

Section 25.6 Electric Safety

29. || The following appliances are connected to a single 120 V,
INT 15 A circuit in a kitchen: a 330 W blender, a 1000 W coffeepot, a 150 W coffee grinder, and a 750 W microwave oven. If these are all turned on at the same time, will they trip the circuit breaker?

30. || John is changing a lightbulb in a lamp. It's a warm summer
BIO evening, and the resistance of his damp skin is only 4000 Ω. While one hand is holding the grounded metal frame of the lamp, the other hand accidentally touches the hot electrode in the base of the socket. What is the current through his torso?

31. || In some countries AC outlets near bathtubs are restricted
BIO to a maximum of 25 V to minimize the chance of dangerous shocks while bathing. A man is in the tub; the lower end of his torso is well grounded, and the skin resistance of his wet, soapy hands is negligible. He reaches out and accidentally touches a live electric wire. What voltage on the wire would produce a dangerous 100 mA current?

32. || If you touch the terminal of a battery, the small area of
BIO contact means that the skin resistance will be relatively large; 50 kΩ is a reasonable value. What current will pass through your body if you touch the two terminals of a 9.0 V battery with your two hands? Will you feel it? Will it be dangerous?

33. ||| Occupational safety experts have developed an alternative
BIO criterion for electrical safety. They have found that shocks lasting less than 3 s will be nonlethal if the product of the voltage drop across the body, the current through the body, and the time ($\leq$3.0 s) that the current flows does not exceed 13.5 V·A·s = 13.5 J. Suppose that one hand of a potential victim is grounded and the other hand touches a voltage source; suppose further that his skin resistance is negligible—a worst-case scenario. Using the criterion above, what is the lowest voltage that will not be lethal for a shock that lasts 1.0 s?

Section 25.7 *RC* Circuits

34. || What is the time constant for the discharge of the capacitor in Figure P25.34?

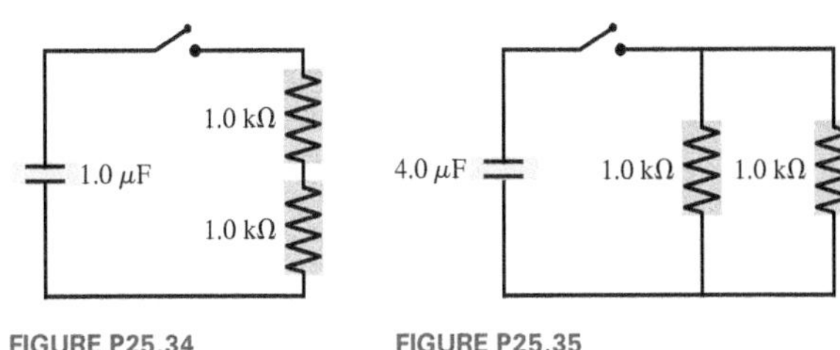

FIGURE P25.34

FIGURE P25.35

35. || What is the time constant for the discharge of the capacitor in Figure P25.35?

36. ||| With the switch open, the potential difference across the capacitor in Figure P25.34 is 10.0 V. After the switch is closed, how long will it take for the potential difference across the capacitor to decrease to 5.0 V?

37. ||| A 10 μF capacitor initially charged to 20 μC is discharged through a 1.0 kΩ resistor. How long does it take to reduce the capacitor's charge to 10 μC?

38. | A capacitor charging circuit consists of a battery, an uncharged 20 μF capacitor, and a 4.0 kΩ resistor. At $t = 0$ s, the switch is closed; 0.15 s later, the current is 0.46 mA. What is the battery's emf?

39. || The switch in Figure P25.39 has been in position a for a long time. It is changed to position b at $t = 0$ s. What are the charge Q on the capacitor and the current I through the resistor (a) immediately after the switch is changed? (b) At $t = 50$ μs? (c) At $t = 200$ μs?

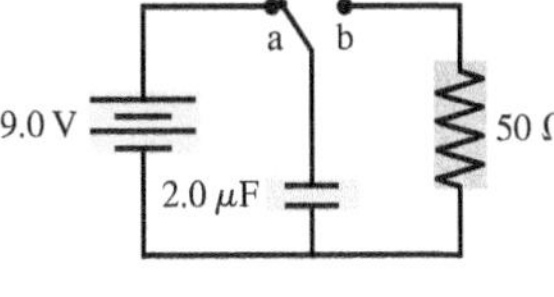

FIGURE P25.39

40. ||| What value resistor will discharge a 1.0 μF capacitor to 10% of its initial voltage in 2.0 ms?

41. ||| A capacitor is discharged through a 100 Ω resistor. The discharge current decreases to 25% of its initial value in 2.5 ms. What is the value of the capacitor?

Section 25.8 Electricity in the Nervous System

42. | Nerve impulse speed can be measured easily in a clinical
BIO setting and is used to diagnose neurological conditions. Two microelectrodes are inserted some distance apart along a long nerve. A voltage pulse to one electrode stimulates the nerve to fire, and the delay time is measured until the action potential is observed at the second electrode. Points on the elbow and wrist, 24 cm apart, are often used.
   a. What is the nerve-impulse speed if the measured delay is 3.2 ms?
   b. How many saltatory jumps are made if the nodes of Ranvier are 1.0 mm apart?

43. | A 9.0-nm-thick cell membrane undergoes an action potential
BIO that follows the curve in the table on page 898. What is the strength of the electric field inside the membrane just before the action potential and at the peak of the depolarization?

44. || A cell membrane has a resistance and a capacitance and thus
BIO a characteristic time constant. What is the time constant of a 9.0-nm-thick membrane surrounding a 0.040-mm-diameter spherical cell?

45. || Axons in the brain have smaller diameters than motor neu-
BIO rons; 0.6 μm is a typical value. Most connections are short range, with an average axon length of about 700 μm. The resistivity of the 7-nm-thick axon membrane is approximately $4 \times 10^7\ \Omega \cdot$ m; the dielectric constant is 5. Model the axon as a cylinder. What are (a) the resistance in MΩ, (b) the capacitance in μF, and (c) the time constant in ms of the axon membrane?

General Problems

46. || How much power is dissipated by each resistor in Figure
INT P25.46?

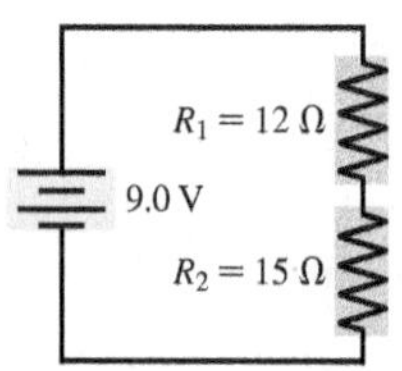

FIGURE P25.46

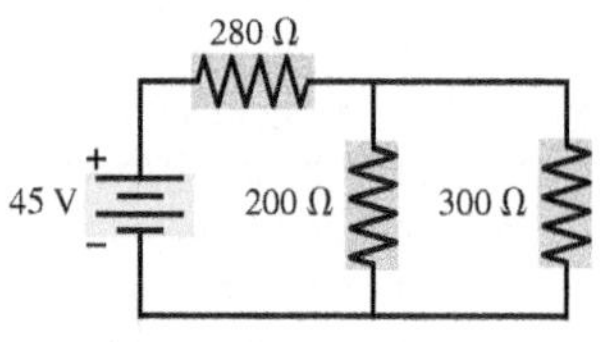

FIGURE P25.47

47. ||| The circuit shown in Figure P25.47 is inside a 15-cm-diameter
INT balloon filled with helium that is kept at a constant pressure of 1.2 atm. How long will it take the balloon's diameter to increase to 16 cm?

48. ||| Two 75 W (120 V) lightbulbs are wired in series, then the
INT combination is connected to a 120 V supply. How much power is dissipated by each bulb?

49. || A real battery is not just an emf. We can model a real 1.5 V
INT battery as a 1.5 V emf in series with a resistor known as the "internal resistance," as shown in Figure P25.49. A typical battery has 1.0 Ω internal resistance due to imperfections that limit current through the battery. When there's no current through the battery, and thus no voltage drop across the internal resistance, the potential difference between its terminals is 1.5 V, the value of the emf. Suppose the terminals of this battery are connected to a 2.0 Ω resistor.
   a. What is the potential difference between the terminals of the battery?
   b. What fraction of the battery's power is dissipated by the internal resistance?

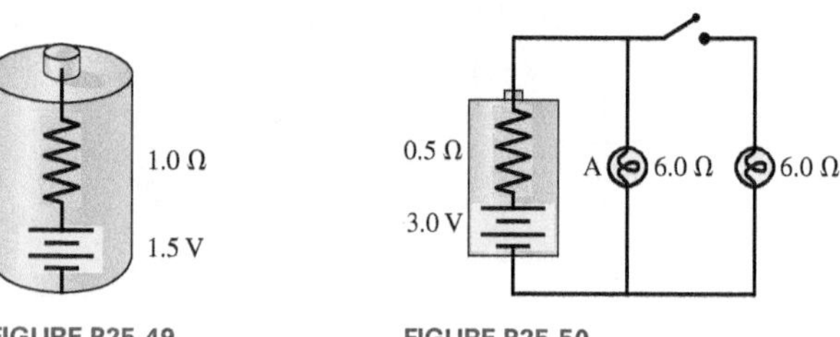

FIGURE P25.49 FIGURE P25.50

50. ||| If the battery in Figure P25.50 were ideal, lightbulb A would not dim when the switch is closed. However, real batteries have a small *internal resistance*, which we can model as the 0.5 Ω resistor shown inside the battery. In this case, the brightness of bulb A changes when the switch is closed.
   a. How much power does bulb A dissipate when the switch is open?
   b. How much power does bulb A dissipate when the switch is closed?

51. || The 10 Ω resistor in Figure P25.51 is dissipating 40 W of
INT power. How much power are the other two resistors dissipating?

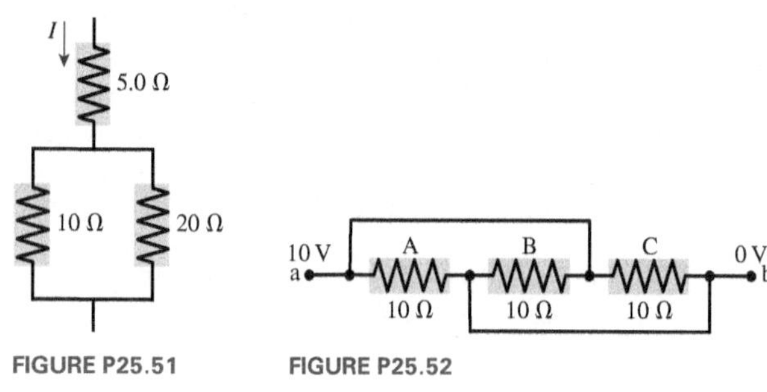

FIGURE P25.51 FIGURE P25.52

52. |||| There's a 10 V potential difference between points a and b in the circuit of Figure P25.52.
   a. What are the currents through resistors A, B, and C?
   b. What is the equivalent resistance between points a and b?

53. ||| What is the current through the battery in Figure P25.53 when the switch is (a) open and (b) closed?

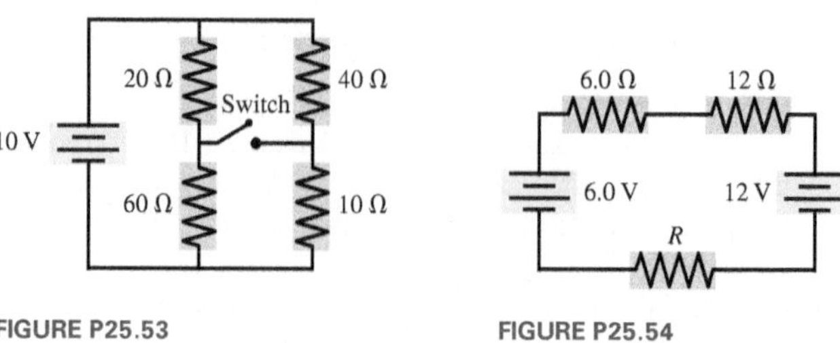

FIGURE P25.53 FIGURE P25.54

54. | There is a current of 0.25 A in the circuit of Figure P25.54.
INT a. What is the direction of the current? Explain.
   b. What is the value of the resistance R?
   c. What is the power dissipated by R?

55. || The ammeter in Figure P25.55 reads 3.0 A. Find I_1, I_2, and $\mathcal{E}$.

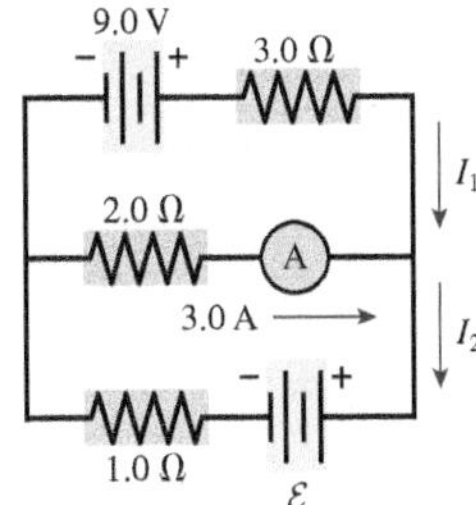

FIGURE P25.55

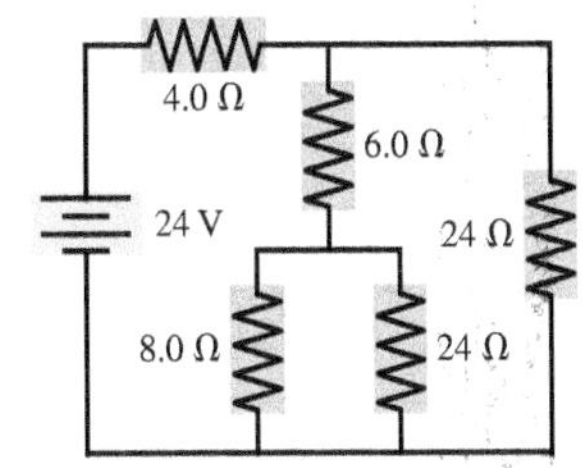

FIGURE P25.56

56. || For the circuit shown in Figure P25.56, find the current through and the potential difference across each resistor. Place your results in a table for ease of reading.

57. || Digital circuits require actions to take place at precise times, so they are controlled by a *clock* that generates a steady sequence of rectangular voltage pulses. One of the most widely used integrated circuits for creating clock pulses is called a 555 timer. Figure P25.57 shows how the timer's output pulses, oscillating between 0 V and 5 V, are controlled with two resistors and a capacitor. The circuit manufacturer tells users that T_H, the time the clock output spends in the high (5 V) state, is $T_H = (R_1 + R_2)C \times \ln 2$. Similarly, the time spent in the low (0 V) state is $T_L = R_2 C \times \ln 2$. You need to design a clock that will oscillate at 10 MHz and will spend 75% of each cycle in the high state. You will be using a 500 pF capacitor. What values do you need to specify for R_1 and R_2?

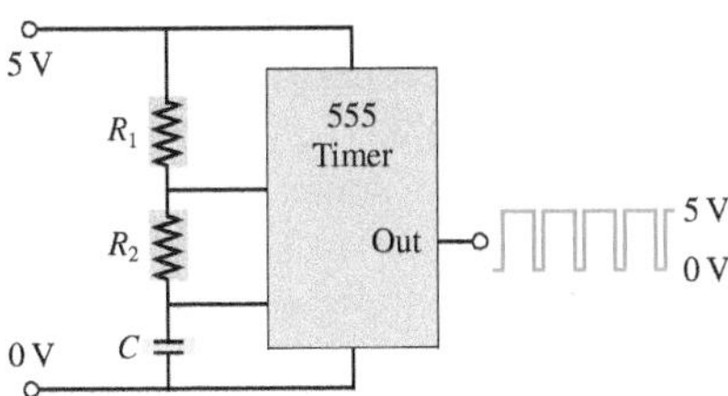

FIGURE P25.57

58. || A 150 μF defibrillator capacitor is charged to 1500 V. When fired through a patient's chest, it loses 95% of its charge in 40 ms. What is the resistance of the patient's chest?
BIO

59. || Large capacitors can hold a potentially dangerous charge long after a circuit has been turned off, so it is important to make sure they are discharged before you touch them. Suppose a 120 μF capacitor from a camera flash unit retains a voltage of 150 V when an unwary student removes it from the camera. If the student accidentally touches the two terminals with her hands, and if the skin resistance of each hand is 1000 Ω, for how long will the current across her chest exceed the danger level of 40 mA?
BIO

60. || Electric eels are known to lift their heads out of the water and touch an animal standing or walking in the water, delivering a painful shock that lasts a few milliseconds. One intrepid scientist allowed a small eel to shock his arm while he was measuring the details. Figure P25.60 shows a model of the scientist's hand in the water. The internal resistance of the eel, the resistance of the arm (including the skin resistance), and the resistance of the return path through the water are shown. Also shown is the resistance of a return path to the water along the exterior of the eel's body.
BIO INT

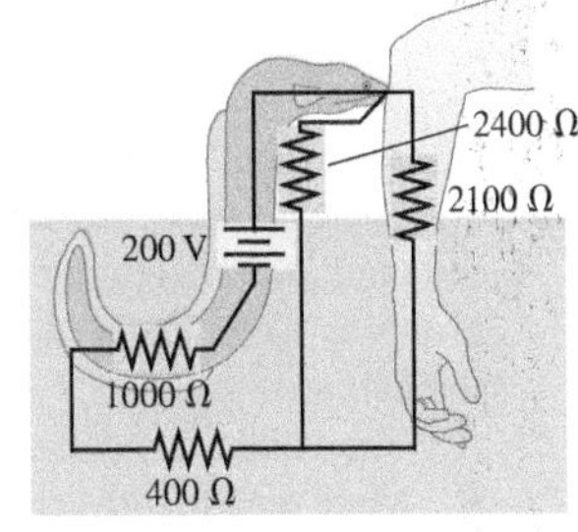

FIGURE P25.60

a. What is the current through the arm?
b. What is the electric power output of the eel during the short duration of the voltage pulse?

61. || A 10 V battery is wired in series with an uncharged capacitor, a resistor, and a switch to make an *RC* circuit. At $t = 0$ s the switch is closed. Figure P25.61 shows the current through the resistor.
a. What is the value of the resistor?
b. What is the value of the capacitor?

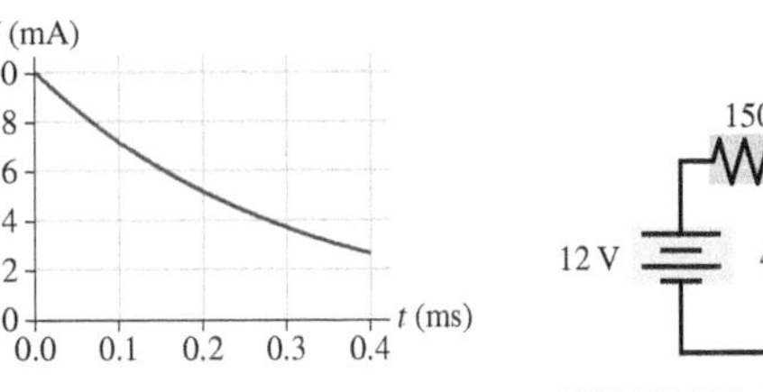

FIGURE P25.61

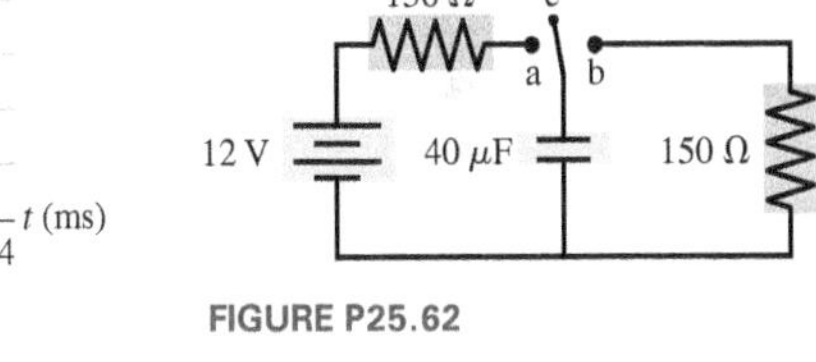

FIGURE P25.62

62. ||| The capacitor in Figure P25.62 is initially uncharged and the switch, in position c, is not connected to either side of the circuit. The switch is now flipped to position a for 10 ms, then to position b for 10 ms, and then brought back to position c. What is the final potential difference across the capacitor?

63. ||| The switch in Figure P25.63 has been closed for a very long time.
a. What is the charge on the capacitor?
b. The switch is opened at $t = 0$ s. At what time has the charge on the capacitor decreased to 10% of its initial value?

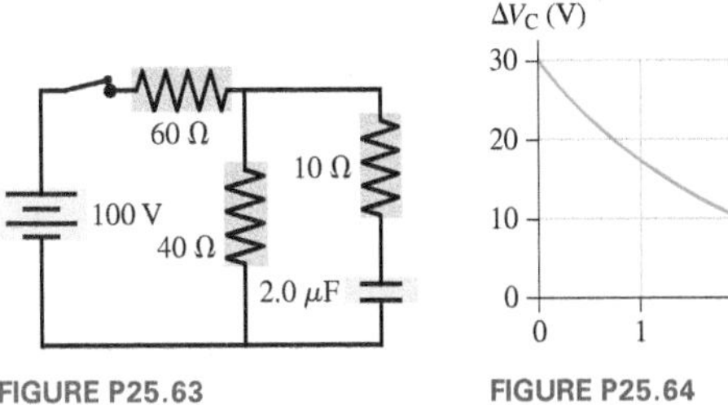

FIGURE P25.63

FIGURE P25.64

64. ||| A 50 μF capacitor that had been charged to 30 V is discharged through a resistor. Figure P25.64 shows the capacitor voltage as a function of time. What is the value of the resistance?

65. ||| A 0.15 F capacitor is charged to 25 V. It is then discharged through a 1.2 kΩ resistor.
INT
a. What is the power dissipated by the resistor just when the discharge is started?
b. What is the total energy dissipated by the resistor during the entire discharge interval?

66. ||| Intermittent windshield wipers use a variable resistor in an RC circuit to set the delay between successive passes of the wipers. A typical circuit is shown in Figure P25.66. When the switch closes, the capacitor (initially uncharged) begins to charge and the potential at point b begins to increase. A sensor measures the potential difference between points a and b, triggering a pass of the wipers when $V_b = V_a$. (Another part of the circuit, not shown, discharges the capacitor at this time so that the cycle can start again.)
   a. What value of the variable resistor will give 12 seconds from the start of a cycle to a pass of the wipers?
   b. To decrease the time, should the variable resistance be increased or decreased?

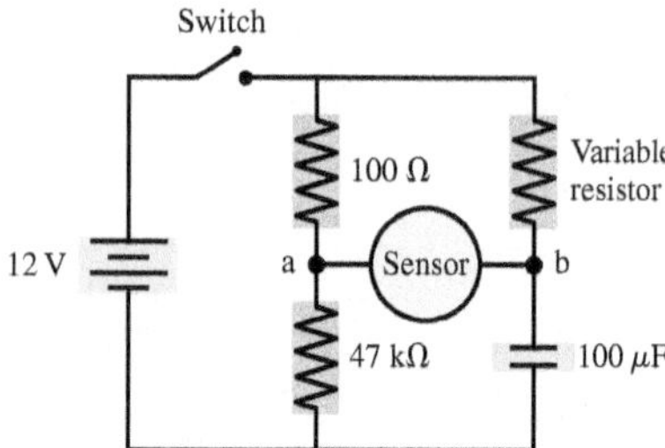

FIGURE P25.66

67. ||| The capacitors in Figure P25.67 are charged and the switch closes at $t = 0$ s. At what time has the current in the 8 Ω resistor decayed to half the value it had immediately after the switch was closed?
INT

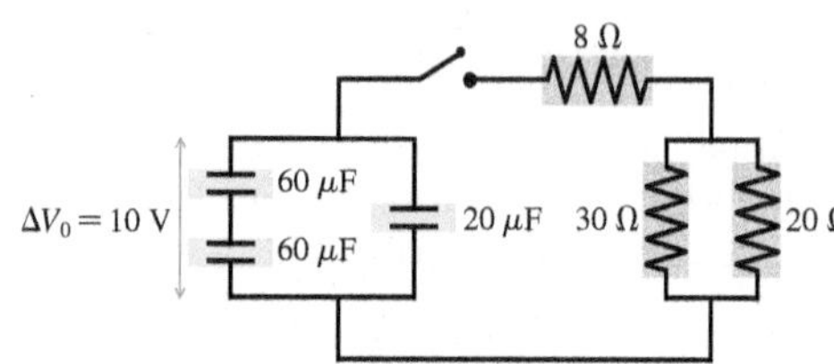

FIGURE P25.67

68. || The giant axon of a squid is 0.5 mm in diameter, 10 cm long, and not myelinated. Unmyelinated cell membranes behave as capacitors with 1 μF of capacitance per square centimeter of membrane area. When the axon is charged to the −70 mV resting potential, what is the energy stored in this capacitance?
BIO INT

69. || A myelinated nerve fiber has a 4.0-μm-diameter cylindrical axon covered by a 1.0-μm-thick layer of myelin. No ions pass through the axon membrane, which is sealed by the myelin, but an action potential at a node causes an ion current to flow along the axis of the axon. The axon is filled with cytosol, whose resistivity was given in Chapter 24 as 0.50 Ω · m. The dielectric constant of myelin is assumed to be 5.0, the same as that of the cell membrane.
BIO
   a. What are the capacitance in pF and the resistance in MΩ of a 1.0-mm-long segment of a myelinated nerve fiber? Use the average radius of the myelin sheath for calculating capacitance.
   b. What is the time constant in μs of this RC circuit?

MCAT-Style Passage Problems

Electric Fish BIO

The voltage produced by a single nerve or muscle cell is quite small, but there are many species of fish that use multiple action potentials in series to produce significant voltages. The electric organs in these fish are composed of specialized disk-shaped cells called *electrocytes*. The cell at rest has the usual potential difference between the inside and the outside, but the net potential difference *across* the cell is zero. An electrocyte is connected to nerve fibers that initially trigger a depolarization in one side of the cell but not the other. For the very short time of this depolarization, there is a net potential difference across the cell, as shown in Figure P25.70. Stacks of these cells connected in series can produce a large total voltage. Each stack can produce a small current; for more total current, more stacks are needed, connected in parallel.

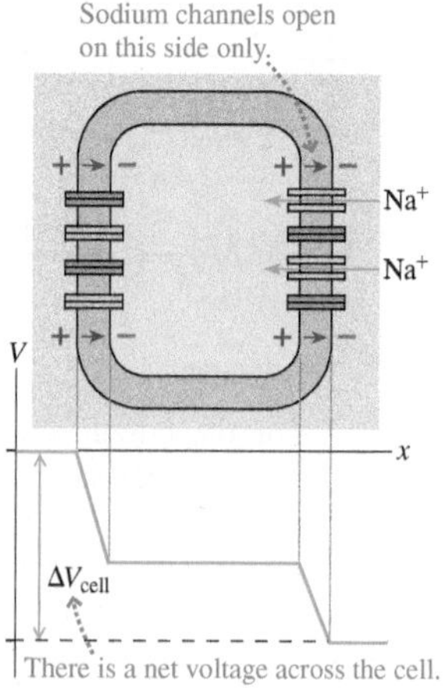

FIGURE P25.70

70. In an electric eel, each electrocyte can develop a voltage of 150 mV for a short time. For a total voltage of 450 V, how many electrocytes must be connected in series?
   A. 300 B. 450 C. 1500 D. 3000
71. An electric eel produces a pulse of current of 0.80 A at a voltage of 500 V. For the short time of the pulse, what is the instantaneous power?
   A. 400 W B. 500 W C. 625 W D. 800 W
72. Electric eels live in fresh water. The torpedo ray is an electric fish that lives in salt water. The electrocytes in the ray are grouped differently than in the eel; each stack of electrocytes has fewer cells, but there are more stacks in parallel. Assuming that the electric eel and the torpedo ray are adapted to their environments, we can assume that
   A. In the lower resistivity of salt water, it is adaptive for an electric fish to provide a higher output current at a lower output voltage.
   B. In the lower resistivity of salt water, it is adaptive for an electric fish to provide a lower output current at a higher output voltage.
   C. In the higher resistivity of salt water, it is adaptive for an electric fish to provide a higher output current at a lower output voltage.
   D. In the higher resistivity of salt water, it is adaptive for an electric fish to provide a lower output current at a higher output voltage.
73. The electric catfish is another electric fish that produces a voltage pulse by means of stacks of electrocytes. As the fish grows in length, the magnitude of the voltage pulse the fish produces grows as well. The best explanation for this change is that, as the fish grows,
   A. The voltage produced by each electrocyte increases.
   B. More electrocytes are added to each stack.
   C. More stacks of electrocytes are added in parallel to the existing stacks.
   D. The thickness of the electrocytes increases.

26 Magnetic Fields and Forces

An aurora occurs when high-energy charged particles from the sun are steered into the upper atmosphere by the earth's magnetic field.

LOOKING AHEAD

Magnetic Fields

Iron filings show the shape of the *magnetic field* around this current-carrying loop of wire.

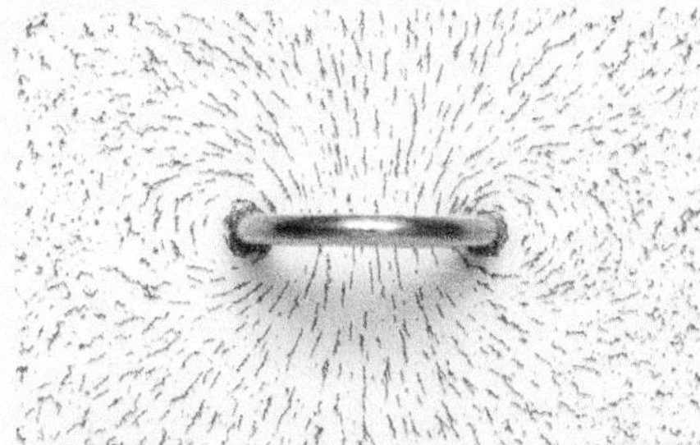

You'll learn how to calculate and use the magnetic fields created when current flows through wires, loops, and coils.

Magnetic Forces

This *mass spectrometer* for carbon dating of materials uses magnetic forces to separate $^{14}C^+$ ions from the more numerous $^{12}C^+$ ions.

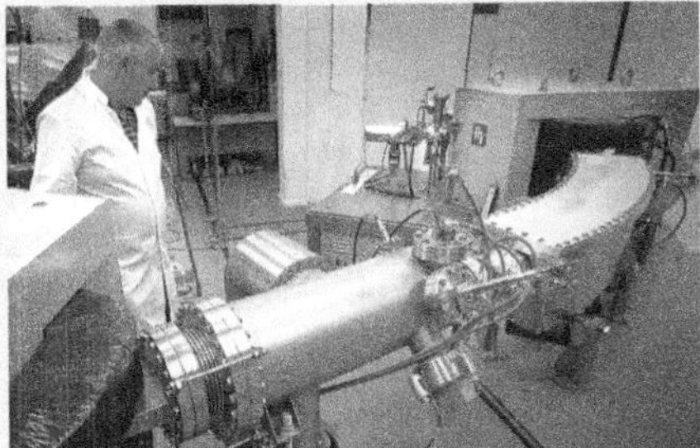

You'll learn to use a *right-hand rule* to determine the magnetic force on a moving charged particle and on a current.

Magnetic Resonance Imaging

Magnetic resonance imaging, or MRI, is a noninvasive technique that produces detailed images of tissues and structures inside the body.

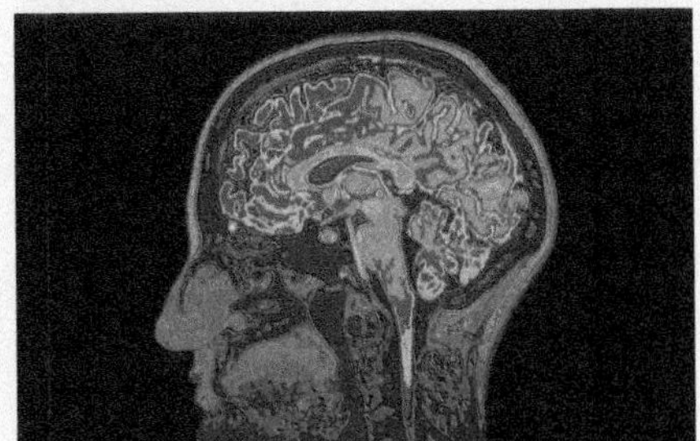

You'll learn how a combination of static and oscillating magnetic fields can measure the density of hydrogen atoms in the body.

GOAL To learn about magnetic fields and how magnetic fields exert forces on currents and moving charges.

PHYSICS AND LIFE

Organisms Use Magnetism in Surprising Ways

A variety of creatures of all sizes, from bacteria to whales, use the earth's magnetic field to navigate. Exactly how they do so is a topic of current research. Magnetism also allows scientists to probe the structures of molecules and physicians to image the body. In particular, nuclear magnetic resonance (NMR) spectroscopy is used to determine molecular structure, including the active structures of proteins in solution, while magnetic resonance imaging (MRI) is a tool to noninvasively image the interior of the body and even, in functional MRI (fMRI), to determine the location of particular types of brain activity. But what is magnetic resonance? What is resonating, and why? Magnetic resonance imaging exploits the remarkable fact that atomic nuclei are subatomic bar magnets that respond to applied magnetic fields. The nature of their response reveals information about the molecular environment that allows both imaging and the determination of molecular structure.

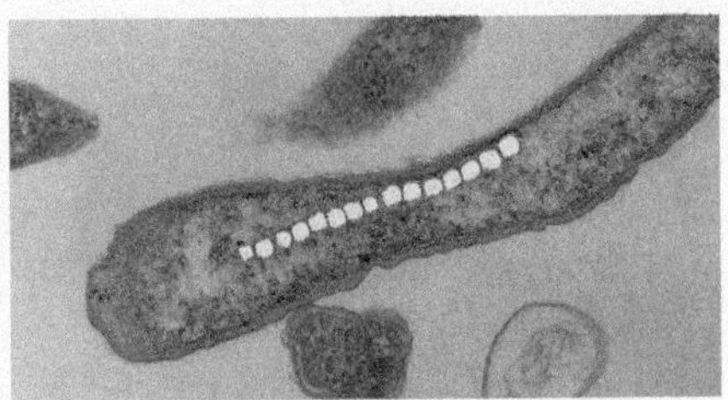

A chain of magnetic nanoparticles inside this bacterium, seen in an electron micrograph, helps it swim along magnetic field lines.

26.1 Magnetism

Magnetism is of profound importance both to living creatures and to medical diagnostics. Organisms ranging from bacteria to reptiles navigate by sensing the earth's magnetic field, a sense known as *magnetoreception.* Magnetic flowmeters are used to measure blood flow, and magnetic resonance imaging is one of the most important diagnostic tools in medicine. We began our investigation of electricity in Chapter 21 by looking at the results of simple experiments with charged rods. Similarly, we'll introduce magnetism by observing some of the interesting properties of *bar magnets,* magnets that have a particularly simple shape.

Discovering magnetism

Experiment 1

If a bar magnet is taped to a piece of cork and allowed to float in a dish of water, it always turns to align itself in an approximate north-south direction. The end of a magnet that points north is called the *north-seeking pole,* or simply the **north pole.** The other end is the **south pole.**

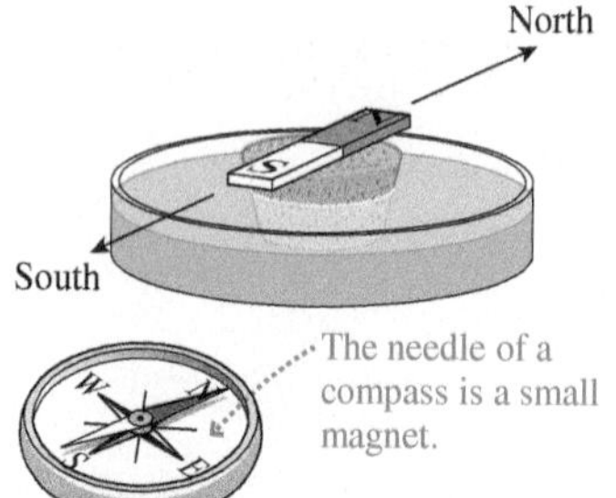

Experiment 2

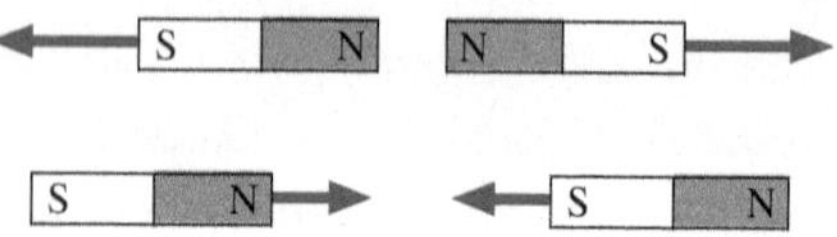

If the north pole of one magnet is brought near the north pole of another magnet, they repel each other. Two south poles also repel each other, but the north pole of one magnet exerts an attractive force on the south pole of another magnet.

Experiment 3

The north pole of a bar magnet attracts one end of a compass needle and repels the other. Apparently the compass needle itself is a little bar magnet with a north pole and a south pole.

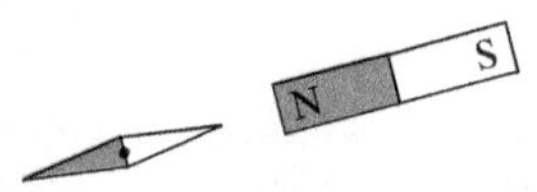

Experiment 4

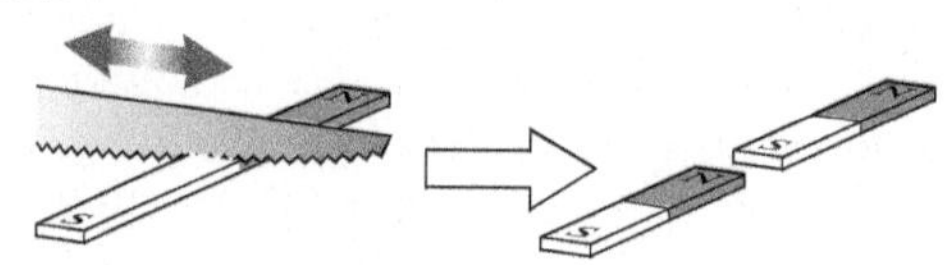

Cutting a bar magnet in half produces two weaker but still complete magnets, each with a north pole and a south pole. No matter how small the magnets are cut, even down to microscopic sizes, each piece remains a complete magnet with two poles.

Experiment 5

Magnets can pick up some objects, such as paper clips, but not all. If an object is attracted to one end of a magnet, it is also attracted to the other end. Most materials, including copper (a penny), aluminum, glass, and plastic, experience no force from a magnet.

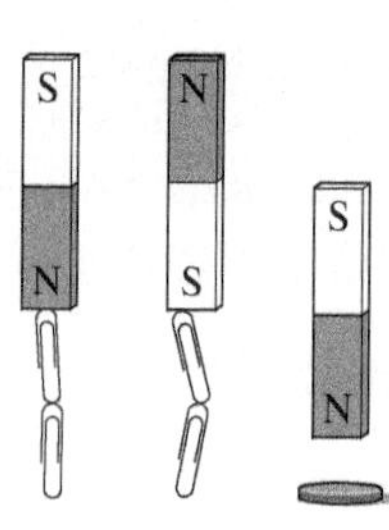

Experiment 6

A magnet does not affect an electroscope. A charged rod exerts a weak *attractive* force on *both* ends of a magnet. However, the force is the same as the force on a metal bar that isn't a magnet, so it is simply a polarization force like the ones we studied in Chapter 21. Other than polarization forces, static charges have *no effects* on magnets.

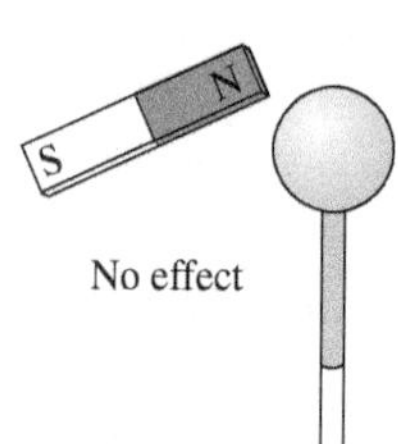

What do these experiments tell us?

- First, magnetism is not the same as electricity. **Magnetic poles and electric charges share some similar behavior, but they are not the same.**
- Magnetism is a long-range force. Paper clips leap upward to a magnet.
- Magnets have two poles, called north and south poles, and thus are **magnetic dipoles.** Two like poles exert repulsive forces on each other; two opposite poles attract. The behavior is *analogous* to electric charges, but, as noted, magnetic poles and electric charges are *not* the same. Unlike charges, isolated north or south poles do not exist.
- The poles of a bar magnet can be identified by using it as a compass. The poles of other magnets can be identified by testing them against a bar magnet. A pole that attracts a known north pole and repels a known south pole must be a south magnetic pole.

- Materials that are attracted to a magnet are called **magnetic materials.** The most common magnetic material is iron. Magnetic materials are attracted to *both* poles of a magnet.

STOP TO THINK 26.1 Does the compass needle rotate clockwise (cw), counterclockwise (ccw), or not at all?

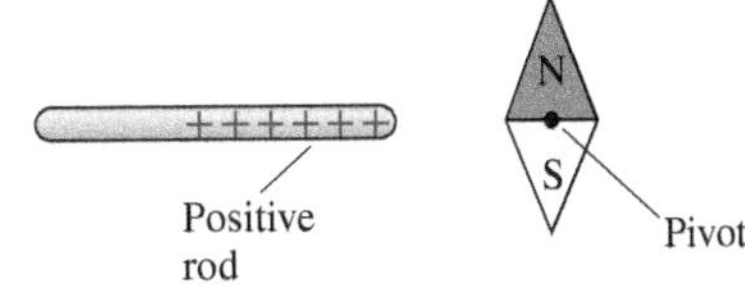
N
S
++++++
Positive rod
Pivot

The Magnetic Field

We introduced the idea of a *field* as a way to understand the long-range electric force. Although this idea appeared rather far-fetched, it turned out to be very useful. We need a similar idea to understand the long-range magnetic force.

Let us define the **magnetic field** $\vec{B}$ as having the following properties:

1. A magnetic field exists at *all* points in space.
2. The magnetic field at each point is a vector. It has both a magnitude, which we call the *magnetic field strength B*, and a direction.
3. The magnetic field exerts forces on magnetic poles. The force on a north pole is parallel to $\vec{B}$; the force on a south pole is opposite $\vec{B}$.

FIGURE 26.1 shows a compass needle in a magnetic field. The field vectors are shown at several points, but keep in mind that the field is present at *all* points in space. A magnetic force is exerted on each of the two poles of the compass, parallel to $\vec{B}$ for the north pole and opposite $\vec{B}$ for the south pole. This pair of opposite forces exerts a torque on the needle, rotating the needle until it is parallel to the magnetic field at that point.

Notice that the north pole of the compass needle, when it reaches the equilibrium position, is in the direction of the magnetic field. Thus we can use a compass needle as a probe of the magnetic field, just as we used a charge as a probe of the electric field. **Magnetic forces align a compass needle parallel to a magnetic field, with the north pole of the compass showing the direction of the magnetic field at that point.**

The SI unit of magnetic field strength is the **tesla,** abbreviated as T. The tesla is defined as

$$1 \text{ tesla} = 1 \text{ T} = 1 \text{ N/A} \cdot \text{m}$$

One tesla is quite a large field; most magnetic fields are a small fraction of a tesla. TABLE 26.1 lists a few magnetic field strengths.

FIGURE 26.1 The magnetic field exerts forces on the poles of a compass needle.

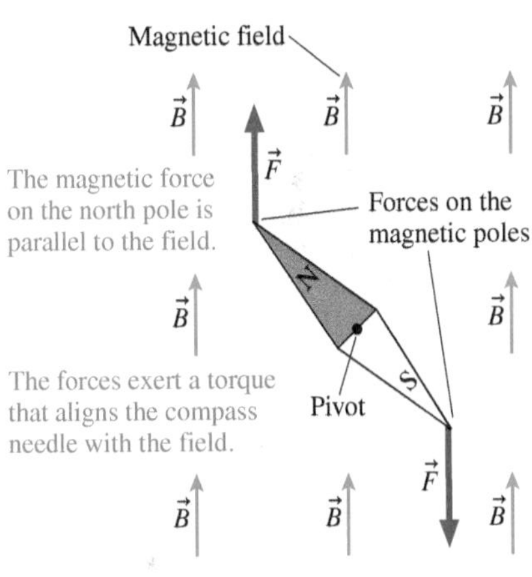
Magnetic field
$\vec{B}$
$\vec{F}$
The magnetic force on the north pole is parallel to the field.
Forces on the magnetic poles
N
S
The forces exert a torque that aligns the compass needle with the field.
Pivot

TABLE 26.1 **Typical magnetic field strengths**

Field source	Field strength (T)
Earth's magnetic field	5×10^{-5}
Refrigerator magnet	0.01
Industrial electromagnet	0.1
Hospital MRI magnet	1 to 3

An Electric Current Creates a Magnetic Field

As electricity began to be seriously studied in the 18th century, some scientists speculated that there might be a connection between electricity and magnetism. Interestingly, the link between electricity and magnetism was discovered *in the midst of a classroom lecture demonstration* in 1819 by the Danish scientist Hans Christian Oersted. Oersted was using a battery—a fairly recent invention—to produce a large current in a wire. By chance, a compass was sitting next to the wire, and Oersted noticed that the current caused the compass needle to turn. In other words, the compass responded as if a magnet had been brought near.

Oersted's discovery that **magnetism is caused by an electric current** is illustrated in FIGURE 26.2 (next page). Compasses placed around a wire all point north if there's no current. But a strong current through the wire causes the compass needles to pivot until they are tangent to a circle around the wire. Part c of the figure demonstrates an important **right-hand rule** that relates the orientation of the compass needles to the direction of the current.

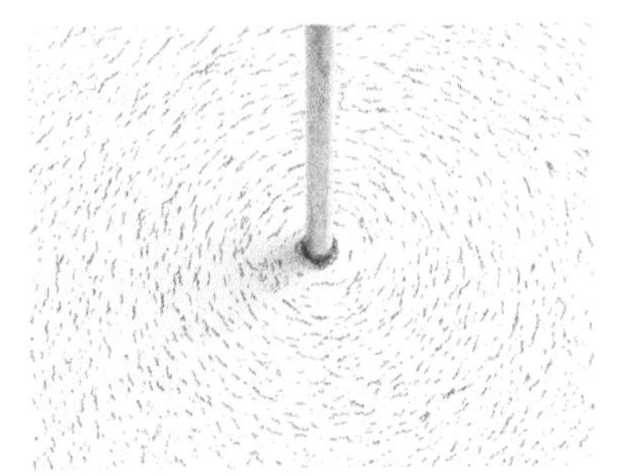

Seeing magnetic fields Iron filings reveal the magnetic field around a current-carrying wire.

FIGURE 26.2 Response of compass needles to a current in a straight wire.

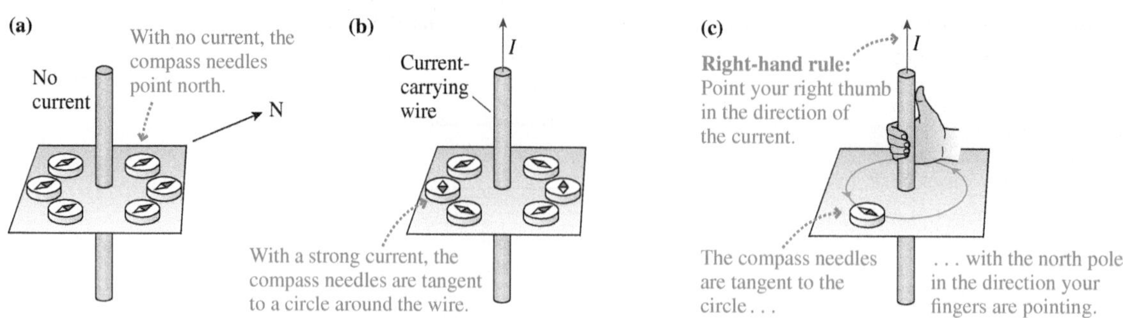
(a)
No current
With no current, the compass needles point north.
N
(b)
I
Current-carrying wire
With a strong current, the compass needles are tangent to a circle around the wire.
(c)
I
Right-hand rule: Point your right thumb in the direction of the current.
The compass needles are tangent to the circle . . .
. . . with the north pole in the direction your fingers are pointing.

FIGURE 26.3 The magnetic field around a current-carrying wire.

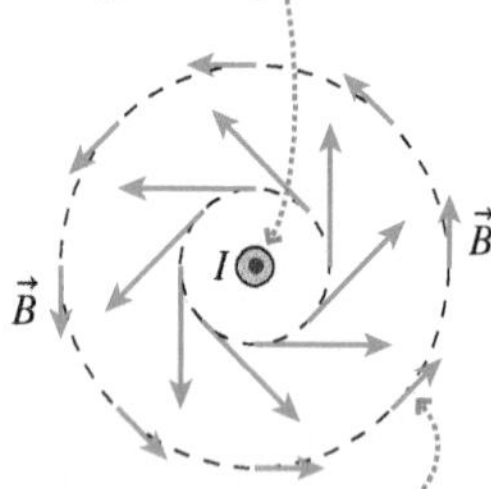
(a) This is a cross section of the wire with the notation showing that the current is coming toward you.
I
$\vec{B}$
$\vec{B}$
The magnetic field *vectors* are tangent to circles around the wire, pointing in the direction given by the right-hand rule.

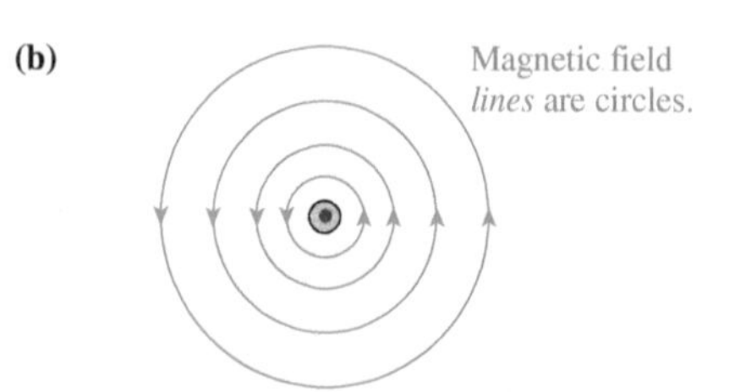
(b)
Magnetic field *lines* are circles.

Because compass needles align with the magnetic field, the magnetic field at each point must be tangent to a circle around the wire. FIGURE 26.3a shows the magnetic field by drawing field vectors. Notice that the field is weaker (shorter vectors) at greater distances from the wire.

Another way to picture the field is with the use of **magnetic field lines.** These are imaginary lines drawn through a region of space so that

- A tangent to a field line is in the direction of the magnetic field, and
- The field lines are closer together where the magnetic field strength is larger.

FIGURE 26.3b shows the magnetic field lines around a current-carrying wire. Notice that magnetic field lines form loops, with no beginning or ending point. This is in contrast to electric field lines, which stop and start on charges. The right-hand rule for fields in Tactics Box 26.1 will help you determine which way the loops circulate.

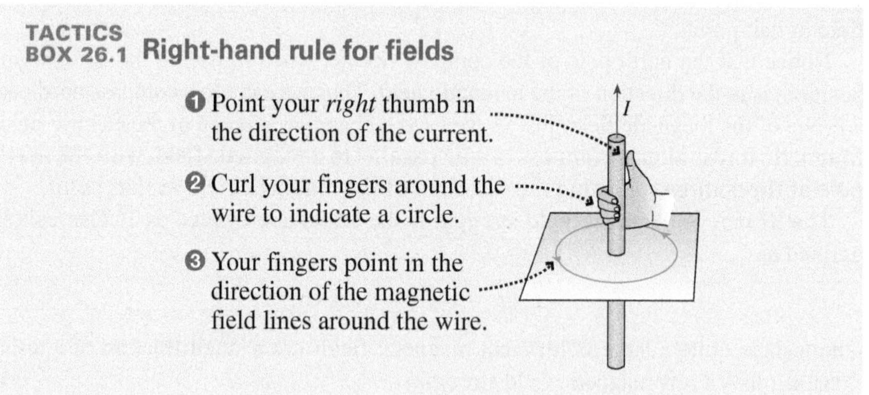
TACTICS BOX 26.1 Right-hand rule for fields
1 Point your *right* thumb in the direction of the current.
2 Curl your fingers around the wire to indicate a circle.
3 Your fingers point in the direction of the magnetic field lines around the wire.
I

FIGURE 26.4 The notation for vectors and currents perpendicular to the plane of the figure.

× × × ×
× × × ×
Vectors going away from you

• • • •
• • • •
Vectors coming toward you

⊗
Current going away from you

⊙
Current coming toward you

Magnetism requires thinking three-dimensionally in a way that electricity does not. But two-dimensional figures such as Figure 26.3 are much easier to draw than the three-dimensional perspective of Figure 26.2. Consequently, we will often need to indicate field vectors or currents that are perpendicular to the plane of the figure. FIGURE 26.4 shows the notation we will use. You can see in Figure 26.3 that the wire's current is coming out of the figure as if we were looking down on Figure 26.2 from above.

Two Kinds of Magnetism?

You might be concerned that we have introduced two kinds of magnetism. We opened this chapter discussing permanent magnets—such as bar magnets—and their forces. Then, without warning, we switched to the magnetic field of a current. It is not at all obvious that the magnetism of a current is the same as that exhibited by stationary chunks of metal called "magnets." Perhaps there are two different types of magnetic forces, one having to do with currents and the other being responsible for permanent magnets. One of the major goals for our study of magnetism is to see that these two quite different ways of producing magnetic effects are really just two different aspects of a *single* magnetic force.

STOP TO THINK 26.2 The magnetic field at position P points

A. Up. B. Down.
C. Toward you. D. Away from you.

26.2 The Magnetic Field of a Current

A current is the source of a magnetic field. Real current-carrying wires, with their twists and turns, have very complex fields. We will emphasize the physical ideas by focusing on three common magnetic fields that are models of more complex fields: the magnetic field of a long, straight, current-carrying wire; the field at the center of a circular loop of current; and the field inside a *solenoid,* a helical coil of wire. The tunnel of an MRI machine is the inside of a solenoid.

The fundamental law of electricity, Coulomb's law, is for the forces between point charges and the electric field of a point charge. The analogous law of magnetism, the **Biot-Savart law** (rhymes with "Leo" and "bazaar"), is for the magnetic field of a tiny piece of current.

FIGURE 26.5 shows a current-carrying wire with a small current segment of length Δx highlighted. The Biot-Savart law gives the magnetic field at distance r from this small current segment:

$$\vec{B}_{\text{current segment}} = \left(\frac{\mu_0}{4\pi}\frac{I\,\Delta x \sin\theta}{r^2}, \text{ direction given by the right-hand rule}\right) \quad (26.1)$$

where θ is the angle between $\vec{r}$ and the current. The constant μ_0, called the **permeability constant,** plays a role in magnetism similar to the role of the permittivity constant ϵ_0 in electricity. Its value is

$$\mu_0 = 1.26 \times 10^{-6}\ \text{T}\cdot\text{m/A}$$

FIGURE 26.5 Finding the magnetic field of a small segment of current in a current-carrying wire.

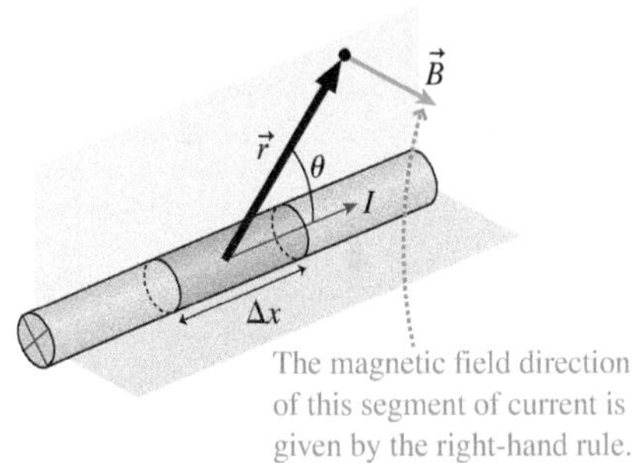

The Biot-Savart law, like Coulomb's law, is an inverse-square law. The Biot-Savart law is somewhat more complex, however, because the magnetic field also depends on the angle between the direction of the current and a line to the point where the field is evaluated.

Tactics Box 26.1 describes how to use the right-hand rule to find the direction of the magnetic field: Point your right thumb in the direction of the current and wrap your fingers around it. FIGURE 26.6 shows that the field vectors $\vec{B}$ are tangent to circles drawn in the current direction. The field gets weaker with distance *and* as angle θ (see Figure 26.5) gets smaller; $\vec{B} = \vec{0}$ directly in front of or behind the current segment where $\theta = 0°$ or $180°$.

The Biot-Savart law is the starting point for calculating all magnetic fields, but we need an additional idea: Magnetic fields, like electric fields, obey the principle of superposition. If different current segments generate magnetic fields $\vec{B}_1$, $\vec{B}_2$, $\vec{B}_3, \ldots$ at a point in space, then the net magnetic field at that point is

$$\vec{B} = \vec{B}_1 + \vec{B}_2 + \vec{B}_3 + \cdots \quad (26.2)$$

Equation 26.2 is the basis for calculating the magnetic fields of several important currents.

FIGURE 26.6 Two views of the magnetic field of a current segment.

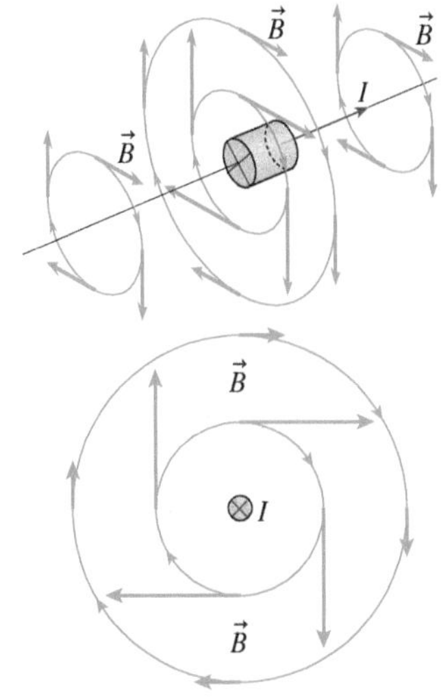

STOP TO THINK 26.3 The current in a current segment is coming toward you. What is the direction of the magnetic field at point P?

P• ⊙

A. Up B. Down C. Left D. Right

The Magnetic Field of a Long, Straight, Current-Carrying Wire

The most basic magnetic field is the field of a very long, straight wire that carries current I. We can use Equation 26.2 to determine this magnetic field. FIGURE 26.7 shows a portion of a current-carrying wire that lies along the x-axis. We want to calculate the magnetic field at point P that is distance y from the wire along the y-axis. The calculation will be similar to the calculations we did in Chapter 22 to find the electric potential of a continuous distribution of charge; we use superposition to form a sum and then convert the sum to an integral.

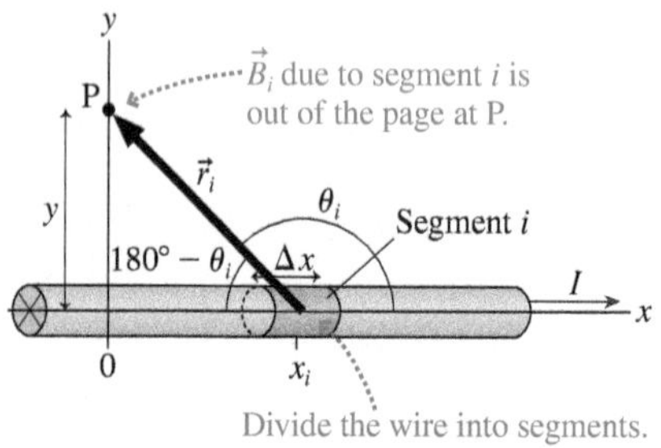

FIGURE 26.7 Calculating the magnetic field of a long, straight, current-carrying wire.

Imagine dividing the wire into a large number of small segments that each have length Δx. We can label the segments with index i. The figure shows a segment at position x_i. You should use the right-hand rule to convince yourself that $\vec{B}_i$, the magnetic field at P due to the current in segment i, points out of the page. This is the field direction no matter where segment i happens to be along the x-axis. Consequently, $\vec{B}$ is tangent to a circle around the wire.

Segment i is at distance $r_i = \sqrt{x_i^2 + y^2}$ from point P. According to the Biot-Savart law, Equation 26.1, the magnetic field at P due to the current in segment i is

$$B_i = \frac{\mu_0}{4\pi}\frac{I\Delta x \sin\theta_i}{x_i^2 + y^2} = \frac{\mu_0}{4\pi}\frac{I\sin\theta_i}{x_i^2 + y^2}\Delta x \tag{26.3}$$

where θ_i is the angle between the direction of the current and the line to point P. Figure 26.7 labels the angle $180° - \theta_i$, and we can use trigonometry to show that

$$\sin(180° - \theta_i) = \frac{\text{far side}}{\text{hypoteneuse}} = \frac{y}{r_i} = \frac{y}{\sqrt{x_i^2 + y^2}} \tag{26.4}$$

But $\sin(180° - \theta_i) = \sin\theta_i$, so we can use Equation 26.4 for $\sin\theta_i$ in Equation 26.3. By doing so, we find that the magnetic field at point P of segment i is

$$B_i = \frac{\mu_0}{4\pi}\frac{I\Delta x \sin\theta_i}{x_i^2 + y^2} = \frac{\mu_0}{4\pi}\frac{Iy}{(x_i^2 + y^2)^{3/2}}\Delta x \tag{26.5}$$

Now we can find the full magnetic field at point P of the current-carrying wire by summing the fields from each of the segments:

$$B_{\text{wire}} = \sum_i B_i = \frac{\mu_0 I y}{4\pi}\sum_i \frac{\Delta x}{(x_i^2 + y^2)^{3/2}} \tag{26.6}$$

The sum becomes an integral in the limit $\Delta x \rightarrow dx$. To sum all the segments in an infinitely long wire, we must set the integration limits at $\pm\infty$:

$$B_{\text{wire}} = \frac{\mu_0 I y}{4\pi}\int_{-\infty}^{\infty}\frac{dx}{(x^2 + y^2)^{3/2}} = \frac{\mu_0 I y}{4\pi}\frac{x}{y^2(x^2 + y^2)^{1/2}}\bigg|_{-\infty}^{\infty} \tag{26.7}$$

The integral in Equation 26.7 is a standard integral that can be found in Appendix A or evaluated with mathematical software. To evaluate the integral at the limits, we note that $x/(x^2 + y^2)^{1/2}$ becomes $x/|x|$ if $x \gg y$, and this is simply ± 1 at $x = \pm\infty$. Consequently,

$$B_{\text{wire}} = \frac{\mu_0}{2\pi}\frac{I}{y} \tag{26.8}$$

Equation 26.8 is the magnitude of the magnetic field; the field direction is determined by the right-hand rule. y is the distance from the wire, which is better represented by r now that we've completed the calculation. With this change, the magnetic field is

$$\vec{B}_{\text{wire}} = \left(\frac{\mu_0}{2\pi} \frac{I}{r}, \text{ tangent to a circle around the wire in the direction of the right-hand rule} \right) \quad (26.9)$$

Magnetic field at distance r from a long, straight wire with current I

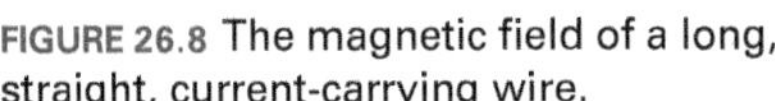
FIGURE 26.8 The magnetic field of a long, straight, current-carrying wire.

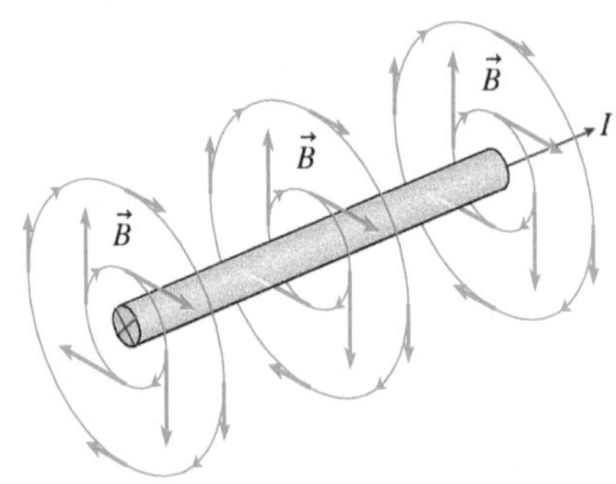

The magnetic field of a long, straight, current-carrying wire is our first important model. FIGURE 26.8 is a three-dimensional visualization of the field.

EXAMPLE 26.1 **The magnetic field strength near a heater wire**

A 1.0-m-long, 1.0-mm-diameter nichrome heater wire is connected to a 12 V battery. What is the magnetic field strength 1.0 cm away from the wire?

PREPARE 1 cm is much less than the 1 m length of the wire, so we will model the wire as infinitely long.

SOLVE The current through the wire is $I = \Delta V_{\text{bat}}/R$, where the wire's resistance R is

$$R = \frac{\rho L}{A} = \frac{\rho L}{\pi r^2} = 1.9\ \Omega$$

The nichrome resistivity $\rho = 1.5 \times 10^{-6}\ \Omega \cdot \text{m}$ was taken from Table 24.1. Thus the current is $I = (12\text{ V})/(1.9\ \Omega) = 6.3$ A. The magnetic field strength at distance $d = 1.0\text{ cm} = 0.010$ m from the wire is

$$B_{\text{wire}} = \frac{\mu_0}{2\pi} \frac{I}{d} = \frac{(1.26 \times 10^{-6}\text{ T} \cdot \text{m/A})(6.3\text{ A})}{2\pi(0.010\text{ m})}$$
$$= 1.3 \times 10^{-4}\text{ T}$$

ASSESS The magnetic field of the wire is slightly more than twice the strength of the earth's magnetic field.

Motors, loudspeakers, metal detectors, and many other devices generate magnetic fields with *coils* of wire. The simplest coil is a single-turn circular loop of wire. A circular loop of wire with a circulating current is called a **current loop.**

EXAMPLE 26.2 **The magnetic field of a current loop**

FIGURE 26.9a shows a current loop, a circular loop of wire with radius R that carries current I. Find the magnetic field strength at the center of the loop.

PREPARE A real coil needs wires to bring the current in and out, but we'll model a current loop as a closed circle of wire with a circulating current I, as shown in FIGURE 26.9b.

FIGURE 26.9 A current loop.

(a) A practical current loop

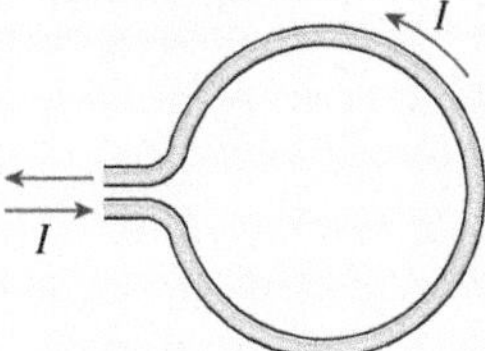

(b) An ideal current loop

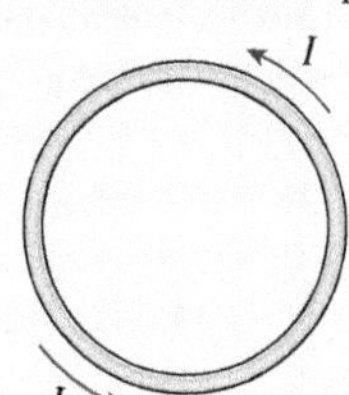

SOLVE FIGURE 26.10 highlights segment i of the wire. This segment is distance $r_i = R$ from the center of the circle, and the angle between $\vec{r}$ and the current is 90°. We can use the Biot-Savart law,

FIGURE 26.10 The magnetic field of a current loop.

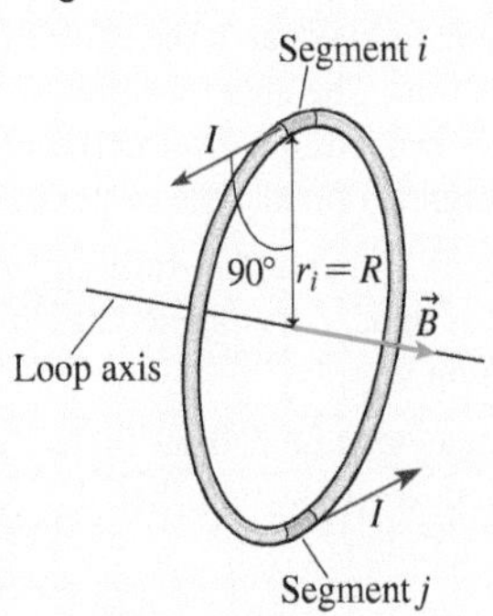

Equation 26.1, to write the magnetic field at the center due to the current in segment i:

$$B_i = \frac{\mu_0}{4\pi} \frac{I \Delta x \sin 90°}{r_i^2} = \frac{\mu_0}{4\pi} \frac{I}{R^2} \Delta x$$

If we use the right-hand rule, with our thumb pointing in the direction of the current, we see that the field points along the axis coming out of the loop.

Continued

Figure 26.10 also highlights a second segment j. If we apply the same reasoning to this segment, we see that segment j creates exactly the same field as segment i—a field with the same strength and pointing in the same direction. In fact, *every* segment around the loop creates an identical field, and the magnetic field at the center is their sum:

$$B_{\text{loop}} = \sum_i B_i = \sum_i \frac{\mu_0}{4\pi}\frac{I}{R^2}\Delta x = \frac{\mu_0}{4\pi}\frac{I}{R^2}\sum_i \Delta x$$

The constants are taken out of the summation, and we're left with a sum that adds the lengths of all the segments around the loop. But that is just the total length of the wire, which is the loop's circumference $2\pi R$. Thus the magnetic field strength at the center of a current loop is

$$B_{\text{loop}} = \frac{\mu_0}{4\pi}\frac{I}{R^2}(2\pi R) = \frac{\mu_0 I}{2R}$$

The direction of the field is along the axis, pointing away from the loop in the direction given by the right-hand rule.

ASSESS If we compare our result to Equation 26.9, the magnetic field of a long, straight, current-carrying wire, we see that the magnetic field strength at the center of a current loop is a factor of π larger than the field strength at distance R from the wire. That makes sense. The segments of a straight wire get farther and farther from the point of interest, and the field strength due to each segment decreases inversely with the square of the distance. In contrast, the segments of a current loop are all equally distant from the center.

In practice, current often passes through a *coil* consisting of N *turns* of wire. If the turns are all very close together, so that the magnetic field of each is essentially the same, then the magnetic field of a coil is N times the magnetic field of a current loop. The magnetic field at the center of a thin N-turn coil, or N-turn current loop, is

$$\vec{B}_{\text{coil}} = \left(\frac{\mu_0}{2}\frac{NI}{R}, \text{ perpendicular to the coil in the direction of the right-hand rule}\right) \tag{26.10}$$

Magnetic field at the center of a thin N-turn coil

This is the second of our key magnetic field models.

EXAMPLE 26.3 Matching the earth's magnetic field

Laboratory studies of how animals use the earth's magnetic field to navigate often begin by observing an animal's behavior in zero magnetic field. One way to create a zero-field region of space is to generate a magnetic field equal to the earth's field but pointing in the opposite direction. The superposition of the two fields is zero. What current is needed in a 5-turn, 10-cm-diameter coil to cancel the earth's magnetic field at the center of the coil?

PREPARE FIGURE 26.11 shows a five-turn coil of wire. We will assume that the coil is thin, so its magnetic field is five times that of a single current loop.

FIGURE 26.11 A coil of wire.

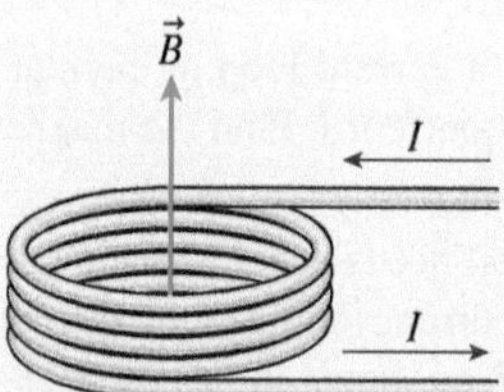

SOLVE The earth's magnetic field, from Table 26.1, is 5×10^{-5} T. We can use Equation 26.10 to find that the current needed to generate a 5×10^{-5} T field is

$$I = \frac{2RB}{\mu_0 N} = \frac{2(0.050\text{ m})(5.0 \times 10^{-5}\text{ T})}{5(1.26 \times 10^{-6}\text{ T}\cdot\text{m/A})} = 0.80\text{ A}$$

ASSESS A 0.80 A current is easily produced. Although there are better ways to cancel the earth's field than using a simple coil, this illustrates the idea.

The Magnetic Field of a Solenoid

In our study of electricity, we made extensive use of the idea of a uniform electric field: a field that is the same at every point in space. We found that two closely spaced, parallel charged plates generate a uniform electric field between them, and this was one reason we focused so much attention on the parallel-plate capacitor.

Similarly, there are many applications of magnetism for which we would like to generate a **uniform magnetic field,** a field having the same magnitude and the same direction at every point within some region of space. Neither a long, straight wire nor a current loop produces a uniform magnetic field.

In practice, a uniform magnetic field is generated with a **solenoid.** A solenoid, shown in FIGURE 26.12, is a helical coil of wire of length L that is formed with N loops—usually called turns—of wire. The same current I passes through each turn as it travels through the coil. Solenoids may have hundreds or thousands of coils, sometimes wrapped in several layers. The photo made with iron filings shows that the magnetic field inside the solenoid is nearly uniform (i.e., the field lines are nearly parallel) and the field outside the solenoid is much weaker.

Our goal of producing a uniform magnetic field can be achieved by increasing the number of coils until we have an *ideal solenoid* that is infinitely long and in which the coils are as close together as possible. **As FIGURE 26.13 shows, the magnetic field inside an ideal solenoid is uniform—the same at every point—and parallel to the axis; the magnetic field outside is zero.** No real solenoid is ideal, but a very uniform magnetic field can be produced near the center of a tightly wound solenoid whose length is much larger than its diameter.

It can be shown that the uniform magnetic field inside a solenoid is

$$\vec{B}_{\text{solenoid}} = \left(\frac{\mu_0 NI}{L}, \text{ parallel to the coil}\right) \tag{26.11}$$

Magnetic field inside a solenoid of length L with N turns

The solenoid is our third important magnetic field model. Measurements that need a uniform magnetic field are often conducted inside a solenoid, which can be built quite large.

FIGURE 26.12 A solenoid and its magnetic field.

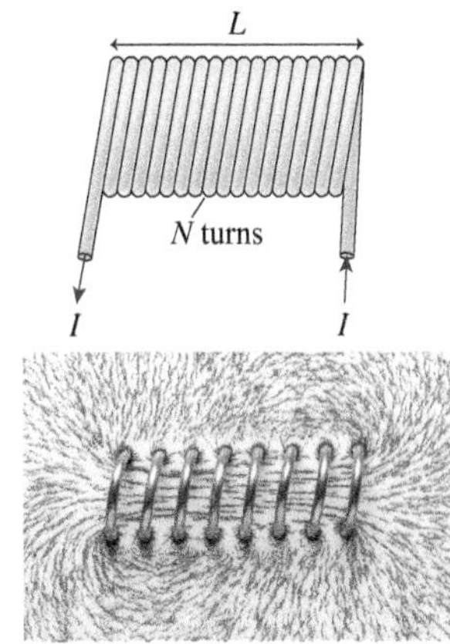

FIGURE 26.13 The magnetic field of an ideal solenoid.

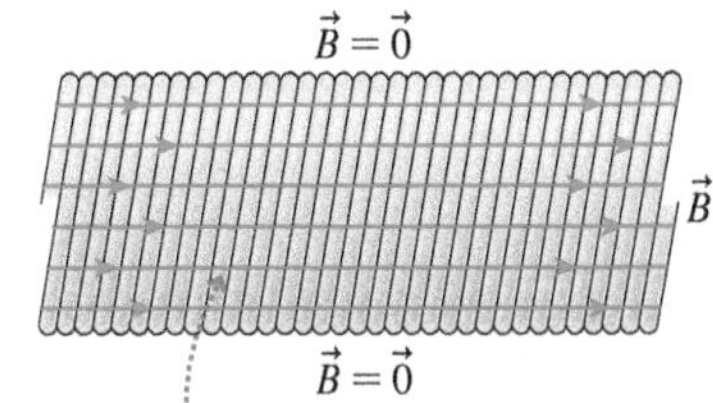

The magnetic field is uniform inside this section of an ideal, infinitely long solenoid. The magnetic field outside the solenoid is zero.

EXAMPLE 26.4 Generating an MRI magnetic field

A 1.0-m-long MRI solenoid generates a 1.2 T magnetic field through the tunnel in which the patient lies. To produce such a large field, the solenoid is wrapped with superconducting wire that can carry a 100 A current. How many turns of wire does the solenoid need?

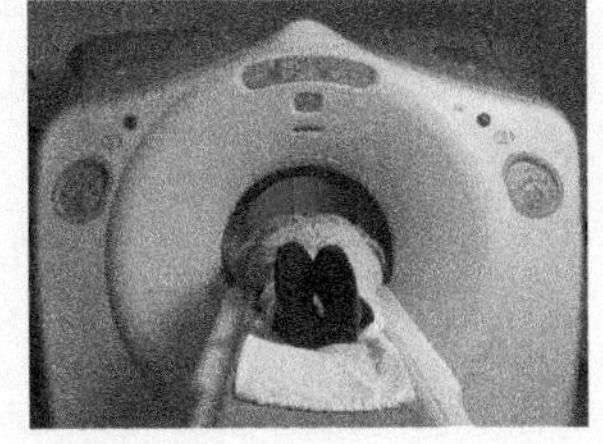

PREPARE Assume that the solenoid is ideal.

SOLVE Generating a magnetic field with a solenoid is a trade-off between current and turns of wire. A larger current requires fewer turns, but the resistance of ordinary wires causes them to overheat if the current is too large. For a superconducting wire that can carry 100 A with no resistance, we can use Equation 26.11 to find the required number of turns:

$$N = \frac{LB}{\mu_0 I} = \frac{(1.0\text{ m})(1.2\text{ T})}{(1.26 \times 10^{-6}\text{ T}\cdot\text{m/A})(100\text{ A})} = 9500 \text{ turns}$$

ASSESS The solenoid coil requires a large number of turns, but that's not surprising for generating a very strong field. If the wires are 1 mm in diameter, there would be 10 layers with approximately 1000 turns per layer.

26.3 Magnetic Dipoles

We were able to calculate the magnetic field at the center of a current loop, but determining the field elsewhere is more difficult. We will be content to use visual representations of the field.

FIGURE 26.14 (next page) shows the full magnetic field of a current loop. This is a field with *rotational symmetry,* so to picture the full three-dimensional field, imagine Figure 26.14a rotated about the axis of the loop. Figure 26.14b shows the magnetic field in the plane of the loop as seen from the right. There is a clear sense, seen in the

FIGURE 26.14 The magnetic field of a current loop.

(a) Cross section through the current loop

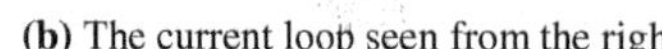

(b) The current loop seen from the right

(c) A photo of iron filings

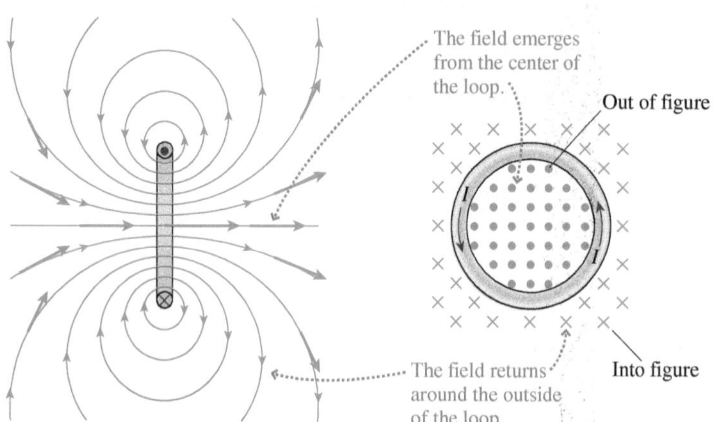

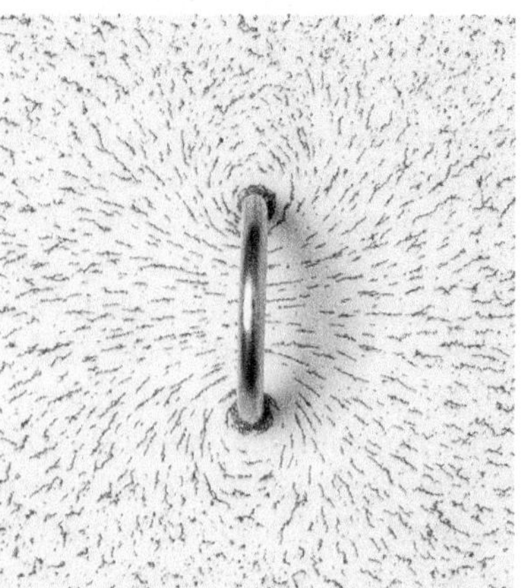

photo of Figure 26.14c, that the magnetic field leaves the loop on one side, "flows" around the outside, then returns to the loop.

There are two versions of the right-hand rule that you can use to determine which way a loop's field points. Try these in Figure 26.14. Being able to quickly ascertain the field direction of a current loop is an important skill.

TACTICS BOX 26.2 Finding the magnetic field direction of a current loop

Use either of the following methods to find the magnetic field direction:

❶ Point your right thumb in the direction of the current at any point on the loop and let your fingers curl through the center of the loop. Your fingers are then pointing in the direction in which $\vec{B}$ leaves the loop.

❷ Curl the fingers of your right hand around the loop in the direction of the current. Your thumb is then pointing in the direction in which $\vec{B}$ leaves the loop.

A Current Loop Is a Magnetic Dipole

A current loop has two distinct sides. Bar magnets also have two distinct sides or ends, so you might wonder if current loops are related to these permanent magnets. Consider the following experiments with a current loop. Notice that we're showing the magnetic field only in the plane of the loop.

Investigating current loops

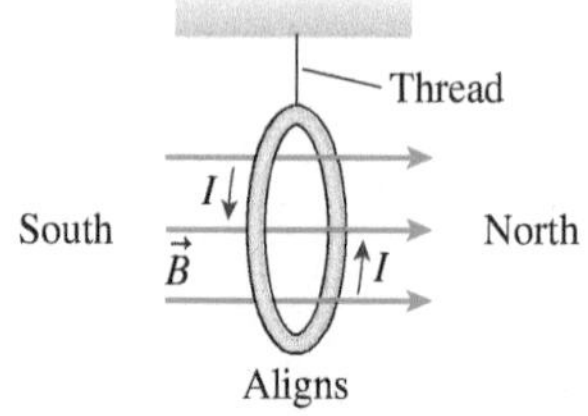

Aligns

A current loop hung by a thread aligns itself with the magnetic field pointing north.

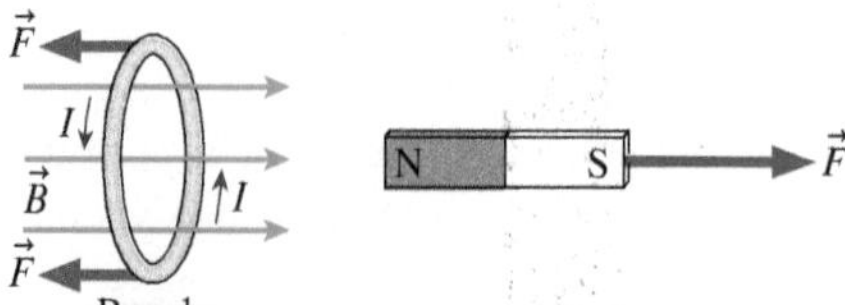

Repels

The north pole of a permanent magnet repels the side of a current loop from which the magnetic field is emerging.

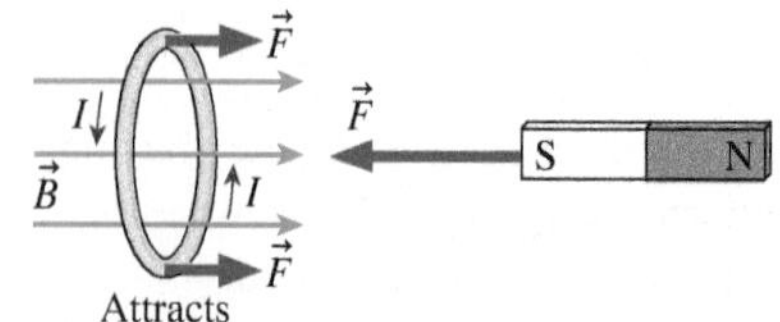

Attracts

The south pole of a permanent magnet attracts the side of a current loop from which the magnetic field is emerging.

These investigations show that **a current loop is a magnet,** just like a permanent magnet. A magnet created by a current in a coil of wire is called an **electromagnet.** An electromagnet picks up small pieces of iron, influences a compass needle, and acts in every way like a permanent magnet.

In fact, FIGURE 26.15 shows that a flat permanent magnet and a current loop generate the same magnetic field—the field of a **magnetic dipole.** For both, **you can identify the north pole as the face or end *from which* the magnetic field emerges.** The magnetic fields of both point *into* the south pole.

NOTE ▶ Magnetic poles, unlike electric charges, are not physical objects but are the regions of a magnet from which the magnetic field emerges and to which it returns. They are usually the regions where the field is strongest. ◀

FIGURE 26.15 A current loop has magnetic poles and generates the same magnetic field—a dipole field—as a flat permanent magnet.

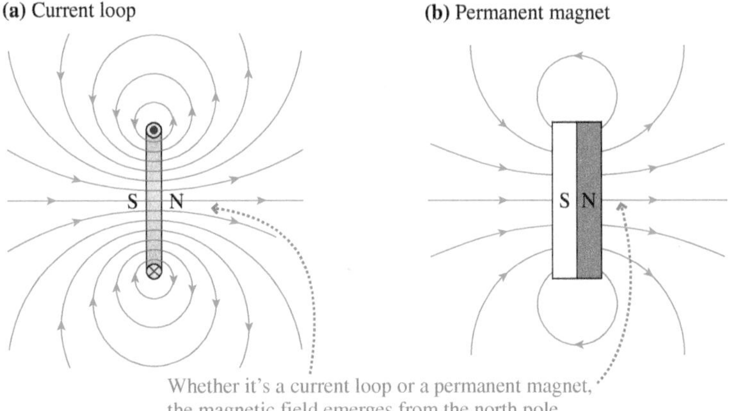

(a) Current loop
(b) Permanent magnet
S
N
S
N
Whether it's a current loop or a permanent magnet, the magnetic field emerges from the north pole.

One of the goals of this chapter is to show that magnetic forces exerted by currents and magnetic forces exerted by permanent magnets are just two different aspects of a single magnetism. This connection between permanent magnets and current loops will turn out to be a big piece of the puzzle.

The Magnetic Dipole Moment

The expression for the electric field of an electric dipole was considerably simplified when we considered the field at distances significantly larger than the size of the charge separation s. We'll let the axis of the dipole be the z-axis. The on-axis field of an electric dipole at distance z when $z \gg s$ was found to be

$$\vec{E}_{\text{dipole}} = \frac{1}{4\pi\epsilon_0}\frac{2\vec{p}}{z^3}$$

where the electric dipole moment $\vec{p} = (qs$, from negative to positive charge).

We can write a similar expression for the on-axis magnetic field of a magnetic dipole. We define the **magnetic dipole moment** of a current loop enclosing area A as

$$\vec{m} = (AI, \text{ from the south pole to the north pole}) \qquad (26.12)$$

We've looked at circular current loops, but the shape of the loop does not matter; the magnetic dipole moment depends on only the enclosed area. The SI units of the magnetic dipole moment are $\text{A} \cdot \text{m}^2$.

The magnetic dipole moment, like the electric dipole moment, is a vector. It has the same direction as the on-axis magnetic field. Thus the right-hand rule for determining the direction of $\vec{B}$ also shows the direction of $\vec{m}$. FIGURE 26.16 shows the magnetic dipole moment of a circular current loop.

FIGURE 26.16 The magnetic dipole moment of a circular current loop.

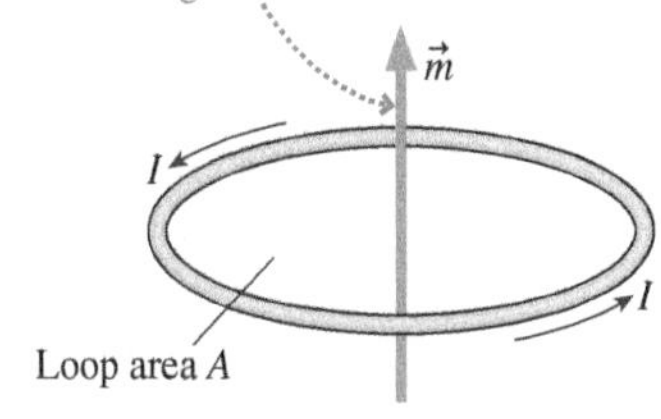

A more advanced calculation, which we will omit, finds that the on-axis field of a magnetic dipole at distances $z \gg R$, the loop size, is

$$\vec{B}_{\text{dipole}} = \frac{\mu_0}{4\pi}\frac{2\vec{m}}{z^3} \tag{26.13}$$

If you compare $\vec{B}_{\text{dipole}}$ to $\vec{E}_{\text{dipole}}$, you can see that the magnetic field of a magnetic dipole is very similar to the electric field of an electric dipole.

A permanent magnet also has a magnetic dipole moment, and its on-axis magnetic field is given by Equation 26.13 when z is much larger than the size of the magnet. Equation 26.13 and laboratory measurements of the on-axis magnetic field can be used to determine a permanent magnet's dipole moment. This magnetic dipole moment cannot be interpreted as the product of a current and an area; instead, it's a property of the atoms that make up the magnet, as we'll discuss shortly.

An especially important magnetic dipole moment is that of an electron. You may recall from chemistry that an electron has a property called *spin* and that the electron's spin is restricted to being either "up" or "down." These two states of the electron are sometimes represented by the *spin quantum number* m_s, which can assume only two values: $m_s = +\frac{1}{2}$ (spin up) and $m_s = -\frac{1}{2}$ (spin down). Other chemistry texts use arrows ↑ and ↓ to indicate the direction of the electron spin. Either way, electron spin plays an important role in the electron configuration of atoms because the Pauli exclusion principle—which we'll discuss in Chapter 29—states that an atomic orbital can hold only two electrons, which must have opposite spins.

The term "spin" may convey the wrong image, an image of a little spinning ball. In fact, the electron is *not* spinning. Instead, the electron has an inherent magnetic dipole moment *as if* it were a spinning ball of charge, but the term is a poetic rather than a literal description. An electron has a mass that allows it to interact with gravitational fields, and a charge that allows it to interact with electric fields. Mass and charge are simply two properties of electrons. Its magnetic dipole moment is a third property, one that allows an electron to interact with a magnetic field.

FIGURE 26.17 The magnetic dipole moment of an electron.

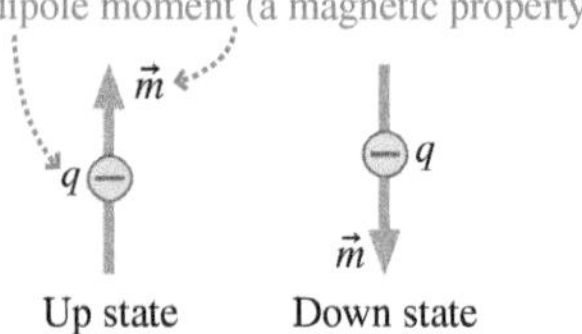
An electron has both a charge (an electric property) and a magnetic dipole moment (a magnetic property).
$\vec{m}$
q
q
$\vec{m}$
Up state
Down state

For reasons associated with quantum physics, the electron's magnetic dipole moment vector can point only up or down along an axis, as illustrated in FIGURE 26.17, and these two orientations give rise to the notion of "spin up" and "spin down" electrons. We'll return to the magnetic dipole moments of electrons and protons later in the chapter when we explore nuclear magnetic resonance and magnetic resonance imaging. For now, we simply note that each electron is a tiny magnet that creates a magnetic field. Consequently, unpaired electrons on neighboring atoms interact magnetically, much like the interaction between two bar magnets. These magnetic interactions are responsible for the features of an NMR spectrum that are used to determine molecular structure.

EXAMPLE 26.5 Measuring current in a superconducting loop

You'll learn in Chapter 27 that a current can be *induced* in a closed loop of wire. If the loop happens to be made of a superconducting material, with zero resistance, the induced current will—in principle—persist forever. The current cannot be measured with an ammeter because any real ammeter has resistance that will quickly stop the current. Instead, physicists measure the persistent current in a superconducting loop by measuring its magnetic field. What is the current in a 3.0-mm-diameter superconducting loop if the axial magnetic field is 9.0 μT at a distance of 2.5 cm?

PREPARE The measurements are made far enough from the loop in comparison to its radius ($z \gg R$) that we can model the loop as a magnetic dipole.

SOLVE The axial magnetic field strength of a dipole is

$$B = \frac{\mu_0}{4\pi}\frac{2m}{z^3} = \frac{\mu_0}{4\pi}\frac{2\pi R^2 I}{z^3} = \frac{\mu_0 R^2 I}{2}\frac{1}{z^3}$$

where we used $m = AI = \pi R^2 I$ for the magnetic dipole moment of a circular loop of radius R. Thus the current is

$$I = \frac{2z^3 B}{\mu_0 R^2} = \frac{2(0.025\text{ m})^3(9.0\times10^{-6}\text{ T})}{(1.26\times10^{-6}\text{ T}\cdot\text{m/A})(0.0015\text{ m})^2} = 99\text{ A}$$

ASSESS This would be a very large current for ordinary wire. An important property of superconducting wires is their ability to carry current that would melt an ordinary wire.

STOP TO THINK 26.4 What is the current direction in this loop? And which side of the loop is the north pole?

A. Current cw; north pole on top
B. Current cw; north pole on bottom
C. Current ccw; north pole on top
D. Current ccw; north pole on bottom

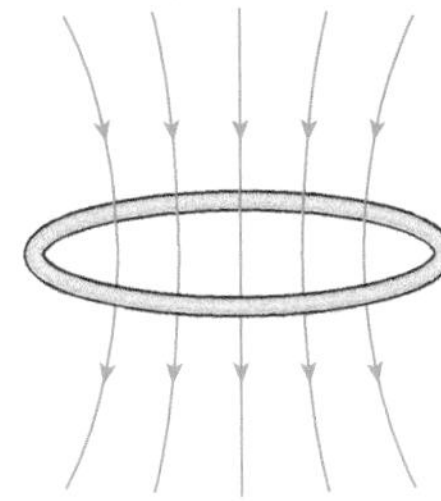

The Magnetic Field of the Earth

The north pole of a compass needle is attracted toward the north pole of the earth. Apparently the earth itself is a large magnet, as shown in FIGURE 26.18a. The reasons for the earth's magnetism are complex, but geophysicists think that the earth's magnetic poles arise from circulating currents in its molten iron core that create a magnetic dipole moment. Two interesting facts about the earth's magnetic field are (1) that the magnetic poles are offset slightly from the geographic poles of the earth's rotation axis, and (2) that the *geomagnetic north pole,* where a compass points, is actually a *south* magnetic pole! You should be able to use what you have learned thus far to convince yourself that this is the case.

FIGURE 26.18 The earth's magnetic field.

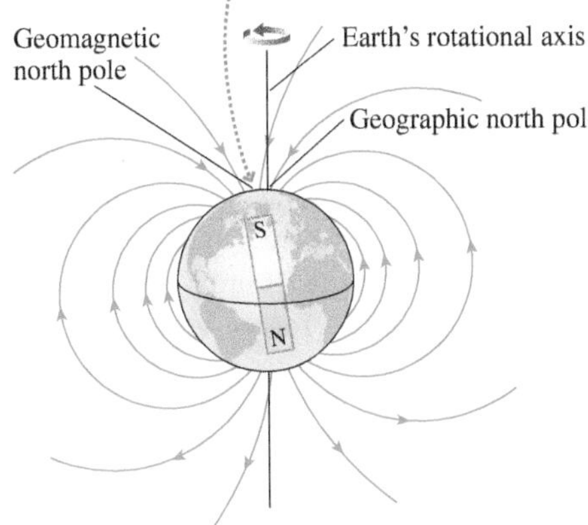
(a)
Geomagnetic north is actually the south magnetic pole of the earth's magnet.
Geomagnetic north pole
Earth's rotational axis
Geographic north pole
S
N

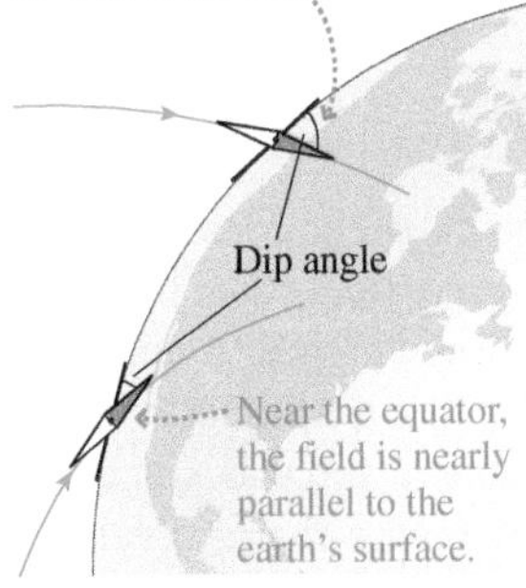
(b)
Near the poles, the field is tipped at a large angle with respect to the earth's surface.
Dip angle
Near the equator, the field is nearly parallel to the earth's surface.

We can see from Figure 26.18a that the earth's magnetic field is very nearly that of a magnetic dipole, which makes sense if the field arises from currents circulating deep in the earth's interior. Measurements find that the strength of the earth's magnetic dipole moment is $m \approx 8 \times 10^{22}\ \text{A} \cdot \text{m}^2$.

Notice that the magnetic field has components both parallel to the ground (horizontal) and perpendicular to the ground (vertical). An ordinary compass responds to only the horizontal component of the field, but a compass needle tilts downward if it is allowed to pivot freely in all directions. The angle of the downward tip—that is, the field's angle below horizontal—is called the **dip angle.** FIGURE 26.18b shows that the dip angle is very small near the equator, where the magnetic field is almost parallel to the surface, but the field becomes almost vertical at locations near the north pole. Sea turtles appear to use the dip angle of the earth's magnetic field to determine their latitude.

Other animals also use the earth's magnetic field to navigate. Some animals, such as birds, seem to have a built-in compass—although biologists are still unsure

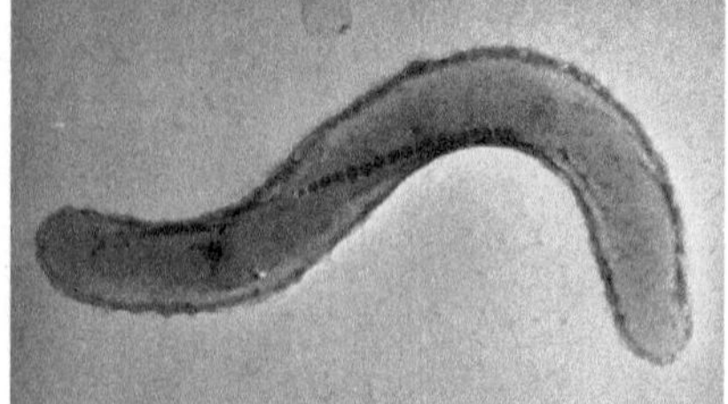

FIGURE 26.19 The black dots are a line of magnetized iron nanoparticles that align a magnetotactic bacterium with the earth's magnetic field.

how it operates—that, along with the angle of the sun, helps them to migrate in the right direction.

Some of the most well studied organisms are *magnetotactic bacteria* that use the earth's field not to distinguish north from south but, by using the dip angle, to tell up from down. These bacteria live in shallow waters where the oxygen concentration is highly stratified. A magnetotactic bacterium contains a line of nanometer-sized microscopic bits of magnetized iron crystals, which can be seen in FIGURE 26.19. A magnetic torque on this chain aligns the bacterium with the earth's magnetic field, as if it were a compass needle, and the bacterium then swims downward along the field lines until it reaches its preferred oxygen concentration.

26.4 The Magnetic Force on a Moving Charge

It's time to switch our attention from how magnetic fields are generated to how magnetic fields exert forces and torques. Oersted discovered that a current passing through a wire causes a magnetic torque to be exerted on a nearby compass needle. Upon hearing of Oersted's discovery, the French scientist André-Marie Ampère, for whom the SI unit of current is named, reasoned that the current was acting like a magnet and, if this were true, that two current-carrying wires should exert magnetic forces on each other.

To find out, Ampère set up two parallel wires that could carry large currents either in the same direction or in opposite (or "antiparallel") directions. FIGURE 26.20 shows the outcome of his experiment. Notice that, for currents, "likes" attract and "opposites" repel. This is the opposite of what would have happened had the wires been charged and thus exerting electric forces on each other.

FIGURE 26.20 The forces between parallel current-carrying wires.

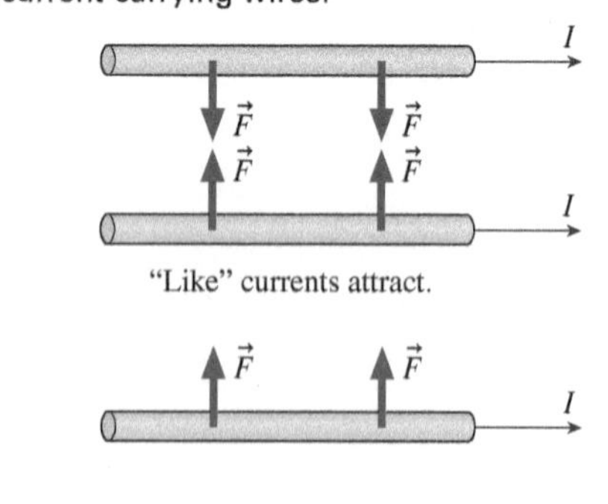

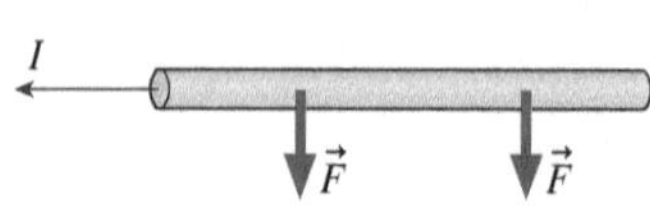

Magnetic Force

The magnetic force between currents has many important applications, from motors to generators. Our investigation of this force begins more simply, with individual moving charges. A current consists of moving charges, so Ampère's experiment implies that **a magnetic field exerts a force on a *moving* charge.** It turns out that the magnetic force is somewhat more complex than the electric force, depending not only on the charge's velocity but also on how the velocity vector is oriented relative to the magnetic field. Consider the following experiments:

Investigating the magnetic force on a charged particle

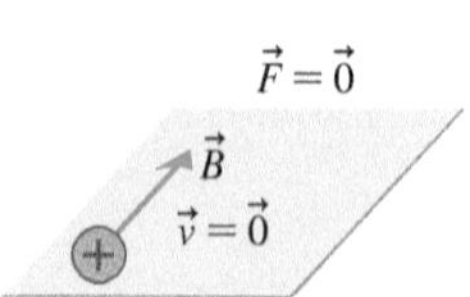

There is no magnetic force on a charged particle at rest.

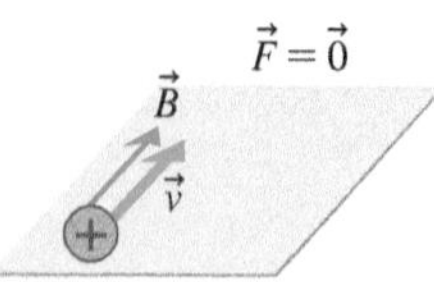

There is no magnetic force on a charged particle moving *parallel* to a magnetic field.

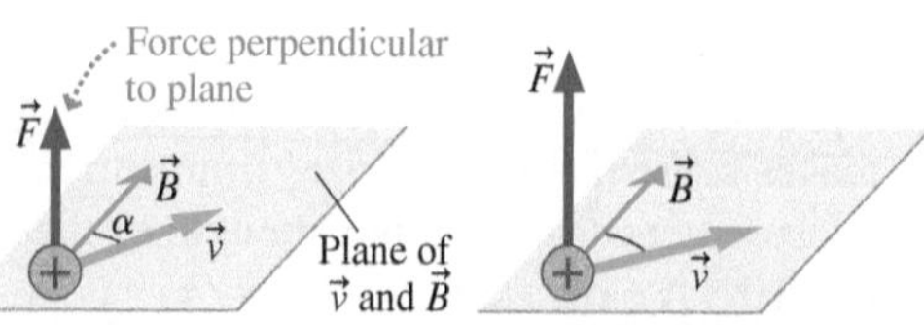

As the angle α between the velocity and the magnetic field increases, the magnetic force also increases. The force is greatest when the angle is 90°. The magnetic force is always perpendicular to the plane containing $\vec{v}$ and $\vec{B}$.

These experiments show that the magnetic force on a charged particle is quite different from the electric force. We see that:

- Only a *moving* charge experiences a magnetic force. There is no magnetic force on a charge at rest ($\vec{v} = \vec{0}$) in a magnetic field.
- There is no force on a charge moving parallel ($\alpha = 0°$) or antiparallel ($\alpha = 180°$) to a magnetic field.

- When there is a force, the force is perpendicular to the plane containing $\vec{v}$ and $\vec{B}$.
- The force is a maximum when $\vec{v}$ is perpendicular to $\vec{B}$.

The magnetic force on a moving charged particle is perpendicular to the plane that contains $\vec{v}$ and $\vec{B}$, but there are two such directions. We can determine the correct force direction with another right-hand rule: the *right-hand rule for forces.*

TACTICS BOX 26.3 Right-hand rule for forces

1. Spread the fingers of your right hand so that your middle finger is perpendicular to your palm.
2. Rotate your hand until your thumb points in the direction of $\vec{v}$ and your index finger points in the direction of $\vec{B}$.
3. Your middle finger now points in the direction of the force $\vec{F}$ on a positive charge. The force on a negative charge points in the opposite direction.

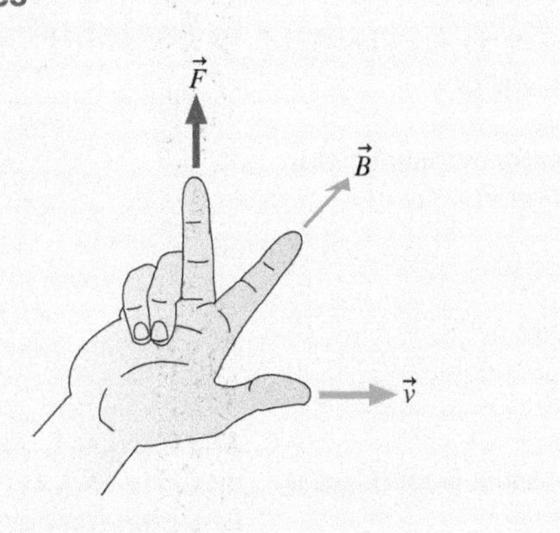

NOTE ▶ The right-hand rule for forces is different from the right-hand rule for finding the direction of a magnetic field. ◀

We can organize all the experimental information about the magnetic force on a moving charged particle in a single equation. The magnetic force on a moving charged particle is

$$\vec{F}_{\text{on } q} = (|q|vB\sin\alpha, \text{ direction of the right-hand rule for forces}) \tag{26.14}$$

Magnetic force on a moving charged particle

where α is the angle between $\vec{v}$ and $\vec{B}$. Equation 26.14 says that the force is zero if $v = 0$. It is also zero if the charge moves parallel to $(\alpha = 0°)$ or opposite to $(\alpha = 180°)$ the field. This is consistent with our observations. The magnetic force reaches a maximum value

$$F_{\text{max}} = |q|vB \tag{26.15}$$

when the charge's motion is perpendicular to the magnetic field $(\alpha = 90°)$.

FIGURE 26.21 shows the relationship among $\vec{v}$, $\vec{B}$, and $\vec{F}$ for four moving charges. (The *source* of the magnetic field isn't shown, only the field itself.) You can see the inherent three-dimensionality of magnetism, with the force perpendicular to both $\vec{v}$ and $\vec{B}$. The magnetic force is very different from the electric force, which is parallel to the electric field.

FIGURE 26.21 Magnetic forces on four moving charges.

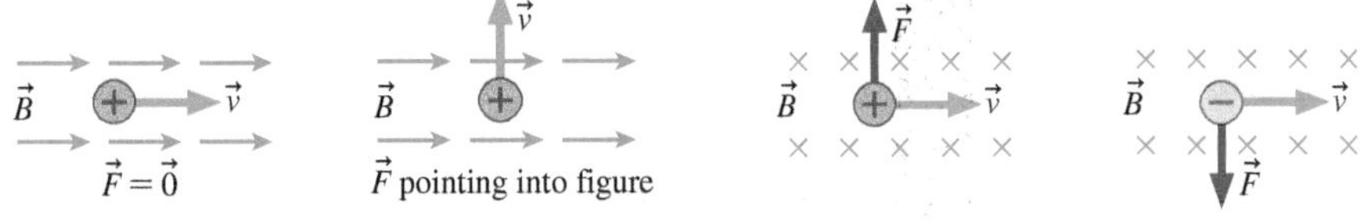

CONCEPTUAL EXAMPLE 26.6 **Earth's magnetic field and a digital compass**

Your smartphone probably has a built-in digital compass. There's no actual pivoting needle, of course. Instead, a sensor in your phone detects the earth's magnetic field by the force it exerts on electrons moving in a conductor. FIGURE 26.22 shows one of these electrons moving to the right when the earth's field points in the direction shown. What is the direction of the magnetic force?

FIGURE 26.22 An electron moving in the earth's magnetic field.

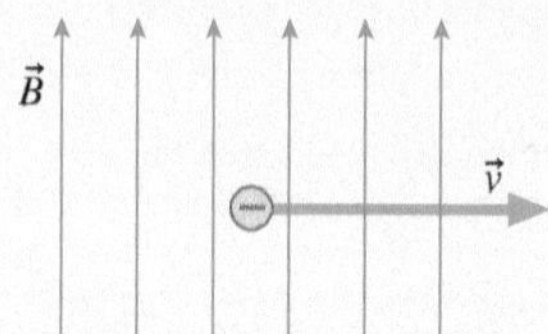

REASON FIGURE 26.23 shows how the right-hand rule for forces is applied to this situation:

- Point your right thumb in the direction of the electron's velocity and your index finger in the direction of the magnetic field.
- Bend your middle finger to be perpendicular to your index finger. Your middle finger, which now points out of the figure, is the direction of the force on a positive charge. But the electron is negative, so the force on the electron is *into* the figure.

FIGURE 26.23 Using the right-hand rule.

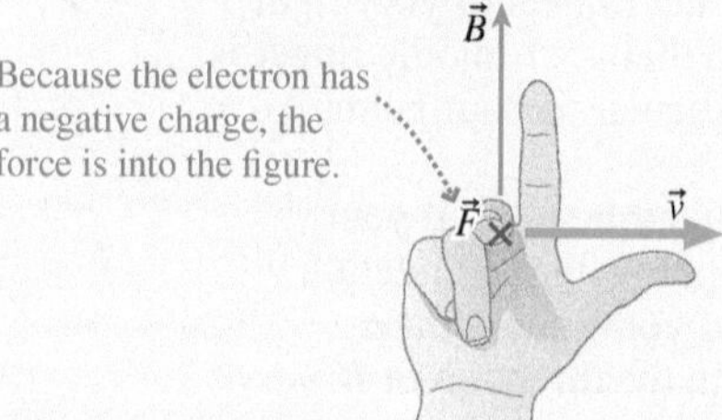

ASSESS The force is perpendicular to both the velocity and the magnetic field, as it must be. The force on an electron is into the figure; the force on a proton would be out of the figure.

EXAMPLE 26.7 **The magnetic force on an electron**

A long wire carries a 10 A current from left to right. An electron 1.0 cm above the wire is traveling to the right at a speed of 1.0×10^7 m/s. What are the magnitude and the direction of the magnetic force on the electron at this instant?

PREPARE FIGURE 26.24 shows the current and an electron moving to the right. The magnetic field is that of a long, straight wire. The right-hand rule for fields tells us that the wire's magnetic field above the wire is out of the figure, so the electron is moving perpendicular to the field.

FIGURE 26.24 An electron moving parallel to a current-carrying wire.

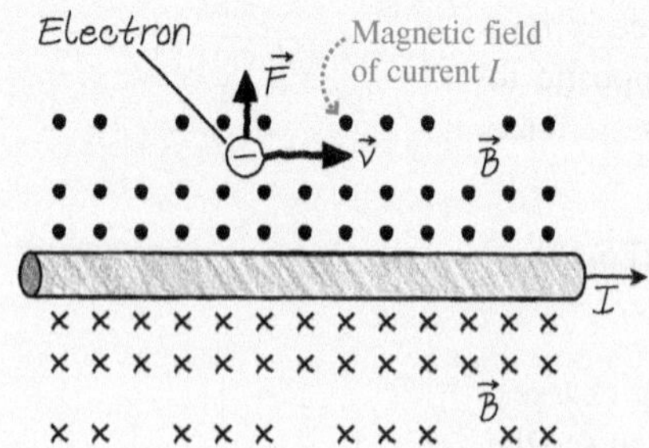

SOLVE If you point your thumb in the direction of $\vec{v}$ and your index finger in the direction of $\vec{B}$, your middle finger points down. But the electron is negative, so the direction of the magnetic force is up, away from the wire. The magnitude of the force is $|q|vB = evB$. The field is at distance $r = 0.010$ m from a long, straight wire:

$$B = \frac{\mu_0 I}{2\pi r} = 2.0 \times 10^{-4}\ \text{T}$$

Thus the magnitude of the force on the electron is

$$F = evB = (1.60 \times 10^{-19}\ \text{C})(1.0 \times 10^7\ \text{m/s})(2.0 \times 10^{-4}\ \text{T})$$
$$= 3.2 \times 10^{-16}\ \text{N}$$

The force on the electron is $\vec{F} = (3.2 \times 10^{-16}\ \text{N, up})$.

ASSESS This force will cause the electron to curve away from the wire.

Cyclotron Motion

Many important applications of magnetism involve the motion of charged particles in a magnetic field. Older television picture tubes use magnetic fields to steer electrons through a vacuum from the electron gun to the screen. Microwave generators, which are used in applications ranging from ovens to radar, use a device called a *magnetron* in which electrons oscillate rapidly in a magnetic field.

You've just seen that there is no force on a charge that has velocity $\vec{v}$ parallel or antiparallel to a magnetic field. Consequently, **a magnetic field has no effect on a charge**

moving parallel or antiparallel to the field. To understand the motion of charged particles in magnetic fields, we need to consider only motion *perpendicular* to the field.

FIGURE 26.25 shows a positive charge q moving with a velocity $\vec{v}$ in a plane that is perpendicular to a *uniform* magnetic field $\vec{B}$. According to the right-hand rule, the magnetic force on this particle is *perpendicular* to the velocity $\vec{v}$. A force that is always perpendicular to $\vec{v}$ changes the *direction* of motion, by deflecting the particle sideways, but it cannot change the particle's speed. Thus **a particle moving perpendicular to a uniform magnetic field undergoes uniform circular motion at constant speed.** This motion is called the **cyclotron motion** of a charged particle in a magnetic field.

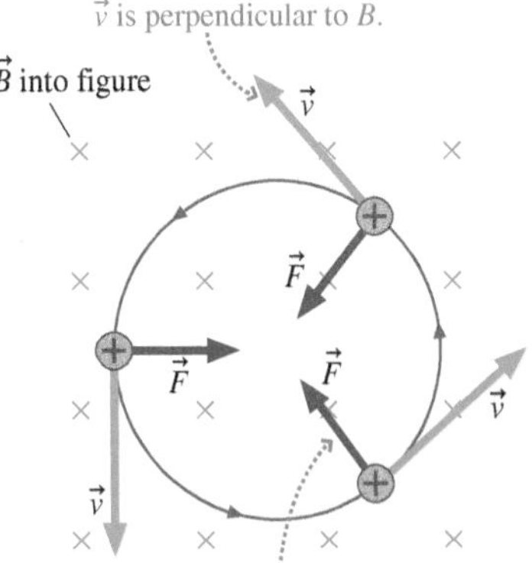

$\vec{v}$ is perpendicular to $\vec{B}$.
$\vec{B}$ into figure
$\vec{v}$
$\vec{F}$
$\vec{F}$
$\vec{F}$
$\vec{v}$
$\vec{v}$
The magnetic force is always perpendicular to $\vec{v}$, causing the particle to move in a circle.

FIGURE 26.25 Cyclotron motion of a charged particle moving in a uniform magnetic field.

NOTE ▶ A negative charge will orbit in the opposite direction from that shown in Figure 26.25 for a positive charge. ◀

You've seen many analogies to cyclotron motion earlier in this text. For a mass moving in a circle at the end of a string, the tension force is always perpendicular to $\vec{v}$. For a satellite moving in a circular orbit, the gravitational force is always perpendicular to $\vec{v}$. Now, for a charged particle moving in a magnetic field, it is the magnetic force of strength $F = qvB$ that points toward the center of the circle and causes the particle to have a centripetal acceleration.

Newton's second law for circular motion, which you learned in Chapter 7, is

$$F = qvB = ma_r = \frac{mv^2}{r} \tag{26.16}$$

Thus the radius of the cyclotron orbit is

$$r_{\text{cyc}} = \frac{mv}{qB} \tag{26.17}$$

The inverse dependence on B indicates that the size of the orbit can be decreased by increasing the magnetic field strength.

We can also determine the frequency of the cyclotron motion. Recall from your earlier study of circular motion that the frequency of revolution f is related to the speed and radius by $f = v/2\pi r$. A rearrangement of Equation 26.17 gives the **cyclotron frequency:**

$$f_{\text{cyc}} = \frac{qB}{2\pi m} \tag{26.18}$$

Cyclotron motion frequency

where the ratio q/m is the particle's *charge-to-mass ratio.* Notice that the cyclotron frequency depends on the charge-to-mass ratio and the magnetic field strength but *not* on the charge's speed.

Going in circles Electrons undergoing circular motion in a magnetic field created by two current loops. You can see the electrons' path because they collide with a low-density gas that then emits light.

EXAMPLE 26.8 The radius of cyclotron motion

In FIGURE 26.26, an electron is accelerated from rest through a potential difference of 500 V, then injected into a uniform magnetic field. Once in the magnetic field, it completes half a revolution in 2.0 ns. What is the radius of its orbit?

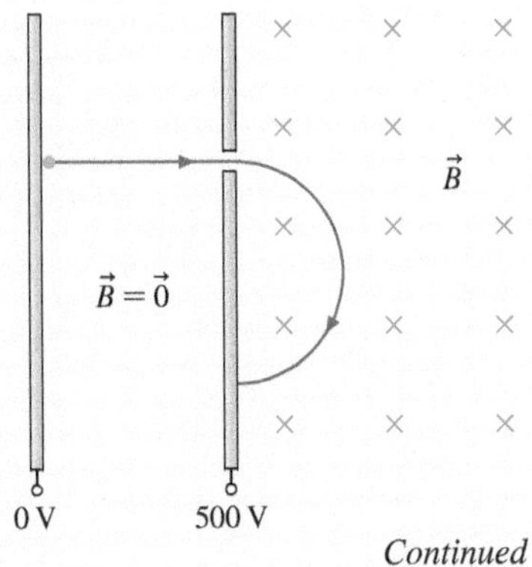

$\vec{B}$
$\vec{B} = \vec{0}$
0 V
500 V

FIGURE 26.26 An electron is accelerated, then injected into a magnetic field.

PREPARE Energy is conserved as the electron is accelerated by the potential difference. The electron then undergoes cyclotron motion in the magnetic field, although it completes only half a revolution before hitting the back of the acceleration electrode.

SOLVE The electron accelerates from rest ($v_i = 0$ m/s) at $V_i = 0$ V to speed v_f at $V_f = 500$ V. We can use conservation of

Continued

energy $K_f + qV_f = K_i + qV_i$ to find the speed v_f with which it enters the magnetic field:

$$\tfrac{1}{2}mv_f^2 + (-e)V_f = 0 + 0$$

$$v_f = \sqrt{\frac{2eV_f}{m}} = \sqrt{\frac{2(1.60 \times 10^{-19}\,\text{C})(500\,\text{V})}{9.11 \times 10^{-31}\,\text{kg}}}$$

$$= 1.33 \times 10^7\,\text{m/s}$$

The cyclotron radius in the magnetic field is $r_{cyc} = mv/eB$, but we first need to determine the field strength. Were it not for the electrode, the electron would undergo circular motion with period $T = 4.0$ ns. Hence the cyclotron frequency is $f = 1/T = 2.5 \times 10^8$ Hz. We can use the cyclotron frequency to determine that the magnetic field strength is

$$B = \frac{2\pi m f_{cyc}}{e} = \frac{2\pi(9.11 \times 10^{-31}\,\text{kg})(2.50 \times 10^8\,\text{Hz})}{1.60 \times 10^{-19}\,\text{C}}$$

$$= 8.94 \times 10^{-3}\,\text{T}$$

Thus the radius of the electron's orbit is

$$r_{cyc} = \frac{mv}{qB} = 8.5 \times 10^{-3}\,\text{m} = 8.5\,\text{mm}$$

ASSESS A 17-mm-diameter orbit is similar to what is seen in the photo just before this example, so this seems to be a typical size for electrons moving in modest magnetic fields.

FIGURE 26.27a shows a more general situation in which the charged particle's velocity $\vec{v}$ is neither parallel nor perpendicular to $\vec{B}$. The component of $\vec{v}$ parallel to $\vec{B}$ is not affected by the field, so the charged particle spirals around the magnetic field lines in a helical trajectory. The radius of the helix is determined by $\vec{v}_\perp$, the component of $\vec{v}$ perpendicular to $\vec{B}$

The motion of charged particles in a magnetic field is responsible for the earth's aurora. High-energy particles and radiation streaming out from the sun, called the *solar wind,* create ions and electrons as they strike molecules high in the atmosphere. Some of these charged particles become trapped in the earth's magnetic field, creating what is known as the *Van Allen radiation belt.*

As FIGURE 26.27b shows, the electrons spiral along the magnetic field lines until the field leads them into the atmosphere. The shape of the earth's magnetic field is such that most electrons enter the atmosphere near the magnetic poles. There they collide with oxygen and nitrogen atoms, exciting the atoms and causing them to emit auroral light, as seen in FIGURE 26.27c.

FIGURE 26.27 In general, charged particles spiral along helical trajectories around the magnetic field lines. This motion is responsible for the earth's aurora.

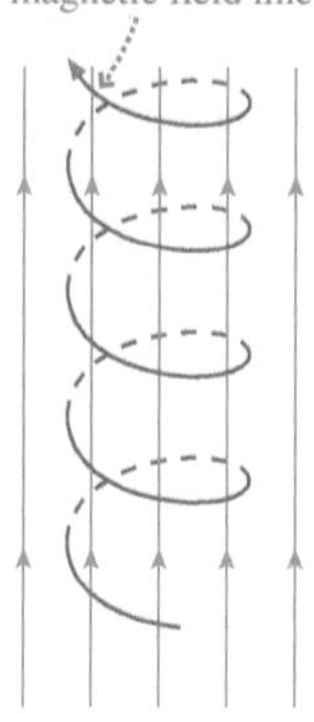
(a) Charged particles spiral around the magnetic field lines.

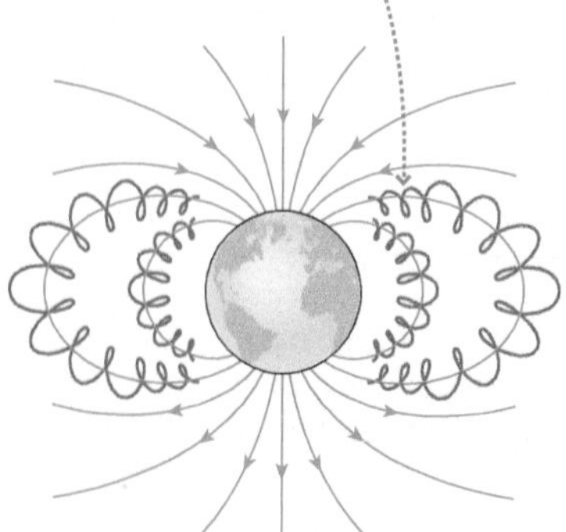
(b) The earth's magnetic field leads particles into the atmosphere near the poles, causing the aurora.

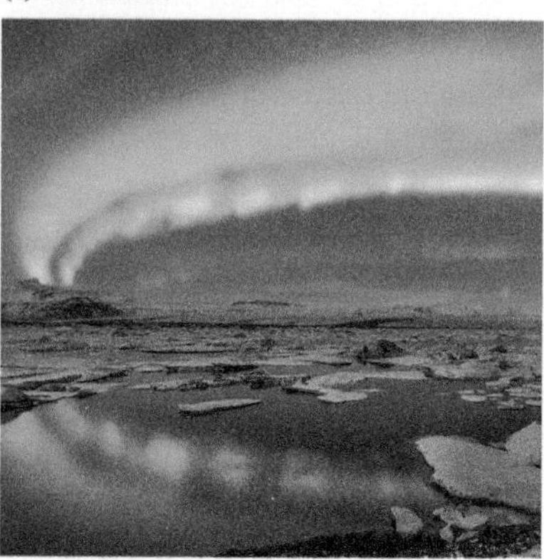
(c) The aurora

STOP TO THINK 26.5 An electron moves perpendicular to a magnetic field. What is the direction of $\vec{B}$?

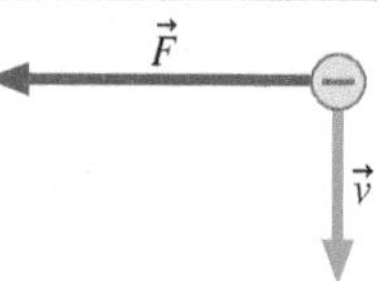
$\vec{F}$
$\vec{v}$

A. Left B. Up C. Into the figure
D. Right E. Down F. Out of the figure

The Mass Spectrometer

Chemists and biochemists who study the structure of molecules use a device called a **mass spectrometer.** There are several different types of mass spectrometers that use different combinations of electric and magnetic forces to separate ions, but all produce similar mass spectra. We'll describe the *magnetic mass spectrometer* shown in FIGURE 26.28.

FIGURE 26.28 A magnetic mass spectrometer.

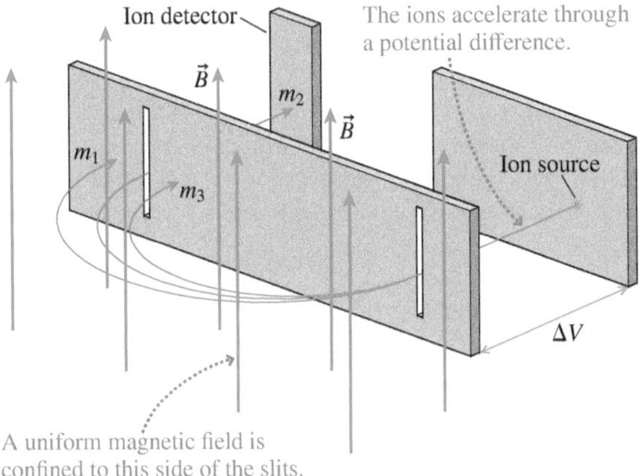

The first step for any mass spectrometer is to create positive ions, usually by bombarding a molecular sample with high-speed electrons. The collisions both ionize the molecules and *fragment* the molecules into smaller ionized pieces. For example, ionization of methanol (CH_3OH) produces the methanol ion CH_3OH^+ as well as fragment molecular ions such as CH_2OH^+, COH^+, and CH_3^+. The fragmentation pattern provides researchers with clues about how the molecule is structured.

Next, the mass spectrometer accelerates the ions through a potential difference ΔV of several hundred volts, and then the ions enter a uniform magnetic field where they undergo cyclotron motion. Ions undergo only half a revolution, however, before either colliding with a barrier or exiting through a narrow slit and being detected by an ion detector. The radius of an ion's cyclotron orbit depends on its mass, so only one type of ion is detected. Figure 26.28 shows ions with mass m_2 being detected; heavier ions with mass m_1 and lighter ions with mass m_3 strike the barrier and are not detected. (We assume that all ions have lost one electron and have charge $q = +e$.)

As the experimenter slowly changes the accelerating voltage ΔV, which alters the speeds of the ions as they enter the magnetic field, ions of different mass are scanned across the exit slit and detected. A record of the ion detection rate as a function of the accelerating voltage is a *mass spectrum.* FIGURE 26.29 is a mass spectrum of methanol. The horizontal axis shows the masses of the ions in amu. Methanol has a molecular

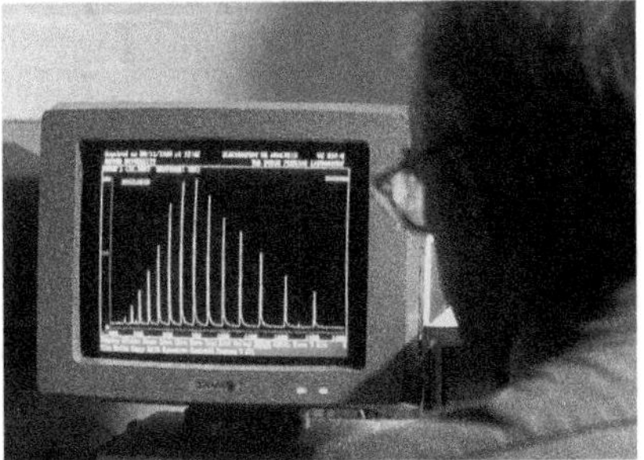

Decoding proteins The researcher is observing the mass spectrum of myoglobin. Mass spectroscopy is used along with other techniques to deduce the structures of biomolecules.

FIGURE 26.29 The mass spectrum of methanol.

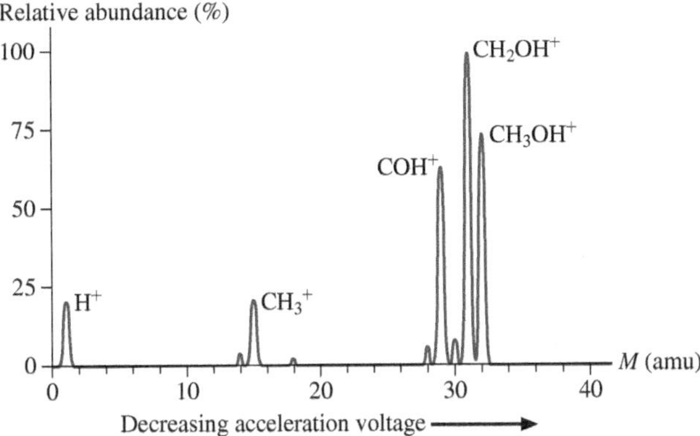

mass of 32 u, and the methanol ion CH_3OH^+ is the heaviest mass in the spectrum. A peak in the spectrum at 15 u corresponds to a fragment with this mass, and CH_3^+ is the only possibility.

EXAMPLE 26.9 Detecting ions

The mass spectrum in Figure 26.29 has a small peak at mass 30 u. This ion is detected when the accelerating voltage is 535 V. At what accelerating voltage is the CH_3OH^+ ion detected?

PREPARE The molecular mass of the CH_3OH^+ ion is 32 u. We need to determine how the mass of the detected ion depends on the accelerating voltage. The ion that is detected has a cyclotron radius $r = R$, where R is half the distance between the entrance slit and the exit slit. We will assume that all ions are singly charged with $q = +e$ and that all start from rest.

SOLVE The cyclotron radius of the detected ions is $R = mv/eB$. The speed v with which the ion enters the magnetic field is determined by the accelerating voltage. We can use conservation of energy $K_f + eV_f = K_i + eV_i$ along with $K_i = 0$ to write

$$\tfrac{1}{2}mv^2 + eV_f = 0 + eV_i$$

Thus

$$\tfrac{1}{2}mv^2 = e(V_i - V_f) = e\Delta V$$

Solving for v, we find

$$v = \left(\frac{2e\Delta V}{m}\right)^{1/2}$$

Thus the cyclotron radius of the detected ions is

$$R = \frac{m}{eB}v = \frac{m}{eB}\left(\frac{2e\Delta V}{m}\right)^{1/2} = \left(\frac{2m\Delta V}{eB^2}\right)^{1/2}$$

We can solve for ΔV to find the accelerating voltage needed to detect ions of mass m:

$$\Delta V = \frac{eR^2B^2}{2m} = \frac{C}{m}$$

where C is a proportionality constant that depends on the design of the mass spectrometer. In other words, the required accelerating voltage is inversely proportional to the mass of the ion to be detected. We don't need to know C if we use ratios:

$$\frac{\Delta V_{32}}{\Delta V_{30}} = \frac{C/32}{C/30} = \frac{30}{32}$$

Consequently, the accelerating voltage needed to detect the CH_3OH^+ ion is

$$\Delta V_{32} = \frac{30}{32}\Delta V_{30} = \frac{30}{32}(535\text{ V}) = 502\text{ V}$$

ASSESS At 535 V, when ions that have a mass of 30 u are detected, the momentum of the heavier CH_3OH^+ ion carries it along a cyclotron orbit with a larger radius. It strikes the barrier at the far side of the exit slit. To be detected, the ion needs a reduced momentum and a smaller orbital radius. That requires a smaller accelerating voltage.

Go with the flow Many scientists and resource managers rely on accurate measurements of stream flows. The easiest way to get a quick measurement of the speed of a river or creek is to use an electromagnetic flowmeter. Water flows between the poles of a strong magnet. Two electrodes measure the resulting potential difference, which is proportional to the flow speed.

Electromagnetic Flowmeters

Blood contains many kinds of ions, such as Na^+ and Cl^-. When blood flows through a vessel, these ions move with the blood. An applied magnetic field exerts a force on these moving charges. We can use this principle to make a completely noninvasive device for measuring the blood flow in an artery: an *electromagnetic flowmeter.*

A flowmeter probe clamped to an artery has two active elements: magnets that apply a strong field across the artery and electrodes that contact the artery on opposite sides, as shown in FIGURE 26.30. The blood flowing in an artery carries a mix of positive and negative ions. Because these ions are in motion, the magnetic field exerts a force

FIGURE 26.30 The operation of an electromagnetic flowmeter.

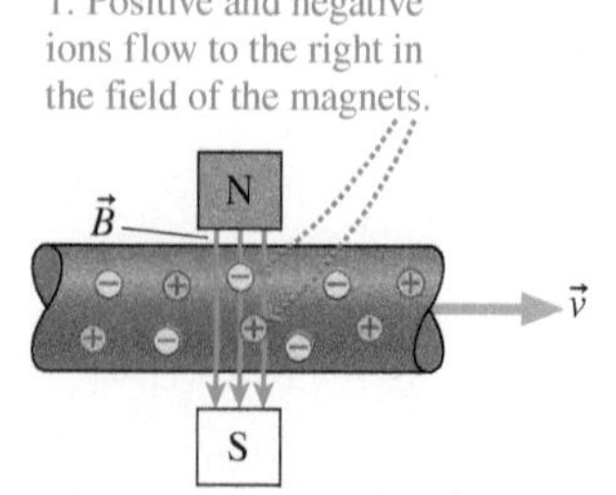
1. Positive and negative ions flow to the right in the field of the magnets.
N
$\vec{B}$
$\vec{v}$
S

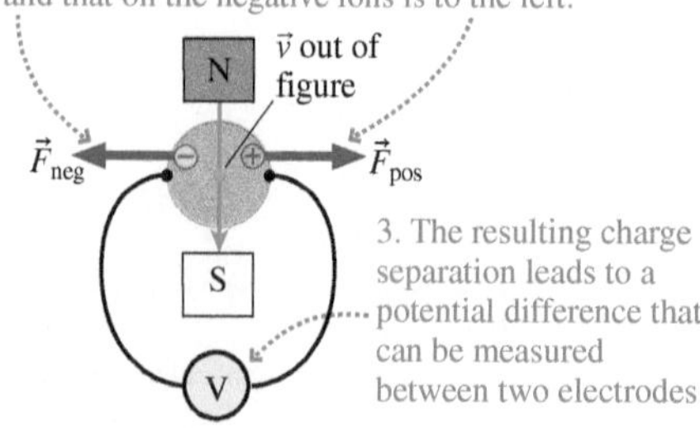
2. An end view shows that the magnetic force on the positive ions is to the right, and that on the negative ions is to the left.
N
$\vec{v}$ out of figure
$\vec{F}_{neg}$
$\vec{F}_{pos}$
S
V
3. The resulting charge separation leads to a potential difference that can be measured between two electrodes.

on them that produces a measurable voltage. We know from Equation 26.14 that the faster the blood's ions are moving, the greater the forces separating the positive and negative ions. The greater the forces, the greater the degree of charge separation and the higher the voltage. The measured voltage is therefore directly proportional to the velocity of the blood.

STOP TO THINK 26.6 These charged particles are traveling in circular orbits with velocities and field directions as noted. Which particles have a negative charge?

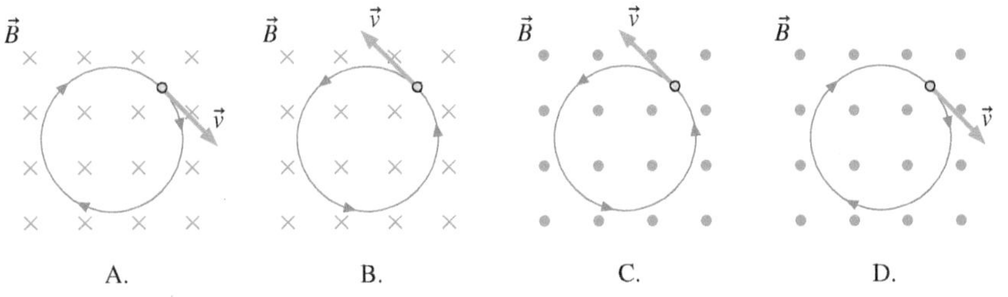
$\vec{B}$
$\vec{v}$
A.
B.
C.
D.

26.5 Magnetic Forces on Current-Carrying Wires

Ampère's observation of magnetic forces between current-carrying wires motivated us to look at the magnetic forces on moving charges. We're now ready to apply that knowledge to Ampère's experiment. As a first step, let us find the force exerted by a uniform magnetic field on a long, straight wire carrying current I through the field. As FIGURE 26.31a shows, there's *no* force on a current-carrying wire *parallel* to a magnetic field. This shouldn't be surprising; it follows from the fact that there is no force on a charged particle moving parallel to $\vec{B}$.

FIGURE 26.31b shows a wire *perpendicular* to the magnetic field. By the right-hand rule, each charge in the current has a force of magnitude qvB directed to the left. Consequently, the entire length of wire within the magnetic field experiences a force to the left, perpendicular to both the current direction and the field direction.

FIGURE 26.31 Magnetic force on a current-carrying wire.

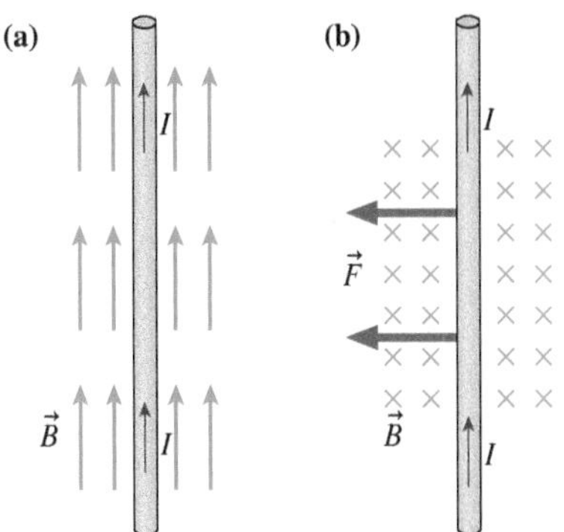
(a)
(b)
I
$\vec{B}$
$\vec{F}$
There's no force on a current parallel to a magnetic field.
There is a magnetic force in the direction of the right-hand rule.

A current is moving charge, and the magnetic force on a current-carrying wire is simply the net magnetic force on all the charge carriers in the wire. FIGURE 26.32 shows a wire carrying current I in which the charge carriers move with speed v. Suppose a small segment of the wire of length Δx contains charge Δq that moves through this segment in time interval Δt. The magnetic force on this charge is $F = \Delta q v B$. If we multiply and divide by Δt, the force on the moving charge in this segment is

$$F = \Delta q v B = \frac{\Delta q}{\Delta t}(v\Delta t)B = I\Delta x B \qquad (26.19)$$

because $\Delta q/\Delta t$ is the current I and $v\Delta t$ is the distance traveled Δx.

If we add the forces on all the small segments in a length l of the wire, we find that the net force on the wire—if it's perpendicular to the magnetic field—is $F = IlB$. More generally, the force on length l of a current-carrying wire is

$$\vec{F}_{\text{wire}} = (IlB \sin\alpha, \text{ direction of the right-hand rule for forces}) \qquad (26.20)$$

Magnetic force on a current-carrying wire of length l

where α is the angle between the wire and $\vec{B}$. A wire perpendicular to $\vec{B}$ has $\alpha = 90°$ and thus $F_{\text{wire}} = IlB$.

FIGURE 26.32 The force on a current is the force on the charge carriers.

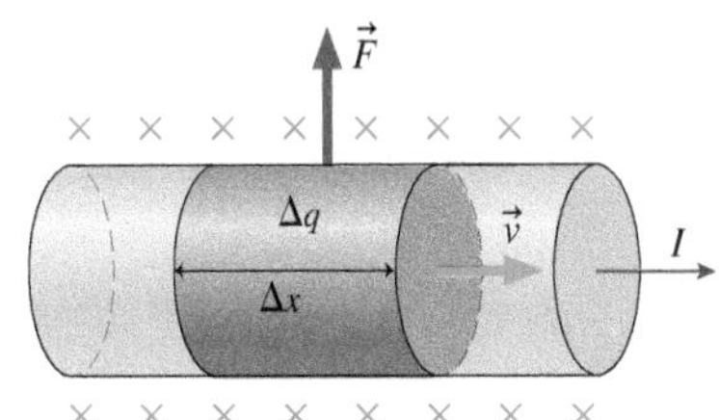
$\vec{F}$
Δq
$\vec{v}$
I
Δx

NOTE ▶ The right-hand rule for forces also applies to a current-carrying wire. Point your right thumb in the direction of the current and your index finger in the direction of $\vec{B}$. Your middle finger is then pointing in the direction of the force $\vec{F}$ on the wire. ◀

EXAMPLE 26.10 Finding the magnetic force on a current-carrying wire

Scientists have studied the aerodynamic forces generated by a flying insect by tethering it to the center of a taut, horizontal wire. As the insect "flies" in a wind tunnel, the insect exerts an upward force on the wire that a researcher can detect by measuring the wire's deflection. First the wire must be calibrated by measuring its deflection when a known force is applied to it. This is done by passing a current through the wire while it is positioned in a magnetic field that is perpendicular to the wire. The researcher requires the force on the wire to point upward with a magnitude of 1.0×10^{-3} N, roughly the mass of a bee. The length of the wire that is in the field is 50 mm, and the wire carries a 75 mA current from left to right. What direction and magnitude of magnetic field will provide the necessary force?

PREPARE A sketch of the situation is shown in FIGURE 26.33a, in which the magnetic force on the wire is shown pointing up, as required. We can use the right-hand rule for forces to find the direction of the field. The wire is perpendicular to the field, so we can use Equation 26.20 with $\alpha = 90°$ for the force on the wire.

FIGURE 26.33 A current-carrying wire in a magnetic field.

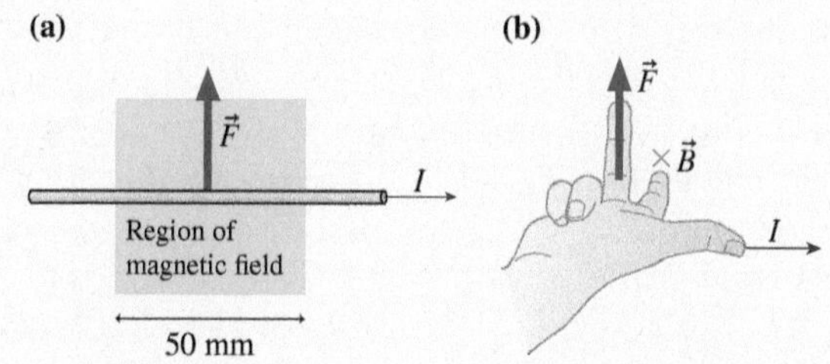

SOLVE As shown in FIGURE 26.33b, the right-hand rule for forces shows that the direction of the magnetic field is into the figure. Solving Equation 26.20 for the magnetic field gives

$$B = \frac{F_{\text{wire}}}{Il} = \frac{1.0 \times 10^{-3}\ \text{N}}{(0.075\ \text{A})(0.050\ \text{m})} = 0.27\ \text{T}$$

ASSESS Table 26.1 shows that 0.27 T is a strong but easily realized magnetic field.

Force Between Two Parallel Wires

Now consider Ampère's experimental arrangement of two parallel wires of length l, distance d apart. FIGURE 26.34a shows the currents I_1 and I_2 in the same direction; FIGURE 26.34b shows the currents in opposite directions. We will assume that the wires are sufficiently long to allow us to use the earlier result for the magnetic field of a long, straight wire: $B = \mu_0 I/2\pi r$.

FIGURE 26.34 Magnetic forces between parallel current-carrying wires.

(a) Currents in same direction (b) Currents in opposite directions

Magnetic field $\vec{B}_2$ created by current I_2

Magnetic field $\vec{B}_1$ created by current I_1

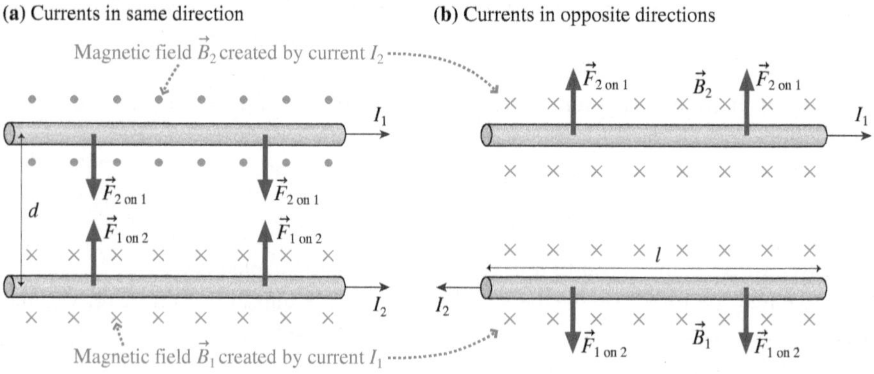

As Figure 26.34a shows, the current I_2 in the lower wire creates a magnetic field $\vec{B}_2$ at the position of the upper wire. $\vec{B}_2$ points out of the figure, perpendicular to current I_1. **It is field $\vec{B}_2$, due to the lower wire, that exerts a magnetic force on the upper wire.** Using the right-hand rule, you can see that the force on the upper wire is downward, thus attracting it toward the lower wire. The field of the lower current is not a uniform field, but it is the *same* at all points along the upper wire because the two wires are parallel. Consequently, we can use the field of a long, straight wire, with $r = d$, to determine the magnetic force exerted by the lower wire on the upper wire:

$$F_{\text{parallel wires}} = I_1 l B_2 = I_1 l \frac{\mu_0 I_2}{2\pi d} = \frac{\mu_0 l I_1 I_2}{2\pi d} \quad (26.21)$$

Magnetic force between two parallel wires

As an exercise, you should convince yourself that the current in the upper wire exerts an upward-directed magnetic force on the lower wire with exactly the same magnitude. You should also convince yourself, using the right-hand rule, that the forces are repulsive and tend to push the wires apart if the two currents are in opposite directions.

Thus two parallel wires exert equal but opposite forces on each other, as required by Newton's third law. **Parallel wires carrying currents in the same direction attract each other; parallel wires carrying currents in opposite directions repel each other.**

EXAMPLE 26.11 A current balance

Two stiff, 50-cm-long, parallel wires are connected at the ends by metal springs. Each spring has an unstretched length of 5.0 cm and a spring constant of 0.025 N/m. The wires push each other apart when a current travels around the loop. How much current is required to stretch the springs to lengths of 6.0 cm?

PREPARE FIGURE 26.35 shows the "circuit." The springs are conductors, allowing a current to travel around the loop. In equilibrium, the repulsive magnetic forces between the wires are balanced by the restoring forces $F_{sp} = k\,\Delta y$ of the springs.

FIGURE 26.35 The current-carrying wires of Example 26.11.

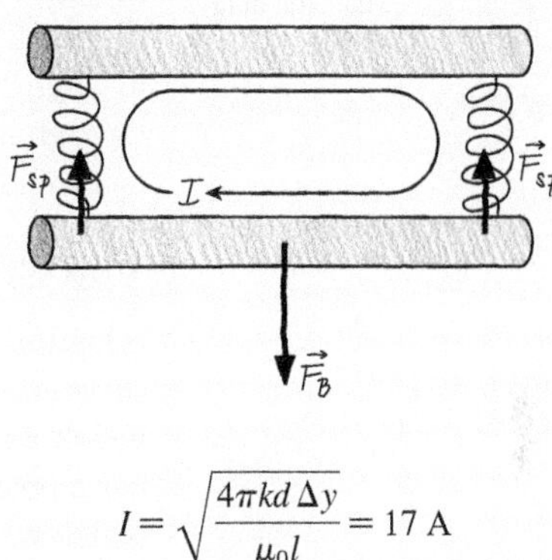

SOLVE Figure 26.35 shows the forces on the lower wire. The net force is zero, hence the magnetic force is $F_B = 2F_{sp}$. The force between the wires is given by Equation 26.21 with $I_1 = I_2 = I$:

$$F_B = \frac{\mu_0 l I^2}{2\pi d} = 2F_{sp} = 2k\,\Delta y$$

where k is the spring constant and $\Delta y = 1.0$ cm is the amount by which each spring stretches. Solving for the current, we find

$$I = \sqrt{\frac{4\pi k d\,\Delta y}{\mu_0 l}} = 17 \text{ A}$$

ASSESS Devices in which a magnetic force balances a mechanical force are called *current balances*. They can be used to make very accurate current measurements.

26.6 Forces and Torques on Magnetic Dipoles

FIGURE 26.36a shows two current loops. Using what we just learned about the forces between parallel and antiparallel currents, you can see that **parallel current loops exert attractive magnetic forces on each other if the currents circulate in the same direction; they repel each other when the currents circulate in opposite directions.**

We can think of these forces in terms of magnetic poles. Recall that the north pole of a current loop is the side from which the magnetic field emerges, which you can determine with the right-hand rule. FIGURE 26.36b shows the north and south magnetic poles of the current loops. When the currents circulate in the same direction, a north and a south pole face each other and exert attractive forces on each other. When the currents circulate in opposite directions, the two like poles repel each other.

Here, at last, we have a real connection to the behavior of magnets that opened our discussion of magnetism—namely, that like poles repel and opposite poles

FIGURE 26.36 Two alternative but equivalent ways to view magnetic forces.

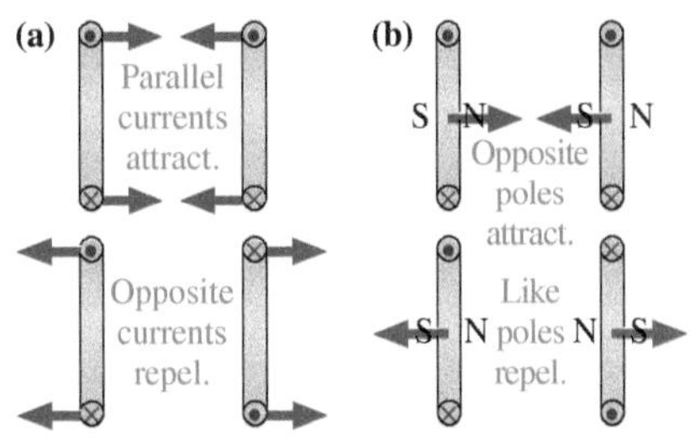

attract. Now we have an *explanation* for this behavior, at least for electromagnets. **Magnetic poles attract or repel because the current in one loop exerts attractive or repulsive magnetic forces on the other current.**

FIGURE 26.37 A uniform magnetic field exerts a torque on a current loop.

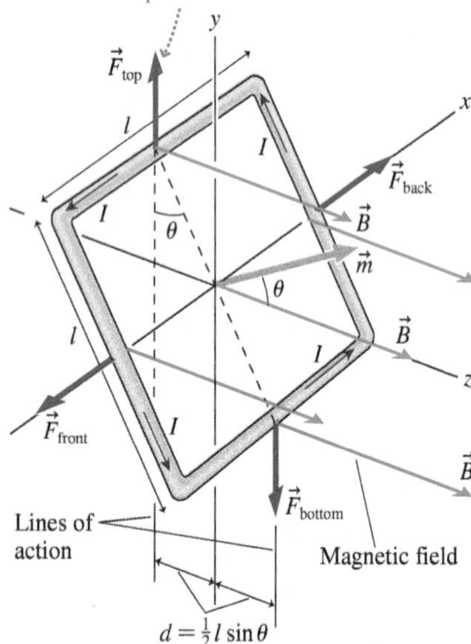

Now let's consider what happens to a current loop in a magnetic field. FIGURE 26.37 shows a square current loop in a uniform magnetic field along the z-axis. As we've learned, the field exerts magnetic forces on the currents in each of the four sides of the loop. Their directions are given by the right-hand rule. Forces $\vec{F}_{front}$ and $\vec{F}_{back}$ are opposite to each other and cancel. Forces $\vec{F}_{top}$ and $\vec{F}_{bottom}$ also add to give no net force, but because $\vec{F}_{top}$ and $\vec{F}_{bottom}$ don't act along the same line they will *rotate* the loop by exerting a torque on it.

Recall that torque is the magnitude of the force F multiplied by the moment arm d, the distance between the pivot point and the line of action. Both forces have the same moment arm $d = \frac{1}{2}l\sin\theta$, hence the torque on the loop—a torque exerted by the magnetic field—is

$$\tau = 2Fd = 2(IlB)\left(\tfrac{1}{2}l\sin\theta\right) = (Il^2)B\sin\theta = mB\sin\theta \tag{26.22}$$

where $m = Il^2 = IA$ is the loop's magnetic dipole moment and θ is the angle between the dipole moment vector $\vec{m}$ and the magnetic field $\vec{B}$. The torque is zero when the magnetic dipole moment $\vec{m}$ is aligned parallel or antiparallel to the magnetic field, and is maximum when $\vec{m}$ is perpendicular to the field.

We derived Equation 26.22 for a square current loop, but the result is valid for any magnetic dipole. What we see is that **a magnetic field exerts a torque on a magnetic dipole.** It is this magnetic torque that causes a compass needle—a magnetic dipole—to align itself with the magnetic field. And magnetic torque spins electric motors through a clever design in which the current through a coil reverses direction every half rotation so that the coil, unlike a compass needle, can never reach equilibrium.

The Energy of a Magnetic Dipole

Equation 26.22 for the magnetic torque on a magnetic dipole has exactly the same structure as the equation for the electric torque on an electric dipole, $\tau = pE\sin\theta$, which we derived in ◀◀ SECTION 21.7. Similarly, a magnetic dipole in a magnetic field has a potential energy that exactly mirrors the potential energy of an electric dipole, which we found in ◀◀ SECTION 22.1:

$$U_{mag} = -mB\cos\theta \tag{26.23}$$

Energy gives us another perspective on the behavior of a magnetic dipole in a magnetic field. We see that the potential energy is a minimum when the magnetic dipole moment $\vec{m}$ is aligned with the magnetic field. If a magnetic dipole has a way to dissipate energy, such as the frictional damping of a compass needle, then it will come to rest aligned with the field because that's the minimum-energy configuration.

STOP TO THINK 26.7 What is the current direction in the loop?

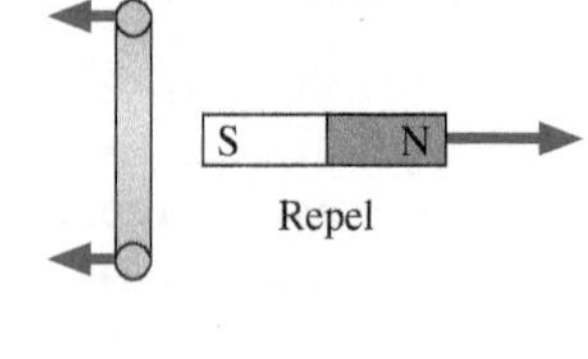

A. Out of the figure at the top of the loop, into the figure at the bottom
B. Out of the figure at the bottom of the loop, into the figure at the top

Ferromagnetism

We started the chapter by looking at permanent magnets. We know that permanent magnets produce a magnetic field, but what is the source of this field? There are no electric currents in these magnets. Why can you make a magnet out of certain materials but not others? Why does a magnet stick to the refrigerator?

Iron, nickel, and cobalt are elements that have very strong magnetic behavior: A chunk of iron (or steel, which is mostly iron) will stick to a magnet, and the chunk can be magnetized so that it is itself a magnet. Other metals—such as aluminum and copper—do not exhibit this property. We call materials that are strongly attracted to magnets and that can be magnetized **ferromagnetic** (from the Latin for iron, *ferrum*).

The key to understanding magnetism at the atomic level is that electrons, as we noted earlier, have an inherent magnetic dipole moment. In other words, each electron is a tiny magnet that generates a tiny magnetic field.

If the magnetic moments of all the electrons in an atom pointed in the same direction, the atom would have a very strong magnetic moment. But this doesn't happen. In atoms that have many electrons, the electrons usually occur in pairs that have magnetic moments in opposite directions. Only the electrons that are unpaired are able to give the atom a net magnetic dipole moment.

Even so, atoms with magnetic dipole moments don't necessarily form a solid with magnetic properties. For most elements, the magnetic dipole moments of the atoms are randomly arranged when the atoms join together to form a solid. As FIGURE 26.38a shows, this random arrangement produces a solid whose net magnetic dipole moment is very close to zero. These solids have no magnetic properties. In contrast, the electron magnetic dipole moments in a ferromagnetic material interact with each other in a way that tends to align the dipole moments in the same direction, as shown in FIGURE 26.38b. The individual dipole moments add to give the material a *macroscopic* magnetic dipole moment. The material has a north and a south magnetic pole, generates a magnetic field, and aligns parallel to an external magnetic field. In other words, it is a magnet—and a very strong one at that.

However, a typical piece of iron is not a strong magnet. It turns out that a piece of iron is made up of small regions, usually smaller than 100 μm, called **magnetic domains.** Each domain is a strong magnet, with all the electron magnetic dipoles aligned. The domains are randomly arranged, however, as shown in FIGURE 26.39, so the larger piece of iron has little or no net magnetic dipole moment. (A permanent magnet, such as a bar magnet, is a ferromagnetic material in which some domains have been induced to grow at the expense of others so that the material does acquire a net magnetic dipole moment.)

FIGURE 26.38 The electron magnetic dipoles in a solid.

(a) A typical solid

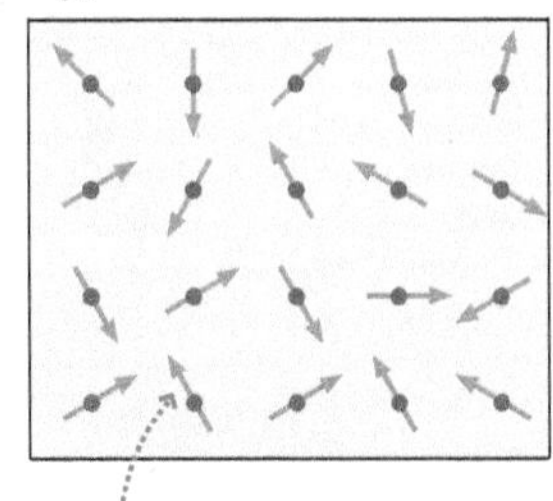

The magnetic dipole moments due to unpaired electrons point in random directions. The solid has no net magnetism.

(b) A ferromagnetic solid

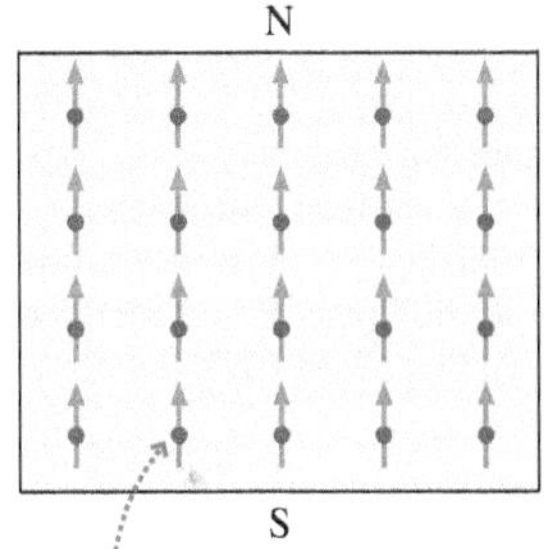

N
S
The electron magnetic dipoles are aligned. The solid has a macroscopic dipole moment with north and south magnetic poles.

FIGURE 26.39 A magnet attracts an unmagnetized piece of iron.

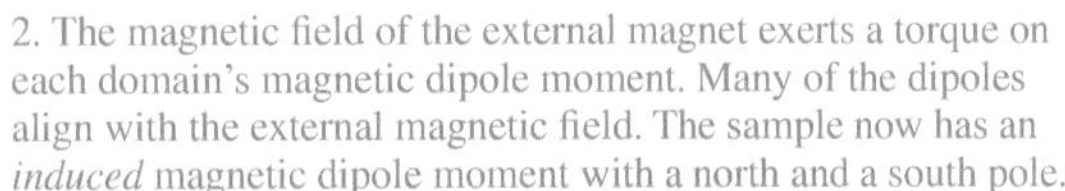

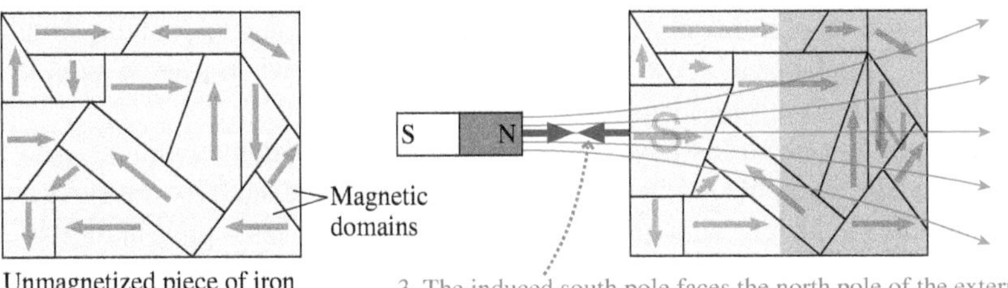

When we bring a magnet near a piece of iron, the magnetic field of the external magnet exerts a torque on the magnetic dipole moment of each domain. The torques cause many of the domains to align their dipole moments with the magnetic field of the magnet. Now the piece of iron has a net magnetic dipole moment that is aligned with the external magnetic field. We say that the iron has an *induced magnetic dipole moment.*

Now we can see how the iron is attracted to the magnet. The induced magnetic dipole moment in the iron has its south pole facing the north pole of the permanent magnet. Opposite poles attract, so the iron is attracted to the magnet, and vice versa. There are several steps in the reasoning, but we can understand how a refrigerator magnet sticks to the refrigerator.

26.7 Magnetic Resonance Imaging

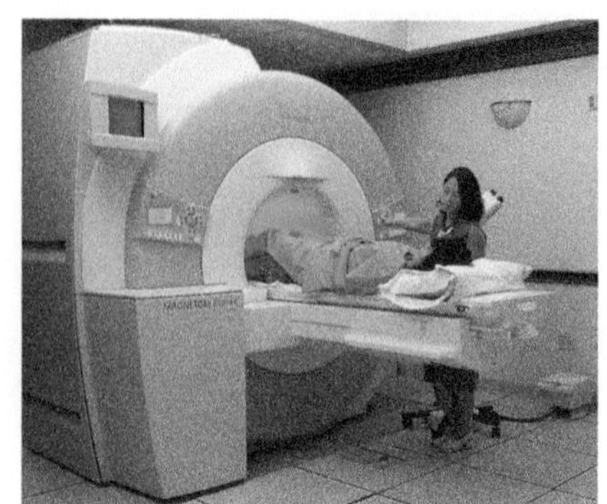

An MRI machine The primary magnetic field of an MRI machine—directed along the length of the patient—is created by a large solenoid that the patient slides into. The most commonly used clinical MRI machines have a field strength of 1.5 T. A field this strong requires an extremely large current. To achieve such currents, the solenoid wires are made from superconducting materials that have zero resistance when cooled by liquid helium to a frigid 4.2 K, barely above absolute zero.

Magnetic resonance imaging, or MRI, is a noninvasive clinical diagnostic tool that provides detailed images of tissues and structures inside the body without exposing the patient to radiation. MRI is an enhancement of **nuclear magnetic resonance,** or NMR, a technique used in organic chemistry to study molecular structure. We're now at a point where we can understand the key ideas of how NMR works and how MRI images are made. There are several key ideas, so we'll start with an overview:

- The patient is placed in a strong, uniform, unchanging magnetic field—the *static field*—that is created by a large solenoid. The alignment of protons—which, like electrons, have a magnetic dipole moment—gives the tissue a net magnetic dipole moment.
- An oscillating magnetic field is applied for a short interval, just a few microseconds. The frequency of this *oscillating field* is adjusted to match the system's natural frequency; this is the resonance in magnetic resonance. The oscillating field rotates the magnetic dipole moment of the tissue by 90°. The rotated magnetic dipole moment generates the signal that is measured in both NMR and MRI.
- Additional current-carrying coils produce a small *gradient field* that varies in strength across the region being imaged. This causes the resonance frequency to vary from point to point and allows responses to be identified with specific locations in the tissue.
- The tissue's rotated magnetic dipole moment then returns to equilibrium—alignment with the static field—with a characteristic *relaxation time* that depends on the tissue type. The contrast between different tissue types in an MRI image is obtained by measuring the relaxation time.

Now let's look at each part of the process in more detail.

Magnetization of the Tissue

Approximately 60% of the atoms in the human body are hydrogen. In a basic MRI image we see the *density* of hydrogen atoms—which varies with the tissue type—as measured by the number of protons that are the nuclei of the hydrogen atoms. A 3 kg sample of tissue, approximately the mass of tissue being imaged, contains roughly 10^{25} protons. Our first task is to determine what happens when a large sample of protons is placed in a strong magnetic field.

We've seen that electrons have spin—an inherent angular momentum—and an associated magnetic dipole moment. Protons also have a spin, and, like an electron, a proton must be in either a "spin up" or a "spin down" state. This representation is adequate for many purposes, but reality is a bit more complex. You may have heard of a surprising feature of quantum physics called the *Heisenberg uncertainty principle,* a topic we'll discuss in Chapter 28. It turns out that a particle's angular momentum cannot be perfectly aligned with a magnetic field because such an alignment would violate the uncertainty principle. Instead, as FIGURE 26.40 shows, the two states of a proton that we'll refer to as *aligned* and *anti-aligned* are states in which the spin and the magnetic dipole moment $\vec{m}$ are at an angle to $\vec{B}$.

FIGURE 26.40 The two possible states of a proton in a magnetic field.

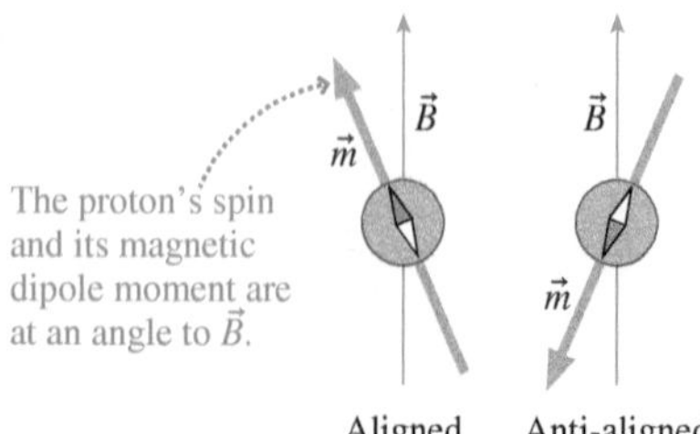

The proton's spin and its magnetic dipole moment are at an angle to $\vec{B}$.
$\vec{m}$
$\vec{B}$
$\vec{B}$
$\vec{m}$
Aligned
Anti-aligned

You've probably seen a spinning top or gyroscope. A top that is not spinning falls over because gravity exerts a torque about the point where the top touches the ground. But a spinning top, as shown in FIGURE 26.41a, stays upright by *precessing* about a vertical axis; that is, the top's rotation axis slowly moves around a circle centered on a vertical (gravitational field) axis. The difference between the nonspinning top and the spinning top is that the spinning top has *angular momentum*. We won't prove it—it's an advanced topic—but a torque on an object with angular momentum causes the object to precess.

A proton in a magnetic field is analogous to a spinning top. The magnetic field exerts a torque on a proton's magnetic dipole moment, but the torque doesn't bring the magnetic dipole moment into alignment with the field. Instead, because the proton has angular momentum—its spin—the torque causes the magnetic dipole moment to *precess* like a spinning top about the direction of the magnetic field, as shown in FIGURE 26.41b.

FIGURE 26.41 Instead of aligning with a magnetic field, the magnetic dipole moment of a proton precesses around the magnetic field in the same way that a spinning top precesses around the gravitational field.

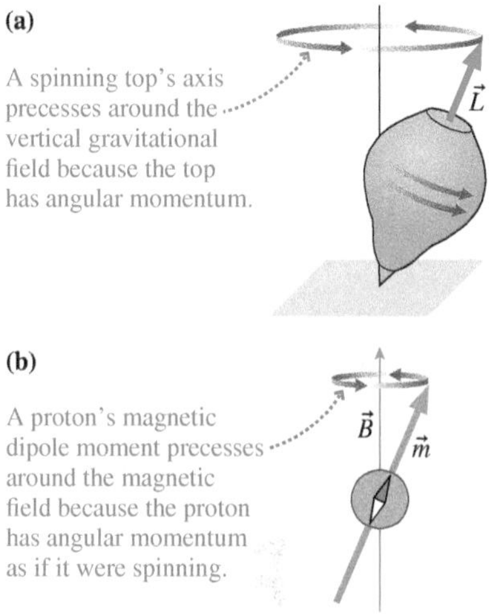

(a) A spinning top's axis precesses around the vertical gravitational field because the top has angular momentum.

(b) A proton's magnetic dipole moment precesses around the magnetic field because the proton has angular momentum as if it were spinning.

FIGURE 26.42 shows the two possible proton states, aligned and anti-aligned, precessing around the magnetic field. We found earlier that the energy of a magnetic dipole in a magnetic field is $U_{mag} = -mB\cos\theta$, so protons in the aligned orientation are in a lower-energy state while anti-aligned protons are in a higher-energy state.

The precession frequency—the number of precessions per second—is proportional to the magnetic field strength and also to the strength of the proton's magnetic dipole moment. A full analysis finds that the precession frequency is

$$f_{precess} = \gamma B \tag{26.24}$$

where $\gamma = 42.6$ MHz/T is the proton's **gyromagnetic ratio.** Clinical MRI machines use magnetic field strengths ranging from 0.5 T to 3 T, so the proton precession frequencies range from ≈ 22 MHz to ≈ 128 MHz. In fact, clinicians often describe the magnet in terms of the **precession** frequency rather than the field strength, as in "This machine uses a 100 MHz magnet."

Suppose a sample with a large number of protons is placed in a uniform magnetic field. The *net* magnetic dipole moment of the sample, which we call its **magnetization** $\vec{M}$, is found by summing the individual dipole moments $\vec{m}_i$, where i labels the protons; that is, $\vec{M} = \Sigma\vec{m}_i$. Every proton in the field precesses at the same frequency $f_{precess} = \gamma B$, but, as FIGURE 26.43a shows, each proton's precession is independent of other protons. That is, the precessions are not coordinated or synchronized. Each $\vec{m}_i$ has a component that is parallel to $\vec{B}$ (its *longitudinal* component) and a component that is perpendicular to $\vec{B}$ (its *transverse* component). The transverse components point in random directions and, when the number of protons is large, sum to zero.

The aligned protons all have the *same* longitudinal component of $\vec{m}$ in the direction of $\vec{B}$. Similarly, the anti-aligned protons all have the same longitudinal component in the opposite direction. As we'll show, the number of aligned protons slightly exceeds the number of anti-aligned protons. Consequently, the sample has a magnetization $\vec{M}$ that is parallel to $\vec{B}$. This is shown in FIGURE 26.43b.

FIGURE 26.42 A proton in a magnetic field has two energy states.

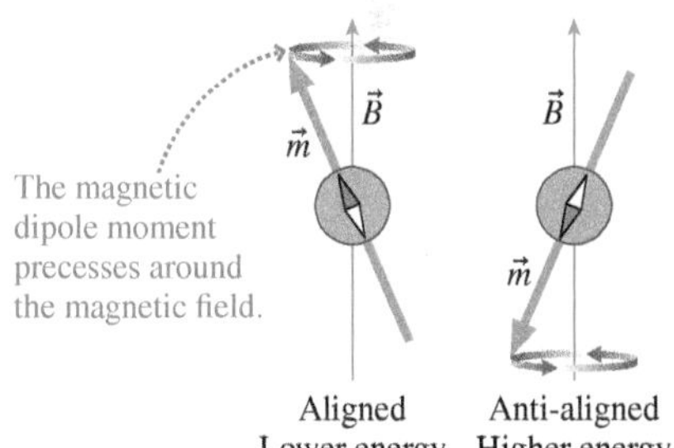
The magnetic dipole moment precesses around the magnetic field.
$\vec{m}$
$\vec{B}$
$\vec{B}$
$\vec{m}$
Aligned
Lower energy
Anti-aligned
Higher energy

FIGURE 26.43 The net magnetization of a sample, which is parallel to the magnetic field, is the sum of all the individual magnetic dipole moments.

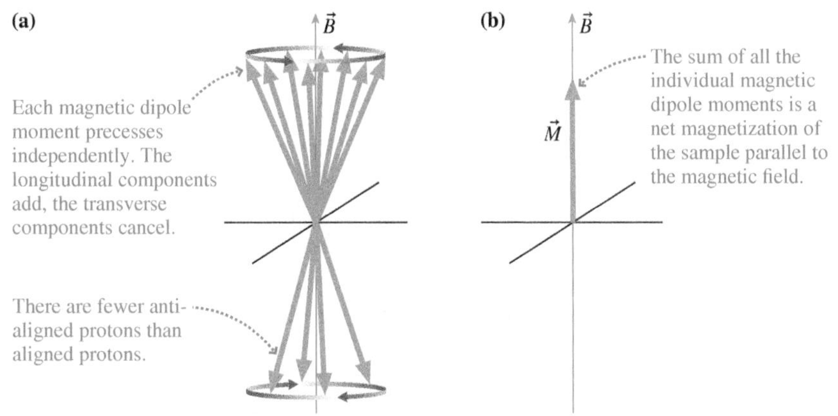
(a)
$\vec{B}$
Each magnetic dipole moment precesses independently. The longitudinal components add, the transverse components cancel.
There are fewer anti-aligned protons than aligned protons.
(b)
$\vec{B}$
$\vec{M}$
The sum of all the individual magnetic dipole moments is a net magnetization of the sample parallel to the magnetic field.

This is our first important conclusion: A tissue sample in a magnetic field develops a net magnetization due to the alignment and precession of a very large number of protons that are the nuclei of hydrogen atoms.

We have one loose end to tie up: How do we know how many protons are in each state? The aligned state has a lower energy and systems seek the lowest energy; thus, we might expect most or all of the protons to be in the lower-energy state, aligned with the magnetic field. The result would be a very large magnetization. On the other hand, if the energy difference ΔE between the aligned and anti-aligned states is much less than $k_B T$, the scale of thermal energy, then thermal interactions would equalize the populations of aligned and anti-aligned protons and the magnetization would be very close to zero.

There are two different but equivalent ways to understand magnetic resonance. We're presenting a *classical* description in terms of precessing magnetic dipoles. There's also a *quantum* description based on energy levels and "spin flips" as protons change energy levels. The quantum description requires an understanding of quantum physics that is not available to us at this point in the text, but we will invoke one quantum idea that you may have learned in chemistry: If there are particles in two energy levels that have separation ΔE, then a photon—a quantum of light—can be absorbed if the photon frequency f satisfies $hf = \Delta E$, where $h = 6.63 \times 10^{-34}\ \text{J}\cdot\text{s}$ is called *Planck's constant.* (We'll explore these ideas in detail in Chapter 28.) For protons in a magnetic field, the relevant frequency is the precession frequency. Thus, the energy difference between the aligned and the anti-aligned states is

$$\Delta E = hf_{\text{precess}} = h\gamma B \tag{26.25}$$

EXAMPLE 26.12 Proton populations in an MRI machine

A clinical MRI machine uses a 1.50 T magnetic field. What are (a) the proton precession frequency and (b) the ratio of anti-aligned protons to aligned protons?

PREPARE We found in ◀◀ SECTION 13.5 that the ratio of the population in a higher energy level to the population in a lower energy level, with an energy difference ΔE, is the Boltzmann factor $e^{-\Delta E/k_B T}$. We will assume a body temperature of 37°C = 310 K.

SOLVE

a. The proton precession frequency is a straightforward calculation:

$$f_{\text{precess}} = \gamma B = (42.6\ \text{MHz/T})(1.50\ \text{T}) = 63.9\ \text{MHz}$$

b. The ratio of the number of higher-energy anti-aligned protons to the number of lower-energy aligned protons is given by the Boltzmann factor:

$$\frac{N_{\text{anti-aligned}}}{N_{\text{aligned}}} = e^{-\Delta E/k_B T}$$

The energy difference between the two states is

$$\Delta E = hf_{\text{precess}} = (6.63 \times 10^{-34}\ \text{J}\cdot\text{s})(63.9 \times 10^{6}\ \text{Hz})$$
$$= 4.24 \times 10^{-26}\ \text{J}$$

Thus the population ratio is

$$\frac{N_{\text{anti-aligned}}}{N_{\text{aligned}}} = \exp\left[-\frac{4.24 \times 10^{-26}\ \text{J}}{(1.38 \times 10^{-23}\ \text{J/K})(310\ \text{K})}\right]$$
$$= \exp(-9.90 \times 10^{-6})$$
$$= 0.999990 = 1 - 1.0 \times 10^{-5}$$

ASSESS The population ratio differs from being exactly 1—equal numbers of aligned and anti-aligned protons—by only 10 part per million. There's *very* little difference in the populations of the two levels. Even so, the slight excess of aligned protons means that the tissue sample has a magnetization, and this makes MRI possible.

Rotation of the Magnetization

A sample of protons in a uniform magnetic field $\vec{B}_0$ has an equilibrium magnetization $\vec{M}$ that is parallel to $\vec{B}_0$. This is shown in FIGURE 26.44a. But a sample of protons in equilibrium is not very useful. NMR and MRI measure how protons respond after being driven *out* of equilibrium.

To understand how a sample's magnetization can be manipulated, it is useful to imagine being in a coordinate system that rotates at the precession frequency. The horses on a merry-go-round move in circles relative to the ground, but they're stationary relative to you when you're on the merry-go-round with them. Similarly,

FIGURE 26.44 An oscillating magnetic field in the transverse plane can rotate the sample's magnetization into the transverse plane.

(a) Laboratory coordinate system

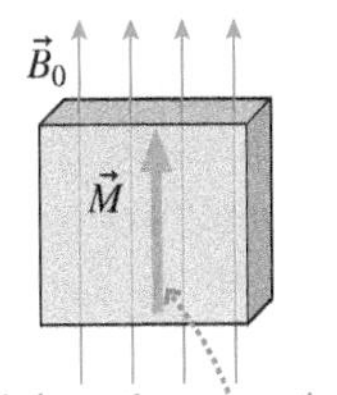

In equilibrium, the magnetization is parallel to the magnetic field.

(b) Laboratory coordinate system

Transverse plane

$\vec{B}_0$

$\vec{M}$

$\vec{B}_1$

The magnetic field in the transverse plane oscillates at the precession frequency f_0.

(c) Rotating coordinate system

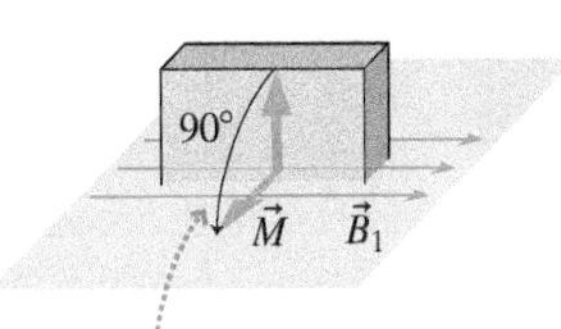

A coordinate system rotating at frequency f_0 sees only a static field $\vec{B}_1$. The magnetization precesses about $\vec{B}_1$. A 90° pulse rotates $\vec{M}$ into the transverse plane.

(d) Laboratory coordinate system

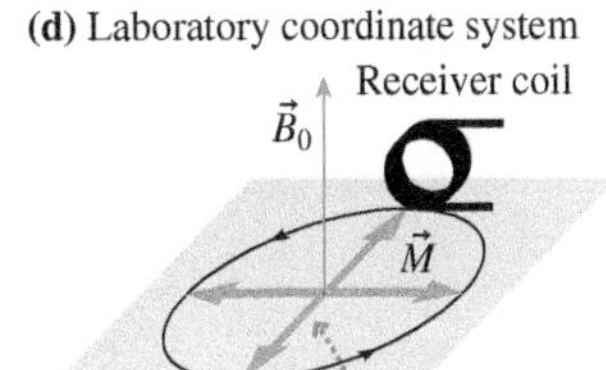

A magnetization in the transverse plane precesses about $\vec{B}_0$ and causes an induced current in the receiver coil.

the protons do *not* precess in a coordinate system that rotates at their precession frequency; their magnetic dipole moments appear frozen in place. Stepping into a coordinate system that rotates at exactly the precession frequency effectively removes the magnetic field $\vec{B}_0$ that causes precession in a laboratory coordinate system.

Suppose we now add a second magnetic field: an *oscillating* magnetic field $\vec{B}_1$ that is perpendicular to $\vec{B}_0$ (i.e., it is in the transverse plane). The frequency of this oscillating field is adjusted to exactly match the proton's precession frequency f_0. This is shown in FIGURE 26.44b. How does this new magnetic field appear in the rotating coordinate system? A bicycle circling a merry-go-round will be at rest relative to you on the merry-go-round if its circular frequency exactly matches the merry-go-round's frequency. Similarly, a magnetic field $\vec{B}_1$ that oscillates in the laboratory coordinate system will *not* oscillate in the rotating coordinate system—it will appear to be a fixed, unchanging field—*if* its frequency exactly matches the coordinate system's rotation frequency. Thus, with $\vec{B}_0$ effectively removed, protons in the rotating coordinate system see only $\vec{B}_1$. And—this is a key idea—if the oscillation frequency exactly matches the precession frequency, the protons see $\vec{B}_1$ as a *static* magnetic field. This is shown in FIGURE 26.44c.

An individual proton's magnetic dipole moment $\vec{m}$ precesses around a magnetic field. Similarly, the sample's magnetization $\vec{M}$ precesses around a magnetic field. The rotating coordinate system has only the static field $\vec{B}_1$, so $\vec{M}$ immediately—in the rotating coordinate system—begins to precess around $\vec{B}_1$ at frequency $f_1 = \gamma B_1$. In other words, **we can rotate the magnetization of the sample by applying a transverse magnetic field that oscillates at the precession frequency.**

You may recall from our study of oscillations in Chapter 15 that a *resonance* occurs when a driving frequency exactly matches a system's natural frequency. If the $\vec{B}_1$ magnetic field oscillates at just the right frequency to rotate the magnetization of a sample of protons, then we have a *magnetic resonance.*

The oscillating field $\vec{B}_1$ is turned on only long enough—typically a few microseconds—to rotate $\vec{M}$ through a 90° angle so that it lies in the transverse plane, perpendicular to $\vec{B}_0$. Then $\vec{B}_1$ is turned off. This short pulse of the oscillating field, called a *90° pulse,* is the key to understanding MRI. A 90° pulse drives the protons out of equilibrium and leaves the magnetization $\vec{M}$ in the transverse plane. Then, when $\vec{B}_1$ is turned off, $\vec{M}$ begins to precess around $\vec{B}_0$ at frequency f_0. This is shown in FIGURE 26.44d.

NOTE ▸ The precession frequency is tens or hundreds of megahertz. Frequencies in this range are called *radio frequencies,* abbreviated RF, because similar frequencies are used for radio broadcasts. Consequently, the oscillating field $\vec{B}_1$ is often called the *RF field.* ◂

It is this macroscopic precession of the sample magnetization $\vec{M}$ around $\vec{B}_0$ that is detected during NMR and MRI. The sample magnetization $\vec{M}$ produces its own magnetic field in the transverse plane, a field that oscillates back and forth as $\vec{M}$

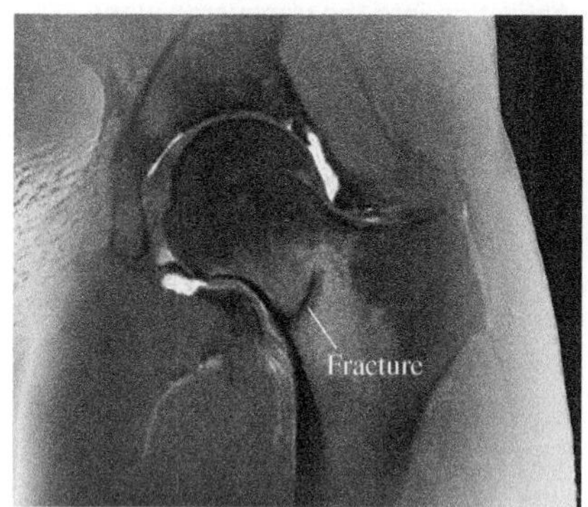

Hip fracture An MRI image is a map of the density of atoms. A greater hydrogen density produces a stronger signal and a brighter image. Different tissues can be distinguished if they have different hydrogen densities. This MRI image shows a hip fracture in a 26-year-old male marathon runner.

processes. This oscillating field is detected by a *receiver coil* whose axis lies in the transverse plane, perpendicular to $\vec{B}_0$. The field that passes through the coil is large when $\vec{M}$ is aligned with the coil, then zero a quarter-cycle later when $\vec{M}$ is perpendicular to the axis of the coil. You'll learn in the next chapter that an oscillating magnetic field inside a coil causes an *induced current* to flow through the coil; that's how many types of radio receivers work. In this case, the sample produces a detectable current in the receiver coil after a 90° pulse as the sample magnetization processes in the transverse plane. This is the essence of how an NMR or MRI signal is generated.

We've assumed that the precession frequency is $f_{\text{precess}} = \gamma B_0$, where $\vec{B}_0$ is the static field of the solenoid. There's actually more to the story. A proton in a sample experiences not only the static field $\vec{B}_0$ but also small magnetic fields created by nearby atoms. These *local fields* arise due to electrons circulating in neighboring atoms and to the magnetic moments of other nuclei. The proton's precession frequency is determined not by $\vec{B}_0$ but by the net magnetic field $\vec{B}_{\text{net}} = \vec{B}_0 + \vec{B}_{\text{local}}$ that it experiences—that is, $f_{\text{precess}} = \gamma B_{\text{net}}$—so local interactions slightly alter the precession frequency. These small shifts in the resonance frequency—only a few parts per million—are what organic chemists measure during NMR spectroscopy because they provide information about the molecular structure of a sample.

The Gradient Field

The strength of the signal—the induced current—is directly proportional to the magnitude of $\vec{M}$, and the magnitude of $\vec{M}$ is directly proportional to the number of hydrogen atoms in the sample because the protons are the nuclei of hydrogen atoms. Different tissues have different densities of hydrogen and would, if isolated, produce different signal strengths. But the technique as we've described it thus far—which includes both NMR and MRI—simultaneously detects all the protons in the sample. To create an image, we need a way to distinguish protons that have different spatial locations within a sample.

FIGURE 26.45 A gradient field provides spatial resolution for the MRI signal.

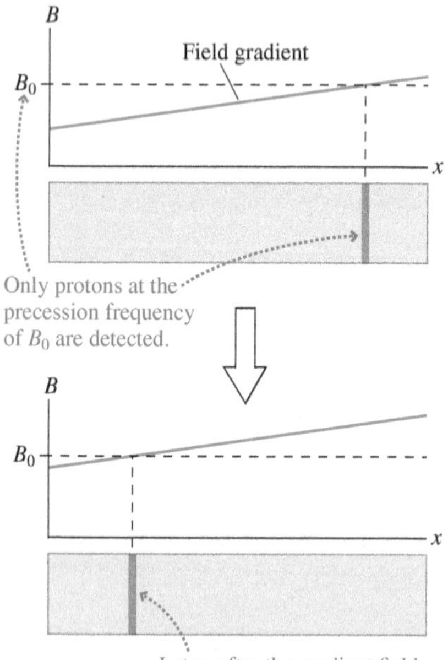

An MRI machine distinguishes protons at different locations by using an additional set of magnet coils to create a *gradient field,* a weak field that changes with position across a sample. FIGURE 26.45 illustrates the idea. The gradient field slightly weakens the external magnetic field on the left side of the sample and slightly strengthens the field on the right side. The apparatus detects only protons that precess at frequency f_0 in a field of strength B_0, so only the protons in a narrow slice of the sample where the field is $\vec{B}_0$ contribute to the signal. Most of the protons in the sample are not in resonance and are not detected.

By slowly increasing the field strength everywhere, the detection point in the sample—the point where the protons are in resonance—moves from right to left across the sample. In other words, gradually strengthening the gradient field scans the detection point across the sample. Computer software then converts the signal strength at each point into an image.

The simple scheme of Figure 26.45 provides only one dimension of imaging. An actual MRI machine uses a more complex gradient field that changes with both space and time to map the proton density point by point throughout a sample, but the principle is the same as in Figure 26.45. The end result, an MRI image, is essentially a map of the density of hydrogen atoms in a cross section of the body.

EXAMPLE 26.13 **Frequency differences in an MRI machine**

A clinical MRI machine with a 3.00 T magnetic field uses a gradient field of 30 mT/m. (a) What is the frequency of the oscillating magnetic field? (b) What is the difference in resonance frequencies at two points separated by 1.0 mm, roughly the spatial resolution of this machine?

PREPARE The proton's precession frequency is $f = \gamma B$, where $\gamma = 42.6$ MHz/T.

SOLVE

a. The oscillating magnetic field is tuned to match the proton's precession frequency before the gradient field is added, when the magnetic field strength is $B_0 = 3.00$ T. This frequency is

$$f_0 = \gamma B_0 = (42.6\text{ MHz/T})(3.00\text{ T}) = 128\text{ MHz}$$

b. With the gradient field applied, two spatially separated points experience slightly different magnetic fields that differ by ΔB. The resonance frequencies at these points differ by $\Delta f = \gamma \Delta B$. With a gradient of 30 mT/m, the change in field strength over 1.0 mm is

$$\Delta B = (30 \times 10^{-3}\ \text{T/m})(1.0 \times 10^{-3}\ \text{m}) = 3.0 \times 10^{-5}\ \text{T}$$

Consequently the resonance frequencies differ by

$$\Delta f = \gamma \Delta B = (42.6 \times 10^{6}\ \text{Hz/T})(3.0 \times 10^{-5}\ \text{T})$$
$$= 1300\ \text{Hz} = 1.3\ \text{kHz}$$

ASSESS The frequencies at these two points differ by only 10 parts per million. To achieve a spatial resolution of 1 mm, the machine has to stimulate and detect protons at the exact resonance frequency of 128 MHz while not stimulating or detecting protons whose resonance frequency differs by only $\approx$1 kHz. In fact, the machine and the electronics can just barely distinguish resonances that are 1 kHz apart, and this is what sets the spatial resolution at $\approx$1 mm. To achieve a smaller spatial resolution requires a finer frequency resolution.

Relaxation and Signal Enhancement

A sample of protons that has had its magnetization rotated to the transverse plane by a 90° pulse creates a signal in the receiver coil as $\vec{M}$ precesses around the magnetic field $\vec{B}_0$. However, the signal strength is seen to decay exponentially with time after the 90° pulse ends. This signal decay is called *relaxation,* and protons undergo two distinct relaxation processes.

First, thermal interactions of the protons with their local environment (i.e., energy exchanges) gradually return the populations of the aligned and anti-aligned states to their equilibrium values. As they do, the longitudinal component of $\vec{M}$—the component parallel to $\vec{B}_0$—increases and the transverse component of $\vec{M}$ that was created by the 90° pulse decreases. This thermal return to equilibrium is called **longitudinal relaxation.** If longitudinal relaxation were the only means of relaxation, the signal would decay with a *time constant* known as T_1.

Second, different protons experience slightly different magnetic fields because nearby nuclei, which have magnetic dipole moments, create a local magnetic field. As a result, different protons in the sample have slightly different precession frequencies. The transverse magnetization $\vec{M}$ is the sum of all the individual magnetic dipole moments, and $\vec{M}$ decays to zero as the precessions of different protons gradually get out of phase with each other. The sample remains out of equilibrium, but near-neighbor interactions among nuclei destroy the synchronized precessions that are required to generate a macroscopic magnetization $\vec{M}$ that can be detected. This process is called **transverse relaxation.** If transverse relaxation were the only means of relaxation, the signal would decay with a time constant known as T_2.

TABLE 26.2 **Proton relaxation times in a 1.5 T magnetic field**

Tissue	T_1 (ms)	T_2 (ms)
Fat	250	80
White matter of cerebrum	780	90
Muscle	870	45
Gray matter of cerebrum	900	100
Cartilage	1100	40
Whole blood (oxygenated)	1300	200
Whole blood (deoxygenated)	1300	50
Synovial fluid	2500	1200

TABLE 26.2 lists some representative values of T_1 and T_2 for different tissues. You can see that there are some large differences. For example, T_1 for muscle is more than three times larger than T_1 for fat, and T_2 for oxygenated blood is four times that of deoxygenated blood. These differences are very useful because there are experimental procedures that distinguish the two forms of relaxation and measure T_1 and T_2.

Computer software can generate an MRI image that emphasizes the signal from tissues that have especially small or especially large values of T_1. For example, fat is much brighter than muscle in an image that emphasizes small values of T_1, whereas fat and muscle look much the same in an image that shows only the hydrogen density. In other words, enhancing the image on the basis of the measured longitudinal relaxation time improves the contrast between some types of tissue. This is called a *T_1-enhanced* or *T_1-weighted image.*

FIGURE 26.46 A T_1-enhanced MRI of a knee reveals severe arthritis.

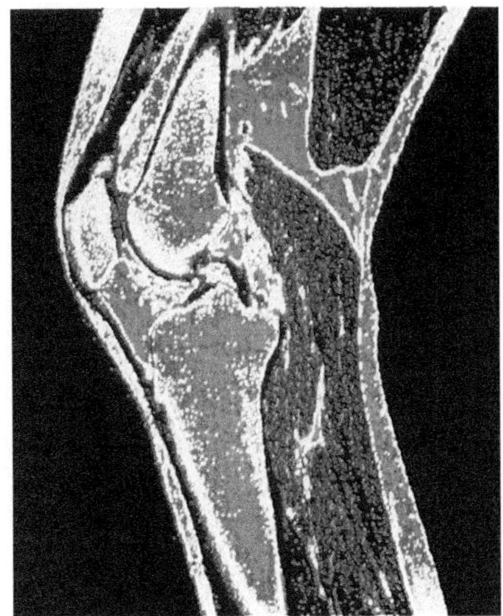

As an example, FIGURE 26.46 is an MRI image of a knee that is T_1-enhanced to emphasize tissue with long relaxation times. This MRI is presented as a "negative" in which strong signals are dark and weak signals bright. Mineral bone has almost no MRI signal because it contains few hydrogen atoms, so it is very bright in the image. Table 26.2 shows that synovial fluid—a viscous fluid found in joints that lubricates the motion of cartilage-covered bones—has a very long T_1 relaxation time.

Consequently, a T_1-enhanced MRI gives significant emphasis to synovial fluid, and its strong signal appears black. Cartilage, with a moderate value of T_1 is somewhat enhanced and is gray.

The left side of the joint shows two gray layers of cartilage separated by a thin black layer of synovial fluid. This side of the joint is healthy. But the right side shows the cartilage of the two bones in contact with no synovial fluid to lubricate the motion. This is an indication of severe arthritis. Cartilage and synovial fluid would appear almost identical in a normal MRI image, but T_1-enhancement allows the two to be distinguished.

MRI images can also be T_2-enhanced to emphasize especially small or especially large values of T_2. For example, T_2 enhancement provides information about blood flow to an organ by distinguishing between oxygenated and deoxygenated blood. T_2 enhancement is also used for MRIs of the brain to distinguish different brain tissues that look almost identical in an unenhanced MRI.

In summary, NMR and MRI, two of the most important tools of modern chemistry and medicine, are based on sophisticated techniques for manipulating and detecting the magnetic dipole moments of protons. The good news is that you now have the knowledge and skills to understand and use these tools.

CHAPTER 26 INTEGRATED EXAMPLE Rotating magnetic bacteria

As we've seen, magnetotactic bacteria contain magnetic crystals that allow them to follow the earth's magnetic field as they swim up and down within the water column. Scientists study magnetotactic bacteria by observing them under a microscope while applying a magnetic field. If the field direction suddenly changes, a torque rotates a bacterium until it is realigned with the new field direction. A bacterium's magnetic dipole moment can be measured by measuring the time it takes to make a 180° rotation after a reversal of the magnetic field.

We learned in Chapter 5 that a swimming microorganism almost immediately reaches a terminal speed in which the viscous drag force balances the applied force. Similarly, a rotating microorganism almost instantly reaches a terminal angular speed in which the torque exerted by drag balances the applied torque. According to Stokes' law, the viscous drag force on a sphere of radius R that is moving with speed v is $D = 6\pi\eta Rv$, where η is the viscosity of the liquid. A similar expression gives the *drag torque* on a sphere of radius R that is rotating with angular speed ω:

$$\tau_{\text{drag}} = 8\pi\eta R^3\omega$$

In one experiment, a bacterium, which can be approximated as a 2-μm-diameter sphere, was observed to reverse direction in 0.5 s when a 1.0 mT magnetic field was reversed. Estimate the magnetic dipole moment of this bacterium. The viscosity of water is 1.0×10^{-3} Pa·s.

PREPARE With strong viscous drag, the bacterium will rotate at an angular speed such that the resisting drag torque always equals the torque on the magnetic dipole moment due to the magnetic field. The torque on a magnetic dipole with dipole moment $\vec{m}$ is $\tau_{\text{mag}} = mB\sin\theta$, where θ is the angle between the field and the dipole. FIGURE 26.47 shows that the magnetic torque is small when the field first reverses and the bacterium is pointing in the wrong direction. The torque is a maximum when $\theta = 90°$ and decreases as the bacterium completes its re-alignment with the magnetic field.

SOLVE At the bacterium's terminal angular speed, the drag torque balances the magnetic torque:

$$\tau_{\text{drag}} = 8\pi\eta R^3\omega = \tau_{\text{mag}} = mB\sin\theta$$

Consequently, the rotational speed when the magnetic dipole moment is at angle θ is

$$\omega = \frac{mB\sin\theta}{8\pi\eta R^3}$$

FIGURE 26.47 The torque on and rotational speed of a bacterium depend on the angle between its magnetic dipole moment and the magnetic field.

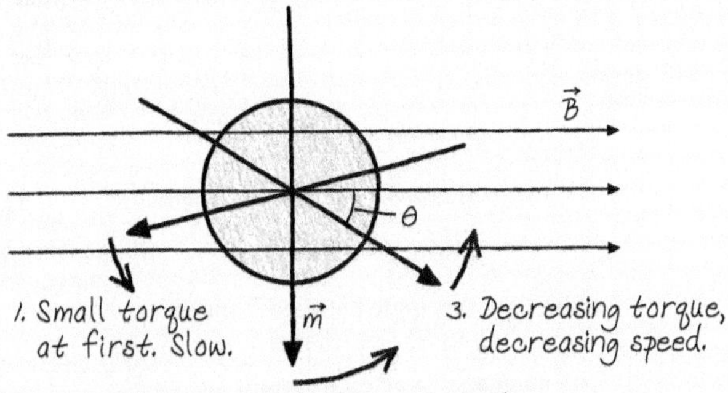

$\vec{B}$
θ
$\vec{m}$
1. Small torque at first. Slow.
2. Biggest torque, fastest speed.
3. Decreasing torque, decreasing speed.

The angular speed varies with the angle, speeding up and then slowing down as θ goes from 180° just after the field reverses to 0° when the bacterium completes its realignment. An exact solution requires a rather difficult integration. For the purpose of an estimate, however, we can replace both $\sin\theta$ and ω with their average values. The value of $\sin\theta$ goes from 0 to 1 and back to 0, with an approximate average value of $\frac{1}{2}$. An approximate average value of ω is the angular speed of an object undergoing uniform rotational motion: $\omega = 2\pi/T$, where T is the period for a complete revolution. The bacterium completes only half of a rotation in a time $\Delta t_{\text{reversal}} = T/2$, so our estimate of the rotational speed is $\omega = \pi/\Delta t_{\text{reversal}}$. If we use these estimates, the expression for ω becomes

$$\frac{\pi}{\Delta t_{\text{reversal}}} = \frac{mB}{16\pi\eta R^3}$$

We can solve this for the magnetic dipole moment:

$$m = \frac{16\pi^2\eta R^3}{B\Delta t_{\text{reversal}}} = \frac{16\pi^2(1.0\times10^{-3}\text{ Pa}\cdot\text{s})(1.0\times10^{-6}\text{ m})^3}{(1.0\times10^{-3}\text{ T})(0.5\text{ s})}$$

$$\approx 3\times10^{-16}\text{ A}\cdot\text{m}^2\times\left(\frac{1\text{ nm}}{10^{-9}\text{ m}}\right)^2 = 300\text{ A}\cdot\text{nm}^2$$

We converted the answer from SI units of $\text{A}\cdot\text{m}^2$ to $\text{A}\cdot\text{nm}^2$, which is more appropriate for expressing the magnetic dipole moments of nanoparticles.

ASSESS This very small magnetic dipole moment is difficult to assess. We can note that a typical magnetotactic bacterium contains roughly 20 magnetite crystals, each approximately 50 nm in size. If each contributes equally to the bacterium's total magnetic dipole moment, then the magnetic dipole moment of each crystal must be about 15 $\text{A}\cdot\text{nm}^2$. This is roughly a factor of 3 larger than measurements of the magnetic properties of nanoparticles. Considering the simplicity of our model, though, to come within a factor of 3 implies that our model has successfully captured the essential features of how magnetotactic bacteria respond to a magnetic field.

SUMMARY

GOAL To learn about magnetic fields and how magnetic fields exert forces on currents and moving charges.

GENERAL PRINCIPLES

Magnetic Fields

Magnetic fields are created by currents.

The magnetic field of a short segment of current is given by the **Biot-Savart law**:

$$B_{\text{current segment}} = \frac{\mu_0}{4\pi}\frac{I\Delta x \sin\theta}{r^2}$$

The magnetic field direction is given by the *right-hand rule for fields.*

The magnetic field of a complete current is found by dividing the current into many small segments and summing their contributions.

Magnetic Forces

The magnetic force on a moving charge is

$$F_{\text{on } q} = |q|vB\sin\alpha$$

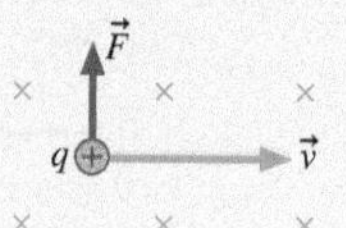

where α is the angle between $\vec{v}$ and $\vec{B}$. There is no force on a charge at rest.

The magnetic force on a current-carrying wire of length l is

$$F_{\text{wire}} = IlB\sin\alpha$$

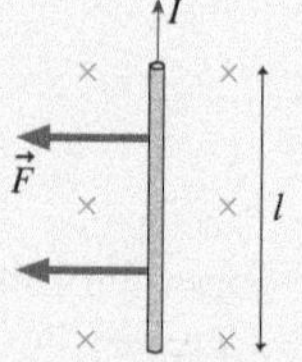

where α is the angle between the wire and $\vec{B}$.

For both, the force direction is given by the *right-hand rule for forces.*

IMPORTANT CONCEPTS

Wire

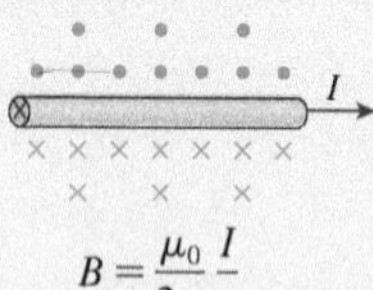

$$B = \frac{\mu_0}{2\pi}\frac{I}{r}$$

Loop or coil

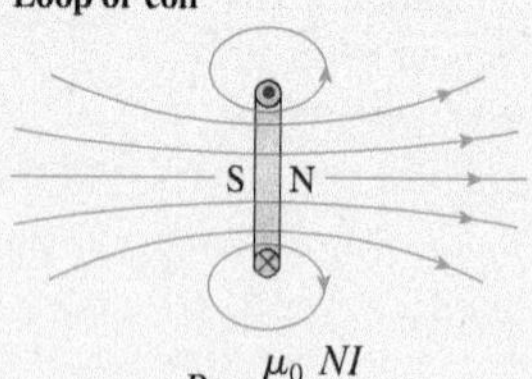

$$B = \frac{\mu_0}{2}\frac{NI}{R}$$

Solenoid

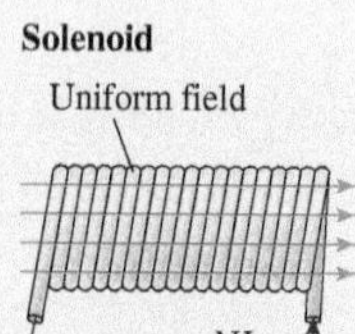

$$B = \frac{\mu_0 NI}{l}$$

Flat magnet

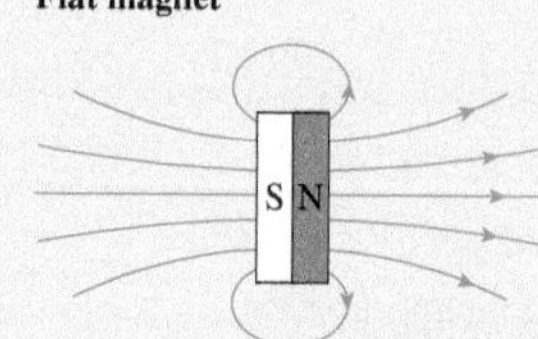

Right-hand rule for fields:

Point your right thumb in the direction of I. Your fingers curl in the direction of $\vec{B}$. For a dipole, $\vec{B}$ emerges from the side that is the north pole.

Charged-particle motion

No force if $\vec{v}$ is parallel to $\vec{B}$.

Cyclotron motion if $\vec{v}$ is perpendicular to $\vec{B}$.

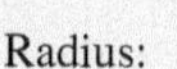

Radius: $r = \dfrac{mv}{qB}$

Frequency: $f = \dfrac{qB}{2\pi m}$

Forces between currents

Parallel currents attract.

Opposite currents repel.

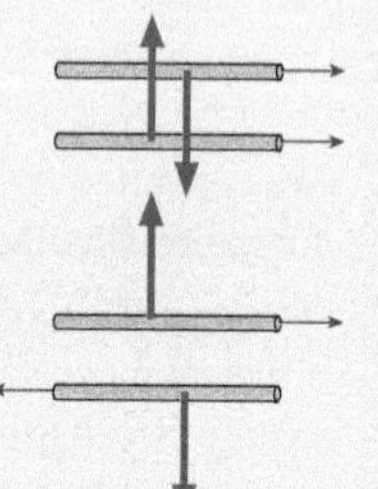

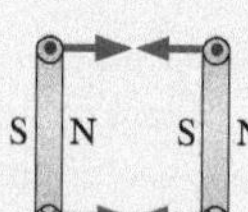

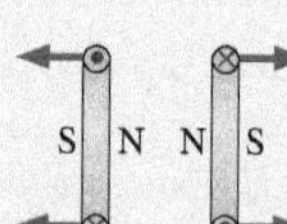

APPLICATIONS

The **magnetic dipole moment** of a current loop of area A is

$$\vec{m} = (IA, \text{from S to N})$$

The torque on a magnetic dipole moment is

$$\tau = mB\sin\theta$$

The energy of a magnetic dipole in a magnetic field is

$$U_{\text{mag}} = -mB\cos\theta$$

Right-hand rule for forces:

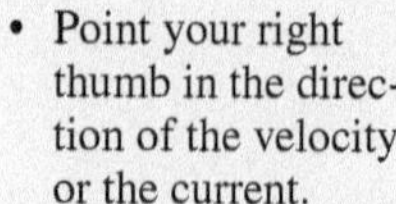

- Point your right thumb in the direction of the velocity or the current.
- Point your right index finger in the direction of the magnetic field.
- Bend your middle finger to be perpendicular to your palm. This is the direction of the force.

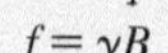

In **magnetic resonance imaging,** protons precess around the magnetic field lines with frequency

$$f = \gamma B$$

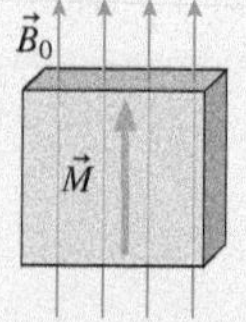
Laboratory coordinates

Initially, the net magnetization $\vec{M}$ points in the direction of $\vec{B}_0$.

A transverse magnetic field $\vec{B}_1$ that oscillates at the precession frequency—a resonance—rotates the magnetization 90° to the transverse plane.

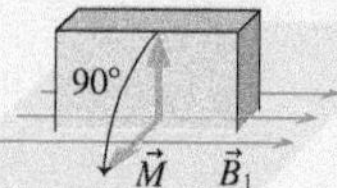

Rotating coordinates

Precession of $\vec{M}$ in the transverse plane is detected with a signal strength proportional to the density of hydrogen atoms.

LEARNING OBJECTIVES After studying this chapter, you should be able to:

- Interpret basic magnetic phenomena. *Conceptual Questions 26.1, 26.2*
- Calculate the magnetic fields of important current distributions. *Conceptual Questions 26.3, 26.4; Problems 26.1, 26.3, 26.5, 26.9, 26.12*
- Determine the poles of a magnetic dipole. *Conceptual Questions 26.5, 26.6; Problems 26.16–26.20*
- Calculate the magnetic force on a moving charged particle. *Conceptual Questions 26.8, 26.11, 26.12; Problems 26.21, 26.23–26.25, 26.29*
- Calculate the magnetic force on a current-carrying wire. *Conceptual Questions 26.13, 26.14; Problems 26.31–26.33, 26.35, 26.36*
- Calculate the torque on a magnetic dipole moment. *Conceptual Question 26.15; Problems 26.37–26.41*
- Explain the physical basis of magnetic resonance imaging. *Problems 26.42–26.46*

STOP TO THINK ANSWERS

Stop to Think 26.1: Not at all. The charge exerts weak, attractive polarization forces on both ends of the compass needle, but in this configuration the forces will balance and have no net effect.

Stop to Think 26.2: C. Point your right thumb in the direction of the current and curl your fingers around the wire.

Stop to Think 26.3: B. The current is coming toward you, so point your right thumb toward yourself and curl your fingers around. Your fingers on the left side are pointing down.

Stop to Think 26.4: B. The right-hand rule gives a downward $\vec{B}$ for a clockwise current. The north pole is on the side from which the field emerges.

Stop to Think 26.5: C. The force is perpendicular to the plane of $\vec{v}$ and $\vec{B}$, so this force requires the field to be either into or out of the figure. The right-hand rule for forces says that the force is to the left if the field is coming out of the figure. But the electron is negative, so it will have a force to the left if the field is into the figure.

Stop to Think 26.6: A, C. The force to produce these circular orbits is directed toward the center of the circle. Using the right-hand rule for forces, we see that this will be true for the situations in A and C if the particles are negatively charged.

Stop to Think 26.7: B. Repulsion indicates that the south pole of the loop is on the right, facing the bar magnet; the north pole is on the left. Then the right-hand rule gives the current direction.

QUESTIONS

Conceptual Questions

1. The lightweight glass sphere in Figure Q26.1 hangs by a thread. The north pole of a bar magnet is brought near the sphere.
   a. Suppose the sphere is electrically neutral. Is it attracted to, repelled by, or not affected by the magnet? Explain.
   b. Answer the same question if the sphere is positively charged.

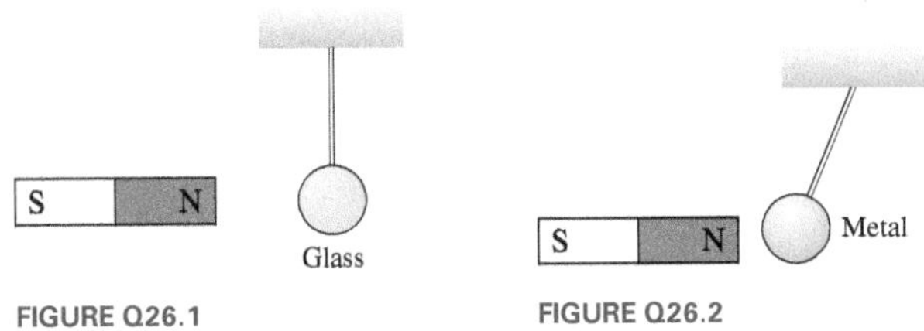

FIGURE Q26.1 FIGURE Q26.2

2. The metal sphere in Figure Q26.2 hangs by a thread. When the north pole of a magnet is brought near, the sphere is strongly attracted to the magnet. Then the magnet is reversed and its south pole is brought near the sphere. How does the sphere respond? Explain.

3. What is the current direction in the wire of Figure Q26.3? Explain.

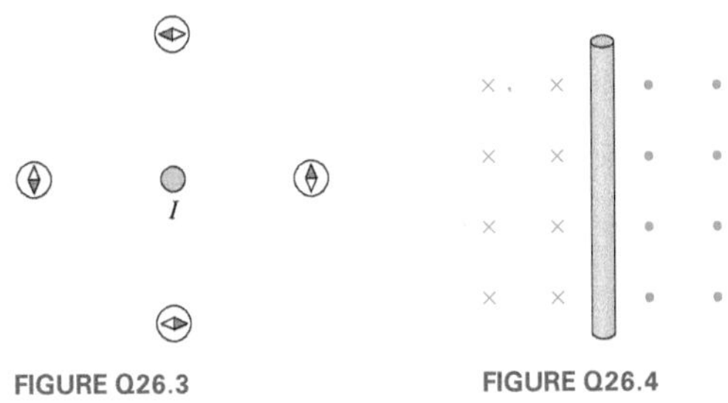

FIGURE Q26.3 FIGURE Q26.4

4. What is the current direction in the wire of Figure Q26.4? Explain.
5. If you were standing directly at the earth's geomagnetic north pole, in what direction would a compass point if it were free to swivel in any direction? Explain.
6. If you took a sample of magnetotactic bacteria from the northern hemisphere to the southern hemisphere, would you expect them to survive? Explain.

Problem difficulty is labeled as | (straightforward) to ||||| (challenging). Problems labeled INT integrate significant material from earlier chapters; BIO are of biological or medical interest; CALC require calculus to solve.

7. One long solenoid is placed inside another solenoid. Both solenoids have the same length and the same number of turns of wire, but the outer solenoid has twice the diameter of the inner solenoid. Each solenoid carries the same current, but the two currents are in opposite directions, as shown in the end view of Figure Q26.7. Does the magnetic field at the center of the inner solenoid point into the figure or out of the figure? Or is it zero? Explain.

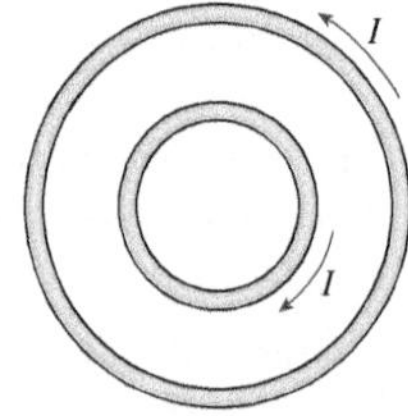
FIGURE Q26.7

8. What is the *initial* direction of deflection for the charged particles entering the magnetic fields shown in Figure Q26.8?

FIGURE Q26.8

9. What is the *initial* direction of deflection for the charged particles entering the magnetic fields shown in Figure Q26.9?

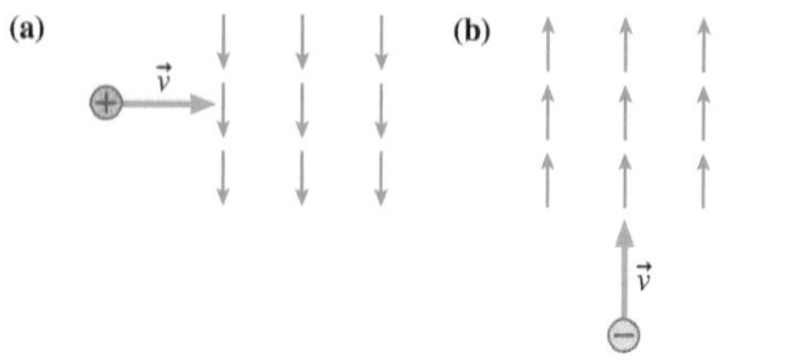
FIGURE Q26.9

10. An electron is moving near a long, current-carrying wire, as shown in Figure Q26.10. What is the direction of the magnetic force on the electron?

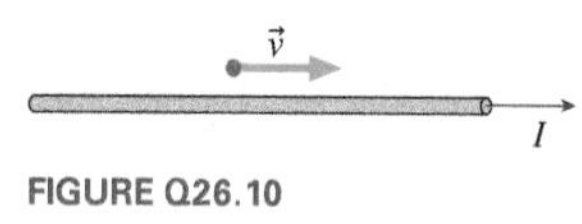
FIGURE Q26.10

11. Three charged particles move in a magnetic field as shown in Figure Q26.11. All the particles have the same mass and the same magnitude of charge. Which particle is moving the fastest? Which particles have positive charges, and which have negative charges?

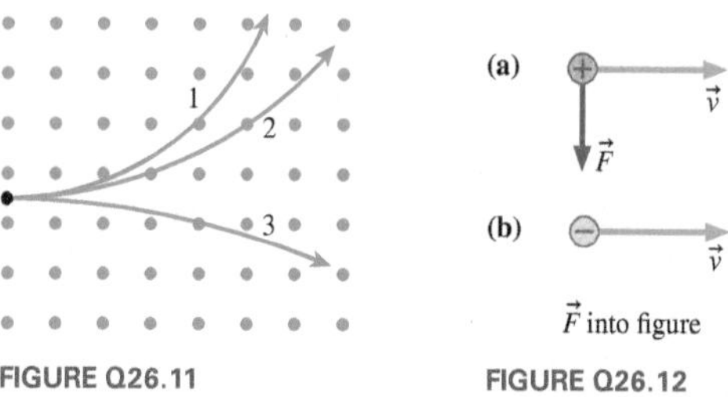
FIGURE Q26.11

FIGURE Q26.12

12. Determine the magnetic field direction that causes the charged particles shown in Figure Q26.12 to experience the indicated magnetic force.

13. A long wire and a square loop lie in the plane of the paper. Both carry a current in the direction shown in Figure Q26.13. Is there a net force on the loop? If so, in which direction? Explain.

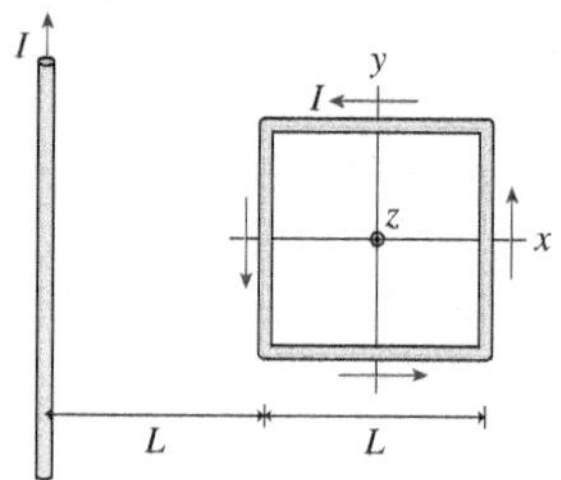
FIGURE Q26.13

14. The south pole of a bar magnet is brought toward the current loop of Figure Q26.14. Does the bar magnet attract, repel, or have no effect on the loop? Explain.

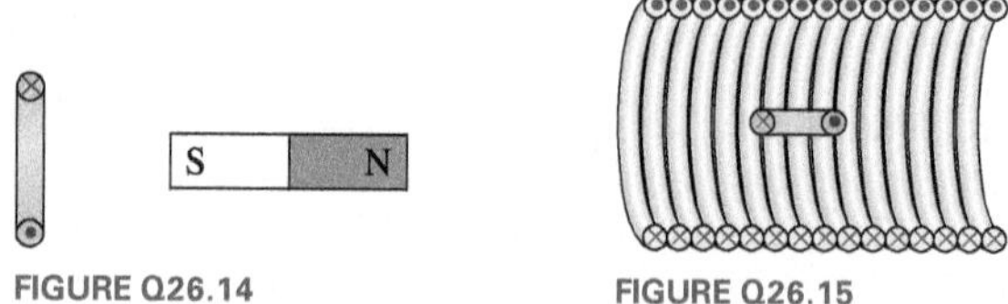
FIGURE Q26.14

FIGURE Q26.15

15. Figure Q26.15 shows a current-carrying solenoid in cross section. A horizontal current loop is at the center of the solenoid. Is there a torque on the current loop? If so, in which direction will the loop rotate? Explain.

Multiple-Choice Questions

16. | Two magnets with unlabeled poles are arranged as shown in Figure Q26.16. A compass points in the direction shown. What are the two poles 1 and 2?
A. 1 = N, 2 = S
B. 1 = N, 2 = N
C. 1 = S, 2 = S
D. 1 = S, 2 = N

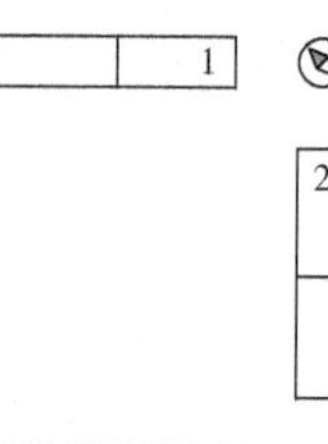
FIGURE Q26.16

17. | Two wires carry equal and opposite currents, as shown in Figure Q26.17. At a point directly between the two wires, the magnetic field is
A. Directed up, toward the top of the figure.
B. Directed down, toward the bottom of the figure.
C. Directed to the left.
D. Directed to the right.
E. Zero.

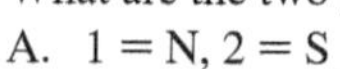
FIGURE Q26.17

18. || If a compass is placed above a current-carrying wire, as in Figure Q26.18, the needle will line up with the field of the wire. Which of the views shows the correct orientation of the needle for the noted current direction?

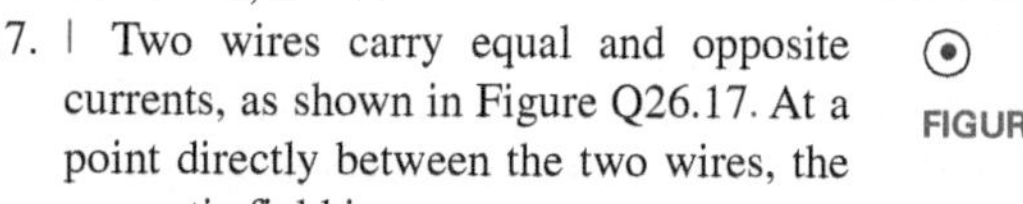
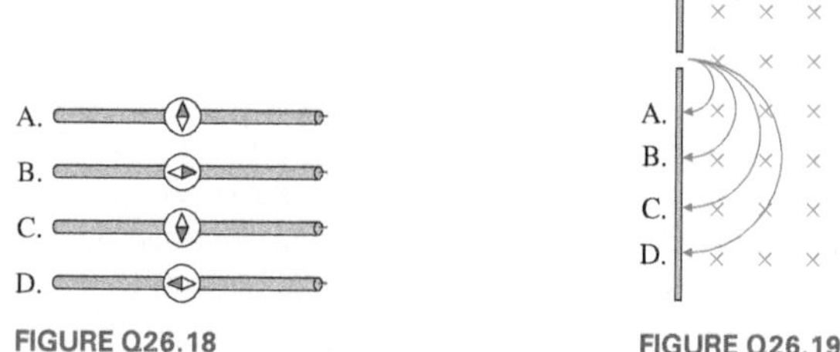
FIGURE Q26.18

FIGURE Q26.19

19. | Figure Q26.19 shows four particles moving to the right as they enter a region of uniform magnetic field, directed into the paper as noted. All particles move at the same speed and have the same charge. Which particle has the largest mass?

20. || A charged particle moves in a circular path in a uniform magnetic field. Which of the following would increase the period of its motion?
    A. Increasing its mass
    B. Increasing its charge
    C. Increasing the field strength
    D. Increasing its speed
21. | Two concentric circular current loops lie in the same plane. The outer loop has twice the diameter of the inner loop. The inner loop carries a 2 A current in a clockwise direction. The magnetic field at the center of the loops is zero. What is the current in the outer loop?
    A. 1 A clockwise B. 1 A counterclockwise
    C. 4 A clockwise D. 4 A counterclockwise
22. || If all of the particles shown in Figure Q26.22 are electrons, what is the direction of the magnetic field that produced the indicated deflection?
    A. Directed up, toward the top of the figure.
    B. Directed down, toward the bottom of the figure.
    C. Out of the plane of the figure.
    D. Into the plane of the figure.

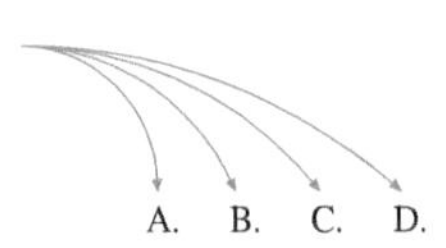

FIGURE Q26.22

A. B. C. D.

FIGURE Q26.23

23. | If two compasses are brought near enough to each other, the magnetic fields of the compasses themselves will be larger than the field of the earth, and the needles will line up with each other. Which of the arrangements of two compasses shown in Figure Q26.23 is a possible stable arrangement?

PROBLEMS

Section 26.1 Magnetism

Section 26.2 The Magnetic Field of a Current

1. | What currents are needed to generate the magnetic field strengths of Table 26.1 at a point 1.0 cm from a long, straight wire?
2. | Although the evidence is weak, there has been concern in re-
BIO cent years over possible health effects from the magnetic fields generated by electric transmission lines. A typical high-voltage transmission line is 20 m above the ground and carries a 200 A current at a potential of 110 kV.
    a. What is the magnetic field strength on the ground directly under such a transmission line?
    b. What percentage is this of the earth's magnetic field of 50 μT?
3. | A biophysics experiment uses a very sensitive magnetic
BIO field probe to determine the current associated with a nerve impulse traveling along an axon. If the peak field strength 1.0 mm from an axon is 8.0 pT, what is the peak current carried by the axon?
4. || The element niobium, which is a metal, is a superconductor (i.e., no electrical resistance) at temperatures below 9 K. However, the superconductivity is destroyed if the magnetic field at the surface of the metal reaches or exceeds 0.10 T. What is the maximum current in a straight, 3.0-mm-diameter superconducting niobium wire?
5. || What should the current I be in the lower wire of Figure P26.5 such that the magnetic field at point P is zero?

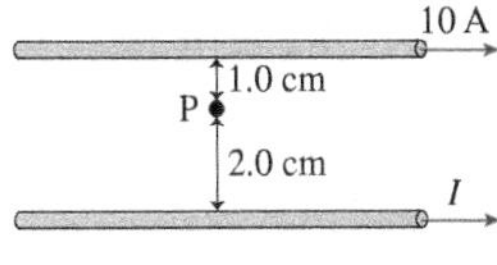

FIGURE P26.5

FIGURE P26.6

6. || What are the magnetic field strength and direction at points 1 to 3 in Figure P26.6?
7. || What are the magnetic field strengths at points 1 to 3 in Figure P26.7? Give your answers as vectors.

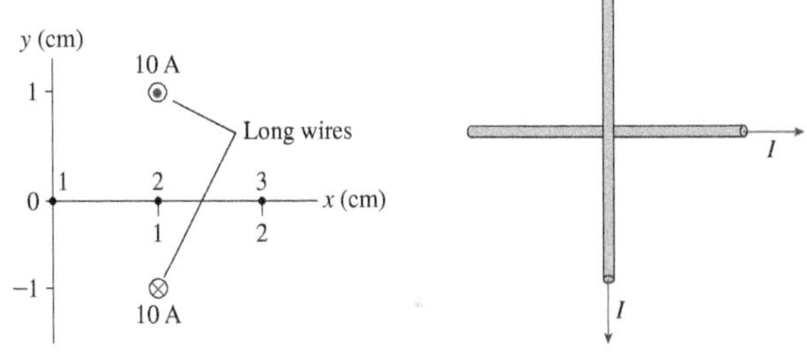

FIGURE P26.7 FIGURE P26.8

8. ||| Two long wires carry 10 A currents in the directions shown in Figure P26.8. One wire is 10 cm above the other. What are the direction and magnitude of the magnetic field at a point halfway between them?
9. || The magnetic field at the center of a 1.0-cm-diameter loop is 2.5 mT.
    a. What is the current in the loop?
    b. A long straight wire carries the same current you found in part a. At what distance from the wire is the magnetic field 2.5 mT?
10. || The wire in Figure P26.10 carries 10 times the current of the loop; the direction of the current in the wire is shown. The magnetic field in the center of the loop is zero. If the radius R of the loop is 1.3 cm,
    a. What is the direction of the current in the loop?
    b. What is the distance d from the wire to the center of the loop?

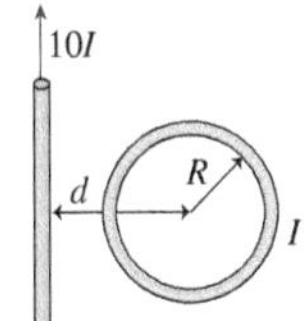

FIGURE P26.10

11. || The magnetic field of the brain has been measured to be ap-
BIO proximately 3.0×10^{-12} T. Although the currents that cause this field are quite complicated, we can get a rough estimate of their size by modeling them as a single circular current loop 16 cm (the width of a typical head) in diameter. What current is needed to produce such a field at the center of the loop?
12. || A solenoid used to produce magnetic fields for research purposes is 2.0 m long, with an inner radius of 30 cm and 1000 turns of wire. When running, the solenoid produces a field of 1.0 T in the center. Given this, how large a current does it carry?

13. || A researcher would like to perform an experiment in zero magnetic field, which means that the field of the earth must be canceled. Suppose the experiment is done inside a solenoid of diameter 1.0 m, length 4.0 m, with a total of 5000 turns of wire. The solenoid is oriented to produce a field that opposes and exactly cancels the 52 μT local value of the earth's field. What current is needed in the solenoid's wire?

14. | BIO Experimental tests have shown that hammerhead sharks can detect magnetic fields. In one such test, 100 turns of wire were wrapped around a 7.0-m-diameter cylindrical shark tank. A magnetic field was created inside the tank when this coil of wire carried a current of 1.5 A. Sharks trained by getting a food reward when the field was present would later unambiguously respond when the field was turned on.
   a. What was the magnetic field strength in the center of the tank due to the current in the coil?
   b. Is the strength of the coil's field at the center of the tank larger or smaller than that of the earth?

15. || BIO Magnetic resonance imaging needs a magnetic field strength of 1.5 T. The solenoid is 1.8 m long and 75 cm in diameter. It is tightly wound with a single layer of 2.0-mm-diameter superconducting wire. What size current is needed?

Section 26.3 Magnetic Dipoles

16. || The on-axis magnetic field strength 10 cm from a small bar magnet is 5.0 μT.
   a. What is the bar magnet's magnetic dipole moment?
   b. What is the on-axis field strength 15 cm from the magnet?

17. || A 100 A current circulates around a 2.0-mm-diameter superconducting ring.
   a. What is the ring's magnetic dipole moment?
   b. What is the on-axis magnetic field strength 5.0 cm from the ring?

18. ||| A small, square loop carries a 25 A current. The on-axis magnetic field strength 50 cm from the loop is 7.5 nT. What is the edge length of the square?

19. || The earth's magnetic dipole moment is $8.0 \times 10^{22}\ \text{A} \cdot \text{m}^2$.
   a. What is the magnetic field strength on the surface of the earth at the earth's north magnetic pole? How does this compare to the value in Table 26.1? You can assume that the current loop is deep inside the earth.
   b. Astronauts discover an earth-size planet without a magnetic field. To create a magnetic field at the north pole with the same strength as earth's, they propose running a current through a wire around the equator. What size current would be needed? The earth's radius is 6370 km.

20. || BIO The heart produces a weak magnetic field that can be used to diagnose certain heart problems. It is a dipole field produced by a current loop in the outer layers of the heart.
   a. It is estimated that the field at the center of the heart is 90 pT. What current must circulate around a 12-cm-diameter loop, about the size of a human heart, to produce this field?
   b. What is the magnitude of the heart's magnetic dipole moment?

Section 26.4 The Magnetic Force on a Moving Charge

21. | A proton moves with a speed of 1.0×10^7 m/s in the direction shown in Figure P26.21. A 0.50 T magnetic field points in the positive x-direction. What are the magnitude and direction of the magnetic force on the proton?

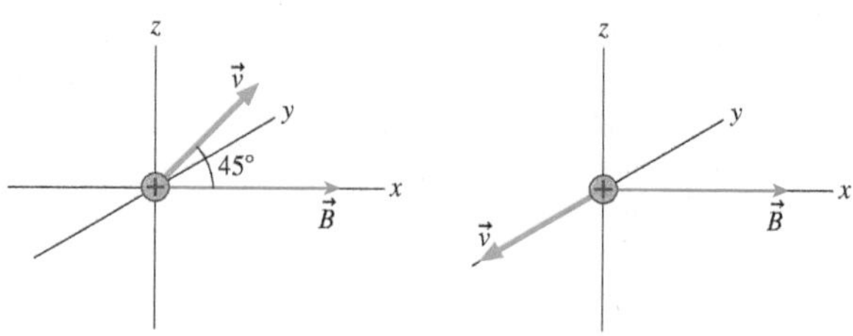

FIGURE P26.21 FIGURE P26.22

22. | A proton moves with a speed of 1.0×10^7 m/s in the direction shown in Figure P26.22. A 0.50 T magnetic field points in the positive x-direction. What are the magnitude and direction of the magnetic force on the proton?

23. || An electron moves with a speed of 1.0×10^7 m/s in the direction shown in Figure P26.23. A 0.50 T magnetic field points in the positive x-direction. What are the magnitude and direction of the magnetic force on the electron?

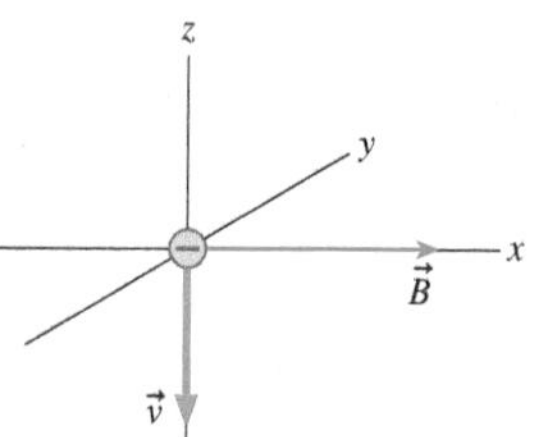

FIGURE P26.23

24. | BIO An electromagnetic flowmeter applies a magnetic field of 0.20 T to blood flowing through a coronary artery at a speed of 15 cm/s. What force is felt by a chlorine ion with a single negative charge?

25. | The aurora is caused when electrons and protons, moving in the earth's magnetic field of $\approx 5.0 \times 10^{-5}$ T, collide with molecules of the atmosphere and cause them to glow. What is the radius of the circular orbit for
   a. An electron with speed 1.0×10^6 m/s?
   b. A proton with speed 5.0×10^4 m/s?

26. | In a simplified model of the hydrogen atom, its electron moves in a circular orbit with a radius of 5.3×10^{-10} m at a frequency of 6.6×10^{15} Hz. What magnetic field would be required to cause an electron to undergo this same motion?

27. | Radio astronomers detect electromagnetic radiation at 45 MHz from an interstellar gas cloud. They suspect this radiation is emitted by electrons spiraling in a magnetic field. What is the magnetic field strength inside the gas cloud?

28. || To five significant figures, what are the cyclotron frequencies in a 3.0000 T magnetic field of the ions (a) O_2^+, (b) N_2^+, and (c) CO^+? The atomic masses are shown in the table; the mass of the missing electron is less than 0.001 u and is not relevant at this level of precision. Although N_2^+ and CO^+ both have a nominal molecular mass of 28, they are easily distinguished by virtue of their slightly different cyclotron frequencies. Use the following constants: $1\ \text{u} = 1.6605 \times 10^{-27}$ kg, $e = 1.6022 \times 10^{-19}$ C.

Atomic masses	
^{12}C	12.000
^{14}N	14.003
^{16}O	15.995

29. || A cyclotron uses a strong magnetic field to keep protons moving in circular orbits. A proton begins in a small-diameter orbit with a low speed, but a weak oscillating electric field gradually accelerates it to higher and higher speeds. This causes the orbit's diameter to slowly increase until the proton exits the machine from a large-diameter orbit. One commercial cyclotron uses a 1.5 T magnetic field to accelerate protons to a final kinetic energy of 1.0×10^{-12} J. What is the diameter of the largest orbit, just before the protons exit the cyclotron?

30. || A 50-cm-long, 10-cm-diameter solenoid creates the uniform magnetic field for an experiment in which electrons undergo cyclotron motion with a frequency of 550 MHz. The solenoid has 2000 turns of wire. What is the current through the solenoid?

Section 26.5 Magnetic Forces on Current-Carrying Wires

31. | What magnetic field strength and direction will levitate the 2.0 g wire in Figure P26.31?

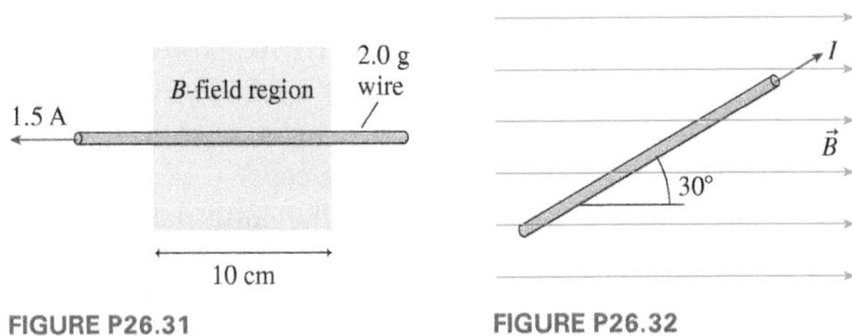

FIGURE P26.31

FIGURE P26.32

32. | A uniform 2.5 T magnetic field points to the right. A 3.0-m-long wire, carrying 15 A, is placed at an angle of 30° to the field, as shown in Figure P26.32. What is the force (magnitude and direction) on the wire?

33. | What is the net force (magnitude and direction) on each wire in Figure P26.33?

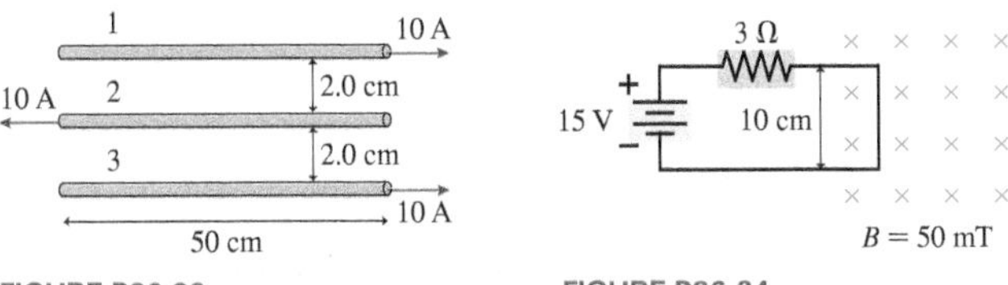

FIGURE P26.33

FIGURE P26.34

34. | The right edge of the circuit in Figure P26.34 extends into
INT a 50 mT uniform magnetic field. What are the magnitude and direction of the net force on the circuit?

35. || The two 10-cm-long parallel wires in Figure P26.35 are sep-
INT arated by 5.0 mm. For what value of the resistor R will the force between the two wires be 5.4×10^{-5} N?

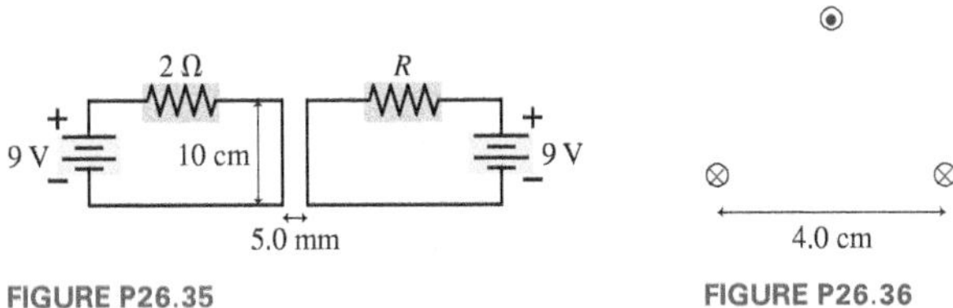

FIGURE P26.35

FIGURE P26.36

36. ||| Figure P26.36 is a cross section through three long wires with
INT linear mass density 50 g/m. They each carry equal currents in the directions shown. The lower two wires are 4.0 cm apart and are attached to a table. What current I will allow the upper wire to "float" so as to form an equilateral triangle with the lower wires?

Section 26.6 Forces and Torques on Magnetic Dipoles

37. || A square current loop 5.0 cm on each side carries a 500 mA current. The loop is in a 1.2 T uniform magnetic field. The axis of the loop, perpendicular to the plane of the loop, is 30° away from the field direction. What is the magnitude of the torque on the current loop?

38. | A small bar magnet experiences a 0.020 N · m torque when the axis of the magnet is at 45° to a 0.10 T magnetic field. What is the magnitude of its magnetic dipole moment?

39. | The magnetic dipole moment of the proton has been measured to be 1.4×10^{-26} A · m². What is the maximum possible torque on a proton in a 1.2 T magnetic field?

40. || What is the magnitude of the torque on the current loop in Figure P26.40?

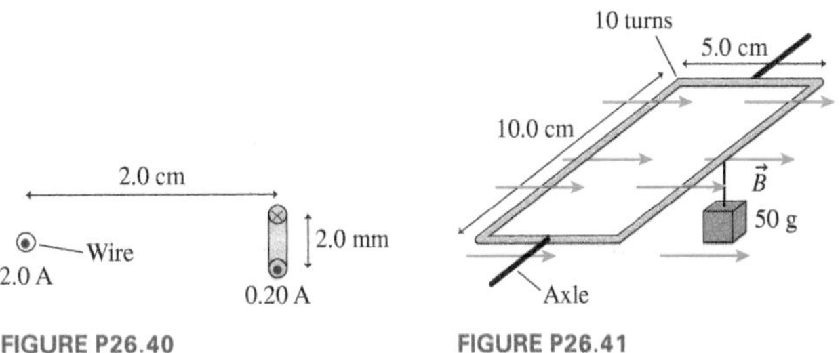

FIGURE P26.40

FIGURE P26.41

41. ||| The 10-turn loop of wire shown in Figure P26.41 lies in a
INT horizontal plane, parallel to a uniform horizontal magnetic field, and carries a 2.0 A current. The loop is free to rotate about a nonmagnetic axle through the center. A 50 g mass hangs from one edge of the loop. What magnetic field strength will prevent the loop from rotating about the axle?

Section 26.7 Magnetic Resonance Imaging

42. || In some facilities, real-time MRI is used to monitor a treat-
BIO ment. This imaging allows the treatment to be adjusted to achieve optimum results. However, the devices that deliver the treatment can themselves produce magnetic fields that interfere with the imaging. In one case, a device produces a 75 μT static field parallel to the 1.5 T field in the MRI machine. By how much does this additional field change the proton's precession frequency?

43. || An NMR spectrometer uses a 7.5 T magnet.
   a. What is the proton precession frequency in the field of the magnet?
   b. Half of the hydrogen atoms in a sample experience only the field of the magnet. The other half experience both the field of the magnetic and a 150 μT field pointing in the same direction that is created by neighboring atoms. The chemist analyzing this sample will find two resonance frequencies, and the frequency difference between them is called the *chemical shift*. What is the chemical shift in this sample?

44. | In MRI, the transverse relaxation time measures how long it
BIO takes precessing protons to get out of phase with each other so that their individual magnetic dipole moments no longer add up to a macroscopic magnetization. In a 1.5 T MRI machine, how many precessions does a proton make in oxygenated whole blood during the transverse relaxation time?

45. I A proton is not the only nucleus that has a magnetic dipole moment. Another is the nucleus of the isotope ^{15}N, which is sometimes imaged in MRI. The gyromagnetic ratio of a ^{15}N nucleus is 10.1% that of a proton. What is the precession frequency of a ^{15}N nucleus in a 1.50 T MRI machine?

46. III The oscillating magnetic field in a 3.0 T MRI machine has an amplitude of 1.0 mT. This field appears static in a coordinate system that is rotating at the precession frequency. What is the duration in μs of a 90° pulse?

General Problems

47. II Point A in Figure P26.47 is 2.0 mm from the wire. What is
INT the magnetic field strength at point A? You can assume that the wire is very long and that all other wires are too far away to contribute to the magnetic field.

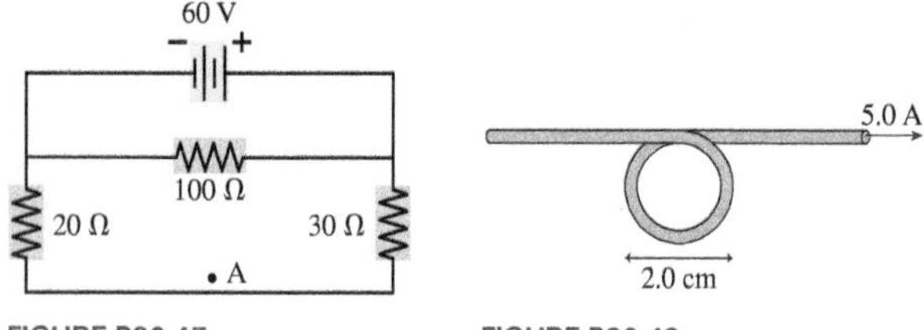

FIGURE P26.47 FIGURE P26.48

48. II What are the strength and direction of the magnetic field at the center of the loop in Figure P26.48?

49. II A scientist measuring the resistivity of a new metal alloy
INT left her ammeter in another lab, but she does have a magnetic field probe. So she creates a 6.5-m-long, 2.0-mm-diameter wire of the material, connects it to a 1.5 V battery, and measures a 3.0 mT magnetic field 1.0 mm from the surface of the wire. What is the material's resistivity?

50. III In Figure P26.50, A and B are long wires perpendicular to the figure. Wire A carries a 10 A current. A compass needle is in equilibrium at the angle shown. What are the magnitude and direction of the current in wire B? Assume that the earth's magnetic field can be neglected.

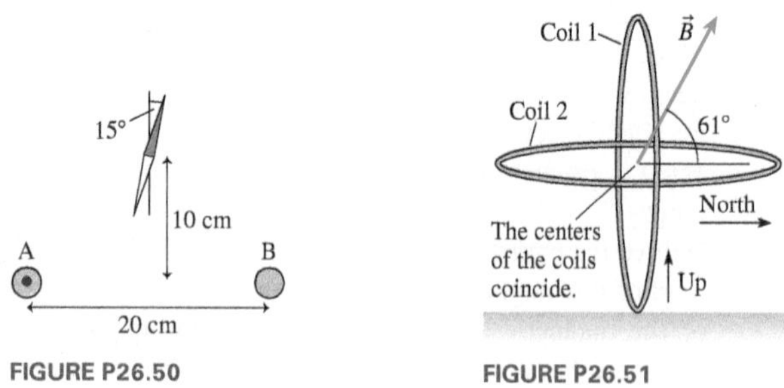

FIGURE P26.50 FIGURE P26.51

51. III Young domestic chickens have the ability to orient them-
BIO selves in the earth's magnetic field. Researchers used a set of two coils to adjust the magnetic field in the chicks' pen. Figure P26.51 shows the two coils, whose centers coincide, seen edge-on. The axis of coil 1 is parallel to the ground and points to the north; the axis of coil 2 is oriented vertically. Each coil has 43 turns and a radius of 1.0 m. At the location of the experiment, the earth's field had a magnitude of 5.6×10^{-5} T and pointed to the north, tilted up from the horizontal by 61°. If the researchers wished to exactly cancel the earth's field, what currents would they need to apply in coil 1 and in coil 2?

52. I The individual magnetized iron particles in magnetotactic
BIO bacteria are typically 25 nm in diameter with a magnetic dipole moment of 30 A · nm^2. What current is needed to create this magnetic dipole moment in a 25-nm-diameter current loop?

53. I Honeybees have a magnetic field sense that is likely based on
BIO iron granules in their abdomens. In one test of bees' ability to sense magnetic fields, bees were placed at the center of a 64-turn, 1.0-m-diameter coil of wire that produced a 65 μT field.
 a. What was the current in the coil?
 b. The field of the coil was perpendicular to the earth's 50 μT magnetic field. What was the strength of the field that the bees experienced?

54. I The magnetic field at the surface of the earth varies with time.
BIO In addition to long-term variations, a relatively strong 7.8 Hz variation is due to currents in the ionosphere. Some investigators are considering the effects of these time-varying fields on living tissue. One team of investigators applied a sinusoidal magnetic field to a cell culture of growing cardiac muscle cells. The cells were at the center of a single 35-mm-diameter loop of wire. The investigators used a 7.8 Hz AC power supply with a peak voltage of 5.0 V in series with a 2.0 kΩ resistor to supply current to the loop. What was the amplitude of the time-varying magnetic field at the position of the cells?

55. II The magnetic field strength at the north pole of a 2.0-cm-diameter, 8-cm-long permanent magnet is 0.10 T. To produce the same field with a solenoid of the same size, carrying a current of 2.0 A, how many turns of wire would you need?

56. II The earth's magnetic field, with a magnetic dipole moment of 8.0×10^{22} A · m^2, is generated by currents within the molten iron of the earth's outer core. Suppose we model the core current as a 3000-km-diameter current loop made from a 1000-km-diameter "wire." The loop diameter is measured from the centers of this very fat wire. What is the current in the current loop?

57. III Bats are capable of navigating using the earth's field—a plus for
BIO an animal that may fly great distances from its roost at night. If, while sleeping during the day, bats are exposed to a field of a similar magnitude but different direction than the earth's field, they are more likely to lose their way during their next lengthy night flight. Suppose you are a researcher doing such an experiment in a location where the earth's field is 50 μT at a 60° angle below horizontal. You make a 50-cm-diameter, 100-turn coil around a roosting box; the sleeping bats are at the center of the coil. You wish to pass a current through the coil to produce a field that, when combined with the earth's field, creates a net field with the same strength and dip angle (60° below horizontal) as the earth's field but with a horizontal component that points south rather than north. What are the proper orientation of the coil and the necessary current?

58. III At the equator, the earth's field is essentially horizontal; near
BIO the north pole, it is nearly vertical. In between, the angle varies. As you move farther north, the dip angle, the angle of the earth's field below horizontal, steadily increases. Green turtles seem to use this dip angle to determine their latitude. Suppose you are a researcher wanting to test this idea. You have gathered green turtle hatchlings from a beach where the magnetic field strength is 50 μT and the dip angle is 56°. You then put the turtles in a 1.2-m-diameter circular tank and monitor the direction in which they swim as you vary the magnetic field in the tank. You change the field by passing a current through a 100-turn horizontal coil wrapped around the tank. This creates a field that adds to that of the earth. What current should you pass through the coil, and in what direction, to produce a net field in the center of the tank that has a dip angle of 62°?

59. |||| You have a 1.0-m-long copper wire. You want to make an N-turn current loop that generates a 1.0 mT magnetic field at the center when the current is 1.0 A. You must use the entire wire. What will be the diameter of your coil?

60. || An electron travels with speed 1.0×10^7 m/s between the two parallel charged plates shown in Figure P26.60. The plates are separated by 1.0 cm and are charged by a 200 V battery. What magnetic field strength and direction will allow the electron to pass between the plates without being deflected?
INT

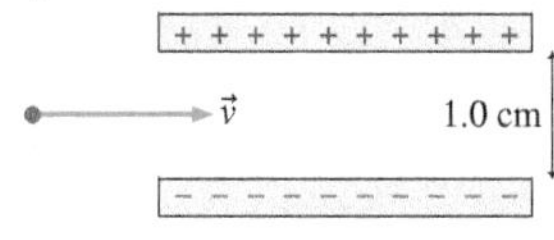

FIGURE P26.60

61. | Typical blood velocities in the coronary arteries range from 10 to 30 cm/s. An electromagnetic flowmeter applies a magnetic field of 0.25 T to a coronary artery with a blood velocity of 15 cm/s. This field exerts a force on ions in the blood, which will separate. The ions will separate until they make an electric field that exactly balances the magnetic force. This electric field produces a voltage that can be measured.
BIO INT
   a. What force is felt by a singly ionized (positive) sodium ion?
   b. Charges in the blood will separate until they produce an electric field that cancels this magnetic force. What will be the resulting electric field?
   c. What voltage will this electric field produce across an artery with a diameter of 3.0 mm?

62. || Irrigation channels that require regular flow monitoring are often equipped with electromagnetic flowmeters in which the magnetic field is produced by horizontal coils embedded in the bottom of the channel. A particular coil has 100 turns and a diameter of 6.0 m. When it's time for a measurement, a 5.0 A current is turned on. The large diameter of the coil means that the field in the water flowing directly above the center of the coil is approximately equal to the field in the center of the coil.
   a. What is the magnitude of the field at the center of the coil?
   b. If the field is directed downward and the water is flowing east, what is the direction of the force on a positive ion in the water above the center of the coil?
   c. If the water is flowing above the center of the coil at 1.5 m/s, what is the magnitude of the force on an ion with a charge $+e$?

63. ||| A mass spectrometer uses a magnetic field to determine the masses of ions. A beam of ions, each with a mass 62 times that of a proton, enters a 0.30 T magnetic field, as shown in Figure P26.63. If the speed of the ions is 1.4×10^5 m/s, at what angle θ do the ions leave the field?

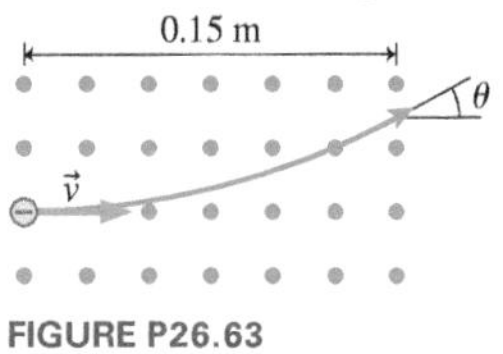

FIGURE P26.63

64. ||| A magnetic mass spectrometer accelerates a CO^+ ion through a 450 V potential difference. The ion then enters a uniform magnetic field. The exit slit in front of the detector is 15 cm away from the entrance slit where ions enter the magnetic field. What magnetic field strength is needed to detect CO^+ ions?
INT

65. ||| Many hospitals use a *cyclotron* to produce radioactive isotopes for nuclear imaging or nuclear medicine. A cyclotron uses a strong magnetic field to keep protons moving in circular orbits. A proton begins in a small-diameter orbit with a low speed, but it is gradually accelerated—with a clever use of electric fields—to higher and higher speeds. This causes the orbit's diameter to gradually increase until the proton exits the machine from a large-diameter orbit. One commercial cyclotron uses a 2.0 T magnetic field.
BIO
   a. How much time does it take a proton to complete one orbit?
   b. What is the kinetic energy of a proton in an initial 1.0-mm-diameter orbit and in a final 1.0-m-diameter orbit?
   c. The proton's acceleration is equivalent to moving through a potential difference $\Delta V = -1000$ V per orbit. How long does it take a proton to move out from its initial orbit to its final orbit?

66. || An antiproton is identical to a proton except it has the opposite charge, $-e$. To study antiprotons, they must be confined in an ultrahigh vacuum because they will annihilate—producing gamma rays—if they come into contact with the protons of ordinary matter. One way of confining antiprotons is to keep them in a magnetic field. Suppose that antiprotons are created with a speed of 1.5×10^4 m/s and then trapped in a 2.0 mT magnetic field. What minimum diameter must the vacuum chamber have to allow these antiprotons to circulate without touching the walls?

67. || The individual magnetized iron particles in magnetotactic bacteria are typically 25 nm in diameter with a magnetic dipole moment of 30 $A \cdot nm^2$.
BIO
   a. What energy is required to rotate one particle from having its magnetic dipole moment parallel to the earth's 50 μT magnetic field to being antiparallel to the earth's field?
   b. What is the thermal energy $k_B T$ at 20°C?
   c. Would one magnetized particle stay aligned with the earth's magnetic field? What about a chain of 10 such particles?

68. ||| Figure P26.68 is an edge view of a 2.0 kg square loop, 2.5 m on each side, with its lower edge resting on a frictionless, horizontal surface. A 25 A current is flowing around the loop in the direction shown. What is the strength of a uniform, horizontal magnetic field for which the loop is in static equilibrium at the angle shown?
INT

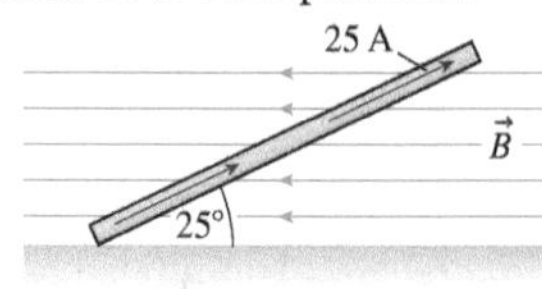

FIGURE P26.68

69. ||| A long, straight wire with a linear mass density of 50 g/m is suspended by threads, as shown in Figure P26.69. There is a uniform magnetic field pointing vertically downward. A 10 A current in the wire experiences a horizontal magnetic force that deflects it to an equilibrium angle of 10°. What is the strength of the magnetic field $\vec{B}$?
INT

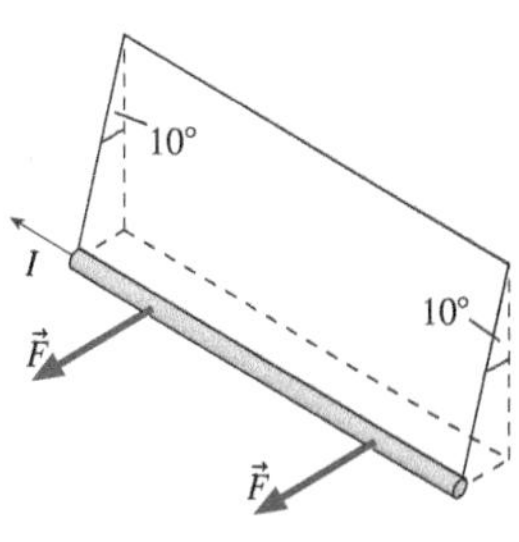

FIGURE P26.69

70. |||| It is shown in more advanced courses that charged particles in circular orbits radiate electromagnetic waves, called *cyclotron radiation*. As a result, a particle undergoing cyclotron motion with speed v is actually losing kinetic energy at the rate
CALC

$$\frac{dK}{dt} = -\left(\frac{\mu_0 q^4}{6\pi c m^2}\right) B^2 v^2$$

How long does it take (a) an electron and (b) a proton to radiate away half its energy while spiraling in a 2.0 T magnetic field?

MCAT-Style Passage Problems

The Velocity Selector

In experiments where all the charged particles in a beam are required to have the same velocity, scientists use a *velocity selector.* A velocity selector has a region of uniform electric and magnetic fields that are perpendicular to each other and perpendicular to the motion of the charged particles. Both the electric and magnetic fields exert a force on the charged particles. If a particle has precisely the right velocity, the two forces exactly cancel and the particle is not deflected. Equating the forces due to the electric field and the magnetic field gives the following equation:

$$qE = qvB$$

Solving for the velocity, we get:

$$v = \frac{E}{B}$$

A particle moving at this velocity will pass through the region of uniform fields with no deflection, as shown in Figure P26.71. For higher or lower velocities than this, the particles will feel a net force and will be deflected. A slit at the end of the region allows only the particles with the correct velocity to pass.

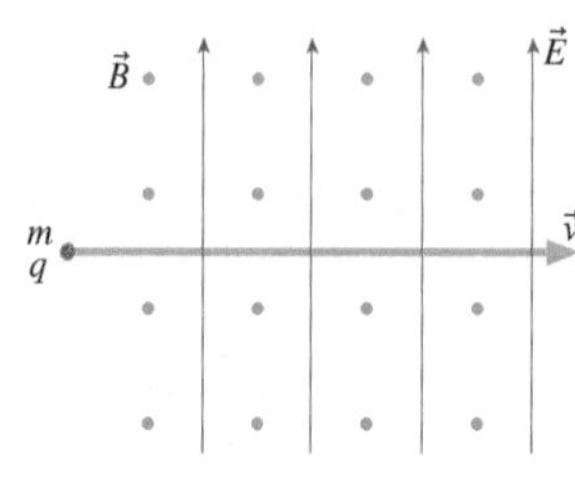

FIGURE P26.71

71. Assuming the particle in Figure P26.71 is positively charged, what are the directions of the forces due to the electric field and to the magnetic field?
    A. The electric force points up, the magnetic force points down.
    B. The electric force points down, the magnetic force points up.
    C. The electric force points out of the figure, the magnetic force points into the figure.
    D. The electric force points into the figure, the magnetic force points out of the figure.
72. How does the kinetic energy of the particle in Figure P26.71 change as it traverses the velocity selector?
    A. The kinetic energy increases.
    B. The kinetic energy does not change.
    C. The kinetic energy decreases.
73. Suppose a particle with twice the velocity of the particle in Figure P26.71 enters the velocity selector. The path of this particle will curve
    A. Upward, toward the top of the figure.
    B. Downward, toward the bottom of the figure.
    C. Out of the plane of the figure.
    D. Into the plane of the figure.
74. Next, a particle with the same mass and velocity as the particle in Figure P26.71 enters the velocity selector. This particle has a charge of $2q$—twice the charge of the particle in Figure P26.71. In this case, we can say that
    A. The force of the electric field on the particle is greater than the force of the magnetic field.
    B. The force of the magnetic field on the particle is greater than the force of the electric field.
    C. The forces of the electric and magnetic fields on the particle are still equal.

Ocean Potentials

The ocean is salty because it contains many dissolved ions. As these charged particles move with the water in strong ocean currents, they feel a force from the earth's magnetic field. Positive and negative charges are separated until an electric field develops that balances this magnetic force. This field produces measurable potential differences that can be monitored by ocean researchers.

The Gulf Stream moves northward off the east coast of the United States at a speed of up to 3.5 m/s. Assume that the current flows at this maximum speed and that the earth's field is 50 μT tipped 60° below horizontal.

75. What is the direction of the magnetic force on a singly ionized negative chlorine ion moving in this ocean current?
    A. East B. West C. Up D. Down
76. What is the magnitude of the force on this ion?
    A. 2.8×10^{-23} N B. 2.4×10^{-23} N
    C. 1.6×10^{-23} N D. 1.4×10^{-23} N
77. What magnitude electric field is necessary to exactly balance this magnetic force?
    A. 1.8×10^{-4} N/C B. 1.5×10^{-4} N/C
    C. 1.0×10^{-4} N/C D. 0.9×10^{-4} N/C
78. The electric field produces a potential difference. If you place one electrode 10 m below the surface of the water, you will measure the greatest potential difference if you place the second electrode
    A. At the surface.
    B. At a depth of 20 m.
    C. At the same depth 10 m to the north.
    D. At the same depth 10 m to the east.

27 Electromagnetic Induction and Electromagnetic Waves

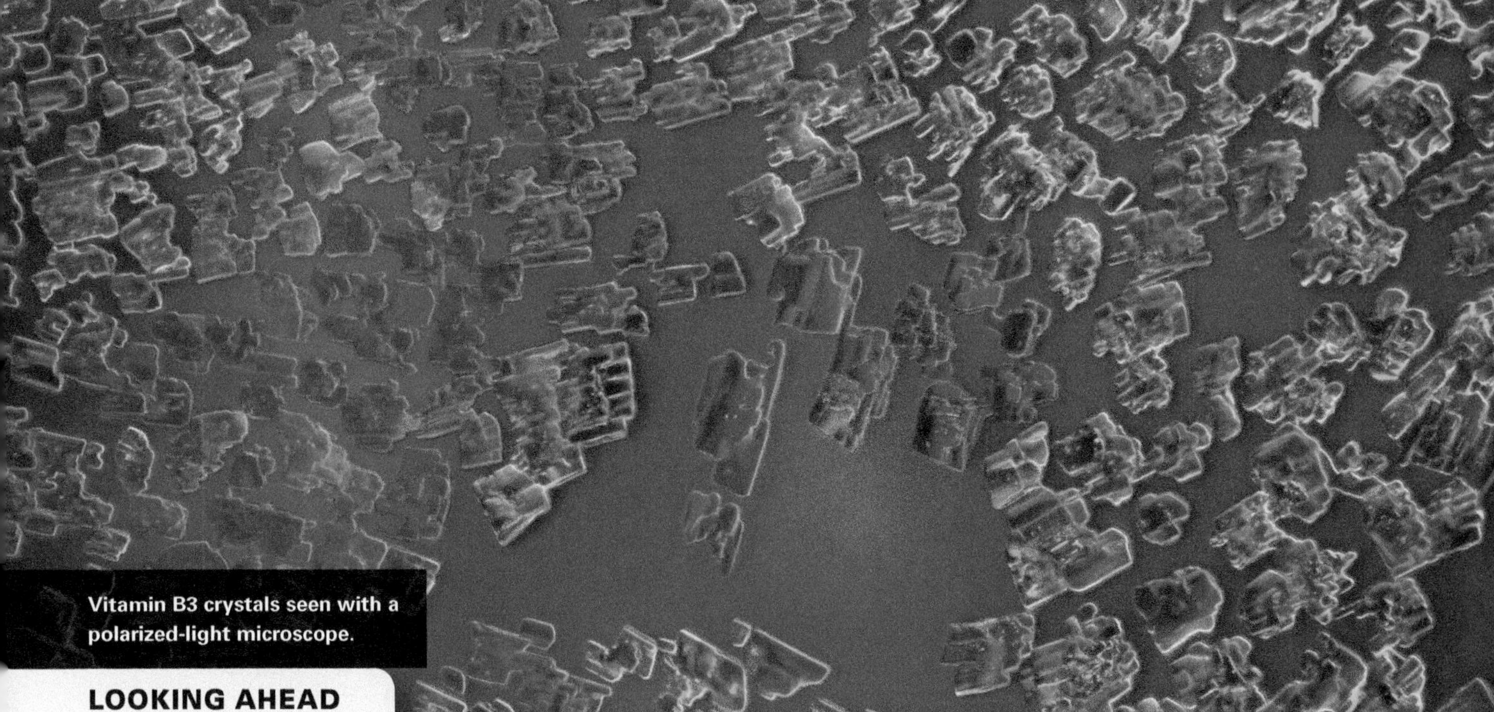

Vitamin B3 crystals seen with a polarized-light microscope.

LOOKING AHEAD

Magnetism and Electricity

These wind turbine blades generate an electric current as a coil of wire rotates in a magnetic field, transforming kinetic energy into electric energy.

You'll continue to explore the close connections between magnetism and electricity.

Induction

Wireless charging—for smartphones, pacemakers, and other medical devices—uses a changing magnetic field to produce an *induced current.*

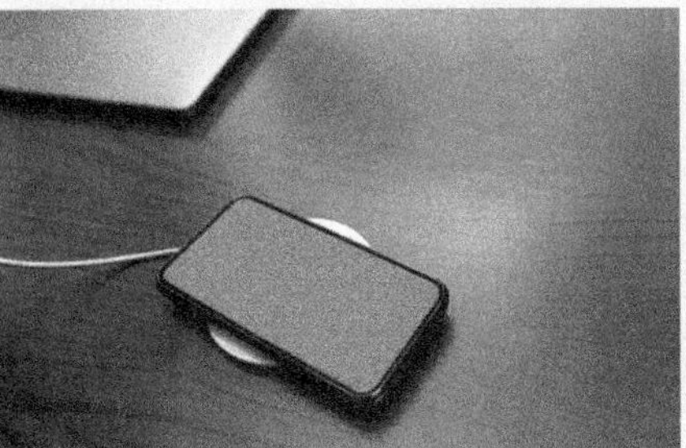

You'll learn to calculate the strengths of induced emfs, induced currents, and induced fields.

Electromagnetic Waves

The antennas on this cell phone tower emit and receive *electromagnetic waves,* self-sustaining oscillations of electric and magnetic fields.

You'll study the properties of electromagnetic waves that range in frequency from radio waves to x rays.

GOAL To understand the nature of electromagnetic induction and electromagnetic waves.

PHYSICS AND LIFE

Electromagnetic Waves Connect Us to the World Around Us

Most animals rely on light, an electromagnetic wave, to perceive their environment, and a wide variety of visual receptors have evolved that enable them to do so. In addition, electromagnetic waves are an essential tool in medical treatment and for the scientific study of living systems. Electromagnetic waves of different wavelengths have distinctive modes of interaction with molecules and tissues, which lead to different applications. For example, microwaves are used for heating, x rays are used for medical imaging and radiation therapy, and polarized visible light is used to determine whether proteins in solution are properly folded. Understanding the properties of electromagnetic waves and the physical processes by which they interact with matter will allow you to make sense of the dizzying variety of techniques, therapies, and animal senses that utilize electromagnetic waves. This chapter, together with Chapters 28 and 29, will develop that understanding.

A marsh marigold seen in ultraviolet light (left, as a bee sees it) and in visible light (right, as we see it).

27.1 Induced Currents

In Chapter 26, we learned that a current can create a magnetic field. As soon as this discovery was widely known, investigators began considering a related question: Can a magnetic field create a current?

FIGURE 27.1 Faraday's discovery of electromagnetic induction.

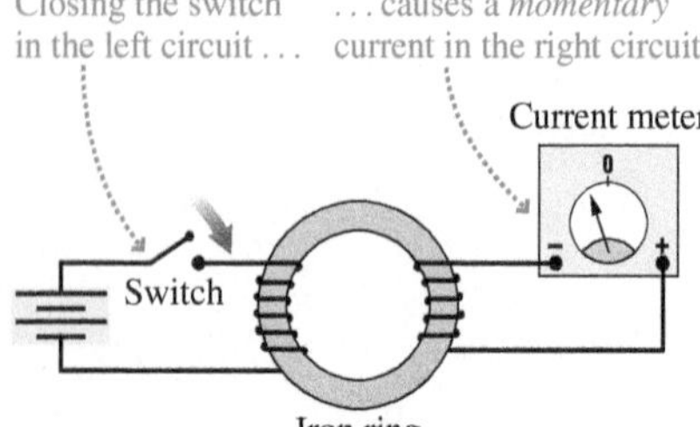
Closing the switch in the left circuit . . .
. . . causes a *momentary* current in the right circuit.
Current meter
Switch
Iron ring

One of the early investigators was Michael Faraday, who in 1831 was experimenting with two coils of wire wrapped around an iron ring, as shown in FIGURE 27.1, when he made a remarkable discovery. He had hoped that the magnetic field generated by a current in the coil on the left would create a magnetic field in the iron, and that the magnetic field in the iron might then somehow produce a current in the circuit on the right.

This technique failed to generate a steady current, but Faraday noticed that the needle of the current meter jumped ever so slightly at the instant when he closed the switch in the circuit on the left. After the switch was closed, the needle immediately returned to zero. Faraday's observation suggested to him that a current was generated only if the magnetic field was *changing* as it passed through the coil. Faraday set out to test this hypothesis through a series of experiments, shown in the following table.

Faraday investigates electromagnetic induction

Faraday placed one coil directly above the other, without the iron ring. There was no current in the lower coil while the switch was held in the open or closed position, but a momentary current appeared whenever the switch was opened or closed.

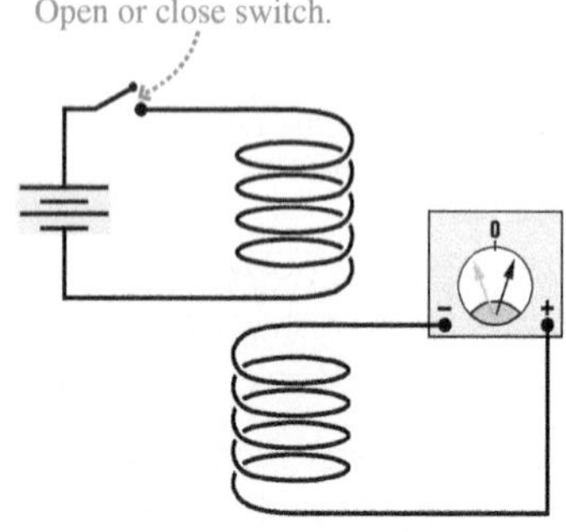
Open or close switch.

Opening or closing the switch creates a momentary current.

Faraday pushed a bar magnet into a coil of wire. This action caused a momentary deflection of the needle in the current meter, although *holding* the magnet inside the coil had no effect. A quick withdrawal of the magnet deflected the needle in the other direction.

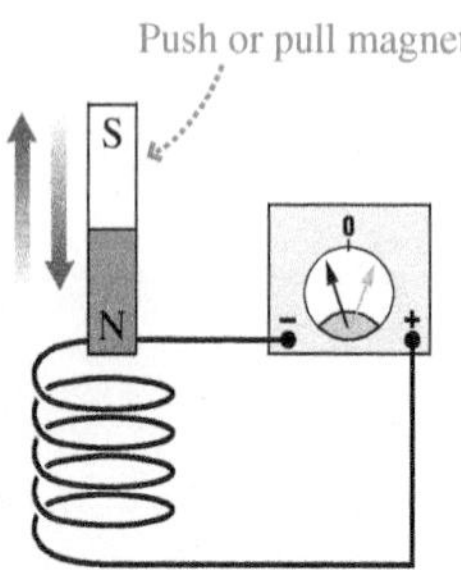
Push or pull magnet.
S
N

Pushing the magnet into the coil or pulling it out creates a momentary current.

Must the magnet move? Faraday created a momentary current by rapidly pulling a coil of wire out of a magnetic field, although there was no current if the coil was stationary in the magnetic field. Pushing the coil *into* the magnet caused the needle to deflect in the opposite direction.

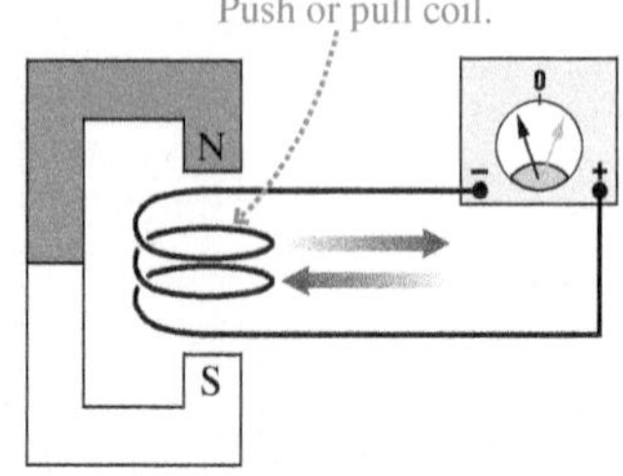
Push or pull coil.
N
S

Pushing the coil into the magnet or pulling it out creates a momentary current.

All of these experiments served to bolster Faraday's hypothesis: **There is a current in a coil of wire if and only if the magnetic field passing through the coil is *changing*.** It makes no difference what causes the magnetic field to change: current stopping or starting in a nearby circuit, moving a magnet through the coil, or moving the coil into and out of a magnet. The effect is the same in all cases. There is no current if the field through the coil is not changing, so it's not the magnetic field itself that is responsible for the current but, instead, it is the *changing of the magnetic field.*

The current in a circuit due to a changing magnetic field is called an **induced current.** Opening the switch or moving the magnet *induces* a current in a nearby circuit. An induced current is not caused by a battery; it is a completely new way to generate a current. The creation of an electric current by a changing magnetic field is our first example of **electromagnetic (EM) induction.**

27.2 Motional emf

In 1996, astronauts on the space shuttle deployed a satellite at the end of a 20-km-long conducting tether. A potential difference of up to 3500 V developed between the shuttle and the satellite as the wire between the two swept through the earth's magnetic field. Why would the motion of a wire in a magnetic field produce such a large voltage? Let's explore the mechanism behind this *motional emf.*

To begin, consider a conductor of length l that moves with velocity $\vec{v}$ through a uniform magnetic field $\vec{B}$, as shown in FIGURE 27.2. The charge carriers inside the conductor—assumed to be positive, as in our definition of current—also move with velocity $\vec{v}$, so they each experience a magnetic force. For simplicity, we will assume that $\vec{v}$ is perpendicular to $\vec{B}$, in which case the magnitude of the force is $F_B = qvB$. This force causes the charge carriers to move. For the geometry of Figure 27.2, the right-hand rule tells us that the positive charges move toward the top of the moving conductor, leaving an excess of negative charge at the bottom.

FIGURE 27.2 The magnetic force on the charge carriers in a moving conductor creates an electric field inside the conductor.

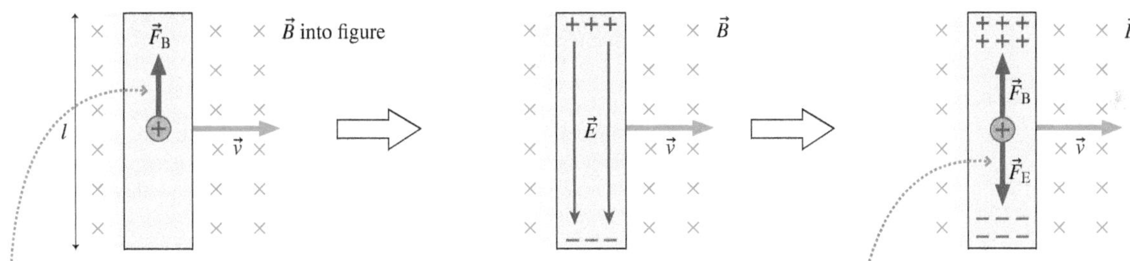
$\vec{F}_B$
$\vec{B}$ into figure
l
$\vec{v}$
+++
$\vec{B}$
$\vec{E}$
$\vec{v}$
− − −
+++
+++
$\vec{B}$
$\vec{F}_B$
$\vec{v}$
$\vec{F}_E$
− − −
− − −
Charge carriers in the conductor experience a force of magnitude $F_B = qvB$. Positive charges are free to move and drift upward.
The resulting charge separation creates an electric field in the conductor. $\vec{E}$ increases as more charge flows.
The charge flow continues until the electric and magnetic forces balance. For a positive charge carrier, the upward magnetic force $\vec{F}_B$ is equal to the downward electric force $\vec{F}_E$.

This motion of the charge carriers cannot continue forever. The separation of the charge carriers creates an electric field. The resulting electric force *opposes* the separation of charge, so the charge separation continues only until the electric force has grown to exactly balance the magnetic force:

$$F_E = qE = F_B = qvB$$

When this balance occurs, the charge carriers experience no net force and thus undergo no further motion. The electric field strength at equilibrium is

$$E = vB \tag{27.1}$$

Thus, **the magnetic force on the charge carriers in a moving conductor creates an electric field $E = vB$ inside the conductor.**

The electric field, in turn, creates an electric potential difference between the two ends of the moving conductor. We found in ◀◀ SECTION 22.6 that the potential difference between two points separated by distance l parallel to a uniform electric field E is $\Delta V = El$. Thus the motion of the wire through a magnetic field *induces* a potential difference

$$\Delta V = vlB \tag{27.2}$$

between the ends of the conductor. The potential difference depends on the strength of the magnetic field and on the wire's speed through the field. This is similar to the action of the electromagnetic flowmeter that we saw in Chapter 26.

There's an important analogy between this potential difference and the potential difference of a battery. FIGURE 27.3a reminds you that a battery uses a nonelectric force—which we called the charge escalator—to separate positive and negative

FIGURE 27.3 Generating an emf.

(a) Chemical reactions separate the charges and cause a potential difference between the ends. This is a chemical emf.

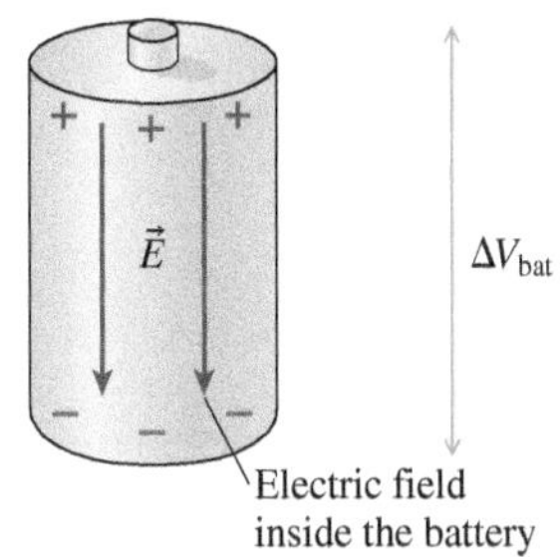
+ + +
$\vec{E}$
ΔV_{bat}
− − −
Electric field inside the battery

(b) Magnetic forces separate the charges and cause a potential difference between the ends. This is a motional emf.

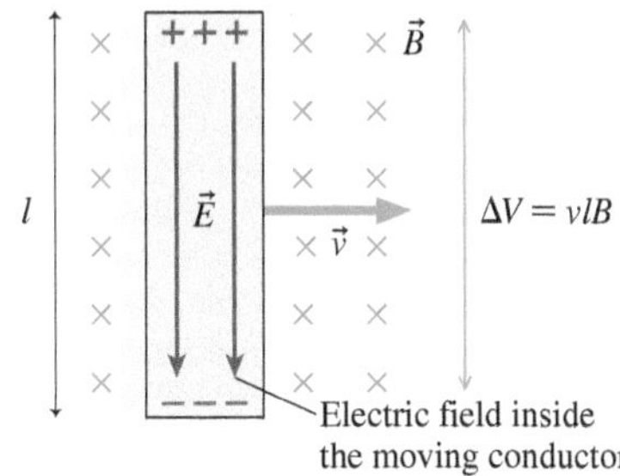
+++
$\vec{B}$
l
$\vec{E}$
$\vec{v}$
$\Delta V = vlB$
− − −
Electric field inside the moving conductor

A head for magnetism? Hammerhead sharks seem to navigate using the earth's magnetic field. It's likely that they detect the *magnetic* field by using their keen *electric* sense, detecting the earth's magnetic field by sensing the motional emf as they move through the water. The width of their oddly shaped heads is an asset because the magnitude of the potential difference is proportional to the length of the moving conductor.

charges. We refer to a battery, where the charges are separated by chemical forces, as a source of *chemical emf.* By contrast, the moving conductor of FIGURE 27.3b separates charges using *magnetic* forces. The charge separation creates a potential difference—an emf—due to the conductor's *motion.* We can thus define the **motional emf** of a conductor of length l moving with velocity $\vec{v}$ perpendicular to a magnetic field $\vec{B}$ to be

$$\mathcal{E} = vlB \tag{27.3}$$

Motional emf of a conductor moving through a magnetic field

EXAMPLE 27.1 Finding the motional emf for an airplane

A Boeing 747 aircraft with a wingspan of 65 m is cruising at 260 m/s over northern Canada, where the magnetic field of the earth (magnitude 5.0×10^{-5} T) is directed straight down. What is the potential difference between the tips of the wings?

PREPARE The wing is a conductor moving through a magnetic field, perpendicular to the field, so there will be a motional emf.

SOLVE The magnetic field is perpendicular to the velocity, so we compute the potential difference using Equation 27.3:

$$\Delta V = vlB = (260 \text{ m/s})(65 \text{ m})(5.0 \times 10^{-5} \text{ T}) = 0.85 \text{ V}$$

ASSESS The earth's magnetic field is small, so the motional emf will be small as well unless the speed and the length are quite large. The tethered satellite generated a much higher voltage due to its much greater speed and the great length of the tether, the moving conductor.

STOP TO THINK 27.1 A square conductor moves through a uniform magnetic field directed out of the figure. Which of the figures shows the correct charge distribution on the conductor?

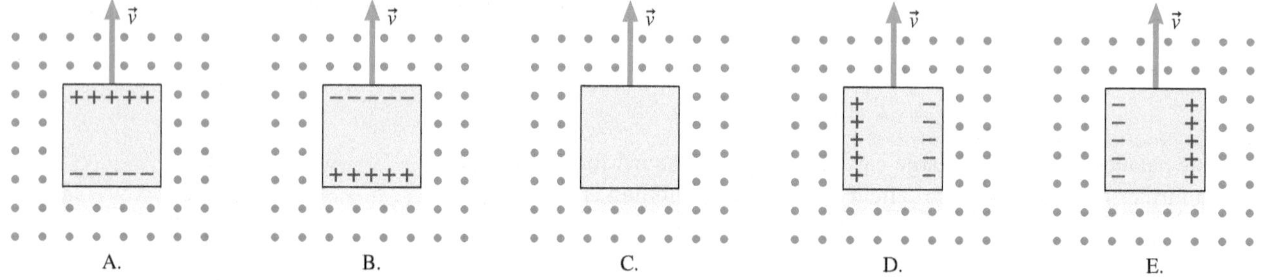

Induced Current in a Circuit

The moving conductor of Figure 27.3 developed an emf, but it couldn't sustain a current because the charges had nowhere to go. We can change this by including the moving conductor in a circuit.

FIGURE 27.4 shows a length of wire with resistance R sliding with speed v along a fixed U-shaped conducting rail. (For simplicity, we will assume that the rail has zero resistance.) The wire and the rail together form a closed conducting loop—a circuit.

Suppose a magnetic field $\vec{B}$ is perpendicular to the plane of the circuit. Charges in the moving wire will be pushed to the ends of the wire by the magnetic force, just as they were in Figure 27.3, but now the charges can continue to flow around the circuit. The moving wire acts like the battery in a circuit.

The current in the circuit is an induced current, due to magnetic forces on moving charges. The total resistance of the circuit is just the resistance R of the moving wire, so the induced current is given by Ohm's law:

$$I = \frac{\mathcal{E}}{R} = \frac{vlB}{R} \tag{27.4}$$

We've assumed that the wire is moving along the rail at constant speed. But we must apply a continuous pulling force $\vec{F}_{pull}$ to make this happen; FIGURE 27.5 shows why. The moving wire, which now carries induced current I, is in a magnetic field. You learned in ◀◀ SECTION 26.5 that a magnetic field exerts a force on a current-carrying wire. According to the right-hand rule, the magnetic force $\vec{F}_{mag}$ on the moving wire points to the left. This "magnetic drag" will cause the wire to slow down and stop *unless* we exert an equal but opposite pulling force $\vec{F}_{pull}$ to keep the wire moving.

NOTE ▶ Think about this carefully. As the wire moves to the right, the magnetic force $\vec{F}_B$ pushes the charge carriers *parallel* to the wire. Their motion, as they continue around the circuit, is the induced current I. Now, because we have a current, a second magnetic force $\vec{F}_{mag}$ enters the picture. This force on the current is *perpendicular* to the wire and acts to slow the wire's motion. ◀

The magnitude of the magnetic force on a current-carrying wire was found in Chapter 26 to be $F_{mag} = IlB$. Using that result, along with Equation 27.4 for the induced current, we find that the force required to pull the wire with a constant speed v is

$$F_{pull} = F_{mag} = IlB = \left(\frac{vlB}{R}\right)lB = \frac{vl^2B^2}{R} \tag{27.5}$$

Energy Considerations

FIGURE 27.6 is another look at the wire moving on a conducting rail. Because a force is needed to pull the wire through the magnetic field at a constant speed, we must do work to keep the wire moving. You learned in Chapter 10 that the power delivered by a force pushing or pulling an object with velocity v is $P = Fv$, so the power provided to the circuit by the force pulling on the wire is

$$P_{input} = F_{pull}v = \frac{v^2l^2B^2}{R} \tag{27.6}$$

This is the rate at which energy is added to the circuit by the pulling force.

But the circuit dissipates energy in the resistance of the circuit. You learned in Chapter 24 that the power dissipated by current I as it passes through resistance R is $P = I^2R$. Equation 27.4 for the induced current I gives us the power dissipated by the circuit of Figure 27.6:

$$P_{dissipated} = I^2R = \frac{v^2l^2B^2}{R} \tag{27.7}$$

Equations 27.6 and 27.7 have identical results. This makes sense: The rate at which work is done on the circuit is exactly balanced by the rate at which energy is dissipated. The fact that our final result is consistent with energy conservation is a good check on our work.

Generators

A device that converts mechanical energy to electric energy is called a **generator.** The example of Figure 27.6 is a simple generator, but it is not very practical. Rather than move a straight wire, it's more practical to rotate a coil of wire, as in FIGURE 27.7 on the next page. The coil is rotated by some external torque, due perhaps to a steam turbine or windmill blades. As the coil rotates, the left edge always moves upward through the magnetic field while the right edge always moves downward. The

FIGURE 27.4 A current is induced in the circuit as the wire moves through a magnetic field.

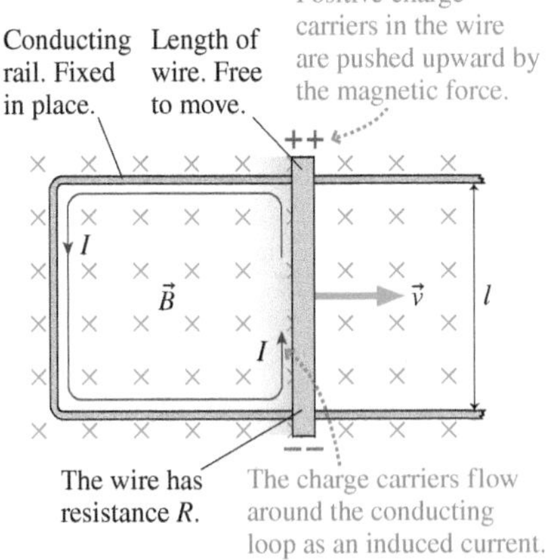
Positive charge carriers in the wire are pushed upward by the magnetic force.
Conducting rail. Fixed in place.
Length of wire. Free to move.
++
I
$\vec{B}$
$\vec{v}$
l
The wire has resistance R.
The charge carriers flow around the conducting loop as an induced current.

FIGURE 27.5 A pulling force is needed to move the wire to the right.

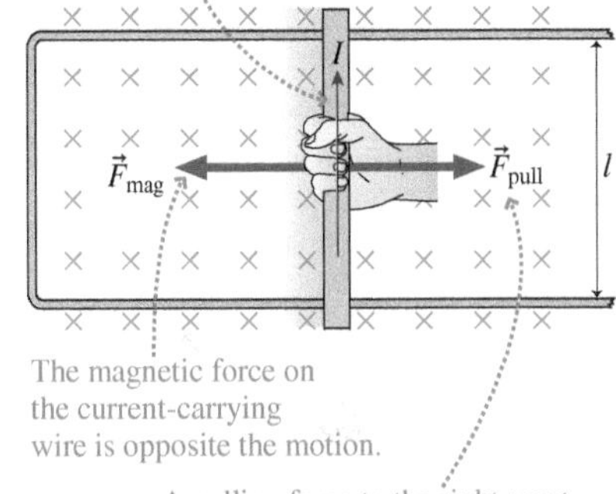
The induced current flows through the moving wire.
I
$\vec{F}_{mag}$
$\vec{F}_{pull}$
l
The magnetic force on the current-carrying wire is opposite the motion.
A pulling force to the right must balance the magnetic force to keep the wire moving at constant speed.

FIGURE 27.6 Power into and out of an induced-current circuit.

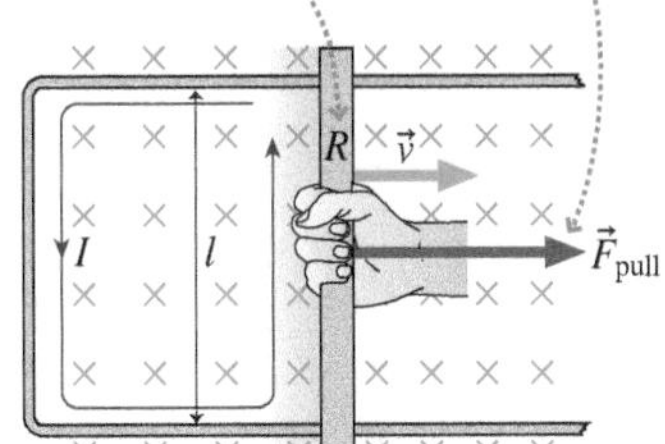
Because there is a current, power is dissipated in the resistance of the wire.
Pulling to the right takes work. This is a power input to the system.
R
$\vec{v}$
I
l
$\vec{F}_{pull}$

FIGURE 27.7 A generator using a rotating loop of wire.

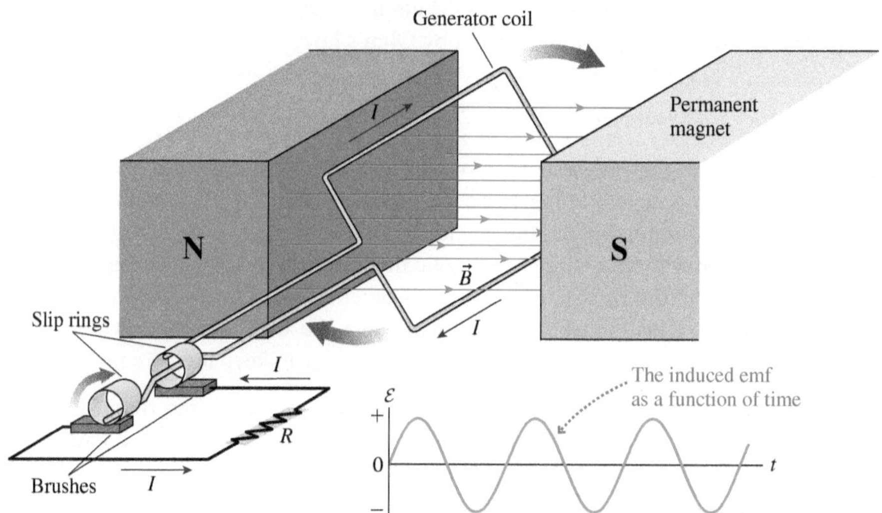

Turning wind into electric power The rotating blades of a wind turbine turn coils of wire in a magnetic field, generating a large emf of around 1000 V. When the blades are running at full power, some 3000 A flow through these coils, which leads to an enormous magnetic drag force on them. The force to overcome this drag comes from the wind itself. The kinetic energy of the wind is transformed into electric energy from the generator.

motion of the wires through the magnetic field induces a current to flow, as noted in the figure. The induced current is removed from the rotating loop by brushes that press up against rotating slip rings. The circuit is completed as shown in the figure.

As the coil in the generator of Figure 27.7 rotates, the amplitude and sign of the emf change, giving a sinusoidal variation of emf as a function of time. Electric power plants use generators of this sort, so the electricity in your house has a varying voltage. The alternating sign of the voltage produces an *alternating current,* so we call such electricity *AC.*

So-called *regenerative braking* in electric cars works much the same way. When your foot is on the accelerator, current from the battery passes through the coil. The coil experiences a magnetic torque and rotates: It's an electric motor that turns the wheels. When you take your foot off the accelerator, the rotating wheels force the coil to keep turning, but now they function as a generator that creates a current. This brakes the car—kinetic energy is being transformed into electric energy—and at the same time recharges the battery.

EXAMPLE 27.2 **Lighting a bulb**

FIGURE 27.8 shows a circuit consisting of a flashlight bulb, rated 3.0 V/1.5 W, and ideal wires with no resistance. The right wire of the circuit, which is 10 cm long, is pulled at constant speed v through a perpendicular magnetic field of strength 0.10 T.

a. What speed must the wire have to light the bulb to full brightness?
b. What force is needed to keep the wire moving?

FIGURE 27.8 Circuit of Example 27.2.

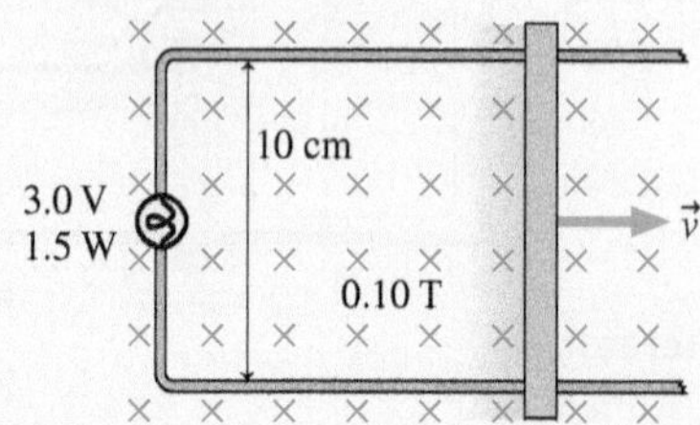

PREPARE We will treat the moving wire as a source of motional emf. The magnetic force on the charge carriers causes a counterclockwise (ccw) induced current.

SOLVE

a. The bulb's rating of 3.0 V/1.5 W means that at full brightness it will dissipate 1.5 W at a potential difference of 3.0 V. Because the power is related to the voltage and current by $P = I\Delta V$, the current causing full brightness is

$$I = \frac{P}{\Delta V} = \frac{1.5\text{ W}}{3.0\text{ V}} = 0.50\text{ A}$$

The bulb's resistance—the total resistance of the circuit—is

$$R = \frac{\Delta V}{I} = \frac{3.0\text{ V}}{0.50\text{ A}} = 6.0\ \Omega$$

Equation 27.4 gives the speed needed to induce this current:

$$v = \frac{IR}{lB} = \frac{(0.50\text{ A})(6.0\ \Omega)}{(0.10\text{ m})(0.10\text{ T})} = 300\text{ m/s}$$

You can confirm from Equation 27.6 that the input power at this speed is 1.5 W.

b. From Equation 27.5, the pulling force must be

$$F_{\text{pull}} = \frac{vl^2B^2}{R} = 5.0 \times 10^{-3}\ \text{N}$$

You can also obtain this result from $F_{\text{pull}} = P/v$.

ASSESS Example 27.1 showed that high speeds are needed to produce significant potential difference. Thus 300 m/s is not surprising. The pulling force is not very large, but even a small force can deliver large amounts of power $P = Fv$ when v is large.

STOP TO THINK 27.2 Is there an induced current in this circuit? If so, what is its direction?

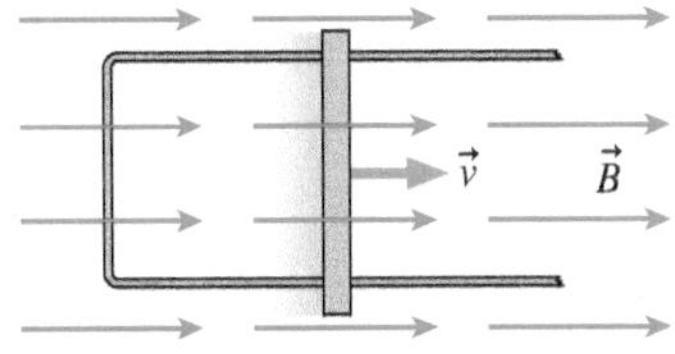

27.3 Magnetic Flux and Lenz's Law

We've begun our exploration of electromagnetic induction by analyzing a circuit in which one wire moves through a magnetic field. You might be wondering what this has to do with Faraday's discovery. Faraday found that a current is induced when the amount of magnetic field passing through a coil or a loop of wire changes. But that's exactly what happens as the slide wire moves down the rail in Figure 27.4! As the circuit expands, more magnetic field passes through the larger loop. It's time to define more clearly what we mean by "the amount of field passing through a loop."

Imagine holding a rectangular loop in front of the fan shown in FIGURE 27.9. The amount of air flowing *through* the loop—the *flux*—depends on the angle of the loop. The flow is maximum if the loop is perpendicular to the flow, zero if the loop is rotated to be parallel to the flow. In general, the amount of air flowing through is proportional to the *effective area* of the loop (i.e., the area facing the fan):

$$A_{\text{eff}} = ab\cos\theta = A\cos\theta \tag{27.8}$$

where $A = ab$ is the area of the loop and θ is the tilt angle of the loop. A loop perpendicular to the flow, with $\theta = 0°$, has $A_{\text{eff}} = A$, the full area of the loop.

We can apply this idea to a magnetic field passing through a loop. FIGURE 27.10 shows a loop of area $A = ab$ in a uniform magnetic field. Think of the field vectors,

FIGURE 27.9 The amount of air flowing through a loop depends on the effective area of the loop.

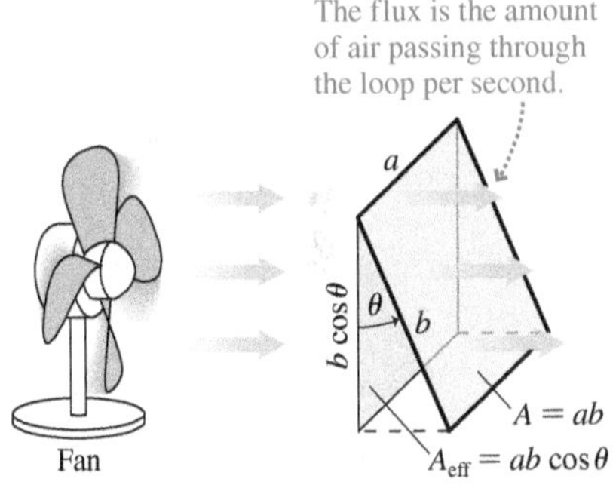

FIGURE 27.10 Magnetic field through a loop that is tilted at various angles.

Loop seen from the side:

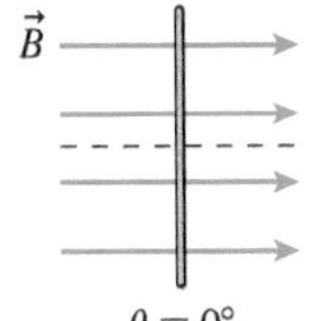

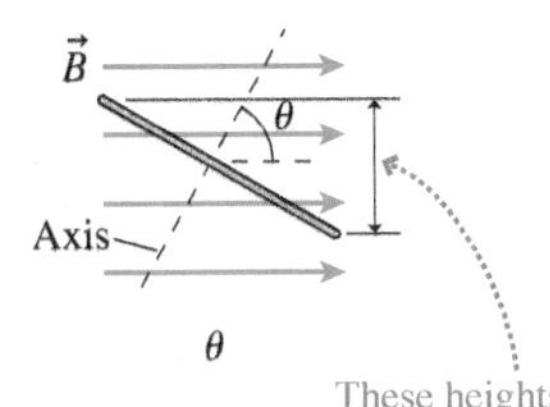

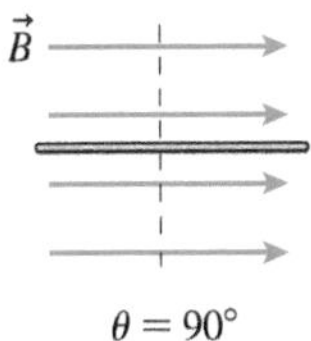

These heights are the same.

Seen in the direction of the magnetic field:

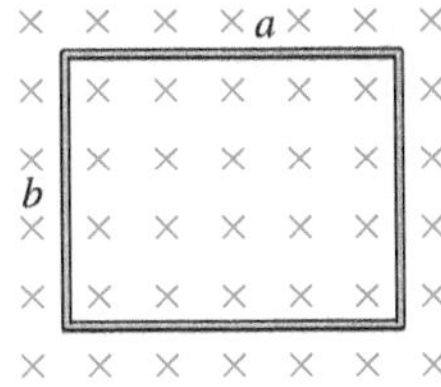

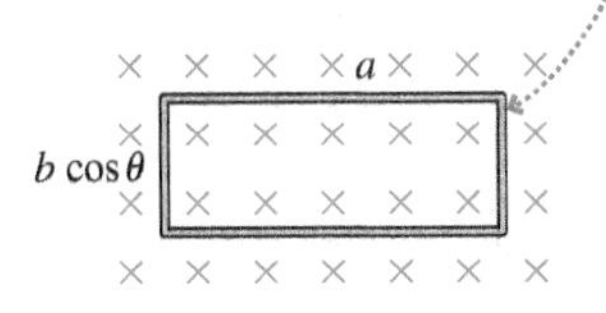

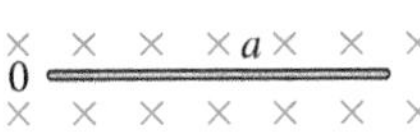

- Loop perpendicular to field.
- Maximum number of arrows pass through.

- Loop rotated through angle θ.
- Fewer arrows pass through.

- Loop rotated 90°.
- No arrows pass through.

seen here from behind, as if they were arrows shot into the figure. The density of arrows (arrows per m^2) is proportional to the strength B of the magnetic field; a stronger field would be represented by arrows packed closer together. The number of arrows passing through a loop of wire depends on two factors:

1. The density of arrows, which is proportional to B, and
2. The effective area $A_{eff} = A\cos\theta$ of the loop.

The angle θ is the angle between the magnetic field and the axis of the loop. The maximum number of arrows passes through the loop when it is perpendicular to the magnetic field ($\theta = 0°$). No arrows pass through the loop if it is tilted 90°.

With this in mind, let's define the **magnetic flux** Φ as

FIGURE 27.11 Definition of magnetic flux.

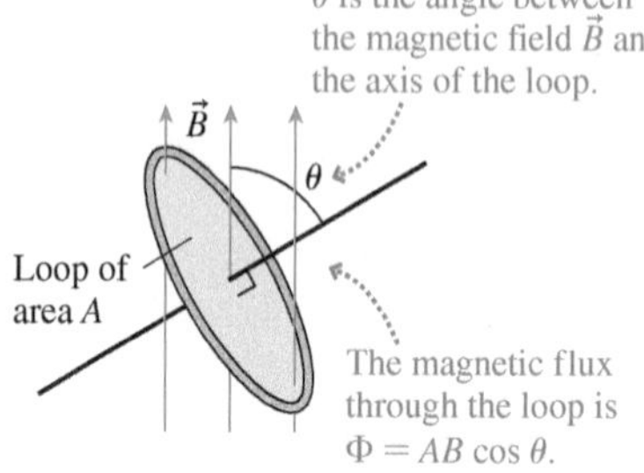
θ is the angle between the magnetic field B and the axis of the loop.
B
θ
Loop of area A
The magnetic flux through the loop is Φ = AB cos θ.

$$\Phi = A_{eff}B = AB\cos\theta \qquad (27.9)$$

Magnetic flux through area A at angle θ to field B

The magnetic flux measures the amount of magnetic field passing through a loop of area A if the loop is tilted at angle θ from the field. The SI unit of magnetic flux is the **weber.** From Equation 27.9 you can see that

$$1\text{ weber} = 1\text{ wb} = 1\text{ T}\cdot\text{m}^2$$

The relationship of Equation 27.9 is illustrated in FIGURE 27.11.

EXAMPLE 27.3 Finding the flux of the earth's magnetic field through a vertical loop

At a particular location, the earth's magnetic field is 50 μT tipped at an angle of 60° below horizontal. A 10-cm-diameter circular loop of wire sits flat on a table. What is the magnetic flux through the loop?

FIGURE 27.12 Finding the flux of the earth's field through a loop.

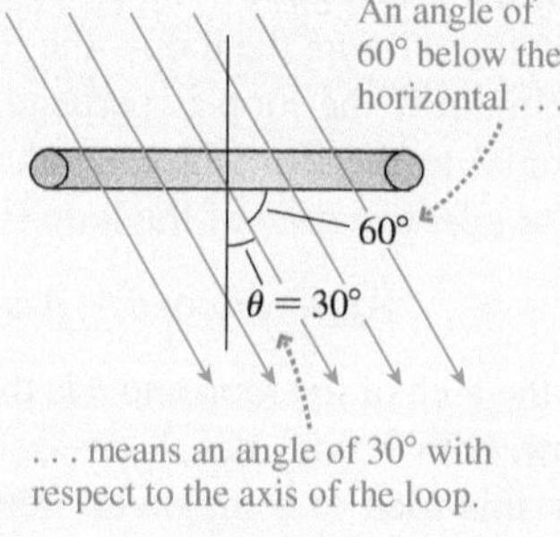
An angle of 60° below the horizontal . . .
60°
θ = 30°
. . . means an angle of 30° with respect to the axis of the loop.

PREPARE FIGURE 27.12 shows the loop and the field of the earth. The field is tipped by 60°, so the angle of the field with respect to the axis of the loop is $\theta = 30°$. The radius of the loop is 5.0 cm, so the area of the loop is $A = \pi r^2 = \pi(0.050\text{ m})^2 = 0.0079\text{ m}^2$.

SOLVE The flux through the loop is then

$$\Phi = AB\cos\theta = (0.0079\text{ m}^2)(50\times10^{-6}\text{ T})\cos 30°$$
$$= 3.4\times10^{-7}\text{ Wb}$$

ASSESS It's a small loop and a weak field, so a very small flux seems reasonable.

FIGURE 27.13 Pushing a bar magnet toward the loop induces a current in the loop.

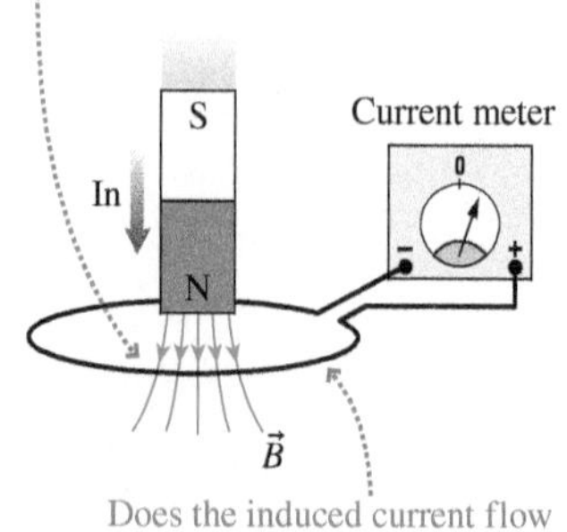
Pushing a bar magnet toward a loop increases the flux through the loop and induces a current to flow.
S
N
In
Current meter
0
B
Does the induced current flow clockwise or counterclockwise?

Lenz's Law

Some of the induction experiments from earlier in the chapter could be explained in terms of motional emf, but others had no motion. What they all have in common, though, is that one way or another the magnetic flux through the coil or loop *changes.* We can summarize all of the discoveries as follows: **Current is induced in a loop of wire when the magnetic flux through the loop changes.**

For example, a momentary current is induced in the loop of FIGURE 27.13 as the bar magnet is pushed toward the loop because the magnetic field through the loop, and hence the flux, increases. Pulling the magnet away from the loop, which decreases the flux, causes the current meter to deflect in the opposite direction. How can we predict the *direction* of the current in the loop?

The German physicist Heinrich Lenz began to study electromagnetic induction after learning of Faraday's discovery. Lenz developed a rule for determining the direction of the induced current. We now call his rule **Lenz's law,** and it can be stated as follows:

> **Lenz's law** There is an induced current in a closed, conducting loop if and only if the magnetic flux through the loop is changing. The direction of the induced current is such that the induced magnetic field opposes the *change* in the flux.

Lenz's law is rather subtle, and it takes some practice to see how to apply it.

NOTE ▶ One difficulty with Lenz's law is the term "flux," from a Latin root meaning "flow." In everyday language, the word "flux" may imply that something is changing. Think of the phrase "The situation is in flux." In physics, "flux" simply means "passes through." A steady magnetic field through a loop creates a steady, *un*changing magnetic flux. ◀

Lenz's law tells us to look for situations where the flux is *changing.* We'll see three ways in which this comes about:

- The magnetic field through the loop changes (increases or decreases).
- The loop changes in area or angle.
- The loop moves into or out of a magnetic field.

We can understand Lenz's law this way: If the flux through a loop changes, a current is induced in a loop. That current generates *its own* magnetic field $\vec{B}_{\text{induced}}$. **It is this induced field that opposes the flux change.** Let's look at an example to clarify what we mean by this statement.

FIGURE 27.14 shows a magnet above a coil of wire. The field of the magnet creates a downward flux through the loop. The three parts of the figure show how the induced current in the loop develops in response to the motion of the magnet and thus the changing flux in the coil.

FIGURE 27.14 The induced current depends on the motion of the magnet.

(a) Magnet at rest

S
N
Current meter
$\vec{B}_{\text{magnet}}$
1. The magnet at rest creates a flux downward through the coil.
2. There is no *change* in flux, so there is no current.

(b) Magnet moving down

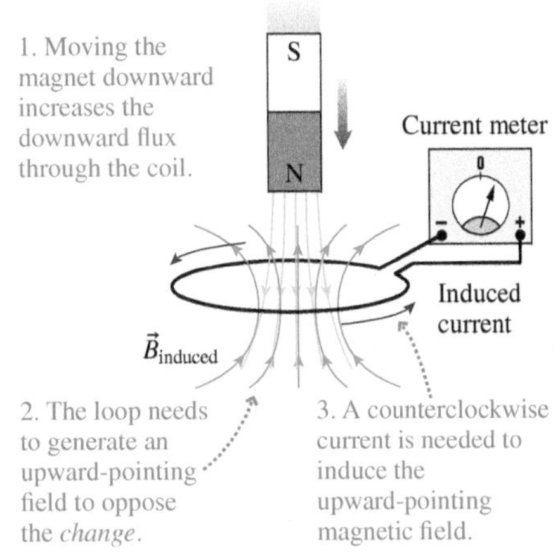

(c) Magnet moving up

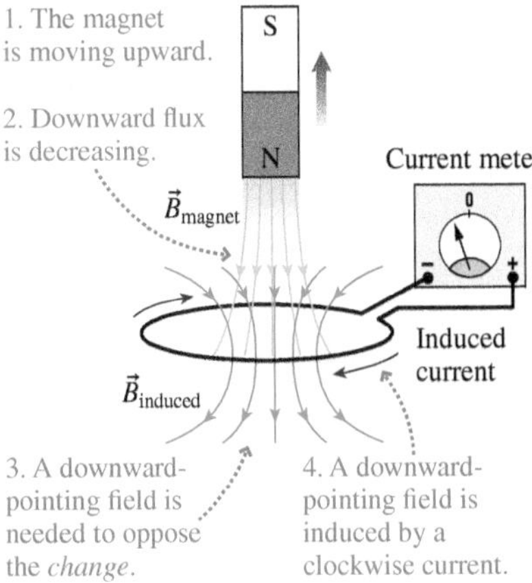

- In part a, the magnet is not moving. There is a flux through the loop, but because there is no *change* in the flux, no current is induced in the loop.
- In part b, the magnet is moving toward the loop, so the downward magnetic flux through the loop increases. According to Lenz's law, the loop will generate a field that opposes this change. To oppose an *increase in the downward flux,* the loop itself needs to generate an *upward-pointing magnetic field.* The induced magnetic field at the center of the loop will point upward if the current is counterclockwise,

according to the right-hand rule you learned in Chapter 26. Thus pushing the north end of a bar magnet toward the loop induces a counterclockwise current around the loop. This induced current ceases as soon as the magnet stops moving.

- Now suppose the bar magnet is pulled back away from the loop, as in part c. There is a downward magnetic flux through the loop, but the flux decreases as the magnet moves away. According to Lenz's law, the induced magnetic field of the loop will oppose this decrease. To oppose a *decrease in the downward flux,* the loop itself needs to generate a *downward-pointing magnetic field.* The induced current is clockwise, opposite the induced current of part b.

The magnetic field of the bar magnet is pointing downward in each part of the figure, but the induced magnetic field can be zero, directed upward, or directed downward depending on the motion of the magnet. It is not the *flux* due to the magnet that the induced current opposes, but the *change in the flux.* This is a subtle but critical distinction.

The approach we took here for using Lenz's law to determine the direction of an induced current is a general one; it's worthwhile to spell out the steps.

TACTICS BOX 27.1 Using Lenz's law

1. Determine the direction of the applied magnetic field. The field must pass through the loop.
2. Determine how the flux is changing. Is it increasing, decreasing, or staying the same?
3. If the flux is steady, there is no induced magnetic field. If the flux is changing, determine the direction of an induced magnetic field that will oppose the change:
   - Increasing flux: The induced magnetic field points opposite to the applied magnetic field.
   - Decreasing flux: The induced magnetic field points in the same direction as the applied magnetic field.
4. Determine the direction of the induced current. Use the right-hand rule to determine the current direction in the loop that generates the induced magnetic field you found in step 3.

CONCEPTUAL EXAMPLE 27.4 Applying Lenz's law 1

The switch in the top circuit of FIGURE 27.15 has been closed for a long time. What happens in the lower loop when the switch is opened?

FIGURE 27.15 Circuits for Example 27.4.

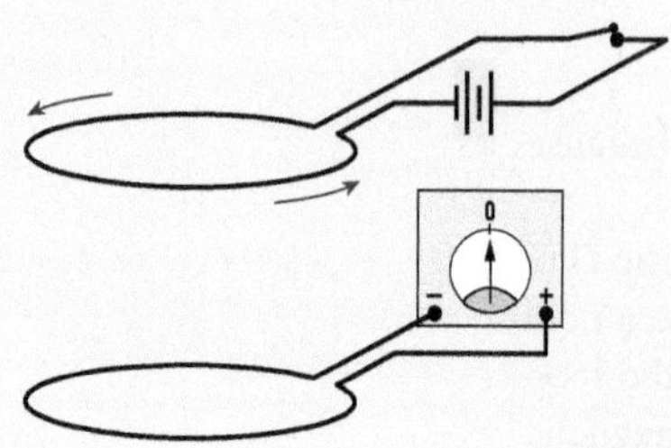

REASON The current in the upper loop creates a magnetic field. This magnetic field produces a flux through the lower loop. When you open a switch, the current doesn't immediately drop to zero; it falls off over a short time. As the current changes in the upper loop, the flux in the lower loop changes.

FIGURE 27.16 shows the four steps of Tactics Box 27.1. Opening the switch induces a counterclockwise current in the lower loop. This is a momentary current, lasting only until the magnetic field of the upper loop drops to zero.

ASSESS The induced current is in the same direction as the original current. This makes sense, because the induced current is opposing the change, a decrease in the current.

FIGURE 27.16 Finding the induced current.

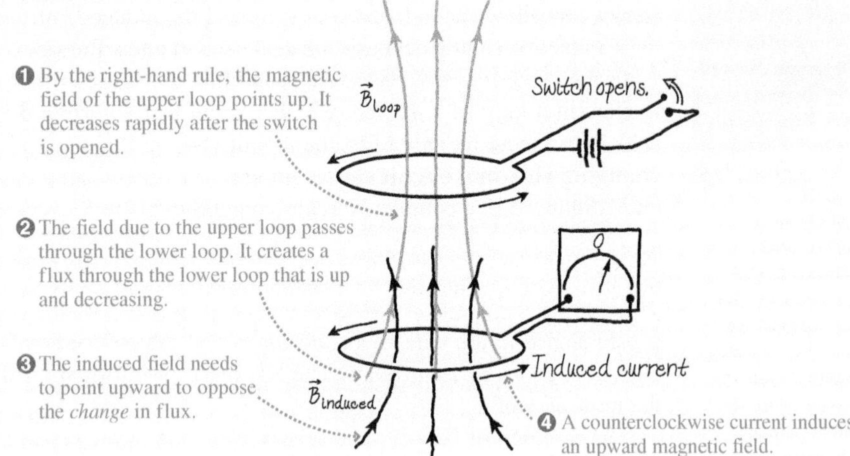
1 By the right-hand rule, the magnetic field of the upper loop points up. It decreases rapidly after the switch is opened.
$\vec{B}_{loop}$
Switch opens.
2 The field due to the upper loop passes through the lower loop. It creates a flux through the lower loop that is up and decreasing.
3 The induced field needs to point upward to oppose the *change* in flux.
$\vec{B}_{induced}$
Induced current
4 A counterclockwise current induces an upward magnetic field.

CONCEPTUAL EXAMPLE 27.5 **Applying Lenz's law 2**

A loop is moved toward a current-carrying wire as shown in FIGURE 27.17. As the wire is moving, is there a clockwise current around the loop, a counterclockwise current, or no current?

REASON FIGURE 27.18 shows that the magnetic field above the wire points into the figure. We learned in Chapter 26 that the magnetic field of a straight, current-carrying wire is proportional to $1/r$, where r is the distance away from the wire, so the field is stronger closer to the wire.

As the loop moves toward the wire, the flux through the loop increases. To oppose the *change* in the flux—the increase into the figure—the magnetic field of the induced current must point out of the figure. Thus, according to the right-hand rule, a counterclockwise current is induced, as shown in Figure 27.18.

ASSESS The loop moves into a region of stronger field. To oppose the increasing flux, the induced field should be opposite the existing field, so our answer makes sense.

FIGURE 27.17 The moving loop.

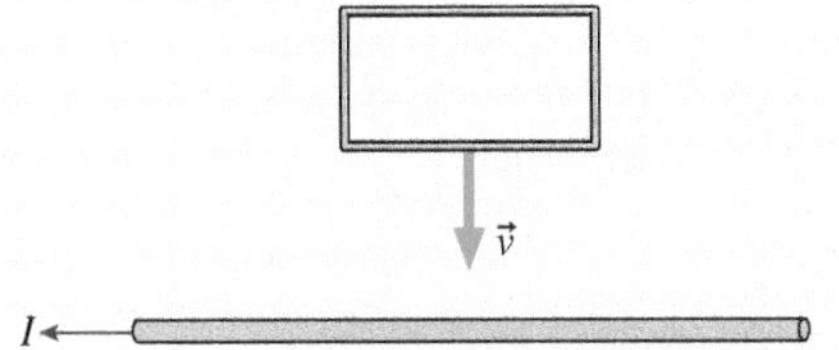
$\vec{v}$
I

FIGURE 27.18 The motion of the loop changes the flux through the loop and induces a current.

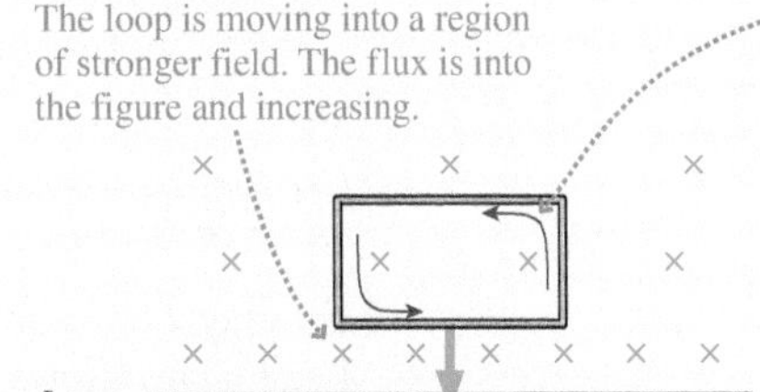
The loop is moving into a region of stronger field. The flux is into the figure and increasing.
The induced current must create a magnetic field out of the figure to oppose the change, so the right-hand rule tells us that the induced current is counterclockwise.
I
$\vec{v}$

STOP TO THINK 27.3 As a coil moves to the right at constant speed, it passes over the north pole of a magnet and then moves beyond it. Which graph best represents the current in the loop for the time of the motion? A counterclockwise current as viewed from above the loop is a positive current, clockwise is a negative current.

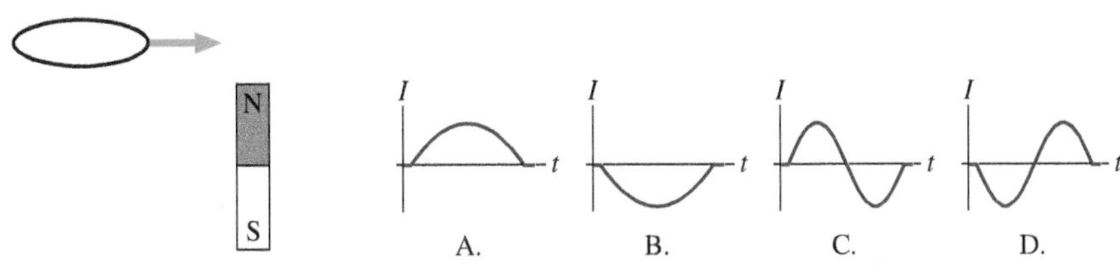
N
S
I
t
A.
B.
C.
D.

Inducing music As you can see in the photo, the pickups on an electric guitar have small disk-shaped magnets that magnetize the nearby steel strings. What you can't see are the small coils below each magnet. As the magnetized strings vibrate, the flux they create in the coils changes rapidly, inducing a changing emf in the coils. When amplified, this emf can drive a loudspeaker, which converts the strings' motion to sound.

27.4 Faraday's Law

A current is induced when the magnetic flux through a conducting loop changes. Lenz's law allows us to find the direction of the induced current. To put electromagnetic induction to practical use, we also need to know the *size* of the induced current.

In the preceding examples, we found that a change in the flux caused a current to flow in a loop of wire. But we also know that a current doesn't just flow by itself—there must be an emf in the circuit to keep the current flowing. Evidently, **a changing flux in a circuit causes an emf,** which we call an **induced emf $\mathcal{E}$.** If this emf is induced in a complete circuit having resistance R, a current

$$I_{\text{induced}} = \frac{\mathcal{E}}{R} \qquad (27.10)$$

is established in the wire as a *consequence* of the induced emf. The direction of the current is given by Lenz's law. The last piece of information we need is the size of the induced emf $\mathcal{E}$.

The research of Faraday and others led to the discovery of the basic law of electromagnetic induction, which we now call **Faraday's law.**

Faraday's law An emf $\mathcal{E}$ is induced in a conducting loop if the magnetic flux through the loop changes. The magnitude of the emf is

$$\mathcal{E} = \left|\frac{d\Phi}{dt}\right| \qquad (27.11)$$

and the direction of the emf is such as to drive an induced current in the direction given by Lenz's law.

In other words, the magnitude of the induced emf is the *rate of change* of the magnetic flux through the loop.

A coil of wire consisting of N turns in a changing magnetic field acts like N batteries in series. The induced emf of each turn of the coil adds, so the induced emf of the entire coil is

$$\mathcal{E}_{\text{coil}} = N\left|\frac{d\Phi_{\text{per turn}}}{dt}\right| \qquad (27.12)$$

PROBLEM-SOLVING STRATEGY 27.1 Electromagnetic induction

Faraday's law allows us to find the *magnitude* of induced emfs and currents; Lenz's law allows us to determine the *direction.*

PREPARE Make simplifying assumptions about wires and magnetic fields. Draw a picture or a circuit diagram. Use Lenz's law to determine the direction of the induced current.

SOLVE The mathematical representation is based on Faraday's law

$$\mathcal{E} = \left|\frac{d\Phi}{dt}\right|$$

For an N-turn coil, multiply by N. The size of the induced current is $I = \mathcal{E}/R$.

ASSESS Check that your result has the correct units, is reasonable, and answers the question.

Let's return to the situation of Figure 27.4, where a wire moves through a magnetic field by sliding on a U-shaped conducting rail. We looked at this problem as an example of motional emf; now, let's look at it using Faraday's law.

EXAMPLE 27.6 Finding the current in an expanding loop

FIGURE 27.19 shows a 10-cm-long wire with resistance 0.050 Ω being pulled along a U-shaped conducting rail at a speed of 1.0 m/s. A 1.5 T magnetic field is perpendicular to the plane of the rail. What are the magnitude and the direction of the induced current? Assume that the conducting rail is an ideal wire.

PREPARE The magnetic field is constant, but the flux through the loop is changing because its area is changing. We'll follow the steps of Problem-Solving Strategy 27.1.

SOLVE The magnetic flux is into the figure and is increasing because the loop's area is increasing. To oppose this change, the induced magnetic field points out of the figure. We can use the right-hand rule for fields to see that the induced current is counterclockwise.

FIGURE 27.19 A wire being pulled along a rail.

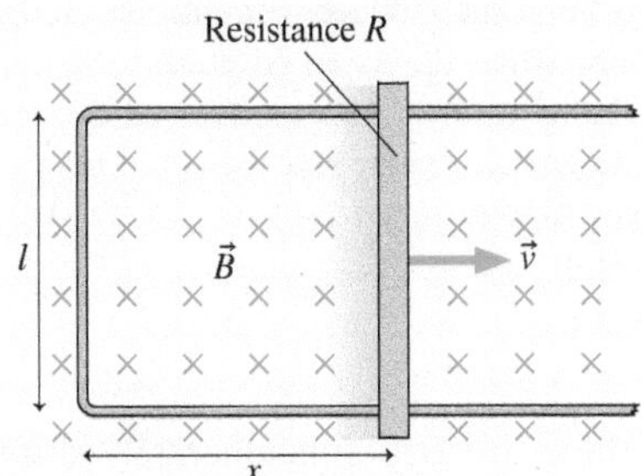

The magnetic field is perpendicular to the plane of the loop, so $\theta = 0°$ and the magnetic flux is $\Phi = AB$. Figure 27.19 shows the sliding wire at distance x from the end of the loop, so the area is $A = xl$ and the flux at this instant of time is

$$\Phi = AB = xlB$$

The flux changes as the wire moves and x increases, and the changing flux induces an emf. Faraday's law for the magnitude of the emf is

$$\mathcal{E} = \left|\frac{d\Phi}{dt}\right| = \left|\frac{d(xlB)}{dt}\right| = lB\left|\frac{dx}{dt}\right|$$

But $|dx/dt|$ is the slide wire's speed v, so the induced emf is

$$\mathcal{E} = vlB$$

The resistance of the loop is simply the resistance of the slide wire, so the induced current is

$$I = \frac{\mathcal{E}}{R} = \frac{vlB}{R} = \frac{(1.0\text{ m/s})(0.10\text{ m})(1.5\text{ T})}{0.050\ \Omega} = 3.0\text{ A}$$

Thus the induced current is 3.0 A counterclockwise.

ASSESS The result $I = vlB/R$ is the same result we found in Section 27.2, where we analyzed the situation by considering the forces on moving charge carriers. Faraday's law is a more general result that includes motional emf but, as the next example shows, also includes situations where there is no motion.

EXAMPLE 27.7 Electromagnetic induction in a solenoid

A 2.0-cm-diameter loop of wire with a resistance of 0.010 Ω is placed in the center of the solenoid seen in FIGURE 27.20a. The solenoid is 4.0 cm in diameter, 20 cm long, and wrapped with 1000 turns of wire. FIGURE 27.20b shows the current through the solenoid as a function of time as the solenoid is "powered up." A positive current is defined to be clockwise (cw) when seen from the left. Find the current in the loop as a function of time and show the result as a graph.

PREPARE The solenoid's length is much greater than its diameter, so we will model it as an infinitely long solenoid. The magnetic field of the solenoid creates a magnetic flux through the loop of wire. The solenoid current is always positive, meaning that it is cw as seen from the left. Consequently, from the right-hand rule, the magnetic field inside the solenoid always points to the right. During the first second, while the solenoid current is increasing, the flux through the loop is to the right and increasing. To oppose the *change* in the flux, the loop's induced magnetic field must point to the left. Thus, again using the right-hand rule, the induced current must flow counterclockwise (ccw) as seen from the left. This is a *negative* current. There's no *change* in the flux for $t > 1$ s, so the induced current is zero.

FIGURE 27.20 A loop inside a solenoid.

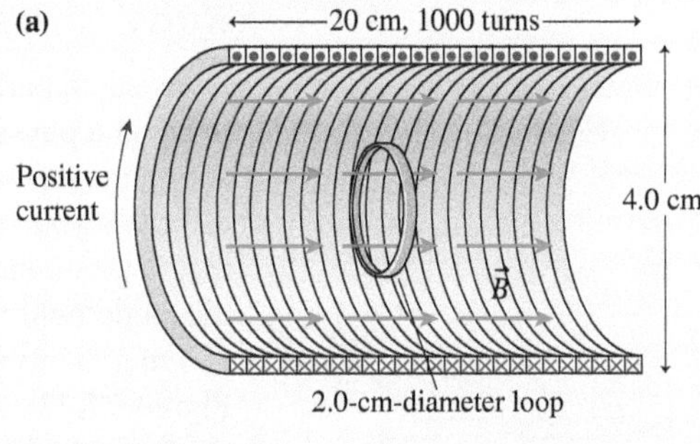

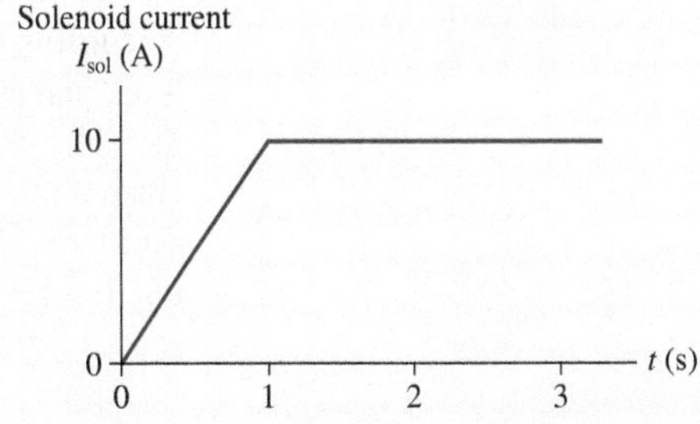

SOLVE Now we're ready to use Faraday's law to find the magnitude of the current. Because the field is uniform inside the solenoid and perpendicular to the loop ($\theta = 0°$), the flux is $\Phi = AB$, where $A = \pi r^2 = 3.14 \times 10^{-4}\text{ m}^2$ is the area of the loop (*not* the area of the solenoid). The field of a long solenoid of length l was found in Chapter 26 to be

$$B = \frac{\mu_0 N I_{sol}}{l}$$

Continued

The flux when the solenoid current is I_{sol} is thus

$$\Phi = \frac{\mu_0 A N I_{sol}}{l}$$

The changing flux creates an induced emf $\mathcal{E}$ that is given by Faraday's law:

$$\mathcal{E} = \left|\frac{d\Phi}{dt}\right| = \frac{\mu_0 A N}{l}\left|\frac{dI_{sol}}{dt}\right| = 2.0 \times 10^{-6}\left|\frac{dI_{sol}}{dt}\right|$$

From the slope of the graph, we find

$$\left|\frac{dI_{sol}}{dt}\right| = \begin{cases} 10 \text{ A/s} & 0.0 \text{ s} < t < 1.0 \text{ s} \\ 0 & 1.0 \text{ s} < t < 3.0 \text{ s} \end{cases}$$

Thus the induced emf is

$$\mathcal{E} = \begin{cases} 2.0 \times 10^{-5} \text{ V} & 0.0 \text{ s} < t < 1.0 \text{ s} \\ 0 \text{ V} & 1.0 \text{ s} < t < 3.0 \text{ s} \end{cases}$$

Finally, the current induced in the loop is

$$I_{loop} = \frac{\mathcal{E}}{R} = \begin{cases} -2.0 \text{ mA} & 0.0 \text{ s} < t < 1.0 \text{ s} \\ 0 \text{ mA} & 1.0 \text{ s} < t < 3.0 \text{ s} \end{cases}$$

where the negative sign comes from Lenz's law. This result is shown in FIGURE 27.21.

FIGURE 27.21 The induced current in the loop.

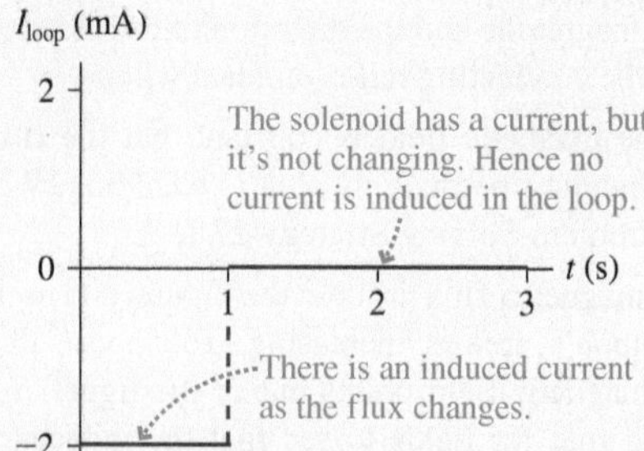
I_{loop} (mA)
2
0
−2
1
2
3
t (s)
The solenoid has a current, but it's not changing. Hence no current is induced in the loop.
There is an induced current as the flux changes.

ASSESS The induced current is proportional to the *rate* at which the solenoid current changes, so a linearly increasing solenoid current gives a constant induced current. The induced current drops to zero not when the solenoid current becomes zero but when it stops changing. Note that an induced current circulating in a loop is a magnetic dipole, and we could refer to this as an induced magnetic dipole.

As these two examples show, there are two fundamentally different ways to change the magnetic flux through a conducting loop:

- The loop can move or expand or rotate, creating a motional emf.
- The magnetic field can change.

Faraday's law tells us that the induced emf is simply the rate of change of the magnetic flux through the loop, *regardless* of what causes the flux to change.

FIGURE 27.22 Eddy currents.

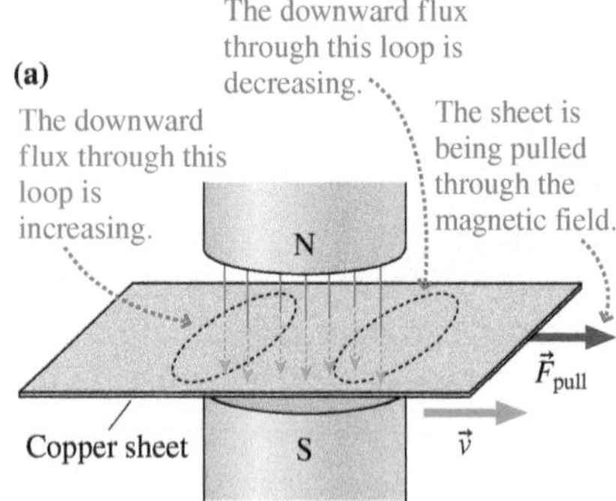
(a)
The downward flux through this loop is decreasing.
The downward flux through this loop is increasing.
The sheet is being pulled through the magnetic field.
N
$\vec{F}_{pull}$
Copper sheet
S
$\vec{v}$

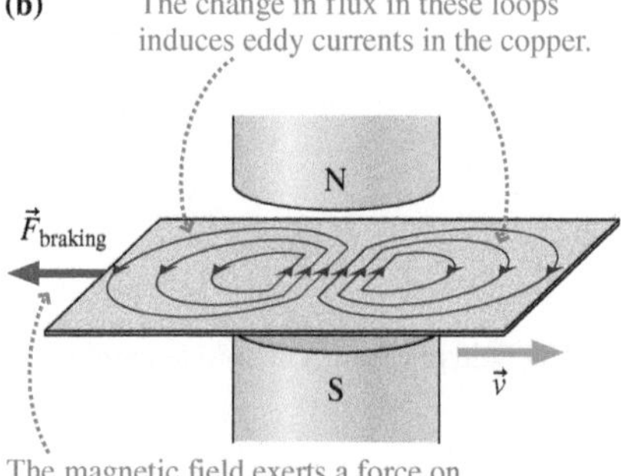
(b)
The change in flux in these loops induces eddy currents in the copper.
N
$\vec{F}_{braking}$
S
$\vec{v}$
The magnetic field exerts a force on the eddy currents, leading to a braking force opposite the motion.

Eddy Currents

Here is a remarkable physics demonstration, perhaps one you'll have an opportunity to try in a physics lab: Take a sheet of copper and place it between the pole tips of a strong magnet, as shown in FIGURE 27.22a. Now, pull the copper sheet out of the magnet as fast as you can. Copper is not a magnetic material and thus is not attracted to the magnet, but, surprisingly, it takes a significant effort to pull the metal through the magnetic field.

Let's analyze this situation to discover the origin of the resisting force. Figure 27.22a shows two "loops" lying entirely inside the metal sheet. The loop on the right is leaving the magnetic field, and the flux through it is decreasing. According to Faraday's law, the flux change will induce a current to flow around this loop, just as in a loop of wire, even though this current does not have a wire to define its path. As a consequence, a clockwise—as given by Lenz's law—"whirlpool" of current (also called an *eddy*) begins to circulate in the metal, as shown in FIGURE 27.22b. Similarly, the loop on the left is entering the field, and the flux through it is increasing. Lenz's law requires this whirlpool of current to circulate the opposite way. These spread-out whirlpools of induced current in a conductor are called **eddy currents.**

Figure 27.22b shows the direction of the eddy currents. Notice that both whirlpools are moving in the same direction as they pass through the center of the magnet. The magnet's field exerts a force on this current. By the right-hand rule, this force is to the left, opposite to the direction of the pull, and thus it acts as a *braking* force. Because of the braking force, **an external force is required to pull a metal through a magnetic field.** If the pulling force ceases, the magnetic braking force quickly causes the metal to decelerate until it stops. No matter which way the metal is moved, the magnetic forces on the eddy currents act to oppose the motion of the metal. *Magnetic braking* uses the braking force associated with eddy currents to slow trains and transit-system vehicles.

Magnetic braking is also used in many laboratory balances, such as the familiar triple-beam balance shown in FIGURE 27.23, to damp out oscillations that occur as the weights are moved. At the end of the beam, a conducting plate passes through a magnetic field. As the beam moves up and down, the magnetic force—which is always directed opposite to the motion—serves to quickly slow the beam. This method is superior to damping based on friction because static friction can cause the beam to stop at a point away from its true equilibrium position. Magnetic damping, however, gets smaller as the beam slows down; when the beam is at rest, there is no magnetic force to disturb equilibrium.

FIGURE 27.23 Magnetic braking of a balance.

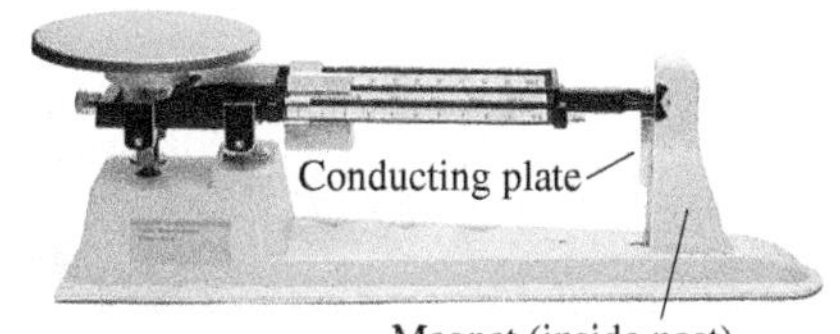

STOP TO THINK 27.4 A loop of wire rests in a region of uniform magnetic field. The magnitude of the field is increasing, thus inducing a current in the loop. Which of the following changes to the situation would make the induced current larger? Choose all that apply.

A. Increase the rate at which the field is increasing.
B. Replace the loop with one of the same resistance but larger diameter.
C. Replace the loop with one of the same resistance but smaller diameter.
D. Orient the loop parallel to the magnetic field.
E. Replace the loop with one of lower resistance.

27.5 Induced Fields

Faraday's law is a tool for calculating the strength of an induced current, but one important piece of the puzzle is still missing. What *causes* the current? That is, what *force* pushes the charges around the loop against the resistive forces of the metal? The agents that exert forces on charges are electric fields and magnetic fields. Magnetic forces are responsible for motional emfs, but magnetic forces cannot explain the current induced in a *stationary* loop by a changing magnetic field.

FIGURE 27.24a shows a conducting loop in an increasing magnetic field. According to Lenz's law, there is an induced current in the counterclockwise direction. Something has to act on the charge carriers to make them move, so we infer that there must be an *electric* field tangent to the loop at all points. This electric field is *caused* by the changing magnetic field and is called an **induced electric field.** The induced electric field is the mechanism that creates a current inside a stationary loop when there's a changing magnetic field.

The conducting loop isn't necessary. The space in which the magnetic field is changing is filled with the pinwheel pattern of induced electric fields shown in FIGURE 27.24b. Charges will move if a conducting path is present, but the induced electric field is there as a direct consequence of the changing magnetic field.

But this is a rather peculiar electric field. All the electric fields we have examined until now have been created by charges. Electric field vectors pointed away from positive charges and toward negative charges. The induced electric field of Figure 27.24b is caused not by charges but by a changing magnetic field.

So it appears that there are two different ways to create an electric field:

1. An electric field is created by positive and negative charges.
2. An electric field is created by a changing magnetic field.

FIGURE 27.24 An induced electric field creates a current in the loop.

(a)

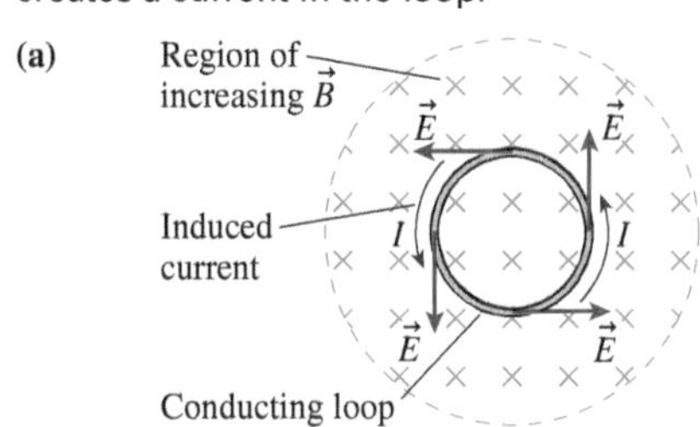

(b)

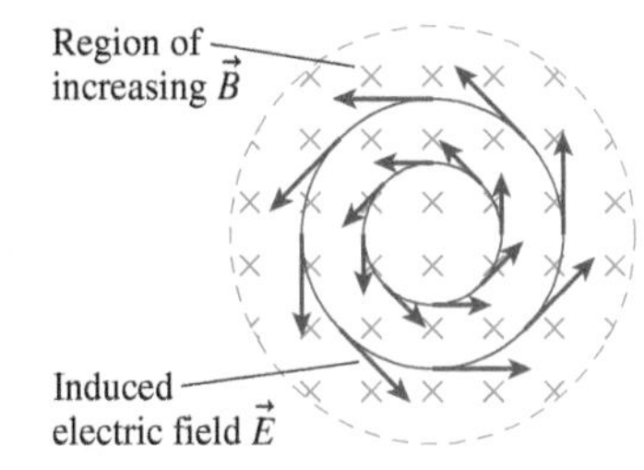

FIGURE 27.25 Two ways to create an electric field.

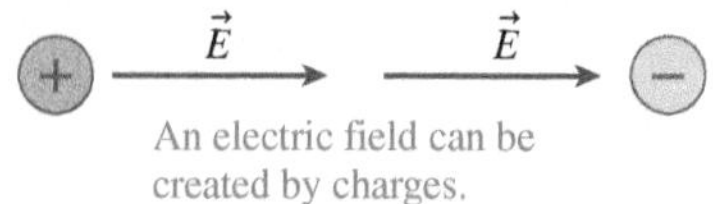

$\vec{E}$
$\vec{E}$
An electric field can be created by charges.

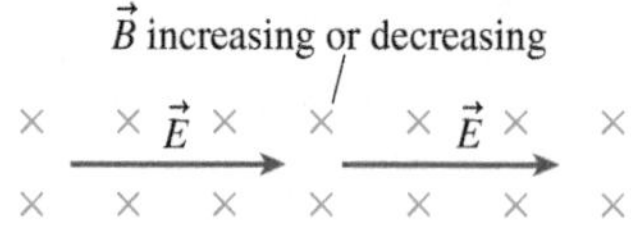

$\vec{B}$ increasing or decreasing
$\vec{E}$
$\vec{E}$
An electric field can also be created by a changing magnetic field.

Both exert a force $\vec{F} = q\vec{E}$ on a charge, and both create a current in a conductor. However, the origins of the fields are very different. FIGURE 27.25 is a quick summary of the two ways to create an electric field.

Calculating the Induced Field

In Chapter 22, we defined the emf $\mathcal{E}$ as the work done *per charge* to separate or move charge; that is,

$$\mathcal{E} = \frac{W}{q} \tag{27.13}$$

In batteries, a familiar source of emf, the work is done by chemical reactions. But the emf in Faraday's law arises when work is done by the force of an induced electric field.

Suppose that a charged particle q travels once around a circular path of radius r in a changing magnetic field. The changing magnetic field creates an induced electric field, and the induced electric field exerts force $\vec{F} = q\vec{E}_{\text{induced}}$ on the charge. This force, like the induced electric field, is tangent to the circle and thus parallel to the particle's displacement at every point. Consequently, this is a situation where W = force $\times$ distance. The distance through which the force acts is the circumference of the circle $2\pi r$, and thus the induced electric field does work

$$W = 2\pi rF = 2\pi rqE_{\text{induced}} \tag{27.14}$$

We can substitute this expression for W into Equation 27.13 to find that the emf created around a circle of radius r by the induced electric field is

$$\mathcal{E} = 2\pi rE_{\text{induced}} \tag{27.15}$$

We also know, from Faraday's law, that $\mathcal{E} = |d\Phi/dt| = |d(AB)/dt|$, where A is the enclosed area. The circular path of the charged particle encloses an unchanging area $A = \pi r^2$. If we equate the emf obtained from Faraday's law to the explicit expression for $\mathcal{E}$ in Equation 27.15, we find

$$\mathcal{E} = \left|\frac{d(AB)}{dt}\right| = \pi r^2\left|\frac{dB}{dt}\right| = 2\pi rE_{\text{induced}} \tag{27.16}$$

Thus the strength of the induced electric field around a circle of radius r is directly proportional to the rate at which the magnetic field changes:

$$E_{\text{induced}} = \frac{r}{2}\left|\frac{dB}{dt}\right| \tag{27.17}$$

Equation 27.17 shows clearly that the induced electric field is created by a *changing* magnetic field. A constant magnetic field with $dB/dt = 0$ gives $E_{\text{induced}} = 0$.

EXAMPLE 27.8 An induced electric field

A 4.0-cm-diameter, 20-cm-long solenoid is wound with 400 turns of wire. The current through the solenoid oscillates at 60 Hz with an amplitude of 2.0 A. What is the maximum strength of the induced electric field inside the solenoid?

PREPARE The electric field lines are concentric circles around the magnetic field lines, as was shown in Figure 27.24b. They reverse direction twice every period as the current oscillates.

SOLVE You learned in Chapter 26 that the magnetic field strength inside a solenoid that has N turns in length L is $B = \mu_0 NI/L$. In this case, the current through the solenoid is $I = I_0 \sin\omega t$, where $I_0 = 2.0$ A is the peak current and $\omega = 2\pi(60 \text{ Hz}) = 377$ rad/s. Thus the induced electric field strength at radius r is

$$E_{\text{induced}} = \frac{r}{2}\left|\frac{dB}{dt}\right| = \frac{r}{2}\frac{d}{dt}\left(\frac{\mu_0 NI_0 \sin\omega t}{L}\right) = \frac{\mu_0 NI_0 r\omega}{2L}\cos\omega t$$

The field strength is maximum at maximum radius ($r = R$) *and* at the instant when $\cos\omega t = 1$. That is,

$$E_{\max} = \frac{\mu_0 NI_0 R\omega}{2L} = 0.019 \text{ V/m}$$

ASSESS This field strength, although not large, is similar to the field strength that the emf of a battery creates in a wire. Hence this induced electric field can drive a substantial induced current through a conducting loop *if* a loop is present. But the induced electric field exists inside the solenoid whether or not there is a conducting loop.

Induced Magnetic Fields

In 1855, less than two years after receiving his undergraduate degree, the Scottish physicist James Clerk Maxwell presented a paper titled "On Faraday's Lines of Force." In this paper, he began to sketch out how Faraday's pictorial ideas about fields could be given a rigorous mathematical basis. Maxwell was troubled by a certain lack of symmetry. Faraday had found that a changing magnetic field creates an induced electric field, an electric field not tied to charges. But what, Maxwell began to wonder, about a changing *electric* field?

To complete the symmetry, Maxwell proposed that a changing electric field creates an **induced magnetic field,** a new kind of magnetic field not tied to the existence of currents. The top half of FIGURE 27.26 shows a region of space where the *electric* field is increasing. This region of space, according to Maxwell, is filled with a pinwheel pattern of induced magnetic fields. The induced magnetic field looks like the induced electric field in the bottom half of the figure, with $\vec{E}$ and $\vec{B}$ interchanged, except that—for technical reasons—the induced $\vec{B}$ points the opposite way from the induced $\vec{E}$.

We won't prove it, but Maxwell's theory predicts that the strength of the induced magnetic field is

$$B_{\text{induced}} = \epsilon_0\mu_0\frac{r}{2}\left|\frac{dE}{dt}\right| \tag{27.18}$$

Equation 27.18 for the induced magnetic field looks very much like Equation 27.17 for the induced electric field, with one exception: the curious combination of the permittivity constant ϵ_0 of electricity and the permeability constant μ_0 of magnetism.

Although there was no experimental evidence at the time that induced magnetic fields exist, Maxwell went ahead and included them in his electromagnetic theory. This was an inspired hunch that was soon vindicated, and the combination $\epsilon_0\mu_0$ played a key role.

FIGURE 27.26 Maxwell hypothesized the existence of induced magnetic fields.

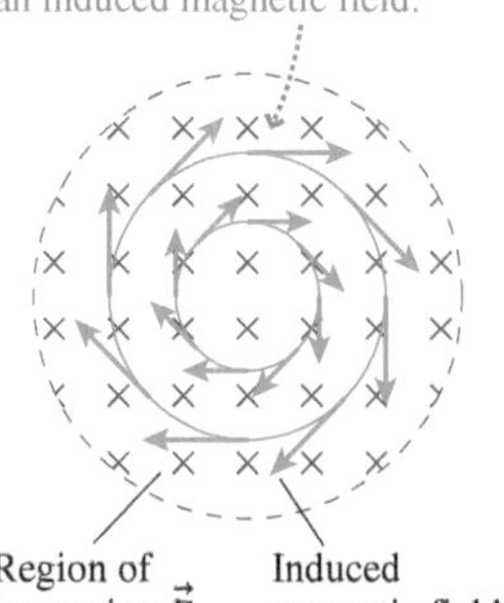

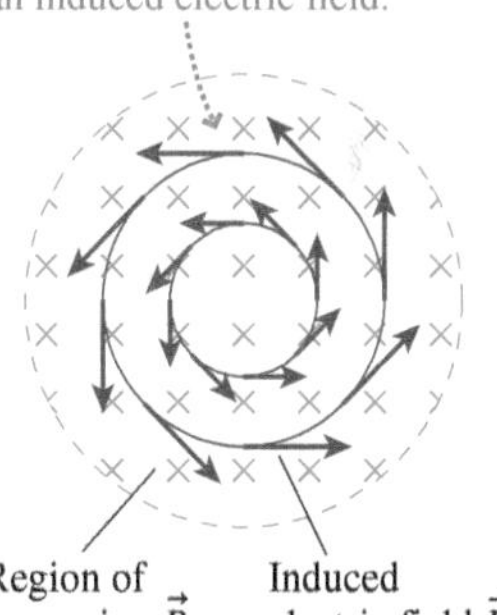

27.6 Electromagnetic Waves

Maxwell observed that an induced magnetic field, if it exists, allows for the possibility of *self-sustaining* electric and magnetic fields. The idea, shown in FIGURE 27.27, is that a changing electric field $\vec{E}$ creates a magnetic field $\vec{B}$, which then changes in just the right way to re-create the electric field, which then changes in just the right way to re-create the magnetic field, and so on. The fields are continually re-created through electromagnetic induction without any reliance on charges or currents.

Maxwell then used his new theory to predict that electric and magnetic fields are self-sustaining, free from charges and current, if they take the form of an **electromagnetic wave.** Furthermore, Maxwell's theory predicts that the wave must travel with speed

$$v_{\text{em}} = \frac{1}{\sqrt{\epsilon_0\mu_0}} \tag{27.19}$$

The values of the constants were known from electrostatic and magnetostatic experiments, and Maxwell computed that an electromagnetic wave, if it existed, would travel with speed $v_{\text{em}} = 3.00 \times 10^8$ m/s.

Maxwell's predicted speed for electromagnetic waves, which comes directly from his theory, is none other than the speed of light! This could be just a coincidence, but Maxwell didn't think so. Making a bold leap, Maxwell concluded that **light is an electromagnetic wave.** It took 25 years for Maxwell's prediction to be tested. In 1886, the German physicist Heinrich Hertz, for whom the unit of frequency is named, discovered how to generate and transmit radio waves, and he soon showed that these waves travel at the speed of light.

FIGURE 27.27 Induced fields can be self-sustaining.

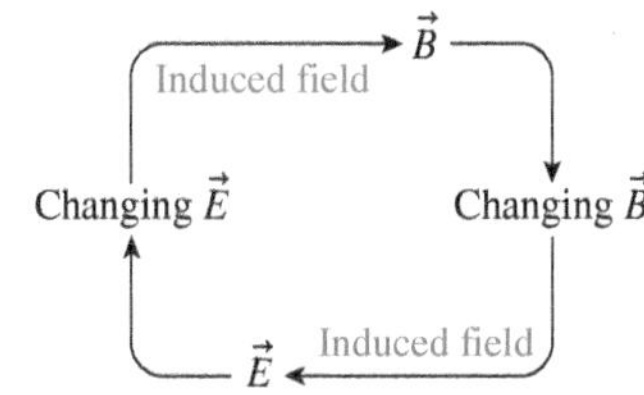

The Structure of an Electromagnetic Wave

FIGURE 27.28 A sinusoidal electromagnetic wave.

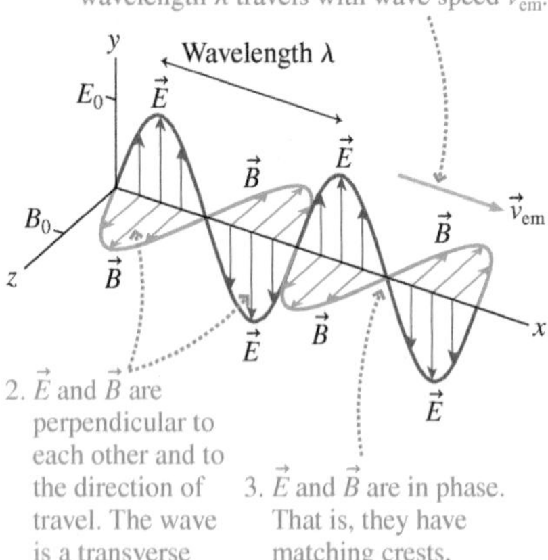
1. A sinusoidal wave with frequency f and wavelength λ travels with wave speed v_{em}.
Wavelength λ
2. $\vec{E}$ and $\vec{B}$ are perpendicular to each other and to the direction of travel. The wave is a transverse wave.
3. $\vec{E}$ and $\vec{B}$ are in phase. That is, they have matching crests, troughs, and zeros.

An electromagnetic wave must have a very specific structure in order to sustain itself and travel through space. FIGURE 27.28 illustrates many of the characteristics of a sinusoidal electromagnetic plane wave that travels along the x-axis. We see that $\vec{E}$ and $\vec{B}$ are perpendicular to each other as well as perpendicular to the direction of travel, so an electromagnetic wave is a *transverse wave*. We also see that $\vec{E}$ and $\vec{B}$ oscillate *in phase,* reaching their peak values E_0 and B_0 at the same instant in each cycle and then passing through zero at the same time.

Figure 27.28 is a useful picture, and one that you will see in many textbooks, but it is a picture that can be misleading if you don't think about it carefully. First and foremost, $\vec{E}$ and $\vec{B}$ are *not* spatial vectors; that is, they don't show how far the wave extends in the y- or z-direction. Instead, these vectors show the field strengths and directions along a single line, the x-axis. An $\vec{E}$ vector pointing in the y-direction means: At this point on the x-axis, where the tail is, this is the direction and strength of the electric field.

Second, this is a plane wave, which, you learned in ◀◀ SECTION 16.5, is a wave for which the fields are the same *everywhere* in any plane perpendicular to $\vec{v}_{em}$. Figure 27.28 shows the fields along only one line. Whatever the fields are doing at a point on the x-axis, they are doing the same thing at every point in the yz-plane that slices the x-axis at that point. FIGURE 27.29 provides a more realistic picture of the electric and magnetic fields of an electromagnetic wave. Figure 27.29a shows that the fields are the same everywhere in a plane that is perpendicular to the direction of travel.

FIGURE 27.29 An electromagnetic wave is a traveling plane wave.

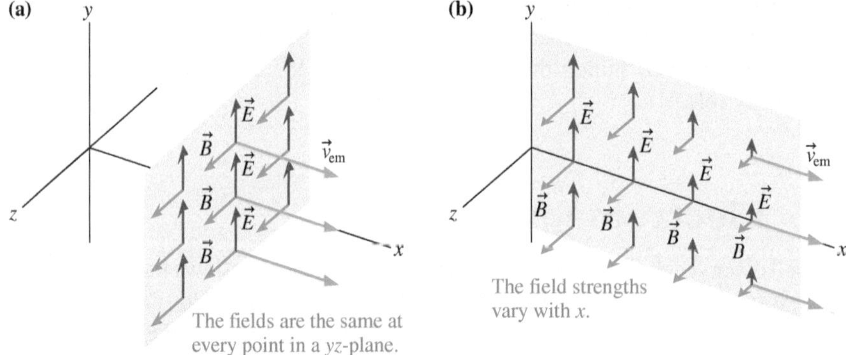
(a)
The fields are the same at every point in a yz-plane.
(b)
The field strengths vary with x.

But a wave is a traveling disturbance, and Figure 27.29b shows that the fields—at one instant of time—*do* change along the x-axis. These changing fields are the disturbance that moves along the x-axis at speed v_{em}. Figure 27.28 is easier to draw than Figure 27.29, but keep in mind that a lot of information is condensed into Figure 27.28.

FIGURE 27.30 The right-hand rule for electromagnetic waves.

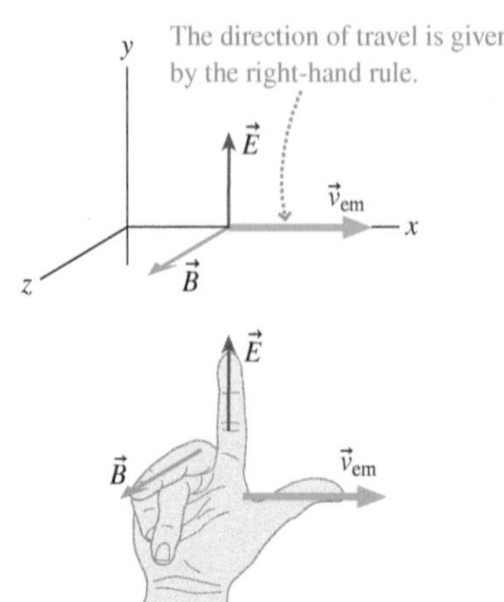
The direction of travel is given by the right-hand rule.

An electromagnetic wave has two additional requirements. First, FIGURE 27.30 shows that the three vectors $\vec{E}$, $\vec{B}$, and $\vec{v}_{em}$ satisfy another right-hand rule: If you point your right index finger in the direction of $\vec{E}$ and, at right angles, your right middle finger in the direction of $\vec{B}$, then your thumb points in the direction of $\vec{v}_{em}$, the direction in which the wave is traveling.

Second, the fields can continually re-create themselves only if there is a specific connection between the field amplitudes E_0 and B_0. Indeed, Maxwell's theory requires that

$$E_0 = cB_0 \tag{27.20}$$

where, as is customary, we use the symbol c for the speed of an electromagnetic wave:

$$c = v_{em} = \frac{1}{\sqrt{\epsilon_0 \mu_0}} = 3.00 \times 10^8 \text{ m/s} \tag{27.21}$$

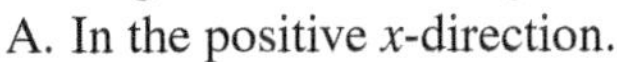

STOP TO THINK 27.5 An electromagnetic wave is traveling in the positive y-direction. The electric field at one instant of time is shown at one position. The magnetic field at this position points

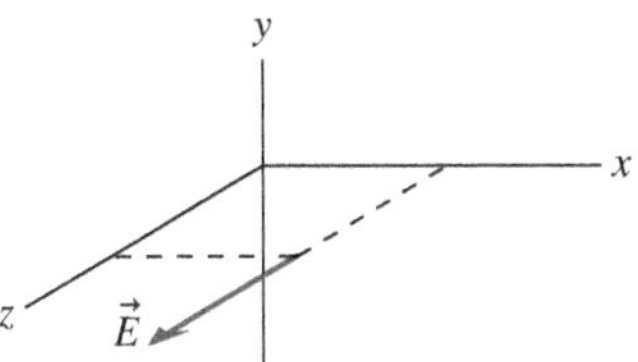

A. In the positive x-direction.
B. In the negative x-direction.
C. In the positive y-direction.
D. In the negative y-direction.
E. Toward the origin.
F. Away from the origin.

Energy and Intensity

Waves transfer energy. Ocean waves erode beaches, sound waves set your eardrums vibrating, and light from the sun warms the earth. We earlier defined the *intensity* of a wave (measured in W/m^2) to be $I = P/A$, where P is the power (energy transferred per second) of a wave that impinges on area A. The intensity of an electromagnetic wave can be written in terms of the amplitude of the oscillating electric field:

$$I = \frac{P}{A} = \frac{1}{2c\mu_0}E_0^2 = \frac{c\epsilon_0}{2}E_0^2 \qquad (27.22)$$

Intensity of an electromagnetic wave with field amplitude E_0

Equation 27.22 relates the intensity of an electromagnetic wave, a quantity that is easily measured, to the amplitude E_0 of the wave's electric field.

The intensity of a plane wave, with constant electric field amplitude E_0, is the same at all points in space. But a plane wave is an idealization; there are no true plane waves in nature. You learned in Chapter 16 that, to conserve energy, the intensity of a wave far from its source decreases with the inverse square of the distance. If a source with power P_{source} emits electromagnetic waves *uniformly* in all directions, the electromagnetic wave intensity at distance r from the source is

$$I = \frac{P_{source}}{4\pi r^2} \qquad (27.23)$$

Equation 27.23 simply expresses the recognition that the energy of the wave is spread over a sphere of surface area $4\pi r^2$.

Solar power The energy from the sun is carried through space by electromagnetic waves—that is, by electric and magnetic fields. This power plant in Spain uses 2650 mirrors across 480 acres to concentrate the sun's power at a tower, where salts are liquefied at 565°C. The thermal energy of this molten salt can be used to power electric generators up to 15 hours later.

EXAMPLE 27.9 **Fields of a cell phone**

A digital cell phone broadcasts a 0.60 W signal at a frequency of 1.9 GHz. What are the amplitudes of the electric and magnetic fields at a distance of 10 cm, about the distance to the center of the user's brain?

PREPARE We'll treat the cell phone as a point source of electromagnetic waves.

SOLVE The intensity of a 0.60 W point source at a distance of 10 cm is

$$I = \frac{P_{source}}{4\pi r^2} = \frac{0.60\ \text{W}}{4\pi(0.10\ \text{m})^2} = 4.78\ \text{W/m}^2$$

We can find the electric field amplitude from the intensity:

$$E_0 = \sqrt{\frac{2I}{c\epsilon_0}} = \sqrt{\frac{2(4.78\ \text{W/m}^2)}{(3.00\times10^8\ \text{m/s})(8.85\times10^{-12}\ \text{C}^2/\text{N}\cdot\text{m}^2)}} = 60\ \text{V/m}$$

The amplitudes of the electric and magnetic fields are related by the speed of light. This allows us to compute

$$B_0 = \frac{E_0}{c} = 2.0\times10^{-7}\ \text{T}$$

ASSESS The electric field amplitude is modest; the magnetic field amplitude is very small. This implies that the interaction of electromagnetic waves with matter is mostly due to the electric field.

Radiation Pressure

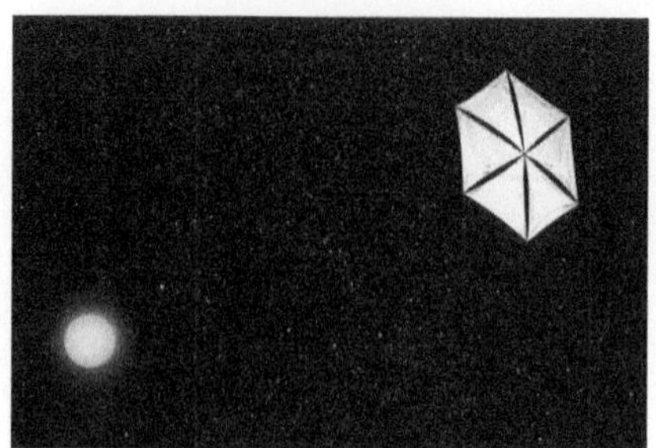

Solar sailing Artist's conception of a future spacecraft powered by radiation pressure from the sun.

Electromagnetic waves transfer not only energy but also momentum. An object gains momentum when it absorbs electromagnetic waves, much as a ball at rest gains momentum when struck by a ball in motion.

Suppose we shine a beam of light on an object that completely absorbs the light energy. If the object absorbs energy during a time interval Δt, its momentum changes by

$$\Delta p = \frac{\text{energy absorbed}}{c}$$

This is a consequence of Maxwell's theory, which we'll state without proof.

The momentum change implies that the light is exerting a force on the object. Newton's second law, in terms of momentum, is $F = \Delta p/\Delta t$. The radiation force due to the beam of light is

$$F = \frac{\Delta p}{\Delta t} = \frac{(\text{energy absorbed})/\Delta t}{c} = \frac{P}{c}$$

where P is the power (joules per second) of the light.

It's more interesting to consider the force exerted on an object per unit area, which is called the **radiation pressure** p_{rad}. The radiation pressure on an object that absorbs all the light is

$$p_{rad} = \frac{F}{A} = \frac{P/A}{c} = \frac{I}{c} \tag{27.24}$$

where I is the intensity of the light wave. The subscript on p_{rad} is important in this context to distinguish the radiation pressure from the momentum p.

EXAMPLE 27.10 Solar sailing

A low-cost way of sending spacecraft to other planets would be to use the radiation pressure on a solar sail. The intensity of the sun's electromagnetic radiation at distances near the earth's orbit is about 1300 W/m^2. What size sail would be needed to accelerate a 10,000 kg spacecraft toward Mars at 0.010 m/s^2?

PREPARE We will assume that the solar sail is perfectly absorbing.

SOLVE The force that will create a 0.010 m/s^2 acceleration is $F = ma = 100$ N. We can use Equation 27.24 to find the sail area that, by absorbing light, will receive a 100 N force from the sun:

$$A = \frac{cF}{I} = \frac{(3.00 \times 10^8 \text{ m/s})(100 \text{ N})}{1300 \text{ W/m}^2} = 2.3 \times 10^7 \text{ m}^2$$

ASSESS If the sail is a square, it would need to be 4.8 km × 4.8 km, or roughly 3 mi × 3 mi. This is large, but not entirely out of the question with thin films that can be unrolled in space. But how will the crew return from Mars?

STOP TO THINK 27.6 The amplitude of the oscillating electric field at your cell phone is 4.0 μV/m when you are 10 km east of the broadcast antenna. What is the electric field amplitude when you are 20 km east of the antenna?

A. 1.0 μV/m
B. 2.0 μV/m
C. 4.0 μV/m
D. There's not enough information to tell.

27.7 Polarization

The plane of the electric field vector $\vec{E}$ and the velocity vector $\vec{v}_{em}$ (the direction of propagation) is called the **plane of polarization** of an electromagnetic wave. FIGURE 27.31 shows two electromagnetic waves moving along the x-axis. The electric field in Figure 27.31a oscillates vertically, so we would say that this wave is *vertically polarized.* Similarly the wave in Figure 27.31b is *horizontally polarized.* Other polarizations are possible, such as a wave polarized 30° away from horizontal.

FIGURE 27.31 The plane of polarization is the plane in which the electric field vector oscillates.

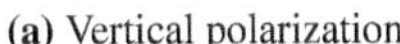
(a) Vertical polarization

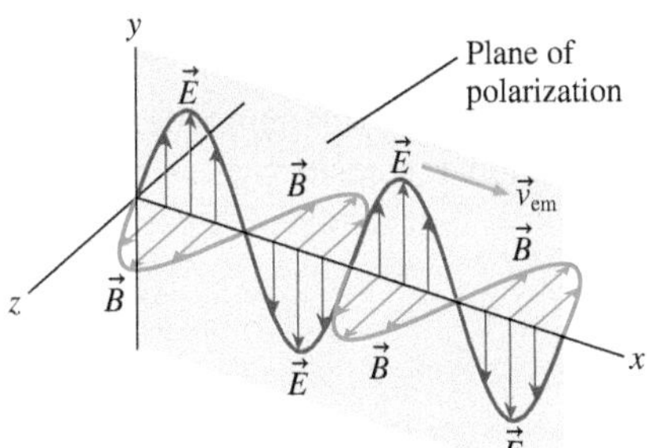
y
Plane of polarization
$\vec{E}$
$\vec{B}$
$\vec{v}_{em}$
z
x

(b) Horizontal polarization

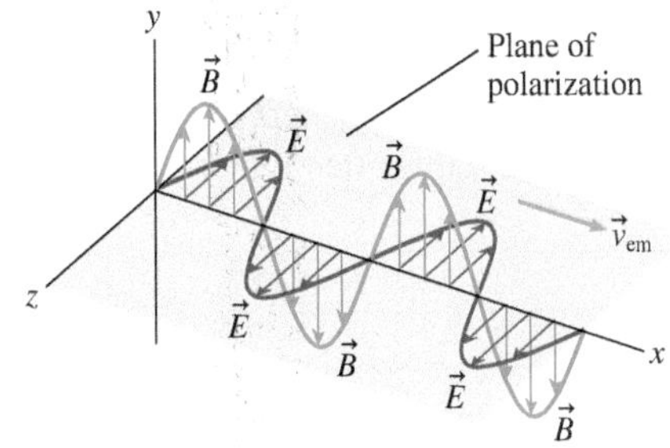
y
Plane of polarization
$\vec{B}$
$\vec{E}$
$\vec{v}_{em}$
z
x

NOTE ▸ This use of the term "polarization" is completely independent of the idea of *charge polarization* that you learned about in Chapter 21. ◂

Some wave sources, such as lasers and radio antennas, emit *polarized* electromagnetic waves with a well-defined plane of polarization. By contrast, most natural sources of electromagnetic radiation are unpolarized, emitting waves whose electric fields oscillate randomly with all possible orientations.

A few natural sources are *partially polarized,* meaning that one direction of polarization is more prominent than others. The light of the sky at right angles to the sun is partially polarized because of how the sun's light scatters from air molecules to create skylight. Bees and other insects make use of this partial polarization to navigate. Light reflected from a flat, horizontal surface, such as a road or the surface of a lake, has a predominantly horizontal polarization. This is the rationale for using polarizing sunglasses.

The most common way of artificially generating polarized visible light is to send unpolarized light through a *polarizing filter.* A typical polarizing filter, as shown in FIGURE 27.32, is a plastic sheet containing very long organic molecules known as polymers. The sheets are formed in such a way that the polymers are all aligned to form a grid, rather like the metal bars in a barbecue grill. The sheet is then chemically treated to make the polymer molecules somewhat conducting.

FIGURE 27.32 A polarizing filter.

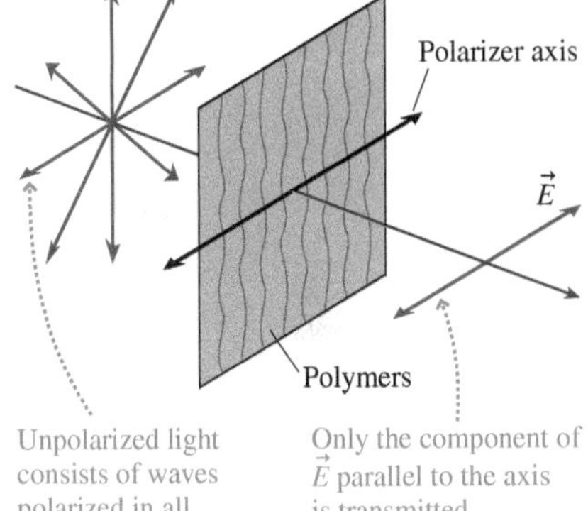
Polarizer axis
$\vec{E}$
Polymers
Unpolarized light consists of waves polarized in all possible directions.
Only the component of $\vec{E}$ parallel to the axis is transmitted.

As a light wave travels through the filter, the component of the electric field oscillating parallel to the polymer grid drives the conduction electrons up and down the molecules. The electrons absorb energy from the light wave, so the parallel component of $\vec{E}$ is absorbed in the filter. But the conduction electrons can't oscillate perpendicular to the molecules, so the component of $\vec{E}$ perpendicular to the polymer grid passes through without absorption. Thus the light wave emerging from a polarizing filter is polarized perpendicular to the polymer grid. The direction of the transmitted polarization is called the *polarizer axis.*

Malus's Law

Suppose a *polarized* light wave of intensity I_0 approaches a polarizing filter. What is the intensity of the light that passes through the filter? FIGURE 27.33 shows that an oscillating electric field can be decomposed into components parallel and perpendicular to the polarizer axis. If we call the polarizer axis the y-axis, then the incident electric field is

$$\vec{E}_{\text{incident}} = (E_x, E_y) = (E_0 \sin\theta,\, E_0 \cos\theta) \tag{27.25}$$

where θ is the angle between the incident plane of polarization and the polarizer axis.

FIGURE 27.33 An incident electric field can be decomposed into components parallel and perpendicular to a polarizer axis.

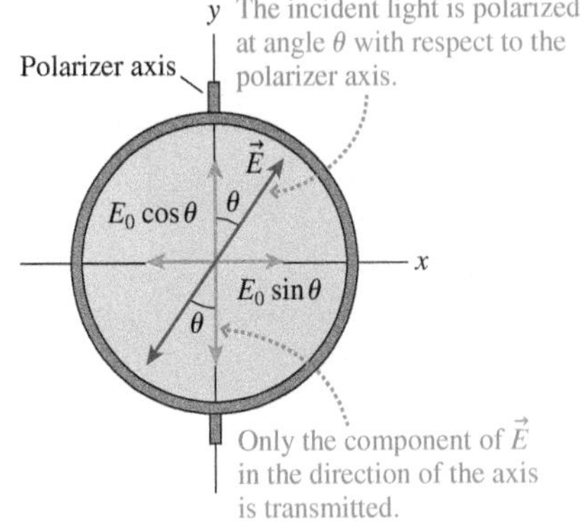
y
The incident light is polarized at angle θ with respect to the polarizer axis.
Polarizer axis
$\vec{E}$
θ
$E_0 \cos\theta$
x
$E_0 \sin\theta$
θ
Only the component of $\vec{E}$ in the direction of the axis is transmitted.

If the polarizer is ideal, meaning that light polarized parallel to the axis is 100% transmitted and light perpendicular to the axis is 100% blocked, then the electric field of the light transmitted by the filter is

$$\vec{E}_{\text{transmitted}} = (E_0 \cos\theta,\ \text{direction of polarizer axis}) \tag{27.26}$$

Because the intensity depends on the square of the electric field amplitude, you can see that the transmitted intensity is related to the incident intensity by

$$I_{\text{transmitted}} = I_0 \cos^2\theta \tag{27.27}$$

Intensity of light transmitted by a polarizing filter

This result, which was discovered experimentally in 1809, is called **Malus's law.**

FIGURE 27.34a shows that Malus's law can be demonstrated with two polarizing filters. The first, called the *polarizer,* is used to produce polarized light of intensity I_0. The second, called the *analyzer,* is rotated by angle θ relative to the polarizer. As the photographs of FIGURE 27.34b show, the transmission of the analyzer is (ideally) 100% when $\theta = 0°$ and steadily decreases to zero when $\theta = 90°$. Two polarizing filters with perpendicular axes, called *crossed polarizers,* block all the light.

FIGURE 27.34 The intensity of the transmitted light depends on the angle between the polarizing filters.

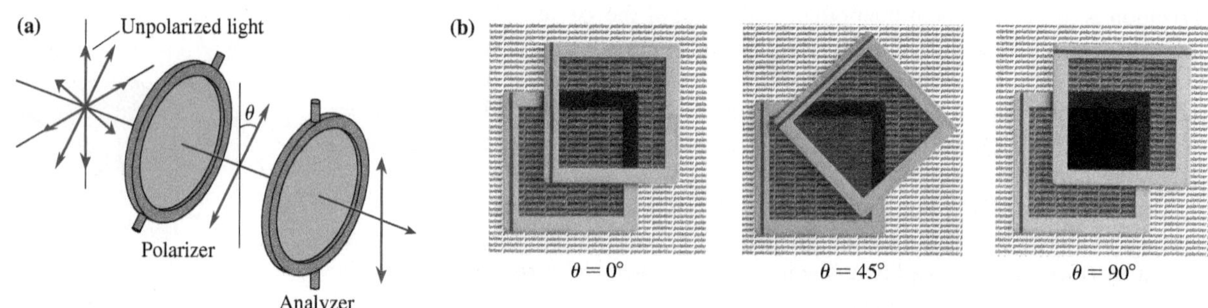

Suppose the light incident on a polarizing filter is *unpolarized,* as is the light incident from the left on the polarizer in Figure 27.34a. The electric field of unpolarized light varies randomly through all possible values of θ. Because the *average* value of $\cos^2\theta$ is $\frac{1}{2}$, the intensity transmitted by a polarizing filter is

$$I_{\text{transmitted}} = \tfrac{1}{2} I_0 \qquad \text{(incident light unpolarized)} \tag{27.28}$$

In other words, a polarizing filter passes 50% of unpolarized light and blocks 50%.

In polarizing sunglasses, the polymer grid is aligned horizontally (when the glasses are in the normal orientation) so that the glasses transmit vertically polarized light. Most natural light is unpolarized, so the glasses reduce the light intensity by 50%. But *glare*—the reflection of the sun and the skylight from roads and other horizontal surfaces—has a strong horizontal polarization. This light is almost completely blocked by a polarizing filter, so the sunglasses "cut glare" without affecting the main scene you wish to see.

You can test whether your sunglasses are polarized by holding them in front of you and rotating them as you look at the glare reflecting from a horizontal surface. Polarizing sunglasses substantially reduce the glare when the glasses are "normal" but not when the glasses are 90° from normal. (You can also test them against a pair of sunglasses known to be polarizing by seeing if all light is blocked when the lenses of the two pairs are crossed.)

Cutting the glare The vertical polarizer blocks the horizontally polarized glare from the surface of the water.

STOP TO THINK 27.7 Unpolarized light of equal intensity is incident on four pairs of polarizing filters. Rank in order, from largest to smallest, the intensities I_A to I_D transmitted through the second polarizer of each pair.

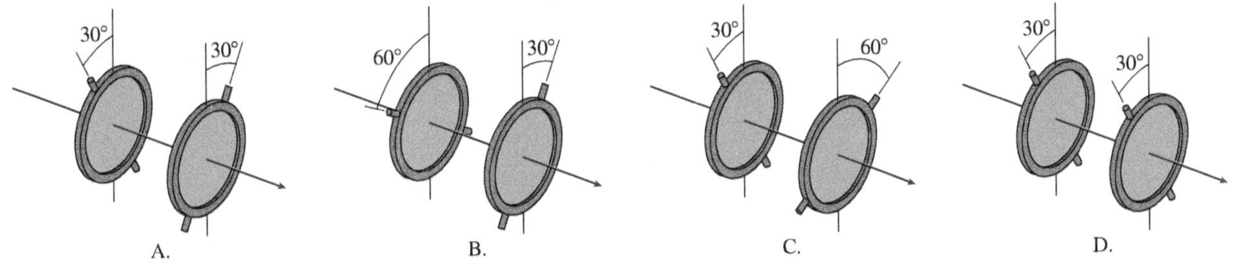

27.8 The Interaction of Electromagnetic Waves with Matter

The oscillating electric and magnetic fields of an electromagnetic wave interact strongly with the electrons in molecules and materials. An electromagnetic wave incident on a material may pass through harmlessly, or it may heat or damage the material. The interaction of electromagnetic waves with living tissue is especially relevant to biology and medicine. We'll take an initial look at these interactions in this section, then expand the ideas in Chapters 28 and 29 after we introduce the photon model of light.

The frequencies of electromagnetic waves span many orders of magnitude, from low-frequency radio waves to high-frequency x rays and gamma rays. The frequencies and wavelengths that our eyes sense—the *visible spectrum*—are a very small slice of the *electromagnetic spectrum* shown in FIGURE 27.35.

FIGURE 27.35 The electromagnetic spectrum.

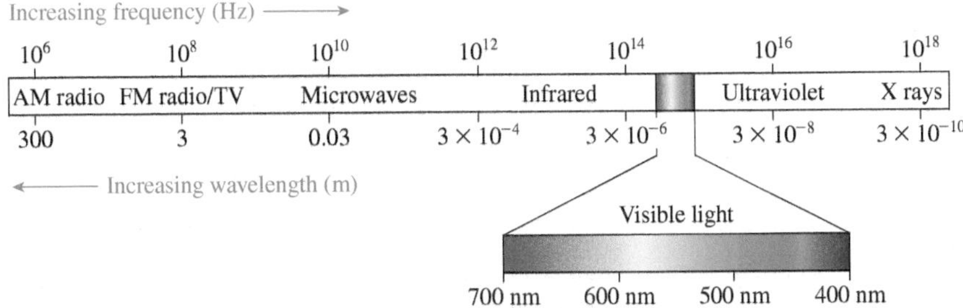

Radio Waves and Microwaves

An electromagnetic wave is self-sustaining, needing no charges or currents to keep it propagating through space. However, charges and currents are needed at the *source* of an electromagnetic wave. Radio waves and microwaves are usually produced by the motion of charged particles in an antenna.

FIGURE 27.36a on the next page reminds you what the electric field of an electric dipole looks like. If the dipole is vertical, the electric field $\vec{E}$ at points along the horizontal axis in the figure is also vertical. Reversing the dipole, by switching the charges, reverses $\vec{E}$. If the charges were to *oscillate* back and forth, switching position at frequency f, then $\vec{E}$ would oscillate in a vertical plane. The changing $\vec{E}$ would then create an induced magnetic field $\vec{B}$, which could then create an $\vec{E}$, which could then create a $\vec{B}$, . . . , and a vertically polarized electromagnetic wave at frequency f would radiate out into space.

This is exactly what an **antenna** does. FIGURE 27.36b shows two metal wires attached to the terminals of an oscillating voltage source. The figure shows an

FIGURE 27.36 An oscillating dipole is an antenna that broadcasts a self-sustaining electromagnetic wave.

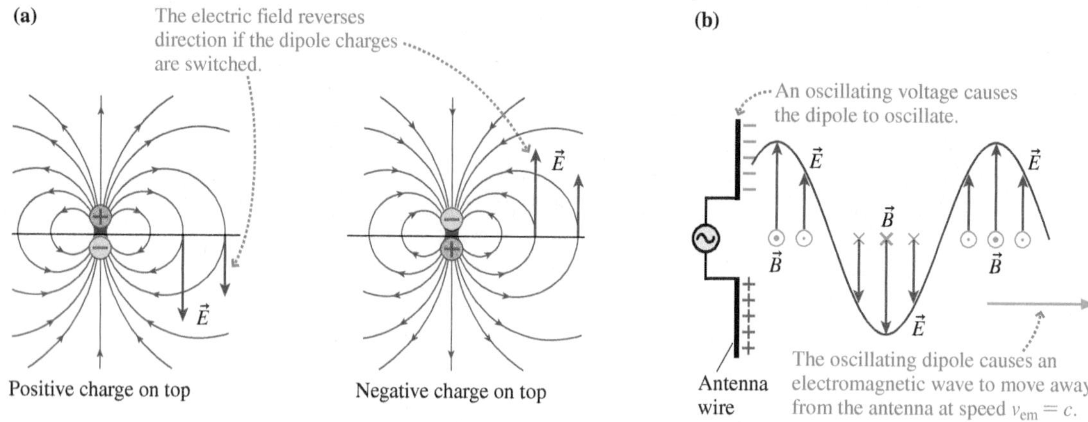
(a)
The electric field reverses direction if the dipole charges are switched.
$\vec{E}$
$\vec{E}$
Positive charge on top
Negative charge on top
(b)
An oscillating voltage causes the dipole to oscillate.
$\vec{E}$
$\vec{B}$
$\vec{E}$
Antenna wire
The oscillating dipole causes an electromagnetic wave to move away from the antenna at speed $v_{em} = c$.

instant when the top wire is negative and the bottom is positive, but these will reverse in half a cycle. The wire is basically an oscillating dipole, and it creates an oscillating electric field. The oscillating $\vec{E}$ induces an oscillating $\vec{B}$, and they take off as an electromagnetic wave at speed $v_{em} = c$. The wave does need oscillating charges as a *wave source,* but once created it is self-sustaining and independent of the source.

Radio waves are *detected* by antennas as well. The electric field of the radio wave drives a current up and down a conductor, producing a potential difference that can be amplified. For best reception, the antenna length should be about $\frac{1}{4}$ of a wavelength. A typical cell phone works at 1.9 GHz, with wavelength $\lambda = c/f = 16$ cm. Thus a cell phone antenna should be about 4 cm long, or about $1\frac{1}{2}$ inches; it is generally hidden inside the phone itself.

CONCEPTUAL EXAMPLE 27.11 Wildlife tracking

The elk shown in the left photo wears a radio collar with a vertical broadcast antenna. A wildlife biologist can search for the signal from the elk's collar by using a receiving antenna. How should the receiving antenna be oriented?

REASON We have seen that an antenna generates a wave that is polarized in the same direction as the antenna itself. As the photo on the right shows, the receiving antenna should be oriented in the direction of the wave's polarization—here, vertically—so that the wave's electric field can drive charges up and down the antenna, creating a measurable potential difference. If the receiving antenna were horizontal, charges would be driven only along the tiny diameter of the antenna wire, creating no potential difference along its length.

ASSESS In reality, tracked animals are always moving, and their antennas may change orientation. To obtain the largest signal, the biologist must adjust the receiving antenna to match.

Radio waves and microwaves interact only weakly with matter. You and everything around you are continuously bathed in electromagnetic waves from radio broadcasts, Wi-Fi transmitters, and cell phones, but those waves pass through you harmlessly.

These low-frequency electromagnetic waves interact with matter primarily by exerting an oscillating torque on molecules, such as water, that have permanent electric dipole moments. As FIGURE 27.37 shows, an oscillating electric field causes an electric dipole to rotate. Molecules that have permanent dipole moments acquire kinetic energy from the wave, then transform this energy into thermal energy as they collide with other molecules. Any heating associated with the low-intensity radio waves and microwaves in our everyday lives is completely unnoticeable.

This is not the case with high-intensity, low-frequency waves. A microwave oven heats food by using very intense 2450 MHz microwaves to drive the rotation of molecules in liquid water. A microwave oven is less effective at defrosting frozen food because the water molecules in ice are not free to rotate. Oils and other nonpolar molecules that lack permanent electric dipole moments are not heated efficiently in a microwave oven.

Diathermy uses intense radio-frequency waves, usually at a frequency of 27 MHz, for heating deep tissue for therapeutic purposes in medicine, especially to increase blood flow and reduce pain. Diathermy transfers energy to water molecules, as in a microwave oven, and it also causes ions to oscillate back and forth until they collide with and transfer energy to other molecules.

FIGURE 27.37 A radio wave interacts with an electric dipole.

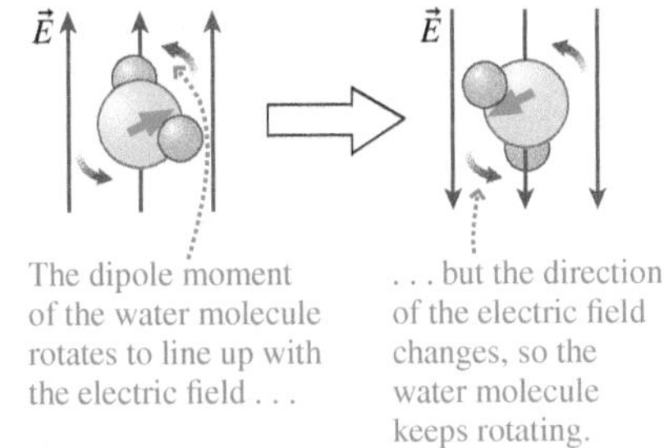

Infrared, Visible Light, and Ultraviolet

Radio waves are produced by oscillating charges in an antenna. At the higher frequencies of infrared, visible light, and ultraviolet, the "antennas" are individual atoms. The continual jostling of thermal energy causes atomic-level oscillations in the charge distributions of molecules or materials, and these oscillations emit *thermal radiation.*

We learned in ◀◀ SECTION 12.5 that thermal radiation is a form of heat transfer between an object and its environment. If a small amount of heat dQ is radiated in a time interval dt by an object with surface area A and absolute temperature T, the *rate* of heat transfer—the radiated power—is

$$\frac{dQ}{dt} = e\sigma A T^4 \tag{27.29}$$

where $\sigma = 5.67 \times 10^{-8}\ \text{W/m}^2 \cdot \text{K}^4$ is the Stefan-Boltzmann constant. The e in Equation 27.29 is the object's emissivity, a measure of its effectiveness at emitting thermal radiation. A perfect emitter, called a *black body,* has $e = 1$.

In Chapter 12 we considered the amount of energy radiated and its dependence on temperature. If you increase the current through an incandescent lightbulb filament, the filament temperature increases and so does the total energy emitted by the bulb, in accordance with Equation 27.29. The three pictures in FIGURE 27.38 show a glowing lightbulb with the filament at successively higher temperatures. We can clearly see an increase in brightness in the sequence of three photographs.

But it's not just the brightness that varies. The *color* of the emitted radiation changes as well. At low temperatures, the light from the bulb is quite red. Looking at the change in color as the temperature of the bulb rises in Figure 27.38, we see that **the spectrum of thermal radiation changes with temperature.**

If we measure the intensity of thermal radiation as a function of wavelength for an object at three temperatures, 3500 K, 4500 K, and 5500 K, the data appear as in FIGURE 27.39. Notice two important features:

- Increasing the temperature increases the intensity at all wavelengths. **Making the object hotter causes it to emit more radiation across the entire spectrum.**
- Increasing the temperature causes the peak intensity to shift to a shorter wavelength. **The higher the temperature, the shorter the wavelength of the peak of the spectrum.**

FIGURE 27.38 The brightness of the bulb varies with the temperature of the filament.

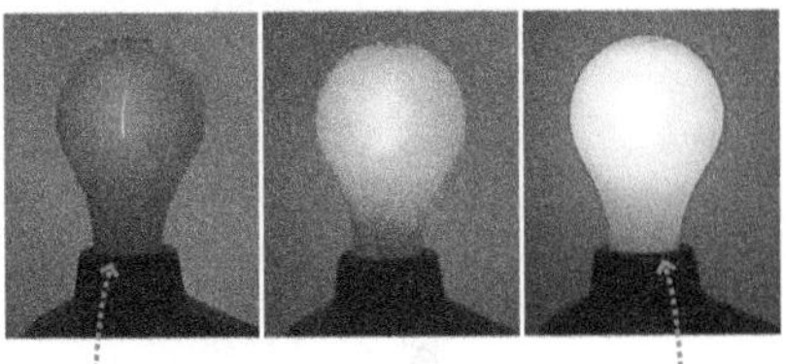

Increasing filament temperature
At lower filament temperatures, the bulb is dim and the light is noticeably reddish.
When the filament is hotter, the bulb is brighter and the light is whiter.

FIGURE 27.39 A thermal emission spectrum depends on the temperature.

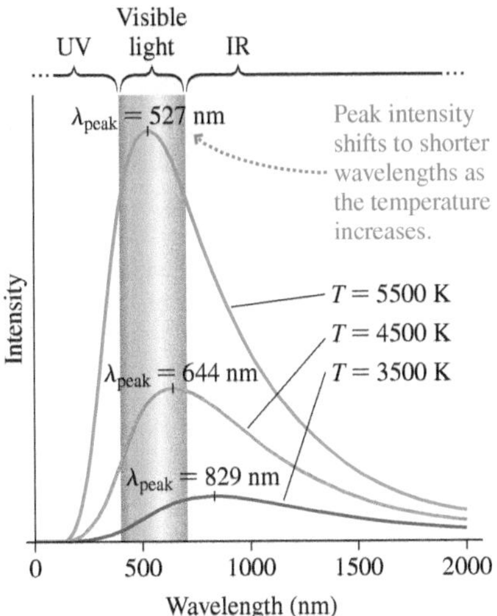

It is this variation of the peak wavelength that causes the change in color of the glowing filament in Figure 27.38. The temperature dependence of the peak wavelength of thermal radiation is known as **Wien's law:**

$$\lambda_{\text{peak}}\,(\text{in nm}) = \frac{2.9\times10^{6}\ \text{nm}\cdot\text{K}}{T\,(\text{in K})} \tag{27.30}$$

Wien's law for the peak wavelength of a thermal emission spectrum

EXAMPLE 27.12 **Finding peak wavelengths of radiating objects**

What are the wavelengths of peak intensity and the corresponding spectral regions for radiating objects at (a) the normal human body temperature of 37°C, (b) the temperature of the filament in an incandescent lamp, 1500°C, and (c) the temperature of the surface of the sun, 5800 K?

PREPARE All of the objects emit thermal radiation, so the peak wavelengths are given by Equation 27.30. We need to convert temperatures to kelvin. The temperature of the human body is $T = 37 + 273 = 310$ K, and the filament temperature is $T = 1500 + 273 = 1773$ K.

SOLVE We use Equation 27.30 to find the wavelengths of peak intensity:

a. $\lambda_{\text{peak}}(\text{body}) = \dfrac{2.9\times10^{6}\ \text{nm}\cdot\text{K}}{310\ \text{K}} = 9.4\times10^{3}\ \text{nm} = 9.4\ \mu\text{m}$

b. $\lambda_{\text{peak}}(\text{filament}) = \dfrac{2.9\times10^{6}\ \text{nm}\cdot\text{K}}{1773\ \text{K}} = 1600\ \text{nm}$

c. $\lambda_{\text{peak}}(\text{sun}) = \dfrac{2.9\times10^{6}\ \text{nm}\cdot\text{K}}{5800\ \text{K}} = 500\ \text{nm}$

The peak of the emission curve at body temperature is far into the infrared region of the spectrum, well below the range of sensitivity of human vision. You don't see someone "glow," although people do indeed emit significant energy in the form of infrared radiation. The sun's emission peaks right in the middle of the visible spectrum, which seems reasonable. Interestingly, most of the energy radiated by an incandescent bulb is *not* visible light. The tail of the emission curve extends into the visible region, but the peak of the emission curve—and most of the emitted energy—is in the infrared region of the spectrum. A 100 W bulb emits only a few watts of visible light. This inefficiency of incandescent lamps is why they are rapidly being replaced by high-efficiency LED lamps.

ASSESS The temperature spans a wide range of absolute temperatures; it makes sense that the wavelengths span an equally wide range.

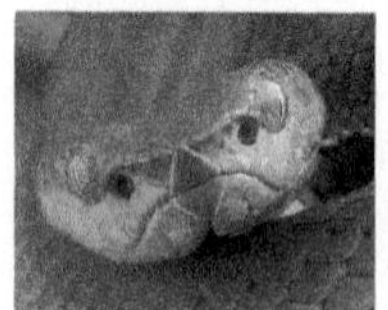
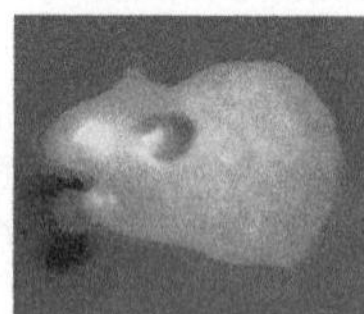

It's the pits . . . Certain snakes—including rattlesnakes and other pit vipers—can hunt in total darkness. The viper in the left photo has pits in front of its eyes. These are a second set of vision organs; they have sensitive tissue at the bottom that allows them to detect the infrared radiation emitted by warm-blooded prey, such as the thermal radiation emitted by the mouse in the thermal image on the right. These snakes need no light to "see" you. You emit an infrared "glow" they can detect.

Infrared radiation and visible light produce effects in tissues similar to the effects of microwaves—heating. You experience this if you sit in the sun or by a heat lamp. However, the absorption is stronger and the penetration less than for microwaves. Infrared radiation reaches a few millimeters into the body, but most visible light is absorbed in the skin by the blood in capillaries. Red light is an exception, and the ability of red light to penetrate 2 or 3 mm is used in clip-on *pulse oximetry* devices to measure blood oxygenation, an application we'll look at in Chapter 29.

In contrast, ultraviolet light has enough energy to cause chemical change, breaking molecular bonds and sometimes ionizing molecules. Ultraviolet radiation is divided into three ranges based on its health effects:

- Ultraviolet A (315–400 nm): Wavelengths in this range damage collagen in the skin and destroy vitamin A. They penetrate deep into the skin to oxidize melanin, thus producing the brown coloration of a suntan.
- Ultraviolet B (280–315 nm): These somewhat shorter wavelengths are absorbed at the surface of the skin, where they can cause sunburn. They can also cause skin cancer by damaging DNA molecules. Ironically, these wavelengths are the most efficient for producing vitamin D. The skin produces this essential

molecule in a process that uses high-energy ultraviolet light to break a bond in a precursor molecule.

- Ultraviolet C (shorter than 280 nm): Wavelengths shorter than 280 nm cause extensive cell damage and death. Ultraviolet C is used in germicidal lamps.

Color Vision

The *cones,* the color-sensitive cells in the retina of the eye, each contain one of three slightly different forms of a light-sensitive photopigment. Each photopigment has a range of wavelengths to which it is sensitive. Our color vision is a result of the differential response of three types of cones that contain three slightly different pigments, shown in FIGURE 27.40.

Humans have three types of cone cells in the eye, mice have two, and chickens four—giving a chicken keener color vision than a human. The three color photopigments that bees possess give them excellent color vision, but a bee's color sense is different from a human's. The peak sensitivities of a bee's photopigments are in the yellow, blue, and ultraviolet regions of the spectrum. A bee can't see the red of a rose, but it is quite sensitive to ultraviolet wavelengths well beyond the range of human vision. Many flowers have a ring of ultraviolet-absorbing pigments near their center that is invisible to humans but helps guide bees to the pollen.

FIGURE 27.40 The sensitivity of different cone cells in the human eye.

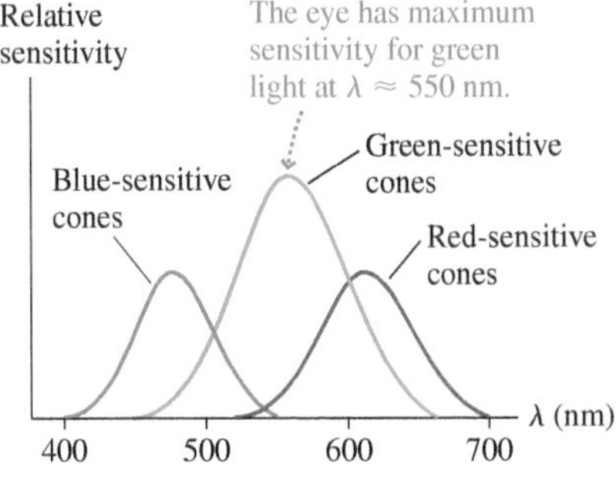

X Rays and Gamma Rays

At the highest energies of the electromagnetic spectrum we find x rays and gamma rays. There is no sharp dividing line between these two regions of the spectrum; the difference is the source of radiation. In fact, both forms of radiation are better described by the photon model of light, which we will discuss in the next chapter, than by the wave model. Generally speaking, high-energy photons emitted by electrons are called x rays. If the source is a nuclear process, we call them gamma rays.

We will look at the production of x rays and gamma rays in atomic and nuclear nuclear processes in Part VI. For now, we will focus on the production of x rays in an x-ray tube, such as the one shown in FIGURE 27.41. X-ray tubes are the source of medical x rays.

FIGURE 27.41 A simple x-ray tube.

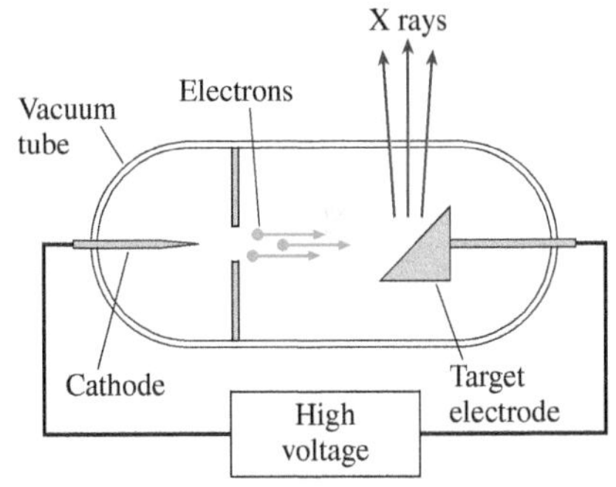

Electrons are emitted from a cathode and accelerated through a potential difference of several thousand volts. The electrons make a sudden stop when they hit a metal target electrode. Maxwell's theory predicts that accelerating charged particles emit electromagnetic waves, and in this case the rapidly decelerating electrons emit x rays—electromagnetic waves with wavelengths shorter than about 10 nm. The x rays pass through a window in the tube and then may be used to produce an image or treat a disease.

X rays and gamma rays (and the short-wavelength part of the ultraviolet spectrum) are **ionizing radiation;** the radiation has sufficient energy to fragment and ionize molecules. When such radiation strikes tissue, the resulting ionization can produce cellular damage. When people speak of "radiation" they often mean "ionizing radiation."

At several points in this chapter we have hinted at places where a full understanding of the phenomena requires some new physics. For example, we mentioned that nuclear processes can give rise to gamma rays. There are other questions that we did not answer, such as why the electromagnetic spectrum of a hot object has the shape that it does. These puzzles began to arise in the late 1800s and early 1900s, and it soon became clear that the physics of Newton and Maxwell was not sufficient to fully describe the nature of matter and energy. Some new rules, some new models, were needed. We will return to these puzzles as we begin to explore the notions of quantum physics in Part VI.

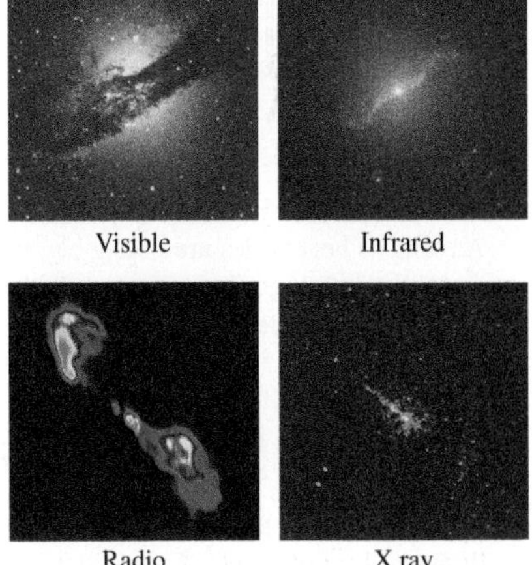

Seeing the universe in a different light These four images of the Centaurus A galaxy have the same magnification and orientation, but they are records of different types of electromagnetic waves. (All but the visible-light image are false-color images.) The visible-light image shows a dark dust lane cutting across the galaxy. In the infrared, this dust lane glows quite brightly—telling us that the dust particles are hot. The radio and x-ray images show jets of matter streaming out of the galaxy's center at right angles to the plane of the galaxy, hinting at the presence of a massive black hole. Views of the cosmos beyond the visible range are important tools of modern astronomy.

STOP TO THINK 27.8 A group of four stars, all the same size, have the four different surface temperatures given below. Which of these stars emits the most red light?

A. 2000 K B. 4000 K C. 7000 K D. 10,000 K

CHAPTER 27 INTEGRATED EXAMPLE Induction heating

Induction heating, which uses an induced current to rapidly heat metal objects to high temperatures, is used to form catheter tips, harden dental drill bits, and create foil seals on plastic bottles in the pharmaceutical industry. To illustrate the idea, consider a copper wire formed into a 4.0 cm × 4.0 cm square loop and placed in a magnetic field—perpendicular to the plane of the loop—that oscillates with 0.010 T amplitude at a frequency of 1000 Hz. What is the wire's initial temperature rise in °C/min?

PREPARE The changing magnetic flux through the loop induces a current that, because of the wire's resistance, heats the wire. Eventually, when the wire gets hot, heat loss through radiation and/or convection will limit the temperature rise, but initially we can consider the temperature change due only to the heating by the current. Assume that the wire's diameter is much less than the 4.0 cm width of the loop.

FIGURE 27.42 shows the copper loop in the magnetic field. The wire's cross-section area A is unknown, but our assumption of a thin wire means that the loop has a well-defined area L^2. Values of copper's resistivity, density, and specific heat are listed in tables at the back of the text. We've used subscripts to distinguish between mass density ρ_{mass} and resistivity ρ_{elec}, a potentially confusing duplication of symbols.

FIGURE 27.42 A copper wire being heated by induction.

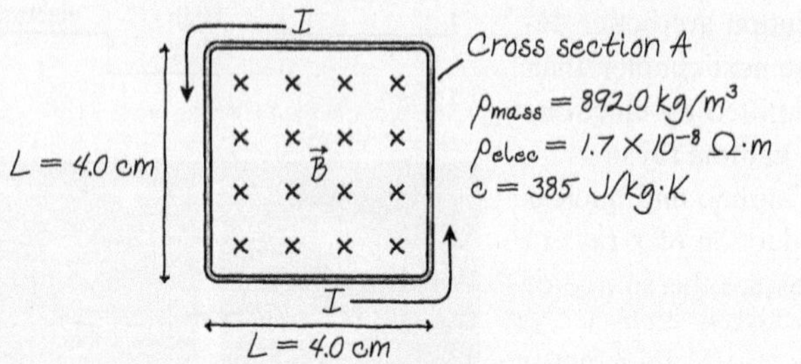

SOLVE Power dissipation by a current, $P = I^2R$, heats the wire. As long as heat losses are negligible, we can use the heating rate and the wire's specific heat c to calculate the rate of temperature change. Our first task is to find the induced current. According to Faraday's law,

$$I = \frac{\mathcal{E}}{R} = \frac{1}{R}\left|\frac{d\Phi}{dt}\right| = \frac{L^2}{R}\left|\frac{dB}{dt}\right|$$

where R is the loop's resistance and $\Phi = L^2B$ is the magnetic flux through a loop of area L^2. The oscillating magnetic field can be written $B = B_0 \cos \omega t$, with $B_0 = 0.010$ T and $\omega = 2\pi \times 1000$ Hz = 6280 rad/s. Thus

$$\frac{dB}{dt} = -\omega B_0 \sin \omega t$$

from which we find that the induced current oscillates as

$$I = \frac{\omega B_0 L^2}{R}\,|\sin \omega t|$$

As the current oscillates, the power dissipation in the wire is

$$P = I^2R = \frac{\omega^2 B_0^{\,2} L^4}{R}\sin^2 \omega t$$

The power dissipation also oscillates, but very rapidly in comparison to a temperature rise that we expect to occur over seconds or minutes. Consequently, we are justified in replacing the oscillating P with its *average* value P_{avg}. Recall that the time average of the function $\sin^2 \omega t$ is $\frac{1}{2}$, a result that can be proven by integration or justified by noticing that a graph of $\sin^2 \omega t$ oscillates symmetrically between 0 and 1. Thus the average power dissipation in the wire is

$$P_{\text{avg}} = \frac{\omega^2 B_0^{\,2} L^4}{2R}$$

Recall that power is the *rate* of energy transfer. In this case, the power dissipated in the wire is the wire's heating rate: $dQ/dt = P_{\text{avg}}$, where here Q is heat, not charge. Using $Q = mc\,\Delta T$, from thermodynamics, we can write

$$\frac{dQ}{dt} = mc\frac{dT}{dt} = P_{\text{avg}} = \frac{\omega^2 B_0^{\,2} L^4}{2R}$$

To complete the calculation, we need the mass and resistance of the wire. The wire's total length is $4L$, and its cross-section area is A. Thus

$$m = \rho_{\text{mass}} V = 4\rho_{\text{mass}} LA$$

$$R = \frac{\rho_{\text{elec}}(4L)}{A} = \frac{4\rho_{\text{elec}} L}{A}$$

Substituting these into the heating equation, we have

$$4\rho_{\text{mass}} LAc\frac{dT}{dt} = \frac{\omega^2 B_0^{\,2} L^3 A}{8\rho_{\text{elec}}}$$

Interestingly, the wire's cross-section area cancels. The wire's temperature initially increases at the rate

$$\frac{dT}{dt} = \frac{\omega^2 B_0^{\,2} L^2}{32\rho_{\text{elec}}\rho_{\text{mass}} c}$$

All the terms on the right-hand side are known. Evaluating, we find

$$\frac{dT}{dt} = 3.3 \text{ K/s} = 200°\text{C/min}$$

ASSESS This is a rapid but realistic temperature rise for a small object, although the rate of increase will slow as the object begins losing heat to the environment through radiation and/or convection. Induction heating can increase an object's temperature by several hundred degrees in a few minutes.

SUMMARY

GOAL To understand the nature of electromagnetic induction and electromagnetic waves.

GENERAL PRINCIPLES

Electromagnetic Induction

Magnetic flux measures the amount of magnetic field passing through a surface:

$$\Phi = AB\cos\theta$$

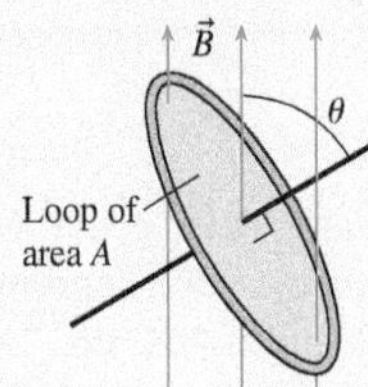

Lenz's law says that there is an induced current in a closed conducting loop if the magnetic flux through the loop is changing. The direction of the induced current is such that the induced magnetic field opposes the *change* in flux.

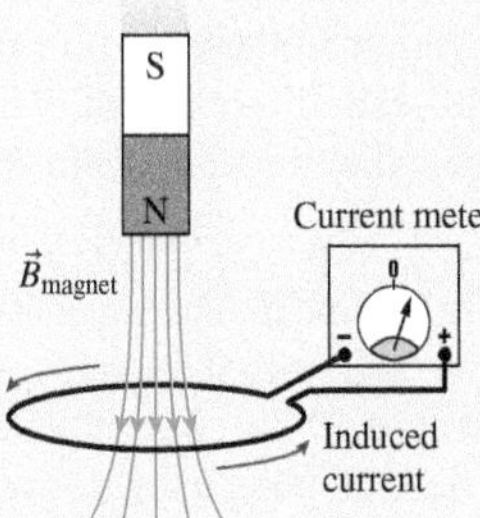

Faraday's law specifies the magnitude of the induced emf around a closed loop:

$$\mathcal{E} = \left|\frac{d\Phi}{dt}\right|$$

Multiply by N for an N-turn coil.

The induced current around a loop with resistance R is

$$I = \frac{\mathcal{E}}{R}$$

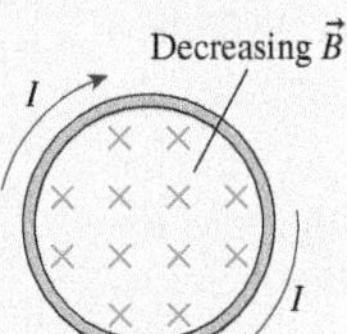

Electromagnetic Waves

An **electromagnetic wave** is a self-sustaining oscillation of electric and magnetic fields.

- The wave is a transverse wave with $\vec{E}$, $\vec{B}$, and $\vec{v}$ mutually perpendicular.
- The wave propagates with speed
$$v_{em} = c = \frac{1}{\sqrt{\epsilon_0\mu_0}} = 3.00 \times 10^8 \text{ m/s}$$
- The wavelength, frequency, and speed are related by
$$c = \lambda f$$
- The amplitudes of the fields are related by
$$E_0 = cB_0$$

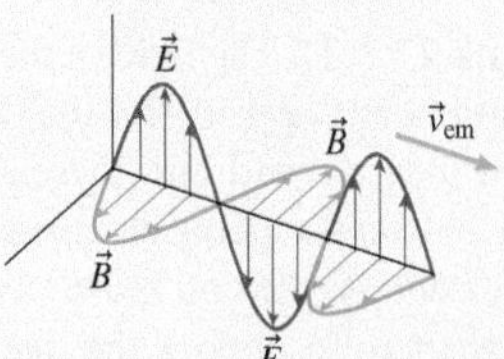

IMPORTANT CONCEPTS

Motional emf

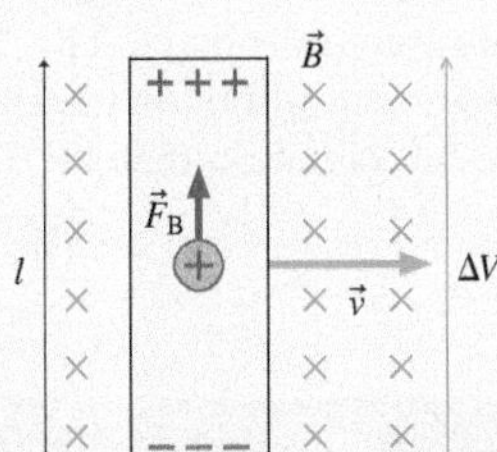

The motion of a conductor through a magnetic field produces a force on the charges. The separation of charges leads to an emf:

$$\mathcal{E} = vlB$$

Induced fields

An **induced electric field** is created by a changing magnetic field.

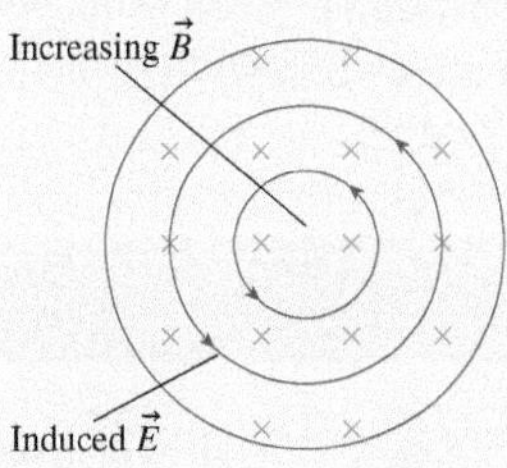

An **induced magnetic field** is created by a changing electric field.

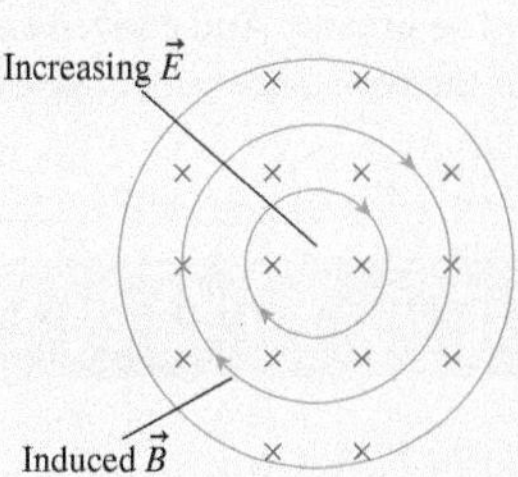

These fields exist independently of charges and currents.

APPLICATIONS

A changing flux in a solid conductor creates **eddy currents.**

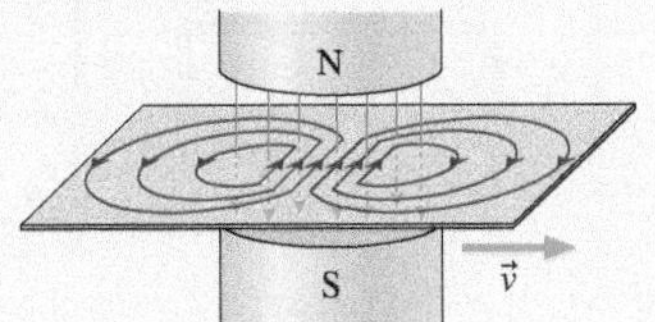

The plane of the electric field of an electromagnetic wave defines its **polarization.** The intensity of polarized light transmitted through a polarizing filter is given by **Malus's law:**

$$I = I_0\cos^2\theta$$

where θ is the angle between the electric field and the polarizer axis.

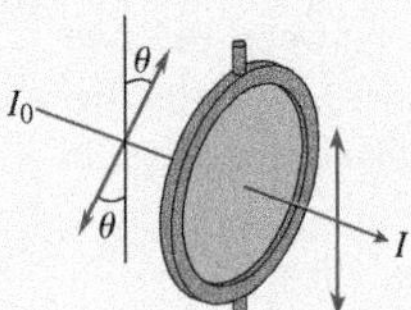

The peak wavelength of a thermal emission spectrum depends on the temperature as given by **Wien's law:**

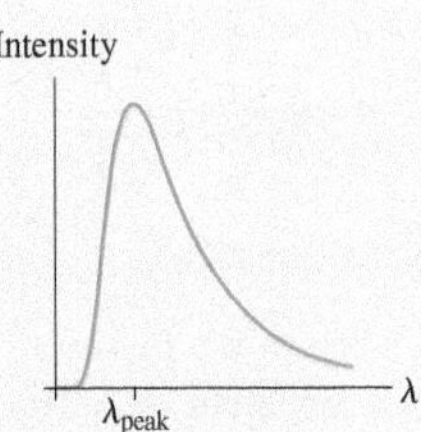

$$\lambda_{peak}\text{ (in nm)} = \frac{2.9 \times 10^6 \text{ nm}\cdot\text{K}}{T}$$

LEARNING OBJECTIVES After studying this chapter, you should be able to:

- Calculate the motional emf of a conductor moving in a magnetic field. *Conceptual Questions 27.1, 27.2; Problems 27.1–27.3, 27.5, 27.8*
- Calculate the magnetic flux through a loop. *Conceptual Question 27.4; Problems 27.9–27.13*
- Determine the direction of an induced current. *Conceptual Questions 27.5, 27.6; Problems 27.14–27.16*
- Calculate the induced emf and induced current of a changing magnetic field. *Conceptual Questions 27.3, 27.8; Problems 27.17–27.19, 27.21, 27.22*
- Calculate the strength of induced electric and magnetic fields. *Conceptual Question 27.10; Problems 27.26–27.29*
- Solve problems about electromagnetic waves. *Conceptual Questions 27.12, 27.13; Problems 27.30–27.33, 27.38*
- Solve problems about polarization and polarizing filters. *Conceptual Question 27.14; Problems 27.39–27.43*
- Apply the wave model of light to the electromagnetic spectrum. *Conceptual Question 27.15; Problems 27.44–27.48*

STOP TO THINK ANSWERS

Stop to Think 27.1: E. According to the right-hand rule, the magnetic force on a positive charge carrier is to the right.

Stop to Think 27.2: No. The charge carriers in the wire move parallel to $\vec{B}$. There's no magnetic force on a charge moving parallel to a magnetic field.

Stop to Think 27.3: D. The field of the bar magnet emerges from the north pole and points upward. As the coil moves toward the pole, the flux through it is upward and increasing. To oppose the increase, the induced field must point downward. This requires a clockwise (negative) current. As the coil moves away from the pole, the upward flux is decreasing. To oppose the decrease, the induced field must point upward. This requires a counterclockwise (positive) current.

Stop to Think 27.4: A, B, E. The induced emf is equal to the rate of change of the flux; the current is proportional to this emf and inversely proportional to the resistance. Increasing the rate of change of the field will increase the rate of change of the flux. If the loop is larger, the flux will be greater, and therefore the rate of change of the flux will also be greater. And a given emf will induce a larger current in the loop if the resistance is decreased.

Stop to Think 27.5: A. To use the right-hand rule, point your right index finger in the direction of $\vec{E}$ and rotate your hand until your thumb—the direction of wave propagation—is up. Your middle finger, the magnetic field direction, points in the positive x-direction.

Stop to Think 27.6: B. The intensity along a line from the antenna decreases inversely with the square of the distance, so the intensity at 20 km is $\frac{1}{4}$ that at 10 km. But the intensity depends on the square of the electric field amplitude, or, conversely, E_0 is proportional to $I^{1/2}$. Thus E_0 at 20 km is $\frac{1}{2}$ that at 10 km.

Stop to Think 27.7: $I_D > I_A > I_B = I_C$. The intensity depends on $\cos^2\theta$, where θ is the angle *between* the axes of the two filters. The filters in D have $\theta = 0°$. The two filters in both B and C are crossed ($\theta = 90°$) and transmit no light at all.

Stop to Think 27.8: D. A hotter object emits more radiation across the *entire* spectrum than a cooler object. The 10,000 K star has its maximum intensity in the ultraviolet region of the spectrum, but it still emits more red radiation than the somewhat cooler stars.

QUESTIONS

Conceptual Questions

1. What is the direction of the induced current in Figure Q27.1?

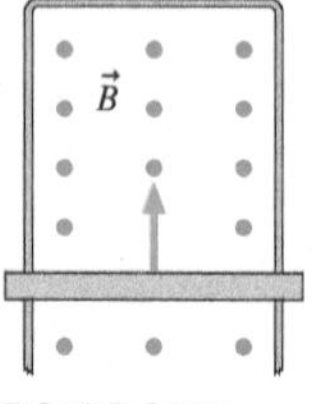

FIGURE Q27.1

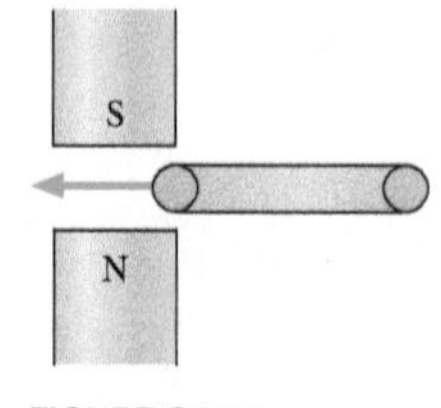

FIGURE Q27.2

2. You want to insert a loop of copper wire between the two permanent magnets in Figure Q27.2. Is there an attractive magnetic force that tends to *pull* the loop in, like a magnet pulls on a paper clip? Or do you need to *push* the loop in against a repulsive force? Explain.
3. The magnetic flux through a loop of wire is zero. Can there be an induced current in the loop at this instant? Explain.
4. Figure Q27.4 shows four different loops in a magnetic field. The numbers indicate the lengths of the sides and the strength of the field. Rank in order the magnetic fluxes Φ_a through Φ_d, from the largest to the smallest. Some may be equal. Explain.

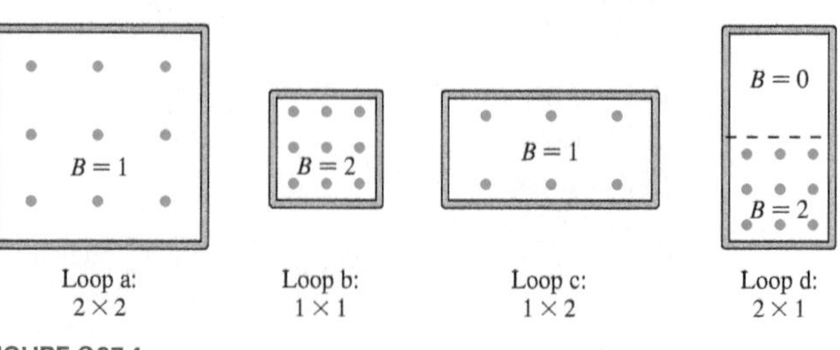

FIGURE Q27.4

Problem difficulty is labeled as | (straightforward) to ||||| (challenging). Problems labeled INT integrate significant material from earlier chapters; BIO are of biological or medical interest; CALC require calculus to solve.

5. Does the loop of wire in Figure Q27.5 have a clockwise current, a counterclockwise current, or no current under the following circumstances? Explain.
   a. The magnetic field points out of the figure and is increasing.
   b. The magnetic field points out of the figure and is constant.
   c. The magnetic field points out of the figure and is decreasing.

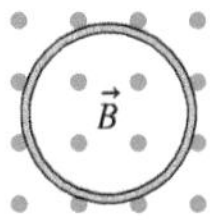

FIGURE Q27.5

FIGURE Q27.6

6. The two loops of wire in Figure Q27.6 are stacked one above the other. Does the upper loop have a clockwise current, a counterclockwise current, or no current at the following times? Explain.
   a. Before the switch is closed.
   b. Immediately after the switch is closed.
   c. Long after the switch is closed.
   d. Immediately after the switch is reopened.
7. Figure Q27.7 shows a bar magnet being pushed toward a conducting loop from below, along the axis of the loop.
   a. What is the current direction in the loop? Explain.
   b. Is there a magnetic force on the loop? If so, in which direction? Explain.
   Hint: A current loop is a magnetic dipole.
   c. Is there a force on the magnet? If so, in which direction?

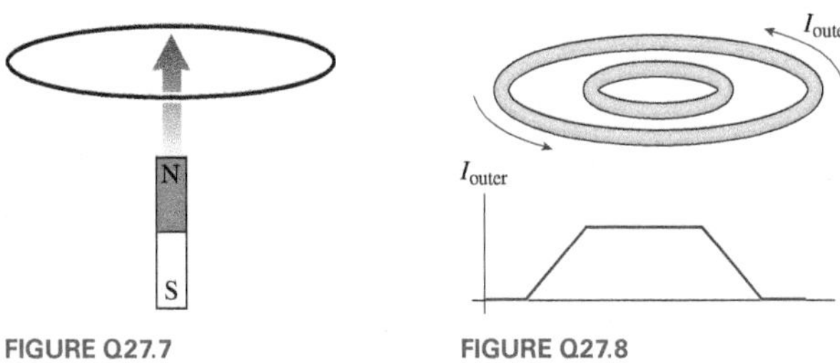

FIGURE Q27.7

FIGURE Q27.8

8. Figure Q27.8 shows two concentric, conducting loops. We will define a counterclockwise current (viewed from above) to be positive, a clockwise current to be negative. The graph shows the current in the outer loop as a function of time. Sketch a graph that shows the induced current in the inner loop. Explain.
9. The current in the straight wire shown in Figure Q27.9 is increasing. Is the induced current in the circular loop clockwise, counterclockwise, or zero?

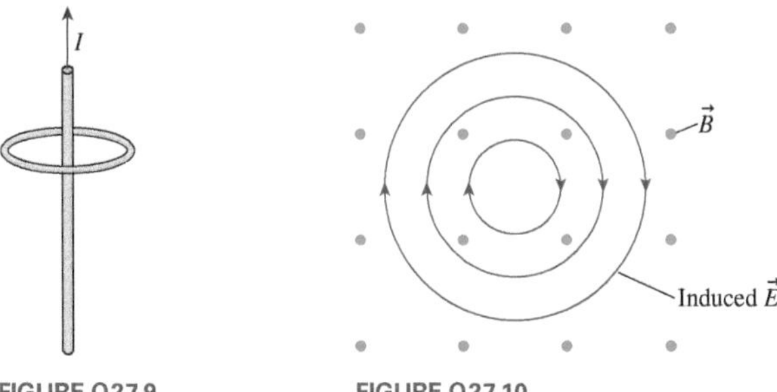

FIGURE Q27.9

FIGURE Q27.10

10. Is the magnetic field strength in Figure Q27.10 increasing, decreasing, or steady? Explain.
11. In what directions are the electromagnetic waves traveling in Figure Q27.11?

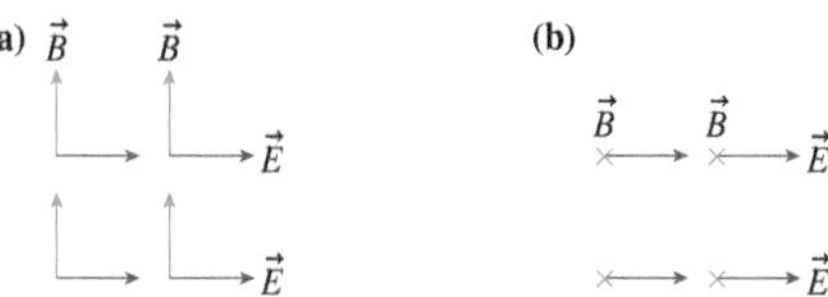

FIGURE Q27.11

12. The intensity of an electromagnetic wave is 10 W/m^2. What will the intensity be if:
    a. The amplitude of the electric field is doubled?
    b. The amplitude of the magnetic field is doubled?
    c. The amplitudes of both the electric and the magnetic fields are doubled?
    d. The frequency is doubled?
13. Two laser beams of equal intensity have wavelengths $\lambda_1 = 500$ nm and $\lambda_2 = 600$ nm. Which laser beam, if either, has the larger electric field amplitude? Explain.
14. A vertically polarized electromagnetic wave passes through the five polarizers in Figure Q27.14. Rank in order, from largest to smallest, the transmitted intensities I_A to I_E.

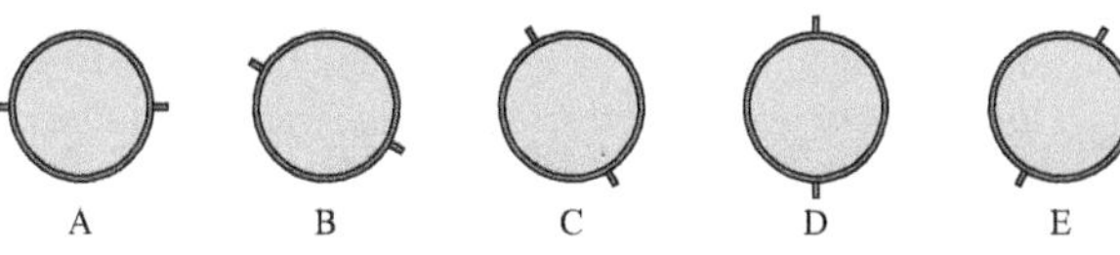

FIGURE Q27.14

15. An object at absolute temperature T emits thermal radiation with peak wavelength λ_0. What is the peak wavelength if the object's temperature is doubled to $2T$?

Multiple-Choice Questions

16. | A circular loop of wire has an area of 0.30 m^2. It is tilted by 45° with respect to a uniform 0.40 T magnetic field. What is the magnetic flux through the loop?
    A. 0.085 T · m^2 B. 0.12 T · m^2 C. 0.38 T · m^2
    D. 0.75 T · m^2 E. 1.3 T · m^2
17. | In Figure Q27.17, a square loop is rotating in the plane of the figure around an axis through its center. A uniform magnetic field is directed into the figure. What is the direction of the induced current in the loop?
    A. Clockwise. B. Counterclockwise.
    C. There is no induced current.

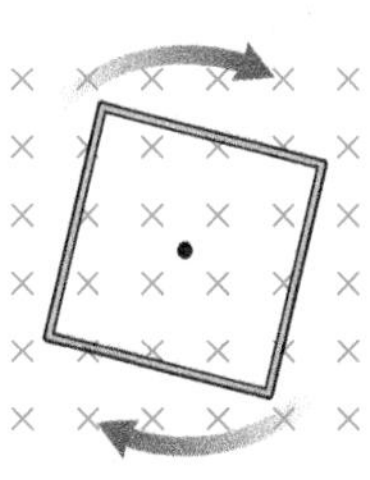

FIGURE Q27.17

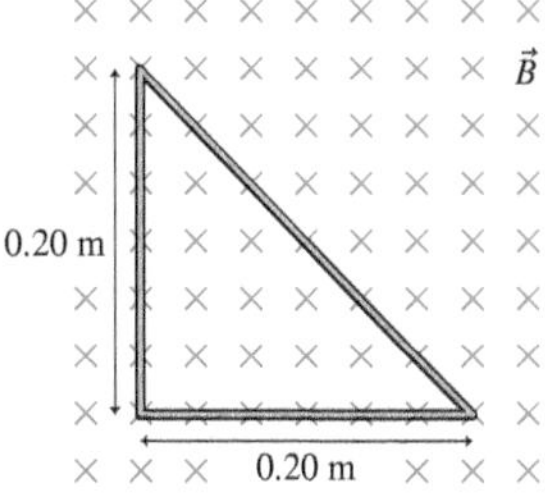

FIGURE Q27.18

18. || Figure Q27.18 shows a triangular loop of wire in a uniform magnetic field. If the field strength changes from 0.30 to 0.10 T in 50 ms, what is the induced emf in the loop?
    A. 0.08 V B. 0.12 V C. 0.16 V D. 0.24 V E. 0.36 V

19. || A diamond-shaped loop of wire is pulled at a constant velocity through a region where the magnetic field is directed into the figure in the left half and is zero in the right half, as shown in Figure Q27.19. As the loop moves from left to right, which graph best represents the induced current in the loop as a function of time? Let a clockwise current be positive and a counterclockwise current be negative.

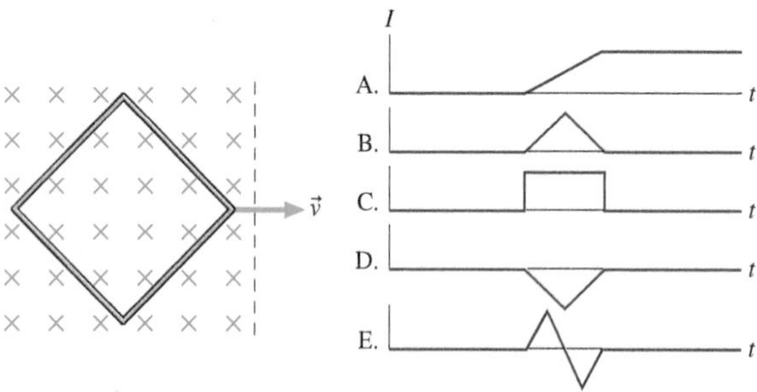

FIGURE Q27.19

20. || A device called a *flip coil* can be used to measure the earth's magnetic field. The coil has 100 turns and an area of $0.010\ m^2$. It is oriented with its plane perpendicular to the earth's magnetic field, then flipped 180° so the field goes through the coil in the opposite direction. The earth's magnetic field is 0.050 mT, and the coil flips over in 0.50 s. What is the average emf induced in the coil during the flip?
A. 0.050 mV
B. 0.10 mV
C. 0.20 mV
D. 1.0 mV

21. | The electromagnetic waves that carry FM radio range in frequency from 87.9 MHz to 107.9 MHz. What is the range of wavelengths of these radio waves?
A. 500–750 nm
B. 0.87–91.08 m
C. 2.78–3.41 m
D. 278–341 m
E. 234–410 km

22. || The beam from a laser is focused with a lens, reducing the area of the beam by a factor of 2. By what factor does the amplitude of the electric field increase?
A. The amplitude does not change.
B. The amplitude increases by a factor of $\sqrt{2}$.
C. The amplitude increases by a factor of 2.
D. The amplitude increases by a factor of $2\sqrt{2}$.
E. The amplitude increases by a factor of 4.

23. || A spacecraft in orbit around the moon measures its altitude by reflecting a pulsed 10 MHz radio signal from the surface. If the spacecraft is 10 km high, what is the time between the emission of the pulse and the detection of the echo?
A. 33 ns
B. 67 ns
C. 33 μs
D. 67 μs

24. | A 6.0 mW vertically polarized laser beam passes through a polarizing filter whose axis is 75° from vertical. What is the laser-beam power after passing through the filter?
A. 0.40 mW
B. 1.0 mW
C. 1.6 mW
D. 5.6 mW

PROBLEMS

Section 27.1 Induced Currents

Section 27.2 Motional emf

1. | A potential difference of 75 mV develops between the ends of a 12-cm-long wire as it moves through a 0.25 T magnetic field. The wire's velocity is perpendicular to the field. What is the wire's speed?

2. | A potential difference of 0.050 V is developed across the 10-cm-long wire of Figure P27.2 as it moves through a magnetic field perpendicular to the figure. What are the strength and direction (in or out) of the magnetic field?

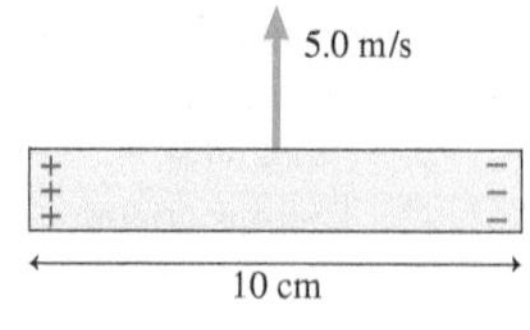

FIGURE P27.2

3. || A scalloped hammerhead shark swims at a steady speed of
BIO 1.5 m/s with its 85-cm-wide head perpendicular to the earth's 50 μT magnetic field. What is the magnitude of the emf induced between the two sides of the shark's head?

4. || A 10-cm-long wire is pulled along a U-shaped conducting rail in a perpendicular magnetic field. The total resistance of the wire and rail is 0.20 Ω. Pulling the wire with a force of 1.0 N causes 4.0 W of power to be dissipated in the circuit.
a. What is the speed of the wire when pulled with a force of 1.0 N?
b. What is the strength of the magnetic field?

5. | Figure P27.5 shows a 15-cm-long metal rod pulled along two frictionless, conducting rails at a constant speed of 3.5 m/s. The rails have negligible resistance, but the rod has a resistance of 0.65 Ω.
a. What is the current induced in the rod?
b. What force is required to keep the rod moving at a constant speed?

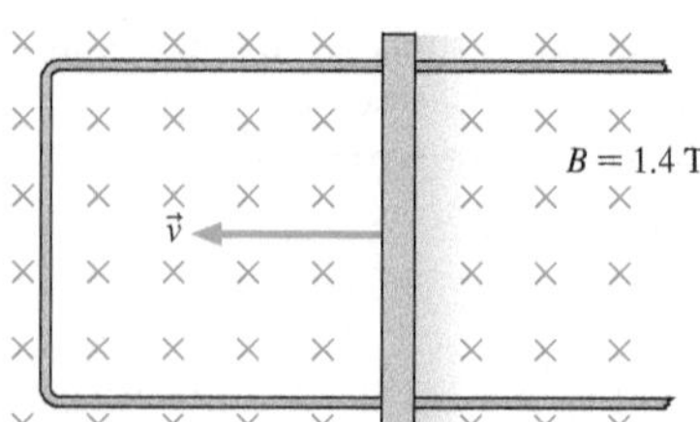

FIGURE P27.5

6. || Two vertical metal rods that are 12 cm apart are connected at the top by a 1.0 Ω resistor. A 0.060 T horizontal magnetic field is perpendicular to the plane of the rods. A 50 g, 12-cm-long, horizontal metal bar is free to slide up and down the rods without friction. Lightweight clips ensure that the bar stays in contact with the rods at all times. The bar is raised to near the top of the rods, then released. What is the terminal speed at which the bar falls?

7. || In the rainy season, the Amazon flows fast and runs deep. In one location, the river is 23 m deep and moves at a speed of 4.0 m/s toward the east. The earth's 50 μT magnetic field is parallel to the ground and directed northward. If the bottom of the river is at 0 V, what is the potential (magnitude and sign) at the surface? Assume that the river contains enough dissolved salts to be a good conductor.

8. || A delivery truck with 2.8-m-high aluminum sides is driving west at 75 km/h in a region where the earth's magnetic field is $\vec{B} = (5.0 \times 10^{-5}\ \text{T, north})$.
   a. What is the potential difference between the top and the bottom of the truck's side panels?
   b. Will the tops of the panels be positive or negative relative to the bottoms?

Section 27.3 Magnetic Flux and Lenz's Law

9. | What is the magnetic flux through the loop shown in Figure P27.9?

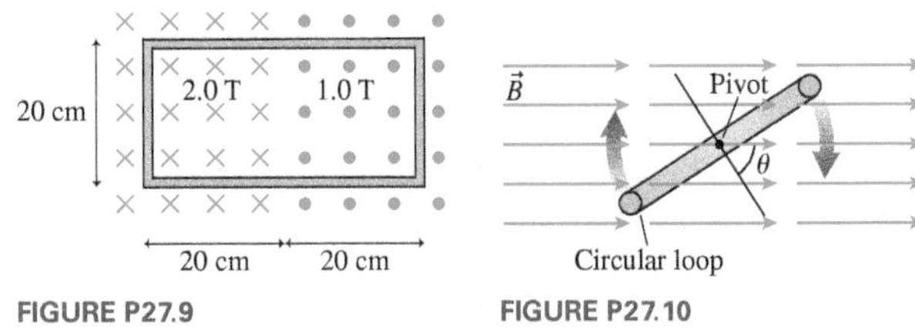

FIGURE P27.9 FIGURE P27.10

10. | Figure P27.10 is an edge-on view of a 10-cm-diameter circular loop rotating in a uniform 0.050 T magnetic field. What is the magnetic flux through the loop when θ is 0°, 30°, 60°, and 90°?

11. || The 2.0-cm-diameter solenoid in Figure P27.11 passes through the center of a 6.0-cm-diameter loop. The magnetic field inside the solenoid is 0.20 T. What is the magnitude of the magnetic flux through the loop (a) when it is perpendicular to the solenoid and (b) when it is tilted at a 60° angle?

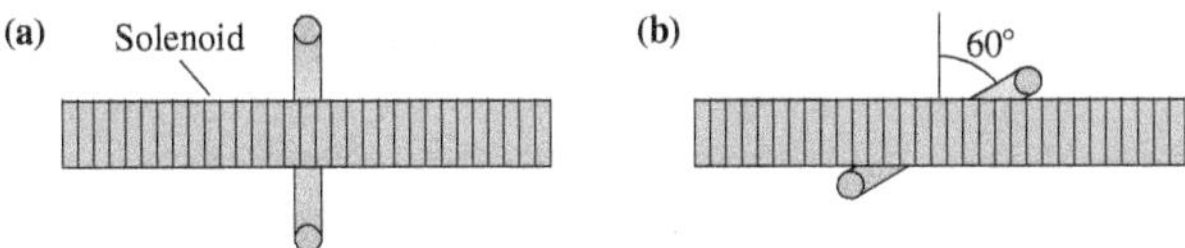

FIGURE P27.11

12. ||| You are given 13.0 m of thin wire. You form the wire into a circular coil with 50 turns. If this coil is placed with its axis parallel to a 0.12 T magnetic field, what is the flux through the coil?

13. || The metal equilateral triangle in Figure P27.13, 20 cm on each side, is halfway into a 0.10 T magnetic field.
   a. What is the magnetic flux through the triangle?
   b. If the magnetic field strength decreases, what is the direction of the induced current in the triangle?

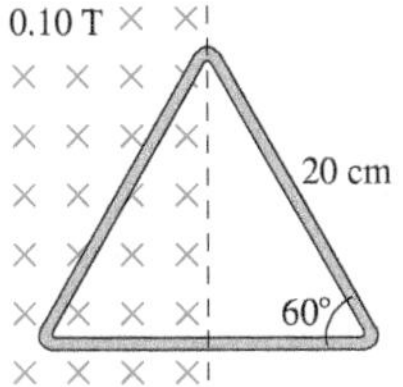

FIGURE P27.13

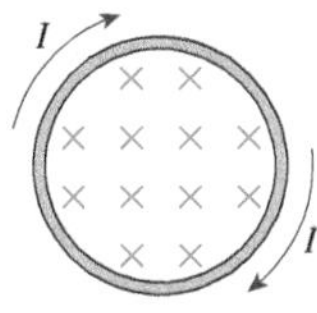

FIGURE P27.14

14. | There is a cw induced current in the conducting loop shown in Figure P27.14. Is the magnetic field inside the loop increasing in strength, decreasing in strength, or steady?

15. | A solenoid is wound as shown in Figure P27.15.
   a. Is there an induced current as magnet 1 is moved away from the solenoid? If so, what is the current direction through resistor R?
   b. Is there an induced current as magnet 2 is moved away from the solenoid? If so, what is the current direction through resistor R?

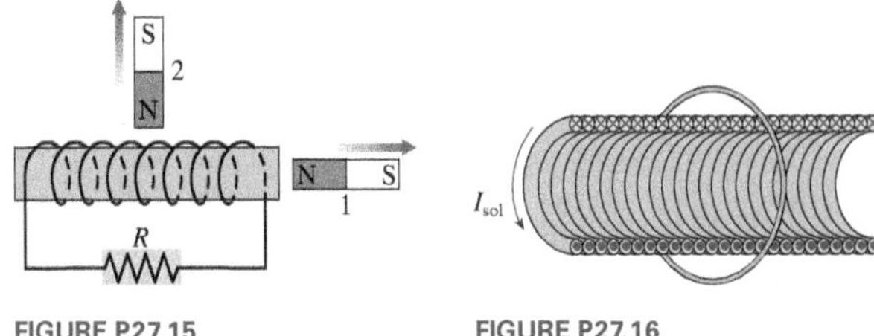

FIGURE P27.15 FIGURE P27.16

16. || The current in the solenoid of Figure P27.16 is increasing. The solenoid is surrounded by a conducting loop. Is there a current in the loop? If so, is the loop current cw or ccw?

Section 27.4 Faraday's Law

17. | Figure P27.17 shows a 10-cm-diameter loop in three different magnetic fields. The loop's resistance is 0.10 Ω. For each case, determine the induced emf, the induced current, and the direction of the current.

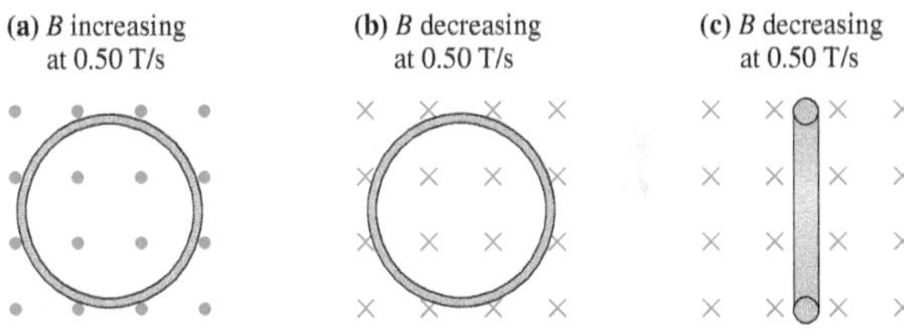

FIGURE P27.17

18. || A 1000-turn coil of wire 2.0 cm in diameter is in a magnetic field that drops from 0.10 T to 0 T in 10 ms. The axis of the coil is parallel to the field. What is the emf of the coil?

19. | Figure P27.19 shows a 100-turn coil of wire of radius 12 cm in a 0.15 T magnetic field. The coil is rotated 90° in 0.30 s, ending up parallel to the field. What is the average emf induced in the coil as it rotates?

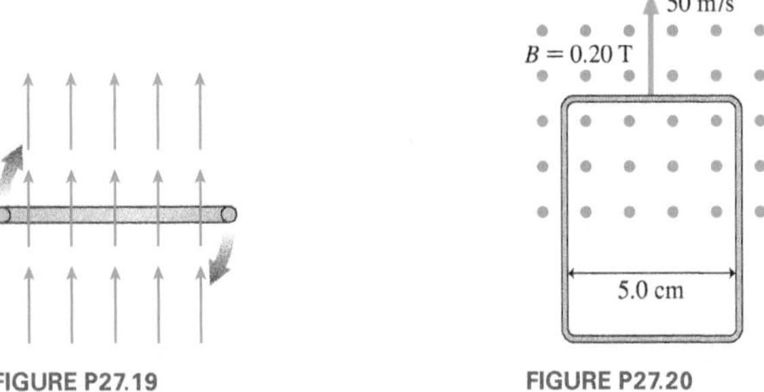

FIGURE P27.19 FIGURE P27.20

20. | The loop in Figure P27.20 is being pushed into the 0.20 T magnetic field at 50 m/s. The resistance of the loop is 0.10 Ω. What are the direction and the magnitude of the current in the loop?

21. || The loop in Figure P27.21 has an induced current as shown. The loop has a resistance of 0.10 Ω. Is the magnetic field strength increasing or decreasing? What is the rate of change of the field, $\Delta B/\Delta t$?

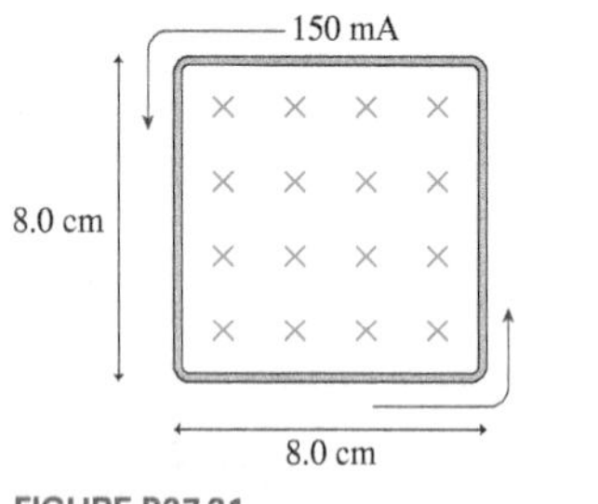

FIGURE P27.21

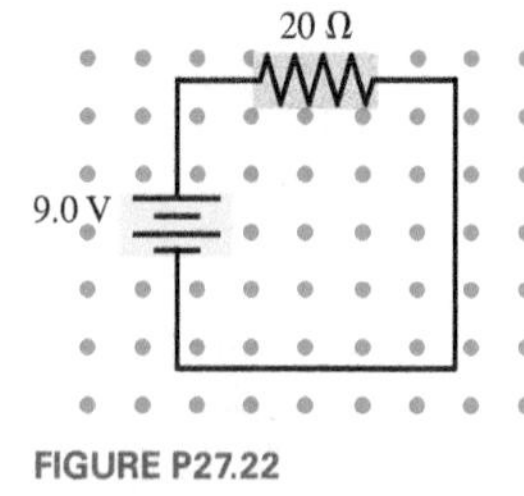

FIGURE P27.22

22. ||| The circuit of Figure P27.22 is a square 20 cm on a side. The magnetic field increases steadily from 0 T to 0.50 T in 10 ms. What is the current in the resistor during this time?

23. || A 20-cm-circumference loop of wire has a resistance of 0.12 Ω. The loop is placed between the poles of an electromagnet, and a field of 0.55 T is switched on in a time of 15 ms. What is the induced current in the loop?

24. || The magnetic field at the earth's surface can vary in response to solar activity. During one intense solar storm, the vertical component of the magnetic field changed by 2.8 μT per minute, causing voltage spikes in large loops of the power grid that knocked out power in parts of Canada. What emf is induced in a square 100 km on a side by this rate of change of field?

25. ||| A 5.0-cm-diameter coil has 20 turns and a resistance of
CALC 0.50 Ω. A magnetic field perpendicular to the coil is $B = 0.020t + 0.010t^2$, where B is in tesla and t is in seconds.
a. Find an expression for the induced current $I(t)$ as a function of time.
b. Evaluate I at $t = 5$ s and $t = 10$ s.

Section 27.5 Induced Fields

26. || Figure P27.26 shows the current as a function of time through a 20-cm-long, 4.0-cm-diameter solenoid with 400 turns. Draw a graph of the induced electric field strength as a function of time at a point 1.0 cm from the axis of the solenoid.

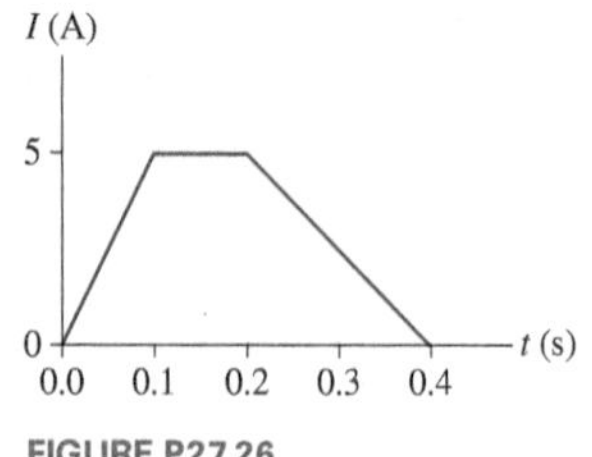

FIGURE P27.26

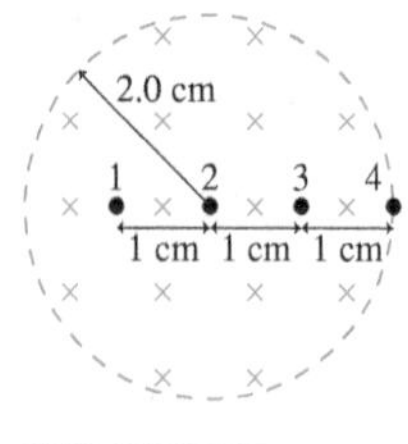

FIGURE P27.27

27. || The magnetic field in Figure P27.27 is decreasing at the rate 0.10 T/s. What is the acceleration (magnitude and direction) of a proton initially at rest at points 1 to 4?

28. || The magnetic field inside a 5.0-cm-diameter solenoid is 2.0 T and decreasing at 4.0 T/s. What is the electric field strength inside the solenoid at a point (a) on the axis and (b) 2.0 cm from the axis?

29. || Scientists studying an anomalous magnetic field find that it is inducing a circular electric field in a plane perpendicular to the magnetic field. The electric field strength 1.5 m from the center of the circle is 4.0 mV/m. At what rate is the magnetic field changing?

Section 27.6 Electromagnetic Waves

30. | What is the electric field amplitude of an electromagnetic wave whose magnetic field amplitude is 2.0 mT?

31. | What is the magnetic field amplitude of an electromagnetic wave whose electric field amplitude is 10 V/m?

32. | Figure P27.32 shows a vertically polarized radio wave of frequency 1.0×10^6 Hz traveling into the figure. The maximum electric field strength is 1000 V/m. What are
a. The maximum magnetic field strength?
b. The magnetic field strength and direction at a point where $\vec{E} = (500 \text{ V/m, down})$?

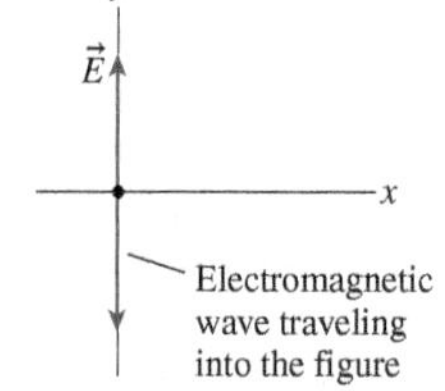

FIGURE P27.32

33. || The maximum allowed leakage of microwave radiation from a microwave oven is 5.0 mW/cm^2. If microwave radiation outside an oven has the maximum value, what is the amplitude of the oscillating electric field?

34. || A typical helium-neon laser found in supermarket checkout scanners emits 633-nm-wavelength light in a 1.0-mm-diameter beam with a power of 1.0 mW. What are the amplitudes of the oscillating electric and magnetic fields in the laser beam?

35. || Biologists often study the patterns of migratory birds by using
BIO radar (1–10 GHz electromagnetic waves) to track their flight. To check whether radar waves influence the birds' flight, researchers tracked the birds visually, both with the radar on and with it off. The 9 GHz radar waves had an intensity of 400 W/m^2 at 250 m. What was the amplitude of the electric field at this distance? (The experiments showed that the radar did not affect the birds.)

36. || At what distance from a 10 mW point source of electromagnetic waves is the electric field amplitude 0.010 V/m?

37. || A radio antenna broadcasts a 1.0 MHz radio wave with 25 kW of power. Assume that the radiation is emitted uniformly in all directions.
a. What is the wave's intensity 30 km from the antenna?
b. What is the electric field amplitude at this distance?

38. || A low-power college radio station broadcasts 10 W of electromagnetic waves. At what distance from the antenna is the electric field amplitude 2.0×10^{-3} V/m, the lower limit at which good reception is possible?

Section 27.7 Polarization

39. | The intensity of a polarized electromagnetic wave is 10 W/m^2. What will be the intensity after passing through a polarizing filter whose axis makes the following angles with the plane of polarization? (a) $\theta = 0°$, (b) $\theta = 30°$, (c) $\theta = 45°$, (d) $\theta = 60°$, (e) $\theta = 90°$.

40. || Only 25% of the intensity of a polarized light wave passes through a polarizing filter. What is the angle between the electric field and the axis of the filter?

41. || A 200 mW vertically polarized laser beam passes through a polarizing filter whose axis is 35° from horizontal. What is the power of the laser beam as it emerges from the filter?

42. ||| A 50 mW laser beam is polarized horizontally. It then passes through two polarizers. The axis of the first polarizer is oriented at 30° from the horizontal, and that of the second is oriented at 60° from the horizontal. What is the power of the transmitted beam?

43. || Unpolarized light with intensity 350 W/m^2 passes first through a polarizing filter with its axis vertical, then through a second polarizing filter. It emerges from the second filter with intensity 131 W/m^2. What is the angle from vertical of the axis of the second polarizing filter?

Section 27.8 The Interaction of Electromagnetic Waves with Matter

44. | The spectrum of a glowing filament has its peak at a wavelength of 1200 nm. What is the temperature of the filament, in °C?

45. | Our sun's 5800 K surface temperature gives a peak wavelength in the middle of the visible spectrum. What is the minimum surface temperature for a star whose emission peaks at some wavelength less than 400 nm—that is, in the ultraviolet?

46. | The hottest ordinary star in our galaxy has a surface temperature of 53,000 K. What is the peak wavelength of its thermal radiation?

47. || The star Sirius is much hotter than the sun, with a peak wavelength of 290 nm compared to the sun's 500 nm. It is also larger, with a diameter 1.7 times that of the sun. By what factor does the energy emitted by Sirius exceed that of the sun?

48. || BIO It is increasingly common to check a person for a fever by measuring the temperature of the forehead with a noncontact infrared thermometer. A person with a normal body temperature has a forehead temperature of about 33°C. This increases to 36°C or higher for a person who has a fever. In principle, an infrared thermometer could measure either the peak wavelength or the radiated power of the thermal emission because both depend on the object's temperature.

   a. What is the percent decrease in the peak wavelength if a person's forehead temperature increases from 33°C to 36°C?
   b. The total radiated power depends on an object's surface area, but the radiated power per unit area depends on only the object's temperature and its emissivity. The emissivity of human skin is 0.97. What is the percent increase in the radiated power per unit area if a person's forehead temperature increases from 33°C to 36°C?
   c. Based on your answers to parts a and b, which of these methods gives a more reliable indication of a fever?

49. ||| BIO INT X rays are generated by accelerating electrons through a large potential difference ΔV and allowing them to strike a metal target. Most of the electron energy is dissipated as heat, but the interaction of the high-speed electrons with the target atoms transforms some of the energy into x rays. The efficiency of the conversion is approximately $(Z \times 10^{-9}\ \text{V}^{-1})\Delta V$, where Z is the atomic number of the target.

   a. A dental x-ray machine has a 60 kV accelerating voltage and a tungsten target. A typical exposure uses an 8.0 mA electron current for $\frac{1}{8}$ s. What is the approximate power of the x-ray emission? The conversion efficiency is not known precisely, so a one-significant-figure answer is appropriate.
   b. What is the x-ray intensity at the jaw, 15 cm from the source? Assume that the x rays are emitted uniformly in all directions from a single point.
   c. A molar tooth can be approximated as a 9-mm-diameter, 20-mm-long cylinder. X rays impinge on a molar from the side, perpendicular to the axis of the cylinder. Approximately what x-ray energy enters a molar during the exposure?

General Problems

50. || Microphones are used to convert sound vibrations to electrical signals. In one kind of microphone, sound waves move a coil of wire up and down in a magnetic field, as shown in Figure P27.50. The coil shown has a diameter of 1.2 cm and has 100 turns; the 0.010 T magnetic field points radially out from the north pole to the south. If the microphone's emf at some instant is 10 mV, what is the speed of the coil?

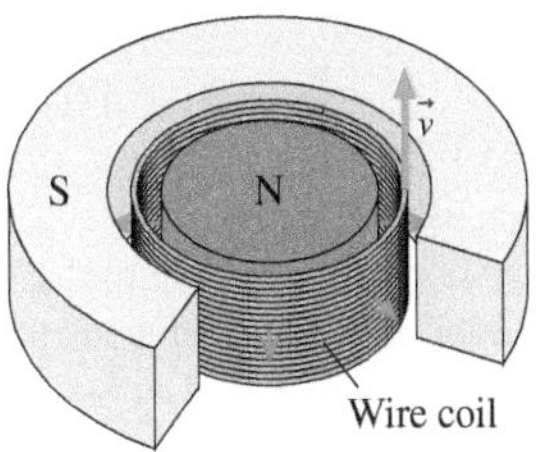

FIGURE P27.50

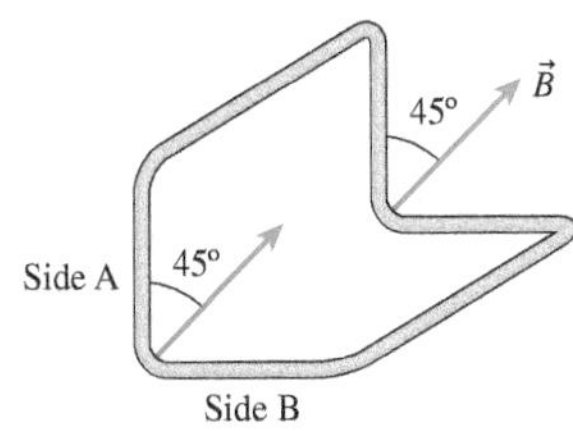

FIGURE P27.51

51. || A square loop of wire, with sides of length 12 cm, is bent by 90° along a line halfway between two of its opposite sides, as shown in Figure P27.51. A uniform magnetic field of 0.85 T is then applied at an angle of 45° from both faces of the bent loop. What is the magnetic flux through the loop?

52. ||| CALC A 1.0 cm × 4.0 cm rectangular loop has its long edge parallel to a long, straight wire. The nearest edge is 1.0 cm from the wire. The wire carries a current of 1.0 A. What is the magnetic flux through the loop?

53. || BIO INT Currents induced by rapid field changes in an MRI solenoid can, in some cases, heat tissues in the body, but under normal circumstances the heating is small. We can do a quick estimate to show this. Consider the "loop" of muscle tissue shown in Figure P27.53. This might be muscle circling the bone of your arm or leg. Muscle tissue is not a great conductor, but current will pass through muscle and so we can consider this a conducting loop with a rather high resistance. Suppose the magnetic field along the axis of the loop drops from 1.6 T to 0 T in 0.30 s, as it might in an MRI solenoid.

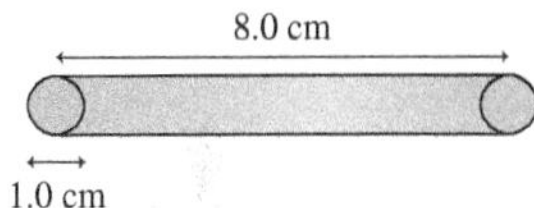

FIGURE P27.53

   a. How much energy is dissipated in the loop?
   b. By how much will the temperature of the tissue increase? Assume that muscle tissue has resistivity 13 Ω · m, density $1.1 \times 10^3\ \text{kg/m}^3$, and specific heat 3600 J/kg · K.

54. || The loop in Figure P27.54 is being pushed into the 0.20 T magnetic field at 50 m/s. The resistance of the loop is 0.10 Ω. What are the direction and magnitude of the current in the loop?

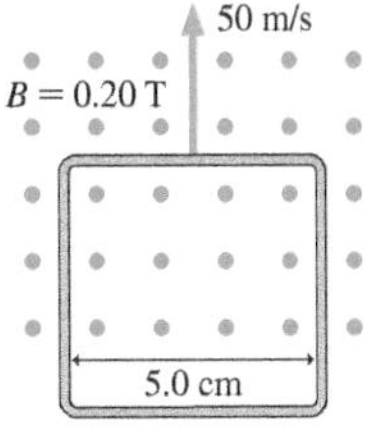

FIGURE P27.54

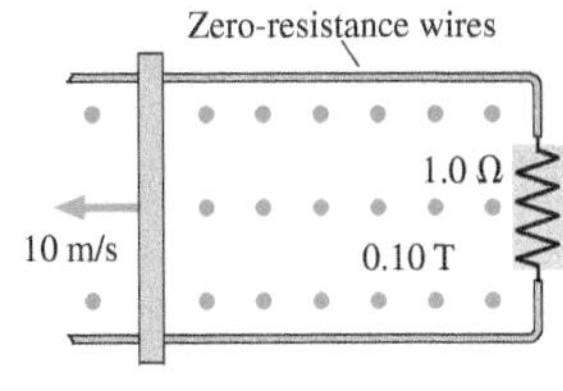

FIGURE P27.55

55. ||| INT The 20-cm-long, zero-resistance wire shown in Figure P27.55 is pulled outward, on zero-resistance rails, at a steady speed of 10 m/s in a 0.10 T magnetic field. On the opposite side, a 1.0 Ω carbon resistor completes the circuit by connecting the two rails. The mass of the resistor is 50 mg.

   a. What is the induced current in the circuit?
   b. How much force is needed to pull the wire at this speed?
   c. How much does the temperature of the carbon increase if the wire is pulled for 10 s? The specific heat of carbon is 710 J/kg · K. Neglect thermal energy transfer out of the resistor.

56. || The 10-cm-wide, zero-resistance wire shown in Figure P27.56 is pushed toward the 2.0 Ω resistor at a steady speed of 0.50 m/s. The magnetic field strength is 0.50 T.
  a. What is the magnitude of the pushing force?
  b. How much power does the pushing force supply to the wire?
  c. What are the direction and magnitude of the induced current?
  d. How much power is dissipated in the resistor?

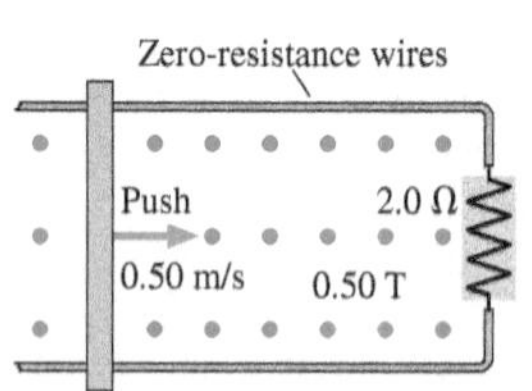

FIGURE P27.56

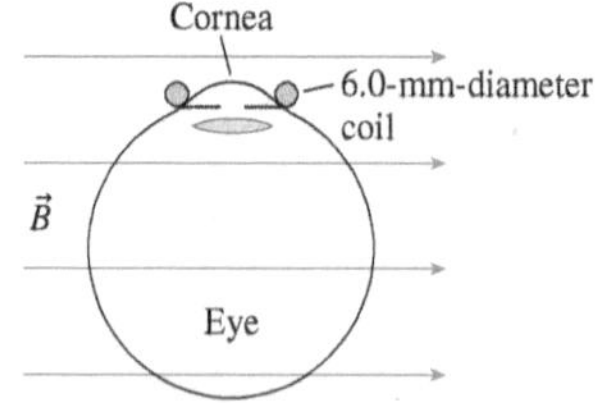

FIGURE P27.57

57. |||| Experiments to study vision often need to track the movements
BIO of a subject's eye. One way of doing so is to have the subject sit in a magnetic field while wearing special contact lenses that have a coil of very fine wire circling the edge. A current is induced in the coil each time the subject rotates his eye. Consider an experiment in which a 20-turn, 6.0-mm-diameter coil of wire circles the subject's cornea while a 1.0 T magnetic field is directed as shown in Figure P27.57. The subject begins by looking straight ahead. What emf is induced in the coil if the subject shifts his gaze by 5.0° in 0.20 s?

58. || Figure P27.58 shows a 4.0-cm-diameter loop with resistance 0.10 Ω around a 2.0-cm-diameter solenoid. The solenoid is 10 cm long, has 100 turns, and carries the current shown in the graph. A positive current is cw when seen from the left. Find the current in the loop at (a) $t = 0.5$ s, (b) $t = 1.5$ s, and (c) $t = 2.5$ s.

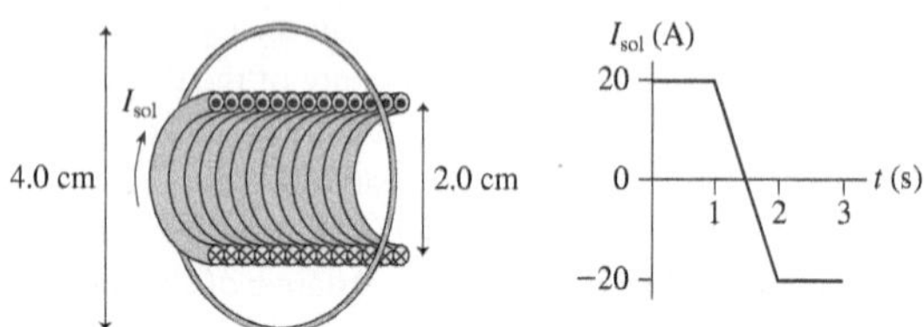

FIGURE P27.58

59. ||| Figure P27.59 shows a 1.0-cm-diameter loop with $R = 0.50\ \Omega$ inside a 2.0-cm-diameter solenoid. The solenoid is 8.0 cm long, has 120 turns, and carries the current shown in the graph. A positive current is cw when seen from the left. Determine the current in the loop at $t = 0.010$ s.

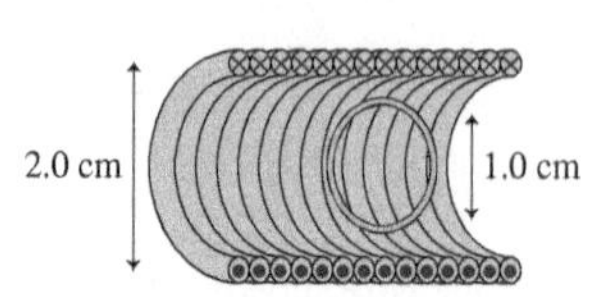

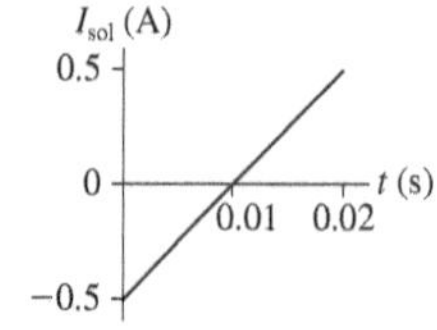

FIGURE P27.59

60. ||| A circular loop made of a flexible conducting wire is shrink-
CALC ing. Its radius as a function of time is $r = (15\text{ cm})e^{-t/\tau}$, where the time constant is $\tau = 2.5$ s. The loop is perpendicular to a 0.50 T magnetic field. What is the induced emf in the loop at (a) $t = 0$ s and (b) $t = \tau$?

61. || Some patients undergoing an MRI report seeing flashes of
BIO light called *magnetophosphenes*. Their exact cause is unknown, but researchers suspect that induced electric fields create small currents in the retina that trigger action potentials. One set of coils in an MRI machine generates a 35 Hz sinusoidally oscillating magnetic field with an amplitude of 15 mT. Assume that the field is perpendicular to the retina, which can be approximated as an 18-mm-diameter circle. What is the maximum strength of the induced electric field around the periphery of the retina?

62. | The magnetic field of an electromagnetic wave in a vacuum
INT is $B_z = (3.00\ \mu\text{T})\sin[(1.00 \times 10^7)x - \omega t]$, where x is in m and t is in s. What are the wave's (a) wavelength, (b) frequency, and (c) electric field amplitude?

63. | The electric field of an electromagnetic wave in a vacuum is
INT $E_y = (20.0\text{ V/m})\cos[(6.28 \times 10^8)x - \omega t]$, where x is in m and t is in s. What are the wave's (a) wavelength, (b) frequency, and (c) magnetic field amplitude?

64. || Satellite TV is broadcast from satellites that orbit 37,000 km above the earth's surface. Their antennas broadcast a 15 kW microwave signal that covers most of North America, an area of about $2.5 \times 10^7\text{ km}^2$.
  a. What is the total power that strikes a 46-cm-diameter ground-based dish antenna?
  b. The dish antenna focuses the incoming wave to a 1.0 cm^2 area. What is the amplitude of the electric field at this focus?

65. || European robins are migratory birds that use the earth's mag-
BIO netic field to orient themselves. There is some evidence that the electromagnetic waves used for cellular phones and other communications interfere with the birds' ability to navigate. Tests have shown that exposure to electromagnetic waves with a magnetic field amplitude of 85 nT can adversely affect a robin's ability to navigate. A typical rural cell phone tower emits 50 W of electromagnetic waves. How far from the tower would a bird need to be to stay below an 85 nT magnetic field threshold?

66. || A LASIK vision correction system uses a laser that emits
BIO 10-ns-long pulses of light, each with 2.5 mJ of energy. The
INT laser is focused to a 0.85-mm-diameter circle. (a) What is the average power of each laser pulse? (b) What is the electric field strength of the laser light at the focus point?

67. || A new cordless phone emits 4.0 mW at 5.8 GHz. The manufacturer claims that the phone has a range of 100 feet. If we assume that the wave spreads out evenly with no obstructions, what is the electric field strength at the base unit 100 feet from the phone?

68. | When the Voyager 2 spacecraft passed Neptune in 1989, it was 4.5×10^9 km from the earth. Its radio transmitter, with which it sent back data and images, broadcast with a mere 21 W of power. Assuming that the transmitter broadcast equally in all directions,
  a. What signal intensity was received on the earth?
  b. What electric field amplitude was detected?

  The received signal was somewhat stronger than your result because the spacecraft used a directional antenna, but not by much.

69. || The maximum electric field strength in air is 3.0 MV/m. Stronger electric fields ionize the air and create a spark. What is the maximum power that can be delivered by a 1.0-cm-diameter laser beam propagating through air?

70. || You've recently read about a chemical laser that generates
INT a 20-cm-diameter, 25 MW laser beam. One day, after physics class, you start to wonder if you could use the radiation pressure from this laser beam to launch small payloads into orbit. To see if this might be feasible, you do a quick calculation of the acceleration of a 20-cm-diameter, 100 kg, perfectly absorbing block. What speed would such a block have if pushed *horizontally* 100 m along a frictionless track by such a laser?

71. || Unpolarized light of intensity I_0 is incident on three polarizing filters. The axis of the first is vertical, that of the second is

45° from vertical, and that of the third is horizontal. What light intensity emerges from the third filter?

72. || Unpolarized light of intensity I_0 is incident on two polarizing filters. The transmitted light intensity is $I_0/10$. What is the angle between the axes of the two filters?

73. || Light reflected from a horizontal surface, such as a road or a lake, has a *partial* horizontal polarization. We can think of the light as a mixture of horizontally polarized light and unpolarized light. Suppose the reflected light from a road surface is 50% polarized and 50% unpolarized. If the light intensity is 200 W/m^2, what is the intensity after the light passes through a polarizing filter whose axis makes an angle from the horizontal of (a) 0°, (b) 45°, and (c) 90°?

74. || The colorful photograph at the beginning of this chapter shows
BIO vitamin B3 crystals seen through a polarized light microscope. Many biological molecules have two possible structural arrangements that are mirror images of each other—often called left-handed and right-handed molecules. When polarized light passes through a sample in which all the molecules are the same, the interaction of the light with the asymmetrical molecules causes the plane of polarization to rotate. The molecules are said to be *optically active.* If polarized light travels distance d through optically active molecules in solution, the plane of polarization rotates through an angle (measured in degrees) $\Delta\phi = \alpha c d$, where c is the concentration in g/L and α is called the *specific rotation.* Specific rotation is a wavelength-dependent characteristic of the molecule.

Naturally occurring fructose is a right-handed sugar, and fructose in solution is optically active. The specific rotation of fructose is $0.92°/[\text{m} \cdot (\text{g/L})]$ for yellow light with a wavelength of 580 nm, $1.16°/[\text{m} \cdot (\text{g/L})]$ for green light with a wavelength of 525 nm. Suppose a 200 g/L solution of fructose is placed in a long tube that has windows at each end. Polarizing filters with perpendicular axes (i.e., crossed polarizers) are placed just outside the windows. White light is incident on the polarizing filter at one end while an observer or camera looks through the polarizing filter at the other end. The observer detects no light coming through the crossed polarizers if the tube is empty. But some light is transmitted if the solution rotates the plane of polarization, and some wavelengths are transmitted better than others. A wavelength has maximum brightness if its plane of polarization is rotated by 90°. For what minimum tube lengths are yellow and green light transmitted through the crossed polarizers with maximum intensity? (Similarly, the different colors in the vitamin B3 photograph correspond to crystals of different thicknesses.)

75. ||| The minimum change in light intensity that is detectable by
BIO the human eye is about 1%. Suppose unpolarized light passes first through a polarizer with a vertical axis, then through a polarizer with its axis at 45°. An observer views the light emerging from the second polarizer. By how much must the angle of the second polarizer axis be increased for the observer to barely perceive a change in the light intensity? A one-significant-figure answer is appropriate.

76. || For radio and microwaves, the depth of penetration into the
BIO human body is proportional to $\lambda^{1/2}$. If 27 MHz radio waves penetrate to a depth of 14 cm, how far do 2.4 GHz microwaves penetrate?

MCAT-Style Passage Problems

Electromagnetic Wave Penetration

Radio waves and microwaves are used in therapy to provide "deep heating" of tissue because the waves penetrate beneath the surface of the body and deposit energy. We define the *penetration depth* as the depth at which the wave intensity has decreased to 37% of its value at the surface. The penetration depth is 15 cm for 27 MHz radio waves. For radio frequencies such as this, the penetration depth is proportional to $\sqrt{\lambda}$, the square root of the wavelength.

77. What is the wavelength of 27 MHz radio waves?
A. 11 m B. 9.0 m C. 0.011 m D. 0.009 m

78. If the frequency of the radio waves is increased, the depth of penetration
A. Increases. B. Does not change. C. Decreases.

79. For 27 MHz radio waves, the wave intensity has been reduced by a factor of 3 at a depth of approximately 15 cm. At this point in the tissue, the electric field amplitude has decreased by a factor of
A. 9 B. $3\sqrt{3}$ C. 3 D. $\sqrt{3}$

The Metal Detector

Metal detectors use induced currents to sense the presence of any metal—not just magnetic materials such as iron. A metal detector, shown in Figure P27.80, consists of two coils: a transmitter coil and a receiver coil. A high-frequency oscillating current in the transmitter coil generates an oscillating magnetic field along the axis and a changing flux through the receiver coil. Consequently, there is an oscillating induced current in the receiver coil.

If a piece of metal is placed between the transmitter and the receiver, the oscillating magnetic field in the metal induces eddy currents in a plane parallel to the transmitter and receiver coils. The receiver coil then responds to the superposition of the transmitter's magnetic field and the magnetic field of the eddy currents. Because the eddy currents attempt to prevent the flux from changing, in accordance with Lenz's law, the net field at the receiver decreases when a piece of metal is inserted between the coils. Electronic circuits detect the current decrease in the receiver coil and set off an alarm.

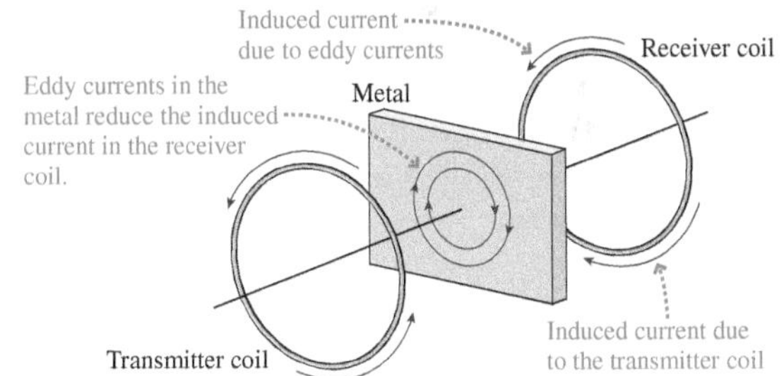

FIGURE P27.80

80. The metal detector will not detect insulators because
A. Insulators block magnetic fields.
B. No eddy current can be produced in an insulator.
C. No emf can be produced in an insulator.
D. An insulator will increase the field at the receiver.

81. A metal detector can detect the presence of metal screws used to repair a broken bone inside the body. This tells us that
A. The screws are made of magnetic materials.
B. The tissues of the body are conducting.
C. The magnetic fields of the device can penetrate the tissues of the body.
D. The screws must be perfectly aligned with the axis of the device.

82. Which of the following changes would *not* produce a larger eddy current in the metal?
A. Increasing the frequency of the oscillating current in the transmitter coil
B. Increasing the magnitude of the oscillating current in the transmitter coil
C. Increasing the resistivity of the metal
D. Decreasing the distance between the metal and the transmitter

PART V SUMMARY

Electricity and Magnetism

FUNDAMENTAL CONCEPTS Charge, like mass, is a fundamental property of matter.

- There are two kinds of charge: positive and negative.
- Charges create electric fields.
- Currents—moving charges—create magnetic fields.
- Charges interact with electric and magnetic fields.
- Fields are also created when other fields change:
  A changing magnetic field induces an electric field.
  A changing electric field induces a magnetic field.
- An electromagnetic field can travel as a wave.

GENERAL PRINCIPLES

Coulomb's law $\vec{E}_{\text{pt charge}} = \left(\dfrac{1}{4\pi\epsilon_0}\dfrac{q}{r^2}, \begin{Bmatrix}\text{away if }+\\ \text{toward if }-\end{Bmatrix}\right)$

Biot-Savart law $\vec{B}_{\text{current}} = \left(\dfrac{\mu_0}{4\pi}\dfrac{I\Delta x \sin\theta}{r^2}, \begin{matrix}\text{direction of}\\ \text{right-hand rule}\end{matrix}\right)$

Faraday's law $\mathcal{E} = |d\Phi/dt|$

Electric force law

$\vec{F}_{\text{elec}} = q\vec{E}$

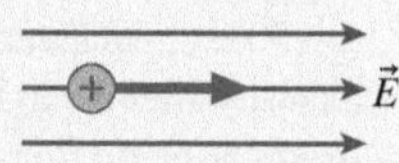

Magnetic force law

$\vec{F}_{\text{mag}} = (|q|vB\sin\alpha$, direction of right-hand rule)

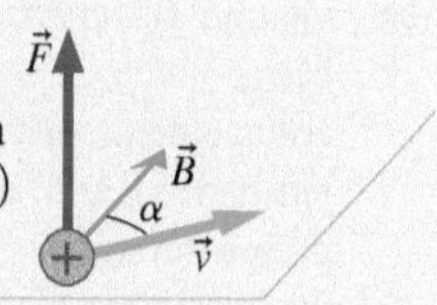

CHARGE MODEL

Two types of charge: positive and negative.

- Like charges repel, opposite charges attract.
- Charge is conserved but can be transferred.

Two types of materials: conductors and insulators.
Neutral objects (no net charge) can be polarized.

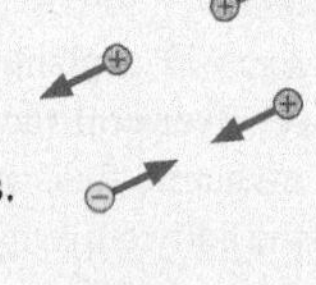

ELECTRIC FIELD MODEL

Charges interact via the electric field.

- Source charges create an electric field.
- Charge q experiences $\vec{F}_{\text{on } q} = q\vec{E}$.

Important electric field models

- Point charge
- Charged sphere
- Parallel-plate capacitor

MAGNETIC FIELD MODELS

Currents both create and interact with magnetic fields.

- Current creates a magnetic field.
- Charge q experiences $F_{\text{on } q} = qvB\sin\alpha$.
- Current I experiences $F_{\text{on } I} = ILB\sin\alpha$.
- Direction given by the right-hand rule.

Important magnetic field models

- Straight, current-carrying wire
- Current loop
- Solenoid

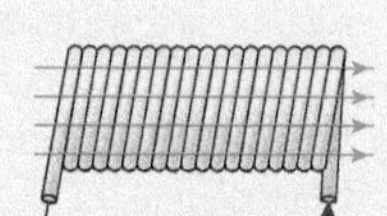

ELECTROMAGNETIC WAVES

An electromagnetic wave is a self-sustaining electromagnetic field.

- $\vec{E}$ and $\vec{B}$ are perpendicular to $\vec{v}_{\text{em}}$.
- Wave speed $v_{\text{em}} = 1/\sqrt{\epsilon_0\mu_0} = c$.

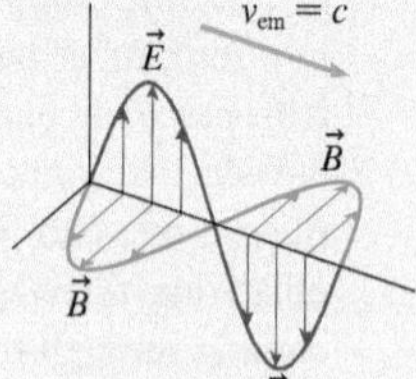

ELECTRIC POTENTIAL

Electric interactions can also be described in terms of the electric potential V.

- A charge q in a potential V has potential energy $U_{\text{elec}} = qV$.
- The mechanical energy $K + U_{\text{elec}}$ of a charged particle is conserved.
- The potential of a point charge is
  $$V_{\text{pt charge}} = q/4\pi\epsilon_0 r$$
- The potential inside a capacitor is
  $$V_{\text{cap}} = (x/d)\,\Delta V_C$$

Field and potential are closely related.

- The field is perpendicular to equipotential surfaces.
- The field points "downhill" toward lower potential.
- The field is
  $$E_x = -dV/dx$$
  with field strength
  $$E \approx \frac{\Delta V}{\Delta x}$$

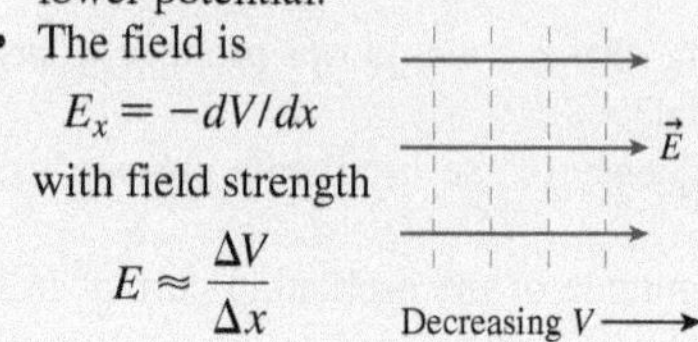

CIRCUITS

emf $\mathcal{E} = W/q$ is the work per charge to separate charge.

A battery is a source of emf.

- $\Delta V_{\text{bat}} = \mathcal{E}$
- ΔV_{bat} creates an electric field in the circuit.
- The electric field pushes charge through the circuit.
- Current is the rate of flow of charge:
  $$I = dQ/dt$$

Capacitors: $Q = C\,\Delta V_C$
Resistors: $I = \Delta V_R/R$ (Ohm's law)
Kirchhoff's laws:

- Loop law: The sum of the potential differences around a loop is zero.
- Junction law: The net current into a junction equals the net current out of the junction.

RC circuits: The time constant for charging and discharging is $\tau = RC$.

INDUCED CURRENTS

Lenz's law: An induced current flows in the direction such that the induced magnetic field opposes the *change* in the magnetic flux.

The magnitude of the induced current is

$$I = \mathcal{E}/R$$

where $\mathcal{E}$ is the emf of Faraday's law.

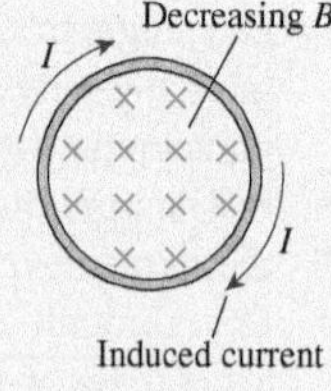

ELECTRICITY IN THE BODY

Fields and potentials in salt water are exponentially screened with Debye length

$$\lambda_D = \sqrt{\kappa\epsilon_0 k_B T/2e^2 c_0}$$

where κ is the dielectric constant and c_0 is the concentration of dissolved ions.

The Nernst potential across a membrane is

$$V_{\text{Nernst}} = (27\text{ mV})\ln(c_{\text{out}}/c_{\text{in}})$$

The membrane potential is about -90 mV for muscle cells and -70 mV for neurons.

The Greenhouse Effect and Global Warming

Electromagnetic waves are real, and we depend on them for our very existence; energy carried by electromagnetic waves from the sun provides the basis for all life on earth. Because of the sun's high surface temperature, it emits most of its thermal radiation in the visible portion of the electromagnetic spectrum. As the figure below shows, the earth's atmosphere is transparent to the visible and near-infrared radiation, so most of this energy travels through the atmosphere and warms the earth's surface.

Although seasons come and go, *on average* the earth's climate is very steady. To maintain this stability, the earth must radiate thermal energy—electromagnetic waves—back into space at exactly the same average rate that it receives energy from the sun. Because the earth is much cooler than the sun, its thermal radiation is long-wavelength infrared radiation that we cannot see. A straightforward calculation using Stefan's law finds that the average temperature of the earth should be $-18°C$, or $0°F$, for the incoming and outgoing radiation to be in balance.

This result is clearly not correct; at this temperature, the entire earth would be covered in snow and ice. The measured global average temperature is actually a balmier $15°C$, or $59°F$. The straightforward calculation fails because it neglects to consider the earth's atmosphere. At visible wavelengths, as the figure shows, the atmosphere has a wide "window" of transparency, but this is not true at the infrared wavelengths of the earth's thermal radiation. The atmosphere lets in the visible radiation from the sun, but the outgoing thermal radiation from the earth sees a much smaller "window." Most of this radiation is absorbed in the atmosphere.

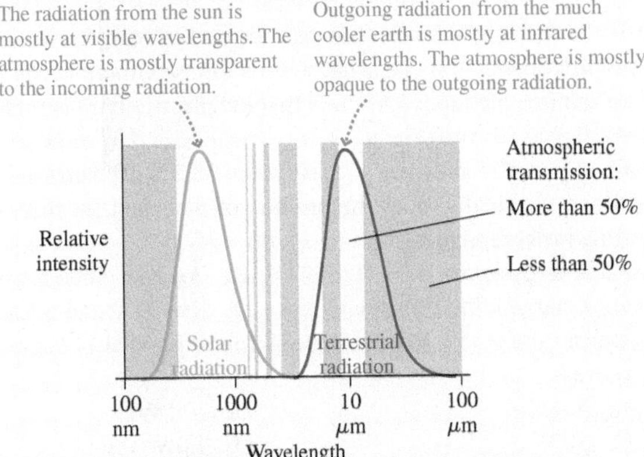

The radiation from the sun is mostly at visible wavelengths. The atmosphere is mostly transparent to the incoming radiation.
Outgoing radiation from the much cooler earth is mostly at infrared wavelengths. The atmosphere is mostly opaque to the outgoing radiation.
Atmospheric transmission:
More than 50%
Less than 50%
Relative intensity
Solar radiation
Terrestrial radiation
100 nm
1000 nm
10 μm
100 μm
Wavelength

Thermal radiation curves for the sun and the earth. The shaded bands show regions for which the atmosphere is transparent (no shading) or opaque (shaded) to electromagnetic radiation.

Because it's easier for visible radiant energy to get in than for infrared to get out, the earth is warmer than it would be without the atmosphere. The additional warming of the earth's surface because of the atmosphere is called the **greenhouse effect**. The greenhouse effect is a natural part of the earth's physics; it has nothing to do with human activities, although it's doubtful any advanced life forms would have evolved without it.

The atmospheric gases most responsible for the greenhouse effect are carbon dioxide and water vapor, both strong absorbers of infrared radiation. These **greenhouse gases** are of concern today because humans, through the burning of fossil fuels (oil, coal, and natural gas), are rapidly increasing the amount of carbon dioxide in the atmosphere. Preserved air samples show that carbon dioxide made up 0.027% of the atmosphere before the industrial revolution. In the last 150 years, human activities have increased the amount of carbon dioxide by nearly 50%, to about 0.040%. By 2050, the carbon dioxide concentration will likely increase to 0.054%, double the pre-industrial value, unless the use of fossil fuels is substantially reduced.

Carbon dioxide is a powerful absorber of infrared radiation. And good absorbers are also good emitters. The carbon dioxide in the atmosphere radiates energy back to the surface of the earth, warming it. Increasing the concentration of carbon dioxide in the atmosphere means more radiation; this increases the average surface temperature of the earth. The net result is **global warming**.

There is strong evidence that the earth has warmed nearly $1°C$ in the last 100 years because of increased greenhouse gases. What happens next? Climate scientists, using sophisticated models of the earth's atmosphere and oceans, calculate that a doubling of the carbon dioxide concentration will likely increase the earth's average temperature by an additional $2°C$ ($\approx 3°F$) to $6°C$ ($\approx 9°F$). There is some uncertainty in these calculations; the earth is a large and complex system. Perhaps the earth will get cloudier as the temperature increases, moderating the increase. Or perhaps the arctic ice cap will melt, making the earth less reflective and leading to an even more dramatic temperature increase.

But the basic physics that leads to the greenhouse effect, and to global warming, is quite straightforward. Carbon dioxide in the atmosphere keeps the earth warm; more carbon dioxide will make it warmer. How much warmer? That's an important question, one that many scientists around the world are attempting to answer with ongoing research. But large or small, change *is* coming. Global warming is one of the most serious challenges facing scientists, engineers, and all citizens in the 21st century.

PART V PROBLEMS

The following questions are related to the passage "The Greenhouse Effect and Global Warming" on the previous page.

1. The intensity of sunlight at the top of the earth's atmosphere is approximately 1400 W/m^2. Mars is about 1.5 times as far from the sun as the earth. What is the approximate intensity of sunlight at the top of Mars's atmosphere?
   A. 930 W/m^2 B. 620 W/m^2
   C. 410 W/m^2 D. 280 W/m^2
2. Averaged over day, night, seasons, and weather conditions, a square meter of the earth's surface receives an average of 240 W of radiant energy from the sun. The average power radiated back to space is
   A. Less than 240 W. B. More than 240 W.
   C. Approximately 240 W.
3. The thermal radiation from the earth's surface peaks at a wavelength of approximately 10 μm. If the surface of the earth warms, this peak will
   A. Shift to a longer wavelength.
   B. Stay the same.
   C. Shift to a shorter wavelength.
4. Biofuels, such as wood, emit carbon dioxide when burned, but they are created by photosynthesis. On average, widespread use of biofuels will likely
   A. Increase the earth's average temperature.
   B. Decrease the earth's average temperature.
   C. Not affect the earth's average temperature.
5. Electromagnetic waves in certain wavelength ranges interact with water molecules because the molecules have a large electric dipole moment. The electric field of the wave
   A. Exerts a net force on the water molecules.
   B. Exerts a net torque on the water molecules.
   C. Exerts a net force and a net torque on the water molecules.

The following passages and associated questions are based on the material of Part V.

Taking an X Ray BIO

X rays are a very penetrating form of electromagnetic radiation. X rays pass through the soft tissue of the body but are largely stopped by bones and other more dense tissues. This makes x rays very useful for medical and dental purposes, as you know.

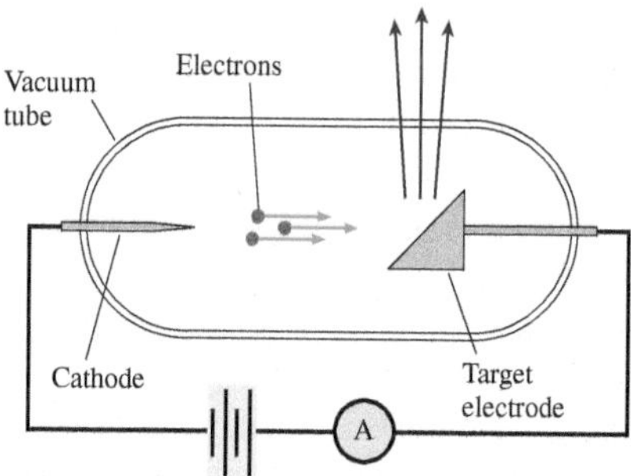

FIGURE V.1

A schematic view of an x-ray tube and a driver circuit is given in Figure V.1. A filament warms the cathode, freeing electrons. These electrons are accelerated by the electric field established by a high-voltage power supply connected between the cathode and a metal target. The electrons accelerate in the direction of the target. The rapid deceleration when they strike the target generates x rays.

An x-ray image is essentially a shadow. X rays are detected (dark in the image) where they pass through soft tissue or around the body, but the detector is unexposed (light in the image) where bones absorb and block x rays.

An x-ray technician uses three parameters to adjust the quality of the image: the accelerating voltage, the current through the tube, and the duration of the exposure. The accelerating voltage determines the energy and wavelength of the x rays, the current through the tube determines the intensity of the x rays, and the exposure time determines the total energy delivered to the target.

In clinical practice, the exposure is characterized by two values: "kVp" and "mAs." kVp is the peak voltage in kV. The value mAs is the product of the current (in mA) and the exposure time (in s) to give a reading in mA · s. This is a measure of the total number of electrons that hit the target and thus the total x-ray energy emitted.

Typical values for a dental x ray are a kVp of 70 (meaning a peak voltage of 70 kV) and mAs of 7.5 (which comes from a current of 10 mA for 0.75 s, for a total of 7.5 mAs). Assume these values in all of the problems that follow.

6. In Figure V.1, what is the direction of the electric field in the region between the cathode and the target electrode?
   A. To the left
   B. To the right
   C. Toward the top of the figure
   D. Toward the bottom of the figure
7. If the distance between the cathode and the target electrode is approximately 1.0 cm, what will be the maximum acceleration of the free electrons? Assume that the electric field is uniform.
   A. 1.2×10^{18} m/s^2 B. 1.2×10^{16} m/s^2
   C. 1.2×10^{15} m/s^2 D. 1.2×10^{12} m/s^2
8. What, physically, does the product of a current (in mA) and a time (in s) represent?
   A. Energy in mJ B. Potential difference in mV
   C. Charge in mC D. Resistance in mΩ
9. During the 0.75 s that the tube is running, what is the electric power?
   A. 7.0 kW B. 700 W
   C. 70 W D. 7.0 W
10. If approximately 1% of the electric energy ends up in the x-ray beam (a typical value), what is the approximate total energy of the x rays emitted?
    A. 500 J B. 50 J
    C. 5 J D. 0.5 J
11. What is the maximum energy of the emitted x-ray photons?
    A. 70×10^3 J
    B. 1.1×10^{-11} J
    C. 1.1×10^{-14} J
    D. 1.6×10^{-18} J

Electric Cars

Electric cars, which have an electric motor powered by large batteries, are increasingly common on our roads. Electric cars are very efficient, and it's very convenient to charge the batteries by plugging into an electric outlet in your house.

But there's a practical problem with such vehicles: the time necessary to recharge the batteries. If you refuel your car with gas at the pump, you add 130 MJ of energy per gallon. If you add 20 gallons, you add a total of 2.6 GJ in about 5 minutes. That's a lot of energy in a short time; the electric system of your house simply can't provide power at this rate.

There's another snag as well. Suppose there were electric filling stations that could provide very high currents to recharge your electric car. Conventional batteries can't recharge very quickly; it would still take longer for a recharge than to refill with gas.

One possible solution is to use capacitors instead of batteries to store energy. Capacitors can be charged much more quickly, and as an added benefit, they can provide energy at a much greater rate—allowing for peppier acceleration. Today's capacitors can't store enough energy to be practical, but future generations may.

12. A typical home's electric system can provide 50 A at a voltage of 220 V. If you had a charger that ran at this full power, approximately how long would it take to charge a battery with the equivalent of the energy in one gallon of gas?
A. 200 min B. 100 min
C. 50 min D. 25 min
13. The electric Tesla Model S has a 400 V battery system that can provide a power of 270 kW. At this peak power, what is the current supplied by the batteries?
A. 75 kA B. 1900 A
C. 680 A D. 75 A
14. The battery of a midsize electric car stores 75 kWh of energy when fully charged. What capacitance is needed in a capacitor that can store this energy if it is charged to 1000 V?
A. 0.15 F B. 15 F
C. 540 F D. 540,000 F
15. One design challenge for a capacitor-powered electric car is that the capacitor voltage decreases as it discharges. Suppose a bank of capacitors is initially charged to 1000 V. What is the voltage when the capacitor bank has provided half of its stored energy?
A. 710 V B. 500 V
C. 250 V D. 100 V

Sensitive Sharks BIO

Sharks have specialized structures called *ampullae of Lorenzini* that can detect electric fields with extraordinary sensitivity; some species can sense fields as small as 10 nV/m. Jelly-filled tubes lead from pores in the skin to the ampullae, which are cavities that contain sensing cells. A typical tube is about 3 mm long with a diameter of 0.2 mm. The jelly has a conductivity of about 0.2 S/m, somewhat higher than that of fat or muscle. Electric fields at the surface of the skin drive a current through the tube; the current is carried by protons, not by electrons. An increase or decrease in the rate at which protons reach the cavity changes the signals from the sensing cells. In fact, some sharks seem able to sense magnetic fields as well by using the ampullae to detect the emf induced as they swim through a magnetic field.

16. 10 nV/m is a very small field. To get a sense of how small, when you rub a balloon on your hair, it acquires a charge of 10 nC. At what approximate distance from the balloon would the field be 10 nV/m? This is an unrealistic, but informative, example.
A. 100 m B. 1 km
C. 10 km D. 100 km
17. If there is an electric field of 10 nV/m along the jelly-filled tube that has the dimensions given in the passage, approximately how many protons pass a given point each second?
A. 40 B. 400
C. 4000 D. 40,000
18. If a shark is swimming due west in a region where the earth's magnetic field is directed to the north, parallel to the surface of the earth, what is the direction of the magnetic force on a proton?
A. North B. South
C. Up D. Down
19. Some researchers think that sharks detect magnetic fields by the motional emf induced between the ends of the tubes as the shark swims. Suppose a shark swims north at 2 m/s in a region where the vertical component of the earth's magnetic field is 35 μT. What is the approximate induced emf? Assume that the tubes, which are in the shark's skin along its sides, are aligned east-west.
A. 200 pV B. 200 nV
C. 200 μV D. 200 mV

Additional Integrated Problems

20. A 20 Ω resistor is connected across a 120 V source. The resistor is then lowered into an insulated beaker, containing 1.0 L of water at 20°C, for 60 s. What is the final temperature of the water?
21. As shown in Figure V.2, a square loop of wire, with a mass of 200 g, is free to pivot about a horizontal axis through one of its sides. A 0.50 T horizontal magnetic field is directed as shown. What current I in the loop, and in what direction, is needed to hold the loop steady in a horizontal plane?

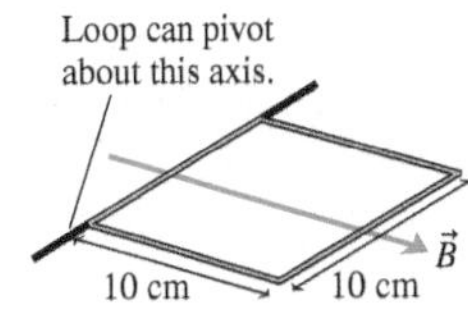

FIGURE V.2

PART VI Modern Physics

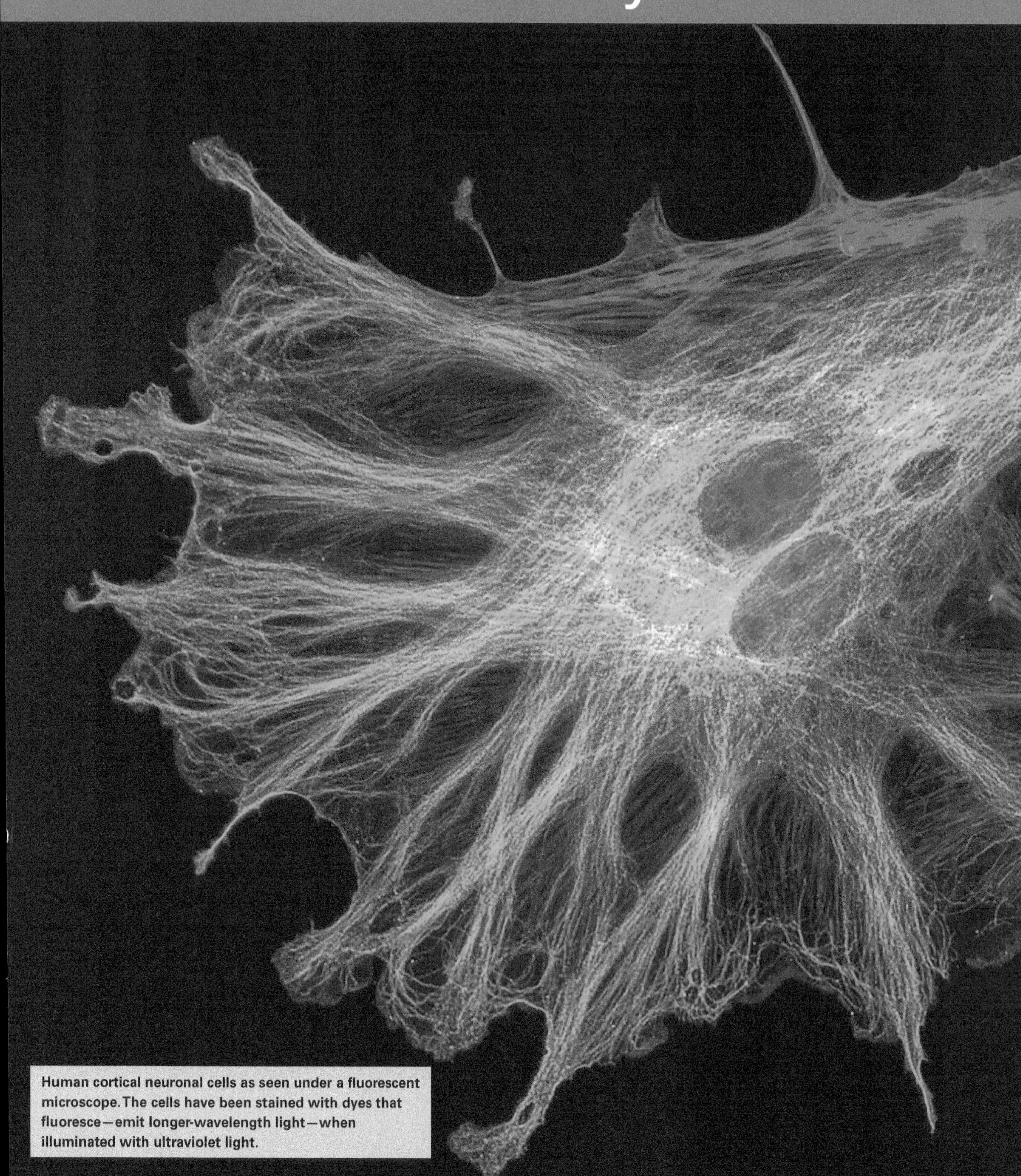

Human cortical neuronal cells as seen under a fluorescent microscope. The cells have been stained with dyes that fluoresce—emit longer-wavelength light—when illuminated with ultraviolet light.

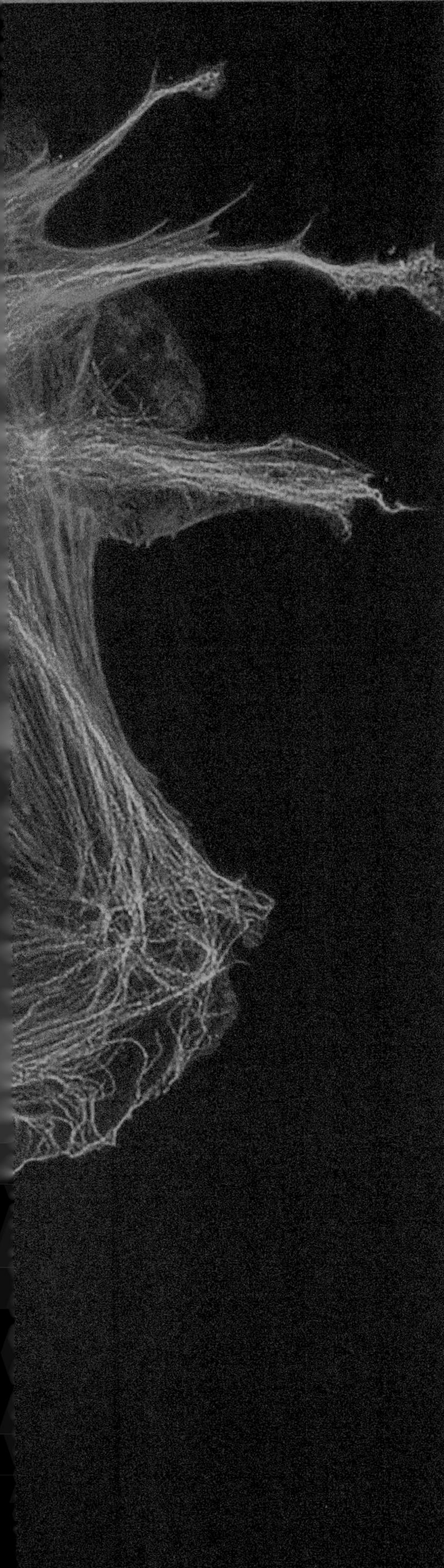

OVERVIEW

New Ways of Looking at the World

Newton's mechanics and Maxwell's electromagnetism are remarkable theories that explain a wide range of physical phenomena. Indeed, physicists in the late 19th century were beginning to think that everything important had been discovered, that only a few loose ends needed to be tied up.

They couldn't have been more wrong. A series of discoveries at the end of the 19th century and beginning of the 20th century profoundly altered our understanding of the universe at the most fundamental level, forcing scientists to reconsider the very nature of space and time and to develop new theories of light and matter. Those new theories laid the foundation for imaging technologies, radiation therapy, our understanding of biochemical reactions, and much more.

Relativity

Albert Einstein and his theory of relativity have become part of our popular culture. Length contraction, black holes, and gravitational waves have completely changed our understanding of the universe. Unfortunately, this text has neither the space nor the time to explore this fascinating theory—with one exception: Einstein's $E = mc^2$, possibly the most famous equation in physics, will enter into our study of nuclear physics to express the idea that matter can be transformed into energy.

Quantum Physics

Light is a wave and atoms are particles. What could seem more sensible and straightforward? Yet both turn out to be much too simplistic. The breakthrough discovery, in 1905, was that light comes in discrete chunks of energy called *photons*. A photon of light is neither a wave nor a particle, although it has aspects of both. Then, less than 20 years later, it was found that electrons can behave like waves. Particles of matter were found to exhibit all the characteristics of waves, such as interference and diffraction.

This odd notion—that there is no clear distinction between particles and waves—is at the heart of a new theory of light and matter called *quantum physics*. We'll see that the wave nature of matter leads to the quantization of energy in atoms and molecules—the key idea upon which modern chemistry has been developed.

Atoms, Nuclei, and Beyond

Throughout this text, we have used the atomic model to explain the properties of matter. But what is an atom? What are its properties? It took quantum physics to answer these questions. Our exploration of the interaction between light and matter will lead to an understanding of how some organisms exhibit bioluminescence and why fluorescence microscopes produce such clear, sharp images.

The word "atom" means "indivisible," and that was the original concept of the atom. But it's not true. As you know, the atom has a tiny core called the *nucleus*. We'll look at what goes on inside the nucleus, why some nuclei undergo radioactive decay, and how nuclear physics has become an indispensable medical tool.

But our questions don't stop there. What's inside a proton? What are the ultimate building blocks of matter? We think we know the answers today, in the 21st century, but no one knows whether tomorrow's discoveries will again upend our understanding of the world around us.

28 Quantum Physics

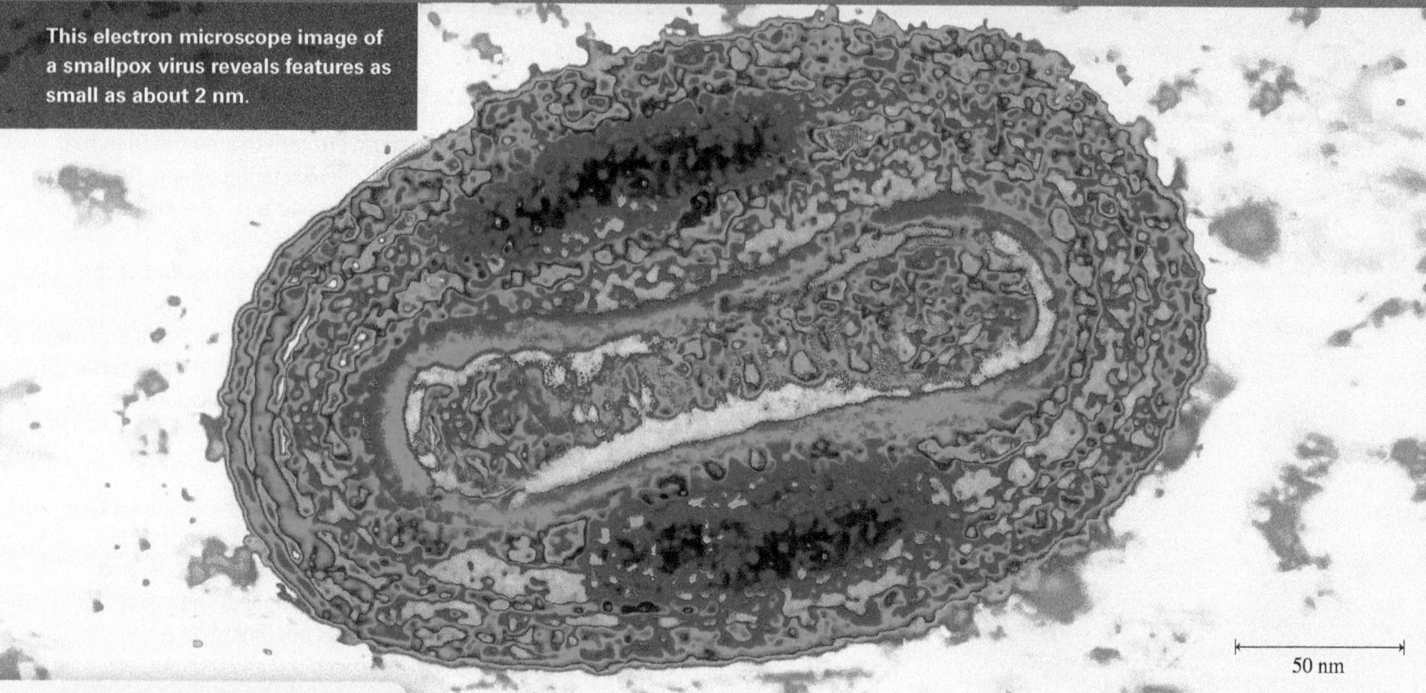

This electron microscope image of a smallpox virus reveals features as small as about 2 nm.

LOOKING AHEAD

Light Has Particle-Like Properties

An interference experiment with extremely low-intensity light shows that light is detected as particle-like spots that form a wave-like pattern.

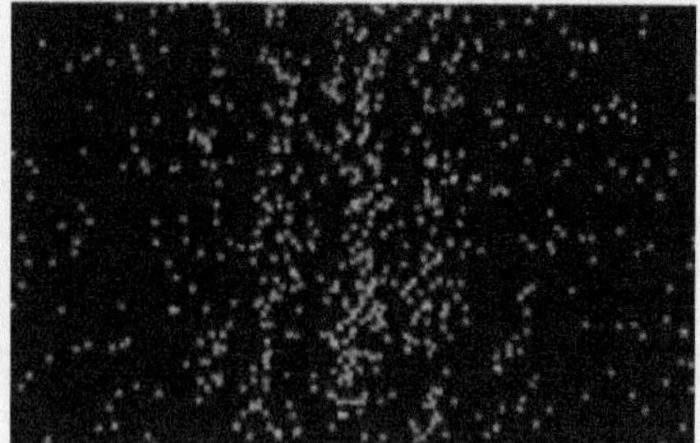

You'll learn that light sometimes exhibits particle-like behavior. The discrete packets of light energy are called *photons*.

Matter Has Wave-Like Properties

This scanning tunneling microscope image of a lipid bilayer shows the individual hydrophilic heads of the molecules.

You'll discover that images such as this are possible because electrons and other atomic particles have wave-like properties.

Energy Is Quantized

A genetically altered mosquito larva expresses the green fluorescent protein and glows green when illuminated with ultraviolet light.

You'll learn that energy is quantized—only certain energies are allowed—and how this affects the interaction of light and matter.

GOAL To understand the quantum behavior of light and matter.

PHYSICS AND LIFE

Exploring the Interactions Between Light and Matter

Many of the tools of biology and medicine—from microscopy to radiation therapy—depend on the interaction of light and matter. Visible light heats an object, ultraviolet radiation breaks bonds, and x rays ionize molecules. Why are the interactions so different? You may know from chemistry that light is made up of discrete packets of energy called *photons*. You'll see that photons are neither particles nor waves, although they have aspects of both. You also may know from chemistry that atoms have only certain allowed energies. You'll see that this is a consequence of the experimental fact that electrons, like photons, have both particle-like and wave-like properties. These observations—discrete energies of light and matter, a blurring of the distinction between particles and waves—are the foundation of quantum physics, the physics of atoms and atomic-level interactions. We'll explore some key ideas of quantum physics in this chapter and then apply them to atoms, molecules, and nuclei in the final two chapters.

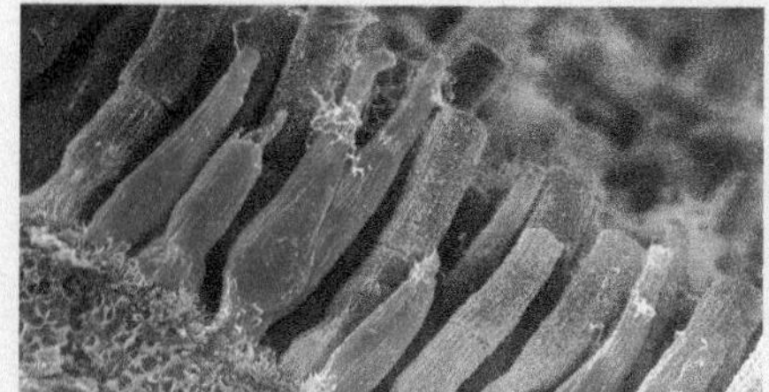

Vision is initiated when specially adapted molecules in rod and cone cells respond to the absorption of photons of light.

28.1 Physics at the Atomic Level

Everything we have studied until this point in the text was known by 1900. Newtonian mechanics, thermodynamics, and electromagnetism form what we call *classical physics*. It is an impressive body of knowledge with immense explanatory power and a vast number of applications.

But several discoveries made around 1900 revealed that classical physics cannot explain the properties of atoms or how atoms interact with light. These discoveries led to *quantum physics*, a new theory of matter and light that is the basis of our current understanding of atoms and molecules. Over the course of the 20th century, insights from quantum physics led to, among other things:

- A modern understanding of chemistry and chemical reactions.
- Lasers and their many uses.
- Computers and other semiconductor devices.
- MRI and other imaging technologies.
- Nuclear medicine

This chapter will introduce the essential ideas of quantum physics—ideas that we will build on in the remaining two chapters as we explore the properties of atoms and nuclei.

Atoms and Electrons

This text has made frequent use of the idea that matter consists of atoms, but we've treated atoms as featureless particles. Now we're ready to study the properties of atoms. How are they constructed? How do atoms interact with one another and with light? Is physics at the atomic level the same Newtonian physics that you're familiar with, or do we need new theories? You may have studied some properties of atoms in chemistry; our goal is to explore the underlying physics.

The idea that matter is made up of discrete, indivisible units is ancient, but that idea was based on philosophical beliefs rather than experimental evidence. John Dalton developed the first scientific theory of atoms in the early 1800s when he noted that the existence of atoms explains why elements combine in simple whole-number ratios to form compounds. According to Dalton, atoms are the fundamental units of nature; they cannot be created, destroyed, or subdivided.

The evidence for atoms grew throughout the 19th century, but it was still assumed that atoms are the ultimate, indivisible particles of matter. Then, in 1897, a series of experiments led the English physicist J. J. Thomson to the discovery that all atoms have smaller, negatively charged constituents that he called *electrons*. Atoms are *not* the ultimate units of matter but, instead, are constructed of smaller, more fundamental pieces. It would take 40 years and the development of quantum physics to fully unravel the issues of atomic structure.

The Electron Volt

The joule is a unit of appropriate size in mechanics and thermodynamics, where we dealt with macroscopic objects, but it is poorly matched to the needs of atomic physics. It will be very useful to have an energy unit appropriate to atomic and nuclear events.

FIGURE 28.1 shows an electron accelerating (in a vacuum) from rest across a parallel-plate capacitor with a 1.0 V potential difference. What is the electron's kinetic energy when it reaches the positive plate? We know from energy conservation that $K_f + qV_f = K_i + qV_i$, where $U_{elec} = qV$ is the electric potential energy. $K_i = 0$ because the electron starts from rest, and the electron's charge is $q = -e$. Thus

$$K_f = -q(V_f - V_i) = -q\,\Delta V = e\,\Delta V = (1.60 \times 10^{-19}\text{ C})(1.0\text{ V})$$
$$= 1.60 \times 10^{-19}\text{ J}$$

FIGURE 28.1 An electron accelerating across a 1 V potential difference gains 1 eV of kinetic energy.

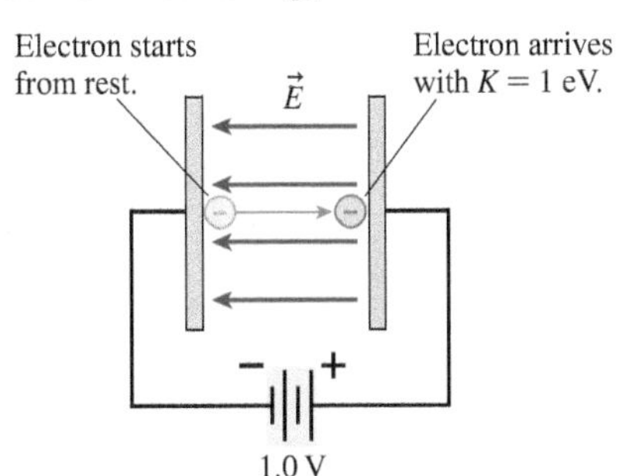
Electron starts from rest.
Electron arrives with $K = 1$ eV.
$\vec{E}$
−
+
1.0 V

Let us define a new unit of energy, called the **electron volt,** as

$$1 \text{ electron volt} = 1 \text{ eV} = 1.60 \times 10^{-19} \text{ J}$$

With this definition, the kinetic energy gained by the electron in our example is

$$K_f = 1 \text{ eV}$$

In other words, **1 electron volt is the kinetic energy gained by an electron (or proton) if it accelerates through a potential difference of 1 volt.**

NOTE ▶ The abbreviation eV uses a lowercase e but an uppercase V. Units of keV (10^3 eV), MeV (10^6 eV), and GeV (10^9 eV) are common. ◀

The electron volt can be a troublesome unit. One difficulty is its unusual name, which looks less like a unit than, say, "meter" or "second." A more significant difficulty is that the name suggests a relationship to volts. But *volts* are units of electric potential, whereas this new unit is a unit of energy! It is crucial to distinguish between the *potential* V, measured in volts, and an *energy* that can be measured either in joules or in electron volts.

NOTE ▶ To reiterate, the electron volt is a unit of *energy,* convertible to joules, and not a unit of potential. Potential is always measured in volts. However, the joule remains the SI unit of energy. It will be useful to express energies in eV, but you *must* convert this energy to joules before doing most calculations. ◀

EXAMPLE 28.1 **The speed of a proton**

Atomic particles are often characterized by their kinetic energy in eV or MeV. For example, proton therapy to treat cancer uses an accelerator to produce a beam of protons with 50 MeV of kinetic energy. What is the speed of a proton? The mass of a proton is 1.67×10^{-27} kg.

PREPARE 50 MeV is 50×10^6 eV. We need to convert this kinetic energy to joules.

SOLVE The proton's kinetic energy is

$$K = \frac{1}{2}mv^2 = (50 \times 10^6 \text{ eV}) \times \frac{1.60 \times 10^{-19} \text{ J}}{1 \text{ eV}} = 8.0 \times 10^{-12} \text{ J}$$

This is a very small energy when expressed in J. We can now use $K = \frac{1}{2}mv^2$ to find the proton speed:

$$v = \sqrt{\frac{2K}{m}} = \sqrt{\frac{2(8.0 \times 10^{-12} \text{ J})}{1.67 \times 10^{-27} \text{ kg}}} = 9.8 \times 10^7 \text{ m/s}$$

ASSESS The proton speed is nearly one-third the speed of light.

28.2 The Photoelectric Effect

It was discovered in the late 1880s that a negatively charged electrode could be discharged by shining ultraviolet light on it. This discovery later caught the attention of J. J. Thomson, who inferred that ultraviolet light causes the electrode to emit electrons until electric neutrality is restored. The emission of electrons from a substance when light strikes its surface is called the **photoelectric effect.** This seemingly obscure discovery soon became a (or maybe *the*) pivotal event that opened the door to the new ideas we discuss in this chapter. We will dwell on this discovery because understanding it is key to understanding quantum physics.

FIGURE 28.2 An experimental device to study the photoelectric effect.

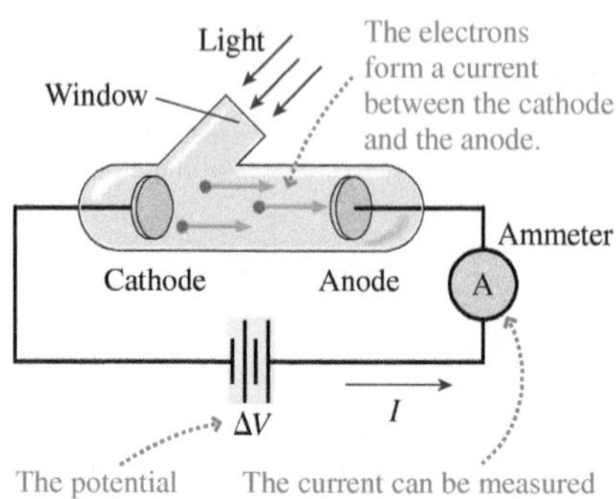
Light
Window
The electrons form a current between the cathode and the anode.
Ammeter
Cathode
Anode
A
ΔV
I
The potential difference can be changed or reversed.
The current can be measured as the potential difference, the light frequency, and the light intensity are varied.

Characteristics of the Photoelectric Effect

FIGURE 28.2 shows an evacuated glass tube with two facing electrodes and a window. When ultraviolet light shines on the cathode, a steady counterclockwise current (clockwise flow of electrons) passes through the ammeter. There are no junctions in this circuit, so the current must be the same all the way around the loop. The current in the space between the cathode and the anode consists of electrons moving freely through space (i.e., not inside a wire) at the *same rate* as the current in the wire.

There is no current if the electrodes are in the dark, so electrons don't spontaneously leap off the cathode. Instead, the light causes electrons to be ejected from the cathode at a steady rate.

The battery in Figure 28.2 establishes an adjustable potential difference ΔV between the two electrodes. With it, we can study how the current I varies as the potential difference and the light's wavelength and intensity are changed. Doing so reveals the following characteristics of the photoelectric effect:

- The current I is directly proportional to the light intensity. If the light intensity is doubled, the current also doubles.
- The current appears without delay when the light is applied.
- Electrons are emitted *only* if the light frequency $f = c/\lambda$ exceeds a **threshold frequency** f_0. This is shown in the graph of FIGURE 28.3a. That is, the photoelectric effect is observed for higher-frequency electromagnetic waves (shorter wavelengths) but not for lower-frequency waves (longer wavelengths).
- The value of the threshold frequency f_0 depends on the type of metal from which the cathode is made.
- If the potential difference ΔV is more than about 1 V positive (anode positive with respect to the cathode), the current changes very little as ΔV is increased. If ΔV is made negative (anode negative with respect to the cathode), by reversing the battery, the current decreases until at some voltage $\Delta V = -V_{stop}$ the current reaches zero. The value of V_{stop} is called the **stopping potential.** This behavior is shown in FIGURE 28.3b.
- The value of V_{stop} is the same for both weak light and intense light. A more intense light causes a larger current, but in both cases the current ceases when $\Delta V = -V_{stop}$.

FIGURE 28.3 The photoelectric current depends on the light frequency f and the battery potential difference ΔV.

(a)

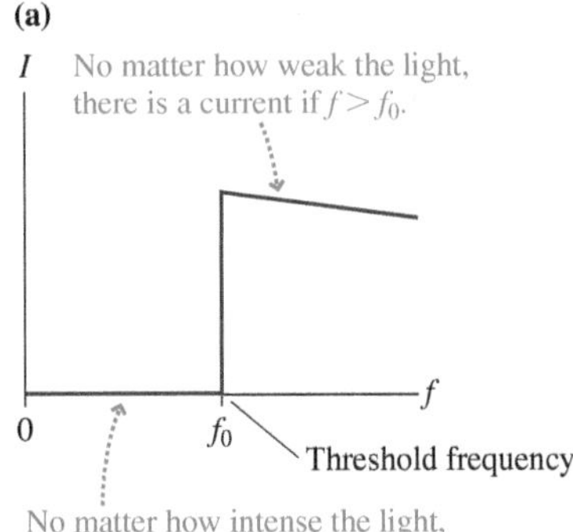
I
No matter how weak the light, there is a current if $f > f_0$.
f
0
f_0
Threshold frequency
No matter how intense the light, there is no current if $f < f_0$.

(b)

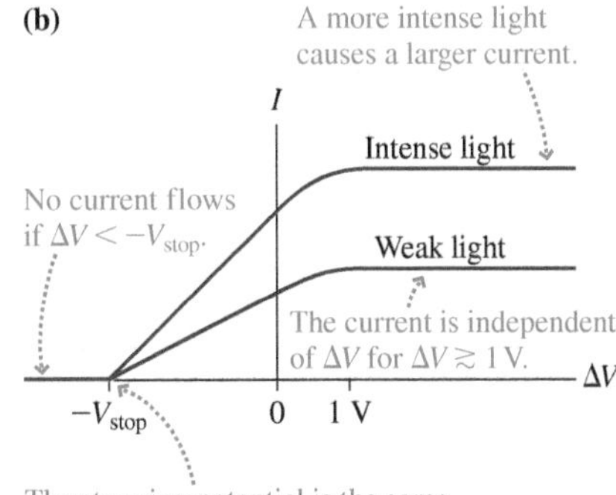
A more intense light causes a larger current.
I
Intense light
No current flows if $\Delta V < -V_{stop}$.
Weak light
The current is independent of ΔV for $\Delta V \gtrsim 1$ V.
ΔV
$-V_{stop}$
0
1 V
The stopping potential is the same for intense light and weak light.

NOTE ▶ We're defining V_{stop} to be a *positive* number. The potential difference that stops the electrons is $\Delta V = -V_{stop}$, with an explicit minus sign. ◀

Understanding the Photoelectric Effect

You learned in Chapter 21 that electrons are the charge carriers in a metal and they move around freely inside like a sea of negatively charged particles. The electrons are bound inside the metal and do not spontaneously spill out of an electrode at room temperature.

A useful analogy, shown in FIGURE 28.4, is the water in a swimming pool. Water molecules do not spontaneously leap out of the pool if the water is calm. To remove a water molecule, you must do *work* on it to lift it upward, against the force of gravity, to the edge of the pool. A minimum energy is needed to extract a water

FIGURE 28.4 A swimming pool analogy of electrons in a metal.

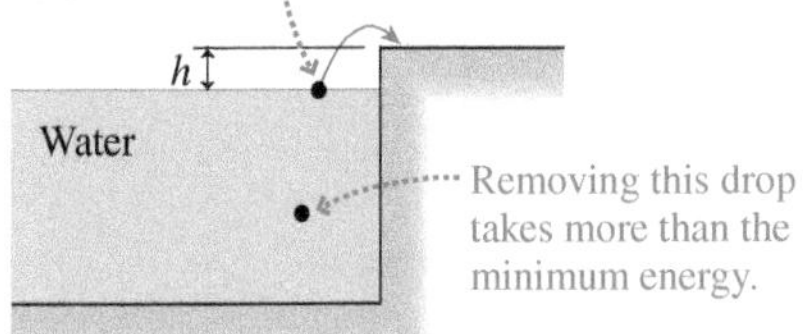
The *minimum* energy to remove a drop of water from the pool is *mgh*.
h
Water
Removing this drop takes more than the minimum energy.

TABLE 28.1 The work functions for some metals

Element	E_0 (eV)
Potassium	2.30
Sodium	2.36
Aluminum	4.28
Tungsten	4.55
Copper	4.65
Iron	4.70
Gold	5.10

molecule—namely, the energy needed to lift a molecule that is right at the surface. Removing a water molecule that is deeper requires more than the minimum energy.

Similarly, a *minimum* energy is needed to free an electron from a metal. To extract an electron, you need to increase its energy until its speed is fast enough to escape. The minimum energy E_0 needed to free an electron is called the **work function** of the metal. Some electrons, like deeper water molecules, may require more energy than E_0 to escape, but all will require *at least* E_0. TABLE 28.1 lists the work functions of some elements. Note that work functions are given in eV.

Now, let's return to the photoelectric-effect experiment of Figure 28.2. When ultraviolet light shines on the cathode, electrons leave with some kinetic energy. An electron with energy E_{elec} inside the metal loses energy ΔE as it escapes, so it emerges as an electron with kinetic energy $K = E_{elec} - \Delta E$. The work function energy E_0 is the *minimum* energy needed to remove an electron, so the *maximum* possible kinetic energy of an ejected electron is

$$K_{max} = E_{elec} - E_0$$

FIGURE 28.5 The effect of different voltages between the anode and cathode.

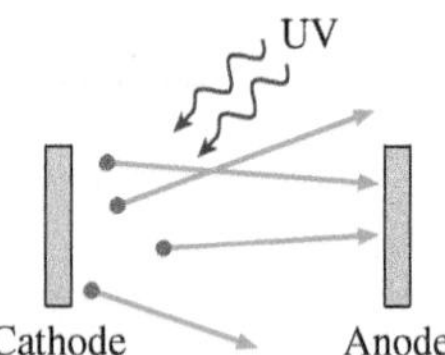

$\Delta V = 0$: The electrons leave the cathode in all directions. Only some reach the anode.

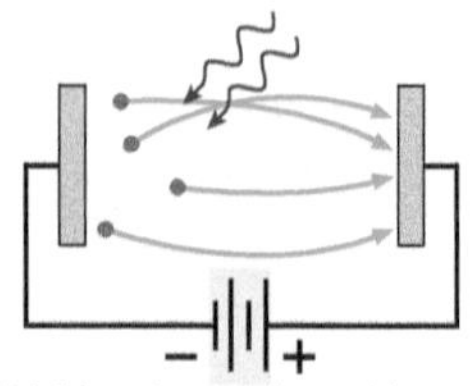

$\Delta V > 0$: Making the anode positive creates an electric field that pushes all the electrons to the anode.

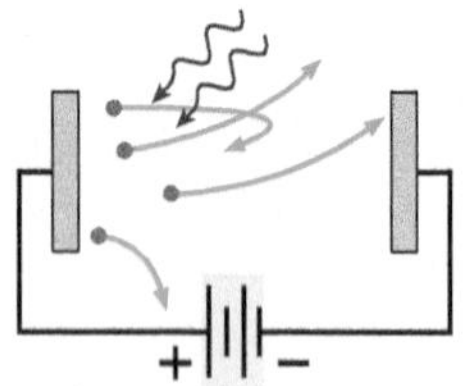

$\Delta V < 0$: Making the anode negative repels the electrons. Only the very fastest make it to the anode.

The electrons, after leaving the cathode, move out in all directions. FIGURE 28.5 shows what happens as the potential difference ΔV between the cathode and the anode is varied:

- If the potential difference between the cathode and the anode is $\Delta V = 0$, there will be no electric field between the plates. Some electrons will reach the anode, creating a measurable current, but many do not.
- If the anode is positive, it attracts *all* of the electrons to the anode. A further increase in ΔV does not cause any more electrons to reach the anode and thus does not cause a further increase in the current I. This is why the curves in Figure 28.3b become horizontal for ΔV more than about 1 V positive.
- If the anode is negative, it repels the electrons. However, an electron leaving the cathode with sufficient kinetic energy can still reach the anode, just as a ball hits the ceiling if you toss it upward with sufficient kinetic energy. A slightly negative anode voltage turns back only the slowest electrons. The current steadily decreases as the anode voltage becomes increasingly negative until, as the left side of Figure 28.3b shows, at the stopping potential, *all* electrons are turned back and the current ceases.

We can use conservation of energy to relate the maximum kinetic energy to the stopping potential. When ΔV is negative, as in the bottom panel of Figure 28.5, electrons are "going uphill," converting kinetic energy to potential energy as they slow down. That is, $\Delta U_{elec} = -e\,\Delta V = -\Delta K$, where we've used $q = -e$ for electrons and ΔK is negative because the electrons are losing kinetic energy. When $\Delta V = -V_{stop}$, where the current ceases, the very fastest electrons, with K_{max}, are being turned back *just* as they reach the anode. They're converting 100% of their kinetic energy into potential energy, so $\Delta K = -K_{max}$. Thus $eV_{stop} = K_{max}$, or

$$V_{stop} = \frac{K_{max}}{e} \tag{28.1}$$

In other words, **measuring the stopping potential tells us the maximum kinetic energy of the electrons.**

Einstein's New Idea

The mere existence of the photoelectric effect is not, as is sometimes assumed, a difficulty for classical physics, and early investigators proposed classical explanations. It was known that a heated metal spontaneously emits electrons, so it was natural to suggest that the light falling on the cathode simply heats it sufficiently to cause electron emission. But this explanation is not consistent with experiment. For light to heat the metal would take time, but, as we've seen, the electrons are emitted without delay when the light is applied.

The heating hypothesis also fails to explain the threshold frequency. If a weak intensity at a frequency just slightly above the threshold can generate a current, then certainly a strong intensity at a frequency just slightly below the threshold should be able to do so—it will heat the metal even more. There is no reason that a slight change in frequency should matter. Yet the experimental evidence shows a sharp frequency threshold, as we've seen.

A new theory that broke with classical ideas about light and energy came in a 1905 paper by Albert Einstein in which he offered an exceedingly simple but amazingly bold idea that explained all of the noted features of the data. Einstein's paper extended the work of the German physicist Max Planck, who had found that he could explain the form of the spectrum of a glowing, incandescent object that we saw in ◀◀ SECTION 27.8 only if he made an unusual assumption. Atoms in a solid oscillate around their equilibrium positions with frequency *f*. The energy of a simple harmonic oscillator depends on the oscillation amplitude and can assume any value.

But to predict the spectrum correctly, Planck had to assume that the oscillating atoms are *not* free to have any possible energy. Instead, the energy of an atom oscillating with frequency *f* had to be one of the specific energies $E = 0, hf, 2hf, 3hf, \ldots,$ where h is a constant. That is, the vibration energies are **quantized.** The constant h, now called **Planck's constant,** is

$$h = 6.63 \times 10^{-34}\ \text{J} \cdot \text{s} = 4.14 \times 10^{-15}\ \text{eV} \cdot \text{s}$$

The first value, with SI units, is the proper one for most calculations, but you will find the second to be useful when energies are expressed in eV.

Einstein was the first to take Planck's idea seriously. Einstein went even further and suggested that **electromagnetic radiation itself is quantized!** That is, light is not really a continuous wave but, instead, arrives in small packets or bundles of energy. Einstein called each packet of energy a **light quantum,** and he postulated that the energy of one light quantum is directly proportional to the frequency of the light. That is, each quantum of light, which is now known as a **photon,** has energy

$$E = hf \qquad (28.2)$$

The energy of a photon, a quantum of light, of frequency *f*

where h is Planck's constant. Higher-frequency light (i.e., light with shorter wavelengths) has higher-energy photons—it is composed of bundles of greater energy. This seemingly simple assumption was completely at odds with the classical understanding of light as a wave, but it allowed Einstein to explain all of the properties of the photoelectric effect.

Einstein's "Miracle Year" Albert Einstein was a little-known young man of 26 in 1905. In that single year, Einstein published three papers on three different topics, each of which would revolutionize physics. One was his initial paper on the theory of relativity. A second paper used statistical mechanics to explain *Brownian motion,* the random motion of small particles suspended in water. It is Einstein's third paper of 1905, on the nature of light, in which we are most interested in this chapter.

Seeing the world in a different light Plants use photosynthesis to convert the energy of light to chemical energy. Photons of visible light have sufficient energy to trigger the necessary molecular transitions, but infrared photons do not, so visible photons are absorbed while infrared photons are reflected. In this infrared photo, infrared photons strongly reflected from the trees' leaves make the trees appear a ghostly white.

EXAMPLE 28.2 **Finding the energy of ultraviolet photons**

Ultraviolet light at 290 nm does 250 times as much cellular damage as an equal intensity of ultraviolet at 310 nm; there is a clear threshold for damage at about 300 nm. What is the energy, in eV, of photons with a wavelength of 300 nm?

PREPARE The energy of a photon is related to its frequency by $E = hf$. Recall from Chapter 16 that the frequency and wavelength of light are related by $f = c/\lambda$.

SOLVE The frequency at wavelength 300 nm is

$$f = \frac{c}{\lambda} = \frac{3.00 \times 10^{8}\ \text{m/s}}{300 \times 10^{-9}\ \text{m}} = 1.00 \times 10^{15}\ \text{Hz}$$

We can now use Equation 28.2 to calculate the energy, using the value of h in eV · s:

$$E = hf = (4.14 \times 10^{-15}\ \text{eV} \cdot \text{s})(1.00 \times 10^{15}\ \text{Hz}) = 4.14\ \text{eV}$$

ASSESS 4.14 eV can be converted to 6.62×10^{-19} J and then, by multiplying by Avogadro's number, to 400 kJ/mol. You may have learned in chemistry that 400 kJ/mol is a fairly typical bond energy. Photons that have a wavelength $\leq$ 300 nm have sufficient energy to disrupt the genetic material in the cell by breaking bonds. Visible light, with lower frequencies, can heat tissue but the photons do not have sufficient energy to break bonds.

Einstein's Postulates

The idea that light is quantized is now widely understood and accepted. But at the time of Einstein's paper, it was a truly revolutionary idea. In his 1905 paper, Einstein framed three postulates about light quanta and their interaction with matter:

1. Light of frequency f consists of discrete quanta, each of energy $E = hf$. Each photon travels at the speed of light c.
2. Light quanta are emitted or absorbed on an all-or-nothing basis. A substance can emit 1 or 2 or 3 quanta, but not 1.5. Similarly, an electron in a metal cannot absorb half a quantum but only an integer number.
3. A light quantum, when absorbed by a metal, delivers its entire energy to *one* electron.

NOTE ▶ These three postulates—that light comes in chunks, that the chunks cannot be divided, and that the energy of one chunk is delivered to one electron—are crucial for understanding the new ideas that will lead to quantum physics. ◀

Let's look at how Einstein's postulates apply to the photoelectric effect. If Einstein is correct, the light shining on the metal is a flow of photons, each of energy $hf = hc/\lambda$. Each photon is absorbed by *one* electron, giving that electron an energy $E_{elec} = hf$. This leads us to several conclusions:

FIGURE 28.6 The ejection of an electron.

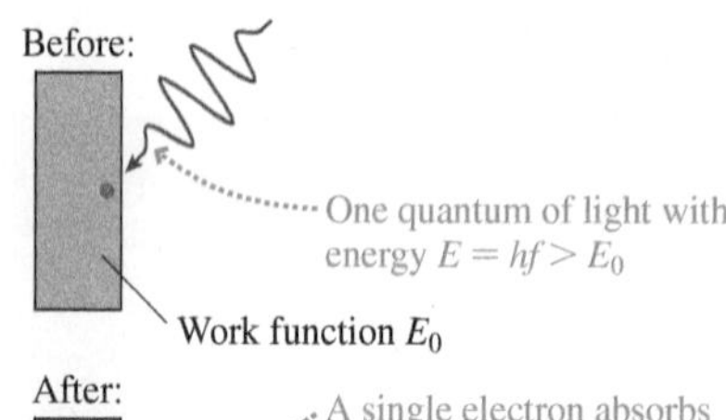
Before:
One quantum of light with energy $E = hf > E_0$
Work function E_0

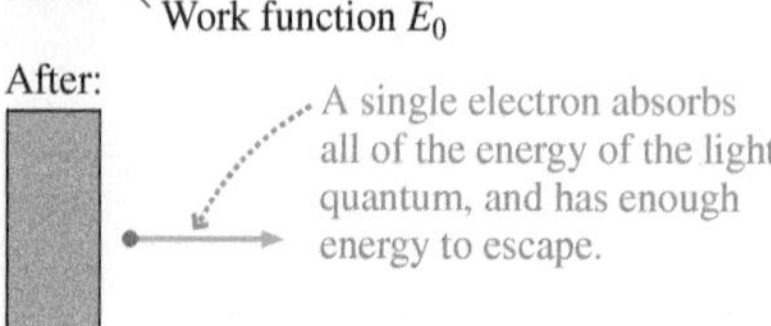
After:
A single electron absorbs all of the energy of the light quantum, and has enough energy to escape.

- An electron that has just absorbed a quantum of light energy has $E_{elec} = hf$. FIGURE 28.6 shows that this electron can escape from the metal if its energy exceeds the work function E_0, or if

$$E_{elec} = hf = \frac{hc}{\lambda} \geq E_0 \tag{28.3}$$

In other words, there is a *threshold frequency*

$$f_0 = \frac{E_0}{h} \tag{28.4}$$

for the ejection of electrons. If f is less than f_0, even by just a small amount, none of the electrons will have sufficient energy to escape no matter how intense the light. But even very weak light with $f \geq f_0$ will give a few electrons sufficient energy to escape **because each photon delivers all of its energy to one electron.** This threshold behavior is exactly what the data show.

- A more intense light delivers a larger number of photons to the surface. These eject a larger number of electrons and cause a larger current, exactly as observed.
- There is a distribution of kinetic energies, because different electrons require different amounts of energy to escape, but the *maximum* kinetic energy is

$$K_{max} = E_{elec} - E_0 = hf - E_0 \tag{28.5}$$

As we noted in Equation 28.1, the stopping potential V_{stop} is a measure of K_{max}. Einstein's theory predicts that the stopping potential is related to the light frequency by

$$V_{stop} = \frac{K_{max}}{e} = \frac{hf - E_0}{e} \tag{28.6}$$

According to Equation 28.6, the stopping potential does *not* depend on the intensity of the light. Both weak light and intense light will have the same stopping potential. This agrees with the data.

- If each photon transfers its energy hf to just one electron, that electron immediately has enough energy to escape. The current should begin instantly, with no delay, exactly as experiments had found.

Ultimately, Einstein's postulates are able to explain all of the observed features of the data for the photoelectric effect, though they require us to think of light in a very different way.

Let's use the swimming pool analogy again to help us visualize the photon model. FIGURE 28.7 shows a pebble being thrown into the pool. The pebble increases the energy of the water, but the increase is shared among all the molecules in the pool. The increase in the water's energy is barely enough to make ripples, not nearly enough to splash water out of the pool. But suppose *all* the pebble's energy could go to *one drop* of water that didn't have to share it. That one drop of water would easily have enough energy to leap out of the pool. Einstein's hypothesis that a light quantum transfers all its energy to one electron is equivalent to the pebble transferring all its energy to one drop of water.

Einstein was awarded the Nobel Prize in 1921 not for his theory of relativity, as many would suppose, but for his explanation of the photoelectric effect. Einstein showed convincingly that energy is quantized and that light, even though it exhibits wave-like interference, comes in the particle-like packets of energy we now call photons. This was the first big step in the development of the theory of quantum physics.

FIGURE 28.7 A pebble transfers energy to the water.

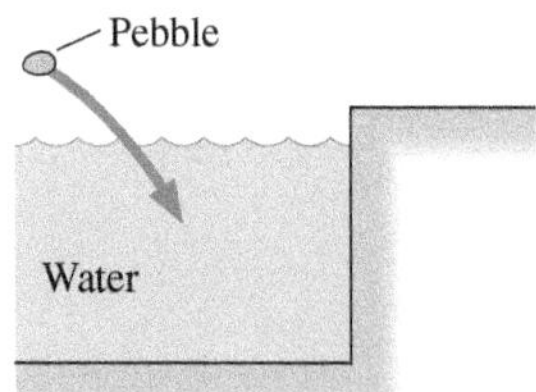

Classically, the energy of the pebble is shared by all the water molecules. One pebble causes only very small waves.

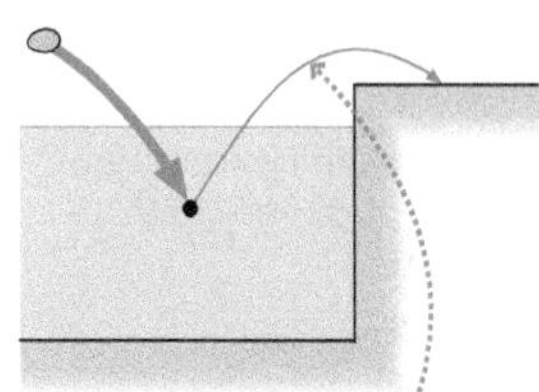

If the pebble could give *all* its energy to one drop, that drop could easily splash out of the pool.

EXAMPLE 28.3 Finding the photoelectric threshold wavelength

What are the threshold wavelengths for electron emission from sodium and from aluminum?

PREPARE From Table 28.1, the work function for sodium is $E_0 = 2.36$ eV and that for aluminum is $E_0 = 4.28$ eV. We can leave these in eV if we use the value of Planck's constant with units of eV · s.

SOLVE With these values, we calculate the threshold wavelengths as

$$\lambda_0 = \frac{hc}{E_0} = \begin{cases} 526 \text{ nm} & \text{sodium} \\ 290 \text{ nm} & \text{aluminum} \end{cases}$$

ASSESS Sodium has a much lower work function, so we expect it to have a much longer threshold wavelength. The photoelectric effect can be observed with sodium for $\lambda < 526$ nm. This includes blue and violet visible light but not red, orange, yellow, or green. Aluminum, with a larger work function, needs ultraviolet wavelengths, $\lambda < 290$ nm.

EXAMPLE 28.4 Determining the maximum electron speed

What are the maximum electron speed and the stopping potential if sodium is illuminated with light of wavelength 300 nm?

PREPARE Each photon gives its energy to one electron. Some of the energy is used to escape the metal; the balance goes to kinetic energy of the electron.

The frequency of the incoming light is $f = c/\lambda = 1.00 \times 10^{15}$ Hz. The energy of the photon, in eV, is thus

$$hf = (4.14 \times 10^{-15} \text{ eV} \cdot \text{s})(1.00 \times 10^{15} \text{ Hz}) = 4.14 \text{ eV}$$

The work function of sodium is $E_0 = 2.36$ eV.

SOLVE The maximum kinetic energy of the electron is given by Equation 28.5:

$$K_{max} = hf - E_0 = 4.14 \text{ eV} - 2.36 \text{ eV} = 1.78 \text{ eV}$$
$$= 2.85 \times 10^{-19} \text{ J}$$

Because $K = \frac{1}{2}mv^2$, where m is the electron's mass, not the mass of the sodium atom, the maximum speed of an electron leaving the cathode is

$$v_{max} = \sqrt{\frac{2K_{max}}{m}} = 7.91 \times 10^5 \text{ m/s}$$

Note that K_{max} must be in J, the SI unit of energy, in order to calculate a speed in m/s.

Now that we know the maximum kinetic energy of the electrons, we can use Equation 28.6 to calculate the stopping potential:

$$V_{stop} = \frac{K_{max}}{e} = 1.78 \text{ V}$$

An anode voltage of -1.39 V will be just sufficient to stop the fastest electrons and thus reduce the current to zero.

ASSESS The stopping potential has the *same numerical value* as K_{max} expressed in eV, which makes sense. An electron with a kinetic energy of 1.78 eV can go "uphill" against a potential difference of 1.78 V, but no more.

STOP TO THINK 28.1 The work functions of metals A, B, and C are 3.0 eV, 4.0 eV, and 5.0 eV, respectively. UV light shines on all three metals, causing electrons to be emitted. Rank in order, from largest to smallest, the stopping voltages for A, B, and C.

28.3 Photons

Einstein showed convincingly that light energy is quantized, but this leaves an important question: Just what *are* photons? To begin our explanation, let's return to the experiment that showed most dramatically the wave nature of light—the double-slit interference experiment. We will make a change, though: We will dramatically lower the light intensity by inserting filters between the light source and the slits. The fringes will be too dim to see with the naked eye, so we will use a detector that can build up an image over time.

FIGURE 28.8 A double-slit experiment performed with light of very low intensity.

(a) Image after a very short time

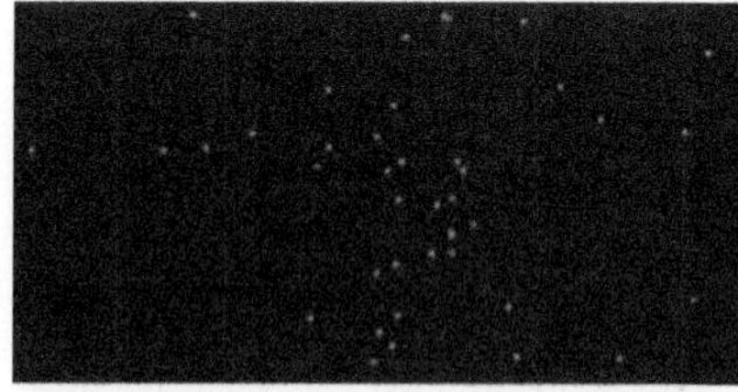

(b) Image after a slightly longer time

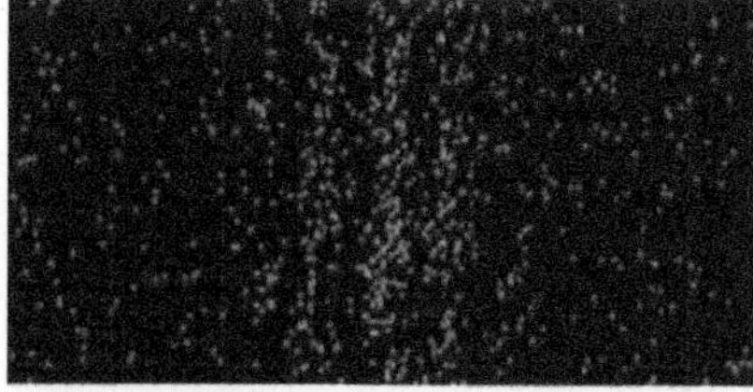

(c) Continuing to build up the image

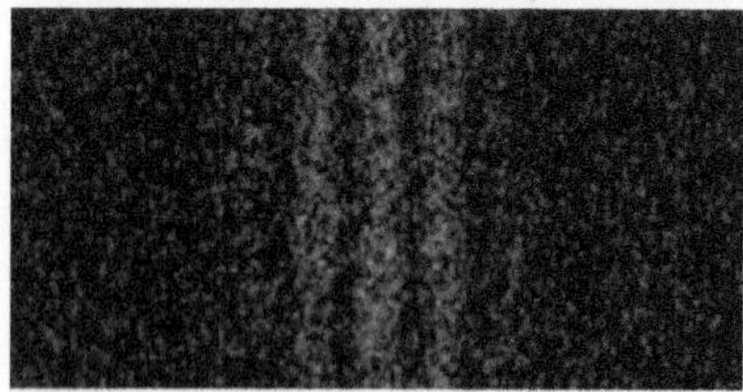

(d) Image after a very long time

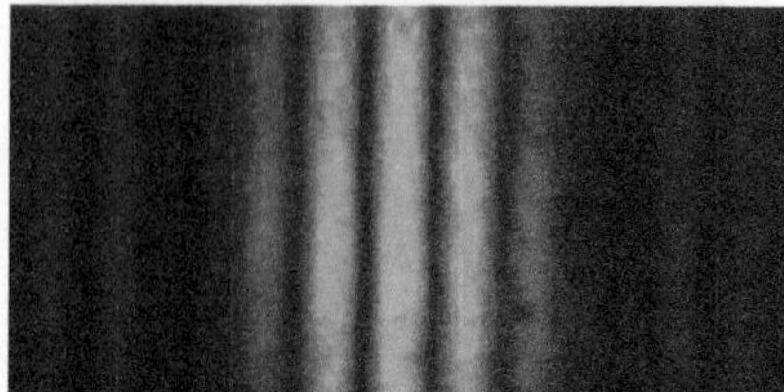

If light is a wave, then lowering the intensity will not change the nature of the interference fringes. It might take a long time to record an image, but the detector will continue to show alternating light and dark bands. However, that's not what happens. FIGURE 28.8 shows the actual outcome at four different times. At early times, contrary to our expectation, the detector shows not dim interference fringes but discrete bright dots. If we didn't know that light is a wave, we would interpret the dots as evidence that light is a stream of some type of particle. The dots arrive one by one, seemingly at random, and each is localized at one point on the detector.

As the detector builds up the image for a longer time, we see that the positions of the dots are not entirely random. They are grouped into bands at *exactly* the positions where we expect to see bright constructive-interference fringes. As the detector continues to gather light, the light and dark fringes become quite distinct. After a long time, the individual dots overlap and the image looks exactly like those we saw in Chapter 18.

The dots of light on the screen, which we'll attribute to the arrival of individual photons, are particle-like, but the overall picture clearly does not mesh with the classical idea of a particle. A classical particle, when faced with a double-slit apparatus, would go through one slit or the other. If light consisted of classical particles, we would see two bright areas on the screen, corresponding to particles that have gone through one or the other slit. Instead, we see particle-like dots forming wave-like interference fringes.

This experiment was performed with a light level so low that only one photon at a time passed through the apparatus. If particle-like photons arrive at the detector in a banded pattern as a consequence of wave-like interference, as Figure 28.8 shows, but if only one photon at a time is passing through the experiment, what is it interfering with? The only possible answer is that the photon is somehow interfering *with itself.* Nothing else is present. But if each photon interferes with itself, rather than with other photons, then each photon, despite the fact that it is a particle-like object, must somehow go through *both* slits! This is something only a wave could do.

This all seems pretty crazy. But crazy or not, this is the way light behaves in real experiments. **Sometimes it exhibits particle-like behavior and sometimes it exhibits wave-like behavior.** The thing we call *light* is stranger and more complex than it first appeared, and there is no way to reconcile these seemingly contradictory behaviors. We have to accept nature as it is, rather than hoping that nature will conform to our expectations. Furthermore, as we will see, this half-wave/half-particle behavior is not restricted to light.

The Photon Model of Light

The photon nature of light is not apparent in our everyday lives; the ray model and the wave model satisfactorily explain phenomena from image formation to diffraction. One reason is that the light sources we're familiar with emit such vast numbers of photons that we're aware of only their wave-like superposition, just as we notice only the roar of a heavy rain on the roof and not the individual raindrops. Only at extremely low intensities does light begin to appear as a stream of individual photons, like the random patter of raindrops when it is barely raining.

Another situation in which the photon perspective is important, even at high intensities, is the interaction of light with matter. Not only is light energy quantized, but so is the energy of atoms and molecules. Atoms, molecules, and even the nuclei of atoms emit and absorb light by emitting and absorbing individual photons. We'll explore these issues in the next two chapters.

Thus we need to add the **photon model** as a third model of light, joining the wave model and the ray model. Each model has a range of validity. The photon model, in which electromagnetic waves with frequency f come in packages of energy $E = hf$, is the appropriate model to describe light of extremely low intensity and to understand how matter emits and absorbs light.

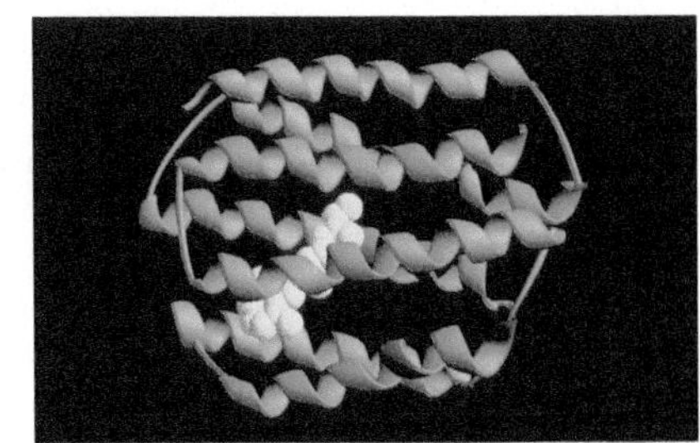

Seeing photons Vision is initiated when specially adapted molecules in the rod and cone cells of the eye respond to the absorption of photons. This image shows a molecule of *rhodopsin* (blue) with a molecule called *retinal* (yellow) nested inside. A single photon of the right energy triggers a transition of the retinal molecule, changing its shape so that it no longer fits inside the rhodopsin "cage." The rhodopsin then changes shape to eject the retinal, and this motion leads to an electrical signal in a nerve fiber.

NOTE ▶ The photon model of light applies to all electromagnetic waves, not just visible light. ◀

In practice, visible and ultraviolet light is more readily characterized by its wavelength than its frequency, so we can use the fundamental relationship for waves $\lambda f = v_{em} = c$ to write the energy of a photon that has wavelength λ:

$$E = \frac{hc}{\lambda} = \frac{1240 \text{ eV} \cdot \text{nm}}{\lambda \text{ in nm}} \tag{28.7}$$

The final expression, which connects the photon energy in eV to the wavelength in nm, comes from using $h = 4.14 \times 10^{-15}$ eV · s. For example, blue light with a wavelength of 460 nm consists of photons of energy $E = (1240 \text{ eV} \cdot \text{nm})/(460 \text{ nm}) = 2.7$ eV, and a laser beam that is a stream of 2.0 eV photons has a red-light wavelength of 620 nm.

To obtain a sense of how many photons are emitted by a typical light source, consider a monochromatic (single color) beam of light that has frequency $f = c/\lambda$. Each photon in the light beam has energy $E_{photon} = hf$, and N photons have energy $E_{light} = Nhf$. We usually measure the *power* of a light beam, or the rate (in joules per second, or watts) at which light energy is delivered. The power is

$$P = \frac{dE_{light}}{dt} = \frac{dN}{dt} hf = Rhf \tag{28.8}$$

where $R = dN/dt$ is the *rate* at which photons arrive or, equivalently, the number of photons per second.

EXAMPLE 28.5 **How many photons per second does a laser emit?**

The 1.0 mW light beam from a laser pointer ($\lambda = 670$ nm) shines on a screen. How many photons strike the screen each second?

PREPARE We'll use the photon model of light.

SOLVE The frequency of the light is

$$f = \frac{c}{\lambda} = \frac{3.00 \times 10^8 \text{ m/s}}{670 \times 10^{-9} \text{ m}} = 4.48 \times 10^{14} \text{ Hz}$$

The laser beam power is 1.0×10^{-3} W, so the rate of photon arrival is

$$R = \frac{P}{hf} = \frac{1.0 \times 10^{-3} \text{ W}}{(6.63 \times 10^{-34} \text{ J} \cdot \text{s})(4.48 \times 10^{14} \text{ Hz})}$$

$$= 3.4 \times 10^{15} \text{ photons/s}$$

ASSESS Each photon carries a small amount of energy, so there must be a huge number of photons per second to produce even this modest power. Our answer makes sense.

CONCEPTUAL EXAMPLE 28.6 Comparing photon rates

A red laser pointer and a green laser pointer have the same power. Which one emits a larger number of photons per second?

REASON Red light has a longer wavelength and thus a lower frequency than green light, so the energy of a photon of red light is less than the energy of a photon of green light. The two pointers emit the same amount of light energy per second. Because the red laser emits light in smaller "chunks" of energy, it must emit more chunks per second to have the same power. The red laser emits more photons each second.

ASSESS This result can seem counterintuitive if you haven't thought hard about the implications of the photon model. Light of different wavelengths is made of photons of different energies, so these two lasers with different wavelengths—though they have the same power—must emit photons at different rates.

Photon Interactions with Matter

Part of Einstein's explanation of the photoelectric effect is that a photon transfers all its energy to *one* electron. This idea is not limited to the photoelectric effect but is also an essential aspect of the photon model of light; that is, photons are emitted and absorbed on an all-or-nothing basis. The details are complex and will be left to other courses, but both vision and photosynthesis begin when a photon is absorbed by a pigment molecule (*rhodopsin* for vision, *chlorophyll* for photosynthesis) and transfers all its energy to that one molecule.

Disinfecting Germicidal lamps were used to disinfect subway cars during the 2020 COVID-19 pandemic. Germicidal lamps, also used in hospitals, have a strong emission at an ultraviolet wavelength of 240 nm. Photons this energetic—5.2 eV—break almost all molecular bonds and kill microorganisms.

You may recall from chemistry that covalent bond energies are typically 300–400 kJ/mol. The energy *per bond* in eV is found by dividing by Avogadro's number and then converting joules to electron volts. By doing so, we discover that covalent bond energies are typically 3 or 4 eV, a nice example of how the electron volt is an appropriate unit for energies at the atomic scale.

Several things can happen when a molecule absorbs light, but one of the most common is for the energy of a photon to be transferred to *one* bond. If the photon energy is less than 3 eV ($\lambda > 400$ nm), the absorbed energy causes the bond to vibrate. Those vibrations are then transferred to other molecules as increased thermal energy. In other words, objects are merely heated by the absorption of visible light.

But for light with wavelength $\lambda < 300$ nm, the photon energy ($E > 4$ eV) exceeds the bond energy of most covalent bonds, so **absorption of short-wavelength ultraviolet light breaks bonds and causes biological damage.** The fact that there's a wavelength threshold for bond breaking—analogous to the wavelength threshold of the photoelectric effect—is a consequence of the photon model of light; the wave model of light does not predict a threshold. Ultraviolet light classified as UVA and UVB, with wavelengths in the 300–400 nm range (photon energies 3–4 eV), is energetic enough to break some but not all bonds; the shorter-wavelength UVC breaks most bonds.

A photon with more than ≈ 10 eV of energy is able to ionize a molecule by transferring its energy to a single electron that then has enough kinetic energy to escape the molecule. Photons this energetic have wavelengths shorter than about 120 nm, so there's another wavelength threshold—an ionization threshold—at ≈ 120 nm. This very-short-wavelength radiation—extreme ultraviolet along with x rays and gamma rays—is *ionizing radiation.*

At the yet higher photon energies of x rays, several keV, additional thresholds are crossed at which photons can eject inner-shell electrons from atoms. The interaction of x rays with tissue during x-ray imaging is very much a photoelectric-effect-like interaction in which the threshold energy for inner-shell ionization plays the role of the work function. We'll have more to say about ionizing radiation in Chapter 30.

STOP TO THINK 28.2 The intensity of a beam of light is increased but the light's frequency is unchanged. Which one (or perhaps more than one) of the following is true?

A. The photons travel faster.
B. Each photon has more energy.
C. There are more photons per second.

28.4 Matter Waves

Prince Louis-Victor de Broglie was a French graduate student in 1924. It had been 19 years since Einstein had shaken the world of physics by introducing photons and thus blurring the distinction between a particle and a wave. As de Broglie thought about these issues, it seemed that nature should have some kind of symmetry. If light waves could have a particle-like nature, why shouldn't material particles have some kind of wave-like nature? In other words, could **matter waves** exist?

With no experimental evidence to go on, de Broglie reasoned by analogy with Einstein's equation $E = hf$ for the photon and with some of the ideas of his theory of relativity. De Broglie determined that *if* a material particle of momentum $p = mv$ has a wave-like nature, its wavelength must be given by

$$\lambda = \frac{h}{p} = \frac{h}{mv} \tag{28.9}$$

De Broglie wavelength for a moving particle

where h is Planck's constant. This wavelength is called the **de Broglie wavelength.**

EXAMPLE 28.7 **Calculating the de Broglie wavelength of an electron**

What is the de Broglie wavelength of an electron with a kinetic energy of 1.0 eV?

PREPARE The wavelength depends on the electron's speed and mass. We can find the speed from the kinetic energy, and then use this to determine the wavelength.

SOLVE An electron with kinetic energy $K = \frac{1}{2}mv^2 = 1.0\text{ eV} = 1.6 \times 10^{-19}\text{ J}$ has speed

$$v = \sqrt{\frac{2K}{m}} = 5.9 \times 10^5\text{ m/s}$$

Although fast by macroscopic standards, the electron gains this speed by accelerating through a potential difference of a mere 1 V. The de Broglie wavelength is

$$\lambda = \frac{h}{mv} = 1.2 \times 10^{-9}\text{ m} = 1.2\text{ nm}$$

ASSESS The electron's wavelength is small, but it is similar to the wavelengths of x rays and larger than the approximately 0.1 nm spacing of atoms in a crystal. We can observe x-ray diffraction, so if an electron has a wave nature, it should be easily observable.

What would it mean for matter—an electron or a proton or a baseball—to have a wavelength? Would it obey the principle of superposition? Would it exhibit diffraction and interference? Surprisingly, **matter exhibits all of the properties that we associate with waves.** For example, FIGURE 28.9 shows the intensity pattern recorded after 50 keV electrons passed through two narrow slits separated by 1.0 μm. The pattern is clearly a double-slit interference pattern, and the spacing of the fringes is exactly as the theory of Chapter 18 would predict for a wavelength given by de Broglie's formula. **The electrons are behaving like waves!**

FIGURE 28.9 A double-slit interference pattern created with electrons.

But if matter waves are real, why don't we see baseballs and other macroscopic objects exhibiting wave-like behavior? The key is the wavelength. We found in Chapter 18 that diffraction, interference, and other wave-like phenomena are observed when the wavelength is comparable to or larger than the size of an opening a wave must pass through. As Example 28.7 just showed, a typical electron wavelength is somewhat larger than the spacing between atoms in a crystal, so we expect to see wave-like behavior as electrons pass through matter or through microscopic slits. But the de Broglie wavelength is inversely proportional to an object's mass, so the wavelengths of macroscopic objects are millions or billions of times smaller than the wavelengths of electrons—vastly smaller than the size of any openings these objects might pass through. The wave nature of macroscopic objects is unimportant and undetectable because their wavelengths are so incredibly small, as the following example shows.

EXAMPLE 28.8 Calculating the de Broglie wavelength of a smoke particle

One of the smallest macroscopic particles we could imagine using for an experiment would be a very small smoke or soot particle. These are $\approx 1\ \mu\text{m}$ in diameter, too small to see with the naked eye and just barely at the limits of resolution of a visible-light microscope. A particle this size has mass $m \approx 10^{-15}$ kg. Estimate the de Broglie wavelength for a 1-μm-diameter particle moving at the very slow speed of 1 mm/s.

SOLVE The particle's momentum is $p = mv \approx 10^{-18}\ \text{kg} \cdot \text{m/s}$. The de Broglie wavelength of a particle with this momentum is

$$\lambda = \frac{h}{p} \approx 7 \times 10^{-16}\ \text{m}$$

ASSESS The wavelength is much, much smaller than the particle itself—much smaller than an individual atom! We don't expect to see this particle exhibiting wave-like behavior.

The example shows that a very small particle moving at a very slow speed has a wavelength that is too small to be of consequence. For larger objects moving at higher speeds, the wavelength is even smaller. A pitched baseball will have a wavelength of about 10^{-34} m, so a batter cannot use the wave nature of the ball as an excuse for not getting a hit. With such unimaginably small wavelengths, it is little wonder that we do not see macroscopic objects exhibiting wave-like behavior.

The Interference and Diffraction of Matter

Though de Broglie made his hypothesis in the absence of experimental data, experimental evidence was soon forthcoming. FIGURES 28.10a and b show diffraction patterns produced by x rays and electrons passing through an aluminum-foil target. The primary observation to make from Figure 28.10 is that **electrons diffract and interfere exactly like x rays.**

FIGURE 28.10 Diffraction patterns produced by x rays, electrons, and neutrons.

(a) Diffraction pattern produced by x rays passing through aluminum.

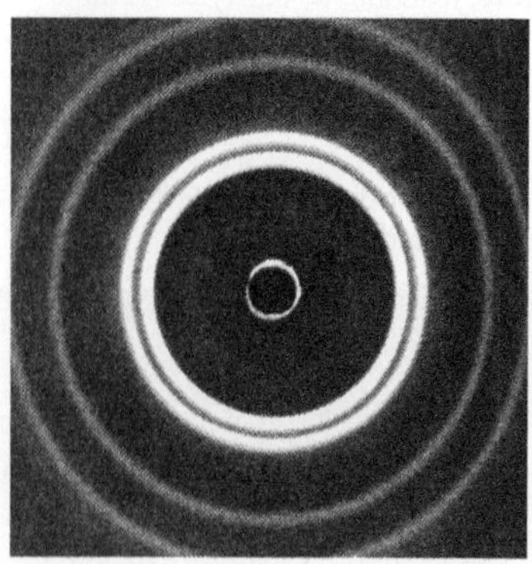

(b) Diffraction pattern produced by electrons passing through aluminum.

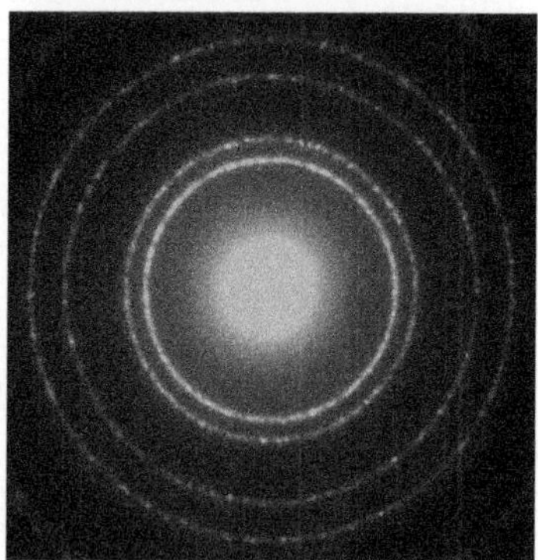

(c) Diffraction pattern produced by neutrons passing through a sodium chloride crystal.

Later experiments demonstrated that de Broglie's hypothesis applies to other material particles as well. Neutrons have a much larger mass than electrons, which tends to decrease their de Broglie wavelength, but it is possible to generate very slow neutrons. The much slower speed compensates for the heavier mass, so neutron wavelengths can be made comparable to electron wavelengths. FIGURE 28.10c shows a neutron diffraction pattern. The pattern appears different than the x-ray and electron patterns because the crystal structure of sodium chloride leads to well-defined diffraction maxima. The appearance of these maxima is a clear indication that a neutron, too, is a matter wave. In recent years it has become possible to observe the interference and diffraction of atoms and even large molecules!

STOP TO THINK 28.3 A beam of electrons, a beam of protons, and a beam of oxygen atoms each pass at the same speed through a 1-μm-wide slit. Which will produce the widest central maximum on a detector behind the slit?

A. The beam of electrons.
B. The beam of protons.
C. The beam of oxygen atoms.
D. All three patterns will be the same.
E. None of the beams will produce a diffraction pattern.

28.5 Energy Is Quantized

De Broglie hypothesized that material particles have wave-like properties, and you've now seen experimental evidence that this must be true. Not only is this bizarre, but the implications are profound.

You learned in ◀◀ SECTION 17.2 that the waves on a string fixed at both ends form standing waves. Wave reflections from both ends create waves traveling in both directions, and the superposition of two oppositely directed waves produces a standing wave. Could we do something like this with particles? Is there such a thing as a "standing matter wave"? In fact, you are probably already familiar with standing matter waves—the atomic electron orbitals that you learned about in chemistry.

Electron standing waves This computer simulation shows the p orbital of an atom. This orbital is an electron standing wave with a clear node at the center.

We'll have more to say about these three-dimensional orbitals in Chapter 29. For now, we'll start our discussion of standing matter waves with a simpler one-dimensional physical system called a "particle in a box." FIGURE 28.11 shows a classical particle—like a little ball—of mass m that bounces back and forth with speed v between the ends of a box of length L. We'll assume that the collisions at the ends are perfectly elastic, with no loss of kinetic energy. Classical physics places no restrictions on the particle's speed or energy.

If matter has wave-like properties, however, perhaps we should consider the "particle" in the box to be not a little ball but a *wave* reflecting back and forth from the ends of the box. The reflections create the standing wave shown in Figure 28.11b. This standing wave is analogous to the standing wave on a string that is tied at both ends.

FIGURE 28.11 A particle confined in a box of length L.

(a) A classical particle bounces back and forth.

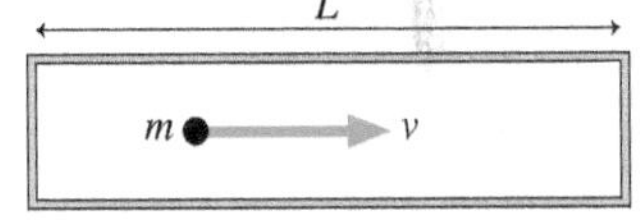

(b) A reflected wave creates a standing wave.

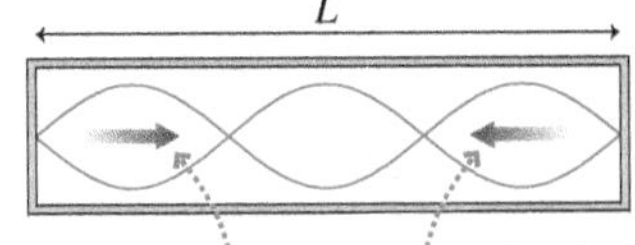

Matter waves travel in both directions.

What can we say about the properties of this standing matter wave? We can use what we know about matter waves and standing waves to make some deductions.

For waves on a string, we saw that there were only certain possible modes. The same will be true for the particle in a box. In Chapter 17, we found that a standing wave of length L must have one of the wavelengths given by

$$\lambda_n = \frac{2L}{n} \qquad n = 1, 2, 3, 4, \ldots \tag{28.10}$$

The wavelength of the particle in a box follows the same formula, but the wave describing the particle must also satisfy the de Broglie condition $\lambda = h/mv$. Equating these two expressions for the wavelength gives

$$\lambda_n = \frac{2L}{n} = \frac{h}{mv} \tag{28.11}$$

Solving Equation 28.11 for the particle's speed, we find

$$v_n = n\left(\frac{h}{2mL}\right) \qquad n = 1, 2, 3, 4, \ldots \tag{28.12}$$

This is a remarkable result; it is telling us that the speed of the particle can have only certain values, the speeds for which the de Broglie wavelength creates a standing wave in the box. Other speeds simply aren't possible. This conclusion is in sharp contrast with classical physics, where a particle can have any speed.

Thus the energy of the particle in a box, which is purely kinetic energy, is

$$E_n = \frac{1}{2}mv_n^2 = n^2\frac{h^2}{8mL^2} \qquad n = 1, 2, 3, 4, \ldots \tag{28.13}$$

Energy levels of a particle in a box of length L

This conclusion is one of the most profound discoveries of physics. Because of the wave nature of matter, **a confined particle can have only certain energies.** This result—that a confined particle can have only discrete values of energy—is called the **quantization** of energy. More informally, we say that energy is *quantized.* The number n is called the **quantum number,** and each value of n characterizes one **energy level** of the particle in the box.

The $n = 1$ energy level is called the **ground state;** it is the state in which the particle has the least possible energy:

$$E_1 = \frac{h^2}{8mL^2} \tag{28.14}$$

The ground-state energy E_1 is analogous to the fundamental frequency f_1 of a standing wave on a string. The particle's other possible energies are

$$E_n = n^2E_1 \tag{28.15}$$

This quantization is in stark contrast to the behavior of classical objects. It would be as if a baseball pitcher could throw a baseball at only 10 m/s, or 20 m/s, or 30 m/s, and so on, but at no speed in between. Baseball speeds aren't quantized, but the energy levels of a confined electron are—a result that has far-reaching implications.

EXAMPLE 28.9 Finding the allowed energies of a confined electron

An electron is confined to a region of space of length 0.19 nm—comparable in size to an atom. What are the first three allowed energies of the electron?

PREPARE We'll model this system as a particle in a box, with a box of length 0.19 nm. The possible energies are given by Equation 28.13.

SOLVE The mass of an electron is $m = 9.11 \times 10^{-31}$ kg. Thus the first allowed energy—the ground-state energy—is

$$E_1 = \frac{h^2}{8mL^2} = 1.7 \times 10^{-18}\ \text{J} = 10\ \text{eV}$$

This is the lowest allowed energy. The next two allowed energies are

$$E_2 = 2^2E_1 = 40\ \text{eV}$$

$$E_3 = 3^2E_1 = 90\ \text{eV}$$

ASSESS You'll see in Chapter 29 that these results are nearly a factor of 10 larger than the energy-level spacings of real atoms, which tells us that this one-dimensional model is *too* simple to describe an atom. Even so, we see that an electron confined to an atomic-sized space needs to be treated as a quantum system, not a classical system.

An atom is certainly more complicated than a simple one-dimensional box, but an electron is "confined" within an atom. Thus the electron orbits must, in some sense, be standing waves, and **the energy of the electrons in an atom must be quantized.** This has important implications for the physics of atomic systems, as we'll see in the next chapter.

EXAMPLE 28.10 Does a virus have quantized energy levels?

The treatment of electrons in atoms must be a quantum treatment, but classical physics still works for baseballs. Where is the dividing line? Suppose we consider a spherical virus, with a diameter of 30 nm, constrained to exist in a long, narrow cell of length 1.0 μm. If we treat the virus as a particle in a box, what is the lowest energy level? Is a quantum treatment necessary for the motion of the virus?

PREPARE We will treat the virus as a particle in a box; the length of the box is the length of the cell.

SOLVE The density of a virus is very close to that of water, so the mass is

$$m = \rho\frac{4}{3}\pi r^3 = (1000\ \text{kg/m}^3)\frac{4}{3}\pi(15\times10^{-9}\ \text{m})^3$$
$$= 1.4\times10^{-20}\ \text{kg}$$

Equation 28.14 gives the lowest energy level as

$$E_1 = \frac{h^2}{8mL^2} = \frac{(6.63\times10^{-34}\ \text{J}\cdot\text{s})^2}{8(1.4\times10^{-20}\ \text{kg})(1.0\times10^{-6}\ \text{m})^2}$$
$$= 3.9\times10^{-36}\ \text{J} = 2.4\times10^{-17}\ \text{eV}$$

This is such an incredibly small amount of energy that there is no hope of distinguishing between energies of E_1 or $4E_1$ or $9E_1$. In principle, the energy is quantized, but the allowed energies are so closely spaced that they will seem to be perfectly continuous. There is no need to use quantum physics to describe the motion of the virus.

ASSESS This result seems reasonable. A virus is small, but at a few million atomic mass units, a typical virus is much larger than electrons and other particles that must be treated with quantum physics.

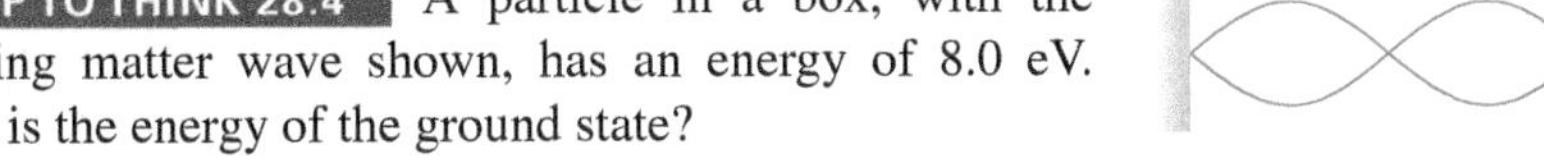

STOP TO THINK 28.4 A particle in a box, with the standing matter wave shown, has an energy of 8.0 eV. What is the energy of the ground state?

A. 1.0 eV B. 2.0 eV C. 4.0 eV D. 8.0 eV

28.6 Energy Levels and Quantum Jumps

Einstein and de Broglie introduced revolutionary new ideas—a blurring of the distinction between waves and particles, and the quantization of energy—but the first to develop a full-blown theory of quantum physics, in 1925, was the Austrian physicist Erwin Schrödinger. Schrödinger's theory is now called *quantum mechanics.* The theory is mathematically complex, so we will be content with looking at some of the key ideas and results. One of the important questions that quantum mechanics answers is How does a quantized system gain or lose energy?

Energy-Level Diagrams

The full theory of quantum mechanics shows that, just as for a particle in a box, the energy of a real physical system, such as an atom, is quantized: Only certain energies are allowed while all other energies are forbidden.

An **energy-level diagram,** like those you may have seen in chemistry, is a useful visual representation of the quantized energies. As an example, FIGURE 28.12a is the energy-level diagram for an electron in a 0.19-nm-long box. We computed these energies in Example 28.9. In an energy-level diagram, the vertical axis represents energy, but the horizontal axis is not a scale. Think of it as a ladder in which the energies are the rungs of the ladder. The lowest rung, with energy E_1, is the ground state. Higher rungs, called **excited states,** are labeled by their quantum numbers, $n = 2, 3, 4, \ldots$. Whether it is a particle in a box, an atom, or the nucleus of an atom, quantum physics requires the system to be on one of the rungs of the ladder.

FIGURE 28.12 Energy levels and quantum jumps for an electron in a 0.19-nm-long box.

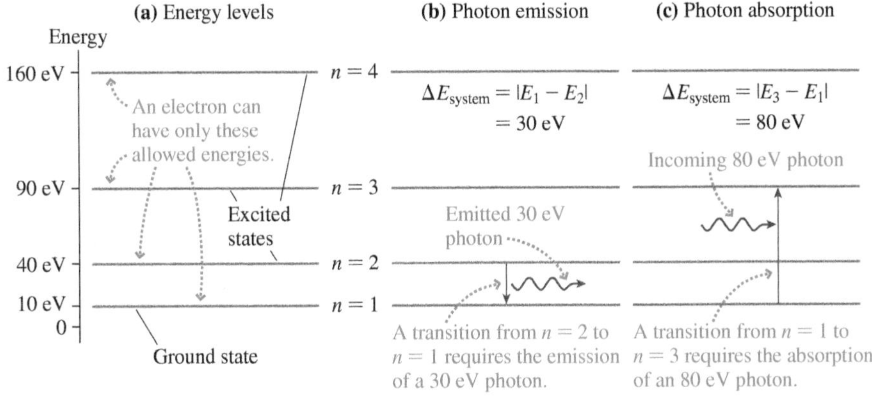

If a quantum system changes from one state to another, its energy changes. One thing that has not changed in quantum physics is the conservation of energy: If a system drops from a higher energy level to a lower, the excess energy ΔE_{system} must go somewhere. In the systems we will consider, this energy generally ends up in the form of an emitted photon. As FIGURE 28.12b shows, a quantum system in energy level E_i that "jumps down" to energy level E_f loses an energy $\Delta E_{\text{system}} = |E_f - E_i|$. This jump corresponds to the emission of a photon of frequency

$$f_{\text{photon}} = \frac{\Delta E_{\text{system}}}{h} \tag{28.16}$$

Conversely, if the system absorbs a photon, it can "jump up" to a higher energy level, as shown in FIGURE 28.12c. In this case, the frequency of the absorbed photon must follow Equation 28.16 as well. Such jumps are called **transitions,** or **quantum jumps.**

Notice that Equation 28.16 links Schrödinger's quantum theory to Einstein's earlier idea about the quantization of light energy. According to Einstein, a photon of frequency f has energy $E_{\text{photon}} = hf$. If a system jumps from an initial state with energy E_i to a final state with *lower* energy E_f, energy will be conserved if the system emits a photon with $E_{\text{photon}} = \Delta E_{\text{system}}$. The photon must have exactly the frequency given by Equation 28.16 if it is to carry away exactly the right amount of energy. As we'll see in the next chapter, these photons form the *emission spectrum* of the quantum system.

Similarly, a system can conserve energy while jumping to a higher-energy state, for which additional energy is needed, by absorbing a photon of frequency $f_{\text{photon}} = \Delta E_{\text{system}}/h$. The photon will not be absorbed unless it has exactly this frequency. The frequencies absorbed in these upward transitions form the system's *absorption spectrum.*

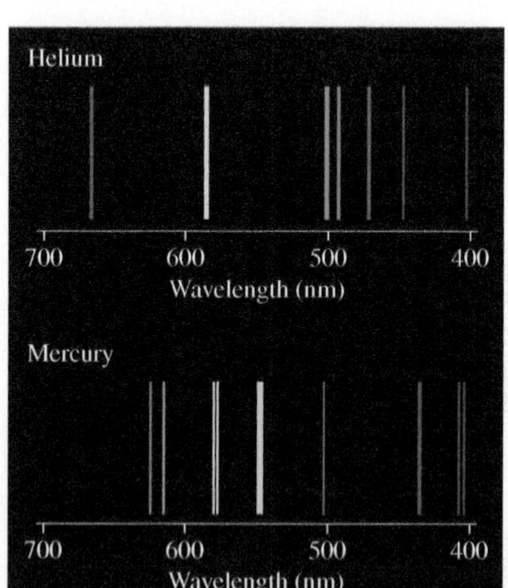

Emission spectra The emission spectra of helium and mercury show discrete wavelengths. Each wavelength in the spectrum corresponds to a quantum jump between two discrete energy levels. The atoms of each element have a unique set of energy levels, so the spectrum is a fingerprint that identifies the element.

Let's summarize what quantum physics has to say about the properties of atomic-level systems:

- **The energies are quantized.** Only certain energies are allowed; all others are forbidden. This is a consequence of the wave-like properties of matter.
- **The ground state is stable.** Quantum systems seek the lowest possible energy state. A particle in an excited state, if left alone, will jump to lower and lower energy states until it reaches the ground state. Once in its ground state, there are no lower energy states to which a particle can jump.
- **Quantum systems emit and absorb a *discrete spectrum* of light.** Only those photons whose frequencies match the energy *intervals* between the allowed energy levels can be emitted or absorbed. Photons of other frequencies cannot be emitted or absorbed without violating energy conservation.

We'll use these ideas in the next two chapters to understand the properties of atoms and nuclei.

EXAMPLE 28.11 **Determining an emission spectrum from quantum states**

An electron in a quantum system has allowed energies $E_1 = 1.0$ eV, $E_2 = 4.0$ eV, and $E_3 = 6.0$ eV. What wavelengths are observed in the emission spectrum of this system?

PREPARE FIGURE 28.13 shows the energy-level diagram for this system. Photons are emitted when the system undergoes a quantum jump from a higher energy level to a lower energy level. There are three possible transitions.

FIGURE 28.13 The system's energy-level diagram and quantum jumps.

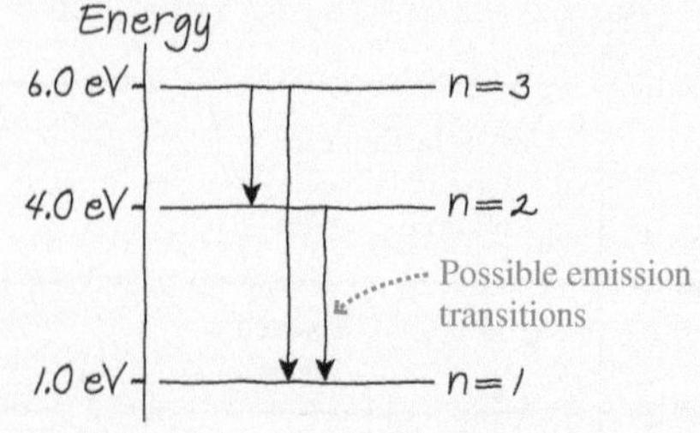

SOLVE This system will emit photons on the $3 \rightarrow 1$, $2 \rightarrow 1$, and $3 \rightarrow 2$ transitions, with $\Delta E_{3\rightarrow1} = 5.0$ eV, $\Delta E_{2\rightarrow1} = 3.0$ eV, and $\Delta E_{3\rightarrow2} = 2.0$ eV. From $f_{\text{photon}} = \Delta E_{\text{system}}/h$ and $\lambda = c/f$, we find that the frequencies and wavelengths in the emission spectrum are

$3 \rightarrow 1$ $\quad f = 5.0\text{ eV}/h = 1.21 \times 10^{15}$ Hz
$\lambda = 250$ nm (ultraviolet)

$2 \rightarrow 1$ $\quad f = 3.0\text{ eV}/h = 7.25 \times 10^{14}$ Hz
$\lambda = 410$ nm (blue)

$3 \rightarrow 2$ $\quad f = 2.0\text{ eV}/h = 4.83 \times 10^{14}$ Hz
$\lambda = 620$ nm (orange)

ASSESS Transitions with a small energy difference, like $3 \rightarrow 2$, correspond to lower photon energies and thus longer wavelengths than transitions with a large energy difference like $3 \rightarrow 1$, as we would expect.

STOP TO THINK 28.5 A photon with a wavelength of 410 nm has energy $E_{\text{photon}} = 3.0$ eV. Do you expect to see a spectral line with $\lambda = 410$ nm in the emission spectrum of the system represented by this energy-level diagram?

Energy
5.0 eV — $n = 3$
3.0 eV — $n = 2$
1.0 eV — $n = 1$

28.7 The Uncertainty Principle

A classical particle that moves along a line—whether it's a dust particle or a baseball—is described by its position x and its velocity v_x. How accurately can we determine the particle's position and velocity? All measurements have experimental uncertainty, but classical physics places no limits on how small those uncertainties can be. Better equipment and better procedures can reduce the uncertainties without limit, and there is no fundamental limit to how well the position and velocity of a classical particle can be known.

One of the strangest aspects of quantum physics is that there *is* an inherent limitation on our knowledge. If we know where a particle is, we can't know exactly how fast it is moving. If we know the particle's velocity, we can't know exactly where it is. This counterintuitive notion is a consequence of the wave nature of matter.

To see where this limitation comes from, FIGURE 28.14 shows a single-slit diffraction experiment carried out with electrons. A classical particle would travel straight through the slit, but, because of the wave nature of electrons, the slit causes electrons to spread out to produce a diffraction pattern. Each electron forms a dot where it hits the detector screen, but the dot pattern replicates the single-slit diffraction pattern that we studied in Chapter 18.

We know something about the electron's vertical position y as it goes through a vertical slit of width a, but our knowledge isn't perfect. We know only that the electron is somewhere within the slit. It could be as much as $\frac{1}{2}a$ to either side of the center line, so we say that our uncertainty in the electron's vertical position at this instant is $\delta y = \frac{1}{2}a$.

NOTE ▸ Some texts write the uncertainty as Δy. We use the Δ notation to indicate the *change* in a quantity; thus Δy is a change of position. To avoid confusion, we'll use a lowercase δ to represent uncertainty. ◂

FIGURE 28.14 A single-slit diffraction experiment with electrons illustrates quantum uncertainty.

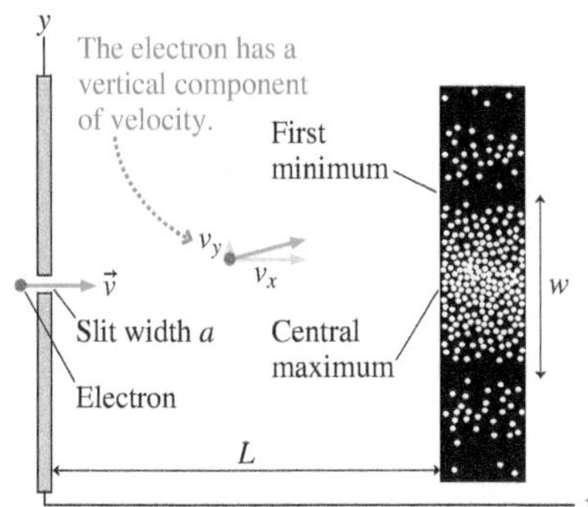

The electrons spread out after passing through the slit to produce the diffraction pattern. An electron that strikes the screen anywhere other than straight ahead must have acquired a vertical component of velocity v_y. We can't predict where any individual electron will land on the screen, so we can't predict its vertical velocity v_y. In other words, there's uncertainty δv_y in our knowledge of v_y for that electron.

Consider an electron that is deflected vertically by Δy. It takes the electron a time interval Δt to move from the slit to the screen, so its vertical velocity after leaving the slit is

$$v_y = \frac{\Delta y}{\Delta t} \tag{28.17}$$

The *uncertainty* in our knowledge of v_y is

$$\delta v_y = \frac{\delta y_{\text{screen}}}{\Delta t} \tag{28.18}$$

where δy_{screen} is our uncertainty in the electron's vertical position when it arrives at the screen.

We learned earlier that the width of the central maximum of the diffraction pattern is $w = 2\lambda L/a$. The electron's vertical position y_{screen} as it reaches the screen could be as much as $\lambda L/a$ to either side of center, so $\delta y_{\text{screen}} = \lambda L/a$ and our uncertainty in v_y is

$$\delta v_y = \frac{\lambda L/a}{\Delta t} = \frac{\lambda}{a}\frac{L}{\Delta t} = \frac{\lambda}{2\delta y} v_x \tag{28.19}$$

We made two substitutions in the last step of Equation 28.19. First, $L/\Delta t$ is the electron's horizontal velocity v_x as it travels distance L to the screen. Second, we found above that $a = 2\delta y$.

If we multiply through by δy, the left-hand side of Equation 28.19 becomes $\delta y\,\delta v_y$. And if we then multiply both sides by the electron mass m, the velocity components v_x and v_y become momentum components $p_x = mv_x$ and $p_y = mv_y$. After doing so, Equation 28.19 is

$$\delta y\,\delta p_y = \frac{\lambda p_x}{2} \tag{28.20}$$

As a final step, we learned earlier that the electron's de Broglie wavelength is $\lambda = h/p \approx h/p_x$. An electron's deflection is extremely small, so there's essentially no difference between p and p_x. If we replace λp_x with Planck's constant h, Equation 28.20 becomes

$$\delta y\,\delta p_y = \frac{h}{2} \tag{28.21}$$

Equation 28.21 tells us that the *product* of the uncertainty in the electron's position y and the uncertainty in its momentum p_y is a constant. We can improve our knowledge of the electron's position by narrowing the slit, thus reducing δy, but doing so spreads the diffraction pattern wider and increases δp_y. In other words, improving our knowledge of where the electron is comes at the expense of *losing* knowledge about its momentum. A wider slit causes less diffraction and has a smaller uncertainty δp_y, but using a wider slit increases the position uncertainty δy.

We've considered a particular experiment, but it is an example of a general principle. For any particle in any experiment, it can be shown in quantum mechanics that the product of the uncertainty δx in the particle's position and the uncertainty δp_x in its momentum has a lower limit:

$$\delta x\,\delta p_x \geq \frac{h}{4\pi} \tag{28.22}$$

Heisenberg uncertainty principle for position and momentum

Equation 28.23 is the **Heisenberg uncertainty principle,** put forward in 1927 by Werner Heisenberg, a key contributor to the theory of quantum mechanics. The product of the uncertainties might be greater than $h/4\pi$, but it can't be less.

Precise definitions of δx and δp_x require the mathematical tools of quantum mechanics, but simple approximations will meet our needs. Consider a particle that is confined within a region of length L centered on x_0. We know that the particle is somewhere in the interval $x_0 \pm \frac{1}{2}L$, so the position uncertainty is $\delta x = \frac{1}{2}L$. A confined particle has, on average, zero momentum, so a momentum uncertainty of δp_x means that the momentum is within the range $\pm\frac{1}{2}\delta p_x$ centered on zero. The most we can say about the particle's velocity v_x is that it must be somewhere in the range $-\frac{1}{2}\delta v_x$ to $+\frac{1}{2}\delta v_x$, where $\delta v_x = \delta p_x/m$. Neither x nor v_x can be known with certainty.

Uncertainties are associated with all experimental measurements, but better procedures and techniques can reduce those uncertainties. Classical physics places no limits on how small the uncertainties can be. A classical particle at any instant of time has an exact position x and an exact momentum p_x, and with sufficient care we can measure both x and p_x with such precision that we can make the product $\delta x\,\delta p_x$ as small as we like. There are no inherent limits to our knowledge.

In the quantum world, it's not so simple. No matter how clever you are, and no matter how good your experiment, you *cannot* measure both x and p_x simultaneously with arbitrarily good precision. Any measurements you make are limited by the condition that $\delta x\,\delta p_x \geq h/4\pi$. **The position and the momentum of a particle are *inherently* uncertain.**

Why? Because of the wave-like nature of matter! The "particle" is spread out in space, so there simply is not a precise value of its position x. Our belief that position and momentum have precise values is tied to our classical concept of a particle. As we revise our ideas of what atomic particles are like, we must also revise our ideas about position and momentum.

Let's revisit particles in a one-dimensional "box," now looking at uncertainties.

EXAMPLE 28.12 **Determining uncertainties**

What constraints does the uncertainty principle place on our knowledge of the world? To get a sense of scale, we'll look at the uncertainty in velocity for a confined electron and for a confined dust particle.

a. What range of velocities might an electron have if confined to a 0.1-nm-wide region, about the size of an atom?

b. A 1-μm-diameter dust particle ($m \approx 10^{-15}$ kg) is confined within a 5-μm-long box. Can we know with certainty if the particle is at rest? If not, within what range is its velocity likely to be found?

PREPARE Neither the electron's nor the dust particle's position can be known with certainty. A particle confined within a region of length L has position uncertainty $\delta x = \frac{1}{2}L$, so the electron has position uncertainty $\delta x = 0.05$ nm and the dust particle has $\delta x = 2.5\ \mu$m.

SOLVE

a. The uncertainty in a particle's velocity is $\delta v_x = \delta p_x/m$. The uncertainty principle relates the momentum uncertainty to the particle's position uncertainty. For the electron, the minimum uncertainty in velocity is

$$\delta v_x = \frac{\delta p_x}{m} = \frac{h}{4\pi m\,\delta x} = 1 \times 10^6 \text{ m/s}$$

Because the *average* velocity is zero (the electron is equally likely to be moving right or left), the most we can say is that the electron's velocity is somewhere in the range -5×10^5 m/s $< v_x < 5 \times 10^5$ m/s. It is simply not possible to know the electron's velocity more precisely than this.

b. We can know with certainty that the dust particle is at rest ($\delta v_x = 0$) only if we have absolutely no knowledge of its position ($\delta x = \infty$). But we know that the particle is in the box ($\delta x = 2.5\ \mu$m), so there must be uncertainty in our knowledge of its velocity. That uncertainty is

$$\delta v_x = \frac{\delta p_x}{m} = \frac{h}{4\pi m\,\delta x} = 2 \times 10^{-14} \text{ m/s}$$

All we know with certainty is that the particle's velocity is somewhere in the range -1×10^{-14} m/s $< v_x <$ 1×10^{-14} m/s.

ASSESS Our uncertainty in the electron's velocity is enormous, a significant fraction of the speed of light. While we can't say with certainty that the dust particle is at rest, knowing that its speed is less than 1×10^{-14} m/s means that the particle is at rest for all practical purposes. At this speed, the particle would take more than 16 years to travel the length of its 5-μm-long box. Quantum physics has profound implications at the atomic scale but that it does not affect how we think about even the smallest macroscopic object.

Wave–Particle Duality

One common theme that has run through this chapter is the idea that, in quantum theory, things we think of as being waves have a particle nature, while things we think of as being particles have a wave nature. What is the true nature of light, or an electron? Are they particles or waves?

The various objects of classical physics are *either* particles *or* waves. There's no middle ground. Planets and baseballs are particles or collections of particles, while sound and light are clearly waves. Particles follow trajectories given by Newton's laws; waves obey the principle of superposition and exhibit interference. This wave–particle dichotomy seemed obvious until physicists encountered irrefutable evidence that light sometimes acts like a particle and, even stranger, that matter sometimes acts like a wave.

You might at first think that light and matter are *both* a wave *and* a particle, but that idea doesn't quite work. The basic definitions of particleness and waviness are mutually exclusive. Two sound waves can pass through each other and can overlap to produce a larger-amplitude sound wave; two baseballs can't. It is more reasonable to conclude that light and matter are *neither* a wave *nor* a particle. At the microscopic scale of atoms and their constituents—a physical scale not directly accessible to our five senses—the classical concepts of particles and waves turn out to be simply too limited to explain the subtleties of nature.

Although matter and light have both wave-like aspects and particle-like aspects, they show us only one face at a time. If we arrange an experiment to measure a wave-like property, such as interference, we find photons and electrons acting like waves, not particles. An experiment to look for particles will find photons and electrons acting like particles, not waves. These two aspects of light and matter are *complementary* to each other, like a two-piece jigsaw puzzle. Neither the wave nor the particle model alone provides an adequate picture of light or matter, but taken together they provide us with a basis for understanding these elusive but most fundamental constituents of nature. This two-sided point of view is called *wave–particle duality.*

For over two hundred years, scientists and nonscientists alike felt that the clockwork universe of Newtonian physics was a fundamental description of reality. But wave–particle duality, along with Einstein's theory of relativity, undermines the basic assumptions of the Newtonian worldview. The certainty and predictability of classical physics have given way to a new understanding of the universe in which chance and uncertainty play key roles—the universe of quantum physics.

STOP TO THINK 28.6 The speeds of an electron and a proton have been measured to the same uncertainty. Which one has a larger uncertainty in position?

A. The proton, because it's more massive.
B. The electron, because it's less massive.
C. The uncertainty in position is the same, because the uncertainty in velocity is the same.

28.8 Applications of Quantum Physics

Quantum physics may be at odds with our intuition about how light and matter should behave, but the predictions of quantum physics have been confirmed in countless experiments. We'll conclude this chapter by looking at two imaging technologies that rest on these remarkable quantum phenomena.

FIGURE 28.15 An energy diagram with a potential-energy barrier for particles that have total energy E_2.

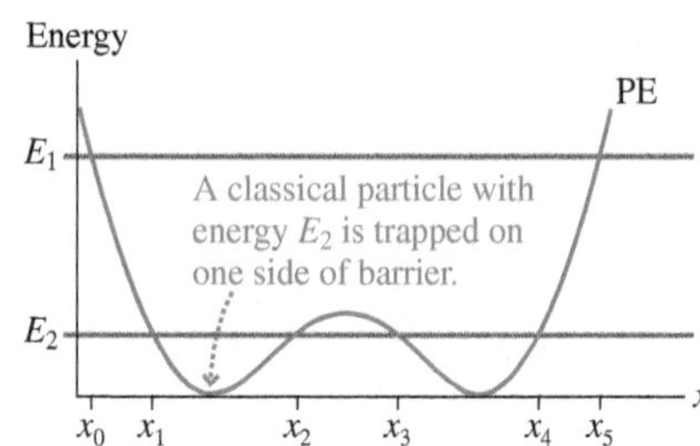

Tunneling and the Scanning Tunneling Microscope

FIGURE 28.15 shows an energy diagram like the ones you learned to analyze in ◀◀ SECTION 11.4. The blue curve is the potential energy U as a particle moves along the x-axis, and the brown lines are two possible total energies $E = K + U$. Recall that the particle's kinetic energy at any point in its motion is the distance $E - U$ from the potential-energy curve to the total energy line.

A particle with total energy E_1 oscillates back and forth between turning points at x_0 and x_5. It slows down each time it passes over the potential-energy "hill," then speeds up as it passes through the potential-energy "valleys."

What about a particle with total energy E_2? One lesson from Chapter 11 was that a particle cannot exist at a position where $E < U$ because that would require negative kinetic energy. A particle with total energy E_2 can oscillate on the left side between x_1 and x_2 or on the right side between x_3 and x_4, but the particle cannot cross the potential-energy barrier from one side to the other.

One of the oddest predictions of quantum physics is that a particle with total energy E_2 *can* move from one side of the potential-energy barrier to the other. A particle initially on the left side of the barrier may, at a later time, be found on the right side. The particle doesn't go *over* the barrier; that would violate energy conservation. Instead, the particle somehow goes *through* the barrier in a process called **tunneling** because it is rather like tunneling through a mountain barrier to get to the other side.

Tunneling is a consequence of the wave-like nature of matter and the uncertainty principle. A classical particle has a well-defined position—it is always on one side of the barrier or the other—but a wave does not. The extent of a quantum particle is not precisely defined—that's what the uncertainty principle tells us—and a particle's waviness extends through the barrier. The particle may seem to be on one side of the barrier, but there's a chance that it will switch to be on the other side. As with everything in quantum physics, the probability that a particle will tunnel from one side of the barrier to the other is meaningful for atomic particles and only if the width of the barrier is comparable to the size of an atom; tunneling is never observed for even the smallest macroscopic particles.

Not only does tunneling actually occur, but it has important practical applications. As an example, we'll look at how a **scanning tunneling microscope,** or STM as it is usually called, creates images of surfaces and molecules at the scale of single atoms. FIGURE 28.16 shows a conducting probe with a *very* sharp tip, just a few atoms wide, positioned just above a surface. The distance between the surface and the probe is about 0.5 nm, only a few atomic diameters. Electrons in the sample are attracted to the positively charged probe, but a classical electron would not be able to cross the gap and no current would flow.

Electrons are not classical particles, however. Electrons, because of their wave-like nature, can tunnel across the gap from the surface to the probe, creating a flow of charge—a current—that can be measured. The probability that an electron will tunnel across the gap is very sensitive to the width of the gap. As the probe is scanned across the surface, the tunneling current increases as the probe passes over an atom, which narrows the gap. Similarly, the current decreases if the probe passes over an atomic-size valley. A map of the surface topography can be constructed from the current-versus-position data as the probe is scanned back and forth.

STM was the first technology that allowed the imaging of individual atoms. The STM image of FIGURE 28.17a on the next page clearly shows the hexagonal arrangement of the individual atoms on the surface of graphite, and in FIGURE 28.17b we see the actual twists of the double helix of DNA. Current research efforts aim to develop DNA-sequencing technologies in which a scanning tunneling microscope "reads" a single strand of DNA.

FIGURE 28.16 The scanning tunneling microscope.

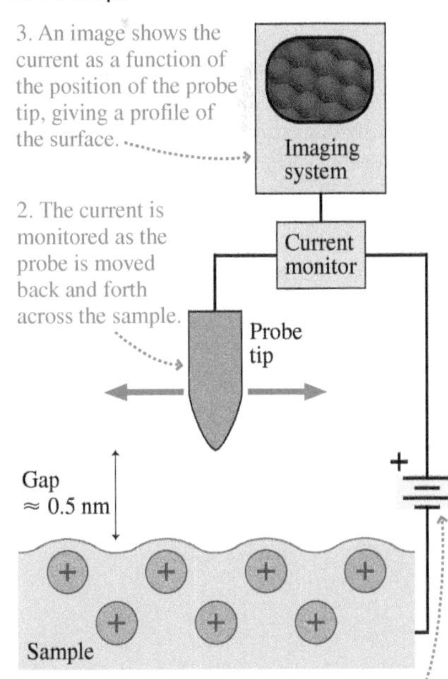
3. An image shows the current as a function of the position of the probe tip, giving a profile of the surface.
Imaging system
2. The current is monitored as the probe is moved back and forth across the sample.
Current monitor
Probe tip
Gap ≈ 0.5 nm
Sample
1. The small positive voltage of the probe causes electrons to tunnel across the narrow gap between the probe tip and the sample.

The Electron Microscope

You learned in ◀◀ SECTION 20.6 that the wave nature of light limits the resolution of an optical microscope—the smallest resolvable separation between two objects—to about half the wavelength of light. The smallest feature that can be resolved with visible light, even with a perfect lens, is about 200 nm. But the image of the smallpox virus at the beginning of this chapter shows features as small as 2 nm, a factor of 1000 better than an optical microscope. This image was made not with light but with electrons.

FIGURE 28.17 STM images.

(a) Surface of graphite

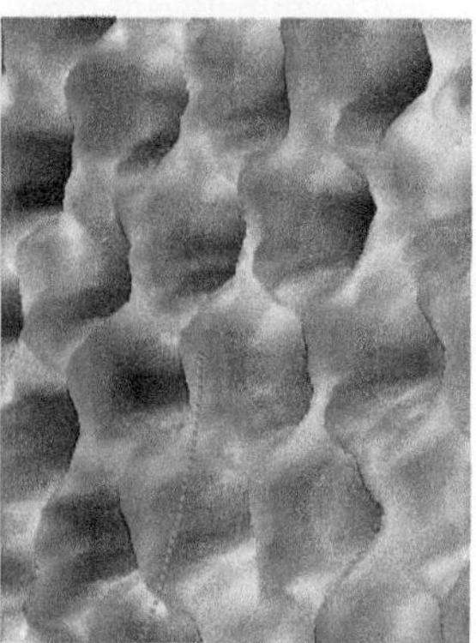

The hexagonal arrangement of atoms is clearly visible.

(b) DNA molecule

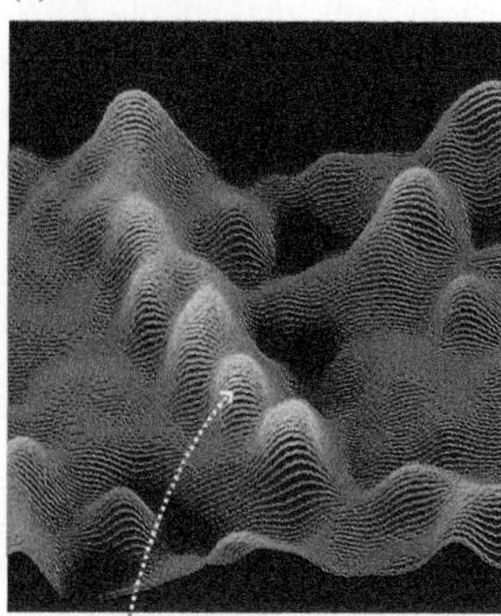

The peaks are the edges of the DNA helix.

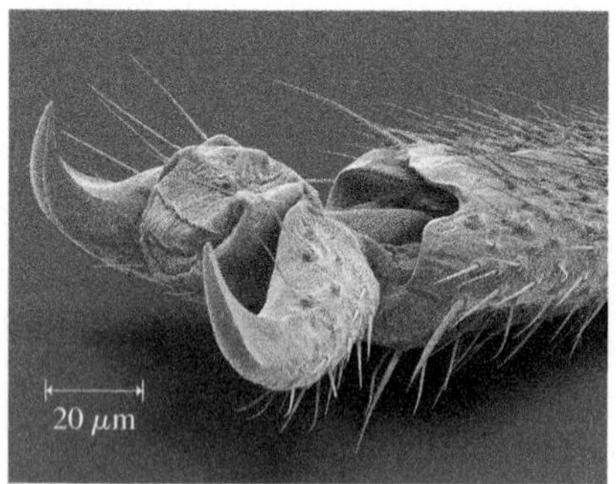

Getting the hook This scanning electron micrograph of an insect foot does not have extraordinarily high resolution—it is comparable to the resolution of an optical microscope—but it provides extraordinarily detailed information about surface features and texture.

An **electron microscope** works much like a light microscope. In the absence of fields, electrons travel through a vacuum in straight lines much like light rays. Electric and magnetic fields bend electron trajectories, much like a surface refracts light, and a suitably designed solenoid can bring electron trajectories together at a focal point; we call this an *electron lens.*

FIGURE 28.18a shows a simplified view of a *transmission electron microscope* (TEM). The source of electrons—equivalent to the light source in an optical microscope—is a hot tungsten filament that emits thermal electrons. The electrons accelerate through a very large potential difference, typically 100 kV, and are then collimated by an electron lens called a *condenser lens* (an optical microscope also has a condenser lens) to illuminate the specimen with a beam of electrons.

Light rays that pass through matter can be transmitted, scattered, or absorbed. The same is true for electrons. Many electrons pass directly through the specimen with little or no scattering. These electrons are collected by the *objective lens,* just as in an optical microscope, and focused onto a detector. Some portions of the specimen absorb electrons or scatter electrons through such large angles that they are not collected by the objective lens; these parts of the specimen are "darker" in the image because fewer electrons reach the detector.

FIGURE 28.18 The electron microscope.

(a) Transmission electron microscope

−
+
100 kV
Electron source
The condenser lens collimates electrons to illuminate the specimen.
Condenser lens
Specimen
Electrons emerging from the specimen are focused onto a detector by the objective lens.
Objective lens
Detector

(b) Scanning electron microscope

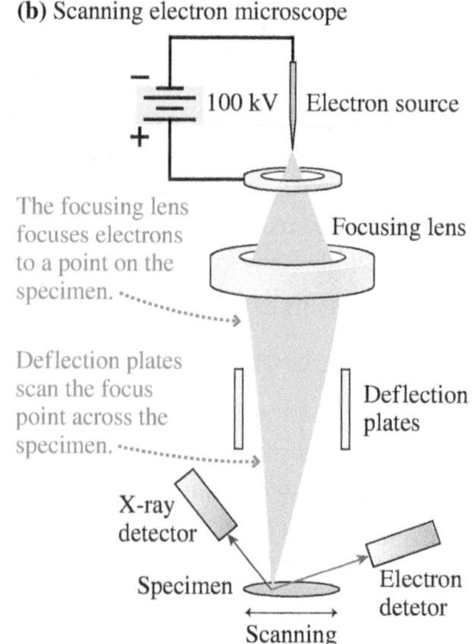

This is purely classical physics; the electrons follow trajectories given by Newton's second law. A carefully designed electron lens can resolve features that are only a few nm in size because electrons that emerge from the specimen at points a few nm apart are focused to distinct points on the detector. But ultimately, just as in a light microscope, the resolution is limited by wave effects. Electrons are not classical point particles; they have a wave-like nature and a de Broglie wavelength.

An electron accelerated through a 100 kV potential difference has 100 keV of kinetic energy. An electron with this much energy has a de Broglie wavelength $\lambda \approx 0.004\text{ nm} = 4\text{ pm}$. The theoretical resolution of an electron microscope, just as with an optical microscope, is about half a wavelength. Thus the best resolution that an electron microscope could achieve is about 2 pm.

The best optical microscopes do reach the theoretical resolution because the best lenses are essentially perfect. That is not yet true for electron lenses; slight imperfections in the lenses prevent electron microscopes from reaching the theoretical limit set by quantum physics. The very best electron microscopes have a resolution of about 50 pm. That is just sufficient to resolve individual atoms, but there's still room for improvement if a clever scientist or engineer can make a better electron lens.

The specimen for a TEM has to be extremely thin; a typically thickness is 100 nm. TEM images are very detailed but also very flat, conveying no information about depth or structure. A different type of electron microscope, the *scanning electron microscope* (SEM), plays a complementary role; it has less resolution but excellent ability to see depth.

FIGURE 28.18b shows that a SEM focuses the electrons to a very small spot directly on the specimen. High-speed electrons have a good chance of passing through a 100-nm-thick specimen, but electrons that strike a thicker specimen deposit their energy in the specimen. Some of the kinetic energy simply increases the specimen's thermal energy, but the impact of high-energy electrons causes a specimen to emit both x rays and *secondary electrons* that are dislodged from atoms in the sample.

A SEM detects either the x rays or the secondary electrons (or sometimes both) as the small spot of focused electrons is scanned back and forth across the specimen, and the image is built up point by point. The spot easily moves over the hills and valleys of the specimen to provide a very dramatic three-dimensional image. A SEM micrograph provides information about the specimen's surface features, in contrast to a TEM image that provides information about the structure and organization of the specimen.

CHAPTER 28 INTEGRATED EXAMPLE A fluorescence experiment

Fluorescence is the absorption of light at one wavelength followed, after excited-state energy is transformed into thermal energy, by the emission of light at a longer wavelength. A common example of fluorescence is the eerie colors when people and objects are illuminated by a "black light," which emits unseen ultraviolet light. In biology, the green fluorescent protein, derived from jellyfish, is used as a marker to image structures or molecules that are otherwise hard to see; it emits green light when illuminated with ultraviolet radiation.

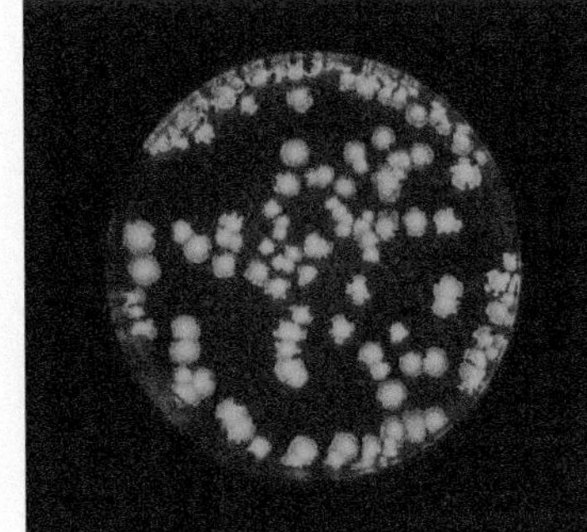

Glowing green *E. coli* bacteria emit a green glow under ultraviolet light after incorporating a plasmid that expresses the green fluorescent protein.

As a simple model of fluorescence, suppose an electron in a 1.50-nm-long box absorbs an ultraviolet photon with a wavelength of 309 nm. After the photon absorption, the electron falls one energy level by transforming some of its energy into thermal energy. This *internal conversion* of energy does not emit a photon. The electron then undergoes a spontaneous quantum jump back to the ground state. What is the wavelength of the photon emitted in this quantum jump?

PREPARE FIGURE 28.19 on the next page shows the process. A system, if otherwise undisturbed, is in its ground state. Thus a photon can be absorbed only if the photon's energy exactly matches the energy *difference* between the ground state ($n = 1$) and an excited state with quantum number n. After excitation, the electron falls

Continued

FIGURE 28.19 The process of fluorescence.

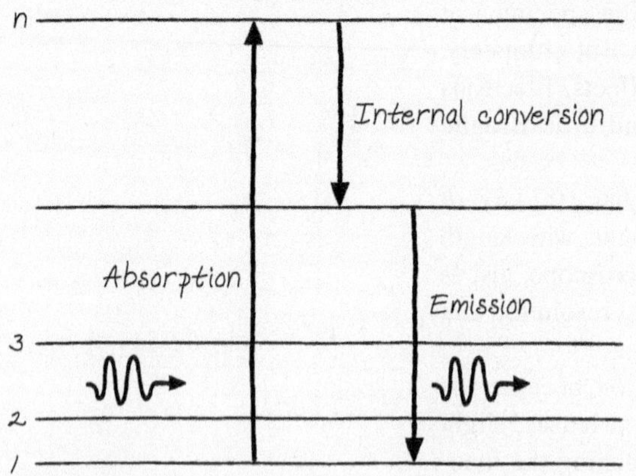

to energy level $n-1$ by transforming some energy into thermal energy. Finally, the electron emits a photon in a $n-1 \rightarrow 1$ transition. The energy levels are those of a particle in a box.

SOLVE The energy levels of a particle in a box are

$$E_n = n^2 E_1$$

where the ground-state energy is

$$E_1 = \frac{h^2}{8mL^2} = \frac{(6.63 \times 10^{-34}\ \text{J} \cdot \text{s})^2}{8(9.11 \times 10^{-31}\ \text{kg})(1.5 \times 10^{-9}\ \text{m})^2}$$

$$= 2.681 \times 10^{-20}\ \text{J} \times \frac{1\ \text{eV}}{1.60 \times 10^{-19}\ \text{J}} = 0.1675\ \text{eV}$$

We'll keep an extra significant figure to avoid round-off error. The energy of the absorbed photon is

$$E_{\text{photon}} = hf = \frac{hc}{\lambda}$$

$$= \frac{(4.14 \times 10^{-15}\ \text{J} \cdot \text{s})(3.00 \times 10^8\ \text{m/s})}{309 \times 10^{-9}\ \text{m}} = 4.019\ \text{eV}$$

The photon can be absorbed, which causes a transition to an excited state n, if

$$\Delta E_{\text{electron}} = n^2 E_1 - E_1 = (n^2 - 1)E_1 = E_{\text{photon}}$$

where the quantum number n must be an integer. We can solve for n:

$$n = \sqrt{\frac{E_{\text{photon}}}{E_1} + 1} = \sqrt{\frac{4.019\ \text{eV}}{0.1675\ \text{eV}} + 1} = 5$$

We see that the absorption of a photon with a wavelength of 309 nm causes a $1 \rightarrow 5$ quantum jump.

The electron then drops to the $n = 4$ energy level without emitting a photon by transforming some of its energy into thermal energy. From there, the electron undergoes a downward $4 \rightarrow 1$ transition that does emit a photon. The electron's energy loss is

$$\Delta E_{\text{electron}} = 4^2 E_1 - 1^2 E_1 = 15E_1 = 2.51\ \text{eV}$$

Energy must be conserved, so this is the energy of the emitted photon. The photon energy is $E_{\text{photon}} = hf = hc/\lambda$, so its wavelength is

$$\lambda = \frac{hc}{E_{\text{photon}}} = \frac{(4.14 \times 10^{-15}\ \text{J} \cdot \text{s})(3.00 \times 10^8\ \text{m/s})}{2.51\ \text{eV}} = 495\ \text{nm}$$

ASSESS 495 nm is a wavelength in the green region of the visible spectrum. In this case, the system absorbs an ultraviolet photon and then fluoresces by emitting a longer-wavelength green photon. In fluorescence, the emitted photon always has a longer wavelength (less energy) than the absorbed photon. Our assumption of an internal conversion of electronic energy into thermal energy is dubious for a particle in a box, but not for molecules where fluorescence actually occurs. A molecule has several modes of storing energy, and some of the electronic energy after photon absorption is easily transformed into vibrational energy of molecular bonds before the molecule undergoes a photon-emitting quantum jump back to the ground state.

SUMMARY

GOAL To understand the quantum behavior of light and matter.

GENERAL PRINCIPLES

Light Has Particle-Like Properties

- The energy of a light wave comes in discrete packets (light quanta) called **photons.**
- For light of frequency f, the energy of each photon is $E = hf$, where h is **Planck's constant.**
- Photons are emitted or absorbed on an all-or-nothing basis.
- When a photon is absorbed, it delivers all its energy to one electron.
- For a beam of light that delivers power P, photons arrive at a rate R such that $P = Rhf$.

Matter Has Wave-Like Properties

- The **de Broglie wavelength** of a particle of mass m is $\lambda = h/p = h/mv$.
- The wave-like nature of matter is seen in the interference and diffraction patterns of electrons, neutrons, and entire atoms.
- Particles can **tunnel** through a potential-energy barrier that a classical particle cannot cross.
- A confined particle sets up a de Broglie standing wave. Standing waves have only certain allowed wavelengths; thus a confined particle has only certain allowed energies. Energy is quantized.

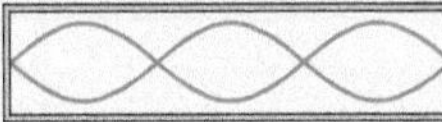

Wave–particle duality

Light and matter are neither particles nor waves but exhibit characteristics of both. An experiment to measure wave-like properties finds wave-like behavior, and an experiment to measure particle-like properties finds particle-like behavior. These two aspects of light and matter are complementary.

Heisenberg uncertainty principle

A particle with wave-like characteristics does not have a precise value of position x or a precise value of momentum p_x. Both are uncertain. The position uncertainty δx and the momentum uncertainty δp_x are related by

$$\delta x\, \delta p_x \geq \frac{h}{4\pi}$$

The better you know the value of one, the less you know about the other.

IMPORTANT CONCEPTS

Photoelectric effect

Light with wavelength λ can eject electrons from a metal only if $\lambda \leq \lambda_0 = hc/E_0$, where E_0 is the metal's **work function.** Electrons will be ejected even if the intensity of the light is very low.

The **stopping potential** that stops even the fastest electrons is

$$V_{stop} = \frac{K_{max}}{e} = \frac{hf - E_0}{e}$$

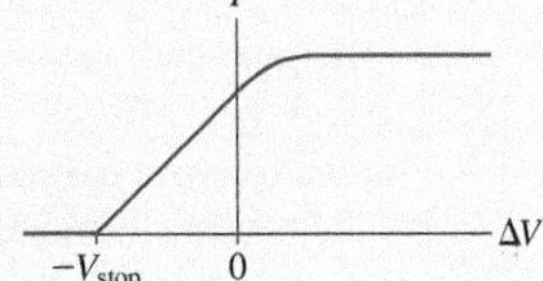

The details of the photoelectric effect could not be explained with classical physics. New models were needed.

Particle in a box

A particle of mass m confined to a one-dimensional box of length L sets up de Broglie standing waves. The allowed energies are

$$E_n = n^2 \frac{h^2}{8mL^2}$$

Energy levels

- Energies are quantized. Only certain energies are allowed.
- The **ground state** is stable. Systems spend most of their time in the ground state.
- Systems emit and absorb photons in **quantum jumps** between energy levels.
- The photon energy must exactly match the energy difference between the energy levels.

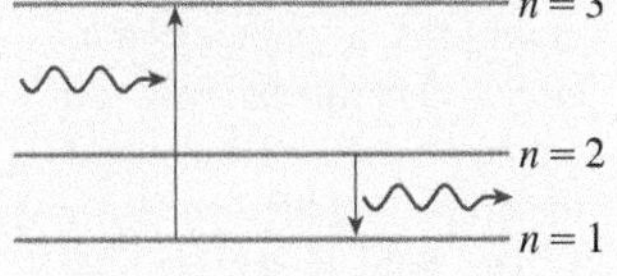

APPLICATIONS

The atomic-scale unit of energy is the **electron volt** (eV). One electron volt is the kinetic energy an electron or proton gains in accelerating through a 1 V potential difference:

$$1\text{ eV} = 1.60 \times 10^{-19}\text{ J}$$

- Particles of matter with de Broglie wavelength λ exhibit interference and diffraction patterns just like those of light with wavelength λ.
- The **scanning tunneling microscope** forms an image by measuring the current as electrons tunnel between the sample and a probe.
- The resolution limit of an **electron microscope** is about half of the de Broglie wavelength of the electrons.

LEARNING OBJECTIVES After studying this chapter, you should be able to:

- Calculate the threshold wavelength and stopping potential for the photoelectric effect. *Conceptual Questions 28.2, 28.4; Problems 28.1–28.3, 28.5, 28.7*
- Apply the photon model of light. *Conceptual Question 28.7; Problems 28.11, 28.12, 28.14, 28.16, 28.22*
- Calculate the de Broglie wavelengths of particles with mass. *Conceptual Questions 28.8, 28.9; Problems 28.26, 28.28–28.30, 28.32*
- Solve problems about the energy levels of a particle in a box. *Conceptual Questions 28.12, 28.13; Problems 28.33–28.35, 28.37, 28.38*
- Solve problems about energy levels and quantum jumps. *Conceptual Question 28.14; Problems 28.39–28.43*
- Apply the Heisenberg uncertainty principle. *Problems 28.44–28.48*
- Solve problems about the electron microscope and the scanning tunneling microscope. *Conceptual Question 28.15; Problems 28.49, 28.50*

STOP TO THINK ANSWERS

Stop to Think 28.1: $(V_{stop})_A > (V_{stop})_B > (V_{stop})_C$. For a given wavelength of light, electrons are ejected faster from metals with smaller work functions because it takes less energy to remove an electron. Faster electrons need a larger negative voltage to stop them.

Stop to Think 28.2: C. Photons always travel at c, and a photon's energy depends on only the light's frequency, not its intensity. Greater intensity means more energy each second, which means more photons.

Stop to Think 28.3: A. The widest diffraction pattern occurs for the largest wavelength. The de Broglie wavelength is inversely proportional to the particle's mass, and so will be largest for the least massive particle.

Stop to Think 28.4: B. The quantum number n is related to the wavelength as $\lambda_n = 2L/n$. The wave shown in the figure has a wavelength equal to L, so for this wave $n = 2$. The lowest energy is the ground state with $n = 1$. According to Equation 28.15, the $n = 1$ state has 1/4 the energy of the $n = 2$ state, or 2.0 eV.

Stop to Think 28.5: No. The energy of an emitted photon is the energy *difference* between two allowed energies. The three possible quantum jumps have energy differences of 2.0 eV, 2.0 eV, and 4.0 eV.

Stop to Think 28.6: B. The position uncertainty is $\delta x = h/4\pi\delta p_x = h/4\pi m\delta v_x$. For particles that have the same velocity uncertainty, a more massive particle will have a smaller position uncertainty.

QUESTIONS

Conceptual Questions

1. Explain why the graphs of Figure 28.3b are mostly horizontal for $\Delta V > 0$.
2. A current is detected in a photoelectric-effect experiment when the cathode is illuminated with green light. Will a current necessarily be detected if the cathode is illuminated with blue light? With red light?
3. Figure Q28.3 is the current-versus-potential-difference graph for a photoelectric-effect experiment with an unknown metal. How might the graph look if:
   a. The light is replaced by an equally intense light with a shorter wavelength? Draw it.
   b. The metal is replaced by a different metal with a larger work function? Draw it.

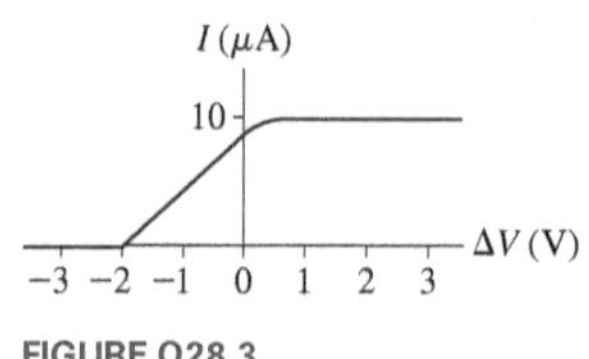

FIGURE Q28.3

4. In a photoelectric-effect experiment, the stopping voltage is 2.0 V when a gold cathode is illuminated with ultraviolet light. Will the stopping voltage increase, decrease, or stay the same if:
   a. The light intensity is doubled?
   b. The wavelength of the light is increased?
   c. The gold cathode is replaced by an aluminum cathode?
5. Metal 1 has a larger work function than metal 2. Both are illuminated with the same short-wavelength ultraviolet light. Do photoelectrons from metal 1 have a higher speed, a lower speed, or the same speed as photoelectrons from metal 2? Explain.
6. An investigator is measuring the current in a photoelectric effect experiment. The cathode is illuminated by light of a single wavelength. What happens to the current if the wavelength of the light is reduced by a factor of two while keeping the intensity constant?

Problem difficulty is labeled as | (straightforward) to ||||| (challenging). Problems labeled INT integrate significant material from earlier chapters; BIO are of biological or medical interest; CALC require calculus to solve.

7. Three laser beams have wavelengths $\lambda_1 = 400$ nm, $\lambda_2 = 600$ nm, and $\lambda_3 = 800$ nm. The power of each laser beam is 1 W.
   a. Rank in order, from largest to smallest, the photon energies E_1, E_2, and E_3 in these three laser beams. Explain.
   b. Rank in order, from largest to smallest, the number of photons per second N_1, N_2, and N_3 delivered by the three laser beams. Explain.
8. Electron 1 is accelerated from rest through a potential difference of 100 V. Electron 2 is accelerated from rest through a potential difference of 200 V. Afterward, which electron has the larger de Broglie wavelength? Explain.
9. An electron and a proton are each accelerated from rest through a potential difference of 100 V. Afterward, which particle has the larger de Broglie wavelength? Explain.
10. When you cool a gas, how does this affect the de Broglie wavelength of the gas atoms?
11. Figure Q28.11 is a simulation of the electrons detected behind two closely spaced slits. Each bright dot represents one electron. How will this pattern change if
    a. The electron-beam intensity is increased?
    b. The electron speed is reduced?
    c. The electrons are replaced by neutrons with the same speed?
    d. The left slit is closed?
    Your answers should consider the number of dots on the screen and the spacing, width, and positions of the fringes.

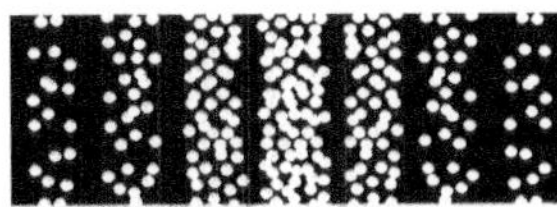
FIGURE Q28.11

12. Figure Q28.12 shows the standing de Broglie wave of a particle in a box.
    a. What is the quantum number?
    b. Can you determine from this picture whether the "classical" particle is moving to the right or to the left? If so, which is it? If not, why not?

FIGURE Q28.12

13. A particle in a box of length L_a has $E_1 = 2$ eV. The same particle in a box of length L_b has $E_2 = 50$ eV. What is the ratio L_a/L_b?
14. Figure Q28.14 shows the energy-level diagram of Element X.
    a. An atom in the ground state absorbs a photon, then emits a photon with a wavelength of 1240 nm. What was the energy of the photon that was absorbed?
    b. An atom in the ground state has a collision with an electron, then emits a photon with a wavelength of 1240 nm. What conclusion can you draw about the initial kinetic energy of the electron?

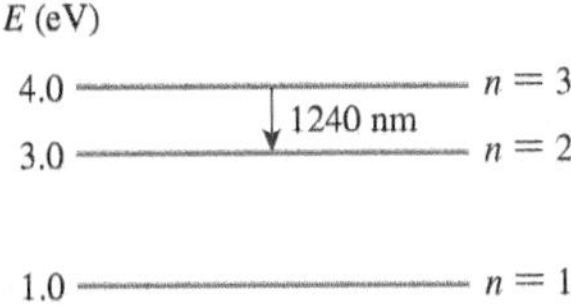

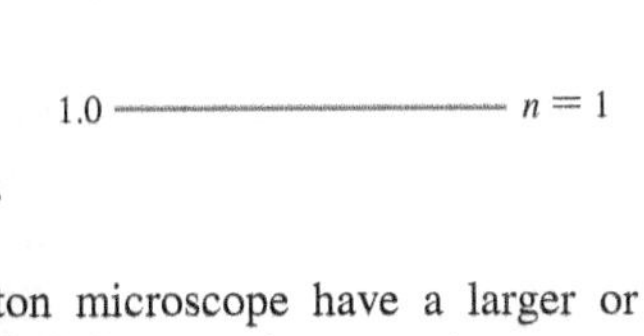

FIGURE Q28.14

15. Would a proton microscope have a larger or smaller theoretical resolution than an electron microscope if both use an accelerating voltage with the same magnitude? Or would their resolutions be the same? Explain.

Multiple-Choice Questions

16. | A light sensor is based on a photodiode that requires a minimum photon energy of 1.7 eV to create mobile electrons. What is the longest wavelength of electromagnetic radiation that the sensor can detect?
    A. 500 nm B. 730 nm
    C. 1200 nm D. 2000 nm
17. || In a photoelectric-effect experiment, the wavelength of the light is decreased while the intensity is held constant. As a result,
    A. There are more electrons.
    B. The electrons are faster.
    C. Both A and B.
    D. Neither A nor B.
18. | In a photoelectric-effect experiment, the intensity of the light is increased while the wavelength, which is below the threshold wavelength, is held constant. As a result,
    A. There are more electrons. B. The electrons are faster.
    C. Both A and B. D. Neither A nor B.
19. | In the photoelectric effect, electrons are never emitted from a metal if the wavelength of the incoming light is above a certain threshold value. This is because
    A. Photons of long-wavelength light don't have enough energy to eject an electron.
    B. The electric field of long-wavelength light does not vibrate the electrons rapidly enough to eject them.
    C. The number of photons in long-wavelength light is too small to eject electrons.
    D. Long-wavelength light does not penetrate far enough into the metal to eject electrons.
20. | Light consisting of 2.7 eV photons is incident on a piece of potassium, which has a work function of 2.3 eV. What is the maximum kinetic energy of the ejected electrons?
    A. 2.3 eV B. 2.7 eV C. 5.0 eV D. 0.4 eV
21. || Visible light has a wavelength of about 500 nm. A typical radio wave has a wavelength of about 1.0 m. How many photons of the radio wave are needed to equal the energy of one photon of visible light?
    A. 2,000 B. 20,000
    C. 200,000 D. 2,000,000
22. | Two radio stations have the same power output from their antennas. One broadcasts AM at a frequency of 1000 kHz and one broadcasts FM at a frequency of 100 MHz. Which statement is true?
    A. The FM station emits more photons per second.
    B. The AM station emits more photons per second.
    C. The two stations emit the same number of photons per second.
23. || An electron is released from rest in a vacuum chamber and allowed to fall a distance of 2.0 m. At the end of the fall, its de Broglie wavelength is
    A. 120 pm. B. 120 nm.
    C. 120 μm. D. 120 mm.
24. || You shoot a beam of electrons through a double slit to make an interference pattern. After noting the properties of the pattern, you then double the speed of the electrons. What effect does this have?
    A. The fringes get closer together.
    B. The fringes get farther apart.
    C. The positions of the fringes do not change.

25. | Photon P in Figure Q28.25 moves an electron from energy level $n = 1$ to energy level $n = 3$. The electron jumps down to $n = 2$, emitting photon Q, and then jumps down to $n = 1$, emitting photon R. The spacing between energy levels is drawn to scale. What is the correct relationship among the wavelengths of the photons?

A. $\lambda_P < \lambda_Q < \lambda_R$
B. $\lambda_R < \lambda_P < \lambda_Q$
C. $\lambda_Q < \lambda_P < \lambda_R$
D. $\lambda_R < \lambda_Q < \lambda_P$

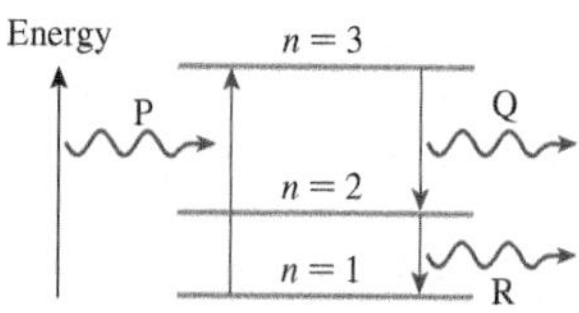

FIGURE Q28.25

PROBLEMS

Section 28.1 Physics at the Atomic Level

Section 28.2 The Photoelectric Effect

1. | Which metals in Table 28.1 exhibit the photoelectric effect for (a) light with $\lambda = 400$ nm and (b) light with $\lambda = 250$ nm?
2. | Electrons are emitted when a metal is illuminated by light with a wavelength less than 388 nm but for no greater wavelength. What is the metal's work function?
3. | Zinc has a work function of 4.3 eV.
   a. What is the longest wavelength of light that will release an electron from a zinc surface?
   b. A 4.7 eV photon strikes the surface and an electron is emitted. What is the maximum possible speed of the electron?
4. | Light with a wavelength of 350 nm shines on a metal surface, which emits electrons. The stopping potential is measured to be 1.25 V.
   a. What is the maximum speed of emitted electrons?
   b. Calculate the work function and identify the metal.
5. || Electrons in a photoelectric-effect experiment emerge from a copper surface with a maximum kinetic energy of 1.10 eV. What is the wavelength of the light?
6. || Potassium and gold cathodes are used in a photoelectric-effect experiment. For each cathode, find:
   a. The threshold frequency
   b. The threshold wavelength
   c. The maximum electron ejection speed if the light has a wavelength of 220 nm
   d. The stopping potential if the wavelength is 220 nm
7. || How many photoelectrons are ejected per second in the experiment represented by the graph of Figure P28.7?

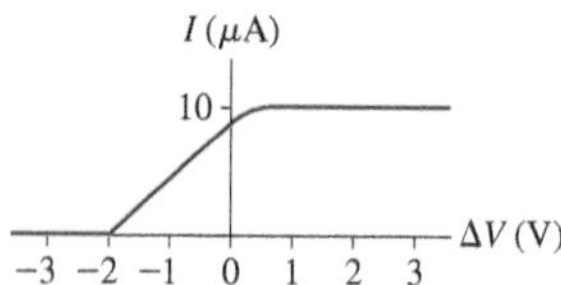

FIGURE P28.7

8. || You need to design a photodetector that can respond to the entire range of visible light. What is the maximum possible work function of the cathode?
9. || A photoelectric-effect experiment finds a stopping potential of 1.93 V when light of 200 nm wavelength is used to illuminate the cathode.
   a. From what metal is the cathode made?
   b. What is the stopping potential if the intensity of the light is doubled?
10. || Light with a wavelength of 375 nm illuminates a metal cathode. The maximum kinetic energy of the emitted electrons is 0.76 eV. What is the longest wavelength of light that will cause electrons to be emitted from this cathode?

Section 28.3 Photons

11. | What is the wavelength, in nm, of a photon with energy (a) 0.30 eV, (b) 3.0 eV, and (c) 30 eV? For each, is this wavelength visible, ultraviolet, or infrared light?
12. | a. Determine the energy, in eV, of a photon with a 550 nm wavelength.
   b. Determine the wavelength of a 7.5 keV x-ray photon.
13. | What is the energy, in eV, of (a) a 450 MHz radio-frequency photon, (b) a visible-light photon with a wavelength of 450 nm, and (c) an x-ray photon with a wavelength of 0.045 nm?
14. | BIO When an ultraviolet photon is absorbed by a molecule of DNA, the photon's energy can be converted into vibrational energy of the molecular bonds. Excessive vibration damages the molecule by causing the bonds to break. Ultraviolet light of wavelength less than 290 nm causes significant damage to DNA; ultraviolet light of longer wavelength causes minimal damage. What is the threshold photon energy, in eV, for DNA damage?
15. | BIO Your eyes have three different types of cones with maximum absorption at 437 nm, 533 nm, and 564 nm. What photon energies correspond to these wavelengths?
16. | BIO Rod cells in the retina of the eye detect light using a photopigment called rhodopsin. 1.8 eV is the lowest photon energy that can trigger a response in rhodopsin. What is the maximum wavelength of electromagnetic radiation that can cause a transition? In what part of the spectrum is this?
17. || BIO The absorption of light by chlorophyll, the pigment in plants that is responsible for converting light to useful energy, is shown in Figure P28.17. Two sharp peaks are visible in the violet and red parts of the spectrum (very little light is absorbed in the green part of the spectrum, which is why leaves appear green). What are the energies in eV of the photons that are absorbed at the two peaks?

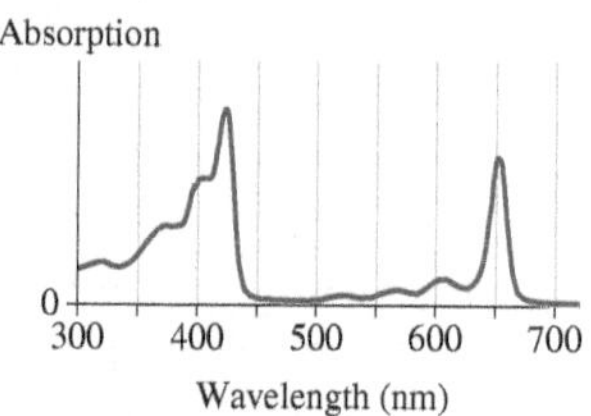

FIGURE P28.17

18. || BIO A firefly glows by the direct conversion of chemical energy to light. The light emitted by a firefly has peak intensity at a wavelength of 550 nm.

   a. What is the minimum chemical energy, in eV, required to generate each photon?
   b. One molecule of ATP provides 0.30 eV of energy when it is metabolized in a cell. What is the minimum number of ATP molecules that must be consumed in the reactions that lead to the emission of one photon of 550 nm light?

19. || One molecule of ATP provides 0.30 eV when it is used to power cellular processes. Photosynthesis in a typical plant requires 8 photons at 550 nm to produce 1 molecule of ATP. What is the overall efficiency of this process? BIO INT

20. | For what wavelength of light does a 100 mW laser deliver 2.50×10^{17} photons per second?

21. | A 100 W incandescent lightbulb emits about 5 W of visible light. (The other 95 W are emitted as infrared radiation or lost as heat to the surroundings.) The average wavelength of the visible light is about 600 nm, so make the simplifying assumption that all the light has this wavelength. How many visible-light photons does the bulb emit per second?

22. || A 193-nm-wavelength UV laser for eye surgery emits a 0.500 mJ pulse. How many photons does the light pulse contain? BIO

23. ||| The human eye can barely detect a star whose intensity at the earth's surface is 1.6×10^{-11} W/m^2. If the dark-adapted eye has a pupil diameter of 7.0 mm, how many photons per second enter the eye from the star? Assume the starlight has a wavelength of 550 nm. BIO

24. || A study of photosynthesis in phytoplankton in the open ocean used short pulses of laser light to trigger photosynthetic reactions. The investigator's system used 0.10 mW pulses of 640 nm laser light, each lasting 200 ps. How many photons were contained in each pulse? BIO

25. || The wavelengths of light emitted by a firefly span the visible spectrum but have maximum intensity near 550 nm. A typical flash lasts for 100 ms and has a power of 1.2 mW. If we assume that all of the light is emitted at the peak-intensity wavelength of 550 nm, how many photons are emitted in one flash? BIO

Section 28.4 Matter Waves

26. || At what speed is an electron's de Broglie wavelength (a) 1.0 nm, (b) 1.0 μm, and (c) 1.0 mm?

27. || What is the de Broglie wavelength of a neutron that has fallen 1.0 m in a vacuum chamber, starting from rest? INT

28. || Through what potential difference must an electron be accelerated from rest to have a de Broglie wavelength of 500 nm? INT

29. || What is the kinetic energy, in eV, of an electron with a de Broglie wavelength of 1.0 nm?

30. || The diameter of an atomic nucleus is about 10 fm ($1 \text{ fm} = 10^{-15}$ m). What is the kinetic energy, in MeV, of a proton with a de Broglie wavelength of 10 fm?

31. | a. What is the de Broglie wavelength of a 150 g baseball with a speed of 30 m/s?
    b. What is the speed of a 150 g baseball with a de Broglie wavelength of 0.20 nm?

32. || What is the de Broglie wavelength of a red blood cell with a mass of 1.00×10^{-11} g that is moving with a speed of 0.400 cm/s? Do we need to be concerned with the wave nature of the blood cells when we describe the flow of blood in the body? BIO

Section 28.5 Energy Is Quantized

33. | What is the quantum number of an electron confined in a 3.0-nm-long one-dimensional box if the electron's de Broglie wavelength is 1.0 nm?

34. || What is the length of a box in which the minimum energy of an electron is 1.5×10^{-18} J?

35. || What is the length of a one-dimensional box in which an electron in the $n = 1$ state has the same energy as a photon with a wavelength of 600 nm?

36. || An electron confined in a one-dimensional box is observed, at different times, to have energies of 12 eV, 27 eV, and 48 eV. What is the length of the box?

37. | The nucleus of a typical atom is 5.0 fm ($1 \text{ fm} = 10^{-15}$ m) in diameter. A very simple model of the nucleus is a one-dimensional box in which protons are confined. Estimate the energy of a proton in the nucleus by finding the first three allowed energies of a proton in a 5.0-fm-long box.

38. || Investigators have created structures consisting of linear chains of ionized atoms on a smooth surface. Electrons are restricted to travel along the chain. The energy levels of the electrons match the results of the particle-in-a-box model. For a 5.0-nm-long chain, what are the energies (in eV) of the first three states?

Section 28.6 Energy Levels and Quantum Jumps

39. | Figure P28.39 is an energy-level diagram for a quantum system. What wavelengths appear in the system's emission spectrum?

$n = 3$ ———— $E_3 = 4.0$ eV

$n = 2$ ———— $E_2 = 1.5$ eV

$n = 1$ ———— $E_1 = 0.0$ eV

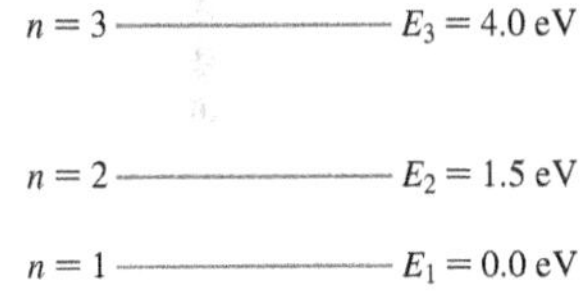

FIGURE P28.39

40. || The allowed energies of a quantum system are 1.0 eV, 2.0 eV, 4.0 eV, and 7.0 eV. What wavelengths appear in the system's emission spectrum?

41. || A quantum system has three energy levels, so three wavelengths appear in its emission spectrum. The shortest observed wavelength is 248 nm; light with a 414 nm wavelength is also observed. What is the third wavelength?

42. || The allowed energies of a quantum system are 0.0 eV, 4.0 eV, and 6.0 eV.
    a. Draw the system's energy-level diagram. Label each level with the energy and the quantum number.
    b. What wavelengths appear in the system's emission spectrum?

43. ||| The color of dyes results from the preferential absorption of certain wavelengths of light. Some dye molecules consist of symmetric pairs of rings joined at the center by a chain of carbon atoms, as shown in Figure P28.43. Electrons of the bonds along the chain of carbon atoms are shared among the atoms in the chain, but are repelled by the nitrogen-containing rings at the end of the chain. These electrons are thus free to move along the chain but not beyond its ends. They look very much like a particle in a one-dimensional box. For the molecule shown, the effective length of the "box" is 0.85 nm. Assuming that the electrons start in the lowest energy state, what are the three longest wavelengths this molecule will absorb?

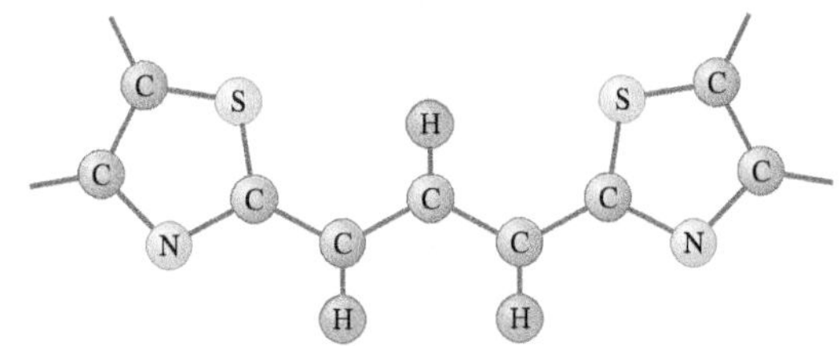

FIGURE P28.43

Section 28.7 The Uncertainty Principle

44. || What is the minimum uncertainty in position, in nm, of an electron whose velocity is known to be between 3.48×10^5 m/s and 3.58×10^5 m/s?

45. || A thin solid barrier in the xy-plane has a 10-μm-diameter circular hole. An electron traveling in the z-direction with $v_x = 0$ m/s passes through the hole. Afterward, is it certain that v_x is still zero? If not, within what range is v_x likely to be?

46. || What is the smallest box in which you can confine an electron if you want to know for certain that the uncertainty in the electron's speed is no more than 20 m/s?

47. ||| A proton is confined within an atomic nucleus of diameter 4.0 fm. Use a one-dimensional model to estimate the smallest range of speeds you might find for a proton in the nucleus.

48. ||| Andrea, whose mass is 50 kg, thinks she's sitting at rest in her 5.0-m-long dorm room as she does her physics homework. Can Andrea be sure she's at rest? If not, within what range is her velocity likely to be?

Section 28.8 Applications of Quantum Physics

49. || The electron microscope image that opens this chapter has a
INT resolution of about 2 nm. If this microscope is ideal, working at the theoretical maximum resolution, through what potential difference are electrons accelerated before they reach the condenser lens?

50. || It is shown in quantum mechanics that the tunneling current measured by a STM decreases exponentially with the width of the gap that electrons must tunnel across: $I = Ce^{-w/\eta}$, where C is a constant, w is the width of the gap, and η is a constant that depends on the work function of the sample. Suppose a STM measures the current I_0 when the probe tip is between two atoms. By what factor does the current increase if the probe tip goes over a 0.10-nm-tall "hill" due to an atom? Assume that η is 0.50 nm, a typical value.

General Problems

51. || In a photoelectric-effect experiment, the wavelength of light shining on an aluminum cathode is decreased from 250 nm to 200 nm. What is the change in the stopping potential?

52. ||| In a photoelectric-effect experiment, the maximum kinetic energy of electrons is 2.8 eV. When the wavelength of the light is increased by 50%, the maximum energy decreases to 1.1 eV. What are (a) the work function of the cathode and (b) the initial wavelength?

53. || Light of constant intensity but varying wavelength was used to illuminate the cathode in a photoelectric-effect experiment. The graph of Figure P28.53 shows how the stopping potential depended on the frequency of the light. What is the work function, in eV, of the cathode?

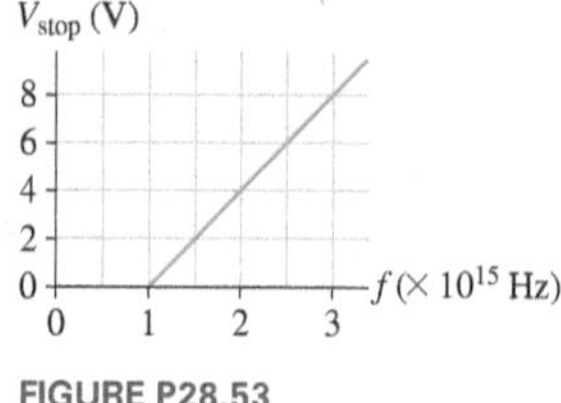

FIGURE P28.53

54. || A typical incandescent lightbulb emits approximately
BIO 3×10^{18} visible-light photons per second. Your eye, when it is fully dark adapted, can barely see the light from an incandescent lightbulb 10 km away. How many photons per second are incident at the image point on your retina? The diameter of a dark-adapted pupil is 6 mm.

55. | Aphids are small insects
BIO that adversely affect many plants that are important for agriculture. One type of aphid has two different colored morphs, green and brown. A team of investigators showed that brown aphids were less sensitive to damaging ultraviolet radiation than green aphids. Individual aphids, which can be modeled as 2.5 mm $\times$ 1.0 mm rectangles, were exposed to ultraviolet light with a 300 nm wavelength and an intensity of 120 W/m^2. If we assume that all photons are absorbed, how many photons did an aphid absorb during a 1.0 h exposure?

56. ||| A red LED (light emitting diode) is connected to a battery;
INT it carries a current. As electrons move through the diode, they jump between states, emitting photons in the process. Assume that each electron that travels through the diode causes the emission of a single 630 nm photon. What current is necessary to produce 5.0 mW of emitted light?

57. ||| In a laser range-finding experiment, a pulse of laser light is fired toward an array of reflecting mirrors left on the moon by Apollo astronauts. By measuring the time it takes for the pulse to travel to the moon, reflect off the mirrors, and return to earth, scientists can calculate the distance to the moon to within a few centimeters. A single mirror receives 0.38 W of power during a 100-ps-long pulse of 532-nm-wavelength laser light. How many photons strike the mirror?

58. || At 510 nm, the wavelength of maximum sensitivity of the
BIO human eye, the dark-adapted eye can sense a 100-ms-long flash of light of total energy 4.0×10^{-17} J. (Weaker flashes of light may be detected, but not reliably.) If 60% of the incident light is lost to reflection and absorption by tissues of the eye, how many photons reach the retina from this flash?

59. || *Dinoflagellates* are single-
BIO cell creatures that float in the world's oceans; many types are bioluminescent. When disturbed by motion in the water, a typical bioluminescent dinoflagellate emits 10^8 photons in a 0.10-s-long flash of light of wavelength 460 nm. What is the power of the flash in watts?

60. ||| Exposure to a sufficient quantity of ultraviolet light will red-
BIO den the skin, producing *erythema*—a sunburn. The amount of exposure necessary to produce this reddening depends on the wavelength. For a 1.0 cm^2 patch of skin, 3.7 mJ of ultraviolet light at a wavelength of 250 nm will produce reddening; at a 300 nm wavelength, 13 mJ are required.
    a. What is the photon energy corresponding to each of these wavelengths?
    b. How many photons does each of these exposures correspond to?
    c. Explain why there is a difference in the number of photons needed to provoke a response in the two cases.

61. || Suppose you need to image the structure of a virus with a
BIO diameter of 50 nm. For a sharp image, the wavelength of the probing wave must be 5.0 nm or less. We have seen that, for imaging such small objects, this short wavelength is obtained by using an electron beam in an electron microscope. Why don't we simply use short-wavelength electromagnetic waves? There's a problem with this approach: As the wavelength gets shorter, the energy of a photon of light gets greater and could damage or destroy the object being studied. Let's compare the energy of a photon and an electron that can provide the same resolution.
   a. For light of wavelength 5.0 nm, what is the energy (in eV) of a single photon? In what part of the electromagnetic spectrum is this?
   b. For an electron with a de Broglie wavelength of 5.0 nm, what is the kinetic energy (in eV)?
62. ||| Some scientists have speculated that the selectivity of ion
BIO channels in cell membranes could be due to quantum interfer-
INT ence, similar to the double-slit interference of electrons seen in Figure 28.9. Whether or not quantum effects could arise depends on the de Broglie wavelength of the ions.
   a. A $^{39}K^+$ ion passes through a 4-nm-long potassium ion channel in about 40 ps. To one significant figure, what is the ion's de Broglie wavelength?
   b. The ion channel's diameter is ≈ 0.3 nm. Suppose that ions undergo circular-aperture diffraction while passing through the channel. What is the width of the central maximum at a distance of 10 nm?
   c. The mean free path of ions in the cell is less than 1 nm. Does it seem likely that any quantum-interference effects could persist for a distance of 10 nm, roughly the thickness of the cell membrane?
63. || Electrons, all with the same speed, pass through a tiny
INT 15-nm-wide slit and create a diffraction pattern on a detector 50 mm behind the slit. What is the electron's kinetic energy, in eV, if the central maximum has a width of 3.3 mm?
64. |||| An experiment was per-
INT formed in which neutrons were shot through two slits spaced 0.10 mm apart and detected 3.5 m behind the slits. Figure P28.64 shows the detector output. Notice the 100 μm scale on the figure. To one significant figure, what was the speed of the neutrons?

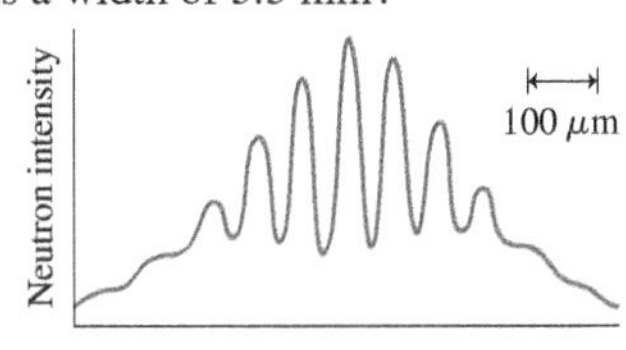

FIGURE P28.64

65. |||| The electron interference pattern of Figure 28.9 was made
INT by shooting electrons with 50 keV of kinetic energy through two slits spaced 1.0 μm apart. The fringes were recorded on a detector 1.0 m behind the slits.
   a. What was the speed of the electrons? (The speed is large enough to justify using relativity, but for simplicity do this as a nonrelativistic calculation.)
   b. Figure 28.9 is greatly magnified. What was the actual spacing on the detector between adjacent bright fringes?
66. ||| What is the length of a box in which the difference between an electron's first and second allowed energies is 1.0×10^{-19} J?
67. || Two adjacent allowed energies of an electron in a one-dimensional box are 2.0 eV and 4.5 eV. What is the length of the box?
68. ||| An electron confined to a one-dimensional box has an energy of 1.28 eV. Another electron in the same box has an energy of 2.88 eV. What is the smallest possible length of the box?
69. |||| An electron confined in a one-dimensional box emits a 200 nm photon in a quantum jump from $n = 2$ to $n = 1$. What is the length of the box?
70. || A proton confined in a one-dimensional box emits a 2.0 MeV gamma-ray photon in a quantum jump from $n = 2$ to $n = 1$. What is the length of the box?
71. || As an electron in a one-dimensional box of length 0.600 nm jumps between two energy levels, a photon of energy 8.36 eV is emitted. What are the quantum numbers of the two levels?

MCAT-Style Passage Problems

Compton Scattering

Further support for the photon model of electromagnetic waves comes from *Compton scattering,* in which x rays scatter from electrons, changing direction and frequency in the process. Classical electromagnetic wave theory cannot explain the change in frequency of the x rays on scattering, but the photon model can.

Suppose an x-ray photon is moving to the right. It has a collision with a slow-moving electron, as in Figure P28.72. The photon transfers energy and momentum to the electron, which recoils at a high speed. The x-ray photon loses energy, and the photon energy formula $E = hf$ tells us that its frequency must decrease. The collision looks very much like the collision between two particles.

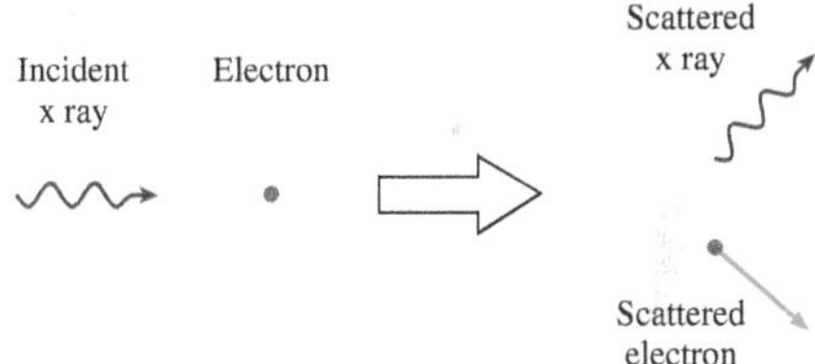

FIGURE P28.72

72. When the x-ray photon scatters from the electron,
   A. Its speed increases.
   B. Its speed decreases.
   C. Its speed stays the same.
73. When the x-ray photon scatters from the electron,
   A. Its wavelength increases.
   B. Its wavelength decreases.
   C. Its wavelength stays the same.
74. When the electron is struck by the x-ray photon,
   A. Its de Broglie wavelength increases.
   B. Its de Broglie wavelength decreases.
   C. Its de Broglie wavelength stays the same.
75. X-ray diffraction can also change the direction of a beam of x rays. Which statement offers the best comparison between Compton scattering and x-ray diffraction?
   A. X-ray diffraction changes the wavelength of x rays; Compton scattering does not.
   B. Compton scattering changes the speed of x rays; x-ray diffraction does not.
   C. X-ray diffraction relies on the particle nature of the x rays; Compton scattering relies on the wave nature.
   D. X-ray diffraction relies on the wave nature of the x rays; Compton scattering relies on the particle nature.

Collisional Excitation

Electrically excited mercury atoms have particularly strong emission of 4.9 eV photons, corresponding to a transition of one of the atom's electrons from a higher energy state to a lower. Mercury vapor absorbs light of this wavelength as well; the energy of the photon moves an electron from the lower state to the higher.

It's also possible to produce an excitation through other means. If an electron with a kinetic energy of 4.9 eV strikes a mercury atom, it can transfer energy, moving an electron in the mercury atom from the lower level to the higher. The electron loses kinetic energy in the process; this is an inelastic collision. Electrons with kinetic energies lower than this transition energy undergo elastic collisions, leaving their kinetic energy unchanged.

This was the idea behind the Franck-Hertz experiment, a classic experiment of early-20th-century physics. The basic setup is illustrated in Figure P28.76a. A tube is filled with mercury vapor. A heated electrode emits slow-moving electrons, and a variable power supply provides a voltage to accelerate the electrons toward a second electrode.

The current varies as the voltage is changed, as shown in Figure P28.76b. As the voltage is increased, this initially leads to an increased current. At a certain point, the electrons have enough energy to excite the mercury atoms, and the collisions lead to a loss in the energy of the moving electrons and a reduction in current, a clear demonstration of the existence of quantized energy levels in the mercury atom.

In modern versions of this experiment performed in student laboratories, the onset of inelastic collisions that excite the mercury atoms leads to a visible glow in the tube. This is illustrated in Figure P28.76a. The position of the glowing gas shows the location in the tube where the accelerating electrons reach the proper energy.

FIGURE P28.76 The setup of the Franck-Hertz experiment and typical data.

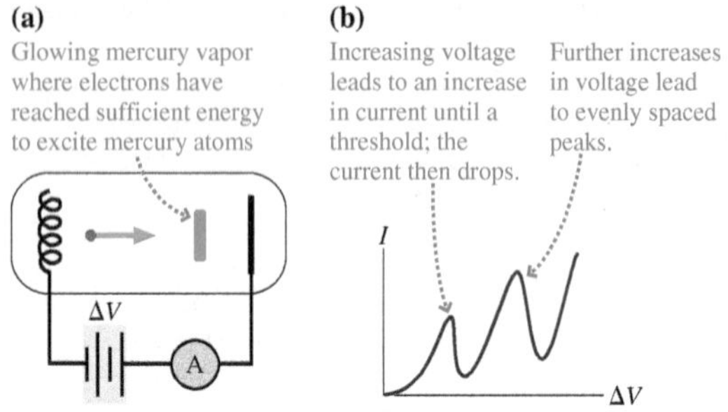

(a)

Glowing mercury vapor where electrons have reached sufficient energy to excite mercury atoms

ΔV

A

(b)

Increasing voltage leads to an increase in current until a threshold; the current then drops.

Further increases in voltage lead to evenly spaced peaks.

I

ΔV

76. If the temperature of the heated electrode is increased, more electrons are emitted. What change does this produce in the graph in Figure P28.76b?
    A. The peaks get taller.
    B. The peaks get shorter.
    C. The peaks occur at higher voltages.
    D. The peaks occur at lower voltages.
77. What is the approximate wavelength of the 4.9 eV photons?
    A. 250 nm
    B. 300 nm
    C. 350 nm
    D. 400 nm
    E. 450 nm
78. Approximately how fast must an electron move to excite a mercury atom in a collision?
    A. 9.3×10^5 m/s
    B. 1.3×10^6 m/s
    C. 1.9×10^6 m/s
    D. 2.6×10^6 m/s
79. If the electrons that lose energy to a mercury atom continue to accelerate, they can acquire enough kinetic energy to excite another atom, leading to a drop in current. At what voltage does the drop in current begin?
    A. 4.9 V
    B. 7.4 V
    C. 9.8 V
    D. 14.7 V

29 Atoms and Molecules

The bright colors of fireworks are due to the photons emitted by excited atoms. Red is from strontium, green is from barium.

LOOKING AHEAD

Spectroscopy

This drone is looking for methane leaks by using an infrared laser to detect specific wavelengths in methane's *absorption spectrum.*

You'll learn how absorption and emission spectra provide clues about the quantum nature of atoms and molecules.

Atomic and Molecular Properties

The sweat and oil in fingerprints are molecules that *fluoresce* with visible light when illuminated with ultraviolet light.

You'll discover that the many ways in which materials absorb and emit light are a consequence of their quantized energy levels.

Lasers

Eye surgery is just one of many applications of lasers in medicine. Lasers slice through tissue, remove stains and lesions, and treat cancer.

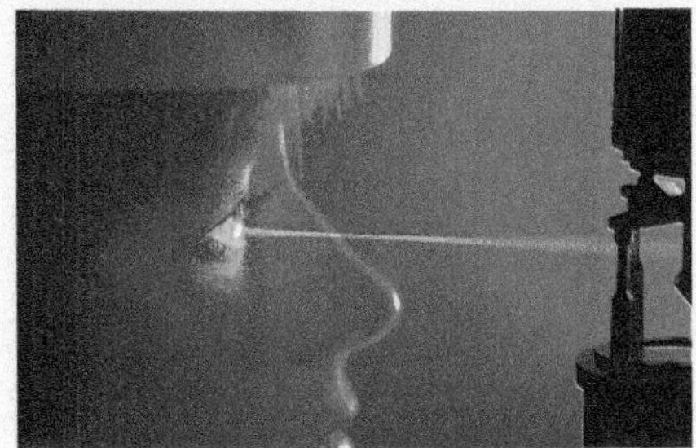

You'll learn that *stimulated emission* allows a suitably prepared material to generate a chain reaction of photon emission.

GOAL To use quantum physics to understand the structure and properties of atoms and molecules.

PHYSICS AND LIFE

Quantum Physics Helps Explain How Biomolecules Respond to Light

How do fireflies glow? What gives dyes and pigments their color? How does light trigger the retinal response of vision and the cellular response of photosynthesis? Why does infrared light warm us but ultraviolet light causes sunburn? All of these phenomena involve the interaction of light and matter, and all are consequences of quantum physics because they depend on the facts that light consists of photons and that a molecule has discrete energy levels. Bioluminescence and fluorescence occur when the arrangement of energy levels allows an excited molecule to dissipate some energy as heat before emitting photons. And the interplay between photon energies and the structure of molecular energy levels explains why infrared, visible, ultraviolet, and x-ray radiation have such profoundly different effects on living tissue.

The summertime flashes of fireflies are an example of *bioluminescence,* the production of light in chemical reactions.

29.1 Spectroscopy

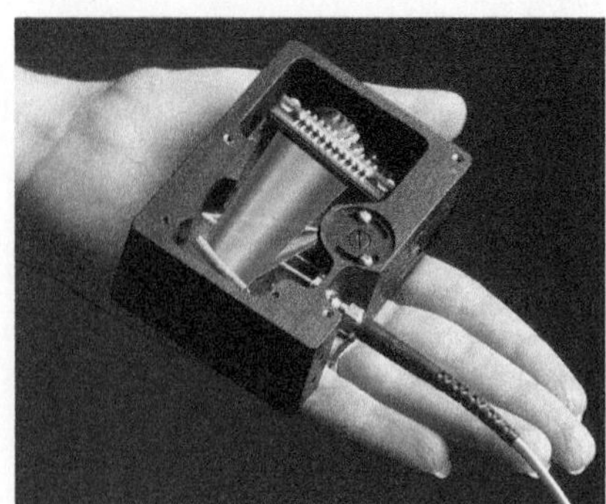

Some modern spectrometers are small enough to hold in your hand. (The rainbow has been added to show the paths that different colors take.)

Chapter 28 introduced a few key ideas from quantum physics, such as matter waves, energy levels, quantum jumps, and photons. These ideas are crucial for understanding atoms and molecules, the topic of this chapter.

Much of our knowledge about atoms and molecules comes from *spectroscopy,* the science of measuring the wavelengths of emitted and absorbed light. The primary tool of spectroscopy is a **spectrometer,** such as the one shown in FIGURE 29.1. The heart of a spectrometer is a diffraction grating that causes different wavelengths of light to diffract at different angles.

Each wavelength is focused to a different position on a detector that is similar to the one in a digital camera. The distinctive pattern of wavelengths emitted by a source of light and recorded on the detector is called the **spectrum** of the light.

As you learned in ◀◀ SECTION 27.8, a hot, self-luminous object, such as the sun or the filament of an incandescent lightbulb, emits a smooth, continuous spectrum in which a rainbow is formed by light being emitted at every possible wavelength. As you learned, this spectrum depends on only the object's temperature and thus contains no information about the atoms that the object is made of.

FIGURE 29.1 A diffraction spectrometer.

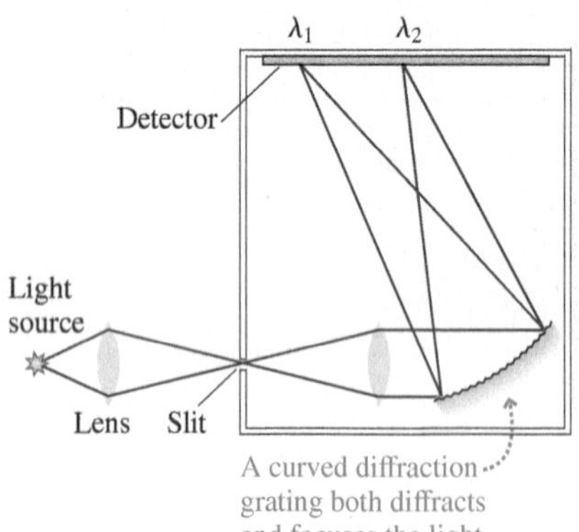

To learn about the light emitted and absorbed by individual atoms, we need to investigate them in the form of a low-pressure gas, where the atoms are far apart and isolated from one another. If a high voltage is applied to two electrodes sealed in a glass tube filled with a low-pressure gas, the gas begins to glow. In contrast to a hot solid object like a lightbulb filament, the light emitted by such a *gas discharge tube* contains only certain discrete, individual wavelengths. Such a spectrum is called an **emission spectrum.**

FIGURE 29.2 shows examples of discrete spectra as they appear on the detector of a spectrometer. Each bright line in a discrete spectrum, called a **spectral line,** represents *one* specific wavelength present in the light emitted by the source. The familiar neon sign is actually a gas discharge tube that contains neon. Such a sign has a reddish-orange color because, as Figure 29.2 shows, nearly all of the wavelengths emitted by neon atoms fall within the wavelength range 600–700 nm that we perceive as orange and red.

FIGURE 29.2 Examples of emission spectra in the visible wavelength range 400–700 nm.

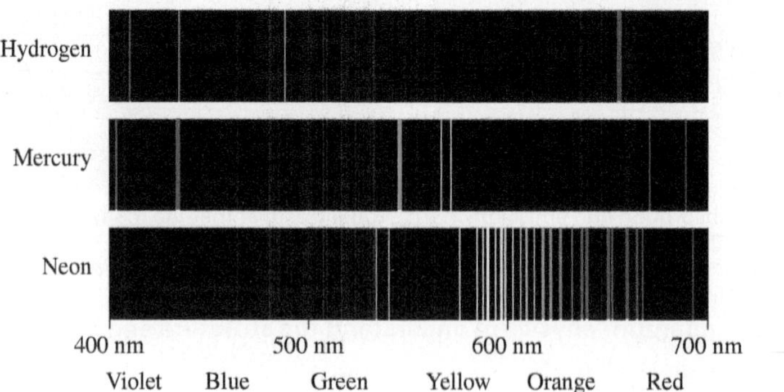

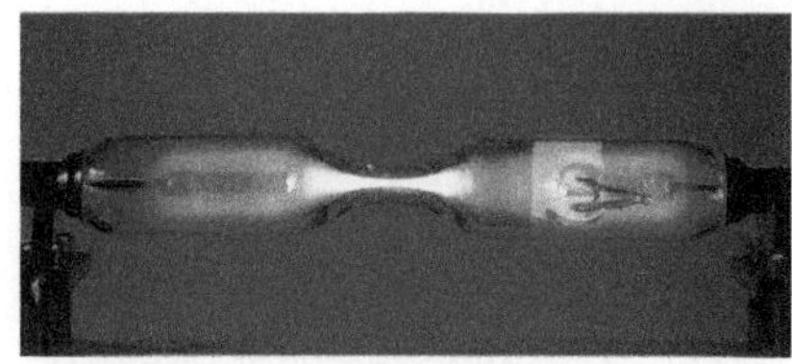

It's a gas A gas discharge tube of the type used to measure atomic spectra. This is the glow of helium.

Figure 29.2 shows that the emission spectra of hydrogen, mercury, and neon look very different from one another. In fact, every element in the periodic table has its own, unique spectrum. The fact that each element emits a unique spectrum means that atomic spectra can be used as "fingerprints" to identify elements. Consequently, atomic spectroscopy is the basis of many contemporary technologies for analyzing the composition of unknown materials, monitoring air pollutants, and studying the atmospheres of the earth and other planets.

Gases can also *absorb* light. To measure an **absorption spectrum,** as shown in FIGURE 29.3a, a white-light source emits a continuous spectrum that, in the absence of a gas, uniformly illuminates the detector. When a sample of gas is placed in the light's path, any wavelengths absorbed by the gas are missing and the detector is dark at that wavelength.

Gases not only emit discrete wavelengths, but also absorb discrete wavelengths. But there is an important difference between the emission spectrum and the absorption spectrum of a gas: **Every wavelength that is absorbed by the gas is also emitted, but *not* every emitted wavelength is absorbed.** The wavelengths in the absorption spectrum appear as a *subset* of the wavelengths in the emission spectrum. As an example, FIGURE 29.3b shows both the absorption and the emission spectra of sodium atoms. All of the absorption wavelengths are prominent in the emission spectrum, but there are many emission lines for which no absorption occurs.

What causes atoms to emit or absorb light? Why a discrete spectrum? Why are some wavelengths emitted but not absorbed? Why is the spectrum emitted by each element different? These questions arose from experiments and explorations in the late 19th and early 20th centuries, but the physics of the time was incapable of providing answers.

In the search for explanations, the first step was to look for patterns in the data. While the spectra of other atoms have dozens or even hundreds of wavelengths, the visible spectrum of hydrogen, between 400 nm and 700 nm, consists of a mere four spectral lines (see Figure 29.2 and TABLE 29.1). If any spectrum could be understood, it should be that of the first element in the periodic table. The breakthrough came in 1885, not by an established and recognized scientist but by a Swiss school teacher, Johann Balmer. Balmer showed that the wavelengths in the hydrogen spectrum could be represented by the simple formula

$$\lambda = \frac{91.1\text{ nm}}{\left(\frac{1}{2^2} - \frac{1}{n^2}\right)} \qquad n = 3, 4, 5, \ldots \tag{29.1}$$

Later experimental evidence, as ultraviolet and infrared spectroscopy developed, showed that Balmer's result could be generalized to

$$\lambda = \frac{91.1\text{ nm}}{\left(\frac{1}{m^2} - \frac{1}{n^2}\right)} \tag{29.2}$$

m can be 1, 2, 3, . . .
n can be any integer greater than m.

Other than at the very highest levels of resolution, Equation 29.2 accurately describes *every* wavelength in the emission spectrum of hydrogen.

Equation 29.2 is what we call *empirical knowledge.* It is an accurate mathematical representation found through experimental evidence, but it does not rest on any physical principles or physical laws. Yet the formula was so simple that it must, everyone agreed, have a simple explanation. It would take 30 years to find it.

STOP TO THINK 29.1 The black lines show the emission or absorption lines observed in two spectra of the same element. Which one is an emission spectrum and which is an absorption spectrum?

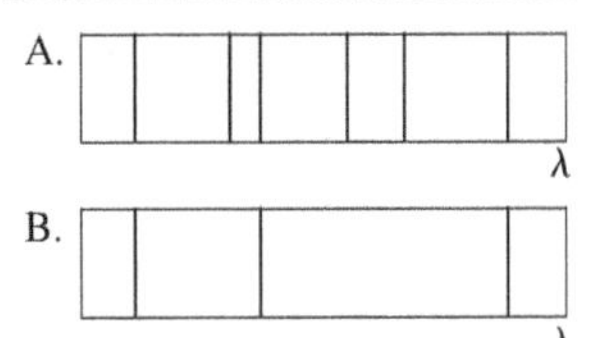

FIGURE 29.3 Measuring an absorption spectrum.

(a) Measuring an absorption spectrum

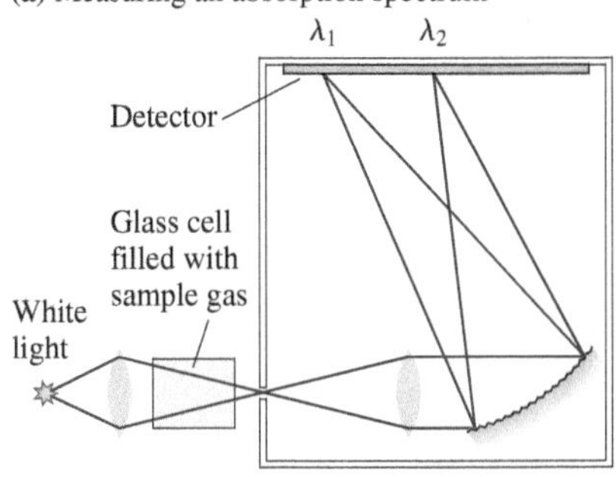

(b) Absorption and emission spectra of sodium

Absorption

Emission

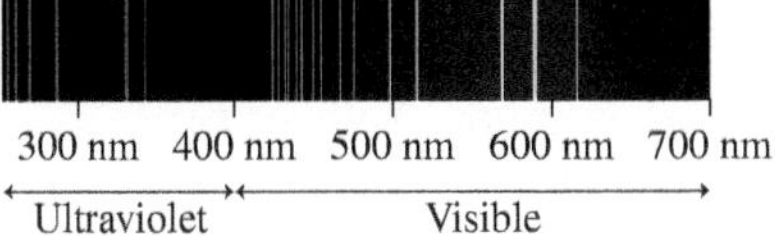

TABLE 29.1 Wavelengths of visible lines in the hydrogen spectrum

656 nm
486 nm
434 nm
410 nm

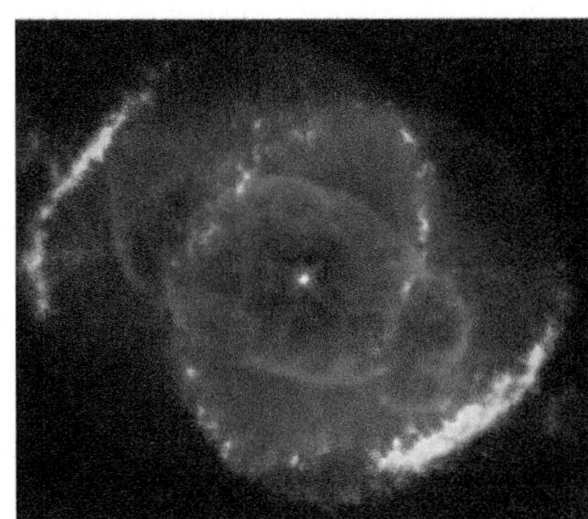

Astronomical colors The red color of this nebula, the Cat's Eye Nebula, is due to the emission of light from hydrogen atoms. The atoms are excited by intense ultraviolet light from the star in the center. They then emit red light, with $\lambda = 656$ nm, as predicted by Equation 29.2 with $m = 2$ and $n = 3$.

29.2 Atoms

The 1897 discovery of the electron by J. J. Thomson had two important implications. First, atoms are not indivisible; they are built of smaller pieces. The electron was the first *subatomic* particle to be discovered. And second, the constituents of the atom are *charged particles*. Hence it seems plausible that the atom must be held together by electric forces.

Almost simultaneously with Thomson's discovery of the electron, the French physicist Henri Becquerel announced his discovery that some new form of "rays" were emitted by crystals of uranium. One of Thomson's former students, Ernest Rutherford, began a study of these new rays and discovered that a uranium crystal actually emits two *different* rays. Beta rays were eventually found to be high-speed electrons emitted by the uranium. Alpha rays (or alpha particles, as we now call them) consist of helium nuclei, with charge $q = +2e$ and mass $m = 6.64 \times 10^{-27}$ kg, emitted at high speed from the sample.

EXAMPLE 29.1 Finding the speed of an alpha particle

The very high speeds of alpha particles make them suitable for experiments that probe the nature of matter. A nucleus ejects an alpha particle with a kinetic energy of 8.3 MeV, a typical energy. How fast is the alpha particle moving?

PREPARE We need to convert the kinetic energy from eV to J:

$$K = 8.3 \times 10^6 \text{ eV} \times \frac{1.60 \times 10^{-19} \text{ J}}{1.00 \text{ eV}} = 1.3 \times 10^{-12} \text{ J}$$

SOLVE Now, using the alpha-particle mass $m = 6.64 \times 10^{-27}$ kg given previously, we can find the speed:

$$K = \frac{1}{2}mv^2 = 1.3 \times 10^{-12} \text{ J}$$

$$v = \sqrt{\frac{2K}{m}} = 2.0 \times 10^7 \text{ m/s}$$

ASSESS This is quite fast, about 7% of the speed of light. We were told to expect a high speed, so this result makes sense.

FIGURE 29.4 Rutherford's experiment to shoot high-speed alpha particles through a thin gold foil.

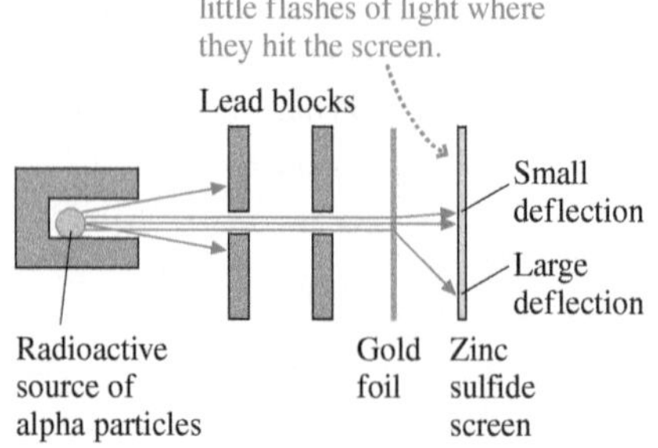

Rutherford soon realized that he could use these high-speed alpha particles as projectiles to probe inside other atoms. In 1909, Rutherford and his students set up the experiment shown in FIGURE 29.4 to shoot alpha particles through very thin metal foils. The alpha particle is charged, and it experiences electric forces from the positive and negative charges of the atoms as it passes through the foil. The amount by which alpha particles are deflected as they pass through the foil provides information about how mass and charge are distributed within the foil's atoms.

It was initially observed that most alpha particles undergo only small deflections. Rutherford then suggested that his students set up the apparatus to see if any alpha particles were deflected at *large* angles. It took only a few days to find the answer. Not only were alpha particles deflected at large angles, but a very few were reflected almost straight backward toward the source!

How can we understand this result? Most alpha particles pass through the outer portions of atoms where forces are weak, so only a small deflection is to be expected. But if an atom has a small positive core, such as the one in FIGURE 29.5, a few of the alpha particles can come very close to the core. Because the electric force varies with the inverse square of the distance, the very large force of this very close approach can cause a large-angle scattering or even a backward deflection of the alpha particle.

FIGURE 29.5 Large-angle scattering of alpha particles suggests that atoms have a small, concentrated positive core.

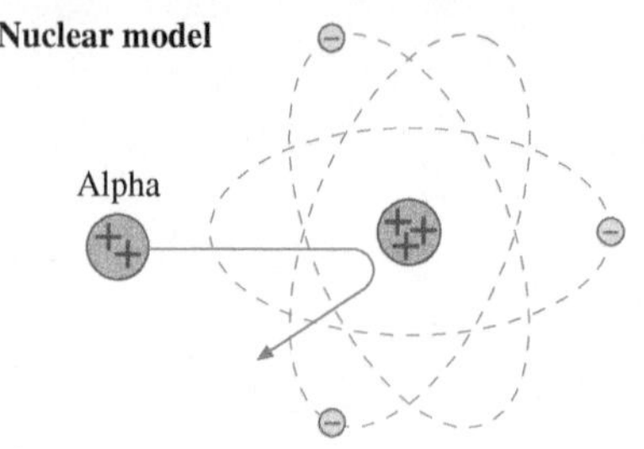

If the atom has a concentrated positive nucleus, some alpha particles will be able to come very close to the nucleus and thus feel a very strong repulsive force.

The discovery of large-angle scattering of alpha particles led Rutherford to envision an atom in which negative electrons orbit a small, massive, positive **nucleus,** rather like a miniature solar system. This is the **nuclear model of the atom.** Further experiments showed that the diameter of the atomic nucleus is $\approx 1 \times 10^{-14}$ m = 10 fm (1 fm = 1 femtometer = 10^{-15} m), a mere 0.01% the diameter of the atom itself. Thus nearly all of the atom is empty space—the void!

EXAMPLE 29.2 **Going for the gold**

An 8.3 MeV alpha particle is shot directly toward the nucleus of a gold atom (atomic number 79). What is the distance of closest approach of the alpha particle to the nucleus?

PREPARE We can solve this problem using conservation of energy; it is similar to problems we solved in Chapter 22. The positively charged alpha particle slows as it approaches the positive nucleus, and it is instantaneously at rest at the point of closest approach. We can ignore the gold atom's electrons. They are spread out over a distance that is very large compared to the size of the nucleus, and their mass is much smaller than that of the alpha particle. The particle can easily push the electrons aside with little change in speed. The gold nucleus is much more massive than the alpha particle, so we will assume that it does not move. Recall that the electric potential near a sphere of charge is the same as near a point charge. FIGURE 29.6 is a pictorial representation. We will assume that the initial distance is so large that the initial potential energy is zero.

FIGURE 29.6 A before-and-after pictorial representation of an alpha particle colliding with a nucleus.

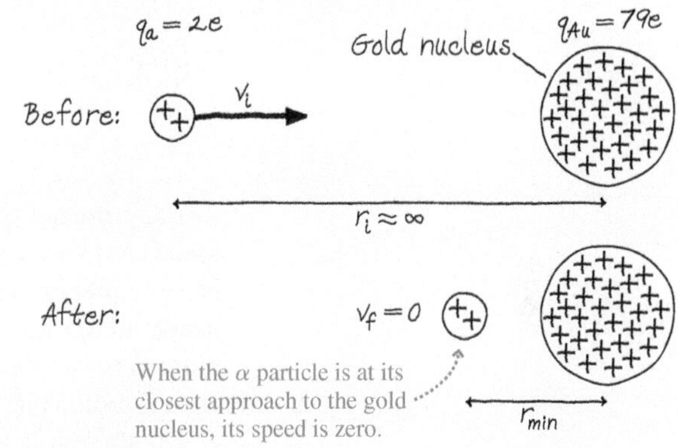

SOLVE Recall, from Chapter 22, that the electric potential near a sphere of charge is

$$V = \frac{1}{4\pi\epsilon_0}\frac{Q}{r}$$

where Q is the charge of the sphere, and r is the distance from the center. The sphere is the gold nucleus, so $Q = q_{Au}$. The electric potential energy at r_{min}, the distance of closest approach, is

$$U_{elec} = q_\alpha V = \frac{1}{4\pi\epsilon_0}\frac{q_\alpha q_{Au}}{r_{min}}$$

where $q_\alpha = 2e$ is the alpha particle's charge. The conservation of energy statement $K_f + U_f = K_i + U_i$ is

$$0 + \frac{1}{4\pi\epsilon_0}\frac{q_\alpha q_{Au}}{r_{min}} = \frac{1}{2}mv_i^2 + 0$$

The solution for r_{min} is

$$r_{min} = \frac{1}{4\pi\epsilon_0}\frac{2q_\alpha q_{Au}}{mv_i^2}$$

The mass of the alpha particle is $m = 6.64 \times 10^{-27}$ kg and its charge is $q_\alpha = 2e = 3.20 \times 10^{-19}$ C. Gold has atomic number 79, so $q_{Au} = 79e = 1.26 \times 10^{-17}$ C. In Example 29.1 we found that an 8.3 MeV alpha particle has speed $v = 2.0 \times 10^7$ m/s. With this information, we can calculate

$$r_{min} = 2.7 \times 10^{-14}\text{ m}$$

ASSESS This distance is much less than the size of the atom, on the scale of the size of the nucleus. The alpha particle will get close enough to the nucleus to experience a very large force, as we expected.

Using the Nuclear Model

The nuclear model of the atom makes it easy to picture atoms and understand such processes as ionization. For example, the **atomic number** of an element, its position in the periodic table of the elements, is the number of orbiting electrons (of a neutral atom) and the number of units of positive charge in the nucleus. The atomic number is represented by Z. Hydrogen, with $Z = 1$, has one electron orbiting a nucleus with charge $+1e$. Helium, with $Z = 2$, has two orbiting electrons and a nucleus with charge $+2e$. Because the orbiting electrons are very light, an x-ray photon or a rapidly moving particle, such as another electron, can knock one of the electrons away, creating a positive *ion*. Removing one electron makes a singly charged ion, with $q_{ion} = +e$. Removing two electrons creates a doubly charged ion, with $q_{ion} = +2e$. This is shown for lithium $(Z = 3)$ in FIGURE 29.7.

FIGURE 29.7 Different ionization stages of the lithium atom $(Z = 3)$.

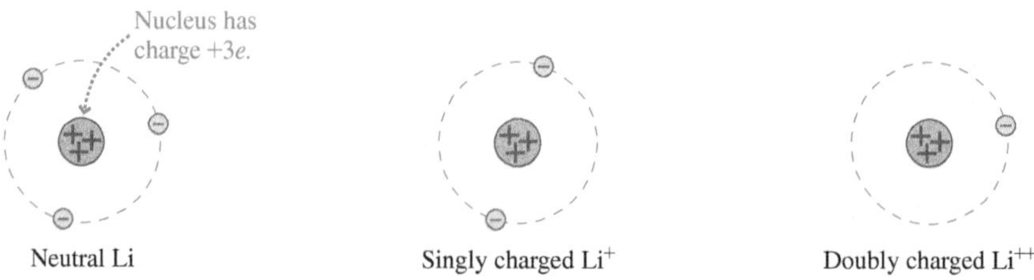

Experiments soon led to the recognition that the positive charge of the nucleus is associated with a positive subatomic particle called the **proton.** The proton's charge is $+e$, equal in magnitude but opposite in sign to the electron's charge. Further, with nearly all the atomic mass associated with the nucleus, the proton is about 1800 times more massive than the electron: $m_p = 1.67 \times 10^{-27}$ kg. Atoms with atomic number Z have Z protons in the nucleus, giving the nucleus charge $+Ze$.

But there was a problem. Helium, with atomic number 2, has twice as many electrons and protons as hydrogen. Lithium, $Z = 3$, has three electrons and protons. If a nucleus contains Z protons to balance the Z orbiting electrons, and if nearly all the atomic mass is contained in the nucleus, then helium should be twice as massive as hydrogen and lithium three times as massive. But it was known from chemistry measurements that helium is *four times* as massive as hydrogen and lithium is *seven times* as massive.

This difficulty was not resolved until the discovery, in 1932, of a third subatomic particle. This particle has essentially the same mass as a proton but *no* electric charge. It is called the **neutron.** Neutrons reside in the nucleus, with the protons, where they contribute to the mass of the atom but not to its charge. As you'll see in Chapter 30, neutrons help provide the "glue" that holds the nucleus together.

We now know that a nucleus contains Z protons plus N neutrons, as shown in FIGURE 29.8, giving the atom a **mass number** $A = Z + N$. The mass number, which is a dimensionless integer, is *not* the same thing as the atomic mass m. But because the proton and neutron masses are both ≈ 1 u, where

$$1 \text{ u} = 1 \text{ atomic mass unit} = 1.66 \times 10^{-27} \text{ kg}$$

the mass number A is *approximately* the mass in atomic mass units.

FIGURE 29.8 The nucleus of an atom contains protons and neutrons.

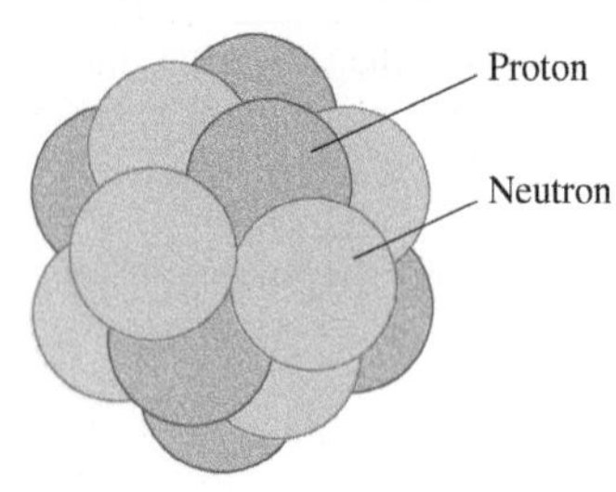

There is a *range* of neutron numbers that happily form a nucleus with Z protons, creating a series of nuclei with the same Z-value (i.e., they are all the same chemical element) but different masses. Such a series of nuclei are called **isotopes.** The notation used to label isotopes is ^{A}Z, where the mass number A is given as a *leading* superscript. For example, the most common isotope of neon has $Z = 10$ protons and $N = 10$ neutrons. Thus it has mass number $A = 20$ and is labeled ^{20}Ne. The neon isotope ^{22}Ne has $Z = 10$ protons (that's what makes it neon) and $N = 12$ neutrons. Helium has the two isotopes shown in FIGURE 29.9. ^{3}He is rare, but it can be isolated and has important uses in scientific research.

FIGURE 29.9 The two isotopes of helium.

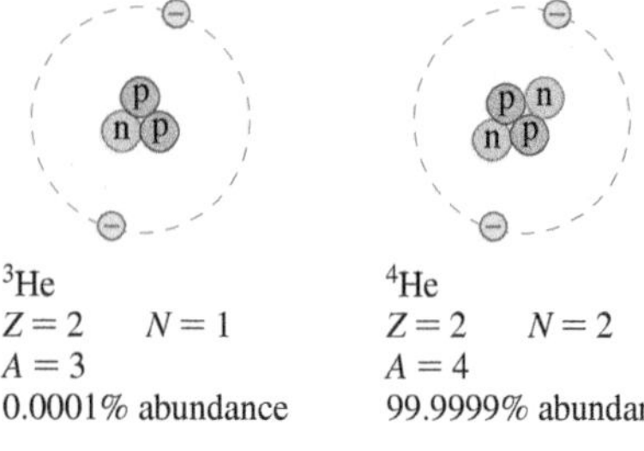

STOP TO THINK 29.2 Carbon is the sixth element in the periodic table. How many protons and how many neutrons are there in a nucleus of the isotope ^{14}C?

The Bohr Model of the Atom

Rutherford's nuclear model was an important step toward understanding atoms, but it had a serious shortcoming. According to Maxwell's theory of electricity and magnetism, the electrons orbiting the nucleus should act as small antennas and radiate electromagnetic waves. As FIGURE 29.10 shows, the energy carried off by the waves would cause the electrons to quickly—in less than a microsecond—spiral into the nucleus! In other words, classical Newtonian mechanics and electromagnetism predict that an atom with electrons orbiting a nucleus would be highly unstable and would immediately self-destruct. This clearly does not happen.

A missing piece of the puzzle, although not recognized as such for a few years, was Einstein's 1905 introduction of light quanta. If light comes in discrete packets of energy, which we now call photons, and if atoms emit and absorb light, what does that imply about the structure of the atoms? This was the question posed by the Danish physicist Niels Bohr.

Bohr wanted to understand how a solar-system–like atom could be stable and not radiate away all its energy. In 1913 he proposed a radically new model of the atom in which he added quantization to Rutherford's nuclear atom. The basic assumptions of the **Bohr model of the atom** are shown in FIGURE 29.11.

FIGURE 29.10 The fate of a Rutherford atom.

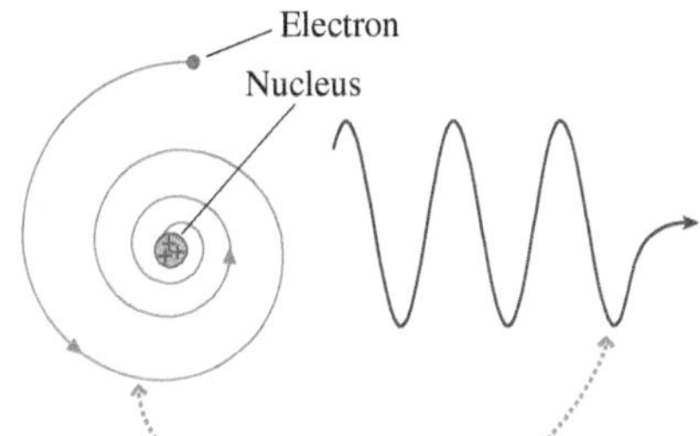

According to classical physics, an electron would spiral into the nucleus while radiating energy as an electromagnetic wave.

FIGURE 29.11 The Bohr model of the atom.

Electrons can exist in only certain allowed orbits.

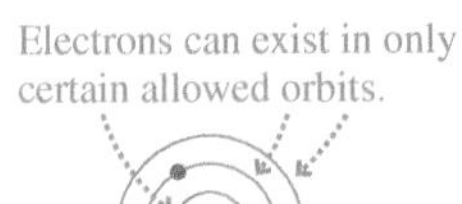

An electron cannot exist here, where there is no allowed orbit.

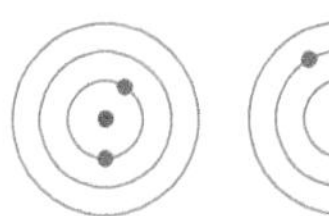

This is one stationary state. This is another stationary state.

The electrons in an atom can exist in only certain *allowed orbits.* A particular arrangement of electrons in these orbits is called a **stationary state.**

Energy-level diagram **Stationary states**

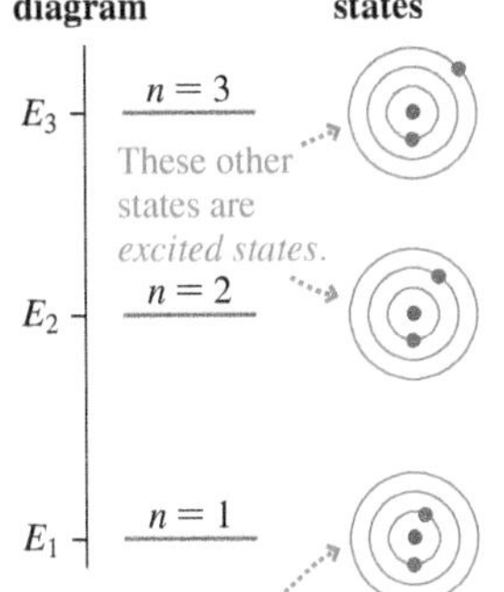

This state, with the lowest energy E_1, is the *ground state.* It is stable and can persist indefinitely.

Each stationary state has a discrete, well-defined energy E_n. That is, atomic energies are *quantized.* The stationary states are labeled by the *quantum number n* in order of increasing energy: $E_1 < E_2 < E_3 < \cdots$.

Photon emission

Excited-state electron

The electron jumps to a lower-energy stationary state and emits a photon.

Photon absorption

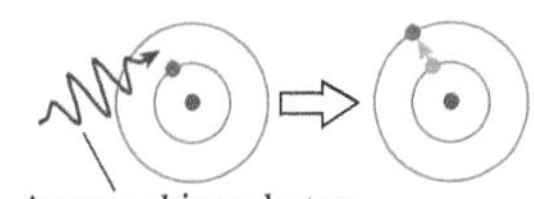

Approaching photon

The electron absorbs the photon and jumps to a higher-energy stationary state.

An atom can undergo a *transition* from one stationary state to another by emitting or absorbing a photon whose energy is exactly equal to the energy difference between the two stationary states.

Collisional excitation

Approaching particle — Particle loses energy.

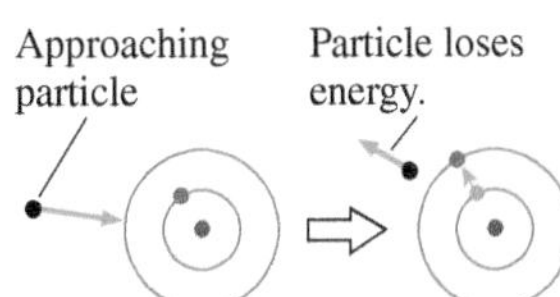

The particle transfers energy to the atom in the collision and excites the atom.

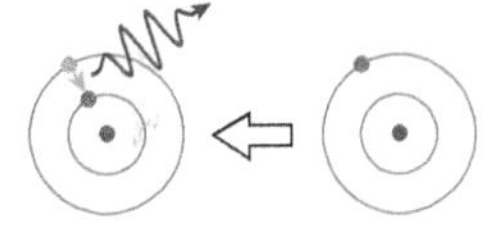

An electron in an excited state soon jumps to lower-energy states, emitting a photon at each jump.

Atoms can also move from a lower-energy state to a higher-energy state by absorbing energy in a collision with an electron or another atom in a process called **collisional excitation.** The excited electrons soon jump down to lower states, eventually ending in the stable ground state.

Einstein proposed the idea that the energy of light is quantized. The Bohr model goes further to propose that the energy of matter—of atoms—is also quantized. The stationary states correspond to the atom's allowed energies. The implications of Bohr's model are profound. In particular:

- **Matter is stable.** Once an atom is in its ground state, there are no states of any lower energy to which it can jump. It can remain in the ground state forever.
- **Atoms emit and absorb a *discrete spectrum.*** When an atom undergoes a transition from an initial state with energy E_i to a final state with energy E_f, conservation of energy requires that it emit or absorb a photon with energy $E_{photon} = \Delta E_{atom} = |E_f - E_i|$. Because $E_{photon} = hf$, this photon must have frequency $f_{photon} = \Delta E_{atom}/h$. Photons of other frequencies cannot be emitted or absorbed without violating energy conservation.

- **Emission spectra can be produced by collisions.** Energy from collisions can kick an atom up to an excited state. The atom then emits photons in a discrete emission spectrum as the electron jumps down to lower-energy states. This is how light is created in a gas discharge tube.
- **Absorption wavelengths are a subset of the wavelengths in the emission spectrum.** Recall that all the lines seen in an absorption spectrum are also seen in emission, but many emission lines are *not* seen in absorption. According to Bohr's model, **most atoms, most of the time, are in their lowest energy state,** the $n = 1$ ground state. Thus the absorption spectrum consists of *only* those transitions such as $1 \rightarrow 2$, $1 \rightarrow 3$, . . . in which the electron jumps from $n = 1$ to a higher value of n by absorbing a photon. Transitions such as $2 \rightarrow 3$ are not observed because there are essentially no atoms in $n = 2$ at any instant of time to do the absorbing. However, atoms that have been excited to the $n = 3$ state by collisions can emit photons corresponding to transitions $3 \rightarrow 1$ *and* $3 \rightarrow 2$. Thus the wavelength corresponding to $\Delta E_{\text{atom}} = E_3 - E_1$ is seen in both emission and absorption, but photons with $\Delta E_{\text{atom}} = E_3 - E_2$ occur in emission only.
- **Each element in the periodic table has a unique spectrum.** The energies of the stationary states are just the energies of the orbiting electrons. Different elements, with different numbers of electrons, have different stable orbits and thus different stationary states. States with different energies will emit and absorb photons of different wavelengths.

Much of this may be familiar to you from chemistry, but it is important to understand *why* atoms act as they do.

EXAMPLE 29.3 Wavelengths in emission and absorption spectra

An atom has stationary states $E_1 = 0.0$ eV, $E_2 = 2.0$ eV, and $E_3 = 5.0$ eV. What wavelengths are observed in the absorption spectrum and in the emission spectrum of this atom?

PREPARE Photons are emitted when an atom undergoes a transition from a higher energy level to a lower energy level. Photons are absorbed in a transition from a lower energy level to a higher energy level. However, most of the atoms are in the $n = 1$ ground state, so the only transitions in the absorption spectrum start from the $n = 1$ state. FIGURE 29.12 shows the energy-level diagram for the atom, with possible transitions noted.

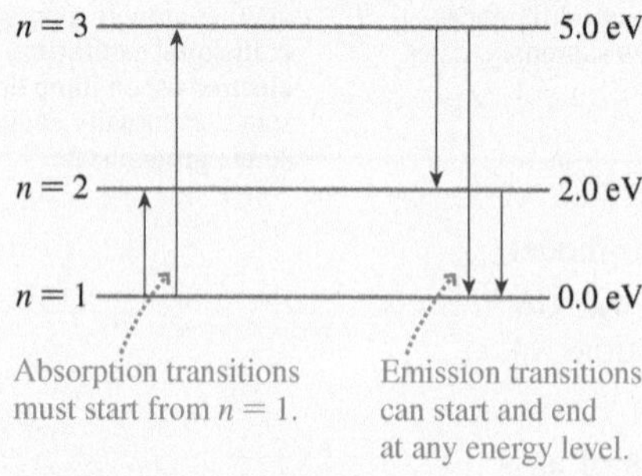

n = 3
5.0 eV
n = 2
2.0 eV
n = 1
0.0 eV
Absorption transitions must start from n = 1.
Emission transitions can start and end at any energy level.

FIGURE 29.12 The atom's energy-level diagram.

SOLVE This atom absorbs photons on the $1 \rightarrow 2$ and $1 \rightarrow 3$ transitions, with $\Delta E_{1\rightarrow 2} = 2.0$ eV and $\Delta E_{1\rightarrow 3} = 5.0$ eV. From $f_{\text{photon}} = \Delta E_{\text{atom}}/h$ and $\lambda = c/f$, we find that the wavelengths in the absorption spectrum are

$$1 \rightarrow 3 \qquad f_{\text{photon}} = 5.0\ \text{eV}/h = 1.2 \times 10^{15}\ \text{Hz}$$
$$\lambda = 250\ \text{nm (ultraviolet)}$$

$$1 \rightarrow 2 \qquad f_{\text{photon}} = 2.0\ \text{eV}/h = 4.8 \times 10^{14}\ \text{Hz}$$
$$\lambda = 620\ \text{nm (orange)}$$

The emission spectrum also has the 620 nm and 250 nm wavelengths due to the $2 \rightarrow 1$ and $3 \rightarrow 1$ transitions. In addition, the emission spectrum contains the $3 \rightarrow 2$ transition with $\Delta E_{3\rightarrow 2} = 3.0$ eV that is *not* seen in absorption because there are too few atoms in the $n = 2$ state to absorb. A similar calculation finds $f_{\text{photon}} = 7.3 \times 10^{14}$ Hz and $\lambda = c/f = 410$ nm. Thus the emission wavelengths are

$$2 \rightarrow 1 \qquad \lambda = 620\ \text{nm (orange)}$$
$$3 \rightarrow 2 \qquad \lambda = 410\ \text{nm (blue)}$$
$$3 \rightarrow 1 \qquad \lambda = 250\ \text{nm (ultraviolet)}$$

STOP TO THINK 29.3 A photon with a wavelength of 410 nm has energy $E_{\text{photon}} = 3.0$ eV. Do you expect to see a spectral line with $\lambda = 410$ nm in the emission spectrum of the atom represented by this energy-level diagram? If so, what transition or transitions will emit it? Do you expect to see a spectral line with $\lambda = 410$ nm in the absorption spectrum? If so, what transition or transitions will absorb it?

$n = 4$ — 6.0 eV
$n = 3$ — 5.0 eV
$n = 2$ — 2.0 eV
$n = 1$ — 0.0 eV

29.3 The Hydrogen Atom

Bohr's hypothesis was a bold new idea, yet there was still one enormous stumbling block: What *are* the stationary states of an atom? Everything in Bohr's model hinges on the existence of these stationary states, of there being only certain electron orbits that are allowed. But nothing in classical physics provides any basis for such orbits. And Bohr's model describes only the *consequences* of having stationary states, not how to find them.

In 1925, a dozen years after Bohr, Erwin Schrödinger developed a very general theory of *quantum* mechanics that can be used to calculate the allowed energy levels (i.e., the stationary states) of any system. We learned that a particle in a box sets up de Broglie one-dimensional standing waves characterized by quantum number n. Schrödinger's theory says, in effect, that the stationary states of an electron in an atom are three-dimensional standing waves characterized by *four* quantum numbers. Quantum mechanics uses advanced mathematics to calculate these stationary states; we will present results without the calculations and focus on the physical interpretation.

We'll look fairly closely at the hydrogen atom, the simplest atom with a single electron, because it is the model for all other elements. A stationary state in the quantum-mechanical hydrogen atom has four key properties, each associated with a quantum number:

- The atom's energy is associated with quantum number n.
- The electron's angular momentum is associated with quantum number l.
- The tilt of the electron's orbit is associated with quantum number m.
- The electron's spin is associated with quantum number m_s.

Let's delve into each of these properties in more depth.

1. The energy of a hydrogen atom is

$$E_n = -\frac{13.60 \text{ eV}}{n^2} \qquad n = 1, 2, 3, \ldots \tag{29.3}$$

The integer n is called the **principal quantum number.** The ground-state energy, with $n = 1$, is -13.60 eV.

It's important to understand why the energies are negative. Earlier we defined the electric potential energy of two point charges to be zero when they are infinitely far apart and no longer interacting. Similarly, a hydrogen atom has zero energy when the proton and electrons are infinitely far apart (no potential energy) and at rest (no kinetic energy). The stationary states are *bound states,* which means that we would have to do work to pull the electron and proton apart, so they have less energy than two separated particles. The absolute value $|E_n|$ is the binding energy of a hydrogen atom in a stationary state with principal quantum number n.

2. The angular momentum L of the electron's orbit, a quantity you learned about in ◀◀ SECTION 8.4, must be one of the values

$$L = \sqrt{l(l+1)}\,\hbar \qquad l = 0, 1, 2, 3, \ldots, n-1 \tag{29.4}$$

The integer l is called the **orbital quantum number.** The quantity $h/2\pi$ occurs so frequently in quantum physics that it is given the special symbol $\hbar$, pronounced "h bar." The units of h and $\hbar$—namely, J · s—are equivalent to the units of angular momentum, kg · m^2/s, so L is an angular momentum.

3. The plane of the electron's orbit can be tilted, but only at certain discrete angles. Each allowed angle is characterized by a quantum number m, which must be one of the values

$$m = -l, -l+1, \ldots, 0, \ldots, l-1, l \tag{29.5}$$

The integer m is called the **magnetic quantum number** because it becomes important when the atom is placed in a magnetic field.

4. The electron's *spin*—discussed later in this section—can point only up or down. These two orientations are described by the **spin quantum number** m_s, which must be one of the values

$$m_s = -\tfrac{1}{2} \text{ or } +\tfrac{1}{2} \tag{29.6}$$

In other words, each stationary state of the hydrogen atom is identified by a quartet of quantum numbers (n, l, m, m_s), and each quantum number is associated with a physical property of the atom.

NOTE ▶ For a hydrogen atom, the energy of a stationary state depends on only the principal quantum number n, not on l, m, or m_s. ◀

EXAMPLE 29.4 Listing quantum numbers

List all possible states of a hydrogen atom that have energy $E = -3.40$ eV.

SOLVE Energy depends on only the principal quantum number n. From Equation 29.3, states with $E = -3.40$ eV have

$$n = \sqrt{\frac{-13.60 \text{ eV}}{-3.40 \text{ eV}}} = 2$$

An atom with principal quantum number $n = 2$ can have either $l = 0$ or $l = 1$, but $l \geq 2$ is ruled out. If $l = 0$, the only possible value for the magnetic quantum number m is $m = 0$. If $l = 1$, then the atom can have $m = -1$, $m = 0$, or $m = +1$. For each of these, the spin quantum number can be $m_s = +\frac{1}{2}$ or $m_s = -\frac{1}{2}$. Thus the possible sets of quantum numbers are

n	l	m	m_s	n	l	m	m_s
2	0	0	$+\frac{1}{2}$	2	0	0	$-\frac{1}{2}$
2	1	1	$+\frac{1}{2}$	2	1	1	$-\frac{1}{2}$
2	1	0	$+\frac{1}{2}$	2	1	0	$-\frac{1}{2}$
2	1	-1	$+\frac{1}{2}$	2	1	-1	$-\frac{1}{2}$

These eight states all have the same energy.

Energy and Angular Momentum Are Quantized

TABLE 29.2 Symbols used to represent quantum number l

l	Symbol
0	s
1	p
2	d
3	f

The energy of the hydrogen atom depends on only the principal quantum number n. For other atoms, however, the allowed energies depend on both n and l. Consequently, it is useful to label the stationary states of an atom by their values of n and l. The lowercase letters shown in TABLE 29.2 are customarily used to represent the various values of quantum number l. Using these symbols, we call the ground state of the hydrogen atom, with $n = 1$ and $l = 0$, the $1s$ state; the $3d$ state has $n = 3$, $l = 2$.

FIGURE 29.13 is an energy-level diagram for the hydrogen atom in which the rows are labeled by n and the columns by l. It was noted in ◀◀ SECTION 28.6 that energy levels are like the rungs of a ladder; now we've given the ladder width as well as height. The left column contains all of the $l = 0$ (or s) states, the next column is the $l = 1$ (or p) states, and so on.

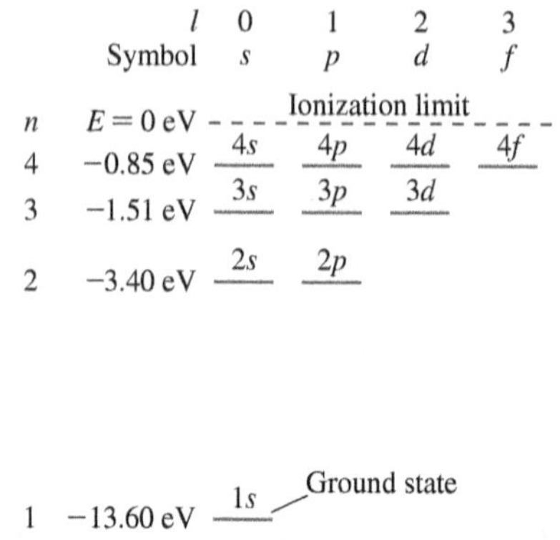

FIGURE 29.13 The energy-level diagram for the hydrogen atom.

As Equation 29.4 shows, the orbital quantum number l of an allowed state must be less than that state's principal quantum number n. For the ground state, with $n = 1$, only $l = 0$ is possible, so that the only $n = 1$ state is the $1s$ state. When $n = 2$, l can be 0 or 1, leading to both a $2s$ and a $2p$ state. For $n = 3$, there are $3s$, $3p$, and $3d$ states; and so on. Figure 29.5 shows only the first few energy levels for each value of l, but there really are an infinite number of levels, as $n \to \infty$, crowding together beneath $E = 0$.

The dashed line at $E = 0$ is called the *ionization limit*. It's the energy of a hydrogen atom in which the electron has been moved infinitely far away to form a H^+ ion. A ground-state hydrogen atom is bound by 13.60 eV, so we would need to provide *at least* 13.60 eV of energy to remove the electron and ionize the atom. Thus $|E_1| = 13.60$ eV is the *ionization energy* of hydrogen. Hydrogen can be ionized by absorbing electromagnetic radiation *if* the photon energy is $E_{photon} \geq 13.60$ eV, corresponding to extreme UV wavelengths less than 91 nm. In other words, electromagnetic radiation with $\lambda < 91$ nm is *ionizing radiation* for hydrogen. The same idea applies to all other elements and molecules. 10 eV is a more typical ionization energy, so electromagnetic radiation with $\lambda < 120$ nm is ionizing radiation for most atoms and molecules.

Classically, the angular momentum L of an orbiting electron can have any value. Not so in quantum mechanics. Equation 29.4 tells us that **the electron's orbital angular momentum is quantized.** The magnitude of the orbital angular momentum must be one of the discrete values

$$L = \sqrt{l(l+1)}\,\hbar = 0, \sqrt{2}\,\hbar, \sqrt{6}\,\hbar, \sqrt{12}\,\hbar, \ldots$$

where l is an integer.

A particularly interesting prediction is that the ground state of hydrogen, with $l = 0$, has *no* angular momentum. A classical particle cannot orbit unless it has angular momentum, but apparently a quantum particle does not have this requirement.

STOP TO THINK 29.4 What are the quantum numbers n and l for a hydrogen atom with $E = -(13.60/9)$ eV and $L = \sqrt{2}\,\hbar$?

The Electron Spin

You learned in ◀◀ SECTION 26.7 that an electron has an inherent magnetic dipole moment—it acts as a tiny bar magnet with a north and a south pole. In association with its magnetic moment, an electron also has an intrinsic *angular momentum* called the *electron spin*. In the early years of quantum mechanics, it was thought that the electron was a very tiny ball of negative charge spinning on its axis, which would give the electron both a magnetic dipole moment and spin angular momentum. However, a spinning ball of charge would violate the laws of relativity and other physical laws. As far as we know today, the electron is truly a point particle that happens to have an intrinsic magnetic dipole moment and angular momentum.

The two possible spin quantum numbers $m_s = \pm\frac{1}{2}$ mean that the electron's intrinsic magnetic dipole points in the $+z$-direction or the $-z$-direction. These two orientations are called *spin up* and *spin down*. It is convenient to picture a little vector that can be drawn $\uparrow$ for a spin-up state and $\downarrow$ for a spin-down state. We will use this notation in the next section.

The Hydrogen Spectrum

The most important experimental evidence that we have about the hydrogen atom is its spectrum, so the primary test of quantum mechanics is whether it correctly predicts the spectrum. FIGURE 29.14 is a simplifed energy-level diagram for the hydrogen atom. The lowest level is the ground state, with $E_1 = -13.60$ eV.

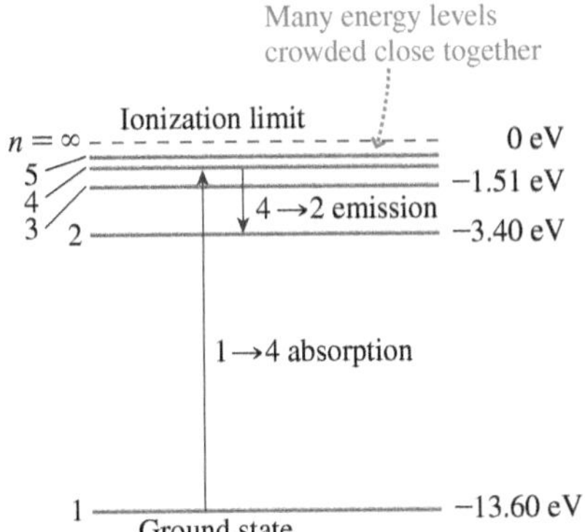

FIGURE 29.14 Photon absorption and emission in the hydrogen atom.

The figure shows a $1 \rightarrow 4$ transition in which a photon is absorbed and a $4 \rightarrow 2$ transition in which a photon is emitted. For two quantum states m and n, where $n > m$ and E_n is the higher-energy state, an atom can *emit* a photon in an $n \rightarrow m$ transition or *absorb* a photon in an $m \rightarrow n$ transition.

According to Bohr's model of atomic quantization, the frequency of the photon emitted in an $n \rightarrow m$ transition is

$$f_{\text{photon}} = \frac{\Delta E_{\text{atom}}}{h} = \frac{E_n - E_m}{h} \tag{29.7}$$

We can use Equation 29.3 for the energies E_n and E_m to predict that the emitted photon has frequency

$$f_{\text{photon}} = \frac{1}{h}\left(\frac{-13.60\text{ eV}}{n^2} - \frac{-13.60\text{ eV}}{m^2}\right) = \frac{13.60\text{ eV}}{h}\left(\frac{1}{m^2} - \frac{1}{n^2}\right)$$

The frequency is a positive number because $m < n$ and thus $1/m^2 > 1/n^2$.

We are more interested in wavelength than frequency because wavelengths are the quantity measured by experiment. The wavelength of the photon emitted in an $n \rightarrow m$ quantum jump is

$$\lambda_{n \rightarrow m} = \frac{c}{f_{\text{photon}}} = \frac{\lambda_0}{\left(\dfrac{1}{m^2} - \dfrac{1}{n^2}\right)} \qquad \begin{array}{l} m = 1, 2, 3, \ldots \\ n = m + 1, m + 2, \ldots \end{array} \tag{29.8}$$

FIGURE 29.15 Transitions producing the Lyman series and the Balmer series of lines in the hydrogen spectrum.

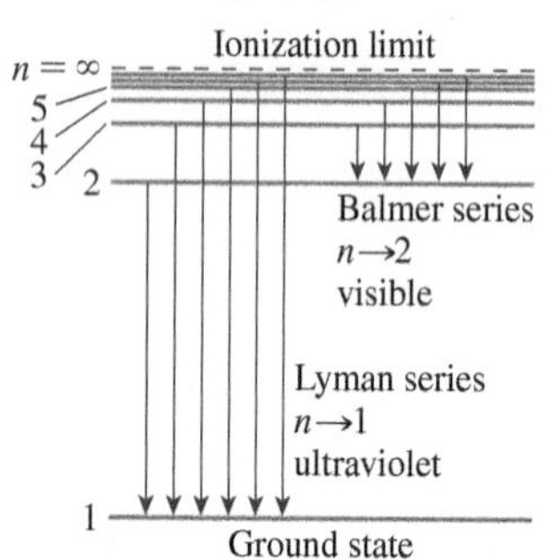

with $\lambda_0 = 91.1$ nm. This should look familiar; it's Equation 29.2 that Balmer discovered empirically.

Unlike previous atomic models, **the quantum-mechanical hydrogen atom correctly predicts the discrete spectrum of the hydrogen atom.** FIGURE 29.15 shows two series of transitions that give rise to wavelengths in the spectrum. The *Balmer series,* consisting of transitions ending on the $m = 2$ state, gives visible wavelengths, and this is the series that Balmer initially analyzed. The *Lyman series,* ending on the $m = 1$ ground state, is in the ultraviolet region of the spectrum and was not measured until later. These series, as well as others in the infrared, are observed in a discharge tube where collisions with electrons excite the atoms upward from the ground state to state n. They then decay downward by emitting photons. Only the Lyman series is observed in the absorption spectrum because, as noted previously, essentially all the atoms in a quiescent gas are in the ground state.

EXAMPLE 29.5 Wavelengths in galactic hydrogen absorption

Whenever astronomers look at distant galaxies, they find that the light has been strongly absorbed at the wavelength of the $1 \rightarrow 2$ transition in the Lyman series of hydrogen. This absorption tells us that interstellar space is filled with vast clouds of hydrogen left over from the Big Bang. What is the wavelength of the $1 \rightarrow 2$ absorption in hydrogen?

PREPARE The wavelengths of the hydrogen spectrum are given by Equation 29.8.

SOLVE Equation 29.8 predicts the *absorption* spectrum of hydrogen if we let $m = 1$. The absorption seen by astronomers is from the ground state of hydrogen $(m = 1)$ to its first excited state $(n = 2)$. The wavelength is

$$\lambda_{1 \rightarrow 2} = \frac{91.1 \text{ nm}}{\left(\dfrac{1}{1^2} - \dfrac{1}{2^2}\right)} = 121 \text{ nm}$$

ASSESS This wavelength is far into the ultraviolet. Ground-based astronomy cannot observe this region of the spectrum because the wavelengths are strongly absorbed by the atmosphere, but with space-based telescopes, first widely used in the 1970s, astronomers see 121 nm absorption in nearly every direction they look.

Electron Clouds and Atomic Orbitals

We've seen that matter has wave-like properties and, as a consequence, a particle of matter does not have a well-defined position. Quantum mechanics deals with this by describing particles in terms of a *wave function.* Rather than predicting a particle's trajectory, the wave function predicts the probability that a particle is found in a certain region of space. For atoms, the region of space in which an electron is likely to be found is often called the **electron cloud.**

FIGURE 29.16 shows electron clouds—you can think of them as three-dimensional standing waves—for the 1s, 2s, and 2p stationary states of hydrogen. These images do *not* imply that the electron itself is smeared out. Instead, because we cannot know exactly where the electron is, these images tell us the probability or likelihood of finding the electron at a particular location. The electron is most likely found in the more darkly shaded regions of space, least likely where the shading is light.

You may have seen pictures like these in chemistry, where these electron clouds are called the 1s, 2s, and 2p *orbitals.* Notice that the s-orbitals are spherically symmetric but that the 2s orbital suggests the idea of *shells,* an important idea in chemistry. These are not shells in a literal sense; instead, the 2s electron cloud tells us that the electron might be found close to the proton *or* farther out, but there is a gap in between where the electron is never found.

FIGURE 29.16 The $1s$, $2s$, and $2p$ orbitals of the hydrogen atom.

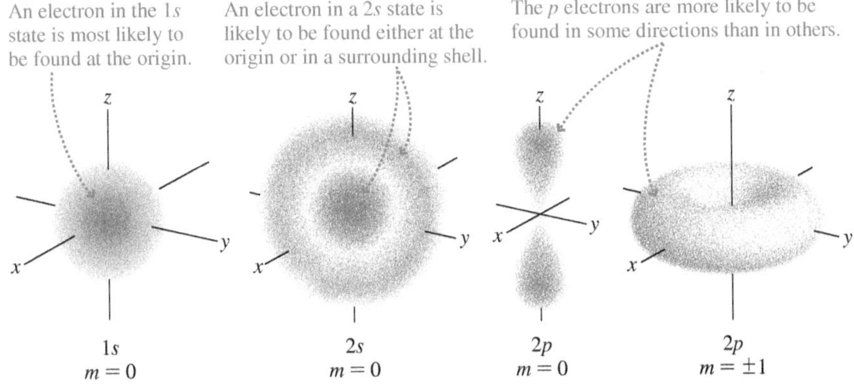

The p-orbitals are interesting because they have directional properties. A p-electron can "reach out" toward a nearby atom to establish a molecular bond. The quantum mechanics of bonding goes beyond what we can study in this text, but the electron-cloud picture of a p-orbital begins to suggest how bonds could form.

29.4 Multi-electron Atoms

The hydrogen atom is unique in that it has only one electron. For all other atoms, *multi-electron atoms,* the electrons are attracted to the positive nucleus *and* repelled by other electrons. A major difference between multi-electron atoms and the simple one-electron hydrogen is that the energy of an electron in a multi-electron atom depends on both quantum numbers *n and l.* Whereas the $2s$ and $2p$ states in hydrogen have the same energy, their energies are different in a multi-electron atom. The difference arises from the electron-electron interactions that do not exist in a single-electron hydrogen atom.

FIGURE 29.17 shows a generic energy-level diagram for the electrons in a multi-electron atom. For comparison, the hydrogen-atom energies are shown on the right edge of the figure. Two features of this diagram are of particular interest:

- For each n, the energy increases as l increases until the maximum-l state has an energy very nearly that of the same n in hydrogen. States with small values of l are significantly lower in energy than the corresponding state in hydrogen.
- As the energy increases, states with different n begin to alternate in energy. For example, the $3s$ and $3p$ states have lower energy than a $4s$ state, but the energy of an electron in a $3d$ state is slightly higher. This has important implications for the structure of the periodic table of the elements.

FIGURE 29.17 An energy-level diagram for electrons in a multi-electron atom.

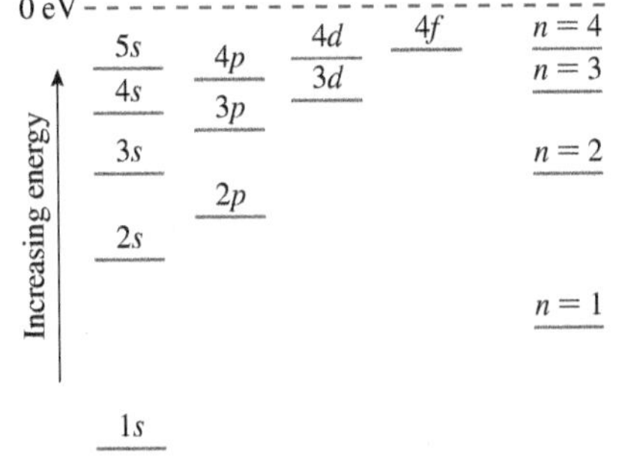

The Pauli Exclusion Principle

By definition, the ground state of a quantum system is the state of lowest energy. What is the ground state of an atom that has Z electrons and Z protons? Because the $1s$ state is the lowest energy state, it seems that the ground state should be one in which all Z electrons are in the $1s$ state. However, this idea is not consistent with the experimental evidence.

In 1925, the Austrian physicist Wolfgang Pauli hypothesized that no two electrons in a quantum system can be in the same quantum state. That is, **no two electrons can have exactly the same set of quantum numbers (n, l, m, m_s).** If one electron is present in a state, it *excludes* all others. This statement is called the **Pauli exclusion principle.** It turns out to be an extremely profound statement about the nature of matter.

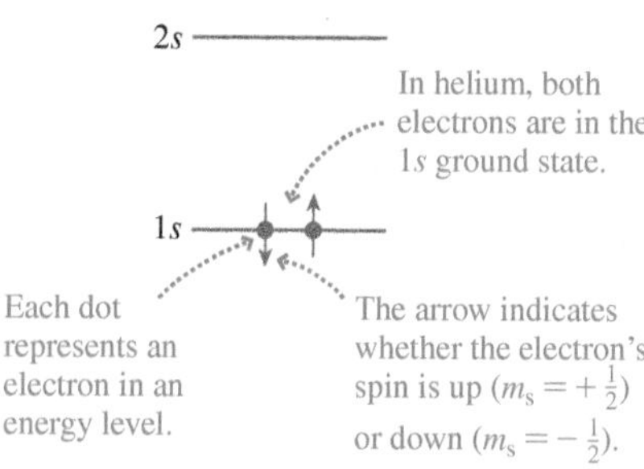
2s
In helium, both electrons are in the 1s ground state.
1s
Each dot represents an electron in an energy level.
The arrow indicates whether the electron's spin is up ($m_s = +\frac{1}{2}$) or down ($m_s = -\frac{1}{2}$).

FIGURE 29.18 The ground state of helium.

The exclusion principle is not applicable to hydrogen, where there is only a single electron, but in helium, with $Z = 2$ electrons, we must make sure that the two electrons are in different quantum states. This is not difficult. For a $1s$ state, with $l = 0$, the only possible value of the magnetic quantum number is $m = 0$. But there are *two* possible values of m_s—namely, $-\frac{1}{2}$ and $+\frac{1}{2}$. If a first electron is in the spin-down $1s$ state $(1, 0, 0, -\frac{1}{2})$, a second $1s$ electron can still be added to the atom as long as it is in the spin-up state $(1, 0, 0, +\frac{1}{2})$. This is shown schematically in FIGURE 29.18, where the dots represent electrons on the rungs of the "energy ladder" and the arrows represent spin down or spin up.

The Pauli exclusion principle does not prevent both electrons of helium from being in the $1s$ state as long as they have opposite values of m_s, so we predict this to be the ground state. A list of an atom's occupied energy levels is called its **electron configuration.** The electron configuration of the helium ground state is written $1s^2$, where the superscript 2 indicates two electrons in the $1s$ energy level.

The states $(1, 0, 0, -\frac{1}{2})$ and $(1, 0, 0, +\frac{1}{2})$ are the only two states with $n = 1$. The ground state of helium has one electron in each of these states, so all the possible $n = 1$ states are filled. Consequently, the electron configuration $1s^2$ is called a **closed shell.**

FIGURE 29.19 The ground state of lithium.

The 1s state can hold only two electrons, so the third electron in lithium must be in the 2s state.

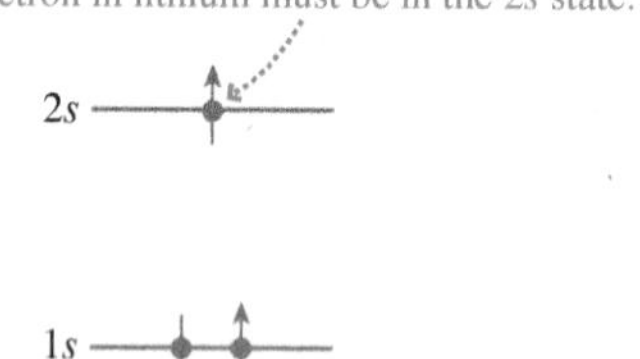
2s
1s

The next element, lithium, has $Z = 3$ electrons. The first two electrons can go into $1s$ states, with opposite values of m_s, but what about the third electron? The $1s^2$ shell is closed, and there are no additional quantum states having $n = 1$. The only option for the third electron is the next energy state, $n = 2$. Figure 29.17 showed that for a multi-electron atom, the next level above the $1s$ level is the $2s$ state, so lithium's third ground-state electron will be $2s$. FIGURE 29.19 shows the electron configuration with the $2s$ electron being spin up, but it could equally well be spin down. The electron configuration for the lithium ground state is written $1s^2 2s$. This indicates two $1s$ electrons and a single $2s$ electron.

The Periodic Table of the Elements

Several 19th-century chemists had noted that some elements have similar chemical properties, but the Russian chemist Dmitri Mendeléev was the first to propose, in 1867, a *periodic* arrangement of the elements based on the regular recurrence of chemical properties. He did so by explicitly pointing out "gaps" where, according to his hypothesis, undiscovered elements should exist. He could then predict the expected properties of the missing elements. The subsequent discovery of these elements verified Mendeléev's organizational scheme, which came to be known as the *periodic table of the elements.*

FIGURE 29.20 The subshells of a multi-electron atom.

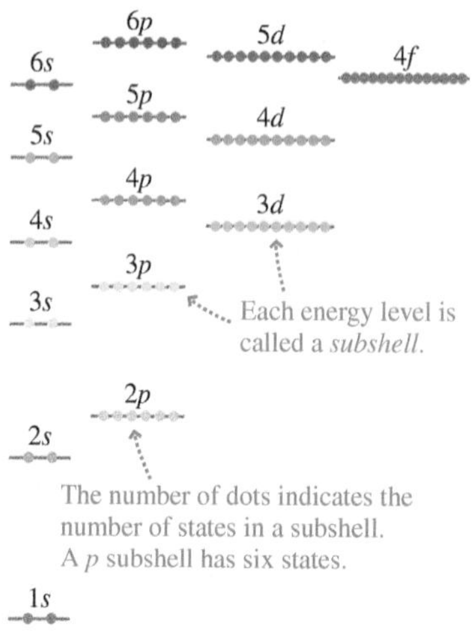

6p 5d 6s 4f 5p 4d 5s 4p 3d 4s 3p 3s 2p 2s 1s

Each energy level is called a *subshell.*

The number of dots indicates the number of states in a subshell. A *p* subshell has six states.

One of the great triumphs of the quantum-mechanical theory of multi-electron atoms is that it explains the structure of the periodic table. We can understand this structure by looking at the energy-level diagram of FIGURE 29.20, which is an expanded version of the energy-level diagram of Figure 29.17. Just as for helium and lithium, atoms with larger values of Z are constructed by placing Z electrons into the lowest energy levels that are consistent with the Pauli exclusion principle.

The s states of helium and lithium can each hold two electrons—one spin up and the other spin down—but the higher-angular-momentum states that will become filled for higher-Z atoms can hold more than two electrons. For each value l of the orbital quantum number, there are $2l + 1$ possible values of the magnetic quantum number m and, for each of these, two possible values of the spin quantum number m_s. Consequently, each energy *level* in Figure 29.20 is actually $2(2l + 1)$ different *states* that, taken together, are called a *subshell.* TABLE 29.3 lists the number of states in each subshell. Each state in a subshell is represented in Figure 29.20 by a colored dot. The dots' colors correspond to the periodic table in FIGURE 29.21, which is color coded to show which subshells are being filled as Z increases.

FIGURE 29.21 The periodic table of the elements. The elements are color coded to the states in the energy-level diagram of Figure 29.20.

Transition elements

Subshells																		
1s	1 H																	2 He
2s, 2p	3 Li	4 Be											5 B	6 C	7 N	8 O	9 F	10 Ne
3s, 3p	11 Na	12 Mg											13 Al	14 Si	15 P	16 S	17 Cl	18 Ar
4s, 3d, 4p	19 K	20 Ca	21 Sc	22 Ti	23 V	24 Cr	25 Mn	26 Fe	27 Co	28 Ni	29 Cu	30 Zn	31 Ga	32 Ge	33 As	34 Se	35 Br	36 Kr
5s, 4d, 5p	37 Rb	38 Sr	39 Y	40 Zr	41 Nb	42 Mo	43 Tc	44 Ru	45 Rh	46 Pd	47 Ag	48 Cd	49 In	50 Sn	51 Sb	52 Te	53 I	54 Xe
6s, 5d, 6p	55 Cs	56 Ba	71 Lu	72 Hf	73 Ta	74 W	75 Re	76 Os	77 Ir	78 Pt	79 Au	80 Hg	81 Tl	82 Pb	83 Bi	84 Po	85 At	86 Rn
7s, 6d, 7p	87 Fr	88 Ra	103 Lr	104 Rf	105 Db	106 Sg	107 Bh	108 Hs	109 Mt	110 Ds	111 Rg	112 Cn	113 Nh	114 Fl	115 Mc	116 Lv	117 Ts	118 Og

Lanthanides (4f)	57 La	58 Ce	59 Pr	60 Nd	61 Pm	62 Sm	63 Eu	64 Gd	65 Tb	66 Dy	67 Ho	68 Er	69 Tm	70 Yb
Actinides (5f)	89 Ac	90 Th	91 Pa	92 U	93 Np	94 Pu	95 Am	96 Cm	97 Bk	98 Cf	99 Es	100 Fm	101 Md	102 No

Inner transition elements

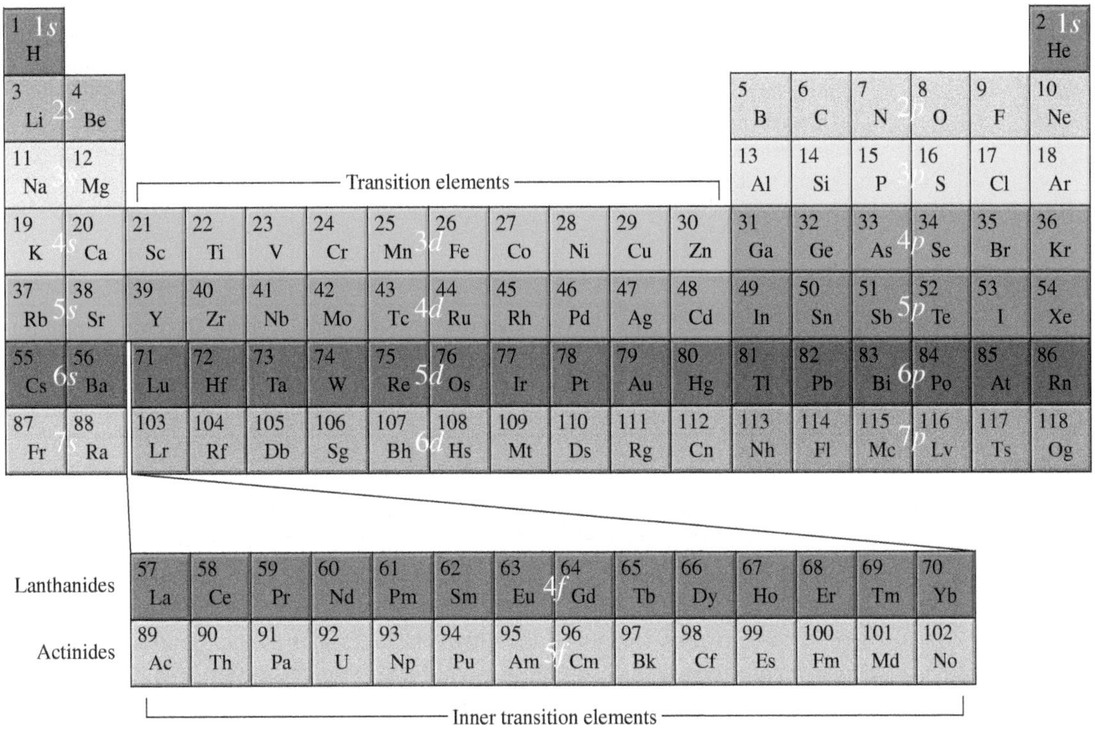

We can use Figure 29.20 to construct the periodic table in Figure 29.21. We've already seen that lithium has two electrons in the $1s$ state and one electron in the $2s$ state. Four-electron beryllium ($Z = 4$) comes next. We see that there is still an empty state in the $2s$ subshell for this fourth electron to occupy, so beryllium closes the $2s$ subshell and has electron configuration $1s^2 2s^2$.

As Z increases further, the next six electrons can each occupy states in the $2p$ subshell. These are the elements boron (B) through neon (Ne), completing the second row of the periodic table. Neon, which completes the $2p$ subshell, has ground-state configuration $1s^2 2s^2 2p^6$. The $3s$ subshell is the next to be filled, leading to the elements sodium and magnesium. Filling the $3p$ subshell gives aluminum through argon, completing the third row of the table.

The fourth row is where the periodic table begins to get complicated. You might expect that once the $3p$ subshell in argon was filled, the $3d$ subshell would start to fill, starting with potassium. But if you look back at the energy-level diagram of Figure 29.20, you can see that the $3d$ state is slightly *higher* in energy than the $4s$ state. Because the ground state is the *lowest energy state* consistent with the Pauli exclusion principle, potassium finds it more favorable to fill a $4s$ state than to fill a $3d$ state. After the $4s$ subshell is filled (at calcium), the 10 *transition elements* from scandium (Sc) through zinc (Zn) fill the 10 states of the $3d$ subshell.

The same pattern applies to the fifth row, where the $5s$, $4d$, and $5p$ subshells fill in succession. In the sixth row, however, after the initial $6s$ states are filled, the $4f$ subshell has the lowest energy, so it begins to fill *before* the $5d$ states. The elements corresponding to the $4f$ subshell, lanthanum through ytterbium, are known as the lanthanides, and they are traditionally drawn as a row separated from the rest of the table. The seventh row follows this same pattern.

Thus the entire periodic table can be built up using our knowledge of the energy-level diagram of a multi-electron atom along with the Pauli exclusion principle.

TABLE 29.3 Number of states in each subshell of an atom

Subshell	l	Number of states
s	0	2
p	1	6
d	2	10
f	3	14

CONCEPTUAL EXAMPLE 29.6 The ground state of arsenic

Predict the ground-state electron configuration of arsenic.

REASON The periodic table shows that arsenic (As) has $Z = 33$, so we must identify the states of 33 electrons. Arsenic is in the fourth row, following the first group of transition elements. Argon $(Z = 18)$ filled the $3p$ subshell, then calcium $(Z = 20)$ filled the $4s$ subshell. The next 10 elements, through zinc $(Z = 30)$, filled the $3d$ subshell. The $4p$ subshell starts filling with gallium $(Z = 31)$, and arsenic is the third element in this group, so it will have three $4p$ electrons. Thus the ground-state configuration of arsenic is

$$1s^2 2s^2 2p^6 3s^2 3p^6 4s^2 3d^{10} 4p^3$$

STOP TO THINK 29.5 Which element has the ground-state electron configuration $1s^2 2s^2 2p^6 3s^2 3p^3$?

A. P B. Al C. B D. Ge

29.5 Excited States and Spectra

The periodic table organizes information about the *ground states* of the elements. These states are chemically most important because most atoms spend most of the time in their ground states. All the chemical ideas of valence, bonding, reactivity, and so on are consequences of these ground-state atomic structures. But the periodic table does not tell us anything about the excited states of atoms. It is the excited states that hold the key to understanding atomic spectra, and that is the topic to which we turn next.

FIGURE 29.22 The $3s$ ground state of the sodium atom and some of the excited states.

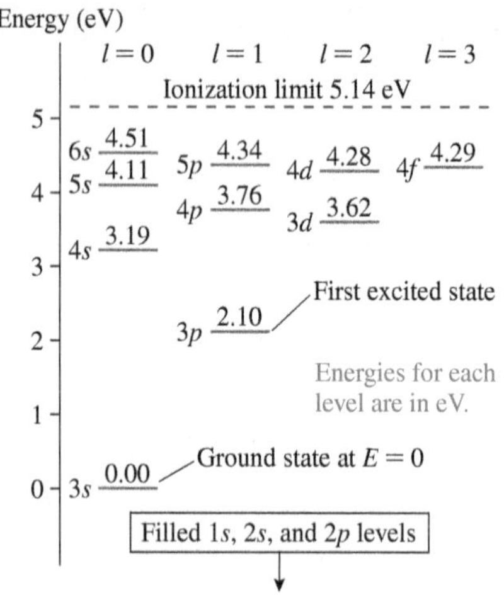

Sodium $(Z = 11)$ is a multi-electron atom that we will use to illustrate excited states. The ground-state electron configuration of sodium is $1s^2 2s^2 2p^6 3s$. The first 10 electrons completely fill the $1s$, $2s$, and $2p$ subshells, creating a *neon core* whose electrons are tightly bound together. The $3s$ electron, however, is a *valence electron* that can be easily excited to higher energy levels. If this electron were excited to the $3p$ state, for instance, then we would write the electron configuration as $1s^2 2s^2 2p^6 3p$.

The excited states of sodium are produced by raising the valence electron to a higher energy level. The electrons in the neon core are unchanged. FIGURE 29.22 is an energy-level diagram showing the ground state and some of the excited states of sodium. The $1s$, $2s$, and $2p$ states of the neon core are not shown on the diagram. These states are filled and unchanging, so only the states available to the valence electron are shown. Notice that the zero of energy has been shifted to the ground state. As we have discovered before, the zero of energy can be located where it is most convenient. With this choice, the excited-state energies tell us how far each state is above the ground state. The ionization limit now occurs at the value of the atom's ionization energy, which is 5.14 eV for sodium.

Left to itself, an atom will be in its lowest-energy ground state. How does an atom get into an excited state? The process of getting it there is called **excitation,** and there are two basic mechanisms: absorption and collision. We'll begin by looking at excitation by absorption.

Excitation by Absorption

FIGURE 29.23 Excitation by photon absorption.

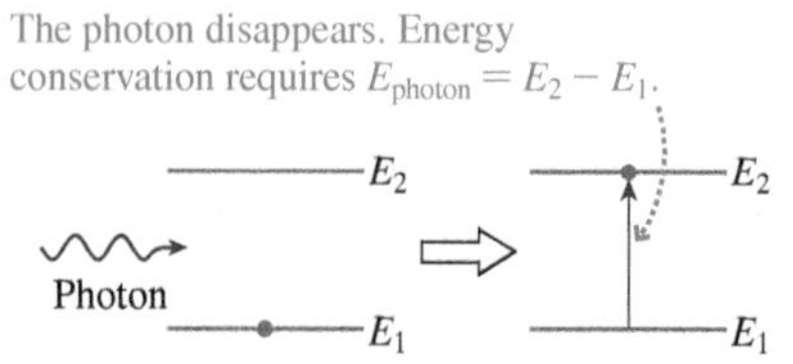

The photon disappears. Energy conservation requires $E_{photon} = E_2 - E_1$.

E_2 E_2 Photon E_1 E_1

One of the postulates of the Bohr model is that an atom can jump from one stationary state, of energy E_1, to a higher-energy state E_2 by absorbing a photon of frequency $f_{photon} = \Delta E_{atom}/h$. This process is shown in FIGURE 29.23. Because we are interested in spectra, it is more useful to write this in terms of the wavelength:

$$\lambda = \frac{c}{f_{photon}} = \frac{hc}{\Delta E_{atom}}$$

Energy transitions are generally expressed in eV and wavelengths in nm. If we express h in eV • s and c in nm/s, we get a version of the above equation that allows us to quickly find wavelengths in nm given a transition energy in eV:

$$\lambda(\text{in nm}) = \frac{1240 \text{ eV} \cdot \text{nm}}{\Delta E_{\text{atom}}(\text{in eV})} \tag{29.9}$$

A quantum-mechanical analysis of how the electrons in an atom interact with a light wave shows that transitions must also satisfy the following **selection rule:** Transitions (either absorption or emission) from a state with orbital quantum number l can occur to only another state whose orbital quantum number differs from the original state by ± 1, or

$$\Delta l = l_2 - l_1 = \pm 1 \tag{29.10}$$

Selection rule for emission and absorption

EXAMPLE 29.7 Analyzing absorption in sodium

What are the two longest wavelengths in the absorption spectrum of sodium? What are the transitions?

PREPARE Figure 29.22 shows the ground state and excited states of the sodium atom. An absorption transition starts from the ground state; the only possible transitions are those for which l changes by ± 1.

SOLVE As Figure 29.22 shows, the sodium ground state is $3s$. Starting from an s state ($l = 0$), the selection rule permits quantum jumps only to p states ($l = 1$). The lowest excited state is the $3p$ state and $3s \rightarrow 3p$ is an allowed transition ($\Delta l = 1$), so this will be the longest wavelength. You can see from the data in Figure 29.22 that $\Delta E_{\text{atom}} = 2.10 \text{ eV} - 0.00 \text{ eV} = 2.10 \text{ eV}$ for this transition. The corresponding wavelength is

$$\lambda = \frac{1240 \text{ eV} \cdot \text{nm}}{2.10 \text{ eV}} = 590 \text{ nm}$$

(Because of rounding, the calculation gives $\lambda = 590$ nm. The experimental value is actually 589 nm.)

The next excited state is $4s$, but a $3s \rightarrow 4s$ transition is not allowed by the selection rule. The next allowed transition is $3s \rightarrow 4p$, with $\Delta E_{\text{atom}} = 3.76$ eV. The wavelength of this transition is

$$\lambda = \frac{1240 \text{ eV} \cdot \text{nm}}{3.76 \text{ eV}} = 330 \text{ nm}$$

ASSESS If you look at the sodium spectrum shown earlier in Figure 29.3b, you will see that 589 nm and 330 nm are, indeed, the two longest wavelengths in the absorption spectrum.

Collisional Excitation

An electron traveling with a speed of 1.0×10^6 m/s has a kinetic energy of 2.85 eV. If this electron collides with a ground-state sodium atom, a portion of its energy can be used to excite the atom to a higher-energy state, such as its $3p$ state. This process is called **collisional excitation** of the atom.

Collisional excitation differs from excitation by absorption in one fundamental way. In absorption, the photon disappears. Consequently, *all* of the photon's energy must be transferred to the atom. Conservation of energy then requires $E_{\text{photon}} = \Delta E_{\text{atom}}$. In contrast, the electron is still present after collisional excitation and can still have some kinetic energy. That is, the electron does *not* have to transfer its entire energy to the atom. If the electron has an incident kinetic energy of 2.85 eV, it could transfer 2.10 eV to the sodium atom, thereby exciting it to the $3p$ state, and still depart the collision with a speed of 5.1×10^5 m/s and 0.75 eV of kinetic energy.

To excite the atom, the incident energy of the electron (or any other matter particle) merely has to *exceed* ΔE_{atom}; that is, $E_{\text{particle}} \geq \Delta E_{\text{atom}}$. There's a threshold energy for exciting the atom, but no upper limit. It is all a matter of energy conservation. FIGURE 29.24 shows the idea graphically.

FIGURE 29.24 Collisional excitation.

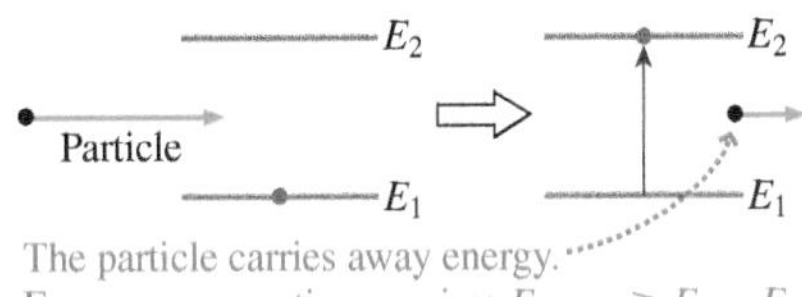
E_2
E_2
Particle
E_1
E_1
The particle carries away energy.
Energy conservation requires $E_{\text{particle}} \geq E_2 - E_1$.

Collisional excitation by electrons is the predominant method of excitation in electrical discharges such as fluorescent lights, street lights, and neon signs. A gas is sealed in a tube at reduced pressure (≈ 1 mm Hg), then a fairly high voltage (≈ 1000 V) between electrodes at the ends of the tube causes the gas to ionize, creating a current in which both ions and electrons are charge carriers. The electrons accelerate in the electric field, gaining several eV of kinetic energy, then transfer some of this energy to the gas atoms upon collision.

NOTE ▶ In contrast to photon absorption, there are no selection rules for collisional excitation. Any state can be excited if the colliding particle has sufficient energy. ◀

CONCEPTUAL EXAMPLE 29.8 **Possible excitation of hydrogen?**

Can an electron with a kinetic energy of 11.4 eV cause a hydrogen atom to emit the prominent red spectral line ($\lambda = 656$ nm, $E_{photon} = 1.89$ eV) in the Balmer series?

REASON The electron must have sufficient energy to excite the upper state of the transition. The electron's energy of 11.4 eV is significantly greater than the 1.89 eV energy of a photon with wavelength 656 nm, but don't confuse the energy of the photon with the energy of the excitation. The red spectral line in the Balmer series is emitted in an $n = 3 \rightarrow 2$ transition with $\Delta E_{atom} = 1.89$ eV, but to cause this emission, the electron must excite an atom from its *ground state,* with $n = 1$, up to the $n = 3$ level. From Figure 29.14, the necessary excitation energy is

$$\Delta E_{atom} = E_3 - E_1 = (-1.51\text{ eV}) - (-13.60\text{ eV}) = 12.09\text{ eV}$$

The electron does *not* have sufficient energy to excite the atom to the state from which the emission would occur.

ASSESS As our discussion of absorption spectra showed, almost all excitations of atoms begin from the ground state. Quantum jumps down in energy, however, can begin and end at any two states allowed by selection rules.

Emission Spectra

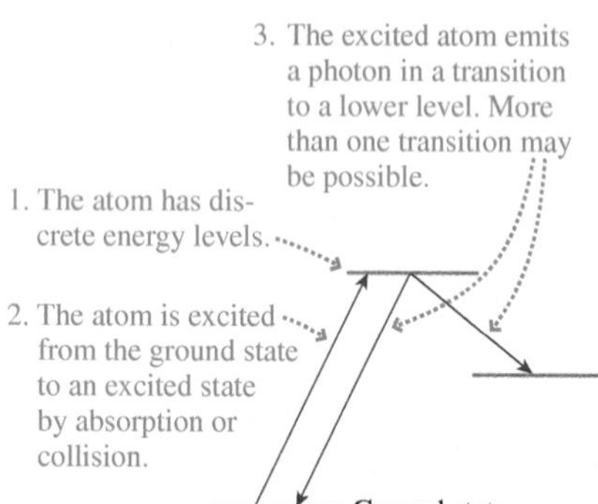

3. The excited atom emits a photon in a transition to a lower level. More than one transition may be possible.
1. The atom has discrete energy levels.
2. The atom is excited from the ground state to an excited state by absorption or collision.
Ground state

FIGURE 29.25 Generation of an emission spectrum.

Understanding emission hinges on the three ideas shown in FIGURE 29.25. Once we have determined the energy levels of an atom, from quantum mechanics, we can immediately predict its emission spectrum.

As an example, FIGURE 29.26a shows some of the transitions and wavelengths observed in the emission spectrum of sodium. This diagram makes the point that each wavelength represents a quantum jump between two well-defined energy levels. Notice that the selection rule $\Delta l = \pm 1$ is obeyed in the sodium spectrum. Atoms in the $5p$ energy level can make transitions to $3s$, $4s$, or $3d$ but *not* to $3p$ or $4p$.

FIGURE 29.26b shows the emission spectrum of sodium as it would be recorded in a spectrometer. (Many of the lines seen in this spectrum start from higher excited states that are not seen in the rather limited energy-level diagram of Figure 29.26a.) By comparing the spectrum to the energy-level diagram, you can recognize that the spectral lines at 589 nm, 330 nm, and 286 nm form a *series* of lines due to possible $np \rightarrow 3s$ transitions. They are the dominant features in the sodium spectrum.

FIGURE 29.26 The emission spectrum of sodium.

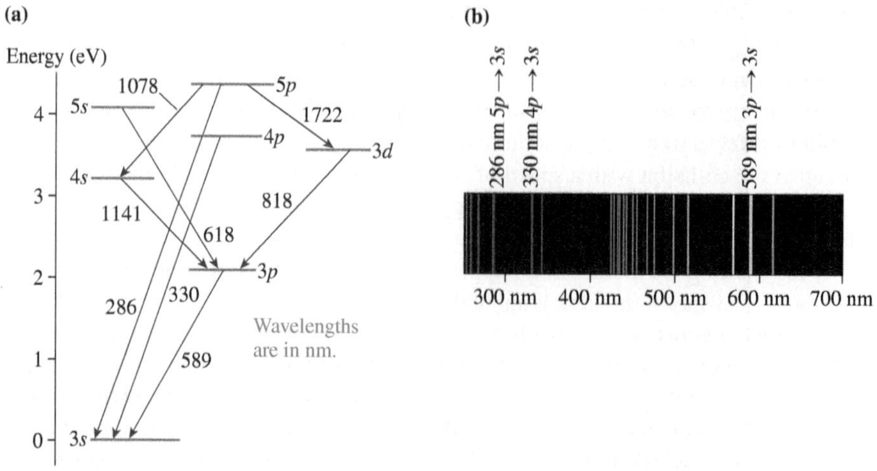

The most obvious visual feature of sodium emission is its bright yellow color, produced by the 589 nm photons emitted in the $3p \rightarrow 3s$ transition. This is the basis of the *flame test* used in chemistry to test for sodium: A sample is held in a Bunsen burner, and a bright yellow glow indicates the presence of sodium. The 589 nm emission is also prominent in the pinkish-yellow glow of the common sodium-vapor street lights. These operate by creating an electrical discharge in sodium vapor. Most sodium-vapor lights use high-pressure lamps to increase their light output. The high pressure, however, causes the formation of Na_2 molecules, and these molecules emit the pinkish portion of the light.

X Rays

Chapter 27 noted that x rays are produced when very-high-speed electrons, accelerated with potential differences of many thousands of volts, crash into metal targets. Rather than exciting the atom's valence electrons, such as happens in a gas discharge tube, these high-speed projectiles are capable of knocking inner-shell electrons out of the target atoms, producing an *inner-shell vacancy*. As FIGURE 29.27 shows for copper atoms, this vacancy is filled when an electron from a higher shell undergoes a quantum jump into the vacancy, emitting a photon in the process.

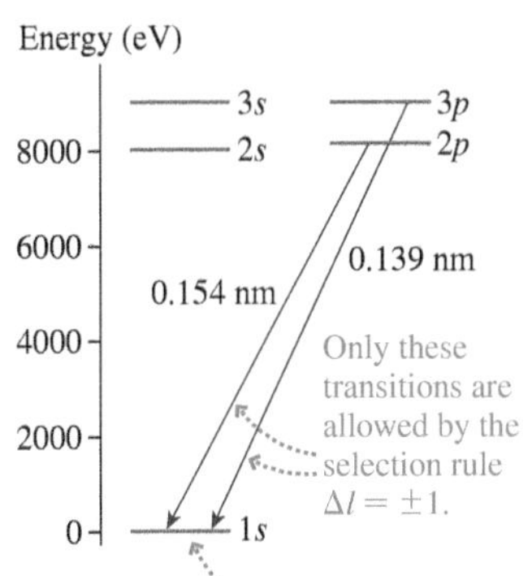

FIGURE 29.27 The generation of x rays from copper atoms.

In heavy elements, such as copper or iron, the energy difference between the inner and outer shells is very large—typically 10 keV. Consequently, the photon has energy $E_{photon} \approx 10$ keV and wavelength $\lambda \approx 0.1$ nm. These high-energy photons are the x rays discovered by Röntgen. X-ray photons are about 10,000 times more energetic than visible-light photons—which makes them ionizing radiation—and the wavelengths are about 10,000 times smaller. Even so, the underlying physics is the same: A photon is emitted when an electron in an atom undergoes a quantum jump.

STOP TO THINK 29.6 In this hypothetical atom, what is the photon energy E_{photon} of the longest-wavelength photons emitted by atoms in the 5*p* state?

A. 1.0 eV
B. 2.0 eV
C. 3.0 eV
D. 4.0 eV

Energy (eV): 5 — 5*s*, 4*d*; 4 — 5*p*; 2 — 4*p*; 1 — 3*d*; 0 — 4*s*

29.6 Molecules

Quantum mechanics applies to molecules just as it does to atoms, but molecules—especially large biomolecules—are more complex because they have internal modes of storing energy. In particular, molecular bonds can *vibrate* as if they were springs, and molecules can *rotate* about their center of mass. In this text, we'll focus on vibrational energies.

It turns out that a vibrating molecular bond, just like a particle in a box, has only certain allowed energies. In other words, the vibrational energy of a molecular bond is quantized. However, the spacing between vibrational energy levels is much less than the spacing between electron energy levels. Consequently, as shown in FIGURE 29.28, (next page) each electron energy level is split into a large number of *vibrational energy levels*. Each vibrational energy level represents a different vibrational amplitude or vibrational mode of the molecule's structure.

The vibrational energy levels are easily distinguished in small molecules, such as O_2 and CO_2, and chemists use *infrared spectroscopy* to observe transitions between one vibrational energy level and another. The infrared spectrum provides

FIGURE 29.28 Energy levels of molecules.

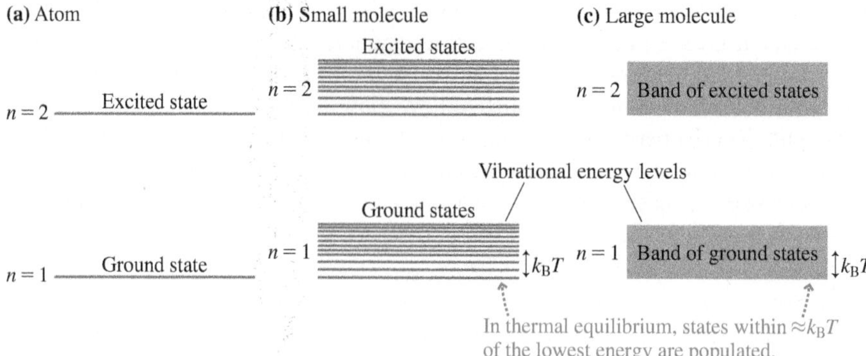

information about the bond structure. In contrast, large biomolecules have such a huge number of vibrational energy levels that they become smeared into a continuous range or *band* of allowed energies; individual vibrational energy levels cannot be distinguished.

Chapter 13 introduced the idea that the probability that an energy level is occupied is given by the Boltzmann factor e^{-E/k_BT}. For atoms, the typical energy-level spacing of a few eV is much greater than the atom's thermal energy: $\Delta E_{atom} \gg k_BT$. Consequently, the equilibrium state of atoms is one in which all the atoms are in the ground state; the population of any excited state is essentially zero.

The electron energy levels of molecules are also spaced several eV apart, so equilibrium finds all the molecules in the electron ground state. However, the spacing between vibrational energy levels is comparable to or, for biomolecules, even less than k_BT. As a result, molecules have anywhere from a few to hundreds of vibrational energy levels that are always populated.

Molecular Absorption Spectra

Atomic spectra are always observed with the atoms in the gas phase, usually at a reduced pressure. The mean time between collisions is long enough that an excited atom has plenty of time to emit a photon of light. Emission spectra can also be observed from small molecules in gas discharge tubes. They are very complex spectra with hundreds of distinct wavelengths that correspond to quantum jumps between all the many combinations of vibrational energy levels in the ground state and the excited state.

FIGURE 29.29 The absorption spectrum and energy levels of chlorophyll.

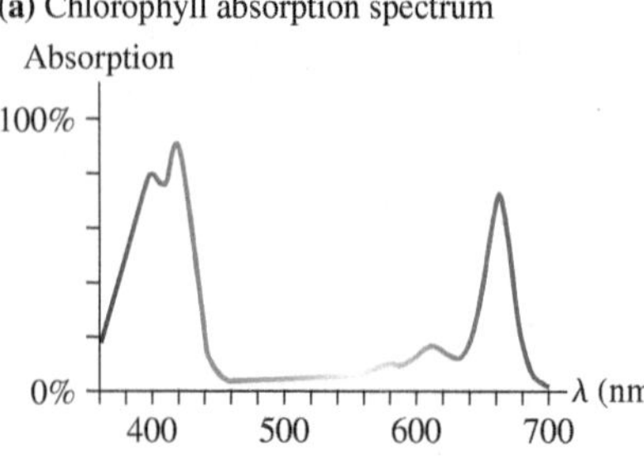

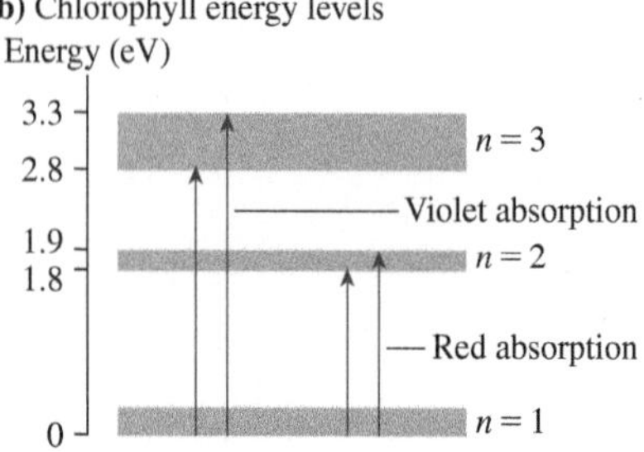

Spectra are still of critical importance in identifying larger molecules; however, larger molecules are invariably observed in solution or, for minerals and pigments, in their solid phase. These molecules absorb light but, with a few exceptions that we'll look at in the next section, are unlikely to emit light. In solution, collisions between molecules are so rapid that an excited molecule gives up its energy to another molecule in a collision before it has time to emit a photon. This *collisional de-excitation* increases the sample's thermal energy instead of producing light. A similar process takes place in most solid materials.

Absorption spectra provide information about the molecule's excited states. As an example, FIGURE 29.29a is the absorption spectrum of chlorophyll, an essential molecule for photosynthesis in green plants. Chlorophyll is a large, complex molecule, so its spectrum does not have discrete spectral lines but, instead, broad absorption features. We see that chlorophyll absorbs violet and ultraviolet light from roughly 380 nm to 440 nm; it also absorbs red light from roughly 640 nm to 680 nm. There is very little absorption between 450 nm and 600 nm.

We can use Equation 29.9, $\lambda(\text{in nm}) = (1240 \text{ eV} \cdot \text{nm})/\Delta E(\text{in eV})$, to calculate the energy corresponding to each of these wavelengths. Doing so allows us to construct the energy diagram of FIGURE 29.29b. We see that there is a first excited-state band ($n = 2$) from 1.8 eV to 1.9 eV and a second excited-state band ($n = 3$) from 2.8 eV to 3.3 eV. The two broad peaks in the absorption spectrum occur due to the absorption of photons that cause transitions from the ground state to these energy bands. There's little or no absorption of photons with energies between 1.9 eV and 2.8 eV because no allowed energies are in this range.

So why are green plants green? Because photons with wavelengths between about 450 nm and 600 nm—the green and yellow region of the visible spectrum—are *not* absorbed. Instead, light with these wavelengths is reflected or, for thin samples, transmitted to give leaves the green color that we see.

In general, the color of liquids and solids is due to the reflection and scattering of photons with energies and wavelengths that are *not* absorbed by the material. Red paint is red because pigment molecules absorb all photons with energies greater than about 2 eV, corresponding to wavelengths shorter than 620 nm. Photons with energy less than 2 eV are not absorbed, so we see reflected and scattered light with wavelengths longer than 620 nm—red light.

FIGURE 29.30 Pulse oximetry is based on the fact that oxygenated hemoglobin molecules and deoxygenated hemoglobin molecules have different molecular absorption spectra.

(a) A pulse oximeter.

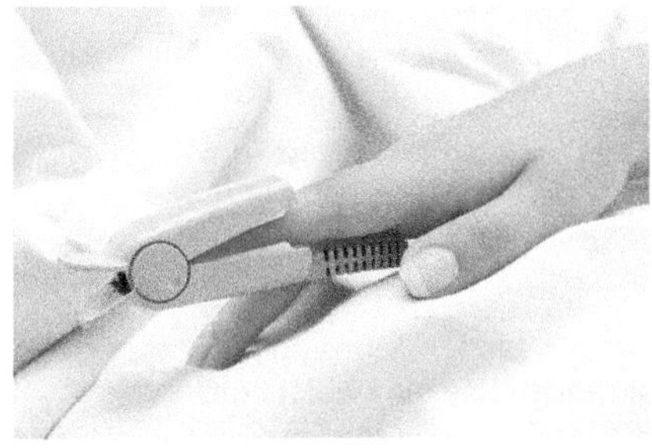

(b) Absorption curves of oxygenated and deoxygenated blood.

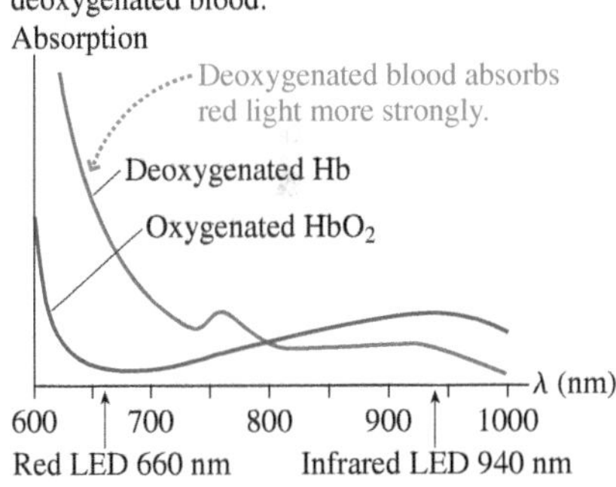

Pulse Oximetry

Pulse oximetry, a noninvasive optical test to measure the oxygen saturation of blood, is a medical application of the molecular absorption of light. Pulse oximetry gained widespread attention during the COVID-19 pandemic of 2020 because a low oxygen-saturation level is an early indicator of the need for medical intervention. FIGURE 29.30a shows a *pulse oximeter* clipped to a patient's index finger. Notice the red light.

The *oxygen saturation* of blood is the percentage of hemoglobin binding sites in the bloodstream that are occupied by an oxygen molecule. The normal oxygen saturation of arterial blood that has just left the heart is 95–100%, which means that almost every hemoglobin binding site is occupied. A lower level of oxygen saturation, called *hypoxia,* is often indicative of underlying medical issues.

Oxygen saturation can be measured optically because oxygenated hemoglobin (HbO_2) and deoxygenated hemoglobin (Hb) have different absorption spectra, as shown in FIGURE 29.30b. All blood strongly absorbs most wavelengths of visible light; only the longest wavelengths are reflected. That is why blood is red. Oxygenated arterial blood reflects wavelengths longer than about 600 nm because the absorption of these wavelengths is low, so arterial blood is a bright cherry red. But deoxygenated venous blood has strong absorption to about 650 nm and reflects only wavelengths longer than about 650 nm, the extreme red end of the visible spectrum. Consequently, deoxygenated venous blood is a dark maroon.

The low absorption of red light by blood means that tissue is fairly transparent to red light. You can demonstrate this by going into a dark room with a mirror and pointing a flashlight into your mouth; you'll see a red glow coming through your cheeks. Your entire fingertip glows red with scattered light if you press a red laser pointer against a fingernail.

Pulse oximetry depends on the fact that red and infrared light can penetrate deeply enough into tissue to pass through a finger. A pulse oximeter uses two light-emitting diodes (LEDs): one that emits red light with a wavelength of 660 nm and a second that emits infrared light with a wavelength of 940 nm. Light from both LEDs passes through the finger and is detected on the opposite side. The red light is more strongly absorbed by deoxygenated blood; the infrared is more strongly absorbed by oxygenated blood. The blood's oxygen saturation can be determined from the ratio I_{660}/I_{940} of detected intensities at the two wavelengths.

Factors other than blood affect the transmission of light through the finger. Skin color, skin condition, and finger thickness all influence the detected intensity. But these factors remain constant, whereas blood flow through the fingertip pulses with each heartbeat. Software that isolates the pulsing component of the light intensity allows the blood signal to be measured while filtering out other contributions to the signal—hence the term "pulse" in pulse oximetry.

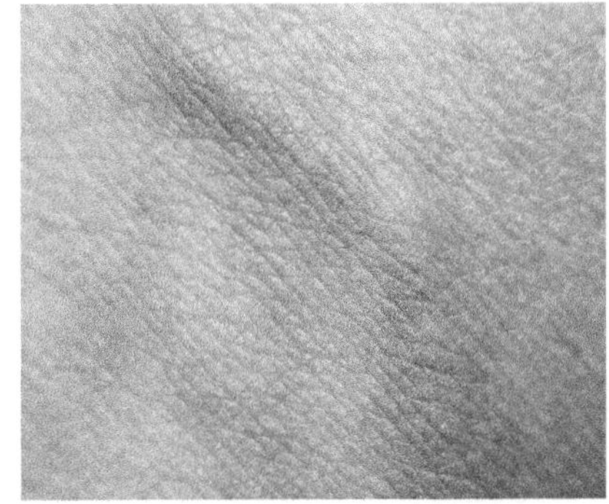

Blue blooded You've probably noticed that veins just below the skin surface of a light-skinned person appear blue. It is *not* because venous blood is blue. Deoxygenated venous blood is darker red than oxygenated arterial blood, but it's still red. And both arteries and veins are red when exposed during surgery. The blueness of veins when seen through skin has two causes, one physical and one perceptual. Physically, the interaction of light with tissue is a complex interplay between the scattering of light by cells and the absorption of light by blood and other molecules. Veins reflect more red light than blue light, as you would expect, but scattering in the tissue causes the skin above a vein to reflect slightly less red light toward your eye than does the nearby skin. Your brain compares the slightly reduced red light from above a vein to the reflected red light from the nearby skin and, somewhat as in an optical illusion, interprets the reduced intensity as a color shift. Seeing "darker" as "bluer" is an issue of perception, not physics.

STOP TO THINK 29.7 A pulse oximeter measures I_{660}/I_{940}, the ratio of the detected intensities at 660 nm and 940 nm. Does a patient with hypoxia have a higher or lower ratio than a person with normal oxygen saturation of the blood?

29.7 Fluorescence and Bioluminescence

As we've seen, molecules in liquids and solids absorb some wavelengths of light much more strongly than other wavelengths. In general, the energy of absorbed light is transformed into thermal energy rather than re-radiated, while light that is not absorbed is scattered or reflected to give the object its color. But there are exceptions.

FIGURE 29.31 A scorpion fluoresces when irradiated by ultraviolet light.

FIGURE 29.31 shows a scorpion that is irradiated by ultraviolet light. Our eyes (and the camera) are not sensitive to ultraviolet light, so the scene would appear completely dark if not for the scorpion. But the scorpion, reddish-gray in daylight, is bright blue. A molecule called β-carboline in the scorpion's exoskeleton is *fluorescing*.

Fluorescence is the emission of longer-wavelength (lower energy) light following the absorption of shorter-wavelength (higher energy) light. In the case of the scorpion, the emission of longer-wavelength blue light follows the absorption of shorter-wavelength ultraviolet light. FIGURE 29.32 illustrates how fluorescence occurs. Some of the absorbed energy is rapidly (≈ 1 ps) transformed into molecular vibrations and from there into thermal energy. This *internal conversion* drops the molecule's energy to the lower edge of the band of excited states. As a result, quantum jumps to the ground state emit photons with lower energy (longer wavelength) than the photons that had been absorbed.

FIGURE 29.32 The process of molecular fluorescence.

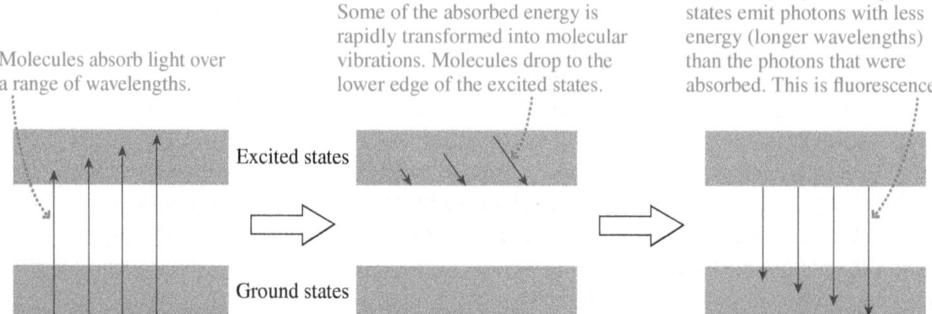

An underwater light show This *Bolinopsis* comb jelly makes its presence known in the dark with several different colors of bioluminescence. About 50% of jellyfish are bioluminescent. Marine biologists are not sure of all the functions of bioluminescence, but for comb jellies, which can flash their bioluminescence on and off, it seems to be part of a defensive tactic to startle predators.

Many biomolecules are fluorescent. A good example is chlorophyll. We noted in the previous section that chlorophyll absorbs violet and red light and that nearly all of the absorbed energy is transformed into thermal energy. However, a small fraction of the molecules that absorb violet light to the $n = 3$ levels in Figure 29.29 undergo an internal conversion of energy, drop to the lower edge of the $n = 2$ band, and then emit longer-wavelength photons in quantum jumps back to the $n = 1$ band. Chlorophyll fluorescence is emitted at deep-red and infrared wavelengths from 680 nm to 750 nm. Fluorescence measurements can be used to determine chlorophyll concentrations; this technology is routinely used in plant physiology and agriculture. Satellite-based observations of chlorophyll fluorescence, which is caused by the absorption of short-wavelength solar radiation, are used to monitor the ecological health of entire ecosystems.

An especially important fluorescent biomolecule is the *green fluorescent protein* (GFP) derived from a species of jellyfish. FIGURE 29.33a shows the absorption and emission spectra of GFP. The protein absorbs blue and violet wavelengths of light ($\lambda < 500$ nm), then fluoresces at green wavelengths (500–550 nm). GFP has become an important tool in cell biology because it can be genetically engineered into organisms and expressed when certain genes are active. For example, FIGURE 29.33b shows the expression of a particular gene in mosquito larvae. The technique of *fluorescence microscopy* was discussed in ◀◀ SECTION 20.6.

The photon emission of fluorescence occurs very quickly, a few ns after the shorter-wavelength photons are absorbed. As a practical matter, fluorescence vanishes instantly when the excitation light is turned off. In a related process, called *phosphorescence,* the emission of light—often very dim—continues after the excitation has been removed, sometimes for hours. Glow-in-the-dark signs, toys, and clock faces are applications of phosphorescence.

Why is phosphorescence a much slower process than fluorescence? Most stable molecules have an even number of electrons, and the electron spins are arranged in up-down pairs so that the molecule has no net spin. Chemists call this a *singlet state.* In some molecules, the internal conversion of energy after a photon has been absorbed causes one electron to flip the direction of its spin. The spins are no longer all paired, so the molecule is in an excited state that has a net spin; chemists call this a *triplet state.* The ground state is a singlet state, so the photon-emitting quantum jump requires the molecule to undergo a spin flip from a triplet state back to a singlet state. The rules of quantum physics strongly discourage transitions that involve a spin flip, so the molecule remains stuck in the excited state for an extended period of time before photon emission occurs. Many minerals exhibit phosphorescence, and you may have seen glowing minerals in a museum display, but biomolecules almost always display fluorescence and not phosphorescence.

FIGURE 29.33 The spectra and use of green fluorescent protein.

(a) Absorption and emission spectra of GFP.

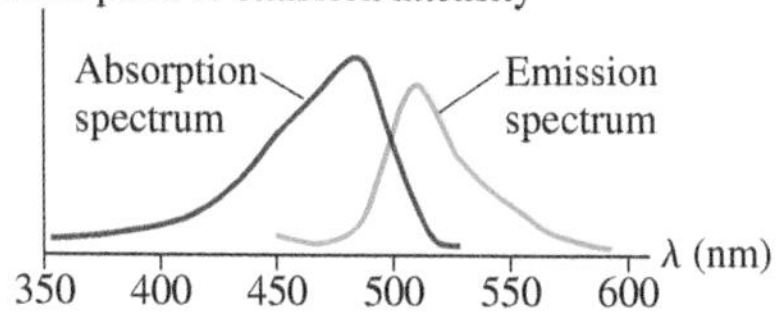

(b) Mosquito larvae expressing GFP.

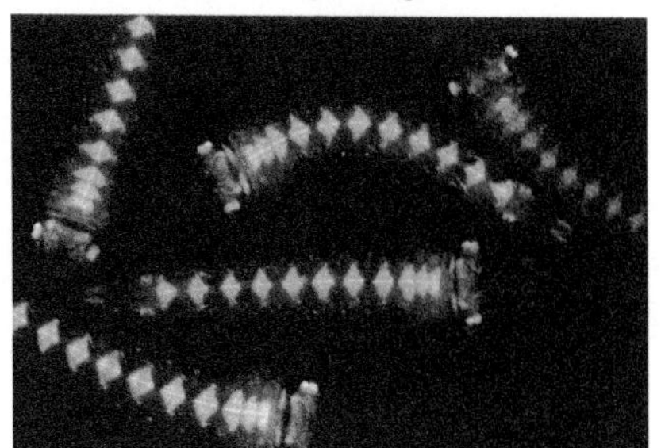

Bioluminescence

One more light-emitting molecular process deserves mention: **bioluminescence,** the soft glow of many marine organisms and a few terrestrial arthropods, such as fireflies. Bioluminescence is a form of *chemiluminescence,* the production of light from a chemical reaction in which a product molecule is created in an excited state from which it emits a photon as it returns to the ground state. Light sticks are a familiar application of chemiluminescence. The emission spectrum of chemiluminescence is very similar to that of fluorescence because in both the quantum jumps occur from the lower edge of the band of excited states, but the excitation process differs.

In organisms, an enzyme called *luciferase* catalyzes a reaction that uses the energy of ATP to oxidize a pigment molecule called *luciferin.* The reaction product *oxyluciferin* is created in an excited state that emits a photon as it transitions to the ground state. Luciferase and luciferin are generic terms, not specific molecules, and different organisms have evolved somewhat different versions. The process is the same, but the light spectrum is species specific. Fireflies emit a yellow light with wavelengths between 550 nm and 650 nm, whereas many marine organisms emit a blue-green light with wavelengths between 450 nm and 550 nm.

29.8 Stimulated Emission and Lasers

We have seen that an atom can jump from a lower-energy level E_1 to a higher-energy level E_2 by absorbing a photon. FIGURE 29.34a illustrates the process. Once in level 2, as shown in FIGURE 29.34b, the atom emits a photon of the same energy as it jumps back to level 1. Because this transition occurs spontaneously, it is called **spontaneous emission.**

In 1917, Einstein proposed a second mechanism by which an atom in state 2 can make a transition to state 1. The left part of FIGURE 29.34c shows a photon approaching an atom in its excited state 2. According to Einstein, if the energy of the photon is exactly equal to the energy difference $E_2 - E_1$ between the two states, this incoming photon can *induce* the atom to make the $2 \rightarrow 1$ transition, emitting a photon in the process. This process is called **stimulated emission.**

FIGURE 29.34 Three types of radiative transitions.

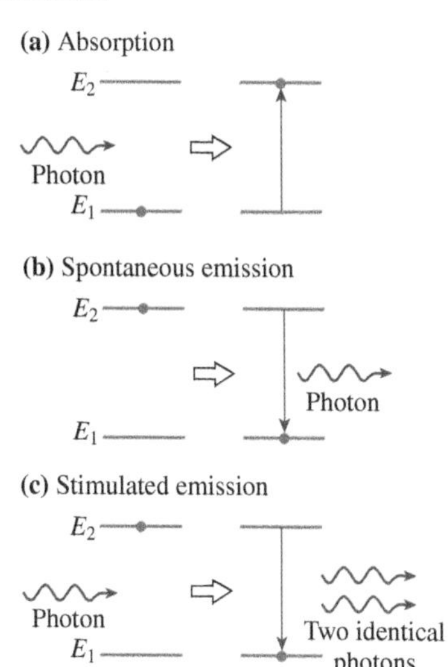

The incident photon is *not* absorbed in the process, so now there are *two* photons. And, interestingly, the emitted photon is *identical* to the incident photon. This means that as the two photons leave the atom they have exactly the same frequency and wavelength, are traveling in exactly the same direction, and are exactly in phase with each other. In other words, **stimulated emission produces a second photon that is an exact clone of the first.**

Lasers

The word **laser** is an acronym for the phrase *light amplification by the stimulated emission of radiation.* But what *is* a laser? Basically it is a device that produces a beam of highly *coherent* and essentially monochromatic (single-color) light as a result of stimulated emission. **Coherent** light is light in which all the electromagnetic waves have the same phase, direction, and amplitude. It is the coherence of a laser beam that allows it to be very tightly focused or to be rapidly modulated for communications.

Let's take a brief look at how a laser works. FIGURE 29.35 represents a system of atoms that have a lower energy level E_1 and a higher energy level E_2. Suppose that there are N_1 atoms in level 1 and N_2 atoms in level 2. Left to themselves, all the atoms would soon end up in level 1 because of the spontaneous emission $2 \rightarrow 1$. To prevent this, we can imagine that some type of excitation mechanism, perhaps an electrical discharge, continuously produces new excited atoms in level 2.

FIGURE 29.35 Energy levels 1 and 2, with populations N_1 and N_2.

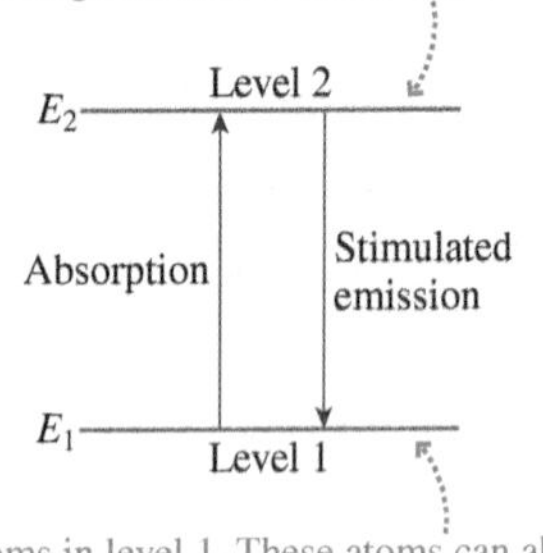
N_2 atoms in level 2. Photons of energy $E_{photon} = E_2 - E_1$ can cause these atoms to undergo stimulated emission.
E_2
Level 2
Absorption
Stimulated emission
E_1
Level 1
N_1 atoms in level 1. These atoms can absorb photons of energy $E_{photon} = E_2 - E_1$.

Let a photon of frequency $f = (E_2 - E_1)/h$ be incident on this group of atoms. Because it has the correct frequency, it could be absorbed by one of the atoms in level 1. Another possibility is that it could cause stimulated emission from one of the level 2 atoms. Ordinarily $N_2 \ll N_1$, so absorption events far outnumber stimulated emission events. Even if a few photons were generated by stimulated emission, they would quickly be absorbed by the vastly larger group of atoms in level 1.

But what if we could somehow arrange to place *every* atom in level 2, making $N_1 = 0$? Then the incident photon, upon encountering its first atom, will cause stimulated emission. Where there was initially one photon of frequency f, now there are two. These will strike two additional excited-state atoms, again causing stimulated emission. Then there will be four photons. As FIGURE 29.36 shows, there will be an *avalanche* of stimulated emission until all N_2 atoms emit a photon of frequency f.

FIGURE 29.36 Stimulated emission creates a chain reaction of photon production in a population of excited atoms.

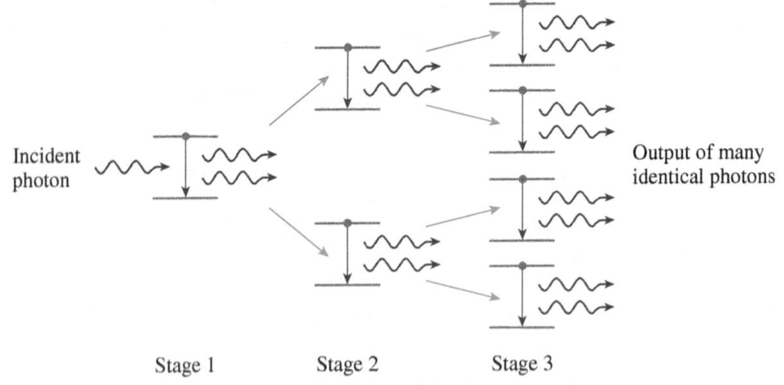
Incident photon
Output of many identical photons
Stage 1
Stage 2
Stage 3

Quite a spectacle A spectacular laser light show depends on three key properties of coherent laser light: It can be very intense, its color is extremely pure, and the laser beam is narrow with little divergence.

In stimulated emission, each emitted photon is *identical* to the incident photon. The avalanche of Figure 29.36 will lead not just to N_2 photons of frequency f, but to N_2 *identical* photons, all traveling together in the same direction with the same phase. If N_2 is a large number, as would be the case in any practical device, the one initial photon will have been *amplified* into a gigantic, coherent pulse of light!

Although the avalanche of Figure 29.36 illustrates the idea most clearly, it is not necessary for every atom to be in level 2 for amplification to occur. All that is needed is to have $N_2 > N_1$ so that stimulated emission exceeds absorption. Such a situation is called a **population inversion.** The stimulated emission is sustained by placing the *lasing medium*—the sample of atoms that emits the light—in an **optical cavity** consisting of two facing mirrors. As FIGURE 29.37 shows, the photons interact repeatedly with the atoms in the medium as they bounce back and forth. This repeated interaction is necessary for the light intensity to build up to a high level. If one of the mirrors is partially transmitting, some of the light emerges as the *laser beam.*

FIGURE 29.37 Lasing takes place in an optical cavity.

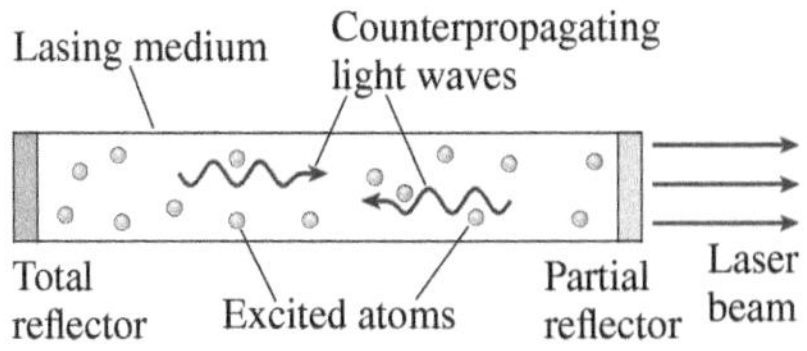

Lasers in Medicine

The invention of lasers was followed almost immediately by medical applications. Even a small-power laser beam can produce a significant amount of very localized *heating* if focused with a lens. More powerful lasers can easily *cut* through tissue by literally vaporizing it, replacing a stainless steel scalpel with a beam of light. Not only can laser surgery be very precise, but it generally involves less bleeding than conventional surgery because the heat of the laser seals the blood vessels and capillaries.

One common medical use of lasers is to remove plaque from artery walls, thus reducing the risk of stroke or heart attack. In this procedure, an optical fiber is threaded through arteries to reach the site. A powerful laser beam is then fired through the fiber to carefully vaporize the plaque. Laser beams traveling through optical fibers are also used to kill cancer cells in *photodynamic therapy.* In this case, light-sensitive chemicals are injected into the bloodstream and are preferentially taken up by cancer cells. The optical fiber is positioned next to the tumor and illuminates it with just the right wavelength to activate the light-sensitive chemicals and kill the cells. These procedures are minimally invasive, and they can reach areas of the body not readily accessible by conventional surgery.

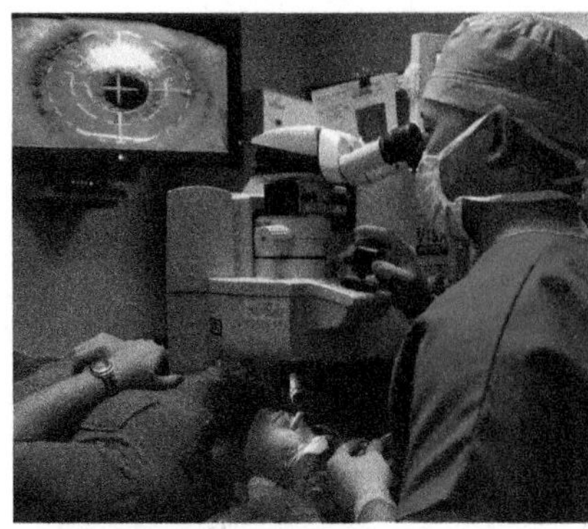

Laser vision Laser-based LASIK surgery can correct for vision defects, such as near- or farsightedness, that result from an incorrect refractive power of the eye. You learned in Chapter 20 that the majority of your eye's focusing power occurs at the surface of the cornea. In LASIK, a special knife first cuts a small, thin flap in the cornea, and this flap is folded out of the way. A computer-controlled ultraviolet laser very carefully vaporizes the underlying corneal tissue to give it the desired shape; then the flap is folded back into place. The procedure takes only a few minutes and requires only a few numbing drops in the eye.

CHAPTER 29 INTEGRATED EXAMPLE Compact fluorescent lighting

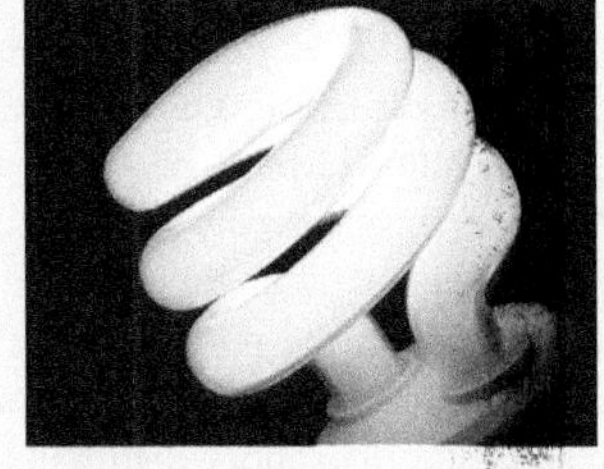

You learned in Chapter 24 how an ordinary incandescent lightbulb works: Current passes through a filament, heating it until it glows white hot. But such bulbs are very inefficient, providing only a few watts of visible light for every 100 W of electric power supplied to the bulb. Most of the power is, instead, converted into thermal energy.

This is why, in recent years, incandescent lighting has given way to other types of lighting: compact fluorescent and LED bulbs. These bulbs are four to six times more efficient than incandescent bulbs at transforming electric energy into visible light. We'll look at the physics of compact fluorescent bulbs.

Inside the glass tube of a fluorescent bulb is a very small amount of mercury, which is in the form of a vapor when the bulb is on. Producing visible light occurs by a three-step process. First, a voltage of about 100 V is applied between electrodes at each end of the tube. This imparts kinetic energy to free electrons in the vapor, causing them to slam into mercury atoms and excite them by collisional excitation. Second, the excited atoms undergo quantum jumps to lower-energy states, many of which emit photons that have ultraviolet (UV) wavelengths. Finally, the UV photons strike a *phosphor* that coats the inside of the tube, causing it to fluoresce with visible light. This is the light you see.

a. A mercury atom, after being collisionally excited by an electron, emits a photon with a wavelength of 185 nm in a quantum jump back to the ground state. Through what minimum distance must an electron accelerate to gain enough kinetic energy to cause this excitation? Assume that the electron starts from rest and the 60-cm-long tube has 120 V applied between its ends.

b. After being collisionally excited, atoms sometimes emit two photons by jumping first from a high energy level to an intermediate level, emitting one photon, and then from this intermediate level to the ground state, emitting a second photon. A mercury atom that has been excited to an energy level 7.72 eV above the ground state emits a photon with a wavelength of 436 nm and then a second photon. What is the wavelength of the second photon?

c. A compact fluorescent bulb is coated with three phosphors that provide red, green, and blue fluorescence. A mix of these three *primary colors* is perceived as white light. FIGURE 29.38 is an energy diagram for the green phosphor. What range of wavelengths does this phosphor emit by fluorescence after absorbing UV light?

PREPARE An electron collisionally excites the mercury atom from its ground state to an excited state, increasing the atom's energy by ΔE_{atom}. Then the atom decays back to the ground state, emitting a 185-nm-wavelength photon. The kinetic energy of the incident electron must equal or exceed ΔE_{atom}; that is, it must equal or exceed the energy of a 185-nm-wavelength photon. The free electrons gain kinetic energy by accelerating through the potential difference inside the tube. We can use conservation of energy to find the distance an electron must travel to gain ΔE_{atom}, the minimum kinetic energy needed to cause an excitation.

In part b, energy conservation requires that the sum of the energies of the two photons must equal the atom's excitation energy of 7.72 eV. If we calculate the energy of the first photon, we can determine the energy and wavelength of the second.

Figure 29.32 showed that quantum jumps during fluorescence all begin at the bottom of the band of excited states but can end anywhere in the lower energy band. This range of energies will give us the range of wavelengths of the emitted photons.

SOLVE

a. The minimum kinetic energy of the electron equals the energy of the 185-nm-wavelength photon that is subsequently emitted. This energy is

$$K_{\text{min}} = E_{\text{photon}} = \frac{hc}{\lambda} = \frac{1240 \text{ eV} \cdot \text{nm}}{185 \text{ nm}} = 6.7 \text{ eV}$$

where we have used the value of hc from Equation 29.9.

FIGURE 29.38 Energy diagram of a compact fluorescent bulb's green phosphor.

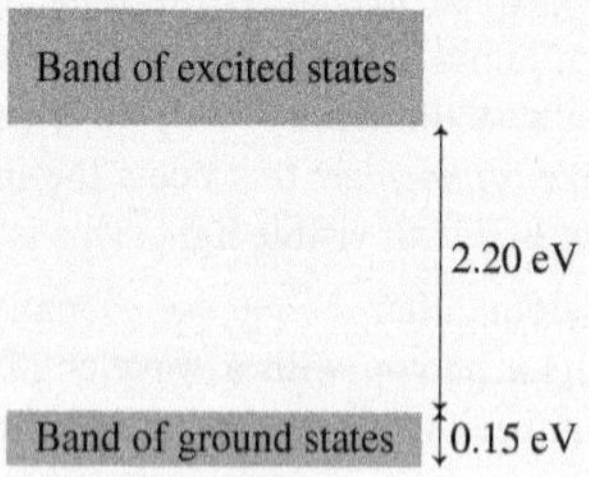

Recall that 1 eV is the kinetic energy gained by an electron as it accelerates through a 1 V potential difference. Here, the electron must gain a kinetic energy of 6.7 eV, so it must accelerate through a potential difference of 6.7 V. The fluorescent tube has a total potential drop of 120 V in 60 cm, or 2.0 V/cm. Thus to accelerate through a 6.7 V potential difference, the electron must travel a distance

$$\Delta x = \frac{6.7 \text{ V}}{2.0 \text{ V/cm}} = 3.4 \text{ cm}$$

b. The first emitted photon has energy

$$E_{\text{photon}} = \frac{hc}{\lambda} = \frac{1240 \text{ eV} \cdot \text{nm}}{436 \text{ nm}} = 2.84 \text{ eV}$$

The energy remaining to the second photon is then 7.72 eV − 2.84 eV = 4.88 eV. The wavelength of this photon is

$$\lambda = \frac{hc}{E_{\text{photon}}} = \frac{1240 \text{ eV} \cdot \text{nm}}{4.88 \text{ eV}} = 254 \text{ nm}$$

c. The energy of the photon emitted during a quantum jump from the bottom of the upper energy band to the top of the lower energy band is 2.20 eV, corresponding to a wavelength of

$$\lambda = \frac{hc}{E_{\text{photon}}} = \frac{1240 \text{ eV} \cdot \text{nm}}{2.20 \text{ eV}} = 564 \text{ nm}$$

The photon energy for a jump to the bottom of the lower energy band is 2.20 eV + 0.15 eV = 2.35 eV. The wavelength of this photon is

$$\lambda = \frac{hc}{E_{\text{photon}}} = \frac{1240 \text{ eV} \cdot \text{nm}}{2.35 \text{ eV}} = 528 \text{ nm}$$

Thus this phosphor, after absorbing UV photons, emits visible light with a wavelength range of 528–564 nm.

ASSESS For part a, it seems reasonable that the electron accelerates for only a small fraction of the tube length before gaining enough energy to collisionally excite an atom. The bulb would be very inefficient if electrons had to travel the full length before colliding with a mercury atom. At the same time, bulbs have to be filled with a low pressure (≈5 mm Hg) of inert gas (usually argon) so that electrons can travel several cm without other losses. The photon-wavelength range in part c is right in the middle of the visible spectrum, appropriate for a green phosphor. Slightly altering the balance between the red and blue phosphors distinguishes a "warm white" bulb from a "cool white" bulb.

SUMMARY

GOAL To use quantum physics to understand the structure and properties of atoms and molecules.

IMPORTANT CONCEPTS

The Structure of an Atom

An atom consists of a very small, positively charged nucleus, surrounded by orbiting electrons.

- The number of protons is the atom's **atomic number** Z.
- The atomic **mass number** A is the number of protons + the number of neutrons.

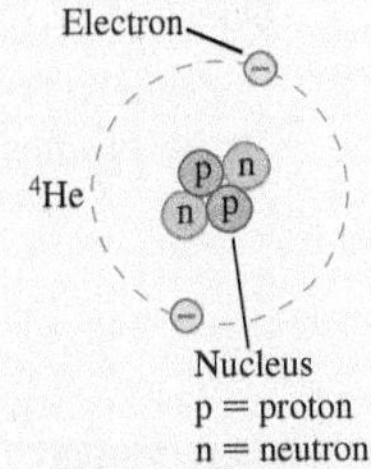

The Bohr Atom

In Bohr's model,

- The atom can exist in only certain **stationary states**. These states correspond to different electron orbits. Each state is numbered by **quantum number** $n = 1, 2, 3, \ldots$.
- Each state has a discrete, well-defined energy E_n.
- The atom can change its energy by undergoing a quantum jump between two states by emitting or absorbing a photon of energy $E_{\text{photon}} = \Delta E_{\text{atom}} = |E_f - E_i|$.

The hydrogen atom

The electron in a hydrogen atom must form a three-dimensional de Broglie standing wave. The conditions for doing so require four quantizations and four quantum numbers.

- The *energy,* principle quantum number n:

$$E_n = \frac{13.60 \text{ eV}}{n^2} \qquad n = 1, 2, 3, \ldots$$

- The *orbital angular momentum,* orbital quantum number l:

$$L = \sqrt{l(l+1)}\hbar \qquad l = 0, 1, 2, \ldots, n-1$$

- The *orbital tilt,* magnetic quantum number m:

$$m = -l, -l+1, \ldots, 0, \ldots, l-1, l$$

- The *electron spin direction,* spin quantum number m_s:

$$m_s = -\tfrac{1}{2} \text{ or } \tfrac{1}{2}$$

The orbital quantum number is usually represented by a letter:

s	$l = 0$	d	$l = 2$
p	$l = 1$	f	$l = 3$

Multi-electron atoms

Each electron in the atom is described by the same four quantum numbers (n, l, m, m_s) used for the electron in the hydrogen atom, but the energy now depends on l as well as n due to electron-electron interactions.

The **Pauli exclusion principle** states that no more than one electron can occupy each quantum state.

According to the rules of quantum numbers, a subshell of the atom characterized by quantum numbers n and l holds $2(2l+1)$ electrons. An s subshell holds 2 electrons, a p subshell 6 electrons, and a d subshell 10 electrons. This explains the structure of the periodic table of the elements.

0 eV
3s 3p 3d
2s 2p
The energy depends on only n.
1s
Hydrogen

0 eV
3p 3d
3s
2p
2s
The energy depends on n and l.
1s
Multi-electron atom

Molecules

In molecules, each electron energy level is split into a large number of vibrational energy levels. For large molecules, the energy levels are smeared into continuous bands of allowed energies. Thus the absorption spectra of large molecules are broad, not discrete.

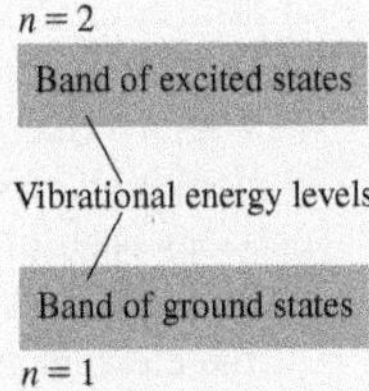

Fluorescence

Fluorescence is the emission of longer-wavelength photons after the absorption of shorter-wavelength photons. Internal conversion of absorbed energy to vibrational energy drops a molecule's energy to the lower edge of the band of excited states so that quantum jumps to the ground state have less energy.

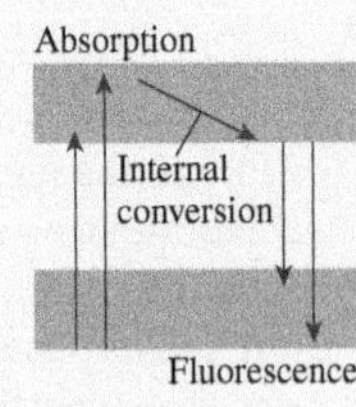

APPLICATIONS

Spectra

Absorption spectra occur from the ground state.

Emission spectra are generated by *excitation* followed by a photon-emitting quantum jump.

- Excitation occurs by absorption of a photon or by collision.
- A quantum jump can occur only if $\Delta l = \pm 1$.

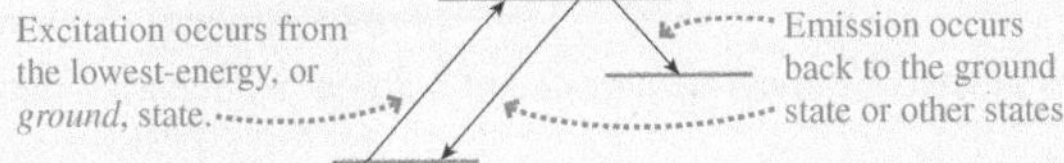

Lasers

A photon with energy $E_{\text{photon}} = E_2 - E_1$ can induce **stimulated emission** of a second photon identical to the first. These photons can then induce more atoms to emit photons. If there are more atoms in level 2 than in level 1—an **inverted population**—this process can cause an avalanche-like buildup of an intense beam of identical photons.

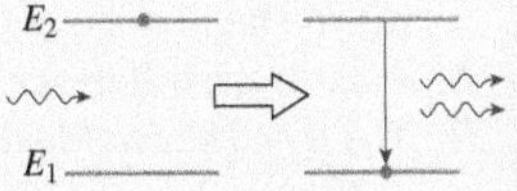

LEARNING OBJECTIVES After studying this chapter, you should be able to:

- Work with and distinguish between emission and absorption spectra. *Conceptual Questions 29.3, 29.4; Problems 29.1–29.3*
- Understand and apply the nuclear model of the atom. *Conceptual Question 29.1; Problems 29.4, 29.6, 29.9, 29.10, 29.13*
- Calculate energy levels and spectra of the hydrogen atom. *Conceptual Questions 29.7, 29.8; Problems 29.18–29.20, 29.22, 29.24*
- Determine the electron configurations of multi-electron atoms. *Conceptual Question 29.10; Problems 29.25–29.27, 29.28, 29.30*
- Solve problems about the spectra of multi-electron atoms. *Conceptual Questions 29.11, 29.12; Problems 29.32–29.36*
- Explain the features of molecular absorption spectra. *Problems 29.37, 29.38*
- Interpret and use the fluorescence of biomolecules. *Conceptual Question 29.15; Problems 29.39–29.42*
- Understand how lasers work. *Conceptual Questions 29.13, 29.14; Problems 29.43–29.47*

STOP TO THINK ANSWERS

Stop to Think 29.1: A is emission, B is absorption. All wavelengths in the absorption spectrum are seen in the emission spectrum, but not all wavelengths in the emission spectrum are seen in the absorption spectrum.

Stop to Think 29.2: 6 protons and 8 neutrons. The number of protons is the atomic number, which is 6. That leaves $14 - 6 = 8$ neutrons.

Stop to Think 29.3: In emission from the $n = 3$ to $n = 2$ transition, but not in absorption. The photon energy has to match the energy *difference* between two energy levels. Absorption is from the ground state, at $E_1 = 0$ eV. There's no energy level at 3 eV to which the atom could jump.

Stop to Think 29.4: $n = 3, l = 1$, or a 3*p* state.

Stop to Think 29.5: A. An inspection of the periodic table in Figure 29.21 shows that the element that has three of the possible six $3p$ states filled is phosphorus (P).

Stop to Think 29.6: C. Emission is a quantum jump to a lower-energy state. The $5p \rightarrow 4p$ transition is not allowed because $\Delta l = 0$ violates the selection rule. The lowest-energy allowed transition is $5p \rightarrow 3d$, with $E_{photon} = \Delta E_{atom} = 3.0$ eV.

Stop to Think 29.7: Lower. A patient with hypoxia has less oxygenated hemoglobin and more deoxygenated hemoglobin. This causes increased absorption at 660 nm and decreased absorption at 940 nm. A pulse oximeter detects *transmitted* light, and there is less transmission at 660 nm, more transmission at 940 nm.

QUESTIONS

Conceptual Questions

1. Can different atoms of the same element have different values of Z? Of A? Explain.
2. A spectral line with a wavelength of 656 nm is a prominent feature in the emission spectrum of hydrogen. Is there a spectral line with a wavelength of 656 nm in the absorption spectrum of hydrogen? Explain.
3. A neon discharge emits a bright reddish-orange spectrum. But a glass tube filled with neon is completely transparent. Why doesn't the neon in the tube absorb orange and red wavelengths?
4. A hypothetical atom has four energy levels. How many lines are in its emission spectrum? In its absorption spectrum?
5. Consider the three hydrogen-atom states 6*p*, 5*d*, and 4*f*. Which has the highest energy?
6. What is the difference between l and L?
7. The $n = 3$ state of hydrogen has $E_3 = -1.51$ eV.
   a. Why is the energy negative?
   b. What is the physical significance of the specific number 1.51 eV?
8. How would you label the hydrogen-atom states with the following (n, l, m) quantum numbers: (a) $(4, 3, 0)$, (b) $(3, 2, 1)$, (c) $(3, 2, -1)$?
9. A hydrogen atom is in a state with principal quantum number $n = 5$. What possible values of the orbital quantum number l could this atom have?
10. Does each of the configurations in Figure Q29.10 represent a possible electron configuration of an element? If so, (i) identify the element and (ii) determine whether this is the ground state or an excited state. If not, why not?

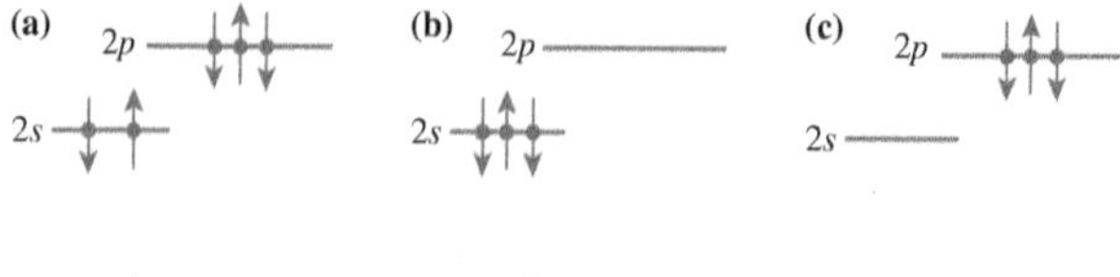

FIGURE Q29.10

Problem difficulty is labeled as | (straightforward) to ||||| (challenging). Problems labeled INT integrate significant material from earlier chapters; BIO are of biological or medical interest; CALC require calculus to solve.

11. An electron is in an f state. Can it undergo a quantum jump to an s state? A p state? A d state? Explain.
12. Figure Q29.12 shows the energy levels of a hypothetical atom.
    a. What *minimum* kinetic energy (in eV) must an electron have to collisionally excite this atom and cause the emission of a photon with a wavelength of 620 nm? Explain.
    b. Can the emission of a photon with a wavelength of 620 nm be caused by the absorption of a 5.0 eV photon?

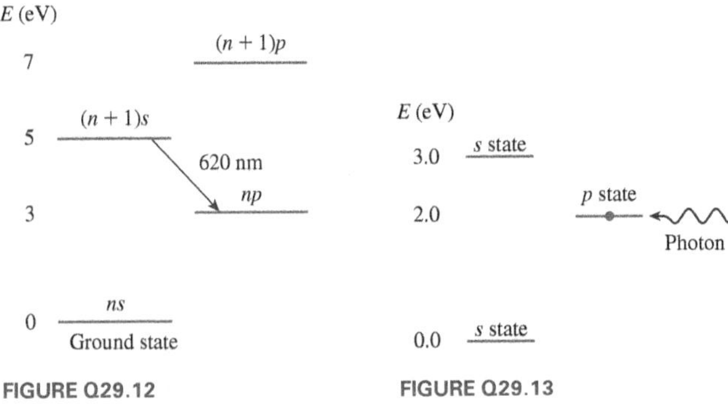

FIGURE Q29.12

FIGURE Q29.13

13. A 2.0 eV photon is incident on an atom in the p state, as shown in the energy-level diagram in Figure Q29.13. Does the atom undergo an absorption transition, a stimulated emission transition, or neither? Explain.
14. A glass tube contains 2×10^{11} atoms, some of which are in the ground state and some of which are excited. Figure Q29.14 shows the populations for the atoms' three energy levels. Is it possible for these atoms to be a laser? If so, on which transition would laser action occur? If not, why not?

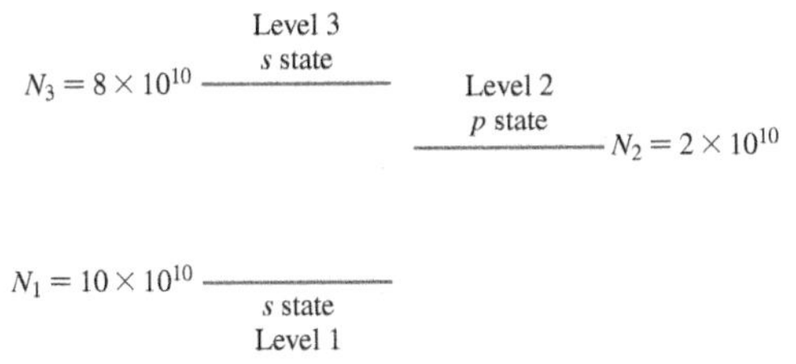

Level 3
s state
$N_3 = 8 \times 10^{10}$
Level 2
p state
$N_2 = 2 \times 10^{10}$
$N_1 = 10 \times 10^{10}$
s state
Level 1

FIGURE Q29.14

15. We usually associate fluorescence with ultraviolet light—objects glow under "black light." But it's possible to excite fluorescence with visible light as well. The light from a green laser pointer causes certain dyes to emit red light. Could the laser pointer cause a dye to emit blue light? Explain.

Multiple-Choice Questions

16. | An atom emits a photon with a wavelength of 275 nm. By how much does the atom's energy change?
    A. 0.72 eV B. 1.06 eV C. 2.29 eV
    D. 3.06 eV E. 4.51 eV
17. || An electron collides with an atom in its ground state. The atom then emits a photon of energy E_{photon}. In this process the *change* ΔE_{elec} in the electron's energy is
    A. Greater than E_{photon}.
    B. Greater than or equal to E_{photon}.
    C. Equal to E_{photon}.
    D. Less than or equal to E_{photon}.
    E. Less than E_{photon}.
18. || How many states are in the $l = 4$ subshell?
    A. 8 B. 9 C. 16 D. 18 E. 22
19. | What is the magnitude of the orbital angular momentum of a hydrogen atom in a $3d$ state?
    A. $2\hbar$ B. $\sqrt{6}\hbar$ C. $3\hbar$ D. $\sqrt{12}\hbar$
20. | A "soft x-ray" photon with an energy of 41.8 eV is absorbed by a hydrogen atom in its ground state, knocking the atom's electron out. What is the speed of the electron as it leaves the atom?
    A. 1.84×10^5 m/s B. 3.08×10^5 m/s
    C. 8.16×10^5 m/s D. 3.15×10^6 m/s
    E. 3.83×10^6 m/s
21. | What is the ground-state electron configuration of calcium ($Z = 20$)?
    A. $1s^2 2s^2 2p^6 3s^2 3p^8$ B. $1s^2 2s^2 2p^6 3s^2 3p^6 4s^1 4p^1$
    C. $1s^2 2s^2 2p^6 3s^2 3p^6 4s^2$ D. $1s^2 2s^2 2p^6 3s^2 3p^6 4p^2$
22. | A fluorescent molecule absorbs a photon with a wavelength of 420 nm and emits a photon with a wavelength of 560 nm. How much energy does the molecule dissipate internally?
    A. 0.74 eV B. 2.21 eV
    C. 2.95 eV D. 5.17 eV
23. | Figure Q29.23 is the energy-level diagram of a hypothetical atom. How many lines are in the atom's emission spectrum?
    A. 2 B. 4 C. 6 D. 10

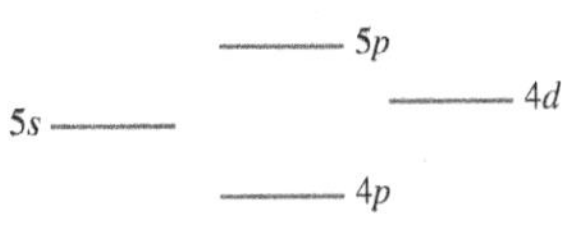

FIGURE Q29.23

PROBLEMS

Section 29.1 Spectroscopy

1. | Table 29.1 shows the wavelengths of the first four lines in the visible spectrum of hydrogen.
    a. What are the values of n and m in Equation 29.2 for these wavelengths in the hydrogen spectrum?
    b. Predict the wavelength of the fifth line in the series.
2. | The wavelengths in the hydrogen spectrum with $m = 1$ form a series of spectral lines called the Lyman series. Calculate the wavelengths of the first four members of the series.
3. | The Paschen series is analogous to the Balmer series, but with $m = 3$. Calculate the wavelengths of the first three members in the Paschen series. What part(s) of the electromagnetic spectrum are these in?

Section 29.2 Atoms

4. | How many electrons, protons, and neutrons are contained in the following atoms or ions: (a) ^{6}Li, (b) $^{13}C^{+}$, and (c) $^{18}O^{++}$?

5. | How many electrons, protons, and neutrons are contained in the following atoms or ions: (a) $^{9}Be^{+}$, (b) ^{12}C, and (c) $^{15}N^{+++}$?
6. | Write the symbol for an atom or ion with
   a. Four electrons, four protons, and five neutrons.
   b. Six electrons, seven protons, and eight neutrons.
7. | Write the symbol for an atom or ion with
   a. Three electrons, three protons, and five neutrons.
   b. Five electrons, six protons, and eight neutrons.
8. | Identify the isotope that is 11 times as heavy as ^{12}C and has 18 times as many protons as ^{6}Li. Give your answer in the form ^{A}S, where S is the symbol for the element. See Appendix D: Atomic and Nuclear Data.
9. | Consider the gold isotope ^{197}Au.
   a. How many electrons, protons, and neutrons are in a neutral ^{197}Au atom?
   b. The gold nucleus has a diameter of 14.0 fm. What is the density of matter in a gold nucleus?
   c. The density of gold is 19,300 kg/m^3. How many times the density of gold is your answer to part b?
10. ||| In a student lab experiment, 5.3 MeV alpha particles from
INT the decay of ^{210}Po are directed at a piece of thin platinum foil. If an alpha particle is directed straight toward the nucleus of a platinum atom, what is the distance of closest approach? How does this compare to the approximately 7 fm radius of the nucleus?
11. ||| In a head-on collision, the closest approach of a 6.24 MeV
INT alpha particle to the center of a nucleus is 6.00 fm. The nucleus is in an atom of what element? Assume the nucleus remains at rest.
12. ||| A 20 MeV alpha particle is fired toward a ^{238}U nucleus. It
INT follows the path shown in Figure P29.12. What is the alpha particle's speed when it is closest to the nucleus, 20 fm from its center? Assume that the nucleus doesn't move.

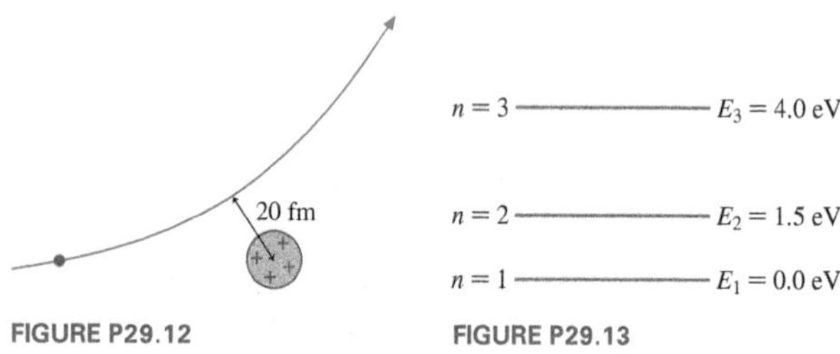

FIGURE P29.12

FIGURE P29.13

13. | Figure P29.13 is an energy-level diagram for a simple atom. What wavelengths appear in the atom's (a) emission spectrum and (b) absorption spectrum?
14. || An electron with 2.0 eV of kinetic energy collides with an atom whose energy-level diagram is shown in Figure P29.13. The electron kicks the atom into an excited state. What is the electron's kinetic energy after the collision?
15. | The allowed energies of a simple atom are 0.0 eV, 4.0 eV, and 6.0 eV.
   a. Draw the atom's energy-level diagram. Label each level with the energy and the principal quantum number.
   b. What wavelengths appear in the atom's emission spectrum?
   c. What wavelengths appear in the atom's absorption spectrum?
16. || An electron with a speed of 5.00×10^6 m/s collides with an atom. The collision excites the atom from its ground state (0 eV) to a state with an energy of 3.80 eV. What is the speed of the electron after the collision?
17. ||| The allowed energies of a simple atom are 0.0 eV, 4.0 eV, and 6.0 eV. An electron traveling at a speed of 1.6×10^6 m/s collisionally excites the atom. What are the minimum and maximum speeds the electron could have after the collision?

Section 29.3 The Hydrogen Atom

18. | List the quantum numbers of (a) all possible $3p$ states and (b) all possible $3d$ states.
19. | When all quantum numbers are considered, how many different quantum states are there for a hydrogen atom with $n = 1$? With $n = 2$? With $n = 3$? List the quantum numbers of each state.
20. | What is the angular momentum of a hydrogen atom in (a) a $4p$ state and (b) a $5f$ state? Give your answers as a multiple of $\hbar$.
21. || The energy of a hydrogen atom is 12.09 eV above its ground-state energy. As a multiple of $\hbar$, what is the largest angular momentum that this atom could have?
22. || A hydrogen atom is in the $5p$ state. Determine (a) its energy, (b) its quantum number l, (c) its angular momentum, and (d) the possible values of its magnetic quantum number m.
23. | A hydrogen atom has orbital angular momentum 3.65×10^{-34} J·s.
   a. What letter (s, p, d, or f) describes the electron?
   b. What is the atom's minimum possible energy? Explain.
24. | Radio astronomers search the universe not for visible light but for long-wavelength radio waves. A spectral feature that is seen in all directions has a wavelength of 21.1 cm. This is an emission from hydrogen atoms, the most abundant atoms in the universe, but it's not from an excited state. An interaction between the small magnetic dipole moments of the electron and the proton causes a very slight energy difference between a spin-up ($m_s = +\frac{1}{2}$) and a spin-down ($m_s = -\frac{1}{2}$) electron in the $1s$ ground state. A photon with a wavelength of 21.1 cm is emitted when a ground-state hydrogen atom undergoes a spin-flip quantum jump from the higher-energy spin-up state to the lower-energy spin-down state. What is the energy separation in μeV between these states?

Section 29.4 Multi-electron Atoms

25. | Predict the ground-state electron configurations of Mg, Sr, and Ba.
26. || Predict the ground-state electron configurations of Si, Ge, and Pb.
27. | If elements beyond $Z = 120$ are ever synthesized, electrons in these heavy atoms will begin filling a g subshell, corresponding to $l = 4$. How many states will be in a g subshell?
28. | Identify the element for each of these electron configurations. Then determine whether this configuration is the ground state or an excited state.
   a. $1s^2 2s^2 2p^5$
   b. $1s^2 2s^2 2p^6 3s^2 3p^6 3d^{10} 4s^2 4p$
29. | Identify the element for each of these electron configurations. Then determine whether this configuration is the ground state or an excited state.
   a. $1s^2 2s^2 2p^6 3s^2 3p^6 4s^2 3d^9$
   b. $1s^2 2s^2 2p^6 3s^2 3p^6 4s^2 3d^{10} 4p^6 5s^2 4d^{10} 5p^6 6s^2 4f^{14} 5d^7$
30. | Explain what is wrong with these electron configurations:
   a. $1s^2 2s^2 2p^8 3s^2 3p^4$
   b. $1s^2 2s^3 2p^4$

Section 29.5 Excited States and Spectra

31. | Hydrogen gas absorbs light of wavelength 103 nm. Afterward, what wavelengths are seen in the emission spectrum?

32. || A $3s \rightarrow 4s$ transition in sodium is not allowed by the selection rule. However, the $4s$ state can be reached in a two-step process in which photon absorption is followed by photon emission. (See Figure 29.22.) What is the longest possible wavelength of the emitted photon?

33. || An electron with a kinetic energy of 3.90 eV collides with a sodium atom. What possible wavelengths of light are subsequently emitted?

34. || What is the electron configuration of the first excited state of lithium?

35. || An electron accelerates through a 12.5 V potential difference,
INT starting from rest, and then collides with a hydrogen atom, exciting the atom to the highest energy level allowed. List all the possible quantum-jump transitions by which the excited atom could emit a photon and the wavelength (in nm) of each.

36. | a. Is a $4p \rightarrow 4s$ transition allowed in sodium? If so, what is its wavelength? If not, why not?
b. Is a $3d \rightarrow 4s$ transition allowed in sodium? If so, what is its wavelength? If not, why not?

Section 29.6 Molecules

Section 29.7 Fluorescence and Bioluminescence

37. || Molecules can vibrate or oscillate like masses connected by springs, and the vibrational energies are quantized. The diatomic molecule HCl has vibrational energy levels with quantum numbers $n = 1, 2, 3, \ldots$. The infrared spectrum of HCl shows absorption lines corresponding to transitions from the $n = 1$ ground state to the excited vibrational states. Transitions to $n = 2$, 3, and 4 occur at wavelengths 3.47 μm, 1.76 μm, and 1.20 μm.
a. What are the energies in eV of the $n = 2$, 3, and 4 energy levels? Assume that the $n = 1$ ground state has $E_1 = 0$ eV.
b. What is the wavelength of the absorption line that corresponds to a transition from the first to the second excited state?

38. ||| Molecules can vibrate or oscillate like masses connected by
INT springs, and the vibrational energies are quantized. The carbon dioxide molecule CO_2 is a linear molecule (O–C–O), and one vibrational mode is the *bending mode* in which the two oxygen atoms move up while the carbon atom moves down, then vice versa half a cycle later. The infrared absorption spectrum of CO_2 has a strong absorption line at a wavelength of 19.0 μm that corresponds to a $n = 1$ to $n = 2$ transition of the bending mode's energy, where v is the vibrational quantum number. What is the equilibrium population ratio N_2/N_1 of these two energy levels for CO_2 molecules at a temperature of 20°C?

39. || Figure P29.39 shows a molecular energy-level diagram. What are the longest and shortest wavelengths in (a) the molecule's absorption spectrum and (b) the molecule's fluorescence spectrum?

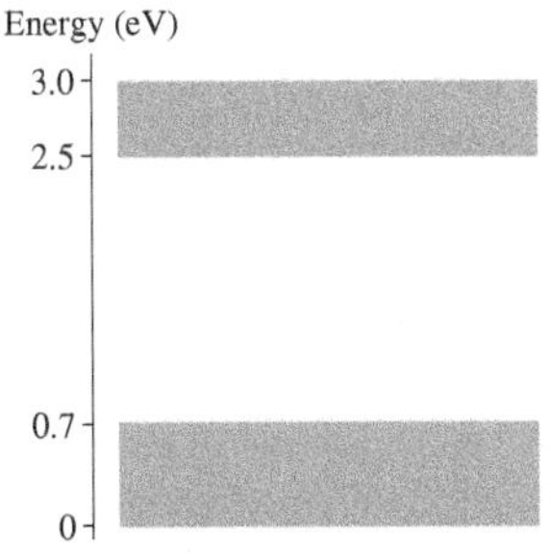

FIGURE P29.39

40. | The molecule whose energy-level diagram is shown in Figure P29.39 is illuminated by 2.7 eV photons. What is the longest wavelength of light that the molecule can emit?

41. || Rhodamine 6G, abbreviated R6G, is an organic dye that is often used in fluorescence microscopy. The absorption spectrum of R6G spans the range 480 nm to 540 nm; its emission spectrum runs from 540 nm to 620 nm.
a. What are the energies in eV of the upper and lower edges of the $n = 2$ band of excited states? Assume that the lowest possible R6G energy is 0 eV.
b. What is the energy in eV of the upper edge of the $n = 1$ band of states?

42. || Not every photon absorbed by a fluorescent molecule leads
BIO to the emission of a lower-energy photon. The ratio of emitted photons to absorbed photons is called the *quantum efficiency*. When excited with 395 nm ultraviolet light, the green fluorescent protein emits 510 nm light with a quantum efficiency of 0.79. What is the *energy efficiency* of the green fluorescent protein for converting ultraviolet light to visible light—that is, the ratio of the energy output to the energy input?

Section 29.8 Stimulated Emission and Lasers

43. || A 1.00 mW helium-neon laser emits a visible laser beam with a wavelength of 633 nm. How many photons are emitted per second?

44. || In LASIK surgery, a laser is used to reshape the cornea of
BIO the eye to improve vision. The laser produces extremely short pulses of light, each containing 1.0 mJ of energy.
a. In each pulse there are 9.7×10^{14} photons. What is the wavelength of the laser?
b. Each pulse lasts only 20 ns. What is the average power delivered to the eye during a pulse?

45. || Port-wine birthmarks, which are caused by malformed capil-
BIO laries close to the skin, can be removed with laser pulses. Laser light of the right color will pass through the skin with little absorption but will be strongly absorbed by oxyhemoglobin in the blood of these capillaries, which destroys them without damaging adjacent tissue. A typical system uses a series of 6900 W pulses of 585 nm laser light. Each pulse lasts 0.45 ms. How many photons are in each pulse?

46. || A ruby laser emits an intense pulse of light that lasts a mere 10 ns. The light has a wavelength of 690 nm, and each pulse has an energy of 500 mJ.
a. How many photons are emitted in each pulse?
b. What is the *rate* of photon emission, in photons per second, during the 10 ns that the laser is "on"?

47. | A laser emits 1.0×10^{19} photons per second from an excited state with energy $E_2 = 1.17$ eV. The lower energy level is $E_1 = 0$ eV.
a. What is the wavelength of this laser?
b. What is the power output of this laser?

General Problems

48. ||| The diameter of an atom is 1.2×10^{-10} m and the diameter of its nucleus is 1.0×10^{-14} m. What percent of the atom's volume is occupied by mass and what percent is empty space?

49. |||| In Example 29.2 it was assumed that the initially stationary
INT gold nucleus would remain motionless during a head-on collision with an 8.3 MeV alpha particle. What is the actual recoil speed of the gold nucleus after that elastic collision? Assume that the mass of a gold nucleus is exactly 50 times the mass of an alpha particle. **Hint:** Review the discussion of perfectly elastic collisions in Chapter 8.

50. ||| Through what potential difference would you need to accel-
INT erate an alpha particle, starting from rest, so that it will just reach the surface of a 15-fm-diameter ^{238}U nucleus?

51. ||| The oxygen nucleus ^{16}O has a radius of 3.0 fm.
INT a. With what speed must a proton be fired toward an oxygen nucleus to have a turning point 1.0 fm from the surface? Assume that the nucleus is heavy enough to remain stationary during the collision.
b. What is the proton's kinetic energy in MeV?

52. ||| A 2.55 eV photon is emitted from a hydrogen atom. What are the values of n and m in Equation 29.2 for the wavelengths of the hydrogen spectrum?

53. || The first three energy levels of the fictitious element X are shown in Figure P29.53.
a. What wavelengths are observed in the absorption spectrum of element X? Give your answers in nm.
b. State whether each of your wavelengths in part a corresponds to ultraviolet, visible, or infrared light.

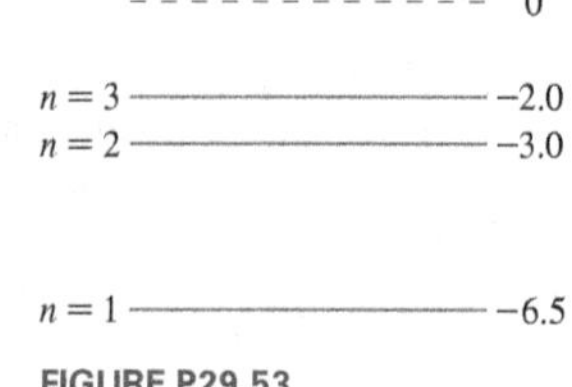

FIGURE P29.53

c. An electron with a speed of 1.4×10^6 m/s collides with an atom of element X. Shortly afterward, the atom emits a 1240 nm photon. What was the electron's speed after the collision? Assume that, because the atom is so much more massive than the electron, the recoil of the atom is negligible.
Hint: The energy of the photon is *not* the energy transferred to the atom in the collision.

54. ||| A simple atom has only two absorption lines, at 250 nm and 600 nm. What is the wavelength of the one line in the emission spectrum that does not appear in the absorption spectrum?

55. || A hydrogen atom in the ground state absorbs a 12.75 eV photon. Immediately after the absorption, the atom undergoes a quantum jump to the next-lowest energy level. What is the wavelength of the photon emitted in this quantum jump?

56. ||| A beam of electrons is incident on a gas of hydrogen atoms.
INT a. What minimum speed must the electrons have to cause the emission of 656 nm light from the $3 \rightarrow 2$ transition of hydrogen?
b. Through what potential difference must the electrons be accelerated to have this speed?

57. ||| Two hydrogen atoms collide head-on. The collision brings
INT both atoms to a halt. Immediately after the collision, both atoms emit a 121.6 nm photon. What was the speed of each atom just before the collision?

58. ||| Germicidal lamps are used to sterilize tools in biological
BIO and medical facilities. One type of germicidal lamp is a low-
INT pressure mercury discharge tube, similar to a fluorescent lightbulb, that has been optimized to emit ultraviolet light with a wavelength of 254 nm. A 15-mm-diameter, 25-cm-long tube emits 4.5 W of ultraviolet light. The mercury vapor pressure inside the tube is 1.0 Pa at the operating temperature of 40°C. On average, how many ultraviolet photons does each mercury atom emit per second?

59. || If N_0 atoms are excited at $t = 0$, perhaps by a very brief laser pulse, the number that remain in the excited state at a later time t is $N = N_0 e^{-t/\tau}$, where τ is called the *lifetime* of the excited state. In other words, the number of excited atoms decays exponentially with time. Suppose 8.0×10^{10} sodium atoms are suddenly excited to the $3p$ state. 6.0×10^{10} photons are emitted by these atoms during the next 24 ns. What is the lifetime of the $3p$ state of sodium?

60. | a. What downward transitions are possible for a sodium atom in the $6s$ state? (See Figure 29.22.)
b. What are the wavelengths of the photons emitted in each of these transitions?

61. || The $5d \rightarrow 3p$ transition in the emission spectrum of sodium (see Figure 29.22) has a wavelength of 499 nm. What is the energy of the $5d$ state?

62. ||| A sodium atom (see Figure 29.26) emits a photon with wavelength 818 nm shortly after being struck by an electron. What minimum speed did the electron have before the collision?

63. | Figure P29.63 shows the first few energy levels of the lithium atom. Make a table showing all the allowed transitions in the emission spectrum. For each transition, indicate
a. The wavelength, in nm.
b. Whether the transition is in the infrared, the visible, or the ultraviolet spectral region.
c. Whether or not the transition would be observed in the lithium absorption spectrum.

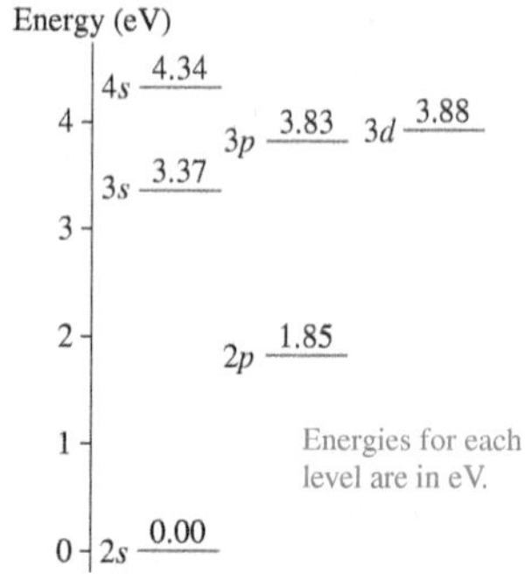

Energy (eV)
4s 4.34; 3p 3.83; 3d 3.88; 3s 3.37; 2p 1.85; 2s 0.00
Energies for each level are in eV.
FIGURE P29.63

64. ||| A simple model can show how pulse oximetry is able to mea-
BIO sure the oxygen saturation of blood. The key idea is that the intensity of light decreases exponentially as it travels through an absorbing medium. This can be written $I = I_0 e^{-\mu x}$, where I_0 is the incident intensity, x is the distance traveled through the medium, and μ is called the *absorption coefficient.* A larger absorption coefficient causes the intensity to decrease more rapidly. At a wavelength of 660 nm, the absorption coefficients of oxygenated hemoglobin HbO_2 and deoxygenated hemoglobin Hb are 2.5 cm^{-1} and 10 cm^{-1}, respectively. Absorption coefficients are customarily given in units of cm^{-1}, so the distance x must be in cm. At a wavelength of 940 nm, the absorption coefficients of HbO_2 and Hb are 7.0 cm^{-1} and 3.0 cm^{-1}. In pulse oximetry, light travels approximately 1.0 cm from the LEDs to the detectors. Assume that absorption by hemoglobin is the only factor that affects the light intensity.
a. Pulse oximetry measures the ratio I_{660}/I_{940} of the detected intensities at the two wavelengths. Assume that both wavelengths enter the finger with the same intensity. What is the ratio if the hemoglobin is completely deoxygenated, corresponding to an oxygen saturation of 0%?
b. What is the ratio if the hemoglobin is completely oxygenated, corresponding to an oxygen saturation of 100%? You can see that the ratio is very sensitive to the oxygen saturation level. The calculation is more complex, but a measured ratio between these two extremes can be used to deduce the fraction of hemoglobin that is oxygenated.

65. |||| In fluorescence microscopy, an important tool in biology, BIO a laser beam is absorbed by target molecules in a sample. These molecules are then imaged by a microscope as they emit longer-wavelength photons in quantum jumps back to lower energy levels, a process known as *fluorescence*. A variation on this technique is *two-photon excitation*. If two photons are absorbed simultaneously, their energies add. Consequently, a molecule that is normally excited by a photon of energy E_{photon} can be excited by the simultaneous absorption of two photons having half as much energy. For this process to be useful, the sample must be irradiated at the very high intensity of at least 10^{32} photons/m$^2\cdot$s. This is achieved by concentrating the laser power into very short pulses (100 fs pulse length) and then focusing the laser beam to a small spot. The laser is fired at the rate of 10^8 pulses each second. Suppose a biologist wants to use two-photon excitation to excite a molecule that in normal fluorescence microscopy would be excited by a laser with a wavelength of 420 nm. If she focuses the laser beam to a 2.0-μm-diameter spot, what minimum energy must each pulse have?

MCAT-Style Passage Problems

Light-Emitting Diodes

Light-emitting diodes, known by the acronym LED, produce the familiar green and red indicator lights used in a wide variety of consumer electronics. LEDs are *semiconductor devices* in which the electrons can exist only in certain energy levels. Much like molecules, the energy levels are packed together close enough to form what appears to be a continuous band of possible energies. Energy supplied to an LED in a circuit excites electrons from a *valence band* into a *conduction band*. An electron can emit a photon by undergoing a quantum jump from a filled state in the conduction band into an empty state in the valence band, as shown in Figure P29.66.

FIGURE P29.66 Energy-level diagram of an LED.

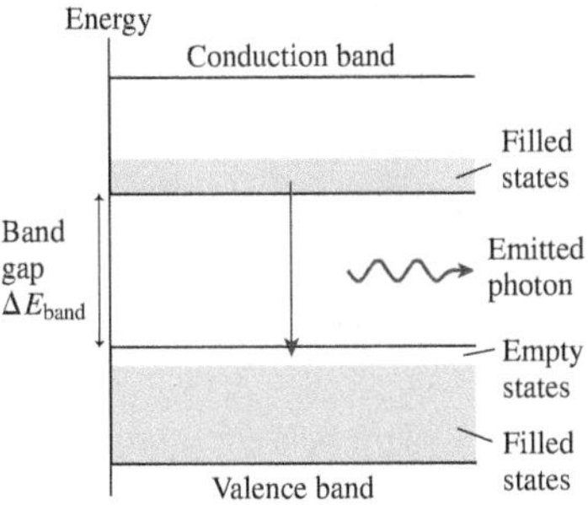

The size of the band gap ΔE_{band} determines the possible energies—and thus the wavelengths—of the emitted photons. Most LEDs emit a narrow range of wavelengths and thus have a distinct color. This makes them well-suited for traffic lights and other applications where a certain color is desired, but it makes them less desirable for general illumination. One way to make a "white" LED is to combine a blue LED with a substance that fluoresces yellow when illuminated with the blue light. The combination of the two colors makes light that appears reasonably white.

66. An LED emits green light. Increasing the size of the band gap could change the color of the emitted light to
    A. Red
    B. Orange
    C. Yellow
    D. Blue
67. Suppose the LED band gap is 2.5 eV, which corresponds to a wavelength of 500 nm. Consider the possible electron transitions in Figure P29.66. 500 nm is the
    A. Maximum wavelength of the LED.
    B. Average wavelength of the LED.
    C. Minimum wavelength of the LED.
68. The same kind of semiconducting material used to make an LED can also be used to convert absorbed light into electric energy, essentially operating as an LED in reverse. In this case, the absorption of a photon causes an electron transition from a filled state in the valence band to an unfilled state in the conduction band. If $\Delta E_{band} = 1.4$ eV, what is the minimum wavelength of electromagnetic radiation that could lead to electric energy output?
    A. 140 nm
    B. 890 nm
    C. 1400 nm
    D. 8900 nm
69. The efficiency of a light source is the percentage of its energy input that gets radiated as visible light. If some of the blue light in an LED is used to cause a fluorescent material to glow,
    A. The overall efficiency of the LED is increased.
    B. The overall efficiency of the LED does not change.
    C. The overall efficiency of the LED decreases.

30 Nuclear Physics

Each of the radial lines is a subatomic particle created in a collision between two gold nuclei, each traveling at 60% the speed of light.

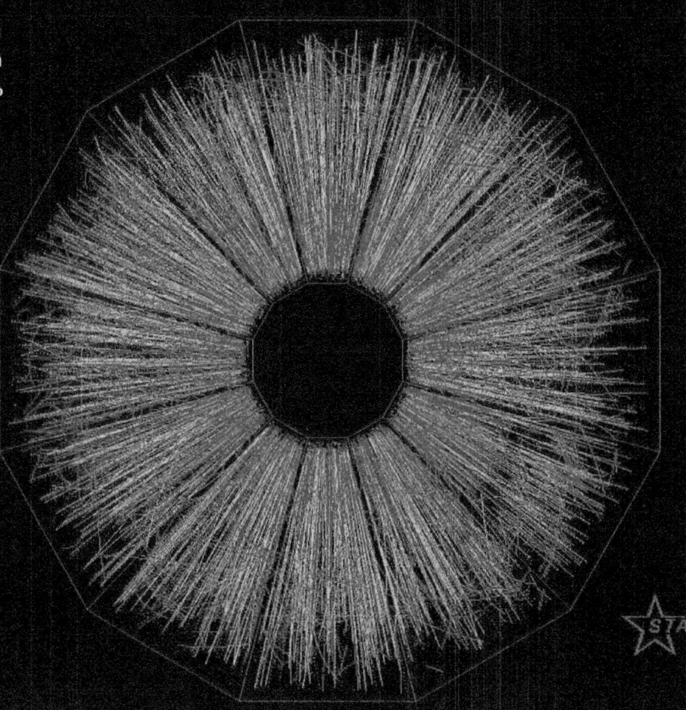

LOOKING AHEAD

Nuclei

A nuclear-powered ice breaker is on its way to the North Pole. *Nuclear fission* reactors power ships and entire cities.

You'll learn about the properties of nuclei and isotopes, and about the energy equivalent of mass, famously expressed as $E = mc^2$.

Radioactive Decay and Half-Lives

An environmental scientist uses a Geiger counter to check for radiation levels at a cleanup site.

You'll learn the ways in which a nucleus can undergo *radioactive decay* and about the *half-life* of radioactive isotopes.

Nuclear Medicine

This image of the liver, showing a tumor on the right side, was recorded by a camera that detects gamma-rays emitted by the isotope ^{99}Tc.

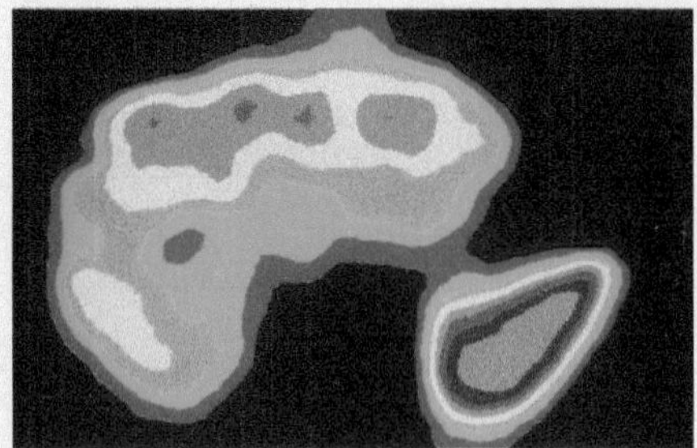

You'll explore how nuclear radiation can harm cells and also how nuclear radiation can be harnessed as a medical tool.

GOAL To learn about the nucleus of the atom and some of the medical applications of nuclear physics.

PHYSICS AND LIFE

Medical Imaging and Cancer Treatment Depend on Nuclear Physics

Ionizing radiation damages biological molecules and tissues by depositing enough energy to break bonds and eject electrons from atoms. This high-energy radiation can be dangerous in uncontrolled situations, but at the same time it is essential in medical imaging and cancer therapy. Effective therapeutic use of radiation requires an understanding of the types of ionizing radiation and the differences in how each interacts with tissue. X rays and gamma rays are very high-energy electromagnetic radiation, but other forms of ionizing radiation consist of high-energy particles; each type differs in how deeply it penetrates and how localized its damage is. Although x rays can be generated with exquisite control in an x-ray tube, gamma rays and many types of particle radiation are produced by random nuclear decays, a fundamentally statistical process. Understanding the particulars of each type of radiation is critical for administering effective and safe therapies.

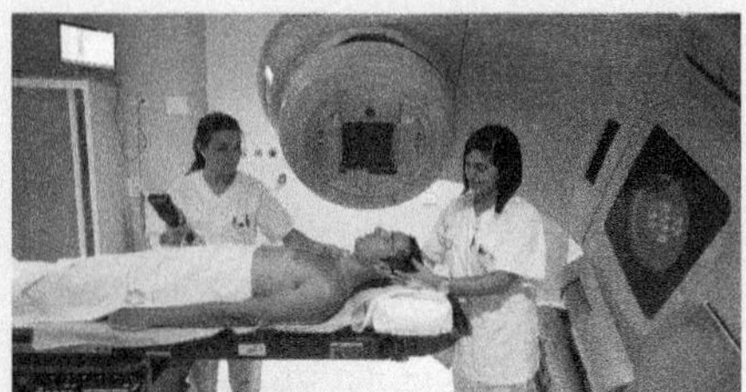

A patient is prepared to undergo proton radiation therapy, in which an intense beam of protons is targeted on a tumor.

30.1 Nuclear Structure

Nuclear reactions and nuclear radiation are like a double-edged sword: They have the potential to be extremely dangerous, but they are also a powerful medical tool for imaging and healing. Our goal for this chapter is to explore how nuclear radiation is emitted and how it interacts with matter, which can then serve as a basis for understanding its safe and beneficial use. We'll start with the structure and properties of the atomic nucleus because the physics of the nucleus determines how and why radiation is emitted.

In 1896, only a year after Röntgen discovered x rays, the French scientist A. H. Becquerel found that uranium crystals also emit some sort of ray that can expose film. Becquerel hoped that he had discovered a new way to generate x rays; instead, he had discovered what became known as *radioactivity.*

Ernest Rutherford began to investigate and soon found not one but three distinct kinds of rays emitted from crystals containing uranium. Not knowing what they were, he named them for their ability to penetrate matter and ionize air. The first, which caused the most ionization and penetrated the least, he called *alpha rays.* The second, with intermediate penetration and ionization, were *beta rays,* and the third, with the least ionization but the largest penetration, became *gamma rays.*

Within a few years, Rutherford was able to show that alpha rays are helium nuclei emitted from the crystal at very high speeds. These became the projectiles that he used in 1909 to probe the structure of the atom. The outcome of that experiment, as you learned in Chapter 29, was Rutherford's discovery that atoms have a very small, dense nucleus at the center.

Rutherford's discovery of the nucleus may have settled the question of atomic structure, but it raised many new issues for scientific research. Foremost among them were:

- What is nuclear matter? What are its properties?
- What holds the nucleus together? Why doesn't the repulsive electrostatic force between protons blow it apart?
- What is the connection between the nucleus and radioactivity?

These questions marked the beginnings of **nuclear physics,** the study of the properties of the atomic nucleus.

The world atom If an atom were the size of the Unisphere in New York City, the nucleus would be only the size of a pea.

Nucleons

The nucleus is a tiny speck in the center of a vastly larger atom. As FIGURE 30.1 shows, the nuclear diameter of roughly 10^{-14} m is only about 1/10,000 the diameter of the atom. Even so, the nucleus is more than 99.9% of the atom's mass. What we call *matter* is overwhelmingly empty space!

The nucleus is composed of two types of particles: *protons* and *neutrons,* which together are referred to as **nucleons.** The role of the neutrons, which have nothing to do with keeping electrons in orbit, is an important issue that we'll address in this chapter. TABLE 30.1 summarizes the basic properties of protons and neutrons.

As you can see, protons and neutrons are virtually identical other than that the proton has one unit of the fundamental charge e whereas the neutron is electrically neutral. The neutron is slightly more massive than the proton, but the difference is very small, only about 0.1%. Notice that the proton and neutron, like the electron, have an *inherent angular momentum* and magnetic moment with spin $\frac{1}{2}$. As a consequence, protons and neutrons obey the Pauli exclusion principle.

The number of protons Z is the element's **atomic number.** In fact, an element is identified by the number of protons in the nucleus, not by the number of orbiting electrons. Electrons are easily added and removed, forming negative and positive ions, but doing so doesn't change the element. The **mass number** A is defined to be $A = Z + N$, where N is the **neutron number.** The mass number is the total number of nucleons in a nucleus.

FIGURE 30.1 The nucleus is a tiny speck within an atom.

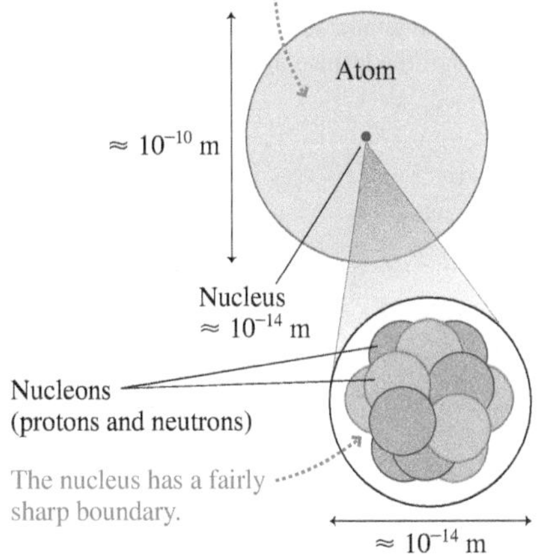
This picture of an atom would need to be 10 m in diameter if it were drawn to the same scale as the dot representing the nucleus.
Atom
≈ 10^{-10} m
Nucleus
≈ 10^{-14} m
Nucleons
(protons and neutrons)
The nucleus has a fairly sharp boundary.
≈ 10^{-14} m

TABLE 30.1 Protons and neutrons

	Proton	Neutron
Number	Z	N
Charge q	$+e$	0
Spin	$\frac{1}{2}$	$\frac{1}{2}$
Mass, in u	1.00728	1.00866

NOTE ▶ The mass number, which is dimensionless, is *not* the same thing as the atomic mass m. We'll look at actual atomic masses later. ◀

Isotopes

It was discovered early in the 20th century that atoms of the same element (same Z) can have different masses. There is a *range* of neutron numbers that happily form a nucleus with Z protons, creating a group of nuclei having the same Z-value (i.e., they are all the same chemical element) but different A-values. The atoms of an element with different values of A are called **isotopes** of that element.

Chemical behavior is determined by the orbiting electrons. All isotopes of one element have the same number of orbiting electrons (if the atoms are electrically neutral) and thus have the same chemical properties, but different isotopes of the same element can have quite different nuclear properties.

The notation used to label isotopes is ^{A}Z, where the mass number A is given as a *leading* superscript. The proton number Z is not specified by an actual number but, equivalently, by the chemical symbol for that element. Hence ordinary carbon, which has six protons and six neutrons in the nucleus, is written ^{12}C and pronounced "carbon twelve." The radioactive form of carbon used in carbon dating is ^{14}C. It has six protons, making it carbon, and eight neutrons.

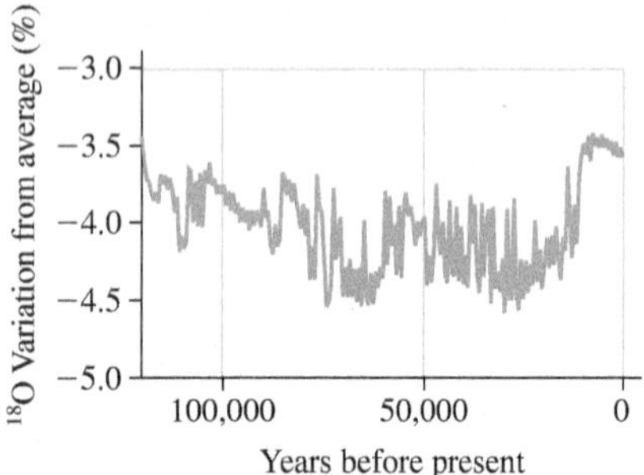

Reading ice cores When water freezes to make snow crystals, the fraction of molecules containing ^{18}O is greater for snow that forms at higher atmospheric temperatures. Snow accumulating over tens of thousands of years has built up a thick ice sheet in Greenland. A core sample of this ice gives a record of the isotopic composition of the snow that fell over this time period. Higher numbers on the graph correspond to higher average temperatures. Broad trends, such as the increase in temperature 20,000 years ago at the end of the last ice age, are clearly seen.

More than 3000 isotopes are known. The majority of these are **radioactive,** meaning that the nucleus is not stable but, after some period of time, will either fragment or emit some kind of subatomic particle in an effort to reach a more stable state. Most of these radioactive isotopes are created by nuclear reactions in the laboratory and have only a fleeting existence. Only 254 isotopes are **stable** (i.e., nonradioactive) and occur in nature. We'll begin to look at the issue of nuclear stability in the next section.

The *naturally occurring* nuclei include the 254 stable isotopes and a handful of radioactive isotopes with such long half-lives, measured in billions of years, that they also occur naturally. The most well-known example of a naturally occurring radioactive isotope is the uranium isotope ^{238}U. For each element, the fraction of naturally occurring nuclei represented by one particular isotope is called the **natural abundance** of that isotope.

Although there are many radioactive isotopes of the element iodine, iodine occurs *naturally* only as ^{127}I. Consequently, we say that the natural abundance of ^{127}I is 100%. Most elements have multiple naturally occurring isotopes. The natural abundance of ^{14}N is 99.6%, meaning that 996 out of every 1000 naturally occurring nitrogen atoms are the isotope ^{14}N. The remaining 0.4% of naturally occurring nitrogen is the isotope ^{15}N, with one extra neutron.

Atomic Mass

You learned in Chapter 12 that atomic masses are specified in terms of the *atomic mass unit* u (equivalent to the *dalton* that is used in chemistry), defined such that the atomic mass of the isotope ^{12}C is exactly 12 u. The conversion to SI units is

$$1 \text{ u} = 1.6605 \times 10^{-27} \text{ kg}$$

The distinction between mass and energy is clear for atoms and for macroscopic objects, but the boundary between mass and energy begins to blur as we move into the realm of nuclear physics and the physics of subatomic particles. Possibly the most famous equation in physics is Einstein's $E = mc^2$, developed as part of his theory of relativity. As we understand physics today, mass and energy are not the same but are *equivalent* in the sense that mass can be transformed into energy and energy can be transformed into mass. Consequently, a careful statement of energy conservation has to include transformations between mass and energy.

In *nuclear reactions,* the mass of the products turns out to be slightly less than the mass of the reactants. The "lost" mass is transformed into energy, which provides the enormous energy release of nuclear reactors and nuclear weapons. A more direct example, shown in FIGURE 30.2, is the collision between an electron and an antielectron, also known as a *positron.* An antielectron has exactly the same mass and spin as an electron, but its charge is $+e$ rather than $-e$. (Physicists discovered in the second half of the 20th century that every fundamental particle has an antiparticle with the opposite charge.)

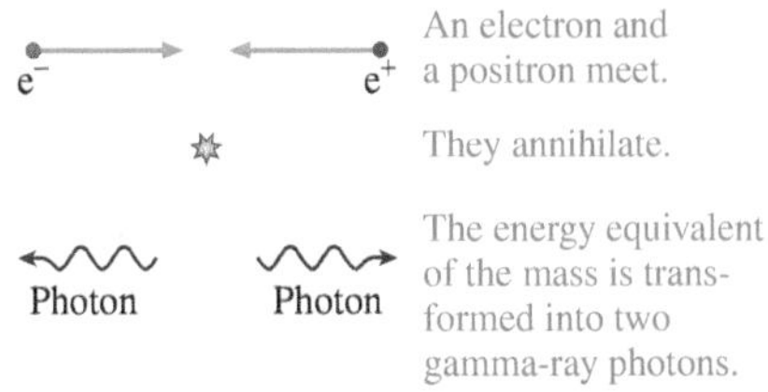

When an electron and a positron collide, they *annihilate*—the particles disappear—and the energy equivalent of their mass is transformed into two very high-energy gamma-ray photons. This is the radiation detected in a PET scan (positron-electron tomography), which we'll discuss later in the chapter.

FIGURE 30.2 The annihilation of an electron and a positron transforms their energy into the energy of two photons.

e⁻
e⁺
An electron and a positron meet.
They annihilate.
Photon
Photon
The energy equivalent of the mass is transformed into two gamma-ray photons.

We can use $E = mc^2$ to calculate the wavelength of the photons. First we need to find the energy equivalent of an electron or positron:

$$\begin{aligned} E = mc^2 &= (9.11 \times 10^{-31}\ \text{kg})(3.00 \times 10^8\ \text{m/s})^2 \\ &= 8.20 \times 10^{-14}\ \text{J} \times \frac{1\ \text{eV}}{1.60 \times 10^{-19}\ \text{J}} = 512\ \text{keV} \end{aligned} \tag{30.1}$$

The energy equivalent of an electron or positron is a factor of 10^5 larger than the typical spacing between atomic energy levels of a few eV. An electron-positron annihilation must conserve energy and momentum, and this is accomplished by the back-to-back emission in opposite directions of two 512 keV photons. The wavelength of each of these very high-energy photons is

$$\lambda = \frac{1240\ \text{eV} \cdot \text{nm}}{512{,}000\ \text{eV}} = 0.0024\ \text{nm} \tag{30.2}$$

This is an unimaginably small wavelength, roughly a factor of 1000 shorter than x-ray wavelengths.

The energy equivalent of 1 u of mass, approximately the mass of a proton, is even larger:

$$\begin{aligned} E &= (1.6605 \times 10^{-27}\ \text{kg})(2.9979 \times 10^8\ \text{m/s})^2 \\ &= 1.4924 \times 10^{-10}\ \text{J} = 931.49\ \text{MeV} \end{aligned} \tag{30.3}$$

Thus the atomic mass unit can be written

$$1\ \text{u} = 931.49\ \text{MeV}/c^2$$

It may seem unusual, but the units MeV/c^2 are units of mass. This will be a useful conversion factor when we need to compute energy equivalents.

NOTE ▶ We're using more significant figures than usual. Many nuclear calculations look for the small difference between two masses that are almost the same. Those two masses must be calculated or specified to four or five significant figures if their difference is to be meaningful. ◀

TABLE 30.2 shows the atomic masses of the electron, the nucleons, and three important light elements. Appendix D contains a more complete list for many isotopes. Notice that the mass of a hydrogen atom is the sum of the masses of a proton and an electron. But a quick calculation shows that the mass of a helium atom (2 protons, 2 neutrons, and 2 electrons) is 0.03038 u less than the sum of the masses of its constituents. The difference is due to the binding energy of the nucleus, a topic we'll look at in Section 30.2.

The isotope ^{2}H is a hydrogen atom in which the nucleus is not simply a proton but a proton and a neutron. Although the isotope is a form of hydrogen, it is called **deuterium.** The natural abundance of deuterium is 0.015%, or about 1 out of every

TABLE 30.2 Some atomic masses

Particle	Symbol	Mass (u)	Mass (MeV/c^2)
Electron	e	0.00055	0.51
Proton	p	1.00728	938.28
Neutron	n	1.00866	939.57
Hydrogen	^{1}H	1.00783	938.79
Deuterium	^{2}H	2.01410	1876.12
Helium	^{4}He	4.00260	3728.40

6700 hydrogen atoms. Water made with deuterium (sometimes written D_2O rather than H_2O) is called *heavy water.*

NOTE ▶ Don't let the name *deuterium* cause you to think this is a different element. Deuterium is an isotope of hydrogen. Chemically, it behaves just like ordinary hydrogen. ◀

The *chemical* atomic mass shown on the periodic table of the elements is the *weighted average* of the atomic masses of all naturally occurring isotopes. For example, chlorine has two stable isotopes: ^{35}Cl, with $m = 34.97$ u, is 75.8% abundant and ^{37}Cl, at 36.97 u, is 24.2% abundant. The average, weighted by abundance, is $0.758 \times 34.97 + 0.242 \times 36.97 = 35.45$. This is the value shown on the periodic table and is the correct value for most chemical calculations, but it is not the mass of any particular isotope of chlorine.

NOTE ▶ The atomic masses of the proton and the neutron are both ≈ 1 u. Consequently, the value of the mass number A is *approximately* the atomic mass in u. The approximation $m \approx A$ u is sufficient in many contexts, such as when we're calculating the masses of atoms in the kinetic theory of gases, but in nuclear physics calculations we almost always need the more accurate mass values that you find in Table 30.2 or Appendix D. ◀

Nuclear Size and Density

FIGURE 30.3 The nuclear radius and volume as a function of A.

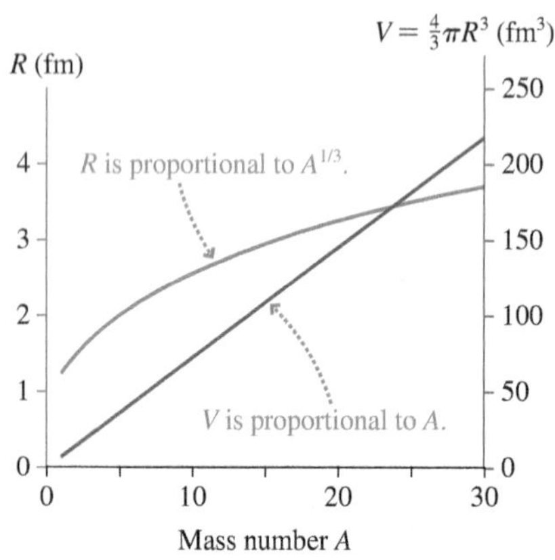

Unlike the atom's electron cloud, which is quite diffuse, the nucleus has a fairly sharp boundary. Experimentally, the radius of a nucleus with mass number A is found to be

$$R = r_0 A^{1/3} \tag{30.4}$$

where $r_0 = 1.2$ fm. Recall that 1 fm = 1 femtometer = 10^{-15} m.

As FIGURE 30.3 shows, the radius is proportional to $A^{1/3}$. Consequently, the volume of the nucleus (proportional to R^3) is directly proportional to A, the number of nucleons. A nucleus with twice as many nucleons will occupy twice as much volume. This finding has three implications:

- Nucleons are incompressible. Adding more nucleons doesn't squeeze the inner nucleons into a smaller volume.
- The nucleons are tightly packed, looking much like the drawing in Figure 30.1.
- Nuclear matter has a constant density.

In fact, we can use Equation 30.4 to calculate the density of nuclear matter. Consider a nucleus with mass number A. Its mass, within 1%, is A atomic mass units. Thus

$$\begin{aligned}\rho_{\text{nuc}} &\approx \frac{A\text{ u}}{\frac{4}{3}\pi R^3} = \frac{A\text{ u}}{\frac{4}{3}\pi r_0^3 A} = \frac{1\text{ u}}{\frac{4}{3}\pi r_0^3} = \frac{1.66\times10^{-27}\text{ kg}}{\frac{4}{3}\pi(1.2\times10^{-15}\text{ m})^3} \\ &= 2.3\times10^{17}\text{ kg/m}^3\end{aligned} \tag{30.5}$$

The fact that A cancels means that **all nuclei have this density.** It is a staggeringly large density, roughly 10^{14} times larger than the density of familiar liquids and solids. One early objection to Rutherford's model of a nuclear atom was that matter simply couldn't have a density this high. Although we have no direct experience with such matter, nuclear matter really is this dense.

FIGURE 30.4 Density profiles of three nuclei.

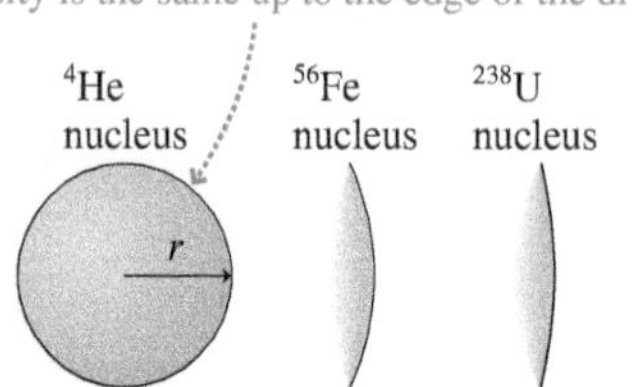

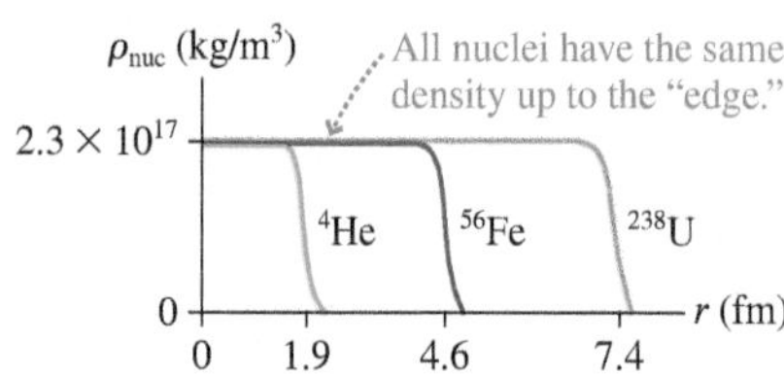

FIGURE 30.4 shows the density profiles of three nuclei. The constant density right to the edge is analogous to that of a drop of incompressible liquid, and, indeed, one successful model of many nuclear properties is called the **liquid-drop model.** Notice that the range of nuclear radii, from small helium to large uranium, is not quite a factor of 4. The fact that ^{56}Fe is a fairly typical atom in the middle of the periodic table is the basis for our earlier assertion that the nuclear diameter is roughly 10^{-14} m, or 10 fm.

STOP TO THINK 30.1 Three electrons orbit a neutral ^{6}Li atom. How many electrons orbit a neutral ^{7}Li atom?

30.2 Nuclear Stability

We've noted that fewer than 10% of the known nuclei are stable (i.e., not radioactive). Because nuclei are characterized by two independent numbers, N and Z, it is useful to show the known nuclei on a plot of neutron number N versus proton number Z. FIGURE 30.5 shows such a plot. Stable nuclei are represented by blue diamonds and unstable, radioactive nuclei by red dots.

FIGURE 30.5 Stable and unstable nuclei shown on a plot of neutron number N versus proton number Z.

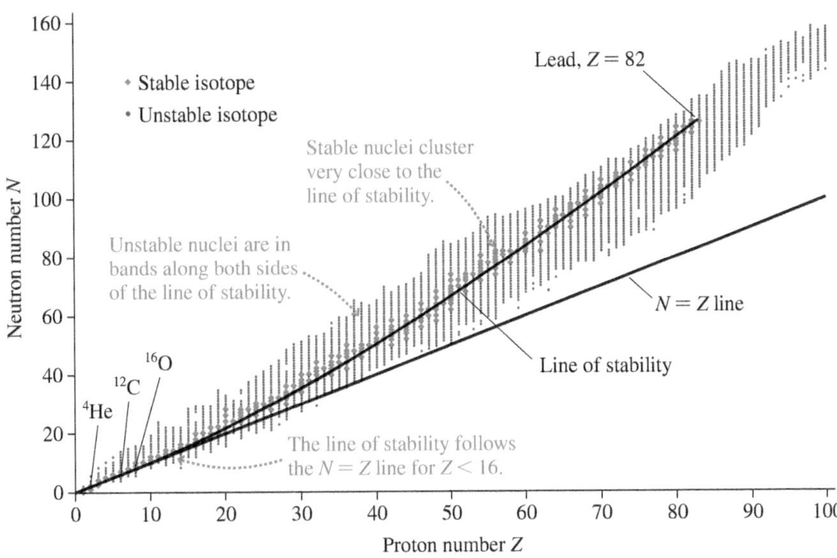

We can make several observations from this graph:

- The stable nuclei cluster very close to the curve called the **line of stability.**
- There are no stable nuclei with $Z > 82$ (lead).
- Unstable nuclei are in bands along both sides of the line of stability.
- The lightest elements, with $Z < 16$, are stable when $N \approx Z$. The familiar elements ^{4}He, ^{12}C, and ^{16}O all have equal numbers of protons and neutrons.
- As Z increases, the number of neutrons needed for stability grows increasingly larger than the number of protons. The N/Z ratio is ≈ 1.2 at $Z = 40$ but has grown to ≈ 1.5 at $Z = 80$.

STOP TO THINK 30.2 The isotopes of an element are found on the plot of Figure 30.5 along

A. A vertical line.
B. A horizontal line.
C. A diagonal line that goes up and to the left.
D. A diagonal line that goes up and to the right.

Binding Energy

A nucleus is a *bound system.* Just as energy is required to break the chemical bonds of stable molecules, we would need to supply energy to disperse the nucleons by breaking the nuclear bonds between them. FIGURE 30.6 shows this idea schematically.

FIGURE 30.6 The nuclear binding energy.

The binding energy is the energy that would be needed to disassemble a nucleus into individual nucleons.

$$B + m_{nuc}c^2 = (Zm_p + Nm_n)c^2$$

Unstable but ubiquitous uranium All the isotopes of uranium are unstable, but one is very long-lived. Half the ^{238}U that was present at the formation of the earth is still around—and it is found all around you at a low concentration in nearly all of the rocks and soil on the earth's surface.

Recall from Chapter 29 that it takes a well-defined amount of energy—the *ionization energy*—to remove an electron from an atom. In much the same way, the energy we would need to supply to a nucleus to disassemble it into individual protons and neutrons is called the **binding energy.** Whereas ionization energies of atoms are only a few eV, the binding energies of nuclei are tens or hundreds of MeV, energies large enough that their mass equivalent is not negligible.

Consider a nucleus with mass m_{nuc}. It is found experimentally that m_{nuc} is *less* than the total mass $Zm_p + Nm_n$ of the Z protons and N neutrons that form the nucleus, where m_p and m_n are the masses of the proton and neutron. That is, the energy equivalent $m_{nuc}c^2$ of the nucleus is less than the energy equivalent $(Zm_p + Nm_n)c^2$ of the individual nucleons. The binding energy B of the nucleus (not the entire atom) is defined as

$$B = (Zm_p + Nm_n - m_{nuc})c^2 \tag{30.6}$$

This is the energy we would need to supply to disassemble the nucleus into its pieces.

The practical difficulty with Equation 30.6 is that laboratory scientists use mass spectroscopy to measure *atomic* masses, not nuclear masses. The atomic mass m_{atom} is m_{nuc} plus the mass Zm_e of Z orbiting electrons. (Strictly speaking, we should allow for the binding energy of the electrons, but these binding energies are roughly a factor of 10^6 smaller than the nuclear binding energies and can be neglected in all but the most precise measurements and calculations.)

Fortunately, we can make Equation 30.6 more useful by adding and subtracting Z electron masses. We begin by writing Equation 30.6 in the equivalent form

$$B = (Zm_p + Zm_e + Nm_n - m_{nuc} - Zm_e)c^2 \tag{30.7}$$

Now $m_{nuc} + Zm_e = m_{atom}$, the atomic mass, and $Zm_p + Zm_e = Z(m_p + m_e) = Zm_H$, where m_H is the mass of a hydrogen *atom*. Finally, we use the conversion factor $1\text{ u} = 931.49\text{ MeV}/c^2$ to write $c^2 = 931.49\text{ MeV/u}$. The binding energy is then

$$B = (Zm_H + Nm_n - m_{atom}) \times (931.49\text{ MeV/u}) \tag{30.8}$$

Binding energy of a nucleus with Z protons and N neutrons

where all three masses are in atomic mass units.

EXAMPLE 30.1 **The binding energy of iron**

What is the binding energy of the ^{56}Fe nucleus?

PREPARE Iron is atomic number 26, so the isotope ^{56}Fe has $Z = 26$ and $N = A - Z = 30$. Appendix D gives the atomic mass of ^{56}Fe as 55.93494 u. The masses of a hydrogen atom and a neutron are listed in Table 30.2 as 1.00783 u and 1.00866 u.

SOLVE The nuclear binding energy is given by Equation 30.8:

$$B = [26(1.00783\text{ u}) + 30(1.00866\text{ u}) - 55.93494\text{ u}] \times 931.49\text{ MeV/u} = 492\text{ MeV}$$

ASSESS The binding energy is extremely large, the energy equivalent of more than half the mass of a proton or a neutron. In other words, the mass of a ^{56}Fe nucleus is less than the combined mass of its 56 protons and neutrons by more than half the mass of a proton. For another comparison, the energy released in the hydrolysis of one ATP molecule is roughly 0.5 eV. Hence the binding energy of a ^{56}Fe nucleus is equivalent to the energy released in the metabolism of a *billion* molecules of ATP.

The nuclear binding energy increases as A increases simply because there are more nuclear bonds. A more useful measure for comparing one nucleus to another is the quantity B/A, called the *binding energy per nucleon.* Iron, with $B = 492$ MeV and $A = 56$, has 8.79 MeV per nucleon. This is the amount of energy, on average, we would need to supply in order to remove *one* nucleon from the nucleus. Nuclei with larger values of B/A are more tightly held together than nuclei with smaller values of B/A.

FIGURE 30.7 The curve of binding energy.

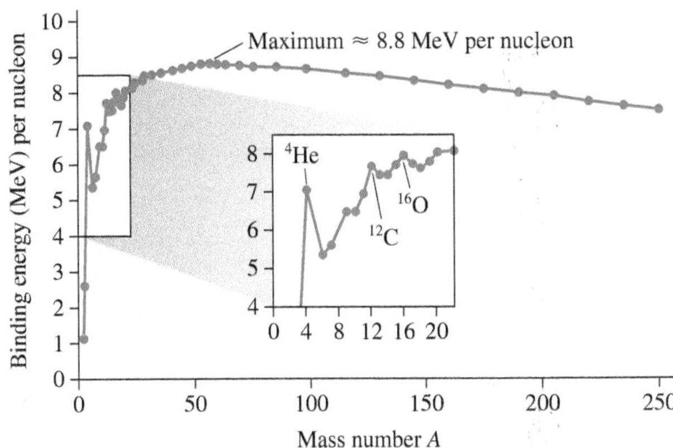

FIGURE 30.7 is a graph of the binding energy per nucleon versus mass number A. The line connecting the points is often called the **curve of binding energy.** This curve has three important features:

- There are peaks in the binding energy curve at $A = 4$, 12, and 16. The one at $A = 4$, corresponding to ^{4}He, is especially pronounced.
- The binding energy per nucleon is *roughly* constant at ≈ 8 MeV per nucleon for $A > 20$. This suggests that, as a nucleus grows, there comes a point where the nuclear bonds are *saturated.* Each nucleon interacts only with its nearest neighbors, the ones it's actually touching. This, in turn, implies that the nuclear force is a *short-range* force.
- The curve has a broad maximum at $A \approx 60$. This will be important for our understanding of radioactivity. In principle, heavier nuclei could become *more* stable (more binding energy per nucleon) by breaking into smaller pieces. Lighter nuclei could become *more* stable by fusing together into larger nuclei. There may not always be a mechanism by which such nuclear transformations can take place, but *if* there is a mechanism, it is energetically favorable for it to occur.

One example of these ideas is *nuclear fission,* a process in which a heavy nucleus—usually an isotope of uranium or plutonium—is induced to fragment into smaller pieces. The fission process that is used to produce energy in nuclear reactors occurs when a slow-moving neutron (n) collides with a ^{235}U nucleus—a nucleus far to the right on the curve of binding energy—to cause the reaction

$$\text{n} + {}^{235}\text{U} \rightarrow {}^{236}\text{U} \rightarrow {}^{90}\text{Sr} + {}^{144}\text{Xe} + 2\text{n} \qquad (30.9)$$

The ^{235}U nucleus absorbs the neutron to become ^{236}U, but ^{236}U is so unstable that it immediately fragments into a ^{90}Sr nucleus, a ^{144}Xe nucleus, and two neutrons. (The fact that the reaction generates extra neutrons, which can then cause additional fissions, is what allows a *chain reaction* to occur.) ^{90}Sr and ^{144}Xe are nearer the peak of the curve of binding energy than the ^{236}U nucleus and thus are more tightly bound, so a great deal of energy is released in this reaction. Fission is analogous to an exothermic reaction in chemistry in which the reaction products are more tightly bound than the reactants, but the energy released per reaction is roughly a factor of 10^6 greater.

EXAMPLE 30.2 **Nuclear fission**

How much energy is released in the nuclear fission reaction in Equation 30.9? A nuclear reactor generates 3.0 GW of power. If Equation 30.9 is the only fission reaction that occurs, and if the reactor generates power 80% of the time (a typical value), how many kilograms of ^{235}U does a reactor "burn" each year?

PREPARE Mass is lost in the fission reaction, and the mass loss Δm is transformed into an amount of energy that we can calculate with $E = mc^2$. The masses of the uranium and strontium isotopes are given in Appendix D; the mass of ^{144}Xe is 143.93851. We'll assume that the kinetic energies are negligible in comparison

Continued

with the energy equivalent of the masses. The reactor generates energy from fission at the rate of 3.0×10^9 J/s, and the number of seconds in a year can be calculated to be 3.15×10^7 s.

SOLVE The reaction has a net gain of one neutron, so the mass loss is

$$\begin{aligned}\Delta m &= m(^{235}\text{U}) - m(^{90}\text{Sr}) - m(^{144}\text{Xe}) - m(n)\\ &= 235.04392\text{ u} - 89.90774\text{ u} - 143.93851\text{ u} - 1.00867\text{ u}\\ &= 0.189\text{ u}\end{aligned}$$

We've seen that $1\text{ u} = 931.49\text{ MeV}/c^2$. This is a useful conversion factor because it avoids having to calculate c^2. The transformation of mass into energy yields

$$E_{\text{fission}} = (\Delta m)c^2 = (0.189\text{ u})(931.49\text{ MeV}/c^2)c^2 = 176\text{ MeV}$$

This is the energy released per reaction.

We can determine the number of fissions required to operate the reactor for a year by calculating the year's energy production in MeV:

$$\begin{aligned}E_{\text{reactor}} &= (0.80)(3.0 \times 10^9\text{ J/s})(3.15 \times 10^7\text{ s}) \times \frac{1\text{ eV}}{1.60 \times 10^{-19}\text{ J}}\\ &= 4.7 \times 10^{29}\text{ MeV}\end{aligned}$$

At 176 MeV per reaction, the number of reactions that occur in 1 yr is

$$N = \frac{E_{\text{reactor}}}{E_{\text{fission}}} = \frac{4.7 \times 10^{29}\text{ MeV}}{176\text{ MeV}} = 2.7 \times 10^{27}\text{ reactions}$$

Each reaction "burns" one ^{235}U nucleus, which has a mass of

$$m(^{235}\text{U}) = 235\text{ u} \times \frac{1.66 \times 10^{-27}\text{ kg}}{1\text{ u}} = 3.9 \times 10^{-25}\text{ kg}$$

The total mass of ^{235}U that undergoes fission is

$$\begin{aligned}M(^{235}\text{U}) &= Nm(^{235}\text{U})\\ &= (2.7 \times 10^{27})(3.9 \times 10^{-25}\text{ kg}) = 1000\text{ kg}\end{aligned}$$

ASSESS We assumed that Equation 30.9 is the only fission reaction. In fact, a ^{236}U nucleus can fragment in several different ways, but the energy yield is very similar for all of them. An online search finds that, indeed, a commercial nuclear reactor uses about 1000 kg of ^{235}U per year. Uranium is so dense that this amount of ^{235}U—enough to power a large city for a year—would occupy a volume of only 50 L, about the volume of your car's gasoline tank. However, such comparisons can be misleading. A nuclear reactor is not fueled with pure ^{235}U. The fuel is a mixture of ^{235}U (which fissions) and the more abundant ^{238}U (which doesn't), and the actual fuel use per year is more like 30,000 kg. Furthermore, the spent fuel—the reaction products—is *extremely* radioactive and will remain so for thousands of years. Dealing safely with even relatively small volumes of spent fuel is a challenge that does not yet have a clear solution.

30.3 The Strong Force

Rutherford's discovery of the atomic nucleus was not immediately accepted by all scientists. Their primary objection was that the protons would blow themselves apart at tremendously high speeds due to the extremely large electrostatic forces between them at a separation of a few femtometers. No known force could hold the nucleus together.

It soon became clear that a previously unknown force of nature operates within the nucleus to hold the nucleons together. This new force has to be stronger than the repulsive electrostatic force; hence it was named the **strong force.** It is also called the *nuclear force.*

The strong force has four important properties:

1. It is an *attractive* force between any two nucleons.
2. It does not act on electrons.
3. It is a *short-range* force, acting only over nuclear distances.
4. Over the range where it acts, it is *stronger* than the electrostatic force that tries to push two protons apart.

FIGURE 30.8 The strong force is the same between any two nucleons.

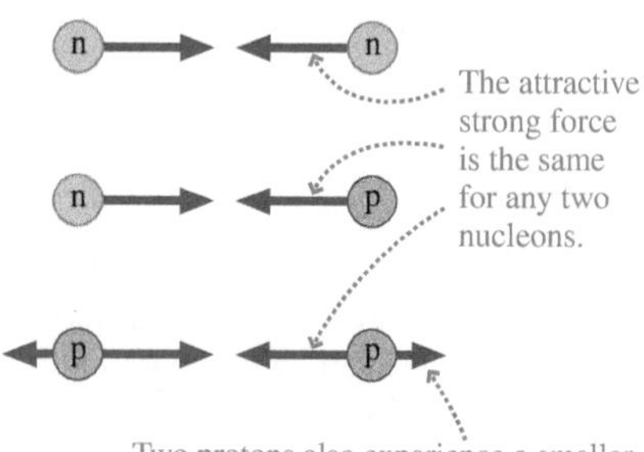
n
n
The attractive strong force is the same for any two nucleons.
n
p
p
p
Two protons also experience a smaller electrostatic repulsive force.

The fact that the strong force is short-range, in contrast to the long-range $1/r^2$ electric, magnetic, and gravitational forces, is apparent from the fact that we see no evidence for nuclear forces outside the nucleus.

FIGURE 30.8 summarizes the three interactions that take place within the nucleus. Whether the strong force between two protons is the same strength as the force between two neutrons or between a proton and a neutron is an important question that can be answered experimentally. The primary means of investigating the strong force is to accelerate a proton to very high speed, using a cyclotron or some other particle accelerator, then to study how the proton is scattered by various target materials.

The conclusion of many decades of research is that the strong force between two nucleons is independent of whether they are protons or neutrons. Charge is the basis for electromagnetic interactions, but it is of no relevance to the strong force. Protons and neutrons are identical as far as nuclear forces are concerned.

Potential Energy

Unfortunately, there's no simple formula to calculate the strong force or the potential energy of two nucleons interacting via the strong force. FIGURE 30.9 is an experimentally determined potential-energy diagram for two interacting nucleons, with r the distance between their centers. The potential-energy minimum at $r \approx 1$ fm is a point of stable equilibrium.

FIGURE 30.9 The potential-energy diagram for two nucleons interacting via the strong force.

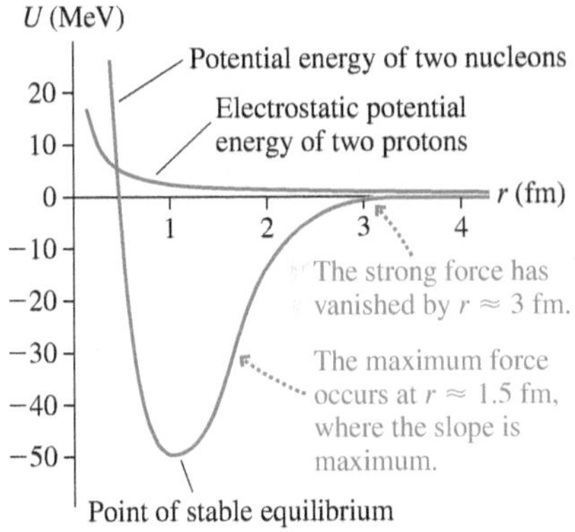

Recall that the force is the negative of the slope of a potential-energy diagram. The steeply rising potential for $r < 1$ fm represents a strongly repulsive force. That is, the nucleon "cores" strongly repel each other if they get too close together. The force is attractive for $r > 1$ fm, where the slope is positive, and it is strongest where the slope is steepest, at $r \approx 1.5$ fm. The strength of the force quickly decreases for $r > 1.5$ fm and is zero for $r > 3$ fm. That is, the strong force represented by this potential energy is effective only over a very short range of distances.

Notice how small the electrostatic energy of two protons is in comparison to the potential energy of the strong force. At $r \approx 1.0$ fm, the point of stable equilibrium, the magnitude of the nuclear potential energy is ≈ 100 times larger than the electrostatic potential energy.

We earlier asked what role neutrons play. Why does a nucleus need neutrons? The answer is related to the short range of the strong force. All protons in the nucleus exert repulsive electrostatic forces on each other, but, because of the short range of the strong force, a proton feels an attractive force only from the very few other nucleons with which it is in close contact. Even though the strong force at its maximum is much larger than the electrostatic force, there wouldn't be enough attractive nuclear bonds for an all-proton nucleus to be stable. Because neutrons participate in the strong force but exert no repulsive forces, **the neutrons provide the extra "glue" that holds the nucleus together.** In small nuclei, where most nucleons are in contact, one neutron per proton is sufficient for stability. Hence small nuclei have $N \approx Z$. But as the nucleus grows, the repulsive force increases faster than the binding energy. More neutrons are needed for stability, causing heavy nuclei to have $N > Z$.

The Shell Model

The potential-energy curve of Figure 30.9 for the strong force is similar in shape to the potential-energy curve of a molecular bond, a curve that you saw in earlier chapters. In classical physics, a molecular bond—an electrostatic interaction between two atoms—oscillates in length about an equilibrium bond length. We saw in Chapter 29 that molecular oscillations are actually quantized, giving rise to discrete vibrational energy levels.

In 1949, Maria Goeppert-Mayer proposed what came to be called the **shell model** of the nucleus. Her model was based on the successful application of quantum mechanics to atoms and molecules, especially the shell model of atoms that explains the periodic table of the elements. The shell model of atoms is a consequence of the Pauli exclusion principle, which allows only one spin-up and one spin-down electron in each energy level. Protons and neutrons also must obey the Pauli exclusion principle, and Goeppert-Mayer recognized that nuclei should also have shells. Nuclear

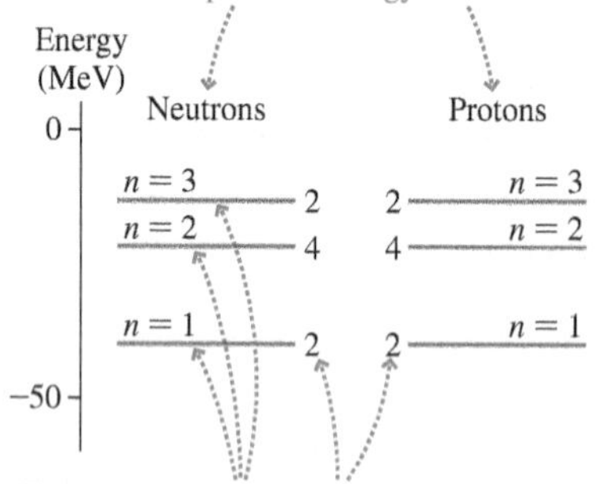

The proton potential energy is nearly identical to the neutron potential energy when Z is small.
Energy (MeV)
Neutrons
Protons
0
−50
n = 3
n = 2
n = 1
2
4
2
These are the first three allowed energy levels. They are spaced several MeV apart.
These are the maximum numbers of protons and neutrons allowed by the Pauli principle.

FIGURE 30.10 The three lowest energy levels of a low-Z nucleus. The neutron energy levels are on the left, the proton energy levels on the right.

interactions are much stronger and more complex than the electrostatic interactions within an atom, however, so nuclear shells don't follow the same simple rules as electron shells. We'll illustrate the basic idea with simple examples.

First consider a low-Z nucleus ($Z \leq 8$). These nuclei have so few protons that we can neglect the proton-proton repulsion and consider only the strong force. We do need to treat proton energy levels and neutron energy levels as distinct, but their energies are the same because the strong force makes no distinction between protons and neutrons.

FIGURE 30.10 is a generic energy-level diagram showing the first three allowed energies of protons and neutrons. Also shown is the maximum number of nucleons that the Pauli principle allows in each. The specific values of the allowed energies vary from nucleus to nucleus, but in all cases the energy levels are separated by several MeV—in contrast to a spacing of a few eV in atoms.

Let's apply these ideas to a series of nuclei with $A = 12$: ^{12}B, ^{12}C, and ^{12}N. These nuclei have 12 nucleons but different combinations of protons and neutrons. **FIGURE 30.11** shows their energy-level diagrams. Look first at ^{12}C, a nucleus with six protons and six neutrons. You can see that exactly six protons are allowed in the $n = 1$ and $n = 2$ energy levels. Likewise for the six neutrons. Thus ^{12}C has a closed $n = 2$ proton shell and a closed $n = 2$ neutron shell.

FIGURE 30.11 Energy levels and occupancy for three $A = 12$ nuclei.

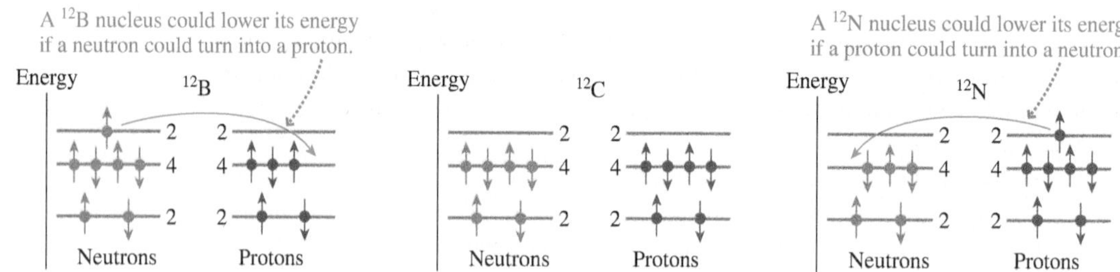

NOTE ► Protons and neutrons are different particles, so the Pauli principle is not violated if a proton and a neutron have the same quantum numbers. ◄

^{12}N has seven protons and five neutrons. The sixth proton fills the $n = 2$ proton shell, so the seventh proton has to go into the $n = 3$ energy level. The $n = 2$ neutron shell has one vacancy because there are only five neutrons. ^{12}B is just the opposite, with the seventh neutron in the $n = 3$ energy level. You can see from the diagrams that the ^{12}B and ^{12}N nuclei have significantly more energy—by several MeV—than ^{12}C.

In atoms, electrons in higher energy levels decay to lower energy levels by emitting a photon as the electron undergoes a quantum jump. That can't happen here because the higher-energy nucleon in ^{12}B is a neutron whereas the vacant lower energy level is that of a proton. But an analogous process could occur *if* a neutron could somehow turn into a proton. And that's exactly what happens! We'll explore the details in Section 30.5, but both ^{12}B and ^{12}N decay into ^{12}C in the process known as *beta decay*.

^{12}C is just one of three low-Z nuclei in which both the proton and neutron shells are full. The other two are ^{4}He (filling both $n = 1$ shells with $Z = 2$, $N = 2$) and ^{16}O (filling both $n = 3$ shells with $Z = 8$, $N = 8$). If the analogy with closed electron shells is valid, these nuclei should be more tightly bound than nuclei with neighboring values of A. And indeed, we've already noted that the curve of binding energy (Figure 30.7) has peaks at $A = 4$, 12, and 16. The shell model of the nucleus satisfactorily explains these peaks. Unfortunately, the shell model quickly becomes much more complex as we go beyond $n = 3$. Heavier nuclei do have closed shells, but there's no evidence for them in the curve of binding energy.

High-Z Nuclei

We can use the shell model to give a qualitative explanation for one more observation, although the details are beyond the scope of this text. FIGURE 30.12 shows the neutron and proton potential-energy wells of a high-Z nucleus. In a nucleus with many protons, the electrostatic potential energy raises the proton energy levels but not the energy levels of the uncharged neutrons. Protons and neutrons now have a different set of energy levels.

As a nucleus is "built," by the addition of protons and neutrons, the proton energy levels and the neutron energy levels must fill to just about the same height. If there were neutrons in energy levels above vacant proton levels, the nucleus would lower its energy by using beta decay to change the neutron into a proton. Similarly, beta decay would change a proton into a neutron if there were a vacant neutron energy level beneath a filled proton level. **The net result of beta decay is to keep the levels on both sides filled to just about the same energy.**

Because the neutron energy levels start at a lower energy, *more neutron states* are available than proton states. Consequently, a high-Z nucleus will have more neutrons than protons. This conclusion is consistent with our observation in Figure 30.5 that $N > Z$ for heavy nuclei.

FIGURE 30.12 The proton energy levels are displaced upward in a high-Z nucleus.

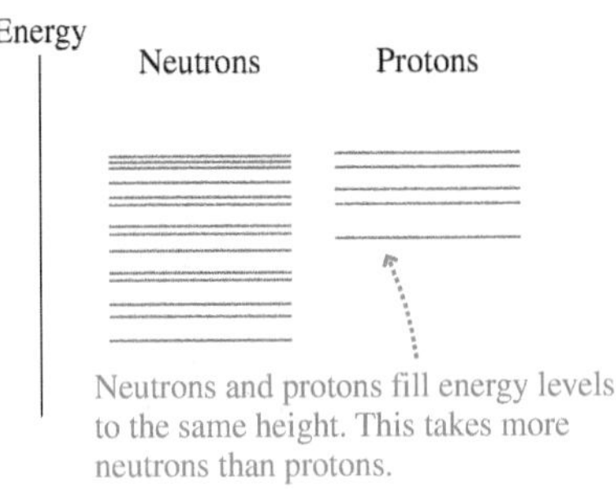
Energy
Neutrons
Protons
Neutrons and protons fill energy levels to the same height. This takes more neutrons than protons.

30.4 Radiation and Radioactivity

Becquerel's 1896 discovery of "rays" from crystals of uranium prompted a burst of activity. In England, J. J. Thomson and Ernest Rutherford worked to identify the unknown rays. Using combinations of electric and magnetic fields, they found three distinct types of radiation. FIGURE 30.13 shows the basic experimental procedure, and TABLE 30.3 summarizes the results.

FIGURE 30.13 Identifying radiation by its deflection in a magnetic field.

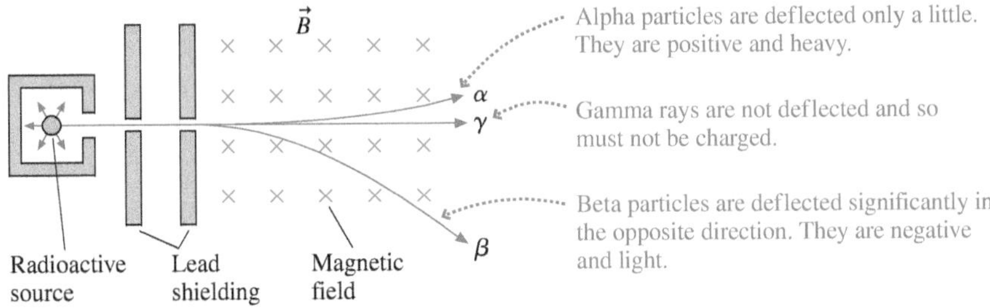
$\vec{B}$
α
γ
β
Alpha particles are deflected only a little. They are positive and heavy.
Gamma rays are not deflected and so must not be charged.
Beta particles are deflected significantly in the opposite direction. They are negative and light.
Radioactive source
Lead shielding
Magnetic field

TABLE 30.3 Three types of radiation

Radiation	Identification	Charge	Stopped by
Alpha, α	^{4}He nucleus	$+2e$	Sheet of paper
Beta, β	Electron	$-e$	Few mm of water
Gamma, γ	High-energy photon	0	Many cm of lead

Within a few years, as Rutherford and others deduced the basic structure of the atom, it became clear that these emissions of radiation were coming from the atomic nucleus. We now define *radioactivity* or **radioactive decay** to be the spontaneous emission of particles or high-energy photons from unstable nuclei as they decay from higher-energy to lower-energy states. Radioactivity has nothing to do with the orbiting valence electrons.

NOTE ▶ The term "radiation" merely means something that is *radiated outward,* similar to the word "radial." Electromagnetic waves are often called "electromagnetic radiation." Infrared waves from a hot object are referred to as "thermal radiation." Thus it was no surprise that these new "rays" were also called radiation. Unfortunately, the general public has come to associate the word "radiation" with *nuclear radiation,* something to be feared. It is important, when you use the term, to be sure you're not conveying a wrong impression to a listener. ◀

Nuclear Decay and Half-Lives

Powered by decay This 10-cm-diameter sphere is made of an oxide of ^{238}Pu, a relatively short-lived ($t_{1/2} = 88$ yr) isotope of plutonium that undergoes alpha decay. The short half-life means that this sphere has a very high activity, so high that the alpha particles heat the sphere enough to make it glow. Radioactive spheres like this one are used to power spacecraft on voyages far from the sun, by using the heat generated by the radioactive decay to produce electric energy. Even after 20 years, *thermoelectric generators* of this kind still produce 85% of their original power.

Rutherford discovered experimentally that for all types of radiation the number of radioactive atoms in a sample decreases exponentially with time. The reason is that radioactive decay—the process in which an unstable nucleus emits a form of radiation—is a *random process.* That is, we can predict only the *probability* that a nucleus will decay, not the exact moment.

To understand why the number decreases exponentially, and to connect experiment to theory, let's assume that the probability that an unstable nucleus decays during a time interval Δt is *independent* of the age of the nucleus. For example, an unstable nucleus that is created when a neutron is absorbed might have a 10% chance of decaying within the first minute, from $t = 0$ min to $t = 1$ min. That means there's a 90% chance that the nucleus *won't* decay during the first minute. If this nucleus happens to survive to $t = 10$ min, our assumption is that it has a 10% chance of decaying during the 1 min interval from $t = 10$ min to $t = 11$ min. This assumption can be justified both theoretically and experimentally.

If Δt is very small, the probability of decay during the time interval Δt is directly proportional to Δt. That is, if the probability of decay in 1 s is 0.1%, it will be 0.2% in 2 s and 0.05% in 0.5 s. (This logic fails if Δt is too large.) We're interested in the limit $\Delta t \rightarrow dt$, for which the concept is valid, so we can write

$$\text{Prob(decay in } \Delta t) = r\,\Delta t \qquad (30.10)$$

where the proportionality constant r is called the **decay rate.** It is a probability *per second,* with units of s^{-1}, and is thus a rate. For example, if a nucleus has a 4% probability of decaying during a 2 s interval, its decay rate is

$$r = \frac{\text{Prob(in } \Delta t)}{\Delta t} = \frac{0.04}{2\text{ s}} = 0.02\text{ s}^{-1}$$

We can interpret this as a decay probability of 2% per s. **The decay rate is a property of the nucleus.** Every unstable nucleus has its own decay rate.

Suppose a sample contains N unstable nuclei that can undergo radioactive decay. During a short time interval, the number that we expect to decay is N multiplied by the probability of decay. Each decay represents the *loss* of a nucleus, so the change in the number of nuclei during Δt is

$$\Delta N = -N \times \text{Prob(decay in } \Delta t) = -Nr\,\Delta t \qquad (30.11)$$

The minus sign shows that the number of undecayed nuclei is decreasing.

If we let $\Delta t \rightarrow dt$, then $\Delta N \rightarrow dN$ and Equation 30.11 becomes

$$dN = -rNdt \qquad (30.12)$$

The symbols are different, but this is exactly the same equation we discovered when calculating how a small particle comes to rest in a viscous liquid and how oscillations die away in a damped oscillator.

We can solve Equation 30.12 by first grouping all the terms involving N on one side of the equation, which gives $dN/N = -rdt$, and then integrating. We integrate from an initial number N_0 at $t = 0$ to the number N at an arbitrary later time t:

$$\int_{N_0}^{N} \frac{dN}{N} = -r\int_0^t dt \qquad (30.13)$$

Both are well-known integrals, and we find that

$$\int_{N_0}^{N}\frac{dN}{N}=\ln N\Big|_{N_0}^{N}=\ln N-\ln N_0=\ln\left(\frac{N}{N_0}\right)=-rt \quad (30.14)$$

We can find the number of still undecayed nuclei at time t by taking the exponential of both sides and then multiplying by N_0:

$$N=N_0e^{-rt} \quad (30.15)$$

Thus the decay of the number of unstable nuclei is another example of an exponential decay.

The quantity $\tau=1/r$ is called the *lifetime* of the nucleus. It is a property of the nucleus. We can write the exponential decay of the number of unstable nuclei in terms of the lifetime as

$$N=N_0e^{-t/\tau} \quad (30.16)$$

Exponential decay of a sample of radioactive nuclei

Equation 30.16 is identical to the equation we found for the amplitude decay of a damped oscillator.

FIGURE 30.14 shows the decrease of N with time. The number of radioactive nuclei decreases from N_0 at $t=0$ to $e^{-1}N_0=0.368N_0$ at time $t=\tau$. In practical terms, the number decreases by roughly two-thirds during one lifetime.

NOTE ▶ An important aspect of exponential decay is that you can choose any instant you wish to be $t=0$. The number of radioactive nuclei present at that instant is N_0. If at one instant you have 10,000 radioactive nuclei whose lifetime is $\tau=10$ min, you'll have roughly 3680 nuclei 10 min later. The fact that you may have had more than 10,000 nuclei earlier isn't relevant. ◀

In practice, it's much easier to measure the time at which half of a sample has decayed than the time at which 36.8% has decayed. Let's define the **half-life** $t_{1/2}$ as the time interval in which half of a sample of radioactive atoms decays. The half-life is shown in Figure 30.14.

FIGURE 30.14 The number of radioactive atoms decreases exponentially with time.

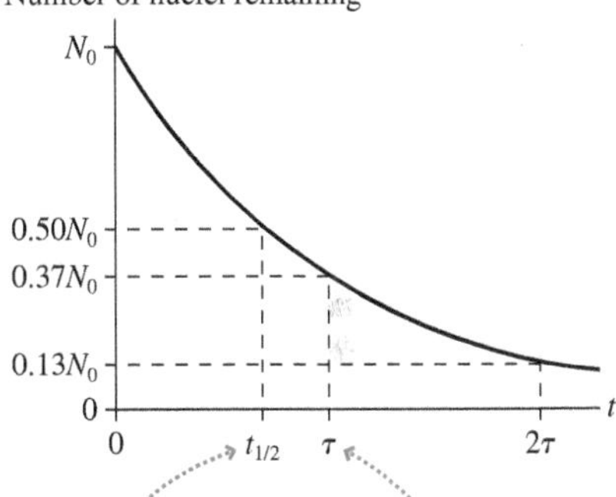
Number of nuclei remaining
N_0
$0.50N_0$
$0.37N_0$
$0.13N_0$
0
0
$t_{1/2}$
τ
2τ
t
The half-life is the time in which half the nuclei decay.
The lifetime is the time at which the number of nuclei is e^{-1}, or 37%, of the initial number.

The half-life is easily related to the lifetime τ because we know, by definition, that $N=\frac{1}{2}N_0$ at $t=t_{1/2}$. Thus, according to Equation 30.16,

$$\frac{N_0}{2}=N_0e^{-t_{1/2}/\tau} \quad (30.17)$$

The N_0 cancels, and we can then take the natural logarithm of both sides to find

$$\ln\left(\frac{1}{2}\right)=-\ln 2=-\frac{t_{1/2}}{\tau} \quad (30.18)$$

With one final rearrangement we have

$$t_{1/2}=\tau\ln 2=0.693\tau \quad (30.19)$$

Equation 30.16 can be written in terms of the half-life as

$$N=N_0\left(\frac{1}{2}\right)^{t/t_{1/2}} \quad (30.20)$$

Radioactive decay in terms of the half-life

Thus $N=N_0/2$ at $t=t_{1/2}$, $N=N_0/4$ at $t=2t_{1/2}$, $N=N_0/8$ at $t=3t_{1/2}$, and so on. **No matter how many nuclei there are, the number decays by half during the next half-life.**

NOTE ▶ Half the nuclei decay during one half-life, but don't fall into the trap of thinking that all will have decayed after two half-lives. ◀

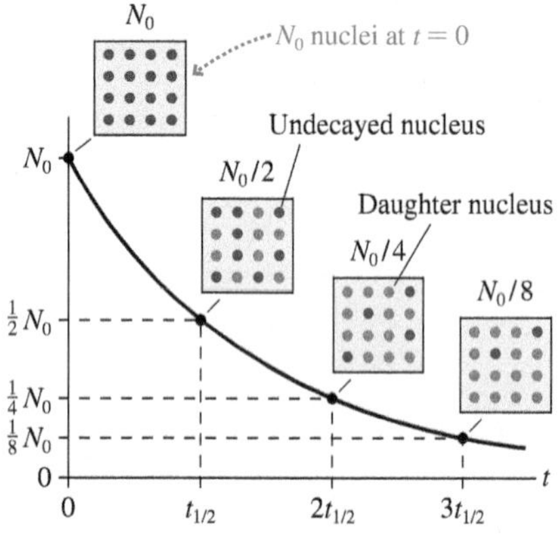

FIGURE 30.15 Half the nuclei decay during each half-life.

FIGURE 30.15 shows the half-life graphically. This figure also conveys two other important ideas:

1. Nuclei don't vanish when they decay. The decayed nuclei have merely become some other kind of nuclei, called the *daughter nuclei.*
2. The decay process is random. We can predict that half the nuclei will decay in one half-life, but we can't predict which ones.

Each radioactive isotope, such as ^{14}C, has its own half-life. That half-life doesn't change with time as a sample decays. If you've flipped a coin 10 times and, against all odds, seen 10 heads, you may feel that a tail is overdue. Nonetheless, the probability that the next flip will be a head is still 50%. After 10 half-lives have gone by, $(1/2)^{10} = 1/1024$ of a radioactive sample is still there. There was nothing special or distinctive about these nuclei, and, despite their longevity, each remaining nucleus has exactly a 50% chance of decay during the next half-life.

EXAMPLE 30.3 The decay of iodine

The iodine isotope ^{131}I, which has an eight-day half-life, is used in nuclear medicine to treat thyroid cancer and other thyroid conditions because nearly all the iodine a person ingests is taken up in the thyroid. A sample of ^{131}I containing 2.00×10^{12} atoms is created in a nuclear reactor.

a. How many ^{131}I atoms remain 36 hours later when the sample is delivered to a hospital?

b. The sample is constantly getting weaker, but it remains usable as long as the number of ^{131}I atoms exceeds 5.0×10^{11}. What is the maximum delay before the sample is no longer usable?

PREPARE The number of ^{131}I atoms decays exponentially.

SOLVE

a. The half-life is $t_{1/2} = 8$ days $= 192$ h. After 36 h have elapsed,

$$N = (2.00 \times 10^{12})\left(\frac{1}{2}\right)^{36\text{ h}/192\text{ h}} = 1.76 \times 10^{12} \text{ nuclei}$$

b. The time after creation at which 5.0×10^{11} ^{131}I atoms remain is given by

$$5.0 \times 10^{11} = 0.50 \times 10^{12} = (2.0 \times 10^{12})\left(\frac{1}{2}\right)^{t/8\text{ days}}$$

To solve for t, we first write this as

$$\frac{0.50}{2.00} = 0.25 = \left(\frac{1}{2}\right)^{t/8\text{ days}}$$

Now we take the logarithm of both sides. Either natural logarithms or base-10 logarithms can be used, but we'll use natural logarithms:

$$\ln(0.25) = -1.39 = \frac{t}{t_{1/2}}\ln(0.5) = -0.693\frac{t}{t_{1/2}}$$

Solving for t gives

$$t = 2.00\,t_{1/2} = 16 \text{ days}$$

ASSESS The weakest usable sample is one-quarter of the initial sample. You saw in Figure 30.15 that a radioactive sample decays to one-quarter of its initial number in 2 half-lives.

Activity

The **activity** R of a radioactive sample is the number of decays per second. This is simply the absolute value of dN/dt, or

$$R = \left|\frac{dN}{dt}\right| = \frac{N}{\tau} = \frac{N_0}{\tau}e^{-t/\tau} = R_0 e^{-t/\tau} = R_0\left(\frac{1}{2}\right)^{t/t_{1/2}} \tag{30.21}$$

where $R_0 = N_0/\tau$ is the activity at $t = 0$. The activity of a sample decreases exponentially along with the number of remaining nuclei.

The SI unit of activity is the **becquerel,** defined as

$$1 \text{ becquerel} = 1 \text{ Bq} = 1 \text{ decay/s or } 1 \text{ s}^{-1}$$

An older unit of activity, but one that continues in widespread use, is the **curie.** The curie was originally defined as the activity of 1 g of radium. Today, the conversion factor is

$$1 \text{ curie} = 1 \text{ Ci} \equiv 3.7 \times 10^{10} \text{ Bq}$$

One curie is a substantial activity. The radioactive samples used in laboratory experiments are typically $\approx 1\ \mu\text{Ci}$ or, equivalently, $\approx 40{,}000$ Bq. These samples can be handled with only minor precautions. Larger sources of radioactivity require lead shielding and special precautions to prevent exposure to high levels of radiation.

EXAMPLE 30.4 **Determining the decay of activity**

A ^{60}Co source used to provide gamma rays to irradiate tumors has an activity of 0.43 Ci. The half-life of ^{60}Co is 5.3 yr.

a. How many ^{60}Co atoms are in the source?
b. What will be the activity of the source after 10 yr?

PREPARE The number of radioactive atoms and the activity decay exponentially. Equation 30.21 relates the number of radioactive atoms in a sample to the sample's activity and half-life as $N = \tau R$.

SOLVE

a. In Bq, the initial activity of the source is

$$R_0 = (0.43\ \text{Ci})\left(\frac{3.7 \times 10^{10}\ \text{Bq}}{1\ \text{Ci}}\right) = 1.59 \times 10^{10}\ \text{Bq}$$

The half-life $t_{1/2}$ in s is

$$t_{1/2} = (5.3\ \text{yr})\left(\frac{3.15 \times 10^{7}\ \text{s}}{1\ \text{yr}}\right) = 1.67 \times 10^{8}\ \text{s}$$

Thus the initial number of ^{60}Co atoms in the source is

$$N_0 = \tau R_0 = \frac{t_{1/2}R_0}{0.693} = \frac{(1.67 \times 10^8\ \text{s})(1.59 \times 10^{10}\ \text{Bq})}{0.693}$$
$$= 3.8 \times 10^{18}\ \text{atoms}$$

b. The decay of activity with time is given by Equation 30.21. After 10 yr, the activity is

$$R = R_0\left(\frac{1}{2}\right)^{t/t_{1/2}} = (0.43\ \text{Ci})\left(\frac{1}{2}\right)^{10\ \text{yr}/5.3\ \text{yr}}$$
$$= 0.12\ \text{Ci}$$

ASSESS N_0 is a large number of atoms, but it is a very small fraction ($\approx 10^{-5}$) of a mole. The source is about 300 μg of ^{60}Co. For part b, 10 years is about two half-lives, so we would expect the activity to fall by about a factor of 4. Our answer is close to this.

Radioactive Dating

Many geological and archeological samples can be dated by measuring the decays of naturally occurring radioactive isotopes. Because we have no way to know N_0, the initial number of radioactive nuclei, radioactive dating depends on the use of ratios.

The most well-known dating technique is carbon dating. The carbon isotope ^{14}C has a half-life of 5730 years, so any ^{14}C present when the earth formed 4.5 billion years ago would long since have decayed away. Nonetheless, ^{14}C is present in atmospheric carbon dioxide because high-energy cosmic rays collide with gas molecules high in the atmosphere. These cosmic rays are energetic enough to create ^{14}C nuclei from nuclear reactions with nitrogen and oxygen nuclei. The creation and decay of ^{14}C have reached a steady state in which the ^{14}C/^{12}C ratio is 1.3×10^{-12}. That is, atmospheric carbon dioxide has ^{14}C at the concentration of 1.3 parts per trillion. As small as this is, it's easily measured by modern chemical techniques.

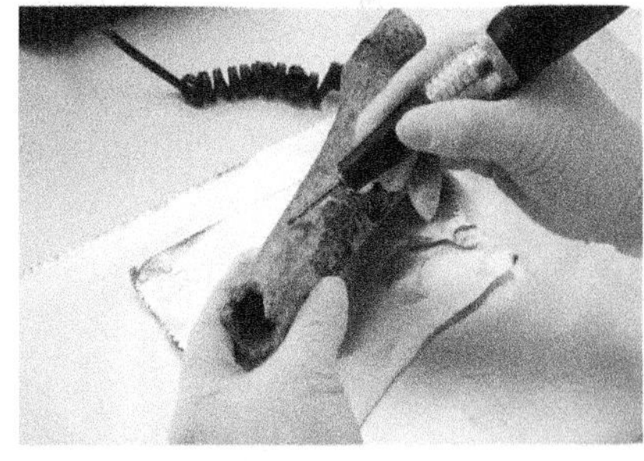

Not speed dating A researcher is extracting a small sample of an ancient bone. She will determine the age of the bone by measuring the ratio of ^{14}C to ^{12}C.

All living organisms constantly exchange carbon dioxide with the atmosphere, so the ^{14}C/^{12}C ratio in living organisms is also 1.3×10^{-12}. When an organism dies, the ^{14}C in its tissue begins to decay and no new ^{14}C is added. Objects are dated by comparing the measured ^{14}C/^{12}C ratio to the 1.3×10^{-12} value of living material.

Carbon dating is used to date skeletons, wood, paper, fur, food material, and anything else made of organic matter. It is quite accurate for ages to about 15,000 years, roughly three half-lives of ^{14}C. Beyond that, the difficulty of measuring such a small ratio and some uncertainties about the cosmic ray flux in the past combine to decrease the accuracy. Even so, items are dated to about 50,000 years with a fair degree of reliability.

Other isotopes with longer half-lives are used to date geological samples. Potassium-argon dating, using ^{40}K with a half-life of 1.25 billion years, is especially useful for dating rocks of volcanic origin.

EXAMPLE 30.5 **Carbon dating**

Archeologists excavating an ancient hunters' camp have recovered a 5.0 g piece of charcoal from a fireplace. Measurements on the sample find that the ^{14}C activity is 0.35 Bq. What is the approximate age of the camp?

PREPARE Charcoal, from burning wood, is almost pure carbon. The number of ^{14}C atoms in the wood has decayed exponentially since the branch fell off a tree. Because wood rots, it is reasonable to assume that there was no significant delay between when the branch fell off the tree and the hunters burned it.

SOLVE The ^{14}C/^{12}C ratio was 1.3×10^{-12} when the branch fell from the tree. We first need to determine the present ratio, then use the known ^{14}C half-life $t_{1/2} = 5730$ years to calculate the time needed to reach the present ratio. The number of ordinary ^{12}C nuclei in the sample is

$$N(^{12}\text{C}) = \left(\frac{5.0\text{ g}}{12\text{ g/mol}}\right)(6.02 \times 10^{23}\text{ atoms/mol})$$

$$= 2.5 \times 10^{23}\text{ nuclei}$$

The number of ^{14}C nuclei can be found from the activity to be $N(^{14}\text{C}) = \tau R = t_{1/2}R/0.693$. After converting the half-life to seconds, $t_{1/2} = 5730$ years $= 1.81 \times 10^{11}$ s, we can compute

$$N(^{14}\text{C}) = \frac{t_{1/2}R}{0.693} = \frac{(1.81 \times 10^{11}\text{ s})(0.35\text{ Bq})}{0.693} = 9.1 \times 10^{10}\text{ nuclei}$$

and the present ^{14}C/^{12}C ratio is $N(^{14}\text{C})/N(^{12}\text{C}) = 0.36 \times 10^{-12}$. Because this ratio has been decaying with a half-life of 5730 years, the time needed to reach the present ratio is found from

$$0.36 \times 10^{-12} = (1.3 \times 10^{-12})\left(\frac{1}{2}\right)^{t/t_{1/2}}$$

To solve for t, we first write this as

$$\frac{0.36}{1.3} = 0.277 = \left(\frac{1}{2}\right)^{t/t_{1/2}}$$

Now we take the logarithm of both sides:

$$\ln(0.277) = -1.28 = \frac{t}{t_{1/2}}\ln(0.5) = -0.693\frac{t}{t_{1/2}}$$

Thus the age of the hunters' camp is

$$t = 1.85t_{1/2} = 11{,}000\text{ years}$$

ASSESS 11,000 years is well within the range for which carbon dating is a reliable tool.

STOP TO THINK 30.3 A sample starts with 1000 radioactive atoms. How many half-lives have elapsed when 750 atoms have decayed?

A. 0.25 B. 1.5 C. 2.0 D. 2.5

30.5 Types of Nuclear Decay

This section will look in more detail at the mechanisms of the three types of radioactive decay.

Alpha Decay

An alpha particle, symbolized as α, is a ^{4}He nucleus, a strongly bound system of two protons and two neutrons. An unstable nucleus that ejects an alpha particle will lose two protons and two neutrons, so we can write the decay as

$$^{A}\text{X}_Z \rightarrow {}^{A-4}\text{Y}_{Z-2} + \alpha + \text{energy} \tag{30.22}$$

Alpha decay of a nucleus

FIGURE 30.16 Alpha decay.

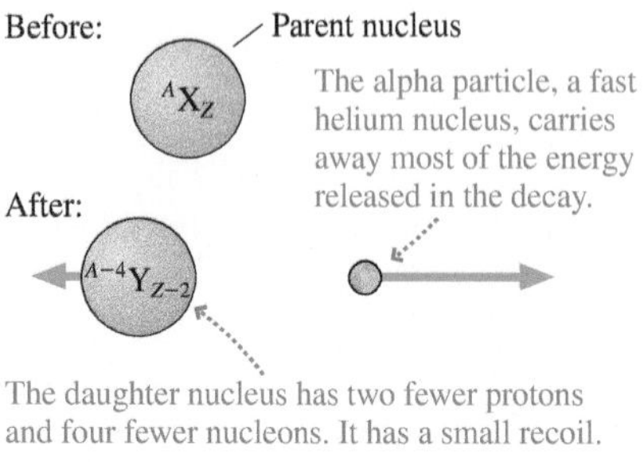

Before: Parent nucleus $^{A}\text{X}_Z$

The alpha particle, a fast helium nucleus, carries away most of the energy released in the decay.

After: $^{A-4}\text{Y}_{Z-2}$

The daughter nucleus has two fewer protons and four fewer nucleons. It has a small recoil.

FIGURE 30.16 shows the alpha-decay process. The original nucleus X is called the **parent nucleus,** and the decay-product nucleus Y is the **daughter nucleus.** This reaction can occur only when the mass of the parent nucleus is greater than the mass of the daughter nucleus plus the mass of an alpha particle. This requirement is met for heavy, high-Z nuclei well above the maximum on the Figure 30.7 curve of binding energy. It is energetically favorable for these nuclei to eject an alpha particle because the daughter nucleus is more tightly bound than the parent nucleus.

For example, the sensing circuit in smoke detectors uses a tiny speck of an isotope of americium, ^{241}Am, with $Z = 95$. ^{241}Am decays by alpha emission with a half-life of 432 yr. The daughter nucleus has two fewer protons, which makes it neptunium with $Z = 93$, and four fewer nucleons, which gives $A = 237$. Thus the decay of ^{241}Am is

$$^{241}\text{Am} \rightarrow {}^{237}\text{Np} + \alpha + \text{energy}$$

Although the mass requirement is based on the nuclear masses, we can express it—as we did the binding energy equation—in terms of atomic masses. The energy released in an alpha decay, essentially all of which goes into the alpha particle's kinetic energy, is

$$\Delta E = (m_{\text{X}} - m_{\text{Y}} - m_{\text{He}})c^2 \approx K_\alpha \tag{30.23}$$

EXAMPLE 30.6 Alpha decay of uranium

The uranium isotope ^{238}U undergoes alpha decay to ^{234}Th. The atomic masses are 238.0508 u for ^{238}U and 234.0436 u for ^{234}Th. What is the kinetic energy, in MeV, of the alpha particle?

PREPARE Essentially all of the energy release ΔE goes into the alpha particle's kinetic energy.

SOLVE The atomic mass of helium is 4.0026 u. Thus

$$K_\alpha = (238.0508\text{ u} - 234.0436\text{ u} - 4.0026\text{ u})c^2$$

$$= \left(0.0046\text{ u} \times \frac{931.5\text{ MeV}/c^2}{1\text{ u}}\right)c^2 = 4.3\text{ MeV}$$

ASSESS This is a typical alpha-particle energy. Notice how the c^2 canceled from the calculation so that we never had to evaluate c^2.

Alpha decay is an example of quantum-mechanical tunneling. The ^{4}He nucleus that forms an alpha particle is so tightly bound that we can think of it as existing "prepackaged" inside the parent nucleus. FIGURE 30.17 shows an approximate potential-energy curve for an alpha particle. The strong force creates a deep potential-energy "well" within the nucleus, but this nuclear potential energy has a sharp cutoff at the edge of the nucleus because the strong force is a short-range force. An alpha particle outside the nucleus experiences a repulsive electric force from the positively charged nucleus, so the external potential energy is that of two like charges.

An alpha particle with $E < 0$ has turning points at $x = \pm R$ where the total energy line crosses the potential-energy curve. These alpha particles are bound particles that move within the nucleus but cannot escape. However, the high *Coulomb barrier* means that there could be an allowed energy level with $E > 0$. A classical particle is bound within the nucleus as long as its total energy line is below the top of the Coulomb barrier because a classical particle cannot be at a position where $E < U$.

But an alpha particle is not a classical particle. It has a delocalized wave-like nature that can extend through the Coulomb barrier. Consequently, an alpha particle in an energy level with $E > 0$ can *tunnel* through the Coulomb barrier and escape. This is how alpha decay occurs. Energy must be conserved, so the kinetic energy of the escaping alpha particle is the height of the energy level above $E = 0$.

The probability that a particle will tunnel through a barrier is *very* sensitive to the width of the barrier, as we noted in our discussion of the scanning tunneling microscope in Chapter 28. The width of the barrier decreases with increasing energy, so an alpha particle with a higher energy level should have a *shorter half-life*—a higher tunneling probability—and escape with *more kinetic energy.* This prediction is in excellent agreement with measured energies and half-lives.

FIGURE 30.17 The potential-energy diagram of an alpha particle in the parent nucleus.

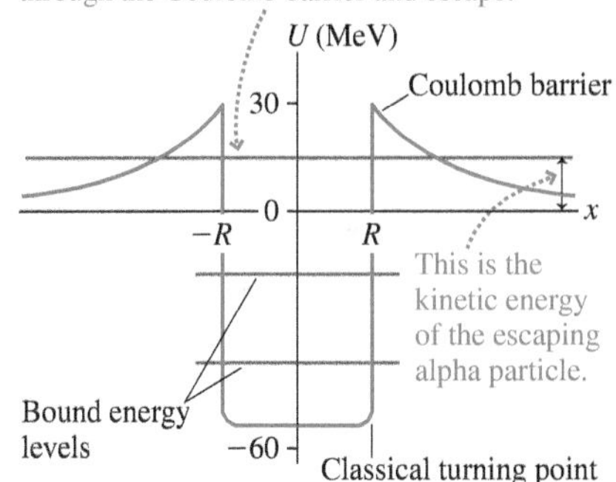
An alpha particle in this energy level can tunnel through the Coulomb barrier and escape.
U (MeV)
Coulomb barrier
30
0
x
−R
R
This is the kinetic energy of the escaping alpha particle.
Bound energy levels
−60
Classical turning point

Beta Decay

Beta decay was initially associated with the emission of an electron e^-, the beta particle. It was later discovered that some nuclei can undergo beta decay by emitting a positron e^+, the antiparticle of the electron, although this decay mode is not as common. A positron is identical to an electron except that it has a positive charge. To be precise, the emission of an electron is called *beta-minus decay* and the emission of a positron is *beta-plus decay.*

A typical example of beta-minus decay occurs in the carbon isotope ^{14}C, which undergoes the beta-decay process $^{14}C \rightarrow {}^{14}N + e^-$. Carbon has $Z = 6$ and nitrogen has $Z = 7$. Because Z increases by 1 but A doesn't change, it appears that a neutron within the nucleus has changed itself into a proton and an electron. That is, the basic beta-minus decay process appears to be

$$n \rightarrow p^+ + e^- \tag{30.24}$$

Indeed, a free neutron turns out *not* to be a stable particle. It decays with a half-life of approximately 10 min into a proton and an electron. This decay is energetically allowed because $m_n > m_p + m_e$. Furthermore, it conserves charge.

Whether a neutron *within* a nucleus can decay depends on the masses of the parent and daughter nuclei. The electron is ejected from the nucleus in beta-minus decay, but the proton is not. Thus the decay process shown in FIGURE 30.18a is

FIGURE 30.18 Beta decay.

(a) Beta-minus decay

Before:

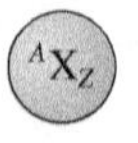

A neutron changes into a proton and an electron. The electron is ejected from the nucleus.

After:

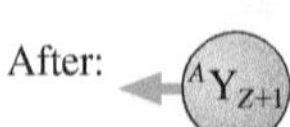

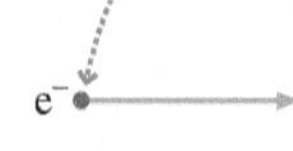

(b) Beta-plus decay

Before:

A proton changes into a neutron and a positron. The positron is ejected from the nucleus.

After:

$$^{A}X_{Z} \rightarrow {}^{A}Y_{Z+1} + e^- + \text{energy} \tag{30.25}$$

Beta-minus decay of a nucleus

Energy is released because the mass decreases in this process, but we have to be careful when calculating the mass loss. Although not explicitly shown in Equation 30.25, the daughter ^{A}Y is actually the ionized atom $^{A}Y^+$ because it gained a proton but didn't gain an orbital electron. Its mass is the atomic mass of ^{A}Y *minus* the mass of an electron. But the full right-hand side of the reaction includes an additional electron, the beta particle, so the total mass of the decay products is simply the atomic mass of ^{A}Y.

Consequently, the energy released in beta-minus decay, based on the mass loss, is

$$\Delta E = (m_X - m_Y)c^2 \tag{30.26}$$

The energy release has to be positive, so we see that **beta-minus decay occurs only if $m_X > m_Y$.** ^{14}C can undergo beta-minus decay to ^{14}N because $m(^{14}C) > m(^{14}N)$. But ^{12}C has the smallest mass of all nuclei with $A = 12$, so ^{12}C is stable and its neutrons cannot decay.

NOTE ► The electron emitted in beta-minus decay has nothing to do with the atom's orbital electrons. The beta particle is created in the nucleus and ejected directly from the nucleus when a neutron is transformed into a proton and an electron. ◄

Beta-plus decay is the conversion of a proton into a neutron and a positron:

$$p^+ \rightarrow n + e^+ \tag{30.27}$$

The full decay process, shown in FIGURE 30.18b, is

$$^{A}X_{Z} \rightarrow {}^{A}Y_{Z-1} + e^+ + \text{energy} \tag{30.28}$$

Beta-plus decay of a nucleus

Beta-plus decay does *not* happen for a free proton because $m_p < m_n$. It *can* happen within a nucleus as long as energy is conserved for the entire nuclear system.

In our earlier discussion of Figure 30.11 we noted that the ^{12}B and ^{12}N nuclei could reach a lower energy state if a proton could change into a neutron, and vice versa. Now we see that such a change can occur if the energy conditions are favorable. And, indeed, ^{12}B undergoes beta-minus decay to ^{12}C while ^{12}N undergoes beta-plus decay to ^{12}C.

In general, beta decay is a process used by nuclei with too many neutrons or too many protons in order to move closer to the line of stability in Figure 30.5. FIGURE 30.19 illustrates this idea for a series of nuclei with $A = 135$. Nuclei with $Z < 56$ have too

FIGURE 30.19 Nuclei move toward the line of stability by undergoing beta decay.

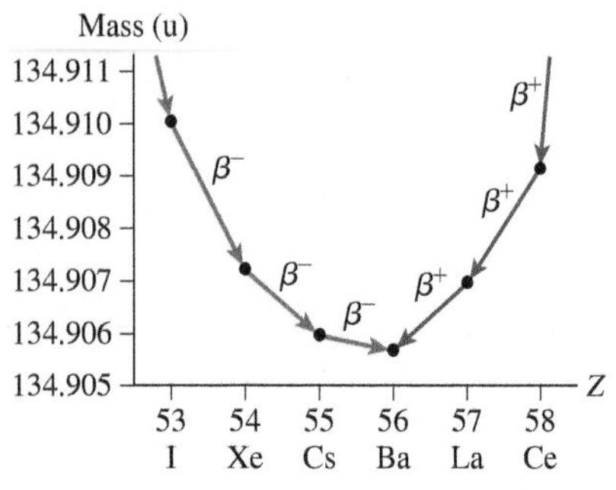

many neutrons and undergo beta-minus decay. Nuclei with $Z > 56$ have too many protons and undergo beta-plus decay. Notice that in each case the daughter nucleus has less mass than its parent. The nucleus ^{135}Ba has the lowest mass in the series and is stable.

A third form of beta decay occurs in some nuclei that have too many protons but not enough mass to undergo beta-plus decay. In this case, a proton changes into a neutron by "capturing" an electron from the innermost shell of orbiting electrons (an $n = 1$ electron). The process is

$$p^+ + \text{orbital } e^- \rightarrow n \tag{30.29}$$

This form of beta decay is called **electron capture,** abbreviated EC. The net result, ${}^{A}X_{Z} \rightarrow {}^{A}Y_{Z-1}$, is the same as beta-plus decay but without the emission of a positron. Electron capture is the only nuclear decay mechanism that involves the orbital electrons.

The Weak Interaction

We've presented beta decay as if it were perfectly normal for one kind of matter to change spontaneously into a completely different kind of matter. For example, it would be energetically favorable for a large truck to spontaneously turn into a Cadillac and a VW Beetle, ejecting the Beetle at high speed. But it doesn't happen.

Once you stop to think of it, the process $n \rightarrow p^+ + e^-$ seems ludicrous, not because it violates mass-energy conservation but because we have no idea *how* a neutron could turn into a proton. Alpha decay may be a strange process because tunneling in general goes against our commonsense notions, but it is a perfectly ordinary quantum-mechanical process. Now we're suggesting that one of the basic building blocks of matter can somehow morph into a different basic building block.

To make matters more confusing, measurements in the 1930s found that beta decay didn't seem to conserve either energy or momentum. Faced with these difficulties, the Italian physicist Enrico Fermi made two bold suggestions:

1. A previously unknown fundamental force of nature is responsible for beta decay. This force, which has come to be known as the **weak interaction,** has the ability to turn a neutron into a proton, and vice versa.
2. The beta-decay process emits a particle that, at that time, had not been detected. This new particle has to be electrically neutral, in order to conserve charge, and it has to be much less massive than an electron. Fermi called it the **neutrino,** meaning "little neutral one." Energy and momentum really are conserved, but the neutrino carries away some of the energy and momentum of the decaying nucleus. Thus experiments that detect only the electron seem to violate conservation laws.

Cold neutrinos The IceCube Neutrino Observatory uses thousands of sensitive photon detectors to monitor a cubic kilometer of ice at the South Pole for faint flashes of light that are emitted when a neutrino occasionally collides with a nucleus in a water molecule. Each year the observatory records about 100 neutrinos from astrophysical sources. The number may be small, but these neutrinos provide an important window into the nuclear processes that take place in extreme events such as supernovae.

The neutrino is represented by the symbol ν, a lowercase Greek nu. The beta-decay processes that Fermi proposed are

$$\begin{aligned} n &\rightarrow p^+ + e^- + \bar{\nu} \\ p^+ &\rightarrow n + e^+ + \nu \end{aligned} \tag{30.30}$$

The symbol $\bar{\nu}$ is an *antineutrino,* although the reason one is a neutrino and the other an antineutrino need not concern us here. FIGURE 30.20 shows that the electron and antineutrino (or positron and neutrino) *share* the energy released in the decay.

FIGURE 30.20 A more accurate picture of beta decay includes neutrinos.

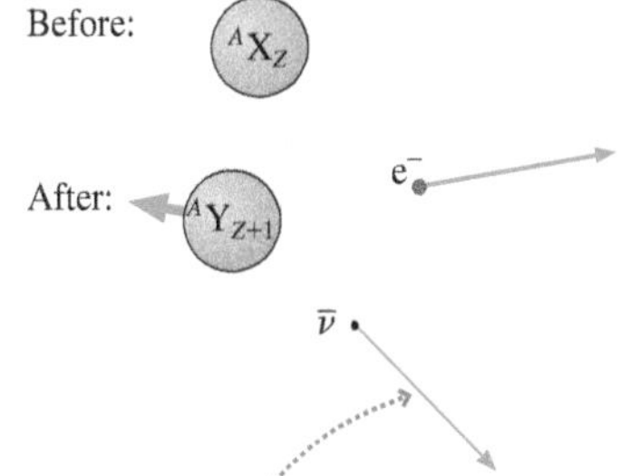

If only the electron and the daughter nucleus are measured, energy and momentum appear not to be conserved. The "missing" energy and momentum are carried away by the undetected antineutrino.

The neutrino interacts with matter so weakly that a neutrino can pass straight through the earth with only a very slight chance of a collision. Trillions of neutrinos created by nuclear fusion reactions in the core of the sun are passing through your body every second. Neutrino interactions are so rare that the first laboratory detection did not occur until 1956, over 20 years after Fermi's proposal.

It was initially thought that the neutrino had not only zero charge but also zero mass. However, experiments in the 1990s showed that the neutrino mass, although very tiny, is not zero. The best current evidence suggests a mass about one-millionth the mass of an electron. Experiments now under way will attempt to determine a more accurate value.

EXAMPLE 30.7 **Beta decay of ^{14}C**

How much energy is released in the beta-minus decay of ^{14}C?

PREPARE The decay is $^{14}C \rightarrow {}^{14}N + e^- + \bar{\nu}$.

SOLVE In Appendix D we find $m(^{14}C) = 14.003\,242$ u and $m(^{14}N) = 14.003\,074$ u. The mass difference is a mere 0.000 168 u, but this is the mass that is converted into the kinetic energy of the escaping particles. The energy released is

$$E = (\Delta m)c^2 = (0.000\,168 \text{ u}) \times (931.5 \text{ MeV/u}) = 0.156 \text{ MeV}$$

ASSESS This energy is shared between the electron and the antineutrino.

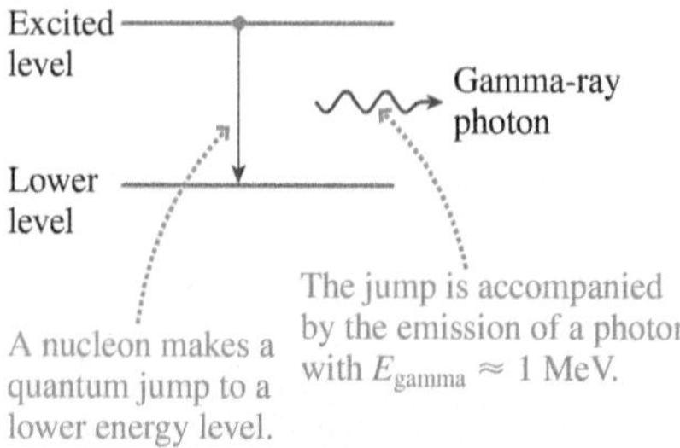

FIGURE 30.21 Gamma decay.

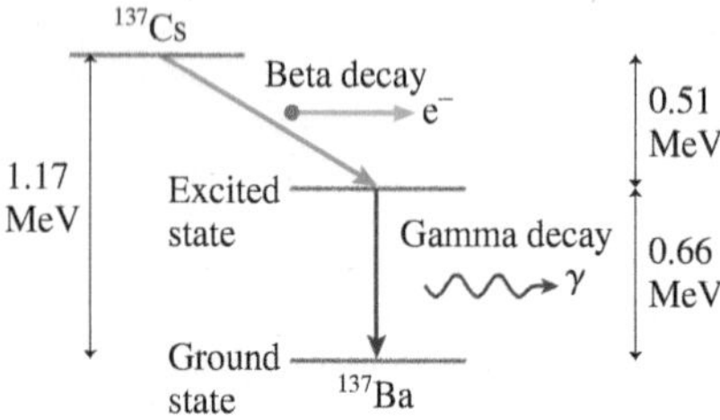

FIGURE 30.22 The decay of ^{137}Cs involves both beta and gamma decay.

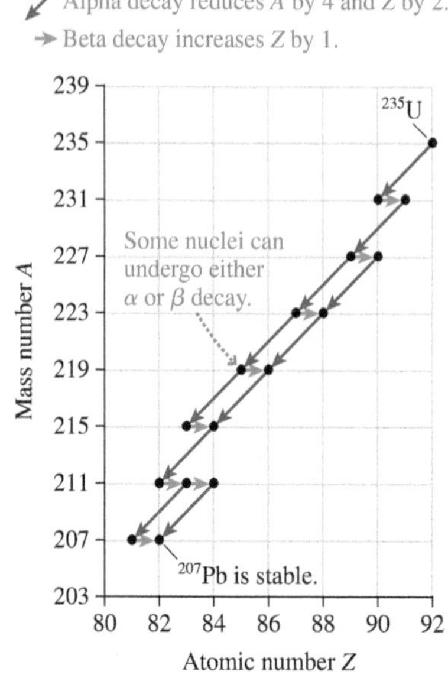

FIGURE 30.23 The decay series of ^{235}U.

Gamma Decay

Gamma decay is the easiest form of nuclear decay to understand. You learned that an atomic system can emit a photon with $E_{photon} = \Delta E_{atom}$ when an electron undergoes a quantum jump from an excited energy level to a lower energy level. Nuclei are no different. A proton or a neutron in an excited nuclear state, such as the one shown in FIGURE 30.21, can undergo a quantum jump to a lower-energy state by emitting a high-energy photon. This is the gamma-decay process.

The spacing between atomic energy levels is only a few eV. Nuclear energy levels, by contrast, are typically 1 MeV apart. Hence gamma-ray photons have $E_{gamma} \approx 1$ MeV. Photons with this much energy have tremendous penetrating power and deposit an extremely large amount of energy at the point where they are finally absorbed.

Nuclei left to themselves are usually in their ground states and thus cannot emit gamma-ray photons. However, alpha and beta decay often leave the daughter nucleus in an excited nuclear state, so gamma emission is usually found to accompany alpha and beta emission.

The cesium isotope ^{137}Cs is a good example. ^{137}Cs is frequently used as a laboratory and medical source of gamma rays, but the actual decay of ^{137}Cs is beta-minus decay to ^{137}Ba. FIGURE 30.22 shows the full process. A ^{137}Cs nucleus undergoes beta-minus decay by emitting an electron and an antineutrino, which share between them a total energy of 0.51 MeV. The half-life for this process is 30 years. This leaves the daughter ^{137}Ba nucleus in an excited state 0.66 MeV above the ground state. The excited Ba nucleus then decays within a few seconds to the ground state by emitting a 0.66 MeV gamma-ray photon. Thus a ^{137}Cs sample *is* a source of gamma-ray photons, but the photons are actually emitted by barium nuclei rather than cesium nuclei.

Decay Series

A radioactive nucleus decays into a daughter nucleus. In many cases, the daughter nucleus is also radioactive and decays to produce its own daughter nucleus. The process continues until reaching a daughter nucleus that is stable. The sequence of isotopes, starting with the original unstable isotope and ending with the stable isotope, is called a **decay series.**

Decay series are especially important for very heavy nuclei. As an example, FIGURE 30.23 shows the decay series of ^{235}U, an isotope of uranium with a 700-million-year half-life. This is a very long time, but it is only about 15% the age of the earth, thus most (but not all) of the ^{235}U nuclei present when the earth was formed have now decayed. There are many unstable nuclei along the way, but all ^{235}U nuclei eventually end as the ^{207}Pb isotope of lead, a stable nucleus.

Notice that some nuclei can decay by either alpha *or* beta decay. Thus there are a variety of paths that a decay can follow, but they all end at the same point.

STOP TO THINK 30.4 The cobalt isotope $^{60}\text{Co}\,(Z = 27)$ decays to the nickel isotope $^{60}\text{Ni}\,(Z = 28)$. The decay process is

A. Alpha decay. B. Beta-minus decay. C. Beta-plus decay.
D. Electron capture. E. Gamma decay.

30.6 The Interaction of Ionizing Radiation with Matter

Electromagnetic waves—from radio waves through ultraviolet radiation—increase an object's thermal energy and temperature when they are absorbed by matter. In contrast with visible-light photon energies of a few eV, the energies of the alpha and beta particles and the gamma-ray photons of nuclear decay are typically 0.1–10 MeV, a factor of roughly 10^6 larger. Rather than simply increasing an object's thermal energy, nuclear radiation *ionizes* matter and breaks molecular bonds. Nuclear radiation (and also x rays, which behave much the same in matter) is called **ionizing radiation.**

Alpha and beta particles interact strongly with matter because they are charged particles. Each collision with an atom or molecule causes ionization—the ejection of electrons—so an alpha or beta particle traveling through matter leaves a trail of ions like the one shown in FIGURE 30.24. Because the ionization energy of an atom or molecule is ≈ 10 eV, a particle with 1 MeV of kinetic energy can ionize $\approx 100{,}000$ atoms or molecules before stopping. The low-mass electrons are scattered to the side, but the more massive positive ions barely move and form the trail.

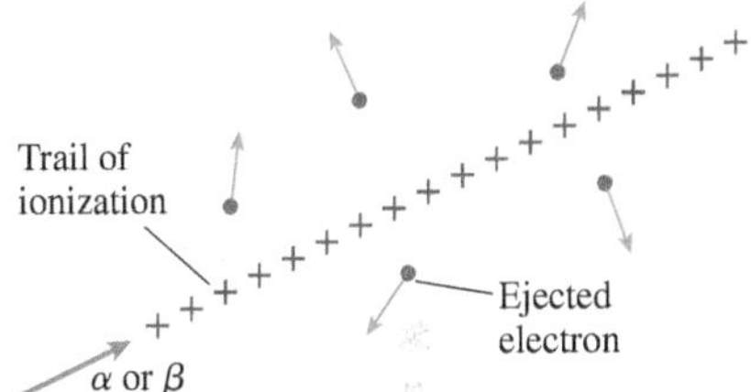

FIGURE 30.24 Alpha and beta particles create a trail of ions as they pass through matter.

For two particles that have different masses but equal kinetic energies $\frac{1}{2}mv^2$, the ratio of their speeds is inversely proportional to the square root of the ratio of their masses. The mass of an alpha particle is nearly 10^4 times that of a beta particle, so the speed of an alpha particle is $\approx 1\%$ that of a beta particle of equal energy. Thus we might expect that an alpha particle will deposit its energy and stop in $\approx 1\%$ the distance of a beta particle. This prediction is validated by the experimental evidence that 1 MeV beta particles penetrate ≈ 5 mm into water or tissue (which is mostly water), but 1 MeV alpha particles penetrate only $\approx 0.05\text{ mm} = 50\ \mu\text{m}$.

Alpha particles from an external source of radiation deposit their energy in the epidermis but cannot penetrate deeper into the human body. Ingested or inhaled sources of alpha particles are a significant health hazard, however, because the decays occur in close proximity to sensitive tissue and the decay energy is deposited in a very small volume. In contrast, beta particles from an external source penetrate deeply enough to cause ionization in the dermis as well as in muscles and organs close to the surface.

X-ray and gamma-ray photons are not charged and interact more weakly with matter than do alpha or beta particles. Consequently, they penetrate much deeper into matter. There are several mechanisms by which very high-energy photons interact with and ionize atoms, but all produce high-energy electrons that cause additional ionization and damage.

When a beam of x-ray or gamma-ray photons travels through matter, the intensity decreases exponentially with distance. That is, the intensity decreases as

$$I = I_0 e^{-\mu x} \tag{30.31}$$

where I_0 is the intensity at the surface ($x = 0$) and μ is the *absorption coefficient.* Measured values of absorption coefficients are typically in units of cm^{-1}, so the distance x has to be in cm. For gamma rays in water, the absorption coefficient is $0.10\ \text{cm}^{-1}$ for 0.5 MeV photons, decreasing to $0.07\ \text{cm}^{-1}$ at 1 MeV and $0.03\ \text{cm}^{-1}$ at 5 MeV.

The beam intensity has decreased to $I_0e^{-1} = 0.37I_0$ when $x = \mu^{-1}$, so μ^{-1} is frequently cited as the *penetration depth.* We can see that the penetration depth of gamma-ray photons in water increases from $\approx$10 cm for 0.5 MeV photons to $\approx$30 cm for 5 MeV photons. A beam of lower-energy gamma rays is mostly absorbed before passing all the way through the body, depositing its energy and causing ionization. Higher-energy gamma rays have less attenuation, and many 5 MeV photons pass entirely through the body without interaction. However, those that do interact deposit 10 times the energy of a 0.5 MeV photon and cause roughly 10 times the ionization and damage.

Radiation Dose

The biological effects of radiation depend on two factors. The first is the physical factor of how much energy is absorbed by the body. The second is the biological factor of how tissue reacts to different forms of radiation.

The **absorbed dose** of radiation is the energy of ionizing radiation absorbed per kilogram of tissue. The SI unit of absorbed dose is the **gray,** abbreviated Gy. It is defined as

$$1 \text{ gray} = 1 \text{ Gy} = 1.00 \text{ J/kg of absorbed energy}$$

The absorbed dose depends only on the energy absorbed, not at all on the type of radiation or on what the absorbing material is.

A 1 Gy dose of gamma rays and a 1 Gy dose of alpha particles have different biological consequences because photons and charged particles interact differently with matter. To account for such differences, the **relative biological effectiveness** (RBE) is defined as the biological effect of a given dose relative to the biological effect of an equal dose of x rays.

TABLE 30.4 shows the relative biological effectiveness of different forms of radiation. Larger values correspond to larger biological effects. Neutrons have a range of values because the biological effect varies with the energy of the particle. Alpha radiation has the largest RBE because, with their small penetration depth, the energy is deposited in the smallest volume.

TABLE 30.4 Relative biological effectiveness of radiation

Radiation type	RBE
X rays	1
Gamma rays	1
Beta particles	1
Neutrons	5–20
Protons	10
Alpha particles	20

The product of the absorbed dose with the RBE is called the **dose equivalent.** Dose equivalent is measured in **sieverts,** abbreviated Sv. To be precise,

$$\text{dose equivalent in Sv} = \text{absorbed dose in Gy} \times \text{RBE}$$

1 Sv of radiation produces the same biological damage regardless of the type of radiation. An older but still widely used unit for dose equivalent is the **rem,** defined as 1 rem = 0.010 Sv. Small radiation doses are measured in millisievert (mSv) or millirem (mrem).

EXAMPLE 30.8 Radiation exposure

A 75 kg laboratory technician working with the radioactive isotope ^{137}Cs receives an accidental 1.0 mSv exposure. ^{137}Cs emits 0.66 MeV gamma-ray photons. How many gamma-ray photons are absorbed in the technician's body?

MODEL The radiation dose is a combination of deposited energy and biological effectiveness. The RBE for gamma rays is 1. Gamma rays are penetrating, so this is a whole-body exposure.

SOLVE The absorbed dose is the dose in Sv divided by the RBE. In this case, because RBE = 1, the dose is 0.0010 Gy = 0.0010 J/kg. This is a whole-body exposure, so the total energy deposited in the technician's body is 0.075 J. The energy of each absorbed photon is 0.66 MeV, but this value must be converted into joules. The number of photons in 0.075 J is

$$N = \frac{0.075 \text{ J}}{(6.6 \times 10^5 \text{ eV/photon})(1.60 \times 10^{-19} \text{ J/eV})}$$
$$= 7.1 \times 10^{11} \text{ photons}$$

ASSESS The energy deposited, 0.075 J, is very small. Radiation does its damage not by thermal effects, which would require substantially more energy, but by ionization.

TABLE 30.5 gives some typical radiation exposures. We are all exposed to a continuous natural background of radiation from cosmic rays (high-energy particles from the sun and astrophysical sources) and from naturally occurring radiation. Much of this is from the radon isotope ^{222}Rn that is part of the decay series of uranium, a trace element in rocks and soil. Radon is an inert gas that seeps into buildings, where it can be inhaled and where its radioactive daughter nuclei attach to dust particles. The radioactive isotope ^{40}K is naturally present at about a 0.01% abundance in the potassium we ingest in food; about 4000 ^{40}K nuclei decay in your body every second. There are large geographical variations in the natural background because of differences in altitude (cosmic-ray radiation is greater at higher altitudes) and in the amount of uranium in the soil.

Medical x rays vary significantly, but most individual x rays give much less than a year's natural exposure to background radiation. However, some imaging technologies, such as CT scans (computed tomography, which uses x rays) and PET scans (positron-emission tomography, which we will discuss in the next section), require a significantly larger dose of radiation. The potential hazards of this dose must be weighed against the procedure's medical benefits.

The question inevitably arises: What is a safe dose? This remains a controversial topic and the subject of ongoing research. The effects of large doses of radiation are easily observed. The effects of small doses are hard to distinguish from other natural and environmental causes. Thus there's no simple or clear definition of a safe dose. A prudent policy is to avoid unnecessary exposure to radiation but not to worry over exposures less than the natural background. It's worth noting that the μCi radioactive sources used in laboratory experiments provide exposures *much* less than the natural background, even if used on a regular basis.

TABLE 30.5 Radiation exposure

Radiation source	Typical exposure (mSv)
PET brain scan	20
CT abdominal scan	10
Natural background (1 year)	3
Mammogram x ray	0.4
Chest x ray	0.2
Dental x ray	0.01

Dosimeters

A *dosimeter* is a device that measures an absorbed dose of radiation. A traditional dosimeter badge contained a piece of film whose level of exposure indicated the radiation dose a person had received. Newer dosimeters are based on solid-state devices. All dosimeters measure only beta and gamma radiation because alpha particles cannot penetrate the cover.

The most common dosimeter in use today is called an OSL dosimeter because it uses a technique called *optically stimulated luminescence.* The detector consists of crystals of aluminum oxide or beryllium oxide that have been "doped" with small amounts of carbon or other elements. The energy levels, shown in FIGURE 30.25a, are similar to the molecular energy levels we examined in ◀◀ SECTION 29.6, with one exception: the presence of *trapped states* between the band of ground states and the band of excited states. (In descriptions of the energy levels of solid crystals, these are called the *valence band* and the *conduction band.*) An electron in a trapped state is unable to jump to the ground state by emitting a photon; as a result, it can remain in the trapped state for an extremely long period of time.

The absorption of a high-energy beta particle or gamma photon in one of these crystals can promote a ground-state electron to a trapped state, where it remains.

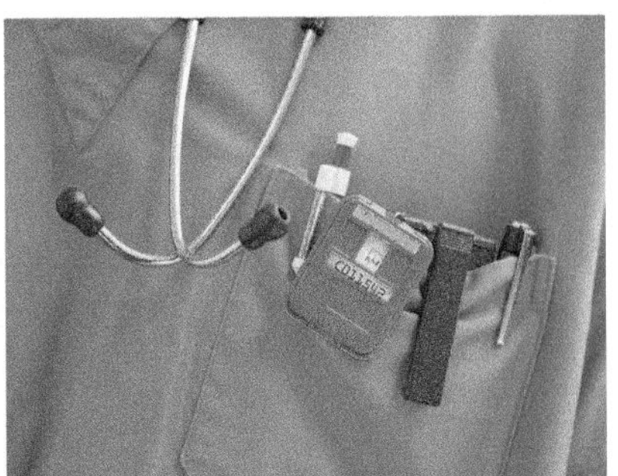

Wearing a badge A personal dosimeter, often called a *dosimeter badge,* is worn by all personnel who work with or around sources of ionizing radiation. Dosimeter badges are typically read once a month.

FIGURE 30.25 Optically stimulated luminescence measures the number of trapped electrons that have been created by ionizing radiation.

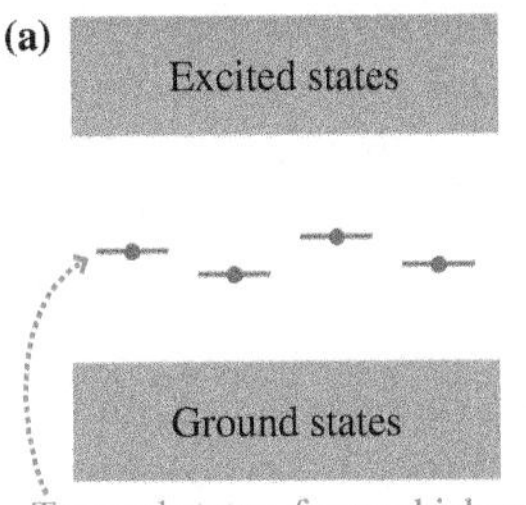
(a)
Excited states
Ground states
Trapped states, from which an electron cannot emit a photon, are excited by ionizing radiation.

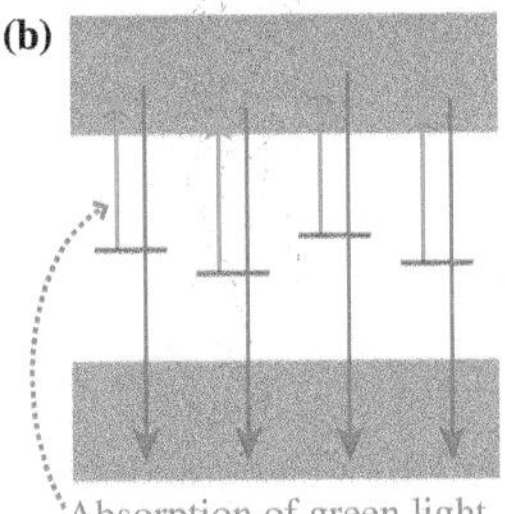
(b)
Absorption of green light is quickly followed by the emission of blue light.

The key idea is that the number of trapped electrons is directly proportional to the absorbed dose, so the absorbed dose can be determined by measuring the number of trapped electrons.

This is done, as shown in in FIGURE 30.25b, by removing the crystal from the dosimeter and inserting it into a machine where it is illuminated with green laser light. A trapped electron is unable to emit a photon, but it can *absorb* a photon and jump upward into the band of excited states. This is called *optical stimulation.* The excited-state electrons then emit photons of shorter-wavelength blue light—*luminescence*—as they quickly make an allowed quantum jump down to the ground state. The absorbed dose is determined from the intensity of blue light that is detected immediately after a pulse of green laser light.

Most dosimeters have small pieces of metal and plastic of different thicknesses—called *filters*—between the front of the dosimeter and the crystal detector. By comparing the blue-light intensity from the portions of the detector that are and are not covered by these filters, and using the known absorption properties of the filters, the dosimeter readout can distinguish between beta and gamma radiation and can provide a rough indication of the energy of the incident particles or photons. This information provides a health safety officer with important clues for identifying the source of any radiation exposure.

STOP TO THINK 30.5 Does a medical professional who wears a dosimeter badge need to keep the badge out of sunlight to avoid an erroneous reading?

30.7 Nuclear Medicine

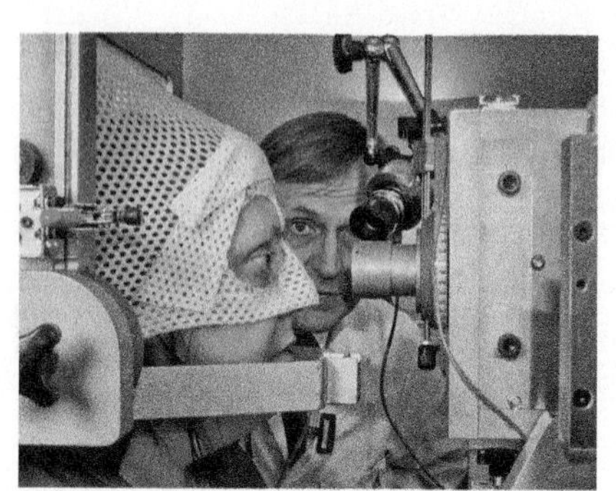

Putting protons to work A patient is being prepared to have a tumor of the eye treated with proton radiation therapy.

Medical imaging of bones was one of the first uses of x rays, 130 years ago. It took several decades for scientists and physicians to learn how radiation interacts with tissue and how to balance its benefits against potential harm, but today nuclear medicine is an essential part of the medical profession. The three main areas of nuclear medicine are diagnostics, imaging, and therapy.

Diagnostics

Radioactive isotopes are used as *tracers* in a number of diagnostic procedures. This technique is based on the fact that all isotopes of an element have identical chemical behavior. As an example, a radioactive isotope of iodine is used in the diagnosis of certain thyroid conditions. Iodine is an essential element in the body, and it concentrates in the thyroid gland. A doctor who suspects a malfunctioning thyroid gland gives the patient a small dose of sodium iodide in which some of the normal ^{127}I atoms have been replaced with ^{131}I. (Sodium iodide, which is harmless, dissolves in water and can simply be drunk.) The ^{131}I isotope, with a half-life of eight days, undergoes beta decay and subsequently emits a gamma-ray photon that can be detected.

The radioactive iodine concentrates inside the thyroid gland within a few hours. The doctor then monitors the gamma-ray photon emissions over the next few days to see how the iodine is being processed within the thyroid and how quickly it is eliminated from the body.

Other important radioactive tracers include the chromium isotope ^{51}Cr, which is taken up by red blood cells and can be used to monitor blood flow, and the xenon isotope ^{133}Xe, which is inhaled to reveal lung functioning. Radioactive tracers are *noninvasive,* meaning that the doctor can monitor the inside of the body without surgery. Animal and plant physiologists also use radioactive tracers, especially isotopes of nitrogen and phosphorous, to study metabolic pathways.

Nuclear Imaging

X-ray images are created with an *external source* of radiation. In contrast, nuclear imaging uses an *internal source*—radiation from decaying isotopes inside the body. And there's another key difference between x rays and nuclear imaging procedures. **An x ray is an image of anatomical structure;** it is excellent for identifying structural problems like broken bones. **Nuclear imaging creates an image of the biological activity of tissues in the body.** For example, nuclear imaging can detect reduced metabolic activity of brain tissue after a stroke.

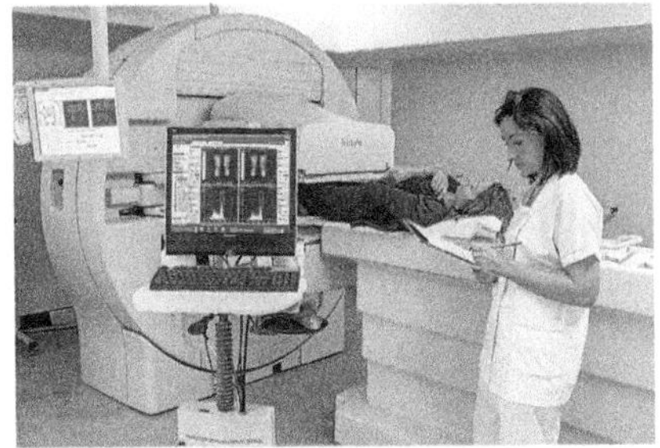

A portrait in gamma The detector is measuring gamma radiation emitted by isotopes taken up by tissues in the woman's knee joints.

Let's look at an example that illustrates the difference between a conventional x-ray image or CT scan and an image made with a nuclear imaging technique. Suppose a doctor suspects a patient has cancerous tissue in the bones. An x ray does not show anything out of the ordinary; the tumors may be too small or may appear similar to normal bone. The doctor then orders a scan with a **gamma camera,** a device that can measure and produce an image from gamma rays emitted within the body.

The patient is given a dose of a phosphorus compound labeled with the gamma-emitter ^{99}Tc. This compound is taken up and retained in bone tissue where active growth is occurring, which could be after a recent injury or because of a tumor. The patient is then scanned with a gamma camera. FIGURE 30.26a shows how the gamma camera can pinpoint the location of the gamma-emitting isotopes in the body and produce an image that reveals their location and intensity. A typical image is shown in FIGURE 30.26b. The bright spots show high concentrations of ^{99}Tc, revealing areas of tumor growth. The tumors may be too small to show up on an x ray, but their activity is easily detected with the gamma camera. With such early detection, the patient's chance of a cure is greatly improved.

FIGURE 30.26 The operation of a gamma camera.

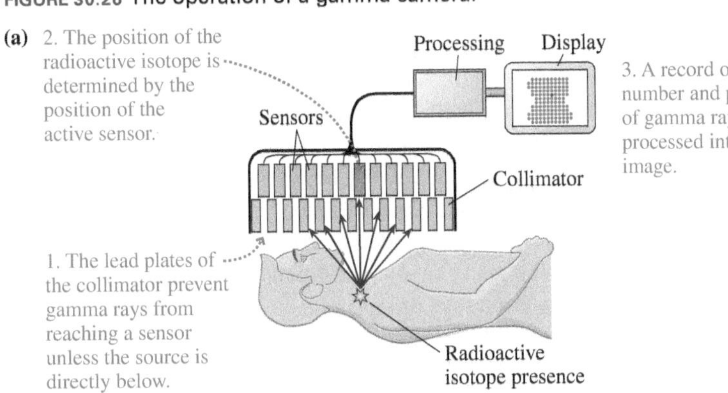
(a)
2. The position of the radioactive isotope is determined by the position of the active sensor.
Processing
Display
3. A record of the number and position of gamma rays is processed into an image.
Sensors
Collimator
1. The lead plates of the collimator prevent gamma rays from reaching a sensor unless the source is directly below.
Radioactive isotope presence

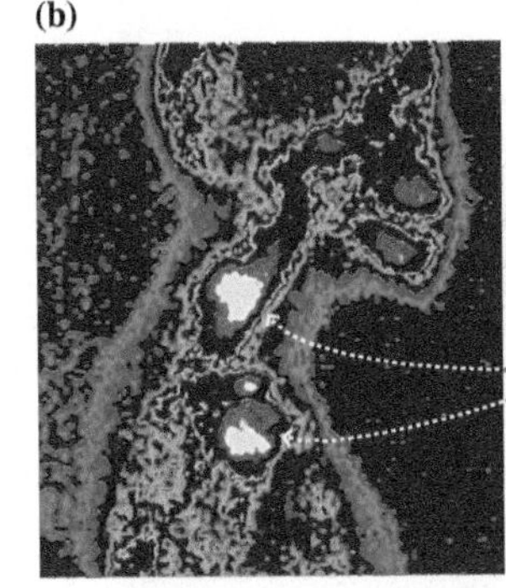
(b)
The bright spots show areas of active tumor growth.

CONCEPTUAL EXAMPLE 30.9 Using radiation to diagnose disease

A patient suspected of having kidney disease is injected with a solution containing molecules that are taken up by healthy kidney tissue. The molecules have been "tagged" with radioactive ^{99}Tc. A gamma camera scan of the patient's abdomen gives the image in FIGURE 30.27. In this image, blue corresponds to the areas of highest activity. Which of the patient's kidneys has reduced function?

FIGURE 30.27 A gamma scan of a patient's kidneys.

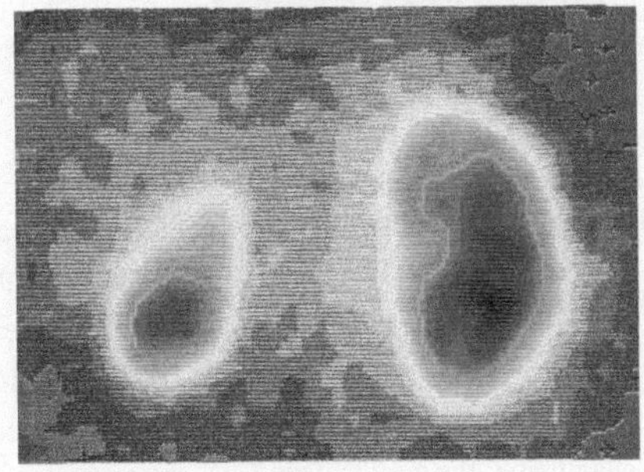

REASON Healthy tissue should show up in blue on the scan because healthy tissue will absorb molecules with the ^{99}Tc attached and will thus emit gamma rays. The kidney imaged on the right shows normal activity throughout; the kidney imaged on the left appears smaller, so it has a smaller volume of healthy tissue. The patient is ill; the problem is with the kidney imaged on the left.

ASSESS Depending on the isotope and how it is taken up by the body, either healthy tissue or damaged tissue could show up on a gamma camera scan.

Positron-Emission Tomography

We have seen that a small number of radioactive isotopes decay by the emission of a positron—beta-plus decay. Such isotopes can be used for an imaging technique known as *positron-emission tomography,* or *PET.* PET is particularly important for imaging the brain.

The imaging process depends on the fact that an electron and a positron annihilate each other when they collide. The particles' mass is transformed into the energy of two gamma-ray photons that are emitted in opposite directions, as shown in FIGURE 30.28a.

FIGURE 30.28 Positron-emission tomography.

(a) When the electron and positron meet . . .

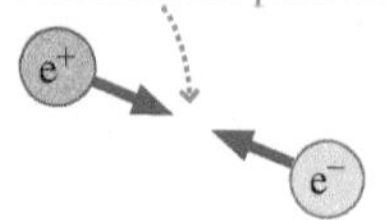
e+
e−
. . . the energy equivalent of their mass is converted into two gamma rays headed in opposite directions.

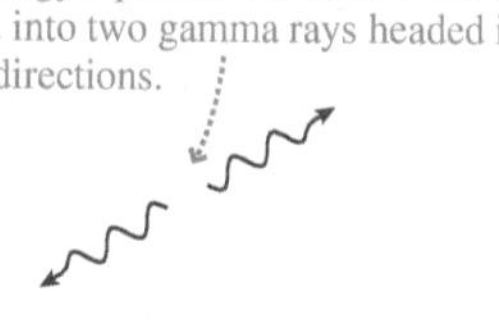

(b) Coincident detection of two gamma rays means that the positron source is along this line.

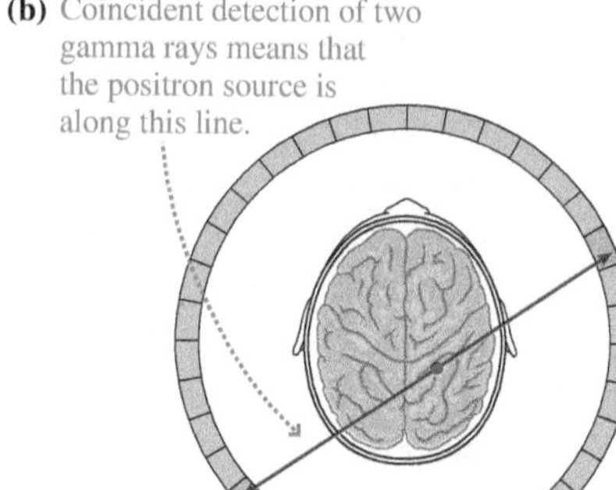

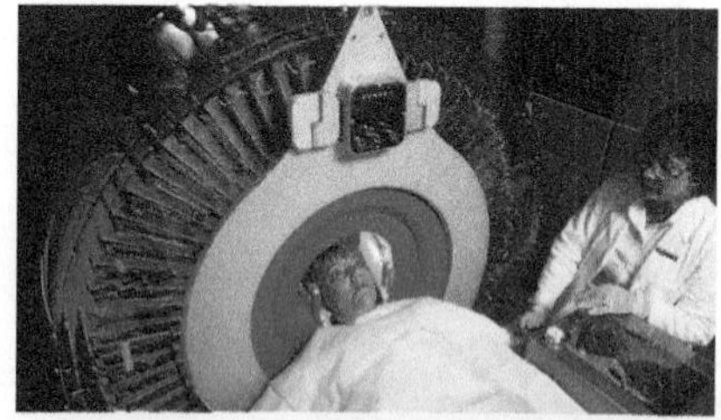

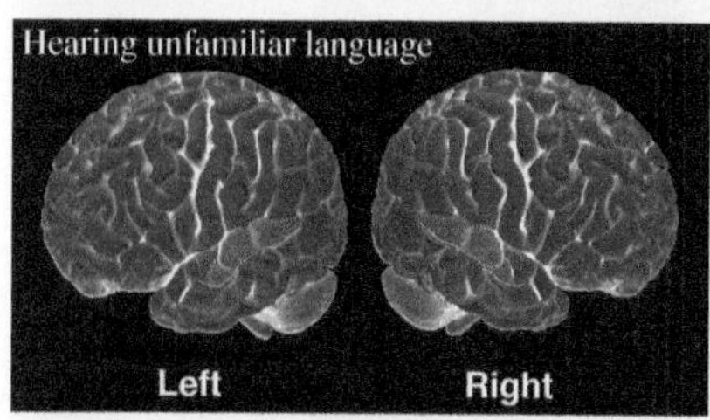
Hearing unfamiliar language
Left
Right

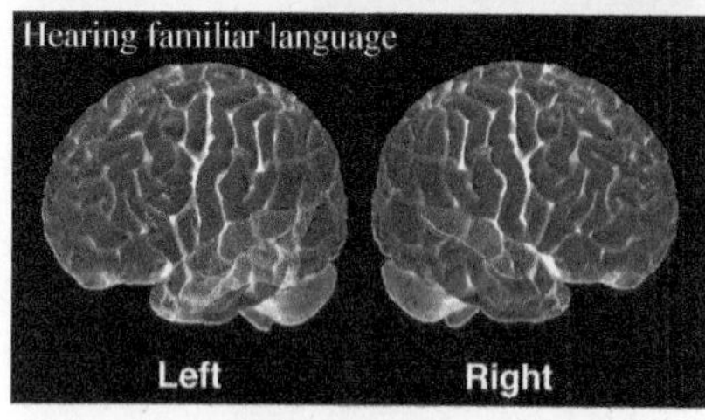
Hearing familiar language
Left
Right

This is your brain on PET The woman in the top photo is undergoing a PET scan not to diagnose disease but to probe the workings of the brain. While undergoing a PET scan, a subject is asked to perform different mental tasks. The lower panels show functional images from a PET scan superimposed on anatomical images from a CT scan. The subject first listened to speech in an unfamiliar language; the active areas of the brain were those responsible for hearing. Next, she listened to speech in a familiar language, resulting in activity in the parts of the brain responsible for speech and comprehension.

Most PET scans use the fluorine isotope ^{18}F, which emits a positron as it undergoes beta-plus decay to ^{18}O with a half-life of 110 minutes. ^{18}F is used to create an analog of glucose called fluorine-18 fluorodeoxyglucose (F-18 FDG). This compound is taken up by tissues in the brain. Areas that are more active are using more glucose, so the F-18 FDG is concentrated in active brain regions. When a fluorine atom in the F-18 FDG decays, the emitted positron immediately collides with a regular electron. The two annihilate to produce two gamma rays that travel out of the brain in opposite directions, as shown in Figure 30.28a.

FIGURE 30.28b shows a patient's head surrounded by a ring of gamma-ray detectors. Because the gamma rays from the positron's annihilation are emitted back to back, simultaneous detection of two gamma rays on opposite sides of the subject indicates that the annihilation occurred somewhere along the line between those detectors. Recording many such pairs of gamma rays shows with great accuracy where the decays are occurring. A full scan will show more activity in regions of the brain where metabolic activity is enhanced, less activity in regions where metabolic activity is depressed. An analysis of these scans can provide a conclusive diagnosis of stroke, injury, or Alzheimer's disease.

Radiation Therapy

Radiation therapy uses ionizing radiation—charged particles or high-energy photons—to destroy cancer cells by breaking the chemical bonds in DNA. Single-strand breaks can often be repaired by enzymes, but a double-strand break is almost always lethal to a cell. A double-strand break is less likely to occur than a single-strand break, so radiation therapy relies on delivering a large absorbed dose of radiation to a tumor.

The primary challenge of radiation therapy is to deliver a lethal dose of radiation to a tumor, which may have an irregular shape, while at the same time minimizing exposure to nearby healthy tissue. One way of doing so is to use a catheter to implant small pellets or "seeds" of radioisotopes directly within or next to the tumor. This technique is called *brachytherapy,* from a Greek root meaning "short" or "short-range."

Brachytherapy uses beta emitters because, as we've seen, the range of beta particles in tissue is a few millimeters—an appropriately short range for irradiating a tumor while limiting the impact on nearby tissue. ^{60}Co, ^{137}Cs, and ^{192}Ir are commonly used isotopes. Brachytherapy requires precise placement of the seeds, and the physician is guided by real-time ultrasound, a CT scan, or other imaging technologies.

A more well-known version of radiation therapy uses external radiation in the form of beams of high-energy photons or charged particles. A beam of radiation irradiates tissue along its path, so the treatment protocol, as shown in FIGURE 30.29, is to rotate the patient or the beam under careful computer control so that the radiation impinges along many different lines that all intersect at the tumor. This protocol delivers a high absorbed dose to the tumor and a much lower dose to surrounding tissue. The radiation treatment plan, which is developed jointly by the physician and a medical physicist, requires an accurate image of the tumor from a CT scan or MRI along with detailed knowledge of how the radiation interacts with the different tissues along each line.

FIGURE 30.29 External-beam radiation therapy is designed to deliver a lethal dose to the tumor without damaging nearby tissue.

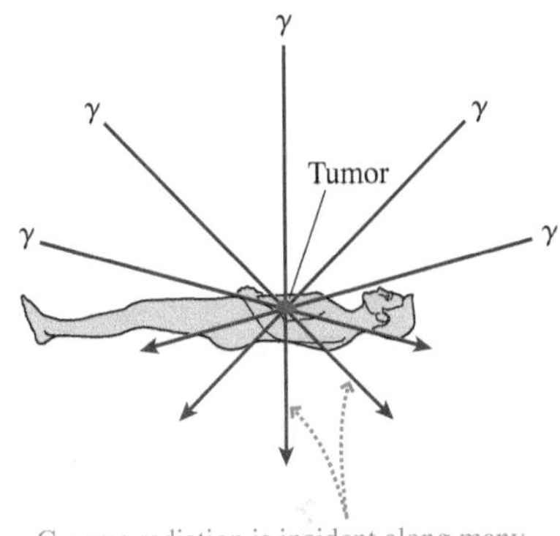

γ
γ
γ
Tumor
γ
γ
Gamma radiation is incident along many lines, all of which intersect the tumor.

External-beam radiation therapy long used the isotope ^{60}Co as a source of gamma-ray photons with energy $\approx$1 MeV. Today, most photon therapy uses x rays generated in an electron linear accelerator—a modern-day version of an x-ray tube. This provides more precise control over the photon energy and the beam shape. Photon energies are typically several MeV, but the value can be adjusted to optimize the treatment.

NOTE ▶ High-energy photons created in a machine are called *x rays* regardless of their energy. Gamma rays are high-energy photons created by nuclear decay or particle-antiparticle annihilation. The multi-MeV x-ray photons used in radiation therapy have higher energy than the gamma-ray photons from most nuclear decays. ◀

One disadvantage of using photons for radiation therapy is that only a small part of the energy is absorbed at the desired depth; the intensity is highest where the beam enters the body, and much of the radiation travels deeper into the body than the target location. It would be preferable to deposit most of the beam energy directly into the tumor.

A newer type of radiation therapy, proton therapy, does exactly this. A proton is a charged particle, and the interaction of a high-energy proton with tissue is similar to that of alpha particles. We've seen that a 1 MeV alpha particle travels only $\approx 50\ \mu$m in tissue. A 1 MeV proton would have a somewhat larger range, although still short, because it is less massive and has only half the charge. But the range of a charged particle increases with energy, and a proton linear accelerator can generate a beam of protons that have tens or hundreds of MeV of kinetic energy. A proton's range in water (the range in tissue is similar) increases from 2 cm at 50 MeV to 8 cm at 100 MeV and 26 cm at 200 MeV. These ranges allow protons to reach anywhere in the body.

Range is not the only factor, however. The electric interaction of a rapidly moving proton with the electrons of atoms is complex, but to a good approximation dE/dx—the particle's energy loss per unit length of travel—is proportional to $1/v^2$. In other words, the energy deposited in tissue per millimeter of travel is initially low when the proton is moving very fast, gradually increases as the proton slows, and becomes very large only at the end of the range as the speed approaches zero.

FIGURE 30.30 The absorbed dose from a beam of 200 MeV protons changes with depth into a patient.

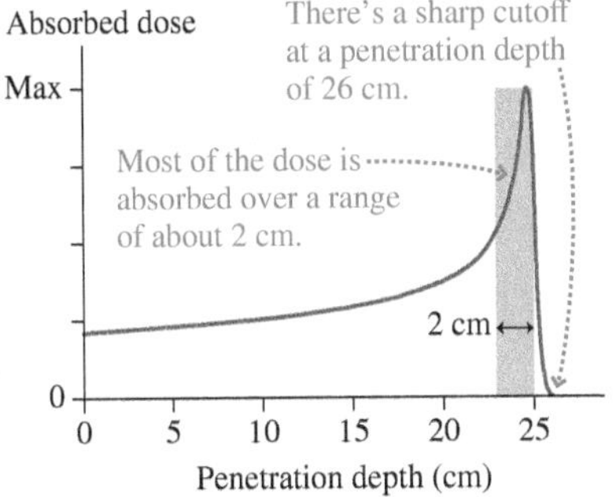

FIGURE 30.30 is a graph of the absorbed dose versus distance into tissue for 200 MeV protons. Notice two important features:

1. The absorbed dose is concentrated into an $\approx$2-cm-wide interval at the end of the proton's range as the proton's speed approaches zero. The dose is significantly lower to tissue that faster protons pass through on their way to the tumor.

2. There's a sharp cutoff at a depth of 26 cm, the depth at which the protons are completely stopped. Tissue beyond 26 cm receives no radiation. This is in contrast to x-ray therapy, where the photon beam is gradually attenuated but does not have a cutoff.

The position of the maximum absorbed dose can be scanned across a tumor by slowly changing the proton energy. By scanning the position of the peak while rotating the proton beam to different angles, a physician can deliver radiation that precisely matches the size and shape of a tumor while sparing nearby healthy tissue.

EXAMPLE 30.10 Using radiation to treat disease

Patients who have prostate cancer can be treated by implanting radioactive "seeds" into the gland, as shown in FIGURE 30.31. For one patient's treatment, the seeds contain the samarium isotope ^{153}Sm, which decays with a half-life of 1.9 days by emitting a 0.81 MeV beta particle. On average, 50% of the decay energy is absorbed by the 24 g prostate gland. If the seeds have an activity of 100 MBq when first implanted, what dose equivalent in Sv does the gland receive during the first day?

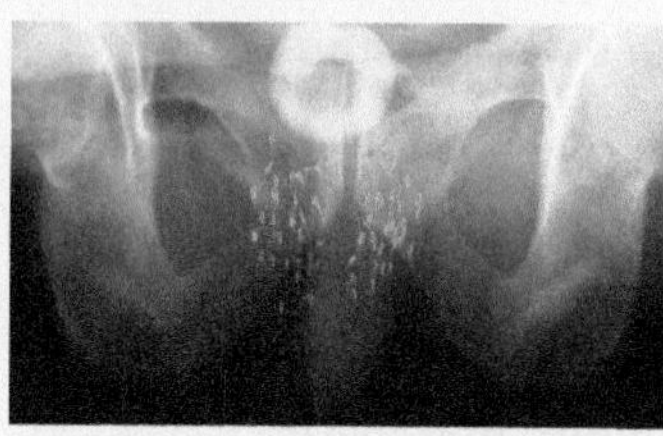

FIGURE 30.31 Radioactive seeds are implanted in the prostate gland to shrink a tumor.

PREPARE To find the dose equivalent, we will need to know how many nuclei have decayed and how much energy each nucleus deposits in the gland. We can find the initial number of nuclei from the activity, as we did in Example 30.4. Then, using the known half-life, we can find the number that remain after one day. The difference between these two numbers is the number of decays in the first day.

We need to convert the half-life from days to seconds:

$$t_{1/2} = (1.9 \text{ days})\left(\frac{8.64 \times 10^4 \text{ s}}{1 \text{ day}}\right) = 1.64 \times 10^5 \text{ s}$$

We also need the absorbed energy in joules. Each atom that decays emits a 0.81 MeV beta particle, and half of this energy is deposited in the gland. In joules, this is

$$E_{\text{decay}} = \frac{1}{2}(0.81 \times 10^6 \text{ eV})(1.60 \times 10^{-19} \text{ J/eV})$$
$$= 6.48 \times 10^{-14} \text{ J}$$

SOLVE The initial number of radioactive nuclei is found from $N_0 = \tau R_0 = (t_{1/2}/\ln 2)\, R_0$:

$$N_0 = \frac{t_{1/2}R_0}{0.693} = \frac{(1.64 \times 10^5 \text{ s})(100 \times 10^6 \text{ decays/s})}{0.693}$$
$$= 2.37 \times 10^{13}$$

After 1.0 day, the number of radioactive nuclei in the sample has decreased to

$$N = N_0\left(\frac{1}{2}\right)^{t/t_{1/2}} = (2.37 \times 10^{13})\left(\frac{1}{2}\right)^{1.0 \text{ day}/1.9 \text{ days}}$$
$$= 1.65 \times 10^{13}$$

The number of decays during the first day is the difference between these numbers:

$$N_0 - N = (2.37 \times 10^{13}) - (1.65 \times 10^{13}) = 7.2 \times 10^{12} \text{ decays}$$

The energy deposited in the gland by these decays is

$$E_{\text{deposit}} = (6.48 \times 10^{-14} \text{ J/decay})(7.2 \times 10^{12} \text{ decays})$$
$$= 0.467 \text{ J}$$

The radiation dose to the 0.024 kg gland is thus

$$\text{dose} = \frac{0.467 \text{ J}}{0.024 \text{ kg}} = 19 \text{ Gy}$$

Beta particles have RBE = 1, so the dose equivalent in Sv is the same:

$$\text{dose equivalent} = 19 \text{ Sv}$$

ASSESS This is a very large dose equivalent, but the goal is to destroy the tumor, so this is a reasonable result.

STOP TO THINK 30.6 A patient ingests a radioactive isotope to treat a tumor. The isotope provides a dose of 0.10 Gy. Which type of radiation will give the highest dose equivalent in Sv?

A. Alpha particles B. Beta particles C. Gamma rays

30.8 The Ultimate Building Blocks of Matter

As we've seen, modeling the nucleus as being made of protons and neutrons allows a description of all of the elements in the periodic table in terms of just three basic particles—protons, neutrons, and electrons. But are protons and neutrons *really* basic building blocks? Molecules are made of atoms. Atoms are made of a cloud of electrons surrounding a positively charged nucleus. The nucleus is composed of protons and neutrons. Where does this process end? Are electrons, protons, and neutrons the basic building blocks of matter, or are they made of still smaller subunits?

This question takes us into the domain of what is known as **particle physics**—the branch of physics that deals with the basic constituents of matter and the relationships among them. Particle physics starts with the constituents of the atom, the proton, neutron, and electron, but there are many other particles below the scale of the atom. We call these particles **subatomic particles.**

Subatomic crash tests The above picture is the record of a collision between an electron and a positron (paths represented by the green arrows) brought to high speeds in a collider. The particles annihilated in the center of a detector that measured the paths of the particles produced in the collision. In this case, the annihilation of the electron and positron created a particle known as a Z boson. The Z boson then quickly decayed into the two jets of particles seen coming out of the detector.

Antiparticles

We've described the positron as the *antiparticle* to the electron. They are antiparticles in the sense that when a positron and an electron meet, they annihilate each other, turning the energy equivalent of their masses into the pure energy of two photons. Mass disappears and light appears in one of the most spectacular confirmations of Einstein's relativity.

Every subatomic particle that has been discovered has an antiparticle twin that has the same mass and the same spin but opposite charge. In addition to positrons, there are antiprotons (with $q = -e$), antineutrons (also neutral, but not the same as regular neutrons), and antimatter versions of all the various subatomic particles we will see. The notation to represent an antiparticle is a bar over the top of the symbol. A proton is represented as p, an antiproton as $\bar{\text{p}}$.

Antiparticles provide interesting opportunities for creating "exotic" subatomic particles. When a particle meets its associated antiparticle, the two annihilate, leaving nothing but their energy behind. This energy must go somewhere. Sometimes it is emitted as gamma-ray photons, but this energy can also be used to create other particles.

The major tool for creating and studying subatomic particles is the *particle collider.* These machines use electric and magnetic fields to accelerate particles and their antiparticles, such as e and $\bar{\text{e}}$, or p and $\bar{\text{p}}$, to speeds very close to the speed of light. These particles then collide head-on. As they collide and annihilate, their mass-energy and kinetic energy combine to produce exotic particles that are not part of ordinary matter. These particles come in a dizzying variety—pions, kaons, lambda particles, sigma particles, and dozens of others—each with its own antiparticle. Most live no more than a trillionth of a second before decaying into other particles.

EXAMPLE 30.11 Determining a possible outcome of a proton-antiproton collision

When a proton and an antiproton annihilate, the resulting energy can be used to create new particles. One possibility is the creation of electrically neutral particles called *neutral pions.* A neutral pion has a rest mass of 135 MeV/c^2. How many neutral pions could be produced in the annihilation of a proton and an antiproton? Assume the proton and antiproton are moving very slowly as they collide.

PREPARE Because the proton and antiproton are moving slowly, with essentially no kinetic energy, the total energy available for creating new particles is the energy equivalent of the masses of the proton and the antiproton. The mass of a proton is given in Table 30.2 as 938 MeV/c^2. The mass of an antiproton is the same.

SOLVE The total energy from the annihilation of a proton and an antiproton is the energy equivalent of their masses:

$$E = (m_{\text{proton}} + m_{\text{antiproton}})c^2 = (938\ \text{MeV}/c^2 + 938\ \text{MeV}/c^2)c^2 = 1876\ \text{MeV}$$

It takes 135 MeV to create a neutral pion. The ratio

$$\frac{\text{energy available}}{\text{energy required to create a pion}} = \frac{1876\ \text{MeV}}{135\ \text{MeV}} = 13.9$$

Continued

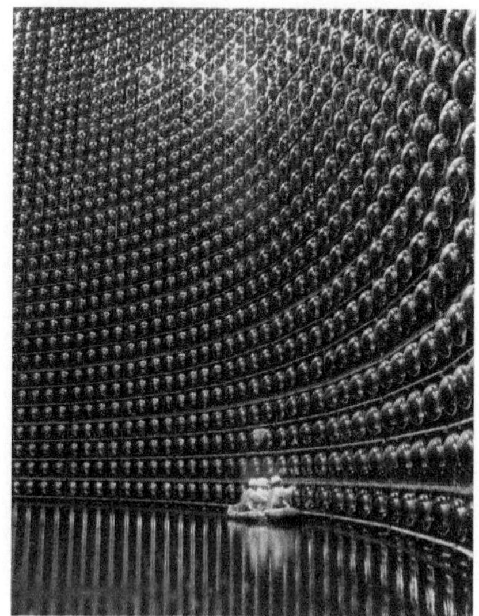

A big detector for a small particle The rubber raft in the photo is floating inside a particle detector designed to measure neutrinos. Neutrinos are so weakly interacting that a neutrino produced in a nuclear reaction in the center of the sun will likely pass through the entire mass of the sun and escape. Of course, the neutrino's weakly interacting nature also means that it is likely to pass right through a detector. The Super Kamiokande experiment in Japan monitors interactions in an enormous volume of water in order to spot a very small number of neutrino interactions.

tells us that we have enough energy to produce 13 neutral pions from this process, but not quite enough to produce 14.

ASSESS Because the mass of a pion is much less than that of a proton or an antiproton, the annihilation of a proton and antiproton can produce many more pions than the number of particles at the start. Though the production of 13 neutral pions is a possible outcome of a proton-antiproton interaction, it is not a likely one. In addition to the conservation of energy, there are many other physical laws that determine what types of particles, and in what quantities, are likely to be produced.

Neutrinos

The most abundant particle in the universe is not the electron, proton, or neutron but the *neutrino,* a neutral, nearly massless particle that interacts only weakly with matter.

It turns out that there are three types of neutrinos. The neutrino involved in beta decay is the *electron neutrino* ν_e; it shows up in processes involving electrons and positrons. The other two, called the *muon neutrino* and the *tau neutrino,* are similarly associated with processes that involve muons and taus, unstable particles that are rather like heavy electrons. Each of these neutrinos has an antiparticle.

The sun is powered by a nuclear fusion process in which hydrogen is converted into helium. Prodigious numbers of neutrinos are created in this process. The number of these neutrinos passing through your body each second is staggering—over 600 trillion. But perhaps even more amazing is that over your entire lifetime, only one or two of these neutrinos will interact with the nuclei of your body; the rest pass through without leaving any trace. Alpha particles can be stopped by a sheet of paper, beta particles by a few centimeters of water. But it would take a piece of lead 1 light-year thick to stop a neutrino!

Quarks

The process of beta decay, in which a neutron can change into a proton, and vice versa, gives a hint that the neutron and the proton are *not* fundamental units but are made of smaller subunits.

There is another reason to imagine such subunits: the existence of dozens of subatomic particles—muons, pions, kaons, omega particles, and so on. Just as the periodic table explains the many different atomic elements in terms of three basic particles, perhaps it is possible to do something similar for this "subatomic zoo."

We now understand protons and neutrons to be composed of smaller charged particles whimsically named **quarks.** The quarks that form protons and neutrons are called **up quarks** and **down quarks,** symbolized as u and d, respectively. The nature of these quarks and the composition of the neutron and the proton are shown in FIGURE 30.32.

FIGURE 30.32 The quark content of the proton and neutron.

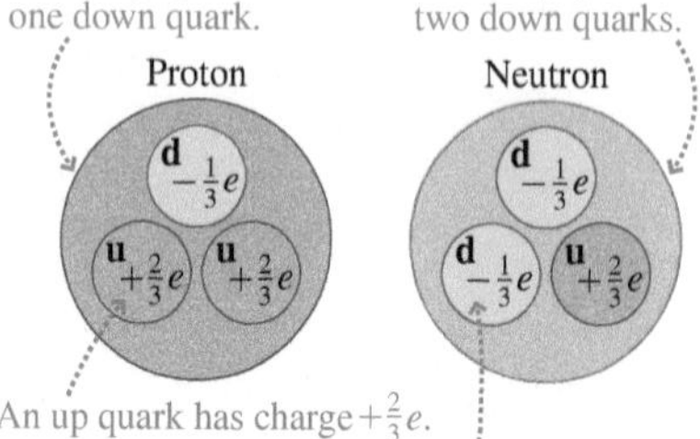

A proton is made of two up quarks and one down quark.
A neutron is made of one up quark and two down quarks.
Proton
Neutron
d $-\frac{1}{3}e$
u $+\frac{2}{3}e$
u $+\frac{2}{3}e$
d $-\frac{1}{3}e$
d $-\frac{1}{3}e$
u $+\frac{2}{3}e$
An up quark has charge $+\frac{2}{3}e$.
A down quark has charge $-\frac{1}{3}e$.

NOTE ▶ It seems surprising that the charges of quarks are *fractions* of e. Don't charges have to be integer multiples of e? It's true that atoms, molecules, and all macroscopic matter must have $q = Ne$ because these entities are constructed from electrons and protons. But no law of nature prevents other types of matter from having other amounts of charge. ◀

A neutron and a proton differ by one quark. Beta decay can now be understood as a process in which a down quark changes to an up quark, or vice versa. Beta-minus decay of a neutron can be written as

$$d \rightarrow u + e^- + \bar{\nu}_e$$

The existence of quarks thus provides an explanation of how a neutron can turn into a proton.

CONCEPTUAL EXAMPLE 30.12 **Quarks and beta-plus decay**

What is the quark description of beta-plus decay?

REASON In beta-plus decay, a proton turns into a neutron, with the emission of a positron and an electron neutrino. To turn a proton into a neutron requires the conversion of an up quark into a down quark; the total reaction is thus

$$u \rightarrow d + e^+ + \nu_e$$

Fundamental Particles

Our current understanding of the truly *fundamental* particles—the ones that cannot be broken down into smaller subunits—is that they come in two basic types: **leptons** (particles like the electron and the neutrino) and quarks (which combine to form particles like the proton and the neutron). The leptons and quarks are listed in TABLE 30.6. A few points are worthy of note:

- Each particle has an associated antiparticle.
- There are three *families* of leptons. The first is the electron and its associated neutrino, and their antiparticles. The other families are based on the muon and the tau, heavier siblings to the electron. Only the electron and positron are stable.
- There are also three families of quarks. The first is the up-down family that makes all "normal" matter. The other families are pairs of heavier quarks that form more exotic particles.

As far as we know, this is where the trail ends. Matter is made of molecules; molecules of atoms; atoms of protons, neutrons, and electrons; protons and neutrons of quarks. Quarks and electrons seem to be truly fundamental. But scientists of the early 20th century thought they were at a stopping point as well—they thought that they knew all of the physics that there was to know. As we've seen over the past few chapters, this was far from true. New tools such as the next generation of particle colliders will certainly provide new discoveries and new surprises.

The early chapters of this text, in which you learned about forces and motion, had very obvious applications to things in your daily life. But in these past few chapters we see that even modern discoveries—discoveries such as antimatter, which may seem like science fiction—can be put to very practical use. As we come to the close of this text, we hope that you have gained an appreciation not only for what physics tells us about the world, but also for the wide range of problems it can be used to solve.

TABLE 30.6 Leptons and quarks

Leptons		Antileptons	
Electron	e^-	Positron	e^+
Electron neutrino	ν_e	Electron antineutrino	$\bar{\nu}_e$
Muon	μ^-	Antimuon	μ^+
Muon neutrino	ν_μ	Muon antineutrino	$\bar{\nu}_\mu$
Tau	τ^-	Antitau	τ^+
Tau neutrino	ν_τ	Tau antineutrino	$\bar{\nu}_\tau$
Quarks		**Antiquarks**	
Up	u	Antiup	$\bar{u}$
Down	d	Antidown	$\bar{d}$
Strange	s	Antistrange	$\bar{s}$
Charm	c	Anticharm	$\bar{c}$
Bottom	b	Antibottom	$\bar{b}$
Top	t	Antitop	$\bar{t}$

CHAPTER 30 INTEGRATED EXAMPLE **A radioactive tracer**

An 85 kg patient swallows a 30 μCi beta emitter that is to be used as a tracer. The isotope's half-life is 5.0 days. The average energy of the beta particles is 0.35 MeV, and they have an RBE (relative biological effectiveness) of 1.5. Ninety percent of the beta particles are stopped inside the patient's body and 10% escape. What total dose equivalent does this patient receive?

PREPARE Beta radiation penetrates the body—enough that 10% of the particles escape—so this is a whole-body exposure. Even the escaping particles probably deposit some energy in the body, but we'll assume that the dose is from only those particles that stop inside the body.

SOLVE The dose equivalent is the absorbed dose in Gy multiplied by the RBE of 1.5. The absorbed dose is the energy absorbed per kilogram of tissue, so we need to find the total energy absorbed from the time the patient swallows the emitter until it has all decayed. The sample's initial activity R_0 is related to the nuclear lifetime τ and the initial

Continued

number of radioactive atoms N_0 by $R_0 = N_0/\tau$. Thus the number of radioactive atoms in the sample, all of which are going to decay and emit a beta particle, is

$$N_0 = \tau R_0 = \frac{t_{1/2}}{\ln 2} R_0$$

In developing this relationship, we used the connection between the lifetime and the half-life.

The initial activity is given in microcuries. Converting to becquerels, we have

$$R_0 = (30 \times 10^{-6}\ \text{Ci}) \times \frac{3.7 \times 10^{10}\ \text{Bq}}{1\ \text{Ci}}$$

$$= 1.1 \times 10^6\ \text{Bq} = 1.1 \times 10^6\ \text{decays/s}$$

The half-life in seconds is

$$t_{1/2} = 5.0\ \text{days} \times \frac{86{,}400\ \text{s}}{1\ \text{day}} = 4.3 \times 10^5\ \text{s}$$

Thus the total number of beta decays over the course of several weeks, as the sample completely decays, is

$$N_0 = \frac{t_{1/2}}{\ln 2} R_0 = \frac{(4.3 \times 10^5\ \text{s})(1.1 \times 10^6\ \text{decays/s})}{\ln 2} = 6.8 \times 10^{11}$$

Ninety percent of these decays deposit, on average, 0.35 MeV in the body, so the absorbed energy is

$$E_{\text{abs}} = (0.90)(6.8 \times 10^{11})\left((3.5 \times 10^5\ \text{eV}) \times \frac{1.60 \times 10^{-19}\ \text{J}}{1\ \text{eV}}\right)$$

$$= 0.034\ \text{J}$$

This is not a lot of energy in an absolute sense, but it is all damaging, ionizing radiation. The absorbed dose is

$$\text{absorbed dose} = \frac{0.034\ \text{J}}{85\ \text{kg}} = 4.0 \times 10^{-4}\ \text{Gy}$$

and thus the dose equivalent is

$$\text{dose equivalent} = 1.5 \times (4.0 \times 10^{-4}\ \text{Gy}) = 0.60\ \text{mSv}$$

ASSESS This dose, typical of many medical uses of radiation, is about 20% of the yearly radiation dose from the natural background. Although one should always avoid unnecessary radiation, this dose would not cause concern if there were a medical reason for it.

SUMMARY

GOAL To learn about the nucleus of the atom and some of the medical applications of nuclear physics.

GENERAL PRINCIPLES

The Nucleus

The nucleus is a small, dense, positive core at the center of an atom.

Z protons, charge $+e$, spin $\frac{1}{2}$
N neutrons, charge 0, spin $\frac{1}{2}$

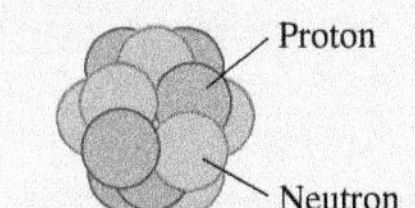

The **mass number** is

$$A = Z + N$$

Isotopes of an element have the same value of Z but different values of N.

The strong force holds nuclei together:

- It acts between any two nucleons.
- It is short-range, <3 fm.

Adding neutrons to a nucleus allows the strong force to overcome the repulsive Coulomb force between protons.

The **binding energy** B of a nucleus depends on the difference in mass between an atom and its constituents:

$$B = (Zm_{\text{H}} + Nm_{\text{n}} - m_{\text{atom}}) \times (931.49\ \text{MeV/u})$$

Nuclear Stability

Most nuclei are not stable. Unstable nuclei undergo **radioactive decay.** Stable nuclei cluster along the **line of stability** in a plot of the isotopes.

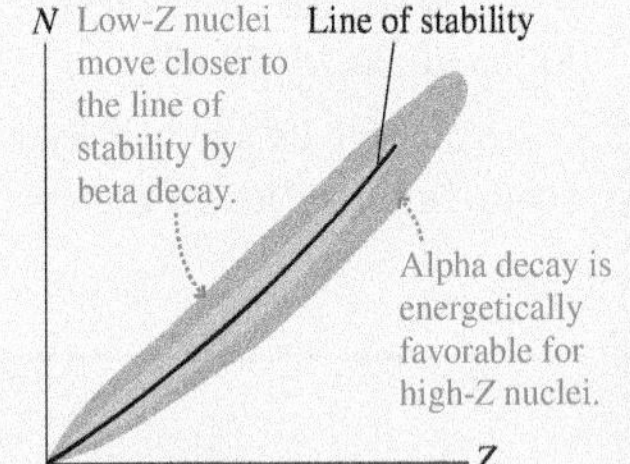

Mechanisms by which unstable nuclei decay:

Decay	Particle	Mechanism	Energy	Penetration
α	^{4}He nucleus	tunneling	few MeV	low
β	e^-	$n \rightarrow p^+ + e^-$	≈ 1 MeV	medium
	e^+	$p^+ \rightarrow n + e^+$	≈ 1 MeV	medium
γ	photon	quantum jump	≈ 1 MeV	high

Alpha and beta decays change the nucleus; the daughter nucleus is a different element.

Alpha decay:

$$^{A}X_{Z} \rightarrow {}^{A-4}Y_{Z-2} + \alpha + \text{energy}$$

Beta-minus decay:

$$^{A}X_{Z} \rightarrow {}^{A}Y_{Z+1} + e^- + \text{energy}$$

IMPORTANT CONCEPTS

The shell model

Nucleons fill nuclear energy levels, similar to how electrons fill energy levels in atoms. A nucleon can sometimes jump to a lower energy level by emitting a beta particle or a gamma-ray photon.

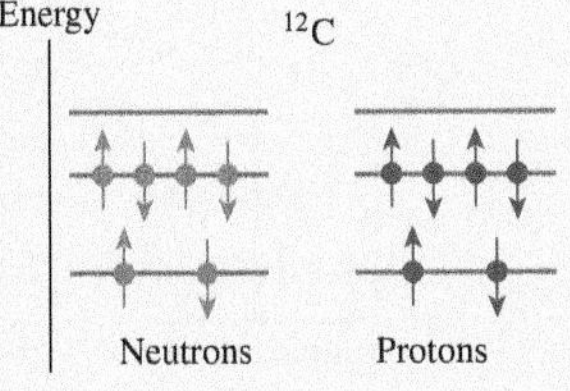

Quarks

Nucleons (and other particles) are made up of quarks. **Quarks** and **leptons** are fundamental particles.

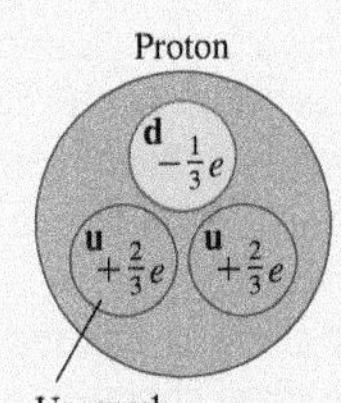

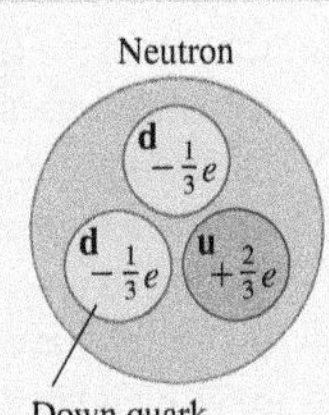

APPLICATIONS

Radioactive decay

The number of still undecayed nuclei decreases exponentially with time:

$$N = N_0 e^{-t/\tau}$$

$$N = N_0 \left(\frac{1}{2}\right)^{t/t_{1/2}}$$

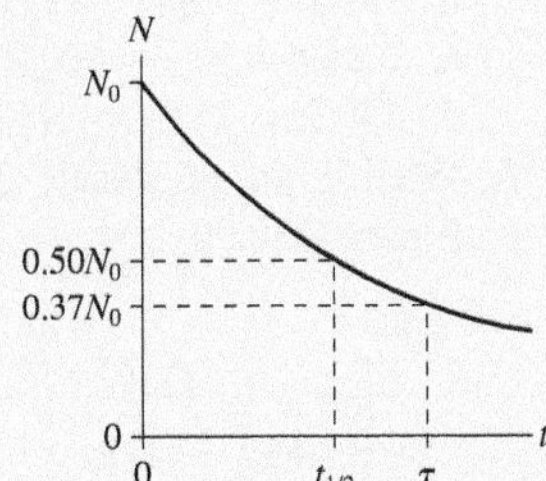

The **half-life**

$$t_{1/2} = \tau \ln 2 = 0.693\tau$$

is the time in which half of any sample decays.

Measuring radiation

The **activity** of a radioactive sample, measured in becquerels, is the number of decays per second. Activity is related to the lifetime by

$$R = \frac{N}{\tau} = \frac{0.693N}{t_{1/2}}$$

The **absorbed dose** is measured in grays, where

$$1\ \text{Gy} = 1.00\ \text{J/kg of absorbed energy}$$

The **relative biological effectiveness** (RBE) is the biological effect of a dose relative to the effect of x rays. The **dose equivalent,** measured in sieverts, is

$$\text{dose equivalent in Sv} = \text{absorbed dose in Gy} \times \text{RBE}$$

LEARNING OBJECTIVES After studying this chapter, you should be able to:

- Describe nuclear structure. *Conceptual Questions 30.1, 30.2; Problems 30.1, 30.3–30.6*
- Calculate nuclear binding energies. *Conceptual Question 30.4; Problems 30.10–30.12, 30.14, 30.15*
- Interpret nuclear energy levels. *Conceptual Question 30.5; Problems 30.16–30.19*
- Calculate activity and half-lives for nuclear decay. *Conceptual Questions 30.8, 30.9; Problems 30.20, 30.22, 30.24, 30.26, 30.29*
- Compare the types and properties of nuclear radiation. *Conceptual Questions 30.6, 30.7; Problems 30.30–30.35*
- Determine the radiation dose for different types of radiation. *Conceptual Questions 30.12, 30.13; Problems 30.36–30.44*
- Identify the major forms of nuclear medicine. *Conceptual Question 30.11; Problems 30.45–30.48*
- Compare the fundamental subatomic particles. *Problems 30.49–30.51*

STOP TO THINK ANSWERS

Stop to Think 30.1: 3. Different isotopes of an element have different numbers of neutrons but the same number of protons. The number of electrons in a neutral atom matches the number of protons.

Stop to Think 30.2: A. The isotopes of an element all have the same proton number Z but different neutron numbers N.

Stop to Think 30.3: C. One-quarter of the atoms are left. This is one-half of one-half, or $(1/2)^2$.

Stop to Think 30.4: B. An increase of Z with no change in A occurs when a neutron changes to a proton and an electron, ejecting the electron.

Stop to Think 30.5: No. A dosimeter responds only to ionizing radiation. The infrared, visible, and near-ultraviolet light from the sun are not ionizing radiation.

Stop to Think 30.6: A. Dose equivalent is the product of dose in Gy (the same for each) and RBE (highest for alpha particles).

QUESTIONS

Conceptual Questions

1. Atom A has a larger atomic mass than atom B. Does this mean that atom A also has a larger atomic number? Explain.
2. Consider the atoms ^{16}O, ^{18}O, ^{18}F, ^{18}Ne, and ^{20}Ne. Some of the following questions may have more than one answer. Give all answers that apply.
   a. Which are isotopes?
   b. Which have the same atomic mass?
   c. Which have the same chemical properties?
   d. Which have the same number of neutrons?
3. Naturally occurring tungsten metal is composed of three stable isotopes: ^{182}W, with a natural abundance of 25%; ^{183}W, abundance 14%; and ^{184}W, abundance 31%. Tungsten also has the highest melting point of any element at 3695 K. If a sample of tungsten were held at its melting point for 10 h, would you expect the natural abundances of its isotopes to change? If so, which would increase the most? Which would decrease the most?
4. a. Is the binding energy of a nucleus with $A = 200$ more than, less than, or equal to the binding energy of a nucleus with $A = 60$? Explain.
   b. Is a nucleus with $A = 200$ more tightly bound, less tightly bound, or bound equally tightly as a nucleus with $A = 60$? Explain.
5. Does each nuclear energy-level diagram in Figure Q30.5 represent a nuclear ground state, an excited nuclear state, or an impossible nucleus? Explain.

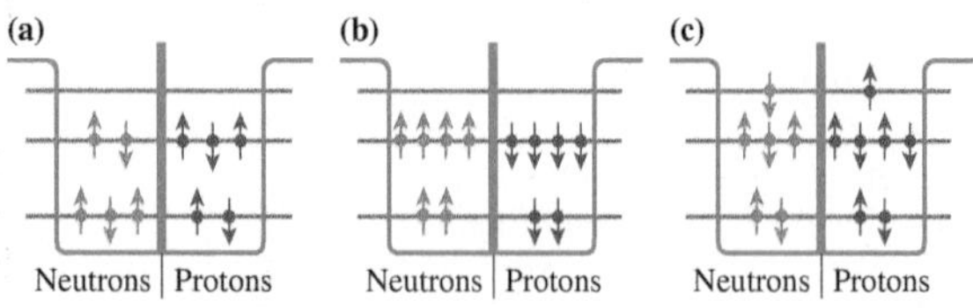

FIGURE Q30.5

6. Are the following decays possible? If not, why not?
   a. $^{232}Th\,(Z=90) \rightarrow {}^{236}U\,(Z=92) + \alpha$
   b. $^{238}Pu\,(Z=94) \rightarrow {}^{236}U\,(Z=92) + \alpha$
   c. $^{11}B\,(Z=5) \rightarrow {}^{11}B\,(Z=5) + \gamma$
   d. $^{33}P\,(Z=15) \rightarrow {}^{32}S\,(Z=16) + e^-$
7. What kind of decay, if any, can occur for the nuclei in Figure Q30.7?

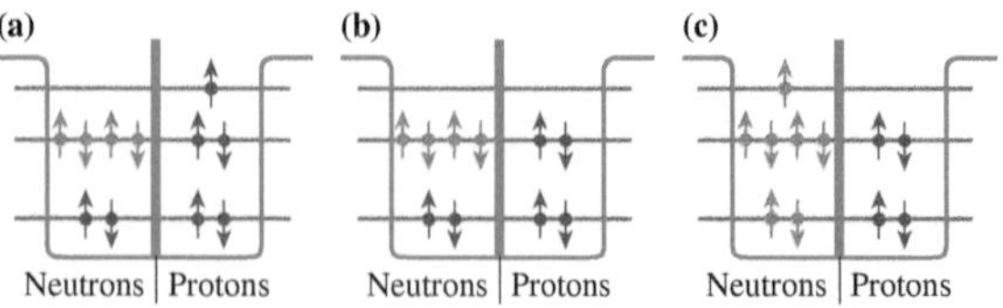

FIGURE Q30.7

Problem difficulty is labeled as | (straightforward) to ||||| (challenging). Problems labeled INT integrate significant material from earlier chapters; BIO are of biological or medical interest; CALC require calculus to solve.

8. Nucleus A decays into nucleus B with a half-life of 10 s. At $t = 0$ s, there are 1000 A nuclei and no B nuclei. At what time will there be 750 B nuclei?
9. A radioactive sample has a half-life of 3 days. Its current activity is 2.0×10^6 Bq. What was its activity 6 days ago?
10. Radiocarbon dating assumes that the abundance of ^{14}C in the environment has been constant. Suppose ^{14}C was less abundant 10,000 years ago than it is today. Would this cause a lab using radiocarbon dating to overestimate or underestimate the age of a 10,000-year-old artifact? (In fact, the abundance of ^{14}C in the environment does vary slightly with time. But the issue has been well studied, and the ages of artifacts are adjusted to compensate for this variation.)
11. The three isotopes ^{212}Po, ^{137}Cs, and ^{90}Sr decay as $^{212}\text{Po} \rightarrow {}^{208}\text{Pb} + \alpha$, $^{137}\text{Cs} \rightarrow {}^{137}\text{Ba} + e^- + \gamma$, and $^{90}\text{Sr} \rightarrow {}^{90}\text{Y} + e^-$. Which of these isotopes would be most useful as a biological tracer? Why?
12. A and B are fresh apples. Apple A is strongly irradiated by nuclear radiation for 1 hour. Apple B is not irradiated. Afterward, in what ways are apples A and B different?
BIO
13. Four radiation doses are as follows: Dose A is 0.10 Gy with an RBE of 1, dose B is 0.20 Gy with an RBE of 1, dose C is 0.10 Gy with an RBE of 2, and dose D is 0.20 Gy with an RBE of 2.
BIO
    a. Rank in order, from largest to smallest, the amount of energy delivered by these four doses.
    b. Rank in order, from largest to smallest, the biological damage caused by these four doses.

Multiple-Choice Questions

14. | ^{74}As is a beta-plus emitter used for locating tumors with PET. What is the daughter nucleus?
BIO
    A. ^{73}As B. ^{74}Ge C. ^{74}Se D. ^{75}As
15. || When uranium fissions, the fission products are radioactive because the nuclei are neutron-rich. What is the most likely decay mode for these nuclei?
    A. Alpha decay B. Beta-minus decay
    C. Beta-plus decay D. Gamma decay
16. || A certain watch's luminous glow is due to zinc sulfide paint that is energized by beta particles given off by *tritium*, the radioactive hydrogen isotope ^{3}H, which has a half-life of 12.3 years. This glow has about 1/4 of its initial brightness. How many years old is the watch?

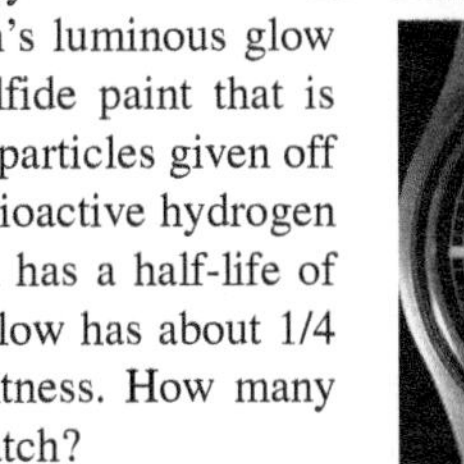

    A. 6 yr B. 12 yr
    C. 25 yr D. 50 yr
17. || What is the unknown isotope in the following fission reaction: $n + {}^{235}\text{U} \rightarrow {}^{131}\text{I} + ? + 3n$
    A. ^{86}Rb B. ^{102}Rb
    C. ^{89}Y D. ^{102}Y
18. | An investigator has 0.010 μg samples of two isotopes of strontium, ^{89}Sr ($t_{1/2} = 51$ days) and ^{90}Sr ($t_{1/2} = 28$ years). The samples contain approximately the same number of atoms. What can you say about the activity of the two samples?
    A. The ^{89}Sr sample has a higher activity.
    B. The ^{90}Sr sample has a higher activity.
    C. The two samples have about the same activity.
19. || The uranium in the earth's crust is 0.7% ^{235}U and 99.3% ^{238}U. Two billion years ago, ^{235}U comprised approximately 3% of the uranium in the earth's crust. This tells you something about the relative half-lives of the two isotopes. Suppose you have a sample of ^{235}U and a sample of ^{238}U, each with exactly the same number of atoms.
    A. The sample of ^{235}U has a higher activity.
    B. The sample of ^{238}U has a higher activity.
    C. The two samples have the same activity.
20. | Suppose you have a 1 g sample of ^{226}Ra, half-life 1600 years. How long will it be until only 0.1 g of radium is left?
    A. 1600 yr B. 3200 yr
    C. 5300 yr D. 16,000 yr
21. | A sample of ^{131}I, half-life 8.0 days, is registering 100 decays per second. How long will be it before the sample registers only 1 decay per second?
    A. 8 days B. 53 days
    C. 80 days D. 800 days
22. | The complete expression for the decay of the radioactive hydrogen isotope *tritium* may be written as $^3\text{H} \rightarrow {}^3\text{He} + X + Y$. The symbols X and Y represent
    A. $X = e^+, Y = \bar{\nu}_e$
    B. $X = e^-, Y = \nu_e$
    C. $X = e^+, Y = \nu_e$
    D. $X = e^-, Y = \bar{\nu}_e$
23. | The quark compositions of the proton and neutron are, respectively, uud and udd, where u is an up quark (charge $+\frac{2}{3}e$) and d is a down quark (charge $-\frac{1}{3}e$). There are also antiup $\bar{u}$ (charge $-\frac{2}{3}e$) and antidown $\bar{d}$ (charge $+\frac{1}{3}e$) quarks. The combination of a quark and an antiquark is called a *meson*. The mesons known as *pions* have the composition $\pi^+ = u\bar{d}$ and $\pi^- = \bar{u}d$. Suppose a proton collides with an antineutron. During such collisions, the various quarks and antiquarks annihilate whenever possible. When the remaining quarks combine to form a single particle, it is a
    A. Proton. B. Neutron.
    C. π^+. D. π^-.

PROBLEMS

Section 30.1 Nuclear Structure

1. | How many protons and how many neutrons are in (a) ^{6}Li, (b) ^{54}Cr, (c) ^{54}Fe, and (d) ^{220}Rn?
2. | How many protons and how many neutrons are in (a) ^{3}He, (b) ^{32}P, (c) ^{32}S, and (d) ^{238}U?
3. | Calculate the nuclear diameters of (a) ^{4}He, (b) ^{56}Fe, and (c) ^{238}U.
4. | Calculate the mass, radius, and density of the nucleus of (a) ^{7}Li and (b) ^{207}Pb. Give all answers in SI units.
5. || Which stable nuclei have a diameter of 7.46 fm?
6. || Use the data in Appendix D to calculate the chemical atomic mass of lithium, to two decimal places.

7. || In *heavy water,* each of the hydrogen atoms is replaced by a deuterium atom. If the density of ordinary water at a given temperature is 1.000 g/cm^3, what is the density of heavy water? Assume that the average distance between the molecules is the same in both cases.

Section 30.2 Nuclear Stability

8. | Use data in Appendix D to make your own chart of stable and unstable nuclei, similar to Figure 30.5, for all nuclei with $Z \leq 8$. Use a blue or black dot to represent stable isotopes, a red dot to represent isotopes that undergo beta-minus decay, and a green dot to represent isotopes that undergo beta-plus decay or electron-capture decay.

9. || a. What is the smallest value of A for which there are two stable nuclei? What are they?
   b. For which values of A less than this are there *no* stable nuclei?

10. || Calculate (in MeV) the total binding energy and the binding energy per nucleon for ^{3}H and for ^{3}He.

11. || Calculate (in MeV) the total binding energy and the binding energy per nucleon for ^{129}I and for ^{129}Xe.

12. || Calculate (in MeV) the binding energy per nucleon for ^{3}He and ^{4}He. Which is more tightly bound?

13. || Calculate (in MeV) the binding energy per nucleon for ^{14}O and ^{16}O. Which is more tightly bound?

14. || When a nucleus of ^{235}U undergoes fission, it breaks into two smaller, more tightly bound fragments. Calculate the binding energy per nucleon for ^{235}U and for the fission product ^{137}Cs.

15. || The sun's energy comes from nuclear fusion reactions in which protons, the nuclei of hydrogen atoms, are squeezed together at very high temperature and pressure to form the nucleus of a helium atom. The process requires three steps, but the overall fusion reaction is

$$4\,^1\text{H} \rightarrow {}^4\text{He} + 2e^- + \text{energy}$$

How much energy is released in this reaction?

Section 30.3 The Strong Force

16. | The strong force between two protons whose centers are
INT 1.5 fm apart is attractive with a magnitude of approximately 1.0×10^4 N. What is the ratio of the magnitude of the strong force to that of the repulsive electrostatic force?

17. || Use the potential-energy diagram in Figure 30.9 to estimate
INT the ratio of the gravitational potential energy to the nuclear potential energy for two neutrons separated by 1.0 fm.

18. | a. Draw energy-level diagrams, similar to Figure 30.11, for all $A = 10$ nuclei listed in Appendix D. Show all the occupied neutron and proton levels.
   b. Which of these nuclei are stable? What is the decay mode of any that are radioactive?

19. | a. Draw energy-level diagrams, similar to Figure 30.11, for all $A = 14$ nuclei listed in Appendix D. Show all the occupied neutron and proton levels.
   b. Which of these nuclei are stable? What is the decay mode of any that are radioactive?

Section 30.4 Radiation and Radioactivity

20. | The barium isotope ^{131}Ba has a half-life of 12 days. A 250 μg sample of ^{131}Ba is prepared. What is the mass of ^{131}Ba after (a) 1 day, (b) 10 days, and (c) 100 days?

21. | The radium isotope ^{226}Ra has a half-life of 1600 years. A sample begins with 1.00×10^{10} ^{226}Ra atoms. How many are left after (a) 200 years, (b) 2000 years, and (c) 20,000 years?

22. || A sample of 1.0×10^{10} atoms that decay by alpha emission has a half-life of 100 min. How many alpha particles are emitted between $t = 50$ min and $t = 200$ min?

23. || The radioactive hydrogen isotope ^{3}H, called *tritium,* has a half-life of 12 years.
   a. What are the decay mode and the daughter nucleus of tritium?
   b. What are the lifetime and the decay rate of tritium?

24. || What is the half-life in days of a radioactive sample with 5.0×10^{15} atoms and an activity of 5.0×10^8 Bq?

25. || The half-life of ^{60}Co is 5.27 years. The activity of a ^{60}Co sample is 3.50×10^9 Bq. What is the mass of the sample?

26. || A 115 mCi radioactive tracer is made in a nuclear reactor.
BIO When it is delivered to a hospital 16 hours later its activity is 95 mCi. The lowest usable level of activity is 10 mCi.
   a. What is the tracer's half-life?
   b. For how long after delivery is the sample usable?

27. || An investigator collects a sample of a radioactive isotope with an activity of 370,000 Bq. 48 hours later, the activity is 120,000 Bq. What is the half-life of the sample?

28. || ^{235}U decays to ^{207}Pb via the decay series shown in Figure 30.23. The first decay in the chain, that of ^{235}U, has a half-life of 7.0×10^8 years. The subsequent decays are much more rapid, so we can take this as the half-life for the complete decay of ^{235}U to ^{207}Pb. Certain minerals exclude lead but not uranium from their crystal structure, so when the minerals form they have no lead, only uranium. As time goes on, the uranium decays to lead, so measuring the ratio of lead atoms to uranium atoms allows investigators to determine the ages of the minerals. If a sample of a mineral contains 3 atoms of ^{207}Pb for every 1 atom of ^{235}U, how many years ago was it formed?

29. || The New Horizons spacecraft, launched in 2006, spent 9.5 yr on its journey to Pluto. The spacecraft generates electric power from the heat produced by the decay of ^{238}Pu, which has a half-life of 88 yr. Each decay emits an alpha particle with an energy of 5.6 MeV. New Horizons was launched with 10 kg of plutonium. How much thermal power was generated by the plutonium (a) at launch and (b) when the spacecraft reached Pluto?

Section 30.5 Types of Nuclear Decay

30. | ^{15}O and ^{131}I are isotopes used in medical imaging. ^{15}O is
BIO a beta-plus emitter, ^{131}I a beta-minus emitter. What are the daughter nuclei of the two decays?

31. | Identify the unknown isotope X in the following decays.
   a. $X \rightarrow {}^{224}\text{Ra} + \alpha$
   b. $X \rightarrow {}^{207}\text{Pb} + e^- + \bar{\nu}$
   c. ${}^7\text{Be} + e^- \rightarrow X + \nu$
   d. $X \rightarrow {}^{60}\text{Ni} + \gamma$

32. | Identify the unknown isotope X in the following decays.
   a. ${}^{230}\text{Th} \rightarrow X + \alpha$
   b. ${}^{35}\text{S} \rightarrow X + e^- + \bar{\nu}$
   c. $X \rightarrow {}^{40}\text{K} + e^+ + \nu$
   d. ${}^{24}\text{Na} \rightarrow {}^{24}\text{Mg} + e^- + \bar{\nu} \rightarrow X + \gamma$

33. ||| An unstable nucleus undergoes alpha decay with the release of 5.52 MeV of energy. The combined mass of the parent and daughter nuclei is 452 u. What was the parent nucleus?

34. || What is the energy (in MeV) released in the alpha decay of ^{239}Pu?

35. | Medical gamma imaging is most often done with the techne-
BIO tium isotope ^{99}Tc, which decays by emitting a gamma-ray photon with energy 140 keV. What is the mass loss of the nucleus, in u, upon emission of this gamma ray?

Section 30.6 The Interaction of Ionizing Radiation with Matter

36. || The absorption coefficient in water is 0.10 cm^{-1} for 0.5 MeV
BIO photons and 0.03 cm^{-1} for 5 MeV photons. Assume that the absorption coefficient in tissue is the same. To one significant figure, in what depth of tissue is 50% of the energy absorbed from a beam of (a) 0.5 MeV x-ray photons and (b) 5 MeV x-ray photons?

37. | A 2 MeV beta particle has a range in tissue of ≈ 1 cm. The
BIO particle loses ≈ 10 eV in each collision with a molecule of the tissue. On average, what is the distance in nm between the ions that are created by the passage of the beta particle?

38. | A passenger on an airplane flying across the Atlantic re-
BIO ceives an extra radiation dose of about 5 μSv per hour from cosmic rays. How many hours of flying would it take in one year for a person to double his or her yearly radiation dose? Assume there are no other significant radiation sources besides natural background.

39. || A 50 kg nuclear plant worker is exposed to 20 mJ of neutron
BIO radiation with an RBE of 10. What is the dose in mSv?

40. || The decay chain of uranium includes radon, a noble gas.
BIO When uranium in the soil decays to radon, it may seep into houses; this can be a significant source of radiation exposure. Most of the exposure comes from the decay products of radon, but some comes from alpha decay of the radon itself. If radon in the air in your home is at the maximum permissible level, the gas in your lungs will have an activity of about 0.22 Bq. Each decay generates an alpha particle with 5.5 MeV of energy, and essentially all that energy is deposited in lung tissue. Over the course of 1 year, what will be the dose equivalent in Sv to the approximately 0.90 kg mass of your lungs?

41. | 1.5 Gy of gamma radiation are directed into a 150 g tumor.
BIO How much energy does the tumor absorb?

42. ||| ^{90}Sr decays with the emission of a 2.8 MeV beta particle.
BIO Strontium is chemically similar to calcium and is taken up by bone. A 75 kg person exposed to waste from a nuclear accident absorbs ^{90}Sr with an activity of 370,000 Bq. Assume that all of this ^{90}Sr ends up in the skeleton. The skeleton forms 17% of the person's body mass. If 50% of the decay energy is absorbed by the skeleton, what dose equivalent in Sv will be received by the person's skeleton in the first month?

43. || Many fruits, vegetables, and spices are treated with
BIO radiation. This irradiation doesn't render the foods radioactive, and it can make perishable items stay fresh for a longer period of time. Papayas don't travel well; after they are picked, they quickly become unusable due to fungal growth. Exposing a 1.5 kg papaya to a total dose of 1.0 kGy of 1.3 MeV gamma rays kills the fungus and allows the papaya to survive its trip to market. How many photons are absorbed in the papaya?

44. |||| A 75 kg patient swallows a 30 μCi beta emitter with a
BIO half-life of 5.0 days, and the radioactive nuclei are quickly distributed throughout his body. The beta particles are emitted with an average energy of 0.35 MeV, 90% of which is absorbed by the body. What dose equivalent does the patient receive in the first week?

Section 30.7 Nuclear Medicine

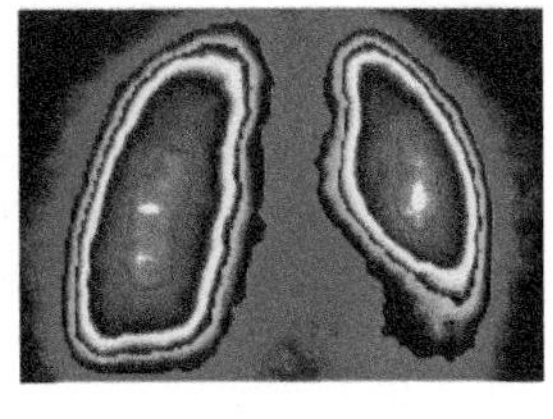

45. || A gamma scan showing the
BIO active volume of a patient's lungs can be created by having a patient breathe the radioactive isotope ^{133}Xe, which undergoes beta-minus decay with a subsequent gamma emission from the daughter nucleus. A typical procedure gives a dose of 3.0 mSv to the lungs. How much energy is deposited in the 1.2 kg mass of a patient's lungs?

46. |||| Certain cancers of the liver can be treated by injecting micro-
BIO scopic glass spheres containing radioactive ^{90}Y into the blood vessels that supply the tumor. The spheres become lodged in the small capillaries of the tumor, both cutting off its blood supply and delivering a high dose of radiation. ^{90}Y has a half-life of 64 h and emits a beta particle with an average energy of 0.89 MeV. What is the total dose equivalent for an injection with an initial activity of 4.0×10^7 Bq if all the energy is deposited in a 40 g tumor?

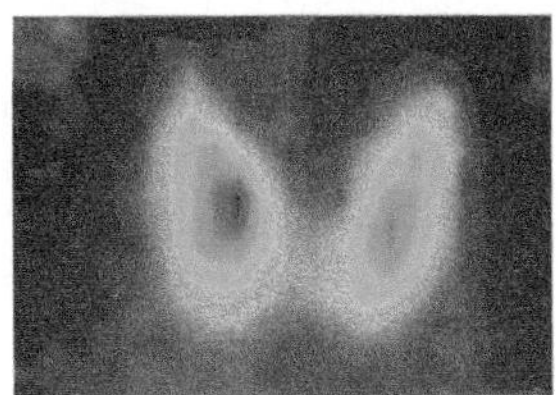

47. || ^{131}I undergoes beta-minus
BIO decay with a subsequent gamma emission from the daughter nucleus. Iodine in the body is almost entirely taken up by the thyroid gland, so a gamma scan using this isotope will show a bright area corresponding to the thyroid gland with the surrounding tissue appearing dark. Because the isotope is concentrated in the gland, so is the radiation dose, most of which results from the beta emission. In a typical procedure, a patient receives 0.050 mCi of ^{131}I. Assume that all of the iodine is absorbed by the 0.15 kg thyroid gland. Each ^{131}I decay produces a 0.97 MeV beta particle. Assume that half the energy of each beta particle is deposited in the gland. What dose equivalent in Sv will the gland receive in the first hour?

48. ||| The rate at which a radioactive tracer is lost from a patient's
BIO body is the rate at which the isotope decays *plus* the rate at which the element is excreted from the body. Medical experiments have shown that stable isotopes of a particular element are excreted with a 6.0 day half-life. A radioactive isotope of the same element has a half-life of 9.0 days. What is the effective half-life of the isotope, in days, in a patient's body?

Section 30.8 The Ultimate Building Blocks of Matter

49. || In a particular beta-minus decay of a free neutron (that is, one not part of an atomic nucleus), the emitted electron has exactly the same kinetic energy as the emitted electron antineutrino. What is the value, in MeV, of that kinetic energy? Assume that the recoiling proton has negligible kinetic energy.

50. || A muon is a lepton that has properties similar to those of an electron, but it is much heavier: $m_{\text{muon}} = 207m_{\text{elec}}$. In addition, muons are unstable; they decay to electrons and neutrinos with a half-life of 1.6 μs. When a muon and an antimuon collide, they annihilate and produce two gamma-ray photons. What is the energy in MeV of each photon produced in such a collision?

51. |||| Positive and negative pions, denoted π^+ and π^-, are antiparticles of each other. Each has a rest mass of 140 MeV/c^2. Suppose a collision between an electron and positron, each with kinetic energy K, produces a π^+, π^- pair. What is the smallest possible value for K?

General Problems

52. || The element gallium has two stable isotopes: ^{69}Ga with an atomic mass of 68.92 u and ^{71}Ga with an atomic mass of 70.92 u. A periodic table shows that the chemical atomic mass of gallium is 69.72 u. What is the percent abundance of ^{69}Ga?
53. || You learned in Chapter 29 that the binding energy of the electron in a hydrogen atom is 13.6 eV.
    a. By how much does the mass decrease when a hydrogen atom is formed from a proton and an electron? Give your answer both in atomic mass units and as a percentage of the mass of the hydrogen atom.
    b. By how much does the mass decrease when a helium nucleus is formed from two protons and two neutrons? Give your answer both in atomic mass units and as a percentage of the mass of the helium nucleus.
    c. Compare your answers to parts a and b. Why do you hear it said that mass is "lost" in nuclear reactions but not in chemical reactions?
54. || Use the graph of binding energy to estimate the total energy released if three ^{4}He nuclei fuse together to form a ^{12}C nucleus.
55. || Could a ^{56}Fe nucleus fission into two ^{28}Al nuclei? Your answer, which should include some calculations, should be based on the curve of binding energy.
56. || What energy (in MeV) alpha particle has a de Broglie wavelength equal to the diameter of a ^{238}U nucleus?
INT
57. || The activity of a sample of the cesium isotope ^{137}Cs, with a half-life of 30 years, is 2.0×10^8 Bq. Many years later, after the sample has fully decayed, how many beta particles will have been emitted?
58. || A 1 Ci source of radiation is a significant source. ^{238}U is an alpha emitter. What mass of ^{238}U has an activity of 1 Ci?
59. || ^{137}Cs is a common product of nuclear fission. Suppose an accident spills 550 mCi of ^{137}Cs in a lab room.
    a. What mass of ^{137}Cs is spilled?
    b. If the spill is not cleaned up, how long will it take until the radiation level drops to an acceptable level, for a room this size, of 25 mCi?
60. || How many half-lives must elapse until (a) 90% and (b) 99% of a radioactive sample of atoms has decayed?
61. || A sample contains radioactive atoms of two types, A and B. Initially there are five times as many A atoms as there are B atoms. Two hours later, the numbers of the two atoms are equal. The half-life of A is 0.50 hour. What is the half-life of B?
62. || Radioactive isotopes often occur together in mixtures. Suppose a 100 g sample contains ^{131}Ba, with a half-life of 12 days, and ^{47}Ca, with a half-life of 4.5 days. If there are initially twice as many calcium atoms as there are barium atoms, what will be the ratio of calcium atoms to barium atoms 2.5 weeks later?
63. || You are assisting in an anthropology lab over the summer by carrying out ^{14}C dating. A graduate student found a bone she believes to be 20,000 years old. You extract the carbon from the bone and prepare an equal-mass sample of carbon from modern organic material. To determine the activity of a sample with the accuracy your supervisor demands, you need to measure the time it takes for 10,000 decays to occur.
    a. The activity of the modern sample is 1.06 Bq. How long does that measurement take?
    b. It turns out that the graduate student's estimate of the bone's age was accurate. How long does it take to measure the activity of the ancient carbon?
64. |||| Wood was excavated from Stonehenge and found, using ^{14}C dating, to be 4500 years old. What is the ^{14}C activity in 1.00 g of carbon taken from this wood?
65. ||| A sample of wood from an archaeological excavation is dated by using a mass spectrometer to measure the fraction of ^{14}C atoms. Suppose 100 atoms of ^{14}C are found for every 1.0×10^{15} atoms of ^{12}C in the sample. What is the wood's age?
66. || Corals take up certain elements from seawater, including uranium but not thorium. After the corals die, the uranium isotopes slowly decay into thorium isotopes. A measurement of the relative fraction of certain isotopes therefore provides a determination of the coral's age. A complicating factor is that the thorium isotopes decay as well. One scheme uses the alpha decay of ^{234}U to ^{230}Th. After a long time, the two species reach an equilibrium in which the number of ^{234}U decays per second (each producing an atom of ^{230}Th) is exactly equal to the number of ^{230}Th decays per second. What is the relative concentration of the two isotopes—the ratio of ^{234}U to ^{230}Th—when this equilibrium is reached?
67. || What is the age in years of a bone in which the ^{14}C/^{12}C ratio is measured to be 1.65×10^{-13}?
68. ||| The smallest ^{14}C/^{12}C ratio that can be reliably measured is about 3.0×10^{-15}, setting a limit on the oldest carbon specimens that can be dated. How old would a sample with this carbon ratio be?
69. || Tumors of the prostate gland are often treated with "seeds" containing radioactive isotopes that provide a localized dose of radiation. One isotope used for such treatment is ^{103}Pd, which decays with a half-life of 17 days by emitting a 21 keV photon. These relatively low-energy photons deposit their energy out to a distance of 1.1 cm, a short range that minimizes the dose to healthy tissue. A typical seed has an activity of 9.0 MBq. If the seed is left in place, what total dose equivalent will it give to the volume of tissue it affects?
BIO
70. || After the 2011 Fukushima nuclear reactor disaster in Japan, people were evacuated from the affected area but not all dogs were. Rescuers later treated the dogs to help them eliminate radioactive nuclei from their bodies. The main isotope of concern was ^{137}Cs, which undergoes beta-minus decay with the subsequent emission of a gamma ray. For one decay, the total energy deposited in tissue is 1.2 MeV. ^{137}Cs has a half-life of 30 years, but, with treatment, this isotope had a biological half-life (the time to eliminate half of the nuclei from the body) of 38 days. One dog had a ^{137}Cs activity of 95 MBq in its 11 kg body.
BIO
    a. Before receiving treatment—and thus no elimination of the nuclei from the body—what dose equivalent did the dog receive in 1 day?
    b. Following treatment, what ^{137}Cs activity remained in the dog's body after 60 days?
71. ||| Uranium has two very long-lived isotopes, ^{235}U and ^{238}U. What were the percent abundances of these two isotopes when the earth was formed 4.5×10^9 yr ago? (The relevant data are given in Appendix D.)
72. || All the very heavy atoms found in the earth were created long ago by nuclear fusion reactions in a supernova, an exploding star. The debris spewed out by the supernova later coalesced to form the sun and the planets of our solar system. Nuclear physics suggests that the uranium isotopes ^{235}U ($t_{1/2} = 7.04 \times 10^8$ yr) and ^{238}U ($t_{1/2} = 4.47 \times 10^9$ yr) should have been created in roughly equal amounts. Today, 99.28% of uranium is ^{238}U and 0.72% is ^{235}U. How long ago did the supernova occur?

73. || What is the total energy (in MeV) released in the beta-minus decay of ^{3}H?

74. || What is the total energy (in MeV) released in the beta decay of a neutron?

75. || A chest x ray uses 10 keV photons with an RBE of 0.85. A
BIO 60 kg person receives a 0.30 mSv dose from one chest x ray that exposes 25% of the patient's body. How many x ray photons are absorbed in the patient's body?

76. || Gamma rays may be used to kill pathogens in ground beef.
BIO One irradiation facility uses a ^{60}C source that has an activity of 1.0×10^6 Ci. ^{60}C undergoes beta decay and then emits two gamma rays, at 1.17 and 1.33 MeV; typically 30% of this gamma-ray energy is absorbed by the meat. The dose required to kill all pathogens present in the beef is 4000 Gy. How many kilograms of meat per hour can be processed in this facility?

77. ||| A 70 kg human body typically contains 140 g of potassium.
BIO Potassium has a chemical atomic mass of 39.1 u and has three naturally occurring isotopes. One of those isotopes, ^{40}K, is radioactive with a half-life of 1.3 billion years and a natural abundance of 0.012%. ^{40}K undergoes beta-minus decay, and each deposits, on average, 1.0 MeV of energy into the body. What yearly dose equivalent in mSv does the typical person receive from the decay of ^{40}K in the body?

78. || Proton therapy with 200 MeV protons uses a 2.0-cm-
BIO INT diameter beam of protons. Half of the proton's energy is deposited in a range of 2.0 cm, centered on the tumor. Assume the tumor's density is that of water. Proton therapy is carried out over several sessions with the goal of delivering a 2.0 Sv dose equivalent per session to this well-defined region.
   a. During one session, how many protons deposit energy in this region?
   b. The proton-beam current is typically 1.0 pA. What is the required exposure time?

79. |||| About 12% of your body mass is carbon; some of this is ra-
BIO dioactive ^{14}C, a beta-emitter. If you absorb 100% of the 49 keV energy of each ^{14}C decay, what dose equivalent in Sv do you receive each year from the ^{14}C in your body? The ^{14}C/^{12}C ratio in living matter is 1.3×10^{-12}.

MCAT-Style Passage Problems

Plutonium-Powered Exploration

The Curiosity rover sent to explore the surface of Mars has an electric generator powered by heat from the radioactive decay of ^{238}Pu, a plutonium isotope that decays by alpha emission with a half-life of 88 years. At the start of the mission, the generator contained 9.6×10^{24} nuclei of ^{238}Pu.

80. What is the daughter nucleus of the decay?
   A. ^{238}Am B. ^{238}Pu
   C. ^{238}Np D. ^{236}Th
   E. ^{234}U

81. What was the approximate activity of the plutonium source at the start of the mission?
   A. 2×10^{21} Bq B. 2×10^{19} Bq
   C. 2×10^{17} Bq D. 2×10^{15} Bq
   E. 2×10^{13} Bq

82. The generator initially provided 125 W of power. If you assume that the power of the generator is proportional to the activity of the plutonium, by approximately what percent did the power output decrease over the first two years of the rover's mission?
   A. 1.5% B. 2.0%
   C. 2.5% D. 3.0%
   E. 3.5%

Dangerous Decays

When a smoker draws in air through a burning cigarette, radioactive isotopes from the decay of radon in the air stick to particles produced by combustion. These particles are concentrated at certain spots in the lungs where the isotopes undergo radioactive decay. Simulations and measurements show that a lifelong smoker is exposed to the decay of 1.0×10^6 atoms of ^{214}Bi at one particularly small patch of tissue at a bifurcation in the lungs. ^{214}Bi undergoes beta-minus decay with a half-life of 20 minutes, but the real worry is the subsequent rapid alpha decay of the ^{214}Po daughter nucleus. All the energy of the 7.7 MeV alpha particles is deposited in a mere 0.014 g of lung tissue, so the absorbed dose is significant, and the RBE of 20 means that the dose equivalent is even larger. The exposure is large enough to lead to a significant increase in the probability of developing malignancies or other abnormalities.

83. What is the daughter nucleus of the ^{214}Po decay?
   A. ^{214}Po B. ^{214}At
   C. ^{212}Pb D. ^{210}Pb

84. If 2000 atoms of ^{214}Bi are deposited in the lungs at one time, how many remain after 1 hour?
   A. 1000 B. 500
   C. 250 D. 125

85. For the data given, what is the approximate lifetime dose equivalent to the small patch of tissue from the decay of ^{214}Po?
   A. 1 Sv B. 2 Sv
   C. 3 Sv D. 4 Sv

PART VI SUMMARY

Modern Physics

FUNDAMENTAL CONCEPTS

Matter and light at the atomic level cannot be explained with classical physics.

- Matter is composed of atoms.
- Atoms have a dense, positively charged nucleus surrounded by negative electrons.
- Light has particle-like properties.
- Matter has wave-like properties.
- The energy of light and matter is quantized.

GENERAL PRINCIPLES

Energy quantization Particles of matter and photons of light have only certain allowed energies.

Heisenberg uncertainty principle $\delta x\, \delta p_x \geq \dfrac{h}{4\pi}$

Pauli exclusion principle No more than one electron or nucleon can occupy the same quantum state.

ATOMIC MODEL

A tiny, dense, positive nucleus is surrounded by negative electrons.

The electrons can occupy only certain stationary states. The lowest-energy stationary state is the ground state.

The nucleus consists of protons and neutrons—called nucleons—held together by the strong force.

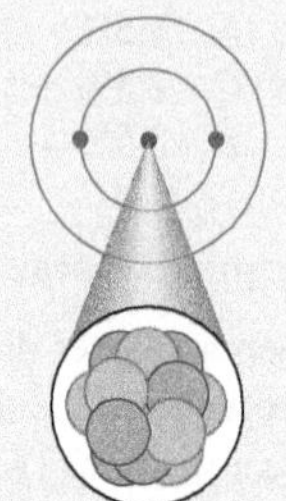

PHOTON MODEL

Light consists of discrete, massless photons that travel in vacuum at the speed of light.

A photon has energy $E = hf$.

Photons are emitted and absorbed on an all-or-nothing basis when a quantum system jumps from one energy level to another.

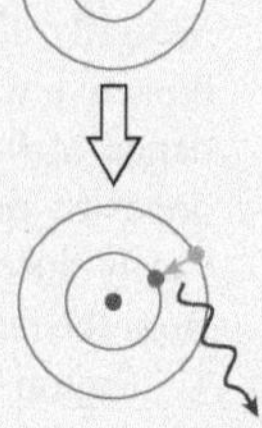

THE HYDROGEN ATOM

The electron in the hydrogen atom is described by the four quantum numbers (n, l, m, m_s). The allowed energies depend on only the principal quantum number n:

$$E_n = -\frac{13.60 \text{ eV}}{n^2}$$

0 eV: 3s, 3p, 3d; 2s, 2p; −13.6 eV: 1s

PARTICLE IN A BOX

A confined particle sets up a de Broglie standing wave. Standing waves have only certain allowed wavelengths, and thus a confined particle has only certain allowed energies:

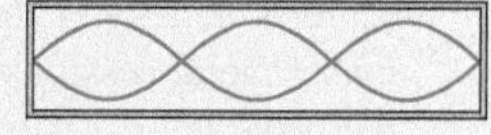

$$E_n = n^2 \frac{h^2}{8mL^2}$$

PHOTOELECTRIC EFFECT

Light with frequency f can eject electrons from a metal only if

$$f \geq f_0 = E_0/h$$

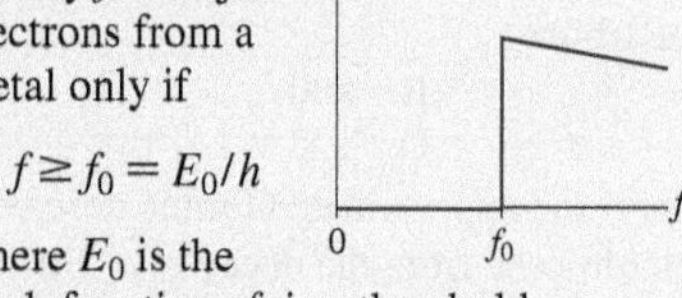

where E_0 is the work function. f_0 is a threshold frequency.

The electron volt is a unit of energy:

$$1 \text{ eV} = 1.60 \times 10^{-19} \text{ J}$$

MULTI-ELECTRON ATOMS

Each electron is described by quantum numbers (n, l, m, m_s). The energy depends on both n and l.

0 eV: 3d; 3p; 3s; 2p; 2s; 1s

The number of electrons in each subshell is determined by the Pauli principle:

2 in s shells, 6 in p shells, 10 in d shells

NUCLEAR SHELL MODEL

Nucleons fill energy levels similar to the way electrons fill energy levels in atoms.

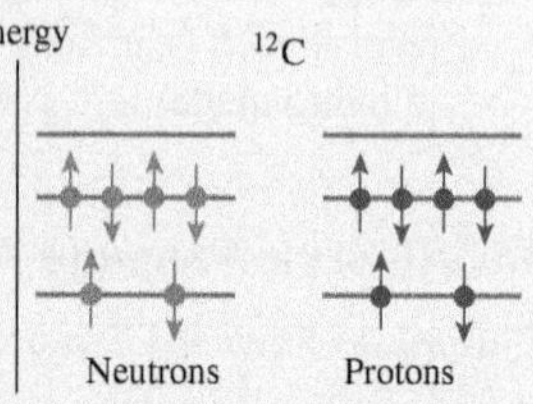

Gamma decay is a quantum jump to a lower energy level.

Beta decay is a quantum jump in which a neutron changes to a proton, or vice versa.

DE BROGLIE WAVELENGTH

A particle of mass m has wavelength

$$\lambda = \frac{h}{p} = \frac{h}{mv}$$

- Electrons, neutrons, and entire atoms can exhibit interference and diffraction.
- Atomic particles can tunnel through a potential-energy barrier that a classical particle cannot cross.
- The position and velocity of a particle are inherently uncertain.

SPECTRA

Absorption spectra occur from the ground state.

Emission spectra are generated by excitation followed by a photon-emitting quantum jump.

- Excitation by photon absorption
- Excitation by collision with a particle

Fluorescence is the emission of longer-wavelength photons after the absorption of shorter-wavelength photons.

RADIOACTIVE DECAY

Unstable nuclei can decay by alpha (^{4}He), beta (e^+ or e^-), or gamma (photon) emission. The number of still undecayed nuclei decreases exponentially with time:

$$N = N_0 e^{-t/\tau} = N_0 \left(\frac{1}{2}\right)^{t/t_{1/2}}$$

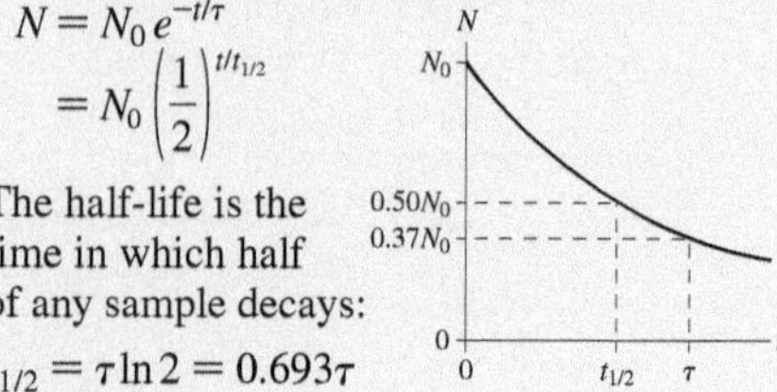

The half-life is the time in which half of any sample decays:

$$t_{1/2} = \tau \ln 2 = 0.693\tau$$

The Physics of Very Cold Atoms

Modern physics is a study of extremes. Relativity deals with the physics of objects traveling at near-light speeds. Quantum mechanics is about the physics of matter and energy at very small scales. Nuclear physics involves energies that dwarf anything dreamed of in previous centuries.

Some of the most remarkable discoveries of recent years are at another extreme—very low temperatures, mere billionths of a degree above absolute zero. Let's look at how such temperatures are achieved and some new physics that emerges.

You learned in Part II that the temperature of a gas depends on the speeds of the atoms in the gas. Suppose we start with atoms at or above room temperature. Cooling the gas means slowing the atoms down. How can we drastically reduce their speeds, bringing them nearly to a halt? The trick is to slow them, thus cooling them, using the interactions between light and atoms that we explored in Chapters 28 and 29.

Photons have momentum, and that momentum is transferred to an atom when a photon is absorbed. Part a of the figure shows an atom moving "upstream" against a laser beam tuned to an atomic transition. Photon absorptions transfer momentum, slowing the atom. Subsequent photon emissions give the atom a "kick," but in random directions, so on average the emissions won't speed up the atom in the same way that the absorptions slow it. A beam of atoms moving "upstream" against a correctly tuned laser beam is slowed down—the "hot" beam of atoms is cooled.

Once a laser cools the atoms, a different configuration of laser beams can trap them. Part b of the figure shows six overlapped laser beams, each tuned slightly *below* the frequency of an atomic absorption line. If an atom tries to leave the overlap region, it will be moving "upstream" against one of the laser beams. The atom will see that laser beam Doppler-shifted to a higher frequency, matching the transition frequency. The atom will then absorb photons from this laser beam, and the resulting kick will nudge it back into the overlap region. The atoms are trapped in what is known as *optical molasses* or, more generally, an *atom trap.* More effective traps can be made by adding magnetic fields. The final cooling of the atoms is by evaporation—letting the more energetic atoms leave the trap.

Ultimately, these techniques produce a diffuse gas of atoms moving at only ≈ 1 mm/s. This corresponds to a nearly unbelievable temperature of just a few nanokelvin—billionths of a degree above absolute zero. This is colder than outer space. The coldest spot in the universe is inside an atom trap in a physics lab.

Once the atoms are cooled, some very remarkable things happen. As we saw in Part VI, all particles, including atoms, have a wave nature. As the atoms slow, their wavelengths increase. In a correctly prepared gas at a low enough temperature, the de Broglie wavelength of an individual atom is larger than the spacing between atoms, and the wave functions of multiple atoms overlap. When this happens, some atoms undergo *Bose-Einstein condensation,* coalescing into one "super atom," with thousands of atoms occupying the same quantum state. The resulting Bose-Einstein condensate illustrates the counterintuitive nature of the quantum world. The atoms in the condensate—that is, their wave functions—are all in the same place at the same time! This is a new and truly bizarre state of matter.

Are there applications for Bose-Einstein condensation? Current talk of atom lasers and other futuristic concepts aside, no one really knows, just as the early architects of quantum mechanics didn't know that their theory would be used to design the chips that power personal computers.

At the start of the 20th century, there was a worry that everything in physics had been discovered. There is no such worry in the 21st century, which promises to be full of wonderful discoveries and remarkable applications. What do you imagine the final chapter of a physics textbook will look like 100 years from now?

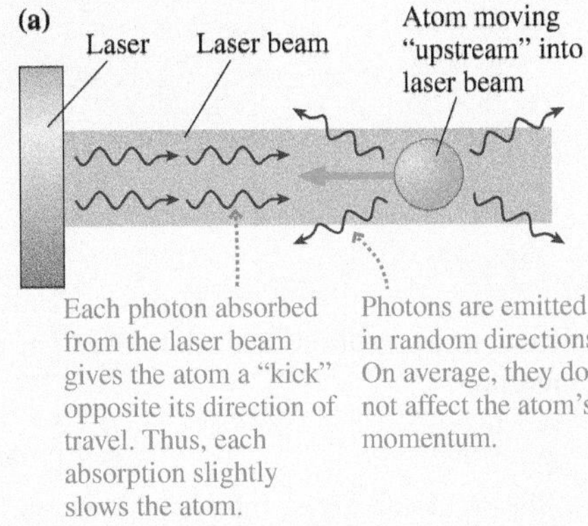
(a)
Laser
Laser beam
Atom moving "upstream" into laser beam
Each photon absorbed from the laser beam gives the atom a "kick" opposite its direction of travel. Thus, each absorption slightly slows the atom.
Photons are emitted in random directions. On average, they do not affect the atom's momentum.

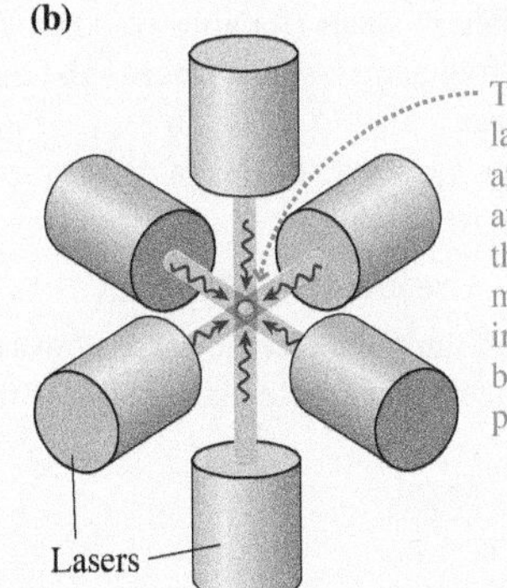
(b)
The intersection of laser beams creates an atom trap. An atom moving out of the trap will be moving upstream into one of the laser beams and will be pushed back.
Lasers

Laser cooling and trapping.

PART VI PROBLEMS

The following questions are related to the passage "The Physics of Very Cold Atoms" on the preceding page.

1. Why is it useful to create an assembly of very slow-moving cold atoms?
   A. The atoms can be more easily observed at slow speeds.
   B. Lowering the temperature this way permits isotopes that normally decay in very short times to persist long enough to be studied.
   C. At low speeds the quantum nature of the atoms becomes more apparent, and new forms of matter emerge.
   D. At low speeds the quantum nature of the atoms becomes less important, and they appear more like classical particles.
2. The momentum of a photon is given by $p = h/\lambda$. Suppose an atom emits a photon. Which of the following photons will give the atom the biggest "kick"—the highest recoil speed?
   A. An infrared photon B. A red-light photon
   C. A blue-light photon D. An ultraviolet photon
3. When an atom moves "upstream" against the photons in a laser beam, the energy of the photons appears to be _______ if the atom were at rest.
   A. Greater than B. Less than C. The same as
4. A gas of cold atoms strongly absorbs light of a specific wavelength. Warming the gas causes the absorption to decrease. Which of the following is the best explanation for this reduction?
   A. Warming the gas changes the atomic energy levels.
   B. Warming the gas causes the atoms to move at higher speeds, so the atoms "see" the photons at larger Doppler shifts.
   C. Warming the gas causes more collisions between the atoms, which affects the absorption of photons.
   D. Warming the gas makes it more opaque to the photons, so fewer enter the gas.
5. A gas of cold atoms starts at a temperature of 100 nK. The average speed of the atoms is then reduced by half. What is the new temperature?
   A. 71 nK B. 50 nK C. 37 nK D. 25 nK
6. Rubidium is often used for the type of experiments noted in the passage. At a speed of 1.0 mm/s, what is the approximate de Broglie wavelength of an atom of ^{87}Rb?
   A. 5 nm B. 50 nm C. 500 nm D. 5000 nm
7. A gas of rubidium atoms and a gas of sodium atoms have been cooled to the same very low temperature. What can we say about the de Broglie wavelengths of typical atoms in the two gases?
   A. The sodium atoms have the longer wavelength.
   B. The wavelengths are the same.
   C. The rubidium atoms have the longer wavelength.

The following passages and associated questions are based on the material of Part VI.

Splitting the Atom

"Splitting" an atom in the process of nuclear fission releases a great deal of energy. If all the atoms in 1 kg of ^{235}U undergo nuclear fission, 8.0×10^{13} J will be released, equal to the energy from burning 2.3×10^{6} kg of coal. What is the source of this energy? Surprisingly, the energy from this nuclear disintegration ultimately comes from the electric potential of the positive charges that make up the nucleus.

The protons in a nucleus exert repulsive forces on each other, but this force is less than the short-range attractive nuclear force. If a nucleus breaks into two smaller nuclei, the nuclear force will hold each of the fragments together, but it won't bind the two positively charged fragments to each other. This is illustrated in Figure VI.1. The two fragments feel a strong repulsive electrostatic force. The charges are large and the distance is small (roughly equal to the sum of the radius of each of the fragments), so the force—and thus the potential energy—is quite large.

In a fission reaction, a neutron causes a nucleus of ^{235}U to split into two smaller nuclei; a typical reaction is

$$\text{n} + {}^{235}\text{U} \rightarrow {}^{87}\text{Br} + {}^{147}\text{La} + \text{neutrons} + \approx 200 \text{ MeV}$$

Right after the nucleus splits, with only the electric force now acting on the two fragments, the electrostatic potential energy of the two fragments is

$$U = \frac{kq_1q_2}{r_1 + r_2}$$

This is the energy that will be released, transformed into kinetic energy, when the fragments fly apart. If we use reasonable estimates for the radii of the two fragments, we compute a value for the energy that is close to the experimentally observed value of 200 MeV for the energy released in the fission reaction. The energy released in this *nuclear* reaction is actually *electric* potential energy.

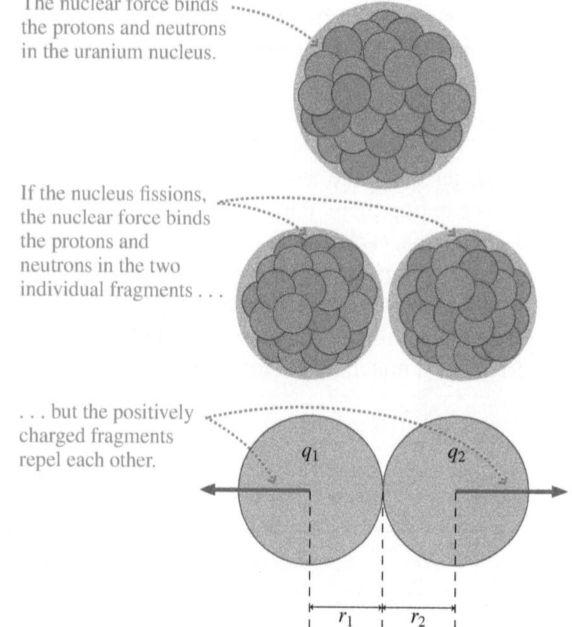

The nuclear force binds the protons and neutrons in the uranium nucleus.
If the nucleus fissions, the nuclear force binds the protons and neutrons in the two individual fragments . . .
. . . but the positively charged fragments repel each other.
q_1
q_2
r_1
r_2

FIGURE VI.1

8. How many neutrons are "left over" in the noted fission reaction?
   A. 1 B. 2
   C. 3 D. 4
9. After a fission event, most of the energy released is in the form of
   A. Emitted beta particles and gamma rays.
   B. Kinetic energy of the emitted neutrons.
   C. Nuclear energy of the two fragments.
   D. Kinetic energy of the two fragments.

10. Suppose the original nucleus is at rest in the fission reaction described previously. If we neglect the momentum of the neutrons, after the two fragments fly apart,
    A. The Br nucleus has more momentum.
    B. The La nucleus has more momentum.
    C. The momentum of the Br nucleus equals that of the La nucleus.
11. Suppose the original nucleus is at rest in the fission reaction noted above. If we neglect the kinetic energy of the neutrons, after the two fragments fly apart,
    A. The Br nucleus has more kinetic energy.
    B. The La nucleus has more kinetic energy.
    C. The kinetic energy of the Br nucleus equals that of the La nucleus.
12. 200 MeV is a typical energy released in a fission reaction. To get a sense for the scale of the energy, if we were to use this energy to create electron-positron pairs, approximately how many pairs could we create?
    A. 50 B. 100 C. 200 D. 400
13. The two fragments of a fission reaction are isotopes that are neutron-rich; each has more neutrons than the stable isotopes for their nuclear species. They will quickly decay to more stable isotopes. What is the most likely decay mode?
    A. Alpha decay B. Beta decay C. Gamma decay

Detecting and Deciphering Radiation

A researcher has placed a sample of radioactive material in an enclosure and blocked all emissions except those that travel in a particular direction, creating a beam of radiation. The beam then passes through a uniform magnetic field, as shown in Figure VI.2, before reaching a bank of detectors. Only three of the detectors record significant signals, so the researcher deduces that the particles coming from the source have taken three different paths, illustrated in the figure, and concludes that the sample is emitting three different kinds of radiation. Assume that the emitted particles move with similar speeds.

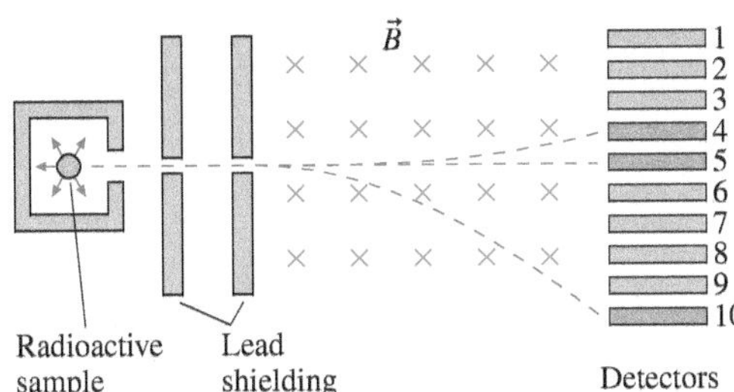

FIGURE VI.2

14. What type of radiation is detected by detector 4?
    A. Alpha B. Beta-minus C. Beta-plus D. Gamma
15. What type of radiation is detected by detector 5?
    A. Alpha B. Beta-minus C. Beta-plus D. Gamma
16. What type of radiation is detected by detector 10?
    A. Alpha B. Beta-minus C. Beta-plus D. Gamma

Double Trouble BIO

Cells that are exposed to radiation are damaged by the resulting ionization. Subsequently exposing the cells to an electric field makes the damage worse by inhibiting repair mechanisms. In one study, investigators were able to dramatically increase the damage to cyanobacteria cells that each received a 2.0 kGy dose of radiation by subsequently exposing them to a 2000 V/m electric field. Cell survival decreased from 70% to 30%. The radiation dose was provided by a ^{60}Co source. ^{60}Co, with a half-life of 5.3 years, emits a 0.3 MeV beta particle, but the radiation dose was provided by the subsequent decay of the daughter nucleus, which decays almost immediately by emitting two gamma-ray photons of average energy 1.3 MeV. Assume that the cells are spherical, with a diameter of 2 μm, and that their density is equal to that of water.

17. What is the daughter nucleus of the ^{60}Co decay?
    A. ^{58}Mn B. ^{60}Fe C. ^{60}Co D. ^{60}Ni
18. For the electric field noted, what was the approximate potential difference across a cell?
    A. 1 mV B. 2 mV C. 3 mV D. 4 mV
19. Approximately how many gamma rays were absorbed in a cell to give the noted dose?
    A. 4 B. 40
    C. 4000 D. 400,000

Additional Integrated Problems

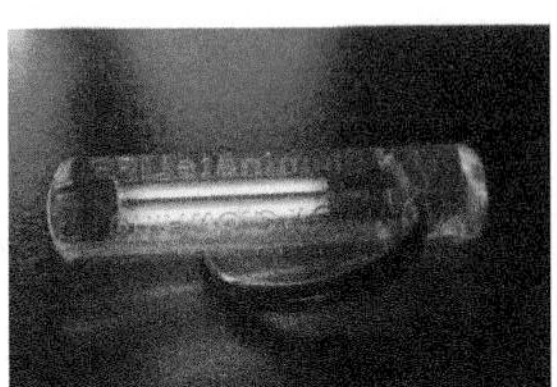

A keychain light powered by the decay of tritium.

20. The glow-in-the-dark dials on some watches and some key-chain lights shine with energy provided by the decay of radioactive tritium, ^{3}H. Tritium is a radioactive isotope of hydrogen with a half-life of 12 years. Each decay emits an electron with an energy of 19 keV. A typical new watch has tritium with a total activity of 15 MBq.
    a. What is the power, in watts, provided by the radioactive decay process?
    b. What will be the activity of the tritium in a watch after 5 years, assuming none escapes?
21. BIO An x-ray tube is powered by a high-voltage supply that delivers 700 W to the tube. The tube converts 1% of this power into x rays of wavelength 0.030 nm.
    a. Approximately how many x-ray photons are emitted per second?
    b. If a 75 kg technician is accidentally exposed to the full power of the x-ray beam for 1.0 s, what dose equivalent in Sv does he receive? Assume that the x-ray energy is distributed over the body, and that 80% of the energy is absorbed.
22. Many speculative plans for spaceships capable of interstellar travel have been developed over the years. Nearly all are powered by the fusion of light nuclei, one of a very few power sources capable of providing the incredibly large energies required. A typical design for a fusion-powered craft has a 1.7×10^6 kg ship brought up to a speed of $0.12c$ using the energy from the fusion of ^{2}H and ^{3}He. Each fusion reaction produces a daughter nucleus and one free proton with a combined kinetic energy of 18 MeV; these high-speed particles are directed backward to create thrust.
    a. What is the kinetic energy of the ship at the noted top speed?
    b. If we assume that 50% of the energy of the fusion reactions goes into the kinetic energy of the ship (a very generous assumption), how many fusion reactions are required to get the ship up to speed?
    c. How many kilograms of ^{2}H and of ^{3}He are required to produce the required number of reactions?

Mathematics Review

Algebra

Using exponents: $a^{-x} = \frac{1}{a^x}$ $\quad a^x a^y = a^{(x+y)}$ $\quad \frac{a^x}{a^y} = a^{(x-y)}$ $\quad (a^x)^y = a^{xy}$

$a^0 = 1$ $\quad a^1 = a$ $\quad a^{1/n} = \sqrt[n]{a}$

Fractions: $\left(\frac{a}{b}\right)\left(\frac{c}{d}\right) = \frac{ac}{bd}$ $\quad \frac{a/b}{c/d} = \frac{ad}{bc}$ $\quad \frac{1}{1/a} = a$

Logarithms: Base 10 logarithms: If $a = 10^x$, then $\log(a) = x$ $\quad \log(10^x) = x$ $\quad 10^{\log(x)} = x$

Natural logarithms: If $a = e^x$, then $\ln(a) = x$ $\quad \ln(e^x) = x$ $\quad e^{\ln(x)} = x$

For both base 10 logarithms or natural logarithms:

$\log(ab) = \log(a) + \log(b)$ $\quad \log\left(\frac{a}{b}\right) = \log(a) - \log(b)$ $\quad \log(a^n) = n\log(a)$

The expression $\ln(a + b)$ cannot be simplified.

Linear equations: The graph of the equation $y = ax + b$ is a straight line. a is the slope of the graph. b is the y-intercept.

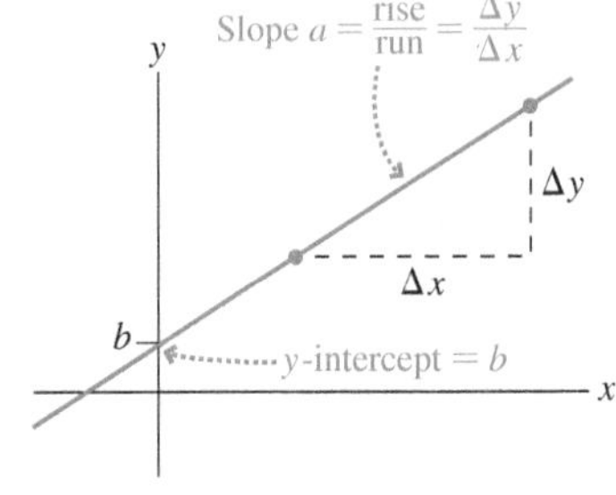

Proportionality: To say that y is proportional to x, written $y \propto x$, means that $y = ax$, where a is a constant. Proportionality is a special case of linearity. A graph of a proportional relationship is a straight line that passes through the origin. If $y \propto x$, then

$$\frac{y_1}{y_2} = \frac{x_1}{x_2}$$

Quadratic equation: The quadratic equation $ax^2 + bx + c = 0$ has the two solutions $x = \frac{-b \pm \sqrt{b^2 - 4ac}}{2a}$.

Geometry and Trigonometry

Area and volume:

Rectangle
$A = ab$

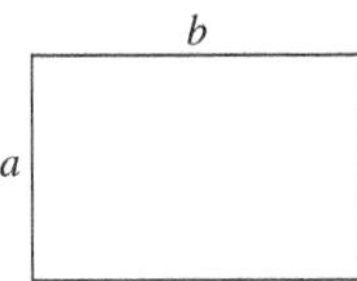

Rectangular box
$V = abc$

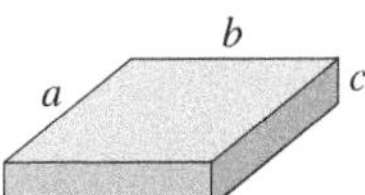

Triangle
$A = \frac{1}{2}ab$

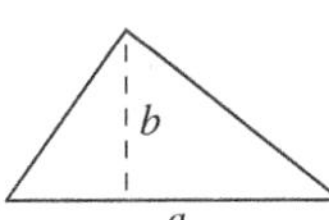

Right circular cylinder
$V = \pi r^2 l$

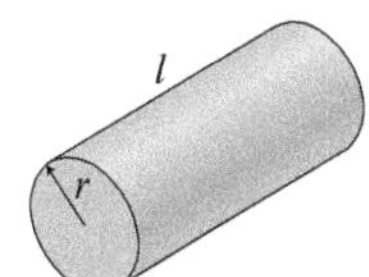

Circle
$C = 2\pi r$
$A = \pi r^2$

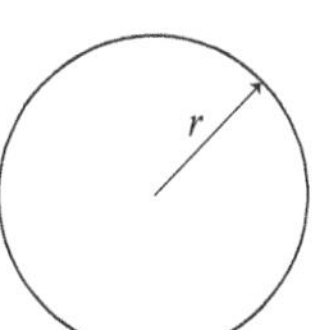

Sphere
$A = 4\pi r^2$
$V = \frac{4}{3}\pi r^3$

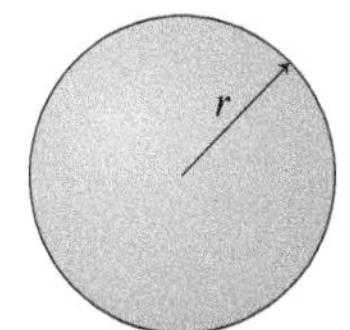

APPENDIX A

Arc length and angle: The angle θ in radians is defined as $\theta = s/r$.
The arc length that spans angle θ is $s = r\theta$.
2π rad $= 360°$

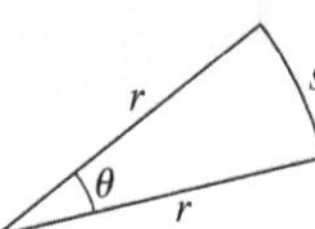

Right triangle: Pythagorean theorem $c = \sqrt{a^2 + b^2}$ or $a^2 + b^2 = c^2$

$$\sin\theta = \frac{b}{c} = \frac{\text{far side}}{\text{hypotenuse}} \qquad \theta = \sin^{-1}\left(\frac{b}{c}\right)$$

$$\cos\theta = \frac{a}{c} = \frac{\text{adjacent side}}{\text{hypotenuse}} \qquad \theta = \cos^{-1}\left(\frac{a}{c}\right)$$

$$\tan\theta = \frac{b}{a} = \frac{\text{far side}}{\text{adjacent side}} \qquad \theta = \tan^{-1}\left(\frac{b}{a}\right)$$

General triangle: $\alpha + \beta + \gamma = 180° = \pi$ rad

Law of cosines $c^2 = a^2 + b^2 - 2ab\cos\gamma$

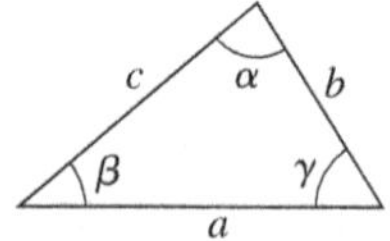

Identities:

$\tan\alpha = \dfrac{\sin\alpha}{\cos\alpha}$ $\qquad$ $\sin^2\alpha + \cos^2\alpha = 1$

$\sin(-\alpha) = -\sin\alpha$ $\qquad$ $\cos(-\alpha) = \cos\alpha$

$\sin(\alpha \pm \beta) = \sin\alpha\cos\beta \pm \cos\alpha\sin\beta$ $\qquad$ $\cos(\alpha \pm \beta) = \cos\alpha\cos\beta \mp \sin\alpha\sin\beta$

$\sin(2\alpha) = 2\sin\alpha\cos\alpha$ $\qquad$ $\cos(2\alpha) = \cos^2\alpha - \sin^2\alpha$

$\sin(\alpha \pm \pi/2) = \pm\cos\alpha$ $\qquad$ $\cos(\alpha \pm \pi/2) = \mp\sin\alpha$

$\sin(\alpha \pm \pi) = -\sin\alpha$ $\qquad$ $\cos(\alpha \pm \pi) = -\cos\alpha$

Graphs: Sine and cosine are functions of the angle θ with a period of 2π rad or $360°$.

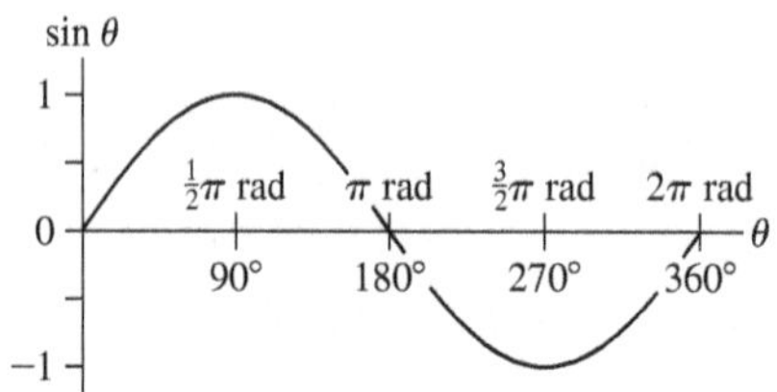

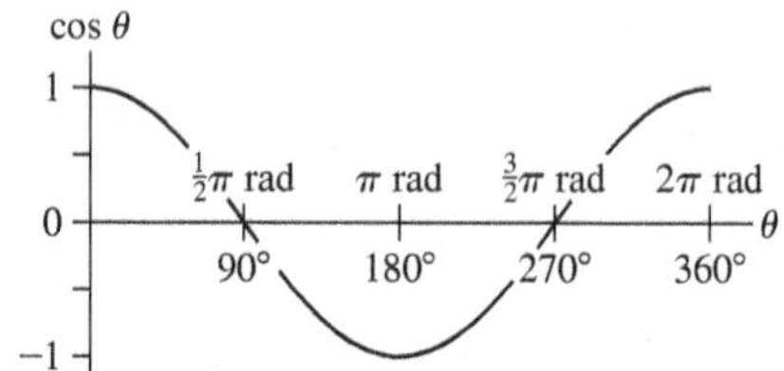

Expansions and Approximations

Binomial approximation: $(1 + x)^n \approx 1 + nx$ if $x \ll 1$

Logarithmic expansion: $\log(1 + x) \approx x$ if $x \ll 1$

Small-angle approximation: If $\theta \ll 1$ rad, then $\sin\theta \approx \tan\theta \approx \theta$ and $\cos\theta \approx 1$.

The small-angle approximation is excellent for $\theta < 5°$ (≈ 0.1 rad) and acceptable up to $\theta \approx 10°$.

Calculus

The letters a and n represent constants in the following derivatives and integrals.

Derivatives

$$\frac{d}{dx}(a) = 0$$

$$\frac{d}{dx}(ax) = a$$

$$\frac{d}{dx}\left(\frac{a}{x}\right) = -\frac{a}{x^2}$$

$$\frac{d}{dx}(ax^n) = anx^{n-1}$$

$$\frac{d}{dx}\left(\ln(ax)\right) = \frac{1}{x}$$

$$\frac{d}{dx}(e^{ax}) = ae^{ax}$$

$$\frac{d}{dx}\left(\sin(ax)\right) = a\cos(ax)$$

$$\frac{d}{dx}\left(\cos(ax)\right) = -a\sin(ax)$$

Integrals

$$\int x\,dx = \frac{1}{2}x^2$$

$$\int x^2\,dx = \frac{1}{3}x^3$$

$$\int \frac{1}{x^2}\,dx = -\frac{1}{x}$$

$$\int x^n\,dx = \frac{x^{n+1}}{n+1} \qquad n \neq -1$$

$$\int \frac{dx}{x} = \ln x$$

$$\int \frac{dx}{a+x} = \ln(a+x)$$

$$\int \frac{x\,dx}{a+x} = x - a\ln(a+x)$$

$$\int \frac{dx}{\sqrt{x^2 \pm a^2}} = \ln\left(x + \sqrt{x^2 \pm a^2}\right)$$

$$\int \frac{x\,dx}{\sqrt{x^2 \pm a^2}} = \sqrt{x^2 \pm a^2}$$

$$\int \frac{dx}{x^2 + a^2} = \frac{1}{a}\tan^{-1}\left(\frac{x}{a}\right)$$

$$\int \frac{dx}{(x^2 + a^2)^2} = \frac{1}{2a^3}\tan^{-1}\left(\frac{x}{a}\right) + \frac{x}{2a^2(x^2 + a^2)}$$

$$\int \frac{dx}{(x^2 \pm a^2)^{3/2}} = \frac{\pm x}{a^2\sqrt{x^2 \pm a^2}}$$

$$\int \frac{x\,dx}{(x^2 \pm a^2)^{3/2}} = -\frac{1}{\sqrt{x^2 \pm a^2}}$$

$$\int e^{ax}\,dx = \frac{1}{a}e^{ax}$$

$$\int xe^{-x}\,dx = -(x+1)e^{-x}$$

$$\int x^2e^{-x}\,dx = -(x^2 + 2x + 2)e^{-x}$$

$$\int \sin(ax)\,dx = -\frac{1}{a}\cos(ax)$$

$$\int \cos(ax)\,dx = \frac{1}{a}\sin(ax)$$

$$\int \sin^2(ax)\,dx = \frac{x}{2} - \frac{\sin(2ax)}{4a}$$

$$\int \cos^2(ax)\,dx = \frac{x}{2} + \frac{\sin(2ax)}{4a}$$

$$\int_0^\infty x^n e^{-ax}\,dx = \frac{n!}{a^{n+1}}$$

$$\int_0^\infty e^{-ax^2}\,dx = \frac{1}{2}\sqrt{\frac{\pi}{a}}$$

Periodic Table of Elements

Key: Atomic number (27), Symbol (Co), Atomic mass (58.9)

Period																		
1	1 H 1.0																	2 He 4.0
2	3 Li 6.9	4 Be 9.0											5 B 10.8	6 C 12.0	7 N 14.0	8 O 16.0	9 F 19.0	10 Ne 20.2
3	11 Na 23.0	12 Mg 24.3											13 Al 27.0	14 Si 28.1	15 P 31.0	16 S 32.1	17 Cl 35.5	18 Ar 39.9
4	19 K 39.1	20 Ca 40.1	21 Sc 45.0	22 Ti 47.9	23 V 50.9	24 Cr 52.0	25 Mn 54.9	26 Fe 55.8	27 Co 58.9	28 Ni 58.7	29 Cu 63.5	30 Zn 65.4	31 Ga 69.7	32 Ge 72.6	33 As 74.9	34 Se 79.0	35 Br 79.9	36 Kr 83.8
5	37 Rb 85.5	38 Sr 87.6	39 Y 88.9	40 Zr 91.2	41 Nb 92.9	42 Mo 95.9	43 Tc [98]	44 Ru 101.1	45 Rh 102.9	46 Pd 106.4	47 Ag 107.9	48 Cd 112.4	49 In 114.8	50 Sn 118.7	51 Sb 121.8	52 Te 127.6	53 I 126.9	54 Xe 131.3
6	55 Cs 132.9	56 Ba 137.3	71 Lu 175.0	72 Hf 178.5	73 Ta 180.9	74 W 183.9	75 Re 186.2	76 Os 190.2	77 Ir 192.2	78 Pt 195.1	79 Au 197.0	80 Hg 200.6	81 Tl 204.4	82 Pb 207.2	83 Bi 209.0	84 Po [209]	85 At [210]	86 Rn [222]
7	87 Fr [223]	88 Ra [226]	103 Lr [262]	104 Rf [267]	105 Db [268]	106 Sg [271]	107 Bh [274]	108 Hs [277]	109 Mt [278]	110 Ds [281]	111 Rg [281]	112 Cn [285]	113 Nh [286]	114 Fl [289]	115 Mc [289]	116 Lv [293]	117 Ts [294]	118 Og [294]

Transition elements: groups 21–30 columns (Sc through Zn and below)

Inner transition elements

Lanthanides 6	57 La 138.9	58 Ce 140.1	59 Pr 140.9	60 Nd 144.2	61 Pm 144.9	62 Sm 150.4	63 Eu 152.0	64 Gd 157.3	65 Tb 158.9	66 Dy 162.5	67 Ho 164.9	68 Er 167.3	69 Tm 168.9	70 Yb 173.0
Actinides 7	89 Ac [227]	90 Th 232.0	91 Pa 231.0	92 U 238.0	93 Np [237]	94 Pu [244]	95 Am [243]	96 Cm [247]	97 Bk [247]	98 Cf [251]	99 Es [252]	100 Fm [257]	101 Md [258]	102 No [259]

An atomic mass in brackets is that of the longest-lived isotope of an element with no stable isotopes.

Preparing for the MCAT

If you are taking this course, there's a good chance you are preparing for a career in the health professions and might well be required to take the Medical College Admission Test, the MCAT. The *Chemical and Physical Foundations of Biological Systems* section of the MCAT assesses your understanding of the concepts of physics by testing your ability to apply these concepts in a biological context. You will be expected to use what you've learned in physics to analyze situations you've never seen before. It is important to realize that the MCAT is an exam of reasoning and critical thinking, not memorization and recall. Study should focus on being able to apply the laws and concepts of physics, not on memorizing equations.

Structure of the MCAT Exam

Most of the test consists of passages of technical information followed by a series of questions based on the passage, much like the MCAT-Style Passage Problems at the end of each chapter in this book. Some details:

- **The passages and the questions are *always* integrated.** That is, understanding the passage and answering the questions will require you to integrate and use knowledge from several different areas of physics.
- **Passages will usually be about topics for which you do not have detailed knowledge.** But, if you read carefully, you'll see that the treatment of the subject is based on information you learned in this course.
- **The test assumes a basic level of physics knowledge.** You'll need to have facility with the central themes and major concepts of physics that you learn in this course, but you won't need detailed knowledge of any particular topic. Any needed details and equations will be provided in the passage.
- **You can't use calculators on the test, so any required math will be reasonably simple.** Being able to quickly estimate an answer with ratio reasoning or with a knowledge of the scale of physical quantities will be a useful skill.
- **The four multiple choice options to each question are all designed to be plausible.** You can't generally weed out the "bad" answers with a quick inspection.
- **The test is given online.** Practicing with Mastering Physics will help you get used to this format.

Preparing for the Test

Because you have used this text as a tool for learning physics, you should use it as a tool as you review for the MCAT exam. Several of the key features of the text will be useful for this, including some that were explicitly designed with the MCAT exam in mind.

As you review the chapters:

- Start with the *Chapter Summaries,* which provide a "big picture" overview of the content. What are the major themes of each chapter? What are the important physics concepts?
- Go through each chapter and review the *Stop to Think* questions. These are a good way to test your understanding of the key concepts.
- In each chapter, the end-of-chapter problems end with one or two *Passage Problems* that are designed to be similar to those on the MCAT exam. They'll give you good practice with the "read a passage, answer questions" format of the MCAT exam.
- The chapter-level Passage Problems usually don't integrate topics that span several chapters—a key feature of the MCAT exam. Turn to the *Part Summaries,* at the end of each of the text's six parts, for integrated passages and problems.

Taking the Test: Reading the Passage

As you read each passage, you'll need to interpret the information presented and connect it with concepts you are familiar with. The example in this appendix is a passage that was written to very closely match the style and substance of an actual MCAT passage. Blue annotations highlight connections you should make as you read. The passage describes a situation (the mechanics and energetics of sled dogs) that you probably haven't seen before, but the basic physics (friction, energy conversion) are principles that you are familiar with. As you read the passage, think about the underlying physics concepts and how they apply to this situation.

Taking the Test: Answering the Questions

The passages on the MCAT exam seem complicated at first, but they are about basic concepts and central themes that you know well. The same is true of the questions; they aren't as difficult as they may seem at first. As with the passage, you should start by identifying the physical concepts that apply in each case. You then proceed by reasoning, using your understanding of these basic concepts to determine the answer to the question. The example illustrates how you might reason your way through the questions associated with the passage on sled dogs.

You Can Answer the Questions in Any Order

The questions test a range of skills and have a range of difficulties. Some questions will involve simple reading comprehension; these are usually straightforward. Others require sophisticated reasoning and some mathematical

Example: Reading and Interpreting a Passage

Reading the Passage

As you read the passage, connect the scenario to examples and problems you've seen before. Think about the basic physical principles that apply.

For travel over snow, a sled with runners that slide on snow is the best way to get around. Snow is slippery, but there is still friction between runners and the ground; the forward force required to pull a sled at a constant speed might be 1/6 of the sled's weight.

As you read this part of the passage, think about the forces involved: There's no net force on a sled moving at a constant speed, so the forward pulling force must be equal to the friction force, which is acting opposite the sled's motion. There are many problems like this in Chapter 5.

Recognize that this is really a statement about the coefficient of kinetic friction.

Sleds are typically pulled by dogs, as in Figure 1. The rope usually makes a slight angle, but you can treat the rope as being parallel to the ground. The pulling force is the tension in the rope.

The force applied to the sled is the tension force in the rope; you're told this.

Figure 1

Sled dogs have great aerobic capacity; a 40 kg dog can provide output power to pull with a 60 N force at 2.2 m/s for hours. The output power is related to force and velocity by $P = F \cdot v$, so they can pull lighter loads at higher speeds.

In the data given here, and the description given above, the sled moves at a constant speed. The net force is zero and the kinetic energy of the sled isn't changing.

Notice that the key equation relating power, force, and velocity is given to you. Equations and constants, will generally be given in the passage. The MCAT is a test of reasoning, not recall.

Doing 100 J of work requires a dog to expend 400 J of metabolic energy; the difference must be exhausted as heat. Given the excellent insulation provided by a dog's fur, this is mostly via evaporation as it pants. At a typical body temperature, the evaporation of 1.0 L of water requires 240,000 J, so this is an effective means of cooling.

The concepts of metabolic energy and energy output were treated in Chapter 11. We see that the efficiency is 25%. 400 J of energy is used by the body; 25% of this, 100 J, is the energy output. This means that 300 J is exhausted as heat.

Chapter 12 discussed means of heat transfer. The information in the passage tells you that the dog's fur limits heat transfer by conduction, convection, and radiation. Evaporation of water by a panting dog is the primary means of cooling.

The data for the energy required to evaporate water are given. If you need such information to answer questions, it will almost certainly be provided.

FIGURE MCAT-EXAM.1 Interpreting a passage.

Example: Answering the Questions

Interpreting the Questions

Think about the physics principles and how they connect to concepts you know and understand. Don't spend time reading through all the answers first. Find a solution, then look at the choices.

You are told that it takes a force that's about 1/6 of the sled's weight to pull it forward. You can estimate the friction coefficient from this information.

1. What is the approximate coefficient of kinetic friction for a sled on snow?

A. 0.35
B. 0.25
C. 0.15
D. 0.05

The pulling force F_{pull} balances the friction force f_k, so the friction force is also 1/6 of the weight: $f_k = w/6$. The model of friction is $f_k = \mu_k n$. The rope is parallel to the ground, so $n \approx w$ and so $f_k \approx \mu_k w$. Thus $\mu_k \approx 1/6 \approx 0.15$.

Assume that the dog's output power is the same for the two cases. This is implied in the passage.

2. A dog pulls a 40 kg sled at a maximum speed of 2 m/s. What is the maximum speed for an 80 kg sled?

A. 2 m/s
B. 1.5 m/s
C. 1.0 m/s
D. 0.5 m/s

Doubling the weight doubles the normal force, which doubles the friction force. This will double the necessary pulling force as well. Given the expression for power in the passage, this means the maximum speed will be halved.

This is a question about energy transformation. What forms of energy are *changing*? Thermal energy is certainly part of the story because some of the stored chemical energy is transformed to thermal energy in the dog's body.

3. As a dog pulls a sled at constant speed, chemical energy in the dog's body is converted to

A. kinetic energy
B. thermal energy
C. kinetic energy and thermal energy
D. kinetic energy and potential energy

Choices A and C are clever distractors—but wrong. It's tempting to include kinetic energy because the sled is in motion, but the kinetic energy is not changing. Friction forces transform any chemical energy the dog supplies to thermal energy, so the correct answer is B.

Increasing the speed increases the power requirement, but energy is not the same as power. A power output P for a time Δt uses energy $E = P \cdot \Delta t$.

4. A dog pulls a sled for a distance of 1.0 km at a speed of 1 m/s, requiring an energy output of 60,000 J. If the dog pulls the sled at 2 m/s, the necessary energy is

A. 240,000 J
B. 120,000 J
C. 60,000 J
D. 30,000 J

Doubling the speed doubles the power requirement P. But at twice the speed, the time needed Δt is halved. These offset each other such that $P \cdot \Delta t$ is unchanged. It's tempting to think that the necessary energy will be doubled, and the MCAT offers you this option as a distractor. It's important to not go for a quick hunch without thinking it through.

The passage tells us that the dog uses 400 J of metabolic energy to do 100 J of work. 300 J, 75%, must be exhausted to the environment. We can assume the same efficiency here.

5. A dog uses 100,000 J of metabolic energy pulling a sled. How much energy must the dog exhaust by panting?

A. 100,000 J
B. 75,000 J
C. 50,000 J
D. 25,000 J

If 75% of the energy must be exhausted to the environment, that's 75,000 J.

manipulations. Start with the easy ones that you can quickly solve. Save the more complex questions for later and skip them if time is short.

Take Steps to Simplify Calculations

You won't be allowed to use a calculator on the exam, so any required math will be reasonably straightforward. Even so, there are some important "shortcuts" to help you rapidly converge on a correct answer choice.

- **Use ratio reasoning.** What's the relationship between the variables involved in a question? You can often use this to deduce the answer with only a very simple calculation, as you've seen many times in the text. For example, suppose you are asked the following question:

  A model rocket is powered by chemical fuel. A student launches a rocket with a small engine containing 1.0 g of combustible fuel. The rocket reaches a speed of 10 m/s. The student then launches the rocket again, using an engine with 4.0 g of fuel. If all other parameters of the launch are kept the same, what final speed would you expect for this second trial?

  This is an energy transformation problem: Chemical energy of the fuel is transformed to the kinetic energy of the rocket. Kinetic energy is related to the speed by $K = \frac{1}{2}mv^2$, so speed is proportional to the square root of kinetic energy: $v \propto K^{1/2}$. The chemical energy—and thus the kinetic energy—in the second trial is increased by a factor of 4, so the speed must increase by a factor of $4^{1/2} = 2$ to 20 m/s.

- **Simplify calculations by rounding numbers.** Round numbers to make calculations more straightforward. Your final result will probably be close enough to choose the correct answer from the list given. For instance, suppose you are asked the following question:

  A ball moving at 2.0 m/s rolls off the edge of a table that's 1.2 m high. How far from the edge of the table does the ball land?

  A. 2 m   B. 1.5 m   C. 1 m   D. 0.5 m

  The vertical motion of the ball is free fall, so the vertical distance fallen by the ball in a time t is $\Delta y = \frac{1}{2}gt^2$. The time to fall 1.2 m is $t = \sqrt{2(1.2\text{ m})/g}$. The free-fall acceleration is very close to 10 m/s^2, and $2(1.2\text{ m}) \approx 2.5\text{ m}$, so the fall time is approximately $t \approx \sqrt{2.5/10} = \sqrt{1/4} = 1/2 = 0.5\text{ s}$. The horizontal motion is constant at 2.0 m/s during this free-fall time, so we expect the ball to land about 1 m away. Our quick calculation shows us that the correct answer must be choice C—no other answer is close.

- **For calculations using values in scientific notation, compute either the first digits or the exponents, not both.** In some cases, a quick calculation can tell you the correct leading digit, and that's all you need to determine the correct answer. In other cases, you'll find possible answers with the same leading digit but very different exponents. In this case, all you need is an order-of-magnitude estimate to decide on the right result.
- **Where possible, use your knowledge of the expected scale of physical quantities to quickly determine the correct answer.** For example, suppose a question asks you to find the photon energy for green light of wavelength 550 nm. If you recall that visible light has photon energies of about 2 eV, or about 3×10^{-19} J, that is probably enough information to allow you to pick out the correct answer without any calculation.

Keep Focused on the Big Picture

The MCAT exam tests your ability to look at a technical passage for which you have some background knowledge and quickly get a sense of what it is saying. Keep this big picture in mind.

- **Don't get bogged down in technical details of the particular situation.** Focus on the basic physics.
- **Don't spend too much time on any one question.** Make an educated guess and move on.
- **Don't get confused by details of notation or terminology.** Different texts and different disciplines use different symbols for physical variables. The meaning should be clear from the context.

Finally, don't forget that the best way to prepare for this or any test is simply to understand the subject. As you prepare for the test, focus your energy on reviewing and refining your knowledge of central topics and techniques. Then practice applying your knowledge by solving problems like you'll see on the actual MCAT.

Atomic and Nuclear Data

Atomic Number (Z)	Element	Symbol	Mass Number (A)	Atomic Mass (u)	Percent Abundance	Decay Mode	Half-Life $t_{1/2}$
0	(Neutron)	n	1	1.008 665		β^-	10.4 min
1	Hydrogen	H	1	1.007 825	99.985	stable	
	Deuterium	D	2	2.014 102	0.015	stable	
	Tritium	T	3	3.016 049		β^-	12.33 yr
2	Helium	He	3	3.016 029	0.000 1	stable	
			4	4.002 602	99.999 9	stable	
			6	6.018 886		β^-	0.81 s
3	Lithium	Li	6	6.015 121	7.50	stable	
			7	7.016 003	92.50	stable	
			8	8.022 486		β^-	0.84 s
4	Beryllium	Be	7	7.016 928		EC	53.3 days
			9	9.012 174	100	stable	
			10	10.013 534		β^-	1.5×10^6 yr
5	Boron	B	10	10.012 936	19.90	stable	
			11	11.009 305	80.10	stable	
			12	12.014 352		β^-	0.020 2 s
6	Carbon	C	10	10.016 854		β^+	19.3 s
			11	11.011 433		β^+	20.4 min
			12	12.000 000	98.90	stable	
			13	13.003 355	1.10	stable	
			14	14.003 242		β^-	5 730 yr
			15	15.010 599		β^-	2.45 s
7	Nitrogen	N	12	12.018 613		β^+	0.011 0 s
			13	13.005 738		β^+	9.96 min
			14	14.003 074	99.63	stable	
			15	15.000 108	0.37	stable	
			16	16.006 100		β^-	7.13 s
			17	17.008 450		β^-	4.17 s
8	Oxygen	O	14	14.008 595		EC	70.6 s
			15	15.003 065		β^+	122 s
			16	15.994 915	99.76	stable	
			17	16.999 132	0.04	stable	
			18	17.999 160	0.20	stable	
			19	19.003 577		β^-	26.9 s
9	Fluorine	F	17	17.002 094		EC	64.5 s
			18	18.000 937		β^+	109.8 min
			19	18.998 404	100	stable	
			20	19.999 982		β^-	11.0 s
10	Neon	Ne	19	19.001 880		β^+	17.2 s
			20	19.992 435	90.48	stable	
			21	20.993 841	0.27	stable	
			22	21.991 383	9.25	stable	

Atomic Number (*Z*)	Element	Symbol	Mass Number (*A*)	Atomic Mass (u)	Percent Abundance	Decay Mode	Half-Life $t_{1/2}$
11	Sodium	Na	22	21.994 434		β^+	2.61 yr
			23	22.989 770	100	stable	
			24	23.990 961		β^-	14.96 hr
12	Magnesium	Mg	24	23.985 042	78.99	stable	
			25	24.985 838	10.00	stable	
			26	25.982 594	11.01	stable	
13	Aluminum	Al	27	26.981 538	100	stable	
			28	27.981 910		β^-	2.24 min
14	Silicon	Si	28	27.976 927	92.23	stable	
			29	28.976 495	4.67	stable	
			30	29.973 770	3.10	stable	
			31	30.975 362		β^-	2.62 hr
15	Phosphorus	P	30	29.978 307		β^+	2.50 min
			31	30.973 762	100	stable	
			32	31.973 908		β^-	14.26 days
16	Sulfur	S	32	31.972 071	95.02	stable	
			33	32.971 459	0.75	stable	
			34	33.967 867	4.21	stable	
			35	34.969 033		β^-	87.5 days
			36	35.967 081	0.02	stable	
17	Chlorine	Cl	35	34.968 853	75.77	stable	
			36	35.968 307		β^-	3.0×10^5 yr
			37	36.965 903	24.23	stable	
18	Argon	Ar	36	35.967 547	0.34	stable	
			38	37.962 732	0.06	stable	
			39	38.964 314		β^-	269 yr
			40	39.962 384	99.60	stable	
			42	41.963 049		β^-	33 yr
19	Potassium	K	39	38.963 708	93.26	stable	
			40	39.964 000	0.01	β^+	1.28×10^9 yr
			41	40.961 827	6.73	stable	
20	Calcium	Ca	40	39.962 591	96.94	stable	
			42	41.958 618	0.64	stable	
			43	42.958 767	0.13	stable	
			44	43.955 481	2.08	stable	
			47	46.954 547		β^-	4.5 days
			48	47.952 534	0.18	stable	
24	Chromium	Cr	50	49.946 047	4.34	stable	
			52	51.940 511	83.79	stable	
			53	52.940 652	9.50	stable	
			54	53.938 883	2.36	stable	
26	Iron	Fe	54	53.939 613	5.9	stable	
			55	54.938 297		EC	2.7 yr
			56	55.934 940	91.72	stable	
			57	56.935 396	2.1	stable	
			58	57.933 278	0.28	stable	

Atomic Number (Z)	Element	Symbol	Mass Number (A)	Atomic Mass (u)	Percent Abundance	Decay Mode	Half-Life $t_{1/2}$
27	Cobalt	Co	59	58.933 198	100	stable	
			60	59.933 820		β^-	5.27 yr
28	Nickel	Ni	58	57.935 346	68.08	stable	
			60	59.930 789	26.22	stable	
			61	60.931 058	1.14	stable	
			62	61.928 346	3.63	stable	
			64	63.927 967	0.92	stable	
29	Copper	Cu	63	62.929 599	69.17	stable	
			65	64.927 791	30.83	stable	
47	Silver	Ag	107	106.905 091	51.84	stable	
			109	108.904 754	48.16	stable	
48	Cadmium	Cd	106	105.906 457	1.25	stable	
			109	108.904 984		EC	462 days
			110	109.903 004	12.49	stable	
			111	110.904 182	12.80	stable	
			112	111.902 760	24.13	stable	
			113	112.904 401	12.22	stable	
			114	113.903 359	28.73	stable	
			116	115.904 755	7.49	stable	
53	Iodine	I	127	126.904 474	100	stable	
			129	128.904 984		β^-	1.6×10^7 yr
			131	130.906 124		β^-	8 days
54	Xenon	Xe	128	127.903 531	1.9	stable	
			129	128.904 779	26.4	stable	
			130	129.903 509	4.1	stable	
			131	130.905 069	21.2	stable	
			132	131.904 141	26.9	stable	
			133	132.905 906		β^-	5.4 days
			134	133.905 394	10.4	stable	
			136	135.907 215	8.9	stable	
55	Cesium	Cs	133	132.905 436	100	stable	
			137	136.907 078		β^-	30 yr
56	Barium	Ba	133	132.905 990		EC	10.5 yr
			134	133.904 492	2.42	stable	
			135	134.905 671	6.59	stable	
			136	135.904 559	7.85	stable	
			137	136.905 816	11.23	stable	
			138	137.905 236	71.70	stable	
79	Gold	Au	197	196.966 543	100	stable	
81	Thallium	Tl	203	202.972 320	29.524	stable	
			205	204.974 400	70.476	stable	
			207	206.977 403		β^-	4.77 min
82	Lead	Pb	204	203.973 020	1.4	stable	
			205	204.974 457		EC	1.5×10^7 yr
			206	205.974 440	24.1	stable	

Atomic Number (Z)	Element	Symbol	Mass Number (A)	Atomic Mass (u)	Percent Abundance	Decay Mode	Half-Life $t_{1/2}$
			207	206.975 871	22.1	stable	
			208	207.976 627	52.4	stable	
			210	209.984 163		α, β^-	22.3 yr
			211	210.988 734		β^-	36.1 min
83	Bismuth	Bi	209	208.980 374	100	α	2×10^{19} yr
			211	210.987 254		α	2.14 min
			215	215.001 836		β^-	7.4 min
84	Polonium	Po	209	208.982 405		α	102 yr
			210	209.982 848		α	138.38 days
			215	214.999 418		α	0.001 8 s
			218	218.008 965		α, β^-	3.10 min
85	Astatine	At	218	218.008 685		α, β^-	1.6 s
			219	219.011 294		α, β^-	0.9 min
86	Radon	Rn	219	219.009 477		α	3.96 s
			220	220.011 369		α	55.6 s
			222	222.017 571		α, β^-	3.823 days
87	Francium	Fr	223	223.019 733		α, β^-	22 min
88	Radium	Ra	223	223.018 499		α	11.43 days
			224	224.020 187		α	3.66 days
			226	226.025 402		α	1 600 yr
			228	228.031 064		β^-	5.75 yr
89	Actinium	Ac	227	227.027 749		α, β^-	21.77 yr
			228	228.031 015		β^-	6.15 hr
90	Thorium	Th	227	227.027 701		α	18.72 days
			228	228.028 716		α	1.913 yr
			229	229.031 757		α	7 300 yr
			230	230.033 127		α	75,000 yr
			231	231.036 299		α, β^-	25.52 hr
			232	232.038 051	100	α	1.40×10^{10} yr
			234	234.043 593		β^-	24.1 days
91	Protactinium	Pa	231	231.035 880		α	32.760 yr
			234	234.043 300		β^-	6.7 hr
92	Uranium	U	233	233.039 630		α	1.59×10^5 yr
			234	234.040 946		α	2.45×10^5 yr
			235	235.043 924	0.72	α	7.04×10^8 yr
			236	236.045 562		α	2.34×10^7 yr
			238	238.050 784	99.28	α	4.47×10^9 yr
93	Neptunium	Np	236	236.046 560		EC	1.15×10^5 yr
			237	237.048 168		α	2.14×10^6 yr
94	Plutonium	Pu	238	238.049 555		α	87.7 yr
			239	239.052 157		α	2.412×10^4 yr
			240	240.053 808		α	6 560 yr
			242	242.058 737		α	3.73×10^6 yr
95	Americium	Am	241	241.056 823		α	432.2 yr
			243	243.061 375		α	7 370 yr

Answers

Chapter 1

Answers to odd-numbered multiple-choice questions

7. C
9. B
10. B

Answers to odd-numbered problems

1. a. 500 b. 0 c. 250
3. a. 0.2 mm b. 1.0 ms c. 2.8 μm
5. 40 s
7. a. 46 ns b. 12 min
9. 32 μm
11. 170 bpm
13. $y \propto \sqrt{x}$
15. The area has been increased by a factor of 1.2.
17. 90 km/h
19. a. 5.2×10^3 μm^3 b. 17 μm
21. a. 20 nm/s b. 3.3 min
23. a. 7.7×10^{13} mammalian cells b. 0.0017 or 0.17%
25. a. 560 hairs/cm^2 b. 0.4 mm
27. There will be approximately 6 times as many small species as large species.
29. 4×10^{-4}
31. A
33. C

Chapter 2

Answers to odd-numbered multiple-choice questions

15. B
17. C
19. A
21. B
23. D

Answers to odd-numbered problems

1.

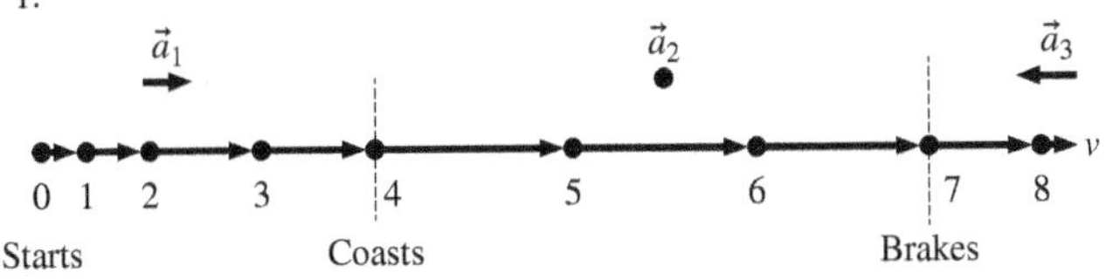

3. a. $x = +2$ mi b. $x = -3$ mi
5. $\Delta x = -23$ mm and $\Delta t = 50$ s
7. 0.46 m/s
9. a. The second 100 m was the fastest. b. 9.88 m/s
11. a. -10 m/s b. -5 m/s c. -10 m/s
13.

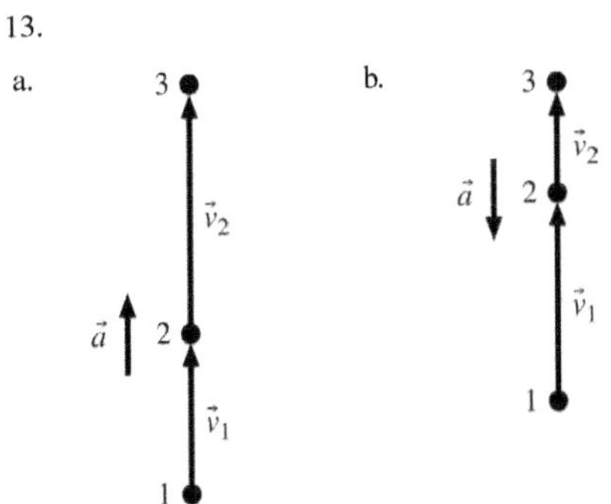

15.

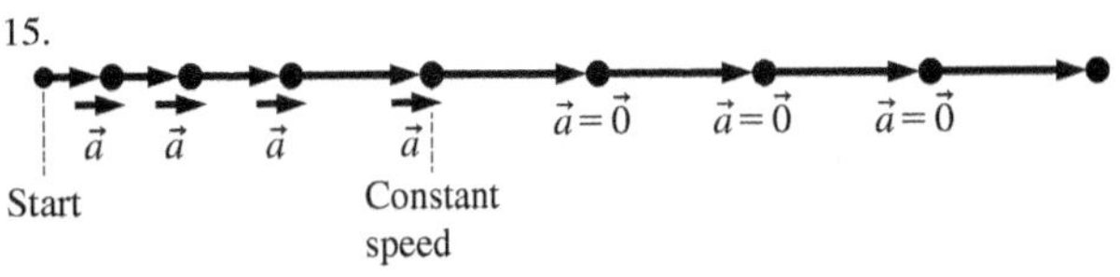

17.

a.

Dot	Time (s)	x (m)
1	0	0
2	2	30
3	4	95
4	6	215
5	8	400
6	10	510
7	12	600
8	14	670
9	16	720

b.

x (m)
100 200 300 400 500 600 700
t (s)
0 2 4 6 8 10 12 14 16

21. a. 9.12×10^{-6} s b. 3.42×10^3 m c. 4.4×10^2 m/s d. 22 m/s
23. a. 3.6×10^3 s b. 8.6×10^4 s c. 3.2×10^7 s
25. a. 3 b. 4 c. 5 d. 3
27. a. $159.31 \times 204.6 = 32{,}590$ b. $5.1125 + 0.67 + 3.2 = 9.0$ c. $7.662 - 7.425 = 0.237$ d. $16.5/3.45 = 4.78$
29. a. 7 m b. 100,000 m c. 30 m/s d. 0.2 m
31. 9.8×10^{-9} m/s $\approx$ 35 μm/h
33. a. 10^{14} cells b. 10^{15} cells. This problem involves estimation, so a range of answers may be correct.

35. 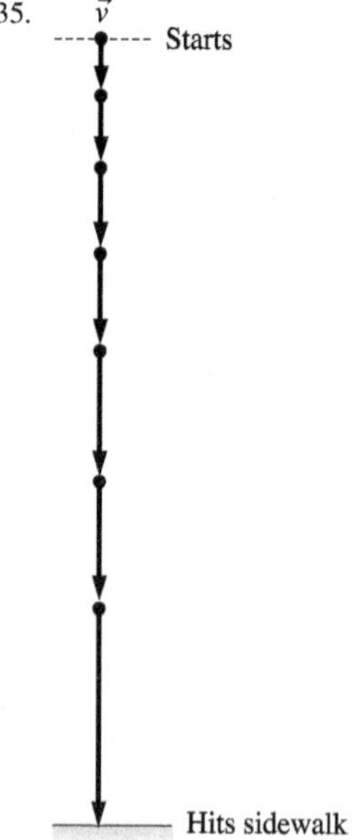

37. 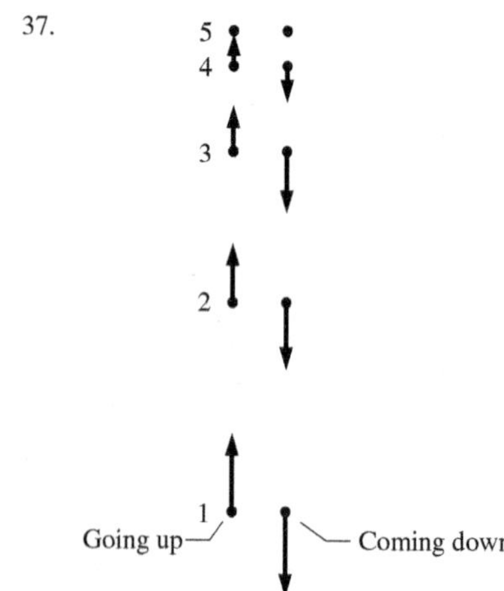

43. 69 mph
45. 1.0×10^{-6} m/s
47. 26 mph
49. 40 ms. This problem involves estimation, so a range of answers may be correct.
51. a. 2×10^{-6} m b. 1×10^{-6} m c. 1×10^{-15} kg d. 1 mm e. 1.7 m
53. a. 3×10^{13} cells b. 10^{8} hemoglobin per red blood cell
55. a. 10 nm b. Dense
57. 9.0 kg/m^3
59. B
61. A

Chapter 3

Answers to odd-numbered multiple-choice questions

15. C
17. C
19. D
21. B
23. B
25. C
27. C

Answers to odd-numbered problems

1. 0.43 s
3. a. 48 mph b. 50 mph
5. 1.71 km
7. 12 m/s
9. a. 7.5 cm b. 4.0 beat
11. a. −4 m/s b. 0 m/s c. 4 m/s
13. a. 1 m/s^2 b. 6 m
15. 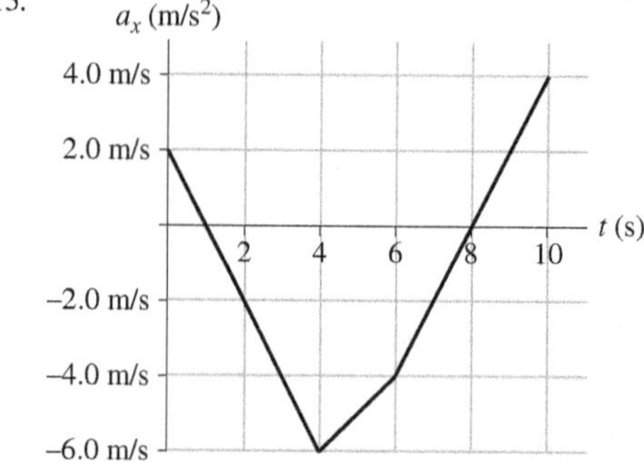

17. There may be some uncertainty in reading the graph. a. $a_x = 5.5$ m/s^2 b. $a_x = 2.3$ m/s^2 c. $a_x = 2.0$ m/s^2
19. Trout
21. 83 m/s^2
23. 0.18 s
25. 52 m
27. 3.3 m
29. a. 33 m b. 25 m/s
31. 0.95 s
33. a. 0.452 s b. 0.565 m
35. a. 10 m/s b. 50°
37. a. 12 m/s b. 11.6 m/s
39. 16 m/s
41. 16 m/s
43. a. 1.7 m/s b. 15 cm
45. a. $\frac{5}{3}$ m b. 2 m/s c. 4 m/s^2
47. 57 mph
49. a. 2.3 m/s^2, 0.23g b. 35 s c. 4.2 km
51. a. 5 m/s^2 b. 0.8 m/s^2 c. $\approx$80 m
53. $k = 2.0$ m/s^3
55. a. 96 m away from the intersection b. $|a_1| = 2.1$ m/s^2 c. 10 s
57. 1 mi
59. 4.6 s
61. a. 2.8 s b. 31 m
63. a. $\approx$1.5 cm b. 52 m/s^2 c. 1.3 m/s
65. a. 6.9 m/s b. 1.7 ms c. 2.4 m
67. 95 m
69. 3.2 s
71. 0.32 m/s^2
73. a. 4.1 s b. Equal speeds
75. 67 m/s^2
77. 6.0 m
79. a. 32.7° b. 97 m c. 81%
81. a. 9.8 m b. 34 m c. 110 m
83. B
85. B

Chapter 4

Answers to odd-numbered multiple-choice questions

15. A
17. A
19. D
21. B
23. C

Answers to odd-numbered problems

1. a. Rear-end collision b. Head-on collision
5.

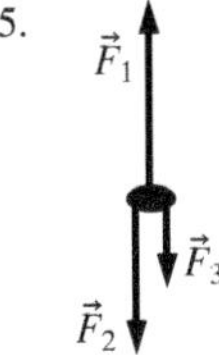

7. $\vec{n}$, $\vec{f}_s$, and $\vec{w}$
9. $\vec{n}$, $\vec{D}$, $\vec{F}_{thrust}$, and $\vec{w}$
11. a. Object B b. Object C c. 3/5
13. a. 16 m/s^2 b. 4.0 m/s^2 c. 8.0 m/s^2 d. 32 m/s^2
15. $m_1 = 0.080$ kg, $m_3 = 0.50$ kg
17. **Motion diagram**

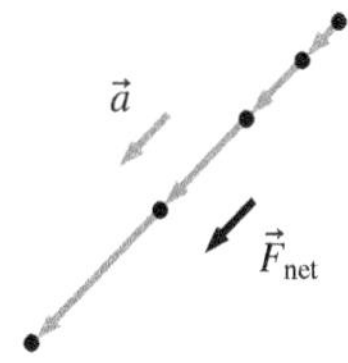

19. a. 1 N b. 2 N
21. 1.5 N
23. $(0.71)g$
25. 6.6 N
27. A possible description is: "An object hangs from a rope and is at rest." Or, "An object hanging from a rope is moving up or down with a constant speed."
29.

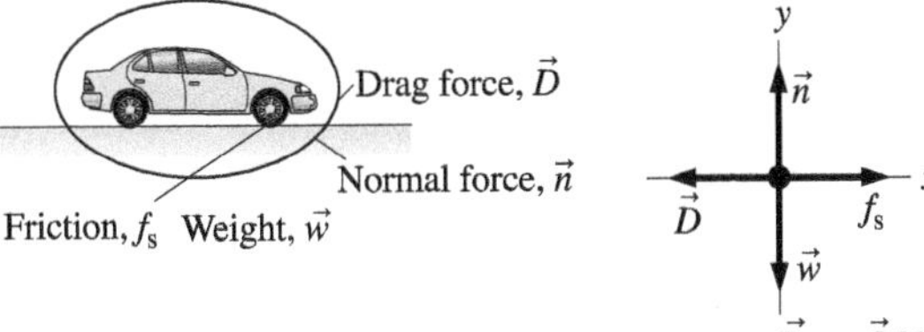

31.

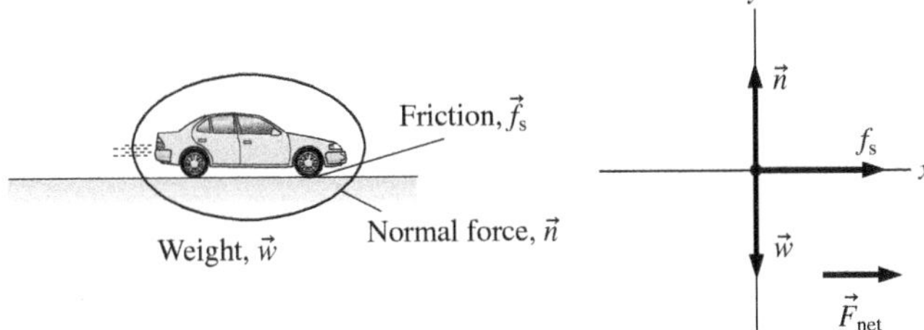

33.

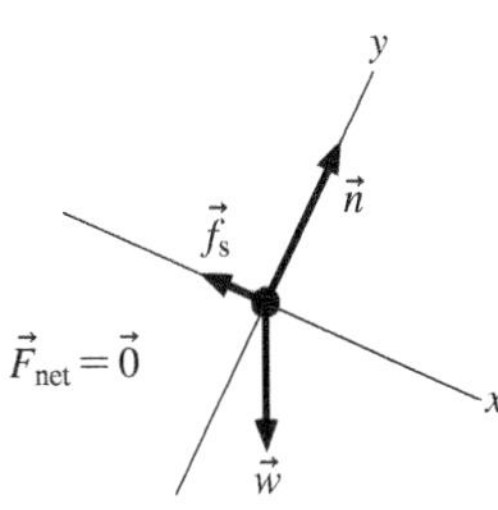

35.

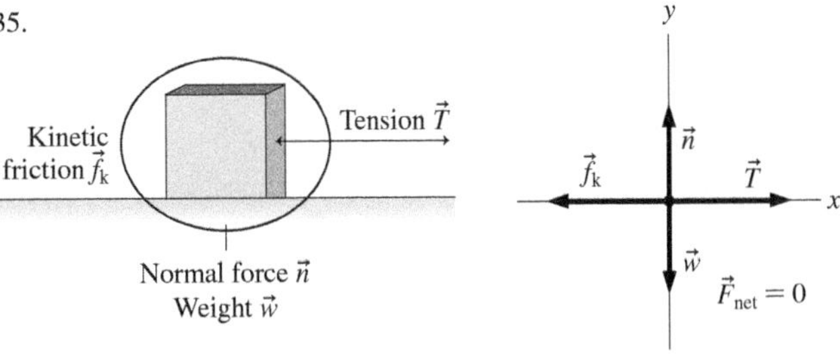

37. Known
$\vec{A}$ $\vec{B}$

Find
$\vec{A}+\vec{B}$ $\vec{A}-\vec{B}$

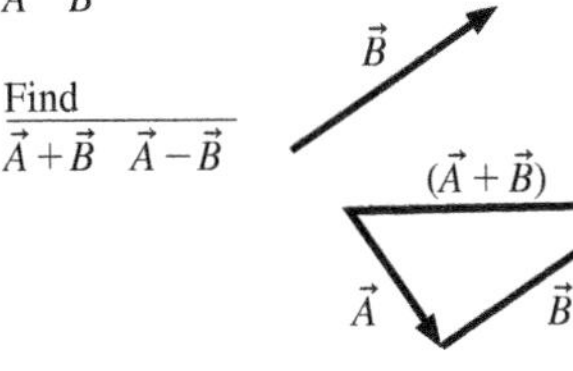

a.

$(\vec{A}-\vec{B})$ $\vec{A}$ $-\vec{B}$

b.

39.

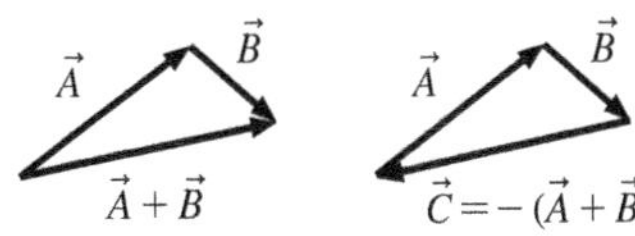

41. 8.0 m
43. 87 m/s
45. a. Known
$d = 2$ km
$\theta = 30°$

Find
d_x d_y

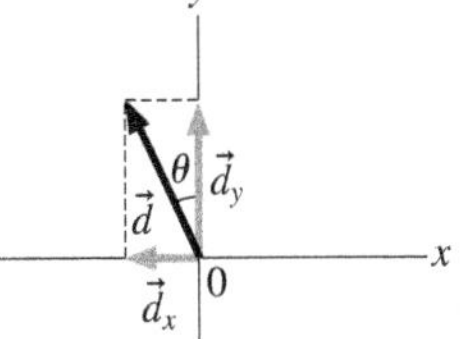

b. Known
$v = 5$ cm/s
$\theta = 90°$

Find
v_x v_y

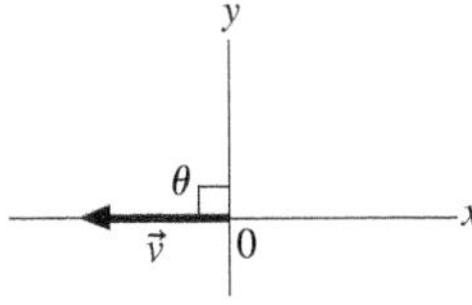

c. Known
$a = 10$ m/s^2
$\theta = 40°$

Find
a_x a_y

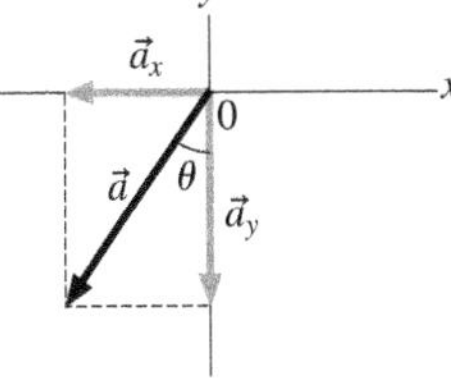

a. $d_x = -1.0$ km, $d_y = 1.7$ km b. $v_x = -5.0$ cm/s, $v_y = 0$ cm/s
c. $a_x = -6.4$ m/s^2, $a_y = -7.7$ m/s^2

47. a. $v = 32$ m/s, $\theta = 72°$

v_y (m/s)

v_x (m/s)

θ

$\vec{v}$

b. $a = 22\text{ m/s}^2, \theta = 27°$

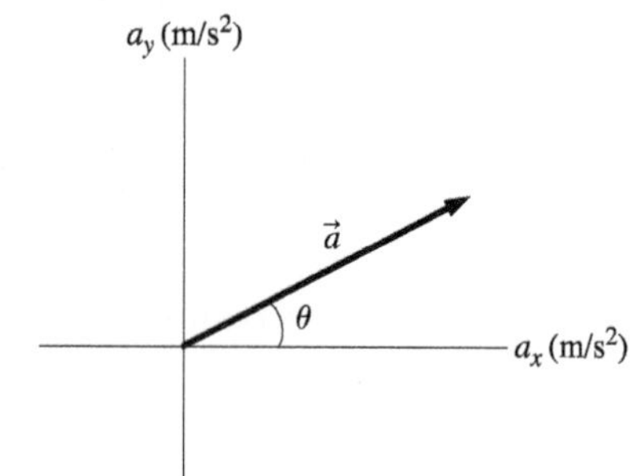

49. No, the tension is 5660 N, so the ropes break.
51. 1.0 m/s^2
53. a. Zero b. 0 N c. 250 N
55.

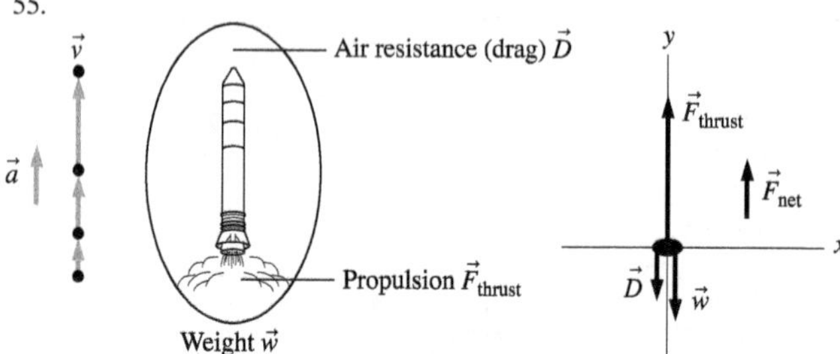

57.

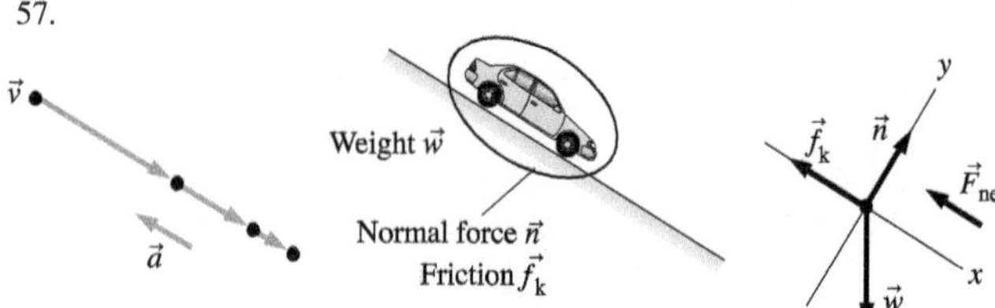

59.

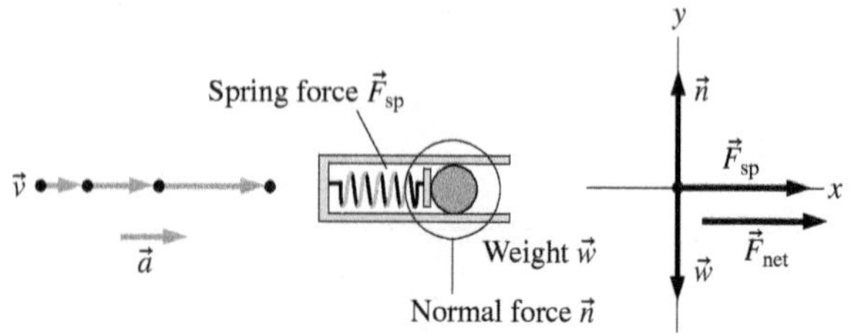

61. $\vec{C} = (-4.8\text{ m}, -2.4\text{ m})$
63. $T_l = 4.1 \times 10^2\text{ N}, T_r = 5.0 \times 10^2\text{ N}$
65. $T_1 = 510\text{ N}$
67. a. $T = 490\text{ N}$ b. $T = 740\text{ N}$
69. $v_{x,f} = 1.2\text{ m/s}$
71. $F_{x,net} = 1.7 \times 10^2\text{ N}$
73. 90 m/s^2
75. C
77. C

Chapter 5

Answers to odd-numbered multiple-choice questions

15. a. B b. D
17. B
19. B
21. A
23. A

Answers to odd-numbered problems

1. a. 539 N b. 89.1 N
3. a. 780 N b. 1600 N
5. a. 780 N b. 1100 N
7. $f_s = 7.8 \times 10^2\text{ N}, n = 4.5 \times 10^2\text{ N}$
9. a. The block will slide the same distance *d*. b. The block will slide a distance of 4*d*.
11. $f_s = 4800\text{ N}$
13. a.

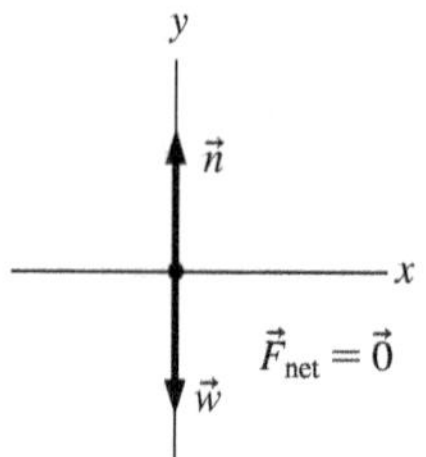

b.

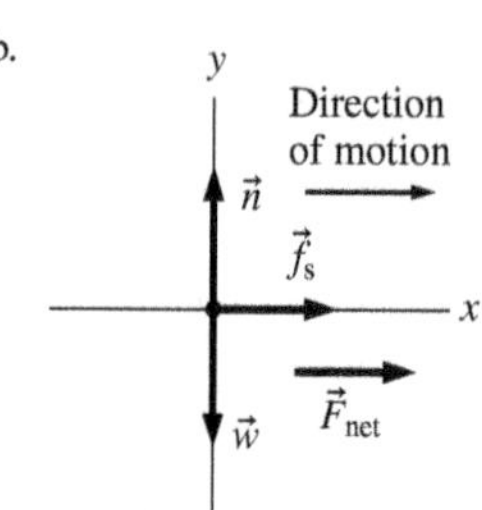

c. $a_{max} = 4.9\text{ m/s}^2$ d. $a = 2.9\text{ m/s}^2$
15. a. 740 N b. 630 N
17. $T = 2.4 \times 10^4\text{ N}$
19. 460 s or about 7.7 min
21. $D = 1.1 \times 10^2\text{ N}$
23. $8.7\ \mu g$
25. a. 0.13 m b. 0.08 m
27. a. $k = 390\text{ N/m}$ b. $y' = -18\text{ cm}$
29. a. $\pi/2$ b. 0 c. $-\frac{2}{3}\pi$ or $4\pi/3$
31. $1.7 \times 10^{-3}\text{ rad/s}$
33. 42 m
35. 5.7 N
37. a. $F_{1\text{ on }2} = F_{2\text{ on }1} = 6.7 \times 10^{-9}\text{ N}$ b. The ratio is 6.8×10^{-9}.
39. 3.9 m/s^2
41. $F_{x,net}(t = 1\text{ s}) = 8\text{ N}, F_{x,net}(t = 4\text{ s}) = 0\text{ N}, F_{x,net}(t = 7\text{ s}) = -12\text{ N}$
43. 2.0 s
45. $1.1 \times 10^3\text{ N}$
47. a. 6800 N b. $1.4 \times 10^6\text{ N}$ c. The force in part a is 11 times the passenger's weight. The force in part b is 2300 times the passenger's weight.
49. a. 1.3 m/s^2 b. 2.0 m/s^2
51. 110 m/s
53. 1900 N or 4.4 times the impala's actual weight
55. The box will stay at rest.
57. $1.0 \times 10^{-3}\text{ m}$
59. 190,000 N
61. 69 g
63. a. 22 m/s b. 1.9 kN, 79 mg c. 2.5 m/s
65. $0.32\ \mu\text{m}$
67. a. 9.8 m/s^2 upward b. 9.8 m/s^2 downward c. 20 m/s^2 upward d. 2.5 m/s e. 0.31 m
69. 0.98 kg
71. $T_1 = 17\text{ N}, T_2 = 27\text{ N}$
73. 99 m
75. 4.08 cm
77. $2.5 \times 10^{-2}\text{ m}$
79. 2400 m
81. 12 cm from the 20 kg sphere
83. B
85. B
87. C

Chapter 6

Answers to odd-numbered multiple-choice questions

13. C
15. D
17. A
19. B
21. D

Answers to odd-numbered problems

1. $-0.20\ \text{N}\cdot\text{m}$
3. 28.3 N
5. 29 N
7. 50 N
9. $-2.1\ \text{N}\cdot\text{m}$
11. a. $32\ \text{N}\cdot\text{m}$ b. $23\ \text{N}\cdot\text{m}$
13. (8.0 cm, 5.0 cm)
15. $n_{\text{right}} = 470$ N and $n_{\text{left}} = 160$ N
17. 0.61
19. 1.6 m from its tail
21. 90 N
23. 1.5 m from the pivot
25. $\theta_{\text{c,empty}} = 35.3°$ and $\theta_{\text{c,full}} = 30.6°$
27. $\theta = 8.9°$
29. 15.0 cm
31. a. 2.5×10^2 N b. 23% the weight of the load
33. 1100 N
35. a. 340 N b. 8.6 times the weight of the arm
37. $1.9 \times 10^{10}\ \text{N/m}^2$
39. $3.7 \times 10^8\ \text{N/m}^2$
41. $3.1 \times 10^8\ \text{N/m}^2$
43. 2.0 kg
45. 7.8×10^{-5} or 0.0078%
47. The woman in high-heeled shoes
49. a. 3.1 mm b. 9.5×10^2 N
51. $-0.94\ \text{N}\cdot\text{m}$
53. 7.5 cm
55. a. $\rho_0 = 2M/(AL)$ b. $x_{\text{cg}} = 2L/3$
57. $F_1 = 750$ N, $F_2 = 1013$ N
59. 350 N
61. a. 2300 N b. 1.0 m
63. a. 0.52 mm b. 0.11 N
65. a. 40 N downward b. 94 N horizontally from the head c. 23°
67. a. 210 N b. 84 N
69. B
71. C
73. C
75. C

Chapter 7

Answers to odd-numbered multiple-choice questions

13. C
15. E
17. B
19. B
21. A

Answers to odd-numbered problems

1. 3.9 m/s
3. a. 0.56 rev/s b. 1.8 s
5. a. 2.3 rotations/s b. 31 m/s c. $450\ \text{m/s}^2$
7. a. 76 m/s b. $100\ \text{m/s}^2$
9. 37 Hz or 2.2×10^3 rpm
11. $0.43\ \text{m/s}^2$
13. a. 1200 N b. 1.8
15. 0.73
17. a. $6 \times 10^3\ \text{m/s}^2$ b. 3.1×10^2 N upward
19. 13 m/s
21. 3.6 N
23. a. $1.8 \times 10^4\ \text{m/s}^2$ b. $4.4 \times 10^3\ \text{m/s}^2$
25. 78%
27. 32 rotations
29. a. 4.7 rad/s b. 1.3 s
31. 3.0 rad
33. 11 m/s
35. $\Delta\theta = 960$ revolutions
37. $7.2 \times 10^{-7}\ \text{kg}\cdot\text{m}^2$
39. 1.8 kg
41. $8.0\ \text{N}\cdot\text{m}$
43. $0.047\ \text{N}\cdot\text{m}$
45. $0.11\ \text{N}\cdot\text{m}$
47. a. 24 m/s relative to the ground b. 3.3×10^2 rpm
49. 35 rotations
51. a. 2π rad/s b. 0 rad/s c. -2π rad/s
53. 1680 km/h or 1000 mph from east to west
55. 19,000 rev
57. 1940 rpm
59. About 12%
61. 0.67
63. 7.4 m/s
65. 69 rpm
67. 54 N
69. $T_1 = 14$ N, $T_2 = 8.3$ N
71. $0.28\ \text{N}\cdot\text{m}$
73. a. $I = \dfrac{M}{2}(R^2 + r^2)$ b. $0.13\ \text{N}\cdot\text{m}$
75. 180 s
77. 0.020 mN, about 170 times the weight of the ant
79. 0.50 s
81. D
83. B
85. A

Chapter 8

Answers to odd-numbered multiple-choice questions

15. C
17. A
19. D
21. C

Answers to odd-numbered problems

1. 75 m/s
3. $J = 2.6\ \text{kg}\cdot\text{m/s}$
5. $J_x = 4.0\ \text{N}\cdot\text{s}$
7. 800 N to the left
9. 0.0 m/s
11. 6.0 m/s to the right
13. $7.5 \times 10^{-10}\ \text{kg}\cdot\text{m/s}$
15. 4.3 m/s
17. 1.4 m/s
19. v_0
21. 8.3×10^{-2} m/s
23. 2.0 mph
25. 8.9 cm/s
27. 0.71 m/s
29. 0.86 m/s and 2.9 m/s
31. 0.20 m/s
33. 3.6 m/s
35. $3v_0$
37. $0.025\ \text{kg}\cdot\text{m}^2\text{/s}$

39. 3.5 times
41. a. 6.4 m/s b. $F_{avg} = 3.6 \times 10^2$ N, which is $612(F_G)_B$.
43. 2.8 m/s
45. $F_x = (12\ \text{kg} \cdot \text{m/s}^3)t$
47. 600 N
49. 5.8 kN
51. 0.50 m/s
53. 31.5 m/s opposite the motion of the squid
55. a. $v_{bullet} = \frac{m+M}{m}\sqrt{2\mu_k g d}$ b. 4.4×10^2 m/s
57. 0.021 m/s
59. a. 8.0 m/s b. $100g$
61. 1.8 km/s
63. 3.0 m/s
65. $(v_{fx})_N = 3.66 \times 10^5$ m/s
67. 400 rpm
69. 18 rev/s, clockwise
71. B
73. C

Chapter 9

Answers to odd-numbered multiple-choice questions

17. A
19. D
21. B
23. B
25. D

Answers to odd-numbered problems

1. 50 mL
3. 16 L
5. 2.4×10^3 kg
7. 1.1×10^3 atm
9. 1.2×10^5 Pa
11. 36 m
13. 8.4 cm
15. Ethyl alcohol
17. 4.3×10^2 N
19. 750 kg/m^3
21. 0.12 N
23. 2.49×10^3 kg/m^3
25. 2.1×10^{-1} Pa · s
27. 3.7×10^{-2} m
29. 6.2×10^{-4} N
31. $h_i - h_f = 1.0$ mm
33. 3.2 m/s
35. 6.8 min
37. $v_1 = 4.0$ m/s, $v_2 = 1.0$ m/s
39. 12 kPa
41. a. 3.1 mm/s b. 94 mm/s
43. 9.6 kPa/m
45. 5.4×10^4 Pa
47. 3.1 kPa or 23 mm Hg
49. a. 1.7×10^{-3} m/s b. 4.7×10^3 Pa
51. 17 Pa
53. 91 mm Hg
55. 8.01%
59. 5%
61. 0.65 m/s
63. a. 42 mN b. 5.9×10^3 balloons
65. 5.2 cm
67. a. 3.8° b. 150 mg
69. 6.0 cm
71. 28 cm
73. a. Lower b. 0.83 kPa c. 7.5×10^4 N, blow out
75. a. 4.9×10^{-4} m^3/s b. 0.25 m/s c. 0.70 m^2
77. The pressure drops 1.8 times as much in the left artery as it does in the right.
79. a. 4.4×10^{-5} L/h or 1.2×10^{-11} m^3/s b. 1.6 mm/s c. 5.028×10^5 Pa d. −9.0 atm
81. a. 0.25 m/s b. 0.25 L/min
83. D
85. B
87. A

Part I

Answers to odd-numbered multiple-choice questions

1. A
3. A
5. C
7. B
9. D
11. A
13. B
15. D
17. A
19. D
21. C

Answers to odd-numbered problems

23. 1.4 s
25. a. 0.26 s b. 9.6 kN

Chapter 10

Answers to odd-numbered multiple-choice questions

15. B
17. A
19. A
21. C

Answers to odd-numbered problems

1. 4.0 m/s
3. The cheetah's kinetic energy is 9.0 times the human's kinetic energy.
5. 67 m/s
7. a. −29 J b. 29 J
9. a. 5.3 J b. 0 J
11. $W_1 = 1.7$ kJ, $W_2 = 1.1$ kJ, $W_3 = -2.0$ kJ
13. $W_{AB} = -4$ J, $W_{BC} = 0$ J, $W_{CD} = 2$ J, $W_{DE} = 2$ J
15. a. 190 J b. −26 kJ c. 26 kJ
17. $v_f(x = 1\ \text{m}) = 8.0$ m/s, $v_f(x = 2\ \text{m}) = 10$ m/s, $v_f(x = 3\ \text{m}) = 11$ m/s
19. 4.1 N/m
21. 32 J
23. a. 3×10^7 N/m^2 b. 24 J
25. 1360 m/s
27. 71%
29. 125 J
31. a. 780 J b. 780 J
33. 14 kJ
35. a. 1.8×10^2 J b. 59 W
37. 42 m^2
39. a. 1.9×10^2 N b. 50 kW, 1.0 kW, 1.5 kW
41. a. $P_{av} = 1.3$ kW b. 18 W/kg
43. 1.5 kW
45. a. 0.24 kJ b. 0.49 W/kg
47. 4.5 MW
49. a. -9.8×10^4 J b. 1.1×10^5 J c. 1.0×10^4 J
51. −28 J
53. a. 2.9 J b. 3.6%
55. $W = kqQd\left(\frac{x_2^2 - x_1^2}{x_1^2 x_2^2}\right)$

57. a. 0.68 kJ b. −0.13 kJ c. 4.9 m/s d. 1.2 m
59. a. 8×10^{21} J b. 7×10^{-8} m/s c. 1×10^{10} J d. 8×10^{21} J
61. 460 W
63. a. 95 W b. 3.8×10^2 W c. 3.2×10^2 Cal
67. $P(t = 0\text{ s}) = 0$ W, $P(t = 0.5\text{ s}) = 1.4$ kW, $P(t = 1.0\text{ s}) = 1.7$ kW, $P(t = 1.5\text{ s}) = 1.5$ kW, $P(t = 2.0\text{ s}) = 1.2$ kW
69. a. 1.1 kW b. 570 W
71. B
73. C

Chapter 11

Answers to odd-numbered multiple-choice questions

17. C
19. C
21. D
23. D
25. C

Answers to odd-numbered problems

1. a. 5 J b. 7 J
3. a. 6.8×10^5 J b. 46 m
5. 6.2 m
7. a. −0.25 kJ b. 2.6×10^5 kg
9. 63 kJ
11. 7.7 m/s
13. 4.2 m/s
15. 0.632 m
17. 18 J
19. 4.0 cm
21. 69 m/s
23. $\sqrt{2}v_0$
25. a. 0.72 J b. 0.67 or 67%
27. $v_B = 3.5$ m/s, $v_C = 2.8$ m/s, $v_D = 4.5$ m/s
29. a. 7.7 m/s b. 10.0 m/s
31. 6.3 m/s
33. 7×10^{-19} J
35. $F_x = 100$ N at $x = 5$ cm, $F_x = 0$ N at $x = 15$ cm and $F_x = -50$ N at $x = 25$ cm
37.

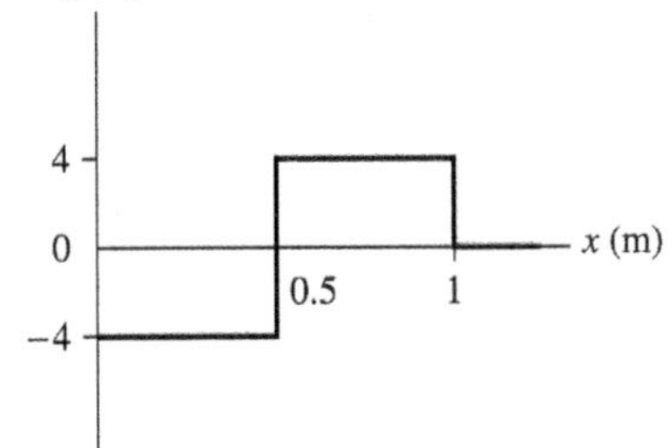

39. 0.16 N
41. K_f is 2 J.
43. 490 Cal
45. Both Jessie and Jaime use 1.4 MJ of energy.
47. a. 3.7×10^2 W b. 5.5×10^2 W
49. 0.20
51. 2.3 m/s
53. a. 2.0 mJ b. 50%
55. 28 m
57. 2.0×10^5 N/m
59. 1.8 m
61. a. 0.51 m b. 0.38 m
63. 41 N/m
65. a. 1.7 m/s b. 1/3
67. a. $x = \dfrac{2A}{B}$ b. Stable
69. $U_x = (-x^3 + \frac{5}{2}x^2 + C)$ J
71. a. 4×10^{24} molecules/hr
 b. 10^{11} molecules/hr
73. a. 1.2×10^3 Cal/mi b. 12 times
75. 260 flights
77. 1200 km
79. B
81. C
83. C
85. A

Chapter 12

Answers to odd-numbered multiple-choice questions

17. A
19. B
21. D
23. B
25. B

Answers to odd-numbered problems

1. 330 K
3. a. $T_Z = 171°$Z b. 944 K
5. 10 J
7. 19 mm
9. 13,100 kg/m^3
11. a. 36°C b. 3000 J
13. 0.98 g
15. 0.27°C
17. 8.7 hours
19. 79 s
21. 14 K
23. 28°C
25. 73.5°C
27. 6 stones
29. 5.8 s
31. a. 150 kW b. 5.8 kW
33. 3800 W
35. a. 28.3 W b. Warm
37. a. 0.050 m^3 b. 1.3 atm
39. a. $1.27V_0$ b. $2.0V_0$
41. a. 1.6 mol b. 9.4×10^{23} molecules c. 7.5×10^{25} m^{-3} d. 200 kPa
43. a. 0.73 atm b. 0.52 atm
45. a. 105.6 kPa b. 42 cm
47. 11.4 psi
49. a. 240 J b. 330 J
51. 60 J
53. a. 31 J b. $\Delta T = 60$°C
55. a. 3.1 atm b. 9.7 L
57. −890.5 kJ/mol
59. a. 36.6 J/mol · K b. 1.41
61. 437°C
63. 4.5×10^{-3} m
65. 16 kJ
67. Aluminum
69. 61 g
71. 250 s or 4.2 min
73. 62%
75. 1.0 mJ
77. 3.8 kg/h
79. 35 psi
81. a. 0.41 L or 4.1×10^{-4} m^3 b. 1.4×10^2 J
83. a. 4.3 L, 610°C b. 3.0 kJ c. 1.0 atm d. −2.3 kJ
85. A
87. B

Chapter 13

Answers to odd-numbered multiple-choice questions

17. C
19. B
21. C

Answers to odd-numbered problems

1. $2.69 \times 10^{25}\ m^{-3}$
3. 0.023 Pa
5. a. $\lambda_2 = \lambda_1 = 300$ nm b. $\lambda \propto V$, $\lambda_2 = 2\lambda_1 = 2(300\text{ nm}) = 600$ nm
7. a. 20 m/s b. 20.2 m/s
9. 283 m/s
11. 2.5 mK
13. 2.12
15. 313°C
17. a. 289 K b. 200 kPa
19. a. 310 nm b. 250 m/s c. 2.1×10^{-22} J
21. a. 1.24×10^{-19} J b. 1.22×10^4 m/s
23. 490 J
25. a. 0.080 K b. 0.048 K c. 0.040 K
27. a. 3.80×10^5 J b. 2.25×10^{-9} m c. $\Delta E_{th} = 0$ J
29. a. Gas B
    b. $E_{Af} = 5200$ J, $E_{Bf} = 7800$ J
31. a. 2.8×10^{-21} J b. 92 K
33. a. $\dfrac{N_5}{N_3} = 0.0080$ b. $\dfrac{N_5}{N_3} = 0.62$
35. 2.3
37. $1.9 \times 10^6\ M^{-1}s^{-1}$
39. 51 kJ/mol
41. 20 hours
43. a. 1.7×10^{-7} s b. 0.12 m/s
45. 1.1×10^{11} molecules, 1.8×10^{-7} μmol/s
47. $p_{N_2} = 13$ kPa, $p_{He} = 1.4 \times 10^5$ Pa
49. a. 1.4×10^5 K b. 9.8×10^3 K c. $m < 9.4$ u d. N_2 cannot, H_2 can
51. 29 J/mol · K
53. a. $v_{\text{most probable}} = \sqrt{\dfrac{2k_BT}{m}}$ b. $\epsilon_{\text{most prob}} = k_BT$
55. a. E_{open} b. 9.3×10^{-21} J
57. 8.5×10^{-20} J
59. 2.2 y
61. a. 15 ms b. 2.1×10^9 molecules
63. C
65. C

Chapter 14

Answers to odd-numbered multiple-choice questions

11. C
13. A
15. B
17. A
19. C

Answers to odd-numbered problems

1. a. 7 macrostates b. The most likely macrostate is 5, with a multiplicity of 4.
3. a. 6 b. 1 c. 5
5. a. $S_1 = 6.9k_B$, $S_2 = 13.8k_B$, $S_3 = 20.7k_B$ b. $S \sim N^1$
7. 0.48 J/K
9. 1.8 kJ/K
11. 0.44 J/K
13. 1.2 J/K
15. 17 J/K
17. a. $9.1R \approx 9R$ b. $\Delta S_{water}/\Delta S_{He} = 3$
19. a. 40 J/K b. −29 J/K c. 11 J/K
21. a. $\Delta H = 0$ b. $\Delta S > 0$ c. $\Delta G = \Delta H - T\Delta S < 0$
23. $T < 462$ K
25. a. −2.90 kJ/mol b. Yes
27. −155 J/K
29. 6.29×10^3 mol
31. a. −8 kJ b. −20 kJ c. −2 kJ
33. 0.80 M
35. 59 atm
37. 9.2 g/L
39. a. 20 m b. 3.4×10^{13} particles/s
41. a. $S_A = 24.5k_B$, $S_B = 6.40k_B$, $S_{tot} = 30.9k_B$ b. $\Delta S_A = (-2.89)k_B$, $\Delta S_B = (3.14)k_B$, $\Delta S_{tot} = (0.25)k_B$ c. With $E_A = E_B = 5$, we find $S_{tot} = 31.3k_B$.
43. 4.0×10^2 K
45. 42 J/K for this one reaction, or 42 J/mol · K
47. 6.5×10^{-3} J/mol · K
49. 22 J/K
51. a. $\Delta G/n - 203.2$ kJ/mol b. Yes, $\Delta G < 0$, so this reaction can occur spontaneously.
53. 21 atm
55. 7.3 atm
57. C

Part II

Answers to odd-numbered multiple-choice questions

1. C
3. A
5. C
7. B
9. C
11. A
13. C
15. D
17. B

Answers to odd-numbered problems

19. 0.96 hp
21. a. $H = \dfrac{k_BT}{mg} = 8.4$ km
    b. $A\int_0^\infty \rho(z)dz = \rho_0 A \int_0^\infty e^{-z/H}dz = -\rho_0 AH(0 - e^{-0/H}) = \rho_0 AH$
    c. $M = 5.2 \times 10^{18}$ kg

Chapter 15

Answers to odd-numbered multiple-choice questions

19. a. B b. C c. B d. B e. B
21. B
23. C

Answers to odd-numbered problems

1. 2.27 ms
3. a. 3.3 s b. 0.30 Hz c. 1.90 rad/s d. 0.25 m e. 0.48 m/s
5. 0.41 s
7. 1.6 mm
9. a. 0.60 m/s b. 0.083 Hz c. 1.1 m
11. a. $\phi_0 = -\frac{1}{3}\pi$ rad b. 13.6 cm/s c. 15.7 cm/s

13.

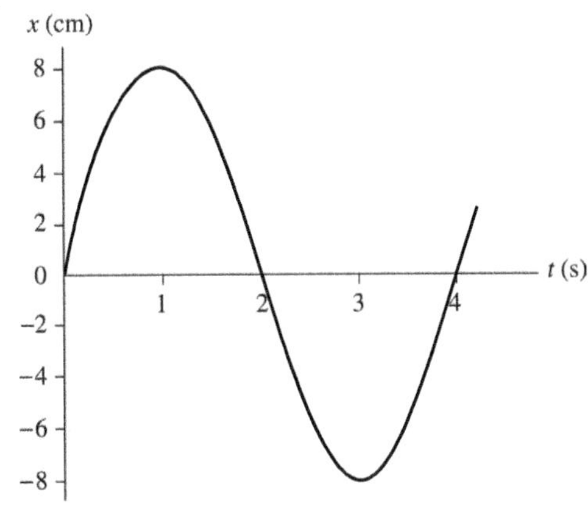

15. $x(t) = (8.0 \text{ cm}) \cos[(\pi \text{ rad/s})t + \pi \text{ rad}]$
17. a. 2.0 cm b. 0.63 s c. 5.0 N/m d. $\phi_0 = -\frac{1}{4}\pi$ rad
e. $x_0 = (2.0 \text{ cm}) \cos(-\frac{1}{4}\pi) = 1.41$ cm and $v_{0x} = -(20 \text{ cm/s}) \sin(-\frac{1}{4}\pi) = 14.1$ cm/s
f. 20 cm/s g. 1.00 mJ h. 1.46 cm/s
19. 5.48 N/m
21. a. 10 cm b. 35 cm/s
23. 4 N/m
25. a. 0.26 m/s^2 b. 2.7% of g
27. a. 2.6×10^{-2} m/s b. 2.0×10^2 m/s^2
29. a. 25 N/m b. 0.90 s c. 0.70 m/s
31. 3.5 Hz
33. 990 m/s^2 or 100g
35. a. $T = T_0 = 4.0$ s b. 5.7 s c. 2.8 s d. $T = 4.0$ s
37. 3.67 m/s^2
39. 54 cm
41. a. 1.2 s b. 97 steps/min
43. 10 oscillations
45. 32 min, additional 22 min
47. a. 5.0 kHz b. 4.8×10^{-3} s
49. 0.079 N/m
51. Speed up
53. a. 1.4 J b. 1.1×10^2 N c. 63%
55. a. 0.42 mHz b. $\phi_0 = \pi/2$ c. 0.010 mM/min
57. 1.02 m/s
59. a. 55 kg b. 0.73 m/s
61. 0.14 s
63. a. 1.1 Hz b. 23 cm c. −4.1 cm
65. 66 rpm
67. 5.0×10^2 m/s
69. 2.8 Hz
71. 0.65 m/s
73. a. 0.25 s b. 0.23 s
75. 2.0 m/s
77. a. 1.1 s b. $\tau = 4.0$ s
79. 21 oscillations
81. D
83. A
85. C
87. B

Chapter 16

Answers to odd-numbered multiple-choice questions

13. A
15. C
17. C
19. D

Answers to odd-numbered problems

1. 141 m/s
3. 2.0 m
5.

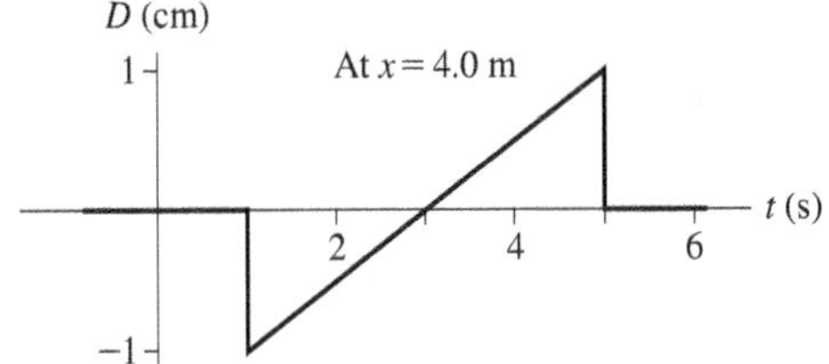

7.

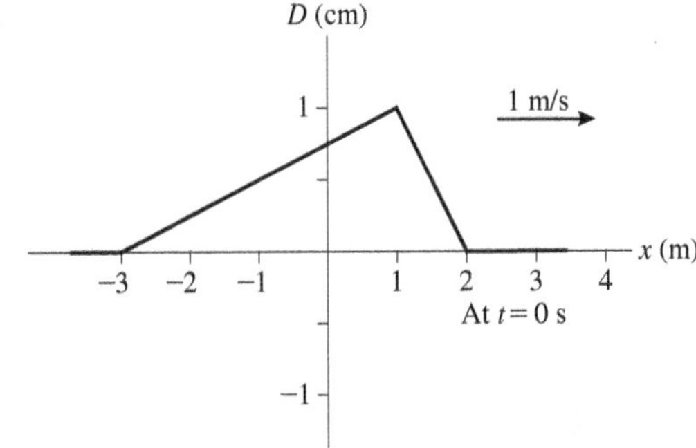

9.

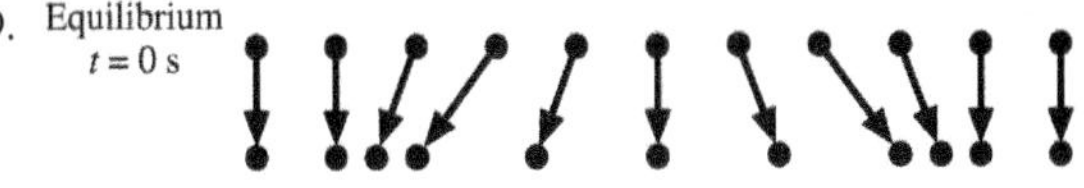

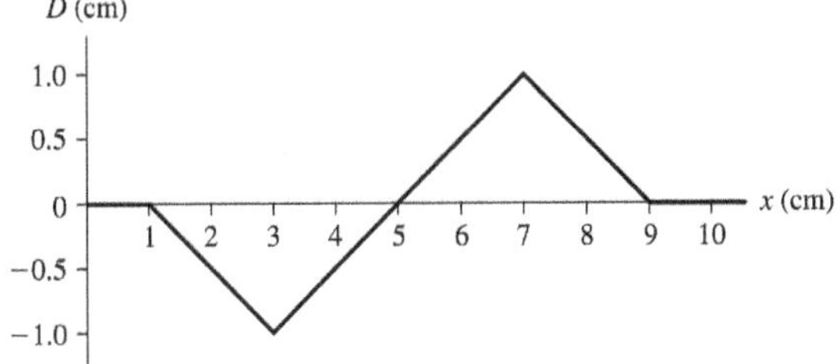

11. a. 3.1 rad/m b. 9.5 m/s
13. a. 11 Hz b. 1.1 m c. 13 m/s
15. $A = 4.0$ cm, $\lambda = 12$ m, $f = 2.0$ Hz, and $\phi_0 = \frac{\pi}{6}$
17. 0.076 s
19. 2.8×10^2 m/s
21. a. 1.5 GHz b. 990 nm
23. a. 2.96 m b. 116 Hz
25. 1500 m/s
27. a. 1.5×10^{-11} s b. 3.4 mm
29. a. $f_{water} = 440$ Hz b. 3.4 m
31. At $r_2 = 3.5$ m, $\phi_2 = \frac{\pi}{2}$ rad and at $r_2 = 4.5$ m, $\phi = \frac{3}{2}\pi$ rad.
33. a. The source is at $x = -2.0$ m. b. $y = 4.6$ m
35. 6.2×10^{-9} J
37. 11 W
39. 36 kW/m^2
41. a. 65 dB b. 105 dB
43. 50 m
45. 1.4 mW/m^2
47. 40 m
49. a. 650 Hz b. 560 Hz
51. 15 m/s
53. −8.0 kHz
55. 1.0 m/s

57. a. 10 m/s b. $\frac{\pi}{6}$ rad or $\frac{5\pi}{6}$ rad

c. $D(x, t) = (1.0 \text{ mm})\sin\left(\frac{2\pi x}{(2.0 \text{ m/s})} + 2\pi(5.0 \text{ Hz})t + \frac{5\pi}{6}\right)$

59. 17.8 kg
61. 790 m
63. 4.5 cm
65. 8×10^8 Hz
67. 130 m/s
69. a. 12.6 N b. 2.00 cm c. 12.8 m/s
71. 1.5 km
73. 2.0×10^{-5} W/m^2
75. 4.7 m
77. 620 Hz, 580 Hz
79. 2.2 cm/s
81. 1.3 m
83. 29 s
85. B
87. C

Chapter 17

Answers to odd-numbered multiple-choice questions

15. A
17. A
19. B
21. D

Answers to odd-numbered problems

1. The displacement at $x = 5.0$ m is always zero, for all the times.
3. 4.0 s
5. 5.9 s
7. 30 Hz
9. a. $\lambda_1 = 4.8$ m, $\lambda_2 = 2.4$ m, $\lambda_3 = 1.6$ m b. 75 Hz
11. $m_A/m_E = 2.24$
13. 0.031 m
15. 4.8 cm from the end
17. a. 8.6 m b. 4.3 m
19. 400 m/s
21. 8.8 Hz
23. 292 Hz
25. a. $f_1 = 2.6 \times 10^2$ Hz b. 1.3 kHz
27. 1.7 m
29. 13.0 cm
31. a. 30 cm b. The distance is closer to 15 cm. c. No
33. a. Open-closed tube
    b. 4.9 cm
35. a. 25 cm b. 25 cm
37. 3.0 m away from the worker
39. a. $\lambda = 80$ cm b. $\Delta\phi_0 = \frac{3\pi}{4}$ rad c. $0.77a$
41. a. Out of phase
    b.

	r_1	r_2	Δr	C/D
P	2λ	3λ	λ	D
Q	3λ	$\frac{3}{2}\lambda$	$\frac{3}{2}\lambda$	C
R	$\frac{5}{2}\lambda$	3λ	$\frac{1}{2}\lambda$	C

43. 8.8×10^2 Hz
45. Maximally destructive
47. 527 Hz
49. 1.26 cm
51. $f_1 = 49$ Hz and $f_2 = 51$ Hz
53. 1.2 m
55. 180 Hz
57. 28.4 cm
59. a. $\lambda = 800$ km b. 2.3×10^{-5} Hz c. 18 m/s
61. 90 N/m
63. 328 m/s
65. 58.3 cm
67. 26 cm, 56 cm, 85 cm
69. a. $\lambda_1 = 20.0$ m, $\lambda_2 = 10.0$ m, $\lambda_3 = 6.67$ m
    b. $v_1 = 5.6$ m/s, $v_2 = 4.0$ m/s, $v_3 = 3.2$ m/s
71. 3.0 m, 5.6 m, 18 m
73. a. 2 Hz b. 2 Hz c. 662 Hz
75. 970 Hz
77. B
79. A

Part III

Answers to odd-numbered multiple-choice questions

1. $v_{light} > v_{earthquake} > v_{sound} > v_{tsunami}$
3. D
5. B
7. B
9. C
11. A
13. A
15. C
17. C
19. C
21. C

Answers to odd-numbered problem

23. 500

Chapter 18

Answers to odd-numbered multiple-choice questions

13. B
15. C
17. C
19. B
21. A

Answers to odd-numbered problems

1. 217 nm
3. 906 nm
5. a. 1.33 b. 87 nm
7. 500 nm
9. 686 nm and 412 nm
11. 0.020 rad or 1.1°; 2.4 cm
13. 0.050 rad
15. 1.2 mm
17. 550 nm
19. 7 bright fringes (3 orders on either side of 1 central maximum)
21. 530 lines per millimeter
23. a. 1.3 m
    b. 7 bright fringes (3 orders on either side of a central maximum)
25. $d_{min} = 2.6 \times 10^{-6}$ m, $d_{max} = 3.4 \times 10^{-6}$ m
27. 0.10 mm
29. 94 μm
31. a. $\left(\frac{a}{\lambda}\right)_{30°} = 2$, b. $\left(\frac{a}{\lambda}\right)_{60°} = 1.15$, c. $\left(\frac{a}{\lambda}\right)_{90°} = 1$
33. No
35. 0.01525 rad = 0.87°
37. 78 cm
39. 180 μm
41. 4
43. 64°
45. 74°, 57°, 34°

47. 1.43
49. 1.4
51. 500 nm
53. 0.286°
55. 0.145 m
57. 1.0 nm
59. 670 lines/mm
61. 16°
63. 500 nm, 550 nm, and 650 nm
65. 0.25 m
67. a. 22.3° b. 16.6°
69. 20 mm
71. a. Double slit b. 0.15 mm
73. 0.1 mm
75. 0.15 nm
77. C

Chapter 19

Answers to odd-numbered multiple-choice questions

15. A
17. C
19. D
21. E
23. A

Answers to odd-numbered problems

1. a. 3.3 ns b. 0.75 m, 0.67 m, 0.46 m
3. 8.00 cm
5. 30°
7. 8 times
9. 1.0 m
11. 55°
13. 1.4
15. 1.5
17. 49° from vertical
19. 76.7°
21. 26.5°
23. 3.2 cm
25. 23 cm
27. 1.52
29. $s' = 30$ cm, real and inverted
31. $s' = -6.7$ cm, virtual and upright
33. $h' = 2.0$ cm
35. $s' = -8.6$ cm, upright; $h' = 1.1$ cm
37. 30 cm, 0.50 cm
39. 40 cm
41. a. 10 cm b. 9.3×10^{-4} m
    c. 1.7×10^{6} W/m^2
43. $s' = 56$ cm, inverted;. $h' = 3.7$ cm
45. $s' = -16$ cm, upright;. $h' = 1.1$ cm
47. a. 180 cm b. 4.7 cm
49. 0.28 mm
51. 1 mm
53. 1.2 m
55. 82.8°
57. 4.7 m
59. a. 87 b. 65 cm c. 43 cm
61. 26°
63. 42°
65. 22.3°
67. 36 cm
69. a. $\dfrac{2nR}{(n-1)}$ b. 9.7 cm, 5.4 cm, 1.6 cm
71. Place 1: 160 cm, 0.5 cm (inverted); place 2: 40 cm, 8.0 cm (inverted)
73. 67 cm
75. 7.7 mm
77. A
79. B

Chapter 20

Answers to odd-numbered multiple-choice questions

13. C
15. C
17. C
19. C
21. D

Answers to odd-numbered problems

1. b. $s'_2 = 49$ cm, $h' = 4.6$ cm
3. b. $s'_2 = -30$ cm, $h' = 6.0$ cm
5. 3.0 mm
7. 6.3 mm
9. 6.0 mm
11. 3.0 mm
13. 60 cm
15. 1.2 cm
17. a. Myopia b. 100 cm
19. 6.3 cm
21. 2.6 cm
23. $s_{near} = 27$ cm, $s_{far} = 50$ cm
25. −2.5 D, 17 cm
27. 1.0 m
29. 67 cm
31. 4.2
33. 4.2 cm
35. −10 cm
37. 2.0 mm
39. 9.2 mm
41. a. 8.0 cm b. 1.2 cm
43. 15 km
45. a. 3.2 μm b. No
47. 0.85 cm
49. 4.9 mm
51. a. 400 nm to 500 nm b. 500 nm to 575 nm
53. a. $L = 14$ cm b. $h' = -1.7$ cm (inverted)
55. $s' = -30$ cm, $h' = -2.7$ cm, inverted
57. 26 cm
59. a. 30 D b. 3.0 D
61. 3.6 mm
63. a. 11 cm b. −160
65. a. 1.3 cm b. 1.2 cm c. 1.7
67. a. 10× objective with NA = 0.30 b. 7.8 mm
69. A
71. B

Part IV

Answers to odd-numbered multiple-choice questions

1. C
3. B
5. A
7. B
9. A
11. B
13. B
15. B
17. A

Answers to odd-numbered problems

19. 35 μm
21. 3.4 m

Chapter 21

Answers to odd-numbered multiple-choice questions

17. A
19. D
21. B
23. A
25. D

Answers to odd-numbered problems

1. a. Electrons removed b. 5.0×10^{10}
3. 1.1×10^8 electrons
5. a. Electrons transferred to the sphere b. 3.1×10^{10}
7. -9.6×10^7 C
9. 1.5×10^7 C/m^3
11. a. 0.90 N b. 0.90 m/s^2
13. a. 3.5×10^{-7} N b. 0.20 m/s^2
15. 10 nC and 15 nC
17. 6.6 m/s^2 downward
19. a. 0 N b. 2.7×10^{-4} N, left
21. $-\frac{4}{9}q$ at $x = \frac{1}{3}L$
23. 400 N/C, 20° above horizontal
25. 0.11 nC
27. 2.1×10^2 m
29. a. 2.3×10^{-6} b. 4.3×10^7N/C, upward
31. a. 2.2×10^4 N/C, 45° above the $+x$-axis
    b. 4.3×10^4 N/C, 90° above the $+x$-axis
    c. 4.3×10^4 N/C, 53° above the $+x$-axis
33. 2500 N/C, horizontal
35. $\vec{E}_{net}$ is 9800 N/C directed 24° below the $+x$-axis.
37. a. 36 N/C b. 18 N/C
39. 4.1 μC/m^2
41. 14 nC, −14 nC
43. 2.7×10^{11}
45. 7.2×10^{-15} N
47. a. 3.6×10^6 N/C b. 8.3×10^5m/s
49. 4.0×10^{-10} C
51. 1.8×10^{-21} N · m counterclockwise
53. 1.4×10^{-13} N · m
55. 100 nC
57. 8.9 kN/m
59. 1.8×10^{-4} N directed horizontally to the right, or 0° counterclockwise from the $+x$-axis
61. 1.0×10^{-3} N directed to the left, or 180° counterclockwise from the $+x$-axis
63. 5.2 m/s
65. 33 nC
67. $\vec{E}_{net} = 1.1 \times 10^5$ N/C, 19° cw from the $+x$-axis
69. 5.2 m/s^2 upward
71. a. −10 cm b. −10 cm
73. a. 3.6×10^3 N/C b. 1.0 cm
75. 0.74 GHz
77. 180 nC
79. 750 nC
81. D
83. A

Chapter 22

Answers to odd-numbered multiple-choice questions

17. B
19. B
21. B
23. A
25. C

Answers to odd-numbered problems

1. 2.7×10^6m/s
3. 25,000 m/s
5. -3.0×10^{-6} C
7. -4.7×10^{-6} J
9. 1.61×10^8 N/C
11. 2.4×10^{-19} J
13. a. 4.4×10^5 m/s b. 1.9×10^7 m/s
15. 9.8×10^7 m/s
17. a. Higher b. 3340 V
19. 3.3 kDa
23. a. 200 V b. 6.3×10^{-10} C
25. b. 2.96×10^5 m/s
27. a. $V_A = V_B = 1.80$ kV, $V_C = 900$ V
    b. $\Delta V_{AB} = 0$ V, $\Delta V_{BC} = -0.90$ kV
29. −1600 V
31. 60 V
33. a. 200 V b. 6.3×10^{-10} C
35. 8.8×10^{-7} C/m^2
37. a. 0.90 J b. 0.90 W
39. 3.0 C
41. a. Point A is at a higher potential than point B.
    b. The potential at A is 70 V higher than at B.
43. 0.050 V/m
45. $\vec{E} = 20$ kV/m, 45° below $-x$-axis
47. Point 1 = 3750 V/m, down; point 2 = 7500 V/m, up
49. −20 V
51. a. 27 V/m b. 3.7 V/m
53. Between electrodes 2 and 3
57. $x = 6.0$ cm, $x = 3.0$ cm
59. 1.5×10^6 m/s
61. -5.1×10^{-19} J
63. 222 nC
65. 0.49 m/s
67. 2.5 cm/s
69. 8.0×10^7 m/s
71. 47 V
73. a. 1000 V b. 1.4×10^{-9}C c. 7.0×10^6m/s
75. a. 560 V/m b. 45 V
77. 1000 V/m 127° cw from the $+x$-axis
79. 2 nC, 4 nC
81. 24°
83. D

Chapter 23

Answers to odd-numbered multiple-choice questions

13. C
15. B
17. C
19. C

Answers to odd-numbered problems

1. 0.20 nF
3. 3.0 V
5. 6.1×10^{-5} C, or 61 μC
7. a. -5.1×10^5 C b. 1.5 F
9. 1400 V
11. 5.3 kJ/kg
13. a. 1.1×10^{-7} J b. 0.71 J/m^3
15. 32 μF
17. 20 μF
19. 37 μF
21. a. 1.9 μF b. 22 μC
23. 1.1×10^{-5} C
25. 2.5 nC
27. a. 12.0 V b. 24 V
29. a. 38 nm b. 7.6 nm c. 0.47 nm

31. a. 1.6×10^{-11} N b. 8.2×10^{-3}
33. a. 1.4 nm b. 2.7×10^{25} ions/m^3 = 0.045 M
35. a. 2×10^{-14} J b. 2×10^5
37. 9.1×10^{-8} M
39. a. $\Delta V_i = 9.0$ V, $Q = \pm 3.2 \times 10^{-11}$ C
b. $\Delta V_f = 18$ V, $Q = \pm 3.2 \times 10^{-11}$ C
41. 13 nF
43. a. 4.1 MJ b. 4.1×10^8 W
45. 22×10^{-6} F
47. 20 V
49. 4 μC, 12 μC, 16 μC; $\Delta V_1 = \Delta V_2 = 1.0$ V, $\Delta V_3 = 8$ V
51. 20 μF
53. 2.4 J
55. a. 83 μC/m^2 b. 14 μC/m^2
57. a. 7.5×10^{-20} J b. 75%
59. a. -1.4×10^{-5} C/m^2 b. 6.7×10^{24} ions/m^3 = 11 mM
61. A
63. D

Chapter 24

Answers to odd-numbered multiple-choice questions

15. C
17. C
19. A
21. B
23. D

Answers to odd-numbered problems

1. 1.9×10^{22}
3. 9.4×10^{13} electrons
5. 3.8×10^{18} ions
7. 2.6 mA to the right
9. 5.6×10^{14} Cl ions
11. 5 A, −2 A
13. 3200 cells
15. 12 V
17. a. 0.70 C b. 7.8 J
19. 8.7 m
21. a. 4.9×10^{-6} S b. 49 μA
23. 4100 Ω
25. 6.4 MΩ
27. 9×10^4 Ω
29. 380 turns
31. 50 Ω
33. 0.50 A
35. 0.43 V/m
37. 9.60 Ω, 12.5 A
39. 0.50 A
41. a. 360 W b. 0.36 J c. 8.0×10^{-4} C
43. a. 11.6 A b. 10.4 Ω
45. a. 50 mA b. 50 mA
47. a. 9.0 Ω b. 19 A c. 3200 W
49. a. 6.3 V b. 0.45 A c. 50 Hz
51. 18 MΩ
53. 33 h
57. 5×10^{-3} Ω · m
59. 2×10^{-11} S
61. 1.1×10^4 ions
63. 20%
65. 0.42
67. 2.5 Ω · m
69. 2.6 m length, 0.24 mm diameter
71. 0.54 Ω
73. 5.6 min
75. 2.5×10^4 C
77. a. 0.36 J b. 0.80 mC
79. Incandescent is \$190; LED is \$34.
81. A
83. B

Chapter 25

Answers to odd-numbered multiple-choice questions

17. B
19. A
21. D
23. D
25. B

Answers to odd-numbered problems

1.

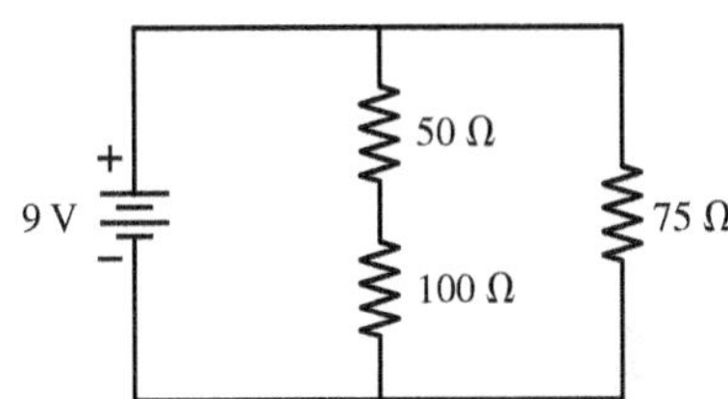

3. 5 A toward the junction (downward)
5. a. $\Delta V_{10} = 5.0$ V, $\Delta V_{20} = 10$ V
b.

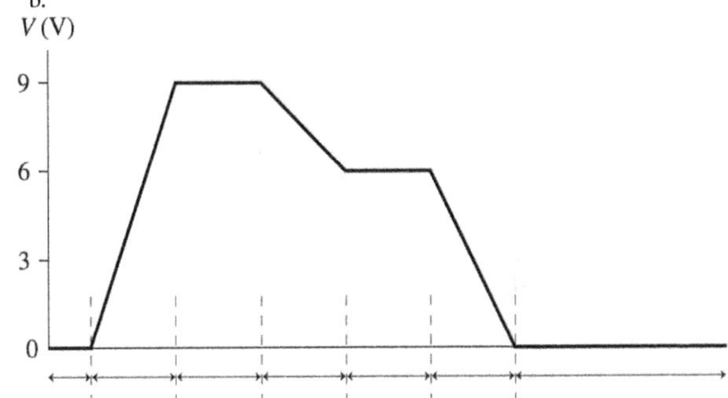

7.

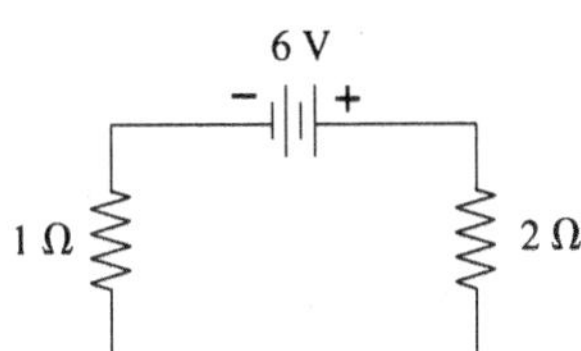

9. 240 Ω
11. 90 Ω
13. 170 Ω
15. 12 Ω
17. 54.5 Ω
19. 34 Ω
21. a. 2.0 A b. 5.0 A
23. $I_1 = 1.0$ A, $I_2 = 2.0$ A, and $E = 15.0$ V
25. Label the top resistor 1 and the other three from left to right 2, 3, and 4.

R	I(A)	ΔV(V)
R_1	1.5	7.5
R_2	0.50	2.5
R_3	0.50	2.5
R_4	0.50	2.5

27. a. For $R_{ph} = 0.56\ k\Omega$, $\Delta V_{1.0\ k\Omega} = 5.8\ V$.
For $R_{ph} = 4.0\ k\Omega$, $\Delta V_{1.0\ k\Omega} = 1.8\ V$.
For $R_{ph} = 20\ k\Omega$, $\Delta V_{1.0 k\Omega} = 0.43\ V$.
b. As the light increases, the voltmeter reading increases.
29. Yes
31. 34 V
33. 94 V
35. 2.0 ms
37. 6.9 ms
39. a. $Q_0 = 18\ \mu C$, $I_0 = 180$ mA
b. $Q = 11\ \mu C$, $I = 110$ mA
c. $Q = 2.4\ \mu C$, $I = 24$ mA
41. 18 μF
43. Before action potential: 7.8 MV/m; peak of depolarization: 4.4 MV/m
45. a. $2 \times 10^8\ \Omega$ b. 8 pF c. 2 ms
47. 14 s
49. a. 1.0 V b. 0.33 or 1/3 of the total
51. $P_{5\ \Omega} = 45$ W, $P_{20\ \Omega} = 20$ W
53. a. 0.33 A b. 0.46 A
55. $I_1 = 5.0$ A, $I_2 = 8.0$ A, $\epsilon = 14$ V
57. $R_1 = 140\ \Omega$, $R_2 = 72\ \Omega$
59. 0.11 s
61. a. 1.0 kΩ b. 3.0×10^{-7} F
63. a. 80 μC b. 0.23 ms
65. a. 0.52 W b. 47 J
67. 0.69 ms
69. a. 0.70 pF, $1.99 \times 10^7\ \Omega$ b. 14 μs
71. A
73. B

Chapter 26

Answers to odd-numbered multiple-choice questions

17. A
19. D
21. D
23. C

Answers to odd-numbered problems

1. 2.5 A, 500 A, 5000 A, 50,000–500,000 A
3. 0.040 μA
5. 20 A
7. $\vec{B}_2 = 4.0 \times 10^{-4}$ T, $+\hat{x}$ direction, $\vec{B}_3 = 2.0 \times 10^{-4}$ T, $+\hat{x}$ direction
9. a. 20 A b. 1.6×10^{-3} m
11. 3.8×10^{-7} A
13. 30 mA
15. 2.4 kA
17. a. $3.1 \times 10^{-4}\ A \cdot m^2$ b. 5.0×10^{-7} T
19. a. 6.2×10^{-5} T b. 6.3×10^8 A
21. $\vec{F} = (5.7 \times 10^{-13}$ N, $+\hat{y}$-direction)
23. $\vec{F} = (8.0 \times 10^{-13}$ N, $+\hat{y}$-direction)
25. a. 11 cm b. 10 m
27. 1.6×10^{-3} T
29. 0.48 m
31. $\vec{B} = (0.13$ T, out of page)
33. $\vec{F}_{on\ 1} = (2.5 \times 10^{-4}$N, up), $\vec{F}_{on\ 2} = \vec{0}$ N, $\vec{F}_{on\ 3} = (2.5 \times 10^{-4}$N, down)
35. 3.0 Ω
37. $7.5 \times 10^{-4}\ N \cdot m$
39. $1.7 \times 10^{-26}\ N \cdot m$
41. 0.12 T
43. a. 320 MHz b. 6.4 kHz
45. 6.45 MHz
47. 0.12 mT
49. $2.4 \times 10^{-8}\ \Omega \cdot m$
51. $I_1 = 1.0$ A, $I_2 = 1.8$ A
53. a. 0.81 A b. 82 μT
55. 3180 turns
57. 0.20 A
59. 1.0 cm
61. a. 6.0×10^{-21} N b. 3.8×10^{-2} V/m c. 1.1×10^{-4} V
63. 31°
65. a. 33 ns b. $K_i = 7.7 \times 10^{-18}$ J, $K_f = 7.7 \times 10^{-12}$ J c. 1.6 ms
67. a. 3.0×10^{-21} J b. 4.0×10^{-21} J c. No, yes
69. $\vec{B} = (8.6$ mT, down)
71. A
73. B
75. A
77. B

Chapter 27

Answers to odd-numbered multiple-choice questions

17. C
19. B
21. C
23. D

Answers to odd-numbered problems

1. 2.5 m/s
3. 64 μV
5. a. 1.1 A b. 0.24 N
7. 4.6 mV
9. 0.040 Wb
11. a. 6.3×10^{-5} Wb b. 6.3×10^{-5} Wb
13. a. 8.7×10^{-4} Wb b. Clockwise
15. a. Yes, left b. No
17. a. 3.9 mV, 39 mA, clockwise b. 3.9 mV, 39 mA, clockwise
c. 0 V, 0 A
19. 2.3 V
21. 2.3 T/s increasing
23. 0.97 A
25. a. $I(t) = 1.6 \times 10^{-3}(1 + t)$ A
b. $I(t = 5\ s) = 9.4 \times 10^{-3}$ A, $I(t = 10\ s) = 1.7 \times 10^{-2}$ A
27. Point 1: $4.8 \times 10^4\ m/s^2$, up; point 2: 0 m/s^2; point 3: $4.8 \times 10^4\ m/s^2$, down; point 4: $9.6 \times 10^4\ m/s^2$, down
29. 5.3 mT/s
31. 3.3×10^{-8} T
33. 0.019 V/m
35. 550 V/m
37. a. $2.2 \times 10^{-6}\ W/m^2$ b. $E_0 = 0.041$ V/m
39. a. 10 W/m^2 b. 7.5 W/m^2 c. 5.0 W/m^2 d. 2.5 W/m^2 e. 0 W/m^2
41. 66 mW
43. 30°
45. 7300 K
47. 26 times
49. a. 2 W b. 7.5 W/m^2 c. 0.17 mJ
51. 8.7×10^{-3} Wb
53. a. 2.17×10^{-2} kg b. 6.6×10^{-11} K
55. a. 0.20 A b. 4.0×10^{-3} N c. 11°C
57. 2.5×10^{-4} V
59. 1.5×10^{-5}A, counterclockwise
61. 0.015 V/m
63. a. 10.0 nm b. 3.00×10^{16} Hz c. 6.67×10^{-8} T
65. 2.1 m
67. 0.016 V/m
69. 9.4×10^7W
71. $\frac{1}{8}I_0$
73. a. 150 W/m^2 b. 100 W/m^2 c. 50 W/m^2
75. 0.3°
77. A
79. D
81. C

Part V

Answers to odd-numbered multiple-choice questions

1. B
3. C
5. B
7. A
9. B
11. C
13. C
15. A
17. B
19. B

Answers to odd-numbered problem

21. 20 A

Chapter 28

Answers to odd-numbered multiple-choice questions

17. B
19. A
21. D
23. C
25. A

Answers to odd-numbered problems

1. a. Sodium and potassium b. All the metals except gold
3. a. 290 nm b. 3.7×10^5 m/s
5. 216 nm
7. 6.25×10^{13} electrons/s
9. a. Aluminum b. 1.93 V
11. a. 4140 nm, infrared b. 414 nm, visible c. 41.4 nm, ultraviolet
13. a. 1.86×10^{-6} eV b. 2.76 eV c. 27.6 keV
15. 2.84 eV, 2.33 eV, 2.20 eV
17. 3.0 eV, 1.9 eV
19. 1.7%
21. 1×10^{19} photons/s
23. 1.7×10^3 photons/s
25. 33×10^{14} photons
27. 90 nm
29. 1.5 eV
31. a. 1.5×10^{-34} m b. 2.2×10^{-23} m/s
33. 6
35. 0.43 nm
37. $E_1 = 1.3 \times 10^{-12}$ J, $E_2 = 5.3 \times 10^{-12}$ J, $E_3 = 1.2 \times 10^{-11}$ J
39. $\lambda = 830$ nm for the $2 \rightarrow 1$ transition
    $\lambda = 500$ nm for the $3 \rightarrow 2$ transition
    $\lambda = 310$ nm for the $3 \rightarrow 1$ transition
41. 618 nm
43. $\lambda_{1\rightarrow 2} = 790$ nm, $\lambda_{1\rightarrow 3} = 300$ nm, $\lambda_{1\rightarrow 4} = 160$ nm
45. No, -18 m/s $\leq v_x \leq 18$ m/s
47. 0 to 2.5×10^7 m/s
49. 90 mV
51. 1.24 V
53. 4.1 eV
55. 1.6×10^{18} photons
57. 1.0×10^8 photons
59. 4.3×10^{-10} W
61. a. 250 eV, x-ray b. 0.060 eV
63. 620 eV
65. a. 1.3×10^8 m/s b. 5.5 μm
67. 0.87 nm
69. 0.43 nm
71. 1 and 3
73. A
75. D
77. A
79. C

Chapter 29

Answers to odd-numbered multiple-choice questions

17. B
19. B
21. C
23. C

Answers to odd-numbered problems

1. a. $n_{656} = 3$, $n_{486} = 4$, $n_{434} = 5$, $n_{410} = 6$ b. 397 nm
3. $\lambda_{4\rightarrow 3} = 1870$ nm, $\lambda_{5\rightarrow 3} = 1280$ nm, $\lambda_{6\rightarrow 3} = 1090$ nm; infrared
5. a. 3 electrons, 4 protons, 5 neutrons
   b. 6 electrons, 6 protons, 6 neutrons
   c. 4 electrons, 7 protons, 8 neutrons
7. a. ^{8}Li b. ^{14}C$^+$
9. a. 79 protons, 79 electrons, 118 neutrons
   b. 2.29×10^{17} kg/m^3
   c. 1.19×10^{13}
11. Aluminum
13. a. $\lambda = 830$ nm for the $2 \rightarrow 1$ transition
       $\lambda = 500$ nm for the $3 \rightarrow 2$ transition
       $\lambda = 310$ nm for the $3 \rightarrow 1$ transition
    b. 830 nm and 310 nm; the $3 \rightarrow 2$ transition is not seen in absorption.
15. a.

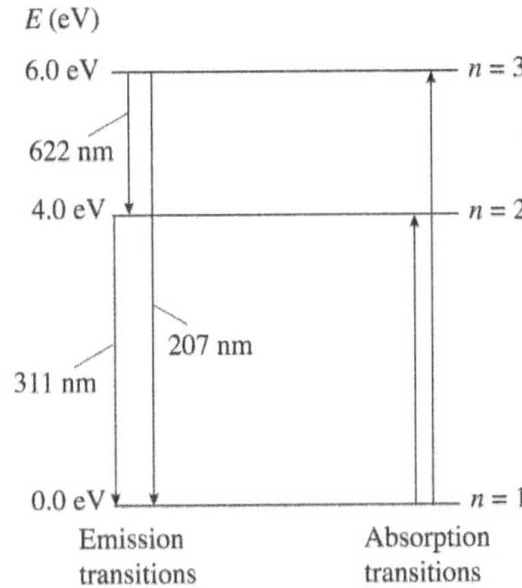

    b. $\lambda_{2\rightarrow 1} = 310$ nm, $\lambda_{3\rightarrow 1} = 210$ nm, $\lambda_{2\rightarrow 1} = 620$ nm
    c. $\lambda_{1\rightarrow 2} = 310$ nm, $\lambda_{1\rightarrow 3} = 210$ nm
17. 6.7×10^5 m/s minimum, 1.1×10^6 m/s maximum
19. For $n = 1$: $(1, 0, 0, \pm\frac{1}{2})$

For $n = 2$:
$(2, 0, 0, \pm\frac{1}{2})$, $(2, 1, -1, \pm\frac{1}{2})$, $(2, 1, 0, \pm\frac{1}{2})$, $(2, 1, 1, \pm\frac{1}{2})$

For $n = 3$:
$(3, 0, 0, \pm\frac{1}{2})$, $(3, 1, -1, \pm\frac{1}{2})$, $(3, 1, 0, \pm\frac{1}{2})$, $(3, 1, 1, \pm\frac{1}{2})$,
$(3, 2, 2, \pm\frac{1}{2})$, $(3, 2, 1, \pm\frac{1}{2})$, $(3, 2, 0, \pm\frac{1}{2})$, $(3, 2, -1, \pm\frac{1}{2})$,
$(3, 2, -2, \pm\frac{1}{2})$

21. $\sqrt{6}\hbar$
23. a. f b. -0.850 eV
25. Mg: $1s^22s^22p^63s^2$; Sr: $1s^22s^22p^63s^23p^64s^23d^{10}4p^65s^2$;
    Ba: $1s^22s^22p^63s^23p^64s^23d^{10}4p^65s^24d^{10}5p^66s^2$
27. 18 states
29. a. Copper b. Iridium
31. 658 nm, 122 nm, 103 nm
33. 330 nm, 590 nm, 816 nm, 1140 nm, 2180 nm, 8860 nm
35.

Transition	E_{photon} (eV)	λ (nm)
$3 \rightarrow 2$	1.89	656
$3 \rightarrow 1$	12.09	102
$2 \rightarrow 1$	10.20	122

37. a. $E_2 = 0.358$ eV, $E_3 = 0.706$ eV, $E_4 = 1.04$ eV b. 3.57 μm
39. a. $\lambda_{largest} = 500$ nm, $\lambda_{smallest} = 410$ nm
    b. $\lambda_{largest} = 690$ nm, $\lambda_{smallest} = 500$ nm
41. a. $E_{lower} = 2.30$ eV, $E_{upper} = 2.59$ eV b. 0.30 eV
43. $3.18 \times 10^{15}\,s^{-1}$
45. 9.1×10^{18} photons
47. a. 1.06 μm b. 1.9 W
49. 7.8×10^5 m/s
51. a. 2.3×10^7 m/s b. 2.9 MeV
53. a. 360 nm, 280 nm b. Both waves are ultraviolet. c. 6.2×10^5 m/s
55. 1870 nm
57. 4.43×10^4 m/s
59. 17 ns
61. 4.59 eV
63.

Transition	a. Wavelength	b. Type	c. Absorption
$2p \to 2s$	670 nm	VIS	Yes
$3s \to 2p$	816 nm	IR	No
$3p \to 2s$	324 nm	UV	Yes
$3p \to 3s$	2696 nm	IR	No
$3d \to 2p$	611 nm	VIS	No
$3d \to 3p$	24,800 nm	IR	No
$4s \to 2p$	498 nm	VIS	No
$4s \to 3p$	2430 nm	IR	No

65. 7.4 pJ
67. A
69. C

Chapter 30

Answers to odd-numbered multiple-choice questions

15. B
17. D
19. A
21. B
23. C

Answers to odd-numbered problems

1. a. 3 protons, 3 neutrons
   b. 24 protons, 30 neutrons
   c. 26 protons, 28 neutrons
   d. 86 protons, 134 neutrons
3. a. 3.8 fm b. 9.2 fm c. 14.9 fm
5. Silicon
7. 1.112 g/cm^3
9. a. ^{36}S and ^{36}Ar b. 5, 8
11. ^{129}I: 1088.1 MeV, 8.44 MeV; ^{129}Xe: 1087.6 MeV, 8.43 MeV
13. ^{14}O: 98.74 MeV, 7.05 MeV; ^{16}O: 127.62 MeV, 7.98 MeV; ^{16}O is slightly more tightly bound.
15. 4.12×10^{-12} J or 25.8 MeV
17. $\dfrac{U_{grav}}{U_{nuclear}} = 2.3 \times 10^{-38}$
19. a.

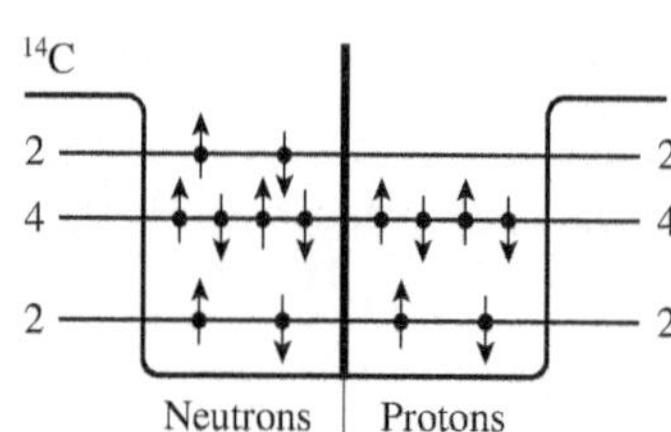

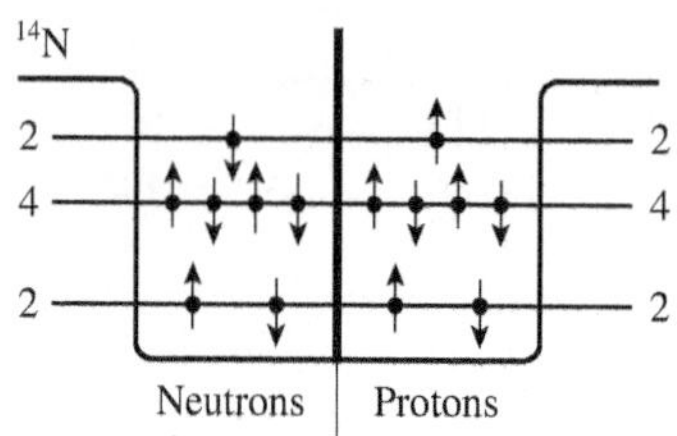

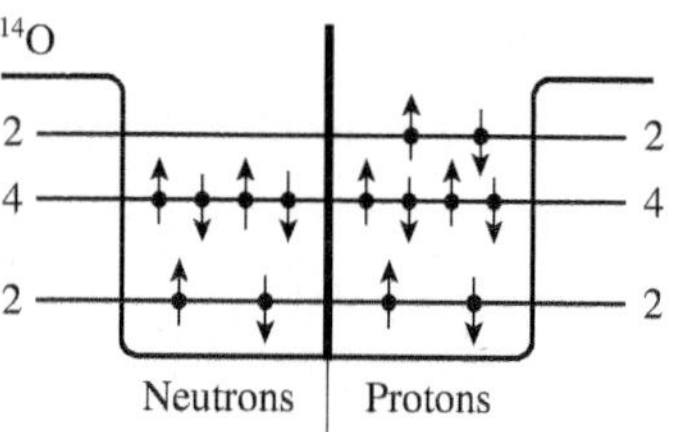

   b. ^{14}N is stable; ^{14}C undergoes beta-minus decay and ^{14}O undergoes beta-plus decay.
21. a. 9.17×10^9 b. 4.20×10^9 c. 1.73×10^6
23. a. $^3H \to {}^3He + \beta^-$ b. $\tau = 5.46 \times 10^8$ s, $r = 1.83 \times 10^{-9}\,s^{-1}$
25. 83.6 μg
27. 30 h
29. a. 5.7 kW b. 5.3 kW
31. a. ^{228}Th b. ^{207}Tl c. ^{7}Li d. ^{60}Ni
33. ^{228}Th
35. 1.50×10^{-4} u
37. 50 nm
39. 4.0 mSv
41. 0.23 J
43. 7.2×10^{15} photons
45. 3.6 mJ
47. 3.4 mSv
49. 0.391 MeV
51. 139 MeV
53. a. 1.46×10^{-8} u, 1.45×10^{-6}% b. 0.0304 u, 0.76%
55. No, ^{56}Fe cannot fission spontaneously.
57. 2.7×10^{17}
59. a. 6.12×10^{-6} kg
    b. 130 yr
61. 1.2 h
63. a. 157 min
    b. 29.5 h
65. 21,000 yr
67. 17,000 yr
69. 11 Sv
71. 23% ^{235}U and 77% ^{238}U
73. 0.0186 MeV
75. 3.3×10^{12}
77. 0.32 mSv
79. 7.4 μSv
81. D
83. D
85. B

Part VI

Answers to odd-numbered multiple-choice questions

1. C
3. A
5. D
7. A
9. D
11. A
13. B
15. D
17. D
19. B

Answers to odd-numbered problems

21. a. 1.1×10^{15} photons/s
    b. 75 mSv

Credits

COVER PHOTO SCIEPRO/Science Photo library/Getty Images

FM **P. v top** Sally Knight/Pearson Education; **p. v second from top** Alicia Aremntrout; **p. v second from bottom** Stuart Field; **p. v bottom** Michael Fripp; **p. xiii** dotshock/Shutterstock; **p. xiv** Yann Hubert/Shutterstock; **p. xv** Zephyr/Sciencesource.

PART I OPENER **P. 2** dotshock/Shutterstock

CHAPTER 1 **P. 4 top** Gregory G. Dimijian/Science source; **p. 4 mid center** Picsfive/Shutterstock; **p. 4 mid right** Viacheslav Nemyrivskyi/123rf; **p. 4 mid right (inset)** Isselee Eric Philippe/123rf; **p. 4 bottom** National Institute of Allergy and Infectious Diseases, National Institutes of Health; **p. 5 top left** Ed Reschke/Getty Images; **p. 5 top center** SciePro/Shutterstock; **p. 5 top right** Caleb Foster/Shutterstock; **p. 5 mid left** Stu Porter/Shutterstock; **p. 5 mid center** ANGELOS ZYMARAS/Alamy Stock Photo; **p. 5 mid right** Dr. David Furness, Keele University/Science Source; **p. 5 bottom left** James Kappernaros/Shutterstock; **p. 5 bottom center** BSIP SA/Alamy Stock Photo; **p. 5 bottom right** Will & Deni McIntyre/Sciencesource; **p. 6** Callista Images/Cultura Creative Ltd/Alamy Stock Photo; **p. 6** Cherly Power/Science Source; **p. 8** Martyn F. Chillmaid/Science source; **p. 12** ALFRED PASIEKA/SCIENCE PHOTO LIBRARY; **p. 14** Eric Isselee/Shutterstock; **p. 14** Four Oaks/Shutterstock; **p. 17** Horizon International Images Limited/Alamy Stock Photo.

CHAPTER 2 **P. 25 top** Jim Zipp/Science source; **p. 25 mid left** faltermaier/vario images/vario images, Inh. Susanne Baumgarten e.K./Alamy Stock Photo; **p. 25 mid center** Martin Meissner/AP/Shutterstock; **p. 25 mid right** Johnny Greig/Science source; **p. 25 bottom** Choksawatdikorn/Shutterstock; **p. 26** Theo Allofs/Getty Images; **p. 26** Paul Prince/Alamy Stock Photo; **p. 26** Richard Megna/Fundamental Photographs; **p. 26** Thegoodly/Room the agency/Alamy Stock Photo; **p. 30 top** Elephant Tracking; **p. 30 bottom** U.S. Air Force; **p. 33** Zumapress; **p. 41** Blaj Gabriel/Shutterstock; **p. 42** NASA; **p. 43** Stuart Field; **p. 43** Stuart Field; **p. 50** Jupiterimages/getty images; **p. 51** Sebastian Kaulitzki/Alamy Stock Photo.

CHAPTER 3 **P. 52 top** Michael Steele/Staff/Getty Images; **p. 52 mid left** jhorrocks/Getty Images; **p. 52 mid center** Steve Bloom Images/Alamy Stock Photo; **p. 52 mid right** EpicStockMedia/Shutterstock; **p. 52 bottom** SIMON SHIM/Shutterstock; **p. 58** Dmitry Ageev/123rf; **p. 63** VINCENT MUNIER/Nature picture library; **p. 65** kuritafsheen/RooM the Agency/Alamy Stock Photo; **p. 69** Ted Kinsman/Sciencesource; **p. 70** Henn Photography/Image Source/Alamy Stock Photo; **p. 71** S.Tuengler - inafrica.de/Alamy Stock Photo; **p. 72** Ted Kinsman/Sciencesource; **p. 73** Richard Megna/Fundamental Photographs; **p. 74** Martin Harvey/Getty Images; **p.75** Gustoimages/Sciencesource; **p. 82** Jim Barber/Shutterstock; **p. 84** KrzysztofOdziomek/Shutterstock; **p. 85** Richard Megna/Fundamental Photographs; **p. 87** EcoPrint/Shutterstock; **p. 87** Mirko Graul/Shutterstock.

CHAPTER 4 **P. 90 top** Nejron Photo/Shutterstock; **p. 90 mid left** MAGNIFIER/Shutterstock; **p. 90 mid center** Air Images/Shutterstock; **p. 90 mid right** Lurii Osadchi/Shutterstock; **p. 90 bottom** Diane Diederich/Shutterstock; **p. 91** NASA; **p. 92 left** National Highway Transportation Safety Agency; **p. 92 center** National Highway Transportation Safety Agency; **p. 92 right** National Highway Transportation Safety Agency; **p. 92 top** Image and Events/Alamy Stock Photo; **p. 92 bottom** Dmitry Kalinovsky/Shutterstock; **p. 93 top** comstock/Getty Images; **p. 93 second from top** Yakov Oskanov/Alamy Stock Photo; **p. 93 second from bottom** Todd Taulman Photography/Shutterstock; **p. 93 bottom** Pearson Education; **p. 95** sportpoint/Shutterstock; **p. 98** D.P. Wilson/FLPA/Sciencesource; **p. 100 top** Richard Megna/Fundamental Photographs; **p. 100 bottom** Diana Eller/123rf; **p. 118** Richard Megna/Fundamental Photographs.

CHAPTER 5 **P. 124 top** Avalon/Photoshot License/Alamy Stock Photo; **p. 124 mid left** 2happy/Alamy Stock Photo; **p. 124 mid center** Arco Images GmbH/Alamy Stock Photo; **p. 124 mid right** dlewis33/Getty Images; **p. 124 bottom** M. I. Walker/Sciencesource; **p. 126** NASA; **p. 127** Erin Patrice O'Brien/Getty Images; **p. 128 top** NASA; **p. 128 bottom** NASA; **p. 132** frank60/Shutterstock; **p. 133** Jari J/Shutterstock; **p. 135 top** Eric Schrader/Pearson Education, Inc; **p. 135 bottom** nulinukas/Shutterstock; **p. 145** Konstantin Novikov/Shutterstock; **p. 152** NASA; **p. 153** NASA; **p. 158** J.D. POOLEY/Associated Press; **p. 161 top** Richard Megna/Fundamental Photographs; **p. 161 bottom** MARTYN HAYHOW/Getty Images.

CHAPTER 6 **P. 163 top** Chris Nash/Getty Images; **p. 163 mid Left** Jonathan Kitchen/DigitalVision/Getty Images; **p. 163 mid center** Larin Andrey/Shutterstock; **p. 163 mid right** Martins Vangs/123rf; **p. 163 bottom** bdavid32/Alamy Stock photo; **p. 164** James Kay/SCPhotos/Alamy Stock Photo; **p. 166 top Left** Travability Images/PhotoAbility/Alamy Stock Photo; **p. 166 top right** Chris Mole/Shutterstock; **p. 166 bottom** Juice Images/Getty Images; **p. 168 top** Corbis/VCG/Getty Images; **p. 168 bottom** Corbis/VCG/Getty Images; **p. 171** Franck Fife/AFP via Getty Images; **p. 178 top** Prof. Stuart Field; **p. 178 bottom left** Richard Megna/Fundamental Photographs; **p. 178 bottom right** Richard Megna/Fundamental Photographs; **p. 179** Dominique Douieb/PhotoAlto/Alamy Stock Photo; **p. 181** Pearson Education; **p. 191 top** UzFoto/Shutterstock; **p. 191 bottom** Kari Marttila/Alamy Stock photo; **p. 194 left** Richie Chan/Shutterstock; **p. 194 right** ratmaner/Shutterstock.

CHAPTER 7 **P. 199 top** fokke baarssen/Shutterstock; **p. 199 mid left** Mikael Damkier/Shutterstock; **p. 199 mid center** DenGuy/Getty Images; **p. 199 mid right** BestPhotoStudio/Shutterstock; **p. 199 bottom** medicalstocks/Shutterstock; **p. 200** James JoneJr/Shutterstock; **p. 200** Everyonephoto Studio/Shutterstock; **p. 204** Jasperimages/Shutterstock; **p. 205** Robert Laberge/Getty Images; **p. 206 top** Matt Perrin Sport/Alamy Stock Photo; **p. 206 center left** Richard Megna/Fundamental Photographs; **p. 206 center right** Richard Megna/Fundamental Photographs; **p. 206 bottom** FPG/Getty Images; **p. 207** Richard Megna/Fundamental Photographs; **p. 213** Prof. Stuart Field; **p. 215** All Canada Photos/Alamy Stock Photo; **p. 219 top** Richard Megna/Fundamental Photographs; **p. 219 bottom** John Mead/Sciencesource; **p. 220** Arco Images Gmbh/Alamy Stock photo; **p. 225 top left** Rafa Irusta/Shutterstock; **p. 225 mid left** Images-USA/Alamy Stock Photo; **p. 225 bottom left** Todd Korol/Reuters; **p. 225 right** Ezra Staff/Getty Images; **p. 226** SCDNR's Southeastern Regional Taxonomic Center (SERTC); **p. 227 top** Hemis/alamy Stock Photo; **p. 227 bottom** dwphotos/Shutterstock; **p. 229 left** Images by J. Edwards/D. Whitaker/M. Laskwaski/A. Acosta; **p. 229 mid left** Images by J. Edwards/D. Whitaker/M. Laskwaski/A. Acosta; **p. 229 mid right** Images by J. Edwards/D. Whitaker/M. Laskwaski/A. Acosta; **p. 229 right** Images by J. Edwards/D. Whitaker/M. Laskwaski/A. Acosta.

CHAPTER 8 **P. 230 top** James Hager/robertharding/Alamy Stock Photo; **p. 230 mid left** Pete Fontaine/Getty Images; **p. 230 mid center** Corepics VOF/Shutterstock; **p. 230 mid right** PEDRO PARDO/Getty Images; **p. 230 bottom** Konstantin Novikov/Shutterstock; **p. 231** Ted Kinsman/Sciencesource; **p. 232** Buiten-Beeld/Alamy Stock Photo; **p. 235** Petr Zurek/Alamy Stock Photo; **p. 236** Gene Blevins/Zumapress/Newscom; **p. 244** Comstock/Getty Images; **p. 244** NASA; **p. 247 left** STOCKFOLIO/Alamy Stock Photo; **p. 247 right** Kostas Tsironis/AP Images.

CHAPTER 9 **P. 256 top** simonkr/Getty Images; **p. 256 mid left** Hero Images Inc./Alamy Stock Photo; **p. 256 mid center** Juancat/Shutterstock; **p. 256 mid right** Henrik Sorensen/Getty Images; **p. 256 bottom** Welcome Collection; **p. 261** Woods Hole Oceanographic Institution/Oceanus Magazine; **p. 264** Joseph P. Sinnot/Fundamental Photographs; **p. 265** Alamy Stock Photo; **p. 269 top** Harald Sund/Getty Images; **p. 269 bottom** vlaru/Shutterstock;

p.270 Vladimir Wrangel/Shutterstock; **p. 271** The Natural History Museum, London/Science Source; **p. 273** mjf99/Shutterstock; **p. 276** Thomas Vogel/Getty Images; **p .279** Erik Mandre/Shutterstock; **p. 280 left** Lisa Kyle Young/Getty Images; **p. 280 right** Don Farrall/Getty Images; **p. 281** Frank Herzog/culture-images GmbH/Alamy Stock Photo; **p. 282** Ed Endicott/Alamy Stock Photo; **p. 283** Zephyr/Science Source; **p. 286** VEM/Sciencesource**; p. 288** Willmott, P.A., et. al. (1997) Flow visualization and unsteady aerodynamics in the flight of the hawkmoth, Manduca sexta. Philosophical Transactions of the Royal Society B, 352(1351), 303–316.; **p. 289** ER Productions Limited/Getty Images; **p. 291** Martin Harvey/Alamy Stock Photo; **p. 292 top** Du Cane Medical Imaging Ltd/Science Source; **p. 292 bottom** David M. Phillips/Science Source; **p. 293** Lillac/Shutterstock; **p.293** Vincent Hazat/PhotoAlto/Alamy Stock Photo; **p. 298** Tom Brakefield/Getty Images; **p. 299** Thomas Vogel/Getty Images; **p. 301** mjf99/Shutterstock; **p. 303** Hartmut Schmidt/imageBROKER/Alamy Stock Photo.

PART I SUMMARY **P. 305** Bill Schoening/Vanessa Harvey/REU Program/NOAA/AURA/NSF; **p. 306 left** Owen Humphreys/Associated Press; **p. 306 right** SPL/Sciencesource.

PART II OPENER **P. 308** Yann Hubert/Shutterstock.

CHAPTER 10 **P. 310 top** Alexander Piragis/Shutterstock; **p. 310 mid left** muratart/Shutterstock; **p. 310 mid center** Lightpoet/Panther Media GmbH/Alamy Stock Photo; **p. 310 bottom** mrbigphoto/Shutterstock; **p. 311 left** Kyodo/Newscom; **p.311 mid left** John Foxx/Getty Images; **p. 311 mid right** Stacey Bates/Shutterstock; **p. 311 right** Timofey Zadvornov/Shutterstock; **p. 312 top left** Alexander Hassenstein/Getty Images; **p. 312 bottom left** numbeos/Getty Images; **p. 312 top right** Vibrant Image Studio/Shutterstock; **p. 312 bottom right** franzfoto.com/Alamy Stock Photo; **p. 322 top** Chris Salvo/Getty Images; **p. 322 bottom** givaga/Shutterstock; **p. 327** ADAM NURKIEWICZ/Getty Images; **p. 328** Juice Images/Getty Images; **p. 332** Adventure Photo/Getty Images; **p. 334** Kristel Segeren/Shutterstock.

CHAPTER 11 **P.335 top** Mitchell Funk/Getty Images; **p. 335 mid left** Strahil Dimitrov/Shutterstock; **p. 335 mid right** Dirima/Shutterstock; **p. 335 bottom** dragasanu/Shutterstock; **p. 337** Andrew Zarivny/Shutterstock; **p. 343** Michael Stubblefield/Alamy Images; **p. 361** Pearson Education, Inc.; **p. 363** SCIENCE PHOTO LIBRARY/Science Photo Library/Alamy Stock Photo; **p. 364 top** Asgeir Helgestad/Nature Picture Library; **p. 364 bottom** Jurgen Freund/Nature Picture Library; **p. 366** Maxim Images archive/Alamy Stock Photo; **p. 367** Charles D. Winters/Science Source; **p. 375** Martin Harvey/Science Source.

CHAPTER 12 **P. 378 top** Gallo Images/Getty Images; **p. 378 mid Left** Steve Hamblin/Alamy Stock Photo; **p. 378 mid center** NHPA/Superstock; **p. 378 mid right** Paul Burns/Getty Images; **p. 378 bottom** vladsilver/Shutterstock; **p. 379** Richard Megna/Fundamental Photographs; **p. 382** Gordon Saunders/Shutterstock; **p. 384** Kenneth B. Storey/Carleton University; **p. 387** Toshi Sasaki/Getty Images; **p. 388** gcoles/Getty Images; **p. 389** John Raoux/Associated Press; **p. 393 left** moodboard/Getty Images; **p. 393 mid Left** Gary S. Settles & Jason Listak/Science Source; **p. 393 mid right** Pascal Goetgheluck/Science Source; **p. 393 right** stockfour/Shutterstock; **p. 395 top** sciencephotos/Alamy Stock Photo; **p. 395 bottom** NASA; **p. 396** Ted Kinsman/Science Source; **p. 397** Gillian Merritt/Alamy Stock Photo; **p. 400** Richard Megna/Fundamental Photographs; **p. 411** Takahashi Outdoors/Shutterstock; **p. 420 top** Fitawoman/Shutterstock; **p.420 bottom** Four Oaks/Shutterstock.

CHAPTER 13 **P. 422 top** Scimat/Science Source; **p. 422 mid left** blue-sea.cz/Shutterstock; **p. 422 mid right** Turtle Rock Scientific/Science Source; **p. 422 bottom** Daniel Schroen, Cell Applications Inc./Science Source; **p. 442** Sirocco/Shutterstock; **p. 443** ihar leichonak/Alamy Stock Photo; **p. 446** Biophoto Associates/Science Source; **p. 447 top** Institue of Molecular Biotechnology; **p. 447 bottom** Robert Hamilton/Alamy Stock Photo; **p. 450** Power and Syred/Science Source; **p. 456** Eye of Science/Science Source; **p. 457** Science Photo Library/Alamy Stock Photo.

CHAPTER 14 **P. 459 top** Design Pics Inc/Alamy Stock Photo; **p. 459 mid left** Warren Price Photography/Shutterstock; **p. 459 mid center** Phil Degginger/Alamy Stock Photo; **p. 459 mid right** gopixa/Shutterstock; **p. 459 bottom** Biomedical Imaging Unit, Southampton General Hospital/Science Source; **p. 465** Randy Knight; **p. 469** HappyRichStudio/Shutterstock; **p. 474** Basement Stock/Alamy Stock Photo; **p. 476** Foodcollection RF/Getty Images; **p. 478** Science History Images/Alamy Stock Photo; **p. 484** Dr Kjaergaard.

PART II SUMMARY **P. 497 top** Morozov Anatoly/Shutterstock; **p. 497 left** IndiaMart; **p. 497 right** Juan Carlos Munoz/NaturePL/Science Source; **p. 498** Egill Bjarnason/Alamy Stock Photo.

PART III OPENER **P. 506** Zephyr/Sciencesource.

CHAPTER 15 **P. 508 top** AiVectors/Shutterstock; **p. 508 mid left** Andrew Lambert Photography/Science Source; **p. 508 mid center** pelfophoto/Shutterstock; **p. 508 mid right** Hank Morgan/Science Source; **p. 508 bottom** Steve Gschmeissner/Science Source; **p. 510 top** Oscar Burriel/Science Source; **p. 510 bottom** Adam Jones/Getty Images; **p. 511** Nastasic/Getty Images; **p. 512** Tom & Pat Leeson/Science Source; **p. 514** Serge Kozak/AGEFotostock; **p. 517** H. Craighead/Cornell Univ; **p. 522** NASA; **p. 526** Antonio Gravante/Shutterstock; **p. 527** Ryzhkov Oleksandr/Shutterstock; **p. 528** MORNING LIGHT/Balan Madhavan/Alamy Stock Photo; **p. 532** Martin Bough/Fundamental Photographs; **p. 533** sirtravelalot/Shutterstock; **p. 536** Leo Mason/Popperfoto/Getty Images; **p. 539** Four Oaks.Shutterstock; **p. 542** Ted Horowitz/Getty Images.

CHAPTER 16 **P. 543 top** ANN CLARKE IMAGES/Getty Images; **p. 543 mid left** UfaBizPhoto/Shutterstock; **p. 543 mid center top** beeboys/Shutterstock; **p. 543 mid center bottom** David Parker/Science Source; **p. 543 mid right** Mike Longhurst/Shutterstock; **p. 543 bottom** Allo4e4ka/Shutterstock; **p. 544** Olga Selyutina/Shutterstock; **p. 545** Adapted from PSSC Physics, 7th edition by Haber-Schaim, Dodge, Gardner, and Shore, Published by Kendall/Hunt, 1991; **p. 547** Aflo Co., Ltd./Alamy Stock Photo; **p. 549** James Gerholdt/Getty Images; **p. 556** yevgeniy11/Shutterstock; **p. 557** Belish/Shutterstock; **p. 558** David Parker/Science Source; **p. 563** Mark Carwardine/Getty Images; **p. 567** NOAA/Associated Press; **p. 569** Chris Gallagher/Science Source; **p. 574** SIHASAKPRACHUM/Shutterstock.

CHAPTER 17 **P. 578 top** Skylines/Shutterstock; **p. 578 mid left** Richard Megna/Fundamental Photographs; **p. 578 mid center** Ysbrand Cosijn/Shutterstock; **p. 578 mid right** Richard Megna/Fundamental Photographs; **p. 578 Bottom** Rasulov/Shutterstock; **p. 580** Renn Sminkey/Creative Digital Vision/Pearson Education Inc.; **p. 582** World History Archive/Alamy Stock Photo; **p. 587** Brian Atkinson/Alamy Stock Photo; **p. 589** David Maiullo; **p. 593** didesign021/Shutterstock; **p. 600** Richard Megna/Fundamental Photographs; **p. 606** Utekhina Anna/Shutterstock; **p. 611** Joe Tucciarone/Science Source; **p. 613** ldutko/Shutterstock.

PART III SUMMARY **P. 617** Courtesy D. Vatvani, Deltares, The Netherlands; **p. 618** Lee Dalton/Alamy Stock Photo; **p. 619** Zoonar/Wolfgang Poelzer/Zoonar GmbH/Alamy Stock Photo.

PART IV OPENER **Pp. 620–621** Larry Lynch.

CHAPTER 18 **P. 622 top** Studio Doros/Alamy Stock Photo; **p. 622 mid left** Ekaterina Sidonskaya/Alamy Stock Photo; **p. 622 mid center** Phil Degginger/Alamy Stock Photo; **p. 622 mid right** sciencephotos/Alamy Stock Photo; **p. 622 bottom right: left** SPL/Sciencesource; **p. 622 bottom right: right** Ericlin1337; **p. 623** Sigitas Baltramaitis/Alamy Stock Photo; **p. 624 left** Richard Megna/Fundamental Photographs; **p. 624 right** Richard Megna/Fundamental Photographs**; p. 625** Stuart Field; **p. 627 top** Pearson Education, Inc.; **p. 627 bottom** Scott Camazine/Alamy Stock Photo; **p. 628 top** National Oceanic and Atmospheric Administration (NOAA)**; p. 628 middle** Terry Oakley/Alamy Stock Photo; **p. 628 bottom** imageBROKER/Alamy Stock Photo; **p. 629 top** Richard Megna/Fundamental Photographs; **p. 629 mid left** Eisele, John; **p. 629 mid right** Ted Kinsman/Science Source; **p. 629 bottom** Adam Pearlstein/Pearson Educations; **p. 630** Nature Picture Library/Alamy Stock Photo; **p. 635 left** Victor Tyakht/Shutterstock; **p. 635 right** Ed Marshall/Alamy Stock Photo; **p. 637** Adam Pearlstein/Pearson Education, Inc.; **p. 638** Melvyn Yeo/ Science Source; **p. 640** Shutterstock; **p. 641 top** Ken Kay/Fundamental Photographs; **p. 641 bottom** Adam Pearlstein/Pearson Education, Inc.; **p. 644** Neil Borden/Science Source; **p. 645** Raymond Gosling/King's College London; **p. 646** Jeff Dahl; **p. 650 top** BorneoRimbawan/Shutterstock; **p. 650 bottom** Raymond Gosling; **p. 651** T. Eidenweil/imageBROKER/Alamy Stock Photo; **p. 652 top** Grigorev Mikhail/Shutterstock; **p. 652 bottom** Steve Gschmeissner/Science Source;**p. 654 left** Arsen Volkov/Alamy

Stock Photo; **p. 654 right** Giraldo, M. A., Parra, J. L., & Stavenga, D. G. (2018). Iridescent colouration of male Anna's hummingbird (*Calypte anna*) caused by multilayered barbules. *Journal of Comparative Physiology* A, 204(12), 965–975; **p. 655 left** Zoonar GmbH/Alamy Stock Photo; **p. 655 top right: left** Vukusic, Pete; **p. 655 top right: right** Vukusic, Pete.

CHAPTER 19 **P. 656 Top** gui00878/Getty Images; **p. 656 mid left** Darren Davis/Alamy Stock Photo; **p. 656 mid center** Turtle Rock Scientific/Science Source; **p. 656 mid right** Smile19/Shutterstock; **p. 656 bottom right** GASTROLAB/ Science Source; **p. 659** Reinhard Dirscherl/Alamy Stock Photo; **p. 661** Tyler Olson/Shutterstock; **p. 662** Pearson Education, Inc.; **p. 663** Brian Jones; **p. 665** Ken Kay/Fundamental Photographs; **p. 666 top** Pete Atkinson/Getty Images; **p. 666 bottom** BSIP SA/Alamy Stock Photo; **p. 667** Pearson Education, Inc.; **p. 668 left** Richard Megna/Fundamental Photographs; **p. 668 right** Richard Megna/Fundamental Photographs; **p. 670** Pearson Education, Inc.; **p. 672 top** pixelklex/Shutterstock; **p. 672 bottom** Andrey Popov/Shutterstock; **p. 678** Yaacov Dagan/Alamy Stock Photo; **p. 679** Solvin Zankl/Alamy Stock Photo; **p. 682** Eric Schrader/Pearson Education, Inc.; **p. 683 top** majeczka/Shutterstock; **p. 683 bottom** BSIP/Science Source; **p. 686** NASA; **p. 691** david hancock/Alamy Stock Photo.

CHAPTER 20 **P. 692 Top** EcoPrint/Shutterstock; **p. 692 mid left** Lee Webb/Alamy Stock Photo; **p. 692 mid center** Stefan Dahl Langstrup/Alamy Stock Photo; **p. 692 mid right** im West/Alamy Stock Photo; **p. 692 bottom right** Jennifer C. Waters/Science Source; **p. 696 top** Richard Megna/Fundamental Photographs; **p. 696 bottom** Murat Baysan/Shutterstock; **p. 697 top** NASA; **p. 697 bottom** Randy Knight; **p. 700** Tetra Images/Alamy Stock Photo; **p. 703 top** Stuart Field; **p. 703 bottom** Stuart Field; **p. 704** Stuart Field; **p. 705** Tetra Images/Alamy Stock Photo; **p. 706** Pearson Education, Inc.; **p. 709 top** Pearson Education, Inc; **p. 709 center** Pearson Education, Inc; **p. 709 bottom** Pearson Education, Inc; **p. 710** James Cavallini/Science Source; **p. 713** DR TORSTEN WITTMANN/Science Source; **p. 714** NCMIR/Tom Deerinck/Science Source; **p. 716** Daniel Hebert/Shutterstock; **p. 722** apilio/Alamy Stock Photo.

PART IV SUMMARY **P. 725 left** Ted Kinsman/Science Source; **p. 725 top right: left** WIM VAN EGMOND/SCIENCE PHOTO LIBRARY; **p. 725 top right: right** M.I. Walker/Science Source.

PART V OPENER **Pp. 728–729** Karen Doody/Stocktrek Images/Getty Images.

CHAPTER 21 **P. 730 Top** scotspencer/Getty Images**; p. 730 mid left** CHARLES D. WINTERS/Science Source; **p. 730 mid center** Leigh Prather/Shutterstock; **p. 730 mid right** Mike Hardiman/Alamy Stock Photo; **p. 730 bottom right** Panu Ruangjan/Alamy Stock Photo; **p. 731** Pearson Education, Inc; **p. 733** YAY Media AS/Alamy Stock Photo; **p. 734** YAY Media AS/Alamy Stock Photo; **p. 736** Geoff Smith/Alamy Stock Photo; **p. 744** Richard Megna/Fundamental Photographs; **p. 745** Thierry GRUN/Alamy Stock Photo; **p. 763** Ondrej Prosicky/Shutterstock.

CHAPTER 22 **P. 770 Top** Nick Everett/Alamy Stock Photo; **p. 770 mid left** Garo/Phanie/Science Source; **p. 770 mid center** lightrain/Shutterstock; **p. 770 bottom right** Paulo Oliveira/Alamy Stock Photo; **p. 779** Aigars Reinholds/Alamy Stock Photo; **p. 785** Luc DIEBOLD/Fotolia; **p. 791** Andrew Lambert Photography/Science Source; **p. 799** Jochen Tack/Alamy Stock Photo; **p. 800** Matt9122/Shutterstock; **p. 801** NASA; **p. 806** Centers for Disease Control and Prevention.

CHAPTER 23 **P. 811 Top** ADAM HART-DAVIS/Science Source; **p. 811 mid left** DR P. MARAZZI/Science Source; **p. 811 mid center** Biochemist Artist/Shutterstock; **p. 811 mid right** GEOFF TOMPKINSON/Science Source; **p. 811 bottom right** HERMANN SCHILLERS, PROF. DR. H. OBERLEITHNER, UNIVERSITY HOSPITAL OF MUENSTER/Science Source; **p. 812** Eric Schrader/Pearson Education, Inc.; **p. 814** Pushish Images/Shutterstock; **p. 816** Dimitri Otis/Getty Images; **p. 836** MIKA Images/Alamy Stock Photo.

CHAPTER 24 **P. 842 Top** ipm/Alamy Stock Photo; **p. 842 mid left** Kenny Williamson/Alamy Stock Photo; **p. 842 mid center** Boris Rabtsevic/Shutterstock; **p. 842 mid right** I MAKE PHOTO 17/Shutterstock; **p. 842 bottom right** Microgen/Shutterstock; **p. 849** Peter Scoones/Nature Picture Library; **p. 852** ethylalkohol/Shutterstock; **p. 853** Dr. Jennifer Mueller; **p. 854 left** pixel shepherd/Alamy Stock Photo; **p. 854 center** Igor S/Shutterstock; **p. 854 right** Brian Jones; **p. 857 top** Brian Jones; **p. 857 bottom** Osmose Utilities Services, Inc; **p. 859** Pearson Education, Inc; **p. 861 top** Rick Friedman/Getty Images; **p. 861 bottom** 1125089601/Shutterstock; **p. 867** Juniors Bildarchiv GmbH/Alamy Stock Photo.

CHAPTER 25 **P. 871 Top** Andrew M. Allport/Shutterstock; **p. 871 mid left** Pavlo Sachek/Fotolia; **p. 871 mid center** Patrick Batchelder/Alamy Stock Photo**; p. 871 mid right** CNRI/Science Source; **p. 871 bottom right** STEVE GSCHMEISSNER/Science Source**; p. 878** Polka Dot/JupiterImages/Alamy Stock Photo; **p. 879** Abelstock/Alamy Stock Photo; **p. 882** Simon Fraser/Science Source**; p. 886 top** Purestock/Alamy Stock Photo; **p. 886 bottom** Purestock/Alamy Stock Photo**; p. 888** Randall Knight; **p. 889** Eric Schrader; **p. 891** SPL/Science Source; **p. 894** STEPHANE DE SAKUTIN/Getty Images.

CHAPTER 26 **P. 907 Top** Denis Belitsky/Shutterstock; **p. 907 mid left** John Eisele/Colorado State University Photography; **p. 907 mid center** James King-Holmes/Alamy Stock Photo; **p. 907 mid right** SpeedKingz/Shutterstock; **p. 907 bottom right** Dennis Kunkel Microscopy/Science Source; **p. 915 top** John Eisele/Colorado State University; **p. 915 bottom** bradwieland/Getty Images; **p. 916** John Eisele/Colorado State University Photography; **p. 920** Atsushi Arakaki et al., "Formation of magnetite by bacteria and its application" *J R Soc Interface*. 2008 September 6; 5(26): 977–999. doi: 10.1098/rsif.2008.0170, **p. 922** Stacy Walsh Rosenstock/Alamy Stock Photo; **p. 923** Richard Megna/Fundamental Photographs; **p. 924** Neutronman/Getty Images; **p. 925** James King-Holmes/Oxford Centre for Molecular Sciences/Science Source; **p. 926** Gado Images/Alamy Stock Photo; **p. 932** Michael Ventura/Alamy Stock Photo; **p. 936** ZEPHYR/Science Source; **p. 937** CNRI/Science Source.

CHAPTER 27 **P. 949 Top** MAREK MIS/Science Source; **p. 949 mid left** Johan Swanepoel/Alamy Stock Photo; **p. 949 mid center** Iryna Veklich/Alamy Stock Photo; **p. 949 mid right** Valmedia/Alamy Stock Photo; **p. 949 bottom right** Cordelia Molloy/Science Source; **p. 952** wildestanimal/Shutterstock; **p. 954** Teun van den Dries/Shutterstock; **p. 960** Steve Tulley/Alamy Stock Photo; **p. 963** Martin Shields/Alamy Stock Photo; **p. 967** Steve Morgan/Alamy Stock Photo; **p. 968** NASA; **p. 970 top left** Richard Megna/Fundamental Photographs; **p. 970 top center** Richard Megna/Fundamental Photographs; **p. 970 top right** Richard Megna/Fundamental Photographs; **p. 970 left top** Richard Megna/Fundamental Photographs; **p. 970 left bottom** Richard Megna/Fundamental Photographs; **p. 972 left** Jim West/Alamy Stock Photo; **p. 972 right** GFC Collection/Alamy Stock Photo; **p. 973 left** Brian Jones; **p. 973 center** Brian Jones; **p. 973 right** ephotocorp/Alamy Stock Photo; **p. 974 left** ephotocorp/Alamy Stock Photo; **p. 974 right** Ted Kinsman/Science Source; **p. 975 top left** NOAA Forecast Systems Laboratory; **p. 975 top right** NASA; **p. 975 bottom left** NASA; **p. 975 bottom right** National Radio Astronomy Observatory (NRAO).

PART VI OPENER **Pp. 990–991** Caleb Foster/Shuttterstock.

CHAPTER 28 **P. 992 Top** James Cavallini/Science Source; **p. 992 mid left** T. L. Dimitrova, A. Weis *American Journal of Physics* 76, 137 (2008); doi: 10.1119/1.2815364 **; p. 992 mid center** PHILIPPE PLAILLY/Science Source**; p. 992 mid right** Sinclair Stammers/Science Source; **p. 992 bottom right** Science History Images/Alamy Stock Photo**; p. 997 top** Bettmann /Getty Images; **p. 997 bottom** RUZvOLD/Zshutterstock; **p. 1000 first** T. L. Dimitrova, A. Weis *American Journal of Physics* 76, 137 (2008); doi: 10.1119/1.2815364; **p. 1000 second** T. L. Dimitrova, A. Weis *American Journal of Physics* 76, 137 (2008); doi: 10.1119/1.2815364**; p. 1000 third** T. L. Dimitrova, A. Weis *American Journal of Physics* 76, 137 (2008); doi: 10.1119/1.2815364; **p. 1000 fourth** T. L. Dimitrova, A. Weis *American Journal of Physics* 76, 137 (2008); doi: 10.1119/1.2815364; **p. 1001** Kenneth Eward/BioGrafx/Science Source; **p. 1002** ITAR-TASS News Agency/Alamy Stock Photo; **p. 1003** Eric Schrader/Pearson Education, Inc.; **p. 1004 left** Copyright 2013 Education Development center, Inc. Reprinted with permission with all other rights reserved, **p. 1004 center** Copyright 2013 Education Development center, Inc. Reprinted with permission with all other rights reserved; **p. 1004 right** Oak Ridge National Laboratory, courtesy AIP ESVA/Science Source**; p. 1005**

Brian Jones; **p. 1008** SPL/Science Source; **p. 1013** Colin Cuthbert/Science Source; **p. 1014 top left** IBM Research/Science Source; **p. 1014 top right** Lawrence Livermore National Laboratory/Science Source; **p. 1014 center** Science Photo Library/Alamy Stock Photo; **p. 1015** Protasov AN/Shutterstock; **p. 1020** Darwin Dale/Science Source; **p. 1022 top** Protasov AN/Shutterstock; **p. 1022 bottom** Solent News/Splash News/Newscom.

CHAPTER 29 **P. 1025 Top** Joshua Davenport/Alamy Stock Photo; **p. 1025 mid left** Heath Consultants, Inc; **p. 1025 mid center** domnitsky/Shutterstock; **p. 1025 mid right** Will & Deni McIntyre/Science Source; **p. 1025 bottom right** Danté Fenolio/Science Source; **p. 1026 top** Ocean Optics; **p. 1026 bottom** Kim Christensen/Shutterstock; **p. 1027** NASA Earth Observing System; **p. 1045 top** Bork/Shutterstock; **p. 1045 bottom** Olga Ovchinnikova/Alamy Stock Photo; **p. 1046 top** Rick & Nora Bowers/Alamy Stock Photo; **p. 1046 bottom** ALEXANDER SEMENOV/Science Source; **p. 1047** Sinclair Stammers/Science Source; **p. 1048** Frankwalker.de/Fotolia; **p. 1049 top** Tech. Sgt. Larry A. Simmons/United States Air Force; **p. 1049 bottom** photastic/Shutterstock.

CHAPTER 30 **P. 1058 Top** Brookhaven National Laboratory; **p. 1058 mid left** Nature Picture Library/Alamy Stock Photo; **p. 1058 mid center** Public Health England/Science Source; **p. 1058 mid right** PROF. J. LEVEILLE/Science Source; **p. 1058 bottom right** agefotostock/Alamy Stock Photo; **p. 1059** Ball and Albanese/Alamy Stock Photo; **p. 1060** British Antarctic Survey/Science Source; **p. 1064** Dan Olsen/Shutterstock; **p. 1070 Los** Alamos National Laboratory; **p. 1073** James King-Holmes/Science Source; **p. 1077** IceCube; **p. 1081** mauritius images GmbH/Alamy Stock Photo; **p. 1082** James King-Holmes/Alamy Stock Photo; **p. 1083 top** Phanie/Alamy Stock Photo; **p. 1083** center CNRI/Science Source; **p. 1083** bottom CNRI/Science Source; **p. 1084 top** Hank Morgan/Science Source; **p. 1084 center** Zephyr/Science Source; **p. 1084 bottom** Zephyr/Science Source; **p. 1086** Chauvaint-Chapon/Phanie/SuperStock; **p. 1087** CERN/Science Source; **p. 1088** Kamioka Observatory; **p. 1093** Robert Gray/Alamy Stock Photo; **p. 1095 top** CENTRE JEAN PERRIN/Science Source; **p. 1095 bottom** CENTRE JEAN PERRIN/Science Source.

PART VI SUMMARY **P. 1101** Bgran, Used under a Creative Commons License, http://creativecommons.org/licenses/by/2.0/deed.en.

Index

A

F

G

J

K

L

M

N

O

P

T

U

V

W

X

Y

Z

Typical Coefficients of Friction

Materials	Static μ_s	Kinetic μ_k	Rolling μ_r
Rubber on concrete	1.00	0.80	0.02
Steel on steel (dry)	0.80	0.60	0.002
Steel on steel (lubricated)	0.10	0.05	
Wood on wood	0.50	0.20	
Wood on snow	0.12	0.06	
Ice on ice	0.10	0.03	

Elastic Properties of Materials

Material	Young's modulus (10^6 N/m^2)	Tensile strength (10^6 N/m^2)
Steel (structural)	200,000	1000
Copper	110,000	220
Concrete (typical)	30,000	5
Compact bone	18,000	140
Wood (Douglas fir)	10,000	75
Dragline silk	10,000	1000
Collagen	1500	100
Elastin	1	2

Properties of Materials

Material	Density (kg/m^3)	Viscosity (Pa · s)	Sound speed (m/s)	Specific Heat (J/kg · K)	Expansion (°C^{-1})	Surface tension (mN/m)	Thermal conductivity (J/kg · K)
Air (1 atm, 0°C)	1.29	1.7×10^{-5}	331				
Air (1 atm, 20°C)	1.20	1.8×10^{-5}	343				0.023
Air (1 atm, 37°C)	1.14	1.9×10^{-5}	353				
Helium (1 atm, 0°C)	0.179	1.9×10^{-5}	970				
Acetic acid	1040	1.1×10^{-3}		2040			
Blood (whole, 37°C)	1060	3.5×10^{-3}				52	
Ethanol (20°C)	790	1.2×10^{-3}	1170	2400		22	0.17
Glycerin	1260	9.5×10^{-1}				63	
Mercury	13,600		1450	140		490	29
Olive oil	910	8.4×10^{-2}	1430	1970		32	0.17
Seawater	1030			3850			
Water (20°C)	1000	1.0×10^{-3}	1480	4190		73	0.61
Water (40°C)	992	7.0×10^{-4}	1530	4190		70	
Aluminum	2700		6420	900	2.3×10^{-5}		240
Copper	8960		4600	385	1.7×10^{-5}		400
Diamond	3520		12,000	510	7.0×10^{-7}		2000
Glass			4540	840	2.7×10^{-5}		1.0
Gold	19,300		3240	129	1.4×10^{-5}		320
Granite	2700		6000	790			
Ice	917		4000	2090	5.5×10^{-5}		1.7
Mammalian body	1005			3400			
Bone			4000				
Fat			1450				0.21
Muscle			1580				0.46
Soft tissue			1540				
Stainless steel	7930		5800	502	1.7×10^{-5}		14

Molar Specific Heats of Gases

Gas	C_V (J/mol · K)	C_P (J/mol · K)
He	12.5	20.8
Ne	12.5	20.8
Ar	12.5	20.8
H_2	20.4	28.7
N_2	20.8	29.1
H_2	20.9	29.1

Phase Changes

Substance	T_m (°C)	L_f (J/kg)	T_b (°C)	L_v (J/kg)
Nitrogen (N_2)	−210	0.26×10^5	−196	1.99×10^5
Ethanol	−114	1.09×10^5	78	8.97×10^5
Mercury	−39	0.11×10^5	357	2.96×10^5
Water	0	3.33×10^5	100	22.6×10^5
Acetic acid	17	1.95×10^5	118	3.95×10^5
Lead	328	0.25×10^5	1750	8.58×10^5